FOOD BIOTECHNOLOGY

Second Edition

FOOD SCIENCE AND TECHNOLOGY

A Series of Monographs, Textbooks, and Reference Books

76. Food Chemistry: Third Edition, *edited by Owen R. Fennema*
77. Handbook of Food Analysis: Volumes 1 and 2, *edited by Leo M. L. Nollet*
78. Computerized Control Systems in the Food Industry, *edited by Gauri S. Mittal*
79. Techniques for Analyzing Food Aroma, *edited by Ray Marsili*
80. Food Proteins and Their Applications, *edited by Srinivasan Damodaran and Alain Paraf*
81. Food Emulsions: Third Edition, Revised and Expanded, *edited by Stig E. Friberg and Kåre Larsson*
82. Nonthermal Preservation of Foods, *Gustavo V. Barbosa-Cánovas, Usha R. Pothakamury, Enrique Palou, and Barry G. Swanson*
83. Milk and Dairy Product Technology, *Edgar Spreer*
84. Applied Dairy Microbiology, *edited by Elmer H. Marth and James L. Steele*
85. Lactic Acid Bacteria: Microbiology and Functional Aspects, Second Edition, Revised and Expanded, *edited by Seppo Salminen and Atte von Wright*
86. Handbook of Vegetable Science and Technology: Production, Composition, Storage, and Processing, *edited by D. K. Salunkhe and S. S. Kadam*
87. Polysaccharide Association Structures in Food, *edited by Reginald H. Walter*
88. Food Lipids: Chemistry, Nutrition, and Biotechnology, *edited by Casimir C. Akoh and David B. Min*
89. Spice Science and Technology, *Kenji Hirasa and Mitsuo Takemasa*
90. Dairy Technology: Principles of Milk Properties and Processes, *P. Walstra, T. J. Geurts, A. Noomen, A. Jellema, and M. A. J. S. van Boekel*
91. Coloring of Food, Drugs, and Cosmetics, *Gisbert Otterstätter*
92. *Listeria*, Listeriosis, and Food Safety: Second Edition, Revised and Expanded, *edited by Elliot T. Ryser and Elmer H. Marth*
93. Complex Carbohydrates in Foods, *edited by Susan Sungsoo Cho, Leon Prosky, and Mark Dreher*

94. Handbook of Food Preservation, *edited by M. Shafiur Rahman*
95. International Food Safety Handbook: Science, International Regulation, and Control, *edited by Kees van der Heijden, Maged Younes, Lawrence Fishbein, and Sanford Miller*
96. Fatty Acids in Foods and Their Health Implications: Second Edition, Revised and Expanded, *edited by Ching Kuang Chow*
97. Seafood Enzymes: Utilization and Influence on Postharvest Seafood Quality, *edited by Norman F. Haard and Benjamin K. Simpson*
98. Safe Handling of Foods, *edited by Jeffrey M. Farber and Ewen C. D. Todd*
99. Handbook of Cereal Science and Technology: Second Edition, Revised and Expanded, *edited by Karel Kulp and Joseph G. Ponte, Jr.*
100. Food Analysis by HPLC: Second Edition, Revised and Expanded, *edited by Leo M. L. Nollet*
101. Surimi and Surimi Seafood, *edited by Jae W. Park*
102. Drug Residues in Foods: Pharmacology, Food Safety, and Analysis, *Nickos A. Botsoglou and Dimitrios J. Fletouris*
103. Seafood and Freshwater Toxins: Pharmacology, Physiology, and Detection, *edited by Luis M. Botana*
104. Handbook of Nutrition and Diet, *Babasaheb B. Desai*
105. Nondestructive Food Evaluation: Techniques to Analyze Properties and Quality, *edited by Sundaram Gunasekaran*
106. Green Tea: Health Benefits and Applications, *Yukihiko Hara*
107. Food Processing Operations Modeling: Design and Analysis, *edited by Joseph Irudayaraj*
108. Wine Microbiology: Science and Technology, *Claudio Delfini and Joseph V. Formica*
109. Handbook of Microwave Technology for Food Applications, *edited by Ashim K. Datta and Ramaswamy C. Anantheswaran*
110. Applied Dairy Microbiology: Second Edition, Revised and Expanded, *edited by Elmer H. Marth and James L. Steele*
111. Transport Properties of Foods, *George D. Saravacos and Zacharias B. Maroulis*
112. Alternative Sweeteners: Third Edition, Revised and Expanded, *edited by Lyn O'Brien Nabors*
113. Handbook of Dietary Fiber, *edited by Susan Sungsoo Cho and Mark L. Dreher*
114. Control of Foodborne Microorganisms, *edited by Vijay K. Juneja and John N. Sofos*
115. Flavor, Fragrance, and Odor Analysis, *edited by Ray Marsili*
116. Food Additives: Second Edition, Revised and Expanded, *edited by A. Larry Branen, P. Michael Davidson, Seppo Salminen, and John H. Thorngate, III*
117. Food Lipids: Chemistry, Nutrition, and Biotechnology: Second Edition, Revised and Expanded, *edited by Casimir C. Akoh and David B. Min*
118. Food Protein Analysis: Quantitative Effects on Processing, *R. K. Owusu-Apenten*
119. Handbook of Food Toxicology, *S. S. Deshpande*
120. Food Plant Sanitation, *edited by Y. H. Hui, Bernard L. Bruinsma, J. Richard Gorham, Wai-Kit Nip, Phillip S. Tong, and Phil Ventresca*
121. Physical Chemistry of Foods, *Pieter Walstra*
122. Handbook of Food Enzymology, *edited by John R. Whitaker, Alphons G. J. Voragen, and Dominic W. S. Wong*
123. Postharvest Physiology and Pathology of Vegetables: Second Edition, Revised and Expanded, *edited by Jerry A. Bartz and Jeffrey K. Brecht*
124. Characterization of Cereals and Flours: Properties, Analysis, and Applications, *edited by Gönül Kaletunç and Kenneth J. Breslauer*
125. International Handbook of Foodborne Pathogens, *edited by Marianne D. Miliotis and Jeffrey W. Bier*
126. Food Process Design, *Zacharias B. Maroulis and George D. Saravacos*
127. Handbook of Dough Fermentations, *edited by Karel Kulp and Klaus Lorenz*

128. Extraction Optimization in Food Engineering, *edited by Constantina Tzia and George Liadakis*
129. Physical Properties of Food Preservation: Second Edition, Revised and Expanded, *Marcus Karel and Daryl B. Lund*
130. Handbook of Vegetable Preservation and Processing, *edited by Y. H. Hui, Sue Ghazala, Dee M. Graham, K. D. Murrell, and Wai-Kit Nip*
131. Handbook of Flavor Characterization: Sensory Analysis, Chemistry, and Physiology, *edited by Kathryn Deibler and Jeannine Delwiche*
132. Food Emulsions: Fourth Edition, Revised and Expanded, *edited by Stig E. Friberg, Kare Larsson, and Johan Sjoblom*
133. Handbook of Frozen Foods, *edited by Y. H. Hui, Paul Cornillon, Isabel Guerrero Legarret, Miang H. Lim, K. D. Murrell, and Wai-Kit Nip*
134. Handbook of Food and Beverage Fermentation Technology, *edited by Y. H. Hui, Lisbeth Meunier-Goddik, Ase Solvejg Hansen, Jytte Josephsen, Wai-Kit Nip, Peggy S. Stanfield, and Fidel Toldrá*
135. Genetic Variation in Taste Sensitivity, *edited by John Prescott and Beverly J. Tepper*
136. Industrialization of Indigenous Fermented Foods: Second Edition, Revised and Expanded, *edited by Keith H. Steinkraus*
137. Vitamin E: Food Chemistry, Composition, and Analysis, *Ronald Eitenmiller and Junsoo Lee*
138. Handbook of Food Analysis: Second Edition, Revised and Expanded, Volumes 1, 2, and 3, *edited by Leo M. L. Nollet*
139. Lactic Acid Bacteria: Microbiological and Functional Aspects: Third Edition, Revised and Expanded, *edited by Seppo Salminen, Atte von Wright, and Arthur Ouwehand*
140. Fat Crystal Networks, *Alejandro G. Marangoni*
141. Novel Food Processing Technologies, *edited by Gustavo V. Barbosa-Cánovas, M. Soledad Tapia, and M. Pilar Cano*
142. Surimi and Surimi Seafood: Second Edition, *edited by Jae W. Park*
143. Food Plant Design, *Antonio Lopez-Gomez; Gustavo V. Barbosa-Cánovas*
144. Engineering Properties of Foods: Third Edition, *edited by M. A. Rao, Syed S.H. Rizvi, and Ashim K. Datta*
145. Antimicrobials in Food: Third Edition, *edited by P. Michael Davidson, John N. Sofos, and A. L. Branen*
146. Encapsulated and Powdered Foods, *edited by Charles Onwulata*
147. Dairy Science and Technology: Second Edition, *Pieter Walstra, Jan T. M. Wouters and Tom J. Geurts*
148. Food Biotechnology, Second Edition, *edited by Kalidas Shetty, Gopinadhan Paliyath, Anthony Pometto and Robert E. Levin*
149. Handbook of Food Science, Technology, and Engineering - 4 Volume Set, *edited by Y. H. Hui*
150. Thermal Food Processing: New Technologies and Quality Issues, *edited by Da-Wen Sun*
151. Aflatoxin and Food Safety, *edited by Hamed K. Abbas*
152. Food Packaging: Principles and Practice, Second Edition, *Gordon L. Robertson*

FOOD BIOTECHNOLOGY

edited by
Kalidas Shetty
Gopinadhan Paliyath
Anthony Pometto
Robert E. Levin

Second Edition

Taylor & Francis
Taylor & Francis Group
Boca Raton London New York

A CRC title, part of the Taylor & Francis imprint, a member of the Taylor & Francis Group, the academic division of T&F Informa plc.

Published in 2006 by
CRC Press
Taylor & Francis Group
6000 Broken Sound Parkway NW, Suite 300
Boca Raton, FL 33487-2742

Printed in the United States of America on acid-free paper
10 9 8 7 6 5 4 3 2

International Standard Book Number-10: 0-8247-5329-1 (Hardcover)
International Standard Book Number-13: 978-0-8247-5329-0 (Hardcover)

Library of Congress Cataloging-in-Publication Data

Catalog record is available from the Library of Congress

T&F informa

Taylor & Francis Group
is the Academic Division of T&F Informa plc.

Visit the Taylor & Francis Web site at
http://www.taylorandfrancis.com

and the CRC Press Web site at
http://www.crcpress.com

Preface

Major challenges facing the world today are not just those of food production and food quality for meeting protein and calorie needs (basic nutritional needs), but also those related to better health. A significant challenge is the outbreak of oxidation-linked disease epidemics caused by calorie sufficiency and excess calories. This nutritional epidemic is occurring not only in the developed world, but also in newly industrialized countries such as China, Brazil, Mexico, and India, which have the most rapidly growing type 2 diabetes problem in the world. Because diabetes is tied to other oxidation-linked diseases such as cardiovascular disease (CVD) and cancer, it will inevitably place a tremendous burden on the emerging health-care systems of these countries. This situation will place further strain and pressure on existing health-care challenges such as treatment of infectious diseases like AIDS, tuberculosis, and foodborne illnesses among lower income populations. In the more developed countries, the continuous and steady advancement of obesity and its subsequent consequences of increases in diabetes, CVD, and cancer are posing additional challenges. All major health challenges, be they of calorie sufficiency- or calorie insufficiency-linked infectious diseases, are directly or indirectly related to diet and environmentally linked disease. Therefore, technologies for disease chemoprevention through dietary modification (reduced calorie intake with more fruits and vegetables and novel ingredients from other food-grade biological and microbial systems) will be very important to help manage these emerging health-care challenges. In addition, advances in food biotechnology must be more nutritionally relevant and must consider the environmental impacts and consequences of food production and consumption.

Thus, with these critical issues in mind, *Food Biotechnology*, 2nd Edition, has been assembled with the hope of being an authoritative, comprehensive, conceptually sound, and highly informative compilation of recent advances in various important areas of food biotechnology. The topics herein deal with bioconversion of food raw materials to processed products, improvement of food quality, food safety, designing of ingredients for functional foods, biochemical advances in traditional fermentation, and, most importantly, they provide an international perspective to the whole field. Biotechnology has become an important tool in recent years, and several scientists across the world are investigating advanced and novel biological, cellular, molecular, and biochemical strategies for improving food production and processing, for enhancing food safety and quality, and for improving from organoleptic to functional aspects of food and food ingredients for better human health. Thus, this volume has amassed diverse topics from appropriate experts in specific areas from across the globe. The book is divided into three sections. The first section deals with food microbiology, the second with plant and animal food applications and functional foods, and the third section deals with food safety, novel bioprocessing, traditional fermentations, and regulatory and patent issues at an international level. In all, there are 70 chapters covering key areas of food biotechnology within the three sections. The first 20 chapters in Section 1, dealing with food microbiology, provide in-depth accounts of basic principles of microbiology, fermentation technologies, aspects of genetic engineering for production of various food ingredients, and several other specialized topics involving microbial systems. Section 2, comprising 27 chapters, is quite diverse and deals with plant tissue culture techniques, genetic engineering of plants and animals, functional food ingredients and their health benefits, probiotics, antibody production for oral vaccines, and several topics on enzyme technologies. Section 3, with 23 chapters, is quite diverse and examines several aspects of food safety issues, bioprocessing, and fermentation biotechnologies used across the globe. In essence, this book has brought together diverse areas of food biotechnology with a strong focus on biochemistry and molecular biology, and it is unique in that respect. This strong molecular- and biochemically based conceptual viewpoint provided by many chapters will form the basis for development of food biotechnology over the next few decades, particularly in the context of designing food ingredients for better health and microbial food safety.

The editors wish to thank all the authors for their outstanding efforts to document and present their research and their conceptual information about their current understanding of this field. Their efforts have particularly advanced our conceptual knowledge with regard to food safety, novel microbial processing, novel applications of plant foods and ingredients, and functional food ingredients.

The editors also would like to thank the staff of Marcel Dekker, CRC, and Taylor and Francis for their help and support in the timely publication of this 2nd edition, and particularly for coordinating the work of the authors of 70 chapters across several countries. All these efforts have advanced the frontiers of food biotechnology and have given it a stronger molecular, metabolic, biochemical, and nutritionally relevant emphasis that is conceptually applicable in any part of the world.

The Editorial Board

Editors

Dr. Kalidas Shetty is a professor of food biotechnology in the Department of Food Science at the University of Massachusetts-Amherst. He received his BS degree from the University of Agricultural Sciences, Bangalore, India, majoring in applied microbiology, and MS and PhD from the University of Idaho, Moscow in microbiology. He then pursued postdoctoral studies in plant biotechnology in Japan (National Institute of Agro-Biological Sciences, Tsukuba Science City) and Canada (University of Guelph) prior to joining the University of Massachusetts in 1993.

Dr. Shetty's research interests focus on redox pathway-linked biochemical regulation of phenolic phytochemicals in food botanicals using novel tissue culture, seed sprout, and fermentation systems. This focus is contributing to innovative advances in the areas of nutraceuticals, functional foods, and food antimicrobial strategies. In particular, the susceptibility of bacterial food pathogens to phenolic phytochemicals at low pH through redox-linked pathways is his major interest in developing new food safety strategies. He has published over 100 manuscripts in peer-reviewed journals and over 25 as invited reviews and in conference proceedings. He holds four United States patents.

Dr. Shetty was appointed as the editor of the journal *Food Biotechnology*, published by Marcel Dekker (now Taylor and Francis). He is also on the editorial board of three additional journals in the areas of food and environmental sciences.

In 2004, Professor Shetty was selected by the U.S. State Department as a Jefferson Science Fellow to advise the Bureau of Economic and Business Affairs on scientific issues as they relate to international diplomacy and international development. This program, administered by the U.S. National Academies, allowed Dr. Shetty to serve as science advisor at the U.S. State Department for 1 year in 2004–2005, and he will continue to serve as science advisor for 5 more years following his return to the University of Massachusetts. Dr. Shetty has widely traveled and has been invited to present lectures and seminars in the areas of food biotechnology, functional foods and dietary phytochemicals, and food safety in over 20 countries in Asia, Europe, and the Americas. In 1998 he was awarded the Asia-Pacific Clinical Nutrition Society Award for his contributions to the area of phytochemicals, functional foods, and human health based on his understanding of Asian food traditions. At the University of Massachusetts he has won the College of Food and Natural Resources Outstanding Teaching Award and a Certificate of Achievement for Outstanding Outreach Contributions.

Dr. Anthony L. Pometto is a professor of industrial microbiology in the Department of Food Science and Human Nutrition at Iowa State University. He received his BS degree in biology from George Mason University, Fairfax, Virginia, and his MS and PhD in bacteriology from the University of Idaho, Moscow, Idaho. Dr. Pometto worked as a full-time scientific aide in the Department of Bacteriology and Biochemistry at the University of Idaho for twelve years. He joined the faculty at Iowa State University in 1988.

Dr. Pometto's research interests focus on microbial degradation of degradable plastics, bioconversion of agricultural commodities into value-added products via fermentation, development of novel bioreactors, production of enzymes for the food industry, and the utilization of food industrial wastes. He has co-authored over 60 peer-reviewed journal articles and over 25 articles as invited reviews, book chapters, and conference proceedings. He is a co-inventor on three United States patents. He is also a member of editorial board of the journal *Food Biotechnology*, published by Marcel Dekker (now Taylor and Francis).

Dr. Pometto became director of the NASA Food Technology Commercial Space Center at Iowa State University in 2000. The Center is associated with the NASA Johnson Space Center, Houston, Texas, which manages all the food systems for the shuttle, International Space Center, and planetary exploration missions. The NASA Food Technology Commercial Space Center at Iowa State University was founded in August 1999 and has the mission to engage industry and academia to develop food products and processes that will benefit NASA and the public. The specific objectives are as follows: (1) to develop food products that meet the shelf life requirements for the shuttle, ISS and the planetary outpost, which are nine months, one year, and five years, respectively; (2) to develop equipment and process technologies to convert the proposed over 15 crops grown on the planetary outpost, Moon or Mars, into safe, edible foods; and (3) to build partnerships with food companies to develop these new food products and processes to make them available for NASA utilization. The space food challenges being addressed by the Center's commercial partners and affiliate faculty are development of new food products, development of new food processing equipment, extending the shelf life of foods, improving and monitoring food safety, packaging of foods, development of food waste management systems, and development of disinfection systems for space travel. For more information, please see the web site http://www.ag.iastate.edu/centers/ftcsc/.

Dr. Pometto has recently been named associate director of the Iowa State University Institute for Food Safety and Security, which was created in 2002 as one of six presidential academic initiatives. Dr. Pometto works with the Institute's director, Dr. Manjit Misra, to bring together the research, education, and outreach components of food safety and security at Iowa State University into one umbrella institute for the purposes of efficient teamwork that is well-positioned among government, industry, and producers.

Dr. Gopinadhan Paliyath is an associate professor at the Department of Plant Agriculture, University of Guelph, Ontario, Canada. Dr. Paliyath has a very broad background in plant science, with a specialization in biochemistry. He obtained his BS. Ed. degree (botany and chemistry) in science education from the University of Mysore, MS degree (botany) from the University of Calicut, and PhD degree (biochemistry) from the Indian Institute of Science, Bangalore. He did postdoctoral work at Washington State University, the University of Waterloo, and the University of Guelph.

The focus of Dr. Paliyath's current research is in the areas of post-harvest biology and technology, functional foods, and nutraceuticals. He is investigating the signal transduction events in response to ethylene and the role of phospholipase D in such events. Various aspects dealing with improvement in fruit and vegetable shelf life and quality, and the efficacy of functional food ingredients are also being investigated. Technologies and products have been developed for enhancing the shelf life and quality of fruits, vegetables, and flowers based on phospholipase D inhibition (US Patent #6,514,914). As well, he is developing novel technologies for the isolation of active nutraceutical fractions and the use of nutraceuticals to enhance the functional food value of processed fruits and vegetables. Phospholipase D inhibition technology for fruit and vegetable preservation has been licensed for commercialization. His current research also includes investigations into the mechanism of action of nutraceuticals (grape and wine polyphenols) as cancer chemopreventive agents.

Dr. Paliyath has authored over 150 research contributions that include peer-reviewed publications (60), patents at various stages (one issued, one on file, one in process), chapters in books (10), and several conference presentations and reports.

Contributors

Motoyasu Adachi
Laboratory of Food Quality Design and Development
Graduate School of Agriculture
Kyoto University
Uji, Japan

Jaime Aguilera
Department of Biotechnology
Instituto de Agroquímica y Tecnología de los Alimentos
Valencia, Spain

Alejandro Amezquita
Safety and Environmental Assurance Centre
Unilever R&D Colworth
Sharnbrook, Bedfordshire, UK

Hákon Örn Birgisson
Islenskar Lyfjarannsóknir - Encode
Reykjavik, Iceland

Linda F. Bisson
Department of Viticulture and Enology
University of California
Davis, California, USA

Hans P. Blaschek
Department of Food Science & Human Nutrition
University of Illinois
Urbana, Illinois, USA

Kathryn J. Boor
Department of Food Science
Cornell University
Ithaca, New York, USA

Mindy M. Brashears
Department of Animal and Food Sciences
Texas Tech University
Lubbock, Texas, USA

Dario Campagnone
Dipartimento di Scienze degli Alimenti
Università di Teramo
Teramo, Italy

Daniel J. Cantliffe
Horticultural Sciences Department
Institute of Food and Agricultural Sciences
University of Florida
Gainesville, Florida, USA

Michele Del Carlo
Dipartimento di Scienze degli Alimenti
Università di Teramo
Teramo, Italy

Tamara Casci
School of Food Biosciences
The University of Reading
Whiteknights, UK

Rod Casey
Department of Metabolic Biology
John Innes Centre
Norwich, UK

Rupsankar Chakrabarti
Central Institute of Fisheries Technology
Visakhapatnam Research Centre
Visakhapatnam, India

Feng Chen
Department of Botany
The University of Hong Kong
Hong Kong, China

Thomas T. Chen
Department of Molecular and Cell Biology
University of Connecticut
Storrs, Connecticut, USA

Pinwen Peter Chiou
Department of Molecular and Cell Biology
University of Connecticut
Storrs, Connecticut, USA

Hanne Risager Christensen
BioCentrum-DTU
Biochemistry and Nutrition
The Technical University of Denmark
Kgs. Lyngby, Denmark

Fergus M. Clydesdale
Department of Food Science
Chenoweth Laboratory
University of Massachusetts
Amherst, Massachusetts, USA

R. Stephen Crespi
British Technology Group
London, UK

Elisabeth Csöregi
Department of Biotechnology
Centre for Chemistry and Chemical
Engineering, Lund University
Lund, Sweden

Ali Demirci
Deptartment of Agricultural and
Biological Engineering
The Hucks Institute of Life Sciences
Pennsylvania State University
University Park, Pennsylvania, USA

Hortense Dodo
Food Biotechnology Laboratory
Department of Food & Animal Sciences
Alabama Agricultural and Mechanical
University
Normal, Alabama, USA

Gilles Feron
Laboratoire de Microbiologie
UMR UB INRA
Ensbana, Dijon, France

K. Helen Fisher
Department of Plant Agriculture
University of Guelph
Vineland, Ontario, Canada

Hanne Frøkiær
BioCentrum-DTU
Biochemistry and Nutrition
The Technical University
of Denmark
Kgs. Lyngby, Denmark

Glenn R. Gibson
School of Food Biosciences
The University of Reading
Whiteknights, UK

Ramon Gonzalez
Departments of Chemical
Engineering and Food Science
& Human Nutrition
Iowa State University
Ames, Iowa, USA

Michael Gray
Department of Food Science
Cornell University
Ithaca, New York, USA

Plinio Guzmán
Departamento de Ingeniería Genética de
Plantas
Centro de Investigación y de Estudios
Avanzados del IPN
Unidad Irapuato
Gto, México

Rajni Hatti-Kaul
Department of Biotechnology
Center for Chemistry and Chemical
Engineering
Lund University
Lund, Sweden

Haijing Hu
Department of Food Science
and Technology
Cornell University
Geneva, New York, USA

Richard K. Hughes
Department of Metabolic Biology
John Innes Centre
Norwich Research Park
Norwich, UK

Divya Jaroni
9440 Forest Glen Dr.
Lincoln, Nebraska, USA

B. Sandesh Kamath
Plant Cell Biotechnology Department
Central Food Technological Research
Institute
Mysore, India

N. Ganesh Karanth
Fermentation Technology and
Bioengineering
Central Food Technological Research
Institute
Mysore, India

Peter L. Keeling
BASF Plant Science
Ames Research
Ames, Iowa, USA

Jenny Khoo
Department of Molecular and Cell
Biology
University of Connecticut
Storrs, Connecticut, USA

Kieran Kilcawley
Teagasc, Dairy Products Research Centre
Fermoy, County Cork, Ireland

Anthony J. Kinney
Crop Genetics Research and Development
DuPont Experimental Station
Wilmington, Delware, USA

Harry H. Klee
Horticultural Sciences Department
Institute of Food and Agricultural
Sciences
University of Florida
Gainesville, Florida, USA

Jeffrey D. Klucinec
BASF Plant Science
Ames Research
Ames, Iowa, USA

Koffi Konan
Food Biotechnology Laboratory
Department of Food & Animal Sciences
Alabama A&M University
Normal, Alabama, USA

Roger A. Korus
Department of Chemical Engineering
University of Idaho
Moscow, Idaho, USA

Parthena Kotzekidou
Department of Food Science and
Technology
Aristotle University of Thessaloniki
Thessaloniki, Greece

Jennifer Kovacs-Nolan
Department of Food Science
University of Guelph
Guelph, Ontario, Canada

Reinhard Krämer
Institute of Biochemistry
University of Köln
Zülpicher, Germany

Hordur G. Kristinsson
Department of Food Science and Human
Nutrition
University of Florida
Gainesville, Florida, USA

Robert E. Levin
Department of Food Science
University of Massachusetts
Amherst, Massachusetts, USA

Yuan-Tong Lin
Department of Food Science
Chenoweth Laboratory
University of Massachusetts,
Amherst, Massachusetts, USA

Susan Lindquist
Whitehead Institute for Biomedical Research
Cambridge, Massachusetts, USA

John E. Linz
Department of Food Science and Human Nutrition
Michigan State University
East Lansing, Michigan, USA

Xue-Jun Liu
Department of Botany
The University of Hong Kong
Hong Kong, China

John R. Lupien
Department of Food Science
University of Massachusetts
Amherst, Massachusetts, USA

Evelyn Mae Tecson-Mendoza
Institute of Plant Breeding, College of Agriculture
University of the Philippines
Los Banos
Laguna, Philippines

V. Maitin
School of Food Biosciences
The University of Reading
Whiteknights, Reading, UK

Nobuyuki Maruyama
Graduate School of Agriculture
Kyoto University
Uji, Japan

Yukie Maruyama
Graduate School of Agriculture
Kyoto University
Uji, Japan

Bo Mattiasson
Department of Biotechnology
Centre for Chemistry and Chemical Engineering
Lund University
Lund, Sweden

Patrick P. McCue
Program in Molecular and Cellular Biology
University of Massachusetts
Amherst, Massachusetts, USA

Lynne McLandsborough
Food Science Department
University of Massachusetts
Amherst, Massachusetts, USA

Michael J. Miller
Department of Food Science
North Carolina State University
Raleigh, North Carolina, USA

Yoshinori Mine
Department of Food Science
University of Guelph
Guelph, Ontario, Canada

Matthew M. Moake
Department of Food Science
and Technology
New York State Agricultural Experiment Station
Cornell University
Geneva, New York, USA

A. Satyanarayan Naidu
N-terminus Research Laboratory
Pomona, California, USA

K. K. Namitha
Plant Cell Biotechnology Department
Central Food Technological Research Institute
Mysore, India

Kendra Kerr Nightingale
Department of Food Science
Cornell University
Ithaca, New York, USA

Mihaela Nistor
Centre for Chemistry and Chemical Engineering
Department of Biotechnology
Lund University
Lund, Sweden

Hector G. Nunez-Palenius
Horticultural Sciences Department
Institute of Food and Agricultural Sciences, University of Florida
Gainesville, Florida, USA

Neftali Ochoa-Alejo
Unit of Biotechnology and Plant Genetic Engineering
Center of Research and Advanced Studies
National Polytechnic Institute
Irapuato, Gto., Mexico

Moustapha Oke
Ontario Ministry of Agriculture and Food
Guelph, Ontario, Canada

N. A. Olasupo
Department of Microbiology
Lagos State University
Ojo, Lagos, Nigeria

Gabriela Olmedo
Departamento de Ingeniería Genética de Plantas
Centro de Investigación y de Estudios Avanzados del IPN
Unidad Irapuato
Gto., México

Daniel J O'Sullivan
Department of Food Science and Nutrition
Center for Microbial and Plant Genomics
Minnesota, USA

Juan Alberto Osuna-Castro
Laboratorio de Biotecnología
Facultad de Ciencias Biológicas y Agropecuarias
Universidad de Colima
Tecoman, Colima, México

Tiago F. Outeiro
Whitehead Institute for Biomedical Research
Cambridge, Massachusetts, USA

Gopinadhan Paliyath
Department of Plant Agriculture
University of Guelph
Guelph, Ontario, Canada

Ashok Pandey
Biotechnology Division
Regional Research Laboratory CSIR
Trivandrum, India

Octavio Paredes-López
Centro de Investigación y de Estudios Avanzados del IPN
Apdo, Gto., México

Socorro Parra
Departamento de Ingeniería Genética de Plantas, Centro de Investigación y de Estudios Avanzados del IPN
Unidad Irapuato, Gto., México

Eugenio Perez-Molphe
Chemistry Department
Center of Basic Sciences
Autonomous University of Aguascalientes
Ags, Mexico

Reena Grittle Pinhero
Department of Food Science
University of Guelph
Guelph, Ontario, Canada

Anthony L. Pometto
Department of Food Science and Human Nutrition, NASA Food Technology Commercial Space Center
Iowa State University
Ames, Iowa, USA

Jose Antonio Prieto
Department of Biotechnology
Instituto de Agroquímica y Tecnología de los
Alimentos, Valencia, Spain

Vernon G. Pursel
U. S. Department of Agriculture
Agricultural Research Service
Beltsville Agricultural Research Center
Biotechnology and Germplasm
Laboratory
Beltsville, Maryland, USA

Nasib Qureshi
United States Department of Agriculture
National Center for Agricultural Utilization
Fermentation/Biotechnology Research
Peoria, Illinois, USA

A. Eugene Raj
Fermentation Technology & Bioengineering
Central Food Technological Research
Institute
Mysore, India

Sumitra Ramachandran
Department of Chemical and Biochemical
Engineering, University Blaise Pascal
Aubiere, France

Rafael Ramirez-Malagon
Institute of Agricultural Sciences
University of Guanajuato
Irapuato, Gto., Mexico

Francisca Randez-Gil
Department of Biotechnology
Instituto de Agroquímica y Tecnología de
los, Alimentos
Valencia, Spain

Reena Randhir
Department of Food Science
Chenoweth Laboratory
University of Massachusetts
Amherst, Massachusetts, USA

E. Rati Rao
Human Resource Development
Department of Food Microbiology
Central Food Technological Research
Institute
Mysore, India

Robert A. Rastall
School of Food Biosciences
The University of Reading
Whiteknights, UK

Colin Ratledge
Department of Biological Sciences
University of Hull
Hull, UK

K. Ravi-Kumar
Department of Food Microbiology,
Central Food Technological Research
Institute
Mysore, India

G.A. Ravishankar
Plant Cell Biotechnology Department
Central Food Technological Research
Institute
Mysore, India

T. Ritu
Plant Cell Biotechnology Department
Central Food Technology Research
Institute
Mysore, India

Louis A. Roberts
Pioneer Valley Life Sciences Institute
Springfield, Massachusetts, USA

T. Roukas
Department of Food Science and
Technology
Aristotle University of Thessaloniki
Thessaloniki, Greece

Gerhard Sandmann
Botanical Institute
J. W. Goethe Universität
Frankfurt, Germany

R. Sarada
Plant Cell Biotechnology Department
Central Food Technological Research
Institute
Mysore, India.

William K. Shaw Jr.
USDA Food Safety and Inspection
Service
Washington, DC, USA

Kalidas Shetty
Department of Food Science
University of Massachusetts
Amherst, Massachusetts, USA

Preethi Shetty
Department of Food Science
Chenoweth Laboratory
University of Massachusetts
Amherst, Massachusetts, USA

Mike A. Singer
Department of Neurology
University of Texas Southwestern Medical Center
Dallas, Texas, USA

Sallie Smith-Schneider
Pioneer Valley Life Sciences Institute
Springfield, Massachusetts, USA

B. Suresh
Plant Cell Biotechnology Department
Central Food Technology Research Institute
Mysore, India

Ian W. Sutherland
Institute of Cell and Molecular Biology
Edinburgh University
Edinburgh, UK

Effie Tsakalidou
Department of Food Science and Technology
Agricultural University of Athens
Iera Odos, Athens, Greece

Joseph Tulpinski
N-terminus Research Laboratory
Pomona, California, USA

S. Umesh-Kumar
Department of Food Microbiology
Central Food Technological Research Institute
Mysore, India

Ragip Unal
N-terminus Research Laboratory
Pomona, California, USA

Shigeru Utsumi
Graduate School of Agriculture
Kyoto University
Uji, Japan

M.C. Varadaraj
Central Food Technological Research Institute
Mysore, India

Dhiraj A. Vattem
Nutritional Biomedicine and Biotechnology FCS Department
Texas State University-San Marcos
San Marcos, Texas, USA

K.S. Venkatesh
Department of Food Microbiology
Central Food Technological Research Institute
Mysore, India

S.V.N. Vijayendra
Department of Food Microbiology
Central Food Technological Research Institute
Mysore, India

Olga Viquez
Food Biotechnology Laboratory
Department of Food and Animal Sciences
Alabama Agricultural and Mechanical University
Normal, Alabama, USA

Y. Waché
Laboratoire de Microbiologie
UMR UB INRA
Ensbana, Dijon, France

Changlu Wang
School of Food Engineering and Biotechnology
Tianjin University of Science and Technology
Tianjin, China

Martin Wiedmann
Department of Food Science
Cornell University
Ithaca, New York, USA

Katy Windham
Department of Food Science
Cornell University
Ithaca, New York, USA

Randy W. Worobo
Department of Food Science and Technology, New York State Agricultural Experiment Station, Cornell University
Geneva, New York, USA

James P. Wynn
Martek Biosciences Corporation
Columbia, Maryland, USA

Yang Zheng
Tianjin C.I. Food Co. Ltd.
Tianjin, China

Zuoxing Zheng
Kraft Foods Research
Global Technology and Quality
Glenview, Illinois, USA

Contents

Section 1

Food Microbiology

1.01

Food Microbiology

Robert E. Levin

CONTENTS

1.1 INTRODUCTION

Food biotechnology integrates biochemistry, chemistry, microbiology, and chemical engineering for the enhanced production of food products. The application of microbiology to food systems encompasses methods involved in the assessment of microbial food safety and the use of microorganisms for the production of foods and beverages, food products, food additives. Microorganisms involved either directly or indirectly with food systems include bacteria, molds, yeasts, and algae. Each of these microbial groups has unique metabolic aspects that are either utilized or circumvented to achieve optimization of various microbial processes.

1.2 GENERAL ASPECTS

1.2.1 Applications of Microbiology to Foods

Ancient Egyptians used fermentation to produce beer and convert grape juice to wine. They also practiced the aerobic conversion of the alcohol in wine to the acetic acid of vinegar, and the leavening of bread. The present practices of using, for example, pectinases for enhanced release of fruit juices from tissue and amylases for the enzymatic modification of starches, are examples involving the indirect application of microorganisms to foods and food components. The production of xanthan gum by the plant pathogenic bacterium *Xanthomonas campestris* for use as a viscosity agent in beverages and semisolid food products is an example of the use of an originally undesirable organism for the production of a desirable food and beverage additive. The use of the mold *Aspergillus niger* to produce high yields of citric acid as a food and beverage acidulant was established in the 1920s and is a classic example of an initial surface culture process that was eventually converted to a submerged aerated process with the use of mutants.

1.2.2 The Nature of Microorganisms

Microscopic organisms are presently divided into three major groups: (1) Eubacteria (bacteria), which lack a discernible nucleus and mitochondria; (2) Archaebacteria (bacteria), which also lack a discernible nucleus and mitochondria; and (3) Eukaryotes (yeasts, molds, algae, and protozoa), which possess both a clearly discernible nucleus and mitochondria, plus filamentous structures known as endothelial reticulum. Mitochondria are self-replicating organelles and contain their own deoxyribonucleic acid (DNA), referred to as mitochondrial DNA. In Eukaryotes, the cytochrome and tricarboxylic acid (TCA) enzymes required for aerobic synthesis of ATP are located in the mitochondrial membrane, while with prokaryotes and Archaebacteria the cytochromes are in the cytoplasmic membrane and the TCA enzymes are in the cytoplasm.

All microorganisms are allocated to a specific group with respect to growth temperature. Obligate psychrophiles are defined as those organisms capable of growth at or near 0°C but not at 20°C. Such organisms usually have a maximum growth temperature of 15–17°C. Psychrotrophic organisms are capable of growth at or near 0°C but exhibit optimum growth at approximately 25°C and are frequently unable to grow at 30°C. Mesophiles exhibit growth from 20–45°C with an optimum growth temperature usually in the range of 30–35°C. Thermophiles exhibit growth in the range of 45–65°C. Hyperthermophiles are organisms from oceanic thermal vents and hot springs that are restricted to growth temperatures from 70–120°C. Hyperthermophiles have not yet been isolated from foods.

1.3 FUNGI

1.3.1 Fungal Cell Walls

The fungal cell wall is composed mainly of carbohydrates together with some protein and lipids. The cytoplasmic membrane, unlike the membrane of bacteria, contains sterols. The most important carbohydrates are mannan, glucan, chitin, and cellulose. The wall of some molds is primarily a chitin–glucan structure, whereas mannan is more predominant in yeasts, resulting in mannan–chitin or mannan–glucan cell wall structures. The digestive juice of the garden snail *Helix pomatia,* available commercially as glusulase, is high in β-1, 3- and β-1, 6-glucanase activity and is frequently used to digest the cell wall of molds and some yeasts. Novozyme 234 (Novo Industries) will yield protoplasts of *Aspergillus* and *Penicillium* (1). Novozyme 234 is notably effective for digesting the cell wall of *Schizosaccharomyces pombe*, while glusulase Nee-154 (DuPont; Endo Laboratories) is used with *Saccharomyces cerevisiae*. Both Novozyme 234 and the yeast lytic enzyme from *Arthrobacter luteus* (ICN Biomedicals), otherwise known as lyticase or zymolase (Sigma), are effective for yielding spheroplasts of *Yarrowia lipolytica* (formerly *Candida lipolytica*) (2). Yeasts and molds harvested from the exponential phase of growth are more sensitive to the activity of these cell wall digesting enzyme preparations than are late exponential or stationary phase cells.

1.3.2 Yeasts

Yeasts can be divided into two metabolic groups: facultative anaerobes and obligate aerobes. The facultative anaerobes are capable of anaerobic growth and fermentative conversion of sugars to ethyl alcohol, CO_2, and cell mass, in addition to the aerobic conversion of sugars to CO_2 and H_2O, and much higher yields of cell mass. Fermentative yeasts such as *S. cerevisiae* (Figure 1.1), when grown in the presence of 3 ppm of the DNA intercolating agent acriflavine, can have their mitochondrial DNA selectively mutated so that mitochondria are eliminated, resulting in obligately fermentative strains unable to utilize oxygen (3). Such strains produce smaller cells than wild-type strains and result in "petite" colonies that are notably reduced in size.

Baker's yeast was originally obtained from the brewing industry; the top yeast *S. cerevisiae* was conveniently skimmed from the top surface of fermentation tanks. During the mid-1800s the brewing industry converted to strains of the bottom-settling yeast *Saccharomyces carlesburgensis,* which precipitated the establishment of the baker's yeast industry. Producing baker's yeast using sucrose derived from molasses requires vigorous aeration of the culture medium so that a maximum amount of carbon flows to cell mass production and not to ethyl alcohol formation. Vigorous aeration of *S. cerevisiae* strains in the presence of an abundant level of carbohydrate (about 3%) results in the metabolic dominance of fermentation and is known as the crabtree effect (4). This in turn results in a significant level of ethyl alcohol and a notably reduced level of cell mass. The baker's yeast industry is able to overcome the crabtree effect using incremental feeding which involves the pulsed addition of molasses to aerated culture tanks, so that at no time does the residual level of sucrose rise above 0.0001%. Thus there is no feedback repression of mitochondria formation caused by elevated levels of sucrose. In this case, derepressed mutants that do not exhibit feedback repression are not used. The yeast *Candida utilis* is facultatively anaerobic; however, under conditions of vigorous aeration and elevated sugar levels the crabtree effect is not observed. Thus, the organism can be conveniently used to convert the lactose in whey and the sugars in sulfite waste liquor to cell mass for use as food and fodder yeast.

All yeasts are capable of utilizing glucose. The utilization of other sugars depends on the species; the spectrum of sugars used constitutes a major criterion for the identity of

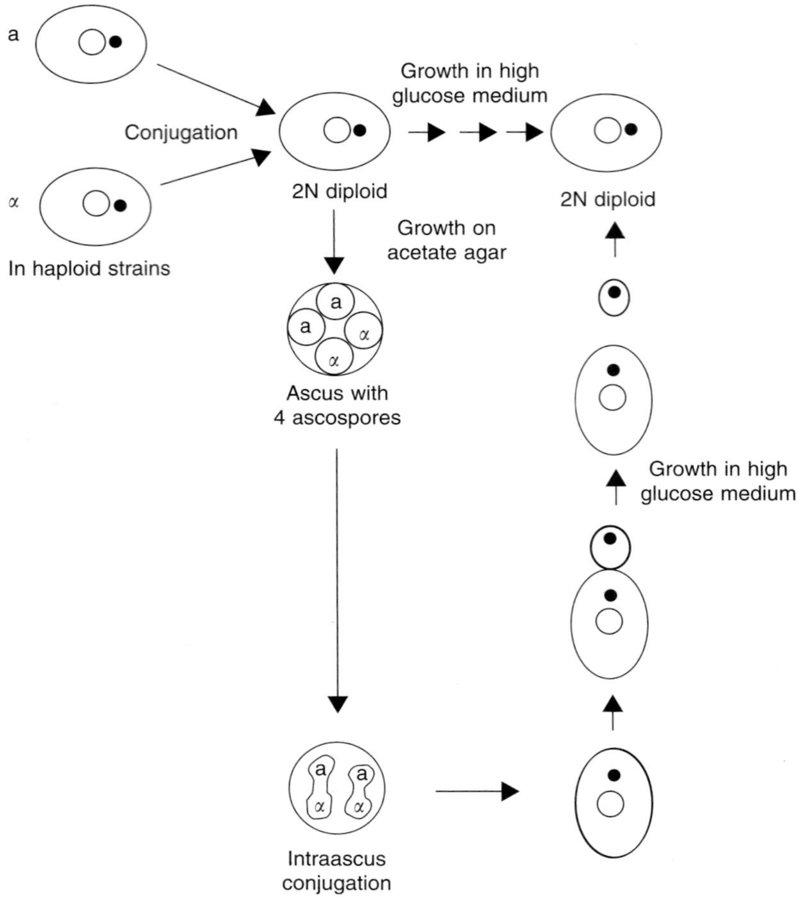

Figure 1.1 Life cycle of *Saccharomyces cerevisiae.*

yeasts. All yeasts are capable of utilizing ammonium sulfate as a sole source of nitrogen. Very few yeasts are capable of utilizing nitrate as a sole nitrogen source. Among ascospore-producing yeasts, the number (1, 4, or 8) and shape of ascospores (spherical, oval, kidney, hat, saturn, needle) in asci constitutes an additional major criterion for genus and species identity. Most yeasts divide by budding; however, members of the strongly fermentative yeast genus *Schizosaccharomyces* divide solely by transverse fission (Figure 1.2).

1.3.3 Molds

In developing mold cultures for the production of food additives, it is important to keep in mind that all molds are obligate aerobes. The maximum production of primary metabolites (e.g., amino acids) and secondary metabolites (e.g., extraceullular enzymes) invariably occurs with wild-type cultures under the condition of static surface growth. This contrasts with submerged cultivation which invariably involves the use of selected mutants. Molds are classified into four classes. The Phycomycetes do not have complete cross walls in their hyphae and therefore exhibit unidirectional protoplasmic streaming (coenocytic movement) or flow throughout their hyphae. Phycomycetes also possess the unifying characteristic of producing aerially borne asexual fruiting structures known as sporangia, with internal sporangiospores borne on a bulblike structure referred to as the

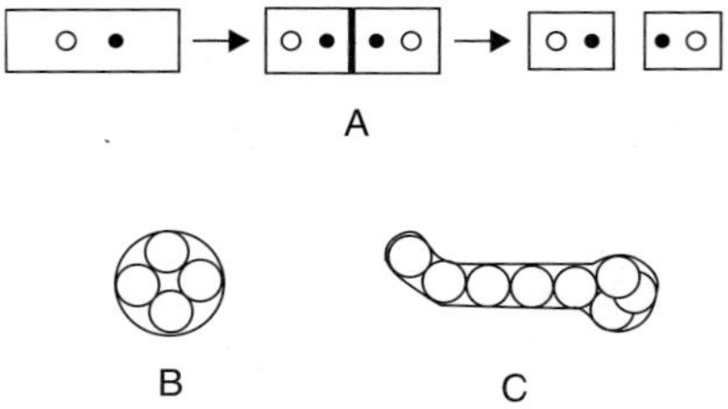

Figure 1.2 Cellular structures and asci of the strongly fermentative yeast genus *Schizosaccharomyces*. (A) Transverse fission of vegetative cells exhibited by all members of the non-budding genus *Schizosaccharomyces*. (B) Ascus containing four ascospores representative of *Schiz. pombe* and *Schiz. versatilis.* (C) Swollen and distended ascus of *Schiz. octosporus* containing eight ascospores.

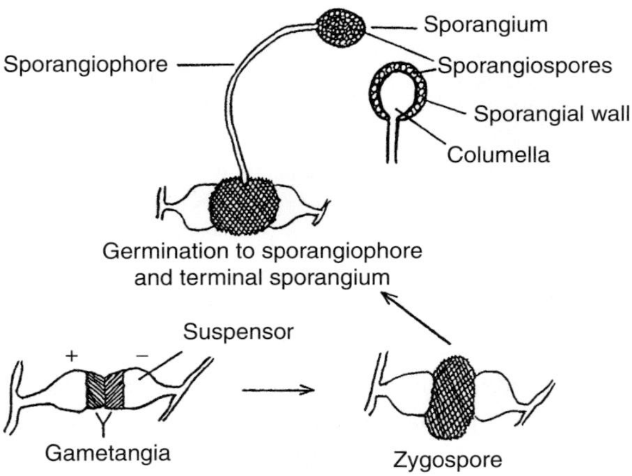

Figure 1.3 Asexual and sexual structures of a typical terrestrial member of the class Phycomycetes.

columella (Figure 1.3). Some, but not all, Phycomycetes produce a sexual spore, known as a zygospore, derived from the fusion of opposite mating types which occurs freely in culture media (Figure 1.3). Color and the microscopic orientation and appearance of these structures are used to establish genera and species. The class Ascomycetes houses fungi (both yeasts and molds) that produce the sexual ascospore. Molds in this class have complete cross walls in their hyphae and therefore do not exhibit protoplasmic streaming. All ascomycete molds produce characteristic conidiospores, which occur in chains or clusters. The characteristic blue-green coloration of members of the genus *Penicillium* (Figure 1.4) is due to the coloration of the long chains of conidiospores borne by all members of this genus. The characteristic coloration (yellow, brown, green) of various species of the genus *Aspergillus* (Figure 1.4) is also due to the coloration of the conidiospores. The major criteria for the establishment of genus and species of this class are the visual coloration of the mass of growth in conjunction with the microscopic appearance and three dimensional orientation of the hyphae and conidiospores. A major distinction between ascomycete yeasts and molds is derived from the fact that yeasts produce "naked" asci and frequently contain four and sometimes eight ascospores, depending on the species. The asci of yeasts occur free in the medium, whereas most ascomycete molds produce asci

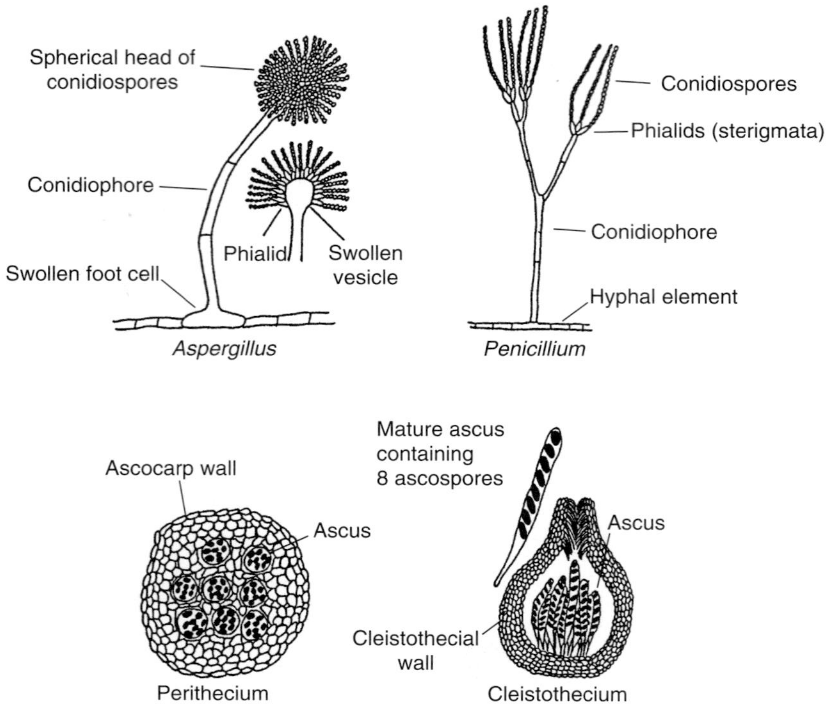

Figure 1.4 Asexual and sexual structures of fungal members of the class Ascomycetes.

with internal ascospores inside a fruiting structure known as a cleistothecium (completely closed) or as a perithecium (open at one end) (Figure 1.4). The class Fungi imperfecti (Figure 1.5), otherwise known as Deuteromycetes, is essentially identical to the Ascomycetes (hyphal crosswalls are present and conidiospores are produced) except that the sexual ascospore is not produced. The class Basidiomycetes houses molds and yeast-like organisms that produce the sexual basidiospore; many also produce conidiospores. Other basidiomycetes produce budding yeastlike cells, which can result in confusing such isolates with true yeasts. The commercial use of molds in various food systems usually involves the harvesting of the asexual sporangiospores or conidiospores for use as inoculum. This allows the density of the inoculum to be based on the precise density or number of spores per unit of volume, which can be readily determined by microscopic count. The use of mycelial mass as an inoculum is more difficult with respect to directly determining the quantity of the cell mass in the inoculum volume, for obvious physical reasons.

1.4 MICROBIAL TAXONOMY

1.4.1 General

After Anton van Leeuwenhoek developed the microscope (circa 1700), Carle Linnaeus developed the binomial system of nomenclature in which each biological entity is allocated to a genus and species. The first letter of the genus designation is always capitalized, the species is entirely lower cased, and both are in italics, e.g., *Penicillium roquefortii, Schizosaccharomyces octosporus, Saccharomyces cerevisiae, Xanthomonas campestris,*

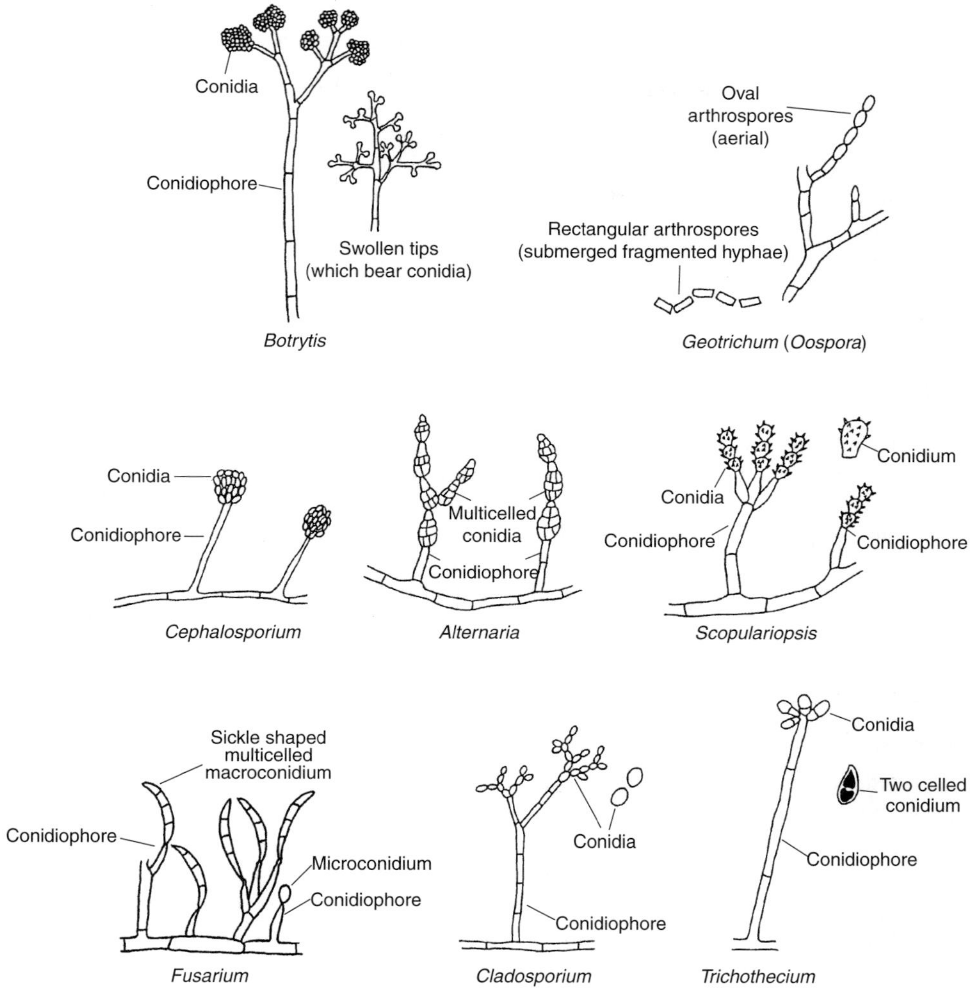

Figure 1.5 Shape and configuration of conidiospores and associated structures of representative members of the class *Fungi imperfecti*.

Lactobacillus acidophilus. The concept of a bacterial genus usually encompasses a well-defined group that is clearly separated from other genera. Interestingly, there is no general agreement on the definition of a genus in bacterial taxonomy (5). The composition of numerous bacterial genera are presently considered to involve a significant level of subjectivity (5).

Species are frequently divided into subspecies, called varieties, serotypes, or biotypes, using the abbreviations "var." or "subsp.", e.g., *Saccharomyces italicus* var. melibiase, *Escherichia coli* subsp. communior. An organism is occasionally found in the literature under several names; e.g., *Candida utilis, Torula utilis, Torulopsis utilis*. Only one of the names is usually correct, the others being synonyms. In this case *Candida utilis* is correct (6). However, when two organisms can be confused in the text resulting from such contractions, e.g., the bacterium *Escherichia coli (E. coli)* vs. the protozoan *Entamoeba coli (E. coli)*, alternate contractions, solely for the purpose of clarity, are then used, e.g., *Esch. coli* vs. *Ent. coli.*

1.4.2 Classification of Bacteria

Bacteria are now classified into two major groups, the Eubacteria and the Archaebacteria (which were formerly grouped under the Protista). The majority of bacteria involved with food systems are Eubacteria. The Archaebacteria presently house the unique halobacteria, which are obligate halophiles and can cause the red tainting of salted fish. All bacteria fall into two convenient groups, those that stain purple with the Gram stain (Gram-positive) and those that stain red with the Gram stain (Gram-negative). In Gram-positive bacteria, there is a bi-lipid membrane between the cell wall and cytoplasm, with the cell wall consisting mainly of peptidoglycan linked with teichoic acids. The cell envelope of Gram-negative bacteria is more complex, consisting of three layers, often referred to collectively as the sacculus. The innermost layer (i.e., the inner cytoplasmic membrane) is adhered to the peptidoglycan layer (otherwise referred to as the murein sacculus (7)), which is covalently linked to elongated lipoprotein molecules. This peptidoglycan–lipoprotein complex partly occupies the periplasmic space between the two (inner and outer) hydrophobic membranes. The outermost layer is an outer membrane consisting of phospholipids, lipopolysaccharides (LPS), and proteins. The LPS content of the sacculus of Gram-negative bacteria constitutes an impermeable barrier to many polar and nonpolar molecules, including dyes and surface-active agents such as bile salts. This difference in permeability to dyes and surface-active agents is used in the selective isolation of Gram-negative organisms with the complete exclusion of Gram-positive organisms, as with the use of MacConkey agar.

There are three general metabolic groups of bacteria: (1) obligate aerobes, (2) facultative anaerobes, and (3) obligate anaerobes. Representative members of each of these groups are found among both the Gram-positive and Gram-negative bacteria.

1.4.3 Serotypes

Serotyping involves the production of antibodies following the injection of a suitable mammal with the microorganism or a specific extract of the organism. If an organism is non-flagellated then serotyping will be based on the somatic antigens. If the organism is flagellated then serotyping may also be based on the flagella antigens. Three antigenic sites are recognized: somatic (O) (German "Ohne") or body, flagella (H) (German "Hauch") or motility, and K (German "Kapsel"), e.g., *Escherichia coli* O157:H7. The O antigens are comprised of the O polysaccharides that are on the surface and are heat stabile. The K and H antigens are heat labile. With whole bacterial cells, agglutination methods are used. With soluble antigens such as toxins, precipitin or gel diffusion assays are used.

1.4.4 Molecular Taxonomy

Each microbial species is presently characterized as having a specific percent molar content of guanosine (G) + cytosine (C) in its DNA. The applicable equation is: 100 mols % = mols % G + mols % C + mols % T + mols % A. Because each guanine nucleotide on one strand of DNA is hydrogen bonded to a cytosine nucleotide on the opposite strand, and because each thymidine nucleotide is hydrogen bonded to a cytosine nucleotide on the opposite strand, the mols % guanine is always equal to the mols % C and the mols % thymine is always equal to the mols % cytosine. By convention, each organism is then defined on the basis its mols % G + C content or GC ratio; 100 mols % = (mols% G + C) + (mols% A + T) or mols % GC = 100 mols % − (mols % A + T). All strains of *S. cerevisiae* are defined as having a molar GC content of 39%. Any yeast strain that deviates significantly from this value cannot be considered *S. cerevisiae*. With unknown isolates, the value of the molar GC content is primarily exclusionary. An unknown organism with a molar GC content of 50% is clearly not *S. cerevisiae*. An unknown isolate with a molar GC content of 39% may be *S. cerevisiae,* but a molar GC content of 39% is not conclusive evidence of identity. Because all biological species fall within the range of about 10% to 90% GC, numerous unrelated species have the same numerical value for their molar GC DNA content. For example, micrococci and all mammals and fish have a molar GC content of 45%.

The practical definition of a species is that it consists of a collection of strains that share many features in common and that differ considerably from other strains (5). A species is presently defined as encompassing strains with approximately 70% or greater DNA–DNA similarity based on DNA strand hybridization, and with 5% or less Tm (Thermal denaturation temperature) for the hybridized strands (8). Phenotypic characteristics should agree with this definition. This corresponds to a 16S ribosomal ribonucleic acid (rRNA) similarity of 98% or higher (9). The nucleotide sequences of rRNA are far more conserved than DNA among various taxonomic groups. rRNA is capable of hybridizing with DNA; however, RNA–DNA hybridization is far less discriminating in terms of recognizing differences between strains of the same species. It is, however, of utility in discerning the difference between two different species of the same genus. Stated more succinctly, DNA–DNA hybridization experiments are used to detect similarities between closely related organisms, whereas RNA–DNA hybridization experiments are used to detect similarities between more distantly related organisms (10). rRNA sequence data is considered more appropriate for determining inter- and intrageneric relationships than for confirming the species identity of an isolate (11). Several groups of organisms have been found to share almost identical 16S rRNA sequences but a DNA–DNA hybridization significantly lower than 70%, indicating that they represent different species (12).

The early classification of microorganisms was based on the utility of their recognition and identification. What emerged with bacteria, however, was a dual system of classification, one based on metabolism and the other on morphology, which are still with us. In 1910, Orla-Jensen proposed that all lactic-acid-producing bacteria (cocci and rods) be housed in the family Lactobacteriacea. In contrast, the family Micrococcaceae houses the various genera of spherical cells or cocci. The bacterial phylogeny that has emerged from molecular sequence data has little in common with these early concepts regarding the morphological relationships of microbial groups (13). Morphology is no longer the guiding principle regarding phylogenetic relationships, in that most characteristics of bacterial morphology are presently regarded as too simple not to have evolved independently in unrelated organisms (13). However, the concept of morphology as a utilitarian character of an organism can be of great value, if, for example, one suspects that a pure culture of a coccus is contaminated and finds rods present.

1.5 METABOLIC CONTROL FOR ENHANCED METABOLITE PRODUCTION

A number of microbial processes in the production of various food additives involve limiting one or more critical nutrients. The submerged production of citric acid by *A. niger* involves limiting both iron and phosphate to achieve maximum yields (14). The production and excretion of maximum amounts of glutamic acid by *Corynebacterium glutamicum* is dependent on cell permeability. Increased permeability can be achieved through biotin deficiency, through oleic acid deficiency in oleic acid auxotrophs, through the addition of saturated fatty acids or penicillin, or by glycerol deficiency in glycerol auxotrophs (15).

1.6 MUTAGENESIS FOR OVERPRODUCTION OF METABOLITES

Increased yields of microbially produced food additives can often be achieved by the selection of overproducing mutants. Such desirable mutants will frequently be produced spontaneously or by a mutagenic agent. There are three fundamental types of mutational events: (1) nucleotide deletions, (2) base-pair substitutions, and (3) gene duplications. Mutants that result from large deletions have the greatest stability. Mutants exhibiting high levels of reversion to wild-type cells are usually derived from base-pair substitution mutations. Alkylating agents are among the most potent direct-acting mutagenic agents.

A variety of methods has been developed for the production of mutants. The frequently used mutagen N-methyl N-nitrosoguanidine (nitrosoguanidine or NTG) functions by forming a methyl group adduct to guanine, and results in multiple mutations. Ethyl methane sulfonate is less lethal and usually results in single mutations. Although yielding large numbers of mutants, such agents usually result in mutations with significant rates of reversion to wild type. Exposure of cells to ultraviolet (UV) irradiation results in approximately equal numbers of both deletions and base-pair substitutions (16). UV irradiation will therefore yield a higher percent of more stable mutants (derived from frame shift mutations) and is the preferred method for mutagenesis. The usual procedure is to expose a cell suspension to UV irradiation to achieve a 50% reduction in viable cells and then to plate out the survivors to isolate mutants. This implies preliminary knowledge of the UV destruction rate. NTG, however, usually produces higher levels of mutagenesis than UV.

Overproducing riboflavin mutants of the mold *Ashbya gossypii* can initially be detected by visually observing the increased intensity of orange coloration of the colonies (17,18). Wild-type isolates of *Pfaffia rhodozyma* produce the red carotenoid pigment astaxanthan which imparts the characteristic pink coloration of salmon tissue, and they are initially a light pink. Sequential mutagenesis has been found to result in a seven- to tenfold increase in pigment production, which is readily detected by visually observing the colonies (19,20). The detection of amino acid overproducing colonies is greatly facilitated by plating overproducing cells (approximately 100 per plate) in the presence of large numbers of cells (approximately 10^8) of an appropriate auxotrophic indicator organism incapable of growth in the agar medium, used because of the absence of the amino acid to be overproduced. Overproduction by a colony is then visually observed and can be semiquantitatively assessed on the basis of the diameter of the zone of growth of the indicator organism surrounding the amino acid overproducing colony.

To use *Brevibacterium ammoniagenes* to overproduce 5′-inosine monophosphate (5′-IMP) by direct fermentation, it is necessary to overcome feedback inhibition due to the synthesis of 5′-adenosine monophosphate (5′-AMP). This can be accomplished by mutationally blocking adenine synthesis while maintaining a low Mn^{++} level, which allows the membrane transit of 5′-IMP into the culture medium (21). The isolation of permeability mutants that allow secretion of 5′-IMP irrespective of Mn concentration has been a significant advance (22). The addition of low levels of adenine to the culture medium for growth is required.

1.7 SELECTIVE CULTIVATION

1.7.1 Selective Enrichment

Microorganisms of public health significance associated with foods are usually present as a small minority of the total microbial population. For both detection and quantitation of such microorganisms, selective enrichment cultivation is usually undertaken. Selective enrichment involves conditions of cultivation that favor the development of the target organism over that of the usual majority of extraneous microorganisms. Selective agents are often derived from the environment in which such organisms of public health significance are found. Most culture media for selective enrichment of *E. coli* make use of the fact that the organism is a common inhabitant of the intestinal tract of mammals and hence is normally in contact with surface-active agents such as bile salts, which are frequently used as selective agents for the organism. Bile salts are notably inhibitory to many Gram-positive organisms. Gram-negative bacteria, in general, are significantly more tolerant of various dyes compared to Gram-positive bacteria, resulting in the incorporation of a variety of dyes for the enrichment cultivation of numerous Gram-negative organisms. This difference in sensitivity to bile salts and dyes is due to the presence of lipopolysaccharide in the external membrane of Gram-negative bacteria that is absent in Gram-positives.

Incubation temperature has also been used successfully with selective chemical agents for the selective cultivation and enumeration of certain bacteria. The use of 45.5°C for the most probable number (MPN) enumeration of *E. coli* using the culture broth known as EC functions selectively in conjunction with the content of 0.15% bile salts no. 3. Selective enrichment of pathogenic vibrios such as *Vibrio vulnificus* and *Vibrio parahaemolyticus* makes use of the fact that they are marine organisms and hence are quite tolerant of the somewhat alkaline pH of seawater, which is usually about 7.8. This tolerance toward an alkaline growth environment extends to a pH of 8.5 to 8.7, which is the usual initial pH range used for enrichment cultivation of these organisms. The natural habitat for *Staphylococcus aureus* is the human skin. The level of sodium chloride on the skin surface during physical exertion under warm weather conditions is often at or near saturation, when one considers the process of evaporation. *S. aureus,* as would be expected, exhibits a significant level of salt tolerance, resulting in the use of enrichment and isolation media for the organism containing 7.5–10% NaCl. NaCl at a level of 7.5% is notably inhibitory to most Gram-positive and Gram-negative bacteria.

A wide variety of antibiotics is now added to media for selective isolation of bacterial pathogens from foods. Yeasts and molds are closely associated with acid fruits. The use of acidified culture media (pH 3.5–5.5) is a long established and highly effective method of inhibiting most bacteria, while allowing unrestricted growth of yeasts and molds.

The successful use of selective enrichment media is predicated on cultivation and enumeration of undamaged cells. Cells of a target organism that have undergone membrane damage through freezing and thawing or heating will often not survive the stress imposed by a selective culture medium originally developed using undamaged cells. The enumeration of damaged cells therefore usually requires initial growth in a nonselective repair medium prior to transfer to a selective medium.

1.7.2 Use of Incubation Temperature for Selective Isolation

Temperature alone can be a highly selective mechanism for the selective isolation of specific groups of organisms. Incubation at 55–65°C will ensure the sole development of obligately thermophilic organisms such as *Bacillus stearothermophilus*. Determination of the number of bacteria on fish generally involves incubation of culture plates at 20°C and resulting development of psychrotrophic organisms that grow optimally at about 25°C and which are also capable of growth at 2°C. By incubating the plates at 10°C, and then replica-plating the resulting colonies to sets of plates incubated at 10 and 20°C, one can readily identify and isolate obligately psychrophilic fishery bacteria, capable of growth from 0–17°C but unable to grow at 20°C (23).

1.7.3 The Use of Mineral Salts Media for the Isolation of Unique Carbon Sources Utilizing Microorganisms

The isolation of microorganisms capable of utilizing unique carbon sources is greatly facilitated if the organism being sought is capable of growth in a strictly glucose–mineral salts medium. Unique carbon and energy sources such as methanol, cholesterol, and naphthalene, can then be added as sole carbon sources with the assurance that only organisms capable of attacking these unique substrates will grow.

1.7.4 Selective Isolation as a Result of Sequential Biochemical Activity

The production of vinegar represents a unique sequence of environmental and biochemical events. The sugars in a fruit juice are first fermented anaerobically by yeasts to ethanol, which is then subjected to vigorous aeration, resulting in the oxidative conversion of ethyl alcohol to acetic acid (usually 4–5%) by resident acetic acid bacteria. Ethanol as an intermediate product will sustain the growth of fewer microorganisms than will glucose because it contains less energy than glucose and is therefore more restrictive with respect to microorganisms capable of attacking it. Acetic acid contains even less energy than ethanol, and, at a level of 4%, results in complete microbial preservation or stability in the absence of oxygen.

The methane fermentation, which is frequently used to digest solid waste materials, involves the initial anaerobic formation of intermediate metabolic products such as ethanol, and lower fatty acids such as acetic acid and butyric acid, by members of the genus *Clostridium* and various facultative anaerobes. These metabolic intermediates are then converted anaerobically to gaseous methane by methane bacteria with the result that the initial solid matter is converted to a gaseous product.

The selective isolation of notably acid tolerant lactic acid bacteria in contrast to moderately acid tolerant lactics can be achieved using the dynamics of the sauerkraut fermentation. Here, 2.5% NaCl is uniformly applied to sliced cabbage in a suitable container. The first brine formed from osmotic expression of water from the tissue by the salt is nearly at a saturation level with respect to NaCl and will not allow microbial growth. As more and more water is drawn from the tissue, along with sugar and other nutrients, the NaCl concentration in the brine becomes progressively less inhibitory. At about 6.5% NaCl coliforms will begin to develop at the initial pH of about 6.5 and will reduce the pH to about 5.5. As these organisms undergo autolysis following their growth, the brine becomes nutritionally enhanced so as to allow a mixture of moderately and notably acid tolerant lactic acid bacteria, that are far more nutritionally fastidious, to initiate growth under anaerobic conditions. The organisms of highest acid tolerance, such as *Lactobacillus plantarum* and *Pediococcus lactis,* will be the last remaining viable organisms by the time a final pH of 3.7 has been reached at an NaCl concentration of about 3.0%.

1.8 ELECTROSTIMULATION OF METABOLITE PRODUCTION

A notably innovative approach toward increasing the microbial production of an end product is electrostimulation. Electrical stimulation of microbial metabolite production is based on the use of an artificial electron carrier such as neutral red, allowing an electric current to indirectly supply the electron-driving force needed to generate a proton-motive force for energy conservation and the electrons needed for growth and end product production (24).

The system involves the use of an electrochemical bioreactor utilizing an anodic chamber and a cathodic chamber divided by a cation selective membrane (24). The cathodic chamber is used as the fermentation vessel. The neutral red serves as a reduced electron carrier and migrates from the anode to the cathode where it enters the cells and couples to a suitable redox enzyme system which accepts the electrons. Electrostimulation has been used to increase L-glutamic acid production by *Brevibacterium flavum* (25), and shows promise with other fermentation processes (26).

1.9 ASPECTS OF MICROBIAL EVOLUTION

One of the foremost examples of microbial evolution and its impact on society involves *E. coli* O157:H7. *E.* is a member of the family Enterobacteriaceae. Molecular phylogeny studies indicate that it is closely related to other mammalian pathogens such as *Vibrio, Shigella, Salmonella,* and *Haemophilus* (27). The enterobacteria are facultative anaerobes, allowing them to live anaereobically as commensulate organisms inside the mammalian intestine and aerobically in various terrestrial environments. Horizontal or lateral gene transfer via sexual recombination, transduction (phage infection), and transformation (direct uptake of naked DNA) is known to occur with strains of *E. coli. Escherichia coli* O157:H7 is an exceptionally lethal example of a harmless commensulate organism (non-pathogenic *E. coli*) that has sequentially acquired a number of discreet genes, each of which enhances human pathogenicity.

O157:H7 strains of *E. coli* are usually described as enterohemolytic, frequently causing hemolytic-uremic syndrome (HUS) with an unusually low infectious dose estimated to be below 100 CFU (28) and a notably high fatality of 10% for those developing HUS (29). Prior to 1982, outbreaks of hemorrhagic colitis and HUS were not reported as attributable to the O157:H7 serotype, this implies that it is a newly emerged pathogen and serotype.

There are presently four recognized phenotypic properties that characterize O157:H7 strains of *E. coli:* (1) inability to ferment sorbitol within 24 hrs, (2) inability to produce β-glucuronidase, (3) poor growth above 44°C, and (4) significantly greater acid tolerance than commensurate strains of *E. coli*. In addition, most strains of *E. coli* O157:H7 produce the shiga-like toxins SLT1 and SLT2, which are encoded for by two independent temperate phages that have presumably previously passed through a strain of *Shigella shiga* with incorporation of the toxin DNA sequences into their genomic DNA, and also harbor a large enterocyte attachment factor (EAF) 60-Mdalt plasmid, designated pO157, that encodes several genes involved in pilus formation and attachment to enterocytes and hemolysin production. The remaining known genes (at least 7) involved in pathogenicity are located in a 35.6 kb chromosomal pathogenicity island termed the locus of enterocyte effacement (30,31,32). The presence of plasmids is known to be associated with the acquisition of virulence factors and antibiotic resistance. In addition, the hypermutable state of O157:H7 strains presumably confers even greater genetic diversity than is present with commensurate strains (33). Inferential conclusions from experimental data indicate that strains of *E. coli* derived from humans are subject to significant levels of gene transfer and genetic recombination (27). Clonal analysis suggests that the immediate ancestor of O157:H7 was a pathogenic clone with a propensity to acquire new virulence factors in nature (34,35). The evolutionary sequence presumably involved has been elegantly presented in detail by Whittam (34). It is quite possible that the selective pressure for the acquisition of such an array of genes enhancing pathogenicity has come about from the development of antibiotic resistant strains.

REFERENCES

1. Collings, A., B. Davis, J. Mills. Factors affecting protoplast release from some mesophilic, thermophilic and thermotolerant species of filamentous fungi using Novozyme 234. *Microbios*, 53:197–210, 1988.
2. Lin, L.F., R.E. Levin. Relative effectiveness of yeast cell wall digesting enzymes on *Yarrowia lipolytica*. *Microbios* 63:109–115, 1990.
3. Ephrussi, B., H. Hottinguer, A.M. Chimenes. Action de l'acriflavine sur les levures: I. la mutation 'petite'. *Ann. Inst. Pasteur* 76:351–367, 1949.
4. De Deken, R.H. The crabtree effect: a regulatory system in yeast. *J. Gen. Microbiol.* 44: 149–156, 1966.
5. Staley, J.T., N.R. Krieg. Classification of prokaryotic organisms: an overview. In: *Bergey's Manual of Systematic Bacteriology*, vol. 1, Holt, J.G., N. R. Krieg, eds., Baltimore: Williams & Wilkins, 1984, pp 1–4.
6. A. Kocková-Kratochvilová. Characteristics of industrial microorganisms. In: *Biotechnology Vol. 1: Microbial Fundamentals,* Rehn, H.-J., G. Reed, eds., Deerfield Beach, FL: Verlag Chemie, 1981, pp 5–71.
7. Park, J.T. The murein sacculus. In: Escherichia coli *and* Salmonella *Cellular and Molecular Biology*, 2nd ed., Vol. 1, Neidhardt, F.C., ed., Washington: ASM Press, 1996, pp 48–57.
8. Wayne, L.G., D.J. Brenner, R.R. Colwell, P.A.D. Grimont, O. Kandler, M.I. Krichevsky, L.H. Moore, W.E.C. Moore, R.G.E. Murray, E. Stackebrandt, M.P. Starr, H.G. Trüper. Report of the ad hoc committee on reconciliation of approaches to bacterial systematics. *Int. J. System. Bacteriol.* 37: 463–464, 1987.
9. Moore, E.R.B., M. Mau, A.K. Arnscheidt, E.C. Bottger, R.A. Hutson, M.D. Collins, Y. van de Peer, R. de Wachter, K.N. Timmis. The determination and comparison of the 16S rRNA gene sequences of species of the genus *Pseudomonas* (sensu stricto) and estimation of the natural intrageneric relationship. *Syst. Appl. Microbiol.* 19:478–492, 1996.
10. Johnson, J.L. Nucleic acids in bacterial classification. In: *Bergey's Manual of Systematic Bacteriology*, vol. 1, Holt, J.G., N. R. Krieg, eds., Baltimore: Williams & Wilkins, 1984, pp 8–11.
11. Fox, G.E., J.D. Wisotzkey, P. Jurtshuk, Jr. How close is close: 16S rRNA sequence identity may not be sufficient to guarantee species identity. *Intl. J. System. Bacteriol.* 42:166–170, 1992.
12. Stackebrandt, E., B.M. Goebel. Taxonomic note: a place for DNA–DNA reassociation and 16SD rRNA sequence analysis in the present species definition in bacteriology. *Intl. J. System. Bacteriol.* 44:846–849, 1994.
13. Woese, C. R. Prokaryote systematics: the evolution of science. In: *The Prokaryotes: A Handbook on the Biology of Bacteria: Ecophysiology, Isolation, Identification, Applications,* 2nd ed., vol I, Balows, A., H.G. Trüper, M. Sworkin, W. Harder, K. Schleifer, eds., New York: Springer-Verlag, 1992, pp 3–17.
14. Lockwood, L.B. Production of organic acids by fermentation. In: *Microbial Technology* vol. 1, Peppler, H.J., D. Perlman, eds., New York: Academic Press, 1979, pp 355–387.
15. Crueger, W., A. Crueger. Amino acids. In: *Biotechnology: A textbook of Industrial Microbiology*, Sunderland, MA: Sinauer Associates, Inc., 1984, pp 133–139.
16. MacPhee, D.G., M.F. Leyden. Effects of caffeine on ultraviolet-induced base-pair substitution and frameshift mutagenesis in *Salmonella. Mutation Res.* 143:1–3, 1985.
17. Pridham, T.G., KB Raper, Studies on variation and mutation in *Ashbya gossypii*. *Mycologia* 44:452–469, 1952.
18. Perlman, D. Microbial process for riboflavin production. In: *Microbial Technology* vol. 1, Peppler, H.J., D. Perlman, eds., New York: Academic Press, 1979, pp 521–527.
19. Gil-Hwan, A., D.B. Schuman, E.A. Johnson. Isolation of *Phaffia rhodozyma* mutants with increased astaxanthan content. *Appl. Environ. Microbiol.* 55:116–124, 1989.
20. Meyers, S.P., G.W. Sanderson. Natural pigments for salmon feeds. *Feed Manage.* November 1992, p. 12, 1992.

21. Furuya, A., S. Abe, S. Kinoshita. Production of nucleic acid-related substances by fermentative processes: XIX, accumulation of 5^1-inosinic acid by a mutant of *Brevibacterium ammoniagone. Appl. Microbiol.* 16:981–987, 1968.
22. Furuya, A., S. Teshiba. Production of 5^1IMP by mutants of *Brevibacterium ammoniogenes.* In: *Genetics of Industrial Microorganisms*, Ikeda, Y., T. Beppu, eds., Tokyo: Kodansha Ltd., 1982, pp 259–263.
23. Makarios-Laham, I., R.E. Levin. Isolation from haddock tissue of psychrophilic bacteria with maximum growth temperatures below 20°C. *Appl. Environ. Microbiol.* 49: 997–998, 1985.
24. Park, D.H., M. Laivenieks, M.V. Guettler, J.K. Jain, J.G. Zeikus. Microbial utilization of electrically reduced neutral red as the sole electron donor for growth and metabolite production. *Appl. Environ. Microbiol.* 65:2912–2917, 1999.
25. Hongho, M., M Iwahara. Application of electro-energized method to L-glutamic acid fermentation. *Agric. Biol. Chem.* 43:2075–2081, 1979.
26. Peguin, S., P. Delorme, G. Goma, P. Soucaille. Enhanced alcohol yields in batch cultures of *Clostridium acetobutylicum* using a three-electrode potentiometric system with methyl viologen as electron carrier. *Biotechnol. Lett.* 16(3):269–274, 1994.
27. Souza, V., A. Castillo, L.E. Eguiarte. The evolutionary ecology *of Escherichia coli. Am. Scientist* 90: 332–341, 2002.
28. Willshaw, G., J. Thirwell, A. Jones, S. Parry, R. Salmon, M. Hickey. Vero cytotoxin producing *Escherichia coli* O157 in beefburgers linked to an outbreak of diarrhea, haemorrhagic colitis and haemolytic uraemic syndrome in Britain. *Lett. Appl. Microbiol.* 19**:** 304–307, 1994.
29. Griffin, P.M.S. Epidemiology of Shiga toxin-producing *Escherichia coli* infections in humans in the United States. In Esherichia Coli *O157H and Other Shiga Toxin-Producing* E. coli *Strains*, Kaper, J.B., A.D. O'Brien, eds., Washington: ASM Press, 1998, pp 15–22.
30. McDaniel, T.K., J.B. Kaper. A cloned pathogenicity island from enteropathogenic *Escherichia coli* confers the attaching and effacing phenotype on *E. coli* K-12. *Mol. Microbiol.* 23:399–407, 1997.
31. Kaper, J.B., S. Elliot, V. Sperandio, N.T. Perna, G.F. Mayhew, F.R. Blattner. Attaching-and-effacing intestinal histophathology and the locus of enterocyte effacement. In: Esherichia Coli *O157H and Other Shiga Toxin-Producing* E. coli *Strains*, Kaper, J.B., A.D. O'Brien, eds., Washington: ASM Press, 1998, pp 163–182.
32. McDaniel, T.K., K.G. Jarvis, M.S. Donnenberg, J.B. Kaper. A genetic locus of enterocyte effacement conserved among diverse enterobacterial pathogens. *Proc. Natl. Acad. Sci. USA* 92:1664–1668, 1995.
33. LeClerc, J.E., B. Li, W.L. Payne, T.A. Cebula. High mutation frequencies among *Escherichia coli* and *Salmonellla* pathogens. *Science* 274: 1208–1211, 1996.
34. Whittam, T.S. Evolution of *Escherichia coli* O157:H7 and other shiga toxin-producing *E. coli* strains. In: Esherichia Coli *O157H and Other Shiga Toxin-Producing* E. coli *Strains*, Kaper, J.B., A.D. O'Brien, eds., Washington: ASM Press, 1998, pp 195–209.
35. Rodriquez, J., I.C.A. Scaletsky, L.C. Campos, T.A.T. Gomes, T.S. Whittam, L.R. Trabulsi. Clonal structure and virulence factors in strains of *Escherichia coli* of the classic serogroup O55. *Infect. Immun.* 64:2680–2686, 1996.

1.02

Principles of Biochemistry and Molecular Biology

Patrick P. McCue and Kalidas Shetty

CONTENTS

2.1 PREFACE

Literally, biochemistry is the study of the chemistry of "life," or, rather, of living organisms. As biochemists, we investigate through experimentation the processes that occur in living organisms. Ever since the first description of cells by the English scientist Robert Hooke in 1663, man has been interested in understanding how cells operate, both alone and in cooperation (as tissues). Desire to investigate objects invisible to the naked eye led to the necessity for scientists to become creative in their design of experiments. Over the years, such ingenuity resulted in the development of technologies that as a whole became the core of a scientific discipline that would soon become known as "biochemistry" by the turn of the nineteenth century.

With the advent of molecular biological technologies adding to the core classical approaches, modern-day biochemistry is comprised of essentially two general areas of experimentation: (1) purification and characterization of macro-scale cellular components via classical approaches, and (2) purification and characterization of micro-scale cellular components via molecular biological approaches. It is difficult to generalize the study of biochemistry into two separate groups of approaches, as researchers typically blend any number of techniques from both approaches into their experimental milieu, but as the aim of this section is to introduce the scientific discipline of biochemistry — its essence, its mode of thinking — it will be easier for the unfamiliar reader if we separate the classical from the modern molecular approaches, if only for the sake of clarity. In so doing, we will review significant aspects of the history of biochemistry, so that the reader may gain an insight into the thinking of the time and better understand how these techniques arose. The scientific range of biochemistry is both vast and complex. In writing this chapter we assume that the reader has studied the general subject in lecture. It is not our aim for this chapter to be all-encompassing, but, rather, for it to give a basic to intermediate understanding of the areas relevant to food biotechnology so that the reader may gain a solid foundation upon which to more fully explore and understand the ideas that will be presented in the succeeding chapters.

2.2 A BRIEF HISTORY OF CLASSICAL BIOCHEMISTRY

The beginnings of biochemistry have their roots in chemists' attempts to define the molecular principles of the world in which we live. In 1803, Englishman John Dalton, a school teacher who studied chemistry in his spare time, postulated the modern atomic theory, in which he stated (1) that all matter must be composed of submicroscopic units named atoms, (2) that each element has its own kind of atom, and (3) that atoms of different elements come together in defined numbers to make the numerous diverse molecules that make up the millions of known substances. In investigating Dalton's theory, chemists found that atoms must obey certain rules in forming molecules and that they are only able to interact with other atoms in certain fixed numbers (i.e., only certain bonds can be formed by certain atoms and in certain numbers).

At that time, it was known that the world that we know and live in was composed of living material (organic) and non-living material (inorganic). Biochemistry as a discipline did not exist yet, but rather was getting its beginnings in an off-shoot of chemistry, known as "chemical physiology." Significant physiological questions remained unanswered, such as what caused the difference between inorganic and organic material and how the processes of living tissue were controlled. With no distinct branch of science to cover this emerging area, investigations were left to adventurous chemists willing or compelled to investigate these intriguing questions. We say "adventurous" in that many of the pioneering scientists were not funded to conduct research in this new field.

In this period of the early nineteenth century, it was believed that only living tissues could perform the reactions of living cells. Beliefs such as this were soon to change. In 1828, Friedrich Wöhler accomplished the synthesis of urea from inorganic matter, proving that reactions of living tissues were not restricted to living tissues, but could be recreated *in vitro* (1). In 1830, proteins were discovered, and chemists found that the formation of proteins from amino acids also obeyed certain rules, similar to as well as dependent upon those of molecule formation by atoms. Discoveries such as these compelled interested scientists to further characterize the cellular machinery.

The idea that reactions of living cells did not need living hosts to occur was advanced further almost 70 years later in 1897, when Edvard and Hans Buchner showed that dead cell extracts can perform reactions of living cells (2). The molecules responsible for performing these reactions were called enzymes. In the late 1800s, Emil Fischer suggested a model mechanism for the action of enzymes that resembled the interaction of a lock and key, in which an enzyme (the lock) interacted with or bound to its substrate (the key), facilitating the subsequent reaction. The model suggested that the enzyme itself would remain intact and unchanged by the reaction. At the turn of the nineteenth century, the field of biochemistry emerged from the area of chemical physiology. The focus of biochemistry was the characterization of the structure and function of enzymes, as well as the elucidation of enzymatic pathways.

As it turns out, food science played a prominent role in the beginnings of biochemistry, as many of the first biochemists investigated various aspects of nutrition. One of the main figures in the establishment of biochemistry as a field of both teaching and research was a British scientist by the name of Frederick Gowland Hopkins, who was interested in understanding the metabolism of living cells. In 1912, Hopkins published a paper in the *Journal of Physiology* in which he demonstrated the importance (discovery) of vitamins and essential amino acids as growth-stimulating agents in the normal diet (3). In 1914, Hopkins was appointed as Cambridge University's first Professor of Biochemistry. Hopkins was knighted in 1925 and received the Nobel Prize in Medicine and Physiology in 1929.

In the early 1900s, biochemistry made another significant advancement when enzymes were proven to be proteins. In 1926, James Sumner and colleagues at Cornell University Medical College demonstrated that urease could be crystallized (4) and, in 1932, that proteolysis of purified urease enzyme results in loss of enzyme activity (5). Though his work was initially met with skepticism, Sumner work was finally accepted, and he received the Nobel Prize in 1946.

The period between the 1940s and 1960s gave witness to arguably some of the most significant achievements in biochemistry, which gave birth to biochemical genetics and led to the emergence of molecular biology. In 1941, studying the metabolism of the orange bread mold *Neurospora crassa* irradiated with X-rays to cause mutations, researchers George Beadle and Edward Tatum demonstrated a 1:1 relationship between a mutation and the absence of a specific enzyme, by identifying mutants defective in specific steps of a metabolic pathway. Assuming that each mutant was defective in a single gene, Beadle and Tatum hypothesized that one gene may contain the genetic information responsible for one enzyme, and that mutants with a defective enzyme carried a defective gene for that enzyme (6). In 1944, building on research from 1928 performed by Frederick Griffith, which demonstrated that heat-killed virulent bacteria contained a "transforming principle (agent)," which could transform live, nonvirulent bacteria into the virulent type, researchers Avery, MacLeod, and McCarty (1944) demonstrated that deoxyribonucleic acid (DNA) was the information-carrying "transforming principle" (7). In 1952, Alfred Hershey and Martha Chase demonstrated that DNA is the material of heredity in experiments that followed the cell fates of radio-labeled DNA and protein (8). In 1953, Watson and Crick reported that the structure of DNA was a double-helix (9). In 1958, Meselon and Stahl demonstrated that DNA replication is semi-conservative (10). In 1961, the existence of messenger ribonucleic acid (mRNA) was postulated by Jacob and Monod (11). Finally, in 1964, the genetic code was elucidated by Nirenberg and Leder using specific trinucleotides (12,13).

2.3 EXPERIMENTAL TECHNIQUES OF CLASSICAL BIOCHEMISTRY

The brief history of classical biochemistry (i.e., pre-molecular biology) related above highlights the progression of the science towards the elucidation of mechanisms (and their constituents) of increasingly smaller sizes from the micro-scale to the molecular. As a number of the "classical" techniques employed in the aforementioned experiments continue to find use alongside the largely molecular techniques in the experimental approaches of modern biochemistry, we will now shift to a brief explanation of the techniques most relevant to the application of biochemistry for food biotechnology. Since the period of time from the 1970s to the present has seen largely the development of experimental techniques relevant to molecular biology, these advances will be covered in the subsequent subsection of this chapter which will discuss relevant molecular biology techniques.

A main preoccupation of biochemistry is the study of enzyme activity (enzymology). For the better part of a century, the primary ways to determine the activity of an enzyme have been (1) to detect the presence or change in concentration of a predicted (or known) reaction product, (2) to follow the incorporation of a radioactively-labeled molecule (such as ^{14}C) in products of a metabolic pathway, or (3) to measure the change in concentration of a required co-factor (such as $NADP^+$) by detecting its absorbance at a certain wavelength of light using a spectrophotometer.

Key to the study of enzyme activity is the isolation of the reaction components (i.e., the enzyme, its substrate(s), and any required co-factors). A quick review of the history of biochemistry will reveal that our understanding of biological processes has increased concomitant to our ability to purify compounds related to or involved in such processes. Once the required reaction components have been isolated, the reaction can be reconstituted *in vitro* and the reaction kinetics investigated. We will introduce and describe two major biochemical approaches that have been (and continue to be) used to purify enzymes (proteins) and other compounds or molecules from crude solutions: chromatography and isoelectric focusing. Both approaches are typically employed subsequent to a salt precipitation procedure to concentrate the protein in a crude sample (with the most common salt used for this procedure being ammonium sulfate). Afterwards, we will discuss spectrophotometry, its history, and its relevance to enzymology.

2.3.1 Principles of Chromatography

2.3.1.1 Column Chromatography

Chromatography is a purification technique by which sample components are separated according to their ability (or inability) to interact with a mobile phase or a stationary phase through which the sample is passed. Although there are many types of chromatography available today, the underlying core separation principle is essentially the same for all of them. One of the most basic and most common chromatographic techniques is column chromatography, in which a glass or plastic tube is filled with a mixture or slurry of a liquid mobile phase (typically a buffer) and a solid stationary phase. The sample, or solute, is applied to the top of the mixture (called the bed) in the vertically-supported tube and is allowed to pass through the two phases either by gravity flow or with pressure applied from an attached pump. The components of the sample distribute themselves between the mobile and stationary phases depending on their affinity for either (which is determined by the inherent chemical properties of each component). Molecules that have little or no affinity for the stationary phase spend more time in the mobile phase, pass quickly through the chromatography system, and are removed, or eluted. Molecules that have affinity for the stationary phase spend more time in the stationary, move more slowly through the system, and are eluted more slowly. The mobile phase that is collected after passing through the chromatography system is called the effluent and is or contains the sample molecules. Usually, the effluent is collected at different times or in fractions (such as for 2 Min periods or in 1–2 mL portions). Typically, fractions which are found to contain the desired components will be combined, or pooled, for further analysis.

In column chromatography, the solid phase is an adsorbing material which contains openings of various sizes (known as mesh sizes) through which the solute may pass. Low mesh sizes (20–50) have a lower number of openings per square inch of packed material and, thus, have a faster flow rate. Higher mesh sizes (200–400) have a higher number of openings per square inch and, thus, have a slower flow rate and are used mainly during high-resolution separations of nearly pure solutes.

2.3.1.2 Other Types of Chromatography

2.3.1.2.1 Ion-Exchange Chromatography Ion-exchange is a form of adsorption chromatography set up similarly to column chromatography in which solute molecules are separated on the basis of their affinity for a charged stationary phase. The column is packed with a stationary phase consisting of a synthetic resin that is tagged with ionic functional groups which electrostatically interact (reversibly) with ionic solutes. Depending on the experimental conditions, the solutes may have a negative or positive charge, or none at all (neutral). Solutes that are neutral, or that have a charge similar to that of the stationary phase, will elute with the buffer. Bound solutes can be released by increasing the ionic strength of the buffer (displacement) or increasing the pH of the buffer (weakening the interaction by charge reduction of the solute or resin).

2.3.1.2.2 Gel Exclusion Chromatography Gel exclusion chromatography is a form of column chromatography in which solute molecules are separated based on their molecular size. The column is packed with a stationary phase consisting of synthetic beads that contain small pores of a controlled size. Solute molecules that are too large to enter the pores will flow quickly through the column. Solute molecules that are small enough in size to enter the pores will remain in the column longer and thus are retarded in their travel through the column. The exclusion limit of the stationary phase defines the molecular mass of the smallest molecule that cannot enter the bead pore. Solute molecules with a molecular size greater than the exclusion limit will pass directly through the column. Gel matrices with an exclusion limit of approximately 6000 Daltons are routinely used for desalting of a protein sample following a salt precipitation procedure.

2.3.1.2.3 Thin-Layer Chromatography (TLC) TLC is performed by (1) coating a glass plate (similar in size to a microscope slide) with a thin layer of silica gel, alumina, or cellulose; (2) spotting a small drop of sample onto the coating and allowing it to dry; (3) and then standing the plate vertically in a shallow dish containing an appropriate solvent. As the solvent front moves vertically through the gel coating by capillary action, the solute components are separated. Samples that are not colored are sometimes detected by radioactive or fluorescent methods, or through treatment with chemicals that develop colors. Unknown sample components are identified by comparing their distances traveled to those of known standards.

2.3.1.2.4 High-Performance Liquid Chromatography (HPLC) HPLC, one of the most common chromatographic techniques performed by food scientists, is an analytical technique well suited to the separation and identification of biological molecules such as proteins, nucleic acids, phenolic compounds, carbohydrates, and lipids. With the development of stationary phases with very small particles sizes and large surface areas, HPLC improves elution rates by applying high pressure to the solvent flow. The result is high resolution of a solute into its individual components in a relatively short period of time (typically between 10–60 Min). Unknown compounds are identified by comparing their retention times (that is, the amount of time required for the compound to elute from the column under specific experimental conditions) to those of known standards, either alone or in combination. The computerized systems in use today are relatively simple to use as they are largely automated and can be controlled by accompanying software.

2.3.2 Principles of Spectrophotometry

Another very common analytical laboratory technique is the measurement of the concentration of a compound, reaction component, or reaction product (or its rate of appearance or disappearance) by measuring its absorbance of a specific wavelength of light. A spectrophotometer is an instrument that can produce light at a selected wavelength, direct it through a sample (typically held in a cuvette), and measure the intensity of light transmitted by the sample. Many types of spectrophotometers exist. A UV/vis spectrophotometer, a type found in most laboratories, can produce light in both the ultraviolet (UV) spectrum (~190–340 nm) and the visible (vis) spectrum (340–800 nm).

The absorbance of a particular compound in a solution can be detected using a spectrophotometer. For example, $NADPH_2$, a co-factor produced by activity of the enzyme glucose-6-phosphate dehydrogenase, absorbs light at the wavelength of 340 nm. Thus, an increase in $NADPH_2$ of a solution can be detected as an increase in the absorbance of the solution at 340 nm. The peak absorbance wavelength has been determined for many substances.

The actual concentration of a chemical can be determined by its absorbance at a given wavelength, such as 340 nm for $NADPH_2$, using the Beer-Lambert Law represented by the mathematical equation $A = \varepsilon\, l\, c$. In this equation, A is the absorbance of a solution at a given wavelength of light. The term ε is a proportionality constant that defines the extent of absorption. It is called the molar extinction coefficient or the molar absorption coefficient, and is typically given in the units cm^{-1} and M^{-1}. The term l is the path length of light through the sample (usually 1 cm), and the term c is the concentration of absorbing material in the sample (usually in moles/liter, or M). Molar extinction coefficients have been determined for a great many compounds at specific wavelengths and environmental temperatures. The chromatographic technique HPLC combines high-pressure column chromatography with a spectrophotometer so as to detect specific chemicals by their characteristic absorption at given light wavelengths as they are eluted from the column and pass through the light beam.

2.4 A BRIEF HISTORY OF MODERN BIOCHEMISTRY – MOLECULAR BIOLOGY

Just as macromolecular entities such as proteins can be detected, purified, and characterized based on their inherent biological properties (i.e., molecular size, ionic charge at certain pHs), so, too, can micro-molecular entities such as nucleic acids (DNA and RNA). In fact, since the core biological properties of DNA and RNA were reported in the 1950s and 1960s, a number of molecular purification and characterization techniques have been developed which take advantage of these properties. The advent of these molecular techniques, first viewed as tools of biochemistry, has led to the emergence of a new and related field of study known as molecular biology. Noting the recent advent of genomics and proteomics, advancements in this field continue to occur at a pace which rivals that of the electronics industry.

Currently, there is much support hailing the discovery of the double helix by Watson and Crick as the beginning of modern biochemistry and molecular biology (9). Others stress that it was the later-proven speculation by Watson and Crick that the pairing arrangement of bases in the double helix implied the existence of an information-carrying genetic code (or "copying mechanism," as they called it). One could just as easily argue that modern biochemistry arose from the work of Hershey and Chase, that demonstrated that DNA was the material of heredity, and shifted much focus away from protein. It would seem to most, though, that the works of Hershey and Chase, Watson and Crick, Messelson and Stahl, Jacob and Monod, and Nirenberg and Leder, as a whole, during 1953–1964, set the foundation for modern biochemistry and molecular biology.

Following the elucidation of the genetic code, the 1970s were witness to a number of important advances in relation to the study of DNA. Right away, the first restriction enzyme was isolated from *Hemophilus influenzae* by Smith and Wilcox (1970) (14). Three years later, Cohen and Boyer utilized restriction enzyme technology in developing DNA cloning and recombinant DNA (15). Ed Southern (1975) developed a DNA detection technique (later named after him) that employed blotting DNA to a membrane and hybridizing radio-labeled DNA probes that were complementary to specific sequences (16). In 1977, Sanger developed a rapid technique for manually sequencing DNA *in vitro* (17).

Similarly, the rapid pace of scientific advancement in microbiology continued during the 1980s and 1990s, and continues right up to the present time. In 1980, the U.S. Supreme Court decreed that life forms, such as genetically-modified organisms, could be patented. This decision facilitated the emergence of numerous biotechnology companies, such as Amgen. The 1980s also saw the invention of the polymerase chain reaction (PCR) technique for amplification of DNA sequences by Kary Mullis (18) and the development of the first genetically-modified crop, a tobacco (19). Soon after, during the late 1980s and early 1990s, Leroy Hood and colleagues developed the first automated DNA sequencer and protocols utilizing fluorescent dyes that could be read automatically (20). In the mid-1990s, J. Craig Venter developed a rapid technique for genomic sequencing known as "shotgun" sequencing, in which genomic DNA is broken into smaller fragments for sequencing and afterwards the resulting sequences are put back in order for further genetic analyses (21). Meanwhile, at Stanford University, Patrick Brown and colleagues were busy developing an automated method for quantitative gene expression, later to be known as the DNA microarray (22). In 1997, the first successfully cloned animal was born in Scotland, a sheep named Dolly (23). And finally, in 2001, after nearly 10 years of tense competition, two research groups, one privately funded and one publicly funded, published the first drafts of the human genome sequence (24,25).

2.5 KEY PRINCIPLES OF BIOCHEMISTRY AND MOLECULAR BIOLOGY

Now that we have an idea as to where biochemistry started and how far our knowledge has come, let us recap the main points that have been learned from the process.

2.5.1 Basic Cellular Biology

Living organisms are complicated, albeit organized, systems. Organization manifests itself as a molecular hierarchy of systems and pathways, occurring as organs (in larger creatures) made up of tissues, which themselves are made up of cells. The chemistry of living cells resembles the chemistry of organic reactions. The constituents of living cells are called biomolecules. Cells contain structures called organelles, which are complex assemblies of large molecules (macromolecules). Both of these components serve functional purposes. Also, living organisms can self-replicate.

Some reactions in the cell require energy to occur. Living systems harness energy from the environments in which they occur. Organisms acquire energy either through photosynthesis (like plants) or through the metabolism of food (like animals), by forming special biomolecules, such as adenosine triphosphate (ATP) or $NADPH_2$, that can serve as chemical sources of stored energy for thermodynamically unfavorable reactions.

All biomolecules contain carbon. The major precursors of biomolecules are water, carbon dioxide, ammonium, nitrate, and nitrogen. Precursors form metabolites. Metabolites are simple, organic compounds that are intermediates in cellular reactions, such as the biosynthesis of amino acids, sugars, nucleotides, and fatty acids. Metabolites are converted into larger macromolecules, such as proteins, polysaccharides, polynucleotides, and lipids. Macromolecules interact to form supramolecular complexes, such as enzyme complexes, ribosomes, chromosomes, and cytoskeletal elements. Organelles are membrane-bound sites of supramolecular complexes, such as the nucleus, mitochondria, and chloroplasts. Prokaryotic cells do not contain organelles. Eukaryotic cells are believed to have originally formed organelles from engulfed bacteria (symbiotic theory).

2.5.2 Macromolecules and Basic Cellular Chemistry

Macromolecules are composed of units that have structural polarity. As such, these molecules are not symmetrical, but rather have ends that are structurally different (a "head" and a "tail"). Some molecules may have a positively-charged end and a negatively-charged end. Structural polarity gives a macromolecule a sense of direction or orientation. Because of this directional nature of their structure, biological macromolecule subunits have the capacity to specify information in their arrangement. Biomolecular interactions, recognition, and environmental conditions are determined by numerous weak forces, such as hydrogen bonds or hydrophobic interactions, as well as structural complementarity.

Enzymes are macromolecules that accelerate (or catalyze) chemical reaction rates by several orders of magnitude, and add a measure of specificity to the reaction (i.e., only select substances may interact with the enzyme). Controlling enzyme activity allows control of metabolic reactions.

2.5.3 Importance of Water, Ionization, and Buffers

As organisms typically contain 70–90% water, water is an indirect participant in biological reactions. The "bent" structure of the water molecule is well-suited to hydrogen-bond formation (with a natural induced dipole charge) and in that formation, water can serve as either a proton (H^+) donor or a proton acceptor. Also, the natural ability of water to form hydration shells around ionic and hydrophobic solutes, as well as ice, influences how various reactions will occur (or not occur). The formation of membranes by fatty acids is influenced by the formation of micelles by amphipathic molecules (both partially hydrophobic and partially hydrophilic) in aqueous solutions.

Water shows a slight tendency to form ions, a property that allows it to conduct electricity. The ions formed from one water molecule are one proton (H^+) and one hydroxyl (OH^-), with the former being immediately hydrated to form a hydronium ion (H_3O^+), although by convention we still refer to the proton concentration in solution and not the hydronium ion concentration.

At equilibrium, the concentration of free protons in water is equal to the concentration of free hydroxyl ions. The product of the two ion concentrations is called the ion product of water (K_w). When a solution is acidic, it contains a higher concentration of free protons and a lower concentration of free hydroxyl ions. Likewise, a basic solution contains more free hydroxyl ions than free protons. To simplify the expression of proton concentrations, the negative logarithm of the free proton concentration ($-\log [H^+]$) has been defined as the "power of hydrogen," or pH, where pH + pOH = 14. For example, the pH of bleach, a strong basic solution, is 12.6, whereas the pH of vinegar, a weak acid, is 2.9. The pH of blood is 7.4. Solutions with a pH of 7 are said to be neutral; that is, neither acid nor base.

Other substances that are able to dissociate into ions in water are called electrolytes, because the addition of their ions to those already present in water increases the ability of the solution to conduct electricity. Substances that dissociate almost completely into ions in water are called strong electrolytes. So-called "strong" acids and bases are examples of strong electrolytes. Similarly, substances that only slightly dissociate into ions in water are called weak electrolytes. Again, so-called weak acids and bases are examples of weak electrolytes. The extent to which a particular substance forms ions in water is called the ionization constant (K_a).

Solutions that resist changes in their pH when acids or bases are added are called buffers. A buffer is most effective when the pH of the solution is within one unit of the pK_a of the buffer solute (i.e., potassium phosphate). Since enzyme activity is very sensitive to changes in pH, maintenance of cellular pH is critical to the survival of living organisms. Intracellular pH is maintained by phosphate and histidine buffer systems. The bicarbonate buffer system controls the pH of blood and extracellular fluids.

2.5.4 Nucleic Acids

Although biochemistry arose as a scientific discipline from a concerted interest in the investigation of proteins and enzymes, modern biochemistry and molecular biology focus arguably more attention on nucleic acids and their role in directing cellular activities. Because nucleic acids encode the blueprints for protein design, thus controlling their activity, we will discuss nucleic acids first, and then proteins.

In the mid 1800s, a monk named Gregor Mendel postulated that each trait of a living organism (in his case, the pea plant) was regulated by tiny physical factors in the cells of the organism (later named "genes") that could be passed on to the next generation (6). As illustrated by the experiments of Hershey and Chase in 1952, nucleic acids contain the material of heredity (8). Later, it was found that, in fact, it is the nucleotide sequences which make up the nucleic acids that contain the genes. Thus, genes composed of specific nucleotide sequences are the material of heredity.

There are two basic types of nucleic acids, deoxyribonucleic acid (DNA) and ribonucleic acid (RNA). Nucleic acids are composed of a five-carbon sugar, a phosphate group, and a nitrogenous base. The sugar in DNA is 2-deoxyribose. The sugar in RNA is ribose. The nitrogenous bases of nucleic acids are derived from either a purine (for adenine and guanine) or a pyrimidine (for cytosine, uracil, and thymine). Adenine, guanine, and cytosine are found in both DNA and RNA. Uracil is found only in RNA, and thymine is found only in DNA. In the DNA double helix, the base adenine (A) in one strand pairs with (i.e., hydrogen bonds to) a thymine base (T) in the opposite strand, while the base guanine (G) pairs with a cytosine (C) base. In DNA, the strands are anti-parallel, which means they run in opposite directions. In a DNA–RNA hybrid complex, the adenine bases in the DNA strand pair with cytosine bases in the RNA strand.

Nucleic acids are linear sequences of nucleotides. Nucleotides are nucleosides bound to a phosphate group through an ester linkage. The term "nucleoside" refers to a nitrogenous base linked to a five-carbon sugar. When bound to a sugar, the base becomes more water-soluble that it normally would be. To create nucleic acids, nucleotides join together to form ordered linear sequences in which the 5′-nucleoside monophosphate of one nucleotide is attached to the free 3′-OH group of the preceding nucleotide. Since the sequence originates at the 5′-end and extends to the 3′-end, base sequences are therefore represented in base shorthand (A, G, C, T, or U) written in a 5′-3′ orientation. The key importance of nucleic acids is in their ability to form ordered linear sequences which allows for nucleic acid sequences to hold [chemical] information, similar to the way words contain information because of the correct orientation of letters in the words.

For most living organisms, genetic information is stored in the form of DNA, although some viruses are known to use RNA. While there are two main types of nucleic acids, there are several main types of RNA, such as messenger RNA (mRNA), ribosomal RNA (rRNA), transfer RNA (tRNA), and, in eukaryotes, small nuclear RNA (snRNA). mRNA is produced during transcription, and acts as a physical messenger to carry the genetic information stored in DNA from the nucleus to the cytosol, where translation activities take place. rRNA molecules are involved in translation activities, along with ribosomes. tRNAs shuttle specific amino acids to the ribosome for protein synthesis. snRNAs join to form ribonucleoproteins that can edit (cut) mRNA sequences before they leave the nucleus.

DNA can be cut, or "cleaved," enzymatically at specific points, by restriction endonucleases. These restriction endonucleases recognize and cut at specific DNA sequences, usually between 4–6 bases in length. One of the first uses of these endonucleases was in restriction mapping of DNA. By this technique, specific sites on DNA molecules can be mapped. The technique involves creating fragments of various lengths, using restriction endonucleases. Then the orientation of the sites is deduced from the patterns formed after separating the fragments by electrophoresis through an agarose gel (explained later).

2.5.5 Proteins

Proteins are composed of one or more peptide chains, folded into specific arrangements. Peptide chains are linear sequences of amino acids. The specific arrangements into which peptide chains join together or fold are largely determined by the amino acid sequences of the peptide chains. There are only 20 naturally occurring amino acids. There are nonpolar, polar uncharged, acidic, and basic amino acids. Consistent with other biomolecules, amino acids are asymmetric. They have a carboxyl group (COO^-) on one end and an amino group (NH_3^+) on the other.

Amino acids can link together through peptide bonds to form linear sequences (polypeptide chains). In a peptide bond, the amino group of one amino acid links to the carboxyl group of another amino acid. One or more polypeptide chains can fold together to form proteins. When notating the sequences of polypeptide chains, amino acid names are typically abbreviated by using three-letter codes.

Proteins generally fold to form the most stable structures possible. In addition to the primary amino acid sequence, other forces influence protein folding and structure, such as hydrogen bonds, hydrophobic interactions, electrostatic interactions, and attractive or repulsive (van der Waals) forces. Sometimes proteins are "helped" into their appropriate conformations by helper proteins called molecular chaperones. A peptide chain folded into its natural conformation is said to be in its native structure. Non-native structures can result when proteins are broken down, or denatured, by cooking, or by treatment with a strong base, such as urea.

The amino acid sequence is the primary (1°) structure of a protein. The secondary structure (2°) of a protein consists of the α-helix or β-pleated sheet conformations formed by the polypeptide chains. Secondary structure largely depends on the amino acid sequences. The tertiary structure (3°) of a protein is the three-dimensional interactions of the secondary structure elements. Quaternary (4°) protein structure refers to the arrangement or interaction of two or more monomer subunits.

2.6 EXPERIMENTAL TECHNIQUES OF MODERN BIOCHEMISTRY AND MOLECULAR BIOLOGY

2.6.1 Detection of Proteins and Nucleic Acids

A great many techniques have been developed for the detection of proteins and nucleic acids. Previously, we discussed chromatography and spectrophotometry. Both of these techniques can be used to detect proteins and nucleic acids. For instance, both can be detected by their absorption of a specific wavelength of UV light. The peak absorption of proteins occurs at 280 nm, whereas the peak absorption of nucleic acids occurs at 260 nm.

2.6.1.1 Principles of Electrophoresis

Probably the most common technique employed in the laboratory to detect proteins or nucleic acids is electrophoresis. Electrophoresis is a technique by which the movement of charged molecules in an electric field can be studied. Most people have likely heard of the term "gel electrophoresis," which employs a buffer-saturated gel matrix as a support medium to which the sample to be analyzed is applied. The migration of the sample through the matrix of the gel is influenced by an applied electric field. Typical forms of gel used are agarose and polyacrylamide. The decision to use one over the other typically depends upon the specific need for a certain resolution power (which is greater for polyacrylamide) or the lack thereof. An agarose gel matrix is usually used in routine electrophoresis of DNA, such as to resolve PCR reaction products. Protein electrophoresis typically uses polyacrylamide.

The electric field is applied such that the positive electrode occurs on the opposite end of the gel to where the sample is applied (usually the top). For DNA electrophoresis, the nucleic acid samples migrate through the gel matrix towards the positive electrode because DNA has an inherent negative charge, due to the accumulation of phosphate groups in the DNA backbone. The charge of proteins, however, largely depends upon the pH of their environment. To resolve this problem, proteins are usually denatured in the presence of sodium dodecyl sulfate or SDS (a type of soap), which coats the denatured protein molecule and gives it an overall negative charge. This type of protein electrophoresis is referred to as SDS-PAGE.

For SDS-PAGE and normal DNA electrophoresis, after the electric field is applied, the molecules migrate through the gel matrix at rates which depend on their molecular size. Thus, larger proteins and oligonucleotides take longer to weave their way through the maze-like matrix, and thus remain "higher" on the gel, or closer to the sample application point. Smaller proteins and oligonucleotides are better able to pass through the matrix and thus are found "lower" or farther down the gel, closer to the positive electrode. Proteins and DNA molecules separated by electrophoresis are typically detected after separation by staining with fluorescent or colorimetric dyes, such as coomassie brilliant blue (for proteins) and ethidium bromide (for DNA).

2.6.2 Southern, Western, and Northern Blotting

Once proteins and DNAs have been separated by their molecular weight, specific molecules can then be detected. However, the gel matrix that is so useful in separating these molecules by their molecular size unfortunately interferes with detection methods that are used to identify specific molecules from a group sample. Further, only a minute amount of protein or DNA is typically present in a band on a gel. In 1975, Ed Southern developed a method known as Southern blotting, in which DNA can be transferred to a nitrocellulose filter (by placing the filter on the gel and allowing the buffer flow from the gel into the filter to carry with it DNA) (16). After DNA is transferred to the filter, the filter can be incubated with radioactively labeled DNA fragments known as probes. The probes will hybridize with complementary DNA sequences, if they are present. The filter then is exposed to x-ray film and any bound probes are detected on the film, in a process called an autoradiogram. This procedure is known as Western blotting when protein is transferred to a filter and Northern blotting when RNA is transferred.

2.6.3 Mass Spectrometry

Mass spectrometry is a powerful tool for accurately detecting and identifying specific proteins separated by 2-dimensional (2D) electrophoresis. First evolving in the early 1990s, the core technique has essentially three steps. In the first step of mass spectrometry, a complex protein sample is separated by 2D electrophoresis. By this method, the protein sample (<u>not</u> treated with SDS) is loaded into one well of a polyacrylamide gel that possesses a pH gradient, usually from 3–10, within the gel matrix. When the proteins in the complex sample travel through the gel during electrophoresis, they stop migrating at the pH at which the net electric charge of the molecule is zero. This pH value is called the isoelectric point or "pI" (1st dimension). After the proteins in the sample are separated according to their pI value, the entire lane of the gel is removed and laid across the top of a new, normal polyacrylamide gel (e.g., without a pH gradient), and the proteins are then further separated by their molecular weight (i.e., size; 2nd dimension) during a second electrophoresis step. During the second step of mass spectrometry, protein bands of interest are then excised from the second gel. Then the gel plug containing the protein-of-interest is digested using the enzyme trypsin. In the third step of the procedure, the peptides that result from the trypsin digest of the protein-of-interest are analyzed by mass spectrometry and the spectra produced by the collision-induced dissociation is searched against mass spectra of peptide sequences contained in a database.

2.6.4 Other Important Molecular Techniques

2.6.4.1 The Polymerase Chain Reaction (PCR)

Probably no other technique developed since the mid-1980s has revolutionized molecular genetics like the polymerase chain reaction (PCR). Developed by Kary Mullis, PCR solved the major problem of analyzing genes that are rare targets in a genome that could contain hundreds of thousands of genes (18). The core premise of PCR exploits certain features of DNA replication by DNA polymerase. First, because DNA polymerase requires a small section of DNA to initiate DNA synthesis, DNA polymerase can be directed to synthesize a specific region of DNA by supplying an oligonucleotide primer that anneals to the template DNA at that point. Second, the specific region of DNA can be amplified by supplying primers that flank the ends of the sequence, heating the reaction mixture to separate the original and newly synthesized strands, and allowing the reaction to occur over and over. In the original protocol, fresh DNA polymerase had to be supplied after each heating step. The method has since been facilitated by the discovery of thermal-stable DNA polymerases, which eliminates the need to add fresh DNA. Because each round or cycle of new DNA synthesis doubles the amount of the target DNA molecule, a typical reaction consisting of 30 cycles can amplify the target sequence to well over 260,000,000 copies.

2.6.4.2 Real-Time PCR

An extension of the basic PCR methodology has been developed to help quantify transcript concentrations without resorting to the more time-consuming Northern blotting analysis. In real-time PCR, mRNA is isolated from a total RNA sample and used as a template for PCR. Using primers conjugated to fluorescent dyes, the efficiency of the PCR reaction can be followed by a computer-assisted instrument that can detect the fluorescent dyes. Since the amount of DNA that is synthesized during the linear portion of the reaction will largely be dependent upon the initial concentration of the template, the initial concentration of specific mRNA templates can be back-calculated using software that accompanies the instrument. Being able to accurately determine mRNA concentrations can help to identify the effect of chemicals or even environmental conditions on the expression of certain genes.

2.6.4.3 Microarrays

With the advent of genomic sequencing came the detection of numerous (hundreds, even thousands) of previously unknown genes in the genomes of various organisms. Now the area of functional genomics, which is the study of gene function through parallel expression measurements of a genome, has emerged. Functional genomics studies have the promising potential of identifying functions for these previously unknown genes. The most common tool to carry out these measurements is the cDNA microarray. The basic technique involves the isolation of mRNA from a biological sample in a normal or conditional state, the conversion of the mRNA into cDNA molecules which contain a fluorescent tag, and the hybridization of these tagged-cDNAs to a microarray which contains tagged-probes for specific genes. The result is something between a few thousand and tens of thousands of measurements of gene expression. The current disadvantage of microarrays is their high cost, but, as the price continues to decrease, the use of microarrays will likely become more and more prevalent, especially in drug discovery research.

REFERENCES

1. Tsipis, C.A., P.A. Karipidis. Mechanism of a chemical classic: quantum chemical investigation of the autocatalyzed reaction of the serendipitous Wöhler synthesis of urea. *J. Am. Chem. Soc.* 125:2307–2318, 2003.
2. Nurse, P. The incredible life and times of biological cells. *Science* 289:1711–1716, 2000.
3. Hopkins, F.G. Feeding experiments illustrating the importance of accessory food factors in normal dietaries. *J. Physiol.* 44:425–460, 1912.
4. Sumner, J.B. The isolation and crystallization of the enzyme urease. *J. Biol. Chem.* 69:435–441, 1926.
5. Sumner, J.B., J.S. Kirk, S.F. Howell. The digestion and inactivation of crystalline urease by pepsin and by papain. *J. Biol. Chem.* 98:543–552, 1932.
6. Leone, F. *Genetics: The Mystery and the Promise*. Blue Ridge Summit, PA: TAB Books, 1992, pp 83–87.
7. Avery, O.T., C.M. Macleod, M. McCarty. Studies of the chemical nature of the substance inducing transformation of pneumococcal types: induction of transformation by a deoxyribonucleic acid fraction isolated from Pneumococcus Type III. *J. Exp. Med.* 79:137–158, 1944.

8. Hershey, A.D., M. Chase. Independent functions of viral protein and nucleic acid in growth of bacteriophage. *J. Gen. Physiol.* 36:39–56, 1952.
9. Watson, J.D., F.H.C. Crick. Molecular structure of nucleic acids; a structure for deoxyribose nucleic acid. *Nature* 171:737–738, 1953.
10. Meselson, M., F.W. Stahl. The replication of DNA in *Escherichia coli*. *Proc. Natl. Acad. Sci. USA* 44:671–682, 1958.
11. Jacob, F., J. Monod. Genetic regulatory mechanisms in the synthesis of proteins. *J. Mol. Biol.* 3:318–356, 1961.
12. Nirenberg, M., P. Leder. RNA codewords and protein synthesis: the effect of trinucleotides upon the binding of sRNA to ribosomes. *Science* 145:1399–1407, 1964.
13. Leder, P., M. Nirenberg. RNA codewords and protein synthesis II: nucleotide sequence of a valine RNA codeword. *Proc. Natl. Acad. Sci. USA* 52:420–427, 1964.
14. Smith, H.O., K.W. Wilcox. A restriction enzyme from *Hemophilus influenzae*: I: purification and general properties. *J. Mol. Biol.* 51:379–391, 1970.
15. Cohen, S.N., A.C. Chang, H.W. Boyer, R.B. Helling. Construction of biologically functional bacterial plasmids *in vitro*. *Proc. Natl. Acad. Sci. USA* 70:3240–3244, 1973.
16. Southern, E.M. Detection of specific sequences among DNA fragments separated by gel electrophoresis. *J. Mol. Biol.* 98:503–517, 1975.
17. Sanger, F., S. Nicklen, A.R. Coulson. DNA sequencing with chain-terminating inhibitors. *Proc. Natl. Acad. Sci. USA* 74:5463–5467, 1977.
18. Mullis, K.B., F.A. Faloona. Specific synthesis of DNA *in vitro* via a polymerase-catalyzed chain reaction. *Methods Enzymol.* 155:335–350, 1987.
19. Abel, P.P., R.S. Nelson, B. De, N. Hoffmann, S.G. Rogers, R.T. Fraley, R.N. Beachy. Delay of disease development in transgenic plants that express the tobacco mosaic virus coat protein gene. *Science* 232:738–743, 1986.
20. Hunkapiller, T., R.J. Kaiser, B.F. Koop, L. Hood. Large-scale and automated DNA sequence determination. *Science* 254:59–67.
21. Venter, V.C., H.O. Smith, L. Hood. A new strategy for genome sequencing. *Nature* 381:364–366, 1996.
22. Schena, M., D. Shalon, R.W. Davis, P.O. Brown. Quantitative monitoring of gene expression patterns with a complementary DNA microarray. *Science* 270:467–470, 1995.
23. Wilmut, I., A.E. Schnieke, J. McWhir, A.J. Kind, K.H. Campbell. Viable offspring derived from fetal and adult mammalian cells. *Nature* 385:810–813, 1997.
24. Venter, J.C., M.D. Adams, E.W. Myers, P.W. Li, R.J. Mural. The sequence of the human genome. *Science* 291(5507): 1304–1351, 2001.
25. International Human Genome Sequencing Consortium. Initial sequencing and analysis of the human genome. *Nature* 409: 860–921, 2001.

1.03

Fermentation Technology and Bioreactor Design

A. Eugene Raj and N.Ganesh Karanth

CONTENTS

3.1 HISTORICAL PERSPECTIVE

Fermentation dates back to prehistoric times when cheese was made in Iraq from cow's and goat's milk (6000 BC). Egyptians first discovered the use of yeast to make leavened bread and wine (4000 BC); Sumerians fermented barley to beer (1750 BC); and Chinese used moldy soybean curds as antibiotics (500 BC) (1). A summary of the fermentation events, reported chronologically, is given in Table 3.1. The first half of the twentieth century saw the development of fermentation processes for the industrial production of alcohol, acetic acid, lactic acid, citric acid, gluconic acid, acetone, butanol, and glycerol. It was during this period that anaerobic fermentations were carried out in submerged culture and

Table 3.1
Historic fermentation events (2)

Application/ Event of Fermentation	Period (AD)
Wine preparation promoted by Roman Emperor	Third century
Fermentation of yeast by Erxleben	1818
Lactic acid fermentation by Pasteur	1860
Fermentation enzymes in yeast by Buchner	1897
Discovery of penicillin by Alexander Fleming	1929
Discovery of other antibiotics	1945–present

aerobic fermentation in surface culture. Continuous demands for increased productivity of aerobic industrial fermentations for baker's yeast and organic acid production led to the use of deep tank vessels sparged with large quantities of sterile air. Development of the penicillin process during the Second World War reaffirmed the development of industrial fermentation. However, the postwar period resulted in a temporary setback due to the growth of the petrochemical industry, and the availability of inexpensive chemicals for cheaper production of organic acids and solvents by chemical synthesis. Toward the end of the twentieth century, a better understanding of molecular biology led to the development of efficient fermentation systems for production of biochemicals. Presently, fermentation technology has gained increased credence due to the power of microbe design biotechnology, perceived hazards to people and the environment of chemical synthesis, and better economics from use of renewable raw materials.

3.2 FERMENTATION IN FOOD BIOTECHNOLOGY

Fermentation is an important part of our lives. The relevance of fermentation to day to day life is evident from the fact that food can be both spoiled and made by fermentation. Many of the foods used for human consumption are fermented foods. Fermentation is one of the oldest techniques used for food preservation. Muscle cells use fermentation to provide energy for a quick response. The oldest food biotechnological processes include the baking of yeast leavened breads, brewing of beer, sake and wine, and production of yogurt and cheese. Biotechnology can improve the baking process with improvements in cereal grains and starter culture through recombinant DNA technology, use of enzymes as processing aids, and application of advanced batch and continuous fermentation technologies (3). Brewing is regarded as a typical example of traditional or old biotechnology, because of its long history (4). Preservation of food remains one of the major objectives of fermentation. In addition, there are other aspects such as wholesomeness, acceptability, and overall quality, which have to be maintained in food fermentations (5).

3.3 TYPES OF FERMENTATION

The fermentation process is mainly divided into two broad categories: submerged fermentation and solid-state fermentation. The former has been readily employed in industries for large scale production of alcohol, organic acids, enzymes, antibiotics, vitamins, and amino acids. Solid-state fermentation has been used for the production of microbial metabolites from fungi, but suffers from limitations of operation at large scales due to operational difficulties.

3.3.1 Submerged Fermentation

Submerged fermentation is the most popularly used technique for the production of a large number of products using a wide range of microorganisms. The medium used for submerged fermentation contains relatively highly processed ingredients. The water activity of the medium is high, making it prone to contamination if asepsis is not maintained. Rheological problems can be encountered at high substrate concentrations. Mass transfer from gas to liquid phase is usually a limiting factor, but due to better mixing, diffusional limitation of nutrients is not encountered in submerged fermentation. Better bioprocess control of fermentation process is possible with the help of online sensors.

3.3.2 Solid-State Fermentation

Solid-state fermentation (SSF) is used for the production of bioproducts from microorganisms under conditions of low moisture content for growth. The medium used for SSF is usually a solid substrate (e.g., rice bran, wheat bran, or grain), which requires no processing. In order to optimize water activity requirements, which are of major importance for growth, it is necessary to take into account the water sorption properties of the solid substrate during the fermentation (6). In view of the low water content, fewer problems due to contamination are observed. The power requirements are lower than submerged fermentation. Inadequate mixing, limitations of nutrient diffusion, metabolic heat accumulation, and ineffective process control renders SSF generally applicable for low value products with less monitoring and control. There exists a potential for conducting SSF on inert substrate supports impregnated with defined media for the production of high value products (7).

3.4 BIOREACTOR CONFIGURATIONS

3.4.1 Submerged Fermentor Systems

The fermentor is the heart of any biochemical process in which microbial, mammalian, or plant cell systems are employed for the economic production of fermentation products. A properly designed fermentor should be used to provide an aseptic, controlled environment to facilitate optimal growth and product formation of a particular cell system. In view of the broad scope of the fermentor for cultivation of microbial, mammalian, or plant cells, it is more commonly referred to as a bioreactor; this term will be used because of its global application.

The efficiency of performance of a bioreactor is dependent on concentration of biomass, maintenance of aseptic conditions, efficient mass and heat transfer, and operation at optimum process conditions. Bioreactors can be classified into three groups based on the type of biochemical process employed (8):

1. Bioreactor with no agitation and aeration (anaerobic processes, e.g., production of wine and beer)
2. Bioreactor with agitation and aeration (aerobic submerged fermentation processes, e.g., production of citric acid and penicillin)
3. Bioreactor with aeration, but no agitation (aerobic solid state fermentation processes, e.g., production of food enzymes)

However, in industrial practice, bioreactors are distinguished by their configuration and design. The common modes of bioreactor configurations are discussed below.

3.4.1.1 Stirred Tank Bioreactor

The most commonly used bioreactor for industrial applications is the conventional stirred tank reactor (STR). The STR offers the advantages of high oxygen transfer rates required for high biomass productivity coupled with low investment and operating costs, which form the basis for any successful aerobic fermentation process. A schematic of a stirred tank bioreactor is shown in Figure 3.1. STRs typically have height to diameter ratios of 1:3 to 1:6. The agitator may be top driven or bottom driven depending on the scale of operation and other operational aspects. The choice of impeller depends on the physical and biological characteristics of the fermentation broth. Usually, a ring-type sparger with perforations is used to supply air to the fermentor. Baffles are provided to avoid vortex formation and improve mixing.

Most fermentation processes use complex medium ingredients like corn steep liquor, molasses, and soybean flour as inexpensive nutritional sources (for carbon and nitrogen), supplemented with vital growth factors (amino acids, proteins, and vitamins) (9). The high turbulence imparted by the impellers in an STR can result in foaming due to the presence of proteinaceous substrates. Although chemical antifoaming agents (silicone or polypropylene glycol) can be added to control the foam, these can have detrimental effects on microbial growth and product recovery. In order to overcome this, mechanical methods of foam suppression such as rakes on the stirrer shaft mounted above the critical

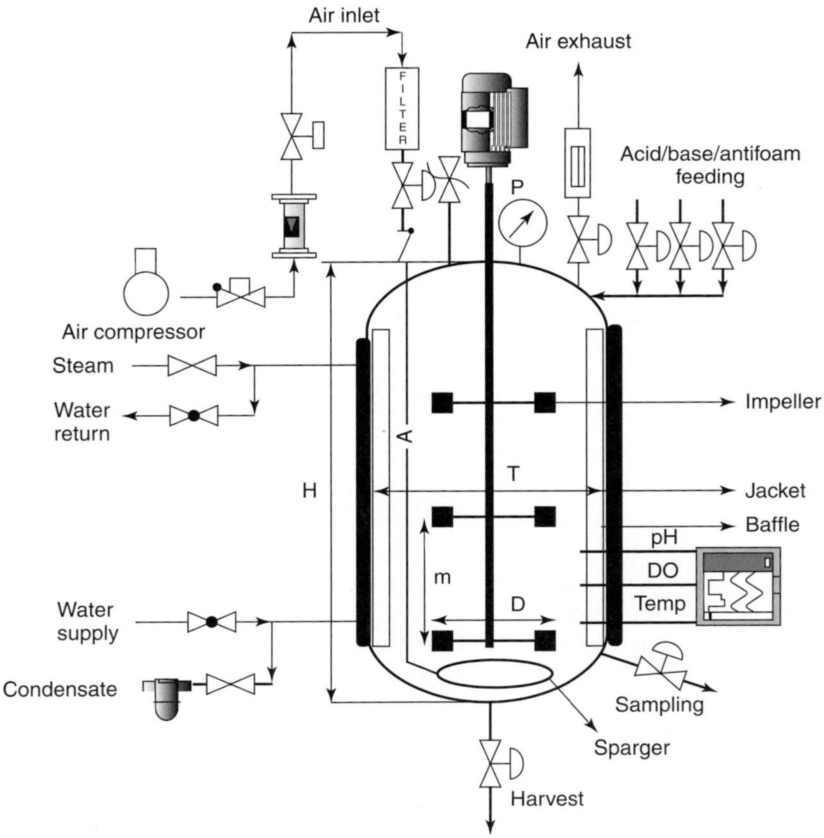

Figure 3.1 Stirred tank bioreactor

level have also been adopted. The emphasis on asepsis of the bioreactor, right from the end of the sterilization cycle to the end of the fermentation, has led to the maintenance of a minimum positive pressure in the fermentor to ensure sterility. A most important aspect of sterility is the point of contact between agitator shaft and vessel, which can be effectively sealed with a lubricated double mechanical seal. The sampling devices and injection ports must be contained in steam sterilizable closures.

3.4.1.2 Air Lift Bioreactor

For fermentations that have low shear and energy requirements, an air lift reactor can be useful. The amount of air required for the fermentation process is usually sufficient to act as the sole source of liquid mixing. In this process, air pumped from the bottom of the reactor creates buoyant, bubbles, which exert a drag on the surrounding fluid. A riser and a "down comer" inside the bioreactor impose a circulating fluid pattern of movement, which provides for oxygenation and mixing of the fermentation broth. The bottlenecks associated with large scale air lift bioreactors are inadequate sterilization, higher capital investment, and aeration requirements. Since mixing in an air lift is solely caused by aeration, the power required for fluid circulation and dispersion can be higher than that needed by an agitator in a stirred tank bioreactor. Air lift bioreactors have also been used for highly viscous fermentations, but the volumetric mass transfer coefficients are predictably low (10). An air lift bioreactor equipped with a draft tube results in better liquid circulation and larger gas to liquid interfacial areas, and gives higher mixing efficiency and oxygen transfer capability (11). A significant example of an air lift bioreactor that has been successfully scaled up to very large commercial levels is the ICI pressure cycle fermentor (12). In this process, methanol is used as substrate to produce single cell protein (SCP) using *Methylophilus methylotrophus* at a production capacity of 60,000 tons per year.

A three phase internal loop gas lift bioreactor was used for experimental studies on beer fermentation (13). This reactor has the advantages of high solid loading and better mass transfer properties than fluidized beds. Low density carriers of yeast cells, such as alginate and carrageenan gel particles, are typically used in three phase gas lift reactors for beer fermentations. The positive attributes of gas lift bioreactors are simple construction, low risk of contamination, easy adjustment and control of the operational parameters, and simple capacity enlargement (14,15). Research studies on air lift bioreactors have been carried out for the production of itaconic acid (16) and ξ-polylysine (17).

3.4.1.3 Fluidized Bed Bioreactor

In the last few decades, there has been a significant increase reported in the application of fluidized bed reactor systems. These have been mainly used for cells that have been immobilized onto particulate matter. This has the advantage that a high density of particles can be used, and that the flow velocity required for the fluidization can be achieved independently of the reactor throughput.

The main advantages of a fluidized bioreactor system as observed in ethanol production from *S.cerevisiae* (18) are superior mass and heat transfer characteristics, very good mixing between the three phases, relatively low energy requirements, and low shear rates (which makes a fluidized bed reactor suitable also for shear sensitive cells such as mammalian and plant cells).

Fluidized bed reactors have been used with cells adsorbed inside the carrier, made either of glass or of ceramics (19–21). The upward feed flow rate in a fluidized bed bioreactor is high enough to provide fluidization of carriers, resulting in improved mixing properties and medium distribution; but this can also induce carrier abrasion and damage.

In addition, fluidization of glass and ceramic carriers may require high medium flow rates that could result in higher pumping costs and eventually cell leakage. Gas liquid solid fluidized bed bioreactors have been employed for production of ligninolytic enzymes (22), treatment of wastewater from refineries (23), and raw wastewater (24).

3.4.1.4 Microcarrier Bioreactor

The majority of mammalian cells need a solid surface such as a microcarrier or a packed bed upon which to grow. The initial idea of culturing anchorage dependent mammalian cells in microcarriers was developed by van Wezel (25). The growth of cells on microcarrier beads depends directly on the surface available for growth up to the point where the microcarrier particles reach sufficient concentration to inhibit the cells and thus reduce cell yield. The microcarriers should have a density between 1.02 and 1.10 kg/m^3 to enable easy suspension in stirred reactors. The toxicity of the support can cause long lag phases, death of the cells in the early stages of development, and limited cell yields. Microcarrier bioreactor systems have been used for cultivation of human fibroblast cells to produce cell mass (26) and in the production of interferon (27). A great advantage of microcarriers is the high surface area for cell growth provided under low shear conditions, while still allowing conventional fermentor equipment to be used. However, bead to bead and bead to impeller collisions, and hydrodynamic shear forces, may cause reduced viability.

3.4.1.5 Membrane Bioreactor

Membrane bioreactors comprising hollow fiber systems have been developed and tested for the growth of mammalian and plant cells, and for the immobilization of bacteria, yeast and enzymes. Hollow fiber reactors have been used in the enzymatic hydrolysis of cellulose (28), penicillin (29), starch (30), hemoglobin (31), protein synthesis (32), and the culture of plant cells (33) and mammalian cells (34).

Hollow fibers can be made from cellulose acetate with a uniform wall matrix, or from acrylic copolymers or polysulphone fibers with asymmetric wall configurations. These hollow fibers have a highly porous surface wall about 70μm thick upon which the cells grow, and a cylindrical lumen of about 200μm in diameter. The surface of the lumen is covered with a thin ultrafiltration layer, which separates the immobilized cells and the lumen. Free diffusion of ions and molecules take place through the ultrafiltration layer, which may have nominal molecular weight cut offs (NMWCO) in the range of 10 kDa to 100 kDa. A hollow fiber bioreactor consists of a cylindrical bundle of a large number of individual fibers, which are held together in a shell and tube-type heat exchanger arrangement. Commercial hollow fiber bioreactor units are available with a luminal surface capacity between 0.01m^2 and 1.0m^2.

The advantages of using a hollow fiber reactor for microbial systems include high density of cell growth, using a perfusion system for simultaneous separation of product and biomass, and biocatalyst regeneration. However, a major disadvantage is the difficulty in monitoring and controlling the growth and metabolism of the culture. Other process constraints associated with microbial hollow fiber reactors are low oxygen transfer rates at high cell density and blockage, and rupture of the membranes due to excessive growth. The accumulation of toxic products in the hollow fiber might also inhibit the metabolic activity of the cell system. Further, the effect of microbial containment on physiology, long term viability, and productivity remains unclear. The technique has been used in the production of lactic acid (35), conversion of L-histidine and biosynthesis of β-galactosidase.

Membrane bioreactors have been used extensively for microbial, plant, and animal cell cultivations (36). A high performance membrane bioreactor has been studied for use in ethanol (37) and organic acid fermentation (38). Continuous fermentation in a membrane bioreactor performed at a very high dilution rate enhanced productivity (39). A double vessel membrane bioreactor was reportedly employed for the production of wine from grape juice (40), wherein the low residual sugar level maintained favored higher wine production compared to a single vessel in continuous fermentation. The production of tissue plasminogen activator (tPA) in microfiltration hollow fiber (MFHF) bioreactors for mammalian cell culture has been reported (41).

3.4.1.6 Photobioreactor

Microalgae have been used successfully, with high productivity compared to higher plants. The high productivity in these systems is due to the high biomass produced in the bioreactor. Microalgae have been used for preparation of vitamins, pigments, antioxidants, and fatty acids, and as feed for aquaculture. The cultivation techniques employed are open systems and closed or semiclosed outdoor photobioreactors. The common photobioreactors used are tubular-type and plate-type reactors (42). The cyanobacterium *Spirulina platensis* has been studied in batch and continuous photobioreactors under varying conditions of incident light energy and nutrient limitations (43).

3.4.1.7 Innovative and Special Bioreactors

3.4.1.7.1 Space Bioreactor The transfer of knowledge from conventional bioreactor technology to microscale space bioreactors for cultivation of cells required in space is an emerging field of space life science research and applications. The first space bioreactors were developed and flown at the end of the last century (44). With the development of an international space station, special attention has been focused on the development of life support systems that allow recycling of expendable materials (i.e., water, air), the treatment of waste byproducts (45), and cultivation of microorganisms, mammalian cells, and tissues for food production. The bioreactors normally used on Earth may not be suitable in space for several reasons:

1. Materials presently utilized for the fabrication of a bioreactor are not acceptable in space for safety and environmental reasons.
2. The space bioreactor equipment has to comply with size, weight, and power requirements.
3. The microgravity conditions in space create a significant hindrance for the operation of a bioreactor due to alteration in the physical factors governing cell sedimentation, nutrient mixing, and byproduct dispersion.
4. Nutrients and oxygen and waste products should be efficiently transported by means of medium exchange, perfusion, or slow mixing, as convection currents are also reduced to near zero.

For these reasons, new types of bioreactors specifically adapted to space investigations had to be developed. Space based bioreactors provide an opportunity to understand how fermentation processes could occur in the absence of gravity in a cell and its surrounding environment. This unique research environment opens new horizons for exploring unconventional bioprocessing techniques (46). Several types of cultivation systems have been designed or are currently under development as given in Table 3.2.

3.4.1.7.2 Tissue Bioreactor Tissue bioreactors are expected to play a dominant role in the production of tissues for healthcare applications. In the twenty-first century,

Table 3.2
Characteristics of the different space cultivation systems reported (47).

Bioreactor	Space Flight	Specifications
Dynamic Cell culture System	Biokosmos 9 (1989), Shuttle (1992)	Cultivation chamber with medium exchange (osmotic pump), no regulation
Swiss Bioreactor (SBR 1)	2 Shuttle flights (1994,1996)	Zero headspace bioreactor for yeast cells, flexible continuous medium exchange (piezo-electric pump), sampling port, flow rate and pressure sensor. pH regulation, online data transfer.

tissue engineering and regenerative medicine are expected to be powerful tools for repairing damaged or diseased tissues and organs, because human donor tissue cannot meet the demand (48). Tissue engineering is the development of biological substitutes to restore, maintain, or improve tissue function (49). A clinically useful bioreactor system will need to be compact and capable of maintaining a large number of cells at relatively high densities over a prolonged period. The cultivation of hepatoma cells used for the production of bioartificial livers has been attempted in tissue bioreactors utilizing fluidized bed and hollow fiber setup (50). With the focus on automation and standardization of tissue manufacture in controlled systems, bioreactors have the potential to reduce production costs, facilitating global use of engineered tissues. The role of bioreactors in processes that are key for the *ex vivo* engineering of three dimensional tissues based on cells and scaffolds include cell seeding of porous scaffolds, nutrition of cells in resulting constructs, and mechanical stimulation of the developing tissues (51). A new challenge for bioprocess engineering in cell culture involves the development of highly sophisticated cultivation techniques for hematopoietic stem cells in novel bioreactors (52,53). The interdisciplinary research associated with tissue engineering will provide a basis for identifying process conditions required for generation of a specific tissue.

3.4.1.7.3 Nuclear Magnetic Resonance Bioreactor Nuclear Magnetic Resonance (NMR) spectroscopy has been used as a tool for noninvasive, real time studies of metabolic processes of cell suspensions in bioreactors (54). The NMR bioreactor is used for online NMR analysis of metabolic reactions in fermentation processes. One of the critical parameters in the evaluation of bioreactors for product formation is the oxygenation state of the cells, which can be determined using NMR (55). An NMR reactor (manufactured by M/s Bioengineering AG, Switzerland) is an autoclavable, miniaturized stirred vessel with ports to measure pH, pO_2, temperature, speed, and aeration, and for the addition of medium. The fully equipped NMR reactor can be inserted into an NMR unit.

3.4.1.7.4 Mass Spectrometer Coupled Bioreactor A minibioreactor with membrane inlet mass spectrometer probe is reported for biological processes with online analysis of volatile compounds such as H_2, CH_4, O_2, N_2, CO_2, ethanol, and methanol (56). The reactor comprises a small, stirred reaction vessel with a thermocouple, a pH probe, agitation, temperature control, and an option for aeration. The reactor is coupled to the mass spectrometer with the help of silicon rubber or fluorohydrocarbon membranes separating the bioreactor from the high vacuum in the mass spectrometer. Volatile compounds are selectively introduced into the mass spectrometer, where they are quantified according to their mass to charge ratio. Typical measurement ranges are from percent to parts per trillion levels in a few seconds to minutes. This bioreactor offers continuous, sensitive detection of small changes in the concentrations of dissolved gases, permitting fast kinetic measurements and in depth metabolic studies.

3.4.1.7.5 Integrated Bioreactor An integrated bioreactor is aimed at improving the productivity of the fermentation process by integration of fermentation and product recovery, for the continuous removal of a potentially inhibitory product. Two phase partitioning bioreactors have a great scope for increasing the productivity of a bioprocess (57). The concept of the two phase partitioning bioreactor can be applied to controlled delivery of a toxic substrate like phenol or benzene dissolved in an organic phase to a cell containing aqueous phase (58). A laboratory scale bioreactor coupling conventional electrodialysis and bipolar membrane electrodialysis was developed for *in situ* product removal and pH control in lactic acid fermentation (59). This electrokinetic process enabled removal of concentrated lactic acid directly from the bioreactor system, enabling good pH control and reduced end product inhibition of glucose catabolism. A membrane distillation integrated bioreactor for ethanol production using *S.cerevisiae* resulted in a higher productivity of 5.3 gm ethanol/dm^3/h, compared to 2.6 gm ethanol/dm^3/h without membrane distillation (60). The morphological and biochemical characteristics of transformed *Nicotiana glauca* roots demand integrated bioreactors for root growth and product extraction. A novel root tube bioreactor separator was used for the increased alkaloid productivity (61).

3.4.2 Solid-State Fermentor Systems

The main difference between submerged and solid-state fermentations is the amount of free liquid in the substrate. Solid-state fermentations (SSF) exhibit a poor conductive gas phase between the particles as compared to submerged fermentation (62,63). The presence of a wide variety of SSF matrices in terms of composition, size of solid substrate, mechanical resistance to air flow, porosity, and water holding capacity renders bioreactor design and control more difficult for the regulation of two important parameters, namely temperature and water content of the solid medium (64). Other factors that influence the bioreactor design are fungal morphological characteristics, resistance to mechanical agitation, and degree of asepsis required for the fermentation process. Solid-state bioreactor systems commonly used are shown in Figure 3.2.

3.4.2.1 Laboratory Scale SSF bioreactor

Small scale SSF equipment can be classified as those without forced aeration and agitation to include Petri dishes, jars, widemouth Erlenmeyer flasks, Roux bottles and roller bottles

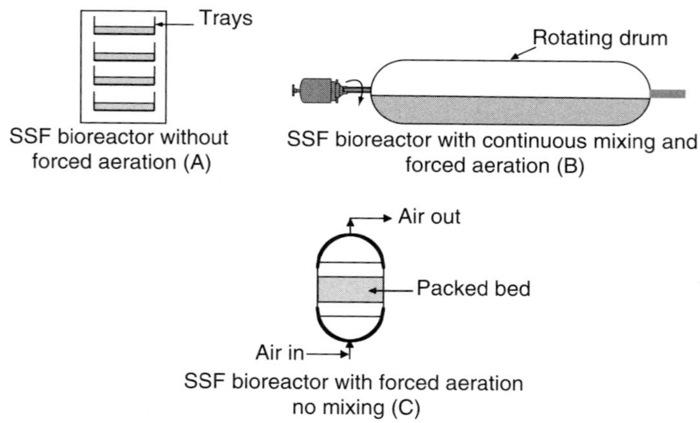

Figure 3.2 Solid state bioreactor systems

(65), and those incorporating continuous agitation of the solid medium such as a rotating drum bioreactor (66), a perforated drum bioreactor (67) and a horizontal paddle mixer. The former are easy to operate in large numbers and commonly used for the screening of substrates or microorganisms for research purposes, while the latter offer the advantage of temperature control due to continuous agitation.

3.4.2.2 Industrial Scale SSF Bioreactor

Industrial scale SSF bioreactors can be built with or without aeration. Those without forced aeration can exhibit limitation of heat and mass transfer as the fermentation progresses, changing the properties of the microorganism involved, particularly in light of associated complexities like heat build up and inadequate oxygen transfer. However, with aeration strategies like circulation of air around the substrate layer or passing air through the substrate layer,, these limitations are reduced to a certain extent.

3.4.2.2.1 SSF Bioreactor Without Forced Aeration On an industrial scale, this bioreactor is generally a tray fermentor; more commonly termed a Koji reactor. The trays containing the solid medium are stacked in tiers and placed in a humidity and temperature controlled chamber (68). This technology has the limitations of not conforming to asepsis conditions, and of high labor requirements. However, it is easily scaled up by the incorporation of additional trays.

3.4.2.2.2 SSF Bioreactor With Forced Aeration and No Mixing In this type of bioreactor, no mechanical agitation is provided, but the medium can be manually agitated *in situ* or it can be transferred into a kneading machine and reloaded into the basket. However, this type of device without agitation is limited by the metabolic heat produced. Considerable temperature gradients can exist within the substrate bed. As the majority of the heat is removed and water is evaporated by forced aeration, the bed dries out, reducing fermentation efficiency. Periodic water addition is required to maintain the moisture content at desired levels (69,70).

Suryanarayan (71) reported an industrial patented bioreactor, Plafractor™, with a capacity to hold 20 kg of wheat bran, constructed by stacking and interconnecting individual modules. The metabolic heat produced during fermentation is removed by conduction. There exist two channels in this bioreactor for ease of operation. The noncommunicating channels deliver cooling and heating fluids sandwiched between two sheets, while the communicating channels deliver fluids for sterilizing (steam or ethylene oxide), for adjusting the moisture and oxygen content, and for extracting the compound of interest after the cultivation. The interior of each module has a mixing arm that revolves about the central axis of the module while rotating.

3.4.2.2.3 SSF Bioreactor With Continuous Mixing and Forced Aeration A rotating drum bioreactor with continuous mixing maximizes the exposure of each substrate particle to a thermostatic air circulating unit in the headspace (72,73). A large reactor, capable of handling 10 kg of steamed wheat bran as substrate, has been reported (74). Large scale use of unagitated SSF is limited by the difficulty in maintaining temperature during the fermentation. However, in a rotating drum bioreactor, efficient heat transfer is possible by convective and evaporative cooling. As the scale of fermentation increases, evaporative cooling becomes significant, because the ratio of the heat produced to the surface area available for convection decreases (75).

The inherent difficulties encountered in the operation of solid-state fermentation systems on a large scale has led to new developments aimed at improving the efficiency of the fermentation process (76,77).

3.5 STAGES IN A FERMENTATION PROCESS

3.5.1 Upstream Processing

The upstream processing in a fermentation process includes preparation of the fermentation medium, sterilization of air and fermentation medium and inoculation of the fermentor.

3.5.1.1 Fermentation Medium

Metabolic activity can only be maintained if the necessary nutrients are available to the cell. A medium is defined as the substance surrounding the cells, which enables the micro-organisms to grow and form products. While the single components of a medium are defined as substrates, generally, only the carbon source is called the substrate. Formulation of proper medium is an essential requirement for the success of a fermentation process.

Medium used for industrial fermentations should have the following criteria (78):

1. Maximum yield and concentration of product or biomass per unit mass of substrate
2. Minimum production of undesirable metabolites
3. Consistent quality of product
4. Minimal problems during sterilization and fermentation

The medium used for fermentation may be classified as defined, complex or technical medium. Defined medium consists only of precisely chemically defined substrates. At laboratory and pilot plant level where physical, chemical, and physiological parameters need to be standardized first, only defined media should be used for studies.

Complex medium is composed of substrates with undefined composition, such as extracts or hydrolysates from waste products, which are cheap substrates commonly used in industrial production. Relatively expensive substrates, such as yeast extract, brain heart infusion, peptone, casamio acids, and tryptone are often used for complex medium. In order to maximize the yield and selectivity of serine alkaline protease production by recombinant *Bacillus subtilis,* based on the amino acid compositions of the enzyme, the fermentation medium was designed with complex nutrients to supply the most needed amino acids to increase the yield (79).

Technical media are used on an industrial scale and are cheaper. Normally, but not necessarily, complex substrates are the main components. Culture medium formulated for the extracellular production of a hybrid ß-1,3-glucanase from *Bacillus* using a recombinant *Escherichia coli* contained lactose, yeast extract, and sodium chloride, and fermentation was carried out in shake flasks and a stirred tank bioreactor (80). The substrate sources can also be derived from industrial waste, and are often highly impure mixtures, requiring pretreatment before they could be used for a fermentation process. Examples are soy meal, whey, fishmeal, malt extract, and sulfite waste liquor. Wastewater from monosodium glutamate production, which contains high levels of chemical oxygen demand (COD), sulphate, and ammoniacal nitrogen at a low pH, has been used as the nitrogen and water source, with sugar beet pulp as the carbon source, for the production of pectinase by solid-state fermentation using *Aspergillus niger* (81).

3.5.1.2 Optimization of Fermentation Medium

Optimization of fermentation medium is done by using an experimental design considering the variables, mostly medium ingredients, that may have a significant influence on the fermentation process. Experimental design for the fermentation process involves the determination of the optimum medium composition, while simultaneously figuring out the best process parameters. Design of experiments (DOE) is a systematic approach to problem solving which is applied to data collection and analysis to obtain information rich data (82). DOE is concerned with carrying out experiments under the constraints of minimum expense of time, costs, and number of runs. A properly designed experiment is more important than detailed statistical analysis. The primary goal of designing an experiment is to obtain valid results using a minimum of effort, time, and resources (83). The limitation of single factor at a time optimization is overcome by taking into account all the significant variables collectively by statistical experimental design using Response Surface Methodology (RSM) (84), which is used to evaluate the relative significance of variables in the presence of complex interactions (85,86).

3.5.1.3 Components of Industrial Fermentation Medium

In industrial processes, the cheapest substrates are used to keep the production costs low. Some of the cheap substrates available include barley, barley malt, blood meal, cane molasses, corn gluten meal, corn meal, corn steep liquor, cotton seed meal, dried distillers' solubles, fish solubles, fishmeal, linseed meal, meat meal, bone meal, oat flour, peanut meal, rice bran, rice flour, soybean meal, wheat flour, whey powder, and yeast hydrolysate (87). The composition of the medium has to reflect the demands necessary for the growth and synthesis of products other than cells. The carbon to nitrogen ratio has been found to be a good indicator for studying the optimum requirements for growth and product formation (88). Generally, media consist of an aqueous solution containing dissolved nutrient salts and a gaseous component (O_2). The different nutrient sources commonly used in fermentation media are shown in Table 3.3.

With the economics of fermentation playing an important role in a successful fermentation process, the efficient conversion of substrate to product is most crucial. In the present stage of advanced biotechnological development, this involves genetic engineering of the strain as well as a practical, focused approach in optimal use of the substrates with the aid of metabolic engineering (89). Metabolic engineering of *S.cerevisiae* for efficient xylose fermentation to ethanol has been reported (90).

Table 3.3
Nutritional sources used in fermentation media (91)

Carbon Source	Nitrogen Source: Organic	Nitrogen Source: Inorganic	Inorganic Elements	Vitamins	Medium Additives
Carbohydrates, Alcohols, Carboxylic acids, Fats, Hydrocarbons, Gaseous substrates.	Urea, Aminoacids, Purines, Pyrimidines, Complex sources such as CSL, dried yeast, protein hydrolysates.	Nitrates, Nitrites, Ammonia, Molecular Nitrogen.	Phosphorus, Sulphur, Magnesium, Potassium, Calcium, Chlorine, Cobalt, Copper, Iron, Manganese, Molybdenum, Zinc.	Thiamine, Riboflavin, Pyridoxin, Biotin, Pantothenic acid, Niacin, Inositol, Choline.	Growth factors, Precursors, Detergents, Antifoaming agents, Anti-microbial agents.

3.5.1.4 Sterilization

Sterilization is a process by which microorganisms are either killed or removed from the material or equipment. Sterilization is necessary to ensure that only the desired microorganism is present to carry out the fermentation, that products are made of predicted quality, that the environment is protected from undesirable contamination, and that deterioration (microbial spoilage) of products is prevented.

3.5.1.4.1 Medium Sterilization Techniques The sterilization techniques applicable for fermentation media sterilization include:

1. Sterilization by high temperature achieved by direct or indirect steam or electric heating, which is the most popular and efficient method.
2. Sterilization of the medium for animal cell culture by membrane filtration with absolute rating of 0.1–0.04 μm.
3. Microwave irradiation, which is used commonly in food industries and has been reported to cause cell death (92).
4. Newer techniques, like high voltage pulses and photosemiconductor powders (93), which involve the rupture of the cell membrane by increasing the transmembrane electric field strength beyond a certain threshold.

3.5.1.4.2 Sterilization by Heat When a medium containing microorganisms is heated above a certain temperature limit, the microorganisms are unable to survive. However, because endospores have higher heat tolerance, inactivation of spores is a good indication of the sterilization process efficiency. In the testing of a heat sterilization process, *Bacillus stearothermophilus* spores, which are the most temperature resistant, are widely used as test organisms (94).

Batch sterilization: Sterilization of medium in the fermentor can be carried out in batch mode by passing steam through the available heat transfer area (jacket or limpet coil) or through direct steam sparging or with electrical heaters. The highest temperature to be operated for batch sterilization of medium is 121°C. The batch heat sterilization is described by a first order kinetics resulting in the equation:

$$N_t / N_o = e^{-kt} \tag{3.1}$$

where N_t is the number of surviving microorganisms at t, N_o is the initial number of microorganisms, k is the specific death rate, and t is time.

Equation 3.1 can be rearranged as:

$$\ln (N_o / N_t) = kt \tag{3.2}$$

The reaction or sterilization rate increases with increase in temperature due to an increase in the reaction rate constant or specific death rate (k). The relationship between temperature (T) and specific death rate (k) can be expressed as:

$$k = A\, e^{-E/RT} \tag{3.3}$$

where A is the Arrhenius constant, E is the activation energy, R is the gas constant and T is the absolute temperature.

Substituting k from Equation 3.3 into Equation 3.2:

$$\ln (N_o/N_t) = A\, e^{-E/RT}.t \tag{3.4}$$

The term "ln (N_o/N_t)" has been used as a design criterion for sterilization (95) and is represented as the Del factor (∇), which is a measure of the fractional reduction in viable cell count produced in a certain temperature and time regime. Thus, we have:

$$\nabla = A\,e^{-E/RT}.t \qquad (3.5)$$

where t is the time required to achieve a certain ∇ value.

Rearranging Equation 3.5:

$$\ln(t) = E/RT + \ln(\nabla/A) \qquad (3.6)$$

From a plot of ln (t) vs 1/T, the thermal death characteristics, namely activation energy (E) and Arrhenius constant (A), can be determined.

The batch sterilization cycle consists of heating, holding, and cooling cycles. The overall Del factor for destruction of cells during the sterilization period is represented as:

$$\nabla_{\text{overall}} = \nabla_{\text{heating}} + \nabla_{\text{holding}} + \nabla_{\text{cooling}} \qquad (3.7)$$

By taking into account the Del factor during heating and cooling cycles in the design of a sterilization process, a minimum holding time is determined.

The advantages of batch sterilization are low capital equipment costs, low risk of contamination, easier manual control, and suitablity for media containing a high proportion of solids.

Continuous sterilization: Continuous sterilization has the advantages of relatively straightforward energy recovery, and lower nutrient degradation of heat labile substances due to shorter holding time. However, a disadvantage is that the presence of solids may give rise to inefficient sterilization, requiring a proper design.

Continuous heat sterilization essentially consists of a heating section using heat exchangers (typically spiral-type plates on an industrial scale), and a holding section where the temperature of the medium is maintained constant at the sterilization temperature. The inlet medium to the continuous sterilizer is preheated using hot fluid from the holding section in the heat exchanger. This process of energy transfer in the heat exchanger between the inlet raw medium and outlet sterilized medium not only cools the sterilized medium effectively, but also results in considerable energy savings. It is important to note that in the presence of solids, the rate at which medium can be heated is limited by the heat transfer rate from the liquid to the solid particles (96).

A continuous sterilizer has the advantage of being more economical in building area. It also offers greater possibility of automatic control, and reduction of sterilization process cycle time, resulting in minimum oversterilization and ease of scale up. Moreover, service requirements for steam and water are constant, compared with the fluctuating demand of batch sterilization. However, it may not be economical for small capacity batch fermentations.

3.5.1.4.3 Sterilization of Air Aerobic fermentation processes require significant quantities of sterile air. The sterilization of air in the strict bacteriological sense means the complete elimination of all viable microorganisms. The removal of bacteria by means of depth filters consisting of granular carbon or fibrous media has been almost universally adopted. In most fermentation systems, a prefilter usually made of cellulose, viscose, or glass wool is installed prior to the absolute filter to remove dust, oil droplets, and moisture from the process air. The most common material of construction of an absolute filter cartridge is pleated PTFE (Poly tetra fluoro ethylene, or Teflon™) membrane, which is hydrophobic and resists wetting. The membrane is intimately supported by pleated material, which prevents from distortion under high hydraulic or gas pressure. The absolute filters have a rating of 0.2–0.01 μm (for total removal of bacteria, viruses, bacteriophage, or spherical particles of that size) and can be sterilized using steam. After air filter sterilization, the filters are held under positive pressure until the end of fermentation. Pharmaceutical grade cartridges are processed with deionized, pyrogen free water and evaluated by bacterial challenge in accordance with international recommendations using standard test organisms (e.g., *Pseudomonas diminuta*, *Acholeplasma laidlawii*).

3.5.1.5 Inoculation

The process of inoculation is the transfer of seed material or inoculum into the fermentor. Inoculation of a laboratory fermentor is generally done using presterilized tubing and a peristaltic pump. However, on a larger scale, inoculum transfer is done by applying a positive pressure on the inoculum fermentor and connecting it aseptically to the production fermentor. The connecting lines are sterilized before being used for transfer of inoculum. Heat susceptible substances such as amino acids and some vitamins must be dissolved in small volumes of water, sterilized by filtration and added separately to the final medium aseptically.

The performance of a fermentation process is dependent largely upon the physiological status of the inoculum (97). It is necessary that the inoculum transfer time be determined experimentally at a laboratory scale and that a standard is set for cultural conditions needed for development of the inoculum. Inoculum transfer is commonly done with vegetatively growing biomass determined by parameters such as turbidity, packed cell volume, dry weight, wet weight, or morphological characteristics (98). Online parameters which are considered for inoculum transfer include pH, dissolved oxygen, and oxygen and carbon dioxide concentration in the exit gas. The effect of inoculum age on the productivity of a secondary metabolite from *Streptomyces* species has been studied (99). The inoculum age at three time intervals, namely early log, log, and declining phases of growth determined online by the carbon dioxide production rates were carried out to determine the best inoculum for fermentation. It was observed that although there was a marginal influence on the biomass formation from the different inoculums, the log phase inoculum gave a significantly higher product concentration compared to the other two in the fermentor.

3.5.2 Fermentation Process

The fermentation process involves actual growth of the microorganism and formation of the product under agitation and aeration, to provide uniform environment and adequate oxygen to the cell for growth, survival, and product formation.

3.5.2.1 Modes of Operation

A fermentation system is usually operated in one of the following modes: batch, fed batch, or continuous fermentation. The choice of the fermentation mode is dependent on the relation of consumption of substrate to biomass and products.

3.5.2.1.1 Batch Fermentation Fermentation processes performed on a batch basis involve a sequence of operations, from development of inoculum from a stock culture, to seed for a production fermentor. The seed and production fermentors are the main concern in fermentation process development. A number of seed stages may be involved, but the production stage is usually performed in a single fermentor (100). The multistage system has the advantages of increased productivity resulting in lower total fermentor volume and the possibility for variation of the environmental conditions from stage to stage. The time required for batch fermentation varies from hours to weeks depending on the conversion attempted and conditions used. Growth rate in batch fermentation is normally uncontrolled and is highest at the start (101).

Productivity of batch fermentation: The productivity of batch fermentation is calculated by the final concentration of biomass or product being produced divided by the complete time of batch, which includes fermentation time and turnaround time (time for emptying, cleaning, sterilizing, and refilling). The batch fermentation setup and productivity for biomass production is shown in Figure 3.3 and Figure 3.4, respectively.

3.5.2.1.2 Continuous Fermentation Continuous fermentation is an open system to maintain cells in a state of balanced growth by continuously adding fresh medium and removing the culture medium at the same rate. Essentially, the two modes of operation for continuous fermentation are chemostats and auxostats. The commonly used auxostats include turbidostats (102), the pHauxostat (103), and the nutristat (104). The above modes of operation have specific control configurations and applications as discussed below.

Chemostat: Presently, the chemostat is the most widely used apparatus for studying microorganisms under constant environmental conditions. It is a continuous fermentation process performed in a Continuous Stirred Tank Reactor (CSTR). A CSTR operates by

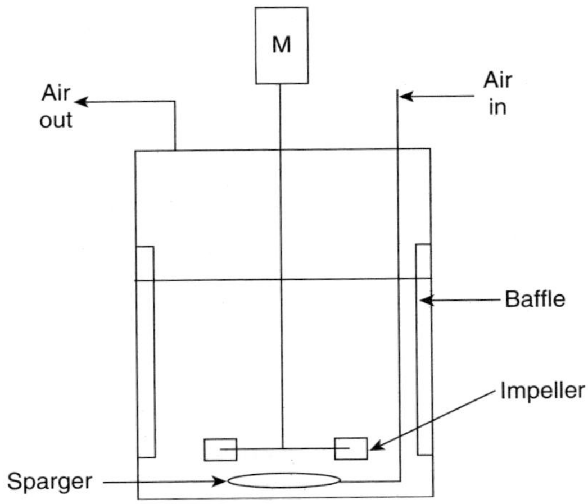

Figure 3.3 Batch fermentation system

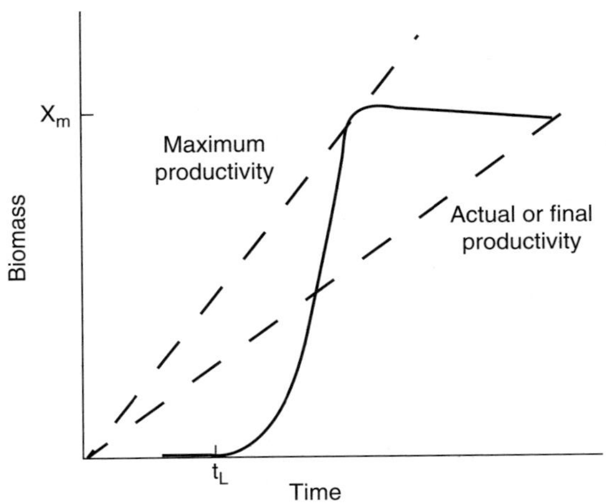

Figure 3.4 Productivity in batch fermentation

maintaining a growth rate through continuously feeding a growth limiting nutrient and withdrawing part of medium at the same rate, thereby achieving steady state growth. The growth limiting nutrient may be carbon, nitrogen, phosphorus, or any other essential nutrient, which influences the specific growth rate. A significant advantage of chemostat mode over batch mode is that by changing the feed rate of growth limiting nutrient, the growth rate can be varied. A schematic of chemostat (single stage) cultivation with and without cell recycle is shown in Figure 3.5 and Figure 3.6 respectively.

Auxostat: An auxostat is a continuous culture technique wherein the dilution rate is regulated based on an indication of the metabolic activity of the culture. A chemostat is essentially used for operation at moderate to low dilution rates, but an auxostat is used at high dilution rates. Population selection pressures in an auxostat lead to cultures that grow rapidly. Practical applications include high rate propagation, destruction of wastes with control at a concentration for maximum rate, open culturing because potential contaminating organisms cannot adapt before washing out, and operation of processes that benefit from careful balance of the ratios of nutrient concentrations. In a pHauxostat, the feed rate is regulated by measurement and control of the pH of the fermentation medium. This can be applied only if there is a change in pH consequent to the growth of microorganism. The pHauxostat has been used for continuous mass cultivation of bacteria for isolation of intracellular products (105,106).

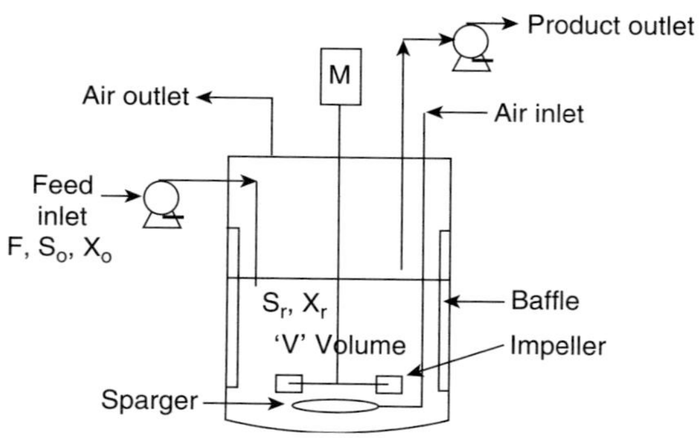

Figure 3.5 Continuous fermentation system

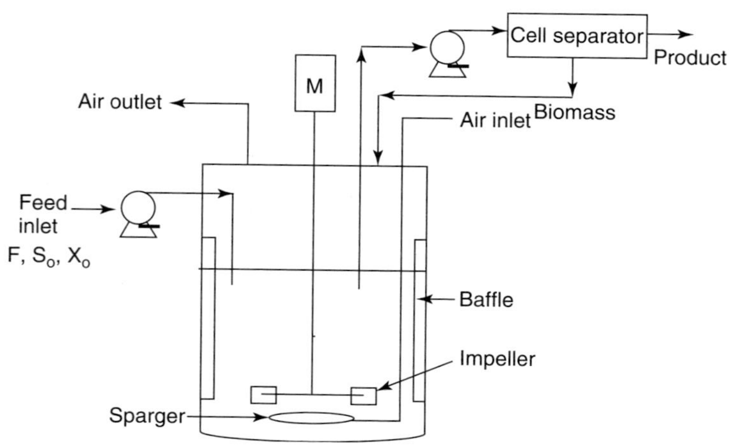

Figure 3.6 Continuous fermentation system with cell recycle

Turbidostat: The turbidostat controls the feed rate depending on the optical density (turbidity) of the fermentation broth, which is proportional to the biomass concentration. In this mode of operation, the culture cannot washout as in a chemostat. This mode of operation is ideal only when operated near maximum growth. The isolation of acid tolerant baker's yeast variants was developed in a turbidostat (107).

Nutristat: The nutristat involves regulation of the feed rate to maintain the residual substrate concentration during fermentation. The use of specific sensors for monitoring the residual substrate level during fermentation is employed. Ion selective electrodes (NH_4^+) have been used for control in the nutristat. However, the lack of reliable and accurate sensors for common substrates is a bottleneck to nutristat operation (108).

Productivity of continuous fermentation: For a continuous fermentation, there is no emptying, cleaning, sterilizing, and refilling component. The productivity of a continuous fermenter is calculated by multiplying the dilution rate (D) by the concentration of product in the outlet stream:

$$\text{Cell productivity} = \text{D.X (kg cells/m}^3\text{/h)} \tag{3.8}$$

$$\text{Product productivity} = \text{D.P (kg product/m}^3\text{/h)} \tag{3.9}$$

The productivity in continuous fermentation for biomass formation is represented in Figure 3.7. The commonly employed commercial applications of continuous culture include baker's yeast, vinegar, gluconic acid, acetone, butanol, and ethanol fermentation systems.

3.5.2.1.3 Fed Batch Fermentation Fed batch fermentation, which is a technique in between batch and continuous fermentation, is a more recent development in industrial fermentation systems. Neither batch nor continuous fermentation is suitable for non growth associated products. In order to produce such products, it is first necessary to build up a high concentration of cells in the growth or batch phase, and then switch the metabolism of the cell to arrest cell growth by feeding product precursors, carbon, and oxygen at a rate sufficient to meet the maintenance and product synthesis requirements. Essentially, fed batch fermentation involves two phases: growth phase and production phase. After the initial growth phase, one or more of the nutrients are supplied to the fermentor while cells and product remain in the fermentor.

The rationale for fed batch fermentation is to match the organism's demand for nutrients by feeding an appropriate amount of that nutrient. To determine the fed batch approach, nutrient demand has to be ascertained. This demand is a function of cell mass and cell yield on the nutrient. Though batch fermentation may be considered simple, fed batch fermentation

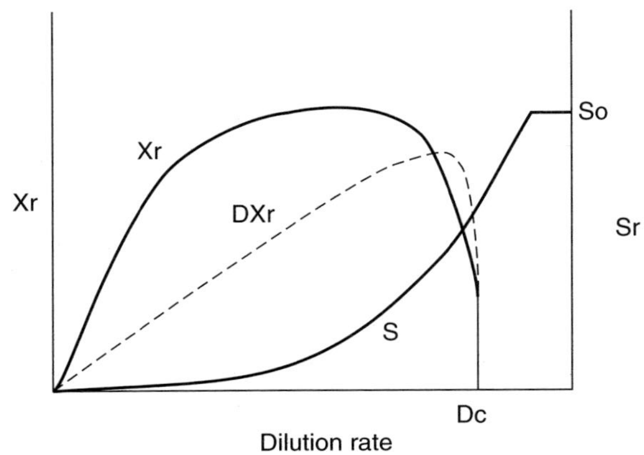

Figure 3.7 Cell productivity in continuous fermanation system

offers the convenience of better control over substrate concentration variations and differentiation of growth, leading to improved overall productivity with essentially the same equipment used for batch fermentation. Fed batch systems can be superior to continuous systems due to problems involving contamination during fermentation.

Fed batch fermentation is well suited for production of compounds during very slow growth where there is no possibility of cell washout. Fed batch fermentation is well suited for producing product or cells when: 1) Substrate is inhibitory and there is a need to maintain low substrate concentration to avoid the cells being inhibited (e.g., citric acid, amylase), and 2) Product or biomass yields at low substrate concentrations are high (e.g., baker's yeast, antibiotic production).

There are two methodologies in the fed batch approach, namely fixed volume fed batch and variable volume fed batch (109). In fixed volume fed batch fermentation, a very concentrated feed nutrient is fed to the fermentor so that there is no appreciable increase in volume. The specific growth rate decreases with time and the biomass increases directly with time. In variable volume fed batch fermentation, there is an increase in volume due to nutrient inflow and no outflow. The specific growth rate is solely dependent on the concentration of the limiting nutrient.

Productivity of fed batch fermentation: The productivity of a fermentation process is better in the fed batch mode compared to a batch mode of operation (110). In batch fermentation of *S. cerevisiae*, a dry cell weight of 10 g/L was obtained, whereas in fed batch mode, with respiratory quotient (RQ) as the control parameter for glucose feed, a final dry cell weight of 31–56 g/L was obtained (111).

3.5.2.2 Agitation

In stirred tank bioreactors, mixing and dispersion of air in the fermentation broth is achieved by mechanical agitation. The presence of impellers on the agitator shaft brings about uniform mixing of microorganisms and nutrients, and dispersion of air in the nutrient solution, resulting in efficient mass and heat transfer. Mixing by aeration or gas movement alone cannot be used in highly viscous systems due to coalescence of the gas bubbles. The heavy mycelial growth produced in antibiotic fermentations is typical of such a condition and hence mechanical agitation is essential for optimum production (112).

A turbine or a propeller, when rotated in a vessel without baffles, imparts circular flow due to swirling of the liquid around the vortex, which renders the motion unconducive for top to bottom mixing. In a baffled tank, on the other hand, the flow pattern may be radial or axial, promoting lateral flow and vertical flow currents, with the application of large amounts of power. Baffles reduce the unproductive tangential velocity component of all impellers to produce a more efficient radial or axial flow. Positioning and design of baffles, along with the placement, type, and design of heating coils has a significant impact on the flow patterns (113).

The types of impellers used in bioreactors are broadly classified based on flow pattern as radial flow (e.g., turbine) and axial flow (e.g., high efficiency impeller).

The Rushton turbine is the most common type of radial flow impeller. It consists of a number of short blades attached to a central shaft. The diameter of a turbine is normally between 30 and 50% of the tank diameter and there are usually between four to six blades. Turbines with flat blades are good for gas dispersion, where the gas is introduced just below the impeller on the axis and is drawn up to the blades and broken up into fine bubbles (114).

An axial flow impeller generates an axial flow pattern with the fluid flowing down the central axis and up on the sides of the tank. This flow pattern is also called down pumping because the impellers circulate the fermentation broth downward against the flow of the rising air, thereby holding the gas in the system longer. In multiphase fermentation systems involving solids, the axial component of the flow due to agitation is useful to keep the solids in suspension (115).

Recent developments in impeller design have led to the emergence of high efficiency impellers like MIG and INTERMIG, which require 25% and 40% less power input to get the same degree of mixing as a turbine impeller. Recently, high solidity hydrofoil impellers with an up pumping mode of operation were reported to increase mass and heat transfer, to and improve blending. Blending, which is less complicated than mass transfer, is not efficient on a large scale compared to small lab scale fermenters due to the formation of dead zones between impellers. Hence, optimal blending is essential for maximum yield and productivity in large scale fermentations (116).

Multiple impeller bioreactors (with impellers fixed at various heights on the central shaft) are becoming increasingly popular due to efficient gas distribution, higher gas phase residence time, increased gas hold up, superior liquid flow (plug flow) characteristics, and lower power consumption per impeller as compared to single impeller systems (117,118). Using both Rushton turbines and high efficiency axial flow impellers on the same shaft is reported to improve fermenter performance with a decrease in the total power consumption (119).

3.5.2.3 Aeration

The cheapest mode of supplying oxygen to the fermentation media is air. The aeration system essentially comprises an air compressor, an air filter (prefilter and sterile filter) and a sparger. The air compressor used for fermentation may be a positive displacement or a nonpositive displacement unit, with the former being a reciprocating or rotary compressor and the latter a centrifugal compressor. The choice of compressor depends upon such factors as the discharge pressure, type of drive, capacity, and cost (120). Equipment sizing of the compressor is usually dependent on the maximum airflow requirement of a full production scale for all the fermentors. Careful scheduling of fermentations can frequently reduce the gap between peak and average demands and result in more economical use of the equipment.

The discharge pressure of the air from the compressor is subjected to pressure drop along the route at predictable locations, such as valves and pipefittings along the supply lines, air filters, the sparger, hydrostatic head, and static pressure, before it reaches the fermentor. The diameter of the supply line is so chosen that the maximum airflow may be enabled without undue pressure loss. A pressure drop of 1–2 p.s.i. per 100 ft of pipe may be used as a first approximation. The first three pressure drops discussed above increase with increasing flow rate, but the hydrostatic head and static pressure are independent of airflow. The hydrostatic head is the depth of nonaerated liquid above the sparger, corrected for density, and usually represents quite a large fraction of the total pressure loss. Air, filtered using sterile air filters, is passed into the fermentor from immediately after sterilization during the cooling cycle until the end of fermentation to maintain a positive pressure and circumvent contamination.

In most aerobic fermentations, peak oxygen consumption occurs over very short periods near the end of fermentation. It is possible to increase the oxygen uptake rate by providing high levels of oxygen transfer rate (OTR). OTR is the rate at which oxygen can be transferred from the air to the fermentation broth. It can be expressed as:

$$\text{OTR} = k_L a\,(c^* - c_L) \tag{3.10}$$

where $k_L a$ is the volumetric mass transfer coefficient, c^* is the dissolved oxygen concentration in equilibrium with the gas phase, and c_L is the dissolved oxygen concentration in the liquid.

The oxygen mass transfer coefficient, k_La, is directly related to power input by aeration and aeration rate by (121):

$$k_La = K\,(P_g/V)^a\,(v_s)^b \tag{3.11}$$

where P_g/V is the gas power input by aeration per unit volume and v_s is the superficial gas velocity.

A high k_La (volumetric mass transfer coefficient) indicates a high oxygen transfer rate (OTR) in the fermentation process. Since oxygen uptake rate is the key for a successful aerobic fermentation, a high k_La is required. However, providing higher levels of oxygen transfer usually requires very powerful drives, which add to mechanical problems and costs, as well as additional heat loads. OTR varies with the nature and the scale of the fermentation. For a $1m^3$ vessel, at an OTR between 250 and 300 mmol/L/h, heat transfer is not a major problem. However, for a $10m^3$ vessel, OTR at the above rates pose significant problems in terms of heat removal from the fermentor, indicationd the need for the use of internal cooling coils and low temperature coolants, which adds considerable operational costs.

3.5.2.4 Process Monitoring and Control

The complexity of metabolic biosynthesis involved in fermentation necessitates the use of process control parameters for monitoring and control to ensure optimum process performance. The process control variables in a fermentation process may be broadly classified as physical, chemical, and biological. The physical variables include temperature, impeller speed, aeration rate, and pressure. The chemical variables include pH, dissolved oxygen, dissolved carbon dioxide, and redox potential. The biological variables include biomass, oxygen uptake rate, carbon dioxide production rate, and respiration quotient. While the physical variables are usually measured by inline sensors forming an integral part of the fermentation equipment, the chemical or process variables are measured by online sensors crucial for the successful operation of a fermentation process. The biological variables are process indicators depending on proper monitoring and control of process variables in a fermentation process. A modern technique for bioprocess monitoring and control uses biosensors. A biosensor is a device that detects, transmits, and records information regarding a physiological or biochemical change resulting in the generation of electronic signals. The main function of a biosensor is integration of a biological component with an electronic transducer to convert the biochemical changes into a quantifiable electrical response. Biosensors make use of a variety of transducers such as electrochemical, optical, acoustic and electronic (122).

The development of miniaturized *in situ* sensors for online monitoring and control of pH and dissolved oxygen has been driven by fermentation processes and is shown in Figure 3.8 (123).

pH: Fermentation at the optimum pH is essential for cell growth and product formation. A sterilizable pH probe which functions based on a potentiometric principle consists of a glass electrode and a reference electrode. The reference electrode system consists of an Ag/AgCl electrode and a KCl electrolyte. A constant voltage of the reference electrode system is provided by the electrolyte. A ceramic diaphragm maintains the electrolytic contact between the reference electrode and fermentation broth. Based on the response from the pH probe to the pH controller, the acid or base pump is triggered for correction of pH. Steam sterilizable glass electrodes have the limitation of low mechanical stability over a period of repeated sterilizations. They can be replaced with optical sensors based on absorbance or fluorescence from pH-sensitive dyes (124).

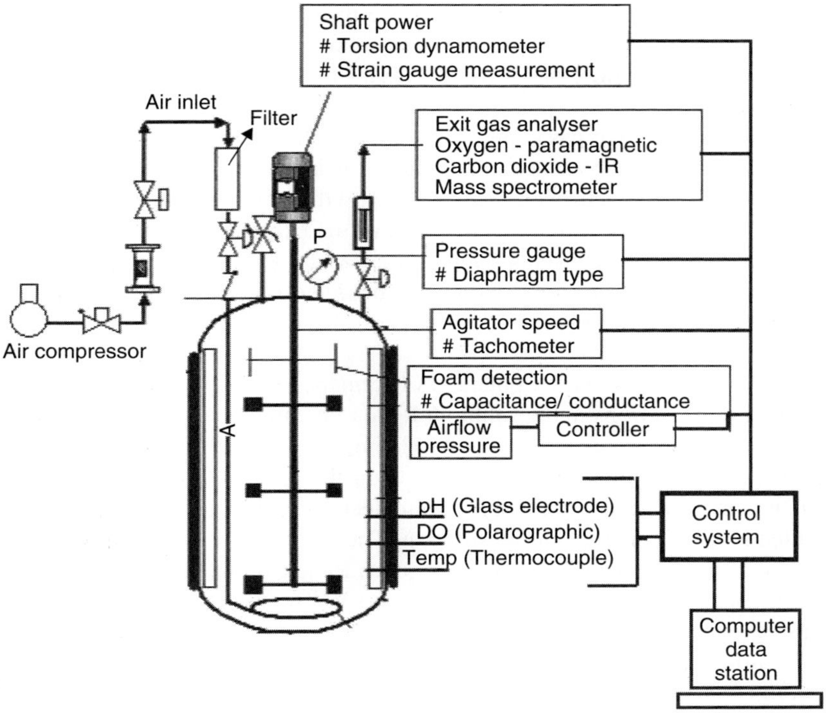

Figure 3.8 Bioprocess control of fermentation process

Dissolved oxygen: Monitoring of the dissolved oxygen (DO) concentration is essential, because sparingly soluble oxygen is one of the limiting substrates for fermentation. Two types of DO electrodes used in fermentation: polarographic and galvanic, with the former more commonly used as it is reliable and resistant to sterilization, while the latter is not widely used due to poor stability. The polarographic or amperometric electrode (−0.7V) needs an external voltage between a Pt cathode and an Ag/AgCl anode. The Pt cathode is in contact with KCl solution encapsulated by an oxygen permeable Teflon membrane. The DO from the broth diffuses across the membrane to be reduced at the Pt electrode surface. The DO concentration in a fermentor can be manipulated by two physical variables: impeller speed and aeration rate. The DO saturation (% pO_2) for a particular fermentation can be controlled individually, sequentially, or simultaneously manipulating the variables based on a feedback from the DO controller. The controller compares the deviation of the signal from the DO electrode with the set point and adjusts the impeller speed controller or the airflow controller for maintenance of DO saturation. Another possibility of DO control in fermentation exists by variation of the oxygen concentration in the gas by application of three proportional valves for controlling the flows of air, oxygen, and nitrogen (125). The steam sterilizable DO electrode (126) is widely used for online bioprocess monitoring of the dissolved oxygen concentration during fermentation. An alternative based on quenching of fluorescence has recently been introduced (127).

Dissolved carbon dioxide: Dissolved carbon dioxide (DCD) is another indicator of the metabolic status of the respiring cell. It is commonly used for animal cell cultivation where DCD concentration helps to maintain the carbon dioxide saturation during the bioprocess. DCD electrode operates on the principle of change in pH of an electrolyte buffer.

DCD sensors with bicarbonate buffers are not sterilizable, and have the limitations of drift and interference from changes in the medium composition. An *in situ* sterilizable DCD sensor with radiometric fluorescent dye HPTS (8-hydroxypyrene-1,3,6-trisulfonic acid) has been introduced (128).

Gas analysis: The exit gas composition from the fermentor, particularly oxygen and carbon dioxide, can be analyzed by a gas analyzer using paramagnetic and infrared detectors respectively. Mass spectrometry can also be used to analyze all gaseous components, including organic volatiles and lower alcohols, of the exhaust gas stream (129). From an oxygen to carbon dioxide gas balance across the fermentor, the oxygen uptake rate (OUR) or carbon dioxide production rate (CPR) can be determined. From a knowledge of the OUR and CPR during the fermentation process, the respiratory quotient (RQ = CPR/OUR) can be determined.

Redox potential: The redox potential of fermentation broth is related to the overall availability of free electrons in the solution. Redox potential operates on the principle that when a noble metal electrode (Pt, Au or Ag) is immersed into a redox system, a potential is developed the magnitude of which depends on the oxidation–reduction concentration ratio. The potential formed at the noble metal electrode is tapped off by a reference electrode, which is also immersed in the solution. The value of the potential is a measure of the concentration of oxidizing or reducing agents. Redox values are found to vary significantly with changing pH and are expressed in mV. The redox potential decreases with increasing pH value of the measured solution. The metabolic activity of microorganisms is strongly influenced by the redox potential of the culture environment. Monitoring and control of the redox potential before inoculation facilitates controlled addition of reducing agents to ensure the proper range for initiation of growth. Control of the redox potential is a vital tool in determining its influence on the metabolic pathways of microorganisms, substrate utilization, or production of specific metabolites (130).

Biomass concentration: Biomass concentration is an important biological variable, which indicates the progress of the fermentation. The common offline cell mass determination methods are based on dry cell weight and packed cell volume. Direct online characterization of cell populations in bioreactors in terms of morphological changes can be determined by *in situ* microscopy equipped with an image analyser (131). Indirect online methods for determination of cell concentration may be based on electrical or optical approach. Electrically, capacitance of the medium measures the fraction of fluid held by polarizable membranes (132). Optically, cell mass is often measured by light absorbance (turbidity) or scattering (nephelometry) continuously in the visible and near infrared ranges (133). The advantages of optical sensors for bioprocess monitoring are that they are very sensitive, give specific and reversible measurements, have a rapid response and versatility, and involve easy maintenance. The disadvantage of the optical method is that scattering and absorption are not linearly dependent on the cell density and are vulnerable to interference from particulates and gas bubbles. A biological receptor, namely an enzyme, microorganism, or antibody that produces an optical signal such as NADH fluorescence (NADH upon UV radiation at 366 nm emits fluoroscence at about 460 nm) in an optical biosensor (134). The signal is converted by an electronic transducer into an electrical signal. Bioluminescent sensors have also been developed, which consist of a bioluminescent enzyme and an optical transducer. Cell mass in animal cell cultivations can also be predicted by software sensors (135). Direct online biomass determination in fermentation processes is limited by the presence of suspended solids. The productivity of a fermentation process depends specifically on the biomass concentration achieved. Maximization of biomass concentration to high cell densities (136) is dependent on operating the fermentor at the optimum controlled process conditions.

Substrate and product concentration: The determination of the exact concentration of substrate and product during fermentation is critical for the maintenance of the optimum feed concentrations required for maximum productivity and efficient substrate utilization.

Flow Injection Analysis (FIA) is a method for online detection of concentration with a suitable detector, which may use amperometry, potentiometry, NADH-fluorescence, chemiluminescence, or UV/Vis spectrophotometry, and involves an analysis time lasting only a few minutes (137). FIA coupled with an immobilized ß-galactosidase biosensor has been reported for monitoring of lactose in a fermentation process (138). Biosensors used for FIA do not need sterile conditions for analysis. FIA systems inject low amounts of samples into the carrier stream and the probability of cell growth and protein precipitation is low because of the high dilution (139). Enzyme based biosensors have been used for monitoring and control of substrates and products involved in glutamate and penicillin fermentation processes (140). Chromatographic techniques like gas chromatography, high performance liquid chromatography, fast protein liquid chromatography, and membrane chromatography have been employed for process monitoring of fermentation processes (141). Recombinant protein product can be determined spectroscopically by attaching the gene for green fluorescent protein (GFP) to the product gene. The product concentration can then be determined by analyzing the fluorescence intensity of the GFP in the culture (142).

Mass spectrometric analysis of gases such as oxygen and carbon dioxide in the exit gas paved the way for determination of dissolved gases such as oxygen, carbon dioxide, methanol, ethanol, butanol, and acetone by the use of appropriate sampling systems (143). The accuracy of analysis is higher than observed with online probes. Sampling for offline analysis is crucial from the point of view of sterility and proper representation of the medium. Many sampling devices based on microfiltration, ultrafiltration, and dialysis have been used in the fermentation industry (144). A powerful analytical tool for noninvasive analyisis of intracellular metabolites is nuclear magnetic resonance (NMR) spectroscopy. *In vivo* determination of metabolites and metabolic processes by NMR spectroscopy serves as a vital tool for metabolic flux analysis and metabolic engineering (145). Fourier-Transform-Near-Infrared (FTNIR) spectroscopy, a standard technique used in the food industry, is now used in biotechnology for online process monitoring (146). Artificial or electronic noses, with their broadband sensing capabilities, prove useful for the online monitoring of both airborne and liquid borne compounds. With the aid of artificial nose technology it may be possible to devise two array sensors to monitor the concentrations of various important components in both the gas and liquid phases in the fermentor, eliminating the need to use a large number of sensors to monitor complex mixtures (147).

3.5.3 Downstream Processing

The products of fermentation are usually found in complex mixtures of dilute solutions and must be concentrated and purified. The separation of the product of interest from the fermentation broth depends on the accumulation of the product, which may be intracellular or extracellular. The typical downstream operations and the unit operations involved in the processing of fermentation broth (148) are:

1. Cell separation (settling, centrifugation, dead end filtration, and cross flow filtration)
2. Cell disruption (high pressure homogenization, wet milling, and lysis)
3. Clarification of extract (centrifugation, extraction, dead end filtration, and cross flow filtration)
4. Enrichment (precipitation, batch adsorption, ultrafiltration, and partition)

5. High resolution techniques (ion exchange, affinity, hydrophobic, gelfiltration, adsorption chromatography, and electrophoresis)
6. Concentration (sterile filtration, diafiltration, ultrafiltration, freeze drying, spray drying, and precipitation).

While downstream processing is an important part of fermentation process development, a detailed discussion on the subject is outside the scope of the chapter and hence not dealt with.

3.6 FERMENTATION SYSTEMS IN PRACTICE

3.6.1 Microbial Cultivation

Traditional microbial biotechnology began during the first World War when the development of acetone, butanol, and glycerol fermentations took place (149). Microbial primary metabolites used in the food and feed industries include: alcohols (ethanol), amino acids (monosodium glutamate, lysine, threonine, phenylalanine, and tryptophan), flavor nucleotides (5′-guanylic acid, 5′-inosinic acid), organic acids (acetic, propionic, succinic, fumaric, and lactic), polyols (glycerol, mannitol, erythritol, and xylitol), polysaccharides (xanthan and gelan), sugars (fructose, ribose, and sorbose), and vitamins (riboflavin, cyanocobalamin, and biotin). The group of microbially produced secondary metabolites important for health and nutrition includes antibiotics, other medicinals, toxins, biopesticides, and animal and plant growth factors (150). The targets uses of antibiotics (the best known group of secondary fermentation metabolites) include DNA replication (actinomycin, bleomycin, and griseofulvin), transcription (rifamycin), translation of 70-S ribosomes (chloramphenicol, tetracycline, lincomycin, erythromycin, and streptomycin), transcription by 80-S ribosomes (cyclohexamide), transcription by 70-S and 80-S ribosomes (puromycin and fusidic acid), and cell wall synthesis (cycloserine, bacitracin, penicillin, cephalosporin and vancomycin) (151).

Over the last three decades, traditional industrial microbiology, using the tools of molecular biology, has led to the development of recombinant organisms aimed at production of high value biopharmaceutical products such as erythropoietin, human growth hormones, and interferons (152). The major microbial hosts for the production of recombinant proteins are *E. coli* (153), *B. subtilis*, *S. cerevisiae*, *Pichia pastoris*, *Hansenula polymorpha* and *Aspergillus niger*. The major thrust of recombinant DNA technology has been in the area of rare mammalian peptides, such as hormones, growth factors, enzymes, antibodies and biological response modifiers (154).

3.6.2 Mammalian Cell Cultivation

Bioreactors developed for mammalian cell culture use a smooth surfaced ceramic for the growth of adherent cells or a porous ceramic for the immobilization of suspension cells. Mammalian cell culture technology has been used for the production of viral vaccines, monoclonal antibodies for diagnostic and therapeutic use, interferons, interleukins, growth factors, and blood regulation proteins. The design of novel bioreactors for mammalian cell culture must take into consideration process constraints such as cell line distribution, product formation, oxygen transfer, nutrient supply, and shear. A viable alternative to scalable mass cell culture is the use of bioreactors with ceramic matrices. Immobilization of adherent and suspension cells on these surfaces has resulted in the favorable production of both cells and cell derived proteins. Cell culture densities ranging from 2×10^6 cells per mL (spinner flask culture) to 1×10^8 cells per mL (high density perfusion reactors) have been obtained (155).

For the cultivation of anchorage dependent animal cells, packed bed bioreactors have been used. Such reactors up to 100 L in size have been successfully operated with BHK 21 C13 monolayer cells for the production of foot-and-mouth disease virus vaccine antigen (156). In a commonly used packed bed bioreactor, glass beads ensure adherence and subsequent colonization *in situ* (157). The use of polyester or polyurethane foam matrices (158) for the culture of hybridomas ensures passive entrapment of the cells in the inert matrix with the added advantages of perfusion of cells with fresh medium, increased process intensity, and improved downstream processing. The first airlift fermentor (1000 L scale) to produce monoclonal antibodies from hybridoma cells in suspension was developed by Celltech (159).

3.6.3 Plant Cell Cultivation

Stirred tank fermentors have also been used for the growth of plant cells for the production of valuable products in single or multiple stages of cultivation. Plant cell culture is carried out by propagation of a callus, which is a mass of undifferentiated cells. To develop a new plant with shoot and root, cells from the callus must be cultured in different media. Desirable metabolites can be extracted from a callus without progression to a shoot and root. Plant cell culture may serve as a good medium for synthesis of metabolites useful in foods, biopesticides, and cosmetics. Plant cell culture has been proposed as an alternative technique for the supply of phytochemical to the normal plantation system (160). The benefits of such a system include a final product of required quality and consistency independent of the variations in plantation due to season, weather, disease, and pest destruction. Cultivation of plant cell suspensions in bioreactors has led to higher productivity of the final product. For instance, *Coleus* cells have been induced to produce rosmarinic acid at a rate of 0.91 g/L/day, which almost compares with microbial fermentation products like penicillin with 1.4-2.1 g/L/day (161). The potential of plant cell cultures for the production of valuable plant secondary metabolites like fragrances, food colors and flavors, and medicinal compounds has long been recognized. A cell suspension culture for the production of an antiinflamatory drug, shikonin, from cells of *Lithospermum erythrorhizon* has been developed on a commercial scale (162).

The main disadvantage of plant cell cultures and immobilized plant cells is the high degree of genetic and biochemical heterogeneity, resulting in process instability and low productivity of metabolites (163). It is possible to grow cultures of plant roots or shoots in the *de novo* synthesis of secondary products.

The industrial application of plant cell suspension cultures has, to date, been limited. Commercialization has essentially been impeded by economic feasibility factors, arising from both biological and engineering considerations. Sensitivity to hydrodynamic shear in bioreactors has generally been attributed to the physical characteristics of the suspended cells. However, shear sensitivity may not be very important (164). Hairy roots formed by plant transformation using *Agrobacterium rhizogenes* could be exploited to produce secondary metabolites (165). The difference between hairy root and cell suspension cultures is shown in Table 3.4 (166).

Due to the extremely slow growth rates and use of rich growth media in plant cell cultures, there is a high risk of microbial contamination, which can be avoided by immobilization of plant cells *in situ* in the reactor. In order to immobilize the plant cell aggregates in the reticulated polyurethane foam matrix, a suspension culture of cells is introduced aseptically into a sterile liquid medium containing the empty foam matrix. The plant cell aggregates are carried from the bulk liquid into the porous foam matrix by the

Table 3.4
Fermentation features of hairy root and plant cell suspension cultures

Fermentation Features	Hairy Root	Plant Cell Suspension
Medium	Simple, no vitamins or hormones	Complex, requires specific vitamins or hormones
Inoculum size	Largely independent, minimal lag	Size dependent , conditioning of medium needed
Genetic stability	Stable	Unstable
Metabolite production	Depends on level and charcateristc of parent plant	Unpredictable
Rheology	Newtonian broth	NonNewtonian broth
Operation	Easy with self immobilization	Requires support, mixing problems
Biomass density and stability	High, easily maintained oxygen transfer	Low, instability due to inefficient
Growth rate	Doubles in 2–7 days	Doubles in 14 days

liquid elements flowing through the open pore structure of the initially empty foam matrices. Once physically entrapped, the cell aggregates grow or adhere to each other to fill the rest of the available space with foam. Some of plant cell culture processes producing valuable products in bioreactor are shown in Table 3.5 (167).

With minor alterations to the conventional stirred tank bioreactor, it is possible to obtain practicable bioreactor systems suitable for use with immobilized plant cell cultures, such as the circulating bed bioreactor and the sheet bioreactor (168). In the former, the bioreactor is initially operated as a packed bed with foam particles held stationary between two retractable stainless steel grids. Once the cell aggregates are fully immobilized, the grids are retracted and the bioreactor is operated as a circulating bed. To achieve efficient immobilization, it is necessary to hold the foam particles stationary in the bioreactor. A modification of the STR with foam matrices in the form of sheets held stationary as vertical baffles around the central impeller is referred to as a sheet bioreactor. Microbial, mammalian, and plant cell cultivation is compared in Table 3.6 (169).

3.7 SCALE UP OF FERMENTATION PROCESSES

Scale up is crucial to the successful development of a fermentation process. Many large scale fermentation processes produced lower yield when compared to laboratory results (170,171). The traditional method for scaling up a fermentation system is based on empirical criteria such as constant power per unit volume, a constant mass transfer coefficient, constant mixing time, or constant impeller tip speed (172). The empirical criteria are subject to lacunae when there is a change in the controlling regime of the process. Another approach aimed at scale up is to scale down with the objective to determine the significant process parameters on a varying production scale in the laboratory (173,174). Due to the numerous pitfalls associated with the scale up of bioreactors and the lack of industrial scale up data available in literature, scale up may still be considered an art rather than science (175).

Scaling up of bioreactors by increasing the number of reactors rather than the size is also one approach, wherein both capital and labor costs increase linearly with increase in scale. The operation of multiple, smaller bioreactors rather than a single, larger bioreactor could be advantageous due to flexibility of operation, ease of start up, inoculum preparation,

Table 3.5
Plant cell cultures producing valuable products in a bioreactor

Plant Cell Line	Product	Mode	Bioreactor
Lithospermum erythrorhizon	Shikonin	Batch	Stirred tank
Coleus blumei	Rosmarinic acid	Fed batch	Stirred tank (Spiral)
Nicotiana tabacum	Geranoil	Batch	Stirred tank

Table 3.6
Fermentation characteristics for microbial, mammalian, and plant cell culture

Fermentation Parameter/Feature	Microbial Cell Culture	Mammalian Cell Culture	Plant Cell Culture
Operating pH range	2–10 depending on the fermentation	Usually between 6.8 and 7.2	Usually between 5.5 and 6.0
Operating temperature	25–50°C depending on the fermentation	Usually between 37 and 42°C	Usually between 25 and 30°C
Aeration rate	Higher rates of 1–2 vvm	Lower rates of up to 0.5 vvm	Lower rates of up to 0.5 vvm
Impeller	High shear impellers (disc turbine)	Low shear impellers (marine/pitched blade)	Low shear impellers (propeller)
Agitation	High speed (up to 1000 rpm) required for viscous broth	Low speed (up to 200 rpm)	Very low speed
Fermentor	Depends on fermentation broth	Membrane bioreactor also used	Air lift fermentor also used
pH control	Acid/alkali addition with peristaltic pump using pH controller	Sparged CO_2/N_2 gas to maintain pH using a four gas controller	Sparged CO_2/N_2 gas to maintain pH using a four gas controller
D.O control	Cascade control with agitation and aeration	Control of air/O_2 using a four gas controller	Control of air/O_2 using a four gas controller
Foam control	Antifoam agents added to reduce foam	Minimize foam using foam elimination chamber	Minimize foam using foam elimination chamber
Shear sensitivity	Insensitive	Sensitive	Sensitive
Specific growth rate	High (doubling time few hours)	Low (doubling time in days)	Low (doubling time in days)

cleaning, and sterilization (176). However, the final justification relies on the benefits from the economies of scale (177,178).

3.8 ASEPSIS IN FERMENTATION PROCESS

Asepsis in biotechnology means freedom from unwanted microorganisms, just as in clinical medicine it means freedom from pathogenic microorganisms (179). However, there do exist many fermentation industries e.g., ethanol, baker's yeast, and vinegar, where asepsis is not a matter of concern. Economic considerations suggest that a contamination probability of 1 in 100 is acceptable for batch fermentations, considering a contamination probability of 1 in 1000 taken into design calculations for a sterilization process (180). Recombinant microorganisms are at a greater risk of being overwhelmed by wild organisms. Cell culture fermentations are susceptible to microbial contamination due to a longer fermentation process and a slow growth rate. However, with use of aseptic procedures large scale animal cell cultures are operated with a contamination rate of about 2% (181).

The sources of contamination in a fermentation process could be attributed to inoculum, nutrient medium, bioreactor system, air or liquid transfer, and mutation. The aseptic procedures necessary to achieve a sterile fermentation process include:

1. A sensitive sterility assessment protocol: The results of conventional sterility checks in nutrient media incubated at optimum conditions may not be possible before sufficient damage is done to the production fermentor. There is a need for accelerated detection of contaminants like use of thioglycolate agar and oxidoreductive indicators (e.g., methylene blue or resazurin) for rapid detection of aerobes and anaerobes in the medium (182).
2. Development of a certified laboratory inoculum: The preparation and propagation of inoculum required for fermentation should be carried out in designated clean rooms. The procedures and equipment used for sterilization of the media and glassware for inoculum preparation should be strictly followed.
3. Efficient sterilization of bioreactor, medium, and air: Proper sterilization of bioreactor, medium, and air is crucial for a successful fermentation batch. Continuous sterilizers have been employed for sterilization of medium required for large scale bioreactors (183). After sterilization of the bioreactor, the system should be held under positive pressure with sterile air to avoid the possibility of external air being drawn due to vacuum formation (184).
4. Asepsis during fermentation: The most common cause of loss of asepsis during fermentation is the depressurization of bioreactor which could be averted by providing the air compressor with automatic change over to a captive power generator and automatic closure of inlet and outlet valves at a fall in the preset pressure or air flow. Foaming is a characteristic feature of gas and liquid operation in presence of surface active agents.
5. Containment of foam: Foaming is perceived as the most probable cause of contamination in fermentation processes, leading to overflows and consequently loss of broth and products. The addition of silicon based antifoaming agents leads to a change in the hydrodynamic and mass transfer characteristics of the bioreactor (185). However, the use of foam breakers has been applied to avoid the need for addition of antifoaming agents, and the necessity of antifoaming agent addition depends on the nature of fermentation (186).
6. Routine and preventive maintenance of bioreactor: The inspection of a bioreactor should cover the shell, dome, agitator seal, air exhaust line, nozzles, sight glasses, O rings, air filters, probes, shaft, impellers, coils, sparger, inoculum, feed, and sampling lines (187).

Asepsis of a bioreactor ultimately rests on the integration of adequate bioreactor design, efficient validation and operating protocols, routine maintenance, and operation by skilled personnel.

3.9 BIOREACTOR DESIGN

The purpose of the fermentor is to provide an aseptic environment in which fermentation can be performed efficiently to achieve process targets. The reliable operation of a fermentation system to achieve process objectives depends on two factors: the fermentor design and the fermentation process. Therefore, rational design of a fermentor requires careful consideration of how it is integrated into the process design. The design should consider from the outset such factors as plant scheduling, space constraints, relationships between fermentor productivity and throughput rates of downstream equipment, containment and validation requirements, utilities requirements, potential interruptions in normal plant operation, overall labor requirements and operating versus capital costs. Consistency, safety, cost, and compliance with statutory requirements usually are of prime concern for production equipment.

3.9.1 Design Criteria

The objective of fermentor design is to produce fermentation systems necessary to build economical production facilities that satisfy well considered performance criteria. Reactor design and scale up considerations are driven by the need to provide the organism optimal conditions for producing the desired product uniformly in the reactor (188).

3.9.1.1 Mechanical Aspects

Mechanical design aspects are important for the successful operation of any fermentation plant. A few design ratios, which have gained significance irrespective of the fermentation system employed, include: vessel diameter to height ratio (should be in the range of 1:3–1:4), and impeller diameter to vessel diameter ratio (should be in the range of 0.3–0.5).

The mechanical design aspects with reference to the geometrical requirements of a fermentor are discussed in Standard Dimensions (Section 3.9.2).

Some practical aspects before getting on to the technical details of vessel design are:

1. Space requirements: The vessel dimensions must be chosen to meet plant space limitations. Poor choice of equipment sizes can cause inordinately high building costs.
2. Transportation: Shopbuilt vessels are usually less expensive and of higher quality than fieldbuilt vessels.
3. Special heads: A hemispherical bottom gives better mixing and less shear in tissue culture vessels than does a standard dished head.

3.9.1.2 Process Aspects

With the lack of adequate information on fermentor design for a particular fermentation process, design has been an applied art rather than a science. It requires a careful consideration of a host of interrelated factors, many of which are risk assessments and value judgments. Therefore, it is imperative that the design engineer applies experience and judgment in reaching reasonable compromises.

The fermentation process guidelines commonly employed without proper consideration of the process aspects do very little to promote good design. These include:

1. Aeration rate: The airflow rate to the fermentor must be generally one volume of air per liquid volume per minute (1 vvm). However, in practice, the aeration

rates depend on oxygen solubility in media, and 1 vvm may be too high for a particular fermentation process when operated at a larger scale.

2. Impeller tip speed: The tip speed of a fermentor impeller must not exceed 7.6 m/s.
3. Arrest of fermentative metabolism: At the end of fermentation, the fermentor broth must be cooled immediately and stored at 4°C, to arrest the fermentative metabolism.
4. The maximum production rate cell mass theory: Process optimization is achieved by obtaining very high cell mass concentrations at very high growth rates.
5. Oxygen transfer rate: The consequences of increasing OTR by increasing air flow rate and agitation could lead to foaming, increased gas holdup (189), higher gas velocities, higher vessel pressure, and oxygen enrichment.
6. Heat transfer rate: Heat transfer usually is the limiting constraint for highly aerobic large scale fermentors. Oxygen and heat transfer rates are coupled closely in aerobic fermentations. The total heat generated during growth (Q_{total}) is given by:

$$Q_{total} = Q_{metabolic} + Q_{mechanical} \tag{3.12}$$

where $Q_{metabolic}$ is the metabolic heat generated and $Q_{mechanical}$ is the heat generated by the agitation required to provide adequate oxygen transfer and mixing.

3.9.2 Standard Dimensions

The standard dimensions of a fermentor are shown in Figure 3.10. The standard geometric configurations in practice are:

1. H/T ratio: The geometric configuration of the fermentor is more commonly denoted by the H/T ratio. The height (H) and diameter (T) actually refers to the height and diameter of the fermentor. The H/T ratios of most stirred tank fermentors range from 3:1 to 4:1 while the ratio can be as high as 10:1 in the case of an air lift fermentor.
2. D/T ratio: The geometric configuration of the impeller is more commonly represented by D/T ratio, i.e., the impeller diameter (D) to tank diameter (T). Impellers with small D/T ratio must be operated at a higher speed in comparison with impellers of large D/T ratio to provide the same power requirement. D/T ratios of 0.3 to 0.5 are commonly used in fermentors. It should be noted that operation at higher impeller speed could result in increased shear rates and consequently may be detrimental for the microorganisms in the fermented broth.
3. Baffle ratio: Baffles play a vital role in improving the aeration efficiency and prevent vortex formation. Baffles should be installed so that there is a gap between the baffle and the vessel wall to minimize the microbial growth on the baffles and walls. The baffle ratio (BR) is given by (190)

$$BR = n_B . w_B / T \tag{3.13}$$

where n_B is the number of baffles installed, w_B is the width of the baffles and T is the diameter of the fermentor.

3.9.3 Jackets and Coils

Jackets are used for the circulation of steam and cooling water during the heating and cooling cycles of sterilization of the fermentor (191). Since microbial reactions are exothermic, the heat produced during fermentations leads to a rise in the temperature of the broth, necessitating the need to maintain the temperature at the optimal value. The heat transfer is dependent on the area of contact surface of the cooling or heating fluid with the fermentor and the temperature gradient of the medium. Normally, steam is used as the heating fluid; and water, chilled water, or chilled brine are used as cooling fluids. The contact surface area of the jacket with the fermentor should be maximal and the pressure drop of the circulating fluids in the jacket should be minimal for better process performance. The layout of the external jackets could be a double jacket, a full pipe, or a half pipe (limpet coil), depending on the required area for heat transfer, the heating or cooling medium, and the circulation velocity (192).

The choice of jacket material usually is not critical, and it usually is best to leave the vessel fabricator some degree of flexibility to allow for his usual shop practices. Some guidelines are:

1. Stainless steel jacket cover made of SS304 to extend vessel life
2. Jacket is appropriately designed to increase vessel strength
3. For vessels less than 5 m^3 providing jacketing at the bottom, dished end may not contribute significantly to heat transfer
4. Jacket cover to the height equivalent to the operating volume of the fermentor
5. Jacket inlet and outlet lines should have end connections (flanged or sanitary coupling) between the external pipes and the jacket
6. Jacket design should favor minimum accumulation of solids to avoid ionic concentrations, which cause corrosion

If internal cooling coils are used, they can not only remove the excess heat from fermentation, but also act as baffles for better mixing. A minimum spacing of 150 mm between coils ensures that they withstand the thermal and mechanical stresses generated during fermentations. Joints between pipes should be butt welded, covered with a welded concentric tube, and lead tested (193).

3.9.4 Safety Codes

The safety of the vessel should be the foremost consideration in vessel design. The vessel must be fabricated in accordance with the Standard Code for unfired pressure vessels (194) and tested at design conditions to insure that the vessel can withstand all forces generated under the specified operating conditions. If operation is required at high operating pressures, one should consider ways to minimize the metal thickness to allow the use of cold rolled sheet rather than plate. This results in better heat transfer, a better interior finish, and a lower price.

3.9.5 Material of Construction

Stainless steel is the more commonly employed material for the fabrication of biotechnology equipment. The selection of the right steel quality in biotechnology is based on a compromise between material costs, availability, and the physical and chemical requirements of the process. Based on the crystal structure, standard chrome steels at room temperature are ferrite based (*alpha* or α iron). When ferric steels are heated to 910°C, the crystal structure is changed to the nonmagnetic austenite (*gamma* or γ iron). The stabilization of the austenite structure at room temperature is enhanced by the addition of austenite formers such as nickel, molybdenum, or carbon. Austenitic steels are the material of choice used as shell plate for fabrication of fermentors. Compared to the nonaustenites, these steels show improved corrosion and heat resistance, and are nonmagnetic, apart from having good tenacity and workability (195).

Iron is the most abundant element in steel and very much prone to corrosion attack. Corrosion in the form of rust can be prevented by the addition of up to 12.5% chromium, which reacts with ambient oxygen to form a passive surface layer of chromium oxide. This chromium oxide layer protects the steel from corrosion. Stainless steels are defined with a maximum corrosion loss of 0.1 mm per year. Carbon in the stainless steel is required for stabilization of the austenite structure. The low carbon steels, SS 304L and SS 316L, are known worldwide as standard steels. Generally, vessels used in biological processes are fabricated with 316 or 316L steel. The vessels widely used in food technology or harvest storage tanks are fabricated with a cheaper and less corrosion resistant steel of grade 304 or 304L. For all other vessel parts not in contact with the product broth, such as jackets, cheaper, nonstainless steel grades are used (196). The commonly used steels in fermentation industry are shown in Table 3.7 (197).

The selection of a vessel material for fabrication should take into consideration:

1. Sensitivity of the organism, particularly eukaryotic cells
2. Extent of vessel corrosion on exposure to fermentation media and utilities (mostly acidic pH)
3. Aseptic operation requires use of SS316, SS316L, SS304, or SS304L, with specification of "L" contributing to an additional 15% vessel cost

Although the above factors govern the choice of material, testing of materials with actual fermentation broth or a growing culture is seldom reported.

3.9.6 Baffles

Baffles are usually welded to the interior of the vessel. Baffles usually take the form of metal strips, roughly one tenth of the vessel diameter in width, extending vertically down the height of the vessel and attached radially to the wall. Removable baffles mean unsealed joints. Baffles should be set off from the wall, to minimize solids build up and to simplify cleaning. Slots between the baffle and the vessel wall prevent the formation of dead spots. The provision of baffles increases the turbulence in the liquid and results in a more efficient utilization of power.

3.9.7 Sparger

A device for introducing air below the liquid level in the fermentor vessel is called a sparger. The use of spargers with very small orifices is more efficient than a single orifice delivering the same volume of air. However, aerobic fermentations, e.g., penicillin, produce large quantities of mycelium and mechanical agitation must be used in order to

Table 3.7
Commonly used steels in fermentation industry

Steel	Carbon (%)	Chromium (%)	Nickel (%)	Molybdenum (%)
SS 304	0.07	18.0	9.5	—
SS 304L	0.03	18.5	11.0	—
SS 316	0.05	17.5	11.0	2.7
SS 316L	0.03	17.5	11.5	2.7

ensure adequate dispersion of air and other nutrients in large scale fermentations. As the degree of agitation is increased, the relative efficiencies of the various types of spargers tend to converge, and at high agitation levels, all spargers give approximately the same performance. The types of spargers in general use may be classified as nozzle spargers, orifice spargers, and porous spargers.

Nozzle spargers consists of either a single nozzle or a cluster of nozzles radiating from a central delivery pipe. Orifice or ring spargers are the type in most common use; essentially a perforated pipe formed in the shape of a ring. The tendency of smaller orifices to block and the high pressure drop are the limiting factors. Porous spargers are of more academic than industrial interest and are used primarily in small scale aeration studies.

The factors relevant for good performance of the sparger include correct placement of the sparger holes in line with the impeller blades, choice of hole diameters to ensure critical orifice at maximum gas flow, and free draining back of the fermentation broth into the vessel. The consideration of gas distribution into the fermentor supports use of a ring sparger over other spargers.

3.9.8 Nozzles and Manways

Nozzle design must take into consideration the following:

1. To ensure aseptic connections to external piping for sizes up to 4 in (10 cm), the best choices are sanitary clamps (3-A), Ingold nozzles, and vacuum seals (Ultraseal or VCO). Above 4 in, welded flanges with flat gaskets are used.
2. In order to facilitate free draining, the nozzle is mounted at an angle of about 5°–15° off the horizontal.
3. For proper and safe cleaning of the nozzles, protrusions inside the vessel should be as minimal as possible.
4. Addition of feed through nozzles to the fermentor to take care that the added liquids do not dribble down the interior surfaces.
5. The use of manways obviates the need for full opening heads on larger vessels. Standard manway diameters range from 12 in to 28 in, usually in 2 in increments. A 16 in manway is the minimum practical size, but 18 in is usually better. The maximum size is determined by the vessel diameter. The manway collar should be kept as short as possible to make it easy to clean.

3.9.9 Piping and Valves

Piping materials: The most commonly used piping materials for biotechnology plants, in order of usage, are stainless steel, thermoplastics (polypropylene, polyethylene, polyvinyldine fluoride), carbon steel, copper, iron, glass, and lined pipe (glass and plastic liners).

The guiding document for engineering, design and installation of Water for Injection (WFI) and other sterile systems is CGMP-LVP (Current Good Manufacturing Practice–Large Volume Parenterals, Part 212.49 of Subpart C, Water and Other Liquid Handling Systems), which was issued by the Food and Drug Administration (FDA) in 1976 as a performance specification, in the absence of a code (198). This document specifically mentions stainless steel as the material of construction for WFI and by inference, sterile services.

In sterile fermentation processes, a great deal of attention is necessary for the layout of lines and the construction of joints. Lines for sterile air and lines for transferring sterile mash are the subject of special care during installation to prevent the formation of pockets where liquid can collect. Adequate slopes, continuous in one direction, have to be given in what would be normally horizontal lines. Loops in lines are best excluded, but if unavoidable have to be provided with drain points so that residual mash and steam condensate can be removed. For purposes of sterilization, steam is introduced wherever possible at the highest point or points in the system, with the steam condensate removed from the lowest points.

Valves used on sterile lines have given cause for thought for a considerable time. For robust industrial processes such as alcohol fermentation, or even for yeast cultivation, the use of the standard type of gate valve is normally acceptable. However, for fermentation systems more prone to contamination, the use of a standard valve has obvious shortcomings. The introduction of a diaphragm capable of standing up to steam for longer periods into the valve has encouraged a gradual changeover to the use of the diaphragm type of valve. Presently, with a diaphragm life of 3–4 months, the use of this type of valve for aseptic applications is justified.

Drain valve: The best drain valve available is a flush mounted diaphragm valve. Designs that include a coupling connected directly to the tank should not be used for any aseptic application. There is no way to keep such a valve clean or free of solids buildup.

Sight Glass: Circular-type sight glasses with sanitary clamp mountings are generally used. These are easy to clean and to replace. The long, rectangular types are much more expensive, and require special modifications to make them self draining.

3.9.10 Steam Locks

A good steam lock assembly should have the following features:

1. Elimination of the chances of dead spots with inefficient sterilization
2. Thorough steaming of sterile lines to be done during non usage to avoid contamination
3. Lines from feed addition tanks should be able to be steamed through their entire lengths at any time after they are connected to the assembly
4. Steam locks to be checked for leakage of steam during steaming
5. Self draining of condensate through steam traps
6. Easy cleaning and maintenance of the steam lock assembly

3.9.11 Welds and Joints

Vessel welds of high quality are required to adhere to code purposes, ensure maximum smoothness and cleanability, and to minimize corrosion problems. Welding for aseptic fermentors should be carried out under an inert gas shield to minimize oxidation and flux residue, and create smoother, pit free welds. The TIG (tungsten inert gas) method is the most acceptable vessel welding technique giving a better finish and not subject to splattering problems compared to the MIG method. It is preferred to use 316L or 304L for multipass welds, to reduce carbide precipitation and consequent pitting near the welds. For single pass welds, one should use 316 or 304 to minimize ferrites in the metal.

A flanged joint for sterile fermentation processes should be constructed with smooth bore continuity throughout the joint. Smooth bore continuity is important, and can be compromised by such factors as tolerances in bore and circularity of commercial tubing, coupled with the tolerances in the bore of such things as slip on welding flanges, and the standard clearance holes for flange bolts. If in addition, the joint is badly cut or badly placed, focal points for lack of sterility with subsequent contamination can easily be produced. In important lines, the use of neck welding flanges is considered justified in spite of their additional cost. The use of welded fittings for branches and reducers, among other fittings, will permit the construction of lines better suited to sterile operations than those normally obtained if the branch is made up by cutting and fitting the tubes.

3.9.12 Surface Treatment and Finish

The surface treatment of a vessel is required for any surface that comes in contact with the product. It is imperative that all stainless steel surfaces are treated and cleaned in a way that prevents corrosion under the operating conditions. Stainless steels are corrosion resistant due to the formation of a microscopically thin, invisible chromium oxide layer, which occurs on clean metal and polished surfaces only. The three main surface treatment methods used are mechanical, chemical, and electrochemical.

Mechanical Surface Treatment: It is possible to improve the surface quality of steel parts in contact with the product by means of mechanical treatment. The surface finish qualities of the vessel exterior are governed more by aesthetic rather than functional attributes. The quality and smoothness of the mechanically treated surfaces are defined by values in terms of arithmetic mean roughness, R_a. Minimal surface qualities of R_a ~0.6 μm (220 grit) are generally accepted and approved in biotechnology. The finest polish or abrasive belt (400 grit) can be used to achieve smoothness values of R_a 0.28–0.34 μm. The best possible smoothness values are achieved with steel grades, SS 304L and SS 316L.

Chemical surface treatment: The two main types of chemical surface treatment are cauterization and passivation. Cauterization is the dissolution of the flaws on the vessel surface with acid mixtures, such as 15–25% (v/v) HNO_3, and 1–8% (v/v) HF, for steel with chromium content above 15.5%. Surface imperfections in the interior of vessel prevent the formation a perfect passive layer. The impurities that require cauterization may be interior surfaces with oxidation tints, welding slag residues, and fine scales or overlaps generated during the working process.

Passivation is used either after cauterization or as a final treatment of ground, brushed, or polished steels with special surface structures. A natural passive layer is invariably formed when stainless steel is exposed to air. Artificial passivation with dilute nitric acid is frequently used for corrosion resistance at critical condition. The oxidizing effect of nitric acid accelerates the formation of a dense passive layer. Reports on passivation show that organic acid chelating agents can produce a more long lasting passive layer (199).

Electrochemical surface treatment: Electropolishing, or mirror polishing, is an electrochemical treatment which smoothes and polishes rough, dull surfaces. The vessel is immersed in an electrolyte and serves as an anode. The quality of the mirror polishing depends mainly on the source material and its mechanical pretreatment. Vessels with good R_a values may sometimes have microscopic surface roughness, which can be smoothed electrolytically. However, there is likelihood that existing imperfections on the material surface may be accentuated after mirror polishing. An electropolished (EP) vessel requires polishing to 320 grit. Finishes commonly used are shown in Table 3.8.

Inspection of Surface Treatment: The quality of the surface and welds after surface treatments are checked with following methods:

1. Visual inspection
2. Detection of ferric contaminations with the ferroxly test
3. Palladium testing to check the surface passivation
4. Detection detergent residues
5. Detection of chloride or sulfur contamination
6. Detection of pickling damage
7. Scanning needle electronic surface measurement

Table 3.8
Finish represented in Grit and Microns

Grit Designation	Microns (μm)
60	6.3
120	3.2
180	1.6
220	0.8
320	0.4
320 EP	< 0.4

3.9.13 Cleanability

The cleaning in place (CIP) system for the fermentor must be designed to ensure that all interior surfaces can be cleaned thoroughly. There should not be any accumulation of solids after CIP treatment. The inside of the fermentor should be smooth, free draining, and allow easy access for maintenance or repair. There should be no nooks and crannies, and all interior joints should be welded (not bolted) if possible. A dished head drains more freely than a flat head, but flat heads may still be used. The size and placement of the manway significantly affects the positioning of spray balls or hose nozzles, and the ease of access for manual cleaning (200).

3.9.14 Sterility Aspects

The design of a fermentor vessel has to meet aseptic conditions before it can be used for fermentation batches. An improper fermentor design may result in a nonsterile batch due to ineffective sterilization of dead volume or leakage.

Dead Volume: Organisms or media that are deposited in dead spaces may lead to subsequent contamination. In air filled cavities, sterilization temperature may not be reached. The installation of temperature probes into the fermentor at critical positions enables the monitoring of the actual temperature to ensure that efficient sterilization occurs.

Leakage: Most fermentation vessels are not hermetically sealed, and have a number of ports into and out of the fermentor. The common points of leakage from the fermentor include improper sealing of the manhole, sight and light glass, probes, inlet and outlet feed streams, mechanical seals and harvest or sampling valves. Leakage from the fermentor vessel may be identified by two methods, namely pressure test and sealing test.

Pressure test: The fermentor is filled to 70% of the total volume and mixed at the operating speed. Air or nitrogen is sparged into the fermentor to achieve a vessel pressure of 1.5 bars. All the ports are tightly secured and fermentor is examined for pressure loss. Care should be taken to maintain the mechanical seal at a positive pressure relative to the fermentor. If there is a drop in pressure, then there is a leakage from the fermentor.

Sealing test: The fermentor is filled with water up to the mechanical seal and a vessel overpressure of 1.5 bars is achieved. The temperature (T_0) and pressure (P_0) are recorded after ten minutes and again after about fifteen hours. If the specific leakage rate (L) in terms of grams of air per hour per meter of sealing length is less than 0.05 gm/h m, the vessel is considered tight; if higher, then the vessel has to be sealed effectively at the leakage points.

3.10 AGITATION SYSTEM DESIGN

Typical agitation equipment consists of the prime mover (usually an electric motor) coupled to the shaft through a reduction gear. Impellers and baffles are fitted to the shaft and vessel, respectively, to give the desired liquid motion. The shaft may enter from the top, side or bottom, and is usually fitted with a mechanical seal at the vessel wall. The number and location of impeller units depends on the vessel. The general practice is to use one impeller for each diameter of depth. The type of impeller is more critical and different designs produce different fluid motions.

In a smooth walled tank, the liquid swirls round in the same general direction as the agitator. As the impeller speed is increased, a vortex is formed and the liquid level at the wall is raised above the average liquid level. This is normally undesirable for the following reasons:

1. Power is wasted in holding up the liquid at the wall
2. The relative speed of the impeller to the liquid is reduced
3. Slight radial movements of the vortex cause the liquid to swirl unevenly, and undesirable side thrusts are set up

3.10.1 Impeller Design

The common impeller designs used for stirred tank reactors are shown in Figure 3.9.

The optimum impeller to tank size ratio for gas or liquid mass transfer depends on the mixing power delivered relative to the gas velocity of the system. At relatively low levels of mixer horsepower, it is seen that relatively large D/T ratios are needed in order to have sufficient impeller pumping capacity to handle the volume of gas passing by the impeller. As the amount of power to the mixer is increased, the optimum D/T is reduced to a relatively small value, approximately 0.15 to 0.2. As the mixer power is increased far beyond that needed to disperse the gas, D/T becomes unimportant because it makes very little difference how the power is distributed. As the impeller diameter is increased, , the maximum shear rate at the same impeller speed is also increased. Thus on scale up, lower average shear rates will be encountered in the stream from the impeller due to lower impeller speeds, but higher maximum shear rates will be encountered due to the higher peripheral velocity of the impeller. The impellers operate at a viscosity range of the fermentation medium as shown in Table 3.9. The impeller spectrum depicting the proportion of flow and shear developed by different impellers is shown in Figure 3.10 (201).

3.10.2 Drive Location

The choice between bottom and top entering drives depends on convenience.

Top drive: Top drive continues to be the most common, because there is no threat of a major spillage as might occur through a bottom seal when subjected to grinding by broth particulates. The risk of a major spill is remote because the leak rate would be quite small, and corrective maintenance could be applied well before any catastrophic event could occur.

Bottom drive: Bottom drives do have some very positive advantages. They have shorter shafts, which means that steady bearings are not necessary in most cases; they have less expensive support structures; they are easier to access and simpler to maintain; and they result in lower overall height. From the point of view of asepsis and of the fermentation per se, there is no real difference between top and bottom entry, assuming proper seal design and maintenance.

3.10.3 Seals

Static Seals: Silicone is the best choice for the head plate and ethylene propylene diene monomer (EPDM) for all other static seals. Teflon has better temperature resistance, but it cold

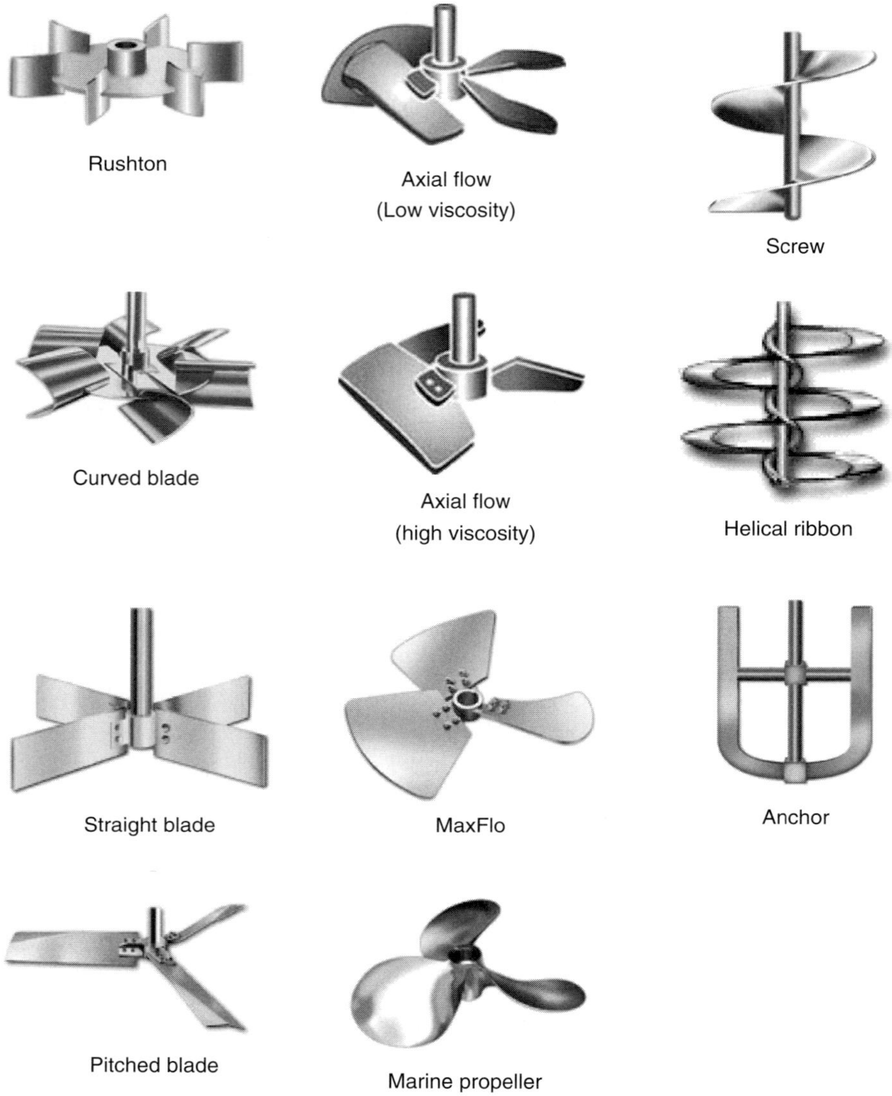

Figure 3.9 Impellers used in stirred tank reactors

flows and does not stretch. Viton™ has good temperature resistance and is quire serviceable, but it hardens.

Rotating Seals: There are two types of rotating seals used in agitated vessels, stuffing box packings and mechanical seals

Stuffing box packings: Stuffing box packings have been used successfully in large fermentors (size 100 m^3 and more), because there is a difficulty in the manufacture of large mechanical seals of size 0.2 m in diameter. The sealing rings in stuffing box packings are compressed in the axial direction first. With subsequent radial deformation, the installed sealing rings are pressed against the stuffing box boring and the spindle.

Mechanical seal: The best choice is a double mechanical seal. Tungsten carbide seats are more expensive than the commonly used ceramic and carbon faces, but they give longer seal life and less maintenance. The static shaft seals should be EPDM, except for hydrocarbon fermentations, for which Viton is recommended. In the conventional design of double mechanical seals, which derives from the chemical process industries, the seals are mounted back to back. Potential problems with this design are that the housing pressure for this design must be kept higher than the vessel pressure. If the top seal leaks, coolant flows into the vessel. Finally, solids can accumulate in the pocket between the vessel flange and the shaft static seal.

The back to back design works very well for chemical reactors, but in chemical reactor design asepsis is not a consideration. For aseptic operation, the top seal is reversed, and the seal housing is operated at atmospheric pressure. This design eliminates pockets where solids can accumulate, does not allow flow into the vessel, compensates automatically for head pressure variations, and results in lower temperature operation of the seals. Steam flows through the housing during sterilization, and sterile steam condensate flows through it during fermentation to provide lubrication and cooling. This design is simpler than the one which must be used for back to back seals.

In addition to the above seals, magnetic couplings have also been used in small fermentors. Magnetic couplings enable noncontact torque transmission to the vessel without any penetration of the vessel. For fermentation with recombinant organisms or mammalian cells, magnetic couplings are usually employed. Since power transmission ceases whenever the design load is exceeded, magnetic couplings are protected against overload. Magnetic couplings require no external bearings; however, the bearing in the reaction chamber may be susceptible to failure due to particulate solids in the form of cells or media.

Table 3.9
Operating range of impellers with respect to viscosity

Impeller Type	Viscosity Range (mPa s)
Turbines	1–50000
Propellers	1–10000
Anchors	100–5000
Paddles	100–50000
Gate anchors	1000–100000
Helical screw	5000–500000
Helical ribbon	10000–5000000

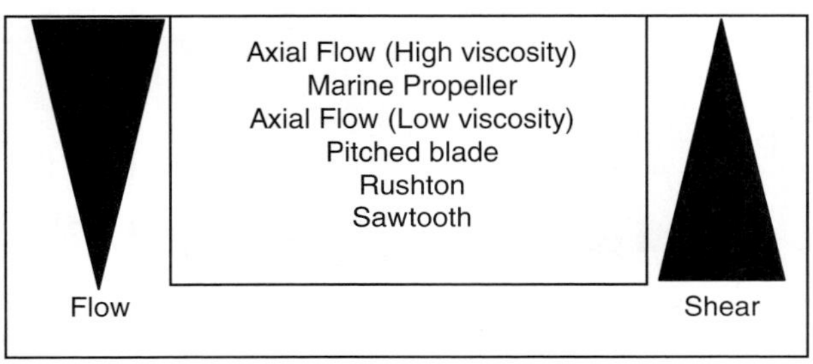

Figure 3.10 Impeller spectrum of different impellers

3.10.4 Power Requirement

The general expression for power consumed by an impeller in an unaerated mixing vessel is given by (202):

$$N_p = K\, N_{Re}^{\ m}.\, N_{Fr}^{\ n} \tag{3.14}$$

where N_p is the power number, which is a constant depending on the impeller type and size, and number of baffles present, N_{Re} and N_{Fr} are the Reynolds and Froude numbers respectively, which describe the kinematic conditions.

The Power: Reynolds and Froude numbers are expressed as:

$$\text{Power number, } N_p = P_o/\rho.N^3.D^5 \tag{3.15}$$

$$\text{Reynolds number, } N_{Re} = D^2.N.\rho\,/\mu \text{ and} \tag{3.16}$$

$$\text{Froude number, } N_{Fr} = D.N^2/g \tag{3.17}$$

where P_o is the power input during unaerated fermentation, D is the impeller diameter, N is the impeller speed, μ is the viscosity, and ρ is the density of the liquid and g is the acceleration due to gravity.

The Reynolds number accounts for the effect of viscosity when viscous forces prevail, while the Froude number accounts for the force of gravity when it plays a part in the motion of the fluid.

The interrelation of impeller speed and diameter, fluid density and viscosity, and physical configuration of the tank and fittings; and consideration of the power necessary to move the impeller have resulted in the formulation of a Rushton plot (203) as shown in Figure 3.11. The Rushton plot is a logarithmic plot of power number (N_p) against the Reynolds number (N_{Re}).

For a particular fermentation system with constant impeller speed, fluid characteristics, and vessel geometry, the Reynolds number for the system, extrapolated to the Y axis in a Rushton Plot can be used to determine the value of N_p. From N_p, P_g/P_o, the ungassed power (P_o) can be calculated.

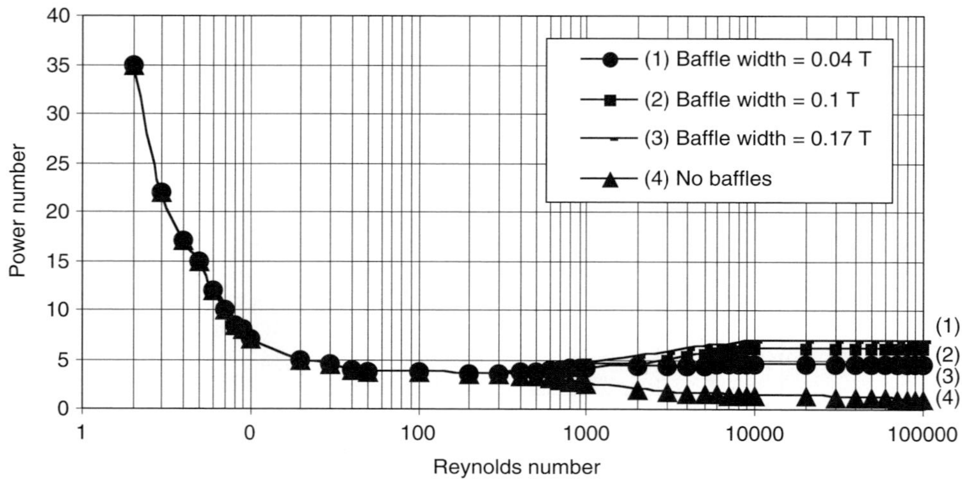

Figure 3.11 Rushton plot for flat blade turbine

The gassed power (P_g) can be determined using correlations expressing the P_g/P_o ratio. The correlation developed by Reuss et al. (204) using dimensional analysis is given by:

$$P_g/P_o = 0.0312.(N_{Re})^{0.064}.(N_{Fr})^{-1.64}.(N_A)^{-0.38}.(T/D)^{0.8} \tag{3.18}$$

where N_A is the aeration or flow number and is represented as Q_g/ND^3 where Qg is the volumetric flow rate of gas and T is the diameter of the vessel.

The power aeration curve (Figure 3.12) which is a plot of P_g/P_o versus the aeration number can be determined for a vessel with liquid height to vessel diameter ratio and stirrer speed (205).

3.11 NATURAL BIOREACTORS USING TRANSGENIC TECHNOLOGY

The biotechnology industry is faced with a shortfall of manufacturing facility for production of recombinant therapeutic proteins. The advent of natural bioreactors in the form of transgenic technology in animals, plants, and insects has the potential for large production capabilities at lower capital and operating costs and posttranslational modifications compared to mammalian cell systems, but is limited by slow production rates and regulatory obstacles.

3.11.1 Animal Bioreactors

Transgenic animal bioreactors have the potential to produce complex, biologically active recombinant proteins, and serve as organ donors for transplantations in humans, in an efficient and economic manner (206,207). Transgenic rabbits expressing human genes have been used as a model for cardiovascular disease, AIDS, and cancer research (208). The recombinant proteins can be produced from the milk of transgenic rabbits not only at lower cost but also on a relatively large scale. Transgenic rabbits, which are a relatively larger mammalian transgenic model compared to transgenic mice, can be used as human disease models and live bioreactors for producing human therapeutic proteins. Transgenic zebra fish have been explored as a useful bioreactor for the expression of living color fluorescent

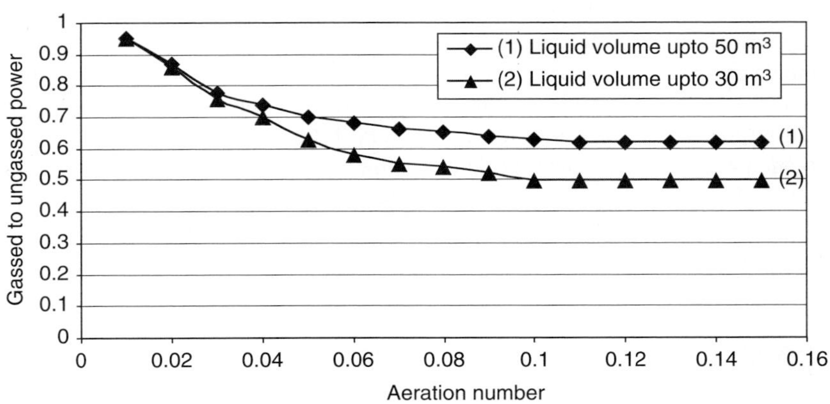

Figure 3.12 Power aeration curve

proteins, namely green fluorescent protein (GFP), yellow fluorescent protein (YFP), and red fluorescent protein (RFP) under a strong muscle specific promoter (209). However, the generation of transgenic animals is perceived as a major problem which limits its applications.

3.11.2 Plant Bioreactors

Transgenic plants are excellent hosts for production of therapeutic proteins in capacities on the order of 10 kg/acre and have been tried out in tobacco, maize, soybean, and alfalfa (210). Transgenic plants have the advantages of accumulation within plants or secretion from roots or leaves, high yields, economic scalability, establishment of permanent lines, containment, and purification (211). Transgenic root systems offer immense potential for introducing additional genes along with the Ri-T-DNA genes for alteration of metabolic pathways and production of useful metabolites (212). The main constraints for commercial exploitation of hairy root cultures for the production of valuable secondary metabolites are the complications associated with scale up. Transgenic potatoes provide a viable alternative to the conventional chemical method for production of high levels of platinose in the storage tissues of the crop (213). Plants can also serve as bioreactors for the production and scale up of functional antibodies used in immunotherapy. Proteins expressed in genetically modified plants (GM-plants) are tested for their potential use as human and animal vaccines (214). To maximize the potential of plants as industrial bioreactors for the production of proteins, efficient expression systems utilizing promoters like cauliflower mosaic virus (CaMV) are widely used (215). From an economical perspective, the production cost of biomolecules using plants is potentially less than that of using a fermentation process, because solar energy, CO_2, and inorganic chemicals are sufficient to drive the plant process; while relatively expensive equipment, substrates, and electric energy are required in fermentation (216).

3.11.3 Insect Cell Bioreactors

Insect cells could also be potential bioreactors and have been cultivated in almost all operation modes of fermentation. The baculovirus insect cell system has been used for the production of high value heterologous proteins, which form an assembly of regular structures called viruslike particles (VLPs) (217). The advantages of the VLPs are that they possess highly immunogenic properties, which have the potential to be used as vaccines. The operation of insect cell expression system is normally by fed batch mode because the baculovirus infection imposes a high metabolic burden on the infected cells thereby inducing cell lysis for the production of VLPs. The significant interdependent variables in the operation of the baculovirus insect system are: multiplication of infection, cell concentration at infection, and harvest time. The scale up of insect cell system to a 25L bioreactor involves consideration of the capability to supply enough oxygen to maintain the high specific oxygen uptake rate during cell infection through sparging of air or O_2 during the cultivation. The cell susceptibility during sparging occurs due to breakage of the bubbles, which can be avoided by the addition of Pluronic F-68. Porcine parvovirus (PPV) is an ethiological agent, which causes reproductive failure of pigs. Using the baculovirus insect cell expression system, production of recombinant self assembling VLPs composed of PPV VP2 capsid protein, which are shown to be highly immunogenic and confer full protection against the disease, was achieved.

3.12 CONCLUSIONS AND FUTURE PERSPECTIVE

Historically, industrial biotechnology has dealt mainly with microbial systems of natural origin. Improvements took place through mutations of the organism by physical and chemical methods on the one hand, and developments and improvements in the design, construction, and operation of fermentors on the other. The production of penicillin represented the first opportunity to apply chemical engineering principles to fermentation technology in large scale aerated, agitated fermentors in batch, continuous, and fed batch mode. Reactor developments took place in terms of mode of conducting the different phases in the reactor, resulting in the application of techniques of air lift, fluidized bed, microcarrier, and membrane bioreactors.

In the recent decades, with the rapid advances in molecular biology and genetic engineering, and the capability to engineer organisms through recombinant DNA technology and introduce foreign genes of desired capability into host organisms, the potential of industrial biotechnology to produce products of economic importance has enormously increased. This has been particularly so for production of high value molecules such as proteins of therapeutic importance through the use of transgenic technology involving animal and plant cells and. Concurrent with these have been the developments in the design of sophisticated bioreactors involving recombinant DNA, microbial, animal, and plant cell systems.

Bioreactor development is dependent on a synthesis of knowledge of the kinetics of the biological reactions, with the flow and mixing processes in the reactor, to achieve an overall mathematical description of the process that can be useful for design and optimization. However, this has not been effectively accomplished, due to reasons such as our inadequate knowledge of biological kinetics, use of complex growth media with undefined compositions, use of reactors with poorly defined flow and mixing characteristics, and the lack of concerted effort toward more systematic and predictive reactor design. Instead, process scale up has proceeded through rather empirical experimental investigations at increasing scale with smaller steps. One reason this has happened is the rapid growth of the ability of biotechnology to produce new products of high commercial value. Other reasons include the high investments involved in systematic experimental investigations, and the tendency of the industry to guard the confidentiality of their results. However, with increased competition in the industry, process optimization and systematic reactor design will definitely become an important factor in the future, particularly for the production of high value products through biotechnology.

REFERENCES

1. Ross, R.P., S. Morgan, C. Hill. Preservation and fermentation: past, present and future. *Intl. J. Food Microbiol.* 79:3–16, 2002.
2. Hulse, J.H. Biotechnologies: past history, present state and future prospects. *Trends Food Sci. Technol.* 15:3–18, 2004.
3. Linko, Y.Y., P. Javanainen, S. Linko. Biotechnology of bread baking. *Trends Food Sci. Technol.* 8:339–344, 1997.
4. Linko, M., A. Haikara, A. Ritala, M. Penttila. Recent advances in the malting and brewing industry. *J. Biotechnol.* 65:85–98, 1998.
5. Holzapfel, W.H. Appropriate starter culture technologies for small-scale fermentation in developing countries. *Int. J. Food Microbiol.* 75:197–212, 2002.
6. Gervais, P., P. Molin. The role of water in solid-state fermentation. *Biochem. Eng. J.* 13:85–101, 2003.
7. Ooijkaas, L.P., F.J. Weber, R.M. Buitelaar, J. Tramper, A. Rinzema. Defined media and inert supports: their potential as solid-state fermentation production systems. *Trends Biotechnol.* 18:356–360, 2000.
8. Verschoor, H. Developments in bioreactors. *Chem. Eng.* 415:39–41, 1985.
9. Cliffe, K. Bioreactors. In: *Biotechnology for engineers: Biological systems in technological processes*, Scragg, A.H., ed., New York: Ellis Horwood, Ltd., 1988, pp 277–301.

10. Kilonzo, P.M., A. Margaritis. The effects of non-Newtonian fermentation broth viscosity and small bubble segregation on oxygen mass transfer in gas-lift bioreactors: a critical review. *Biochem. Eng. J.* 17:27–40, 2004.
11. Fu, C.C., W.T. Wu, S.Y. Lu. Performance of airlift bioreactors with net draft tube. *Enz. Microb. Technol.* 33:332–342, 2003.
12. Zannetti, R. Breathing new life into single-cell protein. *Chem. Eng.* 91:18,1984.
13. Nedovic, V.A., B. Obradovic, I. Leskosek-Cukalovic, G. Vunjak-Novakovic. Immobilized yeast bioreactor systems for brewing: recent achievements. In: *Engineering and Manufacturing for Biotechnology,* Vol. 4, Hofman, M., P. Thonart, eds., Belgium: Kluwer Academic,1993, pp 277–292.
14. Nedovic, V.A., I. Leskosek-Cukalovic, G. Vunjak-Novakovic. Short-time fermentation of beer in an immobilized yeast air-lift bioreactor. In: *Proc. 24th Conv. Inst. Brew.*, Harvey, J. ed., Adelaide: Singapore Winetitles, 1996, pp 245–250.
15. Mensour, N., A. Margaritis, C.L. Briens, H. Pilkington, I. Russel. Application of immobilized yeast cells in the brewing industry. In: *Immobilized Cells: Basics and Applications.* Wijffels, R.H., R.M. Buitelaar, C. Bucke, J. Tramper, eds., Amsterdam: Elsevier, 1996, pp 661–667.
16. Park, Y.S., M. Itida, N. Ohta, M. Okabe. Itaconic acid production using an air-lift bioreactor in repeated batch culture of *Aspergillus terreus. J. Ferment. Bioeng.* 77:329–331, 1994.
17. Kahar, P., K. Kobayashi, T. Iwata, J. Hiraki, M. Kojima, M. Okabe Production of ξ-polylysine in an airlift bioreactor (abr). *J. Biosci. Bioeng.* 93:274–280, 2002.
18. Margaritis, A., D. te Bokkel, M.E. Kashab. Pilot plant production of ethanol using immobilized yeast cells in a novel fluidized bioreactor system. 18th ACS Meeting, Washington, DC, 1983.
19. Shindo, S., H. Sahara, N. Watanabe, S. Koshino. Main fermentations with immobilized yeast using fluidized-bed reactor. In: *Proc. 23rd Conv. Inst. Brew.,* Sydney, 1994, pp 109–113.
20. Umemoto, S., Y. Mitani, K. Shinotsuka. Primary fermentation with immobilized yeast in a fluidized bed reactor. *MBAA Tech. Quart.* 35:58–61, 1998.
21. Tata, M., P. Bower, S. Bromberg, D. Duncombe, J. Fehring, V. Lau, D. Ryder, P. Stassi. Immobilized yeast bioreactor systems for continuous beer fermentation. *Biotechnol. Prog.* 15:105–113, 1999.
22. Moreira, M.T., C. Palma, G. Feijoo, J.M. Lema. Strategies for the continuous production of ligninolytic enzymes in fixed and fluidised bed bioreactors. *J. Biotechnol.* 66:27–39, 1998.
23. Sokol, W. Treatment of refinery wastewater in a three-phase fluidized bed bioreactor with a low-density biomass support. *Biochem. Eng. J.* 15:1–10, 2003.
24. Sokol, W. Operating parameters for a gas-liquid-solid fluidized bed bioreactor with a low-density support. *Biochem. Eng. J.* 8:203–212, 2001.
25. Van Wezel, A.L. Growth of cell strains and primary cells on microcarriers in homogeneous culture. *Nature* 216:64, 1967.
26. Crespi, C.L., T. Imamura, P. Leong, R. J. Fleischaker, H. Brunengraber, W.G. Thilly, D.G. Giard. Microcarrier culture: Applications in biological production and cell biology. *Biotechnol. Bioeng.* 23:2673–2689, 1981.
27. Giard, D.J. Human Interferon production with diploid fibroblat cells grown on microcarriers. Biotechnol. Bioeng. 21:433–442, 1979.
28. Gan, Q., S.J. Allen, G. Taylor. Design and operation of an integrated membrane reactor for enzymatic cellulose hydrolysis. *Biochem. Eng. J.* 12:223–229, 2002.
29. Wenten, I.G., I.N. Widiasa. Enzymatic hollow fiber membrane bioreactor for penicillin hydrolysis. *Desalination* 149:279–285, 2002.
30. Paolucci-Jeanjean, D., M.P. Belleville, G.M. Rios, N. Zakhia. Kinetics of continuous starch hydrolysis in a membrane reactor. *Biochem. Eng. J.* 6:233–238, 2000.
31. Belhocine, D., H. Mokrane, H. Grib, H. Lounici, A. Pauss, N. Mameri Optimization of enzymatic hydrolysis of haemoglobin in a continuous membrane bioreactor. *Chem. Eng. J.* 76:189–96, 2000.
32. Indoles, D.S. Hollow fibre membrane bioreactors using *E.coli* for protein synthesis. *Biotechnol. Bioeng.* 25:2653, 1983.

33. Yang, R.Y.K., O. Bayraktar, H.T. Pu. Plant cell bioreactors with simultaneous electropermeabilization and electrophoresis. *J. Biotechnol.* 100:13–22, 2003.
34. Millward, H.R., B.J. Bellhouse, I.J. Sobey. The vortex wave membrane bioreactor hydrodynamics and mass transfer. *Chem. Eng. J.* 62:175–181, 1996.
35. Moueddeb, H., J. Sanchez, C. Bardot, M. Fick. Membrane bioreactor for lactic acid production. *J. Memb. Sci.* 114:59–71, 1996.
36. Chang, H.N., S. Furusaki. Membrane bioreactors: present and prospects. In: *Advances in Biochemical Engineering Biotechnology,* Vol. 44, Fiechter, A., ed., Berlin: Springer-Verlag, 1991, pp 27–64.
37. Cheryan, M., M.A. Mehaia. A high performance membrane bioreactor for continuous fermentation of lactose to ethanol. *Biotechnol. Letters* 5: 519–524, 1983.
38. Endo, I. A membrane bioreactor. *Membrane* 21:18–22, 1996.
39. Boyaval, P., C. Corre. Continuous fermentation of sweet whey permeate for propionic acid production in a CSTR with UF recycle. *Biotechnol. Lett.* 11:801–806, I987.
40. Takaya, M., N. Matsumoto, H. Yanase. Characterization of membrane bioreactor for dry wine production. *J. Biosci. Bioeng.* 93:240–244, 2002.
41. Tanase, T., Y. Ikeda, K. Iwama, A. Hashimoto, T. Kataoka, Y. Tokushima, T. Kobayash. Comparison of micro-filtration hollow fiber bioreactors for mammalian cell culture. *J. Ferment. Bioeng.* 83:499–501, 1997.
42. Pulz, O., K. Scheibenbogen. Photobioreactors: design and performance with respect to light energy input. In: *Advances in Biochemical Engineering Biotechnology,* vol. 59, Scheper, T., ed., Berlin: Springer-Verlag, 1998, pp 124–152.
43. Cornet, J.F., C.G. Dussap, J.B. Gros. Kinetics and energetics of photosynthetic microorganisms in photobioreactors: application to *Spirulina* growth. In: *Advances in Biochemical Engineering Biotechnology*, Vol. 59, Scheper, T., ed., Berlin: Springer-Verlag, 1998, pp 153–224.
44. Walther, I. Space bioreactors and their applications. *Adv. Space Biol. Med.* 8:197–213, 2002.
45. Wolf, L. Bioregeneration in space. *Adv. Space Biol. Med.* 5:341–356, 1996.
46. Klaus, D.M. Microgravity and its implications for fermentation biotechnology. *Trends Biotechnol.* 16:369–373, 1998.
47. Walther, I., B.V. Schoot, M. Boillat, A. Cogoli. Bioreactors for space: biotechnology of the next century. In: *Engineering and Manufacturing for Biotechnology*, Vol. 4, Hofman, M., P. Thonart, eds., Netherlands: Kluwer Academic, 1993, pp 241–251.
48. Omasaa, T., M. Kishimoto, M. Kawase, K. Yagi. An attempt at decision making in tissue engineering: reactor evaluation using the analytic hierarchy process. *Biochemical Eng. J.* Available online 7th Feb, 2004.
49. Langer, R., J.P. Vacanti. Tissue Engineering. *Science* 260:920–926,1993.
50. Jasmund, I., A. Bader. Bioreactor developments for tissue engineering applications by the example of the bioartificial liver. In: *Advances in Biochemical Engineering Biotechnology*, Vol. 74, Scheper, T., ed., Berlin: Springer-Verlag, 2002, pp 99–109.
51. Martin, I., D. Wendt, M. Heberer. The role of bioreactors in tissue engineering. *Trends Biotechnol.* 22:80–86, 2004.
52. Noll, T., N. Jelinek, S. Schmidt, M. Biselli, C. Wandrey. Cultivation of Hematopoietic stem and progenitor cells: biochemical engineering aspects. In: *Advances in Biochemical Engineering Biotechnology*, Vol. 74, Scheper, T., ed., Berlin: Springer-Verlag, 2002, pp 111–128.
53. Cabrita, G.J.M., B.S. Ferreira, C.L. da Silva, R. Goncalves, G.A. Porada, J.M.S Cabral. Hematopoietic stem cells: from the bone to the bioreactor. *Trends Biotechnol.* 21:233–240, 2003.
54. Melwin, B.K., J.V. Shanks. Influence of aeration on cytoplasmic pH of yeast in an NMR Airlift bioreactor. *Biotechnol. Prog.* 12:257–265, 1996.
55. Zupke, C., B. Foy. Nuclear magnetic resonance analysis of cell metabolism. *Curr. Opin. Biotechnol.* 6:192–197, 1995.
56. Kumar, S., C. Wittmann, E. Heinzle. Minibioreactors. *Biotechnol. Lett.* 26:1–10, 2004.
57. Schugerl, K. Integrated processing of biotechnology products. *Biotechnol. Adv.* 18:581–599, 2000.

58. Malinowski, J.J. Two-phase partitioning bioreactors in fermentation technology. *Biotechnol. Adv.* 19:525–538, 2001.
59. Li, H., R. Mustacchi, C.J. Knowles, W. Skibar, G. Sunderland, I. Dalrympleb, S.A. Jackman. An electrokinetic bioreactor: using direct electric current for enhanced lactic acid fermentation and product recovery. *Tetrahedron* 60:655–661, 2004.
60. Gryta, M. The fermentation process integrated with membrane distillation. *Sep. Purif. Technol.* 24:283–296, 2001.
61. Greens, K.D., N.H. Thomas. An integrated "root tube" bioreactor/separator for transformed root cultures. *J. Ferment. Bioeng.* 81:453–457, 1996.
62. Pandey, A. Solid-state fermentation. *Biochem. Eng. J.* 13:81–84, 2003.
63. Raghavarao, K.S.M.S., T.V. Ranganathan, N.G. Karanth. Some engineering aspects of solid-state fermentation. *Biochem. Eng. J.* 13:127–135, 2003.
64. Durand, A. Bioreactor designs for solid state fermentation *Biochem. Eng. J.* 13:113–125, 2003.
65. Pandey, A., C.R. Soccol, J.A. Rodriguez-Leon, P. Nigam. Aspects of design of fermentor in solid state fermentation. In: *Solid State Fermentation in Biotechnology: Fundamentals and Applications,* Pandey, A., ed., New Delhi: Asiatech Publ., ISBN: 81-87680-06-7, 2001, pp 73–77.
66. Nagel, F.J., J. Tramper, M. Bakker, A. Rinzema. Temperature control in a continuously mixed bioreactor for solid-state fermentation. *Biotechnol. Bioeng.* 72:219–230, 2001.
67. Lonsane, B.K., N.P. Ghildyal, S. Budiatman, S.V. Ramakrishna. Engineering aspects of solid-state fermentation. *Enz. Microbiol. Technol.* 7:258–265, 1985.
68. Chisti, Y. Solid substrate fermentations, enzyme production, food enrichment. In: *Encyclopaedia of Bioprocess Technology: Fermentation, Biocatalysis and Bioseparation*, Vol 5, Flickinger, M.C., S.W. Drew, eds., New York: Wiley, 1999, pp 2446–2462.
69. Durand, A., D. Chereau. A new pilot reactor for solid-state fermentation: application to protein enrichment of sugar beet pulp. *Biotechnol. Bioeng.* 31:476–486, 1988.
70. Sangsurasak, P., D.A. Mitchell. Validation of a model describing two-dimensional heat transfer during solid-state fermentation in packed bed bioreactors. *Biotechnol. Bioeng.* 60:739–749, 1998.
71. Suryanarayan, S. Current industrial practice in solid state fermentations for secondary metabolite production: the Biocon India experience *Biochemical Eng. J.* 13:189–195, 2003.
72. Marsh, A.J., D.M. Stuart, D.A. Mitchell, T. Howes. Characterizing mixing in a rotating drum bioreactor for solid-state fermentation. *Biotechnol. Lett.* 22:473–477, 2000.
73. Kalogeris, E., F. Iniotaki, E. Topakas, P. Christakopoulos, D. Kekos, B.J. Macris. Performance of an intermittent agitation rotating drum type bioreactor for solid-state fermentation of wheat straw. Biores. *Technol.* 86:207–213, 2003.
74. Fung, C.J., D.A. Mitchell. Baffles increase performance of solid state fermentation in rotating drums. *Biotechnol. Tech.* 9:295–298, 1995.
75. Hardin, M.T., T. Howes, D.A. Mitchell. Mass transfer correlations for rotating drum bioreactors. *J. Biotechnol.* 97:89–101, 2002.
76. Robinson, T., P. Nigam. Bioreactor design for protein enrichment of agricultural residues by solid state fermentation. *Biochem. Eng. J.* 13:197–203, 2003.
77. Mitchell, D.A., N. Krieger, D.M. Stuart, A. Pandey. New developments in solid-state fermentation, II: rational approaches to the design, operation and scale-up of bioreactors. *Proc. Biochem.* 35:1211–1225, 2000.
78. Stanbury, P.F., S.J. Whitaker. Media for industrial fermentations. In: *Principles of Fermentation Technology*, 2nd ed., New Delhi: Aditya Books [Original Publisher: Butterworth & Heinemann (1995)], 1997, pp 93–122.
79. Calyk, P., E. Celik, I.E. Telli, C. Oktar, E. Özdemir. Protein-based complex medium design for recombinant serine alkaline protease production. *Enz. Microb. Technol.* 33:975–986, 2003.
80. Beshay, U., H. El-Enshasy, I.M.K. Ismail, H. Moawad, E. Wojciechowska, S. Abd-El-Ghany. b-Glucanase production from genetically modified recombinant *Escherichia coli*: effect of growth substrates and development of a culture medium in shake flasks and stirred tank bioreactor. *Proc. Biochem.* 39:307–313, 2003.

81. Bai, Z.H., H.X. Zhang, H.Y. Qi, X.W. Peng, B.J. Li. Pectinase production by *Aspergillus niger* using wastewater in solid state fermentation for eliciting plant disease resistance. *Biores. Technol.* (In press), 2004.
82. Cornell, J.A. *Experiments with mixtures*, 3rd ed. New York: John Wiley & Sons, 2002.
83. Myers, R.H., D.C. Montgomery. *Response surface methodology: process and product optimization using designed experiments*, 2nd ed. New York: John Wiley & Sons, 2002.
84. Hounsa, C.G., J.M. Aubry, H.C. Dubourguier. Application of factorial and Doehlert design for optimization of pectate lyase production by a recombinant *Escherichia coli. Appl. Microbiol. Biotechnol.* 45:764–770,1996.
85. Rao, K.J., C.H. Kim, S.K. Rhee. Statistical optimization of medium for the production of recombinant hirudin from *Saccharomyces cerevisiae* using response surface methodology. *Proc. Biochem.* 35:639–647, 2000.
86. Dey, G., A. Mitra, R. Banerjee, B.R. Maiti. Enhanced production of amylase by optimization of nutritional constituents using response surface methodology. *Biochem. Eng. J.* 7:227–233, 2001.
87. Wua, J.Z., P.C.K. Cheunga, K.H. Wonga, N.I. Huang. Studies on submerged fermentation of *Pleurotus tuber-regium* (Fr.) singer, part 2: effect of carbon-to-nitrogen ratio of the culture medium on the content and composition of the mycelial dietary fibre. *Food Chem.* 85:101–105, 2004.
88. Atkinson, B., F. Mavituna. Process biotechnology. In: *Biochemical Engineering and Biotechnology Handbook*, 2nd ed., London: Macmillan, 1991, pp 68–73.
89. Aristidou, A., M. Penttila. Metabolic engineering applications to renewable resource utilization. *Curr. Opin. Biotechnol.* 11:187–198, 2000.
90. Kuyper, M., A.A. Winkler, J.P. van Dijken, J.T. Pronk. Minimal metabolic engineering of *Saccharomyces cerevisiae* for efficient anaerobic xylose fermentation: a proof of principle. *FEMS Yeast Res.* 4:655–664, 2004.
91. Zabriskie, D.W., W.B. Armiger, D.H. Phillips, P.A. Albano. *Traders' Guide to Fermentation Media Formulation: Traders Protein*. Memphis: 1999, pp3–20.
92. Jeng, D.K.H., K.A. Kaczmarek, A.G. Woodworth, G. Balasky. Mechanism of microwave sterilization in dry state. *Appl. Environ. Microbiol.* Published by American Society of Microbiology, 54:1330–1333, 1987.
93. Jayaram, S., G.S.P. Castle, A. Margaritis. Kinetics of sterilization of *Lactobacillus brevis* cells by the application of high voltage pulses. *Biotechnol. Bioeng.* 40:1412–1420, 1992.
94. Bes, J. Strategies of sterilization. In: *Operational modes of bioreactors.* 2nd ed. Oxford: Butterworth-Heinemann, 1992, pp 203–234.
95. Deindoerfer, F.H., A.E. Humphrey. Analytical method for calculating heat sterilization times. *Appl. Microbiol.* 7:256–264, 1959.
96. Samsatli, N.J., N. Shah. Optimal design of continuous sterilization networks. *Comp. Chem. Eng.* 19:S95–S100, 1995.
97. Papagianni, M. Fungal morphology and metabolite production in submerged mycelial processes. *Biotechnol. Adv.* 22:189–259, 2004.
98. Hunt, G.R., R.W. Stieber. Inoculum development. In: *Manual of Industrial Microbiology and Biotechnology*, Demain, A.L., N.A. Solomon, eds., Washington: American Society of Microbiology, 1986, pp 32–40. URL:http://www.asm.org/
99. Parton, C., P. Willis. Strain preservation, inoculum preparation and inoculum development. In: *Fermentation: a practical approach*, McNeil, B., L.M. Harvey, eds., Oxford: IRL Press, 1990, pp 39–64.
100. Blakebrough, N. Industrial fermentations. In: *Biochemical and Biological Engineering Science,* Vol. 1, Blakebrough, N., ed., New York: Academic Press, 1967, pp 25–48.
101. Gram, A. Biochemical engineering & industry. *J. Biotechnol.* 59:19–23, 1997.
102. Bryson, V. The turbidostatic selector, part II: a device for automatic isolation of bacterial variants. *Science* 116:48–51, 1952.
103. Martin, G.A., W.P. Hempfling. Method for the regulation of microbial population density during continuous culture at high growth rates. *Arch. Microbiol.* 107:41–47, 1976.

104. Edwards, V.H., R.C. Ko, S.A. Balogh. Dynamics and control of microbial propagators subject to substrate inhibition. *Biotechnol. Bioeng*. 14:939–974, 1972.
105. Oltmann, L.F., G.S. Schoenmaker, W.N.M. Reijnders, A.H. Stouthamer. Modification of the pH-auxostat culture method for the mass cultivation of bacteria. *Biotechnol. Bioeng*. 20:921–925, 1978.
106. Larsson, G., S.O. Enfors, H. Pham. The pH-auxostat as a tool for studying microbial dynamics in continuous fermentation. *Biotechnol. Bioeng*. 36:224–232, 1990.
107. Aarino, T.H., M.L. Suihko, V.S. Kauppinen. Isolation of acetic acid-tolerant baker's yeast variants in a turbidostat. *Appl. Biochem. Biotechnol*. 27:55–63, 1991.
108. Agarwal, P., H.C. Lim. Analyses of various control schemes for continuous bioreactors. *Adv. Biochem. Eng. Biotechnol*. 30:61–90, 1984.
109. Revised edition details: Sinclair, C.G., B. Kristiansen. J.D. Bu'lock. In: *Fermentation Kinetics and Modelling* (*Biotechnology Series*), John Wiley, 1991. London: Open University Press, 1987.
110. Longobardi, G.P. Fed-batch versus batch fermentation. *Bioproc. Eng*. 10:185–194, 1994.
111. Vasavada, A. Improving productivity of heterologous proteins in recombinant *Saccharomyces cerevisiae* fermentations. *Adv. Appl. Microbiol.* 41:25–54, 1995.
112. Oldshue, J.Y. Mixing in fermentation processes. *Annual Reports on Fermentation Processes. Vol. 6,* New York: Academic Press, 1983, pp 75–99.
113. Rushton, J.H., J.Y. Oldshue. Mixing present theory and practice, part I. *Chem. Eng. Prog.* 49:161–168, 1953.
114. Rushton, J.H., J.Y. Oldshue. Mixing present theory and practice, part II. *Chem. Eng. Prog.* 49:267–275, 1953.
115. Ryu, D.Y., J.Y. Oldshue. Reassessment of mixing cost in fermentation processes. *Biotechnol. Bioeng*. 29:621–629, 1977.
116. Coyne, J., P. Kaufman, T.A. Post. Key parameters for consideration in up-pumping technology in fermentation. Proceedings from Interphex East, 1998.
117. Gogate, P.R., A.A.C.M. Beenackers, A.B. Pandit. Multiple-impeller systems with a special emphasis on bioreactors: a critical review. *Biochem. Eng. J*. 6:109–144, 2000.
118. Arjunwadkar, S.J., K. Sarvanan, P.R. Kulkarni, A.B. Pandit. Gas-liquid mass transfer in dual impeller bioreactor. *Biochem. Eng. J*. 1:99–106, 1998.
119. Lally, K.S., J.Y. Oldshue, R.J. Weetman. *Mass transfer and fluid mixing in fermentation.* Presented at American Institute of Chemical Engineers 1985 Annual Meeting, Chicago, Illinois.
120. Walker, J.A.H., H. Holdsworth. Equipment Design. In: *Biochemical Engineering: Unit Processes in Fermentation*, Steel, R., ed., London: Heywood & Company Ltd., 1958, pp 225–273.
121. Kawase, Y., B. Halard, M. Moo-Young. Liquid phase mass transfer coefficients in bioreactors. *Biotechnol. Bioeng*. 39:1133–1140, 1992.
122. Steenkiste, F., K. Baert, D. Debruyker, V. Spiering, B.V.D. Schot, P. Arquint, R. Born, K. Schumann. A microsensor array for biochemical sensing. *Sensors Actuators* B44:409–412, 1997.
123. Harms, P., Y. Kostov, G. Rao. Bioprocess monitoring. *Curr. Opin. Biotechnol.* 13:124–127, 2002.
124. Onken, U., P. Weiland. Control and Optimization. In: *Biotechnology, Vol. 2: Fundamentals of Biochemical Engineering*, Rehm, H.J., G. Reed, eds., Weinheim, Germany: VCH Publishers, 1985, pp 787–806.
125. D'Souza, S.F. Microbial biosensors: review. *Biosensors Bioelectronics* 16:337–353, 2001.
126. Johnson, M.J., J. Borkowski, C. Engblom. Steam sterilizable probes for dissolved oxygen measurement. *Biotechnol. Bioeng*. 67:645–656, 2000.
127. Ocean Optics in Dunedin: FL produces the FOXY probe for dissolved oxygen, as well as several dip probes for pH that could be used to monitor bioprocesses. URL: http://www.oceanoptics.com

128. Pattison, R.N., J. Swamy, B. Mendenhall, C. Hwang, B.T. Frohlich. Measurement and control of dissolved carbon dioxide in mammalian cell culture processes using an *in situ* fiber optic chemical sensor. Biotechnol. Prog. 16:769–774, 2000.
129. Onken, U., R. Buchholz, W. Sittig. Measurement and instrumentation. In: *Biotechnology, Vol. 2: Fundamentals of Biochemical Engineering.* Rehm, H.J., G. Reed, eds., Weinheim, Germany: VCH Publishers, 1985, pp 763–786.
130. Suhr, H., G. Wehnert, K. Schneider, C. Bettner, T. Scholz, P. Gissler, B. Jahne, T. Scheper. *In situ* microscopy for online characterization of cell population in bioreactors. *Biotechnol. Bioeng.* 47:106–116, 1995.
131. Sonnleitner, B., G. Locher, A. Fiechter. Biomass determination. *J. Biotechnol.* 25:5–22, 1992.
132 Harms, P., Y. Kostov, G. Rao. Bioprocess monitoring. *Curr. Opin. Biotechnol.* 13:124–127, 2002.
133. Marose, S., C. Lindemann, R. Ulber, T. Scheper. Optical sensor systems for bioprocess monitoring. *Trends Biotechnol.* 17:30–34, 1999.
134. D'Auria S., J.R. Lakowicz. Enzyme fluorescence as a sensing tool: new perspectives in biotechnology. *Curr. Opin. Biotechnol.* 12:99–104, 2001.
135. Ducommun, P., I. Bolzonella, M. Rhiel, P. Pugeaud, U. von Stockar, I.W. Marison. On-line determination of animal cell concentration. *Biotechnol. Bioeng.* 72:515–522, 2001.
136. Raj, A.E., H.S.S. Kumar, S.U. Kumar, M.C. Misra, N.P. Ghildyal, N.G. Karanth. High cell density fermentation of recombinant *S.cerevisiae* using glycerol. *Biotechnol. Prog.* 18:1130–1132, 2002.
137. Schugerl, K. Progress in monitoring, modeling and control of bioprocesses during the last 20 years. *J. Biotechnol.* 85:149–173, 2001.
138. Ferreira, L.S., M.B. De Souza, J.O. Trierweiler, O. Broxtermann, R.O.M. Folly, B. Hitzmann. Aspects concerning the use of biosensors for process control: experimental and simulation investigations. *Comp. Chem. Eng.* 27:1165–1173, 2003.
139. Ritzka, A., P. Sosnitza, R. Ulber, T. Scheper. Fermentation monitoring and process control. *Curr. Opin. Biotechnol.* 8:160–164, 1997.
140. Liu, J., G. Li. Application of biosensors for diagnostic analysis and bioprocess monitoring. *Sensors Actuators* B65:26–31, 2000.
141. Y.C. Liu, F.S. Wang, W.C. Lee. On-line monitoring and controlling system for fermentation processes. *Biochem. Eng. J.* 7:17–25, 2001.
142. Chae, H.J., M.P. DeLisa, H.J. Cha, W.A. Weigand, G. Rao, W.E. Bentley. Framework for online optimization of recombinant protein expression in high-cell-density *Escherichia coli* cultures using GFP-fusion monitoring. *Biotechnol. Bioeng.* 69:275–285, 2000.
143. Heinzle, E., M. Reuss. *Mass Spectrometry in Biotechnological Process Analysis and Control.* New York: Plenum Press, 1987.
144. Lorenz, T., W. Schmidt, K. Schugerl. Sampling devices in fermentation technology: a review. *Chem. Eng. J.* 35:B15–B22, 1987.
145. Stephanopoulos, G. Metabolic Fluxes and Metabolic Engineering. *Metab. Eng.* 1:1–11, 1999.
146. Kasprow, R.P., A.J. Lange, D.J. Kirvan. Correlation of fermentation yield with yeast extract composition as characterized by near infrared spectroscopy. *Biotechnol. Prog.* 14:318–325, 1998.
147. Dickinson, T.A., J. White, J.S. Kauer, D.R. Walt. Current trends in 'artificial-nose' technology. *Trends Biotechnol.* 16:250–258, 1998.
148. Kula, M.R. Recovery operations. In: *Biotechnology, Vol. 2: Fundamentals of Biochemical Engineering,* Rehm, H.J., G. Reed, eds., Weinheim: VCH Publishers, 1985, pp 725–760.
149. Demain, A.L. Achievements in microbial technology. *Biotechnol. Adv.* 8:291–301, 1990.
150. Demain, A.L. Microbial secondary metabolism: a new theoretical frontier for academia, a new opportunity for industry. In: *Secondary Metabolites: Their Function and Evolution,* Chadwick, D.J., J. Whelan, eds., New York: John Wiley, 1992, pp 3–23.
151. Strohl, W.R. *Biotechnology of Antibiotics*, 2nd ed., New York: Marcel Dekker, 1997.
152. Demain, A.L. Microbial biotechnology. *Trends Biotechnol.* 18:26–30, 2000.

153. Demain, A.L. Contribution of recombinant microbes and their potential. In: *Recombinant Microbes for Industrial and Agricultural Applications.* Murooka, T., T. Imanaka, eds., New York: Marcel-Dekker, 1994, pp 27–46.
154. McCoy, T.A., W. Whittle, E. Conway. A glass helix perfusion chamber for massive growth of cells *in vitro*. *Proc. Soc. Exp. Biol.* 109:235–237, 1962.
155. Whiteside, J.P., R.E. Spier. The scale-up from 0.1 to 100 litres of a unit process system for the production of four strains of FMDV from BHK monolayer cells. *Biotechnol. Bioeng.* 23:551–565, 1981.
156. Felix, H.R., K. Mosbach. Enhanced stability of enzymes in permeabilized and immobilized cells. *Biotechnol. Lett.* 4:181–186, 1982.
157. Mavituna, F., J.M. Park. Growth of immobilized plant cells in reticulate polyurethane foam matrices. *Biotechnol. Lett.* 7:637–640, 1985.
158. Wilkinson, P.J. The development of large-scale production process for tissue culture products. In: *Bioreactors and Biotransformations*, Moody, G.W., P.W. Baker, eds., Essex: Elsevier Science Publishers, 1987, pp 111–120.
159. Swartz, J.R. *Escherichia coli* recombinant DNA technology. In: *Escherichia coli and Salmonella*: *Cellular and Molecular Biology*, Neidhardt, F.C., ed., American Society of Microbiology Press, 1996, pp 1693–1711. URL:http://www.asm.org/
160. Curtin, M.E. Harvesting profitable products from plant tissue culture. *Biotechnology* 1:649–657, 1983.
161. Sahai, O., M. Knuth. Commercializing plant tissue processes: economics, problems and prospects. *Biotechnol. Prog.* 1:1–9, 1985.
162. Scragg, A.H., P. Bond, F. Leckie, R. Gresswell, M.W. Fowler. Growth and product formation by plant cell suspensions cultivated in bioreactors. In: *Bioreactors and Biotransformations*, Moody, G.W., P.W. Baker, eds., Essex: Elsevier Science Publishers, 1987, pp 12–25.
163. Deus, B., M.H. Zenk. Exploitation of plant cells for the production of natural compounds. *Biotechnol. Bioeng.* 24:1965–1974, 1982.
164. Kieran, P.M., P.F. MacLoughlin, D.M. Malone. Plant cell suspension cultures: some engineering considerations. *J. Biotechnol.* 59:39–52, 1997.
165. Hamill, J.D., A.J. Parr, R.J. Robins, M.J.C. Rhodes, N.J. Walton. New routes to plant secondary products. *Biotechnology* 15:800–804, 1987.
166. Wilson, P.D.G., M.G. Hilton, R.J. Robins, M.J.C. Rhodes. Fermentation studies of transformed root cultures. In: *Bioreactors and Biotransformations*, Moody, G.W., P.W. Baker, eds., Essex: Elsevier Science Publishers, 1987, pp 38–51.
167. Wagner, F., H. Vogelmann. Cultivation of plant tissue culture in bioreactors and formation of secondary metabolites. In: *Plant tissue culture and its biotechnological applications,* Barz, W., E. Reinhard, M.H. Zenk, eds., Berlin: Springer-Verlag, 1977, pp 130–146.
168. Mavituna, F., A.K. Wilkinson, P.D. Williams. Production of secondary metabolites by immobilized plant cells in novel bioreactors. In: *Bioreactors and Biotransformations,* Moody, G.W., P.W. Baker, eds., Essex: Elsevier, *Appl. Sci.* 1987, pp 26–37.
169. Vinci, V.A., S.R. Parekh. Mammalian, Microbiol and Plant cells. In: *Handbook of Industrial Cell Culture,* Humana Press, December 2002.
169. Bylund, F., A. Castan, R. Mikkola, A. Veide, G. Larsson. Influence of scale-up on the quality of recombinant human growth hormone. *Biotechnol. Bioeng.* 69:119–128, 2000.
171. George, S., G. Larsson, K. Olsson, S.O. Enfors. Comparison of the baker's yeast process performance in laboratory and production scale. *Bioprocess Eng.* 18:135–142, 1998.
172. Charles, M. Fermentation scale-up: problems and possibilities. *Trends Biotechnol.* 3:124–139, 1985.
173. Bylund, F., F. Guillard, S.O. Enfors, C. Tragardh, G. Larsson. Scale-down of recombinant protein production: a comparative study of scaling performance. *Bioprocess Eng.* 20:377–389, 1999.
174. Oosterhuis, N.M.G., N.W.F. Kossen, A.P.C. Olivier, E.S. Schenk. Scale-down and optimization studies of the gluconic acid fermentation by *Gluconobacter oxydans*. *Biotechnol. Bioeng.* 27:711–720, 1985.
175. Humphrey, A. Shake flask to fermentor: what have we learned? *Biotechnol. Prog.* 14:3–7, 1998.

176. Rouf, S.A., M. Moo-Young, J.M. Scharer, P.L. Douglas. Single versus multiple bioreactor scale-up: economy for high-value products. *Biochemical Eng. J.* 6:25–31, 2000.
177. Horvath, B.E. Mammalian cell culture scale-up: is bigger, better? *Biotechnology* 7:468–469, 1989.
178. Birch, J.R., K. Lambert, P.W. Thompson, A.C. Kenney, L.A. Wood. Antibody production with airlift fermentors. In: *Large Scale Cell Culture Technology,* Lydersen, B.K., ed., Munich: Hanser, 1987.
179. M.C. Sharma, A.K. Gurtu. Asepsis in bioreactors. In: *Advances in Applied Microbiology,* Vol. 39, Neidleman, S., A.I. Laskin, eds., New York: Academic Press, 1993, pp 1–27.
180. Banks, G.T. Scale up of fermentation process. *Topics Enzyme Ferment. Biotechnol.* 3:170–266, 1979.
181. Spier, R. Animal Cells in Culture: Moving into the exponential phase. *Trends Biotechnol.* 6:2–6, 1988.
182. Sutton S.V. W., A. M. Cundell. Microbial identification in the pharmaceutical industry, Pharmacopeial Forum, 30:1884–1894, 2004.
183. Design and analysis of biological reactors, 2nd ed., In: *Biochemical Engineering Fundamentals,* Bailey, J.E., D. F. Ollis, eds., Co.; Singapore: McGraw Hill, 1986, pp 592.
184. Stanbury, P.F., A. Whitaker. Sterilization. In: *Principles of fermentation technology*. New York: Pergamon, 1984.
185. Al-Masry, W.A. Effects of antifoam and scale-up on operation of bioreactors. *Chem. Eng. Process*. 38:197–201, 1999.
186. Ghildyal, N.P., B.K. Lonsane, N.G. Karanth. Foam control in submerged fermentation: State of the art. In: *Adv. Appl. Microbiol.* 33:173–222, 1988.
187. Reisman, H.B. *Economic analysis of fermentation processes.* Boca Raton, FL: CRC Press, 1988.
188. Charles, M., J. Wilson. Fermentor design. In: *Bioprocess Engineering: Systems, Equipment and Facilities,* Lydersen, B.K., N.A. D'Elia, K.L. Nelson, eds., New York: John Wiley, 1994, pp 3–67.
189. Vasconcelos, J.M.T., S.S. Alves. Direct dynamic k_La measurement in viscous fermentation broths: the residual gas hold up problem. *Chem. Eng. J.* 47:B35–B44, 1991.
190. Sukatsch, D.A., A. Dziengel. *Biotechnology: A Handbook of Practical Formulae*. Chicago: Longman, 1987.
191. Behmer, G.J., I.P.T. Moore, A.W. Nienow. Aerated and nonaerated power and mass transfer characteristics of prochem agitators. In: *Biotechnol. Processes scale-up and mixing.* Ho, C.S., J.Y. Oldshue, eds., NY: AIChE, 1987, pp 116–127.
192. Narendranathan, T.J. Designing fermentation equipment. *Chem. Eng*. 5:23–29,1986.
193. Perkowski, C.A. Detection of microscopic leaks in fermentor cooling coil. *Biotechnol. Bioeng*. 26:857–859, 1984.
194. Yokell, S. Understanding the pressure vessel codes. *Chem. Eng*. 93:75–85, 1986.
195. Dillon, C.P., D.W. Rahoi, A.H. Tuthill. Stainless steel for bioprocessing, part 2: classes of alloys. *Bio. Pharm*. 5:32–35, 1992.
196. Demain, A.L., N.A. Solomon. *Manual of Industrial Microbiology and Biotechnology*. Washington: American Society for Microbiology, 1986. URL: http://www.asm.org/
197. Meyer, P. Vessels for Biotechnology. In: *Bioprocess engineering: systems, equipment and facilities*, Lydersen, B.K., N.A. D'Elia, K.L. Nelson, eds., New York: John Wiley, 1994, pp 189–214.
198. Adey, H., M.S. Pollan. Piping and valves for biotechnology. In: *Bioprocess Engineering: Systems, Equipment and Facilities*, Lydersen, B.K., N.A. D'Elia, K.L. Nelson, eds., New York: John Wiley, 1994, pp 213–252.
199. Coleman, D., R. Evans. Fundamentals of passivation and passivity in the pharmaceutical industry. *Pharm. Eng*. 10:43–49, 1990.
200. Seiberling, D.A. Clean-in-place and sterilize-in-place. In: *Aseptic Pharmaceutical Manufacturing*, Olson, W.P., M.J. Groves, eds., Buffalo Grove, NY: Interpharm Press, 1987.
201. Schutte, M., K. Lally, G. Pogal. *Mechanical and process design aspects for fermenter upgrades.* Presented at 6th Annual Bioproc. Eng. Symp. Annaheim, CA, 1992.

202. Rushton, J.H., E.W. Costwich, H.J. Everett. Power characteristics of mixing implellers, part I. *Chemical Eng. Prog*. 46:395–404, 1950.
203. Rushton, J.H., E.W. Costwich, H.J. Everett. Power characteristics of mixing implellers, part II. *Chemical Eng. Prog*. 46:467–476, 1950.
204. Reuss, M., R.K. Bajpai, R. Lenz, H. Niebelschut, Papalexiou. Scale-up strategies based on the interactions of transport and reaction. 6th International Fermentation Symposium, London (Ontario), Canada, July 20–25, 1980.
205. Oosterhuis, N.M.G. Scale-up of Bioreactors a scale-down approach. PhD thesis. Delft, University of Technology, Netherlands, 1984.
206. Dyck, M.K., D. Lacroix, F. Pothier, M.A. Sirard. Making recombinant proteins in animals: different systems, different applications. *Trends Biotechnol*. 21:394–403, 2003.
207. Yang, X., X.C. Tian, Y. Dai, B. Wang. Transgenic farm animals: applications in agriculture and biomedicine. *Biotechnol. Ann. Rev*. 5:269–292, 2000.
208. Fan, J., T. Watanabe. Transgenic rabbits as therapeutic protein bioreactors and using human disease models. *Pharma. Therap*. 99:261–282, 2003.
209. Ghong, Z., H. Wan, T.L. Tay, H. Wang, M. Chen, T. Yan. Development of transgenic fish for ornamental and bioreactor by strong expression of fluorescent proteins in the skeletal muscle. *Biochem. Biophys. Res. Comm.* 308:58–63, 2003.
210. Larrick, J.W., D.W. Thomas. Producing proteins in transgenic plants and animals. *Curr. Opin. Biotechnol*. 12:411–418.
211. Fischer, R., E. Stoger, S. Schillberg, P. Christou, R.M. Twyman. Plant-based production of biopharmaceuticals. *Curr. Opin. Plant Biol*. 7:152–158, 2004.
212. Giri, A., M.L. Narasu. Transgenic hairy roots: recent trends and applications. *Biotechnol. Adv*. 18:1–22, 2000.
213. Bornke, F., M. Hajirezaei, U. Sonnewald. Potato tubers as bioreactors for platinose production. *J. Biotechnol*. 96:119–124, 2002.
214. Sala, F., M.M. Rigano, A. Barbante, B. Basso, A.M. Walmsley, S. Castiglione. Vaccine antigen production in transgenic plants: strategies, gene constructs and perspectives. *Vaccine* 21:803–808, 2003.
215. Yoshida, K., T. Matsui, A. Shinmyo. The plant vesicular transport engineering for the production of recombinant proteins. J. Mol. Cat. B:Enzymatic, 28: 167–171, 2004.
216. Yoshida, K., A. Shinmyo. Transgene expression system in plant, a natural bioreactor. *J. Biosci. Bioeng*. 90:353–362, 2000.
217. Maranga, L., A. Cunha, J. Clemente, P. Cruz, M.J.T. Carrondo. Scale-up of virus-like particles production: effects of sparging, agitation and bioreactor scale on cell growth, infection kinetics and productivity. *J. Biotechnol*. 107:55–64, 2004.

1.04

Process Developments in Solid-State Fermentation for Food Applications

Ashok Pandey and Sumitra Ramachandran

CONTENTS

4.1 INTRODUCTION

Solid-state (substrate) fermentation (SSF) has been defined as the fermentation process occurring in the absence or near absence of free water. SSF processes generally employ a natural raw material as the carbon and energy source. SSF can also employ an inert material as the solid matrix, which requires supplementing a nutrient solution to contain necessary nutrients as well as a carbon source. The solid substrate (matrix), however, must contain enough moisture (1–6). Depending upon the nature of the substrate, the amount of water absorbed could be one or several times more than the dry weight, which leads to relatively high water activity (a_w) on the solid–gas interface, in order to allow a higher rate of the biochemical process. Lower water activity conditions result in low diffusion of nutrients and metabolites, and higher water activity causes compaction of the substrate. Thus, maintenance of an adequate moisture level in the solid matrix, along with suitable water activity, are essential elements for SSF processes (2,4,7–10). Solid substrates should generally have a large surface area per unit volume (in the range of 10^3–10^6 m^2/cm^3), to allow for ready growth on the solid–gas interface (11). Smaller substrate particles provide larger surface area for microbial attack, but pose difficulty in aeration and respiration due to the limitation in interparticle space availability. Larger particles provide better aeration and respiration opportunities, but provide less surface area. In bioprocess optimization, it may sometimes be necessary to use a compromised size of particles (usually a mixed range) to maximize cost effectiveness. For example, wheat bran, which is the most commonly used substrate in SSF, is obtained in two forms, fine and coarse. The former contains particles mostly smaller than 500–600 μ, and the latter contains particles mostly larger than these. Most SSF processes use a mix of these two forms, at different ratios, for optimal production (1,2,12–14).

Solid substrates generally provide a good dwelling environment to the microbial flora comprising bacteria, yeast, and fungi. Among these, filamentous fungi are the best studied for SSF, because, due to their hyphal growth they can not only grow on the surface of the substrate particles, but also penetrate through them. Several agro crops such as cassava and barley, and agro-industrial residues such as wheat bran, rice bran, sugarcane bagasse, cassava bagasse, various oil cakes (e.g., coconut oil cake, palm kernel cake, soybean cake, ground nut oil cake), fruit pulps (e.g., apple pomace), corn cobs, sawdust, seeds (e.g., tamarind, jack fruit), coffee husk and coffee pulp, tea waste, and spent brewing grains are the most often, most commonly used substrates for SSF processes (15–23). During growth on such substrates, hydrolytic exo-enzymes are synthesised by the microorganisms and excreted outside the cells, which create and help in accessing simple products (carbon source and nutrients) by the cells. This in turn promotes biosynthesis and microbial activities.

Apart from these, there are several other important factors which must be considered for development of SSF processes. These include physico-chemical and biological factors such as pH of the medium; temperature and period of incubation; age, size and type of the inoculum; nature of the substrate; and type of microorganism employed.

4.2 SIGNIFICANCE OF SSF

SSF has been considered superior in several aspects to submerged fermentation (SmF). It provides several advantages. It is cost effective due to the use of simple growth and production media comprising agro-industrial residues, and it uses small amounts of water and therefore releases considerably less effluent, thus reducing pollution concerns.

SSF processes are simple, use low volume equipment (lower cost), and are effective in that they provide high product titres (concentrated products). Further, the aeration process (availability of atmospheric oxygen to the substrate) is easier. There is a increased diffusion rate of oxygen into moistened solid substrate, supporting the growth of aerial mycelium. SSF can be effectively used at smaller volumes, which makes it suitable for rural areas.

4.3 SSF PROCESSES FOR FOOD APPLICATIONS

Food fermentation involves the action of microorganisms, depending upon the control of environmental conditions, and should ensure the growth of favourable microbial species for the development of desired sensory qualities in indigenous foods. SSF for food application has been used since the development of human civilization. Many kinds of fermented foods, such as single-cell protein (SCP) probiotics, flavoring products, beverages, pigments, and peptide sweeteners, are produced largely by SSF. In addition to these, several enzymes, organic acids, and exopolysaccharides have been produced using SSF.

Fermented foods can be stored for longer periods and used for food supply during the off season. Fermentation also contributes to the digestibility and enhances the nutritional value of the product. It can also increase fibre digestibility. However, there are chances of accidental contamination of SSF food products by mycotoxins, which are a class of unwanted compounds and could be accidentally present or produced in SSF products. This could be due to the use of contaminated raw materials or due to an accidental contamination of the fermentation with a toxigenic strain. Mycotoxins have been reported to occur as contaminants in many agricultural commodities such as nuts, cottonseed, corn, sorghum, millet, grains, and barley. Blanc et al. (163) reported that about 100 mg/kg citrinin was detected in food quality pigments produced by *Monascus* species.

4.3.1 Historical Developments

Historically, SSF processes have been used since ancient time for food applications. Although it is believed that the discovery of fermentation was purely by chance, food fermentation was developed several thousands of years ago. SSF dates back to 6000 BC when Babylonians made beer from natural yeast. Egyptians used this technique for bread making in 2600 BC, using brewer's yeast. Cheese making with *Penicillium roquefortii* was recorded in Asia before the birth of Christ. *Koji* processing reported to be migrated from China to Japan in the seventh century. Miso, tempeh, tamari sauce, soy sauce, ang-kak, natto, tou-fu-ru, and minchin are some of the other ancient fermented foods known for centuries, which are prepared through SSF. Tempeh and tamari sauce are soybean products, the former is an Indonesian food fermented by *Rhizopus* species and the latter is a Japanese food produced by using *Aspergillus tamari.* Soy sauce, a brown, salty, tangy sauce, is obtained from a sterile mixture of wheat bran and soybean flour, fermented initially by lactic acid bacteria, followed by alcoholic fermentation and ripening. A mash of crushed and steamed soybeans is used as a substrate for miso. Normal fermentation is carried out for one week, followed by two months of ripening. The final product is ground into a paste, which is generally used in combination with other foods.

In the eighteenth century, SSF was used to make vinegar from apple pomace. The beginning of the twentieth century marked the use of SSF for the production of enzymes and organic acids using molds. There were several other developments of SSF technology for nonfood application in subsequent years. The period of the 1970s saw a major focus on production of SCP.

4.3.2 New Developments

SSF gained a new dimension for various food and nonfood applications during the last 15–20 years. In food applications, it has been used particularly for the production of enzymes, organic acids, pigments, SCP, exopolysaccharides, and aroma compounds. Biologically active secondary metabolites such as antibiotics, steroids, biopesticides, and biofertilizers, are some nonfood applications of SSF. Attention is also being given to the development of bioreactors (fermentors) of various kinds, using automation.

4.4 SSF PROCESSES FOR FOOD ENZYMES

Enzymes have become an integral part of human need in day-to-day life, playing a varied role in several industries, particularly the modern food industry. Production of a variety of food products, ranging from baked foods, syrups, and fruit juices to flavoring agents and dairy products, commonly involves the use of several food enzymes. Almost all the microbial enzymes can be produced using SSF systems. Industrially important food enzymes, which include alpha amylase, glucoamylase, lipase, protease, pectinase, inulinase, glutaminase, and tannase, have been widely studied (4,5,19,24–33).

Enzyme production in SSF has often resulted in higher yields in comparison to SmF. Studies conducted to examine the reasons why SSF produced higher enzyme yields than SmF have been unable to explain it fully, although some of the characteristics of SSF provide conditions for the microbes more like the habitat from which they were isolated, and this may be the major reason for higher enzyme production. Most of the SSF processes for the production of food enzymes involve agro-industrial residues as the substrate, although inert materials such as polyurethane foam are also used. These have been mentioned earlier and are also indicated in Table 4.1 (15–18,28,34–37).

4.4.1 Amylases

For starch and glycogen hydrolysis, two major classes of amylases have been identified in the microorganisms: α-amylase and glucoamylase. In addition, β-amylase, which is generally of plant origin, has also been reported from a few microbial sources. α-Amylase (endo-α-1,4-D-glucan glucohydrolase, EC- 3.2.1.1) is an extracellular enzyme, which randomly cleaves the 1,4-α-D-glucosidic linkages in the interior of the amylose chain. Amylases which produce free sugars are termed as *saccharogenic* amylases, and those which liquefy starch without producing free sugars are known as *starch-liquefying* amylases. β-Amylase (α-1,4-glucan maltohydrolase, EC- 3.2.1.2) is an exoacting enzyme, which cleaves the nonreducing chain ends of amylose, amylopectin, and glycogen molecules, yielding maltose. Glucoamylase [also known as amyloglucosidase, glucogenic enzyme, starch glucogenase, and gamma amylase (exo-α-1,4-D-glucan glucanohydrolase, EC- 3.2.1.3)] produces single glucose units from the nonreducing ends of amylose and amylopectin in a stepwise manner. Unlike α- and β-amylase, most glucoamylases (GA) are able to hydrolyse the α-1,6 linkages at the branching points of amylopectin, although at a slower rate than 1,4-linkages. Thus, glucose, maltose, and limit dextrins are the end products of GA starch hydrolysis.

Extensive studies have been done on the production of amylases, in particular for α-amylase and GA in SSF, employing several microorganisms and various kinds of agro-industrial residues, among which starchy substrates have been preferred (Table 4.1) (5,7,12,13,28,32,38,39–50). Reports on microbial production of β-amylase are scanty and very few in SSF. A mutant strain of *Bacillus megaterium,* B6 mutant UN12, was described for the comparative production of beta amylase in SmF and SSF. The starchy wastes used as substrates were from arrowroot, arum, maize, potato, pulse, rice, rice husk, tamarind

Table 4.1
SSF processes for production of food enzymes

Enzymes	Microorganisms	Ref.	Substrates	Ref.
α-amylase	*Aspergillus oryzae*	52	Wheat bran	164
	A. kawachii	165	Spent brewing grain	52
	A. flavus	181	Banana waste	50
	Bacillus subtilis	50	Amaranthus grains	181
	B. licheniformis	164		
	B. coagulans	45		
	Saccharomycopsis capsularis	174		
	Streptomyces megasporium	49		
glucoamylase	*A. niger*	39, 41 42, 44 55, 57	Wheat bran	177
	A. oryzae	61	Corn powder	177
	Trichoderma viride	171	Rice bran	57
	Bacillus stearothermophilus	172	Cassava flour	175
	Saccharomyces cerevisiae	173		
Lipase	*Candida rugosa*	64–66, 70	Coconut oil cake	65
	Penicillium restrictum	68, 72	Olive cake	67
	A. niger	69	Babassu oil cake	68
	Staphylococcus warneri	75	Gingely oil cake	69
	S. xylosus	75		
Protease	*Penicillium citrinum*	85	Rice bran	82
	A. flavus	84	Wheat bran	84
	A. oryzae	170	Steamed rice	168
	A. niger	81		
	Phanerochaete chrysosporium	169	Rape seed meal derived media	167
	Rhizopus oligosporus	78, 79, 82		
	R. oryzae	83		
Pectinase	*A. niger*	88, 94 95, 97	Coffee pulp	87
	A. carbonarius	89	Apple pomace	92
	A. foetidus	93	citrus waste	91
			cocoa pulp	98
			sugarcane bagasse	95
			cranberry and strawberry pomace	99
Glutaminase	*Zygosaccharomyces rouxii*	34	Wheat bran	34
	Vibrio costicola	35		
	Beauvaria sp.	101		
Inulinase	*Staphylococcus* sp.	102	Wheat bran	102
	Kluyveromyces marxianus	102	Chicory roots	103

(*Continued*)

Table 4.1 (*Continued*)

Enzymes	Microorganisms	Ref.	Substrates	Ref.
Tannase	*A. niger*	105	Wheat bran	105
	Aspergillus sp.	166		
	Fusarium sp.	166		
	Trichoderma sp.	166		
	Candida sp.	179		

kernel, cassava, water chestnut, wheat, and wheat bran. Arum and wheat bran gave the highest yields (51).

Although the sources of α-amylases are fairly extensive, the principal commercial preparations are derived from some bacterial and fungal species. These include *Bacillus* sp., *Bacillus amyloliquefaciens, B. coagulans, B. licheniformis, B. megatarium, Aspergillus* sp., *A. niger, A. oryzae, A. kawachii, Aeromonas caviae, Pycnoporus sanguineus,* and *Saccharomycopsis capsularia* (28). The *Bacillus* species is considered the most prolific producer of α-amylase by SSF. *B. amyloliquefaciens* and *B. licheniformis* are considered potent species for thermophilic α-amylase (24,28,45,52,53). Among filamentous fungi, *Thermomyces lanuginosa* was reported to be an excellent producer of α-amylase (54). A strain of *Aspergillus oryzae* was used for α-amylase production using spent brewing grain in SSF and the process was considered economically promising (52). Irrespective of enzyme properties (such as temperature, pH optima, and range), SSFs are typically performed at mesophilic temperatures such as 30°C using agro-industrial residues for 24–96 h.

GA production in SSF was first demonstrated in 1914. The production of GA was carried out in shallow trays with *Aspergillus* sp., *A. awamori, A. niger, A. oryzae,* and *Rhizopus* sp., *R. oligosporus,* under SSF. An extensive study was carried out on the production of GA in solid cultures by a strain of *A. niger* (7,12,13,38–44,55–58). The study included screening of various agro-industrial residues, including wheat bran, rice bran, rice husk, gram flour, wheat flour, corn flour, tea waste, and copra waste, individually and in various combinations. Apart from the substrate particle size, which showed profound impact on fungal growth and activity, substrate moisture and water activity also significantly influenced the enzyme yield. Different types of bioreactors, including flasks, trays, rotary reactors, and columns (vertical and horizontal), were used to evaluate their performance. Enzyme production in trays occurred in optimum quantities in 36 h in comparison to the 96 h typically required in flasks. Supplementation of wheat bran medium with yeast extract increased glucoamylase synthesis by the fungal culture.

Several attempts have been made to compare the GA production in SSF and SmF (59–62). Generally SSF yielded higher enzyme titres. However, contrary to the general findings, *Rhizopus* A-11 showed a 4.6-fold lower GA yield from a conventional SSF on wheat bran medium than the yield in SmF, which used metal ion supplemented medium (60). A similar trend was found with a fungal strain of *A. niger,* which produced higher GA titres in SmF (102 U/ml) than in SSF (66 U/ml) in a shorter period (66 h in comparison to 96 h). In an aqueous biphasic system comprising polystrene glycol (PEG 6000) and potassium phosphate, however, both the GA recovery and yields were twice as high as the control system comprising an aqueous phase only (59).

4.4.2 Lipases

Lipases are essential for the bioconversion of lipids from one organism to another or within the organism itself. Microbial lipases (glycerol ester hydrolases, EC 3.1.1.3) catalyse a wide range of reactions, specifically hydrolysis and interesterification. They also catalyse alcoholysis, acidolysis, esterification, and aminolysis (63). They possess the unique feature of acting as an interface between the aqueous and nonaqueous phase, which distinguishes them from esterases.

Lipases are wide–spread, being found in microbial flora, in the pancreas of mammals such as pigs and humans, and in higher plants such as castor beans and rape seed. Microbial lipases are used for the production of desirable flavor in cheese and other foods, and for the interesterification of fats and oils to produce modified acyl glycerols, which cannot be obtained by conventional esterification (74). The most important species of microorganisms for the synthesis of lipase in SSF include *Candida* species, *Pseudomonas* species, and *Rhizopus* species, with *Candida rugosa* being the most commonly used. *Aspergillus* lipases are highly selective for short chain acids and alcohols (69). *C. rugosa* lipase is more selective for propionic acid, butyric acid, butanol, pentanol, and hexanol (64). The production of flavor esters by the lipases of *Staphylococcus warneri* and *S. xylosus* has been reported (75). *Mucor miehei* and *Rhizopus arrhizus* lipases are more selective for long chain acid and acetates. *M. miehei* lipase is used to prepare gerniol esters, which are important components of fragrances, in a solvent free system (63–72).

Several agro-industrial residues and inert supports have been used to produce lipases in SSF (63–69). These include peanut press cake, coconut oil cake, wheat bran, rice bran, babassu oil cake, olive oil cake, sugarcane bagasse, and amberlite, using various yeast and fungal cultures such as *Candida* sp., *C. rugosa, Neurospora sitophila, A. oryzae, Rhizopus oligosporus, R. delemer, P. candidum,* and *Mucor* sp. Enzyme production in SSF has been reported to be superior in comparison to SmF (yield in SmF was 14 U/ml and in SSF using a polymeric resin, amberlite, as a solid inert substrate, with dextrin as the carbon source, was 96 U/g) (73). Similar observations were reported by Rivera-Munoz et al. (76), who found that the enzyme was more stable. Supplementation of oil cake medium with external carbon or suitable inorganic or organic nitrogen sources resulted in improved enzyme yields (64).

4.4.3 Proteases

Proteolytic enzymes find wide applications in food and other industries. They account for nearly 60% of the industrial market in the enzyme technology. Proteases are produced extracellularly by fungi and bacteria such as *A. oryzae, A. flavus, R. oligosporus, Penicillium citrinum, P. chrysosporium, Bacillus subtilis, B. amyloliquefaciens*, and *Pseudomonas* species (80,85,86). In recent years, different types of proteases such as acid, neutral, and alkaline proteases are being produced by SSF. Proteases production is generally inhibited by carbon sources, indicating the presence of catabolic repression of the biosynthesis. This fact makes it logical to use agro-industrial residues as substrates for proteases production; thus an SSF process becomes imperative. It is interesting to note that although a number of substrates such as wheat bran, rice bran, and oil cakes have been employed for cultivating different microorganisms, wheat bran has been the preferred choice (78–85).

Enzyme production in SSF is generally favoured under partial pressure of carbon dioxide. Studies on the effects of O_2 and CO_2 partial pressure on acid protease production by a strain of *Aspergillus niger* ANH-15 in SSF of wheat bran showed a direct relationship between pressure drop, production of CO_2 and temperature increase (81). Acid protease production was increased when the gas had 4% CO_2 (v/v) and it was directly related with the fungus metabolic activity as represented by the total CO_2 evolved. Acid protease production on rice bran using a strain of *R. oligosporus* by making step changes in the gas environment and temperature during SSF to mimic those changes, which arose during SSF due to mass and heat transfer limitation, showed that a decrease of O_2 concentration from 21 to 0.5% did not alter protease production (82,83).

Comparative study on protease production in SSF and SmF generally showed higher enzyme yields in SSF. For example, a study on acid protease production showed that total protease activity in SSF per gram of wheat bran was equivalent to 100 ml broth (79).

4.4.4 Pectinases

Pectinases are a group of hydrolytic enzymes, usually referred as pectolytic enzymes, which find important applications in the food and beverages industries, in addition to various other industries. Pectolytic enzymes degrade pectic material and reduce the viscosity of a solution, making it easier to handle. They are used industrially in fruit processing to facilitate pressing and to help in the separation of the flocculent precipitate by sedimentation, filtration, or centrifugation in the extraction of the clarified juice (95). Pectinases are used for the elimination of pectin in wine making, coffee and tea processing plants, and maceration of vegetable tissue (93,94). Pectinases comprise hydrolases, lyases, and oxidases, and can be obtained from higher plants, microorganisms, and insects (96). Pectinases can be obtained from several fungi, bacteria (including alkalophilic), actinomycetes, and yeast, such as alkalophilic bacteria, actinomycetes, *Aspergillus versicolor, A. niger, A. awamori,*

Rhizopus stolonifer, Penicillium italicum, P. frequentans, Polyporus squamosus, Thermoascus aurantiacus, Clostridium thermosaccharolyticum, C. felsincum, Tubercularia vulgaris, Sclerotium rolfsii, Erwinia sp., *Erwinia carotovora, Bacillus* sp., *B. subtilis, B. bumilus, B. stearothermophilus, B. polymyxa, Pseudomonas syringae, Rhizoctonia solani, Xanthomonas compestris, Saccharomyces cerevisiae, Kluyveromyces marxianus,* and *Trichosporon penicillatum* (89,93–97,100,101). Extracellular pectinases from *Aspergillus* sp. are of commercial interest and strains of *A. niger, A. carbonarius,* and *A. foetidus* are used. Among these, *A. niger* is considered as the most important for production in SSF. Various agro-industrial residues such as apple pomace (92), citrus waste, orange and lemon peels (91), soy bran (94), sugar cane bagasse (95), wheat bran (97), coffee pulp (87), cocoa pulp, (98) and cranberry and strawberry pomace (99) are used as the substrates. Citrus pulps or peels could be used as inducers for enzyme synthesis (91). In the case of pectinases, generally enzyme yields are higher and the enzyme is said to be more stable (over a wider range of pH and temperature) in SSF in comparison to SmF (88,89,93–97).

4.4.5 L-Glutaminase

L-glutaminase (L-glutamine amidohydrolase - E.C. 3.5.1.2) is an important enzyme deamidating L-glutamine, which plays a major role in the cellular nitrogen metabolism of both prokaryotes and eukaryotes. L-glutaminase is useful in the food industry as it increases the glutamic acid content of the fermented food, thereby imparting a unique flavor. Because the sources for L-glutaminases are limited, the search for potential microbial strains that produce the enzyme in high titres with novel properties for their industrial production is being pursued extensively around the world (31).

L-glutaminase can be produced by many bacterial strains, such as *Escherichia coli, Pseudomonas* sp., *Acinetobacter* sp., *Bacillus* sp., *Proteus morganni, P. vulgaris, Xanthomonas juglandis, Erwinia carotovora, E. aroideae, Serratia marcescens, Enterobacter cloacae,* and *Klebsiella aerogenes.* It can also be produced by *Aerobacter aerogenes*, fungi such as *Aspergillus sojae* and *A. oryzae,* and yeast cultures such as *Hansenula, Cryptococcus, Rhodotorula, Candida scottii, Cryptococcus albidus, C. laurentii, Candida utilis,* and *Torulopsis candida* (31).

One of the major advantages of SSF system for L-glutaminase production lies in cultivation of salt tolerant microbial cultures, which may have better application in food processes involving high salt concentration, such as soy sauce production which requires as high as 20–25% salt concentration. From this point of view, marine microorganisms have been explored and exploited. There seems to be enormous scope for the investigation of deriving new products of economic importance from potential marine microorganisms. Considering the fact that the marine environment, particularly sea water, is saline in nature and chemically closer to human blood plasma, it could be anticipated that this could provide microbial products; in particular, enzymes, that could be safer, having less toxicity and fewer side effects, when used for human application (31).

When SSF was carried out using saline tolerant yeast *Zygosaccharomyces rouxii* with agro-industrial substrates such as coconut oil cake, groundnut oil cake, wheat bran, and sesamum oil cake, 2.2 and 2.17 U enzyme/gram dry substrate (gds) were produced from wheat bran and sesamum oil cake, respectively. Under optimized physico-chemical conditions, enzyme yields increased to 7.5 and 11.61 U/gds for these, respectively, which is 3.5–5 times higher than the initial (without optimization) production (34).

L-glutaminase has also been produced in SSF using inert substrates such as polystyrene (100). A marine organism, *Vibrio costicola,* was used for solid culturing (35,100). Another marine isolate, *Beauveria* species, has also been used for L-glutaminase production (101). However, it was necessary to supplement the fermentation medium with NaCl at high concentrations (~10%) or sea-water when using such cultures. In both cases, enzyme production was growth associated.

4.4.6 Inulinase

Microbial inulinases (2,1-β-D-fructan fructanohydrolase EC 3.2.1.7) are usually inducible and exoacting enzymes, which catalyse the hydrolysis of inulin by splitting off terminal fructosyl units (D-fructose) by cleaving the glycosidic linkages in polymer moiety. They play an important role in the hydrolysis of inulin for its commercial exploitation. Inulin, a polyfructan, consists of linear β-2,1 linked polyfructose chains, terminated by a glucose residue attached through a sucrose-type linkage. It is a reserve carbohydrate in the roots and tubers of plants such as Jerusalem artichoke, chicory, or dahlia. Inulinases catalyse the hydrolysis of inulin to D-fructose (fructose syrup), which has gained an important place in human diets today (30).

Microbial inulinases have mostly been produced in SmF; although the technique of SSF has been recently tried using wheat bran and chicory root as substrates. Indigenously isolated bacterial strains of *Staphylococcus* sp. and a yeast strain of *Kluyveromyces marxianus* were cultivated in SSF, which produced extracellular inulinase. When wheat bran was used as the C source, 16 U of enzymes/g dry fermented substrate were produced. Improvements in process parameters and nutritional conditions led to a 6.5-fold increase in enzyme yields by the bacterial strain in SSF (102,103).

4.4.7 Tannase

Tannase, or tannin acyl hydrolase (EC 3.1.1.20), is an inducible enzyme, which catalyses the breakdown of hydrolysable tannins and gallic acid esters. It transforms tannic acid into glucose and gallic acid. Tannase is used in the industrial processing of fruit juices and coffee flavored soft drinks as a clarifying agent, and in the manufacture of instant tea; it is also used in the production of wine and beer.

Generally tannase is obtained form microbial sources. A wide number of bacterial, fungal, and yeast cultures have been reported to produce the enzyme in SmF or SSF. Among these, *Aspergillus awamori, A. niger, A. oryzae* and *A. japonicus* have been considered as the best producers (104). *Fusarium solani, Rhizopus oryzae, Trichoderma viride,* and *Candida* sp. are other important sources of the enzyme. Tannase production in SSF using polyurethane foam showed superiority of SSF in comparison to SmF. The addition of high tannic acid concentrations increased total tannase activity, whereas the supplementation of glucose proved detrimental (105). SSF is considered to minimize catabolic repression for tannase production.

4.4.8 Recovery of Enzymes from Fermented Matter

Recovery of the enzymes from the fermented matter is a critical factor, which largely affects the cost-effectiveness of the overall process. In SmF, fermented broth is generally subjected to centrifugation or ultrafiltration to remove solid particles and concentrated enzymes, whereas in SSF fermented matter, aqueous extract is prepared to recover the enzyme. Usually distilled water is used for extraction, although buffer solutions or a NaCl solution (0.01M) can also be used. There are several methods which could be used for this purpose. The use of reverse micellar systems for the enzyme extraction has attracted considerable interest, due to their capacity to solubilize specific proteins from the dilute aqueous solutions such as fermented and cell culture media. It represents a new downstream process for the extraction of enzymes without any modification to their conformation. Use of a full forward and backward extraction cycle can remove contaminating matters such as neutral protease, and release of protein can depend on the aqueous phase pH. The temperature of extraction also affects the rate and overall recovery of enzymes. Generally enzymes retain their biological activity fairly well inside the reverse micelles. A work on protease recovery from SSF fermented matter showed that a leaching solution of pH 7 gave the optimum recovery of the enzyme from the fermented matter, although the enzyme had optimum activity at pH 4 (78,79).

4.5 SSF PROCESSES FOR ORGANIC ACIDS

Production of organic acids such as citric acid for food application by SSF has been employed since olden times. For example, citric acid production by SSF (the *Koji* process) was first developed in Japan, and is the simplest production method. SSF can be carried out using several raw materials. Generally, the substrate is moistened to about 70% depending on the water-holding capacity of the substrate. The initial pH is normally adjusted to 4.5–6.0 and the temperature of incubation can vary from 28 to 30°C. One of the important advantages of the SSF process is that the presence of trace elements does not affect citric acid production negatively as it does in SmF (113,119).

Commercially, citric acid is produced mainly using the filamentous fungus *A. niger*, although *Candida* sp. has also been used, employing both molasses- and starch-based media. In SSF, production has been obtained using crops and crop residues such as apple pomace (114), grape pomace (115), coffee husk (116), cassava bagasse, beet molasses (117) pineapple waste (110), and carrot waste (113) as substrates by *A. niger*. Citric acid production is largely dependent on the microorganisms, production techniques, and substrates employed. Generally the addition of methanol increased citric acid production in SSF (118). It has been proposed that overproduction of citric acid was related to an increased glucose flux through glycolysis. At low glucose fluxes, oxalic acid could accumulate. Several studies with different strains of *A. niger* have been made to compare citric acid production in flasks and in different kinds of bioreactors such as trays, packed bed bioreactors (single layered and multilayered), and rotating drums, and varied results have been obtained (106–113,119,120). For example, Tran et al. (109) reported the best production in flasks, and lower yields in tray and rotating drum bioreactors. Lu et al. (111,112) found that a multilayer packed bioreactor improved the mass transfer considerably compared with a single layer packed bed operated under similar conditions. Higher yields were obtained in packed bed than in flasks. In packed bed bioreactors using inert support, heat

Table 4.2
SSF for citric acid production using agro-industrial residues by different strains of Aspergillus niger

Raw Material	Citric Acid
Apple pomace	766–883 g/kg[a]
Grape pomace	413–600 g/kg[a]
Kiwifruit peel	100 g/kg[a]
Cellulose hydrolysate and sugar cane	29 g/kg
Orange waste	46 g/kg
Beet molasses (ca-alginate gel)	35 g/L
Saccharose (sugar cane bagasse)	174 g/kg[b]
Coffee husk	150 g/kg[b]
Carrot waste	29 g/kg[a]
Okara (soy residue)	96 g/kg[a]
Pineapple waste	132–194 g/kg[b]
Glucose (sugarcane bagasse)	21.24 g/L
Kumara (starch containing)	103 g/kg[b]
Mussel processing wastes (polyurethane foams)	300 g/kg
Cassava bagasse	347 g/kg[b]

[a] based on sugar consumed; [b] based on dry matter

removal by conductive mechanism was the least efficient (8.65%) when compared with convective (26.65%) and evaporative (64.7%) (119). Table 4.2 shows SSF for citric acid production using agro-industrial residues by different strains of *A. niger.*

Another important organic acid required for food applications is lactic acid, which has been produced in SSF using fungal as well as bacterial cultures. The commonly employed cultures belong to *Rhizopus* sp. and *Lactobacillus* sp., e.g., *R. oryzae*, *L. paracasei, L. helveticus,* and *L. casei* (121–123). Different crops such as cassava and sweet sorghum, and crop residues such as sugarcane bagasse, sugarcane press-mud, (122) and carrot-processing waste (123) served as the substrate. A comparative study involving fungal strains of *R. oryzae* to evaluate L-(+) lactic acid production in SmF and SSF showed that SSF was superior in production level and productivity. Fermentation yields were 77% (irrespective of media) and yields were 93.8 and 137.0 g/l in SmF and SSF, respectively (121). The productivity was 1.38 g/l per hour in SmF and 1.43 g/l per hour in SSF. In another comparison using *L. paracasei,* lactate concentrations and yields were 88–106 g/l and 91–95% for SmF, and 90g/kg and 91–95% for SSF, respectively, but the time required for SSF was 120–200 h in comparison to 24–32 h in SmF (123).

4.6 PRODUCTION OF AMINO ACIDS

Use of defined SSF media using inert solid support has been considered a very useful method for microbial cultivation for the production of value added products (1,2,3,19,20,25,26,124). The production of L-glutamic acid in SSF with sugarcane bagasse was used as solid inert substrate and a bacterial strain of *Brevibacterium* sp. was used for the fermentation. The study not only showed good L-glutamic acid synthesis in solid culturing but also demonstrated that with media and process parameters manipulation, bacterial strains can also be successfully cultivated in SSF (125). Yield as high as 80 mg glutamic acid per g dry fermented matter was obtained.

4.7 MUSHROOM PRODUCTION

Mushrooms have entered the new era of food technology as a common universally accepted nutritive food. Their commercial cultivation involving SSF has rapidly spread globally, due to their innumerable applications. They are a rich source of protein, carbohydrates, vitamins, and minerals (2,19). Folic acid content in mushrooms has been found to be higher than in liver and spinach. In addition to their nutritive value, mushrooms also possess medicinal properties (1,2,19,20,127–129). They demonstrate antibacterial, antifungal, and antiprotozoal activities, due to the presence of polyacetylene compounds. Today more than 2000 species of mushrooms are known, although only about 20 are cultivated commercially. Button mushrooms, Japanese mark forest mushrooms, Chinese mushrooms, oyster mushrooms, and winter mushrooms are some of the various types of popular mushrooms being cultivated world wide. However, button mushrooms alone account for about 60% of total world production.

Mushroom cultivation involves SSF at three different stages, namely composting, spawn manufacture, and the growth of mushroom on the moist substrate. In order to produce a selective growth medium on which the mycelia colonize and produce fruiting bodies, the compost is prepared by piling up the substrate for a long period of time. During this period, many changes take place, and the resulting substrate differs from the original. Various physico-chemical factors play a vital role in the process. A temperature range of 22–27°C, substrate moisture content of 55–70%, and a pH of 6.0–7.5 are generally considered as the most suitable conditions. Animal manure from horses and chickens, and agro-industrial residues such as wheat straw, paddy straw, barley straw, rice bran, saw dust, banana, maize stover, tannery waste, wool waste, and sugarcane bagasse are used as substrates.

Spawn or inoculum production involves the growth of mycelia of mushroom on cereal grains such as rye, wheat, sorghum, and millet. These support faster mycelial development because of sufficient nutrients availability, and allow easy handling and steadiness during sterilization. The process is carried out with a pure culture. However, application of these substrates increases the costs input; for this reason other raw materials such as sawdust, cereal bran, or other agro-industrial residues have been recommended for spawn preparation. A mixture of sawdust and coffee husk was found quite suitable for spawn preparation for *Agaricus bisporus, Pleurotus* sp., *Lentinus edodus, Flammulina velutipes, and Volvariella volvacea* (127–129). Depending upon the nature of the substrate, optimum conditions were moisture content at 40–60%, a pH of 6.5–7.0, and an incubation temperature of 25°C. The final product spawn should be stored at 2–5°C.

The development of the fruiting body requires a lower temperature than the optimum for mycelial growth; it also requires proper ventilation, which helps in releasing the accumulated carbon dioxide, which retards the formation of fruiting bodies. The crops are harvested at the second or third stage of sporophore development, namely the button or closed-up stage, respectively. At this stage, substrate moisture should be generally higher than previous stages, (i.e., compost formation and spawn preparation). High relative humidity (80–95%) is also a desirable condition in order to control the heat, mass, and gaseous exchange. After harvesting of the fruiting body, the leftover solid residue can be used as manure or as animal feed, depending upon the raw material used as substrate.

4.8 PRODUCTION OF EXOPOLYSACCHARIDES

SSF use in the production of exopolysaccharides such as xanthan and succinoglycan is growing. Xanthan is nontoxic and does not inhibit growth. It is nonsensitizing and does not cause skin or eye irritation. On this basis, xanthan has been approved by the United States Food and Drug Administration (FDA) for use as a food additive without any specific quantity limitations (131). Xanthan gum has been used in a wide variety of foods for a number of important reasons, including emulsion stabilization, temperature stability, compatibility with food ingredients, and pseudoplastic rheological properties (131). For recovering the xanthan, the cells are usually removed first, either by filtration or centrifugation (133). Further purification may include precipitation using water-miscible nonsolvents (isopropanol, ethanol, acetone), addition of certain salts, and pH adjustments (133). One SSF-based process for the production of xanthan gum utilized a bacterial culture of *Xanthomonas campestris* (134). The exopolysaccharide was produced on a number of agro-industrial residues or byproducts such as spent malt grains, apple pomace, grape pomace, citrus peels, and sugar beet molasses (137). With most of the substrates, the gum production was comparable to those obtained with SmF (130). Succinoglycan production by SSF with *Agrobacterium tumefaciens* on various solid substrates, including agar medium, spent malt grains, ivory nut shavings, and grated carrots, impregnated with a nutrient solution, showed the highest yield in static cultivation, reaching 42 g/l of impregnating solution, corresponding to 30 g/kg of wet substrate (135). The polymer production in a tray bioreactor was faster, but the final yield was lower (29 g/l of impregnating solution) (135).

SSF was performed using inert solid particles (spent malt grains) impregnated with a liquid medium. The gum was extracted from the fermented mass with seven volumes of water by shaking at 250rpm for 2h. It was filtered, centrifuged at 4000g for 10 min, followed by ultracentrifugation at 25000g for 20 min. The EPS was precipitated from the supernatant by the addition of 2% KCl, with 2–3 volumes of cold acetone. The precipitated xanthan was dissolved in 2% KCl, precipitated again with acetone, dried at 70°C. and weighed (136). The polymer yields obtained from SSF, as referred to the impregnating liquid volumes, were as follows: 38.8 g/litre xanthan from *Xanthomonas campestris*, 21.8 g/litre succinoglycan from *Rhizobium hedysari* and 20.3 g/litre succinoglycan from *Agrobacterium tumefaciens* PT45. These results made this technique promising as a potential application on the industrial scale. A further advantage of this fermentation process could be in the availability and low cost of substrates, which are obtained as byproducts or wastes from the agricultural or food industry (136). A comparison of SmF and SSF for the production of bacterial exopolysaccharides (EPS) showed that the latter technique yielded 2–4.7 times more polymer than the former on the laboratory scale (136).

4.9 PRODUCTION OF PIGMENTS

Pigments are the normal constituents of the cells or tissues that give color. Pigments play a vital role in the food industry, to make food decorative and appealing. Natural pigments contain provitamin A, and have anticancer activity, and other desirable properties such as stability to heat, light, and pH. There are pigments which are chemo-synthetic counterparts of regular food components, that are referred as natural-identical.

Microbial production of pigments has usually been carried out in SmF, though SSF processes have also been applied. *Candida flareri, C. guilliermundii, Debaromyces subglobosus, Hansenula polymorpha, Saccharomyces, Torulopsis xylinus, Ashbya gossypi* and *Eremothecium ashbyii* have commonly been used for the production of riboflavin, a yellow pigmented B vitamin. *Monascus* pigments have good properties as food colorants possessing reasonable light and chemical stability, tinctorial strength, and water solubility when complexed with appropriate agents.

Monascus pigments may be used as substitutes for traditional food additives, such as nitrites for the preservation of meats (139), and as a potential replacement for synthetic food dyes (181). They have also been used industrially for several years; e.g., as yellow hydrosoluble pigments for candies (140), or red pigments in red rice wine. *Monoscus anka* and *M. pupureus* are cultivated in SSF for red pigment production (Table 4.3). Steamed rice is used as the substrate, although oats, wheat, and barley have also been used. The culturing period is approximately three weeks. Certain sugars, amino acids, and metals have been found important for the production, and yields are typically 10-fold higher in both SSF and SmF. Pigment formation could be inhibited by the presence of glucose in the fermentation medium, but could be increased by limited aeration in SmF. It was also observed that an increase in the partial pressure of CO_2 increases the pigment production (141). For isolating the pigment from fermented matter, simple extract with solvents such as ethanol is effective (1,137,141).

Table 4.3
SSF processes for other food applications

Products	Microorganisms	Ref.	Substrates	Ref.
Citric acid	*A. niger*	108, 110–112 116, 120	Pineapple waste	109, 110
			Apple pomace	114
			Carrot processing waste	113
			Grape pomace	115
			Coffee husk	116
			Beet molasses	117
Lactic acid	*R. oryzae*	121	Sugarcane press mud	122
	Lactobacillus paracasei	123	Sweet sorghum	123
	L.casei	123		
Glutamic acid	*Brevibacterium* sp.	125	Sugarcane bagasse	125
Mushrooms	*Pleurotus* sp	126, 128	Coffee residues	126, 127
	Lentinus edodes	127	Cotton straw	180
Xanthan	*Xanthomonas campestris*	134	Spent malt grains	134
			Citrus pomace	137
			Apple pomace	137
			Citrus peels	137
Succinoglycan	*Agrobacterium tumefaciens*	135	Spent malt grains	135
			Ivory nut shavings	135
Pigments	*Monascus purpureus*	137, 138	Sugarcane bagasse	138
Aroma compounds	*T. viride*	142	Pre gelatinized rice	142
	Kluyveromyces marxianus	148, 149	Sugarcane bagasse	144
	Ceratocystis fimbriata	144, 145	Coffee husk	147
	Rhizopus oryzae	143	Cassava bagasse	145
	Bacillus subtilis	150	Soybeans	150
Vitamins	*Citrobacter freundii*	152	Rice	152
	Klebsiella pneumoniae	152		
Aflatoxin	*A. flavus*	160	Cassava bagasse	160

4.10 PRODUCTION OF AROMA COMPOUNDS

One of the significant applications of SSF in food applications involves production of food aroma compounds. There are two main advantages of SSF as an alternative technology for the production of aroma compounds: (1) release of the products from the microbial membranes is facilitated by the higher concentration in liquid phase, and (2) sometimes the solid substrates or byproducts can be used directly in SSF without any pretreatment of the starting substrates.

SSF has been successfully employed for the production of food aroma compounds using fungal and yeast cultures, such as *Neurospora* sp., *Zygosaccharomyces rouxii*, *Aspergillus* sp., and *Trichoderma viridae* (142), using pregelatinized rice, miso, cellulose fibres, and agar. *Rhizopus oryzae* cultivation of tropical agro-industrial residues results in the production of volatile compounds such as acetaldehyde and 3-methyl butanol (143). *Neurospora* sp. and *T. viride* produce a fruity odour and coconut aroma in SSF with pregelatinized rice and agar medium, respectively (142). Methyl ketones are produced on a commercial scale from coconut oil using *A. niger,* and the yields are as high as about 40% (1,141,143).

Ceratocystis sp. produces a large range of fruity or flower-like aromas (peach, pineapple, banana, citrus, and rose), depending on the strain and cultivation conditions (144). Among these, *C. fimbriata* has been extensively studied for the production of aroma compounds in SSF. Wheat bran, cassava bagasse, coffee husk, and sugarcane bagasse were used as the substrate. Addition of precursors such as urea, leucine, and valine affected growth and aroma production (144–147).

Kluyveromyces marxianus grown on cassava bagasse in SSF using packed bed column bioreactors under different aeration rates produced 11 volatile compounds, out of which nine were identified and two remained unidentified. Ethyl acetate, ethanol, and acetaldehyde were the major compounds produced, with 0.06 l/h/g aeration rate. Maximum TV concentrations were reached at 24 h with 0.06 l/h/g initial dry matter (IDM), and at 40 h for 0.12 l/h/g IDM. At 0.06 l/h/g aeration rate, ethyl acetate and ethanol were the compounds in highest and almost equivalent concentration (~30%) (148,149).

There are some other aroma compounds which can be produced in SSF. These include 2,5-dimethylpyrazine (2,5-DMP) by *Bacillus natto* (150) and tetramethylpyrazine (TTMP) by *B. subtilis* (151) on soybeans in SSF. Results demonstrated the suitability of SSF for their production.

Production of aroma compounds in SSF using naturally occurring substrates offers potential benefits in production of food and fruity aroma compounds for human consumption at low cost. One major difficulty in this regard is the isolation and recovery of compounds produced, especially if the compounds possess lower volatile temperature. A few attempts have been made to trap such compounds in suitable inert materials such as resins by adsorption. However, much remains to be done in this area.

4.11 VITAMINS

SSF has been used for the formation of water soluble vitamins such as vitamin B-12, vitamin B-6, riboflavin, thiamine, nicotinic acid, and nicotinamide. *Rhizopus oligosporus, R. arrhizus*, and *R. stolonifer* formed riboflavin, nicotinic acid, nicotinamide, and vitamin B-6 (153). The final concentrations of these substances depended on the different strains involved and on the fermentation time. Isolates of *R. oligosporus* were generally the best vitamin producers. The molds did not produce physiologically active vitamin B-12.

Citrobacter freundii and *Klebsiella pneumoniae* showed the best capabilities for physiologically active vitamin B-12 production (152,153).

4.12 MYCOTOXINS

Mycotoxins are toxic compounds produced as secondary metabolites by about 350 species of toxigenic fungi, occurring in food commodities and food stuffs such as cereals, soybeans, corn, sorghum, and peanuts. These are pathogenic and their consumption can lead to mycotoxicoses. Cardiac beriberi caused by toxins of *Penicillium* associated with rice, and alimentary toxic aleukia linked with *Fusarium* molds on wheat, millet, and barley are some of the serious problems faced by human beings. Aflatoxin, sterigmatocystin, zearalenone, patulin, ochratoxin, and fumonisin are some of the toxins which could lead to increased incidence of cancer. Aflatoxins B1, B2, G, and G2 are produced by *A. flavus* and *A. parasiticus* in cereal grains such as corn, wheat, sorghum, oats, barley, millet, and rice. Aflatoxin is heat stable at temperature above 212° F. *A. flavus* generally produces toxins at a optimum temperature of 81–86°F and optimum moisture content of 18–24%. Aflatoxin production by *A. flavus* isolates from peanuts, cottonseed, rice, and sorghum showed a higher percentage of aflatoxin-producing strains in peanuts (154). Milk products such as cheese could be contaminated by aflatoxin M-1 when manufactured with milk from dairy cattle that have consumed aflatoxin B-1 contaminated feeds. Zeralenone and deoxnivalenenol are toxins produced by *Fusarium* sp. *Fusarium trincintum,* and some strains of *Fusarium,* produce T-2 and other toxic trichothecenes. T-2 mycotoxin, the only mycotoxin known to have been used as a biological weapon, is highly stable and resistant to UV light destabilization. Citrinin is produced by *P. expansum* in apple juice and, to a lesser extent, in grape juices (155). Frayssinet et al. (156) have discussed methods for the analysis of mycotoxins in foods. They have also outlined techniques for aflatoxins, trichothecenes, zearalenone, fumonisins, and ochratoxin A detection.

Production of mycotoxin in SSF depends on several factors such as temperature, water activity, food sources, and aeration (157). It was reported that aflatoxin production in SSF increased at higher aeration rates in the range of 0.01 to 0.04 ml of air/g of humid corn/min (158). Gonzalez et al. (160) reported that aflatoxin production on cassava bagasse was higher at 29°C, which decreased 8-fold at 35°C. They also reported that at above 40% moisture there was positive effect on aflatoxin production. Maggon et al. (159) reported that in SmF, depletion of the nitrogen or phosphorus source enhanced aflatoxin biosynthesis. A report by Gonzalez et al. (160) stated that when *A. parasiticus* and *A. niger* were grown together in SSF, aflatoxin production was completely inhibited, proving that mycotoxin production is inhibited by competition. Lindenfelser et al. (161) obtained a yield of 2.4g/kg ochratoxin on wheat, using a rotating drum type fermenter. In a study by Saxena et al. (162), it was found that ergosterol could be used as an indicator for potential mycotoxin production, because the concentration of ergosterol was found to be similar to the concentration of ochratoxin A produced by *A. ochraceus* NRRL 3174 and *P. verrucosum* NRRL 3260 cultured on white rice.

REFERENCES

1. Pandey, A., C.R. Soccol, D.A. Mitchell. New developments in solid-state fermentation, I: bioprocesses and products. *Proc. Biochem.* 35(10):1153–1169, 2000.
2. Pandey, A. Recent developments in solid state fermentation. *Proc. Biochem.* 27:109–117, 1992.

3. Pandey, A., C.R. Soccol, J.A. Rodriguez-Leon, P. Nigam. *Solid-State Fermentation in Biotechnology*. New Delhi: Asiatech Publishers Inc, 2001, p 221.
4. Pandey, A., ed. *Solid-State Fermentation*. New Delhi: Wiley Eastern Publishers, 1994, p 183.
5. Pandey. A. Glucoamylase research: an overview. *Starch* 47(11): 439–445, 1995.
6. Pandey, A. Aspects for design of fermentor in solid-state fermentation. *Proc. Biochem.* 26(3):355–361, 1991.
7. Pandey, A., L. Ashakumary, P. Selvakumar, K.S. Vijaylakshmi. Influence of water activity on growth and activity of *Aspergillus niger* NCIM 1245 for enzyme production in solid-state fermentation. *World J. Microbiol. Biotechnol.* 10:485–486, 1994.
8. Acuna-Arguelles, M., M. Gutierrez-Rojas, G. Viniegra-Gonzalez, E. Favela-Torres. Effect of water activity on exo-pectinase production by *Aspergillus niger* CH4 on solid-state fermentation. *Biotechnol. Lett.* 16(1):23–28, 1994.
9. Dorta, B., A. Bosch, J. Arcaas, R. Ertola. Water balance in solid-state fermentation without forced aeration. *Enzyme Microb. Technol.* 16:562–565, 1994.
10. Gervais, P., C. Bazelin. Development of a solid-substrate fermentor allowing the control of the substrate water activity. *Biotechnol. Lett.* 8(3):191–196, 1986.
11. Raimbault. M. General and microbiological aspects of solid substrate fermentation. *Electronic J. Biotechnol.* 1:1–20, 1998.
12. Pandey, A. Effect of particle size of substrate on enzyme production in solid-state fermentation. *Bioresource Technol.* 37(2):169–172, 1991.
13. Pandey, A. Improvements in solid-state fermentation for glucoamylase production. *Biol. Wastes* 34(1):11–19, 1990.
14. Zadrazil, F., A.K. Puniya. Studies on the effect of particle size on solid-state fermentation of sugarcane bagasse into animal feed using white-rot fungi. *Bioresource Technol.* 54:85–87, 1995.
15. Pandey, A., C.R. Soccol, P. Nigam, V.T. Soccol. Biotechnological potential of agro-industrial residues, I: sugarcane bagasse. *Bioresource Technol.* 74(1):69–80, 2000.
16. Pandey, A., C.R. Soccol, P. Nigam, V.T. Soccol, L.P.S. Vandenberghe, R. Mohan. Biotechnological potential of agro-industrial residues, II: cassava bagasse. *Bioresource Technol.* 74(1):81–87, 2000.
17. Pandey, A., C.R. Soccol. Economic utilization of crop residues for value addition: a futuristic approach. *J. Sci. Ind. Res.* 59(1):12–22, 2000.
18. Pandey, A., C.R. Soccol, P. Nigam, D. Brand, R. Mohan, S. Roussos. Biotechnological potential of coffee pulp and coffee husk for bioprocesses. *Biochem. Eng. J.* 6(2):153–162, 2000.
19. Pandey, A., guest ed. Solid-state fermentation. *J. Sci. Ind. Res.* special issue 55:311–482, 1996.
20. Aidoo, K.E., R. Hendry, B.J.B. Wood. Solid-substrate fermentations. *Adv. Appl. Microbiol.* 28:201–237, 1982.
21. Aidoo, K.E., R. Hendry, B.J.B. Wood. Mechanized fermentation systems for the production of experimental soy sauce koji. *J. Food Technol.* 19:389–398, 1984.
22. Soccol, C.R. Biotechnology products from cassava root by solid-state fermentation. *J. Sci. Ind. Res.* 55:358–364, 1996.
23. Banerjee, R., A. Pandey. Bioindustrial applications of sugarcane bagasse: a technological perspective. *Int. Sugar J.* 104(1238):64–67, 2002.
24. Pandey, A., P. Selvakumar, C.R. Soccol, P. Nigam. Solid-state fermentation for the production of industrial enzymes. *Current Sci.* **77**(1):149–162, 1999.
25. Joshi, V.K., A. Pandey, eds. *Biotechnology: Food Fermentation*, Vol 2. New Delhi: Educational Publishers & Distributors, 1999, pp 524–1372.
26. Pandey, A., ed. *Advances in Biotechnology*. New Delhi: Educational Publishers & Distributors, 1998, p 523.
27. Banerjee, R., G. Mukherjee, A. Pandey, A. Sabu. Developments in biotechnology: an overview. *Ind. J. Biotechnol.* 1(1):9–16, 2002.
28. Pandey, A., P. Nigam, C.R. Soccol, D. Singh, V.T. Soccol, R. Mohan. Advances in microbial amylases. *Biotechnol. Appl. Biochem.* 31:135–152, 2000.

29. Pandey, A., S. Benjamin, C.R. Soccol, P. Nigam, N. Krieger, V.T. Soccol. The realm of microbial lipases in biotechnology. *Biotechnol. Appl. Biochem.* 29(2):119–131, 1999.
30. Pandey, A., P. Selvakumar, C.R. Soccol, V.T. Soccol, N. Krieger, J.D. Fontana. Recent developments in microbial inulinases: its production, properties and industrial applications. *Appl. Biochem. Biotechnol.* 81(1):35–52, 1999.
31. Sabu, A., M. Chandrasekaran, A. Pandey. Biopotential of microbial glutaminases. *Chemistry Today/Chimica Oggi* 18(11-12):21–25, 2000.
32. Couri, S., S.D. Terzi, G.A.S. Pinto, S.P. Freitas, A.C.A. da Costa. Hydrolytic enzyme production in solid-state fermentation by *Aspergillus niger* 3T5B8. *Proc. Biochem.* 36(3): 255–261, 2000.
33. Aunstrup, K. Enzymes of industrial interest: traditional products. In: *Annual Report of Fermentation Processes*, Vol. 6, Tsao, G.T., New York: Academic Press, 1983, pp 175–201.
34. Kashyap, P., A. Sabu, G. Szakacs, C.R. Soccol, A. Pandey. Extra-cellular L-glutaminase production by *Zygosaccharomyces rouxii* under solid-state fermentation. *Proc. Biochem.* 38(3):307–312, 2002.
35. Prabhu, G.N., M. Chandrasekaran. Impact of process parameters on L-glutaminase production by marine *Vibrio costicola* in solid state fermentation using polystyrene as an inert support. *Proc. Biochem.* 32:285–289, 1997.
36. Pandey, A., C.R. Soccol. Potential applications of cellulosic residues for the production of bulk chemicals and value added products. In: *Trends in Carbohydrate Chemistry*, Vol. 5, Soni, P.L., V. Kumar, eds., Dehradun, India: Surya International Publications, 1998, pp 83–88.
37. Pandey, A., C.R. Soccol. Bioconversion of biomass: a case study of ligno-cellulosics bioconversions in solid-state fermentation. *Brz. Arch. Biol. Technol.* 41(4):379–390, 1998.
38. Selvakumar, P., L. Ashakumary, A. Pandey. Microbial synthesis of starch saccharifying enzyme in solid-state fermentation. *J. Sci. Ind. Res.* 55(5–6):443–449, 1996.
39. Selvakumar, P., L. Ashakumary, A. Pandey. Biosynthesis of glucoamylase from *Aspergillus niger* by solid-state fermentation using tea waste as the basis of a solid substrate. *Bioresource Technol.* 65:83–85, 1998.
40. Pandey, A., L. Ashakumary, P. Selvakumar. Copra waste: a novel substrate for solid-state fermentation. *Bioresource Technol.* 51:217–220,1995.
41. Pandey, A., L. Ashakumary, P. Selvakumar. Performance of a column bioreactor for glucoamylase synthesis by *Aspergillus niger. Proc. Biochem.* 31:43–46, 1996.
42. Selvakumar, P., L. Ashakumary, L. Helen. A Pandey. Purification and characterization of glucoamylase produced by *Aspergillus niger* in solid state fermentation. *Letts Appl Microbiol* 23:403–406, 1996.
43. Pandey, A., S. Radhakrishnan. Packed-bed column bioreactor for production of enzyme. *Enzyme Microb. Technol.* 14:486–488, 1992.
44. Pandey, A., S. Radhakrishnan. The production of glucoamylase by *Aspergillus niger* NCIM 1245. *Proc. Biochem.* 28:305–309, 1993.
45. Babu, K.R., T. Satyanarayana. Alpha amylase production by thermophilic *Bacillus coagulans* in solid-state fermentation. *Proc. Biochem.* 30:305–309, 1995.
46. Tengerdy, R.P. Solid substrate fermentation for enzyme production. In: *Advances in Biotechnology*, Malik, V.S., P. Sridhar, eds., New Delhi: Educational Publishers & Distributors, 1998, pp 13–16.
47. Pandey, A. Production of glucoamylase in solid-state fermentation. In: *Industrial Microbiology*, Malik, V.S., P. Sridhar, eds., New Delhi: Oxford & IBH Publishing Co., 1992, pp 525–537.
48. Mulimani, V.H., G.N. Patil, J. Ramalingam. Alpha amylase production by solid state fermentation: a new practical approach to biotechnology courses. *Biochem. Edu.* 28:161–163, 2000.
49. Dey, S., S.O. Agarwal. Characterization of a thermostable alpha-amylase from a thermophilic *Streptomyces megasporus* strain SD12. *Ind. J. Biochem. Biophys.* 36:(3)150–157, 1999.
50. Krishna, C., M. Chandrasekarn. Banana waste as substrate for alpha amylase production by *Bacillus subtilis* under solid-substrate fermentation. *Appl. Microbiol. Biotechnol.* 46:106–111, 1996.

51. Ray, R.R., S.C. Jana, G. Nanda. Production of beta amylase from starchy waste by a hyper amylolytic mutant of *Bacillus megaterium*. *Ind. J. Exp. Biol.* 35: 285–288, 1997.
52. Francis, F., A. Sabu, K.M, Nampoothiri, G, Szakacs, A. Pandey. Synthesis of alpha-amylase by *Aspergillus oryzae* in solid-state fermentation. *J. Basic Microbiol.* 42(3): 322–328, 2002.
53. Khire, J.M., A. Pant. Thermostable salt-tolerant amylase from *Bacillus* sp. 64- thermophilic bacterium isolated from hot spring producing halophilic enzyme. *World J. Microbiol. Biotechnol.* 8(2):167–170, 1992.
54. Arnesen, S., S.H. Eriksen, J. Olsen, B. Jensen. Increased production of α-amylase from *Thermomyces lanuginosus* by the addition of Tween 80. *Enzyme Microb. Technol.* 23:249–252, 1998.
55. Pandey, A., L. Ashakumary, P. Selvakumar, K.S. Vijaylakshmi. Effect of yeast extract on glucoamylase synthesis by *Aspergillus niger* in solid-state fermentation. *Ind. J. Microbiol.* 35(4):335–338, 1995.
56. Ashakumary, L., P. Selvakumar, A. Pandey. Column fermentor for solid-state fermentation. In: *Solid-state Fermentation*, Pandey, A., ed., New Delhi: Wiley Eastern Limited, 1994, pp 33–37.
57. Pandey, P., P. Selvakumar, L. Ashakumary. Glucoamylase production by *Aspergillus niger* on rice bran is improved by adding nitrogen sources. *World J. Microbiol. Biotechnol.* 10:348–349, 1994.
58. Mukherjee, P.S., A. Pandey, P. Selvakumar, L. Ashakumary, P. Gurusamy. X-ray diffraction studies on solid-state fermentation for the production of glucoamylase using *Aspergillus niger* NCIM 1248. *J. Sci. Ind. Res.* 57:583–586, 1998.
59. Ramadas, M., O. Holst, B. Mattiasson. Extraction and purification of amyloglucosidase produced by solid state fermentation with *Aspergillus niger*. *Biotechnol. Tech.* 9(12):901–906, 1995.
60. Fujio, Y., H. Morita. Improved glucoamylase production by *Rhizopus* sp. A-11 using metal ion supplemented liquid medium. *J. Ferment. Bioeng.* 82:554–557, 1996.
61. Hata, Y., H. Ishida, Y. Kojima, E.K.A. Ichikawa, K. Suginami, S. Imayasu. Comparison of two glucoamylases produced by *Aspergillus oryzae* in solid-state culture (Koji) and in submerged culture. *J. Ferment. Bioeng.* 84:532–537, 1997.
62. Tani, Y., V. Vongsuvanlert, J. Kumnuanta. Raw cassava starch-digestive glucoamylase of *Aspergillus* sp. N-2 isolated from cassava chips *J. Ferment. Technol.* 64:405–410, 1986.
63. Pandey, A., S. Benjamin, C.R. Soccol, P. Nigam, N. Krieger, V.T. Soccol. The realm of microbial lipases in biotechnology. *Biotechnol. Appl. Biochem.* 29(2):119–131, 1999.
64. Benjamin, S., A. Pandey. *Candida rugosa* lipases: molecular biology and its versatility in biotechnology. *Yeast* 14:1069–1087, 1998.
65. Benjamin, S., A. Pandey. Coconut cake: a potent substrate for production of lipase by *Candida rugosa* in solid-state fermentation. *Acta Biotechnol.* 17:241–251, 1997.
66. Benjamin, S., A. Pandey. Mixed-solid substrate fermentation: a novel process for enhanced lipase production by *Candida rugosa*. *Acta Biotechnol.* 18:315–324, 1998.
67. Cordova, J., M. Nemmaoui, M. Ismaïli-Alaoui, A. Morin, S. Roussos, M. Raimbault, B. Benjilali. Lipase production by solid state fermentation of olive cake and sugar cane bagasse. *J. Mol. Catal. B: Enzymatic* 5:75–78, 1998.
68. Gombert, A.K., A.L. Pinto, L.R. Castilho, D.M.G. Freire. Lipase production by *Penicillium restrictum* in solid-state fermentation using babassu oil cake as substrate. *Proc. Biochem.* 35:85–90, 1999.
69. Kamini, N.R., J.G.S. Mala, R. Puvanakrishnan. Lipase production from *Aspergillus niger* by solid-state fermentation using gingely oil cake. *Proc. Biochem.* 33:505–511, 1998.
70. Benjamin, S., A. Pandey. *Candida rugosa* and its lipases: a retrospect. *J. Sci. Ind. Res.* 57(1):1–9, 1998.
71. Benjamin, S., A. Pandey. Panorama of lipases in bio industry. *Biotechnol. Int. J.* 5(6):11–13, 2000.
72. Castilho, L.R., C.M.S. Polato, E.A. Baruque, G.L. Sant'Anna Jr., D.M.G. Freire. Economic analysis of lipase production by *Penicillium restrictum* in solid-state and submerged fermentations. *Biochem. Eng. J.* 4:239–247, 2000.

73. Christen, P., N. Angeles, G. Corzo, A. Farres, S. Revah. Microbial lipase production on a polymeric resin. *Biotechnol. Tech.* 9(8):597–600, 1995.
74. Novo (1985) Patent. AU – 8432681, 1985.
75. Talon, R., M.C. Montel, J.L.B. Berdague. Production of flavor esters by lipases of *Staphylococcus warneri and S. xylosus*. *Enzyme Microbiol. Technol.* 19(8):620–622, 1996.
76. Rivera-Munoz, G., J.R. Tinoco-Valencia, S. Sanchez, A. Farres. Production of microbial lipases in a solid-state fermentation system. *Biotechnol. Lett.* 13:277–280, 1991.
77. Villegas, E., S. Aubague, L. Alcantara, R. Auria, S. Revah. Solid-state fermentation: acid protease production in controlled CO_2 and O_2 environments. *Biotechnol. Adv.* 11:387–397, 1993.
78. Ikasari, L., D.A. Mitchell. Leaching and characterization of *Rhizopus oligosporus* acid protease from solid-state fermentation. *Enzyme Microb. Technol.* 19:171–175, 1996.
79. Ikasari, L., D.A. Mitchell. Mimicking gas and temperature changes during enzyme production by *Rhizopus oligosporus* in solid-state fermentation. *Biotechnol. Lett.* 20:349–353, 1998.
80. George, S., V. Raju, T.V. Subramanian, K. Jayaraman. Comparative study of protease production in solid substrate fermentation versus submerged fermentation. *Bioproc. Eng.* 16(6):381–382, 1997.
81. Banerjee, R., B.C. Bhattacharya. Some studies on optimization of extraction process for protease production in SSF. *Bioproc. Eng.* 20:485–489, 1999.
82. Ikasari, L., D.A. Mitchell. Protease production by *Rhizopus oligosporus* in solid-state fermentation. *World J. Microbiol. Biotechnol.* 10:320–324, 1994.
83. Aikat, K., B.C. Bhattacharyya. Protease production in solid-state fermentation with liquid medium recycling in a stacked plate reactor and in a packed bed reactor by a local strain of *Rhizopus oryzae*. *Proc. Biochem.* 36:1059–1068, 2001.
84. Malathi, S., R. Chakraborty. Production of alkaline protease by a new *Aspergillus flavus* isolate under solid-substrate fermentation conditions for use as a depilation agent. *Appl. Environ. Microbiol.* 57:712–716, 1991.
85. Germano, S., V.M.G. Lima, C.R. Soccol, R. Pontarolo, G.D. Fontana, N. Krieger. Protease production by *Penicillium citrinum* in solid-state fermentation. In: *Advances in Biotechnology,* Pandey, A., ed., New Delhi: Educational Publishers and Distributors, 1998, pp 59–66.
86. Mitra, P., R. Chakraverty, A.L. Chandra. Production of proteolytic enzymes by solid-state fermentation: an overview. *J. Sci. Ind. Res.* 55, 439–442, 1996.
87. Boccas, F., S. Roussos, M.Gutierrez, L. Serrano, G.G. Viniaegra. Production of pectinase from coffee pulp in solid state fermentation system: selection of wild fungal isolate of high potency by a simple three-step screening technique. *J. Food Sci. Technol.* 31:22–26, 1994.
88. Acuña-Argüelles, M.E., M. Gutiérrez-Rojas, G. Viniegra-González, E. Favela-Torres. Production and properties of three pectinolytic activities produced by *Aspergillus niger* in submerged and solid-state fermentation. *Appl. Microbiol. Biotechnol.* 43:808–814, 1995.
89. Kavitha, R., S. Umesh-Kumar. Genetic improvement of *Aspergillus carbonarius* for pectinase overproduction during solid-state growth. *Biotechnol. Bioeng.* 67(1):121–125, 2000.
90. Rombouts, F.M., W. Pilnik. Enzymes in fruit and vegetable juice technology. *Proc. Biochem.* 18:9–13,1978.
91. Foneseca, M.J.V., S. Said. The pectinase produced by *Tubercularia vulgaris* in submerged culture using pectin or orange-pulp pellets as inducer. *Appl. Microbiol. Biotechnol.* 42:32–35, 1994.
92. Hang, Y.D., E.E.Woodanms. Production of fungal polygalacturonase from apple pomace. *Lebensm. Wiss. U. Technol.* 27:194–196, 1994.
93. Cavalitto, S.F., J.A. Arcas, R.A. Hours. Pectinase production profile of *Aspergillus foetidus* in solid state cultures at different acidities. *Biotechnol. Lett.* 18(3):251–256, 1996.
94. Castilho, L.R., R.A. Medronho, T.L.M. Alves. Production and extraction of pectinases obtained by solid-state fermentation of agro-industrial residues with *Aspergillus niger*. *Bioresource Technol.* 71:45–50, 2000.
95. Acuña-Argüelles, M.E., M. Gutiérrez-Rojas, G. Viniegra-González, E. Favela-Torres. Effect of water activity on exo-pectinase production by *Aspergillus niger* CH4 on solid-state fermentation. *Biotechnol. Lett.* 16:23–28, 1994.

96. Huerta, S., E. Favela-Torres, R. Lopez-Ulibarri, A. Fonseca, G. Viniegra-Gonzalez, M. Gutierrez-Rozas. Absorbed substrate fermentation for pectinase production with *Aspergillus niger. Biotechnol. Tech.* 8:837–842, 1994.
97. Castilho, L.R., T.L.M. Alves, R.A. Medronho. Recovery of pectinolytic enzymes produced by solid state culture of *Aspergillus niger. Proc. Biochem.* 34:181–186, 1999.
98. Schwan, R.F., R.M. Cooper, A.E. Wheals. Endopolygalacturonase secretion by *Kluyveromyces marxianus* and other cocoa pulp-degrading yeasts. *Enzyme Microb. Technol.* 21:234–244, 1997.
99. Zheng, Z., K. Shetty. Solid state production of polygalacturonase by *Lentinus edodes* using fruit processing wastes. *Proc. Biochem.* 35:825–830, 2000.
100. Prabu, G.N., M. Chandrasekaran. Use of inert supports in solid-state fermentation: a polystyrene experiment. In: *Advances in Biotechnology,* Pandey, A., ed., New Delhi: Educational Publishers, 1998, pp 51–58.
101. Sabu, A., T.R. Keerthi, S. Rajeev Kumar, M. Chandrasekaran. L-glutaminase production by marine *Beauvaria* sp. under solid-state fermentation. *Proc. Biochem.* 35:705–710, 2000.
102. Selvakumar, P., A. Pandey. Solid state fermentation for the synthesis of inulinase from *Staphylococcus* sp. and *Kluyveromyces marxianus*. *Proc. Biochem.* 34:851–855, 1999.
103. Pandey, A., P. Kavita, P. Selvakumar. Culture conditions for production of 2-1 beta-D-fructan fructanohydrolase in solid culturing on chicory (*Cichorium intybus*) roots, *Brz. Arch. Biol. Technol.* 41:231–236, 1998.
104. Bhat, T.K., B. Singh, O.P. Sharma. Microbial degradation of tannins: a current perspective. *Biodegredation* 9:343–357, 1998.
105. Aguliar, C.N., C. Augur, E. Favela-Torres, G. Viniegra-Gonzalez. Production of tannase by *Aspergillus niger* Aa-20 in submerged fermentation and solid-state fermentation: influence of glucose and tannic acid. *J. Ind. Microbiol. Biotechnol.* 26:296–302, 2001.
106. Vandenberghe, L.P.S., C.R. Soccol, A. Pandey, J.M. Lebeault. Microbial production of citric acid. *Brz. Arch. Bio. Technol.* 42:262–271, 1999.
107. Grewal, H.S., K.L. Kalra. Fungal production of citric acid. *Biotechnol. Adv.* 3:209–234, 1995.
108. Leangon, S., I.S. Maddox, J.D. Brooks. Influence of the glycolytic rate on production of citric acid and oxalic acid by *Aspergillus niger* in solid state fermentation. *World J. Microbiol. Biotechnol.* 15:493–495, 1999.
109. Tran, C.T., L.I. Sly, D.A. Mitchell. Selection of a strain of *Aspergillus* for the production of citric acid from pineapple waste in solid state fermentation. *World J. Microbiol. Biotechnol.* 14:399–404, 1998.
110. Lima, V.L.A.G., T.L.M. Stamford, A.A. Salgueiro. Citric acid production from pineapple waste by solid state fermentation using *Aspergillus niger*. *Arq. Biol. Technol.* 38:773–783, 1995.
111. Lu, M.Y., I.S. Maddox, J.D. Brooks. Application of a multi-layer packed bed reactor to citric acid production in solid state fermentation using *Aspergillus niger*. *Proc. Biochem.* 33:117–123, 1998.
112. Lu, M.Y., I.S. Maddox, J.D. Brooks. Citric acid production in solid state fermentation in a packed bed reactor using *Aspergillus niger*. *Enzyme Microb. Technol.* 21:392–397, 1997.
113. Garg, N., Y.D. Hang. Microbial production of organic acids from carrot processing waste. *J. Food Sci. Technol.* 32:119–121, 1995.
114. Hang, Y.D., E.E. Woodams. Apple pomace: a novel substrate for microbial production of citric acid. *Biotechnol. Lett.* 6:763–764, 1984.
115. Hang, Y.D., E.E. Woodams. Grape pomace: a novel substrate for microbial production of citric acid. *Biotechnol. Lett.* 7:253–254, 1985.
116. Shankaranand, V.S., B.K. Lonsane. Coffee husk: an inexpensive substrate for production of citric acid by *A. niger* in a solid-state fermentation. *World J. of Microbiol. Biotechnol.* 10:165–168, 1994.
117. Roukas, T. Production of citric acid from beet molasses by immobilised cells of *A. niger. J. Food Sci.* 56:878–880, 1991.

118. Dasgupta, J., S. Nasim, A. Khan, V.C. Vora. Production of citric acid in molasses medium: effect of lower alcohols during fermentation. *J. Microbiol. Biotechnol.* 9:123–125, 1994.
119. Gutierrez-Rojas, M., S.A.A. Hosn, R. Auria, S. Revah, E. Favela-Torres. Heat transfer in citric acid production by solid state fermentation. *Proc. Biochem.* 31:363–369, 1995.
120. Vandenberghe, L.P.S, C.R. Soccol, A. Pandey, J.M. Lebeault. Solid-state fermentation for the synthesis of citric acid by *Aspergillus niger*. *Bioresource Technol.* 74:175–178, 2000.
121. Soccol, C.R., B. Marin, M. Raimbault, J.M. Lebeault. Potential of solid state fermentation for production of L-(+) lactic acid by *Rhizopus oryzae*. *Appl. Microbiol. Biotechnol.* 41:286–290, 1994.
122. Xavier, S., B.K. Lonsane. Sugarcane press-mud as novel and inexpensive substrate for production of lactic acid in a solid state fermentation system. *Appl. Microbiol. Biotechnol.* 41:291–295, 1994.
123. Richter, K., A. Trager. L-(+)-lactic acid from sweet sorghum by submerged and solid state fermentations. *Acta Biotechnol.* 14:367–378, 1994.
124. Ooijkaas, L.P., F.J. Weber, R.M. Buitelaar, J. Tramper, A. Rinzema. Defined media and inert supports: their potential as solid-state fermentation production systems. *Trends Biotechnol.* 18:356–360, 2000.
125. Nampoothiri, K.M., A. Pandey. Solid state fermentation for L-glutamic acid production using *Brevibacterium* sp. *Biotechnol. Lett.* 18(2):199–204, 1996.
126. Fan, L., A. Pandey, C.R. Soccol. Cultivation of *Pleurotus* sp. on coffee residues. In *Proc. 3rd International conference on Mushroom Biology and Mushroom Products & AMGA's 26th National Mushroom Industry Conference*, Broderick, A., T Nair, eds., Sydney, 1999, pp. 301–311.
127. Fan, L., A. Pandey, C.R. Soccol. Growth of *Lentinus edodes* on the coffee industry residues and fruiting body production. In: *Proc. 3rd International conference on Mushroom Biology and Mushroom Products & AMGA's 26th National Mushroom Industry Conference*. Broderick, B., T. Nair, eds., Sydney, 1999, pp. 293–300.
128. Fan, L., A. Pandey, R. Mohan, C.R. Soccol. Use of various coffee industry residues for production of *Pleurotus ostreatus* in solid state fermentation. *Acta Biotechnol.* 20 (1):41–52, 2000.
129. Fan, L., A. Pandey, R. Mohan, C.R. Soccol. Solid-state culturing: an efficient technique to utilize toxic agro-industrial residues. *J. Basic Microbiol.* 40(3):177–187, 2000.
130. Stredansky, M., E. Conti. Xanthan production by solid state fermentation. *Proc. Biochem.* 34:581–587, 1999.
131. Kennedy, J.F., I.J. Bradshaw. Production, properties and applications of xanthan. *Prog. Ind. Microbiol.* 19, 319–371, 1984.
132. Garcia-Ochoa, F., V.E. Santos, J.A. Casas, E. Gomez. Xanthan gum: production/recovery and properties. *Biotechnol. Adv.* 18:549–579, 2000.
133. Flores Candia, J.L., W.D. Deckwer. Xanthan gum. In: *Encyclopedia of Bioprocess Technology: Fermentation, Biocatalysis, and Bioseparation* Vol. 5, Flickinger, M.C., S.W. Drew, eds., New York: Wiley, 1999, pp 2695–2711.
134. Vuyst, L.D., A. Vermeire. Use of industrial medium components for xanthan production by *Xanthomonas campestris* NRRL-B-1459. *Appl. Microbiol. Biotechnol.* 42:187–191, 1994.
135. Stredansky, M., E. Conti. Succinoglycan production by solid-state fermentation with *Agrobacterium tumefaciens*. *Appl. Microbiol. Biotechnol.* 52:332–337, 1999.
136. Stredansky, M., E. Conti, L. Navarini, C. Bertocchi. Production of bacterial exopolysaccharides by solid substrate fermentation. *Process. Biochem.* 34:11–16, 1999.
137. Johns, M.R., D.M. Stuart. Production of pigments by *Monascus purpureus* in solid culture. *J. Ind. Microbiol.* 8:23–28, 1991.
138. Chiu, S.W., S.M. Chan. Production of pigments by *Monascus purpureus* using sugar cane bagasse in roller bottle cultures. *World J. Microbio.l Biotechnol.* 8:68–70, 1992.
139. Fink-Gremmels, J., J. Dresel, L. Leistner, K. von Einstaz. Monascus-extrakten als nitrate alternative beifleischerzeugnissen [Use of Monascus extracts as an alternative to nitrite in meat products]. *Fleischwirtschaft* 71:329–331, 1991.

140. Watanabe, T., A. Yamamoto, S. Nagais, S. Terabe. Separation and determination of *Monascus* yellow pigments for food by micellar electrokinetic chromatography. *Analyt. Sci.* 13:571–575, 1997.
141. Ito, K., K. Yoshida, T. Ishikawa, S. Kobayashi. Volatile compounds produced by the fungus *Aspergillus oryzae* in rice koji and their changes during cultivation. *J. Ferment. Bioeng.* 70:169–172, 1990.
142. Gervais, P., M. Sarrette. Influence of age of mycelium and water activity of the medium on aroma production by *Trichoderma viride* grown on solid substrate. *J. Ferment. Bioeng.* 69:46–50, 1990.
143. Bramorski, A., P. Christen, M. Ramirez, C.R. Soccol, S. Revah. Production of volatile compounds by the edible fungus *Rhizopus oryzae* during solid-state cultivation on tropical agro-industrial substrates. *Biotechnol. Lett.* 20:359–362, 1998.
144. Christen, P., E. Villegas, S. Revah. Growth and aroma production by *Ceratocystis fimbriata* in various fermentation media. *Biotechnol. Lett.* 16(11):1183–1188, 1994.
145. Christen, P., J.C. Meza, S. Revah. Fruity aroma production in solid-state fermentation by *Ceratocystis fimbriata*: influence of the substrate type and the presence of precursors. *Mycol. Res.* 101:911–919, 1997.
146. Meza, J.C., P. Christen, S. Revah. Effect of added amino acids on the production of a fruity aroma by *Ceratocystis fimbriata. Sci. Aliment.* 18:627–636, 1998.
147. Soares, M., P. Christen, A. Pandey, C.R. Soccol. Fruity flavor production by *Ceratocystis fimbriata* grown on coffee husk in solid-state fermentation. *Proc. Biochem.* 35:857–861, 2000.
148. Medeiros, A.B.P., A. Pandey, R.J.S. Freitas, P. Christen, C.R. Soccol. Optimization of the production of aroma compounds by *Kluyveromyces marxianus* in solid-state fermentation using factorial design and response surface methodology. *Biochem. Eng. J.* 6:33–39, 2000.
149. Medeiros, A.B.P., A. Pandey, P. Christen, P.S.G. Fontoura, R.J.S. Freita, C.R. Soccol. Aroma compounds by *Kluyveromyces marxianus* in solid-state fermentation on a packed bed column bioreactor. *World J. Microbiol. Biotechnol.*17:767–771, 2001.
150. Besson, I., C. Creuly, J.B. Gros, C. Larroche. Pyrazine production by *Bacillus subtilis* in solid state fermentation on soybeans. *Appl. Microbiol. Biotechnol.* 47:489–495, 1997.
151. Larroche, C., I. Besson, J.B. Gros. High pyrazine production by *Bacillus subtilis* in solid substrate fermentation on ground soybeans. *Proc. Biochem.* 34:667–674, 1999.
152. Keuth, S., S. Bisping. Vitamin B-12 production by *Citrobacter freundii* or *Klebsiella pneumoniae* during tempeh fermentation and proof of enterotoxin absence by PCR. *Appl. Environ. Microbiol.* 60:1495–1499, 1994.
153. Keuth, S., S Bisping. Formation of vitamins by pure culture of tempe molds and bacteria during the tempe solid substrate fermentation. *J. Appl. Bact.* 75:427–434, 1993.
154. Pitt, J.I., A.D. Hocking. *Fungi and Food Spoilage*, 2nd ed. Gaithersburg, MD: Aspen Publishers, 1997.
155. Vinas, I., J. Dadon, V. Sanchis. Citrinin-producing capacity of *Penicillium expansum* strains from apple packinghouses of Lerida (Spain). *Int. J. Food Microbiol.* 19(2):153–156,1993.
156. Frayssinet, M.C., J.M. Fremy, S. Dragacci, J.L. Multon, eds. *Mycotoxins: Analysis of Food Constituents*. New York: Wiley-VCH, 1997, pp 379–423.
157. Bullernman, L.B., R.L. Buchanan. *J. Food Protect.* 44:701, 1981.
158. Silman, R.W., H.F. Conway, R.A. Anderson, E.B. Bagley. Production aflatoxin in corn by a large-scale solid-substrate fermentation process. *Biotechnol. Bioeng.* 21:1799–1808, 1979.
159. Maggon, K.K., S.K. Gupta, T.A. Venkitasubramanian. Biosynthesis of aflatoxins, *Bact. Rev.* 41:822, 1977.
160. González, J.B., G.M. Rodríguez, A. Tomasini. Environmental and nutritional factors controlling aflatoxin production in cassava solid state fermentation. *J. Ferment. Bioeng.* 70:329–333, 1990.
161. Lindenfelser, L.A., A. Ciegler. Solid-substrate fermentor for ochratoxin A production. *Appl. Microbiol.* 29(3):323–327, 1975.

162. Saxena, J., C. Munimbazi, L.B. Bullermann. Relationship of mould count, ergesterol and ochratoxin A production. *Int. J. Food Microbiol.* 71(1):29–34, 2001.
163. Blanc, P.J., J.P. Laussac, J. Le Bars, P. Le Bars, M.O. Loret, A. Pareilleux, D. Prome, J.C. Prome, A.L. Santerre, G. Goma. Characterization of monascidin from Monascus as citrinin. *Int. J. Food Microbiol.* 27:201–213, 1995.
164. Ramesh, M.V., B.K. Lonsane. Solid-state fermentation for production of high titres of thermostable alpha-amylase with two peaks for pH optima by *B. licheniformis* M27. *Biotechnol. Lett.* 2(1), 49–52, 1989.
165. Sudo, S., T. Ishikawa, K. Sato, T. Oba. Alpha amylase production by *A. kawachii* in solid-state and submerged cultures. *J. Ferment. Bioeng.* 77(5):483–389, 1994.
166. Bajpai, B., S. Patil. Tannin acyl hydrolase (EC 3.1.1.20) activity of *Aspergillus, Penicillium, Fusarium, Trichoderma. World J. Microbiol. Biotechnol.* 12:217–220, 1996.
167. Yeoman, K.H., C. Edwards. Growth of thermophilic bacteria in rapemeal-derived media *J. Appl. Bacteriol.* 77:264, 1994.
168. Narahara, H., Y. Koyama, T. Yoshida, S. Pichangkura, R. Ueda, H. Taguchi. *J. Ferment. Technol.* 60:311, 1982.
169. Datta, A. Purification and characterization of a novel protease from solid substrate cultures of *Phanerochaete chrysosporium. J. Bol. Chem.* 267:728, 1992.
170. Battaglino, R.A., M. Huergo, A.M.R. Pilosuf, G.B. Bartholomai. Culture requirements for the production of protease by *Aspergillus oryzae* in solid state fermentation. *Appl. Microbiol. Biotechnol.* 35:292, 1991.
171. Schellart, J.A., F.M.W. Visser, T. Zandastra, W.J. Middlehoven, Starch degradation by the mould *Trichoderma viride*. I. The mechanism of starch degradation. Antonie Van Leeuwenhoek. *World J. Microbiol. Serial* 42(3):229–238, 1976.
172. Srivastava, R.A.K. Studies on extracellular and intracellular purified amylases from a thermophilic *Bacillus stearothermophilus. Enzyme Microb. Technol.* 6:422, 1984.
173. Janse, B.J.H, I.S. Pretorius. One-step enzymatic hydrolysis of srarch using a recombinant strain of Saccharomyces cerevisiae producing alpha amylase, glucoamylase and pullulanase. *Appl. Microbiol. Biotechnol.* 42:878, 1995.
174. Ghildayal, N.P., M. Ramakrishna, B.K. Lonsane, N.G. Karanth. Efficient and simple extraction of mouldy bran in a pulsed column extractor for recovery of amyloglucosidase in concentrated form. *Proc. Biochem.* 26:235, 1991.
175. Oriol, E., M. Raimbault, S. Roussos, G. Vineiegra-Gonzalez. Water and water activity in solid-state fermentation of cassava starch by *Aspergillus niger. Appl. Microbiol. Biotechnol.* 27:498–503, 1988.
176. Ono, K., S. Shiegeta, S. Oka. *Agric. Biol. Chem.* 52:1701, 1988.
177. Soni, S.K., K.S. Bath, R. Soni. Production of amylases by *Saccharomycopsis capsularis* in solid-state fermentation. *Ind. J. Microbiol.* 36:157–159, 1996.
178. Viswanathan, P., N.R. Surlikar. Production of alpha amylase with *A. flavus* on amaranthus grains by solid-state fermentation. *J. Basic Microbiol.* 41(1):57–64, 2001.
179. Saxena, R.K., P. Sharmila, V.P. Singh. Progress. *Ind. J. Microbiol.* 32:259–270, 1995.
180. Silanikov, N., O. Danai, D. Levanon. Composted cotton straw silage as a substrate for *Pleurotus* sp. cultivation. *Biol. Wastes* 25:219–226, 1988.
181. Juzlova, P., L. Martinkova, V. Kren. Secondary metabolites of the fungus *Monascus:* a review. *J. Industrial Microbiol.* 16:163–170, 1996.

1.05

Metabolic Engineering of Bacteria for Food Ingredients

Ramon Gonzalez

CONTENTS

5.1 INTRODUCTION

Metabolic engineering (ME; see Table 5.1 for a complete list of acronyms used in this chapter) has been defined as the directed improvement of product formation or cellular properties through the modification of specific biochemical reactions or the introduction of new ones with the use of recombinant DNA technology (1,2). Therefore, the *analysis* and *modification* of metabolic pathways is of central importance to ME. The *analytical* part uses experimental and modeling techniques [e.g., labeling experiments, Metabolic Flux Analysis (MFA)], which allow the systematic study of cellular responses (in terms of metabolic fluxes) to genetic and environmental perturbations. This facilitates a rational design of metabolic *modifications*, which are implemented using recombinant DNA technology. Both fields, the *analysis* and the *modification* of metabolic pathways, will be covered in this chapter.

Because of the central role of fermentation in food production and preparation, microorganisms are constantly encountered in these processes. ME can rationally improve the properties of these microorganisms and their efficiency for producing different products. Bacteria occupy a central place among these microorganisms, as either the bacteria *per se* (e.g., as starter cultures and probiotics) or products synthesized by bacteria (i.e., used as biocatalyst in the synthesis of amino acids, organic acids, vitamins, and carbohydrates). This illustrates the tremendous potential of ME for improving the bacteria used by the food industry.

Food additives produced by bacteria can be incorporated into foods as nutritional supplements, flavor enhancers, texturizers, acidulants, preservatives, emulsifiers, surfactants, thickeners, or functional food ingredients. A functional food (also referred to as a nutraceutical or pharmaceutical food) is defined as a food that can beneficially affect one

Table 5.1
Acronyms

Acronym	Definition
AcCoA	Acetyl Coenzyme A
AlaDH	Alanine Dehydrogenase
EPS	Exopolysaccharides
GDH	Glutamate Dehydrogenase
GRAS	Generally Regarded As Safe
LAB	Lactic Acid Bacteria
LAC	Lactic Acid
LDH	Lactate Dehydrogenase
ME	Metabolic Engineering
MFA	Metabolic Flux Analysis
PDH	Pyruvate Dehydrogenase
PEP	Phosphoenolpyruvate
PPC	phosphoenolpyruvate carboxylase
PPP	Pentose Phosphate Pathway
PTS	Phosphotransferase System
PYC	Pyruvate Carboxylase
PYR	Pyruvate
TCA	Tricarboxylic Acid
2,5-DKG	2,5-diketo-D-gluconic Acid
2-KLG	2-keto-L-gulonic Acid

or more targeted bodily functions beyond adequate nutritional effects in a way that improves health and well-being and reduces the risk of disease (3). Functional food ingredients include low calorie sugars, polysaccharides, and vitamins.

Applying ME principles and tools to the production of food ingredients by bacteria has resulted in the efficient production of both native and totally novel products by several cultures, including strains of lactic acid bacteria (LAB), *Escherichia coli*, *Bacillus subtilis*, and *Corynebacterium glutamicum*. These microorganisms have been selected based on a variety of criteria including, availability of genetic tools for cloning, expression, disruption, or replacement of genes (i.e., sequenced genome, promoters, plasmids and cloning vectors, gene transfer methods, and gene expression systems); capacity to metabolize a wide spectrum of carbon sources at high fluxes; rapid growth on inexpensive carbon and nitrogen sources; GRAS (generally regarded as safe) status; resistance to bacteriophage attack; well established physiological knowledge; efficient transport of the final product to the extracellular medium and tolerance to its high levels; and use in large-scale fermentations and production at an industrial level. The examples in this chapter (summarized in Table 5.2) will cover the analysis and modification of central metabolic pathways, biosynthetic pathways, and transport systems involved in producing amino acids, organic acids, vitamins, carbohydrates, bacteriocins, low calorie sugars, and aroma compounds. It is worth noting that many of these products are commercially produced.

5.2 AMINO ACIDS

Amino acids have a great variety of current and potential uses as food, pharmaceuticals, and animal feed. Their main application field is food, where about 50% of the product is applied (4). They can be used as nutritional supplements, flavor enhancers, sweeteners, and in pre- and postoperative nutrition therapy (5). This section will discuss using ME to produce some of these amino acids. Recent progress in this field can be divided into three categories (6): (1) modification of central metabolic pathways, (2) modification of biosynthetic pathways, and (3) modification of transport systems. Figures 5.1, 5.2, and 5.3 summarize some of the examples discussed below.

5.2.1 Modification of Central Metabolic Pathways

From reasoning based on metabolic pathways structure, rerouting a carbon source to produce a desired amino acid should start by increasing the availability of precursor metabolites, energy, and reducing equivalents used in its synthesis. Central metabolic pathways meet these criteria, and therefore engineering central metabolism is essential for the efficient production of amino acids. The analytical tools included in the ME toolbox have played an essential role

Table 5.2
Examples of bacteria that have been engineered for the production of food ingredients

Microorganism(s)	Product (Reference)
Corynebacterium glutamicum	L-tryptophan (16), L-phenylalanine (11,12), L-tyrosine (14), L-threonine (15), L-isoleucine (13), and L-histidine (10)
Lactic acid bacteria	L-alanine (21), Vitamins (50), Exopolysaccharides (57), Bacteriocins (61), and Aroma Compounds (67)
Bacillus subtilis	Vitamins (48) and Lactic Acid (44)
Escherichia coli	Succinate (23,24), Pyruvate (33), Lactate (41,42), and Acetate (45)

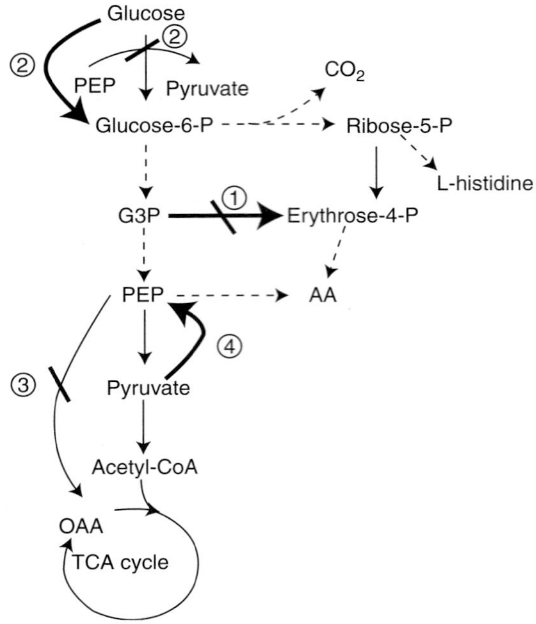

Figure 5.1 Metabolic engineering of central metabolic pathways to increase the synthesis of histidine and aromatic amino acids. Solid and dashed lines represent single and multiple steps, respectively. Solid bars over the arrows represent blocked enzymes, while thick arrows represent amplified enzymes. The following strategies are illustrated. Increasing the production of histidine by increasing the availability of Ribose-5-P, 1- Transketolase-deficient strains. Increasing the production of AA by increasing the supply of erythrose-4-P, 1- Overexpression of transketolases in AA producer. Increasing the production of AA by increasing the availability of PEP, 2- Inactivation of PEP-dependent PTS system for the transport of glucose and amplification of sugar-phosphorylating kinase gene; 3- Inactivation of PEP carboxylase; 4- Amplification of PEP synthase. Abbreviations, AA, aromatic amino acids; G3P, glyceraldehyde-3-P; and PEP, phosphoenolpyruvate.

in elucidating the function of different central pathways and suggesting useful strategies for redirecting carbon flow toward the biosynthesis of amino acids. For example, it has been shown that the pentose phosphate pathway (PPP) supports higher fluxes during the production of L-lysine compared to the production of L-glutamic acid in *C. glutamicum* (7,8). This was attributed to the higher requirements of reducing power (NADPH) in the production of L-lysine. Another example is improving aromatic amino acids and L-histidine production in *C. glutamicum* by increasing the availability of their precursor metabolites, erythrose 4-phosphate and ribose 5-phosphate, respectively, as well as NADPH. This can be done by modifying the flux through the PPP, either by increasing the activity of transketolase (and providing more erythrose 4-phosphate for aromatic amino acids biosynthesis) or by decreasing the activity of transketolase (and providing more ribose 5-phosphate for L-histidine biosynthesis) as shown in Figure 5.1, strategy 1 (6). Both approaches have produced *C. glutamicum* strains with an increased capacity for making aromatic amino acids (9) and L-histidine (10). Figure 5.1 shows additional examples of engineering central metabolic pathways to increase the availability of precursor metabolites used to synthesize aromatic amino acids in *C. glutamicum*. In general, these strategies are based on increasing the availability of erythrose 4-phosphate (strategy 1, Figure 5.1), phosphoenolpyruvate (strategies 2, 3, and 4, Figure 5.1), and ribose-5-P (strategy 1,

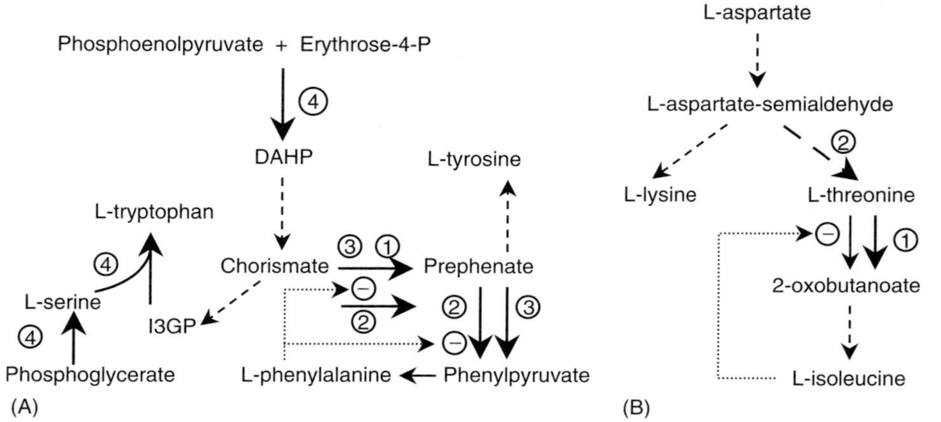

Figure 5.2 Metabolic engineering of biosynthetic pathways to increase the synthesis of aromatic amino acids, threonine, and isoleucine in *C. glutamicum*. See Figure 5.1 for explanation of different types of lines/arrows. In some cases, feedback inhibition has been represented using round dotted lines. Examples illustrated here include amplification of a gene encoding a rate-limiting step, introduction of a heterologous enzyme subjected to a different regulatory mechanism, and redirection of metabolic flux in a branch point (Ikeda (6)). (A) Increasing the synthesis of aromatic amino acids, 1 - Overexpression of chorismate mutase increased L-phenylalanine production; 2- Overexpression of mutated (insensitive to L-phenylalanine) chorismate mutase–prephenate dehydratase from *E. coli* increased the production of L-phenylalanine; 3- Simultaneous amplification of chorismate mutase and prephenate dehydratase resulted in increased production of L-tyrosine and L-phenylalanine; 4- Coexpression of two enzymes catalyzing the initial steps in the biosynthesis of aromatic amino acids (3-deoxy-D-arabino-heptulosonate 7-phosphate synthase) and L-serine (3-phosphoglycerate dehydrogenase) together with tryptophan-biosynthetic enzymes increased tryptophan production. (B) Increasing the synthesis of isoleucine and threonine, 1- Expression of L-isoleucine-insensitive *E. coli* threonine dehydratase (catabolic) enhanced isoleucine production; 2- Amplification of a threonine biosynthetic operon resulted in increased production of threonine in a lysine-producing *C. glutamicum* strain. Abbreviations, DAHP, 3-deoxy-D-arabino-heptulosonate-7-phosphate; I3GP, indole-3-glycerol-phosphate.

Figure 5.1) by inactivating the enzymes involved in their consumption and/or amplifying the enzymes involved in their production (for a review of these strategies, see Reference 6).

5.2.2 Modification of Biosynthetic Pathways

After engineering central metabolic pathways, a sufficient supply of precursor metabolites, energy, and reducing power is ensured, and efforts then need to be focused on engineering biosynthetic pathways that convert precursor metabolites into amino acids. Several strategies have been used to achieve this goal (6), and some of them are illustrated in Figure 5.2. For example, the gene that encodes a rate-limiting enzyme can be amplified, resulting in the release of a bottleneck. Ozaki et al. (11) and Ikeda et al. (12) used this strategy to improve the production of L-phenylalanine in *C. glutamicum*. Overexpression of the gene that encodes chorismate mutase in *C. glutamicum* K38 resulted in a 50% increase in the yield of L-phenylalanine (11). On the other hand, introducing heterologous enzymes subject to different regulatory mechanisms can also result in the release of a bottleneck. For example, overexpressing a mutated (insensitive to L-phenylalanine) *E. coli* gene that encoded the bifunctional enzyme chorismate mutase–prephenate dehydratase in *C. glutamicum* KY10694 led to a 35% increase in the production of L-phenylalanine (12). In addition, expressing the *E. coli* catabolic threonine dehydratase (insensitive to L-isoleucine) in *C. glutamicum*

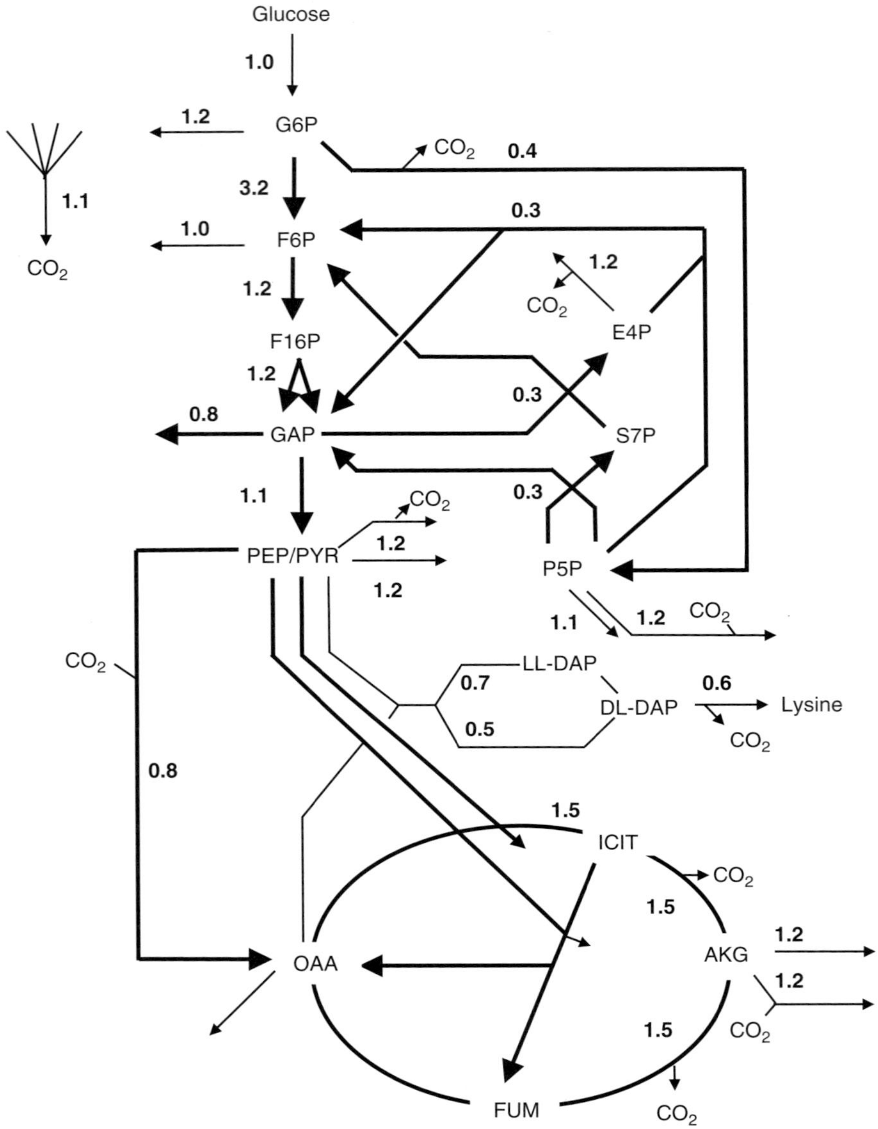

Figure 5.3 Labeling experiments-based MFA of two isogenic glutamate dehydrogenase mutants (homologous, NADPH-dependent, and heterologous, NADH-dependent) of the lysine producer strain *C. glutamicum* MH20-22B. Flux values were converted to flux ratios and expressed as (NADH-dependent mutant)/(NADPH-dependent mutant). Numbers near the thick lines give the estimated net fluxes while those near the thin arrows give the measured fluxes required for biomass synthesis. Adapted from Marx et al. (20). Abbreviations, AKG, α-ketoglutarate; DL-DAP, DL-diaminopimelate; E4P, erythrose-4-P; FUM, fumarate; F16P, fructose-1-6-bisphosphate; F6P, fructose-6-phosphate; GAP, glyceraldehyde-3-phosphate; G6P, glucose-6-phosphate; ICIT, isocitrate; LL-DAP, LL-diaminopimelate; OAA, oxaloacetate; PEP, phosphoenolpyruvate; PYR, pyruvate; P5P, pentose-5-phosphate; S7P, sedoheptulose-7-phosphate.

resulted in increased isoleucine production (13). A second strategy could be the redirection of metabolic flux in a branch point. Ikeda and Katsumata (14) engineered a tryptophan-producing mutant of *C. glutamicum* to produce L-tyrosine or L-phenylalanine in abundance

(26 and 28 g/L, respectively) by overexpressing the branch-point enzymes (chorismate mutase and prephenate dehydratase), catalyzing the conversion of the common intermediate chorismate into tyrosine and phenylalanine. Using a similar approach, Katsumata et al. (15) produced threonine using a lysine-producing *C. glutamicum* strain, by amplifying a threonine biosynthetic operon. Other strategies could include introducing heterologous enzymes that use different cofactors than those used by the native enzyme as well as amplifying the enzyme that catalyzes the steps linking central metabolism and the biosynthetic pathway (6). Ikeda et al. (16) achieved a 61% increase in tryptophan yield (50 g/L of tryptophan) in a tryptophan-producing *C. glutamicum* KY10894 by coexpressing two enzymes catalyzing the initial steps in the biosynthesis of aromatic amino acids (3-deoxy-D-arabino-heptulosonate 7-phosphate synthase) and serine (3-phosphoglycerate dehydrogenase) together with tryptophan biosynthetic enzymes.

5.2.3 Modification of Transport Systems

By altering amino acid transport systems, one could expect to decrease their intracellular concentration and avoid feedback inhibition. The following two examples illustrate that show the significance of this strategy. During the production of L-tryptophan by *C. glutamicum*, the accumulation of this product in the extracellular medium resulted in a backflow into the cells, which produced severe feedback inhibition in the biosynthesis of tryptophan (6). Ikeda and Katsumata (17) solved this problem by creating mutants with lower levels of tryptophan uptake, which resulted in an accumulation of 10 to 20% more tryptophan than in their parent. Another example of transport engineering is the increase in the fermentation yield of cysteine by overexpressing multidrug efflux genes (*mar* genes) in a cysteine-producing strain of *E. coli* (6).

5.2.4 Use of Analytical Tools in the ME Toolbox

In this section, I will give some examples of using MFA [See reference (2) for a detailed description of MFA] to improve the production of specific amino acids, including the two amino acids with the largest production volume worldwide (L-glutamic acid with 800,000 tons/year and L-lysine with 600,000 tons/year).

Glutamate. As with other amino acids, analysis and modification of central metabolic and biosynthetic pathways in glutamate-producing *C. glutamicum* strains have contributed to their improvement. MFA, including estimation of fluxes using labeling experiments, has elucidated the relative contribution of different pathways (e.g., Embden-Meyerhof and hexose monophosphate) under various physiological conditions and genetic backgrounds [(18) and references therein]. For example, MFA has established the relationship between the decline of oxoglutarate dehydrogenase complex and the flux distribution at the metabolic branch point of 2-oxoglutarate glutamate. This suggests that metabolic flux through anaplerotic pathways could be limiting the production of glutamate. Many of these findings have either been verified by genetic approaches or have led to a rational modification of metabolic pathways to improve the production of glutamate [(18) and references therein].

L-lysine. L-lysine production is another example of the application of labeling experiments-based MFA. Very comprehensive approaches have been used to assess all major fluxes in the central metabolism of *C. glutamicum*, which reveal patterns that can be used for designing ME strategies. In a comparison of six metabolic patterns, several metabolic fluxes that depend significantly on the physiological state of the cells were identified [(19) and references therein]. These included the coordinated flux through PPP and the tricarboxylic acid (TCA) cycle, the high capacity for the reoxidation of NADPH, and futile cycles between C3-compounds of glycolysis and C4-compounds of the TCA cycle. As an example, Figure 5.3 shows a comparison of the metabolic flux distribution between two isogenic glutamate dehydrogenase mutants (homologous, NADPH-dependent, and heterologous, NADH-dependent) of the L-lysine producer strain *C. glutamicum* MH20-22B (20). Metabolic flux patterns revealed that the PPP flux was high only for a high demand of NADPH and a low TCA cycle flux. The heterologous, NADH-dependent, glutamate dehydrogenase mutant required more NADH, resulting in an increased TCA cycle flux and then more NADPH supplied by the isocitrate dehydrogenase step in the TCA cycle. This led to a decreased flux of the PPP due to lower NADPH requirements (TCA cycle already produced part of it). The inverse is true for the homologous, NADPH-dependent, glutamate dehydrogenase mutant. There is a higher NADPH requirement, the PPP flux is higher, and the TCA cycle flux is lower.

5.2.5 Use of LAB in the Production of Amino Acids

LAB also has been used to produce different amino acids. *Lactococcus lactis* (21) has been engineered to produce L-alanine as the only end product of fermentation (more than 99%). Rerouting the carbon flux toward alanine was achieved by expressing *Bacillus sphaericus* alanine dehydrogenase (AlaDH) in lactate dehydrogenase (LDH) deficient strains. Finally, stereospecific production ($>$ 99%) of L-alanine was achieved by disrupting the gene encoding alanine racemase.

5.3 MULTIFUNCTIONAL ORGANIC ACIDS

The ability to produce organic acids via fermentation is of great interest to the food industry due to the acids' widespread use as acidulants, food preservatives, beverage ingredients, sweeteners, and flavor enhancers.

Many bacteria, such as *E. coli*, carry out mixed acid fermentation of sugars (e.g., glucose) in which the principal products are formate (or CO_2 and H_2), acetate, lactate, succinate, and ethanol (Figure 5.4) (22). Under these conditions, anaerobic fermentative pathways have two main functions: to produce energy in an anaerobic environment via substrate level phosphorylation, and to provide a source of regeneration of NAD+, closing the redox balance. The aforementioned products can be organized in three groups according to the redox properties of the homofermentative pathway used in their synthesis: (1) net regeneration of redox equivalent by the homofermentative pathway (e.g., lactate, fumarate, and malate), (2) net production of redox equivalents by the homofermentative pathway (e.g., acetate and pyruvate), and (3) net consumption of redox equivalent by the homofermentative pathway (e.g., succinate).

In this section, I will present some examples of ME of *E. coli* for the homofermentative production of some of these organic acids, including succinate, pyruvate, lactate, and acetate. In general, all strategies have been focused on altering the carbon flux ratio at three branch points, phosphoenolpyruvate (PEP), pyruvate (PYR), and acetyl coenzyme A (AcCoA).

5.3.1 Succinic Acid

The production of succinate by several bacterial strains, including the obligate anaerobe *Anaerobiospirullum succiniciproducens*, *Actinobacillus* sp., and *Enterococcus* sp. have been reported. However, ME efforts have focused on engineering *E. coli* to produce succinate as the main fermentation product. In what follows I will summarize those results, which are also schematically represented in Figure 5.5.

E. coli wild-type strains produce succinate only as a minor product of fermentation. Genetic manipulations are required to increase succinate production and reduce byproduct formation. For example, increasing flux at the first step in the succinate branch by

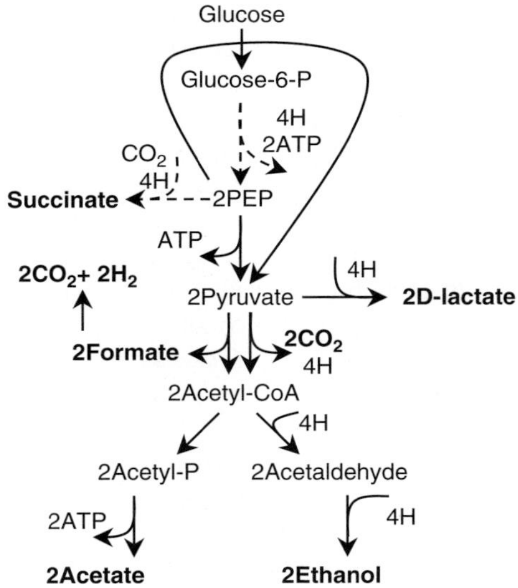

Figure 5.4 Anaerobic fermentation of glucose in *E. coli*. Only relevant central metabolic pathways have been included and it has been considered that glucose is transported into the cell only by the PEP-dependent PTS. Not all steps and metabolites are shown. Solid and broken lines represent single and multiple steps, respectively. Primary fermentation products are indicated in bold. Abbreviations, PEP, phosphoenolpyruvate.

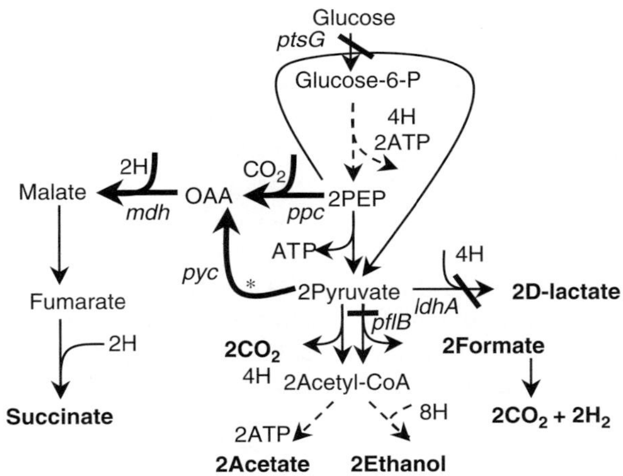

Figure 5.5 Metabolic engineering of *E. coli* to produce succinate. See Figure 5.1 for explanation of different types of lines/arrows. Genes involved in engineered steps are indicated by italics. Primary fermentation products are indicated in bold. Strategies include the blocking (genes *ldhA*, *pflB*, and *ptsG*) and amplification (genes *ppc* and *mdh*) of homologous enzymes as well as the introduction of heterologous enzymes (*R. etli pyc* gene). The asterisk (*) indicates that this step does not exist in *E. coli*. Abbreviations, *ldhA*, gene encoding D-lactate dehydrogenase; *mdh*, gene encoding malate dehydrogenase; PEP, phosphoenolpyruvate; *pflB*, gene encoding pyruvate-formate lyase; *ppc*, gene encoding PEP carboxylase; *ptsG*, gene encoding the enzyme IICB^{Glc} of the PTS system; *pyc*, gene encoding *R. etli* pyruvate carboxylase.

overexpressing PEP carboxylase results in an increase in succinate percentage yield (molar basis) from 12 to 45% (23). PEP is also a required cosubstrate for glucose transport via the phosphotransferase system (PTS) in wild-type *E. coli*. Thus, another approach is to direct pyruvate to the succinate branch. This is achieved by transforming a wild-type *E. coli* strain with plasmid pTrc99A-*pyc*, which expresses *Rhizobium etli* pyruvate carboxylase. This strategy results in an increase in both succinate percentage yield (17%) and productivity (0.17 g/L/h) (24). In order to prevent the accumulation of other undesired products, mutations in genes involved in producing lactate (*ldhA*, encoding lactate dehydrogenase) and formate (*pflB*, encoding pyruvate–formate lyase) were introduced, obtaining the strain *E. coli* NZN111, that grew poorly on glucose under anaerobic conditions. When the gene encoding malic enzyme from *Ascaris suum* was transformed into NZN111, succinate percentage yield and productivity were further increased to 39% and 0.29 g/L/h, respectively (25,26). Expression of *mdh* gene encoding malate dehydrogenase also resulted in improved production of succinate by *E. coli* NZN111 (27).

Donnelly et al. (28) reported an unknown spontaneous chromosomal mutation in NZN111, which permitted anaerobic growth on glucose, and this strain was designated as AFP111. When AFP111 was grown anaerobically under 5% H_2 and 95% CO_2, a succinate percentage yield of 70% and a succinate–acetate molar ratio of 1.97 were obtained. Further improvements of succinate production with this strain included dual phase fermentation (aerobic growth for biomass generation follow by anaerobic growth for succinate production) that resulted in a succinate percentage yield of 99% and a productivity of 0.87 g/L/h (29). AFP111 mutation was mapped to the *ptsG* gene, encoding an enzyme of the PTS (30).

Gokarn et al. (31) found that the expression of *R. etli* pyruvate carboxylase (PYC) in *E. coli* during anaerobic glucose metabolism caused a 2.7-fold increase in succinate concentration, making it the major product by mass. The increase came mainly at the expense of lactate formation. However, in a mutant lacking lactate dehydrogenase activity, expression of PYC resulted in only a 1.7-fold increase in succinate concentration. An accumulation of pyruvate and NADH, metabolites that affect the interconversion of the active and inactive forms of the enzyme pyruvate formate-lyase, may have caused the decreased enhancement of succinate. The same group (32) had previously shown that the presence of the *R. etli pyc* gene in *E. coli* (JCL1242/pTrc99A-pyc) restored the succinate producing ability of *E. coli ppc* null mutants (JCL1242), with PYC competing favorably with both pyruvate formate lyase and lactate dehydrogenase. Flux calculations indicated that during anaerobic metabolism the *pyc*(+) strain had a 34% greater specific glucose consumption rate, a 37% greater specific rate of ATP formation, and a 6% greater specific growth rate than the *ppc*(+) strain. The results demonstrate that when phosphoenolpyruvate carboxylase (PPC) or PYC are expressed, the metabolic network adapts by altering the flux to lactate and the molar ratio of ethanol to acetate formation.

5.3.2 Pyruvic Acid

Several bacterial strains have been used to produce pyruvic acid via the fermentation of different sugars. These include strains of *Escherichia*, *Pseudomonas*, *Enterococcus*, *Acinetobacter*, and *Corynebacterium* (33). The most successful processes (highest yield and concentrations, and shortest fermentation times) are those involving *E. coli* strains. The main strategy used has been limiting pyruvate consumption by pyruvate dehydrogenase complex (PDH) under aerobic conditions. Considering that lipoic acid is a cofactor of PDH, a screen by Yokota et al. (34) identified a lipoic acid auxotroph, *E. coli* W1485*lip2*, which accumulated 25.5 g/L pyruvate from 50 g/L glucose at 32 hours. Pyruvate production was further improved by introducing a mutation in F_1-ATPase, obtaining strain TBLA-1 (35). Strain TBLA-1 exhibited a higher capacity for producing pyruvate, more than 30 g/L of pyruvate were produced from 50 g/L of glucose in only 24 hours. The growth of TBLA-1 decreased to 67% of that in the parent strain, due to lower energy production (TBLA-1 is an F_1-ATPase-defective mutant). The enhanced pyruvate productivity in strain TBLA-1 was thought to be linked to increased activities of some glycolytic enzymes (36). Although there was an increase in glycolytic flux due to a decrease in ATP production (i.e., an F_1-ATPase mutation), the physiological mechanism mediating these changes was not identified. Recently, it was shown that the glycolytic flux in *E. coli* is controlled by the demand for ATP (37). By increasing the ATP hydrolysis, the glycolytic flux was increased by approximately 70%, indicating that glycolytic flux is mainly ($> 75\%$) controlled by reactions hydrolyzing ATP. In light of these results, it is clear that increased pyruvate production in strain TBLA-1 when compared to its wild type *E. coli* W1485*lip2* is due to lower ATP production in TBLA-1, which in turn is driving glycolysis and resulting in higher glycolytic fluxes.

5.3.3 Lactic Acid

Many microorganisms produce D-lactic acid (D-LAC), and some LABs, such as *Lactobacillus bulgaricus*, produce highly pure D-LAC (38). L-LAC has also been produced using other LABs, such as *Lactobacillus helveticus*, *Lactobacillus amylophilus*, and *Lactobacillus delbruekii* (39). Since LABs have complex nutritional requirements and low growth rates, and exhibit incomplete or negligible pentose utilization (40), other bacterial strains have been engineered to produce optically pure D- or L-LAC including *E. coli* (41,42), *Rhizopus orizae* (43), and *B. subtilis* (44). Chang et al. (41) engineered *E. coli* to produce optically pure D- or L-LAC. A *pta* mutant of *E. coli* RR1, which was deficient in the phosphotransacetylase of the Pta-AckA pathway, was found to metabolize glucose to D-LAC and to produce a small amount of succinate byproduct under anaerobic conditions. An additional mutation in *ppc* (encoding PPC) made the mutant produce D-LAC like a homofermentative LAB. In order to produce L-LAC, a nonindigenous fermentation product, an L-lactate dehydrogenase gene from *Lactobacillus casei* was introduced into a *pta ldhA* strain, which lacked phosphotransacetylase and D-LAC dehydrogenase. This recombinant strain was able to metabolize glucose to L-LAC as the major fermentation product, and produced up to 45 g/L of L-LAC. Zhou et al. (42) constructed derivatives of *E. coli* W3110 capable of producing D-LAC in a mineral salts medium. They eliminated competing pathways by chromosomal inactivation of genes encoding fumarate reductase (*frdABCD*), alcohol/aldehyde dehydrogenase (*adhE*), and pyruvate formate lyase (*pflB*). D-LAC production by these new strains approached the theoretical maximum yield of two molecules per glucose molecule and a chemical purity of D-LAC of ~98% with respect to soluble organic compounds. The cell yield and LAC productivity were increased by a further mutation in the acetate kinase gene (*ackA*). The aforementioned ME strategies used to engineer homolactic pathways in *E. coli* are summarized in Figure 5.6.

5.3.4 Acetic Acid

Acetic acid obtained through fermentation is mainly used in the food market (vinegar, meat preservative). Microorganisms currently used in its production are *Saccharomyces cerevisiae*, *Acetobacter aceti*, and *Clostridia* species. *E. coli* W3110 was recently engineered to produce acetic acid from glucose (45). The resulting strain (TC36) converted 60 g/L of glucose into 34 g/L of acetate in 18 h. Strain TC36 was constructed by sequentially assembling deletions that inactivated oxidative phosphorylation (Δ*atpFH*), disrupted the cyclic function of the tricarboxylic acid pathway (Δ*sucA*), and eliminated native fermentation pathways (Δ*focA-pflB*, Δ*frdBC*, Δ*ldhA*, and Δ*adhE*). These mutations minimized the loss of substrate carbon and the oxygen requirement for redox balance. Although TC36 produces only four ATPs per glucose, this strain grows well in a mineral salts medium and

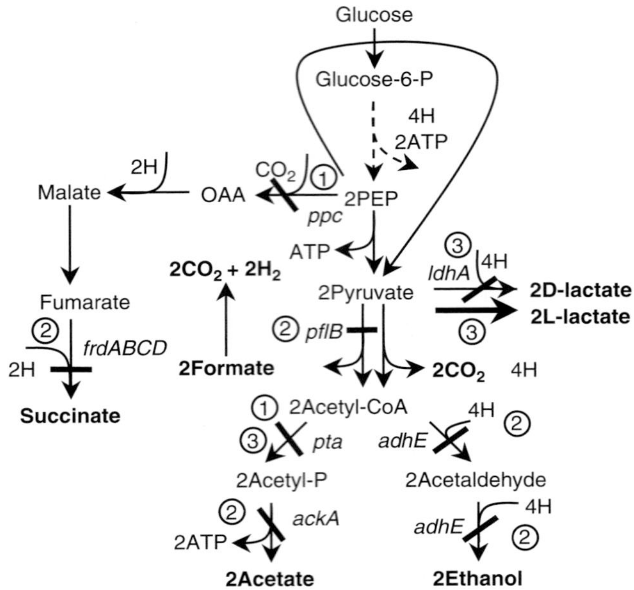

Figure 5.6 Metabolic engineering of *E. coli* to produce D- and L-lactate. See Figure 5.1 for explanation of different types of lines/arrows. Genes involved in engineered steps are indicated by italics. Primary fermentation products are indicated in bold. 1- Engineering a homo-D-lactic pathway by introducing mutations in *pta* and *ppc* genes; 2- Engineering a homo-D-lactic pathway by introducing mutations in *frdBC*, *ackA*, *adhE*, and *pflB* genes; 3- Engineering a homo-L-lactic pathway by introducing mutations in *pta* and *ldhA* genes and expressing an L-lactate dehydrogenase gene from *L. casei*. Abbreviations, *adhE*, gene encoding acetaldehyde/alcohol dehydrogenase; *ackA*, gene encoding acetate kinase; *frdABCD*, operon encoding fumarate reductase; *ldhA*, gene encoding lactate dehydrogenase; PEP, phosphoenolpyruvate; *pflB*, gene encoding pyruvate–formate lyase; *ppc*, gene encoding PEP carboxylase; *pta*, gene encoding phosphotransacetylase.

has no auxotrophic requirement. Glycolytic flux in TC36 (0.3 $\mu mol \cdot min^{-1} \cdot mg^{-1}$ protein) was twice that of the parent. Higher flux was attributed to a deletion of membrane-coupling subunits in $(F_1F_0)H^+$-ATP synthase that inactivated ATP synthesis while retaining cytoplasmic F_1-ATPase activity. The oxygen requirements of this system are energy intensive, so in terms of commercial production of acetate it would be highly desirable to have an anaerobic system able to efficiently produce acetic acid. Our laboratory is currently engineering *E. coli* to produce acetate as the main fermentation product under anaerobic conditions (data not published). The general strategy is based on first eliminating competing pathways (by inactivating correspondent genes) in order to redirect the carbon flux toward the synthesis of acetate, and then transforming the engineered pathway into a viable alternative for cell growth (by introducing modifications that will balance consumption and production of NADH).

5.4 VITAMINS

Vitamins are nutraceuticals or functional foods. They serve as the cofactors of many of the enzymes involved in several metabolic reactions, and so are essential components in the human diet. In this section I will give some examples of using ME to produce three vitamins (riboflavin, folate, and ascorbic acid) in bacterial strains.

5.4.1 Riboflavin (Vitamin B2)

Both *B. subtilis* and *Lac. lactics* have been engineered to produce riboflavin. In *B. subtilis* overexpression of the *rib* genes (involved in the synthesis of riboflavin), which are organized in a cluster (46), is achieved by replacing the two regulated promoters with constitutive ones derived from a phage (47). Although multiple copies of this four gene construction were inserted at two different sites in the genome, a separate overexpression of *ribA* was necessary to reach maximum productivity (48). The *ribA* gene encodes a bifunctional protein with dihydroxy–butanone phosphate synthetase activity at the N-terminal, and GTP cyclohydrolase II activity at the C-terminal. This showed that the initial steps of riboflavin synthesis limited productivity in the last published stage of strain improvement. A metabolic analysis of the wild-type and riboflavin-producing *B. subtilis* strains found that the flux of the oxidative branch of the pentose phosphate pathway was increased in the production strain (49). Therefore, this branch may become rate limiting and new ME strategies may be needed for future strain improvement. *Lac. lactis* is also being engineered to produce riboflavin by overexpressing the gene (*ribA*) encoding the enzyme (GTP cyclohydrolase) catalyzing the first reaction for its synthesis from GTP. This enzyme had been previously reported to limit flux (48) and overexpression of *ribA* resulted in a three-fold increase in riboflavin production (50).

5.4.2 Folate (Vitamin B11)

Lac. lactis has been engineered to produce folate (50). Strategies include the expression of several genes involved in the biosynthesis of folate from GTP, either individually or in combination, and the expression of heterologous γ-glutamyl hydrolase (from humans and rats). These approaches have resulted in three- to six-fold increases in the production of folate and an alteration in the folate spatial distribution (i.e., a shift from mainly intracellular to extracellular accumulation).

5.4.3 L-Ascorbic Acid (Vitamin C)

Almost all biological processes for synthesizing L-ascorbic acid end with the synthesis of 2-keto-L-gulonic acid (2-KLG), which is later converted into ascorbic acid by conventional chemical processing technology (i.e., esterification and lactonization). Therefore, efforts have been focused on engineering strains capable of efficiently producing 2-KLG from different sugars. Two strains of *Gluconobacter oxydans* were engineered to convert sorbitol into 2-KLG. Genes encoding the enzymes sorbitol dehydrogenase and sorbose dehydrogenase were cloned from *G. oxydans* T-100 into *G. oxydans* G624, allowing the production of 2-KLG in three steps (51). An *Erwinia herbicola* has been engineered to produce 2-KLG from glucose in a single fermentation step (52). *E. herbicola* naturally produces 2,5-diketo-D-gluconic acid (2,5-DKG) via glucose oxidation, but lacks the enzyme 2,5-DKG reductase, which can convert 2,5-DKG into 2-KLG. Therefore, *E. herbicola* was engineered by expressing the gene encoding 2,5-DKG reductase from *Corynebacterium* sp., resulting in a recombinant strain capable of converting glucose into 2-KLG in a single fermentation step. This process has been further simplified and 2-KLG concentrations > 120 g/L have been achieved (53).

5.5 CARBOHYDRATES

Several nondigestible polysaccharides have a positive effect on the development of beneficial intestinal microflora, and therefore are known by their function as prebiotics (i.e., nondigestible food ingredients that provide specific nutrients to beneficial bacteria hosted in the colon). Among them are fructo-oligosacharides, galacto-oligosacharides, gluco-oligosacharides, transgalacto-oligosacharides, isomalto-oligosacharides and xylo-oligosacharides (50). LABs are a good source of nondigestible polysaccharides, mainly exopolysacharides (EPS), which are also recognized by their contribution to the texture, mouth feel, taste perception, and stability of the final food product (54). Based on the available genetic information on genes required for EPSs biosynthesis in LABs, heterologous production of EPSs has been achieved in a LAB strain that did not have the capacity to produce them (55). A similar experiment has shown that EPSs with different compositions can be produced from expression of the same cluster of genes due to either selectivity in the export and polymerization system or limitation in the supply of precursors (56). Van Kranenburg et al. (57) have shown that EPS production in *Lac. lactis* can be increased by overexpressing the *epsD* gene, encoding a priming glucosyltransferase. I will give two other examples of using ME strategies to improve EPS production and modify EPS composition in *Lac. lactis*. Looijesteijn et al. (58) showed that overexpression of the *fbp* gene encoding fructose bisphosphatase resulted in increased *Lac. lactis* growth and EPS synthesis using fructose as a sugar source. In addition, Boels et al. (59) studied the UDP-glucose pyrophosphorylase (*galU*) and UDP-galactose epimerase (*galE*) genes of *Lac. lactis* MG1363 to investigate their involvement in the biosynthesis of UDP-glucose and UDP-galactose, which are precursors of glucose- and galactose-containing EPS in *Lac. lactis*. Overexpression of both genes increased the intracellular pools of UDP-glucose and UDP-galactose. These examples clearly illustrate the role ME plays (and will play in the future) in the bacterial production of polysaccharides.

5.6 BACTERIOCINS

One of the most important properties of a starter culture is its capacity to synthesize antimicrobial metabolites, including bacteriocins, which, in turn, can help improve the quality of fermented products (i.e., control of pathogens, extension of shelf life, and improvement of sensory qualities). For example, ability of LAB to act as a preservative in foods is partially attributed to the production of bacteriocins [for a review of bacteriocins produced by LAB with potentials as biopreservatives in foods see O'Sullivan et al. (60)]. Bacteriocins are a diverse group of ribosomally synthesized antimicrobial proteins or peptides, some of which undergo posttranslational modifications (Rodriguez et al. (61)). Bacteriocins can be incorporated into food after they are produced in an individual process or by using live cultures that produce bacteriocins *in situ* in the food. The second option seems to be more cost effective and acceptable (i.e., most strains have a GRAS status). A potential problem associated with the use of bacteriocins in food preservation is the development of resistant populations of pathogenic bacteria. ME plays a central role in solving this problem by enabling the creation of multiple bacteriocin producers (62,63). Horn et al. (64) took advantage of the similarities in the production and secretory systems of pediocin PA-1, from *Pediococcus acidilactici*, and lactococcin A, from *Lac. lactis* bv. diacetylactis WM4. By using the lactococcin A secretory system, they engineered a lactoccocus strain for expressing and secreting pediocin PA-1. This heterologous expression system has been used to coproduce enterocin-A and pediocin PA-1 (62), and pediocin PA-1 and nisin (63). Recently, O'Sullivan et al. (60) have examined the coproduction of lacticin 3147 and lacticin 481. Rodriguez et al. (61) recently published an excellent review of the heterologous production of bacteriocins by LAB. In the future, ME analytical tools (e.g., MFA) could play an important role in improving the heterologous production of bacteriocin in bacteria.

This assertion is based on its contribution to the study of the synthesis of heterologous proteins in yeast (65).

5.7 LOW CALORIE SUGARS

Low calorie sugars, like mannitol, sorbitol, tagatose, and trehalose, contribute to weight loss, which is one reason there is such a high consumer demand for them. Tagatose is a low calorie sweetener that is poorly degraded by the human body. It is considered both a prebiotic and an antiplaque agent (www.tagatose.dk). Recently, *Lac. lactis* has been engineered to produce tagatose by disrupting the *lacC* and/or *lacD* genes, which resulted in the production of either tagatose 6-phosphate or tagatose 1,6-diphosphate (66). Further genetic modifications resulted in the production of tagatose 1,6-diphosphate as the sole end product. Current efforts are focused on the dephosphorylation of tagatose 1,6-diphosphate and excretion of the final product, tagatose.

5.8 AROMA COMPOUNDS

Diacetyl is an important flavor compound in many dairy products like butter, buttermilk, and cheeses (it provides a buttery flavor). *Lac. lactis* has been engineered to synthesize diacetyl as the main fermentation product (67). Efficient diacetyl production resulted from inactivation of the *aldB* gene (encoding α-acetolactate decarboxylase) and overproduction of NADH-oxidase activity as illustrated in Figure 5.7. This approach resulted in the conversion of 80% of the carbon source into diacetyl. Another important group of aroma compounds is obtained from enzymatic degradation of amino acids including isovalerate, isobutyrate, phenylacetate, phenylacetaldehyde, and indole. Rijnen et al. (68) have shown that overexpression of a heterologous glutamate dehydrogenase (GDH) gene in *Lac. lactis*

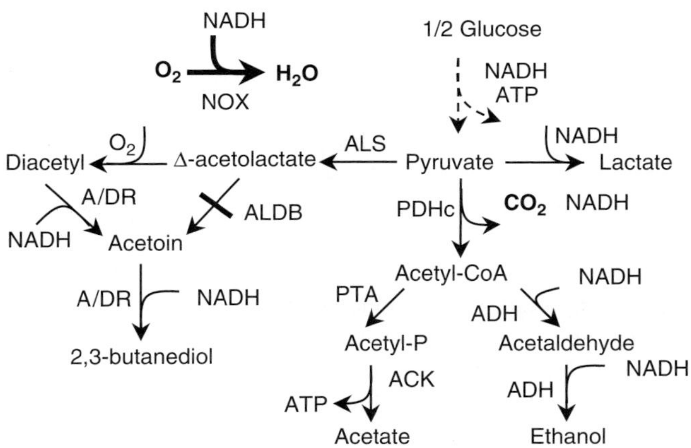

Figure 5.7 Metabolic engineering of *Lac. lactis* to produce diacetyl (ALDB-deficient and NOX overexpressed). Adapted from Hugenholtz et al. (67). See Figure 5.1 for explanation of different types of lines/arrows. Enzymes involved in reactions of interest are indicated. Abbreviations, ACK, acetate kinase; ADH, aldehyde/alcohol dehydrogenase; A/DR, acetoin/diacetyl reductase; ALDB, α-acetolactate decarboxylase; ALS, α-acetolactate synthase; NOX, NADH oxidase; PDHc, pyruvate dehydrogenase complex; PTA, phosphotransacetylase.

increased the conversion of amino acids into aroma compounds. Therefore, instead of adding α-ketoglutarate to cheese, a GDH-producing lactococcal strain could be used to enhance amino acid degradation in aroma compounds.

5.9 FUTURE TRENDS AND POTENTIALS

In the foreseeable future, ME will continue to play a central role in the use of bacteria to produce food ingredients. One potential area for the application of ME is the removal of undesirable sugars present in many foods. A typical example is lactose, which is present in liquid dairy products. Many people are lactose-intolerant and cannot consume products in which lactose is present. Engineering LAB to obtain starter cultures with elevated β-galactosidase activities has been proposed as a way to eliminate lactose (50). Raffinose is another example. Due to the absence of α-galactosidase, it cannot be degraded intestinally by humans. Consuming it causes intestinal disturbances. Engineering LAB to produce high levels of α-galactosidase and using it as starter cultures for removing raffinose has been proposed as a good solution to this problem (50). As a first step, the gene *melA* encoding α-galactosidase from *Lactobacillus plantarum* has been cloned (69) and is being expressed in *Lac. lactis* (50).

Energy metabolism is beginning to attract attention as a new target for ME, especially for engineering central metabolic pathways and transport systems. It was recently shown that an increase in ATP hydrolysis results in an increase in the glycolytic flux of *E. coli* (~70%), indicating that glycolytic flux is mainly (>75%) controlled by reactions hydrolyzing ATP (37). Therefore, one could engineer energy metabolism in order to improve the synthesis of products, such as organic acids, for which the glycolytic flux is a determinant (i.e., the introduction of an independent ATP "sink" would increase the glycolytic flux and the synthesis of the organic acid). One could even think that a similar approach will also be valid for synthesizing secondary metabolites whose precursors are provided by the glycolytic pathway. A second approach points toward a completely different direction, the creation of an energy surplus (in contrast to the energy "sink" created in the previous approach). For example, let us consider that the excretion of a certain product P (e.g., an amino acid) is an energy-requiring process. If a condition of energy surplus is created, cells will use any energy consuming process to dissipate the excess of energy, and therefore the secretion of product P will be stimulated.

The existence of sequenced genomes for several bacteria involved in producing food ingredients (e.g., *Lac. lactis, C. glutamicum, B. subtilis, E. coli*) will dramatically impact ME opportunities. Once genome sequences are available, technologies such as cDNA microarrays can be used to simultaneously monitor and study the expression levels of individual genes at genomic scale (known as the field of transcriptomics). This technology has already played an important role in understanding bacterial metabolism under a great variety of conditions, and in the improvement of numerous bacterial strains (70–74). However, the main contributions from the area of functional genomics are still to come and will result from the combined use of technologies that permit large-scale study of cell functioning and its interaction with the environment (i.e., combined analysis using genomic (DNA), transcriptomic (RNA), proteomic (protein), metabolomic (metabolites), and fluxomic (fluxes/enzymes) information). The era of functional genomics will contribute to an improved understanding of the molecular mechanisms underlying the relationships between microorganism and environment and in that way will establish the link between genotype and phenotype. Thus, in the future, functional genomics is expected to play a large role in the development of bacterial strains for the food industry.

REFERENCES

1. Bailey, J.E. Toward a science of metabolic engineering. *Science* 252:1668–1675, 1991.
2. Stephanopoulos, G., A.A. Aristidou, J. Nielsen. *Metabolic Engineering: Principles and Methodologies*, 1st ed. San Diego, CA: Academic Press, 1998.
3. Mollet, B., I. Rowland. Functional foods: at the frontier between food and pharma. *Curr. Opin. Biotech.* 13:483–485, 2002.
4. Mueller, U., S. Huebner. Economic aspects of amino acids production. *Adv. Biochem. Engin./Biotechnol.* 79:137–170, 2003.
5. Leuchtenberger, W. Amino acids: technical production and use. In: *Biotechnology*, Rehm, H.J., G. Reed, eds., Weinheim, Germany: VCH, 1996, pp 465–502.
6. Ikeda, M. Amino acid production processes. *Adv. Biochem. Engin./Biotechnol.* 79:1–35, 2003.
7. Vallino, J.J., G. Stephanopoulos. Metabolic flux distributions in *Corynebacterium glutamicum* during growth and lysine overproduction. *Biotechnol. Bioeng.* 41:633–646, 1993.
8. Marx, A., A.A. de Graaf, W. Wiechert, L. Eggeling, H. Sahm. Determination of the fluxes in the central metabolism of *Corynebacterium glutamicum* by nuclear magnetic resonance spectroscopy combined with metabolite balancing. *Biotechnol. Bioeng.* 49:111–129, 1996.
9. Ikeda, M., R. Katsumata. Hyperproduction of tryptophan by *Corynebacterium glutamicum* with the modified pentose phosphate pathway. *Appl. Environ. Microbiol.* 65:2497–2502, 1999.
10. Ikeda, M., K. Okamoto, T. Nakano, N. Kamada. Production of metabolite biosynthesized through phosphoribosylpyrophosphoric acid. Japan Patent 12,014,396 A, 2000.
11. Ozaki, A., R. Katsumata, T. Oka, A. Furuya. Cloning of the genes concerned in phenylalanine biosynthesis in *Corynebacterium glutamicum* and its application to breeding of a phenylalanine producing strain. *Agric. Biol. Chem.* 49:2925–2930, 1985.
12. Ikeda, M., A. Ozaki, R. Katsumata. Phenylalanine production by metabolically engineered *Corynebacterium glutamicum* with the pheA gene of *Escherichia coli*. *App. Microb. Biotechnol.* 39:18–323, 1993.
13. Guillouet, S., A.A. Rodal, G. An, P.A. Lessard, A.J. Sinskey. Expression of the *Escherichia coli* catabolic threonine dehydratase in *Corynebacterium glutamicum* and its effect on isoleucine production. *Appl. Environ. Microbiol.* 65:3100–3107, 1999.
14. Ikeda, M., R. Katsumata. Metabolic engineering to produce tyrosine or phenylalanine in a tryptophan producing *Corynebacterium glutamicum* strain. *App. Environ. Microbiol.* 58:781–785, 1992.
15. Katsumata, R., T. Mizukami, Y. Kikuchi, K. Kino. Threonine production by the lysine producing strain of *Corynebacterium glutamicum* with amplified threonine biosynthetic operon. In: *Genetics of Industrial Microorganisms. Proceedings of the Fifth International Symposium on the Genetics of Industrial Microorganisms.* Alacevic, M., D. Hranueli, Z. Toman, eds., Karlovac, Yugoslavia: Ognjen Prica Printing Works, 1987, pp 217–250.
16. Ikeda, M., K. Nakanishi, K. Kino, R Katsumata. Fermentative production of tryptophan by a stable recombinant strain of *Corynebacterium glutamicum* with a modified serine-biosynthethic pathway. *Biosci. Biotechnol. Biochem.* 58:674–678, 1994.
17. Ikeda, M., R. Katsumata. Tryptophan production by transport mutants of *Corynebacterium glutamicum*. *Biosci. Biotechnol. Biochem.* 59:1600–1602, 1995.
18. Kimura, E. Metabolic Engineering of glutamate production. *Adv. Biochem. Engin./Biotechnol.* 79:37–57, 2003.
19. Pfefferle, W., B. Möckel, B. Bathe, A. Marx. Biotechnological manufacture of lysine. *Adv. Biochem. Engin./Biotechnol.* 79:59–112, 2003.
20. Marx, A., B.J. Eikmanns, H. Sahm, A.A. de Graaf, L. Eggeling. Response of the central metabolism in *Corynebacterium glutamicum* to the use of an NADH-dependent glutamate dehydrogenase. *Metabolic Eng.* 1:35–48, 1999.
21. Hols, P., M. Kleerebezem, A.N. Schanck, T. Ferain, J. Hugenholtz, J. Delcour, W.M. de Vos. Conversion of *Lactococcus lactis* from homolactic to homoalanine fermentation through metabolic engineering. *Nat. Biotechnol.* 17:588–592, 1999.

22. Clark, D.P., The fermentation pathways of *Escherichia coli. FEMS Microbiol. Rev.* 63:223-234, 1989.
23. Millard, C.S., Y.P. Chao, J.C. Liao, M.I. Donnelly. Enhanced production of succinic acid by overexpression of phosphoenolpyruvate carboxylase in *Escherichia coli. Appl. Environ. Microbiol.* 62:1808–1810, 1996.
24. Gokarn, R.R., E. Altman, M.A. Eiteman. Metabolic analysis of *Escherichia coli* glucose fermentation in presence of pyruvate carboxylase. *Biotechnol. Lett.* 20:795–798, 1998.
25. Stols, L., M.I. Donnelly. Production of succinic acid through overexpression of NAD^+-dependent malic enzyme in an *Escherichia coli* mutant. *Appl. Environ. Microbiol.* 63:2695–2701, 1997.
26. Stols, L., G. Kulkarni, B.G. Harris, M.I. Donnelly. Expression of *Ascaris suum* malic enzyme in a mutant *Escherichia coli* allows production of succinic acid from glucose. *Appl. Biochem. Biotechnol.* 63–65:153–158, 1997.
27. Boernke, W.E., C.S. Millard, P.W. Stevens, S.N. Kakar, F.J. Stevens, M.I. Donnelly. Stringency of substrate specificity of *Escherichia coli* malate dehydrogenase. *Arch. Biochem.* 322:43–52, 1995.
28. Donnelly, M.I., C.S. Millard, D.P. Clark, M.J. Chen, J.W. Rathke. A novel fermentation pathway in an *Escherichia coli* mutant producing succinic acid, acetic acid, and ethanol. *Appl. Biochem. Biotechnol.* 70–72:187–198, 1998.
29. Nghiem, N.P., M. Donnelly, C.S. Millard, L. Stols. Method for the production of dicarboxylic acids. U.S. patent 5,869,301, 1999.
30. Chatterjee, R.C., S. Millard, K. Champion, D.P. Clark, M.I. Donnelly. Mutation of the *ptsG* gene results in increased production of succinate in fermentation of glucose by *Escherichia coli. Appl. Environ. Microbiol.* 67:148–154, 2001.
31. Gokarn, R.R., J.D. Evans, J.R. Walker, S.A. Martin, M.A. Eiteman, E. Altman. The physiological effects and metabolic alterations caused by the expression of *Rhizobium etli* pyruvate carboxylase in *Escherichia coli. Appl. Microbiol. Biotechnol.* 56:188–195, 2001.
32. Gokarn, R.R., M.A. Eiteman, E. Altman. Metabolic analysis of *Escherichia coli* in the presence and absence of the carboxylating enzymes phosphoenolpyruvate carboxylase and pyruvate carboxylase. *Appl. Environ. Microbiol.* 66:1844–1850, 2000.
33. Li,Y., J. Chen, S.-Y. Lun. Biotechnological production of pyruvic acid. *Appl. Microbiol. Biotechnol.* 57:451–459, 2001.
34. Yokota, A., H. Shimizu, Y. Terasawa, N. Takaoka. Pyruvic acid production by a lipoic acid auxotroph of *Escherichia coli* W1485. *Appl. Microbiol. Biotechnol.* 41:638–643, 1994.
35. Yokota, A., Y. Terasawa, N. Takaoka, H. Shimizu, F. Tomita. Pyruvate production by an F_1-ATPase-defective mutant of *Escherichia coli. Biosci. Biotechnol. Biochem.* 58:2164–2167, 1994.
36. Yokota, A., M. Henmi, N. Takaoka, C. Hayashi, Y. Takezawa, Y. Fukumori, F. Tomita. Enhancement of glucose metabolism in a pyruvic acid-hyperproducing *Escherichia coli* mutant defective in F_1-ATPase activity. *J. Ferment. Bioeng.* 83:132–138, 1997.
37. Koebmann, B.J., H.V. Westerhoff, J.L. Snoep, D. Nilsson, P.R. Jensen. The glycolytic flux in *Escherichia coli* is controlled by the demand for ATP. *J. Bacteriol.* 184:3909–3916, 2002.
38. Benthin, S., J. Villadsen. Production of optically pure D-lactate by Lactobacillus bulgaricus and purification by crystallization and liquid/liquid extraction. Appl Microbiol Biotechnol 42:826–829, 1995.
39. Vickroy, T.B., Lactic acid. In: *Comprehensive Biotechnology: The Principles, Applications, and Regulations of Biotechnology in Industry, Agriculture and Medicine,* Vol. 2. Moo-Young, M., ed., New York: Pergamon Press, 1985, pp 761–776.
40. Stanier, R.Y., J.L. Ingraham, M.L. Wheelis, P.R. Painter. *The Microbial World*, 5th ed., Englewood Cliffs, NJ: Prentice-Hall, 1986, pp 495–504.
41. Chang, D.E., H.C. Jung, J.S. Rhee, J.G. Pan. Homofermentative production of D- or L-lactate in metabolically engineered *Escherichia coli* RR1. *Appl. Environ. Microbiol.* 65:1384–1389, 1999.

42. Zhou, S., T.B. Causey, A. Hasona, K.T. Shanmugam, L.O. Ingram. Production of optically pure D-lactic acid in mineral salts medium by metabolically engineered *Escherichia coli* W3110. *Appl. Environ. Microbiol.* 69:399–407, 2002.
43. Skory, C.D. Isolation and expression of lactate dehydrogenase genes from *Rhizopus oryzae*. *Appl. Environ. Microbiol.* 66:2343–2348, 2000.
44. Ohara, H., H. Okuyama, S. Sawa, Y. Fujii, K. Hiyama. Development of industrial production of high molecular weight poly-L-lactate from renewable resources. *Nippon Kagaku Kaishi* 6:323–331, 2001.
45. Causey, T.B., S. Zhou, K.T. Shanmugam, L.O. Ingram. Inaugural article: engineering the metabolism of *Escherichia coli* W3110 for the conversion of sugar to redox-neutral and oxidized products: homoacetate production. *Proc. Natl. Acad. Sci. USA* 100:825–832, 2003.
46. Mironov, V.N., A.S. Kraev, M.L. Chikindas, B.K. Chernov, A.I. Stepanov, K.G. Skriabin. Functional organization of the riboflavin biosynthesis operon from *Bacillus subtilis* SHgw. *Mol. Gen. Genet.* 242:201–208, 1994.
47. Perkins, J.B., A. Sloma, T. Hermann, K. Theriault, E. Zachgo, T. Erdenberger, N. Hannett, N.P. Chatterjee, V. Williams, G.A. Rufo, R. Hatch, J. Pero. Genetic engineering of *Bacillus subtilis* for the commercial production of riboflavin. *J. Ind. Microbiol. Biotechnol.* 22:8–18, 1999.
48. Humbelin, M., V. Griesser, T. Keller, W. Schurter, M. Haiker, H.-P. Hohmann, H. Ritz, G. Richter, A. Bacher, A.P.G.M. van Loon. GTP cyclohydrolase II and 3,4-dihydroxy-2-butanone 4-phosphate synthase are the rate-limiting enzymes in riboflavin synthesis of an industrial *Bacillus subtilis* strain used for riboflavin production. *J. Ind. Microbiol. Biotechnol.* 22:1–7, 1999.
49. Sauer, U., V. Hatzimankatis, H.P. Hohmann, M. Manneberg, A.P.G.M. van Loon, J.E. Bailey. Physiology and metabolic fluxes of the wild-type and riboflavin-producing *Bacillus subtilis*. *Appl. Environ. Microbiol.* 62:3687–3696, 1996.
50. Hugenholtz, J., W. Sybesma, M.N. Groot, W. Wisselink, V. Ladero, K. Burgess, D. van Sinderen, J. Piard, G. Eggink, E.J. Smid, G. Savoy, F. Sesma, T. Jansen, P. Hols, M. Kleerebezem. Metabolic engineering of lactic acid bacteria for the production of nutraceuticals. *Antonie van Leeuwenhoek* 82:217–235, 2002.
51. Saito, Y., Y. Ishii, H. Hayashi, Y. Imao, T. Akashi, K. Yoshikawa, Y. Noguchi, S. Soeda, M. Yoshida, M. Niwa, J. Hosoda, K. Shimomura. Cloning of genes coding for L-sorbose and L-sorbosone dehydrogenases from *Gluconobacter oxydans* and microbial production of 2-keto-L-gulonate, a precursor of L-ascorbic acid, in a recombinant *G. oxydans* strain. *Appl. Environ. Microbiol.* 63:454–460, 1997.
52. Anderson, S., C.B. Marks, R. Lazarus, J. Miller, S. Stafford, J. Seymour, D. Light, W. Rastetter, D. Estell. Production of 2-keto-L-gulonate, an intermediate in L-Ascorbate synthesis by a genetically modified *Erwinia herbicola*. *Science* 230:144–149, 1985.
53. Chotani, G., T. Dodge, A. Hsu, M. Kumar, R. LaDuca, D. Trimbur, W. Weyler, K. Sanford. The commercial production of chemicals using pathway engineering. *Biochim. Biophys. Acta* 154:3434–455, 2000.
54. Jolly, J., J.F.V. Sebastien, P. Duboc, J.-R. Neeser. Exploiting exopolysaccharides from lactic acid bacteria. *Antonie van Leeuwenhoek* 82:367–374, 2002.
55. Germond, J.E., M. Delley, N. D'Amico, S.J. Vincent. Heterologous expression and characterization of the exopolysaccharide from *Streptococcus thermophilus* Sfi39. *Eur. J. Biochem.* 268:5149–56, 2001.
56. Stingele, F., J.R. Neeser, B. Mollet. Identification and characterization of the eps (exopolysaccharide) gene cluster from *Streptococcus thermophilus* Sfi6. *J. Bacteriol.* 178:1680–90, 1996.
57. van Kranenburg, R., I.C. Boels, M. Kleerebezem, W.M. de Vos. Genetics and engineering of microbial exopolysaccharides for food: approaches for the production of existing and novel polysaccharides. *Curr. Opin. Biotechnol.* 10:498–504, 1999.
58. Looijesteijn, P.J., I.C. Boels, M. Kleerebezem, J. Hugenholtz. Regulation of exopolysaccharide production by *Lactococcus lactis* subsp. cremoris by the sugar source. *Appl. Environ. Microbiol.* 65:5003–5008, 1999.

59. Boels, I.C., R. van Kranenburg, J. Hugenholtz, M. Kleerebezem, W.M. de Vos. Sugar catabolism and its impact on the biosynthesis and engineering of exopolysaccharide production in lactic acid bacteria. *Int. Dairy J.* 11:723–732, 2001.
60. O'Sullivan, L., R.P. Ross, C. Hill. Potential of bacteriocin-producing lactic acid bacteria for improvements in food safety and quality. *Biochimie* 84:593–604, 2002.
61. Rodriguez, J.M., M.I. Martinez, N. Horn, H.M. Dodd. Heterologous production of bacteriocins by lactic acid bacteria. *Int. J. Food Microbiol.* 80:101–116, 2002.
62. Martinez, J.M., J. Kok, J.W. Sanders, P.E. Hernandez. Heterologous coproduction of Enterocin A and Pediocin PA-1 by *Lactococcus lactis*: detection by specific peptide-directed antibodies. *Appl. Environ. Microbiol.* 66:3543–3549, 2000.
63. Horn, N., M.I. Martinez, J.M. Martinez, P.E. Hernandez, M.J. Gasson, J.M. Rodriguez, H. Dodd. Enhanced production of pediocin PA-1 and coproduction of nisin and pediocin PA-1 by *Lactococcus lactis*. *Appl. Environ. Microbiol.* 65:4443–4450, 1999.
64. Horn, N., M.I. Martinez, J.M. Martinez, P.E. Hernandez, M.J. Gasson, J.M. Rodriguez, H. Dodd. Production of pediocin PA-1 by *Lactococcus lactis* using the lactococcin A secretory apparatus. *Appl. Environ. Microbiol.* 64:818–823, 1998.
65. Gonzalez, R., B.A. Andrews, J. Molitor, J.A. Asenjo. Metabolic analysis of the synthesis of high levels of intracellular human SOD in *S. cerevisiae* rhSOD 2060 411 SGA122. *Biotechnol. Bioeng.* 82:152–169, 2003.
66. Hugenholtz, J., E.J. Smid. Nutraceutical production with food-grade microorganisms. *Curr. Opin. Biotechnol.* 13:497–507, 2002.
67. Hugenholtz, J., M. Kleerebezem, M. Starrenburg, J. Delcour, W. de Vos, P. Hols. *Lactococcus lactis* as a cell factory for high-level diacetyl production. *Appl. Environ. Microbiol.* 66:4112–4114, 2000.
68. Rijnen, L., P. Courtin, J.C. Gripon, M. Yvon. Expression of a heterologous glutamate dehydrogenase gene in *Lactococcus lactis* highly improves the conversion of amino acids to aroma compounds. *Appl. Environ. Microbiol.* 66:1354–1359, 2000.
69. Silvestroni, A., C. Connes, F. Sesma, G.S. De Giori, J.C. Piard. Characterization of the melA locus for alpha-galactosidase in *Lactobacillus plantarum*. *Appl. Environ. Microbiol.* 68:5464–5471, 2002.
70. Rhodius, V., T.K. Van Dyk, C. Gross, R.A. LaRossa. Impact of genomic technologies on studies of bacterial gene expression. *Ann. Rev. Microbiol.* 56: 599–624, 2002.
71. Tao, H., R. Gonzalez, A. Martinez, M. Rodriguez, L.O. Ingram, J.F. Preston, K.T. Shanmugam. Engineering a homo-ethanol pathway in *Escherichia coli*: increased glycolytic flux and levels of expression of glycolytic genes during xylose fermentation. *J. Bacteriol.* 183:2979–2988, 2001.
72. Gonzalez, R., H. Tao, S.W. York, K.T. Shanmugam, L.O. Ingram. Global gene expression differences associated with changes in glycolytic flux and growth rate in *Escherichia coli* during the fermentation of glucose and xylose. *Biotechnol. Prog.* 18:6–20, 2002.
73. Gonzalez, R., H. Tao, J.E. Purvis, S.W. York, K.T. Shanmugam, L.O. Ingram. Gene array-based identification of changes that contribute to ethanol tolerance in ethanologenic *Escherichia coli*: comparison of KO11 (parent) to LY01 (resistant mutant). *Biotechnol. Prog.* 19:612–623, 2003.
74. Dharmadi, Y., R. Gonzalez DNA Microarrays: experimental issues, data analysis, and application to bacterial systems. *Biotechnol. Prog.* 20:1309–1324, 2004.

1.06

Technologies Used for Microbial Production of Food Ingredients

Anthony L. Pometto III and Ali Demirci

CONTENTS

6.1 INTRODUCTION

The goal of this chapter is to supplement the industrial microbiology components in the first edition of *Food Biotechnology* by Knorr (1) and to present an updated overview of various technologies currently under investigation and employed by the food industry for the production of microbial food ingredients. For the detailed description of industrial microbiology principals, Demain and Davies *Manual of Industrial Microbiology and Biotechnology*, 2nd edition (2) is highly recommended.

For the food industry, the bottom line is cost. They are selling commodity products (food) with various levels of preconsumer processing. The goal is always to produce the most nutritious and safe product at the lowest possible cost.

For centuries microorganisms have been employed for the production of fermented food products (i.e., cheese, soy sauce, sauerkraut, wine, and bread). The consumption of some live microbial cultures (probiotics) has proven to provide a health benefit to humans and animals (3,4). Some foods containing live cultures are yogurt, buttermilk, and acidophilus milk. These microbial fermented food products also have an extended shelf-life compared to the perishable starting raw material. Thus, microorganisms not only provide a nutritional benefit to humans but act to extend the shelf-life of the food supply.

Microorganisms employed by the food industry include bacteria, yeasts, and molds. These microorganisms have several morphological and physiological differences. Morphologically bacteria are small and difficult to remove, yeasts are larger and will sometime settle out of solution, whereas molds are filamentous and are typically removed by filtration. Physiologically they differ in pH preferences (yeasts and molds prefer a lower pH than bacteria), nutrient requirements (different concentrations and types of nitrogen and other trace elements), growth rates (bacteria grow much faster than yeasts and molds), and more. Thus, different culture media, fermentation methods, and product recovery methods are required depending on the microbial system being cultured.

6.2 MICROORGANISM SELECTION AND DEVELOPMENT

Microorganisms are the biocatalysts that produce and maintain a host of enzymatic pathways that are used to produce the food component of interest. The characteristics of a good industrial microorganism for the production of food ingredients are (1) it must be effective in producing large quantities of a single product, (2) it can be efficiently isolated and purified, (3) it is easy to maintain and cultivate, (4) it is genetically stable, (5) it grows best in an inexpensive culture medium, and (6) it is safe for human consumption. The first step is to isolate the hardiest starter culture possible, then to begin strain improvements via classical mutagenesis or genetic engineering.

A classic example would be the production of L-phenylalanine for the artificial sweetener aspartame (NutraSweet® J.W. Childs Equity Partners II L.P.), which is a dipeptide of L-phenylalanine and L-aspartic acid. When NutraSweet first entered the market in 1981, the L-phenylalanine supply became the bottleneck for production. L-Phenylalanine, L-tyrosine, and L-tryptophan are produced via the shikimic acid pathway in all organisms. To develop a bacterium which over produced L-phenylalanine, first classical chemical mutagenesis of an L-tyrosine auxotroph of *Corynebacterium glutamicum* was employed using -phenylalanine analog resistance in an effort to reduce end product inhibition, and L-tyrosine production (5). Analogs such as *p*-aminophenylalanine, *p*- and *m*-fluorophenylalanine, and β-2-thienylalanine were incorporated into the cellular protein thus poisoning the cell. To combat this poison, surviving mutants must overproduce L-phenylalanine, thus neutralizing the toxic effects of the analogs. This process was repeated several times with mutants resistant to increasing concentrations of analogs. The final analog resistant bacterium selected by Hagino and Nakayama (5) produced 9.5 g/L of L-phenylalanine.

The over producing bacterium was then transformed with plasmids containing L-phenylalanine analog resistant chorismate mutase and prephenate dehydratase genes (6). These are two key enzymes in the shikimic acid pathway for L-phenylalanine production. Except for constitutive enzymes, most enzymes in the cell have a short half life in the cell. Thus, an increase in key enzyme concentrations and residence time in the cell will also increase production. Ozaki et al. (6) transformates produced 19.0 g/L of -Lphenylalanine, thus, illustrating how classical mutagenesis and molecular genetic techniques are employed to further enhance production of some desired metabolites for the food industry.

Another alternative method was whole bioconversion developed by Yamada et al. (7) which produced L-phenylalanine from trans-cinnamic acid via L-phenylalanine ammonia lyase (PAL) reversal in *Rhodotorula glutinis*. In the presence of concentrated ammonium hydroxide, the PAL reversal demonstrated a 70% conversion yield which produced 17.5 g/L of L-phenylalanine. By utilizing a whole cell bioconversion process, no enzyme purification step was needed and the enzyme proved to be more stable within the yeast under the harsh conditions employed. This process was used to produce some of the initial L-phenylalanine used for the production of NutraSweet. Eventually, however, production by the genetically engineered bacterium exceeded the levels in fed-batch fermentation, which did not involve caustic chemicals, and thus, became the method of choice. L-Phenylalanine purification is performed by ion exchange chromatography for all methods.

The genetic stability of cultures requires minimum culture transfers and long term storage capabilities. Fermentation media are inoculated from working cultures which are produced every few months from master cultures depending on the microorganism. The most common procedures for long term storage are freeze drying ($<-18^{\circ}$C) and ultra-low temperature storage ($-70 - -80$°C). Freeze drying requires a cryoprotectant, such as sterile skim milk, followed by freeze drying and vial sealing under a vacuum (2). Sealing under nitrogen gas can also help to stabilize the culture and extend the shelf life. Ultra-low temperature storage is in a rich culture medium with 20% sterile glycerol. Some cultures are sensitive to freeze drying, thus, ultra-low is the most common method employed today, because of long-term culture viability. The risk is loss of electrical power and refrigeration problems.

Suspended cell cultures or spore suspensions are used as inocula for these industrial scale fermentations. Purity is constantly checked until inoculation. For suspended cell inoculation the sequence employed would be culture slant, shake-flask culture, benchtop fermentor, pilot-scale fermentation, then into full scale fermentation. Many fungal fermentations, such as citric acid and soy sauce fermentations, required a suspension of viable fungal spores as the inoculum. These spore suspensions are generated on large agar trays, and then are aseptically transferred into culture bottles suspended in sterile water or saline (2).

Microbial systems frequently constitute efficient mechanisms for the production of nutritionally important ingredients at a relatively low cost, for example, the production of selenomethione in yeast. A slight modification in the yeasts culture medium will force the yeast to substitute the sulfur group in methionine and cysteine with a selenium in standard fed-batch fermentation (8,9). To identify this medium change, Demirci and Pometto (8) developed a gradient delivery unit producing a gradient of sodium selenite or sodium selenate in a continuous bioreactor (Figure 6.1). It has been shown that selenium has several health benefits (10) including a cancer-protective effect (11), and a profound effect on the survival of HIV-infected patients (12).

Furthermore, microbial systems are ideal for the production of essential micronutrients such as amino acids, vitamins, and enzymes, and bulk ingredients such as organic acids and alcohols, whole cell flavor enhancers, and polysaccharides.

6.3 CULTURE MEDIA AND UPSTREAM COMPONENTS

The ideal culture medium will use inexpensive components to supply their complex nutrient requirements. Miller and Churchill (13) provide an excellent summary of inexpensive media components and their makeup. These ingredients are crop, animal, marine or yeast based components. The culture medium alone can represent 30 to 70% of the fermentation production costs. Slight changes in medium micronutrients can have a major impact on the fermentation (14,15). Thus, the food industry demands a consistent product from suppliers of these complex components. Failure to provide a consistent product will eliminate the commercial use. What decides the culture medium makeup? Essentially, it is the nutritional requirements of the microorganism of choice and its ability to biosynthesize essential elements such as amino acids, vitamins, lipids, and carbohydrates. For example, bacteria and yeast are high in protein (40–50%), whereas molds are not (10–25%). Yeast

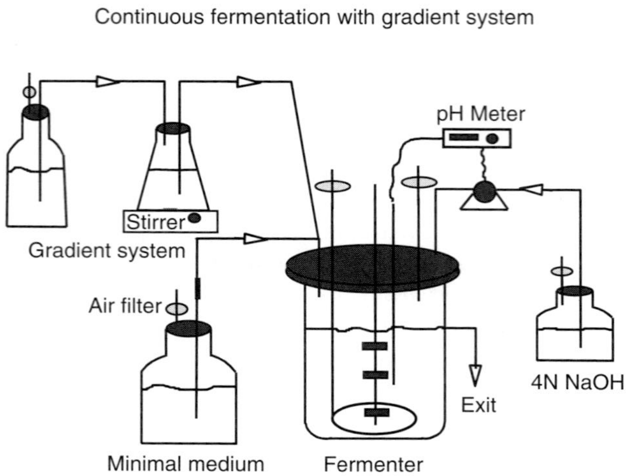

Figure 6.1 Continuous bioreactor gradient delivery system design used for developing the production protocols for selenomethionine production in yeast.

can generally grow in a minimal medium, whereas lactic acid bacteria, essential for the fermented dairy products, require a host of micronutrients to grow.

Each microorganism has different micronutrient requirements. Typically the exact micronutrient which benefits the fermentation that is being supplied by the complex nutrient employed is unknown. Generally it involves specific amino acids, vitamins, trace elements, and lipids. Furthermore, the concentration ratio of carbohydrate to nitrogen and phosphorous has a dramatic impact on microbial growth. Carbon–nitrogen imbalances can result in the production of other byproducts such as extracellular polysaccharide, and fermentation end products such as ethanol. In the case of ethanol production in yeast, excess glucose in the presence of oxygen will direct yeast to produce ethanol. This is called the Crabtree effect (16,17). Yeast cells typically consist of 48% carbon and 8% nitrogen on a dry weight basis. When the C:N ratio is 10:1, yeasts grow aerobically consuming little substrate while producing maximum cell mass, CO_2 and H_2O, but when the C:N ratio is 50:1 yeasts grow anaerobically, consuming much substrate while producing little cell mass, and much CO_2 and ethanol. This difference in yeast growth is also called the Pasteur effect (18).

Why the increase in industrial microbiology fermentation processes over the past 20 years? It is the result of the corn syrup sweetener industry and computer technology. Corn is 70% starch, and when dried to <19%, it can be stored for more than two years. Annually 10 billion bushels of corn are produced in the Unites States of America. Thus, the liquefaction of corn starch to glucose syrups for the production of high fructose corn sweeteners represents a consistent, low cost supply of substrate for most industrial microbiology fermentations. Not only is glucose produced, but customized substrates can be also produced. For example, corn syrups containing 19% dextrose, 14% maltose, 12% maltotriose, and 55% higher saccharides are used to control microbial growth rates and biological heat production in many fermentations. Thus, glucose is the platform chemical used for the microbial production of organic acids, amino acids, vitamins, and more. In fact some food grade fermentation facilities have located adjacent to a corn sweetener facility to permit glucose syrup to be piped directly to their fermentors. Glucose corn syrup is also shipped via truck or rail cars as liquid or dried product.

Furthermore, the advent of computer process controls of industrial scale fermentors has removed many of the fears associated with commercial scale fermentation by providing reliable and easy to operate dissolved oxygen, pH, foam, temperature, and sterilization controls of the process. Sirakaya et al. (19) described fermentation software to monitor and control the utilization fermentation process.

6.4 BIOREACTOR MONITORING SYSTEMS AND DESIGN

The stirred tank bioreactor design is the most common fermentor and consists of agitator, baffles, aeration sparger for aerobic culture growth, sterilizable monitoring probes for pH, dissolved oxygen, temperature, and antifoam, filling and draining ports, and often culture medium sterilization capabilities in the reactor tank. Reactor agitation is essential for temperature control, pH adjustments, oxygen absorption into the liquid medium, overall culture health, and mixing of any required additions of substrate, and nutrients. Typical commercial reactor working volumes for food grade ingredients are 5,000 to 40,000 gallons.

Fermentation health requires real time monitoring system. Microbial growth can be monitored via several methods. The most common method is indirect measurement of biomass by absorbance of the fermentation broth at 620 nm by using a spectrophotometer. The measured absorbance values can then be used to estimate biomass concentrations by using a standard curve. Standard curves are developed by collecting fresh log phase cells via centrifugation, washing cells with water or 0.1 M ammonium acetate pH 7.0 buffers, then serially diluting the pellet (20). Absorbance for each dilution is then determined spectrophotometrically at 620 nm. The actual dry weight biomass (g/l) is determined for each dilution via direct biomass measurement after oven drying of each dilution in preweighed boats at 70°C for 24–36 hr. This needs to be performed in at least replicates of three. By washing biomass with water, any influence of culture medium on dry weights can be eliminated or minimized. Finally, a standard curve can be constructed by plotting absorbance versus actual dry biomass weight (g/l). This method allows for quick, reliable, and easy conversion of absorbance to dry biomass (g/L).

For determining microbial health, viable cell counts can be rapidly performed by using an EPICS XL-MCL flow cytometer (Beckman-Coulter, Miami, FL) in conjunction with the Live/Dead BacLight™ bacterial viability test kit (Molecular Probes, Eugene, OR) (21). BacLight uses a mixture of Syto 9 fluorescence, which is measured as a log FL1 (525 nm) signal, and the propidium iodide fluorescence which is measured as a log FL4 (675 nm) signal. A two color histogram is collected with gating on the bacteria only population from the two parameter light scatter distribution and is used for the analysis of green only (live bacteria), red only (dead bacteria), and both colors (stressed bacteria).

Substrate consumption and product formation rates can be monitored by high pressure liquid chromatography (HPLC) or by membrane bound enzymes biosensors, which requires 20 and 1 min to run, respectively. HPLC analysis is time consuming, but the concentration of multiple metabolites can be monitor simultaneously. HPLC does not provide real time feedback on the health of the fermentation, and it has long sample preparation and run times. In contrast, membrane bound enzyme biosensors such as YSI 2700 select analyzer (Yellow Springs Instruments, Yellow Springs, OH) can analyze a sample in 1 min. However, these units are restricted by the availability of substrate specific oxidases which generate H_2O_2, the measurable product by their electrode. Some compounds currently measurable are glucose, ethanol, maltose, lactic acid, and lysine. Sample preparation is simply filtration (0.45 :m) and dilution with water if the value falls outside the instrument window.

Organic acid production can be continuously monitored via alkali addition rates for pH control. Alkali consumption can be easily monitored by feeding alkali solution from a sterile burette (22). Microbial respiration for aerobic cultures can be continuously monitored via dissolved oxygen probe or the concentration of CO_2 in the exit gas, which can be monitored via off gas analyzers or simply via alkali (4 N NaOH) traps followed by pH titration. However, more technologies are needed to acquire real time measurements of microbial growth, ensuring optimal fermentation time, and product formation in the shortest time possible.

For rapid analysis of biomass, substrate and product concentrations without any sample preparation, Fourier transform mid infrared (FT-MIR) spectroscopy has been successfully utilized for lactic acid (23) and ethanol (24) fermentation. Calibration models have been developed by using principle least square (PLS) and principle component regression (PCR) on suitable spectral wavenumber regions. This calibration model is then used for unknowns. The advantage of this system is that not only no sample preparation required, but it also provides analysis for substrate, product and biomass at the same time. This method can be used for online, real time analysis for monitoring and process control purposes.

For aerobic culture fermentation, house air is generally supplied under pressure. Oxygen transfer into the culture medium depends on the air bubble residence time in the culture medium and bubble size. The smaller the air bubbles the greater the O_2 transfer. Thus, all stirred tank reactors have aeration spargers that generate bubbles right beneath the first agitator blade. The exiting air bubble collides with the standard flat Rashton turbine agitator blade which strikes the bubbles hard and fast as they leave the sparger generating smaller air bubbles for improved oxygen transfer. Many commercial fermentations then follow up the agitator shaft with a series of down draft marine agitator blades, which look like a motor boat propeller (Figure 6.2). This series of down draft marine blades push the air bubble back down as it migrates up the reactor. This increases gas bubble residence time in the liquid medium before exiting out the top. In some fermentations, the rate limiting

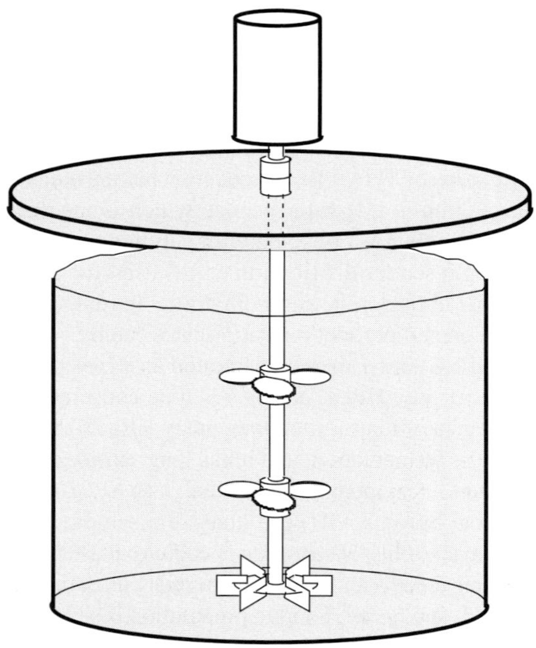

Figure 6.2 Diagram of stirred tank reactor with Rashton turbine agitator blade at the reactor bottom to break up air bubbles followed by a series of down draft marine blades to increase bubble residence time in the culture medium.

substrate will be oxygen. Thus, along with the series of down draft marine agitator blades a supply of pure oxygen may be needed. Pilot scale recombinant *Pichia* fermentations often require pure oxygen supplements to ensure optimal growth.

It is obvious that each microorganism differs in oxygen uptake rate (OUR). Therefore, oxygen transfer rates (OTR) (the rate of oxygen transfer from bubble to fermentation medium) must be equal to or higher than OUR. Maximum OTR can be calculated by using the following equation:

$$\text{OTR (mg } O_2\text{/min)} = k_L a \, C_L^* \, V_L$$

where C_L^* = Equilibrium dissolved oxygen level (mg O_2/L), k_La = Mass transfer coefficient (min−1), and V_L = Liquid volume in tank (L).

Thus, k_La is a critical parameter affecting OTR and the desired dissolved oxygen concentration in a fermentation broth during fermentation, and k_La can be determined experimentally. Briefly, the percentage dissolved oxygen (%DO) level is reduced to almost zero by sparging nitrogen into the culture medium. Then, air or oxygen gas is sparged at the desired temperature, agitation, and aeration conditions. Percent DO values are recorded over time until %DO reaches to saturation level. After converting %DO values into dissolved oxygen concentrations (C_L), plotting ln ($C_L^* - C_L$) versus time gives a straight line with a negative slope which is equal to "$-k_La$". By knowing k_La, one can calculate OTR with the given aeration and agitation. If OTR is less than OUR, some changes can be implemented to improve OTR, such as increasing aeration rate or agitation. Even utilization of various types of propellers can be compared by calculating OTR under each condition.

Agitation and aeration of stirred tank fermentors may also generate foam, which is the entrapment of gas in lipid, polysaccharide, or protein matrix (25). If not controlled, a foam buildup can literally empty the reactor. Mechanical foam breakers are like giant fans which physically break the foam and blow it back down. Mechanical foam breakers have their limitations and ultimately food grade antifoams are employed to control culture broth foam. A list of common antifoams can be found in Hall et al. (25).

Modification to this basic bioreactor design occurs from specific microbial needs. For example, if your production microorganism is sensitive to agitator shear then an air lift bioreactor is employed. Air lift bioreactors consists of a central or an external draft tube whereby air bubbles passing up these tubes generate convection mixing of the medium (26). Many fungal fermentations require the production of fungal balls for maximum product formation. These fungal balls are very sensitive to the shear caused by the agitator blades. For example, *Aspergillus niger* citric acid fermentation requires a defined medium with specific concentrations of trace elements (i.e., copper, manganese, magnesium, iron, zinc, and molybdenum). The bioreactor is typically lined with glass to prevent the addition of any stray trace elements. A fungal spore suspension is used as an inoculum, and mycelium, for optimal production rates, consists of very small solid pellets, or spheres which require a bioreactor with minimum shear. Throughout the entire fermentation period, the minimum dissolved oxygen concentration of 20–25% of saturation is required (27). Thus, tall air lift fermentors are the bioreactors of choice for these types of fermentations.

Biological heat can also be a problem. Bacteria, yeast, and molds will generate different levels of biological heat because their growth rates are so different. All bioreactors require some kind of jacketed cooling and heating system. Also, the time of year and location of the facility will also dictate the level of cooling required. Biological heat is directly related to growth rates. The faster the growth rate the more heat generated. Thus, a rate limiting substrate can be used to control microbial growth. One example is the use of a substrate containing mono, di, oligo, and poly saccharides. For example, a liquefied corn starch described above containing 19% dextrose, 14% maltose, 12% maltotriose, and 55%

higher saccharides is commonly used to control microbial growth rates. Enzymatic hydrolyses of the di, tri, and oligo saccharides will dictate the level of available glucose in the fermentor, and thus, control microbial growth rates.

6.5 FERMENTATION TYPES EMPLOYED: THE ACTUAL PRODUCTION PROCESS

The work horse of the industry is batch and fed-batch fermentation. Batch fermentations are closed fermentations (28). The fermentation sequence starts with medium sterilization, reactor inoculation [1 to 2% (v/v) typically], incubation for complete microbial growth cycle with lag, log, and stationary phases, fermentation termination, draining the reactor for product recovery down stream, cleanup of the reactor, and starting over. Percentage yield is calculated by taking the concentration of the product formed (g/l) divided by the concentration of substrate consumed (g/l) times 100%, whereas productivity is a measure of the product formation rates. It is calculated by dividing the product concentration (g/l) by the fermentation time (hours); thus, it is presented in g/l/hr.

In fed-batch fermentation, additional carbohydrate is supplied to the batch fermentation during the run (28). High carbohydrate concentrations in the initial culture medium are toxic to many microorganisms. Thus, an optimal carbohydrate concentration is employed initially, which permits maximum culture growth to late log phase. When the carbohydrate concentration is reduced to almost zero, additional sterile carbohydrate is injected into the bioreactor to bring the carbohydrate concentration back to the starting level. When this is consumed, the process is repeated until end product inhibition forces the whole bioconversion to stop. Ideally, at the end of the fermentation you will have a product concentration which is three to four times greater than single batch fermentation with no residual carbohydrate. This will generate the highest yield possible. Also, due to increased end product accumulation with each carbohydrate addition, culture production rates will decrease. Thus, the decision as to how many fed-batch fermentations to perform before harvest is based on the desired final product concentration and the optimal fermentor use time. For example, in lactic acid fermentations a final product concentration >120 g/l is desired to enhance product recovery (29). For *Lactobacillus casei* this concentration can only be achieved via fed batch fermentation for a total fermentor run time of eight days.

Continuous fermentations are open fermentations, whereby fresh medium is continually added to the bioreactor, while spent culture medium, cells, and product are continually leaving (28). This fermentation is desired by the industry, because the reactor volume is 10 to 100 times smaller than batch fermentations, a steady stream of fermentation product is produced which will optimize downstream processes, bioconversion rates are always at maximum, operation costs are less, and the system can be fully automated and computer controlled to the point where only two operators are needed to manage the fermentation each day. However, it requires a continuous supply of sterilized or pasteurized culture medium, dilution rates are linked to microbial growth rates and the operational speed of downstream recovery process. Startup is slow, so any facility shut downs have an impact on production, and you are constantly fighting contamination (30). Thus, not all fermentations can be operated this way. The best example is ethanol production for gasohol, which is commonly a continuous fermentation with a four bioreactor train with increasing working volumes in each bioreactor. Thus, the dilution rate is decreasing in each bioreactor over the course of the fermentation. This dilution gradient in the train is critical, because as ethanol builds up in the culture medium, the yeast growth rate slows. Specific growth rate equals dilution rate. Finally a holding tank at the end is used to ensure

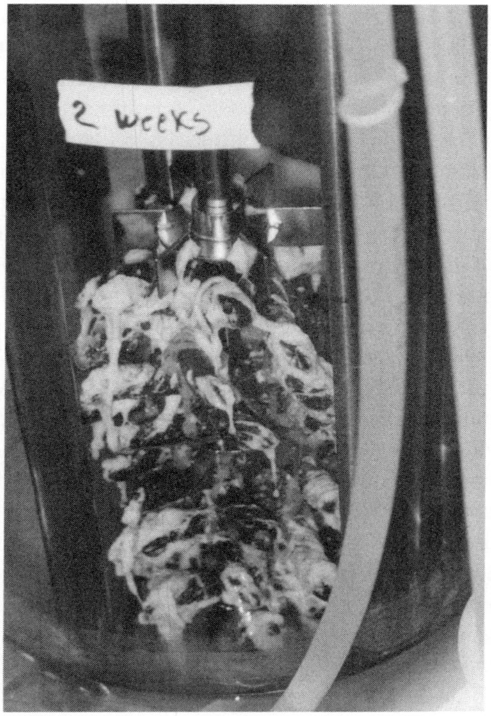

Figure 6.3 Example of *Lactobacillus casei* biofilm development on PCS tubes mounted on the agitator shaft for repeat batch fermentation. PCS blend employed contained 50% (w/w) polypropylene, 35% (w/w) ground soybean hulls, 5% (w/w) bovine albumin, 5% (w/w) Ardamine Z yeast extract, and 5% (w/w) soybean flour.

complete bioconversion of any residual sugars to ethanol. The CO_2 is collected and concentrated, then sold as another valuable byproduct. Fermented beverages (i.e., wine and beer) are still performed in batch.

For some continuous fermentations, an increased concentration of biomass in the reactor is required. This can be achieved by cell recycling, immobilized cell, and biofilm fermentations. Cell recycling reactors employ a filtration unit that allows for the constant bleeding of culture supernatant while retaining biomass (26). However, filtration unit fouling is a problem and must be constantly monitored. This type of operation has found use in the treatment of food processing from some starchy food waste streams (26).

6.6 NOVEL BIOREACTOR DESIGNS

One of the most common forms of immobilized cell bioreactor is entrapment, where high concentrations of cells are trapped in a polymer matrix such as alginate and κ-carrageenin (28,31). Thus, a high cell density is continuously retained in the fermentor while substrate is continuously converted to product. This higher concentration of biocatalysts in the reactor results in higher productivities and yields. The disadvantages of this method are migration of substrate through the matrix to the cell and the migration of product out, potentially high concentrations of product around the cells causing end product inhibition, cell leakage from the polymer matrix due to cell growth, and bead swelling and disintegration over time causing the whole fermentation to be stopped, cleaned, and restarted.

Biofilms are a natural form of cell immobilization in which microorganisms are attached to a solid surface (32). In this bioreactor, cells are continually growing, and sloughing off. Thus, the reactor is a mixture of immobilized and suspended cells. Continuous biofilm fermentations are truly open immobilized cell bioreactors (30). Their operation is equivalent to a suspended cell continuous fermentation with the added advantage of increase biomass concentrations in the bioreactor. Biofilms are typically resistant to harsh conditions, and can tolerate changes in the fermentation feed and conditions. However, not all microorganisms form biofilms. Filamentous microorganisms such as fungi and actinomycetes are natural biofilm formers. For nonfilamentous bacteria to form a biofilm, an extracellular polysaccharide needs to be generated by the bacterium (32).

Some bacteria will form biofilms on any surface such as metal, plastic, and glass. However, certain bacteria, such as lactobacilli, require something to stimulate this biofilm development (33). Plastic composite support (PCS) developed at Iowa State University has proven to stimulate biofilm development of *Lactobacillus casei* (22,34,36), *Zymomonas mobilis* (37,38), *Saccharomyces cerevisiae* (37,38,39), and *Actinobacillus succinogenes* (40). PCS is a high temperature extruded material consisting of at least 50% polypropylene, plus ground soybean hulls, bovine albumin and various culture micronutrients. Soybean hulls keep the extruded product porous due to the release of steam as the PCS leaves the extruder die. Bovine albumin performs as a natural plastizer which protects the temperature sensitive micronutrients. Micronutrients are selected based on the specific cultural requirements for amino acids, vitamins, and lipids. Monosaccharides are avoided due to poor PCS production. For example, the PCS blend for lactobacilli contains 50% (w/w) polypropylene, 35% (w/w) ground dried soybean hulls, 5% (w/w) bovine albumin, 5% (w/w) yeast extract, 5% (w/w) soybean flour, and mineral salts (35). PCS have been evaluated in batch (22), fed-batch (29), and continuous (30) lactic acid fermentations (Figure 6.3). In every application the percentage yields and productivity rates were significantly higher than suspended cell lactic acid fermentations. Furthermore, repeat batch fermentations have operated for more than 1.5 years with virtually no change in percentage yields and productivities. This longevity is attributed to the fact that once a biofilm has established on these customized materials, it will continue to grow as a biofilm. This is supported by the fact that a PCS biofilm reactor washed with concentrated ammonium hydroxide, rinsed with mineral salts solution, and then reinoculated with a fresh culture and medium will reestablish itself overnight. Commercially, the quick vinegar process is the most common biofilm process in current operation which uses wood chips for supports and *Acetobacter aceti* for production (27).

Solid substrate fermentation is when a substrate such as soybeans is ground, inoculated with *Aspergillus oryzae*, then incubated for three days for the production of soy sauce (27). It is a simple fermentation process and commonly used for aerobic fermentation due to its large surface area. Thus, oxygen concentration is high without using any mechanical forced air systems. Solid substrate fermentations require large areas or incubation space. A temperature controlled environment, intermittent monitoring for contamination and quality of starting material is essential for success.

6.7 FUTURE RESEARCH

Research is still needed for isolation of new microbial strains with improved production efficiencies and higher yields. More real time measurements are needed for culture conditions and metabolite formation. Recovery will continue to be the key factor associated with final product purity and cost. As an industry we cannot rely solely on genetic engineering as our method of improving current fermentations. As we have illustrated, there are many other techniques which can be employed to improve productivity and yield, including new inexpensive medium ingredients, more continuous fermentation processes, and new exotic microbial reservoirs in nature and in the food industry.

REFERENCES

1. Knorr, D. *Food Biotechnology*. New York: Marcel Dekker, 1987.
2. Demain, A.L., J.E. Davies. *Manual of industrial Microbiology and Biotechnology, 2nd ed.* Washington: ASM Press, 1999.
3. Hoover, D.G., L.R. Steenson. *Bacteriocins of Lactic Acid Bacteria*. San Diego: Academic Press, Inc., 1993.
4. Tannock, G.W. *Probiotics, A Critical Review*. Norfolk, England: Horizon Scientific Press, 1999.
5. Hagino, H., K. Nakayama. L-Phenylalanine production by analog-resistant mutants of *Corynebacterium glutamicum*. *J. Agric. Biol. Chem.* 38:157–161, 1974.
6. Ozaki, A., R. Katsumata, T. Oka, A. Furuya. Cloning of the genes concerned in phenylalanine biosynthesis in *Corynebacterium glutamicum* and its application to breeding of a phenylalanine producing strain. *J. Agric. Biol. Chem.* 49:2925–2930, 1985.
7. Yamada, S., K. Nabe, N. Izuo, K. Nakamichi, I. Chibata. Production of L-phenylalanine from trans-cinnamic acid with *Rhodotorula glutinis* containing L-phenylalanine ammonia lyase activity. *Appl. Environ. Microbiol.* 42:773–778, 1981.
8. Demirci, A., A.L. Pometto III. Production of organically bound selenium yeast by continuous fermentation. *J. Agric. Food Chem.* 47:2491–2495, 1999.
9. Demirci, A., A.L. Pometto III, D. Cox. Enhanced Organically Bound Selenium Yeast Production by Fed-Batch Fermentation. *J. Agric. Food Chem.* 47:2496–2500, 1999.
10. Burk, R.F. *Selenium in Biology and Human Health*. Heidelburg: Springer-Verlag, 1994.
11. Combs Jr., G.F. Selenium as a cancer-protective agent. In: *The bulletin of Selenium-Tellurium Development Association*. February, 1997, pp 1–4.
12. Bologna, R., F. Indacochea, G. Shor-Posner, E. Mantero-Atienza, M. Grazziutii, M. Sotomayor, M.A. Fletcher, C. Cabrejos, G.B. Scott, M.K. Baum. Selenium and immunity in HIV-1 infected pediatric patients. *J. Nutr. Immunol.* 3:41–49, 1994.
13. Miller, T.L., B.W. Churchill. Substrates for large-scale fermentations. In: *Manual of Industrial Microbiology and Biotechnology,* Demain, A.L., N.A. Solomon, eds., Washington: American Society for Microbiology, 1986, pp 122–136
14. Demirci, A., A.L. Pometto III, B. Lee, P. Hinz. Media evaluation in repeated batch lactic acid fermentation with *Lactobacillus plantarum* and *Lactobacillus casei* subsp. *rhomnosus*. *J. Agric. Food Chem.* 46:4771–4774, 1998.
15. Lee, B., A.L. Pometto III, A. Demirci, P. Hinz. Media evaluation in microbial fermentations for enzyme production. *J. Agric. Food Chem.* 46:4775–4778, 1998.
16. De Deken, R.H. The Crabtree-effect: a regulatory system in yeast. *J. Gen. Microbiol.* 44:149–156, 1966.
17. Postma, E., C. Verduyn, W.A. Scheffers, J.P. van Dijken. Enzymic analysis of the Crabtee effect in glucose-limited chemostat cultures of *Saccharomyces cerevisiae*. *Appl. Environ. Microbiol.* 55:468–477, 1989.
18. Gottschalk, G. *Bacterial Metabolism, 2nd edition*. New York: Springer-Verlag, 1986.
19. Sirakaya, B., J. Gerber, A. Demirci. Fermentation software for bioprocessing: Monitoring growth conditions without programming expertise. *Gene. Eng. News* 21(13):40,71, 2001.
20. Demirci, A., A.L. Pometto III. Enhanced production of D(-)-lactic acid by mutants of *Lactobacillus delbrueckii* ATCC 9649. *J. Indus. Microbiol.* 11:23–28, 1992.
21. Demirci, A., A.L. Pometto III, K.R. Harkins. Rapid Screening of solvents and carrier compounds for lactic acid recovery by emulsion liquid extraction and toxicity on *Lactobacillus casei* (ATCC 11443). *Bioseparation* 7:297–308, 1999.

22. Ho, L.K.G., A.L. Pometto III, P.N. Hinz, J.S. Dickson, A. Demirci. Ingredients selection for plastic composite-supports used in L(+)-lactic acid biofilm fermentation by *Lactobacillus casei* subsp. *rhamnosus*. *Appl. Environ. Microbiol.* 63:2516–2523, 1997.
23. Sivakesava, S., J. Irudayaraj, A. Demirci. Simultaneous determination of multiple components in *Lactobacillus casei* fermentation system using FT-MIR, NIR and FT-Raman spectroscopy. *Process. Biochem.* 37:371–378, 2001a.
24. Sivakesava, S., J. Irudayaraj, A. Demirci. Monitoring a bioprocess for ethanol production using FT-MIR and FT-Raman spectroscopy. *J. Ind. Microbiol. Biotechnol.* 26:185–190, 2001b.
25. Hall, M.J., S.D. Dickinson, R. Pritchard, J.I. Evans. Foams and foam control in fermentation processes, *Prog. Ind. Microbiol.* 12:171–234, 1973.
26. Jin, B., J. van Leeuwen, H.W. Doelle. The influence of geometry on hydrodynamic and mass transfer characteristics in a new external airlift reactor for the cultivation of filamentous fungi. *World J. Microbiol. Biotech.* 15(1):83–90, 1999.
27. Crueger, W., A. Crueger. *A textbook of industrial microbiology, 2nd Ed.* Massachusetts: Sinauer Associates, Inc., 1990.
28. Bjurstrom, E.E. Fermentation systems and processes. In: *Food Biotechnology,* Knorr, D., ed., New York: Marcel Dekker, 1987, pp 193–222.
29. Velázquez, A.C., A.L. Pometto III, K.L.G. Ho, A. Demirci. Evaluation of plastic composite-supports in repeated-fed-batch biofilm lactic acid fermentation by *Lactobacillus casei. J. Appl. Biotech. Microbiol.* 55:434–441, 2001.
30. Cotton, J.C., A.L. Pometto III, J. Gvozdenovic-Jeremic. Continuous lactic acid fermentation using a plastic composite support biofilm reactor. *J. Appl. Biotech. Microbiol.* 57:626–630, 2001.
31. Demain, A.L., N.A. Solomon. *Manual of Industrial Microbiology and Biotechnology*. American Society for Microbiology, 1986.
32. Characklis, W.G. Biofilm processes,. In: *Biofilms,* Charackis, W.G., K.C. Marshall eds., New York: Wiley-Interscience, 1990, p 195–207.
33. Demirci, A., A.L. Pometto III, K.E. Johnson. Lactic acid production in a mixed culture biofilm reactor. *Appl. Environ. Microbiol.* 59:203–207, 1993.
34. Demirci, A., A.L. Pometto III. Repeated-batch fermentation in biofilm reactors with plastic-composite supports for lactic acid production. *Appl. Microbiol. Biotechnol.* 43:585–589, 1995
35. Ho, K.L.G., A.L. Pometto III, P.N. Hinz. Optimization of L (+)-lactic acid production by ring/disc plastic composite-supports through repeated-batch biofilm fermentation. *Appl. Environ. Microbiol.* 63:2533–2542, 1997.
36. Ho, K.L.G., A.L. Pometto III, P.N. Hinz, A. Demirci. Nutrients leaching and end product accumulation in plastic composite-supports for L(+)-lactic acid biofilm fermentation. *Appl. Environ. Microbiol.* 63:2524–2532, 1997.
37. Kunduru, M.R., A.L. Pometto III. Evaluation of plastic composite-supports for enhanced ethanol production in biofilm reactors. *J. Ind. Microbiol.* 16:241–248. 1996.
38. Kunduru, M.R., A.L. Pometto III. Continuous ethanol production by *Zymomonas mobilis* and *Saccharomyces cerevisiae* in biofilm reactors. *J. Ind. Microbiol.* 16:249–256, 1996.
39. Demirci, A., A.L. Pometto III, K.L.G. Ho. Ethanol production by *Saccharomyces cerevisiae* in biofilm reactors. *J. Ind. Microbiol.* 19:299–304, 1997.
40. S.E. Urbance, A.L. Pometto III, A.A. DiSpirito, A. Demirci. Medium evaluation and plastic composite support ingredient selection for biofilm formation and succinic acid production by *Actinobacillus succinogenes. Food Biotechnol.* 17:53–65, 2003.

1.07

Production of Carotenoids by Gene Combination in *Escherichia coli*

Gerhard Sandmann

CONTENTS

7.1 CAROTENOIDS: PROPERTIES, COMMERCIAL ASPECTS, AND BIOLOGICAL FUNCTION IN HUMAN HEALTH

Carotenoids are water-soluble natural pigments of 30–50 carbon atoms. Even shorter structures called apocarotenoids result from their oxidative cleavage. Carotenoids are synthesized de novo by archaea, bacteria, fungi, and higher and lower plants. Animals are supplied with carotenoids from their food and are able to further modify the chemical structure. More than 600 carotenoids are known as intermediates or end products of different biosynthetic branches in various organisms. Most carotenoids consist of 40 carbon atoms; possess an acyclic chain; or carry β-ionone, ε-ionone, or aromatic end groups. Acyclic carotenoids are often modified at C-1 and C-2, e.g., by 1-HO, 1-CH_3O, or 2-keto groups. Typical substitutions of a β-ionone end are 3-hydroxy, 4-keto, and 5,6-epoxy moieties. The 3-hydroxy group can participate in glycosilation or formation of fatty acid esters. The most prominent feature of a carotenoid molecule is the polyene chain. Delocalization of the Π-electrons is responsible for their color and their antioxidative potential. Carotenoids can interact with radical chain reactions and are capable of energy dissipation from photosensitizers as heat. In photosynthesis, carotenoids function as light-harvesting antenna, transferring light energy to chlorophyll. More details on the structure and function of carotenoids can be found in References 1 and 2.

The commercially most important carotenoids are astaxanthin and β-carotene (structures 7.1 and 7.2, Figure 7.1) (3). Most of the carotenoids on the market is produced by chemical synthesis. A chemical process for the synthesis of lycopene (structure 7.3, Figure 7.1) has recently been established. The industrial use of carotenoids involves their application in nutrient supplementation, for pharmaceutical purposes, as food colorants, and in animal feeds. β-Carotene or other carotenoids with an unsubstituted β-ionone group are essential in human nutrition as a provitamin A. Taken up by the body, they are then metabolized to vitamin A, an integral component of the visual process. Lutein and zeaxanthin (structures 4 and 5, Figure 7.1) are constituents of the yellow eye spot, the macula lutea. There, both carotenoids act as protectants, preventing the retina from photodamage (4). Other carotenoids have a less defined role in human health. Evidence is accumulating that various carotenoids stimulate the immune system and play an important role in the prevention of degenerative diseases and cancer (5).

In addition to chemical synthesis, some biological systems for carotenoids are in use (6). Astaxanthin is accumulated in the green alga *Haematococcus pluvialis* or in the fungus

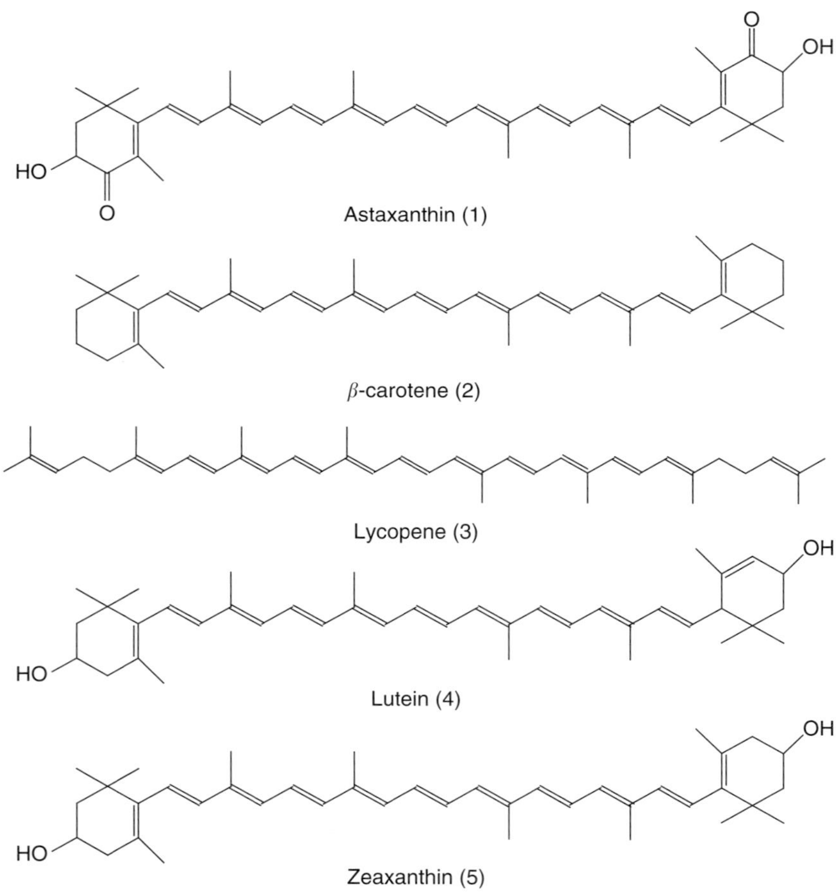

Figure 7.1 Chemical structures of carotenoids of commercial interest or with an effect on human health.

Phaffia rhodozyma (synonymous to *Xanthophyllomyces dendrorhous*). *β*-Carotene is produced by cultivation of the green alga *Dunaliella salina* or the fungus *Blakeslea trispora*. Palm kernel oil is another source for *β*-carotene. Natural sources of other carotenoids are flowers of *Tagetes erecta* for lutein (as fatty acid esters), red fruit of *Capsicum annuum* for capsanthin, and capsorubin or tomato for lycopene.

In recent years, several transgenic microorganisms have been explored as production hosts for carotenoids. These were the fungi *Saccharomyces cerevisiae* and *Candida utilis* as well as the bacteria *Zymomonas mobilis, Agrobacterium tumefaciens,* and *Escherichia coli* (*E. coli*) (7–9). Among them, *E. coli* is the most advanced production system.

7.2 *ESCHERICHIA COLI* AS A CAROTENOID PRODUCTION SYSTEM

Escherichia coli was the first host for functional expression of carotenogenic genes in order to analyze the function of the corresponding enzymes (10,11). Later, this bacterium was very useful in the heterologous overexpression of individual carotenogenic enzymes as functional proteins for their purification and biochemical characterization (12).

Escherichia coli is a very convenient host for heterologous carotenoid production. Because of its fast and easy cultivation in substantial quantity, it can be transformed simultaneously with several plasmids as long as they belong to different incompatibility groups, i.e., possessing different origins of replication. For their stable maintenance in the cells, it is essential that each plasmid carries a different antibiotic resistance. Only when this selection pressure is maintained spontaneous plasmid loss prevented. Several useful plasmids for cotransformation are available. Convenient vectors are pUC-related plasmids with the pMB1 origin of replication and ampicillin resistance, pACYC184 with a p15A origin of replication and chloramphenicol resistance, pRK404 with an RK2 origin of replication and tetracycline resistance, and pBBR1MCS2 with a SC101 origin of replication and kanamycin resistance. These plasmids can be used for expression of individual genes or groups of genes which mediate the formation of certain basic carotenoid structures. For example, one plasmid may carry the genes necessary to obtain certain carotenoid intermediates and others the genes for systematic modifications of the structure (13,14). It should be pointed out that *E. coli* is noncarotenogenic which increases the flexibility to build up a desired carotenoid synthesis pathway. However, *E. coli* has to cope with the drain of prenyl pyrophosphates when carotenogenesis is established. Other plasmids may be used to overexpress certain genes to enhance the *E. coli* metabolic capacity for the supply of carotenoid precursors. Details on the choice of plasmids and growth conditions for carotenoid producing *E. coli* transformants are given in a recent publication (15).

In the majority of bacteria, including *E. coli*, formation of prenyl pyrophosphates, which are the precursors of carotenoids, proceed via a reaction sequence which is different from the mevalonate pathway typically found in fungi and animals (16). It is referred to either as the 1-deoxyxylulose-5-phosphate pathway, because this is the first intermediate, or as the 2-C-methyl-D-erythritol-4-phosphate pathway, because this is the first product which is converted to prenyl pyrophosphate exclusively. To date, most of the reaction steps have been elucidated. Initially, a C_2 unit from pyruvate is condensed to glyceraldehyde by a thiamin-dependent 1-deoxyxylulose-5-phosphate synthase (Figure 7.2). Then, the product is converted to 2-C-methyl-D-erythrol-4-phosphate by 1-deoxyxylulose-5-phosphate reductoisomerase. The following reactions involve the formation of 4-diphosphocytidyl-2-C-methylerythrol in a CTP-dependent reaction and phosphorylation at position 2 by adenosine triphosphate. Subsequently, cytidine-5-phosphate is released and via 2-C-methyl-D-erythritol-2,4-cyclodiphosphate 1-HO-2-methyl-2-butenyl-4-diphosphate is formed.

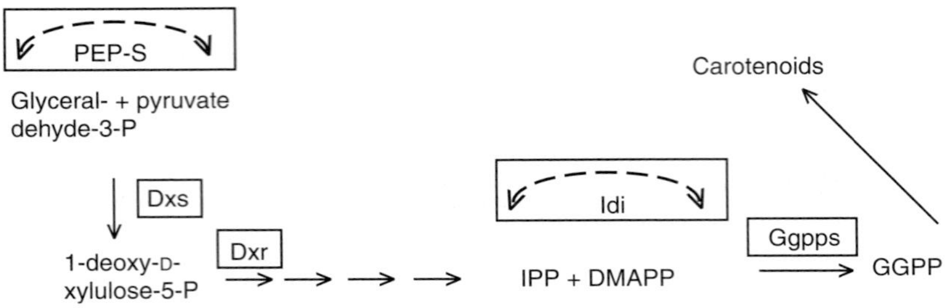

Figure 7.2 Metabolic engineering of the *Escherichia coli* deoxyxylulose-5-phosphate pathway and prenyl pyrophosphate formation for optimum precursor supply. Overexpression of the indicated enzymes either relieve a limitation of the metabolic flow or provide a more favorable balance for two-substrate reactions. PEP-S, phosphoenolpyruvate synthase; Dxs, deoxyxylulose-5-phosphate synthase; Dxr, deoxyxylulose-5-phosphate reducto isomerase; Idi, isopentenyl pyrophosphate isomerase; Gggpps, geranylgeranyl pyrophosphate synthase.

This product may be the branching point for independent routes to isopentenyl pyrophosphate (IPP) and dimethylallyl pyrophosphate (DMAPP). However, the details of the final steps to IPP and DMAPP are not fully understood yet. See Reference 17 for the newest insight into later reactions of this novel pathway.

Precursor supply for carotenogenesis can be increased in *E. coli* by overexpression of limiting enzymes of the deoxyxylulose-5-phosphate pathway and subsequent reactions. Supply of prenyl pyrophosphates and subsequent carotenogenesis was stimulated by transformation with the genes encoding 1-deoxy-D-xylulose-5-phosphate synthase, 1-deoxy-D-xylulose-5-phosphate reductoisomerase, and IPP isomerase under a strong promoter (Figure 7.2). Because geranylgeranyl pyrophosphate is the direct substrate for the formation of the first carotenoid in the pathway and its level is comparably low in *E. coli*, high expression levels of geranylgeranyl pyrophosphate synthase are very important for carotenogenesis (18). Another bottleneck for carotenoid biosynthesis in *E. coli* was relieved in the pathway by overexpressing the gene which encodes phosphoenolpyruvate synthase, a pyruvate consuming enzyme, indicating that the pools of glyceraldehyde 3-phosphate and pyruvate, which both are substrates of 1-deoxy-D-xylulose 5-phosphate synthase, have to be more balanced in the direction of glyceraldehyde 3-phosphate (19). By overexpression of different combinations of the limiting enzymes mentioned above, concentrations of various carotenoids to a final yield of 1.5 mg/g dry weight could be reached (20).

7.3 PRINCIPLES OF CAROTENOID BIOSYNTHESIS

C_{40} carotenoids are synthesized by head-to-head condensation of two molecules of C_{20} geranylgeranyl pyrophosphate catalyzed by the enzyme phytoene synthase (Figure 7.3A). A very similar condensation reaction from C_{15} farnesyl pyrophosphate leads to the formation of C_{30} carotenoids. In the next steps, the conjugated double bond system of C_{30} and C_{40} carotenes is sequentially extended, each time integrating one of the isolated double bonds. Different phytoene desaturases exist with respect to the catalyzed number of desaturation steps. The enzyme Pds from plants inserts only two double bonds symmetrically at positions 11 and 11′ (Figure 7.3B). The bacterial phytoene desaturase from *Rhodobacter capsulatus*, $CrtI_{Rc}$, carries out a 3-step desaturation with an additional

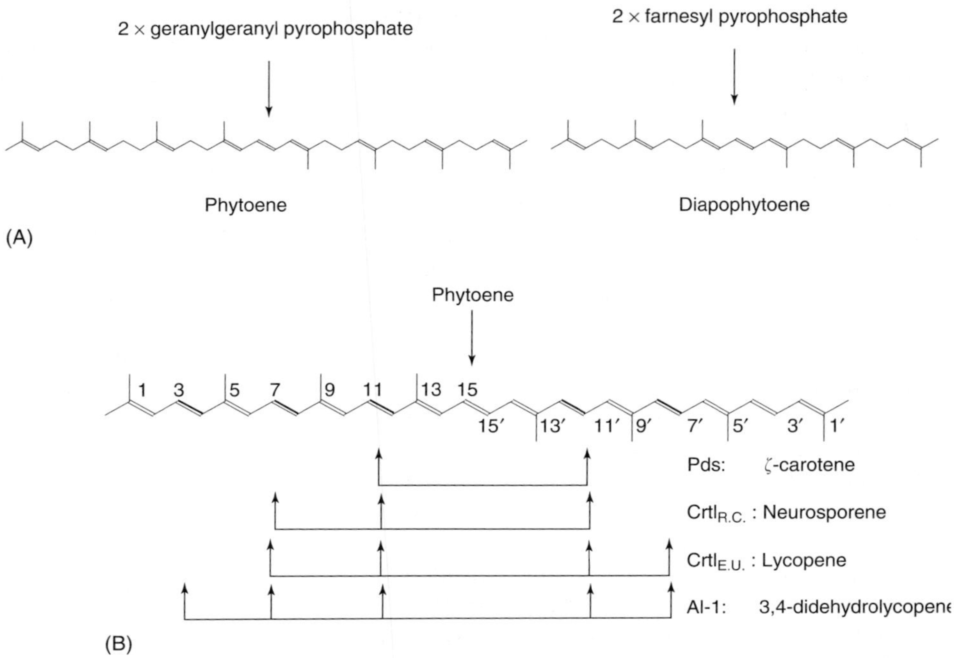

Figure 7.3 Principles of the biosynthesis of acyclic carotenoids with different polyene chains. (A) Establishment of the basic C_{30} and C_{40} carbon chain. (B) Extension of the conjugated double bond system. Double bonds in bold are introduced by the individual phytoene desaturases as indicated by the arrows yielding the specified carotenes.

double bond at position 7. The most abundant bacterial phytoene desaturase like $CrtI_{Eu}$ catalyzes a 4-step desaturation with additional double bonds at position 7 and 7′. The enzyme Al-1 from the fungus *Neurospora crassa* catalyzes a 5-step desaturation with an additional double bond at position 3. These individual desaturation reactions lead to the formation of the products ζ-carotene, neurosporene, lycopene, and 3,4-didehydrolycopene, respectively.

Lycopene can be modified at the terminal double bonds by addition of water, resulting in 1-HO derivatives which can be methylated to 1-CH_3O carotenoids (Figure 7.4). Desaturation at position 3,4 requires the presence of the 1-HO or 1-CH_3O group. These types of carotenoids are typical for photosynthetic bacteria. Finally, a 2-keto group can be introduced under aerobic growth. Another addition to the 1,2 double bond of lycopene is involved in the synthesis of C_{45} and C_{50} carotenoids. This chain elongation at C-2 proceeds via addition of one or two dimethyl allyl cation. Then, the molecule is stabilized by abstraction of a proton from C-17 and C17′.

Cyclization of lycopene ends to ionone groups involves protonation of the 1,2 double bond and the addition of the resulting carbo cation to the 5,6 double bond (Figure 7.4). Stabilization occurred by proton abstraction from either C-6 or C-4 yields β- or ε-rings, respectively. The possible individual substitutions of a β-ionone end group are summarized in Figure 7.4. A cyclic carotenoid may carry a 3-HO, 4-keto, and/or 5,6-epoxy moiety. It should be pointed out that only the combinations of 3-HO and 4-keto, or 3-HO and 5,6-epoxy are known. Details on the carotenogenic pathway can be found in many reviews (1,21,22).

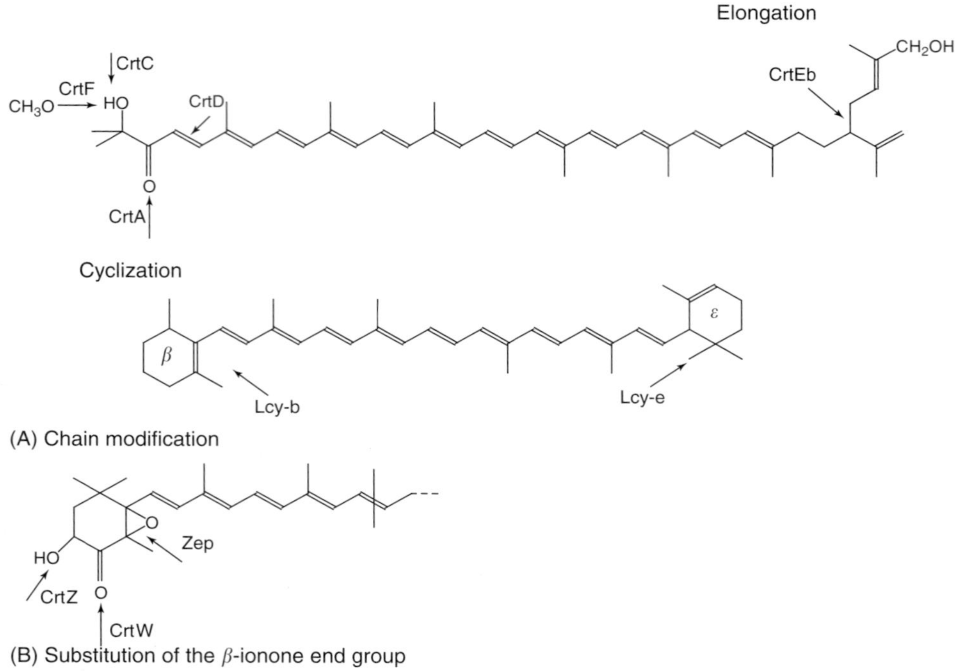

Figure 7.4 Reactions modifying the lycopene molecule (A) or a β-ionone ring (B).

7.4 PRODUCTION OF CAROTENOIDS IN *ESCHERICHIA COLI*: ACHIEVEMENTS AND PERSPECTIVES

Escherichia coli was the first noncarotenogenic organism into which carotenoid biosynthesis was engineered (10,11). Carotenogenic cells were obtained when gene clusters from *Erwinia* species was functionally analyzed. The expression before and after deletions of individual genes lead to the accumulation of the end product zeaxanthin diglucoside and all intermediates like phytoene, lycopene, β-carotene, and zeaxanthin. Many different carotenoids have been produced since, by using genes from many different organisms. Depending on the structure and the amount of carotenoids formed, pigmentation of *E. coli* can be very strong ranging in color from yellow to orange and red.

Genes used for heterologous carotenoid production in *E. coli* originated not only from bacteria but also from fungi, algae, and higher plants. An example is given in Figure 7.5 that demonstrates how genes from all theses organisms were combined in *E. coli,* interacting simultaneously in carotenogenesis. Initially, formation of lycopene is catalyzed by three enzymes encoded by bacterial genes, crtE, crtB, and crtI. Then, a fungal monocyclase Al-2 takes over, forming γ-carotene. Into this molecule, a plant hydroxylase Bhy introduces a 3-hydroxy group. Due to the functionality of all these enzymes in the foreign host *E. coli*, the final product 3-hydroxy-γ-carotene is obtained. The production of this carotenoid is also an example of the synthesis of a carotenoid which is not formed in any of the organisms from which these genes originated. The synthesis of novel carotenoids by combining genes from different organisms only works because the substrate specificity of the enzyme is such that it does not need to recognize the entire substrate molecule, but only certain regions of the molecule which are suitable for conversion. This combinatorial approach can result in the formation of totally novel carotenoids. For example, 1-hydroxylated acyclic

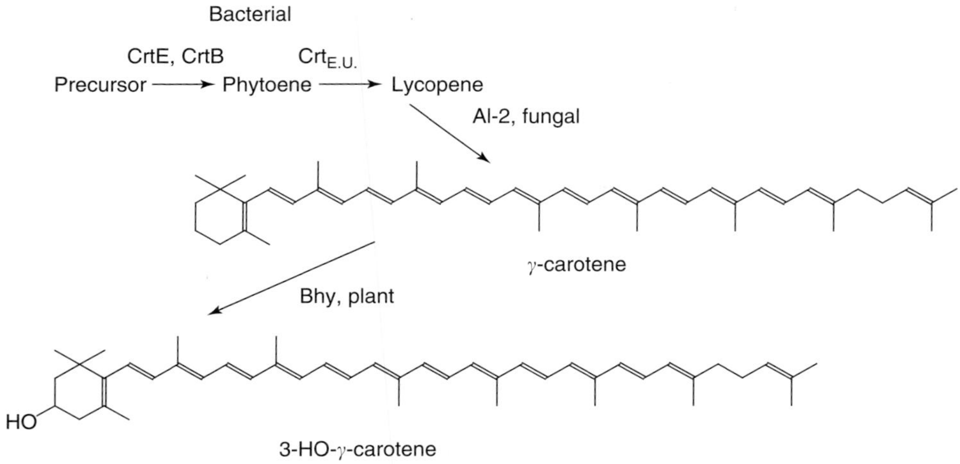

Figure 7.5 Example of heterologous production of the carotenoid 3-HO-γ-carotene by combination of genes from bacteria, fungi and plants.

structures with up to 13 conjugated double bonds were designed with an antioxidative potential which was superior to other related carotenoids (23).

The production of many other carotenoids was successful in the factory *E. coli*. Their synthesis was assembled in a modular way by transformation with the appropriate genes which encode the enzymes responsible for the individual catalytic steps. For this approach, it is important to know the substrate and product specificity of the encoded enzymes. Sometimes, the products of highly homologous genes do not catalyze the same reaction. The 3,4-carotene desaturases $CrtD_{Rg}$ and $CrtD_{Rs}$ from the photosynthetic purple bacteria *Rubrivivax gelatinosus* and *Rhodobacter sphaeroides*, respectively; both enzymes catalyze the conversion of 1-HO-neurosporene and 1-HO-lycopene. However, only the enzyme from *R. gelatinosus* is able to desaturate 1-HO-3′,4′-didehydrolycopene as indicated in Figure 7.6A (24).

A list of genes for the synthesis of different carotenoid structures together with the necessary information on the catalytic function of their products can be found in Reference 25. Table 7.1 comprises the individual carotenoids which can be synthesized in *E. coli*. They include carotenoids of different chain lengths, cyclic and acyclic structures with shorter or longer conjugated double bond systems, unsubstituted hydrocarbons or hydroxy, and keto and epoxy derivatives. All these carotenoids can be produced in *E. coli* in quantities of up to 1–1.5 mg/g dry weight. Many of the carotenoids in Table 7.1 are found in suitable organisms only as biosynthetic intermediates which accumulate only in trace amounts, making it very difficult to extract and purify them. Others have not been detected in biological material before. These carotenoids are marked in bold face. Experimental details for the heterologous production of carotenoids in *E. coli* were recently published (15). Several of the double bonds in a carotenoid molecule can be formed in a cis or trans configuration. For some of the carotenoids produced in *E. coli*, the composition of these geometrical isomers has been determined (26). In general, the all-trans isomer is dominating. In the case of β-carotene, small quantities of 9-, 13-cis, and 9,13,13′-tetracis isomers were also formed. The formation of the major lycopene isomers is determined by the type of gene used to transform *E. coli*. With crtI related genes, mainly all-trans and smaller amounts of 5-, 9- and 13-cis isomers accumulate.

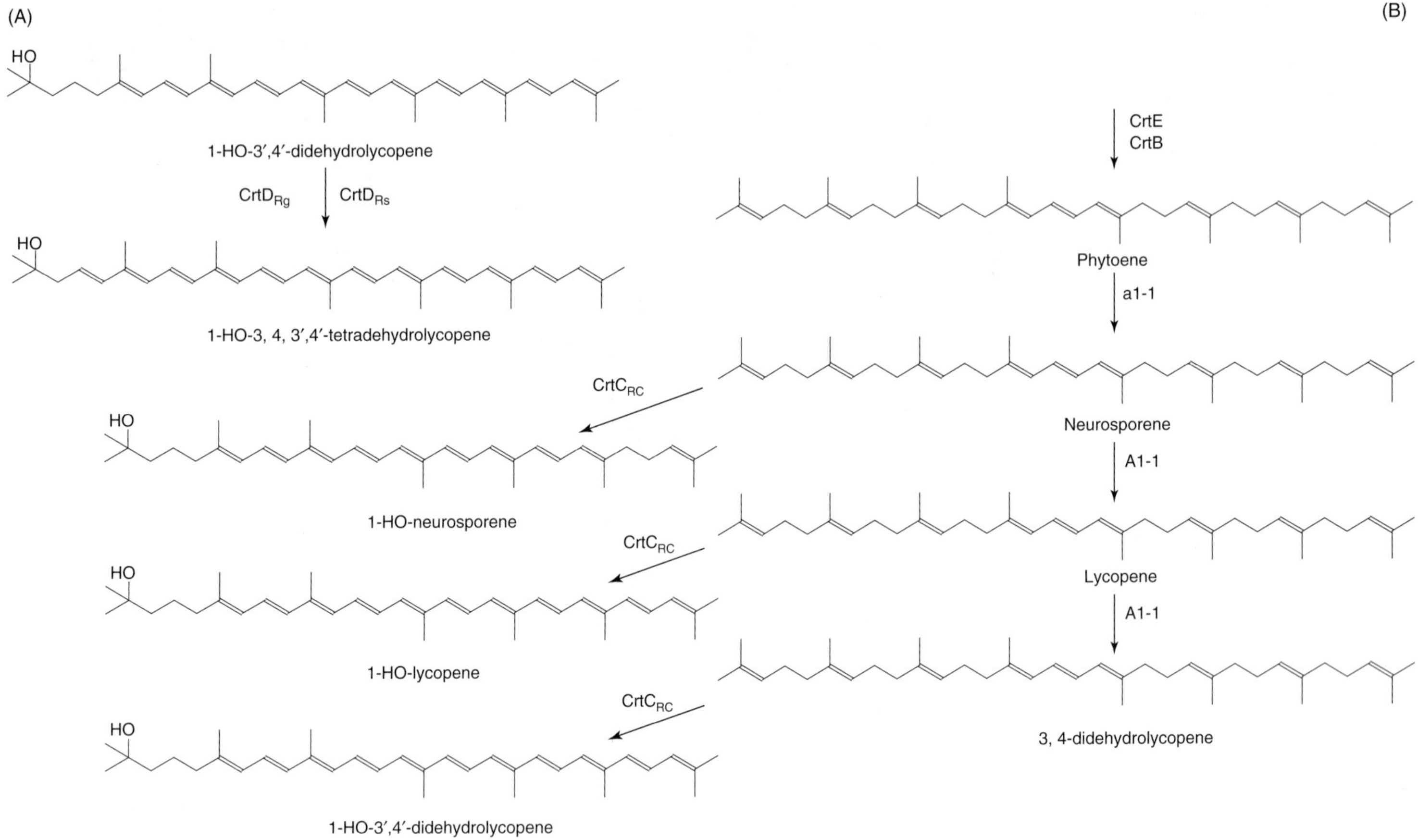

Figure 7.6 Species-specific catalysis of 1-HO-3,4,3′,4′-tetradehydrolycopene formation from 1-HO-3′,4′-didehydrolycopene by enzymes originating from highly homologous genes (A) and an example of intermediate flow into side reactions in the pathway leading to 1-HO-3′,4′-didehydrolycopene (B).

Table 7.1
Gene combinations and production of carotenoids in *E. coli*.

Erwinia:	crtE, crtB,	+ $crtI_{Eu}$, crtY, crtX
		Phytoene, Lycopene, β-Carotene, Zeaxanthin, Zeaxanthinglucosid
	+ Rhodobacter $crtI_{Rc}$, crtC + Erwinia crtY	
		1′,3-$(HO)_2$-γ Carotene§,
	+ Rhodobacter $crtI_{Rc}$ + Erwinia crtY	
		7,8-Dihydro-β-carotene*, 7,8,7′,8′-Tetrahydro-β-carotene*
	+ Rhodobacter/Rubrivivax $crtI_{Rc}$, crtD, crtC	
		1-HO-Lycopene; 1,1′-$(HO)_2$-Lycopene, 1′-HO-γ-Carotene
		Neurosporene, 1-HO-Neurosporene, Demethylsperoidene
		1-HO-3,4-Didehydrolycopene§, 1,1′-$(HO)_2$-3,4-Didehydrolycopene§
		1,1′-$(HO)_2$-3,4,3′,4′-Tetradehydrolycopene§
		7,8-Dihydrozeaxanthin, 3-HO-β-Zeacarotin
	+ Neurospora al-1 + Rhodobacter/Rubrivivax crtC, crtD	
		3,4-Didehydrolycopene, **1-HO-3′,4′-Didehydrolycopene,§**
		1-HO-3,4,3′,4′-Tetradehydrolycopene,§
	+ Erwinia $crtI_{Eu}$, crtY + Synechocystis crtO	
		Echinenone
	+ Erwinia $crtI_{Eu}$, crtY + Haematococcus bkt	
		Canthaxanthin
	+ Synechococcus crtP	
		ζ-Carotene
	+ Erwinia $crtI_{Eu}$ + Corynebacterium crtEb	
		Flavuxanthin, Nonaflavuxanthin
		Staphylococcus crtM
		Diapophytoene
	+ Staphylococcus crtN	
		Diapo-ζ-carotene, Diaponeurosporene
	+ Staphylococcus crtN + Anabaena crtQa	
		Diapolycopene

Note: Carotenoids in bold face have not been found before in biological material; § indicates identification by NMR; others by HPLC, absorbance, and mass spectroscopy.

Source: From Sandmann, G., Chem. Biochem., 3, 629–635, 2002. With permission.

When the crtI gene is replaced by another phytoene desaturase gene crtP together with the ζ-carotene desaturase gene crtQb, prolycopene (=7,9,7′,9′-tetracis) dominates and all-trans lycopene is minor.

In some cases, the end product of an established carotenoid pathway may be much lower than expected. Instead, a mixture of structurally related compound is obtained. This may be the case when the enzymes react, not one after the other in a fixed sequence, but due to their broad substrate specificity in a metabolic web. When, e.g., the desaturase Al-1 and the 1,2-hydratase CrtC interact, not only the end product of the desaturation reaction 3,4-didehydrolycopene but also the intermediates neurosporene and lycopene are modified to the corresponding 1-HO derivatives by the hydratase (Figure 7.6B).

The major advantage of *E. coli* for carotenoid production is the versatility for the generation of very diverse structures. However, compared to the biological carotenoid production systems like *Dunaliella*, *Haematococcus*, *Blakeslea,* and *Phaffia* mentioned before, the yields are lower. Metabolic engineering of the *E. coli* terpenoid biosynthetic reactions led to a considerable increase of carotenoid production and there are indications that precursor supply is not limiting in these engineered strains. The major problem for *E. coli* seems to be carotenoid storage. All the lipophilic carotenoids are sequestered in the cell membranes. Therefore, future activities may aim to extend the carotenoid storage capacity by genetic modification of the density of membranes in *E. coli* cells or by establishing plastoglobuli-like structures.

REFERENCES

1. Goodwin, T.W. *The Biochemistry of the Carotenoids: Vol. I Plants*, 2nd ed. London: Chapman and Hall, 1980.
2. Britton, G. Structure and properties of carotenoids in relation to function. *FASEB J.* 9:1551–1558, 1995.
3. Nonomura, A.M. Industrial biosynthesis of carotenoids, in *Carotenoid Chemistry and Biology*, Krinsky, N.I., M.M. Mathews-Roth, R.F. Taylor, eds. New York: Plenum Press, 1989, pp 365–375.
4. Landrum, J.T., R.A. Bone. Carotenoids in the retina and lens: possible acute and chronic effects on human visible performance. *Arch. Biochem. Biophys*., 385:28–40, 2001.
5. Mayne, S.T. Beta-carotene, carotenoids, and disease prevention in humans. *FASEB J.* 10:690–701, 1996.
6. Nelis, H.J., A.P. de Leenheer. Microbial sources of carotenoid pigments used in foods and feeds. *J. Appl. Bacteriol.* 70:181–191, 1991.
7. Misawa, N., S. Yamano, H. Ikenaga. Production of β-carotene in *Zymomonas mobilis* and *Agrobacterium tumefaciens* by introduction of the biosynthesis genes from *Erwinia uredovora*. *Appl. Environ. Microbiol.* 57:1874–1849, 1991.
8. Yamano, S., T. Ishii, N. Nakagawa, H. Ikenaga, N. Misawa. Metabolic engineering for production of β-carotene and lycopene in *Saccharomyces cerevisiae*. *Biosci. Biotechnol. Biochem.* 58:1112–1114, 1994.
9. Miura, Y., K. Kondo, T. Saito, H. Shimada, P.D. Fraser, N. Misawa. Production of the carotenoids lycopene, β-carotene, and astaxanthin in the food yeast *Candida utilis*. *Appl. Environ. Microbiol.* 64:1226–1229, 1998.
10. Misawa, N., M. Nakagawa, K. Kobayashi, S. Yamano, K. Nakamura, K. Harashima. Elucidation of the *Erwinia uredovora* carotenoid biosynthetic pathway by functional analysis of gene products expressed in *Escherichia coli*. *J. Bacteriol.* 172:6704–6712, 1990.
11. Schnurr, G., A. Schmidt, G. Sandmann. Mapping of a carotenogenic gene cluster from *Erwinia herbicola* and functional identification of its six genes. *FEMS Microbiol. Lett.* 78:157–162, 1991.
12. Sandmann, G. High level expression of carotenogenic genes for enzyme purification and biochemical characterization. *Pure Appl. Chem.* 69:2163–2168, 1997.
13. Sandmann, G., M. Albrecht, G. Schnurr, P. Knörzer, P. Böger. The biotechnological potential and design of novel carotenoids by gene combination in *Escherichia coli*. *Trends Biotechnol.* 17:233–237, 1999.
14. Schmidt-Dannert, C. Engineering novel carotenoids in microorganisms. *Curr. Opin. Biotechnol.* 11:255–261, 2000.
15. Sandmann, G. Combinatorial biosynthesis of novel carotenoids in *Escherichia coli*. In: E. coli *Gene Expression Protocols, Methods in Molecular Biology*, Vol. 205, Vaillantcourt, P.E., ed., Totowa, NJ: Humana Press, 2003, pp 303–314.
16. Lichtenthaler, H.K. The 1-deoxy-D-xylulose-5-phosphate pathway of isoprenoid biosynthesis in plants. *Annu. Rev. Plant Physiol. Plant Mol. Biol.* 50:47–65, 1999.

17. Rhodich, F., S. Hecht, K. Gärtner, P. Adam, C. Krieger, S. Amslinger, D. Arigoni, A. Bacher, W. Eisenreich. Studies on the non-mevalonate terpene biosynthetic pathway: metabolic role of IspH (LytB) protein. *Proc. Natl. Acad. Sci. USA* 99:1158–1163, 2002.
18. Ruther, A., N. Misawa, P. Böger, G. Sandmann. Production of zeaxanthin in *Escherichia coli* transformed with different carotenogenic plasmids. *Appl. Microbiol. Biotechnol.* 48:62–167, 1997.
19. Farmer, W.R., J.C. Liao. Precursor balancing for metabolic engineering of lycopene production in *Escherichia coli. Biotechnol. Prog.* 17:57–83, 2001.
20. Albrecht, M., N. Misawa, G. Sandmann. Metabolic engineering of the terpenoid biosynthetic pathway of *Escherichia coli* for production of the carotenoids β-carotene and zeaxanthin. *Biotechnol. Lett.* 21:791–795, 1999.
21. Cunningham, F.X., E. Gantt. Genes and enzymes of carotenoid biosynthesis in plants. *Annu. Rev. Plant Physiol. Mol. Biol.* 49:557–583, 1998.
22. Sandmann, G. Carotenoid biosynthesis and biotechnological application. *Arch. Biochem. Biophys*. 385:4–12, 2001.
23. Albrecht, M., S. Takaichi, S. Steiger, Z.Y. Wang, G. Sandmann. Novel hydroxycarotenoids with improved antioxidative properties produced by gene combination in *Escherichia coli. Nat. Biotechnol.* 18:843–846, 2000.
24. Steiger, S., S. Takaichi, G. Sandmann. Heterologous production of two novel acyclic carotenoids, 1,1'-dihydroxy-3,4-didehydrolycopene and 1-hydroxy-3,4,3',4' tetradehydrolycopene by combination of the crtC and crtD genes from *Rhodobacter* and *Rubrivivax. J. Biotechnol.* 97:51–58, 2002.
25. Sandmann, G. Combinatorial biosynthesis of carotenoids in a heterologous host: a powerful approach for the biosynthesis of novel structures. *Chem. Biochem.* 3:629–635, 2002.
26. Breitenbach, J., G. Braun, S. Steiger, G. Sandmann. Chromatographic performance on a C_{30}-bonded stationary phase of mono hydroxycarotenoids with variable chain length or degree of desaturation and of lycopene isomers synthesized by different carotene desaturases. *J. Chromatogr.* 936:59–69, 2001.

1.08

Production of Amino Acids: Physiological and Genetic Approaches

Reinhard Krämer

CONTENTS

8.1 INTRODUCTION

Amino acids are the basic constituents of cellular proteins, as well as important nutrients for all living cells. Their importance is closely related to their nutritional value, in particular for those amino acids which are essential for man and animals (lysine, threonine, methionine, tryptophan, phenylalanine, valine, leucine, isoleucine, and, to some extent, arginine and histidine). Amino acids are used for a variety of purposes, depending on their taste, nutritional value, other physiological activities, and chemical properties. Besides their application in human nutrition (flavor enhancers, sweeteners), they are used as feed supplements (animal nutrition), as well as for cosmetics, in the pharmaceutical industry (infusion solutions, dietary food), and as building blocks for chemical synthesis (Table 8.1).

The use of amino acids in food consumes nearly 40% of total amino acid production. Major products used in the food industry are L-glutamate (monosodium glutamate, MSG), L-aspartate, L-phenylalanine, glycine, and L-cysteine. These amino acids are produced by a variety of companies in East Asia, the USA, and Europe (1).

The history of amino acid production is, to a large part, the history of the major producing organism, namely *Corynebacterium glutamicum*. In 1908, K. Ikeda discovered that sodium glutamate was responsible for the specific flavoring component of kelp, a traditional taste enhancing material in Japanese cooking. For a long time, sodium glutamate was commercially isolated from vegetable proteins, because chemical synthesis of glutamate leads to the formation of racemate (racemic acid ester). The sodium salt of the D-isomer, however, is tasteless. In 1957 a screen for bacteria with the capacity to excrete amino acids, in K. Kinoshita and S. Udaka were successful in finding a Gram-positive soil bacterium that was auxotrophic for the vitamin biotin and that spontaneously excreted glutamate under conditions of biotin limitation (2). Since then, a number of bacteria have been isolated that were also characterized by the peculiar capacity of excreting glutamate. Later, they were all reclassified as *C. glutamicum* (3). *Corynebacterium glutamicum*, a soil bacterium, belongs to the group of mycolic acid containing actinomycetes, together with *Nocardia* and *Mycobacteria*. After the successful introduction of *C. glutamicum* for commercial production of large quantities of amino acids, other bacteria, namely *Escherichia coli*, have also successfully been used for this purpose. Interestingly, a bacterium closely related to *C. glutamicum*, namely *Corynebacterium ammoniagenes*, is commercially used for the production of large quantities of nucleotides, another important flavor enhancer in food biotechnology. Recently, the importance of glutamate as a flavor enhancer has been explained by the discovery of special receptor proteins in the human tongue responsible for sensing the taste of glutamate (4).

The majority of amino acids used in the food industry are used as flavor enhancers (mainly L-glutamate, but to some extent also glycine and L-phenylalanine), as sweeteners (L-aspartate and L-phenylalanine, the building blocks for the sweetener aspartame), and for

Table 8.1
Currrent production and application of amino acids. (Data are estimated values for the year 2003)

Amino Acid	Estimated Production (t/y)	Preferred Production Method	Major Use
L-Glutamate	1.100.000	Fermentation	Flavor enhancer
L-Lysine	550.000	Fermentation	Feed additive
D,L-Methionine	550.000	Chemical synthesis	Feed additive
L-Threonine	50.000	Fermentation	Feed additive
Glycine	22.000	Chemical synthesis	Food additive
L-Aspartate	10.000	Enzymatic catalysis	Sweetener (Aspartame)
L-Phenylalanine	10.000	Fermentation	Sweetener (Aspartame)
L-Cysteine	3.000	Enzymatic method	Food additive
L-Arginine	1.000	Fermentation	Pharmaceutical
L-Tryptophan	500	Fermentation	Pharmaceutical
L-Valine	500	Fermentation	Pharmaceutical
L-Leucine	500	Fermentation	Pharmaceutical

a variety of purposes as food additives (e.g., cysteine as a flour additive). In addition, essential amino acids are used as ingredients in dietary food.

In addition to the importance of amino acids as food supplements, there is a huge market for amino acids as additives to animal feed (Table 8.1). The basic goal is to improve the value of animal feed, which typically lacks the essential amino acids lysine, methionine, threonine, and tryptophan as well as other essential amino acids to a minor extent (5,6). The economic driving force for manufacturing and adding amino acids to animal feed is twofold: improvement of the quality of feed leads to a reduction of the quantity needed, which leads to cost reduction; and an enormous reduction of animal waste (manure), because the individual animal will take in as much food as necessary to obtain a sufficient amount of the most limiting nutrients, which, in general, are essential amino acids. The second aspect is thus not only of economic, but also of high ecologic impact. Some aspects of the production of amino acids which are used as feed supplements will also be included in this chapter, because a number of highly relevant and interesting scientific developments in terms of physiology, biochemistry, and molecular approaches have been made in connection with these amino acids, particularly lysine. The results and tools developed in this field will certainly also find their application in the production of amino acids as food additives.

Amino acids which are used in bulk quantities in food and feed applications can be divided into four categories: (1) Those that are mainly synthesized by chemical procedures leading to racemic mixtures. The main example for this class is methionine, because animals (and humans) are able to convert the D-form of this amino acid into the L-form. The achiral amino acid glycine can also be produced by chemical synthesis. (2) Those that are synthesized by enzymatic synthesis, frequently using the enzymes of immobilized bacterial cells. A typical product manufactured by this method is aspartate, for which the enzyme aspartase is used, either from *E. coli* cells immobilized with organic polymers or from *Pseudomonas putida* cells fixed in the natural polymer carrageenan (6). (3) Those that are produced by gene tailored organisms. Glutamate, which is produced by *C. glutamicum*, represents the oldest biotechnological product in this field, and is at the same time the amino acid produced in the largest quantities. (4) The rest of the amino acids, namely lysine, threonine, and phenylalanine, which are produced by using fermentative techniques and microorganisms. The processes for their production substantially differ from those developed for glutamate. In addition, more sophisticated chemical procedures are used in some cases, which lead to stereospecific synthesis of particular amino acids. Most probably, these processes will be replaced in the future by biotechnological techniques based on the use of engineered microorganisms, if the market volume for these amino acids becomes large enough to call for the development of appropriate processes.

The organisms used for biotechnological processes of amino acid production are mainly *C. glutamicum* (glutamate, lysine, aromatic amino acids) and *E. coli* (threonine, aromatic amino acids). A number of other organisms have also been used, e.g., *Serratia marcescens*, a relative to *E. coli*, and others (5,6), these processes, however, are not of high economic impact so far. There is a clear tendency to use only the two major organisms (*C. glutamicum* and *E. coli*) as biotechnological workhorses and, on the basis of the wealth of knowledge on the genetic, biochemical, and in particular physiological level, to adapt their extraordinary capacities to the needs of producing the desired metabolite.

This chapter will have two foci. First, the physiological basis of amino acid excretion in bacteria, a phenomenon that does not seem to be very "physiological" at first view, will be discussed. Second, genetic and biochemical strategies that have been used to improve and to optimize the production of amino acids in a biotechnological scale will be described.

A detailed understanding of the background of amino acid production in terms of physiology, biochemistry, and genetics has proven to be absolutely essential for applying rational strategies to the optimization of amino acid production by metabolic design.

8.2 PHYSIOLOGICAL, BIOCHEMICAL, AND GENETIC BACKGROUND OF AMINO ACID PRODUCTION

It may be argued from a phenomenological point of view that amino acid producing strains can simply be classified into three different categories: wild-type strains able to excrete particular amino acids under specific culture conditions; regulatory mutants from which feedback control of amino acid biosynthesis has been removed; and (3) genetically modified strains in which the biosynthetic capacity has been amplified. These categories may also be combined, as well. This does not, however, fully explain the complexity of the metabolic background behind the peculiar capability to excrete amino acids.

A rational approach to amino acid production is based on a detailed understanding of the particular physiological reasons leading to amino acid excretion. As a matter of fact, amino acid excretion was for a long time interpreted as not being a physiological event, and this perception is still commonly accepted: "The secretion of amino acids serves no purpose for a microbe (7)." Amino acid excretion was thought to result from the impact of biotechnological techniques on the metabolic situation and the integrity of bacterial cells. On the contrary, we now know that amino acid production in most cases has to be rationalized in terms of exploiting and optimizing the physiological property of active amino acid excretion that is already naturally present in many, if not all bacterial organisms (8,9).

Consequently, discussion of the physiological, biochemical, and genetic background of amino acid production necessarily should start by raising and answering the following question. Why would microorganisms, which are generally thought to apply economic strategies in order to survive and to outcompete their neighbors, waste precious compounds such as amino acids, which have to be synthesized at high cost of carbon, nitrogen, and energy? In other words: what is the physiological meaning of the presence of amino acid excretion systems in microorganisms?

Before discussing details of cellular metabolism, one should be aware of a basic misunderstanding. When describing bacterial energy metabolism, F. Harold has conclusively stated: "It must be concluded that the bacterial solution to energy surplus is in general to waste it (10)." Under conditions of carbon and energy surplus, the major survival strategy of bacteria is not the optimization of metabolism in terms of economy and efficiency but in terms of metabolism or growth speed.

On a more detailed level, there are actually at least two plausible reasons that may cause amino acid excretion under physiological conditions (Figure 8.1). They will be explained predominantly in *C. glutamicum*, because of its widespread use in amino acid production. As a soil bacterium, *C. glutamicum* uses all available materials from its natural surroundings as carbon and energy sources. Peptides, derived from decomposing organic matter, are certainly an important source for these needs, and, like most bacteria, *C. glutamicum* harbors efficient uptake systems for peptides. On the other hand, this organism is not very well equipped with efficient degradation systems for various amino acids, e.g., lysine, arginine, branched chain, and aromatic amino acids. If peptides are used as carbon and energy sources, amino acids close to the central metabolism, e.g., glutamate, aspartate, or alanine, are efficiently metabolized. Those amino acids that are not efficiently metabolized, however, will accumulate and will lead to problems due to high internal concentrations. This has been demonstrated by the addition of peptides that contain lysine,

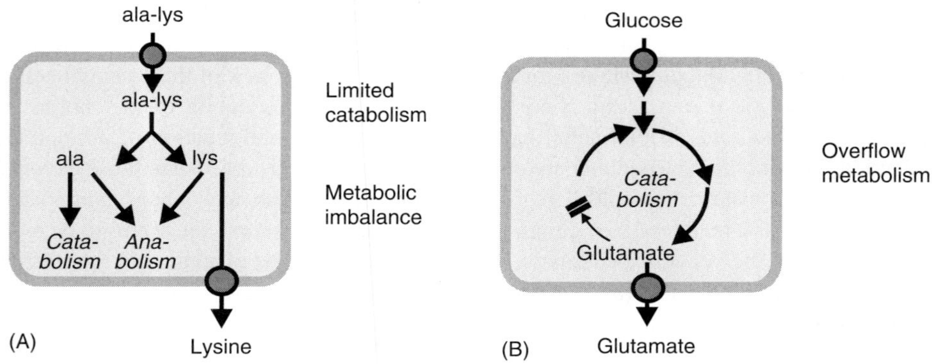

Figure 8.1 Physiological concepts of amino acid excretion. Two major concepts explaining excretion of amino acids based on physiological reasons. (A) Amino acids which are accumulated because of limited catabolic capacity or of a metabolically imbalanced situation are excreted (examples L-lysine in *C. glutamicum*, L-threonine in *E. coli*). (B) Amino acid excretion occurs under conditions of metabolic overflow, i.e., when a metabolic limitation occurs in the presence of a surplus of carbon, nitrogen and energy (example L-glutamate in *C. glutamicum*).

threonine, or isoleucine, as substrates, which leads to high cytoplasmic accumulation (11–13). Even more convincing, it was shown that deletion of the specific lysine excretion system in *C. glutamicum* may cause internal lysine concentrations of more than 1 M as a result of peptide feeding (14). This experiment is actually the most convincing explanation for the physiological necessity of amino acid excretion systems, at least in *C. glutamicum*. This explanation has been designated "metabolic imbalance." In a broader sense, this concept may hold for any imbalanced situation in the cell that can be corrected via export activity. A recent, very interesting example is documented by the discovery of a novel cysteine export system in *E. coli*, the function of which seems to be maintaining redox balance in connection with cytochrome biosynthesis (15).

In addition to lysine, this concept holds for a number of amino acids in *C. glutamicum*, such as threonine and branched chain amino acids, but is not applicable, of course, for glutamate, a metabolite too closely related to the central metabolism. Furthermore, this concept does not easily explain amino acid excretion in other organisms, for example *E. coli*, which is able to degrade those amino acids. Nevertheless, even in those cases it may be an obvious advantage for the bacterial cell to be able to effectively control the steady state concentration of cytoplasmic pools of some amino acids which are not closely related to the central metabolism by efflux, and not only by degradation, in times of a surplus of carbon sources.

In the case of glutamate excretion in *C. glutamicum*, however, a completely different explanation which is based on more general metabolic situations, has been found to apply. It has been studied in a number of microorganisms besides *C. glutamicum*, in particular *E. coli*, and may thus hold for several other cases also (17). If there happens to be an unlimited supply of carbon and nitrogen as well as of energy, and if at the same time some other metabolic limitation occurs which leads to decreased growth or even growth arrest, bacterial cells in general decide not to limit substrate uptake, but to continue the central metabolism at high speed and to waste the surplus of resources instead of switching down their metabolic activity to save nutrients. This situation is called "overflow metabolism" (16), and it is typical for glutamate overproduction in *C. glutamicum* (11). Acetate overproduction in *E. coli*, an unwanted side reaction frequently observed in high cell density fermentation and under oxygen limitation, is caused by the same factors. Limitations leading to overflow mechanisms can naturally occur for different reasons or can be artificially introduced in various ways, including lack of essential metal ions or cofactors. This holds for glutamate production in *C. glutamicum*, where limitation may occur due to lack of the essential cofactor biotin. This type of metabolic condition, leading to overproduction of metabolites in general and amino acids in particular, has also been used for optimization of amino acid production. In principle, introducing auxotrophies for amino acids other than those excreted in production strains leads to controlled growth limitation, and thus may enhance amino acid overproduction. As a matter of fact, glutamate production in *C. glutamicum* seems to be more complicated than the general scenario described above. As will be explained in more detail in section 8.5, in addition to growth limitation, glutamate excretion in *C. glutamicum* is somehow related to alterations in the organism's cell wall, as well as to changes in the activity of cytoplasmic enzymes.

In a number of amino acid production processes, however, these two concepts do not explain the actual situation during fermentation. Under conditions when the flux through central and peripheral pathways is strongly increased due to mutations and engineered enzymatic steps, the pressure for amino acid excretion is much higher as compared to normal physiological situations. Consequently, the basic concepts discussed above still hold as an explanation for the presence of effective excretion transport systems, but they are not applicable to the actual metabolic situation.

Nevertheless, given that amino acids have been shown not to passively cross the permeability barrier of the plasma membrane under production conditions (11,17), it is accepted that most, if not all, amino acid production processes necessarily depend on the presence of active export systems for two reasons. In addition to the trivial argument that most amino acids simply cannot cross the phospholipid bilayer without the help of transport proteins, it is also important to keep the internal amino acid concentration sufficiently low by active extrusion in order to avoid undesired feedback inhibition. This can be achieved only by energy dependent, active export systems, which guarantee that in spite of high rates of biosynthesis as well as increasing external product concentrations, the cytoplasmic amino acid concentration will remain at a relatively low level, compatible with normal physiological situations.

These considerations are also relevant for another situation, which is interesting both from a fundamental and an applied point of view, namely the question of whether amino acid production is coupled to cell growth or not. Whereas this seems to be clear for the basic physiological concepts discussed above, it is no longer obvious for typical production strains. Amino acid excretion as explained by metabolic imbalance may certainly be coupled to growth, but the concept of metabolic overflow at least requires limitation of growth if not growth arrest. On the other hand, with the exception of glutamate production, in typical production strains amino acid excretion is the result of the pressure of an engineered flux increase in the direction of the desired product, and amino acid export just enlists the available efflux systems. Consequently, whether production is coupled to growth or not is more likely to be governed by technical aspects, such as whether the production phase is specifically turned on during fermentation (e.g., threonine in *E. coli*) or not (e.g., lysine in *C. glutamicum*).

For bulk amino acid production, only a small number of microorganisms are used, namely *C. glutamicum* (glutamate, lysine, and aromatic amino acids), *E. coli* (threonine, aromatic amino acids, and cysteine), and, to some extent, *Serratia marcescens* (arginine and histidine). In view of this fact and on the basis of the available models explaining the physiological background of amino acid production, the question arises whether these organisms may have special capacities with respect to amino acid excretion. *Escherichia. coli* is on this list precisely because it is the major source of physiological and biochemical knowledge available. There are already genetic tools at hand that directed the choice of this organism for its use in amino acid production. The same argument holds true for *C. glutamicum* at another level, given that this organism is presumably the best studied bacterium with respect to technical fermentation.

On the other hand, it has frequently been argued that those bacteria would be particularly suitable for the purpose of metabolite production that have a less tightly regulated central and peripheral metabolism. This argument was put forward for *E. coli* and *S. marcescens*, the latter being identified as less tightly regulated. The success of *E. coli* in amino acid production, however, clearly argues against this interpretation. In the same line, *C. glutamicum* is designated as having a less well organized and regulated central and peripheral metabolism as than *E. coli*. There are, for example, three aspartokinase isoenzymes found in *E. coli* at the entrance to the biosynthesis pathway of the aspartate family of amino acids which are separately inhibited by lysine, threonine, and isoleucine, whereas *C. glutamicum* has only one enzyme, which is feedback inhibited in a concerted manner by lysine and threonine. It is, however, not obvious why concerted feedback control should in principle be inferior to control of individual enzymes. Moreover, in the meantime *C. glutamicum* was found to have a number of highly elaborated regulation networks, for example with respect to nitrogen control (9,18), or for the cellular response to osmotic stress (19).*Corynebacterium glutamicum* seems to be especially well equipped in some metabolic control regions, particularly the anaplerotic pathways. A somewhat loose pathway regulation, however, seems in fact to be present in *C. glutamicum* at least in some cases. It has been shown, for example, that simple addition of methionine, which, after being taken up and inhibiting its own biosynthesis, leads to overflow in the direction to lysine biosynthesis, which subsequently causes lysine excretion (20). The argument that less tightly regulated species are more favorable production organisms in general is also contradicted by the fact that *E. coli* seems to be clearly better suited for threonine production than *C. glutamicum*, in spite of the fact that its biosynthesis pathway is much better regulated.

Taken together, the reason for choosing a particular production organism does not seem to be directly derived from its physiological or genetic suitability for this purpose. The choice is obviously driven by the amount of genetic and physiological knowledge available, and by the recognition of a particular organism to be productive (*C. glutamicum*) or easy to handle (*E. coli*).

Nevertheless, *C. glutamicum* and related species seem to be an exception, at least in two aspects: Its physiological capacity to excrete large amounts of glutamic acid already under natural conditions seems to be exceptional, and the peculiar situation that some treatments affecting the state of the cell wall or plasma membrane trigger an increase in the propensity to excrete amino acids has not been reported for any other bacterium. It is important to note that this particular property does not seem to be restricted to glutamate excretion only.

With respect to the central and peripheral metabolism in amino acid producing bacteria, there are a number of specific strategies for metabolic design, which are relevant for metabolite overproduction in general and for amino acid production in particular. Schematically, the metabolic network related to metabolite production can be divided into several individual components (Figure 8.2). Basically, we can discriminate four different major steps being involved: substrate uptake, central metabolic pathways, amino acid biosynthesis, and amino acid excretion. This simple view has to be extended by further important considerations, in particular specific (local) regulatory mechanisms both on the level of gene expression and enzyme activity. Furthermore, the integration into global regulatory networks such as carbon control and nitrogen control has to be considered, as well as the importance of cellular metabolic balances like energy balance and redox balance.

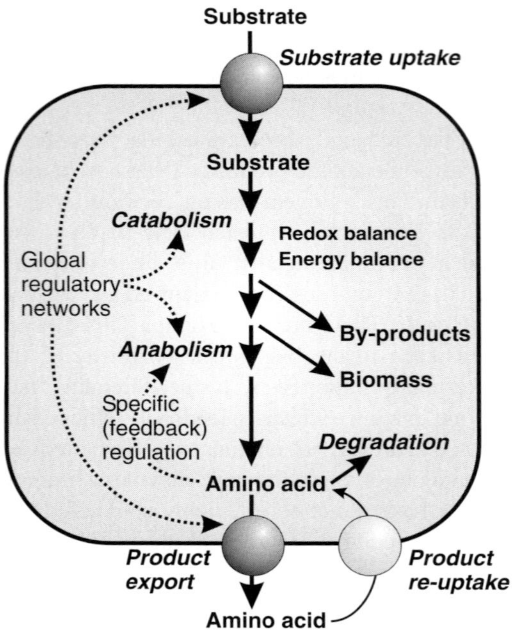

Figure 8.2 Basic steps involved in amino acid production. Summary of major metabolic and regulatory events important for amino acid production by bacteria (see text).

Moreover, specific situations have to be taken into consideration, such as amino acid degradation pathways or product reuptake by amino acid uptake systems.

Most of these aspects will be discussed in detail in section 8.5, in connection with processes used for the production of individual amino acids. This holds for regulatory concepts, as well as for more specific questions like redox balance in the case of lysine, or degradation activity in the case of threonine. The general pathway architecture, i.e., particular reactions as well as general regulation patterns of the major pathways of substrate degradation (e.g., glycolysis, pentose phosphate pathway), central metabolism (e.g., tricarboxylic acid cycle, glyoxylate cycle, anaplerotic reactions), and the common anabolic pathways of amino acid biosynthesis are in principle well known and studied in detail. The special situation of massive overproduction of an anabolic end product, i.e., an amino acid, which is largely different from the usual state under growth conditions, makes it necessary to focus in detail on particular parts of the process. A rational strategy for metabolic design seems to be relatively easy in the case of the final steps, i.e., the anabolic reaction sequences directly leading to the particular amino acid, where a straightforward deregulation or overexpression strategy seems promising. It is, however, obviously not that simple in the case of catabolic and central metabolic pathways involved in the supply of precursor molecules of amino acid biosynthesis, because these pathways also have to fulfil many other purposes in the cell, which may be in conflict with the basic biotechnological goal. This holds in particular for aspects of biomass production (growth), of energy consumption (efficiency), and the balance between desired and undesired pathways (yield).

Many studies have been done on the importance of central metabolic pathways in *C. glutamicum* for the efficient synthesis of products such as lysine or glutamate (21). In these studies, the core significance of anaplerotic reactions within this metabolic node has been demonstrated, making the set of anaplerotic reactions a particularly fascinating target of metabolic engineering. Because of the central position of these reactions, the results obtained are certainly of high relevance for the production of any amino acid the precursors of which are provided by reactions or side reactions of the citric acid cycle. Two precursor metabolites oxaloacetate and pyruvate, which serve as precursors for about 35% of cell biomass and metabolic products, are generated in the anaplerotic node. Several research groups have shown that altogether at least 5 individual enzymes are present in this complex metabolic process. The glyoxylate cycle also fulfils anaplerotic functions (Figure 8.3). By disruption of the gene encoding the phospho*enol*pyruvate carboxylase (22) and by detailed flux analysis using ^{13}C NMR techniques (23,24), it was shown that the pyruvate carboxylase plays a major role in *C. glutamicum*. Altogether, a surprisingly complicated metabolic pattern is found: pyruvate dehydrogenase providing acetyl-CoA for the citric acid cycle; pyruvate carboxylase together with phospho*enol*pyruvate carboxylase supplying oxaloacetate, as well as oxaloacetate decarboxylase, phospho*enol*pyruvate carboxykinase, and the malic enzyme. In addition, the glyoxylate cycle provides C4 precursor compounds. This may not be complete, given that the presence of a pyruvate oxidase was also revealed in the genome of *C. glutamicum*. It may be concluded that the anaplerotic node with its complex scenario of partly counteractive reactions is one of the "secrets" of *C. glutamicum* that explains its exceptional metabolic flexibility and versatility. In addition to obviously being an interesting target for rational metabolic design, it is also an instructive example, in which it has been shown that analysis of the significance of individual enzymes in these processes is not possible by in vitro enzyme tests, nor by genetic studies, but only by applying refined methods of metabolic flux analysis (25–28).

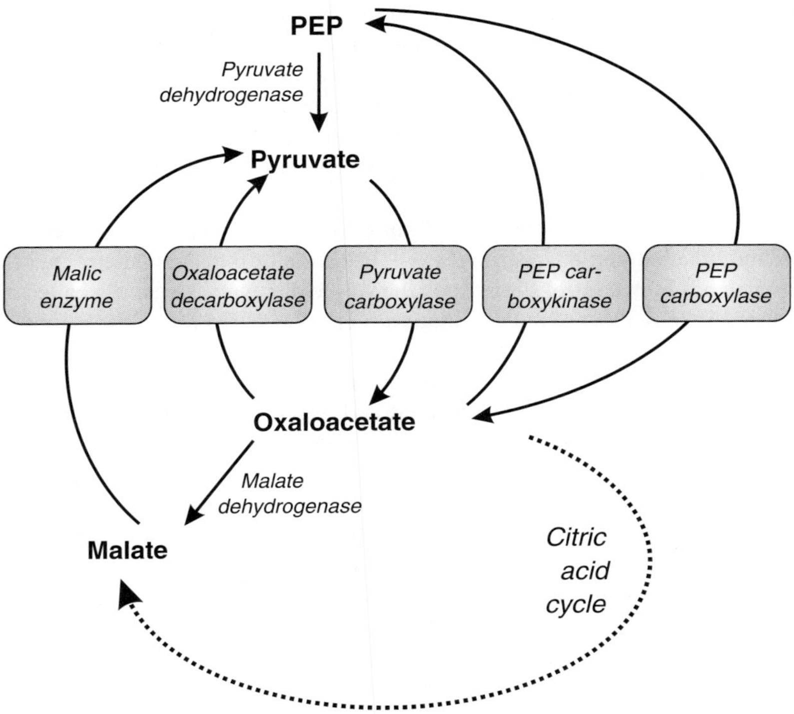

Figure 8.3 Anaplerotic pathways of *C. glutamicum*. Complexity of anabolic reactions in *C. glutamicum*. The figure represents a minimum setup of reactions present (see text).

8.3 PHYSIOLOGICAL, BIOCHEMICAL, AND GENETIC TOOLS USED TO IMPROVE AMINO ACID PRODUCTION

In the past decades, classical approaches including many rounds of mutation, selection, and screening, in some cases combined with molecular biology techniques such as threonine production, have led to impressively efficient strains of *C. glutamicum* and *E. coli* that are used in amino acid production. After introduction of modern genetic tools and sophisticated methods for pathway analysis, the era of rational metabolic design was begun (29,30). Now additional aspects have come into scientific focus, namely substrate uptake, precursor synthesis by central pathways, further steps within the anabolic pathway, redox and energy balance, and last but not least, active product excretion. At the beginning, straightforward molecular approaches such as direct deletion or overexpression of particular genes were, in general, not as successful as conservative methods of strain breeding. This is most likely due to the fact that optimum productivity with respect to a single anabolic end product, i.e., an amino acid, is based on fine tuning of the whole cellular network of pathways and regulation circuits. In order to render modern approaches of metabolic design more efficient than classical procedures, detailed knowledge of the metabolic web, on all possible levels of understanding, is needed. This will include everything from pathway architecture and regulation of the individual enzymes involved, to the sum of carbon, nitrogen, energy and redox fluxes including the construction of the related regulatory circuits. In the genomic and postgenomic era, the full set of information on the level of the genome, the transcriptome, the proteome, and the metabolome becomes more and more available. In the near future it will be fully complemented by dynamic information on metabolic pools and fluxes (the fluxome).

In addition to approaches of optimizing biosynthesis and excretion of amino acids based on the pathway architecture, which are already present in the respective production organisms, metabolic design certainly offers the possibility of introducing new reactions and pathways not so far found in the particular bacterium under study.

As targets for improving a producer organism, all of the different compounds of the metabolic network responsible for amino acid biosynthesis and excretion have to be considered. The vast majority of efforts have been spent on strategies to modify the terminal anabolic pathways (5,31). The most direct approach, which has been applied in all processes except glutamate production, is the amplification of the biosynthetic flux through a reaction which has high control strength on the whole pathway, frequently called a "bottleneck." This may be achieved by deregulation, i.e., release from feedback control by the respective end products of the pathway, a typical result of classical strategies in which high concentrations of substrate analogues were added, giving rise to inhibition insensitive mutant forms of originally feedback inhibited enzymes. Another strategy involves increased or decreased synthesis of enzymes at the corresponding metabolic branch points. An instructive example is lysine production, where a decrease in the dihydrodipicolinate synthase (DapA) increased the flux to lysine. A similar example is redirection of metabolic flux from tryptophan to tyrosine or phenylalanine in *C. glutamicum* (31).

Such strategies can be applied to all other aspects of the metabolic network involved in amino acid production (Figure 8.1). Particularly interesting and successful are the strategies directed toward influencing the connections of anabolic pathways to the central metabolism (precursor supply), decreasing degradation activity, and engineering of transport systems (avoiding futile cycles). Some aspects have not yet been fully exploited, such as optimization of substrate uptake and product excretion, and some fields are not really well understood to a level sufficient to be approached by a rational metabolic design, e.g., global regulatory mechanisms, and energy balance. These will briefly be discussed in section 8.6.

Biochemical and genetic information was successfully used in the case of precursor supply, most prominently concerning anaplerotic reactions. After the set of reactions within the anaplerotic node had been elucidated in *C. glutamicum* (3), the significance of phospho*enol*pyruvate carboxykinase (32) and, in particular, of pyruvate carboxylase, for amino acid production was elucidated. The importance of the latter enzyme for lysine and glutamate production was demonstrated by deletion and overexpression of the corresponding gene (33,34). Further examples of the successful optimization of precursor supply, threonine and aromatic amino acid production in *E. coli* and *C. glutamicum,* are described in section 8.5.

Another obvious strategy involves the inactivation of amino acid degradation pathways that are interfering with high product yield. This is not that important in *C. glutamicum* in view of the limited catabolic capacity of this organism, but it is very important in *E. coli.* An example of this strategy is threonine production in *E. coli.*

Although certainly significant for bacterial metabolism under production conditions, energy and redox balance has not yet been studied in great detail. Because of the obvious need of a sufficient supply of redox equivalents (NADPH) in the case of lysine production, the relation of metabolic flux through glycolysis versus the pentose phosphate pathway has been analyzed using ^{13}C NMR or MS flux analysis (section 8.5). Direct approaches by introducing NADH dependent instead of NADPH dependent enzymes have not yet been successful.

Transport reactions are important in amino acid production and are thus candidates for optimization in three different ways. First, highly active nutrient uptake is a prerequisite for efficient synthesis of amino acids, which, second, have to be actively secreted in order to be accumulated in the surrounding medium. Third, the latter process may be counteracted by the activity of amino acid uptake systems. Sugar uptake via phospho*enol*pyruvate dependent phosphotransferase systems (PTS) has been very well studied in *E. coli* (35), but unfortunately little information is available for *C. glutamicum.* (36–38). PTS systems are highly interesting in connection with amino acid production not only because they are responsible for sugar uptake, but also due to their impact on regulatory networks in the cell with respect to control of carbon metabolism, including various transport systems. Amino acid excretion was not taken into consideration as a relevant aspect of amino acid production for a long time. First on the biochemical level (8), and in recent years by successful molecular identification (39), the discovery of an increasing number of specific and energy dependent amino acid exporters has opened a new world of transport reactions and, at the same time, targets for optimization strategies (Table 8.2). Several new families of transporters are known now: the lysE family (40), the RhtB family, the ThrE family, and the LIV-E family (39). All these families have a number of members found in the genomes of a variety of bacteria. It is interesting that all carriers related to amino acid excretion identified so far belong, with one exception, to the class of secondary carriers, i.e., they are coupled to the electrochemical potential as the driving force. So far, only a single ABC-type, primary carrier has been detected (Table 8.2). It should be mentioned, however, that glutamate excretion was found to depend on the cytoplasmic ATP concentration (41). The first members known were *LysE*, the lysine exporter in *C. glutamicum* (14), *RhtB* and *RhtC* from *E. coli* which are involved in threonine production by this organism (42), *ThrE*, the threonine exporter in *C. glutamicum* (43), and *BrnFE*, the leucine, isoleucine, and valine exporter in *C. glutamicum* (44). Further examples of newly discovered amino acid export systems are the two exporters involved in cysteine and O-acetylserine efflux in *E. coli*, which belong to completely different transporter families from those described above (45,46). The surprisingly large number of homologues found in the genomes of numerous bacterial species indicates that, in clear contrast to the previous perception, amino acid excretion seems to be a common event in the prokaryotic world.

Table 8.2
Amino acid excretion systems identified on the molecular level

Organism	Gene	Transporter Family[1]	Substrate	No. of Putative Transmembrane Segments
C. glutamicum	*lysE*	LysE	L-lysine, L-arginine	5–6
E. coli	*rhtB, rhtC*	RhtB	L-threonine[2], L-homoserine[2], L-homoserine lactone[2]	5–7
E. coli	*ydeD*	PecM	O-acetyl-L-serine, L-cysteine, L-glutamine, L-asparagine	9–10
C. glutamicum	*thrE*	ThrE	L-threonine	10
C. glutamicum	*brnFE*	LIV-E	L-isoleucine, L-leucine, L-valine	7 and 4[3]
E. coli	*yfiK*	RhtB	O-acetyl-L-serine, L-cysteine	5–7
E. coli	*cydDC*	ABC-type[4]	L-cysteine	

[1] The transporter family within the class of secondary transporters is indicated.
[2] Only resistance to the substances listed but no transport was measured.
[3] BrnFE is a transport system consisting of two different subunits.
[4] In contrast to all other transporters listed in Table 8.2, CydDC is an ABC-type transporter.
For references, see text.

It has been demonstrated that amino acid uptake systems may be of significance for amino acid production in two different ways. Tryptophan production by *C. glutamicum* was found to be improved when the uptake system for aromatic amino acids was decreased in its activity. A similar situation was found for threonine production. The advantage of decreasing the activity of amino acid uptake for production may be twofold: a possible energy consuming futile cycle of energy dependent uptake and export can be avoided, and feedback inhibition as well as repression by high internal amino acid concentrations will not occur. For glutamate excretion, it has been shown that the risk of energy consuming futile cycling seems to be avoided by alternative regulation of uptake and excretion systems (47).

In addition to the list of targets for optimization, based on the dissection of metabolic and regulatory events, there are a number of additional aspects to consider. A highly interesting point which frequently is discussed in connection with strain optimization is the presence of auxotrophies for amino acids or related compounds in production strains. Amino acid auxotrophies in strains obtained by classical breeding techniques are in general interpreted as causing release from feedback control or decreasing alternative branches in anabolic metabolism, a conclusive example being lysine production strains (5,6). This explanation, however, is not always conclusive, because frequently auxotrophies for amino acids other than those related to the biosynthesis pathway of the amino acid excreted are found in producer strains. It should be considered that growth limitation may have an effect on productivity as has been discussed in connection with models for overflow metabolism. The disadvantage of the presence of auxotrophies in production strains, i.e., that fermentation media need substantial and defined supplementation with the respective compounds, have been overcome by the introduction of "leaky" strains. A leaky biosynthetic pathway is still functional, but its activity is reduced to a level where it can provide only limited amounts of the required metabolite. Examples are strains with reduced activity of citrate synthase (48) or homoserine dehydrogenase (49). Consequently, the respective amino acid as the end product of the leaky pathway will become growth limiting (metabolic overflow situation) and the cytoplasmic concentrations will be very low (release of feedback control).

A particularly interesting property in this respect is the importance of cell wall integrity for amino acid production in *C. glutamicum*. As discussed in section 8.5 in detail, interfering with cell wall integrity of *C. glutamicum* using a variety of methods causes this organism to excrete glutamate. It is necessary to emphasize, however, that this turns out to be a more general situation, obviously not restricted to glutamate excretion only. It has been shown that modification of the lipid composition (50) as well as the content of mycolic acids in the cell wall (51), also leads to an increase in the excretion of other amino acids. This exceptional property of *C. glutamicum*, however, has not yet been exploited in a rational way, because the molecular basis of these events is still not understood.

A chapter on optimization strategies for amino acid producing strains necessarily should include a brief overview on the modern methods used for this purpose. When developing metabolic concepts of amino acid overproduction, a number of tools from different scientific disciplines have been used and integrated in recent years. These approaches have, of course, been successfully employed also for purposes of rational metabolic design. They include molecular tools on the level of functional genomics, i.e., identification, modification, and controlled expression of genes coding for enzymes involved in amino acid synthesis pathways, in pathways responsible for precursors and supplementary building blocks, and in pathways leading to side products branching off the respective biosynthesis pathway, as well as modulation of the expression of genes coding for regulatory proteins. Analysis of the transcriptome as well as the proteome provides a more detailed understanding of regulatory networks working in the cell. Substantial and successful efforts have been made in *C. glutamicum* in this direction in recent years. With the methods of metabolome analysis, a completely new level of insight into the cell is starting to become available, in particular if its dynamics, i.e., the metabolic fluxes in the cell (fluxome) are also determined. A reasonably well characterized metabolome and the pattern of metabolic fluxes would be a solid basis for the rational understanding of amino acid production and metabolic design. We are, however, far from this situation, because a basic inventory of metabolites (metabolome) of not a single bacterial cell is available.

The genome of *C. glutamicum* has been sequenced independently several times (30), and a number of global studies on the transcriptome are now available (30,52–56). At the same time, proteome analysis has successfully been adapted to *C. glutamicum*, too (57–59), and has recently been extended to phosphoproteome analysis (60).

The metabolome approach, i.e., focusing on metabolic pools and their dynamics, has been applied on a more conceptual level already long before serious efforts at unravelling the metabolome of the cell and the metabolic fluxes have been started. This refers to attempts of identifying and quantifying limiting metabolic steps or "bottlenecks," which was a highly influential concept in the field of amino acid production. Earlier perceptions, which suggested that there would exist one major and dominating "bottleneck" controlling the flux through a particular pathway, were soon proven to be not fully correct or not very useful in application. Certainly, the first entrance step into an amino acid biosynthesis pathway is in general a true bottleneck, because it controls the flux through the consecutive anabolic pathway. In general, however, it is not sufficient to simply deregulate the first control point in order to make possible effective overproduction of a particular metabolic end product. The instructive work of Niederberger et al. (61) may serve as an example, where it has been shown by detailed analysis of aromatic amino acid biosynthesis in yeast that flux control is, in general, distributed among a series of enzymes, and is not confined to a single enzymatic step. There have been a large number of approaches to understand metabolic fluxes on the basis of theoretical considerations as well as experimental measurements (25). Well known, among others, are the concepts of control analysis (62,63) as well as metabolic balancing techniques (64,65). The latter have been extensively applied to studying fluxes in *C. glutamicum* under conditions of amino acid production (21,66–68). These concepts have certainly contributed to understanding the complex situation of interrelated fluxes in cells. The basic problem of most of these methods, however, lies in the fact that in general the complete and exact set of data on metabolite pools, and in particular on their dynamic behavior, is not available to a sufficient extent. Only after application of the sophisticated techniques of nuclear magnetic resonance spectrometry (NMR) (21,26,69) or mass spectrometry (28,67–70) to the analysis of fluxes in *C. glutamicum* was a better understanding of the dynamic aspects of bacterial metabolism provided. In particular, the novel combination of metabolite balancing and MALDI-TOF mass spectrometry seems to be a very useful tool in unravelling dynamic fluxes in bacterial cells (71). For a full understanding, however, a detailed analysis of the metabolome, in addition to the major metabolites which also have been identified in the studies mentioned above, is necessary.

8.4 AMINO ACID PRODUCTION: TECHNICAL ASPECTS

Since the discovery of *C. glutamicum* as a bacterium with the ability to produce large amounts of glutamate directly from sugar and ammonia, new processes of amino acid production were developed, and a new industry started, manufacturing these products by using fermentative techniques. In view of the main goal of this article, namely to understand the physiological, biochemical, and genetic background of amino acid production, it seems valuable to have a brief look at the processes applied for amino acid fermentation. Among many techniques used namely extraction, chemical synthesis, enzymatic synthesis, and fermentation, only the latter, which, except for the production of methionine, is used for all large scale bulk amino acids, is discussed here. It is furthermore obvious that amino acid fermentation is just one step in the whole technical process, followed by downstream processing and wastewater treatment, which are also not discussed here.

Fermentation techniques used can be divided into batch and fed batch cultures, and continuous culture. Industrial amino acid fermentation is mostly performed using the former two procedures. Figure 8.4 shows a typical lysine fermentation procedure with *C. glutamicum.* In general, fed batch processes, in which the substrate is not only provided at the beginning of the process but is added repeatedly during fermentation, lead to higher productivity as compared to batch processes because of several reasons. By lowering the initial concentration of nutrients (sugar), the lag time can be shortened and growth as well as yield can be increased. This is mainly due to osmotic stress caused by high nutrient concentrations (72). Frequently, auxotrophic strains are used. In that case, excess supply of the required nutrient at the beginning of the fermentation may result in a decreased production yield, because of feedback effects caused by the nutrient. Very high growth rates under unlimited supplies of substrate may lead to oxygen limitation in large fermentors, which results in a decreased yield and an increased formation of byproducts, mainly organic acids.

Continuous culture has a number of advantages, as well as disadvantages, in the case of amino acid production. (31,72). The production phase can be controlled much better, and studies have been published where the yields were significantly higher as compared to

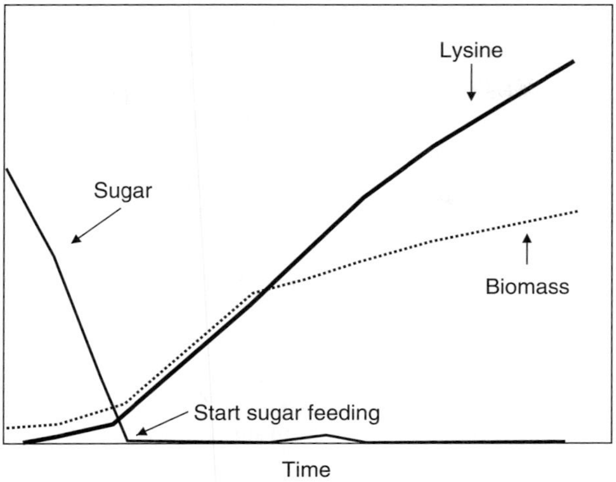

Figure 8.4 Technical l-lysine production process with *C. glutamicum.* Starch hydrolysate or sucrose is used as carbon source and ammonium sulfate as nitrogen source. After consumption of the initial amount of sugar, substrates are added continuously and lysine accumulates up to 170 g/l. The figure was kindly provided by Dr. L. Eggeling, Research Center Jülich.

the same process in fed batch mode (31). As a matter of fact, it has been reported for glutamate production, for example, that the volumetric productivity with 5–8 g L^{-1} h^{-1} of continuous cultivation was at least twice as high as in fed batch fermentations (73). On the other hand, continuous fermentation is technically very demanding, as it requires continuous feeding of sterilized fresh media and sterile air. Furthermore, bacteria often undergo spontaneous mutations which, under the metabolic pressure of amino acid fermentation, may lead to reduced productivity during continuous culture for a long period of time.

It is worth noting that continuous culture is in principle only possible if amino acid excretion occurs during growth. There are several processes in which the production phase is clearly [L-threonine fermentation by *E. coli* (74)] or, to a significant extent (glutamate production by *C. glutamicum*), separated from the growth phase. Because cells are constantly kept in the growth phase in continuous culture, this procedure would not easily be applicable in these cases. Apart from that, continuous culture is a much better tool for studying microbial physiology in response to environmental conditions than batch culture. Consequently, the information obtained from continuous culture experiments is useful for optimizing conditions and feeding strategies in fed batch processes.

8.5 SELECTED EXAMPLES OF AMINO ACIDS

Besides those amino acids which are currently of core interest in the food industry, e.g., glutamate, phenylalanine, aspartate, and cysteine, a number of other amino acids are used for different purposes, namely as feed additives and for the pharmaceutical and cosmetic industry. Some of the latter will be briefly discussed here, because they represent instructive examples of production process development, the physiological and genetic background of which is reasonably well studied.

8.5.1 Glutamate

By far the most important amino acid in food biotechnology, and the first amino acid to be produced by fermentation, is glutamate (monosodium glutamate, MSG). In spite of this fact, and in spite of numerous scientific investigations, it is still less well understood from the physiological and genetic point of view as are, for example, lysine or threonine production.

The historical background of glutamate production, as well as some of its fundamental aspects with respect to the importance of central metabolism and cell wall properties has already been mentioned. For the production of L-glutamate, only *C. glutamicum* is used, although *E. coli* and other organisms are also able to excrete glutamate. Currently, more than 1,000,000 tons of glutamate are produced annually.

The physiological and biochemical background of glutamate production in *C. glutamicum* is highly interesting, and differs significantly from that which is presumably responsible for the excretion of other amino acids as well as from amino acid excretion in other bacteria. L-glutamate is closely related to the central metabolism, synthesized from oxoglutarate by the NADPH dependent glutamate dehydrogenase, which, at the same time, represents the major reaction of nitrogen assimilation under conditions of nitrogen excess in the environment. Under nitrogen limiting conditions, the ATP dependent glutamate synthase glutamate oxoglutarate glutamine aminotransferase system (GS/GOGAT) is predominantly used. The first attempt to explain the property of *C. glutamicum* being an exceptionally good glutamate producer was based on the observation that the activity of the oxoglutarate dehydrogenase (OHDC), i.e., the citric acid cycle reaction following the branching point to glutamate, was surprisingly low. The straightforward explanation of the carbon flow being redirected to glutamate because of the insufficient activity of a major enzyme, thus leading to massive glutamate over synthesis and ultimately to glutamate production, appears convincing.

However, it turned out not to be true, at least in its simple meaning. Later on, it was revealed that *C. glutamicum* in fact possesses significant OHDC activity, which was found to be strongly reduced under conditions which lead to glutamate excretion (75). It has been proven by many investigations studying carbon flux under glutamate production conditions that in fact a flux redirection from oxoglutarate to glutamate is observed under glutamate production conditions (76–78). Moreover, disruption of the OHDC in *C. glutamicum* led to glutamate production comparable to that observed under biotin limited conditions (79). Unfortunately, the reason for the observed reduction of stability and activity of OHDC is not understood so far. It should be mentioned here that another concept has recently been put forward to explain the observation of a reduced flux through the second part of the citric acid cycle, based on the fact that Gram-positive bacteria similar to *C. glutamicum* may need energy input into the succinate dehydrogenase (SDH) reaction. The SDH reaction, which would then depend on the electrochemical proton potential, might possibly not function properly under conditions where the membrane surrounding it is somehow altered, as is suggested after treatments leading to glutamate excretion (80).

For these considerations and their possible consequences in understanding the physiological and biochemical background of glutamate excretion, a critical evaluation of the methods used for inducing glutamate production by *C. glutamicum* is necessary (79). Several different kinds of treatments lead to glutamate excretion. Based on the original observations, glutamate was found to be excreted under conditions of biotin limitation (2). *Corynebacterium glutamicum* requires biotin for growth. Biotin is a cofactor of several enzymes, most importantly of pyruvate carboxylase and acetyl-CoA carboxylase, which is essential for fatty acid synthesis. Glutamate overproduction can also be induced by addition of some kind of detergents, e.g., polyoxyethylene sorbitane monopalmitate (Tween 40) or polyoxyethylene sorbitane monostearate (Tween 60). The reason for induction of glutamate excretion is definitely not simply an increase in membrane permeability, because closely related detergents like the monolaurate (Tween 20) or monooleate esters (Tween 80) are not effective. Consequently, it is assumed that the detergents used exert a regulatory effect rather than a direct increase in permeability. Addition of β-lactam antibiotics (penicillin), which are known to interact in cell wall biosynthesis, leads to glutamate excretion. Strains which are auxotrophic for precursors required in phospholipid biosynthesis, such as glycerol or fatty acids, are able to excrete glutamate into the surrounding medium. The apparently simple method of applying increased temperature also leads to increased glutamate excretion, in particular in the case of temperature sensitive mutants (81). Recently, it was shown that direct changes in the architecture of the cell wall, brought about by disrupting enzymes involved in the synthesis of cell wall components in *C. glutamicum*, such as fatty acids (50) or trehalose (51), lead to increased glutamate excretion, too. In addition to these procedures, also general membrane acting factors, such as addition of local anaesthetics or application of osmotic stress, may induce glutamate excretion in *C. glutamicum* (82,84).

The cell wall of *C. glutamicum* has a very complex structure, consisting of an inner layer composed of cross linked structures of arabinogalactan, peptidoglycan, lipomannan, and lipoarabinogalactan (Figure 8.5). In the middle part, a mycolic acid bilayer structure is present, which, as a true bilayer, is an exception among Gram-positive bacteria, shared only with the other members of the mycolata (*Mycobacteria* and *Nocardia*). The outer layer of the cell wall, finally, is mainly composed of polysaccharides also containing free lipids like trehalose mono- and dimycolates (85,86). It seems to be obvious that all treatments and conditions, inducing glutamate excretion, although they appear to be very diverse, may influence synthesis and composition of the plasma membrane or the cell wall of *C. glutamicum*. Based on this fundamental interpretation, the basic explanation for the trigger being responsible for glutamate excretion was for a long time the concept of an increased leakiness of the membrane or cell wall of *C. glutamicum* (79). In 1989, however, it was shown that glutamate excretion is not due to increased leakiness but, on the contrary, mediated by a specific, energy dependent transport system (8,87). This export system was shown to be responsible for glutamate excretion under conditions of a metabolic overflow situation. Unfortunately, in contrast to lysine, threonine, and isoleucine, the gene or genes encoding this excretion system were not identified until now. The fact that glutamate excretion has been experimentally demonstrated to depend on the presence of a specific export system, however, does not directly help to understand the influence of factors triggering glutamate export or the involvement of the cell wall in this process. This situation led to a revival of the "barrier hypothesis" (82) in which the permeability properties of the cell wall were directly made responsible for the ability to effectively excrete glutamate. All these concepts have to take into account that glutamate uptake systems are located in the plasma membrane, which, under conditions where no glutamate excretion can be observed, take up glutamate with high activity (41,88). Consequently, this requires efficient permeation of glutamate through the cell wall barrier. Thus, glutamate uptake has to take place at the same time that the barrier concept argues for a strictly limited permeability in the direction of efflux, making this concept unlikely for thermodynamic reasons. Furthermore, porin proteins have been identified in the *C. glutamicum* cell wall, providing permeability for hydrophilic solutes (85,89–91).

Concerning a direct involvement of the plasma membrane in glutamate efflux, it has been shown that an increase in passive glutamate permeability is not responsible for glutamate excretion (87). The concept that the observed increase in efflux is correlated with a change in the lipid composition also has to be abandoned, at least in the strict sense that membrane alteration is an essential prerequisite for glutamate export activity (92). It has recently been shown that *C. glutamicum* strains in which the OHDC has been disrupted become efficient glutamate producers without having a changed membrane composition (79). Another concept, related to changes of the physical state of the membrane triggering glutamate excretion, was based on the assumption of a changed viscosity of the cell wall

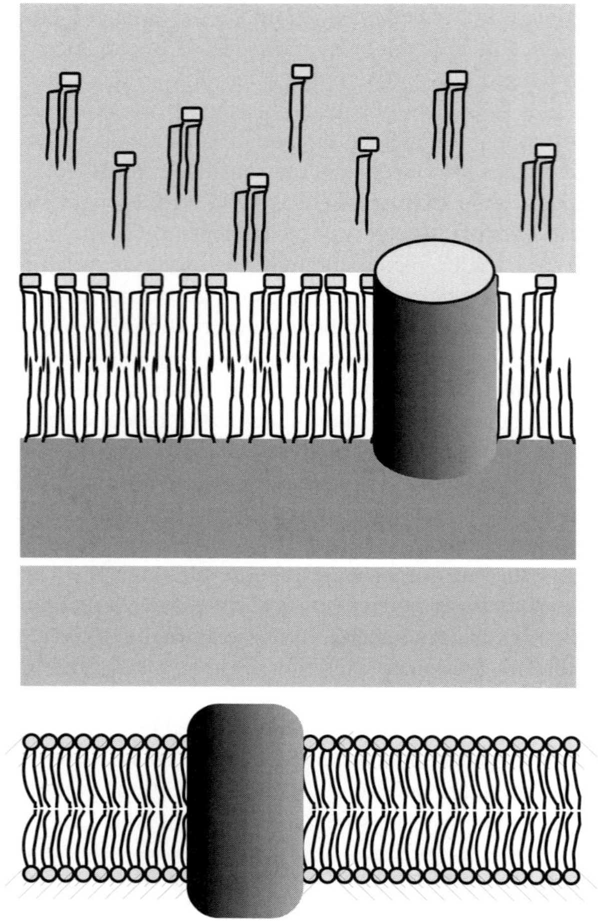

Figure 8.5 Cell wall of *C. glutamicum.* A) Phospholipid bilayer of the plasma membrane with embedded membrane proteins, e.g., amino acid transport systems. B) Peptidoglycan layer. C) Arabinogalactan layer. D) Mycolic acid bilayer of the cell wall. The inner leaflet consists mainly of bound mycolic acids covalently linked to the arabinogalactan layer, whereas free mycolates make up the outer leaflet. Porin channels are embedded in the mycolic acid layer. E) The outer layer mainly composed of polysaccharides also contains trehalose mono- and dimycolates, and in addition, proteins. Drawn after (85).

or plasma membrane under these conditions (93), which, however, was contradicted by results obtained later (82). Nevertheless, the fact remains that the capacity of glutamate excretion seems somehow to be correlated to the state of the membrane or cell wall.

Recently a gene was identified (*dtsR*), which has structural similarity to genes of acetyl-CoA carboxylases, the deletion of which led to fatty acid auxotrophy, a decrease in ODHC activity, and glutamate excretion even in the presence of biotin. Consequently, *dtsR* was proposed to be the real target in detergent treatment as well as under biotin limitation. The future development of this line of research seems to be highly promising.

A possible conceptual way out of this dilemma is based on a hypothesis derived from experiments on the contribution of cell wall biosynthesis to cell division (94, M. Wachi, personal communication). In this concept, the obvious impact on cell wall integrity is thought to trigger a regulatory signal transduced to the cytoplasm via the plasma membrane in *C. glutamicum*, leading to unknown mechanisms resulting in down regulation or inactivation of the oxoglutarate dehydrogenase, and possibly to activation of the glutamate excretion system.

In addition to the impact of cell wall properties on glutamate efflux, particular properties of the central metabolism, particularly the availability of precursor molecules, such as oxaloacetate, in *C. glutamicum* are also of significance for glutamate production. Consequently, both understanding and possible manipulation of the anaplerotic node in *C. glutamicum*, responsible for the supply of oxaloacetate, especially the activity of pyruvate carboxylase, play a pivotal role for effective glutamate production (21,79). The relative contribution of several enzymes being in principle able to contribute to oxaloacetate (the essential precursor compound) supply, namely phospho*enol*pyruvate carboxylase (PEPC), phospho*enol*pyruvate carboxykinase (PEPCk), pyruvate carboxylase (PC), malic enzyme (ME), and the glyoxylate pathway, has been estimated in a number of articles. Most investigations show that PC may possibly be responsible for the greatest contribution to the oxaloacetate pool (24,95,96), whereas the authors of another study using high temperature triggered glutamate excretion came to the conclusion that PEPC may be the most important enzyme (81,96). This result is highly relevant, in particular because PC contains biotin and the latter study was carried out under conditions of biotin limitation.

Flux analysis using NMR spectrometry has turned out to be very helpful in interpreting the metabolic changes related to conditions of active glutamate excretion. The carbon flux distribution within the central metabolism has been evaluated (Figure 8.6). The relative contribution of glycolysis in comparison to the pentose phosphate pathway was found to be increased under conditions of glutamate excretion (37,97), and the bidirectional carbon flux between pyruvate and oxaloacetate was decreased significantly (97). It should be mentioned

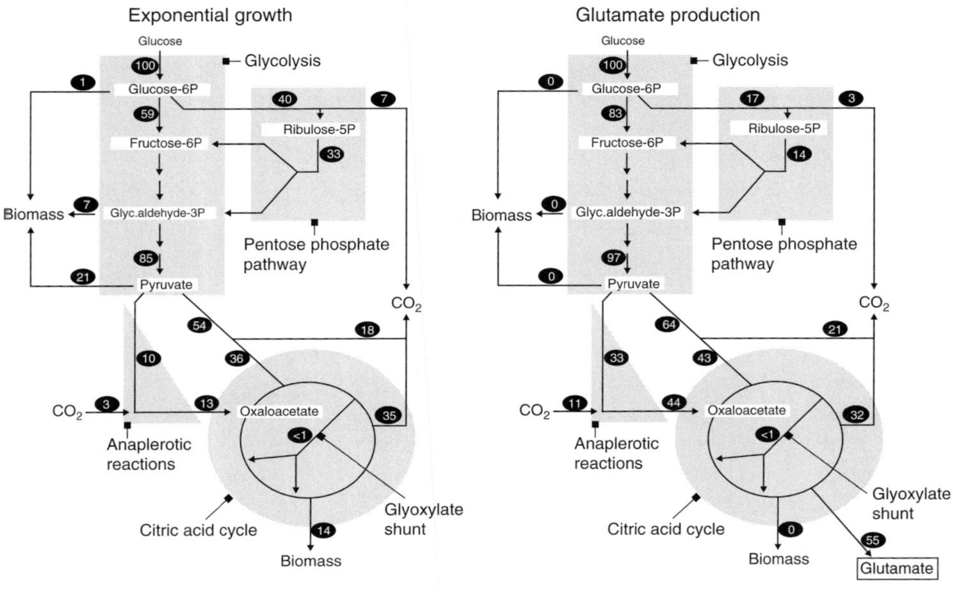

Figure 8.6 Metabolic flux distribution in *C. glutamicum* under conditions of growth and glutamate production. The major metabolic pathways starting from the substrate glucose and leading to biomass, CO_2 and glutamate are represented. The numbers give relative values of metabolic fluxes (see text). Adapted from (112).

that an opposite effect was found for lysine production in both cases. Also, massive flux redirection occurs with respect to anaplerotic reactions and biomass formation (Figure 8.6).

8.5.2 L-phenylalanine

Phenylalanine production is particularly important because of its use in the synthesis of the low calorie sweetener aspartame (methyl ester of the dipeptide L-aspartyl-L-phenylalanine), which is 150–200 times sweeter than sucrose. Besides fermentation, phenylalanine is also produced using chemical synthesis. For microbial production of L-phenylalanine mainly *E. coli* and *C. glutamicum* (and related species) are used (98). Strains have been obtained which are able to produce up to 50g/l of L-phenylalanine (99).

The biosynthetic pathway of aromatic amino acids is divided into a common part leading to chorismate from which the three terminal pathways to L-phenylalanine, L-tyrosine and L-tryptophan branch off (100). This pathway is efficiently regulated in *E. coli* by several different mechanisms both on the level of gene expression and protein activity (Figure 8.7).

Development of phenylalanine production in *E. coli* was based on several strategies to increase the carbon flow of aromatic amino acid biosynthesis. (31,101). The availability of an essential precursor molecule, erythrose-4-phosphate has been increased by modifying the expression of enzymes of the pentose phosphate pathway, namely transketolase (102) and transaldolase. PEP is a key intermediate in the synthesis of aromatic amino acids as well as in many other metabolic reactions. Consequently, efforts have been made to improve the supply of this compound either by direct means (31,101), e.g., by inactivating enzymes that compete for PEP, increasing the flux to PEP, and reducing the use of PEP for sugar uptake via the PTS, or by modulating general networks of carbon control in *E. coli* (103). In addition to increasing the availability of precursors, the specific

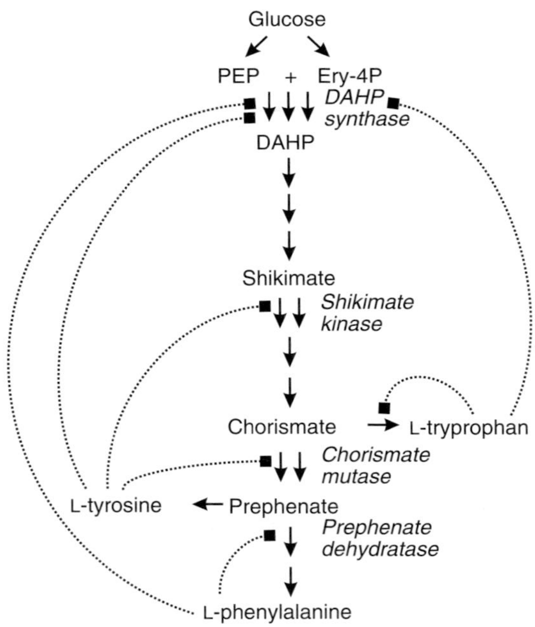

Figure 8.7 Biosynthesis pathways of aromatic amino acids. Simplified version of L-phenylalanine synthesis in *E. coli* including pathways to L-tryptophane and L-tyrosine. The dotted lines indicate regulation mechanisms, both feedback and transcriptional control are active at the indicated points.

pathways responsible for aromatic amino acid biosynthesis have also been successfully engineered. Alleviation of feedback inhibition of 3-deoxy-d-arabino-heptulosonate 7-phosphate synthase (DAHP) synthase isoenzymes by phenylalanine and tyrosine has been achieved, previously through application of appropriate amino acid analogues, and presently by site specific mutation. The genes encoding the enzymes chorismate mutase and prephenate dehydratase have been over expressed to increase the flux toward phenylalanine (101). In general, phenylalanine producers are tyrosine auxotrophs. Consequently, enzymes of the common pathway of aromatic amino acid synthesis are no longer feedback inhibited by this amino acid; moreover, tyrosine accumulation as a result of the shared pathway is prevented. A further beneficial effect is the possibility of a controlled growth limitation during the production phase by appropriate tyrosine feeding.

8.5.3 L-tryptophan

L-tryptophan is an amino acid with increasing potential commercial interest. It is currently mainly used as feed additive, despite its value as an essential amino acid, because of a number of casualties of eosinophilia myalgia syndrome (EMS) in humans which occurred due to the consumption of impure L-tryptophan manufactured by fermentation. L-tryptophan is produced by strains of *E. coli* (104) and *C. glutamicum* (105). In order to obtain effective tryptophan producing organisms, similar alterations of precursor pathways and of the biosynthetic route as described above for phenylalanine production have been carried out. In *E. coli*, it was not necessary to block the route to phenylalanine and tyrosine within the common pathway to aromatic amino acids, which would have led to auxotrophy for these two amino acids (106). Overexpression of the gene encoding the anthranilate synthase (*trpE*), which is the first specific enzyme in the tryptophan synthesis pathway, was sufficient to divert carbon flux to L-tryptophan. In addition, the gene encoding tryptophanase (*trnA*), the enzyme responsible for tryptophan degradation in *E. coli*, was removed in order to optimize tryptophan production (107).

Tryptophan production in *C. glutamicum* is an example where it has been shown that product uptake or reuptake may, in fact, be important in the production process. Engineering a decrease in tryptophan uptake led to a significant enhancement of tryptophan production (108). It was argued that removal of tryptophan uptake results in lower internal tryptophan concentrations and thus reduced feedback effects. Most interestingly, it was shown that tryptophan accumulation by the mutant lacking L-tryptophan uptake was improved mainly in the final part of the fermentation, when high external tryptophan concentrations were accumulated at the outside (31), thus illustrating the advantage of this strategy under conditions of a high backpressure from the external product.

8.5.4 L-lysine

Although lysine is used predominantly as a feed and not as a food additive, it is the best studied example of an amino acid production process in terms of molecular and biochemical aspects. Lysine biosynthesis and production have been intensively investigated with respect to the anabolic pathways involved, pathway regulation, and excretion. Consequently, it may serve here as a paradigm for understanding the tools available and the approaches used in the production of these kinds of amino acids. This holds true for several aspects which seem to be similar in the production of a number of these metabolites, namely pathway fine tuning and regulation, as well as metabolic connection to related pathways; the significance of central pathways for precursor supplementation, in particular the importance of the anaplerotic node; the impact of the cytoplasmic redox and energy balance; the demonstration of how helpful flux analysis may turn out to be in understanding amino acid biosynthesis pathways within the complex cellular metabolic network; and the fundamental demonstration of how important active and energy dependent export systems may be in the course of amino acid production.

Lysine is solely produced by various strains of *C. glutamicum* in fermentative processes, and is used predominantly as a feed additive. Large scale production started in 1958 in Japan. Worldwide production is about 550,000 tons of L-lysine annually (72).

Lysine, like methionine, threonine and isoleucine, is a member of the aspartate family of amino acids. In *C. glutamicum*, the biosynthetic pathway is split into two branches in its lower part (Figure 8.7). The first enzyme of this pathway, aspartokinase, represents the major control point of carbon flow. It is inhibited in *C. glutamicum* by its end products lysine and threonine; consequently, all lysine producing strains contain a deregulated form of this enzyme, which, in general, carries a single amino acid exchange at position 279 of the α-subunit (109). The following were identified as being important in the flux control of this pathway (110,111): (1) deregulation of the aspartokinase (*LysC*); (2) increase of the activity of a feedback resistant aspartokinase, typically by overexpression; (3) increase in the activity of the dihydrodipicolinate synthase (*DapA*), the enzyme situated right after the branching point to the other members of the aspartate family of amino acids; (4) availability of oxaloacetate as precursor for aspartate; (5) availability of NADPH as reducing equivalent: redox balance; and (6) high activity of a specific, energy dependent lysine excretion system *(LysE)*.

Another interesting point about lysine biosynthesis in *C. glutamicum* is that the pathway is split at the level of piperideine-2,6-dicarboxylate (Figure 8.8). In contrast, *E. coli* possesses only the succinylase variant, and some *Bacillus* species only the dehydrogenase variant. One reason for this redundancy is the fact that D,L-diaminopimelate is an essential precursor for cell wall biosynthesis. Only flux analysis using NMR spectrometry revealed that the two pathways are used to highly different extents under differing environmental conditions (112). Because of a low affinity toward its substrate ammonium, the dehydrogenase cannot contribute significantly to lysine biosynthesis at low ammonium concentrations. At high concentrations, however, this pathway is responsible for about 75% of total flux to lysine. If either the succinylase or the dehydrogenase variant is inactivated, lysine production is reduced to about 40%. Consequently, both pathways are functionally combined to ensure the proper supply of D,L-diaminopimelate for the cell.

Last but not least, lysine is an ideal example to demonstrate the importance of the presence of an appropriate excretion system for amino acid production. By 1991, lysine export was shown to depend on the presence of a specific excretion system (11,113). Lysine export is coupled to the electrochemical proton potential, which provides the driving force for its active extrusion from the cytoplasm. Moreover, this system was the first amino acid export system, the encoding gene of which had been identified (14). *LysE*, the lysine excretion carrier, represents a relatively small membrane protein of 25.4 kDa with 5-6 transmembrane domains, presumably being present as a dimer in its active state. If this transport system is deleted, no further export of lysine is possible. Upon addition of lysine containing peptides, the internal lysine concentration reaches toxic values of up to 1 M in this mutant strain of *C. glutamicum*, thus providing an elegant proof for the significance of amino acid excretion carriers in cases of internal amino acid accumulation. *LysE* was the first member of a growing family of transport proteins involved in metabolite export. In addition to the lysine exporter *LysE*, a corresponding transcription factor, named *LysG*, was identified and later characterized (14).

Lysine production is one of the best studied examples of flux analysis in bacteria. Several techniques have been employed for the elucidation of flux distribution in *C. glutamicum* under conditions of lysine production in comparison to normal growth conditions (25). Moreover, methods of flux analysis using ^{13}C-NMR spectrometry

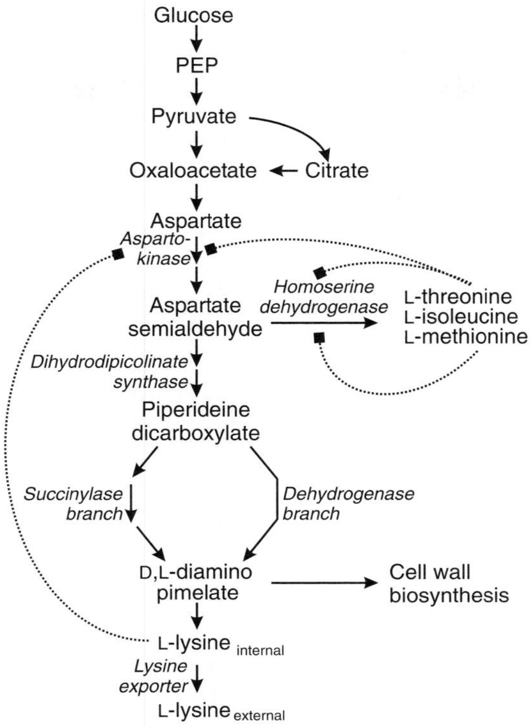

Figure 8.8 Lysine biosynthesis pathway and regulation in *C. glutamicum*. Simplified version of the biosynthesis pathway of the aspartate family of amino acids branches at aspartate semialdehyde to lysine, and to methionine, threonine and isoleucine, respectively. Some reactions of precursor supply are also indicated. The lysine anabolic pathway is split at the level of piperideine-2,6-dicarboxylate into the succinylase branch (4 enzymes, supply of succinyl-CoA and glutamate) and the dehydrogenase branch (1 enzyme, supply of NADPH and NH_4). NADPH is further needed in the steps leading to aspartate semialdehyde and piperideine-2,6-dicarboxylate. Pyruvate is added at the dipicolinate synthase. D,L-diaminopimelate is a precursor of cell wall biosynthesis. Feedback control mechanisms are indicated by dotted lines.

(21,69,95,97,114) as well as MALDI-TOF mass spectrometry (28,67–71) led to a detailed description of the activity of particular enzymes of central and peripheral pathways during lysine production. This made quantification of anaplerotic fluxes, of glycolysis in relation to the pentose phosphate pathway, and the individual share of the two branches of the split lysine pathway possible. Furthermore, particular modifications applied during metabolic engineering approaches could be controlled with respect to their impact on individual fluxes within the reaction network related to lysine production.

An interesting application of this approach concerns the redox balance under production conditions. For several steps within lysine biosynthesis, reducing power in the form of NADPH has to be provided. By using ^{13}C-NMR analysis, the supply of NADPH by the pentose phosphate pathway and by isocitrate dehydrogenase could be calculated. The share of the carbon flux through the pentose phosphate pathway (NADPH generating), in the total flux through the sum of glycolysis (NADH generating), and in the pentose phosphate pathway was found to be significantly increased under lysine production due to the demand of NADPH (68,97). A comparison of this flow and the total flow of NADPH in *C. glutamicum* under conditions of effective lysine production indicated that a limitation by the availability of NADPH is likely. In a recent study in which various lysine producing strains with different lysine yields were systematically compared, an interesting close correlation of lysine yield and the extent of flux through the pentose phosphate pathway as well as the extent of anaplerotic flux was observed (71). The increase of nicotinamide nucleotide transhydrogenase activity in bacteria has also been demonstrated as a successful approach to overcome problems with respect to the redox balance due to a limitation in the NADPH supply (125).

8.5.5 Threonine

Besides lysine, threonine is another member of the aspartate family of amino acids. Like the essential amino acid lysine, threonine is mainly used as a feed additive and its production will thus be described here only briefly. After glutamate and lysine, threonine ranks third in the production volume with about 20,000 tons annually. In contrast to the previous two amino acids, threonine is produced using *E. coli* strains (116). However, studies to improve the application of *C. glutamicum* for threonine production have been reported (117,118). In addition to its commercial significance, threonine is a highly instructive example when reviewing systematic strategies to improve amino acid production. Above and beyond the common features essential for amino acid producing strains, as already discussed for lysine, development of threonine producers is interesting because of two major points. Development of threonine producing strains resembles a straightforward approach in combining classical strain breeding techniques with recombinant DNA technology. It is an example where a broad combination of metabolic changes finally led to the desired result. Among these changes is the introduction of the capacity to use alternate carbon sources as well as the abolishment of amino acid degradation pathways.

In the late 1960s an *E. coli* strain was described which produced threonine because of the presence of a mutant aspartokinase resistant to feedback inhibition by threonine, which was achieved by selection using the threonine analog α-amino-β-hydroxyvaleric acid (119). In the years 1977–1992, a highly productive *E. coli* strain was developed by the group of Debabov (116). The biosynthetic pathway leading to threonine was systematically optimized by several rounds of mutation and selection as well as genetic engineering. In addition to pathway engineering, the production strain was equipped with the capacity to use sucrose as a carbon source, in contrast to wild-type *E. coli* strains which cannot use sucrose. A further significant improvement of threonine production was achieved by reducing threonine degradation. For this purpose, the major threonine degrading enzyme, threonine dehydratase, was inactivated. At the end, a strain producing up to 100 g/L of threonine within 36 h of fermentation was obtained with practically no other products in significant amounts (116).

Threonine is also an example where knowledge on substrate transport in general and on specific excretion systems in particular is available. In *E. coli*, a number of threonine uptake systems are present, the correct contribution of which is not yet understood completely. Interestingly, it has been shown that threonine producer strains have acquired a significantly reduced capacity for threonine uptake in the course of strain construction (74,130). Consequently, expression of the *tdcC* gene encoding an L-threonine uptake system led to decreased threonine production (31). This is another example demonstrating the significance of avoiding product reuptake for effective production strains.

Altogether three genes are assumed to encode for threonine export systems in *E. coli*, namely *rhtA*, *rhtB* and *rhtC* (42), which all belong to a large family of carrier proteins the function of which has not yet been elucidated. The participation of some of these carriers in threonine export, however, has been questioned (120). Although it seems clear that at least one of these transport proteins contributes to threonine export, the situation does not seem to be fully clear yet (39).

Threonine export has been characterized on the molecular level in the case of *C. glutamicum* (43). Also in this case, the recently identified exporter protein *ThrE* again represents the first member of a new transporter family. In the course of the years, there has been considerable effort to optimize *C. glutamicum* with respect to threonine production also (116,117,118), however, the production capacity of these strains never reached that of good *E. coli* strains, due to a relatively low export capacity, as had been found on the biochemical level before (12,121). In agreement with this explanation, in a recent publication it was shown that both reduction of the activity of the major threonine degradation pathway (serine hydroxymethyltransferase, *glyA*) and overexpression of the threonine exporter *ThrE* led in fact to an increase in threonine productivity (118).

8.5.6 Other Amino Acids

A number of amino acids, which are of significant relevance for the food industry, e.g., L-aspartic acid (besides L-phenylalanine necessary for synthesis of aspartame), cysteine, glycine, and alanine, are still produced by chemical synthesis, by enzymatic methods, or by using immobilized cells (6). In recent years, new developments have been described which may lead to introduction of fermentation procedures for the production of at least some of these amino acids.

A new and promising example in this direction was recently reported for cysteine, an amino acid used as a flour additive, which is produced mainly by extraction or enzymatic methods because of the lack of effective producer strains. Cysteine biosynthesis is in general tightly regulated because of its dependence on sulphur metabolism as well as its potential danger of toxic or unwanted side products. Based on recombinant *E. coli* strains, a process leading to significant cysteine production has recently been developed (122). In the development of a cysteine producing strain, care had to be taken not to simply over express the respective genes encoding cysteine biosynthesis enzymes, but to guarantee a coordinated increase in the flux via this end product, in order to avoid accumulation of intermediates which potentially lead to toxic side products. Interestingly, in the effort to optimize cysteine production in *E. coli* strains, two specific exporters related to cysteine efflux were discovered. First, the gene *ydeD* was identified which after over expression significantly increased the efflux of thiazolidine derivatives of cysteine (45). These derivatives are formed intracellularly by the reaction of cysteine with ketones or aldehydes such as pyruvate or glyoxylate. They can be cleaved again in the external medium. The successful screening strategy was based on the property of genes to augment cysteine production in an industrial producer strain. The same strategy led recently to the identification of a further gene, *yfiK*, the product of which had a similar effect on efflux of cysteine in *E. coli* (46). These two transport proteins are both able to excrete O-acetylserine and cysteine, as well as derivatives of it (*YdeD*). Interestingly, a further, completely different cysteine exporter was recently identified in *E. coli*, obviously related to redox homeostasis during cytochrome biosynthesis (15).

8.6 FUTURE DEVELOPMENTS AND PERSPECTIVES

Classical procedures of strain breeding by mutation combined with selection and screening procedures have proven in the past to be powerful tools for providing efficient amino acid producing strains. With easy access to modern molecular techniques, strain improvement on the basis of detailed biochemical knowledge becomes more and more important. The obvious strategy applied previously was directed toward identifying and overcoming metabolic bottleneck reactions. This strategy, however, was frequently not as successful as expected, in particular when it was applied to strains which already had a substantial capacity of amino acid production. Production strains which have evolved through many rounds of mutagenesis and selection are already fine tuned in their metabolic setup for increased production to a significant extent. Genetic manipulation that has strong impact on metabolism, as is normally the case when deleting or strongly over expressing particular genes, frequently leads to disruption of the subtle metabolic balance already obtained in these strains.

In the era of rational metabolic design, we need a combination of data and knowledge from many different fields for understanding bacterial metabolism, to design and to further optimize amino acid producing strains effectively. Process optimization needs an increase in systemic knowledge of how a bacterial cell functions in all relevant aspects. This strategy asks for more direct and accurate methods to analyze and monitor the current status of bacterial cells during fermentation. In the following, a variety of tools presently available for analysis and understanding of the complex network of bacterial metabolism will be reviewed.

The availability of a large number of bacterial genomes, including those of biotechnological relevance such as *E. coli* and *C. glutamicum*, fostered the direct application of genomic information for the design of production strains (30). Genome annotation was helpful for defining the inventory of enzymes and pathways with putative biotechnological significance. Post genome approaches, i.e., transcriptome and proteome analysis, are currently applied with increasing frequency. Several examples have been mentioned previously, such as manipulations within the anaplerotic node of reactions in *C. glutamicum*. A recent, very instructive example is represented by the design of a new production strain utilizing a novel procedure, designated "genome breeding" by the authors (31,123). Based on comparative genome analysis of wild-type and production strains, a set of mutations was identified by which these strains differ. Based on physiological and biochemical knowledge, mutations which were supposed to be relevant for amino acid production were selected and introduced step by step into the wild type strain. The response of these manipulations on amino acid excretion finally led to a production strain with high productivity, but a minimal set of known mutations. In the example mentioned (123) the authors succeeded in creating a lysine producing strain carrying single amino acid replacements in just three enzymes, namely aspartokinase, homoserine dehydrogenase, and pyruvate kinase. The strain obtained showed extremely high lysine productivity, and, due to the fact that growth and glucose consumption was almost as fast as in the wild type, the fermentation period was shortened to nearly half of that normally required. Further developments are to be expected with a systematic application of this approach, exploiting the large variety of available useful mutants which have originated during the long history of classical strain breeding. The aim of this strategy is to combine beneficial genetic properties identified in the genomes of various mutants into newly engineered producing strains (30). Future efforts in this direction necessarily need more input from biochemistry, given that the function of a large number of potentially interesting gene products, in particular transport proteins, are presently not known.

The availability of the genome is the basis for analyzing the transcriptome of a cell, i.e., the sum of mRNA synthesized at a given time. The information obtained with DNA array techniques may be extremely relevant in the near future for strain improvement in two respects. It is the basis for identifying regulons and modulons, sets of genes which are regulated in a concerted manner under particular metabolic states or conditions of environmental stress. This information can then be used to understand connections within local and global regulatory networks, which are relevant to the biosynthetic pathway under study. Instructive examples in connection with central carbon metabolism were recently reported.

DNA microarray analysis led to the finding of the reciprocal regulation of two genes encoding two different glyceraldehyde-3-phosphate dehydrogenase enzymes with different cofactor specificity, in dependence on the flux direction through the central metabolic pathways (54). A combination of microarray analysis and transcriptional analysis of specific genes revealed a surprising complexity of the regulon for acetate utilization in *C. glutamicum* (134).

Furthermore, DNA arrays are in principle a promising future tool for direct process monitoring. Efforts are underway in many labs to establish systems in which DNA array techniques and information on metabolic patterns is combined, in order to be able to follow the metabolic state of a cell suspension in a fermentor as closely as possible to an online situation. If this kind of analytic information were available, the performance of a bacterial culture in a fermenter, in particular in a prolonged fermentation mode, could be significantly improved. Fermentation may be run at the edge of the mass transfer capacity of the fermentor given precise monitoring of the physiological state of the cell (72). This is still far from being realistic due to technical reasons, mainly with respect to the time scale between drawing a sample and obtaining the transcriptome pattern, as well as the routine cost. This kind of monitoring system would be highly attractive for the purpose of keeping bacteria constantly at a metabolic state of optimal performance.

Proteome data will certainly be helpful, too, in the future, in particular if they are combined with information on posttranslational modification of the synthesized proteins, e.g., phosphorylation (60). Posttranslational modifications, which are not accessible by DNA array techniques, are potentially of high interest and may play an important role for regulation networks, in particular in central carbon metabolism. So far, the major effort is still directed toward a complete inventory of the proteins synthesized, mainly due to the fact that some classes of proteins, among them membrane proteins and low abundance proteins such as transcription factors, are still not detectable by 2D gel electrophoresis (57–59). The same holds for another developing tool, namely the investigation of protein to protein interactions, which may become a significant help in unravelling regulatory networks based on direct protein crosstalk.

Certainly a tool of utmost relevance will be metabolome analysis, based on gas chromatography (GC)- or liquid chromatography-mass spectrometry (LC-MS) techniques. As a matter of fact, it is not widely recognized that we still do not have an idea of the total inventory of even the major metabolites in any single bacterial cell. It is obvious that this information would be extremely helpful for understanding the differences between certain metabolic conditions, for example, during fermentation, or for identifying metabolic changes in production strains as compared to the wild type. Moreover, metabolome data represents the basis for sophisticated metabolic flux analysis.

Several examples of the value of metabolic flux analysis combined with metabolite balancing have been described in this chapter. With the introduction of NMR and, in particular, MS techniques into this field, a complete analysis of the dynamic situation of cell metabolism (the fluxome) became, in principle, accessible for routine studies. The evaluation of the significance and value of particular changes in the metabolism of bacterial strains under different metabolic situations seems to be an essential basis for a rational metabolic design. Flux analysis is furthermore a prerequisite for understanding the dynamic situation of cell metabolism including all regulatory connections. Last but not least, it is obvious that the information obtained by various levels of analyses mentioned here can be integrated into a systems biology approach only with the help of powerful bioinformatics tools.

Reaching the aim of an increase in the systemic understanding of cell metabolism, however, does not only require global approaches, but certainly depends to a large extent on additional information from biochemistry and physiology. An instructive example mentioned above is the identification and analysis of specific export proteins to be responsible for amino acid excretion. Furthermore, the architecture of complex regulatory networks cannot be fully elucidated without the help of biochemical techniques.

Although not the topic of this chapter, further development in fermentation techniques is still a major option for improvement of amino acid production, based on detailed knowledge of biochemistry and physiology of the bacterial cell. A potentially promising example concerns the type of fermentation procedure to be used. It has been discussed before that continuous fermentation would in principle provide an interesting method with several advantages, which, however, due to some drawbacks is not yet used for amino acid production. Continuous fermentation needs coupling of growth and amino acid production, which is in fact frequently observed for lysine production by *C. glutamicum*. In a number of procedures, growth and production are artificially uncoupled to some extent, as in for threonine production by *E. coli*. Unfortunately, it is not yet well understood in terms of metabolic, energetic, and regulatory reasons how and to what extent amino acid excretion can be coupled to growth under optimum production conditions.

In the preceding part of this chapter, tools available for future developments in amino acid production have been listed. However, it seems interesting to identify targets for these efforts which seem to be attractive on the basis of current knowledge. To a large extent, this would be *C. glutamicum*, because the wealth of biochemical information available is certainly more impressive for *E. coli*.

In spite of the fact that central and peripheral metabolism in *C. glutamicum* has been studied to a significant extent, this does not hold for sugar metabolism and, in particular, for sugar uptake. Although the flux through glycolysis has been analysed, the regulation of important enzymes, such as fructose bisphosphate kinase, is not yet known. Interestingly, two genes encoding isoenzymes of each the fructose bisphosphate kinase and of the glyceraldehyde-3-phosphate dehydrogenase have been found in the genome of *C. glutamicum* (72). It is well known from *E. coli* that the phospho*enol*pyruvate dependent phosphotransferase systems (PTS) are, besides their significance in sugar uptake, important for regulating other uptake systems and pathways of central carbon metabolism. In several cases, as in for glutamate production, it is not yet known how the cell actually regulates the activity of amino acid excretion systems in dependence of the availability of the carbon and energy supply. It seems to be a plausible hypothesis that sugar uptake via PTS may be involved in this kind of regulation. Moreover, when the carbon flow to amino acid end products in the course of process optimization is further increased, at some point substrate uptake may become limiting in the overall pathway to the product. A further aspect of engineering PTS systems for sugar uptake may be noted here. Efforts have been made to bypass this uptake system in *E. coli* in connection with production of aromatic amino acids, in order to increase the cellular content of phospho*enol*pyruvate, which is both a precursor of aromatic amino acids and the energy donor of the PTS system. In fact, this manipulation resulted in a higher yield of phenylalanine (101), representing another example of transport design, different from that mentioned in connection with reuptake of substrate (tryptophan), as well as in connection with excretion systems.

Certainly a core field of interest in future development will be regulatory networks, including response to stress conditions. Whereas regulation of particular biosynthetic pathways seems to be quite well understood, both in *E. coli* and *C. glutamicum*, this is not true to the same extent for global regulatory mechanism, including the response of the cell to limitation of carbon (carbon control) or nitrogen sources (nitrogen control). Both control systems are well studied in *E. coli* (135), and the latter has also been elucidated in *C. glutamicum* (18,136). Unfortunately, this does not hold for the important network of carbon control in this bacterium. Knowledge on this topic would be furthermore attractive for investigating suspected regulatory interactions between the signal transduction mechanisms of carbon and nitrogen control, which would certainly be essential for a full understanding of the integration of amino acid metabolism in the cellular network. The wealth of regulatory mechanisms as elucidated in *E. coli* to a significant extent by numerous studies is unfortunately not available in *C. glutamicum*. This refers to the variety of sigma factors, transcription factors, two component systems, attenuation mechanisms, RNA stability, posttranslational modifications, enzyme stability, and degradation. A better knowledge of these mechanisms and their significance for amino acid synthesis would also be the basis for the design of appropriate DNA arrays for controlling the status of fermentation processes.

These regulation networks are also relevant for cellular mechanisms of stress response. Bacterial cells are known to possess sophisticated strategies for stress response (heat, osmotic stress, nutrient limitation, and many others) which involve coordinated regulation on the level of gene expression and protein activity. A well studied example is the sigma B regulon in the Gram-positive *Bacillus subtilis* (127). In case of *C. glutamicum*, only the response to osmotic stress (19,128), nitrogen limitation (18,126), and, to some extent, phosphate limitation (129) have been studied in detail so far. Because enforcement of high yield production in fermentors is certainly a kind of stress in terms of growth limitation, pathway rearrangement, concentration of nutrients, and products, both specific and general stress responses will be active under these conditions and need to be further elucidated.

Another field of significant importance, which is still neglected to some extent, is energy metabolism. This is of particular interest for glutamate production in *C. glutamicum*, which seems to be related to a metabolic overflow situation including mechanisms wasting metabolic energy such as futile cycling (130). Information is available on architecture and organization of the respiratory chain and related redox enzymes in *C. glutamicum* (131–136), whereas the role of the core energy converter ATPase and that of possible futile cycles under conditions of metabolic overflow (glutamate production) is not yet clear. Interesting data have recently been published indicating a close relation of ATPase activity and efficiency of glutamate excretion in *C. glutamicum* (137).

In view of an integrative pattern of reactions being involved in amino acid production (Figure 8.1), the presence, activity and regulation of amino acid excretion systems seems also to be highly relevant. Despite the fact that the importance of active amino acid excretion has been elegantly proven in the case of lysine (14,138), increasing knowledge of these systems and their availability for constructing recombinant producing strains has so far not successfully been integrated into application, with just one exception. The possibility has to be taken into consideration, of course, that, at least in the case of lysine excretion, the activity of the exporter may not be limiting even in producer strains. As has been mentioned above, overexpression of a gene coding for a secondary carrier protein has significantly improved export of cysteine derivatives in engineered *E. coli* strains (45). It has recently been reported that for threonine excretion in a recombinant *C. glutamicum* strain, overexpression of the gene encoding the threonine exporter in fact improved threonine accumulation in the medium. (118). The observed positive effect may be due to the fact that the capacity of the amino acid export systems present, and is very limited in both cases reported, which was in fact shown for threonine in *C. glutamicum* (12,121).

In any case, it would be of utmost interest to identify the export system responsible for excretion of glutamate in *C. glutamicum*, which has so far been characterized only at the biochemical level (87,139). Furthermore, the observation that many more products can, in principle, also be observed to be excreted under particular physiological conditions by *C. glutamicum*, *E. coli* and a number of other bacteria, strongly argues for the presence of many more export systems. Because the discovery of some of these systems has already led to the definition of completely new families of transport proteins, more “surprises” of this kind have to be expected.

A truly special property of *C. glutamicum* is the dependence of amino acid excretion on the integrity of the cell wall or plasma membrane. Although in general only discussed in connection with glutamate export, it is important to note that the excretion of other amino acids, such as lysine and proline is also found to be increased when the cell wall is affected . Several different hypotheses for explaining the contribution of the cell wall to amino acid excretion have already been discussed in section 8.5. If a likely explanation were correct, that of a regulatory connection between activity of cell wall synthesis and enzymes of the central metabolism in *C. glutamicum*, it would add an interesting target for further optimization of amino acid excretion for application of molecular and biochemical techniques. Recent publications indicate significant progress toward an improved understanding of cell wall biosynthesis in *C. glutamicum* (140–142).

Finally, after discussing possible tools and targets for future optimization strategies, it is worth briefly discussing more general aims in amino acid production, other than the specific strategies of increasing metabolic fluxes toward the desired end product. In the case of low price products, such as glutamate, lysine, or threonine, future efforts will concentrate on strategies to increase the yield of amino acid production processes. Current yields (g amino acid / g sugar) are 0.45–0.55 for glutamate, 0.4–0.5 for lysine and threonine, and 0.2–0.25 for phenylalanine (31). It is very difficult to calculate true maximum yields, because some additional factors such as ATP stoichiometry and efficiency of energy metabolism are not exactly known. It is obvious, however, that the enormous heat production by bacterial cultures during industrial fermentation due to the high respiration activity argues for a partially uncoupled energy metabolism. Whether this is an intrinsic property of respiratory chain activity and energy metabolism under conditions of amino acid production, or whether it is due to unknown futile cycles is not clear. Consequently, it may in principle be an interesting future target for improvement of efficiency.

Besides improving already available processes in *C. glutamicum* and *E. coli*, efforts on the design of new products or on new processes for available products are a constant challenge for research in amino acid production. The example of L-cysteine production has been mentioned above, where a new process in *E. coli* characterized by novel aspects of strain construction has been reported recently. It may also be possible and attractive in the future to develop a fermentation process for methionine, the only amino acid which is still chemically produced in racemic form in bulk quantities.

For research in microbial biotechnology, as well as for application in industrial processes, a broad basis of physiological, genetic, and biochemical knowledge would be highly attractive. In spite of all classical and modern tools available, we are still far from a complete understanding even of a single bacterial cell in terms of systems biology. The methods applied so far are increasingly well suited for a descriptive understanding; however, they are clearly not detailed enough to be used for predictive statements. Consequently, the final goal of this multidisciplinary work should aim at the "virtual bacterial cell," an aim which is certainly still far from reality, both in the case of *E. coli* and *C. glutamicum*. Due to the combined effort of many groups working on a multilevel analysis of these microorganisms, however, it seems highly probable that a bacterial cell will be the first cell to be understood on that level.

The introduction of techniques of amino acid production by using microorganisms has led to an enormous increase in the beneficial use of these compounds. The application of the molecular tools available will certainly help to overcome remaining limitations, provided these tools can be applied. In that sense, further improvement of amino acid production in the food industry will only be economically relevant if application of genetically engineered bacterial producer strains will not be restricted to an unnecessary extent.

REFERENCES

1. Hodgson, J. The changing bulk biocatalyst market. *Biotechnology* 12:789–790, 1994.
2. Kinoshita, K., S. Udaka, M. Shimono. Studies on the amino acid fermentation, I: production of L-glutamic acid by various microorganisms. *J. Gen. Appl. Microbiol.* 7:193–205, 1957.
3. Liebl, W., M. Ehrmann, W. Ludwig, K.H. Schleifer. Transfer of *Brevibacterium divaricatum* DSM 20297T, "*Brevibacterium flavum*" DSM 20411, "*Brevibacterium lactofermentum*" DSM 20412 and DSM 1412, and *Corynebacterium glutamicum* and their distination by rRNA gene restriction patterns. *Int. J. Syst. Bacteriol.* 41 :225–235, 1991.
4. Nelson, G., J. Chandrashekar, M.A. Hoon, L.X. Feng, G. Zhao, N.J.P. Ryba, C.S. Zuker. An amino-acid taste receptor. *Nature* 416:199–202, 2002.
5. Aida, K., I. Chibata, K. Nakayama, K. Takinami, H. Yamada. *Biotechnology of Amino Acid Production*, Vol. 24. *Progr. Ind. Microbiol.* Amsterdam: Elsevier, 1986.
6. Leuchtenberger, W. Amino acids: technical production and use. In: Biotechnology, Vol. 6, Products of Primary Metabolism, Rehm, H.J. and G., Reed, eds., Weinheim, Germany: VCH Publishers, 1996, pp 455–502.
7. Johnson-Green, P. *Introduction to Food Biotechnology.* Boca Raton: CRC Press, 2002.
8. Kramer, R. Genetic and physiological approaches for the production of amino acids. *J. Biotechnol.* 45:1–21, 1996.
9. Burkovski, A., R. Kramer. Bacterial amino acid transport proteins: occurrence, functions, and significance for biotechnological applications. *Appl. Microbiol. Biotechn.* 58:265–274, 2002.
10. Harold, F. *The Vital Force: A Study of Bioenergetics*. New York: Freeman, 1986, p 163.
11. Broer, S., R. Kramer. Lysine excretion by *Corynebacterium glutamicum*, 1: identification of a specific secretion carrier system. *Eur. J. Biochem.* 202:131–135, 1991.
12. Palmieri, L., D. Berns, R. Kramer, M. Eikmanns. Threonine diffusion and threonine transport in *Corynebacterium glutamicum* and their role in threonine production. *Arch. Microbiol.* 165:48–54, 1996.
13. Hermann, T., R. Kramer. Mechanism and regulation of isoleucine excretion in *Corynebacterium glutamicum. Appl. Environ. Microbiol.* 32:3238–3244, 1996.
14. Vrljic, M., H. Sahm, L. Eggeling. A new type of transporter with a new type of cellular function: L-lysine export from *Corynebacterium glutamicum. Mol. Microbiol.* 22:815–826, 1996.
15. Pittman, M.S., H. Corker, G.H. Wu, M.B. Binet, A.J.G. Moir, R.K. Poole. Cysteine is exported from the *Escherichia coli* cytoplasm by CydDC, an ATP-binding cassette-type transporter required for cytochrome assembly. *J. Biol. Chem.* 277:49841–49849, 2002.
16. Tempest, D.W., O.M. Neijssel. Physiological and energetic aspects of bacterial metabolite overproduction. *FEMS Microbiol. Lett.* 100:169–176, 1992.
17. Kramer, R. Analysis and modeling of substrate uptake and product release by prokaryotic and eukaryotic cells in. In: *Advances in Biochemical Engineering/Biotechnology* , Vol. 54, Scheper, T., ed., Heidelberg: Springer-Verlag, 1996, p 31–74.
18. Burkovski, A. I do it my way: regulation of ammonium uptake and ammonium assimilation in *Corynebacterium glutamicum. Arch. Microbiol.* 179:83–88, 2003.
19. Morbach, S., R. Kramer. Impact of transport processes in the osmotic response of *Corynebacterium glutamicum. J. Biotechnol.* 104:69–75, 2003.
20. Vrljic, M., W. Kronemeyer, H. Sahm, L. Eggeling. Unbalance of L-lysine flux in *Corynebacterium glutamicum* and its use for the isolation of excretion-defective mutants. *J. Bacteriol.* 177:4021–4027, 1995.
21. De Graaf, A.A., L. Eggeling, H. Sahm. Metabolic Engineering for L-Lysine Production by *Corynebacterium glutamicum.* In: *Advances in Biochemical Engineering/Biotechnology* , Vol. 73, Scheper, T. ed., Heidelberg: Springer-Verlag , 2001, p 9–29.
22. Peters-Wendisch, P.G., B.J. Eikmanns, G. Thierbach, B. Bachmann, H. Sahm. Phosphoenolpyruvate carboxylase in *Corynebacterium glutamicum* is dispensable for growth and lysine production. *FEMS Microbiol. Lett.* 112:269–274, 1993.

23. Peters-Wendisch, P.G., V.F. Wendisch, A.A. de Graaf, B.J. Eikmanns, H. Sahm. C-3-carboxylation as an anaplerotic reaction in phosphoenolpyruvate carboxylase-deficient *Corynebacterium glutamicum. Arch. Microbiol.* 165:387–396, 1996.
24. Petersen, S., A.A. de Graaf, L. Eggeling, M. Mollney, W. Wiechert, H. Sahm. *In vivo* quantification of parallel and bidirectional fluxes in the anaplerosis of *Corynebacterium glutamicum. J. Biol. Chem.* 275:35932–35941, 2000.
25. Eggeling, L., H. Sahm, A.A. de Graaf. Quantifying and directing metabolic flux: application to amino acid overproduction. In: *Advances in Biochemical Engineering/Biotechnology* , Vol. 54, Scheper, T., ed., 1996, p 1–30.
26. Wiechert, W., A.A. de Graaf. *In vivo* stationary flux analysis by ^{13}C labeling experiments. In: *Advances in Biochemical Engineering/Biotechnology* , Vol. 54, Scheper, T., ed., 1996, p 109–154.
27. Wiechert, W., M. Mollney, S. Petersen, A.A. de Graaf. A universal framework for ^{13}C metabolic flux analysis. *Metab. Eng.* 3:265–283, 2001.
28. Kromer, J.O., O. Sorgenfrei, K. Klopprogge, E. Heinzle, C. Wittmann. In-depth profiling of lysine-producing *Corynebacterium glutamicum* by combined analysis of transcriptome, metabolome, and fluxome. *J. Bacteriol.* 186:1769–1784, 2004.
29. Kirchner, O., A. Tauch. Tools for genetic engineering in the amino acid-producing bacterium *Corynebacterium glutamicum. J. Biotechnol.* 104(1–3)287–299, 2003.
30. Ikeda, M., S.Nakagawa. The *Corynebacterium glutamicum* genome: features and impacts on biotechnological processes. *Appl. Microbiol. Biotechnol.* 62:99–109, 2003.
31. Ikeda, M. amino acid production processes. In: *Advances in Biochemical Engineering/ Biotechnology* , vol. 79, Scheper, T. ed., Heidelberg: Springer-Verlag, 2003, p 1–35.
32. Riedel, C., D. Rittmann, P. Dangel, B. Mockel, S. Petersen, H. Sahm, B.J. Eikmanns. Characterization of the Phosphoenolpyruvate carboxykinase gene from *Corynebacterium glutamicum* and significance of the enzyme for growth and amino acid production. *J. Mol. Microbiol. Biotechnol.* 3:573–583, 2001.
33. Peters-Wendisch, P., B. Schiel, V.F. Wendisch, E. Katsoulidis, B. Mockel, H. Sahm, B.J. Eikmanns. Pyruvate carboxylase is a major bottleneck for glutamate and lysine production by *Corynebacterium glutamicum. J. Mol. Microbiol. Biotechnol.* 3:295–300, 2001.
34. Koffas, M.A.G., G.Y. Jung, J.C. Aon, G. Stephanopoulos. Effect of pyruvate carboxylase overexpression on the physiology of *Corynebacterium glutamicum. Appl. Environ. Microbiol.* 68:5422–5428, 2002.
35. Postma, P.W., J.W. Lengeler, G.R. Jacobson. Phosphoenolpyruvate: carbohydrate phosphotransferase systems of bacteria. *Microbiol. Rev.* 57:543–594, 1993.
36. Lee, J.K., M.H. Sung, K.H. Yoon, J.H. Yu, T.K. Oh. Nucleotide sequence of the gene encoding the *Corynebacterium glutamicum* mannose enzyme II and analyses of the deduced protein sequence. *FEMS Microbiol. Lett.* 119:137–145, 1994.
37. Dominguez, H., C. Rollin, A. Guyanvarch, J.L. Guerquin-Kern, M. Cocain-Bousquet, N.D. Lindley. Carbon-flux distribution in the central metabolic of *Corynebacterium glutamicum* during growth on fructose, *Eur. J. Biochem.* 254:96–102, 1998.
38. Parche, S., A. Burkovski, G.A. Sprenger, B. Weil, R. Kramer, F. Titgemeyer. *Corynebacterium glutamicum*: a dissection of the PTS. *J. Mol. Microbiol. Biotechnol.* 3:423–428, 2001.
39. Eggeling, L., H. Sahm. New ubiquitours translocators: Amino acid export by *Corynebacterium glutamicum* and *Escherichia coli. Arch. Microbiol.* 180:155–160, 2003.
40. Yen, M.R., Y.H. Tseng, P. Simic, H. Sahm, L. Eggeling, M.H. Saier, Jr. The ubiquitous ThrE family of putative transmembrane amino acid efflux transporters. *Res. Microbiol.* 153:19–25, 2002.
41. Kramer, R., C. Lambert. Uptake of glutamate in *Corynebacterium glutamicum*, 2: evidence for a primary active transport system. *Eur. J. Biochem.* 194:937–944, 1990.
42. Aleshin, V.V., N.P. Zakataeva, V.A. Livshits. A new family of amino-acid-efflux proteins. *Trends Bioch. Sci.* 24:133–135, 1999.
43. Simic, P., H. Sahm, H. Eggeling. L-threonine export: use of peptides to identify a new translocator from *Corynebacterium glutamicum. J. Bacteriol.* 183:5317–5324, 2001.

44. Kennerknecht, N., H. Sahm, M.R. Yen, M. Patek, M.H. Saier, Jr., L. Eggeling. Export of L-isoleucine from *Corynebacterium glutamicum*: a two-gene-encoded member of a new translocator family. *J. Bacteriol.* 184:3947–3956, 2002.
45. Dassler, T., T. Maier, C. Winterhalter, A. Bock. Identification of a major facilitator protein from *Escherichia coli* involved in efflux of metabolites of the cysteine pathway. *Mol. Microbiol.* 36:1101–1112, 2000.
46. Franke, I., A. Resch, T. Dassler, T. Maier, A. Bock. YfiK from *Escherichia coli* promotes export of *O*-Acetylserine and Cysteine. *J. Bacteriol.* 185:1161–1166, 2003.
47. Kramer, R., C. Hoischen. Futile cycling caused by the simultaneous presence of separate transport systems for uptake and secretion of amino acids in *Corynebacterium glutamicum*. In: *Biothermokinetics*, Westerhoff, H, ed., Intercept Publ., 1994, pp 19–26.
48. Shiio, I., H. Ozaki, K. Ujigawa-Takeda. Production of aspartic-acid and lysine by citrate synthase mutants of *Brevibacterium-flavium*. *Agric. Biol. Chem.* 46:101–110, 1982.
49. Shiio, I., K.J. Sano. Microbial production of L-lysine, 2: production by mutants sensitive to threonine or methionine. *Gen. Appl. Microbiol.* 15:267–275, 1969.
50. Nampoothiri, K.M., C. Hoischen, B. Bathe, B. Mockel, W. Pfefferle, K. Krumbach, H. Sahm, L. Eggeling. Expression of genes of lipid synthesis and altered lipid composition modulates L-glutamate efflux of *Corynebacterium glutamicum. Appl. Microbiol. Biotechnol.* 58:89–96, 2002.
51. Wolf, A., R. Kramer, S. Morbach. Three pathways for trehalose metabolism in *Corynebacterium glutamicum* ATCC13032 and their significance in response to osmotic stress. *Mol. Microbiol.* 49:1119–1134, 2003.
52. Loos, A., C. Glanemann, L.B. Willis, X.M. O'Brien, P.A. Lessard, R. Gerstmeir, S. Guillouet, A.J. Sinskey. Development and validation of corynebacterium DNA microarrays. *Appl. Environ. Microbiol.* 67:2310–2318, 2001.
53. Muffler, A., S. Bettermann, M. Haushalter, A. Horlein, U. Neveling, M. Schramm, O. Sorgenfrei. Genome-wide transcription profiling of *Corynebacterium glutamicum* after heat shock and during growth on acetate and glucose. *J. Biotechnol.* 98:255–268, 2002.
54. Hayashi, M., H. Mizoguchi, N. Shiraishi, M. Obayashi, S. Nakagawa, J. Imai, S. Watanabe, T. Ota, M. Ikeda. Transcriptome analysis of acetate metabolism in *Corynebacterium glutamicum* using a newly developed metabolic array. *Biosci. Biotechnol. Biochem.* 66:1337–1344, 2002.
55. Wendisch, V.F. Genome-wide expression analysis in *Corynebacterium glutamicum* using DNA microarrays. *J. Biotechnol.* 104(1-3):273–285, 2003.
56. Huser, A.T., A. Becker, I. Brune, M. Dondrup, J. Kalinowski, J. Plassmeier, A. Puhler, I. Wiegrabe, A. Tauch. Development of a *Corynebacterium glutamicum* DNA microarray and validation by genome-wide expression profiling during growth with propionate as carbon source. *J. Biotechnol.* 106:269–286, 2003.
57. Hermann, T., G.Wersch, E.M. Uhlemann, R. Schmid, A. Burkovski. Mapping and identification of *Corynebacterium glutamicum* proteins by two-dimensional gel electrophoresis and microsequencing. *Electrophoresis* 19:3217–3221, 1998.
58. Hermann, T., W. Pfefferle, C. Baumann, E. Busker, S. Schaffer, M. Bott, H. Sahm, N. Dusch, J. Kalinowski, A. Puhler, A.K. Bendt, R. Kramer, A. Burkovski. Proteome analysis of *Corynebacterium glutamicum*. *Electrophoresis* 22:1712–1723, 2001.
59. Schaffer, S., B.Weil, V.D. Nguyen, G. Dongmann, K. Gunther, M. Nickolaus, T. Hermann, M. Bott. A high-resolution reference map for cytoplasmic and membrane-associated proteins of *Corynebacterium glutamicum*. *Electrophoresis* 22:4404–4422, 2001.
60. Bendt, A.K., A. Burkovski, S. Schaffer, M. Bott, M. Farwick, T. Hermann. Towards a phosphoproteome map of *Corynebacterium glutamicum*. *Proteomics* 3:1637–1646, 2003.
61. Niederberger, P., R. Prasad, G. Miozzari, H. Kacser. A strategy for increasing an *in vivo* flux by genetic manipulations: the trpyptophan system of yeast. *Biochem. J.* 287:473–479, 1992.
62. Kacser, H., J.A. Burns. The control of flux. *Symp. Soc. Exp. Biol.* 27:65–104, 1973.

63. Heinrich, R., T.A. Rapoport. A linear steady-state treatment of enzymatic chains: critique of the crossover theorem and a general procedure to identify interaction sites with an effector. *Eur. J. Biochem.* 42:97–105, 1974.
64. Stephanopoulos, G., J.J.Vallino. Network rigidity and metabolic engineering in metabolite overproduction. *Science* 252:1675–1681, 1991.
65. Vallino, J.J., G. Stephanopoulos. Metabolic flux distributions in *Corynebacterium glutamicum* during growth and lysine overproduction. *Biotechnol. Bioeng.* 41:633–646, 1993.
66. Marx, A., A.A. de Graaf, W. Wiechert, L. Eggeling, H. Sahm. Determination of the fluxes in the central metabolism of *Corynebacterium glutamicum* by nuclear magnetic resonance spectroscopy combined with metabolite balancing. *Biotechnol. Bioeng.* 49:111–129, 1996.
67. Wittmann, C., E. Heinzle. Modeling and experimental design for metabolic flux analysis of lysine-producing *Corynebacteria* by mass spectrometry. *Metab. Eng.* 3:173–191, 2001.
68. Wittmann, C., E. Heinzle. Application of MALDI-TOF MS to lysine-producing *Corynebacterium glutamicum*: a novel approach for metabolic flux analysis. *Eur. J. Biochem.* 268:2441–2455, 2001.
69. Drysch, A., M. El Massoudi, W. Wiechert, A.A. de Graaf, R. Takors. Serial flux mapping of *Corynebacterium glutamicum* during fed-batch L-lysine producing using the sensor reactor approach. *Biotechnol. Bioeng.* 5:497–505, 2004.
70. Kiefer, P., E. Heinzle, O. Zelder, C. Wittmann. Comparative metabolic flux analysis of lysine-producing *Corynebacterium glutamicum* cultured on glucose and fructose. *Appl. Environ. Microbiol.* 70:229–239, 2004.
71. Wittmann, C., E. Heinzle. Genealogy profiling through strain improvement by using metabolic network analysis: metabolic flux genealogy of several generations of lysine-producing corynebacteria. *Appl. Environ. Microbiol.* 68:5843–5859, 2002.
72. Pfefferle, W., B. Mockel, B. Bathe, A. Marx. Biotechnological manufacture of Lysine. In: *Advances in Biochemical Engineering/Biotechnology* , Vol. 79, Scheper, T. ed., Heidelberg: Springer-Verlag, 2003, p 59–112.
73. Koyoma, Y., T. Ishii, Y. Kawahara, Y. Koyoma, E. Shimizu, T. Yoshioka. EU Patent Application EP0844308 A2 980527, 1998.
74. Okamoto, K., K. Kino M. Ikeda. Hyperproduction of L-threonine by an *Escherichia coli* mutant with impaired L-threonine uptake. *Biosci. Biotechnol. Biochem.* 61:1877–1882, 1997.
75. Kawahara, Y., K. Takahashi-Fuke, E. Shimizu, T. Nakamatsu, S. Nakamori. Relationship between the Glutamate production and the activity of 2-Oxoglutarate dehydrogenase in *Brevibacterium lactofermentum. Biosci. Biotech. Biochem.* 61:1109–1112, 1997.
76. Kimura, E., C. Abe, Y. Kawahara, T. Nakamatsu. Molecular cloning of a novel gene, *dtsR*, which rescues the detergent sensitivity of a mutant derived from *Brevibacterium lactofermentum. Biosci. Biotech. Biochem.* 60:1565–1570, 1996.
77. Kimura, E., C. Abe, Y. Kawahara, T. Nakamatsu, H. Tokuda. A *dtsR* gene-disrupted mutant of *Brevibacterium lactofermentum* requires fatty acids for growth and efficiently produces L-glutamate in the presence of an excess of biotin. *Biochem. Biophys. Res. Comm.* 234:157–161, 1997.
78. Kimura, E., C. Yagoshi, Y. Kawahara, T. Ohsumi, T. Nakamatsu. Glutamate overproduction in *Corynebacterium glutamicum* triggered by a decrease in the level of a complex comprising DtsR and a biotin-containing subunit. *Biosci. Biotechnol. Biochem.* 63:1274–1278, 1999.
79. Kimura, E. Metabolic engineering of Glutamate production. In: *Advances in Biochemical Engineering/Biotechnology* , vol. 79, Scheper, T. ed., Heidelburg: Springer-Verlag, 2003, p. 37–57.
80. Schirawski, J., G. Unden. Menaquinone-dependent succinate dehydrogenase of bacteria catalyzes reversed electron transport driven by the proton potential. *Eur. J. Biochem.* 257:210–215, 1998.
81. Delaunay, S., P. Gourdon, P. Lapujade, E. Mailly, E. Oriol, J.M. Engasser, N.D. Lindley, J.L. Goergen. *Enz. Microbial. Technol.* 25:762–768, 1999.

82. Lambert, C., A. Erdmann, M. Eikmanns, R. Kramer. Triggering Glutamate excretion in *Corynebacterium glutamicum* by modulating the membrane state with local anesthetics and osmotic gradients. *Appl. Environ. Microb.* 61:4334–4342, 1995.
83. Eggeling, L., H.Sahm. The cell wall barrier of *Corynebacterium glutamicum* and amino acid efflux. *J. Biosci. Bioeng*. 92:201–213, 2001.
84. Gourdon, P., M. Raherimandimby, H. Dominguez, M. Cocaign-Bousquet, N.D. Lindley. Osmotic stress, glucose transport capacity and consequences for glutamate overproduction in *Corynebacterium glutamicum. J. Biotechnol.* 104:77–85, 2003.
85. Puech, V., M. Chami, A. Lemassu, M.A. Laneelle, B. Schiffler, P. Gounon, N. Bayan, R. Benz, M. Daffé. Structure of the cell envelope of *corynebacteria*: importance of the non-covalently bound lipids in the formation of the cell wall permeability barrier and fracture plane. *Microbiology* 147:1365–1382, 2000.
86. Bayan, N., C. Houssin, M. Chami, G. Leblon. Mycomembrane and S-layer: two important structures of *Corynebacterium glutamicum* cell envelope with promising biotechnology applications. *J. Biotechnol.* 104:55–67, 2003.
87. Hoischen, C., R. Kramer. Evidence for an efflux carrier system involved in the secretion of glutamate by *Corynebacterium glutamicum. Arch. Microbiol.* 151:342–347, 1989.
88. Kronemeyer, W., N. Peekhaus, R. Kramer, H. Sahm, L. Eggeling. Structure of the *gluABCD* cluster encoding the glutamate uptake system of *Corynebacterium glutamicum. J. Bacteriol.* 177:1152–1158, 1995.
89. Lichtinger, T., A. Burkovski, M. Niederweis, R. Kramer, R. Benz. Biochemical and biophysical characterization of the cell wall porin of *Corynebacterium glutamicum*: the channel is formed by a low molecular mass polypeptide. *Biochemistry* 37:15024–15032, 1998.
90. Costa-Riu, N., A. Burkovski, R. Kramer, R. Benz. PorA represents the major cell wall channel of the gram-positive bacterium *Corynebacterium glutamicu. J. Bacteriol.* 185:4779–4786, 2003.
91. Costa-Riu, N., E. Maier, A. Burkovski, R. Kramer, F. Lottspeich, R. Benz. Identification of an anion-specific channel in the cell wall of the Gram-positive bacterium *Corynebacterium glutamicum. Mol. Microbiol.* 50:1295–1308, 2003.
92. Hoischen, C., R. Kramer. Membrane alteration is necessary but not sufficient for effective glutamate secretion in *Corynebacterium glutamicum. J. Bacteriol.* 172:3409–3416, 1990.
93. Duperray, F., D. Jezequel, A. Ghazi, L. Letellier, E. Shechter. Excretion of glutamate from *Corynebacterium glutamicum* triggered by amine surfactans. *Biochim. Biophys. Acta* 1103:250–258, 1992.
94. Kijima, N., D. Goyla, A. Takada, M. Wachi, K. Nagai. Induction of only limited elongation instead of filamentation by inhibition of cell division in *Corynebacterium glutamicum. Appl. Microbiol. Biotechnol.* 50:227–232, 1998.
95. Park, S.M., A.J. Sinskey, G. Stephanopoulos. Metabolic and physiological studies of *Corynebacterium glutamicum* mutants. *Biotechnol. Bioeng*. 55:864–879, 1997.
96. Delaunay, S., D. Uy, M.F. Baucher, J.M. Engasser, A. Guyonvarch, J.L. Goergen. Importance of phosphoenolpyruvate carboxylase of *Corynebacterium glutamicum* during the temperature triggered glutamic acid fermentation. *Metabol. Eng.* 1:334–343, 1999.
97. Marx, A., K. Striegel, A.A. de Graaf, H. Sahm, L. Eggeling. Response of the central metabolism of *Corynebacterium glutamicum* to different flux burdens. *Biotechnol. Bioeng.* 56:168–180, 1997.
98. De Boer, L., L. Dijkhuizen. Microbial and enzymatic processes for L-phenylalanine oproduction. *Adv. Biochem. Eng. Biotechnol.* 41:1–27, 1990.
99. Backmann, K., M.J. O'Connor, A. Maruya, E. Rudd, D. McKay, R. Balakrishnan, M. Radjai, V. DiPasquantonio, D. Shoda, R. Hatch, K. Venkatasubramanian. Genetic engineering of metabolic pathways applied to the production of phenylalanine. *Ann. N. Y. Acad. Sci.* 589:16–24, 1990.
100. Pittard, A.J. Biosynthesis of aromatic amino acids. In: *Escherichia coli and Salmonella, Cellular* and Molecular Biology, Neidhardt, F.C., et al., eds., Washington DC: *Amer. Soc. Microbiol.*, 1996, pp 458–484.

101. Bongaerts, J., M. Kramer, U. Muller, L. Raeven, M. Wubbolts. Metabolic engineering for microbial production of aromatic amino acids and derived compounds. *Metab. Eng.* 3:289–300, 2001.
102. Ikeda, M., R. Katsumata. Hyperproduction of tryptophan by *Corynebacterium glutamicum* with the modified pentose phosphate pathway. *Appl. Environ. Microbiol.* 65:2497–2502, 1999.
103. Tatarko, M., T.Romeo. Disruption of a global regulatory gene to enhance central carbon flux into phenylalanine biosynthesis in *Escherichia coli*. *Curr. Microbiol.* 43:26–32, 2001.
104. Berry, A. Improving production of aromatic compounds in *Escherichia coli* by metabolic engineering. *Trends Biotechnol.* 14:250–256, 1996.
105. Katsumata, R., M. Ikeda. Hyperproduction of tryptophan in *Corynebacterium-glutamicum* by pathway engineering. *Biotechnology* 11:921–925, 1993.
106. LaDuca, R. J., A. Berry, G. Chotani, T. C. Dodge, G. Gosset, F. Valle, J.C. Liao, J. Yong-Xiao, S. D. Power. Metabolic pathway engineering of aromatic compounds. In: *Manual of Industrial Microbiology and Biotechnology*, Demain, A.L., J.E. Davies, eds., Washington DC: *Amer. Soc. Microbiol.*, 1999, pp 605–615.
107. Aiba, S., H. Tsunekawa, T. Imanaka. New approach to tryptophan production by *Escherichia coli*: genetic manipulation of composite plasmids *in vitro*. *Appl. Environ. Microbiol.* 43:289–297, 1982.
108. Ikeda, M., R. Katsumata. Tryptophan production by transport mutants of *Corynebacterium glutamicum*. *Biosci. Biotechnol. Biochem.* 59:1600–1602, 1995.
109. Kalinowski, J., J. Cremer, B. Bachmann, L. Eggeling, H. Sahm, A. Puhler. Genetic and biochemical analysis of the aspartokinase from *Corynebacterium glutamicum*. *Mol. Microbiol.* 5:1197–1204, 1991.
110. Follettie, M.T., O.P. Peoples, C. Agoropoulou, A.J. Sinskey. Gene structure and expression of the *Corynebacterium glutamicum*: molecular cloning, nucleotide sequence, and expression, *J. Bacteriol.* 175:4096–4103, 1993.
111. Jetten, M.S.M., A.J. Sinskey. Recent advances in the physiology and genetics of amino acid-producing bacteria. *Crit. Rev. Biotechnol.* 15:73–103, 1995.
112. Sonntag, K., L. Eggeling, A.A. de Graaf, H. Sahm. Flux partitioning in the split pathway of lysine synthesis in *Corynebacterium glutamicum*. Quantification by 13C- and 1H-NMR spectroscopy. *Eur. J. Biochem.* 213:1321–1325, 1993.
113. Broer, S., R.Kramer. Lysine excretion by *Corynebacterium glutamicum*, 2: Energetics and mechanism of the transport system. *Eur. J. Biochem.* 202:137–143, 1991.
114. Marx, A., B.J. Eikmanns, H. Sahm, A.A. de Graaf, L. Eggeling. Response of the central metabolism in *Corynebacterium glutamicum* to the use of an NADH-dependent glutamate dehydrogenase. *Metabol. Eng.* 1:35–48,1999.
115. Kojima, H., K. Totsuka. Patent Application WO9511985, 1995.
116. Debabov, V.G. The threonine story. In: *Advances in Biochemical Engineering/Biotechnology*, Vol. 79, Scheper, T. ed., Heidelberg: Springer-Verlag, 2003, p 113–136.
117. Ishida, M., K. Sato, K. Hashiguchi, H. Ito, H. Enei, S. Nakamori. High fermentative production of L-threonine from acetate by a *Brevibacterium flavum* stabilized strain transformed with a recombinant plasmid carrying the *Escherichia coli thr* operon. *Biosci. Biotechnol. Biochem.* 57:1755–1756, 1993.
118. Simic, P., J. Willuhn, H. Sahm, L. Eggeling. Identification of *glyA* (encoding Serine Hydr oxymethyltransferase) and its use together with the exporter ThrE to increase L-Threonine accumulation by *Corynebacterium glutamicum*. *Appl. Environ. Microbiol.* 68:3321–3327, 2002.
119. Shiio, I., S. Nakamori. Microbial production of L-threonine, I: production by *Escherichia coli* mutant resistant to alpha-amino-beta-hydroxyvaleric acid. *Agric. Biol. Chem.* 33:1152–1161, 1969.
120. Kruse, D., R. Kramer, L. Eggeling, M. Rieping, W. Pfefferle, J.H. Tchieu, Y.J. Chung, M.H. Saier, A.Burkovski. Influence of threonine exporters on threonine production in *Escherichia coli*. *Appl. Microbiol. Biotechnol.* 59:205–210, 2002.

121. Reinscheid, D.J., W. Kronemeyer, L. Eggeling, B.J. Eikmanns, H. Sahm. Stable expression of HOM-1-THRB in *Corynebacterium glutamicum* and its effect on the carbon flux to threonine and related amino acids. *Appl. Environ. Microbiol.* 60:126–132, 1994.
122. Leinfelder, W., C. Winterhalter. Japan Patent 11 056 381 A, 1999.
123. Ohnishi, J., S. Mitsuhashi, M. Hayashi, S. Ando, H. Yokoi, K. Ochiai, M. Ikeda. A novel methodology employing *Corynebacterium glutamicum* genome information to generate a new L-Lysine-producing mutant. *Appl. Microbiol. Biotechnol.* 58:217–223, 2002.
124. Gerstmeir, R., V.F. Wendisch, S. Schnicke, H. Ruan, M. Farwick, D. Reinscheid, B.J. Eikmanns. Acetate metabolism and its regulation in *Corynebacterium glutamicum. J. Biotechnol.* 104:99–122, 2003.
125. Merrick, M., R.A. Edwards. Nitrogen control in bacteria. *Microbiol. Rev.* 59:604–622, 1995.
126. Burkovski, A. Ammonium assimilation and nitrogen control in *Corynebacterium glutamicum* and its relatives: an example for new regulatory mechanisms in actinomycetes. *FEMS Microbiol. Rev.* 27:617–628, 2003.
127. Hecker, M., U. Volker. General stress response of *Bacillus subtilis* and other bacteria. *Adv. Microb. Physiol.* 44:35–91, 2001.
128. Ronsch, H., R. Kramer, S. Morbach. Impact of osmotic stress on volume regulation, cytoplasmic solute composition and lysine production in *Corynebacterium glutamicum* MH20-22B. *J. Biotechnol.* 104:87–97, 2003.
129. Ishige, T., M. Krause, M. Bott, V.F. Wendisch, H. Sahm. The phosphate starvation stimulon of *Corynebacterium glutamicum* determined by DNA microarray analyses. *J. Bacteriol.* 185:4519–4529, 2003.
130. Russell, J.B., G.M. Cook. Energetics of bacterial growth: balance of anabolic and catabolic reactions. *Microbiol. Rev.* 59:48–62, 1995.
131. Matsushita, K., T. Yamamoto, H. Toyoma, O. Adachi. NADPH oxidase system as a superoxide-generating cyanide-resistant pathway in the respiratory chain of *Corynebacterium glutamicum. Biosci. Biotechnol. Biochem.* 62:1968–1977, 1998.
132. Matsushita, K., A. Otofuji, M. Iwahashi, H. Toyama, O. Adachi. NADH dehydrogenase of *Corynebacterium glutamicum*: purification of an NADH dehydrogenase II homolog able to oxidize NADPH. *FEMS Microbiol. Lett.* 204:271–276, 2001.
133. Kusumoto, K., M. Sakiyama, J. Sakamoto, S. Noguchi, N. Sone. Menaquinol oxidase activity and primary structure of cytochrome bd from the amino acid fermenting bacterium *Corynebacterium glutamicum. Arch. Microbiol.* 173:390–397, 2000.
134. Molenaar, D., M.E. van der Rest, A. Drysch, R. Yucel. Functions of the membrane-associated and cytoplasmic malate dehydrogenases in the citric acid cycle of *Corynebacterium glutamicum. J. Bacteriol.* 182:6884–6891, 2000.
135. Bott, M. A. Niebisch. The respiratory chain of *Corynebacterium glutamicum. J. Biotechnol.* 104, 129–153, 2003.
136. Niebisch, A., M. Bott. Purification of a cytochrome bc-aa3 supercomplex with quinol oxidase activity from *Corynebacterium glutamicum.* Identification of a fourth subunity of cytochrome aa3 oxidase and mutational analysis of diheme cytochrome c1. *J. Biol. Chem.* 278:4339–4346, 2003.
137. Sekine, H., T. Shimada, C. Hayashi, A. Ishiguro, F. Tomita, A. Yokota. H+-ATPase defect in *Corynebacterium glutamicum* abolishes glutamic acid production with enhancement of glucose consumption rate. *Appl. Microbiol. Biotechnol.* 57:534–540, 2001.
138. Broer, S., L. Eggeling, R. Kramer. Strains of *Corynebacterium glutamicum* with different lysine productivities may have different lysine excretion systems. *Appl. Environ. Microbiol.* 59:316–321, 1993.
139. Gutmann, M., C. Hoischen, R. Kramer. Carrier-mediated glutamate secretion by *Corynebacterium glutamicum* under biotin limitation. *Biochim. Biophys. Acta.* 1112:115–123, 1992.
140. Brand, S., K. Niehaus, A. Puhler, J. Kalinowski. Identification and functional analyis of six mycolyltransferase genes of Corynbebacterium glutamicum ATCC 13032. *Arch. Microbiol.* 180:33–44, 2003.

141. De Sousa-D'Auria, C., R. Kacem, V. Puech, M. Tropis, G. Leblon, C. Houssin, M. Daffé. New insights into the biogenesis of the cell envelope of corynebacteria: identification and functional characterization of five new mycolyltransferase genes in *Corynbebacterium glutamicum. FEMS Microbiol. Lett.* 15:35–44, 2003.
142. Portevin, D., C. De Sousa-D'Auria, C. Houssin, C. Grimaldi, M. Chami, M. Daffé, C. Guilhot. A polyketide synthase catalyzes the last condensation step of mycolic acid biosynthesis in mycobacteria and related organisms. *Proc. Natl. Acad. Sci. USA* 101:314–319, 2004.

1.09

Biotechnology of Microbial Polysaccharides in Food

Ian W. Sutherland

CONTENTS

9.1 INTRODUCTION

Polysaccharides are incorporated into foods to alter the rheological properties of the water present and thus change the texture of the product. Most of the polysaccharides used are employed because of their ability to thicken substances or to cause gel formation. They are also used to stabilize foams and emulsions, suspend particulate materials and inhibit or decrease syneresis, as well as increase water retention. Advantage is also taken of the ability of some mixtures of polysaccharides to exhibit synergistic gelling — basically for the two polymers to yield a gel at concentrations of each which will not in themselves form gels. Associated with these readily measurable physical properties are others such as "mouth feel," which are more difficult to define but which also show some correlation with physical properties. In addition, polysaccharides are incorporated because of their capacity to control the texture of foods and to prevent or reduce ice crystal formation in frozen foods; they may also influence the appearance, color, and flavor of prepared foodstuffs. It must also be remembered that many foodstuffs already contain animal or plant polysaccharides such as hyaluronic acid, starch, or pectin. Thus addition of any further polysaccharide or polysaccharides will in all probability involve interactions with these components as well as with proteins, lipids, and other food components. Currently the polysaccharides added to food are sourced from bacteria, algae, and plants, and are exemplified by xanthan and gellan, alginate and carrageenan, and pectin and starch respectively.

The use of microbial polysaccharides in food is also governed by a number of factors unrelated to their physical properties. This includes the observation that humans generally do not metabolize microbial polysaccharides. They are nutritionally inert and nontoxic. Currently there is considerable consumer concern over the use of "food additives," including hydrocolloids. Most regulatory bodies review their lists of permitted food additives at regular intervals, when they may amend the regulations. This may also lead to improved or altered specifications. On the other hand, many of the microbial species, which are currently used in food preparations, synthesize and excrete polysaccharides as well as other products.

Very many bacteria, yeasts, and fungi can produce polysaccharides. While much interest in these polymers is due to their role in infection or adhesion, some of these polymers have proved to be useful industrial products. Dextran was the first microbial polysaccharide to be commercialized and to receive approval for food use. Although it is no longer used for this purpose, several other microbial polymers are now commercial products with a variety of uses. Use of these microbial exopolysaccharides (EPS) in foods is more limited. Only two bacterial polysaccharides are currently extensively employed by the food industry, except in Japan where polymers of this type are regarded as natural products. It must also be remembered that although microbial polysaccharides may be incorporated into foods as food additives, many other EPS are integral products of microorganisms which are involved in the preparation of the food. A number of natural foods contain polysaccharide-producing microorganisms and considerable interest has currently been shown in the lactic acid bacteria (LAB) which are widely used in fermented milk products and other fermented foods. This review will consider both microbial polysaccharides as food additives and those EPS which are present as normal products of microorganisms used in fermented foods.

9.2 NATURAL OCCURRENCE OF MICROBIAL POLYSACCHARIDES IN FOODS

9.2.1 Microbial Polysaccharides Present in Food as Products of Microbial Food Components

A number of the microorganisms that are used in the preparation of foodstuffs are capable of synthesizing extracellular polysaccharides. This is especially true of the lactic acid bacteria (*Streptococcus*, *Lactobacillus* and *Lactococcus* spp.), and a number of structural studies have recently elucidated the nature of some of these polysaccharides. These and other reports can be found in the reviews of De Vuyst and Degeest (1) and Laws et al. (2). Less is yet known about the relationship between structure and physical properties than is the case for Gram-negative bacterial products such as xanthan and gellan (3). As well as the deliberate addition of microbial polysaccharides to food products to obtain specific properties, there are a number of bacterial fermentations in which polysaccharide is produced and is needed to yield a specific type of product. An example of this can be found in certain types of fermented milk product such as yogurts. In some of these, the production of polysaccharide during bacterial growth is claimed to enhance the product, particularly in respect of the body and texture of the product and in its smoothness and mouth feel. This is particularly true of countries such as the Netherlands and France in which the addition of plant or animal stabilizers is prohibited. The polysaccharide from *Streptococcus salivarus thermophilus* is used in this way. Unfortunately, the production of an apparently neutral glycan by these bacteria is unstable and the use of the bacteria tends to lead to a lack of uniformity in the product. Polysaccharide-producing strains of *Lactobacillus delbruekii bulgaricus* are also used for this purpose, strains being available which synthesize from 14 to 400 mg/litre of a culture of viscous polysaccharide in which galactose is the major component. There are also some fermented milk products in which the production of a thick, gel-like texture results from the exopolysaccharide synthesized by the bacteria used. These are traditional products from parts of Finland, which do not have widespread sales elsewhere. Thus, the importance of polysaccharides from mesophilic and thermophilic LAB is recognized in the production and rheological characteristics of fermented milks, but problems have been caused by the instability of products and requirement of many of these bacteria to grow in complex media (4). The textural and other improvements are made by these polysaccharides, even though the actual yields are low. As pointed out by Ruas-Madiedo et al. (5) some of the EPS from LAB may also possess health-promoting effects such as antitumor and immune-modulating activities in addition to possible prebiotic activity and cholesterol-lowering activity (6).

As LAB are recognized food microorganisms, increasing interest has recently been shown in the EPS synthesized by them. It is also accepted that at present EPS derived from these bacteria could not yield significant commercial products (7). Now that a range of polysaccharides from LAB has been studied, it is clear that no common pattern of structures exists, although certain monosaccharides including D-glucose, D-galactose, and L-rhamnose are commonly found (1). While all the heteropolysaccharides are composed of regular repeat units, some carry phosphate groups in addition to their monosaccharide components, while others lack the inorganic residues. Most of the polysaccharides contain the sugars common to other bacterial EPS and as yet few of the rarer monosaccharides have been reported. Among these are galactofuranose found in the pentasaccharide repeat units of the *Lactobacillus rhamnosus* strains C83 EPS (Figure 9.1) (8) and KL37c (9). This monosaccharide is also present in the exopolysaccharide produced by *Streptococcus thermophilus* EU20 (10) as well as *S. thermophilus* strain SY 89 in which it is found together with galactopyranose (Figure 9.2) (11). The majority of EPS from LAB are composed of a small range of common sugars very frequently including D-glucose, D-galactose, and L-rhamnose (2). These are mainly neutral hexoses or methylpentoses, but may also include the corresponding uronic acids and *N*-acetylaminosugars (12). Although L-fucose is common among the EPS of Gram negative bacteria, L-rhamnose is the methylpentose most frequently found in EPS from LAB. Many of the EPS are polyanionic due to the presence of either uronic acids or phosphate groups. The most common uronic acid is D-glucuronic acid, but D-galacturonic acid is also frequently found. The structures are repeat units of 3–7 monosaccharides and although the polymers are essentially linear macromolecules, they

→3)-α-D-Glc*p*-(1→2)-β-D-Gal*f*-(1→6)-α-D-Gal*p*-(1→6)-α-D-Glc*p*-(1→3)-β-D-Gal*f*-(1→

(8,9)

Figure 9.1 The pentasaccharide repeat unit of the *Lactobacillus rhamnosus* strain C83 EPS (8).

→3)-α-D-Glc*p*-(1→3)-β-D-Glc*p*-(1→3)-β-D-Gal*f*-(1→
₁↑⁶
β-D-Gal*p*-1

(11)

Figure 9.2 The tetrasaccharide repeat unit of *Streptococcus thermophilus* strain SY89 containing equimolar amounts of galactopyranose and galactofuranose (11).

β-D-Glc*p* CH_3.COO.
↓1 2↓
6
→ 4)-β-D-Glc*p*-(1→ 3)-β-L- Rha*p*-(1→
1↑3
α-L-Rha*p* (4←*sn*-glycerol 3-phosphate)

(13)

Figure 9.3 The triply branched repeat unit of the EPS of *Lactobacillus sake* 0-1 (13).

→2)-α-L-Rha*p*-(1→2)-α-D-Gal*p*-(1→3)-α-D-Glc*p*-(1→3)-α-D-Gal*p*-(1→3)-α-L-Rha*p*-(1→
↑4
β-D-Gal*p*-(1→4)-β-D-Glc*p*-1

(14)

Figure 9.4 The heptasaccharide repeat of *Lactococcus lactis cremoris* B39 (14).

also very commonly carry branches. Examples of this can be seen in the triply branched repeat unit of *Lactobacillus sake* 0-1 (Figure 9.3) (13) or the heptasaccharide repeat of *Lactococcus lactis cremoris* B39 (Figure 9.4) (14). The branches may be on different monosaccharides, or on the same monosaccharide as is the case for the EPS from *Lactobacillus lactis cremoris* SBT 0495 in which a single D-galactose carries both a rhamnose and a galactose-1-phosphate (15). Strains of the same species may yield several different EPS chemotypes. An isolate of *Lactobacillus delbruekii bulgaricus* from kefir grains yielded a polysaccharide with almost equimolar amounts of D-glucose and D-galactose (16), whereas the EPS from another strain of the same subspecies had a very different structure as can be seen in Figure 9.5 (17). Another strain possessed a heptasaccharide repeating unit in which the ratio of D-galactose: D-glucose: L-rhamnose was 5:1:1 with the structure shown in Figure 9.6 (18). Strain 291 differed again in yielding a pentasaccharide repeat unit containing only glucose and galactose (Figure 9.7) (19).

In addition to monosaccharides, bacterial EPS frequently carry other substituents, which may either be organic or inorganic in nature. The most common organic substituents are either *O*-acetyl groups or pyruvate ketals. The pyruvate ketals are most commonly found attached to D-galactose, D-glucose, or D-mannose residues. It is fairly unusual to find both pyruvate and acetate attached to the same monosaccharide residue. As more structures from Gram-positive species are studied, a number of these have been found to contain pyruvate. One such is EPS from *Lactobacillus rhamnosus* RW-959M in which a galactose residue

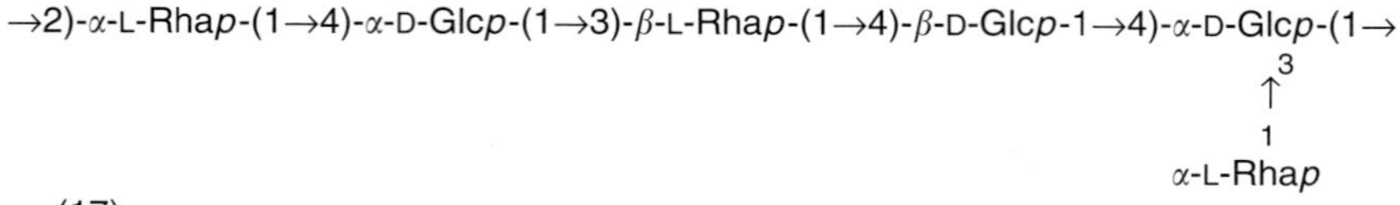

(17).

Figure 9.5 The structure of the EPS from *Lactobacillus delbrueckii bulgaricus* EU23 (17).

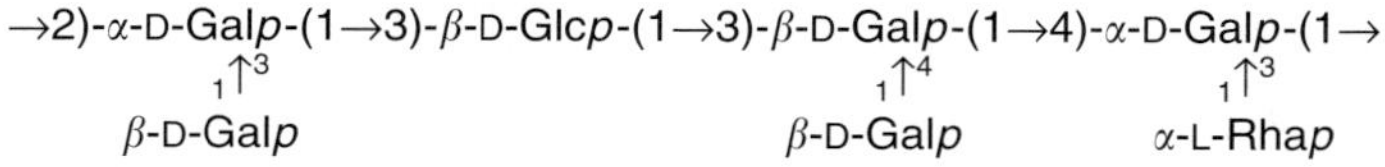

(18)

Figure 9.6 The heptasaccharide structure of the EPS from *Lactobacillus delbrueckii bulgaricus* rr (18).

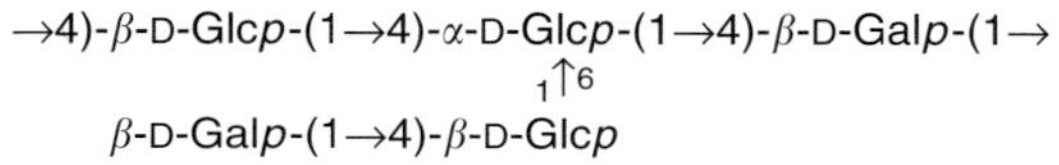

(19)

Figure 9.7 The structure of the EPS from *Lactobacillus delbrueckii bulgaricus* 291 (19).

carried a 4,6-linked ketal (20). In a number of EPS the acyl groups are present in nonstoichiometric amounts. Thus some xanthans contain 0.3 moles of pyruvate per repeat unit. Phosphate groups occur widely among EPS from Gram-positive species including some of the industrially important *Streptococcus* and *Lactobacillus* species, which are employed in the fermentation of milk, yogurt, and cheeses. Quite a few of the extracellular polysaccharides from LAB are neutral polymers as can be seen in Figures 2 and 5. Although some of the LAB can produce homopolymers composed of a single monosaccharide, the majority are heteropolysaccharides. Among the homopolysaccharides synthesized by LAB are glucans and fructans, several of which are produced only when the bacteria are grown in the presence of sucrose. One example is *Lactobacillus reuteri* LB121 which yielded both a fructan and a glucan with masses of 3.5 and 150 kDa respectively (21). Unusually, one strain of *L. lactis lactis* (H414) yielded a galactan. An enzyme from the *L. reuteri* 121 strain was later used to produce a high mass ($>10^7$) fructan with β(2→1) linkages (22). This was thus the first report of bacterial inulin synthesis in a GRAS listed bacterial species (the only other observation of bacterial inulin formation related to a non-GRAS listed species). The strain was also unusual in being capable of forming a highly branched α-D-glucan (23).

It should also be remembered that a number of bacterial strains might have the capacity to form more than one type of polysaccharide, although it is unusual for two or more EPS to be secreted simultaneously. An example of this was *Streptococcus thermophilus* LY03 which yielded two EPS of the same composition, D-glucose and D-galactose in the same molar ratio of 1:4, but differing in mass (24). In contrast to this, strain S3 formed a polymer composed of D-galactose and L-rhamnose in the ratio 2:1 and was also partially acetylated on a D-galactofuranosyl side chain residue (25). The hexasaccharide repeat unit contained three D-galactopyranosyl residues. One of the simplest structures identified to date is the neutral trisaccharide repeat unit from *Lactobacillus* spp. G-77

→6)-α-D-Glc*p*-(1→6)-α-D-Glc*p*-(1→
$_1$↑2
α-D-Glc*p*

(26)

Figure 9.8 The neutral trisaccharide repeat unit from *Lactobacillus* spp. G-77 (26).

(Figure 9.8) (26). This bacterial species was capable of forming two EPS, the other being a 2-substituted (1→3)-*β*-D-glucan, a structure that was also found in the EPS from *Pediococcus damnosus* (27).

Among the most unusual structures determined so far is that for an EPS from *Streptococcus thermophilus* 8S (28). This proved to possess a pentasaccharide repeat unit containing D-galactose: D-glucose: D-ribose: *N*-acetyl-D-galactosamine in the molar ratio 2:1:1:1. In addition an open chain nonionic acid was identified. A substituent of this type has not been reported from other bacterial EPS. Some high mass LAB polymers appear to possess more complex structures with nonstoichiometric amounts of their components. Thus the EPS from *Lactobacillus helveticus* TY1-2 was formed from heptasaccharide repeat units with the molar ratio glucose: galactose: glucosamine 3.0:2.8:0.9 (29). The actual structure was in the form of a trisaccharide backbone with the aminosugar residue carrying both a monosaccharide (a D-galactosyl residue) and a trisaccharide side chain.

9.2.2 Physical Properties

Rheological studies on the EPS from *Propionobacterium acidi-propiionici* DSM4900 revealed shear thinning behavior and suggested that it formed an entangled polymer solution (30), while a polymer from *Streptococcus thermophilus* SY containing rhamnose, glucose, and galactose, was claimed to yield pseudoplastic solutions and have a viscosity higher than xanthan (31). Another LAB EPS, from *L. saké*, had an estimated mass of 4–9 x 10^6 Da and was also claimed to have a higher intrinsic viscosity than xanthan (32). Studies have also been made on the effects of chemical modifications to the EPS from a small number of LAB. Tuinier et al. (33) found that removal of side chain D-galactosyl residues from the EPS of *L. lactis cremoris* polymers increased chain stiffness and thickening efficiency. Deacetylation of the EPS from strain B891 left the chain stiffness unaltered.

In one of the few attempts to model the structure of EPS from *Lactobacillus helveticus*, Faber et al. (34) noted that the EPS from strain 766 exhibited a flexible twisted secondary structure. In their tertiary structure, the polysaccharide chains tended to adopt a random coil conformation.

9.2.3 Production and Synthesis

Yields of most of the EPS from LAB are generally low. However *L. reuteri* LB121 was claimed to form 10g. l^{-1} of two homopolysaccharides (21). Various studies in the laboratory have utilized whey-based media to permit the isolation and characterization of the polysaccharides products from LAB. Thus Ricciardi et al. (31) noted that synthesis of the polymer in *S. thermophilus* was growth associated and obtained yields of 152mg. l^{-1}. Attempts have also been made to improve EPS synthesis through identification and augmentation of key enzymes involved in the earlier stages of carbohydrate metabolism. However this has not always proved successful. In *S. thermophilus*, over expression of *galU* (UDP-glucose pyrophosphorylase) enhanced enzyme activity but not EPS synthesis (35). Only when *galU* and *pgmA* (phosphoglucomutase) were both over expressed were higher yields of EPS obtained. It has also been recognized that in *Lactococcus lactis*, there was competition between EPS synthesis and bacterial growth, with glucose-6-phosphate as the key intermediate metabolite (36). This study indicated possible mechanism by which EPS production might be enhanced through genetic manipulation of the EPS biosynthetic and glycolysis pathways. However a later report from the same researchers (37) revealed that although over expression of the *galU* gene greatly increased sugar nucleotide (UDP-glucose and UDP-galactose) levels, EPS production was not significantly different from the wild type.

9.3 MICROBIAL POLYSACCHARIDES INCORPORATED AS FOOD ADDITIVES

In the US, polysaccharides which are to be used as food additives are subject to Generally Regarded As Safe (GRAS) regulations and must be approved by the USDA. Currently only two microbial polysaccharide products have such approval – xanthan and gellan, although curdlan may also be under evaluation. Previously dextran was also an approved food additive, but it is not currently used in food manufacture. In Japan, a wider view is taken and microbial EPS are regarded as natural products. One thus finds the bacterial product curdlan being used in various foodstuffs, while pullulan (from the fungus *Aureobasidium pullulans*) is also acceptable. In the case of pullulan, the polysaccharide also has potential value as a food packaging material, a role which is also strictly controlled.

9.3.1 Xanthan

Xanthan, a product from the plant pathogenic bacterium *Xanthomonas campestris*, is typically produced copiously as an extracellular slime by the various pathovars of *Xanthomonas campestris* as well as by some other *Xanthomonas* species. Acid hydrolysates of xanthan contain D-glucose, D-mannose, and D-glucuronic acid in the molar ratio 2:2:1. Other components have been reported in polysaccharides from bacterial isolates described as *Xanthomonas* species, but the strains involved have generally been poorly characterized. The primary structure of the polysaccharide is a pentasaccharide repeat unit, effectively a cellulose chain to which trisaccharide side chains are attached at the C-3 position on alternate D-glucosyl residues (38,39) (Figure 9.9). The polysaccharides from several other *Xanthomonas* species seem to share the same composition as that from pathovars of *X. campestris*. Depending on the bacterial strain and on the physiological conditions for bacterial growth, the polysaccharide may carry varying amounts of *O*-acetyl groups on the

→ [-4-β-D-Glc*p*-(1→4)-β-D-Glc*p*-1-]→
$_1$↑^{3}
α-D-Man*p*-O.CO.CH_3
$_1$↑^{2}
α-D-Glc*p*A
$_1$↑^{4}
β-D-Man*p* = Pyr

Typically the internal α-D-mannosyl residue is fully acetylated but only c. 30% of the β-mannosyl termini are ketalated.

(38)

Figure 9.9 The structure of xanthan from *Xanthomonas campestris* (38).

C-6 position of the internal α-D-mannosyl residue and of 4,6-carboxyethylidene (pyruvate ketal) on the side chain terminal β-D-mannosyl residue respectively. Material from some strains (not used for food) carries two acetyl groups on the internal mannosyl residue, and there is some evidence to suggest that certain xanthan preparations may even contain in excess of 2 moles of acetate per repeat unit. However, the material used as a food additive derives from a standard strain (frequently designated NRRL B-1459) and generally contains pyruvate and acetate in approximate molar ratios of 0.3 and 1.0 per pentasaccharide repeat unit.

X. campestris will grow and produce EPS on a wide range of carbon substrates including amino acids, citric acid cycle intermediates, and carbohydrates. Either ammonium salts or amino acids can be used as nitrogen sources. Various ions are needed for bacterial growth and polysaccharide synthesis. Limitation of any of the ions required for substrate uptake or for precursor or polymer synthesis can affect the yield and properties of the EPS. Xanthan formation by *X. campestris* resembles many other bacterial-EPS-producing systems in that polymer production is favored by a high ratio of carbon source/limiting nutrients such as nitrogen. Typically, media for laboratory synthesis of xanthan contain 0.1–0.2% ammonium salt and 2–3% glucose or sucrose. Xanthan can even be produced in fairly good yield when the bacteria are grown in a simple synthetic medium composed of glucose, ammonium sulphate, and salts, but production is improved in the presence of organic nitrogen sources. The quality and the final yield of xanthan may be enhanced by the addition of small amounts of organic acids or of citric acid cycle intermediates such as α-ketoglutaric acid (40). Oxygen is required for growth and for xanthan production, and as the culture viscosity increases as xanthan is formed, oxygen may rapidly become limiting. For satisfactory cultivation in a fermenter, vigorous aeration was essential and fermentation vessels had to be designed to ensure minimal dead space, which would otherwise lead to stagnant areas of culture. The xanthan production process is complicated by mass transfer reactions in the high viscosity broths. Oxygen solubility decreased with increasing xanthan concentration and the diffusion constant of oxygen in dilute solutions of the polysaccharide was reduced relative to water (41). The presence of polysaccharide in the fermentation broth may also affect the availability of carbon substrates such as glucose and sucrose by interaction with these carbohydrates. Consequently, fed-batch culture is usually preferred. Nakajima et al. (42) suggested that in the design of the fermentation vessel, the volume exposed to high shear is of critical importance. The specific rate of xanthan production depended on the volume of the high shear region.

Xanthan is produced throughout growth of *X. campestris* and in the stationary phase. The specific rate of xanthan synthesis was closely related to the bacterial growth rate in batch culture; maximal during exponential growth and minimal during the stationary phase. In this respect, xanthan production resembled the synthesis of several other bacterial exopolysaccharides. Differences could be seen in the viscosity and the acylation of the xanthan synthesized in nutrient limited media during various phases of the growth cycle (43,44). Acetyl CoA and phosphoenolpyruvate may not be readily available during certain stages of growth to permit the complete acylation of the xanthan repeating units. Consequently, the xanthan synthesized in batch culture represents the products of all phases of growth and is possibly a mixture of several molecular types with varying degrees of acylation and varying mass.

9.3.1.1 Industrial Substrates

In the laboratory, pure substrates such as glucose or sucrose are used, whereas in industrial production different substrates are employed. The substrates must be cheap, plentiful carbon sources, which frequently include starch, starch hydrolysates, corn syrup, molasses, glucose, and sucrose (derived from either sugar beet or from sugar cane). It is imperative that they also be acceptable for food use. Optimal synthesis of xanthan requires a balance between utilizable carbon and nitrogen sources. Care must therefore be taken to obtain consistency in yield and product quality when using substrates which may contain nitrogen in addition to carbohydrate. The nutritional versatility of *X. campestris* is clearly a major factor in favor of its use for commercial xanthan production. However, the quality of the xanthan produced from different substrates may vary considerably — the molecular weight and hence the rheological characteristics of xanthan synthesized from glucose or starch may well differ from that formed when whey or other proteinaceous material is employed. The nitrogen sources used for industrial production of xanthan may include yeast hydrolysates, soybean meal, cottonseed flour, distillers' solubles, or casein hydrolysates. In batch culture for industrial production of the polysaccharide, there must be careful control of pH and of the aeration rate. Adequate oxygen transfer may be difficult to achieve unless the fermentation vessel has been carefully designed to ensure that mixing is optimal. To minimize this problem, fed-batch processes may be used. Even so, the conditions used for production and processing must be carefully standardized to ensure that product yield and quality are consistent. In the laboratory, the polymer can be readily prepared in high yield, and then separated from the bacterial cells by high speed centrifugation or by precipitation with quaternary ammonium compounds. As an industrial product, it is manufactured in stirred tank fermenters on a large scale by batch or fed-batch fermentation by a number of commercial companies; normally fermentation proceeds for 3 days at 30°C. In this respect, Linton (45) has claimed that "exopolysaccharide production is a very efficient process" and calculations have shown that the conversion efficiency of substrate to xanthan is very high. After completion of fermentation, the product is subjected to heat treatment to eliminate viable bacteria and to destroy hydrolytic enzymes such as cellulases, amylases, pectinases, and proteases. This treatment also enhanced the rheological properties of xanthan in solution (46). Recovery of the polysaccharide from industrial cultures requires removal or destruction of the cells followed by precipitation with the polar organic solvent isopropanol. To reduce the cost of the process, the solvents are later recovered for reuse by distillation. The fibrous precipitate is dried, milled, and sieved to give material of different mesh sizes. Bacterial cells are difficult to remove from the highly viscous culture broth, although pasteurization may lead to some autolysis and degradation of cell material as well as improving the subsequent separation of cells from the polymer. Polysaccharide recovery in the presence of polar solvents may be also be increased or accelerated by the addition of electrolytes. Subsequent purification can be achieved by a variety of techniques including fractional precipitation and chromatography. Dilute solutions of xanthan may also be subjected to clarification by filtration. The purified xanthan material can then be subjected to various analytical procedures to verify the composition, and to standard techniques for the determination of polysaccharide structure. The procedures used in the production and recovery of xanthan have been reviewed by Garcia-Ochoa et al. (47) and Galindo (46).

Although it is not degraded in the human or animal body, xanthan is biodegradable. It is a substrate for a range of xanthanases (enzymes with endo-1,4-β-D-glucanase activity cleaving the main chain of the polysaccharide) and xanthan lyases. Most crude enzyme preparations contain at least two different types of enzyme activity, although there may also be associated glycosidases (Figure 9.10). Substrates are randomly cleaved to yield oligosaccharides of different sizes; there is accompanying rapid loss of solution viscosity. Most commercial cellulase preparations lack activity against xanthan unless the polysaccharide is in dilute, ion free solution (48,49). The second type of enzyme found associated with the bacterial preparations is a xanthan lyase (4, 5 transeliminase) which cleaves the β-D-mannosyl-D-glucuronic acid linkage of the trisaccharide side chains (50). The activity of the xanthan

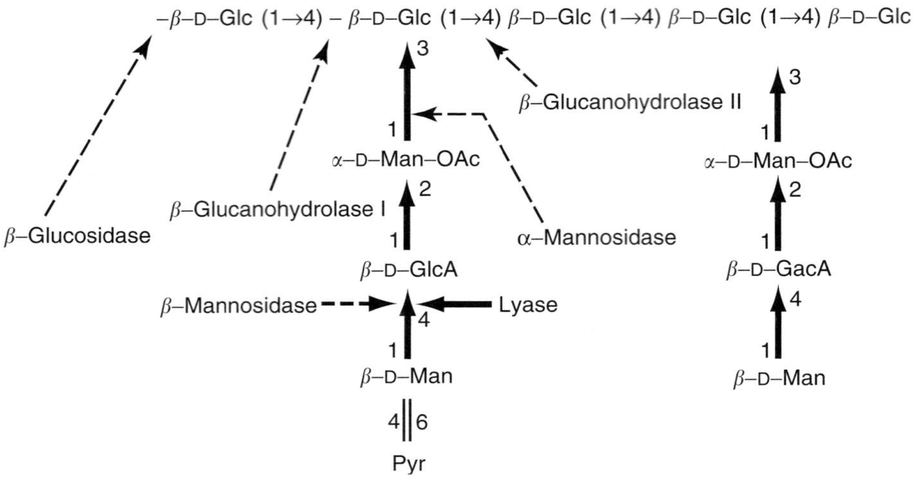

Figure 9.10 Enzymatic degradation of xanthan.

lyase was not greatly influenced by the presence or absence of pyruvate groups on the terminal mannose or of ester-linked acetyl groups on the internal mannose. The enzymes that degrade xanthan are useful laboratory tools for the determination of some of the structurally subtle, acyl modifications found on many xanthan molecules. The fragments generated are more suitable than undegraded polysaccharide for structural studies using fast atom bombardment spectroscopy. They have also proved useful in studies of monoclonal antibody specificity (51). Both the monoclonal antibodies and the enzymes could be used for highly specific quantitation of xanthan in processed foods, as has been suggested recently by Ruijssenaars et al. (52,53), or for the determination of xanthan purity.

9.3.1.2 Biosynthesis of Xanthan

Because of its industrial applications and the very unusual properties of the polysaccharide in aqueous solution, xanthan has been the subject of a large number of structural and physico-chemical studies. These have revealed much information on structure to function relationships in microbial polysaccharides. The biosynthesis of xanthan has also been extensively examined. It shares a common mode of biosynthesis with other bacterial polysaccharides composed of regular repeating units, which are found as cell wall components or extracellular products. Membrane-bound enzymes utilize various activated carbohydrate donors in a tightly regulated sequence to form the polysaccharide on an acceptor molecule. The oligosaccharide repeat units of xanthan are produced by the sequential addition of monosaccharides from the energy rich sugar nucleotides (usually nucleoside diphosphate sugars) to a C_{55} isoprenoid lipid acceptor molecule. At the same time, acyl adornments are added from appropriate activated donors. Thus, the xanthan backbone is formed by the sequential addition of D-glucose-1-phosphate and D-glucose respectively from two moles of UDP-D-glucose. Thereafter, D-mannose and D-glucuronic acid are added from GDP-mannose and UDP-glucuronic acid respectively. Each step requires a specific enzyme and a specific substrate. Absence of the enzyme (or the substrate) inhibits synthesis of the polysaccharide. Depending on the strain used and the physiological conditions under which the bacteria have been grown, and hence on the exact structure of the polymer formed, *O*-acetyl groups are transferred from acetyl CoA to the internal mannose residue, and pyruvate, from phosphoenolpyruvate (PEP), is added to the terminal mannose. This sequence of reactions

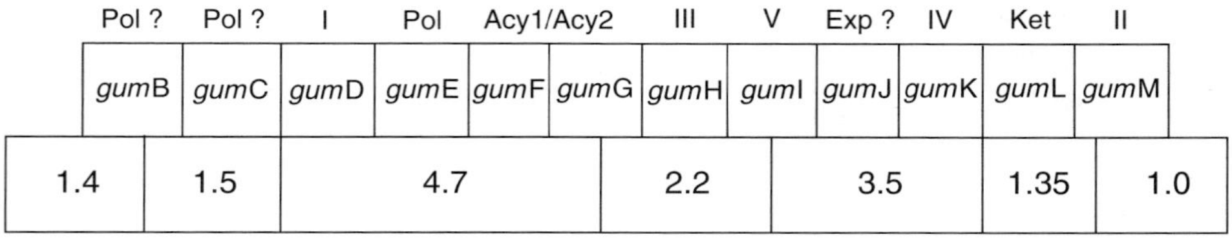

The restriction map of *Xanthomonas campestris.* The *Bam*HI restriction map (bottom) indicates the order and the approximate size (in kilobases) of the fragments of the 16 kb of DNA of the xanthan gene cluster. The genetic map (centre) indicates the twelve separate xanthan genes, designated *gum* B to *gum* M. The biochemical functions (top) indicate the enzymic activity identified for each gene: I–V, transferases I–V; Acy, acetylases (1 and 2); Ket, ketalase; Pol, polymerase; Pol?, possible involvement with polymerisation or export; Exp, post-polymerisation processing or export.

Figure 9.11 Genetic map of the genome segment of *X. campestris* involved in polysaccharide production (54,55).

was elegantly demonstrated by Ielpi and his colleagues (54,55) and is indicated in Figure 9.11. After the pentasaccharide repeat unit has been formed, oligomers are produced by transfer to other lipid intermediates, gradually increasing the size of the carbohydrate chain. Although the structure of the repeating unit is determined by the sequential transfer of the different monosaccharides and acyl groups from their respective donors by highly specific sugar *transferases*, the *polymerase* enzyme responsible for polymerization of the pentasaccharides into a macromolecule of $M_r > 10^6$ is now known to be less specific. This is clear from the ability of mutants to form xanthans defective in their trisaccharide side chains. Such a lack of absolute specificity might account for the natural evolution of bacterial exopolysaccharides such as xanthan, and for the existence of families of structurally related but nonidentical microbial polysaccharides. The final stages of exopolysaccharide secretion from the cytoplasmic membrane passage across the periplasm and outer membrane and finally excretion into the extracellular environment — are less well defined than the earlier biosynthetic steps. The process undoubtedly requires an energy source and may well be analogous to the export of lipopolysaccharide to the outer membrane, in which ATP provides the energy. The genetic system controlling xanthan production in *X. campestris was* elucidated by Vanderslice et al. (56,57). A linear sequence of twelve genes responsible for xanthan production was found. This included seven gene products needed for monosaccharide transfer and for acylation at the lipid intermediate level to form the completely acylated repeat unit (Figure 9.11; Table 9.1). Interestingly, two acetyl transferases were found, one of which appears to function when the pyruvate transferase is defective. The region of the chromosome controlling xanthan synthesis comprised 13 kb. However, other genes, responsible for less well defined functions, may yet be identified in *X. campestris.* The later stages of xanthan biosynthesis presumably include features common to those exopolysaccharide-producing bacteria which have been studied in detail. Thus, in *E. coli*, a region of the chromosome extending to 11.6 kb codes for five proteins, expression of which appears to be necessary for exopolysaccharide production. These proteins *may* form part of a multicomponent system responsible for exporting the polymer from the site of polymerization at the cytoplasmic membrane, to the exterior of the bacterial cell. One of the proteins involved appeared to have attributes common to other ATP-utilizing systems and it may be concerned with excretion (58).

Regulation of xanthan synthesis is not yet well defined. Many of the enzymes needed for the formation of precursors not specifically associated with exopolysaccharide production appear to be under separate control. In *X. campestris*, there are no significant

Table 9.1
Genes Involved in Xanthan Synthesis

Gene products of genes involved in non-specific precursor synthesis:		
UDP-D-glucose pyrophosphorylase		
Gene products of genes involved in specific precursor synthesis:		
GDP-D-mannose pyrophosphorylase		
UDP-D-glucose dehydrogenase		
Gene products of genes involved specifically in synthesis:		
D-Glucose-1-phosphate transferase		(*gumD*)
D-Glucosyl transferase		(*gumM*)
D-Mannosyl transferase 1		(*gumH*)
D-Glucuronosyl transferase		(*gumK*)
D-Mannosyl transferase 2		(*gumI*)
Ketalase		(*gumL*)
Acetylase 1	acetylates the internal mannose	(*gumF*)
Acetylase 2	acetylates the external mannose	(*gumG*)
Polymerase		
(Polysaccharide excretion mechanism)		(*gumB*, *gumC*, *gumE*)

competing demands for substrate as any carbohydrate supplied to the bacteria is metabolized to provide energy, or converted either to essential cell components, wall material, or exopolysaccharide with high efficiency. There are effectively no intracellular storage polymers such as glycogen competing with exopolysaccharide in their synthetic demands on substrate. Xanthan is also the only polysaccharide synthesized by *X. campestris*, unlike several other bacterial species where two or even three extracellular polymers, together with intracellular oligosaccharides, may be produced.

9.3.1.3 Physical Properties of Xanthan

The unusual structure of xanthan as a substituted cellulose confers a number of physical properties to the polymer which are utilized in the food (and other) industries. It is stable at both acid and alkaline pH and forms a pseudoplastic dispersion in water. Relatively low polysaccharide concentrations produce highly viscous solutions and the viscosity does not change greatly on raising the temperature. The solutions are compatible with many other food ingredients and give good flavor release. Currently an estimated 20,000–30,000 metric tons are produced per annum for food and nonfood uses by a number of manufacturers in the US, UK, and elsewhere in the EU, as well as other countries such as China.

The techniques applied to determine the physical attributes of the exopolysaccharide, together with a thorough knowledge of its chemical structure, have enabled the assembly of much information on structure to function relationships pertaining to xanthan and have also provided an excellent model for other microbial exopolysaccharides. Values for the persistence length obtained using techniques such as light scattering and viscometry in dilute solution range from 110–150 nm. A wide range of values proposed for the molecular weight of xanthan has resulted partly from the application of different techniques and also from the tendency of the polysaccharide strands to aggregate in solution. The most consistent results have been obtained using light scattering techniques (Table 9.2). Shatwell (59) noted a range of $0.9–1.6 \times 10^6$ Da for four xanthan preparations with differing acyl substitution from different pathovars grown under standard cultural conditions. Other estimated results ranged between the extremes of 1.1 and 47×10^6 Da. Clearly the very high values must be

Table 9.2
Molecular Mass of Xanthan

Strain	Molecular Weight ($\times 10^6$)
X. campestris pv. *campestris* 646	0.9–1.2
X. campestris pv. *phaseoli* 1128	1.27
X. campestris pv. *phaseoli* 556	1.48
X. campestris pv. *oryzae* PXO61	1.60

Molecular Weights of Xanthans obtained by Static Light Scattering Measurements Results of Shatwell et al.(59).

of dubious accuracy. Measurements of contour length similarly produced a wide range of estimates ($4–120 \times 10^6$). X-ray fiber diffraction studies on xanthan indicated that it formed aligned helices in which there was poor lateral packing. This is consistent with an absence of crystallization and of gelation. The polysaccharide adopted a right handed helical conformation with fivefold symmetry in which the helix pitch was 4.7nm (60). Increase of the temperature of aqueous solutions of xanthan eventually produced a partial melting of the double helix. Xanthan is generally considered to adopt an ordered double helical conformation when in solution at lower temperatures. This conformation causes the apparent stiffness of the polysaccharide at low temperature or in high ionic strength. This stiffness was higher than that estimated for most other macromolecules and is comparable to that of the fungal β-D-glucans which exist as triple helices (61). The conformation was markedly affected by the presence of ions; salt increased the melting temperature. This conformational change, the shift from order to disorder, could be displaced to higher temperatures by the presence of increasing amounts of ions (Figure 9.12) (62). Acyl groups also influence the conformational changes, but light scattering experiments indicated that the inherent stiffness of the polysaccharide molecule is not greatly affected by the pattern of acylation. The presence of acetyl groups on the xanthan structure had a stabilizing effect on the ordered form of the polysaccharide, whereas pyruvate groups had the opposite effect. This could be attributed to intramolecular electrostatic repulsion between pyruvate groups. Conversely, the apolar interactions of the methyl groups of the acetate affected stabilization.

Aqueous solutions of xanthan are highly pseudoplastic whether or not acetyl and ketal substituents are present. The absence of the acyl groups does not normally affect the solubility of xanthans. The viscosity, the degree of pseudoplasticity, and the value for the transition from soft gel to pseudoplastic behavior are all directly related to the polysaccharide concentration. The effects resulting from the addition of salts to the solution also depend on polymer concentration. At ca 0.3% (w/v) xanthan, salts had practically no measurable effect, but at higher polymer concentrations increased viscosity could be observed. This was presumably due to alteration of the macromolecular association. There was probably no significant difference between the viscosity and the degree of acylation. Acetyl and pyruvate contents appeared to have no influence on dilute solution viscosity of xanthan or on the intrinsic viscosity at a given molecular weight. The viscosity of truncated xanthans in which the terminal mannosyl residue is missing was much lower than the wild type material. Surprisingly, the polytrimer, in which the terminal disaccharide is missing, was more viscous than the original xanthan (57,63).

9.3.1.4 Food Usage of Xanthan

Most industrial applications of xanthan derive from its ability to dissolve in hot or cold water to yield high viscosity, pseudoplastic solutions, even at low polysaccharide concentrations. Many of the industrial applications of xanthan were originally in the food industry. It

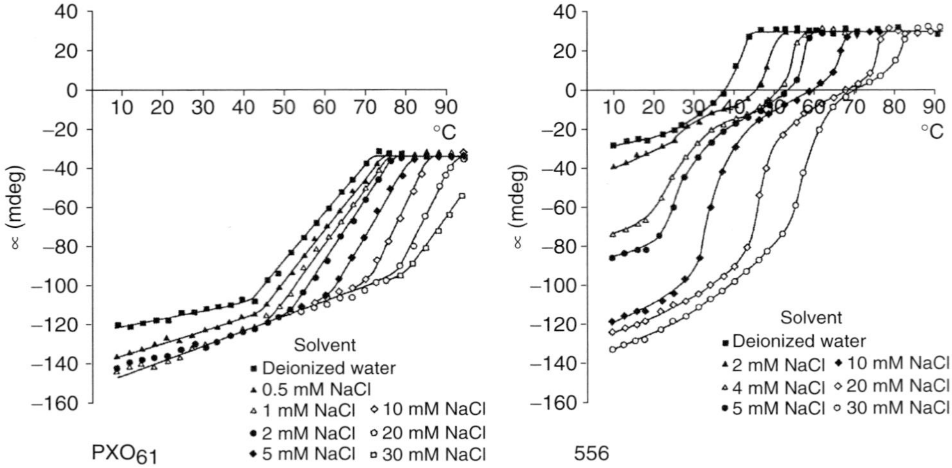

Figure 9.12 Effect of salts on the transition of xanthan from the ordered to the disordered state. Results for two different xanthans (61 - acetylated, non-pyruvylated; 556 - non-acetylated, pyruvylated) monitored by optical rotation measurement. Units mdeg (62).

received approval for food use from the US FDA in 1969 and about 60% of the xanthan currently produced is probably food grade polymer. In many foodstuffs, the useful attributes xanthan include rapid flavor release, good mouth feel and compatibility with other food ingredients including proteins, lipids, and polysaccharides. Salad dressings, relishes, and yogurts to which xanthan is added are of low pH. Xanthan, which is stable over a wide range of pH values, is thus very well suited to such applications. Polysaccharides are added to foods to provide suitable rheological properties during processing and a final appearance which will enhance consumer appeal during purchase and consumption.

Xanthan is incorporated into foods to alter the rheological properties of the water present and thus change the texture of the product. Associated with these readily measurable properties, are others such as mouth feel which are more difficult to define but which also show some correlation with physical properties. Mouth feel has been related to viscosity and, in particular, to nonNewtonian behavior. This also relates to the masking effect of viscosity on the intensity of taste. There is also a specific relationship between the polysaccharide and flavors present in any food. Thus xanthan provides good perception of sweetness and flavor when compared with some nonmicrobial polysaccharides such as gum guar or carboxymethylcellulose. Xanthan is also used for its capacity to control the texture of foods and to prevent or reduce ice crystal formation in frozen foods. As it is compatible with many of the other polysaccharides traditionally used in food, a wide range of textures can be obtained.

Many foods are essentially colloids in which there are complex interactions between the various ingredients. Emulsions of oil in water or water in oil form an important group of food products in which oils or lipids, water, and other ingredients are processed to form sauces, spreads, etc. Some of the emulsion-containing systems are dried and then reconstituted, as in packaged instant soups, desserts, and sauces. Whether used after the initial processing or after reconstitution with water, such emulsions may need to be stabilized with polysaccharides. In this role, xanthan has found many applications in dry mix food products. It can be dispersed in cold or hot water to provide thixotrophic dispersions which can be subsequently heated or refrigerated. The xanthan prevents the constituents from reverting to their original separate phases and ensures the long term stability required for food products.

Some of the foodstuffs into which xanthan is incorporated are of relatively low pH. Xanthan, with its high stability over a wide pH range, is well suited to such applications (64).

Many polysaccharides are employed to provide gelation of foodstuffs. Starch is widely used because of its low cost and its compatibility with many food ingredients, but is not suitable for all food applications. Xanthan on its own only forms very weak gels, but when mixed with various plant galactomannans or glucomannans, and then heated and cooled, it forms thermoreversible gels (*vide infra*).

9.3.1.5 Two Are Better Than One – Synergistic Gels Involving Xanthan

Many food products are formulated using a mixture of two or more polysaccharides to obtain the desired properties after processing (65). The aim may be to reduce the overall content of additive polysaccharides or to form a structured product. Aqueous solutions of xanthan, when mixed with certain plant gluco- or galactomannans, can be induced to form gels, whereas neither component will form a gel on its own. The mixed solution must be heated above the transition temperature of the xanthan in order to denature the xanthan helix and then allowed to cool. X-ray fiber diffraction studies on the mixed gels (66) reveal unique fiber patterns. Previously, gelation of the polymer mixtures had been attributed to the interaction between the helix of the microbial polysaccharide and unsubstituted regions of the plant glucomannan backbone. These results indicated that intermolecular binding could be due to cocrystallization resulting from the stereochemical compatibility between the cellulosic and mannan backbones. Acetyl groups on the xanthan molecules played an important role in inhibiting gelation of the polysaccharide mixtures and increasing the polymer concentrations required (59,62,67).

The most widely used mixture in food manufacture is xanthan and locust bean gum (another widely used food polysaccharide from seeds of the Mediterranean leguminous tree *Ceratonia siliqua*). As pointed out by Copetti et al. (68) concentrated solutions of either xanthan or locust bean gum (LBG) reveal very different properties – xanthan behaves as a weak gel network while LBG represents a hyperentangled macromolecular solution. LBG solutions revealed shear thinning behavior and a Newtonian region at low shear rates, whereas the mixed systems behaved as weak gels. The maximum synergistic effect was observed in a ratio of xanthan: LBG of 1:1. The rheological properties of the mixed gels also depended on the ratio of D-mannose: D-galactose in the LBG (69). LBG with higher mannose–galactose ratios, when mixed with xanthan, showed strongly increased synergism when temperatures were $> 60°C$. The melting points observed also ranged from 40–53°C, again increasing with increased mannose in the LBG. The xanthan-LBG network was formed from xanthan supermolecular strands and was unaffected by the heat treatment or the LBG fraction as addition of LBG failed to influence the xanthan structure. Although these gels are opaque, this is still satisfactory for many food applications where clarity is not required. *Synergistic gels* of this type are employed in a range of foods including spreads and cream cheeses. (Table 9.3).

Mixtures of xanthan with various types of starch have also been examined (70). The pasting peak viscosity of potato starch was reduced in the presence of xanthan, an effect thought to be due to repulsion of the negatively charged bacterial polysaccharide by phosphate groups associated with the starch. The opposite effect, increased viscosity, was seen with xanthan/maize starch mixtures.

9.3.2 Gellan

The polysaccharide (gellan® or gelrite®) was isolated from a bacterial strain initially designated *Auromonas* (*Pseudomonas*) *elodea* but now named *Sphingomonas paucimobilis*

Table 9.3
Food Uses of Synergistic Gels of Xanthan and Glycomannans

Application	Products
Gel formation and stabilization	Cheese and cream cheese
	Dessert gels
	Ice cream
	Milk shakes; milk drinks
	Puddings and pie fillings
Viscosity control	Chocolate drinks
	Cottage cheese dressings
	Instant soups
	Milk shakes
	Ice cream

(71,72). It is an intriguing new polysaccharide of microbial origin with a range of potential food applications, and is the only bacterial polysaccharide other than xanthan to be currently GRAS listed. Like xanthan, it is produced by aerobic submerged fermentation. After production is complete, the viscous broth is pasteurized then treated with mild alkali to remove acyl groups. It is then hot filtered to remove bacterial cell debris. This polysaccharide is an excellent gelling agent providing clear, brittle gels at much lower concentration than agar. Recently a mutant yielding enhanced production of gellan has been reported (73). As well as food use,gellan has been widely employed as a gelling agent in plant biotechnology under the trade name of gelrite or phytogel; it can also be used in place of agar in the formulation of bacterial culture media. Other *Sphingomonas* isolates have yielded a family of polysaccharides which are closely related structurally. Although none form gels, they do yield highly viscous aqueous solutions and have been proposed for various industrial, nonfood uses.

9.3.2.1 Structure

The structure of gellan is a repeating linear tetrasaccharide composed of D-glucose, D-glucuronic acid, and L-rhamnose in a ratio of 2:1:1, but devoid of any side chains (74). In the native material, *O*-acetyl and glyceryl residues are present on the D-glucosyl residue adjacent to D-glucuronic acid (75) (Figure 9.13). As a family of polysaccharides with closely similar carbohydrate structures has now been identified, this has permitted examination of the influence of various substituents and side chains on the physical properties of the polysaccharides (76). One polysaccharide is an exact analogue of gellan, in which the L-rhamnosyl residue has been replaced by L-mannose (77). Most of these polymers possess nonstoichiometric amounts of *O*-acetyl groups, the position of which still remains uncertain in some preparations. None have so far been identified as having food applications.

9.3.2.2 Physical Properties

The commercial gellan product has a mass of ~0.5×10^6 Da. The deacylated polysaccharide forms gels with properties common in some respects to those formed with agar, alginate, or carrageenan. Gelation was dependent on ionic strength as well as on the nature of the cation. The gelation temperature increased with increasing polysaccharide concentration (78). In texture, the gels most resembled those formed with agar or κ-carrageenan. In common with agar gels, marked melting/setting hysteresis was observed. The native structure of the gellan

-3-β-D-Glc*p*-(1→4)-β-D-Glc*p*A-(1→4)-β-D-Glc*p*-(1→4)-α-L-Rha*p*-(1-

(74,75)

Figure 9.13 The structure of gellan (74,75).

polymer that contains *O*-glycerate and ~6% acetate formed weak, elastic, thermoreversible gels. From x-ray fiber diffraction studies, a threefold helical structure with an axial repeat of 2.82 nm has been suggested. The glyceryl and acetyl groups inhibited crystallization of localized regions of the gellan chains (79). Rheological data suggested that in solution, the polysaccharide adopted a locally rigid conformation. The changes observed on deacetylation of the native polysaccharide, resulting in both increased crystallinity and gel stiffness, were interpreted as suggesting that longer sections of gellan contribute to the junction zones formed (80). When the polysaccharide was deacylated, extensive intermolecular association occurred and produced strong, brittle gels closely resembling those obtained from agar. In the crystalline structure, the two left handed threefold helical chains are arranged parallel to one another as an intertwined duplex in which interchain hydrogen bonds stabilize the structure (81). Computer models indicate that an almost fully extended polysaccharide chain of 3_2 double helices packs antiparallel, providing good compatibility with measured x-ray intensities (82). A range of gel textures can be produced through control of the degree of acylation of the polymer. The gels also resembled agar in that they were rigid, brittle, and thermoreversible. They could be reautoclaved and were stable to most commercial enzymes. The polymer behaved as a stiff coil when dissolved in dimethyl sulfoxide (DMSO) but preserved its helical secondary structure in aqueous solutions. The difference between melting and setting temperatures was 45–60°C (64), the gelling temperature increasing with the cation concentration. Unlike the more highly charged macromolecules alginate or pectin, there is marked lack of specificity among the alkaline earth cations, and the temperatures at the midpoint of the sigmoidal transitions (T_m) increased only moderately with ionic strength (83). Gelling and melting temperatures increased when either the gellan or the salt concentration was increased within the concentration range 0.3–2.0% (84), but the values were dependent on the cation used. This was ascribed to the increased number of junction zones produced, and decreased rotational freedom, causing higher heat resistance and increased elastic modulus. The mechanical properties of the gel directly related to the ionic selectivity, with monovalent ions in the order K > Na > Li. Grasdalen and Smidsrød (85) also concluded that gelation of the *A. elodea* subsp. *paucimobilis* polysaccharide occurred in two steps — chain ordering and chain association. Increasing the ion or polysaccharide concentration in the medium raised the gelling and melting point of the polysaccharide gels. The presence of the cations increased the number of junction zones and decreased the rotational freedom. The structures were rendered more heat resistant and the elastic modulus of the gel increased (84). It was also suggested that there was a well defined binding site for Ca^{2+} (86). The *O*-acetyl groups on the native polysaccharide had only a weak effect on the aggregation of the gellan molecules, whereas the L-glyceryl residues were detrimental to crystal packing (81). The potassium salt possessed a crystalline structure in which the L-glyceryl groups prevented the coordinated interactions of ions and carboxylate groups which are required for strong gelation (87). Once deacylated, the coordination of calcium between double helices was so strong that molecular aggregation occurred at very low cation concentrations. The backbone conformation of gellan on reexamination proved to comprise two left handed, threefold helical chains organized in parallel. In aqueous solution, these produced an intertwined duplex in which the chains were displaced half a pitch with respect to each other (88,89).

9.3.2.3 Food Applications

Gellan forms clear, brittle, thermoreversible gels, which give excellent flavor release. The significance of this property in various food applications was indicated by Owen (90). Gellan can also be used in combination with other polysaccharides used by the food industry, although increased concentrations reduced gel strengths. It does, however, induce synergism when blended with gelatine or with gum Arabic, with gel strength increased by 40–60% (64). As pointed out by Gibson (91), blending gellan with other gums or mixtures of gums could produce a wide range of textures ranging from hard and brittle to soft and elastic. The exact texture obtained would depend on the proportion of the different gelling agents. Thus, blending with agar or κ-carrageenan, each of which also form brittle gels, reduced gel strength but retained the brittle texture. The very clear gels formed by low levels of gellan can be expected to have considerable aesthetic as well as practical appeal. Mixtures of gellan and gelatine also provide a range of textures (92). It was suggested that gellan has potential food use where a highly gelled structure or mouth feel was required (Table 9.4) (64).

9.3.3 Curdlan

Different customs exist in different countries. Unlike people in North America and Europe, people in Japan regard microbial polysaccharides as natural products which can be added to food without specific regulatory control. There is thus an upward trend in the use of microbial exopolysaccharides such as curdlan in Japan. Market shortfalls in some of the natural (plant or algal) polymers might lead to increased use of microbial products in their place. This has been exemplified by the fall in the supply of gum arabic in 1985 to 25% of the 1984 level and other shortfalls in the supplies of locust bean gum and guar gum.

Table 9.4
Uses of Gellan

a) Food Application	**Function**	**Concentration Required**
Jellies	Gelling agent in desserts.	0.15–0.2%
Jams	In low calorie spreads and fillings	0.12–0.3
Confectionery		0.8–1.0
Processed foods	Gelling in fruit, vegetables, meats	0.2–0.3%
Processed meats	Texture modification	0.1–1%
Icings	Coating agent	0.05–0.12%
Pie fillings	Texturiser	0.25–0.35%
b) Potential Food Applications of Gellan		
Dairy products	Ice cream, milkshakes, yogurts	
Fabricated foods	Fabricated fruits, meats.	
Icings and frostings	Bakery icings, frostings	
Jams and jellies	Low calorie jams, jellies, fillings	
Pie fillings and puddings	Instant desserts, jellies	
Water-based gels	Aspics	
Pet foods	Gelled pet foods, meat chunks	

Gibson (91)

Gels can be formed by several neutral bacterial polysaccharides. These include the linear β-D-glucan homopolymer, curdlan, which is formed by a number of bacterial species related to the Gram negative bacterium *Agrobacterium radiobacter* and *Rhizobium* (93) as well as *Cellulomonas flavigena* (94,95). Curdlan was approved for food use in Japan, Korea, and Taiwan. It is a β-(1$\rightarrow$3)-linked D-glucan which is insoluble in water, however, when aqueous solutions are heated it forms a gel. It is unusual in that it may either form a thermoreversible (low set) gel with some similarity to agar or, on heating to higher temperatures (greater than 80°C), yield a high set gel which is not normally thermoreversible (96). In the presence of sugars, salt or starch, the gel strength may be lowered but it is unaffected by freezing and thawing, although syneresis is observed. Curdlan is probably unique among biological gums in that a gel is formed when the polysaccharide suspension is heated, and increased heating may even strengthen the gel. The gel thus formed is stable over a wide pH range and is stable to freezing and thawing. From x-ray crystallographic studies, it has been suggested that curdlan forms a triple helical crystalline structure (97). When added to food as a gelling agent, curdlan yields gels intermediate between the elasticity of gelatine and the brittleness of agar (98). Applications of curdlan in food can be seen in Table 9.5.

The different types of gel formed by curdlan under different conditions vary in their specific properties (99), some showing pseudocrystallinity. Differences in syneresis were also noted. When examined by electron microscopy, considerable differences were revealed (100). Some formed fibril units while in others assemblies of these were seen culminating in net-like structures. Curdlan resembles many microbial homopolysaccharides in that it is biodegradable through the action of specific enzymes such as that from a *Bacillus circulans* strain which yields laminaribiose as the sole hydrolysis product (101). However, some of the gel forms are extremely resistant to most β-D-glucanases.

9.4 EXOPOLYSACCHARIDES AS A SOURCE OF FLAVOR COMPONENTS

As many microbial exopolysaccharides contain appreciable amounts of 6-deoxysugars, it has been suggested that the polymers might be used as sources of these sugars (102). The 6-deoxyhexoses in turn, could then be used as intermediates in the synthesis of furaneol and its derivatives. These compounds can be used as flavoring agents for the food industry. Furaneol yields a powerful caramel-like flavor, but when modified by the addition of various short chain fatty acid esters gives a range of either meat or fruit flavors. Furaneol is fairly expensive (~$1–200/kg), and it has been proposed that this cost could be reduced through the use of polysaccharide derived deoxysugars. If this were to be achieved economically, it would require release of the sugars from the polymers by enzymic treatment.

Table 9.5
Food Applications of Curdlan

Food Application	Function	Concentration Required
Jellies	Gelling agent	1–5%
Processed foods	Gelling agent	1–10%
Processed meats	Texture modification	0.1–1%
Sauces	Improved viscosity	0.2–0.7%
Freeze-dried foods	Improved rehydration	0.5–1%

Nakao (96)

Using immobilized enzymes or cells and microbial polysaccharides containing a high percentage of deoxysugars this might be feasible; however, enzymes capable of degrading the appropriate exopolysaccharides to their component sugars have not as yet been described.

9.5 GAZING INTO THE CRYSTAL BALL: THE FUTURE FOR MICROBIAL POLYSACCHARIDES IN FOOD

In the hope that LAB EPS with improved physical properties can be developed, or that strains can be produced with higher yields of EPS, various groups have recently studied the biosynthesis of these polymers. Normal yields of EPS are usually in the range 100–250mg litre^{-1}. (12) Yields are much lower than those obtained for xanthan or gellan. In *Lactococcus lactis* strain NIZO B40, the genes for EPS synthesis were encoded on a 40kb plasmid (103). A 12 kb region contained 14 genes with products having sequence homologies to other gene products involved in EPS, LPS, and teichoic acid biosynthesis. As with production of xanthan and similar EPS, synthesis involved an isoprenoid lipid carrier and genes responsible for transfer of sugars from sugar nucleotides to this carrier were identified. Comparison of the *eps* gene clusters of *S. thermophilus* and *L. lactis* also revealed considerable homology between genes with similar putative functions (103). It also proved possible to introduce the *eps* gene cluster from *Streptococcus thermophilus* Sfi6 into *Lactococcus lactis* MG1363 to produce an altered polysaccharide (104), while eps-encoding genes carried on plasmids from *L. lactis* could be transferred to other strains of the same bacterial species (105). Portions of the gene clusters in LAB revealed high conservation and some similarity to the genes required for capsular EPS synthesis in other Gram positive Streptococci including *S. pneumoniae* and *S. agalactiae*. It has been suggested that successful genetic engineering and targeting of specific genes leading to alteration of the repeat units might beneficially change the physical properties of the polysaccharides. However, low yields might then present problems as was found with EPS from gram negative bacteria including some xanthan mutants (56,57).

The role of specific enzymes involved in carbohydrate metabolism and EPS synthesis was also studied. In *S. thermophilus*, strain differences were observed in the role of UDP-glucose pyrophosphorylase in polysaccharide production (106). In this bacterial species, the amount of EPS produced in synthetic media depended on the carbohydrate source, with highest yields resulting from a combination of lactose and glucose (24). Attempts have also been made to produce EPS from LAB using defined media but at least for *L. delbruekii bulgaricus* various amino acids and vitamins were necessary for both growth and EPS production (107). As pointed out by Jolly et al. (7), a better understanding of structure function relationships of the EPS from LAB is still required, and this must be allied to studies of the interactions with other components present in the food matrix (108). If after achieving such an understanding, polysaccharide structures can be manipulated to yield specific properties, the question remains whether such products would be legislatively and publicly acceptable.

An alternative approach to alter the physical properties of polysaccharides has made use of glycosidases and other enzymes to remove terminal sugars from side chains. Thus, van Casteren et al. (14) employed an enzyme from *Aspergillus aciduleatus* to remove a terminal β-D-galactosyl residue from the disaccharide side chains in a *L. lactis cremoris* polymer (Figure 9.4). Another possible source of novel EPS may be yeasts, several of which already find applications in food either directly or as sources of pigments and flavors. A few species, including *Rhodotorula glutinis* formed highly viscous polyanionic EPS (109) in good yield. The polysaccharide product from strain KCTC7989 formed pseudoplastic solutions in water. There is also the possibility that polysaccharides from suitable nonfood microorganism might be designed to provide very specific physical properties. One candidate for this approach might be acetan from *Acetobacter xylinum*. This polysaccharide resembles xanthan both in its cellulosic backbone and in some of its physical properties (110,111). Mutants with altered chemical structure and differing physical properties have been isolated (112).

9.6 LEGISLATIVE ACCEPTABILITY OF MICROBIAL POLYSACCHARIDES

Before new microbial polysaccharides can be permitted for use as food additives, they must be submitted to a process of approval. In addition to the technological justification for the inclusion of the polymer in foodstuffs, evidence has to be provided of the need to use the polysaccharide and of its safety in use. The producer must also demonstrate that the new additive will benefit the consumer. This covers various categories:

1. The presentation of the foodstuff
2. The need to keep the food wholesome until eaten
3. The extension of dietary choice
4. The need for nutritional supplement
5. Convenience of purchasing, packaging, storage, preparation, and use
6. Economic advantage including longer shelf life or reduced price

Not all these factors will be appropriate in the case of microbial polysaccharides. Any company intending to produce or market the polymer must first embark on an extensive research program. In the U.K., evidence is then presented to the Food Advisory Committee in which there are representatives of consumer and enforcement bodies, the food industry, medicine, and academia. This advises the relevant Government Departments (Health; DEFRA). If the Food Advisory Committee considers that the need has been proved, safety is then examined and is assessed by the Committee on Toxicity of Chemicals in Food, Consumer Products, and Environment. Tests include determination of the LD_{50} in rats, feeding trials to animals through two generations of reproduction, and teratology testing. Tests for dust inhalation must also be performed. The costs for a development programm of this type may well be over \$10 million and 3–6 years may elapse before approval is given. In addition to official approval, there must be consumer acceptance. A major factor in favor of microbial polysaccharides is their lack of digestion by humans and animals. They can thus be used as constituents of low calorie diets as well as other processed foodstuffs. After the Food Advisory Committee has made its recommendation, the polysaccharide still requires approval from the EU if the foodstuffs in which it will be incorporated will be marketed in other Member Countries of the EU. In the U.K., the schedule of "Emulsifiers and Stabilizers in Food Regulations" is promulgated by DEFRA (previously the Ministry of Agriculture, Fisheries, and Food). The EU has attempted to harmonize food additive regulations within its member countries through a "Directive of the Council of Ministers on Emulsifiers, Stabilizers, Thickeners, and Gelling Agents for Food Use." This provides a list of the agents that are fully accepted for use in member states. The compounds are designated in Annex 1 of the EU Directive 80/597/EEC with an appropriate serial number (E400=alginic acid, E406=agar, E415=xanthan, etc). Further details include the foodstuffs in which the use of the emulsifiers is permitted, together with criteria for purity and labeling. The attempted standardization of labeling procedures provides details of trade name, manufacture, designated number, etc. In practice many processed foods are clearly labeled with their contents and consumers can see for themselves if xanthan (or gellan) is one of these.

The World Health Organization (WHO) and Food and Agricultural Organization (FAO), joint expert committee on food additives has the responsibility of proposing Acceptable Daily Intakes. Thus an ADI of 0–10mg/kgm body weight has been established for xanthan. The Committee divided food additives into three categories:

1. Fit for use in food
2. Need to be evaluated
3. Should not be used in foods

After evaluation, detailed specifications for the food additives include identity and purity. Typically, WHO provides data on biological aspects such as ingestion, calorific availability, and digestibility, and on toxicology. Toxicological evaluation includes short term and long term evaluation using a range of different animals. A two year study using rats failed to show any carcinogenic or other toxic effects which could be attributed to xanthan. Tests on human volunteers may also be included; again in the case of xanthan no adverse effects were found although this polysaccharide could effect a slow but significant weight loss in individuals suffering from obesity (113). In the USA, xanthan is permitted as a food additive under regulations controlled by the Food and Drug Administration (FDA). It is on the list of substances GRAS list, being approved in the Federal Register for use as a stabilizer, emulsifier, foam enhancer, thickening, suspending, and bodying agent. Use is permitted in sauces, gravies, and coatings applied to meat and poultry products. It may also be added to cheeses and cheese products, certain milk and cream products, food dressings, table syrups, and frozen desserts. A further use, also requiring FDA approval, is as a component of paper and paperboard intended for use in food packaging and likely to be in contact with the food. Although dextran did receive approval for food use, there is apparently no current food usage of this EPS. Curdlan, which might have a number of food applications, is used in Japan as a natural material, but is not currently permitted in the USA. The other microbial exopolysaccharide which is currently GRAS listed as an approved food additive is gellan.

The definition of EPS, for legislative purposes and for patents, is not particularly satisfactory. They are generally identified by the molar ratio or the composition of their carbohydrate and noncarbohydrate components. More detailed information on structure may be provided, together with properties such as optical rotation, viscosity, physical appearance and ash content. Xanthan is currently described in the National Formulary as: "…a high molecular weight polysaccharide gum produced by a pure culture fermentation of a carbohydrate with *X. campestris*, then purified by recovery with isopropyl alcohol, dried, and milled. It contains D-glucose and D-mannose as the dominant hexose units, along with D-glucuronic acid, and is prepared as the sodium, potassium or calcium salt. It yields not less than 4.2% and not more than 5% carbon dioxide, calculated on the dried basis, corresponding to not less than 91% and not more than 108% xanthan gum. When employed as a food additive, xanthan must conform to the definition provided in the National Formulary (UK) of "…a high molecular weight polysaccharide gum produced by a pure culture fermentation of a carbohydrate with *X. campestris*…". "The food grade polysaccharide must also meet the specifications listed in the Food Chemicals Codex (UK). These include absence of *E. coli* and *Salmonella,* and limits on arsenic (<3ppm), heavy metals (<30 ppm) and isopropanol (<750 ppm). If more exopolysaccharides are proposed for food usage, some may have similar composition and improved methods of definition will be needed. In foods such as yogurt, when addition of xanthan is permitted, it may be difficult using current methodology to distinguish it from polymers synthesized by the LAB present. It is thus surprising that few methods have been developed for the identification and quantification of the EPS once present in food. With the discovery of enzymes specifically acting on the exopolysaccharides, these could well be used to provide highly specific assay procedures for the polymers once they have been incorporated into foods containing other carbohydrate-containing materials. A study by Craston et al. (114) indicated that the gellan lyase from heterologous bacteria (a small pink-pigmented pseudomonad) might prove useful in specific determination of gellan in processed foods. A similar enzyme has been isolated and characterized by Hashimoto et al. (115,116). Their use would have considerable advantages over the tradition assay methods which fail to distinguish between different polysaccharides or carbohydrate-containing material, present in foods. Similarly, pyruvated-mannose-specific xanthan lyase lyases studied by Ruijssenaars et al. (52,53) could be used to quantify xanthan in food as well as determining its quality. Alternatively, techniques based on monoclonal antibodies prepared against xanthan such as those developed by Haaheim and Sutherland (51) could be used. Other enzymes degrading microbial polysaccharides in food, including some of those from LAB, have also been identified but it remains unclear whether they would be as suitable as the xanthan and gellan lyases for quantification of the polymers in foodstuffs (115).

REFERENCES

1. De Vuyst, L., B. Degeest. Heteropolysaccharides from lactic acid bacteria. *FEMS Microbiol. Rev.* 23:153–177, 1999.
2. Laws, A.P., Y. Gu, V.M. Marshall. Biosynthesis, characterization and design of bacterial exopolysaccharides from lactic acid bacteria. *Biotechnol. Adv.* 19:597–625, 2000.
3. Sutherland, I.W. Structure function relationships in microbial exopolysaccharides. *Biotechnol. Adv.* 12:393–448, 1995.
4. Cerning, J. Exocellular polysaccharides produced by lactic acid bacteria. *FEMS Microbiol. Rev.* 87:113–130, 1990.
5. Ruas-Madiedo, P., J. Hugenholtz, P. Zoon. An overview of the functionality of exopolysaccharides produced by lactic acid bacteria. *Int. Dairy J.* 12:163–171, 2002.
6. Nakajima, H., Y. Suzuki, H. Kaizu, T. Hirota. Cholesterol-lowering activity of ropy fermented milk. *J. Food Sci.* 57:1327–1329, 1992.
7. Jolly, L., S.J.F. Vincent, P. Duboc, J.-R. Neeser. Exploiting exopolysaccharides from lactic acid bacteria. *Antonie van Leeuwenhoek* 82:367–374, 2002.
8. Vanhaverbeke, C., C. Bosso, P. Colin-Morel, C. Gey, L. Gamar-Nourani, K. Blondeau, J.M. Simonet, A. Heyraud. Structure of an extracellular polysaccharide produced by *Lactobacillus rhamnosus* strain C83. *Carbohydr. Res.* 314:211–220, 1998.
9. Lipinski, T., C. Jones, X. Lemercinier, A. Korzeniowska-Kowal, M. Strus, J. Rybka, A. Gamian, P.B. Heczko. Structural analysis of the *Lactobacillus rhamnosus* strain KL37C exopolysaccharide. *Carbohydr. Res.* 338:605–609, 2003.
10. Marshall, V.M., H. Dunn, M. Elvin, N. McLay, Y. Gu, A.P. Laws. Structural characterisation of the exopolysaccharide produced by *Streptococcus thermophilus* EU20. *Carbohydr. Res.* 331:413–422, 2001.
11. Marshall, V.M., A.P. Laws, Y. Gu, F. Levander, P. Rådstrom, L. De Vuyst, B. Degeest, F. Vaningelen, H. Dunn, M. Elvin. Exopolysaccharide-producing strains of thermophilic lactic acid bacteria cluster into groups according to their EPS structure. *Lett. Appl. Microbiol.* 32:433–437, 2001.

12. Ricciardi, A., F. Clementi. Exopolysaccharide from lactic acid bacteria: structure, production and tecnological applications. *Ital. J. Food Sci.* 12:23–45, 2000.
13. Robijn, G.W., D.J.C. van den Berg, H. Haas, J.P. Kamerling, J.F.G. Vliegenhart. Determination of the structure of the exopolysaccharide produced by *Lactobacillus saké* O-1. *Carbohydr. Res.* 276:117–136, 1995.
14. Van Casteren, W.H.M., C. Dijkema, H.A. Schols, G. Beldman, A.G.J. Voragen. Structural characterisation and enzymic modification of the exopolysaccharide produced by *Lactococcus lactis* subsp. *cremoris* B39. *Carbohydr. Res.* 324:170–181, 2000.
15. Nakajima, H., T. Hirota, T. Toba, T. Itoh, S. Adachi. Structure of the extracellular polysaccharide from slime-forming *Lactococcus lactis* subsp. *cremoris* SBT 0495. *Carbohydr. Res.* 224:245–253, 1992.
16. Frengova, G.I., E.D. Simova, D.M. Beshkova, Z.I. Simov. Exopolysaccharide produced by lactic acid bacteria of kefir grains. *Z. Naturforsch.(C.)* 57:805–810, 2002.
17. Harding, L.P., V.M. Marshall, M. Elvin, Y. Gu, A.P. Laws. Structural characterisation of a perdeuteriomethylated exopolysaccharide by nmr spectroscopy: characterisation of the novel exopolysaccharide produced by *Lactobacillus delbrueckii* subsp. *bulgaricus* EU23. *Carbohydr. Res.* 338:61–67, 2003.
18. Gruter, M., B.R. Leeflang, J. Kuiper, J.P. Kamerling, J.F.G. Vliegenhart. Structural characterisation of the exopolysaccharide produced by *Lactobacillus*. *Carbohydr. Res.* 239:209–226, 1993.
19. Faber, E.J., J.P. Kamerling, J.F.G. Vliegenhart. Structure of the extracellular polysaccharide produced by *Lactobacillus delbrueckii* subsp. bulgaricus 291. *Carbohydr. Res.* 331:183–194, 2001.
20. Van Casteren, M.R., C. Pau-Roblot, A. Begin, D. Roy. Structure determination of the exopolysaccharide produced by *Lactobacillus rhamnosus* strains RW-9595M. *Biochem. J.* 363:2–17, 2002.
21. Van Geel-Schutten, G.H., E.J. Faber, E. Smit, K. Bonting, M.R. Smith, B. Ten Brink, J.P. Kamerling, J.F.G. Vliegenhart, L. Dijkhuizen. Biochemical and structural characterization of the glucan and fructan exopolysaccharides synthesized by the *Lactobacillus reuteri* wild-type strain and by mutant strains. *Appl. Environ. Microbiol.* 65:3008–3014, 1999.
22. Van Hijum, S.A.F.T., G.H. Van Geel-Schutten, H. Rahaoui, M.J.E.C. van der Maarel, L. Dijkhuizen. Characterization of a novel fructosyltransferase from *Lactobacillus reuteri* that synthesizes high-molecular-weight inulin and inulin oligosaccharides. *Appl. Environ. Microbiol.* 68:4390–4398, 2002.
23. Kralj, S., G.H. Van Geel-Schutten, H. Rahaoui, R.J. Leer, E.J. Faber, M.J.E.C. van der Maarel, L. Dijkhuizen. Molecular characterization of a novel glucosyltransferase from *Lactobacillus reuteri* strain 121 synthesizing a unique, highly branched glucan with α-(1→4) and α-(1→6) glucosidic bonds. *Appl. Environ. Microbiol.* 68:4283–4291, 2002.
24. Degeest, B., L. De Vuyst. Correlation of activities of the enzymes α-phosphoglucomutase, UDP-galactose 4-epimerase and UDP-glucose pyrophosphorylase with exopolysaccharide biosynthesis by *Streptococcus thermophilus* LY03. *Appl. Env. Microbiol.* 66:3519–3527, 2000.
25. Faber, E.J., M.J. van den Haak, J.P. Kamerling, J.F.G. Vliegenhart. Structure of the exopolysaccharide produced by *Streptococcus thermophilus* S3. *Carbohydr. Res.* 331:173–182, 2001.
26. Duenas-Chasco, M.T., M.A. Rodriguez-Carvajal, P. Tejero-Mateo, J.L. Espartero, A. Irastorza-Iribas, A.M. Gil-Serrano. Structural analysis of the exopolysaccharides produced by *Lactobacillus* spp. G-77. *Carbohydr. Res.* 307:125–133, 1998.
27. Duenas-Chasco, M.T., M.A. Rodriguez-Carvajal, P.T. Mateo, G. Franco-Rodriguez, J.L. Espartero, A. Irastorza-Iribas, A.M. Gil-Serrano. Structural analysis of the exopolysaccharide produced by *Pediococcus damnosus* 2.6. *Carbohydr. Res.* 303:453–458, 1997.
28. Faber, E.J., D.J. van Haaster, J.P. Kamerling, J.F.G. Vliegenhart. Characterization of the exopolysaccharide produced by *Streptococcus thermophilus* 8S containing an open-chain nononic acid. *Europ. J. Biochem.* 269:5590–5598, 2002.
29. Yamamoto, Y., S. Murosaki, R. Yamauchi. Structural studies on an exocellular polysaccharide produced by *Lactobacillus helveticus* TY1-2. *Carbohydr. Res.* 261:67–78, 1994.

30. Gorret, N., C.M.G.C. Renard, M.H. Famelart, J.L. Maubois, J.L. Doublier. Rheological characterization of the exopolysaccharide produced by *Propionobacterium acidi-propionici* on milk microfiltrate. *Carbohydr. Polym.* 51:149–158, 2003.
31. Ricciardi, A., E. Parente, M.A. Crudele, F. Zanetti, G. Scolari, I. Mannazzu. Exopolysaccharide production by *Streptococcus thermophilus* SY: production and preliminary characterization of the polymer. *J. Appl. Microbiol.* 92:297–306, 2002.
32. van den Berg, D.J.C., G.W. Robijn, A.C. Janssen, M.L.F. Giuseppin, R. Vreeker, J.P. Kamerling, J.F.G. Vliegenhart, A.T. Ledeboer, C.T. Verrips. Production of a novel extracellular polysaccharide by *Lactobacillus saké* O-! and characterization of the polysaccharide. *Appl. Environ. Microbiol.* 61:2840–2844, 1995.
33. Tuinier, R., W.H.M. van Casteren, E. Looijesteijn, H.A. Schols, A.G.V. Voragen, P. Zoon. Effects of structural modifications on some physical characteristics of exopolysaccharides from *Lactococcus lactis*. *Biopolymers* 59:160–166, 2001.
34. Faber, E.J., J.A. van Kuik, J.P. Kamerling, J.F.G. Vliegenhart. Modeling of the structure in aqueous solution of the exopolysaccharide produced by *Lactobacillus helveticus* 766. *Biopolymers* 63:66–76, 2002b.
35. Levander, F., M. Svensson, P. Rådstrom. Enhanced exopolysaccharide production by metabolic engineering of *Streptococcus thermophilus*. *Appl. Env. Microbiol.* 68:784–790, 2002.
36. Ramos, A., I.C. Boels, W.M. de Vos, H. Santos. Relationship between glycolysis and exopolysaccharide biosynthesis in *Lactococcus lactis*. *Appl. Env. Microbiol.* 67, 33–41, 2001.
37. Boels, I.C., A. Ramos, M. Kleerebezem, W.M. de Vos. Functional analysis of the *Lactococcus lactis galU* and *galE* genes and their impact on sugar nucleotide and exopolysaccharide synthesis. *Appl. Env. Microbiol.* 67:3033–3040, 2001.
38. Jansson, P-E., L. Kenne, B. Lindberg. Structure of the extracellular polysaccharide from *Xanthomonas campestris*. *Carbohydr. Res.* 45:275–282, 1975.
39. Melton, L.D., Mindt, L., Rees, D.A., Sanderson, G.R. Covalent structure of the extracellular polysaccharide from *Xanthomonas campestris*. *Carbohydr. Res.* 46:245–257, 1976.
40. Souw, P. and Demain, A.L. Nutritional studies on xanthan production by *Xanthomonas campestris*. *Appl. Environ. Microbiol.* 37:1186–1192, 1979.
41. Ho, C.S., L.-K. Ju, R.F. Baddour. The anomaly of oxygen diffusion in aqueous xanthan solutions. *Biotechnol. Bioeng.* 32:8–17, 1988.
42. Nakajima, S., H. Funahashi, T. Yoshida. Xanthan gum production in a fermentor with twin impellors. *J. Ferment. Bioengin.* 70:392–397, 1990.
43. Tait, M.I., I.W. Sutherland, A.J. Clarke-Sturman. Effect of growth conditions on the production, composition and viscosity of *Xanthomonas campestris* exopolysaccharide. *J. Gen. Microbiol.* 132:1483–1492, 1986.
44. Casas, J.A., V.E. Santos, F. GarciaOchoa. Xanthan gum production under several operational conditions: molecular structure and rheological properties. *Enzyme Microb. Technol.* 26:282–291, 2000.
45. Linton, J.D. The relationship between metabolite production and the growth efficiency of the producing organism. *FEMS Microbiol. Rev.* 75:1–18, 1990.
46. Galindo, E. Aspects of the process for xanthan production. *Trans. I. Chem. E.* 72:227–237, 1994.
47. Garcia-Ochoa, F., V.E. Santos, J.A. Casas, E. Gómez. Xanthan gum: production, recovery and properties. *Biotechnol. Adv.* 18:549–579, 2002.
48. Rinaudo, M., M. Milas. Enzymic hydrolysis of the bacterial polysaccharide xanthan by cellulase. *Intern. J. Biol. Macromol.* 2:45–48, 1980.
49. Sutherland, I.W. Hydrolysis of unordered xanthan in solution by fungal cellulases. *Carbohydr. Res.* 131:93–104, 1984.
50. Sutherland, I.W. Xanthan lyases: novel enzymes found in various bacterial species. *J. Gen. Microbiol.* 133:3129–3134, 1987.
51. Haaheim, L., I.W. Sutherland. Monoclonal antibodies reacting with the exopolysaccharide from *Xanthomonas campestris*. *J. Gen. Microbiol.*, 135:605–612, 1989.

52. Ruijssenaars, H.J., J.A.M. Debont, S. Hartmans. A pyruvated mannose-specific xanthan lyase involved in xanthan degradation by *Paenibacillus alginolyticus* XL-1. *Appl. Environ. Microbiol.* 65:2446–2452, 1999.
53. Ruijssenaars, H.J., S. Hartmans, J.C. Verdoes. A novel gene encoding xanthan lyase of *Paenobacillus alginolyticus* strain XL-1. *Appl. Environ. Microbiol.* 66:3945–3950, 2000.
54. Ielpi, L., R. Couso, M. Dankert. Lipid-linked intermediates in the biosynthesis of xanthan gum. *FEBS Lett.*, 130:253–256, 1981.
55. Ielpi, L., R.O. Couso, M. Dankert. Xanthan gum biosynthesis acetylation occurs at the prenyl-phospho-sugar stage. *Biochem. Int.*, 6:323–333, 1983.
56. Vanderslice, R.W., D.H. Doherty, M. Capage, M.R. Betlach, R.A. Hassler, N.M. Henderson, J. Ryan-Graniero, M. Tecklenberg. Genetic engineering of polysaccharide structure in *Xanthomonas campestris.* In: *Biomedical and Biotechnicalological Advances in Industrial Polysaccharides*, Crescenzi, V., I.C.M. Dea, S. Paoletti, S. Stivala, I.W. Sutherland, eds., New York: Gordon and Breach, 1989, pp 145–156.
57. Betlach, M.R., M.A. Capage, D.H. Doherty, R.A. Hassler, N.M. Henderson, R.W. Vanderslice, J.D. Marelli, M.B. Ward. Genetically engineered polymers: manipulation of xanthan biosynthesis. In: *Industrial Polysaccharides*, Yalpani, M., ed., Amsterdam: Elsevier, 1987, pp 145–156.
58. Whitfield, C., M.A. Valvano. Biosynthesis and expression of cell-surface polysaccharides in Gram-negative bacteria. *Adv. Microb. Physiol.* 35:135–246, 1993.
59. Shatwell, K.P. The influence of acetyl and pyruvic acid substituents on the solution and interaction properties of xanthan. Ph.D. Thesis, Edinburgh University, 1989.
60. Moorhouse, R., M.D. Walkinshaw, S. Arnott. Xanthan gum - molecular conformation and interactions. *ACS Symp.* 45:90–102, 1977.
61. Norisuye, T., T. Yanaki, H. Fujita. Triple helix of a *Schizophyllum commune* polysaccharide in aqueous solution. *J. Polym. Sci.* 18:547–548, 1980.
62. Shatwell, K.P., I.W. Sutherland, I.C.M. Dea, S.B. Ross-Murphy. The influence of acetyl and pyruvate substituents on the solution properties of xanthan polysaccharide. *Carbohydr. Res.* 206:87–103, 1990.
63. Tait, M.I., I.W. Sutherland. Synthesis and properties of a mutant type of xanthan. *J. Appl. Bacteriol.* 66:457–460, 1989.
64. Kang, K.S., D.J. Pettit. Xanthan, gellan, welan and rhamsan. In: *Industrial Gums*, Whistler, R., J.N. BeMiller, eds., New York: Academic Press, 1993, pp 341–397.
65. Morris, V.J. and Wilde, P.J. Interactions of food biopolymers. *Curr. Opin. Colloid Interface Sci.* 2:567–572, 1997.
66. Morris, V.J., M. Miles. Effect of natural modifications on the functional properties of extracellular bacterial polysacharides. *Int. J. Biol. Macromol.* 8:342–348, 1986.
67. Shatwell, K.P., I.W. Sutherland, S.B. Ross-Murphy, I.C.M. Dea. Influence of the acetyl substituent on the interaction of xanthan with plant polysaccharides, I: xanthan-locust bean gum systems. *Carbohydr. Polym.* 14:29–51, 1991.
68. Copetti, G., M. Grassi, R. Lapasin, S. Pricl. Synergistic gelation of xanthan gum with locust bean gum: a rheological investigation. *Glycoconjugate J.* 14:951–961, 1997.
69. Lundin, L., A. Hermansson. Supermolecular aspects of xanthan-locust bean gum gels based on rheology and electron microscopy. *Carbohydr. Polym.* 26:129–140, 1995.
70. Shi, X., J.N. BeMiller. Effects of food gums on viscosities of starch suspensions during pasting. *Carbohydr. Polym.* 50:7–18, 2002.
71. Pollock, T.J. Gellan related polysaccharides and the genus *Sphingomonas*. *J. Gen. Microbiol.* 139:1939–1945, 1993.
72. Kang, K.S., G.T. Veeder, P.J. Mirrasoul, T. Kaneko, I.W. Cottrell. Agar-like polysaccharide produced by a *Pseudomonas* species: production and basic properties. *Appl. Environ. Microbiol.* 43:1086–1091, 1982.
73. West, T.P. Isolation of a mutant strain of *Pseudomonas* sp. ATCC31461 exhibiting elevated polysaccharide production. *J. Ind. Microbiol. Biotechnol.* 29:185–188, 2002.
74. Jansson, P.-E., B. Lindberg, P.A. Sandford. Structural studies of gellan gum, an extracellular polysaccharide elaborated by *Pseudomonas elodea*. *Carbohydr. Res.* 124:135–139, 1983.

75. Kuo, M., A.J. Mort, A. Dell. Identification and location of L-glycerate, an unusual acyl substituent in gellan gum. *Carbohydr. Res.* 156:173–187, 1986.
76. Cairns, P., M.J. Miles, V.J. Morris. X-ray fibre diffraction studies of members of the gellan family of polysaccharides. *Carbohydr. Polym.* 14:367–371, 1991.
77. O'Neill, M.A., A.G. Darvill, P. Albersheim, K.J. Chou. Structural analysis of an acidic polysaccharide secreted by *Xanthobacter* sp. (ATCC53272). *Carbohydr. Res.* 206:289–296, 1990.
78. Nakamura, K., K. Harada, Y. Tanaka. Viscoelastic properties of aqueous gellan solutions: the effects of concentration on gelation. *Food Hydrocol.* 7:435–447, 1993.
79. Carroll, V., G.R. Chilvers, D. Franklin, M.J. Miles, V.J. Morris. Rheology and microstructure of solutions of the microbial polysaccharide from *Pseudomonas elodea*. *Carbohydr. Res.* 114:181–191 1983.
80. Carroll, V., M.J. Miles, V.J. Morris. Fibre diffraction studies of the extracellular polysaccharide from Pseudomonas elodea. *Intern. J. Biol. Macromol.* 4:432–433, 1982.
81. Chandrasekaran, R., V.G. Thailambal. The influence of calcium ions, acetate and L-glycerate groups on the gellan double helix. *Carbohydr. Polym.* 12:431–442, 1990.
82. Upstill, C., E.D.T. Atkins, P.T. Attwool. Helical conformations of gellan gum. *Int. J. Biol. Macromol.* 8:275–288, 1986.
83. Crescenzi, V., M. Dentini, I.C.M. Dea. The influence of side-chains on the dilute solution properties of three structurally related bacterial anionic polysaccharides. *Carbohydr. Res.* 160:283–302, 1987.
84. Moritaka, H., K. Nishinari, N. Nakahama, H. Fukuba. Effects of potassium chloride and sodium chloride on the thermal properties of gellan gum gels. *Biosci. Biotech. Biochem.* 56:595–599, 1992.
85. Grasdalen, H., O. Smidsrød. Gelation of gellan gum. *Carbohydr. Polym.* 7:371–394, 1988.
86. Larwood, V.L., B.J. Howlin, G.A. Webb. Solvation effects on the conformational behaviour of gellan and calcium ion binding to gellan double helices. *J. Mol. Modeling* 2:175–182, 1996.
87. Chandrasekaran, R., A. Radha, V.G. Thailambal. Roles of potassium ions, acetyl and L-glyceryl groups in the native gellan double helix. *Carbohydr. Res.* 224:1–17 1992.
88. Chandrasekaran, R., A. Radha. Molecular architectures and functional-properties of gellan gum and related polysaccharides. *Trends Food Sci. Technol.* 6:143–148, 1995.
89. Chandrasekaran, R., L.C. Puigjaner, K.L. Joyce, S. Arnott. Cation interactions in gellan: an X-ray study of the potassium salt. *Carbohydr. Res.* 181:23–40, 1988.
90. Owen, G. Gellan quick-setting gum systems. In: *Gums and Stabilisers for the Food Industry* 4, Phillips, G.O., D.J. Wedlock, P.A. Williams, eds., Oxford: IRL Press, 1989, p 173–182.
91. Gibson, W. Gellan. In: *Thickening and Gelling Agents for Food*, Imeson, K., ed., Glasgow: Blackie & Son, 1992, p. 227–250.
92. Wolf, C.L., W.M. LaVelle, R.C. Clark. Gellan gum/gelatine blends. U.S. Patent 4,876,105, 1989.
93. Harada, T., S. Sato, A. Harada. Curdlan. *Bull. Kobe Univ.* 20:143–164, 1987.
94. Buller, C.S. K.C. Voepel. Production and purification of an extracellular polyglucan produced by *Cellulomonas flavigena* strain KU. *J. Ind. Microbiol.* 5:139–146, 1990.
95. Kenyon, W.J., C.S. Buller. Structural analysis of the curdlan-like exopolysaccharide produced by *Cellulomonas flavigena*. *J. Ind. Microb. Biotech.* 29:200–203, 2002.
96. Nakao, Y. Properties and food applications of curdlan. *Agro-Food-Industry Hi-Tech* 8:12–15, 1997.
97. Chuah, C.T., A. Sarko, Y. Deslandes, R.H. Marchessault. Triple helical crystalline structure of curdlan and paramylon hydrates. *Macromolecules* 16:1375–1382, 1983.
98. Kimura, H., S. Moritaka, A. Misaki. Curdlan. *J. Food Sci.* 43, 200–203, 1973.
99. Kanzawa, Y., A. Harada, A. Koreeda, T. Harada, K. Okuyama. Difference of molecular association in two types of curdlan gel. *Carbohydr. Polym.* 10:299–313, 1989.
100. Harada, T., Y. Kanzawa, K. Kanenaga, A. Koreeda, A. Harada. Electron microscopic studies on the ultrastructure of curdlan and other polysaccharides used in foods. *Food Struct.* 10:1–18, 1991.

101. Kanzawa, Y., T. Kurasawa, Y. Kanegae, A. Harada, T. Harada. Purification and properties of a new exo-(1→3)-β-D-glucanase from *Bacillus circulans* YK9 capable of hydrolysing resistant curdlan with formation of only laminaribiose. *Microbiology* 140:637–642, 1994.
102. Graber, M., A. Morin, F. Duchiron, P.F. Monsan. Microbial polysaccharides containing 6-deoxysugars. *Enzyme Microb.Technol.* 10:198–205, 1988.
103. van Kranenburg, R., J.D. Marugg, I.I. Van Swam, N.J. Willem, W.M. de Vos. Molecular characterization of the plasmid-encoded *eps* gene cluster essential for exopolysaccharide biosynthesis in *Lactococcus lactis*. *Mol. Microbiol.* 24:387–397, 1997.
104. Stingele, F., S.J.F. Vincent, E.J. Faber, J.W. Newell, J.P. Kamerling, J.-R. Neeser. Introduction of the exopolysaccharide gene cluster from *Streptococcus thermophilus* Sfi6 into *Lactococcus lactis* MG1363: production and characterization of an altered polysaccharide. *Mol. Microbiol.* 32:1287–1295, 1999.
105. van Kranenburg, R., I.I. Vos, M.K. van Swam, W.M. de Vos. Functional analysis of glycosyltransferase genes from *Lactococcus lactis* and other gram positive cocci: complementation, expression and diversity. *J. Bacteriol.* 181:6347–6353, 1999.
106. Escalante, A., C. Wacherrodarte, M. Garciagaribay, A. Farres. Enzymes involved in carbohydrate metabolism and their role on exopolysaccharide production in *Streptococcus thermophilus*. *J. Appl. Microbiol.* 84:108–114, 1998.
107. Grobben, G.J., I. Chinjoe, V.A. Kitzen, I.C. Boels, F. Boer, J. Sikkema, M.R. Smith, J.A.M. Debont. Enhancement of exopolysaccharide production by *Lactobacillus delbrueckii* subsp. *bulgaricus* NCFB 2772 with a simplified defined medium. *Appl. Environ. Microbiol.* 64:1333–1337, 1998.
108. Duboc, P., B. Mollet. Applications of exopolysaccharides in the dairy industry. *Int. Dairy J.* 11:759–768, 2001.
109. Cho, D.H., H.J. Chae, E.Y. Kim. Synthesis and characterization of a novel extracellular polysaccharide by *Rhodotorula glutinis*. *Appl. Chem. Biotech.* 95:183–193, 2001.
110. Couso, R.O., L. Ielpi, M.A. Dankert. A xanthan gum-like polysaccharide from *Acetobacter xylinum*. *J. Gen. Microbiol.* 133:2133–2135, 1987.
111. Morris, V.J., G.J. Brownsey, P. Cairns, G.R. Chilvers, M.J. Miles. Molecular origins of acetan solution properties. *Intern. J. Biol. Macromol.* 11:326–328, 1989.
112. MacCormick, C.A., J.E. Harris, A.P. Gunning, V.J. Morris. Characterization of a variant of the polysaccharide acetan produced by a mutant of *Acetobacter xylinum* strain CR1/4. *J. Appl. Bacteriol.* 74:196–199, 1993.
113. WHO, 1987.
114. Craston, D.H., P. Farnell, J.M. Francis, S. Gabriac, W. Matthews, M. Saeed, I.W. Sutherland. Determination of gellan gum by capillary electrophoresis and Ce-Ms. *Food Chem.* 73:103–110, 2001.
115. Hashimoto, W., K. Maesaka, N. Sato, S. Kimura, K. Yamamoto, H. Kumagai, K. Murata. Microbial system for polysaccharide depolymerization: enzymatic route for gellan depolymerization by *Bacillus* sp. GL1. *Arch. Biochem. Biophys.* 339:17–23, 1997.
116. Hashimoto, W., E. Kobayashi, H. Nankai, N. Sato, T. Miya, S. Kawai, K. Murata. Unsaturated glucuronyl hydrolase of *Bacillus* sp GL1: novel enzyme prerequisite for metabolism of unsaturated oligosaccharides produced by polysaccharide lyases. *Arch. Biochem. Biophys.* 368:367–374, 1999.
117. Ruijssenaars, H.J., F. Stingele, S. Hartmans. Biodegradability of food-associated extracellular polysaccharides. *Curr. Microbiol.* 40:194–199, 2000.
118. Robijn, G.W., R.C. Gallego, D.J.C. van den Berg, H. Haas, J.P. Kamerling, J.F.G. Vliegenhart. Structural characterization of the exopolysaccharide produced by *Lactobacillus acidophilus* LMG9433 *Carbohydr. Res.* 288:203–218, 1996.
119. Robijn, G.W., A. Imberty, D.J.C. van den Berg, A.M. Ledeboer, J.P. Kamerling, J.F.G. Vliegenhart, S. Perez. Predicting helical structures of the exopolysaccharide produced by *Lactobacillus sake O*-1. *Carbohydr. Res.* 288:57–74, 1996.

1.10

Genetics of Dairy Starter Cultures

Daniel J. O' Sullivan

CONTENTS

10.1 INTRODUCTION

Dairy starter cultures have a long history in the production of fermented milk products dating back several thousand years. These cultures consist primarily of several members of the lactic acid bacteria (LAB), a loosely termed group of bacteria that are arguably the most valuable group of bacteria for humans in terms of health and economic impact. Within the dairy industry, the LAB are involved in the production of a myriad of fermented foods, such as yogurts, sour cream, buttermilk, acidophilus milk, kumiss, kefir, and approximately 2000 different types of cheeses. The workhorse of the dairy starter cultures is undoubtedly *Lactococcus lactis* because of its indispensable role in the production of most cheeses and fermented milks. This bacterium was first described by Louis Pasteur during his studies on soured milks in the 1850s, and was the first bacterium to be isolated from a mixed population in pure culture by Joseph Lister (1). It has since achieved quite a few other firsts among the LAB, being the first one to have genetic tools developed for genetic manipulation, and also the first to have its genome sequenced (2). Currently, genetic tools are available for all pertinent LAB and the genomes of several of them have been deciphered.

While the use of dairy starter cultures dates well before any knowledge of bacteria, knowledge of what microbes are involved is quite recent and is still not known for all dairy starter cultures used. In some cases the old tradition of 'back slopping' some of the previous days product represents the starter culture on a daily basis (3). However, this practice has very limited use in large commercial dairy processing plants today, although undefined starter cultures are still used to a substantial degree. For example, many cheese starter cultures are propagated in milk prior to addition to the cheese vat, but the precise microbial makeup is not known. Interestingly, in the case of cheese starters, the transition from using undefined starters to a defined mix of desirable starter bacteria only occurred to a significant degree over the last 30 years, except in New Zealand, which transitioned decades earlier. Today, most major commercial cheese plants throughout the world now rely on defined starter cultures, as the resulting cheese has a more consistent quality due to the control over the starter bacteria that are present. However, some cheeses still rely heavily on undefined cultures, particularly in Europe, as they can give a better quality product, albeit somewhat inconsistent. Many fermented milks rely totally on undefined starters, such as kefir. Kefir starters are termed kefir grains as the microorganisms are contained in a large grain like mass of protein and carbohydrate, and while the strains are normally undefined, they consist mainly of lactobacilli, lactococci, and yeast (4). Other fermented milk products such as Acidophilus milk, Yakult, and Danone's Actimel or DanActive™ (as marketed in the USA) are sold for the specific strains of LAB they contain.

10.2 TAXONOMY AND IDENTIFICATION OF DAIRY STARTER CULTURES

LAB are the backbone of the dairy culture industry. The main LAB that comprise dairy starters are contained in five genera: *Lactococcus lactis* subsp *lactis* and *cremoris*; *Streptococcus thermophilus*; *Leuconostoc* species *lactis*, *mesenteroides* subsp (*cremoris* and *dextranium*) and *pseudomesenteroides*; *Enterococcus faecalis* and *faecium*; and several species of *Lactobacillus*, most notably, *delbruechii* subsp *bulgaricus*, *casei*, *brevis*, *helveticus*, *rhamnosus*, *acidophilus*, *fermentum*, *curvatus*, *johnsonii*, and *gasseri*. Dairy cultures also contain microorganisms outside the general LAB classification and these include certain bifidobacteria, brevibacteria, propionibacteria, and fungi. These latter three groups of bacteria are from a different taxonomic grouping than the classical LAB, as they have irregular rod shaped cells and a high G + C content in their genome, and group in or close to the *Actinomycetales* branch of high G + C gram positive bacteria, rather than the LAB.

It is essential to be able to accurately and unambiguously identify cultures for their successful application in the diary industry. Previously, identification relied solely on morphological and biochemical profiles, which in some cases can be ambiguous. A 1998 study of commercially available lactobacilli and bifidobacteria showed that many were misclassified at the species level (5). The advent of molecular tools has greatly simplified reliable identification. The most accepted means for typing unknown isolates is by sequence analysis of the16S ribosomal RNA (rRNA). This approach for classifying organisms was first developed by Woese and coworkers (reviewed, 6). The extensive database of rRNA sequences allows detailed studies to be made on the phylogenetic position of unknown isolates. Technically, the procedure is very feasible, as the polymerase chain reaction (PCR) can be used to directly amplify the 16S rRNA genes using primers, which are directed at universally conserved regions at both ends of the gene. The entire PCR amplicon (~1.5 kb) can be directly sequenced and compared to rRNA databases.

While the use of 16S rRNA sequence analysis has greatly facilitated the accurate identification of dairy cultures, it has a limitation in differentiating closely related species of *Bifidobacterium*. This is an important genus for probiotics and its unambiguous identification is critical for its successful use. Within this genus, the rRNA sequence is highly conserved (7) and makes differentiating between very closely related species, such as *B. longum* and *B. infantis*, challenging. To complement the rRNA sequence approach, analysis of other DNA regions, which are not as conserved as the 16S RNA, have been used. However, not every DNA region can be used for this purpose, as it needs to be universally present in bacteria and should be highly conserved to ensure that sequence changes are less influenced by temporary environmental changes. The region between the 16S and 23S rRNA genes, termed the internal transcribed spacer (ITS), has been used for a more detailed analysis of bifidobacteria (7). This molecule is universally present in bacteria, but is not highly conserved (8), and the different ITS regions within the same bacterial cell can exhibit heterogeneity (9). However, the molecule is technically very feasible to obtain as PCR can be used to amplify the molecule directly, using primers for the universally conserved regions within the bordering 16S and 23S rRNA genes. It has been used for further characterizing bifidobacteria and was much more sensitive than rRNA analysis (7).

Another DNA region that has been used for sensitive species differentiation within the genus *Bifidobacterium* is a short internal segment of the *recA* gene. It possesses the important criteria of being universally present in bacteria and being highly conserved. The *recA* gene encodes the RecA protein, which has a central role in recombination, DNA repair and the SOS response (10). Previous studies had established that meaningful bacterial phylogenetic relationships could be obtained by sequence analysis of the RecA protein (11,12). These studies highlighted the possibility that a segment of the *recA* gene might be a useful molecule for phylogenetic analysis within a particular genus. This concept was applied to the genus *Bifidobacterium*, in a study by Kullen et al. (13). The molecule was obtained from both type and intestinal bifidobacteria isolates using PCR with primers directed to universally conserved regions within the *recA* gene. This approach yielded a fragment of ~300 kb, which was sequenced using a single sequence reaction from either end. The phylogenic relationship obtained by sequence analysis of this short segment of the *recA* gene, compared favorably with the analysis from the complete rRNA gene and provided another means for speciation within this genus.

DNA fingerprinting of dairy starter cultures is becoming more common and cultures can be identified if a DNA fingerprinting database is available. This can be very valuable and useful for DNA fingerprinting techniques that are reliable, reproducible, and can differentiate at the strain level. The use of pulse field gel electrophoresis (PFGE) for DNA fingerprinting is the optimum choice for this purpose, as it is very definitive and unambiguous (reviewed, 14). It is however a technically challenging and long procedure, which limits its widespread use. Furthermore, extensive PFGE DNA fingerprinting databases are not yet widely available. A DNA fingerprinting technique that has gained more popularity in industry is ribotyping, as it is automated and results take less that a day (reviewed, 14). The availability of a fully automated RiboPrinter® is frequently an attractive choice because of its technical simplicity and availability of extensive databases for ribotype DNA fingerprints. It is therefore sometimes used in the diary industry to identify cultures. However, it should be noted that DNA fingerprints obtained using ribotyping are more ambiguous for some groups of bacteria than PFGE and can result in inaccurate identification. A recent study of enterococci found ribotyping to be unreliable, suggesting that the technique has limitations for bacterial identification within the LAB (15).

10.3 PLASMID BIOLOGY OF DAIRY STARTER CULTURES

While bacterial plasmids were first observed in the early 1960s (16), they were not detected in the dairy LAB until the early 1970s in the laboratory of Larry McKay at the University of Minnesota (17). Dairy starter lactococci had long been known to loose the ability to grow in milk if they were grown for long periods in laboratory media. This instability in phenotypes prompted McKay's group to search for plasmids, which were known to be readily lost from bacteria if they do not contain essential genes for their current growth environment. Subsequent studies by McKay and other groups found that many phenotypes, essential for milk fermentation by *Lactococcus lactis*, were actually encoded on plasmids. These phenotypes include lactose metabolism, proteinase production, oligopeptide uptake, and bacteriophage defense systems. Other phenotypes such as citrate metabolism, polysaccharide production, and bacteriocin production are also plasmid encoded in many cases.

10.3.1 Role of Plasmids in *L. Lactis*

Strains of *L. lactis* that are used in the dairy industry are originally of plant origin, but have evolved for an extended period in a milk fermentation environment. It is therefore intriguing that these lactococci have a heavy reliance on plasmids for metabolic phenotypes that are essential for growth and survival in their environment (18). This is unusual in bacteria as the basic metabolic genes for growth in a natural habitat are usually chromosomally encoded. Plasmid encoded functions are generally not essential for survival except to confer selective advantages under specific environment conditions. One reason for this unusual reliance on plasmids in lactococci may be to increase the expression of phenotypes that are needed to grow in milk. As multiple genes are involved, it less likely that up mutation in the different gene promoters could achieve the same effect while still being competitive in that environment. Clearly, isolates that can metabolize lactose faster, and obtain amino acids from casein more efficiently, would have a selective advantage over time in an exclusive dairy environment. Evidence for the selective evolution of genes in *L. lactis* to be plasmid encoded for increased phenotypic expression comes from the abortive bacteriophage defense system AbiA. This defense system was localized on a plasmid, pTR2030, from the dairy starter culture *L. lactis* ME2 (19,20). Altering the copy number of A*biA* had a dramatic impact on the AbiA phenotype and inserting it in the chromosome reduced the phenotype essentially below detectable levels (21). While there is undoubtedly selective pressure to maintain some gene systems on plasmids for increased gene expression, it is also likely that some are the result of recent horizontal transfer during adaptation to a milk environment. For example, it has been noted that strains in nature are far less fastidious (22), and adaptation to a protein rich environment would eventually select for strains that utilized the milk casein for its amino acids rather than synthesize them *de novo*.

Plasmids in *L. lactis* range in size from about 2–100 kb (23) and most strain contain several plasmids, sometimes more that 10. They encode all the genes necessary for uptake and metabolism of lactose. Interestingly, the uptake of lactose occurs via a phosphoenolpyruvate (PEP) phosphotransferase system, whereby lactose is phosphorylated and metabolized via phospho-β-galactosidase. While *L. lactis* strains do not contain a

β-galactosidase gene (*lacZ*), they do encode a chromosomally located permease gene that permits the uptake of lactose that can be metabolized if a *lacZ* gene is introduced (24). The ability to uptake oligopeptides from the breakdown of milk casein is also plasmid encoded. A cell wall associated proteinase initially degrades milk casein. Seven different proteinases have been detected in *L. lactis,* with different specificities for casein (25). Some strains only contain a single proteinase type, whereas others have more than one. Subsequent oligopeptides of varying lengths are taken up by the cell and metabolized by peptidases, some of which are chromosomally encoded. Bacteriophage resistance mechanisms are also invariably encoded on plasmids, with few exceptions. These systems are essential for successful fermentation in a milk environment as phage are particularly ubiquitous in this environment. Without effective and multiple phage defense systems, *L. lactis* starter cultures invariably succumb to the prevalent phage, causing a prolonged fermentation time and eventually a failed fermentation. This is especially true for the cheese making industry, which could not exist without defenses to counter phage attack. Currently, the most effective approach for combating phage when using defined starter cultures is to include a number of compatible phage defense systems in each of the strains used, which can impede different stages of the phage developmental cycle. Targeting just a single point of the phage developmental cycle can readily lead to resistant phage appearing, whereas, targeting multiple steps makes it much more unlikely for resistant phage to evolve. This is an analogous situation to fighting HIV infection in humans, where before 1995 only a single drug (AZT) was available to fight the infection by inhibiting its reverse transcriptase, which impeded its genome replication. However, HIV resistance eventually arose upon prolonged usage, limiting its effectiveness. However, in 1995 protease inhibitors were developed which impeded HIV replication at a later stage in its developmental cycle, and using both classes of drugs was described as a ‘one-two punch’ which makes it much more unlikely for resistant viruses to appear. The use of this one-two punch strategy in *L. lactis* starter cultures is effective, but because of the rapid rate of evolution in the microbial world, resistant phage inevitably appear upon prolonged usage. Therefore, a rotation of different defenses in the same starter strain is needed to combat this. This phage defense rotation strategy requires knowledge of the many different defense systems that exist naturally in different strains and has resulted in extensive research of phage defense systems in *L. lactis*. Molecular and phenotypic analysis of lactococcal phage defense systems has revealed four distinct classes: adsorption resistance, whereby the cell produces a polysaccharide material that can block the attachment of phage to the cell; DNA injection blocking, whereby the phage adsorbs to the cell, but is prevented from injecting in its DNA; restriction and modification (R/M) systems, whereby the phage DNA is restricted into segments following injection into the cell; and abortive infection (Abi) systems, whereby the developmental cycle of the phage is prevented at any one of many points following the successful establishment of the phage DNA within the cell (26). It is noteworthy that some strains coevolved combinations of different phage defense systems on a single plasmid highlighting the selective pressure in the milk fermentation environment for the need for multiple defenses to protect *L. lactis* strains. An example of an R/M and Abi system that coevolved together on a native *L. lactis* plasmid, and was characterized in this laboratory (Yang and O’Sullivan, unpublished), is shown in Figure 10.1. This system is the only system uncovered to date, where a methylase gene is required for establishment of both the R/M and Abi systems in a cell, thus intrinsically linking them together. Other plasmid encoding phenotypes in lactococci include uptake and metabolism of citrate (27), which is an essential attribute for effective diacetyl production; resistance to cadmium (28,29); resistance to copper (30); resistance to the bacteriocin nisin (31,32); and production of bacteriocins (33–37).

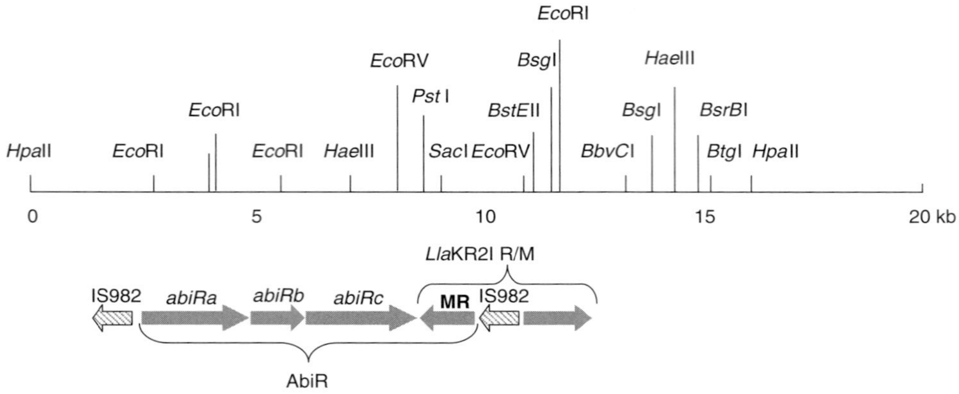

Figure 10.1 Molecular organization of two bacteriophage defense systems, from the native *L. lactis* plasmid pKR223, that are both dependent on a methylase gene for stabilization in *L. lactis*. The methylase protects *L. lactis* from *Lla*KR2I restriction and also from the abortive infection defense system, AbiR.

10.3.2 Role of Plasmids in Other Dairy Starter Cultures

While plasmids play a lesser role in other dairy starter cultures than in *L. lactis*, they do confer a variety of phenotypes in many genera. Many isolates do not contain plasmids and others just contain cryptic plasmids. Plasmids have been detected in some dairy *Leuconostoc* species (38,39). While many *Leuconostoc* do not contain plasmids, and those that do contain mainly cryptic plasmids, some have been shown to encode metabolic phenotypes. Metabolism of lactose via β-galactosidase was linked to a plasmid in *L. mesenteroides* (40) and uptake and metabolism of citrate can also be linked to plasmids (41,42). Recently, diacetyl (acetoin) reductase was located on a plasmid in a dairy *L. pseudomesenteroides* starter culture commonly used for the production cultured buttermilk, sour cream, and ripened cream butter (43).

Dairy enterococci starter cultures are somewhat controversial as many enterococci are pathogenic, including members of the two species relevant to the dairy industry, *E. faecalis* and *E. faecium* (44). However, isolates used by the diary industry are carefully screened for virulence factors. Nevertheless, studies have shown that nonvirulent strains can acquire virulence factors from other enterococci via plasmid transfer (45,46). Indeed, conjugal transfer of tetracycline and vancomycin resistance genes from enterococci to starter *E. faecalis* strains during cheese and sausage fermentations has been demonstrated (47). This illustrates that dairy enterococci have to be carefully monitored for safety purposes. While virulent strains of *E. faecalis* and *E. faecium* can be problematic, members of both these species are prominent in the human large intestine and are considered my many to have probiotic attributes (48). Plasmids are commonly found in both *E. faecalis* and *E. faecium* and while metabolic phenotypes have not been associated with them, they frequently encode bacteriocins (49–53).

Streptococcus thermophilus is the only member of the *Streptococcus* genus used in dairy fermentations. It is a thermophilic starter culture and historically was primarily a yogurt starter culture, but it is increasingly important as a cheese starter. Plasmids are rare in *S. thermophilus*, but are found in some strains and in many cases are cryptic. However, a couple of phenotypes have been associated with *S. thermophilus* plasmids. Some strains have been found to harbor plasmid genes encoding small heat shock proteins (54,55). It is likely that prolonged fermentation pressure at high temperatures provided the selective pressure for acquiring these plasmid genes encoding heat shock proteins via horizontal gene transfer, or reestablishing them on plasmids as a means of increasing expression of this phenotype that is essential for optimum prolonged existence in high temperature environments. Others have been found to harbor R/M systems encoded on plasmids (56,57). Interestingly, R/M systems are the only phage defense systems found in *S. thermophilus* so far, although possible evidence for an Abi system was suggested previously (58). Heterologous expression of the lactococcal Abi system AbiA has been achieved, but expression could only occur at 30ºC and not at the normal growth temperatures of this thermophilic starter (59). It is intriguing that apparently fewer phage defense systems have evolved in *S. thermophilus*, compared to *L. lactis*. One possibility is that many phage defense systems either do not work, or function poorly, at elevated temperatures, as illustrated in the AbiA example (59). This appears to be an evolutionary response to prolonged existence in a stressed environment, where there is a need for rapid evolution to exist more effectively in the stressed environment. Eliminating barriers to DNA uptake is an effective means for speeding up evolution in a species. The finding of plasmid encoded *hsp* genes in *S. thermophilus*, and not in *L. lactis* even though its plasmids have been extensively analyzed over the last 30 years, supports this hypothesis.

The genus *Lactobacillus* covers an extremely broad range of bacteria and the use of lactobacilli in the dairy industry is also very broad as they are used in nearly all applications, from yogurt production, cheese production, fermented milks, and as probiotics. They are also the most extensively found bacterial genus in other food fermentations, such as vegetable fermentations (pickles, sauerkraut, and fermented olives), meat fermentations (summer sausage, salami, and pepperoni), bread fermentations (sourdough), wine malolactic fermentations (with another LAB, *Oenococcus*), and in many fermented foods indigenous to various countries in the world. While plasmids are frequently found in lactobacilli, they are less common in most dairy lactobacilli. The first evidence for plasmids in *Lactobacillus* was from *L. casei* in 1976 (60). As *Lactobacillus* is a very widely studied genus, numerous plasmids have been characterized and numerous phenotypes have been found. Some plasmids have been linked to the metabolism of sugars, including lactose in some strains of *L. acidophilus* and *L. casei*, galactose in *L. acidophilus* and maltose (reviewed, 61). Citrate utilization has been linked to a plasmid in the plant associated *L. plantarum* (62). Unlike lactococci, phage defense plasmids are not common in lactobacilli. However, an R/M plasmid was found in a *L. helveticus* strain (63), and possible evidence for adsorption resistance and Abi plasmids were reported from a nondairy *L. plantarum* isolate (64). Plasmids have also been linked to exopolysaccharide production in some *L. casei* strains (65,66). A plasmid harboring resistance to hops, the broad spectrum antimicrobial plant additive, prominent in beer especially India Pale Ale type beers, was found in *L. brevis* (67). This is a common beer spoilage organism, and the selective pressure for acquiring hop resistance plasmids was likely due to the presence of hops in its environment. Direct evidence for this, was provided recently by Suzuki et al., who demonstrated that *L.* brevis strains readily lost their hop resistant abilities when cultured in a hop free environment (68). Antibiotic resistance plasmids have also been found in lactobacilli, including chloramphenicol resistance in *L. reuteri* (69,70) and *L. acidophilus* (69), erythromycin resistance in *L. reuteri* (71,72) and *L. fermentum* (73), dalfopristin resistance in *L. fermentum* (73), and tetracycline resistance in *L. fermentum* (74) and *L. plantarum* (75). The selective pressure for the acquisition of these antibiotic resistance genes clearly arose from the presence of these antibiotics in their environments. Most of these species are residents of human and animal intestines and the use of antibiotics in humans and animals evidently provided sufficient selective pressure for their acquisition. The one exception is *L. plantarum*, which is not commonly found in human and animal intestines, but is associated with plants.

However, tetracycline is also used in plant agriculture to prevent bacterial infections in plants (76), thus providing the necessary selective pressure for its acquisition by plant associated microbes. Further evidence is provided by the acquisition of tetracycline resistance by other plant associated microbes, such as *Agrobacterium tumefaciens* (77). Numerous plasmids have been associated with the production of bacteriocins including, acidocin B (78), acidocin 8912 (79) and acidocin A (80) from *L. acidophilus*; brevicin 27 from *L. brevis* (81), Curvacin A from *L. curvatus* (82), plantacin 154 from *L. plantarum* (83), and lacticin F from *L. johnsonii* (84,85). The prominence of bacteriocins in *Lactobacillus* suggests that this is a major characteristic utilized by this genus for competing against other bacteria in their natural environment. The ability to attain dominance in a natural habitat is directly related to the ability to outcompete other bacteria. It is noteworthy that microbes that are dominant in neutral pH environments do not rely on bacteriocins to a large extent for competing against other microbes. An interesting example is *Bifidobacterium*, which is a dairy culture of growing importance because of its probiotic attributes. Bifidobacteria are dominant inhabitants of the large intestine, which is largely a neutral pH environment. However, despite extensive searches by various groups throughout the world, only a few strains of *B. bifidum* were found to produce a bacteriocin (86,87,88). This illustrates that this genus uses alternative means to compete against other microbes in the large intestine. A likely reason for this is that dominant microbes in neutral pH environments rely on suppressing their competitors by scavenging for the limiting supply of iron available in neutral pH environments (89). This occurs because iron is extremely insoluble at neutral pH and therefore becomes a limiting, essential growth factor (90). Recently it was demonstrated that dominant bifidobacteria in the human large intestine inhibit the growth of their competitors in neutral pH environments, by secreting potent iron binding compounds that scavenge the limiting supplies of iron thus preventing the growth of their competitors (14,91). This effective means for inhibiting the growth of competitors in their natural habitat is consistent with the absence of bacteriocin production by most bifidobacteria. Lactobacilli on the other hand reside in habitats that are acidic, and in those environments iron is readily soluble and therefore not a limiting factor for growth. It is therefore not surprising that most lactobacilli have evolved the ability to produce bacteriocins to facilitate their dominance in their environment.

The non-LAB dairy cultures, propionibacteria, brevibacteria, and bifidobacteria have not been researched as extensively as the LAB. In all three groups, plasmids are rare, but have been found in some members. The propionibacteria relevant to the dairy industry include *P. jensenii, P. acidipropionici, P. freudenreichii* (subsp *freudenreichii* and *shermanii*), and *P. theonii*. They are important cultures in many cheese fermentations, especially Swiss cheese. They are also of increasing importance as possible probiotic cultures, because of their ability to survive gastric transport and produce propionic acid, ß-galactosidase, bacteriocins, and vitamin B_{12}, as well as attach to intestinal cells (92,93,94). Plasmids have been reported in some strains of *P. jensenii*, *P. acidipropionici,* and *P. freudenreichii*, and are more commonly found in the latter species. While little research has been conducted on plasmid biology in propionibacteria, most plasmids appear to be small and cryptic. Some do contain useful phenotypes, such as encoding bacteriocin production (95,96).

Brevibacteria belong to the coryneform group of bacteria and are important in the dairy industry particularly for surfaced ripened cheeses, such as Limburger, Brick, Muenster, and some blue cheeses. In contrast to all the other dairy cultures, these bacteria only grow in the presence of oxygen and therefore are only found on the surface of cheeses. *B. linens* is the primary species of importance to the dairy industry and is attracting increasing research attention. Brevibacteria are the only bacteria of dairy significance that produce carotenoides, some of which can be aromatic (97). The metabolic capabilities of

B. linens, which include the production of sulphur compounds and pigments, have a major impact on cheese flavor and appearance (98,99). Plasmids are rare in brevibacteria but are sometimes found in *B. levins* (100,101) and also *B. lactofermentum* (102), which is used commercially for the production of amino acids and enzymes (103). However, to date no phenotypes have been attributed to native plasmids in brevibacteria.

Bifidobacteria are increasingly important cultures in the diary industry because of their prominent role in probiotics. It differs from other dairy cultures in that glucose is metabolized exclusively by the fructose-6-phosphate shunt, which utilizes the enzyme fructose-6-phosphoketolase (F6PPK). This is therefore a frequent diagnostic marker for this group of bacteria. Members of the genus *Bifidobacterium* are dominant flora in the large intestine of humans and also in the GI tracts of many farm animals. Studies throughout the twentieth century have consistently substantiated the association of good intestinal health with high bifidobacteria numbers (reviewed, 14). Because of this, bifidobacteria are important probiotic bacteria for humans (104) and farm animals (105), second only to lactobacilli in commercial dominance. The growing interest in bifidobacteria probiotics has greatly increased the research attention of these intriguing bacteria. The species of *Bifidobacterium* of relevance to the dairy industry include *B. longum*, *B. infantis*, *B. bifidum, B. adolescentis, B. breve,* and *B. lactis.* The latter species is a commonly used probiotic and it is intriguing that it is not a normal human inhabitant. It was first isolated in 1997 from fermented milk by Meile et al. (106) and was noted to have a higher tolerance to oxygen than other bifidobacteria. While it was genetically very close to *B. animalis*, it was deemed to have changed sufficiently during adaptation to fermentation environments to warrant a new species name. While the name *B. lactis* is a much more attractive name for a dairy culture, some studies question the justification for a new species name (107,108). The changes that occurred in *B. lactis* during its adaptation to fermentation conditions make it a very resilient strain that can remain viable during processing and storage, longer than other bifidobacteria. These practical reasons contribute to its popularity, but may limit its probiotic properties in the intestine. Plasmids are only found in some species of *Bifidobacterium*, being most common in *B. longum* followed by *B. breve* (109,110,111). It is interesting that most strains of *B. longum* will harbor plasmids, while *B. infantis,* which is very closely related to *B. longum*, does not harbor any plasmids. Plasmids are generally small ($<$ 10 kb) and cryptic. A number have now being characterized at the sequence level (112,113,114). However, phenotypes outside of normal plasmid functions have not been found on bifidobacteria plasmids. One exception is a report of a plasmid linked to production of the bifidocin B bacteriocin in *B. bifidum* (115). It was interesting that the production of bifidocin B was linked to an ~8.0 kb plasmid, while the immunity did not appear to be correlated with it. As bacteriocin production and immunity genes are inherently linked together in bacteria, this would appear to be an unusual setup. However, sequence analysis of the plasmid will be needed to investigate this further.

10.3.3 Plasmid Replication in Dairy Cultures

The need for molecular understanding and genetic manipulation of starter culture phenotypes has resulted in extensive research into plasmid replication mechanisms in these bacteria. Understanding plasmid replication is essential for the development and application of molecular tools. The type of plasmid replication mechanism can influence many crucial aspects of plasmids, including structural and segregational stability (116). While bacterial plasmids are generally circular entities with the cell, some are present in both prokaryotes and eukaryotes that are linear (reviewed, 117). In the dairy cultures, the plasmids found have been invariably circular except for the report of a 48.5 kb linear plasmid in a *Lactobacillus gasseri* strain (118), which has not yet been characterized. Two modes of circular plasmid replication predominate in dairy cultures, rolling circle replication (RCR) and theta replication, with the former being much more common.

The primary distinguishing feature of RCR is the presence of single stranded DNA intermediates that accumulate during replication (116). RCR has been observed for all bacterial genera that are relevant to dairy cultures. The replication origins from RCR plasmids are conserved and depend on the presence of a replication initiation gene (*rep*). This gene encodes a Rep protein that initiates plasmid replication by nicking the DNA at the replication origin, which is a double strand origin termed *ori*, thus permitting replication of one plasmid DNA strand to begin, which is referred to as the leading strand. It also has a ligation role during replication and nicking roles during each subsequent round of replication. The single strand intermediates can be readily seen during replication and can be degraded by S1 nuclease, which is the procedure used to confirm RCR replication in these bacteria. The efficient conversion of single strand DNA into double stranded plasmid DNA is critical for the structural and segregational stability of the plasmid. This is initiated at a single strand origin (*sso*) that permits replication of this strand. Single strand origins tend to form extensive secondary structures, and this can influence plasmid stability. While plasmids utilizing RCR are very common among this group on organisms, and are used frequently as vectors for cloning and other molecular applications, they tend not to be very stable, both structurally and also segregationally. It should be noted that not all RCR plasmids suffer these defects to the same degree, as some are more efficient at converting the single strand intermediates into double strand plasmid DNA, via the *sso*, thus improving their stability. The *sso* therefore has a major role on plasmid stability during replication of RCR plasmids.

The other mode of plasmid replication that has been found among dairy cultures is theta replication, so named after the theta appearance of the plasmid during replication. The absence of single stranded intermediates during replication of these plasmids greatly increases their stability over RCR plasmids in general, and they are therefore more useful as cloning vectors. During theta plasmid replication, DNA nicking is not involved, but an RNA primer is synthesized that can initiate replication of both strands following denaturing the two strands around the origin. This opening of the two DNA strands around the origin resembles the theta symbol. While theta replication is very common for plasmids in gram negative bacteria it is much less frequently observed in gram positives. However, plasmids replicating by this mechanism have been observed, by functional analysis or sequence similarities, for many of the dairy related cultures, including *Lactococcus* (119,120), *Lactobacillus* (121), *Enterococcus* (122), *Streptococcus thermophilus* (123), *Leuconostoc* (124), *Brevibacterium* (101), and *Propionibacterium* (125).

10.4 TOOLS FOR GENETIC MANIPULATION OF DAIRY STARTER CULTURES

The development of molecular tools for dairy cultures began thirty years ago in the 1970s, with the study of plasmids and gene transfer in *Lactococcus lactis* (126) and related pathogenic bacteria (127). Since then, tools have been developed for various purposes for all the bacteria of dairy interest, but are much more sophisticated for the LAB, especially *L. lactis, Enterococcus,* and *Lactobacillus*. Molecular tools are essential for many purposes, including understanding the roles of specific genes in culture phenotypes, understanding how the expression of culture phenotypes are regulated, deletion of undesirable culture features, improvement of current culture phenotypes or expression of new desirable culture phenotypes. Molecular tools available include general cloning vectors, expression systems, and integrative strategies for DNA insertion or knockout.

10.4.1 Cloning Vectors

There is currently a wide range of cloning vectors available for cloning into all dairy bacteria of interest. As *Lactococcus lactis* has received greater research attention than the other dairy cultures, most of them were developed for this bacterium. However, lactococcal vectors are generally applicable for most other LAB as well. While both RCR and theta replication types have been used for cloning vectors, theta replicons tend to make vectors that are less prone to rearrangements and segregational instability. This is particularly the case for large DNA inserts, as the larger the DNA fragment cloned, the more susceptible it is to rearrangements. RCR plasmids however are much more plentiful and tend to be more promiscuous, allowing them to be introduced into a wider range of bacteria (128). Furthermore, there is a need for many different plasmids whose replication mechanisms are compatible, for many applications in cultures. Table 10.1 lists some of the many cloning vectors that have been developed for various starter cultures. Most of the available vectors were developed for *Lactococcus*, but are generally functional in most of the LAB cultures. These greatly simplified research into these cultures, as a selection of useable vectors was already available. Likewise, many of the original vectors used in *Lactococcus* were originally developed for pathogenic streptococci and enterococci. For example, the streptococcal cloning vector pGB301 (145) was the first vector used to clone DNA into *L. lactis* (146). While RCR plasmids predominate the available vectors for dairy cultures, a good selection of theta replicating plasmids is available for most cultures. An exception is *S. thermophilus*, where currently only RCR vectors are used. However, a theta replicating plasmid for this bacterium was recently described which should facilitate the development of stable theta replicating vectors (147).

The need for food grade markers is quickly growing as the there are many potential genetic manipulations that could greatly enhance current dairy cultures. A safe approach, which has growing acceptance, for genetically modifying food microorganisms, is to only use DNA that comes from food grade microorganisms. A summary of the different food grade markers that have been developed for dairy cultures is listed in Table 10.2.

10.4.2 Expression Systems

The ability to be able to express specific genes in dairy cultures to achieve desirable phenotypes is of great interest. The applications of this technology include over expressing existing phenotypes, expressing new enzymes and inhibitory compounds, and expressing surface antigens for immune stimulation in the GI tract. A summary of the different strategies that have been used for expression systems in dairy cultures is given in Table 10.3. While most strategies have been developed for *Lactococcus*, some strategies have been extended to, or developed for, other LAB. The most commonly used expression system for the LAB is the nisin inducible system, which is dependent on the presence of the *nisRK* genes in the expression host, to enable the signal transduction to occur. However, it should also be noted that the *nisA* promoter can also be induced, in the absence of nisin or the *nisRK* genes, during galactose metabolism by the Leloir pathway in *L. lactis* (165). While sophisticated expression systems have not yet been developed for propionibacteria, brevibacteria, and bifidobacteria, basic shuttle vector expression of genes has been achieved. Heterologous genes have been expressed in propionibacteria (166), brevibacteria (167), and bifidobacteria (168,169,170). The genes expressed in bifidobacteria include several tumor suppressors, as these bacteria are believed to be able to grow in the hypoxic regions of solid tumors (168,171).

10.4.3 Integrative Strategies

The ability to be able to integrate DNA into specific sites within the genome is very useful, as this can be used to stabilize a gene or an expression construct in a culture, and also to

Table 10.1
A selection of general cloning vectors available for dairy cultures

Vector	Replication Type	E. coli Shuttle	Markers	Features	Ref.
Developed for *Lactococcus*					
pGK12	RCR	Yes	Erm^r, Cm^r	single promiscuous *ori* low copy[1]; MI[2]	(129)
pNZ18	RCR	Yes	Km^r, Cm^r	single promiscuous *ori* high copy; MI	(130)
pTRKH2	theta	Yes	Em^r	high copy; Blu-W[3]	(131)
pTRKL2	theta	Yes	Em^r	low copy; Blu-W	(131)
p720	RCR	Yes	Cm^r	single promiscuous *ori*	(132)
Developed for *Lactobacillus*					
pJK355	RCR	No	Cm^r	Replicates in *L. lactis* and *B. subtilis*	(133)
pSP1	theta	Yes	Em^r, Cm^r	MI	(134)
pLE16	RCR	Yes	Cm^r	MI	(135)
Developed for *S. thermophilus*					
pND913	RCR	Yes	Em^r, Amp^r	contains a *hsp* gene	(136)
pMEU14'-1	RCR	Yes	Cm^r, Amp^r	MI	(137)
Developed for *Enterococcus*					
pHW800	theta	No	Cm^r	Replicates in *L. lactis* and *Leuconostoc*	(138)
pAM401	theta	Yes	Cm^r, Tet^r	MI	(139)
Developed for *Leuconostoc*					
pFBYC051	theta	Yes	Em^r, Amp^r	Replicates in *L. lactis* and *Lactobacillus*	(124)
Developed for *Propionibacterium*					
pBRESP36A	ND[4]	Yes	Em^r, Amp^r	high stability	(140)
pPK705	ND	Yes	Cm^r, HygB[r5]	Blu-W	(141)
Developed for *Bifidobacterium*					
pDGE7	ND	Yes	Em^r, Cm^r	MI	(142)
pBLES100	ND	Yes	Sp^r, Amp^r	MI	(143)
pRM2	ND	Yes	Sp^r, Amp^r		(144)

[1] refers to the copy number of the plasmid in *L. lactis*;
[2] MI refers to marker inactivation, the ability to clone DNA within an antibiotic resistance gene;
[3] refers to blue/white screening capability in *E. coli*;
[4] not determined;
[5] hygromycin B resistance

delete undesirable DNA regions. It is also possible to integrate a molecular tag that could serve as a unique signature of a particular strain. This process relies on homologous recombination and has been quite successful in many of the dairy relevant bacterial genera. All strategies rely on identifying the DNA region where the insertion is to occur, as this DNA region forms the basis of the integrative vector. If the objective is to delete a gene, then a 5' region and a 3' region of the gene, or its bordering sequences, is cloned at either side of a selection marker. Typically the cloned regions should be > 500 bp to facilitate homologous recombination. A single crossover is first obtained whereby the inserted construct disrupts

Table 10.2
Food grade selection markers available for dairy cultures

Food Grade Marker	Example Vector	Host(s)	Ref.
Melibiose fermentation	pRAF800	*Lactococcus Pediococcus*	(148)
thyA gene[1]	pBSt1	*thyA* mutants of *S. thermophilus*[2]	(149)
Cadmium resistance[3]	pND919	*S. thermophilus*	(150)
Nisin resistance	pFM011	*Lactococcus*	(151,152)
lacF gene	pNZ2104	*lacF* mutants of *Lactococcus*	(153)
ochre suppressor (*supB*)	pFG1	*pyrF* (ochre) mutants of *Lactococcus*[4, 5]	(154)
amber suppressor (*supD*)	pFG200	*pyrF* (amber) mutants of *Lactococcus*[6]	(155)
heat shock protein[7]	pSt04	*S. thermophilus Lactococcus*	(156)
no selection marker[8]	pVEC	*Lactococcus*	(157)

[1] encodes thymidine synthase;
[2] selection must be in a thymidine free medium;
[3] gene isolated from *L. lactis*;
[4] these are pyrimidine auxotrophs and selection must be in pyrimidine-free media;
[5] unstable in many lactococci as SupB suppresses both ochre and amber stop codons, which account for most stop codons in lactococci;
[6] SupD only suppresses amber stop codons which are < 10% used in *Lactococcus*;
[7] selection required using elevated temperatures (52°C for *S. thermophilus* and 42°C for *L. lactis*)
[8] functions by cointroducing a companion plasmid with an Ermr marker and the same replicon as pVEC except lacking a functional recB gene. Selection for Ermr ensures both plasmids are introduced and the Ermr construct is subsequently cured by omitting Erm, due to replicon incompatibility.

Table 10.3
Strategies used for expression systems in dairy cultures

Strategy Used	Features	Expression Host(s)	Ref.
Strong constitutive promoter	P32 promoter pWV01 *ori*	*L. lactis, B. subtilis, E. coli*	(158)
lactose induced promoter	*lacA* promoter	*Lactococcus*	(159)
A lactose induced T7 RNA polymerase	T7 promoter; Lac induced T7 RNA polymerase gene	*Lactococcus*	(160)
Nisin induced	*nisA* promoter; *nisRK* two-component induction system	*Lactococcus Enterococcus Lactobacillus Leuconostoc*	(161,162)
ϕ31 induced	ϕ31 promoter; ϕ31 *ori* low copy number plasmid[1]	*Lactococcus*	(163)
enterocin induction signal	*entA* promoter; *entRK* two-component induction system[2]	*Enterococcus*	(164)

[1] low copy number plasmid facilitates run away replication following infection by ϕ31, [2]induced by the addition of the enterocin induction peptide

the targeted gene. However, in the low G + C LAB, a double crossover readily follows whereby the original gene can be replaced with the constructed disrupted gene.

To achieve homologous recombination in a cell, the construct must first be introduced into the cell and have no means of replicating unless it integrates into the genome of the cell. Therefore, when the cells grow in the presence of the selection agent, integration will have to occur to allow the selection marker gene to be maintained. One strategy for achieving this is to use a construct that does not have a functional replicon, such that when introduced into the cell, integration is the only choice when grown on the selection agent. De Vos and Simons have reviewed a number of constructs that used this approach (172). Unfortunately, this forcible homologous recombination approach is directly dependent of the frequency of gene transfer. A high gene transfer is required, and this is not always possible in *Lactococcus*, not to mid other dairy cultures of interest.

Another strategy is to introduce the construct into to the cell on a permissive replicon, and then, following successful introduction into the cell, disable the plasmid replicon, thereby forcing homologous recombination. This is a much more attractive approach as it negates the need for a very high gene transfer frequency. One way to disable a plasmid replicon is to use a replicon that does not function under all environmental conditions. A common example is temperature sensitivity and a number of these have been used for integrative vectors quite successfully. For example, vectors of the pGhost class have temperature sensitive replicons are have been widely used for gene integration in the LAB (173).

10.5 GENOMICS OF DAIRY CULTURES

Prior to this genomic era, it was only possible to understand a limited number of genes involved in different phenotypes. What the genomics era is bringing is the possibility to know the full complement of genes involved in any phenotype of interest. That would greatly help the efficiency when using cultures for specific purposes. Since the first bacterium genome sequence was published in 1995 (174), there has been an explosion in the number of other microbial sequences that followed. Currently there are more that 600 microbial genome projects either completed or in the process. While obtaining the complete genome sequence of organisms can reveal tremendous insights, determining the biochemical and cellular function of each gene is a necessary subsequent step. This daunting task of the functional analysis of genomes has been greatly enhanced by the advent of microarrays. These are small chips, or glass slides, containing probes corresponding to every gene in a genome. The construction of these high density gene chips occurred from advances in the ability of robotics to spot DNA onto chips and was first used for high density gene profiling in 1995 (175). Early gene chips did have limitations as the signal to background noise level of the hybridization results prevented precise profiling of gene expression. However, recent advances, utilizing multiple oligonucleotide probes per gene, have greatly improved the sensitivity of microarrays to accurately profile gene expression. Hybridizing the chips with RNA from the organism allows one to quantify the expression of all the genes being expressed under specific conditions. This is a necessary step in accurately predicting roles for genes (176).

One disadvantage to the genomics era is the cost of obtaining a fully completed genome sequence, although the cost does continue to decrease. Because of this, many genomes sequenced or being sequenced have industry sponsorship and this can significantly delay the release of the sequence information, due to intellectual property issues. This has also been the case with many dairy culture genome projects. One alternative is to get federal funding such that sequence information would not be complicated by intellectual properties. This was the route taken by US scientists, who formed the Lactic Acid Bacteria Genome Consortium (LABGC), consisting of 10 scientists, and formed a collaboration with the Joint Genome Institute (JGI) to sequence 11 genomes of commercially relevant dairy cultures. This partnership was funded by the Department of Energy to enable the genome project to initiate. Currently all 11 genomes have been completed and are in the process of being published. Unlike genome projects with private funding, the genomes sequenced via the JGI/LABGC will be deposited in public databases upon publication.

Currently, there are three complete genomes from dairy cultures that have been sequenced and published. These are for *Lactococcus lactic* IL1403 (177), *Bifidobacterium longum* NCC2705 (178) and *Lactobacillus johnsonii* NCC533 (179). A fourth genome from *L. plantarum* WCFS1 (180) may also be of interest as it is sometimes speculated to be a potential probiotic culture. The *L. lactis* IL1403 genome was the first one published and it provided an exciting glimpse into the makeup of a lactococcal chromosome and the evolution of this organism. Remnants of a once aerobic existence are evident, and a set of genes analogous to the natural competence genes in *Bacillus*. While analysis of the *L. lactis* IL1403 genome does give intriguing information, it does not shed light on many of the commercially significant phenotypes, as these are largely encoded on plasmids, and IL1403 is a plasmid cured laboratory strain. Therefore, as part of the LABGC effort a commercial strain of *L. lactis* subsp *cremoris* with its full plasmid is being sequenced.

The genome sequence of a strain of *Bifidobacterium longum*, which was a recent human isolate and exhibits characteristics consistent with good competitive abilities in the large intestine, is being compiled in this laboratory as part of the LABGC effort. The comparative analysis of this strain with the other sequenced *B. longum* strain is revealing unique insights into how strains within this species evolve. An interesting feature was that even though the two strains displayed very high sequence identity over much of the chromosome, there was very little colinearity, as demonstrated in Figure 10.2. This was unexpected as comparative genomic analysis of other closely related bacteria have revealed a high degree of colinearity (181). It was also noteworthy that the genome of Strain DJO10A was ~130 kb bigger that strain NCC2705 and that both strains contained significant unique regions. The unique regions of both strains were scattered all across their genomes demonstrating the dynamic nature of strain evolution.

10.6 CONCLUSIONS

Dairy starter cultures are a very unique and exciting group of microbes because of what they do and how they impart human life. Optimum application of specific cultures relies on a fundamental understanding of how the relevant phenotypes are expressed. The current and future endeavors of both bioinformatic and functional genomics of dairy culture's genomes will provide a detailed manual of how every component functions. This will permit intricate metabolic engineering whereby a cultures metabolism can be readily optimized for a specific purpose. It will also facilitate a greater understanding of important criteria for probiotic cultures, which would establish optimum criteria for selection and use of cultures for probiotics.

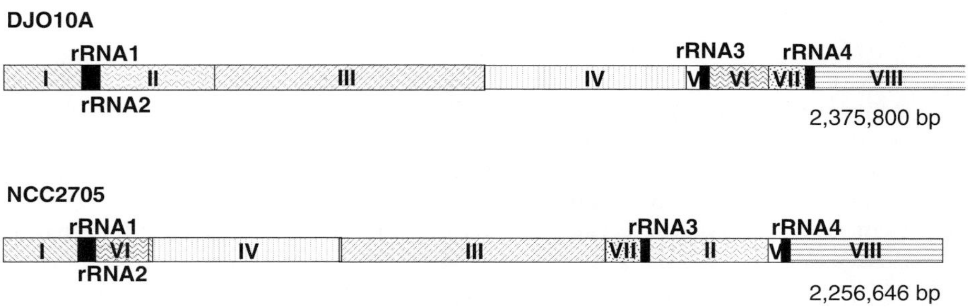

Figure 10.2 Linear maps of the complete genomes of *Bifidobacterium longum* strains DJO10A and NCC2705. Segments I to VIII represent highly identical sequence regions that are not colinear on the genomes.

ADDENDUM

During the publication of this book, the genome sequence of *Bifidobacterium longum* NCC2705, published in 2002, was updated in GenBank. It no longer exhibits the major rearrangements as depicted in Figure 10.2 and discussed in the text. Based on the current deposited sequence for this bacterium, genome rearrangement is not a major factor in *B. Longum* evolution

REFERENCES

1. Lister, J. On lactic fermentation and its bearing on pathology. *Trans. Path. Soc. Lond.* 29:425–467, 1878.
2. Bolotin, A., P. Wincker, S. Mauger, O. Jaillon, K. Malarme, J. Weissenbach, S. Dusko Ehrlich, A. Sorokin. The complete genome sequence of the lactic acid bacterium *Lactococcus lactis* ssp. *lactis* IL1403. *Genome Res.* 11:731–753, 2001.
3. Cogan, T.M. History and taxonomy of starter cultures. In: *Dairy Starter Cultures*, Cogan, T.M., J.P. Accolas, eds., New York: VCH Publishers Inc, 1996, pp 1–24.
4. Farnworth, E.R., I. Mainville. Kefir: a fermented milk product. In: *Handbook of Fermented Functional Foods,* Farnworth, E.R., ed., Boca Raton, FL: CRC Press, 2003, pp 77–112.
5. Klein, G., A. Pack, C. Bonaparte, G. Reuter. Taxonomy and physiology of probiotic lactic acid bacteria. *Int. J. Food Microbiol.* 41:103–125, 1998.
6. Woese, C.R. Bacterial evolution. *Microbiol. Rev*. 51:221–271, 1987.
7. Leblond-Bourget, N., H. Philippe, I. Mangin, B. Decaris. 16S rRNA and 16S to 23S internal transcribed spacer sequence analyses reveal inter- and intraspecific *Bifidobacterium* phylogeny. *Int. J. Syst. Bacteriol.* 56:102–111, 1996.
8. Barry, T., G. Colleran, M. Glennon, L.K. Dunican, F. Gannon. The 16s/23s ribosomal spacer region as a target for DNA probes to identify eubacteria. *PCR Meth. Appl.* 1:51–56, 1991.
9. Garcia-Martinez, J., A. Martinez-Murcia, A. Anton, F. Rodriguez-Valera. Comparison of the small 16S to 23S intergenic spacer region (ISR) of the rRNA operons of some *Escherichia coli* strains of the ECOR collection and *E. coli* K-12. *J. Bacteriol.* 178:6374–6377, 1996.
10. Roca, A.I., M.M. Cox. RecA protein: structure, function, and role in recombinational DNA repair. *Prog. Nucl. Acid Res. Mol. Biol.* 56:129–223, 1997.
11. Eisen, J.A. The RecA protein as a model molecule for molecular systematic studies of bacteria: comparison of trees of RecA's and 16S rRNA's from the same species. *J. Mol. Evol.* 41:1105-1123, 1995.
12. Karlin, S., G.M. Weinstock, V. Brendel. Bacterial classifications derived from RecA protein sequence comparisons. *J. Bacteriol.* 177:6881–6893, 1995.
13. Kullen, M.J., L.J. Brady, D.J. O'Sullivan. Evaluation of using a short region of the *recA* gene for rapid and sensitive speciation of dominant bifidobacteria in the human large intestine. *FEMS Microbiol. Lett.* 154:377–383, 1997.
14. O'Sullivan, D.J. Screening of intestinal microflora for effective probiotic bacteria. *J. Ag. Food Chem.* 49:1751–1760, 2001.
15 Price, C.S., H. Huynh, S. Paule, R.J. Hollis, G.A. Noskin, M.A. Pfaller, L.R. Peterson. Comparison of an automated ribotyping system to restriction endonuclease analysis and pulsed-field gel electrophoresis for differentiating vancomycin-resistant *Enterococcus faecium* isolates. *J. Clin. Microbiol.* 40:1858–1861, 2002.
16. Helinski, D.R., D.B. Clewell. Circular DNA. *Ann. Rev. Biochem.* 40:899–942, 1971.
17. Cords, B.R., L.L. McKay, P. Guerry. Extrachromosomal elements in group N streptococci. *J. Bacteriol.* 117:1149–1152, 1974.
18. McKay, L.L. Functional properties of plasmids in lactic streptococci. *Antoine van Leeuwenhoek* 49:259–274, 1983.
19. Sanders, M.E., P.J. Leonhard, W.D. Sing, T.R. Klaenhammer. Conjugal strategy for construction of fast acid-producing, bacteriophage-resistant lactic streptococci for use in dairy fermentations. *Appl. Environ. Microbiol.* 52:1001–1007, 1986.

20. Hill, C., L.A. Miller, T.R. Klaenhammer. Nucleotide sequence and distribution of the pTR2030 resitance determinat (*hsp*) which aborts bacteriophage infection in lactoccoci. *Appl. Environ. Microbiol.* 56:2255–2258, 1990.
21. Dinsmore, P.K., T.R. Klaenhammer. Phenotypic consequences of altering the copy number of *abiA*, a gene responsible for aborting bacteriophage infections in *Lactococcus lactis*. *Appl. Environ. Microbiol.* 60:1129–1136, 1994.
22. Renault, P. Progress in genetic research of lactic acid bacteria. In: *Lactic Acid Bacteria: Current Advances in Metabolism, Genetics and Applications,* Bozoglu, T.F., B. Ray, eds., Germany: Springer-Verlag, 1996, pp 15–36.
23. Teuber, M. The genus *Lactococcus*. In: *The Genera of Lactic Acid Bacteria, Volume 2,* Wood, B.J.B., W.H. Holzapfel, eds., Glasgow, Scotland: Blackie & Son, 1995, pp 173–234.
24. Chandrapati, S., D.J. O'Sullivan. Nisin independent induction of the *nisA* promoter in *Lactococcus lactis* during growth in lactose or galactose. *FEMS Microbiol. Lett.* 170:191–198, 1999.
25. Kunji, E.R.S., I. Mierau, A. Hagting, B. Poolman, W.N. Konings. The proteolytic systems of lactic acid bacteria. In: *Lactic Acid Bacteria: Genetics, Metabolism and Applications,* Venema, G., J.H.J. Huis, T. Veld, J. Hugenholtz, eds., Dordrecht: Kluwer, 1996, pp 91–125.
26. Allison, G.E., T.R. Klaenhammer. Phage resistance mechanisms in lactic acid bacteria *Int. Dairy J.* 8:207–226, 1998.
27. Defelipe, F.L., C. Magni, D. Demendoza, P. Lopez. Citrate utilization gene cluster of the *Lactococcus lactis* biovar *diacetylactis* - organization and regulation of expression. *Mol. Gen. Genet.* 246:590–599, 1995.
28. Liu, C.Q., N. Khunajakr, L.G. Chia, Y.M. Deng, P. Charoenchai, N.W. Dunn. Genetic analysis of regions involved in replication and cadmium resistance of the plasmid pND302 from *Lactococcus lactis*. *Plasmid* 38:79–90, 1997.
29. Trotter, M., S. Mills, R.P. Ross, G.F. Fitzgerald, A. Coffey. The use of cadmium resistance on the phage-resistance plasmid pNP40 facilitates selection for its horizontal transfer to industrial dairy starter lactococci. *Lett. Appl. Microbiol.* 33:409–414, 2001.
30. Leelawatcharamas, V., L.G. Chia, P. Charoenchai, N. Kunajakr, C.Q. Liu, N.W. Dunn. Plasmid-encoded copper resistance in *Lactococcus lactis*. *Biotechnol. Lett.* 19:639–643, 1997.
31. Froseth, B.R., R.E. Herman, L.L. McKay. Cloning of nisin resistance determinant and replication origin on 7.6 kb *Eco*RI fragment of pNP40 from *Streptococcus lactis* subsp *diacetylactis* DRC3. *Appl. Environ. Microbiol.* 54:5136–2139.
32. Liu, C.Q., M.L. Harvey, N.W. Dunn. Cloning of a gene encoding nisin resistance from *Lactococcus lactis* subsp. *lactis* M189 which is transcribed from an extended -10 promoter. *J. Gen. Appl. Microbiol.* 43:67–73, 1997.
33. Dougherty, B.A., C. Hill, J.F. Weidman, D.R. Richardson. J.C. Venter, R.P. Ross. Sequence and analysis of the 60 kb conjugative, bacteriocin-producing plasmid pMRC01 from *Lactococcus lactis* DPC3147. *Mol. Microbiol.* 29:1029–1038, 1998.
34. Martinez, B., M. Fernandez, J.E. Suarez, A. Rodriguez. Synthesis of lactococcin 972, a bacteriocin produced by *Lactococcus lactis* IPLA 972, depends on the expression of a plasmid encoded bicistronic operon. *Microbiol* 145:3155–3161, 1999.
35. Sanchez, C., A.H. de Rojas, B. Martinez, M.E. Arguelles, J.E. Suarez, A. Rodriguez, B. Mayo. Nucleotide sequence and analysis of pBL1, a bacteriocin-producing plasmid from *Lactococcus lactis* IPLA 972. *Plasmid* 44:239–249, 2000.
36. Martinez-Cuesta, M.C., G. Buist, J. Kok, H.H. Hauge, J. Nissen-Meyer, C. Pelaez, T. Requena. Biological and molecular characterization of a two-peptide lantibiotic produced by *Lactococcus lactis* IFPL105. *J. Appl. Microbiol.* 89:249–260, 2000.
37. Trotter, M., O.E. McAuliffe, G.F. Fitzgerald, C. Hill, R.P. Ross, A. Coffey. Variable bacteriocin production in the commercial starter *Lactococcus lactis* DPC4275 is linked to the formation of the cointegrate plasmid pMRC02. *Appl. Environ. Microbiol.* 70:34–42, 2004.
38. O'Sullivan, T., C. Daly. Plasmid DNA in *Leuconostoc* species. *Irish J. Food Sci. Tech.* 6:206–211, 1982.
39. Coffey, A., A. Harrington, K. Kearney, C. Daly, G.F. Fitzgerald. Nucleotide sequence and structural organization of the small, broad-host range plasmid pCI411 from *Leuconostoc lactis* 533. *Microbiology* 140:2263–2269.

40. Huang, D.Q., H. Prevost, M. Kihal, C. Divies. Instability of plasmid encoding for beta-galactosidase in *Leuconostoc mesenteroides* subsp *mesenteroides*. *J. Basic Microbiol.* 34:23–30, 1994
41. Kihal, M., H. Prevost, M.E. Lhotte, D.Q. Huang, C. Divies. Instability of plasmid-encoded citrate permease in *Leuconostoc*. *Lett. Appl. Microbiol.* 22:219–223, 1996.
42. Martin, M., M.A. Corrales, D. de Mendoza, P. Lopez, C. Magni. Cloning and molecular characterization of the citrate utilization *citMCDEFGRP* cluster of *Leuconostoc paramesenteroides*. *FEMS Microbiol. Lett.* 174:231–238, 1999.
43. Rattray, F.P., D. Myling-Petersen, D. Larsen, D. Nilsson. Plasmid-encoded diacetyl (acetoin) reductase in *Leuconostoc pseudomesenteroides*. *Appl. Environ. Microbiol.* 69:304–311, 2003.
44. Franz, C.M.P.A., M.E. Stiles, K.H. Schleifer, W.H. Holzapfel. Enterococci in foods: a conundrum for food safety. *Int. J. Food Microbiol.* 88:105–122, 2003.
45. Cocconcelli, P.S., D. Cattivelli, S. Gazzola. Gene transfer of vancomycin and tetracycline resistances among *Enterococcus faecalis* during cheese and sausage fermentations. *Int. J. Food Microbiol.* 88:315–323, 2003.
46. Eaton, T.J., M.J. Gasson. Molecular screening of *Enterococcus* virulence determinants and potential for genetic exchange between food and medical isolates. *Appl. Environ. Microbiol.* 67:1628–1635, 2001.
47. Oancea, C., I. Klare, W. Witte, G. Werner. Conjugative transfer of the virulence gene, *esp*, among isolates of *Enterococcus faecium* and *Enterococcus faecalis*. *J. Antimicro. Chemother.* 54:232–235, 2004.
48. Lund, B., I. Adamsson, C. Edlund. Gastrointestinal transit survival of an *Enterococcus faecium* probiotic strain administered with or without vancomycin. *Int. J. Food Microbiol.* 77:109–115, 2002.
49. Balla, E., L.M.T. Dicks, M. Du Toit, M.J. van der Merwe, W.H. Holzapfel. Characterization and cloning of the genes encoding enterocin 1071A and enterocin 1071B, two antimicrobial peptides produced by *Enterococcus faecalis* BFE 1071. *Appl. Environ. Microbiol.* 66:1298–1304, 2000.
50. Ohmomo, S., S. Murata, N. Katayama, S. Nitisinprasart, M. Kobayashi, T. Nakajima, M. Yajima, K. Nakanishi. Purification and some characteristics of enterocin ON-157, a bacteriocin produced by *Enterococcus faecium* NIAI 157. *J. Appl. Microbiol.* 88:81–89, 2000.
51. Folli, C., I. Ramazzina, P. Arcidiaco, M. Stoppini, R. Berni. Purification of bacteriocin AS-48 from an *Enterococcus faecium* strain and analysis of the gene cluster involved in its production. *FEMS Microbiol. Lett.* 221:143–149, 2003.
52. Doi, K., T. Eguchi, S.H. Choi, A. Iwatake, S. Ohmomo, S. Ogata. Isolation of enterocin SE-K4-encoding plasmid and a high enterocin SE-K4 producing strain of *Enterococcus faecalis* K-4. *J. Biosci. Bioeng.* 93:434–436, 2002.
53. Sanchez-Hidalgo, M., M. Maqueda, A. Galvez, H. Abriouel, E. Valdivia, M. Martinez-Bueno. The genes coding for enterocin EJ97 production by *Enterococcus faecalis* EJ97 are located on a conjugative plasmid. *Appl. Environ. Microbiol.* 69:1633–1641, 2003.
54. Somkuti, G.A., D.K. Solaiman, D.H. Steinberg. Structural and functional properties of the *hsp16.4* bearing plasmid pER341 in *Streptococcus thermophilus*. *Plasmid* 40:61–66, 1998.
55. Solow, B.T., G.A. Somkuti. Comparison of low-molecular-weight heat stress proteins encoded on plasmids in different strains of *Streptococcus thermophilus*. *Curr. Microbiol.* 41:177–181, 2000.
56. Solow, B.T., G.A. Somkuti. Molecular properties of *Streptococcus thermophilus* plasmid pER35 encoding a restriction modification system. *Curr. Microbiol.* 42:122–128, 2001.
57. Geis, A., H.A.M. El Demerdash, K.J. Heller. Sequence analysis and characterization of plasmids from *Streptococcus thermophilus*. *Plasmid* 50:53–69, 2003.
58. Larbi, D., B. Decaris, J.M. Simonet. Different bacteriophage resistance mechanisms in *Streptococcus salivarius* ssp. *thermophilus*. *J. Dairy Res.* 59:349–357, 1992.
59. Tangney, M., G.F. Fitzgerald. AbiA, a lactococcal abortive infection mechanism functioning in *Streptococcus thermophilus*. *Appl. Environ. Microbiol.* 68:6388–6391, 2002.
60. Chassey, B.M., E.M. Gibson, A. Giuffrida. Evidence for extrachromosomal elements in *Lactobacillus*. *J. Bacteriol.* 127:1576–1578, 1976.
61. Wang, T.T., B.H. Lee. Plasmids in *Lactobacillus*. *Crit. Rev. Biotechnol.* 17:227–272, 1997.

62. Nakamura, S., T. Miyamoto, M. Izumimito, K Kataoka. Isolation and characterization of citrate metabolism-deficient mutants in *Lactobacillus plantarum* IFO 3070. *Anim. Sci. Technol.* 62:1141–1148, 1991.
63. de los Reyes-Gavilan, C.G., G.K.Y. Limsowkin, L. Sechaud, M. Veaux, J.P. Accolas. Evidence for a plasmid-linked restriction-modification in *Lactobacillus helveticus. Appl. Environ. Microbiol.* 56:3412–3419.
64. Eguchi, T., K. Doi, K. Nishiyama, S. Ohmomo, S. Ogata. Characterization of a phage resistance plasmid, pLKS, of silage-making *Lactobacillus plantarum* NGRI0101. *Biosci. Biotechnol. Biochem.* 64:751–756, 2000.
65. Kojic, M., M. Vujcic, A. Banina, P. Cocconcelli, J. Cerning, L. Topisirovic. Analysis of exopolysaccharide production by *Lactobacillus casei* CG11 isolated from cheese. *Appl. Environ. Microbiol.* 58:4086–4088, 1992.
66. Cerning, J., C.M.G.C. Renard, J.F. Thibault, C. Bouillanne, M. Landon, M. Desmazeaud, L. Topisirovic. Carbon source requirements for exopolysaccharide production by *Lactobacillus casei* CG11 and partial structure analysis of the polymer. *Appl. Environ. Microbiol.* 60:3914–3919, 1994.
67. Sami, M., K. Suzuki, K. Sakamoto, H. Kadokura, K. Kitamoto, K. Yoda. A plasmid pRH45 of *Lactobacillus brevis* confers hop resistance. *J. Gen. Appl. Microbiol.* 44:361–363, 1998.
68. Suzuki, K., M. Koyanagi, H. Yamashita. Isolation of hop-sensitive variants from beer-spoilage *Lactobacillus brevis* strains. *J. Am. Soc. Brewing Chem.* 62:71–74, 2004.
69. Morelli, L., M. Vescovo, V. Bottazzi. Identification of chloramphenicol plasmids in *Lactobacillus reuteri* and *L. acidophilus. Int. J. Microbiol.* 1:1–5, 1983.
70. Lin, Z., F. Fung, C.L. Wu, T.C. Chung. Molecular characterization of a plasmid-borne (pTC82) chloramphenicol resistance determinant (Cat-tc) from *Lactobacillus reuteri* G4. *Plasmid* 36:116–124, 1996.
71. Tannock, G.W., J.B. Luchansky, L. Miller, H. Connell, S. Thode-Anderson, A.A. Mercer, T.R. Klaenhammer. Molecular characterization of a plasmid-borne (pGT633) erythromycin resistance determinant (*ermGT*) from *Lactobacillus reuteri* 100-63. *Plasmid* 31:60–71, 1994.
72. Lin, C.F., T.C. Chung. Cloning of erythromycin-resistance determinants and replication origins from indigenous plasmids of *Lactobacillus reuteri* for potential use in construction of cloning vectors. *Plasmid* 42:31–41, 1999.
73. Gfeller, K.Y., M. Roth, L. Melle, M. Teuber. Sequence and genetic organization of the 19.3-kb erythromycin- and dalfopristin-resistance plasmid pLME300 from *Lactobacillus fermentum* ROT1. *Plasmid* 50:190–201, 2003.
74. Ishiwa, H., S. Iwata. Drug resistance plasmids in *Lactobacillus fermentum. J. Gen. Appl. Microbiol.* 26:71–74, 1980.
75. Danielsen, M. Characterization of the tetracycline resistance plasmid pMD5057 from *Lactobacillus plantarum* 5057 reveals a composite structure. *Plasmid* 48:98–103, 2002.
76. Chopra, I., M. Roberts. Tetracycline antibiotics: mode of action, applications, molecular biology, and epidemiology of bacterial resistance. *Microbiol. Mol. Biol. Rev.* 65:232–260, 2001.
77. Luo, Z.Q., T.E. Clemente, S.K. Farrand. Construction of a derivative of *Agrobacterium tumefaciens* C58 that does not mutate to tetracycline resistance. *Mol. Plant Microbe Interact.* 14:98–103, 2001.
78. Leer, R.J., J.M.B.M. van der Vossen, M. Vangiezen, J.M. Vannoort, P.H. Pouwels. Genetic analysis of acidocin B, a novel bacteriocin produced by *Lactobacillus acidophilus. Microbiology* 141:1629–1635, 1995.
79. Kanatani, K., T. Tahara, M. Oshimura, K. Sano, C. Umezawa. Cloning and nucleotide sequence of the gene for acidocin 8912, a bacteriocin from *Lactobacillus acidophilus* TK8912. *Lett. Appl. Microbiol.* 21:384–386, 1995.
80. Kanatani, K., M. Oshimura, K. Sano. Isolation and characterization of acidocin A and cloning of the bacteriocin gene from *Lactobacillus acidophilus. Appl. Environ. Microbiol.* 61:1061–1067, 1995.
81. Benoit, V., R. Mathis, G. Lefebvre. Characterization of brevicin 27, a bacteriocin synthetized by *Lactobacillus brevis* SB27. *Curr. Microbiol.* 28:53–61, 1994.

82. Tichaczek, P.S., R.F. Vogel, W.P. Hammes. Cloning and sequencing of *curA* encoding curvacin-A, the bacteriocin produced by *Lactobacillus curvatus* LTH1174. *Arch. Microbiol.* 160:279–283, 1993.
83. Kanatani, K., M. Oshimura. Plasmid-associated bacteriocin production by a *Lactobacillus plantarum* strain. *Biosci. Biotechnol. Biochem.* 58:2084–2086, 1994.
84. Muriana, P.M., T.R. Klaenhammer. Conjugal transfer of plasmid-encoded determinants for bacteriocin production and immunity in *Lactobacillus acidophilus* 88*, later reclassified as *L. johnsonii. Appl. Environ. Microbiol.* 53:553–560, 1987.
85. Fremaux, C., C. Ahn, T.R. Klaenhammer. Molecular analysis of the lactacin F operon. *Appl. Environ. Microbiol.* 59:3906–3915, 1993.
86. Anand, S.K., R.A. Srinivasan, L.K. Rao. Antimicrobial activity associated with *Bifidobacterium bifidum. Cult. Dairy Prod. J.* 11:6–7, 1984.
87. Anand, S.K., R.A. Srinivasan, L.K. Rao. Antimicrobial activity associated with *Bifidobacterium bifidum* - II. *Cult. Dairy Prod. J.* 20:21–22, 1985.
88. Yildirim, Z., M.G. Johnson. Characterization and antimicrobial spectrum of bifidocin B, a bacteriocin produced by *Bifidobacterium bifidum* NCFB 1454. *J. Food Protect.* 61:47–51, 1998.
89. O'Sullivan, D.J., F. O'Gara. Traits of fluorescent *Pseudomonas* spp. involved in the suppression of plant root pathogens. *Microbiol. Rev.* 56:662–676, 1992.
90. Neilands, J.B., K. Konopka, B. Schwyn, M. Coy, R.T. Francis, B.H. Paw, A. Bagg. Comparitive biochemistry of microbial iron assimilation. In: *Iron Transport in Microbes, Plants and Animals,* Winkelmann, G., D. van der Helm, J.B. Neilands, eds., New York: VCH Publishers, 1987, pp 3–33.
91. O'Sullivan, D.J. Isolated bifidobacteria that produce siderophores which inhibit growth of *Lactococcus lactis*. US Patent 6,746,672 B2.
92. Zarate, G., A.P. Chaia, G. Oliver. Some characteristics of practical relevance of the beta-galactosidase from potential probiotic strains of *Propionibacterium acidipropionici. Anaerobe* 8:259–267, 2002.
93. Ouwehand, A.C., T. Suomalainen, S. Tolkko, S. Salminen. *In vitro* adhesion of propionic acid bacteria to human intestinal mucus. *Lait* 82:123–130, 2002.
94. Jan, G., P. Leverrier, I. Proudy, N. Roland. Survival and beneficial effects of propionibacteria in the human gut: *in vivo* and *in vitro* investigations. *Lait* 82:131–144, 2002.
95. van Luijk, N., M.P. Stierli, S.M. Schwenninger, C. Herve, G. Dasen, J.P.M. Jore, P.H. Pouwels, M.J. van der Werf , M. Teuber, L. Meile. Genetics and molecular biology of propionibacteria. *Lait* 82:45–57, 2002.
96. Miescher, S., M.P. Stierli, M. Teuber, L. Meile. Propionicin SM1, a bacteriocin from *Propionibacterium jensenii* DF1: isolation and characterization of the protein and its gene. *Sys. Appl. Microbiol.* 23:174–184, 2000.
97. Bikash, C., T. Ghosh, T. Sienkiewicz, K. Krenkel. *Brevibacterium linens*: a useful enzyme producer for cheese: a review. *Milchwissenschaft Milk Sci. Int.* 55:628–632, 2000.
98. Rattray, F.P., P.F. Fox. Aspects of enzymology and biochemical properties of *Brevibacterium linens* relevant to cheese ripening: a review. *J. Dairy Sci.* 82:891–909, 1999.
99. Ummadi, M., B.C. Weimer. Tryptophan catabolism in *Brevibacterium linens* as a potential cheese flavor adjunct. *J. Dairy Sci.* 84:1773–1782, 2001.
100. Leret, V., A. Trautwetter, A. Rince, C. Blanco. pBLA8, from *Brevibacterium linens*, belongs to a Gram-positive subfamily of ColE2-related plasmids. *Microbiology* 144:2827–2836, 1998.
101. Moore, M., C. Svenson, D. Bowling, D. Glenn. Complete nucleotide sequence of a native plasmid from *Brevibacterium linens*. *Plasmid* 49:160–168, 2003.
102. Fernandezgonzalez, C., R.F. Cadenas, M.F. Noirotgros, J.F. Martin, J.A. Gil. Characterization of a region of plasmid pBL1 of *Brevibacterium lactofermentum* involved in replication via the rolling circle model. *J. Bacteriol.* 176:3154–3161, 1994.
103. Deb, J.K., N. Nath. Plasmids of corynebacteria. *FEMS Microbiol. Lett.* 175:11–20, 1999.
104. Sanders, M.E. Probiotics. *Food Technol.* 53:67–77, 1999.

105. Fuller, R. Probiotics for farm animals. In: *Probiotics: A Critical Review*, Tannock, G.W., ed., Norfolk: Horizon Scientific Press, 1999, pp 15–22.
106. Meile, L., W. Ludwig, U. Rueger, C. Gut, P. Kaufmann, G. Dasen, S. Wenger, M. Teuber. *Bifidobacterium lactis* sp. *nov*, a moderately oxygen tolerant species isolated from fermented milk. *Sys. Appl. Microbiol.* 20:57–64, 1997.
107. Cai, Y.M., M. Matsumoto, Y. Benno. *Bifidobacterium lactis* Meile et al. 1997 is a subjective synonym of *Bifidobacterium animalis* (Mitsuoka 1969; Scardovi and Trovatelli 1974). *Microbiol. Immunol.* 44:815–820, 2000.
108. Ventura, M., R. Zink. Rapid identification, differentiation, and proposed new taxonomic classification of *Bifidobacterium lactis*. *Appl. Environ. Microbiol.* 68:6429–6434, 2002.
109. Bourget, N., J.M. Simonet, B. Decaris. Analysis of the genome of 5 *Bifidobacterium breve* strains - plasmid content, pulsed-field gel electrophoresis genome size estimation and rrn loci number. *FEMS Microbiol. Lett.* 110:11–20, 1993.
110. Rossi, M., P. Brigidi, A.G.V.Y. Rodriguez, D. Matteuzzi. Characterization of the plasmid pMB1 from *Bifidobacterium longum* and its use for shuttle vector construction. *Res. Microbiol.* 147:133–143, 1996.
111. Park, M.S., K.H. Lee, G.E. Ji. Isolation and characterization of two plasmids from *Bifidobacterium longum*. *Lett. Appl. Microbiol.* 25:5–7, 1997.
112. O'Riordan, K., G.F. Fitzgerald. Molecular characterisation of a 5.75-kb cryptic plasmid from *Bifidobacterium breve* NCFB2258 and determination of mode of replication. *FEMS Microbiol. Lett.* 174:285–294, 1999.
113. Park, M.S., D.W. Shin, K.H. Lee, G.E. Ji. Sequence analysis of plasmid pKJ50 from *Bifidobacterium longum*. *Microbiology* 145:585–592, 1999.
114. Corneau, N., E. Emond, G. LaPointe. Molecular characterization of three plasmids from *Bifidobacterium longum*. *Plasmid* 51:87–100, 2004.
115. Yildirim, Z., D.K. Winters, M.G. Johnson. Purification, amino acid sequence and mode of action of bifidocin B produced by *Bifidobacterium bifidum* NCFB 1454. *J. Appl. Microbiol.* 86:45–54, 1999.
116. Gruss, A., S.D. Ehrlich. The family of highly interrelated single-stranded deoxyribonucleic acid plasmids. *Microbiol. Rev.* 53:231–241, 1989.
117. Meinhardt, F., R. Schaffrath, M. Larsen. Microbial linear plasmids. *Appl. Microbiol. Biotechnol.* 47:329–336, 1997.
118. Abs El-Osta, Y.G., A.J. Hillier, B.E. Davidson, M. Dobos. Pulsed-field gel electrophoretic analysis of the genome of *Lactobacillus gasseri* ATCC33323, and construction of a physical map. *Electrophoresis* 23:3321–3331, 2002.
119. Kiewiet, R., S. Bron, K. Dejonge, G. Venema, J.F.M.L. Seegers. Theta-replication of the lactococcal plasmid pWVO2. *Mol. Microbiol.* 10:319–327, 1993.
120. Boucher, I., E. Emond, M. Parrot, S. Moineau. DNA sequence analysis of three *Lactococcus lactis* plasmids encoding phage resistance mechanisms. *J. Dairy Sci.* 84:1610–1620, 2001.
121. Pavlova, S.I., A.O. Kilic, L. Topisirovic, N. Miladinov, C. Hatzos, L. Tao. Characterization of a cryptic plasmid from *Lactobacillus fermentum* KC5b and its use for constructing a stable *Lactobacillus* cloning vector. *Plasmid* 47:182–192, 2002.
122. Martinez-Bueno, M., E. Valdivia, A. Galvez, M. Maqueda. pS86, a new theta-replicating plasmid from *Enterococcus faecalis*. *Curr. Microbiol.* 41:257–261, 2000.
123. Turgeon, N., M. Frenette, S. Moineau. Characterization of a theta-replicating plasmid from *Streptococcus thermophilus*. *Plasmid* 51:24–36, 2004.
124. Biet, F., Y. Cenatiempo, C. Fremaux. Identification of a replicon from pTXL1, a small cryptic plasmid from *Leuconostoc mesenteroides* subsp. *mesenteroides* Y110, and development of a food-grade vector. *Appl. Environ. Microbiol.* 68:6451–6456, 2002.
125. Miescher, S., M.P. Stierli, M. Teuber, L. Meile. Propionicin SM1, a bacteriocin from *Propionibacterium jensenii* DF1: isolation and characterization of the protein and its gene. *Syst. Appl. Microbiol.* 23:174–184, 2000.
126. McKay, L.L., K.A. Baldwin. Plasmid distribution and evidence for a proteinase plasmid in *Streptococcus lactis* C2-1. *Appl. Microbiol.* 29:546–548, 1975.

127. Clewell, D.B., Y. Yagi, G.M. Dunny, S.K. Schultz. Characterization of three plasmid deoxyribonucleic acid molecules in a strain of *Streptococcus faecalis*: identification of a plasmid determining erythromycin resistance. *J. Bacteriol.* 117:283–289, 1974.
128. Del Solar, G., R. Giraldo, M.J. Ruiz-Echevarria, M. Espinosa, R. Díaz-Orejas. Replication and control of circular bacterial plasmids. *Microbiol. Mol. Biol. Rev.* 62:434–464, 1998.
129. Kok, J., J.M.B.M. van der Vossen, G. Vennema. Construction of plasmid cloning vectors for lactic streptococci which also replicate in *Bacillus subtilis* and *E. coli*. *Appl. Environ. Microbiol.* 48:726–731, 1984.
130. De Vos, W.M. Gene cloning and expression in lactic streptococci. *FEMS Microbiol. Rev.* 46:281–295, 1987.
131. O'Sullivan, D.J., T.R. Klaenhammer. High and low copy number *Lactococcus* shuttle cloning vectors with features for clone screening. *Gene* 137:227–231, 1993.
132. Espinosa, M. General and specialized vectors derived from pBM02, a new rolling circle replicating plasmid of *Lactococcus lactis*. *Plasmid* 51:265–271, 2004.
133. Klein, J.R., C. Ulrich, R. Plapp. Characterization and sequence analysis of a small cryptic plasmid from *Lactobacillus curvatus* LTH683 and its use for construction of new *Lactobacillus* cloning vectors. *Plasmid* 30:14–29, 1993.
134. Pavlova, S.I., A.O. Kilic, L. Topisirovic, N. Miladinov, C. Hatzos, L. Tao. Characterization of a cryptic plasmid from *Lactobacillus fermentum* KC5b and its use for constructing a stable *Lactobacillus* cloning vector. *Plasmid* 47:182–192, 2002.
135. Chagnaud, P., C.K.N. Chan Kwo Chion, R. Duran, P. Naouri, A. Arnaud, P. Galzy. Construction of a new shuttle vector for *Lactobacillus*. *Can. J. Microbiol.* 38:69–74, 1992.
136. Su, P., K. Jury, G.E. Allison, W.Y. Wong, W.S. Kim, C.Q. Liu, T. Vancov, N.W. Dunn. Cloning vectors for *Streptococcus thermophilus* derived from a native plasmid. *FEMS Microbiol. Lett.* 216:43–47, 2002.
137. Solaiman, D.K.Y., G.A. Somkuti. Characterization of pER371-based *Streptococcus thermophilus Escherichia coli* shuttle vectors. *Biotechnol. Lett.* 19:595–598, 1997.
138. Wyckoff, H.A., M. Barnes, K.O. Gillies, W.E. Sandine. Characterization and sequence analysis of a stable cryptic plasmid from *Enterococcus faecium* 226 and development of a stable cloning vector. *Appl. Environ. Microbiol.* 62:1481–1486, 1996.
139. Wirth, R., F.Y. An, D.B. Clewell. Highly efficient protoplast transformation system for *Streptococcus faecalis* and a new *Escherichia coli-S. faecalis* shuttle vector. *J. Bacteriol.* 165:831–836, 1986.
140. Jore, J.P.M., N. van Luijk, R.G.M. Luiten, M.J. van der Werf, P.H. Pouwels. Efficient transformation system for *Propionibacterium freudenreichii* based on a novel vector. *Appl. Environ. Microbiol.* 67:499–503, 2001.
141. Kiatpapan, P., Y. Hashimoto, H. Nakamura, Y.Z. Piao, H. Ono, M. Yamashita, Y. Murooka. Characterization of pRGO1, a plasmid from *Propionibacterium acidipropionici*, and its use for development of a host-vector system in propionibacteria. *Appl. Environ. Microbiol.* 66:4688–4695, 2000.
142. Rossi, M., P. Brigidi, A.G.V.Y. Rodriguez, D. Matteuzzi. Characterization of the plasmid pmb1 from *Bifidobacterium longum* and its use for shuttle vector construction. *Res. Microbiol.* 147:133–143, 1996.
143. Matsumura, H., A. Takeuchi, Y. Kano. Construction of *Escherichia coli Bifidobacterium longum* shuttle vector transforming *B. longum* 105-A and 108-A. *Biosci. Biotechnol. Biochem.* 61:1211–1212, 1997.
144. Missich, R., B. Sgorbati, D.J. Leblanc. Transformation of *Bifidobacterium longum* with pRM2, a constructed *Escherichia coli B. longum* shuttle vector. *Plasmid* 32:208–211, 1994.
145. Behnke, D., M.S. Gilmore, J.J. Ferretti. Plasmid pGB301, a new multiple resistance streptococcal cloning vehicle and its use in cloning of a gentamicin/kanamycin resistance determinant. *Mol. Gen. Genet.* 182:414–421, 1981.
146. Kondo, J.K., L.L. McKay. Plasmid transformation of *Streptococcus lactis* protoplasts: optimization and use in molecular cloning. *Appl. Environ. Microbiol.* 48:252–259, 1984.

147. Turgeon, N., M. Frenette, S. Moineau. Characterization of a theta-replicating plasmid from *Streptococcus thermophilus*. *Plasmid* 51:24–36, 2004.
148. Boucher, I., M. Parrot, H. Gaudreau, C.P. Champagne, C. Vadeboncoeur, S. Moineau. Novel food-grade plasmid vector based on melibiose fermentation for the genetic engineering of *Lactococcus lactis*. *Appl. Environ. Microbiol.* 68:6152–6161, 2002.
149. Sasaki, Y., Y. Ito, T. Sasaki. *thyA* as a selection marker in construction of food-grade host-vector and integration systems for *Streptococcus thermophilus*. *Appl. Environ. Microbiol.* 70:1858–1864, 2004.
150. Wong, W.Y., P. Su, G.E. Allison, C.Q. Liu, N.W. Dunn. A potential food-grade cloning vector for *Streptococcus thermophilus* that uses cadmium resistance as the selectable marker. *Appl. Environ. Microbiol.* 69:5767–5771, 2003.
151. Froseth, B.R., R.E. Herman, L.L. McKay. Cloning of nisin resistance determinant and replication origin on 7.6-kilobase *Eco*RI fragment of pNP40 from *Streptococcus lactis* subsp. *diacetylactis* DRC3. *Appl. Environ. Microbiol.* 54:2136–2139, 1988.
152. Froseth, B.R., L.L. McKay. Development and application of pFM011 as a possible food-grade cloning vector. *J. Dairy Sci.* 74:1445–1453, 1991.
153. Platteeuw, C., I. van Alen Boerrigter, S. van Schalkwijk, W. de Vos. Food-grade cloning and expression system for *Lactococcus lactis*. *Appl. Environ. Microbiol.* 62:1008–1013, 1996.
154. Dickely, F., D. Nilsson, E.B. Hansen, E. Johansen. Isolation of *Lactococcus lactis* nonsense suppressors and construction of a food-grade cloning vector. *Mol. Microbiol.* 15:839–847, 1995.
155. Sorensen, K.I., R. Larsen, A. Kibenich, M.P. Junge, E. Johansen. A food-grade cloning system for industrial strains of *Lactococcus lactis*. *Appl. Environ. Microbiol.* 66:1253–1258, 2000.
156. El Demerdash, H.A.M., K.J. Heller, A. Geis. Application of the *shsp* gene, encoding a small heat shock protein, as a food-grade selection marker for lactic acid bacteria. *Appl. Environ. Microbiol.* 69:4408–4412, 2003.
157. Emond, E., R.L. Ee, G. Drolet, S. Moineau, G. Lapointe. Molecular characterization of a theta replication plasmid and its use for development of a two-component food-grade cloning system for *Lactococcus lactis*. *Appl. Environ. Microbiol.* 67:1700–1709, 2001.
158. Kok, J. Special purpose vectors for lactococci. In: *Genetics and molecular biology of Streptococci, Lactococci, and Enterococci*, Dunny, G.M., P.P. Cleary, L.L. McKay, eds., Washington: ASM, 1991, pp 97–102.
159. Payne, J., C.A. MacCormick, H.G. Griffin, M.J. Gasson. Exploitation of a chromosomally integrated lactose operon for controlled gene expression in *Lactococcus lactis*. *FEMS Microbiol. Lett.* 136:19–24, 1996.
160. Wells, J.M., P.W. Wilson, P.M. Norton, M.J. Gasson, R.W.F. Lepage. *Lactococcus lactis*: high-level expression of tetanus toxin fragment-C and protection against lethal challenge. *Mol. Microbiol.* 8:1155–1162, 1993.
161. de Ruyter, P.G., O.P. Kuipers, W.M. de Vos. Controlled gene expression systems for *Lactococcus lactis* with the food-grade inducer nisin. *Appl. Environ. Microbiol.* 62:3662–3667, 1996.
162. Kleerebezem, M., M. Beerthuyzen, E.E. Vaughan, W.M. de Vos, O.P. Kuipers. Controlled gene expression systems for lactic acid bacteria: transferable nisin-inducible expression cassettes for *Lactococcus*, *Leuconostoc* and *Lactobacillus*. spp. *Appl. Environ. Microbiol.* 63:4581–4584, 1997.
163. O'Sullivan, D.J., S.A. Walker, S. West, T.R. Klaenhammer. Development of expression vector technology for *Lactococcus lactis* using a lytic phage to trigger explosive plasmid amplification and gene expression. *Bio/Technology* 14:82–87, 1996.
164. R.M. Hickey, D.P. Twomey, R.P. Ross, C. Hill. Potential of the enterocin regulatory system to control expression of heterologous genes in *Enterococcus*. *J. Appl. Microbiol.* 95:390–397, 2003.
165. Chandrapati, S., D.J. O'Sullivan. Characterization of the promoter regions involved in galactose- and nisin-mediated induction of the *nisA* gene in *Lactococcus lactis* ATCC 11454. *Mol. Microbiol.* 46:467–477, 2002.
166. Kiatpapan, P., M. Yamashita, N. Kawaraichi, T. Yasuda, Y. Murooka. Heterologous expression of a gene encoding cholesterol oxidase in probiotic strains of *Lactobacillus plantarum*

and *Propionibacterium freudenreichii* under the control of native promoters. *J. Biosci. Bioeng.* 92:459–465, 2001.
167. Wu, Y.Q., P.H. Jiang, C.S. Fan, J.G. Wang, L. Shang, W.D. Huang. Co-expression of five genes in *E coli* or L phenylalanine in *Brevibacterium flavum. World J. Gastroenterol.* 9:342–346, 2003.
168. K. Yazawa, M. Fujimori, J. Amano, Y. Kano, S. Taniguchi. *Bifidobacterium longum* as a delivery system for cancer gene therapy: selective localization and growth in hypoxic tumors. *Cancer Gene Ther.* 7:269–274, 2000.
169. Nakamura, T., T. Sasaki, M. Fujimori, K. Yazawa, Y. Kano, J. Amano, S. Taniguchi. Cloned cytosine deaminase gene expression of *Bifidobacterium longum* and application to enzyme/prodrug therapy of hypoxic solid tumors. *Biosci. Biotechnol. Biochem.* 66:2362–2366, 2002.
170. Li, X., G.F. Fu, Y.R. Fan, W.H. Liu, X.J. Liu, J.J. Wang, G.X. Xu. *Bifidobacterium adolescentis* as a delivery system of endostatin for cancer gene therapy: selective inhibitor of angiogenesis and hypoxic tumor growth. *Cancer Gene Ther.* 10:105–111, 2003.
171. Kimura, N.T., S. Taniguchi, K. Aoki, T. Baba. Selective localization and growth of *Bifidobacterium bifidum* in mouse tumors following intravenous administration. *Cancer Res.* 40:2061–2068, 1980.
172. DE Vos, W.M., G.F.M. Simons. Gene cloning and expression systems in lactococci. In: *Genetics and Biotechnology of Lactic Acid Bacteria,* Gasson, M.J., W.M. De Vos, eds., Glasgow: Blackie & Son, 1994, pp 53–105.
173. Biswas, I., P. Duwat, S.D. Ehrlich, A. Gruss, T. Hege, P. Langella, Y. Leloir, E. Maguin. Efficient system for genetic modification of lactic bacteria: construction of food grade strains. *Lait* 73:145–151, 1993.
174. Fleischmann, R.D., M.D. Adams, O. White, R.A. Clayton, E.F. Kirkness, A.R. Kerlavage, C.J. Bult, J.F. Tomb, B.A. Dougherty, J.M. Merrick, K. Mckenney, G. Sutton, W. Fitzhugh, C. Fields, J.D. Gocayne, J. Scott, R. Shirley, L.I. Liu, A. Glodek, J.M. Kelley, J.F. Weidman, C.A. Phillips, T. Spriggs, E. Hedblom, M.D. Cotton, J.C. Venter, et al. Whole-genome random sequencing and assembly of *Haemophilus influenzae* Rd. *Science* 269:496–512, 1995.
175. Schena, M., D. Shalon, R.W. Davis, P.O. Brown. Quantitative monitoring of gene expression patterns with a complementary DNA microarray. *Science* 270:467–470, 1995.
176. Noordewier, M.O., P.V. Warren. Gene expression microarrays and the integration of biological knowledge. *Trends Biotechnol.* 19:412–415, 2001.
177. Bolotin, A., P. Wincker, S. Mauger, O. Jaillon, K. Malarme, J. Weissenbach, S.D. Ehrlich, A. Sorokin. The complete genome sequence of the lactic acid bacterium *Lactococcus lactis* ssp *lactis* IL1403. *Genome Res.* 11:731-753, 2001.
178. Schell, M.A., M. Karmirantzou, B. Snel, D. Vilanova, B. Berger, G. Pessi, M.C. Zwahlen, F. Desiere, P. Bork, M. Delley, R.D. Pridmore, F. Arigoni. The genome sequence of *Bifidobacterium longum* reflects its adaptation to the human gastrointestinal tract. *Proc. Natl. Acad. Sci. USA* 99:14422–14427, 2002.
179. Pridmore, R.D., B. Berger, F. Desiere, D. Vilanova, C. Barretto, A.C. Pittet, M.C. Zwahlen, M. Rouvet, E. Altermann, R. Barrangou, B. Mollet, A. Mercenier, T. Klaenhammer, F. Arigoni, M.A. Schell. The genome sequence of the probiotic intestinal bacterium *Lactobacillus johnsonii* NCC 533. *Proc. Natl. Acad. Sci. USA* 101:2512–2517, 2004.
180. Kleerebezem, M., J. Boekhorst, R. van Kranenburg, D. Molenaar, O.P. Kuipers, R. Leer, R. Tarchini, S.A. Peters, H.M. Sandbrink, M.W.E.J. Fiers, W. Stiekema, R.M.K. Lankhorst, P.A. Bron, S.M. Hoffer, M.N.N. Groot, R. Kerkhoven, M. de Vries, B. Ursing, W.M. de Vos, R.J. Siezen. Complete genome sequence of *Lactobacillus plantarum* WCFS1. *Proc. Natl. Acad. Sci. USA* 100:1990–1995, 2003.
181. Glaser, P., L. Frangeul, C. Buchrieser, C. Rusniok, A. Amend, F. Baquero, P. Berche, H. Bloecker, P. Brandt, T. Chakraborty, A. Charbit, F. Chetouani, E. Couve, A. de Daruvar, P. Dehoux, E. Domann, G. Dominguez-Bernal, E. Duchaud, L. Durant, O. Dussurget, K.D. Entian, H. Fsihi, F. Garcia-Del Portillo, P. Garrido, L. Gautier, P. Cossart, et al. Comparative genomics of *Listeria* species. *Science* 294:849–852, 2001.

1.11

Genetic Engineering of Baker's Yeast: Challenges and Outlook

Jose Antonio Prieto, Jaime Aguilera, and Francisca Randez-Gil

CONTENTS

11.1 INTRODUCTION

The bread making process is one of the oldest applications of biotechnology. The term bread defines a great variety of baking products, which vary in formulation, ingredients, and processing conditions. For most of them, however, the use of baker's yeast is fundamental to obtaining a high quality product according to the demands of consumers. Baker's yeast contributes substantially to the flavor and crumb structure of bread, but its primary role is the production of CO_2, via the alcoholic fermentation of sugars, which results in dough leavening.

Commercial baker's yeasts are domesticated strains of *S. cerevisiae* that have been selected and optimized for baking applications. Most of them are homothallic, with a high and irregular degree of ploidy and low sporulation ability (1). In addition, they exhibit chromosomal-length polymorphisms and rearranged chromosomes with multiple translocations (2). These special characteristics are the result of natural selection by adaptative evolution together with the successive application of improvement programs. The polymorphism of baker's yeast reflects, at least in part, selection by continuous long term nutrient limitation. In industrial practice, the yeast is grown aerobically under conditions (fed-batch) in which the supply of the carbohydrate feedstock is limited, to avoid ethanol formation. When a yeast population is cultured under this physiological situation, it undergoes a series of adaptative shifts whereby clones that demonstrate better efficiency of nutrient utilization successively replace one another over time (3). The mechanism underlying this response includes random and nonrandom mutations (4) and unequal crossing over, leading to gene amplification, changes in protein structure, and altered expression patterns (5). When compared with laboratory strains of *S. cerevisiae*, baker's yeasts has more copies of the *SUC* gene, which appears to be amplified and translocated to several chromosomes (6,7). This gene encodes for invertase, the enzyme hydrolyzing sucrose, the main carbon source found in beet and cane molasses used for industrial feeding. Thus, amplification of *SUC* genes endows yeast cells with a higher sucrose hydrolyzing ability, overcoming a limiting step for growth and biomass yield (8). A clear example of this evolutive mechanism is the appearance of multiple tandem duplications of high affinity hexose transporters (*HXT* gene family), in a yeast population grown in a continuous glucose-limited environment (9).

The second factor accounting for many of the characteristics of baker's yeast is the repeated use of improvement strategies that provide a high genetic variability. Strain improvement of baker's yeast has traditionally relied on classical genetic techniques, including random mutagenesis, hybridization, and protoplast fusion. The continuous mating and later segregation of meiotic products in baker's yeasts increase the possibilities of chromosomal reorganizations, both in size and number, leading to polymorphism. This phenomenon is enhanced by the presence of many Ty transposable elements and subtelomeric Y´ regions in the baker's yeast genome (7), a key factor for chromosomal rearrangements (10). This feature could have also favored the selection of traits appreciated in baker's yeast, because the *MAL* loci and the *SUC* genes are located at the telomeres of this organism (11). The selection for broad traits, as robustness, growing rate, or fermentative capacity would also have contributed to the genetic constitution of baker's yeast. Although there is a lack of knowledge of the genes responsible for these properties, they are favored by a high ploidy level, as evidenced by much experimental data. For example, industrial yeasts show higher levels of glycolytic enzymes than wild-type strains (12). Dough-leavening ability also correlates with the activity of the *MAL* genes (13,14), which determine the capacity to transport and hydrolyze maltose, the main sugar than sustains dough fermentation. This is certainly not surprising, because duplication of blocks of genes has been a recurring feature of yeast genome evolution (15,16). Increasing the number of chromosomes will increase the number of favorable genes leading to strains with enhanced fitness, increased maltose utilization capacity, and more vigor (17, 18). Thus, natural adaptation and the selective pressure exerted for decades by producers have led to the transformation, at low cost, of an abundant feedstock to cell biomass of high technological quality.

Nevertheless, there are still important traits in baker's yeast that are far from optimal. Tolerance to various stresses, such as osmosis, freezing, or desiccation, is clearly inadequate. Our limited understanding of the physiological and genetic determinants of these important commercial properties suggests that they are under the control of multiple genes and complex regulatory mechanisms (19, 20). Some of these desired phenotypes could also go against biological design (21). Thus, it is unlikely that classical breeding programs could provide further improvement of these characteristics. Given this scenario, the ability to transform baker's yeast by recombinant DNA technology has opened the possibility to manipulate just a single gene or pathway, without altering other key or valuable genes. Its is also possible to knock out undesirable genes, to engineer the constitutive expression of groups of genes involved in a particular function, or even to switch on or off the transcription of a given gene in a specific technological step. The transfer of heterologous genes to *S. cerevisiae* means that there is now the possibility of using yeast as a cell factory. This has allowed, in the past decade, the development of industrial strains with unsuspected properties, giving a new generation of baker's yeast.

This chapter will collect the available information on the most important advances in the engineering of baker's yeast, from both fundamental and technological aspects. It will also review critically the molecular tools applied for transformation of industrial strains. Information concerning wild *S. cerevisiae* and other industrial yeasts, like brewing or wine strains, will also be considered where it provides knowledge to address unresolved problems or to reveal future perspectives. To date, most of the recombinant approaches have been used for the modification of traits linked to single or a limited number of genes. Thus, the main targets for strain improvement by genetic engineering can be grouped into three categories: (1) extension of substrate range, (2) heterologous enzyme production, and (3) overproduction of essential nutrients.

11.2 MOLECULAR TOOLS

The manipulation of commercial baker's yeast is a difficult task because of the special genetic characteristics of these strains. For instance, the high level of ploidy means that multiple disruption and deletion steps are required to obtain a null mutant. Another consequence of this polyploidy is that most of these strains are prototrophic. Therefore, transformation by complementation of auxotrophic mutations, the most common approach in wild-type *S. cerevisiae*, cannot be employed. On top of this, the molecular tools that could be applied in industrial strains are often limited by legal regulations concerning the commercial exploitation of genetically modified organisms (GMOs). These legal regulations establish, as a criterion of acceptability, the absence of nonessential foreign DNA, and in particular, of vector sequences related to prokaryotic DNAs or antibiotic resistance markers. Additionally, transformants should also (1) keep the same useful properties of the parental strain, (2) have the minimal amount of synthetic DNA linkers, and (3) not express any fusion protein that may arise because of the genomic integration events (22).

11.2.1 Plasmids and Integrative Fragments

The type of cloning vector used for yeast is shuttle plasmids that can also replicate and be selected for, in *E. coli*. They can be grouped into two classes, those that replicate autonomously and those that integrate into a chromosome (23). From all of them, the 2 μm-based plasmids are probably the most popular. Such multicopy plasmids show high transformation frequency and gene expression, which is useful for preliminary tests. However, gene overproduction can have deleterious effects on fermentation performance. In addition, these plasmids tend to be unstable under the absence of selective pressure (23), which is inappropriate for commercial purposes.

To solve this, integration appears as the best strategy, because it offers high stability. As integrative vectors, it is possible to use integrative plasmids, integrative fragments, or vectors based on Ty elements. Integrative plasmids contain an auxotrophic marker, which complements a host mutation. Integration occurs via homologous recombination between the plasmid-borne marker and the chromosomal allele. Nevertheless, this type of vector is not commonly used for industrial yeast, because the lack of auxotrophic markers in these strains.

In contrast, integrative fragments are widely used. These fragments consist of DNA linear fragments, normally generated by restriction or PCR, containing the expression cassette of one or more genes, flanked at both sides by sequences homologous to a target locus in the genome (Figures 11.2, 11.4). In contrast to integrative plasmids, integrative fragments allow multicopy integrations. For this purpose, the strategy is to target for replacement DNA that is reiterated in the genome that is not essential for growth. The locus most commonly employed is a portion of the rDNA unit (24). To ensure a high number of integrations (100–200 copies per cell), Lopes et al. (25) used a *leu2* defective promoter as selection marker. The transformants were highly stable over 70 generations of nonselective growth. Nevertheless, the mitotic stability, as well as the expression level of heterologous genes targeted to rDNA, depends on the size and the nature of the rDNA fragment required for targeting its integration. Lopes et al (26) reported stable maintenance only when the complete plasmid has a size no larger than that of the rDNA unit (9.1 Kb). Several examples of the use of rDNA-targeted integration in industrial strains have been reported (27, 28).

Other interesting integrative fragments used in industrial strains are those based in the δ-integration system described by Sakai et al. (29). This seems to be the most useful system to allow the sequential insertion of multiple cloned genes in yeast. In this vector, the integration is obtained via homologous recombination between the Ty δ-sequences in the yeast chromosome and those subcloned into nonreplicating plasmids (30). This system is chosen to overproduce recombinant glucoamylase in baker's yeast (31).

11.2.2 Dominant Selection Markers

The prototrophic character of industrial baker's yeast implies that genetic manipulation in these strains is based, almost exclusively, on the use of dominant resistance markers. *Saccharomyces* is sensitive to antibiotics such as geneticine, chloramphenicol, cycloheximide, hygromycin B, nourseothricin, or bialaphos. Heterologous enzymes, most of them from bacteria (32–35), can inactivate these antibiotics. Thus, disruption cassettes containing genes which confer resistance to those antibiotics have been developed and widely employed in laboratory strains of *S. cerevisiae*. However, a copy of the resistance marker remains in the genome after the integration event, making this selection system inadequate for industrial strains (22).

In *S. cerevisiae*, the loss of foreign DNA can be promoted by making use of a natural phenomena named "pop-out" or *in vivo* excision. This is an event of intrachromosomic recombination between two identical or very similar sequences (direct repeats), that results in the excision of a single direct repeat copy together with the DNA fragment placed between both direct repeats. Alani et al. (36), who constructed a disruption module containing direct repeats flanking the *URA3* selection marker, first demonstrated the possibilities of this approach. Later, dominant disruption cassettes harboring the *E. coli* gene, which confers resistance to geneticine (*kan*^{r}), flanked by direct repeats of KanMX3 modules (34), were constructed and successfully employed for the repeated use of a single marker. KanMX3-based modules containing genes for hygromycin B, nourseothricin, or bialaphos resistance are also available (35). Furthermore, the efficiency of the *in vivo* excision event has been improved by making use of the loxP-Cre system (37). It has been shown that this a useful tool for industrial strains. Puig and co-workers (38) disrupted the two *URA3* alleles of a wine strain by successive transformation with a kanMX3-based disruption cassette and later excision. Recently, a second set of loxP-Cre plasmids was generated, which are ideally suited to genome functional analysis (39).

An inconvenience of these protocols is that after each excision step, a copy of the direct repeat remains in the genome of the yeast. Due to the requirement of heterologous direct repeats, in order to avoid their homologous recombination with the *S. cerevisiae* DNA, sequences of different sources, like *hisG* from *Salmonella* (36), *LEU2* from *Ashbya gossypii* (34), or loxP from the bacteriophage P1 (37, 40) have been employed. However, for commercial purposes, the use of these exogenous DNA is not convenient.

One possibility to solve this problem would be the use of nonrecombinogenic fragments from yeast species closely related to *Saccharomyces*. Following this approach, several research groups (41, 42); Estruch and Prieto, unpublished results) have constructed food safe grade dominant resistance cassettes containing direct repeats from the *MEL1* gene. This gene, which encodes for α-galactosidase, is present in several *Saccharomyces* species, including some strains currently classified as *S. cerevisiae*, but not in baking strains. These cassettes were used to obtain Ura$^-$ (41) and Trp$^-$ (42) (Estruch and Prieto, unpublished results) commercial baker's yeast strains by consecutive disruption of the corresponding wild-type copies.

A different strategy makes use of the Flp FRT (Flp recombinase recognition target) system from the 2 μm endogenous plasmid of *S. cerevisiae* for multiple gene disruption or heterologous gene expression in industrial yeast. The Flp protein is a yeast recombinase that recognizes FRT sequences. A minimal FRT consists of 34-bp, including two 13-bp inverted repeats surrounding an 8-bp core sequence involved in DNA–DNA pairing during strand exchange in the recombination reaction (43). From *in vitro* (44, 45) and *in vivo* (46) characterization of the Flp FRT recombination model, it is known that DNA heterology between the core regions of two FRT site inhibits their ability to engage in recombination. Utilizing this property, Storici and coworkers (47) designed FRT DNA sequences with different core mutations, which were cloned in direct orientation flanking a dominant marker gene, like KanMx4. Thus, the marker can be easily eliminated by *in vivo* recombination between FRT sequences catalyzed by the endogenous 2 μm plasmid-derived Flp protein.

11.2.3 Homologous Dominant Markers

For food and beverages applications, it is desirable to use dominant markers derived from yeast. Several genes from *S. cerevisiae,* conferring resistance to different harmful substances, have been isolated and tested for the selection of recombinant industrial strains.

The toxic effects of heavy metals ions are countered in *S. cerevisiae* by the *CUP1* gene, which encodes for a metal-binding protein. Cooper resistance was the most popular selection method for industrial yeast strains, because the expression of *CUP1* gene has no significant effects on fermentative behavior (48). Sulfometuron methyl (SM) is an herbicide that blocks yeast growth by inhibition of acetolactate synthase, encoded by *ILV2* (49). In *S. cerevisiae*, 10 phenotypically distinct alleles of *ILV2* have been described. *SMR1-410* (50) and *SMR1B* (51) have been used most widely. Both display mutations affecting the same proline residue, which is required for the binding of SM to the *ILV2* product. In addition, this approach provides a useful site of integration close to the *ILV2* locus, that has been successfully used to generate α-galactosidase- (52) and α-amylase-producing (53) baker's yeast strains. Aureobasidin A (AbA), an antifungal antibiotic produced by *Aureobasidium pullulans*, is highly toxic to fungi, including *S. cerevisiae* and *S. pombe*. Hashida-Okado et al. (54) isolated several dominant mutants to AbA, using the mutated $AUR1^R$ gene from *S. cerevisiae* or the corresponding *AUR1-C* from *S. pombe*. These authors demonstrated the board application of this system for several yeast species and for industrial yeast, including baker's yeast (55). Formaldehyde can also be used as a selecting agent. Indeed, van den Berg and Steensma (56) constructed formaldehyde resistant expression cassettes by placing the gene *SFA1* from *S. cerevisiae*, encoding formaldehyde dehydrogenase (57), under the control of *GPD1* promoter and *CYC1* terminator. This system offers the advantage that formaldehyde is biodegraded to CO_2 during fermentation and therefore its removal during downstream processing is not necessary. The analogue 5,5,5-trifluoro-DL-leucine (TFL), has also been employed in transformation systems. In *S. cerevisiae*, the resistance to this leucine derivative is due to two classes of dominant mutations in the *LEU4* gene (58). Transformation of natural wine yeast strains with linear fragments containing these mutated genes showed a high efficiency, even better than that previously reported for other dominant markers (59). Two *S. cerevisiae* genes, *FZF1-4* and *SSU1,* have been found to exhibit dominant phenotypes useful for selecting transformants in the presence of sulfite. *FZF1-4*, a mutant allele of *FZF1*, a C_2-H_2 zinc finger protein (60), is a positive activator of *SSU1* (61), which encodes for a membrane protein involved in sulfite excretion. An interesting trait of this selection system is the lack of false positives, even at high cell densities (62). This characteristic together with the inexpensive cost of sulfite and its nontoxicity make of this system a good candidate to select transformants from industrial yeast strains.

Recently, a new tool for direct selection of transformants of industrial yeasts, based on the phenotype of multi-drug resistance, has been developed (63, 64). In *S. cerevisiae* this phenotype results from the overexpression of several membrane drug efflux pumps (63, 65). Those genes are under the control of two transcriptional activators Pdr1p and Pdr3p (66, 67). Several dominant point mutations of these effectors enhance its activity (63, 68). Both multicopy and centromeric vectors bearing the mutant *pdr3-9* allele have been used for transformation of laboratory and polyploid industrial strains of baker's yeast (63, 64). Furthermore, *pdr3-9* gene has been also employed as dominant selectable market for natural wine strain transformation (69). These wine strains produce experimental wines of the same quality as their parental natural strain.

11.2.4 Expression Level Promoters

The production of heterologous proteins in *S. cerevisiae* depends on multiple intracellular factors. A fundamental parameter is the level of transcription provided by the promoter (70). The expression of a foreign gene in *Saccharomyces* requires, in most cases, the use of homologous promoters in order to ensure expression and regulation. Thus, strong constitutive promoters are often chosen for efficient protein production (71) The promoters most commonly used are those of the glycolytic genes, phosphoglycerate kinase, *PGK1*, alcohol dehydrogenase and *ADH1*, glyceraldehyde-3-phophate dehydrogenase, *TDH1*, (70) and the promoter of the *ACT1* gene (72). Although those promoters are referred to as constitutive, their transcriptional activity is subjected to regulation, reflecting a particular physiological state of the cell (73, 74). This has been clearly demonstrated in molasses-grown baker's yeast strains by making use of two reporter genes driven by different promoters (75). Expression regulated by the short form of the *ADH1* promoter gave the highest protein yield for cells cultured in molasses medium or flour and water mixtures. Nevertheless, the *ACT1* promoter appeared as the strongest in cells growing actively with sucrose (75). Similar results were reported by Puig et al. (76) when evaluating the use of phase specific gene promoters for the expression of enological enzymes in an industrial wine yeast strain.

Some times, however, the choice of inducible promoters is more useful, like those of the *GAL* genes, where expression is repressed by glucose and induced by galactose, *PHO5*, repressed by inorganic phosphate, *ADH2*, and repressed by glucose or *SUC2* (70). These tightly regulated promoters allow the production of putatively growth inhibitory proteins, because cells can be grown to high densities before the expression of the foreign gene is induced. This strategy also provides a good system for the optimization of fermentation processes for the utilization of lactose by recombinant baker's yeast strains (27). Thus, the choice of a promoter for heterologous gene expression should reflect the particular condition in which the protein production is and the level of recombinant protein required for a desired technological effect.

In this sense, high throughput technologies, such as DNA microarrays and 2D-electrophoresis, are useful tools to provide information on differential gene expressions and regulatory networks (77, 78). Global gene expression analysis during the shift from fermentation to respiration in *S. cerevisiae* (79), or under different stress conditions, such as salt (80), oxidative stress, or thermal stress (81), has identified target genes induced under those physiological conditions. The promoters of these genes are obvious candidates to drive the production of heterologous proteins in different steps of the yeast propagation and bread making process. For instance, the promoter of a strongly induced gene by heat shock could be useful to express a heterologous protein at the baking stage of the bread dough, whereas an osmotically induced gene could provide an ideal promoter for heterologous expression in high sugar dough.

11.3 EXTENSION OF SUBSTRATE RANGE

A blend of beet and cane molasses is commonly used in the commercial propagation of yeast. Sucrose is the main sugar, accounting for more than 40% by weight. Molasses also provides essential nutrients, such as minerals, vitamins, and some amino acids. This composition has made molasses the most widely extended substrate for baker's yeast production. Nevertheless, molasses is not the ideal culture medium for yeast propagation.

Both cane and beet molasses can contain harmful substances that inhibit yeast growth and baking performance. They also contain a great deal of suspended solids that must be removed, at least in part, before yeast production. This pretreatment increases the cost of the process, reducing the product profit margin. In addition, molasses may vary significantly in quality from batch to batch, especially in terms of sugar content. On top of this, the introduction of modern sucrose extraction procedures has reduced the availability of this carbon source for growth. For these reasons, the expression of heterologous genes in baker's yeast has led to the possibility of using new carbon sources and raw materials other than molasses for yeast propagation.

Many examples of substrate range extension are described in the Literature for *S. cerevisiae*. For biomass production, a feedstock having low cost and availability on a large scale is required (82). Consequently, only a few potential carbon sources can be considered for cost effective yeast propagation processes. The practical application of these nonnatural substrates of baker's yeast will require an extensive evaluation of the traits of the newly engineered strains for properties desired in commercial baking.

11.3.1 Melibiose

In addition to sucrose, beet molasses contains up to 8% (by weight) raffinose, which is hydrolyzed by yeast *β*-fructosidase (invertase, EC 3.2.1.26), encoded by the *SUC2* gene in fructose and the disaccharide melibiose (glucose–galactose). Baker's yeast can utilize melibiose after its hydrolysis, but lacks *α*-galactosidase (melibiose, EC 3.2.1.22), which releases the glucose and galactose moieties (Figure 11.1). Thus, the inability of industrial baker's yeast to use melibiose as a carbon source reduces the productivity of the propagation process and increases the organic content of the effluent, which has a high biological demand (BOD).

α-galactosidase is encoded by a *MEL* gene, present in different species of the genus *Saccharomyces*, such as *S. bayanus*, *S. carlsbergensis,* or *S. oleaginosus* (83). Mel$^+$ strains of *S. cerevisiae* have been also isolated, and eleven genes, *MEL1–MEL11*, have been identified in different natural populations (84, 85). Although an accumulation of the *MEL* genes has been found in some specific *S. cerevisiae* populations, most of the *Saccharomyces* spp. strains carry only one *MEL* locus (83, 86), *MEL1* being the best characterized.

Baking strains with complete raffinose utilizing ability were earlier constructed by hybridization with laboratory strains of *S. bayanus* (87). Higher yeast yields (2–3%) on beet molasses were thus obtained, although these industrial hybrids exhibited poor bread dough leavening ability. Mel$^+$ baker's yeast strains were further obtained by cloning and expression of the *MEL1* gene from *S. bayanus* (88, 89). This engineered yeast resulted in an increase in the yield in molasses medium and a significant reduction in the BOD from the factory effluent. However, the presence of bacterial sequences into these industrial

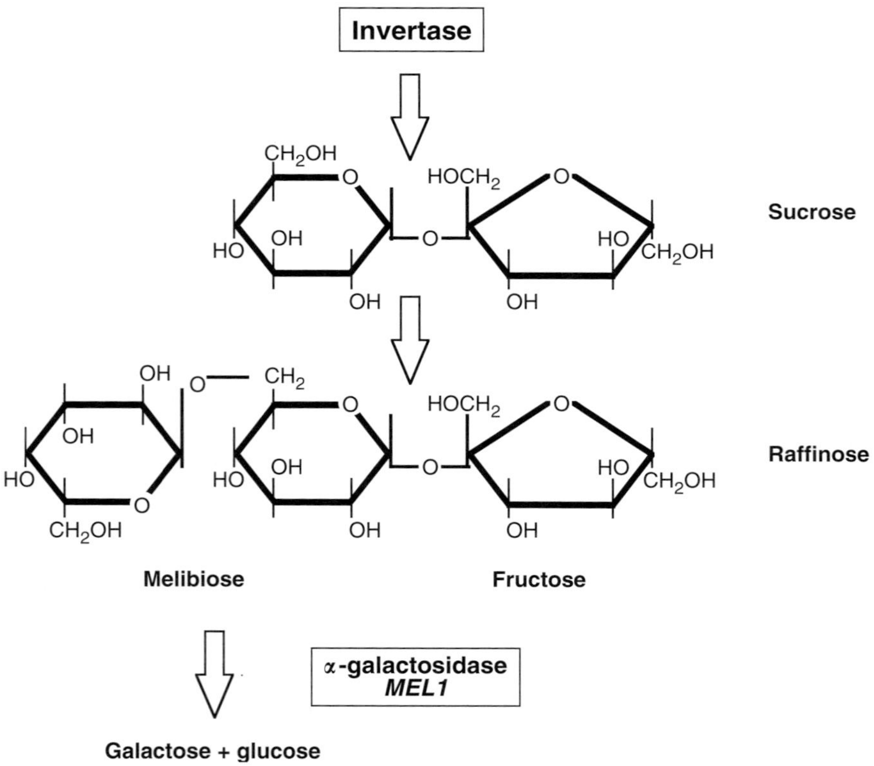

Figure 11.1 Structures of sucrose and raffinose and scheme of their hydrolysis by invertase and *α*-galactosidase. The main hydrolysis products are shown.

strains resulted in a market resistance to their use. A new approach (52) made use of a one step dominant selection procedure to construct stable integrations of *MEL1* devoid of bacterial plasmid DNA in baker's yeast strains (Figure 11.2). These strains were stable, increased the biomass substrate yield up to 8%, and had growth rates similar to those of the untransformed control strains. In a further study, melibiose-utilizing baker's yeast strains produced by genetic engineering and classical breeding were compared (90). The results showed that the strategies render strains with similar fermentative capacity. However, the strains tested differed in genetic background. In addition, biomass yields on molasses under industrial conditions were not determined.

Aiming to find Mel^{+} baker's yeast with optimal performance, Rønnow et al. (91) constructed strains partially relieved of glucose control over galactose metabolism by disruption of *MIG1* and *GAL80*. Mig1p and Gal80p are regulator proteins involved in the glucose repression and galactose induction of the *GAL* genes (92). Overall, these proteins downregulate the synthesis of enzymes involved in the metabolism of galactose and other alternative carbon sources when glucose is present, which means that wild-type strains of baker's yeast consume galactose only when glucose has been exhausted. A simultaneous utilization of both sugars was observed, however, in melibiase-producing *mig1 gal80* mutants (91). Moreover, the recombinant strains showed higher melibiose hydrolysis and galactose uptake rates than the corresponding wild type, resulting in shortened cultivation times on melibiose (93). However, this resulted in a more pronounced respire-fermentative metabolism, with enhanced ethanol yield and less biomass production (91, 93). These

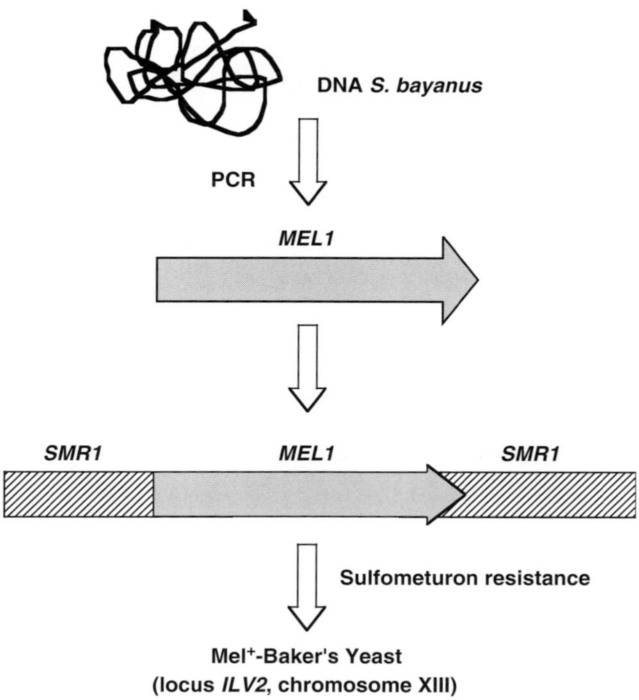

Figure 11.2 Schematic representation of the strategy followed to the selection of stable *MEL1* integrants of baker's yeast strains. Cells were transformed with a PCR-generated DNA fragment containing the *MEL1* gene from *Saccharomyces bayanus* flanked by a *S. cerevisiae SMR1* allele of the *ILV2* gene (chromosome XIII), which confers sulfometuron resistance (50, 52).

derepressed Mel^+ strains could be of particular interest in the production of ethanol, a subproduct of the molasses fermentation, although the reduction in biomass yield makes unlikely their application in the propagation of baker's yeast.

11.3.2 Lactose

Whey is the main byproduct of the cheese industry. In 2001, in the USA alone, this manufacturing process generated approximately 3.7 billion Kg of raw whey (USDA). Whey can be processed for a variety of end uses, like animal feed, human food products, and pharmaceuticals (94). Nevertheless, processing procedures are often expensive (95), and the demand for whey derived products is not enough to cover all whey available. As a result, many dairy facilities often rely on land spreading to dispose of this byproduct (96). This common practice has a great pollution potential due to the high BOD of whey, to which lactose is the major contributor. Consequently, the introduction of new biotransformation methods of lactose and whey into useful products is of great interest.

Different microorganisms, including bacteria, fungi, and yeasts, like *Kluyveromyces lactis* or *K. fragilis*, can metabolize lactose (Figure 11.3). Nevertheless, fermentative transformation of lactose by *S. cerevisiae* would be more profitable, because this organism has special properties that represent a higher benefit (97, 98). *S. cerevisiae* lacks, however, β-galactosidase activity, which releases glucose and galactose, and thus is unable to assimilate the disaccharide. Two different strategies have been followed to endow *S. cerevisiae* with the ability to utilize lactose: (1) release or secretion into the medium of recombinant β-galactosidase, and (2) combined production of a specific lactose transport system coupled with an internal β-galactosidase activity.

Fermentation of lactose by yeast cells overproducing the yeast Gal4p and the β-galactosidase (*lacZ*) from *E. coli* was earlier reported by Porro et al. (99). Expression in a high copy number of *GAL4*, the activating factor of the *GAL* genes (92), induces a certain degree of cell lysis with the consequent release of β-galactosidase into the medium. Physiological studies showed than cells were able to grow on media containing whey lactose, but biomass yield was unsatisfactory. A further improvement of this system was achieved by transforming yeast cells with a multicopy vector containing the *lacZ* gene fused in frame with the *S. cerevisiae var. diastaticus STA2* gene (100). This fusion targeted the enzyme to the yeast periplasm, improving the growth rate and biomass productivity.

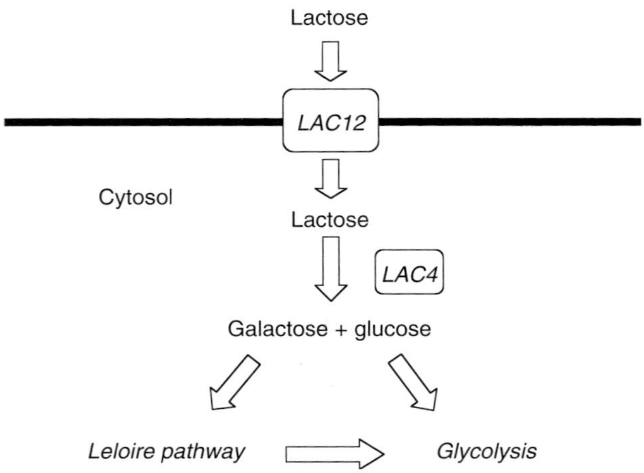

Figure 11.3 Pathway of the transport and catabolism of lactose in *Kluyveromyces*.

Similar results were obtained by expressing an extracellular β-galactosidase, encoded by the *LacA* gene from *A. niger* (101). However, the resulting strains, derived from haploid hosts, showed an unstable Lac^+ phenotype and exhibited a typical diauxic growth (102).

Sreekrishna and Dickson (103) were the first to report the construction of lactose-utilizing strains by transformation of *Saccharomyces* haploid cells with vectors harboring a selectable marker gene, conferring resistance to the antibiotic G418, and the *K. lactis* genes, *LAC4* and *LAC12*, coding for β-galactosidase and lactose permease, respectively (Figure 11.3). Furthermore, they also obtained stable integrative transformants containing both genes and lacking bacterial DNA sequences, as claimed in a U.S. patent (104). This system was later optimized by Rubio-Texeira et al. (105), who reported the construction of a Lac^+ *S. cerevisiae* diploid strain (MRY286), carrying multiple copies of the *K. lactis* genes. In lactose medium, this recombinant strain showed mitotic stability, simultaneous utilization of glucose and galactose, and a respiro-fermentative metabolism characterized by a low ethanol production and a high biomass yield. Following the same strategy, Adam et al. (27) constructed Lac^+ baker's yeast by sequential transformation of a *trp1* auxotrophic recipient strain (106), with linear DNA fragments containing the *LAC12* and *LAC4* genes and flanked by sequences homologous to the rDNA from *Saccharomyces* (Figure 11.4). The resulting strain, L306BY6, was stable under nonselective conditions, and displayed high growth rate in cheese whey with a biomass yield of about 10 g dry weight/liter of culture. Moreover, whey propagated compressed yeast had white color, a characteristic appreciated by bakers, and similar baking properties as molasses-grown commercial baker's yeast. Indeed, measurements of CO_2 production and gassing rate were almost identical for the strains tested (Table 11.1), independently of the culture medium. Bread elaborated with L306BY6 also showed loaf volume and crumb firmness similar to that elaborated with control strains (Table 11.1). Hence, the newly developed strain offers a profitable use for cheese whey and it is a cheap alternative to the traditional procedures of baker's yeast propagation.

11.3.3 Starch

Starch, a cheap raw material of agricultural crops, has been traditionally employed as a source of sugars commonly used in chemical, pharmaceutical, and food industries. The batch hydrolysis of starch into sugar syrups requires its gelatinization by cooking, and two enzymatic steps, liquefaction and saccharification, which involve the use of α-amylase for endo α-1,4 cleavage, and glucoamylase, an exo-acting enzyme able to hydrolyze both,

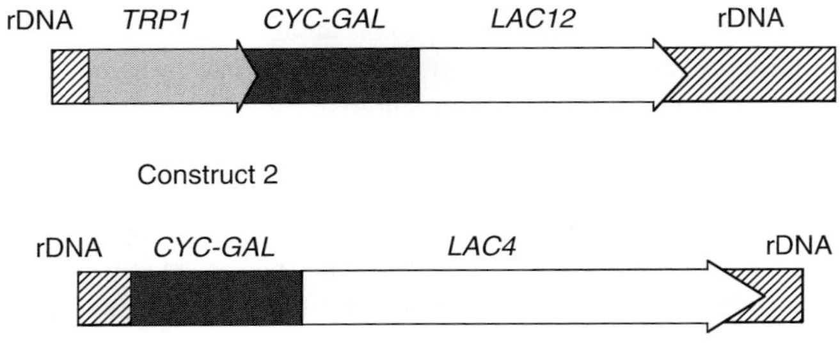

Figure 11.4 DNA arrays (constructs) used in the two sequential yeast transformation steps leading to the construction of the Lac^+ baker's strain L306BY6. Source: Taken from Adam, A.C., J.A. Pricto, M. Rubio-Texeria, J. Polaina, Yeast 15:1299–1305, 1999. Copyright © 1999 John Wiley & Sons Limited. Reproduced with permission.

Table 11.1
CO_2 Production in Bread Dough and Loaf Volume and Firmness of the Corresponding Breads Elaborated with Different Yeast Strains

	Bread Quality Parameters[a]		Gassing Power[b]	
Strain	Loaf Volume (ml)	Firmness (Newtons)	H'm[c] (mm)	CO_2 Production (ml)
L306BY6	412.8 ± 14.3	12.6 ± 1.1	50.2	1120
Commercial strain	394.0 ± 10.3	12.3 ± 2.4	49.1	1076
10a12 (YEplac112)	385.0 ± 16.8	10.1 ± 1.7	51.1	1102

[a]Values are the means ± standard errors, of at least 12 determinations made with loaves from different pieces of bread obtained in two independent baking tests.
[b]CO_2 production was recorded in a Chopin rheofermentometer.
[c]H'm represents the maximum height of the fermented dough piece.
Source: Reference 27. Copyright © 1999 John Wiley & Sons. Reproduced with permission.

α-1,4 and α-1,6 linkages, respectively (107). *Saccharomyces cerevisiae* is unable to hydrolyze starch, but easily ferments its degradation products, glucose and maltose; therefore, the starch-derived sugar syrups can be used as a carbon source for the propagation of baker's yeast. However, because the hydrolysis process requires expensive enzymes, relatively high temperatures (50–80°C), and time-consuming steps, the conversion of starch grains into cell biomass is uneconomic.

Genetically engineered industrial yeast capable of producing starch-degrading enzymes have been constructed, to save some of the cost associated with the traditional methods of starch conversion and to optimize ethanol and single cell protein production. Secretion of α-amylase and glucoamylase activities from *S. cerevisiae* has been done by coexpression of the corresponding genes from amylolytic species, mainly of *Bacillus* and *Aspergillus*, including those from *Schwanniomyces occidentalis* and *Saccharomyces diastaticus* (108–112). These strains perform a single-step hydrolysis of starch with conversion efficiencies greater than 93–95%, although the amylopectin fraction is poorly assimilated. This problem was solved by using a recombinant strain producing α-amlylase, glucoamylase, and pullulanase, a debranching enzyme acting on the α-1,6 linkages (113). The transformed strain showed a high rate of starch hydrolysis and was able to utilize almost 100% of precooked starch. In a different approach, several reports described the development of novel starch-utilizing strains by applying the technique of cell surface engineering. By using the yeast α-agglutinin as an anchor region, Murai et al. (114) were able to target and bind covalently glucoamylase from *Rhizopus oryzae* to the *S. cerevisiae* cell surface. Constructing a laboratory strain that displays glucoamylase and α-amylase on its cell surface yielded a further improvement (115). Although the strain was not evaluated in comparison with other strains expressing secretable enzymes, it is expected that cell-surface engineered cells could maintain the enzyme activity over a longer cultivation period, increasing the efficiency of starch transformation.

Compared to this, the construction of amylolytic baker´s yeast has received less attention, due in part to the difficulty of transforming strains with a polyploid background. Recombinant strains containing the α-amylase gene from *Aspergillus oryzae* or *S. occidentalis* have been generated by employing integrative cassettes targeted to the rDNA or the *ILV2* locus, respectively (28, 53). *Aspergillus oryzae* glucoamylase cDNA has also been expressed in baker's yeast by δ-integration into the Ty1 retrotransposon (31).

However, the ability of these strains to grow on starch as their sole carbon source appears to be quite limited (53), because they lack a complete amylolytic system. Obviously, the most direct approach to this problem would be the coexpression of several amylolytic genes. However, the use of such strains in downstream applications could be problematic, if the high starch-degrading capacity is prolonged during the bread dough fermentation and baking. Excessive starch breakdown results in a moist, sticky, and rubbery breadcrumb and a small volume (22). Nevertheless, this could be solved by an appropriate election of the promoter driving the expression of the heterologous genes. In this way, the application of global expression analysis techniques should help to identify strong promoters showing a differential regulation in the propagation and baking stages.

11.3.4 Lignocellulose

Lignocellulosic wastes are the major potential source of carbon and energy on the earth. Although most of the academic and industrial interest on lignocellulose has been devoted to the bioconversion of this material into fuel (116), improvements regarding pretreatments, enzymatic hydrolysis, fermentation, and process integration should extend its use to the production of different value added commodities. A complex mixture of cellulose, hemicellulose, and lignin are the raw materials of lignocellulose. Total or partial conversion of such complex substrates to yeast fermentable saccharides requires pretreatment to make them capable of enzymatic attack, fractionation into components, and later hydrolysis by a set of several enzymes. Alternatively, lignocellulose hydrolysates can be prepared by strong acid treatment, but this method renders degradation products as hemicellulose-derived acetate and phenolic lignin fragments, which inhibit *S. cerevisiae* growth (117, 118). A detailed description of various pretreatment options, downstream fractionation strategies, and resulting product streams can be found in recent reports (119–123).

Complete enzymatic depolymerization of lignocellulosic materials requires the use of several multicomponent systems, according to the composition and structure of each particular polymer (123, 124). None of these enzymatic complexes is present in *S. cerevisiae*, so far. Consequently, exogenous enzymes must be added to provide a specific conversion of cellulose wastes to sugars. Although these enzymes are commercially available, their high cost has been a major concern in the transformation of lignocellulose to yeast biomass. This can be solved in part by using mixed microbial cultures of *S. cerevisiae* and cellulolytic microorganisms (125, 126), but this strategy is only adequate for ethanol production. Yeast strains genetically engineered for cellulose, hemicellulose, or lignin utilization can overcome most of these problems, rendering them more suitable for industrial fermentation (123, 127). *Saccharomyces cerevisiae* strains able to grow on cellulose and hemicellulose have already emerged. The appearance of lignin-utilizing strains will take longer, because of the complexity of this polymer and because the lignin degradation products are not easily metabolized by *S. cerevisiae*.

11.3.4.1 Cellulose

Cellulose has been traditionally considered the most promising lignocellulosic polymer for yeast cell cultivation, because it consists basically of glucose units, the preferred carbon source for *Saccharomyces*, linked by β-1,4-glycosidic bonds (128). The complete enzymatic degradation of native cellulose requires the synergistic action of cellulases that can be grouped in three main classes, (1) endoglucanases, that release mainly cellooligosaccharides; (2) exoglucanases (cellobiohydrolases), which produce cellobiose from the ends of the cellulose chain; and (3) β-glucosidases, also referred as cellobiose, that hydrolyze the cellobiose disaccharide to D-glucose.

Production of cellulolytic enzymes is common in a large variety of fungi and bacteria, including species from *Trichoderma*, *Aspergillus*, *Penicillium*, *Bacillus,* and *Clostridium* (123, 129). In contrast, *S. cerevisiae* only produces β-1,3-specific glucanases (130), which are unable to degrade cellulose chains. Therefore, recombinant strains of *S. cerevisiae* expressing single genes for heterologous cellulases have been constructed and tested for cellulose bioconversion (127). Although these strains exhibited hydrolytic activity and facilitated the utilization of cellulose wastes, it soon became evident that the effective degradation of cellulose by *Saccharomyces* requires the engineering of a multienzymatic cellulase complex.

With this purpose, van Rensburg et al. (131), obtained a cellulolytic *S. cerevisiae* strain coexpressing the *Phanerochaete chrysosporium* cellobiohydrolase (*CBH1*) and the *Butyrivibrio fibrisolvens* endoglucanase (*END1*) genes. The resulting strain was further improved by incorporating in a multicopy plasmid an expression cassette for *BGL1* and *CEL1*, encoding for cellobiase and cellodextrinase from *Endomyces fibuliger* and *Ruminococcus flavefacies*, respectively (132). The engineered strain was able to degrade efficiently carboxymethylcellulose, laminarin, and barley glucan, and to assimilate cellobiose, but data concerning growth on cellulosic polymers were not provided.

In a similar approach, Cho et al. (133) developed mitotically stable recombinant strains of *S. cerevisiae* producing secretable endo- or exoglucanase and β-glucosidase from *Bacillus*. Expression levels of cellulolytic enzymes were thus significantly increased in comparison with plasmid-based expression systems. Moreover, the resulting transformants showed a significant cell growth on cellulodextrins (133).

Following a different strategy, cell surface engineered yeast strains expressing cell wall anchored β-glucosidase and endoglucanase from *Aspergillus aculeatus* or *Trichoderma reesei* have been successfully constructed (134, 135). This approach tries to mimic the so-called cellulosome structure, a cell surface bound enzymatic complex found in *Clostridium thermocellum* and other species, which promote adhesion to insoluble substrates, degrading efficiently crystalline cellulose (136). Thus, cells of *Saccharomyces* codisplaying cellulolytic enzymes on the cell surface were endowed with the ability to degrade and rapidly assimilate water-soluble cell ooligosaccharides (134) and barley β-glucans (135). Considering the advantages of the cellulosome complex, this system appears to open the possibilities to construct baker's yeast strains able to utilize properly cellulose wastes.

11.3.4.2 Hemicellulose

The hemicellulose fraction of lignocellulosic materials is a complex carbohydrate heteropolymer. Its principal component is xylan, formed by a backbone of β-1,4 linked xylose residues, with L-arabinose and 4-O-methyl-glucuronic acid side chains. Complete degradation of xylan requires the activity of an endo-1,4-β xylanase and a 1,4-β-xylosidase, whereby the monomer xylose is released. In addition, some accessory enzymes breaking down side linkages as α-L-arabinofuranosidases, can act synergistically enhancing the rate and extent of hydrolysis (137). Several genes encoding xylanases and α-L-arabinofuranosidases have been isolated and cloned in *Saccharomyces cerevisiae* laboratory strains with the aim of producing xylan-degrading yeasts (138, 139). Recombinant yeast strains exhibiting both α-L-arabinofuranosidase and β-xylosidase activities have also been constructed (140). Furthermore, the *Aspergillus nidulans* endoxylanase *xlnA*, *xlnB,* and *xlnC* genes, and the *Aspergillus niger* α-L-arabinofuranosidase (*abfB*) gene have been successfully expressed in an industrial baker's yeast strain (141). More recently, La Grange and coworkers (142) have described the construction of a recombinant *S. cerevisiae* strain coexpressing the *A. niger* β-xylosidase (*xlnD*) and the *T. reesei* xylanase II (*xyn2*) genes. Coproduction of these two key enzymes enabled *Saccharomyces* to produce efficiently D-xylose (20 g/liter) from birchwood xylan (50 g/liter), with xylobiose as a minor product.

Despite these advances, the major drawback to converting xylan to cell biomass or other useful products is the inability of *Saccharomyces* to ferment xylose. Numerous efforts have been directed over the last decade toward engineering of *S. cerevisiae* for xylose utilization. In naturally xylose-fermenting yeasts, such as *Pichia stipitis* and *Pachysolen tannophilus*, xylose metabolism begins with its conversion to xylulose in a two step reaction involving xylose reductase, XR, and xylitol dehydrogenase, XDH, encoded by *XYL1* and *XYL2*, respectively. Subsequently, xylulose is phosphorylated by xylulokinase, being further catabolized via the pentose phosphate pathway (Figure 11.5). *S. cerevisiae* displays xylulokinase activity, but lacks the enzymes for xylitol and xylulose formation. For this reason, expression of *P. stipitis XYL1* in *S. cerevisiae* enabled the production of xylitol from xylose, but no growth on this carbon source was observed (143, 144). A significant improvement was obtained by coexpressing *XYL1* and *XYL2* genes from *P. stipitis* (145–147). Indeed, these strains were able to utilize xylose, but the rate of assimilation was poor and took place only in the presence of glucose and under oxygenation conditions. This was attributed to a poor xylose uptake, inefficient xylose to xylulose conversion, starting with an accumulation of xylitol, and the inability of the recombinant strains to route carbon flux trough the pentose phosphate (PP) pathway (145).

Xylose is presumably transported by the same system as glucose, but with a 200-fold lower affinity (148). Consequently, xylose uptake was suggested early as a factor limiting xylose flux in recombinant cells. This appeared to be confirmed by measuring xylose uptake rate in cells of the xylose-utilizing strain TMB3001, cultivated with different xylose–glucose ratios (149). Currently, it is not clear that xylose uptake could have a major impact on xylose utilization. In fact, overexpression of the high affinity glucose transporters of *S. cerevisiae* in the same strain did not result in faster growth on xylose (150).

Xylitol production results in a cofactor imbalance, because xylose reductase uses both NADPH and NADH, while xylitol dehydrogenase utilizes only NAD^+ (Figure 11.5). This was clearly demonstrated by Lidén et al. (151), using a *XYL1*-expressing *S. cerevisiae* mutant unable to produce glycerol. Regenerating NAD^+ could be provided by engineering internal redox sinks, like glycerol (Figure11.5) and succinate synthesis. However, these reactions proceed at the expense of ATP. The redox imbalance could also be overcome by decreasing the XR:XDH activity ratio. Indeed, Walfridsson et al. (152) demonstrated that a strain with much higher specific activity of XDH than of XR metabolized xylose without xylitol formation, but only a small part of the xylose available was consumed. In a different approach, several research groups tried to resolve the redox problem by introducing, instead of the *Pichia* system, bacterial genes for xylose isomerase (XY), an enzyme that transforms xylose into xylulose without a redox cofactor requirement (153). In most of the cases this strategy failed due to the lack of activity of the heterologous protein. Thus, so far only the enzyme from *Thermus thermophilus* has been actively produced in *S. cerevisiae* (154). The recombinant strain displayed ethanol formation from xylose, but no growth was observed. Moreover, xylitol was again the main byproduct. This was ascribed to the presence in *Saccharomyces* of endogenous unspecific aldose reductases. However, deletion of *GRE3*, which encodes the yeast aldose reductase (155), did not improve the xylose metabolism in XY-producing strains (156). Recently, Jeppsson et al. (157) investigated the effects of a flux reduction through the NADPH-producing branch of the PP pathway in a xylose-utilizing strain. Lowering the oxidative PP pathway activity would result in a reduced intracellular level of NADPH, increasing the NADH-dependent reduction of xylose. Although a connection between NADPH levels and xylitol formation was demonstrated, the polyol was still produced. In addition, the recombinant strain showed an 84% lower xylose consumption rate (157).

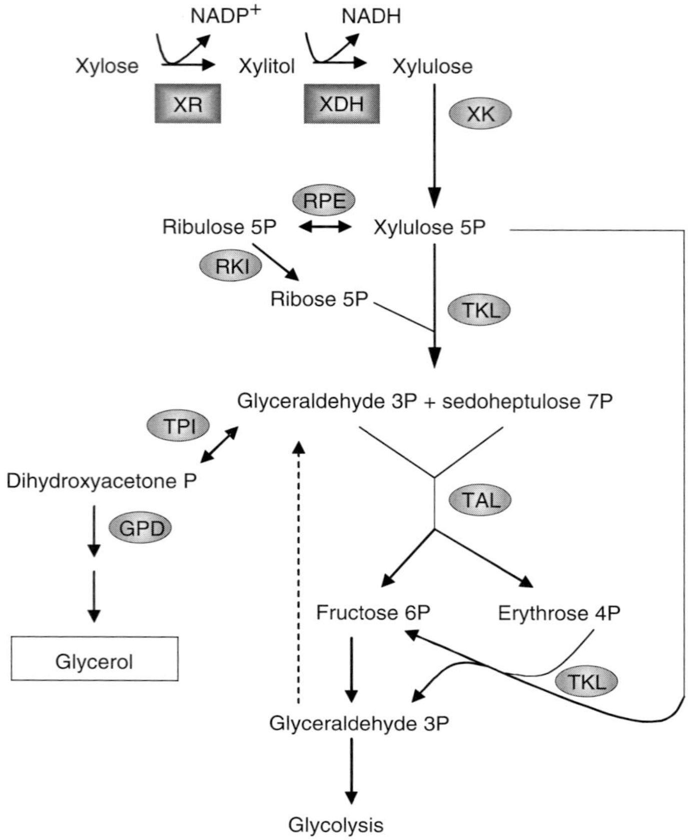

Figure 11.5 Schematic representation of the xylose metabolism through the non-oxidative branch of the pentose phosphate pathway. The main enzymes involved in the pathway, xylulokinase (XK), ribulose-5-phosphate epimerase (RPE), ribose-5-phosphate ketol-isomerase (RKI), transketolase (TKL) and transaldolase (TAL), and related pathways, triosephosphate isomerase (TPI) and glycerol-3-phosphate dehydrogenase (GPD) are shown. The boxes indicate enzymes, xylose reductase (XR) and xylitol dehydrogenase (XDH) that are not present in *S. cerevisiae.*

The capacity of the nonoxidative PP pathway (Figure 11.5) has been traditionally considered as limiting the xylulose consumption in *S. cerevisiae* (158). To investigate this, Walfridsson et al. (159) constructed *XYL1*- and *XYL2*-containing *S. cerevisiae* hosts that overexpress the genes *TKL1* and *TAL1*, encoding transketolase and transaldolase, respectively. These activities catalyze the conversion of two important intermediates of the nonoxidative PP pathway, xylulose-5-phosphate and sedoheptulose-7-phosphate. The study showed that the Tkl1p activity did not control the flux through the PP pathway. Overproduction of Tal1p did enhance growth on xylose, compared with the *XYL1/XYL2* control strain, thus improving biomass yield.

The step catalyzed by xylulokinase (*XKS1*), the third enzyme of the xylose pathway (Figure 11.5), has received special attention. Fermentation studies have been conducted in *S. cerevisiae* strains overexpressing the homologous *XKS1* gene along with the *P. stipitis* *XYL1* and *XYL2* genes (149, 160–163). The results indicated that expression in a high copy number of *XKS1* enhances the specific rate of xylose utilization. In addition, xylose conversion was achieved on xylose as the sole carbon source (161). A significant fraction of the utilized xylose was still excreted as xylitol with biomass as a minor product (160, 163). Moreover, the overexpression of *XKS1* changed the amount and ratios of xylulose-5-phosphate and sedoheptulose-7-phosphate. Thus, both xylulokinase and enzymes downstream of xylulose-5-phosphate appear to limit the flux through the PP pathway in natural *S. cerevisiae* strains. These results demonstrate that the construction by genetic engineering of efficient xylose-growing recombinant yeasts should take into consideration targets in different branches of the metabolic network.

11.4 HETEROLOGOUS ENZYME PRODUCTION

The use of enzymes in the baking industry is widely extended. They modify the flour components, mainly starch, nonstarch polysaccharides, proteins, and lipids, and as a result, the physicochemical properties of the dough and the bread quality parameters (22).

Regular enzyme application in the dough system is carried out by exogenous addition of commercial preparations. Although substantial amounts of these compounds are obtained from recombinant organisms, mainly fungi and bacteria, all of them contain a cocktail of minor enzymatic activities that may negatively affect bread quality. Moreover, these products may act as allergens, producing a high prevalence of occupational hypersensitivity, in terms of dermal and bronchial (asthma) allergies (164–166). Alternatively, recombinant baker's yeast strains producing a technological enzyme can be employed to obtain the same benefits, avoiding airborne allergen pollution at the workplace (167).

11.4.1 Amylases

The amount of fermentable sugars present in flour is only around 1–2% (168). This is clearly insufficient for the optimum growth and gas production of yeast. The fermentation depends thus on the proper production of maltose from damaged starch by the action of wheat β- and α-amylases (169). Regular nongerminated wheat flour contains good levels of β-amylase but little amount of α-amylases. Because of this, wheat flour is normally supplemented with exogenous α-amylase (170).

The reducing sugars produced by the action of α-amylases not only supply the yeast with the substrate for the proper production of gas, but they serve to improve the aroma and color of bread. Furthermore, the addition of α-amylase enhances bread volume, which results in a softer crumb (169,171).

Baker's yeast strains expressing the α-amylase from *Aspergillus oryzae* have been constructed (172). The α-amylase cDNA was placed under the control of the yeast *ACT1* promoter and the resulting transformant produced a recombinant α-amylase with biochemical and kinetic characteristics similar to the native enzyme (173). Dough produced with this transformant gave bread with greater volume, a softer crumb, and a reduced firming rate (172). In addition, the authors showed that a fraction of the enzyme was somehow protected by the yeast structure, so it could remain active for a little bit longer in the oven, as an encapsulated enzyme, resembling then the action of α-amylases of intermediate thermal stability. Therefore, the action of a particular enzyme on its substrate may be modified just by being expressed by yeast.

11.4.2 Xylanases

Wheat flour contains up to 3% nonstarch polysaccharides (pentosanes or hemicelluloses), which play an important role in water absorption and dough rheology. Enzymatic modification of these polysaccharides improves both dough properties and bread quality, increasing volume and retarding the process of "staling" during storage (22, 169).

Using the same strategy as previously described, Monfort et al. (141) expressed different xylanolytic enzymes in industrial baker's yeasts. When transformants producing the *A. nidulans* endoxylanase X24 (*xlnB*) were used in regular straight dough, only a discrete increase in bread volume (5%) was observed. However, when this enzyme and the *A. oryzae* α-amylase were coproduced simultaneously (174), a synergistic effect was observed. The bread elaborated with the new transformant showed a larger volume increase (around 30%) and a softer crumb texture than those prepared with either the *xlnB* overproducing transformant or a control strain. Thus, the combination of xylanase and fungal amylase provided a better overall score on dough and bread quality parameters for the combined exogenous addition of these enzymes (175).

11.4.3 Lipases

Lipases (glycerol ester hydrolases EC 3.1.1.3) catalyze the hydrolysis of triacylglycerols to render mono- and diacylglycerides and free fatty acids. The addition of lipases to the dough improves its rheological properties and stability, resulting in breads with larger volume, softer crumb, and longer shelf life (176). These effects depend on the type of flour and the baking formula, being significant in dough systems without added shortening (175). It has been postulated that the lipase activity would form a kind of *in situ* emulsion by producing mono and diglycerides into the dough (176). However, the total amount of lipids in sound wheat flours is low. Besides, the hydrolysis degree by the lipase action in dough systems is limited. Hence, the amount of mono- and diglycerides produced appears not to be sufficient to explain the effect of the lipase properties on breadmaking (175).

Although the expression of lipases of different origins in laboratory strains of *Saccharomyces cerevisiae* is well documented (177–179), the construction of industrial baker's yeast expressing lipases is now emerging. Monfort et al. (180) reported obtaining of lipolytic baker's yeast able to secrete active lipase by transformation with plasmids containing the *LIP2* gene from *Geotrichum* sp. Recombinant lipase-2 protein exhibited biochemical properties similar to those of the natural enzyme (178). Fermented dough prepared with the recombinant strain rendered bread with higher loaf volume and a more uniform crumb structure than that prepared with control yeast. These effects were strengthened by the addition in the bread dough formulas of a preferment enriched in recombinant lipase 2 (180). Following the same approach, Sánchez et al. (181), described the expression of the *lipA* gene from *Bacillus subtillis* in baker's yeast. Unfortunately, the recombinant strain was unable to target the secretion of the heterologous enzyme into the culture media, even when a signal peptide was fused in frame to *lip*A.

11.4.4 Phytases

The low absorption of iron and zinc from cereal based meals has been ascribed to the high content of phytate (myo-inositol hexaphosphate), which forms insoluble complexes with these minerals at physiological pH values (182). Wheat grain contains endogenous 6-phytase (183), a phosphomonoesterase capable of hydrolyzing phytate to free inorganic phosphate and inositol via mono- to pentaphosphates. This phytase is activated during fermentation but not to such an extent that mineral bioavailability is greatly improved. It has been described as a 3-phytase activity characteristic of microorganisms (184), including baker's yeast (185). To reduce the phytate content of the whole wheat bread to a level considered not to affect mineral absorption, addition of phytase has been proposed (186). Furthermore, fungal phytase increases the specific bread volume and improves crumb texture and bread density (187). This improvement in bread making performance might be associated with the activation of endogenous α-amylase, due to the release of calcium ions from calcium–phytate complexes promoted by phytase activity.

As an alternative to the exogenous addition, phytase (*phyA*) from the fungus *Aspergillus niger* has been biotechnologically expressed in various hosts (188), and a *S. cerevisiae* strain producing an active extracellular phytase is available (189). The baker's yeast secreted phytase was found to be more thermostable than the commercial one, probably due to its high level of glycosylation. According to this, future improvements in commercial baking could be expected from the use of phytase-producing yeasts.

11.4.5 Other Enzymes

Enzyme technology is a major research topic in bread making. The utilization of new enzymes or applications is attracting increasing interest with the possibilities of generating improved products and processes or replacement of chemical substances (emulsifiers and oxidizing agents). In this way, transglutaminases, glucose oxidases, and laccases are a group of interesting enzymes for the baking industry (171, 190–192), because of their ability to oxidize or modify gluten proteins improving dough strength. Transglutaminase is also able to replace emulsifiers (191). This enzyme catalyzes an acyl transfer reaction between lysine and glutamine, stabilizing the gluten structure.

There is a growing interest in the use of some traditional enzymes, like proteases and lipoxygenases (170), because of the possibility of novel applications. For example, the controlled use of proteases with high specificity could increase the level of amino acids and peptides with properties as potential oxidants, taste enhancers, sweeteners, or bitter agents (193). In this scenario, the construction of recombinant strains for new or improved enzymes has started to render promising results. Thus, different pea seed lipoxygenases genes have been characterized and expressed in *S. cerevisiae* (194). Several examples of *S. cerevisiae* producing transglutaminases from different sources have been reported (195). Laccase-encoding cDNA has been isolated from the fungus *Trametes versicolor* and expressed in wild-type *S. cerevisiae* (196) or in recombinant strains overexpressing the protein Sso2p (117), a membrane protein involved in the secretion machinery. The glucose oxidase from *Aspergillus niger* has been also produced in *S. cerevisiae* strains (197).

11.5 OVERPRODUCTION OF ESSENTIAL NUTRIENTS

Human diets are often deficient in amino acids, vitamins, and other substances with high nutritional value. This is especially evident in certain geographic areas and for certain sectors of the population (198). On the other hand, today's lifestyle in developed countries, consumption patterns are evolving dramatically, making it more difficult to maintain a proper diet. In this scenario, bread is considered a good vector to convey effectively the kind of health benefits consumers look for, because it is consumed daily. Thus, several companies are now offering a variety of breads with fiber, ω-3, calcium, vitamins, or prebiotics. Nutritional complements are usually isolated or synthesized by chemical methods, or obtained from high producer organisms, and then purified and employed as additives. As an alternative, metabolic engineering offers us the possibility to convert baker's yeast into a functional ingredient of bread and baked products. Several examples of rerouting metabolic pathways of *S. cerevisiae* for the production of specific compounds are described in the literature (199, 200).

11.5.1 Amino Acids

Cereal flours are deficient in amino acids like threonine, lysine, and triptophan (201). Baker's yeast is also poor in these amino acids (202). In order to palliate this problem, several attempts have been made to improve the amino acid content of the yeast. In general, amino acid biosynthetic routes are regulated by derepression of the pathway coding genes in the absence of the corresponding amino acid, and by feedback enzyme inhibition and gene repression. Amino acid overproducing strains have been isolated by the use of amino acid homologous. These molecules mimic the feedback inhibitory mechanisms of the pathway, and as a consequence, low or no amino acid is produced, causing growth arrest. By means of this method, mutants overproducing threonine, methionine and lysine have been selected from industrial yeast strains (203, 204). Dough fermented with these baker's yeasts produced bakery products with significant higher contents of essential amino acids Lys, Met, and Thr (205). In addition, their use also provided bread with enhanced taste and aroma.

More focused attempts have been made to characterize the threonine-overproducing mutations, and to engineer biosynthetic pathways for further improvements in the threonine content of yeast cells. In *Saccharomyces cerevisiae*, threonine is synthesized from aspartate, through a five step metabolic route (206) (Figure 11.6). The third intermediate of the pathway, homoserine, also contributes to the methionine synthesis pathway. Threonine is the precursor for the production of other amino acids like isoleucine and valine. Threonine production is regulated both at the genetic and enzymatic levels, being the main regulatory point the first enzyme of the pathway, aspartate kinase (Figure 11.6). This enzyme, encoded by the *HOM3* gene, converts aspartate into β-aspartyl-phosphate, and is strongly inhibited by threonine.

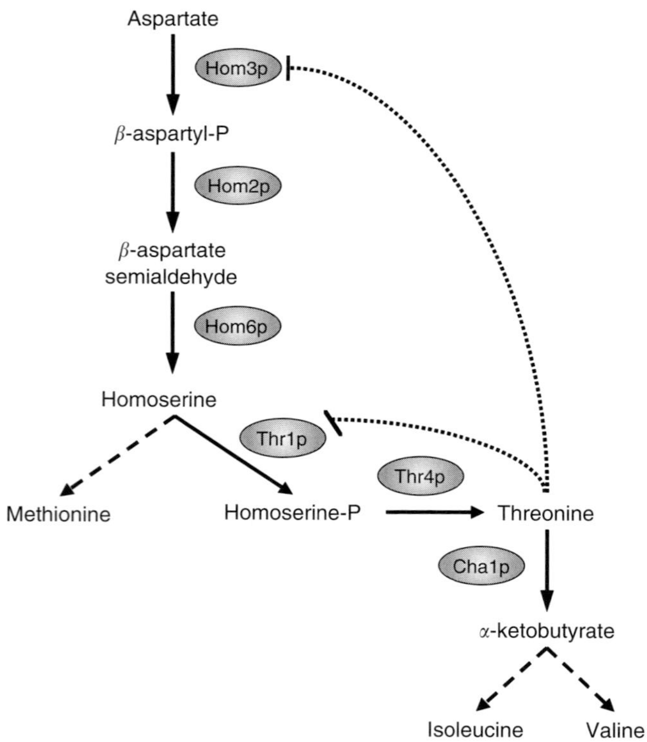

Figure 11.6 Scheme of the threonine biosynthetic pathway. Full arrows indicate direct metabolic steps. Dashed arrows indicate indirect metabolic steps. Dotted, blunt arrows indicate feedback inhibition. Hom3p: aspartate kinase; Hom2p: aspartate-semialdehyde dehydrogenase; Hom6p: homoserine dehydrogenase; Thr1p: homoserine kinase; Thr4p: threonine syntase; Cha1p: threonine deaminase.

The intermediate enzyme homoserine kinase, encoded by *THR1*, also shows threonine inhibition. Thr1p transforms homoserine into homoserine-phosphate, directing homoserine to the synthesis of threonine instead of methionine. Several *HOM3* allelic mutants, impaired in threonine inhibition, have been isolated in laboratory strains (207, 208). All of them overproduce threonine and isoleucine, but not methionine (208), suggesting that only the threonine branch of the pathway is enhanced in these mutants. However, threonine-overproducing yeasts show a growth defect, reaching lower densities that the parental strain (209). This is probably due to interferences between the threonine biosynthetic enzymes and other metabolic routes. To avoid this problem, one of the unrepressible *HOM3* alleles, called *HOM3-R2*, has been expressed under different inducible promoters (209). This approach enabled the authors to induce threonine overproduction using different stimuli. Especially interesting results were obtained with the *GPH1* promoter. This gene, encoding for glycogen phosphorylase, is induced late in the exponential growth phase. Thus, when controlled by *pGPH1*, *HOM3-R2* overexpression only induces threonine production once the exponential growth is over (209), avoiding problems of biomass yield.

Overexpression of other genes involved in threonine synthesis showed good results when combined with the *HOM3-R2* allele, but not by themselves (202). Mutants lacking *CHA1*, which are unable to redirect threonine to the synthesis of isoleucine and valine (Figure 11.6), also showed enhanced threonine content in combination with the *HOM3-R2* mutation (202,209). Therefore, it seems that the inhibitory effect of threonine on the aspartate kinase activity is the key point in the pathway regulation, and threonine cannot be overproduced beyond a threshold value, which is about 350 nmol/mg dry weight (202).

Lysine overproducing yeasts have also been characterized in laboratory strains (210). These mutants are affected in *LYS80*, a general repressor of the genes for the lysine biosynthesis (*LYS* genes), which is upregulated by lysine. Although *LYS80* was first characterized as a specific *LYS* genes repressor, it is actually a general negative regulator of the RAS-cAMP pathway (211). Other lysine overproducing mutants carry altered alleles of the homologous genes *LYS20* and *LYS21*, encoding for the enzymes in the first step of the lysine route (212). On the other hand, the product of the *LYS14* gene is involved in derepression of the *LYS* genes (213). Although there is not a reported work for the moment, *LYS20*, *LYS21*, and *LYS14* genes would be promising targets for the application of genetic engineering techniques, in order to further improve the lysine yeast production in the future.

11.5.2 Vitamins

Deficiencies in one vitamin or another are usual nutritional problems around the world. For example, the risk of deficiency in vitamin A affects up to 2 billion people (198). Although *Saccharomyces cerevisiae* is unable to synthesize this vitamin, heterologous expression of the genes for its synthesis would open new possibilities to produce bread and other baked products enriched in vitamin A. Yamano et al. (214) reported one interesting advance in this direction. These authors engineered *Saccharomyces* to express the whole biosynthetic pathway of the vitamin A precursor, β-carotene, from the Gram-positive bacteria *Erwinia uredovora*. The four genes of this route were situated in a single vector, each one under the control of a different yeast promoter and terminator. The recombinant yeast produced significant amounts of β-carotene. This advance is also relevant because it was the first time that a whole novel secondary metabolic pathway was heterologously expressed in eukaryotic cells.

Other pursued targets for yeast vitamin overproduction is the L-ascorbic acid (L-AA, vitamin C) pathway. Twenty five percent of the worldwide production of L-AA (80,000 tons per year) is used as a food additive (215). This has led to the idea of biotechnologically exploding the microbial production of vitamin C. *Saccharomyces cerevisiae* does not produce L-AA in natural conditions, but does produce its analogue, D-erythroascorbic acid, D-EAA (216). However, when cells are fed with artificial components, as L-galactonolactone (L-GL) or L-galactose (L-gal), L-AA is accumulated (217, 218). This seems to be due to the versatility of the L-EAA pathway enzymes with nonnatural substrates. L-GL is a byproduct of the sugar industry, and L-Gal is known to be produced in plants. Its biosynthetic enzymes might be transferred to yeast. Although this possibility is a recent one, several research groups are taking steps toward an L-AA overproducing yeast (219).

11.5.3 Production of Novel Lipids

Although the body could synthesize most fats *de novo*, certain fatty acids are required as dietary precursors of the most important functional fatty acids. For instance, linoleic acid is considered essential for growth. Similar to classical vitamins deficiencies, essential fatty acids (EFA) deficiency can be overcome by a small amount in the diet. The two classes of EFA (The ω-6 and ω-3 families) may be present in the diet in different proportions to one another. This unbalancing is thought to be the cause of the increased incidence of many chronic diseases (220). Because of this, improving the ω-3 content in our diet should be considered.

Yeast strains that produce novel lipid compounds are currently available. Indeed, *S. cerevisiae* cells harboring *Brassica napus* fatty acid desaturases genes (*FAD2* and *FAD3*), under the control of the fatty acid inducible yeast *PEX11* promoter have been obtained (221). The combined expression yields a reduction in oleic acid content and the appearance of linoleic (ω-6) and linolenic (ω-3) acids (221). In a similar approach, Peyou-Ndi et al. (222) reported the expression of a Δ^{12} desaturase (*FAT2*) from *Caenorhabditis elegans* in *S. cerevisiae*. The recombinant strain produced high levels of polyunsaturated fatty acids. In addition, physiological studies showed a positive correlation between enhanced production of 16:2 and 18:2 acids and membrane fluidity, and tolerance to ethanol and hydrogen peroxide. Furthermore, baker's yeast strains which accumulate high levels of palmitoleic and linoleic acid, have been recently constructed in our laboratory by heterologous expression of the *FAD2-2* and *FAD2-3* genes from *Heliantum* (Rodriguez-Vargas et al., unpublished results).

11.5.4 Prebiotics

Prebiotic oligosaccharides stimulate the growth and colonization of probiotic bacteria; this is, nonpathogenic microorganisms that when ingested are beneficial to health. Consequently, there is currently a great deal of interest in the use of prebiotic oligosaccharides as functional food ingredients (223, 224). There are many known prebiotic oligosaccharides (225), but the market leaders are fructooligosaccharides and inulin. For instance, inulin-containing bread has been recently introduced in the German market (226).

Two different types of enzymes can catalyze the enzymatic synthesis of oligosaccharides: glycosidases (EC 3.2) and glycosyltransferases (EC 2.4). The first group, also denominated glycosynthases, is an engineered retaining glycosidase that can efficiently synthesize oligosaccharides (its substrate), but is unable to hydrolyze them. Currently cloned reported glycosynthases are β-glucosidase from *Agrobacterium* sp. (227), β-glucanase from *Bacillus licheniformis* (228), cellulase (229), and β-glycosidase from *Sulfolobus solfataricus* (230). New glycosynthases may be obtained by identifying and mutating promising candidate enzymes, or by direct evolution of existing glycosidases (231). Baker's yeast expressing these modified enzymes would be an exciting challenge even for the design of anticancer drugs (232).

Fructosyltransferases have been isolated and cloned from various bacterial, fungal, and plant species. Concretely, the inulin production by *S. cerevisiae* transformants expressing the *SFT* gene from *Aspergillus sydowi* (233) is known, as well as the production of 1-kestose by recombinant yeast carrying the enzyme from *Aspergillus foetidus* (234).

Other fructosyltransferases from *Bacillus subtillis* or *Gluconacetobacter diazotrophicus* have also been cloned and expressed in yeast (235, 236). Production of 1-kestose, nystose, and levan by these recombinant yeasts was achieved. Fructosyltransferases (*1-SST* and *1-FFT* encoding respectively sucrose:sucrose 1-fructosyltransferase and fructan:fructan 1-fructosyltransferase) from several *Asteraceae*, like *Cynara scolymus* (237) or *Helianthus tuberosus* (238) could be also expressed in yeast in order to produce long chain inulin molecules. Thus, a new promising perspective in the engineering of baker's yeast would be the direct production of these healthy oligosaccharides.

11.6 PERSPECTIVES

In the past few years, recombinant DNA technology has led to the appearance of new baker's yeast strains with optimized or novel properties, and it is expected that this tool will produce in the near future a huge spectrum of specialized yeasts of high added value. Their introduction in the manufacturing market will produce a dramatic change in formulation, ingredients, or processing conditions currently used in the baking practice, and will provide new end products with enhanced flavor, textures, and shelf life. Right now, it is possible to construct tailor-made baker's yeast strains able to secrete enzyme cocktails that could satisfy almost any baking requirement. As the potential of recombinant gene expression is better understood, the expression of recombinant enzymes could be targeted further to achieve technological benefits in specific baking steps. This would offer a clear advantage over the traditional use of exogenous enzyme improvers. This aim will be encouraged by the introduction of functional genomics approaches in reengineering baker's yeast, leading to the identification of phase regulated gene promoters.

In the future, the major biotechnological challenge in this area will be the significant improvement of fermentative traits, osmotolerance, and cryoresistance of baker's yeast. There is strong evidence that these commercially relevant traits are under the control of genes and cellular functions (21, 167). Again, genome-wide regulation analysis under industrial conditions is the obvious tool to understand how multiple genes are involved in yeast to accomplish metabolic processes determining the performance of baker's yeast. The knowledge, at the molecular level, of metabolic networks could help us to choose transcriptional regulators as targets to control a specific trait. But, we must bear in mind, as recently described by Lee and his coworkers (239), that eukaryotic cellular functions are highly connected through networks of transcriptional regulators that regulate other transcriptional regulators. Keeping a realistic perspective, improvement of these commercial traits will be a difficult task.

Yeast biotechnology will provide benefits not only to yeast producers and bakers, but also to consumers and the public, offering safe and environmentally sound procedures, and healthy and convenient products. Thus, the utilization for yeast biomass production of new raw materials, such as starch, cellulose wastes, or cheese whey, will result in cheaper, cleaner, and more reliable procedures than those traditionally used for baker's yeast propagation. In addition, this technology offers a profitable use for these resources, some of which have a great pollution potential. The use of enzymes as baking additives gives rise to immunoglobulin mediated sensitization and work related asthma, which causes occupational absenteeism in at least 4–25% of the exposed workers. Although this problem could be reduced by the use of encapsulated enzymes or liquid enzyme preparations, the high cost of these particular additives has made their introduction in the baking market difficult. Thus, the use of engineered strains would provide a cost competitive alternative to avoid hypersensitivity to working area allergens.

Nutraceuticals are the fastest-growing segment of today's food industry. Indeed, the current lifestyle generates in the consumer an increased interest in the role of diet in health and health promotion. In addition, the diet in some countries is deficient in several vitamins or essential compounds. Bread, as a diary product, is one of the best candidates to become a functional food. For these reasons, baker's yeast producing vitamin precursors or essential fatty acids would be welcome. In the near future, baker's yeast strains producing some prebiotics will be also a reality. As the knowledge of the gut microbiota increases, it will probably become increasingly important to match the modulatory effects of prebiotics to the microbiota present at specific population. These approaches will clearly increase the added value of baker's yeast, contributing to the profitability of the baking industry and improving quality of life and welfare of consumers.

Despite remarkable progress during the last years and the possibilities that exist today, no genetically modified baker's yeast is on the market. Public acceptance considerations remain the major obstacle to the commercialization of genetically modified industrial yeast strains. Benefits of GMOs are in most cases not perceptible to end consumers, being only a saving of cost or time to yeast producers and bakers. Public awareness of the potential benefits derived from the use of GMOs will increase if high quality, low cost, and new nutritional, hygienic, or environmental advantages are offered. Clear information is required before a product is introduced in the market. Nevertheless, we believe that is only a matter of time before the public is convinced that the benefits outweigh the potential risks.

REFERENCES

1. Evans, I.H. Yeast strains for baking: recent developments. In: *Yeast Technology*, J.F.T. Spencer, D.M. Spencer, eds., Heidelberg: Springer-Verlag, 1990, pp 13–54.
2. Codón, A.C., T. Benitez, M. Korhola. Chromosomal reorganization during meiosis of *Saccharomyces cerevisiae* baker's yeasts. *Curr. Genet.* 32:247–259, 1997.
3. Paquin, C., J. Adams. Frequency of fixation of adaptive mutations is higher in evolving diploid than haploid yeast populations. *Nature* 302:495–500, 1983.
4. Wright, B.E. The effect of the stringent response on mutation rates in *Escherichia coli* K–12. *Mol. Microbiol.* 19:213–219, 1996.
5. Ferea, T.L., D. Botstein, P.O. Brown, R.F. Rosenzweig. Systematic changes in gene expression patterns following adaptive evolution in yeast. *Proc. Natl. Acad. Sci. USA* 96:9721–9726, 1999.
6. Naumov, G.I., E.S. Naumova, E.D. Sancho, M. Korhola. Polymeric *SUC* genes in natural populations of *Saccharomyces cerevisiae*. *FEMS Microbiol. Lett.* 135:31–35, 1996.
7. Codón, A.C., T. Benitez, M. Korhola. Chromosomal polymorphism and adaptation to specific industrial environments of *Saccharomyces* strains. *Appl. Microbiol. Biotechnol.* 49:154–163, 1998.
8. Trivedi, N.B., G.K. Jacbson, W. Tesch. Baker's yeast. *CRC Crit. Rev. Biotechnol.* 24:75–109, 1986.
9. Brown, C.J., K.M. Todd, R.F. Rosenzweig. Multiple duplications of yeast hexose transport genes in response to selection in a glucose-limited environment. *Mol. Biol. Evol.* 15:931–942, 1998.
10. Aguilera, A., H.L. Klein. Chromosome aberrations in simpler eukaryotes. In: *The Causes and Consequences of Chromosomal Aberrations,* Kirsch, I.R., ed., Florida: CRC Press, 1993, pp 51–90.
11. Carlson, M., J.L. Celenza, F.J. Eng. Evolution of the dispersed *SUC* gene family of *Saccharomyces* by rearrangements of chromosome telomeres. *Mol. Cell. Biol.* 5:2894–2902, 1985.

12. van Hoek, P., J.P. van Dijken, J.T. Pronk. Regulation of fermentative capacity and levels of glycolytic enzymes in chemostat cultures of *Saccharomyces cerevisiae. Enzyme Microb. Technol.* 26:724–736, 2000.
13. Higgins, V., M. Braidwood, P. Bell, P. Bissinger, I.W. Dawes, P.V. Attfield. Genetic evidence that high noninduced maltase and maltose permease activities, governed by *MALx3*-encoded transcriptional regulators, determine efficiency of gas production by baker's yeast in unsugared dough. *Appl. Environ. Microbiol.* 65:680–685, 1999.
14. Higgins, V.J., P.J.L. Bell, I.W. Dawes, P.V. Attfield. Generation of a novel *Saccharomyces cerevisiae* strain that exhibits strong maltose utilization and hyperosmotic resistance using nonrecombinant techniques. *Appl. Environ. Microbiol.* 67:4346–4348, 2001.
15. Wolfe, K.H., D.C. Shields. Molecular evidence for an ancient duplication of the entire yeast genome. *Nature* 387:708–713, 1997.
16. Friedman, R., A.L. Hughes. Gene duplication and the structure of eukaryotic genomes. *Genome Res.* 11:373–381, 2001.
17. Adams J., S. Puskas-Rozsa, J. Simlar, C.M. Wilke. Adaptation and major chromosomal changes in populations of *Saccharomyces cerevisiae. Curr. Genet.* 22:13–19, 1992.
18. Bell, P.J.L., V.J. Higgins, P.V. Attfield. Comparison of fermentative capacities of industrial baking and wild-type yeasts of the species *Saccharomyces cerevisiae* in different sugar media. *Lett. Appl. Microbiol.* 32:224–229, 2001.
19. Estruch, F. Stress-controlled transcription factors, stress-induced genes and stress tolerance in budding yeast. *FEMS Microbiol. Rev.* 24:469–486, 2000.
20. Hohmann, S. Osmotic stress signalling and osmoadaptation in yeasts. *Microbiol. Mol. Biol. Rev.* 66:300–372, 2002.
21. Attfield, P.V. Stress tolerance: the key to effective strains of industrial baker's yeast. *Nature Biotechnol.* 15:1351–1357, 1997.
22. Linko, Y.-Y., P. Javanainen, S. Linko. Biotechnology of breadbaking. *Trends Food Sci. Technol.* 8:339–344, 1997.
23. Romanos, M.A., C.A. Scorer, J.J. Clare. Foreign gene expression in yeast: a review. *Yeast* 8:423–488, 1992.
24. Szostac, J.W., R. Wu. Insertion of a genetic marker into the ribosomal DNA of yeast. *Plasmid* 2:536–554, 1979.
25. Lopes, T.S., J. Klootwijk, A.E. Veenstra, P.C. van der Aar, H. van Heerikhuizen, H.A. Raue, R.J. Planta. High-copy-number integration into the ribosomal DNA of *Saccharomyces cerevisiae*: a new vector for high level expression. *Gene* 79:199–206, 1989.
26. Lopes, T.S., I.J. de Wijs, S.I. Steenhauer, J. Verbakel, R.J. Planta. Factors affecting the mitotic stability of high-copy-number integration into the ribosomal DNA of *Saccharomyces cerevisiae. Yeast* 12:467–477, 1996.
27. Adam, A.C., J.A. Prieto, M. Rubio-Texeira, J. Polaina. Construction of a lactose-assimilating strain of baker's yeast. *Yeast* 15:1299–1305, 1999.
28. Nieto, A., J.A. Prieto, P. Sanz. Stable high-copy-number integration of *Aspergillus oryzae* α-amylase cDNA in an industrial baker's yeast strain. *Biotechnol. Prog.* 15:459–466, 1999.
29. Sakai, A., Y. Shimizu, F. Hishinuma. Integration of heterologous genes into the chromosome of *Saccharomyces cerevisiae* using a delta sequence of yeast retrotransposon Ty. *Appl. Microbiol. Biotechnol.* 33:302–306, 1990.
30. Lee, FW., N.A. da Silva. Improved efficiency and stability of multiple cloned gene insertions at the delta sequences of *Saccharomyces cerevisiae. Appl. Microbiol. Biotechnol.* 48:339–345, 1997.
31. Saito, S., Y. Mieno, T. Nagashima, C. Kumagal, K. Kitamoto. Breeding of a new type of baker's yeast by δ-integration for overproduction of glucoamylase using a homothallic yeast. *J. Ferm. Bioeng.* 81:98–103, 1996.
32. Hadfield, C., A.M. Cashmore, P.A. Meacock. An efficient chloramphenicol-resistance marker for *Saccharomyces cerevisiae* and *Escherichia coli. Gene* 45:149–158, 1986.
33. del Pozo, L., D. Abarca, M.G. Claros, A. Jiménez. Cycloheximide resistance as a yeast cloning marker. *Curr. Genet.* 19:353–358, 1991.

34. Wach, A., A. Brachat, R. Pohlmann, P. Philippsen. New heterologous modules for classical or PCR-based gene disruptions in *Saccharomyces cerevisiae*. *Yeast* 10:1793–1808, 1994.
35. Goldstein, A.L., J.H. McCusker. Three new dominant drug resistance cassettes for gene disruption in *Saccharomyces cerevisiae*. *Yeast* 15:1541–1553, 1999.
36. Alani, E., L. Cao, N. Kleckner. A method for gene disruption that allows repeated use of *URA3* selection in the construction of multiply disrupted yeast strains. *Genetics* 116:541–545, 1987.
37. Güldener, U., S. Heck, T. Fiedler, J. Beinhauer, J.H. Hegemann. A new efficient gene disruption cassette for repeated use in budding yeast. *Nucl. Acids Res.* 24:2519–2524, 1996.
38. Puig, S., D. Ramón, J.E. Pérez-Ortin. Optimized method to obtain stable food-safe recombinant wine yeast strains. *J. Agric. Food. Chem.* 46:1689–1693, 1998.
39. Güldener, U., J. Heinisch, G.J. Koehler, D. Voss, J.H. Hegemann. A second set of loxP marker cassettes for Cre-mediated multiple gene knockouts in budding yeast. *Nucl. Acids Res.* 30:e23, 2002.
40. Sauer, B. Functional expression of the cre-lox site-specific recombination system in the yeast *Saccharomyces cerevisiae*. *Mol. Cell. Biol.* 7:2087–2096, 1987.
41. Bauer, J., V. Nacken, A. Loiez. Cassettte de transformation de la levure. EP0994192A2. 2000.
42. Estruch, F., J.A. Prieto. Construction of a Trip commercial baker's yeast strain by using food-safe-grade dominant drug resistance cassettes. *FEMS Yeast Res.* 4:329–338, 2003.
43. Senecoff, J.F., M.M. Cox. Directionality in FLP protein-promoted site-specific recombination is mediated by DNA-DNA pairing. *J. Biol. Chem.* 261:7380–7386, 1986.
44. Senecoff, J.F., P.J. Rossmeissl, M.M. Cox. DNA recognition by the FLP recombinase of the yeast 2 μm plasmid. A mutational analysis of the FLP binding site. *J. Mol. Biol.* 201:405–421, 1988.
45. Dixon, J.E., P.D. Sadowski. Resolution of immobile χ structures by the FLP recombinase of 2 μm plasmid. *J. Mol. Biol.* 243:199–207, 1994.
46. Storici, F., C.V. Bruschi. Molecular engineering with the FRT sequence of the yeast 2 μm plasmid: [cir^0] segregant enrichment by counterselection for 2 μm site-specific recombination. *Gene* 195:245–255, 1997.
47. Storici, F., M. Coglievina, C.V. Bruschi. A 2-μm DNA-based marker recycling system for multiple gene disruption in the yeast *Saccharomyces cerevisiae*. *Yeast* 15:271–283, 1999.
48. Hottiger, T., J. Kuhla, G. Pohlig, P. Fürst, A. Spielmann, M. Garn, S. Haemmerli, J. Heim. 2-μm vectors containing the *Saccharomyces cerevisiae* metallothionein gene as a selectable marker: excellent stability in complex media, and high-level expression of a recombinant protein from a *CUP1*-promoter-controlled expression cassette in *cis*. *Yeast* 11:1–14, 1995.
49. Falco, S.C., K.S. Dumas. Genetic analysis of mutants of *Saccharomyces cerevisiae* resistant to the herbicide sulfometuron methyl. *Genetics* 109:21–35, 1985.
50. Casey, G.P., W. Xiao, G.H. Rank. A convenient selection marker for gene transfer in industrial strains of *Saccharomyces cerevisiae*: *SMR1* encoded resistance to the herbicide sulfometuron methyl. *J. Inst. Brew.* 94:93–97, 1988.
51. Xie, Q., A. Jiménez. Molecular cloning of a novel allele of *SMR1* which determines sulfometuron methyl resistance in *Saccharomyces cerevisiae*. *FEMS Microbiol. Lett.* 137:165–168, 1996.
52. Gasent-Ramírez, J.M., A.C. Codón, T. Benítez. Characterization of genetically transformed *Saccharomyces cerevisiae* baker's yeasts able to metabolize melibiose. *Appl. Environ. Microbiol.* 61:2113–2121, 1995.
53. Marín, D., A. Jiménez, M. Fernández-Lobato. Construction of an efficient amylolytic industrial yeast strain containing DNA exclusively derived from yeast. *FEMS Microbiol. Lett.* 201:249–253, 2001.
54. Hashida-Okado, T., A. Ogawa, M. Endo, R. Yasumoto, K. Takesako. *AUR1*, a novel gene conferring aureobasidin resistance on *Saccharomyces cerevisiae*: a study of defective morphologies in Aur1p-depleted cells. *Mol. Gen. Genet.* 251:236–244, 1996.

55. Hashida-Okado, T., A. Ogawa, I. Kato, K. Takesako. Transformation system for prototrophic industrial yeasts using the *AUR1* gene as a dominant selection marker. *FEBS Lett.* 425:117–122, 1998.
56. van den Berg, M.A., H.Y. Steensma. Expression cassettes for formaldehyde and fluoroacetate resistance, two dominant markers in *Saccharomyces cerevisiae*. *Yeast* 13:551–559, 1997.
57. Wehner, E.P., M. Brendel. Vector YFRp1 allows transformant selection in *Saccharomyces cerevisiae* via resistance to formaldehyde. *Yeast* 9:783–785, 1993.
58. Cavalieri, D., E. Casalone, B. Bendoni, G. Fia, M. Polsinelli, C. Barberio. Trifluoroleucine resistance and regulation of α-isopropyl malate synthase in *Saccharomyces cerevisiae*. *Mol. Gen. Genet.* 261:152–160, 1999.
59. Bendoni, B., D. Cavalieri, E. Casalone, M. Polsinelli, C. Barberio. Trifluoroleucine resistance as a dominant molecular marker in transformation of strains of *Saccharomyces cerevisiae* isolated from wine. *FEMS Microbiol. Lett.* 180:229–233, 1999.
60. Breitwieser, W., C. Price, T. Schuster. Identification of a gene encoding a novel zinc-finger protein in *Saccharomyces cerevisiae*. *Yeast* 9:551–556, 1993.
61. Avram, D., M. Leid, A.T. Bakalinsky. Fzf1p of *S. cerevisiae* is a positive regulator of *SSU1* transcription and its first zinc-finger region is required for DNA binding. *Yeast* 15:473–480, 1999.
62. Park, H., N.I. Lopez, A.T. Bakalinsky. Use of sulfite resistance in *Saccharomyces cerevisiae* as a dominant selectable marker. *Curr. Genet.* 36:339–344, 1999.
63. Nourani, A., D. Papajova, A. Delahodde, C. Jacq, J. Subik. Clustered amino acid substitutions in the yeast transcription regulator Pdr3p increase pleiotropic drug resistance and identify a new central regulatory domain. *Mol. Gen. Genet.* 256:397–405, 1997.
64. Lackova, D., J. Subik. Use of mutated *PDR3* gene as a dominant selectable marker in transformation of prototrophic yeast strains. *Folia. Microb.* 44:171–176, 1999.
65. DeRisi, J., B. van den Hazel, P. Marc, E. Balzi, P. Brown, C. Jacq, A. Goffeau. Genome microarray analysis of transcriptional activation in multidrug resistance yeast mutants. *FEBS Lett.* 470:156–160, 2000.
66. Balzi, E., W. Chen, S. Ulaszewski, E. Capieaux, A. Goffeau. The multidrug resistance gene *PDR1* from *Saccharomyces cerevisiae*. *J. Biol. Chem.* 262:16871–16879, 1987.
67. Delaveau, T., A. Delahodde, E. Carvajal, J. Subik, C. Jacq. *PDR3*, a new yeast regulatory gene, is homologous to *PDR1* and controls the multidrug resistance phenomenon. *Mol. Gen. Genet.* 244:501–511, 1994.
68. Simonics, T., Z. Kozovska, D. Michalkova-Papajova, A. Delahodde, C. Jacq, J. Subik. Isolation and molecular characterization of the carboxy-terminal *pdr3* mutants in *Saccharomyces cerevisiae*. *Curr. Genet.* 38:248–255, 2000.
69. Kozovska, Z., A. Maraz, I. Magyar, J. Subik. Multidrug resistance as a dominant molecular marker in transformation of wine yeast. *J. Biotechnol.* 92:27–35, 2001.
70. Hadfield, C., K.K. Raina, K. Shashi-Menon, R.C. Mount. The expression and performance of cloned genes in yeasts. *Mycol. Res.* 97:897–944, 1993.
71. Nacken, V., T. Achstetter, E. Degryse. Probing the limits of expression levels by varying promoter strength and plasmid copy number in *Saccharomyces cerevisiae*. *Gene* 175:253–260, 1996.
72. Igual, J.C., C. González-Bosch, L. Franco, J.E. Pérez-Ortín. The *POT1* gene for yeast peroxisomal thiolase is subject to three different mechanisms of regulation. *Mol. Microbiol.* 6:1867–1875, 1992.
73. Ruohonen, L., M.K. Aalto, S. Keränen. Modifications to the *ADH1* promoter of *Saccharomyces cerevisiae* for efficient production of heterologous proteins. *J. Biotechnol.* 39:193–203, 1995.
74. Puig, S., J.E. Pérez-Ortín. Expression levels and patterns of glycolytic yeast genes during wine fermentation. *Syst. Appl. Microbiol.* 23:300–303, 2000.
75. Monfort, A., S. Finger, P. Sanz, J.A. Prieto. Evaluation of different promoters for the efficient production of heterologous proteins in baker's yeast. *Biotechnol. Lett.* 21:225–229, 1999.

76. Puig, S., A. Querol, D. Ramón, J.E. Pérez-Ortín. Evaluation of the use of phase-specific gene promoters for the expression of enological enzymes in an industrial wine yeast strain. *Biotechnol. Lett.* 18:887–892, 1996.
77. Kuipers, O.P. Genomics for food biotechnology: prospects of the use of high-throughput technologies for the improvement of food microorganisms. *Curr. Opin. Biotechnol.* 10:511–516, 1999.
78. VanBogelen, R.A., K.D. Greis, R.M. Blumenthal, T.H. Tani, R.G. Matthews. Mapping regulatory networks in microbial cells. *Trends Microbiol.* 7:320–328, 1999.
79. DeRisi, J.L., R.I. Vishwanath, P.O. Brown. Exploring the metabolic and genetic control of gene expression on a genomic scale. *Science* 278:680–686, 1997.
80. Rep, M., M. Krantz, J.M. Thevelein, S. Hohmann. The transcriptional response of *Saccharomyces cerevisiae* to osmotic shock. *J. Biol. Chem.* 275:8290–8300, 2000.
81. Gasch, A.P., P.T. Spellman, C.M. Kao, O. Carmel-Harel, M.B. Eisen, G. Storz, D. Botstein, P.O. Brown. Genomic expression programs in the response of yeast cells to environmental changes. *Mol. Biol. Cell.* 11:4241–4257, 2000.
82. Lynd, L.R., C.E. Wyman, T.U. Gerngross. Biocommodity engineering. *Biotechnol. Prog.* 15:777–793, 1999.
83. Turakainen, H., S. Aho, M. Korhola. *MEL* gene polymorphism in the genus *Saccharomyces*. *Appl. Environ. Microbiol.* 59:2622–2630, 1993.
84. Naumov, G.I., E.S. Naumova, M.P. Korhola. Chromosomal polymorphism of *MEL* genes in some populations of *Saccharomyces cerevisiae*. *FEMS Microbiol. Lett.* 127:41–45, 1995a.
85. Naumov, G.I., E.S. Naumova, H. Turakainen, M.P. Korhola. Identification of the alpha-galactosidase *MEL* genes in some populations of *Saccharomyces cerevisiae*: a new gene *MEL11*. *Genet. Res.* 67:101–108, 1996.
86. Naumov, G.I., E.S. Naumova, E.J. Louis. Genetic mapping of the alpha-galactosidase *MEL* gene family on right and left telomers of *Saccharomyces cerevisiae*. *Yeast* 11:481–483, 1995b.
87. Reed, G., T.W. Nagodawithana. *Yeast Technol.* New York: Van Nostrand Reinhold, 1991.
88. Liljeström-Suominen, P.L., V. Joutsjoki, M. Korhola. Construction of a stable α-galactosidase-producing baker's yeast strain. *Appl. Environ. Microbiol.* 54:245–249, 1988.
89. Liljeström, P.L., R.S. Tubb, M. Korhola. Construction of new alpha-galactosidase producing yeast strains and the industrial application of these strains. US Patent 5,055,401, 1991.
90. Vincent, S.F., P.J.L. Bell, P. Bissinger, K.M.H. Nevalainen. Comparison of melibiose utilizing baker's yeast strains produced by genetic engineering and classical breeding. *Lett. Appl. Microbiol.* 28:148–152, 1999.
91. Rønnow, B., L. Olsson, J. Nielsen, J.D. Mikkelsen. Derepression of galactose metabolism in melibiase producing baker's and distillers' yeast. *J. Biotechnol.* 72:213–228, 1999.
92. Gancedo, J.M. Yeast carbon catabolite repression. *Microbiol. Mol. Biol. Rev.* 62:334–361, 1998.
93. Ostergaard, S., C. Roca, B. Rønnow, J. Nielsen, L. Olsson. Physiological studies in aerobic batch cultivations of *Saccharomyces cerevisiae* strains harboring the *MEL1* gene. *Biotechnol. Bioeng.* 68:252–259, 2000.
94. Yang, S.T., E.M. Silva. Novel products and new technologies for use of a familiar carbohydrate, milk lactose. *J. Dairy. Sci.* 78:2541–2562, 1995.
95. Mawson, A.J. Bioconversions for whey utilization and waste abatement. *Biores. Technol.* 47:195–203, 1994.
96. Matzke, S., W.L. Wendorff. Chloride in cheese manufacturing wastes to be landspread on agricultural land. *Biores. Technol.* 46:251–253, 1994.
97. Rubio-Texeira, M., M. Arévalo-Rodríguez, J.L. Lequerica, J. Polaina. Lactose utilization by *Saccharomyces cerevisiae* strains expressing *Kluyveromyces lactis LAC* genes. *J. Biotechnol.* 84:97–106, 2000.
98. Domingues, L., N. Lima, J.A. Teixeira. Alcohol production from cheese whey permeate using genetically modified flocculent yeast cells. *Biotechnol. Bioeng.* 72:507–514, 2001.
99. Porro, D., E. Martegabni, B.M. Ranzi, L. Alberghina. Lactose/whey utilization and ethanol production by transformed *S. cerevisiae* cells. *Biotechnol. Bioeng.* 39:799–805, 1992.

100. Porro, D., E. Martegabni, B.M. Ranzi, L. Alberghina. Development of high density cultures of engineered *Saccharomyces cerevisiae* cells able to grow on lactose. *Biotechnol. Lett.* 14:1085–1088, 1992.
101. Kumar, V., S. Ramakrishnan, T.T. Teeri, J.K.C. Knowles, B.S. Hartley. *Saccharomyces cerevisiae* cells secreting an *Aspergillus niger* β-galactosidase grow on whey permeate. *Biotechnology* 10:82–85, 1992.
102. Ramakrishnan, S., B.S. Hartley. Fermentation of lactose by yeast cells secreting recombinant fungal lactase. *Appl. Environ. Microbiol.* 59:4230–4235, 1993.
103. Sreekrishna, K., R.C. Dickson. Construction of strains of *Saccharomyces cerevisiae* that grow on lactose. *Proc. Natl. Acad. Sci. USA* 82:7909–7913, 1985.
104. Dickson, R.C., K.K. Sreekrishna. LAC+ *Saccharomyces cerevisiae.* US patent 5,047,340, 1991.
105. Rubio-Texeira, M., J.I. Castrillo, A.C. Adam, U.O. Ugalde, J. Polaina. Highly efficient assimilation of lactose by a metabolically engineered strain of *Saccharomyces cerevisiae. Yeast* 14:827–837, 1998.
106. Randez-Gil, F., P. Sanz. Construction of industrial baker's yeast strains able to assimilate maltose under catabolite repression conditions. *Appl. Microbiol. Biotechnol.* 42:581–586, 1994.
107. Nigam, P., D. Singh. Enzyme and microbial systems involved in starch processing. *Enzyme Microb. Technol.* 17:770–778, 1995.
108. Hollenberg, C.P., A.W.M. Strasser. Improvement of baker's and brewer's yeast by gene technology. *Food Biotechnol.* 4:527–534, 1990.
109. Steyn, A.J., I.S. Pretorius. Co-expression of a *Saccharomyces diastaticus* glucoamylase-encoding gene and a *Bacillus amyloliquefaciens* alpha-amylase-encoding gene in *Saccharomyces cerevisiae. Gene* 100:85–93, 1991.
110. Shibuya, I., G. Tamura, H. Shima, T. Ishikawa, S. Hara. Construction of an alpha-amylase/glucoamylase fusion gene and its expression in *Saccharomyces cerevisiae. Biosci. Biotechnol. Biochem.* 56:884–889, 1992.
111. Luo, J., M. He, W. Li, T. Zhang. Expression and secretion of alpha-amylase and glucoamylase in *Saccharomyces cerevisiae. Chin. J. Biotechnol.* 10:241–248, 1994.
112. De Moraes, L.M., S. Astolfi-Filho, S.G. Oliver. Development of yeast strains for the efficient utilisation of starch: evaluation of constructs that express alpha-amylase and glucoamylase separately or as bifunctional fusion proteins. *Appl. Microbiol. Biotechnol.* 43:1067–1076, 1995.
113. Janse, B.J., I.S. Pretorius. One-step enzymatic hydrolysis of starch using a recombinant strain of *Saccharomyces cerevisiae* producing alpha-amylase, glucoamylase and pullulanase. *Appl. Microbiol. Biotechnol.* 42:878–883, 1995.
114. Murai, T., M. Ueda, M. Yamamura, H. Atomi, Y. Shibasaki, N. Kamasawa, M. Osumi, T. Amachi, A. Tanaka. Construction of a starch-utilizing yeast by cell surface engineering. *Appl. Environ. Microbiol.* 63:1362–1366, 1997.
115. Murai, T., M. Ueda, Y. Shibasaki, N. Kamasawa, M. Osumi, T. Imanaka, A. Tanaka. Development of an arming yeast strain for efficient utilization of starch by co-display of sequential amylolytic enzymes on the cell surface. *Appl. Microbiol. Biotechnol.* 51:65–70, 1999.
116. Wheals, A.E., L.C. Basso, D.M.G. Alves, H.V. Amorim. Fuel ethanol after 25 years. *Trends Biotechnol.* 17:482–487, 1999.
117. Larsson, S., E. Palmqvist, B. Hahn-Hägerdal, C. Tenborg, K. Stenberg, G. Zacchi, N.-O. Nilvebrant. The generation of inhibitors during dilute acid hydrolysis of softwood. *Enzyme Microb. Technol.* 24:151–159, 1999.
118. Larsson, S., P. Cassland, L.J. Jönsson. Development of a *Saccharomyces cerevisiae* strain with enhanced resistance to phenolic fermentation inhibitors in lignocellulose hydrolysates by heterologous expression of laccase. *Appl. Environ. Microbiol.* 67:1163–1170, 2001.
119. Chandrakant, P., V.S. Bisaria. Simultaneous bioconversion of cellulose and hemicellulose to ethanol. *Crit. Rev. Biotechnol.* 18:295–331, 1998.
120. Aristidou, A., M. Penttilä. Metabolic engineering applications to renewable resource utilization. *Curr. Opin. Biotechnol.* 11:187–198, 2000.

121. Mielenz, J.R. Ethanol production from biomass: technology and commercialization status. *Curr. Opin. Microbiol.* 4:324–329, 2001.
122. Galbe, M., G. Zacchi. A review of the production of ethanol from softwood. *Appl Microbiol Biotechnol* 59:618–628, 2002.
123. Lynd, L.R., P.J. Weimer, W.H. Van Zyl, I.S. Pretorius. Microbial cellulose utilization: fundamentals and biotechnology. *Microbiol. Mol. Biol. Rev.* 66:506–577, 2002.
124. Pérez, J., J. Muñoz-Dorado, T. de la Rubia, J. Martínez. Biodegradation and biological treatments of cellulose, hemicellulose and lignin: an overview. *Int. Microbiol.* 5:53–63, 2002.
125. Chadha, B.S., S.S. Kanwar, H.S. Garcha. Simultaneous saccharification and fermentation of rice straw into ethanol. *Acta. Microbiol. Immunol. Hung.* 42:71–75, 1995.
126. Mamma, D., D. Koullas, G. Fountoukidis, D. Kekos, B.J. Macris, E. Koukios. Bioethanol from sweet sorghum: simultaneous saccharification and fermentation of carbohydrates by a mixed microbial culture. *Process. Biochem.* 31:377–381, 1996.
127. Zaldivar, J., J. Nielsen, L. Olsson. Fuel ethanol production from lignocellulose: a challenge for metabolic engineering and process integration. *Appl. Microbiol. Biotechnol.* 56:17–34, 2001.
128. Himmel, M.E., M.F. Ruth, C.E. Wyman. Cellulase for commodity products from cellulosic biomass. *Curr. Opin. Biotechnol.* 10:358–364, 1999.
129. Kuhad, R.C., A. Singh, K.E. Eriksson. Microorganisms and enzymes involved in the degradation of plant fiber cell walls. *Adv. Biochem. Eng. Biotechnol.* 57:45–125, 1997.
130. Larriba, G., E. Andaluz, R. Gueva, R.D. Basco. Molecular biology of yeast exoglucanases. *FEMS Microbiol. Lett.* 125:121–126, 1995.
131. van Rensburg, P., W.H. van Zyl, I.S. Pretorius. Co-expression of a *Phanerochaete chrysosporium* cellobiohydrolase gene and a *Butyrivibrio fibrisolvens* endo-β-1,4-glucanase gene in *Saccharomyces cerevisiae*. *Curr. Genet.* 30:246–250, 1996.
132. van Rensburg, P., W.H. van Zyl, I.S. Pretorius. Engineering yeast for efficient cellulose degradation. *Yeast* 14:67–76, 1998.
133. Cho, K.M., Y.J. Yoo, H.S. Kang. δ-Integration of endo/exo-glucanase and β-glucosidase genes into the yeast chromosomes for direct conversion of cellulose to ethanol. *Enzyme Microb. Technol.* 25:23–30, 1999.
134. Murai, T., M. Ueda, T. Kawaguchi, M. Arai, A. Tanaka. Assimilation of cellooligosaccharides by a cell surface-engineered yeast expressing β-glucosidase and carboxymethylcellulase from *Aspergillus aculeatus*. *Appl. Environ. Microbiol.* 64:4857–4861, 1998.
135. Fujita, Y., S. Takahashi, M. Ueda, A. Tanaka, H. Okada, Y. Morikawa, T. Kawaguchi, M. Arai, H. Fukuda, A. Kondo. Direct and efficient production of ethanol from cellulosic material with a yeast strain displaying cellulolytic enzymes. *Appl. Environ. Microbiol.* 68:5136–5141, 2002.
136. Shoham, Y., R. Lamed, E.A. Bayer. The cellulosome concept as an efficient microbial strategy for the degradation of insoluble polysaccharides. *Trends Microbiol.* 7:275–281, 1999.
137. Kirk, K., D. Cullen. Enzymology and molecular genetics of wood degradation by white rot fungi. In: *Environmental Friendly Technologies for Pulp and Paper Industry*. Young, R.A., M. Akhtar, eds. New York: Wiley Interscience, 1998, pp 273–307.
138. Pérez-González, J.A., L.H. de Graaff, J. Visser, D. Ramón. Molecular cloning and expression in *Saccharomyces cerevisiae* of two *Aspergillus nidulans* xylanase genes. *Appl. Environ. Microbiol.* 62:2179–2182, 1996.
139. La Grange, D.C., I.S. Pretorius, W.H. van Zyl. Cloning of the *Bacillus pumilus* beta-xylosidase gene (*xynB*) and its expression in *Saccharomyces cerevisiae*. *Appl. Microbiol. Biotechnol.* 47:262–266, 1997.
140. Margolles-Clark, E., M. Tenkanen, T. Nakari-Setala, M. Penttilä. Cloning of genes encoding alpha-L-arabinofuranosidase and beta-xylosidase from *Trichoderma reesei* by expression in *Saccharomyces cerevisiae*. *Appl. Environ. Microbiol.* 62:3840–3846, 1996.
141. Monfort, A., A. Blasco, J.A. Prieto, P. Sanz. Construction of baker's yeast strains that secrete different xylanolytic enzymes and their use in bread making. *J. Cereal Sci.* 26:195–199, 1997.
142. La Grange, D.C., I.S. Pretorius, M. Claeyssens, W.H. Van Zyl. Degradation of xylan to D-xylose by recombinant *Saccharomyces cerevisiae* coexpressing the *Aspergillus niger*

β-xylosidase (*xlnD*) and the *Trichoderma reesei* xylanase II (*xyn2*) genes. *Appl. Environ. Microbiol.* 67:5512–5519, 2001.

143. Hallborn, J., M. Walfridsson, U. Airaksinen, H. Ojamo, B. Hahn-Hägerdal, M. Penttilä, S. Keränen. Xylitol production by recombinant *Saccharomyces cerevisiae*. *Bio/Technology* 9:1090–1095, 1991.
144. Amore, R., P. Kötter, C. Kuster, M. Ciriacy, C.P. Hollenberg. Cloning and expression in *Saccharomyces cerevisiae* of the NAD(P)H-dependent xylose reductase-encoding gene (*XYL1*) from the xylose-assimilating yeast *Pichia stipitis*. *Gene* 109:89–97, 1991.
145. Kötter, P., M. Ciriacy. Xylose fermentation by *Saccharomyces cerevisiae*. *Appl. Microbiol. Biotechnol.* 38:776–783, 1993.
146. Tantirungkij, M., N. Nakashima, T. Seki, T. Yoshida. Construction of xylose-assimilating *Saccharomyces cerevisiae*. *J. Ferment. Bioeng*. 75:83–88, 1993.
147. Hallborn, J., M. Penttilä, H. Ojamo, M. Walfridsson, U. Airaksinen, S. Këranen, B. Hahn-Hägerdal. Recombinant yeasts containing the DNA sequences coding for xylose reductase and xylitol dehydrogenase enzymes. PCT/AU647104. 1994.
148. Busturia, A., R. Lagunas. Catabolite inactivation of the glucose transport system in *Saccharomyces cerevisiae*. *J. Gen. Microbiol.* 132:379–385, 1986.
149. Eliasson, A., C. Christensson, C.F. Wahlbom, B. Hahn-Hägerdal. Anaerobic xylose fermentation by recombinant *Saccharomyces cerevisiae* carrying *XYL1*, *XYL2*, and *XKS1* in mineral medium chemostat cultures. *Appl. Environ. Microbiol.* 66:3381–3386, 2000.
150. Hamacher, T., J. Becker, M. Gárdonyi, B. Hahn-Hägerdal, E. Boles. Characterization of the xylose-transporting properties of yeast hexose transporters and their influence on xylose utilization. *Microbiology* 148:2783–2788, 2002.
151. Lidén, G., M. Walfridsson, R. Ansell, M. Anderlund, L. Adler, B. Hahn-Hägerdal. A glycerol-3-phosphate dehydrogenase-deficient mutant of *Saccharomyces cerevisiae* expressing the heterologous *XYL1* gene. *Appl. Environ. Microbiol.* 62:3894–3896, 1996.
152. Walfridsson, M., M. Anderlund, X. Bao, B. Hahn-Hägerdal. Expression of different levels of enzymes from the *Pichia stipitis XYL1* and *XYL2* genes in *Saccharomyces cerevisiae* and its effects on product formation during xylose utilization. *Appl. Microbiol. Biotechnol.* 48:218–224, 1997.
153. Hahn-Hägerdal, B., C.F. Wahlbom, M. Gárdonyi, W.H. van Zyl, R.R. Cordero Otero, L.J. Jonsson. Metabolic engineering of *Saccharomyces cerevisiae* for xylose utilization. *Adv. Biochem. Eng. Biotechnol.* 73:53–84, 2001.
154. Walfridsson, M., X. Bao, M. Anderlund, G. Lilius, L. Bülow, B. Hahn-Hägerdal. Ethanolic fermentation of xylose with *Saccharomyces cerevisiae* harboring the *Thermus thermophilus xylA* gene, which expresses an active xylose (glucose) isomerase. *Appl. Environ. Microbiol.* 62:4648–4651, 1996.
155. Aguilera, J., J.A. Prieto. The *Saccharomyces cerevisiae* aldose reductase is implied in the metabolism of methylglyoxal in response to stress conditions. *Curr. Genet.* 39:273–283, 2001.
156. Träff, K.L., R.R. Otero Cordero, W.H. van Zyl, B. Hahn-Hägerdal. Deletion of the *GRE3* aldose reductase gene and its influence on xylose metabolism in recombinant strains of *Saccharomyces cerevisiae* expressing the xylA and XKS1 genes. *Appl. Environ. Microbiol.* 67:5668–5674, 2001.
157. Jeppson, M., B. Johansson, B. Hahn-Hägerdal, M.F. Gorwa-Grauslund. Reduced oxidative pentose phosphate pathway flux in recombinant xylose-utilizing *Saccharomyces cerevisiae* strains improves the ethanol yield from xylose. *Appl. Environ. Microbiol.* 68:1604–1609, 2002.
158. Senac, T., B. Hahn-Hägerdal. Intermediary metabolite concentrations in xylulose- and glucose-fermenting *Saccharomyces cerevisiae* cells. *Appl. Environ. Microbiol.* 56:120–126, 1989.
159. Walfridsson, M., J. Hallborn, M. Penttilä, S. Keränen, B. Hahn-Hägerdal. Xylose-metabolizing *Saccharomyces cerevisiae* strains overexpressing the *TKL1* and *TAL1* genes encoding the pentose phosphate pathway enzymes transketolase and transaldolase. *Appl. Environ. Microbiol.* 61:4184–4190, 1995.
160. Ho, N.W.Y., Z. Chen, A.P. Brainard. Genetically engineered *Saccharomyces* yeast capable of effective cofermentation of glucose and xylose. *Appl. Environ. Microbiol.* 64:1852–1859, 1998.

161. Toivari, M.H., A. Aristidou, L. Ruohonen, M. Penttilä. Conversion of xylose to ethanol by recombinant *Saccharomyces cerevisiae:* importance of xylulokinase (*XKS1*) and oxygen availability. *Metab. Eng.* 3:236–249, 2001.
162. Johansson, B., C. Christensson, T. Hobley, B. Hahn-Hägerdal. Xylulokinase overexpression in two strains of *Saccharomyces cerevisiae* also expressing xylose reductase and xylitol dehydrogenase and its effect on fermentation of xylose and lignocellulosic hydrolysate. *Appl. Environ. Microbiol.* 67:4249–4255, 2001.
163. Zaldivar, J., A. Borges, B. Johansson, H.P. Smits, S.G. Villas-Boas, J. Nielsen, L. Olsson. Fermentation performance and intracellular metabolite patterns in laboratory and industrial xylose-fermenting *Saccharomyces cerevisiae*. *Appl. Microbiol. Biotechnol.* 59:436–442, 2002.
164. Baur, X., I. Sander, A. Posch, M. Raulf-Heimsoth. Baker's asthma due to the enzyme xylanase — a new occupational allergen. *Clin. Exp. Allerg.* 28:1591–1593, 1998.
165. Merget, R., I. Sander, M. Raulf-Heimsoth, X Baur. Baker's asthma due to xylanase and cellulase without sensitization to alpha-amylase and only weak sensitization to flour. *Int. Arch. Allerg. Immunol.* 124:502–505, 2001.
166. Quirce, S., M. Fernandez-Nieto, B. Bartolomé, C. Bombin, M. Cuevas, J. Sastre. Glucoamylase: another fungal enzyme associated with baker's asthma. *Ann. Allerg. Asthma Immunol.* 89:197–202, 2002.
167. Randez-Gil, F., P. Sanz, J.A. Prieto. Engineering baker's yeast: room for improvement. *Trends Biotechnol.* 17:237–244, 1999.
168. Oliver, S.G. Classical yeast technology. In: *Saccharomyces*: *Biotechnology Handbooks*, vol. 4, Tuite, M.F., S.G. Oliver, eds. London: Plenum Press, 1991, pp 213–248.
169. Matthewson, P.R. Enzymatic activity during bread baking. *Cereal Foods World* 45:98–101, 2000.
170. Kulp, K. Enzymes as dough improvers. In: *Advances in baking technology,* Kamel, B.S., C.E. Stauffer, eds. New York: VCH Publishers, 1993, pp 152–178.
171. Taylor, J. Not a stale subject. *Food Rev.* 25:19–21, 1998.
172. Randez-Gil, F., J.A. Prieto, A. Murcia, P. Sanz. Construction of baker's yeast strains that secrete *Aspergillus oryzae* α-amylase and their use in bread making. *J. Cereal Sci.* 21:185–193, 1995.
173. Randez-Gil, F., P. Sanz. Expression of *Aspergillus oryzae* α-amylase gene in *Saccharomyces cerevisiae*. *FEMS Microbiol. Lett.* 112:119–124, 1993.
174. Monfort, A., A. Blasco, J.A. Prieto, P. Sanz. Combined expresión of *Aspergillus nidulans* endoxilanase X24 and *Aspergillus oryzae* α-amylase in industrial baker's yeast and their use in bread making. *Appl. Environ. Microbiol.* 62:3712–3715, 1996.
175. Si, J.Q. Synergistic effect of enzymes for breadbaking. *Cereal Foods World* 42:802–807, 1997.
176. Olesen, T., Q.J. Si, V. Donelyan. Use of lipase in baking. PCT/DK 93/00274. 1994.
177. Yamaguchi, S., T. Mase, K. Takeuchi. Secretion of mono- and diacylglycerol lipase from *Penicillium camembartii* U-150 by *Saccharomyces cerevisiae* and site-directed mutagenesis of the putative catalytic sites of the lipase. *Biosci. Biotecnol. Biochem.* 56:315–319, 1992.
178. Bertolini, M.C., J.D. Schrag, M. Cygler, E. Ziomek, D.Y. Thomas, T. Vernet. Expression and characterization of *Geotrichum candidum* lipase I gene. Comparison of specificity profile with lipase II. *Eur. J. Biochem.* 228:863–869, 1995.
179. Nagao, T., Y Shimada, A. Sugihara, Y. Tominaga. C-terminal peptide of *Fusarium heterosporum* lipase is necessary for its increasing thermostability. *J. Biochem.* 124:1124–1129, 1998.
180. Monfort, A., A. Blasco, P. Sanz, J.A. Prieto. Expresión of *LIP1* and *LIP2* genes from *Geotrichum* sp. in baker's yeast strains and their application to the bread making process. *J. Agri. Food Chem.* 47:803–808, 1999.
181. Sánchez, M., N. Prim, F. Randez-Gil, F.I.J. Pastor, P. Diaz. Engineering of baker's yeasts, *E. coli* and *Bacillus* hosts for the production of *Bacillus subtilis* lipase A. *Biotechnol. Bioeng.* 78:339–345, 2002.
182. Graf, E.. *Phytic Acid: Chemistry and Applications*. Minneapolis, MN: Pilatus Press, 1986, pp 1–21.

183. Reddy, N.R., S.K. Sathe. *Food Phytates*. Boca Raton, FL: CRC Press LLC, 2002, pp 107–126.
184. Pandey, A., G. Szakacs, C.R. Soccol, J.A. Rodriguez-Leon, V.T. Soccol. Production, purification and properties of microbial phytases. *Bioresource Technol.* 77:203–214, 2001.
185. Türk, M., A.-S. Sandberg, N.-G. Carlsson, T. Andlid. Inositol hexaphosphate hydrolysis by baker's yeast: capacity, kinetics, and degradation products. *J. Agric. Food Chem.* 48:100–104, 2000.
186. Türk, M., A.-S. Sandberg. Phytate degradation during breadmaking: effect of phytase addition. *J. Cereal Chem.* 15:281–294, 1992.
187. Haros, M., C.M. Rosell, C. Benedito. Use of fungal phytase to improve breadmaking performance of whole wheat bread. *J. Agric. Food. Chem.* 49:5450–5454, 2001.
188. Lei, X.G., C.H. Stahl. Biotechnological development of effective phytases for mineral nutrition and environmental protection. *Appl. Microbiol. Biotechnol.* 57:474–481, 2001.
189. Han, Y., D.B. Wilson, X.G. Lei. Expression of an *Aspergillus niger* phytase gene (*phyA*) in *Saccharomyces cerevisiae*. *Appl. Environ. Microbiol.* 65:1915–1918, 1999.
190. van Oort, M.G.. Oxidases in baking. A review of the uses of oxidases in bread making. *Int. Food Ingred.* 4:42–44, 47, 1996.
191. Schmitt, M., G. Jungschaffer. Transglutaminase for baking. *Proceedings of the IX Meeting on Industrial Applications of Enzymes*, Barcelona, 2001, p 157.
192. van der Plas, A.G. Reducing and oxidizing agents in the bread process. *Proceedings of the IX Meeting on Industrial Applications of Enzymes*, Barcelona, 2001, p 11.
193. Clemente, A. Enzymatic protein hydrolysates in human nutrition. *Trends Food Sci. Technol.* 11:254–262, 2000.
194. Knust, B., D. von Wettstein. Expression and secretion of pea-seed lipoxygenase isoenzymes in *Saccharomyces cerevisiae*. *Appl. Microbiol. Biotechnol.* 37:342–351, 1992.
195. Schu, P., M. Reith. Evaluation of different preparation parameters for the production and cryopreservation of seed cultures with recombinant *Saccharomyces cerevisiae*. *Cryobiology* 32:379–388, 1995.
196. Cassland, P., L.J. Jonsson. Characterization of a gene encoding *Trametes versicolor* laccase A and improved heterologous expression in *Saccharomyces cerevisiae* by decreased cultivation temperature. *Appl. Microbiol. Biotechnol.* 52:393–400, 1999.
197. Park, E.H., Y.M. Shin, Y.Y. Lim, T.H. Kwon, D.H. Kim, M.S. Yang. Expression of glucose oxidase by using recombinant yeast. *J. Biotechnol.* 81:35–44, 2000.
198. Ramakrishnan, U. Prevalence of micronutrient malnutrition worldwide. *Nutr. Rev.* 60: S46–52, 2002.
199. Stephanopoulos, G. Metabolic fluxes and metabolic engineering. *Metab. Eng.* 1:1–11, 1999.
200. Ostergaard, S., L. Olsson, J. Nielsen. Metabolic engineering of *Saccharomyces cerevisiae*. *Microbiol. Mol. Rev.* 64:34–50, 2000.
201. Kamel, B.S., C.E. Stauffer. *Advances in baking technology*. New York: Blackie & Sons, 1993, pp 38–87.
202. Farfán, M.J., I.L. Calderón. Enrichment of threonine content in *Saccharomyces cerevisiae* by pathway engineering. *Enzyme Microb. Technol.* 26:763–770, 2000.
203. Martínez-Force, E., T. Benítez. Selection of amino-acid overproducer yeast mutants. *Curr. Genet.* 21:191–196, 1992.
204. Gasent-Ramírez, J.M., T. Benítez. Lysine-overproducing mutants of *Saccharomyces cerevisiae* baker's yeast isolated in continuous culture. *Appl. Environ. Microbiol.* 63: 4800–4806, 1997.
205. Rincón, A.M., T. Benítez. Improved organoleptic and nutritive properties of bakery products supplemented with amino acid overproducing *Saccharomyces cerevisiae* yeasts. *J. Agric. Food Chem.* 49:1861–1866, 2001.
206. Jones, E.W., G.R. Fink. Regulation of amino acids and nucleotide biosynthesis in yeast. In: *The Molecular Biology of the Yeast Saccharomyces*, Vol. 2, Strathern, J.N., E.W. Jones, J.R. Broach, eds., Cold Spring Harbor, New York: Cold Spring Harbor Laboratory Press, 1982, pp 181–299.

207. Delgado, M.A., J. Guerrero, J. Conde. Genetic and biochemical study of threonine-overproducing mutants of *Saccharomyces cerevisiae*. *Mol. Cell. Biol.* 2:731–736, 1982.
208. Ramos, C., I.L. Calderón. Overproduction of threonine by *Saccharomyces cerevisiae* mutants resistant to hydroxynorvaline. *Appl. Environ. Microbiol.* 58:1677–1682, 1992.
209. Farfán, M.J., L. Aparicio, I.L. Calderón. Threonine overproduction in yeast strains carrying the *HOM3-R2* mutant allele under the control of different inducible promoters. *Appl. Environ. Micriobiol.* 65:110–116, 1999.
210. Ramos, F., J.-M. Wiame. Mutation affecting the specific regulatory control of lysine biosynthetic enzymes in *Saccharomyces cerevisiae*. *Mol. Gen. Genet.* 200:291–294, 1985.
211. Feller, A., F. Ramos, A. Pièrard, E. Dubois. Lys80p of *Saccharomyces cerevisiae*, previously proposed as a specific repressor of *LYS* genes, is a pleiotropic regulatory factor identical to Mks1p. *Yeast* 13:1337–1346, 1997.
212. Feller, A., F. Ramos, A. Pièrard, E. Dubois. In *Saccharomyces cerevisiae*, feedback inhibition of homocitrate synthase isoenzymes by lysine modulates the activation of *LYS* gene expression by Lys14p. *Eur. J. Biochem.* 261:163–170, 1999.
213. Ramos, F., E. Dubois, A. Pièrard. Control of enzyme synthesis in the lysine biosynthesis pathway of *Saccharomyces cerevisiae*. Evidence for a regulatory role of gene LYS14. *Eur. J. Biochem.* 171:171–176, 1988.
214. Yamano, S., T. Ishii, M. Nakagawa, H. Ikenaga, N. Misawa. Metabolic engineering for production of β-carotene and lycopene in *Saccharomyces cerevisiae*. *Biosci. Biotech. Biochem.* 58:1112–1114, 1994.
215. Hancock, R.D., R. Viola. The use of micro-organisms for L-ascorbic acid production: current status and future perspectives. *Appl. Microbiol. Biotechnol.* 56:567–576, 2001.
216. Huh, W.K., B.H. Lee, S.T. Kim, Y.R. Kim, G.E. Rhie, Y.W. Baek, C.S. Hwang, J.S. Lee, S.O. Kang. D-erythroascorbic acid is an important antioxidant molecule in *Saccharomyces cerevisiae*. *Mol. Microbiol.* 30:895–903, 1998.
217. Onofri, S., E. Poerio, P. Serangeli, F. Tosi, I. Garuccio, O. Arrigoni. Influence of L-galactonic and γ-lactone on ascorbate production in some yeast. *Antonie van Leeuwenhoek* 71:277–280, 1997.
218. Hancock, R.D., J.R. Galpin, R. Viola. Biosynthesis of L-ascorbic acid (vitamin C) in *Saccharomyces cerevisiae*. *FEMS Microbiol. Lett.* 186:245–250, 2000.
219. Hancock, R.D., R. Viola. Biotechnological approaches for L-ascorbic acid production. *Trends Biotechnol.* 20:299–305, 2002.
220. Simopoulos, A.P. ω-3 fatty acids in health and disease and in growth and development. *Am. J. Clin. Nutr.* 54:438–463, 1991.
221. Dyer, J.M., D.C. Chapital, J.W. Kuan, R.T. Mullen, A.B. Pepperman. Metabolic engineering of *Saccharomyces cerevisiae* for production of novel lipid compounds. *Appl. Microbiol. Biotechnol.* 59:224–230, 2002.
222. Peyou-Ndi, M.M., J.L. Watts, J. Browse. Identification and characterization of an animal Δ^{12} fatty acid desaturase gene by heterologous expression in *Saccharomyces cerevisiae*. *Arch. Biochem. Biophys.* 376:399–408, 2000.
223. Gibson, G.R., R. Fuller. Aspects of *in vitro* and *in vivo* research approaches directed toward identifying probiotics and prebiotics for human use. *J. Nutr.* 130:391S–395S, 2000.
224. Hammes, W.P., C. Hertel. Research approaches for pre- and probiotics: challenges and outlook. Food Res. Int. 35:165–170, 2002.
225. Playne, M.J., R. Crittenden. Commercially available oligosaccharides. *Br. Int. Dairy Fed.* 313:10–22, 1996.
226. Walter, T. Bread goes prebiotic. *Int. Food. Ingred.* 2:20–21, 1999.
227. Mayer, C., D.L. Jakeman, M. Mah, G. Karjala, L. Gal, R.A.J. Warren, S.G. Withers. Direct evolution of new glycosynthases from *Agrobacterium* β-glucosidase: a general screen to detect enzymes for oligosaccharide synthesis. *Chem. Biol.* 8:437–443, 2001.
228. Malet, C., A. Planas. From β-glucanase to β-glucansynthase: glycosyl transfer to α-glycosyl fluorides catalyzed by a mutant endoglucanase lacking its catalytic nucleophile. *FEBS Lett.* 440:208–212, 1998.

229. Fort, S., V. Boyer, L. Greffe, G. Davies, O. Moroz, L. Christiansen, M. Schulein, S. Cottaz, H. Driguez. Highly efficient synthesis of (1,4)-oligo- and -polysaccharides using a mutant cellulase. *J. Am. Chem. Soc.* 122:5429–5437, 2000.
230. Trincone, A., G. Perugino, M. Rossi, M. Moracci. A novel thermophilic glycosynthase that effects branching glycosylation. *Bioorg. Med. Chem. Lett.* 10:365–368, 2000.
231. Arnold, F.H., A.A. Volkov. Directed evolution of biocatalysts. *Curr. Opin. Chem. Biol.* 3:54–59, 1999.
232. Danishefsky, S.J., J.R. Allen. From the laboratory to the clinic: A retrospective on fully synthetic carbohydrate-based anticancer vaccines. *Angew Chem. Int. Ed. Engl.* 39:837–863, 2000.
233. Heyer, A.G., R. Wendenburg. Gene cloning and functional characterization by heterologous expression of the fructosyltransferase of *Aspergillus sydowi* IAM 2544. *Appl. Environ. Microbiol.* 67:363–370, 2001.
234. Rehm, J., L. Willmitzer, A Heyer. Production of 1-kestose in transgenic yeast expressing a fructosyltransferase from *Aspergillus foetidus*. *J. Bacteriol.* 180:1305–1310, 1998.
235. Scotti, P.A., R. Chambert, M.F. Petit-Glatron. Extracellular levansucrase of *Bacillus subtilis* produced in yeast remains in the cell in its precursor form. *Yeast* 10:29–38, 1994.
236. Trujillo, L.E., J.G. Arrieta, F. Dafhnis, J. García, J. Valdés, Y. Tambara, M. Pérez, L. Hernández. Fructo-oligosaccharides production by the *Gluconacetobacter diazotrophicus* levansucrase expressed in the methylotrophic yeast *Pichia pastoris*. *Enzyme Microb. Technol.* 28:139–144, 2001.
237. van den Ende, W., A. Michiels, D. van Wonterghem, R. Vergauwen, A. van Laere. Cloning, developmental, and tissue-specific expression of sucrose:sucrose 1-fructosyl transferase from *Taraxacum officinale*: fructan localization in roots. *Plant Physiol.* 123:71–80, 2000.
238. van der Meer, I.M., A.J. Koops, J.C. Hakkert, A.J. van Tunen. Cloning of the fructan biosynthesis pathway of Jerusalem artichoke. *Plant J.* 15:489–500, 1998.
239. Lee, T.I., N.J. Rinaldi, F. Robert, D.T. Odom, Z. Bar-Joseph, G.K. Gerber, N.M. Hannett, C.T. Harbison, C.M. Thompson, I. Simon, J. Zeitlinger, E.G. Jennings, H.L. Murray, D.B. Gordon, B. Ren, J.J. Wyrich, J.-B. Tagne, T.L. Volkert, E. Fraenkel, D.K. Gifford, R.A. Young. Transcriptional regulatory networks in *Saccharomyces cerevisiae*. *Science* 298:799–804, 2002.

1.12

The Biotechnology of Wine Yeast

Linda F. Bisson

CONTENTS

12.1 INTRODUCTION

The production of wine from grapes is one of the world's oldest biotechnological processes. The alcoholic fermentation, that is, the conversion of grape sugars to ethanol, is conducted by yeasts of the genus *Saccharomyces*. In addition to wide-ranging roles in beverage and food processing, *Saccharomyces* is also a premier research organism because of its genetic tractability. Its genome has been sequenced, and an extensive array of molecular technologies has been developed for the genetic manipulation of this organism. In spite of the availability of these molecular tools, genetically engineered yeast strains are not yet in use in the commercial wine industry. Anxiety over consumer acceptance is one reason genetically modified strains are not employed, but there are other concerns as well.

Wine production is unique in many respects. The wine production process is not conducted under sterile conditions, and any modified organism has the potential to become an enduring resident of the winery flora. Further, wine production is both science and art. There is concern that many of the proposed genetic modifications of yeast are designed to correct processing errors or incompetence of the winemaker rather than being useful tools to enhance the artistry of wine making. In spite of these concerns, the efficacy of genetic modification of wine yeast has been demonstrated.

12.1.1 The Winemaking Process

Wine production is a simple process (Figure 12.1). Following harvest, grapes are crushed. The crushed grape aggregate, called must, is used either directly for fermentation (red wines) or pressed to release the juice (white wines). The juice or must is then fermented in wood, stainless steel, or concrete vats, with or without temperature control. Since the

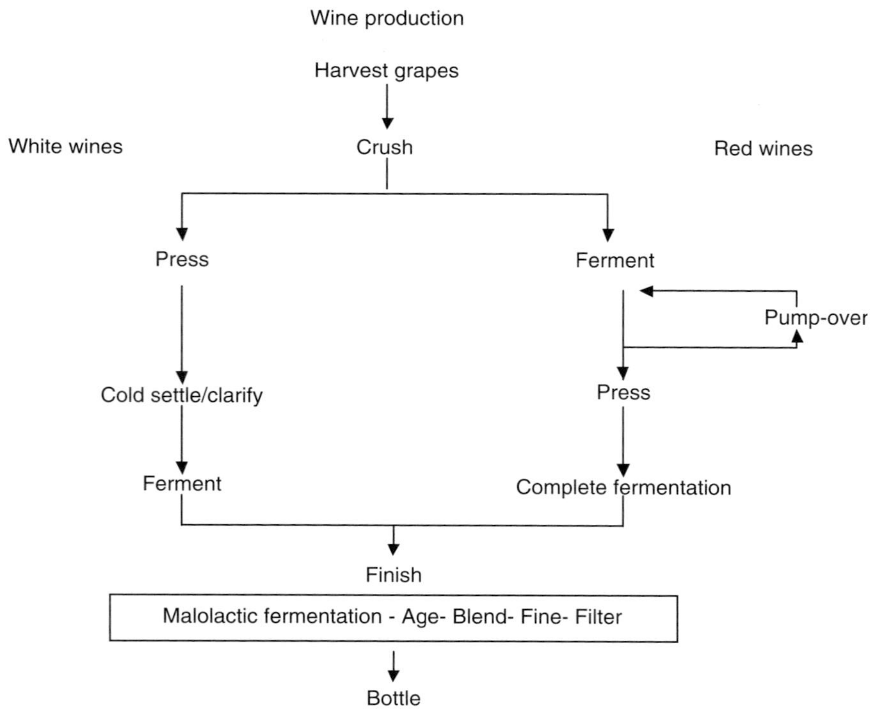

Figure 12.1 Schematic representation of wine production.

red pigments of grape, the anthocyanins, are located in the skin cells, there is a further difference between red wine and white wine production. Red wines are produced from fermentations conducted on the skins to extract the pigments and other materials present in the skin. In contrast, in white wine production the juice is separated from the skins prior to fermentation. In red wine fermentations the grape skins and other insoluble material form a cap on the surface of the wine once fermentation initiates and carbon dioxide is being evolved. In order to extract skin cell components, it is necessary to submerge or bathe the cap, using the partially fermented juice. The accumulating ethanol enhances the extraction of the skin cell components as well as inhibiting the growth of aerobes that may be on the exposed surface of the cap. The cap is either submerged in the tank (a process called "punch down"), or fermenting juice from the bottom of the tank is pumped over the cap on the surface ("pump over"). Although various mixing strategies are used, none are typically performed under sterile or anaerobic conditions in red wine production. The type and frequency of mixing affect oxygen availability for the yeast and the natures and numbers of other microbes present.

The fermentations may be inoculated with selected yeast or left to ferment by the native flora present on the grapes and in the winery. In either case, the fermentation is eventually dominated by *Saccharomyces*. This yeast is a single celled eukaryotic fungus that reproduces by budding. Two species of *Saccharomyces*, *S. cerevisiae* and *S. bayanus*, are commonly found in wine fermentations. It is important to note that there is some confusion in the nomenclature of wine yeast strains, specifically in the designation "*S. bayanus*". True *S. bayanus* (*Saccharomyces bayanus uvarum*) is quite distinct from *S. cerevisiae* (1,2) displaying only 80% sequence identity in coding regions and 62% identity in noncoding regions (1). Many strains used commercially in the wine industry that have been labeled as *S. bayanus* are actually *S. cerevisiae bayanus*. *Saccharomyces cerevisiae cerevisiae*, *S. cerevisiae bayanus* and *S. bayanus uvarum* can all be found in fermenting grape juice or must. Unfortunately, the designation "*S. bayanus*" has been used in the wine literature for both *S. bayanus uvarum* and *S. cerevisiae bayanus*. It is therefore important to understand which organism is actually the subject of the study.

Depending upon the style, the wine either is immediately ready for bottling following completion of fermentation or undergoes postfermentation processing such as aging in wood or stainless steel, a secondary bacterial fermentation, fining to remove undesirable components, or filtration to remove hazes and precipitates. Finally, the wines may be blended to achieve greater complexity and style. The last step in the process is bottling, which may be done under sterile conditions but frequently is not. One of the goals of wine production is to achieve microbial stability of the wine prior to bottling so that sterile conditions are not required for preservation of the wine.

There are several key issues with respect to this process that impact the development of genetically modified strains. First, it is important to reemphasize that it is not a sterile process. It takes advantage of the ability of *Saccharomyces* to dominate a mixed culture fermentation and produce sufficient ethanol to inhibit other microbes. If a genetic modification impacts the competitiveness of the yeast strain, the modified organism will be quickly dominated during the fermentation by unmodified native strains. Thus, any genetic modification will have to have little or no impact on the cell's ability to conduct and dominate a fermentation.

Second, numerous parameters affect the relative numbers and persistence of the wild microbial flora (3,4,5). These factors may be intrinsic to the juice or must (nutrient composition, presence of antimicrobial compounds, length of time the grapes are held postharvest, and nature of the grape bioflora) or extrinsic (length of time juice is in contact with the skins, temperature of contact and of fermentation, nature of the fermenting vat, juice supplementation practices, inoculation strategies, juice and must processing strategies, use of antimicrobial agents, and winery sanitation practices). The nutritional content of grapes varies from excessive to marginal: by varietal, by vineyard practices, and by growing season. Some of these differences can be mitigated by addition of nutrients to the fermentation. However, a genetically modified strain will have to retain the ability to dominate a fermentation under quite diverse environmental and growing conditions.

Third, there is no one standard practice or recipe for the production of wine, as processing strategies vary by the style and producer, but also within a given winery, as alternate practices may be employed for wines of the same nominal variety and overall style that are targeted to different market segments. If a genetic modification is tailored for one style, the winemaker must make certain that no problems will occur if that strain is present throughout the winery.

Fourth, with a few notable exceptions, there is no consensus among producers internationally or even regionally as to what traits a yeast strain should be engineered to possess. What one region or winemaker may find desirable may be undesired by another. Finally, given the potential of lateral gene transfer (6), any genetic change that would impose a high risk if acquired by the wild flora should not be considered.

12.1.2 Genetic Characteristics of *Saccharomyces*

Saccharomyces cerevisiae cells contain 16 linear chromosomes. The sequence of the genome indicates that approximately 6000 functional genes are present (7). Roughly two thirds of the genes have been assigned a function or putative function, based on the phenotype of mutations or on homology to proteins of known functions (8). There are several gene families, mostly encoding plasma membrane proteins. Systematic mutational analyses have been conducted with null mutations generated for each open reading frame (putative gene) (9). It is known which mutations are lethal. Global screening of this set of mutations is defining the phenotypes of the nonlethal mutations. Within the near future the functional role(s) of every gene will be known. The nature and kinds of protein–protein interactions are also being delineated for each gene product (10,11). The generation of specific genetic modifications and determination of the effect of those modifications on the global biological properties of *Saccharomyces* will be routine.

Saccharomyces can grow vegetatively as haploid or diploid cells. Haploids are able to mate yielding a diploid cell and diploids are able to sporulate yielding haploid spores (Figure 12.2). There are two mating types, designated "a" and "α". Mating type is specified by the information at the mating type or *MAT* locus. Haploid strains may be either *MATa* or *MAT*α. Sporulation of diploid cells leads to the formation of four spores, two of the "a" mating type (inheriting the *MATa* locus) and two of the "α" mating type (inheriting the *MAT*α locus). Spores are formed within a cell or ascus; thus *Saccharomyces* is an ascomycete. Because there are four spores in an ascus, the ascus is called a "tetrad". Tetrads can be dissected, allowing each individual spore to give rise to a colony. The recovery of all four products of a single meiotic division allows sophisticated genetic analyses to be performed.

Yeast strains may be homothallic or heterothallic (12). Homothallic yeast strains are self fertile. An individual haploid cell is able to yield a diploid population. This is because mother cells (cells that have produced at least one bud or daughter) can switch mating type and thus mate with their own daughter cells (Figure 12.3). Heterothallic strains do not have this property, and so stably propagate as haploids, and are only able to mate with a haploid cell of the opposite mating type. Heterothallism is conferred by the loss of a single gene, the *HO* gene that encodes a specific endonuclease required for switching the information at the mating type locus (12).

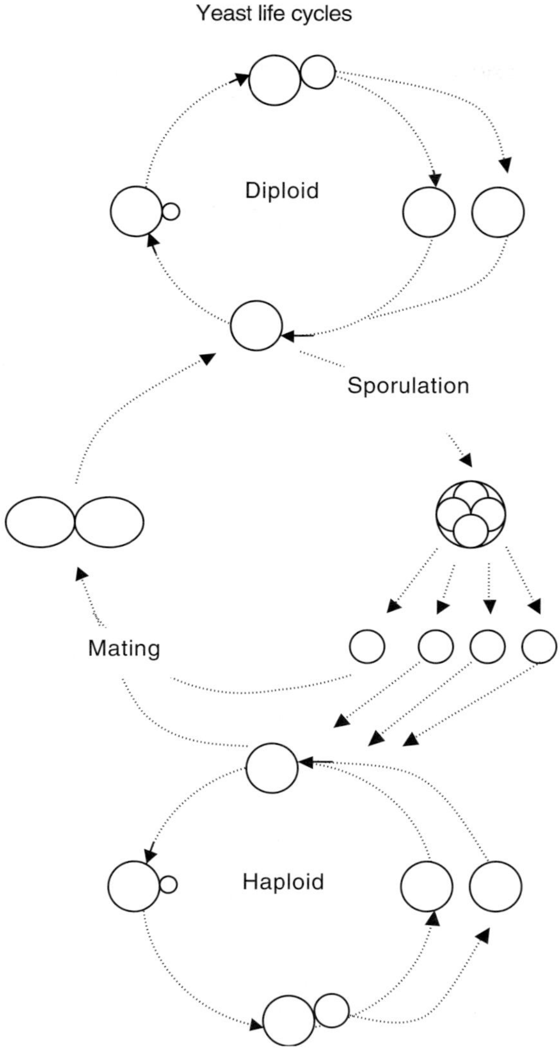

Figure 12.2 The life cycles of *Saccharomyces*. This yeast can grow vegetatively by budding as either a diploid or haploid. Diploids may sporulate or undergo meiotic division to yield four spores that may then grow vegetatively as haploids or mate to form a diploid. Diploids may grow vegetatively as diploids or sporulate.

12.1.2.1 Laboratory, Native and Industrial Strains

Genetic inheritance in *Saccharomyces* has been extensively studied under laboratory conditions. Laboratory strains were selected, however, for their genetic tractability — effectiveness of mutagenesis, stability of mutations, efficiency of sporulation and mating, and predictability of growth under laboratory conditions. To facilitate genetic analysis, most laboratory strains are heterothallic, allowing the researcher to control the genotype of the cells involved in genetic crosses. These strains are genetically stable, displaying low rates of spontaneous mutation. *Saccharomyces* is most commonly found in nature as a minor resident on the surface of grapes (3,4,5,13,14). It can be isolated from other types of rotting fruit and is transferred by insect vectors. *Saccharomyces* is also found as part

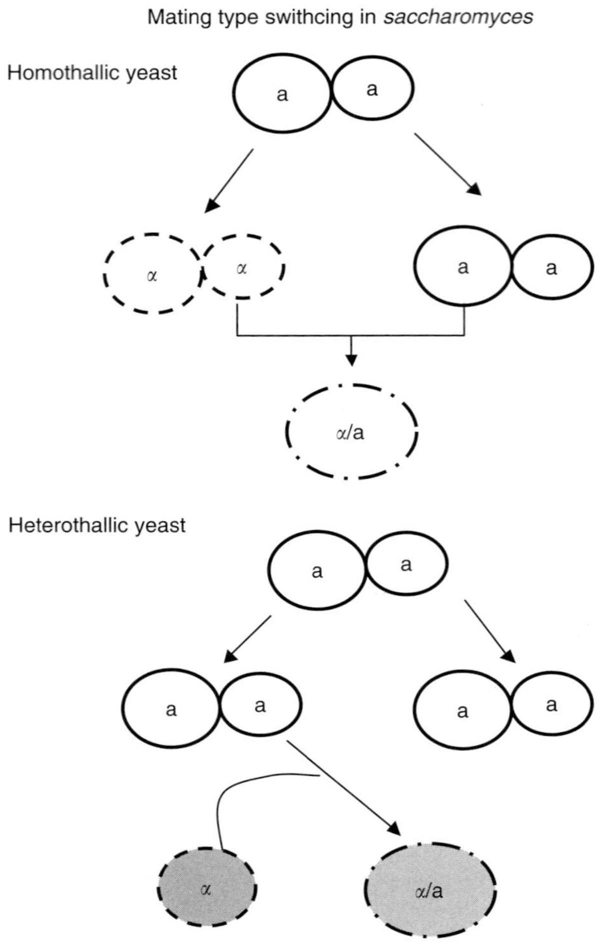

Figure 12.3 Homothallic strains of *Saccharomyces* are able to switch mating type following division. Mother and daughter cells may then mate. Heterothallic yeasts are not able to switch mating type and can only mate when a cell of the opposite mating type is present in the area [reviewed in (12)].

of the bioflora of wineries (3,5). Studies suggest that it is the winery bioflora strains that generally dominate the juice or must fermentation rather than the strains from the vineyard (13). Native isolates of *Saccharomyces* are typically homothallic diploids (15,16). In contrast to laboratory strains, native isolates accumulate mutations during normal growth, displaying higher rates of spontaneous mutation. Sporulation efficiency varies among the native isolates.

Industrial strains are also generally homothallic, but often contain multiple aneuploidies, being monosomic or trisomic for some chromosomes (17,18,19). Industrial strains are defined as those grown under commercial conditions to generate "active dry" inocula. These strains must be able to retain viability following the dehydration–rehydration procedure, and are frequently maintained on media distinct from grape juice. The conditions under which commercial strains are produced are likely responsible for the genomic variation obtained. Chromosome rearrangements are also common in industrial strains. Some commercial strains are not diploid, but display a higher ploidy. The genetic diversity of industrial strains limits sporulation efficiency and spore viability, and poses challenges to the genetic manipulation of these strains. Although chromosomal rearrangements occur in native strains of *Saccharomyces*, the native isolates are generally all normal diploids. However, these strains have usually accumulated several heterozygosities (20).

12.1.2.2 Genomic Instability of Yeast Strains

Commercial strains of *Saccharomyces* have been obtained following physiological characterization of native isolates from growing regions all over the world. Successful candidates are chosen primarily on the basis of their ability to be produced commercially and secondarily on their winemaking attributes. Genetic analysis of these strains suggests that native isolates from the same region are generally related (21). In one analysis of 13 commercial isolates, sporulation and spore viability varied dramatically (20). Of ten genetic traits analyzed, strains contained from zero to seven heterozygosities. Mitotic recombination and chromosomal rearrangements appear to occur at higher frequencies among the native populations as compared to laboratory strains (17,18,19,20,21). Analysis of a single isolate indicated the existence of mutations affecting global regulatory processes. Transcriptome analysis indicated that approximately 6% of the genes displayed differences in expression patterns in spores obtained from the original isolate (22). Thus, there is considerable genetic divergence among native strains of *Saccharomyces* isolated from vineyards and from wineries that have not employed commercial inocula.

Karyotype analysis has demonstrated significant chromosome polymorphisms that exist in wine yeast strains. Different studies have reached the conclusion that these polymorphisms arise due to recombination events between homologous Ty elements on different chromosomes (23). There is variety in the number and location of these Ty elements across yeast strains (17). Previously these elements were regarded as "junk" DNA, but it is now thought that they play an important role in generating genetic diversity among native populations (23).

12.1.3 Role of Indigenous Yeasts and Bacteria in Wine Production

The flavors and aromas of commercial wines are contributed in part by the grape and by the biological activities of the organisms present. Various yeast and bacterial species (and in some wine styles, filamentous fungi), may be present during grape must and juice fermentation and may impact the final characteristics of the wine (reviewed in (24,25,26)). In addition to *Saccharomyces*, several other genera of yeast are commonly found: *Hanseniaspora* (*Kloeckera*), *Metschnikowia*, *Candida*, and *Pichia.* Various factors affect the presence and persistence of the non-*Saccharomyces* yeasts, such as temperature of fermentation, nutrient addition practices, aeration, skin contact, the nature of the *Saccharomyces* strain used, and inoculation practices. There are other important yeast genera in wine production, the spoilage yeasts: *Brettanomyces*, *Zygosaccharomyces*, and other species of *Candida* and *Pichia.* Under ideal conditions, the yeasts present prior to and during fermentation strip the juice or must of nutrients, limiting the ability of spoilage yeasts to grow.

Several bacterial species are also present. The lactic acid bacteria can persist under wine production conditions and are responsible for the conversion of grape malate to lactate, the "malo-lactic" fermentation. They also impart other characteristics to the wine, such as the buttery note of diacetyl. The principal lactic acid bacterium found in wine is *Oenococcus oeni* (formerly known as *Leuconostoc oenos*) (24,26). *Oenococcus oeni* is able to thrive over the range of low pH values commonly found in grape must and juice (pH 3.2–3.9). Other lactic acid bacteria, members of the genera *Lactobacillus* and *Pediococcus*, can be found in wine if the pH values are greater than 3.5. These bacteria can grow under anaerobic conditions and in the presence of high ethanol. These organisms, depending upon the spectrum of end products produced and their desirability, may cause bacterial spoilage. *Acetobacter* is also an important spoilage organism. This bacterium produces large quantities of acetic acid and can turn wine into vinegar. It is an obligate aerobe, so its growth is confined to the exposed surface of the wine. It can be easily controlled by limiting the development of a headspace over the surface of the wine.

The skin of the grape teems with other aerobic filamentous fungi and bacteria, but these organisms are quickly eliminated by the anaerobic conditions rapidly established in the must or juice by microbial and enzymatic activity. The role these organisms play in wine fermentation is secondary and is due to impacts on grape composition and on the microflora of the berry.

12.1.4 Historical Aspects of Wine Yeast Biotechnology: Uses of Classical Breeding and Selection

Specific strains of *Saccharomyces* have been selected and used as inocula for decades. The brewing industry was the first to develop pure culture inocula (27). Interestingly, resistance to the use of pure culture inocula persisted in the wine industry for diametrically opposed reasons. Some felt the regional character of the wine was derived from the regional character and diversity of the native flora, whereas others felt that any yeast would do, as the yeast played no role in the final characteristics of the wine. Neither group perceived a need for pure culture inocula. Analysis of the properties of oenological wine strains demonstrated that significant variation existed and showed that some strains produced more off characters or were more prone to fermentation arrest than others. These observations fueled interest in use of known or well characterized strains to assure the commercial acceptability of the wine. In wine production, inocula may be one of two types: native flora, where one tank is inoculated by juice from a tank that has initiated fermentation, or a commercial preparation.

Commercial strains have been selected on the basis of specific attributes: ethanol tolerance, sulfur dioxide tolerance, predictability of fermentation behavior, fermentation to dryness (reduction of sugar below 0.2%), absence of production of spoilage characters (such as hydrogen sulfide), production of specific desirable esters, ability to dominate diverse fermentation conditions, tolerance of other microbes, and neutrality (minimal impact on grape varietal character). These strains are derived from native isolates following evaluation of the physiological characteristics of those isolates. Commercial preparations do not necessarily contain a single strain. Some vary due to spontaneous genomic rearrangements having occurred during production of the strains. Others are deliberate mixtures of different strains to assure that at least one strain will be able to complete the fermentation and to add to the complexity of yeast characters produced during the fermentation.

Several classical genetic techniques have been applied to the improvement of industrial strains of *Saccharomyces*. These include classic breeding of strains relying on genetic crosses between unrelated haploid strains, forced mass matings to detect rare events, and mutation and selection for desired phenotypes (28,29). Protoplast fusion has also been employed. In this case, cell walls of two different strains are digested, the strains are placed in contact under conditions that promote fusion of their plasma membranes, and the hybrid strains isolated (30). Cytoduction has been used to transfer cytoplasmically inherited genetic information between strains. Karyogamy (nuclear fusion) fails in *kar1* mutants of *Saccharomyces* (31,32). Crosses between haploid *kar1* mutants result in failure to form a diploid and in the creation of hybrid nuclei. Each of these techniques has been used to generate strains for use in the brewing, distilling, and baking yeast industries. In contrast, there has been little effort to genetically construct wine yeast strains using classical breeding methods. Instead, the genetic diversity of native isolates has provided ample material from which to select wine yeasts expressing specific traits.

12.2 GENETIC ENGINEERING TECHNOLOGY FOR YEAST

The discovery over two decades ago that yeast could be transformed as easily as bacteria, allowing stable modification of the genome, revolutionized the genetic analysis of *Saccharomyces*. It was then possible to isolate any gene of interest in the test tube and transfer the wild type or mutated allele to yeast. A second important discovery was the ease with which yeast genes could be disrupted, allowing mutations to be made at any known locus. As a consequence, any chromosomal allele could be replaced with any other allele. Combined with the strengths of classical genetic analysis, these technologies allowed unprecedented manipulation of the genome of a eukaryotic organism. Applications of these technologies to the genetic alteration of industrial strains soon followed their development in laboratory yeast.

12.2.1 Molecular Genetic Tools

Several transformation strategies have been developed for *Saccharomyces*, leading to the stable incorporation of introduced DNA into the yeast genome (reviewed in (33)). The first transformation protocol developed for yeast relied on the enzyme glusulase (snail gut enzyme) to digest the vegetative cell wall. This same preparation had been used for years to dissect the ascal wall, allowing isolation of the spores of an individual tetrad. The spheroplasts formed can translocate external DNA. Cells are then maintained under conditions conducive to the regeneration of the cell wall. The uptake of extracellular DNA can also be achieved by the use of salts such as lithium or cesium. This method is technically easy, but differences in permeabilization and sensitivity to the toxic effects of the ions render it challenging to use for some strains. Electroporation, use of an electrical pulse to drive the uptake of DNA, can also be employed with yeast, as can a "gene gun" or biolistic transformation. In our experience, each of these techniques is equally effective in the transformation of native isolates of *Saccharomyces*. Success in the transformation of industrial strains varies, probably as a consequence of their aneuploid nature rather than the specific technique used.

Several different plasmid vectors have been developed for yeast transformation (Table 12.1). These vectors differ in their replications of origin, mode of inheritance, selectable markers, and nature of the cloning site. The two principal modes of inheritance of yeast plasmids are: (1) reliance on the presence of a centromere on the plasmid or of the sites required for replication, and (2) segregation of the yeast 2-micron plasmid. Other plasmids, the integrating vectors, are designed for integration into a yeast chromosome and do not contain a separate inheritance region. Origins of replication may be chromosomal autonomous replication sequence (ARS) sites, or replication origins for the 2-micron plasmid. There are ARS vector plasmids that only contain an ARS sequence but no region of inheritance. These plasmids are transmitted to progeny cells in a passive manner, simply by virtue of being in the right place at the right time of nuclear division. They are the least stable of the yeast vectors.

Recently, a set of interchangeable or "cassette" vectors has been developed (34). A unique restriction site flanks each functional region. In addition to selectable markers, these plasmids also have counter selectable markers, markers that can be selected against, leading to recovery of strains that have lost the plasmid. Cassettes of functional regions are available, allowing easy manipulation of plasmid features.

Linear fragments of DNA can also be efficiently transformed into yeast cells and stabilized by insertion into the yeast genome. If the ends of the linear fragment are homologous to chromosomal sequences, homologous recombination will lead to replacement of the genomic sequence with the one introduced on the fragment (Figure 12.4). Thus, it is possible to replace alleles of any desired gene with another allele, or to simply disrupt the

Table 12.1
Options for Construction of Yeast Vectors

Inheritance Element	Origin of Replication	Selectable Marker	Counter-Selectable Marker	Multiple Cloning Site Fusion Options
2 micron circle	2 micron circle	**Metabolic Genes:**	*URA3*	None
CEN	ARS	*ADE1, 2*	*LYS2*	Promoter
		URA3	*MET15*	Terminator
		LYS1, 2	*PKA*	Promoter and Terminator
		TRP1	*GAL10-GIN11*	
		HIS3, 4		
		LEU2		
		MET2, 15		
		Resistance Markers:		
		Blasticidin		
		Bialaphos		
		Cerulinin		
		Copper		
		Cycloheximide		
		Formaldehyde		
		Fluoroacetate		
		Hygromycin B		
		Kanamycin		
		Multidrug resistance		
		Nourseothricin		
		Sulfur dioxide		

(31,34,36,42,43,44,45,46,47,48,49,50,51,52,53,54; reviewed in 33,56)

sequence, creating a null strain for that gene. Multiple replacements of the same locus can occur. The "gamma deletion" method, originally developed by Sikorski and Hieter (35), can be readily replaced by other constructs (36). This method relies on homology introduced by inserted plasmid sequences to direct subsequent recombination events.

Plasmids designed to facilitate exchange or swap of markers have also been developed (37). This allows the researcher to replace markers that have been used in strain construction for other markers while leaving the original null or knock out mutation intact. Double fusion PCR can also be used to create specific unique constructs that can be efficiently transformed into yeast (Figure 12.5) (38). In this technique, a region of interest is fused to flanking regions that can then be used to direct integration of the construct into the genome by homology. Several related adaptamer-mediated PCR methods have been employed in the genetic engineering of *Saccharomyces* (39).

12.2.2 Selectable Markers

Successful DNA transfer is dependent upon the detection of the transformants. Numerous markers have been developed in *Saccharomyces* for the selection of transformants (reviewed in (33)). These markers are used for positive selection, that is, a selection for the marker gene. If the marker is present, then, in theory, sois the rest of the construct, but this should always be independently confirmed. Many of these markers rely on restoration of auxotrophic mutations to prototrophy. Although quite effective in laboratory strains, the

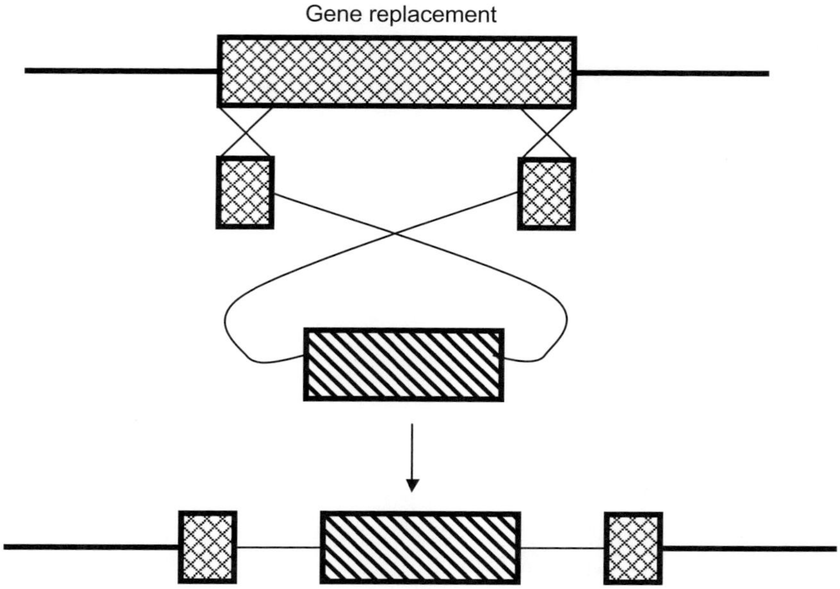

Figure 12.4 Gene replacement in *Saccharomyces*. A linear fragment containing homology to chromosomal sequences on the ends of the fragment will undergo homologous recombination with the chromosomal locus leading to replacement of the information at the locus with that from the fragment [reviewed in (33)].

use of these strategies in the genetic modification of industrial strains is limited by the lack of availability of specific mutations and by the need to maintain the competitive nature of the organism in the natural environment.

Several alternate selectable markers have been developed that are more suitable for wine production. The most commonly used marker is G14 (kanamycin) resistance (40). The *kanMX* marker is a hybrid of the kan^r transposon Tn903 with the promoter and terminator sequences of the *TEF* gene of *Ashbya gossypii* directing expression (41). This gene construct confers resistance to G418 in a variety of fungi. An advantage of this gene is the lack of homology to any sequences within *Saccharomyces* resulting in efficient disruption of the target locus. There are problems with the use of G418 resistance in wild and industrial strains of *Saccharomyces*. Native resistance to this antibiotic varies, and it is important to determine the level of sensitivity of the strain targeted for genetic modification.

The *Aspergillus* blasticidin S deaminase has also been used as a selectable marker in *Saccharomyces* (42). The presence of this gene confers blasticidin resistance. The dehH1 gene from *Moraxella* encodes fluoroacetate dehalogenase and has been used as a selectable marker in *Saccharomyces* (43). Genes conferring resistance to hygromycin B nourseothricin and bialaphos have also been used as dominant selectable markers in *S. cerevisiae* (44). Other drug resistance markers have also been developed (reviewed in 26). The majority have the same two drawbacks, first that they are encoded by heterologous DNA and second that spontaneous drug resistance appears with varying frequency across the native yeast populations.

Because of a perceived reluctance of acceptance of yeast strains carrying bacterial or other nonyeast DNA in the food supply, researchers have explored the development of alternate selectable markers that would originate solely from yeast. Mutants resistant to the antimicrobial agent sulfur dioxide, which is used in wine production, have been isolated

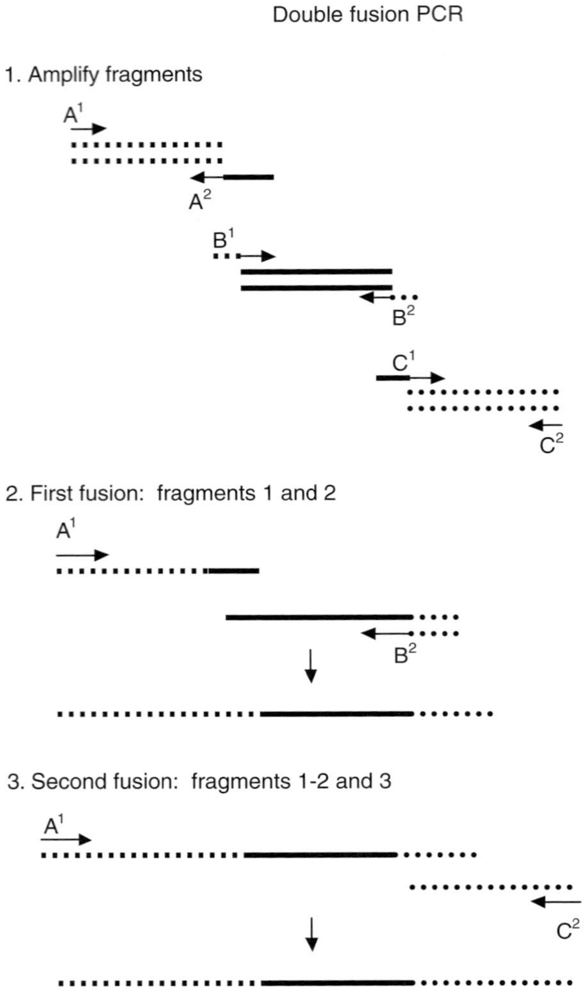

Figure 12.5 Diagram of double fusion PCR (38). Primers are designed that contain regions of homology to the target fusion fragment on one end (Primers A^2, B^1, B^2 and C^1 in the diagram). This results in overlapping homology on the ends of the PCR fragments generated from the primers. A fusion PCR reaction is then done using the flanking primers for each of the two fragments. This can be repeated sequentially to produced double fusion products that can be used for gene replacement in *Saccharomyces*.

and the dominant mutant alleles cloned. These mutant alleles, encoded by the *SSU1* and *FZF1* genes, have been used as selectable markers in *Saccharomyces* transformations (45). The *SFA1* gene of *Saccharomyces,* which encodes formaldehyde dehydrogenase, confers resistance to formaldehyde when over expressed (46). It can also be used as a dominant selectable marker. *CUP1*, encoding metallothionein, can also be used as a selectable marker, as this gene will confer resistance to copper (47). Similarly, mutations of the *PDR3* gene, which is involved in multidrug resistance, can also be used as a dominant selectable marker (48). The *YAP1* gene encodes an AP-1-like transcription factor and is involved in stress adaptation in yeast (49). Over expression of this gene has been shown to lead to resistance to cerulinin, an inhibitor of fatty acid synthesis, and to cyclohexamide, an inhibitor of protein synthesis. An advantage of these genes as selectable markers is the resistance conferred to more than one drug, facilitating the confirmation that transformation rather than mutation to spontaneous drug resistance has occurred.

Counter selectable markers are also important in strain construction. These are makers that confer a distinct phenotype when lost, one that can be both selected for and easily detected. This is also called a negative selection, as the presence of the gene is being selected against. The availability of such markers allows for the engineering of mutations requiring sequential recombination events. The most common counter selectable markers are based upon resistance to inhibitory analogs of metabolites of nucleic acid or amino acid biosynthesis. Resistance to the inhibitory compound occurs upon loss of the function of a specific gene involved in biosynthesis. If auxotrophic mutations are available, these genes also can be selected for, by restoration of growth in the absence of the specific nucleotide base or amino acid. The most frequently used gene is *URA3*. The *URA3* gene encodes orotodine-5'-monophosphate (OMP) decarboxylase, and this enzyme is required for the conversion of 5 fluro-orotic acid to 5 fluoro-uracil, which is toxic to yeast cells (50). *URA3*, therefore, confers sensitivity to 5 fluoro-orotic acid. A positive selection exists for *URA3*, the conversion of *ura3* mutants to uracil prototrophy, as well as a negative selection, resistance to 5 fluoro-orotic acid, for the conversion of *URA3* strains to *ura3*. The *LYS2* and *MET15* genes can be used similarly. *LYS2* confers sensitivity to α aminoadipic acid (51), and *MET15* to methyl mercury (52). In addition to this counter selectivity, *met15* mutants also become darkly pigmented in the presence of lead ions due to the formation of lead sulfide (53). A colony color screen can therefore also be used to distinguish *MET15* wild type from *met15* mutant cells.

The shortcoming of these counter selectable markers with respect to industrial strains is the need to have auxotrophic mutations in the strain background. Such auxotrophies will likely confer a growth disadvantage to the strain and would have to be corrected subsequent to their use to generate the construct of interest. If needed, the auxotrophic marker would best be introduced by transformation rather than by mutagenesis, to avoid the potential for cryptic background mutations. Once the desired gene has been incorporated into the genome, a second event to replace the mutated gene with a wild type allele will have to occur.

Counter selectable markers not requiring the presence of specific mutations in the strain background have also been developed. The *GIN11* region, which contains part of a subtelomeric region, is growth inhibitory when over expressed. This region has been placed under the control of the *GAL10* promoter (54). In the presence of galactose, the promoter is induced and growth inhibition occurs. The presence of the *GAL10–GIN11* fusion can be selected against by requiring growth in the presence of galactose. Similarly, the *PKA3* gene, which encodes the catalytic subunit of the cAMP-dependent protein kinase, also is growth inhibitory when over expressed. This gene has also been placed under the control of regulatable promoters (34). Counter selection occurs by isolation of those strains able to grow under conditions of induction of the *PKA3* gene. Just about any gene that confers a phenotype of growth arrest when over expressed can be used as a counter selectable marker. However, one shortcoming of these constructs is the absence of a positive selection in the first place. This can be compensated for by having two independent markers on the same construct, one for which a positive selection exists and one for which there is a negative selection. Industrial strain construction strategies will need to involve multiple steps of selection and counter selection if the use of auxotrophic mutations is to be avoided.

12.2.3 Strain Construction Strategies

There are two different goals for the construction of genetically modified wine yeast strains: the alteration of an inherent property of the strain, or the expression of a novel or heterologous activity. The first goal is essentially a swapping of one allele for another, either a null allele or one mutated to confer a different regulatory or catalytic property. Ideally, genetic modification of an allele should not leave any heterologous DNA as part of the construction, should be stably inherited and not borne on plasmids or other elements that may facilitate lateral gene transfer, should not require the strain be auxotrophic or subjected to mutagenesis, should be readily detectable and identifiable in mixed populations, and should not interfere with any other biological processes that may confer a disadvantage to the organism in its natural environment. Heterologous gene expression should likewise be stably inherited and minimize opportunities for lateral gene transfer.

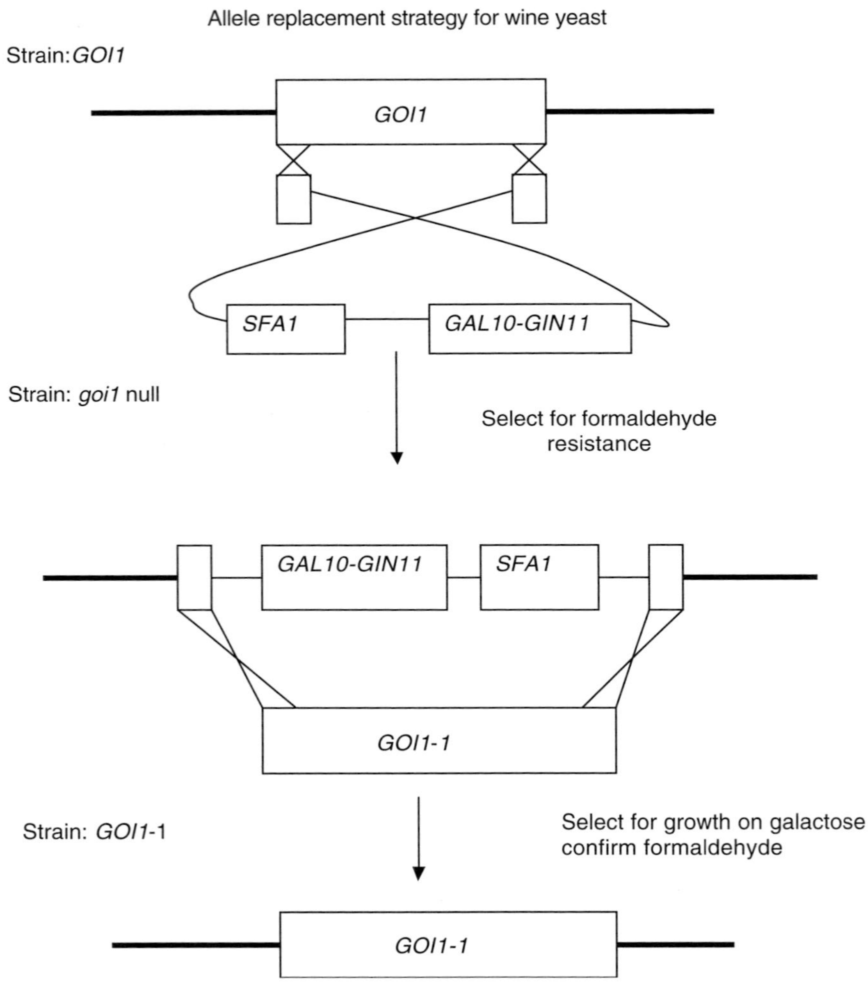

Figure 12.6 A strategy for allele replacement in *Saccharomyces*. A hypothetical gene, *GOI1*, "*G*ene *O*f *I*nterest" is shown. Both selectable and counter-selectable markers must be present. The allele is replaced in a two-step process. First, a null allele is created, in this case by selecting for fomaldehyde resistance. The null construct also contains a counter selectable marker, the *GAL10-GIN11* fusion (54) in this case. The null allele is then replaced with the desired allele, by a second homologous recombination event and selection against the counter-selectable maker, by requiring growth in the presence of galactose. In this example, only *Saccharomyces* markers are used. If double fusion PCR were employed, only *Saccharomyces* DNA sequences need be used throughout.

The amount of foreign DNA should be kept to a minimum, and it is very important that the full impact of the incorporation of the novel trait into the genome be known.

A sample strategy for allele replacement is shown in Figure 12.6. In this case a null mutation of the target gene is first constructed using a fragment containing both a dominant selectable marker and a marker that can be counter selected. Transformants expressing the dominant marker are then isolated and the location of the marker confirmed using PCR, Southern analysis, or another mapping technique. This allele can then be replaced with the altered allele by repeating the transformation and selecting against the counter selectable marker. The location of the construct should be confirmed by PCR analysis followed by DNA sequencing.

Heterologous gene expression requires the presence of efficient promoter and terminator sequences to allow transcription of the gene (Figure 12.7). These promoters may be constitutive or regulated. A series of promoters displaying expression at different times of grape juice fermentation have been identified and can be used to time expression to the desired physiological window (55). Promoters have also been designed that express target genes to varying levels (56) allowing adjustment of the level of expression of a heterologous gene. These genes should be targeted to a nonessential region of the genome. Targeting the heterologous constructs to the *HO* gene resulting in conversion of the strain to heterothallism has been proposed as a benign gene disruption (57). This is not expected to confer any disadvantage during growth and fermentation. However, there is a concern that this may lead to spread of genes through mating with other partners.

The lack of an impact on other metabolic pathways can be confirmed by conducting a functional genomic analysis of the modified strains (59,60). Transcriptome technologies allow display of global gene expression patterns, and the impact of the modification on gene expression can be determined (59,61,62,63,64). Proteome analysis can further demonstrate the absence of an effect on the protein constitution of a cell (60,65).

12.3 IMPROVEMENT OF OENOLOGICAL TRAITS OF YEAST STRAINS

The genetic engineering of wine yeast can be divided into two broad categories: modifications that serve to enhance existing or intrinsic properties of wine strains and those that aim to introduce novel traits (Figure 12.8, Figure 12.9). The modification of existing traits may take the form of (1) the *in vitro* creation of a specific allele, (2) isolation of a specific

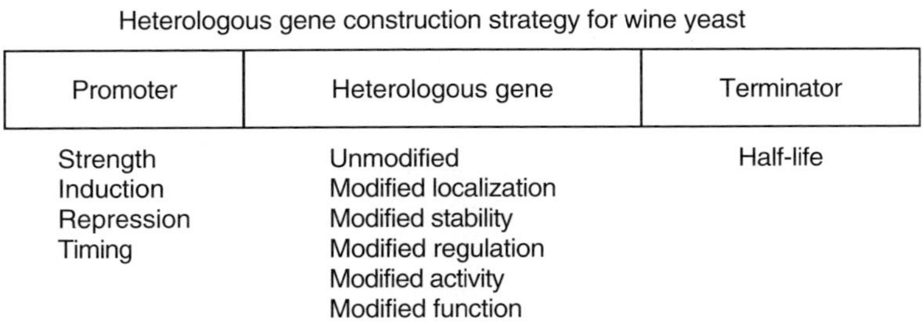

Figure 12.7 Options for the construction of an expression cassette for a heterologous or novel function. The promoter and terminator regions may be selected based on the desired regulation and level of expression (55,56). The heterologous gene may be used unmodified or altered in one or more properties to produce a more effective gene product.

Genetic engineering of intrinsic traits of wine yeast

Fermentation performance	Enhance: Stress tolerance Rate of fermentation Substrate utilization Competitiveness
Off-character production	Reduce: Sulfur volatiles Acetate/acids Aldehydes Higher alcohols Phenolic derivatives
Aroma character production	Increase: Ester formation Glycerol production Lactone production Terpene production S-conjugates release Autolysis flavors production Reduce: Esterase
Production performance	Enhance: Flocculation Sedimentation Hydrolase activity Tannin reduction Mannoprotein release/production
Healthfulness improvement	Decrease: Ethyl carbamate Increase: Vitamin production Pesticide scavenging Metal ion scavenging

Figure 12.8 Proposed intrinsic property targets of genetic modification in wine strains of *Saccharomyces*.

allele from another strain or different species of *Saccharomyces* followed by allele replacement, (3) alteration of promoter or terminator sequences to change the pattern of regulation of the gene or protein, or (4) the construction of a novel hybrid protein that will display altered activity or posttranslational regulation. In some cases a null allele may be desired to eliminate an unwanted activity. Null alleles may impose strong selective pressure for suppression or other compensating alterations of the genome, so the outcome of the presence of these mutations in the genetic background must be clearly understood. Furthermore, our understanding of gene function derives almost exclusively from analysis of laboratory strains, strains that generally are not capable of growth in the natural environment. It is necessary therefore to determine if a given gene, thought to be dispensable for laboratory growth, is dispensable in the natural environment. Alteration of existing traits is largely considered to be a benign genetic modification, and one that does not pose any risks to the

Genetic engineering of extrinsic traits of wine yeast

Aroma/flavor composition	Enhance: Malate degradation Terpene production Glycosidic cleavage of precursors Novel flavors production
Production performance	Express: Pectinase Cellulase Bacteriocins
Healthfulness of product	Enhance: Resveratrol production Protection of antioxidants Degradation of biogenic Amines

Figure 12.9 Proposed extrinsic or novel properties to be expressed in wine strains of *Saccharomyces*.

food supply. Since these same alterations could occur in the wild population, the threat of negative environmental effects due to lateral gene transfer is minimal.

The creation of recombinant strains, strains expressing DNA elements not normally found in *Saccharomyces* or in yeast in general, causes more concern among consumers and producers. The risks of creation of a strain expressing heterologous traits depend of course upon the trait and the source of the DNA. There is a greater threat to the producer: a modified strain possessing an activity that is not uniformly desired in all fermentations, and the appearance of the strain and the unwanted characters in other wine lots. Cross contamination of wine production lots is the norm, and any organism present in one fermentation has the potential to be present in all fermentations.

12.3.1 Fermentation Performance

There are several existing traits of *Saccharomyces* that it may be desirable to modify by genetic engineering. These fall into four categories: (1) the adjustment of traits associated with fermentative activity and growth, (2) the elimination of the production of negative traits or off characters, (3) the stimulation of the production of positive aroma and flavor characters by the yeast, and (4) the modification of yeast activities that impact grape aroma and flavor or wine stability and processing.

With respect to fermentation performance, the goal is to generate strains that are able to complete the fermentation; that is, consume the available sugar in spite of the appearance of hostile environmental conditions. Yeasts are subjected to numerous stress factors during fermentation of grape juice: the initial high osmolarity (22–26% sugar or more); the generation of high ethanol concentrations (11–17%), the low pH (3.0–3.9), competition from other organisms, the lack of nutrients, extremes of temperature, and the presence of inhibitory compounds such as acetate. These stress factors have two major impacts on yeast biology during wine fermentation. If the strain is not able to adapt to these conditions, fermentation will slow and can be arrested if conditions are severe enough (66). These arrested, or "stuck" fermentations, are notoriously difficult to restart, and lead to losses in production. In addition, the presence of stress in the environment may influence the spectrum of end products produced by the organism, leading to the appearance of negative characters in the wine. As with stuck fermentations, some of these characters are very difficult to remove or correct. Successful removal is often at the cost of stripping other desirable characters from the wine. Not surprisingly, the objectives for improvement of yeast stress response are to enhance osmotolerance, thermotolerance, tolerance to ethanol, and tolerance to the presence of inhibitory microbial compounds (reviewed in (26)). To define the genes that should be modified, researchers are actively investigating the role of various stress proteins under oenological conditions (reviewed in (67)). Genes identified as being highly expressed under specific stress conditions will be evaluated for their role in stress resistance, and for the effects of altered patterns of regulation of the stress response gene. Once this work is complete, heartier industrial strains will likely be brought to the marketplace.

One of the main causes of fermentation arrest is nutrient deficiencies, particularly of nitrogen. Nitrogen limitation of fermentation can be addressed by nutrient addition, as diammonium phosphate and other nitrogen containing nutrients can be added to the fermentation (25). There are two problems with this practice. First, it may leave excess nutrient levels in the wine, encouraging spoilage organisms and necessitating sterile bottling of the wine. Second, the addition of nitrogen sources will impact the formation of esters derived from amino acid catabolism, which can be important aroma characters in the wine. To address these issues, genetic alteration of nitrogen utilization is being investigated (68,69). Mutants defective in key regulatory factors controlling nitrogen assimilation are being evaluated for their impact on fermentation capacity.

Microarray or transcriptome analyses have been applied to wine strains (64,70,71,72). These studies are identifying genes and pathways important to normal fermentation progression and ethanol stress resistance. There is a remarkable consistency in the transcriptional profiles revealed in spite of the use of different strains, different growth conditions, and different array technologies. These analyses are providing many potential genetic targets of modification to create superior production strains.

The transcriptome array technology can also be used to aid in strain construction and characterization. The DNA complement of a strain can also be analyzed using these arrays. In this case, the array is blotted directly with labeled genomic DNA (73). In the case of oligonucleotide arrays, divergence from the original sequence of the laboratory strain of *S. cerevisiae* can be identified as a failure to hybridize to that particular oligonucleotide, with hybridization to other oligonucleotides still occurring. Allele diversity can then be determined for the strains under investigation. Crosses can be done and this same technique used to define regions of variation or interest that are not represented by a specific oligonucleotide. Once the allele differences have been determined, a hybrid diploid strain containing both of the different alleles can be created and the phenotype of each allele in the same genetic background can then be evaluated using the technique of reciprocal hemizygosity (73). In this technique, each allele is inactivated via a null mutation, leaving only one allele functional. This is done for both alleles, and the biological properties of the two strains compared to each other as well as to the original parent. Any differences in phenotype would be due to the specific allele that is being expressed. This allows the researcher to definitively determine the role of each individual allele in the biological property under investigation, while still retaining the diploid nature of the original strain. This technique can also be used to analyze quantitative traits in *Saccharomyces*. Quantitative traits are those that display a distinct dosage effect phenomenon, meaning that the phenotype is not plus or minus, but a differing degree of expression. In the case of wine strains, hydrogen sulfide production is an example of such a trait. Strains produce varying amounts of this compound in response to identical environmental conditions. In other words, strains are not producing or nonproducing, but are high, moderate, low or nonproducers of hydrogen sulfide. The quantitative nature of the trait is due to the involvement of several different genes. Expression of yeast aroma compounds is similarly dependent upon multiple genes, the concerted action of which confers the quantitative nature of the phenotype. Use of quantitative trait locus analysis in combination with reciprocal hemizygosity should provide important information on construction strategies for more complex characteristics of wine yeast strains.

12.3.2 Loss of Negative Characters

Another important area of investigation in the construction of improved commercial strains is the elimination of negative characters; that is, those that are undesirable from a sensory perspective, in the finished wine. These undesirable compounds fall into several groups: the sulfur volatiles, volatile acids, higher alcohols, aldehydes, volatile phenol derivatives, and esters. Sulfur volatiles include hydrogen sulfide (H_2S) and other sulfides, thiols, and mercaptans. These compounds derive from the reduction of sulfate for biosynthesis (H_2S) or from the degradation of sulfur containing amino acids as nitrogen sources (25). Inactivation of the gene *MXR1*, which encodes methionine sulfoxide reductase, prevents formation of dimethyl sulfide (74), a compound found in wine that has a rotten clam aroma. The production of these characters could also be eliminated by mutation of the genes responsible for their appearance (75), which would create strains that were methionine or cysteine auxotrophs. It is unlikely such strains would be competitive against non-auxotrophic strains, and supplementation with the sulfur-containing amino acids may lead to the formation of off characters by other non-*Saccharomyces* organisms.

Similar concerns exist for the inhibition of the production of acetate, other acids, and aldehydes. These compounds are likely produced under conditions requiring balancing of the oxidation/reduction status of the cytoplasm. Preventing this balancing will likely impact the competitiveness of the strain in the natural environment. In the case of organic acids, disruption of the tricarboxylic acid (TCA) cycle altered the levels of TCA acids produced by the yeast (76). Mutation of the *FUM1* gene resulted in loss of respiratory ability, but the strains were still able to ferment (76). However, the competitiveness of these strains versus wild type strains was not evaluated.

Volatile phenols may be produced by yeast during fermentation, likely as a means of resistance to the inhibitory effects of specific phenolic compounds. However, these compounds confer negative aroma characters that are not desired in the wine. Strains carrying mutations blocking phenol modification have been isolated (77). Alteration of the ability to eliminate inhibitory compounds would obviously be contraindicated in the presence of these compounds, if the strain were to survive. Thus, while the targets of gene disruption to eliminate off character production are obvious, such genetic modifications would confer a biological disadvantage to the organism. This is one of the challenges of genetically engineering a strain that must retain full competitiveness against wild organisms of the same species.

An alternative, although more time-consuming, strategy is to evaluate the genetic basis of loss or reduction in synthesis of these negative traits that occurs throughout the native yeast populations. The fact that these strains are still competent to complete a fermentation indicates that they have dealt with the loss of the ability to create a negative character in a manner not affecting competitiveness in the native environment.

12.3.3 Enhanced Production of Positive Characters

Just as there are undesirable yeast traits, there are also positive yeast traits some winemakers would like to see enhanced. Yeast esters, derived from amino acid degradation, confer generic fruity and floral characters to a wine as well as some yeast specific notes, such as toasty. Currently these aromas may accompany the production of off characters also derived from amino acid degradation, so the goal is to increase the concentration of the positive characters without increasing the negative. The level of expression of these compounds must also be controlled. At low levels the esters add to the complexity of a wine, but at high levels they can dominate the overall profile and decrease the perception of complexity. Increased expression of some catabolic amino acid reactions should lead to enhanced ester formation. Over expression of alcohol acetyltransferase (*ATF1*), one of the enzymatic activities responsible for ester formation, increased the production of fruity esters in wine strains (78,79). Ester production was further enhanced by mutative loss of the *IAH1* gene that encodes a yeast esterase (80).

Another character thought to have beneficial sensory effects is glycerol. Glycerol is believed to positively impact mouth feel and to add a certain element of sweetness to the wine. Strains of *Saccharomyces* have been genetically engineered to increase glycerol production (81,82). These strains have other problems — the increase in production of glycerol disturbs the redox balance of the cells such that more acetic acid is produced (81,82). Additional modification is therefore required to reduce or prevent acetic acid production (83,84). These strains have been used in pure culture to generate wines of elevated glycerol content; however, their effectiveness and persistence under industrial conditions is not clear.

The winemaking technique of "sur lies" aging refers to aging the wine for an extended period of time on the yeast lees. During this time, which is on the order of several months, the yeast cells undergo autolysis; that is, they self degrade, due to the release of enzymes from the yeast vacuole. The activity of these enzymes on both wine and yeast components alters the composition of the wine, generating a more "silky" mouth feel and specific sur lies characters. Construction of yeast strains that more quickly or more readily generate these positive characters would be desirable.

Yeast strains that express enzymatic activities enhancing grape flavor and aroma characters are also being developed and tested. These strains express higher levels of innate activities impacting flavor compound production, or contain mutations of native genes leading to the release of metabolic intermediates that possess an aromatic character (78,80,85,86,87). Mutation of the *ERG20* gene, which is involved in sterol biosynthesis, leads to the release of terpenes by yeast (88). Terpenes are responsible for the characteristic odors of Muscat grapes. The use of these strains confers a Muscat note to the wines.

Finally, the role of *Saccharomyces* in the formation of volatile thiols from S-cysteine conjugate precursors has recently been established (29,89). These compounds are important aroma constituents of white wines (90). Classical breeding strategies have been used to develop hybrid strains between *S. cerevisiae* and *S. bayanus uvarum* producing high levels of volatile thiols (29). Once the mechanism by which these strains yield high volatile thiol contents have been elucidated, targeted gene constructions can be conducted to transfer this property to any desired commercial strain.

It is unclear how well accepted these strains will be by the public. There is the concern that such yeasts will compensate for poor harvesting or processing decisions on the part of the winemaker and will modify the regional characteristics of the fruit to the point of obscurity. Many wines are marketed by the special characters imparted to the grapes by virtue of the growing conditions: the interplay of soil, geography, and climate. If this is altered significantly by the yeast strain used, regional wines might not be so highly prized in the marketplace.

12.3.4 Facilitation of Wine Processing and Enhancement of Wine Organoleptic Properties

Many yeast strains are naturally flocculent and will clump during growth in liquid media. Flocculent wine yeasts have been developed to allow better settling of the yeast lees from the wine. Introduction of enhanced flocculation does not appear to affect fermentation performance or competitiveness with other yeast strains (91). Flocculating yeasts are particularly important in sparkling wine production following the secondary alcoholic fermentation that occurs in the bottle. The process of riddling settles the yeast cake in the tip of the bottle so that it can be removed. If the yeast strain more readily clumps, a more complete reduction of yeast biomass can result.

Researchers are also generating recombinant strains of *Saccharomyces* that can facilitate wine processing or the attainment of wine stability. Proteins present in wine postfermentation can denature, leading to the formation of a visible haze. Yeast strains expressing vacuolar acid protease externally can decrease wine protein content, thereby reducing haze (92). However, certain peptides confer a bitter character, so it is important to assess the bitterness level of protease treated wines. To date, enhanced expression of the *Saccharomyces* native proteases has been attempted. Expression of heterologous proteases can be tricky, as the protease has to remain inactive until release into the external environment so as not to damage the internal constituents of the cell.

Wine tannins are responsible for astringency. Different yeast strains appear to remove tannins from the wine at differing rates and to different extents. Although the mechanism of tannin removal by yeast is not known, once this process becomes better understood, genetic manipulation of yeast may allow alteration of tannin binding activity and permit adjustment of the natural tannin content, and therefore astringency, of a wine.

Yeast mannoproteins, constituents of the yeast cell surface, have been shown to be released during fermentation and to facilitate stabilization of the wine against the appearance of a protein haze (93). Again, the mechanism of action of yeast mannoproteins in haze reduction is not known, but once it is elucidated, strains with enhanced or targeted mannoprotein release may be developed.

Other genetic alterations that are being explored are designed to make wine safer for human consumption or to enrich its nutritional content; for example, efforts to genetically engineer strains that reduce ethyl carbamate levels in the finished wine. Ethyl carbamate is a naturally occurring carcinogen formed in wine and other fermented products spontaneously from the reaction of carbamyl group-containing compounds such as urea and ethanol (94,95,96). The objective of this research is to design strains that do not release carbamyl donors such as urea to the medium during fermentation.

It is also possible to engineer strains that would elevate the nutritional value of wine by increasing the content of micronutrients. This is being explored by the creation of strains that more efficiently sequester vitamins and minerals from the environment and that then release those components upon cell death or autolysis. Construction of vitamin over producing strains is also under consideration. Such strains have been successfully developed for the brewing industry. However, it is important to remember that *Saccharomyces* shares the grape juice fermentation with a host of other organisms, and the excess vitamin may simply feed a neighbor removed along with *Saccharomyces* during filtration and not remain in the wine at the time of bottling.

12.4 NOVEL TRAITS

There are several enzymatic activities that would be of benefit if expressed in wine yeast during the fermentation. These novel activities impact grape flavor and aroma or facilitate the processing or stability of the wine. A relatively recent suggestion is to generate yeast strains that will enhance the antioxidant character or potential of a wine.

12.4.1 Flavor Production

There are three goals of the creation of more flavorful yeast: (1) the use of yeast to enhance the natural grape flavors and aromas, (2) the creation of strains that produce novel flavors not currently found in the grape, and (3) the construction of yeast strains that will carry out some of the important reactions associated with the malolactic fermentation. Many grape aroma compounds are present in the fruit in the form of nonvolatile precursors, or glycosidically bound molecules (reviewed in (79)). The glycoside moiety holds the compound in solution. Only the free compounds (nonglycosidically bound) are aromatic. Unbound or free terpenes impart floral and intense fruity notes to grapes and grape juice. The attachment of a glycoside group prevents volatilization and therefore detection of these compounds. Bound terpenes are hydrolyzed over time thereby maintaining the aroma characteristics of the fruit. The grape glycosidase are inhibited by glucose and are therefore not active in juice or during the early stages of fermentation. Once fermentation has been completed and the sugar content decreased, these enzymes can slowly hydrolyze the precursor compounds over time. Yeast strains expressing a fungal β(1-4) endoglucanase have been generated (98,99,100,101,102). These strains result in the release of more intense fruity character in the wine. However, the use of these strains may impact the long term aging potential of the wine, and lead to the loss of too many characters too quickly.

Malate and tartrate represent the two major acid types found in grapes. In cooler growing regions, malate levels and therefore acidity and sourness may be unacceptably high at the time of harvest. The lactic acid bacteria can metabolize the dicarboxylic acid malate to the monocarboxylic lactate and thereby reduce acidity (25). Yeast and the lactic acid bacteria compete for nutrients during fermentation and produce compounds inhibitory toward the other species, and it can be difficult to obtain both the alcoholic fermentation and malolactic conversion in the same wine. The lactic acid bacteria produce a spectrum of organoleptic characters in addition to the reduction in sourness. Some of these traits, such as the cream and buttery notes, are desired in some wines, while other traits (the mousy and other animal characters) are not.

Several teams of researchers have constructed strains of *Saccharomyces* that will metabolize malate thus reducing acidity without the need for bacterial malolactic fermentation. One early strategy was to express the bacterial malolactic enzyme in yeast (103,104,105).

A different strategy based upon creating strains of *Saccharomyces* capable of metabolizing malate to ethanol has also been pursued (106,107). This strategy takes advantage of the ability of the yeast *Schizosaccharomyces pombe* to degrade malate to ethanol. Malate is first converted to pyruvate, which is subsequently degraded to ethanol and CO_2. Although *S. cerevisiae* possesses the same enzymatic activities, the substrate affinity of the *Saccharomyces* malic enzyme for malate is too low to result in sufficient degradation of malate. Further, *Saccharomyces* does not possess an efficient malate permease. The genes encoding the *S. pombe* malate permease (*mae1*) and the malic enzyme (*mae2*) were both transformed into wine strains of *S. cerevisiae* (106). The recombinant strain efficiently degraded malate to ethanol (107).

12.4.2 Facilitation of the Production Process

Similarly to protein hazes, polysaccharides can form visible cloudiness in wine. Yeast strains possessing enzymatic activities to break down these complex polysaccharides are under development. Similarly, wine filterability is improved by degradation of grape polysaccharides. Strains of *Saccharomyces* with pectinase or cellulase activity have been constructed that offer an alternative to the use of commercial enzyme preparations (108, 109,110,111,112,113,114,115,116). The advantage of a recombinant wine strain would be the elimination of any potentially undesirable side activity that is present in the commercial preparations. However, an enzyme addition may be easier to control and monitor than an activity produced by a recombinant strain.

Wine strains of *Saccharomyces* producing bacteriocins have also been constructed (117). Bacteriocin production by yeast strains offers many advantages over the current practice of use of sulfur dioxide. Individuals deficient in sulfite oxidase display a strong hypersensitivity to sulfite. In severe cases, exposure can lead to death. The role of sulfite in elimination of unwanted bacteria can be effectively carried out by bacteriocins. In addition, bacteriocins show more selectivity than sulfite and can be used to inhibit specific classes of bacteria, not harming those that are desired in the wine.

12.4.3 Improved Healthfulness of Product

Finally, efforts are underway to generate recombinant yeast strains that will increase the levels of important phytochemicals such as resveratrol (118,119). The reductive environment created by the yeast during fermentation is important in the retention of these compounds in their bioactive form. The aim of this work is to increase the level of bioactive phenolic compounds that have been shown to have positive dietary outcomes in the prevention of disease. Such engineered strains may be of more value to the population as a whole than those leading to subtle differences in wine aroma and flavor or the elimination of economic loss for producers. Also, yeast may be used to remove other compounds in the wine for which there are health concerns such as the biogenic amines (histamine, tyramine and tryptamine), implicated as causative agents of wine headache, that can be produced by lactic acid bacteria (120). It may be possible to produce strains of *S. cerevisiae* that will lead to reduction in the levels of the precursors of these compounds.

12.5 FUTURE PERSPECTIVE

The ability to alter the genetic composition of wine strains of *Saccharomyces* appears limitless. A significant amount of information is accumulating on the physiology of this organism under native environment or grape juice fermentation conditions providing researchers with a fertile database upon which to improve inherent traits or guide the construction of strains expressing novel activities. This review covered some of the current constructions that are underway in several laboratories. Other desirable modifications will emerge as we learn more about the nature of wine flavor and aroma. Genetic engineering of wine yeast poses some unusual challenges, such as the need to maintain strain competitiveness against wild types of the same species and against other organisms, the inability to disturb stress tolerance, and the necessity to sustain patterns of metabolites produced. In essence, the modification needs to have little to no impact on the other biological properties of the organism. Because of this fact, most of the changes to wine yeast are expected to be benign and well characterized, with no risk to the consumer.

In addition to modification of *Saccharomyces*, alteration of other yeasts, those present during the fermentation and those with the potential to spoil the finished wine, can also be explored. Many of these yeasts play important roles in wine aroma and flavor. They also produce off characters that are undesirable. Engineering of strains with enhanced production of positive traits and reduced formation of undesired characters would be beneficial to the wine industry.

REFERENCES

1. Kellis, M., N. Patterson, M. Endrizzi, B. Birren, E.S. Lander. Sequencing and comparison of yeast species to identify genes and regulatory elements. *Nature* 423:241–254, 2003.
2. Cliften, P., P. Sudarsanam, A. Desikan, L. Fulton, B. Fulton, J. Majors, R. Waterston, B.A. Cohen, M. Johnston. Finding functional features in *Saccharomyces* genomes by phylogenetic footprinting. *Science* 301:71–76, 2003.
3. Bisson, L.F., R.E. Kunkee. Microbial interactions during wine production. In: *Mixed Cultures in Biotechnology*, Zeikus, G., E.A. Johnson, eds., New York: McGraw-Hill, 1991, pp 37–68.
4. Pretorius, I.S., T.J. Van der Westhuizen, O.P.H. Augustyn. Yeast biodiversity in vineyards and wineries and its importance to the South African wine industry. *S. Afr. J. Enol. Vitic.* 20:61–74, 1999.
5. Lachance, M.A., W.T. Stramer. *Ecology and yeasts*. In: *The Yeasts: A Taxonomic Study*, Kurtzman, C.P., J.W. Fell, eds., Amsterdam: Elsevier, 1998, pp 21–30.
6. Marinoni, G., M. Manuel, R.F. Peterson, J. Hvidtfeldt, P. Sulo, J. Piskur. Horizontal transfer of genetic material among *Saccharomyces* yeasts. *J. Bacteriol.* 181:6488–6496, 1999.
7. Goffeau, A., B.G. Barrell, H. Bussey, R.W. Davis, B. Dujon, H. Feldmann, F. Galibert, J.D. Hoheisel, C. Jacq, M. Johnston, E.J. Louis, H.W. Mewes, Y. Murakami, P. Philippsen, H. Tettelin, S.G. Oliver. Life with 6000 genes. *Science* 274:546–547, 1996.
8. Wu, L.F., T.R. Hughes, A.P. Davierwala, M.D. Robinson, R. Stoughton, S.J. Altschuler. Large-scale production of *Saccharomyces cerevisiae* gene function using overlapping transcriptional clusters. *Nat.Genet.* 31:255–265, 2002.
9. Entian, K.D., P. Kotter. Yeast mutant and plasmid collections. In: *Yeast Gene Analysis*, Brown, A.J.P., M.F. Tuite, eds., San Diego: Academic Press, 1998, pp 431–449.
10. Uetz, P., L. Giot, G. Cagney, T.A. Mansfield, R.S. Judson, J.R. Knight, D. Lockshon, V. Narayan, M. Srinivasan, P. Pochart, A. Qureshi-Emili, Y. Li, B. Godwin, D. Conover, T. Kalbfleisch, G. Vijayadamodar, M. Yang, M. Johnston, S. Fields, J.M. Rothberg. A comprehensive analysis of protein-protein interactions in *Saccharomyces cerevisiae. Nature* 403:623–627, 2000.
11. Lecrenier, N., F. Foury, A. Goffeau. Two-hybrid systematic screening of the yeast proteome. *Bio. Essays* 20:1–6, 1998.
12. Herskowitz, I., Y. Oshima. Control of cell type in *Saccharomyces cerevisiae*: Mating type and mating-type interconversion. In: *The Molecular Biology of the Yeast* Saccharomyces*: Life Cycle and Inheritance*, Strathern, J.N., E.W. Jones, J.R. Broach, eds., Cold Spring Harbor, New York: Cold Spring Harbor Laboratory Press, 1981, pp 181–209.
13. Vaughan, A., A. Martini. Facts, myths and legends on the prime industrial microorganism. *J. Ind. Microbiol.* 14:514–522, 1995.
14. Mortimer, R., M. Polsinelli. On the origins of wine yeast. *Res. Microbiol.* 150:199–204, 1999.
15. Miklos, I., T. Varga, A. Nagy, M. Sipiczki. Genome instability and chromosome rearrangements in a heterothallic wine yeast. *J. Basic Microbiol.* 37:345–354, 1997.
16. Puig, S., A. Querol, E. Barrio, J.E. Perez-Ortin. Mitotic recombination and genetic changes in *Saccharomyces cerevisiae* during wine fermentation. *Appl. Environ. Microbiol.* 66:2057–2061, 2000.
17. Rachidi, N., P. Barre, B. Blondin. Multiple Ty-mediated chromosomal translocations lead to karyotype changes in a wine strain of *Saccharomyces cerevisiae. Mol. Gen. Genet.* 261:841–850, 1999.
18. Codon, A.C., T. Benitez, M. Korhola. Chromosomal polymorphism and adaptation to specific industrial environments of *Saccharomyces* strains. *Appl. Microbiol. Biotechnol.* 49:154–163, 1998.
19. Winzeler, E.A., D.R. Richards, A.R. Conway, A.L. Goldstein, S. Kalman, M.J. McCullough, J.H. McCusker, D.A. Stevens, L. Wodicka, D.J. Lockhart, R.W. Davis. Direct allelic variation scanning of the yeast genome. *Science* 281:1194–1197, 1998.

20. Johnston, J.R., C. Baccari, and R.K. Mortimer. Genotypic characterization of strains of commercial wine yeasts by tetrad analysis. *Res. Microbiol.* 151:583–590, 2000
21. Querol, A., D. Ramon, The application of molecular techniques in wine microbiology. *Trends Food Sci. Tech.* 7:73–78, 1996.
22. Cavalieri, D., J.P. Townsend, D.L. Hartl. Manifold anomalies in gene expression in a vineyard isolate of *Saccharomyces cerevisiae* revealed by DNA microarray analysis. *Proc. Natl. Acad. Sci. USA* 97:12369–12374, 2000.
23. Kidwell, M.G., D.R. Lisch. Transposable elements and host genome evolution. *Trends Ecol. Evol.* 15:95–99, 2000.
24. Benitez, T., J.M. Gasent-Ramirez, F. Castrejon, A.C. Codon. Development of new strains for the food industry. *Biotechnol. Prog.* 12:149–163, 1996.
25. Boulton, R.B., V.L. Singleton, L.F. Bisson, R.E. Kunkee. *Principles and Practices of Winemaking*, 1st ed. New York: Chapman & Hall, 1996.
26. Pretorius, I.S. Tailoring wine yeast for the new millennium: novel approaches to the ancient art of winemaking. *Yeast.* 16:675–729, 2000.
27. Enari, T.M. Genetic modification of food and beverage yeast. *Ann. N.Y. Acad. Sci.* 646:181–192, 1991.
28. Thornton, R.J. An introduction to the genetics of industrial yeasts. *Proceedings of the Alko Symposium on Industrial Yeast Genetics, Helsinki*, 1987, pp. 11–25.
29. Masneuf, I., M.L. Murat, G.I. Naumov, T. Tominaga, D. Dubourdieu. Hybrids *Saccharomyces cerevisiae* X *Saccharomyces bayanus* var. *uvarum* having a high liberating ability of some sulfur varietal aromas of *Vitis vinifera* Sauvignon blanc wines. *J. Int. Sci. Vigne. Vin.* 36:205–212, 2002.
30. Spencer, J.F.T., D.M. Spencer, N. Reynolds. Protoplast fusion for the improvement of industrial yeasts, *Proceedings of the Alko Symposium on Industrial Yeast Genetics, Helsinki*, 1987, 27–42.
31. Cox, B.S. Genetic analysis in *Saccharomyces cerevisiae* and *Schizosaccharomyces pombe.* In: *The Yeasts, Vol. 6: Yeast Genetics*, Wheals, A.E., A.H. Rose, eds., London: Academic Press, 1995, pp 7–67.
32. Georgieva, B., R. Rothstein. *kar*-mediated plasmid transfer between yeast strains: alternative to traditional transformation methods. In: *Guide to Yeast Genetics and Molecular and Cellular Biology, Part B: Methods in Enzymology*, Vol. 350, Guthrie, C., G.R. Fink, eds., New York: Academic Press, 2002, pp 278–289.
33. Rose, M.D. Modern and postmodern genetics in *Saccharomyces cerevisiae*. In: *The Yeasts, Vol. 6: Yeast Genetics*, Wheals, A.E., A.H. Rose, eds., London: Academic Press, 1995, pp 69–120.
34. Olesen, K., P.F. Johannesen, L. Hoffman, S.B. Sorensen, C. Gjermansen, J. Hansen. The pYC plasmids, a series of cassette-based yeast plasmid vectors providing means of counter-selection. *Yeast* 16:1035–1043, 2000.
35. Sikorski, R.S., P. Hieter. A system of shuttle vectors and yeast host strains designed for efficient manipulation of DNA in *Saccharomyces cerevisiae. Mol. Cell Biol.* 11:19–27, 1989.
36. Vidal, M., R.F. Gaber. Selectable marker replacement in *Saccharomyces cerevisiae. Yeast* 10:141–149, 1994.
37. Voth, W.P., Y.W. Jiang, D.J. Stillman. New 'marker swap' plasmids for converting selectable markers on budding yeast gene disruptions and plasmids, *Yeast.* 20:985–993, 2003.
38. Amberg, D.C., D. Botstein, E.M. Beasley. Precise gene disruption in *Saccharomyces cerevisiae* by double fusion polymerase chain reaction. *Yeast* 11:1275–1280, 1995.
39. Reid, R.J.D., M. Lisby, R. Rothstein. Cloning-free genome alterations in *Saccharomyces cerevisiae.* In: *Guide to Yeast Genetics and Molecular and Cellular Biology, Part B: Methods in Enzymology*, Vol. 350, Guthrie, C., G.R. Fink, eds., New York: Academic Press, 2002, pp 258–277.
40. Wach, A., A. Bracht, C. Rebischung, S. Steiner, K. Pokorni, S. Tetteesen, P. Philippsen. PCR-based gene targeting in *Saccharomyces cerevisiae.* In *Yeast Gene Analysis*, Brown, A.J.P., M.F. Tuite, eds., San Diego: Academic Press, 1998, pp 67–81.

41. Steiner, S., P. Philippsen. Sequence and promoter analysis of the highly expressed TEF gene of the filamentous fungus *Ashbya gossypii*. *Mol. Gen. Genet.* 242:263–271, 1994.
42. Fukuda, H., Y. Kizaki. A new transformation system of *Saccharomyces cerevisiae* with blasticidin S deaminase gene. *Biotechnol. Lett.* 21:969–971, 1999.
43. Van Den Berg, M.A., H.Y. Steensma. Expression cassette for formaldehyde and fluoroacetate resistance, two dominant markers in *Saccharomyces cerevisiae*. *Yeast* 13:551–559, 1997.
44. Goldstein, A.L., J.H. McCusker. Three new drug resistance cassettes for gene disruption in *Saccharomyces cerevisiae*. *Yeast* 15:1541–1553, 1999.
45. Park, H., N.I. Lopez, A.T. Bakalinsky. Use of sulfite resistance in *Saccharomyces cerevisiae* as a dominant selectable marker. *Curr. Genet.* 36:339–344, 1999.
46. Schmidt, M., A. Coemer, M. Greg, M. Brendel. Genetic engineering of baker's and wine yeasts using formaldehyde hyperresistance-mediating plasmids, *Brz. J. Med. Biol. Res.* 30:1407–1414, 1997.
47. Hottiger, T., J. Kuhla, G. Pohlig, P. Fuerst, A. Spielmann, M. Garn, S. Haemmerli, J. Heim. 2-mu vectors containing the *Saccharomyces cerevisiae* metallothinein gene a selectable marker: excellent stability in complex media, and high-level expression of a recombinant protein from a *CUP1*-promoter controlled expression cassette in cis. *Yeast.* 11:1–14, 1995.
48. Kozovska, Z., A. Maraz, I. Magyar, J. Subik. Multidrug resistance as a dominant molecular marker in transformation of wine yeast. *J. Biotechnol.* 92:27–35, 2001.
49. Akada, R., Y. Shimizu, Y. Mastsushita, H. Kawahata, Y. Nishizawa. Use of a *YAP1* overexpression cassette conferring specific resistance to cerulenin and cycloheximide as an efficient selectable maker in the yeast *Saccharomyces cerevisiae*. *Yeast* 19:17–28, 2002.
50. Boeke, J.D., F. Lacroute, G.R. Fink. A positive selection for mutants lacking orotidine-5'-phosphate decarboxylase activity in yeast: 5-fluoro-orotic acid resistance. *Mol. Gen. Genet.* 197:345–346, 1984.
51. Chattoo, B.B., F. Sherman, D.A. Azubalis, T.A. Fjellstedt, D. Mehnert, M. Ogur. Selection of *lys2* mutants of the yeast *Saccharomyces cerevisiae* by the utilization of α-aminoadipate. *Genetics* 93:51–65, 1979.
52. Singh, A., F. Sherman. Association of methionine requirement with methyl mercury resistance mutants of yeast. *Nature* 247:227–229, 1974.
53. Cost, G.J., J.D. Boeke. A useful colony colour phenotype associated with the yeast selectable/counter-selectable marker *MET15*. *Yeast* 12:939–941, 1996.
54. Akada, R., I. Hirosawa, M. Kawahata, H. Hoshida, Y. Nishizawa. Sets of integrating plasmids and gene disruption cassettes containing improved counter-selectable makers designed for repeated use in budding yeast. *Yeast* 19:393–402, 2000.
55. Puig, S., A. Querol, D. Ramon, J.E. Perez-Ortin. Evaluation of the use of phase-specific gene promoters for the expression of enological enzymes in an industrial wine yeast strain. *Biotechnol. Lett.* 18:887–892, 1996.
56. Funk, M., R. Niedenthal, D. Mamberg, K. Brinkmann, V. Ronicke, T. Henkel. Vector systems for the heterologous expression of proteins in *Saccharomyces cerevisiae*. In: *Guide to Yeast Genetics and Molecular and Cellular Biology, Part B: Methods in Enzymology*, Vol. 350, Guthrie, C., G.R. Fink, eds., New York: Academic Press, 2002, pp 248–257.
57. VanZyl, W.H., E.J. Lodolo, M. Gericke. Conversion of homothallic yeast to heterothallism through HO gene disruption. *Genetics* 23:290–294, 1993.
58. Schena, A., D. Shalon, R.W. Davis, P.O. Brown. Quantitative monitoring of gene expression patters with a complementary DNA microarray. *Science* 270:467–470, 1995.
59. DeRisi, J.L., V.R. Iyer, P.O. Brown. Exploring the metabolic and genetic control of gene expression on a genomic scale. *Science* 278:680–686, 1997.
60. Futcher, B., G.I. Latter, P. Monardo, C.S. McLaughlin, J.I. Garrels. A sampling of the yeast proteome. *Mol. Cell. Biol.* 19:7357–7368, 1999.
61. Gasch, A.P., P.T. Spellman, C.M. Kao, O. Carmel-Huvel, M.B. Eisen, G. Storz, D Botstein, P.O. Brown. Genomic expression programs in the response of yeast cells to environmental changes. *Mol. Biol. Cell.* 11:4241–4257, 2000.

62. Kang, J.J., R.M. Watson, M.E. Fisher, R. Higuchi, D.H. Gelfand, M.J. Holland. Transcript quantitation in total yeast cellular RNA using kinetic PCR. *Nucl. Acids. Res.* 28:1–8, 2000.
63. DeRisi, J., B. van den Hazel, P. Marc, E. Balzi, P. Brown, C. Jacq, A. Goffeau. Genome microarray analysis of transcriptional activation in multidrug resistance yeast mutants. *FEBS Lett.* 470:156–160, 2000.
64. Backhus, L.E., J. DeRisi, L.F Bisson, Functional genomic analysis of a commercial wine strain of *Saccharomyces cerevisiae* under differing nitrogen conditions. *FEMS Yeast Res.* 1:111–125, 2001.
65. Shevchenko, A., O.N. Jensen, A.V. Podtelejnikov, F. Sagliocco, M. Wilm, O. Vorm, P. Mortensen, H. Boucherie, M. Mann. Linking genome and proteome by mass spectrometry: large scale identification of yeast proteins from two dimensional gels. *Proc. Natl. Acad. Sci. USA* 93:14440–14445, 1996.
66. Bisson, L.F. Stuck and sluggish fermentations. *Am. J. Enol. Vitic.* 50:1–13, 1999.
67. Bauer, F.F., I.S. Pretorius. Yeast stress response and fermentation efficiency: how to survive the making of wine: a review. *S. Afr. J. Enol. Vitic.* 21:5–26, 2000.
68. Salmon, J.M., P. Barre. Improvement of nitrogen assimilation and fermentation kinetics under enological conditions by derepression of alternative nitrogen-assimilatory pathways in an industrial *Saccharomyces cerevisiae* strain. *Appl. Environ. Microbiol.* 64:3831–3837, 1998.
69. Martin, O., M.C. Brandriss, G. Schneider, A.T. Bakalinsky. Improved anaerobic use of arginine by *Saccharomyces cerevisiae*. *Appl. Environ. Microbiol.* 69:1623–1628, 2003.
70. Marks, V.D., G.K. van der Merwe, H.J.J. van Vuuren. Transcriptional profiling of wine yeast in fermenting grape juice: regulatory effect of diammonium phosphate. *FEMS Yeast Res.* 3:269–287, 2003.
71. Erasmus, D.J., G.K. van der Merwe, H.J.J. van Vuuren. Genome-wide expression analysis: metabolic adaptation of *Saccharomyces cerevisiae* to high sugar stress. *FEMS Yeast Res.* 3:375–399, 2003.
72. Rossignol, T., L. Dulau, A. Julien, B. Blondin. Genome-wide monitoring of wine yeast gene expression during alcoholic fermentation. *Yeast* 20:1369–1385, 2003.
73. Steinmetz, L.M., H. Sinha, D.R. Richards, J.I. Spiegleman, P.J. Oefner, J.H. McCusker, R.W. Davis, Dissecting the architecture of a quantitative trait locus in yeast, *Nature* 416:326–330, 2002.
74. Hansen, J. Inactivation of *MXR1* abolishes formation of dimethyl sulfide from dimethyl sulfoxide in *Saccharomyces cerevisiae*. *Appl. Environ. Microbiol.* 65:3915–3919, 1999.
75. Jiranek, V., P. Langridge, P.A. Henschke. Determination of sulphite reductase activity and its response to assimilable nitrogen status in a commercial *Saccharomyces cerevisiae* wine yeast. *J. Appl. Bacteriol.* 81:329–336, 1996.
76. Magarifuchi, T., K. Goto, I. Kuniyasu, M. Tade, G. Tamura. Effect of yeast fumarase gene (*FUM1*) disruption on production of malic, fumaric and succinic acids in sake mash. *J. Ferm. Bioeng.* 80:355–362, 1995.
77. Van Wyk, C.J., I.M. Rogers. A "phenolic" off-odour in white table wines: Causes and methods to diminish its occurrence. *S. Afr. J. Enol. Vitic.* 31:52–57, 2000.
78. Lilly, M., M.G. Lambrechts, I.S. Pretorius. Effect of increased yeast alcohol acetyltransferase activity on flavor profiles of wine and distillates. *Appl. Environ. Microbiol.* 66:744–753, 2000.
79. Verstrepen, K.J., S.D.M. Van Laere, B.M.P. Vanderhagen, G. Derdelinckx, J.P. Dufour, I.S. Pretorius, J. Winderickx, J.M. Thevelein, F.R. Delvaux. Expression levels of the yeast alcohol acetyltransferase genes *AFT1*, *Lg-AFT1* and *AFT2* control the formation of a broad range of volatile esters. *Appl. Environ. Microbiol.* 69:5228–5237, 2003.
80. Fukuda, K., N. Yamamoto, Y. Kiyokawa, T. Yanagiuchi, Y. Wakai, Y. Kitamoto, Y. Inoue, A. Kimura. Balance of activities of alcohol acetyltransferase and esterase in *Saccharomyces cerevisiae* is important for production of isoamyl acetate. *Appl. Environ. Microbiol.* 64:4076–4078, 1998.
81. Michnick, S., J.L. Roustan, F. Remise, P. Barre, S. Dequin. Modulation of glycerol and ethanol yields during alcoholic fermentation in *Saccharomyces cerevisiae* strains overexpressed

or disrupted for *GPD1* encoding glycerol 3-phosphate dehydrogenase. *Yeast* 13:783–793, 1997.

82. Remiz, F., J.L. Roustan, J.M. Sablayrolles, P. Barre, S. Dequin. Glycerol overproduction by engineered *Saccharomyces cerevisiae* wine yeast strains leads to substantial changes in by-product formation and to stimulation of fermentation rate in stationary phase. *Appl. Environ. Microbiol.* 65:143–149, 1999.
83. Eglinton, J.M., A.J. Heinrich, A.P. Pollnitz, P. Langridge, P.A. Henschke, M. de Barros Lopes, Decreasing acetic acid accumulation by a glycerol over-producing strain of *Saccharomyces cerevisiae* by deleting the *ALD6* aldehyde dehydrogenase gene. *Yeast* 19:295–301, 2002.
84. Remize, F., E. Andrieu, S. Dequin. Engineering of the pyruvate dehydrogenase bypass in *Saccharomyces cerevisiae*: role of the cytosolic Mg^{2+} and mitochondrial K^{+} acetaldehyde dehydrogenases Ald6p and Ald4p in acetate formation during alcoholic fermentation. *Appl. Environ. Microbiol.* 66:3151–3159, 2000.
85. Bardi, L., C. Crivelli, M. Marzona. Esterase activity and release of ethyl esters of medium chain fatty acids by *Saccharomyces cerevisiae* during anaerobic growth. *Can. J. Microbiol.* 44:1171–1176, 1998.
86. Bardi, L., C. Cocito, M. Marzona, *Saccharomyces cerevisiae* cell fatty acid composition and release during fermentation without aeration and in absence of exogenous lipids. *Int. J. Food Microbiol.* 47:133–140, 1999.
87. Ganga, M.A., F. Pinaga, S. Valles, D. Ramon, A. Querol. Aroma improving microvinification processes by use of a recombinant yeast strain expressing the *Aspergillus nidulans xlnA* gene. *Int. J. Food Microbiol.* 47:171–178, 1999.
88. Javelot, C., P. Griard, B. Colonna-Ceccaldi, B. Vladescu. Introduction of terpene-producing ability in a wine strain of *Saccharomyces cerevisiae*. *J. Biotechnol.* 21:239–252, 1991.
89. Murat, M.L., I. Masneuf, P. Darriet, V. Lavigne, T. Tominaga, D. Dubourdieu. Effect of the *Saccharomyces cerevisiae* yeast strains on the liberation of volatile thiols in Sauvignon blanc wines. *Am. J. Enol. Vitic.* 52:136–139, 2000.
90. Tominaga, T., R. Baltenweck-Guyot, C. Peyrot des Gachons, D. Dubourdieu. Contribution of volatile thiols to the aromas of white wines made from several *Vitis vinifera* grape varieties. *Am. J. Enol. Vitic.* 51:178–181, 2000.
91. Shinohara, T., S. Mamiya, F. Yanagida. Introduction of flocculation property into wine yeasts (*Saccharomyces cerevisiae*) by hybridization. *J. Ferm. Bioeng.* 83:96–101, 1997.
92. Lagace, L.S., L.F. Bisson. Survey of yeast acid proteases for effectiveness of wine haze reduction. *Am. J. Enol. Vitic.* 41:147–155, 1990.
93. Dupin, I.V.S., B.M. McKinnon, C. Ryan, M. Boulay, A.J. Markides, G.P. Jones, P.J. Williams, E. Waters. *Saccharomyces cerevisiae* mannoproteins that protect wine from haze: their release during fermentation and lees contact and a proposal for their mechanism of action. *J. Agric. Food Chem.* 48:3098–3105, 2000.
94. Ough, C.S. Ethyl carbamate in fermented beverages and foods, I: naturally occurring ethyl carbamate. *J. Agric. Food Chem.* 24:323–327, 1976.
95. Ough, C.S., E.A. Crowell, B.R. Gutlove. Carbamyl compound reactions with ethanol. *Am. J. Enol. Vitic.* 39:139–142, 1988.
96. Monteiro, F.F., E.K. Trousdale, L.F. Bisson. Ethyl carbamate formation in wine: Use of radioactively labeled precursors to demonstrate the involvement of urea. *Am. J. Enol. Vitic.* 40:1–8, 1989.
97. Ebeler, S.E. Analytical chemistry: unlocking the secrets of wine flavor. *Food Rev. Int.* 17:45–64, 2001.
98. Perez-Gonzalez, J.A., R. Gonzalez, A. Querol, J. Sendra, D. Ramon. Construction of a recombinant wine yeast strain expressing β-(1,4)-endoglucanase and its use in microvinification processes. *Appl. Environ. Microbiol.* 59:2801–2806, 1993.
99. Van Rensburg, P., W.H. Van Zyl, I.S. Pretorius. Co-expression of a *Phanerochaete chrysosporium* cellobiohydrolase gene and a *Butyrivibrio fibrisolvens* endo-β-1,4-glucanase gene in *Saccharomyces cerevisiae*. *Curr. Genet.* 30:246–250, 1996.

100. Van Rensburg, P., W.H. Van Zyl, I.S. Pretorius. Over-expression of the *Saccharomyces cerevisiae* exo-β-1,3-glucanase gene together with the *Bacillus subtilis* endo-β-1,3-1,4 glucanase gene and the *Butyrivibrio fibrisolvens* endo-β-1,4-glucanase gene in yeast. *J. Biotechnol.* 55:43–53, 1997.
101. Van Rensburg, P., W.H. Van Zyl, I.S. Pretorius. Engineering yeast for efficient cellulose degradation. *Yeast* 14:67–76, 1998.
102. Manzanares, P., M. Orejas, J.V. Gil, L.H. de Graaff, J. Visser, D. Ramon. Construction of a genetically modified wine yeast strain expressing the *Aspergillus aculeatus rhaA* gene, encoding an α-L-rhamosidase of enological interest. *Appl. Environ. Microbiol.* 69:7558–7562, 2003.
103. Williams, S.A., R.A. Hodges, T.L. Strike, R. Snow, R.E. Kunkee. Cloning the gene for the malolactic fermentation of wine from *Lactobacillus delbruekii* in *Escherichia coli* and yeasts. *Appl. Environ. Microbiol.* 47:288–293, 1988.
104. Ansanay, V., S. Dequin, B. Blondin, P. Barre. Cloning, sequence and expression of the gene encoding the malolactic enzyme from *Lactococcus lactis*. *FEBS Lett.* 332:74–80, 1993.
105. Denayrolles, M., M. Aigle, A. Lonvaud-Funel. Functional expression in *Saccharomyces cerevisiae* of the *Lactococcus lactis mleS* gene encoding the malolactic enzyme. *FEMS Microbiol. Lett.* 125:37–44, 1993.
106. Volschenk, H., M. Vilijoen, J. Grobler, B. Petzold, F. Bauer, R.E. Subden, R.A. Young, A. Lonvaud, M. Denayrolles, H.J.J. Van Vuuren, *Engineering pathways for malate degradation in Saccharomyces cerevisiae. Nature Biotech.* 15:253–257, 1997.
107. Volschenk, H., M. Viljoen, J. Grobler, F. Bauer, A. Lonvaud-Funel, M. Denayrolles, R.E. Subden, H.J.J. Van Vuuren. Malolactic fermentation in grape must by a genetically engineered strain of *Saccharomyces cerevisiae*. *Am. J. Enol. Vitic.* 48:193–197, 1997.
108. Cantwell, B, G. Brazil, J. Hurley, D. McConnell. Expression of the gene for the endo-β-1,3-1,3- glucanase from *Bacillus subtilis* in *Saccharomyces cerevisiae* from *CYC1* and *ADH1* promoters. *Curr. Genet.* 11:65–70, 1986.
109. Skipper, N.A., R.P. Bozzato, D. Velter, R.W. Davis, R. Wong, J.E. Hopper. Use of the melibiase promoter and signal peptide to express a bacterial cellulase from yeast. In: *Biological Research in Industrial Yeast*, Stewart, G.G., P.R. Russell , R.D. Klein, R.R. Hiebsch, eds., Boca Raton Florida: CRC Press 1987, pp 137–148.
110. Laing, E., S. Pretorius. Co-expression of an *Erwinia chrysanthemi* pectate lyase-encoding gene (pelE) and an *Erwinia carotovora* poly-galacturonase-encoding gene (peh1) in *Saccharomyces cerevisiae*. *Appl. Microbiol. Biotech.* 39:181–188, 1993.
111. Ooi, T., K. Micamiguchi, T. Kawaguchi, H. Okada, S. Murao, M. Arai. Expression of the cellulase (F1-CMCase) gene of *Aspergilllus aculeatus* in *Saccharomyces cerevisiae*. *Biotechnol. Biochem.* 58:954–956, 1994.
112. Crowhurst, M.D., E.H.A. Rikkevrink. The pectin lyase encoding gene (pn1) family from *Glomerella cingulata*: characterization of pn1A and its expression in yeast. *Gene* 142:141–146, 1994.
113. Crous, J.M., I.S. Pretorius, W.H. Van Zyl. Cloning and expression of an *Aspergillus kawachii* endo 1,4-β-xylanase gene in *Saccharomyces cerevisiae*. *Curr. Genet.* 28:467–473, 1995.
114. Lang, C., A.C. Looman. Efficient expression and secretion of *Aspergillus niger* RH5344 polygalacturonase in *Saccharomyces cerevisiae*. *Appl. Microbiol. Biotechnol.* 44:147–156, 1995.
115. LaGrange, D.C., I.S. Pretorius, W.H. Van Zyl. Expression of the *Trichoderma reesei* β-xylanase gene (XYN2) in *Saccharomyces cerevisiae*. *Appl. Environ. Microbiol.* 62:1036–1044, 1996.
116. Luttig, M., I.S. Pretorius, W.H. Van Zyl. Cloning of two β-xylanase-encoding genes from *Aspergillus niger* and their expression in *Saccharomyces cerevisiae*. *Biotech. Lett.* 19:411–415, 1997.
117. Schoeman, H., M.A. Vivier, M. Du Toit, L.M.T. Dicks, I.S. Pretorius. The development of bactericidal yeast strains by expressing the *Pediococcus acidilactici* pediocin gene (pedA) in *Saccharomyces cerevisiae*. *Yeast* 15:647–656, 1999.

118. Gonzalez-Candelas, L., J.V. Gil, P.R. Lamuela-Raventos, D. Ramon. The use of transgeneic yeast expressing a gene encoding a glycosyl-hydrolase as a tool to increase resveratrol content in wine. *Int. J. Food Microbiol.* 59:179–183, 2000.
119. Becker, J.V.W., G.O. Armstrong, M.J. van der Merwe, M.G. Lambrechts, M.A. Vivier, I.S. Pretorius. Metabolic engineering of *Saccharomyces cerevisiae* for the synthesis of the wine-related antioxidant resveratrol. *FEMS Yeast Res.* 4:79–85, 2003.
120. Guerrini, S., S. Mangani, L. Granchi, M. Vincenzini. Biogenic amine production by *Oenococcus oeni*. *Curr. Microbiol.* 44:374–378, 2002.

1.13

Stress Tolerance, Metabolism, and Development: The Many Flavors of Trehalose

Mike A. Singer, Tiago F. Outeiro, and Susan Lindquist

CONTENTS

13.1 INTRODUCTION

Trehalose has a remarkable capacity to protect cells and their component macromolecules against a daunting array of environmental stresses – from extreme heat (1,2) to freezing cold (3), from anoxia (4) to oxidative damage (5), from osmotic stress (6) to withering dehydration (7). The disaccharide is one of a group of unrelated small molecules known alternatively as compatible solutes (osmolytes) or chemical chaperones — the former because even high concentrations do not inhibit enzyme activity (8), and the latter in recognition of their stabilizing effects on proteins (9). Their properties read like the pages of a superhero comic, and indeed harnessing them has long been largely a province of the imagination, making investigation in this area an exciting frontier in food biotechnology.

The discovery of trehalose was itself the product of efforts in food biotechnology. Working in Germany in the early nineteenth century, H.A.L. Wiggers studied ergot, a toxic fungus which contaminates rye and other cereal crops. Wiggers reported among its constituents a new sugar, later named trehalose. (For further background on the colorful history of trehalose, please see (10) and (11).) Trehalose has since been identified in organisms across an extraordinarily broad spectrum of taxonomy, including bacteria, fungi, plants, insects, and invertebrates (12). We focus here, after a brief review of its structure and proposed mechanisms of function, on the role of trehalose in the budding yeast *Saccharomyces cerevisiae*. In addition to its great commercial importance in baking and brewing, *S. cerevisiae* is a well characterized model system offering convenient and powerful techniques for genetic manipulation. Application of these methods has provided significant insight into complex molecular process even far removed from yeast itself, including cancer (13), aging (14), and neurodegenerative disorders (15–19).

Work by many investigators has revealed that protection from stress is but one aspect of the emerging multifaceted role of trehalose in yeast. Not only synthesis of the sugar, but also its degradation is necessary to optimize survival of extreme heat. Equally intriguing is the finding that the enzymes and products in the trehalose pathway critically influence fundamental cellular metabolism, gene expression, and development. Similar recent discoveries in *Arabidopsis* and *Drosophila* offer the compelling prospect that this link between the mundane and the extreme may be highly conserved, with significant implications for biology and biotechnology.

13.2 STRUCTURE AND PROTECTIVE MECHANISMS

The structure of trehalose is elegantly simple: two monomers of D-glucose joined at their alpha-carbons. This flexible glycosidic bond, together with the absence of internal hydrogen bonds, yields a supple molecule thought to conform readily to the irregular polar groups of proteins (20). While the bond bends, it does not easily break: the 1 kcal/mol linkage is highly resilient, enabling the molecule to withstand a wide range of temperature and pH conditions (21).

Importantly, the bond links the reactive (anomeric) carbon atoms of the two glucose monomers, preventing them from interacting with other molecules and thereby rendering trehalose among the most chemically inert sugars (12,22). One significant advantage is that trehalose consequently does not participate in the Maillard (browning) reaction (23,24). This process is actually a complex series of reactions by which sugars and amino acids are mutually degraded; the end result is multiple products which produce a variety of flavors and aromas. Although desirable in many areas of food biotechnology, the Maillard reaction is an anathema to others, because it is thought to be a key cause of damage to proteins during storage (25).

Early experiments with restriction endonucleases dramatically demonstrated the extraordinary protective capacity of trehalose. In a routine familiar to anyone employing molecular biology techniques, these enzymes are carefully stored in freezers and even when briefly removed for use, are maintained in freezer blocks to preserve their activity. If dried in a trehalose containing solution, however, these normally labile enzymes retained activity upon reconstitution, even after extended storage at room temperature or after being heated to 70°C (8,26,27). In addition to stabilizing native proteins, trehalose has also been found to prevent aggregation of denatured polypeptides, a function previously thought to require molecular chaperones (28) (Figure 13.1).

The protective effects of trehalose shield not only proteins, but other macromolecules as well. Lipid vesicles, which are potentially important tools for drug delivery, fare poorly under storage conditions of freezing or dehydration: rather than crystallize, they aggregate and fuse, resulting in leakage of their contents. If freeze dried in the presence of trehalose,

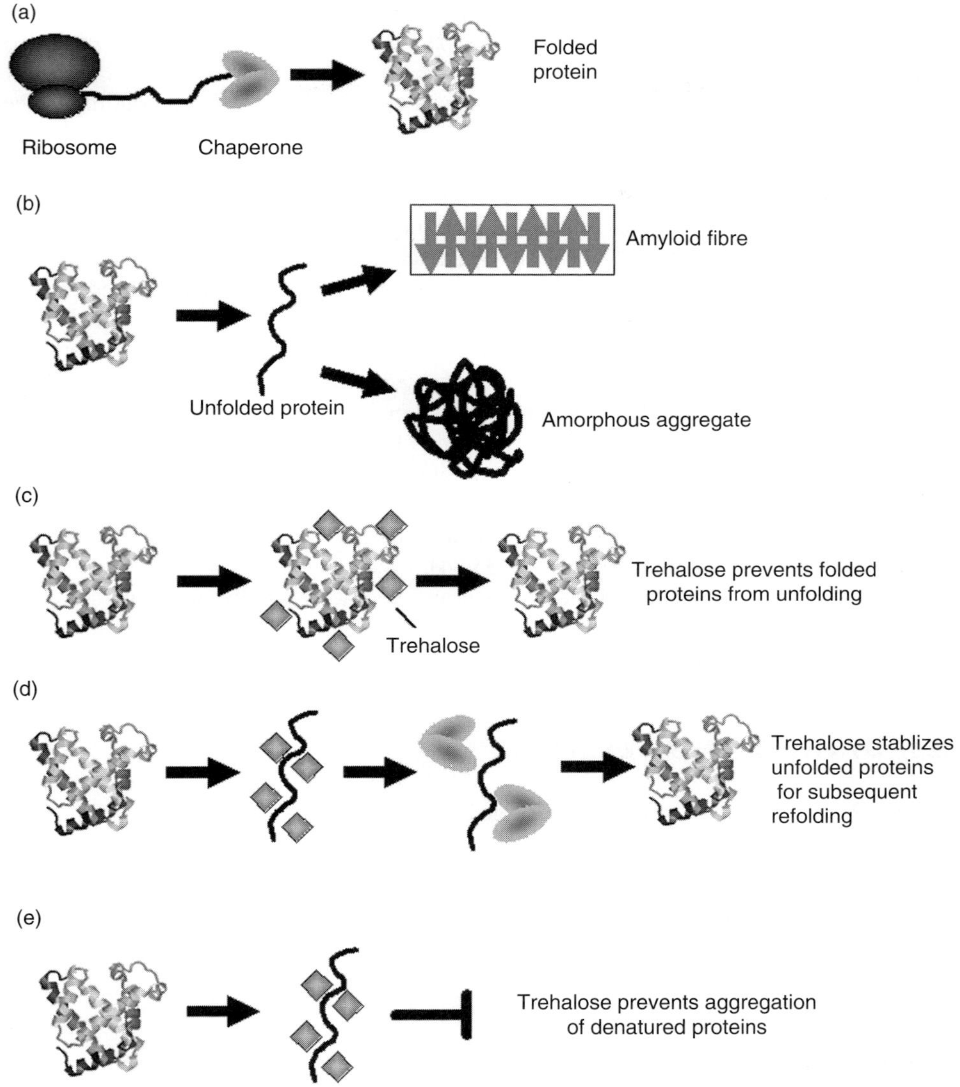

Figure 13.1 Model for trehalose-mediated protein stabilization in cells: (a) Protein folding under normal conditions. (b) Undesirable associations of denatured proteins. (c-e) Roles of trehalose during heat shock.

however, the vesicles remain intact upon rehydration (21). Carbohydrates in the form of starches become disordered during cooking, but may subsequently associate and precipitate, thereby contributing to staling. This undesirable process, termed retrogradation, can be significantly inhibited by trehalose (29). The damaging effects of radiation on DNA are well known, causing structural alterations and strand breaks. These effects, thought to be caused by hydroxide radicals generated by the decomposition of water molecules, are inhibited by trehalose, which may act as a scavenger of those highly reactive species (30).

The mechanisms by which trehalose protects from other forms of damage, however, are less readily apparent. Two general hypotheses have been proposed to explain how trehalose stabilizes macromolecules and prevents their nonproductive associations: water replacement, and the formation of a glass (20). An important factor is the remarkable concentrations of trehalose which can accumulate, reaching in yeast, for example, up to 0.5 *M*, or 20% of the dry weight of the cells (2).

The water replacement hypothesis emphasizes the role of hydrogen bonding in the protective process. According to this model, macromolecules are stabilized by hydrogen bonds provided by the water that normally surrounds them. During freezing or dehydration, these water molecules are removed. Trehalose may stabilize substrates either by ordering and maintaining the surrounding water, or by interacting with polar regions to substitute as water is removed (20). The finding that trehalose stabilizes denatured proteins suggests that it may also form nonpolar associations, preventing normally buried regions of unfolded proteins from interacting as well (28).

Glass refers a liquid of such high viscosity that it is capable of slowing or even halting chemical reactions (23). Sugars in solution can adopt the glass state (vitrify), holding enclosed macromolecules either in their native state or a conformation which returns to the native form upon rehydration. Consistent with this hypothesis, trehalose readily forms a glass which resists crystallization or separation of phases, and restricts mobility of macromolecules (31,32). Addition of substances such as borate, cations, and glucose may provide increased stabilization; such compounds naturally occurring in cells likely contribute to trehalose mediated stabilization *in vivo* (33,34).

The water replacement and glass transition hypotheses are by no means mutually exclusive, and likely are interrelated. Indeed, studies of dehydration tolerance indicate that both hydrogen bonding and formation of the glass state are necessary (23).

13.3 SYNTHESIS IN *S. CEREVISIAE*

Under normal conditions, trehalose is formed in the cytosol of *S. cerevisiae* through the efforts of a quartet of proteins acting together (2,35,36) (Figure 13.2). The reaction proceeds

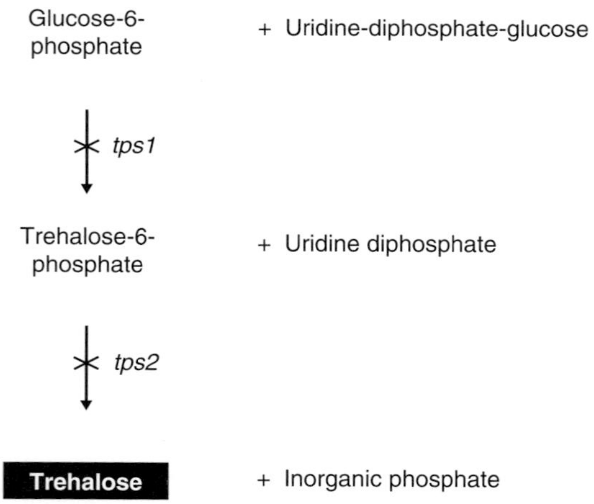

Figure 13.2 Biosynthesis of trehalose: *S. cerevisiae* produces trehalose in the cytosol through the sequential action of the Tps1 and Tps2 proteins. Steps at which trehalose synthesis is blocked in the absence of these proteins are indicated. Tps1 and Tps2 associate physically to form the trehalose-synthase complex, together with Tsl1 and Tps3. The latter components appear to regulate the activity of the complex. (Adapted from Singer, M.A., S. Lindquist, *Trends Biotechnol.* 16:460–468, 1998).

in two steps: UDP-glucose is joined to glucose-6-phosphate (G6P) to form trehalose-6-phosphate (T6P). The phosphate molecule is then cleaved, yielding trehalose (37,38).

The components of the trehalose synthase complex are Tps1, Tps2, Tps3, and Tsl1 (36). Though 35% identical over a length of about 500 amino acids, their functions are distinct (39). Tps1 catalyzes formation of trehalose-6-phosphate (T6P) by uniting the two glucose moieties. The phosphate is then cleaved by Tps2 to yield trehalose and recycle inorganic phosphate. Tps3 and Tsl1 appear to act as regulators, in a manner that remains enigmatic. All members are present constitutively, with expression further induced by heat (40,41), diauxic shift and stationary phase (42), and upon DNA damage (40).Their stoichiometry is not yet known; gel filtration studies estimate the size of the complex as 600–800 kD, indicating that one or more of the subunits is present in multiple copies (36,44).

Tps1 (trehalose phosphate synthase), which facilitates the first step in synthesis, is the smallest subunit (56 kD) (45,46). Mutants lacking Tps1 are unable to produce trehalose, and are severely impaired in thermotolerance (47). Despite their sequence similarity, the other components of the complex cannot substitute for its absence, leading to speculation that the identical regions may instead promote physical association (36).

Upon fractionation of extracts, the other subunits are primarily found in complex, whereas Tps1 is present both as part of the complex and as a free monomer. Free Tps1 is functional, although its catalytic activity is diminished (36). These properties are all the more intriguing given another, as yet unexplained finding: mutants lacking Tps1 are unable to grow on glucose (48,49). We return to this link between trehalose and metabolism more extensively in what follows.

Tps2 (102 kD) carries out the final step in trehalose synthesis, dephosphorylation of trehalose-6-P to yield trehalose (60). In contrast to *tps1* mutants, cells in which *Tps2* is deleted are still able to accumulate trehalose, indicating that one or more other cellular phosphatases are to some extent able to hydrolyze the phosphate group from trehalose-6-P (36).

Tps3 (115 kD) and Tsl1 (the largest subunit, 123 kD) appear to exert a regulatory and stabilizing effect, in a manner that remains poorly understood (40). The former, discovered by sequence homology, is coregulated with Tps1 and Tps2 (35). Deletion of either *Tps3* or *Tsl1* individually destabilizes the complex, yielding more degradation products and free Tps1. In the absence of both genes, no evidence of the complex is found (36). The individual mutants also show diminished trehalose–phosphate synthase and phosphatase activity, an effect that similarly is more pronounced in the absence of both proteins (36).

13.4 DEGRADATION

Like its synthesis, degradation of trehalose in *S. cerevisiae* also poses some puzzling complexities. Hydrolysis is catalyzed by Nth1 (neutral trehalase), an 80 kD cytoplasmic protein which functions as a dimer (45,51,52). Cells lacking this protein grow normally in media containing trehalose, but are unable to degrade intracellular trehalose (53).

Paradoxically, expression of Nth1 is induced by heat, and mutants lacking the enzyme are impaired in their recovery from heat shock (53–56). That is, both synthesis and degradation of trehalose are important for surviving exposure to extreme heat. As the functions of trehalose in thermotolerance, detailed in what follows, have become clear, these seemingly counterintuitive findings can now be explained: trehalose stabilizes native and denatured proteins during stress, but that same process may interfere with the refolding of denatured proteins afterward. Consequently, the continued presence of trehalose may impede return to normal growth.

Nth1 activity is high in log-phase cells, and decreased in stationary phase, circumstances corresponding respectively to low and high levels of intracellular trehalose (52,53). *In vivo*, neutral trehalase is activated by stimuli such as glucose, which triggers germination of spores (52,57). Nth1 is tightly regulated through phosphorylation, particularly by protein kinase A, which activates the enzyme. Nth1 also has a putative calcium binding site, and *in vitro* is stimulated both by calcium and magnesium.

A second gene, *NTH2*, was identified in *S. cerevisiae* by sequence comparison and is 77% identical to *NTH1* (58). Also cytoplasmic and heat inducible, its function is not clear: neither deletion nor overexpression of *NTH2* affects intracellular trehalase activity or trehalose levels, but mutants are impaired in recovery from heat shock, the more so if both *NTH2* and *NTH1* are deleted (59) (Figure 13.3).

A third trehalase enzyme, Ath1 (218 kD) or acid trehalase, is also present in yeast. This heavily glycosylated protein has no homology to Nth1 or Nth2, and does not function in thermotolerance. Instead, it is found in vacuoles, consistent with its optimum pH of 4.5 (53,60). Mutants lacking Ath1 are unable to grow on exogenous trehalose (53). In nature, many fungi utilize trehalose as a carbon source, because it accumulates in insects and microbes and therefore is often present at high levels in decaying soil. Expression and activity of Ath1 is induced by addition of trehalose, and increases in stationary phase (56,61,62). Unlike its cytoplasmic counterparts, Ath1 is activated neither by phosphorylation nor by calcium or magnesium (60,63).

Acid trehalase (Ath1) is thought to be secreted into the periplasm, and then internalized into vacuoles together with exogenous trehalose. Support for this model comes from

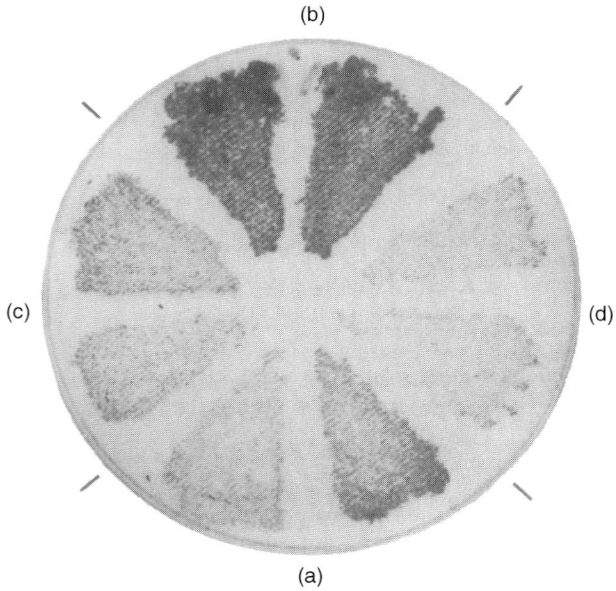

Figure 13.3 Poor growth of the *nth1*, *ath1,* and *nth1ath1* double mutants on glycerol: (a) Two independent *nth1* spores (no neutral trehalase activity), but wild-type for the ATH1 gene. (b) Two independent spores (no acid trehalase activity), but wild-type for NTH1. (c) Two independent spores, wild-type for NTH1 and ATH1. (d) Two independent spores with both *nth1* and *ath1* mutations (neither neutral nor acid trehalase activity). The cells were grown on a YEPD plate for 23 days at 30°C, replica plated onto a YEPGlycerol plate, and then incubated at 30°C for 5 days. (Reproduced from Nwaka, S., B. Mechler, M. Destruelle, H. Holzer, *FEBS Lett.* 360:286–290, 1995).

the finding that Ath1 is present in the periplasm of cells overexpressing the protein, and from the observation that *end1* mutants, in which the endocytosis pathway is impaired, are unable to grow on trehalose (52).

13.5 TREHALOSE AND METABOLISM

As noted earlier, an intriguing aspect of trehalose in *S. cerevisiae* is its relationship to glycolysis, one of the fundamental pathways of cellular metabolism. For reasons that remain elusive, mutants lacking Tps1 cannot regulate the influx of glucose, and become flooded with the sugar. The glucose is rapidly phosphorylated for entry into glycolysis, depleting intracellular phosphate and adenosine triphosphate (ATP) and leading to cell death (48).

Teusnick et al. suggest that this lethal circumstance arises because of the "turbo" architecture of the glycolytic pathway, in which ATP is first invested to promote the breakdown of glucose, and only then are returns obtained by subsequent metabolism. A sudden increase in glucose levels, therefore, can paralyze this process, causing ATP to be depleted faster than it can be replenished (64).

Three models have been proposed to explain the role of the trehalose pathway in glycolysis (49). The first posits that Tps1 regulates glucose influx by production of trehalose-6-phosphate, which in turn inhibits the activity of hexokinases and thereby prevents the uncontrolled phosphorylation of glucose. This hypothesis is based on the finding that regulation of glycolysis and growth on glucose can be restored by mutation of Hxk2, the primary enzyme responsible for phosphorylating intracellular glucose (65). Further support comes from the *in vitro* observation that Hxk2 activity is inhibited by T6P (66,67).

Aside from the genetic finding with Hxk2, results *in vivo* are less straightforward. Expressing the *Escherichia coli* Tps1 homolog otsA restores T6P levels, yet only partially suppresses deregulation of glycolysis (68). Overexpression of Hxk2 in wild–type *S. cerevisiae* does not interfere with growth on glucose (69), and even a broad range of T6P levels had only a limited effect on the accumulation of glucose-6-phosphate in cells (70). In all likelihood, then, Tps1 probably influences glycolysis *in vivo* through multiple mechanisms.

Such mechanisms could include one or both of the other proposed models (49). The second model proposes that by utilizing G6P for synthesis of trehalose, the trehalose pathway enables cells to recover phosphate and regenerate ATP. The absence of a growth defect for *tps2* mutants on glucose may be explained by the ability of nonspecfic phosphatases to compensate in removing the phosphate from T6P. Finally, the Tps1 protein itself may have, in addition to its role in trehalose synthesis, a distinct regulatory function to limit influx of glucose into glycolysis. This latter suggestion is supported by the finding noted earlier, of free Tps1 separate from the trehalose synthase complex (36,49,71).

13.6 TREHALOSE IN HEAT SHOCK

Until recently, it was thought that trehalose, like glycogen, served as a carbohydrate storage reserve for cells under nutrient deprivation. The two compounds indeed have similarities: both are found in high concentrations under conditions when food is scarce, such as stationary phase and in spores, and correspondingly, both are largely absent when it is abundant, as during exponential growth (72–74). Further consideration of the data, however, together with key additional observations, indicated that trehalose functions primarily to help cells withstand not starvation, but stress (1,2).

An energy storage molecule would be expected to be produced when nutrients are abundant, in anticipation of future scarcity. Glycogen is indeed synthesized under such circumstances, during the logarithmic phase of growth. Cells only produce trehalose, however, once they are already entering stationary phase and food is no longer readily available (74).

Similarly, energy stores would normally be consumed after environmental resources are depleted. Cellular glycogen is utilized during maturation of ascospores (72). In contrast, trehalose is not used, and remains present in mature spores. During long term incubation in stationary phase, *S. cerevisiae* first turn to glycogen, and only draw on trehalose after the glycogen has been depleted and cells begin to die of starvation (2,74).

These anomalies became even more compelling in the context of several striking experimental results. Although trehalose was shown *in vitro* to stabilize proteins under adverse conditions, it was not thought to do so *in vivo*, because the concentrations required are much higher than were found in cells. That view changed following the discovery that trehalose in *S. cerevisiae* is present exclusively in the nucleus and cytoplasm and not, for example, in vacuoles, which occupy a large fraction of the intracellular space (54). This key finding indicated that the concentration in those areas was actually on the order of 0.5 M, consistent with levels observed to protect macromolecules *in vitro* (2,8).

Additional *in vivo* support came from the observation that trehalose accumulated in *S. cerevisiae*, as well as in other fungi, prokaryotes, and nematodes, under conditions of environmental stress, such as oxidation, heat, desiccation, and ethanol. Moreover, elevated intracellular levels of trehalose correlated with increased thermotolerance (Figure 13.4) in yeast spores, as well as cells in stationary phase and in certain mutants impaired in the cyclic AMP signaling pathway (1,75).

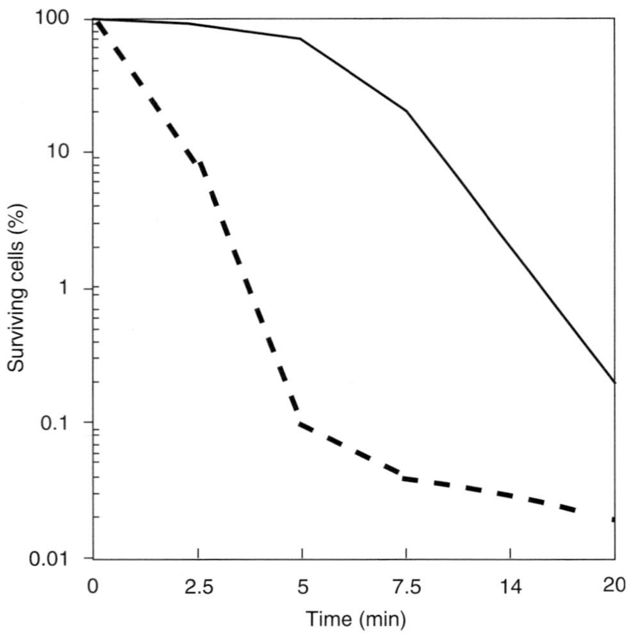

Figure 13.4 Thermotolerance of wild-type and T*ps1* *S. cerevisiae* strains in stationary phase: Cultures of wild-type (—) and Tps1 (- -) cells were grown to stationary phase. Thermotolerance was measured as survival following incubation at 54°C for the times indicated. (Adapted from De Virgilio, C., T. Hottiger, J. Dominguez, T. Boller, A. Wiemken, *Eur. J. Biochem.* 219:179–186, 1994).

An important caveat, of course, was that these conditions also stimulate increased synthesis of heat shock proteins (HSPs) (76). Also, trehalose is degraded very rapidly after a stress has passed, unlike HSPs, which are long lived (62,75). The crucial test of this hypothesis could finally be performed once the *Tps1* gene was successfully cloned. Indeed, mutation of the gene resulted in dramatically diminished survival of extreme heat – to an extent similar to that of cells lacking Hsp104, the key HSP mediator of thermotolerance in yeast (47,77,78).

Further experiments established that synthesis of HSPs was unchanged in *Tps1* mutants, confirming that the thermotolerance deficit was due to the absence of trehalose (28). That trehalose exerted its powerful effects through stabilization of proteins *in vivo* was demonstrated using two different heat sensitive reporter proteins. Both showed diminished activity in T*ps1* cells compared to wild–type at elevated temperatures (28).

Two additional, surprising, results emerged from this work. First, trehalose also prevented aggregation of proteins that were already denatured – a function previously thought to be carried out exclusively by HSPs. Second, the continued presence of trehalose impaired reactivation of those denatured proteins (28). Other experiments correspondingly demonstrated that cells unable to degrade trehalose were impaired in recovery from heat shock relative to their wild-type counterparts (53,55,56).

As noted, the current model for trehalose function in stress tolerance resolves these seemingly paradoxical findings: Trehalose stabilizes proteins in their native state and prevents them from denaturing during stress. Together with HSPs, trehalose also keeps those proteins that do unfold from forming nonproductive and potentially toxic aggregates. Trehalose must be rapidly removed, however, because its continued presence may interfere with degradation or chaperone mediated refolding of denatured proteins (11).

13.7 TREHALOSE IN PLANTS AND *DROSOPHILA*

With the exception of the "resurrection plants", *Selaginella lepidophylla* and *Myrothamnus flabellifolius*, which utilize trehalose to survive extreme desiccation, the disaccharide was thought to be largely absent in plants (79). At the same time, however, extending the protective effects of trehalose to plants had been a sought after goal of work in stress tolerance (80,81). A number of exciting recent discoveries have dispelled the notion that trehalose is rare in plants. Equally striking is the discovery that the same basic parameters which characterize trehalose in *S. cerevisiae* are also seen in plants, and now in *Drosophila* as well: trehalose enhances resistance to environmental stress; accumulation of trehalose can be detrimental during growth; and finally, components and intermediates of the trehalose synthesis process are intimately connected to metabolism and gene expression.

The finding that trehalose is widespread in plants was made serendipitously: in the course of efforts to engineer trehalose production in tobacco and potato, the control plants were found to contain trace levels of the sugar (81). Other work demonstrated that treatment of *Arabidopsis* with the trehalase inhibitor validomycin A resulted in significant accumulation of trehalose (79,82). TPS and TPP homologs have since been identified in rice, soybean, and tomato, and more recently, a host of related genes, as well as a putative trehalose, have been found in *Arabidopsis* (71).

The *Arabidopsis* sequences cluster into two groups (71). The first subfamily, members of which most resemble Tps1, contains four genes which are 63–80% identical in sequence. One gene, designated *AtTps1*, restores the ability of *S. cerevisiae tps1* mutants to produce trehalose (67). The *AtTps1* sequence includes a 100 amino acid N-terminal region not present in yeast Tps1 or the other plant Tps1 homologs. This sequence is homologous to segments of the yeast regulator protein Tsl1 and likewise appears to have a regulatory function: when deleted, activity of the enzyme increases (83). Members of the second subfamily are most similar to Tps2. This group shares 54–83% sequence identity, and all members contain a C-terminal sequence common to phosphatases (71).

Overexpression of yeast and *E. coli* trehalose synthesis enzymes in transgenic plants indeed enhanced resistance to stress (82,84,85). As with *S. cerevisiae nth1* mutants, however, this benefit was not without cost: the transgenic plants displayed altered metabolism and morphological growth defects, such as stunted growth and lancet shaped leaves (82,86).

Studies in *Arabidopsis* revealed that *AtTps1* is expressed at low levels in all tissues examined, and is strongly upregulated during embryo maturation (seed development) (87). Perhaps the most dramatic parallels to yeast were seen in a key study in which the gene was mutated. Although the thermotolerance properties of the mutant have not yet been reported, the effects on development were manifest, resulting in recessive embryonic lethality (87). Development in higher plants such as *Arabidopsis* proceeds in three overlapping stages. Cells first divide and differentiate, establishing the pattern of the embryo. Next, cells expand in size and accumulate storage reserves. Finally, the embryo desiccates and growth halts. The *AtTps1* mutant plants were unable to proceed past the torpedo stage, at the transition between pattern formation and growth (87).

Under normal conditions, sucrose levels rise dramatically in *Arabidopsis* embryos at this juncture, as does expression of AtTps1. The increase is thought to be required both for producing storage reserves and for influencing gene expression to control this developmental transition. The latter suggestion is supported by abnormalities in expression of the seed maturation marker genes *At2S2* and *AtOLEOSN2* seen in the mutant.

As mentioned earlier, *S. cerevisiae tps1* mutants die in the presence of glucose. The sugar enters the cells and is phosphorylated in an unrestricted manner, depleting cellular phosphate and ATP stores. Moreover, Tps1 is also required for carbon catabolites to control gene expression (49). These cells are, however, able to grow on carbon sources such as galactose. As discussed earlier, growth on glucose can be restored by mutating the gene encoding Hxk2, the enzyme primarily responsible for phosphorylating glucose for entry into glycolysis (65). Just as *S. cerevisiae tps1* mutants can be rescued by restricting entry of glucose into glycolysis, *AtTps1* mutants similarly are able to proceed with development *in vitro* if sucrose levels are reduced. Unlike in yeast, growth could not be restored in these cells by targeting Hxk2 (here by antisense methods or treatment with the hexokinase inhibitor glucosamine), nor is *Arabidopsis* Hxk2 inhibited by trehalose-6-phosphate (87). The reason for this failure is not clear, but several explanations have been proposed. The *HXK* gene family in *Arabidopsis* is not well characterized, so another isoform (or perhaps an entirely distinct kinase) may be the critical target. Alternatively, sucrose can enter glycolysis through more than one route in plants, so its catabolism at this developmental stage may proceed by a different pathway, such as that utilizing sucrose synthase, which does not require phosphorylation of glucose (87,88). Finally, it is also possible that AtTps1 acts directly or via trehalose to influence development and gene expression. Based on their finding that the *E. coli* trehalose synthase gene can complement the *AtTps1* mutation, Schluepmann et al. favor the view that the mutant phenotype results from the absence of activity of AtTps1, rather than a regulatory property of the protein (89).

The likelihood that the protean roles of the trehalose pathway are broadly conserved was underscored by exciting work in *Drosophila* paralleling these findings. *Drosophila* is unusually resistant to anoxic injury, a property of considerable medical interest because of its potential to help limit damage to the brain in patients following stroke or cardiac arrest. By overexpressing the *Drosophila Tps1* gene, Chen et al. succeeded in making that organism even more resistant. Strikingly, deletion of the gene or its overexpression under certain conditions led to developmental abnormalities or death of the developing flies (4).

13.8 APPLICATIONS IN BIOTECHNOLOGY

Efforts to elucidate the role of trehalose in adverse conditions and in growth and metabolism have advanced concurrently with the drive to harness its protective properties in food biotechnology. That work, which has primarily involved genetic engineering and exogenous use of the sugar, also holds outstanding promise broadly in industry and medicine.

Utilization of trehalose as an additive for food preservation or storage has recently become feasible through the development of new techniques enabling economical large scale production from starch (24,29). The sugar's mild sweetness — 45% that of sucrose (29) — and "masking effect" of bitter-tasting compounds, as well as lack of an aftertaste or laxative effect, enhance its versatility as a contributor to the texture and balance of flavors in foods (90). It is not easily degraded during processing, and its resistance to taking up moisture reduces caking in mixtures with other ingredients (29). The protective effects of trehalose prolong the shelf life of starch-containing products by preventing retrogradation, as noted; similarly, the stabilization of proteins and lipids during freezing and drying extends the shelf life of products containing eggs or meat (29).

Genetic methods have successfully been employed to improve the freeze tolerance of frozen dough (91,92). As mentioned earlier, tobacco and potato plants modified to produce trehalose display enhanced tolerance to a diverse range of stresses, including salt, drought, and low temperatures (79). Considerable potential also holds for the use of such methods to promote proper folding and prevent aggregation of recombinant proteins synthesized in *E. coli* or other organisms, for production of foods or pharmaceuticals (11). Trehalose has already been shown to help preserve vaccines and liposomes, as well as organs in preparation for transplant (30,90). The novel finding that disruption of *Tps2* in *Candida* leads to diminished infectivity of that organism suggests that the trehalose pathway may be an attractive target for antibiotics (93,94). The same may also hold true for pesticides: diminished infectivity and pathogenicity was similarly seen for trehalose mutants of the rice blast fungus *Magnaporthe grisea* (95).

Finally, trehalose may be helpful in the treatment and prevention of disorders resulting from misfolding or aggregation of proteins (9,96,97). Expression of α-synuclein, a protein involved in Parkinson's disease, is toxic to yeast. In a genetic screen to identify modulators of this toxicity, Willingham et al. discovered that cells lacking Tsl1 were more susceptible to the deleterious effects of α-synuclein (98). Another recent study, involving a transgenic mouse model of Huntington's disease, demonstrated that mice fed trehalose dosed water developed far fewer polyglutamine aggregates (one of the hallmarks of Huntington's disease pathology) than did untreated controls (99). The trehalose group also lived longer and performed better in coordination tests, suggesting a correlation between the decreased number of aggregates and improvements in the Huntington's-like symptoms.

13.9 PERSPECTIVES

Overall, three general principles are emerging from work on trehalose in *S. cerevisiae* and its striking correlates in plants and *Drosophila*: (1) trehalose has a protective role *in vivo* as well as *in vitro* in stress tolerance; (2) the same properties which can protect macromolecules may also be detrimental as environmental circumstances change; and (3) a critical but still elusive connection exists between the trehalose pathway and cellular metabolism, gene expression, and development. It is constructive to consider these principles in the broader context of stress tolerance mechanisms, particularly other organic solutes and the heat shock proteins (HSPs).

Regarding the former, trehalose is not unique in its protective effects. Rather, similar properties are shared by a host of other small molecules, such as the amino acids proline and betaine and the sugars sucrose and sorbitol (8,28). The diversity in nature of these "compatible solutes" or "chemical chaperones" is intriguing. It may reflect propensities to stabilize different sensitive structures in different organisms, or perhaps regulatory and metabolic functions (such as the interaction of trehalose with glycolysis) that make certain osmolytes of greater utility in particular organisms. Thus, a vital connection to cell biology and organismal development may exist more generally with compatible solutes, providing another unifying characteristic and potentially powerful tool in food biotechnology.

A paradigm for such an association between fundamental cellular processes and stress tolerance already exists, with the heat shock proteins. While a subgroup are expressed only under adverse environmental conditions, many others, such as the members of the Hsp70 and Hsp40 families, have variants which play essential roles in protein folding and translocation under normal circumstances (100). Even Hsp104, an *S. cerevisiae* heat shock protein initially thought to function exclusively in survival of extreme stress (75,77), subsequently was found to have a key role at normal temperatures for propagation of *[PSI$^+$]*, *[RNQ$^+$]*, and *[URE3]* (101). Also known as yeast prions, these unconventional genetic elements represent alternative but stable conformations of a particular protein, which propagate heritable information through maintenance of a particular conformation (101–03).

This finding and other work in HSPs highlight the subtleties of cellular adaptation to environmental challenges. *Drosophila* utilize Hsp70 as their primary protective mechanism in response to heat shock. In tissue culture cells and embryos, that protein is sequestered into granules with astonishing speed after return to normal conditions. As with trehalose, its removal likely occurs because of interference with protein folding and progress of development (104,105). A similar explanation may underlie the puzzling finding that although *AtTps1* expression has been found in all *Arabidopsis* tissues examined, the disaccharide itself under normal conditions is detectable only at low levels, if at all (88).

As these incongruities continue to inspire further efforts, it is clear that in the nearly two centuries since its discovery, trehalose has indeed come full circle – from a component of the toxin ergot contaminating rye, to a versatile and valuable tool – in a transcendent irony with great promise for food biotechnology.

REFERENCES

1. Van Laere, A. Trehalose, reserve and/or stress metabolite? *FEMS Microbiol. Rev.* 63:201–210, 1989.
2. Wiemken, A. Trehalose in yeast, stress protectant rather than reserve carbohydrate. *Antonie Van Leeuwenhoek* 58:209–217, 1990.
3. Diniz-Mendes L., E. Bernardes, P.S. de Araujo, A.D. Panek, V.M. Paschoalin. Preservation of frozen yeast cells by trehalose. *Biotechnol. Bioeng.* 65:572–578, 1999.
4. Chen, Q., E. Ma, K.L. Behar, T. Xu, G.G. Haddad. Role of trehalose phosphate synthase in anoxia tolerance and development in *Drosophila melanogaster*. *J. Biol. Chem.* 277:3274–3279, 2002.
5. Benaroudj, N., D.H. Lee, A.L. Goldberg. Trehalose accumulation during cellular stress protects cells and cellular proteins from damage by oxygen radicals. *J. Biol. Chem.* 276:24261–24267, 2001.

6. Hounsa, C.G., E.V. Brandt, J. Thevelein, S. Hohmann, B.A. Prior. Role of trehalose in survival of *Saccharomyces cerevisiae* under osmotic stress *Microbiology* 144(3): 671–680, 1998.
7. Holmstrum, K.-O., E. Mantyla, B. Welin, A. Mandal, E.T. Plava, O.E. Tunnela, J. Londesborough. Drought tolerance in tobacco. *Nature* 379:683–684, 1996.
8. Hottiger, T., C. De Virgilio, M. N. Hall, T. Boller, A. Wiemken. The role of trehalose synthesis for the acquisition of thermotolerance in yeast, II: physiological concentrations of trehalose increase the thermal stability of proteins *in vitro. Eur. J. Biochem.* 219:187–193, 1994.
9. Tatzelt, J., S.B. Prusiner, W.J. Welch. Chemical chaperones interfere with the formation of scrapie prion protein. *EMBO J.* 15:6363–6373, 1996.
10. Harding, T.S. History of trehalose, its discovery and methods of preparation. *Sugar* 25:476–478, 1923.
11. Singer, M.A., S. Lindquist. Thermotolerance in *Saccharomyces cerevisiae*: the Yin and Yang of trehalose. *Trends Biotechnol.* 16:460–468, 1998.
12. Elbein, A.D. The metabolism of alpha, alpha-trehalose. *Adv. Carbohydr. Chem. Biochem.* 30:227–256, 1974.
13. Bjornsti, M.A. Cancer therapeutics in yeast. *Cancer Cell.* 2:267–273, 2002.
14. Haber, J.E. Aging: the sins of the parents. *Curr. Biol.* 13:R843–R845, 2003.
15. Walberg, M.W. Applicability of yeast genetics to neurologic disease. *Arch. Neurol.* 57:1129–1134, 2000.
16. Outeiro, T.F., S. Lindquist. Yeast cells provide insight into alpha-synuclein biology and pathobiology. *Science* 302:1772–1775, 2003.
17. Muchowski, P.J., G. Schaffar, A. Sittler, E.E. Wanker, MK. Hayer-Hartl, F.U. Hartl. Hsp70 and hsp40 chaperones can inhibit self-assembly of polyglutamine proteins into amyloid-like fibrils *Proc. Natl. Acad. Sci. USA* 97:7841–7846, 2000.
18. Meriin, A.B., X. Zhang, X. He, G.P. Newnam, Y.O. Chernoff, M.Y. Sherman. Huntingtin toxicity in yeast model depends on polyglutamine aggregation mediated by a prion-like protein Rnq1. *J. Cell. Biol.* 157:997–1004, 2002.
19. Krobitsch, S., S. Lindquist. Aggregation of huntingtin in yeast varies with the length of the polyglutamine expansion and the expression of chaperone proteins. *Proc. Natl. Acad. Sci. USA* 97:1589–1594, 2000.
20. Colaco, C.A.L.S., B. Roser. Trehalose: a multifunctional additive for food preservation. In: *Food Packaging and Preservation,* Mathlouthi, M., ed., London: Blackie Professional, 1995, pp 123–140.
21. Paiva, C.L., A.D. Panek. Biotechnological applications of the disaccharide trehalose. *Biotechnol. Annu. Rev*. 2:293–314, 1996.
22. Birch, G.G. Trehaloses. *Adv. Carbohydr. Chem.* 18:201–225, 1963.
23. Crowe, J.H., J.F. Carpenter, L.M. Crowe. The role of vitrification in anhydrobiosis. *Annu. Rev. Physiol.* 60:73–103, 1998.
24. Schiraldi, C., I. Di Lernia, M. De Rosa. Trehalose production: exploiting novel approaches. *Trends Biotechnol.* 20:420–425, 2002.
25. Li, S., T.W. Patapoff, D. Overcashier, C. Hsu, T.H. Nguyen, R.T. Borchardt. Effects of reducing sugars on the chemical stability of human relaxin in the lyophilized state *J. Pharm. Sci.* 85:873–877, 1996.
26. Colaco, C., S. Sen, M. Thangavelu, S. Pinder, B. Roser. Extraordinary stability of enzymes dried in trehalose: simplified. *Mol. Biol. Biotechnol. NY* 10:1007–1011, 1992.
27. Uritani, M., M. Takai, K. Yoshinaga. Protective effect of disaccharides on restriction endonucleases during drying under vacuum. *J. Biochem.* 117:774–779, 1995.
28. Singer, M.A., S. Lindquist. Multiple effects of trehalose on protein folding *in vitro* and *in vivo. Mol. Cell.* 1:639–648, 1998.
29. Richards, A.B., S. Krakowka, L.B. Dexter, H. Schmid, A.P. Wolterbeek, D.H. Waalkens-Berendsen, A. Shigoyuki, M. Kurimoto. Trehalose: a review of properties, history of use and human tolerance, and results of multiple safety studies. *Food Chem. Toxicol.* 40:871–898, 2002.

30. Yoshinaga, K., H. Yoshioka, H. Kurosaki, M. Hirasawa, M. Uritani, K. Hasegawa. Protection by trehalose of DNA from radiation damage. *Biosci. Biotechnol. Biochem.* 61:160–161, 1997.
31. Crowe, J.H., L.M. Crowe, A.E. Oliver, N. Tsvetkova, W. Wolkers, F. Tablin. The trehalose myth revisited: introduction to a symposium on stabilization of cells in the dry state. *Cryobiology* 43:89–105, 2001.
32. Sun, W.Q., P. Davidson. Protein inactivation in amorphous sucrose and trehalose matrices: effects of phase separation and crystallization. *Biochim. Biophys. Acta* 1425:235–244, 1998.
33. Miller, D.P., R.E. Anderson, J.J. de Pablo. Stabilization of lactate dehydrogenase following freeze thawing and vacuum-drying in the presence of trehalose and borate. *Pharm. Res.* 15:1215–1221, 1998.
34. Mazzobre, M.F., M. Del Pilar Buera. Combined effects of trehalose and cations on the thermal resistance of beta-galactosidase in freeze-dried systems. *Biochim. Biophys. Acta* 1473:337–344, 1999.
35. Reinders, A., N. Burckert, S. Hohmann, J.M. Thevelein, T. Boller, A. Wiemken, C. De Virgilio. Structural analysis of the subunits of the trehalose-6-phosphate synthase/phosphatase complex in *Saccharomyces cerevisiae* and their function during heat shock. *Mol. Microbiol.* 24:687–695, 1997.
36. Bell, W., W. Sun, S. Hohmann, S. Wera, A. Reinders, C. De Virgilio, A. Wiemken, J.M. Thevelein. Composition and functional analysis of the *Saccharomyces cerevisiae* trehalose synthase complex. *J. Biol. Che*m. 273:33311–33319, 1998.
37. Cabib, E., L.F. Leloir. The biosynthesis of trehalose phosphate. *J. Biol. Chem.* 231:259–275, 1958.
38. Leloir, L.F., E. Cabib. The enzymic sythesis of trehalose phosphate. *J. Am. Chem. Soc.* 75:5445–5446, 1953.
39. Francois, J., J. L. Parrou. Reserve carbohydrates metabolism in the yeast *Saccharomyces cerevisiae*. *FEMS Microbiol. Rev*. 25:125-145 (2001).
40. Vuorio, O.E., N. Kalkkinen, J. Londesborough. Cloning of two related genes encoding the 56-kDa and 123-kDa subunits of trehalose synthase from the yeast *Saccharomyces cerevisiae. Eur. J. Biochem.* 216:849–861, 1993.
41. Winderickx, J., J.H. de Winde, M. Crauwels, A. Hino, S. Hohmann, P. Van Dijck, J.M. Thevelein. Regulation of genes encoding subunits of the trehalose synthase complex in *Saccharomyces cerevisiae*: novel variations of STRE-mediated transcription control? *Mol. Gen. Genet*. 252:470–482, 1996.
42. DeRisi, J.L., V.R. Lyer, P.O. Brown. Exploring the metabolic and genetic control of gene expression on a genomic scale. *Science* 278:680–686, 1997.
43. Gasch, A.P., M. Huang, S. Metzner, D. Botstein, S. J. Elledge, P. O. Brown. Genomic expression responses to DNA-damaging agents and the regulatory role of the yeast ATR homolog Mec1p. *Mol. Biol. Cell.* 12:2987–3003, 2001.
44. Londesborough, J., O.E. Vuorio. Purification of trehalose synthase from baker's yeast: its temperature-dependent activation by fructose 6-phosphate and inhibition by phosphate. *Eur. J. Biochem.* 216:841–848, 1993.
45. Londesborough, J., O. Vuorio. Trehalose-6-phosphate synthase/phosphatase complex from bakers' yeast: purification of a proteolytically activated form. *J. Gen. Microbiol*. 137(2):323-330, 1991.
46. Vandercammen, A., J. Francois, H.G. Hers. Characterization of trehalose-6-phosphate synthase and trehalose-6-phosphate phosphatase of *Saccharomyces cerevisiae*. *Eur. J. Biochem.* 182:613–620, 1989.
47. De Virgilio, C., T. Hottiger, J. Dominguez, T. Boller, A. Wiemken. The role of trehalose synthesis for the acquisition of thermotolerance in yeast, I: genetic evidence that trehalose is a thermoprotectant. *Eur. J. Biochem.* 219:179–186, 1994.
48. Van Aelst, L., S. Hohmann, B. Bulaya, W. de Koning, L. Sierkstra, M. J. Neves, K. Luyten, R. Alijo, J. Ramos, P. Coccetti. Molecular cloning of a gene involved in glucose sensing in the yeast *Saccharomyces cerevisiae*. *Mol. Microbiol.* 8:927–943, 1993.

49. Thevelein, J.M., S. Hohmann. Trehalose synthase: guard to the gate of glycolysis in yeast? *Trends Biochem. Sci.* 20:3–10, 1995.
50. De Virgilio, C., N. Burckert, W. Bell, P. Jeno, T. Boller, A. Wiemken. Disruption of TPS2, the gene encoding the 100-kDa subunit of the trehalose-6-phosphate synthase/phosphatase complex in *Saccharomyces cerevisiae*, causes accumulation of trehalose-6-phosphate and loss of trehalose-6-phosphate phosphatase activity. *Eur. J. Biochem.* 212:315–323, 1993.
51. App, H., H. Holzer. Purification and characterization of neutral trehalase from the yeast ABYS1 mutant. *J. Biol. Chem.* 264:17583–17588, 1989.
52. Nwaka, S., H. Holzer. Molecular biology of trehalose and the trehalases in the yeast *Saccharomyces cerevisiae*. *Prog. Nucleic Acid Res. Mol. Biol.* 58:197–237, 1998.
53. Nwaka, S., B. Mechler, M. Destruelle, H. Holzer. Phenotypic features of trehalase mutants in *Saccharomyces cerevisiae*. *FEBS Lett.* 360:286–290, 1995.
54. Keller, F., M. Schellenberg, A. Wiemken. Localization of trehalase in vacuoles and of trehalose in the cytosol of yeast (*Saccharomyces cerevisiae*). *Arch. Microbiol.* 131:298–301, 1982.
55. Wera, S., E. De Schrijver, I. Geyskens, S. Nwaka, J.M. Thevelein. Opposite roles of trehalase activity in heat-shock recovery and heat-shock survival in *Saccharomyces cerevisiae*. *Biochem. J.* 343(3):621–626, 1999.
56. Nwaka, S., M. Kopp, H. Holzer. Expression and function of the trehalase genes NTH1 and YBR0106 in *Saccharomyces cerevisiae*. *J. Biol. Chem.* 270:10193–10198, 1995.
57. Jorge, J.A., M.L. Polizeli, J.M. Thevelein, H.F. Terenzi. Trehalases and trehalose hydrolysis in fungi. *FEMS Microbiol. Lett.* 154:165–171, 1997.
58. Wolfe, K.H., A.J. Lohan. Sequence around the centromere of *Saccharomyces cerevisiae* chromosome II: similarity of CEN2 to CEN4. *Yeast* 10(A):S41–46, 1994.
59. Nwaka, S., B. Mechler, H. Holzer. Deletion of the ATH1 gene in *Saccharomyces cerevisiae* prevents growth on trehalose. *FEBS Lett.* 386:235–238, 1996.
60. Londesborough, J., K. Varimo. Characterization of two trehalases in baker's yeast. *Biochem. J.* 219:511–518, 1984.
61. Destruelle, M., H. Holzer, D.J. Klionsky. Isolation and characterization of a novel yeast gene, ATH1, that is required for vacuolar acid trehalase activity. *Yeast* 11:1015-1025, 1995.
62. Hottiger, T., P. Schmutz, A. Wiemken. Heat-induced accumulation and futile cycling of trehalose in *Saccharomyces cerevisiae*. *J. Bacteriol.* 169:5518–5522, 1987.
63. Mittenbuhler, K., H. Holzer. Purification and characterization of acid trehalase from the yeast suc2 mutant. *J. Biol. Chem.* 263:8537–8543, 1988.
64. Teusink, B., M.C. Walsh, K. van Dam, H.V. Westerhoff. The danger of metabolic pathways with turbo design. *Trends Biochem. Sci.* 23:162–169, 1998.
65. Hohmann, S., M.J. Neves, W. de Koning, R. Alijo, J. Ramos, J.M. Thevelein. The growth and signalling defects of the ggs1 (fdp1/byp1) deletion mutant on glucose are suppressed by a deletion of the gene encoding hexokinase PII. *Curr. Genet.* 23: 281-289, 1993.
66. Hohmann, S., W. Bell, M.J. Neves, D. Valckx, J.M. Thevelein. Evidence for trehalose-6-phosphate-dependent and -independent mechanisms in the control of sugar influx into yeast glycolysis. *Mol. Microbiol.* 20: 981-991, 1996.
67. Blazquez, M.A., R. Lagunas, C. Gancedo, and J. M. Gancedo. Trehalose-6-phosphate, a new regulator of yeast glycolysis that inhibits hexokinases. *FEBS Lett.* 329:51–54, 1993.
68. Bonini, B.M., C. Van Vaeck, C. Larsson, L. Gustafsson, P. Ma, J. Winderickx, P. Van Dijck, J.M. Thevelein. Expression of *Escherichia coli* otsA in a *Saccharomyces cerevisiae* Tps1 mutant restores trehalose 6-phosphate levels and partly restores growth and fermentation with glucose and control of glucose influx into glycolysis. *Biochem. J.* 350(1):261–268, 2000.
69. Ernandes, J.R., C. De Meirsman, F. Rolland, J. Winderickx, J. de Winde, R. L. Brandao, J.M. Thevelein. During the initiation of fermentation overexpression of hexokinase PII in yeast transiently causes a similar deregulation of glycolysis as deletion of Tps1. *Yeast* 14:255–269, 1998.
70. van Vaeck, C., S. Wera, P. van Dijck, J.M. Thevelein. Analysis and modification of trehalose 6-phosphate levels in the yeast *Saccharomyces cerevisiae* with the use of *Bacillus subtilis* phosphotrehalase. *Biochem. J.* 353:157–162, 2001.

71. Leyman, B., P. Van Dijck, J.M. Thevelein. An unexpected plethora of trehalose biosynthesis genes in *Arabidopsis thaliana*. *Trends Plant Sci.* 6: 510–513, 2001.
72. Kane, S.M., R. Roth. Carbohydrate metabolism during ascospore development in yeast. *J. Bacteriol.* 118:8–14, 1974.
73. Thevelein, J.M. Regulation of trehalose mobilization in fungi. *Microbiol. Rev.* 48:42–59, 1984.
74. Lillie, S.H., J.R. Pringle. Reserve carbohydrate metabolism in *Saccharomyces cerevisiae*: responses to nutrient limitation. *J. Bacteriol.* 143:1384–1394, 1980.
75. Attfield, P.V. Trehalose accumulates in *Saccharomyces cerevisiae* during exposure to agents that induce heat shock response. *FEBS Lett.* 225:259–263, 1987.
76. Sanchez, Y., J. Taulien, K.A. Borkovich, S. Lindquist. Hsp104 is required for tolerance to many forms of stress. *Embo J.* 11:2357–2364, 1992.
77. Elliott, B., R.S. Haltiwanger, B. Futcher. Synergy between trehalose and Hsp104 for thermotolerance in *Saccharomyces cerevisiae*. *Genetics* 144:923–933, 1996.
78. Sanchez, Y., S.L. Lindquist. HSP104 required for induced thermotolerance. *Science* 248:1112–1115, 1990.
79. Muller, J., R.A. Aeschbacher, A. Wingler, T. Boller, and A. Wiemken. Trehalose and trehalase in Arabidopsis. *Plant Physiol.* 125:1086–1093, 2001.
80. Penna, S. Building stress tolerance through over-producing trehalose in transgenic plants. *Trends Plant Sci.* 8:355–357, 2003.
81. Goddijn, O.J., K. van Dun. Trehalose metabolism in plants. *Trends Plant Sci.* 4:315–319, 1999.
82. Goddijn, O.J., T.C. Verwoerd, E. Voogd, R.W. Krutwagen, P.T. de Graaf, K. van Dun, J. Poels, A.S. Ponstein, B. Damm, J. Pen. Inhibition of trehalase activity enhances trehalose accumulation in transgenic plants. *Plant Physiol.* 113:181–190, 1997.
83. Van Dijck, P., J.O. Mascorro-Gallardo, M. De Bus, K. Royackers, G. Iturriaga, J.M. Thevelein. Truncation of *Arabidopsis thaliana* and *Selaginella lepidophylla* trehalose-6-phosphate synthase unlocks high catalytic activity and supports high trehalose levels on expression in yeast. *Biochem. J.* 366:63–71, 2002.
84. Romero, C., J.M. Belles, J.L. Vaya, R. Serrano, F.A. Culianez-Macia. Expression of the yeast trehalose-6-phosphate synthase gene in transgenic tobacco plants: pleiotropic phenotypes include drought tolerance. *Planta* 201:293–297, 1997.
85. Garg, A.K., J.K. Kim, T.G. Owens, A.P. Ranwala, Y.D. Choi, L.V. Kochian, R.J. Wu. Trehalose accumulation in rice plants confers high tolerance levels to different abiotic stresses. *Proc. Natl. Acad. Sci. USA* 99:15898–15903, 2002.
86. Muller, J., T. Boller, A. Wiemken. Effects of validamycin A, a potent trehalase inhibitor, and phytohormones on trehalose metabolism in roots and nodules of soybean and cowpea. *Planta* 197:362–368, 1995.
87. Eastmond, P.J., A.J. van Dijken, M. Spielman, A. Kerr, A.F. Tissier, H.G. Dickinson, J.D. Jones, S.C. Smeekens, I.A. Graham. Trehalose-6-phosphate synthase 1, which catalyses the first step in trehalose synthesis, is essential for Arabidopsis embryo maturation. *Plant J.* 29:225–235, 2002.
88. Eastmond, P.J., I.A. Graham. Trehalose metabolism: a regulatory role for trehalose-6-phosphate? *Curr. Opin. Plant Biol.* 6: 231–235, 2003.
89. Schluepmann, H., T. Pellny, A. van Dijken, S. Smeekens, M. Paul, Trehalose 6-phosphate is indispensable for carbohydrate utilization and growth in Arabidopsis thaliana. *Proc. Natl. Acad. Sci. USA* 100:6849–6854, 2003.
90. Richards, A.B., L.B. Dexter. Trehalose. In: *Alternative Sweeteners*, 3rd ed., Nabors, L.O., ed., New York: Marcel Dekker, 2001, pp 423–462.
91. Hirasawa, R., K. Yokoigawa, Y. Isobe, H. Kawai. Improving the freeze tolerance of baker's yeast by loading with trehalose. *Biosci. Biotechnol. Biochem.* 65:522–526, 2001.
92. Shima, J., A. Hino, C. Yamada-Iyo, Y. Suzuki, R. Nakajima, H. Watanabe, K. Mori, H. Takano. Stress tolerance in doughs of *Saccharomyces cerevisiae* trehalase mutants derived from commercial baker's yeast. *Appl. Environ. Microbiol.* 65:2841–2846, 1999.

93. Van Dijck, P., L. De Rop, K. Szlufcik, E. Van Ael, J.M. Thevelein. Disruption of the *Candida albicans* TPS2 gene encoding trehalose-6-phosphate phosphatase decreases infectivity without affecting hypha formation. *Infect. Immun.* 70:1772–1782, 2002.
94. Zaragoza, O., C. de Virgilio, J. Ponton, C. Gancedo. Disruption in *Candida albicans* of the TPS2 gene encoding trehalose-6-phosphate phosphatase affects cell integrity and decreases infectivity. *Microbiology* 148:1281–1290, 2002.
95. Foster, A.J., J.M. Jenkinson, N.J. Talbot. Trehalose synthesis and metabolism are required at different stages of plant infection by *Magnaporthe grisea*. *EMBO J.* 22:225–235, 2003.
96. Cohen, F.E., J.W. Kelly. Therapeutic approaches to protein-misfolding diseases. *Nature* 426:905–909, 2003.
97. Perlmutter, D.H. Chemical chaperones: a pharmacological strategy for disorders of protein folding and trafficking. *Pediatr. Res.* 52:832–836, 2002.
98. Willingham, S., T.F. Outeiro, M.J. DeVit, S.L. Lindquist, P.J. Muchowski. Yeast genes that enhance the toxicity of a mutant huntingtin fragment or alpha-synuclein. *Science* 302:1769–1772, 2003.
99. Tanaka, M., Y. Machida, S. Niu, T. Ikeda, N.R. Jana, H. Doi, M. Kurosawa, M. Nekooki, N. Nukina. Trehalose alleviates polyglutamine-mediated pathology in a mouse model of Huntington's disease. *Nat. Med.* 2004.
100. Parsell, D.A., A.S. Kowal, M.A. Singer, S. Lindquist. Protein disaggregation mediated by heat-shock protein Hsp104. *Nature* 372:475-478, 1994.
101. Uptain, S.M., S. Lindquist. Prions as protein-based genetic elements. *Annu. Rev. Microbiol.* 56:703–741, 2002.
102. Wickner, R.B., K.L. Taylor, H.K. Edskes, M.L. Maddelein, H. Moriyama, and B.T. Roberts. Prions of yeast as heritable amyloidoses. *J. Struct. Biol.* 130:310–322, 2000.
103. Wickner, R.B., H.K. Edskes, B.T. Roberts, M. Pierce, U. Baxa. Prions of yeast as epigenetic phenomena: high protein "copy number" inducing protein "silencing". *Adv. Genet.* 46:485–525, 2002.
104. Feder, J.H., J.M. Rossi, J. Solomon, N. Solomon, S. Lindquist. The consequences of expressing hsp70 in Drosophila cells at normal temperatures. *Genes Dev.* 6:1402–1413, 1992.
105. Welte M.A., J.M. Tetrault, R.P. Dellavalle, S.L. Lindquist. A new method for manipulating transgenes: engineering heat tolerance in a complex, multicellular organism. *Curr. Biol.* 3:842–853, 1993.

1.14

Production of Pectinases and Utilization in Food Processing

K.S. Venkatesh and S. Umesh-Kumar

CONTENTS

14.1 INTRODUCTION

Raw materials of plant origin are processed in a variety of ways for the production of food. Modern food processing uses enzyme preparations to improve pulp and juice yields from fruits, in wine preparation to increase the color and quality of the wine, in tea manufacture for flavor release, in oil extraction from olives, and in steeping maize for the recovery of sugar. In each of these processes, pectin degradation is an important step.

Pectin is a complex colloidal carbohydrate found in plants. Pectin is found in the middle lamella and primary cell walls of higher plants and is made of anhydrogalacturonic acid units. The carboxyl groups of galacturonic acid in pectin may be partly esterified by methyl groups or partly or totally neutralized by one or more bases. Pectic substances are produced in plants from a water insoluble parent pectic substance called

Protopectin. Pectinic acid is a colloidal polygalacturonic acid containing a methyl ester group.

Under suitable conditions, pectinic acids form gels with sugars and acids. If the methoxyl content is low in pectinic acid, gel formation can also occur with ions. Pectin contains pectinic acids of varying methoxyl ester content. This complex polysaccharide contains "smooth regions" of α-1, 4-linked D-galacturonic acid residues as the backbone, sometimes acetylated at *O*-2 or *O*-3 and methoxylated at *O*-6. In the "hairy regions", a xylogalacturonan consists of a D-xylose substituted galacturonan backbone. Similarly, rhamnogalacturonan and arabinogalacturonan contain rhamnose and arabinose respectively. Certain plant tissues show further complexity in their pectins. Sugar beet and apple pectins contain ferrulic acid as terminal residues attached to *O*-5 of arabinose residues, or *O*-2 of galactose residues. Side chains of rhamnogalacturonan also contain uncommon sugars such as 2-*O*-methoxy-L-fucose and 3-Deoxy-D-manno-2-octulonic acid. These variations in pectin structures are attributes of protective mechanism that the plants have evolved to protect against invading microorganisms.

The presence of such a complex heteropolysaccharide as pectin in the plant world necessitated identification of pectinases useful for food processing. Generally, the enzymes that act on pectic substances can be divided into two categories. The depolymerizing enzymes like polygalacturonase or pectin hydrolase and lyases act by hydrolytic or transeliminating cleavage of glycosydic bonds with endo- or exosplitting action, to break down the pectic acid or pectin. The saponifying enzymes, commonly referred to as pectin methyl esterases, catalyze the deesterification of the methoxyl group of pectin to make the molecule susceptible to hydrolytic enzymes. Several pectinases like rhamnogalacturonase, arabinofuronosidase, and xylogalacturonan hydrolases have found specific applications in food processing industries due to their ability to act on hairy regions of the pectin.

Aspergillus rhamnogalacturonase and xylogalacturonase hydrolytically cleave the main chain of rhamnogalacturonan and xylogalacturonan, respectively. Arabinofuranosidases hydrolyse both α-1,5-and α-1,3-glycosydic bonds of arabinan. The activity of these enzymes is hindered by the presence of acetyl residues in the pectin main chain. Hence, for efficient hydrolysis of sugar backbone, several enzymes called acetyl esterases have been characterized from *Aspergillus* species. All the hydrolyzing enzymes cleave the substrate through an inverting mechanism (1). The characteristics of the enzymes that degrade pectin are described in Table 14.1.

In food processing industries, hydrolases have wide application. In previous years, polygalacturonases and pectin methyl esterases were identified as enzymes that degrade pectin. Because pectin contains methylated polygalacturonic acid, which is not susceptible to polygalacturonase activity, predeesterification of the methyl groups by pectin methyl esterase is important for the activity of the pectin hydrolase and polygalacturonase. Recent advances in the elucidation of the pectin structure, specifically the hairy regions and their resistance to the activity of pectinases, led to the discovery of newer enzymes that found application in food processing. The advent of molecular biological techniques has also resulted in selective production of these enzymes industrially.

14.2 PRODUCTION OF PECTINASES

Though pectinases are produced by bacteria, mold, and yeasts, most of the commercial enzymes are produced from the mold *Aspergillus* (34). Typical commercial preparations of polygalacturonase from *Aspergillus* show different forms of the enzyme that cross react with a single antibody raised against one of the polygalacturonase (Figure 14.1).

Table 14.1
Characteristics of Pectic Enzymes Produced by Microorganisms

Enzyme	Organisms	Molecular Weight (kDa)	pH Optima	Ref.
Endopolygalacturonase	*Colletotrichum lindemuthianum*	38	4.2 & 5.4	2
Endopolygalacturonase	*Rhizopus stolonifer*	60	4.6–4.8	3
Endopolygalacturonase (PG-III)	*A. niger*		5.0	4
Exopolygalacturonase (PG-I)			5.0	
Exopolygalacturonase (PG-II)			5.0	
Endopolygalacturonase (Endo-I)	*A. niger*	55	4.9	5
Endopolygalacturonase (Endo-II)		38	4.8	
Endopolygalacturonase (Endo-IIIA)		57	4.3	
Endopolygalacturonase (Endo-IIIB)		57	4.5	
Endopolygalacturonase (Endo-IV)		59	4.8	
Endopolygalacturonase (EndoPG)	*A. niger*	35.5	4.5	6
Endopolygalacturonase (EndoPG)	*C. lindemuthianum*	42	5.2	7
Endopolygalacturonase (PG-A)	*A. niger*	35	4.0	8
Endopolygalacturonase (PG-B)		35	5.0	8
Endopolygalacturonase (PG-D)		51	4.2	9
Endopolygalacturonase	*A. oryzae*		4.0	10
Endopolygalacturonase (PG-A)		41	5.0	11
Endopolygalacturonase (PG-B)		39	5.0	11
Endopolygalacturonase (EndoPG-1)	*A. niger*	35	4.1	12
Endopolygalacturonase (EndoPG-2)		80	3.8	
Endopolygalacturonase (EndoPG-I)	*Kluveromyces fragilis*	46	4.0–5.0	13
Endopolygalacturonase (EndoPG-II)		50	4.0–5.0	
Endopolygalacturonase (EndoPG-III)		30	4.0–5.0	
Endopolygalacturonase (PG-I)	*A. carbonarius*	61	4.0	14
Endopolygalacturonase (PG-II)		42	4.1	
Endopolygalacturonase (PG-III)		47	4.3	
Endopolygalacturonase (EndoPG-II)	*A. niger*	35	4.5	15
Endopolygalacturonase (Endo-Pgl1P)	*Saccharomyces cerevisiae*	42	4.5	16
Endopolygalacturonases	*K. marxianus*	45, 42, 39, 36	5.0	17
Endoxylogalacturonan hydrolase	*A. tubingensis*		5.0	18
Exopolygalacturonase	*A. aculeatus*	42		19
Exopolygalacturonase (ExoPG-I)	*A. niger*	66	3.8	20
Exopolygalacturonase (ExoPG-II)		63	4.5	
Exopolygalacturonase (ExoPG-I)	*A. alliaceus*	40	3.5	21
Exopolygalacturonase (ExoPG-2)		40	6.0	
Pectate lyase (PEL-A)	*A. nidulans*	40		22
Pectates lyase (Ply-A)	*A. niger*	43	7.5–8.5	23
Pectin lyase	*A. joponicus*	32	5.2	24
Pectin lyase (PL-I)	*A. niger*	37.5		25

(Continued)

Table 14.1 (*Continued*)

Enzyme	Organisms	Molecular Weight (kDa)	pH Optima	Ref.
Pectin lyase (PL-II)		37.5		25
Pectin lyase (PL-B)		40	8.5–9.0	26
Pectin methyl esterase (PME1)	*A. aculeatus*	43	4.6	27
Rhamnogalacturonan hydrolases	*A. aculeatus*			
RhgA		51	3.0–4.0	28
Rgase A		59	3.5	29
Rhamnogalacturonan rhamnohydrolase (RG-RH)		84	4	30
Rhamnogalacturonan galacturonohydrolase RG-GH		66	4	31
Rhamnogalacturonan Hydrolase A (RHG-A), Rhamnogalacturonan	*A. niger*	72, 70	3.6, 4.1	32
Hydrolase B (RHG-B) Rhamnogalacturonan lyase (Rgase-B)	*A. aculeatus*	55	6.0	29
Rhamnogalacturonase acetylesterase (RGAE)	*A. aculeatus*	26.7	6	33

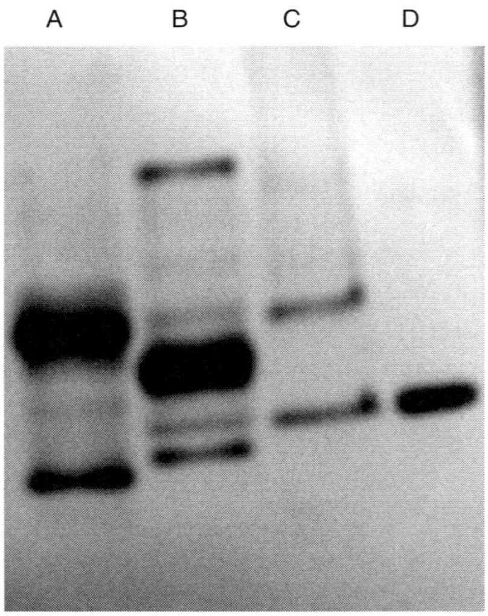

Figure 14.1 Different isoforms of polygalacturonase of *Aspergillus* identified by antibody cross reactivity on western blot reaction. The polygalacturonases of *Aspergillus niger* obtained from two commercial samples [A] & [B] were found to cross react with the antibody raised to a polygalacturonase purified from *A. carbonarius* culture filtrates [D]. Antibody reactivity of another polygalacturonase produced by *A. carbonarius* is also shown [C].

Differences in the molecular mass of polygalacturonases within the *Aspergillus* are also known (34) and it is believed that highly active polygalacturonase is produced by *A. carbonarius* (14). Given that degradation of plant pectin involves activity by several enzymes [Figure 14.2(a), 14.2(b)], commercial preparations generally contain more than one pectinolytic enzyme. Industrially, pectinases are produced by both solid-state fermentation (SSF) and submerged fermentation (SmF) methods using *Aspergillus* sp. Generally, SSF has been considered more suitable for obtaining a particular composition of the enzyme mixture with higher enzyme yields (35,36,37).

14.2.1 Submerged Fermentation

In submerged fermentation using *Aspergillus*, designing of a balanced fermentation medium has been found to be critical. Production of extracellular pectinase by this mold was shown to be induced by the presence of pectic material in the culture medium (38). Pectin is a high molecular weight polysaccharide. Enzyme induction is apparently due to low levels of basal constitutive activity produced by the organism that degrades the polymeric substrate. This results in low molecular weight products that serve as inducers in the cell to promote cell growth and pectinase production (39). General inducers added to the growth medium include sources rich in pectin like beet pulp, citrus peel, or apple pomace (38,40). The medium contains nitrogen, in the form of ammonium salts, corn steep liquor, distillers soluble yeast extract, gelatin, or casein. Mineral supplementation is also provided in the form of inorganic salts to achieve higher growth rates.

In SmF, for the production of pectinase, the type of pectin inducers used influence the pectic enzyme produced by the culture. When an alkali pretreated lemon peel was used, reduced pectin esterase activity was detected in the culture broth (41). In these culture filtrates, at least four times higher endopolygalacturonase activity was estimated. Extensive studies on polygalacturonase production from *A. niger* revealed that the productivity of this enzyme is a function of both temperature and cultivation medium (42). When the fungus was grown in a defined medium containing purified pectin as a carbon source, production of polygalacturonase was most efficient at 18°C and insignificant at 30°C. Higher enzyme activity was also obtained at acidic pH around 3.0 (37,43–49). In *Rhizopus nigricans,* production of endopolygalacturonase was found to be highest at pH 3.0 (50).

Catabolic repression of pectinase production in *A. niger* (44,51) defines a two stage fermentation system for obtaining higher enzyme yields. In such a system, the mycelia are developed using readily metabolizable sugars such as molasses and starch hydrolysates in the first stage. Enzyme synthesis is induced in grown mycelium in the second stage by feeding pectin containing substrate as a carbon source (43).

Mutation of *Aspergillus* spores and selection of mycelium that formed aggregates, such as clumps and pellets, were found to produce a higher concentration of pectinases in SmF (52). The addition of potassium hexocyanoferrate, which affected the morphology of *A. niger* to form small compact pellets with smooth surfaces, was reported to increase enzyme production, reducing viscosity of culture broth and improving the fermentation process for enzyme production (53).

14.2.2 Solid-State Fermentation

Solid-state fermentation (SSF) has been adopted as a fermentation technique for pectinase production, apparently because cellular regulations that affect enzyme production are absent when fungi are grown in solid-state substrates. Low water activity (a_w) in the substrate enhanced enzyme secretion during SSF (54). Reports in literature suggest use of wheat bran, cassava fibrous waste, citrus pulp waste, sugar cane bagasse, coffee pulp, and orange peel as substrates for the production of pectic enzymes using *Aspergillus* (54–60).

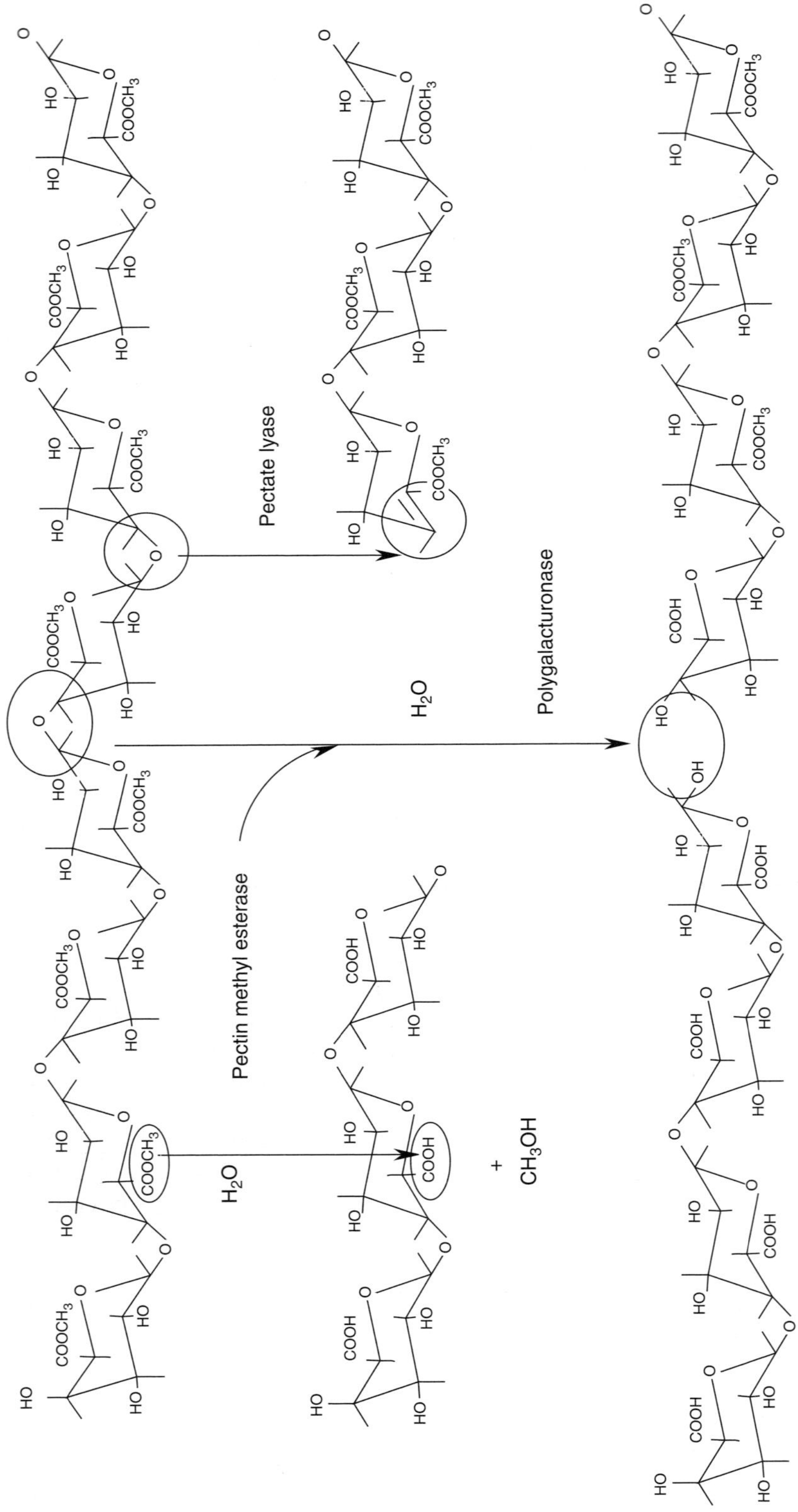

Figure 14.2(a) Pectin main chain structure and enzymes involved in degradation.

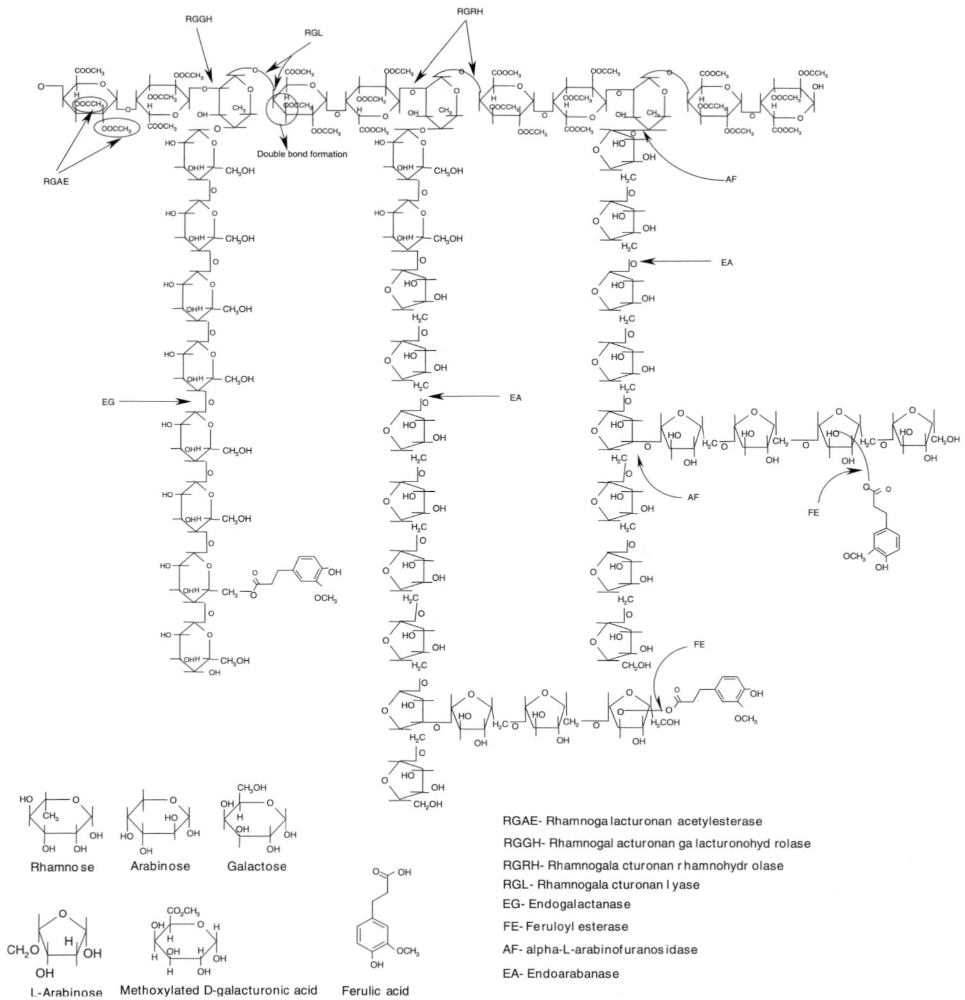

Figure 14.2(b) Structure of hairy region of pectin and pectic enzymes involved in the degradation.

In SSF, an initial acidic pH of the medium increased enzyme yields. In the authors' laboratory, a simple wheat bran based medium has been routinely used for the production of pectinase using *A. carbonarius* by SSF. The bran medium, containing 40% moisture, was acidified with 2N HCl before sterilization. Spores or actively growing mycelia in liquid culture were used to inoculate the substrate. An assay of the enzyme extracted from fermented substrate using 0.1 M of an acetate buffer of pH 4.3 detected maximum pectinases in bran medium after 60–72 h growth.

Glucose catabolic repression has been the major constraint in pectinase production in SmF by a wild-type strain of *Aspergillus* (43). Though in such strains, SSF cultivation gives greater enzyme yields than SmF, mutational selection to avoid regulation by glucose can make SmF an ideal choice of fermentation for pectinase production. This is because the cost inputs are less in SmF, with the added advantage of the absence of an extraction procedure. In SSF, the enzyme is usually dilute, unlike in SmF, due to the involvement of an extraction step using 1:3 to 1:5 ratios of fermented bran and buffer (61). Recently a strain of *A. carbonarius* that constitutively produced polygalacturonase in SmF was isolated by ultra violet radiation induced mutation by the authors.

14.2.3 Strain Improvement

According to the type of fermentation system adopted for pectinase production using *Aspergillus*, several strain improvement strategies have been described.

In SSF, low a_w and efficient colonization of the substrate apparently dictated yields of pectinase. Hence, *dgr* (2-Deoxy D-glucose resistant) mutants selected for tolerance to low a_w were found to produce higher pectinase when grown in solid-state culture (59). Increased production of pectinase by such mutants was reasoned to be due to modification in the phospholipid fatty acid saturation in the membrane, causing increased permeability of pectinase into culture media.

Observations on differential pectinase production, according to mycelial structure, resulted in the isolation of slow growing mutant strains of *Aspergillus* that formed aggregated mycelia (52). This mutant, when grown in solid-state, produced more pectinase compared to the parent strain.

Inter- and intraspecific fusion of *Aspergillus* protoplasts have been attempted, to obtain strains producing higher pectinase activities during solid-state growth (62,63). Though the results were not encouraging, claims have been made that useful isoforms of polygalacturonases have been produced. One of the reports described fusion of protoplasts of *A. niger* with *A. flaviceps*. A strain isolated by fusion techniques showed increased enzyme production, growth, and specific activity (64).

In SSF, sporulation was identified as a constraint for industrial production of pectinase (55,62). Mutants of *A. carbonarius*, deformed in cell differentiation, were isolated as nonsporulant strains after UV radiation of spores of the parent strain. The nonsporulant mutant showed higher polygalacturonase production by SSF. In fungi, sporulation indicates cessation of growth. Sporulation has several disadvantages in industrial pectinase production by SSF because it contaminates the fermentation section and sometimes causes allergic reactions in the personnel. Absence of sporulation, in addition to improving pectinase production due to vegetative mycelial growth of the fungus, also avoided inherent difficulties of SSF described previously (62).

Cellular regulation of pectinase production by glucose (glucose catabolic repression) was shown to affect pectinase production by *Aspergillus* in SmF (65). In the authors' laboratory, glucose catabolic repression of polygalacturonase production by *A. carbonarius* was overcome by isolating specific mutants of the fungus after inducing mutation with UV radiation. The mutants were isolated by temperature selection at 42°C in a medium containing both pectin and glucose analogue 2-Deoxy-D-glucose, and the pectinase producing mutant was identified using ruthenium red or *O*-toluidine blue (66). The mutants so isolated produced constitutive polygalacturonase with activities about 200-fold higher than the parent strain, in a simple medium containing starch hydrolysates (67). Similar results on hyperendopolygalacturonase-producing mutants of *Aspergillus* strains have been obtained by inducing mutation with N-methyl-N'-nitro-N-nitrosoguanidine (NTG) and the selection of strains resistant to catabolic repression (68).

14.3 GENES OF PECTIN DEGRADING ENZYMES

Promoter deletion analysis of one of the genes (*pga*II) identified the region 5'-TYATTGGTGGA-3' as important for high level gene expression. This region had high similarity to the promoter *pgaA* of *A. niger*. Expression of most of the pectinase genes was reported to be induced by D-galacturonic acid. The induction of several other carbohydrases like arabinofuronosidase, endoarabinase, β-galactosidase, and α-glucuronidase by galacturonic acid and glucuronic acid suggested a general system of gene expression in these carbohydrases (34). Galacturonic acid did not induce the *pgaB* gene, the promoter of which does not contain the 5'-TYATTGGTGGA-3', and this further augmented the importance of the promoter region in pectinase gene expression (69). A hexanucleotide sequence CCCTGA was identified in the promoters of several pectin lyase-encoding genes. Characterization of several pectinase genes including the gene of polygalacturonase has resulted in cloning and overexpression. These genes are described in Table 14.2.

14.4 PECTINASES OF *ASPERGILLUS*

Several enzymes involved in the degradation of pectin have been purified and their gene sequences obtained. Studies on the expression patterns of the lyase and hydrolase genes of *A. niger* showed differences in the expression patterns (78). The crystal structure of *A. aculeatus* polygalacturonase glycoprotein revealed one *N*- and ten *O*-glycosylation sites. The enzyme folds into a right handed parallel β-helix with the long T_1 loop containing a catalytic tyrosine residue. This residue was significantly different from that found in *A. niger* polygalacturonase. All three of the polygalacturonases analyzed had six subsites. It is hypothesized that the long T_1 loop of *A. niger* polygalacturonases undergos a conformational change upon binding of the substrate to bring the tyrosine residue close to subsite 1 (79).

Polygalacturonase I and II (PG-I and PG-II) were identified as major enzymes produced by *A. niger* (5,34). Probes designed based on the NH_2-terminal sequence of PG-I and PG-II and screening genomic DNA libraries of *A. niger* isolated five new genes. The products of these genes were also detected in *A. nidulans* transformants by western blot analysis, using antibodies raised against PG-I. The sequence of one of the polygalacturonases (*PGC*) identified three introns in the genomic DNA. The product of the gene, a 383 amino acid preproprotein was reportedly cleaved after a pair of basic amino acids. In the upstream region of the *A. niger* polygalacturonase gene, a conserved sequence CCAAT apparently represented a binding site for the regulatory protein (74).

Sequence analysis of *pga*A and *pga*B identified one intron in the former and two introns in the latter. Apparently, these enzymes are also synthesized as preproenzymes with the prosequences cleaved by a KEX2-like dibasic peptidase after the lysine and arginine residues (77). The N-terminal amino acid sequence of several polygalacturonases of

Table 14.2
Cloned genes encoding for pectinases in *Aspergillus*

Organism	Genes	Enzyme	Ref.
A. niger	*pel*A,	Pectin lyase-A	70
	*pel*B	Pectin lyase-B	71
A. niger,	*pga*II	Polygalacturonase-II	72
A. tubigensis			73
A. niger	*pga*I	Polygalacturonase-I	74
A. niger	*pga*C	Polygalacturonase-C	74
A. niger	*pme*	Pectin methyl esterase	75
A. nidulans	*pel*A	Pectin lyase-A	22
A. nidulans	*pel*D	Pectate lyase-D	76
A. niger	*prepro PG*II	Prepropolygalacturonase-II	77

Aspergillus identified three different enzymes. However, polygalacturonase III produced by *A. niger* and *A. oryzae* were similar (79). Sequence analysis of polygalacturonase identified 27 amino acids characteristic of a secretary signal sequence in NH_2-terminal, which is cleaved off in the mature protein (73,80).

A similar processing of rhamnogalacturonan acetyl esterase (RGAE) from *A. aculeatus* was described. The *rha*1 cDNA was also found to encode a signal peptide of 17 amino acids. The amino acid sequence of this enzyme did not show significant similarities with other pectinases, suggesting it as a novel enzyme representing a new family of esterases (33). Analysis of the endoxylogalacturonan hydrolase (xghA) of *A. tubingensis* cDNA revealed the presence of an 18 amino acid signal sequence in the NH_2-terminus. Comparison of the xghA amino acid sequence with sequences in the European Molecular Biology Laboratory (EMBL) data library revealed homology to polygalacturonase sequences of prokaryotes, fungi, and plants and to the *rhg*A and *rhg*B sequences of *A. aculeatus* and *A. niger* (18). The pectin methyl esterase protein of *A. aculeatus* had a molecular mass of 43 kDa as estimated on SDS-PAGE gels. The cDNA gene sequence revealed an apparent signal sequence of 17 amino acids with a typical signal cleave site between Ala-17 and Ala-18 (27).

Three polygalacturonases of molecular mass 61-, 42- and 47-kDa were purified and characterized from *A. carbonarius.* Of these enzymes, the polygalacturonase of molecular mass 42 kDa was highly active on a citrus polygalacturonate substrate with specific activity as high as 7000 U/mg of protein (14). Further studies on this enzyme in the authors' laboratory showed that the N-terminal amino acid sequence Gly-Ser-Cys-Thr-Phe of the protein was identical to the N-terminal sequence of polygalacturonase II of *A. tubingensis* (81), endopolygalacturonase III of *Sclerotinia sclerotiorum* (82,83), and polygalacturonase II of *A. niger* (84). Homology to polygalacturonase of *A. niger* (72) and *A. oryzae* (48) was also found in the four N-terminal amino acids. The polygalacturonases of *A. niger* and *A. oryzae* differed from *A. carbonarius* polygalacturonase in the first N-terminal amino acid. While Asp constituted the N-terminal amino acid in *A. niger* and *A. oryzae,* Gly was the N-terminal amino acid in *A. carbonarius.*

The structure of rhamnogalacturonan hydrolases A and B from *A. niger* predicted the occurrences of three introns at the same positions as identified in *A. aculeatus rhg*A. A high degree of homology was also found between rhamnogalacturonan hydrolase proteins of *A. niger* and *A. aculeatus*. Rhamnogalacturonan hydrolase B showed about 72% identity with rhamnogalacturonan hydrolase A protein of *A. aculeatus* (32).

14.5 CLONING AND EXPRESSION OF PECTINASE ENZYMES

Genomic DNA coding pectinase protein contains introns. Hence, cloning strategies used cDNA in heterologous hosts for expression of various enzymes degrading pectin. The protein expression in homologous hosts was also achieved by cloning genomic DNA of pectic enzymes. Enzyme overproduction using recombinant genes described multicopy expression systems, gene fusions, and expression under the control of strong promoters.

The promoters *pgaA* and *pgaB*, the two constitutively expressed endopolygalacturonases of *A. niger*, were cloned in *A. niger* plasmid pGW635 carrying the *pyr*A gene that encoded orotidine – 5'-phosphate decarboxylase as a selection marker for restoration of uridine prototrophy. The multicopy transformants were used for overproduction of the enzyme. For expression, promoter gene fusions were made in plasmids harboring *pki*A promoter. Standard assays performed indicated overproduction of both the enzymes. Product analysis using polygalacturonate revealed a random cleavage pattern for both the enzymes (8).

The heterologous expression of *A. niger* endopolygalacturonase in *Saccharomyces cerevisiae* yeast has been reported with technological advantages. Endopolygalacturonase cDNA was expressed under the control of the alcohol dehydrogenase (*ADH*I) promoter in yeast — the *Escherichia coli* shuttle vector. Because this yeast secreted only a few of its own proteins in culture fluid, heterologous gene expression yielded polygalacturonase with a high specific activity. It was observed that the enzyme secretion was efficiently directed by the fungal leader sequence and the processing of the enzyme occurred at the same site of yeast as in *Aspergillus*. A shorter version of yeast *ADH*I promoter enhanced the expression level. The yeast *ADH*I promoter is regulated by glucose, and glucose is essential for efficient induction. Hence, the removal of the upstream region of the promoter avoided this regulation. Removal of bacterial sequences from the expression vector increased the plasmid stability and plasmid copy. This led to an increased yield of heterologous protein. The yeast derived enzyme was more glycosylated than that obtained from the mold and showed the same biochemical properties (6).

The cloning strategy and heterologous gene expression of the polygalacturonase gene of *A. awamori* in *S. cerevisiae* involved the preparation of a cDNA probe using the N-terminal amino acid sequence "Ser-Thr-Cys-Thr-Phe-Thr". The gene was isolated from the genomic DNA by Polymerase Chain Reaction (PCR) and identified by southern hybridization. The gene library was prepared by inserting the *Hind* III digested fragment of DNA into pUC118. The transformed *E. coli* DH5α were identified by hybridization and the plasmid extracted from positive clones was used to isolate the polygalacturonase gene. The introns were excised by two step PCR and ligation. The "intron free" polygalacturonase gene was inserted between the phosphoglycerate kinase (*pgk*) promoter and terminator in a yeast expression plasmid containing the *LEU-2* gene as a selection marker. The transformed yeast reportedly secreted the enzyme whose molecular weight (41 kDa) corresponded with the native enzyme. Based on heterologous gene expression studies it was presumed that the glycosylation mechanism in the secretion path of *S. cerevisiae* was similar to that in fungi (81).

The rhamnogalacturonan hydrolase genes (*rghA* and *rghB*) identified in *A. niger* using the genomic rhamnogalacturonan hydrolase gene of *A. aculeatus* as a probe were cloned, and the proteins expressed in *A. niger*. For this study, a genomic library of *A. niger* was screened for the presence of the rhamnogalacturonan hydrolase gene. The expression of the proteins was achieved by putting the genes under the control of the endoxylanase A gene (*exlA*) promoter of *A. awamori*. Despite multicopy insertions, western blot analysis showed very little expression of the two proteins in *A. niger*. Transcription analysis for the mRNAs also showed very little expression. Multicopy integration of the gene in *A. awamori* improved protein expression with yields of 3 and 40 mg of purified rhamnogalacturonan hydrolase A and B, respectively, from one liter of culture fluid (32).

An *Aspergillus tubingensis* expression library constructed in *Kluyveromyces lactis* identified xylogalacturonan hydrolyzing activities in some recombinants that encoded a xylogalacturonan hydrolase containing 406 amino acids. This protein was used to define substrate activity, which appeared to be specific for xylose substituted galacturonic acid backbone (18).

A full length cDNA encoding the pectin methyl esterase isolated from the filamentous fungus *A. aculeatus* was cloned and expressed in *A. oryzae* in fungal expression vector pHD464 for expression of the protein. The expression host, which does not produce endogenous pectin methyl esterase or polygalacturonase, heterologously expressed the pectin methyl esterase of *A. aculeatus*. Though the amino acid sequence of recombinant enzyme showed homology with the authentic enzyme, it was found to be processed differently at the N-terminal by signal peptidase. The authentic enzyme was processed at the

21st amino acid by *A. aculeatus* whereas the heterologous host processed the recombinant enzyme at the 18th amino acid at the N-terminal and it had more glycosylation than the authentic enzyme (27).

A DNA probe constructed based on the N-terminal sequence of rhamnogalacturonan acetylesterase (rha1) of *A. aculeatus* was used to isolate cDNA, encoding the 250 amino acids of precursor protein with a 17 amino acid sequence of signal peptide from the cDNA library. The sequence was expressed in *A. oryzae*, a strain that did not have rhamnogalacturonan acetylesterase activity. Subcloning the cDNA behind the *A. oryzae* α-amylase promoter and expressing the resultant in *A. oryzae* yielded high level production of *rha1* (33).

A heterologously expressed glycoprotein, rhamnogalacturonan acetyl esterase from *A. aculeatus* in *A. oryzae*, was used to elucidate the structure and function of this new family of hydrolase. It was found that the rhamnogalacturonan acetylesterase protein folded into an $\alpha/\beta/\alpha$ structure, with the active site as an open cleft containing "Ser-His-Asp". The sequence similarities and active site residues suggested that this enzyme, with seven other proteins, formed a new hydrolase family, the SGNH hydrolase family ,which included the carbohydrate esterase family 12 (85,86).

14.6 REGULATION OF PECTINASE PRODUCTION IN *ASPERGILLUS*

Aspergilli that produce pectic enzymes are known to grow over a wide pH range. In *A. kawachi*, it was demonstrated that the pH of the culture broth influenced the type of polygalacturonase produced by the fungus (48). Thus, a role for the pH regulatory protein (PacC) was identified in pectinase gene regulation.

Generally, pH regulation is mediated by *pac* genes, *pac*C, *pad*A, *pal*B, *pal*C, *pal*F, *pal*H, and *pal*I. Mutations that mimic acidity, alkalinity, and neutrality gene expressions have been obtained. This suggested direct involvement of a key transcription factor, which is a product of *pacC*, in the regulation of gene expression by ambient pH. The major arabinofuronosidase gene is reportedly regulated by PacC (87). Evidence suggesting acidic pH induction of polygalacturonase production in *A. carbonarius* (Figure 14.3) was also obtained recently in the authors' laboratory. Hence, regulation of pectinase gene expression apparently is controlled by pH, as in the genes encoding several cell wall degrading enzymes like xylanases and cellulases (88,89).

14.7 APPLICATIONS

Pectinases are enzymes involved in the degradation of the plant cell wall. Hence, these enzymes are applied to provide a good alternative to chemical processing. Figures 14.4, 14.5, and 14.6 describe examples of pectinase applications in industrial processing of fruits. Major applications involve their wide spread use in the beverage industry, due to their ability to improve pressing of fruits and clarification of fruit juices. In wine production, pectinases are used to improve the juice and color yields, enhancing the clarity of the wine. Pectinases, when applied along with other cell wall degrading enzymes, result in increased yields of the oil that can be extracted from lemon peel and olive fruits. Other applications are in the production of carrot puree and neutraceutically important oligouronides, the monosaccharides which are the building blocks of pectin polymer. Arabinose, which is the precursor for α-fructose and L-glucose, released from the pectin polymer arabinogalacturonan (commonly found in apples) can be used as a noncalorific sweetener

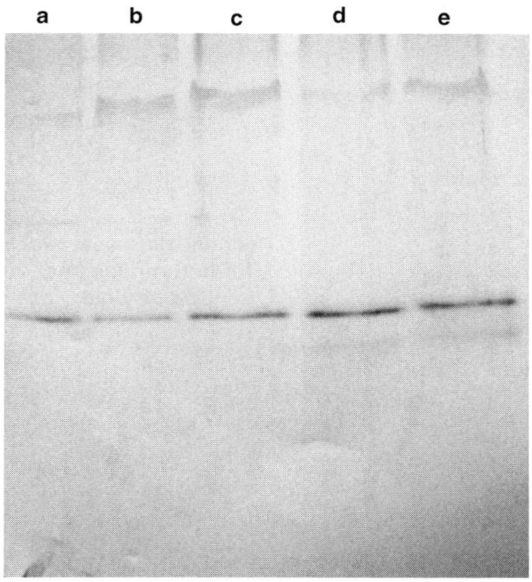

Figure 14.3 Effect of pH on polygalacturonase production in *A. carbonarius*. The fungus was grown at pH 5.0 (a), 4.5 (b), 4.0 (c), 3.5 (d), and 3.0 (e). in a medium buffered to the required pH with 0.25M citrate phosphate buffer. Polygalacturonase was quantitated by the intensity of the reaction in Western Blots after SDS PAGE with identical protein concentrations. Antibody used for this reaction was raised in rabbit using the purified protein (Figure 14.1, column D).

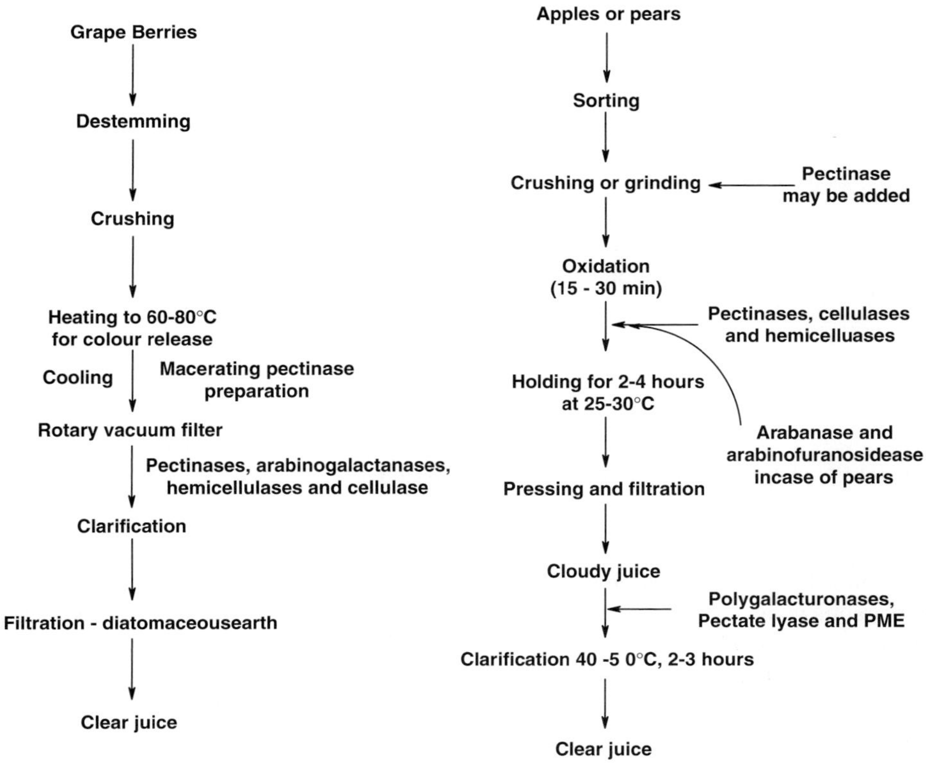

Figure 14.4 Application of pectinases in the processing of fruits for juice production.

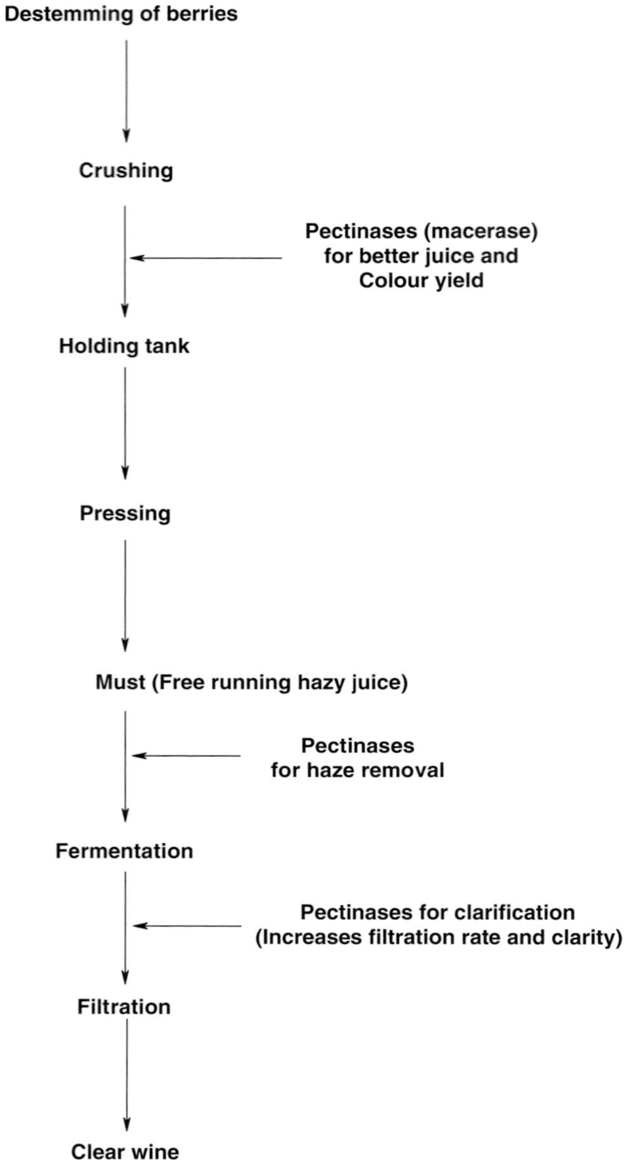

Figure 14.5 Application of pectinases in wine production.

in low calorie foods. 5-Deoxy-L-arabinose has been reported to have anti-Parkinson properties (90). Furaneol can be obtained by transformation of arabinose, the sugar obtained from arabinogalacturonan. Polygalacturonases that produce galacturonic acid make a good substrate for L-ascorbic acid. Galacturonic acid can also be used to produce surface active agents by esterification with fatty acids. Furaneol is used in caramel as fruit flavor (91). Furaneol can be obtained by chemical transformation of rhamnose released from rhamnogalacturonan by rhamnogalacturonan hydrolase. Coffee beans contain pectin in their mucilage coat and pulp. Use of pectinases along with cellulases and hemicellulases accelerate the removal of mucilage coat during coffee processing. Likewise, in the manufacture of tea, the degradation of tea leaf pectin helps in tea fermentation by improving the foam

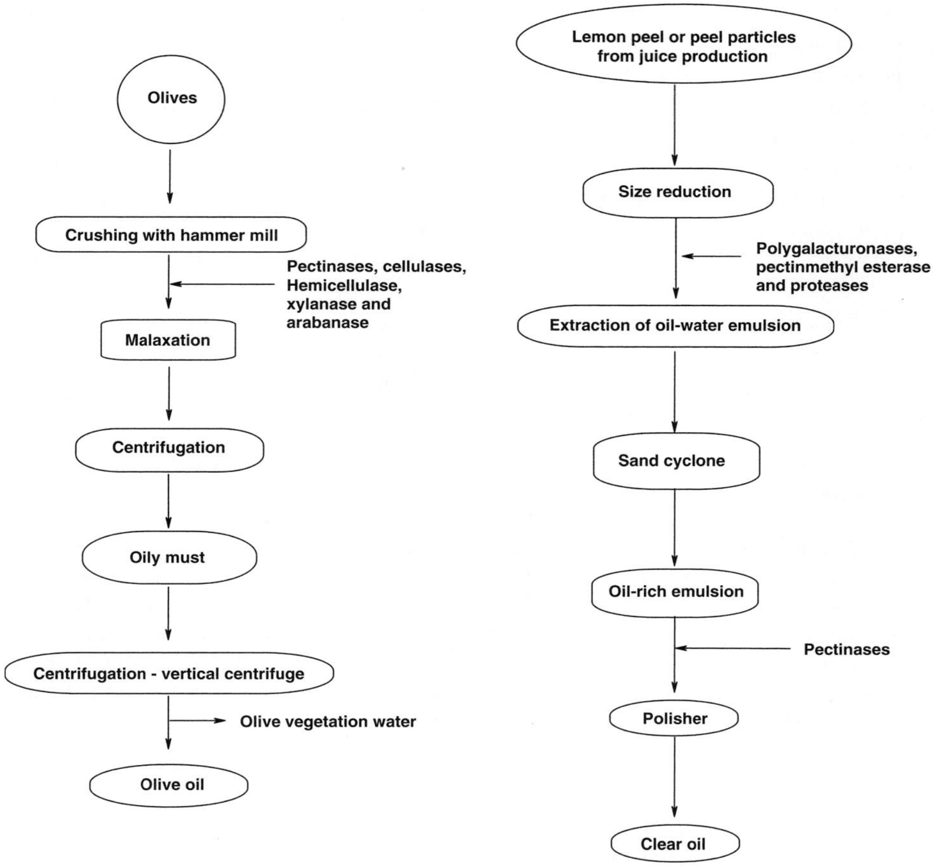

Figure 14.6 Application of pectinases in oil extraction.

forming property in plant tea powders. Application of pectinases describes processing of plant materials. Recently, *in vivo* modification of plant polysaccharides has been reported using *Aspergillus* enzymes. Transgenic plants containing the *A. aculeatus* endogalacturonase encoding gene have been used to reduce the galactosyl content of rhamnogalacturonan I by 70% (92).

Bioactive food ingredients using pectin are emerging as food products. The oligogalacturonides produced by enzymatic degradation of pectin serve as nondigestible health promoting compounds analogous to the biologically active fructo-oligosaccharide molecules. These oligosaccharides are termed as "prebiotic" because they support the growth of beneficial intestinal bacteria (93).

Pectinases also find application in papermaking; wastewater treatment, especially from citrus processing industries; vetting and degumming of fiber; and for the production of protoplast useful in plant breeding by genetic manipulation.

14.8 PERSPECTIVES

Microbial pectinases have been used to identify variation in pectin polymers of plant tissues. These studies have resulted in the characterization of newer microbial enzymes useful in food processing industries. The structural features of pectinases, the amino acid sequence of the enzyme protein, and their properties and characteristics are emerging as important contributions. These may predict changes in enzyme functionalities. Research in this direction can decipher the problem of haze and precipitation often encountered by fruit juice industries due to the conversion of the soluble L-arabanan to less soluble (1→5)-L-arabanan by the unwanted enzymes acting on arabanan. Apparently this requires enzymes that have greater activity on (1→5) linkages of arabanan than on (1→3) linkages. A possible solution to this problem is protein engineering, because we now know that the two hydrolytic activities reside in a single enzyme.

REFERENCES

1. Biely, B., J. Benen, K. Heinrichova, H.C.M. Kester, J. Visser. Inversion of configuration during hydrolysis of α-1,4-galacturonidic linkage by three polygalacturonases. *FEBS Lett.* 382:249–255, 1996.
2. Keon, J.P.R., G. Waksman. Common amino acid domain among endopolygalacturonases of *Ascomycetes* fungi. *Appl. Environ. Microbiol.* 56:2522–2528, 1990.
3. Trescott, A.S., J Tampion. Properties of the endopolygalacturonase secreted by *Rhizopus stolonifer*. *J. Gen. Microbiol.* 80:401–409, 1974.
4. Behere, A., V. Satyanarayan, S.R. Padwal Desai. Separation and limited characterization of three polygalacturonases of *Aspergillus niger*. *Enzyme Microb. Technol.* 15:158–161, 1993.
5. Kester, H.C.M., J. Visser. Purification and characterization of polygalacturonases produced by the hyphal fungus *Aspergillus niger*. *Biotechnol. Appl. Biochem.* 12:150–160, 1990.
6. Lang, C., A.C. Looman. Efficient expression and secretion of *Aspergillus niger* RH5344 polygalacturonase in *Saccharomyces cerevisiae*. *Appl. Microbiol. Biotechnol.* 44:147–156, 1995.
7. Hugouvieux, V., S. Centis, C. Lafitte, M. Esquerre-Tugaye. Induction by α-L-rhamnose of endopolygalacturonase gene expression in *Colletotrichum lindemuthianum*. *Appl. Environ. Microbiol.* 63:2287–2292, 1997.
8. Parenicova, L., J.A.E. Benen, H.C.M. Kester, J. Visser. *pgaA* and *pgaB* encode two constituently expressed endopolygalacturonases of *Aspergillus niger*. *Biochem. J.* 345:637–644, 2000.
9. Parenicova, L., H.C.M. Kester, J.A.E. Benen, J. Visser. Characterization of a novel endopolygalacturonase from *Aspergillus niger* with unique kinetic properties. *FEBS Lett.* 467:333–336, 2000.
10. Ueda, S., Y. Fujio, J.Y. Lim. Production and some properties of pectic enzymes from *Aspergillus oryzae* A-3. *J. Appl. Biochem.* 4:524–532, 1982.
11. Kitamoto, N., J. Matsui, Y. Kawai, A. Kato, S. Yoshino, K. Ohmiya, N. Tsukagoshi. Utilization of the *TEL1*-alpha gene (*TEF1*) promoter for expression of polygalacturonase genes *pgaA* and *pgaB* in *Aspergillus oryzae*. *Appl. Micorbiol. Biotechnol.* 50:85–92, 1998.
12. Cooke, R.D., C.E.M. Ferber, L. Ganakasabapathy. Purification and characterization of polygalacturonases from a commercial *Aspergillus niger* preparation. *Biochem. Biophys. Acta* 452:440–451, 1976.
13. Inque, S., Y. Nagamatsu, C. Hatanaka. Preparation of cross-linked pectate and its application to the purification of endopolygalacturonase of *Kluyveromyces fragilis*. *Agric. Biol. Chem.* 48:633–640, 1984.
14. Devi, N.A., A.G.A. Rao. Fractionation, purification and preliminary characterization of polygalacturonases produced by *Aspergillus carbonarius*. *Enzyme Microb. Technol.* 18:59–65, 1996.
15. van Santen, Y., J.A.E. Benen, K.H. Schroter, K.H. Kalk, S. Armand, J. Visser, B.W. Dijkstra. 1.68 A^0crystal structure of endopolygalacturonase II from *Aspergillus niger* and identification of active site residues by site directed mutagenesis. *J. Biol. Chem.* 274:30474–30480, 1999.

16. Gainvors, A., N. Nedjaoum, S. Gognies, M. Muzart, M. Nedjma, A. Bearbi. Purification and characterization of acidic endopolygalacturonase encoded by the *PGL1-1* gene from *Saccharomyces cerevisiae*. *FEMS Microbiol. Lett.* 183:131–135, 2000.
17. Schwan, R.F., R.M. Cooper, A.E. Wheals. Endopolygalacturonase secretion by *Kluyveromyces marxianus* and other cocoa pulp degrading yeasts. *Enzyme Microb. Technol.* 21:234–244, 1997.
18. van der Vlugt-Bergmans, C.J.B., P.J.A. Meeuwsen, A.G.J. Voragen, A.J.J. van Ooyen. Endo-xylogalacturonan hydrolase, a novel pectinolytic enzyme. *Appl. Environ. Microbiol.* 66:36–41, 2000.
19. Beldman, G., L.A.M. van den Brock, H.A. Schol, M.J.F. Searle-van Leeuwen, K.M.J van Laere, A.G.J. Voragen. An exogalacturonase from *Aspergillus aculeatus* able to degrade xylogalacturonan. *Biotechnol. Lett.* 18:707–712, 1996.
20. Hara, T., J.Y. Lim, Y. Fujio, S. Ueda. Purification and some properties of exopolygalacturonase from *Aspergillus niger* cultured in the medium containing Satsuma mandarin peel. *Nippon Shokuhin Kogyo Gakkaishi* 31:581–586, 1984.
21. Mikhailova, R.V., L.I. Sapunova, A.G. Lobanok. Three polygalacturonases constitutively synthesized by *Aspergillus alliaceus. World J. Microbiol. Biotechnol.* 11:330–332, 1995.
22. Dean, R.A., W.E. Timberlake. Regulation of the *Aspergillus nidulans* pectate lyase gene (*pel*A). *Plant Cell* 1:275–284, 1989.
23. Benen, J.A.E., H.C.M. Kester, L. Parenicova, J. Visser. Characterization of *Aspergillus niger* pectate lyase A. *Biochemistry* 39:15563–15569, 2000.
24. Ishii, S., Y. Yokotsuka. Purification and properties of pectin lyase from *Aspergillus joponicus. Agric. Biol. Chem.* 39:313–321, 1975.
25. van Houdenhoven, F.E.A. Studies on pectin lyase. PhD thesis, Wageningen University, Wageningen, The Netherlands, 1975.
26. Kester, H.C.M., J. Visser. Purification and characterization of pectin lyase B, a novel pectinolytic enzyme from *Aspergillus niger. FEMS Microbiol. Lett.* 120:63–68, 1994.
27. Chistgau, S., L.V. Kofod, T. Halkier, L.N. Adersen, M. Hockauf, K. Dorreich, H. Dalboge, S. Kauppinen. Pectin methyl esterase from *Aspergillus aculeatus*: expression cloning in yeast and characterization of the recombinant enzyme. *Biochem. J.* 319:705–712, 1996.
28. Schols, H.A., C.J.M. Geraeds, M.F. Searle-van Leeuwen, F.J.M. Kormelink, A.G.J. Voragen. Rhamnogalacturonase: a novel enzyme that degrades the hairy regions of pectin. *Carbohydr. Res.* 206:104–115, 1990.
29. Kofod, L.V., S. Kauppinen, S. Christgau, L.N. Adersen, H.P. Heldt-Hangen, K. Derreich, H. Dalboge. Cloning and characterization of two structurally and functionally divergent rhamnogalacturonase from *Aspergillus aculeatus. J. Biol. Chem.* 269:29182–29189, 1994.
30. Mutter, M., G. Beldman, H.A. Schols, A.G.J. Voragen. Rhamnogalacturanan α-L-rhamnopyronosylhydrolase: a novel enzyme specific for the terminal non reducing rhamnosyl unit in rhamnogalacturonan region of pectin. *Plant Physiol.* 106:241–250, 1994.
31. Mutter, M., G. Beldman, S.M. Pitson, H.A. Schols, A.G.J. Voragen. Rhamnogalacturonan α-D-galactopyranosyluronohydrolase: an enzyme that specifically removes the terminal non reducing galacturonosyl residue in rhamnogalacturonan region of pectin. *Plant Physiol.* 117:153–163, 1994.
32. Suykerbuyk, M.E.G., H.C.M. Kester, P.J. Scheap, H. Stam, W. Muster, J. Visser. Cloning and characterization of two rhamnogalacturonan hydrolase genes from *Aspergillus niger. Appl. Environ. Microbiol.* 63:2507–2515, 1997.
33. Kauppinen, S., S. Christgau, L.V. Kofod, T. Halkier, K. Dorreirch, H. Dalboge. Molecular cloning and characterization of a rhamnogalacturonan acetyl esterase from *Aspergillus aculeatus. J. Bio. Chem.* 270:27172–27178, 1995.
34. de Vries, R.P., J. Visser. *Aspergillus* enzymes involved in degradation of plant cell wall polysaccharides. *Microbiol. Mol. Biol. Rev.* 65:497–522, 2001.
35. Ward, O.P. Hydrolytic enzymes. In: *Comprehensive Biotechnology,* Blanch, H.W., S. Drew, D.I. Wang, eds., Oxford: Pergamon Press, 1985, Vol. 3, pp 819–835.

36. Rombouts, F.M., W. Pilnik. Microbial enzymes and bioconversions. In: *Economic Microbiology,* Rose, A.H., ed., London: Academic Press, 1980, Vol. 5, pp 227–252.
37. Pilnik, W., F.M. Rombouts. Pectic enzymes. In: *Enzymes and Food Processing,* Birch, G.G., N. Blakebrough, K.J. Parker, eds., London: Applied Science, 1981, pp 105–128.
38. Fogarty, W.M., C.T. Kelly. Pectic enzymes. In: *Microbial Enzymes and Biotechnology,* Fogarty, W.M., ed., London: Applied Science, 1983, pp 131–182.
39. Leone, G., J. van den Heuvel. Regulation by carbohydrates of the sequential *in vitro* production of pectic enzymes by *Botrytis cinerea. Can. J. Bot.* 65:2133–2141, 1987.
40. Kilara, A. Enzymes and their uses in processed apple industry: a review. *Process Biochem.* 6:35–41, 1982.
41. Maldonado, M.C., A. Navarro, D.A.S. Callieri. Production of pectinases by *Aspergillus sp.* using deferentially pretreated lemon peel as the carbon source. *Biotechnol. Lett.* 8:501–504, 1986.
42. Bailey, M.J. Effect of temperature on polygalacturonase production by *Aspergillus niger. Enzyme Microb. Tech.* 12:622–624, 1990.
43. Aguilar, G., C. Huitron. Application of fed-batch cultures in the production of extracellular pectinases by *Aspergillus sp. Enzyme Microb. Tech.* 9:541–546, 1986.
44. Galiotou-Panayotou, M., M. Kapantai, O. Kalantai. Growth conditions of *Aspergillus* sp. ATUM-3842 for polygalacturonase production. *Appl. Microbiol. Biotechnol.* 47:425–429, 1997.
45. Aguilar, G., C. Huitron. Constitutive exo-pectinase produced by *Aspergillus* sp. CH-Y-1043 on different carbon sources. *Biotech. Lett.* 12:655–660, 1990.
46. Baracat, M.C., M.C.D. Vanetti, E.F. de Araujo, D.O. Silva. Growth conditions of a pectinolytic *Aspergillus fumigatus* for degumming of natural fibers. *Biotechnol. Lett.* 13:693–696, 1991.
47. Cavalitto, S.F., J.A. Arcas, R.A. Hours. Pectinase production profile of *Aspergillus foetidus* in solid-state cultures at different acidities. *Biotechnol. Lett.* 18:251–256, 1996.
48. Kojima, Y., T. Sakamoto, M. Kishida, T. Sakai, H. Kawasaki. Acidic condition inducible polygalacturonase of *Aspergillus kawachii. J. Mol. Catal. B Enzymatic* 6:351–357, 1999.
49. Aguilar, G., A.B. Trejo, J.M. Garcia, C. Huitron. Influence of pH on endo- and exo-pectinase production by *Aspergillus* sp. CH-Y-1043. *Can. J. Microbiol.* 37:912–917, 1991.
50. Ros, J.M., D. Saura, M.C. Salmeron, J. Laencina. Culture pH effect on the endopolygalacturonase production from *Rhizopus nigricans. Biotechnol. Lett.* 14:565–566, 1992.
51. Torakazu, T., K. Hirokasu, S. Atsuhiko, E. Toshio. Inhibition of accumulation of polygalacturonase forming activity during catabolic repression in *Aspergillus niger. J. Ferment. Technol.* 53:409–412, 1975.
52. Leuchtenberger, A., G. Mayer. Changed pectinase synthesis by aggregated mycelium of same *Aspergillus niger* mutants. *Enzyme Microb. Technol.* 14:18–22, 1992.
53. Gibbs, P.A., R.J. Sevour, F. Schmid. Growth of filamentous fungi in submerged culture: problems and possible solutions. *Crit. Rev. Biotechnol.* 20:17–48, 2000.
54. Acuna-Arguelles, M., M.G. Rojas, G.V. Gonzalez, E.F. Torres. Effect of water activity on exopectinase production by *Aspergillus niger* CH4 on solid-state fermentation. *Biotechnol. Lett.* 16:23–28, 1994.
55. Kavitha, R. Studies on polygalacturonase production of *Aspergillus* species and genetic characterization of a polygalacturonase overproducing mutant. PhD thesis, University of Mysore, India, 2001.
56. Budiatman, S., B.K. Lonsane. Casava fibrous waste residue: a substitute to wheat bran in solid-state fermentation. *Biotechnol. Lett.* 9:597–600, 1987.
57. Siessere, V., S. Said. Pectic enzymes production in solid-state fermentations using citrus pulp pellets by *Thalaromyces flavus, Tubercularia vulgaris* and *Penicillium charlessi. Biotechnol. Lett.* 11:343–344, 1989.
58. Arguelles, M.A., M.G. Rojas, G.V. Gonzalez, E.F. Torres. Production and properties of three pectinolytic activities produced by *Aspergillus niger* in submerged and solid-state fermentation. *Appl. Microbiol. Biotechnol.* 43:808–814, 1995.

59. Antier, P., A. Minjares, S. Roussos, M. Raimbault, G. Viniegra-Gonzalez. Pectinase hyper-producing mutants of *Aspergillus niger* C28B25 for solid-state fermentation of coffee pulp. *Enzyme Microb. Technol.* 15:254–260, 1993.
60. Ismail, A.S. Utilization of orange peels for the production of multienzyme complex by some fungal strains. *Process. Biochem.* 31:645–650, 1996.
61. Cestilho, L.R., R.A. Medronho, T.L.M. Alves. Production and extraction of pectinases obtained by solid-state fermentation of agroindustrial residues with *Aspergillus niger*. *Bioresource Technol.* 71:45–50, 2000.
62. Kavitha, R., S. Umesh-Kumar. Genetic improvement of *Aspergillus carbonarius* for pectin-ase overproduction during solid-state growth. *Biotechnol. Bioeng.* 67:121–125, 2000.
63. Loera, O., J. Aguirre, G. Viniegra-Gonzalez. Pectinase production by a diploid construct from two *Aspergillus niger* overproducing mutants. *Enzyme Microb. Technol.* 25:103–108, 1999.
64. Solis, S., M.E. Flores, C. Huitron. Improvement of pectinase production by interspecific hybrids of *Aspergillus* strains. *Lett. Appl. Microbiol.* 24:77–81, 1997.
65. Solis-Pereira, S., E. Favela-Torres, G. Viniegra-Genzalez, M. Gutierrez-Rojas. Effects of different carbon sources on the synthesis of pectinase by *Aspergillus niger* in submerged and solid-state fermentation. *Appl. Microbiol. Biotechnol.* 39:36–41, 1993.
66. McKeon, T.A. Activity stain for polygalacturonase. *J. Chromatogr.* 455:376–381, 1988.
67. Raj, A.E., K.S. Venkatesh, K. Ravikumar, S. Umeshkumar, M.C. Misra, N.P. Ghildyal, N.G. Karanth, M.K. Gowthaman. An improved process for the production of pectinase enzyme from *Aspergillus carbonarius*. Indian Patent (CSIR) No.222/DEL/02, 2002.
68. Solis, S., M.E. Flores, C. Huitron. Isolation of endopolygalacturonase hyper producing mutants of *Aspergillus* sp. CH-Y-1043. *Biotechnol. Lett.* 12:751–756, 1990.
69. Visser, J., H.J. Bussink, C. Witteveen. Gene expression in filamentous fungi: expression of pectinases and glucose oxidase in *Aspergillus niger*. In: *Gene Expressions in Recombinant Microorganisms,* Smith, A., ed., New York: Marcel Dekker, 1994, pp 241–308.
70. Kuster-van Someren, M.A., H.A.M. Harmsen, H.C.M. Kester, J. Visser. Structure of the *Aspergillus niger pel*A gene and its expression in *A. niger* and *A. nidulans*. *Curr. Genet.* 20:293–299, 1991.
71. Kuster-van Someren, M.A., M. Flipphi, L. de Graaff, H. van den Broeck, H. Kester, A. Hinnen, J. Visser. Characterization of the *Aspergillus niger pel*B gene: structure and regulation of the expression. *Mol. Gen. Genet.* 234:113–120, 1992.
72. Bussink, H.J.D., F.P. Buxton, J. Visser. Expression and sequence comparison of the *Aspergillus niger* and *Aspergillus tubingensis* genes encoding polygalacturonase II. *Curr. Genet.* 19:467–474, 1991.
73. Ruttkowski, E., R. Labitzke, N.Y. Khanh, E. Loffler, M. Gottschalk, K.D. Jany. Cloning and DNA sequence analysis of polygalacturonase cDNA from *Aspergillus niger* RH5344. *Biochem. Biophys. Acta* 1087:104–106, 1990.
74. Bussink, H.J.D., F.P. Buxton, B.A. Fraaye, L.H. de Graaff, J. Visser. The polygalacturonases of *Aspergillus niger* are encoded by a family of diverged genes. *Eur. J. Biochem.* 48:83–90, 1992.
75. Khanh, N.Q., E. Ruttkowski, K. Leidinger, H. Albrecht, H. Gottschalk. Characterization and expression of a genomic pectin methyl esterase encoding gene *Aspergillus*. *Gene* 106:71–77, 1991.
76. Gysler, C., J.A.M. Harmsen, H.C.M. Kester, J. Visser, J. Heim. Isolation and structure of pectin lyase D-encoding gene from *Aspergillus niger*. *Gene* 89:101–108, 1990.
77. Bussink, H.J.D., H.C.M. Kester, J. Visser. Molecular cloning, nucleotide sequence and expression of the gene encoding prepropolygalacturonase II of *Aspergillus niger*. *FEBS Lett.* 273:127–130, 1990.
78. Bussink, H.J.D., J.P. van den Hombergh, T.W. van den Ijesel, J. Visser. Characterization of polygalacturonase overproducing *Aspergillus niger* transformants. *Appl. Microbiol. Biotechnol.* 37:324–329, 1992.

79. Cho, S.W., S. Lee, W. Shin. The x-ray structure of *Aspergillus aculeatus* polygalacturonase and a modeled structure of the polygalacturonase-octagalacturonate complex. *J. Mol. Biol.* 314:863–878, 2001.
80. Bussink, H.J.D., K.B. Brouner, L.H. de Graaff, H.C.M. Kester, J. Visser. Identification and characterization of a second polygalacturonase gene *Aspergillus niger*. *Curr. Genet.* 20:301–307, 1991.
81. Nagai, M., A. Oxawa, T. Katsuragi, H. Kawasaki. Cloning and heterologous expression of gene encoding a polygalacturonase from *Aspergillus awamori*. *Biosci. Biotechnol. Biochem.* 64:1580–1587, 2000.
82. Reymond, P., G. Deleage, C. Rascle, M. Fevre. Cloning and sequence analysis of a polygalacturonase encoding gene from the phytopathogenic fungus *Sclerotinia sclerotiorum*. *Gene* 146:233–237, 1994.
83. Fraissinet-Tachet, L., P. Reymond-Cotton, M. Fevre. Characterization of multigene family encoding an endopolygalacturonase in *Sclerotinia sclerotiorum*. *Curr. Genet.* 29:96–99, 1995.
84. Ruttkowski, E., N.R. Khan, F.J. Wientjes, M. Gottschalk. Characterization of a polygalacturonase gene of *Aspergillus niger* RH 5344. *Mol. Microbiol.* 5:1353–1361, 1991.
85. Molegaard, A., S. Kauppinen, S. Larsen. Rhamnogalacturonan acetylesterase elucidates the structure and function of a new family of hydrolases. *Structure* 8:373–383, 2000.
86. Molegaard, A., J.F.W. Petersen, S. Kauppinen, H. Dalboge, A.H. Johnsen, J.C.N. Poulsen, S. Larsen. Crystallization and preliminary x-ray diffraction studies of the heterogeneously glycosylated enzyme rhamnogalacturonan acetylesterase from *Aspergillus aculeatus*. *Acta Crystallogr.* D54:1026–1029, 1998.
87. Gielkens, M.M.C., L. Gonzales-Candelas, P. Sanchez-Torres, P.J.I. van de Vondervoort, L.H. de Graaf, J. Visser. The *abf*B gene encoding the major α-L-arabinofuranosidase of *Aspergillus nidulans*: nucleotide sequence, regulation and construction of a disrupted strain. *Microbiology* 145:735–741, 1999.
88. MacCabe, A.P., M. Orejas, J.A. Perz-Gonzalez, D. Ramon. Opposite patterns of expression of two *Aspergillus nidulans* xylanases genes with respect to ambient pH. *J. Bacteriol.* 180:1331–1333, 1998.
89. Stewart, J.C., J.B. Parry. Factors influencing the production of cellulase by *Aspergillus fumigatus* (Fresenius). *J. Gen. Microbiol.* 125:33–39, 1981.
90. Vogel, M. Alternative utilization of sugar beet pulp. *Zuckerindustrie* 116:265–270, 1991.
91. Wong, C.H., F.P. Mazemod, G.M. Whiteside. Chemical and enzymatic synthesis of 6-deoxyhexoses: conversion to 2,5-dimethyl-4-hydroxy-2,3-dihydrofuran-3-one (Furaneol) and analogues. *J. Org. Chem.* 48:3493–3497, 1983.
92. Sorensen, S.O., M. Pauly, M. Bush, M. Skjot, M.C. McCann, B. Borkhardt, P. Ulvoskov. Pectin engineering: modification of potato pectin by *in vivo* expression of endo-1,4-β-D-galacturonase. *Proc. Natl. Acad. Sci. USA* 97:7639–7644, 2000.
93. Voragen, A.G.J. Technological aspects of functional food-related carbohydrates. *Trends Food Sci. Technol.* 9:328–335, 1998.

1.15

Biotechnology of Citric Acid Production

T. Roukas

CONTENTS

15.1 INTRODUCTION

Citric acid (2-hydroxy-1,2,3-propanetricarboxylic acid), a tricarboxylic acid, has the following structure:

$$
\begin{array}{ccc}
 & \overset{1}{C}H_2COOH & \\
 & | & \\
OH- & \overset{2}{C} & -COOH \\
 & | & \\
 & \overset{3}{C}H_2COOH &
\end{array}
$$

It is naturally found in fruits such as lemon, orange, pineapple, plum, and pear; in the seeds of different vegetables; and in animal bone, muscle, and blood (1,2). Citric acid was first commercially obtained from lemon juice, crystallized in 1784 by Carl Wilhelm Scheele, a Swedish chemist (2,3). Citric acid produced from fruits is known as "natural citric acid," in contrast to that produced by microorganisms (3). Although citric acid is currently being chemically synthesized, there is no chemical method that is superior to microbial fermentation. In 1893 Wehmer investigated citric acid production from sucrose with strains of *Mucor* and *Penicillium* (1). This experiment was not successful and was abandoned in 1903. Difficulties included the selection of proper microorganisms, microbial degeneration, contamination, long fermentation time, high plant construction cost, and an insufficient spread between raw material cost and the price of finished product (4). Wehmer reported that until 1910 no successful citric acid fermentation process was carried out. Until 1920 all commercially produced citric acid came from lemon or lime juice (5). In 1917 Currie described the first production of citric acid with *Aspergillus niger*. He reported that different strains of *A. niger*, grown in low pH medium supplemented with 15% (w/v) sucrose plus nutrients, converted 55% of the sugar to citric acid (4). Citric acid was isolated from the fermentation broth as insoluble calcium citrate by adding calcium hydroxide, followed by regeneration of the acid using sulfuric acid. Citric acid crystallizes into two forms: anhydrous and monohydrous. The anhydrous form is produced at a temperature exceeding 36.6°C while the monohydrate is produced at temperatures of up to 36.0°C (1,5). By 1930 a number of manufacturers used the submerged *A. niger* fermentation process for the production of citric acid from sucrose (5,6). In 1952, America Miles Company (Miles Laboratories, Inc., Elkhart, IN) used the submerged fermentation process in a pilot program on a plant scale to produce citric acid. Little information exists about the processes they used. From the results claimed in their patents, however, it is clear that they used pretreated molasses with a cation exchanger as a substrate deficient in phosphate, iron, and manganese

(3,4). After 1952, many countries used glucose or beet and cane molasses as substrates to produce citric acid. Among the microorganisms producing citric acid, strains of *A. niger* have been used, but more recently, some yeast strains (*Yarrowia, Hansenula, Pichia, Torula*) have also been employed. These strains, which were isolated from soil or plant material, were further improved by mutation (7). Recently, different strains of yeasts have been used instead of *A. niger* for the production of citric acid (8–11). Today, most citric acid used in food and other industries is produced by microbial fermentation.

Two methods are commonly used for the production of citric acid by *A. niger*: surface fermentation and submerged culture fermentation. Beet and cane molasses were for many years the preferred substrates for the production of citric acid. The production of citric acid from n-paraffins by different strains of *Candida* or other yeasts has been developed (8,11–15), but due to production of isocitric acid during fermentation as well as the prohibitive price of petroleum products, this method has proved uneconomical and has never been used commercially (7,10,11).

The current annual world production of citric acid among 35 countries is ~600,000 tons (16). Among these, the two largest manufacturing areas are Western Europe followed by the USA. However, the USA is still a net importer of citric acid, while Western Europe is a net exporter (11). Citric acid is used in the food and beverage industries (70%), in pharmaceuticals (12%), and in other industries and applications (18%) (3,17–19). It is also used for metal cleaning, as an acidulant in soft drinks and confectionary, and can serve as a buffer and pH stabilizer, as well as a fat antioxidant in various food products (2,7). In the food industry it is used as a preservative in that it prevents color and flavor deterioration, inhibits oxidation, protects ascorbic acid from oxidation, inactivates oxidative enzymes, acts as an antioxidant, and is an emulsifier in dairy products. Citric acid is used in the preparation of soft drinks, desserts, jams, jellies, candies, wines, and frozen fruits. It also provides a tart taste in soft drinks. In the pharmaceutical industry, it is used in syrups, astringents, and effervescent tablets and powders. It is also used in blood transfusions. In the chemical industry, its uses include water conditioning, metal pickling, and used as a foam inhibitor for vinyl sheeting and polyester resin. Also, it is used as a nontoxic plasticizer to make plastic film (2). Citric acid can be used to adjust pH levels that are neutral or high, thus permitting its wide application in industry in areas such as electroplating, leather tanning, and reactivation of oil wells where pores of the sand face have become clogged with iron (20). Finally, citric acid is used in the detergent industry to replace phosphates and in the removal of sulfur in stack gases and other industrial facilities (3).

15.2 MICROORGANISMS USED FOR THE PRODUCTION OF CITRIC ACID

15.2.1 Fungi

The selection of a suitable fungus strain is critical to the production of citric acid, because he fungus plays a central role in the process. A strain used in an industrial scale operation should have long term stability, high sporulation, good growth in the substrates, a short fermentation time, resistance to other microorganisms, and produce a high concentration of the acid (70.0–100.0 g/L) in different fermentation systems. Citric acid's production stability is difficult to maintain because of spontaneous mutation and autolysis of the strains during fermentation. These problems can be avoided by periodic reisolation from single spores, storage at low temperature (4.0–7.0°C), and avoidance of "foggy" patches of sterile mycelium (fungal mat lacking spores) when mass transfers are made (6).

Over the years, many microorganisms have been used for the production of citric acid. However, *A. niger* still remains the microorganism of choice in industrial scale production. Other microorganisms, such as yeasts and bacteria, produce potentially large amounts of citric acid (50.0–70.0 g/L), but until recently they have not been used for commercial production (3). There is general agreement that only selected strains of *A. niger* are useful citric acid producers because they can be handled easily, are inexpensive, and give high and consistent yields, thereby making the process economical. Other species of *Aspergillus* which have been found to accumulate citric acid include strains of *A. awamori, A. fenicis, A. fonsecaeus, A. luchensis, A. fumaricus, A. wentii, A. saitoi, A. usami, A. phoenicus, A. lanosus, A. foetidus,* and *A. flavus* as well as some strains of *Penicillium* such as *P. janthinellum, P. simplicissimum,* and *P. restrictum* (3,6,21–24). Fungi that produce citric acid include strains of *Trichoderma viride, Mucor piriformis, Ustulina vulgaris,* and species of *Botrytis, Ascochyta, Absidia, Talaromyces, Acremonium,* and *Eupenicillium* (3,6). The yields of citric acid obtained with these strains are lower than those obtained with strains of *A. niger.* It is generally recognized that *A. niger* consists of a large group of strains that differ from one another in their morphology and biochemical characteristics. The differences reported include: color of spores and mycelium, size and quantity of spores, mycelium size, substrate utilization, fermentation time, and ability to produce citric acid from different substrates (6,25–27). Industrial scale production by fungi has been carried out from chemically defined media or beet and cane molasses using surface or submerged fermentation. Today, almost all citric acid produced by fermentation is manufactured by strains of *A. niger* in submerged culture.

15.2.2 Yeasts

Although *A. niger* is the traditional producer of citric acid, in the past 30 years researchers have been attracted to the use of yeasts as citric acid producers (9). Yeasts have some advantages compared to *A. niger* strains. The fermentation time is short (half the time of *A. niger* fermentation) and thus, productivity is higher. Yeast strains are insensitive to molasses variations and can be used for developing a continuous process (3,5). Also, yeasts are not only more tolerant of contamination, but are capable of metabolizing high initial sugar or n-alkane concentrations (100.0–150.0 or 40.0–60.0 g/L, respectively), with rapid growth resulting in high productivity rates. Moreover, they have a greater tolerance for metal ions, thus allowing the use of less refined substrates. These capabilities can lead to significant reductions in substrate and waste treatment costs and in product recovery costs (10,11). The advantages of using n-alkanes over carbohydrates are potentially lower costs for substrates and higher citric acid concentrations (11). The major disadvantage of using yeasts is the production of isocitric acid during fermentation. The amount of isocitric acid produced depends on the yeast strains used, the chemical composition of the substrates, the fermentation system, and generally the conditions under which fermentation takes place. To overcome this problem, some methods have been developed to reduce the amount of isocitric acid produced. For instance, the utilization of mutants that are sensitive to fluoroacetate results in the production of very small amounts of isocitric acid (5).

The variety of yeasts that produce citric acid belong to the genus *Candida, Saccharomycopsis, Hansenula, Pichia, Debaryomyces, Torulopsis, Kloeckera, Trichosporon, Torula, Rhodotorula, Sporobolomyces, Endomyces, Nocardia, Nematospora, Saccharomyces,* and *Zygosaccharomyces* (3). Among these, strains of *Candida* are widely used for the production of citric acid. These strains include *C. lipolytica, C. tropicalis, C. zeylanoides, C. fibrae, C. intermedia, C. parapsilosis, C. petrophilum, C. subtropicalis, C. oleophila, C. hitachinica, C. citrica, C. guilliermondii,* and *C. sucrosa* (3,8–15,28–39). Recently, the production of citric acid by immobilized and mutant strains of *C. lipolytica* has been reported (40–51). The mechanism by which *C. lipolytica* produces a high concentration of citric acid is unclear. In general, unlimited growth of *C. lipolytica* in a rich medium results in low citric acid production. However, citric acid can be produced in increased amounts if yeast growth is properly restricted during the acid producing phase of fermentation (11). Production by yeasts, carried out in aerobic and agitated fermentation broth and at temperatures of 25–35°C, depended on the yeast strains and equipment used. The medium consisted of either glucose, molasses, n-alkanes, n-paraffins, methanol, butanol, ethanol, C_{12-16} alcohols, acetate, fatty acids, and natural oils, or fats supplemented with nitrogen sources such as $(NH_4)_2SO_4$, NH_4NO_3, or NH_4Cl, KH_2PO_4, $MgSO_4{\bullet}7H_2O$, $CaCO_3$, thiamin hydrochloride, vitamin B complex, and trace elements such as Fe^{2+}, Zn^{2+}, Mn^{2+}, and Cu^{2+} (3,9–14,28–39). The medium was inoculated with 10–12% (v/v) of *C. lipolytica* (culture aged 48 h). Crolla and Kennedy (11) reported that excess inoculum concentration leads to high biomass concentration (20.0 g/L) and lower citric acid production (25.0 g/L), while low amounts of inoculum lead to long fermentation times.

Citric acid production by yeasts in the laboratory employs two phases: (1) a preliminary growth phase on a complete medium, followed by (2) a production phase without nitrogen sources. In some cases the medium is supplemented with limited amounts of nitrogen (1.0 g/L) in order to keep cellular activity at an acceptable level (10). In industrial scale production with *Y. lipolytica*, three steps have been identified: (1) the exponential growth phase, (2) the citrate production lag phase, and (3) the linear production phase. The production phase is connected to a preliminary reduction in intracellular nitrogen content (10). Fermentation runs from 3 to 6 days, with pH controlled at 4.5–6.5. Citric acid concentrations varied from 20 to −60 g/L,, depending on the strain used, the substrate (synthetic or byproduct), the fermentation system (surface culture or submerged fermentation), and the general conditions under which fermentation took place (initial sugar concentration, pH, and temperature). The yeast based process for citric acid production was maintained at neutral pH (5). Thus, the citrate produced was a sodium or calcium salt, which complicated recovery. In order to produce citric acid by yeast in industrial scale, further research is needed in selecting a high citrate producing strain and in optimizing its metabolic pathways as well as in understanding the best operating conditions for minimizing the production of isocitric acid (10).

15.2.3 Bacteria

Little information is available on the production of citrate by bacteria, as most of the references in the literature are to patents (3). Bacteria generally include *Bacillus, Brevibacterium, Arthrobacter, Corynebacterium, Klebsiella, Aerobacter, Pseudomonas,* and *Micrococcus.* Among these, *B. subtilis, B. licheniformis, B. flavum,* and *A. paraffinens* are the most promising (3,52). Kapoor et al. (3) reported that *B. licheniformis* grown in medium containing glucose, urea, calcium carbonate, and ammonium sulfate or glutamate (pH 7.0) produced 42.0 g/L of citric acid. Also, strains of *A. paraffinens, Corynebacterium* sp., and *Bacillus* sp. yielded maximum citric acid concentrations of 28–40 g/L when they were grown in media containing dodecane or a mixture of C_{12}–C_{14} paraffins (3). In all cases the fermentation was aerobic at 30–37°C for 2–5 days, depending on the strain and the composition of medium used. Generally, citric acid production by bacteria was 50–100% lower than that by fungi or yeasts. Nonetheless, these reports have opened a new avenue for the production of citric acid by bacteria.

15.3 BIOSYNTHESIS OF CITRIC ACID

The metabolic pathways by which *A. niger* converts glucose or sucrose to citric acid are outlined in Figure 15.1. Citric acid biosynthesis involves both the Embden–Meyerhof–Parnas

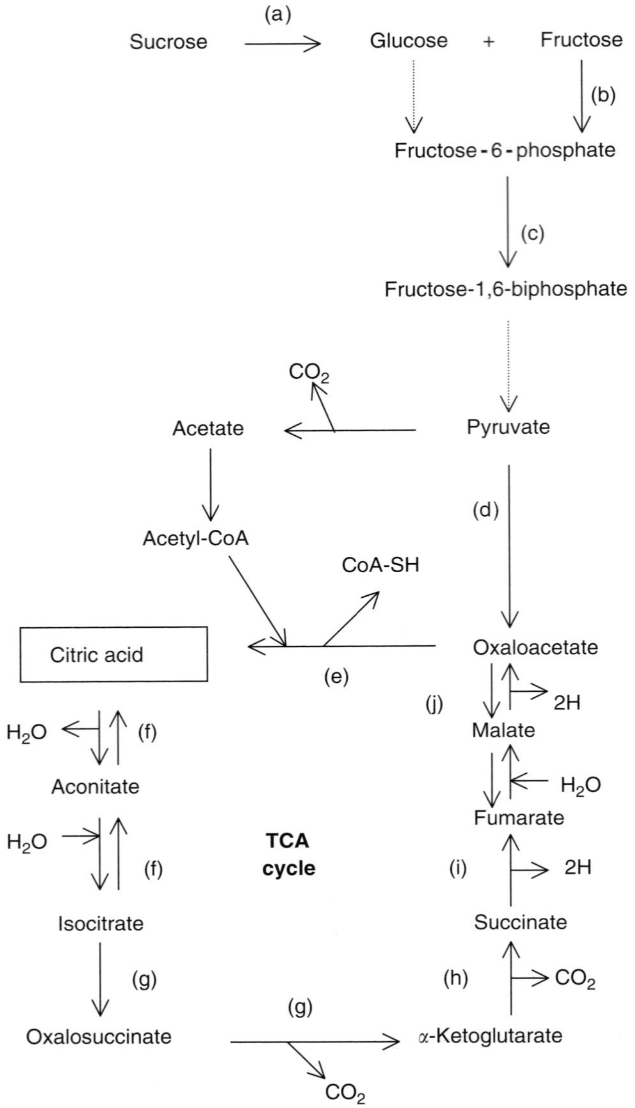

Figure 15.1 Biosynthesis of citric acid from sucrose by *A. niger*. (a) Invertase; (b) hexokinase; (c) phosphofructokinase; (d) pyruvate carboxylase; (e) citrate synthase; (f) aconitase; (g) isocitrate dehydrogenase; (h) α-ketoglutarate dehydrogenase; (i) succinic dehydrogenase; (j) malate dehydrogenase. (From Milsom, P.E., *Food Biotechnology*: 1, King, R.D., P.S.J. Cheetham, eds., London: Elsevier Applied Science, 1987, pp 273–308).

(EMP) pathway and the tricarboxylic acid (TCA) cycle (3,5,7,53–55). Citric acid is formed by condensation of acetyl-coenzyme A with oxaloacetic acid. Experiments with $C^{14}O_2$ indicated that CO_2 was the source of the a-carboxyl group of oxaloacetate (54). Woronick and Johnson (54) and Kapoor et al. (3) reported that *A. niger* contains two different systems for synthesizing C_4 dicarboxylic acids by condensing CO_2 with 3-carbon units. One of these systems seems to be identical to the phosphoenol pyruvate carboxylase kinase system:

$$\text{Phosphoenol pyruvate} + \text{ADP} + CO_2 \rightleftharpoons \text{Oxaloacetate} + \text{ATP} \qquad (15.1)$$

The second system requires pyruvate and ATP to produce oxaloacetate, and the reaction catalyzes by pyruvate carboxylase:

$$\text{Pyruvate} + CO_2 + H_2O + \text{ATP} \rightleftharpoons \text{Oxaloacetate} + \text{ADP} + P_i \qquad (15.2)$$

The major radioactive products were aspartate, malate, and fumarate, and were found to be present in the same ratios (1.6:4.3:1) as in the phosphoenol pyruvate plus ADP system (54). The citrate is metabolized via isocitrate to oxalosuccinate, which loses one CO_2 to yield a-ketoglutarate, which, in turn, loses a further molecule of CO_2 to form succinate. It is further metabolized via fumarate and malate to regenerate oxaloacetate (5). Smith and Anderson (55) studied the activities of the most important enzymes of the EMP and pentose phosphate (PP) pathways and found that the enzyme activities showed both EMP and PP pathways were present and operative at all stages of conidiophore development, but that the relative activities of the pathways differed. The highest activity of the EMP pathway occurred during the initial period of vegetative growth before any signs of morphological change. High EMP activity and high 6-C/1-C ratios were obtained in a medium that did not support conidiophore initiation but allowed vegetative growth (55). Generally, during conidiophore differentiation, PP pathway enzymes were higher in activity (0.35 units/mg protein) than EMP pathway enzymes (0.12 units/mg protein). The direct oxidation of glucose through the PP pathway may be of considerable importance during conidiophore development in *A. niger.*

Because *A. niger* produces both invertase and hexokinase (HXK), which convert sucrose to fructose-6-phosphate, sucrose can be used as a source for the production of citric acid. Verhoff and Spradlin (56) reported that only 10% of the glucose is available for the production of mycelia and CO_2, both of which occur primarily during the growth phase early in fermentation. Thus, practically all the glucose entering into the cell during the citric acid production phase must be converted to citric acid. However, Nowakowska and Sokotowski (57) found that the conversion yield of glucose to citric acid, mycelium, and CO_2 was 79.0%, 10.0%, and 11.0%, respectively. Steinbock et al. (58) reported high HXK activity when *A. niger* was grown in media containing high concentrations of glucose or sucrose (200.0–250.0 g/L). The specific activity of HXK was ~40 and 35 mmol/min/mg protein with glucose and fructose as substrates, respectively. Also, glucose affinity was about tenfold higher than for fructose. Finally, they found that the synthesis of very high HXK activities counteracted citrate inhibition, thereby guaranteeing a high glycolytic flux during citric acid accumulation. The key enzymes that were responsible for the biosynthesis of citric acid from sucrose by *A. niger* were phosphofructokinase (PFK), pyruvate kinase (PKI), pyruvate carboxylase, citrate synthase (CS), aconitase (ACH), and isocitric dehydrogenase (ICDH). Phosphofructokinase is an essential enzyme for the conversion of glucose or fructose to pyruvate. Inhibition of PFK by citrate is nullified by an accumulation of NH_4^+ in the cells. This state results from manganese deficiency, which also adversely affects the synthesis or activity of ACH and ICDH (7). Phosphofructokinase inhibition by citrate produces a measured feedback control over the whole system (5). Phosphofructokinase inhibition was observed in low citric acid producing strains, but was not observed in high yielding strains. Moreover, elevated levels of ammonium ion in the cells relieve the inhibition of PFK by citrate.

Pyruvate carboxylase catalyzes the reaction of pyruvate to oxaloacetate. Smith and McIntosh (59), who studied the effect of dilution rate on PFK and PKI activity, and found that these enzyme activities varied directly with dilution rate under glucose limitation in chemostat culture. Jaklitsch et al. (60) reported that pyruvate carboxylase was found predominantly or exclusively in cytosol. Also, low level activity of this enzyme was present in the mitochondria during the microbial log phase. Hence, intracellular distribution of pyruvate carboxylase may be influenced by culture conditions. Thus, pyruvate carboxylase is present in both the mitochondrion and the cytosol, with the proportion in the cytosol rising as culture age increases (60). When *A. niger* was cultivated in media containing glucose, sucrose or fructose stimulate citric acid, which also led to increased intracellular levels of fructose 2,6-disphosphate, a potent activator of PFK. Ruijter et al. (61) studied the effect of overexpression of PFK and PKI in citric acid producing *A. niger* and found that PFK or PKI, or both simultaneously, did not significantly increase citric acid production. Data suggest that overexpression of PKI has a negative effect on citric acid production. Increased PFK and PKI did not influence the activities of other enzymes in the pathway, nor did they change intermediary metabolite levels. Strains of *A. niger* produced high concentrations of PFK (0.84 units/mg protein), while the level of fructose 2,6-disphosphate was reduced almost twofold when wild-type *A. niger* was used. Furthermore, a reduction in the fructose 2,6-disphosphate level decreased significantly (30–40%) the specific activity of PFK. Generally, PFK and PKI did not contribute in a major way to flux control in the biosynthetic route involved in the formation of citric acid (61). With use of mutant strains of *A. niger* during increased citric acid production, the activities of HXK and PFK appeared to be twofold higher than with the use of the parental strain. However, improved citric acid production was not due to higher PFK activity but to increased HXK activity (61). In the case of citric acid production by yeasts, the effect of overproduction of PFK, PKI, and a number of other glycolytic enzymes on glycolytic flux was investigated. The results showed that none of the enzymes increased glycolytic flux. One exception was a small increase in glycolytic flux upon overexpression of PFK (61).

Because pyruvate dehydrogenase and CS are present only in the mitochondria, the synthesis of citrate from pyruvate is localized in mitochondrion. Szczodrak (62) studied the activity of selected enzymes of the TCA cycle in *A. niger* mycelium and found that CS activity was maintained at almost the same level or increased slightly until the maximum concentration of citric acid was reached (60.0 g/L). During citrate decomposition, the decrease in CS activity was very high compared to its initial activity. Ruijter et al. (63) reported that overproduction of CS did not increase the rate of citric acid production by *A. niger,* suggesting that CS contributes little to flux control in the pathway involved in citric acid biosynthesis. No significant changes were observed in the activity of other enzymes relevant to citric acid production, such as glycolytic enzymes, pyruvate carboxylase, malate dehydrogenase, and pyruvate dehydrogenase. Furthermore, overproduction of CS did not cause significant changes in the levels of intermediary metabolites (L-malate, oxaloacetate, and ATP). The activities of ACH and both dehydrogenases (NAD-ICDH and NADP-ICDH) decreased significantly with an increase in citric acid concentration (3,5,60,64,65). The activity of ACH was highest in the growth phase of the microorganism. Experiments showed that the activities of the above enzymes never disappeared completely during citric acid production by *A. niger* (62). Decrease in the concentration of citric acid is accompanied by a distinct increase in the activity of the enzymes responsible for the degradation of this acid (ACH and ICDH) as well as a large decrease in CS activity (30–40%). Based on the above observations it was concluded that citric acid production is affected by a larger inhibition of ACH and ICDH activity rather than by an increase in CS activity. Milsom (5) reported that NAD-ICDH was present in the cytosol, and two NADP-specific enzymes (NADP-ICDH and $NADP^+$-specific ICDH) were present in the mitochondria. Generally, mitochondrial enzymes were inhibited by normal concentrations of citric acid (60.0–70.0 g/L) while the cytoplasmic enzyme was not so inhibited (5). Moreover, Legisa and Kidric (66) found that glycerol slowly diffuses out of the cells and possibly into the mitochondria. Because mitochondrial $NADP^+$-ICDH is inhibited by glycerol, citric acid begins to accumulate in the cells. Other workers (67) reported that

$NADP^+$-specific ICDH activity was inhibited by citrate, but was not affected by other metabolites of the TCA cycle. Inhibition of this enzyme was caused by physiological concentrations of citric acid. Experiments showed that a close relationship existed between the citrate production and $NADP^+$-specific ICDH of mitochondria. Thus, $NADP^+$-specific ICDH was one of the key enzymes involved in citric acid production by *A. niger.*

La Nauze (68) and Szczodrak and Ilczuk (69) studied the effect of iron on the activity of ACH and found that a 20-fold increase in the concentration of iron doubled the activity of ACH, while citric acid accumulation was only decreased by 25%. Kubicek and Rohr (70) found that concentrations of Fe^{3+} required to decrease citric acid production were at least 1000-fold higher than those of Mn^{2+}, and it is probable that Mn^{2+} impurities in the Fe^{3+} source caused the actual inhibition. Finally, the results showed that inhibition of ACH was not necessary for the accumulation of citric acid.

Roukas (65) studied the influence of impeller speed on citric acid production and selected enzyme activities of the TCA cycle. The fermentation was carried out in a 12 L stirred tank fermentor (S.K. Fermenters Ltd., Manchester, UK) with a working volume of 9 L and a temperature of 30°C. The fermentor consisted of a glass vessel with stainless steel end plates and had four equally spaced vertical baffles. The aeration rate was 1 vvm (volume of air per volume of substrate per minute) and agitation was provided by a six flat blade turbine operating at 300–600 rpm. The impeller was located 5 cm above the bottom of the vessel. Roukas found that the activity of CS decreased by 80% with an increase of agitation speed from 300 to 600 rpm, while the activity of ACH, NAD-ICDH, and NADP-ICDH increased by 100% with the same increase in agitation speed. The lowest ACH and ICDH activities were obtained at the same time as maximum concentrations of citric acid (20.0–27.0 g/L) were obtained. The above results show that increase in agitation speed resulted in changes in the activity of enzymes, which are connected to citric acid accumulation. This was due to the change of dissolved oxygen concentration, biomass concentration, viscosity, and pH with increases of speed of agitation (65).

Another important enzyme of the TCA cycle is α-ketoglutarate dehydrogenase (AKGDH), which catalyzes the oxidation of α-ketoglutarate to succinate. It was found that the activity of this enzyme is inhibited by NADH, oxaloacetate, succinate, and *cis*-aconitate (71). Monovalent cations including NH_4^+ were inhibitory at high concentrations (>20 mM). The highest enzyme activity (0.065 units/mg protein) was found in mycelia of *A. niger* grown in media supplemented with glucose-NH_4^+ or glucose-peptone. Kubicek and Rohr (72) reported that during the log phase of *A. niger* the activity of almost all the enzymes of TCA cycle, with the exception of 2-oxo acids, was high. During the induction phase of citrate accumulation, malate, fumarate, and isocitrate decreased, whereas pyruvate, oxaloacetate, and citrate increased. The interrelations of the momentary concentration intermediates mainly demonstrate a lack of activity of 2-oxoglutarate dehydrogenase, representing a block in the TCA cycle concomitant with a strongly operating glycolysis as a prerequisite for citrate accumulation. Under conditions of manganese deficiency, it has been observed that the citric acid production stage is characterized by a significant increase in the concentration of pyruvate and oxaloacetate (100%), apparently causing increased activity of citric acid biosynthesis (50–100%) (72). Heinrich and Rehm (73) observed that in a stirred tank fermentor and in a fixed bed reactor, a high concentration of gluconic acid occurred immediately after nitrogen exhaustion during citric acid production by free, and immobilized, *A. niger* cells, respectively. Because manganese is a necessary supplement for glucose oxidase (GOD) formation, manganese from the stainless steel parts of the vessel was responsible for both gluconic acid production and small citric acid yields. After maximum concentration of gluconic acid, the production rate of gluconic acid decreased rapidly at low pH values. This decrease may be explained by the fact that gluconolactone was formed at low pH values.

At pH 5.5 or above, such a decrease was prevented by the activity of lactonase converting gluconolactone to free gluconic acid. High gluconolactone concentrations perhaps lead to feedback inhibition or, by equilibrium, to a decreasing conversion rate (73). Mischak et al. (74) studied the formation and location of GOD during citric acid production by *A. niger* and reported that GOD could be induced *de novo* by a shift in pH from 1.7 to 5.5. Induction required the intracellular presence of either glucose or glucose-6-phosphate. Glucose oxydase produced was rapidly secreted into the fermentation broth.

The role of intracellular cyclic AMP in the control of citric acid production by *A. niger* has been investigated (75). It was found that citric acid accumulation was only observed in media containing high sugar concentrations and low manganese concentrations. When sucrose was substituted by other monosaccharides or disaccharides, similar intracellular concentrations of cyclic AMP were observed. However, citric acid accumulation was only significant with sucrose, glucose, and fructose. Thus, it is clear that the intracellular level of cyclic AMP is not related to the accumulation of citric acid by *A. niger* and is not affected by manganese deficiency. Also, the addition of cyclic AMP to the medium resulted in a drastic alteration in the morphology of *A. niger*, from filamentous to pellet growth (75).

Another two enzymes (the cytosolic $NADP^+$-specific glycerol dehydrogenase and a mitochondrial $NADP^+$-specific glycerol dehydrogenase) were detected in *A. niger* mycelium (76). It was found that these enzymes were regulated by bicarbonate or CO_2. A change in carbohydrate metabolism from PP pathway to glycolytic flux was detected during the early stages of the growth of *A. niger.* The decrease of PP metabolism was shown to be due to inhibition of 6-phosphogluconate dehydrogenase (6-PGDH) by intracellular citrate (77). The excretion of citric acid by *A. niger* in the early stages of fermentation is connected with glycerol accumulation from the intermediates stages of the PP pathway. Glycerol inhibits mitochondrial $NADP^+$-specific ICDH, causing citrate accumulation. Citrate, by inhibiting 6-PGDH, stops glucose degradation via the PP pathway, which, in turn, results in cessation of glycerol synthesis. When glycerol disappears there is already enough citrate present in the cells to keep the mitochondrial $NADP^+$-specific ICDH blocked. For rapid citric acid accumulation, uninhibited glycolytic flux is required. This process is dependent on 6-PFK activity, but the mechanism of its activation still remains unexplained (77). Other researchers (78) found that *A. niger* accumulates a considerable amount of polyols during citric acid production. Moreover, these authors reported that during the first phase of citric acid accumulation, more sugar is taken up than is used in the production of biomass, CO_2, and citric acid. In contrast, during later phases of fermentation more citric acid, CO_2, and biomass are formed than sugar uptake. Accumulation of polyols as byproducts of citric acid fermentation has also been observed with yeast, but in this case mannitol and arabitol were the main accumulated polyols (78).

Recently, studies by Torres (79,80) showed that the most important step controlling the glycolytic flux occurs prior to formation glucose 6-phosphate such as during glucose uptake or glucose phosphorylation. In addition, *A. niger* contains a single hexokinase enzyme which shares a significant part in the control of the rate of citric acid accumulation. Several studies have been carried out on glycolysis regulation and the stimulation of its flux rate. The presence of high concentrations of glucose or sucrose ($\geq$50 g/L) causes a rise in the intracellular concentration of Fru-2,6-P_2, a strong activator of glycolysis. This may be explained by the fact that poor regulation of 6-phosphofructo-2-kinase (PFK2) in *A. niger* mycelium is carried out, which suggests that the metabolic steps prior to PFK1 and PFK2 are of major importance for glycolytic flux and citric acid accumulation (81). Moreover, Torres et al. (81) found that *A. niger* contains a single, high affinity glucose transporter when grown in low glucose concentrations (1% w/v), but forms an additional low affinity transporter when grown in high glucose concentrations (15% w/v). Both glucose transporters exhibit decreased activities at low pH values, and are inhibited by citric acid. However, the activity of the low affinity transporter is much less affected by these conditions. The results showed that the low affinity glucose transporter has a major function in glucose catabolism under conditions of excess extracellular glucose, because it was specifically formed only under these conditions. However, the fact that this carrier was also less sensitive to a low pH, and that the inhibition by extracellular citric acid was not strong enough to significantly reduce its activity, suggested a role in citric acid accumulation (81). Hence, despite the importance of the low affinity glucose transporter for the rate of glycolysis and citric acid accumulation, hexokinase activity appears to contribute most significantly to the control of the glycolytic flux to citric acid. Finally, Torres et al. (81) concluded that the low affinity glucose transporter takes part in the mechanism by which *A. niger* responds to high extracellular glucose concentrations that lead to citric acid accumulation.

Other workers (82) investigated the relationship between specific glucose uptake rate and glucose concentration during citric acid production by *A. niger.* In shake flask cultures, airlift bioreactors, and stirred tank fermentors, they found a simple relationship between glucose concentration and specific uptake rate which is independent of fermentation parameters. However, in a glucostat sfermentation system (the concentration of glucose is kept constant while the other fermentations parameters are not controlled), no significant relationship between the specific glucose uptake rate and glucose concentration was observed. Finally, Wayman and Mattey (82) found a simple diffusion model that fit all of the observed data and explained the relationship between specific uptake rate and glucose concentration, which should not exist under carrier saturated conditions. It also reflected the known effects of trehalose-6-phosphate on the flux through glycolysis that produced citric acid, as the values obtained for the diffusive flux constant were lower than expected, with glucose concentrations >400 mM. Because simple diffusion is an inevitable physical process, it is not capable of being regulated directly by the microorganism. This fact may, in itself, account for the dramatic overproduction of citric acid under the conditions of this process. The simple nature of this mechanism also explains the similarity of the uptake relationships from different sources despite the use of different strains and growing conditions (82).

Alvarez-Vasquez et al. (83) studied the metabolism of citric acid production by *A. niger* and found that when the total enzyme concentration was increased two times higher than the basal value, the citric acid production rate was capable of being increased by more than 12-fold, and even larger values were attainable if the total enzyme concentration was allowed to increase even more (up to 50-fold when the total enzyme concentration rose up to tenfold the basal value). Also, Alvarez-Vasquez et al. (83) investigated the metabolic optimization of a model of *A. niger* metabolism under conditions of citric acid production. The results showed that a minimum of 13 enzymes responsible for the biosynthesis of citric acid were needed to increase significantly the production of the acid.

Recently, Guebel and Torres (84) studied optimization of citric acid production by *A. niger* through a metabolic flux balance model and suggested three strategies to increase citric acid production: (1) increasing the glucose uptake rate; (2) decreasing the biosynthesis rate of byproducts (i.e., polyols); and (3) decreasing fluxes diverting mass from the pathway leading to citrate precursors, namely, the biomass synthesis and the TCA cycle. However, the results showed that the synthesis of citric acid described not only a carbon consuming process but also an energy consuming one. Consequently, the biosynthesis of the citric acid would be improved either by a higher ATP (3.7 mmol/g/h) availability from oxidative phosphorylation, or by a diminished activity of H^+/ATPases (1.8 mmol/g/h). The influence of both glucose uptake rate and polyol excretion has already been discussed in relation to carbon availability. Also, the results showed that the TCA cycle would have a dual effect: a larger and positive effect as a source of FADH and ATP, and a quantitatively less important but negative effect because of carbon loss through production of CO_2 (84). During citric acid production by *A. niger*, a significant amount of CO_2 is produced. It can be generated by the hexose monophosphate (HMP) cycle, the TCA cycle, and the 4-aminobutirate cycle. The results of Guebel and Torres (84) showed that increases in mitochondrial phosphate carrier activity enhance citric acid rates of production. The main factor underlying this effect is the enhancement of ATP availability. Thus, while the activity of cytoplasmatic H^+-ATPase is diminished, the mitochondrial H^+- ATPase is more active. Because the mitochondrial pump is energetically most efficient, a net ATP saving can be expected. Other perturbations, based on the blocking of some diverting fluxes (the HMP pathway or the excretion polyols, mainly glycerol and arabitol), have been shown to have no effect on citric acid production. Finally, Guebel and Torres (84) concluded that the HMP pathway accounts for 16% of the glucose input, the TCA cycle for 13%, and citric acid biosynthesis for the remaining 71%. This profile implies an operative glycerol-P shuttle. It recycles 93% of the cytosolic glycerol-P to cytosolic dihydroxy acetone phosphate (DHAP), thus coupling the transformation of cytosolic NADH to mitochondrial FADH. The GABA cycle and $NH^+{}_4/NH_3$ are not significant during citric acid production by *A. niger.* Specific citric acid productivity can be increased by 45% if glucose influx is duplicated.

Kapoor et al. (3) reported a hypothetical pathway for the production of citric acid from n-alkanes by yeasts (Figure 15.2). n-Alkanes are converted to propionyl-CoA and acetyl-CoA through β-oxidation. Propionyl-Coa condenses with oxaloacetate to form methyl citrate. This is isomerized to methyl isocitrate which is further cleaved to succinate and pyruvate. Succinate is oxidized via the TCA cycle to oxaloacetate, which is then recycled. Finally, carboxylation of pyruvate to oxaloacetate and condensation with acetyl-CoA forms citric acid. Moreover, acetyl-CoA through the glyoxylate cycle converts to oxaloacetate, which condenses with acetyl-CoA again to yield citric acid. The most important problem during citric acid production by yeasts is the production of isocitric acid. Thus, the activities of NAD-ICDH and NADP-ICDH play a significant role in the biosynthesis of citrate by yeasts. Media supplemented with low concentrations of potassium ferrocyanide or quinaldinic acid, or media that are deficient in thiamin, decrease the activity of isocitrate dehydrogenase and increase the accumulation of citric acid (3,12). When n-alkanes are used as substrates for the production of citric acid by *Yarrowia lipolytica,* citric and isocitric acid are generated at the end of the exponential growth phase when nitrogen in the fermentation broth becomes limited (12–14). Simultaneously, the activities of AMP, ADP, and NAD-ICDH decrease significantly. This prevents the oxidation of citrate to isocitrate, while the activity of isocitrate lyase is not inhibited. During the growth phase, CO_2 production and oxygen consumption rates increase, approximately in parallel with the cell production rate. They then decrease with a decline in cell concentration and reach constant values during the production phase (12). Aiba and Matsuoka (13) studied the production of citric acid from n-alkanes by *Yarrowia lipolytica* in nitrogen limited chemostat culture and found that the ratio of citrate to isocitrate remained virtually unchanged (about 1.2). This signifies that ACH must have an appreciable activity to establish a dynamic equilibrium between intracellular pools of citrate and isocitrate. The results showed that there was a relatively high specific activity of ACH comparable to that of CS. Also, a large increase in the ratio of citrate to isocitrate was achieved with an ACH leaky mutant of *Yarrowia lipolytica.* Aiba and Matsuoka (13) found that the specific activity of isocitrate lyase (ICL) decreased from 184.0 to 64.6 μmoles/min with the decrease in specific growth rate from 0.073 to 0,012 h^{-1} , while the specific activities of CS, NAD-ICDH, and NADP-ICDH were not affected. In addition, the activity of CS was always higher than the activities of NAD-ICDH and NADP-ICDH, and ICL controlled the production of citric

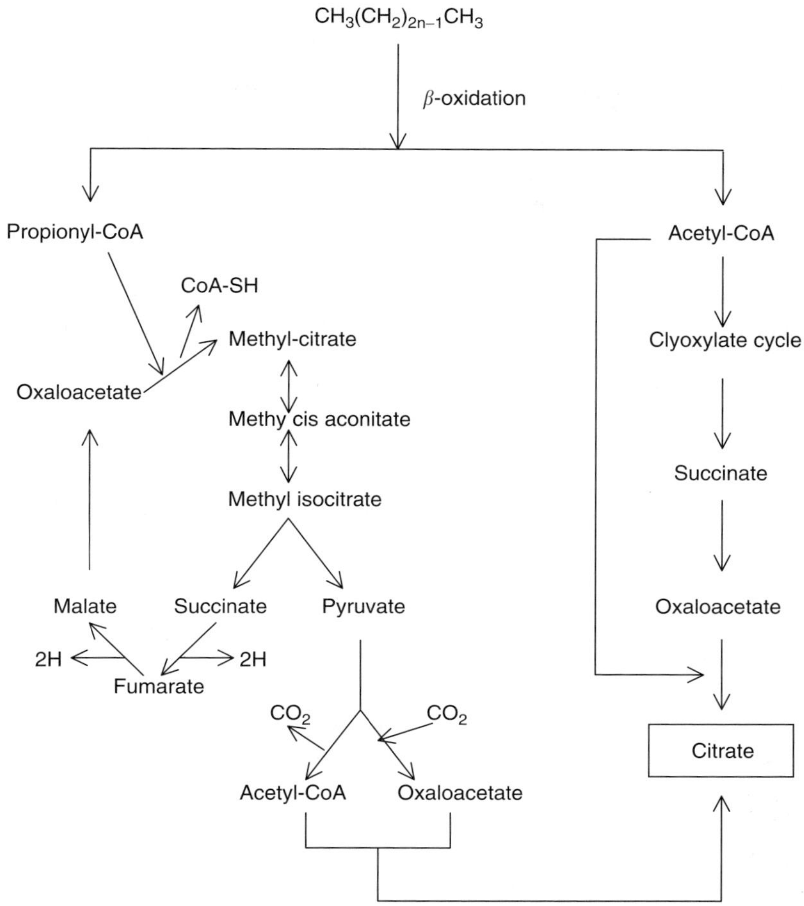

Figure 15.2 Biosynthesis of citric acid from n-alkanes by yeasts. (From Kapoor, K.K., K. Chaudhary, P. Tauro, *Prescott and Dunn`s Industrial Microbiology*, Reed, G., ed.. UK: MacMillan Publishers Ltd, 1983, pp 709–747).

and isocitric acid. Finally, Kapoor et al. (3) concluded that excretion of citric and isocitric acids probably occurred due to other physiological changes such as alteration of cell permeability in relation to these acids.

15.4 FACTORS AFFECTING CITRIC ACID PRODUCTION

15.4.1 Nutrients

The composition of the media used for the production of citric acid by *A. niger* depends on the strain of the microorganism used and the type of process. Generally, strains that can use one carbon source efficiently fail to show good acid production when cultured in a medium containing another (3). *Aspergillus niger* grows well in media containing carbohydrates (sucrose, glucose, fructose, maltose, mannose, and starch), nitrogen (as ammonium or nitrate ions), phosphate, low amounts of potassium, magnesium, sulfate, and trace metals such as iron, manganese, zinc, and copper. It has been found that citric acid yields are much higher when *A. niger* strains are grown in simple synthetic media rather than in complex media. The initial sugar concentration plays an important role. The highest citric acid concentrations were observed in cultures grown at high initial sugar concentrations (15–20% w/v) (3,27,85,86). Further increase of sugar concentration (i.e., 250 g/L)) resulted in a decrease of acid concentration by 15%. The decreased concentration of acid encountered with the highest concentration treatment was probably due to osmotic effects. It has been reported that above a critical substrate concentration (150 g/L), decreased water activity and the onset of plasmolysis combine to cause a decrease in the rates of fermentation and product concentration (27,86). Also, initial sugar concentration (>20%) leads to low sugar utilization, making the process uneconomical, while a low concentration of sugar (<5%) leads to low yields of citric acid and a high accumulation of oxalic acid (3,27,86). Precultivation of *A. niger* at 1% (w/v) sucrose and transference to a 14% concentration of various other sugars (sucrose, maltose, glucose, mannose, and fructose) induced citric acid accumulation. This accumulation can be blocked by the addition of cyclohexamide, an inhibitor of *de novo* protein synthesis. However, precultivation of *A. niger* at high sucrose concentration (14%) and subsequent transfer to the same concentrations of various other carbohydrates led to the formation of citrate. These results indicate that high concentrations of certain carbon sources are required for high citric acid yields (60–70%) because they induce the appropriate metabolic imbalance required for acidogenesis (85).

In addition to sugar concentration, the source of carbohydrate has been shown to have a marked effect on citric acid production by *A. niger.* Xu et al. (85) found that sucrose and maltose were better carbon sources for citric acid production by *A. niger* than glucose and fructose. Hossain et al. (87) studied the effect of the sugar source on citric acid production by *A. niger* and found that sucrose was the most favorable source, followed by glucose and fructose, and then lactose. They also observed that while the fungus grew in medium containing galactose as the carbon source, no citric acid was produced. Strong relationships were observed between citric acid production and the activities of enzymes that were responsible for the biosynthesis of the acid.

When sucrose, glucose, or fructose was the sugar source pyruvate carboxylase activity was high (1050, 600, and 500 μmole/min/mg protein, respectively), while 2-oxoglutarate dehydrogenase activity was not detected. However, when galactose was the sugar source, pyruvate carboxylase activity was low (20 μmole/min/mg protein), but 2-oxoglutarate dehydrogenase activity was high (45 μmole/min/mg protein). The activities of ACH showed a strong relationship with the citric acid production rates during fermentation. The activity was highest (2200 μmole/min/mg protein) when sucrose was the sugar source and lowest (150 μmole/min/mg protein) with galactose. The activity of the enzyme decreased during fermentation except in the presence of galactose, where activity increased. Similar effects were seen with both NAD-ICDH and NADP-ICDH. In conclusion, the results demonstrate that the nature of the sugar source has a strong effect on citric acid production. These findings can be related to the activities of certain enzymes. The 2-oxoglutarate dehydrogenase appears to be the key enzyme for the accumulation of citric acid, and it is repressed during growth on glucose and fructose, but not on galactose (87).

In addition to carbohydrates, nitrogen and the phosphate concentrations have a strong influence on citric acid production. Generally, a nitrogen or phosphate concentration less than 0.2% (w/v) in the medium appears to be adequate. High concentrations of nitrogen (>0.8 g/L) resulted in a reduction by 100% in citric acid production (3–6,88–92). Kristiansen and Sinclair (88) studied production of citric acid in continuous submerged culture and reported that citric acid is produced by cells that accumulate carbon under nitrogen limitation. The nitrogen in the feed is accounted for by the new cells growing at the hyphal tip. The older cells suffer nitrogen limitation and consequently store carbon and produce citric acid. The transformation of basic cells into storage cells is probably brought about by the streaming of cytoplasm. In general, the effect of nitrogen on citric acid production by *A. niger* can be described as follows (88). The number of cells formed increases with nitrogen concentration and streaming increases with the number of cells formed. Thus, at low nitrogen concentration few cells are formed and with little or no streaming citric acid production is low. When nitrogen concentration increases, the rate of formation of storage cells increases and more citric acid is produced. The production of citric acid is carried out in the mitochondria, which may be carried along toward the nonproducing tip of hyphae not suffering nitrogen limitation if the streaming becomes too pronounced, thereby causing citric acid concentration to drop. This process offers a possible explanation for the drastic reduction in citric acid production at high nitrogen levels. Experiments performed in batch fermentation under phosphate limited growth conditions showed that citric acid yield was inversely related to excessive nitrogen concentration in the medium (89). Results from chemostat culture confirmed a negative relationship between citric acid production and both specific growth rate and nitrogen consumption rate. Dawson and Maddox (89) concluded that under phosphate limited growth conditions, citric acid production by *A. niger* is possibly subject to nitrogen catabolite repression. Also, a fed batch fermentation process operated under nitrogen and phosphate limitation is efficient and economical. Moreover, Choe and Yoo (90) studied the production of citric acid using batch and fed batch culture, and found that citric acid biosynthesis increased more significantly in the fed batch system (40 g/L) than in the batch culture (21 g/L). Other workers (91) reported that not only the timing of the addition of ammonium ions is important, but the concentration of nitrogen must be considered. Optimum addition time of $(NH_4)_2SO_4$ was in the range of 40 and 75 h, and concentrations of the above nitrogen source between 0.25 and 0.5 g/L yielded a 12% increase in maximum citric acid concentration relative to standard fermentation. The concentration of phosphate in the medium is very important to the growth of *A. niger* and the biosynthesis of citric acid (92). A high concentration of phosphate (>2.0 g/L) promotes more growth of the microorganism and less acid production. As in the case of nitrogen, the production of citric acid starts after exhaustion of phosphorus compounds (3). Finally, the maintenance of low levels of phosphorus (<1.0 g/L) also becomes a key factor, at least in cultures developed on natural carbohydrate sources, an effect that has been related to the presence of trace metals in the medium. In addition to the above nutrients, *A. niger* needs amino acids and vitamins for growth and for production of citric acid. Lal and Srivastava (93) studied the effect of amino acids on citric acid production by *A. niger.* They found that the presence of glutamic acid and aspartic acid stimulated citric acid production to the extent of 79.6% and 76.7%, respectively. Lysine was effective in increasing yield by 62%. However, serine could not influence the yield (50.4%) to a greater extent, while the effect of cysteine was found to be detrimental. Hamissa et al. (34), however, found that the addition of certain amino acids to the medium completely inhibited citric acid biosynthesis by *C. lipolytica* Y 1095. When the medium was supplied with vitamins such as thiamine, nicotinic acid, and nicotinamide, concentration of citric acid increased. The optimum concentration of thiamine favoring the yield was 6.0 mg/L.

Trace elements play a significant role. High concentrations of trace metals (5.0 mg/L) decrease concentration of citric acid while low concentrations (1.0 mg/L) improve production of the acid (3–6,23,34,94–102). Clark et al. (94,95) studied the effect of potassium ferrocyanide on the chemical composition of molasses used for the production of citric acid. They found that the removal of trace metals (Fe^{2+}, Mn^{2+}, Zn^{2+}, Cu^{2+}, Ca^{2+}, Mg^{2+}) from molasses resulted in a significant increase in citric acid production. Banik (96) studied the effect of minerals on citric acid production by *A. niger* and observed that K_2HPO_4

and $MgSO_4$•$7H_2O$ were required at concentrations of 0.1 and 0.02%, respectively. The optimum level of each of the trace elements Fe^{2+}, Mn^{2+}, and Zn^{2+} was 1.0 mg/L , whereas NaCl and $CaCl_2$ at lower concentrations had no effect. Other trace elements such as Cu^{2+}, Co^{2+}, and Mo^{2+} had an adverse effect, while Ni and V were without effect. Kubicek and Rohr (97) reported that a comparison of citric acid fermentations in manganese deficient and manganese containing media showed that manganese strongly influenced idiophase metabolism. In the presence of manganese, cell growth increases, sugar consumption is diminished, and acidogenesis decreases drastically. An investigation of the key enzymes of glycolysis, the PP pathway, the TCA cycle, nitrogen metabolism, and gluconeogenesis indicated that manganese deficiency was accompanied by a repression of anabolic and TCA cycle enzymes, with the exception of citrate synthase. In general, manganese deficiency mainly affects the operation of biosynthetic reactions in *A. niger,* thus leading to an overflow of citric acid. Production of citric acid from molasses by *A. niger* in surface and submerged fermentation is highly sensitive to the concentration of trace metals in the molasses. However, studies on citric acid production from wheat bran in solid-state fermentation showed no such influence despite high concentrations of minerals in the wheat bran. In contrast, when the medium was supplemented with iron, manganese, copper, zinc, magnesium, and phosphorus, the concentration of citric acid was increased by 100% (99). When the mineral ions were used individually at optimum levels the concentration of citric acid increased by 1.4–1.9 times. However, only iron, copper, and zinc were effective in combination. These results indicate that the advantage of solid-state fermentation in tolerating high concentrations of elements in the production of citric acid is probably due to concentration gradients and impaired transport of the metal ions in conditions of limited water availability (99). Addition of 0.5 g/L $CaCl_2$ to the fermentation medium decreased the biomass dry weight by 35% whereas the uptake rates of phosphate and sucrose and the concentration of citric acid increased by 15, 35, and 50%, respectively (100). Tsekova et al. (101) found that copper and cadmium inhibited the growth of *A. niger* and the production of citric acid. The activity of ACH and both NAD-ICDH and NADP-ICDH was strongly inhibited by copper. The contents of DNA and proteins in the cells decreased, but the contents of lipids and polysaccharides increased considerably in the presence of both heavy metals. However, Haq et al. (102) reported that the addition of 2.0×10^{-5} M $CuSO_4$ to the molasses solution increased concentration of citric acid (92.0 g/L) by *A. niger* GCB-47. This means that the effect of heavy metals on citric acid production depends on the strain used and the composition of the medium. Finally, Hamissa et al. (34) studied the effect of micronutrients such as $FeSO_4$•$7H_2O$, $PbSO_4$, $CuSO_4$•$5H_2O$, $MnSO_4$•$4H_2O$, $ZnSO_4$•$7H_2O$, $CoCl_2$•H_2O, and $(NH_4)_6$•Mo_7O_{24}•$4H_2O$ on citric acid production by *C. lipolytica* Y 1095. They found that $FeSO_4$ •$7H_2O$ and $MnSO_4$•$4H_2O$ were more suitable than the other micronutrients for citric acid production. Generally, the tolerance of different strains to trace elements varies so widely that it is necessary to determine the optimum concentrations of such ions in order to avoid the inhibitory effects caused when these cations are present in toxic concentrations (103).

15.4.2 Inhibitors and Stimulants

When certain inhibitors are added to the medium, more specific product may be produced or a metabolic intermediate that is normally metabolized is accumulated instead. In most cases the inhibitor is effective in increasing the concentration of the desired product and reducing the yield of undesired related products. Inhibitors have also been used to affect the structure of the cell wall and increase its permeability for release of metabolites (104).

Stimulants are chemical substances that increase the concentration of the product when added to the medium. The most important stimulants used for improving citric acid yield by

A. niger are methanol and ethanol (3–6,27,86,103,105–111). These chemicals have been found to retard growth, delay sporulation, and increase (by 30–50%) citric acid production.

Roukas and Liakopoulou-Kyriakides (27), Roukas (86), and Roukas and Kotzekidou (103,105) studied the effect of methanol on citric acid production from fig extract, carob pod extract, brewery wastes, and date syrup in surface fermentation by *A. niger* and found that the addition of 4% (v/v) methanol in the above media increased the concentration of citric acid by 50%, 41%, 17%, and 60%, respectively. Moreover, Roukas (106,107) studied the production of citric acid from carob pod and fig by *A. niger* in solid-state fermentation and found that the addition of 6% (w/w) methanol into the media increased the concentration of citric acid by 62% and 50%, respectively. In all cases methanol was added on the third day of fermentation. The observed increases in citric acid concentration show that methanol has a profound effect on the metabolism of sugars by *A. niger.* It was reported that the effect of methanol is at the cell permeability level; it allows citrate to be excreted from the cell. The cell then responds by increasing its citrate production via repression of 2-oxoglutarate dehydrogenase in an attempt to maintain an adequate intracellular level of the metabolite (86,108). In addition, the high stimulation effect of methanol can be attributed to inhibition of spore formation and an increase of microorganism tolerance to high levels of minerals contained in the media (105).

Other workers (111) reported that the addition of methanol to the production medium remarkably depressed cellular protein synthesis without inhibiting nitrogen uptake, thus causing an increase of amino acids, peptides, and low molecular mass protein pooled in the mycelium, especially at the early stage of cultivation. It was also found that addition of methanol to the medium affected the activity of some enzymes of the TCA cycle. Saha et al. (109) studied the effect of methanol and ethanol on citric acid production by *A. niger* using a magnetic rotating biological drum contactor. They found that ethanol promoted citric acid production as well as productivity. Jianlong and Ping (111) reported that the addition of ethanol resulted in a twofold increase in CS activity, and a 75% decrease in ACH activity. The activities of other TCA cycle enzymes increased slightly (5–10%). Increase in citric acid accumulation with the addition of ethanol might be due to the slow degradation of citric acid consequent to reduction in aconitase activity. Both continuous and repeated batch fermentations could be easily carried out using a magnetic drum contactor. In continuous fermentation, methanol and ethanol increased citric acid concentration by 82% and 85%, respectively, in comparison to 49% when no ethanol was used. The highest citric acid concentration and productivity in the continuous culture was attained when ethanol was added during fermentation.

Other stimulants such as fats and oils significantly increased the production of citric acid. It was reported that fats and oils act as carbon sources and are broken down to glycerol and fatty acids. The latter enter the TCA cycle via glycolysis, whereas the former enter directly by the formation of acetyl-CoA (111). Roukas and Kotzekidou (103) found that the addition of sodium monofluoroacetate to brewery wastes resulted in decreased citric acid yield. This chemical is toxic to the mold and does not stimulate citric acid (111).

Addition of some inhibitors such as calcium fluoride, sodium fluoride, potassium fluoride, hydrogen peroxide, naphthaquinone, methylene blue, sodium malonate, potassium ferricyanide, iodoacetate, sodium azide, and sodium arsenate to the different media increased (by 30–40%) citric acid concentration (111,112). Hossain et al. (113) observed that the presence of galactose caused inhibition of citric acid production and also reduced the rate of glucose utilization. Enzyme analyses using mycelial cell free extracts indicated that galactose interfered with glucose repression of the key enzyme 2-oxoglutarate dehydrogenase. Legisa and Mattey (114) studied the production of citric acid by *A. niger* and found that the production of 1% glycerol at the early stage of fermentation caused inhibition of the $NADP^+$-specific ICDH, which led to the accumulation of citric acid. Also, citric acid concentration increased when the medium was supplemented with cyclic AMP at a concentration of $\geq 10^{-6}$ M. Adenosine, ATP, or cGMP stimulate citric acid production at a concentration of 10^{-3} M, while ADP strongly inhibits citric acid accumulation. Addition of thiophylline along with cAMP increases the effect of cAMP. It is suggested that citric acid production in fungi perhaps results from an abnormal cAMP metabolism (3). Kirimura et al. (115) reported that in *A. niger*, a cyanide (CN)- and antimycin A insensitive and salicylhydroxamic acid (SHAM) sensitive respiratory pathway exists beside the cytochrome pathway and is catalyzed by the alternative oxidase (AOX). When *A. niger* WU-2223L was cultivated with antimycin A or methanol, both citric acid concentration and productivity increased with an increase of both the activity of AOX and the rate of CN insensitive and SHAM sensitive respiration. However, when *A. niger* WU-2223L was cultivated with SHAM, an inhibitor of AOX, CN insensitive and SHAM sensitive respiration were not detected; citric acid concentration and productivity drastically decreased, although mycelia growth was not affected. These results clearly indicate that CN insensitive and SHAM sensitive respiration, catalyzed by AOX and localized in the mitochondria, contributed to citric acid production by *A. niger* (115).

Recently, Jianlong and Ping (111) reported a new stimulator that increased citric acid production by *A. niger* to about 1.1- to 1.7-fold. They observed that the addition of sodium phytate to the fermentation medium at a concentration of 1% increased citric acid production and substrate utilization and slightly inhibited the growth of the microorganism. Several key enzymes involved in the TCA cycle were affected by addition of phytate to the medium (for instance, sodium phytate depressed the activity of NADP-ICDH and activated pyruvate carboxylase, hence favoring the accumulation of citric acid).

15.4.3 Inoculum

To prepare inoculum, *A. niger* is grown in standard media for molds. It is usually cultivated on potato dextrose agar (PDA) slants or in petri dishes at 28–30°C for 3–5 days. The spores obtained are suspended in sterile water containing 0.1% Tween 80. Inoculation is carried out using spores of *A. niger* or pregrown mycelia. When mycelia pellets are used, they are grown in submerged fermentation for 2–3 days in medium that has the same composition as the production medium. The production medium is then inoculated at a concentration of 5–10% (v/v). When spores are used as inoculum, the substrate is inoculated with 0.5–1.0% (v/v) of the inoculum to give a final concentration of $\sim 1.0 \times 10^6$ spores/mL (3,27,86). In solid-state fermentation, the medium is inoculated with 5% (v/w) of the inoculum containing 1.0×10^8 spores/mL to give a concentration of 0.5×10^7 spores/g wet substrate (106,107).

15.4.4 Fermentation Time

The optimum time for the maximum production of citric acid depends on the strain used, the chemical composition of the medium, the fermentation system, and generally, the conditions under which fermentation takes place. In the surface culture, fermentation time is usually completed in 10–20 days, while in the submerged culture incubation time is much shorter (5–10 days). In solid-state fermentation the fermentation time depends strongly on the amount of inoculum used, the moisture content of the substrate, the initial pH, the temperature, and the particle size of the medium (e.g., carob pod or fig) (106,107). In this case, the maximum citric acid concentration is between 10 and 15 days.

15.4.5 Temperature

Asperillus niger and other fungi used in the production of citric acid from synthetic media or molasses in submerged fermentation have an optimum temperature between 25 and

30°C (3,116). However, Roukas (86) studied the effect of temperature on citric acid production from carob pod extract by *A. niger* and found no significant differences between cultures grown at 25 and 40°C in terms of citric acid concentration (67.7 and 70.0 g/L, respectively) and citric acid productivity (3.22 and 3.33 g/L/h, respectively). Also, no significant differences were noted in biomass yield, specific biomass production rate, and specific sugar uptake rate among cultures grown at 25 and 30°C. Increasing the fermentation temperature from 25 to 40°C significantly affected specific citric acid production rate (0.157 to 0.111 g citric acid/g biomass dry weight/d) and biomass dry weight (20.5 to 30.0 g/L). Other kinetic parameters, namely, biomass dry weight, biomass yield, specific biomass production rate, and fermentation efficiency, were at their maximum at 40°C (86). In addition, this author studied the effect of temperature on citric acid production from carob pods and figs in solid state fermentation and observed that the maximum citric acid concentration (176 and 64 g/kg dry carob pod or fig, respectively) was achieved at 30°C. Very similar results were obtained by Szewczyk and Myszka (117) who found that the optimum temperature for growth of *A. niger* in solid-state fermentation was 26–34°C. At temperatures >34°C, the maintenance coefficient decreases as the temperature increases. This reflects a deactivation effect of temperature on endogenous metabolism. Perlman and Sih (6) reported that a two stage process, where *A. niger* was incubated at 28–30°C for 2–3 days and then for a week at 20°C, resulted in higher yields than when the temperature was maintained at 28–30°C.

15.4.6 pH

One important factor that affects the biosynthesis of citric acid is the initial pH of the medium. When *A. niger* is used for the production of citric acid, the initial pH is dependent on the medium employed. In synthetic media the initial pH of the medium is usually adjusted to 2.5–3.5, while in the case of molasses the initial pH must be neutral or slightly acidic in order for germination and growth of the microorganism to occur. The pH of the medium is adjusted with HCl, H_2SO_4, or NaOH. The pH of the substrate decreases during fermentation (from 7.0 to 2.0) due to the production of citric acid and other acids generated in the TCA cycle (27,86,106,107). Roukas and Alichanidis (118) studied the effect of initial pH on citric acid production from molasses using surface fermentation. They found that the best results were obtained using an initial pH of 6.5 without control during fermentation. Adjustments of pH of the fermentation medium at appropriate times have also been made to affect the change from the growth to the production phase (2). In this process the initial pH of the substrate was within the range 4–6. After the desired amount of mycelial growth was obtained, the pH was lowered to stimulate citric acid production. When yeasts are used for the production of citric acid, the pH is often adjusted to 6.0–7.0 by addition of lime, calcium carbonate, or sodium hydroxide.

15.4.7 Aeration and Agitation

Aspergillus niger is an aerobic microorganism and therefore requires oxygen. Aerating and agitating the fermentation broth normally satisfies the oxygen demand of a fermentation process. The effect of agitation and aeration on citric acid production in submerged fermentation is extremely important for the fermentation's successful progress. Agitation is important for adequate mixing, mass transfer, and heat transfer. It not only assists mass transfer between the different phases present in the culture, but also maintains homogeneous chemical and physical conditions in the culture by continuous mixing. Agitation creates shear forces, which affect microorganisms in several ways, causing morphological changes, variation in their growth and product formation, and also damage to cell structures.

Aeration can be beneficial to growth and performance of microbial cells by improving mass transfer characteristics with respect to substrate, product and byproduct, and oxygen. The morphology of the microorganism strongly influences product formation, because it affects broth rheology and consequently the mass and heat transfer capabilities of the fermentation broth (119). Three techniques are used for the aeration of the substrate during citric acid production: (1) the shake flask culture, where Erlenmeyer flasks are placed on reciprocating or rotary shakers (200–250 rpm); (2) surface culture, where oxygen diffuses through the surface, which is maintained in a static condition; and (3) submerged culture, in which filter sterile air or oxygen is pumped into the medium from the bottom of the reactor. Industrial production of various metabolites by filamentous fungi is susceptible to regulation by the dissolved oxygen tension (DOT) of the medium (120). In this case, the critical DOT for growth and DOT for product formation are distinct parameters; in general, the latter is higher by 25–30%. Kubicek et al. (120) conclude that citric acid accumulation is favored by increasing the DOT of the fermentation broth (40–150 mbar). Furthermore, this demonstrates that oxygen acts as a direct regulator of citric acid accumulation because a response was obtained within only minutes after altering the oxygen supply. These changes in citric acid production were accompanied by changes in the oxygen uptake rate, whereas carbon dioxide formation remained relatively unaffected, which suggests that there is a regulatory linkage between respiration and citric acid accumulation. However, the lack of coordination in the behavior of adenine nucleotides and citric acid synthesis makes it probable that the oxygen effect is not exerted via the adenine nucleotides (120). Also, these authors reported a critical DOT of 9–10% of air saturation and 12–13% of air saturation for growth and production phases, respectively, during citric acid production by *A. niger.* Moreover, they found that the concentration of citric acid steadily increased between DOT values of 20% and 75% of saturation. Dawson et al. (121) studied the effect of interruptions to the air supply on citric acid production by *A. niger.* Interruptions were carried out by treating the cultures with nitrogen gas or by stopping the air supply and impeller speed. Provided that the measured DOT of the culture did not fall below 20% saturation, no gross effect was obtained on the performance of the culture. In contrast, when the DOT fell to zero, both mycelial growth and citric acid production were inhibited.

Roukas (65) studied the influence of impeller speed on citric acid production and found that the highest citric acid concentration (28.0 g/L) was obtained in culture agitated at lower speed (300 rpm). The concentration of dissolved oxygen increased with the increase of speed of agitation. In the cultures agitated at 300, 400, 500, and 600 rpm the concentration of dissolved oxygen from the second to the 12th day remained at ~22%, 30%, 40%, and 50% of the initial saturation level, respectively. It fell rapidly during the first 2 days of fermentation after which it increased more slowly due to the rapid increase of biomass concentration (6.0–9.0 g/L) observed at the same time.

Because of significant oxygen requirements during citric acid production and the relatively low solubility of oxygen in water, Jianlong (17) studied the production of citric acid by *A. niger* using n-dodecane as an oxygen vector. The advantage of using the oxygen vector in fermentation is that it increases the oxygen transfer rate from the gas phase to the cells of the microorganism without the need for an extra energy supply. In addition, oxygen vectors can act as surface active agents to lower the surface tension of water and increase the gaseous specific interfacial area. Jianlong (17) found that the addition of 5% (v/v) n-dodecane to the medium enhanced citric acid accumulation by 40%, significantly increased the volumetric oxygen transfer coefficient ($K_L a$) (130 h^{-1}), reduced the residual sugar concentration by 15–100%, stimulated mycelial growth (25–90%), and did not affect the activity of citrate synthase. Generally, the use of n-dodecane as an oxygen vector is of great potential interest for a number of fermentation processes, particularly those that depend strongly on oxygen concentration. It is biologically inert, has no toxicity, is readily available commercially, is inexpensive, and can be reused (17).

15.5 MORPHOLOGY OF *ASPERGILLUS NIGER* AND RHEOLOGICAL PROPERTIES OF THE FERMENTATION BROTH DURING CITRIC ACID PRODUCTION

The *A. niger* group consists, among others, of the fungi we commonly call black mold. *Aspergillus niger* grown on Czapek agar at 25°C forms colonies attaining a diameter of 4–5 cm within 7 days. They consist of a compact white or yellow basal felt with a dense layer of dark brown to black conidiophores (122). Conidial heads are globose, radiate, or (as they grow) split into several columns of conidial chains. Conidiophore stipes are smooth walled and hyaline, but often brown in color. They arise from the substratum, varying from 200 μm to several millimeters in length, and 10–20 μm in diameter. Vesicles are globose to subglobose, colorless to yellowish brown, and fertile over the whole surface. Phialides are borne on metulae that are 7.0–9.5 × 3–4 μm in size. Metulae are hyaline to brown, often septate, and 15–25 μm long by 4.5–6.0 μm wide. Conidia are globose to subglobose, 3.5–5.0 μm in diameter, brown, and ornamented with irregular warts, spines, and ridges (122). In submerged culture, the morphological characteristics of *A. niger* have been correlated with high yields of citric acid. The most desired morphological characteristics are as follows: abnormally short, stubby, forked, bulbous mycelium; numerous swollen, oval to spherical shaped cells well distributed throughout the mycelial structure; mycelial structures showing granulation and numerous vacuoles or refractile bodies; an absence of normal reproductive bodies; and formations of compact aggregates or colonies having gross granular appearance, with cross sections under 0.5 mm and averaging about 0.1 mm. The individual pellets are 0.2–0.5 mm in diameter, with precipitated matter enmeshed in the pellet; the lateral hyphae are short, thickened, vacuolated, and granular, with short club like branches (6).

Jernejc et al. (123) studied the composition of the mycelium of *A. niger* and found that the total lipids in mycelia grown in the presence of manganese were up to two times higher than in conditions favoring citric acid accumulation. The concentrations of phospholipids and neutral lipids were lower under manganese deficiency. Regarding phospholipid composition, phosphatidylethanolamine and diphosphatidylglycerol prevailed under citric acid accumulating conditions, whereas under restricted citric acid biosynthesis, phosphatidylcholine was present in larger amounts. Among the fatty acids contained in mycelium, unsaturated acids predominated in mycelia of *A. niger.* The ratio of saturated (palmitic and stearic) to unsaturated (oleic and linoleic) acids was always lower in accumulating conditions of citric acid. In this case, the amount of linoleic acid was higher compared to the amount of palmitic and stearic acid. Moreover, trace levels of myristic and palmitoleic acid, and an odd chain marguaric acid were detected during fermentation (123).

In another work (124), it was reported that phosphatidylcholine and sterols play an important role in membrane permeability. *Aspergillus niger* mycelia grown in conditions hindering citric acid accumulation contained higher amounts of sterols, with ergosterol as the main component and six other sterols representing minor components. In submerged fermentation, *A. niger* forms pellets or mycelium during fermentation, depending on the type of inoculum used. Inoculating the production medium with spores or preculture of pellets causes *A. niger* to form filamentous suspension or pellets, respectively. The most important advantage afforded by a pellet suspension is a significant decrease in viscosity compared with a filamentous suspension, which considerably enhances the desirable mixing and mass transfer properties of the suspension. However, in the case of pellets, growth of microorganism and acidity decrease due to limited diffusion of oxygen and other nutrients into the interior of the pellets (125). Citric acid production requires high dissolved oxygen concentration (30–40% saturated) which may adversely affect the fermentative capacity of hyphae and pellet formations. In general, it is accepted that pellet formation commences through aggregation of conidia during germination and that development progresses at low stirrer speeds and high growth rates (125).

Gomez et al. (125) studied the effect of inoculation methods and aeration conditions on the morphology of *A. niger.* They found that inoculation of the production medium with spores gave rise to cultures with a filamentous morphology at low agitation speeds (450 rpm) and either long filaments or a mixture of filaments with a small proportion of fluffy, loose pellets at high impeller speeds (1000 rpm). In contrast, inoculation with a preculture of pellets gave rise to cultures of small, compact, smooth pellets of uniform size. Culture morphology was either uniform, consisting solely of such pellets, or else the pellets were mixed with larger, gelatinous pellets at high or low impeller speeds, respectively, during the first two days of fermentation. These results suggest that the morphology of *A. niger* is sensitive to aeration conditions employed in the initial phase of fermentation. Generally, slow agitation speed during germination and the beginnings of vegetative growth of the spores and high agitation speed during the exponential growth phase of the microorganism favor the formation of small, compact, and smooth pellets (125). A relationship between culture morphology and citric acid production was observed during fermentation. In general, the cultures of standard pellets yielded higher levels of citric acid than did the cultures consisting of filaments (125). However, Paul et al. (126) reported that citric acid titer was higher when the morphology was filamentous or pelleted culture. Moreover, these authors reported that the citric acid titer obtained from a culture with small pellets was higher than that obtained from a culture containing large pellets with extensive filamentous annular regions.

Papagianni et al. (127–130) studied the morphology of *A. niger* in batch and fed batch culture using a stirred tank fermentor or a tubular loop bioreactor. They reported that the most important factors influencing the morphology of *A. niger* in submerged fermentation are size of inoculum, type and concentration of carbon source, nitrogen and phosphate limitation, trace metals, pH, dissolved oxygen tension, aeration, and agitation. The following morphological parameters were measured using an image analysis method: mean perimeter of clumps (P1), mean perimeter of the core of clumps (P2), mean length of filaments (L), and mean diameter of filaments (d). In batch culture, during the first phase of fermentation, the length of filaments decreased with a decrease of initial glucose concentration. The mean length and specific growth rate was found to increase for the first 2 days of fermentation with decreasing initial glucose concentrations, indicating a link between specific growth rate and branching frequency. Also, the perimeter of the core of clumps changed with initial glucose concentration. The shape of clumps changed because at the end of the fermentation, the ratio P1/P2 increased as the concentration of glucose decreased. The reduction of clumps at low glucose concentrations could be explained by the observation that the filaments became highly vacuolated and fragmentation occurred, keeping the size of the filaments small (129). In fed batch culture, the main observations on morphology of the microorganism were similar with those of batch culture. In this fermentation system, the mean length of filaments decreased with decreasing glucose levels. Comparing the mean values of P1 and L from batch and fed batch culture, it was found that mean values were smaller in fed batch culture than in batch culture. Finally, Papagianni et al. (129) concluded that the morphology of *A. niger*, as determined by clump perimeter and hyphal length, was found to be related to specific growth rate and thus was indirectly linked to sugar concentration. The large clumps with long filaments increased viscosity of the broth to the point at which oxygen became a limiting nutrient. A small increase in phosphate concentration in the medium caused a drastic change in morphology. The clumps lost their compact structure and the length of filaments increased during fermentation. However, at lower phosphate concentrations, after early growth phase, the length of the filaments remained the same until the end of fermentation.

Among trace metals, only the effect of manganese concentration has been shown to influence the morphology of *A. niger.* Paul et al. (126) reported that manganese deficiency in the medium influenced the development of pellet morphologies of *A. niger* in submerged culture. Addition of manganese to the medium caused an undesirable change in the morphology from a pellet like form to a filamentous form (126). Other morphological changes such as prevention of clumping, absence of swollen cells, and reduced diameter of filaments took place during incubation. In spore inoculated fermentations a high initial spore concentration usually tended to produce dispersed growth while a low one favored pellet formation. However, evidence has been presented that pellets can arise even at high spore concentrations due to spore agglomeration during the early stages of germination (126). The perimeter of the clumps, the core perimeter of the clumps, and the length of filaments increase by 90%, 15%, and 50%, respectively, with increases of pH from 2.1 to 4.5. No unidirectional response takes place for the diameter of filaments. In addition to the small size of clumps and small length of filaments at low pH values, an unusually high number of swollen cells and tips occurs in the mycelium (126,127).

Paul, et al. (126) studied the effect of agitation speed on the morphology of *A. niger* and found that with low agitation (300 rpm) the pellets of an initial mean equivalent diameter of 1.4 mm grew larger with fermentation time, until the agitation was increased, at 96 h, to a very high level of 800 rpm. The pellets then fragmented rapidly from 3.2 to 1.6 mm in diameter. In that case, both citric acid yield (0.49 g/g glucose) and cell growth were very low. The glucose uptake rate, oxygen uptake rate, and carbon dioxide production rate were also low compared to nonpelleted fermentation, although all showed a sharp rise after fragmentation of the pellets. The high agitation speed of the second pelleted fermentation (500–800 rpm) retarded continued pellet size increases. The mean equivalent diameter at inoculation was 1.4 mm and rose to only 1.5 mm toward the end of fermentation. Both citric acid yield and cell growth were higher than those obtained from fermentation with large pellets but were still lower than in the dispersed morphology case.

Papagianni, et al. (127,128) studied the relationship between *A. niger* morphology and citric acid production in two reactor systems with different configurations, a tubular loop bioreactor and a stirred tank fermentor. They found that in both reactors citric acid production and morphology were dependent on agitation speed. The length of the filaments was shown to be the most important parameter related to citric acid production in both bioreactors. However, for the same concentration of citric acid produced, the morphology of the microorganism grown in the tubular loop reactor differed significantly from that grown in the stirred tank fermentor. The perimeter of the clumps decreased as circulation frequency through the pump increased in the loop bioreactor and as agitation speed increased in the stirred tank fermentor. In the loop bioreactor, citric acid production appeared to increase sharply with hyphal diameter, while in the stirred tank reactor this relationship was not as clear. Intensive agitation conditions transformed the filaments from long, thin, and almost unbranched to short and thick, with many branches. At high agitation intensity, fragmentation of the mycelia was carried out in both reactors and resulted in the presence of free filaments in the fermentation broth. These free filaments formed new and smaller clumps. This cycle of fragmentation and regrowth kept the average size of clumps small (127).

Roukas (65) studied the effect of agitation speed on *A. niger* morphology and found that at agitator speeds ≥300 rpm, pellets were formed within 24 to 48 h of inoculation, which crushed later into pieces of mycelia in which the hyphae formed a homogeneous suspension dispersed through the fermentation broth. The biomass concentration increased by 25% with an increase in speed of agitation (300–600 rpm). This was due to the crushing of the pellets into small pieces of mycelium, resulting in the surface of the mycelium having much better contact with nutrients and oxygen. In another work, Papagianni, et al. (130) observed that under intensive agitation conditions and during early fermentation stages, the characteristics observed were increased specific growth rates and hyphal branching, along with low vacuolation levels. These were followed by fragmentation of the highly vacuolated parts of filaments and regrowth at later stages of the fermentation, a process that did not increase biosynthetic activity. In fed batch culture, increased vacuolation (162.5 μm) and low specific production rates were observed at low glucose levels (17.0 g/L). The results indicate that vacuolation weakened the hyphae, and low glucose levels created conditions that favored fragmentation and made the mycelium more susceptible to it when exposed to increased agitation. Generally, Papagianni, et al. (128) reported a close link between morphology and productivity, with both dependent on "relative mixing time," whether in a stirred tank or a tubular loop bioreactor.

Other workers (131) studied the effect of cAMP on the morphology of *A. niger* and the production of citric acid. They found that high levels of intracellular and extracellular cAMP in the later stages of fermentation were associated with those treatments that yielded pellets of intermediate size and the highest citric acid concentration. It was suggested that cAMP stimulates conidial aggregation under certain conditions in *A. niger,* and that such conidial aggregation may influence pellet formation. It is perhaps not surprising that intermediate sized pellets have a higher cAMP content than larger pellets, because the center of large pellets is hollow or contains physiologically inactive mycelium. It is possible that the intermediate pellet size is optimal for a range of physiological activities, not just citric acid production, and that the high rate of citric acid production obtained is a result of an optimum metabolic state in the mycelium which exhibits this morphological form (131). The morphology of *A. niger* influences the activity of enzymes that are responsible for biosynthesis of citric acid. Thus, glyoxylate and several intermediates of the TCA cycle also promoted conidiation in the presence of ammonium. The activities of two glutamate dehydrogenases (NAD-GDH and NADP-specific GDH) and of aspartate and alanine aminotransferases varied during growth as a function of the stage of the life cycle and of the growth medium (132). It was also found that the activities of key enzymes of EMP and PP pathways were present and operative at all stages of conidiophore development, but that the relative activities of the pathways differed. The highest activity of the EMP pathway occurred during the initial period of vegetative growth before any signs of morphological change. High EMP activity and high 6-C:1-C ratios were obtained in a medium that did not support conidiophore initiation but allowed vegetative growth to occur. In contrast, high activity of the PP pathway and very low 6-C:1-C ratios accompanied each stage of conidiophore development (133).

During citric acid production, the fermentation broth consists of liquid medium, biomass, citric acid, and any product produced by the microorganism. Thus, the rheology of fermentation broth is affected by the composition of the original medium, the morphology of biomass, the concentration of microbial products, the mixing qualities (impeller speed and working volume), and the dissolved oxygen concentration (134,135). The morphology of *A. niger* in submerged culture varies between the pelleted and the filamentous form, depending on strain and culture conditions (130,136–138). Fungal morphology affects the rheological properties of the fermentation broth and has a strong effect on bioreactor transport phenomena relating to problems of mass transfer, heat transport, and energy consumption. The rheological behavior of fermentation broth is also of great importance for the design, scale up, and operation of reactors (134). In the filamentous form, the hyphae are dispersed throughout the medium, thus rheological behavior is highly nonnewtonian and extremely viscous, so that agitation and aeration become difficult. The pellet growth form consists of branched hyphae intertwined to a stable aggregate, and this growth form causes less viscous fluids than the filamentous form (136,137). The main disadvantage of the pelleted form, however, is that the center of the pellet is subject to autolysis, due to oxygen limitation, and does not participate in product synthesis (130).

Among the factors affecting the morphology of *A. niger,* agitation has been studied extensively and shown to account for significant changes in morphology and productivity of citric acid. Roukas (65) studied the effect of impeller speed on shear dependent viscosity and found that shear dependent viscosity decreased by 25% with increase in speed of agitation from 300 to 600 rpm. This may be explained by the fact that filamentous pellets were crushed into smaller pieces by increased agitation speeds (600 rpm) as a result of shear stress between the fermentation broth and the impeller. In general, the reduction in viscosity with increasing stirred speed indicated that culture morphology is the major determinant of rheological properties of the fermentation broth (65).

Berovic, et al. (138) studied the rheological properties of fermentation broth during citric acid production by *A. niger* in a stirred tank fermentor. They observed that during the first period of fermentation (36 h after inoculation), the fermentation broth was characterized by nearly newtonian behavior, showing a high power law index (n = 1.0–0.95), a low consistency index (K = 5–10 mPas^n), and a high volumetric oxygen transfer coefficient (K_{la} = 156 h^{-1}). During the second fermentation period the change in *A. niger* morphology to spherical pellets and the increase in biomass (18.5 g/L) changed the rheological behavior of the fermentation broth to a pseudoplastic one, reducing the power law index (n = 0.45) and the volumetric oxygen transfer coefficient (K_{la} = 28 h^{-1}) and increasing the consistency index (K = 130 mPas^n). The power law index and the consistency index were evaluated from a log-log plot of the equation $n_a = K\gamma^{n-1}$ where n_a is the apparent viscosity and γ is the shear rate. The equation can be written as $\log n_a = \log K + (n-1)\log \gamma$. In the curve formed by plotting the data $\log n_a - \log \gamma$, the intercept is $\log K$ and the slope is equal to $n-1$ (138). Generally, it was observed that a submerged citric acid fermentation broth shows nonnewtonian rheological behavior. The most significant changes in rheological behavior occurred during the period of maximal citric acid production at the time of exponential increase in biomass. It seems that the pellet growth form represents resistance to shear stress in the system, causing pseudoplastic rheology and significantly reducing the volumetric oxygen transfer coefficient (138).

15.6 FERMENTATION PROCESSES FOR CITRIC ACID PRODUCTION BY FREE AND IMMOBILIZED MICROBIAL CELLS

15.6.1 Surface Fermentation

Surface fermentation was the first fermentation system used for the production of citric acid on an industrial scale. Nowadays fermentation in deep tanks, the so called submerged process, is also employed. Only 20% of world production of citric acid is through use of the surface fermentation process using molasses as raw material. In this process, *A. niger* forms a mycelium layer on the liquid surface of the aluminum or stainless steel trays. These trays are stacked in fermentation rooms supplied with filtered air which serves both to supply oxygen and to control the temperature of fermentation (5,7). The air supply for the chambers is sterilized by passage through a 2 in. thick cotton filter impregnated with salicylic acid, then passed through a water spray and heaters to bring it to 40% humidity at 30°C. The air supply is 0.25 vvm (4). The trays are 2 × 2.5 m in size and 12 cm deep. They are sterilized and filled to a depth of 8 cm with the medium, inoculated with spores of *A. niger*, and incubated 9–12 days at 28–30°C. After the maximum concentration of citric acid is reached, the mycelium is separated from the fermentation broth by filtration. The biomass is washed with water to remove citric acid and the washings are added to the main liquor. The citric acid contained in the solution is precipitated as calcium citrate. The fermentation chambers are sterilized by washing with 1% NaOH, then with water, then with 6% formaldehyde. Finally, sulfur dioxide is blown into the chambers with the airstream (4).

In recent years, several methods for citric acid production have been proposed, and various continuous or multistage processes have been patented. Despite various advances in process development, surface and submerged batch culture techniques are still being used. However, in the batch technique, a significant amount of productive biomass is discarded each time. Cell recycling techniques have advantages over batch techniques because the biomass is conserved, batch downtime is mostly eliminated, and productivity is often increased. These techniques make use of the fact that citric acid production occurs in cells not in the active stage of growth (139).

Roukas and Alichanidis (139) studied the production of citric acid from beet molasses at a varying pH profile using cell recycling of *A. niger.* Fermentation was carried out in Erlenmeyer conical flasks containing 100 mL of molasses solution (initial sugar concentration of 14%). For the first cycle, the pH of the medium was adjusted to 6.5. The flasks were incubated at 30°C, as in surface fermentation. When the maximum concentration of citric acid (65.0 g/L) was achieved, the original medium was aseptically withdrawn from each flask and aseptically replaced with 100 mL of fresh sterile medium, with an initial pH of 6.5, 4.5, or 3.0.This cycle was repeated until the productivity of citric acid was decreased (six cycles). The experiments showed that best results in terms of citric acid concentration (75.0 g/L), yield (0.62 g citric acid/g sugar utilized), productivity (0.3 g/L/h), and specific citric acid productivity (0.0071 g citric acid/g biomass dry weight/h) were obtained with a substrate pH of 3.0. Moreover, the authors concluded that the most significant advantages for citric acid production from beet molasses, at pH 3.0, using cell recycle surface fermentation techniques, are as follows: (1) increases in citric acid production and productivity; (2) energy savings from washing and resterilization of equipment; and (3) possible cost savings from repeated use (up to six times) of the culture.

In another relevant work, Roukas (140) studied the production of citric acid from carob pod extract by cell recycling of *A. niger.* It was found that citric acid concentration (85.5 g/L) and productivity (0.16 g/L/h) remained constant for 42 and 60 days, respectively.

15.6.2 Submerged Fermentation

In submerged fermentation, the microorganism is grown in the fermentation broth. Fermentation is carried out in shake flasks, aerated stainless steel tanks, agitated reactors, sparged towers, a loop bioreactor, a bubble column, a tower fermentor, a disk fermentor, and a rotating disk contactor or trickle flow fermentor (141–149). Among these fermentors, the tower fermentor is the most economical. This fermentor, having a proportion from 1:4 to 1:6 (diameter:height), was found to be especially suitable. Air is supplied from the bottom of the column via a distribution system. Oxygen transfer by rising air bubbles also ensures a thorough mixing in the fermentor. Thus, an agitator, which needs additional energy and makes the fermentor system much more complicated, is not necessary (1). The use of oxygen instead of air to improve oxygen transfer has been applied in unagitated vessels. Foaming of the broth is counteracted by mechanical or chemical foam breakers (7).

Some cell recycle reactors consist of a mechanical device that separates the cells from the liquid, returning the cells to the fermentor. A hollow fiber filter is connected to the fermentor by silicon tubing and a peristaltic pump. Such reactors conserve substrate, because cell mass is conserved. They allow continuous operation and increased productivity by eliminating batch downtime and allow high cell densities. Enzminger and Asenjo (37) studied the production of citric acid from glucose using an aerobic continuous stirred tank bioreactor with cell recycling of *S. lipolytica.* They observed that in batch fermentation, under nitrogen limited conditions, stability of citric acid biosynthesis (75.0 g/L) and excretion was constant over a period of 700 h. In addition, cell concentration (26.0 g/L) and citric acid productivity (1.16 g/L/h) were constant over 200 h of operation under conditions of cell recycling.

Tongwen and Weihua (150) studied the effect of cell configurations on energy consumption and the electroacidification parameters in citric acid production by bipolar membrane electrodialysis. Three basic cell arrangements, type I: A (anion membrane)-C (cation membrane)-BP (bipolar membrane)-A-C, type II: C-BP-C, and type III: BP-A-C-BP, were used. Type I generated acid citrate by acidification with acid sulfate produced from the dissociation of sodium sulfate, and type II produced acid citrate by replacing Na^+ with H^+ generated at bipolar membrane electrodialysis. Type III generated acid citrate by directly splitting sodium citrate. From energy consumption, current efficiency, and concentration of the produced acid citrate, it was suggested that type II seems to be a favorable cell configuration for the production of citric acid. Other workers (141,147) studied the production of citric acid from synthetic medium by *S. lipolytica* and *A. niger* using a stirred tank reactor and a disk fermentor. They found that maximum citric acid concentration was 95.0 and 80.0 g/L, respectively.

15.6.3 Comparison of Surface Fermentation and Submerged Fermentation

Surface fermentation is easy to control and to implement. It needs no aeration or agitation of the fermentation broth, so it needs no instrumentation for aeration and agitation. The separation of citric acid from the mycelium is easy because the microorganism is not dispersing into the medium (151). Only the temperature and humidity of the fermentation chamber need controlling. It can be used easily in small plants as well as in third world countries. With surface fermentation, the fermentation broth is concentrated due to a high evaporation rate during fermentation. Thus, expenses and losses during recovery and purification are low (152). However, surface fermentation has the following disadvantages: Building investment costs are high. Personnel expenses are high in developed industrial countries with extremely high wages. Fermentation time is long and therefore productivity is low (152).

Submerged fermentation is favored over surface fermentation for the following reasons: lower total investment costs; higher yields of citric acid; improved process control; reduced fermentation time; reduced floor space requirements; lower labor costs; simpler operations; and easier maintenance of aseptic conditions on an industrial scale (1,4,152). However, submerged fermentation has some disadvantages compared to surface fermentation: expenses for equipment are higher; consumption of electrical energy is higher; and the process is very sensitive to short interruptions or breakdowns in aeration and vulnerable to infections, which result not only in losses of yield, but also in a total breakdown of respective batches.

Shierholt (152) concludes that surface fermentation is theoretically superior to submerged fermentation. Some essential criteria for industrial management of these two processes are ground requirements as well as the number of and costs for assistants to be employed.

15.6.4 Continuous Culture

In continuous culture the substrate is fed to the fermentor at a constant rate, and the culture is harvested at the same rate, so that the culture volume remains constant. Continuous culture is superior to batch culture in productivity, uniformity of operation, and ease of automation but is more susceptible to contamination (104). Very little published information is available on the production of citric acid by free cells of *A. niger* in continuous culture (59,88,153,154). Kristiansen and Sinclair (88) investigated the production of citric acid from synthetic medium by *A. niger* in single stage continuous culture operated under nitrogen limiting conditions at dilution rates between 0.04 to 0.21 h^{-1}. Citric acid concentration increased rapidly as the dilution rate decreased and appears to be critically dependent on the pH of the fermentation broth and the nitrogen concentration in the feed. The maximum citric acid productivity (0.43 g/L/h) was observed at a dilution rate of ~0.075 h^{-1}. The biomass dry weight, which was independent of pH in this range, did not increase significantly until the dilution rate fell below 0.1 h^{-1}. The optimum pH range for the maximum citric acid productivity was ~3.4. The growth rate was not affected by dissolved oxygen concentration at a dilution rate between 0.07 and 0.3 h^{-1}. The production of citric acid is strongly dependent on culture conditions, especially pH. It appears that pH affects the enzymes that are active in degrading the substrate or the permeability of the cell membrane to substrate and product. The pH does not appear to directly influence the biosynthesis of citric acid (88). Dawson et al. (154) studied the production of citric acid from synthetic medium in continuous culture and found that maximum citric acid yield and production rates were observed at low dilution rates (0.017 h^{-1}) and high dissolved oxygen tension values (90% of saturation). Roukas and Harvey (153) studied the effect of pH on production of citric and gluconic acid from beet molasses using continuous culture and found that at pH values >2.5 gluconic acid was the major product, while at low pH values (1.8–2.0), citric acid was the predominant product. The optimum specific activities of CS, ACH, NAD-ICDH, and NADP-ICDH occurred at pH 4.0 and of glucose oxidase at pH 5.0.

15.6.5 Fed Batch Culture

Fed batch culture is a batch culture fed continuously or sequentially with substrate without the removal of fermentation broth. Compared to conventional batch culture, fed batch culture has several advantages including a very low concentration of residual sugars, higher dissolved oxygen in the medium, decreased fermentation time, higher productivity, and reduced toxic effects of the medium components that occur at high concentrations (155).

Dawson et al. (154) studied the production of citric acid by *A. niger* in batch, chemostat, and fed batch cultures. They found that in fed batch culture, the maintenance of active growth resulted in improved citric acid yield and production rates when compared with batch culture. However, the specific citric acid production rate in fed batch culture was still less than the rate observed in chemostat culture. This was probably due to a higher dissolved oxygen tension value or a higher dilution rate used in the chemostat. To obtain high dissolved oxygen tension values in fed batch culture, where biomass concentration is high, it may be necessary to use pure oxygen rather than air or provide significantly greater power input. Finally, the authors found that in fed batch culture, citric acid productivity was two times higher than that of a conventional batch culture. These results agree with those of Dawson, et al. (89) and Choe and Yoo (90). In addition, Dawson, et al. (89) observed that a fed batch fermentation process operated under nitrogen and phosphate limitation is an efficient and economic process for citric acid production.

15.6.6 Solid-State Fermentation

Solid-state fermentation (SSF) is generally defined by the growth of the microorganism in a low water activity environment on an insoluble material that acts both as a physical support and a source of nutrients (156,157). However, it is not necessary to combine the role of support and substrate, but rather to reproduce the conditions of low water activity and high oxygen transference by using a nutritionally inert material soaked with a nutrient solution (156). In recent years, SSF technology has gained more attention. Solid-state fermentation offers numerous advantages for the production of microbial products. It requires less economic investment, an important aspect when methods for waste treatment are being encouraged (156). It has lower energy requirements and produces less wastewater than submerged fermentation (19,157). In addition, increasing environmental concerns regarding the disposal of solid wastes and their use as substrates for the commercial production of microbial metabolites is becoming an attractive proposition (157). Generally, SSF for citrate production is equal or superior to surface or submerged fermentation. Given the above advantages of this technology, use of this fermentation system is expected to increase (158).

For commercial implementation of SSF, the tray reactor is currently the most popular technology. It is simple but has some disadvantages such as a poor air supply. It also requires a large area and thus poses serious handling problems on a commercial scale (18,157). Moreover, in tray bioreactors the transport of heat and mass is limited by convection, conduction, and diffusion, and the gradients quickly become steep, placing a low limit on the design thickness of the solid substrate bed (18). Thus, other types of reactors are being investigated, including the packed bed reactor (18,157,159,160). A packed bed reactor consist of a glass column of 700 mm in length and 48 cm in diameter (157). Dry air from a cylinder is passed through a rotameter, a cotton wool filter, and then a solution of 20% (w/v) Na0H to remove carbon dioxide. It is then adjusted to 29–30°C and 95–100% RH as it passes through the humidifier. The moist air enters the reactor at the top to prevent lifting of the substrate, which can cause serious blockage in the reactor during the later period of operation. The air exiting the substrates bed is passed through a condenser (0°C) and then a water trap to strip its moisture (157). The substrate is inoculated prior to installation in the sterile reactor by mixing with a spore suspension at a rate of 2.5×10^{-4} spores/g wet weight of substrate. The most important fermentation parameters influencing the production of citric acid by *A. niger* in SSF are inoculum size, moisture, pH, temperature, particle size, aeration rate, and bed height (18,106,107,161).

Lu et al. (157,159) studied the production of citric acid from kumara, a starch containing root crop, by *A. niger* in SSF using a single layer and multilayer packed bed reactor. They found that maximum citric acid productivity to be 0.82 g citrate kg^{-1} wet weight kumara h^{-1}. This amount was twice that observed in flask culture. In this respect, the results in the packed bed reactor were superior to those of flask culture, probably because of improved aeration. In addition, the multilayer packed bed reactor yielded a marked increase in citric acid concentration compared to the single layer reactor, presumably because of improved mass transfer.

Other workers (18,160) studied the production of citric acid from apple pomace by *A. niger* in a packed bed reactor. It was found that the maximum citric acid concentration was124–130 g of citric acid per kilogram of dry apple pomace. This amount of citric acid was obtained at an aeration rate of 0.8 L/min, a bed height of 10 cm, particle sizes of 0.6–2.33 mm, and a moisture content of 78% (w/w) (18).

Pintado et al. (156) studied the production of citric acid from mussel processing effluents by different strains of *A. niger* in SSF using polyurethane foam particles soaked with the culture medium. The results showed that in comparison with submerged fermentation, the strains with requirements of N and P seemed to be disfavored in SSF. The ones with low requirements were favored and showed a bigger tolerance to a surplus of both nutrients. The maximum citric acid concentration (46.5 g/L) was obtained in culture grown in SSF. Another relevant work (162) examined the possibility of monitoring citric acid production by *A. niger* in SSF on sugarcane bagasse. This study was conducted by measuring CO_2 and O_2 concentration in exhaust gases using an automatic sampler connected to a gas chromatograph and a data acquisition system. A relationship between citric acid accumulation and a decrease in CO_2 production was reported. Other workers (163) examined the production of citric acid at high glucose concentration (400 g/L) by *A. niger* in SSF on Amberlite as an inert support. They observed that maximum citric acid concentration (94.5 g/L) and maximum productivity (1.3 g/L/h) were obtained without inhibition related to the presence of heavy metals at high concentrations.

15.6.7 Immobilized Cells

In the past few years, immobilization of microbial cells has been studied with greater interest, and immobilized cells have been used for the production of organic acids, amino acids, antibiotics, enzymes, and other compounds. The production of citric acid by immobilized *A. niger* or yeasts cells compared with free cell systems has several advantages. The cells can be reused for long periods and transferred simply by draining the supernatant and replacing with fresh medium. The fermentation process can be controlled more easily. A less expensive fermentor design is required. Continuous fermentation takes place at a high dilution rate without washout and with higher product yield. Working at dilution rates greater than the growth rate of contaminating microorganisms also helps to overcome infection problems. Immobilized microbial cells are more stable than free cells. In the case of *A. niger,* the immobilization process generally leads to a consistent decrease in medium viscosity, thus enhancing nutrient and oxygen transfers which makes repeated batch and continuous processes possible (164–167). However, immobilized cell systems have some problems such as possible metabolic changes, a need to ensure diffusion of substrates and products, contamination of the medium with free cells, the cost of the immobilization matrix, and, in the case of *A. niger*, a layer of the mycelium forms outside of beads, which prevents the diffusion of substrate and air into the beads (164,165).

Fermentation systems used for the production of citric acid by immobilized microbial cells are the batch and repeated batch culture systems, and the continuous culture system. Fermentation is usually carried out in a bubble column reactor, where high cell concentrations and thus productivity values can be attained. Furthermore, the construction cost of the bioreactor is low, and these reactors can be readily used in continuous flow mode. Consequently, they can be maintained for long periods of time (44). In recent years, citric acid has been produced on a laboratory scale with *A. niger* or yeast cells immobilized in calcium alginate beads (5,41,43,164,166,168–176), k-carrageenan (5,41,168), polyacrylamide gel (40,177), polyurethane foam (41,167,178), cellulose microfibrils (179), and sawdust (44). When the media contains large amounts of nitrogen compounds, the production of citric acid by immobilized *A. niger* cells in calcium alginate gels is carried out in two phases. In the first phase, gel bead immobilized spores or hyphae are precultivated in medium containing 200 mg/L NH_4NO_3. In this case, a peripheral layer of the mycelium is grown on the gel bead. In the second phase, the growth medium is replaced with the production medium, which is supplemented with a limited concentration of NH_4NO_3 (150 mg/L). This limitation of the nitrogen source induces a higher mycelium growth in the center of the beads (169). Eikmeier and Rehm (169) reported that in the case of immobilized *A. niger* cells the development of the mycelium layer near the gel bead surface becomes overloaded at high NH_4NO_3

concentrations (>400.0 mg/L), and free cells occur in the medium. Moreover, these densely packed and actively metabolizing peripheral mycelia build up a critical diffusion resistance which may overcome the normal diffusional limitation of the alginate bead. This might explain the enormous decrease of acid production of the immobilized *A. niger* cells at high nitrogen concentrations (169). In fermentation systems with nitrogen limitations, the immobilized *A. niger* cells produce more citric acid than the free pellets, perhaps due to the fact that the immobilized microorganisms grow from a large amount of spores, whereby the formation of free pellets depends on a limited spore inoculum. Given this fact, immobilized mycelia are characterized by thin or just germinated young hyphae which have a higher metabolic activity than submerged cultivated free cells (169). Using an air lift fermentor, Eikmeier and Rehm (169) found that the outgrowth of free mycelia into the medium could be provided by increasing the particle volume:medium volume ratio, by which means the productivity increased twofold, over that obtained in shaking culture. In general, high nitrogen concentrations (400.0 mg/L) cause a peripheral development of the fungus and outgrowing of the cells, whereas at low concentrations of NH_4NO_3 (200.0 mg/L), *A. niger* forms thin mycelia which grow through the whole interior of the particles, with no outgrowing cells. The immobilization system with low nitrogen concentrations and fungal growth inside the particles seems to be favorable for fermentation of hydrophilic substrates such as molasses and sucrose, without contamination of the media by free cells (168).

Honecker et al. (174) studied the influence of sucrose concentration and phosphate limitation on citric acid production by immobilized cells of *A. niger.* They found that immobilized cells of *A. niger* needed a lower initial sucrose concentration (120.0 g/L) than free cells in order to obtain maximal yields of citric acid (50.0 g/L). However, high sucrose concentrations (240.0 g/L) led to reduced yields (40.0 g/L) and increased polyol formation (glycerol, erythritol, arabitol). These results agree with those of Bisping, et al. (175) who studied the formation of citric acid and polyols by immobilized cells of *A. niger.* Continuous culture with media containing low sugar concentrations prevented the formation of polyols. The change from nitrogen limited to phosphate limited precultivation of immobilized spores significantly increased the productivity of citric acid. Bayraktar and Mehmetoglu (166) found that immobilized cells of *A. niger* require a 2-day preactivation period at a 0.05 g/L NH_4NO_3 concentration. Maximum citric acid concentration was obtained with medium containing 0.01 g/L of NH_4NO_3.

The rate of citric acid production in the nitrogenous medium was 33% higher when oxygen was used instead of air during the production phase. In the nonnitrogenous production medium, citric acid production was not significantly influenced when oxygen was used instead of air. Similar results were observed by Vaija and Linko (170) who studied the production of citric acid by *A. niger* cells entrapped in calcium alginate beads in continuous culture. They observed that oxygen transport was the limiting factor when air was used as the oxygen source. The effectiveness factor (n) was estimated to be between 0.1 and 0.2 with air and 0.6 and 0.95 with pure oxygen. The use of pure oxygen appeared to be the only effective way to reduce oxygen transport limitations. The average volumetric productivity under optimal conditions was 0.3–0.5 g/L/h, while the specific productivity at the beginning of the process was about 0.025 g/g/h, decreasing slowly to about 0.01 g/g/h after 33 days of fermentation.

In another similar work (166), citric acid production was tested in a packed bed bioreactor, and an increase in productivity by a factor of 22 was achieved compared to the batch system. Other workers (173) studied the production of citric acid by immobilized *A. niger* cells in calcium alginate beads using semicontinuous and continuous culture. They reported that citric acid production could better be prolonged by semicontinuous cultivation with medium exchange every 7 or 14 days, respectively. After 32 days, the remaining activity in semicontinuous culture was 1.4-fold higher than in comparable batch experiments. Similar improvements were obtained with a continuous process at a dilution rate of 0.125 h^{-1}, whereby medium efflux kept completely free of detaching mycelia.

Tsay and To (171) examined the production of citric acid from synthetic medium by *A. niger* cells entrapped in calcium alginate gels using batch and repeated batch fermentation. They found that maximum citric acid concentration (77.0 g/L) was obtained at an initial sucrose concentration of 14%, an initial pH of 3.3–4.8, and a temperature of 35°C. The final pH was maintained lower than 2.0; such a low pH value is advantageous in that high yields of citric acid are favored, oxalic acid formation is suppressed, and the danger of contaminations is minimized. Under the same conditions with a batch wise culture, immobilized cells of *A. niger* retained the ability to produce citric acid up to 31 days without marked activity loss when the microorganism was grown on chemically defined medium in surface fermentation. In contrast, free cells maintained a shorter acid producing phase (17 days). This indicated that immobilized cells could retain enzyme activities for long periods, which may be due to a change of the chemical composition of cells (proteins, lipids, DNA, RNA, inorganic substances, saccharides) and the activity of some key enzymes (164). Gupta and Sharma (176) studied the production of citric acid from sugarcane molasses using a combination of submerged calcium alginate immobilized and surface stabilized culture (free cells) of *A. niger* in a continuous flow horizontal bioreactor. They observed that citric acid productivity depended on the dilution rate, with an optimum value of 0.015 h^{-1}. Presaturation of fermentation medium with sterile air, in addition to surface aeration, before feeding to the bioreactor enhanced citric acid productivity. The highest citric acid concentration, productivity, and yield was 110.0 g/L, 1.7 g/L/h, and 91.0%, respectively. The fermentation system was continuously used for 30 days without any apparent loss in citric acid productivity.

Roukas (164) immobilized cells of *A. niger* in calcium alginate beads and studied the production of citric acid from beet molasses in a shake flask and glass bioreactor. It was found that the production of citric acid from molasses by immobilized *A. niger* cells presented the problem of outgrowing of free cells in the medium, which was due to large amounts of nitrogen compounds contained in molasses. To remove the nitrogen compounds, the molasses solution containing 14% initial sugars was treated with a cation and an anion exchange resin. The pH of the solution was adjusted to 3.0, and this substrate (production medium) was used for the production of citric acid. Fermentation was carried out in two phases: immobilized *A. niger* spores were grown on molasses medium, pH 6.5, containing 20 g/L sucrose and 0.6 g/L nitrogen compounds, and the above growth medium was replaced by production medium. In one experiment, 20 g of calcium alginate gel beads entrapping the fungi spores were added in 500 mL Erlenmeyer flasks containing 100 mL of growth medium. The flasks were shaken on a rotary shaker at 250 rpm for 4 days at 30°C. Thus, immobilized growing cells were obtained. The gel beads were then washed with sterile physiological salt solution and 100 mL of production medium was added to the flasks. When a maximum citric acid concentration was achieved, the original production medium was removed, the gel beads were washed twice with sterile distilled water, and fresh production medium was added to the flasks. The above batch fermentation was repeated four times. In a second experiment, calcium alginate gel beads were packed in a glass bioreactor containing the production medium. (The gel beads were incubated in growth medium as described above prior to being added to the bioreactor.) Incubation was carried out under aeration at rates of 0.5, 1.5, and 2.5 vvm at 30°C. The results showed that for the production of citric acid, shaking culture was a better fermentation system than the bioreactor. Maximum citric acid concentration (35.0 g/L) was observed from immobilized *A. niger* cells in shake flasks after 28 days of fermentation. In repeated batch fermentations,

A. niger cells entrapped in calcium alginate gel beads retained their ability to produce citric acid for up to 84 days. In the glass bioreactor, citric acid concentration increased with increased aeration rate, from 0.5 to 2.5 vvm, but was not significantly increased when aeration rate increased beyond 1.5 vvm. This may be explained by the fact that the higher the aeration rate, the better the growth of cells into the beads, and when the aeration rate increased, the diffusion coefficient of substrate into the beads increased. At an aeration rate of 0.5, 1.5, and 2.5 vvm, the maximum concentration of citric acid was 6.0, 22.0, and 27.0 g/L, respectively, after 28 days of incubation.

Horitsu et al. (177) immobilized *A. niger* cells in polyacrylamide gels and studied the effect of aeration volume, oxygen tension, and shape of immobilized cells on citric acid production from sucrose in a glass bioreactor. They observed that in the case of immobilized cells, the rate of citric acid production was ~2.4 times higher than of free cells. This fact indicates that immobilized *A. niger* cells retained their activity to produce citric acid for a long period of time, whereas in the case of free cells their activities are difficult to retain because autolysis occurs under conditions where they are supplied minimum nutrients. Increasing the aeration rate from 0.7 vvm to 2.8 vvm increased the concentration of citric acid 1.4-fold, and then a falloff occurred. The rate of citric acid production using only oxygen was ~1.7 times higher than that of air. This means that immobilized cells require more oxygen tension for citric acid production. The effect of shape of immobilized cells on citric acid production showed that the rate of production with the $4 \times 4 \times 1$ mm slice was ~1.4 times higher than that of the 4 mm^3 cube. In this case, the rate of citric acid production with slices aerated at 1.4 vvm of O_2 was ~2.1 times higher than that of a cube at 1.4 vvm of air. The above results show that the higher the oxygen tension, the more citric acid is produced; and the greater the surface of immobilized particle, the more citric acid is produced because the surface grown mycelia have much better contact with the nutrient and oxygen supply. In a glass bioreactor, the maximum rate of citric acid was 39.1 mg/h per 40 g gels, and the period of half life was found to be 105 days. However, in a tower fermentor the maximum rate of citric acid and the period of half life were 96.6 mg/h per 80 g gels and 96 h, respectively.

Other workers (178) immobilized *A. niger* cells on polyurethane foam and examined the production of citric acid from synthetic medium in a bubble column. They reported that most of the adsorbed cells remained on the support and, as a result, high oxygen tension was maintained during the reactor operation. However, uncontrolled growth of the pellets made continuous reactor operation difficult. The maximum citric acid productivity (0.135 g/L/h) was observed from 15% foam particle concentration. This productivity of immobilized cells was almost the same as that of free cells. The oxygen level dropped to half saturation in 5 days in the immobilized cell culture, in contrast to 2 days in the free cell culture. Jianlong (167) used a rotating biological contactor (RBC) to produce citric acid by immobilized *A. niger* cells on polyurethane foam (PUF). The reactor was a 2.0 L fermentor with 1.0 L working volume. The reactor contained five plastic disks (120 cm^2/disk). Polyurethane foam, 60 cm^2 per side, was attached to both sides of each disk. The disks, with a total surface area of 600 cm^2, were connected to a single stainless steel shaft, which had bearings at both ends and was rotated by a variable speed motor using a reducing gear. The disks were placed in the fermentor containing growth medium. Mycelial fungi showed a strong tendency to grow on PUF to form the biofilm and immobilized on the surface of PUF. During fermentation, air was flowed into and out of the upper portion of air space at an air flow rate of 1.2 vvm. The disks were rotated and the biofilms were exposed alternately to the fermentative medium and the air space. The reactor was operated in a batch mode. During incubation, with slow rotation of the plastic disks (10 rpm), the germinated spores with hyphae became attached to the surface of PUF upon contact. They grew on and around the disks, eventually covering the entire surface of the PUF, and formed a mycelial biofilm. During fermentation, the disks were rotated at 20 rpm, slightly faster than during the microbial growth phase. The slower rotational speed used during the biofilm formation phase was to allow the mycelia to attach to the surface of PUF without much friction. The results showed that maximum citric acid concentration (86.0 g/L) was obtained from an initial average glucose of 120 g/L, which is a higher concentration than that in the stirred tank fermentor. Fermentation time was 4 days, which is much shorter than that in the stirred tank fermentor. The productivity of citric acid obtained with the RBC-PUF system (0.896 g/L/h) was almost three times higher than that obtained with a stirred tank fermentor (0.33 g/L/h). The immobilized biofilm was active for over 8 cycle periods of citric acid production with repetitive use without loss of bioactivity. In general, the most significant advantage of the RBC-PUF system for citric acid production is an increase of volumetric productivity and the reusability of fungal mycelia.

Sankpal et al. (179) immobilized *A. niger* cells by adsorption onto cellulose microfibrils of a porous fabric and used them in batch, fed batch, and continuous production of citric acid. The support used for immobilization of *A. niger* and the mode of fermentation have played important roles in enhancing productivity of citric acid due to the different morphology achieved under conditions of submerged and surface modes of fermentations. In submerged fermentation, the mycelia form a thick mat of microfibrils whereas in the surface mode, a protruding mycelial structure was seen in a cobweblike network of free filaments between the microfibrils of fabric. These microfibrils also show lumens, which may play an important role in medium flow and interfacial interactions during medium movement. Immobilized cells showed a great variability in morphology, which is an important consideration in heat and mass transfer. Interface interaction and metabolite excretion are the major advantages in immobilization techniques. Highly improved interface interaction between mycelia and oxygen, as well as substrate, improves the diffusion of substrate; continuous removal of the metabolite formed also contributes to improved productivity. In the fed batch mode using a recycle reactor, the dissolved oxygen concentration of the system was maintained at 20 mg/L using oxygen enriched air. This improved volumetric productivity to 1.85 g/L/h of citric acid, representing a 15-fold increase over results obtained simultaneously using a shake flask and a 1.6-fold increase over a conventional aerated batch reactor. An overall specific production rate of citric acid of 0.147 and 0.208 g/g/h was achieved using cane juice and sucrose, respectively. In continuous culture, a medium containing 50 g/L of sucrose was allowed to drip through the fabric support at a residence time of 20 h. As a result of interface interaction, a citric acid productivity of 2.08 g/L/h was achieved for 26 days without any significant loss of productivity.

Several workers have investigated the production of citric acid by entrapped yeast cells in different matrixes. Maddox and Kingston (40) immobilized *Y. lipolytica* cells in polyacrylamide gel to produce citric acid from glucose. It was found that the production rate was 50 mg/L/h. In comparison with free cells in a stirred tank fermentor, this production rate of citric acid was low. Nonetheless, the immobilized cells of the yeast retained their activity to produce citric acid for 45 days, and the gel particles were able to be stored at 4°C for 14 days without loss of activity. Kautola et al. (41) studied the production of citric acid by immobilized *Y. lipolytica* cells in calcium alginate beads, k-carrageenan, polyurethane gel, nylon web, and polyurethane foam using repeated batch culture. The highest citric acid productivity (155 mg/L/h) was observed with calcium alginate beads in the first batch. A decrease in bead diameter from 5 to 6 mm to 2 to 3 mm increased volumetric citric acid productivity threefold. In an air lift bioreactor the highest citric acid productivity of 120 mg/L/h with a product concentration of 16.4 g/L obtained from cells entrapped in k-carrageenan gel. Rymowicz et al. (43) investigated the production of citric acid by *Y. lipolytica* cells entrapped in calcium alginate beads in a repeated batch and continuous air lift bioreactor. The highest citric acid concentration (39 g/L) was obtained in medium containing 150 g/l glucose, 0.105 g/l potassium dihydrogen phosphate, 0.84 g/l magnesium sulphate, and 21 mg/l copper sulphate. The production of citric acid was further improved by hardening the alginate carrier beads with glutaraldehyde, and by activation of the immobilized biocatalyst in a nutrient solution. In continuous air lift bioreactor with a height to diameter ratio of 3.0, the highest citric acid productivity (350 mg/L/h) was achieved at a dilution rate of 0.023 h^{-1}.

Tisnadjaja et al. (44) immobilized *C. guilliermondii* cells by adsorption onto sawdust and used in a bubble column reactor for the continuous production of citric acid. The results showed that at a dilution rate of 0.21 h^{-1} in a nitrogen limited medium containing glucose, a productivity of 0.24 g/l/h has been achieved which is twice that observed in a batch fermentor culture using free cells. The corresponding specific production rate of citric acid was 0.024 g/g biomass/h , while the yield was 0.1 g/g glucose utilized. These latter values were lower than those observed using free cells. In conclusion, the technique of immobilization of the yeast cells onto sawdust is cheap, simple to operate, and allows reactor productivities comparable to those reported using other immobilization techniques.

15.7 SUBSTRATES USED FOR THE PRODUCTION OF CITRIC ACID

15.7.1 Chemically Defined Media

Chemically defined media consist of (g/L) sucrose 100–140; NH_4NO_3 1.5–2.5 ; KH_2PO_4 0.25–1.0; $MgSO_4.7H_2O$ 0.25; $CuSO_4.5H_2O$ 40.0 mg/L, and $K_4Fe(CN)_6$ 6.0 mg/L (180,181). The pH of the medium is adjusted to 2.5–5.0. It was found that the maximum citric acid concentration was 110.0 g/L when *A. niger* was grown in synthetic medium in shake flask culture (180). Qazi et al. (181) studied the production of citric acid from synthetic medium in a pilot study (500 L and 2.6 m^3 stirred tank fermentor) and found a maximum concentration of citric acid of 60.0 and 80.0 g/L, respectively. Maddox et al. (182) studied the production of citric acid from glucose, mannose, galactose, xylose, arabinose, or a mixture of the above substances by *A. niger* or *S. lipolytica* and found that the maximum concentration of citric acid ranged between 5.0 and 27.0 g/L. Also, they reported that *S. lipolytica* failed to assimilate xylose, arabinose, and galactose, while *A. niger* did not metabolize galactose. Asenjo, et al. (35) investigated the production of citric acid (6.0 g/L) from cellulose hydrolysate by *C. guilliermondii.* Kirimura et al. (183) studied the production of citric acid from xylan and xylan hydrolysate by *A. niger* Yang no. 2 strain. The microorganism produced 72.4, 52.6, 51.6, and 39.6 g/L of citric acid from xylose, arabinose, xylan hydrolysate, and xylan, respectively. Drysdale and McKay (184) studied the production of citric acid from inulin in surface fermentation and found that citric acid yields can be improved by airflow over the surface of the fermentation, but yields from inulin were 20–30% lower than from sucrose, the traditional commercial substrate.

15.7.2 Molasses

Molasses is a byproduct of the sugar industry and readily available at relatively low cost. It contains water, sugars (sucrose, invert sugar, ~50% w/w), nitrogen compounds (betaine, glutamine, asparagine, leucine, isoleucine, alanine, valine, glycine, and nitrogen as nitrates and nitrites), organic acids, and heavy metals such as iron, zinc, copper, manganese, magnesium, and calcium (185). High heavy metal concentrations in molasses solution cause a critical problem during fermentation as they inhibit the growth of microorganisms, influence substrate pH, and are involved in the inactivation of enzymes associated with product biosynthesis. To overcome this problem, molasses solution containing 14% (w/w) initial sugars is treated with cation exchange resin, sulfuric acid, tricalcium phosphate, potassium ferrocyanide, or EDTA. Among these chemical methods, potassium ferrocyanide treatment is the most common used. The maximum concentration of citric acid produced from molasses by *A. niger* in surface and submerged culture ranged between 60.0 and 110.0 g/L (186–190). Hamissa and Radwan (191) and Qazi et al. (181) studied the production of citric acid from cane molasses by *A. niger* in a pilot study using surface or submerged fermentation and found a maximum concentration of citric acid of 60.8 and 67.0 g/L, respectively. Garg and Sharma (192) examined the production of citric acid from cane molasses by cell recycling of *A. niger* in surface fermentation. It was observed that the rate of citric acid production was doubled, reducing fermentation time in half compared to the normal single cycle batch submerged or surface fermentation process. About 80% of the sugar was converted to citric acid in 5 days of batch fermentation, and three batches were carried out with the same fungal mat without any significant loss of productivity. Jianlong et al. (193) developed a novel method of citric acid production from beet molasses in which an anion exchange resin packed column was connected to a fermentor for separation of citric acid from fermentation broth. The results indicated that, as compared with a conventional batch, the new fermentation technique increased citric acid productivity and sugar conversion from 0.338 g/L/h and 82.2% to 0.543 g/L/h and 94.8%, respectively.

Jianlong (194) improved the production of citric acid from beet molasses by *A. niger* with the addition of phytate to the medium. Phytate is an important plant constituent and can be found in the seeds of cereals and legumes. The effect of phytate addition was found to be dependent on concentration and the stage of fermentation at which phytate was added. When added at the beginning of incubation, the optimal concentration of phytate in the medium for citric acid production was 10.0 g/L, which resulted in a ~3.1-fold increase in citric acid accumulation. Addition of 16.0 g/L phytate to the medium, after 3 days incubation, gave the maximal citric acid concentration, which was ~2.4-fold higher than the control experiment.

Adham (195) reported that natural oils (almond, caster, maize, nigella, olive, peanut, soybean, sunflower) added at concentrations of 2 and 4% (v/v) to beet molasses medium caused a considerable increase in citric acid yield by *A. niger.* Among these natural oils, best citric acid yields were obtained with olive, sunflower, and maize oils at 4% doses. Maximum citric acid concentration (72.8 g/L) was achieved in the medium supplemented with 4% olive oil after 12 days of incubation using surface fermentation.

15.7.3 Cereal Constituents

Starch is the main constituent of cereals. The production of citric acid from different sources of starch (corn, potato, cassava, yam bean) has been reported (196–200). Generally, in the starch medium, the critical factors of citric acid production are the aeration efficiency of the medium and the amylase formation of the strain (196). Mourya and Jauhri (197) tested 17 strains of *A. niger* for their capacity to produce citric acid using starch hydrolysate as a substrate. The most efficient strain, *A. niger* ITCC-605, was selected for further improvement in citric acid content by mutation. In this case, the maximum citric acid concentration was 64.7 g/kg of glucose consumed. Bolach et al. (198) reported that liquefied starch (DE <25) was a better substrate for the production of citric acid than saccharified starch (DE >80). Tan, et al. (199) found that a typical medium containing 2% cassava starch yielded 5.4 g/L biomass dry weight and 74.8 g/L organic acids (citric, malic, gluconic, succinic, fumaric). Sarangbin and Watanapokasin (200) investigated the production of citric acid from yam bean starch by a protease negative mutant strain of *A. niger.* They found a maximum citric acid concentration of 106.0 g/L, whereas the parental strain produced 58.0 g/L from 140.0 g/L of soluble starch in semisolid culture after 5 days of fermentation. Esuoso et al. (201) reported the production of citric acid from imumu *Cyperus esculentus* and maize *Zea mays* by *A. niger.* Imumu, although a weed, has a tuber rich in carbohydrate. It is widely distributed and constitutes one of the world's worst weeds. Proximate analysis reveals a high percentage of starch, with pockets of sugar, protein, and lipids. It was observed that the maximum citric acid concentration (14.3 g/L) was obtained in unagitated and defatted cultures compared to agitated and undefatted cultures, respectively. Hang and Woodams (202,203) studied the production of citric acid from corncobs and corn husks by *A. niger.* Corncobs and corn husks are important byproducts of the sweet corn processing industry that have been either used as animal feed or are returned to the harvested field. Methanol had a significant effect on the production of citric acid from corncobs. Of the four cultures examined, *A. niger* NRRL 2001 was found to produce the highest amount of citric acid (250 g/kg dry corncobs) after 72 h of incubation at 30°C in the presence of 3% methanol. The yield of citric acid was over 50% based on the amount of sugar consumed. Combined treatments of corn husks with dilute NaOH and Rapidase Pomaliq (a commercial apple juice processing enzyme preparation) significantly enhanced the yield of citric acid. Under favorable conditions (pretreated with 0.5 M NaOH, followed by 120 h fermentation at 30°C in the presence of Rapidase Pomaliq and 3% methanol), the yield of citric acid was 260.0 g/kg of dry matter of corn husks. Yuguo et al. (204) investigated citric acid production from dried sweet potato mash with its dregs by *A. niger* in an external loop airlift bioreactor and in a stirred tank fermentor. They found that an average of 10.6 g/L of citric acid was obtained in the external loop airlift bioreactor under an airflow rate of 1.3 vvm and liquid volume of 8.5 L as compared to 9.6 g/L of citric acid in the 10 L stirred tank fermentor at an agitation rate of 200 rpm, airflow rate at 1.0 vvm, and liquid volume of 6.5 L.

Khare et al. (205) reported the production of citric acid from okara (soy residue) in SSF using a cellulolytic *A. terreus* and citric acid producing *A. niger.* Okara or soy residue is generated as a byproduct from soy milk and tofu production. The medium supplemented with ammonium sulfate (0.1% N) when fermented by *A. niger* with simultaneous saccharification using *A. terreus* at pH 8.3 and incubation temperature of 30°C resulted in the production of 5.1 g citric acid/100 g dry solids.

Other workers (158) produced citric acid from kumara (*Ipomoea batatas*), taro (*Colacasia esculenta*), and potato (*Solanum tuberosum*) by *A. niger* in SSF. Raw tubers were peeled, heated at 121°C for 30 min, and blended into a paste. The fermentation was carried out in 250 mL conical flasks containing 40 g of paste, inoculated with 10^8 spores per flask and incubated at 30°C as surface fermentation. The results showed that kumara and taro were excellent substrates for citric acid production, while potato was a poor medium, although it supported profuse fungal growth. The maximum citric acid concentrations were 69.0, 66.0, and 3.0 g/kg wet weight of the substrate from kumara, taro, and potato, respectively, after 6 days of fermentation. The optimum moisture content of the kumara for citrate production was ≥65% (w/w), while metal ions were shown not to be inhibitory to the process.

15.7.4 Fruit Extracts

In past years, a considerable interest has been shown in using agricultural products such as dates, carob pods, and figs for the production of citric acid by *A. niger* (27,86,105–107,206,207). Date syrup is produced from dates by boiling the dates in water, filtering the solution, and concentrating the filtrate under vacuum. The extract of the date syrup contains large amounts of heavy metals which are removed by treatment of syrup with cation exchange resin, sulfuric acid, tricalcium phosphate, potassium ferrocyanide, and EDTA (105,206,207). Al-Obaidi and Berry (207) applied a filtration technique to produce citric acid from date syrup. In this process the medium was exchanged using an external filtration loop in which only a small percentage of the medium passing through the upper chamber of the filter was removed at any one pass. In this way cells were maintained suspended in the growth medium throughout the filtration process. Using this technique, the maximum citric acid concentration and the yield were 102.0 g/L and 72.2%, respectively. Roukas and Kotzekidou (105) studied the production of citric acid from date syrup and found that the maximum citric acid concentration (55.0 g/L), citric acid yield (50.0%), and sugar utilization (73.0%) were obtained in medium treated with 2% tricalcium phosphate. The optimum pH for citric acid production was 6.5. The addition of 4% (v/v) methanol in the date syrup solution increased the concentration of citric acid by 63.5%. Roukas (86,106) investigated the production of citric acid from carob pod extract and carob pod using surface culture and SSF, respectively. After removing the seeds, carob pod residue, or kibble, was chopped into small particles which were mixed with water at a ratio of solid to liquid of 1:4. The mixture was shaken on a rotary shaker incubator at 250 rpm for 2 h at 70°C in order to extract the sugars from the kibble. The extract was centrifuged at 4000 $\times$ g for 15 min, and the supernatant (production medium) was used for the production of citric acid. In the case of SSF, the kibble was chopped into small particles and pulverized in a Waring Blendor at high speed. The pulverized particles were oven dried overnight at 70°C and passed through sieves with a pore size of 0.5, 1.2, 2.5, and 5.0 mm. The medium was moistened with the appropriate amount of distilled water to contain 55, 60, 65, and 70% moisture. The results showed that the maximum citric acid concentration (85.5 g/L) was achieved at an initial sugar concentration of 200 g/L, pH of 6.5, and a temperature of 30°C. In solid-state fermentation, the highest citric acid concentration (176.0 g/kg dry pod) was obtained at a particle size of 0.5 mm, moisture level of 65%, pH of 6.5, and temperature of 30°C. The external addition of 6% (w/w) methanol into the substrate increased the concentration of citric acid by 50%. In other relevant studies, Roukas (27,107) reported the production of citric acid from figs by *A. niger* in surface and SSF. Figs were chopped into small particles which were mixed with distilled water (solid/liquid ratio 1:2), and the mixture was maintained at room temperature for 24 h in order to extract the sugars from the figs. In SSF, the particles of figs were moistened with an appropriate amount of distilled water to contain 60, 65, 70, 75, or 80% moisture. The maximum citric acid concentration was 20.5 g/L in the surface culture, whereas in the SSF the highest citric acid concentration (64.0 g/kg dry figs) was obtained at a moisture level of 75%, initial pH 7.0, temperature 30°C, and fermentation time 15 days.

15.7.5 Hydrocarbons

The production of citric acid from hydrocarbons (n-paraffins, n-alkanes, a-olefins) by the yeast *Y. lipolytica* has been described (8,11–15,29–31,36). Paraffins used for the production of citric acid were usually composed of 12–19 carbon atoms. The medium containing n-paraffins (50.0 g/L) was supplemented with (g/L) the following: NH_4NO_3 5.0; KH_2PO_4 4.0; $MgSO{\bullet}7H_2O$ 2.0; $FeSO_4{\bullet}7H_2O$ 0.4; $MnSO_4{\bullet}7H_2O$ 0.05; and thiamine•HCl 0.0005. The n-paraffins mixture had the following composition (in %): C_{12} 0.10; C_{13} 3.50; C_{14} 21.50; C_{15} 28.70; C_{16} 25.30; C_{17} 18.20; C_{18} 2.30; C_{19} 0.15; isoparaffins 0.25; and aromatics 0.01 (12). The disadvantage of citric acid production from hydrocarbons by yeasts is the formation of isocitric acid during fermentation. Aiba and Matsuoka (13) suggested that the production of the above acids is controlled by the enzyme isocitrate lyase (ICL). Treton, et al. (29) gave several explanations for the nonconsumption of isocitric acid observed in the presence of citric acid: (1) If the same permease mediates the uptake of both citric and isocitric acids, one could expect that phenomena of competitive inhibitions would occur to the prejudice of isocitric acid uptake. Actually, citric acid is always in excess over isocitric acid. (2) There is also the possibility that the carriers might be different, as has already been shown with different microorganisms. Citric acid could act as a repressor or an inhibitor of the transport system specific for isocitric acid. The maximum concentration of citric and isocitric acid produced from n-paraffins by *Y. lipolytica* ranged between 55.0 and 65.0 and 30.0 and 35.0 g/L, respectively (12,29). Terasawa et al. (31) studied the production of citric acid from *l*-tetradecane and *n*-tetradecane and found that the amount of total citrates (citrate and isocitrate) was 40.0 and 48.0 g/L, respectively. Crolla and Kennedy (11) studied the optimization of citric acid production from n-paraffin by *C. lipolytica.* They concluded that the optimal initial concentrations of inoculum, n-paraffin, and ferric nitrate for maximum citric acid production were 10–12%, 10–15%, and 10 mg/L, respectively, at temperatures of 26–30°C. Furukawa and Ogino (15) investigated the production of citric acid from n-paraffins by *S. lipolytica* using a semicontinuous cell recycle system in order to prolong the effective production phase, to minimize product inhibition, and to shorten the lag phase preceding the production phase. They found that the lag phase was shortened and the overall productivity of citric acid was maintained at a higher rate (0.92 g/L/h) than that with the batch culture (0.58 g/L/h), although productivity during the effective phase decreased gradually.

15.7.6 Agricultural Wastes

Recently, research has been directed to finding ways of utilizing agricultural wastes for the production of citric acid. Hang and Woodams (25,208–210) studied the production of citric acid from apple and grape pomace by *A. niger* in SSF. Apple pomace is the residue left from juice extraction and constitutes about 25% of the weight of fresh fruit. Disposal is by trucking to farms for soil application. This procedure is very costly and also poses a serious environmental problem. Grape pomace is rich in carbohydrates, but its nitrogen and phosphorus contents are low. It is dumped despite increasing disposal problems and efforts at byproduct utilization. It was found that maximum citric acid concentrations were 300 g/kg dry apple pomace at moisture content of 65–75% and 90 g/kg grape pomace in the presence of 3% methanol. In another study, Hang et al. (211) described the production of citric acid from kiwifruit peel by *A. niger.* Kiwifruit peel is a byproduct resulting from the manufacture of kiwifruit into nectars or slices and represents nearly 10–16% of the weight of the original fruit, depending on the peeling method used. Its current disposal poses considerable economic and environmental problems. The highest citric acid concentration (100 g/kg dry kiwifruit peel) was obtained in the presence of 2% methanol at 30°C in 4 days. Potvin et al. (38) studied the production of citric acid from kraft black liquor by *C. tropicalis.* Kraft black liquor is the residual liquor from the kraft pulping process. It also contains significant amounts of acetic, formic, lactic, and other carboxylic acids. The maximum concentration of citric acid was 8.0 g/L. The utilization of immobilized cells improved the process parameter and enhanced fermentation capabilities. Kumagai et al. (212) produced citric acid from mandarin orange waste in SSF. It was found that maximum citric acid concentration was obtained in medium containing 6% (w/v) of mandarin orange peel and 11 mL of citrus molasses (14% total sugars content). The average yield of citric acid after 3 days of fermentation was 55–65%. Tran and Mitchell (213) and Usami and Fukutomi (214) studied the production of citric acid from pineapple waste by *A. niger* using SSF. It was found that the highest concentration of citric acid was 160.0 g/kg dry pineapple waste with a moisture content of 70% in the presence of 3% methanol. This represents a yield of 62.4% based on the sugar consumed (213). Kiel et al. (215) examined the production of citric acid in two stage fermentations. However, mycelium obtained from surface cultures in cotton waste medium yielded more citric acid when transferred to sucrose containing media than when directly inoculated to sucrose containing media. It was concluded that cotton waste can be used for the growth of the microorganism and for increasing yields of citric acid fermentation by *A. niger.* Hossain and Brooks (216) studied the production of citric acid from whey permeate by a mutant strain of *A. niger.* In this case, a maximum citric acid concentration of 8.3 g/L was observed. Supplementation of the permeate with lactose to a final concentration of 140 g/L increased the production of citric acid to 14.8 g/L. The addition of methanol at a concentration of 3% (v/v) increased the production of citric acid to 25 g/L. The natural pH of the whey permeate (pH 4.5) was the most suitable for the process, and pH control during fermentation was unnecessary. Hang et al. (21,217) and Roukas and Kotzekidou (26,103) investigated the production of citric acid from brewery wastes by *A. niger.* Spent grain liquor is the liquor resulting from the spent grain recovery process in a brewing plant. It contains total reducing sugars as glucose, 25.7 g/L, and has a pH of 6.0. Lager tank sediment is the sludge obtained after wort fermentation and lagering of beer. It is centrifuged at 4,500 X *g* for 15 min and the supernatant is used as fermentation medium. The liquor contains total reducing sugars as glucose, 31.5 g/L, and has a pH of 4.4. The highest concentrations of citric acid produced from spent grain liquor and lager tank sediment were 19.0 and 11.5 g/L, respectively. Vandenberghe et al. (19) reported the production of citric acid from three different agroindustrial wastes, sugar cane bagasse, coffee husk, and cassava bagasse by *A. niger* in SSF. The results showed that cassava bagasse gave the highest yield of citric acid among the tested substrates. Citric acid production reached a maximum (88.0 g/kg dry matter) when fermentation was carried out with cassava bagasse having an initial moisture of 62% at 26°C for 120 h.

15.8 GENETIC ENGINEERING OF MICROORGANISMS USED FOR THE PRODUCTION OF CITRIC ACID

There is little information about mutagenic techniques applied to the improvement of microorganisms used for the production of citric acid. The main techniques are old and involve the treatment of microorganisms with UV, gamma rays, nitrogen mustard, UV nitrous acid, UV ethyleneamine, UV nitrosoguanidine, and UV N-nitroso-N-methyl urea (3,197,200,218–232). Mutations were carried out in some strains of *A. niger* and *Y. lipolytica.* The irradiation facility used was Cobalt-60 gamma chamber 4000 A. Spore suspensions were exposed to 2.0–3.0 KGy at ambient temperature. In the case of UV irradiation, fresh conidia of *A. niger* are subjected to shortwave UV irradiation for 15 min. The conidial suspension is inoculated in substrate containing 30.0 g /L of glucose and 5.0 g/L of 2-deoxyglucose (MMG5DG medium) and cultivated aerobically by shaking. After 48 h, the liquid is filtered on a sintered glass filter with a pore size of 20–30 μm to remove ungerminated conidia. The washed mycelia are resuspended in distilled water and then spread on MMG5DG agar medium. After cultivation for 2 days, growing colonies are transferred onto Koji-extract agar (KEA) medium. The conidia are inoculated on MMG5DG agar plates and growing colonies are picked up and maintained on KEA slants. Conidia of these strains are inoculated on medium containing 30 g/L glucose and 30 g/L cellobiose (MMC medium) and cultivated for 9 days. Colonies showing good growth on MMG5DG and MMC are selected and evaluated for citric acid production (220). The mutant strains of *A. niger* have some advantages over the parent strains: (1) They produce higher concentrations of citric acid. (2) Fermentation time is shorter. (3) They increase metabolic flux through the pathway leading to citric acid formation. (4) There is a direct increase of the flux through the main pathway (i.e., by overproduction of the enzymes involved in the production of the acid). (5) Finally, they tolerate high concentrations of heavy metals and produce large amounts of citric acid when they are grown in molasses solution (3).

Mutant strains of *A. niger* have some disadvantages as well. During the screening of mutants for high citric acid concentration producers, there is a lack of a precise and quick method by which citrate producing strains can be easily selected. The variability of the strains is decreased after repeated mutagen treatments (237). Mourya and Jauhri (197) studied the production of citric acid from starch hydrolysate by a mutant strain of *A. niger*. It was found that it produced 130 times more citric acid than the parent. For further increases in citric acid production, the medium contained 15% (glucose equivalent) supplemented with NH_4NO_3 0.25%; KH_2PO_4 0.15%; nicotinic acid 0.0001%; and had an initial pH of 2.0. In this case, the mutant strain produced 490 g of citric acid/kg of glucose consumed at 30°C in 8 days of fermentation. Sarangbin and Watanapokasin (200) reported the production of citric acid from yam bean starch using a protease negative mutant strain of *A. niger.* These strains were obtained by UV irradiation of the parent strain. Using a haloselection medium, the authors selected a number of mutants with decreased extracellular protease activity. In addition, when 17 g per plate of rod shaped yam bean was used instead of soluble starch, the protease negative mutant strain produced 490 g/L of citric acid, which is ~1.5 times higher than the parent strain and showed enhanced extracellular glucoamylase production. An investigation was undertaken by Das (218) where parasexual crosses were made between strains of *A. niger.* The mutagenic treatment was carried out with UV irradiation (2537 A) and 1% N-nitroso-N-methyl urea (NMU). A heterozygous diploid was derived from the two auxotrophic mutants and produced segregants, including parental haploids and a recombinant. Kirimura et al. (219) and Sarangbin et al. (220) studied the production of citric acid by 2-deoxyglucose-resistant mutant strains of *A. niger.* It was found that the mutant strain showed faster glucose consumption and growth rate, and fermentation time was shortened from 9 to 6 days. Moreover, in semisolid culture, when the mutant strain was grown in medium containing cellobiose it produced 1.6 times more citric acid than the parent and showed enhanced β-glucosidase production. These results indicate that mutant strains are insensitive to catabolite repression. A study was made of the effects of various lipids on citric acid production by an *A. niger* mutant strain (221). Fatty acids with <15 carbon atoms inhibited growth and no citric acid was produced. Natural oils with a high content of unsaturated fatty acids, when added at 2% (v/v) to suitable media, increased citric acid yield ~20%. This may be explained by the fact that unsaturated lipids act as alternative hydrogen acceptors to oxygen during fermentation and thus improve the production of citric acid (221). Schreferl, et al. (222) reported that mutant strains of *A. niger* with reduced citrate control of carbohydrate catabolism (cic mutants) grow faster than the parent strain on media containing 5% (w/v) citrate. The mutant tolerated a higher intracellular citrate concentration (114.0 g/L) than the parent strain (85.0 g/L). Also, one mutant (cic-7/3) contained phosphofructokinase activity significantly less sensitive toward citrate than the enzyme from the parent strain. When this mutant was grown under citrate accumulating conditions, acidogenesis was far less sensitive to inhibition by Mn^{2+} than in the parent strain. Some cic mutants showed altered citrate inhibition of NADP-specific isocitrate dehydrogenase. Ghosh and Banik (224) studied the effect of chemical nutrients on aconitase activity during citric acid fermentation by a mutant strain of *A. niger.* The results showed that Fe^{2+} stimulated aconitase activity while Co^{2+} (up to 5.0 mg/L) and Ni^{2+} (10 mg/L) strongly depressed enzyme activity. Boric acid (0.5 mg/L) had a very good stimulatory response toward citric acid production. Steinbock, et al. (226)

investigated the properties of a single hexokinase from citric acid accumulating *A. niger.* They isolated mutant strains of *A. niger* by UV irradiation, which tolerated 5.0 g/L of 2-deoxyglucose in the presence of an equal concentration of fructose. These strains showed decreased growth rate and activity of hexokinase during growth on glucose. It was concluded that the synthesis of very high hexokinase activities may counteract citrate inhibition, thereby guaranteeing a high glycolytic flux during citric acid accumulation.

Islam, et al. (227) produced citric acid from cane molasses by gamma ray induced mutants of *A. niger.* Molasses contains many organic and inorganic metallic ions that inhibit the growth of microorganism and citric acid production. These are usually removed by chemical treatment before fermentation. It was found that two mutant strains of *A. niger* showed a good ability to produce citric acid without pretreatment of molasses. Thus, other workers (228–230) studied the production of citric acid from different carbohydrate media by gamma ray induced mutants of *A. niger* and by the parasexual cycle. The results indicated that improved reproducibility of a citric acid producing *A. niger* strain can be obtained after mutagen treatment or diploidization (228). Citric acid yield was found to be sevenfold higher compared to the parent strain (229). A natural isolate of *A. niger,* and two of its second step mutants, grown on sucrose medium gave maximum citric acid yields of 34.0, 70.0, and 126.0 g/L, respectively. A combination of two sugars, at 50% each, improved citric acid yield for sucrose:glucose, glucose:sorbitol, glucose:xylose, and xylose:sorbitol combinations with the mutant strains. Inclusion of galactose in combinations decreased citric acid yield (230).

Chaudhary et al. (231) and Roy and Das (232) studied the production of citric acid by nitrosoguanidine induced (NTG) mutants of *A. niger.* Lower concentrations of NTG (0.5 g/L) were more effective in producing high citric acid yielding mutants. Of 100 mutants selected for preliminary screening, 12 were found to be high citric acid producers. Mutant strains gave 1.5–2.0 times higher citric acid yield than parent strains. Promper et al. (233), however, found that a mutant strain of *A. niger* accumulated much more citric acid intracellularly; it excreted less citrate than the parental strain. Under growth conditions that led to uncontrolled catabolic flux through glycolysis, a dramatic catabolic overflow occurred in the mutant strain. The parent strain, under these conditions, excreted large amounts of citrate to moderate the intracellular catabolic overflow.

Watanabe, et al. (234) investigated the production of citric acid from cellulose hydrolysate by a 2-deoxyglucose resistant mutant strain of *A. niger.* The hydrolysate prepared in acetate buffer was concentrated to contain 150.0 g/L of reducing sugars and was used as a medium for fermentation. Parent and mutant strains produced 77.9 and 98.7 g/L of citric acid, respectively, in 9 days of incubation. When the hydrolysate was prepared in citrate buffer and used as a medium, parent and mutant strains produced 92.2 and 102.3 g/L of citric acid, respectively, after 3 days of fermentation.

Yeasts, particularly the strains of the genus *Yarrowia,* are used for the production of citric acid from different substrates. In these fermentations, a mixture of citric and isocitric acid is produced. The development of mutant strains that accumulate only citric acid has been attempted. Good et al. (45) examined the production of citric acid from canola oil by a mutant strain of *S. lipolytica.* It was isolated after mutagenesis of the parent strain with NTG and had an improved citric to isocitric acid ratio. The maximum citric and isocitric acid concentration was 152.3 and 38.5 g/L, respectively. Ermakova, et al. (46) and Finogenova, et al. (47) studied the properties of *C. lipolytica* mutants with the modified glyoxylate cycle and their ability to produce citric and isocitric acid. They found that the key enzyme citrate synthase had the highest activity in both parent and mutant strains grown on glucose or hexadecane. NAD-dependent isocitrate dehydrogenase had the minimum activity. The mutant strain had actively functioning enzymes for both anaplerotic pathways—pyruvate carboxylase, isocitrate lyase, and malate synthase. The low activities of both isocitrate lyase and pyruvate carboxylase in the mutant strain showed that this strain has an additional pathway for oxalacetic acid synthesis during the assimilation of n-alkane. The mutant with high isocitrate lyase and low aconitate hydratase activities was shown to synthesize citric acid almost exclusively when grown on glucose, glycerol, hexadecane, acetate, or ethanol. However, the mutant with low isocitrate lyase and high aconitate hydratase activities produced primarily isocitric acid in media with n-alkanes. Tani, et al. (48) produced citric acid from methanol by a fluoroacetate resistant mutant of *Candida* sp. Y-1. This mutant strain had lower aconitase activity than the wild strain and produced 4.6 g/L citric acid from methanol after 4 days of incubation. Four commercial strains and two mutants of the yeast *Y. lipolytica* were screened using batch fermentation (49). The mutant strain induced with UV irradiation was found to be the most suitable for citric acid production from glucose hydrol (39.9% glucose and 2.1% other sugars), a byproduct of glucose production from potato starch. In this case, the highest citric acid concentration (100.0 g/L) was obtained after 80 h of fermentation.

Finogenova et al. (50) studied the biosynthesis of citric and isocitric acids by the wild-type and mutant strains of *C. lipolytica.* They found that when the mutant strains cultivated in media containing acetate, ethanol, glycerol, glucose, or hexadecane, supersynthesis of the acids started after complete consumption of the nitrogen source and resulted in a delay of culture growth. In another relevant work (51), it was observed that comparisons of enzyme activities among parent and mutant strains showed that mutant strains possess higher citrate synthase activity and lower aconitate hydratase activity than the parent, implying that the altered activities may favorably affect citric acid production of mutant strains.

Other workers (225,235,236) studied the production of citric acid by a diploid strain of *A. niger.* The parent strain was treated with UV irradiation which produced auxotrophic mutants. These mutants were less productive than the parent. All possible crosses in pairwise combinations were carried out between these auxotrophs, and three heterokaryons. Finally, one heterozygous diploid was isolated from each. These heterokaryons and diploids showed improved productivity when compared with their corresponding parents, except for one diploid. The citric acid concentration of the diploid strain was 1.2–1.5 times higher than the parent strain.

Martinkova et al. (237) reported that the disadvantages of mutagenesis and parasexual hybridization (low frequency of hybrid formation and incompatibility of the strains) could be avoided when using protoplast fusion. Auxotrophic strains of *A. niger* were obtained from citric acid producing strains of the fungus after UV irradiation. Protoplasts were isolated from young hyphae of auxotrophic strains after treatment with snail enzyme and then treated with polyethylene glycol (30%, w/v) in a Ca^{2+} solution (10 mmol/L) at pH 9.0 The frequency of the heterokaryons was 0.67%. Prototrophic heterozygous spores were isolated from a heterokaryon with the frequency of 1.2×10^{-6}. Citric acid production in the best heterozygous strains was about 15% higher than that of the parent strain.

Production of citric acid by *A. niger* depends on a high flux through glycolysis. Arisan-Atac et al. (238) investigated the possibility of flux control by trehalose 6-phosphate synthase A (T6PSA), an inhibitor of hexokinase. In order to study the relevance of T6PSA inhibition of hexokinase on the glycolytic flux and citric acid accumulation, they used a recombinant strain of *A. niger* carrying multiple copies of *ggsA* by cotransformation. The results showed that these strains produced citric acid with similar yields on 1.0 or 2.5% (w/v) sucrose. At 5–14% sucrose, the *ggsA* disrupted strain initiated citric acid accumulation earlier, whereas the multicopy strain showed the reverse effect. This means that while the disruption of the *ggsA* gene did not increase the product yield, it reproducibly decreased the time required to reach maximum citric acid concentration. These results demonstrate for the first time the possibility of improving citric acid fermentation by a recombinant method. Previous attempts to amplify genes encoding glycolytic enzymes have been unsuccessful. This can be explained by the fact that glycolytic flow is subject to tight control by several metabolites that antagonize the effect of gene amplifications (238).

Verdoes, et al. (239) reported the cloning of the *nicB* gene of *A. niger*. They describe an improved gene cloning strategy by complementation of mutant alleles in *A. niger*. The strategy was based on the use of a fungal autonomously replicating vector, pAB4-ARp1. This vector was constructed by the introduction of a sequence involved in autonomous replication of an apical membrane antigen 1 (AMA1) into a *pyrG* integrative vector, pAB4-1. With vector pAB4-ARpl, a 10- to 100-fold increase in transformation frequency was obtained, as compared to pAB4-1. The use of pAB4-ARp1 in gene cloning was demonstrated by the complementation of two linkage group VII-specif *A. niger* mutants. Complementation of a lysF mutant was achieved by cotransformation of pAB4-ARp1 with total genomic *A. niger* DNA (instant banK). A nicB-deficient *A. niger* was complemented by cotransformation with pAB4-ARp1 and an *A. niger* cosmid library. The complementing DNA was re-isolated from a Nic^+ transformant by transforming *E. coli* with total genomic DNA of this transformant. A complementary DNA (cDNA) fragment encoding the mitochondrial alternative oxidase, the enzyme responsible for cyanide insensitive and salicylhydroxamic acid sensitive respiration, from the citric acid producing fungus *A. niger* was cloned and expressed in *E. coli* as a host strain (240). Synthetic primers were designed from the conserved nucleotide sequences of the alternative oxidase genes from higher plants and a yeast. The 210-bp DNA fragment was amplified by PCR with these primers using chromosomal DNA of *A. niger* as a template, and was employed to screen a cDNA library of the culture. One full length cDNA clone of 1.2 kb was obtained and was sequenced to reveal that the clone contained an open reading frame (ORF-AOX1) encoding a polypeptide of 351 amino acids. In the 5'-terminus region of the ORF-AOX1, a mitochondrial targeting motif was found. The whole ORF-AOX1 was ligated to plasmid pkk223-3 to construct the expression vector pkAOX1 and showed that cyanide insensitive and SHAM sensitive respiration, and expression were increased approximately twofold by the addition of isopropyl-beta-D-thiogalactopyranside (IPTG). These results indicated that the ORF-AOX1 encodes an alternative oxidase of *A. niger* (240).

15.9 RECOVERY OF CITRIC ACID

The fermentation broth obtained either from surface or submerged fermentation is filtered to remove mycelia or cells and other suspended impurities. Citric acid is precipitated from the filtrate as calcium citrate by the addition of lime slurry at 95°C for 1 h or 50°C for 20 min (3). The precipitate is washed to remove soluble impurities and treated with sulfuric acid to precipitate calcium sulfate and regenerate the citric acid. The solution is then filtered to remove $CaSO_4$. The liquid is decolorized with charcoal and ion exchangers. The purified solution is concentrated by evaporation and run into low temperature crystallizers. Finally, the crystals are removed by centrifugation. Citric acid is marketed as an anhydrous crystalline chemical, as a monohydrate or as a crystalline sodium salt (241). A general flow sheet for the production and purification of citric acid from molasses by *A. niger* is given in Figure 15.3.

Citric acid may be extracted from the fermentation broth using solvent extraction methods. Solvents employed include 2-butanol, tributyl phosphate, secondary or tertiary amines or a mixture of n-octyl alcohol, synthetic isoparaffinic petroleum hydrocarbons,

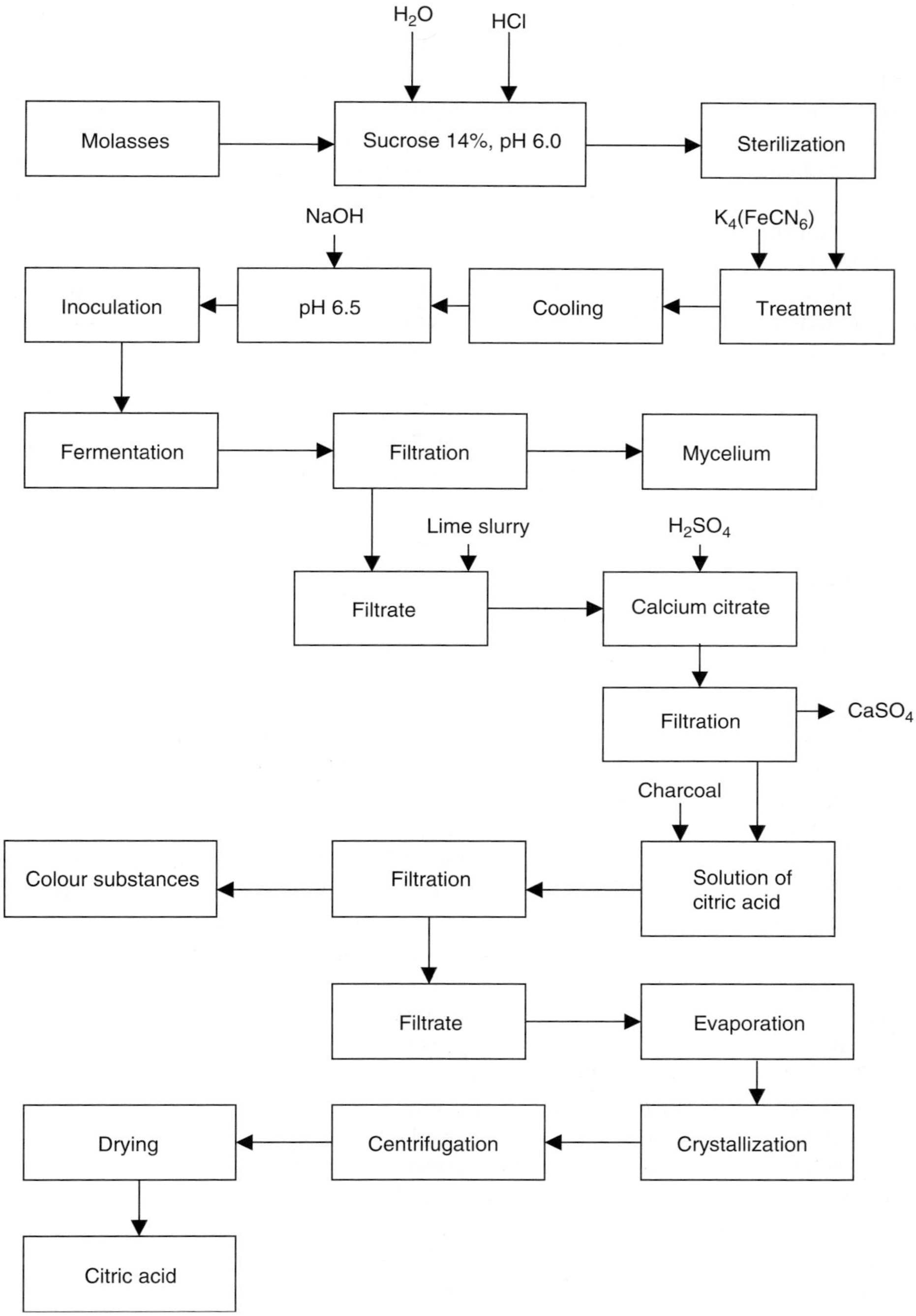

Figure 15.3 Schematic diagram of the production and purification of citric acid from molasses by *A. niger* (From Atkinson, B., F. Mavituna. *Biochemical Engineering and Biotechnology Handbook*. Hong Kong: The Nature Press, 1985, pp 1033–1036).

and tridodecylamine (3,5,7). The citric acid is extracted into the solvent at a low temperature and reextracted into water at a higher temperature. The solution containing the acid is then concentrated and crystallized. Kapoor et al. (3) reported a process in which citric acid is extracted from the fermentation broth by two phase extraction using tridecylamine or triisononylamine and ester, ketone, or ethanol. Citric acid salts are precipitated by extracting the mixed solvent extracts with aqueous solutions of ammonia, alkalihydroxides, carbonates, or bicarbonates. Kilic et al. (16) investigated a method for the extraction of citric acid in the presence of corn oil and Hostarex A327 in oleyl alcohol. The addition of corn oil reduced solvent toxicity. These authors combined the extraction method with the effect of potassium ferrocyanide on citric acid production. They found that citric acid concentration increased with increasing $K_4Fe(CN)_6$ concentration. Using the response surface methodology, they found that 0.02–0.10% (w/v) potassium ferrocyanide, the ratio of organic phase volume to the aqueous phase volume (V_{or}/V_{aq} = 0.5–1.5), 0–20% (v/v) of corn oil concentration, and 10–50% (v/v) of Hostarex A327 content in oleyl alcohol were the critical components. Optimum citric acid concentration was predicted under these optimum extraction conditions: $K_4Fe(CN)_6$ of 0.06% (w/v), V_{or}/V_{aq} of 1.25, corn oil concentration of 13.7% (v/v), and Hostarex A327 content in oleyl alcohol of 33% (v/v). Citric acid production by extractive fermentation was increased ~40% with respect to the control run.

15.10 CONCLUSIONS

Citric acid, a tricarboxylic acid, is used in the food and beverage industries, in pharmaceuticals, and in other industrial applications. It is produced by fungi, yeasts, and bacteria. *Aspergillus niger* strains dominate other microorganisms producing citric acid on an industrial scale. They can be handled easily, are resistant to fermentation conditions, grow in various raw materials and agricultural wastes, and give high and consistent yields of citric acid by utilization of inexpensive substrates, thereby making the fermentation process economical. Citric acid production takes place during exponential and stationary growth phases. The strains used for citric acid production need major elements such as carbon, nitrogen, phosphorus, and sulfur in addition to various trace elements and stimulants. Also, fermentation parameters such as fermentation time, initial sugar concentration, pH, temperature, agitation, and aeration play a profound role in yield. The fermentation systems used for citric acid production are surface fermentation, submerged fermentation, continuous culture, fed batch culture, solid-state fermentation, and immobilized microbial cells. Among these systems only surface and submerged fermentation are used on an industrial scale. The substrates used include chemically defined media, molasses, agricultural wastes from the food and beverage industry, hydrocarbons, and oils. Currently, molasses is the superior medium for the production of citric acid on a pilot plant scale. Understanding of basic biochemical pathways and genetic mechanisms in microorganisms is useful for improving the organisms used for citric acid production. Thus, recent research has been focused on immobilized microbial cells and genetically improved microorganisms. These new methods aim to improve citric acid production by obtaining shorter fermentation time, by using raw materials without any pretreatment, and by finding strains that will produce high citric acid yields. The discovery of yeasts that produce higher amounts of citric acid than fungi in shorter fermentation time has been accompanied by the disadvantage of the production of isocitric acid. Mutant strains of the yeasts producing low amounts of isocitric acid have been evaluated in bench and pilot studies. Generally, while the new methods are promising, they are far from adequate, and further studies are required before a commercial process can be envisaged.

REFERENCES

1. Sodeck, G., J. Modl, J. Kominek, W. Salzburn. Production of citric acid according to the submerged fermentation process. *Process. Biochem.* October/November: 9–11, 1981.
2. Smith, J.E., A. Nowakowska-Waszczuk, J.G. Anderdson. Organic acid production by mycelial fungi. *Ind. Aspects Biochem.* 297–317, 1974.
3. Kapoor, K.K., K. Chaudhary, P. Tauro. Citric acid. In: *Prescott and Dunn's Industrial Microbiology*, Reed, G., ed.. UK: MacMillan Publishers Ltd, 1983, pp 709–747.
4. Johnson, M.V. The citric acid fermentation. In: *Industrial Microbiology*, S Prescott, S., C.G. Dunn, eds., New York: McGraw-Hill, 1949, pp 420–445.
5. Milsom, P.E. Organic acids by fermentation, especially citric acid. In: *Food Biotechnology*: 1, King, R.D., P.S.J. Cheetham, eds., London: Elsevier Applied Science, 1987, pp 273–308.
6. Perlman, D., C.J. Sih. Fungal sythesis of citric, fumaric and itoconic acids. In: *Progress in Industrial Microbiology*,Vol. 2. New York: Interscience Publishers, 1960, pp 169–194.
7. Meers, J.L., P.E. Milsom. Organic acids and amino acids. In: *Basic Biotechnology*, Bu'lock, J., B. Kristiansen, eds., London: Academic Press, 1987, pp 359–383.
8. Wojtatowicz, M., G.L. Marchin, L.E. Erickson. Attempts to improve strain A-101 of *Yarrowia lipolytica* for citric acid production from n-paraffins. *Process. Biochem.* 28:453–460, 1993.
9. Arzumanov, T.E., I.A. Sidorov, N.V. Shishkanova, T.V. Finogenova. Mathematical modeling of citric acid production by repeated batch culture. *Enzyme Microb. Technol.* 26:826–833, 2000.
10. Antonucci, S., M. Bravi, R. Bubicco, A. Di Michele, N. Verdone. Selectivity in citric acid production by *Yarrowia lipolytica*. *Enzyme Microb. Technol.* 28:189–195, 2001.
11. Crolla, A., K.J. Kennedy. Optimization of citric acid production from *Candida lipolytica Y-1095* using n-paraffin. *J. Biotechnol.* 89:27–40, 2001.
12. Marchal, R., O. Chaude, M. Metche. Production of citric acid from n-paraffins by *Saccharomyces lipolytica*: kinetics and balance of fermentation. *Eur. J. Appl. Microbiol.* 4:111–123, 1977.
13. Aiba, S., M Matsuoka. Citrate production from n-alkane by *Candida lipolytica* in reference to carbon fluxes *in vivo*. *Eur. J. Appl. Microbiol.* 5:247–261, 1978.
14. Moresi, M., D. Cimarelli, G. Gasparrini, G. Liuzzo, R. Marinelli. Kinetics of citric acid fermentation from n-paraffins by yeasts. *J. Chem. Tech. Biotechnol.* 30:266–277, 1980.
15. Furukawa, T., T. Ogino. Citric acid production from n-paraffins by use of a semicontinuous cell-recycle system. *J. Ferment. Technol.* 60:377–380, 1982.
16. Kilic, M., E. Bayraktar, S. Ates, U. Mehmetoglu. Investigation of extractive citric acid fermentation using response-surface methodology. *Process. Biochem.* 37:759–767, 2002.
17. Jianlong, W. Enhancement of citric acid production by *Aspergillus niger* using n-dodecane as an oxygen vector. *Process. Biochem.* 35:1079–1083, 2000.
18. Shojaosadati, S.A., V. Babaeipour. Citric acid production from apple pomace in multi-layer packed bed solid-state bioreactor. *Process. Biochem.* 37:909–914, 2002.
19. Vandenberghe, L.P.S., C.R. Soccol, A. Pandey, J.M. Lebault. Solid-state fermentation for the sythesis of citric acid by *Aspergillus niger*. *Biores. Tech.* 74:175–178, 2000.
20. L.B. Lockwood. Production of organic acids. In: Microbial Technology, Vol. 1, Peppler, H.J., D Perlman, eds., New York: Academic Press, 1979, pp 356–367.
21. Hang, Y.D., D.F. Splittstoesser, E.E. Woodams, R.M. Sherman. Citric acid fermentation of brewery waste. *J. Food. Sci.* 42:383–384, 1977.
22. Chen, H.C. Response-surface methodology for optimizing citric acid fermentation by *Aspergillus foetidus*. *Process. Biochem.* 29:399–405, 1994.
23. Majolli, M.V., S.N. Aguirre. Effect of trace metals on cell morphology, enzyme activation, and production of citric acid in a strain of *Aspergillus wentii*. *Revista Argentina de Microbiologica* 31:65–71, 1999.

24. Franz, A., W. Burgstaller, B. Muller, F. Schinner. Influence of medium components and metabolic inhibitors on citric acid production by *Penicillium simplicissimum*. *J. General Microbiol.* 139:2101–2107, 1993.
25. Hang, Y.D., E.E. Woodams. Grape pomace: a novel substrate for microbial production of citric acid. *Biotechnol. Letters*. 7:253–254, 1985.
26. Roukas, T., P. Kotzekidou. Production of citric acid from brewery wastes by surface fermentation using *Aspergillus niger. J. Food Sci.* 51:225–228, 1986.
27. Roukas, T., M. Liakopoulou-Kyriakides. Optimization study for the production of citric and gluconic acid from fig water extract by *Aspergillus niger* in surface fermentation. *Food Biotechnol.* 16:17–28, 2002.
28. Marchal, R., J.P. Vandecasteele, M. Metche. Regulation of the central metabolism in relation to citric acid production in *Saccharomyces lipolytica*. *Arch. Microbiol.* 113:99–104, 1977.
29. Treton, B., M.T. Le Dall, H Heslot. Excretion of citric and isocitric acids by the yeast *Saccharomyces lipolytica*. *Eur. J. Appl. Microbiol. Biotechnol.* 6:67–77, 1978.
30. Behrens, U., E. Weissbrodt, W. Lehmann. Kinetics of citric acid production in *Candida lipolytica*. *Zeitschrift fur Allgemeine Mikrobiologie* 18:549–558, 1978.
31. Terasawa, M., T. Nagata, J. Takahashi. Utilization of α-olefins by yeasts, part 1: production of citric acid from 1-tetradecene and n-tetradecene. *Agric. Chem. Soc. Jap.* 53:227–232, 1979.
32. Omar, S.H., H.J. Rehm. Physiology and metabolism of two alkane oxidizing and citric acid producing strains of *Candida parapsilosis*, I: activities of the tricarboxylic and glyoxylate cycle enzymes. *Eur. J. Appl. Microbiol. Biotechnol.* 11:35–41, 1980.
33. Omar, S.H., H.J. Rehm. Physiology and metabolism of two alkane oxidizing and citric acid producing strains of *Candida parapsilosis*, II: effect of carbon sources on morphology, lipid content, and fatty acid composition, on enzymes of the citric acid and glyoxylate cycles and catalase. *Eur. J. Appl. Microbiol. Biotechnol.* 11:42–49, 1980.
34. Hamissa, F.A., A.Z. Abou-Zeid, A.A. Radwan. Influence of micronutrients on citric acid production by *Candida lipolytica* (Y 1095). *Zentralbl. Bakteriol. Naturwiss.* 135:332–338, 1980.
35. Asenjo, J.A., J. Szuhay, D. Chiu. Growth and citric acid production by *Candida guilliermondii* using a cellulose substrate. *Biotechnol. Bioeng. Symp.* 12:111–120, 1982.
36. Furukawa, T., T. Ogino. Citric acid production from n-paraffins by use of a semicontinuous cell-recycle system. *J. Ferment. Technol.* 60:377–380, 1982.
37. Enzminger, J.D., J.A. Asenjo. Use of cell recycle in the aerobic fermentative production of citric acid by yeast. *Biotechnol. Lett.* 8:7–12, 1986.
38. Potvin, J., M. Desrochers, Y. Arcand. Fermentation of kraft black liquor for the production of citric acid by *Candida tropicalis*. *Apl. Microbiol. Biotechnol.* 28:350–355, 1988.
39. Rane, K.D., K.A. Sims. Oxygen uptake and citric acid production by *Candida lipolytica* Y 1095. *Biotechnol. Bioeng.* 43:131–137, 1994.
40. Maddox, I.S., P.J. Kingston. Use of immobilized cells of the yeast, *Saccharomyces lipolytica*, for the production of citric acid. *Biotechnol. Lett.* 5:795–798, 1983.
41. Kautola, H., W. Rymowicz, Y.Y. Linko, P. Linko. Production of citric acid with immobilized *Yarrowia lipolytica*. *Appl. Microbiol. Biotechnol.* 35:447–449, 1991.
42. Linko, P., Y.Y. Linko. Industrial applications of immobilized cells. *CRC Crit. Rev. Biotechnol.* 1:289–338, 1984.
43. Rymowicz, W., H. Kautola, M. Wojtatowicz, Y.Y. Linko, P. Linko. Studies on citric acid production with immobilized *Yarrowia lipolytica* in repeated batch and continuous air-lift bioreactors. *Appl. Microbiol. Biotechnol.* 39:1–4, 1993.
44. Tisnadjaja, D., N.A. Gutierrez, I.S. Maddox. Citric acid production in a bubble column reactor using cells of the yeast *Candida guilliermondii* immobilized by adsorption onto sawdust. *Enzym. Microb. Technol.* 19:343–347, 1996.
45. Good, D.W., R. Droniuk, G.R. Lawford, J.E. Fein. Isolation and characterization of a *Saccharomycopsis lipolytica* mutant showing increased production of citric acid from canola oil. *Can. J. Microbiol.* 31:436–440, 1985.
46. Ermakova, I.T., N.V. Shishkanova, O.F. Melnikova, T.V. Finogenova. Properties of *Candida lipolytica* mutants with the modified glyoxylate cycle and their ability to produce citric

and isocitric acid, I: physiological, biochemical, and cytological characteristics of mutants grown on glucose or hexadecane. *Appl. Microbiol. Biotechnol.* 23:372–377, 1986.
47. Finogenova, T.V., N.V. Shishkanova, I.T. Ermakova, I.A. Katavea. Properties of *Candida lipolytica* mutants with the modified glyoxylate cycle and their ability to produce citric and isocitric acid, II: synthesis of citric and isocitric acid by *C. lipolytica* mutants and peculiarities of their enzyme systems. *Appl. Microbiol. Biotechnol.* 23:378–383, 1986.
48. Tani, Y., Y. Sakai, S.G. Chou. Production of citric acid from methanol by a fluoroacetate-resistant mutant of *Candida* sp. Y-1. *Appl. Microbiol. Biotechnol.* 34:5–9, 1990.
49. Wojtatowicz, M., W. Rymowicz, H. Kautola. Comparison of different strains of the yeast *Yarrowia lipolytica* for citric acid production from glucose hydrol. *Appl. Biochem. Biotechnol.* 31:165–174, 1991.
50. Finogenova, T.V., A.V. Grinchak, V.I. Illarionova, N.V. Shishkanova. Biosynthesis of citric and isocitric acids by the wild type and mutant strains of *Candida lipolytica* in media containing different carbon sources. *Prikl. Biokhim. Mikrobiol.* 15:811–816, 1979.
51. Furukawa, T., T. Ogino, T. Matsuyoshi. Fermentative production of citric acid from n-paraffins by *Saccharomyces lipolytica*. *J. Ferment. Technol.* 60:281–286, 1982.
52. Hanson, R.S., D.P. Cox. Effect of different nutritional conditions on the sythesis of tricarboxylic acid cycle enzymes. *J. Bacteriol.* 93:1777–1787, 1967.
53. Stern, J.R., B. Shapiro, E.R. Stadtman, S. Ochoa. Enzymatic synthesis of citric acid, III: reversibility and mechanism. *J. Biol. Chem.* 193:703–720, 1951.
54. Woronick, C.L., M.J. Johnson. Carbon dioxide fixation by cell-free extracts of *Aspergillus niger*. *J. Biol. Chem.* 235:9–15, 1960.
55. Ng, W.S., J.E. Smith, J.G. Anderson. Changes in carbon catabolic pathways during synchronous development of conidiosphores of *Aspergillus niger*. *J. Gen. Microb.* 71:495–504, 1972.
56. Verhoff, F.H., J.E. Spradlin. Mass and energy balance analysis of metabolic pathways applied to citric acid production by *Aspergillus niger*. *Biotechnol. Bioeng.* 18:425–432, 1976.
57. Nowakowska-Waszczuk, A., A. Sokolowski. Application of carbon balance to submerged citric acid production by *Aspergillus niger*. *Appl. Microbiol. Biotechnol.* 26:363–364, 1987.
58. Steinbock, F., S. Choojun, I. Held, M. Roehr, C.P. Kubicek. Characterization and regulatory properties of a single hexokinase from the citric acid accumulating fungus *Aspergillus niger*. *Biochim. Biophys. Acta* 1200:215–223, 1994.
59. Ng, A.M.L., J.E. Smith, A.F. McIntosh. Influence of dilution rate on enzyme synthesis in *Aspergillus niger* in continuous culture. *J. Gen. Microb.* 81:425–434, 1974.
60. Jaklitsch, W.M., C.P. Kubricek, M.C. Scrutton. Intracellular location of enzymes involved in citrate production by *Aspergillus niger*. *Can. J. Microbiol.* 37:823–827, 1991.
61. Rujiter, G.J.G., H. Panneman, J. Visser. Overexpression of phosphofructokinase and pyruvate kinase in citric acid producing *Aspergillus niger*. *Biochim. Biophys. Acta* 1334:317–326, 1997.
62. Szczodrak, J. Biosynthesis of citric acid in relation to the activity of selected enzymes of the Krebs cycle in *Aspergillus niger* mycelium. *Eur. J. Appl. Microbiol. Biotechnol.* 13:107–112, 1981.
63. Ruijter, G.J.G., H. Panneman, D.B. Xu, J. Visser. Properties of *Aspergillus niger* citrate synthase and effects of *cit*A overexpression on citric acid production. *FEMS Microbiol. Lett.* 184:35–40, 2000.
64. Agrawal, P.K., C.S. Bhatt, L. Viswanathan. Studies on some enzymes relevant to citric acid production by *Aspergillus niger*. *Enzym. Microb. Technol.* 5:369–372, 1983.
65. Roukas, T. Influence of impeller speed on citric acid production and selected enzyme activities of the TCA cycle. *J. Ind. Microbiol.* 7:221–226, 1991.
66. Legisa, M., J. Kidric. Initiation of citric acid accumulation in the early stages of *Aspergillus niger* growth. *Appl. Microbiol. Biotechnol.* 31:453–457, 1989.
67. Bowes, I., M. Mattey. A study of mitochondrial $NADP^+$- specific isocitrate dehydrogenase from selected strains of *Aspergillus niger*. *FEMS Microbiol. Lett.* 7:323–325, 1980.
68. La Nauze, J.M. Aconitase and isocitric dehydrogenases of *Aspergillus niger* in relation to citric acid production. *J. Gen. Microbiol.* 44:73–81, 1966.

69. Szczodrak, J., Z. Ilczuk. Effect of iron on the activity of aconitate hydratase and synthesis of citric acid by *Aspergillus niger*. *Z. Mikrobiol*. 140:567–574, 1985.
70. Kubicek, C.P., M. Rohr. Aconitase and citric acid fermentation by *Aspergillus niger*. *Appl. Environ. Microbiol*. 50:1336–1338, 1985.
71. Meixner-Monori, B., C.P. Kubicek, A. Habison, E.M. Kubicek-Pranz, M. Rohr. Presence and regulation of the α-ketoglutarate dehydrogenase multienzyme complex in the filamentous fungus *Aspergillus niger*. *J. Bacteriol*. 161:265–271, 1985.
72. Kubicek, C.P., M. Rohr. The role of the tricarboxylic acid cycle in citric acid accumulation by *Aspergillus niger*. *Eur. J. Appl. Microbiol. Biotechnol*. 5:263–271, 1978.
73. Heinrich, M., H.J. Rehm. Formation of gluconic acid at low pH values by free and immobilized *Aspergillus niger* cells during citric acid fermentation. *Eur. J. Appl. Microbiol. Biotechnol*. 15:88–92, 1982.
74. Mischak, H., C.P. Kubicek, M. Rohr. Formation and location of glucose oxidase in citric acid producing mycelia of *Aspergillus niger*. *Appl. Microbiol. Biotechnol*. 21:27–31, 1985.
75. Xu, D.B., M. Rohr, C.P. Kubicek. *Aspergillus niger* cyclic AMP levels are not influenced by manganese deficiency and do not correlate with citric acid accumulation. *Appl. Microbiol. Biotechnol*. 32:124–128, 1989.
76. Legisa, M., M. Mattey. Glycerol synthesis by *Aspergillus niger* under citric acid accumulating conditions. *Enzym. Microb. Technol*. 8:607–609, 1986.
77. Legisa, M., M. Mattey. Citrate regulation of the change in carbohydrate degradation during the initial phase of the citric acid production by *Aspergillus niger*. *Enzym. Microb. Technol*. 10:33–36, 1988.
78. Roh, M., C.P. Kubicek, O. Zehentgruber, R. Orthofer. Accumulation and partial re-consumption of polyols during citric acid fermentation by *Aspergillus niger*. *Appl. Microbiol. Biotechnol*. 27:235–239, 1987.
79. Torres, N.V. Modeling approach to control of carbohydrate metabolism during citric acid accumulation by *Aspergillus niger*, I: model definition and stability of the steady state. *Biotechnol. Bioeng*. 44:104–111, 1994.
80. Torres, N.V. Modeling approach to control of carbohydrate metabolism during citric acid accumulation by *Aspergillus niger*, II: sensitivity analysis. *Biotechnol Bioeng* 44:112–118, 1994.
81. Torres, N.V., J.M. Riol-Cimas, M. Wolschek, C.P. Kubicek. Glucose transport of *Aspergillus niger*: the low affinity carrier is only formed during growth on high glucose concentrations. *Appl. Microbiol. Biotechnol*. 44:790–794, 1996.
82. Wayman, F.M., M. Mattey. Simple diffusion is the primary mechanism for glucose uptake during the production phase of the *Aspergillus niger* citric acid process. *Biotechnol. Bioeng*. 67:451–456, 2000.
83. Alvarez-Vazquez, F., C. Gonzales Alcon, N.V. Torres. Metabolism of citric acid production by *Aspergillus niger*:model definition, steady state analysis and constrained optimization of citric acid production rate. *Biotechnol. Bioeng*. 70:82–108, 2000.
84. Guebel, D.V., N.V. Torres Darias. Optimization of the citric acid production by *Aspergillus niger* through a metabolic flux balance model. *Elect. J. Biotechnol*., 2001.
85. Xu, D.B., C.P. Madrid, M. Rohr, C.P. Kubicek. The influence of type and concentration of the carbon source on production of citric acid by *Aspergillus niger*. *Appl. Microbiol. Biotechnol*. 30:553–558, 1989.
86. Roukas, T. Carob pod: a new substrate for citric acid production by *Aspergillus niger*. *Appl. Biochem. Biotechnol*. 74:43–53, 1998.
87. Hossain, M., J.D. Brooks, I.S. Maddox. The effect of the sugar source on citric acid production by *Aspergillus niger*. *Appl. Microbiol. Biotechnol*. 19:393–397, 1984.
88. Kristiansen, B., C.G. Sinclair. Production of citric acid in continuous culture. *Biotechnol. Bioeng*. 21:297–315, 1979.
89. Dawson, M.W., I.S. Maddox, J.D. Brooks. Evidence for nitrogen catabolite repression during citric acid production by *Aspergillus niger* under phosphate-limited growth conditions. *Biotechnol. Bioeng*. 33:1500–1504, 1989.

90. Choe, J., Y.J. Yoo. Effect of ammonium ion concentration and application to fed-batch culture for overproduction of citric acid. *J. Ferment. Bioeng.* 72:106–109, 1991.
91. Yigitoglu, M., B. McNeil. Ammonium ion and citric acid supplementation in batch cultures of *Aspergillus niger* B 60. *Biotechnol. Lett.* 14:831–836, 1992.
92. Pintado, J., M.A. Murado, M.P. Gonzales, J. Miron, L. Pastrana. Joint effect of nitrogen and phosphorus concentrations on citric acid production by different strains of *Aspergillus niger* grown on an effluent. *Biotechnnol. Lett.* 15:1157–1162, 1993.
93. Lal, D.N., A.S. Srivastava. Effect of amino acids on microbial production of citric acid by *Aspergillus niger*. *Z. Mikrobiol.* 137:31–35, 1982.
94. Clark, D.S., K. Ito, P. Tymchuk. Effect of potassium ferrocyanide on the chemical composition of molasses mash used in the citric acid fermentation. *Biotechnol. Bioeng.* 7:269–278, 1965.
95. Clark, D.S., K. Ito, H. Horitsu. Effect of manganese and other heavy metals on submerged citric acid fermentation of molasses. *Biotechnol. Bioeng.* 8:465–471, 1966.
96. Banik, A.K. Mineral nutrition of *Aspergillus niger* for citric acid production. *Folia Mikrobiol.* 21:139–143, 1976.
97. Kubicek, C.P., M. Rohr. Influence of manganese on enzyme synthesis and citric acid accumulation in *Aspergillus niger*. *Eur. J. Appl. Microbiol.* 4:167–175, 1977.
98. Bowes, I., M. Mattey. The effect of manganese and magnesium ions on mitochondrial $NADP^+$ - dependent isocitrate dehydrogenase from *Aspergillus niger*. *FEMS Microbiol. Lett.* 6:219–222, 1979.
99. Shankaranand, V.S., B.K. Lonsane. Ability of *Aspergillus niger* to tolerate metal ions and minerals in a solid-state fermentation system for the production of citric acid. *Process. Biochem.* 29:29–7, 1994.
100. Pera, L.M., D.A. Callieri. Influence of calcium on fungal growth, hyphal morphology and citric acid production in *Aspergillus niger*. *Folia Mikrobiol.* 42:551–556, 1997.
101. Tsekova, K., D. Dentchev, D. Todorova. Effect of cadmium and copper on the production of citric acid by *Aspergillus niger*. Folia Mikrobiol. 45:331–334, 2000.
102. Haq, I.U., S. Ali, M.A. Qadeer, J. Iqdal. Effect of copper ions on mould morphology and citric acid productivity by *Aspergillus niger* using molasses based medium. *Process. Biochem.* 37:1085–1090, 2002.
103. Roukas, T., P. Kotzekidou. Influence of some trace metals and stimulants on citric acid production from brewery waste by *Aspergillus niger*. *Enzym. Microb. Technol.* 9:291–294, 1987.
104. Stanbury, P.F., A. Whitaker. *Principles of Fermentation Technology*. New York: Pergamon Press, 1984.
105. Roukas, T., P. Kotzekidou. Pretreatment of date syrup to increase citric acid production. *Enzym. Microb. Technol.* 21:273–276, 1997.
106. Roukas, T. Citric acid production from carob pod by solid-state fermentation. *Enzym. Microb. Technol.* 24:54–59, 1999.
107. Roukas, T. Citric and gluconic acid production from fig by *Aspergillus niger* using solid-state fermentation. *J. Ind. Microbiol. Biotechnol.* 25:298–304, 2000.
108. Maddox, I.S., M. Hossain, J.D. Brooks. The effect of methanol on citric acid production from galactose by *Aspergillus niger*. *Appl. Microbiol. Biotechnol.* 23:203–205, 1986.
109. Saha,M.L., Y. Sakai, F Takahashi. Citric acid fermentation by magnetic drum contactor: use of methanol and ethanol for higher production. *J. Biosci. Bioeng.* 87:394–396, 1999.
110. Pintado, J., M.P. Gonzales, M.A. Murado. Interactions between pretreatment and nutrient concentrations of mussel processing effluents for citric acid production. *Enzym. Microb. Technol.* 20:544–549, 1997.
111. Jianlong, W., L. Ping. Phytate as a stimulator of citric acid production by *Aspergillus niger*. *Process. Biochem.* 33:313–316, 1998.
112. Agrawal, P.K., C.S. Bhatt, L. Viswanathan. Effect of some metabolic inhibitors on citric acid production by *Aspergillus niger*. *Enzym. Microb. Technol.* 5:373–376, 1983.
113. Hossain, M., J.D. Brooks, I.S. Maddox. Galactose inhibition of citric acid production from glucose by *Aspergillus niger*. *Appl. Microbiol. Biotechnol.* 22:98–102, 1985.

114. Legisa, M., M. Mattey. Glycerol as an initiator of citric acid accumulation in *Aspergillus niger*. *Enzym. Microb. Technol*. 8:258–259, 1986.
115. Kirimura, K., M. Yoda, H. Shimizu, S. Sugano, M. Mizuno, K. Kino, S. Usami. Contribution of cyanide-insensitive respiratory pathway,catalyzed by the alternative oxidase, to citric acid production in *Aspergillus niger*. *Biosci. Biotechnol. Biochem*. 64:2034–2039, 2000.
116. Rokosu, A.A., C.A. Anenih. Effects of various conditions on the production of citric acid during fermentation of molasses by *Aspergillus niger*. Enzym. Microb. Technol. 2:61–62, 1980.
117. Szewczyk, K.W., L. Myszka. The effect of temperature on the growth of *Aspergillus niger* in solid state fermentation. *Bioprocess. Eng*. 10:123–126, 1994.
118. Roukas, T., E. Alichanidis. The effect of pH on the production of citric acid from beet molasses by surface fermentation. *8th International Biotechnology Symposium*, Paris, 1988, p 218.
119. Mantzouridou, F., T. Roukas, P. Kotzekidou. Effect of the aeration rate and agitation speed on β-carotene production and morphology of *Blakeslea trispora* in a stirrer tank reactor: mathematical modeling. *Biochem. Eng. J*. 10:123–135, 2002.
120. Kubicek, C.P., O. Zehentgruber, H. El-Kalak, M. Rohr. Regulation of citric acid production by oxygen: effect of dissolved oxygen tension on adenylate levels and respiration in *Aspergillus niger*. *Eur. J. Appl. Microbiol. Biotechnol*. 9:101–115, 1980.
121. Dawson, M.W., I.S. Maddox, J.D. Brooks. Effect of interruptions to the air supply on citric acid production by *Aspergillus niger*. *Enzym. Microb. Technol*. 8:37–40, 1986.
122. Samson, R.A., E.S. Hoekstra, C.A.N. Van Oorschot. *Introduction to Food-Borne Fungi*, 2nd ed., Baarn: Centraalbureau Voor Schimmelcultures, 1984.
123. Jernejc, K., M. Vendramin, A. Cimerman. Lipid composition of *Aspergillus niger* in citric acid accumulating and nonaccumulating conditions. *Enzym. Microb. Technol*. 11:452–456, 1989.
124. Jernejc, K., A. Cimerman, M. Vendramin, A. Perdih. Lipids of a citric acid producing *Aspergillus niger* strain grown in copper and in manganese supplemented media. *Appl. Microbiol. Biotechnol*. 32:699–703, 1990.
125. Gomez, R., I. Schnabel, J. Garrido. Pellet growth and citric acid yield of *Aspergillus niger* 110. *Enzym. Microb. Technol*. 10:188–191, 1988.
126. Paul, G.C., M.A. Priede, C.R. Thomas. Relationship between morphology and citric acid production in submerged *Aspergillus niger* fermentation. *Biochem. Eng. J*. 3:121–129, 1999.
127. Papagianni, M., M. Mattey, B. Kristiansen. Morphology and citric acid production of *Aspergillus niger* PM 1. *Biotechnol. Lett*. 16:929–934, 1994.
128. Papagianni, M., M. Mattey, B. Kristiansen. Citric acid production and morphology of *Aspergillus niger* as functions of the mixing intensity in a stirred tank and a tubular loop bioreactor. *Biochem. Eng. J*. 2:197–205, 1998.
129. Papagianni, M., M. Mattey, B. Kristiansen. The influence of glucose concentration on citric acid production and morphology of *Aspergillus niger* in batch and fed batch culture. Enzym. Microb. Technol. 25:710–717, 1999.
130. Papagianni, M., M. Mattey, B. Kristiansen. Hyphal vacuolation and fragmentation in batch and fed-batch culture of *Aspergillus niger* and its relation to citric acid production. *Process. Biochem*. 35:359–366, 1999.
131. Al Obaidi, Z.S., D.R. Berry. cAMP concentration, morphological differentiation and citric acid production in *Aspergillus niger*. *Biotechnol. Lett*. 2:5–10, 1980.
132. Galbraith, J.C., J.E. Smith. Sporulation of *Aspergillus niger* in submerged liquid culture. J. Gen. Microbiol. 59:31–45, 1969.
133. Ng, W.S., J.E. Smith, J.G. Anderson. Changes in carbon catabolic pathways during synchronous development of conidiophores of *Aspergillus niger*. J. Gen. Microbiol. 71:495–504, 1972.
134. Roukas, T. Rheological properties of pullulan fermentation broth in a stirred tank fermentor. *Food Biotechnol*. 13:255–266, 1999.
135. Olsvik, E.S., B. Kristiansen. On-line rheological measurements and control in fungal fermentations. *Biotechnol. Bioeng*. 40:375–387, 1992.
136. Olsvik, E., B. Kristiansen. Rheology of filamentous fermentations. *Biotechnol. Adv*. 12:1–39, 1994.

137. Mitard, A., J.B. Riba. Rheological properties of *Aspergillus niger* pellet suspensions. *Appl. Microbiol. Biotechnol.* 25:245–249, 1986.
138. Berovic, M., A. Cimerman, W. Steiner, T. Koloini. Submerged citric acid fermentation: rheological properties of *Aspergillus niger* broth in a stirred tank reactor. Appl. Microbiol. Biotechnol. 34:579–581, 1991.
139. Roukas, T., E. Alichanidis. Citric acid production from beet molasses by cell recycle of *Aspergillus niger*. *J. Ind. Microbiol.* 7:71–74, 1991.
140. Roukas, T. Citric acid production from carob pod extract by cell recycle of *Aspergillus niger* ATCC 9142. *Food Biotechnol.* 12:91–104, 1998.
141. Briffaud, J., M. Engasser. Citric acid production from glucose, I: growth and excretion kinetics in a stirred fermentor. *Biotechnol. Bioeng.* 21:2083–2092, 1979.
142. Roehr, M., O. Zehentgruber, C.P. Kubicek. Kinetics of biomass formation and citric acid production by *Aspergillus niger* on pilot plant scale. *Biotechnol. Bioeng.* 23:2433–2445, 1981.
143. M. Berovic, A. Cimerman. Redox potential in submerged citric acid fermentation. *Eur. J. Appl. Microbiol. Biotechnol.* 16:185–188, 1982.
144. Khan, K.H., S.S. Shaukat. Citric acid production with mixed strains of *Aspergillus niger* in submerged culture. *Acta Microbiolog. Hung.* 37:9–13, 1990.
145. Allen, D.G., C.W. Robinson. Hydrodynamics and mass transfer in *Aspergillus niger* fermentations in bubble column and loop bioreactors. *Biotechnol. Bioeng.* 34:731–740, 1989.
146. Blain, J.A., J.G. Anderson, J.R. Todd, M. Divers. Cultivation of filamentous fungi in the disc fermentor. *Biotechnol. Lett.* 146:269–274, 1979.
147. Anderson, J.G., J.A. Blain, M. Divers, J.R. Todd. Use of the disc fermentor to examine production of citric acid by *Aspergillus niger*. *Biotechnol. Lett.* 2:99–104, 1980.
148. Sakurai, A., H. Imai. Effect of operational conditions on the rate of citric acid production by rotating disc contactor using *Aspergillus niger*. *J. Ferment. Bioeng.* 73:251–254, 1992.
149. Briffaud, J., M. Engasser. Citric acid production from glucose. II: growth and excretion kinetics in a trickle-flow fermentor. *Biotechnol. Bioeng.* 21:2093–2111, 1979.
150. Tongwen, X., Y. Weihua. Effect of cell configurations on the performance of citric acid production by a bipolar membrane electrodialysis. *J. Memb. Sci.* 5224:1–9, 2002.
151. Sakurai, A., H. Imai, T. Ejiri, K. Endoh, S. Usami. Citric acid production by surface culture using *Aspergillus niger*: kinetics and simulation. *J. Ferment. Bioeng.* 72:15–19, 1991.
152. Shierholt, J. Fermentation processes for the production of citric acid. *Process. Biochem.* 12:20–21, 1977.
153. Roukas, T., L. Harvey. The effect of pH on production of citric and gluconic acid from beet molasses using continuous culture. *Biotechnol. Lett.* 10:289–294, 1988.
154. Dawson, M.W., I.S. Maddox, I.F. Boag, J.D. Brooks. Application of fed-batch culture to citric acid production by *Aspergillus niger*: the effects of dilution rate and dissolved oxygen tension. *Biotechnol. Bioeng.* 32:220–226, 1988.
155. Roukas, T. Ethanol production from non-sterilized beet molasses by free and immobilized *Saccharomyces cerevisiae* cells using fed-batch culture. *J. Food Eng.* 27:87–96, 1996.
156. Pintado, J., A. Torrado, M.P. Gonzales, M.A. Murado. Optimization of nutrient concentration for citric acid production by solid-state culture of *Aspergillus niger* on polyurethane foams. *Enzym. Microb. Technol.* 23:149–156, 1998.
157. Lu, M.Y., I.S. Maddox, J.D. Brooks. Application of a multi-layer packed-bed reactor to citric acid production in solid-state fermentation using *Aspergillus niger*. *Process. Biochem.* 33:117–123, 1998.
158. Lu, M.Y., I.S. Maddox, J.D. Brooks. Citric acid production by *Aspergillus niger* in solid-substrate fermentation. *Biores. Technol.* 54:235–239, 1995.
159. Lu, M., J.D. Brooks, I.S. Maddox. Citric acid fermentation by solid-state fermentation in a packed-bed reactor using *Aspergillus niger*. *Enzym. Microb. Technol.* 21:392–397, 1997.
160. Hang, Y.D. Microbial production of citric acid in fixed-bed column bioreactors. *Biotechnol. Lett.* 10:421–426, 1988.
161. Oriol, E., B. Schettino, G. Viniegra-Gonzales, M. Raimbault. Solid-state culture of *Aspergillus niger* on support. *J. Ferment. Technol.* 66:57–62, 1988.

162. Pintado, J., B.K. Lonsane, I. Gaime-Perraud, S. Roussos. On-line monitoring of citric acid production in solid-state culture by respirometry. *Process. Biochem.* 33:513–518, 1998.
163. Gutierrez-Rojas, M., J. Cordova, R. Auria, S. Revah, E. Favela-Torres. Citric acid and poyols production by *Aspergillus niger* at high glucose concentration in solid-state fermentation on inert support. *Biotechnol. Lett.* 17:219–224, 1995.
164. Roukas, T. Production of citric acid from beet molasses by immobilized cells of *Aspergillus niger*. *J. Food Sci.* 56:878–880, 1991.
165. Nunez, M.J., J.M. Lema. Cell immobilization :application to alcohol production. *Enzym. Microb. Technol.* 9:642–651, 1987.
166. Bayraktar, E., U. Mehmetoglu. Production of citric acid using immobilized conidia of *Aspergillus niger*. *Appl. Biochem. Biotechnol.* 87:117–125, 2000.
167. Jianlong, W. Production of citric acid by immobilized Aspergillus niger using a rotating biological contactor (RBC). *Biores. Technol.* 75:245–247, 2000.
168. Eikmeier, H., F. Westmeier, H.J. Rehm. Morphological development of *Aspergillus niger* immobilized in Ca-alginate and K-carrageenan. *Appl. Microbiol. Biotechnol.* 19:53–57, 1984.
169. Eikmeier, H., H.J. Rehm. Production of citric acid with immobilized *Aspergillus niger*. *Appl. Microbiol. Biotechnol.* 20:365–370, 1984.
170. Vaija, J., P. Linko. Continuous citric acid production by immobilized *Aspergillus niger*: reactor performance and fermentation kinetics. *J. Mol. Cataly.* 38:237–253, 1986.
171. Tsay, S.S., K.Y. To. Citric acid production using immobilized conidia of *Aspergillus niger* TMB 2022. *Biotechnol. Bioeng.* 29:297–304, 1987.
172. Eikmeier, H., H.J. Rehm. Stability of calcium alginate during citric acid production of immobilized *Aspergillus niger*. *Appl. Microbiol. Biotechnol.* 26:105–111, 1987.
173. Eikmeier, H., H.J. Rehm. Semicontinuous and continuous production of citric acid with immobilized cells of *Aspergillus niger*. *Z. Natur. C. J. Biosci.* 42:408–413, 1987.
174. Honecker, S., B. Bisping, Z. Yang, H.J. Rehm. Influence of sucrose concentration and phosphate limitation on citric acid production by immobilized cells of *Aspergillus niger*. *Appl. Microbiol. Biotechnol.* 31:17–24, 1989.
175. Bisping, B., H. Hellfors, S. Honecker, H.J. Rehm. Formation of citric acid and poyols by immobilized cells of *Aspergillus niger*. *Food Biotechnol.* 4:17–23, 1990.
176. Gupta, S. C.B. Sharma. Continuous production of citric acid from sugar cane molasses using a combination of submerged immobilized and surface stabilized cultures of *Aspergillus niger* KCU 520. *Biotechnol. Lett.* 16:599–604, 1994.
177. Horitsu, H., S. Adachi, Y. Takahashi, K. Kawai, Y. Kawano. Production of citric acid by *Aspergillus niger* immobilized in polyacrylamide gels. *Appl. Microbiol. Biotechnol.* 22:8–12, 1985.
178. Lee, Y.H., C.W. Lee, H.N. Chang. Citric acid production by *Aspergillus niger* immobilized on polyurethane foam. *Appl. Microbiol. Biotechnol.* 30:141–143, 1989.
179. Sankpal, N.V., A.P. Joshi, B.D. Kulkarni. Citric acid production by *Aspergillus niger* immobilized on cellulose microfibrils: influence of morphology and fermenter conditions on productivity. *Process. Biochem.* 36:1129–1139, 2001.
180. Jernejc, K., A. Cimerman, A. Perdih. Citric acid production in chemically defined media by *Aspergillus niger*. *Eur. J. Appl. Microbiol. Biotechnol.* 14:29–33, 1982.
181. Qazi, G.N., C.N. Gaind, S.K. Chaturvedi, C.L. Chorpa, M. Trager, U. Onken. Pilot-scale citric acid production with *Aspergillus niger* under several conditions. J. Ferment. Bioeng. 69:72–74, 1990.
182. Maddox, S., K. Spencer, J.M. Greenwood, M.W. Dawson, J.D. Brooks. Production of citric acid from sugars present in wood hemicellulose using *Aspergillus niger* and *Saccharomyces lipolytica*. *Biotechnol. Lett.* 7:815–818, 1985.
183. Kirimura, K., T. Watanabe, T. Sunagawa, S. Usami. Citric acid production from xylan and xylan hydrolysate by semi-solid culture of *Aspergillus niger*. *Biosci. Biotechnol. Biochem.* 63:226–228, 1999.
184. Drysdale, C.R., A.M. McKay. Citric acid production by *Aspergillus niger* in surface culture of inulin. *Lett. Appl. Microbiol.* 20:252–254, 1995.

185. Hungerford, E.H. Molasses. In: *Beet Sugar Technology*, 2nd ed., RA McGinnis, R.A., ed., Colorado: Beet Sugar Development Foundation, 1971, pp 579–587.
186. Clark, D.S., C.P. Lentz. Submerged citric acid fermentation of beet molasses in tank-type fermentors. *Biotechnol. Bioeng*. 5:193–199, 1963.
187. Chaudhary, K., S. Ethiraj, K. Lakshminarayana, P. Tauro. Citric acid production from Indian cane molasses by *Aspergillus niger* under solid-state fermentation conditions. *J. Ferment. Technol*. 56:554–557, 1978.
188. Berovic, M., A. Cimerman. Foaming in submerged citric acid fermentation on beet molasses. *Eur. J. Appl. Microbiol. Biotechnol*. 7:313–319, 1979.
189. Ilczuk, Z. Attempts at improving citric fermentation on molasses solutions. *Eur. J. Appl. Microbiol. Biotechnol*. 17:69–72, 1983.
190. Kundu, S., T. Panda, S.K. Majumdar, B. Guha, K.K. Bandyopadhyay. Pretreatment of Indian cane molasses for increased production of citric acid. *Biotechnol. Bioeng*. 26:1114–1121, 1984.
191. Hamisa, F.A., A. Radwan. Production of citric acid from cane molasses on a semi-pilot scale. *J. Gen. Appl. Microbiol*. 23:325–329, 1977.
192. Garg, K., C.B. Sharma. Repeated batch production of citric acid from sugar cane molasses using recycled solid-state surface culture of *Aspergillus niger*. *Biotechnol. Lett*. 13:913–916, 1991.
193. Jianlong, W., W. Xianghua, Z. Ding. Production of citric acid from molasses integrated with *in situ* product separation by ion-exchange resin adsorption. *Biores. Technol*. 75:231–234, 2000.
194. Jianlong, W. Improvement of citric acid production by *Aspergillus niger* with addition of phytate to beet molasses. *Biores. Technol*. 65:243–245, 1998.
195. Adham. N.Z. Attempts at improving citric acid fermentation by *Aspergillus niger* in beet molasses medium. *Biores. Technol*. 84:97–100, 2002.
196. Nguyen, T.K., L. Martinkova, L. Seichert, F. Machek. Citric acid production by *Aspergillus niger* using media containing low concentrations of glucose or corn starch. *Folia Mikrobiol*. 37:433–441, 1992.
197. Mourya, S., K.S. Jauhri. Production of citric acid from starch hydrolysate by *Aspergillus niger*. Microbiol. Res. 155:37–44, 2000.
198. Bolach, E., W. Lesniak, J. Ziobrowski. Starch hydrolysis in citric acid fermentation. *Acta Alimentaria Polonica* 11:95–101, 1985.
199. Tan, K.H., L.B. Ferguson, C. Carlton. Conversion of cassava starch to biomass carbohydrates and acids by *Aspergillus niger*. *J. Appl. Biochem*. 6:80–90, 1984.
200. Sarangbin, S., Y. Watanapokasin. Yam bean starch: a novel substrate for citric acid production by the protease-negative mutant strain of *Aspergillus niger*. *Carboh. Polym*. 38:219–224, 1999.
201. Esuoso, K.O., R.A. Oderinde, J.I. Okogun. Citric acid production from imumu *Cyperus esculentus* and maize *Zea mays*. *J. Ferment. Bioeng*. 71:200–202, 1991.
202. Hang, Y.D., E.E. Woodams. Production of citric acid from corncobs by *Aspergillus niger*. *Biores. Technol*. 65:251–253, 1998.
203. Hang, Y.D., E.E. Woodams. Corn husks: a potential substrate for production of citric acid by *Aspergillus niger*. *Lebensm. Wiss. U. Technol*. 33:520–521, 2000.
204. Yuogo, Z., W. Zhao, C. Xiaolong. Citric acid production from the mash of dried sweet potato with its dregs by *Aspergillus niger* in an external-loop airlift bioreactor. *Process. Biochem*. 35:237–242, 1999.
205. Khare, S.K., K. Jha, A.P. Gandhi. Citric acid production from okara (soy-residue) by solid-state fermentation. *Biores. Tech*. 54:323–325, 1995.
206. Al-Obaidi, Z., D.R. Berry. The use of deionised date syrup as a substrate for citric acid fermentation. *Biotechnol. Lett*. 1:153–158, 1979.
207. Al-Obaidi, Z., D.R. Berry. Extended production of citric acid using an exchange filtration technique. *Biotechnol. Lett*. 1:221–224, 1979.
208. Hang, Y.D., E.E. Woodams. Apple pomace: a potential substrate for citric acid production by *Aspergillus niger*. *Biotechnol. Lett*. 6:763–764, 1984.

209. Hang, Y.D., E.E. Woodams. Effect of substrate moisture content on fungal production of citric acid in a solid-state fermentation system. *Biotechnol. Lett.* 9:183–186, 1987.
210. Hang, Y.D., E.E. Woodams. Utilization of grape pomace for citric acid production by solid-state fermentation. *Am. J. Enol. Vitic.* 37:141–142, 1986.
211. Hang, Y.D., B.S. Luh, E.E. Woodams. Microbial production of citric acid by solid-state fermentation of kiwifruit peel. *J. Food. Sci.* 52:226–227, 1987.
212. Kumagai, K., S. Usami, S. Hattori. Citric acid production from mandarin orange waste by solid state culture of *Aspergillus niger*. *Hakkokogaku* 59:461–464, 1981.
213. Tran, C.T., D.A. Mitchell. Pineapple waste: a novel substrate for citric acid production by solid-state fermentation. *Biotechnol. Lett.* 17:1107–1110, 1995.
214. Usami, S., N. Fukutomi. Citric acid production by solid-state fermentation method using sugar cane bagasse and concentrated liquor of pineapple waste. *Hakkokogaku* 55:44–50, 1977.
215. Kiel, H., R. Gurin, Y. Henis. Citric acid fermentation by *Aspergillus niger* on low sugar concentrations and cotton waste. *Appl. Environ. Microbiol.* 42:1–4, 1981.
216. Hossain, M., J.D. Brooks. Production of citric acid from whey permeate by fermentation using *Aspergillus niger*. *N.Z. J. Dairy Sci. Tech.* 18:161–168, 1983.
217. Hang, Y.D., D.F. Splittstoesser, E.E. Woodams. Utilization of brewery spent grain liquor by *Aspergillus niger*. *Appl. Microbiol.* 30:879–880, 1975.
218. Das, A. Parasexual hybridisation and citric acid production by *Aspergillus niger*. *Eur. J. Appl. Microbiol. Biotechnol.* 9:117–119, 1980.
219. Kirimura, K., S. Sarangbin, S. Rugsaseel, S. Usami. Citric acid production by 2-deoxyglucose-resistant mutant strains of *Aspergillus niger*. *Appl. Microbiol. Biotechnol.* 36:573–577, 1992.
220. Sarangbin, S., K. Kirimura, S. Usami. Citric acid production from cellobiose by 2-deoxy-glucose-resistant mutant strains of *Aspergillus niger* in semi-solid culture. *Appl. Microbiol. Biotechnol.* 40:206–210, 1993.
221. Millis, N.F., B.H. Trumpy, B.M. Palmer. The effect of lipids on citric acid production by an *Aspergillus niger* mutant. *J. Gen. Microbiol.* 30:365–379, 1963.
222. Schreferl, G., C.P. Kubicek, M. Rohr. Inhibition of citric acid accumulation by manganese ions in *Aspergillus niger* mutants with reduced citrate control of phosphofructokinase. *J. Bacteriol.* 165:1019–1022, 1986.
223. Pelechova, J., L. Petrova, E. Ujcova, L. Martinkova. Selection of a hyperproducing strain of *Aspergillus niger* for biosynthesis of citric acid on unusual carbon substrates. Folia Mikrobiol. 35:138–142, 1990.
224. Ghosh, P., A.K. Banik. Effect of chemical nutrients on aconitase activity during citric acid fermentation by a mutant strain of *Aspergillus niger*. *Acta Microbiol. Polon.* 47:253–260, 1998.
225. Das, A., P. Roy. Improved production of citric acid by a diploid strain of *Aspergillus niger*. *Can. J. Microbiol.* 24:622–625, 1978.
226. Steinbock, F., S. Choojun, I. Held, M. Roehr, C.P. Kubicek. Characterization and regulatory properties of a single hexokinase from the citric acid accumulating fungus *Aspergillus niger*. *Biochim. Biophys. Acta* 1200:215–223, 1994.
227. Islam, M.S., R. Begum, N. Choudhury. Semi-pilot scale production of citric acid in cane molasses by gamma-ray induced mutants of *Aspergillus niger*. *Enzym. Microb. Technol.* 8:469–471, 1986.
228. R Bonatelli, R., Jr., J.L. De Azevedo. Improved reproducibility of citric acid *production* in *Aspergillus niger*. *Biotechnol. Lett.* 4:761–766, 1982.
229. Islam, M.S., R. Begum, N. Choudhury. Semi-pilot scale studies on citric acid fermentation by a gamma-ray induced mutant of *Aspergillus niger*. *Biotechnol. Lett.* 6:431–434, 1984.
230. Begum, A.A., N. Choudhury, M.S. Islam. Citric acid fermentation by gamma ray induced mutants of *Aspergillus niger* in different carbohydrate media. *J. Ferment. Bioeng.* 70:286–288, 1990.
231. Chaudhary, K., K. Lakshminarayana, I.K. Dev, S.R. Vyas. Nitrosoguanidine induced mutation of *Aspergillus niger* for obtaining high citric acid producing mutants. *Ind. J. Microbiol.* 14:42–43, 1974.

232. Roy, P., A. Das. The mutagenic action of N-methyl N`-nitro-N-nitrosoguanidine on *Aspergillus niger* in relation to citric acid production. *Sci. Cult.* 43:461–463, 1977.
233. Promper, C., R. Schneider, H. Weiss. The role of the proton-pumping and alternative respiratory chain NADH: ubiquinone oxidoreductases in overflow catabolism of *Aspergillus niger. J. Biochem.* 216:223–230, 1993.
234. Watanabe, T., A. Suzuki, H. Nakagawa, K. Kirimura, S. Usami. Citric acid production from cellulose hydrolysate by a 2-deoxyglucose-resistant mutant strain of *Aspergillus niger.* Biores. Technol. 66:271–274, 1998.
235. Shcherbakova, E.I., M.N. Rezvaia. Formation of diploids by *Aspergillus niger* and their biosythesis of citric acid. *Mikrobiologia* 46:1064–1069, 1977.
236. Kirimura, K., I. Nakajima, S.P. Lee, S. Kawabe, S. Usami. Citric acid production by the diploid strains of *Aspergillus niger* obtained by protoplast fusion. *Appl. Microbiol. Biotechnol.* 27:504–506, 1988.
237. Martinkova, L., M. Musilkova, E. Ujcova, F. Machek, L. Seichert. Protoplast fusion in *Aspergillus niger* strains accumulating citric acid. *Folia Mikrobiol.* 35:143–148, 1990.
238. Arisan-Atac, I., M.F. Wolschek, C.P. Kubicek. Trehalose-6-phosphate synthase A affects citrate accumulation by *Aspergillus niger* under conditions of high glycolytic flux. *FEMS Microbiol. Lett.* 140:77–83, 1996.
239. Verdoes, J.C., P.J. Punt, P. van der Berg, F. Debets, A.H. Stouthamer, C.A.M.J.J. van den Hondel. Characterization of an efficient gene cloning strategy for *Aspergillus niger* based on an autonomously replicating plasmid: cloning of the *nicB* gene of *A.niger. Gene* 146:159–165, 1994.
240. Kirimura, K., M. Yoda, S. Usami. Cloning and expression of the cDNA encoding an alternative oxidase gene from *Aspergillus niger* WU-2223L. *Curr. Genet.* 34:472–477, 1999.
241. Atkinson, B., F. Mavituna. *Biochemical Engineering and Biotechnology Handbook.* Hong Kong: The Nature Press, 1985, pp 1033–1036.

1.16

Microbial Biotechnology of Food Flavor Production

G. Feron and Y. Waché

CONTENTS

16.1 INTRODUCTION

Traditional fermentations use the ability of microorganisms to transform raw materials in complex food products. This process of biotransformation involves a wide range of biochemical reactions that depend on both the microorganism used and its direct environment. Since the 1980s, biotechnologists have tended to exploit this biological capacity in order to produce high value added products, especially aroma or flavor compounds. During the last twenty years, a large number of reviews and contributions discussing the possible industrial applications of microbial volatiles have been published (1–13). In spite of many advances in cell biology, enzymology, genetic engineering, and process control, and considering the high demand and commercial value of natural flavors, to date about 15% of natural flavoring molecules are produced as pure compounds by biotechnological means. This is principally explained by economic constraints that limit the industrial scaling up of many laboratory processes. However, the recent development of genetic engineering tools as well as new investigations on the direct control of the biocatalyst environment via redox potential or hydrophobicity and polarity opens novel and innovating developments in this field. Such developments may help to overcome the limits encountered and widen the spectrum of biotechnologically accessible flavor compounds.

16.2 MARKET OF FLAVOR: COST AND LEGACY CONSIDERATON

Perfumes and flavors appeared in early Egyptian times via plant extracts and derivatives. However, authors believe the foundation of the modern flavor industry was established in 1843 with the synthesis of methyl salicylate, followed by cinnamic aldehyde in 1856, and benzaldehyde in 1863. Then came the pioneering explosion for the flavor industry: the synthesis of vanillin, still the key ingredient in flavor creativity, in 1872.

How large is this industry? The total global market is worth 10.9 billion euros (€) and grows by 4–5% per year. There are approximately 1800 registered global flavor houses but there are probably an additional 1500 flavor companies spread over China and India as well as a few other countries. However, there are 10 international companies that command 90% of the whole market of perfumes and flavors, and they are principally split between the USA and Europe.

About 10,000 volatiles have been mentioned in food products (6) that are representative of the complexity of the investigations in this field. To date, 2000 synthetic flavors are available on the market and about 400 natural flavors. It is generally considered that about 15–20% of these flavors may be produced by biotechnological means (13).

Beverages and cooked products represent around 70% of the total use of flavor additives. However, if we consider only natural flavors, up to 90% are used in beverages and only 20% in sweets and candies. This can be explained by the cost of flavoring but also the consumer target.

The use of natural or synthetic flavors depends partly on the legislation applied in the different countries, principally in the US and in Europe, which are both considered the principal producers and users of flavors in manufactured products. To summarize briefly, in the USA, the system that has been adopted since 1965 is the positive list. The flavoring substances are evaluated and classified as GRAS (generally recognized as safe) on the basis of long usage without toxicity or by undertaking the safety evaluation of novel flavoring ingredients. The only alternative designations for the consumers are "natural flavor" and "artificial flavor," the latter being avoided wherever possible. The term "natural flavor" or "natural flavoring" is defined in Title 21 of the Code of Federal Regulations (14). In contrast, in Europe, the system is a combination of positive and negative lists and is now being changed to a total positive list. Three designations are possible: "natural 'X' flavor" if the amount of the flavoring substance is greater than 90% from 'X' ('X' referring to the named source); "natural flavor" if the flavoring component consists of exclusively natural flavoring substances, including "bioflavoring"; "'X' flavor" or "'X' flavor(s)" if nature identical flavoring or artificial substances are used alone or in combinations.

If we consider the cost ratio between natural and artificial flavor (Table 16.1), one can see that, depending on the flavor family, the price of natural flavoring substances is from 10 to 100 times more than the corresponding synthetic flavor. As an example, synthetic vanilla is sold at about 15 €/kg whereas vanilla extract is ~4000 €/kg. These cost considerations help explain the preferential use of nature identical flavor in the European Food Industry.

16.3 TRADITIONAL FERMENTATIONS AND *DE NOVO* SYNTHESIS OF FLAVOR

The origins of fermented food go back many thousands of years with evidence of consumption of fermented products in China 6000 years ago. In terms of total production and consumption, the three major groups of fermented foods are beverages (primarily wine and beer), dairy foods, and cereals (15), but many other raw materials such as meat, fish, legumes, and fruits are used in fermentation processes.

Fermented foods are those foods which have been subjected to the action of microorganisms or enzymes so that desirable biochemical changes of the raw materials cause significant modifications to the food. The biogenesis of flavor is one of the main modifications that appear during the fermentation process and that contributes largely to the organoleptic characteristics of the corresponding food. The development of flavor during fermentation involves a variety of biochemical pathways using the main three classes of raw materials: glucoside or carbons, lipids, and proteins.

16.3.1 Metabolism

16.3.1.1 Carbon Metabolism

During fermentation, the source of carbon is essentially used for the growth of the microflora that are naturally present in the raw material or that have been added for the improvement of

Table 16.1
Cost ratio between different chemical classes of natural flavoring substances and their natural identical correspondence (G. Feron, personal data, 2003).

Class of Aroma	Cost Ratio : Natural/ Natural Identical
Lactones	20 – 25
Esters	10 – 20
Acids	10
Alcohols	25 – 30
Aldehydes	100 or more

the process. For this reason the diversity of flavor directly resulting from glucose catabolism is not very broad. Glucose is oxidized to pyruvate via the Emden-Meyerhof pathway of glycolysis. As summarized in Figure 16.1, pyruvate can be metabolized to some major flavor compounds such as diacetyl, acetoine, 2-3 butanediol, ethanol and acetaldehyde. Diacetyl, acetoine, and 2-3 butanediol contribute principally to the butter odor characteristic of many dairy products, but in this case they come from the metabolism of lactose and citrate present in milk. Diacetyl is also an important flavor compound in fermented beverages, especially beer and wine. Ethanol is principally synthesized during anaerobic fermentation and significantly contributes to the organoleptic properties of alcoholic beverages. Ethanol is synthesized from acetaldehyde, which when diluted possesses an apple like flavor, but at higher concentrations gives a characteristic pungent and penetrating odor that is not well appreciated in the fermented foods.

16.3.1.2 Lipid Metabolism

Unlike glucose metabolism, the oxidation and/or hydrolysis of lipids can lead to an important source of flavors (Figure 16.2).

Through the lipolysis of triglycerides, a broad variety of fatty acids can be released. Short chain fatty acids contribute directly to the flavoring of fermented products with aromatic notes such as vinegar (acetic acid), goat (4-ethyloctanoic acid), or rancid (butyric, octanoic, and decanoic acid).

After lipolysis, free medium and long chain fatty acids can be oxidized leading to different groups of flavors: methyl ketones, lactones, aldehydes, and alcohols. Methyl ketones are responsible for the fruity, musty, and blue cheese notes of cheeses, other dairy products, and fermented meats such as dried sausage. Methyl ketones are synthesized through the β-oxidation pathway of medium or short chain fatty acids (16) (Figure 16.3). After the third step of β-oxidation, β-acetoacyl-CoA is released from the cycle. This acyl-CoA is hydrolyzed with a thiolase and then decarboxylated, leading to methyl ketone characteristically possessing an odd number of carbon atoms. Some microorganisms such as

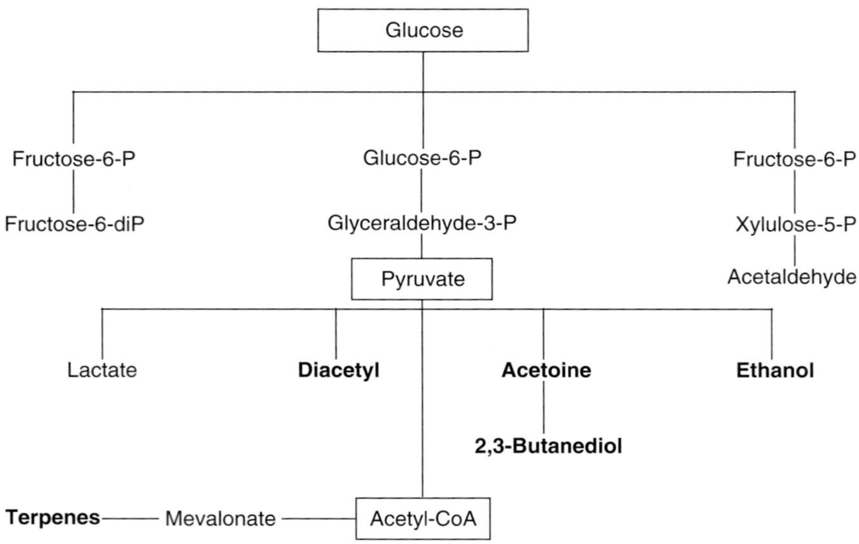

Figure 16.1 Formation of flavor compounds from glucose/carbon metabolism.

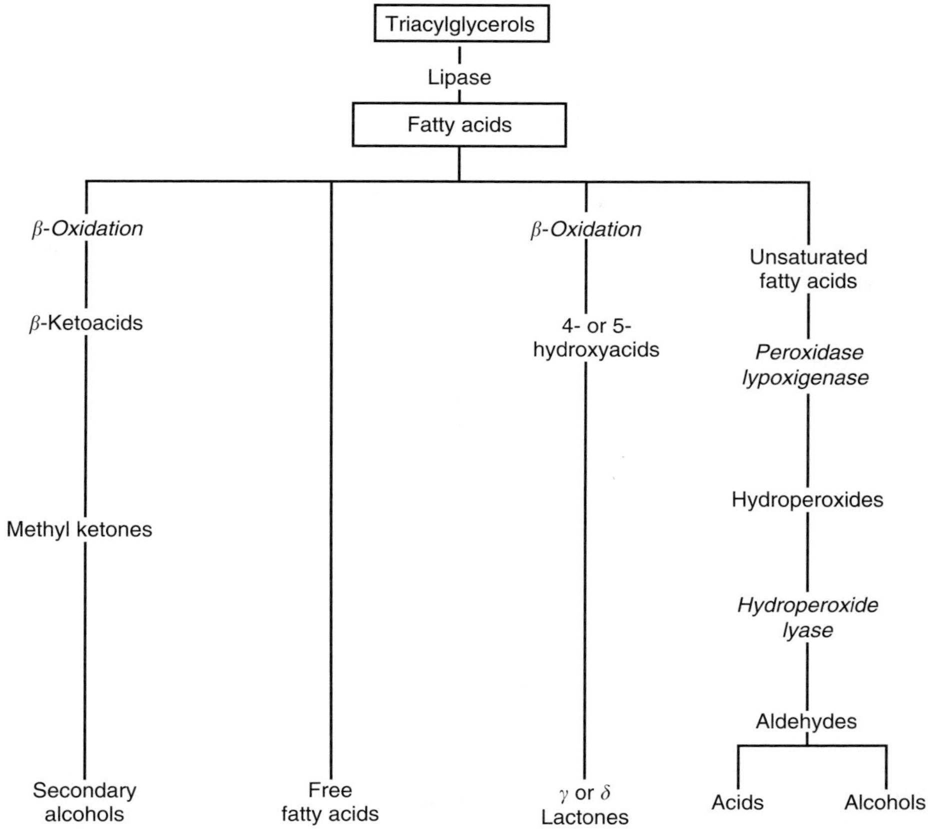

Figure 16.2 Formation of flavor compounds from lipid metabolism (adapted from Molimard, P., H.E. Spinnler, *J. Dairy Sci.* 79:169–184, 1996).

Penicillium camemberti use this metabolic pathway in order to detoxify the medium against high concentrations of lipids. This pathway is interesting because of the use of one CoA instead of two for a complete degradation of the fatty acids. β-oxidation is particularly important in the fermentation of raw material rich in fats and lipids; in some cases 60% of the carbonyl compounds produced are methyl ketones (17). Methyl ketones can be further reduced leading to the formation of different alcohols such as 2-pentanol, 2-heptanol, or 2-nonanol.

Even lactones come principally from the β-oxidation of hydroxy fatty acids. Hydroxy fatty acids can be naturally present in triglycerides or can be generated from the oxidation of polyunsaturated fatty acids by the action of lipoxygenases or hydratases. They are characterized by very pronounced fruity notes such as peach, apricot, and coconut. They are present in many products and are particularly appreciated in beverages. As an example, 21 different lactones have been identified in white wine (18,19). The metabolic pathway for odd lactones is not well understood in spite of their flavoring importance. Because odd fatty acids are not present naturally in high quantity and diversity, the corresponding lactones cannot come from β-oxidation. However, in 1978, Tressl (20) proposed another oxidative pathway for the synthesis of γ-nonalactone during beer fermentation (Figure 16.4).

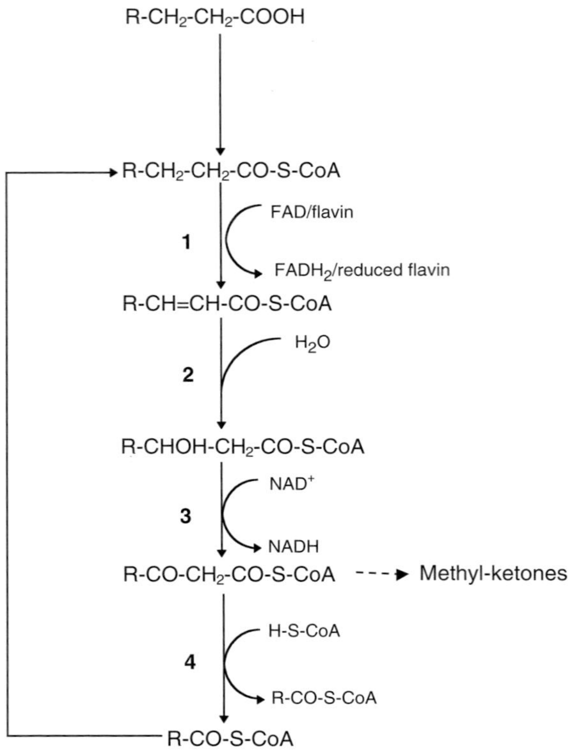

Figure 16.3 β-Oxidation cycle and methyl-ketone synthesis. Individual enzyme reactions: (a) fatty acyl-CoA dehydrogenase (mitochondria) or oxidase (peroxisomes); (b) 2,3-enoyl-CoA hydratase; (c) 3-hydroxyacyl-CoA dehydrogenase; (d) 3-oxoacyl-CoA thiolase. [FAD : Flavin Adenine Dinucleotide, NAD(H) : Nicotinamide Adenine Dinucleotide (reduced form), R= $CH_3(CH_2)_n$].

The intrachain oxidation of fatty acids can also lead to the synthesis of different alcohols and aldehydes. In particular, linoleic and linolenic acids are precursors of eight carbon aroma compounds, particularly 3,1-octenol (mushroom), 2,1-octenol, 1,5-octa-3-dienol, and 1,5-octa-1-dienol. This oxidation is catalyzed by lipoxygenases and hydroperoxide lyases with different specificities (Figure 16.2).

16.3.1.3 Amino Acid Metabolism

Proteolytic activities lead to the formation of peptides that are degraded by aminopeptidase and carboxypeptidase activities. Amino acids coming from this degradation are subjected to transamination, dehydrogenation, decarboxylation, and reduction reactions that directly contribute to the flavoring of fermented foods (17,21) (Figure 16.5).

Different classes of aroma compounds are synthesized in amino acid metabolism, such as alcohols, sulfur compounds, amines, and some organic acids. Recent work showed that, in dairy products as well as in fermented meat products, the synthesis of such compounds depends on lactic bacterial transaminases activities showing different substrate specificities (22–26). Two transaminases have been isolated and cloned. One is active on branched chain fatty acids and after a series of decarboxylation or dehydrogenation reactions leads to the formation of alcohols such as 2/3-methylbutanol, 2-methylpropanol, and the corresponding aldehydes, which are globally characterized by fruity and malty flavors.

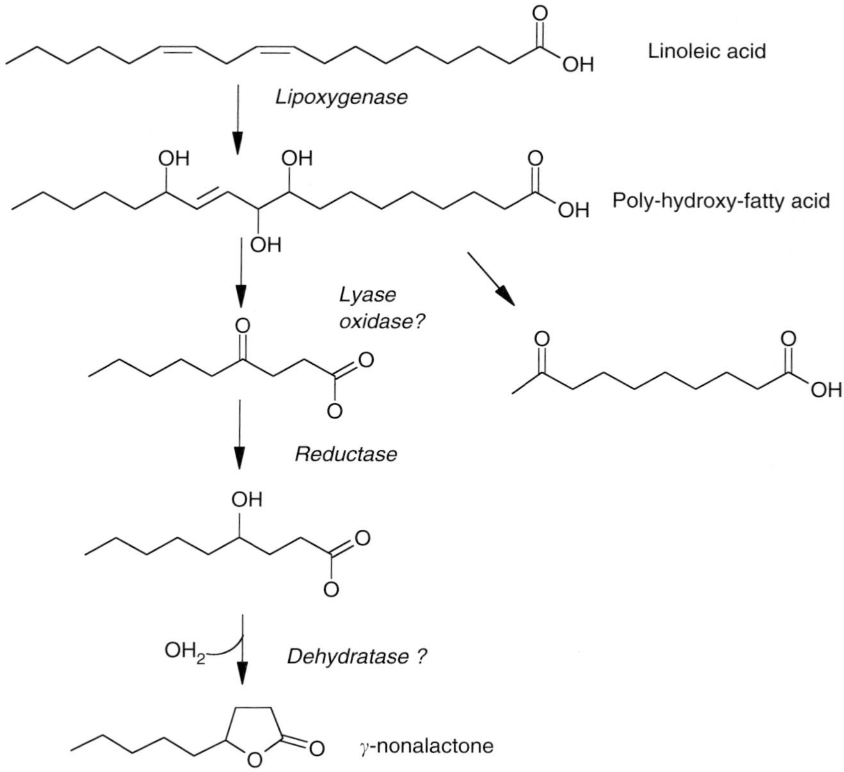

Figure 16.4 Proposed metabolic pathway for the synthesis of γ-nonalactone from lipid oxidation (adapted from Tressl, R., M. Apetz, R. Arrieta, K.-G. Grünewald, *Flavor of Foods and Beverages*, London: Academic Press, 1978, pp 145–168).

Further steps of oxidation can generate short chain fatty acids (isovaleric, 2-methyl butyric, and isobutyric acids) that are responsible for rancid, fecal, putrid, and sweaty flavors (17).

The second transaminase is active mainly on aromatic amino acids such as phenylalanine, tyrosine, and tryptophan leading to the formation of floral notes. For example, the degradation of phenylalanine can generate a broad variety of flavor compounds such as benzaldehyde (almond), cinnamaldehyde (cinnamon), veratraldehyde (woody, vanilla like), or phenyl ethanol (rose) (Figure 16.6) (7).

The degradation of methionine is directly responsible of the synthesis of sulfur compounds in fermented food products such as cheese and beer (Figure 16.7) (27,28). Sulfur compounds are flavors of major importance. They can be found in many different products in very low quantity (mainly at a ppb level). However, their detection threshold is also very low (from 0.004 ppb for isopentenylmercaptan to 1.2 ppm for methionol). Some of the main compounds described in the literature are dimethylsulfide, dimethyldisulfide, dimethyltrisulfide, methional, methionol and methanethiol. Their odors are often described as onion, garlic, radish, and cabbage, but also as excrement or putrefaction. For this reason, sulfur compounds play an important role in consumer preferences and choices.

Pyrazines are synthesized from threonine through a series of biochemical reactions (reduction, decarboxylation, and hydration) (29). Pyrazines are typical aroma components of heated foodstuffs, with a characteristic roasty nutty flavor or Maillard flavor. They are particularly found in beverages such as wine and beer, but also in meat products.

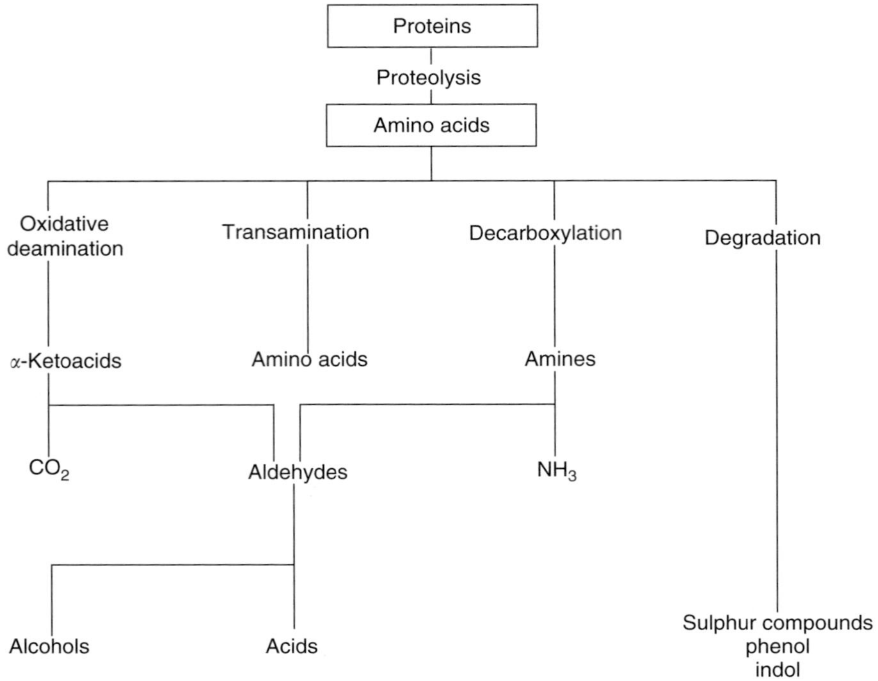

Figure 16.5 Formation of flavor compounds from protein and amino acid metabolisms (adapted from Molimard, P., H.E. Spinnler, *J. Dairy Sci.* 79:169–184, 1996).

16.3.1.4 Cometabolism

Esterification reactions occur between the different alcohols and acids synthesized from lipids and carbon metabolism enabling the formation of a large diversity of ester compounds. Esters are very important flavors in all types of fermented food. There is a great diversity of ester compounds. The most common are the acetates formed from an alcohol and acetyl-CoA. The esterification reactions can be catalyzed by carboxylesterases, characterized by a wide range of substrates, and arylesterases. The aromatic notes used to characterize esters are fruity [pineapple (ethylacetate, ethylpropanoate and ethylbutanoate), banana (isoamylacetate), apricot (isoamylpropanoate, ethyloctanoate), and pear (isoamylacetate)], floral [rose (2-phenylethylacetate)], honey (phenylethylbutyrate) or wine (ethyloctanoate) (17,21).

16.3.1.5 Conclusion

All of the aspects of microbial metabolism previously cited contribute to the flavoring of fermented foods. The large variety of such products observed all around the world depends principally on the characteristics of the raw material, the biocatalysts, the conditions and processes of fermentation and the conditions of storage. Hence, the results typically come from the combination of many different metabolic activities. As an example, Table 16.2 shows the different flavors found in dry sausage. The entire metabolism of the different microorganisms present or added to the meat is expressed and, except for the flavor coming from the spices added to the meat, all classes of compounds described in the previous chapter have been identified in dried sausage (30).

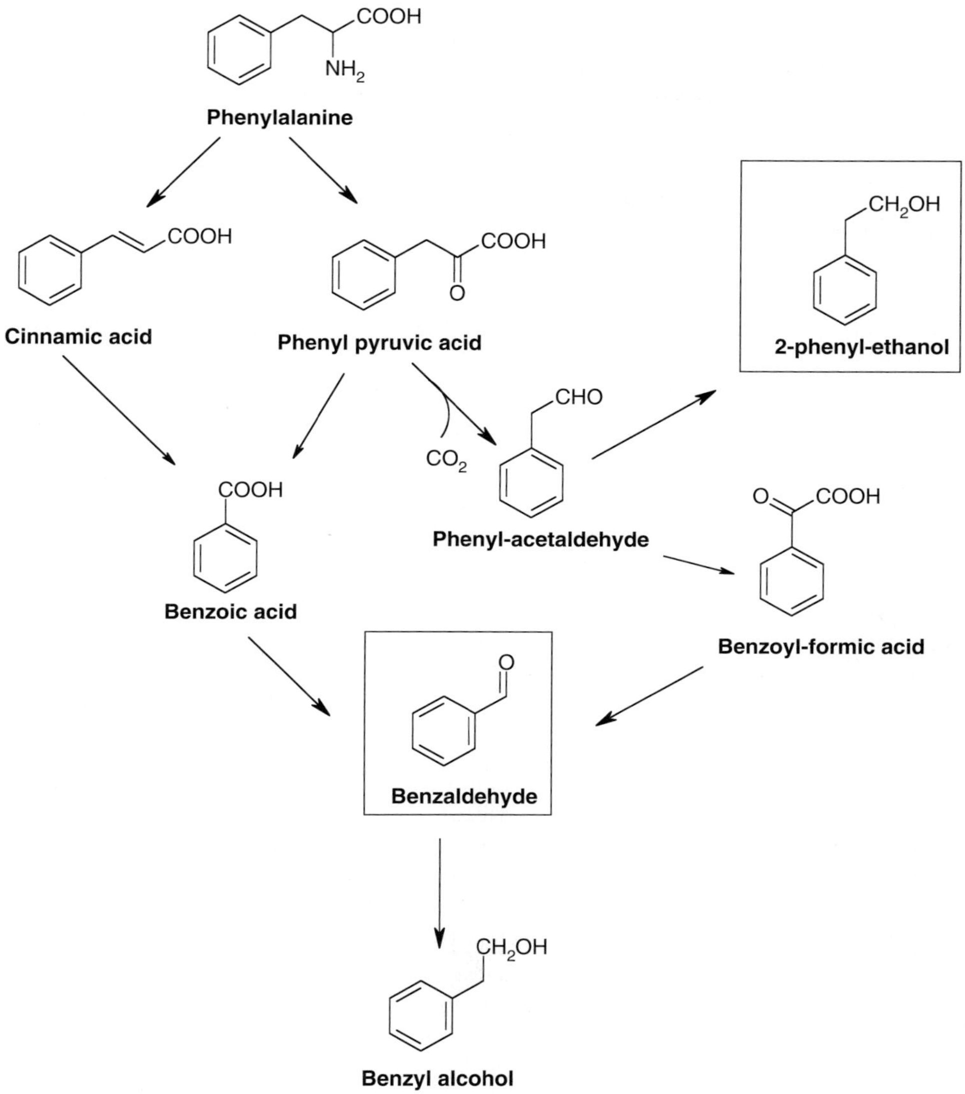

Figure 16.6 Proposed metabolic pathways for the biotechnological synthesis of benzaldehyde and β-phenethylalcohol (adapted from Feron, G., P. Bonnarme, A. Durand, *Trends Food Sci. Technol.* 7:285–293, 1996).

In parallel to traditional fermentations, many microorganisms not commonly used for the elaboration of food are able to produce many different flavor compounds *de novo*. These microorganisms can be also used by the industry for the production of pure natural flavor compounds. This is the purpose of the following sections.

16.3.2 *De Novo* Synthesis of Flavors

It has been known for a long time that some microorganisms can generate pleasant odors (31). Odor description has therefore been used for many years as a parameter of taxonomic classification for some microorganisms, especially wood destroying fungi (32). In 1992, an

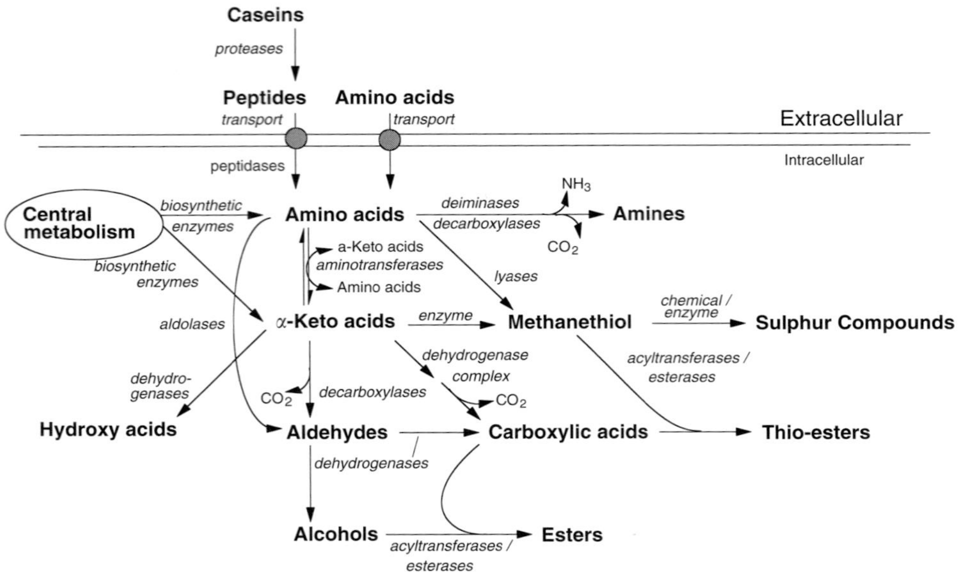

Figure 16.7 Potential routes for conversion of amino acids during cheese ripening (after Engels, W., E. Floris, R. van Kranenburg, J. van Hylckama Vlieg, W. Noordman, G. Smit. *Proceedings of the 10th Weurman Flavor Research Symposium*, Beaune, 2002, pp 326–331).

excellent review listing the flavor producing microorganisms was published by Janssens (4). A total of 130 microorganisms were cited in this paper (Table 16.3) and concerned mainly mold and fungi (72 genera and species described). This is principally due to the large biodiversity that occurs among the world of filamentous fungi and especially in the ascomycetes and basidiomycetes orders. As an example the latter can produce a wide variety of phenolic compounds such as benzaldehyde, anisaldehyde or veratraldehyde (6) (Table 16.4), as well as lactone compounds, from the γ-butyrolactone (C4) to the γ-dodecalactone (C12) (33); and also a large variety of terpenes. Moreover, this diversity can be enhanced by modifying simple constituents of the culture medium as carbon or nitrogen source. An example can be found in the genus *Ceratocystis* which is able to produce a high diversity of aroma compounds (34–36). Depending on the composition of the culture medium, we can enlarge the variety of accessible flavor compounds such as fruity or banana flavor (isoamyl acetate), weak potato (unknown compound) or peach flavor (γ-decalactone) (37).

Yeasts are also able to produce *de novo* flavors. One of the well known examples is *Sporidiobolus salmonicolor* which produces a high variety of lactones (38) such as γ-decalactone or delta lactones. Other species such as *Kluyveromyces lactis* or *Williopsis saturnus* are also capable of synthesizing large amounts of terpenes or fruity ester flavors (39).

Because of their limited genome, bacteria do not show the same biodiversity for the production of flavor. However one particular example can be found in recent works (40) showing that strains of *Bacillus cereus* were able to produce different pyrazines such as 2,5-dimethylpyrazine, 2,6-dimethylpyrazine or trimethylpyrazine. This production depended on the bacteria strains, the temperature, and the culture medium. The concentrations did not reach more than 4 mg/l of medium.

De novo synthesis of flavor can be considered as a good indication of the flavors that could be potentially accessible by biological means. However the production levels are

Table 16.2
Chemical classes of volatile constituents of dry sausage, odor descriptions and metabolic origins (adapted from Berdague, J.L., P. Monteil, M.C. Montel, R. Talon, *Meat Sci.* 35:275–287, 1993).

Volatile Compounds	Fragrant Descriptors	Metabolic Origin
Alcanes- Alcenes		
Pentane, heptane, 2-octene	None	Lipid oxidation
Terpenes		
Limonene, b-pinene	Fruity, lemon	Pepper
Aldehydes		
Pentanal, hexanal, heptanal, methylated aldehydes, benzaldehyde	Grassy, fatty, rancid, fruity, roasty, bitter almond	Lipid oxidation and amino acid metabolism
Alcohols		
Ethanol, methyl-propanol, butanediol, hexanol, 1-octen-3-ol	Piquant, ethereal, fatty, buttery, fruity, mushroom, flowery	Sugar fermentation, lipid oxidation, amino acid metabolism
Ketones		
Methyl-pentanone, pentanone, heptanone, octanone, nonanone, pentanedione, butanedione	Grassy, fruity, blue cheese, buttery, fatty, milky	Lipid oxidation, sugar fermentation
Carboxylics acids		
Acetic acid, methyl-buta/propanoic acids, buta/penta/hexanoic acids	Vinegar, "feet", "dirty socket", rancid, acid, blue cheese	Sugar fermentation, lipid oxidation, amino acid metabolism
Esters		
Methyl/ethyl-acetate, ethyl-butanoate, ethyl-propanoate, ethyl-3/2-methylbutanoate	Soft, ethereal, fruity, "green", banana, pine-apple, acide	Sugar fermentation, lipid oxidation, amino acid metabolism

very poor, and thus constitute a limit for industrial exploitation. For this reason, biotechnologists have focused on bioconversion processes that offer more economic advantages.

16.4 BIOCONVERSION

Starting from specific substrates, microorganisms are able to catalyze many different reactions such as oxidations, reductions, dehydrations, hydrolytic reactions, and other metabolic reactions in order to produce food flavors. In contrast to *de novo* synthesis, the concentrations of molecules produced are high enough to permit industrial exploitation of the bioproduction process. This potential to exploit the diversity of life for the production of added value compounds has led to a strong interest, even if the applications have not been as numerous as we have expected.

Price is one of the major obstacles for the development of bioflavors, because cost is strongly related to production volume, which is generally low. From the 400 compounds of bioorigin available, a substantial part are produced in very small quantities (less than 10 kg/year) as impact molecules. It is generally admitted that an average consumption of 100 kg/year of an additive seems to be the minimum necessary to expect to find a market for a bioflavor.

Table 16.3
Synthetic survey of flavor, de novo produced from microorganisms (summarized from Janssens, L., H.L.D. Pooter, N.M. Schamp, E.J. Vandamme, *Proc. Biochem.* 27:195–215, 1992). Numbers in () correspond to the number of different strains in a class or in a genus cited in the review.

Mould (72)	**Class of Aroma**
Aspergillus (3)	Acids, alcools, terpenes
Cerastocystis (9)	Terpenes, esters, alcohols
Fusarium (2)	Terpenes, lactones
Geotrichum (3)	Esters, lactones
Trichoderma (5)	Terpenes, esters, lactones
Levures (36)	
Dipodascus (3)	Esters, alcools
Hansenula (4)	Esters, alcools
Kluyveromyces (3)	Terpenes, esters, alcohols
Sporobolomyces (2)	Lactones
Saccharomyces (8)	Terpenes, lactones, esters, alcohols.
Bacteria (22)	
Clostridium (2)	Esters, alcohols, acids
Corynebacterium (2)	Pyrazines
Pseudomonas (6)	Esters, pyrazines
Streptomyces (4)	Terpenes, pyrazines

Table 16.5 summarizes the production cost evaluation as proposed by Delest in 1995 (41), and is based on a combination of production size, concentrations and duration. The first part of this table presents a high precursor cost with a low production size for the bioflavor. Considering the different parameters that must be taken into account, the finding of cheap raw material is one of the main limits encountered when a process for the bioproduction of flavor is investigated.

In spite of the different technical and economical constraints encountered, different processes have been described the last twenty years, as shown in Table 16.6. Some of them have led to commercially available natural flavors. Because of the industrial confidentiality commonly observed in this field, this table is not exhaustive. In fact about 80–100 flavor compounds coming from biotechnology are considered to be on the market (13).

16.4.1 Vanillin

Vanillin is considered the most widely used flavoring agent in food products with a market of 12000 t/yr for artificial vanillin (15 €/kg) and 50 t/yr for vanilla extracts (4,000 €/kg) (42). Between these two sources of aroma, a third possibility was envisaged given that the development of biotechnology permits the access to a biovanillin with an estimated market of 5,000 t/yr and an average price of 800–1,000 €/kg.

Biovanillin has long been considered as the Holy Grail of biotechnologists and flavoring companies with an impressive quantity of processes, papers, reviews and patents published on the subject (43). Figure 16.8 summarizes briefly the most important routes investigated in this field.

Table 16. 4
Volatile aromatics compounds produced by filamentous fungi (adapted from Berger, R.G., Berlin: Springer-Verlag, 1995, and Lapadatescu, C., PhD dissertation, Université de Bourgogne, France, Dijon, 1999).

Mould	Aroma
Agaricus bisporus	benzaldehyde, phenyl acetaldehyde
Ascoidea hylecoeti	2-phenyl ethanol
Bjerkandera adusta	anisaldehyde, veratraldehyde
Camarophyllus virgineus	anisaldehyde
Hebeloma saccariolens	2-amino benzaldehyde
Hyanellum suaueolens	coumarine
Inocybe sp.	methylcinnamate
Ischnoderma benzoinum	benzaldehyde, anisaldehyde
Lentinus sp.	benzyl acetate, methyl anisate, methyl cinnamate
Hycoacia vda	methyl acetophenone, alcool methyl benzylique, *p*-tolualdehyde
Nidula sp.	Cinnamic acid derivatives, raspberry
Phanerochaete chrysosporium	veratraldehyde
Phellinus sp.	methylbenzoate, salicylate
Pleurotus euosmus	coumarine
Poria sp.	Cinnamic and anthranilic acid derivatives
Pycnoporus cinnabarinus	methylanthranilate, vanilline
Sirodesmium diversum	*p*-hydroxybenzaldehyde
Stereum subpilatum	methylcoumarate
Trametes sp.	anisaldehyde, methylphenylacetate
Tyromyces sambuceus	benzaldehyde, ethylbenzoate

Table 16.5
Production cost evaluation of biotechnological compounds (adapted from Delest, P., Paris: INRA, 1995, pp. 13–19.)

Raw Materials (€/kg)	120	15	1.5
Fermentation/Bioconversion (€/kg)	280	80	4
Recovery/purification (€/kg)	80	30	2.5
Total (€/kg)	460	120	8
Assumptions Concentrations (g/l)	10	60	50
Duration (Days)	2	1	2
Reactor Size (m3)	1	5	100

The first involves the transformation of natural stilbene as isorhapontin to vanillin via an oxidation reaction catalyzed by a stilbene dioxygenase. This process has led to numerous patents (5).

The most promising route may be the transformation of eugenol, via coniferaldehyde or ferulic acid, to vanillin, because eugenol is an inexpensive and commercially available natural raw material. However, production is still very low, mainly due to numerous side reactions. An alternative solution could come from the use of isoeugenol which provides better production rates, but natural isoeugenol is not available in large volume,

Table 16.6
Principal natural flavoring substances produced by biotechnological means

Compounds	Aromatic Note	Estimated Market T/Yr (1997)	Precursor	Biocatalyst (Not Exhaustive)	Process	Concentration g/l	Commercial Availability	Ref.
Vanilline	Vanilla	5000	Ferulic acid	*Amycolatopsis sp.* *Pycnoporus sp.*	Liquid	11	Yes	(42,129, G. Feron, personal data, 1997)
γ-decalactone	Fruity/peach	5–10	Ricinoleic acid	*Yarrowia lypolitica* *Sporidiobolus sp.*	Liquid	10–15	Yes	(131)
δ-decalactone	Nuts/milk/ coconut/fruity	3.5	Massoia lactonez Other ?	*Saccharomyces cerevisiae* Various filamentous fungi	Liquid	> 1	Yes	(132,133, G. Feron, personal data, 1997)
γ-dodecalactone	Peach/musk	0.5	Oleic acid	Strain 458 + *Saccharomyces cerevisiae*	Liquid	> 5 ?	Yes	(56,57, G. Feron, personal data, 1997)
δ-dodecalactone	Butter	?	5-ketododecanoic acid	*Saccharomyces cerevisiae*	Liquid	< 1	Yes	(132,133, G. Feron, personal data, 1997)
γ-octalactone	Coconut	0.6–1	Esters of octanoic acid	*Mucor sp.*	Liquid	~ 11	Yes	(134, G. Feron, personal data, 1997)
Alcohols, aldehydes : Cis-3-hexenol Trans-2-hexenal, hexanal…	« leaf alcohol », fruity	5–10	Poly-unsaturated fatty acids	Soybean lipoxygenase	Liquid and HPL	2–5	Yes	(67)
Raspberry-ketone	Raspberry	5–20	Betuloside	Enzyme (glycosidase) + yeast (*S. cerevisiae*)	Liquid	1	No	(46,47, 130, G. Feron, personal data, 1997)
2-phenylethanol	Rose, honey	0.5–1	Phenyl-alanine	*Pichia sp., S. cervisiae, Kluyveromyces sp.*	Liquid	2–3	Yes	(54)
Benzaldehyde	Almond	115	Phenyl-alanine	*Bjerkandera adusta, Pycnoporus cinabarini*	Liquid	< 1	No	(135)
Sotolon	Curry	?	OH-isoleucine	*M. morganii*	Liquid	?	?	(136,137)
Pyrazine	Roasty, nutty	?	Threonine	*Bacillus subtilis*	Solid	2–3	?	(29)
Methyl-ketones	Blue cheese notes	?	Coconut fat	*Aspergillus niger*	Solid	? (Yield = 42%)	yes	(4)

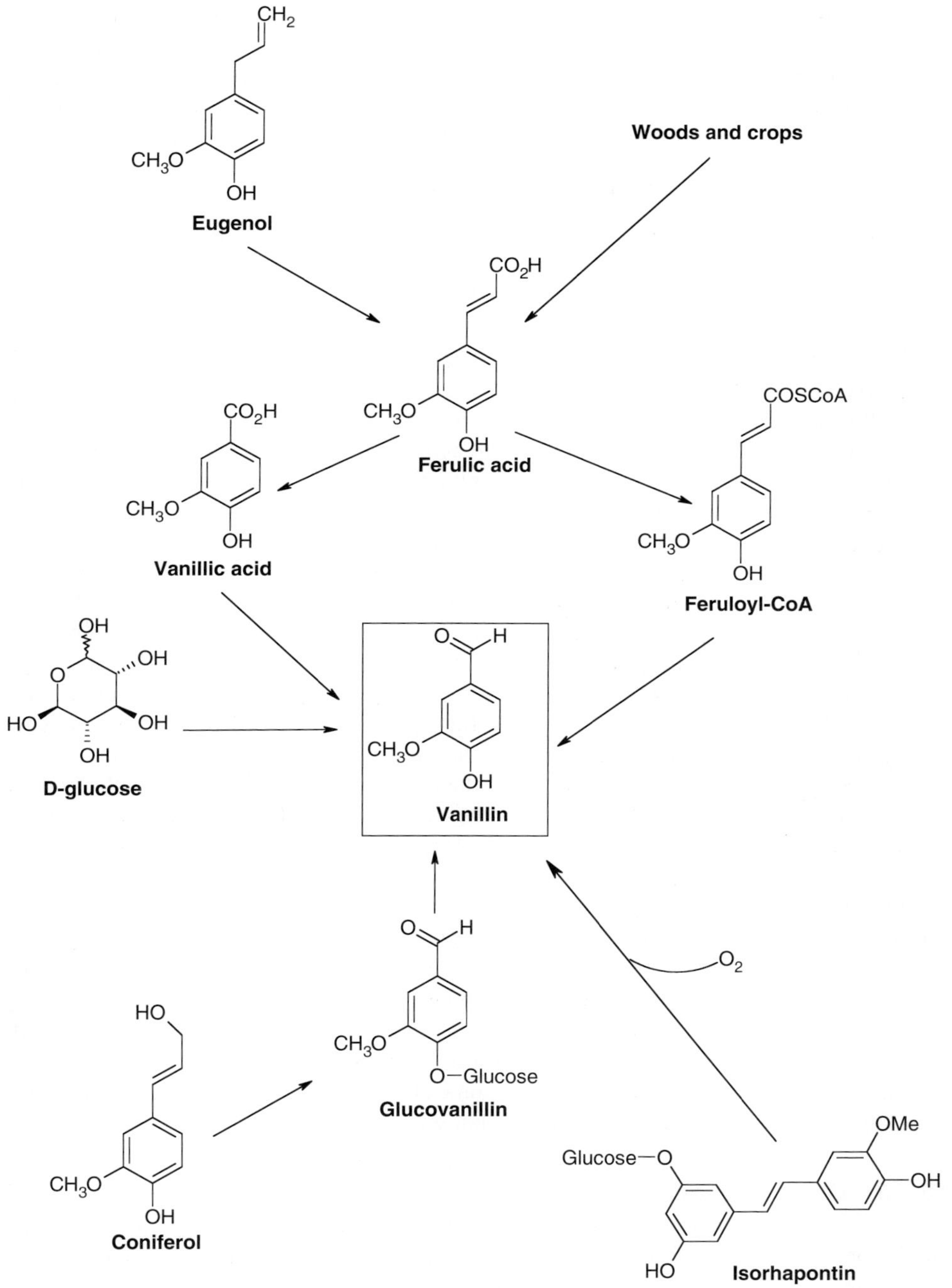

Figure 16.8 Proposed metabolic pathways for the biotechnological synthesis of vanillin (adapted from Feron, G., P. Bonnarme, A. Durand, *Trends Food Sci. Technol.* 7:285–293, 1996; Priefert, H., J. Rabenhorst, A. Steinbüchel, *Appl. Microbiol. Biotechnol.* 56:296-314, 2001; and Walton, N.J., A. Narbad, C. Faulds, G. Williamson, *Curr. Opinion Biotech.* 11:490–496, 2000).

which limits its use for vanillin production. In the 1990s, studies tended to focus on the direct use of ferulic acid. This acid is one of the constituents of various grasses and crops, and is also a product of the microbial oxidation of lignin. For these reasons it constitutes an excellent precursor. However its recovery as pure precursor was difficult, which was one of the main limits for its use. The first and most important advance in the field was published in 1997 with the work of Rabenhorst (for H&R company) (44) that described a two step process, one step for the bioconversion of eugenol to ferulic acid by a novel *Pseudomonas* species and the second step describing the conversion of ferulic acid to vanillin by *Amycolatopsis* species. The yield of the bioconversion attained was 78% with a final concentration of 11.5 g/L. In parallel with Rabenhorst's work, the development during the ame decade of molecular tools and genetic engineering opened the possibility of overcoming the different drawbacks mentioned, and led, very recently, to the development of industrial processes (Section 16.4.6.1).

16.4.2 Raspberry Ketone or Frambinone

Raspberry ketone (para-hydroxyphenyl-butan-2-one) is the key compound of raspberry flavor. It is present in very low concentration in the berry (from 0.1 to 2.0 ppm). Its threshold of detection is around 1 to 10 ppb. As an additive in food products, it is used at concentrations from 5 to 50 ppm. Unlike vanilla flavor, natural raspberry ketone is not available because of the high price of extracting it from the corresponding fruit (45). From a financial point of view, natural raspberry ketone occupies the second place after vanillin with a total theoretical market comprising between 6 and 10 millions euros.

If biovanillin was considered as the Holy Grail for some biotechnologists, frambinone (raspberry ketone) is more like the Lost Ark. The difficulty comes principally from the absence of a natural cheap precursor close to the final product and available at enough quantity. However, in spite of these limits, some attempts to produce bioframbinone have been described in the literature. They are summarized in Figure 16.9. The first one concerns the bioconversion of betuligenol, the chosen precursor of frambinone, and betuloside, its glucoside. The later is found in rhododendron, birch, alder, yew, maple, and fir. In the patented processes (46,47), the chosen betuloside source was silver birch, especially the bark. After extraction from the bark, the betuloside was hydrolyzed with a β-glucosidase and then oxidized using yeast alcohol dehydrogenase, which appears to be an elegant and innovative bioprocess. To our knowledge there is not an industrial application yet, mostly due to technological constraints and costs.

Another interesting route was proposed by Whitehead in 1998 (48). The approach consists of exploring the possibility of developing a process close to the chemical pathway of synthesis, that is, condensation of 4-hydroxy-benzaldehyde with acetone. The main difficulty results in the finding of a biocatalyst able to condense, via an aldolase reaction, acetone and a phenol compound. However, some aldolases of biological origin are capable of acting on a large spectrum of substrates, but with variable catalytic constants (49).

16.4.3 Other Phenolic Compounds

As described in the Figure 16.6 and in section 3.1.3, phenylalanine is the source of different flavor compounds in traditional fermented products. As phenylalanine is an abundant and cheap compound, it can be also used as natural substrate for the biosynthesis of natural benzaldehyde and natural β-phenethyl alcohol.

16.4.3.1 Benzaldehyde

In quantity, benzaldehyde is the second most important molecule, after vanillin, for its use in the flavor and fragrance industries. Natural benzaldehyde is used as an ingredient in cherry and other natural fruit flavors. It is generally extracted from fruit kernels such as

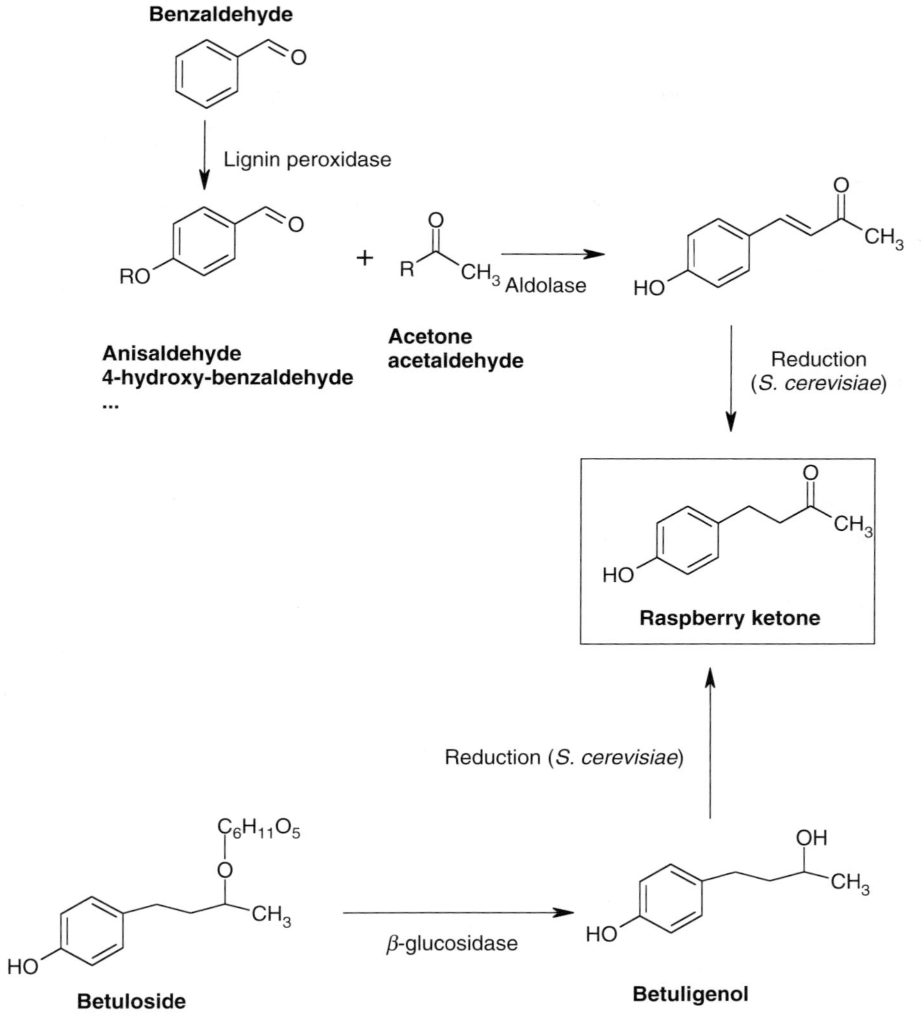

Figure 16.9 Proposed metabolic pathways for the biotechnological synthesis of raspberry ketone (adapted from Dumont, B., P. Hugueny, J.M. Belin, French Patent Applications FR 2 724 666 A1, 1994; Whitehead, I.M, *Food Tech.* 52:40–46, 1998; and Hugueny, P., B. Dumont, F. Ropert, J.-M. Belin, *Bioflavor 95*: INRA Editions, 1995, pp 269–273).

apricots. Natural benzaldehyde has a market of ~20 t/yr and a price of ~240 €/kg. Benzaldehyde obtained from natural cinnamaldehyde can be purchased for 100 €/kg with an estimated market of more than 100 t/yr. However its classification as "natural" remains questionable in view of European Union rules.

The fermentation of phenylalanine offers an alternative route for the biosynthesis of natural benzaldehyde. This route has been studied in a mutant of *Pseudomonas putida* (50) and in the white rot fungi *Polyporus tuberaster* (51). However, production via a fermentation process could become commercially acceptable only if sufficient yields can be obtained, which is not the case as yet. One limitation to scaling up the process is the toxicity of the product (52).

16.4.3.2 2-phenylethanol

After vanillin and benzaldehyde, 2-phenylethanol shows a sizeable market. Its bioproduction is of industrial interest because its price is ~1000 €/kg with an average market of 0.5–1 t/yr (53,54). The production of 2-phenylethanol can be done by harnessing the Ehrlich pathway of different yeast species by bioconversion of L-phenylalanine. As observed for the production of benzaldehyde, above a certain concentration attained, 2-phenylethanol is highly toxic for the biocatalyst. *In situ* product removal appears to be necessary to overcome toxicity and inhibition of the reaction of bioconversion.

16.4.4 Lactones and Methyl Ketones

Lactones constitute an important class of value added bioflavoring additives (from 6000€ to 1200€ with an average market of 5T/yr). They are used in many flavoring preparations like butter or fruit. The general strategy used for the production of biolactones consists of the reduction of the aliphatic chain of hydroxy fatty acids carried out via the β-oxydation system as described in Figure 16.10 with the catabolism of castor oil. After a number of cycles of β-oxydation, the CoA ester of the reduced hydroxyl fatty acid is released from the β-oxidation complex. Usually, the precursor used for the production of a lactone is a hydroxy fatty acid. Depending upon the position of the hydroxy group on the aliphatic chain, this acid could lead to the formation of a γ-lactone or a δ-lactone. In fact, the only natural precursor available in sufficient quantity and at a cheap enough price is ricinoleic acid [(*R*12)-hydroxy-(Z9)-($C_{18:1}$)], the major fatty acid (90%) of castor oil, which is obtained from the plant *Ricinus communis*. In this case, the added value obtained by bioconversion of the product (γ-decalactone) is sufficient for commercial exploitation of the bioconversion system, as is evident from

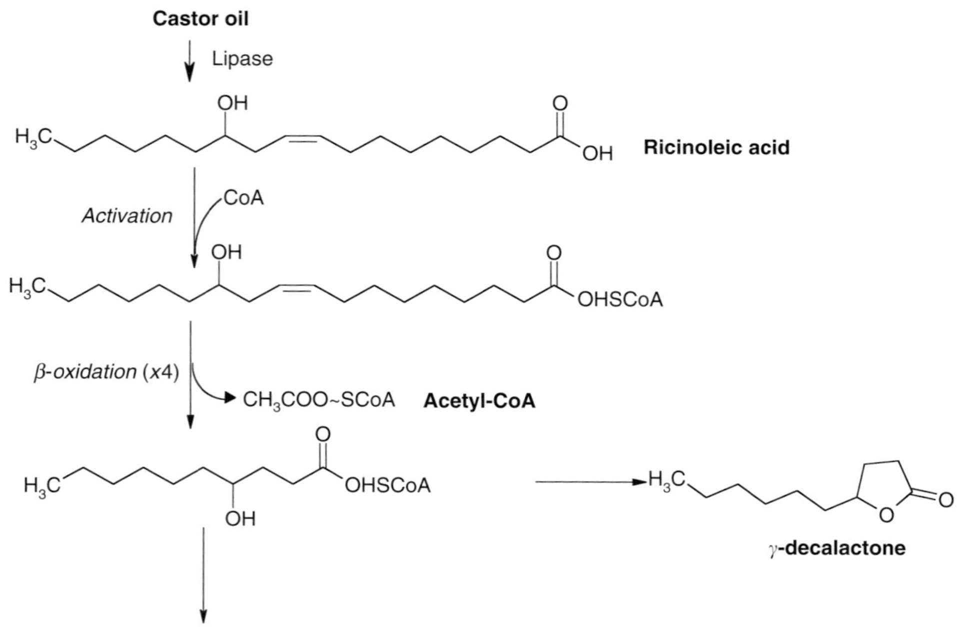

Figure 16.10 β-oxidation of ricinoleic acid and 4-hydroxy-decanoic acid lactone synthesis. Original pathway as deduced from the works of Hutton and Stumpf in plant (Hutton, D., P.K. Stumpf, *Arch. Biochem. Biophys.* 142:48–60, 1971) and Okui *et al.* in *Candida sp.* (Okui, S., M. Uchiyama, M. Mizugaki, *J. Biochem.* 54:536–540, 1963).

the numerous patents on this production system (33). To a lesser extent, 4-decanolide can be obtained from coriolic acid and δ-decalactone can be produced by *Saccharomyces cerevisiae* from 3,11-dihydroxymyristic acid, which is present in Jalap resin (55).

Because there are few natural sources of natural hydroxyl fatty acids, processes have been developed for their biotechnological production, such as the microbial production of 10-hydroxystearic acid for the synthesis of γ-dodecalactone (56,57). Another promising possibility for the future might be the use of isolated enzymes that act on natural polyunsaturated fatty acids, as has been successfully applied using lipoxygenases (58). In this case it might be possible to obtain new lactones. However, in some cases, the production yield is closely related to the structure of the hydroxy fatty acids used as precursors, and this selective affinity has not yet been explained. As described in § 4.6.1., novel knowledge obtained with genetic engineering on producing species can help to understand these physiological particularities.

Methyl ketones are synthesized from the β-oxidation of short chain fatty acids (Figure 16.3, Section 16.3.1.2), with their typical blue cheese aroma and they are commonly used for the flavoring of many dairy products, but are also used in soups, meat products, cooked food, and dressings (59). The major studies have focused on the use of spores of *Penicillium* spp. or *Aspergillus* spp. with oil and fats from plant origin, essentially coconut. The innovative aspect of the work was the development of solid-state fermentation (SSF) processes that are particularly adapted to filamentous fungi cultures. This technique was developed at an industrial and commercial level by the British company, Stafford Specialty Ingredients (4). To our knowledge, this is the only example of the production of bioflavor by SSF at an industrial scale. The production of pyrazine from threonine (29) has recently been reported with the same mode of culture, but industrial exploitation of the results is not mentioned.

16.4.5 Green Notes

Volatile aldehydes and alcohols are key compounds in the fresh and green sensorial notes of vegetables and fruits (60). They are produced by plants in response to various stresses, and they play a major role in plant defense mechanisms (61). The pathway is summarized in Figure 16.11. Polyunsaturated octadecanoic acid is oxidized to 9-, 10-, or 13-hydroperoxides depending on the specificity of the lipoxygenase catalyst, and these compounds are then cleaved by hydroperoxide lyase into C6- C9- or C10-aldehydes, which can be reduced to the corresponding alcohols by yeast alcohol dehydrogenase. This metabolic pathway is a typical plant pathway, which has been mimicked in order to produce natural aldehydes and alcohols (62–65). The highest demand from the flavor industry concerns *cis*-3-hexenol and *trans*-2-hexenal (66).

The worldwide market for natural green notes is estimated to be 5–10 metric tons per year with an average price of 3,000 €/kg. With the growing demand for these natural flavoring compounds, efficient and viable industrial scale processes have been developed, starting from readily available polyunsaturated fatty acids (oleic, linoleic, and linolenic acids) and using plant and microbial systems, essentially soybean lipoxygenase and hydroperoxyde lyase from different plant origins, then alcohol dehydrogenase from *S. cerevisiae* (67). Presently, some flavoring companies produce up to 50% of their green notes compounds by these biotechnological processes.

16.4.6 Prospects

The previous paragraphs illustrate different approaches and processes that have been developed for the production of natural bioflavor. However, in spite of many advances in this field, many applications have shown some limits for commercial exploitation. This is

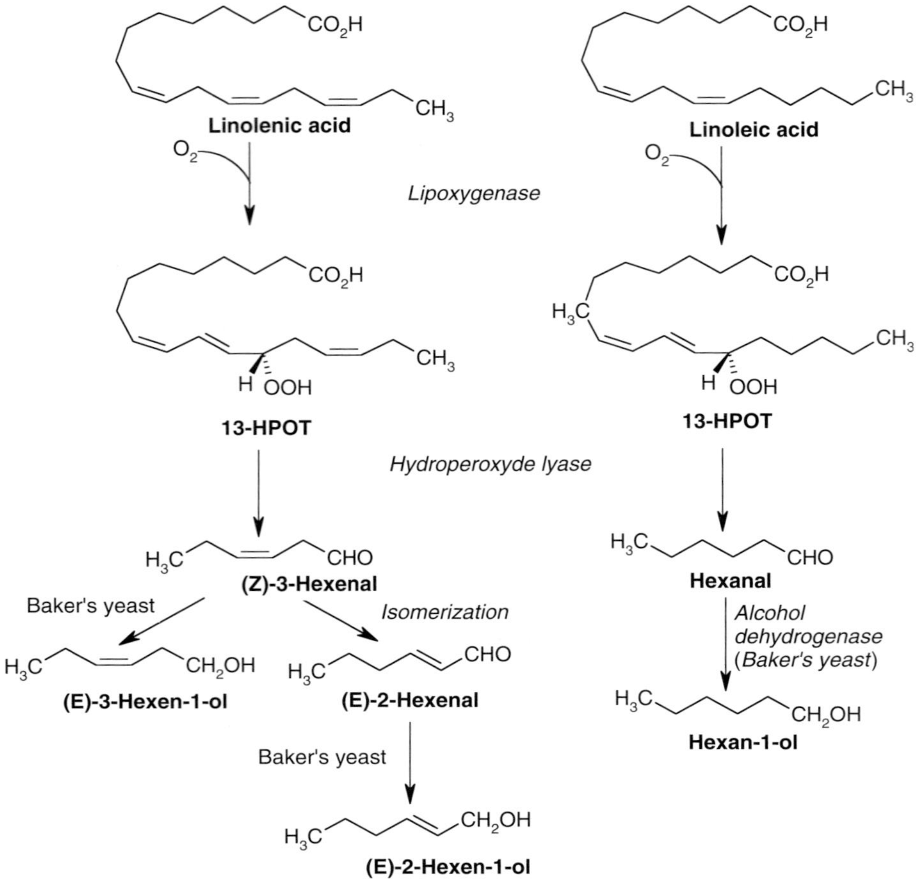

Figure 16.11 Degradation of polyunsaturated C18-acids in the lipoxygenase/hydroperoxide lyase pathway and synthesis of green notes (adapted from Muller, B.L., C. Dean, I.M. Whitehead, *Bioflavor 95*: INRA Editions, 1995, pp 339–344).

mainly due to the lack of basic information on the metabolism of the related compounds. Nowadays, research tends to focus on better control of the different metabolic pathways in order to preferentially orient the fluxes of these pathways to the desired compounds. The following sections present some recent and innovative studies that have been used for this purpose and that could appear as new prospects for the development of the field.

16.4.6.1 Genetic Engineering

Natural synthesis of flavor compounds occurs in tissues where both low concentrations of substrates and weakly active enzymes are present. As described above, biotechnologists try to mimic Nature by mixing tissues or cells possessing high concentrations of substrates and very active or highly specific enzymes. In the genetic era it is expected that they also propose to increase the substrate concentration or the catalyst activity, or even to modify this activity through genetic engineering. As a matter of fact, Nature itself continuously modifies genomes of all species, resulting in the diversity present on Earth. For instance, ricinoleic acid, the substrate for the synthesis of γ-decalactone, is only present at trace levels in most species, but it constitutes 70 to 80% of hydrolyzed castor oil. To explain this specificity, it can be noted that the substitution of only four amino acids can change a 12-desaturase activity into a 12-hydroxylase one (68).

However, genetic engineering is not so easily understandable by consumers, and the public perception of biotechnology and especially of genetic engineering is, particularly in Europe, rather weak (69). It is true that this perception is partly due to the lack of understanding (69) but also to the philosophical, social, and economical dimensions of the subject. When a gene from an equatorial plant (produced by the third world agriculture) is added to a microbial cell to enable production of exotic flavors in developed countries' reactors, two different questions arise; one about the unnaturalness of the obtained product, and the other about the behavior of rich countries toward poor ones. However, the use of microorganisms to produce aroma compounds in closed reactors is not comparable to open air culture and the resulting flavor, which contains no genetically modified DNA or protein, may not even require specific labeling in the new European rules. Nevertheless, except for the production of specific compounds, genetic engineering in the production of aroma compounds is not significantly booming. This can be explained by the still developing legislative framework in Europe [firms dare not yet utilize genetically modified organisms (GMOs) as catalysts to produce natural products] and the fact that rules are not so constraining in North America. In the USA, for a product to be labeled "Natural" often only requires one "natural" step, which can be completed by the use of classical (nonmicrobial) chemical techniques. In this moving context, it is obvious that many aroma producers develop genetic engineering processes without publishing them (although some processes are patented). This part of the chapter will focus on the production of some compounds that will illustrate the various approaches employed. The main strategies will be summarized in the conclusion of this section.

16.4.6.1.1 Diacetyl Diacetyl is derived from pyruvate through an α-acetolactate degradation pathway (Figure 16.12). Although the oxidative decarboxylation can be catalyzed by enzymes, it occurs mainly spontaneously in media in the presence of metallic catalysts or with a suitable electron acceptor (70,71). In *Lactococcus lactis* strains, the main pathway results in the accumulation of lactic acid (homolactic). Other pathways can convert pyruvate to other end products including acetic acid, ethanol, diacetyl, acetoin, and 2,3-butanediol (heterolactic or mixed acid fermentation). As the acetolactate decarboxylase enzyme (coded by *aldB*) converts acetolactate to acetoin without enabling the production of diacetyl, this enzyme was the first target to increase production (72). However, production with the *aldB* mutant strain was not so high because the homolactic pathway was dominant. To significantly increase the yields required a rerouting to the heterolactic pathway. As shown in Figure 16.12, the homolactic pathway consumes NADH,H^+. The first step of the heterolactic pathway is redox neutral (produces NAD^+). The importance of the redox balance of cofactors led Lopez de Felipe et al. (73) to modify it by adding an NADH,H^+ oxidase coding gene in order to direct the fluxes into the mixed acid pathway. This was done with success giving rise to a high amount of acetoin, but required the combined disruption of *aldB* to significantly increase the production of diacetyl (74) as, in adequate conditions, oxidative decarboxylation could take place giving rise to diacetyl instead of acetoin (75). A similar strategy could be followed using a classical mutagenesis approach selecting mutants deficient in acetolactate decarboxylase and possessing a low lactate dehydrogenase activity. With that two step mutagenesis approach, mutants of *Lactococcus lactis* subsp. *lactis* biovar Diacetylactis were obtained by Monnet et al. (76) in a nongenetically modified way, but the resulting strains were very sensitive to aerobic conditions and required catalase or yeast extract to grow in the presence of oxygen. In all these cases, the objectives were not so much the production of the aroma compound diacetyl, but the utilization of dairy strains overproducing it in milk products.

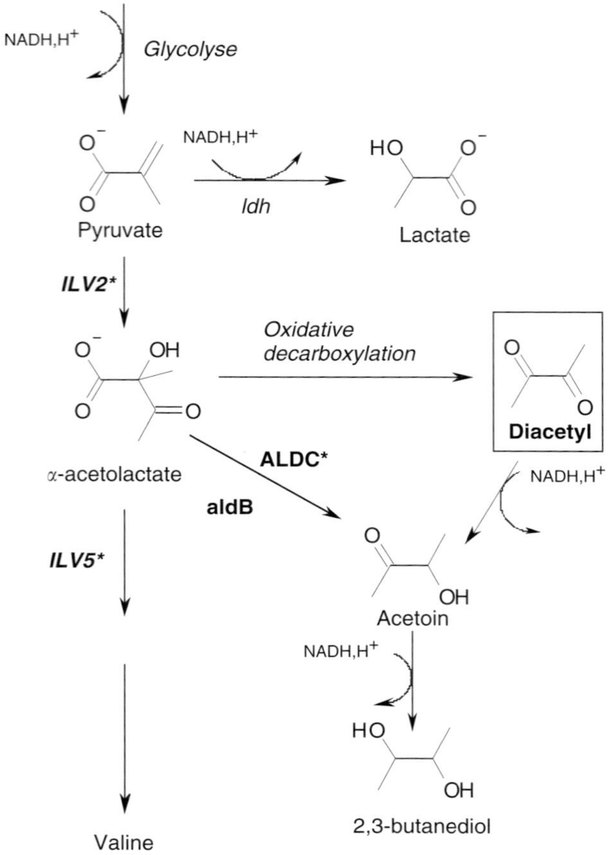

Figure 16.12 Schematic representation of pyruvate metabolism in *L. lactis* (a) and *S. cerevisiae**(b). (a): Pyruvate resulting from the glycolysis pathway can be converted into lactate or into α-acetolactate which is readily decarboxylated into diacetyl (through an oxidative decarboxylation) or, after catalysis of acetolactate decarboxylase (encoded by *aldB*) into acetoin and 2,3-butanediol. (b): Pyruvate is converted to the unstable acetolactate by an acetohydroxyacid synthetase (encoded by *ILV2*) and later to valine (using the reductoisomerase *ILV5*). Some acetolactate can be degraded into diacetyl and therefore, an heterologous acetolactate decarboxylase (ALDC) can be added to favor the pathway leading to acetoin.

It can be noted that in other products such as beer, diacetyl is an important off flavor which can be detected at rather low concentrations (below 0.02 to 0.10 mg/L). Genetic engineering tools have been used to decrease its production in brewer's yeast. In yeast, the metabolic pathways are different from what they are in lactic acid bacteria, and α-acetolactate is an intermediate in the biosynthesis of valine (Figure 16.12). Diacetyl, when formed, can be reduced in yeast into acetoin and 2,3-butanediol, but this requires sometimes several weeks (77). The various strategies have been described by Hansen and Kielland-Brandt (78). Approaches interfering with the valine pathway have been attempted through *ILV2* inactivation or *ILV5* overexpression (79,80). Rerouting from α-acetolactate to acetoin was also tried through the expression of *E. aerogenes* α-acetolactate decarboxylase gene (81).

16.4.6.1.2 Vanillin and Precursor Compounds The production of vanillin or its precursors is probably the field that gave rise to the highest number of patents for genetically modified strains. Several strategies are used to increase the synthesis of vanillin or of its precursors, and to decrease degradation with genetic manipulations or natural mutations. Most are derived from the microbial pathways described previously, but some authors have also proposed genetic construction enabling new pathways (from glucose) (82).

Because vanillin is very reactive, it exhibits a toxic effect to most microorganisms, which convert it rapidly to less toxic compounds (43). The first approach was to select mutants from the producing strain that are unable to degrade vanillin. The genetic approach of disrupting a target gene such as the one encoding vanillin dehydrogenase (83) resulted in the realization that many nonspecific dehydrogenases were able to convert vanillin (43). Thus, more classical strategies consisting in the isolation of strains on a screening medium were used to select strains unable to grow on vanillin (84), or unable to degrade coniferyl aldehyde (85). Transposon mutants were also used to identify the disrupted genes (86). It was thus observed that mutants disrupted for *vanA* and *vanB* were unable to degrade vanillic acid. These genes code for components of the vanillate-*O*-demethylase which catalyzes the first step of vanillate catabolism.

Other strategies using heterologous DNA were also used with the expression of a phenolic stilbene dioxygenase from *P. paucimobilis* in *E. coli* (87), of a ferulic acid decarboxylase from *S. cerevisiae* W3 in a strain used in the production of shochu, a Japanese distilled liquor (88), or even of the pathway from glucose to vanillin in *E. coli* KL7 (89).

16.4.6.1.3 Biotransformation of Methyl Ricinoleate into γ-decalactone Fewer patented processes have been disclosed for the genetic engineering of γ-decalactone, but several approaches have been attempted, mainly with the yeast *Yarrowia lipolytica*, an organism possessing several families of lipid degrading enzymes (90) for which genetic tools have been developed. The use of this species to produce lactones has been reviewed recently (91).

As shown previously (Section 16.4.4), the biotransformation pathway includes the introduction of a C18 hydroxy acid into the β-oxidation pathway and its exit at the C10-level for the lactonization step (spontaneous). Two main reactions can cause a decrease in the yields: the further β-oxidation of the C10-precursor, or the degradation of the lactone through the same pathway and the exit of the pathway at another stage of the β-oxidation cycle resulting in 3-hydroxy-γ-decalactone.

16.4.6.1.3.1 Acyl-CoA Oxidases of Y. lipolytica Conserved nucleotides blocks from acyl-CoA oxidases encoding genes of *S. cerevisiae* (*POX1*-M27515), *C. maltosa* (*AOX1* and *POX1*-X06721 and D21228) and *C. tropicalis* (*PXP4*, *PXP5* and *PXP2*-M12160, M12161, and P18259) were used to amplify fragments of the genes encoding *Y. lipolytica* acyl-CoA oxidases. Five genes (*POX1* to *POX5*) coding for five acyl-CoA oxidases (Aox1p to Aox5p) have been detected in the yeast *Y. lipolytica* (92). Aox2p and Aox3p are respectively long and short chain specific, whereas Aox1p does not exhibit any activity and Aox4p and Aox5p are active on a wide range of substrates (93).

16.4.6.1.3.2 Blockage of β-oxidation on Short Chain Compounds For yeast species possessing several forms of acyl-CoA oxidases (*C. tropicalis, C. maltosa,* and *Y. lipolytica*), chain length specificities have been observed. The strategy then consists in deleting genes encoding short chain specific enzymes to maintain good activity between C18 and C10 and to block it for shorter acyl-CoA. This requires a good knowledge of the characteristics of the enzymes, which is not always easy to acquire. For *Y. lipolytica*, these characteristics have been obtained by studying how mutant strains grow on methyl ricinoleate, and how γ-decalactone accumulates (94,95). This resulted in the construction of a Δ*pox2pox3pox5* strain with a multicopy insertion of *POX2* (96) but, as this construction was still able to degrade γ-decalactone, the *pox4* gene was also deleted giving rise to a strain able to produce 10 times more γ-decalactone than the wild type and unable to degrade it (97).

16.4.6.1.3.3 Decreasing Acyl-CoA Oxidase Activity in Order to Modify the Fluxes Knowledge of how the global acyl-CoA oxidase was targeted to the peroxisome (where the β-oxidation takes place) was necessary to improve the construction. This was explained by Titorenko et al. (98) who observed that acyl-CoA oxidase was targeted as a heteropentamer including each Aoxp and that Aox2p or Aox3p was necessary for a good assembly of the polymer. Without these Aoxp, the acyl-CoA oxidase activity was weak.

Because *Y. lipolytica* was able to produce more 3-hydroxy-γ-decalactone than γ-decalactone, it was proposed that the acyl-CoA oxidase activity was higher in the β-oxidation flux than the 3-hydroxyacyl-CoA dehydrogenase. Attempts to decrease the acyl-CoA activity in order to increase the production of γ-decalactone were made with success (99). A similar strategy of decreasing β-oxidation fluxes have been carried out with a completely different approach. A process has been proposed using high cell concentrations of a nongrowing Po1d, a *leu⁻ura⁻*derived strain (100).

16.4.6.1.3.4 Acyl-CoA Oxidase Activity Design The current work on acyl-CoA oxidases and lactone production deals with the selectivity of these enzymes. The activity of Aox2p and Aox3p have been characterized with *pox2* and *pox3* genes expressed in *E. coli* (101,102) and protein shuffling experiments between these two forms are being carried out (103).

16.4.6.1.4 Aldehydes (Green Notes) Although biotechnology uses the natural pathway utilizing naturally concentrated sources of substrate (vegetable oil) and enzymes (e.g., soy bean flour) (Section 16.4.5), there is a scaling up problem resulting from the instability of hydroperoxide lyase. Whether this enzyme is sensitive to hydroperoxides or to the reaction products is not known, but its activity rapidly decreases when employed, suggesting a suicidal behavior of the enzyme (65). Moreover, crude extracts are often associated with isomerase activities converting *cis*-3-hexenal into *trans*-2-hexenal (60). As many hydroperoxide lyase encoding genes have been characterized, cloned, and expressed in microorganisms, new processes involving lipoxygenase produced hydroperoxides and microorganisms containing plant hydroperoxide lyase and alcohol dehydrogenase have been proposed. Hydroperoxide lyase exhibits specificities resulting in the production of the highly desired hexenal from the hydroperoxide of linolenic acid or to the more common hexanal from the hydroperoxide of linoleic acid. However, expression in a microbial cell may modify the activity. In a mixture of hydroperoxides, recombinant hydroperoxide lyase from alfalfa expressed in *E. coli* results in a good ratio of *cis*-3-hexenal whereas the expression of this enzyme from green bell pepper in *Y. lipolytica* gives rise to high ratios of hexanal (104), although its specificity in green bell pepper or when expressed in *E. coli* results in the unsaturated aldehyde (105).

Other hydroperoxide lyases have been cloned and expressed in microorganisms: from tomato fruits (106), *Arabidopsis thaliana* (107), melon (108), guava (109). This resulted in several processes of production of green notes which were disclosed very recently (110–113). The wound responding regulation system of the hydroperoxide lyase can be used (114) or, more conveniently, another expression system such as the lipid inducible *POX*-promoter of *Y. lipolytica* (104).

16.4.6.1.5 Conclusion From the set of examples described above, the approaches can be divided into three main strategies (Table 16.7). The first one consists in stopping a degradation pathway in the producing strain. This can require a thorough knowledge of the pathway or simply a good screening medium. In this case, the simple deletion of a target gene often shows that another degradation pathway can be active: the inactivation of the gene coding for vanillin dehydrogenase decreases vanillin dehydrogenation only slightly, showing that several enzymes can catalyze this reaction; and the disruption of *pox3* and *pox5* showed that another acyl-CoA oxidase enzyme (encoded by *POX4*) was active on γ-decalactone (96,97). This step can usually be overcome using traditional mutation procedures that do not give rise to "genetically modified organisms" (76).

A second strategy, which necessitates a good global knowledge of the metabolic fluxes, consists in modifying these fluxes. Lopez de Felipe et al. (73,115) had the original approach to modify the cofactor redox equilibrium inside the cell by expressing a gene coding for an NADH,H^+ oxidase thus changing the metabolism of lactococci from homolactic to heterolactic. Once again, Monnet et al. (76) had a similar result by blocking the homolactic pathway through a classical mutagenesis strategy, rerouting substrates to the mixed acid pathway showing that a lot can be done within the nongenetically modified framework, as long as adequate screening conditions can be used. Another attempt to modify the flux inside a metabolic pathway consisted in decreasing the acyl-CoA oxidase activity within the β-oxidation pathway to favor the exit of the pathway at this stage and decrease the exit at the dehydrogenase step (99).

Eventually, the third possibility is to add a heterologous enzyme or pathway inside an organism that is well known and for which genetic tools have been developed. Some examples (see Table 16.7) have been described for the production of vanillin from glucose in *E. coli* or the production of hexenal from hydroperoxides in various organisms. These strategies are often more complex than initially thought, and problems of expression or of interference with other pathways are common.

Table 16.7
Examples illustrating the main approaches used in the genetic engineering of aroma compounds biosynthesis

Compound	Target	Organism	Mutation	Ref.
Blockage of the degradation of the compound of interest				
Diacetyl	*aldB*	*L. lactis*	GMO	(72,74)
	combined with redox balance		GMO	
Diacetyl	acetolactate decarboxylase	*L. lactis*	mutagenesis	(76)
γ-decalactone	*pox3pox4pox5*	*Y. lipolytica*	GMO	(97)
Vanillin	*vdh*	*P. fluorescens*	GMO	(83,138)
Vanillin	*vanA, vanB* (transposon mutant)	*P. fluorescens* BF13	GMO	(86,138)
Vanillin	degradation of coniferyl aldehyde	*Pseudomonas* sp.	mutagenesis	(85)
Vanillin	growth on vanillin	*Pseudomonas* sp.	mutagenesis	(84)
Modification of fluxes/rerouting				
Diacetyl	redox state: NADH oxidase	*L. lactis*	GMO	(73)
γ-decalactone	β-oxidation fluxes/*ura⁻, leu⁻*	*Y. lipolytica*	GMO	(100)
γ-decalactone	β-oxidation fluxes/*pox*	*Y. lipolytica*	GMO	(99)
diacetyl	rerouting	*L. lactis*	mutagenesis	(76)
Heterologous gene or pathway				
green notes	hydroperoxide lyase	*E. coli, S. cerevisiae, Y. lipolytica*	GMO	(104,110-113)
Vanillin	ferulic acid decarboxylase	*S. cerevisiae* W3	GMO	(88)
Vanillin	stilbene dioxygenase	*P. paucimobilis* to *E. coli*	GMO	(87)
Vanillin	pathway from glucose to vanillin	*E. coli* KL7	GMO	(89)
Directed evolution/protein shuffling				
γ-decalactone	acyl-CoA oxidase L7	*Y. lipolytica*	GMO	(103)

In conclusion, genetic engineering of the production of aroma compounds is often complex and requires not only a thorough knowledge of the pathway but also of the fluxes inside the cell. For that reason, genetic tools can be completed by the use of flux modelling systems such as Metabolic Control Analysis strategies (116).

16.4.6.2 Biocatalyst Environment

16.4.6.2.1 Oxydoreduction Potential Considerable efforts have been made to understand how cellular metabolism reacts to variations in pH, temperature, pressure, and osmotic pressure. However, little information is available on the response of microorganisms to changes in external oxidoreduction potential (ORP). Early studies suggested that each species or strain has a preferred redox potential range, within which growth is possible (117,118). Certain enzymes activities are controlled by ORP, and modification of this parameter can alter metabolic balances during anaerobic growth of *Escherichia coli* (119). This is also true for the synthesis and the proportion of amino acids synthesized by *Clostridium glutamicum* (120). More recently, Bespalov et al. (121) have shown that *E. coli* can sense the medium's ORP and swim to a preferred ORP niche by means of redox taxis.

As for the pH, ORP is an intrinsic parameter of all media, because they contain at least one molecule able to move from one redox state to another. Oxidoreduction potential can be considered as the average electron availability of a medium. Biochemical reactions are often redox reactions occurring in an aqueous environment, and protons play important roles in electron exchange reactions. Consequently, pH and ORP are not independent parameters. Each medium has its own ORP value, and the different redox coupling system equilibria depend on this value. Moreover it is possible to manipulate this value by the addition of molecules (such as cysteine and ascorbate) as well as by gas mixing (O_2 and H_2).

This control of ORP was first performed to increase the rate of production of some metabolites, such as xylitol in the yeast *Candida* (122). Very recently it has been shown that this physicochemical parameter also affects the functioning of yeast for the production of some flavors. In particular, in view of the importance of ORP in the regulation of β-oxidation flux (123,124), the sensitivity of the genus *Sporidiobolus* to a modification of external ORP by using gas (O_2 or H_2) or molecules {DTT, $K_3[Fe(CN)_6]$}, with regard to their ability to synthesize 4-decanolide, was studied in our laboratory [125]). Results indicated that the synthesis of lactone was highly improved under an ORP reduced to 50 mV. This production is 4 times higher than that of control (ORP 300 mV). However, when the ORP dropped to –150 mV, no lactone production was observed (Figure 16.13). This was directly related to the functioning of the β-oxidation system via a modification of the cofactor level (NAD^+:NADH ratio) and the enzyme activities of the yeast (126). To our knowledge, this was the first example of the effect of external ORP modifications on the production of a flavoring substance.

Basically, these different studies led to the question of the defining of an optimal ORP (as it is already done for pH) for the functioning of major enzymatic systems and metabolic pathways (Figure 16.14), and especially those leading to the synthesis of flavor compounds. Moreover, in terms of application, ORP seems to appear also as an essential environmental parameter that has to be taken into account during fermentation processes as well as in traditional fermented food technology.

16.4.6.2.2 Polarity Flavor compounds are mainly hydrophobic compounds. For this reason, their toxicity is often one of the limits encountered in the scaling up of the process. Above a certain concentration of aroma, the biocatalytic activity is strongly inhibited. To minimize the disadvantage of substrate and product toxicity, studies based on solvent resistant organisms have been developed. In particular, some investigations have focused

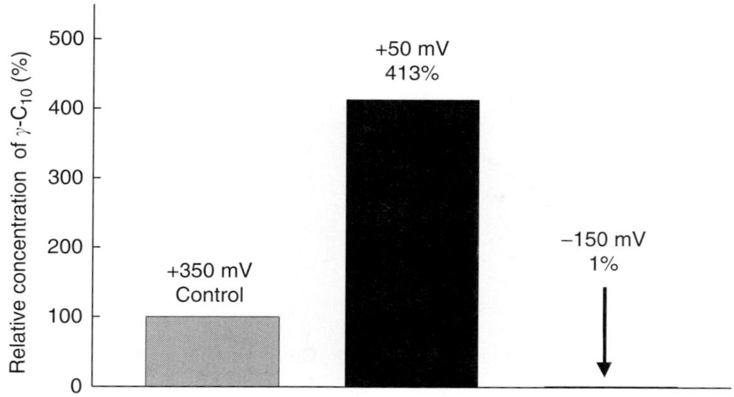

Figure 16.13 Relative maximum amount of γ-decalactone produced by yeast (*Sporidiobolus ruinenii* in this example) under different oxydo-reduction potential conditions (from Feron, G., X.-D. Wang, C. Viel, C. Perrin, G. Mauvais, R. Cachon, C. Diviès, *Proceedings of the 10th Weurman Flavor Research Symposium*, Beaune, 2002, pp 389–392).

on the biotransformation of terpenes by a particular strain of *Pseudomonas alcaligines* able to growth on limonene. From different mutants obtained on this strain, it is possible to better understand the enzymes involved in terpenes biotransformation (127) for the production of food grade flavoring products such as α-terpineol or carvone, for example.

Beside the toxicity aspect, the hydrophobicity of flavor compounds frequently implies their channeling through enzymatic complexes, making their recovery difficult when they are not the terminal product of a metabolic pathway. Based on the works of Spinnler et al. (128), the possibility to modify extracellular polarity by the use of organic solvents and lypophilic compounds can be an interesting way to modify catalytic and equilibrium constants inside a pathway and, as a consequence, the recovery of interesting intermediaries that are not released generally outside the biocatalyst (Figure 16.14). The feasibility of such a strategy was demonstrated by studying the catabolism of ricinoleic for the recovery of γ-decalactone with the yeast *Sporidiobolus ruinenii* grown in the presence of decane (128). As a result, the corresponding 4-hydroxy-acid intermediates released during the β-oxidation of (*R*)-ricinoleic acid have been detected with a final accumulation of γ-decalactone up to 1 g/L.

In spite of the difficulty to find and select microorganism able to growth on organic solvents, these results open a very innovative area particularly adapted to the production of "bioflavor".

16.5 GENERAL CONCLUSION

As a general conclusion, one must considered that the biotechnological production of microbial flavors is an industrial reality that will tend to be further developed in the future. In fact, the exponential progress of results acquired on the understanding of metabolic pathways involved in the synthesis of the desired compounds will permit better control of the flavoring metabolism of the biocatalysts, either in biotechnological or in traditional fermentation processes. The sole limits that could occur on this development might come from legal considerations, but also from food consumer acceptance of genetically modified microbial (GMO) products. Presently, as it was observed for the vague chemophobia of the past, the use of flavor from GMO is totally avoided in the food industry, especially in Europe.

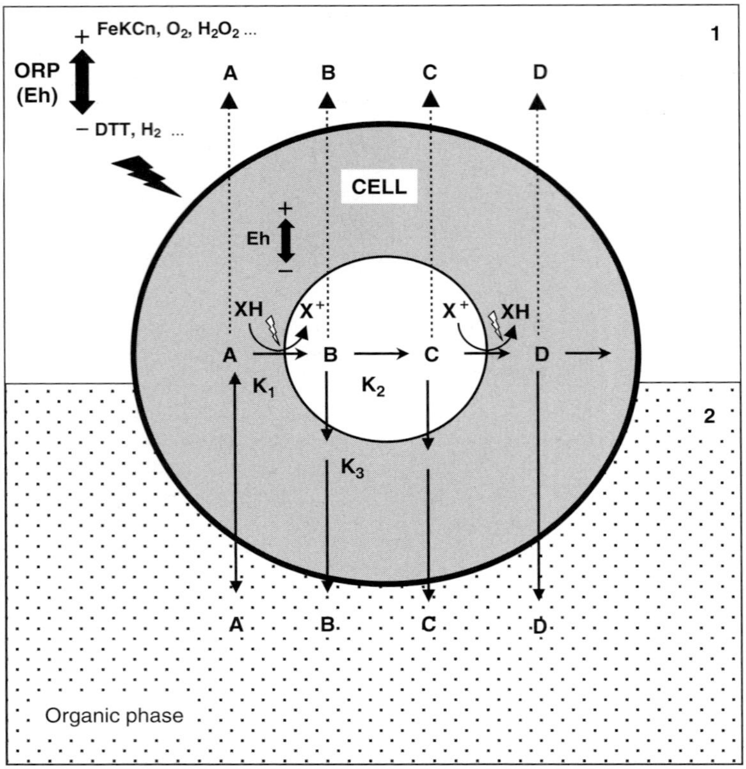

Figure 16.14 Strategy to orientate metabolic pathways leading to flavor compounds in a cell with the modification of external ORP (Eh) [Figure 16.14(1)] or extracellular polarity/hydrophobicity [Figure 16.14(2)] by the use of organic solvents. In the first case, external modification of Eh would affect internal Eh, modifying internal redox balance and metabolic flux. In the second case, hydrophobic intermediates (flavors B, C, and D), formed by a sequence of intracellular reactions, would cross the cell membrane and partition favorably in the organic phase, where they should accumulate (adapted from Spinnler, H.E., C. Ginies, J.A. Khan, E.N. Vulfson, *Proc. Natl. Acad. Sci. USA* 93:3373–3376, 1996).

Even questioning the character "natural" and "safe" of GMO, especially in the case of interspecies transfers, genetic engineering still constitutes one of the most promising possibilities either as a tool for production of novel and safe additives or creating material for discovering new biocatalytic capacities.

REFERENCES

1. Maga, J.A. The potential of certain fungi as sources for natural flavor compounds. *Chem. Sens Flav.* 2:255–262, 1976.
2. Bell, E.R., E.B. White. The potential of biotechnology for the production of flavors and colours for the food industry. *Int. Ind. Biotech.* 9:20–26, 1989.
3. Murray, W.D., S.J.B. Duff. Biosynthesis of natural food flavors. In: *Food Enzymology,* Fox, P.F., ed., London: Elsevier, 1991, pp 105–141.
4. Janssens, L., H.L.D. Pooter, N.M. Schamp, E.J. Vandamme. Production of flavors by microorganisms. *Proc. Biochem.* 27:195–215, 1992.
5. Hagedorn, S., B. Kaphammer. Microbial biocatalysis in the generation of flavor and fragrance chemicals. *Annu. Rev. Microbiol.* 48:773–800, 1994.

6. Berger, R.G. *Aroma Biotechnology*. Berlin: Springer-Verlag, 1995.
7. Feron, G., P. Bonnarme, A. Durand. Prospects for the microbial production of food flavors. *Trends Food Sci. Technol.* 7:285–293, 1996.
8. R.G. Berger, A. Böker, M. Fisher, J. Taubert. Microbial flavors. In: *Flavor Chemistry : 30 Years of Progress,* Academic, K., ed., New York: Plenum Publishers, 1999, pp 229–238.
9. Krings, U., R.G. Berger. Biotechnological production of flavors and fragrances. *Appl. Microbiol. Biotechnol.* 49:1–8, 1998.
10. Van Den Bremt, K., G. Gasarasi, F. Delvaux, H. Derdelinckx, H. Verachtert. Bioflavoring by refermentation, I: general introduction: natural flavors for food and beverages. *Cerevisia* 24:31–39, 1999.
11. Dufossé, L., C. Blin-Perrin, G. Feron. Production of flavors by microbial cultures. In: *11th World Congress of Food Science and Technology*, Seoul (Korea), 2001.
12. Dufossé, L., C. Blin-Perrin, I. Souchon, G. Feron. Microbial production of flavors for the food industry: a case study on the production of gamma-decalactone, the key compound of peach flavor, by the yeasts *Sporidiobolus* sp. *Food Sci. Tech.* 11:192–202, 2002.
13. R.G. Berger, U. Krings, H. Zorn. Biotechnological flavor generation. In: *Food Flavor Technology,* A.J. Taylor, ed., Sheffield: Sheffield Academic Press Ltd, 2002, pp 60–104.
14. Knights, J. Flavor legislation. In: Taylor, A.J., ed., *Food Flavor Technology.* Sheffield: Sheffield Academic Press Ltd, 2002, pp 277–295.
15. Campbell-Platt, G. Fermented foods: a world perspective. *Food Res. Int.* 27:253–257, 1994.
16. Baltazar, M.F., F.M. Dickinson, C. Ratledge. Oxidation of medium-chain acyl-CoA esters by extracts of *Aspergillus niger*: enzymology and characterization of intermediates by HPLC. *Microb. UK* 145:271–278, 1999.
17. Molimard, P., H.E. Spinnler. Compounds involved in the flavor of surface mold-ripened cheeses: origins and properties. *J. Dairy Sci.* 79:169–184, 1996.
18. Dufossé, L., A. Latrasse, H.E. Spinnler. Importance des lactones dans les arômes alimentaires: structure, distribution, propriétés sensorielles. *Sci. Alim.* 9:427–454, 1994.
19. Maarse, H., M. Gessner. *Volatile Compounds in Food: Qualitative and Quantitative Data*, 6th ed., Utrecht: TNO-CIVO Food Analysis Institute, 1988.
20. Tressl, R., M. Apetz, R. Arrieta, K.-G. Grünewald. Formation of lactones and terpenoids by microorganisms. In: *Flavor of Foods and Beverages*, Charalambous, Inglett, ed., London: Academic Press, 1978, pp 145–168.
21. Marilley, L., M.G. Casey. Flavors of cheese products: metabolic pathways, analytical tools and identification of producing strains. *Int. J. Food Microbiol.* 90:139–159, 2004.
22. Banks, J.M., M. Yvon, J.C. Gripon, M.A. de la Fuente, E.Y. Brechany, A.G. Williams, D.D. Muir. Enhancement of amino acid catabolism in cheddar cheese using [alpha]-ketoglutarate: amino acid degradation in relation to volatile compounds and aroma character. *Int. Dairy J.* 11:235–243, 2001.
23. Bonnarme, P., K. Arfi, C. Dury, S. Helinck, M. Yvon, H.-E. Spinnler. Sulfur compound production by *Geotrichum candidum* from -methionine: importance of the transamination step. *FEMS Microbiol. Lett.* 205:247–252, 2001.
24. Yvon, M., L. Rijnen. Cheese flavor formation by amino acid catabolism. *Int. Dairy J.* 11:185–201, 2001.
25. Kieronczyk, A., S. Skeie, T. Langsrud, D. Le Bars, M. Yvon. The nature of aroma compounds produced in a cheese model by glutamate dehydrogenase positive *Lactobacillus* INF15D depends on its relative aminotransferase activities towards the different amino acids. *Int. Dairy J.* 14: 227–235, 2004.
26. Rijnen, L., M. Yvon, R. van Kranenburg, P. Courtin, A. Verheul, E. Chambellon, G. Smit. Lactococcal aminotransferases AraT and BcaT are key enzymes for the formation of aroma compounds from amino acids in cheese. *Int. Dairy J.* 13:805–812, 2003.
27. Engels, W., E. Floris, R. van Kranenburg, J. van Hylckama Vlieg, W. Noordman, G. Smit. Enzymatic conversion of methionine results in the formation of sulphur compounds. *Proceedings of the 10th Weurman Flavor Research Symposium*, Beaune, 2002, pp 326–331.

28. Perpète, P., L. Gijs, S. Collin. Methionine: a key amino acid for flavor biosynthesis in beer. In: *Brewing Yeast Fermentation Performance*, Smart, K., ed. Oxford: Blackwell, 2003.
29. Larroche, C., I. Besson, J.-B. Gros. High pyrazine production by *Bacillus subtilis* in solid substrate fermentation on ground soybeans. *Proc. Biochem.* 34:667–674, 1999.
30. Berdague, J.L., P. Monteil, M.C. Montel, R. Talon. Effects of starter cultures on the formation of flavor compounds in dry sausage. *Meat Sci.* 35:275–287, 1993.
31. Omelianski, V.L. Aroma-producing microorganisms. *J. Bacteriol.* 8:393–419, 1923.
32. Nobles, M.K. Identification of cultures of wood-inhabiting hymenomycetes. *Can. J. Botany.* 43:1097–1139, 1965.
33. Dufossé, L. Production de lactones par des levures appartenant au genre *Sporidiobolus*. PhD dissertation, Université de Bourgogne, Dijon (France), 1993.
34. Christen, P., J.C. Meza, S. Revah. Fruity aroma production in solid state fermentation by *Ceratocystis fimbriata*: influence of the substrate type and the presence of precursors. *Mycol. Res.* 101:911–919, 1997.
35. Hubbal, J.A., R.P. Collins. A study of factors affecting the synthesis of terpenes by *Ceratocystis variospora*. *Mycologia* 10:117–129, 1978.
36. Lanza, E., K.W. Ko, J.K. Palmer. Aroma production by cultures of *Ceratocystis moniliformis*. *J. Agric. Food Chem.* 24:1247–1250, 1976.
37. Crouzet, J. La biogenese des arômes et leur évolution. *Proceedings of Colloque CPCIA*, Massy-France, 1988.
38. Tahara, S., K. Fujiwara, H. Ishizaka, J. Mizutani, Y. Obata. Gamma decalactone: one of constituents of volatiles in cultured broth of *Sporobolomyces odorus*. *Agric. Biol. Chem.* 36:2585–2587, 1972.
39. E.J. Vandamme, W. Soetaert. Bioflavors and fragrances via fermentation and biocatalysis. *J. Chem. Tech. Biotech.* 77:1323–1332, 2002.
40. Demyttenaere, J., S. Macura, N. De Kimpe, R. Verhé. Production of pyrazines and 2-acetyl-1-pyrroline by *Bacillus cereus* strains. *Proceedings of the 10th Weurman Flavor Research Symposium*, Beaune, 2002, pp 344–349.
41. Delest, P. Natural flavors: biotech limited... or unlimited? In: *Bioflavor '95,* Etiévant, P., P. Schreier, ed., Paris: INRA, 1995, pp 13–19.
42. Lomascolo, A., C. Stentelaire, M. Asther, L. Lesage-Meessen. Basidiomycetes as new biotechnological tools to generate natural aromatic flavors for the food industry. *TIBTECH* 17:282–289, 1999.
43. Priefert, H., J. Rabenhorst, A. Steinbüchel. Biotechnological production of vanillin. *Appl. Microbiol. Biotechnol.* 56:296–314, 2001.
44. Rabenhorst, J., R. Hopp. Verfahren zur herstellung von vanillin und dafür geeignete mikroorganism. German Patent Applications GE 19532317 A1, 1997.
45. Böker, A., M. Fisher, R.G. Berger. Rasberry ketone from submerged cultured cells of the basidiomycete *Nidula niveo-tometosa*. *Biotech. Progress* 17:568–572, 2001.
46. Falconnier, B., N. Godard, L. Attard, P. Girard. Bioconversion de cétone framboise. French Patent Applications FR 2 776 301 A1, 1999.
47. Dumont, B., P. Hugueny, J.M. Belin. Production par bioconversion de cétone framboise. French Patent Applications FR 2 724 666 A1, 1994.
48. Whitehead, I.M. Challenges to biocatalysis from flavor chemistry. *Food Tech.* 52:40–46, 1998.
49. DeSantis, G., J. Liu, D.P. Clark, A. Heine, I.A. Wilson, C.H. Wong. Structure-based mutagenesis approaches toward expanding the substrate specificity of D-2-Deoxyribose-5-phosphate aldolase. *Bioorg. Med. Chem.* 11:43–52, 2003.
50. S.D. Geusz, D.M. Anderson. Process of using bacteria that metabolize phenylacetate through mandelate. US Patent Application US 5151353, 1992.
51. Kawabe, T., H. Morita. Production of benzaldehyde and benzyl alcohol by the mushroom *Polyporus tuberaster* k2606. *J. Agric. Food Chem.* 42:2556–2560, 1994.
52. Lapadatescu, C. Production de benzaldéhyde naturel par bioconversion. PhD dissertation, Université de Bourgogne, France, 1999.

53. Bonnarme, P., G. Feron, A. Durand. The biotechnology of flavors. *Eur. Food Drink Rev.* Autumn: 67–70, 1996.
54. Etschmann, M.M., W. Bluemke, D. Sell, J. Schrader. Biotechnological production of 2-phenylethanol. *Appl. Microbiol. Biotechnol.* 59:1–8, 2002.
55. Gatfield, I. La production d'arômes au moyen d'enzymes et de microorganisms. *Contact* 1:3–9, 1996.
56. Gocho, S., T. Yamada. Microbial manufacture of γ-dodecalactone from 10-hydroxy-stearic acid. Japanese Patent Applications JP 03198 787, 1991.
57. Gocho, S., N. Tabogami, M. Inagaki, C. Kawabata, T. Komai. Biotransformation of oleic acid to optically active γ-dodecalactone. *Biosci. Biotech. Biochem.* 59:1571–1572, 1995.
58. Cardillo, R., C. Fuganti, M. Barbeni, P. Cabella, P.A. Guarda, A. Gianna. Procédé de production microbiologique des gamma- et delta-lactones. European Patent Applications EP 0 412 880, 1991.
59. Chalier, P., J. Crouzet. Methyl ketone production from copra oil by *Penicillium roqueforti* spores. *Food Chem.* 63:447–451, 1998.
60. Nordermeer, M.A., W. Van Der Goot, A.J. Van Kooij, J.W. Veldsink, G.A. Veldink, J.F.G. Vliengenthart. Development of a biocatalytic process for the production of C6-aldehydes from vegetable oils by soybean lipoxygenase and recombinant hydroperoxide lyase. *J. Agric. Food Chem.* 50:4270–4274, 2002.
61. Blée, E. Impact of phyto-oxylipins in plant defense. *Trends Plant Sci.* 7:315–321, 2002.
62. Garger, S.J., R.B. Holtz, M.J. McCulloch, H.F. Phillips, R.K. Teague. Method for providing green note compounds. World Patent Applications WO 9526413, 2001.
63. Brunerie, P., Y. Koziet. Procédé de production de cis-3-hexénol naturel à partir d'acides gras insaturés. French Patent Applications FR 95 06761, 1995.
64. JM Belin, J.M., B. Dumont, F. Ropert. Method for the enzymatic preparation of aromas, particularly ionons and C6 to C10 aldehydes. World Patent Applications WO 94 08028, 1994.
65. Rehbock, B., D. Gansser, R.G. Berger. Efficient generation of 2*E*-hexenal by a hydroperoxide lyase from mung bean seedlings. *Food Chem.* 63:161–165, 1998.
66. Gatfield, I.L. Biotechnological production of natural flavor materials. In: *Flavor Chemistry, Thirty Years of Progress,* Teranishi, R., E.L. Wick, I. Hornstein, eds., New York: Plenum Press, 1999, pp 211–227.
67. Muller, B.L., C. Dean, I.M. Whitehead. The industrial use of plant enzymes for the production of natural "green note" flavor compounds. In: *Bioflavor '95.* Etiévant, P., P. Schreier, eds., INRA Editions, 1995, pp 339–344.
68. Broun, P., J. Shanklin, E. Whittle, C. Somerville. Catalytic plasticity of fatty acid modification enzymes underlying chemical diversity of plant lipids. *Science* 282:1315–1317, 1998.
69. Smith, J. Public perception of biotechnology. In: *Basic Biotechnology*, Ratledge, C., B. Kristiansen, eds., Cambridge: Cambridge University Press, 2001.
70. De Man, J.C. The formation of diacetyl and acetoin from alpha-acetolactic acid. *Recueil* 78:480–486, 1959.
71. Park, S.H., R. Xing, W.B. Whitman. Nonenzymatic acetolactate oxidation to diacetyl by flavin, nicotinamide and quinone coenzymes. *Biochim. Biophys. Acta* 1245:366–370, 1995.
72. Swindell, S.R., K.H. Benson, H.G. Griffin, P. Renault, S.D. Ehrlich, M.J. Gasson. Genetic manipulation of the pathway for diacetyl metabolism in *Lactococcus lactis*. *Appl. Environ. Microbiol.* 62:2641–2643, 1996.
73. Lopez de Felipe, F., M. Kleerebezem, W.M. de Vos, J. Huenholtz. Cofactor engineering: a novel approach to metabolic engineering in *Lactococcus lactis* by controlled expression of NADH oxidase. *J. Bacteriol.* 180:3804–3808, 1998.
74. Hugenholtz, J., M. Kleerebezem, M. Starrenburg, J. Delcour, W. De Vos, P. Hols. *Lactococcus lactis* as a cell factory for high-level diacetyl production. *Appl. Environ. Microbiol.* 66:4112–4114, 2000.
75. Monnet, C., P. Schmitt, C. Diviès. Diacetyl production in milk by an alpha-acetolactic acid accumulating strain of *Lactococcus lactis* ssp. *lactis* biovar. *diacetylactis*. *J. Dairy Sci.* 77:2916–2924, 1994.

76. Monnet, C., F. Aymes, G. Corrieu. Diacetyl and alpha-acetolactate overproduction by *Lactococcus lactis* subsp. *lactis* biovar *diacetylactis* mutants that are deficient in alpha-acetolactate decarboxylase and have a low lactate dehydrogenase activity. *Appl. Environ. Microbiol.* 66:5518–5520, 2000.
77. Dequin, S. The potential of genetic engineering for improving brewing, wine-making and baking-yeasts. *Appl. Microbiol. Biotechnol.* 56:577–588, 2001.
78. Hansen, J., M. Kielland-Brandt. Brewer's yeast. In: *Yeast Sugar Metabolism, Biochemistry, Genetics, Biotechnology and Applications,* Zimmerman, F.K., E.K.D., eds., New York: Technomic Publishing, 1997, pp 503–526.
79. Gjermansen, C., T. Nilsson-Tillgren, J.G. Petersen, M.C. Kielland-Brandt, P. Sigsgaard, S Holmberg. Towards diacetyl-less brewer's yeast: influence of *ilv2* and *ilv5* mutations. *J. Basic Microbiol.* 28:175–183, 1988.
80. Villanueba, K.D., E. Goossens, C.A. Masschelein. Subthreshold vicinal diketone levels in lager brewing yeast fermentations by means of ILV5 gene amplification. *J. Am. Soc. Brew. Chem.* 48:111–114, 1990.
81. Ehrlich, S., P. Godon, J.-J. Renault. Nucleic acid coding for an alpha-acetolactate decarboxylase and applications. World Patent Applications WO 9408019, 1994.
82. Frost, J. Synthesis of vanillin from a carbon source. World Patent Applications WO 0017319, 2000.
83. Overhage, J., H. Priefert, J. Rabenhorst, A. Steinbüchel. Biotransformation of eugenol to vanillin by a mutant of *Pseudomonas* sp. strain HR199 constructed by disruption of the vanillin dehydrogenase (*vdh*) gene. *Appl. Microbiol. Biotechnol.* 52:820–828, 1999.
84. Washisu, S., T. Aida, N. Hashimoto. Production of vanillin and its related compound by fermentation, Production of vanillin and its related compound by fermentation. Japanese Patent Applications JP 5227980, 1993.
85. Cooper, B. Preparation of coniferaldehyde by a microorganism. US Patent Applications US 4874701, 1989.
86. Civolani, C., P. Barghini, A.R. Roncetti, M. Ruzzi, A. Schiesser. Bioconversion of ferulic acid into vanillic acid by means of a vanillate-negative mutant of *Pseudomonas fluorescens* strain BF13. *Appl. Environ. Microbiol.* 66:2311–2317, 2000.
87. Kamoda, S., M. Samejima. Cloning of a ligno stilbene-alpha, beta-dioxygenase gene from *Pseudomonas paucimobilis* TMY1009-expression in *E. coli. Agric. Biol. Chem.* 55:1411–1412, 1991.
88. Ago, S., Y. Kikuchi. Ferulic acid decarboxylase. US Patent Applications US 5955137, 1998.
89. Berry, A. Improving production of aromatic compounds in *Escherichia coli* by metabolic engineering. *Trends Biotechnol.* 14:250–256, 1996.
90. Barth, G., C. Gaillardin. *Yarrowia lipolytica*. In: *Nonconventional Yeasts in Biotechnology, a Handbook.* Berlin: Springer-Verlag, 1996, pp 314–388.
91. Waché, Y., M. Aguedo, J.-M. Nicaud, J.-M. Belin. Catabolism of hydroxyacids and production of lactones by the yeast *Yarrowia lipolytica*. *Appl. Microbiol. Biotechnol.* 61:393–404, 2003.
92. Wang, H., M.-T. Le Dall, Y. Waché, C. Laroche, J.-M. Belin, J.-M. Nicaud. Cloning, sequencing and characterization of five genes coding for Acyl-CoA oxidase isozymes in the yeast *Yarrowia lipolytica*. *Cell Biochem. Biophys.* 31:165–174, 1999.
93. Wang, H., M.-T. Le Dall, Y. Waché, C. Laroche, J.-M. Belin, C. Gaillardin, J.-M. Nicaud. Evaluation of acyl CoA oxidase (Aox) isozyme function in the n-alkane-assimilating yeast *Yarrowia lipolytica*. *J. Bacteriol.* 181:5140–5148, 1999.
94. Waché, Y., Y. Pagot, J.-M. Nicaud, J.-M. Belin. Acyl-CoA oxidase, a key step for lactone production by *Yarrowia lypolitica*. *J. Molec. Catal. B: Enzym.* 5:165–169, 1998.
95. Waché, Y., C. Laroche, K. Bergmark, C. Moller-Andersen, M. Aguedo, M.-T. Le Dall, H. Wang, J.-M. Nicaud, J.-M. Belin. Involvement of acyl-CoA oxidase isozymes in biotransformation of methyl ricinoleate into gamma-decalactone by *Yarrowia lipolytica*. *Appl. Environ. Microbiol.* 66:1233–1236, 2000.

96. Waché, Y., M. Aguedo, M.-T. LeDall, J.-M. Nicaud, J.-M. Belin. Optimization of *Yarrowia lipolytica*'s beta-oxidation pathway for lactones production. *J. Molec. Catal. B: Enzym.* 19-20:347–351, 2002.
97. Groguenin, A., Y. Waché, E. Escamilla Garcia, M. Aguedo, F. Husson, M. LeDall, J. Nicaud, J. Belin. Genetic engineering of the beta-oxidation pathway in the yeast *Yarrowia lipolytica* to increase the production of aroma compounds. *J. Molec. Catal. B*: 28: 75–79, 2004.
98. Titorenko, V.I., H. Chan, R.A. Rachubinski. Fusion of small peroxisomal vesicles *in vitro* reconstructs an early step in the *in vivo* multistep peroxisome assembly pathway of *Yarrowia lipolytica. J. Cell Biol.* 148:29–43, 2000.
99. Waché, Y., M. Aguedo, A. Choquet, I. Gatfield, J.-M. Nicaud, J.-M. Belin. Role of beta-oxidation enzymes in the production of gamma-decalactones from methyl ricinoleate. *Appl. Environ. Microbiol.* 67:5700–5704, 2001.
100. Nicaud, J.-M., J.-M. Belin, Y. Pagot, A. Endrizzi-Joran. Bio-conversion of substrate with microbe auxotrophic for compound in medium deficient in this compound. French Patent Applications FR 2734843, 1996.
101. Luo, Y.S., H.J. Wang, K.V. Gopalan, D.K. Srivastava, J.M. Nicaud, T. Chardot. Purification and characterization of the recombinant form of Acyl CoA oxidase 3 from the yeast *Yarrowia lipolytica. Arch. Biochem. Biophys.* 384:1–8, 2000.
102. Luo, Y.S., J.M. Nicaud, P.P. Van Veldhoven, T. Chardot. The acyl-CoA oxidases from the yeast *Yarrowia lipolytica*: characterization of Aox2p. *Arch. Biochem. Biophys.* 407:32–38, 2002.
103. Mlícková, K., Y. Luo, S. D'Andrea, P. Pec, T. Chardot, J.-M. Nicaud. Acyl-CoA oxidase, a key step for lipid accumulation in the yeast *Yarrowia lipolytica. J. Molec. Catal. B.*, 28: 81–85, 2004.
104. Bourel, G., J.M. Nicaud, B. Nthangeni, P. Santiago-Gomez, J.M. Belin, F. Husson. Fatty acid hydroperoxide lyase of green bell pepper: cloning in *Yarrowia lipolytica* and biogenesis of volatile aldehydes. *Enz. Microb. Technol.*, 35: 293–299, 2004.
105. Matsui, K., T. Shibutani, T. Hase, T. Kajiwara. Bell pepper fruit acid hydroperoxide lyase is a cytochrome P450 (CYP74B). *FEBS Lett.* 394:21–24, 1996.
106. Matsui, K., C. Miyahara, J. Wilkinson, B. Hiatt, V. Knauf, T. Kajiwara. Fatty acid hydroperoxide lyase in tomato fruits: cloning and properties of a recombinant enzyme expressed in *Escherichia coli. Biosci. Biotechnol. Biochem.* 64:1189–1196, 2000.
107. Matsui, K., J. Wilkinson, B. Hiatt, V. Knauf, T. Kajiwara. Molecular cloning and expression of *Arabidopsis* fatty acid hydroperoxide lyase. *Plant Cell Physiol.* 40:477–481, 1999.
108. Tijet, N., C. Schneider, B.L. Muller, A.R. Brash. Biogenesis of volatile aldehydes from fatty acid hydroperoxides: molecular cloning of a hydroperoxide lyase (CYP74C) with specificity for both the 9- and 13-hydroperoxides of linoleic and linolenic acids. *Arch. Biochem. Biophys.* 386:281–299, 2001.
109. Tijet, N., U. Waspi, D.J. Gaskin, P. Hunziker, B.L. Muller, E.N. Vulfson, A Slusarenko, A.R. Brash, I.M. Whitehead. Purification, molecular cloning, and expression of the gene encoding fatty acid 13-hydroperoxide lyase from guava fruit (*Psidium guajava*). *Lipids* 35:709–720, 2000.
110. Feusner, I., E. Berndt. Preparation of C9 aldehydes, alcohols and their esters, for use as aroma and flavoring agents, from unsaturated fatty acids, comprises modifying specific activities in e.g. plants. World Patent Applications WO 0194606, 2001.
111. Häusler, A., N. Silke, K. Lerch, A. Muheim. Hydroperoxide lyase. US Patent Applications US 6238898, 2001.
112. Slusarenko, A.J., I.M. Whitehead, D.J.H. Gaskin, N. Tijet, A.R. Brash. Guava (*Psidium guajava*) 13-hydroperoxide lyase and uses thereof. World Patent Applications WO 9958648, 2001.
113. Whitehead, I.M., A. Brash, N. Tijet. Muskmelon (*Cucumis melo*) hydroperoxide lyase and uses thereof. World Patent Applications WO 0173075, 2001.
114. Gleddie, S., D. Hegedus, U. Schafer, D. Brown, N. Bate. Hydrogen peroxide lyase regulatory region. World Patent Applications WO 0250291, 2002.

115. Lopez de Felipe, F., J. Hugenholtz. Pyruvate flux distribution in NADH-oxidase-overproducing *Lactococcus lactis* strain as a function of culture conditions. *FEMS Microbiol. Lett.* 179:461–466, 1999.
116. Kleerebezem, M., I.C. Boels, M.N. Groot, I. Mierau, W. Sybesma, J. Hugenholtz. Metabolic engineering of *Lactococcus lactis*: the impact of genomics and metabolic modelling. *J. Biotechnol.* 98:199–213, 2002.
117. Riondet, C. Effet du potentiel d'oxydoréduction et du pH sur le métabolisme de *Escherichia coli*: rôle de ces paramètres sur la force proton motrice, sur la survie au choc thermique et sur l'orientation des flux de carbone. PhD dissertation, Université de Bourgogne, Dijon (France), 1999.
118. Riondet, C., R. Cachon, Y. Waché, G. Alcaraz, C. Diviès. Changes in the proton-motive force in *Escherichia coli* in response to external oxidoreduction potential. *Eur. J. Biochem.* 262: 595–599, 1999.
119. Winpenny, J.W.T., D.K. Necklen. The redox environment and microbial physiology. *Biochim. Biophys. Acta* 253:352–359, 1971.
120. Kwong, S.M.C., G. Rao. Effect of reducing agents in an aerobic amino acid fermentation. *Biotech. Bioeng.* 40:851–857, 1992.
121. Bespalov, V.A., I.B. Zhulin, B.L. Taylor. Behavioral responses of *Escherichia coli* to changes in redox potential. *Proc. Natl. Acad. Sci. USA* 93:10084–10089, 1996.
122. Oh, D.-K., S.-Y. Kim, J.-H. Kim. Increase of xylitol production rate by controlling redox potential in *Candida parapsilosis*. *Biotech. Bioeng.* 58:440–444, 1998.
123. George, S.M., L.C.C. Richardson, I.E. Pol, M.W. Peck. Effect of oxygen concentration and redox potential on recovery of sublethally heat-damaged cells of *Escherichia coli O157:H7*, *Salmonella enteritidis* and *Listeria monocytogenes*. *J. Appl. Microbiol.* 84: 903–909, 1998.
124. Eaton, S., D.M. Turnbull, K. Bartlett. Redox control of β-oxidation in rat liver mitochondria. *Eur. J. Biochem.* 220:671–681, 1994.
125. Wang, X.D., G. Mauvais, R. Cachon, C. Diviès, G. Feron. Addition of the reducing agent dithiotreitol improves 4-decanolide synthesis by the genus *Sporidiobolus*. *J. Biosci. Bioeng.* 90:338–340, 2000.
126. Feron, G., X.-D. Wang, C. Viel, C. Perrin, G. Mauvais, R. Cachon, C. Diviès. Influence of a reducing environment on the bioconversion of ricinoleic acid to gamma decalactone by yeast from the genus *Sporidiobolus:* cellular and enzymatic approaches. *Proceedings of the 10th Weurman Flavor Research Symposium*, Beaune, 2002, pp 389–392.
127. Teunissen, M.J., J.A.M. De Bont. Will terpenes be of any significance in future biotechnology? In: *Bioflavor '95.* Etiévant, P., P. Schreier, eds., INRA Editions, 1995, pp 329–337.
128. Spinnler, H.E., C. Ginies, J. A. Khan, E.N. Vulfson. Analysis of metabolic pathways by the growth of cells in the presence of organic solvents. *Proc. Natl. Acad. Sci. USA* 93:3373–3376, 1996.
129. Walton, N.J., A. Narbad, C. Faulds, G. Williamson. Novel approaches to the biosynthesis of vanillin. *Curr. Opinion Biotech.* 11:490–496, 2000.
130. Hugueny, P., B. Dumont, F. Ropert, J.-M. Belin. The raspberry ketone, a biotechnological way for its production. In: *Bioflavor '95.* Etiévant, P., P. Schreier, eds., INRA Editions, 1995, pp 269–273.
131. Spinnler, H.E., L. Dufossé, I. Souchon, A. Latrasse, C. Piffaut-Juffard. Production de gamma-decalactone par bioconversion. French Patent Applications FR 2705971, 1994.
132. Van der Schaft, P.H., N. Terburg, S. Van Den Bosch, A.M. Cohen. Microbiol production of natural δ-decalactone and δ-dodecalactone from the corresponding unsaturated lactones in massoi bark oil. *Appl. Microbiol. Biotechnol.* 36:712–716, 1992.
133. Van der Schaft, P.H., W.H. De Laat. Natural delta lactones and process of the production thereof. European Patent Applications EP 0.425.001 A1, 1991.
134. Page, G.V., R.G. Eilerman. Process for the preparation of gamma and delta lactones. World Patent Applications WO 89/12104, 1989.
135. Lapadatescu, C., G. Feron, C. Vergoignan, A. Djian, A. Durand, P. Bonnarme. Influence of cell immobilization on the production of benzaldehyde and benzyl alcohol by the white-rot fungi

Bjerkandera adusta, *Ischnoderma benzoinum* and *Dichomitus squalens*. *Appl. Microbiol. Biotechnol.* 47:708–714, 1997.

136. Blank, I., J. Lin, L.B. Fay, R. Fumeaux. Formation of 4,5-dimethyl-3-hydroxy-2(5)-furanone (sotolon) from 4-hydroxy-L-isoleucine. In: *Bioflavor '95*. Etiévant, P., P. Schreier, eds., INRA Editions, 1995, pp 385–388.
137. Blank, I., J. Lin, R. Fumeaux, D.H. Welti, L.B. Fay. Formation of 3-hydroxy-4,5-dimethyl-2(5H)-furanone (Sotolone) from 4-hydroxy-l-isoleucine and 3-amino-4,5-dimethyl-3,4-dihydro-2(5H)-furanone. *J. Agric. Food Chem.* 44:1851–1856, 1996.
138. Overhage, J., H. Priefert, J. Rabenhorst, A. Steinbüchel. Construction of production strains for producing substituted phenols by specifically inactivating genes of the eugenol and ferulic acid catabolism. World Patent Applications WO 0026355, 2000.

1.17

Microbial Production of Oils and Fats

James P. Wynn and Colin Ratledge

CONTENTS

17.1 INTRODUCTION

17.1.1 The Concept of Single Cell Oils

Human beings have used microorganisms as sources of foods and food ingredients since ancient times. We have been consuming a wide variety of fermented beverages, including beers and wines, and numerous fermented foods, such as cheeses, yogurts, salamis, sauerkraut, and tempeh, since almost the very beginnings of civilization. Extensive use of microorganisms for food production has occurred in many civilizations, and the safe consumption of microorganisms themselves, as well as the foods that they have produced, is now well established as a global activity.This is evidenced by the many chapters in this volume that deal with the various aspects of microbial technologies in the production of a wide variety of foods.

The use of isolated components from microbial cultures in foods also has a long history. These range from vinegar (i.e., acetic acid, which is, of course, a product again almost as old as civilization) to citric acid, numerous amino acids, and polysaccharide gums. Of more recent appearance has been the advent of microbial proteins in which the whole organism (or almost so) is the object of consumption. The first uses of this technology came in the development of food yeast (the Torula process), which was devised as a means of converting surplus sugar and molasses in the West Indies in the 1940s and 1950s to a biomass that could be used as an animal feed supplement. (The nutritional value added to the biomass was derived by the yeast converting the sugar and ammonia, supplied as the sole nitrogen source, into proteins, which then were eaten by the animals.) Of major significance in the use of microorganisms for food was the advent of various Single Cell Protein (SCP) processes that further developed the original concept of the Torula process for use with other microorganisms and other substrates. In particular, alkanes were (and are) produced by the petroleum industry in vast amounts almost as a waste material during the refinement of crude petroleum oil. This led to major advances in the technology of large scale microbial cultivations with, eventually, fermentors up to 1500 m^3 being built to produce large amounts of SCP. Such fermentation plants could easily generate upward of 5000 tons (dry weight) of biomass per year on a site about the size of three or four football fields. With the use of such large scale production facilities came economies of scale: yeast biomass derived from alkanes could be produced for as little as $150–200 per ton in the 1970s.

Against this background of being able to produce large amounts of microbial biomass as a source of protein food, the current ideas of producing oils and fats using microorganisms were conceived (1). The name Single Cell Oils (SCO) was coined (2) to be in keeping with the concept of SCP and to indicate the distinctiveness of this material. The term SCO was meant to denote those oils that could be extracted from a microorganism, which would approximate an animal or plant oil and be suitable for use in foods or in other products where oils and fats from more conventional sources were normally used.

The concept of producing SCOs as alternatives to plant oils, though, was far from novel and the quest to produce them has a long history [see (3–5) for reviews covering this period]. Initial attempts to use various yeasts and filamentous fungi as potential sources of oil were probably started in the first two decades of the twentieth century. By the 1930s, serious attempts were being made in Europe and the U.S. to develop these processes. During World War II, the first attempts to produce microbial oils on an industrial scale were developed in Germany, with Henkel, as one of the major oils and fats companies, being a key instigator. Production details are somewhat scant, but it appears that the fat-containing yeast produced was only used as a feed material for army horses! But as the German army at the time had over a million horses, this may not have been an inconsiderable amount or an insignificant contribution to animal nutrition. However, the oil within the yeast does not appear to have been extracted, though technically it could have been. Clearly, if the entire material was only going to be used for animal nutrition, then there was no point in extracting the oil and then feeding it together with the residual high protein microbial biomass to the horses; rather, just feed the entire high fat biomass.

The importance of this early research in Germany and elsewhere is not to be underestimated. It laid down several important principles that hold good today and that can be summarized as follows:

1. Not all microorganisms can accumulate lipid. Those that do so above an arbitrary limit of about 20% of the dry cell biomass are termed oleaginous (6).

2. The main organisms that accumulate lipid are yeasts and fungi. Bacteria are not good producers.
3. Some microorganisms can accumulate more lipid than others.
4. Oil contents of a few microorganisms can reach up to 70% (w/w) of the dry biomass.
5. Lipid accumulation in a microorganism is triggered when a nutrient (usually nitrogen) in the growth medium becomes exhausted and the surplus carbon source (usually glucose) continues to be assimilated by the microorganism, which is converted into a storage oil or fat.
6. The oils that are accumulated by microorganisms are the same type as are found in seed oil plants (i.e., predominantly triacylglycerols and the esterified fatty acids).

17.1.2 Oleaginous Microorganisms and Lipid Accumulation

The range of fatty acids that microorganisms produce is extensive (7) and is almost as diverse as those from plant sources. However, if we restrict ourselves to the oleaginous organisms, that is those accumulating more than about 20% (w/w dry wt) oil, then the range of fatty acids is much more restricted and, almost without exception, are the same as those found in plant and animal oils and fats. A selection of fatty acids from various oleaginous yeasts, molds, and some algae is given in Tables 17.1–17.3. More complete lists of fatty acids found in the oleaginous microorganisms can be found in previous reviews of the authors (8–10).

Table 17.1
Lipid Contents and Fatty Acid Profiles of some Selected Oleaginous Yeasts[a]

Yeast Species[b]	Maximum Lipid Content [% (w/w)]	Major Fatty Acyl Residues [Relative % (w/w)] 16:00	16:1	18:0	18:1	18:2	18:3	Others
Cryptococcus albidus var. *aerius*	65	12	1	3	73	12	-	
Crypt. albidus var. *albidus*	65	16	trace	3	56	-	3	21:0 (7%) 22:0 (12%)
Crypt. curvatus D[a]	58	32	-	15	44	8	-	
Lipomyces starkeyi	63	34	6	5	51	3		
Lip. tetrasporus	67	31	4	15	43	6	1	
Rhodosporidium toruloides	66	18	3	3	66	-	-	23:0 (3%) 24:0 (6%)
Rhodotorula glutinis	72	37	1	3	47	8	-	
Rta. graminis	36	30	2	12	36	15	4	
Trichosporon beigelii[b]	45	12	-	22	50	12	-	
Waltomyces lipofer[c]	64	37	4	7	48	3	-	
Williopsis saturnus[d]	28	16	16	-	45	16	5	
Yarrowia lipolytica[e]	36	11	6	1	28	51	1	

[a] originally known as *Candida curvata*, then *Apiotrichum curvatum*
[b] formerly *Trichosporon cutaneum*
[c] formerly *Lipomyces lipofer*
[d] formerly *Hansenula saturnus*
[e] formerly *Candida lipolytica* and also *Saccharomycopsis lipolytica*; only some strains may be oleaginous
Source: Adapted from References 8–10

Table 17.2
Fatty Acid Profiles of Lipid from Selected Molds

	Major Fatty Acyl Groups [Relative % (w/w)] of							
	14:0	16:0	18:0	18:1	18:2	18:3		
						(n-6)	(n-3)	Others
Lower fungi								
Conidiobolus nanodes	1	23	15	25	1	4	-	20:1 (13%) 22:1 (8%) 20:4 (4%)
Cunninghamella japonica	trace	16	14	48	14	8	-	-
Entomophthora coronata	31	9	2	14	2	1	-	12:0 (40%)
Mortierella alpina	-	19	8	28	9	8	-	20:3 (7%) 20:4 (21%)
Mort. elongata	-	7	2	18	12	25	-	20:4 (16%) 20:5 (15%)
Mort. isabellina	1	29	3	55	3	3	-	-
Mucor alpina-peyron	10	15	7	30	9	1	-	20:0 (8%) 20:3 (6%) 20:4 (5%)
Pythium ultimum	[illegible]	15	2	20	16	1	-	20:1 (4%) 20:4 (15%) 20:5 (12%)
Pyth. irregulare	8	17	2	14	18	-	-	20:1 (5%) 20:4 (11%) 20:5 (14%)
Rhizopus arrhizus	19	18	6	22	10	12	-	
Ascomycetes								
Aspergillus terreus	2	23	trace	14	40	-	21	
Fusarium oxysporum	trace	17	8	20	46	-	5	
Pellicularia practicola	trace	8	2	11	72	-	2	
Penicillium spinulosum	-	15	7	42	31	-	1	
Higher fungi								
Cladosporium herbarum	trace	31	12	35	18	-	1	
Claviceps purpurea	trace	23	2	19	8	-	-	12-HO-18:1 (42%)
Tolyposporium ehrenbergii	1	7	5	81	2	-	-	

Source: Adapted from References 8–10.

As can be seen from the tables, the composition of these fatty acids covers the type and range of fatty acids seen in many of the commercial plant oils though some microorganisms, particularly species of molds and algae, clearly produce some polyunsaturated fatty acids (PUFA) that are of a longer chain length and are even more unsaturated than those that occur in plants. It is these microorganisms that have become increasingly of more interest as the demands for PUFAs as dietary supplements have increased over the past decade. These aspects of microbial oil production have then become the cornerstone for the very recent developments in this area and therefore form the main focus of this chapter.

Table 17.3
Fatty Acid Profiles of Selected Microalgae

	Major Fatty Acyl Residues in Lipids [Relative % (w/w)]											
	14:00	**16:00**	**16:1 (n-7)**	**18:1 (n-9)**	**18:2 (n-6)**	**18:3 (n-6)**	**18:3 (n-3)**	**20:3 (n-6)**	**20:4 (n-6)**	**20:5 (n-7)**	**22:6 (n-3)**	**Others**
Prokaryotes	8	63	2	4	9	12	-	-	-	-	-	
Spirulina maxima												
S. platensis	1	26	5	23	10	21	-	-	-	-	-	
Eukaryotes											-	
Chlorella minutissima	12	13	21	1	2	-	-	-	3	45	-	
Chlorella vulgaris	-	16	2	58	9	-	14	-	-	-	-	
Crypthecodium cohnii	16	16	1	21	1	-	-	-	-	-	40	
Isochrysis galbana	12	10	11	3	2	-	-	-	<1	25	11	18:4, 11%
Monodus subterraneus	-	19	10	5	2	<1	-	-	14	34	-	
Nannochloropsis oculata	4	15	22	3	1	-	-	1	4	38	-	
Nannochloropsis sp.	5	14	21	4	3	-	-	-	7	38	-	
Phaeodactylum tricornutum	-	10	21	1	4	1	-	-	1	33	4	
Porphyridium cruentum	-	30	5	<1	5	1	-	<1	16	-	-	

Source: Adapted from References 8–10.

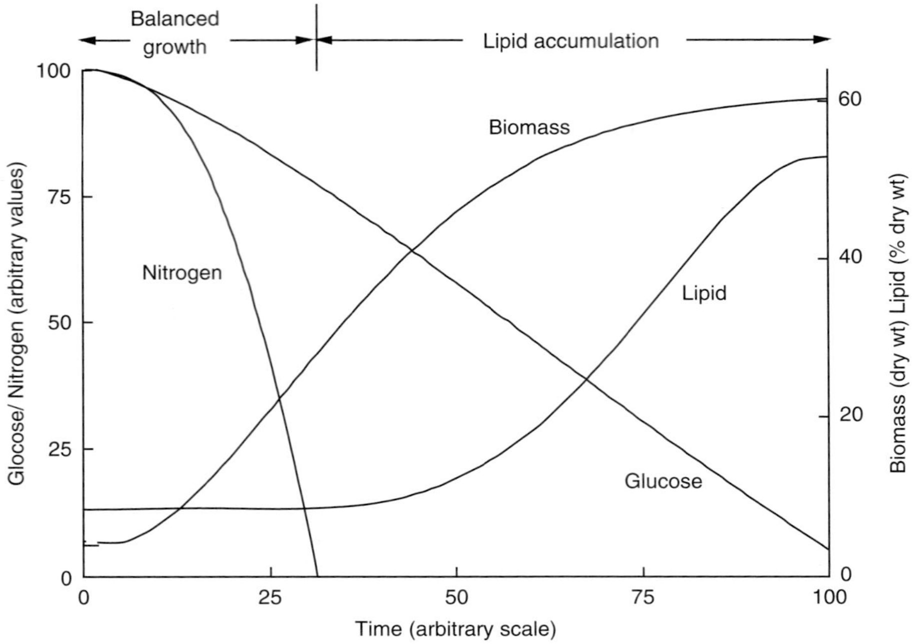

Figure 17.1 The course of lipid accumulation by a typical oleaginous microorganism. The concentration of nitrogen (usually NH_3) in the medium is adjusted in the initial formulation so that it becomes exhausted after about the first 24 to 36 hours' growth.

In most cases, though not in every one, the accumulation of lipid in an oleaginous species of microorganism follows a similar pattern (Figure 17.1). In essence, the microorganism of choice is grown in a culture medium that is high in carbon (usually glucose or sucrose) and low in nitrogen (usually this is an ammonium salt though urea is used occasionally) so that after the initial phase of balanced growth, where all nutrients are in excess, the organism exhausts the supply of nitrogen but continues to assimilate the carbon source. The absence of nitrogen from the growth medium stops cells from dividing and continuing to grow as the nitrogen is obviously needed for both protein and nucleic acid synthesis, both of which are essential for the creation of new cells. The continued assimilation of the carbon source, however, means that the cells must do something with it; oleaginous microorganisms convert it into oils and fats, though other organisms might convert it into other storage materials, such as polysaccharides (11), or even produce large amounts of metabolites, such as citric acid (12). The process of lipid accumulation (see Figure 17.1) continues until the cells reach a personal limit of obesity. Some cells may continue to produce oil until they are physically unable to accumulate any more (Figure 17.2). Other microorganisms, still being classed as oleaginous, stop lipid accumulation at some apparent limit that may vary from 20 to 25% (w/w dry wt) up to 60% or so. This limit is evidently genetically predetermined and is a characteristic of each individual organism. The reason why microorganisms accumulate lipid but then limit their production of it resides in the complexities of the biochemistry and genetics of these cells. This area has recently been reviewed by the present authors (13), and that article should be consulted for further information regarding the biochemistry and molecular biology of lipid accumulation in microorganisms and for an explanation for the limits to lipid production by individual microorganisms.

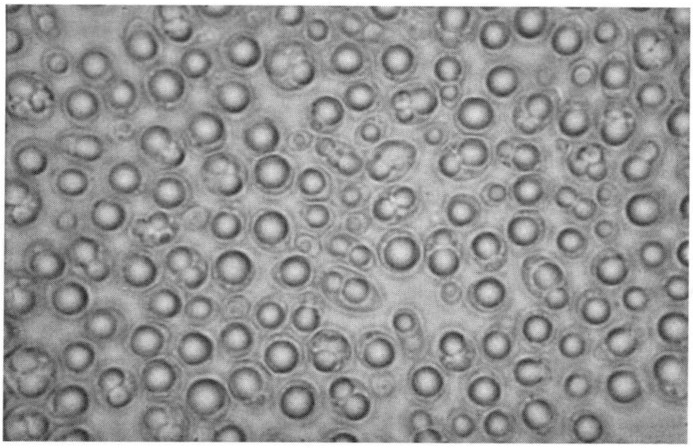

Figure 17.2 Cells of an oleaginous yeast, *Cryptococcus curvatus* (formerly *Candida curvata* D), showing the accumulation of lipid in the form of discrete intracellular droplets.

17.2 GENERAL PRINCIPLES, ECONOMICS AND PROCESSES FOR SCO PRODUCTION

The intrinsic cheapness of the commodity oils and fats obtained from agriculture, where the costs are usually between $400 and $800 per ton (see www.fas/usda/gov/oilseed), means that any potential biotechnology SCO process must aim at producing speciality oils and fats that would be valued at considerably higher prices. The biochemical efficiency of converting sugar into triacylglycerol indicates that to produce one ton of microbial oil, three tons of sugar substrate will be needed. A further one to two tons of sugar will also be needed to provide the remainder of the (oil free) biomass. As sugar costs about $300 per ton at world prices, the costs of substrate alone will be about $1500 to produce one ton of SCO. Add to this the cost of fermentation, which even on the largest scale will cost a minimum of $2000 per ton of oil generated, we already have a production cost that greatly exceeds the price of all commodity plant oils. Further costs will be incurred in the extraction and refining that must be done to produce a high quality, food grade oil. Hence even the cheapest SCO is not likely to cost less than $5000 per ton and, if a slow growing microorganism is used with fastidious growth requirements or with a low oil yield, then these costs could easily double or triple.

The exact costs of SCO production by fermentation technology, though, are difficult to calculate in general as there are so many variables that need to be taken into account. Apart from the costs of the process itself (Figure 17.3) and the costs of the substrate and other direct costs, there could also be other costs incurred, for example, in the disposal of the waste fermentation broth. Further revenue must be included in the eventual selling price of the SCO to cover all the capital outlay that will have been needed to customize the fermentation plant, or even to build a new one on a green field site, and also to cover the costs of all the research that will have been done to develop the process, as well as the final marketing, sales, and distribution of the oil.

While it may be difficult to estimate the cost of producing an SCO beyond the ballpark figures indicated above, we nonetheless can indicate the type of process that will be used. Figure 17.3 shows a typical large scale fermentation process of the type that would be used for SCO production. An appropriate microbial culture is initially grown through shake-flask culture, then placed into small (10–100 L) stirred fermentors, and finally

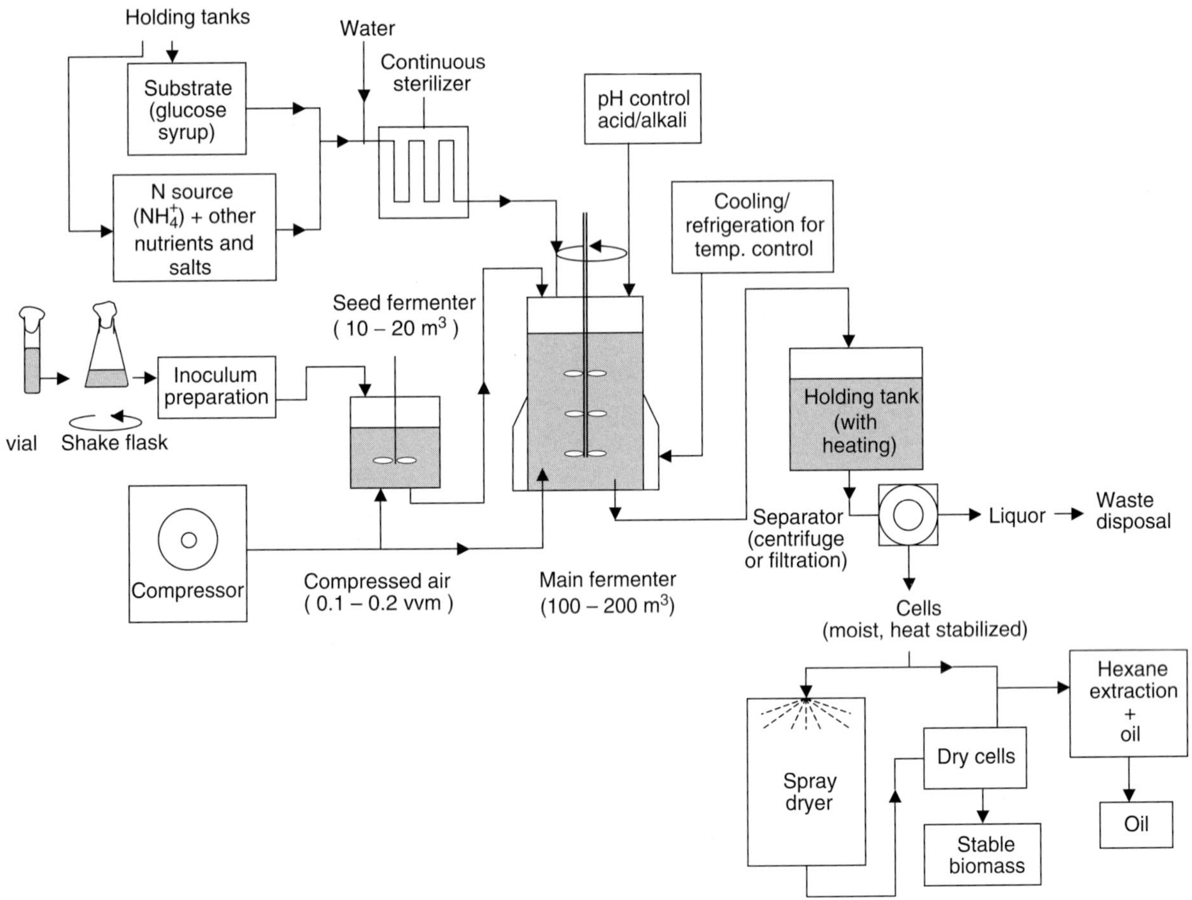

Figure 17.3 A diagrammatic presentation of a fermentation system used for the production of Single Cell Oils. (see text, Section 17.2).

placed in a large seed fermentor that is usually about 10% of the volume of the final fermentor to be used. The use of a large inoculum volume means that the growth time of the organism in the final fermentor is minimized; that is, the productivity (measured as g oil produced per unit time per unit volume of fermentor) is maximized.

Growth of the SCO-producing organism will probably follow the pattern given in Figure 17.1. All the major parameters, such as temperature, pH, and dissolved oxygen, are carefully measured and controlled during the fermentation process. Temperature control is critical and in most fermentations which, of course, are exothermic, this requires continuous cooling using a water supply rather than expensive refrigeration. For this reason, fermentations that run at 30°C or higher are usually preferred.

The sterility of the systems is also of paramount importance, and there are frequent checks at all stages of both the inoculum preparation as well as the final fermentation process itself that the culture remains axenic. Contamination usually means that the individual production run has to be aborted but, fortunately, this is usually a rare event.

In the final hours of the fermentation, when the oil content of the cells is reaching its maximum, the amount of residual substrate (e.g., glucose or sucrose) may be exhausted. Although this may seem to be good practice, in that all the added substrate will have been consumed, this situation can stimulate the cells to use the lipid that they have just accumulated. What happens is that oil accumulation occurs as a means of the cells storing excess carbon as a reserve material. However, when the substrate is exhausted, the cells then sense that they are starving and instigate a survival mechanism in which the stored reserve of oil is now mobilized for the cells' own benefit. Consequently, if this happens in the fermentation process, not only will the final oil yield diminish but, more importantly, the switch from the cells using glucose to using oil as a substrate induces the formation of lipases. Lipases are induced as the first step in oil utilization, and their presence can be detrimental to the final quality of the oil.

Even if the glucose or other substrate is not completely exhausted at the end of growth and at the time of harvesting the cells, the final process of cell harvesting (Figure 17.3) means that any residual substrate will be removed from the cells at this stage. The cells in the absence of glucose, whether they are being harvested by filtration or centrifugation, still respond to the stress of there being no carbon source by inducing the formation of lipases; they still sense the starvation stress and begin to utilize their oil reserves. If the activity of the induced lipases is not checked, this will inevitably lead to loss of valuable oil but will also cause free fatty acids to be liberated. These will not be easily removed and, indeed, may remain in the final oil to detract from its final appearance, flavor, and general characteristics. Induction of lipases in the cells, either during the final hours of the fermentation process, or during the harvesting of the cells, therefore has to be prevented. This is most easily achieved by heating the fermentation broth in the final hours to a point where the lipases are destroyed or the mechanism of inducing the lipases is prevented by heat denaturation of the protein-synthesizing ribosomes. In practice, heating the fermentation broth to no more than 60°C is normally regarded as sufficient.

It is usually more economical if the final fermentation broth is removed as quickly as possible from the fermentor as this speeds up the turnaround time before the fermentor can be cleaned, refilled, and reinoculated, and the process repeated. To achieve this fast turnaround, the fermentor is usually discharged into a holding tank (see Figure 17.3) which can, as an alternative to heating the fermentor, be heated to inactivate the lipases and stabilize the oil. Harvesting of the cells from the holding tank can then take place over a much longer period than harvesting directly from the fermentor itself — perhaps taking up to 48 hours depending on the size of the centrifuges or rotary drum filtration units available.

The harvested cells are still moist. They can be completely stabilized by flash drying (e.g., spray drying) in which case the dry cells can be either stored until required or sold if there is a market for an oil-rich microbial biomass. Alternatively, the moist cells can be directly extracted using hexane (which is a permitted solvent for oil extraction). The process of oil extraction either from the moist cells, or from the spray dried cells, is the same as is used with the extraction of oils from plant seeds.

Usually for SCO extraction, a small scale extraction unit either must be built or a suitable existing commercial unit has to be identified. Such small extraction units are used commercially in the extraction of various high value, low volume, speciality oils or in the extraction of essential oils from various plants seeds and tissues. A final deodorizing step together with a refinement step then completes the oil extraction and production. [Further information regarding the extraction of various SCOs from organisms being grown commercially is given in (14).]

The overall aim is to produce a clear, bright oil with a minimum of coloring; though some yellowness is acceptable, a dark, brown oil would not be regarded as desirable. The various characteristics of the oil need to be carefully evaluated, chief among them being the overall stability of the oil. Some food grade, antioxidant material can be added to ensure a high stability of the oil, though in practice many microbial oils have been found to be very stable to oxidation because of the presence of natural antioxidants that have been co-extracted with the oil.

17.3 SCO PRODUCTION PROCESSES

Because of the high cost of producing microbial oils, most recent work has concentrated on the production of various oils rich in polyunsaturated fatty acids (PUFA) as these are the most expensive oils. There is, however, one exception, and that has been the pursuit of producing a possible cocoa butter equivalent fat using yeasts.

17.3.1 Process for the Production of a Cocoa Butter Equivalent Fat

In the mid 1980s the price of cocoa butter reached an all time high of more than $8000 per ton. This made the possibility of producing a facsimile fat — known as a cocoa butter equivalent (CBE) fat — using yeast technology somewhat attractive. However, cocoa butter is a triacylglycerol in which the three component fatty acyl groups are palmitic acid (16:0), oleic acid (18:1), and stearic acid (18:0) (Table 17.4). Unfortunately, most microbial oils are low in their contents of stearic acid (see Table 17.1) and therefore modification of the composition of the yeast fat was necessary if a CBE was to be produced. Several approaches were tried to achieve this end, and these have been reviewed in four separate chapters in ref. 15. The most successful work was led by Henk Smit of the Free University of Amsterdam, Netherlands, using *Cryptococcus curvatus* D and involving the deletion of the gene that codes for the Δ9–desaturase, which converts stearate into oleate (16). This genetic deletion led to the mutant accumulating 50% of its total fatty acids as stearic acid (Table 17.4) and a greatly diminished amount of oleate. So complete was the genetic deletion that, for the yeast then to grow, it was necessary to include a small amount of oleic acid in the growth medium. Subsequent modification of the mutation led to the isolation of a partial revertant that could now grow without the addition of oleate but which nevertheless still accumulated a large amount of stearic acid. Unfortunately significant amounts of linoleic acid (18:2) still occurred in the final fat, which detracted from the final quality of the yeast CBE. Also it was found that when this final mutant (R22.72) was grown in

Table 17.4
Formation of a Cocoa Butter Substitute by the Yeast *Cryptococcus curvatus* D[a]

	Major Fatty Acids (Rel. % w/w)					
	16:0	18:0	18:1	18:2	18:3	20:0
Cocoa butter	23–30	32–37	30–37	2–4	-	-
Yeast						
Wild type	17	12	55	8	2	1
Wild type[b]	18	24	48	3	1	2
Ufa 33[c]	20	50	6	11	4	4
R22.72[d]	16	43	27	7	1	2

[a] Formerly known as *Candida curvatus* and also *Apiotrichum curvatum*
[b] Yeast grown with restricted supply of O_2 (see Reference 16)
[c] Mutant lacking Δ9-desaturase and requiring oleic acid to grow (see Reference 17)
[d] Partial revertant of mutant Ufa 33 no longer requiring oleic acid for growth (see Reference 17)
Source: Adapted from References 8–10.

large scale fermentations, it was not particularly stable and the quality of the final fat was now below that achieved in small laboratory cultivations.

The simplest and possibly the most successful approach to achieving an increase in the stearate content of the yeast fat was that used by Julian Davies in New Zealand, which was to grow the yeast with a deficiency of oxygen (17). This required trials to be done in 100 m^3 fermentors in order to decrease the aeration rate sufficiently to cause the desaturases to stop working. (All fatty acid desaturases require molecular oxygen for their reaction.) When this was done, the content of stearate was increased to 24% of the total fatty acids (Table 17.4). The low aeration rate in the fermentor also diminished the contents of 18:2 and 18:3 in the yeast fat thus adding to the overall attractiveness of the product.

The properties of the final yeast CBE produced in this manner were sufficiently good for it to be considered suitable for use alongside more conventionally produced CBEs. However, by the time the work had been completed the price of cocoa butter had fallen to about $3000/ton forcing down the price of the CBE fat to under $2000/ton. Even though the proposed process intended to use the waste lactose arising from cheese creameries in New Zealand, where there are vast amounts of it available throughout the year, and could therefore almost claim to be using a substrate of zero cost, the process was still deemed to be uneconomical.

Although this work was ultimately unsuccessful, it did show that SCOs could be tailor-made to a desirable specification by appropriate genetic manipulations of an amenable organism, such as *C. curvatus*. Furthermore, careful manipulation of the fermentation conditions, in this case using a very low aeration rate, could also help change the fatty acid profile of the organism toward the most desirable configuration. These principles would therefore apply to other organisms being used for other SCOs and illustrate one of the strengths of the biotechnological approach to oil production, namely to take an individual organism and then to optimize both oil quality and production by careful selection of the organism and control of its growth conditions.

17.3.2 Polyunsaturated Fatty Acids (PUFAs)

Before covering the present SCO processes for the production of various PUFAs, we thought that it would be helpful if we explained the reasons behind the current upsurge of interest in these lipids.

PUFAs are now in considerable demand mainly as dietary supplements and are regarded as nutraceutical materials. The major sources of these fatty acids are given in Table 17.5 together with their biological roles. PUFAs fall into two main categories: the ω-6 or n-6, (sometimes given just as n6), and the ω-3 (n-3 or n3) series, depending on the position of the final double bond in the acyl chain relative to the terminal methyl group (Figure 17.4).

17.3.2.1 Dietetic Significance of Long Chain Polyunsaturated Fatty Acids

Although fat in general is very much seen as a harmful element in the human diet of course, in moderation, fat is a perfectly healthy food component and certain dietary fats are actually dietetically essential for human health. Biological membranes are composed of a lipid bilayer and several groups of cell signalling molecules (prostaglandins, leukotrienes, and eicosanoids) that are essential for cell function are synthesized from lipid precursors. Although animal cells are capable of synthesizing fat *de novo* from glucose certain specific fatty acids (the basic building blocks of lipids) cannot be produced *de novo* by animal cells. This is a result of cells lacking certain enzyme activities involved in polyunsaturated fatty biosynthesis.

The biosynthesis of the two major groups of polyunsaturated fatty acids, the n-6 and n-3 series, are shown in Figure 17.4. These two groups are synthesized, respectively, from linoleic acid [18:2(n-6)] and α-linolenic acid [18:3(n-3)]. Both linoleic acid and α-linolenic acid are essential fatty acids, as they cannot be synthesized *de novo* in animals which lack the Δ12 desaturase required for 18:2 biosynthesis from oleic acid and the Δ12 and Δ15 desaturases required for 18:3(n-3) biosynthesis from oleic acid. As a result of this metabolic deficiency, animals require sources of both 18:2 and 18:3(n-3) in their diet. Fortunately, both these essential fatty acids are synthesized by plants, which almost without exception possess a Δ12 desaturase and a Δ15 desaturase. As a result, a dietary deficiency of either of these fatty acids is unheard of in nature and difficult to obtain even under laboratory conditions. (Even if one eats a minimum of plant derived oils, these essential fatty acids are still present in animal fats as, invariably, the animals that we eat are herbivores and will have derived their own linoleic and linolenic acids from their plant diet.)

Once taken up by animal cells, both 18:2 and 18:3(n-3) can be further elongated and desaturated to the full compliment of long chain polyunsaturated fatty acids (LC-PUFA), i.e., fatty acids containing 20 or more carbon atoms and three or more double bonds, (Figure 17.4). It should be noted that although animal cells can synthesize LC-PUFA from the essential fatty acid precursors, plant cells are unable to produce LC-PUFA and so dietary sources of LC-PUFA are traditionally animal oils (especially fish oils).

Several of the LC-PUFA, of both the n-3 and n-6 class, play important roles in cell function. LC-PUFA are key components in cell membranes, playing a role in regulating the fluidity of membranes and therefore the activity of many membrane associated processes. Indeed two LC-PUFA, arachidonic acid [20:4(n-6)], and docosahexaenoic acid [DHA, 22:6(n-3)] are highly enriched in mammalian brain and nervous tissue where they play an important role in brain and neural development, especially in neonates (19). The role of DHA in brain development has been highlighted by reports that a diet rich in fish oils containing high levels of DHA may have played a defining role in the evolution of mankind in terms of brain size and therefore intelligence (19).

Certain LC-PUFA are the precursors of a number of cell signalling compounds that are vital for many vital processes from inflammatory response to blood clotting. Arachidonic acid [20:4(n-6)] and dihomolinolenic acid [20:3(n-6)] are key n-6 precursors of the group 2 and group 1 series of prostanoids while eicosapentaenoic acid [20:5(n-3)],

Table 17.5
The Major Sources and Biological Functions of the Major PUFAs* and Their Applications

Sources	n-6 Family		n-3 Family		
	GLA (18:3; 6,9,12)	**ARA (20:4; 5,8,11,14)**	**ALA (18:3; 9,12,15)**	**EPA (20:5; 5,8,11,14,17)**	**DHA (20:6; 4,7,10,13,16,19)**
Plants and animals	Plants: *Oenothera, Borago, Ornithogalum* spp.	Fish: *Brevoortia, Clupea, Sardina* spp., animal tissues	Plants: *Brassica, Glycine, Linum* spp.	Minor component of tissues	Fish: *Brevoortia, Engraulis, Sardina, Scomber* spp.
Microorganisms	Fungi: *Mucor, Mortierella.* Algae: *Chlorella and Spirulina* spp.	Fungi: *Pythium, Mortierella* spp. Algae: *Porphyridium* spp. Mosses: *Rhytidiadelphus, Brachythecium, Erthynchium* spp.	Algae: *Chlorella* spp.	Fungi: *Mortierella, Phythium* spp. Algae: *Chlorella, Monodus, Porphyidium, Nannochloropsis, Cryptoleura, Schizymenia, Navicula* spp. Mosses: *Brachythecium, Eurhynchium, Scleropodium* spp. Bacteria: *Rhodopseudomonas, Shewanella* spp, *Photobacterium*	Fungi: *Thraustochytrium, Entomophthora* spp. Algae: *Gonyaulax, Gyrodinium, Cryptheconidium* Bacteria: *Colwellia, Moritella (Vibrio) marinus*
Major dietary sources	Evening primrose seed oil, from linoleic acid	Liver, brain, egg yolk lecithin	Vegetable oils (such as soy, linseed, pumpkin seed oils), leafy vegetables	Salmon, tuna, sardines, cod, shellfish, algae	Cold water fish, shellfish, algae
Biological functions	Intermediates in the biosynthesis of other fatty acids, precursor of prostaglandins PGE1	Major component of most membrane phospholipids, precursor of prostaglandin PGE2	Minor component of tissues	Precursor of prostaglandin PGE3, prevents thrombosis, inhibit the production of n-6 eicosanoids, modulates immune function	Present in the retina and the grey matter of brain of mammals, influences on the visual and neural development of infants
Nutritional values and applications	Dietary supplement intake for several disorders such as eczema, use in cosmetic	Ingredient in various infant formulae along with DHA	Beneficial effects on blood lipid profiles, to reduce the risk of coronary heart disease, arthritis, inflammation, hypertension, psoriasis, other autoimmune disorders and cancer, nutraceutical additives for processed food, DHA incorporated into infant formulae for improvement of vision and memory.		

*ARA: Arachidonic acid; ALA: α-linolenic acid; DHA: docodahexaeneoic acid; EPA: eicosapentaenoic acid; GLA: γ-linolenic acid.
Source: From Reference 18. With permission.

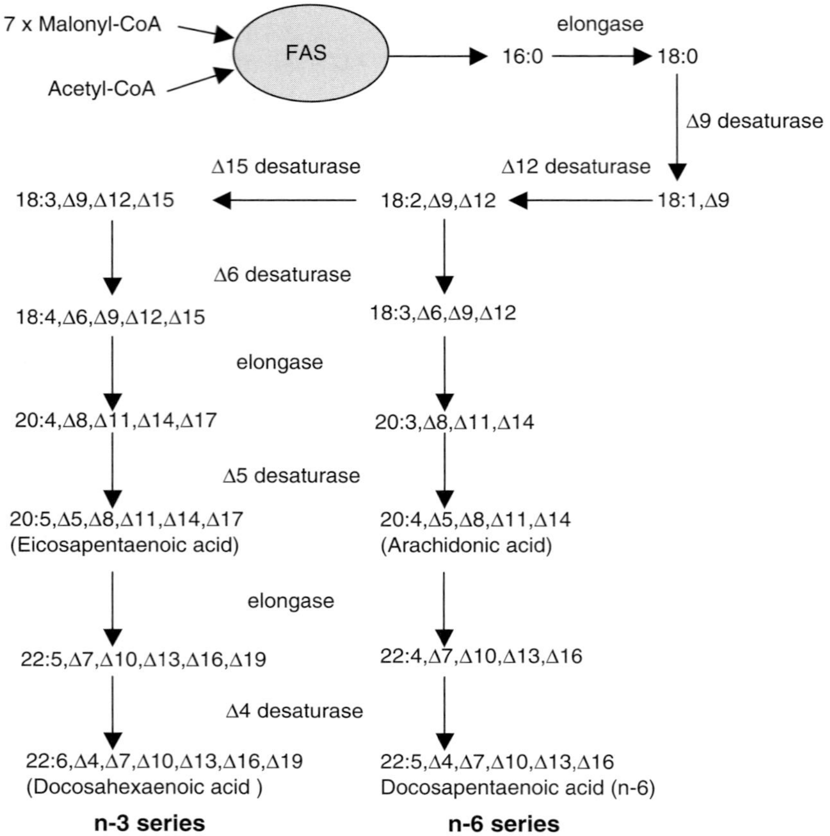

Figure 17.4 Diagram showing the routes of *de novo* biosynthesis of polyunsaturated fatty acids of the n-3 and n-6 series.

another n-3 LC-PUFA, is the precursor of the group 3 series of prostanoids (20). As the group 3 series of prostanoids often act antagonistically to the group 1 and 2 series and, because the enzyme activities that produce the prostanoids from their respective n-3 and n-6 precursors are the same, the ratio of n-3 and n-6 fatty acids in the diet can have far reaching physiological implications.

The current Western European/USA-type diet is becoming increasingly deficient in LC-PUFA of the n-3 class. This is a result of the decline in the consumption of food types rich in these fatty acids. Good sources include oily fish and certain animal organs (e.g., liver) that are losing favor with today's consumer. As a result there is an interest in fortifying foodstuffs with n-3 fatty acids and various attempts have been made to this end either by direct addition to foodstuffs (such as bread) or by feeding farm animals on a diet rich in n-3 fatty acids in the form of LC-PUFA themselves (fish oil) or as α-linolenic acid (flaxseed oil). Neither of these approaches is completely satisfactory, however. The inclusion of fish oil in animal diets or in food products directly can impart a distinct fishy taste that is generally not well received by consumers and has been noted to cause farm animals to go off their feed (21). Flaxseed oil is not associated necessarily with off-flavors and, although it increases the n-3 fatty acid content of the food, it does not boost the n-3 LC-PUFA levels which is the desired outcome. Although animals, including man, can synthesize n-3 LC-PUFA from α-linolenic acid, the rate at which this occurs is generally slow (20) and several factors, including age, and hormonal balance, can decrease these conversions further still. Therefore additional supplementation of these key fatty acids is frequently thought to be desirable.

17.3.2.2 Processes for the Production of Various PUFAs

Of particular importance are: γ-linolenic acid [GLA, octadecatrienoic acid, 18:3 (6,9,12), or abbreviated to 18:3(n-6)], arachidonic acid [ARA, eicosatetraenoic acid, 20:4 (5,8,11,14), 20:4(n-6)], eicosapentaenoic acid [EPA, 20:5 (5,8,11,14,17), 20:5(n-3)], and docosahexaenoic acid [DHA, 22:6 (4,7,10,13,16,19), 22:6(n-3)]. Of these four PUFAs, only GLA can be obtained from plant sources (Table 17.5); all the others must either be obtained from animal sources or by using SCO technology. When animal sources are used for PUFA production, the source material is often in short supply or extensive purification is needed as with the production of purified EPA and DHA starting with fish oils that contain a mixture of these PUFAs. Consequently, the overall costs to production are very high thereby making the microbial route to their production an attractive and economic alternative.

17.3.2.2.1 γ-Linolenic Acid-rich SCO The first commercial microbial oil was produced in 1985 and was rich in γ-linolenic acid [GLA, 18:3(n-6)]. The oil was aimed at being an alternative source of the oil from seeds of the evening primrose (*Oenothera biennis*). Evening primrose oil was, at that time, considered to be useful in the treatment of multiple sclerosis though these claims have since been discounted. The oil was also useful for the treatment of a number of other disorders and today is still sold as an over-the-counter nutraceutical for the relief of premenstrual tension. It is also thought to be useful for the treatment of eczema, especially in children. A considerable amount of clinical and medical interest still exists with this PUFA and further information regarding the current roles of GLA can be found in two recent monographs on the topic (22,23).

Against this background, the search for an alternative source to evening primrose oil began in one of the present authors' (CR) laboratory in the late 1970s by examining fungi belonging to the class known as the *Mucorales*. Prior work (24,25) had indicated that GLA occurred in most, if not all, species in this group and, after a search of more than 300 species and strains, one organism was identified as being the most productive. This was *Mucor circinelloides*, also known as *Mucor javanicus*. Initial toxicity trials of the oil showed that it was without any detrimental effect on a variety of cells and whole animals. Furthermore, as the fungus itself has a long association with the fermented food known as tempeh, clearance for the use of the oil, coming from what was a Generally Recognized As Safe (GRAS) organism, quickly followed. Sales of the oil began in 1985 and lasted until 1990. It was sold under the trade name of Oil of Javanicus and a small market for it was quickly established. Approximately 5 to 8 tons a year of the oil was produced using fermentors up to 220 m^3. The company producing the oil was J. & E. Sturge Ltd at Selby, North Yorkshire, UK. This company normally produced citric acid by fermentation technology and, for the production of the GLA-SCO, one of the existing fermentation vessels was employed without modification although the medium being used was one with a high C:N ratio in order to engender lipid accumulation (Figure 17.1). The production process is shown in broad outline in Figure 17.3.

The specifications of the oil were superior to those of evening primrose oil in almost every respect (Table 17.6). With 18–20% GLA, the SCO had more than double the content of GLA of evening primrose oil, it has less residual herbicide and pesticide materials (though the amounts found in evening primrose oil were much less than the maximum permitted levels) and had exceptional long term stability due to the presence of natural

Table 17.6
Fatty Acid Profiles of GLA-Rich Oils from Plants and Fungi used in Industrial Fermentations

	Relative % (w/w) Major Fatty Acids in Neutral Lipids (= Triacylglycerol Oil)							
	16:0	**16:1**	**18:0**	**18:1**	**18:2**	**18:3 (n-6)**	**18:3 (n-3)**	**20:1**
Evening primrose	6	-	2	8	75	8	0.2	0.2
Borage[a]	10	-	4	16	40	22	0.5	4.5
Mucor circinelloides[b]	22	1	6	40	11	18	-	-
Mortierella isabellina[c]	27	1	6	44	12	8	-	0.4

[a] Also contains 22:1 (~2.5%) and 24:1 (~1.5%)
[b] Production organism used by J & E Sturge Ltd, UK; oil content of biomass 25%
[c] Production organism used by Idemitzu Ltd, Japan; oil content of biomass ~50%

antioxidants that were coextracted with the oil from the fungal cells. The fermentation process could, moreover, guarantee an unchanging specification of the oil which could be available at any time of the year. However, and perhaps not too surprising, the oil was difficult to market as there was an obvious reluctance on behalf of the public to purchase what was a microbial oil. Marketing strategies therefore very carefully avoided using words such as "fungus", "mold", or even "microorganism," though "biotechnology" was considered to be acceptable.

The arrival of this "new" oil in the market place had an immediate effect in decreasing the price of evening primrose and thus the initial, apparent profitability of the biotechnology process was therefore not fully realized. Further, another plant source of GLA had been developed during the second half of the 1980s which was borage (*Borago officinalis*). This oil, which was sold under the name of starflower oil, had an even higher content of GLA than Oil of Javanicus (Table 17.6) and could be produced for less than the cost of evening primrose oil as this plant was an annual crop and not a biennial crop like evening primrose. It was thus an immediate and strong competitor to the GLA-SCO.

Of telling significance in the final economics of oil production was the status of the various oils in the eyes of the European Union Agricultural Committee. Both evening primrose and borage were regarded as non-food crops and, as such, then enjoyed financial subsidies from the EU for their cultivation. (Round 1 to the plants.) Next, the fermentation process had to buy the sugar being used in the production of the GLA-SCO at European prices and not at World price, which was about half the EU price because of trade tariffs being imposed. With approximately 10 tons of sugar being needed to produce one ton of oil, this was Round 2 to the plants and, indeed, it was "game, set, and match." Production of Oil of Javanicus ceased in 1990 as it could no longer compete against borage oil.

Another process for the production of GLA using fermentation technology was developed in the late 1980s by Idemitzu Ltd in Japan. While the process in the UK had used a species of *Mucor* that contained 18–20% of its total fatty acids as GLA, the Japanese process, which used *Mortierella isabellina* another member of the Mucorales group, settled for a species that had only about 8% GLA in its oil (Table 17.6). However, the oil content of the Japanese organism was about 50% of the biomass whereas that of the UK organism was only 25%. Thus although more oil was obtained in the Japanese process, the GLA content was not sufficiently high to compete against the two major plant sources. Sales of the Japanese oil did not achieve a significant part of the market and the oil no longer appears to be available. Some details of the process have appeared (26) but information as to the size of the production system and the amounts produced has not been disclosed.

Although no longer in production, Oil of Javanicus still represents one of the best sources of oil from which to purify GLA itself, if such a pure fatty acid should ever be needed. The very low content of linoleic acid (18:2) in the oil (Table 17.6) means that fractionation of the fatty acids, using a technique such as urea adduct formation, would be relatively simple in comparison with trying to purify it from either evening primrose or borage oils which have much higher contents of 18:2. However, at the time of writing, no such demand for a high purity GLA has been expressed and thus the fermentation process must now languish until some favorable economic circumstances arise that would justify its resurrection.

17.3.2.2.2 Arachidonic Acid (ARA) rich SCO The prospect of producing an oil rich in ARA from a microbial source was first investigated in the mid 1960s. Initial interest was due to a misapprehension that this fatty acid was a potential chicken flavor additive. With this as the driving force, a small research group led by Bob Shaw at Unilever Ltd, UK, initiated a study of ARA production by microbial sources (24,25) thereby identifying a number of potentially useful organisms. Another early, and nondietetic, application for a microbial oil rich in ARA was as a cosmetic additive with the Lion Corp of Japan developing and patenting a process in 1988 for the production of an ARA-rich oil (27).

17.3.2.2.2.1 Microbial Sources of ARA-rich SCO Many different micro-organisms have been examined for their capacity to synthesize ARA, however, only two – both fungi (*Mortierella alpina* and *Pythium* sp) have been seriously considered for commercial production (28). Subsequent development has, though, concentrated solely on *Mt. alpina* as this organism is now widely considered the most productive species (29,30). *Mt. alpina* is the fungus used in all the current commercial ARA-rich SCO processes.

Mortierella alpina is a ubiquitous soil fungus that has been isolated from soils all over the world. There are at least 37 different isolates of *Mt. alpina* that have been deposited in the world's major culture collections (though there maybe some overlap in these collections). All the off-the-shelf strains have been exhaustively examined for their capacity to synthesize arachidonic acid and the most productive strain is reported to be ATCC 32222 (31,32). However the production organisms used commercially are proprietary strains that have been isolated either directly from the environment or via strain selection (30,33). It should be noted that strain selection with *Mt. alpina* is far from straightforward as stimulating sporulation in this fungus, which is a prerequisite to strain improvement, is difficult if not impossible for many strains. One notable exception is the 1S-4 strain, which sporulates freely, and was isolated by Prof. S. Shimizu's group. It is now used commercially by Suntory Corp. in Japan.

17.3.2.2.2.2 Commercial Production of ARA-Rich SCO There are currently at least three commercial processes operating to produce ARA-rich SCOs. Two are operating in the Far East. In Japan, Suntory Corp., one of the pioneers of ARA-rich SCO technology continues to produce ARA-rich SCO using technology developed by the group of Prof. S. Shimizu. The process employed by Suntory is relatively well documented and basically follows the steps outlined in Figure 17.3. In China, Wuhan Alking Bioengineering Co. Ltd. (Wuhan City) have been producing ARA-rich SCO since 2001 using a process that is far less well reported. Both of these processes, while operated on a commercial basis providing ARA-rich SCO for infant formula supplementation and as health supplements, do not constitute a major global source of ARA-rich SCO. That honor goes to the ARA-rich SCO produced by DSM Co. (formerly Gist-brocades) in Italy under contract for Martek Biosciences Corp. Their process is responsible for >95% of the current ARA-rich SCO produced and will be even more important when a second plant (in Belvidere, NJ, USA) comes on-line to meet the current demand for ARA-rich SCO which cannot be met by the Italian plant alone.

This ARA-rich oil (sold under the trade name ARASCO™ as part of the infant formula additive Formulaid™) is produced using *Mortierella alpina* (Table 17.7) in a process that shares many features with the process operated by Suntory. However, the strains used in the two competing processes differ and, so inevitably, do the details of the processes, which are optimized for the specific production strain employed. Both though approximate to the production outline given in Figure 17.3.

A cryovial of a certified stock of the production organism (a proprietary strain) is thawed and cultivated in shake flasks, the medium used is relatively simple containing simple salts, glucose and yeast extract. After the cells have grown to a designated density, cultures are used to inoculate a set of increasingly large fermentation vessels so as to maintain an inoculum volume of approximately 5–10% (v/v). The final vessels are in the region of 50 to 100 m^3 and are nitrogen fed, as the amount of total nitrogen needed to support the final cell density cannot be added into the initial medium without suppressing the mass transfer capacity of the fermentation vessels and creating problems maintaining the dissolved O_2 concentration above a critical value (PUFA biosynthesis being an aerobic process). The nitrogen feeding regimes allow very high biomass concentrations to be achieved. Cell densities in excess of 50 g/l have been reported (34) and it must be assumed that the commercial process exceeds this value. Very high cell densities are, of course, required for all SCO processes as the product is intracellular. As a consequence, the fermentation has to support these high cell densities and maintain sufficiently high oxygen transfer rates within the tank to support not only growth but also the O_2-dependent desaturase enzymes that are required for PUFA biosynthesis.

A key factor in obtaining very high cell densities, while not exceeding the O_2 transfer capacity of the fermentation vessel, is culture morphology. As a filamentous fungus, *Mt. alpina* is capable of growth as either dispersed hyphae yielding a culture that resembles porridge (oatmeal) or as discrete pellets (of varying sizes). The morphology observed is interdependent on a number of factors including strain, nitrogen source, dissolved O_2, and metal ion concentrations (34,35). While dispersed hyphae are most productive at low cell densities, due to the ease of O_2 transfer to the hyphae, at high cell densities the viscosity of the medium becomes problematic (in terms of mass transfer and physical mixing of the broth). As a result, dispersed biomass is not suitable for commercial production. In contrast, pellet formation can be suboptimal for ARA production as the interior of the pellet can rapidly become nutrient and O_2-limited. The optimum biomass morphology is a compromise between different factors and appears to be very small pellets. Small pellets minimize the problems associated with nutrient transfer across the pellet radius whilst having the benefit of reducing culture viscosity to promote high final cell densities.

Once the cell density has reached a suitable level (and the cell lipid is of adequate quantity and quality – the total fatty acids contain in excess of 42% ARA) the biomass is removed from the fermentation vessel and harvested. The harvesting involves dewatering the biomass, using either a continuous centrifuge and a screw press, or a filter press, and then drying the biomass prior to extraction of the cell oil. Drying of the biomass is a critical step as drying must be achieved rapidly to avoid spoilage of the cell lipid (either due to contamination of the wet cake with other microorganisms or breakdown of the oil, *via* β-oxidation by the fungus itself). However, heating the biomass to a substantial degree (see also Section 7.2) also compromises the quality of the oil obtained due to oxidation of the highly unsaturated lipid.

Early attempts to remove oil from the fungal biomass (and those still used by Suntory) involved a disruption step designed to release the cell lipid and make it more accessible to hexane extraction (36). Although disruption increased extraction efficiency, the downside was the retention in the extracted oil of colloidal particulate

Table 17.7
Fatty Acid Profiles of Microbial Oils Rich in Arachidonic Acid and Docosahexaenoic Acid that are Produced Commercially

	Fatty Acid Composition (Rel. % w/w)													
Oil	**12:0**	**14:0**	**16:0**	**16:1**	**18:0**	**18:1**	**18:2 (n-6)**	**18:3 (n-6)**	**18:3 (n-3)**	**20:3 (n-6)**	**20:4 (n-6)**	**22:5 (n-6)**	**22:6 (n-3)**	**24:0**
ARASCO™[a]	-	0.4	8	0	11	14	7	4	-	4	49	-	0	1
DHASCO™[b]	4	20	18	2	0.4	15	0.6	-	-	-	-	-	39	-
Schizochytrium oil[c]	-	13	29	12	1	1	2	-	3	1	-	12	25	-

[a] Production organism: *Mortierella alpina*
[b] Production organism: *Crypthecodinium cohnii*
[c] Production organism: *Schizochytrium* sp.
NB: ARASCO™ and DHASCO™ are registered trade names of Martek Corp. Inc.

matter which complicated the downstream treatment of the oil. A more recent adaptation has been an extrusion process that yields pelletized biomass (37) that allows relatively efficient oil extraction without the need for extensive cell disruption and the problems associated with this. The biomass pelletization process also contributes significantly to oil stability.

Another process that can greatly increase the quality of the crude oil obtained is pasteurization of the culture broth at the termination of the fermentation (see Section 17.2). Pasteurization has two benefits, it decreases the risk of a fast growing microbial contaminate (bacteria or yeast) becoming established in the culture broth during harvest and it also ensures that the *Mt. alpina* biomass is non-viable and that all enzymatic activity has ceased (37). Non-viable *Mt. alpina* biomass is advantageous as it minimizes the risk of triacylglycerol hydrolysis by endogenous enzymes during the harvest period (21,37).

The extracted oil is refined, bleached, and deodorized using methods developed for vegetable oils, resulting in translucent yellow oil with a distinctive but not strong flavor (the authors' personal experience). Despite the high level of unsaturation the oil is remarkably resilient to oxidation suggesting the presence in the extracted and processed oil of naturally occurring antioxidants.

17.3.2.2.2.3 DHA-rich SCO Once the beneficial properties of DHA, in terms of neonate neural and visual development, were realized a clean fermentable source of DHA was an obvious target for biotechnologists. Although fish oils were commonly available containing significant quantities of DHA, these oils had certain disadvantages that made their inclusion in infant formula unadvisable. Most significant of these was the occurrence in fish oils of another n-3 LC-PUFA, eicosapentaenoic acid [EPA, 20:5(n-3)]. Although this LC-PUFA has certain potentially attractive therapeutic properties (see below) it is contra-indicated for infant nutrition as it is associated with growth retardation (38). As a result of this (and due to many fish oils now containing low levels of environmental pollutants) the inclusion of fish oil in infant formula is not considered advisable and has not obtained GRAS status from the Food and Drug Agency (FDA) of the USA for this application, thereby barring its inclusion in infant milk formulae in the USA.

A number of marine microorganisms that produce DHA are known (Table 17.3) but these were considered little more than of academic interest as they were hard to cultivate to the high densities with substantial lipid reserves (39–41) as would be required for commercial SCO production. Two competing American companies, OmegaTech (Boulder, CO), and Martek Biosciences (Columbia, MD), which became a single company in 2002 when Martek acquired OmegaTech, did, however, develop high density cultivation systems for some of these marine microbes and these cultivation systems are the basis for the current DHA-rich SCOs (see Table 17.7).

17.3.2.2.2.4 DHA Production Organisms As mentioned, a number of marine microorganisms are known to produce DHA (substantial DHA biosynthesis is strongly associated with marine environments for reasons yet to be fully explained). These organisms include both microalgae and, surprisingly, bacteria (Table 17.5). Bacteria for a long while were considered unable to synthesize PUFA (42) but this property is now found in many marine bacteria and in bacteria isolated from intestines of fish. (Indeed, many people would consider that the LC-PUFAs of fish are derived either from these intestinal bacteria or are derived from ingested algae and are not synthesized by the fish themselves.)

Two heterotrophic microalgae, a dinoflagellate (*Crypthecodinium cohnii*), and a stramenopile (*Schizochytrium* – formerly classified as a "fungoid protist") have been selected as production organisms for commercial DHA production. Both of these organisms produce the majority of their LC-PUFA as DHA (Table 17.7). In *C. cohnii*, DHA (up to 50% of total fatty acids) corresponds to >95% of all PUFA with only 28:8(n-3) being detectable at >1% of total fatty acids (43). The fatty acid profile of *Schizochytrium* is also relatively simple [one of the criteria on which this strain was selected, see (21)] but does contain detectable amounts of other PUFA belonging to both the n-3 and n-6 pathways. Of most significance was the presence of the n-6 PUFA, docosapentaenoic acid [DPA n-6, 22:5(n-6)], at approximately one third the level of DHA (43). As the appearance of this fatty acid in humans was associated with a deficiency of DHA (and of n-3 PUFA in general), it was thought that this was a fatty acid that should be avoided in infant formula. Although the evidence to support this assumption was far from unequivocal, this doubt was sufficient to preclude the DHA-rich SCO from *Schizochytrium* from being considered for inclusion in infant formula. As a result, while the *C. cohnii* oil was developed as an infant formula additive, the *Schizochytrium* oil was developed as a general food additive and as an animal food ingredient aimed at increasing the amount of DHA in the human diet in the form of meat and eggs (21). Since April 2002, Martek and OmegaTech have been a single company (the former acquiring the latter) and both DHA-rich oils remain under development.

A third production organism is now being employed by Nutinova, a subsidiary of Celanese, in Germany to produce other DHA-rich SCO. This oil has a fatty acid composition similar to the *Schizochytrium* oil (see Table 17.7) although the phylogeny of the production organism is the subject of some conjecture and debate. The commercial significance of this oil is likewise unclear though the indications are that the product will be launched in 2004 under the trade name of DHActive. A similar oil is also thought to be near commercialisation in Japan by Nakase-Suntory.

17.3.2.2.2.5 Commercial Production of DHA-rich SCO Although many microalgae that produce DHA are photosynthetic (Table 17.3) the species selected for industrial development are heterotrophic. This is a key attribute as it allows cultivation in large fermentation tanks (up to 200 m^3) at high cell density. A photosynthetic production system would be hindered by the requirement for cellular access to light which severely limits cell densities and causes such processes to be many times more expensive to operate than heterotrophic systems. A recent economic analysis calculated that photosynthetic cultivation of microalgae for LC-PUFA production was at least an order of magnitude too expensive to compete with highly purified fish oils (45).

Both the *C. cohnii* and *Schizochytrium* processes operate in a similar fashion (as shown in general outline in Figure 17.3). In both cases, a cryopreserved stock of the production organism, a proprietary strain obtained by classical strain selection to have a range of desirable attributes, is thawed and cultivated in shake-flask culture. As with the ARASCO™ protocol, the culture is scaled through a seed train to maintain a suitable inoculum volume of about 5–10% for each step in the scale up process. The culture medium throughout the seed train is kept constant so as to avoid stressing the culture and slowing culture growth. The culture media used have been extensively modified and bear very little resemblance to the high chloride-containing seawater-based media that appear in the literature and are used in academic studies (39,46). One reason for this is the corrosive nature of the high chloride concentrations in sea water.

The marine ancestry of the DHA-producing microalgae was a problem for large scale production as these organisms would only grow in high salt (NaCl) conditions. High Cl^- concentrations, while not a problem in glass shake-flasks or even laboratory scale glass fermentation vessels, is a major concern in production scale stainless steel fermentors. Seawater contains 18–19000 ppm Cl^- whereas standard stainless steel (type 304) can withstand high Cl^- concentrations for a short while it is corroded at relatively low Cl^- concentrations if exposure is prolonged. A more resistant, and expensive, stainless steel (type 316) is often used for industrial fermentation tanks but this can only withstand Cl^- concentrations up to 1000 ppm. To avoid problems with corrosion of fermentation tanks and downstream equipment, strains capable of growth in low saline conditions have been selected (21) and media formulations have been developed with other ions replacing Cl^- (and containing a Cl^- concentration of >250 ppm) that sustain growth of these organisms (47).

Although previously thought to be difficult to grow at high density, a combination of strain selection and media optimization means that these production organisms can now be cultivated at densities in excess of 100 g/l (44,46). These cultures also are high in oil (>20% w/w dry wt) and rich in DHA (>25% total fatty acids). Once optimal biomass and oil yields have been obtained, the biomass is harvested by continuous centrifugation and the cake spray dried to yield a powder that, after disruption, can relatively efficiently be extracted with hexane (Figure 17.3). As with ARA-SCO, the crude oil is then processed by refining, bleaching and de-odorizing in a process similar to that used by the vegetable oil industry. Although the resultant translucent orange oil is stable (the color is a result of β-carotene content that probably contributes to the oil stability), vitamin E is added to further increase the stability of the oil to a level where it is stable at room temperature for a minimum of 12 months. Again, as one of us (JPW) can attest through personal experience, the DHA-rich SCO (both DHASCO™ from *C. cohnii,* and the *Shizochytrium*-derived DHA-rich SCO) are remarkably palatable oils that lack the fishy taste associated with purified fish oil unless they are very badly abused.

17.4 SAFETY OF SINGLE CELL OILS

The arrival of SCO on the food scene is a relatively recent event – the first commercial SCO (Oil of Javanicus) not being produced commercially until 1985 (see Section 17.3.2.2.1) – well after regulations and regulatory bodies relating to food safety were in place. Therefore the safety of these food products had to be determined and proven to the regulatory authorities and the general public. The safety of Oil of Javanicus was relatively easy to confirm as both the producing organism (*Mucor circinelloides*) and the active ingredient (γ-linolenic acid) have long been part of the human diet and were therefore deemed safe by historical association with food products.

γ-Linolenic acid was the PUFA responsible for the therapeutic activity of evening primrose oil, the high value plant oil that Oil of Javanicus was developed to compete against. Evening primrose oil has been taken for centuries as a folk remedy for a number of ailments, the most well documented of which is premenstrual syndrome. Indeed, its reputed efficacy against general illnesses was such that it was also known as "King's Cure All."

The filamentous fungus *Mucor circinelloides* (the production organism for Oil of Javanicus) has a long documented association with human foods, being the organism used in the fermentative production of the oriental food tempeh. This historical connection and evidence of long term (hundreds if not thousands of years) human consumption without reported adverse effects was instrumental in proving the safety of Oil of Javanicus, and was one of the major factors (other than γ-linolenic acid productivity) involved in selection of *Mucor circinelloides* as the production organism of choice. Having a whole organism that was historically demonstrated non-toxogenic made it fairly easy to argue that a cellular constituent of that organism (the oil) would also be safe for human consumption. Likewise, as the methods used to extract and process the oil from *Mucor circinelloides* were essentially identical to those used to prepare vegetable oils for human consumption (14), it was evident that the processing of the fungal biomass to release the oil would not introduce any deleterious qualities to the final product.

Unlike Oil of Javanicus, the safety of ARA-rich SCO from Mt alpina could not be inferred by the association of the producing organism with any traditional foodstuff. Furthermore, there were even reports suggesting the potential harmful effects of an increase in the dietary intake of arachidonic acid (48). As a result extensive safety data had to be generated before ARASCO™ could be considered safe for human consumption, especially as its primary application is as an ingredient in baby formula. As a result ARA-rich SCO, along with DHASCO™, are probably the most extensively safety tested edible oils currently on the market. The safety of ARASCO™ has been reported by numerous authors who have assessed this oil from a number of different perspectives. The safety of ARASCO™ (the ARA-rich SCO produced by Martek Biosciences Corp) was even the topic of an entire symposium at the AOCS meeting in Indianapolis in 1996 (49).

For some years the inclusion of high levels of ARA in the human diet was considered undesirable, due to potential adverse effects on blood clotting (48). However the apparent lack of toxicity in animal models prompted a major study at the Western Human Nutrition Research Center, San Francisco, USA, of the effect of dietary supplementation with ARASCO™. During an intensive and prolonged study (in which human volunteers lived in a "metabolic ward" to decrease the effect of external influences) the inclusion of 1.5 g ARA/day (3 g ARASCO™/day) did not cause any significant adverse effects on any of the test individuals. Factors examined included blood coagulation, immune response, and the production of arachidonic acid derived signal molecules. The conclusion of the study was that the inclusion of 3 g/day ARASCO™ did not have any toxic effects on adult humans and that this oil should be considered safe as a human food ingredient (50–53).

Further studies have been undertaken in Europe to determine the safety of ARASCO™ as a nutritional supplement specifically in relation to its intended use in infant formula. A review of the literature was carried out to determine if the production organism (*Mort. alpina*) was, or had ever been, associated with human disease (33). This study concluded that this fungus had never been unequivocally linked to human disease (due largely to the fact it is unable to grow at human body temperature) and that there were no reports of mycotoxin production. As *Mt. alpina* is a ubiquitous soil fungus and therefore in common contact with humans it was argued that it was clearly a completely benign organism and should be considered safe for the production of human food material.

Animal studies employing rats as a mammalian model system have failed to demonstrate acute toxicity of ARASCO™ at levels up to 20 ml/kg body weight, (corresponding to a 75 kg human taking 1.5 l of ARASCO™ in a single dose). The only negative finding was that test animals experienced diarrhea the day following administration! Likewise, subchronic studies using a range of ARASCO™ doses in the rat model found that while >2 g/kg body weight/day caused some physiological changes in terms of blood lipid composition and organ weights. These changes are associated with high-lipid diets in general and are not specific to ARASCO™ consumption in particular. No obvious pathological changes were seen (i.e., none of the animals became ill). Furthermore, reproductive function was unaffected and the pups from the ARASCO™-fed group did not demonstrate any ill effects (54,55).

This detailed examination of ARASCO™, in terms of the producing organism and the SCO derived from it, demonstrated unequivocally that this SCO presents no greater danger to infant and human health than any other typical vegetable oil. It should also be pointed out that most of today's commodity and speciality plant oils have never undergone the extensive scrutiny to which SCOs have been subjected. We assume that such esoteric oils as walnut oil, macadamia oil, and sesame seed oil are intrinsically safe, as indeed they probably are; but these have never gone through the extensive examination that SCO have had to.

The DHA-rich SCO from *C. cohnii*, like Oil of Javanicus, does not contain a LC-PUFA with questionable toxicological effects. Diets very rich in DHA are documented for several human populations – this DHA is invariably derived from ingestion of sea fish and sea mammals and has never been associated with any adverse effects. Indeed it was the supposed health benefits of the diets enjoyed by populations of Iceland and Japan (who have a very low incidence of cardiovascular disease, despite a high fat diet, and a low incidence of premature infant deliveries) that first highlighted the potential therapeutic effects of DHA. Ironically, the high intake of sea fish is now being suggested to represent a health risk to these same populations due to the presence in fish meat and especially oil of pollutants released into the marine environment by man (56,57).

As a result of the accepted inherent safety of an oil rich in DHA, the safety issues related to DHA-rich SCO were restricted to a demonstration of the safety of the microbially derived oil. In many studies, fish oils have been used as a negative control (i.e., "safe" oil) with which to compare DHA-rich SCO. The safety of DHASCO™, the DHA-rich SCO that is used on a commercial scale as a infant formula additive as part of the DHASCO™/ARASCO™ blend, has been comprehensively studied both by the manufacturer (Martek Biosciences Corp.), and by formulae manufacturers. These studies have looked at the possible *in vitro* effects of DHASCO™ to test for mutagenic and clastogenic (chromosome damaging) effects, animal studies and human feeding studies (Tables 17.8, 17.9). Animal studies using rats as a model system have checked acute toxicity at doses up to the maximum dose physically possible to administer (~20 g/kg body wt). Additionally, sub-acute studies, where DHASCO™ was fed at several times the anticipated human doses for prolonged periods (up to 90 days), have also studied the effect of DHA supplementation of female animals prior to and during pregnancy as well as to the offspring. None of these studies found any indication of toxicological problems and the LD_{50} for DHASCO™ (if one exists) was found to exceed the maximum possible dose.

Table 17.8
In Vitro and Animal (*in Vivo*) Studies on the Safety of DHASCO™ Produced by Martek Biosciences Corporation

Study	Max DHASCO™ Dose	Model System Employed	Conclusion	Ref.
Ames mutagenicity test	5 mg/plate	Salmonella typhimurium	Not mutagenic	58
Forward mutation test	5 mg/plate	Mouse lymphoma cells	Not mutagenic	58
Chromosomal aberration test	5 mg/plate	Chinese hamster ovary cells	Not clastogenic	58
Acute toxicity	20 g/kg body wt	rat	No deaths	59
Subchronic toxicity				
28 day	1.25 g/kg body wt/day	rat	Not toxic	59,62
90 day	1.25 g/kg body wt/day	rat	Not toxic	60
90 day, *in utero* exposure	3 g/kg body wt/day	rat	Not toxic	61,62
Developmental toxicity test	1.25 g/kg body wt/day	rat	Not toxic	58

Table 17.9
Human Safety Studies on the Safety of DHASCO™ Produced by Martek Biosciences Corporation

Dose (g/day)	Duration (weeks)	No. of Subjects per Group (Experimental and Control)	Conclusion	Ref.
4	15	14	No effect on platelet aggregation/homeostatic factors	63
4.5	6	12 vegetarians	No effect on thrombogenic factors	64
15	13	6	No change in blood clotting or immune paramters	65,66
1.5 – 7.3	2	8	No significant adverse effects	67
0.75 – 1.5	4	12–15	No serious adverse effects	68

Human studies (Table 17.9) comparing DHASCO™ alongside equivalent doses of DHA from fish oil or meat likewise have failed to detect any adverse effect associated with ingestion of DHASCO™; the most serious reported adverse effect being "fishy burps."

The safety assessments carried out on DHA-rich SCO from *Schizochytrium,* ingested either as extracted oil or as intact microbial biomass, have been similarly rigorous to those carried out on DHASCO™, i.e., from *C. cohnii* (69–73). Likewise these studies have observed a lack of toxicity at the genetic level when administered at either single high (acute) or prolonged lower (sub-acute) doses.

The extensive safety evaluation of both DHA-rich SCO and ARA-rich SCO was a necessary prerequisite to the release of these "novel" foods on to the open market, particularly as their major application was infant nutrition. However, as the intrinsic safety of microbe-derived oils has now been demonstrated, it seems likely that the next generation of SCOs will not have to undergo the same high level of toxicological testing. As microbial oils have been shown to be no more toxic than oils from traditional sources it should be sufficient in the future to demonstrate the lack of pathogenicity and toxigenicity of the producing organism. This should decrease the development costs of future products significantly.

One of the first beneficiaries of being a second generation SCO looks to be the DHAactive™ product developed by Celanase/Nutrinova. The safety of their product has been evaluated to a large degree by confirming its equivalence to previously available microalgal oils (i.e., DHASCO™ and DHAGold™) (74).

Whilst discussing the safety of SCO it is worthwhile noting that the safety of traditional high PUFA oils should not be tacitly accepted – although it often is. Fish oils (a major source of EPA and DHA-rich oil) have potential hazards themselves, these include the potential for contamination with environmental pollutants (dioxins, heavy metals), which have to be carefully removed by manufacturers prior to sale. Furthermore, fish oils can be very rich in the oil-soluble vitamins A and D (21) which, although beneficial in low doses, can lead to liver and kidney problems with excessive or prolonged exposure. Finally, while any possible microbial contamination of the biomass used for SCO production is closely monitored (sterility checks are made daily in production fermentors and any contaminated batches are discarded), no such monitoring (as far as the authors are aware) occurs with the fish bodies and livers that are extracted for fish oils. As fish livers have undoubtedly been handled and exposed to the atmosphere for a considerable time between harvesting of the fish and oil extraction, the microbial contamination of this material can only be guessed at!

17.5 COMMERCIAL SIGNIFICANCE OF SCOS

Although the first SCO appeared on the market in 1985 it was not until the twenty-first century that SCO have really become a successful commercial commodity. This commercial success has been based on the inclusion of a mix of ARA-rich SCO (ARASCO™), and DHA-rich SCO (DHASCO™) in infant formulae in many countries in Europe, Australasia, and the Far East. However, the market was somewhat limited due to the inclusion of the DHA/ARA-SCO blend in only formula designed for premature babies, and to the initial refusal of the FDA to allow inclusion of the DHA/ARA-SCO in infant formulae in the U.S. The breakthrough came in May 2001 when the FDA finally gave GRAS status to the DHA/ARA-SCO for its inclusion in infant formulae in the USA. Since February 2002 when the first formulae containing SCO were available in the US market, the SCO-fortified formulae have captured over 50% of US formulae sales. As a result of this success, >95% of all SCO production globally is destined for formula use; the size of the market is now limited by the supply, rather than demand, of both DHA-rich, and ARA-rich SCOs. Martek Biosciences acquired a new fermentation plant in October 2003 in an attempt to keep up with demand and DSM, the producers of ARASCO for Martek, have also exceeded the capacity of their first production plant and have recently commissioned a new plant in Belvidere (NJ, USA).

It is calculated that the production of SCO in 2003 will exceed 500 tonnes (Figure 17.5), equalling or even exceeding the total SCO produced cumulatively between 1985 and 2002 (75). Production and consumption of ARA-rich SCO, and DHA-rich SCO should continue to expand for at least the next five years based solely on the increased use of these products in infant formula. It should be noted that although the current use of SCO is rapidly expanding, this expansion represents, to a very large degree, the success of a single product, a DHA/ARA-SCO blend (Formulaid™) for one specific application (infant formula). If SCO production is to continue to grow beyond the next 5 to10 years, then new products and new markets must be developed to sustain this growth.

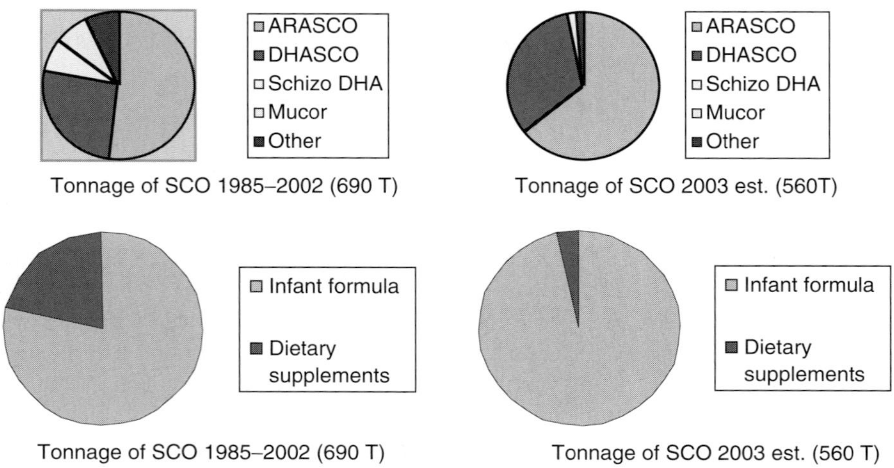

Figure 17.5 Production and usages of Single Cell Oils (SCO), 1985–2003. Top diagrams (left and right): production tonnages of the various SCOs; bottom diagrams (left and right): proportion of SCOs used for infant formulae and as dietary supplements (75).

REFERENCES

1. Whitworth, D.A., C. Ratledge. Microorganisms as a potential source of oils and fats. *Proc. Biochem.* 9(9):14–22, 1974.
2. Ratledge, C. Microbial production of oils and fats. In: *Food from Waste*, Birch, G.G., K.J. Palmer, J.T. Worgan, eds., London: *Applied Science* 1976, pp 98–113.
3. Lundin, H. Fat synthesis by microorganisms and its possible applications in industry. *J. Inst. Brew.* 56:17–28, 1950.
4. Woodbine, M. Microbial fat: microorganisms as potential fat producers. *Prog. Ind. Microbiol.* 1:179–245, 1959.
5. Ratledge, C. Biotechnology as applied to the oils and fats industry. *Fette Sifen Anstrichmittel* 86:379–389, 1984.
6. Thorpe, R.F., C. Ratledge. Fatty acid distribution in triglycerides of yeasts grown on glucose or *n*-alkanes. *J. Gen. Microbiol.* 72:151–163, 1972.
7. Ratledge, C., S.G. Wilkinson. *Microbial Lipids,* vols. 1, 2. London: Academic Press, 1989, 1990.
8. Ratledge, C. Yeasts, moulds, algae and bacteria as sources of lipids. In: *Tehcnological Advances in Improved and Alternative Sources of Lipids,* Kamel, B.S., Y. Kakuda, eds., Glasgow: Blackie & Sons, 1994, pp 235–291.
9. Ratledge, C. Microbial lipids. In: *Biotechnology, vol. 7: Products of Secondary Metabolism,* 2nd ed., Rehm, H.J., R. Reed, A. Puhler, P. Stadler, H. Kleinhauf, H. von Dohren, eds., Weinheim, Germany: VCH, pp 133–197.
10. Ratledge, C. Microorganisms as sources of polyunsaturated fatty acids. In: *Structured and Modified Lipids*, Gunstone, F.D., ed., New York: Marcel-Dekker, 2001, pp 351–399.
11. Anderson, A.J., J.P. Wynn. Microbial polyhydroxyalkanoates, polysaccharides and lipids. In: *Basic Biotechnology,* 2nd ed., Ratledge, C., B. Kristiansen, eds., London: Cambridge University Press, 2001, pp 325–348.
12. Kubicek, C.P. Organic acids. In: *Basic Biotechnology,* 2nd ed., Ratledge, C., B. Kristiansen, eds., London: Cambridge University Press, 2001, pp 305–324.
13. Ratledge, C., J.P. Wynn. The biochemistry and molecular biology of lipid accumulation in oleaginous microorganisms. *Adv. Appl. Microbiol.* 51: 1–51, 2002.
14. Ratledge, C., H. Streekstra, Z. Cohen. Processing Aspects of Single Cell Oils. In: *Nutritionally Enhanced Edible Oil Processing,* Dunford, N.T., H.B. Dunford, eds., Champaign, IL: AOCS Press, 2004, in press.
15. Kyle, D.J., C. Ratledge. *Industrial Applications of Single Cell Oils*. Champaign, IL: AOCS Press, 1992.
16. Davies, J. Scale up of yeast technology. In: *Industrial Applications of Single Cell Oils,* Kyle, D.J., C. Ratledge, Champaign, IL: AOCS Press, 1992, pp 196–218.
17. Smit, H., A. Ykema, E.C. Verbree, I.I.G.S. Verwoert, M.M. Kater. Production of cocoa butter equivalents by yeast mutants. In *Industrial Applications of Single Cell Oils,* Kyle, D.J., C. Ratledge, Champaign, IL: AOCS Press, 1992, pp 185–195.
18. Zhou, B. Studies on the biochemistry of malic enzyme in *Mucor circinelloides*. MSc Thesis, University of Hull, UK, 2004.
19. Broadhurst, C.L., Y. Wang, M.A. Crawford, S.C. Cunnane, J.E. Parkington, W.F. Schmidt. Brain-specific lipids from marine, lacustrine, or terrestrial food resources: potential impact on early *Homo sapiens*. *Comp. Biochem. Physiol. B. Biochem. Mol. Biol.* 131: 653–673, 2002.
20. Hornstra, G. Essential fatty acids in mothers and their neonates. *Am. J. Clin. Nutr.* 71: 1262S-1269S, 2000.
21. Barclay, W.R. Process for the heterotrophic production of products with high concentrations of omega-3 highly unsaturated fatty acids. Patent No. WO9107498, 1991.
22. Huang, Y.S., D.E. Mills. *γ-Linolenic Acid: Metabolism and Its Roles in Nutrition and Medicine.* Champaign, IL: AOCS Press, 1996.

23. Huang, Y.S., A. Ziboh. *γ-Linolenic Acid: Recent Advances in Biotechnology & Clinical Applications.* Champaign, IL: AOCS, 2001.
24. Shaw, R. The occurrence of gamma-linolenic acid in fungi. *Biochim. Biophys. Acta* 98: 230–237, 1965.
25. Shaw, R. The polyunsaturated fatty acids of microorganisms. *Adv. Lipid Res.* 4:107–174, 1966.
26. Nakahara, T., T. Yokocki, Y. Kamisaka, O. Suzuki. Gamma-linolenic acid from genus *Mortierella*. In: *Industrial Applications of Single Cell Oils,* Kyle, D.J., C. Ratledge, Champaign, IL: AOCS Press, 1992, pp 61–97.
27. Lion Corporation. External preparation for skin. Japanese patent No. 64–38007, 1988.
28. Kyle, D.J. Arachidonic acid and methods for the production and use thereof. Patent No. WO0213086, 1992.
29. Stredanska, S., J. Sajbador. Oligounsaturated fatty acid production by selected strains of micromycetes. *Folia. Microbiol.* 37:357–359, 1992.
30. Shinmen, Y., S. Shimizu, K. Akimoto, H. Kawashima, H. Yamada. Production of arachidonic acid by *Mortierella* fungi: selection of a potent producer and optimization of culture conditions for large scale. *Appl. Microbiol. Biotechnol.* 31:11–16, 1989.
31. Bajpai, P., P.K. Bajpai, O.P. Ward. Eicosapentaenoic acid (EPA) production by *Mortierella alpina* ATCC 32222. *Appl. Biochem. Biotechnol.* 31: 267–272, 1991.
32. Singh, A., O.P. Ward. Production of high yields of arachidonic acid in a fed batch system by *Mortierella alpina* ATCC 32222. *Appl. Microbiol. Biotechnol.* 48: 1–5, 1997.
33. Streekstra, H. On the safety of *Mortierella alpina* for the production of food ingredients, such as arachidonic acid. *J. Biotechnol.* 56:153–165, 1997.
34. Higashiyama, K., T. Yagushi, K. Akimoto, S. Fujikawa, S. Shimizu. Effects of mineral addition on the growth morphology of and arachidonic acid production by *Mortierella alpina* 1S-4. *J. Am. Oil Chem. Soc.* 75:1815–1819, 1998.
35. Park, E.Y., Y. Koike, K. Higashiyama, S. Fujikawa, M. Okabe. Effect of nitrogen source on mycelial morphology and arachidonic acid production in cultures of *Mortierella alpina. J. Biosci. Bioeng.* 88: 61–67, 1999.
36. Yamada, H., S. Shimizu, Y. Shinman, K. Akimoto, H. Kawashima, S. Jareonkitmongkol. Production of dihomo-γ-linolenic acid, arachidonic acid and eicosapentaenoic acid by filamentous fungi. In: *Industrial Applications of Single Cell Oils,* Kyle, D.J., C. Ratledge, Champaign, IL: AOCS Press, 1992, p 118–138.
37. Wolf, J.H., A. Schaaf, H.L. Bijl, J.M.J. Visser. Preparation of microbial fatty acid containing oil from pasteurized biomass. Patent No. US6441208, 2001.
38. Kyle, D.J., S.E. Reeb, V.J. Silcotte. Docosahexaenoic acid, methods for its production and compounds containing the same. Patent No. WO91/11918, 1991.
39. Tuttle, R.C., A.R. Loeblich. An optimal growth medium for the dinoflagellate *Crypthecodinium cohnii. Phycologia* 14:1–8, 1975.
40. Beach, D.H., G.G. Holz. Environmental influences on the docosahexaenate content of the triacylgycerols and phosphatidylcholine of a heterotrophic marine dinoflagellate *Crypthecodinium cohnii. Biochim. Biophys. Acta* 316:56–65, 1973.
41. Kendrick, A., C. Ratledge. Lipids in selected molds grown for the production of n-3 and n-6 polyunsaturated fatty acids. *Lipids* 27:15–20, 1992.
42. Russell, N.J., D.S. Nichols. Polyunsaturated fatty acids in marine bacteria: a dogma rewritten. *Microbiology* 145:767–779, 1999.
43. Van Pelt, C., M.-C. Huang, C.L. Tshanz, J.T. Brenna. An octaene fatty acid, 4,7,10,13,16,19,22,25-octacosaoctaenoic acid (28:8n-3), found in marine oils. *J. Lipid Res.* 40:1501–1505, 1999.
44. Barclay, W.R., P. Mirrasoul, G.T. Weeder, T. Kaneko, D. Dimasi, J. Hansen, C.M. Reuker, R.B. Bailey. Enhanced production of lipids containing polyenoic afatty acids by high density cultures of eukaryotic microbes in fermentors. Patent WO0154510, 2001.
45. Molina Grima, E., E.-H. Elardi, F.G. Acien Fernandez, A. Robles Medina, Y. Chisti. Recovery of microalgal biomass and metabolites: process options and economics. *Biotechnol. Adv.* 20:491–515, 2003.

46. De Swaaf, M., L. Sijtsma, J.T. Pronk. High-cell-density fed-batch cultivation of the docosahexaenoic acid producing marine alga *Crypthecodinium cohnii*. *Biotechnol. Bioeng.* 81:666–672, 2003.
47. Barclay, W.R. Process for growing *Thraustochytrium* and *Schizochytrium* using non-chloride salts to produce a microfloral biomass having omega-3 highly unsaturated fatty acids. U.S. Patent 5,340,742, 1994.
48. Seyberth, H.W., O. Oelz, T. Kenedy, B.J. Sweetman, A. Danon, J.C. Frolich, H. Heimberg, J.A. Oates. Increased arachidonate in lipids after administration to man. *Clin. Pharmacol. Therapeu.* 18:521–529, 1975.
49. Anon. A human dietary arachidonic acid supplementation study conducted in a metabolic unit. *Lipids* 32:413–456, 1997.
50. Nelson, G.J., D.S. Kelley, E.A. Emken, S.D. Phinney, D. Kyle, A. Ferretti. A human dietary arachidonic acid supplementation study conducted in a metabolic research unit: rationale and design. *Lipids* 32:415–420, 1997.
51. Nelson, G.J., P.C. Schmidt, G. Bartolini, D.S. Kelley, D. Kyle. The effect of dietary arachidonic acid on platelet function, platelet fatty acid composition, and blood coagulation in humans. *Lipids* 32:421–425, 1997.
52. Nelson, G.J., P.C. Schmidt, G. Bartolini, D.S. Kelley, S.D. Phinney, D. Kyle, S. Silbermann, E.J. Schaefer. The effect of dietary arachidonic acid on plasma lipoprotein distributions, apoproteins, blood lipid levels and tissue fatty acid compositions in humans. *Lipids* 32:427–433, 1997.
53. Kelley, D.S., P.C. Taylor, G.J. Nelson, P.C. Schmidt, B.C. Mackey, D. Kyle. Effects of dietary arachidonic acid on human immune response. *Lipids* 32:449–456, 1997.
54. Hempenius, R.A., J.M.H. Van Delft, M. Prinzen, B.A.R. Lina. Preliminary safety assessment of an arachidonic acid-enriched oil derived from *Mortierella alpina*: summary of toxicological data. Food Chem. Toxicol. 35: 573–581, 1997.
55. Hempenius, R.A., B.A.R. Lina, R.C. Haggit. Evaluation of a subchronic (13 week) oral toxicological study, preceded by an *in utero* exposure phase, with arachidonic acid oil derived from *Mortierella alpina* in rats. *Food Chem. Toxicol.* 38:127–139, 2000.
56. Grandjean, P., P. Weihe. Arachidonic acid status during pregnancy is associated with polychlorinated biphenyl exposure. *Am. J. Clin. Nutr.* 77:715–719, 2003.
57. Falandysz, J., S. Tanabe, R. Tatsukawa. Most toxic and highly bioaccumulative PCB congeners in cod-liver oil of Baltic origin processed in Poland during the 1970s and 1980s, their TEQ-values and possible intake. *Sci. Total Environ.* 145:207–212, 1994.
58. Arterburn, L.M., K.D. Boswell, T. Lawlor, M.A. Cifone, H. Murli, D.J. Kyle. *In vitro* genotoxicity testing of ARASCO and DHASCO oils. Food Chem. Toxicol. 38: 971–976, 2000.
59. Boswell, K., E.-K. Koskelo, L. Carl, S. Glaza, D.J. Hensen, K.D. Williams, D.J. Kyle. Preclinical evaluation of single-cell oils that are highly enriched with arachidonic acid and docosahexaenoic acid. Food Chem. Toxicol. 34:585–593, 1996.
60. Arterburn, L.M., K.D. Boswell, E. Koskelo, S.L. Kassner, C. Kelly, D.J. Kyle. A combined subchronic (90-day) toxicity and neurotoxicity study of a single-cell source of docosahexaenoic acid triglyceride (DHASCO oil). *Food Chem. Toxicol.* 38: 35–49, 2000.
61. Burns, R.A., G.J. Wibert, D.A. Diersen-Schade, C.M. Kelly. Evaluation of single-cell sources of docosahexaenoic acid and arachidonic acid: 3-month rat oral safety study with an *in utero* phase. *Food Chem. Toxicol.* 37:23–36, 1999.
62. Arterburn, L.M., K.D. Boswell, S.M. Henwood, D.J. Kyle. A developmental safety study in rats using DHA- and ARA-rich single-cell oils. *Food Chem. Toxicol.* 38:763–771, 2000.
63. Agren, J.J., S. Vaisanen, O. Hanninen, A.D. Muller, G. Hornstra. Hemostatic factors and platelet aggregation after a fish-enriched diet or fish oil or docosahexaenoic acid supplementation. *Prostaglandins Leukot. Essent. Fatty Acids* 57:419–421, 1997.
64. Conquer, J.A., B.J. Holub. Supplementation with an algae source of docosahexaenoic acid increases (n-3) fatty acid status and alters selected risk factors for heart disease in vegetarian subjects. *J. Nutr.* 126:3032–3039, 1996.

65. Nelson, G.J., P.S. Schmidt, G.L. Bartolini, D.S. Kelley, D. Kyle. The effect of dietary docosahexaenoic acid on platelet function, platelet fatty acid composition, and blood coagulation in humans. *Lipids* 32:1129–1136, 1997.
66. Kelley, D.S., P.C. Taylor, G.J. Nelson, B.E. Mackey. Dietary docosahexaenoic acid and immunocompetence in young healthy men. *Lipids* 33:559–566, 1998.
67. Innis, S.M., J.W. Hansen. Plasma fatty acid responses, metabolic effects and safety of microalgal and fungal oils rich in arachidonic and docosahexaenoic acids in healthy adults. *Am. J. Clin. Nutr.* 64:159–167, 1996.
68. Otto, S.J., A.C. van Houwelingen, G. Hornstra. The effect of different supplements containing docosahexaenoic acid on plasma and erythrocyte fatty acids of healthy non-pregnant women. *Nutr. Res.* 20:917–927, 2000.
69. Abril, R., J. Garrret, S.G. Zeller, W.J. Sander, R.W. Mast. Safety assessment of DHA-rich microalgae from *Schizochytrium* sp, part V: target animal safety/toxicity study in growing swine. *Regul. Toxicol. Pharmacol.* 37:73–82, 2003.
70. Hammond, B.G., D.A. Mayhew, M.W. Naylor, F.A. Rueker, R.W. Mast, W.J. Sander. Safety assessment of DHA-rich microalgae from *Schizochytrium* sp. *Regul. Toxicol. Pharmacol.* 33:192–204, 2001.
71. Hammond, B.G., D.A. Mayhew, J.F. Holson, M.D. Nemec, R.W. Mast, W.J. Sander, Safety assessment of DHA-rich microalgae from *Schizochytrium* sp. *Regul. Toxicol. Pharmacol.* 33:205–217, 2001b.
72. Hammond, B.G., Mayhew, D.A., K. Robinson, R.W. Mast, W.J. Sander. Safety assessment of DHA-rich microalgae from *Schizochytrium* sp. *Regul. Toxicol. Pharmacol.* 33:356–362, 2001.
73. Hammond, B.G., D.A. Mayhew, L.D. Kier, R.W. Mast, W.J. Sander. Safety assessment of DHA-rich microalgae from *Schizochytrium* sp. *Regul. Toxicol. Pharmacol.* 35:255–265, 2002.
74. Kroes, R., E.J. Schaefer, R.A. Squire, G.M. Williams. A review of the safety of DHA45-oil. *Food Chem. Toxicol.* 41:1433–1446, 2003.
75. Wynn, J.P., C. Ratledge. Microbial Oils and Fats. In: *Bailey's Encyclopedia of Industrial Oil and Fat Products*, 6th ed., Shahidi, F., ed., Hoboken, NJ: J. Wiley, in press.

1.18

Potential Uses of Cyanobacterial Polysaccharides in the Food Industry

Xue-Jun Liu and Feng Chen

CONTENTS

18.1 INTRODUCTION

In the past two decades, algal biotechnology has successfully employed advanced techniques matching economic development in the exploitation of novel biochemicals in algae. Humankind has benefited from the progress of algal biotechnology in alleviating the problems resulting from a phenomenal increase in world population and the concurrent scarcity of food resources to meet surging demands. This is witnessed in the application of microalgae (including cyanobacteria) and algae-like organisms in myriad areas of agriculture, aquaculture, environmental management, and pharmaceutical and food industries. The use of algal biotechnology for the production of a number of valuable biochemicals has become a firm reality, including the commercial production of β-carotene from the halophilous green algae *Dunaliella*, biomass as health food from *Spirulina* (*Arthrospira*), polyunsaturated fatty acids (PUFAs) from certain flagellates and diatoms, phycobiliproteins from certain red algae and cyanobacteria, and astaxanthin from *Haemotococcus* (1,2).

The tendency of market demand to switch from synthetic to natural production has been recognized as the driving force for this vigorously developing biotechnology area (3). Some mature technologies have been applied directly to industry, while the others (e.g., recombinant DNA techniques, modern separation techniques) have been blended with conventional biotechnology, resulting in the birth of new research areas, new processes, and new products. Among them, the production of polysaccharides from the microalgae cyanobacteria is of significant importance, as reflected by the steadily increasing number of scientific publications concerning current research and progress pertinent to their huge potentials in various food and biotechnological industries (Table 18.1).

Polysaccharides are extensively used as gelling agents, stabilizers, thickeners, and emulsifiers, mainly in food products as well as in paints, photographic films, and pharmaceuticals, and in tertiary oil recovery (3–5). Traditionally, seaweeds, especially the red macroalgae, are the main sources of algal polysaccharides. However, seaweeds are usually harvested from their natural habitats (3), and production of seaweeds is extremely susceptible to seasonality and improper harvesting, which lead to inconsistent supply. This phenomenon has indicated the need to search for new and constant sources of algal polysaccharides. As a result, cyanobacteria have been widely investigated for the production of polysaccharides due to their high contents of the products and versatility in industry.

This review summarizes current knowledge of cyanobacterial polysaccharides with special emphasis on the properties pertinent to their applications in food, and explores commercial potentials of cyanobacterial polysaccharides in the food industry. Extensive information on the ecological implications and the application in wastewater treatment of cyanobacterial polysaccharides can be found in Guiseley (6), Adhikary (7), and Vilchez et al. (8). Red and brown algae, the main source of polysaccharides and their polysaccharide production potential have been reviewed by Arad (3), and Siddhanta and Murthy (9). The application of algal polysaccharides in medicine has been reported and discussed by Witvrouw and DeClercq (10), and Schaeffer and Krylov (11).

Table 18.1
Numbers of publications concerning traditional and cyanobacterial polysaccharides in different databases

	Polysaccharides[1]		Microbial Polysaccharides[2]	Cyanobacterial Polysaccharides
Year	ISI[3]	FSTA[4]	ISI	ISI
1981–1985	127	244	17	1
1986–1990	137	343	27	3
1991–1995	543	638	104	13
1996–2000	659	804	138	28
2001–2004	4567	2121	238	33

[1] Polysaccharides from various sources
[2] Key words are put as microbial and bacterial polysaccharides
[3] ISI Web of Science
[4] Food Science and Technology Abstracts

18.2 THE ALGAE CYANOBACTERIA

Cyanobacteria are conventionally called blue-green algae. Cyanobacteria are a diverse group of obligate photoautotrophs (~2,000 known species), as reflected by their widespread occurrence (except for those environments at lower pH), frequent abundance, and morphological diversity. These O_2-producing photosynthesizing prokaryotes are ultrastructurally and biochemically similar to both eubacteria and the chloroplasts of microalgae or higher plants (12,13), and thus possess both bacterial and algal attributes with high flexibility in their metabolisms. Typically, water acts as the electron donor during photosynthesis, leading to the production of oxygen, although several species are able to shift between H_2O and H_2S as electron donors (14). The cells of cyanobacteria contain only unstacked photosynthetic lamellae (thylakoids), where the electron transport reactions of photosynthesis and respiration occur. Cyanobacteria possess the ability to synthesize chlorophyll *a* and certain types of hydrophilic phycobiliproteins, including phycocyanin, phycoerythrin, and allophycocyanin (15). Nucleotide base sequence data of 16S and 5S rRNA have suggested that cyanobacteria are phylogenetically coherent with the Gram-negative eubacteria (16).

Much of the success of cyanobacteria, in view of their ecological and physiological adaptation to their corresponding environments, is undoubtedly a consequence of a number of features widespread in the group, in comparison to their counterparts in other taxa (13). First, cyanobacteria show considerable morphological diversity. They may be unicellular (e.g., *Chroococcus, Microcystis*) or filamentous (normally composed of vegetative cells, e.g., *Anabaena, Nostoc*), and they may form macroscopic or microscopic colonies. Unicells may divide in one, two, or three planes. Filaments may branch or not. Maintenance of their diverse colonial structures is aided by the presence of exopolysaccharides, such as slimes or a firm sheath. During their life cycle, cyanobacteria form highly specialized cells, namely, heterocyst and akinete, which help the organisms in the process of nitrogen fixation and to adapt to stress environments, respectively. Different cell forms (i.e., vegetative cells, proheterocysts, heterocysts, akinetes, germlings) alternate in a regular manner during their life spans. The discrepancy of these characteristics in different taxa provides the basis for morphologically based taxonomy. For instance, the presence or absence of a heterocyst is an important feature separating genera (17). Cyanobacteria exist in a wide range of ecological habitats because of several traits that allow them to survive in marginal environments. These include their tolerance to high temperature, high UV irradiance, desiccation, and free sulfide, as well as their ability to utilize low photon flux density and low levels of inorganic carbon, and to fix molecular nitrogen (18).

Given their diversity in ecophysiology and their obvious importance in a wide variety of ecosystems, cyanobacteria show a metabolic diversity while acclimatizing to the environments. Consequently, from a biotechnological point of view, they are a group of organisms that produce a wide spectrum of fine chemicals, or renewed energy, naturally or artificially. Indeed, cyanobacteria are nature's gift to mankind. First, their oxygenic photosynthesis provides a unique means to utilize cheap substrates (e.g., CO_2, H_2O) and solar energy for the primary production of organic compounds and many biotechnological products. Meanwhile, N_2-fixing cyanobacteria can utilize sunlight as their sole source of energy for the fixation of carbon and nitrogen, and thus make potential biofertilizers. There is renewed interest in the role of biological N_2 fixation by cyanobacteria. Recently cyanobacteria have been utilized as a source of nitrogen fertilizers in some developing countries to reduce the dependency of agriculture on fossil fuel, which, on the other hand, would lead to the reduced emission of CO_2 into the atmosphere (19). Cyanobacteria, like other microalgae, provide O_2 for BOD removal and incorporate nutrients such as nitrogen and phosphorus into biomass, thereby reducing the eutrophication potential of receiving waters. The immobilized cyanobacteria can also remove heavy metals, nutrients, and toxic organic compounds due to their affinity for these contaminants (13). Cyanobacteria serve as a food source in various parts of the world; in the past, the main focus has been on the production of single cell protein (SCP), although other applications that require mass cultivation are apparent. For instance, the cyanobacterium *Spirulina platensis* is grown commercially and the biomass is used as a health food worldwide. It is also reported that cyanobacteria are a rich source of type II restriction endonucleases, some of which are currently marketed. Other products from cyanobacteria include amino acids and pigments that can be used as food colorants and diagnostic probes (13). It is worth noting that some hydrophilic pigments have been marketed. Phycobiliproteins, especially phycoerythrin, are stable pigments that can be stored for a long period of time, and show high fluorescence which is not quenched by a number of biomolecules, and so could be used as fluorescent probes (20). Other biochemicals may include animal cell growth stimulants, anti-HIV pharmaceuticals, and cyanobacterial toxins (13).

Recently, cyanobacterial polysaccharides have received much attention of bacteriologists, phycologists, and biotechnologists for the purpose of industrial-scaled applications of this fascinating and promising food ingredient. They stand out among fine chemicals not only for multiple functions in their habitats but also for huge commercial potentials.

18.3 POSSIBLE ROLES OF CYANOBACTREIAL POLYSACCHARIDES

In Bergey's classification system, the polysaccharide-producing cyanobacteria belong to all 5 subsections (17). In totality, 33 genera in 5 subsections have demonstrated the ability to release polysaccharides during their life cycles (Table 18.2).

The terminology associated with mucilaginous layers in cyanobacteria is quite dubious and confusing. As pointed out by Bertocchi et al. (21), the structural and functional differences between different sublayers of those mucilaginous layers are very arbitrary, and thus are often not evident in the literature. These structures are defined by Bertocchi et al. (21), further confirmed by De Phillipis and Vincenzini (22), and slightly modified by Adhikary (7). In general, the external envelope layers are divided into three structured or unstructured layers or zones (sheaths, slimes, and capsules), and can be observed under the microscope using Nomarski differential interference contrast photomicrography. The sheath is defined as a thin, electron-dense layer that loosely surrounds cells or cell groups, The layer is usually visible by light microscopy without staining. The capsule generally consists of a thick and slimy layer intimately associated with the cell surface, with sharp outlines, and is structurally coherent to exclude particles. Slime refers to the mucilaginous material dispersed around the organism but not reflecting the shape of the cells (22). The schematic diagram is shown in Figure 18.1.

The functioning of cyanobacterial polysaccharides has been extensively studied for both the ecologically and biotechnologically relevant purposes. The obvious function is to serve as a protective boundary between the cell and its immediate environment (22). For some species of the genus *Nostoc*, these mucilaginous layers equip them to overcome environmental stresses due to desiccation or low water activity in arid or semiarid regions and saline environments. It is known that the secreted glycan provides a repository for water, thereby acting as a buffer between cells and the atmosphere, and representing the key component of the mechanism to tolerate desiccation (23). Likewise, a role as buffer compounds for accumulation and slow release of water has been suggested for polysaccharides

Table 18.2
Polysaccharide-producing cyanobacteria in Bergey's classification system (17)

Subsections (No. of Genus)	No. of Polysaccharide-producing Genera	Genus Names
I (14)	10	*Chamaesiphon, Chroococcus, Cyanobium, Cyanotheca, Gloeobacter, Gloeocapsa, Gloeotheca, Microcystis, Synechococcus* (including former *Anacystis*), *Synechocystis* (including former *Aphanocapsa*)
II (7)	6	*Dermocarpella, Stanieria* (including former *Dermocarpa*), *Xenococcus, Chroococcidiopsis, Myxosarcina, Pleurocapsa*
III (17)	6	*Geitlerinemia* (including former *Phormidium*), *Lyngbya, Microcoleus, Oscillatoria, Pseudoanabaena, Spirulina*
IV (12)	9	*Anabaena, Anabaenopsis, Cyanospira, Cyclindrospermum, Nodularia, Nostoc, Scytonema, Calothrix, Tolypothrix* (including former *Microchaete*)
V (6)	2	*Chlorogloeopsis* (including former *Mastigocladus*), *Fischerella*

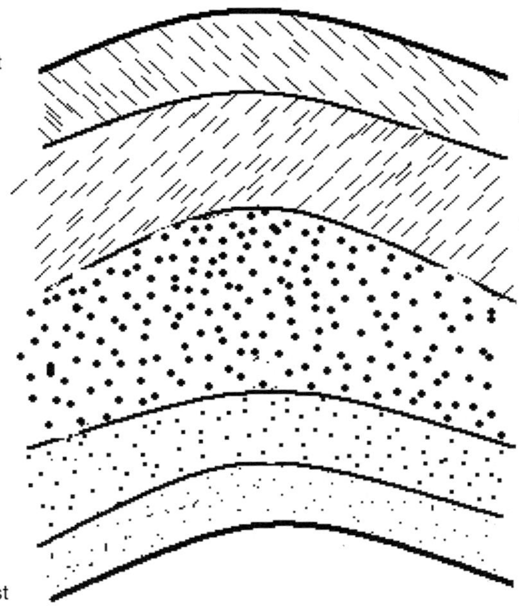

Figure 18.1 Schematic diagram of cell surface in cyanobacteria. [Reproduced from Bertocchi et al. (2[illegible]) and Adhikary (7) with slight modifications.]

produced by *Chroococcidiopsis* strains (24). It is suggested that for *Phormidium autumnale*, the release of extracellular polysaccharides enables the algal cells to survive very strong grazing pressure, thus making themselves less preferred foods than other microalgae, which are devoid of capsules (25). In some species of *Nostoc*, the outermost slimy shrouds that trap trichomes have been suggested to facilitate the homogeneous dispersion of trichomes into the culture broth for improving light utilization and nutrient uptake, which results in the decrease in metabolic energy (26). It is also proposed that cyanobacterial polysaccharides can take a special role in certain important metabolic processes. In *Nostoc cordubensis*, the thick mucilaginous envelope surrounding the heterocysts appears to be essential for the protection of nitrogenase activity from inactivation by oxygen during the process of nitrogen fixation (27). In addition, because of their anionic substituents, cyanobacterial polysaccharides play an important role in handling metal ions when exposed to environments with high concentrations of metallic ions. They help increase the availability of essential metals (e.g., Fe, Mn) by binding to these metals to accumulate and release them slowly to be taken up. For the case of nonessential or toxic heavy metals (e.g., Ag, Pb, Hg), cyanobacterial polysaccharides have a tendency to sequester the metals in order to alleviate toxicity (28,29). For terrestrial or symbiotic species, it is suggested that polysaccharides may act as an adhesive for cyanobacterial cells. For some desert species, microbial crust cohesion is mainly attributed to algal aggregates, which is supposed to be due to the vigorous secretion of polysaccharides to survive wind force, as well as both high relative air humidities and low moisture content (30,31). The release of polysaccharides by the cyanobiont *Anabaena azollae* has been considered essential for the attachment of this cyanobacterium to the earth surface and to the cavities of the host plant *Azolla filiculoides* (32). It has been discovered that the synthesis of extracellular polysaccharides, UV-absorbing compounds, and antioxidants is a unique strategy to counteract UV damage on proteins, DNA, and membranes (33).

From an ecological point of view, polysaccharide-releasing cyanobacteria can tolerate high concentrations of heavy metals by binding them with living cells or dry cells, and consequently mitigating the toxicity of the heavy metals. One recent study shows the nontoxic cyanobacterium *Gloeotheca magna* can be used to remove cadmium and manganese (34).

Increasing attentions have focused upon the application of cyanobacterial monosaccharides and polysaccharides in medical antiviral therapy (10,11). Experiments have proved that the presence of a sulfate group is necessary for multiple antiviral activities (such as anti-HIV and anti-HSV), and the efficacy increases with the degree of sulfation. Studies using both sulfated and nonsulfated homo- and heteropolysaccharides isolated from algae or other sources, or synthesized, have revealed the mechanisms of binding of drugs to the viron, and the mechanisms of viral binding to host cells. Given the few classes of compounds investigated, however, the pharmacology of most compounds from algae and cyanobacteria with antiretroviral activity is still unknown (11).

18.4 CORRELATION BETWEEN CHEMICAL COMPOSITION AND PHYSICAL PROPERTIES OF CYANOBACTERIAL POLYSACCHARIDES

In general, cyanobacterial polysaccharides fall into the category of heteropolysaccharides; they possess repeating units in which several different monosaccharides are present. Ten of 15 different monosaccharides have been found in the polysaccharides of the thermophilic cyanobacterium *Mastigocladus laminosus* (35), which can be further divided into 4 categories according to their chemical structures: (1) the hexoses (glucose, galactose, and mannose); (2) the pentoses (ribose, xylose, and arabinose); (3) the deoxyhexoses (fucose and rhamnose); and (4) the acidic hexoses (glucouronic and galactoronic acids) (22). In a few cases, some monosaccharides with methyl or amino functional groups have been reported, such as amino sugars (e.g., N-acetyl-gluco- and galactosamine), methylated sugars (e.g., 3-0-methyl-pentose, 3-0-methyl deoxyhexose, 4-0-methyl hexose), and other rare sugars (e.g., 4-0-[1-carboxyethyl] mannose) (35). Glucose is the most frequently found monosaccharide (in more than 90% of cyanobacterial polysaccharides), followed by galactose, mannose, and rhamnose, respectively and xylose is the most common among the pentoses (21). The molar ratio varies from species to species; and at times within the same species if the growth conditions are altered. In most cases, glucose is also the most abundant monosaccharide. Sometimes, arabinose, galactose, or fucose are also reported to be present at higher concentrations than glucose. Ribose has only been found in small number of polymers (~9% of the polysaccharides analyzed). Furthermore, in most cyanobacteria, uronic acids are the important constituents. The presence of acidic sugars in the macromolecules accounts for the anionic nature of almost all the polysaccharides studied so far (4,36). The possibility of simultaneous presence of two acidic monosaccharides is about 50% (22). Quantitatively, the relative contribution of glucouronic acid and galactouronic acid to total acidic sugars differs among species (7,22). Comparison of monosaccharide composition (in terms of molar ratio) between several intensively studied cyanobacterial polysaccharides is shown in Table 18.3.

Polysaccharides in cyanobacteria differ markedly from those of bacterial origin. First, most cyanobacterial polysaccharides are characterized by an anionic nature and many of them contain two different uronic acids, a feature rarely found in the polymers released by strains belonging to other microbial groups (42). Second, most polysaccharides synthesized by cyanobacteria are quite complex, either in the number of the monosaccharides or monomers, or in diversity of linkage types, resulting in even more complex repeating units and a broad range of possible macromolecular structures (42,43). As emphasized by De Philippis and Vincenzini (22), the qualitative and quantitative analyses of the monosaccharides are strongly affected by the conditions employed for hydrolysis of the polymers, and thus the moderate hydrolysis should be applied to the analytical practice of cyanobacterial polysaccharides. The most accessible carbohydrates are uronic acids. Briefly, incomplete hydrolysis due to the glycosidic linkage stabilized by the carboxyl group may result in an underestimate of the contents of constituent sugars, while excessive hydrolysis may lead to degradation or lactonization to some extent (44,45,46).

The presence of sulfate groups has been found in some prokaryotic polysaccharides (4). However, the occurrence of phosphate residues, frequently found in polysaccharides of many other bacterial groups, has never been investigated in cyanobacterial polysaccharides, although phosphate-containing polysaccharides are attracting much interest because of their possible immunological significance (42).

Another important feature that contributes to the physico-chemical properties of the polysaccharides is the presence of a polypeptide moiety or other nonsaccharidic components such as organic substituents (e.g., acetyl, pyruvyl, succinyl groups), or inorganic ones (e.g., sulfate or phosphate groups) (42). However, systematic investigations of the presence of these nonsaccharidic components in cyanobacterial polysaccharides have only been undertaken recently, and as a result, data available are very limited (22). Large amounts of the same amino acids have also been found in polypeptide moiety of the polysaccharide produced by *N. calcicola*, which are found to contribute significantly to the hydrophobicity of the macromolecule (47). On the other hand, the removal of the proteinaceous moiety drastically reduces the viscosity of aqueous solutions and the adhesive capacity of the polysaccharides produced by some microalgae. A significant presence of ester-linked acetyl groups, as well as deoxy-sugars like fucose and rhamnose, may give an appreciable contribution to the emulsifying properties of polysaccharides, owing to a certain lipophilic character introduced by these small molecules in the macromolecules which

Table 18.3
Comparison of monosaccharide composition (in terms of molar ratio) among several intensively studied cyanobacterial polysaccharides

	Monosaccharides											
Strains	**Ara**	**Fuc**	**Gal**	**Glc**	**Man**	**Rha**	**Rib**	**Xyl**	**GalA**	**GlcA**	**UrA**	**Ref.**
Cyanothece sp. 16Som2		1.6	2.4	6.8	4.8			2.9	2.0	1.0		37
Aphanocapsa halophytica MN11		26.5	1.5	12.5	7.5	1.0		1.5				38
Cyanothece sp. PE 13		3.8	11.6	22.0	5.9	1.0	0.3	5.9			20.9	54
Microcystis flos-aquae C3-40			1.0	1.0	3.0	3.0		2.0	43.0			55
Synechocystis sp. PCC 6714	5.5	2.1	6.0	34.8	3.8	2.8		2.8			16.7	56
Synechocystis sp. PCC 6803		6.0	1.0	6.7	3.9	3.6		3.5			16.4	56
Oscillatoria amphibian PCC 7105			16.0	33.0	21.7	4.5		12.3			6.7	57
Phormidium foveolarum C52	0.6	0.5	3.4	43.0	15.3	1.5		5.5			29.4	57
Spirulina platensis		0.7	2.7	2.0		0.3		1.3			40.0	51
Cyanospira capsulata ATCC43193	1.0	1.0		1.0	1.0				2.0			58
Nostoc calciola 79WA01	1.0	2.8	3.8	5.9	1.7	1.0		6.1	3.0	2.8		47
N. commune UTEX584	1.6	1.7	6.5	2.0	1.3	1.0		2.8	4.0	6.7		47
N. flagelliforme (natural colony)	3.3		21.0	45.7	7.3			23.2				59
N. flagelliforme (laboratory culture)			21.6	40.0	15.2					23.3		59
N. commune (natural colony)			21.8	50.2	1.1			27.0				59
N. commune (laboratory culture)	12.8	4.9	21.2	7.1	14.9	2.1		43.5				59

otherwise would be highly hydrophilic. However, some polysaccharides containing high levels of rhamnose do not show any particular emulsifying activity. Acetyl groups may hinder cation binding, but may also facilitate gel swelling, as in the case of alginate gel beads, which contribute to the stabilization of the ordered form of the polymers as reported for the xanthan structure (42). The charge on these molecules is conferred either through the presence of uronic acids residues (frequently glucouronic acid or pyruvate), linked as a ketal to a neutral sugar (36). Polysaccharides characterized by high concentrations of charged components (e.g., uronic acids, sulfate or phosphate groups, pyruvate ketals) usually form stable gels in the presence of metallic ions, and are most promising for the removal of toxic metals from polluted waters. However, the mere determination of charged groups is not enough for anticipating the actual metal binding capacity of a polymer, because depending on the conformational structure of the macromolecules, some of the charged groups are hardly accessible for the ions. Indeed, the ion uptake depends on both charge density and charge distribution on the polymers (42). Charged groups also significantly contribute to polysaccharide solubility in water and improve the ability of the macromolecules to bind water molecules. Consequently, the viscosity of their aqueous solutions increases, closely related to the effective volume occupied by the macromolecules (48). On the other hand, the protective role against desiccation suggested for microbial polysaccharides is particularly effective as the presence of acidic components in the macromolecules increases their water retaining capacity (49).

A large number of studies are available on the rheological properties of aqueous solutions of cyanobacterial polysaccharides, most of which consider that the viscosity of polysaccharides would depend on shear rate. All the polysaccharides show pseudoplastic behavior. However, if their viscosity dependence on shear rate is compared with that of xanthan solutions at the same concentration, many differences become evident. The polysaccharide produced by *Synechococcus* BG0011 shows a more marked shear thinning behavior but lower values of viscosity than xanthan gum at the same concentration (50). Polysaccharides of *Spirulina platensis* show pseudoplastic behavior at very low concentrations and are also characterized by a significant decrease of viscosity with increasing ionic strength of the solution (51). A rather wide range of rheological behaviors has been observed in aqueous solutions of the polysaccharides produced by *Cyanotheca* strains or *Cyanospira capsulata* (22). Indeed, flow properties highly differentiate these polymers, some of which show a behavior quite similar to or even better than xanthan gum, while some others show a more marked shear thinning behavior, a property that could be of particular interest for certain special applications (4). Other comparative data between xanthan gum and the polysaccharide from *Cyanospira capsulata*, with regard to the dependence of viscosity on pH, NaCl concentration, and temperature, have been demonstrated (22), suggesting a very promising behavior of these cyanobacterial polymers.

18.5 FACTORS INFLUENCING PRODUCTION OF CYANOBACTERIAL POLYSACCHARIDES

Researches have suggested that each algal strain should be carefully tested in order to envisage the right culture strategies aimed at optimizing polysaccharide production. Because most cyanobacterial strains have only been tested in small culture devices and in studies not oriented to maximizing its production, they attain only moderate levels of polysaccharides except *Cyanospira capsulata* and some *Cyanotheca* strains (52,53). The comparison of the biomass productivity between cyanobacteria and red microalgae in different culture systems are listed in Table 18.4.

Table 18.4
Comparison of productivity between cyanobacteria and red microalgae in different culture modes

Polysaccharide Sources	Flask	Column	Fermentor	Pond	Ref.
Cyanobacteria					
Cyanospira capsulata			116 mg L^{-1} D^{-1}	144 mg L^{-1} D^{-1}	37
Aphanocapsa halophytica MN11	32 mg L^{-1} D^{-1}				38
Nostoc insulare 54.79	47.0 mg L^{-1} D^{-1}		18.4 mg L^{-1} D^{-1}		39
	36 mg L^{-1} D^{-1} (for 7 D culture),				22
Anabaena flos-aquae A37	20 mg L^{-1} D^{-1} (for 12 D culture)	46.4 mg L^{-1} D^{-1}			
Red microalgae					
Porphiridium sp.				55-75, 133 mg L^{-1} D^{-1}	22
Botryococcus braunii	130-145 mg L^{-1} D^{-1}				40
Bacteria					
Xanthomonas campestris	7–10 g L^{-1} D^{-1}				41

There is seemingly an absence of a common behavior of polysaccharide release among the polysaccharide-producing cyanobacteria described so far. Gantar et al. report that the mucilaginous sheath of *Nostoc* 2S9B is mainly synthesized in the aseriate stage of the developmental cycle and then released as empty shells, at the same instance when homogonia are liberated (60). Some other species, *Cyanothece* sp. BH68K, *N. calciola*, and *Phormidium* sp. J-1, show a significant release of polysaccharides starting from the late exponential growth phase, suggesting that cyanobacteria produce polysaccharides as a typical secondary metabolite (22). On the contrary, the polysaccharides released may be regarded as a primary metabolite evident in the fact that polysaccharide production parallels biomass production (51). This result is consistent with that of Mehta and Vaidya (61), in which the highest rates of polysaccharide synthesis and release are achieved by young cultures in a *Nostoc* strain. The kinetics of polysaccharide release have been considered as a complex dynamic equilibrium among different processes, i.e., trichome elongation and akinete germination (62). It has been concluded that the polysaccharide synthesis is mainly directed toward the formation of the capsule, and as opposed processes, i.e., trichome fragmentation and akinete differentiation, which causes the release of the polymer into the culture medium. There are also reports that the sugar composition varies quantitatively as well as qualitatively with the age of the culture (56).

In cyanobacterial cultures, the production of polysaccharides appears to depend on the C:N ratio. A change in light intensity, temperature, and the concentrations of sulfur, iron, phosphate, and potassium also affect polysaccharide production (22). By means of nutrient starvation, especially under nitrate and sulfate limitations, *Synechococcus* produce a large amount of exopolysaccharides (7). However, the exudation of polysaccharides under special culture conditions is demonstrated only for phycocyanin-rich cultures, suggesting different metabolic pathways for phycocyanin and phycoerythrin-rich *Synechococcus* species. More important, the formation of substituents of the polysaccharides, which may play a crucial role in determining the physicochemical properties of the cyanobacterial polysaccharides, can be manipulated by changing the growth conditions (35,46).

Available studies on enhancing polysaccharide release by means of optimizing the culture conditions mainly focus on the effect of nitrogen limitation or starvation. The response to nitrogen starvation, however, is species-specific in many cases and depends on the nitrogen sources used (63,64). In some nitrogen-fixing species, the production of polysaccharides is reported to be associated with carbon metabolic flux or an impairment of balanced growth both from a nutritional point of view and from physicochemical parameters (35,52). Nicolaus et al. reports that the dramatic decrease in polysaccharide yield occurs when the strain *Phormidium* sp. is grown with a light–dark cycle, in the absence of aeration and phosphorus (65). Likewise, the increase in P and N in the medium, and the absence of combined nitrogen, have a similar effect on the yield, giving rise to a one third decrease with respect to the amount obtained under standard growth conditions. For *Spirulina*, on the other hand, lowering the nitrogen content and nitrogen deprivation causes a strong increase in the total amount of polysaccharides even in the absence of phosphate (65). This finding has been confirmed in polysaccharide analyses of three species of *Nostoc* (59), indicating that combined nitrogen in the media has qualitative influence on the composition of extracellular polysaccharide, but not on intracellular polysaccharides. It is noticeable that the conclusions drawn from different researchers working on different species or strains are often confusing and sometimes even contradictory (65). The ambiguity in polysaccharide production is explained as the differences in physiological roles played by the exocellular polysaccharides in different strains (22). For instance, when the shortage of metallic ions stimulates the enhanced secretion of polysaccharides, the function of the polymer as a chelating agent for cations essential for cell life may be more pronounced. Likewise, when nitrogen limitation or starvation stimulates polysaccharide release, the polymer may act as a product of overflow metabolism which is exuded to allow cells to get rid of the carbon excesses (66). Another explanation is that nitrogen starvation or limitation causes a decrease in photosynthetic pigments, leading to the decrease in photosynthetic efficiency, and meanwhile, the response of microalgae to nitrogen deficiency is to degrade N-containing compounds inside the cells, giving rise to the decrease in nitrogen content and increase in carbohydrates and fatty acids (35).

Another approach to increasing the amount of polysaccharide release is to supplement the algal culture with intermediate metabolites such as glyoxylate in glycometabolism. This approach has proven promising when applied to *Cyanospira capsulata*, leading to the net enhancement of the final yield of cyanobacterial polysaccharide without affecting the growth rate (52).

18.6 PERSPECTIVES FOR COMMERCIALIZATION OF CYANOBACTERIAL POLYSACCHARIDES IN THE FOOD INDUSTRY

Polysaccharides are derived primarily from the red and brown algae, and higher plants, which have been widely employed in the food industry. These polysaccharides serve as emulsifying agents, gelling agents, thickening agents, and stabilizers (4). Microorganisms are a highly promising renewable source for the production of polysaccharides. Microbial polysaccharides offer greater advantages over the polysaccharides derived from other sources because they are obtained from cultures of selected microbial strains with a high and stable yield in a controllable environment. Some cyanobacterial polysaccharides even possess a superior property to *Xanthomonas* polysaccharides (Figure 18.2). Recent studies have indicated the interesting properties of these polymers, which make them suitable for forming stable gels, fibers, films, and liquid crystals; stabilizing suspensions and emulsions; enhancing viscosity of aqueous solutions; and serving as flocculants. All these features make them particularly suitable for use in the food industry, although they can be used equally well in the other industries (Table 18.5). The assessment of their potential use for various industrial purposes or specific application is quite an arduous task. In order to envisage the possible applications of these polymers, one should refer to some recent reviews on properties and uses of polysaccharides from the other microbial sources (4,5).

One of the most important prerequisites of a polysaccharide, which determines many of the properties generally considered useful for its industrial utilization (i.e., high viscosity of its aqueous solutions, capability of forming gels with good tensile strength, stabilizing emulsion), is that it possesses, together with an adequate composition and structure, a high molecular weight (MW) (69). In this respect, only seven cyanobacterial polysaccharides have been tested, five of which show MWs higher than 1000 kDa. The highest MW values, in the range of 1400–1900 kDa, have been recorded for the polysaccharides produced by *Cyanospira capsulata* (46,62).

Physical properties of certain polysaccharides must be seriously considered before putting them into use in the food industry. The important properties of cyanobacterial polysaccharides include high viscosity at a low concentration, high pseudoplasticity (shear thinning), yield value and suspending power, and stable viscosity toward salts, and a wide range of temperature and pH (70). In order to provide a scientific basis for industrial applications, future research should focus on the following: (1) alteration of rheological properties in different aqueous solutions, the change in texture of the products, and the correlations between chemical composition and physical properties, especially the ability

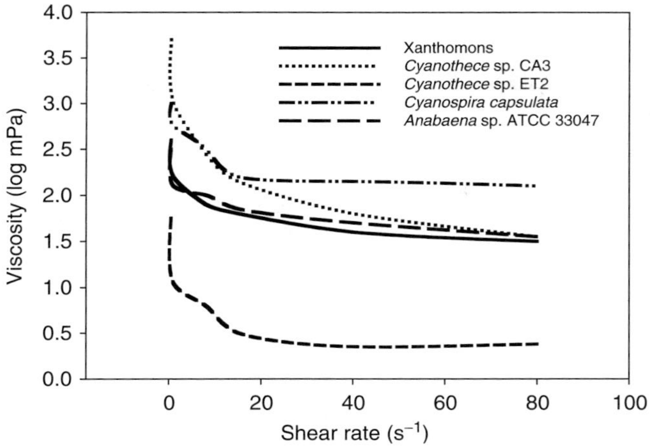

Figure 18.2 Comparison of viscosity dependence of aqueous solutions of polysaccharides produced by *Xanthomonas* and some cyanobacterial species/strains (0.2% w/v ratio for *Anabaena* sp., for other strains, 0.1%). Sources: De Philippis et al. (54,67), Moreno et al. (68).

Table 18.5
Potential applications of cyanobacterial polysaccharides

Applications	Physico-chemical Properties
Thickening agents	Excellent viscosity stability
Dietary fiber	The ability to stabilize suspensions and emulsions
Antiviral drugs	S-containing and thus inhibition of virus infection
An essential ingredient to formulate new functional foods	P-containing and thus improvement of immune systems
Food additives to improve texture and water-holding capacity of other foods	Rheological properties
Toothpaste and cosmetics as suspendants and emulsifiers	Pseudoplasticity
Environmental engineering, such as the removal of contaminants	Charged components
Textile industry	Flocculating activity

to thicken or to cause gel formation; (2) mixtures of several polysaccharides to exhibit synergistic gelling; (3) possible interaction of polysaccharides with food components such as proteins, lipids, and metallic ions, and (4) the stability over a wide range of physical factors, such as pH, temperature, and light (42). However, judging from the limited information available, it is still not possible to generalize an overall picture in physical attributes, either based on taxonomy or morphology or ecological habitat of the cyanobacteria of interest. The reason is that these physical characteristics are highly species or strain specific (22).

From culturing, harvesting, and processing to final marketing of biotechnological products, many procedures are involved, and thus there is room for improvement. As a first step in the production of cyanobacterial polysaccharides, development of a cost effective cultivation process is vitally important. Although the use of a tubular photobioreactor system has high capital costs, they may be offset by significantly high productivities and reduced costs for harvest due to high cell density (71). Major improvement in cultivation techniques is needed to ensure the maximal conversion of substrate to biotechnological products. It is desirable to develop a heterotrophic fermentation process for polysaccharide production because light is no longer required (i.e., not a limiting factor) in heterotrophic culture, and consequently the cell density and productivity can be greatly enhanced on a large scale (1). The use of heterotrophic fermentation to produce xanthan gum by other bacteria has been well established, and much of the expertise and knowledge on fermentation production of xanthan gum can be adopted for developing an industrial fermentation process for polysaccharides using cyanobacteria as producing organisms. Equally importantly, complementary to basic researches on structure–function investigation, physico-chemical properties, and the condition of optimal polysaccharide production, the improvement of the intrinsic capacities of the cultivated strains may lead to significant future developments in exploitation of cyanobacterial polysaccharides for industrial applications. This can be achieved either by the isolation and selection of spontaneous mutants displaying appropriate characteristics, or by modification of some physico-chemical attributes through genetic engineering using genomic information implicated in polysaccharide production (35), because for some polysaccharide producing cyanobacteria, most genes controlling the enzymatic syntheses have been decoded and published (Tanaka, Cyanobase, http://www.kazusa.or.jp/cyano/). Some cyanobacteria (e.g., *Synechocystis*) have been regarded as model organisms in genetic manipulation. A recent study of an obligate photoautotrophic fatty-acid-producing microalga has demonstrated the possibility of introducing a human glucose transporter gene into the microalga so that the alga can grow heterotrophically on glucose, indicating conventional fermentation technology may be used for growing microalgae on a commercial scale (72).

The present knowledge of cyanobacterial polysaccharides suggests that these polysaccharides may cover a broad range of complex chemical structures and physical properties. It is reasonable to anticipate that further studies on polysaccharide-producing cyanobacteria will lead to the discovery of new polymers possessing properties different from those of existing available polymers.

In the future, the relations between structures and properties of most cyanobacterial polysaccharides still have to be elucidated in detail in order to determine proper application for each polysaccharide. It has to be stressed that the development of any microbial polysaccharide into a viable biotechnological product requires a multidisciplinary effort, which combines many different areas of expertise (i.e., microbiology, biochemistry, chemistry, genetics, engineering). There is a pressing need to fill the existing gap between current knowledge and exploiting cyanobacteria for the production of useful polysaccharides (67).

18.7 SUMMARY

Cyanobacteria are a diverse group of prokaryotes performing oxygenic photosynthesis, which are widely distributed in different habitats (from aquatic to terrestrial environments and to Arctic and Antarctic environments). Due to their physiological adaptation to the environments, cyanobacteria exhibit a great diversity in metabolic pathways during growth. Therefore, they are versatile in the production of a wide range of bioactive materials as both primary and secondary metabolites, among which polysaccharides have received much attention both in academia and industry. Technically, cyanobacterial polysaccharides are readily recovered from the culture solutions. In view of their chemical composition and conformational structure, cyanobacterial polysaccharides are heteropolysaccharides, being composed of 6–10 monosaccharides (e.g., neutral sugars, deoxy sugars, uronic acids, amino sugars, methylated sugars). The peculiar features of these polymers in comparison with those of other microbial origin are characterized by an anionic nature, many possessing two different uronic acids and are complex either in the number of the monosaccharides or monomers or diversity of linkage types, resulting in even more complex repeating units and a broad range of possible macromolecular structures. Another important feature that contributes to the physico-chemical properties of cyanobacterial polysaccharides is the presence of a polypeptide moiety or other nonsaccharidic components such as organic (e.g., acetyl, pyruvyl, succinyl group) or inorganic (e.g., sulfate or phosphate groups) substituents, which modify or determine important physical properties in food. These structural and conformational features give rise to a great variety of physical and chemical properties of cyanobacterial polysaccharides. Previous studies have revealed that cyanobacterial polysaccharides possess unique physico-chemical properties compared with the existing polymer source (xanthan), of primarily bacterial origin. There is an increasing appeal from both the scientific community and the food industry to introduce these biopolymers into the market. In the future, concerted efforts are needed to fill the existing gap between current knowledge and that is needed in order to scale up the production of cyanobacterial polysaccharides, making them comparable to existing commercially available polysaccharides. Studies on cyanobacterial polysaccharides under diverse environmental or cultural conditions will enable an enhanced understanding of cyanobacterial metabolic processes which is necessary to develop new industrially profitable products for human uses. In conclusion, the use of cyanobacterial polysaccharides for the food industry is still in its infancy, the opportunity is enormous, and once scale up procedures are successfully developed, the opportunity will become a reality.

ACKNOWLEDGMENT

The authors are grateful to the support of CRCG (The University of Hong Kong Committee on Research and Conference Grants). The authors would also like to extend their special thanks to Dr. R. De Philippis at the Universita' degli Studi di Firenze (Italy) for providing valuable references, and Mr. S. C. Chang for improving the early draft of this review.

REFERENCES

1. Chen, F. High cell density culture of microalgae in heterotrophic growth. *Trends Biotechnol.* 14:421–426, 1996.
2. Chen, F., Y. Jiang. *Algae and their Biotechnological Potential.* Dordrecht: Kluwer, 2001, pp 1–316.
3. Arad, S. (M.). Polysaccharides of red microalgae. In: *Chemicals from Microalgae,* Cohen, Z., ed.. London: Taylor & Francis, 1999, pp 282–291.
4. Sutherland, I.W. *Biotechnology of Microbial Exopolysaccharides, Cambridge Studies in Biotechnology*, Vol. 9. London: Cambridge University Press, 1990, pp 1–163.
5. Sutherland, I.W. Polysaccharides from microorganisms, plants and animals. In: *Polysaccharides I: Polysaccharides from Prokaryotes,* Vandamme, E.J., S. De Baets, A. Steinbuchel, eds., Weinheim, Germany: Wiley-VCH Verlag GmbH, 2002, pp 1–19.
6. Guiseley, K.B. Chemical and physical properties of algal polysaccharides used for cell immobilization. *Enzyme Microb. Technol.* 11:706–716, 1989.

7. Adhikary, S.P. Polysaccharides from mucilaginous envelope layers of cyanobacteria and their ecological significance. *J. Sci. Ind. Res.* 57:454–466, 1998.
8. Vilchez, C., I. Garbayo, M.V. Lobato, J.M. Vega. Microalgae-mediated chemicals production and waste removal. *Enzyme Microb. Technol.* 20:562–572, 1997.
9. Siddhanta, A.K., A.S.K. Murthy. Bioactive polysaccharides from marine brown algae (phaeophyceae). *J. Ind. Chem. Soc.* 78:431–437, 2001.
10. Witvrouw, M., E. DeClercq. Sulfated polysaccharides extracted from sea algae as potential antiviral drugs. *Gen. Pharmacol.* 29:497–511, 1997.
11. Schaeffer, D.J., V.S. Krylov. Anti-HIV activity of extracts and compounds from algae and cyanobacteria. *Ecotox. Environ. Saf.* 45:208–227, 2000.
12. Doolittle, W.F. Molecular evolution. In: *The Biology of Cyanobacteria,* Carr, N.G., B.A. Whitton, eds., Berkeley: University of California Press, 1982, pp 307–331.
13. Whitton, B.A. Diversity, ecology and taxonomy of the cyanobacteria. In: *Photosynthetic Prokaryotes,* Mann, N.H., N.G. Carr, eds., New York: Plenum Press, 1992, pp 1–51.
14. Lem, N.W., B.R. Glick. Biotechnological uses of cyanobacteria. *Biotechnol. Adv.* 3:195–208, 1985.
15. Glazer, A.N. Phycobilisome: a macromolecular complex optimized for light energy transfer. *Biochim. Biophys. Acta* 768:29–51, 1984.
16. Woese, C.R. Bacterial evolution. *Microbiol Rev.* 51:221–271, 1987.
17. Castenholz, R.W., B.X. Phylum. Cyanobacteria: Oxygenic photosynthetic bacteria. In: *Bergey's Manual of Systematic Bacteriology*, Vol.1, 2nd ed., Boone, D., R.W. Castenholz, eds., New York: Springer-Verlag, 2001, pp 474–599.
18. Whitton, B.A., M. Potts. Introduction to the cyanobacteria. In: *The Ecology of Cyanobacteria: Their Diversity in Time and Space,* Whitton, B.A., M. Potts, eds., Dordrecht: Kluwer Academic Publishers, 2000, pp 1–11.
19. Vaishampayan, A., R.P. Sinha, D.P. Häder, T. Dey, A.K. Gupta, U. Bhan, A.L. Rao. Cyanobacterial biofertilizers in rice agriculture. *Bot. Rev.* 67:453–516, 2001.
20. Glazer, A.N., L. Stryer. Phycofluor probes. *Trends Biochem. Sci.* 8:423–427, 1984.
21. Bertocchi, C., L. Navarini, A. Cesaro, M. Anastasio. Polysaccharides from cyanobacteria. *Carbohydr. Polymers* 12:127–153, 1990.
22. De Philippis, R., M. Vincenzini. Exocellular polysaccharides from cyanobacteria and their possible applications. *FEMS Microbiol. Rev.* 22:151–175, 1998.
23. Hill, D.R., A. Peat, M. Potts. Biochemistry and structure of the glycan secreted by desiccation-tolerant *Nostoc commune* (cyanobacteria). *Protoplasma* 182:126–148, 1994.
24. Grilli-Caiola, M., D. Billi, E.I. Friedmann. Effect of desiccation on envelopes of the cyanobacterium *Chroococcidiopsis* sp. (*Chroococcales). Eur. J. Phycol.* 31:97–105, 1996.
25. Pajdak-Stós, A., E. Fialkowska, J. Fyda. *Phormidium autumnale* (cyanobacteria) defense against three ciliates grazer species. *Aqua. Microb. Ecol.* 23:237–244, 2001.
26. Martin, T.J., J.T. Wyatt. Extracellular investments in blue-green algae with particular emphasis on genus *Nostoc*. *J. Phycol.* 10:204–210, 1974.
27. Prosperi, C.H. A cyanophytes capable of fixing nitrogen under high levels of oxygen. *J. Phycol.* 30:222–224, 1994.
28. Bender, J., S. Rodriguez-Eaton, U.M. Ekanemesang, P. Phillips. Characterization of metal-binding bioflocculants produced by the cyanobacterial component of mixed microbial mats. *Appl. Environ. Microbiol.* 60:2311–2315, 1994.
29. Parker, D.L., B.R. Schram, J.L. Plude, R.E. Moore. Effects of metal cations on the viscosity of a pectin-like capsular polysaccharide from the cyanobacterium *Microcystis flos-aquae* C3-40. *Appl. Environ. Microbiol.* 62:1208–1213, 1996.
30. Hu, C.X., Y.D. Liu, L.R. Song, D.L. Zhang. Effects of desert soil algae on the stabilization of fine sands. *J. Appl. Phycol.* 12:281–292, 2002.
31. Mazor, G., G.J. Kidron, A. Vonshak, A. Abeliovich. The role of cyanobacterial exopolysaccharides in structuring desert microbial crusts. *FEMS Microbiol. Ecol.* 21:121–130, 1996.
32. Robins, R.J., D.O. Hall, D.J. Shi, R.J. Turner, M.J.C. Rhodes. Mucilage acts to adhere cyanobacteria and cultured plant cells to biological and inert surfaces. *FEMS Microbiol. Lett.* 34:155–160, 1986.

33. Ehling-Schulz, M., S. Scherer. UV protection in cyanobacteria. *Euro. J. Phycol.* 34:329–338, 1999.
34. Mohamed, Z.A. Removal of cadmium and manganese by a non-toxic strain of the freshwater cyanobacterium *Gloeotheca magna. Water Res.* 35:4405–4409, 2001.
35. Morvan, H., V. Gloaguen, L. Vebret, F. Joset, L. Hoffmann. Structure-function investigations on capsular polymers as a necessary step for new biotechnological applications: the case of the cyanobacterium *Mastigocladus laminosus. Plant Physiol. Biochem.* 35:671–683, 1997.
36. Sutherland, I.W. Industrially useful microbial polysaccharides. *Microbiol. Sci.* 3:5–9, 1986.
37. De Philippis, R., M.C. Margheri, E. Pelosi, S. Ventura. Exopolysaccharide production by a unicellular cyanobacterium isolated from a hypersaline habitat. *J. Appl. Phycol.* 5:387–394, 1993.
38. Sudo, H., J. Grant Burgess, H. Takamasa, N. Nakamura, T. Matsunaga. Sulphated exopolysaccharide produced by the halophilic cyanobacterium *Aphanocapsa halophytica. Curr. Microbiol.* 30:219–222, 1995.
39. Fischer, D., U.G. Schlosser, P. Pohl. Exopolysaccharide production by cyanobacteria grown in closed photobioreactors and immobilized using white cotton toweling. *J. Appl. Phycol.* 9:205–213, 1997.
40. Lupi, F.M., H.M.L. Fernandes, I. Sa-Correia, J.M. Novais. Temperature profiles of cellular growth and exopolysaccharides synthesis by *Botryococcus braunii* Kutz. UC58. *J. Appl. Phycol.* 3:35–42, 1991.
41. Linton, J.D., S.G. Ash, L. Huybrechts. Microbial polysaccharides. In: *Biomaterials: Novel Materials from Biological Sources,* Byrom, D., ed., New York: Stockton Press, 1991, pp 215–261.
42. Sutherland, I.W. Structure-function relationships in microbial exopolysaccharides. *Biotechnol. Adv.* 12:393–448, 1994.
43. Atkins, E.D.T. Biomolecular structures of naturally occurring carbohydrate polymers. *Int. J. Biol. Macromol.* 8:323–329, 1986.
44. Matulewicz, C.M., E.E. Percival, H. Weigel. Water-soluble polysaccharides of artarctic and cultured *Phormidium* species of cyanophyceae. *Phytochemistry* 23:103–105, 1984.
45. Low, C.S.F., D.C. White. Regulation of external polymer production in benthic microbial communities. In: *Microbial Mats: Physiological Ecology of Benthic Microbial Communities*, Cohen, Y., E. Rosenberg, eds., Washington, DC: American Society for Microbiology, 1989, pp 228–238.
46. Cesáro, A., G. Liut, C. Bertocchi, L. Navarini, R. Urbani. Physicochemical properties of the exocellular polysaccharides from *Cyanospira capsulata. Int. J. Biol. Macromol.* 12:79–84, 1991.
47. Flaibani, A., Y. Olsen, T.J. Painter. Polysaccharides in desert reclamation: composition of exocellular proteoglycan complexes produced by filamentous blue-green and cellular green edaphic algae. *Carbohydr. Res.* 190:235–248, 1989.
48. Indegaard, M., K. Ostgaard. Polysaccharides for food and pharmaceutical uses. In: *Seaweed Resources in Europe: Uses and Potentials.*, Guiry, M.D., G. Blunden, eds., Hoboken, NJ: John Wiley and Sons, 1991, pp 169–183.
49. Rees, D.A. Polysaccharide gels: a molecular view. *Chem. Ind.* 19:630–636, 1972.
50. Phlips, E.J., C. Zeman, P. Hansen. Growth, photosynthesis, nitrogen fixation and carbohydrate production by a unicellular cyanobacterium, *Synechococcus* sp. *cyanophyta. J. Appl. Phycol.* 1:137–145, 1989.
51. Filali-Mouhim, R., J.F. Cornet, T. Fontaine, B. Fournet, G. Dubertret. Production, isolation and preliminary characterization of the exopolysaccharides of the cyanobacterium *Spirulina platensis. Biotechnol. Lett.* 15:567–572, 1993.
52. De Philippis, R., C. Sili, G. Tassinato, M. Vincenzini. Effects of growth conditions on exopolysaccharides production by *Cyanospira capsulate. Biores. Technol.* 38:101–104, 1991.
53. De Philippis, R., M.C. Margheri, C. Sili, M. Vincenzini. Cyanobacteria: a promising group of exopolysaccharides producers. *Proceedings of 2nd European Workshop: "Biotechnology of Microalgae"*, IGV Institut fur Getreideverarbeitung GmbH, Bergholz-Rehbrucke, 1995, pp 78–81.

54. De Philippis, R., M.C. Margheri, R. Materassi, M. Vincenzini. Potential of unicellular cyanobacteria from saline environments as exopolysaccharide producers. *Appl. Environ. Microbiol.* 64:1130–1132, 1998.
55. Plude, J.L., D.L. Parker, O.J. Schommer, R.J. Timmerman, S.A. Hagstrom, J.M. Joers, R. Hnasko. Chemical characterization of polysaccharide from the slime layer of the cyanobacterium *Microcystis flos-aquae* C3-40. *Appl. Environ. Microbiol.* 57:1696–1700.
56. Panoff, J., B. Priem, H. Morvan, F. Joset. Sulphated exopolysaccharides produced by two unicellular strains of cyanobacteria, *Synechocystis* PCC 6803 and 6714. *Arch. Microbiol.* 150:558–563, 1988.
57. Gloaguen, V., H. Morvan, L. Hoffmann. Release and capsular polysaccharides of *Oscillatoriaceae (Cyanophyceae, Cyanobacteria)*. *Arch. Hydrobiol. (Alg. Stud.)* 78:53–69, 1995.
58. Vincenzini, M., R. De Philippis, C. Sili, R. Materassi. Studies on exopolysaccharides release by diazotrophic batch culture of *Cyanospira capsulata*. *Appl. Microbiol. Biotechnol.* 34:392–396, 1990.
59. Huang, Z.B., Y.D. Liu, B.S. Paulsen, D. Klaveness. Studies on polysaccharides from three edible species of *Nostoc* (cyanobacteria) with different colony morphologies: comparison of monosaccharide compositions and viscosities of polysaccharides from field colonies and suspension cultures. *J. Phycol.* 34:962–968, 1998.
60. Gantar, M., P. Powell, N.W. Kerby, I.W. Sutherland. Role of extracellular polysaccharide in the colonization of wheat (*Triticum vulgare* L.) roots by N_2-fixing cyanobacteria. *Biol. Fertil. Soils* 19:41–48, 1995.
61. Mehta, V.B., B.S. Vaidya. Cellular and extracellular polysaccharides of the blue-green alga *Nostoc*. *J. Exp. Bot.* 29:1423–1430, 1978.
62. Vincenzini, M., R. De Philippis, C. Sili, R. Materassi. A novel exopolysaccharide from a filamentous cyanobacterium: production, chemical characterization and rheological properties. In: *Novel Biodegradable Microbial Polymers*, Dawes, E.A., ed., Dordrecht: Kluwer Academic Publishers, 1990, pp 295–310.
63. Lama, L., B. Nicolaus, V. Calandrelli, M.C. Manca, I. Romano, A. Gambacorta. Effect of growth conditions on endo- and exopolymer biosynthesis in *Anabaena cylindrica* 10C. *Phytochemistry* 42:655–659, 1996.
64. Tischer, R.G., E.B. Davis. The effect of various nitrogen sources upon the production of exocellular polysaccharide by the blue-green alga *Anabaena flos-aquae* A-37. *J. Exp. Bot.* 22:546–551, 1971.
65. Nicolaus, B., A. Panico, L. Lama, I. Romano, M.C. Manca, A. De Giulio, A. Gambacorta. Chemical composition and production of exopolysaccharides from representative members of heterocystous and non-heterocystous cyanobacteria. *Phytochemistry* 52:639–647, 1999.
66. Dawes, E.A. *Microbial Energetics*. Glasgow: Blackie & Son, 1986, pp 1–187.
67. De Philippis, R., C. Sili, R. Paperi, M. Vincenzini. Exopolysaccharide-producing cyanobacteria and their possible exploitation: a review. *J. Appl. Phycol.* 13:293–299, 2001.
68. Moreno, J., M.A. Vargas, J.M. Madiedo, J. Munoz, J. Rivas, M.G. Guerrero. Chemical and rheological properties of an extracellular polysaccharide produced by the cyanobacterium *Anabaena* sp. ATCC 33047. *Biotechnol. Bioengin.* 67:283–290, 2000.
69. Shepherd, R., J. Rockey, I.W. Sutherland, S. Roller. Novel bioemulsifiers from microorganisms for use in foods. *J. Biotechnol.* 40:207–217, 1995.
70. Sanford, P.A., I.W. Cottrell, D.J. Pettitt. Microbial polysaccharides: new products and their commercial applications. *Pure Appl. Chem.* 56:879–892, 1984.
71. Borowitza, M.A. Algal biotechnology products and processes: matching science and economics. *J. Appl. Phycol.* 4:267–279, 1992.
72. Zaslavskaia, L.A., J.C. Lippmeier, C. Shih, D. Ehrhardt, A.R. Grossman, K.E. Apt. Trophic conversion of an obligate photoautotrophic organism through metabolic engineering. *Science* 292:2073–2075, 2001.

1.19

Food Applications of Algae

G.A. Ravishankar, R. Sarada, B. Sandesh Kamath, and K.K. Namitha

CONTENTS

19.1 HISTORICAL PERSPECTIVES

Algae have been long recognized as a heterogeneous group of organisms, ranging from prokaryotic blue greens to distantly related eukaryotic lines. Their growth modes range from phototrophy through photoheterotrophy to heterotrophy. Each growth mode in turn can be either obligate or facultative (1). The genetic diversity of microalgae is evident still further in their ecological diversity and ubiquitous distribution. Different species can grow in water ranging from fresh water to hypersaline environments; from the water surface to the limits of photic zone, which is often 200–300 meters; either free floating or attached (2).

This great genetic diversity of algae leads to a great variety of products. Microalgae have a wide range of physiological and biochemical characteristics, many of which are rare or absent in other taxonomic groups. For example, unlike higher plants and animals and most prokaryotic microorganisms, some algae synthesize long chain polyunsaturated fatty acids (LC-PUFAs), such as eicosapentaenoic acid (EPA) and docosahexaenoic acid (DHA) (3). Some species of microalgae can efficiently synthesize intra- or extracellular polysaccharides, some of which contain unusual monomer sugars. For example, some algae synthesize extracellular polysaccharides containing substantial quantities of D-galactose, while others use L-galactose or a mixture of the two isomers (4).

Microalgae such as *Spirulina* have also had a long history in human nutrition. *Spirulina platensis* was consumed by the native population of the subSaharan region of Kanem, northeast of lake Chad and Belgium (5,6). The dried mats of *S. platensis* in the shape of biscuits or flakes were sold as djhe in the markets of the capital city of Fort Lamy (known as N'Djamena today). Djhe was used to prepare the sauces for the millet meal eaten by the Kanembou population. The spiral shape of the trichomes and the presence of abundant gas vacuoles resulted in the formation of large floating mats of *S. platensis* that the wind brought to the shores and sun dried (7).

The potentiality of *Spirulina* as a food for humans was exploited in ancient Mexico as well. Historical records reveal that at the time of conquest of Mexico by Spain, biscuits analogous to djhe were sold in the Mexican markets (8). The biscuits were prepared, as in the case of djhe, by sun drying mats of microalgae (*S. maxima*) that grows abundantly in the highly alkaline waters of Lake Texcoco near what is now Mexico City where first large scale factory for the production of *Spirulina* operates.

Since the 1950s, microalgae have been a major focus of attention as potential sources of protein, liquid fuels, and fine chemicals (9–12). Although initial work with microalgae focused mainly on protein production for human and animal nutrition (13), economic realities have meant that the emphasis today is on high value fine chemicals. There are now well established microalgae production plants in various countries, processing several different species of microalgae, including *Chlorella* (Taiwan, Australia), *Spirulina* (U.S., China, India, Mexico), *Dunaliella* (U.S., India, Australia), and *Haematococcus* (U.S., India).

Marine macroalgae (seaweeds) are grouped under three broad groups, brown, red, and green, based on pigmentation. These are referred to as Phaeophyceae, Rhodophyceae, and Chlorophyceae respectively. They have a long history of exploitation and cultivation. They are used directly as food or are processed to yield products used as food additives, pharmaceuticals, and industrial chemicals (14).

Over the last 50 years, the main products from macroalgae have been the phycocolloids, mainly agar and carrageenans (15,16). These phycocolloids have been essential to the development of modern microbiology and biotechnology (17). Since prehistoric times coastal human populations must have collected and eaten a variety of seaweeds. What is known about previous and existent uses of algae as sea vegetables has been compiled by Levring et al. (18) and by Chapman and Chapman (14). For example, a green sea lettuce (*Ulva lactuca*) was consumed in Brittany and elsewhere and esteemed for curing scurvy. The red alga, *Chondrus crispus* (Irish moss), is still commercially available as food in Ireland and is used by the people living in the folk coasts of Latin America, where it is not uncommon to find red algae such as *Gracilaria* as cushuro on the menu. Only in Japan, however, has human consumption of seaweeds developed into a real cult. Seaweed fishing has been described in the literature since 274 A.D (14). Approximately 20 kinds of sea vegetables are normal items on the Japanese shopping list. They provide up to 5–10% of Japanese food requirements.

Seaweeds have been used as food since the fourth century in Japan and the sixth century in China. Today these two countries, along with Republic of Korea, are among the largest consumers of seaweeds as food. In the past, the requirements were met by harvesting the supply of seaweeds from natural (wild) stocks. However, increasing demand over the last fifty years and a better understanding of the life cycle of these seaweeds has led to the development of cultivation industries, which produce more than 90% of market demand today. China is the largest producer of edible seaweeds, growing about 5 million tons, followed by the Republic of Korea with about 800,000 wet tons and Japan with about 600,000 wet tons. The main edible seaweeds include *Laminaria*, *Undaria*, and *Hizikia* in the brown seaweed category; *Porphyra* and *Palmaria* in the red seaweed category; and *Monostroma* and *Enteromorpha* in the green seaweed category. These seaweeds are being extensively farmed in countries such as China, the Philippines, Indonesia, Japan, the U.S., France, and Chile, to meet the ever-increasing demand (19–21).

19.2 NUTRITIONAL VALUE OF MICRO- AND MACROALGAE

Micro- and macroalgae are rich in protein, polysaccharides, carbohydrates, lipids, amino acids, trace elements, and vitamins (22). Table 19.1 gives the chemical composition of *Spirulina, Chlorella,* and *Scenedesmus*.

Table 19.1
Chemical Composition of Algae

Component	*Spirulina* (Percentage Dry Weight Basis)	*Chlorella*	*Scenedesmus*
Protein	50–60	40–50	50–55
Lipids	2–3	8–12	8–14
Carbohydrates	15–20	12–16	10–15
Fibre	5–8	6–8	10–12
Ash	10–12	8–10	6–8
Nucelic acid	5–7	6–8	4–6
Moisture	5–8	5–8	5–7

Source: Reference 23

19.2.1 Proteins

Various studies on the chemical composition of the biomass of *S. platensis* and *S. maxima* document an unusually high protein content, comprising up to 62–65% of its dry weight; *Chlorella* has been found to contain 58% protein. Such a high level of protein in *Spirulina* is uncommon even in the microbial world, being surpassed only by certain bacteria like *Cellulomonas*, which are recorded to have protein levels above 80% of their dry weight (24,25). However, the high protein content in these bacteria is accompanied by high nucleic acid content, which is medically undesirable. Excess nucleic acids (26,27) result in uric acid accumulation due to purine catabolism, which may in turn lead to pathological conditions such as gout. In *Spirulina* the concentration of nucleic acid is always below 5% of the dry weight and is thus advantageous.

Research conducted on *Spirulina* by the World Health Organization and scientists in the United States, France, West Germany, Mexico, Vietnam, and Japan confirm that it has a blend of nutrients that no other single plant source can provide. The protein content of *Spirulina* is higher than that of any other food (28,29). The net protein utilization (NPU) value of *Spirulina* is comparable to that of other vegetarian sources, indicating good amino acid composition, digestibility, and biological value. The amino acid profile of *Spirulina* protein is impressive. Indeed, with the exception of cysteine and lysine, which are present in quantities somewhat lower than in standard proteins, all the other essential amino acids are present in adequate concentrations (30–34). The potential use of *Spirulina* as a source of protein for human consumption has been widely recognized (35). The essential amino acids which it produces (Table 19.2) are of value in designer foods. The recommended intake of *Spirulina* based on the recommended daily allowance (RDA) requirements is given in Table 19.3. The protein content of seaweeds varies. Protein comprises 5–11% of the dry matter of brown algae. Red algae have a higher protein content ranging from 30–40% of dry matter; green algae can have up to 20%.

19.2.2 Minerals

The mineral makeup of *Spirulina* is attractive, as well. The iron level of *Spirulina* is 12 times higher than that of any other food. *Spirulina* is also rich in magnesium, potassium (Table 19.4), and other trace elements (29). *Spirulina*, being particularly rich in iron and calcium, is good for blood rejuvenation and the healthy functioning of bones and teeth.

Seaweed contains enormous amounts of mineral elements, which account for up to 36% of its dry mass. The mineral macronutrients include sodium, calcium, magnesium

Table 19.2
Amino-Acid Profile in *Spirulina*

Essential Amino Acids	Per 10 gms	% Total
Isoleucine	350 mg	5.6
Leucine	540 mg	8.7
Lysine	290 mg	4.7
Methionine	140 nmg	2.3
Phenylalanine	280 mg	4.5
Threonine	320 mg	5.2
Tryptophan	90 mg	1.5
Valine	400 mg	6.5
Nonessential Amino Acids		
Alanine	470 mg	7.6
Arginine	430 mg	6.9
Aspartic acid	610 mg	9.8
Cystine	60 mg	1.0
Glutamic acid	910 mg	14.6
Glycine	320 mg	5.2
Histidine	100 mg	1.6
Proline	270 mg	4.3
Serine	320 mg	5.2
Tyrosine	300 mg	4.8
Total amino acids	6200 mg	100%

Source: Reference 29

Table 19.3
Adult Essential amino acid (EAA) requirements provided by *Spirulina*

Essential Amino Acids	Recommended Daily Allowance(g/day)	*Spirulina* gm/10 gm	% RDA
Isoleucine	0.84	0.35	42
Leucine	1.12	0.54	48
Lysine	0.84	0.29	35
Methionine	0.70	0.20	29
Phenylalanine	1.12	0.58	52
Threonine	0.56	0.32	43
Valine	0.98	0.40	41

Source: Reference 29

potassium, chlorine, sulfur, and phosphorous. The micronutrients include iodine, zinc, copper, selenium, molybdenum, fluoride, manganese, boron, nickel, and cobalt (36).

Among the micronutrients, seaweeds have a high proportion of iodine content. The iodine content in brown algae ranges from 1500–8000 ppm in dry kelps and 500–1000 ppm in dry rockweed; in red and green algae, the iodine content ranges from 100–300 ppm in dry seaweeds. The recommended daily intake of iodine is 150 mg/day. One gram of dried brown algae provides 5–8 mg of iodine and one gram of red algae (purple nori)

Table 19.4
Different minerals found in *Spirulina*

Minerals	Per 10 gms	RDA	% RDA
Calcium	100 mg	1000 mg	10
Iron	15 mg	18 mg	83
Zinc	300 μg	15 mg	2
Phosphorous	90 mg	1000 mg	9
Magnesium	40 mg	400 mg	10
Copper	120 μg	2 mg	6
Sodium	60 mg	2–5 gm	1
Potassium	160 mg	6 gm	3
Manganese	500 μg	3 mg	17
Selenium	2 μg	100 μg	2

Source: Reference 29

provides 1–3 mg of iodine. Besides iodine, seaweed is one of the richest sources of calcium, the content ranging from 4–7% of dry matter (37).

19.2.3 Vitamins

Seaweeds, like green vegetables, contain all kinds of vitamins and are a natural source of vitamins to humans. Among the group B vitamins, seaweeds contain vitamin B_{12}, which is particularly recommended for treatment of the effects of aging, chronic fatigue syndrome, and anemia. *Chlorella* contains vitamins C, E, B, and A. *Spirulina* is the richest source of B_{12} and the daily ingestion of 1 g of *Spirulina* would be enough to meet the daily requirement of B_{12} (29,38). Apart from vitamin B_{12}, *Spirulina* has an excellent blend of other vitamins, including vitamins A, B_1, B_2, B_6, E, and H (biotin) (Table 19.5). It contains 21% of the RDA of thiamin and riboflavin (29). *Spirulina* has a β-carotene (provitamin A) content (0.1% of the dry weight) which is 20 times higher than that of carrots. Its folic acid and Vitamin B_{12} contents make it a good therapeutic food for anemia.

Algae provide a good source of vitamin C (39). The levels of vitamin C range between 500–3000 mg/kg of dry matter for green and brown algae, whereas red algae contains 100–800 mg/kg. Vitamin C is of interest in human nutrition because it strengthens the immune defense system, activates the intestinal absorption of iron, controls the formation of conjunctive tissue, acts in trapping free radicals, and regenerates vitamin E (37).

The brown seaweeds contain higher levels of vitamin E than the green and red seaweeds. Vitamin E is important in human nutrition, because, due to its antioxidant activity, it inhibits the oxidation of low density lipoproteins. It also plays an important role in the arachidonic acid chain, by inhibiting the formation of prostaglandins and thromboxan. Brown algae also contain alpha, beta, and gamma tocopherol; the green and red algae contain only alpha tocopherol. Gamma and alpha tocopherol have been shown to increase the production of nitric oxide and nitric oxide synthase activity and also help in prevention of cardiovascular disease (40).

19.2.4 Lipids

Microalgae contain significant quantities of lipids, with a composition similar to that of vegetable oils (Table 19.6). Under certain conditions, microalgae have been reported to contain lipids up to 85% of their dry weight (41,42). The lipid content in general ranges between 20 and 40% of the dry weight. At least one microalga, *Botryococcus braunii*, produces large amounts of C_{15} and C_{33} hydrocarbons (43).

Table 19.5
Different Vitamins present in *Spirulina*

Vitamins	Per 10 gms	Recommended Daily Allowance	% RDA
Vitamin A (β-carotene)	23000 IU	5000 IU	460
Vitamin B_1 (Thiamine)	0.31 mg	1.5 mg	21
Vitamin B_2 (Riboflavin)	0.35 mg	1.7 mg	21
Vitamin B_3 (Niacin)	1.46 mg	20.0 mg	7
Vitamin B_6 (Pyridoxine)	80.00 μg	2.0 mg	4
Vitamin B_{12} (Cyanocobalamine)	32.00 μg	6.0 μg	533
Vitamin E (α-tocoferol)	1.00 IU	30.0 IU	3
Folacin	1.00 μg	400.0 μg	0.04
Panthothenic acid	10.00 μg	10.0 mg	1
Biotin	0.50 μg	-	-
Inositol	6.40 mg	-	-

Source: Reference 29

Table 19.6
Lipids from Microalgae

Microalgae	Lipid	Ref.
Porphyridium cruentum	Eicosapentaenoic acid	49
	Arachidonic acid	59
Porphyridium aeruginosa	Arachidonic acid	53
	Palmitic acid	
	Oleic acid	
	Linoleic acid	
Ochromonas danica	γ-Linolenic acid	60
Monodus subterraneous	Eicosapentaenoic acid	61
Euglena gracilis	α-Linolenic acid	62
Ulothrix aequalis	α-Linolenic acid	63
Lauderia borealis	Eicosapentaenoic acid	64
Phaeodactylum tricornutum	Eicosapentaenoic acid	65
Chlorella minutissima	Eicosapentaenoic acid	61
Nannochloropsis oculata	Eicosapentaenoic acid	65

The lipids of microalgae are generally esters of glycerol and fatty acids with a chain length of C_{14} to C_{22}. They may be either saturated or unsaturated. Some blue green algae, especially the filamentous species, tend to have large quantities of polyunsaturated fatty acids (25–60% of the total) (44–48).

The lipids of some algal species are rich in essential fatty acids such as the C_{18} linoleic (18: 2ω3) and γ-linolenic (18: 3ω3) acids and their C_{20} derivatives, eicosapentaenoic acid (20: 5ω3), and arachidonic acid (20: 4ω6). These fatty acids are an essential component of the diet of humans and animals and are becoming important feed additives in aquaculture.

The eukaryotic algae have a predominance of saturated and monosaturated fatty acids (49). Triglycerides are the most common storage lipids and may constitute up to 80% of the total lipid fraction (50,51). Besides the triglycerides, the other major algal lipids are sulphoquinovosyl diglyceride (SL), monogalactosyl diglyceride (MGDG), digalactosyl diglyceride (DGDG), lecithin, phosphatidyl glycerol, and phosphatidyl inositol (52).

The red unicellular *Porphyridium cruentum* and *P. aeruginosa* have high levels of arachidonic acid (20:4) as well as palmitic, oleic (18:1), and linoleic acids (Table 19.6) (49,53). The chrysophytes have relatively high contents of 22:6 and 22:4 acids (54,55) and the dinophytes have significant quantities of highly unsaturated 18:4 and 22:6 acids, as well as the saturated 16:0 and 20:0 acids (54,56). The major unsaponifiable lipid in the *Chlorophyceae* and *Cyanophyceae* is n-heptadecane (57,58), with sterols being more common in the green algae.

Spirulina has yet another merit in its lipid composition, being cholesterol free and rich in polyunsaturated essential fatty acids (66), which is desirable for treating or preventing conditions like atherosclerosis, obesity, and blood pressure. Additionally, this alga is rich in γ-linolenic acid (Table 19.7), a precursor for prostaglandin biosynthesis, and thus has even further medical implications (29). In short, the composition of *Spirulina* makes it a low calorie, low fat, cholesterol free source of protein.

Although seaweeds have a low fat content, ranging from 1–5% of dry weight, they have a higher proportion of essential fatty acids than the green plants (land plants). Green algae have a higher oleic and α-linoleic acid content and red and brown algae are rich in EPA and arachidonic acid content (67).

19.3 ALGAE AS A SOURCE OF NUTRACEUTICALS

The very great diversity of the microalgae is also reflected in their chemical composition, and the microalgae are, therefore, potential sources of an almost unlimited range of chemicals (Table 19.8). The main chemicals currently being commercialized or under consideration for commercial extraction are carotenoids, phycobilins, fatty acids, polysaccharides, vitamins, sterols, and biologically active molecules for use in human and animal health (43,68–70). Selection of promising algae and their products requires evaluation of the suitability of the alga for mass culture, the value of the product, its concentration in the alga, the size of the market for the products, and existing and future alternative sources.

19.3.1 Carotenoids

The carotenoids, β-carotene and astaxanthin, are attractive algal products because they command a high market price (\$600 kg^{-1} for β-carotene; \$3000 kg^{-1} for astaxanthin)

Table 19.7
Essential fatty acid composition of *Spirulina*

Fatty Acids	mg/10 gms	% Total
C14:0 Myristic	1 mg	0.2
C16:0 Palmitic	244 mg	45.0
C16:1 Palmitoleic	33 mg	5.6
C17:0 Heptadecanoic	3 mg	0.3
C18:0 Stearic	8 mg	1.4
C18:1 Oleic	12 mg	2.2
C18:2 Linoleic	97 mg	17.9
C18:3 γ-linolenic	135 mg	24.9
C20:0 others	14 mg	2.5

Source: Reference 29

Table 19.8
Products from Microalgae

Product	Source
Carotenoids	
β-carotene	*Dunaliella salina*
astaxanthin	*Haematococcus pluvialis*
canthaxanthin	Some chlorophytes and mutants of *Dunaliella*
lutein	
Other Pigments	
Phycocyanin	*Spirulina*, other blue-green algae
Phycoerythrin	*Porphyra, Porphyridium* and other red algae
Fatty acids	
Eicosapentaenoic acid (EPA)	Diatoms, *Porphyridium*, prymenesiophytes, cryptomonads
Docosahexaenoic acid (DHA)	Prymenesiophytes, cryptomonads
Polysaccharides	
Many Polysaccharides	Mainly blue-green algae, *Porphyridium*
Sterols	
Many Sterols	Many species
Vitamins	
Tocopherol	Brown algae
Vitamin B_{12}	Various algae
β-carotene (pro-vitamin A)	*Dunaliella salina*
Bioactive compounds	
Anti-fungal, anti-bacterial, anti-viral, antineoplastic, pharmacologically active compounds	Many species
Biomass	
Health Food	*Spirulina, Chlorella*

Source: Reference 71

(121, www.healthfoods.com) and they are present in high concentrations in some algal cells. Carotenoids are powerful antioxidants. Recent studies have shown a correlation between a diet rich in carotenoids and a diminishing risk of some diseases. For example, a diet rich in β-carotene and lycopene helps diminish cardiovascular disease and cancers; one rich in lutein and zeaxanthin helps reduce opthamalogical diseases. Brown seaweeds are particularly rich in carotenoids, especially in fucoxanthin, β-carotene, and violaxathin. The main carotenoids of red algae are β-carotene and α-carotene and their dihydroxylated derivatives, lutein and zeaxanthin. The carotenoid composition of green algae is similar to that of higher plants; the main carotenoids are β-carotene, lutein, violaxathin, antheraxanthin, zeaxanthin, and neoxanthin. The antioxidant properties of algal carotenoids, and the role they play in preventing many pathologies linked to oxidative stress, have been demonstrated by Okuzumi et al. (72) and Yan et al. (73).

19.3.2 Fatty Acids

Apart from the carotenoids, the long chain polyunsaturated fatty acids — arachidonic acid (AA; 20: 4n-6), eicosapentaenoic acid (EPA; 20: 5n-3), and docosahexaenoic acid (DHA; 22: 6n-3) — are of great interest. For example, the red unicellular *Porphyridium cruentum* can contain more than 30% of its total fatty acids as AA and equally high concentrations of EPA (74,75); in the diatom *Phaeodactylum* more than 35% of the total fatty acids can be EPA (76). DHA, on the other hand, is more abundant in prymnesiophytes such as *Isochrysis*, *Pavlova*, and in the cryptomonad *Chroomonas salina* (77).

Not only are there very great variations in the fatty acid composition of different species, but the total fatty acid content also varies. Furthermore, the fatty acid content and composition can be manipulated by altering culture conditions (78,79). Genetic enhancement of the production of desirable fatty acids also remains a possibility for the future.

A major barrier to the commercialization of algal fatty acids is the fact that no single species is a distinctly better producer than another, although *Phaeodactylum tricornutum*, *Porphyridium cruentum*, and *Nannochloropsis salina* are the most likely candidate organisms. The production of pure fatty acids or algal lipids for pharmaceutical use requires the ability to extract the lipids and purify the fatty acids effectively.

These fatty acids, especially the omega-3-fatty acids EPA and DHA, have a function in the prevention of cardiovascular disease and rheumatic arthritis, and are also apparently essential in infant nutrition (80–82). They are also important in the treatment of atherosclerosis, cancer, rheumatoid arthritis, psoriasis, ocular diseases, hypertension, and diseases of old age such as Alzheimer's disease and age related macular degeneration (83,84). Estimates of the potential U.S. market for these fatty acids range up to $790 million and, although fish oil is presently the main source of EPA and DHA, the supply of suitable fish oils is limited and the quality variable. Cultured microalgae could be a more reliable source.

These essential fatty acids are also important in the nutrition of aquaculture species (85), and the demand for high quality algal feed for this purpose is increasing. Most hatcheries grow their own algae on site, because live feeds are preferred. This small scale cultivation is generally unreliable, and the cultures often collapse. This is mainly due to the very primitive culture systems used, and the lack of experience of the hatchery staff. If a preserved microalgal product of reliable composition could be produced at a reasonable cost, this could overcome some of these problems. One such product of heterotrophically grown *Tetraselmis* was available for some time and was effective as a feed (86). The main problem with this product was the cost, and the fact that the composition of the alga was not optimal for the nutrition of the aquaculture species.

19.3.3 Phycobilins

Phycocyanin and phycoerythrin can be easily isolated from red or blue green algae (87). Highly purified phycobilins are being used in immunofluorescent labeling (88,89). Although the value for these high purity phycobilins is very high ($3.00–17.00/mg) (www.phycobiliprotein.com), the overall market size is small and unlikely to increase significantly. An alternative use for the phycobilins is as a natural food color. This is obtained as protein extract and marketed as Lina Blue (a brand name) (90), as an edible dye especially for food products such as ice cream and yogurt. Phycocyanin is used for this purpose in Japan. In recent years (91,92) the hepatoprotective, antiinflammatory, and antiarthritic properties of phycocyanin have been reported. Phycocyanin is also used as natural dye by the cosmetics industry. There is a potentially large market for the phycobilins; however, it will require extensive marketing and the compounds will have to be registered with the appropriate regulatory authorities in various countries. Unfortunately, this will take some time and will be an expensive process.

19.3.4 Bioactive Compounds

One area of interest in recent years is the production, by microalgae, of biologically active substances, with possible applications in human and animal health. Although screening of microalgae has been limited so far, compounds with antineoplastic (93,94), antibacterial (95–97), antifungal (98–100), and antiviral (101,102) activities have been isolated from microalgae, mainly blue green algae, and also macroalgae (Table 19.9). Other pharmacologically active substances have also been found. If eventually found to be suitable for medical use it is likely that many of these will be manufactured by synthesis. However, the complexity of some of these molecules makes production by algal culture a potential possibility.

19.4 INDUSTRIAL PRODUCTION PROCESSES

19.4.1 *Chlorella*

Chlorella is a unicellular, fresh water, green microalga, one of the earliest organisms used in applied research. Its potential use as a source of single cell protein was realized early by the Taiwanese. In 1951 the first *Chlorella* pilot plant was constructed and operated for the Carnegie Institution by Arthur D. Little, Inc. at Cambridge Mass. In 1960s, the large scale production of this alga was started in Taiwan (10). Another series of laboratory and pilot plant studies followed in Japan under the guidance of Dr. Himoshi Tamiya after his stay at Stanford. His strong working group at the famous Tokugawa Institute in Tokyo concentrated on *Chlorella* research, as did a study conducted in Israel. The results of this first boom of applied research on microalgae were published in the classic and still fascinating report edited by Burlew (9), which contains an interesting article on the use of natural algal suspension consisting mainly of *Chlorella* for the nutrition and apparently successful treatment of leprosy patients (113).

19.4.1.1 Nutritive Value

Chlorella is considered a nutritional powerhouse. It is richer in chlorophyll than *Spirulina* or other blue green algae, contains nearly 60% protein, and contains all of the essential

Table 19.9
Bioactive compounds from seaweeds

Seaweed	Bioactive Compound	Application	Ref.
Eucheuma	Carrageenan	Anticonstipation agent	103
Digenia simplex	Kainic acid and its isomer allokainic acid	Antihelminthic compounds	104
Phyllophora nervosa	An alkaloid hordenine	Sedative	105
Laminaria angustata	Aminoacid Laminine	Antihypertensive compound	106
Porphyra yezoeusis	γ-butyrobetaine	Hypocholesterolemic compounds	107
Fucus vesiculosus	Fucans	Blood anticoagulant compounds	108
Porphyra tenera	Porphyrosin	Antiulcer compound	109
Enteromorpha and *Ulva*	Acrylic acid	Antimicrobial	110
Dictyopteris zonaroides	Tetrabromoheptanone	Antimicrobial	111
Bonnemaisonia hemifera	4 polymers of polyhalo 3-butene-2-one	Antimicrobial	111
Discodermia kiiensis	Discodermin A-D	Cardiac activity	112

amino acids. It contains β-carotene, B complex, vitamin C, E, and K, minerals, carbohydrates, fatty acids, and lipoic acid. The major *Chlorella* production in the world is achieved in Taiwan; 10–15% of the total is produced in green houses in Japan. The main products of *Chlorella* available in the health food market are in the form of biomass dry powder, compressed pills, and *Chlorella* extract, marketed under different brand names. The cost of production of *Chlorella* extract ranges from $10 to $15/kg dry weight biomass (114). Over the last 20 years, there has been no significant breakthrough in the technology utilization, probably because of the knowledge of diversified uses of *Spirulina* (29).

19.4.1.2 Beneficial Effects

Chlorella is known for its ability to stimulate the growth of beneficial bacteria in the intestines and detoxify chemicals and heavy metals from the body. It has been reported to stimulate interferon production and increase the activity of B and T cells, thus enhancing the body's natural defense system. It has also been found to accelerate healing, improve digestion, protect against radiation, prevent degenerative diseases, help in the treatment of fungal infections such as *C. albicans*, relieve arthritis pain, and, due to its nutritional content, aid in the success of numerous weight loss programs.

19.4.2 *Spirulina*

Spirulina is a multicellular, filamentous blue green alga recognized as a unique health food throughout the world because of its nutrient density. The nutrient medium used for cultivation is simple and highly alkaline (pH 9.5–11). Growing in high alkaline media where other microorganisms do not survive is a great advantage in allowing mass production of *Spirulina* under outdoor conditions. The production centers of *Spirulina* are spread throughout the world including India (114).

19.4.2.1 Nutritive Value

Besides its high protein content (55–65%), *Spirulina* has also been reported to be a good source of β-carotene (precursor of vitamin A), B vitamins, pigments-phycobilins, chlorophyll, minerals, and γ-linolenic acid, an important nutraceutical.

19.4.2.2 Beneficial Effects

Spirulina has shown to possess cholesterol-reducing effect. The Harvard University School of Dental Medicine showed reduced oral cancer cells when *Spirulina* extracts were used (115). In Japan, the reports showed that *Spirulina* reduced kidney nephrotoxicity from mercury and three pharmaceutical drugs in experimental rats (116). *Spirulina* may also have positive effects against diabetes and may help to reduce high blood pressure (117). Research in Japan showed that *Spirulina* increased *Lactobacillus* in rats threefold over a control group (118). In France, pharmaceutical compounds containing *Spirulina* accelerated wound healing (30). A study in Mexico also hinted at the beneficial effect when *Spirulina* was incorporated in the diet of undernourished children and adults. In Togo, rapid recovery of malnourished infants through the use of *Spirulina* was reported in a village clinic. In India, large scale studies with preschool children showed carotenes in *Spirulina* helped the children recover from symptoms of vitamin A deficiency (119). In Romania, studies by Fica et al. (120) showed that when *Spirulina* tablets were prescribed to patients with nutritional deficiencies, the patients gained weight and their health improved. In China, *Spirulina* has been prescribed as a baby-nourishing ingredient in baked barley sprouts. When given *Spirulina*, 27 out of 30 children aged from 2 to 6 years recovered in a short period from bad appetite, night sweats, diarrhea, and constipation. In 1989, the National Cancer Institute announced that chemicals from blue green algae were found to be remarkably active against the AIDS virus and 100 types of cancer (101). The sulfolipid portions of the glycolipids in *Spirulina* can prevent viruses from either attaching to or penetrating into cells, resulting in prevention of viral infection.

19.4.3 *Dunaliella*

β-carotene from the alga *Dunaliella salina* was the first high value algal product commercialized (121) and there are now major producers in Australia (Western Biotechnology Ltd; Betatene Ltd), the U.S. (Microbio Resources Inc; Cyanotech), India (Parry Agro), and Israel (Naturebeta). The Australian producers use very large (5 ha or greater) shallow ponds without any means of mixing (122,123), whereas in the U.S., India, and Israeli, producers use paddle wheel mixed raceway ponds to grow the algae. The β-carotene is sold mainly as an oil extract, as a suspension in vegetable oil, or as a dried algal powder (121).

The *Dunaliella* β-carotene process has several advantages:

1. *Dunaliella salina* is the most halophilic eukaryote known, with an optimum salinity of about 22% w/v NaCl for growth, and an ability to withstand NaCl concentrations >30% and high temperatures. This means that extensive open air culture of the alga is possible.
2. The cells contain up to 14% of dry weight as β-carotene.
3. This β-carotene has a high sale price and a large and growing market.

In the market, *Dunaliella* products are available in dried form or extracted form. *Dunaliella* is harvested by different means of concentration and centrifugation, followed by spray drying. The final product in powder or pill form is considered to be of highest quality (124). The product in this form is mainly aimed at the health food market for direct human consumption. The price of the product is based on the β-carotene content; it can be as high as \$2000/kg of β-carotene. The extracted form of the product is obtained from *Dunaliella* wet biomass harvested by flocculation. The β-carotene is extracted into vegetable oil and the product thus obtained can be applied as a food dye and a provitamin additive for human consumption, for fish and poultry feed, or in the cosmetics industry as an additive to sunscreen products. The price of this product will be comparatively lower and depends on the β-carotene concentration. The size of the market for natural β-carotene from *Dunaliella* is difficult to estimate, but the trend and demand globally is increasing.

The disadvantages and threats to the algal process are:

1. The biomass in the growth ponds rarely exceeds $1g^{-1}$ and the algal cells have a density almost identical to that of the growth medium; this makes harvesting a complicated and expensive process.
2. Most of the world markets are supplied by synthetically manufactured β-carotene, and the production capacity of the synthetic product could be increased relatively easily.

The major competitive advantage of *Dunaliella* β-carotene over the synthetic product is that the algal product is natural. There are also some suggestions that the natural algal mixtures of the cis and trans isomers of β-carotene are more biologically active than the synthetically produced pure trans β-carotene (125–127). Furthermore, the increasing evidence for a preventive action of β-carotene against certain cancers, particularly epithelial cancers (128–130) should lead to an increase in the demand for β-carotene.

The algal β-carotene producer still needs to continue improving the algal culture process, so as to increase the β-carotene productivity by increasing growth rate and reliability, and by improving the β-carotene content of the cells. This requires improvements in the design of the production plants, a better understanding of *Dunaliella* physiology, and better algal strains, which can be obtained by either strain selection or by mutagenesis (131) and by genetic engineering.

19.4.4 *Haematococcus*

The production of the ketocarotenoid astaxanthin using the freshwater alga *Haematococcus pluvialis*, is also very attractive, but has fewer advantages than the *Dunaliella* β-carotene process. First, *H. pluvialis* is a freshwater alga, and thus open air culture is extremely difficult due to contamination by many other undesirable algal species. *Haematococcus pluvialis* culture will require closed culture systems such as tubular photo bioreactors. Although some pilot scale units have been tested (132,133), these systems require further refinements. However, recently completely closed photo bioreactors with artificial light, and a combination of closed photo bioreactors and open culture ponds are being used for *Haematococcus* cultivation (134,135). Second, unlike *Dunaliella, H. pluvialis* changes from a motile, flagellated cell to a nonmotile, thick walled aplanospore during the growth cycle (136,137); the astaxanthin is contained in the aplanospore. This means that the physical properties (density, settling rate, cell fragility) and nutrient requirements of the cells change during the culture process, and this alters the optimum conditions for growth and carotenoid accumulation during the growth cycle (138). Third, the content of astaxanthin in the aplanospores is about 1–2 % of dry weight and their thick wall requires physical breakage before the astaxanthin can either be extracted or be available to organisms consuming the alga (139).

On the other hand, the high value of astaxanthin, and the rapid growth of its main market, the aquaculture industry, means that the potential markets are very promising. The main competing product is synthetic astaxanthin, and once again, the algal process has the advantage of producing a natural product. The only natural alternatives are crustacean meal, which is in limited supply and only has low astaxanthin content, and the yeast *Phaffia rhodozyma,* which also has much lower astaxanthin content than the alga (140).

The development of a commercially viable algal astaxanthin process requires the development of an effective closed culture system and the selection (either from nature or by mutagenesis) of strains of *Haematococcus* with higher astaxanthin content and an ability to tolerate higher temperatures than the wild strains.

19.4.4.1 Astaxanthin as a Nutraceutical

Most of the microalgae products like *Spirulina* sp. and *Aphanizomenon flosaquae* are used as nutraceuticals. The medicinal benefits of nutraceuticals depend upon high quality manufacturing and effective control of production environments. Astaxanthin, a powerful bioactive antioxidant, has demonstrated efficacy in animal or human models against Alzheimer's and Parkinson's diseases, and macular degeneration (141), a cause of blindness in the U.S. Astaxanthin also ameliorates the effects of LDL, protects against cancer, and repairs cell damage caused by lack of oxygen, hence a potent nutraceutical.

19.4.4.2 Astaxanthin as Antioxidant

Astaxanthin has been shown to be a powerful quencher of singlet oxygen activity in *in vitro* studies (142,143). Astaxanthin has stronger antioxidant activity (10 times higher than β-caroten e) and is more than 500 times as effective as α-tocopherol. Astaxanthin has been purported to be the super vitamin E (134,143–146). The antioxidant property has been demonstrated in a number of biological membranes (147–149). Astaxanthin has preventive effects against aflatoxin B1 carcinogenicity (150). Kobayashi and Sakamoto (151) showed antioxidative activity of astaxanthin under both hydrophobic and hydrophilic conditions, while Kobayashi et al. (152) reported *in vivo* activity against super oxide anion radicals using a whole cell assay system. Astaxanthin also has strong activity as an inhibitor of lipid peroxidation mediated by active forms of oxygen (143,153). The strong antioxidative activities of astaxanthin suggest its potential as a photoprotectant against UV irradiation (154,155). Preparations containing Astaxanthin for the prevention of light induced aging of the skin have been developed by Suzuki et al. (156,157).

19.4.4.3 Astaxanthin for Health

Mammals lack the ability to synthesize astaxanthin, or to convert dietary astaxanthin into vitamin A. Unlike β-carotene, astaxanthin has no provitamin activity in these animals (158). Astaxanthin has been shown both *in vitro* and in a study with human subjects to be effective for the prevention of the oxidation of low density protein (159), suggesting that it can be used to prevent arteriosclerosis, coronary artery disease, and ischemic brain development. A number of astaxanthin health products are under study (159,160). Studies on rats have shown no toxicity of astaxanthin even in diets containing 400 ppm astaxanthin (161). Dietary administration of astaxanthin has proved to significantly inhibit carcinogenesis in the mouse urinary bladder, rat oral cavity, and rat colon (162–164). In addition, it is reported to induce xenobiotic-metabolizing enzymes in rat liver (150). Astaxanthin has been shown to enhance *in vitro* antibody production by mouse spleen cells stimulated with sheep red blood cells (145,165) and in human blood cells *in vitro* (158,166). Further it has not exhibited any mutagenicity in *in vitro* study at doses up to 14.4 mg/day for two weeks (159). Yamashita (167) has reported an antiinflammatory effect of astaxanthin when administered together with aspirin. An oral preparation has been developed by Alejung and Wadstroem (168) for the treatment of helicobacter infections of the mammalian gastrointestinal tract. There is strong evidence to suggest that astaxanthin modulates the humoral and nonhumoral immune system. It enhances the release of interleukin-1 and tumor necrosis factor α in mice, with greater effect than canthaxanthin or β-carotene, and has the greatest cytokinin inducing activity (169). Astaxanthin has a significant enhancing action on the production of immunoglobulin A, M, and G, and on T-helper cell antibody production, even when suboptimal amounts of antigen are present (158,166). Consequently, at the initial stage of a pathogen invasion, doses of a particular antigen may be suboptimal for eliciting an effective immune reaction; astaxanthin appears to enhance this response. Astaxanthin might also be useful in preventing age related macular degeneration that causes irreversible blindness.

19.4.4.4 Astaxanthin for Salmon and Trout Feeds

The predominant source of carotenoids for salmonids has been synthetic astaxanthin, which has been used for pigmentation for the last 20 years, since FDA approval in 1996. Natural sources of astaxanthin for commercially raised salmonids have been utilized, which include processed crustacean waste from krill, shrimp, crab, and crawfish. However, crustacean waste products contain high amounts of moisture, ash, and chitin, which limits the use in salmonids feeds. Another natural source, *Phaffia rhodozyma,* requires a large amount of feed for sufficient pigmentation, leading to higher ash contents. The efficiency of dietary astaxanthin using microalgae for flesh pigmentation of Atlantic salmon and rainbow trout has been demonstrated by Torrison et al. (170) and Storebakken (171). For salmon, astaxanthin is even considered a vitamin that is essential for the proper development and survival of juveniles (172). Choubert and Heinrich (173) showed that feeing rainbow trout with algae up to 6% of the diet had no major effect on growth or mortalities. Thus, the algae was concluded to be a safe and effective source of pigment. Astaxanthin has been used to enhance the immune response of fish and shrimp for maximum survival and growth. Natural microalgal astaxanthin has shown superior bioefficacy over the synthetic form. Full approval in Japan has been received for the use of astaxanthin as a pigment in feeds and foods; registration for approval is in progress for the United States, the European community, and Canada. An amount of 25–100 ppm of carotenoids in the final feed has been considered to give desired pigmentation in various salmonid species (173).

In poultry astaxanthin has shown to reduce the mortality of chicks by 50%, and to reduce *Vibrio* infections in eggs, thereby improving the nutritional value of eggs, especially among European consumers. However, the livestock feed market for astaxanthin, which is presently small, may grow to a size comparable to the market for synthetic pigments, which is estimated at $185 million. The largest market for astaxanthin, aquaculture, constituting 24% of total global fisheries production, is currently valued at $35 billion per annum and is expected to grow to $49 billion by 2010.

Limited studies have been carried out on dietary astaxanthin intake by humans. In a study reported by Miki et al. (159), astaxanthin was tested to protect low density lipoprotein from oxidation; 3.6–14.4 mg/day of an astaxanthin-containing drink was administered over a period of 2 weeks. Progressively slowed LDL oxidation with increasing doses of astaxanthin was observed and no ill effects were reported.

Sterlie et al. (174,175) reported that when 100 mg of synthetic astaxanthin in olive oil containing meal was given to male volunteers, maximum plasma concentration of 1.24 mg/L astaxanthin was observed in the first 6 hours postprandially. The relative concentration of total astaxanthin in HDL decreased compared to the other lipoprotein fractions in the 72 hour study.

Based on a study conducted with 40 healthy volunteers, Lignell (176) reported the effect of astaxanthin on mammalian muscle function. Volunteers received 1 capsule of 4 mg astaxanthin each morning in association with food. No significant difference was observed between the treatment and placebo group in any physical parameters measured.

The effect of dietary astaxanthin on the health of humans as studied by Aquasearch (177) on 33 volunteers consuming daily 3.85 mg (low dose) and 19.25 mg (high dose) for a period of 29 days indicated no ill effects or toxicity due to consumption of astaxanthin as analyzed by medical and clinical parameters.

Tso and Lam (146) suggested astaxanthin could be useful for the prevention and treatment of neuronal damage associated with age related macular degeneration, and it may also be effective at treating ischemic reperfusion injury, Alzheimer's disease, Parkinson's disease, spinal cord injuries, and other types of central nervous system injuries. Astaxanthin was found to easily cross the blood–brain barrier and did not form crystals in the eye.

19.4.5 *Porphyra*

Porphyra, commonly known as nori or purple laver, is the largest source of food from red seaweeds in the world. Nori has been a staple in the diet of most Asian countries, such as Japan, China, and the Republic of Korea.

19.4.5.1 Nutritive Value

Porphyra contains significant amount of proteins, vitamins, and minerals. The vitamin C content in *Porphyra* is greater than in raw oranges. It has vitamin A content comparable to spinach. The free and proteinaceous amino acids in nori are similar to that of vegetables. *Porphyra* contains high amount of arginine, which is generally present in animal protein. The characteristic taste of nori is caused by the large amounts of three amino acids: alanine, glutamic acid, and glycine. It also contains an abundant amount of taurine, which is essential for liver activity in preventing the occurrence of gallstone disease and to control blood cholesterol levels. Nori also contains large amounts of essential trace elements, such as zinc, needed for certain enzyme functions (20,178).

19.4.5.2 Cultivation Methods

Porphyra cultivation dates back as far as 300 years ago, when it was collected by hand from natural resources. Actual cultivation began during the 1960s when the life cycle of *Porphyra* was clearly understood. Detailed information on its life cycle and methods of cultivation can be found in Oohusa (179) and Mumford and Muira (180).

19.4.5.3 Food Uses

Porphyra is dried and processed into thin purplish black sheets. Nori is mainly used as a luxury food. It is commonly used in Japanese sushi, wrapped around a small handful of boiled rice topped with a slice of raw fish. Nori can be incorporated into soy sauce and can be used as a raw material for jam and wine. In China, it is mostly used in soups and for seasoning fried foods. In the Republic of Korea, it has similar uses to Japan, except that the popular snack with beer is Hoshinori, which is nori quickly fried in a pan with a little oil (181).

19.4.6 *Enteromorpha*

Enteromorpha is commonly known as aonori or green laver. Its life history involves an alternation of generations, both generations having the same appearance of long, tubular filaments.

19.4.6.1 Nutritive Value

Enteromorpha contains about 20 percent protein, little fat, low sodium, and high iron and calcium. Its vitamin B group content is generally higher than most vegetables, and while its vitamin A is high, it is only half of that found in spinach.

19.4.6.2 Cultivation Methods

Enteromorpha is cultivated in Japan and the Republic of Korea. It is also found in other parts of the world including Europe and North America. For cultivation, rope nets are seeded with spores by submerging them in areas where *Enteromorpha* is growing naturally, usually attached to rocks. Harvesting can be done 2–3 times during the growing period, either by hand picking from the nets or by machine (182). The quantities of *Enteromorpha* harvested in Japan and the Republic of Korea are 1400 and 1038 tons/yr respectively (www.surialink.com).

19.4.6.3 Food Uses

Enteromorpha is lightly roasted to improve the flavor and powdered. The powder can be used as a condiment in soups and foods. The algae can be crushed into small pieces and used as a garnish. More information on its life cycle and cultivation methods, along with descriptions and useful illustrations, can be found in Ohno (183).

19.4.7 *Eucheuma*

Eucheuma is a red algae belonging to the family *areschougiaceae*, mainly used as a source of carrageenan. The Philippines is the leading producer and exporter of *Eucheuma* raw material. It accounts for more than 70% of the world's production of carrageenan (184). In India, experimental field cultivation of *Eucheuma* has been successfully done by the Central Salt and Marine Chemicals Research Institute, Bhavnagar (185). The cultivation methods of *Eucheuma* and the uses of carrageenan are explained in the next section.

19.5 ALGAE BASED INDUSTRIAL PRODUCTS OF FOOD PROCESSING

Seaweeds have been commercially exploited for hydrocolloids since 1658, when the gelling properties of agar, extracted with hot water from red seaweed, were first discovered in Japan. Various red and brown seaweeds are used to produce three hydrocolloids, namely agar, alginate, and carrageenan (Table 19.10). A hydrocolloid is a noncrystalline substance with very large molecules which dissolves in water to give a thickened (viscous) solution. Industrial uses of seaweed extracts expanded rapidly after the Second World War, which were subsequently limited by the availability of raw materials. Research into seaweed life cycles has led to the development of cultivation industries that now supply a high proportion of the raw material for some hydrocolloids. Today, approximately 1 million tons of wet seaweed are harvested and extracted to produce these three hydrocolloids. Total hydrocolloid production is about 55,000 tons, with a value of $585 million (181).

19.5.1 Agar

The main seaweeds used for agar extraction are *Gelidium* and *Gracilaria*.

19.5.1.1 Cultivation Methods

Gelidium is a small slow growing plant; cultivation in ponds and tanks is possible but not economically feasible. One exception is a Canadian company in Vancouver, Marine Bioproducts International Corp., which claims to be growing consistent, high quality *Gelidium* from which high grade agar and agarose products are obtained (www.marbio.com).

Table 19.10
Sources of Seaweeds for Hydrocolloids

Hydrocolloids	Seaweeds
Agar	*Gelidium*
	Gracilaria
Alginate	*Ascophyllum*
	Durvillaea
	Ecklonia
	Lessonia
	Laminaria
	Macrocystis
Carrageenan	*Kappaphycus alvarezii*
	Eucheuma denticulatum
	Chondrus crispus
	Betaphycus gelatinum

Source: Reference 181

Gracilaria cultivation is widespread and several methods are used. It can be grown vegetatively in open waters on the bottom of bays, estuaries, or reef flats; on lines, ropes or nets; in ponds; or in tanks (186–188). The harvesting methods for *Gelidium* and *Gracilaria* are given in Table 19.11.

19.5.1.2 Extraction of Agar

Agar is used because of its ability to form gels, and the unique properties of these gels. Agar dissolves in boiling water and when cooled, forms a gel between 32° and 43°C, depending on the seaweed source of agar (189). The procedure for the extraction of agar from seaweeds is shown in Figure 19.1.

19.5.1.3 Food Applications

- About 90% of the agar produced is used for food applications. The U.S. Food and Drug Administration has classified agar as GRAS (Generally recognized as safe).
- Agar can be used as a stabilizer and thickener in pie fillings, icings, and meringues in the baked goods industry.
- Some agars, especially those extracted from *Gracilaria chilensis,* are used in confections with very high sugar content like fruit candies.
- Agar is a popular component of jellies in Asian countries and is used in gelled meat and fish products because of its high melting temperature and gel strength.
- Agar has been used to stabilize sherbets and ices in combination with other gums.
- Agar is used to improve texture of dairy products like cream cheese and yogurt and to clarify wines.

Further information on the uses of agar in food applications can be found in Armisen (190) and Armisen and Galatas (189,191).

19.5.2 Alginate

The main seaweeds used for alginate production are given in Table 19.10.

19.5.2.1 Cultivation Methods

None of the alginate producers are cultivated, since the costs of harvesting and transporting are expensive compared to alginate production. The harvesting methods used for alginophytes are given in Table 19.12.

Table 19.11
Harvesting methods for Agarophytes

Agarophytes	Harvesting Methods
Gelidium	A high percentage of harvest is by gathering of storm-cast seaweed by dragging a net.
Gracilaria	Fully grown Gracilaria resists water movement and eventually break off. These broken pieces drift and are collected by nets or picked up after they wash onto the shore.

Source: Reference 181.

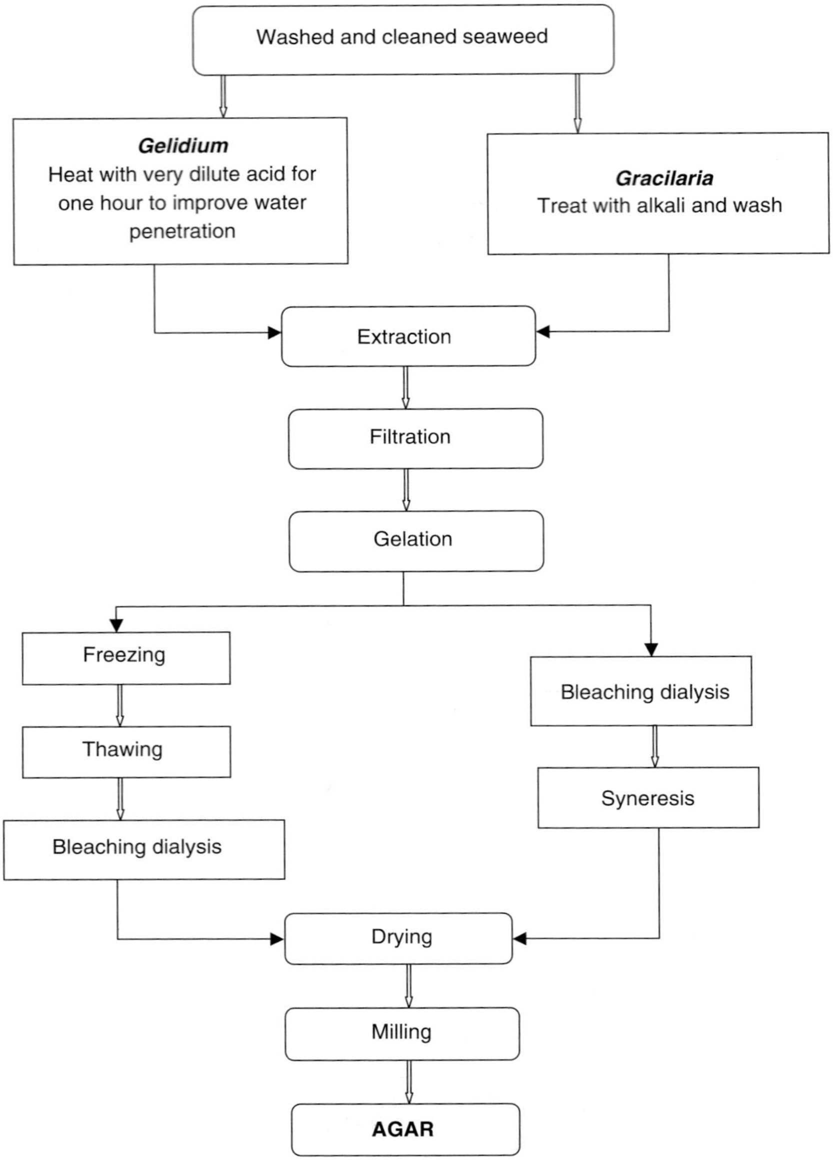

Figure 19.1 Flow chart for the production of agar (Source: Reference 181)

19.5.2.2 Extraction of Alginate

Alginate is present in the cell walls of brown seaweeds and is partly responsible for the flexibility of the seaweed. The methods used to extract alginate (192) from seaweeds are shown in Figure 19.2. The main applications of alginate are in thickening aqueous solutions and forming gels.

19.5.2.3 Food Applications

- Alginates are used to stabilize ice cream; addition of alginate reduces the formation of ice crystals during freezing and gives a smooth product.

Table 19.12
Harvesting methods for wild Alginophytes

Alginophytes	Harvesting Methods
Ascophyllum	Mechanically harvested using a flat-bottom vessel, which has rotating cutters to cut the seaweed. These are then collected with the help of nets.
Durvillaea	Collected only from storm-cast material.
Ecklonia	Collected from beach-cast seaweed.
Laminaria	Harvested using a mechanical device, the scoubidou, mounted on a boat

Source: Reference 181

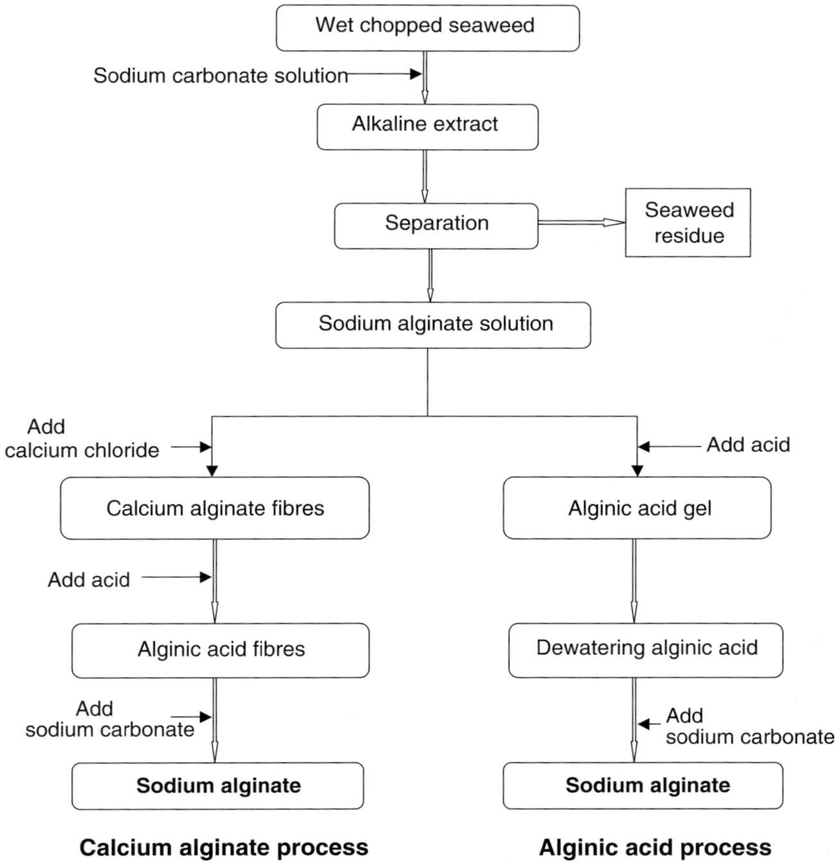

Figure 19.2 Flow chart for the production of sodium alginate (Source: Reference 181)

- Propylene glycol alginate at low concentrations provides stable, long lasting beer foam.
- Even though a large number of agents are used for the removal of off colors and to clarify wine, addition of sodium alginate is found to be effective in more difficult cases.
- Alginate–calcium mixtures are used to make edible dessert jellies; because they do not melt, alginate jellies have a different, firmer mouth feel when compared to gelatin jellies.

- Alginate gels are used in restructured or reformed food products such as restructured meat.
- Calcium alginate films and coatings have also been used to preserve frozen fish, which protects the fish from air.

19.5.2.4 Other Uses

- Calcium alginate is used to immobilize enzymes or cells.
- Good quality stable fibers produced from mixed salts of sodium and calcium alginate and processed into nonwoven fabric are used in wound dressings as they have wound healing and haemostatic properties.
- Alginic acid has also been used in some dietary foods, such as biscuits.
- Recently, oral controlled release systems involving alginate microspheres sometimes coated with chitosan to improve mechanical strength have been tested as a way of delivering various drugs.
- Crude alginate is used as a binder in salmon and other fish feeds.

19.5.3 Carrageenan

The main seaweeds used for carrageenan production are *Kappaphycus alvarezzi*, *Eucheuma denticulatum*, and *Betaphycus gelatinum.* About 120,000 dry tons/yr of *K. alvarezzi* and 30,000 dry tons/yr of *E. denticulatum* are harvested, mainly from the Philippines, Indonesia, and Tanzania.

19.5.3.1 Cultivation Methods

Kappaphycus and *Eucheuma* are cultivated by the same methods, the most popular being the fixed, off bottom line method and the floating raft method (193).

The fixed, off bottom line method is the simplest one. A suitable site is chosen and two wooden stakes are driven into the bottom about 5–10 m apart. A monofilament nylon line or a polypropylene rope 20–30 cm above the sea bottom is stretched between the stakes. Seaweed pieces (50–100 g) are tied to the line. If the site is suitable and the farming is maintained regularly, seaweed reaches 10 times its original size in 6–8 weeks, when it can be harvested.

The second method of cultivation is the floating raft method. It is suitable for protected areas where water current is weak or where water is too deep for fixed bottom lines. A floating construction is used to suspend the seaweed about 50 cm below the surface. A timber frame made of bamboo or mango timber is used with 3 mm polypropylene ropes stretched parallel in one direction between the timbers. The seedlings are tied to ropes and the raft is anchored to the bottom. Regular maintenance is required. At the time of harvest, the entire raft is removed and used as a drying rack. The harvesting methods for other carrageenophytes are given in Table 19.13.

More information on *Kappaphycus* and *Eucheuma* cultivation with useful illustrations can be found in Foscarini and Prakash (194).

There are several carrageenans, differing in their chemical structure and properties. The carrageenans of commercial interest are called iota, kappa, and lambda (Table 19.14). Their uses are related to their ability to form thick solution or gels and they vary, as will be seen.

19.5.3.2 Extraction of Carrageenan

The procedure for the extraction of carrageenan is given in Figure 19.3.

Table 19.13
Harvesting methods for other carrageenophytes

Carrageenophytes	Harvesting Methods
Chondrus crispus	Collected by hand rakes using small boats.
Betaphycus gelatinum	Harvested during may to September, where it is pulled by hand during low tide.
Gigartina canaliculata	During low tide, seaweeds are pulled from rocks by hand.
Sarcothalia crispata	Hooks are used to drag the seaweeds off the rocks.

Source: Reference 181

Table 19.14
Properties of different Carrageenans

Carrageenan	Properties
Iota	Elastic gels formed with calcium salts.
	Clear gel with no bleeding of liquid.
	Gels freeze/thaw stable.
Kappa	Strong, rigid gel, formed with potassium salts.
	Brittle gel forms with calcium salts.
	Slightly opaque gel becomes clear with sugar addition.
Lambda	No gel formation, forms high viscosity solutions.

Source: Reference 181

19.5.3.3 Food Applications

- Kappa carrageenan at 0.01–0.04% added to cottage cheese and to ice cream prevents separation of whey.
- Lambda carrageenan or a mixture of carrageenans helps improve liquid coffee whiteners by preventing separation of fat at 0.2–0.3% levels.
- Combining various carrageenans with locust bean gum, konjac flour, and starch provides a variety of melting and nonmelting gels and gel textures which can be used as suitable substitutes for gelatin.
- Mixtures of kappa and iota carrageenan are used in low or noncalorie fruit jellies.
- A combination of carrageenan and xanthan gum can be used in low oil or no oil salad dressings to help suspend herbs.
- Adding phosphates and carrageenan to low fat products helps prevent the loss of tenderness, juiciness, and flavor of such products.
- Seaweed flour (kappa carrageenan) and locust bean gum are used with meat pieces in canned pet foods.
- Apart from these applications, carrageenan gels provide a good medium for immobilizing enzymes and whole cells.

More information on applications of carrageenan can be found in Stanley (195), Therkelsen, (196) and Nussinovitch (197).

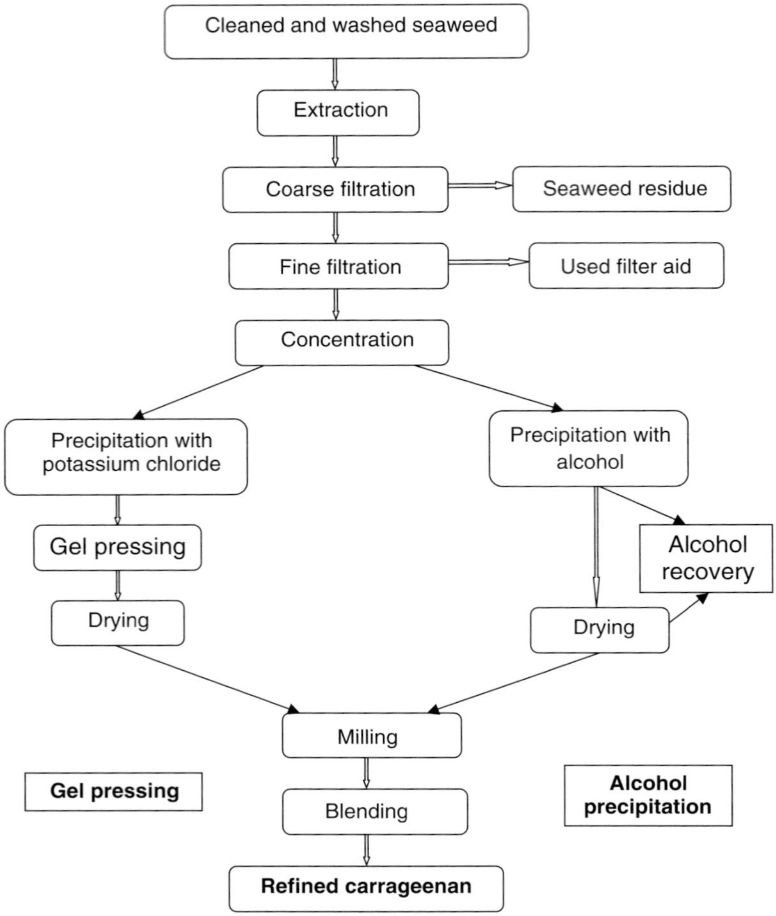

Figure 19.3 Flow chart for the production of refined carrageenan (Source: Reference 181)

19.6 SUMMARY AND FUTURE PROSPECTS

Forms of algae have been used from time immemorial as an alternative source of food. Several civilizations have adopted this efficiently, and developed recipes, which helped human beings in meeting the nutritional needs for their wellbeing. Algae were often used as a famine food by virtue of its availability under drought conditions. Persistent consumption from generation to generation established the safety of such algal food materials. The need for ensuring nutritional security has lead to the evaluation, from the point of view of essential micro- and macronutrients, of algal resources, both from fresh water and marine forms. The uniqueness of algae as a source of protein, vitamins, minerals, and essential fatty acids and health promoting substances, as well as the advancement of food science and nutrition, have thrown light on the versatility of algal forms in meeting nutritional and nutraceutical requirements of the human beings. The food chain in the ecosystem involves utilization of algal forms, which are primary producers, by the depending animal forms for their food needs. This has been extrapolated to develop a number of feed formulations for fish, crustacean, and poultry farming, which is now a flourishing industry providing fortified feed ingredients mainly in the fish and meat farming sector. From the 1970s it was recognized that the large scale cultivation of micro and macro algal forms could lead to controlled production systems, with high quality and safety, that would provide standard food material in abundant quantity for human needs. This lead to the development of large scale production ponds for the microalgae *Chlorella*, *Spirulina*, *Dunaliella,* and *Haematococcus,* leading to the annual production of over 20,000 tons of biomass which have been used as food supplements. While *Chlorella* has been mainly used in health foods, *Spirulina* is used both as a health food and as a source of nutraceuticals such as β-carotene, vitamins, and minerals in therapeutic formulations. The health promoting qualities of algae are also being exploited for developing newer beneficial formulations for targeted applications. Microalgae such as *Dunaliella* and *Haematococcus* are also cultivated as a rich source of carotenoids, especially β-carotene and astaxanthin. Undoubtedly these algal forms are the richest source of these important dietary carotenoids, which are essential for human physiology and disease prevention. Macroalgal forms such as *Porphyra*, *Enteromorpha,* and *Eucheuma* are directly used as food, and their potential applications have not been fully exploited. Their cultivation is restricted to certain regions of the world, mainly in East Asian countries. However, the use of macroalgae for gums and thickening agents is well recognized and utilized by the food processing industries.

The newer potentials of algal forms are in the area of health foods, designer foods, and functional foods. There is need for continued effort to unravel the health benefits of algal constituents and to develop standard specifications for the products derived from them. including scientific validation of their efficacy. Designer foods for specific needs such as pediatric or geriatric formulations, including specialty foods for women and children, may be looked at from the point of view of nutritional security, especially in the developing world, where there is a demanding nutritional need. Foods for various human disorders and for the prevention of disease such as cancer, atherosclerosis, hypertension, and immunological disorders are being addressed through algal supplementation. It is necessary to ensure the quality and quantities of such identified algae and algal products. Therefore, there is need for continued research from the biochemical, nutritional, nutraceutical, pharmaceutical, and medical oriented angle. Also there is need for development of large scale production systems with the involvement of engineers for both biomass production and downstream processing. Despite the potential of algae, adequate attention has not been given for their utility and their exploitation for food uses. Future years will certainly witness tremendous advancements in both algal cultivation and utilization, especially for the health food sector.

REFERENCES

1. Kalpan, D., A.E. Richmond, Z. Dubinsky, S. Aaronson. Algal nutrition. In: *Handbook of Microalgal Mass Culture*, Richmond, A., ed., Boca Raton, FL: CRC Press, 1986, pp 147–198.
2. Lee, R.E. *Phycology*, 2nd ed. Cambridge: Cambridge University Press, 1989, p 170.
3. Pohl, P. *Marine Biology: An Ecological Approach*. New York: Harper and Row, 1982, p 446.
4. O'colla, P.S. Mucilages. In: *Physiology and Biochemistry of Algae*, Lewin, R.A., ed., New York: Academic Press, 1962, pp 337–356.
5. Dangeard, P. Sru une algue bleue alimentaire pour I'homme *Arthospira platensis* (Nordst.) Gomont. *Actes Soc Linn Boreaux Extr Proces-verbaux* 91:39–41, 1940.
6. Leonard, J., P. Compere. *Spirulina platensis* (Gom.) Geitl., algue bleue de grande valeur alimentaire par sa richesse en proteins. *Bull. Jard. Bot. Nat. Belg*. 37:1–23, 1967.
7. Delpeuch, F., A. Jospeh, C. Cavelier. Conmsommation alimentaire et apport nutritionnel des algues bleues (*Oscillatoria platensis*) chez quelques populations du kanem (Tchad). *Ann. Nutr. Aliment*. 29:497–516, 1976.
8. Farrar, W.V. Tecuitlatl: a glimpse of Aztec food technology. *Nature* 211:341–342, 1966.

9. Burlew, J.S. *Algae Culture: from Laboratory to Pilot Plant*. Publication 600, Washington, DC: Carnegie Institution of Washington, 1953, pp 335–357.
10. Richmond, A. *CRC Handbook of Microalgal Mass Culture*. Boca Raton, FL: CRC Press, 1986, p 528.
11. Borowitzka, M.A., L.J. Borowitzka. *Micro-algal Biotechnology*. Cambridge: Cambridge University Press, 1988, p 466.
12. Cresswell, R.C., T.A.V. Rees, N. Shah. *Algal and Cyanobacterial Biotechnology*. Harlow, Essex: Longman Scientific and Technical Publishers, 1989, pp 341–362.
13. Venkataramam, L.V., E.W. Becker. In: *Biotechnology and Utilization of Algae- The Indian Experience,* New Delhi: Department of Science and Technology, 1985, pp 121–174.
14. Chapman, V.J., D.J. Chapman. *Seaweeds and their uses*, 3rd ed. London: Chapman & Hall, 1980, p 334.
15. McCandless, E.L. Polysaccharides of the seaweeds. In: *The Biology of Seaweeds,* Lobban, C.S., M.J. Wynne, eds., Oxford: Blackwell Scientific, 1981, pp 559–588.
16. Glicksman, M. Utilization of seaweed hydrocolloids in the food industry. *Hydrobiologia* 151/152:31–47, 1987.
17. Renn, D.W. Seaweeds and biotechnology: inseparable companions. *Hydrobiologia* 204: 7–13, 1990.
18. Levring, T., H.A. Hoppe, O.J. Schmid. Marine Algae. In: *A Survey of Research and Utilization.* Hamburg: Cram, de Gruyter and Co., 1969, pp 42.
19. Tseng, C.K. Commercial cultivation. In: *The Biology of Seaweeds*, Lobban, C.S., M.J. Wynne, eds., Oxford: Blackwell Scientific, 1981, pp 680–725.
20. Nisizawa, K., H. Noda, R. Kikuchi, T. Watanabe. The main seaweed foods of Japan. *Hydrobiologia* 151/152:5–29, 1987.
21. Tseng, C.K., X.G. Fei. Macroalgal commercialization in the orient. *Hydrobiologia* 151/152:167–172, 1987.
22. Waaland, J.R. Commercial utilization. In: *The Biology of Seaweeds.* Lobban, C.S., M.J. Wynne, eds., Oxford: Blackwell Scientific, 1981, pp 726–741.
23. Mahadevaswamy, M. Problems of contamination in outdoor cultures in the alga *Scenedesmus acutus* for utilization of a single cell protein. M. Sc Thesis (by Research), University of Mysore, Mysore, India, 1980.
24. Aaronson, S., Z. Dubinsky. Mass production and microalgae. *Experentia* 38:36–40, 1982.
25. Litchfield, J.H. Single-cell proteins. *Science* 219:740–746, 1983.
26. Aaronson, S., T. Berner, Z. Dubinsky. Microalgae as a source of chemicals and natural products. In: *Algae Biomass, Production and Use,* Shelef, G., C.J. Soeder, eds., Amsterdam: Elseiver/North Holland, 1980, pp 575–601.
27. Matelese, R.I. The physiology of single-cell protein (SCP) production. In: *Microbial Technology: Current state, future prospects, 29th Symp. Soc. Gen. Microbiol.*, Cambridge: Cambridge Univ. Press, 1979, pp 29–52.
28. Cifferi, O. *Spirulina*, the edible microorganism. *Microbial. Rev.* 47:551–578, 1983.
29. Henrickson, R. *Earth Food Spirulina*. Laguna Beach, California: Ronore Enterprises Inc, 1989, pp 23–42.
30. Clement, G. Wound treating medicaments containing algae, Fr.M.5279 (Int.cl.A61), Institut Francais du Petrol, 11 Sept., 1967.
31. Fidanza, F., A. Maurizi. Caratteristiche nutrizionali *in vitro* delle biomass di *Spirulina*. In: *Prospettive della coltura di Spirulina in Italia,* Materassi, R., ed., Rome: Firenze-Accademia dei Georgofili, 1980, pp 179–193.
32. Salcedo-Olavarrieta, N., M.M. Ortega, M.E. Marin-Garcia, C. Zavala-Moreno. Study of edible algae of the valley of Mexico. III. Comparative study of amino acids. *Rev. Lat.-Am. Microbiol.* 20:215–217, 1978.
33. Santillan, C. Mass production of *Spirulina. Experentia* 38:40–43, 1982.
34. Paoletti, C., M. Vencenzini, F. Bocci, R. Materrasi. Composizione biochimica generale delle biomasse di *Spirulina platensis* e *S. maxima.* In: *Prospettive della Coltura di Spirulina in Italia,* Materasssi, R., ed., Rome: Firenze-Accademia dei Georgofili, 1980, pp 111–125.

35. Jaya, V.T., M.L. Scarnio, M.A. Spandoni. Caratteristiche nutrizionali *in vivo* di *Spirulina maxima.* In: *Prospettive della coltura di Spirulina in Italia,* Materassi, R., ed., Rome: Firenze-Accademia dei Georgofili, 1980, pp 195–203.
36. Ito, K., K. Hori. Seaweed: Chemical composition and potential food uses. *Food Rev. Int.* 5(1):101–144, 1989.
37. Burtin, P. Nutritional value of seaweeds. *Electron. J. Environ. Agric. Food Chem.* 2(4), 2003.
38. Watanabe, F., S. Takenaka, H. Katsura, S.A.M. Zakir Hussain Masumimder, K. Abe, Y. Tamura, Y. Nakano. Dried green and purple lavers (Nori) contain substantial amounts of biologically active Vitamin B12 but less of dietary iodine relative to other edible seaweeds. *J. Agric. Food. Che.* 47:2341–2343, 1999.
39. Qasim, R., S. Barkati. Ascorbic acid and dehydro ascorbic acid contents of marine algal species from Karachi. *Pakisthan J. Sci. Ind. Res.* 28(2):129–133, 1985.
40. Solibami, V.J., S.Y. Kamat. Distribution of tocopherol (Vitamin E) in marine algae from Goa, west coast of India. *Ind. J. Marine Sci.* 14:228–229, 1985.
41. H.A. Spoeher, H.W. Milner. The chemical composition of *Chlorella*: effect of environment conditions. *Plant Physiol.* 24:120–149, 1949.
42. Iwamoto, H., G. Yonekawa, T. Asai. Fat synthesis in unicellular algae, I: culture conditions for fat accumulation in *Chlorella cells. Bull. Agric. Chem. Soc.* 19:240–246, 1955.
43. Borowitzka, M.A. Fats, oils and hydrocarbons. In: *Microalgal Biotechnology,* Borowitzka, M.A., L.J. Borowitzka, eds., Cambridge: Cambridge University Press, 1988, pp 257–287.
44. Parker, P.L., C. Van Baalen, L. Maurer. Fatty acids in eleven species of blue green algae: geochemical significance. *Science* 155:707–708, 1976.
45. Holton, R.W., H.H. Blecker. Fatty acids in blue-green algae. In: *Properties and products of algae.* J.E. Zaick, ed., New York: Plenum Press, 1972, pp 117–127.
46. Kenyon, C.N., R. Rippka, R.Y. Stainer. Fatty acid composition and physiological properties of some filamentous blue green algae. *Archiv. fur Mikrobiologie* 83:216–236, 1972.
47. Nichols, B.W. Lipid composition and metabolism. In: *The Biology of Blue Green Algae.* Carr, N.G., B.A. Whitton, eds., Oxford: Blackwell, 1973, pp 144–161.
48. Zepke, H.D., E. Heinz, A. Radunz, M. Linscheid, R. Pesch. Combination and positional distribution of fatty acids in lipids from blue green algae. *Arch. Microbiol.* 119:157–162, 1978.
49. Wood, B.J.B. Fatty acids and saponifiable lipids. In: *Algal Physiology and Biochemistry,* W.D.P. Stewart, ed., Oxford: Oxford University Press, 1974, pp 236–265.
50. Klyachko-Gurvich, G.L. Changes in the content and composition of triglyceride fatty acid during restoration of *Chlorella pyrenoidosa* cells after nitrogen starvation. *Soviet Plant Physiol.* 21:611–618, 1974.
51. Tornabene, T.G., G. Holzer, S. Lien, N. Burris. Lipid composition of the nitrogen starved green alga *Neochloris oleoabundans. Enzyme Microbiol. Technol.* 5:435–440, 1983.
52. Werner, D. Silicate metabolism. In: *Biology of Diatom*, Werner, O., ed., Berkeley: University of California Press, 1977, pp 110–131.
53. Yurieva, M.I., A.A. Temnykh, L.M. Gostrenko, V.N. Akulin. Fatty acid composition of alga *Porphyridium cruentum. Biologya Morya* 6:45–48, 1984.
54. Cheucas, L., J.P. Riley. Component fatty acids of the total lipids of some marine phytoplankton. *J. Mar. Biol. Asso. UK* 49:97–116, 1969.
55. Ben-Amotz, A., T.G. Tornabene, W.H. Thomas. Chemical profiles of selected species of microalgae with emphasis on lipids. *J. Phycol.* 21:72–81, 1985.
56. Harrington, G.B., D.H. Beach, J.E. Dunham, G.G. Holz. The polyunsaturated fatty acids of marine dinoflagellates. *J. Protozool.* 17:13–219, 1970.
57. Paoletti, C., P. Pushparaj, G. Florenzano, P. Capella, G. Lercker. Unsaponifiable matter of green and blue green algal lipids as a factor of biochemical differentiation of their biomasses, I: total unsaponifiable and hydrocarbon fractions. *Lipids* 11:258–265, 1976.
58. Paoletti, C., P. Pushparaj, G. Florenzano, P. Capella, G. Lercker. Unsaponifiable matter of green and blue green algal lipids as a factor of biochemical differentiation of their biomasses, II: terpenic alcohol and sterol fractions. *Lipids* 11:266–271, 1976.

59. Sada, E., S. Katoh, A. Kheirolomoon, H. Yokoi. Effects of light intensity on the growth rate of the red alga *Porphyridium cruentum*. *J. Ferment. Bioeng*. 67:135–137, 1989.
60. Nichols, B.W., R.S. Appleby. The distribution and biosynthesis of arachidonic acid in algae. *Phytochemistry* 8:1907, 1969.
61. Vazhappilly, R., F. Chen. Eicosapentanoic acid and docosahexaenoic acid production potential of microalgae and their heterotrophic growth. *JAOCS* 75:393–397, 1998.
62. Pohl, P., H. Wagner. Control of fatty acid and lipid biosynthesis in *Euglena gracilis* by ammonia, light and DCMU. *Z. Naturforsch. Teil. B*. 72:53–61, 1972.
63. Holton, R.W., H.H. Blecker, T.S. Stevens. Fatty acids in blue-green algae: possible relation to phylogenetic position. *Science* 160:545, 1968.
64. Beach, D.H., G.W. Harrington, G.G. Holz, Jr. The polyunsaturated fatty acids of marine and fresh water cryptomonads. *J. Protzools*. 17:501, 1970.
65. Tonon, T., D. Harvey, T.R. Larson, I.A. Graham. Long chain polyunsaturated fatty acid production and partitioning to triacylglycerols in four microalgae. *Phytochemistry* 61:15–24, 2002.
66. Hudson, B.J.F., I.G. Kari. The lipids of alga *Spirulina*. *J. Sci. Food Agric*. 25:759–763, 1974.
67. Fleurence, J., G. Gutbier, S. Mabeau, C. Leray. Fatty acids from 11 marine macro algae of the French Brittany coast. *J. Appl. Phycol*. 6:527–532, 1994.
68. Borowitzka, M.A. Vitamins and fine chemicals. In: *Microalgal Biotechnology*. Borowitzka, M.A., L.J. Borowitzka, eds., Cambridge: Cambridge University Press, 1988, pp 153–196.
69. Borowitzka M.A. Microalgae as sources of essential fatty acids. *Aust. J. Biotech*. 1:58–62, 1988.
70. Moore, R.E., M.L. Patterson, W.W. Carmichael. New pharmaceuticals from cultured blue–green algae. In: *Biomedical importance of marine organisms,* Fautin, D.G., ed., San Francisco: California Academy of Sciences, 1988, pp 143–150.
71. Borowitzka, M.A. Products from algae. In: *Algal Biotechnology in the Asia-Pacific Region,* Moi, P.S., L.Y. Kun, M.A. Borowitzka, B.A. Whitton, eds., Kuala Lumpur: Institute of Advanced Studies, University of Malaya, 1994, pp 5–15.
72. Okuzumi, J., T. Tkahashi, T. Yamane, Y. Kitao, M. Inagake, K. Ohya, H. Nishino, Y. Tanaka. Inhibitory effects of fucoxanthin, a natural carotenoid, pm N-ethyl-N^1-nitro-N-nitrosoguanidine-induced mouse duodenal carcinogenesis. *Cancer Lett.* 68:159–168, 1993.
73. Yan, X., Y. Chuda, M. Sizmki, T. Nagata. Fucoxanthin as a major antioxidant in *Hijikia fusiformis* a common edible seaweed. *Biosci. Biotechnol. Biochem*. 63:605–607, 1999.
74. Cohen, Z., A. Vanshak, S. Boussiba, A. Richmond. The effect of temperature and cell concentration on the fatty acid composition of outdoor cultures of *Porphyridium cruentum*. In: *Algal Biotechnology,* T. Stadler, J. Mollion, M.C. Verdus, Y. Karamanos, H. Morvan, D. Christiaen, eds., London: Elsevier Applied Science, 1988, pp 421–429.
75. Cohen, Z., S. Cohen. Preparation of eicosopentaenoic acid (EPA) concentrate from *Porphyridium cruentum*. *JAOCS* 68:16–19, 1991.
76. Veloso, V., A. Reis, L. Gouvi, H.L. Fermamdes, J.A. Empis, J.M. Novais. Lipid production by *Phaeodactylum tricornutum*. *Bioresource Technol*. 38:115–119, 1991.
77. Volkman, J.K., S.W. Jeffrey, P.D. Nichols, G.I. Rogers, C.D. Garland. Fatty acid and lipid composition of 10 species of micro algae used in mariculture. *J. Exp. Mar. Biol. Ecol.* 128:219–240, 1989.
78. Sukenik, A., Y. Carmeli. Regulation of eicosapentaenoic acid content by light in the marine eustigmatophyte *Nannochloropsis* sp. In: *Current Topics in Marine Biotechnology*. Miyachi, S., I. Karube, Y. Ishida, eds., Tokyo: The Japanese Society for Marine Biotechnology, 1990, pp 127–130.
79. Yongmanitchai, W., O.P. Ward. Growth of and Omega-3 fatty acid production by *Phaeodactylum tricornutum* under different culture conditions. *Appl. Env. Microbiol.* 57:419–425, 1991.
80. Koletzko, B., E. Schnidt, H.J. Brener, M. Haug, G. Harzer. Effects of dietary long-chain polyunsaturated fatty acids on the essential fatty acid status of premature infants. *Eur. J. Pediatr*. 148:669–675, 1989.

81. Simopoulos, A.P. Summary of the NACO advanced research workshop on dietary W3 and W6 fatty acids: biological effects and nutritional essentiality. *J. Nutr.* 119:521–528, 1989.
82. Gibson, R.A., M. Makrides. Long-chain polyunsaturated fatty acids in breast milk: are they essential? *Adv. Exp. Med. Biol.* 501:375–383, 2001.
83. Drevon, C.A., I. Baksaas, H.E. Krokan. *Omega-3 Fatty Acids: Metabolism and Biological Effects*. Basel, Switzerland: Birkhauser Verlag AG, 1993.
84. Simopoulos, A.P. Essential fatty acids in health and chronic disease. *Am. J. Clin. Nutr.* 70:560s–569s, 1999.
85. Brown, M.R., S.W. Jeffrey, C.D. Garland. Nutritional aspects of microalgae used in mariculture: a literature review. *CSIRO Marine Laboratories Report*, 205:1–43, 1989.
86. Laing, I., C.G. Verdugo. Nutritional value of spray dried *Tetraselmis succica* for juvenile bivalues. *Aquaculture* 92:207–218, 1991.
87. Herrera, A., S. Boussiva, V. Napoleone, A. Holberg. Recovery of C- phycocyanin from the cyanobacterium *Spirulina maxima*. *J. Appl. Phycol.* 1:325–331, 1989.
88. Kronick, M.N., P.D. Grossman. Immunoassay techniques with fluorescent phycobiliprotein conjugates. *Clin. Chem.* 29:1582–1586, 1983.
89. Wilson, M.R., S. Crowley, G.A. Odgers, L. Show. Immunofluorescent labeling using covalent linked anti phycoerythrin antibodies and phycoerythrin polymers. *Cytometry* 12:373, 1991.
90. Dainippon Ink and Chemicals Inc. Cosmetics containing water soluble phycocyanin. Japanese Patent 79–138755, 1981.
91. Vadiraja, B.B., K.M. Madyastha. C-Phycocyanin: a potent peroxyl radical scavenger *in vivo* and *in vitro*. *Biochem. Biophys. Res. Comm.* 275:20–25, 2000.
92. Reddy, C. M., B.B. Vadiraja, G. Kiranmai, M.N. Reddy, P. Reddanna, K.M. Madyastha. Selective inhibition of cyclooxygenase-2 by C-phycocyanin, a biliprotein from *Spirulina platensis*. *Biochem. Biophys. Res. Comm.* 277:599–603, 2000.
93. Barchi, J.J., R.E. Moore, G.M.L. Patterson. Antiphycin and 20, 21-didehydroacutiphycin, new antieo plastic agents from the cyanophyte *Oscillatoria acutissima*. *J. Am. Chem. Soc.* 106:8193–8197, 1984.
94. Patterson, G.M.L., C.L. Boldwin, C.M. Bohis, F.R. Caplan, H. Karuso, L.K. Larsen, I.A. Levina, R.E. Moore, C.S. Nelson, K.D. Tschappat, G.D. Tuang, E. Furusawa, S. Furusawa, T.R. Nortea, R.B. Raybourne. Antineoplastic activity of cultured blue-green algae (cyanophyta). *J. Phycol.* 27:530–536, 1991.
95. Reichelt, J.L., M.A. Borowitzka. Antibiotics from algae: results of a large scale-screening programme. *Hydrobiologia* 116/117:158–168, 1984.
96. Kellam, S.J., J.M. Walker. Antibacterial activity from marine micro algae. *Brit. Phycol. J.* 24:191–194, 1989.
97. Miura, Y., T. Matsunga. Antibiotics production from marine microalgae. In: *Current Topics in Marine Biotechnology,* Miyachi, S., I. Karube, Y. Ishida, eds., Tokyo: The Japanese Society for Marine Biotechnology, 1990, pp 189–190.
98. Ishibashi, M., R.E. Moore, G.M.L. Patterson, C. Xu, J. Clardy. Scytophycins, cytotoxic and antimycotic agents from the cyanophyte *Scytonema pseudohofmanni*. *J. Org. Chem.* 51:5300–5306, 1986.
99. Decaire, G.Z., M.S. Decano, M.C.Z. Demule, D.R. Dehalperin. Antimycotic products from the cyanobacterium *Nostoc muscorum* against *Rhizoctonia solani*. *Phyton* 51:1–4, 1990.
100. Bonjouklain, R., T.A. Smitka, L.E. Doolin, R.M. Molloy, M. Denbono, S.A. Shaffer, R.E. Moore, J.B. Stewart, G.M.L. Patterson. Tjipanazoles, new antifungal agent from the blue–green alga *Tolypothrix tjipanasensis*. *Tetrahedron* 47:7739–7750, 1991.
101. Gustafson, K.R., J.H. Cardellina, R.W. Fuller, O.S. Weislow, R.F. Kise, K.M. Snade, K.M.L. Patterson, M.R. Boyd. AIDS-antiviral sulfolipids from cyanobacteria (blue-green algae). *J. Nat. Cancer Inst.* 81:254–1258, 1989.
102. Knubel, G., L.K. Larsen, R.E. Moore, I.A. Levine, G.M.L. Patterson. Cytotoxin, antiviral indolocarbazoles from a blue-green alga belonging to the Nostocaceae. *J. Antibiot.* 43:1236–1239, 1990.
103. Arasaki, S., T. Arasaki. *Vegetables from the sea.* Tokyo: Japan Publications Inc., 1983.

104. Murakami, S., T. Takemoto, Y. Shimizu. Studies on the effective principles of *Diagenea simplex* Aq., I: separation of the effective fraction by liquid chromatography. *J. Pharm. Soc. Japan* 73:1026–1028, 1953.
105. Guven, K.C., A. Bora, G. Sunan. Hordenine from the alga *Phyllophora nervosa*. *Phytochem* 9:1893, 1970.
106. Takemoto T., K. Diago, N. Takagi. Hypotensive constituents of marine algae. I. A new basic amino acid 'Laminine' and other basic constituents isolated from *Laminaria angustata*. *J. Pharm. Soc. Jpn.* 84:1176, 1964.
107. Kaneda, T., S. Abe. *Bull. Jpn. Soc. Sci. Fishes* 39:239, 1973.
108. Bernardi, G., G.F. Springer. Properties of highly purified Fucan. *J. Biol. Chem.* 237:75, 1962.
109. Sakagami, Y. Isolation of porphyrosin and verucoysin from edible seaweeds. In: *Biochemistry of marine algae and their application*, Tokyo: Japan Society of Science and Fish, 1983, pp 90–100.
110. Sieburth, J.M. Acrylic acid, an antibiotic principle in phaeocystis blooms in Antarctic waters. *Science* 132:676–677, 1960.
111. Kokate, C.K., A.P. Purohit, S.B. Gokhale. Marine drug. In: *Pharmacognosy.* Pune: Nirali Prakashan, 2000, pp 478–486.
112. Ohshima, T. Recovery and use of nutraceutical products from marine resources. *Food T. Technol*. 52(6):50–54, 1998.
113. Soeder, C.J. An historical outline of applied algology. In: *CRC Handbook of Microalgal Massculture,* Richmond, A., ed., Boca Raton, FL: CRC Press, 1986, pp 29–30.
114. Venkataraman, L.V., N. Bhagyalakshmi, G.A. Ravishankar. Commercial production of micro and macroalgae-problems and potentials. *Ind. J. Microbiol*. 35(1):1–19, 1995.
115. Babu, M. Evaluation of chemoprevention of oral cancer with *Spirulina*. *Nutr. Cancer* 24:197–202, 1995
116. Yamane, Y. The effect of *Spirulina* on nephrotoxicity in rats. Annual symposium of the Pharmaceutical Society of Japan, April 15, 1988.
117. Takai, Y. Effects of *Spirulina* on caecum content in rats. Japan: Chiba Hygiene College Bulletin, Vol 5, No.2, Feb 1987.
118. Takai, Y. Effect of water-soluble and water insoluble fractions of *Spirulina* over serum lipids and glucose resistance of rats. *J. Jap. Soc. Nutr. Food Sci.* 4:73–77, 1991.
119. Seshadri, C.V. Large-scale nutritional supplementation with *Spirulina* alga. All India project, MCRC, Madras, 1993.
120. Fica, V. Observations on the utilization of *Spirulina* as an adjuvant nutritive factor in treating diseases accompanied by a nutritional deficiency, Clinica 11 Medicala, Spitalui clinic Municipiului Bucuresti. *Med Interna* 36(3), 1984.
121. Borowitzka, L.J., M.A. Borowitzka. β-carotene (provitamin A) production with algae. In: *Biotechnology of Vitamins, Pigments and Growth Factors.* Vandamme, E.J., ed., London: Elseiver Applied Science, 1989, pp 15–26.
122. Curtain, C.C., S.M. West, L. Schipalius. Manufacture of β-carotene from the salt lake alga *Dunaliella salina*; the scientific and technical background. *Aust. J. Biotechnol.* 1:51–57, 1987.
123. Borowitzka, L.J., M.A. Borowitzka. Commercial production of β-carotene by *Dunaliella salina* in open ponds. *Bull. Mar. Sci.* 47:244–252, 1990.
124. Vonshak, A. Recent advances in microalgal biotechnology. *Biotech. Adv.* 8:709–727, 1990.
125. Ben-Amotz, A., A. Lers, M. Avron. Stereoisomers of β-carotene and phytoene in the alga *Dunaliella bardawil. Plant Physiol.* 86:1286–1291, 1988.
126. Mokady, S., M. Avron, A. Ben-Amortz. Accumulation in chick livers of 9-cis versus all-trans β-carotene. *J. Nutr.* 120:889–892, 1990.
127. Stich, H.F., B. Mathew, R. Sankaranarayanan, M.K. Nair. Remission of precancerous lesions in the oral cavity of tobacco chewers and maintenance of the protective effect of beta-carotene or vitamin-A. *Am. J. Clin. Nutr.* 53:S298–S304, 1991.
128. Lambert, L.A., W.H. Koch, W.G. Wamer, A. Kornhauser. Antitumor activity in skin of Skh and Sencar mice by 2 dietary β-carotene formulations. *Nutr. Cancer* 13:213–221, 1990.

129. Harris, R.W.C., T.J.A. Key, P.B. Silcocks, D. Bull, N.J. Wald. A case-control study of dietary carotene in men with lung cancer and in men with other epithelial cancers. *Nutr. Cancer* 15:63–68, 1991.
130. Stahelin, H.B., K.F. Gey, M. Eichhjolzer, E. Ludin. *β*-carotene and cancer prevention: the Basel study. *Am. J. Clin. Nutr.* 53:S265–S269, 1991.
131. Shaish, A., A. Ben-Amotz, M. Avron. Production and selection of high *β*-carotene mutants of *Dunaliella bardawil* (Chlorophyta). *J. Phycol.* 27:652-656, 1991.
132. Chaumont, D., C. Thepenier, C. Gudin, C. Junjas. Scaling up a tubular photoreactor for continuous culture of *Porphyridium cruentum* from laboratory to pilot plant (1981–1987). In: *Algal Biotechnology,* Stadler, T., J. Mollion, M.C. Verdus, Y. Karamonos, H. Morvan, D. Christian, eds., London: Elseiver Applied Science, 1988, pp 199–208.
133. Borowitzka, L.J., M.A. Borowitzka. Industrial production: methods and economics. In: *Algal and Cyanobacterial Biotechnology,* Cresswell, R.C., T.A.V. Rees, N. Shah, eds., Harlow, Essex: Longman Scientific and Technical Publishers, 1989, pp 294–316.
134. Lorenz, R.T., G.R. Cysewski. Commercial potential for *Haematococcus* microalgae as a natural source of astaxanthin. *TIBTECH* 18:160–167, 2000.
135. Fabregas, J., A. Otero, A, Maseda, A. Dominguez. Two-stage cultures for the production of Astaxanthin from *Haematococcus pluvialis*. *J. Biotech.* 89:65–71, 2000.
136. Borowitzka, M.A., J.M. Huisman, A. Osborn. Culture of the astaxanthin-producing green alga *Haematococcus pluvialis,* I: effects of nutrients on growth and cell type. *J. Appl. Phycol.* 3:295–304, 1991.
137. Kobayashi, M., T. Kakizono, S. Nagai. Astaxanthin production by a green alga. *Haematococcus pluvialis* accompanied with morphological changes in acetate media. *J. Ferment. Bioeng.* 71:335–339, 1991.
138. Sarada, R., T. Usha, G.A. Ravishankar. Influence of stress on astaxanthin production in *Haematococcus pluvialis* grown under different culture conditions. *Process. Biochem.* 37:623–627, 2002.
139. Sommer, T.R., W.T. Pott, N.M. Morrissy. Utilization of microalgal astaxanthin by Rainbow Trout (*Oncorhynchus mykiss*). *Aquaculture* 94:79–88, 1991.
140. An, G.H., D.B. Schuman, E.A. Johnson. Isolation of *Phaffia rhodozyma* mutants with increased astaxanthin content. *Appl. Envir. Microbiol.* 55:116–124, 1989.
141. Snodderly, D.M. Evidence for protection against age-related macular degeneration by carotenoids and antioxidant vitamins. *Am. J. Clin. Nutr.* 62:1448S–1461S, 1995.
142. DiMascio, P., T.P.A. Devasagayam, S. Kaiser, H. Sies. Carotenoids, tocopherols and thiols as biological singlet molecular oxygen quenchers. *Trans. Biochem. Soc.* 18:1054–1056, 1990.
143. Miki, W. Biological functions and activities of animal carotenoids. *Pure Appl. Chem.* 63:141–146, 1991.
144. Palozza, P., N.I. Krinsky. *Antioxidant effects of carotenoids* in vivo *and* in vitro*: an overview.* Packer, L., ed., San Diego: Academic Press, 1992.
145. Jyonouchi, H., L. Zhang, M. Gross, Y. Tomita. Immunomodulating actions of carotenoids: enhancement of *in vivo* and *in vitro* antibody production to T-dependent antigens. *Nutr. Cancer* 21:47–58, 1994.
146. Tso, M.O., T.T. Lam. Method of retarding and ameliorating central nervous system and eye damage, US patent # 5527533. Board of trustees of the University of Illinois, United States of America, 1996.
147. Kurashige, M., E. Okimasu, M. Inoue, K. Utsumi. Inhibition of oxidative injury of biological membranes by astaxanthin. *Physiol. Chem. Phys. Med. NMR* 22:27–38, 1990.
148. Oshima, S., F. Ojima, H. Sakamoto, Y. Ishiguro, J. Terao. Inhibitory effect of beta-carotene and astaxanthin on photosensitized oxidation of phospholipid bilayers. *J. Nutr. Sci. Vitaminol.* 39:607–615, 1993.
149. Nakagawa, K., S.D. Kang, D.K. Park, G.J. Handelman, T. Miyazawa. Inhibition by beta-carotene and astaxanthin of NADPH-dependent microsomal phospholipid peroxidation. *J. Nutr. Sci. Vitaminol.* 43:345–355, 1997.

150. Gradelet, S., P. Astorg, J. LeClerc, J. Chevalier, M.F. Vernevaut, M.H. Siess. Effects of canthaxanthin, astaxanthin, lycopene and lutein on liver xenobiotic-metabolizing enzymes in the rat. *Xenobiotica* 26:49–63, 1996.
151. Kobayashi, M., Y. Sakamoto. Singlet oxygen quenching ability of astaxanthin esters from the green alga *Haematococcus pluvialis*. *Biotech Lett.* 21:265–269, 1999.
152. Kobayashi, M., T. Kakizono, N. Nishio, S. Nagai, Y. Kurimura, Y. Tsuji. Antioxidant role of astaxanthin in the green alga *Haematococcus pluvialis*. *Appl. Microbiol. Biotechnol.* 48:351–356, 1997.
153. Ranby, B., J.F. Rabek. *Singlet Oxygen*. Chichester, England: Wiley Europe Ltd., 1978.
154. Savouré, N., G. Briand, M.C. Amory-Touz, A. Combre, M. Maudet, M. Nicol. Vitamin A status and metabolism of cutaneous polyamines in the hairless mouse after UV irradiation: action of β-carotene and astaxanthin. *Internat. J. Vit. Nutr. Res.* 65:79–86, 1995.
155. Black, H. Radical interception by carotenoids and effects on UV carcinogenesis. *Nutr. Cancer* 31:212–217, 1998.
156. Suzuki, K., H. Masaki, M. Takei. External preparation for skin. Japanese Patent #08073312, 1996 [in Japanese].
157. Suzuki, K., H. Masaki, M Takei. External preparation for skin. Japanese Patent #08073311, 1996 [in Japanese].
158. Jyonouchi, H., S. Sun, M. Gross. Effect of carotenoids on *in vitro* immunoglobulin production by human peripheral blood mononuclear cells: Astaxanthin, a carotenoid without vitamin A activity, enhances *in vitro* immunoglobulin production in response to a T-dependent stimulant and antigen. *Nutr. Cancer* 23:171–183, 1995.
159. Miki, W., K. Hosada, K. Kondo, H. Itakura. Astaxanthin – containing drink. Patent application number 10155459. Japanese Patent office. 16 June 1998.
160. Murillo, E. Efecto hipercolesterolémico de la cant Astaxanthinantina y la Astaxanthin en ratas. *Arch. Latinoamericanos Nutr*. 42:409–413, 1992.
161. Nishikawa, Y., Y. Minenaka, M. Ichimura. Physiological and biochemical effects of carotenoid (beta-carotene and astaxanthin) on rat. *Koshien Daigaku Kiyo* 25:19–25, 1997 [in Japanese].
162. Tanaka, T., Y. Morishita, M. Suzui, T. Kojima, A. Okumura, H. Mori. Chemoprevention of mouse urinary bladder carcinogenesis by the naturally occurring carotenoid astaxanthin. *Carcinogenesis* 15:15–19, 1994.
163. Tanaka, T., H. Makita, M. Ohnishi, H. Mori, K. Satoh, A. Hara. Chemoprevention of rat oral carcinogenesis by naturally occurring xanthophylls, astaxanthin and canthaxanthin. *Cancer Res.* 55:4059–4064, 1995.
164. Tanaka, T., T. Kawamori, M. Ohnishi, H. Makita, H. Mori, K. Satoh, A. Hara. Suppression of azomethane-induced rat colon carcinogenesis by dietary administration of naturally occurring xanthophylls astaxanthin and canthaxanthin during the postinitiation phase. *Carcinogenesis* 16:2957–2963, 1995.
165. Jyonouchi, H., R.J. Hill, Y. Tomita, R.A. Good. Studies of immunomodulating actions of carotenoids, I: effects of beta-carotene and astaxanthin on murine lymphocyte functions and cell surface marker expression in *in vitro* culture system. *Nutr. Cancer* 16:93–105, 1991.
166. Jyonouchi, H., S. Sun, Y. Tomita, M.D. Gross. Astaxanthin, a carotenoid without vitamin A activity, augments antibody responses in cultures including T-helper cell clones and suboptimal doses of antigen. *J. Nutr*. 124:2483–2492, 1995.
167. Yamashita, E., Anti-inflammatory agent, Japanese Patent #07300421 [in Japanese], 1995.
168. Alejung, P., T. Wadstroem., Oral preparation for treatment of *Helicobacter* sp. infections: comprises xanthophylls, especially astaxanthin esterified with a fatty acid and derived from the alga *Haematococcus* sp., World Patent #9837874, 1998.
169. Okai, Y., K. Higashi-Okai. Possible immunomodulating activities of carotenoids in *in vitro* cell culture experiments. *Int. J. Immunopharmacol*. 18:753–758, 1996.
170. Torrissen, O.J., W.H. Hardy, K.D. Shearer. Pigmentation of salmonids-carotenoid deposition and metabolism. *Rev. Aquat. Sci.* 1:209–227, 1989.
171. Storebakken, T. Krill as a potential feed source for salmonids. *Aquaculture* 70:193, 1988.

172. Christiansen, R., O. Lie, O.J. Torrissen. Growth and survival of Atlantic Salmon, *Salmo salar* L, fed different dietary levels of astaxanthin: first-feeding fry. *Aquaculture Nutr.* 1:189–198, 1995.
173. Choubert, G., O. Heinrich. Carotenoid pigments of green alga *Haematococcus pluvialis*: assay on rainbow trout *Oncorhynchus mykiss*, pigmentation in comparison with synthetic astaxanthin and canthaxanthin. *Aquaculture* 112:217–226, 1993.
174. Sterlie, M., B. Bjerkeng, S. Liaaen-Jensen. Blood appearance and distribution of astaxanthin E/Z isomers among plasma lipoproteins in humans administered a single meal with astaxanthin. *Abstracts of the Twelfth International Carotenoid Symposium, Cairns, Australia, Abstract 2A- 13*, 18–23 July, 1999, p 72.
175. Sterlie, M., B. Bjerkeng, S. Liaaen-Jensen. On bioavailability and deposition of bent Z – isomers of astaxanthin, *Proceedings of the First International Congress on Pigments in Food Technology*, Sevilla, Spain, 24–26 March, 1999, pp 157–161.
176. Lignell, A. Medicament for improvement of duration of muscle function or treatment of muscle disorders or diseases, AstaCarotene AB, Sweden, Patent Cooperation Treaty application # 9911251.1999.
177. Aquasearch, Inc., *Haematococcus pluvialis* and astaxanthin safety for human consumption, Technical Report TR., 3005, 001, 1999.
178. Fleurence, J. Seaweed proteins: biochemical, nutritional aspects and potential uses. *Trends Food Sci. Technol.* 10:25–28, 1999.
179. Oohusa, T. The cultivation of *Porphyra* "nori". In: *Monostroma and Enteromorpha "Aanori"*. Ohno, M., A.T. Critchley, eds., Yokosuka, Japan: Japan International Coorperation Agency, 1993, pp 57–73.
180. Mumford, T.F., A. Muira. *Porphyra* as food cultivation and economics. In: *Algae and Human Affairs,* Lembi, C.A., R. Waaland, eds., Cambridge: Cambridge University Press, 1988, pp 87–117.
181. McHugh, D.J. *A Guide to Seaweed Industry.* Rome: FAO Fisheries Technical Paper 441, 2003.
182. Ohno, M., D.B. Largo. The seaweed resources of Japan. In: *The Seaweed Resources of World,* Critchley, A.T., M. Ohno, D.B. Largo, R.D. Gillespie, eds., Yokosuka, Japan: Japan International Coorperation Agency, 1998, pp 1–14.
183. Ohno, M. Cultivation of the green alga. In: *Monostroma and Enteromorpha "Aanori",* Ohno, M., Critchley, eds., 1993, pp 51–56.
184. Trono, G.C. In: *Seaweed Culture in the Asia-Pacific Region,* Bangkok, Thailand: RAPA Publication, FAO Regional Office for Asia and the Pacific, 1987, pp 41.
185. Ghosh, P.K., K. Eswaran, O.P. Mairh. Experimental field cultivation of *Kappaphycus alvarezzi* (Doty) Doty ex. P. Silva at Mandapam region. *Seaweed Res. Util.* 24(1):67–72, 2002.
186. Buschmann, A.H., R. Westermeier, C.A. Retamales. Cultivation of *Gracilaria* on the seabottom in southern Chile: a review. *J. Appl. Phycol.* 7:291–301, 1995.
187. Buschman, A.H., M. Troell, N. Kautsky. Integrated algal farming: a review. *Cah. Biol. Mar.* 42:83–90, 2001.
188. Friedlander, M., I. Levy. Culivation of *Gracilaria* in outdoor tanks and ponds. *J. Appl. Phycol.* 7:315–324, 1995.
189. Armisen, R., F. Galatas. Agar. In: *Handbook of Hydrocolloids,* Phillips, G., P. Williams, eds., Boca Raton, FL: CRC press, 2000, pp 21–40.
190. Armisen, R. Agar. In: *Thickening and Gelling Agents for Food*, 2nd Edn., Imeson, A., ed., London: Blackie Academic and Professional, imprint of Chapmann and Hall 1997, pp 1–21.
191. Armisen, R., F. Galatas. Production, properties and uses of agar. In: *FAO*, 1987, pp 1–57.
192. McHugh, D.J. Worldwide distribution of commercial resources of seaweeds including *Gelidium. Hydrobiologia* 221:19–29, 1991.
193. Doty, M.S. The production and use of *Eucheuma.* In: *FAO*, 1987, pp 123–164.
194. Foscarini, R., J. Prakash. *Handbook on Eucheuma seaweed cultivation in Fiji.* Ministry of Primary Industries, Suva (Fiji), Fisheries Div., (Field document of FAO project RAS/116/

JPN., South Pacific Aqualculture Development FAO Library Accession No: 306175), 1990, pp 42.
195. Stanley, N. Production properties and uses of carrageenan. In: *FAO*, 1987a, pp 116–146.
196. Therkelsen, G.H. Carrageenan. In: *Industrial Gums,* R.L. Whistler, J.N. Bemiller, eds., San Diego, CA: Academic Press, 1993, pp 145–180.
197. Nussinovitch, A. *Hydrocolloid Applications: Gum Technology in the Food and Other Industries.* New York: Blackie Academic & Professional, 1997, pp 354.

1.20

Butanol Production from Agricultural Biomass

Nasib Qureshi and Hans P. Blaschek

CONTENTS

20.1 INTRODUCTION

The process of producing butanol by fermentation was first discovered by Pasteur in 1861 (1). Butanol fermentation, also called acetone butanol ethanol (ABE) or solvent fermentation, is one of the oldest fermentation processes, second only to ethanol (1). The typical ratio of acetone to butanol to ethanol in the final product is usually 3:6:1 with maximum concentration of total solvents being 20 g L^{-1} when using traditional strains and traditional batch fermentation processes. The low concentration of solvents is caused by end product inhibition and results in a high cost for solvent recovery using distillation. During the early part of the twentieth century this fermentation was commercially viable. However, in the 1950s and 1960s butanol and acetone produced by fermentation were unable to compete economically with petrochemically produced solvents.

Butanol, being an important industrial chemical, is produced chemically. Currently, it is produced using either the oxo process starting from propylene (with H_2 and CO over a rhodium catalyst) or the aldol process starting from acetaldehyde (2). Acetone, a coproduct of the butanol fermentation, is produced chemically either by the cumene hydroperoxide process, or the catalytic dehydrogenation of isopropanol (3). In either case, these synthetic routes have proven to be economically advantageous in comparison to the fermentation-based processes. For this reason all butanol/acetone fermentation facilities around the world have ceased operation.

In 1996 the worldwide annual production of butanol and acetone was 2.49×10^9 kg and 2.10×10^9 kg, respectively. The total production of acetone and butanol was achieved by chemical processes using petroleum-based raw materials. Since 1990, production of butanol has been constant in the USA at 1.17×10^9 kg, while worldwide butanol production has fluctuated slightly. Acetone and butanol production by regions of the world is shown in Table 20.1 (4).

Butanol has several applications in the chemical industry and as a fuel (Table 20.2). When used as a fuel butanol can contribute to clean air by reducing emissions such as

Table 20.1
World production of butanol and acetone by region

Region	Butanol (kg)	Acetone (kg)
United States	1.17×10^9	1.20×10^9
North America (Total)	1.17×10^9	1.20×10^9
South America	5.12×10^7	8.52×10^7
Europe	8.43×10^8	3.31×10^8
Asia	4.30×10^8	4.76×10^8
Total	2.49×10^9	2.10×10^9

Reproduced with permission from Springer-Verlag GmbH & Co. KG. (N Qureshi, HP Blaschek. ABE production from corn: a recent economic evaluation. *J. Ind. Microbiol. Biotechnol.* 27:292–297, 2001; Table 1)

Table 20.2
Applications of butanol and acetone

Butanol	
Manufacture of dibutyl phthalate (precursor), butyl acetate (precursor), butyl acrylate (as a latex), glycol ethers, methyl butanol, and amine resins. Is an excellent fuel; used in plastic industry as a feedstock chemical; a solvent in the manufacture of oil, pharmaceuticals, and perfumes; as a solvency enhancer in the formation of nitrocellulose lacquers.	
Acetone	
Solvent for paints, lacquers, resins, nitrocellulose, varnishes, various processing, and cellulose acetate. Use in the manufacture of: methyl isobutyl ketone, methacrylates, methyl butanol, methyl isobutylcarbinol, isophorone, and diacetone alcohol.	

Reproduced with permission from Springer-Verlag GmbH & Co. KG. (N Qureshi, HP Blaschek. ABE production from corn: a recent economic evaluation. *J. Ind. Microbiol. Biotechnol.* 27:292–297, 2001; Table 2)

unburned hydrocarbons in tail pipe exhaust. Butanol has research and motor octane values of 113 and 94 compared to 111 and 92 for ethanol (5). Some of the advantages of using butanol as a fuel have been reported in the literature (5), including a vapor pressure for pure butanol of 0.63 psi as compared to 2.25 psi for ethanol, and a heat of vaporization of 141.3 k cal kg^{-1} for butanol as compared to 204.1 k cal kg^{-1} for ethanol. Additionally, the high boiling point (118°C) and comparatively low vapor pressure of butanol may enhance cold starting. Butanol is more miscible with gasoline and diesel fuel and is less miscible with water, than ethanol. It is currently used as a feedstock chemical in the plastic industry and as a food grade extractant in food and flavor industries (Table 20.2). Because of the potential for carcinogen carryover, the use of petroleum-derived butanol is not desirable for food-based applications (6).

Oil price increases in early 1970s resulted in a revival of research activities on a number of fermentations, including butanol, ethanol, and 2,3-butanediol, with a long range view toward reducing dependence on foreign oil. As a consequence, research has focused on developing technologies to produce fuels and chemicals from easily and widely produced renewable resources including agricultural biomass. A mixed degree of success was achieved in various countries when bioconversion programs involving alcohol fuel production were examined. This resulted in partially depressed crude oil market prices, which made it clear that fuel alcohol process development research should now be directed toward resolving problems associated with the efficiency of both new and traditional processes. By 1975, Brazil had introduced its ethanol production program (7) followed in 1977 by the USA (8). By 1982 many countries had introduced similar programs, but the subsequent reduction in oil prices led to diminishing interest in fermentation derived ethanol.

In order to make butanol fermentation an economically viable option, various laboratories around the world identified the following factors to impact the economics of butanol fermentation: high cost of substrate; low product concentration (<20 g L^{-1}); low reactor productivities (< 0.5 g $L^{-1}h^{-1}$); low ABE yields (0.29–0.33); and an escalated cost for butanol recovery by distillation, which is the classical approach for solvent recovery. Additional factors such as bioreactor costs, interest rates on borrowed capital, and the rate of return on the investment were also identified as factors which affect the price of fuels derived from renewable resources (9–13).

Because corn is a major crop in the Midwest region of the United States, *Clostridium beijerinckii* BA101 (previously called *C. acetobutylicum*) was developed to hydrolyze and utilize corn starch, and accumulate higher concentrations of butanol in the fermentation broth (6,14,15). *Clostridium beijerinckii* BA101 was developed from *C. beijerinckii* 8052. The developed culture is able to hydrolyze starch into glucose faster than it is able to utilize the glucose (16). Although most butanol producing cultures secrete amylases (starch hydrolyzing enzymes), *C. beijerinckii* BA101 has an enhanced capability to hydrolyze starch. In addition, *C. beijerinckii* BA101 produces a higher concentration of total solvents (25–33 g L^{-1}) under optimized fermentation conditions (15). In addition to corn, other renewable and economically viable substrates such as agricultural wastes, including food and fruit processing industry wastes, soy molasses (17), starch-based packaging materials (16), corn fiber hydrolysate, molasses, whey permeate, and cellulosic biomass hydrolysate can potentially be used with *C. beijerinckii* BA101 and *C. acetobutylicum* in order to produce solvents. In addition to the development of new microbial strains, techniques for improved production rates, energy efficient solvent recovery, and continuous fermentations have been developed. These developments have occurred with an eye toward using butanol in internal combustion engines (18,19). With these developments in butanol fermentation and downstream recovery we have reached a stage where fermentation derived butanol is able to compete economically with petroleum-based and chemically manufactured butanol. While the reader may be surprised to find a chapter on butanol production in a text on Food Biotechnology, the fact that agricultural food crops and derived carbohydrates obtained from those crops can be used to produce this valuable chemical by fermentation making this an appropriate topic of this book. The various problems and solutions which have been associated with this fermentation will be discussed in this chapter.

20.2 SUBSTRATES

20.2.1 Corn/Degermed Corn

Of the many substrates outlined here, two have been used on an industrial scale: corn (also called maize) and molasses. Corn contains approximately 71% starch which can be easily hydrolyzed by the butanol producing cultures. The molecular biology and genetics of substrate utilization, including starch, in *Clostridia* has been published elsewhere (20). This capability on the part of the cultures eliminates the need for a hydrolysis step which is essential for ethanol producing strains. The composition of corn is shown in Table 20.3 (21). During the ABE fermentation starch and glucose are utilized while corn fiber, protein, and oil are left behind depending on the process used for processing the corn. Pentosans and cellulosic fibers can be utilized only after hydrolysis, either by enzymatic

Table 20.3
Average composition of corn

Components	Average Composition (%)
Starch	71.1
Total sugars as glucose	2.6
Protein	9.5
Oil	4.3
Pentosans (in corn fiber)	6.2
Cellulose (in corn fiber)	3.3
Ash	1.4

Source: Reference 21

treatment, or dilute acid treatment, or a combination of both. The use of corn as substrate for this fermentation is well documented (1,22–24).

Degermed corn is obtained after removal of the germ, which contains oil. The oil content of corn is approximately 4.3% and its recovery results in substantial byproduct credit for this fermentation. However, a number of extra processing steps (soaking, grinding, fine grinding, sieving, and centrifugation) are involved in the germ recovery process as compared to the dry milling process (25). For this reason, capital costs associated with producing degermed corn are higher, which offsets the byproduct credit. Hence, corn dry milling is preferred over wet corn milling. This is consistent with the current trend in the USA toward establishment of more dry mills. Another drawback of corn wet milling is that during soaking of the corn nutrients are leached out, making the corn mash deficient in nutrients for fermentation. Hence, nutrient supplementation is required in order to produce butanol from degermed corn (26).

20.2.2 Molasses

Molasses as a fermentation substrate has many advantages over corn including easy handling and processing. Molasses contains approximately 50% sucrose. Sucrose can be hydrolyzed by solventogenic cultures during the fermentation to produce butanol. There are several types of molasses, namely blackstrap, invert (high test), and beet molasses (27). Until the early 1980s a butanol fermentation plant was operational in South Africa [National Chemical Products (NCP) Germiston, South Africa]. The process was terminated due to a shortage of molasses which was caused by drought.

20.2.3 Whey Permeate

In the past, whey permeate was considered to be a waste product of the dairy industry; however, new uses for whey permeate have changed that impression so that now whey permeate is considered a byproduct. Whey permeate has numerous applications in the fermentation industry which have been documented in the literature (28). The lactose content in whey permeate, which is the carbon source for fermentation, varies from process to process. Typically, the lactose content in whey permeate is in the range of 44–50 g kg^{-1}, making it a suitable substrate for the butanol fermentation. This is because butanol fermentation limits substrate concentration to <60 g L^{-1} due to product toxicity. For fermentations other than butanol (i.e., ethanol or 2,3-butanediol) supplementation with additional carbohydrate is required. For details on the composition of whey permeate see Maddox et al. (28). Various types of whey permeates have been used for the production of butanol using *C. acetobutylicum,* suggesting that it is a good substrate for this fermentation. Both *C. acetobutylicum* P262, and *C. beijerinckii* BA101 have been used for butanol production from whey permeate. It should be noted that this substrate is a rich source of minerals and hence further supplementation of minerals to the fermentation medium is not required.

20.2.4 Other Butanol Fermentation Substrates

The high cost of substrates, including corn, molasses, and whey permeate, has been identified as a major factor affecting the economic viability of butanol production by fermentation (1,12,29). In order to find a solution to this problem a number of carbohydrate containing alternative substrates have been investigated. Butanol producing cultures are able to utilize a wide variety of carbohydrates such as sucrose, glucose, fructose, mannose, lactose, dextrin, starch, pentoses (xylose, arabinose), and inulin (1). Use of these carbohydrates by butanol producing cultures makes it possible to ferment byproducts and agricultural wastes to butanol. The details of some of these substrates are presented in what follows.

20.2.4.1 Soy Molasses

Spray dried soy molasses (SDSM) is a byproduct of the soy processing industry and contains 746 g kg^{-1} of carbohydrate, of which 434 g kg^{-1} is fermentable (17). The sugars that are present in soy molasses include glucose, sucrose, fructose, pinitol, raffinose, verbascose, melibiose, and stachyose (17). The fermentable sugars include glucose, sucrose, fructose, and galactose. It was determined that *C. beijerinckii* BA101 was not able to ferment raffinose, pinitol, melibiose, verbascose, and stachyose due to its inability to hydrolyze the α 1-6 glycosidic bond present in these sugars. In order to be able to use these sugars, a strain of *C. beijerinckii* BA101 should be developed with a capability of hydrolyzing α 1-6 glycosidic bonds. Development of such a strain would result in the more complete utilization of sugars present in SDSM. Figure 20.1 shows a diagram of an oligosaccharide present in soy molasses.

20.2.4.2 Agricultural Wastes

Examples of agricultural wastes include waste corn or contaminated corn, and fruit industry waste such as apple and pear drops, and apple pomace (1,16,30). A mixture of cracked corn, apple drops, and starch-containing packing peanuts (peanut shape packing material) was successfully fermented without adjusting the pH or supplementing with nutrients (16). The fruits used in this study showed significant signs of decay suggesting that such orchard wastes can be used. Apple pomace contains approximately 10% sugars which can be fermented to butanol (1). The residue which is obtained after sugar fermentation can be hydrolyzed and subsequently fermented to butanol as well. Another source of fermentation substrate is corn fiber, which can be hydrolyzed, using either enzymes or acid or both, into simple sugars followed by fermentation. Approximately 4.7×10^6 dry tons of corn fiber are produced in the USA annually (31). At the time of preparation of this manuscript authors were examining the hydrolysis of corn fiber and fermentation to butanol. Other agricultural biomasses such as corn stover (corn stalk), wheat straw, rice husks, and various grasses can become substrates for this fermentation as well. It should be noted that sugars derived from the hydrolysis of these lignocellulosics can be utilized by the solventogenic *Clostridia* (1).

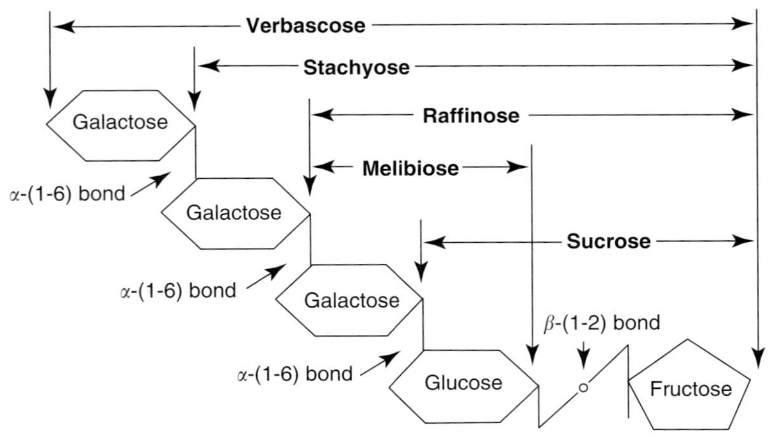

Figure 20.1 Predominant soy molasses oligosaccharides. [Reproduced with permission from Springer-Verlag GmbH & Co. KG (N Qureshi, A Lolas, HP Blaschek. Soy molasses as fermentation substrate for production of butanol using *Clostridium beijerinckii* BA101. *J. Ind. Microbiol. Biotechnol.* 26: 290–295, 2001; Figure 1)]

In the USA a significant amount of such wastes is available, which could become potential substrates for butanol fermentation and could replace conventional substrates such as corn. Direct utilization of cellulosic waste using single or mixed cultures is another possible approach for utilizing these economically available substrates for butanol fermentation. However, keeping in view the size and objective of this chapter, it is not possible to describe those processes or cultures here. For preliminary information on such processes and cultures, see Jones and Woods (1).

20.2.4.3 Jerusalem Artichokes

Jerusalem artichoke is an agricultural crop which contains the carbohydrate inulin. Use of inulin requires hydrolysis using either dilute acid or the enzyme inulinase prior to fermentation (1). Following hydrolysis, Jerusalem artichokes have been used as a fermentation substrate for producing solvents (32). The hydrolysates were prepared using inulinase and fermentation resulted in the production of 23–24 g L^{-1} solvents in 36 h.

20.2.4.4 Potatoes

Because *Clostridia* possess strong amylolytic activity (33), substrates other than corn have been considered for this fermentation. Potatoes and potato wastes are economically viable substrates which could be used for butanol fermentation. Although it has been suggested that this substrate has viscosity problems (34), both hydrolyzed and unhydrolyzed potato waste has been successfully used for butanol production using *C. acetobutylicum* or *C. beijerinckii* strains (35). An initial potato starch concentration of 45–48 g L^{-1} was used in the fermentation thereby producing up to 12 g L^{-1} ABE (35). These results were obtained with unhydrolyzed potatoes. Potato starch hydrolysis with α-amylase and amyloglucosidase prior to the fermentation resulted in the production of 11.4 and 10.4 g L^{-1} total ABE, respectively. This suggested that hydrolysis prior to fermentation was not beneficial to the fermentation. Other examples of successful butanol production from potato include that of Grobben et al. (36) who used *C. acetobutylicum* DSM 1731.

Various additional substrates that can be used for the production of butanol include sulfite waste liquors (a byproduct of the paper industry), sugar beet, wheat, rye, and oats. Sulfite waste liquor contains glucose, xylose, and arabinose which are fermentable by *Clostridia*. At the time of preparation of this chapter research is being planned on the use of other economically viable substrates such as CO_2 and distillers dry grain solids (DDGS). CO_2 is a byproduct of the ethanol and butanol fermentations. Similarly, DDGS is obtained as a byproduct of the corn dry milling ethanol industry and could be used as a substrate for butanol fermentation following hydrolysis of the cellulosics. In the USA approximately 5.10×10^9 kg of DDGS is produced each year, from which 1.93×10^9 kg of hexoses and pentoses can be obtained following hydrolysis.

20.3 UPSTREAM PROCESSING OF SUBSTRATES

The raw materials mentioned previously may require upstream processing depending upon the process being used for fermentation. Molasses requires dilution and removal of insoluble solids; whey permeate requires concentration, possibly by reverse osmosis; and corn requires removing bran, germ, and oil; milling; sieving; centrifugation; and saccharification. Table 20.4 shows the upstream process steps which may be required for the production of butanol from various substrates. The most economically viable substrate is cellulosic biomass, which requires prior hydrolysis either by dilute acid treatment or

Table 20.4
Upstream processing steps that may be required for butanol production from various substrates.

Corn	
Wet Milling (germ removal): Steeping, grinding, fine grinding, sieving, and centrifugation	
Dry Milling: Grinding, mash cooking	
Molasses	
Dilution and sediment removal by centrifugation	
Whey Permeate	
For batch fermentation no upstream processing is required	
For integrated fermentation and product removal process: lactose concentration possibly by reverse osmosis is required	
Soy Molasses	
Dilution, removal of sediment and supplementation with additional carbon source is required	
Jerusalem Artichokes	
Cooking, mashing, sediment removal, hydrolysis with inulinase and dilution	
Corn Fiber	
Steam explosion, acid or enzymatic hydrolysis and possible removal of inhibitors	
Potato	
Cooking, mashing, sediment removal and dilution	
Agricultural Waste/Biomass	
Steam explosion, hydrolysis by dilute acid or enzymatic method	
CO_2	
Due to slow dissolution rate in aqueous solutions, rate of reaction would be slow	
Distillers Dry Grain Solids (DDGS)	
Steam explosion, hydrolysis with dilute acid or enzymes	

enzymatic treatment or both. The acid hydrolysis method is comparatively faster than the enzymatic method; however, acid hydrolysate contains sugar degradation products which may inhibit the fermentation. Removal of these inhibitors is essential for good productivity in the fermentation. Jerusalem artichokes require hydrolysis prior to fermentation by using the enzyme inulinase.

20.4 MICROBIAL CULTURES

There are a number of cultures capable of producing significant amounts of solvents from various carbohydrates. The strain that was used for the production of solvents in large scale was *Clostridium acetobutylicum* P262. Unfortunately the plant (South Africa) was shut down due to molasses shortages caused by drought. The other strains that have been studied in the past include *C. acetobutylicum* ATCC 824, *C. acetobutylicum* NRRL B643, *C. acetobutylicum* B18, *C. beijerinckii* 8052, *C. beijerinckii* BA101, and *C. beijerinckii* LMD 27.6.

Various different species of butanol producing *Clostridia* are recognized based on the ratio and type of solvents produced. *Clostridium butylicum* produces isopropanol rather than acetone. It has been reported that *C. aurantibutyricum* produces all three solvents including acetone, isopropanol, and butanol. *Clostridium tetanomorphum* produces equal amounts of ethanol and butanol. This culture does not produce any acetone or isopropanol 1).

A significant amount of research efforts has been made to genetically improve some of the existing cultures or isolate new solventogenic strains. Montoya et al. (37) isolated strains that produced 24.2 to 29.1 g L^{-1} total solvents. *Clostridium beijerinckii* BA101 is a strain developed from *C. beijerinckii* 8052. *Clostridium beijerinckii* BA101 has been reported to accumulate 23.5-33 g L^{-1} solvents (15,38). Attempts have been made to enhance solvent production capability of *C. acetobutylicum* ATCC 824. The newly developed strain is reported to produce 25.6 to 30 g L^{-1} solvents (39). Note that none of the parental strains produced > 20 g L^{-1} solvents.

20.5 GENETICS OF CLOSTRIDIUM BEIJERINCKII

The hyper-amylolytic *C. beijerinckii* BA101 mutant was generated using N-methyl-N-nitro-N-nitrosoguanidine (MNNG), together with selective enrichment on the nonmetabolizable glucose analog 2-deoxyglucose (14). Amylolytic enzyme production by *C. beijerinckii* BA101 was 1.8- and 2.5-fold higher than that of the parent strain grown in starch and glucose, respectively. Dramatically elevated levels of butanol and acetone resulted in higher butanol and total solvent yields for hyper-amylolytic *C. beijerinckii* BA101 relative to the NCIMB 8052 parent strain when grown in P2 medium containing either 6% glucose or STAR-DRI 5 maltodextrin (6). The *C. beijerinckii* BA101 strain consistently produced on the order of 19 g L^{-1} butanol in 20 L batch fermentations. This represents an increase of more than 100% in butanol concentration by the BA101 strain when compared to the parent NCIMB 8052 strain. The kinetics of butanol production over time also indicated a more rapid rate of butanol production by *C. beijerinckii* BA101 in semidefined P2 medium containing glucose or maltodextrin. The lower levels of butyric and acetic acids produced over the course of the fermentation carried out by *C. beijerinckii* BA101 is consistent with an enhanced capacity for uptake and recycling of these acids. A low acid producing mutant of *C. acetobutylicum* designated B18 was generated following ethyl methyl sulfonate treatment (40). This mutant utilized glucose (up to 8%) more completely and accumulated less acid in batch fermentations. Enzyme assays indicated that the B18 mutant has twofold higher butanol dehydrogenase and butyraldehyde dehydrogenase activities. *C. beijerinckii* BA101 also appears to more completely utilize carbohydrate when compared to the *C. beijerinckii* NCIMB 8052 strain (6). This corresponds with the higher total butanol and solvent production by the *C. beijerinckii* BA101 strain when grown in either glucose or maltodextrin. Carbon balance following fermentation by *C. beijerinckii* NCIMB 8052 and BA101 indicates sufficient carbon is available for the twofold increase in butanol concentration observed during *C. beijerinckii* BA101 fermentations (6).

Two reports have suggested that the instability of certain solventogenic genes (*ctfA*: acetoacetyl-CoA; ctfB: acetate/butyrate CoA transferase; *aad*: aldehyde/alchohol dehydrogenase; and *adc*: acetoacetate decarboxylase) may be the cause of strain degeneration in *C. acetobutylicum* (41,42). Specifically, the genes for butanol and acetone formation in *C. acetobutylicum* ATCC 824 were found to reside on a large 210-kb (pSOL1) plasmid whose loss led to degeneration of this strain (42). Eight genes concerned with solventogenesis in *C. beijerinckii* NCIMB 8052 were found at three different locations on the genome (43). In *C. beijerinckii* NCIMB 8052, new evidence has demonstrated that the *ctfA* gene is localized on the chromosome in a solvent operon downstream from the *ald* gene and upstream from the *ctfB* and *adc* genes (44). Chen and Blaschek (44) reported that the degeneration of *C. beijerinckii* NCIMB 8052 does not involve loss of the *ctfA/B* genes, and that degeneration is prevented by the addition of sodium acetate into the medium. It has also been demonstrated that the addition of sodium acetate increased levels of solvent production by *C. beijerinckii* NCIMB 8052 and BA101 (15,44).

Chen and Blaschek (45) also demonstrated that the specific activities and mRNA expression levels of Coenzyme-A transferase, acetoacetate decarboxylase and butyraldehyde dehydrogenase were elevated in *C. beijerinckii* BA 101 when compared to *C. beijerinckii* NCIMB 8052. Harris et al. (39) found that strain PJC4BK (pTAAD), a mutant that carries a second copy of the *C. acetobutylicum* alcohol dehydrogenase gene, produced more ethanol than the wild type strain. Similarly, increasing the dosage of *sol* operon genes in *C. beijerinckii* BA 101 may lead to increased butanol production.

Although these key enzymes in the solvent forming pathway of the hyper-butanol producing *C. beijerinckii* BA101 mutant have previously been shown to have elevated levels of activity and expression (15,44), these findings do not fully explain the metabolic activities associated with *C. beijerinckii* BA101. Because *C. beijerinckii* BA101 was produced using chemical mutagenesis (14), the molecular basis for why this strain produces enhanced levels of acetone and butanol (relative to the *C. beijerinckii* NCIMB 8052 wild type strain) is not fully understood.

Subsequently, much of the focus of our lab has been to characterize the hyper-solvent producing mutant *C. beijerinckii* BA101. Studies have shown that the mutant exhibits higher levels of expression and activities for *ptb* and the genes of the *sol* operon when compared to the wild type *C. beijerinckii* NCIMB 8052 (45). The *sol* operon and the *ptb/buk* operon from *C. beijerinckii* NCIMB 8052 and *C. beijerinckii* BA101 have been cloned and sequenced. The sequences of the structural genes from the *sol* operon, *ald*, *ctfA*, *cftB* and *adc*, have been compared and results show that the two strains are identical with respect to these genes. In addition, the regulatory regions of the *sol* operons from *C. beijerinckii* BA101 and *C. beijerinckii* NCIMB 8052 have been compared, and no differences have been found. It has been proposed that the *sol* operon is controlled directly by Spo0A (46), however, we have found no consensus sequences for Spo0A boxes in the upstream region of the *sol* operons in these organisms.

To investigate the expression of the cloned *sol* operon, constructs of the shuttle vectors pTJ1 and pTJ2 containing the *sol* operon from *C. beijerinckii* BA101 were introduced into the degenerate mutant *C. beijerinckii* SA2 (47) via electrotransformation. We have successfully restored butanol production in this strain with the constructs pTJ1*sol* and pTJ2*sol*. These plasmids were derived from the CAK1-based phagmid (48). However, southern hybridization data suggests that the replicative plasmids pTJ1*sol* and pTJ2*sol* have undergone integration events. To investigate the effect of downregulating the acidogenic enzymes phosphotransbutyrylase (PTB) and butyrate kinase (BK), we produced two nonreplicative plasmid constructs containing internal fragments of the acidogenic genes *ptb* and *buk* and electrotransformed *C. beijerinckii* BA101. Southern hybridization analysis of the putative integrational mutants showed that the constructs had not integrated at the *ptb* and *buk* target sites but elsewhere in the chromosome. Fermentation analysis is being carried out in order to see if these recombination-induced variants of *C. beijerinckii* BA101 show improved solvent production characteristics.

In addition to the study of the acidogenic and solventogenic genes in *C. beijerinckii* BA101, studies into the sugar transport by this strain have also been carried out. The increased glucose utilization by *C. beijerinckii* BA101 has been reported previously in our laboratory (6). It was observed that *C. beijerinckii* BA101 utilizes an ATP-dependent phosphorylation system to take up glucose at a rate double that of *C. beijerinckii* NCIMB 8052. This appears to allow *C. beijerinckii* BA101 to transport and utilize glucose during the solventogenic stage (49).

20.6 SUGAR TRANSPORT

Saccharolytic *Clostridia* have the capability of utilizing a wide spectrum of carbon sources for growth and are transported into the cell by a variety of mechanisms. Phosphoenolpyruvate (PEP) dependent phosphotransferase system (PTS) is the predominant transport mechanism for carbohydrates in *Clostridia* (50). The clostridial PTS has been shown to be functionally related to the PTS present in other bacteria. However, not all carbohydrates are accumulated by *Clostridia* via the PTS. By analogy with other bacterial transport systems, accumulation of some carbohydrates is likely to be driven by ion gradients or by ATP hydrolysis. Comparatively few studies examining non-PTS based transport systems have been undertaken with obligate anaerobes. Recent work in our laboratory provided evidence for an alternative glucose transport system in *C. beijerinckii* NCIMB 8052 and BA101 (49).

Glucose uptake and accumulation by *Clostridium beijerinckii* BA101, a butanol hyper-producing mutant, was examined during various stages of growth. Glucose uptake in *C. beijerinckii* BA101 was repressed 20% by 2-deoxyglucose (2-DG) and 25% by mannose, while glucose uptake in *C. beijerinckii* 8052 was repressed 52% and 28% by these sugars, respectively. We confirmed the presence of PTS associated with cell free extracts of *C. beijerinckii* BA101 by glucose phosphorylation by PEP. The PTS activity associated with *C. beijerinckii* BA101 was 50% of that observed for *C. beijerinckii* 8052. *Clostridium beijerinckii* BA101 also demonstrated lower PTS activity for fructose and glucitol (sorbitol). Glucose phosphorylation by cell free extracts derived from both *C. beijerinckii* BA101 and 8052 was also dependent on the presence of ATP, consistent with the presence of glucokinase activity in *C. beijerinckii* extracts. ATP-dependent glucose phosphorylation was predominant during the solventogenic stage when PEP-dependent glucose phosphorylation was dramatically repressed. A nearly twofold greater ATP-dependent phosphorylation rate was observed for solventogenic stage *C. beijerinckii* BA101 than for solventogenic stage *C. beijerinckii* 8052. These results suggest that *C. beijerinckii* BA101 is defective in PTS activity and that *C. beijerinckii* BA101 compensates for this defect with enhanced glucokinase activity resulting in an ability to transport and utilize glucose during the solventogenic stage.

Glucose uptake by both acidogenic and solventogenic stage *C. beijerinckii* BA101 was inhibited by proton conductors, ATPase inhibitors and non-PTS substrates in contrast with what was observed for acidogenic stage *C. beijerinckii* 8052. However, it was observed that these compounds were also potent inhibitors of glucose uptake by solventogenic *C. beijerinckii* 8052. ATP-dependent glucose phosphorylation was dramatically induced during solventogenic stage when PEP-dependent glucose phosphorylation was decreased. Approximately a twofold greater ATP-dependent phosphorylation rate was observed during all growth stages for *C. beijerinckii* BA101 versus *C. beijerinckii* 8052. These results suggest that an alternative glucose transport mechanism is present in *C. beijerinckii,* which is predominant during the solventogenic stage (51). The alternative transport mechanism together with enhanced glucokinase activity may allow *C. beijerinckii* BA101 to more completely utilize glucose in spite of a defective PTS being associated with this strain.

20.7 FERMENTATION

20.7.1 Batch Process

The most commonly studied process for butanol production is batch fermentation. As mentioned in the substrates section, a number of carbohydrates can be used for butanol production. The upstream processing for fermentation depends upon the carbohydrate used for fermentation. The most commonly used substrates for butanol fermentation are molasses, corn, or glucose derived from corn.

The production of butanol on an industrial scale was carried out using large fermenters ranging in capacity from 200,000 to 800,000 L. The industrial process used 8–10% corn mash which was cooked for 90 min at 130–133°C. Corn contains approximately 71% starch. A schematic flow sheet for butanol production from corn is shown in Figure 20.2. Butanol production is a biphasic fermentation where acetic acid and butyric acid are produced during acidogenic phase followed by their conversion into acetone and butanol. During the acidogenic phase the pH drops due to acid production and subsequently rises during solvent production. Figure 20.3 shows the results of a typical batch fermentation process.

The use of molasses offers many advantages over using corn, including presence of essential vitamins and micronutrients (27). In industrial processes beet, invert, and blackstrap molasses were diluted to give a fermentation sugar concentration of 50 to 75 g L^{-1}, most commonly 60 g L^{-1}. The molasses solution was sterilized at 107–120°C for 15–60 min followed by adding organic and inorganic nitrogen, phosphorus and buffering chemicals. During cooling, nitrogen or carbon dioxide is swept across surface to keep the medium anaerobic. After inoculation the medium is sparged with these gases to mix the inoculum. The final solvent concentration ranged from 12 to 20 g L^{-1}. The yield of solvent is usually low, at 0.29–0.33. Distillation has been the method of choice to recover butanol; however, during the last two decades a number of alternative techniques have been investigated for the economical recovery of butanol, which will be discussed in the recovery section.

The solvent concentration that can be produced in a batch process depends upon the culture used. As mentioned in the Microbial Cultures Section, ABE on the order of 25–33 g L^{-1} can be produced in a batch process depending upon the culture used. Using newly developed *C. beijerinckii* BA101 an ABE yield of 0.45 can be obtained (6).

20.7.2 Fed-Batch Fermentation

The fed-batch fermentation is an industrial technique which is applied to processes where a high substrate concentration is toxic to the culture. In such a situation the reactor is initiated in a batch mode with low substrate concentration and low volume of medium (usually less than half the volume of the fermenter). As the substrate is utilized by the culture,

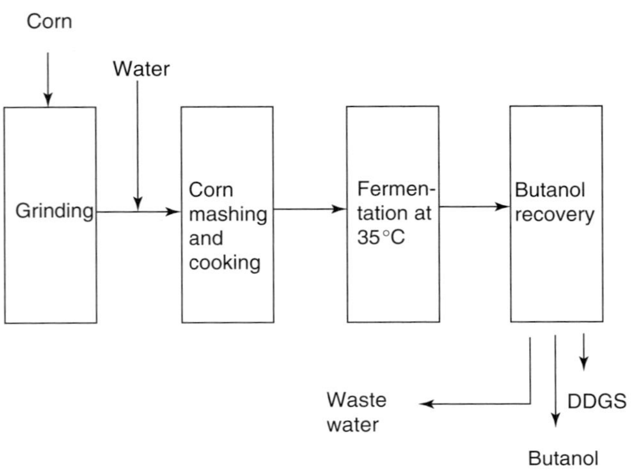

Figure 20.2 A schematic diagram of butanol production from corn by fermentation.

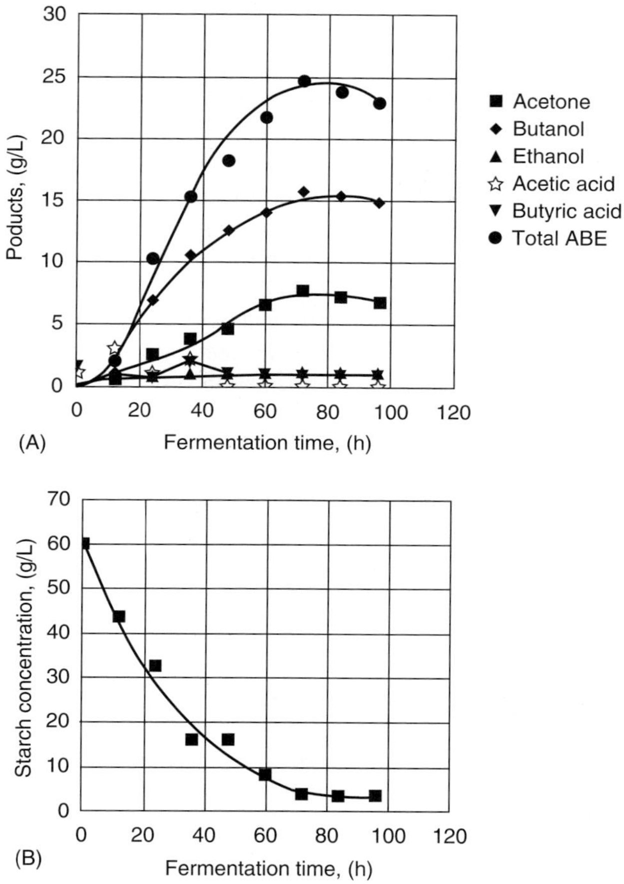

Figure 20.3 Production of butanol from corn starch by *C. beijerinckii* BA101. [Reproduced with permission from Springer-Verlag GmbH & Co. KG (TW Jesse, TC Ezeji, N Qureshi, HP Blaschek. Production of butanol from starch-based waste packing peanuts and agricultural waste. *J. Ind. Microbiol. Biotechnol.* 29: 117–123, 2002; Figure 1)]

it is replaced by adding a concentrated substrate solution at a slow rate, thereby keeping the substrate concentration in the fermenter below a level toxic to the culture. When using this approach, the culture volume increases over time. The culture is harvested when liquid volume is approximately 75% of the volume of reactor. Because butanol is toxic to *C. acetobutylicum* or *C. beijerinckii* cells, the fed-batch fermentation technique cannot be applied in this case unless one of the novel product recovery techniques is applied for simultaneous separation of the product.

20.7.3 Continuous Fermentation

The continuous culture technique may be used to improve reactor productivity and study the physiology of the culture in steady state. A number of studies exist for the continuous fermentation of butanol. Because of the production of fluctuating levels of solvents and complexity of butanol fermentation, the use of a single stage continuous reactor does not seem to be practical at the industrial scale. In continuous culture, a serious problem exists in that solvent production may not be stable for long time periods and ultimately declines over time, with a concomitant increase in acid production. In a single stage continuous system high reactor productivity may be obtained; however this occurs at the expense of low product concentration when compared to that achieved in a batch process. In a single stage continuous reactor 15.9 g L^{-1} total solvents was produced at a dilution rate of 0.1 h^{-1} resulting in a productivity of 1.5 g L^{-1} h^{-1} (52). In order to improve the productivity the dilution rate was increased to 0.22 h^{-1}, increasing productivity to 2.55 g L^{-1} h^{-1}. However, the product concentration decreased to 12 g L^{-1}. In a more recent study a continuous fermentation was run in order to study the effect of fermentation gases and dilution rate on solvent production (53). In this system a productivity of 0.58 g $L^{-1}h^{-1}$ was obtained at a dilution rate of 0.07 h^{-1}, producing 8.3 g L^{-1} total solvents. In order to improve reactor productivity Mulchandani and Volesky (54) used a single stage spin filter perfusion bioreactor. In the perfusion bioreactor a maximum productivity of 1.14 g $L^{-1}h^{-1}$ was obtained, however, solvent concentration fluctuated over time.

Two or more multistage continuous fermentation systems have been investigated in the attempt to reduce fluctuations and increase solvent concentration in the product stream (55–57). This is done by allowing growth, acid production, and solvent production to occur in separate bioreactors. In a two stage system, Bahl et al. (55) reported a solvent concentration of 18.2 g L^{-1} using *C. acetobutylicum* DSM 1731 which is comparable to the solvent concentration in a batch reactor. This type of multistage reactor system (7–11 fermentors in series) was successfully tested at the pilot plant and full scale plant level in Russia (then Soviet Union) (56).

20.8 IMMOBILIZED CELL BIOREACTORS

Highly productive immobilized cell bioreactors have been developed for the production of butanol. In these reactors productivities as high as 6.5–15.8 g L^{-1} h^{-1} have been achieved as compared to $<$0.50 g L^{-1} h^{-1} in batch reactors. Due to space limitations we can not describe further details of these reactors; however, a brief overview will be presented in section 11 (Butanol Fermentation Limitations).

20.9 RECOVERY

20.9.1 Distillation

The cost of recovery of butanol is high because its concentration in the fermentation broth is low due to product inhibition. In addition to the low product concentration, the boiling point of butanol is higher than water (118°C). The usual concentration of total solvents in the fermentation broth is 20–30 g L^{-1}, of which butanol is only about 13–18 g L^{-1}. This makes butanol recovery by distillation energy intensive. Phillips and Humphrey (13) evaluated the economics of butanol removal from the fermentation broth using distillation and demonstrated that, as the concentration of butanol is increased from 10 to 40 g L^{-1}, energy saving by a factor of several orders of magnitude may be achieved. At 10 g L^{-1} feed butanol the ratio of tons of oil used for fuel to the ratio of tons of 100% recovered butanol is 1.5, while at 40 g L^{-1} feed butanol concentration this ratio is 0.25. This suggests that tremendous energy savings can be achieved if the butanol concentration in the fermentation broth can be increased from 10 to 25 g L^{-1}.

In order to economize butanol recovery, a number of recovery techniques have been investigated including *in situ* gas stripping, pervaporation, liquid–liquid extraction, perstraction, and reverse osmosis. Details of these techniques have been described elsewhere (58–59).

Any of these techniques can be applied for *in situ* butanol removal, removing butanol simultaneously as it is produced. The objective is to prevent the butanol concentration from increasing above the tolerance level of the culture. The product is subsequently recovered either by condensation (gas stripping and pervaporation) or by distillation (extraction).

20.9.2 Gas Stripping

Gas stripping is a simple technique for recovering butanol (and acetone and ethanol) from the fermentation broth (60). Nitrogen or fermentation gases (CO_2, H_2) are bubbled through the fermentation broth followed by passing the gas through a condenser. As the gas is bubbled through the fermenter it captures solvents which are condensed in the condenser followed by collection in a receiver. Once the solvents are condensed the gas is recycled back to the fermenter to capture more solvents. This process continues until all the sugar in the fermenter is utilized by the culture. In some cases a separate stripper can be used to strip off solvents followed by recycling the stripper effluent that is low in solvents. Figure 20.4 shows a typical schematic diagram of solvent removal by gas stripping. Gas stripping has been successfully applied to remove solvents from batch (61), fed-batch (62), fluidized bed (63), and continuous reactors (64,65). In addition to removal of solvents, a concentrated sugar solution was fed to the reactor to reduce the volume of process streams and economize the butanol production process (62,63,65).

20.9.3 Pervaporation

Pervaporation is a membrane-based process which is used to remove solvents from fermentation broth by using a selective membrane (58,59). The liquids or solvents diffuse through a solid membrane leaving behind nutrients, sugar, and microbial cells. The concentration of solvents across the membrane depends upon membrane selectivity, which is a function of feed solvent concentration, and membrane composition. Application of pervaporation to batch butanol fermentation has been described by Groot et al. (66), Larrayoz and Puigjaner (67), and Qureshi and Blaschek (68). Pervaporation has also been applied for the removal of butanol from the fermentation broth in fed-batch reactors (69,70). In fed-batch reactors a concentrated sugar solution is used to reduce the process stream volume. It is interesting to note that acids did not diffuse through the membranes used by these authors. Qureshi et al. (71) used a polypropylene membrane through which acids diffused, though at high acid concentration in the fermentation broth.

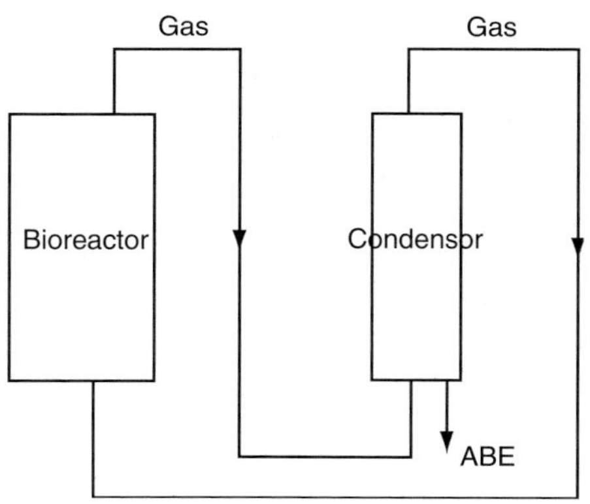

Figure 20.4 A schematic diagram of butanol production by fermentation and recovery by gas stripping.

The use of a liquid membrane, rather than a solid membrane, has been described by Matsumura et al. (72). The oleyl alcohol containing liquid membrane was supported on a microporous 25 mm thick flat polypropylene sheet. The liquids which diffused through the membrane showed a selectivity of 180 as compared to silicone membrane selectivity of approximately 45. It was estimated that if this pervaporation membrane was used as a pretreatment process for butanol separation, the energy requirements would be only 10% of that of conventional distillation. Unfortunately, this was not a stable membrane.

In order to develop a stable and high selectivity membrane, silicalite, an adsorbent was included in a silicone membrane (73). The selectivity of the silicone-silicalite membrane was improved from approximately 40 to 209. The working life of the membrane is 3 years. The membrane was used with both butanol model solutions and fermentation broths. A comparison of various membranes suggested that this membrane may be superior to other membranes (74).

20.9.4 Liquid–Liquid Extraction

Liquid–liquid extraction is another technique which can be used to remove ABE (acetone, butanol, and ethanol) from the fermentation broth. In this process an extraction solvent is in contact with the fermentation broth (58). Butanol, acetone, and ethanol are diffused into the extraction solvent and recovered by back-extraction into another extraction solvent or by distillation (75). The basic requirements have been presented by Maddox (58), Mattiasson and Larsson (76) and Ennis et al. (77). Some of the requirements for extractive butanol fermentation are:

1. Nontoxic to the producing organism
2. High partition coefficient for the fermentation products
3. Immiscible and nonemulsion forming with the fermentation broth
4. Inexpensive and easily available
5. Can be sterilized and does not pose health hazards.

An early report on the application of extractive butanol fermentation was that described by Wang et al. (78) who used corn oil as an extraction solvent and reported an increase in butanol production in repeated batch fermentations. Since that time there have been many reports on the use of various extraction solvents for extractive butanol fermentation (58). Of the many extraction solvents reported in literature, oleyl alcohol appears to meet some of the requirements, and it has been a subject of a number of investigations (75,79,80). Roffler et al. (80) were successful producing butanol in an extractive fed-batch fermentation using a sugar solution containing 339 g L^{-1} glucose.

Extractant toxicity is a major problem with extractive fermentations. In order to avoid the toxicity problem brought about by the extraction solvent, some investigators have used a membrane to separate the extraction solvent from the cell culture (81,82). Eckert and Schugerl (82) used a microfiltration unit in order to separate bacterial cells producing butanol from the extraction solvent (decanol). In this process, a continuous fermentation cell recycle system was used. The fermentation broth was circulated through the membrane and the cells were returned to the fermenter, while the permeate was extracted with decanol to remove the butanol. In this system a productivity of 3.08 g L^{-1} h^{-1} was achieved. In cell recycle systems, without extraction, productivities as high as 6.5 g L^{-1} h^{-1} have been achieved. Compared to this productivity a productivity of 3.08 g $L^{-1}h^{-1}$ is low, however, it is difficult to compare the two systems unless the biomass concentration and fermentation parameters are presented.

Another approach for reducing the toxicity and improving the partition coefficient has been to mix a high partition coefficient, high toxicity extractant with a low partition coefficient and low toxicity extractant (83). The resultant mixture is an extractant with a high partition coefficient and low toxicity. At this time research is underway in the authors' laboratory on the extractive recovery of butanol from the fermentation broth of *C. beijerinckii* BA101. In integrated extractive fermentations, oleyl alcohol forms an emulsion when agitated at high speed to improve rate of removal of butanol.

20.9.5 Other Recovery Techniques

The prominent recovery techniques that have been used in the ABE fermentation have been presented. In addition to these techniques, adsorption (58,84), reverse osmosis (85), and aqueous two phase extraction (58,76,86) have also been used; with, however, lower success due to limitations associated with these techniques.

20.10 BUTANOL FERMENTATION LIMITATIONS

The major problems with the butanol fermentation are product inhibition due to solvent toxicity, use of dilute sugar solutions, low ABE yield, low reactor productivity, and uneconomical product recovery. In the butanol fermentation a maximum solvent concentration of 25–30 g L^{-1} is rarely attained (69). The fermentation broth contains approximately 970–980 gL^{-1} water to be removed in order to separate the solvents. Therefore, simultaneous recovery should be an essential part of this fermentation process using one of the novel techniques mentioned. Use of one of the novel product recovery techniques allows for concentrated feed streams in addition to efficient product recovery, thus benefiting the fermentation process. An alternative approach would be to develop a superior culture which could produce solvents at higher concentrations. However, there has been limited success with this approach leaving *in situ* or in line removal as the only viable options. In addition, in line or *in situ* recovery techniques eliminate the use of dilute feed solutions. Those approaches have been applied in batch (68,87), fed-batch (69), and continuous (63,88) fermentations.

Another major problem associated with the ABE fermentation is low ABE yield. In most fermentations, an ABE yield of 0.3–0.45 g g^{-1} is achieved with the release of the rest of the carbon (approximately 53%) to the environment in the form of CO_2 gas. In addition, H_2 gas is also released, which is a loss of energy. It is suggested that H_2 be separated from CO_2 using a membrane and used as a fuel. Currently, the authors are investigating the conversion of CO_2 gas to butanol to improve solvent yield. Low butanol yield adversely affects butanol price (Figure 20.5).

Another problem with the ABE fermentation has been the low reactor productivity when batch fermentation is employed. In a batch reactor, productivity on the order of <0.50 g L^{-1} h^{-1} is achieved. In order to improve reactor productivity, continuous immobilized cell reactors (63,89–91) or cell recycle reactors (57) have been developed. In these reactors, productivities as high as 6.5 to 15.8 g L^{-1} h^{-1} have been achieved as compared to batch reactor productivity of approximately 0.50 g L^{-1} h^{-1} or less.

Substrate cost is another factor which significantly influences the price of butanol (Figure 20.6) (29). For this exercise a corn price of \$73 ton^{-1} was taken into consideration as a base case scenario. A cheaper substrate would likely reduce the cost of butanol. However, this also depends on upstream processing of the substrate and the processes that can be applied for fermentation and recovery.

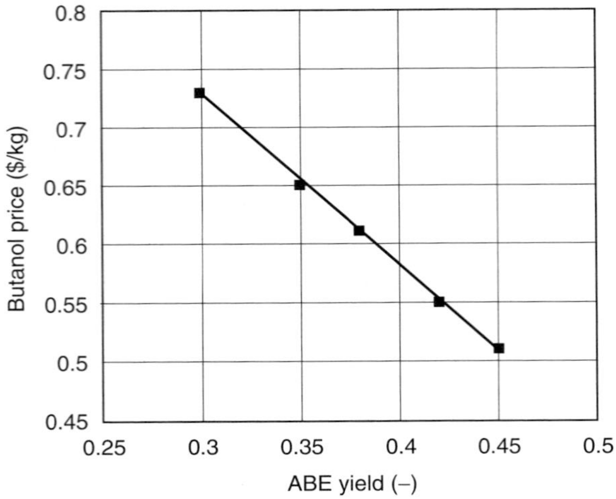

Figure 20.5 Effect of ABE yield on butanol price. [Reproduced with permission from Institution of Chemical Engineers (N Qureshi, HP Blaschek. Economics of butanol fermentation using hyper-butanol producing *Clostridium beijerinckii* BA101. Trans IChemE, Part C, 78:139–144, 2000; Figure 3)]

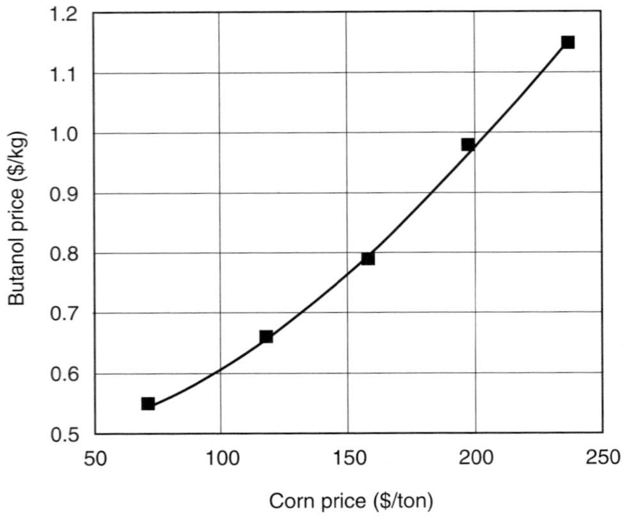

Figure 20.6 Effect of corn price on butanol price. [Reproduced with permission from Institution of Chemical Engineers (N Qureshi, HP Blaschek. Economics of butanol fermentation using hyper-butanol producing *Clostridium beijerinckii* BA101. Trans IChemE, Part C, 78:139–144, 2000; Figure 2)]

The economic recovery of butanol plays a major role in determining the price of butanol (29,92). However, at this stage it appears that integrated product recovery plays an important role in the economics of the butanol fermentation.

20.11 INTEGRATION OF FERMENTATION AND RECOVERY

The major problems associated with butanol fermentation have been uneconomical product recovery, and the use of dilute sugar solution (usually 60 g L^{-1} for batch fermentation), which results in a large reactor and process stream volumes thereby making butanol fermentation noncompetitive when compared to butanol obtained from petrochemical sources. Simultaneous recovery and fermentation when using novel recovery techniques have solved these problems because concentrated sugar solutions (130–350 g L^{-1}) can be used in combination with economic product recovery systems (62,63,68–71). Product recovery techniques such as gas stripping, pervaporation, and liquid–liquid extraction have been applied to high productivity reactors (63,93,94) thereby making butanol fermentation and recovery much more attractive. These reactors offer the advantages of high productivity, use of concentrated sugar solutions, reduction in reactor size, and efficient recovery of product. With these developments only a few problems remain with the butanol fermentation (i.e., low ABE yield). It is anticipated that with the recent developments in integrated fermentation and recovery, the butanol fermentation would be closer to commercialization. Using batch fermentation and distillative recovery the butanol price is estimated to be \$0.55 kg^{-1}, while using integrated technology it is estimated to be \$0.12–0.37 kg^{-1} which is competitive to petrochemically derived butanol. Figure 20.7 shows a schematic diagram of integrated butanol fermentation and recovery process. In a continuous butanol fermentation and product recovery experiment 460 g L^{-1} (reactor volume) total ABE was produced from 1163g L^{-1} (reactor volume) total sugar as opposed to 20–25 g L^{-1} ABE from 60 g L^{-1} sugar in a nonintegrated batch fermentation. Integration of fermentation and product recovery makes the butanol fermentation economically more attractive.

20.12 PROBLEMS OF COMMERCIALIZATION AND FUTURE DEVELOPMENTS

In a recent independent study on butanol commercialization conducted at the College of Commerce at the University of Illinois (Urbana, IL), it was determined that fermentatively produced butanol is competitive with petrochemically derived butanol. This is because of the use of integrated process technology in combination with *C. beijerinckii* BA101 which can produce 25–33 g L^{-1} total ABE in batch process. However, it is anticipated that petrochemical industries would reduce price of butanol in an attempt to prevent the fermentative production of butanol from being successful. Currently, petrochemical industries have

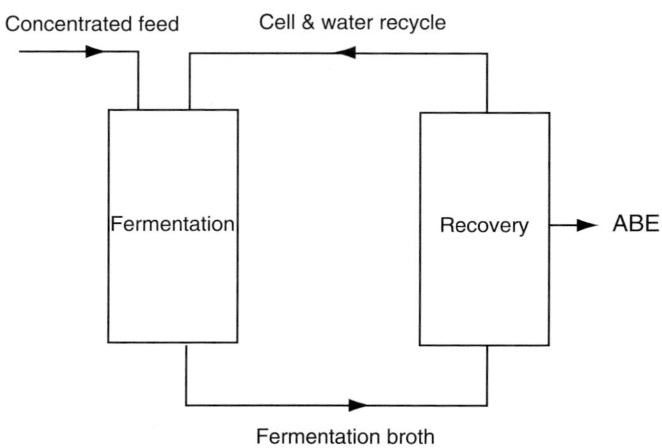

Figure 20.7 A schematic diagram of integrated production of butanol.

a monopoly with respect to butanol market. We suggest that researchers should focus on the following problems.

Although butanol fermentation is more economically competitive today than it was two decades ago, certain problems still remain to be resolved. These include the development of a superior culture that is able to produce elevated levels of butanol. Along these lines attempts have been made, but with limited success. *Clostridium beijerinckii* BA101 and *C. acetobutylicum* PJC4BK are examples of two strains that have been modified to produce elevated levels of solvents. However, the maximum level of solvents that these strains can produce is on the order of 25–33 g L^{-1} which is still low from a product recovery point of view. Another problem is the low ABE yield. Due to the release of CO_2 and H_2 approximately 53% of the substrate is lost in the form of these gases. An improvement in yield would further economize the production of butanol. Alternately, CO_2 could be captured and converted to acetone and butanol. At this time authors are studying the conversion of these gases to solvents.

Another problem with butanol fermentation is the inability of these cultures to use the sugars derived from economically available substrates such as corn fiber hydrolysate (95). As with corn fiber hydrolysate, it is anticipated that sugars derived from hydrolysed corn stalks, wheat straw, and rice husk could not be utilized without pretreatment of these substrates, which would add to the cost. In order to meet these challenges new strains capable of utilizing agricultural biomass derived sugars should be developed. Alternately, economic methods capable of removing inhibitors from the hydrolysates should be developed. Simultaneous saccharification and fermentation is another approach that should be considered.

20.13 BUTANOL FERMENTATION WASTE AND BYPRODUCTS

In the traditional butanol fermentation plants a number of byproducts and waste streams were generated, including a large waste water stream, waste cell mass, and CO_2 and H_2 gases. However, due to recent developments in bioreactor design, process integration, and recovery technology the size of some of these streams has been reduced significantly. As mentioned, the application of process integration allows for the use of concentrated sugar solutions thereby enabling the recycle of water and a reduction in waste disposal streams. Recycling the waste water stream allows for economic benefits in addition to being environmentally friendly. Use of highly productive reactors permits biomass (cell mass) reuse, thereby reducing waste biomass. However, the biomass that is generated can be used as an animal feed. Hydrogen gas can also be separated from the mixture of CO_2 and H_2 and used as an excellent fuel. CO_2 can be recycled either as a substrate (which is under investigation) or sold as a gas for other applications.

ACKNOWLEDGMENTS

N. Qureshi and H. Blaschek would like to thank Michael A. Cotta (United States Department of Agriculture, National Center for Agricultural Utilization Research, Peoria, IL) for reading this chapter and making valuable suggestions.

REFERENCES

1. Jones, D.T., D.R. Woods. Acetone-Butanol fermentation revisited. *Microbiol. Rev.* 50:484–524, 1986.

2. Sherman, P.D. *Kirk-Othmer Encyclopedia of Chemical Technology*, 3rd ed, Vol. 4. New York: John Wiley, 1979, pp 338–328.
3. Nelson, D.L., B.P. Webb. *Kirk-Othmer Encyclopedia of Chemical Technology*, 3rd ed, Vol. 1. New York: John Wiley, 1979, pp 179–190.
4. *Industrial Commodity Statistics*. New York: United Nations, 1998, pp 416–423.
5. Ladisch, M.R. Fermentation-derived butanol and scenarios for its uses in energy-related applications. *Enzyme Microbial Technol.* 13:280–283, 1991.
6. Formanek, J., R. Mackie, H.P. Blaschek. Enhanced butanol production by *Clostridium beijerinckii* BA101 grown in semidefined P2 medium containing 6 percent maltodextrin or glucose. *Appl. Environ. Microbiol.* 63:2306–2310, 1997.
7. Carioca, J.O.B., H.L. Arora, A.S. Khan. Biomass conversion program in Brazil. *Adv. Biochem. Eng.* 20:153–162, 1981.
8. Philips, J.A., A.E. Humphrey. Microbial production of energy: liquid fuels. In: *Biotechnology and Bioprocessing Engineering*, Ghose, T.K., ed., New Delhi: United India Press, 1985, pp 157–186.
9. Qureshi, N., H.P. Blaschek. Recent advances in ABE fermentation: hyper-butanol producing *Clostridium beijerinckii* BA101. *J. Ind. Microbiol. Biotechnol.* 27:287–291, 2001.
10. Manderson, G.J., K. Spencer, A.H.J. Paterson, N. Qureshi, D.E. Janssen. Price sensitivity of bioethanol produced in New Zealand from *Pinus radiata* wood. *Energy Sources* 11:135–150, 1989.
11. Qureshi, N., G.J. Manderson. Bioconversion of renewable resources into ethanol: an economic evaluation of selected hydrolysis, fermentation, and membrane technologies. *Energy Sources* 17:241–265, 1995.
12. Qureshi, N., I.S. Maddox. Application of novel technology to the ABE fermentation process: an economic analysis. *Appl. Biochem. Biotechnol.* 34–35:441–448, 1992.
13. Phillips, J.A., A.E. Humphrey. An overview of process technology for the production of liquid fuels and chemical feedstocks via fermentation. In: *Organic Chemicals from Biomass*, Wise, D.L., ed., Menlo Park, CA: Benjamin Cummings, 1983, pp 249–304.
14. Annous, B.A., H.P. Blaschek. Isolation and characterization of *Clostridium acetobutylicum* mutants with enhanced amylolytic activity. *Appl. Env. Microbiol.* 57:2544–2548, 1991.
15. Chen, C.K., H.P. Blaschek. Acetate enhances solvent production and prevents degeneration in *Clostridium beijerinckii* BA101. *Appl. Microbiol. Biotechnol.* 52:170–173, 1999.
16. Jesse, T.W., T.C. Ezeji, N. Qureshi, H.P. Blaschek. Production of butanol from starch-based packing peanuts and agricultural waste. *J. Ind. Microbiol. Biotechnol.* 29:117–123, 2002.
17. Qureshi, N., A. Lolas, H.P. Blaschek. Soy molasses as fermentation substrate for production of butanol using *Clostridium beijerinckii* BA101. *J. Ind. Microbiol. Biotechnol.* 26:290–295, 2001.
18. Schoutens, G.H., W.J. Groot. Economic feasibility of the production of isopropanol-butanol-ethanol fuel from whey permeate. *Proc. Biochem.* 20:117–121, 1985.
19. Schoutens, G.H., W.J. Groot, J.B.W. Hoebeek. Application of isopropanol-butanol-ethanol mixtures as an engine fuel. *Proc. Biochem.* 21:30, 1986.
20. Bronnenmeier, K., W.L. Staudenbauer. Molecular biology and genetics of substrate utilization in clostridia. In: *The Clostridia and Biotechnology*, Woods. D.R., ed., Oxford: Butterworth-Heinemann, 1993, pp 261–309.
21. Watson, S.A., P.E. Ramstad. *Corn Chemistry and Technology*. St. Paul, Minnesota: American Association of Cereal Chemists, Inc., 1987, pp 72.
22. Beesch, S.C. Acetone-butanol fermentation of sugars. *Eng. Proc. Dev.* 44:1677–1682, 1952.
23. Spivey, M.J. The acetone/butanol/ethanol fermentation. *Process. Biochem.* 13:2–5, 1978.
24. Walton, M.T., J.L. Martin. Production of butanol-acetone by fermentation. In: *Microbial Technology*, 2nd ed., Vol. I. New York: Academic Press, 1979, pp 187–209.
25. Singh, V., S.R. Eckhoff. Effect of soak time, soak temperature, and lactic acid on germ recovery parameters. *Cereal Chem.* 73:716–720, 1996.
26. Campos, E.J., N. Qureshi, H.P. Blaschek. Production of acetone butanol ethanol from degermed corn using *Clostridium beijerinckii* BA101. *Appl. Biochem. Biotechnol.* 98–100:553–561, 2002.

27. Paturau, J.M. *By-products of Cane Sugar Industry: An Introduction To Their Industrialized Utilization*, 3rd ed. New York: Elsevier, 1989, pp 265–285.
28. Maddox, I.S., N. Qureshi, N.A. Gutierrez. Utilization of whey by clostridia and process technology. In: *The Clostridia and Biotechnology*, Woods. D.R., ed., Oxford: Butterworth-Heinemann, 1993, pp 343–369.
29. Qureshi, N., H.P. Blaschek. Economics of butanol fermentation using hyper-butanol producing *Clostridium beijerinckii* BA101. *Trans IChemE (Transactions of Chemical Engineers), Part C* 78:139–144, 2000.
30. Voget, C.E., C.F. Mignone, R.J. Ertola. Butanol production from apple pomance. *Biotechnol. Lett.* 7:43–46, 1985.
31. Grohmann, K., R.J. Bothast. Saccharification of corn fiber by combined treatment with dilute sulphuric acid and enzymes. *Proc. Biochem.* 32:405–415, 1997.
32. Marchal, R., D. Blanchet, J.P. Vandecasteele. Industrial optimization of acetone-butanol fermentation: a study of the utilization of Jerusalem artichokes. *Appl. Microbiol. Biotechnol.* 23:92–98, 1985.
33. Jones, D.T., S. Keis. Origins and relationships of industrial solvent-producing clostridial strains. *FEMS Microbiol. Rev.* 17:223–232, 1995.
34. Linden, J.C., A.R. Moreira, T.G. Lenz. Acetone and butanol. In: *Comprehensive Biotechnology*, Blanch, H.W., S. Drew, D.J.C. Wang, eds., Oxford: Pergamon Press, 1985, pp 915–931.
35. Gutierrez, N.A., I.S. Maddox, K.C. Schuster, H. Swoboda, J.R. Gapes. Strain comparison and medium preparation for the acetone-butanol-ethanol (ABE) fermentation process using a substrate of potato. *Bioresource Technol.* 66:263–265, 1998.
36. Grobben, N.G., G. Eggink, E.P. Cuperus, H.J. Huizing. Production of acetone, butanol and ethanol (ABE) from potato wastes: fermentation with integrated membrane extraction. *Appl. Microbiol. Biotechnol.* 39:494–498, 1993.
37. Montoya, D., S. Spitia, E. Silva, W.H. Schwarz. Isolation of mesophilic solvent-producing clostridia from Colombian sources: physiological characterization, solvent production and polysaccharide hydrolysis. *J. Biotechnol.* 79:117–126, 2000.
38. Parekh, M., J. Formanek, H.P. Blaschek. Pilot-scale production of butanol by *Clostridium beijerinckii* BA101 using a low cost fermentation medium based on corn steep water. *Appl. Microbiol. Biotechnol.* 51:152–157, 1999.
39. Harris, L.M., R.P. Desai, N.E. Welker, E.T. Papoutsakis. Characterization of recombinant strains of the *Clostridium acetobutylicum* butyrate kinase inactivation mutant: need for new phenomenological models for solventogenesis and butanol inhibition? *Biotechnol. Bioeng.* 67:1–11, 2000.
40. Rogers, P., N. Palosaari. *Clostridium acetobutylicum* mutants that produce butyraldehyde and altered quantities of solvents. *Appl. Environ. Microbiol.* 53:2761–2766, 1987.
41. Stim-Herndon, K.P., R. Nair, E.T. Papoutsakis, G.N. Bennet. Analysis of degenerate variants of *Clostridium acetobutylicum* ATCC 824. *Anaerobe* 2:11–18, 1996.
42. Cornillot, E.R., V. Nair, E.T. Papoutsakis, P. Soucaille. The genes for butanol and acetone formation in *Clostridium acetobutylicum* ATCC 824 reside on a large plasmid whose loss leads to degeneration of the strain. *J. Bacteriol.* 179:5442–5447, 1997.
43. Wilkinson, S.R., M. Young. Targeted integration of genes into the *Clostridium acetobutylicum* chromosome. *Microbiology* 140:89–95, 1994.
44. Chen, C.K., H.P. Blaschek. Effect of acetate on molecular and physiological aspects of *Clostridium beijerinckii* NCIMB 8052 solvent production and strain degeneration. *Appl. Env. Microbiol.* 65(2):499–505, 1999.
45. Chen, C.K., H.P. Blaschek. Examination of physiological and molecular factors involved in enhanced solvent production by *Clostridium beijerinckii* BA101. *Appl. Environ. Microbiol.* 65:2269–2271, 1999.
46. Ravagnani, A., K. Jennert, E. Steiner, R. Grunberg, J.R. Jefferies, S.R.Wilkinson, D.I. Young, E.C. Tidswell, D.P. Brown, P. Youngman, J.G. Morris, M. Young. Spo0A directly controls the switch from acid to solvent production in solvent-forming clostridia. *Mol. Microbiol.* 37(5):1172–1185, 2000.

47. Baer, S., H.P. Blaschek, T.L. Smith. Effect of butanol challenge and temperature on the lipid composition and membrane fluidity of butanol-tolerant *Clostridium acetobutylicum*. *Appl. Env. Microbiol.* 53:2854–2861, 1987.
48. Li, Y., H.P. Blaschek. Molecular characterization and utilization of the CAK1 filamentous viruslike particle derived from *Clostridium beijerinckii*. *J. Ind. Microbiol. Biotechnol.* 28:118–126, 2002.
49. Lee, J., W.J. Mitchell, H. Blaschek. Glucose uptake in *Clostridium beijerinckii* NCIMB 8052 and the solvent-hyperproducing mutant BA101. *Appl. Environ. Microbiol.* 57(11):5035–5031, 2001.
50. Mitchell, W.J. Physiology of carbohydrate to solvent conversion by clostridia. *Adv. Microbiol. Physiol.* 39:31–130, 1998.
51. Lee, J., W.J. Mitchell, M. Tangney, H.P. Blaschek. Evidence for the presence of an alternative glucose transport system in *Clostridium beijerinckii* NCIMB 8052 and the solvent hyper-producing BA101 mutant. 71:3384–3387, 2005.
52. Leung, J.C.Y., D.I.C. Wang. Production of acetone and butanol by *Clostridium acetobutylicum* in continuous culture using free cells and immobilized cells. *Proc. 2nd World Cong. Chem. Eng.* 1:348–352, 1981.
53. Mollah, A.H., D.C. Stuckey. The influence of H_2, CO_2 and dilution rate on the continuous fermentation of acetone-butanol. *Appl. Microbiol. Biotechnol.* 37:533–538, 1992.
54. Mulchandani, A., B. Volesky. Production of acetone-butanol-ethanol by *Clostridium acetobutylicum* using a spin filter perfusion bioreactor. *J. Biotechnol.* 34:51–60, 1994.
55. Bahl, H., W. Andersch, G. Gottschalk. Continuous production of acetone and butanol by *Clostridium acetobutylicum* in a two-stage phosphate limited chemostat. *Eur. J. Appl. Microbiol. Biotechnol.* 15:201–205, 1982.
56. Yarovenko, V.L. Principles of the continuous alcohol and butanol-acetone fermentation processes. In: *Continuous cultivation of microorganisms: 2nd Symposium*, Malek, I., ed., Prague: Czechoslovakian Academy of Sciences, pp 205–217, 1964.
57. Afschar, A.S., H. Biebl, K. Schaller, K. Schugerl. Production of acetone and butanol by *Clostridium acetobutylicum* in continuous culture with cell recycle. *Appl. Microbiol. Biotechnol.* 22:394–398, 1985.
58. Maddox, I.S. The acetone-butanol-ethanol fermentation: recent progress in technology. *Biotechnol. Genet. Eng. Rev.* 7:190–220, 1989.
59. Groot, W.J., R.G.J.M. van der Lans, K.Ch.A.M. Luyben. Technologies for butanol recovery integrated with fermentations. *Proc. Biochem.* 27:61–75, 1992.
60. Qureshi, N., H.P. Blaschek. Recovery of butanol from fermentation broth by gas stripping. *Renewable Energy: An International Journal* 22:557–564, 2001.
61. Ennis, B.M., C.T. Marshall, I.S. Maddox, A.H.J. Paterson. Continuous product recovery by *in-situ* gas stripping/condensation during solvent production from whey permeate using *Clostridium acetobutylicum*. *Biotechnol. Lett.* 8:725–730, 1986.
62. Maddox, I.S., N. Qureshi, K. Roberts-Thompson. Production of acetone-butanol-ethanol from concentrated substrates using *Clostridium acetobutylicum* in an integrated fermentation-product removal process. *Proc. Biochem.* 30:209–215, 1995.
63. Qureshi, N., I.S. Maddox. Integration of continuous production and recovery of solvents from whey permeate: use of immobilized cells of *Clostridium acetobutylicum* in fluidized bed bioreactor coupled with gas stripping. *Bioproc. Eng.* 6:63–69, 1991.
64. Groot, W.J., R.G.J.M. van der Lans, K.Ch.A.M. Luyben. Batch and continuous butanol fermentation with free cells: integration with product recovery by gas stripping. *Appl. Microbiol. Biotechnol.* 32:305–308, 1989.
65. Ezeji, T.C., N. Qureshi, H.P. Blaschek. Continuous production of butanol using *Clostridium beijerinckii* BA101 and recovery by gas stripping. *Clostridium 2002: The Seventh International Conference and Workshop on Regulation of Metabolism, Genetics and Development of the Solvent and Acid Forming Clostridia*. Rostock, Germany, September 19–21, 2002.
66. Groot, W.J., C.E. van den Oever, N.W.F. Kossen. Pervaporation for simultaneous product recovery in the butanol/isopropanol batch fermentation. *Biotechnol. Lett.* 6:709–714, 1984.

67. Larrayoz, M.A., L. Puigjaner. Study of butanol extraction through pervaporation in acetobutylic fermentation. *Biotechnol. Bioeng*. 30:692–696, 1987.
68. Qureshi, N., H.P. Blaschek. Production of acetone-butanol-ethanol (ABE) by a hyper-butanol producing mutant strain of *Clostridium beijerinckii* BA101 and recovery by pervaporation. *Biotechnol. Prog*. 15:594–602, 1999.
69. Qureshi, N., H.P. Blaschek. Butanol production using hyper-butanol producing mutant strain of *Clostridium beijerinckii* BA101 and recovery by pervaporation. *Appl. Biochem. Biotechnol*. 84:225–235, 2000.
70. Qureshi, N., M.M. Meagher, J. Huang, R.W. Hutkins. Acetone butanol ethanol (ABE) recovery by pervaporation using silicalite-silicone composite membrane from fed-batch reactor of *Clostridium acetobutylicum. J. Membrane Sci*. 187:93–102, 2002.
71. Qureshi, N., I.S. Maddox, A. Friedl. Application of continuous substrate feeding to the ABE fermentation: relief of product inhibition using extraction, perstraction, stripping and pervaporation. *Biotechnol. Prog*. 8:382–390, 1992.
72. Matsumura, M., H. Kataoka, M. Sueki, K. Araki. Energy saving effect of pervaporation using oleyl alcohol liquid membrane in butanol purification. *Bioproc. Eng*. 3:93–100, 1988.
73. Qureshi, N., M.M. Meagher, R.W. Hutkins. Recovery of butanol from model solutions and fermentation broth using a silicalite/silicone membrane. *J. Membrane Sci*. 158:115–125, 1999.
74. Qureshi, N., H.P. Blaschek. Butanol recovery from model solution/fermentation broth by pervaporation: evaluation of membrane performance. *Biomass and Bioenergy* 17:175–184, 1999.
75 Roffler, S.R., H.W. Blanch, C.R. Wilke. *In-situ* recovery of butanol during fermentation, part I: batch extractive fermentation. *Bioproc. Eng*. 2:1–12, 1987.
76. Mattiasson, B., M. Larsson. Extractive bioconversions with emphasis on solvent production. *Biotechnol. Genet. Eng. Rev*. 3:137–174, 1985.
77. Ennis, B.M., N.A. Gutierrez, I.S. Maddox. The acetone-butanol-ethanol fermentation: a current assessment. *Process Biochem* 21:131–147, 1986.
78. Wang, D.I.C., C.L. Cooney, A.L. Demain, R.F. Gomez, A.J. Sinskey. Production of acetone and butanol by fermentation. *MIT Quarterly report to the US Department of Energy*, COG-4198-9, 141–149, 1979.
79. Taya, M., S. Ishii, T. Kobayashi. Monitoring and control for extractive fermentation of *Clostridium acetobutylicum. J. Ferment. Technol*. 63:181–187, 1985.
80. Roffler, S.R., H.W. Blanch, C.R. Wilke. *In-situ* recovery of butanol during fermentation, part 2: fed-batch extractive fermentation. *Bioproc. Eng*. 2:181–190, 1987.
81. Traxler, R.W., E.M. Wood, J. Mayer, M.P. Wilson. Extractive fermentation for the production of butanol. *Dev. Ind. Microbiol*. 26: 519–525, 1985.
82. Eckert, G., K. Schugerl. Continuous acetone-butanol production with direct product removal. *Appl. Microbiol. Biotechnol*. 27:221–228, 1987.
83. Evans, P.J., H.Y. Wang. Enhancement of butanol formation by *Clostridium acetobutylicum* in the presence of decanol-oleyl alcohol mixed extractants. *Appl. Environ. Microbiol.* 54:1662–1667, 1988.
84. Ennis, B.M., N. Qureshi, I.S. Maddox. In-line toxic product removal during solvent production by continuous fermentation using immobilized *Clostridium acetobutylicum. Enzyme Microbial Technol*. 9:672–675, 1987.
85. Garcia, A., E.L. Ianotti, J.L. Fisher. Butanol fermentation liquor production and separation by reverse osmosis. *Biotechnol. Bioeng*. 28:785–791, 1986.
86. Larsson, M., B. Mattiasson. Novel process technology for biotechnological solvent production. *Chem. Ind*. 12:428–431, 1984.
87. Ezeji, T.C., N. Qureshi, H.P. Blaschek. Production of butanol by *Clostridium beijerinckii* BA101and *in-situ* recovery by gas stripping. *World J. Microbiol. Biotechnol*., 19:595–603 2003.
88. Lienhardt, J., J. Schripsema, N. Qureshi, H.P. Blaschek. Acetone butanol ethanol (ABE) production in an immobilized cell reactor of *Clostridium beijerinckii* BA101 and recovery by pervaporation. *Appl. Biochem. Biotechnol.* 98:591–598, 2002.

89. Qureshi, N., I.S. Maddox. Continuous solvent production from whey permeate using cells of *Clostridium acetobutylicum* immobilized by adsorption onto bonechar. *Enzyme Microbial Technol*. 9:668–671, 1987.
90. Qureshi, N., I.S. Maddox. Reactor design for the ABE fermentation using cells of *Clostridium acetobutylicum* immobilized by adsorption onto bonechar. *Bioproc. Eng*. 3:69–72, 1988.
91. Qureshi, N., J. Schripsema, J. Lienhardt, H.P. Blaschek. Continuous solvent production by *Clostridium beijerinckii* BA101 immobilized by adsorption onto brick. *World J. Microbiol. Biotechnol.* 16:377–382, 2000.
92. Qureshi, N., H.P. Blaschek. Evaluation of recent advances in butanol fermentation, upstream, and downstream processing. *Bioproc. Biosystems Eng*. 24:219–226, 2001.
93. Qureshi, N., I.S. Maddox. Continuous production of acetone butanol ethanol using immobilized cells of *Clostridium acetobutylicum* and integration with product removal by liquid-liquid extraction. *J. Ferment*. Bioeng. 80:185–189, 1995.
94. Friedl, A., N. Qureshi, I.S. Maddox. Continuous Acetone-Butanol-Ethanol (ABE) fermentation using immobilized cells of *Clostridium acetobutylicum* in a packed bed reactor and integration with product removal by pervaporation. *Biotechnol. Bioeng*. 38:518–527, 1991.
95. Ebener, J., N. Qureshi, H.P. Blaschek. Corn fiber hydrolysis and fermentation to butanol using *Clostridium beijerinckii* BA101. In: *Proceedings of 25th Biotechnology Symposium on Fuels and Chemicals*, Colorado, 65, 2003.

Section 2

Plant and Animal Food Applications and Functional Foods

2.01

Methods in Plant Tissue Culture

Hector G. Nunez-Palenius, Daniel J. Cantliffe, Harry H. Klee, Neftali Ochoa-Alejo, Rafael Ramirez-Malagon, and Eugenio Perez-Molphe

CONTENTS

1.1 INTRODUCTION

For thousands of years people have benefited from the things provided by the higher plants, things such as food, building materials, dyes, fragrances, and medicines. As human beings domesticated plants, they were able to take advantage of plant genetic resources, leading to assorted varieties and cultivars of plants that helped feed, cloth, and shelter people. Nevertheless, world population has risen considerably during the last century, compromising worldwide food reserves. Agriculture is expected to feed an increasing human population, forecast to reach 8 billion by 2020 (1).

Plant biotechnology could present feasible alternatives which could greatly help overcome the risks of food shortages. Within the plant biotechnology discipline, plant tissue culture methods have played an essential role, allowing the development of transgenic plants with a number of desirable agronomic and food traits. What is plant tissue culture? The common understanding is that the term "plant tissue culture" generally refers to *in vitro* cultivation, on nutrient media, of any plant component, under a sterile environment. The plant component may be a single cell, tissue, or an organ. The result may be an entire *de novo* regenerated plant, or the utilization of the plant tissue culture technique to further research in plant biology.

1.2 HISTORY

The first recorded attempt at plant tissue culture was in 1902, when Haberlandt (2) placed epidermal, pith parenchyma, and palisade cells of monocotyledons in culture containing various sugars, inorganic salts, and amino acids. None of the cells divided. The first successful attempt at tissue culture was recorded in 1904, when Hannig (3), using similar media, was able to grow whole plants from embryos.

During the next two decades, work centered on improving embryo culture techniques. It was found that several complex additives, like yeast extract, coconut milk, and peptone, greatly stimulated growth. In 1934, White (4) placed excised tomato root tips in culture. When yeast extract was added to the medium, the roots were able to divide and grow indefinitely.

In 1936, Brown and Gardener found that tumors, similar to the tumors induced by crown gall bacteria, could be mimicked by exogenous auxin applications. This information was quickly applied to tissue culture. In 1939, Gautheret (5), Nobecourt (6), and White (7) each independently demonstrated that auxin indoleacetic acid, when incorporated into culture media, stimulated the production of undifferentiated callus. This was the first time hormones were added to tissue culture media. In 1956, Miller and others (8) discovered that the active ingredient in coconut milk and yeast extract which promoted growth in culture was a new class of hormones, called cytokinin. Skoog and Miller (9) further reported in 1957 that the auxin–cytokinin balance was very important for growth of cells in culture. If auxin levels were relatively high and cytokinin level relatively low, undifferentiated callus growth occurred. High levels of cytokinin relative to auxin induced the appearance of differentiated shoots. Thus, it became possible to regenerate a whole plant from a single undifferentiated cell.

In 1962, Murashige and Skoog (10) published the details of a distinct medium with high NH_4 levels, for tobacco cells; a paper most commonly cited in publications regarding tissue culture. This medium has given excellent results with both monocot and dicot species. A further advance came in 1964 when Guha and Maheshwari (11) placed Datura anthers in culture. From these anthers, embryoids formed which developed into haploid plants. Tissue cultures of haploid plants were a major advance for genetic studies. Vasil and Hildebrandt (12) were able to regenerate a whole plant from single tobacco cells, using a micro-culture method with fresh medium, enriched with coconut water.

In 1971, Takebe et al. (13) demonstrated that it was possible to regenerate whole plants from protoplasts. Protoplasts are plant cells without cell walls, usually produced by incubating tissue in crude cellulolytic preparations. The first somatic hybrid plant was produced in 1972, when Carlson and colleagues (14) fused protoplasts from two tobacco species and regenerated the hybrid cell into a whole plant.

In 1974, Zaenen et al. (15) discovered that a plasmid (Ti plasmid) in *Agrobacterium tumefaciens* was responsible for inducing tumors in crown gall-infected plants. Through the use of mutants, Ooms et al. (16) demonstrated that genes on the Ti plasmid affected a host plant's auxin–cytokinin balance, showing that one gene controlled auxin metabolism and other the cytokinin metabolism. In 1983, the first example of the expression of a foreign protein in a plant cell was demonstrated by Fraley and others (17). They inserted into the Ti plasmid a gene for kanamycin resistance. Resultant petunia and tobacco cells infected with this hybrid Ti plasmid were resistant to the bactericide kanamycin. In 1984 Horsch et al. (18) reported that the transformed cells were regenerated into whole plants and that the kanamycin-resistant gene was inherited as a single gene dominant trait. Genetic engineering was no longer a dream, thanks to plant tissue culture methods.

Throughout the last 70 years, plant tissue culture methods have evolved dramatically. Many techniques and methods have been developed since Haberlandt's early attempts. This technology has now been used for plant propagation, embryo culture, haploid production, germplasm conservation, mutant production, hybridization, or cell fusion, developmental studies, and metabolic studies, including secondary metabolites. Tissue culture technology has had a tremendous impact in both basic and applied plant science.

1.3 LABORATORY OPERATIONS

1.3.1 Laboratory Facilities

To design any type of facility for plant tissue culture operations, either small or large, the most important consideration must be organization and sanitation. The success of sanitation procedures requires proper work techniques by the operator as well as the correct design of the facilities used for tissue culture. Street (19) has recommended that individual module facilities should be prearranged as in a production line. As Wetherell (20), and Kyte (21) pointed out, any plant tissue culture facility must have three main basic areas (laboratory, transfer room, and culture room) and must be designed to provide proper aseptic conditions. Windows and doors should be sliding, in order to avoid air currents that could carry spores, vegetative cells, or other reproductive structures of bacteria or fungi. On the other hand, several authors (22,23) have considered plant tissue culture facilities equipped with up to 10 different rooms. Explicit plant tissue culture laboratory designs, material, tools, and equipment can be found in Street (19), Wetherell (20), Kyte (21), George (23), and Gamborg and Phillips (24).

1.3.2 Sterile Techniques

As previously pointed out, the preservation of sterile conditions is critical for successful tissue plant culture procedures. All media, instruments, tools, and culture vessels must be sterilized to remove all contamination. Contamination may be due to bacteria, yeast, virus, or fungi. Nevertheless, in some cases even mites [(20), RM Rangel, personal communication, 1997] or thrips (24) can thrive in the moist environment of plant tissue cultures, causing disastrous contaminations.

Sterilized plant material, nutrient media, equipment, and tools must be handled using strict aseptic practices. Culture manipulation could be carried out on an open laboratory bench, but this is a very unsafe course of action because of increased chances for contamination. For this reason culture manipulations are done in a transfer room with air flow cabinets (Figure 1.1). The transfer room is an undersized space within the building, which is sterilized by ultraviolet irradiation lamps, used when workers are not present. Air is

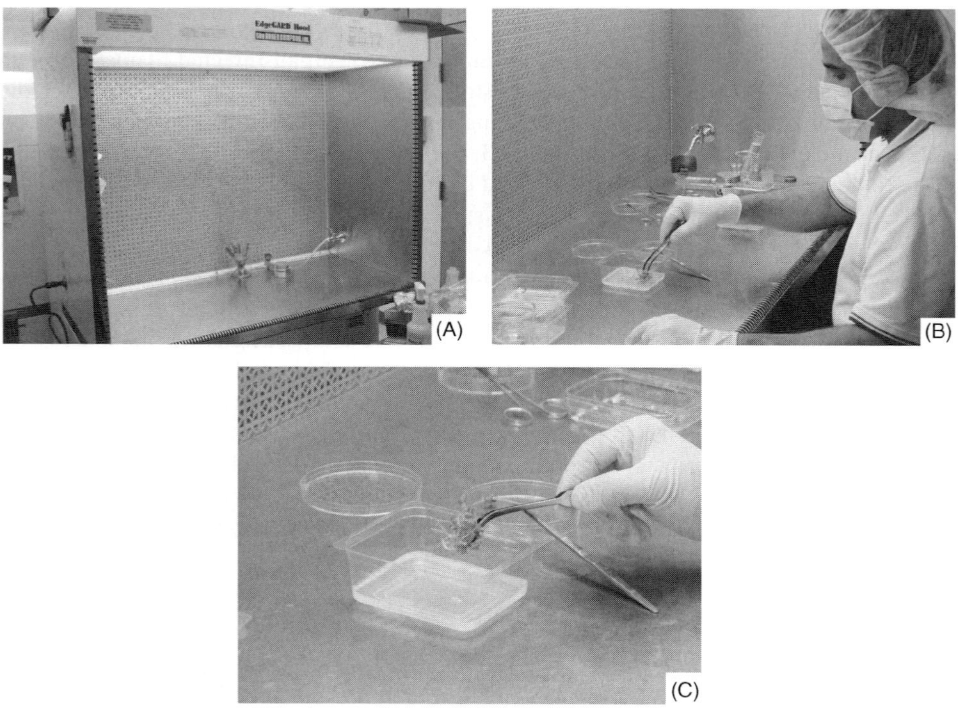

Figure 1.1 Laminar air-flow cabinet for *in virto* culture. **A**, ready for use. **B** and **C**, in use for strawberry micropropagation. (**B** and **C** courtesy of Yousef Dlaigan).

cleaned via a filter system. Laminar air flow cabinets are more common and more popular than transfer rooms, mostly because they are less expensive, easier to maintain, and safer, as they do not require UV lamps. There are two basic types of laminar air-flow cabinets, both equipped with a high-efficiency particulate air (HEPA) filter; vertical-air-flow and horizontal-air-flow. The former model seems to provide a better and more continuously sterile environment (25).

Methods of sterilization may be classified as dry heat, moist heat, radiation, microwaves, membrane filtration, and chemical. These sterilization procedures can be applied in sterilization of media and containers and in preparation of axenic or sterile explants (26).

Dry heat is used most with metal instruments, glassware, and other nonorganic materials able to tolerate high (180°C) temperatures, with no presence of moisture (Figure 1.2A). On the other hand, moist heat, steam sterilization, and autoclaving are commonly used to sterilize

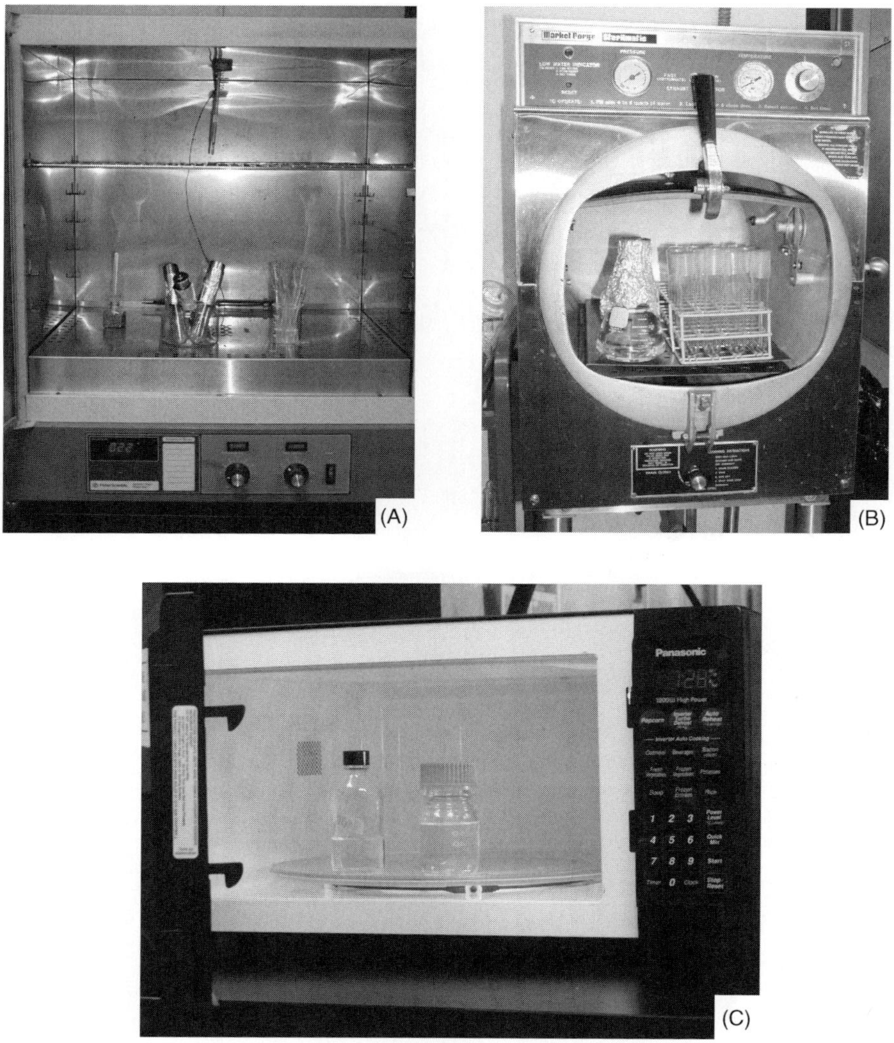

Figure 1.2 **A** Oven for dry-heat sterilization. **B** Autoclave for moist-heat sterilization. **C** Media-microwave sterilization.

nutrient media, distilled water, paper products, heat tolerant plastic ware, and contaminated material. A constant pressure of 1.06 kg cm^{-2} (105 kPa) and 121°C (250°F) for 15–20 minutes usually kills most living tissue (Figure 1.2B). Gamma (γ)-irradiation (radiation) is most frequently used with plastic materials, using a viable irradiation plant. Media microwave sterilization (Figure 1.2C) can be completed utilizing commercial microwaves, though the correct conditions for obtaining sterility have to be empirically obtained (27). Membrane filtration or microfiltration is a simple procedure easily achieved using Millipore membranes of 0.025–0.45 μm pore size and a syringe (Figure 1.3A). This method is generally useful for thermosensitive substances, such as vitamins, plant growth regulators, and complex organic additives, among others. Chemicals may be used on plant materials. Ethanol (70–80% concentration) is used for flaming metallic tools and to disinfect the air flow cabinet (Figure 1.3B).

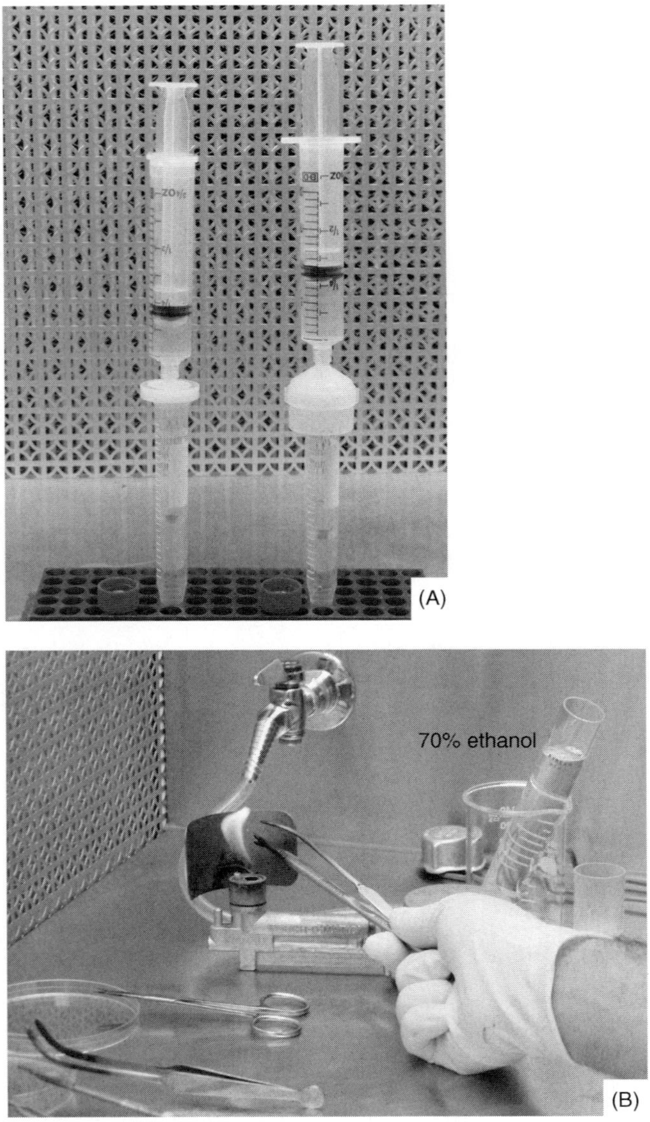

Figure 1.3 **A** Membrane sterilization by microfiltration. **B** flaming metallic tools.

Plant or explant surface sterilization may be achieved using a number of chemical substances. Table 1.1 depicts several chemicals regularly used in plant material surface sterilization, although the most appropriate chemical agent, its concentration, the sequential combination of agents, and the length of sterilization must be determined empirically for different plant materials.

1.3.3 Media Composition, Preparation, and Handling

Plant cells, tissues, and organs only have normal *in vitro* growth when supplied with nutrients and the appropriate environmental conditions. Moreover, different cells, tissues, and organs will have varying nutrient requirements: thus, there is no one single medium competent to sustain growth for all types of *in vitro* cultures. As a result, the best nutrient media for any specific cell, tissue, or organ generally has to be determined by trial and error. Nonetheless, related plant cell genotypes will have similar needs, reducing the size of preliminary experiments. Numerous formulations have been published for several types of tissues and are able to sustain good *in vitro* plant growth with diverse tissues.

A suitable plant culture medium must contain at least the known essential inorganic salts (macro- and micronutrients) in proper balance, as well as a carbon source for energy, frequently sucrose. In addition, the medium should contain water soluble vitamins and one or more plant growth regulators, because plant cells are heterotrophic when they are cultured *in vitro*. Organic nitrogen, organic acids, and complex extracts such as coconut water, casein hydrolysate, yeast extract, tomato juice, orange juice, or malt extract, are frequently added to the medium. But these substances are added for specific purposes, such as in ovule and

Table 1.1
Surface Sterilizing Chemicals for Plant Tissue Culture

Chemical	Concentration	Time Length (min)
Sodium hypochlorite[1]	0.9–4%	5–40
Calcium hypochlorite[1]	9–12%	5–40
Hydrogen peroxide	5–15%	5–30
Bromine water	1–2%	2–10
Silver nitrate	1%	5–30
Mercuric chloride[2]	0.1–1%	0.5–30
Ethanol	70–95%	0.5–5
Chlorine gas[3]	--	30–180
Tincture of iodine	0.005%	15–20
Chloramine B	3%	5–20
Chloramine T	5%	10–30
Potassium permanganate	0.01%	4–32
Halozone[4]	0.04%	30
Zephiran[5]	0.01–0.1%	5–30
Antibiotics	4–500 mg l^{-1}	30–240
Fungicides	6*	6*

[1]w/v solution. It equates to 5–40% v/v of a commercial bleach solution.
[2]Extremelly poison and environmentally hazardous
[3]Caution, very toxic!
[4]*p*-carboxybenzenesulphodichloramide
[5]Benzalkonium chloride
[6]*Follow manufacturer instructions

anther cultures and plant specific media. The pH range for most plant tissue culture media is usually around 5–6. Maintaining proper pH is very important for *in vitro* culture because it determines the solubility and availability of the mineral ions and modifies gel properties of the agar. Any pH adjustments have to be done before adding the gelling agent (e.g., agar, agarose, gellan gum). These gelling agents are obligatory for semisolid media preparation. Owen et al. (29) claimed that culture medium pH is influenced by basal medium, carbohydrate source, gelling agent, activated charcoal, and medium storage method. When certain pH levels have to be kept constant through culture, a 2-Morpholinoethanesulfonic acid (MES) buffer can be added, following previously reported protocols (28).

Optimization for plant growth and regeneration of specific cell and tissues from diverse species might require adjustments to current media, rather than new formulations of nutrient media. Table 1.2 lists the composition of the most commonly used plant tissue culture media. Furthermore, Leifert et al. (30) has published an extensive and valuable review on mineral and carbohydrate nutrition of plant cell and tissues cultures.

Inorganic nitrogen is most commonly supplied as nitrate, but urea or ammonium ions may be used as well. Nitrogen, as an organic source, is provided as amino acid (glycine, asparagine, glutamine, aspartic acid, glutamic acid, and cysteine, among others), or casein-hydrolyzed. The nitrogen source has an important effect on plant cell metabolism and developmental response (26,31) For instance, an elevated level of ammonium and nitrate are used for callus culture, low concentrations of mineral salts are needed for shoot and root cultures, and a half strength medium favors cell differentiation and plant regeneration.

As previously pointed out, sucrose is the most commonly used carbon source in culture media. However, fructose and glucose can support similar or higher cell growth rates than sucrose in certain species such as *Catharanthus roseus* cells (32). Moreover, various lines and mutant rice cells are capable of using mannose, ribose, galactose, sorbitol, or even acetate as the only energy source (33).

Plants normally are competent to synthesize all required vitamins, but plant cells in culture are dependent on an exogenous source. Because of that, several vitamins may be included to the media. Thiamine is by far the most common vitamin added, because it is an essential cofactor in carbohydrate metabolism and is directly involved in the biosynthesis of some amino acids. However nicotinic acid, pyridoxine, folic acid, calcium pantothenate, choline chloride, *p*-amino-benzoic acid, and biotin are added as supplements, as well. *Myo*-inositol is an additional organic compound that functions as a vitamin (complex B) and it can be found in most of the media. Ascorbic acid (vitamin C) and citric acid, which are antioxidants, are added to media in certain cases in order to avoid excessive tissue oxidation or browning reactions.

Most plant *in vitro* cultures require exogenous plant growth regulators or plant hormones for differentiation and growth. Hormones can maintain or arrest growth *in* vitro. As growth begins, endogenous plant hormone levels of the explant will eventually start to be synthesized. As growth requirements change during culture, tissues are subcultured and hormone content can be altered. There are six main types of plant growth regulators: auxins, cytokinins, gibberellins, abscisic acid, brassinosteroids, and ethylene. Table 1.3 depicts some of most commonly used plant regulators in cell, tissue, and organ culture. Auxins and cytokinins are most widely used. An appropriate balance of these two hormones is required to induce either shoots (caulogenesis) or roots (rhizogenesis).

Plant tissue culture nutrient media can be obtained by preparing it from "stock solutions" or by buying basic commercial media from commercial companies. Once the medium is made, it is preferable to let it stand for 2–4 days at 25°C to watch for possible contamination. Media storage for long periods should be at 0–4°C, in order to avoid nutrient degradation and potential microorganism growth.

Table 1.2
Chemical Composition of Several Widely Used Tissue Culture Media (mg l^{-1})

Constituents	MS	LM	SH	White's	Heller's	B_5	Nitsch's	N_6
Inorganic								
NH_4NO_3	1650	1650	—	—	—	—	720	—
$NH_4H_2PO_4$	—	—	300	—	—	—	—	—
KNO_3	1900	1900	2500	80	—	2500	950	2830
$CaCl_2{\cdot}2H_2O$	440	440	200	—	75	150	—	166
$CaCl_2$	—	—	—	—	—	—	166	—
$MgSO_4{\cdot}7H_2O$	370	370	400	750	250	250	185	185
KH_2PO_4	170	170	—	—	—	—	68	400
$(NH_4)_2SO_4$	—	—	—	—	—	134	—	463
$Ca(NO_3)_2{\cdot}4H_2O$	—	—	—	300	—	—	—	—
$NaNO_3$	—	—	—	—	600	—	—	—
Na_2SO_4	—	—	—	200	—	—	—	—
$NaH_2PO_4{\cdot}H_2O$	—	—	—	19	125	150	—	—
KCl	—	—	—	65	750	—	—	—
KI	0.83	0.83	1	0.75	0.01	0.75	—	0.8
H_3BO_3	6.2	6.2	5	1.5	1	3	10	1.6
$MnSO_4{\cdot}4H_2O$	22.3	22.3	—	5	0.1	—	25	4.4
$MnSO_4{\cdot}H_2O$	—	—	10	—	—	10	—	—
$ZnSO_4{\cdot}7H_2O$	8.6	8.6	1	3	1	2	10	1.5
$Na_2MoO_4{\cdot}2H_2O$	0.25	0.25	0.1	—	—	0.25	0.25	—
MoO_3	—	—	—	0.001	—	—	—	—
$CuSO_4{\cdot}5H_2O$	0.025	0.025	0.2	0.01	0.3	0.025	0.025	—
$CoCl_2{\cdot}6H_2O$	0.025	0.025	0.1	—	—	0.025	—	—
$AlCl_3$	—	—	—	—	0.03	—	—	—
$NiCl_2{\cdot}6H2O$	—	—	—	—	0.03	—	—	—

(*Continued*)

Table 1.2 *(Continued)*

Constituents	MS	LM	SH	White's	Heller's	B_5	Nitsch's	N_6
$FeCl_3$·6H2O	—	—	—	—	1	—	—	—
$Fe_2(SO_4)_3$	—	—	—	2.5	—	—	—	—
$FeSO_4$·7H2O	27.8	27.8	20	—	—	—	27.8	27.8
Na_2·EDTA·$2H_2O$	37.3	37.3	15	—	—	—	37.3	37.3
Sequest. 330Fe	—	—	—	—	—	28	—	—
Organic								
Myo-inositol	100	100	1000	—	—	100	100	—
Nicotinic acid	0.5	—	5	0.05	—	1	5	0.5
Pyridoxine-HCl	0.5	—	0.5	0.01	—	1	0.5	0.5
Thiamine-HCl	0.1	0.4	5	0.01	—	10	0.5	1
Glycine	2	—	—	3	—	—	2	2
Folic acid	—	—	—	—	—	—	0.5	—
Biotin	—	—	—	—	—	—	0.05	—
Sucrose	3%	3%	3%	2%	3%	2%	2%	5%
pH	5.7–5.8	5.6	5.9	5.6	5.8	5.7	5.5	5.8

Note: Plant growth regulators and nutrient mixtures described by particular authors are not reported here.

MS (Murashige & Skoog. 1962. Physiol. Plant., 15, 473)

LM (Linsmaier & Skoog. 1965. Physiol. Plant., 18, 100)

SH (Schenk & Hildebrand. 1972. Can. J. Bot. 50, 199)

White's (White. 1963. The Cultivation of Animal and Plant Cells, Second Edition. The Ronald Press Company, New York)

Heller's (Heller. 1953. *Ann. Sic. Nat Biol. Veg.* 14:1–233)

B_5 (Gamborg *et al.* 1968. Exp. Cell Res., 50, 151)

Nitsch's (Nitsch & Nitsch. 1969. Science. 163, 85)

N_6 (Chu *et al.* 1975. Scientia Sinic. 18, 659)

Table 1.3
Plant Growth Regulators Chemicals Commonly Used for Plant Tissue Culture

Name	**Abbreviation**
Auxins	
3-indolyl-acetic acid	IAA
2,4-dichlorophenoxyacetic acid	2,4-D
3-indolebutiric acid	IBA
1-naphthaleneacetic acid	NAA
2-naphthyloxyacetic acid	NOA
4-chlorophenoxyacetic acid	4-CPA
p-chlorophenoxyacetic acid	PCPA
2-methyl-4-chlorphenoxyacetic acid	MCPA
2,4,5,-trichlorophenoxyacetic acid	2,4,5-T
3,6-dichloroanisic acid	dicamba
4-amino-3,5,6-trichloropicolinic acid	picloram
Cytokinins	
4-hydroxy-3-methyl-*trans*-2-butenylaminopurine	Zeatin
6-(4-hydroxy-3-methyl-*trans*-2-butenyl)aminopurine	Dihydrozeatin
6-(γ,γ-dimethylallylamino)purine	2-iP
Benzyladenine (6-benzylaminopurine)	BA
Adenine sulphate	AS
6-furfurylaminopurine	Kinetin
N-(2-chloro-4-pyridyl)-*N*'-phenylurea	CPPU
N-(phenyl)-*N*'-1,2,3-thiadiazol-5-ylurea	Thidiazuron (TDZ)
Gibberellins	
Gibberellic acid A_3	GA_3
Gibberellic acid A_4	GA_4
Gibberellic acid A_7	GA_7
Abscisic acid	
cis-(+) abscisic acid	ABA
Brassinosteroids	
24-epibrassinolide	24-EB
24-epicastasterone	24-EC
Ethylene inhibitors	
Silver thiosulphate	AgS_2O_3
Aminoethoxyvinylglycine	AVG
2,5-norbornadiene	NBD
Cobalt chloride	$CoCl_2$

As a general rule, nutrient media are sterilized in an autoclave (Figure 1.2B). Biondi and Thorpe (34) have recommended increasing sterilization time as media volume increases. For vessels having 10 to 20 ml of media, 20 min is adequate, while for flasks with 1 l of media, 40 min are needed because larger volumes heat more slowly. However, media oversterilization must be avoided, as this can result in degradation of media constituents and caramelization of sugars.

Sources of contamination for media may include the culture vessel, the medium itself, the explant, the environment of the transfer area, instruments and tools, and conditions in the culture room. All these sources have to be rigorously checked and meticulously sterilized to avoid contamination. All contaminated media have to be properly sterilized and then discarded.

1.3.4 Explant Selection and Culture Conditions

Any plant organ, tissue, and cell may be used as explant for *in vitro* culture. Nonetheless, some plant materials are more suitable to be used as explants, because they are naturally less contaminated and they have a higher propensity to induce sustained growth *in vitro*. Explants may come from either field grown or *in vitro* plants. Elite healthy young plants in a very active growing phase are the most suitable source of explant. Roots and underground tissues are highly likely to have contamination. Thus, these organs are not the best source for explants, unless they can be disinfected or grown in sterilized media. Adventitious or axillary buds and shoots, cotyledons, hypocotyls, and stems are common sources of explants.

Etiolated plants tend to give less contaminated explants, compared with normal plants. This could be because bacteria and fungi do not have appropriate conditions to reproduce in dark or dry conditions. Shoots from active growing plants are less contaminated than older tissues; in addition rough surfaces are more susceptible to contaminants. In summary, vigorous and healthy field- or greenhouse growing plants will give cleaner and less problematic explants. Undoubtedly, the best plant sources for explants are those which were grown in *in vitro* conditions, because plants are completely axenic. These *in vitro* cultures may be obtained from seeds. They are more easily sterilized under rigorous conditions without damage; mostly because they bear seed coats.

Plant cell and tissue cultures have several requirements for optimal growth and culture maintenance. The main physical necessities are temperature, light, and humidity. These factors can be accomplished using an incubator, plant growth chamber, or plant culture room. Temperature is frequently regulated between 22–30°C; nevertheless particular plant species and culture types will have specific temperature requirements. In general, tropical species will need higher temperatures than temperate-grown plants. Plant tissue light requirements may be affected by light irradiance, light quality, and photoperiod. Light conditions can regulate plant morphogenesis. Light irradiance is expressed as $\mu molm^{-2}s^{-1}$ and, in general culture growth is best between 40 and 110 $\mu molm^{-2}s^{-1}$. Light quality is related to the electromagnetic spectrum source, which can be different depending on the type of lamp used at the growth chamber. Finally, the length of light and dark periods (photoperiod) would generally be set at 6–18 hours of light, and the complementary dark period to give a total 24 h cycle. To achieve proper photoperiod, lights are attached to timers so that the length of the photoperiod can be controlled automatically. Normally, cool white or daylight fluorescent light lamps can provide the light irradiance and quality needed, although some specific plant cultures require incandescent lamps.

Relative humidity (RH) is a way to assess the amount of water vapor contained in a gaseous atmosphere. The RH is not easy to manage inside the plant tissue vessel or container, mostly because the container has to be sealed to avoid contamination, therefore reducing the gas exchange. Nonetheless, culture containers may be sealed with Parafilm™ or a wrapping film material, allowing gas exchange between the exterior and interior of the vessel. This prevents media contamination and desiccation. Temperature will determine the RH above the medium within the culture vessel. When air temperature is equal to media temperature, and the container is tightly sealed, RH will be approximately 98–99%. The type of closure will have an effect on intra-vessel RH. Short et al. (35) proved that a closed screw-top lid gave 100% of RH, and a partially closed lid yielded 80% RH. A cotton wool plug gave 70% RH, cling-film wrap gave 60% RH, and filter paper led to 50% RH.

1.4 DIFFERENTIATION AND MORPHOGENESIS

Single plant cells can give rise to multicellular whole plants when they are cultured *in vitro* under proper conditions. This characteristic is known as totipotency. Schleiden (36) and Schwann (37) independently stated the basis of cellular theory and postulated that the cell is capable of autonomy and that it is totipotent. F.C. Steward (38) was the first to show totipotency of plant cells from tissue culture when he cultured whole carrot plants from single cells. Taking into account that all cells of a multicellular organism derive from an egg cell, theoretically they have the same genetic information and the same capacity to generate a whole organism. However, during the developmental process, different kinds of cells with different morphological and structured characteristics are generated, and each plays a specific function in the plant. This means that cells became differentiated, and the process by which they acquire function is known as differentiation. Plant cells undergo differentiation to generate tissues, organs, and eventually whole plants; this phenomenon is called morphogenesis.

1.4.1 Organogenesis (Direct and Indirect)

Plant cells can divide and dedifferentiate or redifferentiate when they are cultured *in vitro*. When plant cells are cultured, they can change their differentiated state or the developmental pattern to give rise to new forms and structures (morphogenesis). Two main pathways of morphogenesis can be recognized in *in vitro* cultures: organogenesis and embryogenesis. Organogenesis implies *de novo* organ formation (e.g., roots, buds, shoots, flowers, tubers) from cells, tissues, or organs. Organogenesis can occur from differentiated cells, tissues, or organs, and from non-differentiated cells or tissues (callus). In the first case, organ formation is referred as direct organogenesis, whereas organ formation in the second case is known as indirect organogenesis.

Organogenesis occurs in some cell or tissue cultures as a response to manipulation of exogenous levels of phytohormones or growth regulators. Skoog and Miller (9) demonstrated for the first time that the type of organogenesis was controlled by the balance of auxin and cytokinin in the tissue culture medium. Media relatively high in auxin and low in cytokinin induced roots, whereas media relatively low in auxin and high in cytokinin induced shoots. These results are very clear in the case of tobacco and many other plant species, but are not the rule, and for this reason, the *in vitro* multiplication of any plant usually requires special adjustment.

Organogenesis is the most common pathway to micropropagate plants. Micropropagation techniques allow the maintenance of characters and yield populations of plants identical to the individual used initially; in other words, this allows cloning elite material. Maintenance of genetic stability requires caution and care must be taken to avoid culture media inducing genetic variability.

1.4.2 Somatic Embryogenesis (Direct and Indirect)

The ability of somatic plant cells in culture to regenerate entire organisms, by a process resembling normal sexual embryogenesis, is one of the most remarkable features of plants. Somatic or adventive embryogenesis is the process by which somatic cells develop into differentiated plants through characteristic embryological stages without fusion of gametes. In contrast to organogenesis, a bipolar structure with a root–shoot axis is formed. This process occurs naturally in a wide range of species. Embryogenesis can be induced directly from cells of an organized structure such as a stem, leaf, flower, tubers, or root segments, among others (direct embryogenesis), or from non-differentiated tissue cultures (callus), or cell suspensions (indirect embryogenesis).

The earliest published reports of *in vitro* somatic embryogenesis were in 1958 in carrot plants (38,39). Our current understanding of the process of *in vitro* somatic embryogenesis has arisen from a variety of experimental approaches, using a number of plant systems (40), of which the carrot (*Daucus carota*) has been the most important. In general, somatic embryogenesis can be divided into two phases: the induction of cells with embryogenic competence, and the subsequent development of the embryogenic masses into embryos. These phases are sometimes referred to as primary and secondary cultures, and the culture media used are similarly referred to as primary and secondary. Somatic embryogenesis has also been divided into the following stages: early growth, embryo maturation, and germination, or conversion.

The term embryogenic cell is restricted to those cells that have completed the transition from a somatic state to one in which no further exogenously applied stimuli, such as the application of growth regulators, are necessary to produce a somatic embryo (41). The cells that are in a transitional state and have started to become embryogenic, but still require exogenously applied stimuli, are identified as competent cells (42,43). During this somatic-to-embryogenic transition, cells have to dedifferentiate, activate their cell division cycle, and reorganize their physiology, metabolism, and gene expression patterns (44). Acquisition of embryogenic competence largely relies on dedifferentiation, a process whereby existing transcriptional and translational profiles are erased or altered in order to allow cells to set a new developmental program. The activation of cell division is required to maintain the dedifferentiated cell fate, as well as for embryogenic differentiation. This is not only true for those embryogenic systems with embryogenic callus (indirect somatic embryogenesis), but also for those where somatic embryos develop on primary explants without an intervening callus formation (direct embryogenesis).

Three factors have been shown to be important in the production of somatic embryos *in vitro*: the explant, the culture medium, and the culture environment (31,45). Although somatic embryogenesis can be induced from a range of tissues, immature or young explants are generally more responsive, with the developmental stage of the explant often being a prime factor. The genotype is also important. Although a wide range of basal media can be used for the induction of somatic embryogenesis, the MS medium (10) has been the most commonly used culture medium. Generally, primary culture requires an auxin, of which 2,4-dichlorophenoxyacetic acid (2,4-D) and naphthalene acetic acid (NAA) have been successfully used. During secondary culture, a lower level of auxin is needed, but there is sometimes the requirement for other hormones, particularly cytokinin and abscisic acid. The culture environment also influences somatic embryogenesis. It can occur under a variety of light dark regimes, but in general, darkness may be better. Osmotic stress can enhance somatic embryogenesis (46,47).

Somatic embryogenesis provides a unique experimental model to understand the molecular and cellular bases of developmental plasticity in plants (48). In the last few years, the application of modern experimental techniques has resulted in the accumulation of novel data on the molecular genetics and molecular biology of somatic embryogenesis (44,49,50). Genes differentially expressed during the early and late stages of somatic embryogenesis have been reported (50). The recent identification of genes that are markers of switch in cell fate or are themselves capable of inducing embryogenic development in somatic cells has opened up new approaches to the question of developmental flexibility and determination in plants (44).

1.4.3 Somaclonal Variation

Spontaneous genetic variation and mutation breeding have traditionally been used to develop new plant varieties. Spontaneous genetic variation usually occurs at a frequency too low to be used in most breeding programs. Mutation breeding, on the other hand, is difficult to control. Usually seeds, consisting of a shoot apex with hundreds of cells, are mutagenized. The resultant plants are usually mosaics and require several generations to stabilize. In addition, at mutation frequencies that are high enough to isolate useful traits, such modified plants usually contain additional deleterious changes and require significant breeding effort to generate new, improved varieties.

It was observed several decades ago that plants regenerated from cultured somatic cells were not genetically uniform but showed significant genetic variability. Larkin and Scowcroft (51) advanced the term "somaclonal variation" to denote such variation, and summarized many of its pertinent manifestations as they occur in cultured plant cells and regenerated plants. Evans et al. (52) extended the definition to somaclonal and gametoclonal variation to refer to those variations observed in regenerated plants from cultured somatic cells or from gametophytes, respectively. Variation has been detected in cultured cells for morphological and biochemical characteristics, and chromosome number and structure (53,54). The genetic variation observed in the regenerated plants appears to result from both preexisting variation in the explant donor tissue as well as from variation induced by the *in vitro* culture (55–57). The frequency of genetic change is significantly higher for somaclonal variation (1–3%) than that observed for spontaneous genetic changes (58). Therefore, this variation can be used to recover at high frequency natural genetic variability from existing crop cultivars.

Somaclonal variation can arise at the cell level by modification in the chromosomes, plastids, or mitochondria (59,60). Variation in regenerated plants can be genetic (heritable) or epigenetic. Genetic changes are stable and expressed in the progeny of regenerated plants after sexual cycle, whereas epigenetic changes are usually induced by culture conditions and are not useful for crop improvement because they are unstable even through asexual propagation (61). The term epigenetic has become used to indicate variation in phenotypic expression that is perpetuated by cloning but does not involve permanent changes in the genotype. All of the ontogenetic and phenological changes that occur in the development of an individual plant can be considered as epigenetic (62). Polyploidy is the most frequently observed chromosomal abnormality, but also chromosome rearrangements (translocations, deletions, inversions) can occur among regenerated plants (63,64). Mitotic crossing over could also account for some of the variation detected in regenerated plants. Single gene mutations and organelle gene mutations have also been detected in regenerated plants (64). Participations of retrotransposons as the mechanism for mutations induced by tissue culture has been proposed by Hirochika et al. (65). Gene amplification and changes in sequence copy number can also account for the observed *in vitro* genetic variation (66). Differences in DNA methylation in cell cultures or plant somaclones have been demonstrated (67–69). Molecular biology techniques have been used quite often to investigate genetic polymorphism in regenerated plants (70–74).

Somaclonal variation depends on the preexisting variation (mosaicism, chimeras) and on *in vitro* induced variation (culture medium composition, cultivar source of explants, age of the cultivar, length of time *in vitro*, proliferation rate, selection pressure, and cultural conditions) (58). Callus, cells suspensions, and protoplast cultures are the systems of choice for the recovery of mutants and variants. Growth regulators such as 2,4-D and benzyladenine (BA) have been implicated in the induction of variability (58).

In vitro genetic variation has been useful for the isolation of a number of variant cell lines resistant to different drugs, antibiotics, amino acid analogs, herbicides, and abiotic and biotic factors, among others (75,76). This genetic variation has also been used to generate new plant cultivars (58) and somaclonal plants with interesting agronomic traits like resistance against pathogens (77).

In summary, somaclonal variation is an important source of genetic variation that can be used for the isolation and recovery of variant and mutant cell lines or plants with characteristics of interest.

1.5 PLANT CELL, TISSUE, AND ORGAN CULTURES

1.5.1 Initiation and Maintenance of Callus

Callus may be defined as an amorphous plant cellular aggregation, slightly differentiated, and attaining rapid proliferation. Almost all plant tissues response to callus induction, and it is known that successful callus initiation is largely dependant on genotype and the influence of exogenously supplied growth regulators. The type of growth regulator requirement and its concentration in the medium depend strongly on species and potential endogenous hormone content of the explant (78). In natural conditions, callus appears as a healing mechanism when the plant has been wounded, or after a tumour is induced by some phytopathogenic organisms like *Agrobacterium tumefaciens*. Natural development of callus tissue is a consequence of a change in the level of endogenous growth hormones, usually auxin and cytokinin. This type of callus may be easily cultured *in vitro* when suitable concentrations of needed growth regulators are applied.

Calli from all kinds of flowering plants have been induced with emphasis on dicotyledoneous materials, although monocotyledonous plant callus have been also induced and cultured (79–82).

Callus is composed of cells with similar characteristics to those of parenchyma cells, though cytological studies show that it is a heterogeneous tissue in cellular composition. When critically examined, callus culture is not a homogeneous mass of cells, because it is usually made up of two types of tissue: differentiated and nondifferentiated (78). The characteristics of a callus tissue *in vitro* depend on the type and quantity of growth regulators used; on organ, species, and source; and on the physical incubation conditions. One of the most important properties of the callus tissue is "friability." This is the ability of cells to break apart from clumps and to disperse. In general, high auxin concentrations favour callus friability, whereas presence of cytokinins tends to produce calli with hard consistency. Friability is desired for cell suspension culture.

With a suitable culture medium, almost any plant tissue with a high number of viable cells is able to produce callus tissue. It is known that only a few cells of the initial explant give rise to callus tissue; usually these are located on the inoculum surface or adjacent to wounds produced during explant handling. For the induction of callus tissue, culture medium such as described by Murashige and Skoog (10) or Gamborg et al. (83), generally semisolid, and with a suitable concentration of auxins and cytokinins, is needed. The induction of callus takes place when the amount of auxin is greater than cytokinin. It has been observed that auxins are callus inducers, whereas cytokinins only favor cell proliferation. Nevertheless, the requirements for growth regulators vary, depending on the tissue and species cultured. Auxins commonly used for callus induction are 2,4-D, which is usually active at 10^{-5}–10^{-7} M; indol-3-acetic acid, (IAA), at concentrations from 10^{-5} to 10^{-10} M; and naphthaleneacetic acid (NAA), that can be used at somewhat higher concentrations than IAA. The most common cytokinins are kinetin and benzyladenine (BA).

Callus proliferation has a high multiplication rate, thus subcultures every 3–8 weeks, depending on the species, are needed to maintain conditions suitable for callus growth. If subcultures are not performed with the appropriate frequency, nutrient depletion and medium desiccation appears, as well as a potentially toxic metabolite accumulation that may induce tissue necrosis. For the callus subculture, only healthy tissue, with no dehydration or necrosis, should be taken and transferred to fresh culture medium. This media may be the same as used for induction medium, though in some cases it is desirable to reduce the concentration of auxin, and to add supplements such as hydrolysed casein and malt extract, which in several species help maintain the constant callus growth. Some calli need cytokinin to maintain cell proliferation, whereas with other calli this is not necessary, because natural endogenous cytokinin levels are sufficient to support proliferation. Additionally, complex additives such as coconut milk, yeast extract, or casein hydrolysate may be required.

In order to have more precise information about *in vitro* callus proliferation and the effects of the growth regulators or other additives, it is important to know callus growth rate. The simplest method is to measure fresh weight at the time of inoculation and at the end of culture. Weighing does not damage tissue; nevertheless, final weight may be influenced by callus hydration levels, and therefore it may not give precise biomass growth information. Alternatively, it is possible to use dry weight as a more precise measurement of biomass increment, unfortunately tissues are killed by this method. Other methods to determine the rate of tissue proliferation are determination of cell number per weight unit and mitotic index.

Callus of many species may be maintained indefinitely with appropriate subcultures; nevertheless, long term maintenance of callus may favor induction of genetic and epigenetic changes such as chromosomal aberrations, mutations, and different ploidy types that, as a whole, give rise to genetic variability in these cells, for what it is widely accepted that most long term callus cultures are chromosomally unstable (78). These changes generally cause a progressive loss in the ability of callus to be morphogenetically competent. Because of regeneration of organs or complete plants is arrested, it is necessary be aware that plant regeneration may not be achieved from long term callus proliferation. Therefore, for a callus maintenance program it may be best to control calli proliferation, and check frequently the morphogenetic competence by induction of plantlets (Figure 1.1C).

Callus tissues have several applications in plant tissue culture, due to their easy maintenance in controlled conditions and their rapid proliferation. Thus callus cultures are used for cellular metabolism studies, production of secondary metabolites, phytotoxicology, and cellular ultra structure. Callus cultures are useful to study cellular mechanisms that induce a response to different types of stress such as water deficit, salinity, extreme temperatures, and pathogen attack. Cell suspension cultures usually start from callus tissues and are intermediate steps for plant regeneration through organogenesis or somatic embryogenesis. Organogenic callus tissue can lead to the formation of adventitious shoots. This process is known as indirect organogenesis. Cell suspension as well as callus can give rise to embryogenic tissue, usually from auxin rich medium. Development of the somatic embryos starts as soon as the auxins are eliminated from the medium. Not every callus tissue has an organogenetic or embryogenetic potential; therefore, plant regeneration from a callus tissue is not always possible, mainly due to the genetic and epigenetic variations that take place in callus cells. It is possible to regenerate plants that are phenotypically different from the original plant material. This makes callus useful for breeding programs where natural genetic variability of some species is limited.

1.5.2 Initiation and Maintenance of Suspension Cultures

A cell suspension culture consists of dispersed cells and small aggregates growing in moving liquid media. It is normally initiated by transferring pieces of undifferentiated, friable calli to a liquid medium which is continuously agitated by a suitable device. Suspension cultures can be started from seedlings, imbibed embryos, or leaves by mechanical methods.

Leaves, soft tissues, or other plant segment can be gently ground and can be broken up in a hand held glass homogenizer. The homogenate, containing intact living cells, dead cells, and cell debris, is cleared by filtration and centrifugation and then transferred to a moving liquid medium. An enzymatic method for isolation of single cells by the use of pectinases, which digest the pectin wall, has also been employed for numerous plant species.

Cultures are initiated by placing freshly cut sections of plant organs on a solidified nutrient medium containing growth regulators. On this medium, the explant exhibits callusing, which is separated from the parent explant and transferred to a fresh medium to build up reasonable amount of callus tissue. This callus is transferred to a liquid medium and agitated to raise fine suspension of cells (78). The transferred callus initiates a fragmentation period forming small aggregates that produce free cells and new cell groups. However, it is almost impossible to establish a suspension culture formed by only individual cells. The plant cell culture requires a minimum size inoculum to be able to grow satisfactorily (84). An appropriate inoculum for chilli pepper (*Capsicum annuum* L.) cell suspensions is 5 mg dry weight per ml of culture medium (75).

In regard to culture growing, development is a sigmoid behaviour that includes five characteristic growing phases:

1. Lag phase: a preparation period before the starting of an accelerated division of plant cells
2. Log phase, or logarithmic phase: characterized by an exponential increase of the cell number due to an accelerated cell division
3. Linear phase: characterized by a linear increase of cell population, which increases fresh and dry weight of the culture
4. Deceleration phase: characterized by a gradual deceleration of cellular division rate, due to media depletion and cellular residues accumulation
5. Stationary phase: cellular division ceases

Cell suspension cultures usually require more frequent subcultures compared to callus culture and should be subcultured before they reach the deceleration phase to avoid loss of cell viability. Cell suspensions are grown in liquid media which has several advantages. All the cells are uniformly in touch with the media therefore, nutrient gradients are not formed. In theory, cell suspension culture may be initiated from any explant inoculated in a liquid agitated medium; nevertheless greater success is obtained when initiation is made from callus tissue. Cell cultures should be agitated or subjected to forced aeration to allow an adequate gas exchange (85,86). Container size and the liquid medium volume are important factors to obtain adequate aeration. Culture medium volume should be 20% of the flask volume. Generally Erlenmeyer flasks mounted on laboratory shakers are used for culturing. Specialized equipment has been designed for plant cell suspension cultures such as the platform orbital shakers, spinning cultures, stirred cultures, continuous culture systems, and airlift reactors (19,87,88). Platform (orbital) shakers are widely used for the initiation and serial propagation of plant cell suspension culture (Figure 1.4). They should have a variable speed control (30–150 rpm) and the stroke should be in the range of 4–8 cm/orbital motion. Besides aeration, agitation of media on a shaker serves two purposes: it exerts a mild pressure on cell aggregates, breaking them into smaller clumps and single cells, and agitation maintains uniform distribution of cell and cell clumps in the medium.

A number of bioreactor variables can easily be monitored on line, sensors for *in situ* monitoring of the two most critical variables for plant suspension cultures: biomass concentration and aggregate size are available (89–91). As pointed out previously there are

Figure 1.4 A platform orbital shaker for the aeration of liquid cultures. This shaker has clips for 250 ml culture flasks.

several ways to characterize biomass concentration in plant suspension cultures, but dry weight measurements are often considered to be the most meaningful and reliable. However this method requires sampling the bioreactor and takes 1–3 days to obtain a measurement, minimizing the measurement's usefulness for on line control. Similarly, aggregate size distribution is often measured by gently screening a sample from the bioreactor using a series of standard sieves and obtaining dry weight of the biomass retained on each screen. The mechanical sieving process has the potential to give size distribution. Thus, *in situ* techniques to monitor biomass concentration (dry weight l^{-1}) and aggregate size for plant suspension cultures need to be developed.

When small reactors or shaking flasks are used to grow cell suspension cultures, routine operations are relatively easy. With large scale operations involving aerobic high cell density plant cells, maintenance becomes more complex. In many cases scale up of plant cell suspensions has been accompanied by a reduction in system productivity (92). The process has to be scaled up in one or more steps, because shake flasks and bioreactors have completely different geometry, mixing, and gas regime characteristics (88). Problems, such as non uniform mixing, hydrodynamic shear damage of fragile plant cells, and low rates of oxygen to cell mass transfer, are potential barriers to process development. Cell damage can be avoided by operating the stirred tank bioreactor at low mixing speeds.

Despite the fact that oxygen requirements of plant cells are comparatively lower to microorganisms, high cell densities may lower the oxygen mass transfer coefficient, which in turn may cause oxygen starvation at low mixing speeds. Oxygen supply is known to affect both cell growth and the production of metabolites (93,94). Generally, growth rate increases when oxygen supply rise until it reaches a maximum. However, further increase of oxygen supply by increasing aeration rate or stirrer speed can decrease growth rate. This is not due to an oxygen increase *per se*, but to an increase in hydrodynamic shear and carbon dioxide stripping (95).

Oxygen starvation may be overcome using techniques proposed by Farrés and Kallio (96), who, using the binary vector pBVHb, on an LBA-4404 *Agrobacterium tumefaciens* strain, overcame oxygen starvation in tobacco suspension cultures by making cells capable of expressing active Vitreoscilla-hemoglobin protein (VHb). VHb-expressing cultures do not have a significant lag phase and can grow to higher cell densities relative to controls.

No significant differences in growth rate, biomass yield on sucrose, or ethanol production between VHb-expressing cell lines and controls were found.

Calli from a wide range of explants (cotyledons, hypocotyls, and leaves) and different media composition [Murashige and Skoog (10), Gamborg et al. (83), and Linsmaier and Skoog (97)] have been used with success for cell suspension induction. Vitamins, inositol, sucrose, and growth regulators, especially auxin (2,4-D at concentrations of 1–5 μM), should be added for ensuring cell division. The addition of 1 μM kinetin may be beneficial. The culture medium for suspension culture should be such that it maintains superior growth of cells (78).

Ozeki and Takeda (98), studying the regulation of phenylalanine ammonia lyase genes in carrot suspension cultured cells, established that plant cell cultures are often used as model systems to investigate regulation of secondary metabolism by environmental and developmental stimuli, because the cells can be grown in a stable and homogenous environment with respect to temperature, light, and nutrients. This reduces the complexity of the experimental system as compared to intact plants, and facilitates analysis of regulatory mechanisms.

1.5.3 Bioreactors, Immobilized Plant Cells, and Secondary Metabolites

The relevance of a bioreactor culture technique for plant micropropagation either via organogenesis or somatic embryogenesis has previously been established (91,99). This is a way to reduce production cost by scaling up and automation. For instance, Harrell and Cantliffe (100) developed a machine vision based measurement system which noninvasively quantifies the population characteristics of a heterogeneous plant tissue culture suspension grown in a bioreactor. Likewise, Harrell et al. (90) demonstrated the aseptic and automated harvest of somatic embryos from a bioreactor packed with *Ipomoea batatas* suspension culture. Using a machine vision system, which emulated the selection criteria of an experienced biologist, these authors were capable to classify somatic embryos as harvestable or nonharvestable, without affecting the aseptic conditions of the culture. The efficiency rate of this system was as high as 99%. On the other hand, periodic immersion liquid culture using ebb and flood system and column-type bubble bioreactors were found to be suitable for micropropagation of plants via the organogenetic pathway. Whereas balloon-type bubble bioreactors proved better performance for micropropagation via somatic embryogenesis with less shear stress on cultured cells (99).

Plant cell and tissue cultures have enormous prospects as a substitute method for the production of secondary metabolites of commercial value, which normally are extracted from cultivated or wild plants. In general, *in vitro* cultures accumulate lower amounts of secondary metabolites than whole plants. Nevertheless, a number of *in vitro* cultures have been able to yield higher levels of phytochemicals than cultivated or wild plants (101,102). To overcome the low secondary metabolite yield several strategies have been suggested, such as optimization of culture conditions, selection and improving of high-producing strains, addition of bioprecursors, biotransformation systems, elicitor treatment, metabolic engineering, and cellular immobilization (102–110).

Rosevear (111) has defined the immobilization technique as an artificial system which confines a catalytically active cell or enzyme within a reactor, and prevents its entry into the mobile phase which carries substrate and product. In 1966, Mosbach and Mosbach (112) reported for the first time a plant cell immobilization process, using cells of the lichen *Umbilicaria pustulata* entrapped in polyacrylamide gel. After that achievement, extensive information on immobilized plant cells, describing characteristics, properties, techniques, and commercial potentialities of that system have been published (113).

In the last 18 years, the appearance of recombinant DNA technology has opened the opportunity to change the expression and repression of genes which are related directly to biosynthesis, transport, accumulation, and degradation of secondary metabolites (114–117). In other words, it is now possible to regulate pathways that lead to phytochemical substance accumulation. Bailey (118) has defined metabolic engineering as the improvement of cellular activities by manipulation of enzymatic, transport, and regulatory functions of the cell with the use of recombinant DNA technology. This metabolic engineering approach can be used either in a plant or plant cell culture, and successful examples have been reported for monoterpene, lignin, carotenoid, flavonoid, and alkaloid biosynthesis (102,106,119).

As previously pointed out, a number of approaches have been developed to improve the use of the bioreactor culture technique for plant micropropagation, as well as to increase the secondary metabolite accumulation in plant cell cultures. However, not all of those approaches have been completely successful. Possibly that goal may be the result of the combination of improved bioreactor technologies, enhanced cell, tissue, and organ culture protocols and an exceptional metabolic engineering approach.

1.5.4 Root Cultures

Roots are very important organs for plant survival, because they have vital functions for absorption, and transport of water and nutrients. Plant roots have been valuable economically as food sources and medicines. *In vitro* culture of isolated roots was one of the most important successes within plant tissue culture. The advances were made by White in 1934 (4), who successfully cultured tomato roots in completely artificial conditions. The original radicle lines obtained by White were grown for 25 years in a liquid chemically defined medium, and were the initial *in vitro* cultures of plant organs (120).

In vitro culture of roots is a system of great interest for the study of metabolic processes that take place in those organs, as well those that are related to the metabolism of carbohydrates, vitamins, growth regulators, nutrient absorption, and translocation (121,122). Likewise, these types of cultures have been used in studies of differentiation, root development, and in the interaction of these organs with other organisms, such as Rhizobium, and parasitic nematodes. Another very interesting application of root culture is the synthesis of a large number of secondary compounds of interest that are produced by this tissue, and the potential production of them (123,124).

The response of roots to *in vitro* culture depends on factors such as species, culture medium, and environmental conditions of the culture. Roberts and Street (125) have classified root cultures into three different classes, depending on their response to exogenously applied auxins. In a group, which includes those of tomato, the addition of auxins has an inhibitory effect on growth or at best no effect. In the second group, auxins stimulate growth when they are added in suitable concentrations. In the last group, radicle cultures are entirely dependent on an external auxin source to be able to grow. Later, Butcher and Street (126) compiled a list with all root cultures published, and concluded that auxins are needed to support a suitable growth of radicular tissues. On the other hand, variation has been observed related to the time during which the roots may be growing *in vitro*.

Three groups of species are defined:

1. Roots that may be cultured indefinitely *in vitro*, needing only the medium and the appropriate subcultures. The roots of tomato, Datura, and other species have this type of behavior.
2. Roots that grow and develop *in vitro* during variable periods of time, but that do not achieve indefinite growth. In all cases, sooner or later, the rate of growth

gradually decreases, lateral root formation is inhibited, and finally the culture dies. This is the case with some leguminous and cereal roots.

3. Roots that up to now have not been cultured *in vitro*. To this group belong many woody species.

Related to the nourishing medium to the *in vitro* root cultures, several have been tested with a different degree of success, but the most widely used are those derivatives of the original White's medium, although some species responded better to Murashige and Skoog (10) or Gambor et al. (83) media. Sucrose is the principal carbon source used; nevertheless, some cultures of monocotyledons plants have a good response to glucose. Carbohydrate levels are between 1.5 and 2% (w/v), because major concentrations cause metabolic root alterations. Liquid media is used more frequently, stationary, or under agitation, though in some cases it is possible to use a semisolid culture medium.

In general, exogenous application of growth plant regulators must be carefully done, because some auxins combined with cytokinins may induce a rapid root dedifferentiation to turn into cellular suspensions. In certain cases, it has been observed that gibberellins strengthen growth (122).

A variant of root culture that has recently reached enormous importance is the root culture transformed by *Agrobacterium rhizogenes*. This subject is widely covered in section 1.6.3.2 of this chapter. The future work of *in vitro* transformed root culture should be based on the *Agrobacterium rhizogenes* genetic knowledge, and on the biochemistry and genetics of plants, in order to spread the range of plants that may be transformed and to increase both the rates of growth and the production of useful substances.

1.5.5 Shoot Cultures

Plant tissue culture techniques include the culture of any type of organ plant; but, among them, shoots are the organ preferred for commercial micropropagation. In general, herbaceous plant species are more amenable for micropropagation than woody plants because the latter may be more dramatically affected by seasonal changes that may have a profound effect on the physiology of tissues and organs. For example, dormant tissues or organs dissected from winter plants do not respond to *in vitro* culture as well as those collected during the spring. Shoot propagation is a general term used, but it includes apical and lateral meristems, as well as apical and lateral apexes. *In vitro* clonal propagation means the production of a plant population derived from a single individual by asexual reproduction. Plant propagation is derived from the plant's ability to attain a constant grow. However, not all plant cells are able to achieve steady division under normal conditions. As a result, plant growth requires the presence of a special type of tissue that always preserves the dividing capacity. This type of tissue is named meristem or meristematic tissue, and is made up of small, spherical, nondifferentiated cells, with a high relation of nucleus to cytoplasm, very small vacuoles, thin cellular walls, and nondifferentiated plastids.

The axillary and apical meristems inside the lateral and terminal buds are the natural vegetative growing points with capacity of new shoot formation. That capacity is maintained when they are established and cultured *in vitro*. Thus, the culture of terminal or lateral buds is an easy way for obtaining new shoots, which can be rooted to produce new healthy and vigorous plants. Therefore, this propagation system is based on a new shoot formation system from preexisting meristems, which does not imply cellular dedifferentiation, redifferentiation phenomena such as those that take place in the organogenesis, and somatic embryogenesis. The meristem and bud culture is a regeneration system less complex than organogenesis or somatic embryogenesis.

Shoot tips or meristem *in vitro* culture, is an ideal system for clonal propagation, because it provides the maximum genetic stability, which does not always is obtained from organogenesis or somatic embryogenesis. The system is relatively simple, and is very similar for many species. Thus, suitable conditions needed for propagation can be easily found. Nevertheless, bud propagation is slower and less productive than organogenesis and somatic embryogenesis. As an advantage, the meristem culture system alone or combined with thermotherapy or chemotherapy, allows rapid development of pathogen- and virus-free plants. This has an enormous importance especially in vegetative propagated species, because this is the easy way to obtain healthy plant material.

This methodology has been successfully applied to *in vitro* germplasm conservation by means of cryopreservation. Meristem or bud culture facilitates the transport of plant material without regulatory phytosanitary problems.

Micropropagation has several disadvantages, including high tissue sensibility to surface sterilization methods, elevated amounts of phenolic compounds which are produced during cutting, mostly in woody plants, and induced tissue vitrification in some of the cultured species.

According to the type of explant used, shoot culture technique may be divided into apex, axillary bud, and meristem culture.

1.5.5.1 Apex Culture

For culture of apexes, apical explants of 4–10 mm length are used. Besides the meristem, they contain several bud primordia as well as vascular differentiated tissue. This method is simple due to the manipulation of bigger explants and to the higher probability of their survival. Nevertheless, this method is not appropriate when pathogen free plants are needed, for what it is important to start with clean material from donor plants cultured under greenhouse conditions. Regarding the culture media needed for this type of propagation, Murashige and Skoog (10) or Gamborg et al. (83) media normally are used. Cytokinins are used with concentrations from 0.5 to 10 mg l^{-1}, favoring a multiple shoot proliferation from a single bud; although auxins in a small amount (0.01–1 mg l^{-1}) are used as well. Figure 1.5 illustrates *Photinia* X *Fraseri in vitro* plants cultured in MS medium.

1.5.5.2 Axillary Bud Culture

Explant size is not critical in this system, because it carries one or more axillary buds, from which shoot production shall take place. This is the simplest method for plant micropropagation, and it may be coupled to other systems to increase the material (40). For example, once a pathogen-free plant is obtained, it may be propagated through axillary bud system and after several cycles, the number of plants will be increased exponentially (127). Figure 1.6 depicts axillary bud propagation of *Agave tequilana* cactus plant which is used to obtain the Tequila liquor.

1.5.5.3 Meristem

For meristem culture, explants must be isolated by dissection under a stereomicroscope. The isolated explant is very small, usually less than 1 mm in diameter, and consists of the apical dome and a small number of leaf primordia, excluding the vascular differentiated tissue. To work and isolate so tiny explant presents certain difficulties. The explant size reduces the possibility of its successful *in vitro* establishment, and it needs a more complex culture media. Nevertheless, as the explant size is smaller, the probability of being pathogen free is increased. The main use of this type of culture is the production of plants free of virus or any pathogenic agent (128). This method may be combined with some type of

Figure 1.5 *Photinia X Fraseri in vitro* plants. They were propagated by mean of apex culture. Supplemented with 5 mg l^{-1} BA, exhibiting a shoot propagation rate of eight plants per explant within a lapse time of 45 days.

Figure 1.6 Axillary propagation of *Agava tequilana* depicting many axillary shoots (12 shoots per original explant).

thermo- or chemotherapy to assure pathogen liberation. Thermotherapy consists of incubation of whole plants or fragments to high temperatures before the isolation of meristem. Typical treatments are 7–9 days to 40°C, or 16–168 days to 32°C. After thermotherapy treatments, plants must be analyzed by serological or molecular methods, or by electronic microscopy to confirm the absence of specific pathogens.

Ramírez-Malagón et al. (129) studied the elimination of viruses from two commercial garlic varieties using heat or chemical treatment and meristematic dissection. Growing explants were subjected to ELISA tests using potyvirus antisera. Thermotherapy killed 70% of explants; however, 64% of the surviving explants were seronegative. Chemotherapy treatment did not destroy the explants, but only 33% of them were seronegative. Field performance of virus-free plants was evaluated. *In vitro* plants had longer stems, higher fresh and dry weight per bulb, and increased yield in comparison with commercial lines (Figure 1.7).

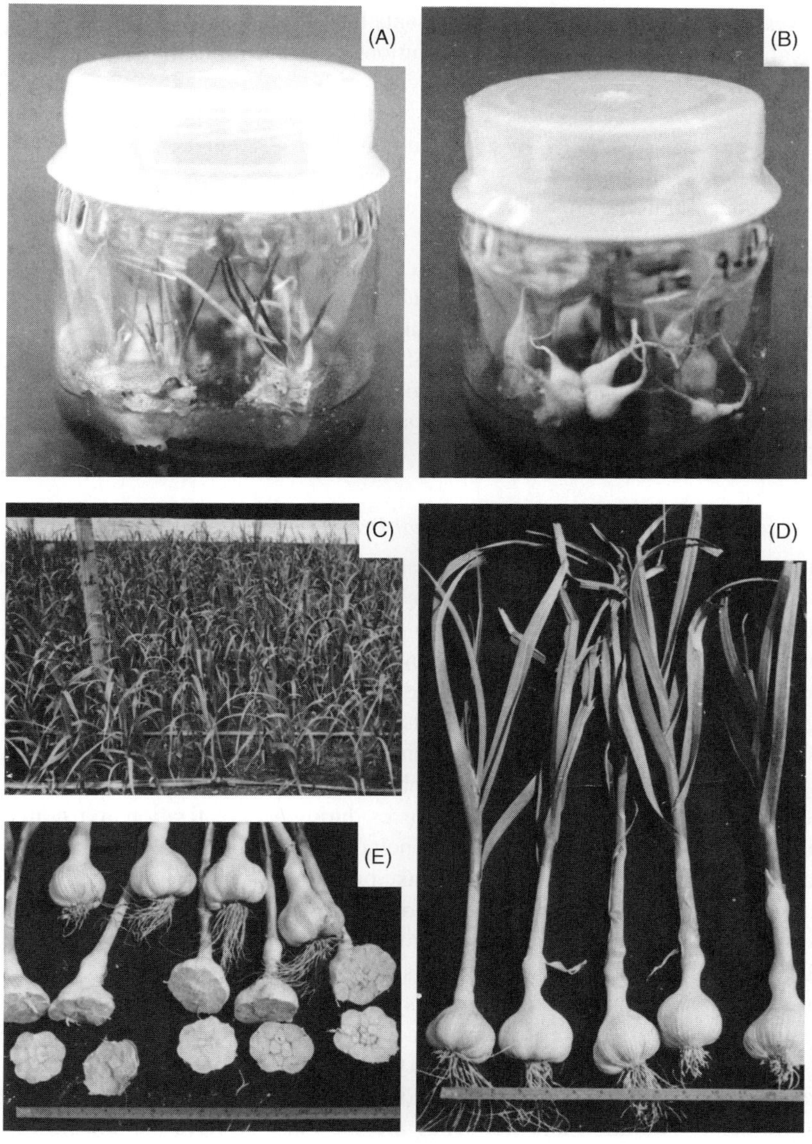

Figure 1.7 Garlic propagated plants from meristem culture. **A**. *In vitro*-growing garlic plant after meristematic termotherapy or chemotherapy treatments. **B**. *In vitro* clove development. **C**, **D**, and **E**. Garlic plants and cloves harvested after virus elimination and field cultivation.

1.5.5.4 Acclimatization

Acclimatization is defined as the process by which an organism adapts to environmental changes. *In vitro* regenerated plants have several characteristics which complicated their adaptation to external environment. High relative humidity inside *in vitro* containers combined with a rapid cell proliferation by hormone action, induce formation of plants lacking of normal systems to avoid water loss. *In vitro* plants have a thin cuticle, wider stomata, and an atrophied closure mechanism. Therefore, the acclimatization process of *in vitro* plants must be done as gradual as possible, providing a slow reduction in relative humidity. In general, plants from *in vitro* propagation have to be cultured under greenhouse conditions provided with mist application system, with a different misting distance according to diverse species. Different antitranspirants have been tested to protect plants during this critical period, among them glycerol solutions or liquid waxes have been used, but the results not always have justified their use.

During *in vitro* culture plants do not perform a normal photosynthesis, and their carbon requirements are satisfied by the culture medium. Leaf anatomy from *in vivo* growing plants, differs from *in vitro* growing counterpart. The last ones are thinner, with less chlorophyll concentration. Therefore, transference of *in vitro* plants to soil must be done step by step, with slow light increments and slow relative humidity decrements.

In vitro plants are incubated under axenic environment consequently they have not developed their natural resistances against pathogens. During the first adaptation step, it is advisable to work under very clean conditions. Likewise, it is advisable to eliminate any culture medium residue adhered to the plant, because, due to its high sucrose content, it might favor microorganism growth. Evidence of successful plant adaptation is the development of new leaves after the transference to soil. At this time, watering should be reduced and light increased.

1.5.5.5 Current Approaches for Mass Clonal Propagation

Current micropropagation techniques require a large number of small containers, semisolid media, and aseptic division of plant tissues by hand. Also, plant micropropagation involves periodic transfers of plant material to fresh media due to exhaustion of nutrients in the medium and because of continuous tissue growth and proliferation, which is quickly limited by the size of the culture container. Gelling products are not inert and complicate automation. Labor generally accounts for 40–60% of production costs. Cutting and planting represent the most expensive part of micropropagation even in countries with low incomes. Furthermore, cleaning, filling, and handling of a large number of containers also adds to the final cost of the process. Other major costs result from losses during acclimatization and stem and root hyperhydricity, or vitrification. It has been concluded that commercial application of micropropagation for various species would only take place if new technologies were available to automate procedures, and if acclimatization protocols were improved (130). High production costs generally limit the commercial use of micropropagation to products with a very high unit value, such as ornamentals, foliage plants, and selected fruit crops.

In recent years, temporary immersion systems for plant micropropagation have been investigated as a means of overcoming some problems for *in vitro* mass propagation of plants (131). The positive effects of temporary immersion on micropropagation are recommended for shoot proliferation and microcuttings, microtuberization, and somatic embryogenesis. Immersion duration and frequency are the most decisive parameters for system efficiency. The parameters most involved in reducing production costs by temporary immersion systems include: the drastic reduction in work, reduction in shelving area, reduction in number of containers used, and better biological yields. Scaling up somatic embryogenesis and shoot proliferation procedures involving temporary immersion systems in order to commercialize this process are now taking place for different plant species (131).

1.5.6 Isolation and Culture of Protoplasts

By definition, a protoplast is a plant cell coming from plasmolysed tissues and deprived of its cell wall. Protoplasts may be obtained by physical or enzymatic methods from almost any kind of plant tissue. For several species, it is possible to regenerate whole plants using just one single protoplast. Isolation and culture of protoplasts are made to pursue several different objectives, such as somatic hybridization, genetic transformation, and modelling systems for physiological and molecular studies (132,133). These cells, lacking a cellular wall, may easily allow the passage of organelles and different particles such as virus and naked DNA, by relatively simple methods like electroporation and microinjection. On the other hand, a naked plant cell may be the solution to harvest useful secondary metabolites, because cell walls limit the mass transfer to the culturing medium (134). Additional information on somatic hybridization and genetic transformation using protoplasts can be found in section 1.6 of this chapter (genetic manipulation).

Although protoplast isolation may be done by mechanical methods, the most important is the enzymatic one. This is based on the enzymatic degradation of the cellular wall components, which allows the protoplast obtaining without damaging the tissue. Enzymes that are used for this purpose belong to three different groups: cellulases, hemicellulases, and pectinases. In general mixtures of two types are used (135). Explicit and detailed protoplast isolation protocols can be found in Thomas and Davey (136), Gamborg et al. (137), Bhojwani and Razdan (138), Kyte (139), Collin and Edwards (140), Augé and Boccon-Gibod (141), and Dodds and Roberts (142).

A protoplast is the most unique system for whole plant regeneration from a single cell, presenting great potential involving several developmental pathways, such as suppression of events leading to cell death, active oxygen species generation, and induction of cell division (133).

1.5.7 Anther, Pollen, and Ovule Cultures

One of the most interesting applications of plant tissue culture is undoubtedly the generation of haploid and double haploid (DH) plants through anther, microspore, and ovule culture. Culture of these specialized organs was started by Guha and Maheswari (11) who were able to induced embryogenesis while culturing anthers from *Datura innoxia*. The successful culture of anthers and pollen of approximately 137 plant species belonging to 25 families, the majority of them solanaceae or gramineae have been achieved (143). Regeneration of complete plants in these systems is through direct or indirect embryogenesis induction.

If anthers or microspores are used as explants, the regeneration process is named androgenesis. Using this technique is possible to obtain homozygosity in one generation, reducing the numerous cycles of inbreeding necessary in conventional pure line breeding systems (144). The main applications of haploid plants generated *in vitro* culture are: studies of quantitative genetics and gene interactions, estimation of the gene number that take part in the polygenic characteristics, production of isogenic lines to be used in breeding programs, and isolation of homozygous recessive characteristics not easily achieved by conventional methods. By means of traditional breeding methods, to obtain one isogenic line can take from five to six generations, while by haploid plant systems followed by a dihaploidization process; it is possible to obtain isogenic lines in a very short period of time. There are additional advantages to this process: saving time from select plant material generation to marketing, selection efficiency increases due to an increment of additive genetic variation, absence of variation among the segregating families, and decrease in effects of variation due to the environment (145).

The dihaploids obtained directly from the regenerated plants may be 30–60%, and the percentages of haploid plants up to 40–70% (146). The most frequent event is that the percentage of dihaploids obtained by breeders does not exceed 30% of the regenerated plants. Several workers are looking for increments of the number of DH plants by different treatments, such as increasing the concentration of mannitol in the culture media (147). Other authors consider more suitable treatments with colchicine to increase the number of DH plants (146,148,149).

Among the factors influencing haploid or double haploid regeneration are: physiology of donor plant, physical factors of culture, developmental stage of pollen, genotype, and culture media.

1.5.7.1 Physiology of Donor Plants

Age and physiology of donor plants have an influence on anther or microspore *in vitro* culture. Donor plants must be free of any type of environmental stress, phytopathological problem, or entomological problem. It is known that vigor of the donor plant is a requisite for a suitable response (150–156). For wheat anther culture, a temperature of 32°C is recommend if the donor plants are field cultivated, yet 30°C is an adequate temperature if donor plants are cultured in a greenhouse. Interestingly, anthers from field plants had a better *in vitro* regeneration response than those from greenhouse (150,154). Kristiansen and Andersen (153), working on *Capsicum annuum,* found that 26.4°C was the best temperature for incubation. Embryo formation was not affected by photoperiod, instead, age of the donor plant was important because a decline response was obtained when donor plant age was increased. Donor plant chemical pretreatments may affect anther response. Wassom et al. (156), on maize established different responses after abscisic acid (ABA), ancymidol, and gibberellic acid (GA) treatments. Other chemical pretreatment to donor plants might positively influence callus, and green plant formation. It consists in spraying the donor plant at different stages around meiosis with a chemical hybridization agent, such as hybrex^{TM} or fenridazon–potassium, which have gametocidal effects (157–159).

1.5.7.2 Physical Factors of Culture

Cold or heat treatment to the anthers, either before or after *in vitro* culture has a main influence on regeneration response. Cold treatment induces a synchronous effect on pollen grain development, delays degeneration of microspores, and accelerates nuclear division during induction process. High temperature during early stages of embryogenesis is associated with protein synthesis. The effect of a temperature treatment has not been established yet, and available reports are contradictory (160). On the other hand, several reports about temperature androgenic response conclude that genotype, rather than cold or heat treatment is the response inductor (160–162).

1.5.7.3 Developmental Stage of Pollen

Pollen grain development stage is determinant for *in vitro* embryogenic response. Pollen grain must be collected before the dinucleated stage. Novák (163), and Heberle-Bors (158) established that the middle or late uninucleate stage are ideal for haploid and dihaploid embryogenesis.

1.5.7.4 Genotype

Genotype is the critical factor in anther *in vitro* culture, and it is considered the main factor that affects anther culture response and embryoid induction (160,164). These responses are dependent not only on genetics, but also on the interaction of environment–genotype (158). Ouyang et al. (150) and Jones and Petolino (151) confirmed that interaction, and reported that even plants from the same genotype gave different culture responses if the donor plants are grown under different conditions. Moreover, anthers from the same genotype and grown in the same field conditions, but in different years, sometimes require different culture temperature. This complicates a single recipe for successful anther culture response, because for every genotype it is necessary to study its particular environmental, physical, and chemical requirements.

1.5.7.5 Culture Media

Media commonly used for anther and microspore culture are Murashige and Skoog medium (10), Nitsch medium (165), Potato-2 medium (166), and N6 medium (167). Media may be gelled by agar, gellan gum, or agarose. In order to prevent inhibitory substances from agar, liquid medium may be used or 10% Ficoll™ added, which avoids that anther sink into the medium (168–170). Trotier et al. (171) suggested the use of anther supports like membrane rafts. Different sucrose levels have been used in anther culture. Chu and Hill (172) and Saidi et al. (173) reported a concentration of 9% sucrose for wheat anther culture. Alemanno and Guiderdoni (146) established a sucrose addition of 60 g l^{-1} for *Oryza sativa*. Furthermore, Barnabás et al. (174) suggested a sucrose supplementation of 120 g l^{-1} in maize. Ferrie et al. (175) found for *Brassica oleracea* that the best sucrose concentration was 170 g l^{-1}. Culturing *Cucurbita pepo* anthers, Metwally et al. (176) recommended 150 g l^{-1} of sucrose.

Regarding plant growth regulators, 2,4-dichlorophenoxiacetic acid (2,4-D) is more frequently used, with concentrations from 1.5 to 2 mg l^{-1} mainly for anther or microspores from monocotyledon plants (151,152,169,172,173,177–179). For dicotyledon plants, 2,4-D concentrations are from 0.01 to 1 mg l^{-1}.

Other components included in the anther culture media are silver nitrate (180), copper sulphate at 10 mg l^{-1} (149), and activated charcoal (181). Beneficial effects of amino acids on androgenesis have been reported. Nitsch (165) attributes a positive effect of serine for *Nicotiana tabacum* anther cultures. Maheswari et al. (182) and Henry and DeBuyser (183) established that asparagine and glutamine are promoters of androgenesis. Nonetheless, Xu and Sunderland (184) and Hoekstra et al. (185) reported an inhibitory effect of glutamine on barley anther cultures.

1.5.7.6 Microspore Culture

Microspore culture may be considered just a variant of anther culture. In fact, several researchers (174,186,187) reported that anther culture is a microspore culture due to the fact that almost all dicots shed microspores into medium several days after anther incubation. All the factors reviewed for anther culture are applicable for microspore culture.

1.5.7.7 Ovule Culture

Another possible route for generation of haploid plants is culture of nonfertilized ovules. As for anther culture, gynogenesis depends upon genotype, physiological status of donor plant, type of explant, culture medium, and their interactions. Ovule culture has been much more complicated than anther or microspore culture process, and in spite of many efforts trying to improve the haploid induction technique, the yield of gynogenic embryos and haploid plants obtained is frequently low (188–190). In spite of that, several Cucurbitaceae ovule cultures have shown similar responses to those from anther culture. In *Cucurbita pepo,* Metwally et al (176) has produced 8.8 and 9.1 embryos per 100 cultured ovules when 2,4-D was used, at a dosage of 1 or 5 mg l^{-1}. For *Cucumis sativus*, Gémes-Juhász et al. (191) have reported an 18.4% of gynogenesis frequency with 7.1% of plant regeneration. For *Cucumis melo*, Lotfi et al. (192) reported the regeneration of 175 haploid plants using gynogenesis, which were subjected to colchicine treatments, and induced 10 dihaploid and 100 mixoploid plants. These authors applied 0.05 mg l^{-1} α-naphtaleneacetic acid, and 0.2 mg l^{-1} benzyladenine as plant growth regulators.

1.5.8 Germplasm Storage

The main goal of constituting germplasm banks is to have easily available material for crop improvement, and to preserve local and introduced materials. Likewise, banks constitute an ideal means for maintaining useful lines for commercial hybrids, and for preserving valuable germplasm threatened by overcollection and severe perturbation of a habitat. There are many germplasm banks based on plant seed storage, although if genetic material is difficult to preserve as seed collection, botanic gardens or orchards are established as field gene banks. This system requires a large land area, considerable labor, and a sizeable capital investment. Because of this, the development of alternative conservation techniques is desirable. During the past few decades, different *in vitro* conservation methods have been developed as substitutes for field gene banks and have been widely employed depending on storage duration required (193). *In vitro* techniques compared to traditional methods, offer different germplasm storage based mainly on vegetative plant material preservation using cold storage, chemicals inhibitors, or cryopreservation techniques.

It is possible to preserve *in vitro* active collections by subculturing them with normal culture traits, but the success of that method depends greatly on plant species. Many species support a continued subculture for several years without any apparent change, such as strawberries, potato, *Photinia*, African violet, and gerbera (194). But others, such as cucumber, and maybe all cucurbits, lose their regeneration capacity through two to three years, and apparently it is not possible to use cold storage technique, because their survival depends on relatively hot temperatures (195). A number of advantages for *in vitro* preservation may be listed, such as the small space needed for storage, easy shipping, year round availability, maintenance of disease-free status, and quick propagation and distribution if materials are requested by users (196).

For several species there are *in vitro* conservation techniques based on modified incubation parameters. Strawberry preservation studies suggests that shoot culture storage has to be done under temperatures of 4°C, subculturing shoots over filter paper bridges in liquid medium and adding fresh medium every three months (197). Reed and Humer (194) studied in the strawberry the influence of benzyladenine in the storage medium, prestorage acclimatization, and exposure to a photoperiod during storage. They concluded that strawberry preservation at 4°C for 9–24 months intervals can be obtained before they require propagation.

Light is another physical factor studied for long term plant storage. It is important for plants to preserve photosynthetic capabilities and their absence implies a rapid elongation for many plants, nevertheless when this factor is combined with cold for plant material storage different responses are observed depending on genotype studied. Prune shoots had a positive and better response to a photoperiod of 16 h combined with a temperature of 4–8°C, compared with similar cold treatment in darkness (198). Romano and Martins-Louçao (199) were able to culture Cork oak (*Quercus suber* L.) explants on multiplication medium at 5±1°C without a subculture for more than 22 months. They evaluated light and dark conditions for explant survival finding that light negatively affected culture viability. On the contrary, 50% of the cultures survived after two years of cold storage in dark incubation, and multiplication rate and rooting of cold storage plants was similar to controls.

It is important to asses the genetic stability of *in vitro* cultures, mainly after prolonged storage periods, because one of the objectives of germplasm storage is to preserve the genetical identity. A genetic stability study after one year of cold storage was conduced using apple plants tissues (193). These authors established that all stored apple shoots survived one year at 4°C, having a 12 h photoperiod of 11 μmol m^{-2} s^{-1} light irradiance. By means of an amplified fragment length polymorphism study (AFLP), they did not detect any differences in DNA fragment patterns using 20 primer combinations, while a methylation sensitive amplified polymorphism (MSAP) assay demonstrated that cold stored plants had developed methylation changes compared to controls. This methylation event is important because change in DNA methylation status may be a response to stresses associated with *in vitro* conservation, and further investigation is necessary to determine the physiology and genetic role of DNA methylation change for *in vitro* cultured species. This phenomenon has been found in several species such as potato, apple, and *Swietenia macrophylla*, (193,200,201).

Cryopreservation is a harder treatment to preserve *in vitro* plant cell, tissue, and organ cultures. When material to be preserved is a shoot apex collection, liquid nitrogen can be used as a suitable method to maintain viability and phenotypic and genetic stability, thereby ensuring long term germplasm storage. Many studies have emphasized that cold pretreatments of shoot cultures of hardy species improves the resistance of shoot tips to freezing in liquid nitrogen (202,203). Although high levels of shoot regrowth have been recorded with techniques such as vitrification and encapsulation–dehydration, none of them could suppress the deleterious effect of cold hardening pretreatment in shoot cultures (204). However, with similar encapsulation–dehydration technique, Sales et al. (205) successfully cryopreserved *Digitalis obscura* shoots. Their best treatment consisted in cold hardening of *in vitro* cultures before sucrose treatment, recovering up to 86% of the treated explants. Nevertheless, after applying the same treatment with diverse genotypes of the same spices, they found that cryopreservation responses mainly depend on genotype. More information on germplasm storage using *in vitro* techniques can be found in Westcott et al. (206), Kartha (207), Bhojwani and Razdan (208), Brennan et al. (209), Dodds and Roberts (210), Ashmore (211), Drew (212), and Collin and Edwards (213).

In vitro germplasm storage offers an alternative to traditional storage methods. It is a tool for different biological studies, and at same time it is a method to maintain for long time worthy research materials, mostly those which are vegetatively propagated. Although for commercial application, qualified personnel and minimal equipment is needed, undoubtedly plant tissue culture is offering an alternative to germplasm storage that is increasing as different techniques are developed, and many others are under study.

1.6 GENETIC MANIPULATION

Plant genetic improvement by conventional methods is mostly limited by narrow available genetic variability and by natural barriers of crossing. One of the most interesting applications of modern plant biotechnology is the possibility to genetically manipulate plants, in order to confer desirable agronomic traits, to carry out studies on genetics and metabolism, and to obtain plants with new characteristics.

This genetic manipulation can be made through two different methods: somatic hybridization and genetic transformation. The former is used to produce somatic hybrids which are produced by fusion of two protoplasts isolated from different plant sources. This technique allows mixing two genomes of diverse origin, and obtaining plant hybrids that could not be generated using traditional methods. The second method is the direct introduction of genes to the plant cell genome, also called genetic transformation. This process requires first the delivery of foreign DNA to the nucleus of a plant cell, in a manner that minimizes damage using transformation methods. This includes the insertion, integration, expression, and inheritance of new foreign DNA. In a second step, fertile transgenic plants are regenerated starting from those transformed cells, using *in vitro* plant culture methods.

Plant genetic transformation is now a routine technique in numerous laboratories around the world, because it offers a valuable tool to accelerate crop genetic improvement, by transferring genes which confer resistance to pests, diseases, herbicides, and environmental stress, as well as quality traits such as improved postharvest storage, nutritional content, flavor, or color. Some examples of plant improvement using genetic transformation are depicted in Table 1.4. Additionally, genetic transformation can be used to study

Table 1.4
Examples of some Phenotypic Traits Improved by Plant Genetic Transformation.

Phenotypic Trait	Transgene(s)	Ref.
Altered lignification in forest trees	*COMT* (caffeic acid 5-hydroxyferulic acid O-methyltransferase gene,depressed)	(245)
Bacterial diseases resistance	Bacterial detoxification gene	(249)
	Antibacterial protein genes from insects	(250)
Cold tolerance	*gpat* Glycerol 3-phosphate acyltransferase gene for fatty acid unsaturation	(251)
	Anti-freeze gene from fish	(252)
	afp (carrot antifreeze protein)	(253)
Enhanced fruit quality	*Spe2* (yeast S-adenosylmethionine decarboxylase gene)	(254)
Enhanced growth	*vhb* (gene for *Vitreoscilla* hemoglobin)	(255)
Fungal diseases resistance	Bean chitinase gene	(256)
Herbicide resistance	EPSPS (glyphosate resistance)	(257)
High bioavailable iron-content plants	Soybean ferritin gene	(258)
Increased rooting capacity	*rolABC* from *Agrobacterium rhizogenes*	(259)
Insect resistance	*Bt* gene from *Bacillus thuringiensis*	(260)
	Cowpea trypsin inhibitor gene	(261)
Phytoremediation (use of plants for removal and detoxification of environmental pollutants)	Human cytochrome P450 (for oxidize a wide range of organic pollutants)	(262)
	merA9 (Bacterial organomercurial lyase for mercury phytoremediation)	(263)
	onr (PETN reductase for detoxification of explosives)	(264)
Rootstock transformation to alter scion growth habits	*rolABC* from *Agrobacterium rhizogenes*	(265)
Salt tolerance	*mtlD* Mannitol-1 phosphate dehydrogenase gene (mannitol metabolism)	(266)
Virus resistance	Viral coat protein genes	(267)
	Viral satellite RNA (cDNA copy)	(268)
	Pokeweed antiviral protein gene (a ribosome inhibiting protein)	(269)
Water stress tolerance	*p5cs* Δ^1-Pyrroline 5 carboxylate synthetase gene for proline biosynthesis	(270)

the organization, function, and evolution of plant genes, and their expression systems. Irrespective of the transformation method used, evidences indicate that foreign DNA is integrated at random into plant genome, and that the transgene is inherited according to established genetic rules. Foreign genes introduced to plant cells genome, also called transgenes, are usually chimeric assemblies of cloned DNA from diverse organisms.

A typical transgene consists of a coding sequence with 5′ and 3′ untranslated regions, including the promoter and the sequences for poly A addition that function in plants (214). In order to select and recover transformed cells, tissues, or plants, selective marker genes have been developed which confer resistance to an antibiotic or herbicide. Presence of these genes gives to transformed cells the ability to grow on media containing a toxic compound for untransformed cells. Examples of these marker genes are *nptII* that confers kanamycin resistance, *hpt* that confers hygromycin resistance, and *csr1-1* that confers chlorsulfuron resistance (215). In addition, a number of reporter genes are available, encoding enzymes or proteins that can be assayed or observed easily in plant tissues, facilitating the detection of transformation events during gene transfer experiments. The gene for the β-Glucuronidase (*uidA* or *gus*), which originates from *E. coli*, is the most widely used reporter gene in plant transformation protocols (216). Green fluorescent protein (GFP) is another interesting reporter gene, responsible for the jellyfish *Aequorea victoria* green bioluminescence. The intense fluorescence of GFP is due to the chromophore composed nature of modified amino acids within the polypeptide. Formation of fluorescent chromophore is species independent and apparently does not require any additional factors. Because the gene product is easily detectable *in vivo* in transformed tissues by its intense fluorescence, the GFP has become a unique reporter gene (217).

Usually, marker and reporter genes have no agronomic or nutritional value, and their presence in the plant genome can generate some plant grower and consumer concerns. For this reason, the development of transformation systems that produce marker free transgenic plants is desirable and has been intensely researched (218). Introduction of multiple genes for metabolic engineering applications that require simultaneous and coordinated expression of genes encoding multiple steps in a metabolic pathway is also under study (219).

1.6.1 Somatic Hybridization

Protoplasts are useful for a somatic hybridization technique, giving place for somatic hybrid cells, which are originated from the fusion of two protoplasts from different species. These somatic hybrids may be able to attain the posterior regeneration of a whole plant. Somatic hybridization enables the transfer of desired properties across species boundaries and avoids sexual incompatibilities. Protoplast fusion is not a spontaneous event, because they have a membrane with a negative peripheric charge, and thus they have the tendency to mutually repulse. Nevertheless, it is possible to use a chemical or physical fusion method to overcome that repulse. Chemical fusion consists on placing the protoplasts in medium that alter the characteristics of their membrane surfaces, such as $NaNO_3$, a high calcium concentration (Ca^{2+}), and polyethylene–glycol (PEG). These compounds change the plasma membrane's properties from having a negatively peripheric charge to a positively one. On the other hand, the most important physical method is electrofusion (220). A derivative of somatic hybridization technique is the production of cybrids. For the production of somatic hybrids or cybrids, is essential to have an efficient method for regeneration of whole plants starting from protoplasts.

In a comparative study of both methods using pea protoplasts, De Filippis et al. (221) reported that PEG induced about 15% of somatic hybrids, and that over 50% of the initial protoplast mix was lost or damaged. Additionally, PEG treated protoplasts had decreased viability and respiratory O_2 consumption. In presence of light, PEG also induced a marked decrease in O_2 evolution, rate of CO_2 fixation, and protein and chlorophyll levels. Membrane damage was present in PEG treated protoplasts, as evident by large amounts of efflux of intracellular potassium (>70%) and CO_2 fixation products (>40%). On the other hand, electrofusion protocol produced 20–30% fusion products, lower protoplast membrane damage, and normal respiratory O_2 consumption. Under light, O_2 evolution and CO_2 fixation were close to, and in some cases higher than, freshly isolated protoplasts. Protein and chlorophyll content of electrofused protoplasts were close to the controls, and the loss of intracellular K was about 10%.

Nuclei fusion may also take place after protoplast fusion, and this is when true somatic hybrids are obtained. An interesting variation of protoplast fusion is when nuclei of one population of protoplasts to be fused are irradiated with x-rays or gamma rays. Resulting hybrids have the nuclei of one cellular line, and the cytoplasm of both lines, including mitochondria and plastid genomes. This type of cytoplasmic hybrid is called a cybrid.

In its beginning, somatic hybridization was considered as a technique that would revolutionize agricultural research. Nevertheless, difficulties in protoplast isolation, culture, and regeneration, and elevated ploidy levels in the resulting somatic hybrids, as well as the emergence of genetic transformation techniques, lead to limited use of this procedure. Despite this fact, this technique continues being an important tool in the improvement of some species. For example, citrus somatic hybridization and cybridization has become an integral part of several programs of citrus improvement (222). Applications of these techniques to citrus rootstock improvement include the production of allotetraploid hybrids that combine complementary diploid rootstocks, and to combine sexually incompatible or difficult to hybridize genera that possess traits of interest (222).

1.6.2 Direct Transformation

1.6.2.1 Gene Transfer to Protoplast

Uptake of exogenous DNA into plant protoplasts can be promoted by various treatments. Those more frequently used, as in somatic hybridization, are polyethylene–glycol (PEG) and electroporation (223). These methods do not require biological vectors and the DNA uptake is a physical process.

For regeneration of whole transgenic plants, it is essential an efficient regeneration system starting from protoplasts. This is one of the main obstacles for this transformation method, because plant regeneration from protoplasts always is a complex process that depends on many parameters and not easily achieved. Use of direct gene transfer to protoplasts as a method for production of whole transgenic plants is being taken over by biolistic or *Agrobacterium*-based transformation methods. However, this method maintains its importance in transient expression studies, which neither require integration of transforming DNA into genome nor regeneration of whole plants.

1.6.2.2 Gene Transfer Using Biolistics

Biolistics is a transformation method based on gene delivery into intact plants cells or organized tissues by high velocity micro-projectiles coated with DNA. These projectiles are tungsten or gold particles 1–2 μm in diameter. A number of devices for accelerating microprojectiles have been described (224). Acceleration is achieved either by gun powder charge, capacitance discharge through a drop of water, or the sudden release of compressed air, nitrogen, or helium. Development of this technique has allowed the transformation of many plant species considered recalcitrant, because the biolistic process appears to be effective regardless of species or tissue type (225,226).

Biolistic transformation process is influenced by several parameters such as origin and stage of the tissue; cell survival after bombardment; and the material, size, propulsion velocity, impact distribution, and DNA-carrying capacity of the particles. It is important to target cells that are actively dividing, as these have the greatest capacity for foreign DNA integration. Tissue culture conditions and media that induce the division of cells with high morphogenic competence will facilitate the transformation and recovery of transgenic plants (227).

Biolistic transformation is also possible for chloroplasts (228), and is a valuable method for transient gene expression studies. The disadvantages of the biolistic process are that it requires special instrumentation and the fact that it is a direct gene transfer method, and therefore has no direct effect on chromosomal integration mechanism. For this reason biolistics can not guarantee the stable transformation of bombarded cells.

1.6.2.3 Other Direct Transformation Systems

Microinjection, DNA introduction mediated by silicon carbide, and tissue electroporation are alternative direct transformation methods. Microinjection is a direct physical method for introducing DNA or any other kind of molecule into single plant cells using an optical microscope and micromanipulators. This is one of the most difficult gene transfer methods, but it has some advantages such as control of the quantity of DNA injected, and the possibility to observe and control the cells before, during, and after DNA transfer (229).

The direct delivery of DNA to plant cells using silicon carbide whiskers is a simple and inexpensive option for the production of transgenic plants. Cells are mixed abruptly with plasmid DNA in the presence of silicon carbide whiskers. The resultant collisions between cells and needle-like whiskers result in cell penetration and DNA delivery (230). Electroporation-mediated DNA transfer into intact plant cells and regeneration of whole maize transgenic plants were reported by D´Haullin et al. (231). This demonstrates that it is not indispensable to eliminate the cell wall to introduce DNA by electroporation.

1.6.3 Agrobacterium-Mediated Transformation

1.6.3.1 Agrobacterium tumefaciens

Agrobacterium tumefaciens is a Gram-negative soil bacterium that induces tumors called crown galls on many plant species. This is due to the natural capacity of this bacterium to introduce a DNA segment of its large tumor-inducing plasmid (Ti) into plant cells. The transferred DNA, called T-DNA, is integrated into plant cell genome via recombination. Products of virulence (*vir*) genes located on a nontransferred segment of the Ti plasmid are responsible for excision of T-DNA for transfer into a plant nucleus and for chromosomal integration in recipient cell. The *vir* genes are tightly regulated, so their expression occurs only in the presence of wounded plant cells, the targets of infection. This is because the product of *vir A* gene detects small phenolic compounds, like acetosyringone and others, released by wounded cells. Genes transferred to plant genome, located in the T-DNA, encoded enzymes involved in the production of plant growth regulators like auxin and cytokinin, and tumor-specific novel amino acid derivatives called opines. The first ones, called oncogenic genes, alter the development patterns of the plant tissue forming tumors, while the products of the second kind of genes are used by *Agrobacterium tumefaciens* as sources of carbon and nitrogen. The type of opine produced is used to classify the *Agrobacterium* strains.

Plant cells containing the T-DNA wild genes cannot regenerate whole plants. However, deletion of the oncogenic genes from the T-DNA does not interfere with the transfer process to plant cells. This is because only two repeated regions of 25 bp, located in the ends of T-DNA, called border sequences, are indispensable for the transfer process. These border sequences are recognized by endonucleases encoded by *vir* genes and they are the only *cis* elements necessary to direct T-DNA transfer. Plant cells transformed with T-DNA lacking oncogenes are able to regenerate complete plants. Therefore, by using such disarmed Ti plasmids it is possible to obtain fertile transgenic plants. Genes of interest can be introduced into plants by linking them to the disarmed T-DNA via recombination or by cloning them between border sequences in an independent plasmid called a binary vector. Any DNA between the border sequences will be transferred to the plant cell genome.

Efficiency of T-DNA transfer by *A. tumefaciens* can vary dramatically for different plant species, cultivars, or target tissues, and the principal limitation of this genetic transformation system is the host range of the bacteria (232,233). A typical transformation protocol using *A. tumefaciens* involves a 24–72 h cocultivation of bacteria (containing the plasmid with foreign genes) with plant tissues. This is followed by culture of explants on shoot induction medium with the appropriate antibiotic or herbicide (234) to select transformed cells, and one antibiotic to eliminate residual *Agrobacterium*. Modifications of this approach have been adapted to several plant species.

1.6.3.2 Agrobacterium rhizogenes

Agrobacterium rhizogenes is a soil bacterium responsible for development of hairy root disease on a range of dicotyledonous plants. The hairy root phenotype is caused by transferring to the plant genome, a DNA sequence from the Ri plasmid that encodes functions which induce root formation, in a manner similar to the development of crown gall disease by *A. tumefaciens*. Transformed roots are easily distinguished by their rapid growth on hormone-free media and plagiotropic root development. These roots can be cultivated *in vitro* easily and are an important tool for the study and production of plant secondary metabolites. *Agrobacterium rhizogenes* will also transfer in *trans* the T-DNA of binary vectors, thereby enabling the production of hairy roots containing foreign genes. Tissues transformed by wild genes of *A. rhizogenes* are able to regenerate fertile plants, which is important because regeneration from tissues containing wild genes of *A. tumefaciens* is a very rare event. However, due to independent insertion of Ri T-DNA and binary vector T-DNA, segregation in subsequent generations can produce plants free of wild genes of *A. rhizogenes*. Therefore, *A. rhizogenes* can be used as an alternative transformation system for the generation of plants transformed with foreign genes, including agronomical useful traits.

Transgenic plants are usually obtained after transferring transformed roots segments to a shoot induction media (235,236). Sometimes plants regenerated from hairy roots exhibit an altered phenotype characterized by several morphological changes including wrinkled leaves, shortened internodes, reduced apical dominance, reduced fertility, altered flowering, and plagiotropic roots (237).

Protocols for hairy roots establishment have been created and described in detail (238,239). Frequently two types of problems may arise trying to establish hairy root cultures, one relates to the difficulty of transforming certain plant species, and another concerns specifically the complication presented as to when transformed roots should be removed from the original tissue to establish the appropriate cultures. The common range of *Agrobacterium rhizogenes* hosts is limited to dicotyledoneus plants (240), although within that group there are recalcitrant woody species difficult to transform. Fortunately, it is possible to increase the host range of different bacterial strains introducing additional copies of virulent active genes or carrying out a double plant material transformation (241,242).

The most important characteristics of roots transformed by *Agrobacterium rhizogenes* is their major growth rate in comparison to normal roots and their complete independence of exogenous growth regulators to induce growth. The major growth arises from several characteristics, such as main lateral root formation, high linear extension speed, secondary diameter increase as consequence of expansion and cellular differentiation, profusion of radical hair, and geotropism absence (238). Growing speed of *in vitro* transformed root cultures changes widely depending on plant species (243), and culture conditions as well (244).

Culture of transgenic roots is increasing especially in the field of study and production of secondary metabolites. As it was pointed out, White grew root cultures as early as 1934, but these culture systems were not used, mainly due to their poor growth rate. With the discovery of transformed roots, the disadvantage of scanty growth was overcome. Therefore, in this *in vitro* system of isolated plant organs, growth and expression of secondary metabolism biosynthetic route were compatible. On the other hand, root biosynthetic capacity is also maintained in transformed roots, consequently they become attractive as systems for the study and production of secondary metabolites that are synthesized and accumulated specifically in the roots. Plant tissue transformation protocols of *Agrobacterium rhizogenes* infection technique may be used for a wide number of industrial interest metabolite producing species (245,246). Transformed root cultures have certain characteristics that make them appropriated to be cultured in large fermenters to produce at commercial level several secondary metabolites (247).

1.7 CONCLUDING REMARKS

Plant cell, tissue, and organ culture methods have evolved considerably throughout the last 100 years, since Haberlandt's first attempts to culture *in vitro* plant fragments. It is interesting in this regard to note that these methodologies have conveniently been utilized as research systems to study at the molecular, genetical, biochemical, cellular, physiological, and morphological levels several processes that take place in plants. The outcome of those studies have helped enormously and complemented a better understanding of plant biology function. Therefore, *in vitro* culture of plants has had an important role in basic plant science. Alternatively, plant cell, tissue, and organ culture have also been applied to the commercial production and propagation (micropropagation) of elite plants, as well as to obtain pathogen-free plants. In addition, these *in vitro* systems have been used in the production of secondary metabolites at industrial stage.

On the other hand, not all plant tissue culture techniques have had the same success rate. Initial techniques and observable facts related to *in vitro* culture, such as somatic hybridization and somaclonal variation, were considered at first as powerful tools for plant improvement. However, ambiguous and unexpected results, as well as many drawbacks, such as lack of genetic control in those systems, have posed somaclonal variation and other tissue culture techniques' expectatives in a moderate context. Nevertheless, as previously noted in this chapter, numerous plant *in vitro* methodologies are still in use by scientists, and new technologies have recently emerged, increasing the role of *in vitro* tissue culture on plant science. Furthermore, plant cell, tissue, and organ techniques have played a key role in the advance of plant transformation protocols. Without plant *in vitro* regeneration techniques, GMO crop plants could not be obtained at all. It is important to emphasize that the combination of plant *in vitro* techniques and genetic engineering can make significant contributions to the development of food biotechnology.

In summary, tissue culture technology has had, and will have, a remarkable impact on and relevance to both basic and applied plant science. Consequently, it is expected that these plant *in vitro* methods will contribute to the advance and improvement of food biotechnology science.

REFERENCES

1. F.A.O., Biotechnology in food and agriculture. (www.fao.org).
2. Haberlandt, G. Culturversuche mit isolierten Pflanzenzellen. *Sitzungsber Kaizer Akad. Wiss.* 111:69–92, 1902.
3. Hannig, E. Zur Physiologie pflanzlicher Embryonen. *Bot. Z.* 62:45–80, 1904.
4. White, P.R. Potentially unlimited growth of excised tomato root tips in a liquid medium. *Plant Physiol.* 9:585–600, 1934.
5. White, P.R. Potentially unlimited growth of excised plant callus in an artificial nutrient. *Am. J. Bot.* 26:59–64, 1939.
6. Gautheret, R.J. Sur la possibilité de réalizer la culture indéfinie des tissus de tubercules de carrote. *C. R. Acad. Sci.* 208:118–121, 1939.
7. Nobécourt, P. Sur la perennite de l´augmentation de volume des cultures de tissus vegetaux. *C. R. Soc. Biol.* 130:1270–1271, 1939.
8. Miller, C., F. Skoog, F. Okumura, F. Saltza, F. Strong. Isolation, structure and synthesis of kinetin, a substance promoting cell division. *J. Am. Chem. Soc.* 78:1375–1380, 1956.
9. Skoog, F., C. Miller. Chemical regulation of growth and organ formation in plant tissues cultured *in vitro*. *Symp. Soc. Exp. Biol.* 11:118–130, 1957.
10. Murashige, T., F. Skoog. A revised medium for rapid growth and bioassays with tobacco tissue cultures. *Physiol. Plant.* 15:473–479, 1962.
11. Guha, S., S. Maheshwari. *In vitro* production of embryos from anthers of *Datura*. *Nature* 204:497–501, 1964.
12. Vasil, V., A. C. Hildebrandt. Differentiation of tobacco plants from single, isolated cells in microcultures. *Science* 150:889–892, 1965.
13. Takebe, I., G. Labib, G. Melchers. Regeneration of whole plants from isolated mesophyll protoplasts of tobacco. *Naturwissenschaften* 58:318–320, 1971.
14. Carlson, P., H. Smith, R. Dearing. Parasexual interspecific plant hybridization. *Proc. Natl. Acad. Sci. USA* 69:2292–2294, 1972.
15. Zaenen, I., N. Larebeke, H. Teuchy, M. Montagu, J. Shell. Super coiled DNA in crown-gall inducing *Agrobacterium* strains. *J. Mol. Biol.* 56:109–127, 1974.
16. Ooms, O., P. Hooykaas, G. Moolenaar, R. Schileroot. Crown gall plant tumours of abnormal morphology induced by *Agrobacterium* carrying mutated octapine Ti plasmids. *Gene* 14:33–50, 1981.
17. Fraley, R., S. Rogers, R. Horsch, P. Sanders, J. Flick, S. Adams, M. Bittner, L. Brand, C. Fink, J. Fry, G. Galluppi, S. Goldberg, N. Hoffman, S. Woo. Expression of bacterial genes in plant cells. *Proc. Natl. Acad. Sci. USA* 80:4803–4807.
18. Horsch, R., R. Fraley, S. Rogers, P. Sanders, A. Lloyd, N. Hoffmann. Inheritance of functional foreign genes in plants. *Science* 223:496–498. 1984.
19. Street, H.E. Laboratory Organization. In: *Plant Tissue and Cell Culture*, 2nd ed., Street, H.E., ed. Berkely: University of California Press, 1977, pp 11–30.
20. Wetherell, D.F. *Introduction to* In Vitro *Propagation*. New York: Avery Publishing Group, 1982, pp 7–29.
21. Kyte, L. *Plants from Test Tubes*. Portland, OR: Timber Press, Inc., 1987, pp 20–35.
22. Boccon-Gibod, J. The Technology of *In Vitro* Culture. In: H Vidalei, ed. In Vitro *Culture and Its Applications in Horticulture*. New Delhi, Lebanon, NH: Science Publishers, 1995, pp 38–64.
23. George, E.F. *Plant Propagation by Tissue Culture: Part 1 The Technology*. 2nd ed. Britain, Edington: Exegetics Limited, 1993, pp 95–116.

24. Gamborg, O.L., G.C. Phillips, eds., *Plant Cell, Tissue and Organ Culture: fundamental methods*. Berlin; New York: Springer, 1995, pp 3–20.
25. FASTER. Laminar flow cabinets and systems - fume hoods. http://www.faster-air.com/lab/lab.htm
26. Nuñez-Palenius, H.G., N. Ochoa-Alejo. *In vitro* mass cultivation of cells and tissues. In: *Molecular Biotechnology for Plant Food Production*, Paredes-López, O., ed. Lancaster: Technomic Pub. Co., 1999, pp 89–130.
27. Tisserat, B., D. Jones, P.D. Galletta. Microwave sterilization of plant tissue culture media. *HortScience*. 27:358–361.
28. Parfitt, D.E., A.A. Almehdi, L.N. Bloksberg. Use of organic buffers in plant tissue culture systems. *Sci. Hortic*. 36:157–163, 1988.
29. Owen, H.R., D. Wengerd, A.R. Miller. Culture medium pH is influenced by basal medium, carbohydrate source, gelling agent, activated charcoal, and medium storage method. *Plant Cell Rep*. 10:583–586.
30. Leifert, C., K.P. Murphy, P.J. Lumsden. Mineral and carbohydrate nutrition of plant cell and tissue cultures. *Critic. Rev. Plant Sci*. 14:83–109, 1995.
31. Chée, R.P., D.I. Leskovar, D.J. Cantliffe. Optimizing embryogenic callus and embryo growth of a synthetic seed system for sweetpotato by varying media nutrient concentrations. *J. Amer. Soc. Hort. Sci*. 117:663–667.
32. Sagishima, K., K. Kubota, H. Ashihara. Uptake and metabolism of sugars by suspension-cultured *Catharanthus roseus* cells. *Ann. Bot*. 64:185–193, 1989.
33. Lee, T.K., W.S. Lee. Diauxic growth in rice suspension cells grown on mixed carbon sources of acetate and glucose. *Plant Physiol*. 110:465–470, 1996.
34. Biondi, S., T.A. Thorpe. Requirements for a tissue culture facility. In: *Plant Tissue Culture: Methods and Applications in Agriculture,* Thorpe, T.A., ed., New York: Academic Press, 1981, pp 1–20.
35. Short, K.C., J. Warburton, A.V. Robert. *In vitro* hardening of cultured cauliflower and chrysanthemum plantlets to humidity. *Acta. Hort*. 212:329–334, 1987.
36. Schleiden, M.J. Beiträge zur phytogenesis. *Müller Arch. Anat. Physiol*. 3:137–176, 1838.
37. Schwann, T.H. Mikroskopische untersuchungen über die übereinstimmung in der struktur und dem wachstum der thiere und pflanzen. Nr. 176:Oswalds Berlin, 1839.
38. Steward, F.C., M.O. Mapes, K. Mears. Growth and organized development of cultured cells, II: organization in cultures grown from freely suspended cells. *Am. J. Bot*. 45:705–708, 1958.
39. Reinert, J. Utersuchungen uber die morphogenese an gewebekulture. *Ber. Dtsch. Bot. Ges*. 71:15, 1958.
40. Schultheis, J.R., D.J. Cantliffe, H.H. Bryan. Early plant growth and yield of sweetpotato grown from seed, vegetative cuttings, and somatic embryos. *J. Amer. Soc. Hort. Sci*. 119:1104–1111, 1994.
41. De Jong, A.J., E.D.L. Schmidt, S.C. De Vries. Early events in higher-plant embryogenesis. *Plant Mol. Biol*. 22:367–377, 1993.
42. Finstad, K., D.C.W. Brown, K. Joy. Characterization of competence during induction of somatic embryogenesis in alfalfa tissue culture. *Plant Cell. Tiss. Org. Cult*. 34:125–132, 1993.
43. Toonen, M.A.J., T. Hendriks, E.D.L. Schmidt, H.A. Verhoeven, A. Van Kammen, S.C. De Vries. Description of somatic-embryo-forming single cells in carrot suspension cultures employing video cell tracking. *Plantarum* 194:565–572, 1994.
44. Fehér, A., T.P. Pasternak, D. Dudits. Transition of somatic plant cells to an embryogenic state. *Plant Cell Tiss. Org. Cult*. 74:201–228, 2003.
45. Chée, R.P., D.J. Cantliffe. Embryo development from discrete cell aggregates in *Ipomoea batatas* (L.) Lam. in response to structural polarity. In Vitro *Cell. Dev. Biol*. 25:757–760.
46. Kamada, H., W. Ishikawa, H. Saga, H. Harada. Induction of somatic embryogenesis in carrot by osmotic stress. *Plant. Tiss. Cult. Lett*. 10:38–44, 1993.
47. Litz, R.E. Effect of osmotic stress on somatic embryogenesis in *Carica papaya* suspension cultures. *J. Am. Soc. Hortic. Sci*. 111:969–972, 1986.

48. Komamine, A. Mechanism of SE in cell cultures: physiology, biochemistry and molecular biology. In Vitro *Cell. Dev. Biol.* 28:11–14, 1992.
49. Meinke, D.W. Molecular genetics of plant embryogenesis. *Ann. Rev. Plant. Physiol. Plant Mol. Biol.* 46:369–394, 1995.
50. Rojas-Herrera, R., F. Quiróz-Figueroa, L. Sánchez-Teyer, V.M. Loyola-Vargas. Molecular analysis of somatic embryogenesis: an overview. *Physiol. Mol. Biol. Plants* 8:171–184, 2002.
51. Larkin, P.J., W.R. Scowcroft. Somaclonal variation: a novel source of variability from cell cultures for plant improvement. *Theor. Appl. Genet.* 60:197–214, 1981.
52. Evans, D.A., W.R. Sharp, H.P. Medina-Filho. Somaclonal and gametoclonal variation. *Am. J. Bot.* 71:759–774, 1984.
53. Evans, D.A. Somaclonal variation: genetic basis and breeding applications. *Trends Genet.* 5:46–50, 1989.
54. Orton, T.J. Experimental approaches to the study of somaclonal variation. *Plant Mol. Biol. Rep.* 1:67–76, 1983.
55. Duncan, R.R. Tissue culture-induced variation and crop improvement. *Adv. Agron.* 58:201–240, 1997.
56. Gill, D.E., L. Chao, S.L. Perkins, J.B. Wolf. Genetic mosaicism in plants and animals. *Ann. Rev. Ecol. Syst.* 26:423–444, 1995.
57. van den Bulk, R.W., H.J.M Löffler, W.H. Lindhout, M. Koornneef. Somaclonal variation in tomato: effect of explant source and comparison with chemical mutagenesis. *Theor. Appl. Gene.* 80:817–825, 1990.
58. Skirvin, R.M., K.D. McPheeters, M. Norton. Sources and frequency of somaclonal variation. *Hort. Sci.* 29:1232–1237, 1994.
59. Chowdhury, M.K.U., G.W. Schaeffer, R.L. Smith, L.R. DeBonte, B.F. Matthews. Mitochondrial DNA variation in long-term tissue cultured rice lines. *Theor. Appl. Genet.* 80:81–88, 1990.
60. Rani, V., K.P. Singh, B. Shiran, S. Nandy, S. Goel, R.M. Devarumath, H.L. Sreenat, S.N. Raina. Evidence for nuclear and mitochondrial genome organizations among high-frequency somatic embryogenesis-derived plants of allotretaploid *Coffea arabiga* L. (Rubiaceae). *Plant Cell. Rep.* 19:1013–1020, 2000.
61. Kaeppler, S.M., H.F. Kaeppler, Y. Rhee. Epigenetic aspect of somaclonal variation in plants. *Plant Mol. Biol.* 43:179–188, 2000.
62. Kester, D.E. The clone in horticulture. *Hort. Sci.* 18:831–837, 1983.
63. Sinden, S.L., J.F. Shepard. Variation at the cellular level. *Hort. Sci.* 18:837–840, 1983.
64. Lee, M., R.L. Phillips. The chromosomal basis of somaclonal variation. *Ann. Rev. Plant Mol. Biol.* 39:413–437, 1988.
65. Hirochika, H., K. Sugimoto, Y. Otsuki, H. Tsugawa, M. Kanda. Retrotransposons of rice involved in mutations induced by tissue culture. *Proc. Natl. Acad. Sci. USA* 93:7783–7788, 1996.
66. Goldsbrough, P.B., E.M. Hatch, B. Huang, W.G. Kosinski, W.E. Dyer, K.M. Herrmann, S.C. Weller. Gene amplification in glyphosate tolerant tobacco cells. *Plant Sci.* 72:53–62, 1990.
67. Arnholdt-Schmitt, B., S. Herterich, K.-H. Neumann. Physiological aspects of genome variability in tissue culture, I: growth phase-dependent differential DNA methylation of the carrot genome (*Daucus carota* L.) during primary culture. *Theor. Appl. Genet.* 91:809–815, 1995.
68. Diaz-Sala, C., M. Rey, A. Boronat, R. Besford, R. Rodriguez. Variations in the DNA methylation and polypeptide patterns of adult hazel (*Corylus avellana* L.) associated with sequencial *in vitro* subcultures. *Plant Cell. Rep.* 15:218–221, 1995.
69. Sabbah, S., M. Raise, M. Tal. Methylation of DNA in NaCl-adapted cells of potato. *Plant Cell. Rep.* 14:467–470, 1995.
70. Al-Zahim, M.A., B.V. Ford-Lloyd, H.J. Newbury. Detection of somaclonal variation in garlic (*Allium sativum* L.) using RAPD and cytological analysis. *Plant Cell. Rep.* 18:473–477, 1999.
71. Chowdari, K.V., W. Ramakrishna, S.A. Tamahankar, R.R. Hendre, V.S. Gupta, N.A. Sahasrabudhe, P.K. Ranjekar. Identification of minor DNA variations in rice somaclonal variants. *Plant Cell. Rep.* 18:55–58, 1998.

72. Fourré, J.-L., P. Berger, L. Niquet, P. André. Somatic embryogenesis and somaclonal variation in Norway spruce: morphogenetic, cytogenetic and molecular approaches. *Theor. Appl. Genet.* 94:159–169, 1997.
73. Müller, E., P.T.H. Brown, S. Hartke, H. Lörz. DNA variation in tissue culture-derived rice plants. *Theor. Appl. Genet.* 80:673–679, 1990.
74. Vendrame, W.A., G. Kochert, H.Y. Wetzstein. AFLP analysis of variation in pecan somatic embryos. *Plant Cell. Rep.* 18:853–857, 1999.
75. Salgado-Garciglia, R., N. Ochoa-Alejo. Increased capsaicin content in PFP-resistant cells of chili pepper (*Capsicum annuum* L.). *Plant Cell. Rep.* 8:617–620, 1990.
76. Santos-Diaz, M.S., N. Ochoa-Alejo. PEG-tolerant cell clones of chili pepper (*Capsicum annuum* L.): Growth, osmotic potentials and solute accumulation. *Plant Cell. Tissue Org. Cult.* 37:1–8, 1994.
77. Shepard, J.F. Protoplasts as a source of disease resistance in plants. *Ann. Rev. Phytopathol.* 19:145–155, 1981.
78. Chawla, H.S. *Introduction to Plant Biotechnology*, 2nd ed., Enfield, NH: Science Publishers, 2002, pp 57–73.
79. Ramírez-Malagón, R., N. Ochoa-Alejo. Adventitious shoot formation and plant regeneration from tissues of tomatillo (*Physalis ixocarpa* Brot). *Plant Cell. Tissue Org. Cult.* 25:185–188, 1991.
80. Koch, M., Z. Tamani, R. Salomón. Improved regeneration of shoots from garlic callus. *Hort. Sci.* 30:378, 1995.
81. Machado, M.F.P.S., A.J. Prioli. Micropropagation of *Cereus peruvianus* Mill. (Cactaceae), by areole activation. In Vitro *Cell. Dev. Biol. Plant* 32:199–203, 1996.
82. Myers, J.M., P.V. Simon. Regeneration of garlic callus as affected by clonal variation, plant growth regulators and culture conditions over time. *Plant Cell Rep* 19:32–36, 1999.
83. Gamborg, O.L., R.A. Miller, K. Ojima. Nutrient requirements of suspension cultures of soybean root cells. *Exp. Cell. Res.* 50:151–158, 1968.
84. Manoharan, K., A. Gnanam. Growth stimulation by conditioned medium and spermidine in low-density suspension cultures of rice. *Plant Cell. Physiol.* 33:1243–1246, 1992.
85. Perata, P., A. Alpi. Ethanol metabolism in suspension cultured carrot cells. *Physiol. Plant* 82:103–108, 1991.
86. Mirjalili, N., J.C. Linden. Gas composition effects on suspension cultures of *Taxus cuspidata*. *Biotechnol. Bioeng.* 48:123–132, 1995.
87. Fischer, U., U.J. Santore, W. Huseman, W. Bartz. Semicontinuous cultivation of photoautotrophic cell suspension cultures in a 20 l airlift-reactor. *Plant Cell. Tissue Org. Cult.* 38:123–134, 1994.
88. Hoopen, H.J.G., W.M.V Gulic, J.E. Schlatmann, P.R.H. Moreno, J.L. Vinke, J.J. Heijnen, R. Verpoorte. Ajmalicine production by cell cultures of *Catharanthus roseus*: from shake flask to bioreactor. *Plant Cell. Tissue Org. Cult.* 38:85–91, 1994.
89. Harrell, R.C., M. Bieniek, D.J. Cantliffe. Noninvasive evaluation of somatic embryogenesis. *Biotechnol. Bioeng.* 39:378–383, 1992.
90. Harrell, R.C., M. Bieniek, C.F. Hood, R. Munilla, D.J. Cantliffe. Automated, *in vitro* harvest of somatic embryos. *Plant Cell. Tissue Org. Cult.* 39:171–183, 1994.
91. Cantliffe, D.J. Bioreactor technology in plant cloning. *Proceedings of the Fourth International Symposium on* In Vitro *Culture and Horticultural Breeding*, Finland, 2001, pp 345–351.
92. Pan, Z.W., H.Q. Wang, J.J. Zhong. Scale up study on suspension cultures of *Taxus chinensis* cells for production of taxane diterpene. *Enzime Microb. Technol.* 27:714–723, 2000.
93. Gao, J.W., J.M. Lee. Effect of oxygen-supply on the suspension-culture of genetically modified tobacco cells. *Biotechnol. Prog.* 8:285–290, 1992.
94. Huang, S.Y., C.J. Chou. Effect of gaseous composition on cell growth and secondary metabolite production in suspension culture of *Stizolobium hassjoo* cells. *Bioprocess. Eng.* 23:585–593, 2000.
95. Ho, C.H., K.A. Henderson, G.L. Rorrer. Cell-damage and oxygen mass-transfer during cultivation of *Nicotiana Tabacum* in stirred-tank bioreactor. *Biotechnol. Prog.* 11:140–145, 1995.

96. Farrés, J., P. Kallio. Improved cell growth in tobacco suspension cultures expressing Vitreoscilla hemoglobin. *Biotechnol. Prog.* 18:229–233, 2002.
97. Linsmaier, E.M., F. Skoog. Organic growth factor requirements of tobacco tissue cultures. *Physiol. Plant.* 18:100–127, 1965.
98. Ozeki, Y., J. Takeda. Regulation of phenylalanine ammonia-lyase genes in carrot suspension cultured cells. *Plant Cell. Tissue Org. Cult.* 38:221–225, 1994.
99. Paek, K.Y., E.J. Hahn, S.H. Son. Application of bioreactors for large-scale micropropagation systems of plants. In Vitro *Cell. Dev. Biol. Plant* 37:149–157, 2001.
100. Harrell, R.C., D.J. Cantliffe. Automated evaluation of somatic embryogenesis in sweet potato by machine vision. In: *Cell Culture and Somatic Cell Genetics of Plants, Vol. 8:Cell Culture in Phytochemistry*, Constabel, F., I.K. Vasil, eds., New York: Academic Press, 1991, pp 179–195.
101. Stöckigt, J., P. Orbitz, H. Falkenhagen, R. Lutterbach, S. Endress. Natural products and enzymes from plant cell cultures. *Plant Cell. Tissue Org. Cult.* 43:97–109, 1995.
102. Bourgaud, F., A. Gravot, S. Milesi, E. Gontier. Production of plant secondary metabolites: a historical perspective. *Plant Sci.* 161:839–851, 2001.
103. Misawa, M. *Plant Tissue Culture: An Alternative for Production of Useful Metabolite. FAO Agricultural Services Bulletin No. 108.* Rome: Food and Agriculture Organization of the United Nations, 1994, pp. 65–78.
104. Vanek, T., I. Valterova, R. Vanková, T. Vaisar. Biotransformation of (-)-limonene using *Solanum aviculare* and *Dioscorea deltoidea* immobilized plant cells. *Biotechnol. Lett.* 21:625–628, 1999.
105. Ishihara, K., H. Hamada, T. Hirata, N. Nakajima. Biotransformation using plant cultured cells. *J. Mol. Catal. B- Enzym.* 23:145–170, 2003.
106. Verpoorte, R., R. van der Heijden, H.J.G. ten Hoopen, J. Memelink. Metabolic engineering of plant secondary metabolite pathways for the production of fine chemicals. *Biotechnol. Lett.* 21:467–479, 1999.
107. Choi, H.J., B.Y. Tao, M.R. Okos. Enhancement of secondary metabolite production by immobilized *Gossypium arboreum* cells. *Biotechnol. Prog.* 11:306–311, 1995.
108. Gillet, F., C. Roisin, M.A. Flinaux, A. Jacquin-Dubreuil, J.N. Barbotin, J.E. Nava-Saucedo. Immobilization of *Nicotiana tabacum* cell suspensions within calcium alginate gel beads for the production of enhanced amounts of scopolin. *Enzyme Microb. Tech.* 26:229–234, 2000.
109. Mera, N., H. Aoyagi, F. DiCosmo, H. Tanaka. Production of cell wall accumulative enzymes using immobilized protoplasts of *Catharanthus roseus* in agarose gel. *Biotechnol. Lett.* 25:1687–1693, 2003.
110. Yuana, M.J., W. Dignum, R. Verpoorte. Glucosylation of exogenous vanillin by plant cell cultures. *Plant Cell. Tissue Org. Cult.* 69:177–182, 2002.
111. A. Rosevear. Immobilized biocatalyst: a critical review. *J. Chem. Tech. Biotechnol.* 34B:127–134, 1984.
112. Mosbach, K., R. Mosbach. Entrapment of enzymes and microorganisms in synthetic cross-linked polymers and their application in column techniques. *Acta Chem. Scand.* 20:2807–2810, 1966.
113. Yeoman, M.M. Techniques, characteristics, properties, and commercial potential of immobilized plant cells. In: *Cell Culture and Somatic Cell Genetics of Plants*, Vol. 4, Constabel, F., I.K. Vasil, eds., New York: Academic Press, 1987, pp 197–216.
114. Bohmert, K., I. Balbo, J. Kopka, V. Mittendorf, C. Nawrath, Y. Poirier, G. Tischendorf, R.N. Trethewey, L. Willmitzer. Transgenic Arabidopsis *plants can accumulate polyhydroxybutyrate to up to 4% of their fresh weight. Plantarum* 211:841–845, 2000.
115. Hoa, T.T., S. Al-Babili, P. Schaub, I. Potrykus, P. Beyer. Golden Indica and Japonica rice lines amenable to deregulation. *Plant Physiol.* 133:161–169, 2003.
116. Van Eenennaam, A.L., K. Lincoln, T.P. Durrett, H.E. Valentin, C.K. Shewmaker, G.M. Thorne, J. Jiang, S.R. Baszis, C.K. Levering, E.D. Aasen, M. Hao, J.C. Stein, S.R. Norris, R.L. Last. Engineering vitamin E content: from Arabidopsis mutant to soy oil. *Plant Cell.* 15:3007–3019, 2003.

117. Siebert, M., S. Sommer, S. Li, Z. Wang, K. Severin, L. Heide. Genetic engineering of plant secondary metabolism. Accumulation of 4-hydroxybenzoate glucosides as a result of expression of the bacterial ubiC gene in tobacco. *Plant Physiol.* 112:811–819, 1996.
118. Bailey, J.E., Toward a science of metabolic engineering. *Science* 252:1668–1675, 1991.
119. Mahmoud, S.S., R.B. Croteau. Strategies for transgenic manipulation of monoterpene biosynthesis in plants. *Trends Plant Sci.* 7:366–373, 2002.
120. Dodds, J.H., L.W. Roberts. *Experiments in Plant Tissue Culture*. New York: Cambrige University Press, 1995, pp 117–125.
121. Oksman-Caldentey, K.M.N., L. Sevon, L. Vanhala, R. Hiltunen. Effect of nitrogen and sucrose on the primary and secondary metabolism of transformed root cultures of *Hyoscyamus muticus*. *Plant Cell. Tissue Org. Cult.* 38:263–272, 1994.
122. Rhodes, M.J.C., A.J. Parr, A. Giuletti, E.L.H. Aird. Influence of exogenous hormones on the growth and secondary metabolite formation in transformed roots. *Plant Cell. Tissue Org. Cult.* 38:143–151, 1994.
123. Flores, H.M., M.W. Hoy, J.J. Picard. Secondary metabolites from root cultures. *Trends Biotech.* 9:19–26, 1987.
124. Toivonen, L. Utilization of hairy root cultures for production of secondary metabolites. *Biotech. Progr.* 9:12–20, 1993.
125. Roberts, E.H., H. Street. The continuous culture of excised roots. *Physiol. Planta.* 8:238–262, 1955.
126. Buchter, D.N., H.E. Street. Excised root culture. *Bot. Rev.* 30:523–586, 1964.
127. Evans, N.E. Micropropagation: axillary bud multiplication. In: *Methods in Molecular Biology, Vol 6:Plant Cell and Tissue Culture*, Pollard, J.F., J.M. Walker, eds., Totowa, NJ: The Humana Press, 1990, pp 93–104.
128. Grout, B.W.W. Meristhem tip culture. In: *Methods in Molecular Biology, Vol 6: Plant Cell and Tissue Culture*, Pollard, J.F., J.M. Walker, eds., Totowa, NJ: The Humana Press, 1990, pp 81–91.
129. Ramírez-Malagón, R., L. Perez-Moreno, A. Borodanenko, J.G. Salinas-Gonzalez. Heat thermal, chemical and meristhematic elimination of *Potyvirus* from garlic (*Allium sativum*). *Sixth International Workshop on Allium White Rot*, University of Guanajuato, Irapuato, Gto, México, 1998, pp 1–7.
130. Kitto, S.L. Commercial micropropagation. *Hort. Sci. 32*:1012–1014, 1997.
131. Etienne, H., M. Berthouly. Temporary immersion systems in plant micropropagation. *Plant Cell. Tiss. Org. Cult.* 69:215–231, 2002.
132. Staken, E.B., T.H. Iversen. Effect of simulated and real weightless on early regneration stages of *Brasica napus* protoplasts. In Vitro *Cell. Dev. Bio. Plant.* 36:312–318, 2000.
133. Papadakis, A.K., A. Roubelakis-Angelakis. Oxidative stress could be responsible for the recalcitrance of plant protoplasts. *Plant Physiol. Biochem.* 40:549–559, 2002.
134. Aoyagi, H., H. Tanaka. Development of simple methods for preparation of yeast plant protoplasts immobilized in alginate gel beads. *Biotechnol. Tech.* 13:253–258, 1999.
135. Endress, R. *Plant Cell Biotechnology*. Berlin: Springer-Verlag, 1994, pp 353.
136. Thomas, E., M.R. Davey. *From Single Cells to Plants*. London: Wykeham Publications, 1975, pp 76–97.
137. Gamborg, O.L., J.P. Shyluk, E.A. Shahin. Isolation, Fusion, and Culture of Plant Protoplasts. In: *Plant Tissue Culture*, Thorpe, T.A., ed., New York: Academic Press, 1981, pp 115–153.
138. Bhojwani, S.S., M.M. Razdan. *Plant Tissue Culture: Theory and Practice*. New York: Elsevier, 1983, pp 237–260.
139. Kyte, L. *Plants from Test Tubes*. Portland, OR: Timber Press, 1987, pp 82–83.
140. Collin, H.A., S. Edwards. *Plant Cell Culture*. New York: Springer-Verlag, 1998, pp 71–81.
141. Augé, R., J. Boccon-Gibod. Applications in Horticulture. In: In Vitro *Culture and Its Applications in Horticulture*. Lebanon, NH: Science Publishers, 1995, pp 87–93.
142. Dodds, J.H., L.W. Roberts. *Experiments in Plant Tissue Culture*. New York: Cambrige University Press, 1995, pp 167–182.

143. Bhojwani, S.S., M.K. Razdan. *Plant Tissue Culture: Theory and Practice*, rev. ed. Amsterdam: Elsevier, 1996, pp 37–56.
144. Zheng, M.Y. Microspore culture in wheat (*Triticum aestivum*) doubled haploid production via induced embryogenesis. *Plant Cell. Tissue Org. Cult.* 73:213–230, 2003.
145. Snape, J.W. Doubled haploid breeding: theoretical basis and practical applications. In: *Plant Biotechnology 1985–1988*, International Maize and Wheat Improvement Center: México, 1989, pp 19–30.
146. Alemanno, L., E. Guiderdoni. Increased doubled haploid plant regeneration from rice (*Oriza sativa* L.) anthers cultured on colchicine supplemented media. *Plant Cell. Rep.* 13:432–436, 1994.
147. Cistué, L., A. Ramos, A.M. Castillo, I. Romagosa. Production of large number of doubled haploid plants from barley anthers pretreated with high concentrations of mannitol. *Plant Cell. Rep.* 13:709–712, 1994.
148. Barnabás, B., P.L. Pfahler, G. Kovacs. Direct effect of colchicine on the microspore embryogenesis to produce dihaploid plants in wheat. *Theor. Appl. Genet.* 81:675–678, 1991.
149. Ghaemi, M., A. Sarrafi, G. Alibert. The effects of silver nitrate, colchicine, cupric sulfate and genotype on the production of embryoids from anthers of tetraploid wheat (*Triticum turgidum*). *Plant Cell. Tissue Org. Cult.* 36:355–359, 1994.
150. Ouyangm, J.W, D.G. He, G.H. Feng, S.E. Jia. The response of anther culture temperature varies with growth conditions of anther-donor plants. *Plant Sci.* 49:145–148, 1987.
151. Jones, A.M., J.F. Petolino. Effects of donor plant genotype and growth environment on anther culture of soft-red winter wheat (*Triticum aestivum* L.). *Plant Cell. Tissue Org. Cult.* 8:215–223, 1987.
152. Simmonds, J. Improved androgenesis of winter cultivars of *Triticum aestivum* L., in response to low temperature treatment of donor plants. *Plant Sci.* 65:225–231, 1989.
153. Kristiansen, K., S.B. Andersen. Effects of donor plant temperature, photoperiod, and age on anther culture response of *Capsicum annuum* L. *Euphytica* 67:105–109, 1993.
154. Lu, C.S., H.C. Sharma, H.W. Ohm. Wheat anther culture: effect of genotype and environmental conditions. *Plant Cell. Tissue Org. Cult.* 24:233–236, 1991.
155. Orshinsky, B.R., R.S. Sadasivaiah. Effect of plant growth conditions, plating density, and genotype on anther culture response of soft white spring wheat hybrids. *Plant Cell. Rep.* 16:758–762, 1997.
156. Wassom, J.J., C. Mei, T.R. Rocheford, J.M. Widholm. Interaction of environment and ABA and GA treatments on maize anther culture response. *Plant Cell. Tissue Org. Cult.* 64:69–72, 2001.
157. Pikard, E., C. Hours, S. Gregoire, T.H. Phan, J.P. Meunier. Significant improvement of androgenetic happloid and doubled haploid induction from wheat plants treated with a chemical hybridization agent. *Theor. Appl. Genet.* 74:289–297, 1987.
158. Heberle-Bors, E. *In vitro* haploid formation from pollen: a critical review. *Theor. Appl. Genet.* 71:361–374, 1985.
159. Hewstone, O.N., C. Nich, M.C. Hewstone, S.C. Muñoz. Uso del gametocida hybrex para aumentar la androgenesis en trigo. *Agricultura Técnica (Chile)* 52:101–104, 1992.
160. Xinias, I.N., I.A. Zamani, E. Gouli-Vavdinoudi, D.G. Roupakias. Effect of cold pretreatment and incubation temperature on bread wheat (*Triticum aestivum* L.) anther culture. *Cereal Res. Commn.* 29:331–338, 2001.
161. Callenberg, E.K., L.B. Johansson. The effect of starch incubation temperature in anther culture of potato. *Plant Cell. Tissue Org. Cult.* 32:27–34, 1993.
162. Kiviharju E., E. Pehu. The effect of cold and heat pretreatments on anther culture response of *Avena sativa* and *A. sterilis*. *Plant Cell. Tissue Org. Cult.* 54:97–104, 1998.
163. Novák, F.J., Induction of haploid callus in anther cultures of Capsicum sp. *Z. Pflanzenzuchtg.* 72:46–54, 1974.
164. Lazar, M.D., G.W. Schaeffer, P.S. Baezinger. The effects of interactions of culture environment with genotype on wheat (*Triticum aestivum*) anther culture response. *Plant Cell. Rep.* 8:525–529, 1990.

165. Nitsch, C. Culture of isolated pollen on Synthetic Medium. *Comptes Rendus Hebdomadaires Des Seances De L Academie Des Sciences Serie D*. 278D:1031–1034, 1974.
166. Chuang, C.C., T.W. Ouyang, H. Chia, S.M. Chou, C.K. Chang. A set of potato medium for wheat anther culture. *Proceedings of Symposium on Plant Tissue Culture, Beijing, 1978*, pp 51–56.
167. Chu, C.C. The N_6 medium and its application to anther culture of cereal crops. *Proceedings of Symposium on Plant Tissue Culture, Beijing 1978*, pp 43–50.
168. Kao, K. Plant formation from barley anther cultures with ficoll medium. *Z. Pflanzenphysiol.* 103:437–443, 1981.
169. Armstrong, T.A., S.G. Metz, P.N. Mascia. Two regenerated systems for the production of haploid plants from wheat anther culture. *Plant Sci.* 51:231–327, 1987.
170. Yuan, H.M., V.D. Keppenne, P.S. Baezinger, T. Berke, G.H. Liang. Effect of genotype and medium on wheat (*Triticum aestivum* L.) anther culture. *Plant Cell Tissue Org. Cult.* 21:253–258, 1990.
171. Trotier, M.C., J. Collin, A. Comeau. Comparison of media for their aptitude in wheat anther culture. *Plant Cell Tissue Org. Cult.* 35:59–67, 1993.
172. Chu, C.C., R.D. Hill. An improved anther culture method for obtaining higher frequency of pollen embryoids in *Triticum aestivum* L. *Plant Sci.* 55:175–181, 1988.
173. Saidi, N., S. Cherkaoui, A. Chlyah, H. Chlyah. Embryo formation and regeneration in *Triticum turgidum* ssp. *Durum* anther culture. *Plant Cell Tissue Org. Cult.* 51:27–33, 1997.
174. Barnabás, B., B. Obert, G. Kovács. Colchicine, an efficient genome-doubling agent for maize (*Zea mays* L.) microspores cultured *en anthero*. *Plant Cell Rep.* 18:858–862, 1999.
175. Ferrie, A.M.R., D.C. Taylor, S.L. MacCkenzie, W.A. Keller. Microspore embryogenesis of high sn-2 erucic acid *Brassica oleracea* germplasm. *Plant Cell Tissue Org. Cult.* 57:79–84, 1999.
176. Metwally, E.I., S.A. Moustafa, B.I. El-Sawy, T.A. Shalaby. Haploid plantlets derived by anther culture of *Cucurbita pepo. Plant Cell Tissue Org. Cult.* 52:171–176, 1998.
177. Ting, Y.C., M. Yu, W.Z. Zheng. Improved anther culture of Maize (*Zea mays*). *Plant Sci Let* 23:139–145, 1981.
178. Ball, S.T., H.P. Zhou, C.F. Konzak. Influence of 2,4-D, IAA, and duration of callus induction in anther cultures of spring wheat. *Plant Sci.* 90:195–200, 1993.
179. Mandal, A.B., A.K. Bandyopadhyay. *In vitro* anther culture response in indica rice hybrids. *Cereal Res. Comm.* 25:891–896, 1997.
180. Beyer, E.M. A potent inhibitor of ethylene action in plants. *Plant Physiol.* 58:268–271, 1976.
181. Nervo, G., G. Carannante, M.T. Azzimonti, G.L. Rotino. Use of anther culture method in pepper breeding: factors affecting plantlets production. In: *Current Issues in Plant Molecular and Cellular Biology*, Terzi, M., et al., eds., Dordrecht: Kluwer, 1995, pp 155–160.
182. Maheswari, S.C., A.K. Tiagy, K. Malhotra. Induction of haploid from pollen grains of angiosperms: the current status. *Theor. Appl. Genet.* 58:193–206, 1980.
183. Henry, Y., J. De Buyser. Float culture of wheat anthers. *Theor. Appl. Genet.* 60:77–79, 1981.
184. Xu, Z.H., N. Sunderland. Glutamine, inositol and conditioning factor in the production of barley pollen callus *in vitro*. *Plant Sci. Lett.* 23:161–168, 1981.
185. Hoekstra, S., M.H. Van Zijderveld, J.D. Louwerse, F. Heidekamp, F. VanDerMark. Anther and microspore culture of *Hordeum vulgare* cv. Igri. *Plant Sci.* 86:89–96, 1992.
186. Barnabás, B., G. Kovács, A. Hegedüs, S. Erdei, G. Horváth. Regeneration of doubled haploid plants from *in vitro* selected microspores to improve aluminium tolerance in wheat. *Plant Physiol.* 156:217–222, 2000.
187. Bárány, I., P.S. Testillano, J. Mitykó, M.C. Risueño. The switch of the microspore developmental program in Capsicum involves HSP70 expression and leads to the production of haploid plants. *Int. J. Dev. Biol.* 45 (S1):S39–S40, 2001.
188. Champion, B., M.T. Azzimonti. Evolution of ploidy level in haploid plants of onion (*Allium cepa* L.) obtained through *in vitro* gynogenesis. *4th Allium Symposium, Eucarpia, Institute of Horticultural Research*, Wellesbourne, Warwick, UK, 1988, pp 85–89.

189. Martínez, L.E., C.B. Agüero, C.R. Galmarini. Obtention of haploid plants by ovaries and ovules culture in onion (*Allium cepa L.*). *Acta Horticulturae.443*: 447–453, 1997.
190. Bohanec, B., M. Jakse, Variation of gynogenic response among long-day onion (*Allium cepa* L.) accessions. *Plant Cell Rep.* 18:737–742, 1999.
191. Gémes-Juhász, A., P. Balogh, A. Ferenczy. Effect of optimal stage of female gametophyte and heat treatment on *in vitro* gynogenesis induction in cucumber (*Cucumis sativus* L.). *Plant Cell Rep.* 21:105–111, 2002.
192. Lotfi, M., A.R. Alan, M.J. Henning, M.M. Jahn. Production of haploid and doubled haploid plants of melon (*Cucumis melo* L.) for use in breeding for multiple virus resistance. *Plant Cell Rep.* 21:1121–1128, 2003.
193. Hao, Y.J., X.X. Deng. Genetically stable regeneration of apple plants from slow growth. *Plant Cell Tissue Org. Cult.* 72:253–260, 2003.
194. Reed, B.M., K. Humer. Conservation of germplasm of strawberry (*Fragaria* species). In: *Biotechnology in agriculture and forestry, cryopreservation of plant germplasm I*, Vol. 32., Bajaj, Y.P.S., ed., Berlin:Springer-Verlag, 1995, pp 354–370.
195. Niemirowicz-Szczytt, K., N.M. Faris, M. Rucinska, V. Nikolova. Conservation and storage of a haploid cucumber (*Cucumis sativus* L.) collection under *in vitro* conditions. *Plant Cell. Rep.* 19:311–314, 2000.
196. Towill, L.E. Germplasm preservation. In:*Plant Tissue Culture Concepts and Laboratory Exercises*, Trigiano, R.N., D.J. Gray, eds., Boca Raton, FL: CRR Press, 1996, pp 291–296.
197. Mullin, R.H., D.E. Schlegel. Cold storage maintenance of strawberry meristhem plantlets. *HortScience* 11:100–101, 1976.
198. Marino, G., P. Rosati, F. Sagrati. Storage of *in vitro* cultures of *Prunus* rootstocks. *Plant Cell Tissue Org. Cult.* 5:73–78, 1985.
199. Romano, A., M.A. Martins-Louçao. *In vitro* cold storage of cork oak shoots cultures. *Plant Cell Tissue Org. Cult.* 59:155–157, 1999.
200. Harding, K. The methylation status of DNA derived from potato plants recovered from slow growth. *Plant Cell. Tissue Org. Cult.* 37:31–38, 1994.
201. Harding, K., M. Marzalina, B. Krishnapillay, N.A.N. Zaimah, M.N. Normah, E.E. Benson. Molecular stability assessment of trees regenerated from cryopreserved mahogany (*Swietenia macrophylla*) seed germplasm using non-radioactive techniques to examine the chromatin structure and DNA methylation status of the ribosomal RNA genes. *J. Trop. For. Sci.* 12:149–163, 2000.
202. Dereuddre, J., M. Galerne, C. Gazeau. Effects du saccharose sur la rèsistance à la congèlation dans l'azote liquide (-196ºC) de meristèmes d'œillet (*Dianthus caryophyllus* L.) cultivès *in vitro*. *CR Acad. Sci. Paris Ser.* 3 304:485–488, 1987.
203. Brison, M., M.T. De Boucaud, F. Dosba. Cryopreservation of *in vitro* grown shoot tips of two interspecific Prunus rootstock. *Plant Science* 105:235–242, 1995.
204. Paul, H., G. Daigny, B.S. Sanguan-Norreel. Cryopreservation of apple (*Malus x domestica* Borkh.) shoot tips following encapsulation-dehydration or encapsulation-vitrification. *Plant Cell Rep.* 19:768–774, 2000.
205. Sales, E., S.G. Nebauer, I. Arrillaga, J. Segura. Cryopreservation of *Digitalis obscura* selected genotypes by encapsulation-dehydration. *Planta. Med.* 67:833–838, 2001.
206. Westcott, R.J., G.G. Henshaw, B.W.W. Grout, W.M. Roca. Tissue culture methods and germplasm storage in potato. *Acta Hort. (ISHS)* 78:45–50, 1977.
207. Kartha, K.K. Meristem culture and cryopreservation:methods and applications. In:*Plant Tissue Culture: Methods and Applications in Agriculture*, Thorpe, T.A., ed., New York: Academic Press, 1981, pp 181–211.
208. Bhojwani, S.S., M.M. Razdan. *Plant Tissue Culture: Theory and Practice*. New York: Elsevier, 1983, pp 373–385.
209. Brennan, R.M., S. Millam, D. Davidson, A. Wilshin. Establishment of an *in vitro* Ribes germplasm collection and preliminary investigations into long-term low temperature germplasm storage. *Acta Hort. (ISHS)* 280:109–112, 1990.

210. Dodds, J.H., L.W. Roberts. *Experiments in Plant Tissue Culture*. New York:Cambrige University Press, 1995, pp 195–203.
211. Ashmore, S.E. *Status Report on the Development and Application of* In Vitro *Techniques for the Conservation and Use of Plant Genetic Resources*. Rome:International Plant Genetic Resources Institute, 1997.
212. Drew, R.A. *The Application of Biotechnology to the Conservation and Improvement of Tropical and Subtropical Fruit Species*. Rome:Seed and Plant Genetic Resources Service: Food and Agriculture Organization of the United Nations, 1997.
213. Collin, H.A., S. Edwards. *Plant Cell Culture*. New York:Springer-Verlag, 1998, pp 83–90.
214. Miki, B. Transgene expression and control. *In Vitro Cell. Dev. Biol. Plant.* 38:139–145, 2002.
215. Guerineau, F. Tools for expressing foreign genes in plants. In:*Methods in Molecular Biology, Vol. 49:Plant Gene Transfer and Expression Protocols*, Jones, H., ed., Totowa, NJ, Humana Press Inc. 1995. pp 1–32.
216. Jefferson, R.A., T.A. Kavanagh, M.W. Bevan. *gus* fusions: β-glucuronidase as a sensitive and versatile gene fusion marker in higher plants. *EMBO J.* 6:3901–3907, 1987.
217. Prasher, D.C. Using GFP to see the light. *TIG* 11:320–323, 1995.
218. Hare, P.D., N.H. Chua. Excision of selectable marker genes from transgenic plants. *Nat. Biotechnol.* 20:575–580, 2002.
219. Hunt, A.G., I.B. Maiti. Strategies for expressing multiple foreign genes in plants as polycistronic constructs. *In Vitro Cell. Dev. Biol. Plant.* 37:313–320, 2001.
220. R. Endress. *Plant Cell Biotechnology*. Berlin:Springer-Verlag, 1994, pp 270–308.
221. De Filippis, L.F., R. Hampp, H. Ziegler. Membrane permeability changes and ultrstructural abnormalities observed during protoplast fusion. *J. Plant Physiol.* 156:628–634, 2000.
222. Grosser, J.W., P. Ollitrault, O. Olivares-Fuster. Somatic hybridization in citrus:an effective tool to facilitate variety improvement. *In Vitro Cell. Dev. Biol. Plant* 36:434–449, 2000.
223. Songstad, D.D., D.A. Somers, R.J. Griesbach. Advances in alternative DNA delivery techniques. *Plant Cell Tiss. Org. Cult.* 40:1–15, 1995.
224. McCabe, D.E., J.R. Wong, C. Sautter, H. Morikawa, K. Chiba, K. Irifune, A. Iida, M. Seki, M. Nishihara, T. Tanaka, T. Yamashita, N. Asakura. Particle bombardment:instrumentation and designs. In:*Particle Bombardment Technology for Gene Transfer*, Yang, N.-S., P. Christou, eds., New York:Oxford University Press, 1994, pp 39–69.
225. Christou, P. Transformation technology. *Trends Plant Sci.* 1:423–431, 1996.
226. Fischer, R., C. Vaquero-Martin, M. Sack, J. Drossard, N. Emans, U. Commandeur. Towards molecular farming in the future:transient protein expression in plants. *Biotechnol. Appl. Biochem.* 30:113–116, 1999.
227. Sanford, J.C. Biolistic plant transformation. *Physiol. Plant.* 79:206–209, 1990.
228. Svab, Z., P. Hajdukiewicz, P. Maliga. Stable transformation of plastids in higher plants. *Proc. Natl. Acad. Sci. USA* 87:8526–8530, 1990.
229. Neuhaus, G. Microinjection into plant cells:methodology and applications. In:*Gene Transfer to Plants*, Potrykus, I., G. Spangenberg, eds., Berlin:Springer-Verlag, 1995, pp 173–175.
230. Kaeppler, H.F., W. Gu, D.A. Somers, H.W. Rines, A.F. Cockburn. Silicon carbide fiber-mediated DNA delivery into plant cells. *Plant Cell Rep.* 9:415–418, 1990.
231. Haullin, K.D., E. Bonne, M. DeBeukeler, J. Leemans. Transgenic maize plants by tissue electroporation. *Plant Cell.* 4:1495–1505, 1992.
232. Zupan, J.R., P. Zambryski. Transfer of T-DNA from *Agrobacterium* to the plant cell. *Plant Physiol.* 107:1041–1047, 1995.
233. Gelvin, S.B. *Agrobacterium* and plant genes involved in T-DNA transfer and integration. *Annu. Rev. Plant Physiol. Plant Mol. Biol.* 51:223–256, 2000.
234. Torres, A.C., R.T. Nagata, R.J. Ferl, T.A. Bewick, D.J. Cantliffe. *In vitro* assay selection of glyphosate resistance in lettuce. *J. Amer. Soc. Hort. Sci.* 124:86–89, 1999.
235. Tepfer, D. Genetic transformation using *Agrobacterium rhizogenes*. *Physiol. Plant.* 79:140–146, 1990.

236. Christey, M.C. Use of Ri-mediated transformation for production of transgenic plants. *In Vitro Cell Dev. Biol. Plant* 37:687–700, 2001.
237. Tepfer, D. Ri T-DNA from *Agrobacterium rhizogenes*:a source of genes having applications in rhizosphere biology and plant development, ecology and evolution. In:*Plant-Microbe Interactions:Molecular and Genetic Perspectives,* Vol. 3, Kosuge, T., E.W. Nester, eds., New York:McGraw-Hill, pp 294–342, 1989.
238. Hamill, J.D., A.J. Parr, M.J.C. Rhodes, J.R. Robins, N.J. Walton. New routes to plant secondary products. *Biotechnology* 5:800–804, 1987.
239. Rhodes, M.J.C., R.J. Robins, J.D. Hamill, A.J. Parr, N.J. Walton. Secondary products formation using *Agrobacterium rhizogenes* transformed hairy roots cultures. *TCA Newsletter* 53:2–15, 1987.
240. De Cleene, M., D. Ley. The host range infectious hairy-root. *Botanical Review* 47:147–194, 1981.
241. Phythoud, K.V.P., V.P. Sinkar, E.W. Nester, M.P. Gordon. Increased virulence of *Agrobacterium rhyzogenes* conferred by the vir region of pTi Bo542:application to genetic engineering of poplar. *Biotechnology* 5:1323–1327, 1987.
242. Jaziri, M., K. Yoshimatsu, J. Homs, K. Shimomura. Traits of transgenic *Atropa belladonna* doubly transformed with different *Agrobacterium rhyzogenes* strains. *Plant Cell Tissue Org. Cult.* 38:257–262, 1994.
243. Hilton, M.G., P.D.G. Wilson, R.J. Robins, M.J.C. Rhodes. Transformed root cultures: fermentations aspects. In:*Manipulating Secondary Metabolism in Culture*, Robins, R.J., M.J.C. Rhodes, eds., Cambridge: Cambridge University Press, 1988, pp 239–246.
244. Yonemitsu, H., K. Shimorura, M. Sakate, S. Mochida, M. Tanaka, T. Endo, A. Kaji. Lobeline production by hairy root cultures of *Lobelia inflata* L *Plant Cell Rep.* 9:307–310, 1990.
245. Robins, R.J., P. Bachmann, A.C.J. Peerless, S. Rabot. Sterification reactions in the biosynthesis of tropane alkaloids in transformed root cultures. *Plant Cell Tissue Org. Cult.* 38:241–247, 1994.
246. Vázquez-Flota, F.O., M.L. Moreno-Valenzuela, J. Miranda-Ham, C. Waldron, E.B. Murphy, J.L. Roberts, G.D. Gustafson, S.L. Armour, S.K. Malcom. Resistance to higromycin B: a new marker for plant transformation studies. *Plant Mol. Biol.* 5:103–108, 1985.
247. Scheidegger, A. Plant biotechnology goes commercial in Japan. *Trends Biotech.* 8:197–198, 1990.
248. Van Doorsselaere, J., M. Baucher, E. Chognot, B. Chabbert, M.T. Tollier, M. Petit-Conil, J.C. Leplé, G. Pilate, D. Cornu, B. Monties, M. Van Montagu, D. Inzé, W. Boerjan, L. Jouanin. A novel lignin in poplar trees with a reduced caffeic acid 5-hydroxyferulic acid O-methyltransferase activity. *Plant J.* 8:855–864, 1995.
249. Anzai, H., K. Yoneyama, I. Yamaguchi. Transgenic tobacco resistant to a bacterial disease by the detoxification of a pathogenic toxin. *Mol. Gen. Genet.* 219:492–494, 1989.
250. Jaynes, J.M., P. Nagpala, L. Destefano-Beltran, J.H. Huang, T. Kim, T. Denny, S. Cetiner. Expression of a cecropin B litic peptide analog in transgenic tobacco confers enhanced resistance to bacterial wilt caused by *Pseudomonas solanacearum*. *Plant Sci.* 89:43–53, 1993.
251. Murata, N., O. Ishizaki-Nishizawa, S. Higashi, H. Hayashi, Y. Tasaka, I. Nishida. Genetically engineered alteration in the chilling sensitivity of plants. *Nature* 356:710–713, 1992.
252. Georges, F., M. Saleem, A.J. Cutler. Design and cloning of a synthetic gene for the flounder antifreeze protein and its expression in plant cells. *Gene* 91:159–165, 1990.
253. Fan, Y., B. Liu, H. Wang, S. Wang, J. Wang. Cloning of an antifreeze protein gene from carrot and its influence on cold tolerance in transgenic tobacco plants. *Plant Cell Rep.* 21:296–301, 2002.
254. Mehta, R.A., T. Cassol, N. Li, N. Ali, A.K. Handa, A.K. Mattoo. Engineered polyamine accumulation in tomato enhances phytonutrient content, juice quality and vine life. *Nat. Biotech.* 20:613–618, 2002.
255. Holmberg, N., G. Lilius, J.E. Bailey, L. Bülow. Transgenic tobacco expressing *Vitreoscilla* hemoglobin exhibits enhanced growth and altered metabolite production. *Nat. Biotechnol.* 15:244–247, 1997.

256. Broglie, K., I. Chet, M. Holliday, R. Cressman, P. Biddle, S. Knowlton, J. Mauvals, R. Broglie. Transgenic plants with enhanced resistance to the fungal pathogen *Rhizoctonia solani*. *Science* 254:1194–1197, 1991.
257. Nagata, R.T., J.A. Dusky, R.J. Ferl, A.C. Torres, D.J. Cantliffe. Evaluation of glyphosate resistance in transgenic lettuce. *J. Amer. Soc. Hort. Sci.* 125:669–672, 2000.
258. Goto, F., T. Yoshihara, N. Shigemoto, S. Toki, F. Takaiwa. Iron accumulation in tobacco plants expressing soybean ferritin gene. *Transgenic Res.* 7:173–180, 1998.
259. Rugini, E., A. Pellegrineschi, M. Mencuccini, D. Mariotti. Increase of rooting ability in the woody species kiwi (*Actinidia deliciosa*) by transformation with *Agrobacterium rhizogenes* rol genes. *Plant Cell Rep.* 10:291–295, 1991.
260. Vaeck, M., A. Reynaerts, H. Hofte, S. Jansens, M. De Beuckeleer, C. Dean, M. Zabeau, M. Van Montagu, J. Leemans. Transgenic plants protected from insect attack. *Nature* 328:33–37, 1987.
261. Hilder, V.A., A.M.R. Gatehouse, S.E. Sheerman, R.F. Barker, D. Boulter. A novel mechanism of insect resistance engineered in transgenic plants. *Nature* 330:160–163, 1987.
262. Doty, S.L., T.Q. Shang, A.M. Wilson, J. Tangen, A.D. Westergreen, L.A. Newman, S.D. Strand, M.P. Gordon. Enhanced metabolism of halogenated hydrocarbons in transgenic plants containing mammalian cytochrome P450 2E1. *Proc. Natl. Acad. Sci. USA* 98:6287–6291, 2000.
263. Rugh, C.L., H.D. Wilde, N.M. Stack, D.M. Thompson, A.O. Summers, R.B. Meagher. Mercuric ion reduction and resistance in transgenic *Arabidopsis thaliana* plants using a modified bacterial *merA* gene. *Proc. Natl. Acad. Sci. USA* 93:3182–3187, 1996.
264. French, C.E., S.J. Rosser, G.J. Davies, S. Nicklin, N.C. Bruce. Biodegradation of explosives by transgenic plants expressing pentaerythritol tetranitrate reductase. *Nat. Biotechnol.* 17:491–494, 1999.
265. Rugini, E., G. Caricato, M. Muganu, C. Taratofolo, M. Camilli, C. Camilli. Genetic stability and agronomic evaluation of six-year-old transgenic kiwiplants for *rolABC* and *rolB* genes. *Acta Hort.* 447:609–610, 1997.
266. Tarczynski, M.C., R.G. Jensen, H.J. Bohnert. Stress protection of transgenic tobacco by production of osmolyte mannitol. *Science* 259:508–510, 1993.
267. Powell-Abel, P., R.S. Nelson, N. Hoffmann, S.G. Rogers, R. Fraley, R.N. Beachy. Delayed disease development in transgenic tobacco that expresses the tobacco mosaic virus coat protein gene. *Science* 232:738–743, 1986.
268. Paek, K., K.W. Hahn. Development of virus resistance in transgenic tobacco plants that express the cucumber mosaic virus satellite RNA. *Mol. Cell.* 1:295–300, 1991.
269. Lodge, J.K., W.K. Kaniewski, N.E. Tumer. Broad-spectrum virus resistance in transgenic plants expressing pokeweed antiviral protein. *Proc. Natl. Acad. Sci. USA* 90:7089–7093, 1993.
270. Kishor, K.P.K., Z. Hong, G.H. Miao, C.A.A. Hu, D.P.S. Verma. Overexpression of Δ^1-Pyrroline 5 carboxylate synthetase increases proline production and confers osmotolerance in transgenic plants. *Plant Physiol.* 108:1387–1394, 1995.

2.02

Clonal Screening and Sprout Based Bioprocessing of Phenolic Phytochemicals for Functional Foods

Kalidas Shetty, Fergus M. Clydesdale, and Dhiraj A. Vattem

CONTENTS

2.1 PHENOLIC PHYTOCHEMICAL INGREDIENTS AND BENEFITS

Phenolic phytochemicals are secondary metabolites synthesized by plants to protect themselves against biological and environmental stresses such as pathogen attack or high energy radiation exposure (1,2). These compounds involved in the plant defense response are one of the most abundant classes of phytochemicals and are also invariably important components of our diets (3,4,5). Commonly consumed fruits such as apples, bananas, grapes, and several types of berries and their beverages are examples of plant foods as sufficiently rich sources of phenolic phytochemicals. Similar phytochemicals in our diet are also obtained from diverse commonly consumed vegetables such as tomato, cabbage, and onions to grains such as cereals and millets as well as legumes such as soybean, common beans, mung beans, fava beans, and peas, depending on the specific regions of the world (4,5,6). In addition many different types of herbs and spices containing phenolic phytochemicals are widely consumed through the diet. Therefore, there are many different types of phenolic phytochemicals from the diet that are mediators of different biological functions for health and wellbeing. The profile of phenolic phytochemicals is often a characteristic of that plant species and is a result of the evolutionary pressures experienced by that species (5,6,7). The most abundant phenolic compounds in fruits are flavonols and flavonoids. Flavonoids, isoflavonoids, phenolic acids, and tannins are important phenolic phytochemicals in many legumes (5,6,7). Biphenyls such as rosmarinic acid are common in many herbs (1).

Chemically phenolic phytochemicals refers to a wide range of chemical compounds containing at least one aromatic ring and a hydroxyl substituent. Metabolic processing in plants after their synthesis results in chemical variations in basic phenolic structure (8). More than 8000 different phenolic structures, categorized into 10 subclasses, have been identified and are a result of differences in substituent groups and linkages. Structurally different phenolic phytochemicals having distinct properties range from simple molecules (e.g., phenolic acids with a single ring structure) to biphenyls and flavanoids having two or three phenolic rings (9,10). Polyphenols that contain >10 phenolic groups are another abundant group of phenolic phytochemicals in fruits and vegetables. Proanthocyanidins, tannins, and their derivatives phlorotannins are often referred to as polyphenols (9,10).

Phenolic phytochemicals are synthesized in plants via a common biosynthetic pathway and derive precursors from the shikimate–phenylpropanoid and the acetate–malonate (polyketide) pathways (1,11). Cinnamic acid, coumaric acid, and caffeic acid, and their derivatives are widespread in fruits, vegetables, and herbs and are derived primarily from the shikimate/phenylpropanoid pathway (1). Oxidative modifications of side chains produce benzoic acid derivatives, which include protocatechuic acid and its positional isomer gentisic acid. Fruits and legumes are especially rich sources of another group of phenolic phytochemicals called flavonoids and iso-lavonoids, which constitute the most abundant group of phenolic phytochemicals derived from the phenylpropanoid–acetate–malonate pathway (2,11).

Flavonoids are subdivided into several families such as flavonols, flavones, flavanols, isoflavones, and antocyanidins, which are formed as a result of hydroxylation, methylation, isoprenylation, dimerization, and glycosylation of the substituents in the aromatic rings (2,11). Phenolic phytochemicals are often esterified with sugars and other chemicals such as quinic acid to increase their solubility and to prevent their enzymatic and chemical degradation. Esterification also helps to target the phenolics to specific parts of the plant (11). Phenolic phytochemicals esterified via their hydroxyl groups to sugars are called glycosides. The sugar most commonly involved in esterification is glucose. However, the glycosides of phenolics with galactose, sucrose, and rhamnose are also found in some plant species (11).

2.2 BIOLOGICAL FUNCTION OF PHENOLIC PHYTOCHEMICALS

Reactive oxygen species (ROS) are now associated with manifestation of several oxidation-linked diseases such as cancer, cardiovascular diseases (CVD) and diabetes; epidemiological studies indicate that diets rich in fruits and vegetables are associated with lower incidences of such oxidation linked diseases. These disease protective effects of fruits and vegetables are now linked to the presence of antioxidant vitamins and phenolic phytochemicals having antioxidant activity, which support the body's antioxidant defense system (12–15). This has led to an interest in the use of diet as a potential tool for the control of these oxidative diseases (5,16–18). This is further supported by recent *in vitro* and clinical studies which have shown that lack of physical activity, exposure to environmental toxins, and consumption of diets rich in carbohydrate and fats induced oxidative stress, which was decreased by consuming fruits, vegetables, and their products (19–26).

Phenolic phytochemicals with antioxidant properties are now widely thought to be the principle components in fruits, vegetables, and herbs that have these beneficial effects. Most phenolic phytochemicals that have positive effect on health are believed to be functioning by countering the effects of reactive oxygen species (ROS) species generated during cellular metabolism. Consumption of natural dietary antioxidants from fruits, vegetables, and herbs has been shown to directly enhance scavenging of ROS, prevent the formation of ROS, and enhance the function of the antioxidant defense response mediated by Glutathione, Ascorbate, superoxide dismutase (SOD), catalase (CAT), and glutathione–s–transferase (GST) interface (15–20).

Specifically, oxidation of biological macromolecules as a result of free radical damage has now been strongly associated with development of many physiological conditions that can manifest into disease (27–31). The first and widely accepted mode of action of these phenolic phytochemicals in managing oxidation stress related diseases is due to the direct involvement of the phenolic phytochemicals in quenching the free radicals from biological systems. Phenolic phytochemicals, due to their phenolic ring and hydroxyl substituents, can function as effective antioxidants due to their ability to quench free electrons and chelate metal ions that are responsible for generating free radicals (32). Phenolic antioxidants can therefore scavenge the harmful free radicals and thus inhibit their oxidative reactions with vital biological molecules (32).

Emerging research into the biological functionality of phenolic phytochemicals also strongly suggests their ability to modulate cellular physiology both at the biochemical/physiological and at molecular level. Structural similarities of phenolic phytochemicals with several key biological effectors and signal molecules have been suggested to be involved in induction and repression of gene expression or activation and deactivation of proteins, enzymes, and transcription factors of key metabolic pathways (27,33–35). They are believed to be able to critically modulate cellular homeostasis as a result of their physiochemical properties such as size, molecular weight, partial hydrophobicity, and ability to modulate acidity at biological pH through enzyme (dehydrogenases) coupled reactions. As a consequence of many modes of action of phenolic phytochemicals they have been shown to have several different functions. Potential anticarcinogenic and antimutagenic properties of phenolic phytochemicals such as gallic acid, caffeic acid, ferulic acid, catechin, quercetin, and resveratrol have been described in several studies (36–38). It is believed that phenolic phytochemicals might interfere in several of the steps that lead to the development of malignant tumors, including, inactivating carcinogens, inhibiting the expression of mutant genes (39). Many studies have also shown that these phenolic phytochemicals can repress the activity of enzymes such as cytochrome P 450 (CYP) class of enzymes involved in the activation of procarcinogens. Their protective functions in liver against carbon tetrachloride toxicity (40) have shown that phenolic phytochemicals also decrease the carcinogenic potential of a mutagen and can activate enzymatic systems (Phase II) involved in the detoxification of xenobiotics (41). Antioxidant properties of the phenolic phytochemicals linked to their ability to quench free radicals has been shown to prevent oxidative damage to the DNA which has been shown to be important in the age related development of some cancers (42). Phenolic phytochemicals have been shown to inhibit the formation of skin tumors induced by 7, 12-dimethyl-benz(a) anthracene in mice (43). Skin tumors in mice, and development of preneoplastic lesions in rat mammary gland tissue in cultures in the presence of carcinogens, were inhibited by resveratrol which is an important biphenyl found in wine (44,45).

Ability of phenolic phytochemicals is preventing of cardiovascular diseases (CVD) has been well described by epidemiological studies. The "French paradox" describes a famous study linking the lower incidences of CVD in the population consuming wine as part of their regular diet (46). Recent research has revealed that these beneficial effects of wine are due to the presence of a biologically active phenolic phytochemical "resveratrol." Inhibiting of LDL oxidation (47) and preventing platelet aggregation (48) are now believed to be the mechanisms by which resveratrol and other phenolic antioxidants prevent development of CVD. Phenolic phytochemicals have also been able to reduce blood pressure and have antithrombotic and antiinflammatory effects (48,49). Phenolic phytochemicals have also been shown to inhibit the activity of alpha-amylase and alpha-glucosidase which are responsible for postprandial increase in blood glucose level, which has been implicated in the manifestation of type-II diabetes and associated cardiovascular diseases (50,51).

In addition to managing oxidation linked diseases, immune modulatory activities of phenolic phytochemicals such as antiallergic (52) properties as a result of suppressing the hypersensitive immune response have also been defined. Antiinflammatory responses mediated by suppression of the TNF-alpha mediated proinflammatory pathways were also shown to be mediated by phenolic phytochemicals (53). Several studies have shown phenolic phytochemicals to have antibacterial, antiulcer, antiviral, and antifungal properties (54–57) and therefore are being implicated in management of infectious diseases.

2.3 PHYTOCHEMICALS AND FUNCTIONAL FOOD

The use of foods and food components such as phenolic phytochemicals as medicine has had an extraordinarily long history in the East and is still practiced successfully in many countries in Asia with Ayurvedic Medicine in India being one of many examples. However, the origin of the modern concept of "functional foods" as a separate and government regulated category of foods is quite new and was first developed in Japan in the 1980s. Faced with inflationary health care costs, the Japanese government instituted a regulatory system to approve certain foods with documented health benefits in order to improve health. These foods have a special seal and are known as "Foods for Specified Health Use" (FOSHU) (1). According to a recent report, as of September 2001, 271 food products had FOSHU status in Japan (58).

Interestingly, with this exception, there is no international, nor in most cases national, agreement on how functional foods should be defined nor is there a legal or regulatory definition. Functional foods seem to be regarded simply as foods that have physiological and psychological effects beyond traditional nutritional effects (59). In spite of the lack of legal or regulatory status consumers are extremely interested in the potential "medicinal" effects of food because they are becoming more and more convinced that their health and quality of life is, in part, a controllable gift, and they want to be an active participant in the process of this control. Added to this desire is a health care system, or lack thereof, which is increasingly perceived as distant, cold, and uncaring along with an explosion in emerging science which points to nontraditional health benefits of foods, food components, and plants (botanicals). This does not imply that consumers wish to give up traditional food, medicine, or health care but means that other options are being offered to the consumer, that form a continuum of treatment. These options are attractive to the consumer but they are also extremely attractive to governments who are being overwhelmed by the cost of health care and to growers and the food industry as a source of new value added products, which fulfill consumer demand and have the potential to create a healthier and higher quality of life.

The concept of functional foods is thus a key component in the new paradigm for health care. This paradigm is unique in that it appeals to not only consumers but governments and industries as well. The paradigm is illustrated graphically in Table 2.1, Table 2.2, and Table 2.3, which have been modified from Clydesdale (60). Each table shows some of the disease states that may be linked to diet and an approximation of the percentage of the population at risk. For these disease states three traditional options are shown "Established Public Health Programs," "Emerging/Controversial Public Health Recommendations," and "Clinical Research." In Table 2.1 the scope of therapy to achieve these treatments is shown as a continuum from individualized treatment of the disease for critical care patients to reduction of risk of the disease for those responding to public health programs. In Table 2.2 the level of health professional involvement in order to

Table 2.1
Disease States Linked to Diet and Current Scope of Therapy

Reported Problem	Nutrient Deficiency	Dental Caries	Obesity	Cardiovascular Disease	Cancer	Osteoporosis	Critical Care
% of Population At risk	100%	25%	33%	25%	20%	8%	<5%
	ESTABLISHED PUBLIC HEALTH PROGRAM		**EMERGING/CONTROVERSIAL PUBLIC HEALTH RECOMMENDATIONS**				**CLINICAL RESEARCH**
Scope of Therapy	← **REDUCTION OF RISK**				**TREATMENT OF DISEASE** →		

Table 2.2
Disease States Linked to Diet with Health Professional Involvement

Reported Problem	Nutrient Deficiency	Dental Caries	Obesity	Cardiovascular Disease	Cancer	Osteoporosis	Critical Care
% of Population At risk	100%	25%	33%	25%	20%	8%	<5%
	ESTABLISHED PUBLIC HEALTH PROGRAM		**EMERGING/CONTROVERSIAL PUBLIC HEALTH RECOMMENDATIONS**				**CLINICAL RESEARCH**
Scope of Therapy	← **REDUCTION OF RISK**				**TREATMENT OF DISEASE** →		
Level of Health Professional Involvement	← **LOW**						**HIGH** →

Table 2.3
Disease States Linked to Diet and Scope of Functional Foods

Reported Problem	Nutrient Deficiency	Dental Caries	Obesity	Cardiovascular Disease	Cancer	Osteoporosis	Critical Care
% of Population At risk	100%	25%	33%	25%	20%	8%	<5%

	ESTABLISHED PUBLIC HEALTH PROGRAM	EMERGING/CONTROVERSIAL PUBLIC HEALTH RECOMMENDATIONS	CLINICAL RESEARCH
Scope of Therapy	REDUCTION OF RISK ←		→ TREATMENT OF DISEASE
Level of Health Professional Involvement	LOW ←		→ HIGH
Scope of Delivery Options	FUNCTIONAL FOODS AND BEVERAGES ←	HERBS SUPPLEMENTS MEDICAL FOODS	→ DRUGS
Individual Participation	HIGH ←		→ LOW
Cost	LOW ←		→ HIGH

achieve these therapies is also shown as a continuum from a high level for individual treatment to a low level for reduction of risk utilizing established public health programs. In the latter case there is an important educational component even though health professional involvement is low compared to the rest of the continuum. In Table 2.3 several other continua are overlapped on those in Table 2.2. With the advent of functional foods it can be seen that the scope of delivery options has been enlarged. The options continuum begins with drugs at the high cost, low consumer participation end, and then moves through medical foods, supplements, herbs, and finally functional foods and beverages at the low cost, high consumer participation preventative end of the spectrum. Functional foods and beverages are becoming, therefore, a part of established public health programs to reduce the risk of specific diseases.

The ultimate success of functional foods will depend on the ability to deliver bioactive healthful components from plants in a predictable and assured manner after they have been proven scientifically to be efficacious in reducing the risk of disease. In addition to the availability of bioactives such as phenolic phytochemicals, we will also need to insure that the foods we create with these components will provide a stable environment for them so they can deliver their physiological benefits as well as being sensorially pleasing, convenient, safe, and affordable.

It has been proposed that the effectiveness of a public health measure can be defined in the following equation:

$$\text{Effectiveness} = \text{Efficacy} \times \text{Compliance} \tag{2.1}$$

Therefore it is essential that functional foods do not compromise flavor or convenience to insure continued use (compliance) by consumers.

Functional foods are intended to be eaten as part of a healthy diet and in so doing reduce the risk of disease in the future along with the symptoms of disease which destroy our quality of life.

The other part of the equation, efficacy, requires sound science in order to validate a food's effectiveness. The question always arises as to the type of science and how much is necessary to validate effectiveness and thus make a claim of some sort for the food. The Keystone National Policy Dialogue on Food, Nutrition, and Health (61) noted that three types of evidence were considered by the FDA in assessing the validity of health claims:

Epidemiology: clinical data derived from observational epoidemiologic studies assessing associations between food substances and disease

Biologic mechanisms: data derived from chemical, cellular, or animal models investigating plausible mechanisms of action of food substances

Intervention trials: controlled assessment of clinical food substance interventions in the human population

The FDA felt that these combinations of data met their "significant scientific agreement" standard of proof. It should be mentioned that the "significant scientific agreement" standard has been called into question and FDA is facing legal challenges in the landmark "Pearson vs. Shalala" dietary supplement claims case. A Federal appeals court directed FDA, based on first amendment rulings, to allow certain health claims accompanied by FDA approved label disclaimers, even though they had failed to meet FDA's "significant scientific agreement" standard (62). This issue has not yet been fully resolved.

A slightly different approach to validate efficacy has been suggested in a consensus document for Europe (63): "The design and development of functional foods is a key issue, as well as a scientific challenge, which should rely on basic scientific knowledge relevant to target functions and their possible modulation by food components. Functional foods themselves are not universal, and a food based approach would have to be influenced by local considerations. In contrast, a science based approach to functional food is universal and, because of this, is very suitable for a panEuropean approach. The function driven approach has the science base as its foundation — in order to gain a broader understanding of the interactions between diet and health. Emphasis is then put on the importance of the effects of food components on well identified and well characterized target functions in the body that are relevant to health issues, rather than solely on reduction of disease risk."

In this case the European position suggests validation of the efficacy of functional foods on functions in the body rather than disease states as is the case with the FDA. In both cases the need for government accepted biomarkers (biological markers of disease) is implicit and unfortunately there is no standard procedure to follow to have biomarkers approved. Hopefully this situation will be rectified in the not to distance future.

One might argue that the three types of evidence suggested by the FDA would be valid for proving the efficacy of a food to reduce the risk of disease or to affect a function in the body that is relevant to health. Thus, in the U.S. evidence from epidemiology, biological mechanisms and clinical trials could be used to validate both health claims and structure-function claims with the difference being the end point in question. In health claims the end point would be disease or an accepted biomarker of disease and in structure function claims the end point would be the specific structure or function in question.

It is probably fair to say that the majority of countries who have policies on functional foods would be within the parameters of evidence set by the U.S., Europe, and Japan (59). Functional foods, however, may be viewed more broadly than the preceding discussion indicates. For instance the American Dietetic Association (ADA) takes a very broad view in a position statement issued in 1999 (58). ADA stated that functional foods are "any potentially healthful food or food ingredient that may provide a health benefit beyond the traditional nutrients it contains." In addition they added that functional foods may be whole, fortified, enriched, or enhanced foods, to have a beneficial effect on health. A functional food would have to be consumed as part of a varied diet on a regular basis, at effective levels, and it is likely that all foods are functional at some physiological level.

Indeed, Milner (64) has asked "What is a functional food? I have never identified a nonfunctional food. I do not think anyone can truly describe one. All foods should be functional under the right circumstances." He goes on to use diet colas as an example of being functional for one who wishes to avoid calories, to increase palatability or to use the caffeine in it as a stimulant. All this simply means is that both consumers and those working on functional foods should have a definition in mind because most governments do not currently have one. In fact as noted by Clydesdale (59) it might be more productive to focus on what might ethically and scientifically be said about foods and what health claims might be made, rather than on the concept of functional foods themselves. If the discussion becomes involved with terminology, one can easily become lost in a morass of terms, such as functional foods, nutraceuticals, phytochemicals, natural remedies, bio-chemo-preventatives, medical foods, dietary supplements, and foods for special medical purposes that may or may not be regulated, depending on the country.

Further, we should be clear that our focus is on foods and not on drugs and that we make a clear difference between the two. In the U.S., the terms prevention and cure are generally ascribed, by statute, to drugs. Indeed, one of the U.S. statutory drug definitions relies on intended use whereby a drug is any article intended for use in the diagnosis, cure, mitigation, treatment, or prevention of disease in man (65). Therefore, potential health claims for foods or beverages that ascribe prevention, mitigation, or cure relationship between the compound and the disease would therefore, be drug claims. Thus food and beverage claims should be limited to a level that ascribes risk reduction or improved biological function (60). Foods, in general, will have a future benefit rather than an immediate benefit like drugs. Because of this they rarely have side effects as their level of bioactivity is lower and therefore long term rather than high and short term like drugs. Foods therefore have a presumption of safety for the entire population, at every age, gender, and ethnic group, which drugs do not. The benefit/risk of drugs is used as part of their regulatory approval in the U.S. whereas the regulation of foods cannot consider benefit/risk assessments.

If the bioactive component of a food botanical was found to be drug-like in its function then a functional food utilizing the bioactive would become a drug if a claim for prevention or cure of a disease was made. This would mean that there would not be a presumption of safety and testing under the drug statutes would have to be conducted.

As noted previously the European Commission has taken a different approach (63) in which they select and discuss key target functions in detail (66). Based on this they concluded that key target functions should (63):

play a major role in maintaining an improved state of health and wellbeing or reduction of risk of disease
appropriate markers should be available and feasible
potential opportunities should exist for modulation by candidate food components

Such reasoning would lend scientific credibility to structure function claims and this along with the scientific rigor required for health claims in the U.S. would provide for a meaningful measure of efficacy for use in the equation:

$$\text{Effectiveness} = \text{Efficacy} \times \text{Compliance} \tag{2.2}$$

Food is a wonderful vehicle for bioactive ingredients such as phenolic phytochemicals because the consumer finds little difficulty with compliance as long as the food is appealing; tastes, smells, and looks good; has good storage stability; is convenient; and is reasonably priced. However, one should not gloss over the importance of food as a vehicle because the technology to create an optimum vehicle for bioactive ingredients is as critical as the science and technology to create physiologically important bioactive compounds. Food, the vehicle which will carry these bioactive compounds must provide an environment which maintains their stability, bioactivity, and bioavailability during preparation and storage such that the bioactive ingredients are consumed at a known and efficacious level by the consumer.

It is important that consumers understand that claims for functional foods are based on sound science. This is at times made difficult by the different types of claims, differences in the laws which govern foods and supplements, and the wide diversity of terminology mentioned previously. However, as society becomes more accustomed to foods with specific biological properties these issues should and must be resolved. The science to provide proof of quality, safety, and efficacy is critical to the future success of functional foods. As Gruenwald (67) notes: the key market for herbal medicines is Germany, followed by France, because these countries have the highest sales but also the most rigid requirements regarding quality, safety, and efficacy.

Understanding the physiological benefits of foods and their ingredients has the potential to provide significant advances in public health, not only from the benefits provided by the bioactive ingredients but also from the impetus provided to consumers to take some responsibility for their own health. Certainly sales of supplements botanicals and functional foods have been growing with functional foods showing the largest rate of growth in the U.S. since 2000 (68). This, however, may be somewhat deceptive as the greatest growth has been in sports bars, beverages, and meal supplements, many of which are legally classified as supplements in the U.S. rather than functional foods. Nevertheless, the future has great potential and new foods will be developed as science allows us to uncover the bioactivity of the foods and plants that we eat.

2.4 RELEVANCE OF PHENOLIC ANTIOXIDANTS FOR FUNCTIONAL FOOD AND COMPARATIVE METABOLIC BIOLOGY CONSIDERATIONS

It is clear that food plants are excellent sources of phenolic phytochemicals, especially as bioactives with antioxidant property. As is evident, phenolic antioxidants from dietary sources have a history of use in food preservation, however, many increasingly have therapeutic and disease prevention applications (69–72). Therefore, understanding the nutritional and the disease protective role of dietary phytochemicals and particularly phenolic antioxidants is an important scientific agenda well into the foreseeable future (73). This disease protective role pf phytochemicals is becoming more significant at a time when the importance of in the prevention of oxidation linked chronic diseases is gaining rapid recognition globally. Therefore, disease prevention and management through the diet can be considered an effective tool to improve health and reduce the increasing health care costs for these oxidation linked chronic diseases, especially in low income countries.

As discussed earlier, phenolic phytochemicals have been associated with antioxidative action in biological systems, acting as scavengers of singlet oxygen and free radicals (74–77). Recent studies have indicated a role for phenolics from food plants in human health and, in particular, cancer (76,78). Phenolic phytochemicals (i.e., phenylpropanoids) serve as effective antioxidants due to their ability to donate hydrogen from hydroxyl groups positioned along the aromatic ring to terminate free radical oxidation of lipids and other biomolecules (79). Phenolic antioxidants, therefore, short circuit destructive chain reactions that ultimately degrade cellular membranes. Examples of food based plant phenolics that are used as antioxidant and antiinflammatory compounds are curcumin from *Curcuma longa* (80–82), *Curcuma mangga* (83), and *Zingiber cassumunar* (84), and rosmarinic acid from *Rosmarinus officinalis* (72,85). Examples of phenolics with cancer chemopreventive potential are curcumin from *Curcuma longa* (80,86–89), isoflavonoids from *Glycine max* (90–92), and galanigin from *Origanum vulgare* (93). Other examples of plant phenolics with medicinal uses include lithospermic acid from *Lithospermum* sp. as antigonadotropic agent (94), salvianolic acid from *Salvia miltiorrhiza* as an antiulcer agent (95), proanthocyanidins from cranberry to combat urinary tract infections (96,97), thymol from *Thymus vulgaris* for anticaries (98), and anethole from *Pimpinella anisum* as an antifungal agent (99).

We have targeted enhanced production of oxidation disease relevant plant biphenyl metabolites such as rosmarinic acid, resveratrol, ellagic acid, and curcumin using novel tissue culture and bioprocessing approaches. Other phenolic phytochemicals also targeted are flavonoids, quercetin, myrcetin, scopoletin, and isoflavonoids. Among simple phenolics, there is major interest in the overexpression of L-tyrosine and L-DOPA from legumes in a high phenolic antioxidant background (100,101). Rosmarinic acid has been targeted from clonal herbs (1,69) for its antiinflammatory and antioxidant properties (85,102,103). Resveratrol has shown antioxidant and cancer chemopreventive properties (104,105) and its overproduction has been targeted from several fruits using solid-state bioprocessing (106,107). Ellagic acid has been targeted for antioxidant and cancer chemopreventive properties (108,109) and has been similarly targeted via solid-state bioprocessing from fruits and fruit processing byproducts (106). Extensive studies have shown cancer chemopreventive and antioxidant properties for *Curcuma longa* and its major active compound, curcumin, (81,86,110) and the developmental and elicitor stress mediated overexpression of curcumin is being investigated.

The emergence of dietary and medicinal applications for phenolic phytochemicals, harnessing their antioxidant and antimicrobial properties in human health and wellness has sound rationale. As stress damage on the cellular level appears similar among eukaryotes, it is logical to suspect that there may be similarities in the mechanism for cellular stress mediation between eukaryotic species. Plant adaptation to biotic and abiotic stress involves the stimulation of protective secondary metabolite pathways (111–113) that results in the biosynthesis of phenolic antioxidants. Studies indicate that plants exposed to ozone responded with increased transcript levels of enzymes in the phenylpropanoid and lignin pathways (114). Increase in plant heat tolerance is related to the accumulation of phenolic metabolites and heat shock proteins that act as chaperones during hyperthermia (115).

Phenolics and specific phenolic-like salicylic acid levels increase in response to infection, acting as defense compounds, or to serve as precursors for the synthesis of lignin, suberin, and other polyphenolic barriers (116). Antimicrobial phenolics called phytoalexins are synthesized around the site of infection during pathogen attack and, along with other simple phenolic metabolites, are believed to be part of a signaling process that results in systemic acquired resistance (111–113). Many phenylpropanoid compounds such as flavonoids, isoflavonoids, anthocyanins, and polyphenols are induced in response to wounding (117), nutritional stress (118), cold stress (119), and high visible light (120). UV irradiation induces light-absorbing flavonoids and sinapate esters in *Arabidopsis* to block radiation and protect DNA from dimerization or cleavage (121). In general, the initiation of the stress response arises from certain changes in the intracellular medium (122) that transmit the stress induced signal to cellular modulating systems and results in changes in cytosolic calcium levels, proton potential as a long distance signal (123), and low molecular weight proteins (124). Stress can also initiate free radical generating processes and shift the cellular equilibrium toward lipid peroxidation (125). It is believed that the shift in prooxidant–antioxidant equilibrium is a primary nonspecific event in the development of the general stress response (126). Therefore, protective phenolic antioxidants involved in such secondary metabolite linked stress responses in food plant species has potential as a source of therapeutic and disease-preventing functional ingredients for oxidation disease linked diet (high carbohydrate and high fat diets) and environment (physical, chemical, and biological) influenced chronic disease problems (69).

2.5 CLONAL SCREENING OF PHENOLIC SYNTHESIS IN HERB SHOOT CULTURE SYSTEMS

The hypothesis that the biosynthesis of plant phenolic metabolites is linked to the proline linked pentose–phosphate pathway (1,69; Figure 2.1) was developed based on the role of the proline linked pentose–phosphate pathway in regulation of purine metabolism in mammalian systems (127). Proline is synthesized by a series of reduction reactions from glutamate (69). In this sequence, P5C and proline known to be metabolic regulators function as a redox couple (127,128). During respiration, oxidation reactions produce hydride ions, which augment reduction of P5C to proline in the cytosol. Proline can then enter mitochondria through proline dehydrogenase (129) and support oxidative phosphorylation (alternative to NADH from Krebs/TCA cycle). This is important because shunting the TCA cycle toward proline synthesis likely deregulates normal NADH synthesis. The reduction of P5C in the cytosol provides $NADP^+$, which is the cofactor for glucose-6-phosphate dehydrogenase (G6PDH), an enzyme that catalyzes the rate-limiting step of the pentose–phosphate pathway. Proline synthesis is therefore hypothesized and has been partly shown to both regulate and stimulate pentose–phosphate pathway activity in erythrocytes (130) and cultured fibroblasts (131) when P5C is converted to proline. This was shown to stimulate purine metabolism via ribose-5-phosphate, which affects cellular physiology and therefore function (127,132).

Based on several studies, Shetty (1,69) proposed a model that the proline linked pentose–phosphate pathway (PLPPP) could be the critical control point (CCP) of shikimate and phenylpropanoid pathways and hypothesized that stress linked modulation of PLPPP could lead to the stimulation of phenolic phytochemicals (1,69; Figure 2.1). Using this model, proline, proline precursors, and proline analogs were effectively utilized to stimulate total phenolic content and a specific phenolic metabolite, rosmarinic acid (133,134). Further, it was shown that proline, proline precursors, and proline analogs

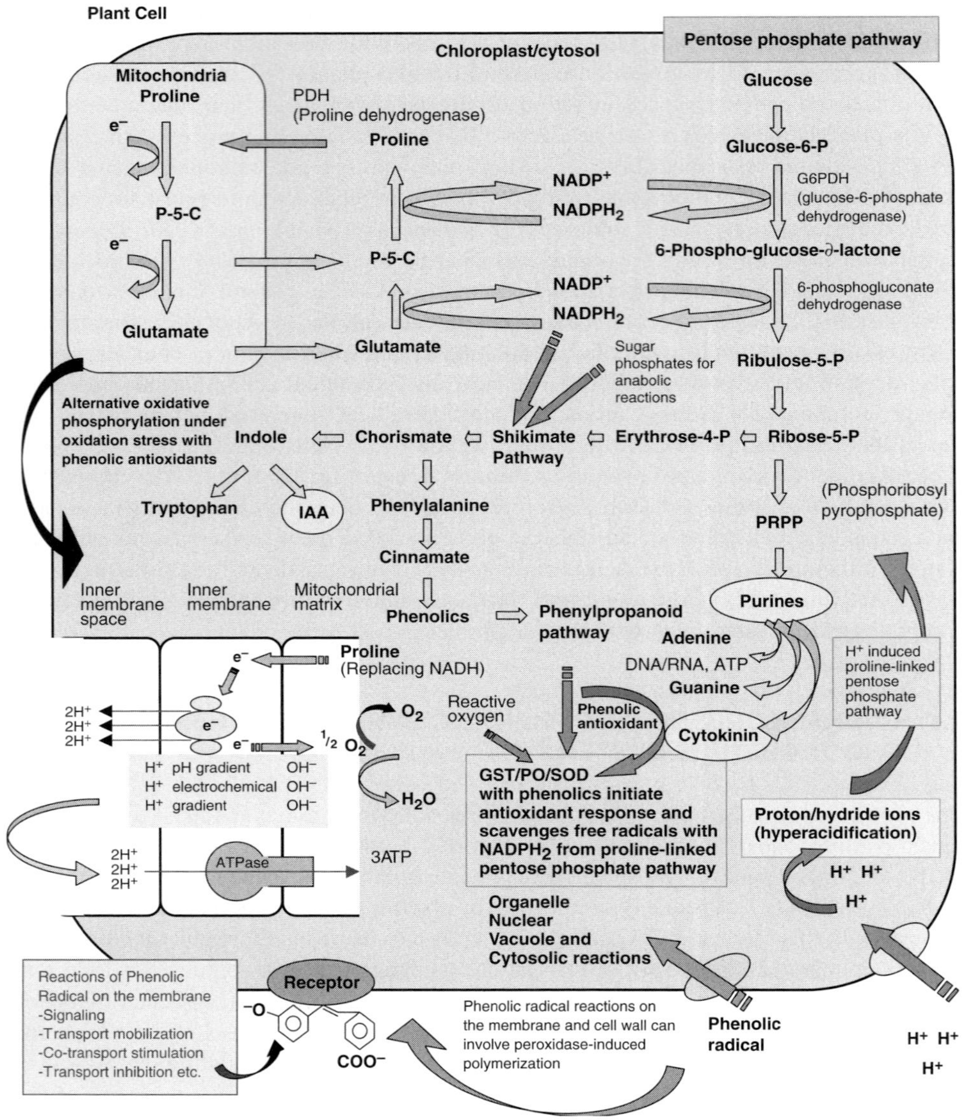

Figure 2.1 Model for the role of proline linked pentose–phosphate pathway in phenolic synthesis and its utilization for screening high phenolic clonal lines of single seed genetic origin.

stimulated somatic embryogenesis in anise, which correlated with increased phenolic content (135). It was also established that during *Pseudomonas* mediated stimulation of total phenolic and rosmarinic acid (RA), proline content was stimulated in oregano clonal shoot cultures (136). Therefore, it was proposed that $NADPH_2$ demand for proline synthesis during response to microbial interaction and proline analog treatment (1,69), may reduce the cytosolic $NADPH_2$/ $NADP^+$ ratio, which should activate G6PDH (137,138). Therefore, deregulation of the pentose–phosphate pathway by proline analog and microbial induced proline synthesis may provide the excess erythrose-4-phosphate (E4P) for shikimate and, therefore, the phenylpropanoid pathways (69). At the same time, proline and P5C could serve as superior reducing equivalents (RE), alternative to NADH (from Krebs/TCA cycle) to support increased oxidative phosphorylation (ATP synthesis) in the mitochondria during the stress response (69,127,128).

The proline analog, azetidine-2-carboxylate (A2C), is an inhibitor of proline dehydrogenase (139). It is also known to inhibit differentiation in Leydig cells of rat fetal testis, which can be overcome by exogenous proline addition (140). Another analog, hydroxyproline, is a competitive inhibitor of proline for incorporation into proteins. According to the model of Shetty (1,69), either analog, at low levels, should deregulate proline synthesis from feedback inhibition and stimulate proline synthesis (1,69). This would then allow the proline linked pentose–phosphate pathway to be activated for $NADPH_2$ synthesis, and concomitantly drive metabolic flux toward E4P for biosynthesis of shikimate and phenylpropanoid metabolites, including RA. Proline could also serve as a RE for ATP synthesis via mitochondrial membrane associated proline dehydrogenase (129,69).

Using this rationale for the mode of action of proline analogs and links to the pentose–phosphate pathway, high RA-producing, shoot based clonal lines originating from a single heterozygous seed among a heterogeneous bulk seed population of lavender, spearmint, and thyme have been screened and isolated based on tolerance to the proline analog, A2C, and a novel *Pseudomonas* sp. isolated from oregano (Figure 2.2; 141–143). This strategy for screening and selection of high RA clonal lines is also based on the model that proline linked pentose phosphate pathway is critical for driving metabolic flux (i.e., E4P) toward shikimate and phenylpropanoid pathways (Figure 2.1). Any clonal line with a deregulated proline synthesis pathway should have an overexpressed pentose phosphate pathway which allows excess metabolic flux to drive shikimate and phenylpropanoid pathway toward total phenolic and RA synthesis (69). Similarly, such proline overexpressing clonal lines should be more tolerant to proline analog, A-2-C. If the metabolic flux to RA is overexpressed, it is likely to be stimulated in response to *Pseudomonas* sp (69). Therefore,

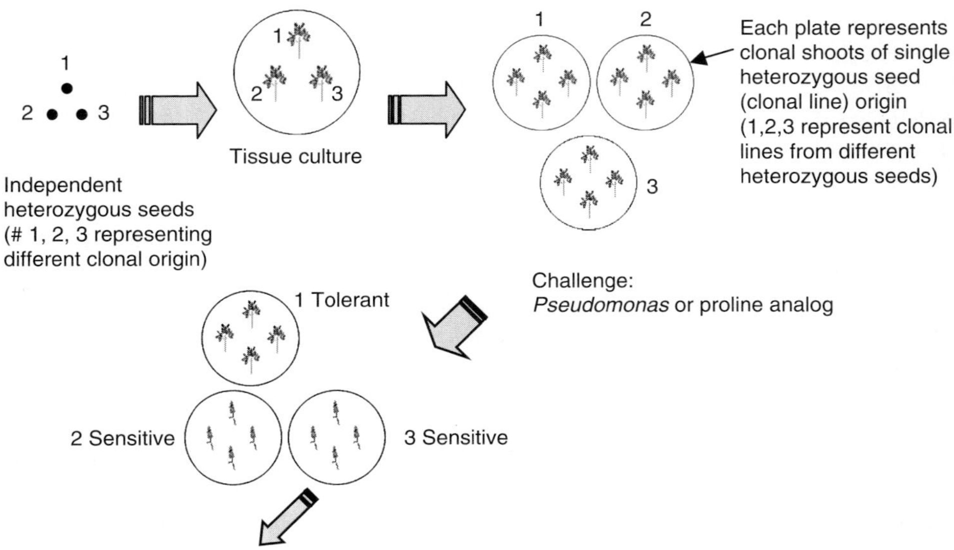

Figure 2.2 Isolation of clonal lines of single seed origin based bacterial or proline analog tolerance. Clonal lines can be used for investigations on Rosmarinic acid inducibility and endogenous antioxidant enzyme response in various clonal lines under various nutritional conditions. Further, high phenolic clonal lines can be used for development of ingredient profiles of single seed clonal origin.

such a clonal line is equally likely to be tolerant to *Pseudomonas* sp. Further, such a clonal line should also exhibit high proline oxidation and RA content in response to A2C and *Pseudomonas* sp. The rationale for this model is based on the role of the pentose–phosphate pathway in driving ribose-5-phosphate toward purine metabolism in cancer cells (127), differentiating animal tissues (140), and plant tissues (144). The hypothesis of this model is that the same metabolic flux from over expression of proline linked pentose–phosphate pathway regulates the interconversion of ribose-5-phosphate to E4P driving shikimate pathway (69). Shikimate pathway flux is critical for both auxin and phenylpropanoid biosynthesis, including RA (69). This hypothesis has been strengthened by preliminary results in which RA biosynthesis in several oregano clonal lines was significantly stimulated by exogenous addition of proline analog (e.g., A2C) and ornithine (133,134). The same clonal lines are also tolerant to *Pseudomonas* sp. and respond to the bacterium by increasing RA and proline biosynthesis (136,145). High RA-producing clonal lines selected by approaches based on this model (141–143,145–147) are being targeted for preliminary characterization of the key enzymes in phenolic synthesis and large scale production of such phenolics using tissue culture generated clonal lines in large scale field production systems. This strategy for investigation and stimulation of phenolic biosynthesis in clonal plant systems using PLPPP as the CCP can be the foundation for designing dietary phenolic phytochemicals from cross-pollinating, heterogeneous species for functional foods (1,69).

2.6 PHENOLIC SYNTHESIS IN SEED SPROUTS

Preliminary results (1,69,72,133,134,148) have provided empirical evidence for a link between proline biosynthesis and oxidation, as well as stimulation of G6PDH. In light mediated sprout studies in pea (*Pisum sativum*), acetylsalicylic acid in combination with fish protein hydrolysates (a potential source of proline precursors) stimulated phenolic content and guaiacol peroxidase (GPX) activity during early germination with corresponding higher levels of proline and G6PDH activity (149). In parallel light mediated studies in pea, low pH and salicylic acid treatments stimulated increased phenolic content and tissue rigidity. Similarly, there was concomitant stimulation of G6PDH and proline (148). This work supported the hypothesis that pentose–phosphate pathway stimulation may be linked to proline biosynthesis and that modulation of a proton linked redox cycle may also be operating through proline linked pentose–phosphate pathway (148). In dark germinated studies in pea, high cytokinin-containing anise root extracts stimulated phenolic content and antioxidant activity, which correlated with increased proline content but inversely with G6PDH activity (150). In further dark germination studies in mung bean (*Vigna radiata*), dietary grade microbial polysaccharide treatments stimulated phenolic content and enzyme activity, G6PDH and GPX compared to controls (151), with concomitant stimulation of proline content. In addition, specific elicitors xanthan gum, yeast extract, and yeast glucan stimulated antioxidant activity. In additional studies, oregano phenolic extracts were used as elicitors to stimulate phenolic content during dark germination of mung bean. Again, increased phenolic content corresponded to an increase in activity of G6PDH and GPX and phenolic related antioxidant activity were also stimulated (152). In studies with dark germinated fava bean, support for the hypothesis that stimulation of proline linked pentose–phosphate pathway would stimulate phenolic metabolism under elicitor and stress response was probed (69). In polysaccharide elicitor studies, gellan gum stimulated fava bean total phenolic content ninefold in late stages of germination with a corresponding increase in proline content and GPX activity, although the effect on antioxidant and G6PDH activity was inconclusive (100). In the same fava bean system, UV mediated stimulation of phenolic content in dark germinated fava bean

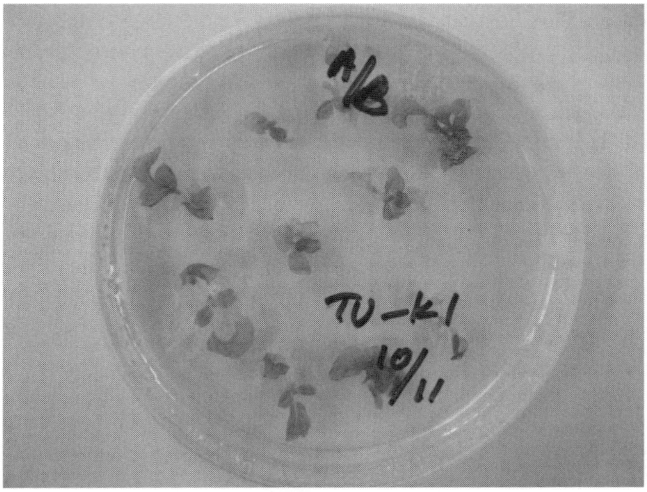

Figure 2.3 High phenolic clonal shoot culture of oregano (*Origanum vulgare*) with genetic origin from single heterozygous seed.

sprouts indicated a positive correlation to G6PDH and GPX activities with a concomitant increase in proline content (153). It was further confirmed that proline analog, A2C also stimulated phenolic content in fava bean with positive correlation to G6PDH and GPX activities as well as proline content (154). Similar to studies in clonal shoot cultures of thyme (133) and oregano (134), the proline analog mediated studies in fava bean confirmed that proline overexpression was not only possible, but involved stimulating G6PDH and therefore likely diverted the pentose–phosphate pathway toward phenylpropanoid biosynthesis. The late stage stimulation of phenolic content and GPX activity in response to microwave mediated thermal stress in dark germinated fava bean strongly correlated with stimulation of free radical scavenging activity of free phenolics as measured by the quenching of 1,1-diphenyl-2-picrylhydrazyl (DPPH) free radical assay and stimulation of superoxide dismutase (SOD) activity (155). Therefore, using elicitor and physical stress in early stages of seed sprouting could be effectively used to stimulate key phenolic phytochemicals for design of functional foods. In designing appropriate phenolic phytochemical extracts diverting carbohydrate and protein metabolic flux through the proposed CCP, proline linked pentose–phosphate pathway (PLPPP) has substantial potential. This CCP link to PLPPP could be the basis for a more effective Systems Biology approach for metabolic control of functional phenolic phytochemicals and appropriate design of phytochemicals from different seed sprout systems for different oxidation linked disease targets.

2.7 SUMMARY OF STRATEGIES AND IMPLICATIONS

In an age where diet and environmental related oxidation and infectious diseases are on the rise, the role of dietary phenolic phytochemicals in preventive management of diseases is becoming important. Strategies to improve functional phytochemicals that addresses the problem of consistency have to be developed. Within this framework we have developed strategies for isolating clonal dietary herbs of single seed genetic origin with a specific profile of phenolic phytochemicals and also stress elicitation based enhancement of phenolic phytochemicals during legume seed sprouting. The rationale for development of these techniques for phenolic phytochemical enhancement for functional food design is based on harnessing the potential of proline linked pentose–phosphate pathway (PLPPP) as the critical control point (CCP) in clonal shoots of single seed genetic origin such as herbs from the family Lamiaceae and seed sprouts in self-pollinating species such as various legumes. This strategy can be extended to develop foods with better phenolic phytochemical profile and functionality. Further it can be extended to develop functional foods and supplements with consistent ingredient profiles targeted against a disease condition. This concept is now being extended to specifically isolate antioxidants for diverse disease conditions, antimicrobials against bacterial pathogens, phytochemicals for diabetes management, angiotensin converting enzyme inhibitors for hypertension management, L-DOPA for Parkinson's management, dietary cyclooxygenase (COX-2 inhibitors) for inflammatory diseases and isoflavones for women's health.

REFERENCES

1. Shetty, K. Biotechnology to harness the benefits of dietary phenolics: focus on *Lamiaceae*. *Asia Pac. J. Clin. Nutr*. 6:162–171, 1997.
2. Briskin, D.P. Medicinal plants and phytomedicines. Linking plant biochemistry and physiology to human health. *Plant Physiol.* 124:507–14, 2000.
3. Mann, J. *Oxford Chemistry Series: Secondary Metabolism*. Oxford: Clarendon Press, 1978.
4. Bravo, L. Phenolic phytochemicals: chemistry, dietary sources, metabolism, and nutritional significance. *Nutr. Rev*. 56:317–333, 1998.
5. Crozier, A., J. Burns, A.A. Aziz, A.J. Stewart, H.S. Rabiasz, G.I. Jenkins, C.A. Edwards, M.E.J. Lean. Antioxidant flavonols from fruits, vegetables and beverages: measurements and bioavailability. *Biol. Res*. 33:79–88, 2000.
6. Urquiaga, I., F. Leighton. Plant Polyphenol Antioxidants and Oxidative Stress. *Biol. Res.* 33:55–64, 2000.
7. Moure, A., J.M. Cruz D. Franco, J.M. Domínguez, J. Sineiro, H. Domínguez, M.H. Núñez, J.C. Parajó. Natural antioxidants from residual sources. *Food Chem.* 72(2):145–171, 2001.
8. Bors, W., C. Michel. Chemistry of the antioxidant effect of polyphenols. *Ann. NY Acad. Sci.* 957:57–69, 2002.
9. Harborne, J.B. Plant phenolics. In: *Encyclopedia of Plant Physiology*, Bella, E.A., B.V. Charlwood, eds, Heidelburg: Springer-Verlag, 8:329–395, 1980.
10. Haslam, E. *Practical Polyphenolics: From Structure to Molecular Recognition and Physiological Action.* Cambridge: Cambridge University Press, 1998.
11. Strack, D. Phenolic metabolism. In: *Plant Biochemistry,* Dey, P.M., J.B. Harborne, eds., San Diego: Academic Press, 1997, pp 387–416.
12. Jakus, V. The role of free radicals, oxidative stress and antioxidant systems in diabetic vascular disease. *Bratisl. Lek. Listy.* 101(10):541–551, 2000.
13. Barbaste, M., B. Berke, M. Dumas, S. Soulet, J.C. Delaunay, C. Castagnino, V. Arnaudinaud, C. Cheze, J. Vercauteren. Dietary antioxidants, peroxidation and cardiovascular risks. *J. Nutr. Health Aging*. 6(3):209–23, 2002.
14. Gerber, M., C. Astre, C. Segala, M. Saintot, J. Scali, J. Simony-Lafontaine, J. Grenier, H. Pujol. Tumor progression and oxidant-antioxidant status. *Cancer Lett.* 19:114(1–2): 211–214, 1997.
15. Block, G., B. Patterson, A. Subar. Fruit, vegetables, and cancer prevention: a review of the epidemiological evidence. *Nutr. Cancer* 18:1–29, 1992.
16. Serdula, M.K., M.A.H. Byers, E. Simoes, J.M. Mendlein, R.J. Coates. The association between fruit and vegetable intake and chronic disease risk factors. *Epidemiology* 7:161–165, 1996.

17. Tapiero, H., K.D. Tew, G.N. Ba, G. Mathe. Polyphenols: do they play a role in the prevention of human pathologies? *Biomed. Pharmacother.* 56(4):200–207, 2002.
18. Duthie, G.G., P.T. Gardner, J.A. Kyle. Plant polyphenols: are they the new magic bullet? *Proc. Nutr. Soc.* 62(3):599–603, 2003.
19. Lundberg, A.C., A. Åkesson, B. Åkesson. Dietary intake and nutritional status in patients with systemic sclerosis. *Ann. Rheum. Dis.* 51:1143–1148, 1992.
20. Yoshioka, A., Y. Miyachi, S. Imamura, Y. Niwa. Anti-oxidant effects of retinoids on inflammatory skin diseases. *Arch. Dermatol. Res.* 278:177–183, 1986.
21. Wilks, R., F. Bennett, T. Forrester, N. Mcfarlane-Anderson. Chronic diseases: the new epidemic. *West Ind. Med. J.* 47(4):40–44, 1998.
22. Leighton, F., A. Cuevas, V. Guasch, D.D. Perez, P. StrobelA. San Martin, U. Urzua, M.S. Diez, R. Foncea, O. Castillo, C. Mizon, M.A. Espinoza, I. Urquiaga, J. Rozowski, A. Maiz, A. Germain. Plasma polyphenols and antioxidants, oxidative DNA damage and endothelial function in a diet and wine intervention study in humans. *Drugs Exp. Clin. Res.* 25(2–3):133–41, 1999.
23. Gillman, M.W., L.A. Cupples, D. Gagnon, B.M. Posner, R.C. Ellison, W.P. Castelli, P.A. Wolf. Protective effect of fruits and vegetables on development of stroke in men. *J. Am. Med. Assoc.* 273:1113–1117, 1995.
24. Joshipura, K.J., A. Ascherio, J.E. Manson, M.J. Stampfer, E.B. Rimm, F.E. Speizer, C.H. Hennekens, D. Spiegelman, W.C. Willett. Fruit and vegetable intake in relation to risk of ischemic stroke. *J. Amer. Med. Asso.* 282:1233–1239, 1999.
25. Cox, B.D., M.J. Whichelow, A.T. Prevost. Seasonal consumption of salad vegetables and fresh fruit in relation to the development of cardiovascular disease and cancer. *Public Health Nutr.* 3:19–29, 2000.
26. Strandhagen, E., P.O. Hansson, I. Bosaeus, B. Isaksson, H. Eriksson. High fruit intake may reduce mortality among middle-aged and elderly men: the study of men born in 1913. *Eur. J. Clin. Nutr.* 54:337–341, 2000.
27. Droge, W. Free radicals in the physiological control of cell function. *Physiol. Rev.* 82(1): 47–95, 2002.
28. Morel, Y., R., Barouki Repression of gene expression by oxidative stress. *Biochem J.* 342(3):481–96, 1999.
29. Parke, A.L., C. Ioannides, D.F.V. Lewis, D.V. Parke. Molecular pathology of drugs: disease interaction in chronic autoimmune inflammatory diseases. *Inflammopharmacology* 1:3–36, 1991.
30. Schwarz, K.B. Oxidative stress during viral infection: a review. *Free Radic. Biol. Med.* 21(5):641–9, 1996.
31. Offen, D., P.M. Beart, N.S. Cheung, C.J. Pascoe, A. Hochman, S. Gorodin, E. Melamed, R. Bernard, O. Bernard. Transgenic mice expressing human Bcl-2 in their neurons are resistant to 6-hydroxydopamine and 1-methyl-4-phenyl-1,2,3,6-tetrahydropyridine neurotoxicity. *Proc. Natl. Acad. Sci.* 95:5789–5794, 1998.
32. Rice-Evans, C.A., N.J. Miller, G. Paganga. Antioxidant properties of phenolic compounds. *Trends Plant Sci.* 2:152–159, 1997.
33. Lusini, L., S.A. Tripodi, R. Rossi, F. Giannerini, D. Giustarini, M.T. del Vecchio, G. Barbanti, M. Cintorino, P. Tosi, P. Di Simplicio. Altered glutathione anti-oxidant metabolism during tumor progression in human renal-cell carcinoma. *Int. J. Cancer* 91(1):55–59, 2001.
34. Mates, J.M., F. Sanchez-Jimenez. Antioxidant enzymes and their implications in pathophysiologic processes. *Front. Biosci.* 4D:339–345, 1999.
35. Mates, J.M., C. Perez-Gomez, I. Nunez de Castro. Antioxidant enzymes and human diseases. *Clin. Biochem.* 32(8):595–603, 1999.
36. Yamada, J., Y. Tomita. Antimutagenic activity of caffeic acid and related compounds. *Biosci. Biotechnol. Biochem.* 60(2):328–329, 1996.
37. Mitscher, L. A., H. Telikepalli, E. McGhee, D.M. Shankel. Natural antimutagenic agents. *Mutat. Res.* 350:143–152, 1996.

38. Uenobe, F., S. Nakamura, M. Miyazawa. Antimutagenic effect of resveratrol against Trp-P-1. *Mutat. Res.* 373:197–200, 1997.
39. Kuroda, Y., T. Inoue. Antimutagenesis by factors affecting DNA repair in bacteria. *Mutat. Res.* 202(2):387–391, 1988.
40. Kanai, S., H. Okano. Mechanism of protective effects of sumac gall extract and gallic acid on progression of CC14-induced acute liver injury in rats. *Am. J. Chin. Med.* 26:333–341, 1998.
41. Bravo, L. Phenolic phytochemicals: chemistry, dietary sources, metabolism, and nutritional significance. *Nutr. Rev.* 56:317–333, 1998.
42. Halliwell, B. Establishing the significance and optimal intake of dietary antioxidants: the biomarker concept. *Nutr. Rev.* 57:104–113, 1999.
43. Kaul, A., K.I. Khanduja. Polyphenols inhibit promotional phase of tumorigenesis: relevance of superoxide of superoxide radicals. *Nurt. Cancer* 32:81–85, 1998.
44. Clifford, A.J., S.E. Ebeler, J.D. Ebeler, N.D. Bills, S.H. Hinrichs, P.-L. Teissedre, A.L. Waterhouse. Delayed tumor onset in transgenic mice fed an amino acid-based diet supplemented with red wine solids. *Am. J. Clin. Nutr.* 64:748–756, 1996.
45. Jang, M., L. Cai, G.O. Udeani, K.V. Slowing, C. Thomas, C.W.W. Beecher, H.H.S. Fong, N.R. Farnsworth, A.D. Kinghorn, R.G. Mehta, R.C. Moon, J.M. Pezzuto. Cancer chemopreventive activity of resveratrol, a natural product derived from grapes. *Science* 275:218–220, 1997.
46. Ferrieres, J. The French paradox: lessons for other countries. *Heart* 90(1):107–111, 2004.
47. Frankel, E.N, J. Kanner, J.B. German, E. Parks, J.E. Kinsella. Inhibition of oxidation of human low-density lipoprotein by phenolic substances in red wine. *Lancet* 341:454–457, 1993.
48. Gerritsen, M.E., W.W. Carley, G.E. Ranges, C.-P. Shen, S.A. Phan, G.F. Ligon, C.A. Perry. Flavonoids inhibit cytokine-induced endothelial cell adhesion protein gene expression. *Am. J. Pathol.* 147:278–292, 1995.
49. Muldoon, M.F., S.B. Kritchevsky. Flavonoids and heart disease. *BMJ* 312(7029):458–459, 1996.
50. McCue, P., K. Shetty. Inhibitory effects of rosmarinic acid extracts on porcine pancreatic amylase *in vitro*. *Asia Pac. J. Clin. Nutr.* 13(1):101–106, 2004.
51. Andlauer, W., P. Furst. Special characteristics of non-nutrient food constituents of plants: phytochemicals: introductory lecture. *Int. J. Vitam. Nutr. Res.* 73(2):55–62, 2003.
52. Noguchi, Y., K. Fukuda, A. Matsushima, D. Haishi, M. Hiroto, Y. Kodera, H. Nishimura, Y. Inada. Inhibition of Df-protease associated with allergic diseases by polyphenol. *J. Agric. Food Chem.* 47:2969–2972, 1999.
53. Ma, Q., K. Kinneer. Chemoprotection by phenolic antioxidants. Inhibition of tumor necrosis factor alpha induction in macrophages. *J. Biol. Chem.* 277(4):2477–2484, 2002.
54. Chung, K.T., T.Y. Wong, C.I. Wei, Y.W. Huang, Y. Lin. Tannins and human health: a review. *Crit. Rev. Food Sci. Nutr.* 38:421–464, 1998.
55. Ikken, Y., P. Morales, A. Martínez, M.L. Marín, A.I. Haza, M.I. Cambero. Antimutagenic effect of fruit and vegetable ethanolic extracts against N-nitrosamines evaluated by the Ames test. *J. Agric. Food Chem.* 47:3257–3264, 1999.
56. Vilegas, W., M. Sanomimiya, L. Rastrelli, C. Pizza. Isolation and structure elucidation of two new flavonoid glycosides from the infusion of *Maytenus aquifolium* leaves. Evaluation of the antiulcer activity of the infusion. *J. Agric. Food Chem.* 47:403–406, 1999.
57. Abram, V., M. Donko. Tentative identification of polyphenols in *Sempervivum tectorum* and assessment of the antimicrobial activity of *Sempervivum* L. *J. Agric. Food Chem.* 47:485–489, 1999.
58. Meister, K. *Facts About Functional Foods*. New York: A report by the American Council on Science and Health, 2002, pp 5–29.
59. Clydesdale, F. M. A proposal for the establishment of scientific criteria for health claims for functional foods. *Nutr. Rev.* 55:413–422, 1997.
60. Clydesdale, F. M. Science, education technology: new frontiers for health. *Crit. Rev. Fd. Sci. Nutr.* 38:397–419, 1998.

61. Keystone Center. *The Final Report of the Keystone National Policy Dialogue on Food, Nutrition and Health,* Keystone, CO and Washington, DC, 1996, pp 33–56.
62. FDA seeks public comment after first amendment rulings. *Food Chemical News,* Washington, DC. 2002:44, pp. 1, 26.
63. Diplock, A.T., P.J. Aggett, M. Ashwell, F. Bornet, E.B. Fern, M.B. Roberfroid. Scientific concepts of functional foods in Europe: consensus document. *Brit. J. Nutr.* 81(1):S1–S27, 1999.
64. Milner, J.A. Expanded definition of nutrients. *Proceedings of the Ceres Forum on What is a Nutrient?: Defining the Food – Drug Continuum.* Georgetown University Center for Food and Nutrition Policy, Washington, DC. 1999, pp 50–55.
65. 21 U.S.C. 1994, 321 (g).
66. Bellisle, F., A.T. Diplock, G. Hornstra, B. Koletzko, M. Roberfroid, S. Salminen, W.H.M. Saris. Functional food science in Europe. *Brit. J. Nutr.* 80(1):S1–S193, 1998.
67. Gruenwald, J. The future of herbal medicines in Europe. *Nutraceuticals World* 5(6):26, 2002.
68. Gruenwald, J., F. Herzberg. The situation of major dietary supplement markets. *Nutracos.* 1(3):18021, 2002.
69. Shetty, K., M.L. Wahlqvist. A model for the role of proline-linked pentose phosphate pathway in phenolic phytochemical biosynthesis and mechanism of action for human health and environmental applications; a Review. *Asia Pac. J. Clin. Nutr.* 13:1–24, 2004.
70. Shetty, K., Labbe, R.L. Food-borne pathogens, health and role dietary phytochemicals. *Asia Pac. J. Clin. Nutr.* 7:270–276, 1998.
71. Shetty, K. Phytochemicals: biotechnology of phenolic phytochemicals for food preservatives and functional food applications. In: *Wiley Encyclopedia of Food Science and Technology,* 2nd ed., Francis, F.J., ed., New York: Wiley Interscience, 1999, pp 1901–1909.
72. Shetty, K. Biosynthesis of rosmarinic acid and applications in medicine. *J. Herbs Spices Med. Plants* 8:161–181, 2001.
73. Shetty, K., P. McCue. Phenolic antioxidant biosynthesis in plants for functional food application: integration of systems biology and biotechnological approaches. *Food Biotechnol.* 17:67–97, 2003.
74. Rice-Evans, C.A., N.J. Miller, P.G. Bolwell, P.M. Bramley, J.B. Pridham. The relative antioxidant activities of plant-derived polyphenolic flavonoids. *Free Rad. Res.* 22:375–383, 1995.
75. Hertog, M.G.L., P.C.H. Hollman, M.B. Katan. Content of potentially anticarcinogenic flavonoids of 28 vegetables and 9 fruits commonly consumed in the Netherlands. *J. Agric. Food Chem.* 40:2379–2383, 1992.
76. Hertog, M.G.L., D. Kromhout, C. Aravanis, H. Blackburn, R. Buzina, F. Fidanza, S. Giampaoli, A. Jansen, A. Menotti, S. Nedeljkovic, M. Pekkarinen, B.S. Simic, H. Toshima, E.J.M. Feskens, P.C.H. Hollman, M.B. Kattan. Flavonoid intake and long-term risk of coronary heart disease and cancer in the seven countries study. *Arch. Intern. Med.* 155:381–386, 1995.
77. Jorgensen, L.V., H.L. Madsen, M.K. Thomsen, L.O. Dragsted, L.H. Skibsted. Regulation of phenolic antioxidants from phenoxyl radicals: an ESR and electrochemical study of antioxidant hierarchy. *Free Rad. Res.* 30:207–220, 1999.
78. Paganga, G., N. Miller, C.A. Rice-Evans. The polyphenolic contents of fruits and vegetables and their antioxidant activities: what does a serving constitute? *Free Rad. Res.* 30:153–162, 1999.
79. Foti, M., M. Piattelli, V. Amico, G. Ruberto. Antioxidant activity of phenolic meroditerpenoids from marine algae. *J. Photochem. Photobiolo.* 26:159–164, 1994.
80. Huang, M.T., T. Lysz, T. Ferraro, A.H. Conney. Inhibitory effects of curcumin on tumor promotion and arachidonic acid metabolism in mouse epidermis. In: *Cancer chemoprevention,* Wattenberg, L., M. Lipkin, C.W. Boone, G.J. Kellof, eds., Boca Raton, FL: CRC Press, 1992, pp 375–391.
81. Osawa, T., Y. Sugiyama, M. Inayoshi, S. Kawakishi. Antioxidant activity of tetrahydrocurcuminoids. *Biosci. Biotechnol. Biochem.* 59:1609–1612, 1995.

82. Lim, G.P., T. Chu, F. Yang, W. Beech, S.A. Frauschy, G.M. Cole. The curry spice curcumin reducces oxidative damage and amyloid pathology in an Alzheimer transgenic mouse. *J. Neurosci*. 21:8370–8377, 2001.
83. Jitoe, A., T. Masuda, G.P. Tengah, D.N. Suprapta, I.W. Gara, N. Nakatani. Antioxidant activity of tropical ginger extracts and analysis of the contained curcuminoids. *J. Agric. Food Chem.* 40:1337–1340, 1992.
84. Masuda, T., A. Jitoe. Antioxidative and anti-inflammatory compounds from tropical gingers: isolation, structure determination, and activities of cassumunins A, B and C, new complex curcuminoids from *Zingiber cassumunar*. *J. Agric. Food Chem.* 42:1850–1856, 1994.
85. Peake, P.W., B.A. Pussel, P. Martyn, V. Timmermans, J.A. Charlesworth. The inhibitory effect of rosmarinic acid on complement involves the C5 convertase. *Int. J. Immunopharmac.* 13:853–857, 1991.
86. Ruby, A.J., G. Kuttan, K.D. Babu, K.N. Rajasekharan, R. Kuttan. Anti-tumour and antioxidant activity of natural curcuminoids. *Cancer Lett*. 94:79–83, 1995.
87. Singh, S.V., X. Hu, S.K. Srivastava, M. Singh, H. Xia, J.L. Orchard, H.A. Zaren. Mechanism of inhibition of benzo[a] pyrene-induced forestomach cancer in mice by dietary curcumin. *Cacinogenesis* 19:1357–1360, 1998.
88. Khar, A., A.M. Ali, B.V. Pardhasaradhi, Z. Begum, R. Anjum. Antitumor activity of curcumin is mediated through the induction of apoptosis in AK-5 tumor cells. *FEBS Lett.* 445:165–168, 1999.
89. Inano, H., M. Onoda, N. Inafuku, M. Kubota, Y. Kamada, T. Osawa, H. Kobayashi, K. Wakabayashi. Potent preventive action of curcumin on radiation-induced initiation of mammary tumorigenesis in rats. *Carcinogenesis* 21:1835–1841, 2000.
90. Hutchins, A.M., J.L. Slavin, J.W. Lampe. Urinary isoflavonoid phytoestrogen and lignan excretion after consumption of fermented and unfermented soy products. *J. Am. Dietetic Assoc.* 95:545–551, 1995.
91. Messina, M.J. Legumes and soybeans: overview of their nutritional and health effects. *Am. J. Clin. Nutr.* 70:439S–450S, 1999.
92. Nagata, C., N. Takatsuka, N. Kawakami, H. Shimizu. A prospective cohort study of soy product intake and stomach cancer death. *Br. J. Cancer* 87:31–36, 2002.
93. Kanazawa, K., H. Kawasaki, K. Samejima, H. Ashida, G. Danno. Specific desmutagens and (antimutagens) in oregano against a dietary carcinogen, Trp-P-2 are galagin and quercetin. *J. Agric. Food Chem.* 43:404–409, 1995.
94. Winterhoff, H., H.G. Gumbinger, H. Sourgens. On the antiogonadotropic activity of *Lithospermum* and *Lycopus* species and some of their phenolic constituents. *Planta. Medica.* 54:101–106, 1988.
95. Harbone, J.B., H. Baxter. Phenylpropanoids. In: *Phytochenical Dictionary: A Handbook of Bioactive Compounds from Plants*, Harbone, J.B., H. Baxter, eds., London: Taylor and Francis, 1993, pp 472–488.
96. Howell, A.B., N. Vorsa, A. Der Marderosian, L.Y. Foo. Inhibition of adherence of P-fimbriated *Escherichia coli* to uroepithelial-cell surfaces by proanthocyanidin extracts from cranberries. *N. Eng. J. Med.* 339(15):1085–1086, 1998.
97. Howell, A.B., B. Foxman. Cranberry juice and adhesion of antibiotic-resistant uropathogens. *JAMA* 287(23):3082–3083, 2002.
98. Guggenheim, S., S. Shapiro. The action of thymol on oral bacteria. *Oral Microbiol. Immunol.* 10:241–246, 1995.
99. Himejima, M., I. Kubo. Fungicidal activity of polygodial in combination with anethol and indole against *Candida albicans*. *J. Agric. Food Chem.* 41:1776–1779, 1993.
100. Shetty, P., M.T. Atallah, K. Shetty. Enhancement of total phenolic, L-DOPA and proline contents in germinating fava bean (*Vicia faba*) in response to bacterial elicitors. *Food Biotechnol.* 15:47–67, 2001.
101. Randhir, R., P. Shetty, K. Shetty. L-DOPA and total phenolic stimulation in dark geriminated fava bean in response to peptide and phytochemical elicitors. *Process Biochem.* 37:1247–1256, 2002a.

102. Engleberger, W., U. Hadding, E. Etschenberg, E. Graf, S. Leyck, J. Winkelmann, M.J. Parnham. Rosmarinic acid: A new inhibitor of complement C3 – convertase with anti-inflammatory activity. *Intl. J. Immunopharmac.* 10:729–737, 1988.
103. Kuhnt, M., A. Probstle, H. Rimpler, A. Bauer, M. Heinrich. Biological and pharmacological activities and further constituents of *Hyptis verticillata. Planta. Medica.* 61:227–232, 1995.
104. Fremont, L. Biological effects of resveratrol. *Life Sci.* 66:663–673, 2000.
105. Pinto, M.C., J.A. Garcia-Barrado, P. Macias. Resveratrol is a potent inhibitor of the dioxygenase activity of lipoxygenase. *J. Agric. Food Chem.* 47:4842–4846, 1999.
106. Vattem, D.A., K. Shetty. Solid-state production of phenolic antioxidants from cranberry pomace by *Rhizopus oligosporus. Food Biotechnol.* 16:189–210, 2002.
107. Vattem, D.A., K. Shetty. Ellagic acid production and phenolic antioxidant activity in cranberry pomace mediated by *Lentinus edodes* using solid-state system. *Process Biochem.* 39:367–379, 2003.
108. Singh, K., A.K. Khanna, R. Chander. Protective effect of ellagic acid on t-butyl hydroperoxide induced lipid peroxidation in isolated rat hepatocytes. *Ind. J. Exp. Biol.* 37:939–943, 1999.
109. Narayanan, B.A., G.G. Re. IGF-II down regulation associated cell cycle arrest in colon cancer cells exposed to phenolic antioxidant ellagic acid. *Anticancer Res.* 21:359–364, 2001.
110. Quiles, J.L., M.D. Mesa, C.L. Ramirez-Tortosa, C.M. Aguilera, M. Battino, M.C. Ramirez-Tortosa. *Curcuma longa* extract supplementation reduces oxidative stress and attenuates aortic fatty streak development in rabbits. *Arterioscler. Thromb. Vasc. Biol.* 22:1225–1231, 2002.
111. Dixon, R.A., M.J. Harrison, C.J. Lamb. Early events in the activation of plant defense responses. *Ann. Rev. Phytopath.* 32:479–501, 1994.
112. Dixon, R.A., N. Paiva. Stress-induced phenylpropanoid metabolism. *Plant Cell.* 7:1085–1097, 1995.
113. Rhodes, J.M., L.S.C. Wooltorton. The biosynthesis of phenolic compounds in wounded plant storage tissues. In: *Biochemistry of wounded plant tissues,* Kahl, G., ed., Berlin: W de Gruyter, 1978, pp –286.
114. Brooker, F.L., J.E. Miller. Phenylpropanoid metabolism and phenolic composition of soybean [Glycine max (L) Merr.] leaves following exposure to ozone. *J. Exp. Bot.* 49:1191–1202, 1998.
115. Zimmerman, Y.Z., P.R. Cohill. Heat shock and thermotolerance in plant and animal embryogenesis. *New Biol.* 3:641–650, 1991.
116. Yalpani, N., A.J. Enyedi, J. Leon, I. Raskin. Ultraviolet light and ozone stimulate accumulation of salicylic acid, pathogenesis related proteins and virus resistance in tobacco. *Planta* 193:372–376, 1994.
117. Hahlbrock, K., D. Scheel. Physiology and molecular biology of phenylpropanoid metabolism. *Plant Mol. Biol.* 40:347–369, 1989.
118. Graham, T.L. Flavanoid and isoflavanoid distribution in developing soybean seedling tissue and in seed and root exudates. *Plant Physiol.* 95:594–603, 1991.
119. Christie, P.J., M.R. Alfenito, V. Walbot. Impact of low temperature stress on general phenylpropanoid and anthocyanin pathways: enhancement of transcript abundance and anthocyanin pigmentation in maize seedlings *Planta* 194:541–549, 1994.
120. Beggs, C.J., K. Kuhn, R. Bocker, E. Wellmann. Phytochrome induced flavonoid biosynthesis in mustard *Sinapsis alba* L cotyledons: enzymatic control and differential regulation of anthocyanin and quercitin formation. *Planta* 172:121–126, 1987.
121. Lois, R., B.B. Buchanan. Severe sensitivity to ultraviolet light radiation in *Arabidopsis* mutant deficient in flavanoid accumulation: mechanisms of UV-resistance in *Arabidopsis. Planta* 194:504–509, 1994.
122. Kurganova, L.N., A.P. Veselov, T.A. Goncharova, Y.V. Sinitsyna. Lipid peroxidation and antioxidant system protection against heat shock in pea (*Pisum sativum* L.) chloroplasts. *Fiziol. Rast. Russ. J. Plant Physiol., Engl. Transl.* 44:725–730, 1997.

123. Retivin, V., V. Opritov, S.B. Fedulina. Generation of action potential induces preadaptation of *Cucurbita pepo* L. stem tissues to freezing injury. *Fiziol. Rast. Russ. J. Plant Physiol. Engl. Transl.* 44:499–510, 1997.
124. Kuznetsov, V.V., N.V. Veststenko. Synthesis of heat shock proteins and their contribution to the survival of intact cucumber plants exposed to hyperthermia. *Fiziol. Rast. Russ. J. Plant Physiol. Engl. Transl.* 41:374–380, 1994.
125. Baraboi, V.A. Mechanisms of stress and lipid peroxidation. *Usp. Sovr. Biol.* 11:923–933, 1991.
126. Kurganova, L.N., A.P. Veselov, Y.V. Sinitsina, E.A. Elikova. Lipid peroxidation products as possible mediators of heat stress response in plants. *Exp. J. Plant Physiol.* 46:181–185, 1999.
127. Phang, J.M. The regulatory functions of proline and pyrroline-5-carboxylic acid. *Curr. Topics Cell. Reg.* 25:91–132, 1985.
128. Hagedorn, C.H., J.M. Phang. Transfer of reducing equivalents into mitochondria by the interconversions of proline and δ-pyrroline-5-carboxylate. *Arch. Biochem. Biphys.* 225:95–101, 1983.
129. Rayapati, J.P., C.R. Stewart. Solubilization of a proline dehydrogenase from maize (*Zea mays* L.) mitochondria. *Plant Physiol.* 95:787–791, 1991.
130. Yeh, G.C., J.M. Phang. The function of pyrroline-5-carboxylate reductase in human erythrocytes. *Biochem. Biophys. Res. Commun.* 94:450–457, 1980.
131. Phang, J.M., S.J. Downing, G.C. Yeh, R.J. Smith, J.A. Williams. Stimulation of hexosemonophosphate-pentose pathway by δ-pyrroline-5-carboxylic acid in human fibroblasts. *Biochem. Biophys. Res. Commun.* 87:363–370, 1979.
132. Hagedorn, C.H., J.M. Phang. Catalytic transfer of hydride ions from NADPH to oxygen by the interconversions of proline and δ-pyrroline-5-carboxylate. *Arch. Biochem. Biphys.* 248:166–174, 1986.
133. Kwok, D., K. Shetty. Effect of proline and proline analogs on total phenolic and rosmarinic acid levels in shoot clones of thyme (*Thymus vulgaris* L.). *J. Food Biochem.* 22:37–51, 1998.
134. Yang, R., K. Shetty. Stimulation of rosmarinic acid in shoot cultures of oregano (*Origanum vulgare*) clonal line in response to proline, proline analog and proline precursors. *J. Agric. Food Chem.* 46:2888–2893, 1998.
135. Bela, J., K. Shetty. Somatic embryogenesis in anise (*Pimpinella anisum* L.): the effect of proline on embryogenic callus formation and ABA on advanced embryo development. *J. Food Biochem.* 23:17–32.
136. Perry, P.L., K. Shetty. A model for involvement of proline during *Pseudomonas*-mediated stimulation of rosmarinic acid. *Food Biotechnol.* 13:137–154, 1999.
137. Lendzian, K.J. Modulation of glucose-6-phosphate dehydrogenase by NADPH, NADP+ and dithiothreitol at variable NADPH/NADP+ ratios in an illuminated reconstituted spinach (*Spinacia oleracea* L.) choroplast system. *Planta* 148:1–6, 1980.
138. Copeland, L., J.F. Turner. The regulation of glycolysis and the pentose-phosphate pathway. In: *The Biochemistry of Plants*, Vol. 11, Stumpf, F., E.E. Conn, eds., New York: Academic Press, 1987, pp 107–125.
139. Elthon, T.E., C.R. Stewart. Effects of proline analog L-thiazolidine-4-carboxylic acid on proline metabolism. *Plant Physiol.* 74:213–218, 1984.
140. Jost, A., S. Perlman, O. Valentino, M. Castinier, R. Scholler, S. Magre. Experimental control of the differentiation of Leydig cells in the rat fetal testis. *Proc. Natl. Acad. Sci. USA.* 85:8094–8097, 1988.
141. Al-Amier, H., B.M.M. Mansour, N. Toaima, R.A. Korus, K. Shetty. Tissue culture-based screening for selection of high biomass and phenolic-producing clonal lines of lavender using *Pseudomonas* and azetidine-2-carboxylate. *J. Agric. Food Chem.* 47:2937–2943, 1999.
142. Al-Amier, H., B.M.M. Mansour, N. Toaima, R.A. Korus, K. Shetty. Screening of high biomass and phenolic-producing clonal lines of spearmint in tissue culture using *Pseudomnas* and azetidine-2-carboxylate. *Food Biotechnol.* 13:227–253, 1999.

143. Al-Amier, H.A., B.M.M. Mansour, N. Toaima, L. Craker, K. Shetty. Tissue culture for phenolics and rosmarinic acid in thyme. *J. Herbs Spices Med. Plant* 8:31–42, 2001.
144. Kohl, D.H., K.R. Schubert, M.B. Carter, C.H. Hagdorn, G. Shearer. Proline metabolism in N_2-fixing root nodules: energy transfer and regulation of purine synthesis. *Proc. Natl. Acad. Sci. USA* 85:2036–2040, 1988.
145. Eguchi, Y., O.F. Curtis, K. Shetty. Interaction of hyperhydricity-preventing *Pseudomonas* spp. with oregano *(Origanum vulgare)* and selection of high rosmarinic acid-producing clones. *Food Biotechnol.* 10:191–202, 1996.
146. Shetty, K., T.L. Carpenter, D. Kwok, O.F. Curtis, T.L. Potter. Selection of high phenolics-containing clones of thyme (*Thymus vulgaris* L.) using *Pseudomonas* spp. *J. Agric. Food Chem.* 44:3408–3411, 1996.
147. Yang, R., O.F. Curtis, K. Shetty. Selection of high rosmarinic acid-producing clonal lines of rosemary (*Rosmarinus officinalis*) via tissue culture using *Pseudomonas* sp. *Food Biotechnol.* 11:73–88, 1997.
148. McCue, P., Z. Zheng, J.L. Pinkham, K. Shetty. A model for enhanced pea seedling vigor following low pH and salicylic acid treatments. *Process Biochem.* 35:603–613, 2000.
149. Andarwulan, N., K. Shetty. Improvement of pea (*Pisum sativum*) seed vigor by fish protein hydrolysates in combination with acetyl salicylic acid. *Process Biochem.* 35:159–165, 1999.
150. Duval, B., K. Shetty. The stimulation of phenolics and antioxidant activity in pea (*Pisum sativum*) elicited by genetically transformed anise root extract. *J. Food Biochem.* 25:361–377, 2001.
151. McCue, P., K. Shetty. A biochemical analysis of mungbean (*Vigna radiata*) response to microbial polysaccharides and potential phenolic-enhnacing effects for nutraceutical applications. *Food Biotechnol.* 6:57–79, 2002.
152. McCue, P., K. Shetty. Clonal herbal extracts as elicitors of phenolic synthesis in dark-germinated mungbeans for improving nutritional value with implications for food safety. *J. Food Biochem.* 26:209–232, 2002.
153. Shetty, P., M.T. Atallah, K. Shetty. Effects of UV treatment on the proline-linked pentose phosphate pathway for phenolics and L-DOPA synthesis in dark germinated *Vicia faba*. *Process Biochem.* 37:1285–1295, 2002.
154. Shetty, P., M.T. Atallah, K. Shetty. Stimulation of total phenolics, L-DOPA and antioxidant activity through proline-linked pentose phosphate pathway in response to proline and its analog in germinating fava beans (*Vicia faba*). *Process Biochem.* 38:1707–1717, 2003.
155. Randhir, R., K. Shetty. Microwave-induced stimulation of L-DOPA, phenolics and antioxidant activity in fava bean (*Vicia faba*) for Parkinson's diet. *Process Biochem.* 39:1775–1785, 2004.

2.03

Genomic Basics for Food Improvement

Gabriela Olmedo, Socorro Parra, and Plinio Guzmán

CONTENTS

3.1 INTRODUCTION

One ultimate goal of many plant scientists is to gain in depth knowledge of the function of each gene in a given plant. Understanding how genes coordinate growth and development of a living being is of great importance in many areas of science and technology. In the last decade, sequencing projects turned out myriads of genes. Because traditional approaches involve studying genes one at a time, and given that there are at least 20,000 genes in any given plant species, understanding the functions of all these genes requires new strategies. In the last few years, several technological advances for deciphering and manipulating the genetic information revolutionized the field of genetics, giving birth to a new area in biological sciences: genomics. Genomics provides the strategies for the simultaneous study of hundreds or thousands of genes. Using microarrays, for instance, the expression of many genes can be studied in parallel. Furthermore, it is hoped that genomics can be complemented with proteomics to also assess the function of each gene's product.

Genomics deals with the molecular cloning and characterization of entire genomes. It studies the organization and function of the genetic information, and its goal in a given species is to determine the sequence of the complete genome and to establish the function of all genes contained in it. The basic understanding of how a microorganism, a worm, an insect, a plant, a fish, or a mammal functions provides invaluable information for biotechnological developments. Drugs can be manufactured using new approaches, and early detection of a number of diseases can be envisioned. Crop and livestock productivity as well as food quality and safety can be expected to improve. Because genomics deciphers the structure of the genome it will also impact our knowledge of the evolution of species.

Two major areas of genomic research are considered here: the structural and the functional analysis of the genome. The structural analysis of the genome, known as structural genomics, establishes the composition and organization of the whole genome; and the functional analysis of the genome, known as functional genomics, studies the expression of the genetic elements and searches for the function of every gene product. Pioneer genomic projects have been accomplished in several of the research model systems and current projects are now being established in a wide range of organisms, including crops of nutritional importance. Structural and functional genomics have already generated information from some of the model systems, but most of the genomic projects are still at the structural analysis stage. In plant biotechnology the weed *Arabidopsis thaliana*, as the main model systems in dicots, is driving the genomic field. The sequence of its genome has been completed and tools for the functional analysis of its genes are now available. Likewise, the draft of the rice genome has been created, providing a first panoramic view of the genome of this very important food crop. This chapter reviews the existing knowledge of the approaches and tools used in the genomic analysis of plants.

3.2 WHOLE GENOME SEQUENCING OF PLANT GENOMES

Deciphering the basic composition of a genome is a crucial step in the understanding of its function as a whole. A central procedure in genomics is DNA sequencing; this has undergone several improvements since first described by Sanger in the 1970s (1). The automation of routine processes for template preparation (the DNA to be sequenced) and the development of sophisticated machines have speeded up DNA sequence procedures; now the determination of vast amounts of DNA sequence is possible in much less time and at much less cost than a decade ago.

The complete sequence of a genome is obtained by first determining the DNA sequence of small segments and then deducing the position of each one in a linear order. One approach begins with the cloning of fragments representing the entire genome into vectors with high cloning capabilities, usually greater than 150.000 base pairs long. These are the Yeast Artificial Chromosome vectors (YACs), manipulated in yeast, and the Bacterial Artificial Chromosomes (BACs) and P1 Artificial Chromosome vectors (PACs) maintained as *Escherichia coli* clones (2). Overlapping clones (clones containing segments of identical sequence) are then arranged into contigs, a series of contiguous clones, to generate a physical map of the organism. Next, 500–1000 base pair long fragments are obtained from each of the clones and subcloned into vectors for the subsequent determination of the DNA sequence. The subclones are sequenced randomly, and overlapping sequences are used to assemble the complete original clone *in silico*. Another approach is to start to sequence random clones from the beginning. In this approach, a library of small size clones (1000 base pair long inserts) is generated and sequenced randomly, and then the entire genome is assembled from overlapping sequences. The first output is the draft sequence that is used to assemble the original clone or genome. However, to ensure data accuracy, additional sequencing is performed, resulting in 5–8 sequences from each individual clone. This process clarifies uncertainties in the sequence and helps find most of the overlapping sequences between clones. There are regions of the genome of plants and mammals that cannot be cloned or that are very unstable. These regions are not sequenced and are not present in the description of the complete sequence of a genome. These are the heterochromatic regions of the chromosome, including centromeres, telomeres, and regions rich in repeated sequences.

Whole genome sequencing projects have been completed in several organisms, ranging from small bacterial genomes to large mammalian and plant genomes. In eukaryotes, the genomes from three model organisms were the first to be completed: the yeast *Saccharomyces cerevisiae* (3), the worm *Caenorhabditis elegans* (4), and the insect *Drosophila melanogaster* (5). To date there are 261 genomes that have been sequenced: 21 archaeal, 207 bacterial, and 33 from eukaryotic organisms (see http://wit.integratedgenomics.com/GOLD/). The first plant genome to be sequenced was that of *Arabidopsis thaliana* (6), and more recently two draft sequences of the rice genome were reported (7,8). The completion of the sequence of a genome is only the beginning of new challenges in biological research. *Arabidopsis thaliana* is a weed of the Brassicaceae family that does not have any agriculturally important value. However, this weed provides the resources to understand basic phenomena in plant growth and development. One result of a complete genome is establishing the exact number of genes in an organisms and the map location of each. This permits the identification of single copy genes as well as the composition of gene families. The next challenge is to establish the role for all the genes encoded in the genome. The genome of *Arabidopsis* will be the reference book for other plant genomes; comparative analysis will help with gene identification and to establish gene function in crop species.

A landmark in plant genomics is the availability of the DNA sequences of the *Arabidopsis* and rice genomes. These two species possess the smallest known genomes among plants (see Table 3.1), and represent two important lineages in the plant kingdom: monocots and dicots. The availability of the complete sequences of the *Arabidopsis thaliana* and rice genomes permits researchers to inquire for the first time about the number of conserved genes between plant species (see Figure 3.1). There are 25,000 genes encoded in *Arabidopsis* and 32,000 to 55,000 in rice. Analysis of the presence of *Arabidopsis* genes in rice revealed that 85% of predicted genes are also present in rice, and about 12% of these with high homology, suggesting that they are important genes for cell functioning. For another 30% of the conserved genes there is no information in databases, indicating potential genes to be discovered. Plant specific genes can also be enlisted (see Figure 3.1); these genes are present in both *Arabidopsis* and rice genomes but absent in other eukaryotic genomes like *S. cerevisiae, D. melanogaster,* and *C. elegans*. Another group of genes that may reveal specific genes for dicot plants are those that do not have homologues in

Table 3.1
Genome Size of Some Plant Genomes

Organism	Genome Size (Mega Base Pairs)
Arabidopsis thaliana	125
Oriza sativa	430
Lotus japonicus	450
Medicago truncatula	500
Gossypium hirsutum L cv. Maxxa	2,118
Hordeum vulgare var. distichum	5,000
Zea mays	5,000
Avena sativa	16,000
Triticum aestivum	16,000

Data obtained from: http://wit.integratedgenomics.com/GOLD

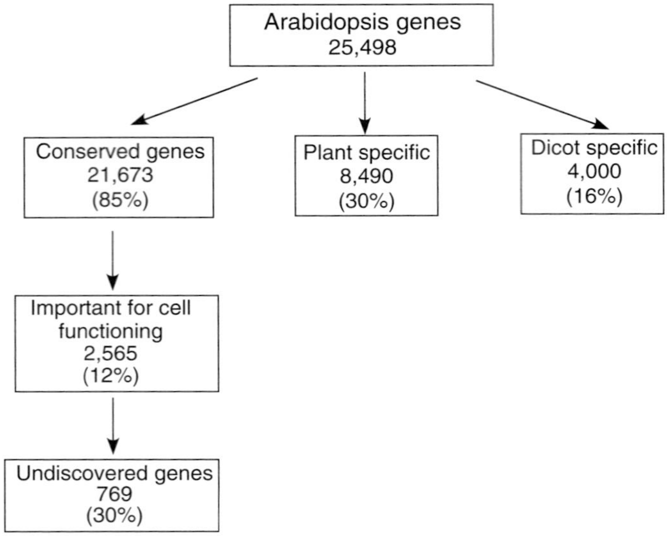

Figure 3.1 Conservation of genes between two plant species: *Arabidopsis* and rice. Conserved genes are those present in both plant species; genes important for cell functioning are those that show a high homology between them; undiscovered genes have no annotation in databases; plant specific genes are present in both species but absent in other sequenced genomes; dicot specific genes are those only present in *Arabidopsis*. Data adapted from (7).

rice or other eukaryotes (see Figure 3.1). Additionally, the availability of the rice sequence also permits confirmation of observations previously only inferred in *Arabidopsis*; there are genes that are neither encoded in *Arabidopsis* nor in rice. Examples of this are the genes encoding the nuclear steroid receptors, Notch/Lin12, p53, Hedgehog, and Wingless/Ent. Conversely, genes that are found in high numbers in *Arabidopsis* when compared to other eukaryotes are also found in high numbers in the rice genome. These are families of RING zinc fingers and F-box containing proteins. Interestingly, these two classes of genes are involved in the regulation of protein turnover in other organisms, suggesting that such large gene families have a function particular to plants, possibly participating in some stages of plant growth and development (7).

Comparative analysis between *Arabidopsis* and rice can also be applied to address specific questions. An example is the comparison of the number of transcription factors between both species. Transcriptional regulation of gene expression is intrinsic to most of the biological phenomena in all living organisms. Differential expression of sets of genes during development is controlled by transcription factors, and basic changes in transcription factors may give rise to diversity among species. The *Arabidopsis* and rice genomes encode for at least 1533 and 1306 predicted transcription factors, respectively, and show a similar number of members in each of the classes (7, 9); an example of these are displayed in Table 3.2. Almost 45% of these regulatory proteins constitute families specific to plants, such as the AP2/ERF/RAV and WRKY families. Others are found in higher numbers in plants than in nonplant eukaryotes, such as Myb and MADS factors. In general plant transcription factors show sequence conservation with other eukaryotes primarily in the DNA binding domains.

The next complete plant genomes to be sequenced may be those of two legume species, *Lotus japonicus* and *Medicago truncatula* (10). *Lotus japonicus* is a model for forage legumes with a genome size of 450 Mb. *Medicago truncatula* is a model system for studying crop species relative to alfalfa; its genome is similar to *L. japonicus* in size, 500 Mb. Pilot sequencing

Table 3.2
Transcription factors in Arabidopsis and Rice

Transcription Factor Family	Arabidopsis	Rice
Ap2/ERF/RAV	144	143
WRKY	72	83
MYB	190	156
MADS box	82	71
C2H 2zinc finger	105	125
Dof	36	21
BZip	81	75

Data adapted from (7, 9)

projects have been initiated for *M. truncatula* using shotgun cloning strategies, and the entire sequence of the chloroplast genome has been completed. Legumes have a key role in food production, with many of them having been the source of very nutritious seeds for many years. Another significant feature is that legumes benefit from symbioses with nitrogen-fixing rhizobia and mycorrhiza when growing in soils limited in nutrients. Legumes also have unique pathways for secondary metabolite biosynthesis.

Not all the plant genomes have the size and simplicity of *Arabidopsis* (Table 3.1). Genome sizes in most crops show immense variations; for instance, the genome size of oats and wheat are estimated to be in the 16,000 Mb range, 5 times larger than that of the human genome. In addition, wheat is hexaploid and oat is an allopolyploid of three genomes. The size difference does not correlate with the number of genes. Enormous sizes are caused by the invasion of the genome with transposable elements and repetitive sequences. These repeats impair sequence analysis, making it impossible to assemble overlapping clones. Nevertheless, there are at least 16 genome projects ongoing in plants. Genome sequencing is very revealing, because regulatory elements located outside transcribed regions are contained in the sequence. Likewise, genes transcribed at very low levels are obtained that would be difficult to obtain by other means. A futher problem is how to obtain sequence information from complex plant genomes. An original approach has been described in maize (11). It was shown that repeats are preferably modified by methylation. By selecting clones in methylation-restrictive *E. coli* hosts, or by using methylation-sensitive restriction enzymes to generate libraries, the clones containing methylated DNA can be circumvented. This approach enriches the number of nonmethylated sequences that are likely to correspond to genes.

3.3 LARGE SCALE EST SEQUENCING PROJECTS IN CROP SPECIES

As mentioned previously, genomes are complex and possess not only coding genes but repetitive and structural elements as well. The approach of some sequencing projects is to sequence coding genes exclusively. Coding genes give rise to mRNAs that can be separated and converted into cDNAs. ESTs (expressed sequence tags) are partial DNA sequences obtained from either the 5′ or 3′ ends of cDNAs. ESTs provide very helpful data on the nature of coding regions of genes and constitute a basic tool for the assembly of high density cDNA microarrays for global expression studies. A shortcoming of ESTs is that the cDNAs lack DNA sequence information encoded beyond the transcribed region of the gene; consequently knowledge about regulatory elements that are not encoded in the transcript cannot be obtained. Likewise, genes that contain large untranslated regions make it difficult to discover genes, because sequence comparison is less feasible.

Considering the size and complexity of most crop genomes, large scale ESTs sequencing projects have been an effective alternative for getting insights into their genomes. A list of current EST projects, including those of plants important for human nutrition, is displayed in Table 3.3. The rising number of sequences in data bases is astonishing; more than 100,000 ESTs are available for some crops, such as oats, soybeans, and corn; and more than 400,000 ESTs are reported for wheat (Table 3.3). EST collections are usually obtained from cDNA libraries constructed from specific tissues or from tissues representing specific developmental stages, or from different treatments such as response to pathogens or to abiotic stresses. This diverse array of treatments provides invaluable information on the pattern and level of expression of genes in an organism. For example, 21 libraries are the sources of the current EST collection in maize (12). These libraries were prepared from various developmental stages: embryos collected at different days after pollination; immature apical meristem tissue; whole premeiotic anthers and pollen shed; from diverse organs: tassel, kernel, silk, husk, root, and leaf; and under diverse conditions: adult silk infected with the fungus *Fusarium* and salt-stressed seedlings and shoots.

Table 3.3
Number of ESTs and Database Entries for Plant Species

	ESTs	Nucleotide Entries	Protein Entries
Triticum aestivum (wheat)	415642	418106	2177
Hordeum vulgare (barley)	339034	35341	1239
Glycine max (soybean)	308564	320244	2168
Zea mays (corn)	198523	546188	4850
Medicago truncatula (barrel medic)	180939	185548	227
Lycopersicon esculentum (tomato)	148566	162415	2252
Solanum tuberosum (potato)	94423	95874	1661
Sorghum bicolor (sorghum)	84712	120337	517
Pinus taeda (loblolly pine)	60226	60721	184
Populus tremula × *Populus tremuloide*	56013	56190	61
Gossypium arboreum (tree cotton)	38894	38953	39
Lotus japonicus	33124	80015	381
Mesembryanthemum crystallinum (common ice plant)	25446	25689	355
Sorghum propinquum	21387	27470	1
Beta vulgaris (beet)	19617	14	17
Gossypium hirsutum (upland cotton)	10725	13867	733
Secale cereale (rye)	8930	9221	222
Lycopersicon pennellii	8346	8377	21
Lycopersicon hirsutum	2504	2535	20
Marchantia polymorpha (liverwort)	1415	1515	710
Medicago sativa (alfalfa)	805	1363	679
Avena sativa (oat)	501	730	190

Data obtained from: www.ncbi.nlm.nih.gov/PMGifs/Genomes/PlantList.html

Since the early days of DNA sequencing it was noted that the genomes of eukaryotes share conserved sequences. The comparative analysis of complete genomes, known as comparative genomics, confirmed that different genomes have a substantial quantity of conserved genes. Such genes are mostly involved in basic cellular processes. Comparison of ESTs in databases facilitates the detection of conserved sequences among diverse and unrelated organisms. This fact increases the possibility of accurate annotation of unknown cDNA clones; that is, to be able to classify a gene only by analyzing the information of its predicted coding product. Percentages of maize ESTs that correspond to protein homologues in other plants is shown in Table 3.4. As expected, the greatest number of matches is from *Arabidopsis* and rice, given that their complete genome sequence is known. Comparative genomics between *Arabidopsis* genes and wheat EST sequences shows that about 50% of *Arabidopsis* genes match 2 or more wheat ESTs (Figure 3.2) (13). About 75% of the matching genes are annotated with a description, many of them corresponding to genes with a possible function in general processes. This comparison is useful as an initial categorization to distinguish genes involved in species–specific processes. These

Table 3.4
Percentages of Maize EST Protein Homologs by Organism

Organism	Percentage
Arabidopsis thaliana	42.7
Oryza sativa	23.4
All remaining species	5.8
Triticum aestivum	1.2
Hordeum vulgare	1.1
Sorghum bicolor	0.2

Data adapted from (12)

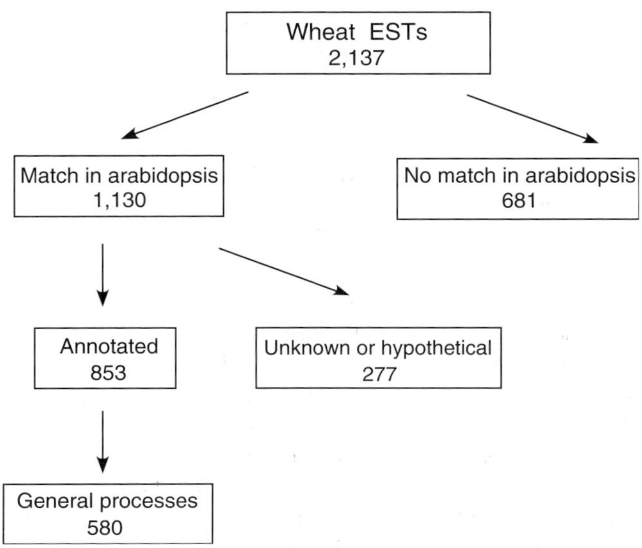

Figure 3.2 Comparative genomics: *Arabidopsis* as a tool to annotate wheat genes. Flow diagram showing the outcome of comparing wheat ESTs to *Arabidopsis* genes. See description in the text. Data adapted from (13).

genes may be those that do not have a match in both genomes. The rapid increase in the number of sequences in the databases will greatly improve the annotation process of genes from plants of nutritional importance.

3.4 GENOME WIDE ANALYSIS OF GENE EXPRESSION USING DNA MICROARRAYS

The analysis of the presence and the measurement of the level of different mRNAs during specific growth or developmental conditions is a method that provides useful information about the function of a gene. The availability of large collections of genes from several organisms and of complete genome sequences has prompted the development of technology to facilitate the analysis of the expression of thousands of genes at the same time. This technology is known as DNA microarrays, also referred to as DNA chips, and one of its uses is to monitor global expression patterns (14). DNA microarrays consist of DNA samples bound to a solid surface and arrayed in a defined order so that the identity of each is known. These arrayed samples are then hybridized to RNA that has been labeled with fluorescent dyes to allow the detection of the DNA sample represented in the array corresponding to the expressed RNA.

There are basically two different DNA microarray methods for monitoring gene expression. One depends on the use of cDNA clones and the other on the use of oligonucleotides. Amplification of cDNA clones in a size range of 200 to 2000 base pairs by PCR is the most common source of DNA. These cDNA clones are usually obtained form EST sequencing projects and primers from vector sequences can direct the amplification of the cDNA inserts. Amplification of genes using genomic DNA as a template is also a source of DNA to be tested. In this case the genomic sequence has to be known to be able to design gene specific primers. The second option makes use of synthetic oligonucleotides 60 to 80 nucleotides long. This alternative has the advantage that the oligonucleotides can be designed to be gene specific, which helps in the analysis of gene families, as it increases the possibility of getting a differential hybridization among its members. This approach obviously needs previous knowledge of the gene sequences to be able to design the specific oligonucleotides to be spotted.

The supports for spotting the DNA samples in the microarrays are either glass or nylon. When DNA samples are mechanically spotted at a high density on a glass slide using a robotic spotter, these are known as glass microarrays. Nylon is also used for spotting microarrays but it has some disadvantages that increase the experimental error. The nylon membrane can shrink, the degree of hybridization of each DNA sample adhered to the membrane is variable, and, as in glass, the amount of DNA that binds to the membrane is not always constant; nevertheless nylon is a good choice for analysis of gene expression on a small scale.

There is an alternative method to spotting, which is the direct synthesis of oligonucleotides on a modified glass surface by a photolithographic technique. In this technique, 25 nucleotid-long oligonucleotides with a long linker arm are directly synthesized on a surface (15). Two types of oligonucleotides are designed and placed on the same chip. One type perfectly matches the target sequence, and is called the perfect match (PM); the other is a corresponding oligonucleotide that displays a single nucleotide difference in the center of the sequence, called the mismatch (MM). For each one of the genes on the array, 10–20 PM oligonucleotides and their corresponding MM oligonucleotides are synthesized. These oligonucleotides include some that are designed from nonconserved regions of the genes, usually the 3′ untranslated region (UTR). This method has several advantages: the long linker arm allows each base to be equally accessible to hybridization, thus the hybridization kinetics is similar to that in solution; and very high density and uniform microarrays can be manufactured, reducing the background fluorescence and producing high quality and reproducible data.

Glass microarray experiments from either spotted cDNA or synthetic oligonucleotides are simultaneously hybridized with two mRNA samples obtained from two conditions or sources that will be compared (see Figure 3.3). Each mRNA sample is labeled by reverse

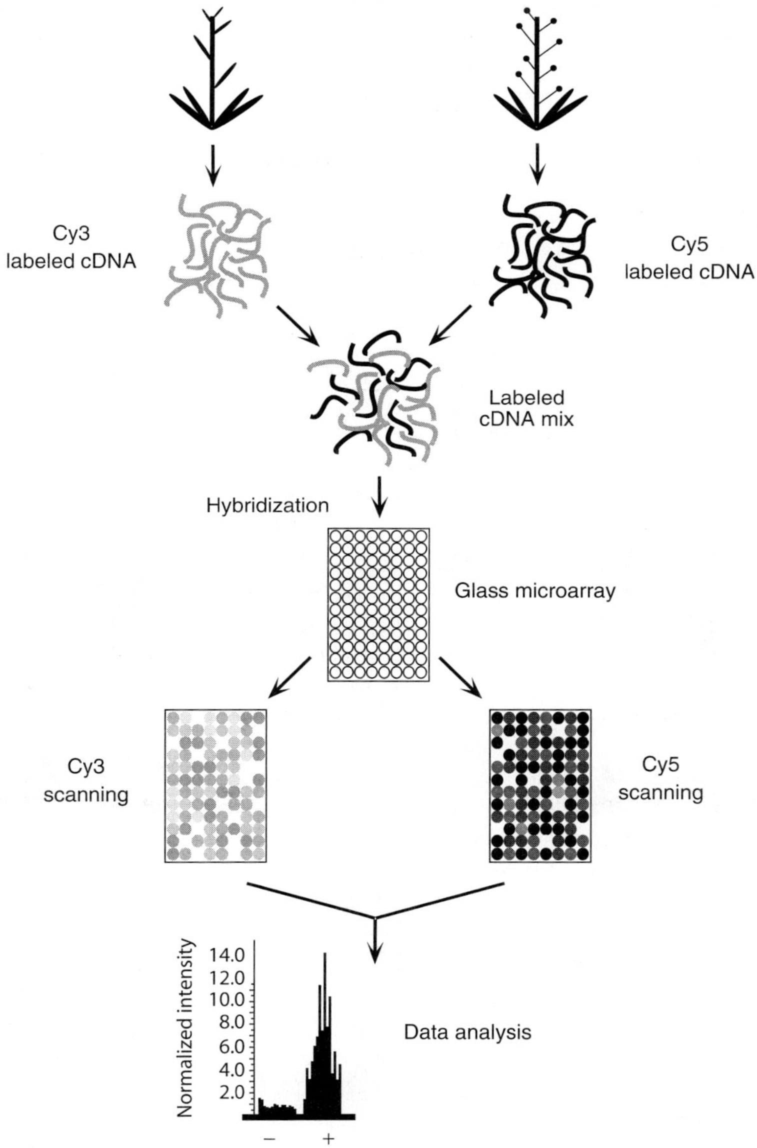

Figure 3.3 Glass microarrays. RNA is isolated from each sample (plant or tissue) and labeled by means of reverse transcription. One sample is labeled in the presence of Cy3, a green-fluorescent dye (depicted in gray and tones of light gray), and the other in the presence of Cy5, a red-fluorescent dye (depicted in black and dark gray). Samples are then mixed and used to hybridize a single microarray. After hybridization and washing, the fluorescent intensities for Cy3 and Cy5 are independently scanned and compared. The data is evaluated using software designed for microarray analysis.

transcription into cDNA, incorporating during the reaction a different fluorescent dye: cyanine 5 (Cy5), a red fluorescent dye, or cyanine 3 (Cy3), a green fluorescent dye. The fluorescence is determined separately for each fluorophore at each one of the spots using a scanner. The ratio of the measurements of intensity between these two fluorophores is used to compare transcript expression levels (16). For high density photolithographic oligonucleotide microarray experiments, however, a single labeled sample is used for each hybridization (see Figure 3.4). The method employs the synthesis of a biotinylated antisense copy of the mRNA (cRNA) followed by staining using streptavidin coupled to phycoerythrin; chips

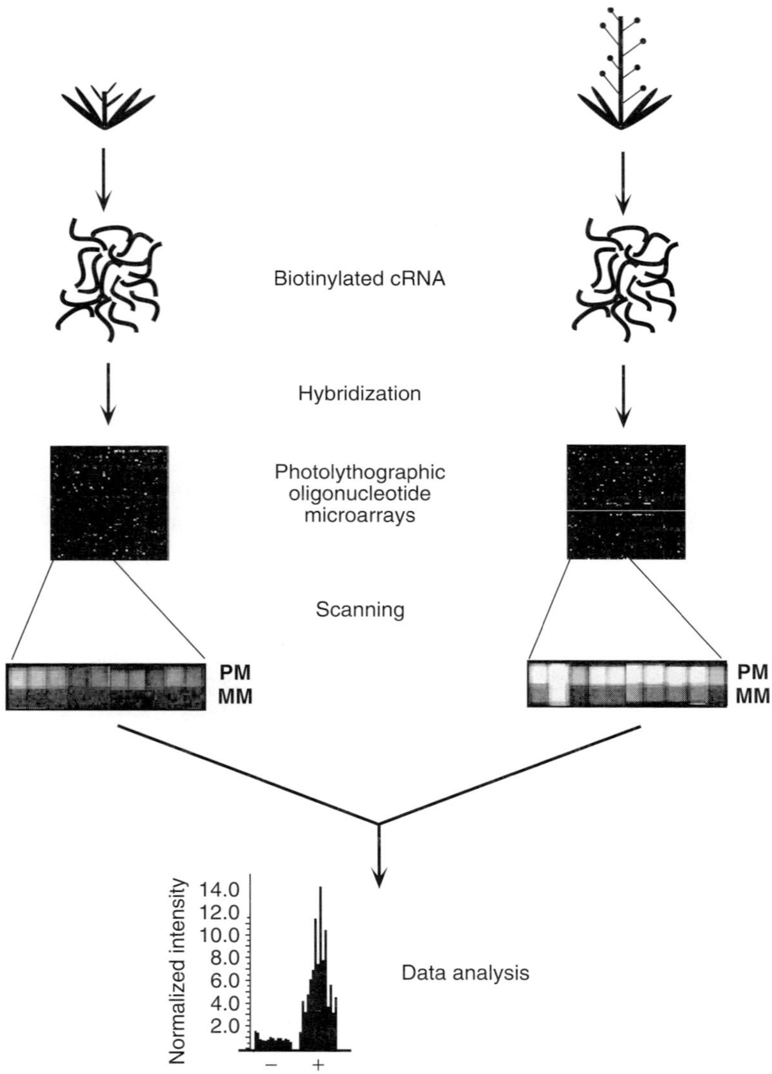

Figure 3.4 Photolithographic oligonucleotide microarrays. RNA is isolated from each sample and labeled in the presence of biotinylated ribonucleotides by means of T7-polymerase. The resulting cRNA is amplified and fragmented into small segments. The fragmented cRNA is used to hybridize a single microarray. After hybridization the microarray is stained with streptavidin coupled to phycoerythrin and then scanned. The data is evaluated using software designed for microarray analysis.

are then scanned and the measurements compared with the results of other chips from the experiment (15). In this case the hybridization occurs from RNA to DNA, whereas in spotted cDNA or oligonucleotide microarrays the hybridization occurs from DNA to DNA.

A very important aspect of microarrays is the design of a "good" experiment. When planning a microarray experiment it is necessary to program enough controls, and to include experimental and biological replicas, because hybridizations are very sensitive to experimental errors. Likewise, validation of the data by an independent method such as RT-PCR, Northern blot, or real time PCR is advisable. Because microarrays assess the expression of thousands of genes in a single experiment, statistical analysis is needed for the examination of the intensities of the spots; several software products are now available to perform such tasks (17).

The use of microarray technology for global gene expression studies is in its infancy in most of the crop species of nutritional importance. cDNA microarrays are preferred for most species because nonredundant clones obtained from EST projects can be readily employed to design the array. cDNA microarrays have been used in strawberries to study gene expression patterns during fruit ripening. Interestingly, an important gene involved in flavor biogenesis was identified after the analysis of the data (18). The global expression analysis unveiled novel processes connected to ripening such as the development of the vascular system, oxidative stress, and auxin-regulated gene expression (19). The strawberry is not a real fruit; it is the expansion of the receptacle carrying in the surface of the epidermis the achenes, the real fruit. In an attempt to establish specific gene expression profiles, DNA microarrays were also used to compare gene expression between receptacle and achene (20). Another example is cacao, where a genomic project that includes microarray analysis has been initiated. Microarrays of 997 cDNAs have been hybridized to samples isolated from leaves or beans from five different cacao genotypes. This project will contribute to the understanding of disease resistance, because pests cause major devastation of cocoa plants, and will seek new traits important for cocoa production (21). cDNA microarrays are also being developed for the model legumes. An array of 18,144 nonredundant ESTs has been used to analyze up regulated and down regulated genes during early stages of nodule formation in *Lotus*. Likewise, an array of 8,000 cDNAs from *M. truncatula* has been developed to analyze the gene expression under several conditions (10).

Significant advances in microarray technology have been achieved in model plant species. A variety of arrays have been generated for *Arabidopsis*, some of them including almost all the genes encoded in the genome. cDNA microarrays have been used to analyze expression profiles during seed development (22), light stress (23), defense response, drought, cold, and high salinity stress conditions (24), among others. Three types of microarrays for *Arabidopsis* are now commercially available. One is an oligonucleotide array of 14,800 genes represented by oligonucleotides 60 bases long (www.chem.agilent.com). Genes for the array were selected considering availability of the annotation and accuracy on gene prediction. A collection of 26,029 oligonucleotides 70 nucleotides long is also available (www.qiagen.com). The genes for this array were obtained from experimental data, ESTs, and predicted transcripts. To generate the array, the collection has to be spotted on glass slides using a robotic arrayer. Photolithographic oligonucleotide microarrays representing approximately 24,000 *Arabidopsis* genes were generated using eleven pairs of oligonucleotides for each transcript (www.affymetrix.com). A previous version of this microarray containing 8,256 genes was used to analyze gene expression profiles in several experiments, including responses to gravity (25), wounding, biotic and abiotic stresses, and response to hormones (26–29).

Large scale expression analysis in rice is also under way. cDNA microarrays from 10,243 cDNA sequences have been constructed from EST sequencing projects and about

30% of these genes have been annotated (www.jircas.affrc.go.jp). To assemble the array, each gene is represented by a full length cDNA insert and by a segment of its 3'UTR. The addition of the 3'UTR is useful in some cases to resolve gene specific expression between members of gene families. This microarray was used to obtain expression profiles for low temperature, high salt, nutrient deprivation, oxidative stress, and hormone response, among others. A custom whole genome photolithographic oligonucleotide microarray was made for rice and is under evaluation (www.tmri.org). It contains 50,119 rice genes selected by gene prediction; 43% of them with high evidence, 25% with some evidence and 32% with low evidence. In the design of this array, ten perfect match oligonucleotides for each gene were included and the mismatch oligonucleotides (MM, see Figure 3.4) were eliminated.

cDNA microarrays have been also used in maize to obtain gene expression profiles. An array of 5,376 unique genes was designed and the expression from thirteen samples from seven organs assessed (30). To discern hybridization occurring between closely related members of gene families, oligonucleotides of 45 nucleotides were simultaneously included in the microarray for members of some gene families. Almost all genes were expressed in all organs with little variation between samples; 7% of the genes displayed at least a fourfold difference in level of expression in one of the organs and 37% displayed differences between two and fourfold. Interestingly, from the analysis of the gene expression profiles novel patterns of expression and organ hierarchies were established. It was inferred that gene expression in silks, immature ears, and embryos is closely related, and that it is different from vegetative tissues. Photolithographic oligonucleotide microarrays representing approximately 1,400 maize selected genes have been also manufactured (31). These arrays were tried out to establish gene expression profiles of endosperm in wild-type maize and in the mutants altered in storage protein synthesis, opaque2 and floury2; these experiments are the basis of the understanding of the mutant phenotypes.

New technologies for microarray analysis have been recently tested in plants. Laser-capture microdissection (LCM), a high resolution technique for gene expression analysis, was used in maize (32). LCM has been widely used in mammals to isolate specific cell types and to obtain global profiles from such cells. LCM was recently successfully tested in maize to isolate epidermal cells and vascular tissues to obtain a gene expression profile for each of these cell types in a cDNA microarray. Another approach combines genome wide genetic analysis and global gene expression profiling strategies in the same experiment (see Figure 3.5). In this strategy, termed genetical genomics, DNA fingerprinting and microarray analysis are performed on each individual of a segregating population. This procedure has been tested with human, mouse, and maize samples. In maize both QTLs (quantitative trait loci) and a microarray of 24,473 60-nucleotide-long oligonucleotides were used. By these means it is possible to identify gene expression patterns associated with a specific genetic trait.

The correlation of the expression of a gene with a particular response is not sufficient to definitely assign a function to it. A powerful strategy to prove that a gene has a certain function is to actually mutate the gene and observe the effect caused by its disruption. Reverse genetic tools have been successfully established for *Arabidopsis*. Gene knockouts caused by insertional mutagenesis have been instrumental to establish gene function in plants. Transfer T-DNA of *Agrobacterium tumefaciens* and transposable elements have been adapted as insertional mutagens. In various cases a direct correlation of a knockout phenotype with gene function can be made. Large collections of insertion lines compiling more than 100,000 insertions exactly positioned within the genome are available for *Arabidopsis* (http://signal.salk.edu/tabout.html) (33). These lines were generated by T-DNA insertion; the position of a single insertion for each line was determined by sequencing the genomic region flanking the insertion. Lines harboring T-DNAs on genes of interest can be

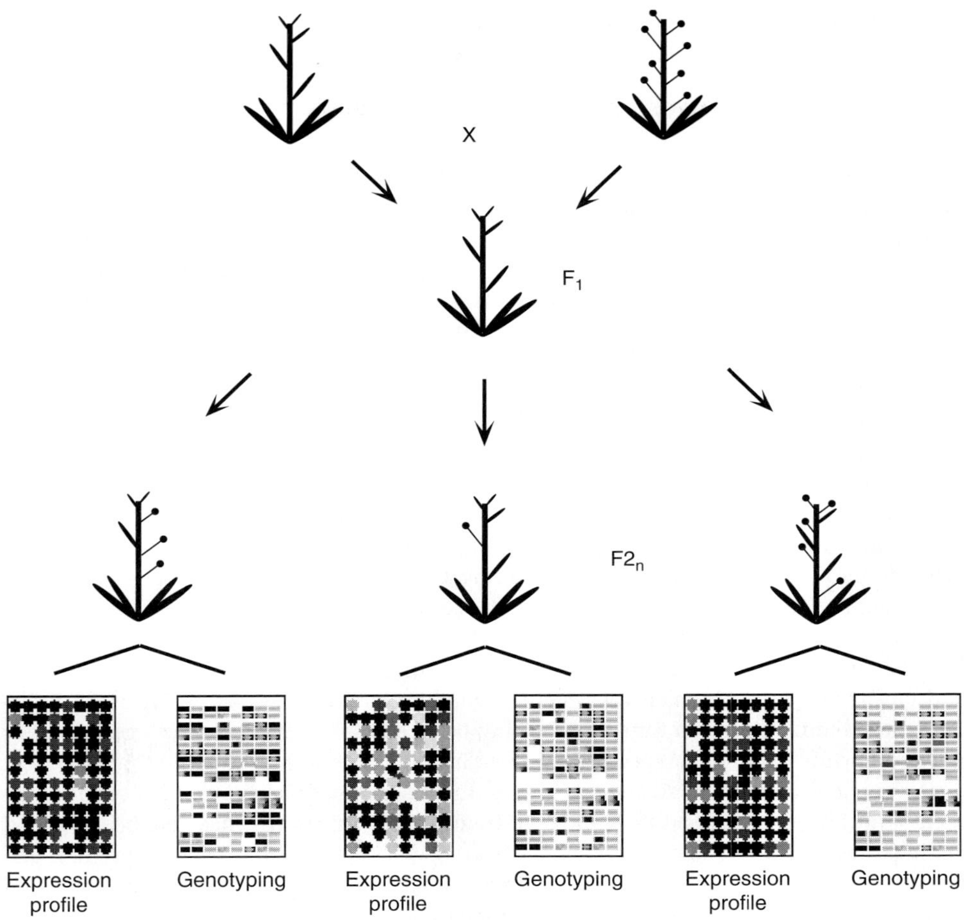

Figure 3.5 Genetical genomics approach. The F1 generation of the cross of two plant lines is generated, and subsequently, an F2 population (sufficiently large to carry out statistical analysis) is obtained. By correlating phenotypical variation with molecular genotyping, any trait that is subject to quantification can be tracked down to a chromosomal location. Likewise, variation in gene expression for each individual in the population can be monitored. Thus, specific chromosomal locations that correlate with altered gene expression are revealed. In each individual, the left panel represents expression profiles generated by microarray analysis and the right panel shows genotyping analysis.

readily identified in databases. Although these collections of lines are valuable tools for gene discover in plants, they have some limitations; members of gene families often do not display scorable phenotypes, hindering gene function prediction.

Because a T-DNA insertional library represents a random library, the estimated 25,000 genes in *Arabidopsis* (with a fourfold redundancy) require approximately 100,000 insertional mutants to cover most of the genome. In principle, the number of insertional mutants should be proportional to the genome size, not the number of genes. If the process of generating insertional mutants is random, and, if on average there is a gene every 5 kb in the plant genome, then one needs 24,000 mutants for *Arabidopsis*, which has a genome size of 1.2×10^8 bp. To allow a margin of safety, obtaining a tenfold redundancy means 240,000 mutants. Using the same calculations, one needs 800,000 mutants for rice (genome size 4×10^8 bp) and 4,800,000 for maize (genome size 2.4×10^9 bp).

3.5 PROTEOMICS: SYSTEMATIC AND HIGH THROUGHPUT CHARACTERIZATION OF GENE PRODUCTS

Analysis of mRNA transcripts on microarrays has provided information regarding which genes are expressed in cells under a given set of experimental conditions, yielding clues as to which proteins are involved in certain pathways. However, differences in the half-lives of RNA and proteins, as well as posttranslational modifications important to protein function (such as phosphorylation, glycosylation, or cleavage of an N- or C-terminal peptide) prevent mRNA profiles from being perfectly correlated to the actual protein profiles. In a given organism, the set of gene products encoded in the genome are called the proteome and "proteomics" would be the counterpart to genomics, aimed at the high throughput characterization of the proteome.

Protein analysis at a global scale is more technically demanding than genomics. For instance, mRNAs are chemically homogeneous molecules and thus a single method is useful for the identification of every mRNA. Protein separation, in contrast, is complex, given the different sizes and physical–chemical properties of proteins. The systematic analysis of expressed proteins requires three key technologies: the first is used for separating the proteins, and is most commonly two dimensional polyacrylamide electrophoresis; the second is a high throughput method of protein identification, which is afforded by mass spectrometry; and the third comprises the bioinformatics tools for the analysis and storage of data.

Proteomics is in a very early stage of advance in plant species and most of the work is now being carried out in *A. thaliana*. Many *A. thaliana* proteins have been characterized and localized by a two dimensional gel system. The usefulness of the analysis of differences in protein expression for distinguishing between closely related genotypes has been pointed out for many plant species (34,35). This approach was used to classify different developmental mutants of *Arabidopsis*, with the discrimination obtained being in agreement with phenotypical and physiological traits (36). *Oryza sativa* has also been studied at the proteome level. The rice genome has been used as a source for mapping proteins and linking genomic with proteomic data for function analysis (37–39). There is an increasing number of reports dealing with the use of proteomic techniques to identify proteins from specific tissues and developmental stages, such as rice anther young microspore (40), proteins related to abiotic stress (41), and proteins expressed during plant–microbe interactions in rice (42). Protein reference maps of subproteomes of different plant species obtained by two dimensional gel electrophoresis are expected to become a central tool for cataloguing and analyzing plant proteomes.

Most proteins do not function alone; rather they are components of protein complexes and networks. In many cases, knowing the proteins with which an unknown protein interacts helps to understand its function. Recently, three methods (the yeast two-hybrid system, protein microarrays, and mass spectrometry) have reached a proteomic scale to examine protein networks (43). Since its invention, the yeast two-hybrid system has been successfully used to identify protein–protein interactions of proteins from several organisms, including plants. This system is based on the reassembly of an active transcription factor in yeast by the interaction of two domains carried by two different recombinant proteins (each of which is therefore a hybrid protein). The hybrids are each coupled either to the DNA binding domain or to the activation domain of a yeast transcription factor. This reassembled transcription factor triggers the expression of a gene, usually a reporter gene that is easy to detect or to select for. Automation of the process and the generation of arrays of yeast strains expressing fusion proteins have permitted the analysis of thousands of interactions in a single experiment. A network of 2,358 interactions among 1,548 proteins has been established in the yeast *S. cerevisiae*. It was possible to predict the functional category for 72% of the 1,393 known yeast proteins and to infer a possible function for 364 unknown proteins (44). Likewise, maps of protein–protein interaction networks during vulval development and DNA damage response in *C. elegans* have been obtained (45,46).

Mass spectrometry is a strategy that is also used to identify protein–protein interactions. A protein of interest is the affinity reagent used to purify a protein complex that binds to it. Compared with two-hybrid and chip-based approaches, this strategy has the advantage that the fully processed and modified protein is used for the binding, that the interactions take place in the native environment and cellular location, and that multicomponent complexes can be isolated and analyzed in a single operation (47). Mass spectrometry-based protein interaction experiments have three essential components: presentation of the protein used to trap the interacting proteins, affinity purification of the complex, and analysis of the bound proteins. Ideally, endogenous proteins can serve as trap or bait if an antibody or other reagent exists that allows specific isolation of the protein with its bound partners. A more generic strategy is to tag the proteins of interest with a sequence readily recognized by an antibody specific for the tag.

With the advent of high throughput molecular biology, it is now possible to prepare large, normalized collections of cloned genes. Recently, protein arrays are emerging as a new tool for the profiling and functional characterization of recombinant proteins encoded by globally or differentially expressed cDNA clones. This technology implies the cloning of large numbers of cDNA clones in appropriate expression vectors or the throughput subcloning of open reading frames (48). Protein arrays (or protein chips) are constructed by spotting many hundreds of capture molecules onto a solid surface, which can then be used to interrogate a particular sample for protein content. A critical first step in generating these arrays has been the development of general methods for arraying a genomic set of proteins on a solid surface without denaturing the proteins, and at high enough density for detection of activity. It was demonstrated that a single protein could be detected at high resolution on a single glass slide in the midst of 10,799 identical spots of another protein, showing the huge potential of protein microarrays for parallel biochemical analysis (49).

Peptides can also be analyzed using microarrays. Synthesis of peptide microarrays may become more practical with the development of methods for *in situ* synthesis of high density peptide microarrays, using photolithography. Carbohydrate and small molecule microarrays have also shown great potential for characterizing proteins to small molecule binding activities (43). Another type of microarray is the analytical microarray, in which a genomic set of protein-specific ligands such as antibodies, nucleic acid aptamers, or chemical probes are spotted on a microarray, and then the levels of different proteins in an extract are quantified in parallel by binding protein extracts to the microarray. Protein microarray analysis in plants is at an early stage of development. However, work is in progress to gain access to this revolutionary technology; for example, using the ORF approach for cloning different *Arabidopsis* genes into a specialized *E. coli* expression vector, for the synthesis of recombinant proteins to generate *Arabidopsis* protein chips for future applications (48).

3.6 IMPACT OF GENOMICS ON NUTRITION

Scientific innovation in biotechnology holds great promise for meeting the food needs of a growing world population and for protecting human health through novel strategies to improve the nutritional qualities of plant crops. In this century, there is urgent need to increase food production without expansion of cultivable lands. The increase in food supply is possible by reducing the losses caused by pests and diseases, changing environments and postharvest losses. Genomic sciences can assist both "classical" and transgenic approaches to improve crops. These technologies will also impact areas such as food quality and safety.

The goal of genomic techniques is to identify the key biomolecules controlling a specific biological process; theoretically if all mRNAs (genomics), proteins (proteomics), or metabolites (metabolomics) that respond to environmental changes are identified; subsequently, differences and potential links between the data can be identified through bioinformatics. Thus, the major tasks for genomics research over at least the next 25–50 years will be to correlate data on the DNA sequence of each gene and its protein product with biological functions. One of the postgenome challenges will be interpreting the vast amount of new data. Scientific knowledge obtained on the biology of plant cells, tissues, and organization pathways is expected to increase tremendously, and thus novel strategies and new technologies are expected to arise.

The main impact that genomics techniques will have on classical breeding is to accelerate it and make it more accurate. Transgenic technology is not ready to handle the important quantitative traits, and genetic transformation of some crops is still challenging. Molecular techniques offer a more effective way to manipulate germ plasm via marker-assisted selection for elucidating genetic bases of traits with complex inheritances, identifying optimal germ plasm for specific breeding objectives and monitoring response to selection and breeding progress. The knowledge of the genome organization and the assignment of special functions will facilitate the identification of plant material with a desired combination of traits.

Genetic engineering involves taking genes from one species and inserting them into another in an attempt to transfer a desired trait or character. Scientists have introduced genes taken from bacteria, viruses, insects, animals, or even humans into plants. Crops resistant to pests or weeds, more tolerant to salt or drought, or that produce foods that are tastier or more nutritious have been genetically engineered. Plants have been also modified to produce specific compounds, such as industrial oils, plastics, enzymes, and even drugs and vaccines. There is abundant literature that refers to the techniques involved in producing transgenics (50). Transgenic technology will benefit by incorporating the desired gene(s) identified by genome analysis. Moreover, transgenic technology will itself develop into a more precise and controlled technology. EST and microarray technologies are being used to identify promoters that meet specific quantitative, temporal, tissue specific, or cell specific gene expression requirements for a particular trait. Synthetic promoters and novel transcription factors are being designed to increase gene expression levels. Efforts are also directed at stabilizing and enhancing gene expression and to reducing effects related to the insertion site of a transgene. Chemically regulated promoter systems have also been used to regulate gene expression (51). Additionally, because there is so much concern about possible unintended effects of transgenic plants, the use of genomics, proteomics, and metabolomics provides an ample, nonbiased screen to assess any possible changes associated with the recombinant plant.

Genomics has the potential to greatly enrich the availability and quality of food supplies. Thousands of different plant species are grown for food, and many more plants that are not even cultivated are known to produce secondary metabolites of medicinal, nutritive, or commercial interest. It is not feasible, however, to carry out the sequencing and expression profiling of every crop of interest. The challenge to agricultural plant scientists is to exploit the databases being generated for rice, maize, and *Arabidopsis* toward the genetic improvement of nonmodel crop species.

From a dietary standpoint, rice alone is the main staple food for half of the world population, while pearl millet, sorghum, maize, or wheat sustain an important portion of the food requirements of the remaining half. With the genomic sequence of rice already available, the next logical step is to capitalize on this information using the rice plant as a model system to develop strategies and tools for quickly isolating genes in cereals, deciphering their function, and determining their potential role in improving food security and nutritional value. As genes become associated with specific functions or traits in rice, it will be feasible to search the database to identify orthologs governing the same trait in other cereals. To this end, a powerful and so far underutilized approach is offered by comparative genomics based on gene colinearity (52). The comparative analysis across cereals of mutants and QTLs influencing key traits will allow us to identify a number of genomic targets for the implementation of more detailed studies profiling the proteome and the metabolome (53). Collectively, these studies will lead to the identification and cloning of a number of genes controlling the traits of interest, many of which will target different types of abiotic and biotic stresses, growth, and composition of the grain.

Genomics is already having great impact in providing "fingerprinting" of plant materials that may be used to determine, from the species and cultivars of given plants, to the quality and authenticity of foods, whether a given plant product is or derives from a GMO. These will be easily and accurately determined with microarray techniques. Functional proteomics will aid the identification of marker proteins that can be used as a screening tool, such as indicators of plant conditions and ageing, indicators of genomic traits (e.g., disease resistance, bending of the stem of flowers, abiotic plant or root stress, plant growth), and quality indicators (e.g., for screening of raw materials or development of agrochemicals).

Microbial genomics is an area necessarily linked to food production. Novel molecular genetic methods are being applied to detect, identify, and classify mycotoxigenic fungi, plant pathogens, food-borne pathogenic bacteria, food spoilage microorganisms, fungi, and bacteria used for biological control of agricultural pests, and microorganisms used for conversion of raw agricultural materials into value added products. Whole genome information can be used to improve industrial and food-related microorganisms. In many instances analyzing the components of the growth media to make secondary metabolite production a more reproducible process can identify critical parameters. Also, in complex fermentations containing different microorganisms and complex substrates, quality of the proteome or metabolome of the starter culture can be used to predict the quality of the fermented end product. For example, as ripening is often a long process, enormous cost savings can be achieved if "bad" starter culture batches can be identified and eliminated beforehand.

In nutritional genomics a main focus is the identification and isolation of genes that are required for the synthesis and accumulation of a target compound desired in human diet. Research has mainly concentrated on vitamins (like folate and vitamins E, B6, and A) and minerals (like iron, calcium, selenium, and iodine). In the past, breeding programs largely overlooked the micronutrient composition and density of crops. For many plant secondary metabolites, including those of nutritional importance to humans, gene identification by classical biochemical approaches has given way to molecular genetic approaches. For example *Arabidopsis* mutants exhibiting altered production of carotenoids, flavanoids, tocopherols, and ascorbic acid have been used to establish the genetic basis for their synthesis. Research for health-promoting secondary metabolites, such as anticarcenogenic or infection-inhibiting compounds, has not reached the state of development of micronutrients because the best known plants produce those compounds in very small amounts or not at all (54).

Finally, for agribusiness companies, insect genomics are another point of interest. Insect genomics will be very useful for the development of new chemical pesticides or other biological control agents for insects. Just as *Arabidopsis* is the model system of plant genomics, *D. melanogaster* represents the model for insect genomics. The whole sequence of the genome is known (5) and tools to manipulate gene expression have been developed.

REFERENCES

1. Sanger, F., S. Nicklen, A.R. Coulson. DNA sequencing with chain-terminating inhibitors. *Proc. Natl. Acad. Sci. USA* 74:5463–5467, 1977.
2. Monaco, A.P., Z. Larin. YACs, BACs, PACs and MACs: artificial chromosomes as research tools. *Trends Biotechnol.* 12:280–286, 1994.
3. Mewes, H.W., K. Albermann, M. Bahr, D. Frishman, A. Gleissner, J. Hani, K. Heumann, K. Kleine, A. Maierl, S.G. Oliver, F. Pfeiffer, A. Zollner. Overview of the yeast genome. *Nature* 387:7–65, 1997.
4. The C. elegans Sequencing Consortium. Genome sequence of the nematode *C. elegans*: a platform for investigating biology. *Science* 282:2012–2018, 1998.
5. Adams, M.D., S.E. Celniker, R.A. Holt, C.A. Evans, J.D. Gocayne, P.G. Amanatides, S.E. Scherer, P.W. Li, R.A. Hoskins, R.F. Galle, R.A. George, S.E. Lewis, S. Richards, M. Ashburner, S.N. Henderson, G.G. Sutton, J.R. Wortman, M.D. Yandell, Q. Zhang, L.X. Chen, R.C. Brandon, Y.H. Rogers, R.G. Blazej, M. Champe, B.D. Pfeiffer, K.H. Wan, C. Doyle, E.G. Baxter, G. Helt, C.R. Nelson, G.L. Gabor, J.F. Abril, A. Agbayani, H.J. An, C. Andrews-Pfannkoch, D. Baldwin, R.M. Ballew, A. Basu, J. Baxendale, L. Bayraktaroglu, E.M. Beasley, K.Y. Beeson, P.V. Benos, B.P. Berman, D. Bhandari, S. Bolshakov, D. Borkova, M.R. Botchan, J. Bouck, P. Brokstein, P. Brottier, K.C. Burtis, D.A. Busam, H. Butler, E. Cadieu, A. Center, I. Chandra, J.M. Cherry, S. Cawley, C. Dahlke, L.B. Davenport, P. Davies, B. de Pablos, A. Delcher, Z. Deng, A.D. Mays, I. Dew, S.M. Dietz, K. Dodson, L.E. Doup, M. Downes, S. Dugan-Rocha, B.C. Dunkov, P. Dunn, K.J. Durbin, C.C. Evangelista, C. Ferraz, S. Ferriera, W. Fleischmann, C. Fosler, A.E. Gabrielian, N.S. Garg, W.M. Gelbart, K. Glasser, A. Glodek, F. Gong, J.H. Gorrell, Z. Gu, P. Guan, M. Harris, N.L. Harris, D. Harvey, T.J. Heiman, J.R. Hernandez, J. Houck, D. Hostin, K.A. Houston, T.J. Howland, M.H. Wei, C. Ibegwam, M. Jalali, F. Kalush, G.H. Karpen, Z. Ke, J.A. Kennison, K.A. Ketchum, B.E. Kimmel, C.D. Kodira, C. Kraft, S. Kravitz, D. Kulp, Z. Lai, P. Lasko, Y. Lei, A.A. Levitsky, J. Li, Z. Li, Y. Liang, X. Lin, X. Liu, B. Mattei, T.C. McIntosh, M.P. McLeod, D. McPherson, G. Merkulov, N.V. Milshina, C. Mobarry, J. Morris, A. Moshrefi, S.M. Mount, M. Moy, B. Murphy, L. Murphy, D.M. Muzny, D.L. Nelson, D.R. Nelson, K.A. Nelson, K. Nixon, D.R. Nusskern, J.M. Pacleb, M. Palazzolo, G.S. Pittman, S. Pan, J. Pollard, V. Puri, M.G. Reese, K. Reinert, K. Remington, R.D. Saunders, F. Scheeler, H. Shen, B.C. Shue, I. Siden-Kiamos, M. Simpson, M.P. Skupski, T. Smith, E. Spier, A.C. Spradling, M. Stapleton, R. Strong, E. Sun, R. Svirskas, C. Tector, R. Turner, E. Venter, A.H. Wang, X. Wang, Z.Y. Wang, DA Wassarman, G.M. Weinstock, J. Weissenbach, S.M. Williams, T. Woodage, K.C. Worley, D. Wu, S. Yang, Q.A. Yao, J. Ye, R.F. Yeh, J.S. Zaveri, M. Zhan, G. Zhang, Q. Zhao, L. Zheng, X.H. Zheng, F.N. Zhong, W. Zhong, X. Zhou, S. Zhu, X. Zhu, H.O. Smith, R.A. Gibbs, E.W. Myers, G.M. Rubin, J.C. Venter. The genome sequence of *Drosophila melanogaster. Science* 287:2185–2195, 2000.
6. The Arabidopsis Genome Initiative. Analysis of the genome sequence of the flowering plant *Arabidopsis thaliana. Nature* 408:796–815, 2000.
7. Goff, S.A., D Ricke, T.H. Lan, G. Presting, R. Wang, M. Dunn, J. Glazebrook, A. Sessions, P. Oeller, H. Varma, D. Hadley, D. Hutchison, C. Martin, F. Katagiri, B.M. Lange, T. Moughamer, Y. Xia, P. Budworth, J. Zhong, T. Miguel, U. Paszkowski, S. Zhang, M. Colbert, W.L. Sun, L. Chen, B. Cooper, S. Park, T.C. Wood, L. Mao, P. Quail, R. Wing, R. Dean, Y. Yu, A. Zharkikh, R. Shen, S. Sahasrabudhe, A. Thomas, R. Cannings, A. Gutin, D. Pruss, J. Reid, S. Tavtigian, J. Mitchell, G. Eldredge, T. Scholl, R.M. Miller, S. Bhatnagar, N. Adey, T. Rubano, N. Tusneem, R. Robinson, J. Feldhaus, T. Macalma, A. Oliphant, S. Briggs. A draft sequence of the rice genome (*Oryza sativa* L. ssp. japonica). *Science* 296:92–100, 2002.
8. Yu, J., S. Hu, J. Wang, G.K. Wong, S. Li, B. Liu, Y. Deng, L. Dai, Y. Zhou, X. Zhang, M. Cao, J. Liu, J. Sun, J. Tang, Y. Chen, X. Huang, W. Lin, C. Ye, W. Tong, L. Cong, J. Geng, Y. Han, L. Li, W. Li, G. Hu, J. Li, Z. Liu, Q. Qi, T. Li, X. Wang, H. Lu, T. Wu, M. Zhu, P. Ni, H. Han, W. Dong, X. Ren, X. Feng, P. Cui, X. Li, H. Wang, X. Xu, W. Zhai, Z. Xu, J. Zhang, S. He, J. Xu, K. Zhang, X. Zheng, J. Dong, W. Zeng, L. Tao, J. Ye, J. Tan, X. Chen,

J. He, D. Liu, W. Tian, C. Tian, H. Xia, Q. Bao, G. Li, H. Gao, T. Cao, W. Zhao, P. Li, W. Chen, Y. Zhang, J. Hu, S. Liu, J. Yang, G. Zhang, Y. Xiong, Z. Li, L. Mao, C. Zhou, Z. Zhu, R. Chen, B. Hao, W. Zheng, S. Chen, W. Guo, M. Tao, L. Zhu, L. Yuan, H. Yang. A draft sequence of the rice genome (*Oryza sativa* L. ssp. indica). *Science* 296:79–92, 2002.

9. Riechmann, J.L., J. Heard, G. Martin, L. Reuber, C. Jiang, J. Keddie, L. Adam, O. Pineda, O.J. Ratcliffe, R.R. Samaha, R. Creelman, M. Pilgrim, P. Broun, J.Z. Zhang, D. Ghandehari, B.K. Sherman, G. Yu. *Arabidopsis* transcription factors: genome-wide comparative analysis among eukaryotes. *Science* 290:2105–2110, 2000.
10. VandenBosch, K.A., G. Stacey. Summaries of legume genomics projects from around the globe: community resources for crops and models. *Plant Physiol.* 131:840–865, 2003.
11. Rabinowicz, P.D., K. Schutz, N. Dedhia, C. Yordan, L.D. Parnell, L. Stein, W.R. McCombie, R.A. Martienssen. Differential methylation of genes and retrotransposons facilitates shotgun sequencing of the maize genome. *Nat. Genet.* 23:305–308, 1999.
12. Lunde, C.F., D.J. Morrow, L.M. Roy, V. Walbot. Progress in maize gene discovery: a project update. *Funct. Integr. Genomics* 3:25–32, 2003.
13. Clarke, B., M. Lambrecht, S.Y. Rhee. *Arabidopsis* genomic information for interpreting wheat EST sequences. *Funct. Integr. Genomics* 3:33–38, 2003.
14. Schena, M., D. Shalon, R.W. Davis, P.O. Brown. Quantitative monitoring of gene expression patterns with a complementary DNA micro array. *Science* 270:467–470, 1995.
15. McGall, G.H., F.C. Christians. High-density genechip oligonucleotide probe arrays. *Adv. Biochem. Eng. Biotechnol.* 77:21–42, 2002.
16. Lou, X.J., M. Schena, F.T. Horrigan, R.M. Lawn, R.W. Davis. Expression monitoring using cDNA micro arrays: a general protocol. *Methods Mol. Biol.* 175:323–340, 2001.
17. Yang, Y.H., T. Speed. Design issues for cDNA micro array experiments. *Nat. Rev. Genet.* 3:579–588, 2002.
18. Aharoni, A., L.C. Keizer, H.J. Bouwmeester, Z. Sun, M. Alvarez-Huerta, H.A. Verhoeven, J. Blaas, A.M. van Houwelingen, R.C. De Vos, H. van der Voet, R.C. Jansen, M. Guis, J. Mol, R.W. Davis, M. Schena, A.J. van Tunen, A.P. O'Connell. Identification of the SAAT gene involved in strawberry flavor biogenesis by use of DNA micro arrays. *Plant Cell.* 12:647–662, 2000.
19. Aharoni, A., L.C. Keizer, H.C. Van Den Broeck, R. Blanco-Portales, J. Munoz-Blanco, G. Bois, P. Smit, R.C. De Vos, A.P. O'Connell. Novel insight into vascular, stress, and auxin-dependent and -independent gene expression programs in strawberry, a non-climacteric fruit. *Plant Physiol.* 129:1019–1031, 2002.
20. Aharoni, A., A.P. O'Connell. Gene expression analysis of strawberry achene and receptacle maturation using DNA micro arrays. *J. Exp. Bot.* 53:2073–2087, 2002.
21. Jones, P.G., D. Allaway, D.M. Gilmour, C. Harris, D. Rankin, E.R Retzel, C.A. Jones. Gene discovery and micro array analysis of cacao (*Theobroma cacao L.*) varieties. *Planta* 216:255–264, 2002.
22. Ruuska, S.A., T. Girke, C. Benning, J.B. Ohlrogge. Contrapuntal networks of gene expression during *Arabidopsis* seed filling. *Plant Cell.* 14:1191–1206, 2002.
23. Rossel, J.B., I.W. Wilson, B.J. Pogson. Global changes in gene expression in response to high light in *Arabidopsis. Plant Physiol.* 130:1109–1120, 2002.
24. Seki, M., M. Narusaka, J. Ishida, T. Nanjo, M. Fujita, Y. Oono, A. Kamiya, M. Nakajima, A. Enju, T. Sakurai, M. Satou, K. Akiyama, T. Taji, K. Yamaguchi-Shinozaki, P. Carninci, J. Kawai, Y. Hayashizaki, K. Shinozaki. Monitoring the expression profiles of 7000 *Arabidopsis* genes under drought, cold and high-salinity stresses using a full-length cDNA micro array. *Plant* J. 31:279–292, 2002.
25. Moseyko, N., T. Zhu, H.S. Chang, X. Wang, L.J. Feldman. Transcription profiling of the early gravitropic response in *Arabidopsis* using high-density oligonucleotide probe micro arrays. *Plant Physiol.* 130:720–728, 2002.
26. Cheong, Y.H., H.S. Chang, R. Gupta, X. Wang, T. Zhu, S. Luan. Transcriptional profiling reveals novel interactions between wounding, pathogen, abiotic stress, and hormonal responses in *Arabidopsis. Plant Physiol.* 129:661–677, 2002.

27. Fowler, S., M.F. Thomashow. *Arabidopsis* transcriptome profiling indicates that multiple regulatory pathways are activated during cold acclimation in addition to the CBF cold response pathway. *Plant Cell.* 14:1675–1690, 2002.
28. Tao, Y., Z. Xie, W. Chen, J. Glazebrook, H.S. Chang, B. Han, T. Zhu, G. Zou, F. Katagiri. Quantitative nature of *Arabidopsis* responses during compatible and incompatible interactions with the bacterial pathogen Pseudomonas syringae. *Plant Cell.* 15:317–330, 2003.
29. Goda, H., Y. Shimada, T. Asami, S. Fujioka, S. Yoshida. Micro array analysis of brassinosteroid-regulated genes in *Arabidopsis. Plant Physiol.* 130:1319–1334, 2002.
30. Cho, Y., J. Fernandes, S.H. Kim, V. Walbot. Gene-expression profile comparisons distinguish seven organs of maize. *Genome Biol.* 3:1–16, 2002.
31. Hunter, B.G., M.K. Beatty, G.W. Singletary, B.R. Hamaker, B.P. Dilkes, B.A. Larkins, R. Jung. Maize opaque endosperm mutations create extensive changes in patterns of gene expression. *Plant Cell.* 14:2591–2612, 2002.
32. Nakazono, M., F. Qiu, L.A. Borsuk, P.S. Schnable. Laser-capture microdissection, a tool for the global analysis of gene expression in specific plant cell types: identification of genes expressed differentially in epidermal cells or vascular tissues of maize. *Plant Cell.* 15:583–596, 2003.
33. Sessions, A., E. Burke, G. Presting, G. Aux, J. McElver, D. Patton, B. Dietrich, P. Ho, J. Bacwaden, C. Ko, J.D. Clarke, D. Cotton, D. Bullis, J. Snell, T. Miguel, D. Hutchison, B. Kimmerly, T. Mitzel, F. Katagiri, J. Glazebrook, M. Law, S.A. Goff. A high-throughput *Arabidopsis* reverse genetics system. *Plant Cell* 14:2985–2994, 2002.
34. Thiellement, H., N. Bahrman, C. Damerval, C. Plomion, M. Rossignol, V. Santoni, D. de Vienne, M. Zivy. Proteomics for genetic and physiological studies in plants. *Electrophoresis* 20:2013–2026, 1999.
35. Thiellement, H., M. Zivy, C. Plomion. Combining proteomic and genetic studies in plants. *J. Chromatogr. B. Analyt. Technol. Biomed. Life Sci.* 782:137–149, 2002.
36. Santoni, V., M. Delarue, M. Caboche, C. Bellini. A comparison of two-dimensional electrophoresis data with phenotypical traits in *Arabidopsis* leads to the identification of a mutant (cri1) that accumulates cytokinins. *Planta* 202:62–69, 1997.
37. Hirano, H. Screening of rice genes from the cDNA catalog using the data obtained by protein sequencing. *J. Protein Chem.* 16:533–536, 1997.
38. Komatsu, S., A. Muhammad, R. Rakwal. Separation and characterization of proteins from green and etiolated shoots of rice (*Oryza sativa* L.): towards a rice proteome. *Electrophoresis* 20:630–636, 1999.
39. Tsugita, A., M. Kamo, T. Kawakami, Y. Ohki. Two-dimensional electrophoresis of plant proteins and standardization of gel patterns. *Electrophoresis* 17:855–865, 1996.
40. Imin, N., T. Kerim, J.J. Weinman, B.G. Rolfe. Characterisation of rice anther proteins expressed at the young microspore stage. *Proteomics* 1:1149–1161, 2001.
41. Rakwal, R., G.K. Agrawal, M. Yonekura. Separation of proteins from stressed rice (*Oryza sativa* L.) leaf tissues by two-dimensional polyacrylamide gel electrophoresis: induction of pathogenesis-related and cellular protectant proteins by jasmonic acid, UV irradiation and copper chloride. *Electrophoresis* 20:3472–3478, 1999.
42. Konishi, H., K. Ishiguro, S. Komatsu. A proteomics approach towards understanding blast fungus infection of rice grown under different levels of nitrogen fertilization. *Proteomics* 1:1162–1171, 2001.
43. Phizicky, E., P.I. Bastiaens, H. Zhu, M. Snyder, S. Fields. Protein analysis on a proteomic scale. *Nature* 422:208–215, 2003.
44. Schwikowski, B., P. Uetz, S. Fields. A network of protein-protein interactions in yeast. *Nat. Biotechnol.* 18:1257–1261, 2000.
45. Walhout, A.J., R. Sordella, X. Lu, J.L. Hartley, G.F. Temple, M.A. Brasch, N. Thierry-Mieg, M. Vidal. Protein interaction mapping in C. elegans using proteins involved in vulval development. *Science* 287:116–122, 2000.
46. Boulton, S.J., A. Gartner, J. Reboul, P. Vaglio, N. Dyson, D.E. Hill, M. Vidal. Combined functional genomic maps of the C. elegans DNA damage response. *Science* 295:127–131, 2002.

47. Aebersold, R., M. Mann. Mass spectrometry-based proteomics. *Nature* 422:198–207, 2003.
48. Kersten, B., L. Burkle, E.J. Kuhn, P. Giavalisco, Z. Konthur, A. Lueking, G. Walter, H. Eickhoff, U. Schneider. Large-scale plant proteomics. *Plant Mol. Biol.* 48:133–141, 2002.
49. MacBeath, G., S.L. Schreiber. Printing proteins as micro arrays for high-throughput function determination. *Science* 289:1760–1763, 2000.
50. Stewart, S.N. *Transgenic Plants: Current Innovations and Future Trends.* Knoxville: Sci Press, 2003.
51. Bohner, S., I.I. Lenk, M. Rieping, M. Herold, C. Gatz. Technical advance: transcriptional activator TGV mediates dexamethasone-inducible and tetracycline-inactivatable gene expression. *Plant J.* 19:87–95, 1999.
52. Laurie, D.A., K.M. Devos. Trends in comparative genetics and their potential impacts on wheat and barley research. *Plant Mol. Biol.* 48:729–740, 2002.
53. Consoli, L., A. Lefevre, M. Zivy, D. de Vienne, C. Damerval. QTL analysis of proteome and transcriptome variations for dissecting the genetic architecture of complex traits in maize. *Plant Mol. Biol.* 48:575–581, 2002.
54. DellaPenna, D. Nutritional genomics: manipulating plant micronutrients to improve human health. *Science* 285:375–379, 1999.

2.04

Molecular Design of Soybean Proteins for Enhanced Food Quality

Nobuyuki Maruyama, Evelyn Mae Tecson-Mendoza, Yukie Maruyama, Motoyasu Adachi, and Shigeru Utsumi

CONTENTS

4.1 INTRODUCTION

Generally, the protein contents of seeds are high; those of legume seeds and cereal seeds range from ~20% to 40% and from ~7% to 15%, respectively (1). Protein contents of soybean seeds are especially high, and those of some cultivars are more than 40%. Thus, they play an important role as a protein resource for human and domestic animals. In addition, soybean seeds are oil seeds, with oil content comprising ~20% of total weight. The annual production of soybean seeds in the world is ~180 million tons with more than 80% being used for oil expression. The amount of proteins in the residues after oil expression reaches around 58 million tons. The residues after oil expression have a history of safe use as food material for many years. Most of them are used as feed for domestic animals and as fertilizer. They are very cheap (U.S. $1.2/10kg). Even isolated soybean proteins extracted from the residues are cheap (700–850 yen/kg or U.S. $6–8/kg).

Fuji Oil in Japan and Dupont in U.S.A. have endeavored to develop new foods for which soybean proteins are utilized, but their applications have been limited due to inadequate physicochemical properties of soybean proteins. Food proteins should have specific nutritional quality, palatability, physicochemical functions, safety, economical efficiency, and physiological properties as food material (1). Nutritional quality is a property determined by the amino acids, especially the essential amino acids, of the protein. Soybean proteins are deficient in sulfur containing amino acids. Nutritional quality is also influenced by the digestibility of the protein. On the other hand, palatability is determined by taste and sensory quality (texture and hardness). Generally, animal proteins are superior to plant proteins in terms of palatability.

Physicochemical properties are properties such as gelation, emulsification, foaming, and water absorption properties and are determined by structural features of soybean proteins. Thus, the physicochemical properties unique to soybean make it possible to produce traditional foods such as tofu, koori-tofu, and ganmodoki from soybeans, and to use soybean proteins as a modifier for chilled food and sausage.

The physiological property of a protein can help maintain and enhance human health. It is well known that ingestion of 6 g of soybean proteins per day lowers serum cholesterol levels in humans (2). Recently, it has been reported that soybean proteins also lower serum triglyceride levels in humans (3). Soybean proteins, therefore, are capable of improving lipid metabolism. The soybean Bowman-Birk protease inhibitor has been reported to be capable of suppressing cancer of the colon (4), prostrate cancer (5,6), oral cancers, and cancer of the head and neck (7). A soybean peptide called *lunasin* has been isolated and shown to have antimitotic activity (8,9) and is a possible chemopreventive agent (10). While some soybean proteins may have excellent physiological or biological activities, in general, the nutritional quality and physicochemical functions of soybean proteins are inferior to those of animal proteins.

Recently, in Japan and in many developed countries, health problems related to obesity, such as hypertension and heart disease, have increased. Aging of the population has also progressed. In many developing countries, the problems of lack of food and protein malnutrition exist, and legumes like soybeans are the major source of proteins. Therefore, it is desirable to develop foods that will address these problems. It is estimated that world population will be about 8.4 billion around 2070, and food production at the present rate will be insufficient to feed the growing population. Therefore, this situation demands an increase of food production at a worldwide scale.

Soybeans could serve as a major source of proteins for the world's growing population. Soybean proteins have excellent physiological properties, are cheap, and can be produced in large quantities. Further, their physicochemical and functional properties can be improved. Protein engineering provides a means to improve the properties of soybean proteins by modifying their primary structures through gene manipulation. The ultimate objective, of course, is to develop crops in which soybean proteins with enhanced food quality accumulate. As a requisite, modification by protein engineering for enhanced quality must not disturb the folding ability of proteins. If the modified proteins cannot fold correctly, they may be degraded during the processes of biosynthesis, transportation and accumulation, and may therefore not accumulate in seeds (11,12). Thus, it is most important to understand the structures of soybean proteins.

This review will discuss the structures and improvement of nutritional quality, physicochemical functions, and physiological properties of soybean proteins, with emphasis on seed storage proteins, by protein engineering.

4.2 SOYBEAN PROTEINS

Most soybean proteins are seed storage proteins. Globulins, salt soluble proteins, are the primary storage proteins. The globulins of legume seeds are classified into two types according to their sedimentation coefficients; the 7S globulin and 11S globulin (13). Both account for approximately 80% of soybean proteins. The ratio of 11S and 7S globulins varies from 0.5 to 1.7 among soybean cultivars (14). This ratio determines the nutritional quality and physicochemical functions of soybean proteins. Aside from the 7S and 11S globulins, γ-conglycinin and basic 7S globulin also exist in soybean proteins.

4.2.1 β-Conglycinin

The 7S globulin of soybean is called β-conglycinin. β-conglycinin is a glycoprotein with a trimeric structure composed of three subunits and has a molecular mass of 150–200 kDa. The constituent subunits of β-conglycinin are α, α', and β. The α and α' subunits are composed of an extension region and a core region, while the β subunit consists of only the core region (15). The core regions of three subunits exhibit high absolute homologies (90.4%, 76.2%, and 75.5% between α and α', between α and β, and between α' and β, respectively). The extension regions of the α and α' subunits exhibit lower absolute homologies (57.3%) and a highly acidic property. The α and α' subunits are glycosylated at two sites (α, Asn199 and Asn455; α', Asn215 and Asn471), and the β subunit is glycosylated at one site (Asn328) (16). The N-glycosylation site of the β subunit corresponds to the latter site of the α and α' subunits. The α and α' subunits and the β subunit are synthesized on polysomes as preproform and preform, respectively. Their signal peptides are removed cotranslationaly in the endoplasmic reticulum, and the resultant subunits assemble into trimers. The trimers are targeted from the endoplasmic reticulum to protein storage vacuoles. Finally, the propeptides of the α and α' subunits are removed, resulting in the mature forms. Molecular heterogeneity of a subunit composition is present (17,18). In other words, ten molecular species having different subunit compositions with random combinations exist (homotrimers; $\alpha3$, $\alpha'3$, and $\beta3$; heterotrimers; $\alpha2\beta1$, $\alpha1\beta2$, $\alpha'2\beta1$, $\alpha'1\beta2$, $\alpha2\alpha'1$, $\alpha1\alpha'2$, and $\alpha1\beta1\alpha'1$).

4.2.2 Glycinin

The 11S globulin is called glycinin. Glycinin is a simple protein composed of six subunits, with a molecular mass of 300–400 kDa. Some of them have a potential glycosylation site, but none of them is glycosylated. The constituent subunits are A1aB1b, A1bB2, A2B1a,

A3B4, and A5A4B3. Five subunits are classified into two groups according to their amino acid sequences: group I (A1aB1b, A1bB2 and A2B1a) and group II (A3B4 and A5A4B3). The sequence identity in each group and between groups is about 80% and 45%, respectively. Each subunit is composed of an acidic polypeptide with acidic pI and a molecular mass of 30–40 kDa, and a basic polypeptide with basic pI and a molecular mass of 18–22 kDa. The acidic and basic polypeptides are linked together by a disulfide bond. It is considered that many molecular species having different subunit compositions with random combinations exist (19). The processes of synthesis, assembly, and accumulation of glycinin are shown in Figure 4.1. The constituent subunits are synthesized as a single polypeptide precursor, the preproglycinin. The signal peptide is removed cotranslationaly in the endoplasmic reticulum and the resultant proglycinin assembles into a trimer. Proglycinins are sorted into a protein storage vacuole and a specific posttranslational cleavage occurs. The cleavage results in a mature form of the glycinin. In contrast to β-conglycinin, none of the peptides is removed by this processing.

4.3 THREE DIMENSIONAL STRUCTURES OF SOYBEAN PROTEINS

4.3.1 Crystallization

X-ray crystallography is a suitable technique to elucidate three dimensional structures of proteins of high molecular masses. X-ray crystallography needs high quality crystals. However, it is difficult to obtain crystals of samples prepared from normal soybean seeds because of the molecular heterogeneity of both β-conglycinin and glycinin. Both proteins from soybean seeds had not been crystallized. To exclude the influence of molecular

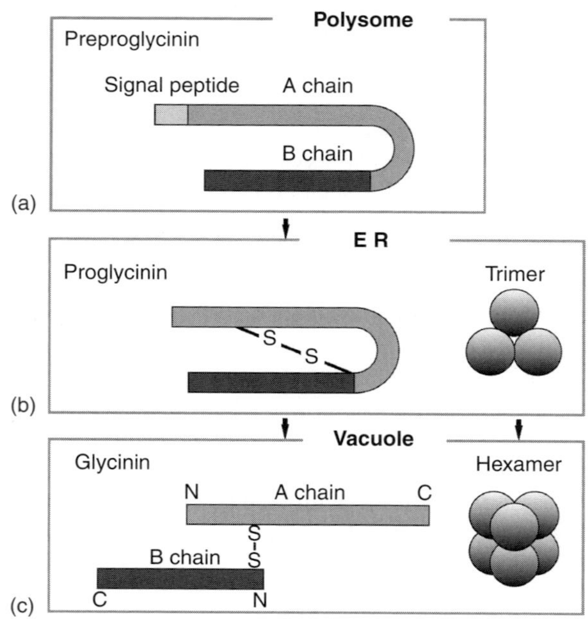

Figure 4.1 Schematic representation of processes of (a) biosynthesis, (b) assembly, and (c) transportation of soybean glycinin.

heterogeneity, we purified the storage proteins by constructing an *Escherichia coli* expression system and by using soybean seeds of mutant cultivars containing β-conglycinin or glycinin composed of a single or limited subunits.

Molecular species composed of a single subunit can be prepared easily by means of an *E. coli* expression system, because the cDNA of a single subunit is expressed in *E. coli*. Thus, we constructed an *E. coli* expression system for deletion mutants of the α and α′ subunits lacking the extension regions (α_{core} and α'_{core}) in addition to individual normal subunits of β-conglycinin (α, α′ and β) and A1aB1b and A3B4 subunits of glycinin. The pET-21d was used as an expression vector. *Escherichia coli* does not contain enzymes for the processing of the pro regions of the α and α′ subunits and a specific posttranslational cleavage of the proglycinins. Therefore, we expressed the α and α′ subunits without the pro regions. As a result, mature forms of the α and α′ subunits and proglycinin A1aB1b and A3B4 were prepared by means of the *E. coli* expression system. Meanwhile, researchers of the Ministry of Agriculture, Forestry, and Fisheries of Japan have developed mutant soybean cultivars containing β-conglycinin lacking the α subunit, the α′ subunit or both subunits (20,21). Further, mutant soybean cultivars containing glycinin composed of only the A3B4 subunit were developed (22,23). As a collaborative effort, we prepared native homotrimers of individual subunits of β-conglycinin (α3, α′3 and β3) and A3B4 homohexamer (glycinin A3B4) from seeds of these mutant cultivars.

Crystallization of the purified glycinin and β-conglycinin from the mutant cultivars was done by means of vapor diffusion, dialysis, and a batch method under various conditions such as pH, protein concentration, and temperature using polyethyleneglycol, 2-methyl-2,4-pentanediole and ammonium sulfate as a precipitant. The crystals of the recombinant and native β homotrimer (β3), the recombinant α'_{core} homotrimer (α'_{core}3), the recombinant A1aB1b homotrimer (proglycinin A1aB1b), the recombinant A3B4 homotrimer (proglycinin A3B4), and the native glycinin A3B4 were suitable for X-ray crystallography for analysis of their three dimensional structures.

4.3.2 X-ray Crystallography

When we started X-ray crystallography of soybean proteins, the crystal structures of the 11S globulin had not been reported; only those of French bean phaseolin and jack bean canavalin, 7S globulins, had been determined (24–26).

4.3.2.1 β-Conglycinin

We tried to determine the crystal structures of the native and recombinant β3 by the molecular replacement method using the protein structure models of phaseolin and canavalin as a search model. As a result, we determined the three dimensional structure of β3 by molecular replacement using canavalin as a search model (27) [Figure 4.2]. The β monomers consist of aminoterminal and carboxyterminal modules which are very similar to each other and are related by a pseudodiad axis. Each module of the β monomer is subdivided into a β-barrel and an α-helix domain. The superposition of the models of the native and recombinant β monomers shows a root mean square deviation (RMSD) of 0.43–0.51 Å for 343 common Cα atoms within 2.0 Å. This result indicates that the N-linked glycan does not influence the final structure of β3. Four regions including N- and C-termini were not included in the models of all the recombinant and native β monomers, because electron density maps in these regions were too thin to trace the correct sequence. These are likely to be disordered on the molecular surface [Figure 4.3(a)]. Comparison of sequences between 7S globulins by Wright (28) showed conserved regions (amino acid identity is high) and variable regions (amino acid identity is low), and it was noted that five variable

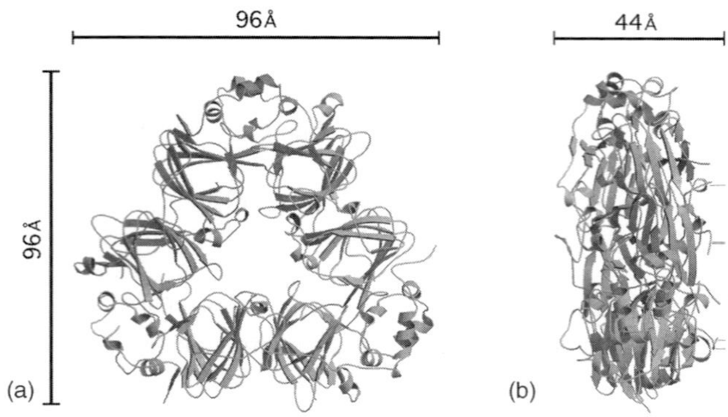

Figure 4.2 Ribbon diagrams of the three dimensional structures of (a) β-conglycinin and (b) β-homotrimer. Numbers indicate dimensions. (From: Maruyama, N., M. Adachi, K. Takahashi, K. Yagasaki, M. Kohno, Y. Takenaka, E. Okuda, S. Nakagawa, B. Mikami, S. Utsumi, *Eur. J. Biochem.* 268:3595–3604, 2001.)

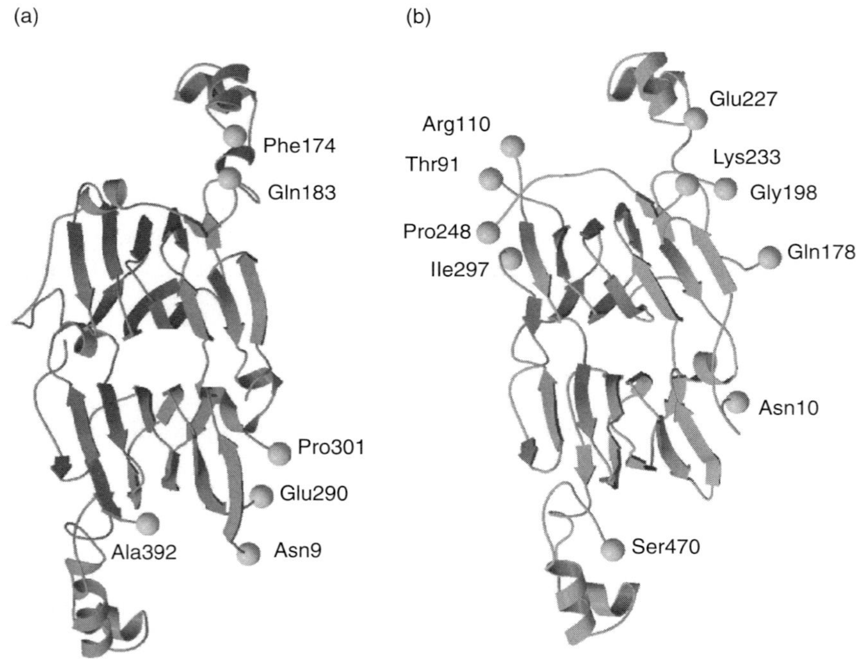

Figure 4.3 Disordered regions of soybean proteins. Disordered regions of β-conglycinin β (a) and of glycinin A1aB1b (b). The termini of the disordered regions are indicated by balls.

regions exist. In fact, there are only four variable regions, because one of them was identified by using an incorrect sequence of the α' subunit (15). Therefore, the disordered regions in the β subunit are consistent with the four variable regions. An amino acid replacement is liable to occur in variable regions, because they are disordered regions.

$\alpha3$ and $\alpha'3$ among the three homotrimers of β-conglycinin formed only small crystals, but $\alpha_{core}3$ and $\alpha'_{core}3$ with no extension region formed crystals suitable for X-ray crystallography. Therefore, the presence of the extension region probably perturbs the formation of crystals of good quality. We determined the three dimensional structure of $\alpha'_{core}3$ by the molecular replacement using the model of α3 (unpublished data). The scaffold of $\alpha'_{core}3$ was identical to that of β3. All α atoms of the α'_{core} monomer and homotrimer could be superimposed on corresponding atoms of the β monomer and homotrimer with a small RMSD of 0.6 and 0.7 Å, respectively. These values indicate that scaffolds of these two proteins are very similar to each other.

4.3.2.2 Glycinin

The amino acid identity between 7S and 11S globulins is only 15%, but they do exhibit partially significant identity. Therefore, it is considered that these two globulins are evolutionarily related. The 11S globulin has a hexameric structure, but proglycinin has a trimeric structure. In other words, there is a possibility that pro forms of 11S globulins have a structure analogous to the 7S globulin. Thus, we tried to determine the structure of proglycinin A1aB1b by molecular replacement using the model of β-conglycinin β3, canavalin, and phaseolin but failed. Finally, we successfully determined the structure of proglycinin A1aB1b by the multiple isomorphous replacement method using heavy atom derivatives (29). Our results showed that the structure of proglycinin A1aB1b is similar to that of β-conglycinin β (Figure 4.4). Superposition of the structure of proglycinin A1aB1b on that of β-conglycinin β exposes the high similarity between the proteins (Figure 4.5). A least squares fit of a protomer between them produced an RMSD of 1.35 Å, and that of β-barrel in a protomer between them was 1.2 Å. Further, RMSDs of the comparable Cα atoms of β-barrel domains in both N- and C-terminal modules between proglycinin A1aB1b and β-conglycinin β were approximately 0.9 Å, whereas those of α-helix domains in N- and C-terminal modules were 1.35 and 1.39 Å, respectively. This indicates that the scaffold of proglycinin A1aB1b is very similar to that of β-conglycinin β, but that the configurations between a β-barrel and an α-helix domain in protomers are slightly different.

All six disordered regions existing in proglycinin A1aB1b protomer are located on the molecular surface similarly to β-conglycinin [Figure 4.3(b)]. Amino acid sequences have been previously aligned among various 11S globulins from legumes and nonlegumes, and five variable regions have been proposed (14). Five of the six disordered regions correspond to the five variable regions. The only disordered region that does not correspond to the variable region is the shortest one. The second disordered region from the C-terminus is the

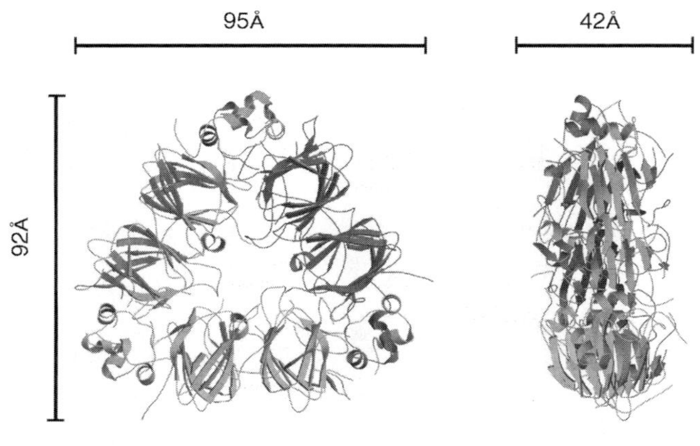

Figure 4.4 Ribbon diagram of the three dimensional structure of A1aB1b homotrimer. The view in the left diagram is depicted with the threefold symmetry axis running perpendicular to the paper, whereas the depiction on the right is related to the view on the left by a rotation of 90°. (From: Adachi, M., Y. Takenaka, A.B. Gidamis, B. Mikami, S. Utsumi., *J. Mol. Biol.* 305:291–305, 2001.)

longest, and the region of A1aB1b is composed of 48 residues. This region is called the hypervariable region, because both an amino acid sequence and a number of an amino acid in this region are the most variable among the five.

The three dimensional structure of the native glycinin A3B4 homohexamer is determined by the molecular replacement using the model of proglycinin A1aB1b (Figure 4.6) (30). As a result, a trimer of a proglycinin assembles into a hexamer by stacking of faces in which the processing site for a mature form exists.

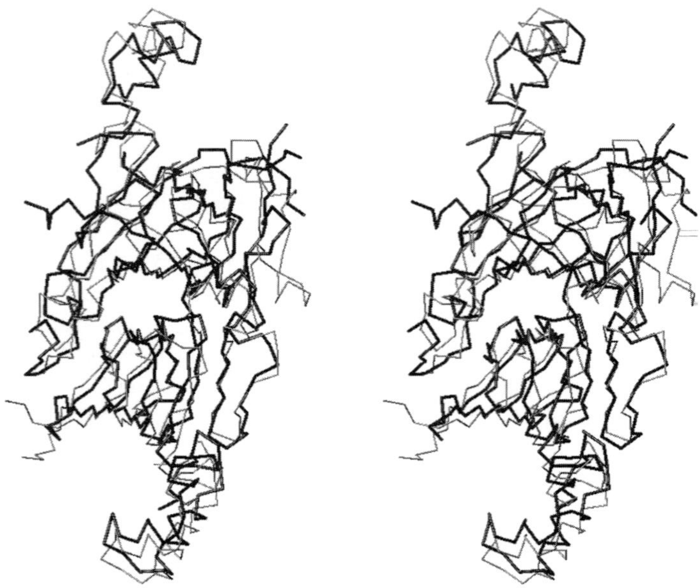

Figure 4.5 Superposition of the Cα trace of the β-conglycinin β and proglycinin A1aB1b. The Cα traces of β-conglycinin β and proglycinin A1aB1b are shown in gray and black lines, respectively.

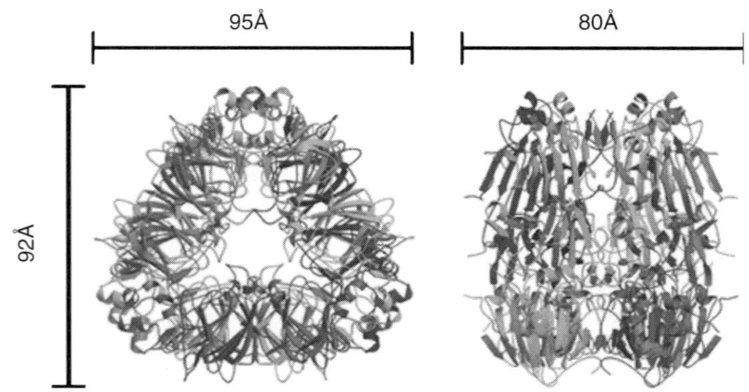

Figure 4.6 Ribbon diagram of the three dimensional structure of mature A3B4 homohexamer. The view in the left diagram is depicted with the threefold symmetry axis running perpendicular to the paper. The depiction on the right is related to the view on the left by a rotation of 90°. Numbers indicate dimensions. (From: Adachi, M., J. Kanamori, T. Masuda, K. Yagasaki, K. Kitamura, M. Bunzo, S. Utsumi. *Proc. Natl. Acad. Sci. USA*, 100:7395–7400, 2003.

4.3.3 Cupin Superfamily

The 7S and 11S globulins are classified into the cupin superfamily (31,32).

Cupin comes from "cupa," the Latin term for a small barrel. The characteristic cupin domain comprises two conserved motifs, $G(x)_5HxH(x)_{3,4}E(x)_6G$ and $G(x)_5PxG(x)_3N$ (31,32). Based on structural features, cupin proteins are classified as either a monocupin having one cupin motif per one monomer, and a bicupin having two cupin motifs per one monomer. The 7S and 11S globulins are bicupins. Cupin proteins are also known to have diverse functions. For example, in addition to seed storage proteins, an oxalate oxidase and a sucrose binding protein belong to this group (33,34).

In the three dimensional structure of the cupin protein phaseolin, which was determined at the early stage of research of the cupin superfamily, the cupin motifs are located at the C and D strands and at the G and H strands of the β-barrel (Figure 4.7). The length between motifs varies largely among species of proteins. Later, the three dimensional structures of many cupin proteins were shown to have a jellyroll type β-barrel structure. Further, it was elucidated that in addition to a tertiary structure, a quaternary structure is common among cupin proteins. Three dimensional structures of many cupin proteins indicate that an arabinose binding domain of AraC transcription factor from *E. coli*, a dimeric dTDP-4-dehydrorhamnose 3,5-epimerase from *Salmonella enterica*, a hexameric germin from barley and an oxalate decarboxylase from *Bacillus subtilis* are very similar to the domain, protomer, and trimer of proglycinin and β-conglycinin, and the hexamer of glycinin, respectively (27,29,33,34) (Figure 4.7).

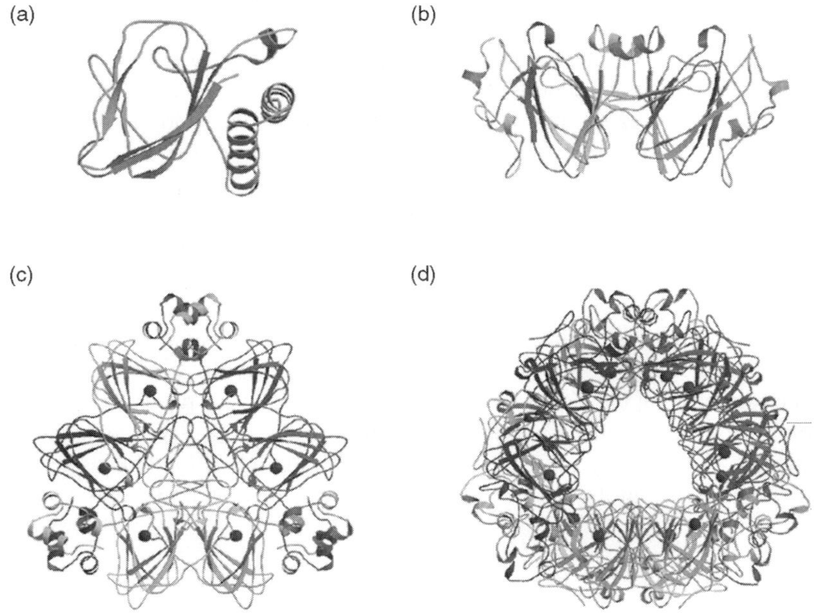

Figure 4.7 Ribbon diagrams of the three dimensional structures of cupin superfamily. (a) arabinose-binding domain of AraC transcription factor from *E. coli* (PDB code: 2ARA); (b) dimeric dTDP-4-dehydrorhamnose 3,5-epimerase from *Salmonella enterica* (1DZR); (c) hexameric germin from barley (1FI2); (d) oxalate decarboxylase from *Bacillus subtilis* (IJ58). The cupin motifs in (b) are shown in dark gray. Manganese ions are indicated by balls in (c) and (d). (From: Dunwell, J.M., A. Culham, C.E. Carter, C.R. Sosa-Aguirre, P.W. Goodenough. *Trends Biochem. Sci.* 26:740–746, 2001; and Anand, R., P.C. Dorrestein, C. Kinsland, T.P. Begley, S.E. Ealick. *Biochemistry* 41:7659–7669, 2002.)

The cupin fold is evolutionally conserved, although amino acid sequences are not conserved except the cupin motifs. This indicates that the cupin motif is essential for the formation and maintenance of the structures of glycinin and β-conglycinin.

4.4 STRUCTURAL FEATURES OF SOYBEAN PROTEINS

4.4.1 β-Conglycinin

We analyzed the structural features of the native homotrimers ($\alpha 3$, $\alpha' 3$, and $\beta 3$) of the individual subunits of β-conglycinin prepared from seeds of mutant cultivars containing β-conglycinin with limited subunit compositions, the recombinant homotrimers ($\alpha 3$, $\alpha' 3$, $\beta 3$, $\alpha_{core}3$, and $\alpha'_{core}3$) of individual subunits, and the α_{core} and α'_{core} of β-conglycinin by means of *E. coli* expression system (35–37).

On a sucrose density gradient centrifugation, three recombinant homotrimers sedimented at similar positions. However, the recombinant $\alpha 3$ and $\alpha' 3$ eluted faster than did the recombinant $\beta 3$ on a gel filtration chromatography, similarly to the native glycinin hexamer. These indicate that the extension regions largely contribute to the dimensions of $\alpha 3$ and $\alpha' 3$. Therefore, the structural models indicated in Figure 4.8 can be proposed. The order of thermal stability of the recombinant and native homotrimers was $\alpha 3 < \alpha' 3 < \beta 3$, and $\beta 3$ was the most stable among the three subunits. On the other hand, the $\alpha_{core}3$ and $\alpha'_{core}3$ exhibited similar thermal stability to those of $\alpha 3$ and $\alpha' 3$, respectively. These results indicate that the extension regions and the glycans do not affect the thermal stability, and that the core regions determine the thermal stability of the individual subunits of β-conglycinin. Further, an analysis of the thermal stability of the heterotrimers of β-conglycinin indicated that thermal stabilities of most heterotrimers are determined by that of the constituent subunit which has the lower thermal stability (38). On the other hand, the order of surface hydrophobicity was $\alpha 3 > \alpha' 3 \geq \beta 3$, and dependent on that of the core regions. Surface hydrophobicities of the heterotrimers were the arithmetic mean of those of constituent subunits. The extension regions and the glycans largely affect solubility under lower ionic strength, but not under higher ionic strength. The solubility of the trimer containing even one α or α' subunit is much higher than that of the $\beta 3$, $\alpha_{core}3$ and $\alpha'_{core}3$, and close to that of the $\alpha 3$ and $\alpha' 3$, indicating that the extension regions contribute largely to the solubility.

Because the thermal stability of the core region determines that of the subunit, we tried to analyze the structural factors related to thermal stability by comparing the three dimensional structure of the $\alpha'_{core}3$ with that of $\beta 3$ in detail. Thus, five structural factors that can account for the result that $\beta 3$ is more thermostable than $\alpha'_{core}3$ were elucidated (39).

First, protein packs tightly but some cavities are often observed inside the molecule, resulting in a decrease in its thermal stability (40). We calculated the total cavity volume

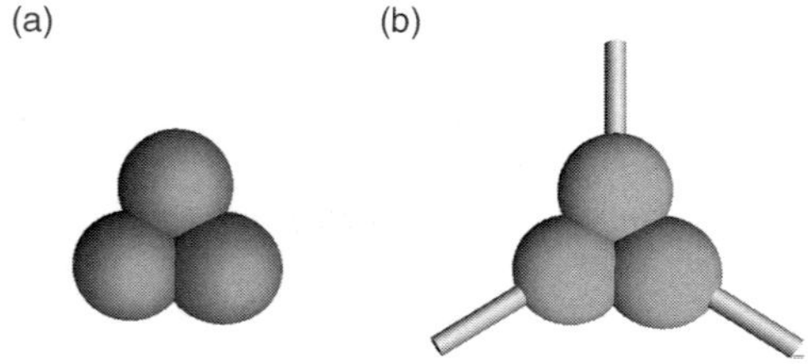

Figure 4.8 Structure models of soybean β-conglycinin. (a) $\beta 3$, (b) $\alpha 3$ or $\alpha' 3$. The core and the extension regions are shown in ball and stick, respectively.

of the $\alpha'_{core}3$ (172 Å^3) to be larger than that of $\beta3$ (79 Å^3), indicating that $\beta3$ is more tightly packed than $\alpha'_{core}3$. The main cavities are located in the β-barrels of the N- and C-terminal modules. The cavities of the C-terminal β-barrels of $\beta3$ and $\alpha'_{core}3$ exhibit a prominent difference between them. Twelve residues among 15 residues lining the cavity were common between the β and α'_{core}, and three residues were different. In other words, Phe233, Leu262, and Phe325 of β were substituted with Leu376, Phe405, and Val468 of α'_{core}, respectively. Because of the compensating substitutions of Leu376 and Phe405 in the α'_{core} to Phe and Leu in β, respectively, a substitution of Val468 in the α'_{core} to Phe in β contributes mainly to the difference in the cavity volume of the C-terminal β-barrel.

Second, because long ion pair networks have been found in several enzymes from hyperthermophilic bacterium (41,42), they are thought to be a common mechanism to stabilize the protein of hyperthermophiles (43). Sequence alignment of β and α'_{core} indicates that some of the charged residues in β are substituted with noncharged residues in the α'_{core}. These residues exist in the α-helix domain of the C-terminal module, and form a cluster at the intermonomer interface. The length of an ion pair network of $\beta3$ was longer than that of the $\alpha'_{core}3$, because E353, R359, E362, R363, R379, K383, and R385 of β were replaced with S496, S502, P505, S506, N522, S526, and S528 of the α'_{core}, respectively.

Third, the hydration of nonpolar groups apparently destabilizes proteins (44). Experimental analysis of surface hydrophobicity using a hydrophobic column demonstrated that the molecular surface of $\alpha'_{core}3$ was more hydrophobic than that of $\beta3$. Further, a comparison of three dimensional structures between $\beta3$ and $\alpha'_{core}3$ indicated that the ratio of hydrophobic atoms (carbon and sulfur) to all atoms of solvent accessible surface of $\alpha'_{core}3$ was higher than that of $\beta3$.

Fourth, the number of proline residues, which stabilize the protein structure by decreasing the entropy of denatured structure, is 57 in the $\alpha'_{core}3$ compared to 63 in $\beta3$. According to Matthews et al. (45), this difference contributes 4.8 kcal/mol to the ΔG. Therefore, the difference in the number of proline residues may affect the thermal stability.

Fifth, shorter loops are one of stabilizing factors in thermophilic bacteria (46). The loop between helix 3 and strand J' is five residues shorter in β than that in α'_{core}. The electron densities of this loop region were broken in all monomers of the $\alpha'_{core}3$ and two of $\beta3$, but appeared in one monomer of $\beta3$, giving the clearest density map among the three monomers. Because the resolution of $\beta3$ is lower, this may suggest relatively rigid conformation of this loop region in the β subunit. The length of the loop is also considered one of the important factors related to the structural stability of the cupin superfamily to which β-conglycinin belongs.

As described, we investigated possible factors for the difference in the thermal stability between the $\alpha'_{core}3$ and $\beta3$ through the comparison of three dimensional structures. More hydrogen bonds were observed in each module of $\alpha'_{core}3$, which suggests more stable packing of $\alpha'_{core}3$, and is not consistent with the experimental data. This difference should be more than compensated by the accumulation of the effects of many factors which account for less thermal stability of the $\alpha'_{core}3$, such as larger cavity volumes, lack of short ion pair networks, higher surface hydrophobicity, lower number of proline residues, and a longer loop. We think that $\beta3$ would be stabilized more than the $\alpha'_{core}3$ by the sum of the contributions of each of these factors.

4.4.2 Glycinin

Researchers of the Ministry of Agriculture, Forestry, and Fisheries of Japan have developed mutant soybean cultivars containing glycinin of limited subunit compositions, such as only group I, only group II, and only A5A4B3, in addition to only A3B4 (22,23). In collaboration with them, we prepared glycinin (group I, group II, A3B4, and A5A4B3) from these mutant cultivars and analyzed their structural features (unpublished data). On gel filtration chromatography, subunits having the longer hypervariable region eluted faster. In other words, A5A4B3 and group II eluted faster than A3B4 and group I did, respectively. These results indicate that the variable regions influence the dimensions of glycinin similarly to the extension region of β-conglycinin. The thermal stability of A3B4 (87.2°C) was slightly lower than those of the other subunits (92.9–95.0°C), but group II containing A3B4 exhibited thermal stability similar to the others. Therefore, the effect of subunit compositions on the thermal stability is not significant. The order of surface hydrophobicity on a hydrophobic column chromatography was group I $<$ 11S $<$ A3B4 $<$ group II $<$ A5A4B3. Thus, the surface hydrophobicity of glycinin comprised of several subunits is the arithmetic mean of those of individual subunits. Further, we confirmed that the molecular surfaces of the three dimensional structures of recombinant proglycinin A1aB1b and native glycinin A3B4 corresponded to the difference of surface hydrophobicity on a hydrophobic column chromatography. However, it is difficult to discuss the difference of the thermal stability in detail, because we have only determined the three dimensional structure of A3B4 among native glycinins.

4.5 IMPROVEMENT OF NUTRITIONAL QUALITY BY PROTEIN ENGINEERING

There are two approaches to improve the nutritional quality of soybean proteins: fortify the content of a limiting amino acid such as amino acids containing sulfur, and increase the digestibility of the proteins.

4.5.1 Increasing Sulfur Containing Amino Acids

In grain legumes such as soybean, methionine is the primary limiting essential amino acid. The biological value of legume protein is limited to 55–75% of that of animal protein because of this limitation or imbalance in amino acid composition (47). Several ways to increase the content of sulfur containing amino acids have been cited (48): (1) the insertion and replacement of amino acid residues at appropriate sites in a storage protein; (2) increasing the level of endogenous methionine rich protein; (3) introduction of heterologous genes for methionine rich protein; and (4) manipulating key enzymes in the biosynthetic pathway. A description of the work done in our laboratory using the first strategy, protein engineering, follows.

Both glycinin and β-conglycinin have variable regions which are less conservative. Thus, it may be possible to replace and insert amino acids in the variable regions to enhance nutritional quality.

At the start of our research to improve glycinin by protein engineering, no data for the three dimensional structures of glycinin were available. Therefore, we attempted to modify the variable regions based on the alignment of amino acids. Five variable regions were named I, II, III, IV, and V from the N-terminus [Figure 4.9(a)]. We designed modified glycinins with five contiguous Met residues inserted into variable region IV or V (IV+4Met and V+4Met). Further, we also designed a modified glycinin whose five contiguous Glu in the variable region IV were substituted with five Met and one Leu [IV(Met)]. These modified glycinins were produced by means of an *E. coli* expression system and their folding ability was analyzed. All of them were accumulated as soluble proteins in *E. coli* at a high level, and each one self assembled. Because the modified glycinins exhibit solubility as a globulin, we concluded that the modified glycinins fold correctly. Moreover, the three modified glycinins can be crystallized, and IV+4Met and

V+4Met can be accumulated in tobacco seeds, rice seeds, and potato tubers in a way similar to the intact molecule (49–53). In other words, an insertion and a replacement in variable regions could be useful in improving the nutritional quality of glycinin.

It is difficult to insert amino acids in conserved regions, but it is possible to replace amino acids based on three dimensional structures. For example, Figure 4.10(a) shows one β-strand in the β-barrel of C-terminal module of A1aB1b. This strand contains a hydrophobic peptide, Val-Ile-Leu-Val. Based on molecular modeling, we considered that it is possible to replace these four amino acids with four Met residues [Figure 4.10(b)]. We have not examined the folding ability of this modified glycinin yet, but we think that it probably can fold correctly.

4.5.2 Increasing Digestibility

There are two approaches to improving digestibility: by introducing a cleavable site for digestive enzymes, and by destabilizing a structure. If aromatic residues recognized by chymotrypsin are introduced inside the molecule, however, its structures may be stabilized

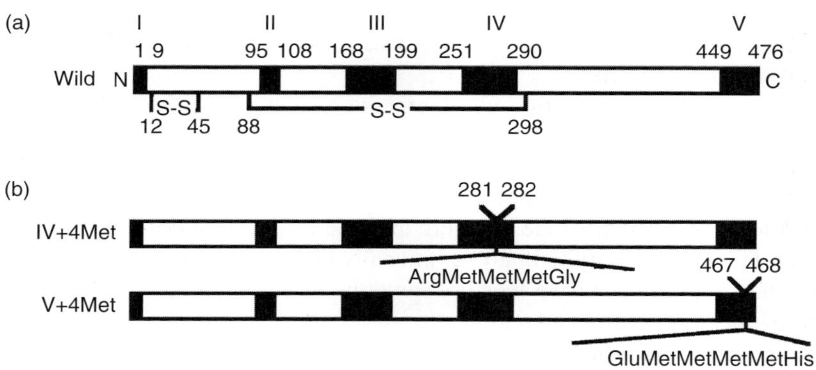

Figure 4.9 Schematic representation of the wild and modified proglycinin A1aB1b. (a) wild type, (b) tetramethionine-inserted proglycinin. White and black areas indicate the conserved and variables regions, respectively. The number of the residues from the N-terminus for the variable regions (I–V) are shown above the alignment. S-S indicates a disulfide bond.

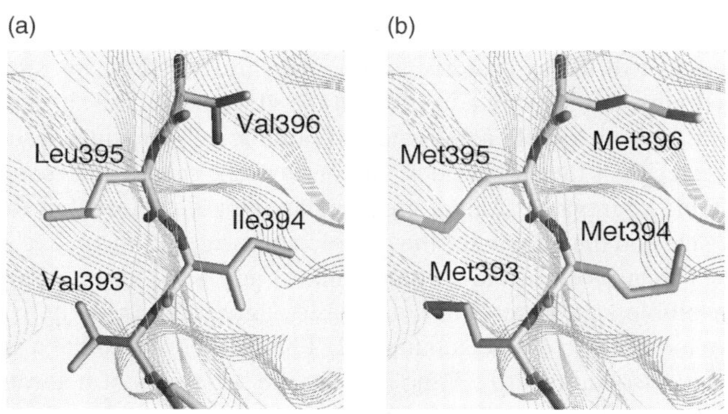

Figure 4.10 Fortification of Met into β-barrel of the proglycinin A1aB1b. (a) wild type; (b) methionine-introduced proglycinin. Backbone and side chains are indicated by bond models.

by the increased hydrophobicity inside the molecules, resulting in the lowering of digestibility. On the other hand, introduction of lysine and arginine residues recognized by trypsin inside the molecule may probably have a large influence on digestibility, because they not only add the site for cleavage, but also destabilize the structure due to the introduction of a charged residue inside the molecule. Most of the salt bridges inside molecules contribute largely to structural stability. Therefore, a substitution of an acidic residue which forms a salt bridge inside the molecule with another amino acid destabilizes a structure while keeping the site for trypsin. In glycinin A1aB1b, a salt bridge between Asp157-Arg161 exists inside the molecule. We replaced Asp157 with Ala (D157A) (54). This mutant, produced in *E. coli*, was found to fold correctly. Furthermore, we confirmed that D157A formed crystals suitable for X-ray crystallography. The T_m value of D157A was 15°C lower than that of the wild type and its sensitivity to chymotrypsin increased.

4.6 IMPROVEMENT IN PHYSICOCHEMICAL FUNCTIONS BY PROTEIN ENGINEERING

4.6.1 Structure to Physicochemical Function Relationships of Soybean Proteins

The physicochemical functions of soybean proteins are determined by solubility, degree of hydrophobicity, and hydrophilicity of a molecular surface. Their distribution and balance, structural stability, exposure of hydrophobic regions under denaturation, and interactions between the same and the different molecular species also influence physiochemical function. Therefore, elucidation of the structure to physicochemical function relationships of soybean proteins is necessary to improve rationally the physicochemical functions of soybean proteins. Such studies have mostly been done in our laboratory using mutant cultivars, and by means of protein engineering.

We prepared homotrimers of the individual subunits of β-conglycinin and analyzed their physicochemical functions (35–38). The order of emulsifying ability was $\alpha3 > \alpha'3 > \beta3$ in the cases of both the recombinant and native homotrimers. In comparison with the homotrimers lacking the extension region, the order of emulsifying ability was $\alpha_{core}3 > \alpha'_{core}3 > \beta3$, although the emulsifying abilities of $\alpha'_{core}3$ and $\alpha_{core}3$ were lower than those of $\alpha'3$ and $\alpha3$, respectively. This order was related to those of thermal stability and surface hydrophobicity. The heterotrimers containing two α or α' subunits ($\alpha2\beta1$, $\alpha'2\beta1$) exhibited emulsifying ability similar to those of $\alpha3$ or $\alpha'3$, respectively. The heterotrimers containing one α or α' subunit ($\alpha1\beta2$, $\alpha'1\beta2$) exhibited emulsifying ability similar to that of $\beta3$. Therefore, the number of the extension regions, structural stability, and surface hydrophobicity contribute to the emulsifying abilities, but the glycans do not. The native and recombinant $\alpha3$ and $\alpha'3$ formed soluble aggregates without accompanying insoluble aggregates by heating at greater than their T_m values. In contrast, the native and recombinant $\beta3$ did not form the soluble aggregates at all at greater than their T_m values, but formed insoluble aggregates. The size and the amount of the soluble aggregates of the native $\alpha3$ and $\alpha'3$ were smaller and fewer than those of the recombinant $\alpha3$ and $\alpha'3$, respectively, but $\alpha'_{core}3$ and $\alpha_{core}3$ formed insoluble aggregates without soluble aggregates similarly to that of $\beta3$. The heterotrimers containing two α or α' subunit ($\alpha2\beta1$ and $\alpha'2\beta1$) exhibited a similar tendency about heat induced associations to $\alpha3$ and $\alpha'3$, but the behaviors of heterotrimers containing two β subunits ($\alpha1\beta2$ and $\alpha'1\beta2$) resembled that of $\beta3$ rather than those of $\alpha3$ and $\beta3$. These results indicate that the presence and the number of the extension regions contribute to the heat induced association, but the glycans prevent such heat induced association.

Similarly, we measured the emulsifying ability of glycinin prepared from mutant cultivars having subunit compositions different from a normal one (unpublished data). Group II exhibited higher emulsifying ability than that of group I, and that of glycinin containing all subunits (normal glycinin) was intermediate between those of group I and II. A5A4B3 exhibited slightly higher ability than A3B4 did, and the ability of group II was intermediate between those of A3B4 and A5A4B3. These results indicated that A5A4B3 has the highest emulsifying ability among all subunits of glycinin, and that the heterohexamer exhibits the arithmetic mean of the abilities of constituent subunits. Variable region IV (the hypervariable region), the longest in five variable regions of glycinin, is rich in negative charges. The length of this region is different among subunits of glycinin, and that of group II, especially A5A4B3, is longer than that of group I. Therefore, it can be considered that the length of the hypervariable region is largely related to the emulsifying ability of glycinin. This resembles the effect of the extension region of β-conglycinin on the emulsifying ability. However, the order of the emulsifying ability of glycinins did not correlate with those of the surface hydrophobicity and the structural stability, in contrast to β-conglycinin. In other words, the degree of the contribution by each factor is different among the species of proteins and subunits.

To exhibit excellent emulsifying ability, proteins need the affinity for both oil and water at their interface. Therefore, flexibility and amphiphatic properties are important factors for emulsifying ability. In fact, it is pointed out that in many proteins surface hydrophobicity and structural stability are related to emulsifying ability. The contribution of surface hydrophobicity and structural stability of glycinin to its emulsifying ability is different from that of β-conglycinin, although the presence of large hydrophilic regions contributes to the emulsifying ability of both glycinin and β-conglycinin. Elucidation of the three dimensional structures of all subunits of glycinin and β-conglycinin can probably shed light on this difference. It is expected that destabilization of a structure and strengthening of hydrophilicity or hydrophobicity can improve emulsifying ability. However, the effect is probably different between glycinin and β-conglycinin.

The subunit compositions of glycinins vary among cultivars (55). Mori et al. analyzed heat induced gel forming ability of glycinins prepared from five cultivars containing different subunit compositions (56). As a result, glycinins which have larger amounts of A3B4 formed harder gel. Pseudoglycinins were reconstituted from A1aB1b, A2B1a, and A3B4 isolated on a chromatography in the presence of urea, and their heat induced gel forming abilities were analyzed (57,58). The results showed that the hardness of the gels is directly proportional to the amount of A3B4. The number of cysteine residues in A3B4 is less than those of A1aB1b and A2B1a. The topology of free sulfhydryl groups of A3B4 is different from those of A1aB1b and A2B1a. Further, A3B4 has lower structural stability than A1aB1b and A2B1a. Changes in conformation and SH/S-S interchange reactions are essential for heat induced gel forming ability of glycinin. Therefore, it can be considered that the reactivity of SH/S-S interchange of A3B4 is higher than those of A1aB1b and A2B1a due to the difference in the topology of the free sulfhydryl group and lower structural stability. As a result, A3B4 can form hard gels with fine network structures. Therefore, we can consider that a destabilization of a structure and modification of the topology of free sulfhydryl groups are an approach to improving gel forming ability of glycinin. In terms of fineness of network, the addition of a number of SH and S-S could be a strategy. We cannot describe the mechanism of gel forming ability of β-conglycinin, because analysis is still insufficient.

4.6.2 Improvement in the Physicochemical Functions of Glycinin

Variable regions of glycinin are rich in hydrophilic amino acids. Therefore, deletions of these regions can change the balance between hydrophilicity and hydrophobicity of glycinin.

There is a possibility that the deletions destabilize the molecule, because the regions exist in the intact molecule. So we designed mutants of glycinin lacking each of the variable regions (ΔI, ΔII, ΔIII, ΔIV, ΔV, ΔV36 and ΔV8) [Figure 4.9(a)] (49). Existing restriction enzyme sites were used to construct the mutants because the use of PCR was not yet popular during the time of these experiments. Therefore, some parts of conserved regions were also deleted, and glycinin lacking C-terminal 8 or 36 residues (ΔV8, ΔV36) has two extra amino acids, Leu and Asn, at its C-terminus.

If a peptide composed of four Met residues is introduced into the variable region [Figure 4.9(b)], a small hydrophobic patch is formed. This may result in structure destabilization because a small hydrophobic patch is introduced into a hydrophilic region. Therefore, it is expected that this mutation improves the protein's emulsifying ability and gel forming ability in addition to nutritional quality.

Glycinin has two disulfide bonds in each constituent subunit: one is an intraacidic chain bond (Cys12-Cys45 in A1aB1b) and another links the acidic and basic chains (Cys88-Cys298 in A1aB1b). These disulfide bonds can be confirmed in the three dimensional structure of A1aB1b (29). Because the corresponding cysteine residues are conserved in all 11S globulins whose primary structures have been determined, it is considered that these two disulfide bonds are also conserved in all 11S globulins (1). In case of the subunits of group I, it is assumed that one more disulfide bond (C271-C278 in A1aB1b) exists. However, we cannot confirm this disulfide bond in the three dimensional structure, because these residues are located in the disordered region. If one of the cysteine residues forming a disulfide bond is replaced by other amino acids, a free sulfhydryl group arises. The lack of a disulfide bond probably induces a destabilization of a structure. It is therefore expected that the emulsifying ability and gel forming ability are improved by a disruption of a disulfide bond. To test this hypothesis, mutants lacking one disulfide bond (C12G and C88S) and both disulfide bonds (C12G/C88S) were designed [Figure 4.11(b)] (59).

On the other hand, Cys53 and Cys377 in A1aB1b are free sulfhydryl groups. If an amino acid near the site of Cys53 or Cys377 is replaced by a cysteine residue, a new disulfide bond can be rationally introduced, considering the directions of the side chains of the replaced, the cysteine residues, and the distance of Cα between them shown by a simulated model. On the other hand, if two amino acids are replaced by cysteine residues by a similar method, a new disulfide bond can be introduced. Two and three mutants containing a new sulfhydryl residue and a new disulfide bond were designed, respectively (60).

These mutants were expressed in *E. coli*. Solubilities, self assemblies, and stabilities of expressed proteins indicated that ΔI, ΔV8, C88S, C12G/C88S, R161C, P248C, N116C, P248C, and N116C/P248C, in addition to IV+4Met, V+4Met. and D157A can fold correctly. In other words, modifications of the variable regions and the conserved regions based on three dimensional structure can be acceptable. Further, it was elucidated that the two disulfide bonds conserved among 11S globulins are not essential for the formation and the maintenance of the conformation of glycinin. However, it was difficult to purify C12G/C88S, because it was susceptible to attack of proteinase of *E. coli* during purification under a low ionic strength. This indicates that the removal of both disulfide bonds makes the conformation of C12G/C88S less compact than that of normal glycinin under low ionic strength.

We confirmed the expected introduction of a new free sulfhydryl group and a new disulfide bond by measuring the content of the free sulfhydryl group of the modified glycinins after blocking a free sulfhydryl group (60).

We analyzed the physicochemical functions of the modified glycinins, which can be prepared at a large scale. We compared the heat induced gel forming ability of ΔI, ΔV8, IV+4Met, V+4Met, C12G, C88S, R161C, F163C, N116C, P248C, and N116C/P248C with those of native glycinin and normal proglycinin A1aB1b. Proglycinin A1aB1b exhibits

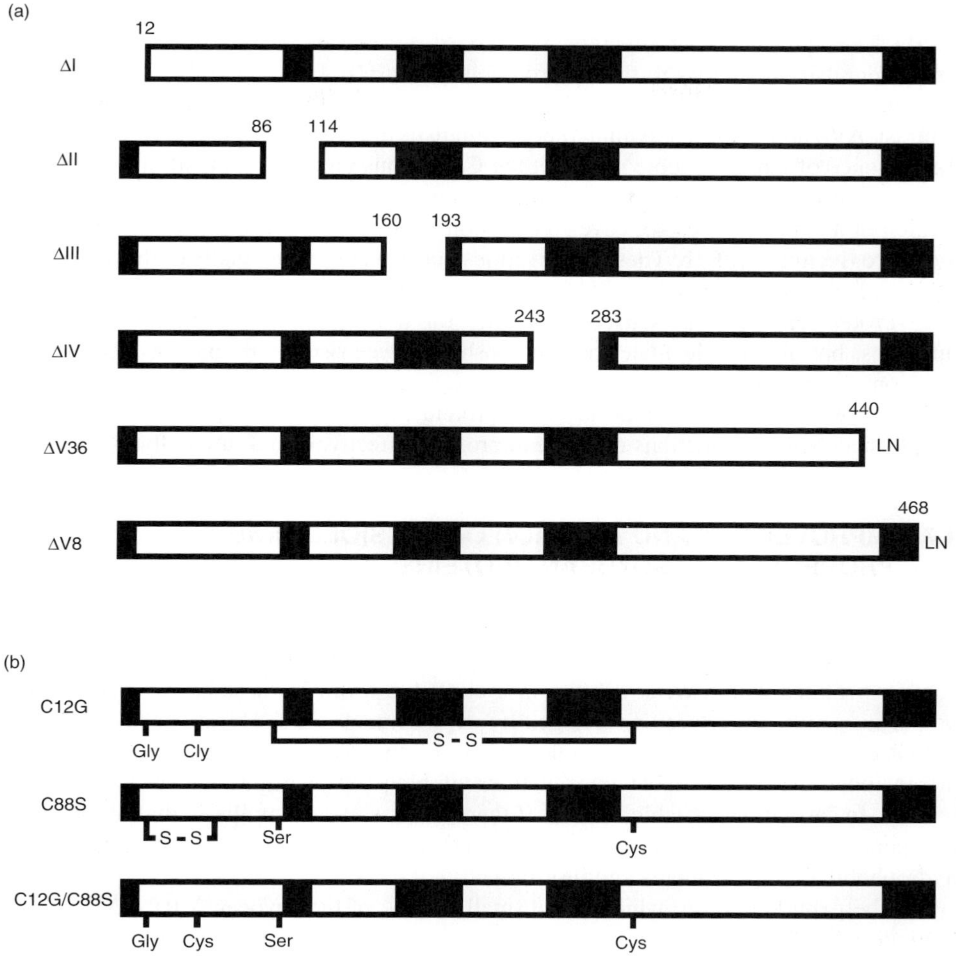

Figure 4.11 Schematic representation of modified proglycinin A1aB1b. (a) variable region-deleted mutants; (b) disulfide bond-deleted mutants. Numbers and S-S indicate the residue number from N-terminus and a disulfide bond, respectively. (From: Kim, C.S., S. Kamiya, T. Sato, S. Utsumi, M. Kito, *Protein Eng*. 3:725–731, 1990.)

gel forming ability and gel hardness similar to those of the native glycinin. On the other hand, ΔI, IV+4Met, V+4Met, C88S, F163C, N116C, P248C, and N116C/P248C formed harder gels than the normal one did. C88S and N116C/P248C showed the most prominent results. On the other hand, R161C formed harder gels than normal at protein concentrations higher than 7%, but formed gels with gel hardness similar to that of normal to one at protein concentration lower than 7%. C12G formed gels with gel hardness similar to that of normal glycinin at higher protein concentration, but could not form gel at a protein concentration lower than 5.6%. These results indicate: (1) the disulfide bond between Cys12 and Cys45 plays an important role in the initiation of a disulfide exchange reaction for a gelation; (2) introduction of a new sulfhydryl group and a disulfide bond is useful for improvement of heat induced gel forming ability and gel hardness; (3) the reactivity of the disulfide bond between Cys12 and Cys45 of ΔI, where its N-terminus is Cys12, increases; and (4) enhancement of surface hydrophobicity may improve gel forming ability.

We compared the emulsifying ability of ΔI, ΔV8, IV+4Met, V+4Met, C12G, C88S, and D157A with that of proglycinin A1aB1b. ΔI, IV+4Met, C12G, and C88S exhibited similar emulsifying ability to that of proglycinin A1aB1b. D157A exhibited worse emulsifying ability than proglycinin A1aB1b did (the details will be described elsewhere). In contrast, ΔV8 and V+4Met exhibited better emulsifying ability. These results indicate: (1) the presence of a hydrophobic region at the C-terminus increases emulsifying ability; (2) the effect of increasing hydrophobicity is different among the mutants; and (3) a destabilization of a structure does not have an effect on the improvement of the emulsifying ability of proglycinin A1aB1b. These observations correspond to the results of the analyses of glycinin having different subunit compositions described in the previous section.

Protein engineering is useful not only for the improvement of physicochemical functions, but also to elucidate the relationship between structure and physicochemical function. Therefore, the elucidation of the relationship between structure and physicochemical function at a molecular level, by protein engineering, will enable us to improve the physicochemical functions of soybean proteins effectively and rationally.

4.7 IMPROVEMENT AND ADDITION OF PHYSIOLOGICAL PROPERTIES OF SOYBEAN PROTEINS

Many kinds of bioactive peptides have been isolated from enzymatic digests of food proteins (61–64). Some proteins contain such peptides in a noncleavable form, and some contain peptides with a weak activity in a cleavable form. In these cases, physiological properties can be introduced or changed by replacement of a few amino acids. It is easy to replace a few amino acids in variable regions. Further, simulation models based on three dimensional structures enable introductions of bioactive peptides into even conserved regions. In case of the variable regions, if there is no sequence similar to that of the bioactive peptide, it is possible to introduce bioactive peptides that are not long and highly hydrophobic by replacements and insertions.

The peptide lunasin, which is the small subunit of the soybean MRP GM2S-1, has antimitotic properties and is a potential chemopreventive agent (8,10). As earlier mentioned, the soybean Bowman-Birk protease inhibitor has been shown to suppress various types of cancer: colon cancer (4), prostrate cancer (5,6), oral cancers, and cancers of the head and neck (7).

In this section, we review the research status of the improvement of physiological properties of soybean protein.

4.7.1 Fortifying the Bile Acid Binding Ability

It is well known that soybean proteins have a cholesterol lowering effect in human serum (65,66). Several mechanisms for the regulation of serum cholesterol have been proposed. As a major one, an undigested insoluble fraction of soybean protein interacts with cholesterols and bile acids in the intestine, and they are excreted into an easing (67,68). Most probably, hydrophobic interactions play an important role in these interactions. Makino et al. found that the 114–161 residues in A1aB1b, one of the peptides generated by trypsin digestion of glycinin, have a bile acid binding ability (69). On the other hand, Iwami et al. confirmed that peptides generated by hydrolysis of mild acids contain residues having highly hydrophobic properties derived from β-conglycinin, and these peptides actually bind bile acids (70).

The sequence of 114–161 residues in A1aB1b found by Makino et al. is very similar to those of the corresponding regions of the other glycinin subunits, because the region is located in the conserved region. Further, the cleavage sites for trypsin exist at both sides of the corresponding regions of the other subunits. Makino et al. confirmed that a peptide generated from A2B1a also has bile acid binding activity. However, the activity of peptides generated from any of the subunits except A1aB1b and A2B1a has not been identified. There is a possibility that the peptide derived from A1bB2 was not detected because the sequence of the corresponding region of A1bB2 is very similar to that of A1aB1b, and the amount of A1bB2 in seeds is smaller than those of A1aB1b and A2B1a. All normal cultivars examined so far contain A3B4, whereas some normal cultivars of soybean do not contain A5A4B3. Therefore, Makino et al. probably used a cultivar containing no A5A4B3, and the activity of the corresponding region in A3B4 is weaker than those of A1aB1b and A2B1a.

Our experiments using an affinity column conjugated with cholic acid, one of the bile acids, supports this idea (71). By replacing the corresponding region in A3B4 with the region of 114–161 residues in A1aB1b, the bile acid binding ability can be increased in A3B4. Recently, we constructed deletion mutants of A1a and evaluated their bile acid binding ability by affinity column and their dissociation constants by fluorescence measurement (71). As a result, VAWWMY, the most hydrophobic sequence in the region of 114–161 residues of A1aB1b, was found to be very important in binding bile acids. We can easily fortify the bile acid binding ability of food proteins by using this peptide. For example, the sequences of A3B4 and A5A4B3 corresponding to VAWWMY of A1aB1b are VPYWTY and VPYWTY, respectively. Therefore, a replacement of three amino acids will probably enable the addition a bile acid binding ability into them. However, the folding ability of the mutants in these cases must also be considered, because the amino acids to be replaced are located inside the β-barrel. As described in the section on improvement of nutritional quality, the folding ability of the mutants can be evaluated by simulations before constructing and preparing them. On the other hand, the bile acid binding sequence (VAWWMY) can be introduced into variable regions easily, because it is easy to insert and replace amino acids in these regions. Thus, we introduced the bile acid binding sequence into two sites (M1 and M2) in the variable region IV (hypervariable region) of A1aB1b. The dissociation constants of M1 and M2 for sodium cholate were calculated by fluorescence measurements to be 30 mM and 37 mM, respectively, and that of A1a was 49 mM. These results indicate that the bile acid binding activity of A1a can be fortified as expected.

The bile acid binding peptide sequence can probably be introduced into β-conglycinin by a similar method, because β-conglycinin also has variable regions. Both glycinin and β-conglycinin have β-barrels rich in hydrophobic amino acids. Therefore, the bile acid binding ability of soybean proteins may be fortified by introduction of the bile acid binding sequence (VAWWMY) into the β-barrel based on simulation modeling.

4.7.2 Addition of Phagocytosis Stimulating Activity

Recently, Yoshikawa et al. isolated a phagocytosis stimulating peptide from trypsin digests of soybean proteins (72). This peptide, named soymetide, is derived from the α' subunit of β-conglycinin, and its primary sequence is MITLAIPVNKPGR. Met at its N-terminus is essential for the activity and MITL, the four residues from N-terminus, is the shortest peptide exhibiting phagocytosis stimulation. When the Thr, third residue from the N-terminus of the soymetide, was replaced by Phe or Trp, the activities of the modified peptides greatly increased (Thr $<$ Phe $<$ Trp) (72). Although the regions of the α and β subunits corresponding to the soymetide are highly conserved among the three subunits, and the cleavage sites for trypsin exist at both sides of the regions, their digests by trypsin do not exhibit the activity. The reason is that the residues of the α and β subunits corresponding to the

N-terminal Met in the soymetide, which is essential for the activity, are Leu and Ile, respectively. It was also found that Lys124 in the β subunit was not favorable for phagocytosis stimulation (72).

We attempted to introduce the phagocytosis stimulating peptide sequence into the β subunit by replacing Ile122 and Lys124 of the wild type with Met and Thr/Phe/Trp (I122M/K124T, I122M/K124F and I122M/K124W), respectively (73). These residues are located in the β barrel of the core region and their side chains are inside the molecule. Therefore, there is a possibility that the introduction of the phagocytosis stimulating peptide sequence into the β subunit might induce the loss of the folding ability by causing unfavorable contacts, especially in the case of the replacement of Lys with Phe or Trp, which have side chains bigger than that of Lys.

At first, we simulated the models of the three mutants using the three dimensional structure of the β subunit by the programs Insight II and Discover to confirm whether these mutants could fold correctly or not. The RMSD for the entire Cα atoms between the starting and simulated monomer models was around 0.67 Å for all the mutants. The distance of Cα atoms between the wild type and the simulated models at the positions 122 and 124 were 0.47–0.49 and 0.29–0.50 Å in all the mutants, respectively. These values mean that the position of the backbone scarcely moved. Furthermore, no unfavorable van der Waal's interactions between the side chains of the replaced residues and those of their neighboring residues were observed. Therefore, we concluded that all the mutants could fold correctly. To confirm our assumption that all the mutants could fold correctly based on the molecular modeling, we characterized the structural features of the mutants expressed in *E. coli* by circular dichroism measurement, differential scanning calorimetry, and gel filtration column chromatography. No significant difference in circular dichroism spectra was observed between the wild type and the mutants. This result indicates that the mutations for introducing the phagocytosis stimulating peptide sequence into the β subunit have little effect on the secondary structure. T_m values of all the mutants measured by differential scanning calorimetry were 1.9–3.1°C lower than that of the wild type. Although there is a hydrogen bond between Lys124 and Tyr109 in the β barrel of the wild type, all the mutants lost this hydrogen bond by the replacement of Lys124. Therefore, the loss of the hydrogen bond of the mutants might induce a slight decrease in T_m values. In gel filtration chromatography, all the mutants eluted similarly to the wild type. The results of circular dichroism measurement, differential scanning calorimetry, and gel filtration chromatography indicate that all the mutants folded correctly similar to the wild type. Further, we determined the crystal structure of I122M/K124W to investigate the effect of mutations in detail, because Trp was biggest among the introduced residues (Figure 4.12). The distances of Cα between Ile and Met (residue number 122) and between Lys and Trp (residue number 124) were 0.48 Å and 0.17 Å, respectively. Furthermore, no unfavorable van der Waal's interactions between the side chains of the replaced residues and those of their neighboring residues were observed. These results indicate that the replacement has little influence on the backbone structures and that the assumption about the conformation of I122M/K124W from the simulated model is correct. Further, all the mutants exhibited phagocytosis stimulating activity in the order of I122M/K124T $<$ I122M/K124F $<$ I122M/K124W as expected, and the wild type did not (Figure 4.13). The results confirm that we could introduce the phagocytosis stimulating peptide sequence into the β subunit with the correct folding.

When bioactive peptides are introduced into food proteins, a simulation based on three dimensional structures is a powerful tool to estimate the folding ability of the mutants. In the future, the accumulation of data on three dimensional structures of higher resolutions and on the folding abilities of mutants will make it possible to introduce bioactive peptide sequences exactly into food proteins by protein engineering.

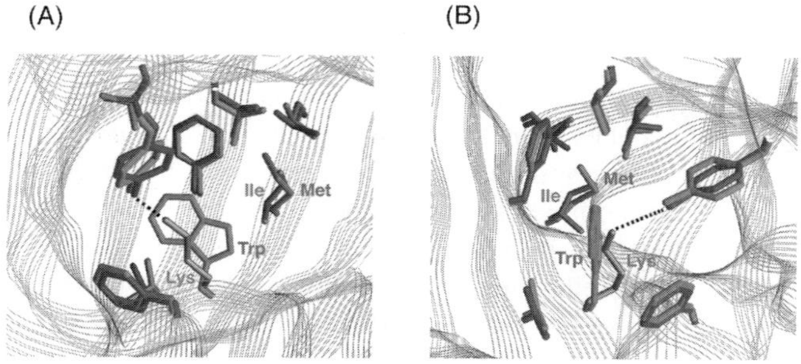

Figure 4.12 Structural comparison of the mutation site of I122M/K124W with corresponding site of the wild type. The Cα traces of wild type and I122M/K124W are represented by lines, respectively. The view in panel (b) is related to that depicted in panel (a) by a rotation of 90°. Dotted line indicates a hydrogen bond. (From: Maruyama, N., Y. Maruyama, T. Tsuruki, E. Okuda, M. Yoshikawa, S. Utsumi, Biochim. Biophys. Acta 1648:99–104, 2003.)

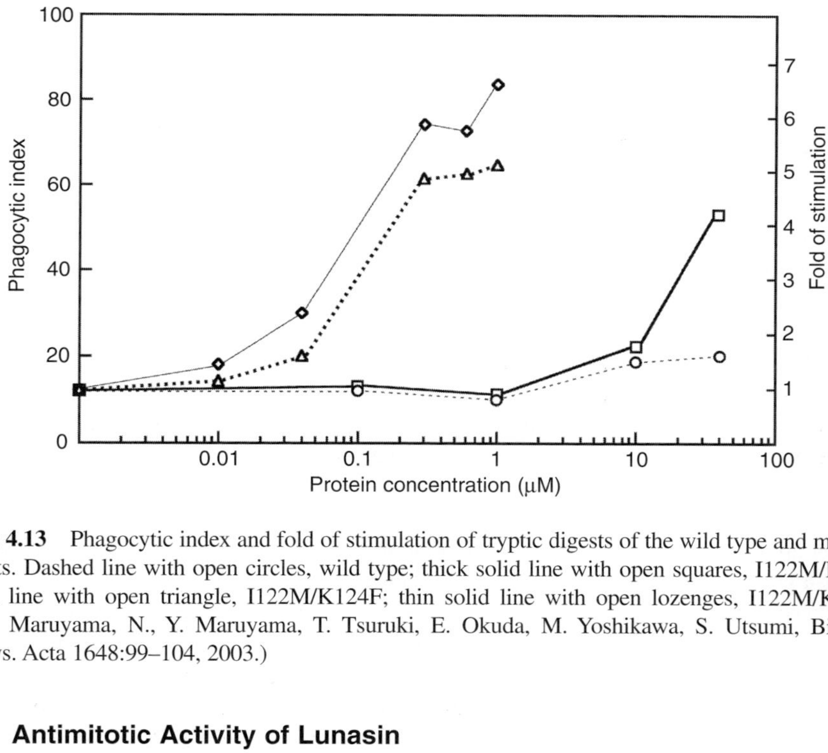

Figure 4.13 Phagocytic index and fold of stimulation of tryptic digests of the wild type and mutant β subunits. Dashed line with open circles, wild type; thick solid line with open squares, I122M/K124T; dashed line with open triangle, I122M/K124F; thin solid line with open lozenges, I122M/K124W. (From: Maruyama, N., Y. Maruyama, T. Tsuruki, E. Okuda, M. Yoshikawa, S. Utsumi, Biochim. Biophys. Acta 1648:99–104, 2003.)

4.7.3 Antimitotic Activity of Lunasin

Lunasin is the small subunit peptide of a seed specific methionine rich 2S albumin (Gm2S-1) (74). It consists of 43 amino acids, and contains the cell adhesion motif Arg-Gly-Asp (RGD) followed by eight aspartic acid residues at its C-terminal end. Lunasin was found to arrest mitosis, leading to cell death, when the lunasin gene was transfected and expressed in mammalian cells (8). In addition to arresting cell division, the lunasin peptide caused abnormal spindle fiber elongation, chromosomal fragmentation, and cell lysis when transiently transfected into murine embryo fibroblast, murine hepatoma, and human breast cancer cells. Deletion of the polyaspartyl tail abolished the biological activity of lunasin. The antimitotic activity of lunasin is therefore ascribed to the binding of its polyaspartyl tail carboxyl end to regions of hypoacetylated chromatin. Galvez et al. (10) have also shown that exogenous application of the peptide inhibits chemical carcinogen induced transformation of murine fibroblast cells to cancerous foci. When applied on the skin, lunasin (250 μg/week) reduces skin tumor incidence by 70%, tumor yield per mouse significantly, and delays tumor appearance by 2 weeks compared to the positive control.

It was estimated that commercial soy products contain about 5.48 mg lunasin/g of protein (defatted soy flour) to 16.52 mg of lunasin/g protein (soy concentrate) (10). However, it is not known whether ingestion of soy products at the FDA recommended daily consumption of 25 g of soy protein (75) will be adequate for chemoprevention. Bioavailability studies of natural and recombinant lunasin are needed to determine the physiological doses of lunasin that will be effective for preventing cancer. At present, the biological role of the methionine rich protein GM2S-1 or lunasin in soybean seeds is not known. Its allergenic potential has also not been established. Lunasin with identical structure and anti mitotic properties has also been isolated from barley (76).

Lunasin is presently produced either by recombinant DNA technology or chemical synthesis (10). It may be possible to add this biologically active peptide to one of the storage proteins of soybean based on three dimensional structure to prevent or minimize changes in structural conformation as described earlier. This can possibly increase the amount of lunasin in the soybean seeds.

4.8 FUTURE TRENDS

Understanding and use of three dimensional structures have made it possible for us to increase or add specific nutritional properties, physicochemical functions, and physiological properties to soybean proteins more accurately and rationally by protein engineering. The nutritional quality of soybean proteins can be improved by adding sulfur containing amino acids, while physiological properties can be enhanced or added by the introduction of bioactive peptide sequences. However, to improve physicochemical function, which is most important for usage as a food material, requires the elucidation of the relationship between structure and physicochemical function in detail. This can be done to a certain extent by analysis of a molecular species consisting of a single subunit prepared by means of an *E. coli* expression system and a limited number of subunits by using the mutant cultivars. However, this is still insufficient, as the relationship between the structure and the physicochemical function are different among a species of proteins. It still remains to be elucidated whether there is a general rule on structure and physicochemical function.

Protein engineering is a powerful tool for describing these subjects. If we can elucidate the relationship between structures and physicochemical functions of soybean proteins at a molecular level by protein engineering in the near future, it is expected that we can develop soybean proteins with excellent physicochemical functions in addition to nutritional and physiological properties. This will contribute to solving the food and health problems of the twenty-first century.

REFERENCES

1. Utsumi, S. Plant food protein engineering. In: *Advances in Food and Nutrition Research*, Kinsella, J.E., ed., San Diego, CA: Academic Press, 1992, Vol. 36, pp 89–208.
2. Kito, M., T. Moriyama, Y. Kimura, H. Kambara. Changes in plasma lipid levels in young healthy volunteers by adding an extruder-cooked soy protein to conventional meals. *Biosci. Biotechnol. Biochem.* 57:354–355, 1993.

3. Aoyama, T., M. Kohno, T. Saito, K. Fukui, K. Takamatsu, T. Yamamoto, Y. Hashimoto, M. Hirotsuka, M. Kito. Reduction by phytate-reduced soybean beta-conglycinin of plasma triglyceride level of young and adult rats. *Biosci. Biotechnol. Biochem.* 65:1071–1075, 2001.
4. Kennedy, A.R., P.C. Billings, X.S. Wan, P.M. Newberne. Effects of Bowman-Birk inhibitor on rat colon carcinogenesis. *Nutr. Cancer* 43:174–186, 2002.
5. Kennedy, A.R., X.S. Wan. Effects of the Bowman-Birk inhibitor on growth, invasion, and clonogenic survival of human prostate epithelial cells and prostate cancer cells. *Prostate* 50:125–133, 2002.
6. Malkowicz, S.B., W.G. McKenna, D.J. Vaughn, X.S. Wan, K.J. Propert, K. Rockwell, S.H. Marks, A.J. Wein, A.R. Kennedy. Effects of Bowman-Birk inhibitor concentrate (BBIC) in patients with benign prostatic hyperplasia. *Prostate* 48:16–28, 2001.
7. Meyskens, F.L. Development of Bowman-Birk inhibitor for chemoprevention of oral head and neck cancer. *Ann. N.Y. Acad. Sci.* 952:116–123, 2001.
8. Galvez, A.F., B.O. de Lumen. A soybean cDNA encoding a chromatin-binding peptide inhibits mitosis of mammalian cells. *Nat. Biotechnol.* 17:495–500, 1999.
9. de Lumen, B.O., A.F. Galvez, M.J. Revilleza, D.C. Krenz. Molecular strategies to improve the nutritional quality of legume proteins. *Adv. Exp. Med. Biol.* 464:117–126, 1999.
10. Galvez A.F., N.Chen, J. Macasieb, B.O. de Lumen. Chemopreventive property of a soybean peptide (lunasin) that binds to deacetylated histones and inhibits acetylation. *Cancer Res.* 61:7473–7478, 2001.
11. Utsumi, S., T. Katsube, T. Ishige, F. Takaiwa. Molecular design of soybean glycinins with enhanced food qualities and development of crops producing such glycinins. In: *Food Proteins and Lipids*, Damodaran, S., ed., New York: Plenum Press, 1997, pp 1–15.
12. Katsube, T., N. Maruyama, F. Takaiwa, S. Utsumi. Food protein engineering of soybean proteins and the development of soy-rice. In: *Engineering Crop Plants for Industrial End Uses*, Shewry, P.R., J.A. Napier, P. Davis, eds., London: Portland Press, 1998, pp 65–76.
13. Derbyshire, E., D.J. Wright, D. Boulter. Legumin and vicilin, storage proteins of legume seeds. *Phytochemistry* 15:3–24, 1976.
14. Wright, D.J. The seed globulins: part II. In: *Developments in Food Proteins: 5*, Hudson, B.J.F., ed., London: Elsevier, 1987, pp 81–157.
15. Maruyama, N., T. Katsube, Y. Wada, M.H. Oh, A.P. Barba de la Rosa, E. Okuda, S. Nakagawa, S. Utsumi. The roles of the N-linked glycans and extension regions of soybean β-conglycinin in folding, assembly and structural features. *Eur. J. Biochem.* 258:854–862, 1998.
16. Utsumi, S., Y. Matsumura, T. Mori. Structure-function relationships of soy proteins. In: *Food Proteins and Their Applications*, Damodaran, S., A. Paraf, eds., New York: Marcel Dekker, 1997, pp 257–291.
17. Thanh, V.H., K. Shibasaki. Heterogeneity of beta-conglycinin. *Biochim. Biophys. Acta* 439:326–338, 1976.
18. Thanh, V.H., K. Shibasaki. Major proteins of soybean seeds: subunit structure of β-conglycinin. *J. Agric. Food Chem.* 26:692–695, 1978.
19. Utsumi, S., H. Inaba, T. Mori. Heterogeneity of soybean glycinin. *Phytochemistry* 20:585–589, 1981.
20. Takahashi, K., H. Banba, A. Kikuchi, M. Ito, S. Nakamura. An induced mutant line lacking the α subunit of β-conglycinin in soybean [*Glycine max* (L.) Merril]. *Breed. Sci.* 44:65–66, 1994.
21. Takahashi, K., Y. Mizuno, S. Yumoto, K. Kitamura, S. Nakamura. Inheritance of the α-subunit deficiency of β-conglycinin in soybean [*Glycine max* (L.) Merril] line induced by γ-ray irradiation. *Breed. Sci.* 46:251–255, 1996.
22. Yagasaki, K., N. Kaizuma, K. Kitamura. Inheritance of glycinin subunits and characterization of glycinin molecules lacking the subunits in soybean [*Glycine max* (L.) Merril]. *Breed. Sci.* 46:11–15, 1996.
23. Yagasaki, K., T. Takagi, M. Sasaki, K. Kitamura. Biochemical characterization of soybean protein consisting of different subunits of glycinin. *J. Agric. Food Chem.* 45:656–660, 1997.
24. Lawrence, M.C., T. Izard, M. Beuchat, R.J. Blagrobe, P.M. Colman. Structure of phaseolin at 2.2 Å resolution: implications for a common vicilin/legumin structure and the genetic engineering of seed storage proteins. *J. Mol. Biol.* 238:748–776, 1994.

25. Ko, T.-P., J.D. Ng, A. McPherson. The three-dimensional structure of canavalin from jack bean (*Canavalia ensiformis*). *Plant Physiol.* 101:729–744, 1993.
26. Lawrence, M.C., E. Suzuki, J.N. Varghese, P.C. Davis, A. Van Donkelaar, P.A. Tulloch, P.M. Colman. The three-dimensional structure of the seed storage protein phaseolin at 3 Å resolution. *EMBO J.* 9:9–15, 1990.
27. Maruyama, N., M. Adachi, K. Takahashi, K. Yagasaki, M. Kohno, Y. Takenaka, E. Okuda, S. Nakagawa, B. Mikami, S. Utsumi. Crystal structures of recombinant and native soybean β-conglycinin β homotrimers. *Eur. J. Biochem.* 268:3595–3604, 2001.
28. Wright, D.J. The seed globulins: part II. In: *Developments in Food Proteins: 6.* London: Elsevier, 1987, pp 119–178.
29. Adachi, M., Y. Takenaka, A.B. Gidamis, B. Mikami, S. Utsumi. Crystal structure of soybean proglycinin A1aB1b homotrimer. *J. Mol. Biol.* 305:291–305, 2001.
30. Adachi, M., J. Kanamori, T. Masuda, K. Yagasaki, K. Kitamura, M. Bunzo, S. Utsumi. Crystal structure of soybean 11S globulin: glycinin A3B4 homohexamer. *Proc. Natl. Acad. Sci. USA*, 100:7395–7400, 2003.
31. Dunwell, J.M. Cupins: a new superfamily of functionally diverse proteins that include germins and plant seed storage proteins. *Biotechnol. Genet. Eng. Rev.* 15:1–32, 1998.
32. Dunwell, J.M., P.J. Gane. Microbial relatives of seed storage proteins: conservation of motifs in a functionally diverse superfamily of enzymes. *J. Mol. Evol.* 46:147–154, 1998.
33. Dunwell, J.M., A. Culham, C.E. Carter, C.R. Sosa-Aguirre, P.W. Goodenough. Evolution of functional diversity in the cupin superfamily. *Trends Biochem. Sci.* 26:740–746, 2001.
34. Anand, R., P.C. Dorrestein, C. Kinsland, T.P. Begley, S.E. Ealick. Structure of oxalate decarboxylase from *Bacillus subtilis* at 1.75 Å resolution. *Biochemistry* 41:7659–7669, 2002.
35. Maruyama, N., R. Sato, Y. Wada, Y. Matsumura, H. Goto, E. Okuda, S. Nakagawa, S. Utsumi. Structure-physicochemical function relationships of soybean β-conglycinin constituent subunits. *J. Agric. Food Chem.* 47:5278–5284, 1999.
36. Utsumi, S., N. Maruyama, R. Satoh, M. Adachi. Structure-function relationships of soybean proteins revealed by using recombinant systems. *Enzyme Microb. Tech.* 30:284–288, 2002.
37. Maruyama, N., M.S Mohamad Ramlan, K. Takahashi, K. Yagasaki, H. Goto, N. Hontani, S. Nakagawa, S. Utsumi. The effect of the N-linked glycans on structural features and physicochemical functions of soybean β-conglycinin homotrimers. *J. Am. Oil Chem. Soc.* 79:139–144, 2002.
38. Maruyama, N., M.S. Mohamad Ramlan, K. Takahashi, K. Yagasaki, H. Goto, N. Hontani, S. Nakagawa, S. Utsumi. Structure-physicochemical function relationships of soybean β-conglycinin heterotrimers. *J. Agric. Food. Chem.* 50:4323–4326, 2002.
39. Maruyama, Y., N. Maruyama, B. Mikami, S. Utsumi. Structure of the core region of the soybean β-conglycinin α' subunit. *Acta Cryst.* D60:289–297, 2004.
40. Pace, C.N. Contribution of the hydrophobic effect to globular protein stability. *J. Mol. Biol.* 226:29–35, 1992.
41. Yip, K.S., T.J. Stillman, K.L. Britton, P.J. Artymiuk, P.J. Baker, S.E. Sedelnikova, P.C. Engel, A. Pasquo, R. Chiaraluce, V. Consalvi. The structure of *Pyrococcus furiosus* glutamate dehydrogenase reveals a key role for ion-pair networks in maintaining enzyme stability at extreme temperatures. *Structure* 3:1147–1158, 1995.
42. Lim, J.H., Y.G. Yu, Y.S. Han, S. Cho, B.Y. Ahn, S.H. Kim, Y. Cho. The crystal structure of an Fe-superoxide dismutase from the hyperthermophile *Aquifex pyrophilus* at 1.9 Å resolution: structural basis for thermostability. *J. Mol. Biol.* 270:259–274, 1997.
43. Karshikoff, A., R. Ladenstein. Ion pairs and the thermotolerance of proteins from hyperthermophiles: a "traffic rule" for hot roads. *Trends Biochem. Sci.* 26:550–556, 2001.
44. Spassov, V.Z., A.D. Karshikoff, R. Ladensten. The optimization of protein-solvent interactions: thermostability and the role of hydrophobic and electrostatic interactions. *Protein Sci.* 4:1516–1527, 1995.
45. Matthews, B.W., H. Nicholson, W.J. Becktel. Enhanced protein thermostability from site-directed mutations that decrease the entropy of unfolding. *Proc. Natl. Acad. Sci. USA* 84:6663–6667, 1987.

46. Gerhard, V., W. Stefanie, A. Patrick. Protein thermal stability, hydrogen bonds, and ion pairs. *J. Mol. Biol.* 269:631–643, 1997.
47. Müntz, K., V. Christov, G. Saalbach, I. Saalbach, D. Waddell, T. Pickardt, O. Schieder, T. Wustenhagen. Genetic engineering for high methionine grain legumes. *Nahrung* 42:125–127, 1998.
48. de Lumen, B.O., H. Uchimiya. Molecular strategies to enhance the nutritional quality of legume protein: an update. *AgBiotech. News Info.* 9:53N–58N, 1997.
49. Kim, C.S., S. Kamiya, T. Sato, S. Utsumi, M. Kito. Improvement of nutritional value and functional properties of soybean glycinin by protein engineering. *Protein Eng.* 3:725–731, 1990.
50. Utsumi, S., S. Kitagawa, T. Katsube, I.J. Kang, A.B. Gidamis, F. Takaiwa, M. Kito. Synthesis, processing and accumulation of modified glycinins of soybean in the seeds, leaves and stems of transgenic tobacco. *Plant Sci.* 92:191–202, 1993.
51. Takaiwa, F., T. Katsube, S. Kitagawa, T. Hisaga, M. Kito, S. Utsumi. High level accumulation of soybean glycinin in vacuole-derived protein bodies in the endosperm tissue of transgenic tobacco seed. *Plant Sci.* 111:39–49, 1995.
52. Utsumi, S., S. Kitagawa, T. Katsube, T. Higasa, M. Kito, F. Takaiwa, T. Ishige. Expression and accumulation of normal and modified soybean glycinins in potato tubers. *Plant Sci.* 102:181–188, 1994.
53. Katsube, T., N. Kurisaka, M. Ogawa, N. Maruyama, R. Ohtsuka, S. Utsumi, F. Takaiwa. Accumulation of soybean glycinin and its assembly with the glutelins in rice. *Plant Physiol.* 120:1063–1074, 1999.
54. Okuda, E., M. Adachi, B. Mikami, S. Utsumi. Significance of the salt bridge Asp157-Arg161 for structural stability of soybean proglycinin A1aB1b homotrimer. [In preparation]
55. Mori, T., S. Utsumi, H. Inaba, K. Kitamura, K. Harada. Differences in subunit composition of glycinin among soybean cultivars. *J. Agric. Food. Chem.* 29:20–23, 1981.
56. Nakamura, T., S. Utsumi, K. Kitamura, K. Harada, T. Mori. Cultivar differences in gelling characteristics of soybean glycinin. *J. Agric. Food Chem.* 32:647–651, 1984.
57. Mori, T., T. Nakamura, S. Utsumi. Formation of pseudoglycinins and their gel hardness. *J. Agric. Food Chem.* 30:828–831, 1982.
58. Nakamura, T., S. Utsumi, T. Mori. Formation of pseudoglycinins from intermediary subunits of glycinin and their gel properties and network structure. *Agric. Biol. Chem.* 49:2733–2740, 1985.
59. Utsumi, S., A.B. Gidamis, J. Kanamori, I.J. Kang, M. Kito. Effect of deletion of disulfide bonds by protein engineering on the conformation and functional properties of soybean proglycinin. *J. Agric. Food Chem.* 41:687–691, 1993.
60. Adachi, M., E. Okuda, Y. Kaneda, A. Hashimoto, A.D. Shutov, C. Becker, K. M ntz, S. Utsumi. Crystal structures and structural stabilities of the disulfide bond-deficient soybean proglycinin mutants C12G and C88S. *J. Agric. Food Chem.*, 51:4633–4639, 2003.
61. Fujita, H., R. Sasaki, M. Yoshikawa. Potentiation of the antihypertensive activity of orally administered ovokinin, a vasorelaxing peptide derived from ovalbumin, by emulsification in egg phosphatidylcholine. *Biosci. Biotechnol. Biochem.* 59:2344–2345, 1995.
62. Takahashi, M., S. Moriguchi, M. Ikeno, S. Kono, K. Ohata, H. Usui, K. Kurahashi, R. Sasaki, M. Yoshikawa. Studies on the ileum-contracting mechanism and identification as a complement C3a receptor agonist of oryzatensin, a bioactive peptide derived from rice albumin. *Peptides* 17:5–12, 1996.
63. Fujita, H., H. Usui, K. Kurahashi, M. Yoshikawa. Isolation and characterization of ovokinin, a bradykinin B1 agonist peptide derived from ovalbumin. *Peptides* 16:785–790, 1995.
64. Fujita, H., M. Yoshikawa. LKPNM: a prodrug type ACE-inhibitory peptide derived from fish protein. *Immunopharmacology* 44:123–127, 1999.
65. Nagata, C., N. Takatsuka, Y. Kurisu, H. Shimizu. Decreased serum total cholesterol concentration is associated with high intake of soy products in Japanese men and women. *J. Nutr.* 128:209–213, 1998.
66. Wong, W.W., E.O. Smith, J.E. Stuff, D.L. Hachey, W.C. Heird, H.J. Pownell. Cholesterol-lowering effect of soy protein in normocholesterolemic and hypercholesterolemic men. *Am. J. Clin. Nutr.* 68:1385S–1389S, 1998.

67. Sugano, M., Y. Yamada, K. Yoshida, Y. Hashimoto, T. Matsuo, M. Kishimoto. The hypocholesteroletic action of the undigested fraction of soybean protein in rats. *Atherosclerosis* 72:115–122, 1988.
68. Sugano, M., S. Goto, Y. Yamada, K. Yoshida, Y. Hashimoto, T. Matsuo, M. Kimoto. Cholesterol-lowering activity of various undigested fractions of soybean protein in rats. *J. Nutr.* 120:977–985, 1990.
69. Makino, S., H. Nakashima, K. Minami, R. Moriyama, S. Takano. Bile acid-binding protein from soybean seed: isolation, partial characterization and insulin-stimulating activity. *Agric. Biol. Chem.* 52:803–809, 1988.
70. Kato, N., K. Iwami. Resistant protein: its existence and function beneficial to health. *J. Nutr. Sci. Vitaminol. (Tokyo)* 48:1–5, 2002.
71. Choi, S.K., M. Adachi, S. Utsumi. Identification of the bile acid-binding region in the soy glycinin A1aB1b subunit. *Biosci. Biotechnol. Biochem.* 66:2395–2401, 2002.
72. Tsuruki, T., K. Kishi, M. Takahashi, M. Tanaka, T. Matsukawa, M. Yoshikawa. Soymetide, an immunostimulating peptide derived from soybean β-conglycinin, is an fMLP agonist. *FEBS Lett.* 540:206–210, 2003.
73. Maruyama, N., Y. Maruyama, T. Tsuruki, E. Okuda, M. Yoshikawa, S. Utsumi. Creation of soybean β-conglycinin β with strong phagocytosis-stimulating activity. *Biochim. Biophys. Acta* 1648:99–104, 2003.
74. Galvez, A.F., M.J.R. Revilleza, B.O. de Lumen. A novel methionine-rich protein from soybean cotyledon: cloning and characterization of cDNA (Accession No. AF005030) (PGR97-103). *Plant Physiol.* 114:1567, 1997.
75. FDA Talk Paper. *FDA approves new health claim for soy protein and coronary heart disease: FDA, United States Department of Heath and Human Services, October 26, 1999.* Washington DC: United States Government Printing Office, 1999.
76. Jeong, H.J., Y. Lam, B.O. de Lumen. Barley lunasin suppresses ras-induced colony formation and inhibits core histone acetylation in mammalian cells. *J. Agric. Food Chem.* 50(21):5903–5908, 2002.

2.05

Genetic Modification of Plant Starches for Food Applications

Jeffrey D. Klucinec and Peter L. Keeling

CONTENTS

5.1 INTRODUCTION

Starch is a unique natural material, valued for its uses in food, feed, and industry. It is found in higher plants, mosses, ferns, and some microorganisms where it serves as an important store of energy. In higher plants, starch is deposited as transitory starch in leaves and as storage starch in specialized storage organs such as seeds or tubers. Starch is also an important component of many fruit crops such as apple, pear, melon, banana, and tomato. Storage starch is one of the main components of cereal grain (seeds) harvested from crops like wheat, maize, oats, barley, sorghum, and rice as well as of tubers harvested from crops like cassava, yam, and potato. Whether in its native state in grain or tubers, or in isolated granular form, starch is a convenient stable material, cheap to produce, suitable for long term storage without spoilage, convenient for high volume transport, and an important source of calories, retaining functional properties for use in many potential product applications. Grain and tubers are often used directly for animal feed or human food, with little or no processing, such as cooked whole cereal grain or potatoes. Cereal grain is also ground or milled to make flour or meal, which is subsequently mixed with other ingredients and cooked to make breads and pastries. Alternatively, starch may be extracted from the storage organs and the purified starch used as a key functional ingredient added to foodstuffs such as pie fillings, puddings, soups, sauces, gravies, coatings, candies, confectionary products, yogurts, and other dairy products. Extracted starch also has many nonfood industrial uses, such as paper sizing aids, textile sizing aids, molded plastics, ceramics, dye carriers, or suspension aids. Globally, starch is an essential commodity providing most (~80%) of the worlds calories. This vital commodity supply comes from just six different plant species: three cereal crops (rice, maize, and wheat) and three tuber crops (potato, yam, and cassava).

As a result of advances in genetics and biochemistry, we have discovered much about how starch is synthesized in crop plants. Furthermore we have also unraveled the biochemical and genetic basis of some useful natural genetic variations that affect starch synthesis and consequently starch structure and functionality. Some of these variants are already commercially exploited. Examples include variants that accumulate less starch and more sugar (e.g., sweet peas, sweet corn, sweet potato) and others that cook to form clear sols rather than opaque gels (e.g., waxy corn, waxy rice, waxy wheat) and yet others that are useful industrially (e.g., amylose extender corn), and finally others valued for imparting stickiness when cooked (e.g., indica vs. japonica rice). Further progress in this area depends upon improvements in our understanding of the relationship between starch synthetic genes and enzymes, starch structure and functionality. Thus, by linking these findings with further advances in our understanding of the genes required for starch synthesis, an opportunity has appeared for us to make starches with increased usefulness and value.

This paper seeks to pull together the disciplines of biochemistry, genetics, biotechnology, and food technology of plant starches. First we will review current knowledge of starch structure and how starch is synthesized in plants. The primary focus of this review will be storage starch because this provides the main source of food starch today. Next we will summarize the effects on starch composition, physical properties, and functionality due to genetic modifications that cause changes in starch biosynthetic enzymes. Finally we will focus on food applications that might benefit from genetic modifications of crop plants and discuss future opportunities coming from traditional plant breeding and modern biotechnology.

5.2 STRUCTURE

Physically, after extraction and drying, normal starch is a white powder consisting of a mixture of amylose and amylopectin in semicrystalline granules. Starch granules are microscopic structures approximately 0.5 to 100 μm in diameter. In shape, they are spherical, elliptical, or polyhedral. The size and morphology of starch granules is characteristic of the organ and species in which they are produced (102). Starch granules appear rather similar in size and morphology with and without amylose. Under most environmental conditions, starch granules can be considered moderately inert with little capacity to hold water. These characteristics of starch granules make them ideal vessels for storage and shipping, whether in grain or tubers or from processed isolated starch.

Chemically, starch is classified as a complex carbohydrate and is a mixture of two polymers of glucose: amylose and amylopectin. Amylose is a generally linear α-1,4 glucan which is sometimes lightly branched with α-1,6-glycosidic linkages. Amylopectin is normally higher in molecular weight than amylose. It is also an α-1,4 glucan, but is highly branched with α-1,6-glycosidic linkages. The proportions on a dry weight basis of amylose and amylopectin in starches isolated from storage tissues like potato tubers or cereal grain is normally between 20 to 30 percent amylose and 70 to 80 percent amylopectin. In addition to amylose and amylopectin, granules contain small quantities of protein and lipid. Between species there is variation in the structure of amylopectin (104), the size and structure of amylose (84,207,209,211,213), and the nature and amounts of proteins (77,78,155) and lipids (154,214). Because starch physical behavior is dependent on all of these components (55,70,104,116–119) there are specific uses of starches from different species. In addition, within a given species, rare examples have been found of grains, tubers, or roots producing starches that deviate from the typical amylose to amylopectin ratio or have altered amylopectin structure. These plants have been selected because of their unique cooking behavior due to their unusual starch composition that confers unique properties to the crop storage organ. Some of these natural variants are now cultivated on a commercial scale.

Examination of starch structure began over 60 years ago (191). The first widely accepted model shows starch as a branched structure with alternating regions of higher branching density separated by more lightly branched regions (110). A more widely accepted model shows amylopectin arranged in alternating clusters (178,179). Based on the chain length profile of debranched amylopectin and a refinement of the cluster model, the amylopectin chains were categorized into type A, B1, B2, B3, B4 and a single C chain (82). Recently, three refinements for the different modes of interconnection of the amylopectin clusters were presented (82,216) (See Figure 5.1). In starch granules, some of the chains of amylopectin are believed to be associated with one another through hydrogen bonding, forming double helices. The double helices either form higher ordered crystalline structures or may exist independently of crystalline order. The double helices are oriented radially within the granule, with the reducing ends of the chains oriented toward the center or hilum of each granule. Within the granule, crystalline regions, often referred to as growth rings, are separated in a radial fashion from each other by amorphous regions. The crystalline regions are further subdivided into amorphous and crystalline lamellae, which have a periodicity between clusters of approximately 9 nm (105). The branch points in amylopectin are believed to be the primary component of the amorphous lamellae, with the ordered amylopectin side chain double helices clustered in the crystalline lamellae. Differences in the internal chain lengths of amylopectin affect starch crystallinity (163). Important new insights into how amylopectin chain architecture may affect packing have been advocated based on small angle x-ray scattering studies and analogies with liquid crystals (230–233). Using these models it is possible to discuss the mechanisms and kinetics of interchain associations in the context of visualizing starch as a liquid crystalline polymer having different degrees of crystalline order depending on physical conditions.

Amylose contributes to the overall crystallinity of normal starch through the formation of crystalline complexes of amylose with lipids and, it is believed, through participation

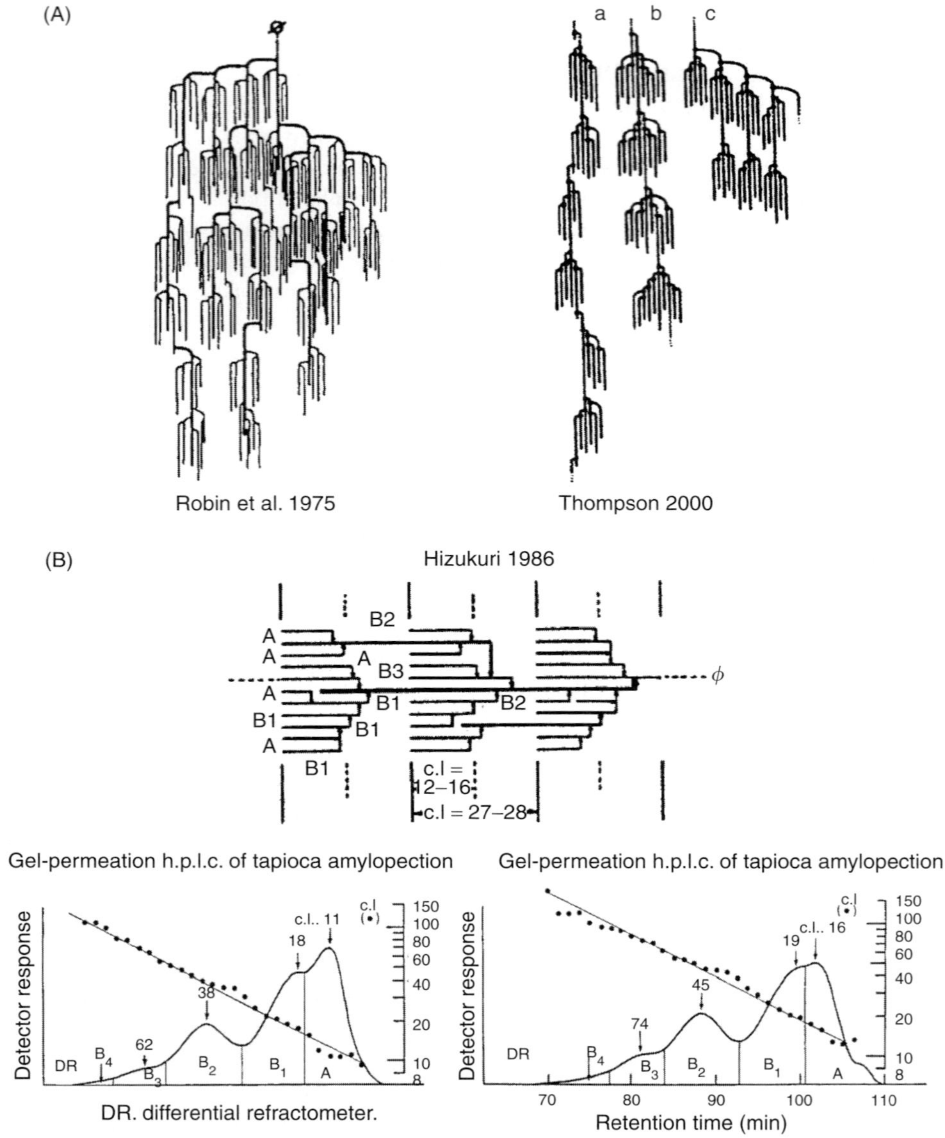

Figure 5.1 (A) Mode of Interconnection of clusters (B) Cluster models of amylopectin

in some starch double helices. However, although amylose readily forms double helices in solution, it readily leaches out of hydrated starch at granule gelatinization temperatures (68). This observation shows that amylose in starch granules does not associate in the same manner inside granules as it does outside of the granule. This observation additionally raises some intriguing questions about the nature of amylose synthesis and the kinetic trapping of amylose within starch granules.

While normal starch can be readily separated into two components, amylose and amylopectin, studies of high amylose starches have revealed what some authors describe as a third starch component, sometimes termed intermediate starch. For normal starches containing ~25% amylose, the results from quantification by iodine inclusion complex formation

(11,139,181,127), or precipitation in the presence of complexing agents (191,192,126,240,241) are in general agreement. However, for high amylose starches, these methods give different results. Thus, from detailed observations, high amylose starches were initially considered to contain either normal amylopectin and short chain linear amylose (1,2,75) or longer chain amylopectin compared with amylopectin from normal starch (148,149). Liquid chromatographic studies (140) have confirmed the latter. Additional studies of high amylose starches by differential precipitation with 1-butanol combined with Sepharose 2B chromatography (245) revealed an inability to clearly separate amylose from amylopectin. Further fractionation revealed high molecular weight material in what traditionally was the amylose fraction (8,9,209). Later studies showed that this material can be removed by repeated precipitation with 1-butanol and was most likely contaminating amylopectin (209,213), while removal of the low molecular weight material in the amylopectin fraction using differential precipitation techniques has not been successful. In summary, starch from high amylose mutants appears to contain a significant amount of an amylopectin-like component having an altered architecture. This intermediate starch component is characterized by:

1. An inability to precipitate with 1-butanol
2. An ability to eluete within the same molecular size range as amylose
3. An ability to bind iodine and having a lambda maximum between amylose and amylopectin

Estimates of intermediate material defined in this way have exceeded 55% (w/w) of the total starch of high amylose mutants (10).

Other workers (116,209,239) have defined intermediate starch to be the material that has the ability to precipitate in the presence of some complexing agents (e.g., 2-nitropropane) or mixtures of complexing agents (1-butanol with isoamyl alcohol) but not others (e.g., 1-butanol, 1-nitropropane). This type of intermediate material obtained by differential precipitation is a relatively small proportion of normal starches (<10%) (116,209,220,239) irrespective of the amylose content of the high amylose starch. The molecules have been considered by some as amylopectin molecules with long external chains and a limited capacity to form clathrate complexes and precipitate (116,209,220,239). It has also been suggested (209) that this type of intermediate material could be a mixture of amylopectin and a small amount of contaminating amylose, which might be the case if the amylopectin from high amylose starch has an overall molecular size similar to that of amylose from normal starch (116).

Amylopectin molecules within granules are believed to be organized radially, with the long C-chain innermost (84). As a result of high magnification microscopy studies it was proposed that the radially oriented amylopectin clusters are organized into super helices, which may relate to the blocklets seen in microscopy studies (166). In turn the super helices may be responsible for the formation of concentric spherical rings. Although these growth rings are a characteristic of all starch granules, the mechanisms determining their formation are still not well understood (60,170). Recently, Bertoft proposed a bidirectional backbone model, where the super helix could be organized so that the longer amylopectin chains are oriented in line with the super helix, while the amylopectin clusters may be oriented radially (18) (Figure 5.2).

X-ray crystallography has shown that there are three distinct types of crystalline order in starch: A-type, B-type and V-type. V-type crystallinity is often associated with crystalline packing of amylose lipid complexes. The A-type and B-type starches differ in the organization of helices: A-type crystals are densely packed hexagonal arrays of double helices, B-type crystals, though also double helices packed in a hexagonal array, are unlike A-type

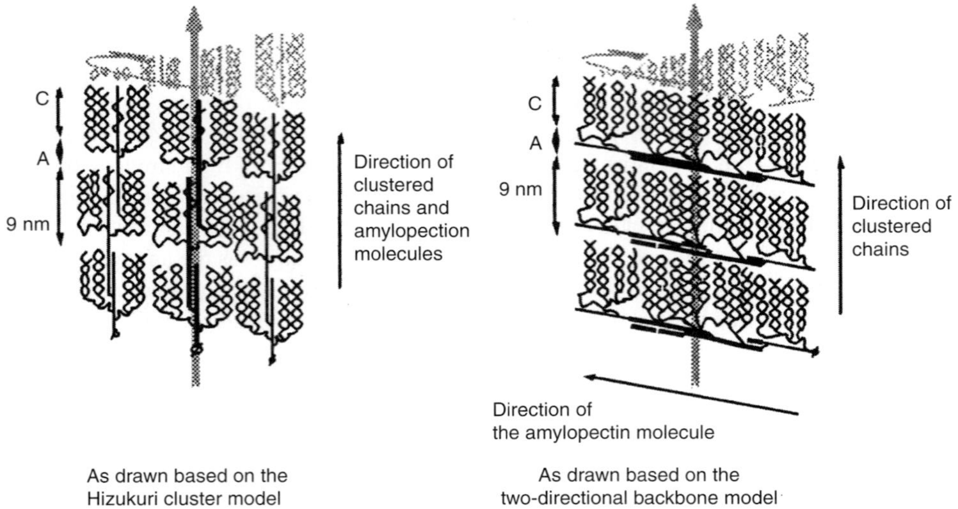

Figure 5.2 The super helix model as drawn in the Hizukuri cluster model and modified based on the two-directional backbone model.

crystals because they have an open central cavity which results in an increased water content (92). C-type starches are a mixture of A-type and B-type crystallinity (22,27,69).

5.3 SYNTHESIS

For an overview of starch synthesis, the reader is referred to recent reviews (99,151,218). In general terms, it is important to note that storage organs are composed of individual starch storing cells and each cell contains several subcellular compartments: including particularly the cytosol and amyloplast compartments. Sucrose, made in the leaves, is transported to the storage organ where it is imported into the cytosolic compartment of each cell. A well characterized pathway (see Figure 5.3) of starch synthesis achieves the enzymatic conversion of sucrose to starch. The first part of this pathway is localized in the cytosol, while the final steps of starch synthesis are located in a specialized subcellular compartment called the amyloplast.

In the cytosol, the glucosyl and fructosyl moieties of sucrose are converted into sugar phosphates. One of these types of hexose phosphates, glucose-1-phosphate is converted into ADP-glucose by the enzyme ADP-glucose pyrophosphorylase (AGPase), the first committed step in starch synthesis. In the endosperm of monocot crops, AGPase is located predominantly in the cytosol, whereas in the storage organs of dicot crops, AGPase is located predominantly in the amyloplast. Thus in monocot crops, ADP-glucose is transported into the amyloplast, while in dicot crops hexose phosphate is imported to support ADP-glucose synthesis inside the amyloplast. The conversion of ADP-glucose to starch is performed by several enzymes, which include, but may not be limited to, soluble starch synthase (SS), granule bound starch synthase (GBSS), starch branching enzymes (SBE), and isoamylase (ISA). Other enzymes that also may be involved in starch synthesis include phosphorylase (PHO), pullulanase (PU) and disproportionating enzyme (DisPE). GBSS is involved primarily in amylose biosynthesis and SSs, SBEs, and ISAs are involved in the biosynthesis of amylopectin.

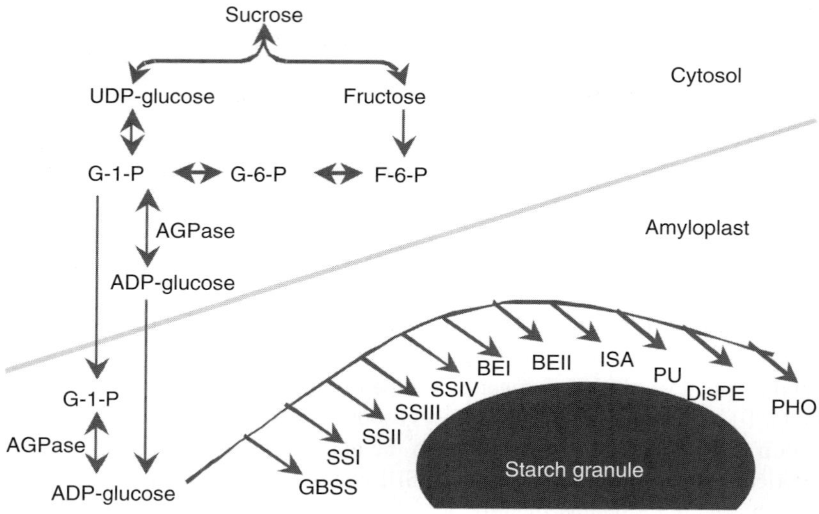

Figure 5.3 Pathway of starch biosynthesis

Across species, well conserved families of genes encode the enzymes in the pathway. Thus, diverse plant species possess similar sets of multiple forms (isoforms) of each enzyme. The conservation of isoforms suggests that each has a unique role in the process of starch synthesis. This idea has been largely confirmed in several plant species by studies of genetic modifications in which particular isoforms have been overexpressed or reduced.

5.3.1 Granule Bound Starch Synthase

Granule bound starch synthase (GBSS) is a member of the glycosyltransferase class of enzymes more generally known as starch synthases. GBSS is more formally named ADPglucose:1,4-α-D-glucan-4-α-D-glucosyltransferase (E.C. 2.4.1.21). In storage organs there are several starch synthase proteins associated with starch. However, only one isoform (GBSSI) is found predominantly associated with starch, and based on studies of amylose free mutants which lack the GBSSI protein, this isoform appears to be exclusively responsible for amylose synthesis in maize (200) and other plant species (44,88). There is a similar isoform in leaves (229). GBSSI has been identified, mapped and cloned from several species (218). The activity of GBSSI is correlated with the product of the waxy, lam, and amf locus of cereals, pea, and potato respectively (44,88,89,160,200).

As well as associating with starch granules, another key attribute of GBSSI is its ability to elongate a growing glucan chain processively. This means that the enzyme does not necessarily dissociate from the glucan chain after the addition of the first glucose but can remain associated with it to add further glucose units before finally dissociating. Soluble starch synthases enzymes in comparison add single glucose units per encounter with the glucan chain (distributive elongation). This processive mechanism of elongation by GBSSI is consistent with the proposed role of GBSSI in amylose synthesis (45,46). One of the more intriguing aspects of the function of GBSSI is that in isolated granules, its amylose synthesizing activity appears to be dependent on the presence of low concentrations of maltooligosaccharides (45,46). These experiments with isolated granules also showed that GBSSI can elongate glucan chains within amylopectin, a finding that may explain why mutants lacking GBSSI have altered amylopectin structure as well as lacking amylase (83,172,251).

5.3.2 Soluble Starch Synthase

Soluble starch synthase (SS) isoforms are all members of the glycosyltransferase class of enzymes. SS is more formally named ADPglucose:1,4-α-D-glucan-4-α-D-glucosyltransferase (E.C. 2.4.1.21). Like GBSSI, the SS enzymes are also associated with starch granules, but are also present in the stroma of the amyloplast, leading to their classification as soluble starch synthase. SS catalyses the transfer of glucose from ADP-glucose to an acceptor glucan chain, by a mechanism that has been described as distributive (46).

Four SS forms have been identified, mapped and cloned from several species (218). Although the various isoforms are believed to be present in all starch synthesizing cells, they appear to have different relative activities in different species and tissues. SS is believed to be primarily responsible for amylopectin synthesis, and there is evidence that each isoform is responsible for the synthesis of different chain lengths. SSI has been cloned and mapped from several species, but the only known mutant type described so far is in rice (159). SSI appears to be primarily responsible for synthesizing shorter chains of amylopectin based on biochemical studies of the isolated enzyme (38) and structural studies of the rice mutant (159). The SSII and SSIII isoforms appear to be involved in synthesizing the intermediate chains of amylopectin, as evidenced by changes observed with mutant and transgenic rice (40,63,221). The precise role of SSIV remains to be elucidated, and it may even be revealed that SSIV is a subclass of the SSIII isoform. All of these classifications are consistent with evidence obtained from starch isolated from potatoes transformed with antisense constructs to the potato enzymes (51,68).

5.3.3 Branching Enzyme

Starch branching enzyme (SBE) is a member of the glycosyltransferase class of enzymes. The branching enzymes are more formally and collectively named 1,4-α-D-glucan:1,4-α-D-glucan 6-α-D-(1,4-α-D-glucano)-transferase (E.C. 2.4.1.18). Like the SS enzymes, the SBEs appear to have a weak association with the starch granule as evidenced by the fact that these enzymes may be found entrapped inside starch granules. SBEs catalyze the interchain cleavage of a glucan chain with subsequent bonding of the cleaved portion to the parent glucan chain. SBEs act by cutting an α-1,4 linkage forming a new linkage between this cleaved glucan chain and an adjacent glucan chain via an α-1,6 linkage. There appear to be at least two major classes of highly conserved SBEs across plant species (SBEI and SBEII). Thus, like the soluble starch synthases, SBEs appear to have a distinct role in the mechanism of amylopectin structure development. The SBEs must also play some role in amylose synthesis, as amylose is lightly branched. All known SBEs have been mapped, cloned, and sequenced from several plant species. SBEIIb appears to be primarily responsible for transferring longer chains of amylopectin based on biochemical studies of the isolated enzyme (76,212) and structural studies of the maize and rice mutant (162,205,212). The role of SBEI and SBEIIa is less clear as mutants of SBEI in rice have reduced intermediate and long chains (159)while in maize the chain lengths are unaffected in mutants of SBEI and SBEIIa (20,21).

5.3.4 Debranching Enzyme

Debranching enzymes are members of the amylase class of enzymes. There appear to be at least four debranching enzymes in the storage organs: three isoforms of Isoamylase (ISA) and one Pullulanase (PU). Although ISA is known to catalyze the selective cleavage of α-1,6 linkages in branched glucans, their precise catalytic role in starch granule biosynthesis remains unresolved. Plants lacking ISAI accumulate higher levels of sucrose, more starch granules and phytoglycogen and have reduced amounts of starch, strongly indicating that ISAI is required for normal starch biosynthesis (28,29,59,122). Some studies have highlighted the involvement of ISA in starch granule initiation, probably mediated by degradation of soluble glucans which would otherwise initiate granule formation (28,29,122). The role of ISAII and ISAIII in starch biosynthesis remains unresolved at the present time, as no mutants are known to exist. What little is known about their catalytic properties has led to the proposal that each isoform has retained a specific role (90). Studies in maize of the limit dextrinase/pullulanase (PUI) class of debranching enzymes has successfully identified a mutant, which on its own appears to have minimal effects on starch deposition (13,47).

5.4 MODIFICATIONS

Most of the maize starch mutants affecting starch structure were discovered before the genes and enzymes responsible for starch synthesis were known. Their names are therefore based upon the phenotypic changes observed in that storage organ or on the properties of the starch, rather than that of the gene or enzyme affected by the mutation. For example, numerous mutant phenotypes have been reported for maize (73,36) and several phenotypes (e.g., waxy, amylose extender, dull, shrunken, sugary-2, and sugary) have been described extensively with regard to their effects on carbohydrate composition and response to genetic background, allelic dosage, or interaction with other mutations (42,64,65,87). Examination of maize kernels with differing starch phenotypes has been instrumental in determining which enzymes are required for starch synthesis in this storage organ (24,194). Mutations that are responsible for most of the abnormal starch phenotypes have been located in genes encoding starch synthetic enzymes. Related isoforms, for which there are as yet no mutants available, have also been identified and characterized. Table 5.1 summarizes the enzymes

Table 5.1
Summary of individual isoforms of enzymes linked with specific mutants and the effects of these mutations on starch structure (similar genetic modifications affecting these starch biosynthetic enzymes exist in other species).

Enzyme and Specific Isoform	Mutant (Maize)	Effect on Starch
Granule Bound Starch Synthase		
GBSSI	Waxy (*wx*)	Low amylose content.
Soluble Starch Synthase		
SSI	None known	Not known
SSII	Sugary2 (*su2*)	Lacks intermediate glucan chains in amylopectin
SSIII	Dull1 (*du*)	Lacks longer glucan chains in amylopectin
SSIV	None known	Not known
Branching Enzyme		
BEI	BEI (*be1*)	None/Minimal
BEIIb	Amylose extender (*ae*)	High amylose content.
Debranching Enzyme		
ISAI	Sugary1 (*su1*)	Forms compound granules and phytoglycogen
ISAII	None known	Not known
ISAIII	None known	Not known
PUI	Pullulanase1 (*pu1*)	None/Minimal

required for normal starch synthesis in maize kernels, the associated mutant loci and the starch phenotypes of these mutants.

As a result of the great progress made in genetics in the last decade, in addition to the more traditional modifications used in plant breeding, it has become possible to modify or engineer starch synthesis enzymes using modern biotechnology. Collectively this is opening an oppurtunity to consider using all the tools of breeding, genetics and biotechnology to genetically control starch deposition in plants so as to enhance starch yield and starch quality. Thus, instead of crudely controlling enzyme activity by selecting certain mutants or selecting among different alleles of starch synthesis enzymes using plant breeding, it is possible to more precisely reduce the expression or activity of certain enzymes (such as by using antisense or RNAi technology). Alternatively the enzymes may be overexpressed: an opportunity that is unavailable using conventional plant breeding. What is additionally exciting about enzyme overexpression is that novel enzyme isoforms can be selected in order to act in concert with the existing starch pathway enzymes. With full sequencing of many homologous genes from different plants, we have begun to be able to identify domains within enzymes, which impart particular catalytic properties. This reclassification of enzymes into domain classes is an important new facet of our time, and is generating new understandings of the origins, mode of action and potential for genetic modification of different enzymes. The starch created from such work has the potential to be extremely novel and valuable. However, while the opportunities seem limitless, we are constrained by our limited understanding of the links between starch structural changes and functionality. Furthermore we are significantly constrained by our understanding of what determines the specific catalytic properties of the enzymes.

In the following sections, we will expand upon the properties and roles of the best studied of the enzymes of starch synthesis and describe the effects on the properties of starch of genetic modifications that eliminate or modify the genes encoding them.

5.4.1 Modification of GBSSI Activity

Modifications in granule bound starch synthase activity result in changes in amylose content or amylose structure. In maize, the GBSSI mutant phenotype was named waxy because the intact seed has a waxy phenotypic appearance, unlike normal seed that has a shiny and glossy appearance (194,236). No other enzyme appears to be involved in determining amylose content, as the waxy mutants of cereals (96,157), the amf mutants of potato (88), the lam mutants of pea (44), and GBSSI antisense lines of potato (4,228) show either a reduction or elimination of GBSS activity and a specific reduction of amylose in starch from tubers. Some studies have attempted to restore the production of amylose in amylose free potato plants by transforming the plants with genes for GBSSI enzymes produced by other plants. Between 3.5% and 13% amylose was restored to amylose free mutants of potato by transformation with the cassava GBSSI enzyme (183). Similarly, amylose free potatoes transformed with pea GBSSI isoforms resulted in potatoes with amylose contents of between 0.8% and 1%. Like, the other low amylose potatoes and pea starch, heterogeneity in amylose and amylopectin content was observed within the granules: the granules stained with iodine revealed amylose in concentric rings or having blue-staining granule cores (52).

Extensive work, initially in Japan, has identified wheat mutants lacking GBSSI activity (143,144,145,156,158,255). Miura and Sugawara (144) showed that substitution of genes producing functional GBSSI enzyme with the null alleles could result in starches with a 22 to 23% amylose content rather than the 25.5% amylose content of the normal control. Likewise, Miura et al. have shown that elimination of an active GBSSI enzyme at 2 of the 3 loci in wheat endosperm results in a wheat starch that has an amylose content of 16% to 21% (143). Thus, the presence of a functional GBSSI enzyme from a single locus pair is sufficient to produce a starch with an amylose content of at least 16%. Others (164) have shown that low amylose wheat starches having amylose content between 14.1 and 16.7% can be created through ethyl methanesulphonate (EMS) mutagenesis of the seeds. Amylose free wheat starches were created using triple null combinations of GBSSI mutants (255) and using mutagenesis of a double null wheat known as Ike to generate a nonnull wheat (WO09815621) which stained red when stained with iodine.

Two functional Wx alleles of rice exist: Wx^a, which produces a large amount of amylose, and Wx^b, which produces a smaller amount of amylose. Studies of the effects of the two alleles on the gene expression at the waxy locus in rice (187) showed that the Wx^b allele resulted in an ineffective production of GBSSI enzyme and amylose in japonica rice, while the Wx^a allele produced larger quantities of GBSSI enzyme and amylose in indica rice (186). On a specific activity basis, other authors have shown that the Wx^a allele was less effective in the production of amylose than the Wx^b allele based on analysis of 40 rice varieties (226). It was observed that for two wild-type rice alleles, Wx^a and Wx^b, Wx^b had a GBSS expression tenfold lower than Wx^a at the protein and mRNA levels (97). The decrease in the expression of Wx^b compared to Wx^a was the result of a point mutation within the genetic sequence for the normal rice enzyme (Wx^a allele). The Wx^b allele resulted in the synthesis of a 3.4 kilobase pair mRNA transcript compared to a 2.3 kilobase pair mRNA transcript for Wx^a as a result of the inclusion of an intron into the mRNA sequence as a result of the point mutation (97). Amylose produced from rice plants was related to the ability of the plant to excise the intron 1 from the mRNA sequence (235). Plants expressing high levels of mature mRNA (without intron 1) and no pre-mRNA (containing intron 1) produced the highest levels of GBSSI protein and the highest levels of amylose (20.0 to 27.8% amylose). These were all indica species. With more balanced expression of mature and pre-mRNA, lower levels of GBSSI protein and amylose were observed (6.7 to 16.0% amylose). Both indica and japonica species were within this group. When all of the mRNA contained intron 1 and no mature mRNA was observed, no GBSSI protein was observed and no amylose was detected (235). This pattern relating amylose content to mature mRNA with properly excised intron 1 could be applied across 31 different rice cultivars (235). Thus, based on extensive studies (97,198,235), low amylose rice appears to be the result of a decrease in the amount of normal GBSSI through a mutation which results in problems with mRNA processing rather than due to a mutation in the mature mRNA sequence. However, some differences in the behavior of Wx^a and Wx^b may be present in different rice species.

In recent years a number of patents have been filed covering genetic modifications of granule bound starch synthase in plants, utilizing knowledge gained after cloning, sequencing and transforming plants with the GBSSIs. Examples include WO9211376, US5365016, WO09827212, WO028052, WO02018606, and WO04078983.

5.4.2 Modification of SS Activity

Modifications in starch synthase activity result in changes in amylopectin content or amylopectin structure. Starch synthase may have either a subtle or profound impact on the starch, depending on the activity or inactivity of a specific isoform on the structure and composition of the starch. Although SSI is a relatively minor isoform in potato, it is the predominant isoform of SS in the cereals. Of the starch synthases, the effects of SSIIa and SSIII on starch structure and composition are the best elucidated, especially in maize. Mutants lacking the putative SSIV isoform have not yet been reported. The first reported SSI mutant was in rice (159) and no mutants have yet been reported in other plant species. In rice, the SSI mutant has only a minor effect on starch content and quality. In maize, SSIIa maps to a locus known as Sugary-2 which when mutated produces the sugary-2 mutant. Starch from mutants lacking SSIIa lack certain intermediate amylopectin chains in maize, pea and rice (31,40,221). Wheat, null for the SSIIa enzyme in each of the A, B, and D genomes, has an increase in starch chains with a DP below 10, an elevated amylose content, and a poor x-ray diffraction pattern (246). Barley null for the SSIIa enzyme produces a starch with an apparent amylose content between 50 and 70% of the dry weight of the starch, with a concurrent increase in the proportion of short chains through a DP of 35 (150). The structural gene for SSIII in maize is known to be the Dull1 locus which when mutated gives the dull1 mutant. Starch from the dull1 mutant is relatively lacking in longer amylopectin chains (62). In potato the simultaneous antisense inhibition of SSII and SSIII resulted in a grossly modified amylopectin (51,136), with yet further changes in structure if GBSSI as well as SSII and SSIII are inhibited (108).

In recent years a number of patents have been filed covering genetic modifications of starch synthase in plants, utilizing knowledge gained after cloning, sequencing and transforming plants with the starch synthase enzymes. Examples include WO9409144, US6013861, US5824790, JP06070779, US6130367, WO9720936, EP779363, WO09726362, WO09744472, WO09745545, WO9844780, WO9924575, WO9966050, WO006755, WO031274, and WO112826.

5.4.3 Modification of SBE Activity

Modifications in branching enzyme activity result in changes in apparent amylose content which are believed to be due to an increase in amylose and to changes in amylopectin structure. The first naturally high amylose starch was reported 50 years ago when Vinyard and Bear successfully found a maize endosperm mutant termed *amylose extender* (*ae*) (12,227). Despite extensive research, kernels containing exclusively amylose as the reserve starch polysaccharide have never been found (197,201). The relationship between the Ae mutation and SBE activity was proposed nearly 30 years ago (25). We now know that the structural gene for SBEIIb is the *Ae* locus (205), which when mutated produces *ae* mutant starch having an amylose content of up to 50%. Although the biochemical basis of maize starches having amylose contents above 50% clearly requires the *ae* mutant (SBEIIb), the biochemical basis of the additional increases over 50% amylose is not clear at the present time (201). Furthermore, it does not appear as though SBI is involved, as SBI mutants do not have increased amylose alone or in combination with SBEIIb mutants (20,250). Similarly, SBEIIa mutants (21) had no detectable effects on seed starch structure.

Irrespective of the means of defining or partitioning *ae* starch into amylose and amylopectin components, it is clear that the relatively clear demarcation between amylose and amylopectin existing for normal starch is blurred for *ae* starch, giving rise to a third type of starch termed intermediate starch. Additional evidence for this comes from material collected as amylopectin from *wxae* intermediate starch which has no GBSSI activity and hence produces no true amylose, and has reduced amylopectin branching (61,104,116,135,196,254). Further, the beta-amylolysis limits for *wx* starch, *wxae* starch, and the amylopectins from normal maize and *ae* starches are all between 55 and 61% (259, 260). Thus, both the average exterior chain length and average interior chains length are proportionally longer for the amylopectin from *ae*-containing starches and the double mutant *wxae* starch than for *wx* starch and the amylopectin for normal maize starch. The combination of the absent GBSSI and BEIIb activity in the *wxae* double mutant produces an amylopectin which has sufficient linearity to have apparent amylose content approximately 20% as measured by iodine binding (25).

SBEIIb mutants also exist in rice (162) but there appear to be pleiotropic effects that complicate directly linking SBE activity with increased amylose. Rice SBEI mutants have also been found (189) which, like maize, do not have increased amylose content and do not further increase amylose when combined with SBEIIb mutants. High amylose rice starches appear to have amylose contents between 30 and 50% of the weight of the starch (226). However, with the variability of amylose content of what may be considered normal rice varieties (19,173,210) and the recent implication that other starch biosynthetic enzymes affect the amylose content of such normal rice starch (221) with a few exceptions (146,147,225), it is difficult to know whether deficiencies in SBE are explicitly responsible for some of these high amylose rice starches. In such instances, the rice starch granules have the characteristic changes in starch properties, amylose properties, and amylopectin structure seen with high amylose maize starches (6,225,249). Similar to rice, high amylose barley starches with an amylose content between 30 and 45% of the weight of the starch clearly exist (184,203,252), although it is unknown whether these starches are a result of down regulation of SBE activity or are a result of changes in the expression and activity of other enzymes.

Reports of high amylose potato starch obtained through transgenic down regulation of multiple starch branching enzymes have been published recently (85,107,182,193). Work on development of a high amylose potato starch has been occurring for at least the past 10 years. Amylose contents as high as commercially available maize starches have been obtained in potato with a decrease in the overall molecular weight distribution.

In recent years a number of patents have been filed covering genetic modifications of branching enzymes in plants, utilizing knowledge gained after cloning, sequencing and transforming plants with the SBEs. Examples include US6013861, WO9211375, WO9214827, WO09507355, WO9634968, WO9722703, WO9720040, WO9820145, WO9837214, WO9837213, WO9914314, WO9964562, WO015810, WO031282, WO022140, JP0621767, JP06098656, JP05317057, and JP04088987.

5.4.4 Modification of DeBE Activity

Modifications in debranching enzyme activity can result in significant changes in starch granule structure. ISAI mutants accumulate starch in compound instead of simple starch granules (compound refers to amyloplasts containing many small granules, while simple refers to amyloplasts containing one major granule) and sometimes also accumulate phytoglycogen (a highly branched nongranular storage product). Mutants known in ISAI include the sugary1 locus in maize and sugary of rice (100,122) and in barley by lines named *Riso17* and *Notch-2* (28). Similar results were observed using antisense technology to reduce ISAI activity in rice (59). In potato where antisense constructs for ISAI and ISAII were combined, the tubers accumulated large numbers of small granules (28,29,122). Mutations in PUI have been identified (47), but effects on starch content are minimal. Modifications in the other isoamylase (ISAII and ISAIII) have not yet been identified, although there is some evidence that they each play distinct roles in starch synthesis (90). The maize sugary mutants are important because they are one of the main sources of producing sweet corn.

In recent years a number of patents have been filed covering genetic modifications of debranching enzymes in plants, utilizing knowledge gained after cloning, sequencing and transforming plants with the ISA and PU enzymes. Examples include WO9202614, WO09504826, WO09619581, US5750876, WO9603513, US5912413, WO09732985, WO09742328, WO9850562, WO09906575, WO9912950, WO09958690, and WO0001796.

5.4.5 Modification of Multiple Pathway Enzymes

By eliminating multiple starch biosynthesis enzymes, other alterations of the starch biosynthetic pathway can occur resulting in even more novel starches. Several patents exist on the creation and use of such starches (US4428972, 4615888, 4767849, 4789557, 4789738, 4801470, 5009911, and 5482560). More recently, several patents and published applications have described the production and utilization of heterozygous combinations of mutations in the starch biosynthetic pathway to obtain commercially useful starches (WO9535026, US5356655, US5502270, and 5516939). The production of many of these starches involves the use of double or triple mutant plants. Due to the number of mutations required to sufficiently alter the starch (at least 2 or 3 within a single plant) many of these starches are difficult and costly to produce commercially, so many of these starches from plants with mutations in the starch biosynthetic pathway are uncompetitive with chemically modified starches. Further, these combinations of two or more mutations, whether they are combined homozygously or heterozygously in the plant endosperm, rely on the alteration of the structure of amylopectin from normal or waxy starch.

5.5 FUNCTIONALITY

The origin of wide angle x-ray scattering (WAXS) diffraction patterns from amylose fibers was determined to be due to the crystalline packing of double helices of starch chains (242,243). This work was further developed to explain the crystallinity of starch granules. When starch granules are heated sufficiently in excess water, the WAXS diffraction pattern is lost (257) and an endothermic transition is observed (39). Double helical order is additionally lost during this endothermic event, and it is this loss and not the loss of crystalline order that is believed to be responsible for the endothermic event (39). This loss of granular, crystalline, and double helical order is called gelatinization, and it is irreversible with respect to the three dimensional granule organization (7). Subsequent to gelatinization, starch chains of both amylose and amylopectin may organize into new double helical and crystalline structures in a process called retrogradation (7,70,176).

In its native granular form mixed with water, starch makes a high solids mixture, useful in making batters and doughs. During heating, starch granules swell, associating with water up to 30-fold its dry weight, to form a hydrated starch. This process is reversible provided temperatures and pressures are lower than those that lead to disruption of the organized intramolecular glucan chain associations within the granule. Above these temperatures and pressures, the granules will undergo irreversible changes and glucan chains will dissociate, permitting further swelling. With additional heating and applied shear forces, swollen hydrated granules will eventually collapse to form an unorganized paste of starch molecules. Such hydration, with heating, results in a thickening effect, imparting texture and structure. In functional terms, this process of starch granule swelling and dissociation is known as gelatinization (7,18,214,238). The glucan polymers of a gelatinized starch are able to associate with each other or other components of food to impart additional character to the food system including stickiness, tackiness or a rubbery texture (258). In addition, amylose readily leaches out of the granule during gelatinization and interacts with the other food components (17,53,165). Under mild cooking conditions, amylopectin does not so readily leach from the granule and hence is enriched inside the gelatinized granule. However, in more extreme cooking conditions, the amylopectin is much more dispersed, resulting in products with a tacky or sticky character.

Once a cooked starch is allowed to cool, the glucan chains begin to reassociate in a process resulting in a change in the functional characteristics, including a decrease in paste clarity and gelation of the paste. In functional terms, this process of reorganization is known as retrogradation (7,70,176). Normal starches are generally recognized for their ability to retrograde within hours (175) and usually form opaque gels (41). Retrogradation of amylopectin gels occurs during days or weeks of storage. As a consequence of their different molecular weights and chain length profiles, the rates of retrogradation of amylose and amylopectin are not the same. The rapid setting of the structure of breads is believed to be due to rapid (within seconds or minutes) amylose retrogradation to form a network structure. Starch functionality is therefore a consequence of the degree of gelatinization and is influenced by retrogradation, time, temperature, concentration, and the presence of other food components or additives. In addition, modifying starch using chemical, enzymatic, or physical treatments alters and extends its functional properties.

Measuring starch functionality and applying it to food applications is problematic because the results are subject to extrapolation to systems and processes which are far more complex than laboratory testing is able to emulate. Analytical instruments (e.g., Differential Scanning Calorimetry, DSC) are frequently used to quantify the temperature range and amount of energy needed to melt crystalline starch. The amounts and molecular size of amylose and amylopectin may be measured using gel permeation or size exclusion chromatography. Granule size and shape are measured microscopically or by light diffraction techniques, and granule viscosity is measured using various rheometers. Rheological measurements may include various temperature and shear programs that attempt to mimic thermal treatments, pumping, and shearing forces that occur during food processing. Such measurements of texture provide information on adhesiveness, cohesiveness, yield stress, viscous flow, and rigidity of starch sols and gels.

5.5.1 Amylopectin Retrogradation

Because retrogradation manifests itself in many ways and over many different spatial scales, establishing relationships between the structural orders and the physical properties of the starch are complicated. Thus, a number of techniques are required for such an assessment. NMR has been used to probe the most fundamental physical order of the starch: double helical order (71), and x-ray diffraction has been used to examine the crystallinity of starches (114). However, neither of these two measures of starch order adequately represents the rheological properties of a cooked starch paste because rheology is dependent on not just the order of the starch but also the larger scale interrelationships of this order with the covalent structure of the starch molecules themselves. There are additional problems that make relationships between low level molecular order and larger scale order difficult to establish: residual granular order (41,141,153) or polymer incompatibility (49,66,112) may result in macromolecular heterogeneity in the system. To avoid these complications, many investigators interested in starch retrogradation have examined the behavior of waxy-type starches (16,30,57,134,135,144,154,175,176,195,253). Likewise, the properties of amylopectin free amylose have been studied either by preparing synthetic amylose using glucose-1-phosphate and the enzyme phosphorylase (35,68,70,180), or by studying highly pure amylose fractionated from the native starch mixtures of amylose and amylopectin (116,126,209,239). Amylopectin retrogradation has been examined by multiple techniques including DSC (57,95,137,169,196,254), WAXS (175,176,217), turbidimetry (98,176), and rheology (16,30,55,111,117,176,253).

The length of a double helix, whether in a starch granule or in a retrograded starch, is limited by the length of the external chains participating in the double helix. Linear chain lengths of amylopectin are generally shorter than the DP 40 to DP 70 lengths observed for amylose double helices (103,129,138). Amylopectin double helices are generally observed to dissociate below 70°C by DSC (39) when the starch is in excess water (<30% starch). The DSC endotherm of retrograded amylopectins containing a higher proportion of longer chains than normal amylopectin (i.e., amylopectin from some mutant starches) may extend above 100°C. The endotherms from these mutant starches are also broader than an endotherm observed for retrograded amylopectin from normal starch, perhaps indicating that amylopectin with longer external chains has an ability to form a broader distribution of double helical lengths than amylopectin with shorter chains. The absolute lower limit for the length of a stable double helix appears to be six anhydroglucose residues (69), the length required to complete one turn of a chain in a double helix (242,243). Practically, the lower limit appears to be closer to ten anhydroglucose units, given that in systems of a pure oligosaccharide, chains with a length (or degree of polymerization, DP) less than ten anhydroglucose units could not self associate into stable double helices and crystallize (69). However, this minimum appears to be somewhat dependent on the lengths of other chains available for double helix formation, because in mixed systems of short chain oligosaccharides chains longer than a DP of 6 may form double helices with longer chains.

As with double helical and crystalline formation with oligosaccharides, the ability for amylopectin chains to retrograde is closely related to the average external chain length (ECL) of the molecules. With stubs of DP 1, 2, or 3 anhydroglucose residues as external chains, β-limit dextrins of amylopectin do not retrograde after one month of storage at 4°C (176). Shortening the average ECL by just three anhydroglucose units (from 14 to 11) by alpha-amylase was sufficient to reduce the enthalpy and solid character (assessed by DSC and pulsed NMR, respectively) to 10% of those of the native starch (12,244). Chromatography of the modified starches was instructive, because it showed that the component chains of the starch do not all need to be equally shortened to have an effect on starch retrogradation: some chains were reduce to a DP less than 5, with the remainder remaining largely unchanged (12,244).

Examination of starches with intact chains (135,196,254) has indicated that starches with high proportions of chains below a DP of 11 to 12 have a slower rate of retrogradation than those starches with fewer short chains, irrespective of concentration. The retrogradation rate of starches with higher proportions of shorter chains is also more concentration dependent than for those starches with fewer short chains (57,135). However, conclusions from these studies must be tempered by any possible confounding effects of additional changes in the fine structure of the amylopectin (e.g., branching pattern).

5.5.2 Amylose Gelation

Amylose retrogradation has been monitored by the development of supramolecular organization using turbidimetric and rheological methods (35,48,56,70,141,142). Gels of amylose have been suggested to develop within minutes through the retrogradation of amylose into double helices (35,70) or through an initial phase separation of amorphous amylose, followed by retrogradation into double helices in these concentrated amylose regions. The aggregation event is followed by an additional retrogradation event: the development of double helical crystallites within hours (91,128,141,142). Increasing the chain length and concentration of the amylose both have a destabilizing effect on amorphous amylose in solution (70).

Destabilization of amylose may lead to retrogradation in the form of gelation, precipitation, or both. The relatively long, stable double helices of amylose result in more thermally stable gels than gels of amylopectin. In the formation of amylose gels, one outcome of amylose retrogradation, gel development is understood to be related to the ability of a single molecule to interact in double helical junction zones which bind molecules together into a three dimensional network structure, and the subsequent ability for the double helices to either orient or aggregate once formed (70,128). For synthetic, completely linear amylose, the network formation is possible for amylose molecules with a DP above 100 because these molecules are able to interact with more than one additional molecule to form a network (70). Precipitation, the other outcome of amylose retrogradation, is favored over gelation for amylose molecules below a DP of 250, gelation is favored over precipitation for amylose molecules above a DP of 1100, and both network forming (gelation) and breaking (precipitation) retrogradation phenomena are similarly favored for amylose molecules with a DP of 440 and 660 (35). The concentration of amylose also influences whether amylose will gel or precipitate. Higher concentrations of amylose result in an increased frequency for chain to chain interactions, which would both increase the rate of gel development and decrease the time required for the onset of a three dimensional network.

Amylose gels prepared in water have been described as phase separated systems consisting of a solvent rich phase and a filamentous polymer rich phase consisting of bundles of amylose chains (74,91,129,222). Portions of amylose chains that participate in the double helices residing in the crystalline regions of amylose gels have been estimated to range in length between DP 40 and DP 70 (103,128,137,138). As measured by DSC, the double helices of amylose appear to dissociate over a very broad temperature range (~50°C) with an endothermic peak temperature between 120°C and 160°C (23,37,117,129,202).

5.5.3 Gelation of Mixtures of Amylose and Amylopectin

Understanding the retrogradation behavior of normal starches, considered as mixtures of amylose and amylopectin, is even more complex than understanding amylose or amylopectin retrogradation in isolation. This is due to the possibilities for phase separation of the two types of molecules and for interactions between amylose and amylopectin to develop.

Cooked normal starches have been considered phase separated systems of swollen granules consisting of amylopectin trapped within a continuous gel matrix of leached amylose (141,153,167). In these systems, the amylose and amylopectin behaviors appear to be relatively independent of each other: the amylose forms a thermally irreversible (at <100°C) gel network within hours, followed by a thermally reversible increase in gel elastic modulus attributable to an increase in granule rigidity due to amylopectin retrogradation (113,141,167). In the absence of granular order, amylose and amylopectin mixtures have been shown to phase separate over time (66,111). Examination of gels of artificial mixed systems of amylose and amylopectin showed that over a small increase (10–15% w/w) in the amylose content of gels (from 25 to 40% amylose on a total starch basis), the gels developed the behavior of pure amylose gels in their rate of turbidity development (49), their elastic modulus, and their susceptibility to enzymatic and acid hydrolysis (128). Both groups suggested that a phase inversion from an amylopectin to an amylose continuous phase at an amylose to amylopectin ratio of about 30:70 or about 17:83 was the cause of the observed differences in gel characteristics at high and low amylose to amylopectin ratios. The differences may have been with the starches and concentrations used, with the methods of preparation of both the amylose and amylopectin prior to mixing, and the method of mixing the amylose and amylopectin. From examination of the fracture of gels of jet cooked high amylose maize starches prepared between temperatures of 121°C and 166°C, it was hypothesized (33)that higher processing temperatures resulted in more homogeneous mixtures of amylose and amylopectin than processing at lower temperatures, and that the more homogeneous mixtures would take longer to phase separate and gel during cooling at a constant rate. The delay in phase separation during cooling was thought to result in phase separation and gelation at lower temperatures, resulting in weaker gels due to the more rapid formation of a less perfect gel network.

Others have suggested that amylose and amylopectin interact more extensively than suggested by the relatively independent behavior of mixtures of phase separated amylose and amylopectin (101,117,168,190) and that the short chains of branched molecules interfere with the molecular association of amylose molecules (23,117). In gels prepared from reconstituted mixtures of amylose and amylopectin, the chain length distribution of the amylopectin, the size of the amylopectin, and the amylose content appear to influence the formation and properties of starch gels. From examination of gels formed by waxy maize starch (considered amylopectin) and oligosaccharide chains with a DP of 21 or 35, linear chains as short as DP 21 appear sufficient to form junction zones with multiple amylopectin chains resulting in a gel (190). Acetylation, substitution of the hydroxyl groups of the starch with structure-inhibiting moieties, also appear to destroy the gel-forming ability of the amylopectin, and the addition of amylose instead of the oligosaccharides appears to result in stronger gels. Substitution was suggested to limit associations preventing gelation, but was not sufficiently high to prevent gelation in the presence of amylose or short, linear chains (190). Using a variety of amylopectin sources with different chain lengths and a consistent source of amylose and ratio of amylose to amylopectin ratio (20:80), the strongest gels appear to form from the mixtures containing the amylopectin with the longest chains (101). Using 25:75 or 50:50 (w/w) mixtures of amylose with either *wx* starch or small, medium, or large dextrins prepared from waxy maize starch, gels prepared with the medium sized dextrins, irrespective of amylose content, formed weaker gels than gels prepared from the other dextrins or the *wx* starch (168). Examination of starches by DSC shows that amylose appears to retrograde independently when the starch is heated to temperatures below 140°C, but does not do so when heated to temperatures above 160°C in an excess water environment (23). Further, dynamic shear rheology of these same starches indicated that the length of these double helices formed between amylose and amylopectin was dictated by the length of the external chains of the amylopectin: when gels prepared from mixtures of completely dispersed starches were heated, the decrease in the elastic modulus paralleled the decrease in the elastic modulus for the amylopectin isolated from the starch (117).

5.6 APPLICATIONS

As well as providing essential calories, starches from different crops play an important role in foods, such as improving processing, shelf life, consistency and appearance by providing texture and thickening capability for suspending solids. There are myriads of applications of starches in foods and these uses have been extended due to application work by food technologists using physical and chemical modifications (designated by E-numbers in the EU). These include oxidation (E1404, E1451), monostarch phosphorylation (E1410), distarch phosphorylation (E1412), acetylation (E1414, E1420, E1422), hydroxypropylation (E1440, E1442), and octenylsuccination (E1450). In this section, we will focus on starch applications of various genetic modifications of the major crop plants and how these modifications impact the uses of these starches in foods. In particular we will emphasize modifications in amylose and amylopectin content and molecular weight, amylopectin chain length, starch granule size and morphology, crystal structure and phosphate content. We will conclude with a discussion of new opportunities for enhanced starches resulting from new genetic modifications of plants.

5.6.1 Amylose Free Starch

The most widely used amylose free starches are obtained from waxy maize and rice and in all cases are considered non-GM natural variants of normal maize and rice. Staining starch with iodine readily identifies amylose free types: normal starch will stain blue or purple whereas the starch produced from amylose free plants will stain red or brown or brownish red in color. Amylose free maize starches are generally considered to contain zero or almost zero amylose. Amylose free rice starches have been shown to contain between 0 and 3% amylose, though collectively these starches are referred to as amylose free or waxy rice starches (109,173,185). With these amylose free rice starches, it has been assumed that the differing cooking and paste properties are due to differences in the structure of the amylopectin of the starch rather than variations in the low levels of amylose of the starch (234). The effects of amylose and other molecular and compositional characteristics of rice starches on rice (19,34) or rice starch properties remain unclear (125).

Amylose free starches have useful functionality that has encouraged their commercial development. They are considered useful as water binders, viscosity builders, and texturizers in food as well as in industrial applications (171). However, these starches are less resistant to shear, acid, and high temperatures than are normal starches, and extended cooking results in stringy, cohesive pastes. Amylose free starches are generally recognized for their improved transparency after processing compared to normal starches (41) and have better freeze and thaw stability compared to normal starches once cooked (171,237). They are also recognized for their improved long term storage capability as they require weeks to gel if they could be considered to gel at all (15,253). To correct for some of the negative paste attributes of amylose free starch, such as poor stability to temperature, shear and acid and undesirable paste quality, most amylose free starches are chemically modified by substitution, cross linking, or both (171,238,256).

In recent years there has been significant interest in developing amylose free starches in other crops so as to take further advantage of any species specific qualities of the starch produced from that species. However, although mutants are readily found in some plant species (such as maize, rice, and barley) this is more difficult in other plant species where there are multiple copies of each gene (such as the polyploid species like potato, oats, and wheat). In the case of wheat, the advances have been made by screening for mutants (non-GM), while in potato the waxy types were made using biotechnology (GM). Applications of amylose free wheat and potato are at present still being developed. However, the most likely applications for amylose free potato include the paper, adhesive, textile, and packing industries. In the EU, certain GM varieties of modified amylose free (high amylopectin) potato have not been approved thusfar (http://europa.eu.int/comm/food/fs/sc/scp/out24_en.html), while others appear to be in the approval process (http://europa.eu.int/comm/food/fs/sc/scp/out129_gmo_en.pdf). Amylose free wheats are finding applications in foods such as noodles and baked goods including breads (5,247).

5.6.2 Low Amylose Starch

Although staining starch with iodine readily identifies amylose free starches (161), care must be taken when using this quantitatively to adequately account for the iodine binding capacity of the amylopectin (120). For example, amylopectin from an amylose free plant having an inactive GBSSI enzyme might appear to contain 5% amylose based on iodine binding, blue value, measurement (116). After carefully considering these potential problems certain low amylose starches have been identified. For example, in the early 1940s, a waxy maize mutant (wx^a) was discovered in two exotic Argentinean small seeded flint varieties that contained a starch that had an amylose content of 2.4% and stained a pale violet color with iodine (26). Additionally, the amylose content of the starch increased from 0% (full waxy) to 0.65% to 1.3% to 2.4% (full *wx*^{a}) with increasing dose of the *wx*^{a} allele (26,204). With these same crosses, the viscosity of starch pastes decreased with increasing dose of the *wx*^{a} allele. The *wx*^{a} allele was described as resulting in a 95% reduction in the amount of GBSSI protein produced and a starch with a low amylose content (174, 204,261).

Wheat starches have been produced with amylose contents of about 7.5% and 13.5% by crossing normal wheat with amylose free wheat (188). Peak viscosities of all starches differed by less than 20% of the peak viscosity of the amylose free wheat starch, with the low amylose starches having a higher peak viscosity than both normal and waxy wheat starch. The gelatinization temperatures and enthalpy were highest for waxy wheats and decreased in the order waxy > 13.5%, amylose wheat > 7.5%, and amylose wheat > normal wheat starch. The retrogradation temperatures and enthalpy were insignificantly different for amylose free wheat, normal wheat, or any of the low amylose wheat starches. Starch granules extracted from a wheat strain derived from mutagenized Tanikei A6099 had an apparent amylose content of 1.6% and stained dark brown with dark cores compared to red-staining waxy wheat starch with 0.4% apparent amylose (115). The Tanikei A6099 mutant wheat starch had a higher initial pasting stability than an amylose free wheat starch (0.4% amylose). However, the viscosity of the low amylose starch paste decreased dramatically, to the same viscosity as the amylose free wheat, during continued cooking and remained at the same viscosity as amylose free wheat after cooking. The mutagenized Tanikei A6099 wheat is known to produce a mutant GBSSI enzyme (115,248).

By screening microtubers from plants exposed to x-ray radiation an amylose free mutant of potato was identified (88). Potato starches considered amylose free have been shown to have an amylose content varying from 0 to 7.9% (183,204,223,224). The amylose free potatoes were null for the GBSSI enzyme. Staining with iodine gave varied results: sometimes the starch stained red and other times reddish brown and blue. These results are indicative of a heterogeneous mixture of amylose free starch and amylose containing starch of unknown quality within the potato tuber. In further attempts to understand the link between function and activity of GBSSI in potato, antisense transgenic plants having amylose contents between 3.0 and 8% were produced (123,224,228). These tubers had both blue and red brown staining portions (228) again indicative of heterogeneous mixtures of amylose free starch and amylose containing starch of unknown quality. Others (123) observed additional heterogeneity at a granule level, with starch granules having blue cores surrounded by a red brown shell of starch. The size of the blue core appeared to be correlated with the amylose content of the starch. Starch extracted from plants produced from crosses between an amylose free potato and a normal potato had no linear correlation between GBSS activity and amylose content (58). Additionally it was observed that the swelling and rheological properties of the granules could not be clearly linked with amylose content (183).

The waxy barley starches have been shown to contain up to approximately 5% apparent amylose (215). However, this apparent amylose is due to a mixture of starch granules within the barley seeds. The amylose content of the granules typically ranges from an undetectable level up to approximately 10%, with the granules closest to the surface of the seed having the highest amylose content (3). Recent work with waxy barley starch (with amylose contents up to 6.44%) shows delayed peak viscosity development and viscosity varying during cooking under shear (132). Additionally, all of the waxy barley starches began to develop viscosity at a similar time and temperature in the cooking process.

Low amylose rice starches have been shown to have amylose contents between 7 and 15% (124). Shimada et al. (199) produced several antisense rice plants with starch having an amylose contents between 6 and 13%. The iodine staining qualities of these starch granules were not reported. Further, any cooking properties of the starches, the elastic properties and gelling abilities of pastes and the gel properties of gels produced from these low amylose rice starches produced by transgenic rice plants are unknown.

5.6.3 High Amylose Starch

Commercialized high amylose maize starch is a result of down regulation of the maize BEIIb enzyme, utilizing the *amylose extender* mutation. High amylose maize starches bring important differentiated properties for food applications, having amylose contents between 30 and 90% of the weight of the starch. The starch imparts gelling ability to the food system, improving adhesion to water impermeable surfaces and altering product texture. These starches also have improved film forming ability and improved fat impermeability compared to the normal starch counterparts. High amylose starches provide firmness, extend cooking times and increase the crispiness of coatings.

Because of their high resistance to processing and subsequent digestion, in addition to their rapid retrogradation if gelatinization occurs, high amylose maize starches are also being utilized as a source of resistant starch in a number of food products. Reviews (43,79,93) provide an overview of resistant starch and its valued physiological benefits in foods. Commercialized resistant starches include unprocessed starches from different botanical sources (e.g., green banana, legumes, potato) as well as more highly processed forms of high amylose starches. High amylose starches from other sources are also being advocated for their resistant starch properties.

The most widely available high amylose starches originate from maize using the *amylose extender* mutation that results in a starch having 40–50% amylose. As a result of extensive breeding and selection work higher amylose contents (up to 90%) have been achieved (201). High amylose starches are also available in barley and rice.

5.6.4 Amylopectin Chain Length

As already stated, care must be taken when using the iodine binding assay for quantifying amylose content. In a more extreme example of this problem, the combination of the absent GBSSI activity and the *ae* mutation produces an amylose free maize starch that stains blue or purple and appears to have an amylose content of 15–26% (194). This is because the *ae* mutation causes a decrease in starch branching enzyme activity (24), which results in the formation of long chain amylopectin (104), which will itself stain blue with iodine. Other examples of this come from genetic modifications that affect amylopectin structure.

Loss of SSIIa activity in maize results in a starch with an amylose content near 40% of the total starch weight. This starch develops viscosity very slowly at temperatures above 90°C and forms stable gels that strengthen only very slowly compared to normal starch gels (31). The loss of SSIIa activity resulting in elevated amylose contents highlights the often observed disconnect between the viscosity development of the starch and the thermally detected gelatinization of the starch. Despite the resistance of the starch to develop viscosity, likely a consequence of the elevated amylose content, loss of SSIIa activity results in a decrease in the gelatinization temperature range of the starch from about 70°C to 80°C for normal starch to approximately 55°C to 65°C (31,21). Examination of the chain length distribution of the maize starch in combination with the absence of amylose, as a result of the inactivity of GBSSI, indicates that the starch has a elevated proportion of short component chains below a DP of 30 compared to normal starch (212) and additionally an elevated proportion of very short component chains below a DP of 10 compared to normal starch (94,134). This high proportion of very short chains imparts a decreased tendency to retrograde compared to normal starch (134,135).

Recently, potatoes have been engineered to eliminate both SSIIa and GBSSI activity, resulting in an increase in short chains below a DP of 14 compared with normal starch (108). This had the benefit of a decrease in the tendency for the starch to retrograde after cooking which may have implications for improved freeze and thaw tolerance. Potatoes engineered to have reduced SSIII activity had decreased amounts of chains longer than DP 17 (60).

The SSIIa enzyme of rice has been implicated as one of the major enzymes that affects whether the grain is of the indica-type vs. the japonica-type (221). Indica-type rices have been long known to have higher kernel integrity and higher granule stability than japonica-type rices (131). The properties of indica rice starches have been attributed to their higher proportion of longer chains than japonica-type rice starches.

It has been known for some time that the inactivation of SSIII in the dull mutant of maize results in an elevation in the apparent amylose content to 30 to 40% (81). Beyond this, a number of chromatographic studies on dull starches, including those additionally lacking GBSSI enzyme activity, indicate that the chain distribution of maize starch is only slightly different from normal starch, with some elevation in the chains with a DP less than 30 to 50 (94,196,254). The beta-limit dextrins, starches with the exterior chains digested to stubs of maltose to maltotriose, provide a more complete picture of the changes observed with elimination of SSIII in maize. In this case the lengths of the residual chains containing all of the branching are considerably shorter when SSIII is absent during synthesis. It has been suggested (135) that this change has implications for the retrogradation of starch produced in the absence of SSIII. Thus, during gelatinization and then continued heating after gelatinization, the starch chains are less able to randomly orient themselves compared with starches having longer spacing between branch points (i.e., normal or waxy starch). Thus, the retrogradation rate of *du wx* starch is less concentration dependent than *wx* starch despite the similar chain length distributions of the two starches after debranching.

5.6.5 Novel Starches

Recent developments in biotechnology are opening ways of making novel starches in plants. Thus starches obtained from different crops having varied functionalities due to differences in composition may now theoretically be readily transferred into crops from one or another different plant species. In recent years a number of patents have been filed covering novel genetic modifications of starch in plants. Examples include WO09711188, WO008184, WO09961580, WO9814601, WO9404692, WO09601904, WO09729186, WO014246, WO9924593, WO9839460, WO047727, WO011192, WO9929875, WO9747806, WO9747807, WO9747808, and WO9822604. In general we can consider making changes in any aspect of amylose/amylopectin, phosphate, protein, phospholipids, crystallinity, gelation, and pasting characteristics, flavor and starch granule morphology. One merely has to understand the genetic basis of the differences.

One interesting example of a cross species difference that may be exploited is phosphate content in potato. Recent work (177) has shown that starch phosphate content may be attributed to a novel glucan water dikinase enzyme. This discovery opens up the possibility of creating novel genetically modified starches having varied phosphate contents. Although it is too early to conclude whether there would be useful applications in food, possibilities include changing starch digestibility or starch viscosity after cooking. Another interesting character in one species that is not available in others is associated with the genes responsible for endosperm texture in some cereal crops. In this case the puroindoline genes, described as PinA and PinB are interesting candidates (14,72,86,152). Transforming the wheat genes into rice successfully created the softness trait in the seed (121). At this time it is still too early to say how useful this character will be in food applications.

Although starch granule morphology varies extensively across species (102) and starch from different species is highly valued for certain applications, there are few specific reports that granule morphology is the attribute desired for valuable food applications. Furthermore, although there have been some reports of progress in understanding the mechanisms that control and influence starch granule size (28,29,122), for granule morphology there is little that can be defined as clear enough to arouse interest in a genetic approach. Similarly, flavor can be considered to vary extensively across species and a bland flavor of (e.g., tapioca starch) is valued over certain cereal starches that impart a mealy character. However, as with granule morphology, our understanding of the genetic components determining this character is rather poor, making flavor a difficult target for genetic modification at the present time. Thus in these cases it is difficult to see a way of using such differences in granule morphology and flavor to exploit that character in another species.

Another novel concept has come from bifunctional domains involving cellulose and starch, which may enable developments with novel biomaterials (130). Another possibility is to explore incorporation of inulins with starch (80). Even more novel possibilities to consider include adding new functionality to starch using gene constructs containing fusion proteins (106). The fusion proteins can be proteins, which might affect starch properties or enzymes, or other bioactive proteins.

ACKNOWLEDGMENTS

The authors wish to thank Dr Kay Denyer for comments and constructive criticism of the manuscript. All starch structure diagrams kindly supplied by Dr Eric Bertoft (with permission).

REFERENCES

1. Adkins, G.K., C.T. Greenwood. Studies on starches of high amylose-content, part VIII: the effect of low temperature on the interaction of amylomaize starch with iodine: a unique characterization. *Carbohydr. Res.* 3:152–156, 1966.
2. Adkins, G.K., C.T. Greenwood. Studies on starches of high amylose content, part X: an improved method for fractionation of maize and amylomaize starches by complex formation from aqueous dispersion after pretreatment with methulsuphoxide. *Carbohydr. Res.* 11:217–224, 1969.
3. Anderson, J.M. Characterisation of starch from inner and peripheral parts of normal and waxy barley kernels. *J. Cereal Sci.* 30:165–171, 1999.
4. Andersson, M., A. Trifonova, A. Andersson, M. Johansson, L. Bulow, P. Hofvander. A novel selection system for potato transformation using a mutated AHAS gene. *Plant Cell Rep.* 22:261–267, 2004.
5. Araki, E., H. Miura, S. Sawada. Differential effects of the null alleles at the three *Wx* loci on the starch-pasting properties of wheat. *Theor. Appl. Gen.* 100:1113–1120, 2000.
6. Asaoka, M., K. Okuno, Y. Sugimoto, M. Yano, T. Omura, H. Fuwa. Characterization of endosperm starch from high-amylose mutants of rice (*Oryza sativa* L.). *Starch* 38:114–117, 1986.
7. Atwell, W., L.F. Hood, D.R. Lineback, E. Varriano-Martson, H.F. Zobel. The terminology and methodology associated with basic starch phenomena. *Cereal Foods World* 33:306–311, 1988.
8. Baba, T., M. Arai, T. Yamamoto, T. Itoh. Some structural features of amylomaize starch. *Phytochemistry* 21:2291–2296, 1982.
9. Baba T, Y. Arai. Structural characterization of amylopectin and intermediate material in amylomaize starch granules. *Agric. Biol. Chem.* 48:1763–1775, 1984.
10. Baba, T., R. Uemura, M. Hiroto, Y. Arai. Structural features of amylomaize starch: components of amylon 70 starch. *J. Jpn. Soc. Starch Sci.* 34:213–217, 1987.
11. Bates, F.L., D. French, R.E. Rundle. Amylose and amylopectin content of starches determined by their iodine complex formation. *J. Am. Chem. Soc.* 65:142–148, 1943.
12. Bear, R., M.L. Vineyard, M.M. MacMasters, W.L. Deatherage. Development of "amylomaize": corn hybrids with high amylose starch, II: results of breeding efforts. *Agron. J.* 50:598–602, 1958.
13. Beatty, M.K., A. Rahman, H.P. Cao, W. Woodman, M. Lee, A. Myers, M. James. Purification and molecular genetic characterization of ZPU1, a pullulanase-type starch-debranching enzyme from maize1. *Plant Physiol.* 119:255–266, 1999.

14. Beecher, B., A.D. Bettge, E. Smidansky, M. Giroux. Expression of wild-type pinB sequence in transgenic wheat complements a hard phenotype. *Theor. Appl. Genet.* 105:870–877, 2002.
15. Beliaderis, C.G. Characterization of starch networks by small strain dynamic oscillatory rheometry. In: *Developments in Carbohydrate Chemistry,* Alexander, R.J., H.F. Zobel, eds., St Paul: American Association of Cereal Chemists, 1992, p 103.
16. Bello-Pérez, L.A., O. Paredez-López. Starch and amylopectin: rheological behavior of gels. *Starch/Stärke* 46:411–413, 1994.
17. Bello, A.B., R.D. Waniska, M.H. Gomez, L.W. Rooney. Starch solubilization and retrogradation during preparation of To (a food jell) from different sorghum varieties. *Cereal Chem.* 72:80–84, 1995.
18. Bertoft, E. On the nature of categories of chains in amylopectin and their connection to the super helix model. *Carbohydr. Polym.* 57:211–234, 2004.
19. Bett-Garber, K.L., E.T. Champagne, A.M. McClung, K.A. Moldenhauer, S.D. Linscombe, and K.S. McKenzie. Categorizing rice cultivars based on cluster analysis of amylose content, protein content, and sensory attributes. *Cereal Chem.* 78:551–558, 2001.
20. Blauth, S.L., K.N. Kim, J.D. Klucinec, J.C. Shannon, D.B. Thompson, M.J. Guilitinan. Identification of *Mutator* insertional mutants of starch-branching enzyme 1 (SBEI) in *Zea mays* L. *Plant Mol. Biol.* 48:287–297, 2002.
21. Blauth, S.L., Y. Yao, J.D. Klucinec, J.C. Shannon, D.B. Thompson, M.J. Guilitinan. Identification of *Mutator* insertional mutants of starch-branching enzyme 2a in corn. *Plant Physiol.* 125:1396–1405, 2001.
22. Bogracheva, T.Y., T. Wang, V.J. Morris, S. Ring, C. Hedley. The granular structure of C-type pea starch and its role in gelatinisation. *Biopolymers* 45:323–332, 1998.
23. Boltz, K.W., D.B. Thompson. Initial heating temperature and native lipid affects ordering of amylose during cooling of high-amylose starches. *Cereal Chem.* 76:204–211, 1999.
24. Boyer, C.D. Soluble starch synthases and starch branching enzymes from developing seeds of sorghum. *Phytochemistry* 24:15–18, 1985.
25. Boyer, C., D.L. Garwood, J. Shannon. Interaction of the amylose-extender and waxy mutants of maize. *J. Heredity* 67:209–214, 1976.
26. Brimhall, B., G.F. Sprague, and J.E. Sass. A new waxy allel in corn and its effect on the properties of endosperm starch. *J. Am. Soc. Agron.* 37:937–944, 1945.
27. Buleon, A., C. Gerard, C. Riekel, R. Vuong, H. Chanzy. Details of the ultrastructure of C-starch granules revealed by synchrotron microfocus mapping. *Macromolecules* 31:6605–6610, 1998.
28. Burton, R.A., H. Jenner, L. Carrangis, B. Fahy, G.B. Fincher, C. Hylton, D.A. Laurie, M. Parker, D. Waite, S. Van Wegen, T. Verhoeven, K. Denyer. Starch granule initiation and growth are altered in barley mutants that lack isoamylase activity. *Plant J.* 31:97–112, 2002.
29. Bustos, R.,B. Fahy, C. Hylton, R. Seale, N.M. Nebane, A. Edwards, C. Martin, A.M. Smith. Starch granule initiation is controlled by a heteromultimeric isoamylase in potato tubers. *Proc. Nat. Acad. Sci. USA* 101:2215–2220, 2004.
30. Cameron, R.E., C.M. Durrani, A.M. Donald. Gelation of amylopectin without long range order. *Starch/Stärke* 46:285–287, 1994.
31. Campbell, M.R., L.M. Pollak, P.J. White. Dosage effect at the sugary-2 locus on maize starch structure and function. *Cereal Chem.* 71:464–468, 1994.
32. Campbell, M.R., L.M. Pollak, P.J. White. Interaction of two sugary-1 alleles (su1 and su1st) with sugary-2 (su2) on characteristics of maize starch. *Starch/Stärke* 48:391–395, 1996.
33. Case, S.E., T. Capitani, J.K. Whaley, Y.C. Shi, P. Trzasko, R. Jeffcoat, H.B. Goldfarb. Influence of amylose-amylopectin ratio on gel properties. *J. Cereal Sci.* 27:301–314, 1998.
34. Champagne, E.T., K.L. Bett, B.T. Vinyard, A.M. McClung, F.E. Barton, K. Moldenhauer, S. Linscombe, K. McKenzie. Correlation between cooked rice texture and rapid visco analyser measurements. *Cereal Chem.* 76:764–771, 1999.
35. Clark, A.H., M.J. Gidley, R.K. Richardson, S.B. Ross-Murphy. Rheological studies of aqueous amylose gels: the effect of chain length and concentration on gel modulus. *Macromolecules* 22:346–351, 1989.

36. Coe, E. The genetics of corn. In: *Corn and Corn Improvement*, Sprague, G.W., J.W. Dudley, eds., Madison: American Society of Agronomy, 1988.
37. Colonna, P., C.F. Morris. Behavior of amylose in binary DMSO-water mixtures. *Food Hydrocolloids* 1:573–574, 1987.
38. Commuri, P.D., P.L. Keeling. Chain-length specificities of maize starch synthase I enzyme: studies of glucan affinity and catalytic properties. *Plant J.* 25:475–486, 2001.
39. Cooke, D., M. Gidley. Loss of crystalline and molecular order during starch gelatinization: origin of the enthalpic transition. *Carbohydr. Res.* 227:103–112, 1992.
40. Craig, J., J.R. Lloyd, K. Tomlinson, L. Barber, A. Edwards, T.L. Wang, C. Martin, C.L. Hedley, A.M. Smith. Mutations in the gene encoding starch synthase II profoundly alter amylopectin structure in pea embryos. *Plant Cell.* 10:413–426, 1998.
41. Craig, S.A.S., C.C. Maningat, P. Seib, R.C. Hoseney. Starch paste clarity. *Cereal Chem.* 66:173–182, 1989.
42. Creech, R.G. Genetic control of carbohydrate synthesis in maize endosperm. *Genetics* 52:1175–1186, 1965.
43. Cummings, J.H., E.R. Beatty, S.M. Kingman, S.A. Bingham, H.N. Englyst. Digestion and physiological properties of resistant starch in the human large bowel. *Br. J. Nutr.* 75:733–747, 1995.
44. Denyer, K., C.M. Hylton, C.F. Jenner, A.M. Smith. Identification of multiple isoforms of soluble and granule bound starch synthase in developing wheat endosperm. *Planta* 196:256–265, 1995.
45. Denyer, K., D. Waite, A. Edwards, C. Martin, A.M. Smith. Interaction with amylopectin influences the ability of granule-bound starch synthase I to elongate malto-oligosaccharides. *Biochem. J.* 342:647–653, 1999.
46. Denyer, K., D. Waite, S. Motawia, B.L. Moller, A.M. Smith. Granule-bound starch synthase I in isolated starch granules elongates malto-oligosaccharides processively. *Biochem. J.* 340:183–191, 1999.
47. Dinges, J.R., C. Colleoni, M. James, A. Myers. Purification and Molecular Genetic Characterization of ZPU1, a Pullulanase-Type Starch-Debranching Enzyme from Maize1. *Plant Cell* 15:666–680, 2003.
48. Doublier, J.L., I. Coté, G. Llamas, M.L. Meur. Effect of thermal history on amylose gelation. *Prog. Coll. Polymer Sci.* 90:61–65, 1992.
49. Doublier, J.L., G. Llamas. A rheological description of amylose-amylopectin mixtures. In: *Food Colloids and Polymers: Stability and Mechanical Properties*, Doublier, J.L., G. Llamas, eds., Cambridge: The Royal Society of Chemistry, 1993, pp 138–146.
50. Echt, C., D. Schwartz. The *wx* locus is the structural gene for the *Wx* protein. *Maize Genetics Cooperation Newsletter* 55:8–9, 1981.
51. Edwards, A., D.C. Fulton, C.M. Hylton, S.A. Jobling, M. Gidley, U. Rössner, C. Martin, A.M. Smith. A combined reduction in activity of starch synthases II and III of potato has novel effects on the starch of tubers. *Plant J.* 17:251–261, 1999.
52. Edwards, A., J.P. Vincken, L.C.J.M. Suurs, R.G.F. Visser, S. Zeeman, A. Smith, C. Martin. Discrete forms of amylose are synthesized by isoforms of GBSSl in pea. *Plant Cell* 14:1767–1785, 2002.
53. Eerlingen, R.C., H. Jacobs, K. Block, J.A. Delcour. Effects of hydrothermal treatments on the rheological properties of potato starch. *Carbohydr. Res.* 297:347–356, 1997.
54. Eliasson, A.C., H.R. Kim. Changes in rheological properties of hydroxypropyl potato starch pastes during freeze-thaw treatments, I: a rheological approach for evaluation of freeze-thaw stability. *J. Texture Studies* 23:279–295, 1992.
55. Eliasson, A.C., H.R. Kim. A dynamic rheological method to study the interaction between starch and lipids. *J. Rheol.* 39:1519–1534, 1995.
56. Ellis, H.S., S. Ring. A study of some factors influencing amylose gelation. *Carbohydr. Polymers* 5:201–213, 1985.
57. Fisher, D.K., D.B. Thompson. Retrogradation of maize starch after thermal treatment within and above the gelatinization temperature range. *Cereal Chem.* 74:344–351, 1957.

58. Flipse, E., C.J.M. Keetels, E. Jacobsen, R.G.F. Visser. The dosage effect of the wild type GBSS allele is linear for GBSS activity but not for amylose content: Absence of amylose has a distinct influence on the physico-chemical properties of starch. *Theor. Appl. Genet.* 92:121–127, 1996.
59. Fujita, N., A. Kubo, D. Suh, K.S. Wong, J.-L. Jane, K. Ozawa, F. Takaiwa, Y. Inaba, Y. Nakamura. Antisense inhibition of isoamylase alters the structure of amylopectin and the physicochemical properties of starch in rice endosperm. *Plant Cell Physiol.* 44:607–618, 2003.
60. Fulton, D., A. Edwards, E. Pilling, H.L. Robinson, B. Fahy, R. Seale, L. Kato, A.M. Donald, P. Geigenberger, C. Martin, A.M. Smith. Role of granule bound starch synthase in determination of amylopectin structure and starch granule morphology in potato. *J. Biol. Chem.* 277:10834–10841, 2002.
61. Fuwa, H., D.V. Glover, K. Miyaura, N. Inouchi, Y. Konishi, Y. Sugimoto. Chain length distribution of amylopectins of double- and triple-mutants containing the waxy gene in the inbred Oh43 maize background. *Starch/Stärke* 39:295–298, 1987.
62. Gao, M., D.K. Fisher, K.N. Kim, J.C. Shannon, M.J. Guiltinan. Independent genetic control of maize starch-branching enzymes IIa and IIb: isolation and characterization of a *Sbe2a* cDNA. *Plant Physiol.* 114:69–78, 1997.
63. Gao, M., J. Wanat, P.S. Stinard, M.G. James, A.M. Myers. Characterization of *dull1*, a maize gene coding for a novel starch synthase. *Plant Cell* 10:399–412, 1998.
64. Garwood, D.L., J.C. Shannon, R.G. Creech. Starches of endosperms possessing different alleles at the *amylose-extender* locus in *Zea mays* L. *Cereal Chem.* 53:355–364, 1976.
65. Garwood, D.L., S.F. Vanderslice. Carbohydrate composition of alleles at the sugary locus in maize. *Crop Sci.* 22:367–371, 1982.
66. German, M.L., A.L. Blumenfeld, Y.V. Guenin, V.P. Yuryev, V.B. Tolstoguzov. Structure formation in systems containing amylose. *Carbohydr. Polym.* 18:27–34, 1992.
67. Gidley, M. Factors affecting the crystalline type (A-C) of native starches and model compounds: a rationalisation of observed effects in terms of polymorphic structures. *Carbohydr. Res.* 161:301–304, 1987.
68. Gidley, M. Molecular mechanisms underlying amylose aggregation and gelation. *Macromolecules* 22:351–358, 1989.
69. Gidley, M., C. Bulpin. Crystallisation of malto-oogisaccharides as models of the crystalline forms of starch: minimum chain-length requirement for the formation of double helices. *Carbohydr. Res.* 161:291–300, 1987.
70. Gidley, M., C. Bulpin. Aggregation of amylose in aqueous systems: the effect of chain length on phase behavior and aggregation kinetics. *Macromolecules* 22:341–346, 1989.
71. Gidley, M.J., S.M. Bociek. Molecular organization in starches: a 13C CP/MAS NMR study. *J. Am. Chem. Soc.* 107:7040–7044, 1985.
72. Giroux, M., C.F. Morris. Wheat grain hardness results from highly conserved mutations in the friabilin components puroindoline a and b. *Proc. Natl. Acad. Sci. USA* 95:6262–6266, 1998.
73. Glover, D.V., E.T. Mertz. Corn. In: Nutritional quality of cereal grains. Genetic and agronomic improvement (Olson, R.A., Frey, K.J., eds). Agronomy No. 28. American Society of Agronomy, Madison, pp 183–336, 1987.
74. Goodfellow, B.J., R.H. Wilson. A fourier transform IR study of the gelation of amylose and amylopectin. *Biopolymers* 30:1183–1189, 1990.
75. Greenwood, C.T., S. McKenzie. Studies on starches of high amylose-content, Part IV. The fractionation of amylomaize starch; a study of the branched component. *Carbohydr. Res.* 3:7–13, 1966.
76. Guan, H.P., J. Preiss. Differentiation of the properties of the branching isozymes from maize (*Zea mays*). *Plant Physiol.* 102:1269–1273, 1993.
77. Han, X.-Z., O.H. Campanella, H. Guan, P.L. Keeling, B.R. Hamaker. Influence of maize starch granule-associated protein on the rheological properties of starch pastes, I: large deformation measurements of paste properties. *Carbohydr. Polym.* 49:323–330, 2002.

78. Han, X.-Z., O.H. Campanella, H. Guan, P.L. Keeling, B.R. Hamaker. Influence of maize starch granule-associated protein on the rheological properties of starch pastes, II: dynamic measurements of viscoelastic properties of starch pastes. *Carbohydr. Polym.* 49:315–321, 2002.
79. Haralampu, S.G. Resistant starch: a review of the physical properties and biological impact of RS3. *Carbohydr. Polym.* 41:285–292, 2000.
80. Hellwege, E.M., S. Czapla, A. Jahnke, L. Willmitzer, A.G. Heyer. Transgenic potato (*Solanum tuberosum*) tubers synthesize the full spectrum of inulin molecules naturally occurring in globe artichoke (*Cynara scolymus*) roots. *Proc. Natl. Acad. Sci. USA* 97:8699–8704, 2000.
81. Helm, J.L., V.L. Ferguson, M.S. Zuber. Interaction of dosage effects on amylose content of corn at the *Du* and *Wx* loci. *J. Heredity* 60:259–260, 1969.
82. Hizukuri, S. Polymodal distribution of the chain lengths of amylopectins and its significance. *Carbohydr. Res.* 147:342–347, 1986.
83. Hizukuri, S., Y. Takeda, N. Maruta. Molecular structure of rice starches. *Carbohydr. Res.* 189:227–235, 1989.
84. Hizukuri, S., Y. Takeda, M. Yasuda, A. Suzuki. Multi-branched nature of amylose and the action of de-branching enzymes. *Carbohydr. Res.* 94:205–213, 1981.
85. Hofvander, P., M. Andersson, C. Larsson, H. Larsson. Field performance and starch characteristics of high-amylose potatoes obtained by antisense gene targeting of two branching enzymes. *Plant Biotechnol. J.* 2:311–320, 2004.
86. Hogg, A.C., T. Sripo, B. Beecher, J.M. Martin, M.J. Giroux. Wheat puroindolines interact to form friabilin and control wheat grain hardness. *Theor. Appl. Genet.* 108:1089–1097, 2004.
87. Holder, D.G., D.V. Glover, J.C. Shannon. Interaction of shrunken-2 with five other carbohydrate genes in corn endosperm. *Crop Sci.* 14:643–646, 1997.
88. Hovenkamp-Hermelink, J.H.M., E. Jacobsen, A.S. Ponstein, R. Visser, G.H. Vos-Scheperkeuter, B. Witholt, J.N. de Vries, E.W. Bijmolt, W.J. Feenstra. Isolation of an amylose-free starch mutant of the potato (*Solanum tuberosum* L.). *Theor. Appl. Genet.* 75:217–221, 1987.
89. Hseih, J.S. Genetic studies of the Wx gene of sorghum (*Sorghum bicolor* [L.] Moench). *Crop Sci.* 14:643–646, 1988.
90. Hussain, H., A. Mant, R. Seale, S. Zeeman, E. Hinchliffe, A. Edwards, C. Hylton, S. Bornemann, A.M. Smith, C. Martin, R. Bustos. Three isoforms of isoamylase contribute different catalytic properties for the debranching of potato glucans. *Plant Cell* 15:133–149, 2003.
91. I'Anson, K.J., M.J. Miles, V.J. Morris, S. Ring. A study of amylose gelation using a synchrotron x-ray source. *Carbohydr. Polym.* 8:45–53, 1988.
92. Imberty, A., A. Buleon, V. Tran, S. Perez. Recent advances in knowledge of starch structure. *Starch/Stärke* 43:375–384, 1991.
93. Björck, I. Starch: Nutritional Aspects. In: *Carbohydrates in Food*, Eliasson, A.C., ed., New York: Marcel Dekker, 1996.
94. Inouchi, N., D.V. Glover, H. Fuwa. Chain length distribution of amylopectins of several single mutants and the normal counterpart, and sugary-1 phytoglycogen in maize (*Zea mays* L.). *Starch/Stärke* 39:259–266, 1987.
95. Inouchi, N., D.V. Glover, H. Fuwa. Structure and physicochemical properties of endosperm starches of a waxy allelic series and their respective normal counterparts in the inbred Oh43 maize background. *Starch/Stärke* 47:421–426, 1995.
96. Ishikawa, N., J. Ishihara, M. Itoh. Artificial induction and characterization of amylose-free mutants of barley. *Barley Genet. News* 24:4953, 1994.
97. Isshiki, M., K. Morino, M. Nakajima, R. Okagaki, S.R. Wessler, T. Isawa, K. Shimamoto. A naturally occurring functional allele of the rice *waxy* locus has a GT to TT mutation at the 5' splice site of the first intron. *Plant J.* 15:133–138, 1998.
98. Jacobson, M.R., J.N. BeMiller. Method for determining the rate and extent of accelerated starch retrogradation. *Cereal Chem.* 75:22–29, 1998.
99. James, M., K. Denyer, A. Myers. Starch synthesis in the cereal endosperm. *Curr. Opin. in Plant Biol.* 6:215-222, 2003.

100. James, M.G., D.S. Robertson, A.M. Myers. Characterization of the maize gene sugary1, a determinant of starch composition in kernels. *Plant Cell* 7:417–429, 1995.
101. Jane, J.-L., J.F. Chen. Effect of amylose molecular size and amylopectin branch chain length on paste properties of starch. *Cereal Chem.* 69:60–65, 1992.
102. Jane, J.-L., T. Kasemsuwan, S. Leas, I. Ames, H.F. Zobel, I.L. Darien, J.F. Robyt. Anthology of starch granule morphology by scanning electron microscopy. *Starch/Stärke* 46:121–129, 1994.
103. Jane, J.-L., J.F. Robyt. Structure studies of amylose-V complexes and retrogradaded amylose by action of alpha amylases, and a new method for preparing amylodextrins. *Carbohydr. Res.* 132:105–118, 1984.
104. Jane, J.L., Y.Y. Chen, L.F. Lee, A.E. McPherson, K.-S. Wong, M. Radosavljevic, T. Kasemsuwan. Effects of amylopectin branch chain length and amylose content on the gelatinization and pasting properties of starch. *Cereal Chem.* 76:629–637, 1999.
105. Jenkins, P.J., R.E. Cameron, A.M. Donald. A universal feature in the structure of starch granules from different botanical sources. *Starch/Stärke* 45:417–420, 1993.
106. Ji, Q., J.P. Vincken, L.C. Suurs, R.G.F. Visser. Microbial starch-binding domains as a tool for targeting proteins to granules during starch biosynthesis. *Plant Mol. Biol.* 51:789–801, 2003.
107. Jobling, S.A., G.P. Schwall, R.J. Westcott, C.M. Sidebottom, M. Debet, M.J. Gidley, R. Jeffcoat, R. Safford. A minor form of starch branching enzyme in potato (*Solanum tuberosum* L.) tubers has a major effect on starch structure: cloning and characterisation of multiple forms of *SBE A*. *Plant J.* 18:163–171, 1999.
108. Jobling, S.A., R.J. Westcott, A. Tayal, R. Jeffcoat, G.P. Schwall. Production of freeze-thaw stable potato starch by antisense inhibition of three starch synthase genes. *Nat. Biotechnol.* 20:295–299, 2002.
109. Juliano, B.O. M.B. Nazareno, N.B. Ramos. Properties of waxy and isogenic nonwaxy rices differing in starch gelatinization temperature. *J. Agric. Food Chem.* 17:1364–1369, 1969.
110. Kainuma, K., D. French. Nägeli amylodextrin and its relationship to starch granule structure, I: preparation and properties of amylodextrins from various starch types. *Biopolymers* 10:1673–1680, 1971.
111. Kalichevski, M.T., P.D. Orford, S. Ring. The retrogradation and gelation of amylopectins from various botanical sources. *Carbohydr. Res.* 198:49–55, 1990.
112. Kalichevski, M.T., S. Ring. Incompatability of amylose and amylopectin in aqueous solution. *Carbohydr. Res.* 162:323–328, 1987.
113. Keetels, C.J.A.M., T.V. Vilet, P. Walstra. Gelation and retrogradation of concentrated starch systems, 2: retrogradation. *Food Hydrocolloids* 10:355–362, 1996.
114. Kim, J.O., W.S. Kim, M.S. Shin. A comparative study on retrogradation of rice starch gels by DSC, X-ray and α-amylase methods. *Starch/Stärke* 49:71–75, 1997.
115. Kiribuchi-Otobe. *Cereal Chem.* 75:671–672, 1998.
116. Klucinec, J.D., D.B. Thompson. Fractionation of high-amylose maize starches by differential alcohol precipitation and chromatography of the fractions. *Cereal Chem.* 75:887–896, 1998.
117. Klucinec, J.D., D.B. Thompson. Amylose and amylopectin interact in retrogradation of dispersed high-amylose starches. *Cereal Chem.* 76:282–291, 1999.
118. Klucinec, J.D., D.B. Thompson. Amylopectin nature and amylose-to-amylopectin ratio as influences on the behaviours of gels of dispersed starch. *Cereal Chem.* 79:24–35, 2002.
119. Klucinec, J.D., D.B. Thompson. Structure of amylopectins from ae-containing maize starches. *Cereal Chem.* 79:19–23, 2002.
120. Knutson, C.A., M.J. Grive. Rapid method for estimation of amylose in maize starches. *Cereal Chem.* 71:469–471, 1994.
121. Krishnamurthy, K., M.J. Giroux. Expression of wheat puroindoline genes in transgenic rice enhances grain softness. *Nat. Biotechnol.* 19:162–166, 2001.
122. Kubo, A., N. Fujita, K. Harada, T. Matsuda, H. Satoh, Y. Nakamura. The starch-debranching enzymes isoamylase and pullulanase are both involved in amylopectin biosynthesis in rice endosperm. *Plant Physiol.* 121:399–409, 1999.

123. Kuipers, A.G.J., E. Jacobsen, R.G.F. Visser. Formation and deposition of amylose in the potato tuber starch granule are affected by the reduction of granule-bound starch synthase gene expression. *Plant Cell* 6:43–52, 1994.
124. Kumar, I., G.S. Khush. Inheritance of amylose content in rice (*Oryza sativa* L.). *Euphytica* 38:261–269, 1988.
125. Lai, V.M.-F., S. Lu, C.Y. Lii. Molecular characteristics influencing retrogradation kinetics of rice amylopectins. *Cereal Chem.* 77:272–278, 2000.
126. Lansky, S., M. Kooi, T.J. Schoch. Properties of the fractions and linear subfractions from various starches. *J. Am. Chem. Soc.* 71:4066–4075, 1949.
127. Larson, B.L., K.A. Gilles, R. Jenness. Amperometric method for determining the sorption of iodine by starch. *Anal. Chem.* 25:802–804, 1953.
128. Leloup, V.M., P. Colonna, A. Buleon. Influence of amylose-amylopectin ratio on gel properties. *J. Cereal Sci.* 13:1–13, 1991.
129. Leloup, V.M., P. Colonna, S.G. Ring, K. Roberts, B. Wells. Microstructure of amylose gels. *Carbohydr. Polym.* 18:189–197, 1992.
130. Levy, I., T. Paldi, O. Shoseyov. Engineering a bifunctional starch-cellulose cross-bridge protein. *Biopolymers* 25:1841–1849, 2004.
131. Lii, C.-Y., Y.-Y. Shao, K.-H. Tseng. Gelation mechanism and rheological properties of rice starch. *Cereal Chem.* 72:393–400, 1995.
132. Li, J.H., T. Vasanthan, B. Rossnagel, R. Hoover. Starch from hull-less barley: II. Thermal, rheological, and acid hydrolysis characteristics. *Food Chem.* 74:407–415, 2001.
133. Liu, H., L. Ramsden, H. Corke. Physical properties and enzymatic digestibility of acetylated *ae*, *wx*, and normal maize starch. *Carbohydr. Polym.* 34:283–287, 1997.
134. Liu, Q., D.B. Thompson. Effects of moisture content and different gelatinization heating temperatures on retrogradation of waxy-type maize starches. *Carbohydr. Res.* 314:221–235, 1998.
135. Liu Q., D.B. Thompson. Retrogradation of *du wx* and *su2 wx* maize starches after different gelatinization heat treatments. *Cereal Chem.* 75:868–874, 1998.
136. Lloyd, J.R., V. Landschütze, J. Kossmann. Simultaneous antisense inhibition of two starch-synthase isoforms in potato tubers leads to accumulation of grossly modified amylopectin. *Biochem. J.* 338:515–521, 1999.
137. Lu, S., L.N. Chen, C.-Y. Lii. Correlations between the fine structure, physicochemical properties, and retrogradation of amylopectins from Taiwain rice varieties. *Cereal. Chem.* 74:34–39, 1997.
138. Lu, T.-J., J.-L. Jane, P.L. Keeling Temperature effect on retrogradation rate and crystalline structure of amylose. *Carbohydr. Polymers* 33:19–26, 1997.
139. McCready, R.M., W.Z. Hassid. The separation and quantitative estimation of amylose and amylopectin in potato starch. *J. Am. Chem. Soc.* 65:1154–1157, 1943.
140. Mercier, R. The fine structure of corn starches of various amylose-percentage: waxy, normal, and amylomaize. *Starch/Stärke* 25:78–83, 1973.
141. Miles, M.J., V.J. Morris, P.D. Orford, S. Ring. The roles of amylose and amylopectin in the gelation and retrogradation of starch. *Carbohydr. Res.* 135:271–281, 1985.
142. Miles, M.J., V.J. Morris, S. Ring. Some recent observations on the retrogradation of amylose. *Carbohydr. Polym.* 4:73–77, 1984.
143. Miura, H., E. Araki, S. Tarui. Amylose synthesis capacity of the three *Wx* genes of wheat cv. Chinese Spring. *Euphytica* 108:91–95, 1999.
144. Miura, H., A. Sugawara. Dosage effects of the three Wx genes on amylose synthesis in wheat endosperm. *Theor. Appl. Genet.* 93:1066–1070, 1996.
145. Miura, H., S. Tanii, T. Nakamura, N. Watanabe. Genetic control of amylose content in wheat endosperm starch and differential effects of three Wx genes. *Theor. Appl. Genet.* 89:276-280, 1994.
146. Mizuno, K., T. Kawasaki, H. Shimada, H. Satoh, E. Kobayashi, S. Okumura, Y. Arai, T. Baba. Alteration of the structural properties of starch components by the lack of an isoform of starch branching enzyme in rice seeds. *J. Biol. Chem.* 268:19084–19091, 1993.

147. Mizuno, K., K. Kimura, Y. Arai, T. Kawasaki, H. Shimada, T. Baba. Starch branching enzymes from immature rice seeds. *J. Biochem.* 112:643–651, 1992.
148. Montgomery, E.M., K.R. Sexon, F.R. Senti. High-amylose corn starch fractions. *Starch/Stärke* 13:215–222, 1961.
149. Montgomery, E.M., K.R. Sexon, F.R. Senti. Physical properties and chemical structure of high-amylose corn starch fractions. *Starch/Stärke* 16:345–351, 1964.
150. Morell, M.K., B. Kosar-Hashemi, M. Cmiel, M.S. Samuel, P. Chandler, S. Rahman, A. Buleon, I.L. Batey, Z. Li. Barley *sex6* mutants lack starch synthase IIa activity and contain a starch with novel properties. *Plant J.* 34:173–185, 2003.
151. Morell, M.K., Z. Li, S. Rahman. Starch biosynthesis in the small grained cereals: wheat and barley. In: *Starch: Advances in Structure and Function,* Barsby, T., A.M. Donald, P.J. Frazier, eds., Cambridge: The Royal Society of Chemistry, 2001, pp 129–137.
152. Morris, C.F. Puroindolines: the molecular genetic basis of wheat grain hardness. *Plant Mol. Biol.* 48:633–647, 2002.
153. Morris. V.J. Starch gelation and retrogradation. *Trends Food Sci. Technol.* 1:2–6, 1990.
154. Morrison, W.R. Lipids in cereal starches: a review. *J. Cereal Sci.* 8:1–15, 1988.
155. Mu-Forster, C., B.P. Wasserman. Surface localization of zein storage proteins in starch granules from maize endosperm: proteolytic removal by thermolysin and *in vitro* cross-linking of granule-associated polypeptides. *Plant Physiol.* 116:1563–1571, 1996.
156. Murai, J., T. Taira, D. Ohta. Isolation and characterization of the three *Waxy* genes encoding the granule-bound starch synthase in hexaploid wheat. *Gene* 234:71–79, 1999.
157. Murata, K., T. Sagiyama, T. Akazawa. Enzymic mechanism of starch synthesis in glutinous rice grains. *Biochem. Biophys. Res. Commun.* 18:371–376, 1965.
158. Nakamura, T., M. Yamamori, H. Hirano, S. Hidaka. Identification of 3 Wx proteins in wheat (*Triticum aestivum* L). *Biochem. Genet.* 31:75–86, 1993.
159. Nakamura, Y. Towards a better understanding of the metabolic system for amylopectin biosynthesis in plants: rice endosperm as a model tissue. *Plant Cell Physiol.* 43:718–725, 2002.
160. Nakamura, Y., M. Yamamori, H. Hirano, S. Hidaka, Y. Nagamura. Production of waxy (amylose-free) wheats. *Mol. Gen. Genet.* 248:253–259, 1995.
161. Neuffer, G. *In Mutants of Maize*. Cold Spring Harbor, NY: Cold Spring Harbor Laboratory Press, 1997, p 298.
162. Nishi, A., Y. Nakamura, N. Tanaka, H. Satoh. Biochemical and genetic analysis of the effects of *amylose-extender* mutation in rice endosperm. *Plant Physiol.* 127:459–472, 2001.
163. O'Sullivan, A.C., S. Perez. The relationship between internal chain length of amylopectin and crystallininty in starch. *Biopolymers* 50:381–390, 1999.
164. Oda, S., C. Kiribuchi, H. Seko. A bread wheat mutant with low amylose content induced by ethyl methanesulphonate. *Jpn. J. Breed.* 42:151–154, 1992.
165. Ong, M.H., J.M.V. Blanshard. Texture determinations of cooked, parboiled rice, II: physicochemical properties and leaching behaviour of rice. *J. Cereal Sci.* 21:261–269, 1995.
166. Oostergetel, G.T., E.F.J. Vanbruggen. The crystalline domains in potato starch granules are arranged in a helical fashion. *Carbohydr. Polym.* 21:7–12, 1993.
167. Orford, P.D., S. Ring, V. Carroll, M.J. Miles, V.J. Morris. The effect of concentration and botanical source on the gelation and retrogradation of starch. *J. Sci. Food Agric.* 39:177, 1987.
168. Paravouri, P., R. Manelius, T. Suortti, E. Bertoft, K. Autio. Effects of enzymatically modified amylopectin on the rheological properties of amylose-amylopectin mixed gels. *Food Hydrocolloids* 11:471–477, 1997.
169. Paredez-López, O., L.A. Bello-Pérez, M.G. López. Amylopectin: structural, gelatinization and retrogradation studies. *Food Chem.* 50:411–417, 1994.
170. Pilling, E., A.M. Smith. Growth ring formation in the starch granules of potato tubers. *Plant Physiol.* 1323:365–371, 2003.
171. Reddy, I., P.A. Seib. Modified waxy wheat starch compared to modified waxy corn starch. *J. Cereal Sci.* 31:25–39, 2000.
172. Reddy, R.K., Z.S. Ali, K.R. Bhattacharya. The fine structure of rice-starch amylopectin and its relation to the texture of cooked rice. *Carbohydr. Polym.* 22:267–275, 1993.

173. Reyes, A.C., E.L. Albano, V.P. Briones, B.O. Juliano. Varietal differences in physicochemical properties of rice starch and its fractions. *J. Agric. Food Chem.* 13:438–442, 1965.
174 Echt, C., D. Schwartz. The *wx* locus is the structural gene for the *Wx* protein. Maize Genetics Cooperation Newsletter 55:8-9, 1981.
175. Ring, S. Some studies on starch gelation. *Starch/Stärke* 37:80–83, 1985.
176. Ring, S., P. Colonna, K.J. I' Anson, M.T. Kalichevski, M.J. Miles, V.J. Morris, P.D. Orford. The gelation and crystallization of amylopectin. *Carbohydr. Res.* 162:277–293, 1987.
177. Ritte, G., J.R. Lloyd, N. Eckermann, A. Rottmann, J. Kossmann, M. Steup. The starch-related R1 protein is an α-glucan, water dikinase. *Proc. Natl. Acad. Sci. USA* 99:7166–7171, 2002.
178. Robin, J.P., C. Mercier, R. Charbonierre, A. Guilbot. Lintnerized starches: gel filtration and enzymatic studies of insoluble residues from prolonged acid treatment of potato starch. *Cereal Chem.* 51:389–406, 1974.
179. Robin, J.P., C. Mercier, F. Duprat, R. Charbonierre, A. Guilbot. Amidons lintnerises: etudes chromatographique et enzymatique des residus insolubles provenant de l'hydrolyse chlorhydrique d'amidons de cereales, en particulier de mais cireux. *Starch/Stärke* 27:36–45, 1975.
180. Roger, P., P. Colonna. The influence of chain length on the hydrodynamic behaviour of amylose. *Carbohydr. Res.* 227:73–83, 1992.
181. Rundle, R.E., R.R. Baldwin. The configuration of starch and the starch-iodine complex, I: The dichroism of flow of starch-iodine solutions. *J. Am. Chem. Soc.* 65:554–558, 1943.
182. Safford, R., S.A. Jobling, C. Sidebottom, R.J. Westcott, D. Cooke, K.J. Tober, B. Strongitharm, A.L. Russell, M. Gidley. Consequences of antisense RNA inhibition of starch branching enzyme activity on properties of potato starch. *Carbohydr. Polym.* 35:155–168, 1998.
183. Salehuzzaman, S.N.I.M., J.P. Vincken, M. Van de Wal, I. Straatman-Engelen, E. Jacobsen, R.G.F. Visser. Expression of a cassava granule-bound starch synthase gene in the amylose-free potato only partially restores amylose content. *Plant Cell Environ.* 22:1311–1318, 1999.
184. Salomonsson, A.C., B. Sundberg. Amylose content and chain profile of amylopectin from normal, high amylose and waxy barleys. *Starch/Stärke* 46:325–328, 1994.
185. Sanchez, P.C., B.O. Juliano, V.T. Laude, C.M. Perez. Nonwaxy rice for tapuy (rice wine) production. *Cereal Chem.* 65:240–243, 1988.
186. Sano, Y. Differential regulation of waxy gene expression in rice endosperm. *Theor. Appl. Genet.* 68:467–471, 1984.
187. Sano, Y., M. Katsumata, K. Okuno. Genetic studies of speciation in cultivated rice. 5. Inter- and intraspecific differentiation in the waxy gene expression of rice. *Euphytica* 35:1–9, 1986.
188. Sasaki, T., T. Yasui, J. Matsuki. Effect of amylose content on gelatinization, retrogradation, and pasting of starches from waxy and nonwaxy wheat and their F1 seeds. *Cereal Chem.* 77:58–63, 2000.
189. Satoh, H., A. Nishi, K. Yamishita, Y. Takemoto, Y. Tanaka, Y. Hosaka, Y. Sakurai, N. Fujita, Y. Nakamura. Starch-branching enzyme I-deficient mutation specifically affects the structure and properties of starch in rice endosperm. *Plant Physiol.* 133:1111–1121, 2004.
190. Schierbaum, F., W. Vorwerg, B. Kettlitz, F. Reuther. Interaction of linear and branched polysaccharides in starch gelling. *Die Nahrung* 30:1047–1049, 1986.
191. Schoch, T.J. Fractionation of starch by selective precipitation with butanol. *J. Am. Chem. Soc.* 64:2957–2961, 1942.
192. Schoch, T.J. Preparation of starch and the starch fractions. In: *Methods in Enzymology III*, Colowick, S.P., N.O. Kaplan, eds., New York: Academic Press, 1954, pp 5–17.
193. Schwall, G.P., R. Safford, R.J. Westcott, R. Jeffcoat, A. Tayal, Y.C. Shi, M.J. Gidley, S.A. Jobling. Production of very-high-amylose potato starch by inhibition of SBE A and B. *Nat. Biotechnol.* 18:551–554, 2000.
194. Shannon, J.C., D.L. Garwood. Genetics and physiology of starch development. In: *Starch: Chemistry and Technology*, ed. 2, Whistler, R.L., J.N. BeMiller, E.F. Paschall, eds., New York: Academic Press, 1984, pp 25–86.

195. Shi, Y.C., P. Seib. The structure of four waxy starches related to gelatinization and retrogradation. *Carbohydr. Res.* 227:131–145, 1992.
196. Shi, Y.C., P. Seib. Fine structure of maize starches from four *wx*-containing genotypes of the W64A inbred line in relation to gelatinization and retrogradation. *Carbohydr. Polym.* 26:141–147, 1995.
197. Shi Y.C., P. Seib. Molecular structure of a low-amylopectin starch and other high-amylose maize starches. *J. Cereal Sci.* 27:289–299, 1998.
198. Shimada, H., Y. Tada, T. Kawasaki, T. Fugimura. Antisense regulation of the rice waxy gene expression using a PCR-amplified fragment of the rice genome reduces the amylose content in grain starch.*Theor. Appl. Genet.* 86:665–672, 1993.
199. Shimada, H., Y. Tada, T. Kawasaki, T. Fujimura. Antisense regulation of the rice waxy gene expression using a PCR-amplified fragment of the rice genome reduces the amylose content in grain starch. *Theor. Appl. Genet.* 86:665–672, 1993.
200. Shure, M., S.R. Wessler, N. Federoff. Molecular identification and isolation of the waxy locus in maize. *Cell* 225–233, 1983.
201. Sidebottom, C., M. Kirkland, B. Strongitharm, R. Jeffcoat. Characterization of the difference of starch branching enzyme activities in normal and low-amylopectin maize during kernel development. *J. Cereal Sci.* 27:279–287, 1998.
202. Sievert, D., P. Wursch. Thermal behavior of potato amylose and enzyme-resistant starch from maize. *Cereal Chem.* 70:333–338, 1993.
203. Song, Y., J.-L. Jane. Characterization of barley starches from waxy, normal and high amylose varieties. *Carbohydr. Polym.* 41:365–377, 2000.
204. Sprague, G.F., M.T. Jenkins. The development of waxy corn for industrial use. *Iowa State Coll. J. Sci.* 22:205–213, 1948.
205. Stinard, P.S., D.S. Robertson, P.S. Schnable. Genetic isolation, cloning and analysis of a mutator-induced, dominant antimorph of the maize amylose extender1 locus. *Plant Cell* 5:1555–1566, 1993.
206. Sun, C.X., P. Sathish, S. Ahlandsberg, C. Jansson. The two genes encoding starch-branching enzymes IIa and IIb are differentially expressed in barley. *Plant Physiol.* 118:37–49, 1998.
207. Takeda, C., Y. Takeda, S. Hizukuri. Structure of amylomaize amylose. *Cereal Chem.* 66:22–25, 1989.
208. Takeda, Y., H.P. Guan, J. Preiss. Branching of amylose by the branching isoenzymes of maize endosperm. *Carbohydr. Res.* 240:253–263, 1993.
209. Takeda, Y., S. Hizukuri, B.O. Juliano. Purification and structure of amylose from rice starch. *Carbohydr. Res.* 148:299–308, 1986.
210. Takeda, Y., S. Hizukuri, B.O. Juliano. Structures of rice amylopectins with low and high affinities for iodine. *Carbohydr. Res.* 168:79–88, 1987.
211. Takeda, Y., S. Hizukuri, C. Takeda, A. Suzuki. Structures of branched molecules of amyloses of various origins, and molar fractions of branched and unbranched molecules. *Carbohydr. Res.* 165:139–145, 1987.
212. Takeda, Y., Preiss, J. Structures of B90 (sugary) and W64A (normal) maize starches. *Carbohydr. Res.* 240:265–275, 1993.
213. Takeda, Y., K. Shirakawa, S. Hizukuri. Examination of the purity and structure of amylose by gel-permeation chromatography. *Carbohydr. Res.* 132:83–92, 1984.
214. Tester, R.F., W.R. Morrison. Swelling and gelatinization of cereal starches, I: effects of amylopectin, amylose, and lipids. *Cereal Chem.* 67:551–557, 1990.
215. Tester, R.F., W.R. Morrison. Swelling and gelatinization of cereal starches, 3: some properties of waxy and normal nonwaxy barley starches. *Cereal Chem.* 69:654–658, 1992.
216. Thompson, D.B. On the non-random nature of amylopectin branching. *Carbohydr. Polym.* 43:223–239, 2000.
217. Thompson, D.B., J.M.V. Blanshard. Retrogradation of selected *wx*-containing maize starches. *Cereal Foods World* 40:670, 1995.
218. Tomlinson, K., K. Denyer. Starch synthesis in cereal grain. *Adv. Bot. Res.* 40:1–61, 2003.

219. Topping, D.L., M.K. Morell, R.A. King, Z. Li, A.R. Bird, M. Noakes. Resistant starch and health – Himalaya 292, a novel barley cultivar to deliver benefits to consumers. *Starch/Stärke* 55:539–545, 2003.
220. Tziotis, A., K. Seetharaman, K.S. Wong, J.D. Klucinec, J.-L. Jane, P. White. Structural properties of starch fractions isolated from normal and mutant corn genotypes by using different methods. *Cereal Chem.* 2004.
221. Umemoto, T., M. Yano, H. Satoh, J.M. Short, A. Shamura, Y. Nakamura. Mapping of a gene responsible for the difference in amylopectin structure between japonica-type and indica-type rice varieties. *Theor. Appl. Genet.* 104:1–8, 2002.
222. Vallêra, A.M., M.M. Cruz, S.G. Ring, F. Boue. The structure of amylose gels. *J. Phys.* 6:301–320, 1994.
223. Hovenkamp, J.H.M., E. Jacobsen, A.S. Ponstein, R.G.F. Visser, G.H. Vos-Scheperkeuter, E.W. Bijmolt, J.N. de Vries, B. Witholt, W.J. Feenstra. Isolation of an amylose-free starch mutant of the potato (*Solanum tuberosum* L.). *Theor. Appl. Genet.* 75:217–221, 1991.
224. Van der Leij F.R., R.G.F. Visser, K. Oosterhaven, D.A.M. van der Kop, E. Jacobsen, W.J. Feenstra. Complementation of the amylose-free starch mutant of potato (*Solanum tuberosum* L.) by the gene encoding granule-bound starch synthase. *Theor. Appl. Genet.* 82:289–295, 1991.
225. Villareal, C.P., S. Hizukuri, B.O. Juliano. Amylopectin staling of cooked milled rices and properties of amylopectin and amylose. *Cereal Chem.* 74:163–167, 1997.
226. Villareal, C.P., B.O. Juliano. Comparative levels of waxy gene product of endosperm starch granules of different rice ecotypes. *Starch/Stärke* 41:369–371, 1989.
227. Vinyard, M.L., R. Bear. Amylose content. *Maize Genet. Coop. Newsl.* 26:5, 1952.
228. Visser R.G.F., I. Somhorst, G.J. Kuipers, N.J. Ruys, W.J. Feenstra, E. Jacobsen. Inhibition of the expression of the gene for granule-bound starch synthase in potato by antisense constructs. *Mol. Gen. Genet.* 225:289–296, 1991.
229. Vrinten, P.L., T. Nakamura. Wheat granule-bound starch synthase I and II are encoded by separate genes that are expressed in different tissues. *Plant Physiol.* 122:255–263, 2000.
230. Waigh, T.A., A.M. Donald, F. Heidelbach, C. Riekel, M.J. Gidley. Analysis of the native structure of starch granules with small angle x-ray microfocus scattering. *Biopolymers* 49:91–105, 1999.
231. Waigh, T.A., M. Gidley, B.U. Komanshek, A.M. Donald. The phase transformations in starch during gelatinisation:a liquid crystalline approach. *Carbohydr. Res.* 328:165–176, 2000.
232. Waigh, T.A., K.L. Kato, A.M. Donald, M. Gidley, C. Riekel. Side chain liquid crystalline models for starch. *Starch/Stärke* 52:450–460, 2000.
233. Waigh, T.A., P.A. Perry, C. Riekel, M. Gidley, A.M. Donald. Chrial side chain liquid crystalline properties of starch. *Macromolecules* 31:7980–7904, 1998.
234. Wang, Y.J., L.F. Wang. Structures of four waxy rice starches in relation to thermal, pasting and textural properties. *Cereal Chem.* 79:252–256, 2002.
235. Wang, Z.Y., F.Q. Zheng, G.Z. Shen, J.P. Gao, D.P. Snustad, M.G. Li, J.L. Zhang, M.M. Hong. The amylose content in rice endosperm is related to the post-transcriptional regulation of the *waxy* gene. *Plant J.* 7:613–622, 1995.
236. Weatherwax, P. A rare carbohydrate in waxy maize. *Genetics* 7:568–572, 1922.
237. Whistler, R.L., J.N. BeMiller. Starch. In: *Carbohydrate Chemistry for Food Scientists,* Whistler, R.L., J.N. BeMiller, eds., St. Paul: Eagan Press, 1997, p 146.
238. Whistler, R.L., R.M. Daniel. Carbohydrates in feed chemistry. In: *Food Chemistry,* Fennema, O.R., ed., New York: Marcel Dekker, 1985, pp 118–120.
239. Whistler, R.L., W.M. Doane. Characterization of intermediary fractions of high-amylose corn starches. *Cereal Chem.* 38:251–255, 1961.
240. Whistler, R.L., G.E. Hilbert. Separation of amylose and amylopectin by certain nitroparaffins. *J. Am. Chem. Soc.* 67:1161–1165, 1945.
241. Wilson, E.J., T.J. Schoch, C.S. Hudson. The action of *macerans* amylase on the fractions from starch. *J. Am. Chem. Soc.* 65:1380–1383, 1943.

242. Wu, H.C., A. Sarko. The double-helical molecular structure of crystalline A-amylose. *Carbohydr. Res.* 61:7–25, 1978.
243. Wu, H.C., A. Sarko. The double-helical molecular structure of crystalline B-amylose. *Carbohydr Res* 61:27–40, 1978.
244. Würsch, P, D. Gumy. Inhibition of amylopectin retrogradation by partial beta-amylolysis. *Carbohydr Res* 256:129–137, 1994.
245. Yamada, T., M. Taki. Fractionation of maize starch by gel-chromatography. *Starch/Stärke* 28:374–377, 1976.
246. Yamamori, M., S. Fujita, K. Hayakawa, J. Matsuki, T. Yasui. Genetic eliminiation of a starch granule protein, SGP-1, of wheat generates an altered starch with apparent high amylose. *Theor. Appl. Genet.* 101:21–29, 2000.
247. Yamamori, M., N.T. Quynh. Differential effects of Wx-A1,-B1 and-D1 protein deficiencies on apparent amylose content and starch pasting properties in common wheat. *Theor. Appl. Genet.* 100:32–38, 2000.
248. Yanagisawa, T., C., Kiribuchi-Otobe, H. Yoshida. An alanine to threonine change in the Wx-D1 protein reduces GBSS I activity in waxy mutant wheat. *Euphytica* 121:209–214, 2001.
249. Yano, M., K. Okuno, J. Kawakami, H. Satoh, T. Omura. High amylose mutants of rice, *Oryza sativa* L. *Theor. Appl. Genet.* 69:253–257, 1985.
250. Yao, Y., D.B. Thompson, M.J. Guilitinan. Maize starch-branching enzyme isoforms and amylopectin structure: in the absence of starch-branching enzyme IIb, the further absence of starch-branching enzyme Ia leads to increased branching. *Plant Physiol.* 136:3515–3523, 2004.
251. Yeh, J.Y., D.L. Garwood, J. Shannon. Characterisation of starch from maize endosperm mutants. *Starch/Stärke* 33:222–230, 1981.
252. Yoshimoto Y., J. Tashiro, T. Takenouchi, Y. Takeda. Molecular structure and some physicochemical properties of high-amylose barley starches. *Cereal Chem.* 77:279–285, 2000.
253. Yuan, R.C., D.B. Thompson. Rheological and thermal properties of aged starch pastes from three waxy maize genotypes. *Cereal Chem.* 75:117–123, 1998.
254. Yuan, R.C., D.B. Thompson, C.D. Boyer. Fine structure of amylopectin in relation to gelatinization and retrogradation behavior of maize starches from 3 wx containing genotypes in 2 inbred lines. *Cereal Chem.* 70:81–89, 1993.
255. Zhao, X.C., P.J. Sharp. Production of all eight genotypes of null alleles at 'waxy' loci in bread wheat, *Triticum aestivum* L. *Plant Breed.* 117:488–490. 1998.
256. Zheng, G.H., H.L. Han, R.S. Bhatty. Functional properties of cross-linked and hydroxypropylated waxy hull-less barley starches. *Cereal Chem.* 76:182–188, 1999.
257. Zobel, H.F., S.N. Young , L.A. Rocca. Starch gelatinization: an x-ray diffraction study. *Cereal Chem.* 65:443–446, 1988.
258. Slade, L., H. Levine. Glass transitions and water-food structure interactions. In: *Advances in Food and Nutrition Research,* Kinsella, J.E., S.L. Taylor, eds., San Diego, CA: Academic Press, 1995, pp 103–269.
259. Yun, S.-H. N.K. Matheson. Structures of the amylopectins of waxy, normal, amylose-extender, and wx:ae genotypes and of the phytoglycogen of maize. *Carbohydr. Res.* 243:307–321, 1993.
260. Hizukuri, S. Starch:analytical aspects. Pgs. 347–429 in: Carbohydrates in Food. Eliasson, A.-C., ed., Marcel Dekker, New York, 1996.
261. Sprague, G.F., M.T. Jenkins. The development of waxy corn for industrial use. Iowa State College Journal of Science 22:205-213, 1948.

2.06

Bioprocessing of Starch Using Enzyme Technology

K. Ravi-Kumar and S. Umesh-Kumar

CONTENTS

6.1 INTRODUCTION

Starch is a polymer of glucose produced by the plants as their principal food reserve: polysaccharide. Starch is stored in seeds such as cereals and millets, and tubers such as tapioca (cassava) and potato. Starch is a major source of carbohydrates for human and animal diets. Starches, or their liquefied products, are found in almost every processed food. In the U.S.A, fructose syrups are the primary sweeteners.

In its native state, starch is crystalline; the shape of the granules varies according to the source of starch. Because starch granules have the property to swell reversibly in cold water, they are used in industrial applications to loosen the starch protein in grain matrix. Starch irreversibly swells, forming a paste; when boiled in water this process is called gelatinization. The result increases the viscosity of a food product.

Structurally, starch can be separated into two polymers called amylose and amylopectin. Amylose is a linear polymer of D-glucose units linked together by α-1,4-glucosidic linkages, whereas amylopectin has branching at α-1,6 positions of starch. The concentration of amylose and amylopectin in each plant varies. Waxy starches from maize contain 2% amylose and 98% amylopectin, whereas most starches contain 15–30% amylose. A starch granule exists in a ring structure of amylose and amylopectin extending from the hilum toward the edge of the granule.

Starch hydrolases act either by cleaving α-1,4 and α-1,6 glucosidic bonds at random positions, releasing oligosaccharides of different chain lengths called dextrins or from the nonreducing ends of the starch molecules, successively releasing D-glucose (Figure 6.1). Smaller chain lengths of starch molecules such as disaccharides and oligosaccharides are also substrates for specific enzymes. Hence, enzyme processing of starch defines processes for the preparation of food components, substrates for the production of beverages, and industrial alcohol using microorganisms. The processing also defines technologies for modified starch components used in paper, textile, soaps, cosmetic, and pharmaceutical industries.

6.2 ENZYMES INVOLVED IN STARCH DEGRADATION

Microorganisms are the major source for starch hydrolases, generally called amylases (Table 6.1). The number of amylases produced by microorganisms in culture can be easily identified by performing enzyme zymograms after electrophoresis (Figure 6.2) in starch incorporated sodium dodecyl sulfate containing polyacrylamide gels (SDS-PAGE) under nonreducing condition (18). Amylases are classified according to the specific glucosidic bond they cleave as α-1,4-glucanases or α-1,6-glucanases. Endoglucanases act on interior bonds of starch while exoglucanases cleave the bonds successively from nonreducing ends of starch. Activities of amylases result in smaller molecules called dextrins, disaccharides, and monosaccharides. Glycosyl transferases are enzymes that synthesize cyclic molecules from starch.

6.2.1 α-Amylase (1,4-α-D-glucan Glucanohydrolase EC 3.2.1.1)

α-Amylases are α-1,4-endoglucanases that rapidly decrease starch viscosity, resulting in oligosaccharides. Some of the α-amylases produce higher concentrations of mono- and disaccharides; they are classified as saccharifing α-amylases. α-Amylases that reduce starch viscosity by producing precursor products for mono- and disaccharides are called liquefying enzymes. Though bond specificity for this enzyme is for α-1,4 linkages, some enzymes acting on α-1,6 linkages of starch molecule have also been reported (19). End products of the reaction result in oligosaccharides with α-configuration at the first carbon. Several of these enzymes, acting at temperatures above 60°C (normal gelatinization temperature of starch molecules), have been identified and characterized from bacterial sources. The enzyme most stable to high temperature, acting at temperatures above 100°C, has been described from *B. licheniformis* by Novo industries (20).

α-Amylases of bacterial origin are extracellular and act at near neutral pH. However, the pH optimum of an enzyme from *B. stearothermophilus* was reported to be dependent on temperature (10). α-Amylases active at pH as high as 9.0 and 11 have been described from alkalophilic *Bacillus spp* (21).

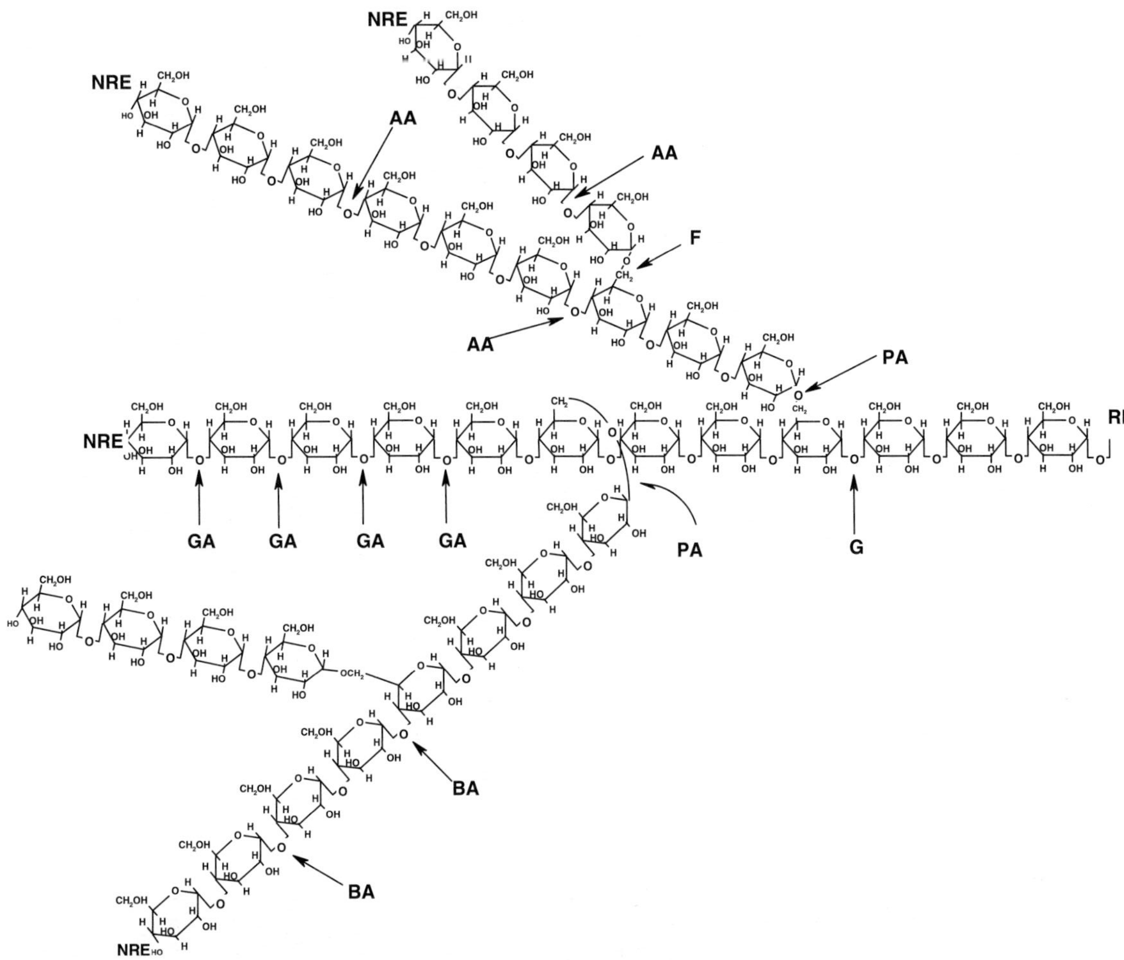

Figure 6.1 Action of different amylases on starch molecule
AA – α-Amylase (Endo 1,4-α-D-glucan glucanohydrolase), GA – Glucoamylase (Exo 1,4-α-D-glucan glucanohydrolase), BA – β-Amylase (Exo 1,4-α-D-glucan maltohydrolase), PA – Pullulanase (Endo α-dextrin 6-glucanohydrolase), RE – Reducing end, NRE – Non reducing end, F – α-1,6-glucosidic linkage, G – α-1,4-glucosidic linkage.

Table 6.1
Amylases of microbial origin

Microbial Source	Amylase Produced[a]	Activity on Starch	Ref.
Aspergillus awamori	Glucoamylase	Exo	1, 2
	α-Glucosidase	Exo	1
A. niger	α-Amylase	Endo	3
	Glucoamylase	Exo	4
	α-Glucosidase	Exo	5
A. oryzae	Taka-Amylase	Endo	6
	Glucoamylase	Exo	7
Bacillus amyloliquefaciens	α-Amylase	Endo	8
B. licheniformis	α-Amylase	Endo	9
B. stearothermophilus	α-Amylase	Endo	10
B. subtilis	α-Amylase	Endo	11
B. megaterium	α-Amylase	Endo	12
B. acidopullulyticus	Pullulanase	Endo	13
Clostridium thermosulfurogens	β-Amylase	Exo	14
Chalara paradoxa	α-Amylase	Endo	15
Lactobacillus amylovorus	α-Amylase	Endo	16
Thermomonospora viridis	Maltogenic amylase	Endo	17

[a] α-Amylases, α-Glucosidases, Taka-amylase, β-Amylase and Maltogenic amylase cleave α-1,4-glucosidic linkages; Pullulanase cleaves α-1,6-glucosidic linkages; Glucoamylases cleave both α-1,4 and α-1,6-glucosidic linkages.

α-Amylase stability at high temperatures requires calcium ions for maintaining the protein conformation at high temperatures (22). However, for *B. licheniformis* thermostable α-amylase, the calcium requirement is low (23). Inhibition of the enzyme activity by calcium ions has also been reported (24). Modeling of *Bacillus* α-amylase suggested conserved regions for calcium binding and the amino acids between 116 and 127 are involved in calcium binding in *B. subtilis* α-amylase (25). In *B. amyloliquefaciens,* calcium is bound outside the active center (26). In *B. subtilis* Taka-amylase, calcium is bound to the SH group of cysteine (27). α-Amylases of bacterial origin contain tryptophan, tyrosine, and histidine in their active sites (28).

The three dimensional structure of Taka-amylase (α-amylase of *Aspergillus oryzae*) and *Bacillus* α-amylase revealed that seven glucose units of the substrate are bound by at least two amino acids on the surface of the molecule. In this reaction, calcium binds close to the active center, apparently to stabilize the cleft in the active center (28). The catalytic site of Taka-amylase contains conserved aspartic acid at positions 206 and 297 of the protein chain. Protein engineering of *B. stearothermophilus* α-amylase revealed that aspartic acid is in the catalytic site at 331 position (28).

6.2.2 Glucoamylase or Amyloglucosidase (1,4-α-D-glucan Glucanohydrolase EC 3.2.1.3)

This enzyme catalyses release of glucose from the nonreducing ends of starch, dextrins, and maltose. Glucoamylases occur widely in microorganisms and plants with filamentous fungi as the major source of the enzyme (29). *Rhizopus* glucoamylase hydrolyses both α-1,6 and α-1,4 linkages of starch molecule with equal efficiency compared to the *A. niger* glucoamylase, which has higher affinity for α-1,4 linkages of starch (30).

Glucoamylases occur as multiple forms in several fungi. Because all fungal glucoamylases are glycoproteins, the difference in molecular mass was attributed to varying

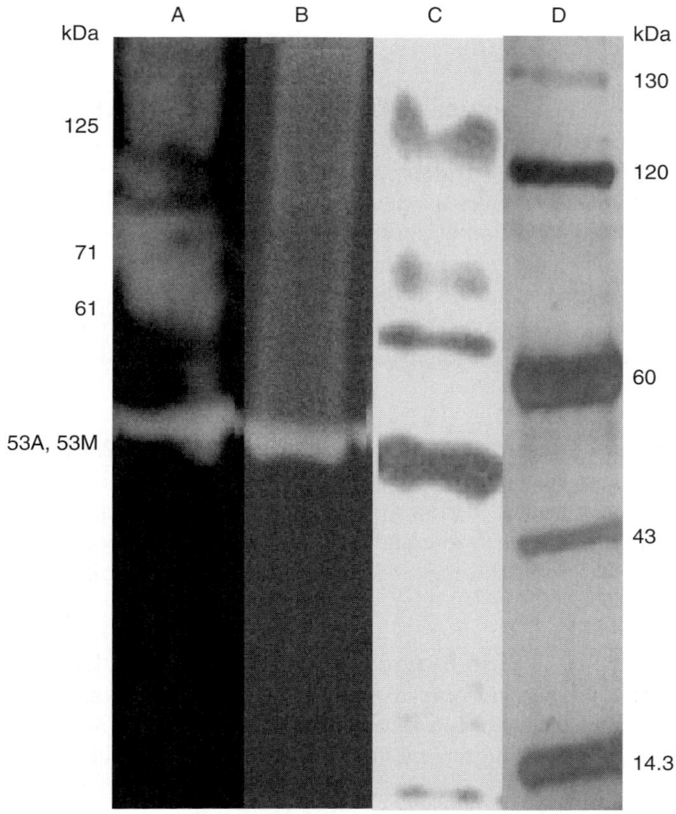

Figure 6.2 Amylases of *A. niger* identified as zymogram reactions after SDS-PAGE performed under nonreducing conditions in 10% acrylamide gels. Haloes correspond to enzyme activity as determined after iodine addition. A. Zymogram at pH 4.3; B. Zymogram at pH 6.0; C. Corresponding silver stained amylase proteins; D. Protein Standards. 125-Starch Hydrolyzing enzyme; 71 and 61-Glucoamylases; 53A-α-amylase (active at pH 6.0); 53M-Taka amylase (active at pH 4.3).

amounts of carbohydrate moieties. The glucoamylases of molecular mass 71 and 61 kDa from *Aspergillus niger,* referred to as G1 and G2, have been extensively studied. It was initially believed that there were two glucoamylases because of differential splicing of mRNA encoded by a single gene that contained a number of intervening sequences (31,32). But it was shown that the 61 kDa G2 protein came from a proteolytic splitting off of a 10 kDa peptide from the 71 kDa G1 protein C-terminus (33). The activity differences of G1 and G2 on raw starch were determined to be the C-terminus peptide. Recently it was shown that a 125 kDa precursor starch hydrolysing enzyme is the major enzyme in *A. niger* and this enzyme is apparently processed by proteases to give rise to 71-, 61-kDa glucoamylases and 53 kDa Taka amylase enzyme (18).

Glucoamylases of fungal origin contain about 28% carbohydrate, which is responsible for the maintenance of the enzyme conformation (28). The carbohydrate is *O*-glycosidically linked to serine and threonine of the protein. Glucoamylases have optimum activity at acidic pH and act at temperatures around 60°C. Tryptophan residues have been proposed to be essential for enzyme activity (34). Biochemical analysis after site directed mutagenesis of glucoamylase gene indicated that tryptophan 120 is essential for transitional state stabilization (34). Glutamic acid 179 and 400 were identified as general catalytic acid and base catalyst respectively (35,36).

6.2.3 β-Amylase (1,4-α-D-glucan Maltohydrolase, EC 3.2.1.2)

Occurrence of β-amylase is common in plants and perhaps the best characterized enzymes of microbial origin are of *Clostridium thermosulfurogenes* (14) and *B. polymyxa* (37). β-Amylase hydrolytically cleaves the penultimate α-1,4 bond at the nonreducing ends of starch and causes the production of anomeric β-maltose. Because the enzyme cannot act on α-1,6 linkages of starch, it also produces β-limit dextrins.

Plant β-amylases are not thermostable. The enzyme of the thermophilic anaerobe *C. thermosulfurogenes* is thermostable, maintaining its activity at 80°C for two hours in the presence of 5% starch (28). Cysteine residues seem to be essential for both activity and conformation of β-amylase.

6.2.4 Isoamylase (Glycogen 6-glucanohydrolase, EC 3.2.1.68)

This enzyme predominantly degrades α-1,6-glucosidic linkages of amylopectin, glycogen, dextrins, and oligosaccharides. Its low affinity to short chains of pullulan makes this substrate less susceptible to the enzyme activity. Isoamylases have been characterized from *Bacillus spp*. (38,39). Molecular mass of the enzyme from microbial sources range from 65–121 kDa.

6.2.5 Pullulanase (α-dextrin 6-glucanohydrolase EC 3.2.1.4)

Very few organisms produce this enzyme that hydrolyse α-1,6 linkages of pullulan. Their molecular mass range from 80–145 kDa. Pullulanase of *C. thermohydrosulfuricum* and *B. acidopullulyticus* are thermostable with optimum activities in acidic pH (4.9–6.0).

6.2.6 α-Glucosidase, (α-D-glucoside Glucohydrolase EC 3.2.1.20)

α-Glucosidases are exoacting enzymes that catalyze the splitting of α-glucosyl residue from the nonreducing terminals of substrates to liberate α-glucose. Typically they are called maltases, because they hydrolyse maltose to glucose.

Fungal α-glucosidase has a wide substrate specificity degrading starch, dextrins, and maltose (5). The enzyme purified and crystallized from *A. niger* was a glycoprotein of molecular mass of about 125 kDa as estimated by SDS-Polyacrylamide gel electrophoresis. The activities of this enzyme resembled that of glucoamylase.

6.2.7 Glucose Isomerase or Xylose Isomerase (EC 5.3.1.5)

A range of microorganisms like *Streptomyces*, *Bacillus,* and *Arthrobacter* normally produces glucose isomerase or xylose isomerase intracellularly. Glucose isomerase generally act at 60°C, isomerising glucose to fructose. Because its affinity for glucose is low, a concentrated solution of substrate is used for isomerization reaction. The molecular mass of the enzyme range from 80 kDa in the case of *Actinoplanes missouriensis* to 157 kDa in the case of *Streptomyces spp*. This enzyme is strongly inhibited by Ca^{+2} and Mn^{+2}. However, it requires magnesium for activity and cobalt for maintenance of stability.

6.3 STARCH PROCESSING

Starch based derivatives are produced industrially for different applications. In its unmodified form, starch obtained from cassava is used in puddings, fruit fillings, and biscuits. Acid modified starches such as the dextrins find applications in baby food formulations.

They have also been used in the paper and textile industries. Hydrolyzed starches form substrates for sweeteners and brewing industries.

In the early years, starch was processed using concentrated acid at high temperature and pressure. By this procedure, the yield of products was inconsistent. Starch loss due to retrogradation (spontaneous precipitation of starch) was also a major constraint in the acid hydrolysis process. Advent of enzyme reactions and their suitability for application in starch processing has nearly replaced the acid hydrolysis processing of starch. Discovery of thermostable amylases and specificity of the enzymatic reactions upgraded industrial processing of starch, Product yields were in order obtain products consistent with defined end points, thereby saving on. This saved processing cost by reducing starch loss due to retrogradation.

The primary step in starch hydrolysis for processing (Figure 6.3) is its gelatinization at high temperatures followed by a thinning reaction. Discovery of bacterial α-amylases like Termamyl from *B. licheniformis,* which was stable even at temperatures as high as 104°C, completely replaced acid hydrolysis of starch, making enzyme technology a viable process. Use of this enzyme permitted, processing of even 30% starch slurries in steam jet cookers. Being a liquefying enzyme, the dextrins produced due to the activity of thermostable α-amylase constituted nearly 70% of the product. From these simple starch molecules, cyclodextrins can be created with the enzyme cyclodextrin glycosyl transferase. Cyclodextrins find applications as carriers of fragile vitamins and flavors. Only enzymatic synthesis of cyclodextrins is possible, because no chemical means exist for making this class of substance.

Two and three stage enzyme processing of starch hydrolysed with thermostable α-amylase defined clear protocols for industrial production of glucose, maltose, fructose, and limit dextrins.

6.3.1 Production of Sweeteners and Confectionaries

Enzymatically hydrolysed starch has been used in the manufacture of syrups (glucose, maltose, and fructose), baby food formulations (dextrins), and in confectionaries (maltose and β-limit dextrins).

Glucose is produced from starch in two stages. In the first stage, starch is gelatinized at temperatures ranging from 105–110°C and liquefied to reduce viscosity using thermostable α-amylase to evolve α-limit dextrins. The enzyme reaction is performed at pH approximately 7.0 to maintain the enzyme stability. In the second stage, glucoamylase or β-amylase enzymes are used to produce glucose and maltose syrups from the dextrins respectively. Glucoamylases and β-amylases act at acidic pH (3.5–5.0 and 4.8–6.5 respectively) and at temperatures around 60°C. Hence, the dextrin solution evolved from the thinning reaction is acidified to the required pH and cooled prior to enzyme action. *Aspergillus* is the major source for commercial glucoamylase. Because the enzyme is slow acting on α-1,6-linkages of dextrins, pullulanase is added along with glucoamylase to improve enzyme efficiency to evolve glucose syrups containing 95–96% glucose in shorter periods of time (40). Glucoamylases are also known to catalyze reversible condensation of glucose into isomaltose at higher concentrations of glucose. Hence, technologies describing glucose syrup production define a very low concentration of glucoamylase to prevent reversible reactions (40).

In the manufacture of maltose syrups, the dextrins obtained from gelatinized starch after α-amylase treatment is further hydrolyzed using thermostable β-amylase. Products of this enzyme activity evolve β-maltose and β-limit dextrins. α-1,6-Glucosidic bonds of dextrins are resistant to β-amylase activity. Simultaneous treatment of the dextrins with β-amylase and pullulanase improves the yield of maltose, but the process efficiency is reduced due to condensation reaction of maltose to form branched tetra-, penta-, and

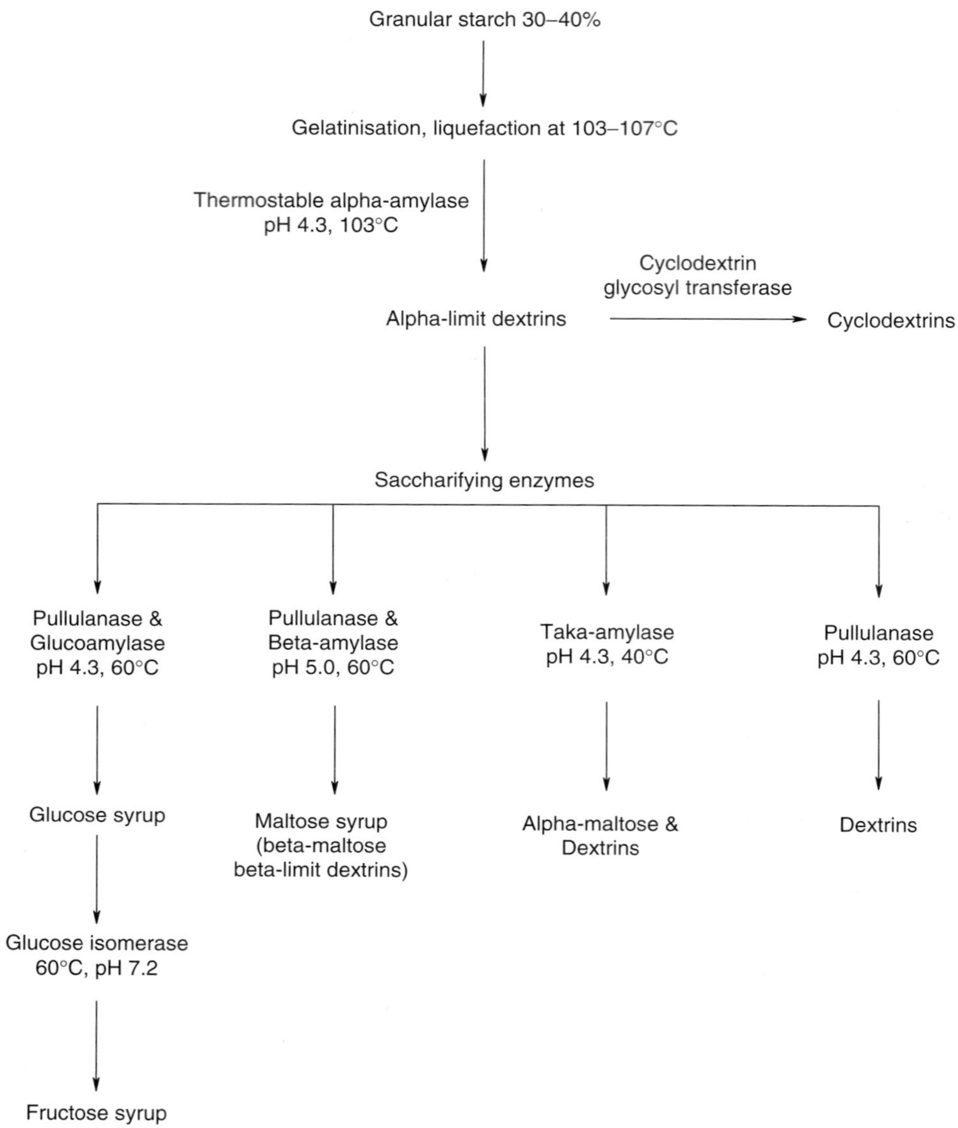

Figure 6.3 General processing steps of starch using enzymes.

hexa- saccharides. Maltose syrups are nonhygroscopic and are used in hard candies to avoid stickiness. These syrups normally contain 45–60% maltose, 3–5% glucose, and 37–55% oligosaccharides. Because maltose syrups are characterized by mild sweetness, defined enzyme technologies have attempted to make extra high maltose syrups by decreasing substrate concentrations and by using new pullulanase that lack reversible activities. In foods, rather than sweetness, the primary importance for maltose syrups is for their functional properties, such as low hygroscopicity, low viscosity in solution, resistance to crystallization, reduced browning capacity, and good heat stability. These properties of maltose syrups have applications as moisture conditioners, crystallization inhibitors, stabilizers, carriers, and bulking agents (41).

Production of high fructose syrup involves a three stage processing of starch. The first two stages of starch processing to generate glucose have been described above. In the third stage of processing, isomerization of glucose to fructose is achieved using glucose isomerase. The enzyme obtained from *B. coagulans* is most widely used (42). Calcium ions, the potent inhibitor of glucose isomerase are acquired as impurities from raw materials. Though calcium ions are important for the stability of α-amylase used to liquefy starch, they have to be removed completely prior to the isomerization reaction. Hence, the glucose syrup obtained after glucoamylase treatment is passed through ion resin columns to remove calcium. To make the process economical, glucose isomerase immobilized by cross linking with glutaraldehyde is used in column reactors to achieve isomerization. By this procedure, the need for the enzyme is considerably reduced and the process is made amenable to microprocessor control. Different immobilization procedures are described to retain high enzyme activity, as described in Table 6.2. Glucose isomerases are optimally active at pH 7.5–7.8. Hence, prior to conversion of glucose to its isomer fructose, the pH of the glucose syrup is adjusted generally with soda ash or sodium hydroxide. Because isomerization reaction does not result in high fructose syrups, industries employ chromatography columns to partially separate glucose. This gives the industries the flexibility to sell syrups with different levels of fructose. This is important because fructose is sweeter than glucose and syrups that carry a broad range of sweetness. The variety of sweeteners assists in the manufacture of many food products (43).

6.3.2 *In situ* Starch Modifications for Food and Beverage Preparations

Amylolytic enzymes are widely used in modifying texture in breads. They are an important application in biochemically modifying barley to substitute malt in the brewing process for the manufacture of beer. Staling, that makes bread hard and brittle, is thought to be due to reversion of starch granules from soluble to insoluble form (retrogradation). In bread making, thermostable α-amylase reduces the retrogradation of starch to improve the shelf life of breads. α-Amylases, due to their stability at high temperatures, are able to work after the baking process, breaking the starch to smaller length dextrins, preventing retrogradation and increasing the shelf life of bread. Also, production of dextrins from starch reduces high dough viscosity in bread and accelerates the fermentation process by yeast, thereby improving the crust formation. In brewing of beer, use of amylase, apart from thinning the mash and accelerating fermentation, also modifies barley biochemically to substitute for malt. In beer, malt has an impact on flavor and is usually used as source of β-amylase enzyme. β-Amylases of microbial origin can modify β-glucans in the barley

Table 6.2
Immobilized glucose isomerase used for fructose syrups

Immobilization Method	Enzyme Activity [b]
Whole cells	
Cross-linked *Streptomyces spp.*	121
S. violaceoniger in gelatine fiber	260
Bacillus coagulans cross-linked	200
Purified/crude enzyme	
Actinoplanes missouriensis on alumina	1350
Enzyme bound to DEAE-cellulose	1487

[b]Activity IU/g corresponds to micro moles of fructose formed min^{-1}

to lower the wort viscosity for improving flavor. Activity of this enzyme also releases maltose, which serves as a nutrient for yeast in the fermentation process.

6.4 CURRENT STATUS

Though microorganisms have been identified, which produce amylases that find direct application in starch processing industries, the demand and requirement to improve processing parameters necessitated further developments in this area. Overproduction of amylases by recombinant DNA technologies and modification of the enzyme property by protein engineering have reduced the cost of producing enzymes as well as producing enzymes designed for a particular application.

Most cloned α-amylase genes have been expressed using their own promoters in bacteria and fungi. High level production using cloned genes has been obtained for α-amylase in *B. amyloliquefaciens* and *A. oryzae* and β-amylase in *A. schirousami*. Glucoamylase gene has been found expressed in *Saccharomyces cerevisiae* under the control of enolase (*ENO1*) promoter (44). This study described construction of the yeast strain that hydrolyzed and simultaneously fermented starch, because of the presence of *A. awamori* glucoamylase gene.

Insight into carbohydrate binding properties of amylases and their mechanisms of action acquired by a combination of x-ray crystallography, chemical modification, inhibition studies, and site directed mutagenesis, explained functionality of amylases for specific action patterns. Replacement of histidine 238 with aspartic acid in α-amylase showed thermostability reduction but the mutant protein derived hydrolyzed starch to give rise to a different oligodextrin profile (45). In another approach, replacement of histidine 133 with tyrosine resulted in α-amylase that was more stable (45). The tryptophan 178 to arginine, and serine 119 and to tyrosine mutations in *A. niger* glucoamylase increased the cleavage of α-1,6 glucosidic bonds of starch without significantly decreasing activity toward α-1,4 bonds (45).

The extreme thermostability of *B. licheniformis* α-amylase was attributed to electrostatic interactions involving few specific lysines (46). Stabilization against irreversible thermal inactivation was achieved in a chimeric gene constructed using *B. amyloliquefaciens* and *B. licheniformis* α-amylase genes and by deleting arginine 176 and glycine 177, substituting lysine 269 with alanine and asparagine 266 with aspartic acid (47). Truncation of amylolytic enzymes has also led to the understanding of starch granule binding function. Hence, protein engineering of amylases through chemical modifications and site directed mutagenesis described the art of changing enzyme properties without losing catalytic efficiency (48). Selectivity of glucoamylase for increasing glucose yields by site directed mutagenesis of the glucose catalytic domain (49–52), modified the dimension of the active site just enough to prevent the prevalent byproduct, isomaltose, from binding the active site without affecting the binding of maltose. This was achieved by stiffening the α-helices of the protein by mutating glycine to alanine or producing disulphide bonds by substituting cysteine on adjacent loops that restricted the movement of the active site so that it did not accommodate isomaltose.

Thermostability of glucoamylase was also found to increase when amino acid substitutions were made by site directed mutagenesis. Significantly, replacement of glycine 137 with alanine was reported to stiffen an α-helix that improved the enzyme thermostability at temperatures over 70°C. This was because the mutation retarded general unfolding of the enzyme (53). Several other mutants obtained by site directed mutagenesis have also shown substantial improvement in thermostability and activity of glucoamylase (54).

While several problems of industrial processing of starch using enzymes were solved through isolation and engineering of amylases, still this technology is affected by low thermostability of enzymes like β-amylase, glucoamylase, pullulanase, and glucose isomerase. Hence there is a continuous search for highly thermostable enzymes from microbes occurring in extreme environments (55).

pH variations described for the production of different types of sweeteners is also an important consideration in starch processing. Thermostable α-amylase of *B. licheniformis*

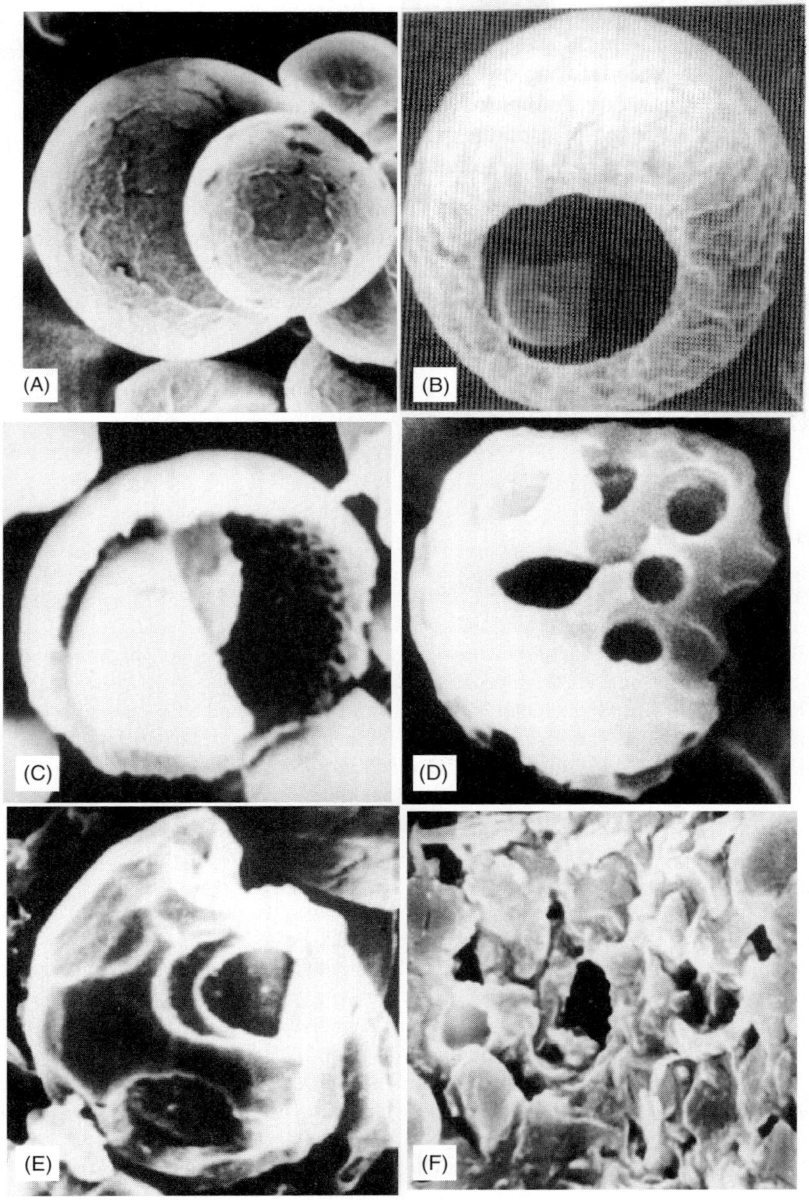

Figure 6.4 Granular starch hydrolysis by amylases of A. niger
(A) Starch granules, (B) Pit formation and (C) pit enlargement due to Glucoamylase activity. (D) Multiple pitting, (E) Facial Fracture, (F) Flaking are reactions due to the activity of Starch Hydrolyzing Enzyme (18).

that has an optimum pH range of 5.5 to 6.0 has been reported (9). Liquefaction of starch at low pH using this enzyme was claimed to prevent the formation of maltulose (4-α-D-Glucopyranosyl-D-fructose) precursors. In the manufacture of glucose syrups, maltulose precursors are hydrolyzed to maltulose by glucoamylase, reducing the yield of glucose.

Starch processing as described involves high energy inputs, because gelatinization of granular starch at high temperature is a primary requisite. Degradation of crystalline granular starch by amylolytic enzymes (Figure 6.4) has been widely reported. Structural studies on glucoamylases have characterized a separate globular domain that has the capacity to bind starch granules. These studies and the three dimensional structures of the enzyme proteins have identified the structure — function relationships of the enzyme and substrate — that have revealed clues for rational engineering of enzymic properties for granular starch saccharification. Granular starch hydrolysing glucoamylase of *Rhizopus* has been cloned and expressed in yeast (56). Thus, enzymatic degradation of granular starch to make syrups and sweeteners is not a distant possibility.

It has been reported that in order to improve starch processing, a transgenic potato crop has been developed that produces fructose and glucose when mashed and treated at 65°C for 45 minutes (57). This was achieved by expressing chimeric gene construct coding for *Bacillus stearothermophilus* α-amylase and *Thermus thermophilus* glucose isomerase in potatoes. The possibility of starch modification in plants apparently will revolutionize the existing technologies because it can be a solution to many problems encountered by the starch processing industries that use enzymes.

REFERENCES

1. Yamasaki, Y., Y. Suzuki, J. Ozawa. Three forms of α-glucosidase and a glucoamylase from *Aspergillus awamori*. *Agric. Biol. Chem.* 41:2149–2161, 1977.
2. Bhella, R.S., I. Altosaar. Production of multiple forms of glucoamylase in *Aspergillus awamori*. *Biochem. Cell. Biol.* 65:762–765, 1987.
3. Ramasesh, N., K.R. Sreekantiah, V.S. Murthy. Purification and characterization of a thermophilic α-amylase of *Aspergillus niger* van Tieghem. *Starch* 34:274–279, 1982.
4. Lineback, D.R., I.J. Russell, C. Rasmussen. Two forms of the glucoamylase of *Aspergillus niger*. *Arch. Biochem. Biophys.* 134:539–553, 1969.
5. Kita, A., H. Matsui, A. Somoto, A. Kimura, M. Takata, S. Chiba. Substrate specificity and subsite affinities of crystalline α-glucosidase from *Aspergillus niger*. *Agric. Biol. Chem.* 55:2327–2335, 1991.
6. Matsuura, Y., M. Kusonoki, W. Harada, M. Kakudo. Structure and possible catatlytic residues of Taka- amylaseA. *J. Biochem.* 95:697–702, 1984.
7. Saha, B.C., T. Mitsue, S. Ueda. Glucoamylase produced by submerged culture of *Aspergillus oryzae*. *Starch* 31:307–314, 1979.
8. Milner, J.A., J.D. Martin, A. Smith. Two-stage inocula for the production of alpha-amylase by *Bacillus amyloliquefaciens*. *Enzyme Microbiol. Technol.* 21:382–386, 1997.
9. Antrim, R.L., B.A. Solheim, L. Solheim, R. Meadows, A.L. Auterinen, J. Cunefare, S. Karppelin. A new *Bacillus licheniformis* alpha-amylase capable of low pH liquefaction. *Starch* 43:355–360, 1991.
10. Ogasahara, K., A. Imanishi, T. Isemura. Studies on thermophilic α-amylase from *Bacilus stearothermophilus*, I: some general and physico-chemical properties of thermophilic α-amy lase. *J. Biochem.* 67:65–75, 1970.
11. Ohdan, K., T. Kuriki, H. Kaneko, J. Shimada, T. Takada, Z. Fujimoto, H. Mizuno, S. Okada. Characteristics of two forms of α-amylases and structural implication. *Appl. Environ. Microbiol.* 65:4652–4658, 1999.

12. Brumm, P.J., R.E. Hebeda, W.M. Teague. Purification and characterization of the commercialized, cloned *Bacillus megaterium* α-amylase, part I: purification and hydrolytic properties. *Starch* 43:315–319, 1991.
13. Lappalainen, A., M.L. Nikupaavola, T. Suortti, K. Poutanen. Purification and characterization of *Bacillus acidopullulyticus* pullulanase for enzymatic starch modification. *Starch* 43:477–482, 1991.
14. Hyun, H.H., J.G. Zeikus. General biochemical characterization of thermostable extracellular β-amylase from *Clostridium thermosulfurogens*. *Appl. Environ. Microbiol.* 49:1162–1167, 1985.
15. Monma, M., Y. Yamamoto, N. Kagei, K. Kainuma. Raw starch digestion by α-amylase and glucoamylase from *Chalara paradoxa*. *Starch* 41:382–385, 1989.
16. Sanoja, R.R., J. Morlon-Guyot, J. Jore, J. Pintado, N. Juge, J.P. Guyot. Comparative characterization of complete and truncated forms of *Lactobacillus amylovorus* α-amylase and role of the c-terminal direct repeats in raw-starch binding. *Appl. Environ. Microbiol.* 66:3350–3356, 2000.
17. Takahashi, K., A. Totsuka, T. Nakakuki, N. Nakamura. Production and application of a maltogenic amylase by a strain of *Thermomonospora viridis* TF-35. *Starch* 44:96–101, 1992.
18. Dubey, A.K., C. Suresh, R. Kavitha, N.G. Karanth, S. Umesh-Kumar. Evidence that the glucoamylases and α-amylase secreted by *Aspergillus niger* are proteolytically processed products of a precursor enzyme. *FEBS Lett.* 471:251–255, 2000.
19. Sakano, Y., M. Sano, T. Kobayashi. Hydrolysis of α-1,6 glucosidic linkages by α-amylases. *Agric. Biol. Chem.* 49:3041–3043, 1985.
20. Rosendal, P., B.H. Nielsen, N.K. Lange. Stability of bacterial alpha-amylase in the starch liquefaction process. *Starch* 31:368–372, 1979.
21. Kelly, C.T., W.M. Fogarty. Microbial alkaline enzymes. *Process. Biochem.* 11(6):3–9, 1976.
22. Klibanov, A.M. Stabilization of enzymes against thermal inactivation. *Adv. Appl. Microbiol.* 29:1–28, 1983.
23. Chiang, J.P., J.E. Alter, M. Sternberg. Purification and characterization of a thermostable α-amylase from *B. licheniformis*. *Starch* 31:86–92, 1979.
24. Umesh-Kumar, S., F. Rehana, K. Nand. Production of an extracellular thermostable calcium-inhibited α-amylase by *Bacillus licheniformis* MY 10. *Enzyme Microbiol. Technol.* 12:714–716, 1990.
25. Rogers, J.C. Conserved amino acid sequence domains in α-amylases from plants, mammals and bacteria. *Biochem. Biophys. Res. Commun.* 128:470–476, 1985.
26. Hsiu, J., E.H. Fischer, E.A. Stein. α-Amylases as calcium-metalloenzymes, II: calcium and the catalytic activity. *Biochemistry* 3:61–66, 1964.
27. Toda, H., I. Kato, K. Narita. Correlation of the masked sulfhydryl group with the essential calcium in Taka-amylase A. *J. Biochem.* 63:295–301, 1968.
28. Vihinen, M., P. Mantsala. Microbial amylolytic enzymes. *Critic. Rev. Biochem. Mol. Biol.* 24:329–418, 1989.
29. Fogarty, W.M., C.T. Kelly. Amylases amyloglucosidases and related glucanases. In: *Microbial Enzymes and Bioconversions: Economic Microbiology,* Rose, A.H., ed., New York: Academic Press, 1980, pp 115–170.
30. Dubey, A.K. Genetic improvement of *A. niger* for the production of a novel starch hydrolysing enzyme. Ph.D. Thesis, University of Mysore, India, 1999.
31. Boel, E., I. Hjort, B. Svensson, F. Norris, K.E. Norris, N.P. Fill. Glucoamylases G1 and G2 from *Aspergillus niger* are synthesized from two different but closely related mRNAs. *EMBO J.* 3:1097–1102, 1984.
32. Boel, E., M.T. Hansen, I. Hjort, J. Hoegh, N.P. Fill. Two different types of intervening sequences in the glucoamylase gene from *Aspergillus niger*. *EMBO J.* 3:1581–1585, 1984.
33. Svensson, B., K. Larsen, A. Gunnarsson. Characterization of a glucoamylase G2 from *Aspergilus niger*. *Eur. J. Biochem.* 154:497–502, 1986.
34. Olsen, K., U. Christensen, M.R. Sierks, B. Svensson. Reaction mechanisms of Trp 120: Phe and wild type glucoamylases from *A. niger* interactions with maltooligodextrins and acarbose. *Biochemistry* 32:9686–9693, 1993.

35. Ly, H.D., S.G. Withers. Mutagenesis of glycosidases. *Annu. Rev. Biochem.* 68:487–522, 1999.
36. Chiba, S. Molecular mechanism in α-glucosidase and glucoamylase. *Biosci. Biotechnol. Biochem.* 61:1233–1239, 1997.
37. Robyt, J., D. French. Purification and action pattern of an amylase from *B. polymyxa*. *Arch. Biochem. Biophys.* 104:338–345, 1964.
38. Maitin, V., R. Kavitha, S. Umesh-Kumar. Properties of extracellular amylase purified from a *Bacillus* species. *World J. Microbiol. Biotechnol.* 17:823–826, 2001.
39. Urlaub, H., G. Wober. Identification of isoamylase a glycogen-debranching enzyme from *Bacillus amyloliquefaciens*. *FEBS Lett.* 57:1–4, 1975.
40. Norman, B.E. A novel debranching enzyme for application in the glucose syrup industry. *Starch* 34:340–346, 1982.
41. Saha, B.C., J.G. Zeikus. Biotechnology of maltose syrup production. *Process. Biochem.* 22:78–82, 1987.
42. Bucke, C. Glucose-transforming enzymes. In: *Microbial Enzymes and Biotechnology,* Fogarty, W.M., ed., London: Applied Science, 1983, pp 93–129.
43. Boyce, C.O.L. *Novo's Hand Book of Practical Biotechnology.* Copenhagen: Novo industries A/S, 1986, pp 35–41.
44. Innis, M.A., M.J. Holland, P.C. McCabe, G.E. Cole, V.P. Wittman, R. Tal, K.W.K. Watt, D.H. Gelfand, J.P. Holland, J.H. Meade. Expression glycosylation and secretion of an *Aspergillus* glucoamylase by *Saccharomyces cerevisiae*. *Science* 228:21–26, 1985.
45. Svensson, B., M. Soggard. Protein engineering of amylases. *Biochem. Soc. Trans.* 20:34–42, 1992.
46. Tomazic, S.J., A. Klibanov. Why is one *Bacillus* α-amylase more resistant against irreversible thermoinactivation than another? *J. Biol.Chem.* 263:3092–3096, 1988.
47. Suzuki, Y., N. Ito, T. Yuuki, H. Yamagata, S. Udaka. Amino acid residues stabilizing a *Bacillus* α-amylase against irreversible thermoinactivation. *J. Biol. Chem.* 264:18933–18938, 1989.
48. Gottschalk, T.E., H.P. Fierobe, E. Mirgorodskaya, A.J. Clarke, D. Tull, B.W. Sigurskjold, T. Christensen, N. Payre, T.P. Frandsen, N. Juge, K.A. McGuire, S. Cottaz, P. Roepstorff, H. Driguez, G. Williamson, B. Svensson. Structure function and protein engineering of starch-degrading enzymes. *Biochem. Soc. Trans.* 26:198–204, 1998.
49. Fang, T.Y., P.M. Coutinho, P.J. Reilly, C. Ford. Mutations to alter *A. awamori* glucoamylase selectivity, I: Tyr 48 Phe 49→Trp, Tyr 116→Trp, Tyr 175→Phe, Arg 241→Lys, Ser 411→ Ala and Ser 411→Gly. *Protein Eng.* 11:119–126, 1998.
50. Fang, T.Y., R.B. Honzatko, P.J. Reilly, C. Ford. Mutations to alter *A. awamori* glucoamylase selectivity, II: mutations at positions 119 and 121. *Protein Eng.* 11:127–133, 1998.
51. Liu, H.L., P.M. Coutinho, C. Ford, P.J. Reilly. Mutations to alter *A. awamori* glucoamylase selectivity, III: Asn 20→Cys/Ala 27→Cys/Ala 27→Pro, Ser 30→Pro, Lys 108→Met, Gly 137→Ala, 311–314 loop, Tyr 312→Trp and Ser 436→*Protein Eng.* 11:389–398, 1998.
52. Liu, H.L., C. Ford, P.J. Reilly. Mutations to alter *A. awamori* glucoamylase selectivity, IV: Asn 20→Cys/Ala 27→Cys, Ser 30→Pro, Gly 137→Ala, 311–314 loop, Ser 411→Ala and Ser 436→*Protein Eng.* 12:163–172, 1999.
53. Chen, H.M., Y. Li, T. Panda, F.U. Buehler, C. Ford, P.J. Reilly. Effect of replacing helical Glycine residues with alanines on reversible and irreversible stability and production of *A. awamori* glucoamylase. *Protein Eng.* 9:499–505, 1996.
54. Reilly, P.J. Protein engineering of glucoamylase to improve industrial performance: a review. *Starch* 51:269–274, 1999.
55. Leveque, E., S. Janecek, B. Haye, A. Belarbi. Thermophilic archeal amylolytic enzymes. *Enzyme Microbiol. Technol.* 26:3–14, 2000.
56. Ashikari, T., N. Nakamura, Y. Tanaka, N. Kiuchi, Y. Shibano,T. Tanaka, T. Amachi, H. Yoshizumi. *Rhizopus* raw-starch-degrading glucoamylase: its cloning and expression in yeast. *Agric. Biol. Chem.* 50:957–964, 1986.
57. Beaujean, A., C. Ducrocq-Assaf, R.S. Sangwan, G. Lilius, L. Bulow, B.S. Sangwan-Norreel. Engineering direct fructose production in processed potato tubers by expressing a bifunctional α-amylase/glucose isomerase gene complex. *Biotechnol. Bioeng.* 70:9–16, 2000.

2.07

Genetic Modification of Plant Oils for Food Uses

Anthony J. Kinney

CONTENTS

7.1 INTRODUCTION

Almost 100 million metric tons of vegetable oil are produced annually, and most of it is used for food purposes. The majority of edible vegetable oil in the world (70 million metric tons) is derived from four crops: soybean, canola, palm and sunflower. About half finds its way (after refining, bleaching, and deodorizing) directly to the retail market in the form of bottled oil for cooking and salads (1). These oils tend to be rich in linoleic acid, an omega-6 polyunsaturated fatty acid, and for this reason the ratio of omega-6 to omega-3 fatty acids in the Western diet is greater than 10 moles of omega-6 to one of omega-3. Most medical experts now believe that this ratio should be about three to one, and suggest a reduction in omega-6 fatty acid consumption or an increased intake of omega-3 fatty acids (2,3).

The half of edible vegetable oil production that is not bottled directly is used by the food industry for the manufacture of shortening, confections, and margarine, and as such is usually hydrogenated to a greater or lesser degree depending upon the application (1,4). Oils are hydrogenated to remove oxidatively unstable polyunsaturated fatty acids, and thus increase the shelf life of oil containing food products. Oils are also hydrogenated to achieve solid fat functionality, by increasing the saturated fatty acid content and by the introduction of *trans* fats, which have a much higher melting point than their *cis* isomers. The relative abundance of saturated and monounsaturated (*cis* and *trans*) fatty acids in the oil will determine the solid fat content (SFC) curve of the oil and thus its food application. A developing understanding of the negative health effects associated with the consumption of *trans* fatty acids is leading food manufacturers to consider alternatives to hydrogenated oils, such as fractionation and transesterification, for many food applications (5).

In recent years the importance in the diet of relatively modest concentrations of bioactive fatty acids, such as long chain omega-3 and conjugated fatty acids, especially conjugated linoleic acid (CLA), has been recognized. Currently the main dietary sources of these fatty acids are dairy products (for CLA) and fish oils (for long chain omega-3 fatty acids). Consumption of these fatty acids has been strongly correlated with a reduced risk for coronary disease and a number of other positive health benefits (2,3). There is a potential for including these bioactive fatty acids in the diet as food ingredients, although their widespread use is currently limited by the high cost of purifying and refining them to a point where they will not negatively impact the flavor of the food to which they are added. Food ingredient quality refined fish oil, for example, can cost as much as $50 per kilogram.

Biotechnology has the potential to impact both the low added value, high volume commodity oil market and the low volume, high added value specialty oil market. Changing the relative abundance of saturated, monounsaturated, and polyunsaturated fatty acids in seed oils by manipulating existing biosynthetic pathways could result in oils that have desired health and functional properties without hydrogenation, fractionation, or other modifications (6). Adding new pathways for the synthesis of conjugated or long chain polyunsaturated fatty acids represents a new direction for food oil biotechnology, and could result in lower cost, higher quality replacements for fish and microbial oils as health promoting food ingredients.

Many of the technical challenges of both these areas have been met, and the feasibility of producing a broad range of novel plant oils in the major oilseed crops has been demonstrated in both model plants and oilseed crops. Some of these oils (such as lauric acid rich canola and high oleic acid soybean oils) have even been commercialized in limited quantities. It is not expected, however, that any of these new oils will make a commercial impact until the end of the current decade. It is hard to imagine any of the major edible plant oils (with the exception of gourmet, organic, and artisanal oils) that will not have been improved, by the middle of this century, by the use of biotechnology.

7.2 FRYING OILS

The most important property of frying oils is resistance to oxidation and polymerization (4). Oxidative stability is usually measured by either the active oxygen method (AOM) or the oxidative stability index (OSI). In both cases the higher the number, the more stable the oil to oxidation and polymerization, and the two measurements normally correlate with each other. A typical refined, bleached, deodorized (RBD) vegetable oil will have an AOM in the 15–25 hour range (4). This is fine for home frying that is done in small batches and often discarded after a single use. For the kind of heavy duty, continuous frying done in the food industry, however, far more stable oils are needed, with AOMs of around 100–200 hours (4,6,7). This stability is usually achieved by either partial hydrogenation or a combination of partial hydrogenation, transesterification, and fractionation. Partial hydrogenation will remove most of the polyunsaturated fatty acids and replace them with *cis* and *trans* monounsaturated fatty acids. There is now a substantial body of evidence linking the consumption of *trans* fatty acids to coronary heart disease, and food manufacturers will soon be required to label the *trans* content of packaged foods (8). Thus there is an incentive to remove polyunsaturated fatty acids from oils by means other than hydrogenation.

The polyunsaturated fatty acids linoleic and linolenic acid are synthesized in the developing oilseed by the action of two types of desaturase enzyme. A delta-12 desaturase inserts a second double bond into oleic acid while it is esterified to the membrane lipid phosphatidylcholine (9). This enzyme is encoded by the Fad 2 gene in plants. Linolenic acid is synthesized by the insertion of a third double bond into linoleic acid by an omega-3 desaturase, encoded by the Fad 3 gene. Again, this reaction occurs while the fatty acid is esterified to a membrane lipid. Because this synthesis of linolenic acid from oleic acid is sequential, blocking the first step by silencing the Fad 2 gene should theoretically result in blocking the formation of polyunsaturated fatty acids and result in an increase in oleic acid. This generally proves to be the case, although most oilseed plants have two or three Fad 2 genes. To achieve the maximum oleic acid content all three may have to be silenced in a seed specific manner. A small amount of polyunsaturated found in seed oils is actually synthesized in a separate pathway located in the plastid. Nevertheless, silencing of Fad 2 genes has resulted in oils with an oleic acid content up to 90%. High oleic soybean oil (HOS), which has an oleic acid content of about 85% and less than 5% total polyunsaturated fatty acids, has an AOM of 150 hours. Polymer formation in HOS oil during frying is very low, equivalent to heavy duty, hydrogenated shortening (7).

Interestingly, high oleic soybean oil has a higher AOM than other modified vegetable oils with a similar oleic and polyunsaturate content. This is most likely because soybean oil is naturally richer in tocopherols than other plant oils, and the form is predominantly *gamma*-tocopherol. When, for example, high oleic sunflower oil is stripped of tocopherols, and tocopherols are added back to the final content and composition of soybean oil, the AOM of the sunflower increases close to that of high oleic soy (7). Most of the genes that control tocopherol formation and the degree of tocol methylation have been cloned (10–12). It is now theoretically possible, therefore, to combine genetic modifications in fatty acid desaturation with modifications of tocopherol metabolism to produce very high stability frying oils from most oilseed species.

Because of concerns surrounding the consumption of genetically enhanced foods and the advent of novel, cheaper transesterification techniques, the commercial production of transgenic high oleic acid oils, even in the US, has been limited. The increasing demand for *trans* free oils created by the new labeling requirements and increased public awareness of *trans* related health issues is likely to reverse this situation. Consequently, in the near future genetically enhanced high oleic oils should become more widely available.

7.3 OILS FOR BAKING AND CONFECTIONARY APPLICATIONS

The bulk of the hydrogenated vegetable oil produced in the US is used as shortening for baking applications (4). These oils were originally developed to replace animal fats such as lard and butter. They typically need a high oxidative stability (AOM between 100 and 200 hours), a high melting point (35–60°C) and usually a SFC curve that is basically flat or very gently sloping between 10 and 40°C. The actual SFC curve requirement depends upon the application, although SFC curve requirements for piecrusts, puff pastry, and baker's margarine would be very similar (4). Consumer margarine applications, which represent about 5% of hydrogenated vegetable oil produced in the US, require very similar SFC curves and properties. Hydrogenated vegetable oils are also used for confectionary coatings, cookie fillers, and wafer fillers. These particular oils, however, need to be solid at room temperature but liquid at body temperature, and thus have a very different SFC requirement from baking and margarine oils (4).

Producing modified plant oils with these two very different types of SFC curves can be done with similar technical approaches. Margarine type baking oils may be replicated by oils with a higher stearic acid and higher oleic acid content than regular plant oils. The coating, confectionary, and filler-type fats can be mimicked by fractionating these same oils to enrich for particular triacylglycerol species. Oils with a confectionary-type SFC curve can also be made by engineering seeds to produce structured triacylglycerols that are also rich in certain medium chain fatty acids.

High oleic, high stearic acid oils can be made by combining the Fad 2 silencing effect described with a gene for increased stearic acid. There are a number of possible approaches to increase stearic acid in oilseeds. All of these approaches involve providing more favorable flux conditions for 18:0-ACP leaving the cytoplasm (where it becomes a substrate for triacylglycerol [TAG] biosynthesis) than for it being desaturated to 18:1-ACP.

Stearic acid is the predominant fatty acid (45–55%) in the seed oil of *Garcinia mangostana* (mangosteen). This is result of the activity of an unusual FatA-type thioesterase, *GarmFatA1*, which has low activity toward 16:0-ACP but high activity toward 18:0 and 18:1-ACPs (13). When a cDNA encoding this FatA was expressed in seeds of canola, it was able to siphon away large amounts of plastidal 18:0-ACP from the acyl-ACP desaturase and transfer it to the cytoplasm as stearoyl-CoA. The resultant stearate accumulation in the seed oil was increased from 1% to 20% of the total fatty acids (13).

It is also possible to slow flux through the acyl-ACP desaturation step to allow the endogenous FatB-type thioesterase to release more stearic acid from 18:0-ACP into cytoplasm. This can be achieved by silencing the gene for one of the number of SAD genes in oilseeds that encode 18:0-ACP desaturases. Knocking out a single gene of the acyl-ACP desaturase gene family usually results in a stearic acid content of around 25% of total fatty acids (6,14).

Because, in both these cases, flux into 18:1 has been reduced, the final oleic acid content of high stearate, high oleic combinations is usually in the 55–60% range (6,9). Vegetable oils which contain 20–25% stearic acid and 55–60% oleic acid have a unique SFC functionality that makes them suitable for baking and margarine applications. This functionality may be further extended by fractionation of these oils.

It has been possible to make oils suitable for chocolate substitutes, confectionary coatings, and cookie fillings by modifying fatty acid synthesis to produce shorter chain fatty acids, from 8:0 to 12:0 carbons in length. A 12:0-ACP FatB-type thioesterase was isolated from California bay laurel. Expressing this FatB-type thioesterase in canola resulted in seed oil rich in lauric acid, up to 50% of the total fatty acids (15). The initial intent of producing this oil (which has the trade name Laurical ®) was for nonfood uses, such as soaps. Once the oil had been produced and its functionality measured, it was also found to be suitable for various food applications (16). Commercial production of this oil has been limited, however, by the availability of relatively inexpensive coconut oil. Nevertheless, the invention of Laurical was a major milestone along the road to more exotic food oils (16).

Subsequently, cDNAs encoding 8:0-ACP, 10:ACP, and 14:0-ACP FatB-type thioesterases have been isolated from *Cuphea* species with oils rich in the corresponding acyl chains. Expression of these cDNAs in canola resulted in oil rich in these medium chain fatty acids (18).

The functionality of all these medium chain oils is limited by the positions of the fatty acids on the triaclyglycerol molecule. In the bay thioesterase oil, for example, the lauric acid is found exclusively at the sn-1 and sn-3 positions. Coexpression in canola seeds of a coconut 12:0-CoA Lysophosphatidic acid acyltransferase, along with the bay thioesterase, facilitated efficient laurate deposition at the sn-2 position resulting in the accumulation of lauric acid at all three positions of the seed triacylglycerol (19).

The cloning and expression of acyl transferases has paved the way for the production of triacylglycerols (TAGs) structured with specific fatty acids at different positions on the glycerol backbone. The utility of these structured triacylglycerols (STAGs) extends beyond the functional properties related to their SFC curve and their nutritional importance is more profound than their bulk fatty acid profile might suggest (20,21). For example, STAGS with medium chain fatty acids at positions sn-1 and sn-3 of the glycerol backbone and an essential fatty acid at the sn-2 position have been useful in the diets of patients suffering from malabsorbtion because they provide both essential fatty acids and a source of energy equivalent to soluble sugars. The medium chain fatty acids are absorbed as free fatty acids, independently of acylcarnitine, and hence faster and more efficiently than longer chain fatty acids. They thus provide a rapid source of energy without having the side effects of high sucrose supplements (21).

In another example, most of the palmitic acid in human milk is at the sn-2 position of triacylglycerol. Enriching infant formula with TAGs having palmitic acid at this position results in a food that more closely resembles natural sources and results in a number of health benefits for the infant, including reduced calcium loss (21,22).

The utility of foods containing STAGs is currently limited by the very high cost of producing the structured molecule itself. By combining the fatty acid pathway manipulation techniques described with the expression of specific acyltransferases it should be possible to directly produce relatively inexpensive plant oils enriched in specific STAGs for various food and medical purposes.

7.4 FATTY ACIDS FOR FOOD INGREDIENTS

There is an increasing body of evidence supporting the conclusion that the fatty acids in the food we eat have a major influence on our physical and mental health (23). This is particularly the case with the essential omega-6 and omega-3 polyunsaturated fatty acids (PUFAs) linoleic and linolenic acids, which are the precursors of the long chain PUFAs arachidonic acid (ARA) and eicosopentaenoic acid (EPA) in humans. In turn these long chain PUFAS are the direct precursors of bioactive eicosanoids (prostaglandins, prostacyclins, thromboxanes, and leukotrienes) and the bioactive omega-3 fatty acid docosohexaenoic acid (DHA). The eicosanoids mediate many functions in the human body including the inflammatory response, the induction of blood clotting, the regulation of mental functions such as the sleep/wake cycle, and the regulation of blood pressure (24). The balance of omega-6 and omega-3 derived eicosanoids is important in the body because those derived from omega-6 ARA are proinflammatory and those derived from omega-3 EPA are antiinflammatory (24). DHA is an important component of mammalian brain membranes and has been shown to play a role in the cognitive development of infants and the mental health of adults (25–27).

Because of dietary changes over the last century many populations, particularly those in the U.S. and Europe, now consume a high proportion of omega-6 fatty acids, mostly in the form of linoleic acid. The main source of omega-3 fatty acids in human diets is fish, the consumption of which is not consistent across any population (23). Thus many people are thought to have suboptimal concentrations of long chain omega-3 fatty acids in their bodies (28). It is no surprise, therefore, that consumption of omega-3 fatty acids, particularly the long chain PUFAs EPA and DHA, has been associated with many positive health benefits such as the prevention of cardiovascular disease in adults, improved cognitive development in infants (25–27), and a greatly reduced risk of sudden cardiac arrest, one of the major causes of death in the U.S. (29).

It has been calculated that adding between 800mg and 1000mg of EPA per day to the human diet would redress the omega-6:omega-3 balance in the body and lead to substantially improved health (2). This is equivalent to eating a 4–12 oz serving of fresh tuna every day, depending on the EPA/DHA content of the fish itself (2). It is very difficult to persuade people to change drastically their dietary habits, particularly if the change involves eating more of a strongly flavored food such as fish. In addition, the world supply of fish (either wild or farmed, because farmed fish need fish oil from the wild to sustain them) is limited, environmentally sensitive, and subject to periodic issues of contamination with mercury and other pollutants (30).

Biotechnology has the potential to provide a sustainable source of high quality and relatively inexpensive long chain omega-3 PUFAs that could be added to many different foods, from salad dressings to high protein nutrition bars. This can be achieved by adding a long chain PUFA biosynthetic pathway to the existing fatty acid biosynthetic pathway in developing oilseeds and thus producing a PUFA enriched plant oil that could be used as a food ingredient.

There are a number of potential pathways that could be added, and a number of potential sources for the pathway genes (31). Fish make few if any long chain PUFAs, obtaining all they need from marine plankton. Over the past few years a wide range of PUFA pathways have been discovered in microbes. Certain marine prokaryotes, such as *Shewanella* species, *Moritella* species, and *Photobacter profundum*, contain significant quantities of either EPA or DHA which they synthesize via a polyketide synthase (PKS) enzyme complex (32). Some eukaryotic microbes, such as many thraustochytrids, also contain a PUFA synthesizing PKS (31,32). It is likely that PUFA PKSs make EPA or DHA from malonyl-CoA by a cyclical extension of the fatty acid chain with isomerization and enoyl reduction occurring in some of the cycles.

Many other eukaryotic microbes synthesize EPA and DHA using an aerobic pathway similar to that of humans, involving desaturases and fatty acid elongating enzymes. In these pathways EPA or DHA is made by the elongation and aerobic desaturation of linoleic and linolenic acids (31).

The PUFA synthesizing PKS was first identified in the EPA rich prokaryote *Shewanella*. The five open reading frames necessary for reconstituting this system in *E. coli* had homology to type I and II PKS components and to type II fatty acid synthase enzymes. A similar PKS from the DHA rich eukaryotic microbe *Schizochytrium* was found to contain three open reading frames (ORFs) with domains homologous to those in *Shewanella* (32). When expressed with a 4′-phosphopantetheinyl transferase, to activate the PKS ACP, the *Schizochytrium* PKS was able to synthesize DHA in *E. coli*. Current efforts are underway to engineer *Schizochytrium* PKS into oilseeds crops such as safflower and thus provide a renewable source of DHA in the seed oil (33).

In humans EPA is synthesized by the delta-6 desaturation of linolenic acid, and delta-6 elongation of the resulting stearidonic acid, followed by delta-5 desaturation to EPA. The same enzymes also convert linoleic acid to ARA (31). This pathway is also used by a wide range of microbes including fungi (e.g., *Mortierella alpina* and *Saprolegnia diclina*), mosses (e.g., *Physcomitrella patens*) and microalgae (e.g., *Phaeodactylum tricornutum*). A number of other freshwater and marine microbes (*Isochrysis galbana, Euglena gracillus*) have been shown to use an alternative aerobic pathway of EPA biosynthesis that begins with the delta-9 elongation of linoleic acid. This is followed by delta-8 and delta-6 desaturations to yield ARA, and then an omega-3 (delta-17) desaturation to EPA (34).

The synthesis of DHA from EPA in humans is complex, involving two elongation steps, a delta-6 desaturation, and β-oxidation in the peroxisome to DHA (31). Microbial DHA synthesis follows a simpler route. EPA is elongated to docosopentaenoic acid (DPA) which is then desaturated by a delta-4 desaturase to DHA.

There has also been some success in engineering both the delta-8 and the delta-6 aerobic pathways for EPA biosynthesis into plants. The delta-8 pathway was reconstructed in the model oilseed plant *Arabidopsis*, by the heterologous expression of an *I. galbana* delta-9 elongase, a *Euglena* delta-8 desaturase, and an *M. alpina* delta-5 desaturase. Significant accumulation of both ARA and EPA (7% and 3% respectively) was observed in *Arabidopsis* leaves, although the seed content of these long chain PUFAs was not reported (35). A similar EPA content was observed in flax seeds containing a delta-6 pathway transferred from *Physcomitrella patens* (36). Flax was chosen because it has a naturally abundant linolenic acid content, thus encouraging the formation of long chain omega-3, rather than omega-6, fatty acids (37). Another approach to omega-3 enrichment is to add extra omega-3 desaturase genes along with the long chain PUFA pathway (38,39). Transforming soybeans with a delta-6 desaturase and delta-6 elongase from *M. alpina*, a delta-5 desaturase from *S. diclina* and two extra omega-3 desaturases (under the control of various seed specific promoters) resulted in soybean oil containing up to 20% EPA and no ARA (40). It has also been possible to include two more genes, encoding a C20-elongase and a delta-4 desaturase, to the set for this aerobic pathway in transgenic plants, resulting in the production of small quantities of DHA in seed oils (36,40).

All of these results are very encouraging regarding future plant derived sources of long chain omega-3 fatty acids although considerable progress still needs to be made in optimizing pathway expression, producing crops with good agronomics and in stabilizing the new, highly unsaturated oil during production (to prevent oxidation).

Another bioactive fatty acid that is attracting some interest from the biotechnology world is conjugated linoleic acid (CLA). This fatty acid is normally found in dairy products, and has attracted this interest because of its reported effects on reducing body fat mass and increasing lean body mass (41). Other beneficial health related effects of CLA that have been reported in the literature include anticarcinogenic effects, and beneficial effects on the prevention of coronary heart disease, as well as modulation of the immune system (42). Most of the research on CLA, however, has been done with animals and there is not yet a scientific consensus about the human benefits of CLA consumption (42). Part of the problem may be that CLA is really a collective term signifying a group of different geometrical and positional isomers of linoleic acid and it is still unclear which isomers are the most active forms (43,44). Most natural food sources of CLA consist predominantly of the *cis-9 trans-11* isomer, while commercial CLA preparations are usually a mixture of *cis-9 trans-11* and *trans-10 cis-12* isomers. Consequently, most research has focused on these two isomers, although it is quite possible that other isomers and other conjugated fatty acids may also have some biological activity. Only very recently have researchers begun to examine the effect of purified CLA isomers in humans, and in these studies both *cis-9 trans-11* and *trans-10 cis-12* isomers have been shown to have some biological effects on blood lipids and immune function (44,45).

Attempts have been made to provide new food sources of CLA by enriching seed oils with conjugated fatty acids. One approach to this is expressing a polyenoic *cis-trans* fatty acid isomerase in developing oilseeds to convert newly synthesized linoleic acid to CLA before it is incorporated into oil (46). Isomerase genes have been cloned from algal (*Ptilota filicina*) and bacterial (*Propionium, Bacterium acnes*) sources, and attempts made to express these genes in plant cells (46,47). The substrates for these isomerases, however, are free fatty acids, which are only present in trace amounts in plant cells. Thus only trace amounts of CLA have been observed when isomerases were expressed in plants (47). Because most fatty acids in plant cells are part of complex lipids such as phospholipids, galactolipids and triacylglycerol, it seems likely that coexpressing an isomerase in developing oil seeds with some kind of lipase might be necessary to observe useful quantities of CLA in plant oils.

Another approach to making significant quantities of conjugated fatty acids in plant oils is to use plant enzymes that operate on esterified fatty acids. Dienoic conjugated fatty acids, such as CLA, are only present in plants in trace amounts although trienoic fatty acids are abundant in some plant species. The trienoic isomer of linolenic acid, calendic acid (*trans-8, trans-10, cis-12* CLnA), is found in high relative abundance (over 65% of total fatty acids) in the seeds of the pot marigold *Calendula officinalis* (48). This conjugated linolenic acid isomer contains the *trans-10 cis-12* double bonds thought to be involved in the biological activity of commercial CLA. The *trans-8, trans-10, cis-12* CLnA is synthesized in the developing marigold seed from linoleic acid, while the fatty acid is esterified to the membrane lipid phosphatidylcholine (48). The synthesis is catalyzed by a delta-12 oleic acid desaturase related enzyme encoded by a member of the Fad 2 gene family. When the marigold Fad 2 conjugase was expressed in soybean seeds, significant concentrations of calendic acid (15–20%) were observed in the seed oil (48).

Similar delta-12 oleic acid desaturase related enzymes catalyze the synthesis of other linolenic acid isomers, such as α-eleostearic acid (*cis-9, trans-11, trans-13* CLnA) in Chinese bitter melon (*Momordica charantia*) seeds (49) and trichosanic acid (*cis-9, trans-11, cis-13* CLnA) in *Trichosanthes kirilowii* and *Punica granatum* seeds (50). Again, it is possible to produce these other CLnA isomers in plant oils by expressing the Fad 2-related gene encoding the specific linoleic acid conjugase in the developing seeds of an oilseed plant (49,50).

It is not yet known how the biological activities of various CLnA isomers compare with those of CLA isomers, although there have been a number of provocative recent studies that have even raised the suggestion that dietary CLnA may be more biologically potent than CLA (51–54). For example, it has been shown that eleostearic acid prepared from both pomegranate seed and bitter melon seeds had very potent anticarcinogenic effects in rats (51,52). In another study, chemically synthesized eleostearic acid was found to be substantially more effective at suppressing tumors in rats than was CLA (53). Other studies in rats have found calendic acid to be more effective than CLA at increasing lean body mass and reducing adipose tissue weight (54).

The biotechnology of bioactive food lipids such as long chain omega-3 PUFAs and conjugated fatty acids is still in its infancy. Nevertheless, the ability to produce these fatty acids in plant oils has now been demonstrated and has the potential to impact the long term physical and mental health of human populations in a significant and positive way.

REFERENCES

1. *USDA Economic Research Service Oil Crops Situation and Outlook Yearbook*, U.S. Department of Agriculture, Springfield, VA, October, 2003.
2. Kris-Etherton, P.M., W.S. Harris, L.J. Appel. American Heart Association Nutrition Committee: Fish consumption, fish oil, omega-3 fatty acids, and cardiovascular disease. *Circulation* 106:2747–2757, 2002.
3. Wijendran, V., Hayes, K.C. Dietary n-6 and n-3 fatty acid balance and cardiovascular health. *Annu. Rev. Nutr.* 24:597–615, 2004.
4. Stauffer, C. *Fats and Oil*. St. Paul: Eagen Press, 1996.

5. Ascherio, A. Epidemiologic studies on dietary fats and coronary heart disease *Am. J. Med.* 113(9)B:9S–12S, 2002.
6. Kinney, A.J. Development of genetically engineered soybean oils for food applications. *J. Food Lipids* 3:209–273, 1996.
7. Kinney, A.J., Knowlton, S. Designer oils: the high oleic soybean. In: *Genetic Modification in the Food Industry*, Roller, S., S. Harlander, eds. Glasgow: Blackie & Son, 1998.
8. Food and Drug Administration, HHS. Food labeling: trans fatty acids in nutrition labeling, nutrient content claims, and health claims, final rule. *Fed. Regist.* 68:1433–1506, 2003.
9. Voelker, T., A.J. Kinney. Variations in the biosynthesis of seed-storage lipids. *Annu. Rev. Plant Physiol. Plant. Mol. Biol.* 52:335–361, 2001.
10. Van Eenennaam, A.L., K. Lincoln, T.P. Durrett, H.E. Valentin, C.K. Shewmaker, G.M. Thorne, J. Jiang, S.R. Baszis, C.K. Levering, E.D. Aasen, M. Hao, J.C. Stein, S.R. Norris, R.L. Last. Engineering vitamin E content: from Arabidopsis mutant to soy oil. *Plant Cell* 15:3007–3019, 2003.
11. Cahoon, E.B., S.E. Hall, K.G. Ripp, T.S. Ganzke, W.D. Hitz, S.J. Coughlan. Metabolic redesign of vitamin E biosynthesis in plants for tocotrienol production and increased antioxidant content. *Nat. Biotechnol.* 21:1082–1087, 2003.
12. Cheng, Z., S. Sattler, H. Maeda, Y. Sakuragi, D.A. Bryant, D. DellaPenna. Highly divergent methyltransferases catalyze a conserved reaction in tocopherol and plastoquinone synthesis in cyanobacteria and photosynthetic eukaryotes. *Plant Cell* 15:2343–2356, 2003.
13. Hawkins, D.J., J.C. Kridl. Characterization of acyl-ACP thioesterases of mangosteen (*Garcinia mangostana*) seed and high levels of stearate production in transgenic canola. *Plant J.* 13:743–752, 1998.
14. Knutzon, D.S., G.A. Thompson, S.E. Radke, W.B. Johnson, V.C. Knauf, J.C. Kridl. Modification of Brassica seed oil by antisense expression of a stearoyl-acyl carrier protein desaturase gene. *Proc. Natl. Acad. Sci. USA* 89:2624–2628, 1992.
15. Voelker, T.A., A.C. Worrell, L. Anderson, J. Bleibaum, C. Fan, D.J. Hawkins, S.E. Radke, H.M. Davies. Fatty acid biosynthesis redirected to medium chains in transgenic oilseed plants. *Science* 257:72–74, 1992.
16. Del Vecchio, A.J. High laurate canola. *Inform* 7:230–243, 1996.
17. Dehesh, K., A. Jones, D.S. Knutzon, T.A. Voelker. Production of high levels of 8:0 and 10:0 fatty acids in transgenic canola by overexpression of Ch FatB2, a thioesterase cDNA from *Cuphea hookeriana*. *Plant J.* 9:167–172, 1996.
18. Voelker, T.A., A. Jones, A.M. Cranmer, H.M. Davies, D.S. Knutzon. Broad-range and binary-range acyl-acyl-carrier protein thioesterases suggest an alternative mechanism for medium-chain production in seeds. *Plant Physiol.* 114:669–677, 1997.
19. Knutzon, D.S., T.R. Hayes, A. Wyrick, H. Xiong, H. Maelor Davies, T.A. Voelker. Lysophosphatidic acid acyltransferase from coconut endosperm mediates the insertion of laurate at the sn-2 position of triacylglycerols in lauric rapeseed oil and can increase total laurate levels. *Plant Physiol.* 120:739–746, 1999.
20. Mu, H., C.E. Hoy. The digestion of dietary triacylglycerols. *Prog. Lipid Res.* 43:105–133, 2003.
21. Mu, H., C.E. Hoy. Intestinal absorbtion of specific structured triacylglycerols. *J. Lipid Res.* 42:792–798, 2001.
22. Tomarelli, R.M., B.J. Meyer, J.R. Weaber, F.W. Bernhart. Effect of positional distribution on the absorption of the fatty acids of human milk and infant formulas. *J Nutr.* 95:583–590, 1968.
23. Lands, W.E. Essential fatty acids in the foods we eat have a subtle but powerful influence on hundreds of different processes in the life and death of humans *Nutr. Metab. Cardiovasc. Dis.* 13:154–164, 2003.
24. Yaqoob, P. Fatty acids and the immune system: from basic science to clinical applications. *Proc. Nutr. Soc.* 63:89–104, 2003.
25. Willatts, P., Forsyth, J.S. The role of long-chain polyunsaturated fatty acids in infant cognitive development. *Prostaglandins Leukot. Essent. Fatty Acids* 63:95–100, 2000.

26. Iribarren, C., J.H. Markovitz, D.R. Jacobs, Jr., P.J. Schreiner, M. Daviglus, J.R. Hibbeln. Dietary intake of n-3, n-6 fatty acids and fish: relationship with hostility in young adults – the CARDIA study. *Eur. J. Clin. Nutr.* 58:24–31, 2004.
27. Stoll, A.L., K.E. Damico, B.P. Daly, W.E. Severus, L.B. Marangell. Methodological considerations in clinical studies of omega 3 fatty acids in major depression and bipolar disorder. *World Rev. Nutr. Diet.* 88:58-67, 2001.
28. Harris, W.S., C. Von Schacky. The Omega-3 Index: a new risk factor for death from coronary heart disease? *Prev Med.* 39:212–220, 2004.
29. Richter, W.O. Long-chain omega-3 fatty acids from fish reduce sudden cardiac death in patients with coronary heart disease. *Eur. J. Med. Res.* 8:332–336, 2003.
30. Naylor, R.L., R.J. Goldburg, J.H. Primavera, N. Kautsky, M.C. Beveridge, J. Clay, C. Folke, J. Lubchenco, H. Mooney, M. Troell. Effect of aquaculture on world fish supplies. *Nature* 405:1017–1024, 2000.
31. Wallis, J.G., J.L. Watts, J. Browse. Polyunsaturated fatty acid synthesis: what will they think of next? *Trends Biochem Sci.* 27:467–473, 2002.
32. Metz, J.G., P. Roessler, D. Facciotti, C. Levering, F. Dittrich, M. Lassner, R. Valentine, K. Lardizabal, F. Domergue, A. Yamada, K. Yazawa, V. Knauf, J. Browse. Production of polyunsaturated fatty acids by polyketide synthases in both prokaryotes and eukaryotes. *Science* 293:290–293, 2001.
33. Metz, J.G. PUFA synthases: characterization, distribution and prospects for PUFA production in heterologous systems. *6th Congress of the International Society for the Study of Fatty Acids and Lipids, Brighton, UK, Abstract book*, 2004, p 41.
34. Sayanova, O.V., J.A. Napier. Eicosapentaenoic acid: biosynthetic routes and the potential for synthesis in transgenic plants. *Phytochemistry* 65:147–158, 2004.
35. Qi, B., T. Fraser, S. Mugford, G. Dobson, O. Sayanova, J. Butler, J.A. Napier, A.K. Stobart, C.M. Lazarus. *Nat Biotechnol.* 22:739–745, 2004.
36. Heinz, E. Biosynthesis of VLCPUFAs in transgenic oilseeds. *16th International Plant Lipid Symposium, Budapest, Hungary, Abstract book*, 2004, p 8.
37. Drexler, H., P. Spiekermann, A. Meyer, F. Domergue, T. Zank, P. Sperling, A. Abbadi, E. Heinz. Metabolic engineering of fatty acids for breeding of new oilseed crops: strategies, problems and first results. *J. Plant Physiol.* 160:779–802, 2003.
38. Spychalla, J.P., A.J. Kinney, J. Browse. Identification of an animal omega-3 fatty acid desaturase by heterologous expression in *Arabidopsis*. *Proc. Natl. Acad. Sci. USA.* 94:1142–1147, 1997.
39. Pereira, S.L., Y.S. Huang, E.G. Bobik, A.J. Kinney, K.L. Stecca, J.C. Packer, P. Mukerji. A novel omega3-fatty acid desaturase involved in the biosynthesis of eicosapentaenoic acid. *Biochem. J.* 378:665–671, 2004.
40. Cahoon, E.B., H.G. Damude, W.D. Hitz, A.J. Kinney. Production of very long chain polyunsaturated fatty acids in oilseed plants. U.S. Patent Application US20040049813, 2004.
41. Jeukendrup, A.E., S. Aldred. Fat supplementation, health, and endurance performance. *Nutrition* 20:678–688, 2004.
42. Rainer, L., C.J. Heiss. Conjugated linoleic acid: health implications and effects on body composition *J. Am Diet Assoc.* 104:963–968, 2004.
43. Wang, Y., P.J. Jones. Dietary conjugated linoleic acid and body composition. *Am. J. Clin. Nutr.* 79(6):1153S–1158S, 2004.
44. O'Shea, M., J. Bassaganya-Riera, I.C. Mohede. Immunomodulatory properties of conjugated linoleic acid. *Am. J. Clin. Nutr.* 79(6):1199S–1206S, 2004.
45. Burdge, G.C., B. Lupoli, J.J. Russell, S. Tricon, S. Kew, T. Banerjee, K.J. Shingfield, D.E. Beever, R.F. Grimble, C.M. Williams, P. Yaqoob, P.C. Calder. Incorporation of cis-9,trans-11 or trans-10,cis-12 conjugated linoleic acid into plasma and cellular lipids in healthy men. *J. Lipid Res.* 45:736–771, 2004.
46. Zhen, W., L. Yuan, J. Metz. Nucleic acid sequences encoding polyenoic fatty acid isomerase and uses thereof, International Patent Application WO2001009296, 2002.

47. Hornung, E. Production of conjugated fatty acids in plants with bacterial double bond isomerases: biosynthesis of VLCPUFAs in transgenic oilseeds, *16th International Plant Lipid Symposium, Budapest, Hungary,* Abstract book, 2004, p 27.
48. Cahoon, E.B., K.G. Ripp, S.E. Hall, A.J. Kinney. Formation of conjugated delta8,delta10-double bonds by delta12-oleic-acid desaturase-related enzymes: biosynthetic origin of calendic acid. *J. Biol. Chem.* 276:2637–2643, 2001.
49. Cahoon, E.B., T.J. Carlson, K.G. Ripp, B.J. Schweiger, G.A. Cook, S.E. Hall, A.J. Kinney. Biosynthetic origin of conjugated double bonds: production of fatty acid components of high-value drying oils in transgenic soybean embryos. *Proc. Natl. Acad. Sci. USA* 96:12935–12940, 1999.
50. Iwabuchi, M., J. Kohno-Murase, J. Imamura. Delta 12-oleate desaturase-related enzymes associated with formation of conjugated trans-delta 11, cis-delta 13 double bonds. *J. Biol. Chem.* 278:4603–4610, 2003.
51. Kohno, H., R. Suzuki, Y. Yasui, M. Hosokawa, K. Miyashita, T. Tanaka. Pomegranate seed oil rich in conjugated linolenic acid suppresses chemically induced colon carcinogenesis in rats. *Cancer Sci.* 95:481–486, 2004.
52. Kohno, H., Y. Yasui, R. Suzuki, M. Hosokawa, K. Miyashita, T. Tanaka Dietary seed oil rich in conjugated linolenic acid from bitter melon inhibits azoxymethane-induced rat colon carcinogenesis through elevation of colonic PPARgamma expression and alteration of lipid composition. *Int. J. Cancer* 20(110):896–901, 2004.
53. Tsuzuki, T., Y. Tokuyama, M. Igarashi, T. Miyazawa. Tumor growth suppression by alpha-eleostearic acid, a linolenic acid isomer with a conjugated triene system, via lipid peroxidation. *Carcinogenesis* 25:1417–1425, 2004.
54. Koba, K., A. Akahoshi, M. Yamasaki, K. Tanaka, K. Yamada, T. Iwata, T. Kamegai, K. Tsutsumi, M. Sugano. Dietary conjugated linolenic acid in relation to CLA differently modifies body fat mass and serum and liver lipid levels in rats. *Lipids* 37:330–343, 2002.

2.08

Molecular Biotechnology for Nutraceutical Enrichment of Food Crops: The Case of Minerals and Vitamins

Octavio Paredes-López and Juan Alberto Osuna-Castro

CONTENTS

8.1 MICRONUTRIENTS: ALLEVIATING NUTRITIONAL DISORDERS BY NUTRACEUTICALS

Ever since the discovery of vitamins, there has been a great interest in deciphering the effects of food constituents on human health. The requirements for a healthy life with a balanced metabolism (metabolic homeostasis) and preservation of the body cell mass require a protein balance having all the essential amino acids, fatty acids, carbohydrates, and micronutrients (1,2).

Approximately 40 micronutrients (the vitamins, essential minerals, and other compounds needed in small amounts for normal metabolism) are required in human diet (Tables 8.1, 8.2). Obtaining maximum health and life span requires metabolic harmony. For each micronutrient, metabolic harmony requires an optimal intake (i.e., to give maximal life span); deficiency distorts metabolism in numerous and complicated ways, many of which may lead to DNA damage (3–7). Many micronutrient minerals and vitamins act as substrates or cofactors in key DNA maintenance reactions, and the exact concentration of these in the cell may be critical. Suboptimal levels of key micronutrients will thus lead to impaired activity of enzymes needed for genomic stability in man, producing effects similar to inherited genetic disorders or exposure to carcinogens (Tables 8.1, 8.2) (3–7). In fact, a deficiency of some micronutrients (folic acid, vitamin B12, vitamin B6, niacin, vitamin C, vitamin E, selenium, iron, or zinc) appears to mimic exposure to radiation or chemicals by causing single and double strand breaks in DNA, oxidative lesions, or both (4–12). Chromosomal aberrations such as double strand breaks are a strong predictive factor for human cancer.

In adults, superoxide, hydrogen peroxide, and hydroxy radicals (produced by some 10^{10} free radicals per cell each day) can potentially cause in the order of 10^6 mutational alterations of DNA per cell per day. Fortunately, the activities of these dangerous mutagens are countered by antioxidants, DNA repair, the removal of persistent alterations by apoptosis (normally a cellular program is activated, that causes the cell to self destruct), differentiation, necrosis, and activation of the immune system so that only about one mutation per cell per day persists (13). By old age, many mutations have accumulated because the repair system is not working properly, and as a consequence, cancer may occur. However, the increased consumption of dietary antioxidants such as vitamins C and E, quercetin (flavonoid), and carotenoids (zeaxanthin and lycopene) can diminish DNA oxidation and therefore, less cancer incidence is presented (3,13–18).

The U.S. Recommended Dietary Allowances (RDAs) for micronutrients refer to daily levels of intake of essential nutrients, at or above defined minimum values, judged by the Food and Nutrition Board to be adequate to meet the known nutrients needed for practically all healthy persons and to prevent deficiency diseases. The level of daily requirement for each nutrient varies with age, sex, and physiological status (e.g., pregnancy, lactation, disease related stress) (Tables 8.1, 8.2) (19). RDAs are established by estimating the requirement for the absorbed nutrient, adjusting for incomplete utilization of the ingested nutrient, and incorporating a safety factor to account for variability among individuals (19).

Also, now it is known that vitamins, minerals, and other dietary components consumed at varying levels, higher than RDAs, are significant contributors to the reduction of risks of chronic diseases such as cancers, cardiovascular diseases, and degenerative diseases associated with aging (Alzheimer's, premature aging). Their role in maintaining human genomic stability is likely to be critical to alleviating the classical nutritional deficiency diseases which include scurvy (vitamin C), anemia (folic acid, iron), and pellagra (niacin) (Tables 8.1, 8.2) (4–7,17,20,21).

While deficiencies of dietary energy (i.e., calories) and protein currently affect more than 800 million people in food insecure regions, incredibly, micronutrient malnutrition, known as "hidden hunger", now afflicts over 40% of the world population, especially the most vulnerable being in many developing nations (resource-poor women, infants, and children), and a surprisingly large amount of people in the advanced countries, where food diversity, abundance, and supply are excellent, but also poor food eating habits are common and in both cases the numbers are rising (4,22–25). Even though micronutrients are needed in minute quantities (i.e., micrograms to milligrams per day), they have tremendous impact on human health and well-being (Tables 8.1, 8.2) (22,26,27). Insufficient dietary intake of these microcomponents impairs the functions of the brain, the immune and reproductive systems, and energy metabolism (26,27).

At present, deficiencies of iron, vitamin A and iodine are of the most concern to the community and healthcare systems. Unfortunately, for zinc there is a lack of a simple, quick, and cheap clinical screening test for determining marginal zinc deficiency in humans. Specialists in zinc nutrition suggest that its deficiency presents similar problems as caused by a deficiency of iron (10,22,24,25,28). Iron, zinc, and vitamin A all play important roles in brain development, and when deficiencies of these micronutrients are manifest during pregnancy, or even for up to two years postpartum, permanent damage to offspring is possible. Thereafter, further loss of cognitive ability is found (10,22,24–31).

Iodine deficiency is the greatest single cause of preventable brain damage and mental retardation in the world today. More than 2 billion people around world live in iodine deficiency environments. Deficiency in iodine occurring in late infancy and childhood have been demonstrated to produce mental retardation, delayed motor development, and stunted growth, occurrence of neuromuscular disorders, and speech and hearing defects. Even mild iodine deficiency has been reported to decrease intelligence quotients by 10–15 points (22,31).

Rapid growth during fetal development, infancy, childhood, and adolescence demand greater amount of micronutrients, when their deficiency can cause developmental abnormalities (22,25). In addition to multiple micronutrient deficiencies (those of vitamin A and C, decreased iron absorption), interactions between micronutrient deficient conditions and infectious diseases may become significant. These effects can complicate health care efforts to control various diseases such as vitamin A deficiency and measles, and increased severity of measles leading to vitamin A deficiency blindness and death. Malaria and hookworm infections are also associated with Fe deficiency anemia (22,25,30,32).

Plants synthesize and accumulate an astonishingly diverse array of vitamins, and nutraceuticals that have health-promoting properties (33,34). Many naturally occurring compounds have health promoting or disease preventing properties beyond the mere provision of nutrients for basic nutrition (1,34–36). In this way, plants not only constitute the base of the human food chain but they are also important means to improve human health and well-being, with the exception of vitamin B12 and D. A diverse and well-balanced plant-based diet, that includes mixed sources of grains, fruits, and vegetables, can ensure the proper micronutrient nutrition and health at all stages of the life cycle (Tables 8.1, 8.2) (18,20,21,37–39). It is known that seeds are good sources of lipid-soluble vitamins, but tend to have low levels of bioavailable iron, zinc, and calcium, whereas leafy vegetables can supply most minerals and vitamins. Fruits provide water-soluble vitamins and several types of carotenoids but generally are minor sources of certain minerals. Thus, concentrations of very few individual plant foods are able to supply all the daily recommended intake of any micronutrient in an average or reasonable serving size (20–22,37–39). Unfortunately, many people do not consume a sufficiently diverse diet. In fact, many low-income families from the developing world subsist with a simple diet formed mainly of a staple food (i.e., maize, soybean, beans, wheat, tubers, root) that are poor sources of many micronutrients (2,16,20–25,27,37–39). Heavy and monotonous consumption of cereal-based foods with low concentrations and diminished bioavailability of iron and zinc has been considered a major reason for the current widespread deficiency of both minerals in developing countries (16,20–25,27,37–39). Furthermore, even in developed nations, the average intake of fruits and vegetables may fall below official recommendations (19,21).

Table 8.1
Minerals: Dietary allowances per day and safe upper intake limit values, sources, deficiency and importance on human genomic stability.

				Deficiency	Deficiency	Deficiency	
Minerals	**Maximum Adult RDA**$^{\Phi}$	**Safe Upper Intake Limits (Relative to RDA)**$^{\theta}$	**Predominant Food Sources**	**Some Classical Nutritional Disorders and Diseases**	**DNA Damage (Negative Effecton Genomic Stability)**	**New Health Effects**	**Functions on Genomic Stability**
Selenium	70 μg	13X	Seeds	Keshan disease, an endemic cardiomyopathy in China	DNA Oxidation (radiation mimic)	Prostate cancer, increased risk for heart diseases and other type of cancer	Retardation of oxidative damage to DNA, proteins and lipids, modulation of cellular events critical in cell growth inhibition and multi-step carcinogenesis process, cofactor of glutathione peroxidase
Iodine	150 μg	13X	Iodized salt, sea foods, plants and animal grown in areas where soil iodine is not depleted	Goiter, mental retardation, brain damage and reproductive failure	—	—	—
Chromium	50–200 μg	1X	Various	Rare	—	—	—
Molybdenum	75–250 μg	1X	Seeds	Rare	—	—	—
Copper	1.5–3 mg	1X	Organ meats, sea foods, nuts, seeds	Rare	—	—	Cofactor of cuproenzymes of antioxidant system
Fluoride	1.5–4 mg	1X	Aerial tissues	Problems with bone and teeth	—	—	—

Manganese	2.–5 mg	1X	Whole seeds, fruit	Extremely rare	—	—	—
Zinc	15 mg	1X	Meat, eggs, nut, seeds	Growth retardation, delayed skeletal and sexual maturity, dermatitis, diarrhea, alopecia and defects in immune system function with decreased resistance to infections	Elevated DNA Oxidation (radiation mimic) DNA breaks and increased chromosome damage rate	Immune and brain dysfunction, cancer	Co-factor of enzymes: Cu/Zn superoxide dismutase, endonuclease IV, zinc fingers
Iron	13–15 mg	5X	Meat, seeds, leafy vegetables	Nutritional anemia, problem pregnancies, stunted growth, tiredness and diminished food-energy conversion, poor work performance	DNA breaks, radiation mimics	Immune and brain dysfunction with lower resistance to infections and long-term impairment of neural motor development, and mental function, cancer	Fe-containing enzymes that are particularly important for gene regulation
Magnesium	350 mg	1X	Whole grains, nuts, green leafy vegetables	Rare, but when it occurs there are some problems with bone structures	—	—	—
Sodium	500 mg	5X	Common salt	Rare, but excesive intakes may lead to hypertension	—	—	
Chloride	750 mg	5X	Various	Rare	—	—	—

(Continued)

Table 8.1 (*Continued*)

				Deficiency	Deficiency	Deficiency	
Minerals	**Maximum Adult RDA**[Φ]	**Safe Upper Intake Limits (Relative to RDA)**[θ]	**Predominant Food Sources**	**Some Classical Nutritional Disorders and Diseases**	**DNA Damage (Negative Effecton Genomic Stability)**	**New Health Effects**	**Functions on Genomic Stability**
Phosphorus	1200 mg	2X	Ubiquitous, animal products tend to be good sources	Rare due to presence in virtually all foods	—	—	—
Calcium	1200 mg	2X	Milk products, green leafy vegetables, tofu, fish bones	Osteoporosis in elderly, rickets	—	—	—
Potassium	2000 mg	9X	Fruits, vegetables, meats	Rare	—	—	—

[Φ] Recommended daily dietary allowances. RDA values given mean that highest either for male or female adult, but not for pregnant or lactating women. [θ] The concept assumes that there is individual variation in both requirement for the minerals and tolerance for elevated intake.

Sources: References 3–7, 9–11, 16, 19–22, 24–29, 31, 37, 43, 46.

Table 8.2
Vitamins: Dietary allowances per day and safe upper intake limit values, sources, deficiency and importance on human genomic stability.

				Deficiency/Traditional Functions	Deficiency	Deficiency	
Vitamins	**Maximum Adult RDA**$^{\Phi}$	**Safe Upper Intake Limits (Relative to RDA)**$^{\theta}$	**Predominant Food Sources**	**Classical Nutritional Disorders and Diseases**	**DNA Damage (Negative Effects on Genomic Stability)**	**New Health Effects**	**Role on Genomic Stability**
Fat Soluble							
Vitamin D (Chole-calciferol)	10 μg	4X	Milk and dairy products	Rickets, whose common symptoms include weak and misshapen bone, bowlegged and poor muscle tone; Required to incorporate calcium, phosphate and magnesium into bones	—	Colorectal cancer and adenoma, increase cell proliferation	—
Vitamin K	80 μg	375X	Leafy vegetables (spinach, cauliflower, cabbage), tomato and some vegetable oils	Elevated hemorrhage due to reduced prothrombin levels, particularly in premature or anoxic infants	—	—	—
Vitamin A	1 mg RE$^{\Psi}$	5X (retinol); 100X (β-carotene)	Some pigmented vegetables and fruits	Poor night vision, eye lesions, and in severe cases, permanent blindness; elevated illness and death from infections	—	—	

(*Continued*)

Table 8.2 (*Continued*)

				Deficiency/Traditional Functions	Deficiency	Deficiency	
Vitamins	**Maximum Adult RDA**$^{\Phi}$	**Safe Upper Intake Limits (Relative to RDA)**$^{\theta}$	**Predominant Food Sources**	**Classical Nutritional Disorders and Diseases**	**DNA Damage (Negative Effects on Genomic Stability)**	**New Health Effects**	**Role on Genomic Stability**
Fat Soluble							
Vitamin E	10 mg α-TE $^{\vartheta}$	100X	Oilseeds, leafy vegetables	Increased fragility in red cells and hemolysis, as well as impaired sensation and neuromuscular activity	Increased baseline level of DNA strand breaks, chromosome breaks and oxidative DNA damage and lipid radicals adduct on DNA	Heart disease, colon cancer, immune system impairment	Prevention of lipid and DNA (radiation mimic) oxidation
Water Soluble							
Vitamin B12 (Cyano-cobalamin)	2 μg	500X	Fish, dairy products, meat	Megaloblastic anemia where proper maturation of red blood cells is hampered, among other symptoms such as anorexia, intestinal discomfort, depression and some neurological problems	Chromosome breaks	Neuronal damage also see folate	Maintaining methylation level in DNA

Biotin	30–100 μg	300X	Seeds, animal products	Rare	—	—	—
Folate	200 μg	50X	Fruits, leafy vegetables, legumes including dark and dried beans	Megaloblastic anemia, weight loss, nausea, confusion, irritability and dementia are typical.	Uracil misincorporation into DNA provoking DNA breakage (DNA hypomethylation), chromosome breaks (radiation mimic)	Colorectal cancer, most heart diseases, altered risks of birth defects (neural tube), brain dysfunction (Down's syndrome), tumorigenesis, acute lymphoblastic leukemia in children	Maintaining methylation level in DNA, efficient recycling of folate
Vitamin B1 (Thiamine)	1.5 mg	67X	Crops	Beriberi, symptoms include weak muscles, fatigue, depression, irritability, weight loss, with frequency cardiovascular complications and paralysis	—	—	—
Vitamin B2 (Riboflavin)	1.7 mg	—	Leafy vegetables (i.e., broccoli), cereals, liver, beef and cheese	Nouse dermatitis, altered coloration and texture of lips and tongue, fatigue and sensitive and vascularized eyes, bloodshot	—	—	—

(*Continued*)

Table 8.2 (*Continued*)

Vitamins	Maximum Adult RDA$^{\Phi}$	Safe Upper Intake Limits (Relative to RDA)$^{\theta}$	Predominant Food Sources	Deficiency/Traditional Functions: Classical Nutritional Disorders and Diseases	Deficiency: DNA Damage (Negative Effects on Genomic Stability)	Deficiency: New Health Effects	Role on Genomic Stability
Vitamin B6 (Pyridoxin)	2 mg	125X	Whole grains, meat	Vomiting, irritability, weakness, ataxia	Chromosome breaks	Cancer of prostate, and lung, See folate	See folate
Pantothenic acid	4–7 mg	150X	Cereal seeds, eggs, meat, milk, fresh vegetables	—	—	—	—
Vitamin B3 (Niacin)	19 mg NE*	150X	Leafy vegetables, crops	Pellagra, characterized by skin lesions, diarrhea, and mental apathy	Disables DNA repair, (augmented percentage of unrepaired nicks in DNA); High levels of chromosome fractures, mutagen sensitivity	Neurological problems; memory loss	Substrate of poly (ADP-ribose) polymerase enzyme, that cleaves DNA and rejoins cut ends, and telomere length maintenance, component of DNA involved in poly ADP-ribose

							defense against DNA strand fractures
Vitamin C	60 mg	16X	Citrus fruits, vegetables	Irritability, growth delay, anemia, poor wound healing, increased tendency to bleed, and susceptibility to infections; weak cartilages and tenderness in the legs are typical symptoms of scurvy	See vitamin E	Cataract, See vitamin E	See vitamin E

[Φ] Recommended daily dietary allowances. RDA values given mean that highest either for male or female adult, but not for pregnant or lactating women. θ The concept assumes that there is individual variation in both requirement for the vitamin and tolerance for elevated intake. [Ψ] Vitamin A activity is expressed in retinol equivalent (RE). One RE represents 1 mg of all-*trans*-retinol or 6 mg of all-*trans*-β-carotene. [ϑ] one α-tocopherol equivalent (TE) is similar to 1 mg of (R, R, R)-α-tocopherol. [*] One mg NE (niacin equivalent) represents 1 mg of dietary tryptophan.

Sources: References 3–8,12–14, 16–22, 24, 25, 31, 37, 38, 42, 45, 92.

The world population is expected to grow from 6 billion to around 10 billion by 2050. Virtually all of the anticipated increase in population will occur in Africa, Latin America, and Asia (23–25,30,31). These regions already face serious problems of food production and access, water scarcity, nutritional disorders, and malnutrition of macronutrients and micronutrients (23–25,30,31). Among the undernourished, children under five years of age are particularly at risk. A high percentage of children are underweight, and low birth weight becomes an important factor in child malnutrition and premature death (23–25,30). Notably, countries that have dramatically reduced the incidence of malnutrition have also concurrently reduced their birth rates dramatically over the last century.

Therefore, to ensure an adequate dietary intake of all essential micronutrients and to augment the consumption of various nutraceuticals by people around the world, great and intelligent efforts are needed to improve the nutritional and nutraceutical quality of agricultural produce, with regard to their micronutrient composition, concentration as well as their biological availability (16,17,27,38–41).

8.2 MICRONUTRIENT BIOAVAILABILITY AND APPROACHES TO REDUCE MICRONUTRIENT MALNUTRITION

The mineral and vitamin content of food crops should be considered at the same time with its bioavailability (20,42,43). The total content, or absolute concentration, of a given nutrient in a food is not always a reliable indicator of its useful nutritional quality, because not all of the nutrients in food are absorbed (20,42,43).

The human nutritionists employ the concept of bioavailability of micronutrients, which refers to the percentage of a given nutrient that is potentially available for absorption from an ingested food, and once absorbed, utilizable for normal physiological functions and even for storage in the body (25,40,44). Bioconversion is defined as the fraction of a bioavailable micronutrient (i.e., absorbed provitamin A carotenoids) that is converted to the active form of a nutrient (retinol) (44,45).

The digestive processes help to release and solubilize nutrients so they can diffuse out of the bulk food matrix into the enterocytes of the human intestine. Micronutrients can be present in several chemical species in plant foods and their quantities may vary according to growth environment, plant species and genotype, and cultural practices utilized to grow the plant (21,22,24,43,46,47). These chemical forms may have characteristically distinct solubilities and reactivities related to other plant components and other meal constituents; absorption then is a function of physicochemical properties and form of micronutrients (i.e., free, complexed, charged, dispersible). The absorption and transport of nutrients may be influenced by the presence of inhibitory (i.e., antinutrients) and promotive (i.e., promoters that can enhance absorption and utilization) substances in the food matrix (Table 8.3) (21,22,24,32,39,43,46,47).

As a result, for a complete and realistic description of the nutritional adequacy of food from dietary micronutrient viewpoint, three factors should be assessed: (1) the concentration of the micronutrients at the time of consumption; (2) the nature of chemical species of the micronutrients present; and (3) the bioavailability of these forms of the micronutrient as they exist in the meal consumed (21,22,24,32,39,42,43,46,47).

The bioavailability of normal micronutrients from plant sources ranges from less than 5% for some minerals such as calcium and iron, over 90% for sodium and potassium (20,21,43). The three most common and widely recognized strategies for reducing micronutrient mineral and vitamin malnutrition include: (1) supplementation with pharmaceutical preparations, (2) fortification, (3) and dietary food diversification (16,21,22,31,39–43,48–50).

For instance, iron and zinc supplementation has been useful in developing countries for rapid improvement of the status of both minerals in deficient individuals, but this approach is relatively expensive and often has poor compliance, particularly with medicinal Fe because of unpleasant side effects (22,43).

Commercial fortification of foods is familiar to most of us. Minerals and vitamins are added to a particular food vehicle during its processing, well after the foods have left the farm and before they are distributed through various marketing channels for consumer purchase and consumption (31). Thus, fortification refers to incorporation of nutrients at superior levels than those existing in the original (i.e., unprocessed) (42,48,49). This may include adding nutrients not normally associated with the food. Food fortification is usually the best long term strategy, but its effectiveness depends on the compatibility of fortified foods with local culinary habits. Iron is the most difficult mineral to incorporate into foods (due to technical problems related to the choice of suitable iron compounds) and to ensure adequate absorption (22,43,48,49). For example, those with relatively high bioavailability, which are water-soluble iron compounds such as ferrous sulfate, usually provoke unacceptable color and flavor changes in food. When water-soluble iron compounds are incorporated into cereal flour, they often cause rancidity; and in low grade salt, they rapidly lead to undesirable color formation. Furthermore, other compounds which are organoleptically inert, such as elemental iron powders, are so poorly absorbed as to be of little or no nutritional value (22,43,48,49).

Both supplementation and fortification have treated the symptoms of micronutrient malnutrition rather than the underlying causes. While many of these interventions have been successful in the short term, these strategies have proved to be prohibitively expensive in marginal economies, unsustainable and incapable of reaching all the individuals affected. Ironically, those segments of population at high risk (poor women, infants, and children) usually live in remote places either from a clinic and health care professionals or do not have ready access to processed and fortified foods (22,24,30,43,48,49). Finally, food diversification may also be quite difficult, but not impossible, in developing countries for economic, social, or traditional reasons, and people who avoid milk products in developed nations normally do so because they dislike or are intolerant to dairy products (16,20,38).

Within the agricultural community, conventional plant breeding efforts during the green revolution were focused on productivity and efficiency of cropping systems (mostly cereal crops), being greatly successful in supplying enough calories and macronutrients to prevent the threatening global starvation and shortage of foods predicted at the beginning of 1960s (37,51). Remarkably, the nutrition community has never embraced agriculture as a key tool to be used to fighting "hidden hunger"; paradoxically the green revolution may have contributed to some unforeseen negative consequences on human nutrition and health referent to micronutrient malnutrition (22,25,37). Thus, in some developing countries the rise in micronutrient deficiencies may be associated to changes in the patterns in cultivation toward cereals, which was paralleled by a significant decreased per capita production of traditional edible legumes that are a much richer source of micronutrients than cereals, especially after cereals have been milled and polished before consumption (22,23,25).

Plant mineral and vitamin concentrations vary among plant sources (i.e., species, cultivars) and within plant tissues (i.e., leafy structures against seeds); thereby demonstrating the existence of genetic variability, which can contribute to the plant's ability to acquire, sequester, synthesize, and accumulate micronutrients (16,21,22,33,34,37,38,52). In contrast to macronutrients which can represent up to 30 to 50% in dry weight in some tissues (g/100 g portion of food), individually, minerals and vitamins when present constitute a very tiny quantity ranging from a microgram to a milligram per 100 g of food (less than 0.1%) (2,16,21,22,33,34,37,38,52). Therefore, significant quantitative and qualitative

Table 8.3
Antinutritional and promoter substances that reduce and enhance, respectively, the bioavailability of some important micronutrients for human nutrition and health, and even genomic stability, and their food distribution.

Antinutrient(s)	Micronutrient(s) Negatively Affected (Less Bioavailability)	Common Dietary Food Sources of Antinutrient(s)	Promoter(s) [Enhancer(s)]	Micronutrient(s) Positively Affected (Improved Bioavailability)	Common Dietary Food Sources of Promoter(s)
Some tannins and other polyphenols	Iron, zinc	Pigmented beans, sorghum, coffee, tea	Meat factors (cysteine-rich polypeptides)	Iron, zinc	Animal meats (i.e., pork, beef, fish)
Hemagglutinins (i.e., lectins)	Iron, zinc	Wheat and most legumes	Some free and essential amino acids (i.e., lysine, histidine, sulfur-containing)	Iron and/or zinc	Animal meats (i.e., pork, beef, fish)
Fiber (i.e., cellulose, cutin, hemicellullose, lignin)	Iron, zinc	Whole-cereal-based products	Some organic acids and/or their salts (i.e., ascorbic acid, citrate, fumarate, malate)	Iron and/or zinc	Several fresh fruits and vegetables
Toxic heavy metals (i.e., Cd, Hg, Pb, Ag)	Iron, zinc	Contamined leafy vegetables, tubers and grains (i.e., Cd in rice) from metal-polluted soils	Plant ferritin	Iron	Legume crops, leafy vegetables

Phytic acid or phytin	Iron, zinc, magnesium, calcium	Whole legume and cereal crops	Hemoglobin	Iron	Animal red meats
Oxalic acid	Calcium	Spinach	β-Carotene, vitamin A	Iron	Green and orange vegetables, red palm oil, yellow maize
—	—	—	Iron, zinc	Vitamin A	Animal meats
—	—	—	Selenium	Iodine	Sea foods, organ meats, tropical nuts and cereals, amounts vary depending on soil levels
—	—	—	Vitamin E (α-tocopherol)	Vitamin A	Vegetable oils, green leafy vegetables

Sources: Adapted from References 22, 24, 27.

changes are feasible for minor plant components. Genetic manipulation to augment the levels of these components would require minimal diversion of precursors and some modifications in the plant's ability to store or sequester the target micronutrients (16,33,34,52,53). Interestingly, the tolerable upper intake levels in humans for minerals fluctuate between 1- and 13-fold in their RDA values, whereas those for vitamins are higher than those for minerals, and, as it has been reviewed earlier, micronutrient intakes superior to RDAs, but not higher than those tolerable levels for humans, provide health benefits including genomic stability (Tables 8.1, 8.2) (3–7).

All this raises the opportunity for the molecular biotechnology as an emerging and powerful approach with the potential for improving nutritional quality of food plants, altering the composition, content, and bioavailability of the existing micronutrients [i.e., modifying chemical forms of the stored micronutrient, removing (or reducing the level of) antinutritional compounds, or elevating the amount of promoter substances] (Table 8.3), or accumulating novel and bioavailable minerals and vitamins in edible parts (i.e., the endosperm of cereals), which usually lack these components (18,33,34,41,49,50).

8.3 MINERALS

In developing nations, cereal grains such as wheat, maize, rice, and sorghum, and some legumes such as common bean and soybean, are the primary and cheap sources of essential minerals as iron, zinc, and calcium (20,37,43). Minerals with chemical similarities can compete for transport proteins or other uptake mechanisms, as well as for chelating organic substances, hindering absorption (Table 8.3) (32,54,55). In fact, it has also been suggested that antinutritional factors that interfere with proper nutrient absorption and bioavailability account for a large proportion of world wide micronutrient deficiencies (40,55).

Thus, increasing the amount of bioavailable micronutrients in plant foods for human consumption by molecular biotechnology is a challenge that is not only important for developing countries, but also for many industrialized countries. Theoretically, it could be achieved by increasing the total level of micronutrients in the edible part of staple crops, such as cereals and pulses, while simultaneously increasing the concentration of compounds which promote their uptake, for example ascorbic acid, and by decreasing the concentration of chemicals that inhibit their absorption, such as phytic acid or some phenolic compounds (Table 8.3) (22,41,43,50).

8.3.1 Iron

Iron is both an essential micronutrient and a potential toxicant to cells; as such, it requires a highly sophisticated, coordinated, and complex set of regulatory mechanisms to meet the demands of cells as well as prevent excess accumulation (29). The human body requires Fe for the synthesis of the oxygen transport proteins hemoglobin and myoglobin and for the formation of heme enzymes and other Fe-containing enzymes that are particularly important for energy production, gene regulation, immune defense, regulation of cell growth and differentiation and thyroid function (Table 8.1) (28,29,43). The body normally regulates Fe absorption so as to replace the obligatory iron losses of about 1–1.5 mg per day. Thus, the body must be economical in its handling of iron, for example, when a red blood cell dies, its iron is reutilized, and excess level of iron can be stored by a specially designed protein, ferritin, which is used at times of increased iron metabolic requirements (28). In spite of these ingenious physiological approaches, iron deficiency is estimated to affect around 30% of the world population, making iron by far the most deficient nutrient worldwide. In general, the etiology of iron deficiency can be viewed as a negative balance between iron intake and iron loss. Whenever there is a rapid growth, as occurs during infancy, early childhood, adolescence, and pregnancy, positive iron balance is difficult to maintain (9,28). The blood volume expands in parallel with growth, with a corresponding increase in iron requirement (9).

Dietary iron is constituted by heme iron (animal origin) and nonheme iron (inorganic salts mainly from vegetal sources); the first one is absorbed by a distinct route and more efficiently than nonheme iron (Table 8.1). Usually, nonheme iron bioavailability is very low (less than 5–10%) due to its poor solubility and interaction with other diet components known as antinutrients (28,43). Inadequate absorption of this mineral will first lead to the mobilization of storage iron, and finally to lower hemoglobin levels or anemia (29). Iron deficiency is the most common cause of anemia and is usually due to inadequate dietary intake of bioavailable iron and/or excessive loss due to physiological conditions of parasitic infections (30,32). For example, dietary iron sources in developing countries consist mainly of nonheme iron. Because cereal and legume staples are rich in phytic acid, a potent inhibitor of mineral absorption, and in addition, the intake of foods that enhance nonheme iron absorption such as fruits, vegetables, and animal muscle tissues is often limited, these conditions may serve as major factors responsible for the anemia (Table 8.3) (21,22,43,47,49,55). The major consequences of this deficiency are poor pregnancy outcome, including increased mortality of mothers and children, reduced psychomotor and mental development in infants, decreased immune functions, tiredness, and poor work performance (25,26,30–32).

Increasing ferritin, the natural iron store protein, in food crops, has been suggested as an approach to raise iron levels and bioavailability. Ferritin is a multimeric iron storage protein, composed of 24 subunits, and has a molecular structure highly conserved among plants, animals, and bacteria (56,57). This protein is capable of storing up to 4500 Fe atoms in its central cavity, which are nontoxic, biologically available, and releases them when iron is required for metabolic functions. In fact, recent studies have demonstrated that iron from animal and plant ferritin can successfully be utilized by anemic rats and humans (56,58,59).

While staple food, such as corn and wheat flours are usually fortified with iron, rice grains present much hard problems and challenges. In addition, whole brown rice is barely consumed, and its commercial milling (polishing) produces considerable loss of micronutrients, up to 30% and 67% for zinc and iron, respectively, by eliminating its outer layers where these metals are accumulated (27).

With the aim of increasing the iron content and its bioavailability in rice, two different research groups have overexpressed a ferritin gene into its endosperm isolated from either common bean or soybean (Table 8.4) (60–62). In both cases the plant ferritin is produced at high levels and correctly accumulated in the cereal endosperm. Notably, the iron content of bean ferritin rice is 22.1 μg/g dry weight whereas soybean ferritin rice stores up to 31.8 μg/g dry weight, resulting in two- and threefold greater levels, respectively, than that of the corresponding untransformed crop (10–11 μg/g dry weight). A two to three times extra iron enrichment in ferritin in transgenic grains would appear to be of nutritional significance. In fact, a daily consumption of about 300 g of the iron-rich rice by an adult would be sufficient to provide 50–75% of the daily adult requirements for this mineral, which is about 13–15 mg (Table 8.1).

Recently, Vansconcelos et al. (63), a third distinct research group, also reported the expression of soybean ferritin gene, driven by the endosperm-specific glutelin promoter, leading to higher iron accumulation in transgenic *indica* rice seed than control grains, even after commercial milling. They selected as target the *indica* rice line IR68144-3B-2-2-3, an elite line, which presents high tolerance for tungro virus and an excellent grain quality, good yield and resistance for growing in mineral-poor soils, and high iron level in the crop (15–17 μg/g untransformed brown rice). Transgenic rice lines were obtained containing as much as 71 μg iron/g dry weight unpolished rice. This accounts for a 4.4-fold increase in iron compared with that of the control; a two- to threefold extra iron content of transformed rice with plant ferritins would already be of nutritional relevance as noted earlier (Tables 8.1, 8.4) (60–62). But when the iron levels of the rice grains (untransformed and transgenic) were assessed after the seeds were polished, they indicated that the highest iron content of transgenic lines ranged from 19–37 μg/g milled rice, versus control material of only 10 μg/g milled rice. This is the first report which shows that after commercial milling the iron concentration remains higher than that of that milled negative control, and even that of untransformed brown crop. These results with transgenic rice expressing a ferritin from either soybean or bean would imply that low iron concentration in food seeds may not result from low iron availability for transport, but rather from a lack of sequestering and storing capacity in the seeds.

In order to explore and test the potential benefit of iron-improved transgenic rice incorporating soybean ferritin in its edible tissue (60), a standard hemoglobin depletion bioassay was employed with anemic rats followed by complete diets having equivalent quantities of either iron as $FeSO_4$ (a popular compound used in anemic human beings in medical treatments) or bioengineered ferritin rice. Iron-rich rice diet was as effective as the diet containing $FeSO_4$, and it was shown that full recovery of anemia in rats occurred after 28 days of treatment with any of the iron sources (Table 8.4) (64).

It is generally agreed that nutrients are effectively utilized from breast milk and that breast-fed infants possess a lower prevalence of infections than those fed with commercial formula. Breast milk not only provides the infant with a well balanced supply of nutrients, but also several unique components that facilitate nutrient digestion and absorption, protection against pathogenic microorganisms, and promotion of healthy growth and development. It is believed that those benefits are due in part to milk proteins (41,65,66). One of such bioactive proteins is lactoferrin. Lactoferrin is an 80 kDa iron-binding glycoprotein belonging to the transferrin family and is found in elevated levels (1–2 g/l) in human milk. Proposed biological activities for this protein include antimicrobial properties, regulation and facilitating iron absorption, immune system modulation, cellular growth activity, and antivirus and anticancer activities (41).

Rice was used as an useful bioreactor to produce, in its edible endosperm, recombinant human lactoferrin to infant food because it presents a low allergenicity, and is likely a vehicle safer than transgenic microorganisms or animals. Therefore, a human milk lactoferrin linked to a rice glutelin 1 promoter was inserted into rice cells and a very high expression level was reached in a large scale field trial (5 g of recombinant human lactoferrin per kilogram of dehusked transgenic rice), being stable for four generations. In fact, the boosting expression of lactoferrin in rice endosperm turned this cereal grains pink, as a consequence of iron bound to lactoferrin (65). The gross nutrient composition of transgenic cereal was similar to that of nontransformed rice, except for a twofold increase in iron content (negative control, 8.7 μg/g dehusked rice; transgenic rice, 19.3 μg/g dehusked rice) probably because each molecule of lactoferrin is able to bind two Fe^{3+} ions (Table 8.4) (65).

Additionally, the lactoferrin purified from transgenic rice exhibited similar pI, antimicrobial activity against a human pathogen (i.e., inhibition of growth of enteropathogenic *Escherichia coli*, one of the most common causes of diarrhea in infants and children) and bind and release iron capacity at acidic gastric pH as those of native human lactoferrin (Table 8.4) (65,66). Lactoferrin-rich rice crops (as lactoferrin is bioactive and therefore, has the ability to store iron in a bioavailable manner) can be incorporated directly into infant formula or baby foods, and even to be consumed by people at any age; without purification

Table 8.4
Manipulation of selected micronutrients for human nutrition and nutraceutical uses by molecular biotechnology.

Micronutrient	Molecular Approach	Improved Plant	Results and Comments	Ref.
Mineral				
Iron	Expression of bean-ferritin- or soybean-ferritin	Rice	Both bean-ferritin- and soybean-ferritin-rice presented high levels of stored iron, in their edible endosperm being of nutritional and nutraceutical significance for fighting against anemia disease and other iron-related deficiencies. In fact, anemic rat fed with iron-rich transgenic rice had a full recovery from that disease, and thus a food with enhanced iron bioavailability.	60–63.
	Production of human milk lactoferrin	Rice	Overaccumulation of lactoferrin in rice endosperm, which turned pink; transgenic protein showed similar antimicrobial activity against a human pathogen and both bind capacity and release iron at acidic gastric pH as native human lactoferrin; also immune system modulation, cellular growth activity, antivirus and anticancer properties are expected	65, 66.
Phosphorus	Expression of microbial or plant phytase	Canola, soybean, rice, maize, wheat	Phytic acid reduced amount in crops; significant improvement in essential mineral bioavailability such as iron, calcium and zinc and in protein digestibility are expected. Low-phytic-acid maize enhanced iron absorption in humans that consumed tortillas whereas pigs fed with this cereal exhibited a higher phosphorus absorption	43, 47, 61, 68–71.
Vitamin				
Vitamin A	Production of carotenoid biosynthetic enzymes	Canola, rice, tomato	Enhanced content of provitamin A and other nutraceutical carotenoids such zeaxanthin	79–82.
Vitamin E	γ-Tocopherol methyltransferase expression	*Arabidopsis*	Increased levels of α-tocopherol, higher vitamin E activity, nutraceutical	90.
Vitamin C	Expression of rat L-gulono-γ-lactone oxidase or strawberry D-galacturonic acid reductase	Lettuce, *Arabidopsis*	Improved antioxidant and nutraceutical properties and iron bioavailability; reduction, or elimination of bisulfite utilization to prevent browning in leaf lettuce	94, 95.

this milk protein provides a convenient advantage over other heterologous protein expression systems (65,66).

These achievements have demonstrated the feasibility of producing ferritin or lactoferrin in a very important food crop as rice, as a cheap and good source of bioavailable iron (Table 8.4). However, further investigations are now needed to show biological usefulness in the human diet and their contribution to a solution to global problems of iron deficiency.

8.3.2 Zinc

Most of the zinc in the human body is in the bones and muscles. This mineral acts as a stabilizer of the structures of membranes and cellular components. Its biochemical function is as an essential component of a large number of Zn-dependent enzymes, especially in the synthesis and degradation of biomacromolecules such as carbohydrates, proteins, lipids, and nucleic acids, as well as wound healing (43). Because of a significant portion of cellular zinc is found in the nucleus, it seems that mechanistically this metal is participating in processes which include stabilization of chromatin structure, DNA replication and transcription by the activity of transcription factors and DNA and RNA polymerases, as well as playing a key role in DNA repair and programmed cell death (Table 8.1) (10).

These features give Zn an essential and unique role for healthy growth and development of human beings. In fact, its deficiency reduces appetite, growth, sexual maturity, and the immune defense system (22,30,43). Meat and seafood are good sources of Zn for people in industrialized nations, providing up to 70% of their requirements. However, its deficiency has just recently taken dimensions as a serious public health problem (10).

8.3.3 Calcium

Calcium is required for the normal growth and development of the bones. It accumulates at the rate of about 150 mg per day during human skeletal growth until genetically predetermined peak bone mass is reached in the early twenties. Bone mass is then stable until about 50 years of age in men or before menopause in women (43). After that time, Ca balance becomes negative and bone is lost from all skeletal sites, which is related with a marked rise in fracture rate in both sexes, but particularly in women. So, adequate Ca intake during adolescence is critical. Osteoporosis in old age is characterized by a microarchitectural deterioration of bone tissue, leading to increased bone fragility as well as to an increased risk of fractures (Table 8.1) (22,46). Also, during early childhood this mineral is of a significant concern as its deficiency can cause rickets (22).

Besides its structural role in humans, Ca plays major regulatory functions in several biochemical and physiological processes such as blood clotting, muscle contraction, cell division, transmission of nerve impulses, enzyme activity, cell membrane function and hormone secretion (37,46). Milk and dairy products are the most important sources of calcium, but again these foods are scarce in less developed countries; in Mexico maize tortillas are a good and important Ca source for great part of its population (20,39).

8.3.4 Phytic Acid

Seeds normally accumulate severalfold more phosphorus than that which is needed to support basic cellular functions. In the normal seed this excess P is incorporated into a single small molecule referred to as phytic acid (*myo*-inositol-1,2,3,4,5,6-hexa*kis*phosphate). Grain crops typically contain about 10 mg of phytic acid per gram of dry seed weight, representing about 60–85% of total P (47). Once synthesized, most phytic acid is deposited as a mixed phytate or phytin salt of K and Mg, although phytates also contain other mineral cations such as Fe and Zn. During germination, phytate salts are broken down by the action of phytases, releasing P and *myo*-inositol for use by the growing seedling. Phytic acid is a polyanion at physiological pH and an effective chelator of nutritionally important mineral cations (Ca, Zn, Mg, and Fe) (Table 8.3) (22,32,40,47,54,55,67). Once consumed in human foods or animal feeds, phytic acid binds to these minerals in the intestinal tract to form mixed and unavailable salts that are largely excreted. This phenomenon contributes to mineral deficiency in human populations (40). In addition, diverse investigations suggest that phytic acid may also react with proteins making them partially unavailable for human absorption (22,67). On the other hand, phytic acid can readily be degraded in cereal and legume foods by the addition of exogenous phytases either during food processing or during digestion, increasing mineral absorption dramatically (22,43,47,50). Other interesting approach that has been proposed is for an *in vivo* decrease of phytic acid levels through raising phytase activity in crops using genetic engineering. Also, it is important that the enzyme should be able to withstand the cooking temperatures that occur during food preparation. Thus, highly important agronomic crops such as canola, soybean, rice, wheat, and even tobacco seeds have been successfully transformed with fungal phytase genes, driving their expression by either constitutive or seed-specific promoters (Table 8.4) (43,50,61,68,69). Significant enhancement on mineral bioavailability is expected in transgenic crops synthesizing and accumulating functional microbial phytases. On the other hand, consumption of natural low phytic-acid mutant maize improved iron absorption in humans fed with maize-based diets, for example tortillas prepared with this type of maize. The phytic acid level was reduced significantly representing only 35% of that found in wild-type maize and outstandingly, absorption of iron from transgenic tortillas was nearly 50% greater than from normal tortillas (Table 8.4) (70). Moreover, when pigs were fed with low phytic-acid maize, a higher P bioavailability of up to 62% than that obtained from the nonmutant, wild-type grain, was observed (71).

8.4 VITAMINS AND NUTRACEUTICALS

Vitamins are defined as a diverse group of food-based essential organic substances (relatively small molecules but comparable in size to amino acids or sugars) that are not synthesized by the human body, but by plants and microorganisms. Therefore, vitamins are nutritionally essential micronutrient for humans and function *in vivo* in several ways, including: (1) as coenzymes or their precursors (niacin, thiamin, riboflavin, biotin, pantothenic acid, vitamin B6, vitamin B12, and foliate); (2) in specialized functions such as vitamin A in vision and ascorbate in distinct hydroxylation reactions; and (3) as components of the antioxidative defense systems (vitamins C and E and some carotenoids), and as factors involved in human genetic regulation and genomic stability (folic acid, vitamin B12, vitamin B6, niacin, vitamin C, vitamin E, and vitamin D) (Table 8.2) (5,6,42).

Vitamins present in a food source, once taken up by the body, are dissolved either in water or fat (37,42). As a consequence vitamins are classified on the basis of their solubility as water-soluble or fat-soluble vitamins. Vitamins and the chemical structures of each group are presented in Figures 8.1 and 8.2.

8.4.1 Carotenoids as Food Pigment and Provitamin A

The term carotenoids summarizes a class of structurally related compounds, which are mainly found in plants, algae, and several lower organisms, bacteria, and fungi. At present, more than 600 different carotenoids have been identified (72). Saffron, pepper, leaves, and red palm oil possessing carotenoids as their main color components, have been exploited as food colors for several centuries. The color of carotenoids, together with beneficial properties such as vitamin A precursor and antioxidant activity, has led to their wide application in the food industry. They have been used for pigmentation of margarine, butter, fruit juices and

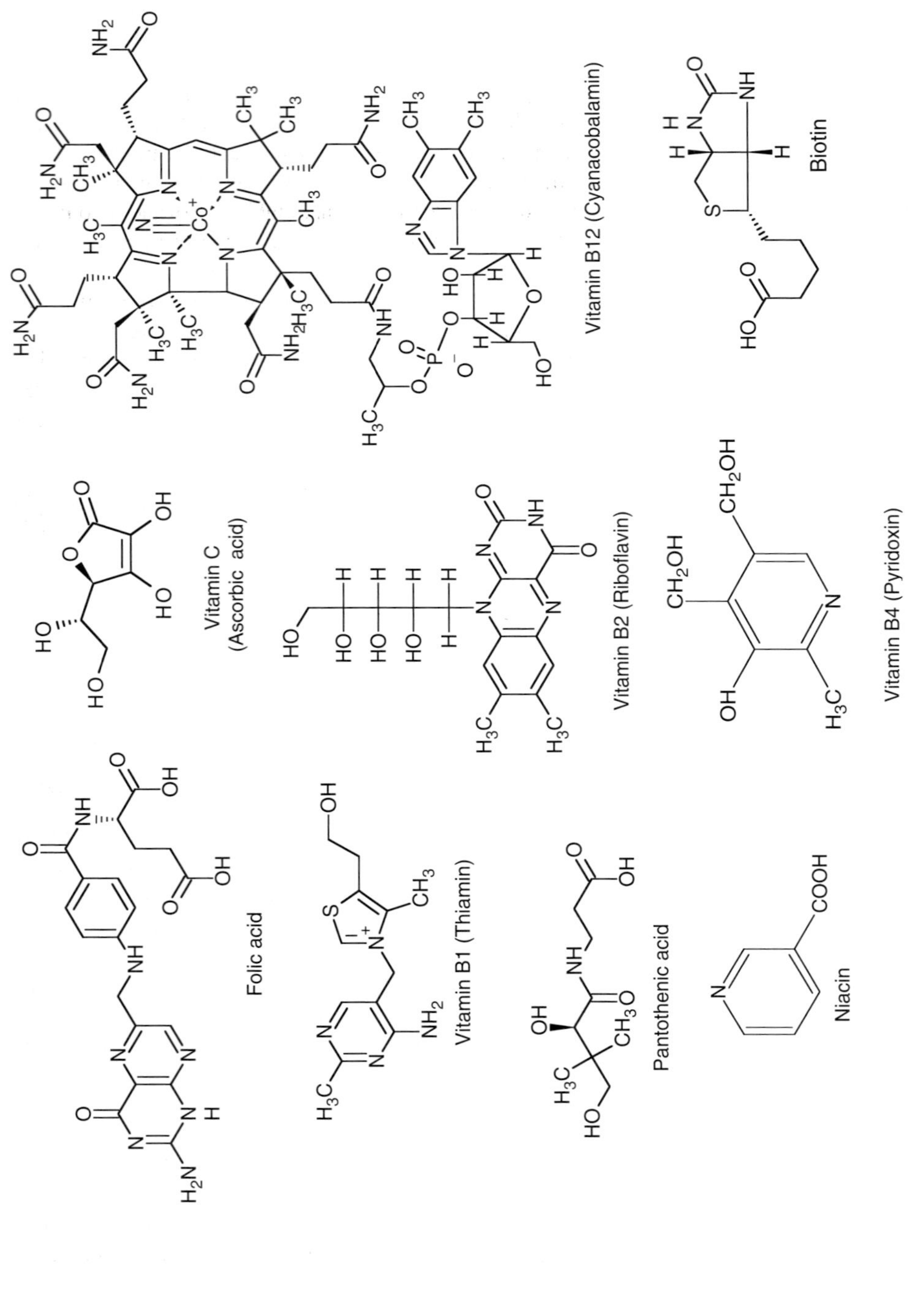

Figure 8.1 Chemical structures of water soluble vitamins.

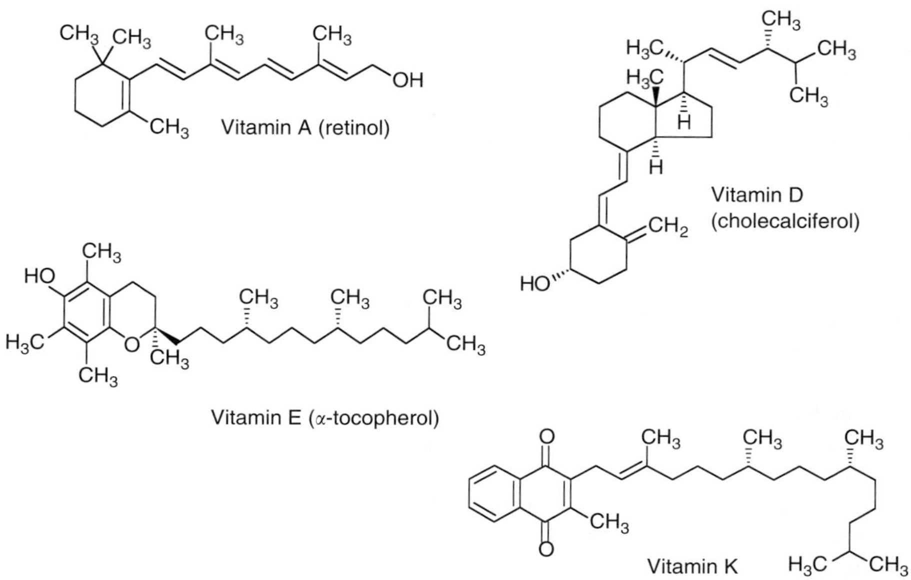

Figure 8.2 Chemical structures of fat soluble vitamins.

beverages, canned soups, dairy and related products, desserts and mixes, preserves and syrups, sugar and flour confectionery, salad dressings, meat, pasta and egg products (17,73).

Carotenoids form one of the most important classes of plant pigments and play a crucial role in defining the quality parameters of fruits and vegetables. Their function as antioxidants in plants in the prevention of photo oxidative damage shows interesting parallels with their potential roles as antioxidants in foods and humans (72). Carotenoids as food colorants take advantage of their very high stability to pH and reducing agents (ascorbic acid); their colors fluctuate from yellow to red, including orange, with variations of brown and purple (17).

Epidemiological studies suggest that the onset of chronic disease states such as coronary heart disease, certain cancers and macular degeneration can be reduced by high dietary intakes of carotenoid-rich foods (72–74). Lycopene has been shown to prevent the incidence of prostate cancer, while zeaxanthin and lutein offer protection against the occurrence of age-related macular degeneration. In addition, β-carotene is the most potent provitamin A carotenoid; α-carotene and β-cryptoxanthin also possess provitamin A activity, but to a lesser extent than β-carotene. Their deficiency can result in blindness and premature death (Table 8.2) (17,74).

Marigold (*Tagetes erecta* L.) varieties range in petal color from white to dark orange. This pigmentation of flowers is due to the massive synthesis and accumulation of carotenoids during petal development. The dark orange varieties possess concentrations of carotenoids that are 20-fold greater than those found in ripe tomato fruit (75). For this reason, marigolds are grown commercially as an important source of carotenoids and used as an animal feed supplement (73).

In our laboratory, Delgado-Vargas et al. (76) found that sunlight illumination produces a favorable equilibrium toward all *trans*-lutein isomers. Also, *trans*-isomers showed increased red hues, and the poultry fed with sunlight-illuminated marigold meal (lutein as its main carotenoid) showed a better egg yolk-pigmentation than controls. These authors also suggested that other components different from carotenoids may participate in pigmentation efficiency.

Also, the antimutagenicity activity of carotenoids from Aztec marigold has been evaluated. It was concluded that lutein was the compound with a higher activity in marigold extracts; however, the mixture of carotenoids in the marigold extract had higher antimutagenicity activity. In addition, it was suggested that lutein and 1-nitropyrene (mutagen) formed an extracellular complex that limits the bioavailability of 1-nitropyrene and consequently its mutagenicity (77).

8.4.1.1 Carotenogenesis in Tomato

In general, carotenoids are compounds comprised of eight isoprenoid units whose order is inverted at the center of the molecule. Formally, all carotenoids can be considered as lycopene ($C_{40}H_{56}$) derivatives by reactions involving hydrogenation, dehydrogenation, cyclization of the ends of the molecules, or oxidation (Figure 8.3A), which in addition to the number and location of conjugated double bonds inside their structure, influence biological action (antioxidant properties) and pigmentation efficiencies of colorants in food systems (17,74,76,78).

Based on their chemical composition, carotenoids are subdivided into two groups. Those which contain only carbon and hydrogen atoms are collectively assigned as carotene (β-carotene, α-carotene, and lycopene). The majorities of natural carotenoids have at least one oxygen function, such as keto, hydroxy, or epoxy groups and are referred to as xanthophylls (oxy-carotenoids) (17,72,74,76,78). The first step in the carotenoid biosynthetic pathway is the head-to-head condensation of two molecules of geranylgeranyl pyrophosphate (GGPP) to produce the colorless intermediate phytoene (Figure 8.3A) (17,74,75,78,79). This reaction is catalyzed by the enzyme phytoene synthase (PSY). Introduction of four double bonds convert phytoene to red lycopene. In plants, two enzymes carry out those transformations, phytoene desaturase (PDS), and ζ-carotene desaturase (ZDS). Each desaturase catalyzes two symmetric dehydrogenation steps; thus, the first enzyme catalyzes the conversion of ζ-carotene via phytofluene (phytoene to phytofluene to ζ-carotene) whereas the second transforms ζ-carotene to lycopene through neurosporene. In contrast, the bacterial phytoene desaturase (ctrl) is capable of introducing the four double bonds converting phytoene to lycopene. Subsequently, the ends of the linear lycopene can be cyclized by lycopene cyclases, resulting in the formation of β-carotene and α-carotene. Hydroxylation reactions carried out by hydroxylases [β-carotene hydroxylase (β-*Chy*) and ε-hydroxylase (ε-*Chy*)] produce the xanthophylls β-cryptoxanthin, zeaxanthin, and lutein (Figure 8.3A) (17,74,75,78,79).

Vitamin A refers to a family of essential, fat-soluble dietary compounds required for vision, growth, reproduction, cell proliferation, cell differentiation, and the integrity of the immune system (Table 8.2) (45). The most potent vitamin A compound, all-*trans*-retinol, is able to reverse signs and symptoms of vitamin A deficiency (44,45).

All vitamin A is originally derived from carotenoids, and all green plants produce carotenoids as β-carotene in their tissues as essential photosynthetic pigments. But for many staple crops, to have nutritional value from a micronutrient viewpoint, the carotenoids must also be produced in the nonphotosynthetic edible tissues consumed by humans; however traditional breeding methods have had little success in producing staple crops containing a high vitamin A concentration (17,22).

Research on carotenoid metabolic engineering in plants has focused on increasing nutritional quality, and it includes: (1) introducing variation in carotenoid products in tomato; (2) production of higher levels of preexisting carotenoids in canola; and (3) accumulation of carotenoids in normally carotenoid-free tissues, such as rice endosperm (Table 8.4) (78–83).

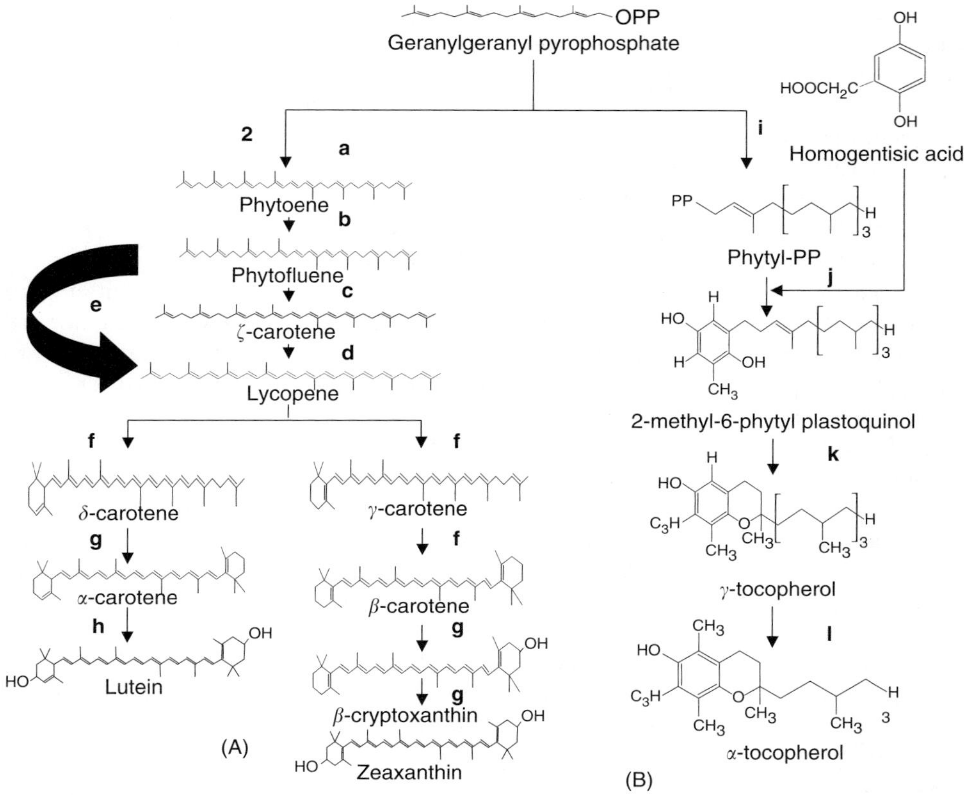

Figure 8.3 Summary of carotenoid (A) and vitamin E (B) biosynthesis pathway in plants. Enzymes carrying out specific reactions (a-h, for carotenoids and i-l, for vitamin E); are named as: a, phytoene synthase (PSY); b and c, phytoene desaturase (PDS); d, ζ-carotene desaturase (ZDS); e, the formation of lycopene through a bacterial phytoene desaturase (ctrl) is indicated with a black thick arrow; f, lycopene ε-cyclase, g, β-carotene hydroxylase (*β-Chy*); h, ε-hydroxylase (*ε-Chy*); i, hydrogenation enzyme; j, phytyl/prenyl transferase; k, methyl transferase 1 and cyclase; l, tocopherol γ-methyltransferases (γ-TMT). Adapted from Ref. 17, 21, 74, 75, 78, 79, 90.

Tomato fruit and its processed products are the principal dietary sources of lycopene, and also useful for their β-carotene content in minor amounts in the red varieties. At the green stage, tomato fruit has a carotenoid content similar to leaves (essentially β-carotene, lutein, and violaxanthin). But during fruit ripening, the genes mediating lycopene synthesis are up regulated whereas those controlling its cyclization are down regulated leading to over accumulation of lycopene in ripe fruit (up to 90% of total carotenoid, followed by β-carotene and only traces of lutein) (Figure 8.3A) (79).

Because the activity of the enzyme phytoene synthase (PSY) increases during tomato fruit ripening and when lycopene deposition occurs, it has been a preferred target to modify carotenoid profiles and levels in the fruit (Figure 8.3A). The constitutive expression of tomato PSY-1 in transgenic tomato plants led to dwarfism (plants with reduced size), due to redirecting GGPP from the gibberellins biosynthetic pathway into the carotenoid pathway, which reduced the amount of those hormones in the plants. Transgenic fruits produced lycopene earlier in development, but the final levels in the ripe fruits were lower than those in control samples, due to silencing of the endogenous gene (Table 8.4) (84). On the other hand,

Romer and coworkers (81) achieved a significant increment in the levels of β-carotene (provitamin A) in transgenic tomatoes by manipulating their desaturation activity. They produced transgenic lines overexpressing a PDS from *Erwinia uredovora* bacterium using a constitutive promoter; an enzyme capable of producing lycopene from phytoene (Figure 8.3A). Although total carotenoid concentration was not affected, ripe tomato fruits from transformed plants showed a threefold increase in β-carotene levels, accounting for up to 45% of total carotenoids. Tomato with improved provitamin A quality possessed about 5 mg/g fresh weight of all-*trans*-β-carotene (800 retinal equivalents), which satisfy 42% of the RDA, in contrast to normal fruit satisfying only 23% (Tables 8.2, 8.4).

Tomato is one of most productive based on carotenoid production per unit of cultivated area. By metabolic engineering, the large pool of red lycopene could be converted into high value added downstream carotenoids such as xanthophylls, which are an important type of target nutraceuticals, because of their antioxidant properties, their chemical stability and the difficulty in their chemical synthesis (79).

Recently, Dharmapuri et al. (79) reported the hyperexpression in tomato of an *Arabidopsis* β-lycopene cyclase (β-*Lcy*) gene with and without pepper β-carotene hydroxylase (β-*Chy)* gene under the control of the fruit-specific PD*S* promoter (Figure 8.3A). They found that the color of the ripe fruit varied from complete red as in wild-type tomato (by natural lycopene accumulation) to red-orange or to complete orange for transgenic tomatoes expressing only recombinant β-lycopene cyclase, or both recombinant β-lycopene cyclase and β-carotene hydroxylase enzymes, suggesting significant changes in carotenoid composition. The transformed fruits showed up to a 12-fold increase in β-carotene content with respect to their untransformed parental line. The transgenic tomato (containing genes β-*Lcy* + β-*Chy)* accumulated β-carotene as much as 63 μg/g fresh weight as compared to 5 μg/g fresh weight produced by the untransformed fruit. Notably, modified tomato also stored very good levels of both nutraceutical xanthophylls; β-cryptoxanthin (11 μg/g fresh weight) and zeaxanthin (13 μg/g fresh weight), but they are not detectable in control parent fruit or transformed tomato with only β-*Lcy* gene (Table 8.4). These results demonstrate that β-carotene pool can be converted into xanthophylls by the overexpression of pepper β-*Chy* and thus adding nutraceutical and commercial value to tomato.

8.4.1.2 Carotenogenesis in Other Food Crops

Another successful genetic manipulation of carotenogenesis has been reported in canola (*Brassica napus*). The bacterial (*Erwinia uredovora*) phytoene synthase with a plastid targeting sequence was overexpressed in a seed-specific manner using a napin promoter from rapeseed (Figure 8.3A) (80). Transgenic embryos were visibly orange, as compared with those green from control seeds, and the mature seed exhibited up to 50-fold more carotenoids, mainly α- and β-carotene, both presenting activities of provitamin A. Carotenoid-rich transgenic canola seed reached up to 1.6 mg per gram of fresh weight and produced an oil with 2 mg of carotenoids per gram of oil (Tables 8.2, 8.4).

Worldwide, vitamin A deficiency causes visible eye damage in around 3 million preschool children and up to 500 thousand of those children become partially or totally blind each year because of this deficiency, and approximately two thirds of them die within months of going blind. The estimates of the subclinical prevalence of vitamin A deficiency range between 100 and 200 million across the world. In developing countries several clinical trials have shown that vitamin A capsules can reduce mortality rates among preschool children by 23%, whereas improved vitamin A nutrition could prevent 1.2 million deaths annually among children aged 1–4 years (31). In addition, because vitamin A deficiency is common among vast populations of Asia, Africa, and South America, whose principal source of food is rice, engineering these crops to produce provitamin A-type carotenoids is of great importance (72,74).

Rice is generally consumed in its polished version, as commercial milling removes the oil-rich aleurone layer which becomes rancid during storage mainly in tropical and subtropical regions. Mature rice (*Oryza sativa*) endosperm is capable of synthesizing and accumulating GGPP but completely lacks carotenoids (Figure 8.3A) (85). In rice endosperm, the additional enzymatic activities needed to produce β-carotene was genetically engineered (82). To achieve this, four plant enzymes are necessary, but alternatively the number may be reduced to three as a bacterial carotene desaturase catalyzes the introduction of four double bonds required to produce lycopene (Figure 8.3A). Therefore three new genes (a plant phytoene synthase gene from daffodil, a bacterial phytoene desaturase gene from *Erwinia uredovora* and a lycopene β-cyclase also from daffodil plant) were transferred to rice. The first and third genes were driven by the endosperm-specific promoter of the rice glutelin gene, whereas a constitutive CaMV 35S promoter was used for the phytoene desaturase gene. Interestingly, transgenic rice seeds exhibited a beautiful golden yellow color after milling and increased accumulation of β-carotene in edible endosperm; as well as the xanthophylls, zeaxanthin, and lutein were formed to some extent, resulting in a carotenoid qualitative profile somehow analogous to that of green leaves. However, golden rice showed a higher proportion of β-carotene in their endosperm than that of the other two carotenoids, with a maximum amount of 1.6 μg/g dry weight; however, this quantity only represents 1–2% of carotenoid concentration in transgenic rapeseed (80) (Table 8.4). Nevertheless, it is noteworthy that in a typical Asian diet (about 300 g of rice per day), provitamin A-rich golden rice could provide nearly the full daily vitamin A requirement (85) (Table 8.2). It has been suggested that β-carotene accumulated in golden rice endosperm can be converted to retinol easier than β-carotene in vegetables, where provitamin A is converted to retinol at a rate equivalent to 26 to 1, due to the physicochemical properties of endosperm matrix (31,85). This significant accomplishment in plant biotechnology of improving the nutritional value of rice with beneficial carotenoids can benefit human nutrition and health (83). A future project aims to join iron-rich rice lines expressing ferritin with golden rice lines because it is known that provitamin A improves the iron bioavailability (Table 8.3) (85).

Nopal cactus (its young cladodes are called nopalitos) is low in calories and since Mesoamerican times has been eaten as a vegetable and fruit source in Mexico. At present, the economic and social importance of nopal is not only because large areas are covered with wild and commercial species in all arid and semiarid Mexican regions, but also because of its remarkable nutritional and nutraceutical qualities. For example, the intake of broiled nopalitos improved glucose control in adult people with noninsulin-dependent diabetes mellitus; it is also known that nopal diminishes human cholesterol levels (35,86). On the other hand, nopal is a good source of dietary fiber, calcium, iron, zinc, and vitamin C, and thereby can also be used in treatments against scurvy. However, although β-carotene is accumulated in nopal, its amount is very low; in fact its provitamin A level depends on nopalito development stage and nopal variety (35,86).

In our laboratory, Paredes-López et al. (87) have developed a genetic transformation system for nopal cactus (*Opuntia* sp.) through *Agrobacterium tumefaciens*, obtaining regenerated transgenic nopal plants. Thus, it has been proposed to use the nopal plant as a bioreactor for improving the production and storage of provitamin A by introducing additional genes for carotenoid biosynthesis to obtain β-carotene (P. Garcia-Saucedo, O. Paredes-López, personal communication, 2004). Nopal could become an important, cheap, and accessible source of vitamin A, and other nutraceuticals such as lutein or zeaxanthin for a great part of the population in Mexico.

8.4.2 Vitamin E

The intense research efforts which have surrounded vitamin E, a lipid-soluble antioxidant, support the hypothesis that preventing free radical-mediated tissue damage, for example to cellular lipids, proteins, or DNA, may play a key role in decreasing or delaying the pathogenesis of a variety of degenerative diseases such as cardiovascular disease, cancer, inflammatory diseases, neurological disorders, cataract, and age-related macular degeneration, and a decline in the immune system function (Table 8.2) (8,21). It has been suggested that vitamin E supplementation of 100 to 400 international units (IU) or around 250 mg of α-tocopherol per day may help reduce the magnitude of occurrence of such health disorders. Vitamin E is represented by a family of structurally related compounds, eight of which are known to occur in nature, being isolated from vegetable oils and other plant materials. The eight naturally occurring compounds are α-, β-, γ-, and δ-tocopherol (which differ only in the number and position of methyl substituents on the aromatic ring), and α- β-, γ-, and δ-tocotrienols (8,33). Tocotrienols differ from the corresponding tocopherols in that the isoprenoid side chain is unsaturated at C3′, C7′, and C11′. The phenolic hydroxyl group is key for the antioxidant activity of vitamin E, as donation of hydrogen from this group stabilizes free radicals (Figure 8.2). The presence of at least one methyl group on the aromatic ring is also critical. α-tocopherol, with three methyl groups, is the most biologically active of all homologues occurring in nature as a single isomer, followed by β-tocopherol, γ-tocopherol, and δ-tocopherol (8,88). Changes in the isoprenoid side chain also influence vitamin E activity; this biological activity is defined in terms of equivalents of α-tocopherol (α-TE). (R, R, R)-α-tocopherol has an activity of 1 α-tocopherol (α-TE) equivalent per milligram of compound. The activities of (R, R, R)-β-, (R, R, R)-γ-, and (R, R, R)-δ-tocopherols are 0.5, 0.1 and 0.03 per mg of compound, respectively (8,21). Of the tocotrienols, only α-tocotrienol has significant biological activity (0.3 mg α-TE/mg). Lengthening or shortening the side chain results in a progressive loss of vitamin E activity. The previous information is based on all tocopherols which are absorbed to similar extents during digestion, however single (R, R, R)-α-tocopherol is successfully stored and distributed in the entire body, while the other species are not processed with the same efficiency. It has been estimated that one α-TE molecule is capable of protecting 2000 phospholipids (8,88).

The physiological role of vitamin E centers on its ability to react with and quench free radicals in cell membranes and other lipid environments, thereby preventing polyunsaturated fatty acids (PUFAs) from being damaged by lipid oxidation. An imbalance in the production of free radicals and the natural protective system of antioxidants may lead to oxidized products, able to harm tissue; in fact tissue damage due to free radicals has been associated to several human chronic diseases (Table 8.2) (4–8).

Refined and processed foods are usually exposed to light, heat, or metal ions that can cause structural degradation of their constituent lipids by triggering the process of lipid oxidation (42). The rate of lipid oxidation in a food depends on the concentration and type of PUFAs it contains, the amount and effectiveness of the antioxidants present in the food and the heating, processing and storage conditions to which it is subjected to. Tocopherols and tocotrienols are the most important natural antioxidants in fats and oils, acting as primary or chain breaking antioxidants by converting lipid radicals to more stable products (8,42). At normal oxygen pressure, the major lipid radical is the peroxyl radical (ROO°) which can be converted to a hydroperoxide (ROOH) by proton donors such as tocopherols and tocotrienols. The hydrogen is donated from their phenolic groups, stabilizing the radicals and stopping the propagation phase of the oxidative chain reaction.

Supplementation of α-tocopherol has been demonstrated to positively affect sensorial quality of meat as well as saving money (i.e., in the U.S. beef industry meat, color degradation provokes losses up to $1 billion each year). In poultry, the high tocopherol level increased the stability of its meat; whereas in the case of pork and beef, vitamin E protects against rancid flavor, odor, and discoloration improving shelf life of packaged meat (89).

However, traditional plant breeding and food processing technologies have not concerned themselves with maximizing the levels of tocopherols in the human diets and even in diets for domestic animals used for meat, and the supplementation is necessary both for nutritional reasons and for the protection of fat-rich foods against oxidative rancidity (8,42). Unfortunately, synthetic α-tocopherol used as supplement is a complex mixture of stereo-isomers with less biological activity than natural single (R, R, R)-α-tocopherol (16,21,88). Significant changes in the α-tocopherol levels of major edible crops are necessary because there is a growing body of evidence to suggest that the dietary intake of vitamin E is insufficient to protect against the long term health risks associated with oxidative stress (8,16,21,88,89). Normally, the tocopherol composition of cultivated sunflower (*Helianthus annuus* L.) seed is primarily α-tocopherol, 95–100% of the total tocopherol pool. However, two mutant sunflower lines have been identified with tocopherol compositions of 95% γ-tocopherol/5% α-tocopherol, and 50% β-tocopherol/50% α-tocopherol. Although these presumed tocopherol methylation mutants showed severe alterations in their tocopherol profiles in seeds, their overall levels do not differ significantly from those of wild-type sunflower (8,89). These results suggest that it should be possible to alter the tocopherol profile of different crop species by manipulating the expression of one or both tocopherol methyltransferases (TMT), without having a detrimental effect on the total tocopherol pool size (Figure 8.3B). An exquisite and noteworthy research in the context of increasing the overall level of vitamin E activity available to consumers from plant foods was carried out by Shintani and DellaPenna (90). By overexpressing γ-TMT (Figure 8.3B), it was possible to increase α-tocopherol content in the *Arabidopsis* seed to about 85–95% of the total tocopherol, as compared with levels of 1.1% α-tocopherol and 97% γ-tocopherol in the untransformed seeds. Transgenic *Arabidopsis* showed a vitamin E activity about nine times greater than the negative control (Table 8.4). The authors speculate that if γ-TMT activity is limiting in commercially important oilseed crops such as soybean, corn, and canola, all of which have low γ-tocopherol to α-tocopherol ratios, overexpressing the γ-TMT gene in these crops should also elevate α-tocopherol amounts and improve their nutraceutical value (8,16,21,33,34,89).

8.4.3 Vitamin C

Vitamin C is used in large scale as an antioxidant in food, animal feed, beverages, pharmaceutical formulations, and cosmetic applications (91). This water-soluble vitamin, defined as L-ascorbic acid (L-AA), structurally is one of the simplest vitamins (Figure 8.1); its oxidation product is termed dehydroascorbate. It is related to the C6 sugars, being the aldono-1,4-lactone of a hexonic acid (L-galactonic or L-gulonic acid) and contains an enediol group on carbons 2 and 3 (92). In animal metabolism, including that of humans, the biological functions of L-AA are centered on its antioxidant properties and on its role to modulate a number of important enzymatic reactions. Thus, generally it acts as an enzyme cofactor, free radical scavenger and donor and acceptor in electron transfer reactions. For example, L-AA is required for collagen synthesis, and consequently in the formation and maintenance of cartilage, bones, gums, skin, teeth, and wound healing. In fact, the Fe-dioxygenases involved in collagen biosynthesis need L-AA for maximal activity, where the function of L-AA is to keep the transition metal ion centers of these enzymes in a reduced form. In the disease scurvy, which is known to be the result of vitamin C deficiency, its symptoms are directly related with the inadequate collagen formation (Table 8.2) (12,92). This micronutrient is also crucial for the normal, and enhanced, functioning of immune system, and is required for carnitine synthesis. There is now strong evidence to link high intake dietary vitamin C with reduced risk for several oxidative stress-associated diseases such as cardiovascular diseases, various types of cancers, aging, neurodegenerative diseases, and cataract formation. Cataracts appear to be due to the oxidation of lens protein, and antioxidants such as vitamin C and E, and carotenoids seem to protect against cataracts and macular degeneration of eye in rodents and humans. On the other hand, increased oxidative damage from low vitamin C intake, chronic inflammation, smoking, or radiation, together with elevated levels of uracil in DNA, would be expected to lead to more double strand (chromosome) breaks in individuals who are deficient in both folate and antioxidant (Table 8.2) (4–7,12).

However, there is also large body literature on supplementation studies with vitamin C in humans using biomarkers of oxidative damage to DNA, lipids (its oxidation releases mutagenic aldehydes), and protein. Some studies suggest that blood cell saturation occurs at about 100 mg vitamin C/day and the evidence suggests that this level minimizes DNA damage. Both experimental and epidemiological data support that vitamin C provides protection against stomach cancer, a result that is plausible because of the role of oxidative damage from inflammation by *Helicobacter pylori* infection, which is the main risk factor for stomach cancer (4–7,12). Unfortunately, the plasma levels of L-AA in large sections of the population around the world are suboptimal for those health benefic effects of this vitamin; in fact about 15% of the population consumes less than half the RDA (60 mg/day) of ascorbate (Table 8.2).

It is thought that L-AA secreted in gastric juices in animals enhances the absorption of iron from plant foods through two mechanisms: by forming Fe(III) complexes and by reducing the less soluble Fe^{3+} to the more soluble and bioavailable Fe^{2+} valence state (Table 8.3) (12,43).

Plants and most animals (i.e., rats, dogs, cats) can synthesize their own vitamin C, but a few mammalian species, including primates, human beings, and guinea pigs have lost this capability, and thus entirely depend upon dietary sources to meet needs for this vital micronutrient. This deficiency has been localized to a lack of the terminal flavo-enzyme L-gulono-1,4-lactone oxidase (L-gulono-γ-lactone oxidase, [GuLO]); the gene encoding it was found in the human genome, but was not expressed due to the accumulation of various mutations. In vitamin C-producing animals, GuLO catalyzes the final reaction in the L-AA route corresponding to the oxidation of L-gulono-1,4-lactone, whereas in plants the enzyme L-galactono-1,4-lactone dehydrogenase employs L-galactono-1,4-lactone as a substrate for carrying out the terminal step in the vitamin C production (Figure 8.4) (92–95).

Vitamin C is the single most important specialty chemical manufactured in the world. The current world market of ascorbic acid is 60,000 to 70,000 metric tons each year and generates annual revenues in excess of US$ 500 million (91). But its industrial production is a lengthy procedure involving microbial fermentation and diverse chemical steps (38,91). Until quite recently, little focus has been given to improving the L-AA content of plant foods, either in terms of the amounts present in commercial crop varieties or in minimizing losses prior to consumption. Notably, plants and animal possess different pathway for synthesizing L-AA; the expression in transgenic lettuce plants of an animal cDNA encoding a rat GuLO under the control of CaMV 35S promoter led to accumulation up to seven times more L-AA than untransformed crops (the basal levels of L-AA varied among the three unmodified lettuce cultivars from 0.36–0.58 μmol/g fresh weight) (Table 8.4) (95). In food science and technology, vitamin C as well as bisulfites are used to prevent oxidation in peaches, potato chips, apples, potatoes, peanut butter, beer, fat, and oils (20,42). Therefore, in the future, L-AA-rich transgenic lettuce may diminish the commercial application of bisulfite to avoid browning of its leaf, as well as enhancing the nutritional and nutraceutical value of this food vegetable. A recent report also showed that the L-AA content of *Arabidopsis thaliana* (untransformed plants have a vitamin C level of about 2 μmol/g fresh weight) was increased two- and threefold by hyperexpression of a D-galacturonic acid reductase gene from strawberry

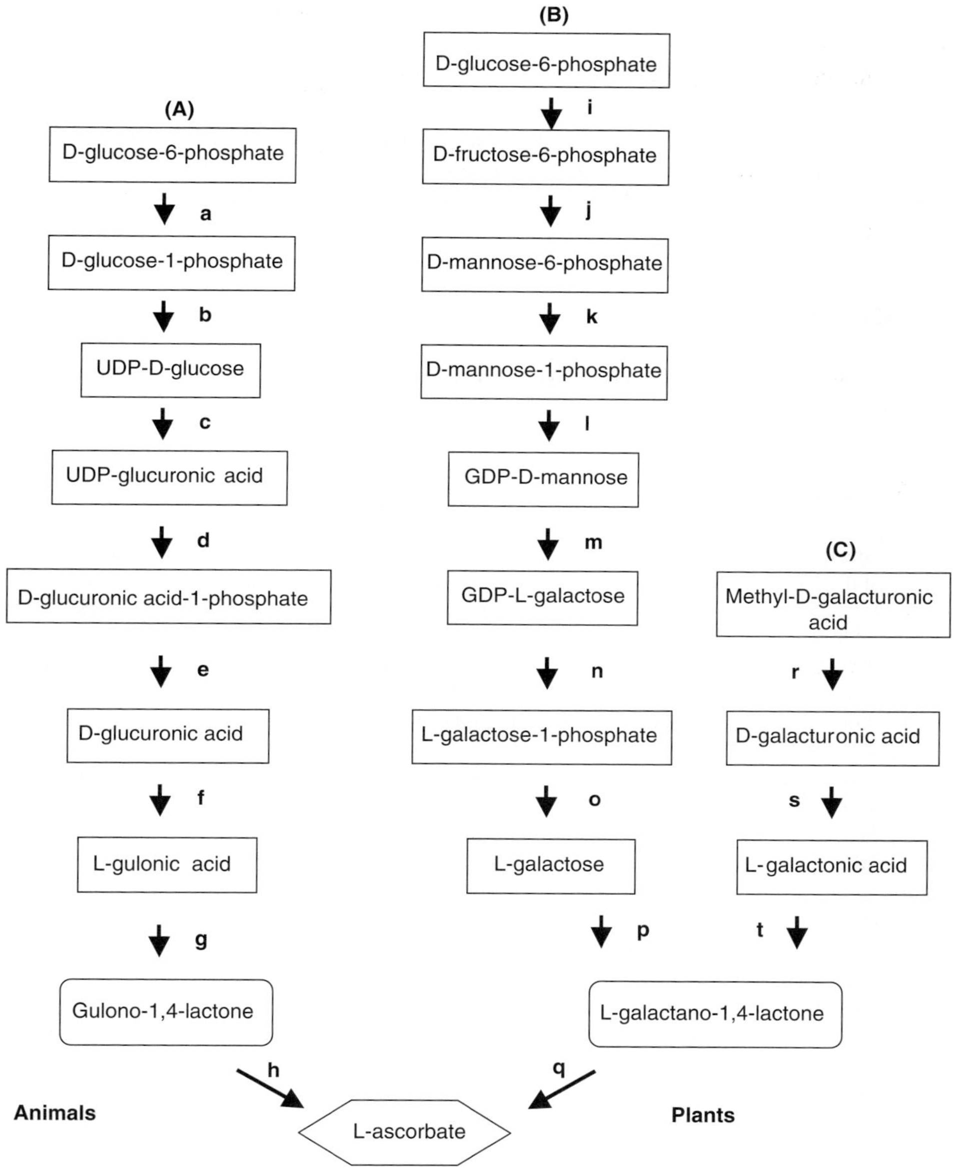

Figure 8.4 Proposed pathways for L-ascorbic acid biosynthesis in animals (A) and plants (B and C). Enzymes catalyzing the individuals reactions (a-h, for animals; i-t, for plants) are given next: a, phosphoglucomutase; b, UDP-glucose pyrophosphorylase; c, UDP-glucose dehydrogenase; d, glucuronate-1-phosphate uridylyltransferase; e, glucurono kinase; f, glucuronate reductase; g, aldono-lactonase; h, L-gulono-1,4-lactone oxidase, GuLO; i, glucose-6-phosphate isomerase; j, mannose-6-phosphate isomerase; k, phosphomannomutase; l, GDP-mannose pyrophosphorylase; m, GDP-mannose-3,5-epimerase; n, phosphodiesterase; o, sugar phosphatase; p, L-galactose-1- dehydrogenase; q, L-galactono-1,4-lactone dehydrogenase, r, methylesterase, s, D-galacturonate reductase; t, aldono-lactonase. Adapted from Ref. 92–95.

(Table 8.4) (94). This gene encodes the enzyme D-galacturonate reductase that converts D-galacturonic acid into L-galactonic acid, which is readily transformed to L-galactono-1,4-lactone, the immediate precursor of L-AA (Figure 8.4C).

The previous works demonstrate the possibility that the basal content of minerals and vitamins in important food crops can be augmented by metabolic engineering, and therefore raise the realistic possibility that such increases may substantially benefit vulnerable populations in their daily dietary intakes, without the need for fortification or for a change in dietary habits as a whole.

ACKNOWLEDGMENTS

We acknowledge partial financial support from National Council of Science and Technology (CONACYT).

REFERENCES

1. Blackburn, G.L. Pasteur's quadrant and malnutrition. *Nature* 409:397–401, 2001.
2. Willet, W.C., M.J. Stampfer. Rebuilding the food pyramid. *Sci. Am.* 288:64–71, 2003.
3. Lachance, P.A. Overview of key nutrients: micronutrient aspects. *Nutr. Rev.* 56:S34–S39, 1998.
4. Ames, B.N. DNA damage from micronutrient deficiencies is likely to be a major cause of cancer. *Mut. Res.* 475:7–20, 2001.
5. Ames, B.N., P. Wakimoto. Are vitamin and mineral deficiencies a major cancer risk? *Nat. Rev.* 2:694–704, 2002.
6. Fenech, M., L.R. Ferguson. Vitamins/minerals and genomic stability in humans. *Mut. Res.* 475:1–6, 2001.
7. Fenech, M. Micronutrients and genomic stability: a new paradigm for recommended dietary allowances (RDAs). *Food Chem. Toxicol.* 40:1113–1117, 2002.
8. Bramley, P.M., I. Elmadfa, A. Kafatos, F.J. Kelly, Y. Manios, H.E. Roxborough, W. Shuch, P.J. Sheehy, K.H. Wagner. Vitamin E. *J. Sci. Food Agric.* 80:913–938, 2000.
9. De Freitas, J.M., R. Meneghini. Iron and its sensitive balance in the cell. *Mut. Res.* 475: 153–159, 2001.
10. Dreosti, I.E. Zinc and gene. *Mut. Res.* 475:161–167, 2001.
11. El-Bayoumy, K. The protective role of selenium on genetic damage and on cancer. *Mut. Res.* 475:123–139, 2001.
12. Halliwell, B. Vitamin C and genomic stability. *Mut. Res.* 475:29–35, 2001.
13. Trewavas, A., D. Stewart. Paradoxical effects of chemicals in the diet on health. *Curr. Opin. Plant Biol.* 6:185–190, 2003.
14. Collins, A.R. Carotenoids and genomic stability. *Mut. Res.* 475:21–28, 2001.
15. Kaur, C., H.C. Kapoor. Antioxidants in fruits and vegetables: the millennium's health. *Int. J. Food Sci. Technol.* 36:703–725, 2001.
16. DellaPenna, D. Nutritional genomics: manipulating plant micronutrients to improve human health. *Science* 285:375–379, 1999.
17. Delgado-Vargas, F., O. Paredes-López. *Natural Pigments for Food and Nutraceutical Uses*. Boca Raton, FL: CRC Press, 2003.
18. Smil, V. Magic beans. *Nature* 407:567, 2000.
19. National Research Council (U.S.), Food and Nutrition Board. *Recommended Dietary Allowances*, 10th ed. Washington: National Academy Press, 1989.
20. Guzmán-Maldonado, S.H., O. Paredes-López. Biotechnology for the improvement of nutritional quality of food crop plants. In: *Molecular Biotechnology for Plant Food Production*, Paredes-López, O., ed., Boca Raton, FL: CRC Press, 1999, pp 553–620.

21. Grusak, M.A., D. DellaPenna. Improving the nutrient composition of plants to enhance human nutrition and health. *Annu. Rev. Plant Physiol. Plant Mol. Biol.* 50:133–161, 1999.
22. Graham, R.D., R.M. Welch, H.E. Bouis. Addressing micronutrient malnutrition through enhancing the nutritional quality of staple foods: principles, perspectives and knowledge gaps. *Adv. Agron.* 70:77–142, 2001.
23. Cakmak, I. Plant nutrition research: priorities to meet human needs for food in sustainable ways. *Plant Soil* 247:3–24, 2002.
24. Welch, R.M. Breeding strategies for biofortified staple plant foods to reduce micronutrient malnutrition globally. *J. Nutr.* 132:495S–499S, 2002.
25. Welch, R.M., R.D. Graham. Breeding crops for enhanced micronutrient content. *Plant Soil* 245:205–214, 2002.
26. Grantham-McGregor, D.M., C.C. Ani. The role of micronutrients in psychomotor and cognitive development. *Brit. Med. Bull.* 55:511–527, 1999.
27. Welch, R.M. The impact of mineral nutrients in food crops on global human health. *Plant Soil*, 247:83–90, 2002.
28. Hallberg, I. Perspectives of nutritional iron deficiency. *Annu. Rev. Nutr.* 21:1–21, 2001.
29. Beard, J.L. Iron biology in immune function, muscle metabolism and neuronal functioning. *J. Nutr.* 131:568S–580S, 2001.
30. Bhan, M.J., H. Sommerfelt, T. Strand. Micronutrient deficiency in children. *Brit. J. Nutr.* 85:199S–203S, 2001.
31. Bouis, H.E. Plant breeding: a new tool for fighting micronutrient malnutrition. *J. Nutr.* 132:491S–494S, 2002.
32. Sandström, B. Micronutrient interactions: effects on absorption and bioavailability. *Brit. J. Nutr.* 85:181S–185S, 2001.
33. MA Grusak, M.A. Genomics-assisted plant improvement to benefit human nutrition and health. *Trends Plant. Sci.* 4:164–166, 1999.
34. Grusak, M.A. Phytochemicals in plants: genomics-assisted plant improvement for nutritional and health benefits. *Curr. Opin. Biotechnol.* 13:508–511, 2002.
35. Guzmán-Maldonado, S.H., O. Paredes-López. Functional products of indigenous plants to Latin America: amaranth, quinoa, common beans and botanicals. In: *Functional Foods: Biochemical and Processing Aspects,* G Mazza, ed., Boca Raton, FL: CRC Press 1978, pp 293–328.
36. Elliot, R., T.J. Ong. Nutritional genomics. *BMJ* 324:1438–1442, 2002.
37. Chrispeels, M.J., D.E. Sadava. *Plants, Genes, and Agriculture*. Sudbury, MA: Jones & Bartlett Publishers, Inc, 1994.
38. Osuna-Castro, J.A., O. Paredes-López. Introduction to molecular food biotechnology. In: *Food Science and Food Biotechnology,* Gutiérrez-López, G.F., G.V. Barbosa-Canovas, eds, Boca Raton FL: CRC Press, 2003, pp 1–62.
39. Osuna-Castro, J.A., O. Paredes-López. Mejoramiento de características y calidad alimentarias y nutracéuticas de plantas mediante biotecnología molecular; algunos ejemplos. In: *Fundamentos y Casos Exitosos de la Biotecnología,* Bolívar, F., ed. México: El Colegio Nacional, 2004, pp. 451–503.
40. King, J.A. Biotechnology: a solution for improving nutrient bioavailability. *Int. J. Vitam. Nutr. Res.* 72:7–12, 2002.
41. Lönnerdal, B. Genetically modified plants for improved trace element nutrition. *J. Nutr.* 133:1490S–1493S, 2003.
42. Gregory, J.F. III. Vitamins, 3rd ed. In: *Food Chemistry*, Fennema, O.R., ed., New York: Marcel Dekker, Inc., 1996, pp 531–616.
43. Frossard, E., M. Bucher, M. Machler, A. Mozafar, R. Hurrell. Potential for increasing the content and bioavailability of Fe, Zn and Ca in plants for human nutrition. *J. Sci. Food Agric.* 80:861–879, 2000.
44. Van Lieshout, M., C.E. West, R.B. Van Breemen. Isotopic tracer techniques for studying the bioavailability and bioefficacy of dietary carotenoids, particularly β-carotene, in humans. *Am. J. Clin. Nutr.* 77:12–28, 2003.

45. Dawson, M.I. The importance of vitamin A in nutrition. *Curr. Pharm. Des.* 6:311–325, 2000.
46. Miller, D.D. Minerals, 3rd ed. In: *Food Chemistry*, Fennema, O.R., ed. New York: Marcel Dekker, Inc., 1996, pp 617–649.
47. Raboy, V. Seeds for a better future: 'low phytate' grains help to overcome malnutrition and reduce pollution. *Trends Plant. Sci.* 6:458–462, 2001.
48. Hurrell, R.F. Fortification: overcoming technical and practical barriers. *J. Nutr.* 132:806S–812S, 2002.
49. King, J.A. Evaluating the impact of plant biofortification on human nutrition. *J. Nutr.* 132:511S–513S, 2002.
50. Lucca, P., R. Hurrell, I. Potrykus. Approaches to improving the bioavailability and level of iron in rice seeds. *J. Sci. Food Agric.* 81:828–834, 2001.
51. NE Borlaug,N.E. Ending world hunger: the promise of biotechnology and the threat of antiscience zealotry. *Plant Physiol.* 124:487–490, 2000.
52. Grusak, M.A. Enhancing mineral content in plant food products. *J. Am. Coll. Nutr.* 21:178S–183S, 2002.
53. Kochian, L.V., D.F. Garvin. Agricultural approaches to improving phytonutrient content in plants: an overview. *Nutr. Rev.* 2:S13–S18, 1999.
54. Carbonaro, M., G. Grant, M. Mattera, A. Aguzzi, A. Pusztai. Investigation of the mechanisms affecting Cu and Fe bioavailability from legumes. *Biol. Trace Elem. Res.* 84:181–196, 2001.
55. Sandberg, A.-S. Bioavailability of minerals in legumes. *Brit. J. Nutr.* 88:281S–285S, 2002.
56. Masuda, F., F. Goto, T. Yoshihara. A novel plant ferritin subunit from soybean that is related to a mechanism in iron release. *J. Biol. Chem.* 276:19575–19579, 2001.
57. Theil, E.C. Ferritin: at the crossroads of iron and oxygen metabolism. *J. Nutr.* 133:1549S–1553S, 2003.
58. Skikne, B., D. Fonz, S.R. Lynch, J.D. Cook. Bovine ferritin iron bioavailability in man. *Eur. J. Clin. Invest.* 27:228–233, 1997.
59. Murray-Kolb, L., R. Welch, E.C. Theil, J.L. Beard. Women with low iron stores absorb iron from soybeans. *Am. J. Clin. Nutr.* 77:180–184, 2003.
60. Goto, F., T. Yoshihara, N. Shigemoto, S. Toki, F. Takaiwa. Iron fortification of rice seed by the soybean ferritin gene. *Nat. Biotechnol.* 17:282–286, 1999.
61. Lucca, P., R. Hurrell, I. Potrykus. Genetic engineering approaches to improve the bioavailability and the level of iron in rice grains. *Theor. Appl. Genet.* 102:392–397, 2001.
62. Lucca, P., R. Hurrell, I. Potrykus. Fighting iron deficiency anemia with iron-rich rice. *J. Am. Coll. Nutr.* 21:184S–190S, 2002.
63. Vasconcelos, M., K. Datta, N. Oliva, M. Khalekuzzaman, L. Torrizo, S. Krishnan, M. Oliveira, F. Goto, S.K. Datta. Enhanced iron and zinc accumulation in transgenic rice with the *ferritin* gene. *Plant Sci.* 164:371–378, 2003.
64. Murray-Kolb, L., F. Takaiwa, F. Goto, T. Yoshihara, E. Theil, J.L. Beard. Transgenic rice is a source of iron for iron-depleted rats. *J. Nutr.* 132:957–960, 2002.
65. Nandi, S., Y.A. Suzuki, J. Huang, D. Yalda, P. Pham, L. Wu, G. Bartley, N. Huang, B. Lönnerdal. Expression of human lactoferrin in transgenic rice grains for the application in infant formula. *Plant Sci.* 163:713–722, 2002.
66. Suzuki, Y.A., S.L. Kelleher, D. Yalda, L. Wu, J. Huang, N. Huang, B. Lönnerdal. Expression, characterization and biologic activity of recombinant human lactoferrin in rice. *J. Pediatr. Gastroenterol. Nutr.*, 36:190–199, 2003.
67. Guzmán-Maldonado, S.H., J. Acosta-Gallegos, O. Paredes-López. Protein and mineral content of a novel collection of wild and weedy common bean (*Phaseolus vulgaris* L). *J. Sci. Food Agric.* 80:1874–1881, 2000.
68. Tramper, J. Modern biotechnology: food for thought. In: *Food Biotechnology*, Bielecki, S., J. Tramper, J. Polak, eds., Amsterdam: Elsevier Science BV, 2000, pp 3–12.
69. Ponstein, A.S., J.B. Bade, T.C. Verwoerd, L. Molendijk, J. Storms, R.F. Beudeker, J. Pen. Stable expression of phytase (*phyA*) in canola (*Brassica napus*) seeds: towards a commercial product. *Mol. Breed.* 10:31–44, 2002.

70. Mendoza, C., F.E. Viteri, B. Lonnerdal, K.A. Young, V. Raboy, K.H. Brown. Effect of genetically modified, low-phytic acid maize on absorption of iron from tortillas. *Am. J. Clin. Nutr.* 68:1123–1127, 1998.
71. Golovan, S.P., R.G. Meidinger, A. Ajakaiye, M. Cottrill, M.Z. Wiederkehr, D.J. Barney, C. Plante, J.W. Pollard, M.Z. Fan, M.A. Hayes, J. Laursen, J.P. Hjorth, R.R. Hacker, J.P. Phillips, C.W. Forsberg. Pigs expressing salivary phytase produce low-phosphorus manure. *Nat. Biotechnol.* 19:741-745, 2001.
72. Van den Berg, H., R. Faulks, H.F. Granado, J. Hirschberg, B. Olmedilla, G. Sandmann, S. Southon, W. Stahl. The potential for the improvement of carotenoid levels in foods and the likely systemic effects. *J. Sci. Food Agric.* 80:880–912, 2000.
73. Delgado-Vargas, F., A.R. Jiménez, O. Paredes-López. Natural pigments: carotenoids, anthocyanins, and betalains: characteristics, biosynthesis, processing, and stability. *Crit. Rev. Food Sci. Nutr.* 40:173–289, 2000.
74. Fraser, P.D., S. Romer, J.W. Kiano, C.A. Shipton, P.A. Mills, R. Drake, W. Schuch, P.M. Bramley. Elevation of carotenoids in tomato by genetic manipulation. *J. Sci. Food Agric.* 81:822–827, 2001.
75. Moehs, C.P., L. Tian, K.W. Osteryoung, D. DellaPenna. Analysis of carotenoid biosynthetic gene expression during marigold petal development. *Plant Mol. Biol.* 45:281–293, 2001.
76. Delgado-Vargas, F., O. Paredes-López, E. Avila-González. Effects of sunlight illumination of marigold flower meals on egg yolk pigmentation. *J. Agric. Food Chem.* 46:698–706, 1998.
77. González-de-Mejía, E., G. Loarca-Piña, M. Ramos-Gómez. Antimutagenicity of xanthophylls present in Aztec marigold (*Tagetes erecta*) against 1-nitropyrene. *Mut. Res.* 389 :219–226, 1997.
78. Sandmann, G. Genetic manipulation of carotenoid biosynthesis: Strategies, problems and achievements. *Trends Plant Sci.* 6:14–17, 2001.
79. Dharmapuri, S., C. Rosati, P. Pallara, R. Aquilani, F. Bouvier, B. Camara, G. Guiliano. Metabolic engineering of xanthophyll content in tomato fruits. *FEBS Lett.* 519:30–34, 2002.
80. Shewmaker, C.K., J.A. Sheehy, M. Daley, S. Colburn, D.Y. Ke. Seed specific overexpression of phytoene synthase: increase in carotenoids and other metabolic effects. *Plant J.* 20:401–412, 1999.
81. Romer, S., P.D. Fraser, J.W. Kiano, C.A. Shipton, N. Misawa, W. Schuch, P.M. Bramley. Elevation of the provitamin A content of transgenic tomato plants. *Nat. Biotechnol.* 18 :666-669, 2000.
82. Ye, X., S. Al-Babili, A. Klöti, J. Zhang, P. Lucca, P. Beyer, I. Potrykus. Engineering provitamin A (β-carotene) biosynthetic pathway into (carotenoid-free) rice endosperm. *Science* 287:303–305, 2000.
83. Potrykus, I. Golden rice and beyond. *Plant Physiol.* 125:1157–1161, 2001.
84. Fray, R.G., A. Wallace, P.D. Fraser, D. Valero, P. Hedden, P.M. Bramley, D. Grierson. Constitutive expression of a fruit phytoene synthase gene in transgenic tomatoes causes dwarfism by redirecting metabolites from the gibberellin pathway. *Plant J.* 8:693–701, 1995.
85. Beyer, P., S. Al-Babili, X. Ye, P. Lucca, P. Schaub, R. Welch, I. Potrykus. Golden rice: introducing the β-carotene biosynthesis pathway into rice endosperm by genetic engineering to defeat vitamin A deficiency. *J. Nutr.* 132:506S–510S, 2002.
86. Mizrahi, Y., A. Nerd, P. Nobel. Cacti as crops. *Horticul. Rev.* 18:291–321, 1997.
87. Paredes-López, O., H. Silos, J.L. Cabrera, Q. Rascón. Método para la transformación genética y regeneración de plantas transgénicas de nopal. Patente en trámite. México, D.F.: Agosto 9, 2000.
88. Hirschberg, J. Production of high-value compounds: carotenoids and vitamin E. *Curr. Opin. Biotechnol.* 10:186–191, 1999.
89. Rocheford, T.R., J.C. Wong, C.O. Egesel, R.J. Lambert. Enhancement of vitamin E levels in corn. *J. Am. Coll. Nutr.* 21:191S–198S, 2002.
90. Shintani, D., D. DellaPenna. Elevating the vitamin E content of plants through metabolic engineering. *Science*, 282:2098–2100, 1998.

91. Chotani, G., T. Dodge, A. Hsu, M. Kumar, R. LaDuca, D. Trimbur, W. Weyler, K. Sanford. The commercial production of chemicals using pathway engineering. *Biochim. Biophys. Acta*, 1543:434–455, 2000.
92. Davey, M.W., M. Van Montagu, D. Inze, M. Sanmartin, A. Kanellis, N. Smirnoff, I.J. Benzei, J.J. Strain, D. Favell, J. Fletcher. Plant L-ascorbic acid: chemistry, function, metabolism, bioavailabilty and effect of processing. *J. Sci. Food Agric.* 80:825–860, 2000.
93. Wheeler, G.L., M.A. Jones, N. Smirnoff. The biosynthetic pathway of vitamin C in higher plants. *Nature* 393:365–369, 1998.
94. Agius, F., R. González-Lamothe, J.L. Caballero, J. Muñoz-Blanco, M.A. Botella, V. Valpuesta. Engineering increased vitamin C levels in plants by overexpression of a D-galacturonic acid reductase. *Nat. Biotechnol.* 21:177–181, 2003.
95. Jain, A.K., CL Nessler. Metabolic engineering of an alternative pathway for ascorbic acid biosynthesis in plants. *Mol. Breed.* 6:73–78, 2000.

2.09

Potential Health Benefits of Soybean Isoflavonoids and Related Phenolic Antioxidants

Patrick P. McCue and Kalidas Shetty

CONTENTS

9.1 INTRODUCTION

Soybeans are a well-known dietary staple of Asian countries such as Japan and are now consumed worldwide (1). Unlike string beans or snap peas, soybeans cannot be eaten raw, because they contain trypsin inhibitors, which can disrupt digestion activities in the stomach, leading to cramping and associated discomfort. Soybeans are usually fermented to produce distinct cultural or ethnic foods or food ingredients, or sometimes are sprouted for use in salads, as are mungbean and alfalfa sprouts. In Japan, soybeans are sometimes blanched (i.e., boiled for about five minutes), cooled, and then consumed readily with beer as a snack. The Japanese also process soybeans to produce a protein rich meat substitute, tofu. In Indonesia, soybeans are fermented by a food grade fungus of the *Rhizopus* species to produce an alternative protein rich meat substitute called tempeh (1).

9.2 CONSUMPTION OF SOYBEAN AND REDUCED INCIDENCE OF DISEASE

East Asian populations that regularly consume soybeans as a part of their dietary intake seem to have lower incidences of cancers and oxidation linked diseases of old age than are prevalent in Western populations. Numerous epidemiological studies have demonstrated an association between the consumption of soybeans and improved health, particularly as a reduced risk for cancers or diseases, such as breast cancer, cardiovascular disease, and atherosclerosis (2–8). Consumption of soy foods has also been associated with a reduced risk of prostate cancer (9,10).

Although soybean protein was first suspected to potentiate the health-promoting benefits of soybean consumption, these properties have more recently been linked to the biological activities of a specific group of phenolic compounds, found mainly in soybeans, known as isoflavonoids (5). While the chemopreventive properties of purified and synthetic isoflavonoids have been heavily investigated, a fermented soybean extract was recently shown to perform better at reducing the incidence of mammary tumor risk than a similar mixture of its constituent isoflavonoids, suggesting that the food background may play a positive role in the chemopreventive actions associated with soybean consumption, in addition to that of isoflavonoids (11–16).

9.2.1 Soybean Isoflavonoids

Isoflavonoids are a unique subgroup of the flavanoids, one of the largest classes of plant phenolics, with more than 5000 compounds currently identified. Isoflavonoids are found mainly in soybeans, and possess a chemical structure that is similar to the hormone estrogen (1). The chief isoflavonoids found in soybeans are genistein and daidzein. Because their structures resemble estrogen and they can interact with the estrogen receptor, soybean isoflavonoids are sometimes referred to as phytoestrogens (17).

Isoflavonoids are flavanoid variants in which the location of one of the phenolic rings is shifted. As diphenolic secondary metabolites, isoflavonoids are synthesized from products of the shikimic acid and malonyl pathways in the fusion of a phenylpropanoid with three malonyl CoA residues (1). Isoflavonoid content in soybeans ranges from 0.14 to 1.53 mg/g, and in soy flour from 1.3 to 1.98 mg/g (18). The Japanese are estimated to consume 25–100 mg of isoflavonoids per day (19). Chinese women are estimated to consume 39 mg of isoflavonoids per day (20). The consumption of isoflavonoids in Western diets is much lower, at less than 1 mg/day in the U.S. and U.K. (18,21).

Phenolic compounds normally occur as glucoside bound moieties called glycones (22,23). However, it is the aglycone (glucoside free) form that is metabolically active (24). After consumption, probiotic enzymes in the intestine cleave the glycoside moieties from glycone isoflavonoids and release the biologically active health-promoting aglycone isoflavonoids. Aglycone phenolic compounds possess higher antioxidant activity and are absorbed faster in the intestines than glucoside bound forms (23,25,26). Interestingly, fermented soy foods are rich in phenolic aglycones due to microbial bioprocessing during fermentation (27,28). However, once inside the bloodstream, biologically active aglycone genistein travels to the liver were it is converted back into an inactive glycone (β-glucuronide) (24). Cellular glucuronidases must remove the glycone moiety before genistein can exert its biological activity (24).

Isoflavonoids have been well studied and possess numerous biological activities (1). For example, genistein possesses inhibitory activity against topoisomerase II, tyrosine kinase, NF-κB, cancer cell proliferation, and nonoxidative pentose–phosphate pathway ribose synthesis in cancer cells (29–33). Many of the health-promoting benefits of isoflavonoids have been linked to the ability of phenolics to serve as antioxidants (34–37).

9.2.2 Major Bioactivities of Soybean Isoflavonoids

9.2.2.1 Phytoestrogenic and Postmenopausal Activity

The structure of soybean isoflavonoids is uniquely similar to that of estrogen (17) and may account for their weak ability to act as agonists at estrogen receptors (38). Many have speculated that soybean isoflavonoids may be useful for the treatment of somatic, mood, and cognitive disturbances associated with the onset of menopause (39). Diet supplementation with soybean phytoestrogens has been reported to ameliorate hot flashes and other symptoms of menopause (40–43).

Soybean isoflavonoids may also have potential in natural chemoprevention therapies against long term health problems associated with menopause, particularly for osteoporosis (44–47). After menopause, the ovaries stop producing estrogen. Because estrogen positively affects the metabolism of calcium, lack of sufficient estrogen can lead to bone loss and osteoporosis (48). Hormone replacement therapy (HRT) can reduce bone loss and the risk of osteoporosis in postmenopausal women, but unfortunately appears to also increase the risk for certain estrogen linked cancers (49–51).

Current osteoporosis prevention research is focused on the development of estrogen-like compounds (selective estrogen receptor modulators, or SERMs) that can selectively act against bone loss without causing negative estrogenic action against the uterus (52). The soybean isoflavonoid genistein has shown SERM activity in ovariectomized mice (53). When provided at optimal dosages, soybean isoflavonoids (especially genistein and daidzein) have been shown to improve bone mass and reduce bone resorption (54,55).

9.2.2.2 Cancer Chemoprevention

Soybean isoflavonoids also possess various biological activities that may help to explain the cancer chemopreventive properties associated with the consumption of soybean foods (3,8,49,56). In *in vitro* studies, daidzein was reported to activate the catalase promoter, to stimulate caspase-3 and apoptosis, and to down regulate the activities of Bcl-2 and Bcl-xL (57,58). Genistein can stimulate p53, antioxidant enzyme activities, BRCA2, caspase-3 and apoptosis, and chloride efflux (59–63). Genistein has also been reported to suppress activation of NF-κB, matrix metalloproteinases, lipogenesis, and COX-2 (31,64–66).

The exact mechanism by which these compounds exert their chemopreventive properties is not yet clear.

9.2.2.3 Prevention of Cardiovascular Disease

Soybean consumption has also been linked to a reduced risk for cardiovascular disease (47). Addition of soybean to foods has been shown to result in reduced cholesterol (67). In 1999, the US Food and Drug Administration reported that the consumption of soy protein as part of a healthy diet could help reduce the risk of coronary heart disease by lowering blood cholesterol levels (68). Soy protein isolates typically contain soybean isoflavonoids, which are believed to be largely responsible for the health benefits assigned to soy protein. Related herbal flavanoids prevented *in vitro* platelet aggregation and *in vivo* thrombogenesis in mouse arteries (69). Inclusion of isoflavonoid rich soybean in diets was also reported to protect against coronary heart disease by causing reductions in blood lipids, oxidized LDL, homocysteine, and blood pressure (7).

9.2.3 Approaches Toward Isoflavone Enrichment of Soybean

9.2.3.1 Genetic Modulation of Soybean

Throughout recorded history, man has used conventional breeding and selection techniques to improve crop species for desired traits. When modern genetic engineering techniques became available, agricultural scientists sought to improve crop species' phenolic content through the use of genetic technologies. At first, progress was slow, as knowledge of the biosynthetic pathways responsible for producing beneficial phenolic phytochemicals was limited.

One of the most studied pathways is the anthocyanin biosynthesis pathway, as phenotypic changes in flower color aided genetic analysis and metabolic understanding. Study of anthocyanin biosynthesis has also aided in the understanding of isoflavonoid biosynthesis as both pathways share a dependence on substrate flux through the flavanoid biosynthetic pathway. The isoflavonoid biosynthetic pathway is now almost completely characterized and genetic manipulation techniques have matured enough that it is now possible to alter synthesis at many different stages (70).

Knowledge of key enzymes involved in isoflavonoid biosynthesis in legumes led to attempts to engineer isoflavonoid biosynthesis in nonlegumes through genetic manipulation, in order to expand the delivery of dietary isoflavonoids as well as to develop new sources for their isolation (1). Unfortunately, initial attempts to incorporate key enzymes involved in isoflavonoid synthesis, such as soybean isoflavone synthase (IFS) and alfalfa chalcone isomerase in *Arabidopsis*, corn, and tobacco resulted in little to no formation of genistein or daidzein, the major isoflavonoids produced by soybean (71–73). Significant accumulation of isoflavonoids in nonlegumes was thought to be hindered by limited activity of the introduced IFS enzyme, by precursor pool limitations, and by competition (flux partitioning) between IFS and other enzymes that use the flavanoid naringenin as a substrate (73,74). More recent attempts to engineer flavanoids in bacteria and increased isoflavonoid content in soybean have been more successful, with the latter largely by coengineering the suppression of the naringenin-utilizing enzyme flavanone-3-hydroxylase to block competing pathways (72,75).

Dietary safety of genetically engineered foods remains a major concern of potential consumers. Although genetically modified (GM) foods appear no more harmful than conventionally produced foods, concerns remain as to the safety of the newly added DNA, its gene product, the overall safety of the rest of the food, the potential toxicology of the expressed protein, potential changes in allergenicity, changes in nutrient composition, unintended effects that could give rise to allergens or cause toxicity, and the safety of antibiotic resistance marker encoded proteins included with the transgene (76).

9.2.3.2 Enrichment of Soybean Isoflavone Content via Nongenetic Approaches

The technological challenge of stably introducing a foreign gene into a food crop and having that gene product function as desired, the problems of controlling substrate flux partitioning to drive a desired biosynthetic pathway, and the potential risks posed by transgenic food crops are troublesome issues that have underscored the need for continued research on and development of nongenetic approaches for the enrichment of isoflavonoids in soybean foods and food ingredients (1). Major nongenetic approaches for increasing phenolic content in dietary plants include bioprocessing of soybean substrates and stimulation of the plant defense responses, which is known to result in the stimulation of phenolic synthesis. Both of these approaches could potentially be used to increase isoflavonoid content in soybeans.

Bioprocessing of various plant based foods by dietary fungus is a technology that has been used throughout history in the context of producing fermented foods, such as tempeh. Currently, this technology is being utilized in conjunction with specific dietary fungi to produce certain desired products. In this context, fungal bioprocessing (also known as solid-state fermentation) has been employed to enrich various solid food substrates such as grape pomace, cornmeal, mango, date, wheat bran, and wine for products such as protein, C/N ratio, β-glucan, and pectinase (77–81). The use of dietary fungal bioprocessing of fruit and legume food substrates for enrichment of aglycone phenolic antioxidants such as ellagic acid in cranberry and isoflavonoids from soybeans has been reported (27,28,82–84). Enhanced isoflavonoid content in soybeans and soybean meal following bioprocessing by *Aspergillus* species has also been reported (85,86). The number of different microbial species and substrates available for isoflavonoid and other phenolic enrichment by fungal bioprocessing is likely to grow as knowledge of dietary microbial species increases.

One of the results of elicitor mediated activation of the plant defense system is an increase in phenolic secondary metabolite biosynthesis (87,88). Exogenous application of salicylic acid, a phenolic metabolite thought to act as a chemical signal in the defense response system, can stimulate phenolic content in peas (89). Similarly, pure dietary phenolics and phenolic rich extracts have been reported to stimulate phenolic content in legumes (90–96). Further, application of bacteria, bacterial polysaccharides, and UV and microwave radiation (all of which are known inducers of plant defense responses) have been shown to stimulate plant phenolic content (96–99). The type of phenolics elicited by activation of the defense response is largely determined by the nature of the treated plant (i.e., phenolic profiles vary from plant to plant). Therefore, application of plant defense response elicitors to soybean may potentially stimulate higher isoflavonoid content without the need for genetic manipulation.

9.2.4 Toward a Model Mechanism for Action of Soybean Isoflavonoids and Related Phenolic Antioxidants against Cancer

9.2.4.1 Metabolism of Dietary Isoflavonoids

As stated earlier, genistein and daidzein occur in soybeans as glucoside conjugates that must be converted into aglycones to be metabolically active (22,24). In humans, this can be performed by intestinal flora. Interestingly, fermented soy foods are rich in phenolic aglycones due to microbial bioprocessing during fermentation, which may allow for rapid intestinal absorption upon consumption (27,28).

9.2.4.2 Control of Energy Metabolism and Oxidative Stress in a Healthy Cell

In cells, the production of energy adenosine triphosphate (ATP) occurs in mitochondria by reduced nicotinamide adenine dinucleotide (NADH)-mediated oxidative phosphorylation (oxPHOS) (100). The tricarboxylic acid (TCA) cycle produces NADH to support mitochondrial ATP synthesis. Although some reactive oxygen species (ROS) are generated during mitochondrial oxPHOS, cells possess an extensive antioxidant response system which operates to scavenge ROS and protect cellular components from oxidative damage (100,101).

Antioxidant enzymes play a key role in cellular antioxidant response (100). Chief among these enzymes are superoxide dismutase (SOD) and catalase. SOD converts superoxide (O_2^-) into hydrogen peroxide (H_2O_2), which is less reactive. Catalase converts H_2O_2 into water (H_2O) and oxygen (O_2). Manganese superoxide dismutase (MnSOD) occurs within mitochondria to protect the organelle from oxidative damage, while copper–zinc SOD (CuZnSOD) and catalase occur in the cytosol, both to protect cytosolic bound cellular components from oxidative damage and to maintain a proper redox environment, because redox imbalances can activate certain cellular activities (102).

It has been proposed that phenolic antioxidants may have chemopreventive potential through modulation of the antioxidant enzyme response through the proline linked pentose–phosphate pathway (103). Exogenous antioxidants, such as dietary plant phenolic compounds, have been shown to scavenge ROS in cells *in vitro* and may help protect cells against oxidative damage *in vivo* (103,104).

9.2.4.3 Control of Energy Metabolism and Oxidative Stress in a Tumorigenic Cell

Many of the diseases for which a reduced risk of incidence has been associated with soy food consumption are oxidation linked diseases, such as cancer and cardiovascular disease (100). Oxidation linked diseases have been linked to a general breakdown in the regulation of cellular activities (such as growth or energy production), an accumulation of ROS such as superoxide and hydrogen peroxide, a cellular redox imbalance, and accumulated oxidative damage in normal cells (106).

Evidence indicates that, similar to healthy cells, tumor cells obtain much of their ATP for energy requirements via NADH linked mitochondrial oxPHOS (107). However, in tumorigenic cells, energy generation is inefficient as mitochondrial respiration activities are defective compared to healthy cells (108–110). Interestingly, in many cancer cells the activity of glucose-6-phosphate dehydrogenase (G6PDH), the key regulatory enzyme of the oxPPP, is reduced by up to 90%, while the nonoxidative pentose–phosphate pathway (nonoxPPP) flux is increased, possibly to support higher glycolytic flux toward the TCA cycle to support additional demand for NADH (111,112).

In addition to an abnormal energy metabolism, tumor cells possess a reduced antioxidant response system. Catalase and CuZnSOD activities, important for controlling cytosolic ROS levels, are decreased in numerous cancer cell lines (113,114). Low activity of antioxidant enzymes leaves cancer cells particularly susceptible to increased oxidative damage upon ROS accumulation, and eventually apoptosis (cell death) or necrosis (113,114).

9.2.5 Stress Response and the Proline Linked Pentose–Phosphate Pathway

Healthy cells possess a mechanism that couples increased mitochondrial ATP synthesis to increased glucose flux through the oxPPP for biosynthetic substrates (glucose phosphates, reduced nicotinamide adenine dinucleotide phosphate ($NADPH_2$) that support cellular energy demands during times of stress (100). In this mechanism, mitochondrial ATP generation and oxPPP activity are coupled through the biosynthesis of proline (115). Proline biosynthesis provides a mechanism for the transfer of reducing equivalents from $NADPH_2$ into mitochondria (via proline oxidation) and is linked to glucose oxidation in the oxPPP by $NADPH_2$ turnover, a coupling known as the proline linked pentose–phosphate pathway (PL-PPP) (116). Notably, in this mechanism mitochondrial oxPHOS switches its dependence on NADH to proline as a source of reducing equivalents to support ATP synthesis while maintaining cellular $NADPH_2$ biosynthesis for anabolic reactions. Increased proline metabolism has been shown to stimulate oxPPP activity via NADP mediated redox regulation (117).

As the activity of several antioxidant enzymes depends on the availability of $NADPH_2$, activity of the cellular antioxidant response system in stressed cells may depend directly upon flux through the oxPPP and, therefore, indirectly upon the activity of a functional proline cycling mechanism (100). Increased oxPPP activity has been shown to protect cells against H_2O_2 and NO stresses (118,119). Similarly, a phagocyte derived ROS increase occurs during the immune response and is followed by an increase in G6PDH activity (120). G6PDH and oxPPP activity protected cells from oxidant and radiation induced apoptosis (121). Recently, antioxidant enzymes were found to be essential for protecting cells against ROS mediated damage (122).

9.3 A HYPOTHETICAL MODEL FOR THE CANCER CHEMOPREVENTIVE ACTION OF THE SOYBEAN ISOFLAVONE GENISTEIN

The mechanistic coupling of mitochondrial ATP generation and oxPPP through proline metabolism that supports normal cells during stress and which appears to be dysfunctional in many tumorigenic cells provides a foundation for a possible model mechanism for the chemopreventive action of phenolic antioxidants, such as genistein, against certain cancers (100).

When soybean is consumed, its main phenolic phytochemical, the isoflavone genistein, is converted by intestinal flora bioprocessing and subsequent hepatic activities into a β-glucuronide conjugate. In an interesting twist of fate, the expression of β-glucuronidase is low or not detected in normal tissues, but is high in tumors (123). Therefore, in tissues that take up the genistein-β-glucuronide from the bloodstream, free and biologically active genistein is most likely to occur in a higher amount in tumor cells than in healthy cells due to the selective nature of β-glucuronidase expression in these tissues.

Once activated by glucuronidase, free genistein may stimulate proline metabolism through the activation of proline dehydrogenase (PDH) via p53. Genistein can induce p53 expression in colorectal cancer cells (124). The mechanism for stimulation of p53 by genistein is not clear, but other dietary antioxidants can activate p53 by a ref1 dependent redox mechanism (125,126). The transcription factor p53 can activate PDH, as well as ROS and apoptosis in cancer cells (127–129).

Stimulation of mitochondrial PDH in the tumor cell by phenolic antioxidants such as genistein should cause a demand for proline (117) that would have several major metabolic effects: (1) increased mitochondrial oxPHOS supported by proline oxidation would generate increased ROS that could leak into the cytosol and damage essential cellular components; (2) proline would be shunted to the mitochondria and away from collagen biosynthesis, potentially crippling tumor growth, expansion, and proliferation activities; and (3) energy metabolism would be redirected toward proline mediated mitochondrial ATP synthesis and away from TCA linked, NADH mediated oxPHOS via activity of the PL-PPP. An occurrence of these metabolic effects would be uniquely detrimental to the functioning of a tumor cell because, by all reported indications, normal cellular metabolism, including the antioxidant response system, in tumor cells is dysfunctional at mitochondrial and cytosolic levels.

First, increasing the activity of PDH would drive mitochondrial oxPHOS toward ATP synthesis, and produce increased ROS levels in the process. In a normal cell, increased ROS production by increased mitochondrial activity would likely be countered by activation of the cellular antioxidant enzyme response system, but in cancer cells antioxidant enzymes appear to be less active and may be less able to adequately protect tumor cells from oxidative damage, and potentially promote apoptosis (109,110,122,130). Because tumor cell mitochondria are already dysfunctional, increased respiration activities activity driven by genistein stimulated PDH activity should produce increased amounts of cytosolic ROS into the cytosol, endangering numerous essential cytosolic components such as proteins and organelles. MnSOD is expressed at low levels in normal cells, but at high levels in tumors, perhaps in response to high ROS levels (131). However, expression of catalase, glutathione peroxidase (GPx), and CuZnSOD is greatly diminished, which may facilitate accumulation of excessive oxygen radicals and oxidative damage in tumor cells (109,110). Although dietary antioxidants, such as curcumin, ascorbic acid, and flavanoids, have been shown to stimulate antioxidants and important phase II antioxidant enzymes such as SOD and catalase, genistein did not stimulate catalase or SOD in prostate cancer (132–135). In fact, genistein and soy isoflavone extracts have been shown to stimulate caspase-3 and apoptosis in cancer cells (58,62). Thus, phenolic antioxidants such as genistein may stimulate apoptosis linked activities that support the generation and accumulation of oxygen radicals and oxidative damage in tumor cells. The action of genistein could be further enhanced synergistically by other soluble phenolics from a food system. This is further supported by recent evidence that fermented soymilk and whole soy extracts inhibit tumor growth better than genistein alone (16,136,137).

Second, the stimulation of PDH activity by genistein and synergistic phenolic profiles would cause a metabolic demand for proline as a reductant to support mitochondrial oxPHOS (energy production) and may derail tumor growth supporting collagen biosynthesis by redirecting a necessary substrate (e.g., proline). Genistein has been shown to inhibit the growth and proliferation of cancer cells *in vitro* (32). If genistein activates of mitochondrial PDH and the proline cycle as part of a key stress response mechanism, proline cycle demand for proline may have higher cellular priority than collagen biosynthesis during times of stress (117).

Finally, and perhaps most importantly, induction of mitochondrial PDH activity and proline cycling by genistein would shift cellular energy metabolism in the tumor cell away from NADH linked mitochondrial oxPHOS to a proline linked mitochondrial oxPHOS system (100). Dysfunctional tumor cell mitochondria may produce increased ROS generation via increased mitochondrial respiration that may lead to mitochondrial membrane damage, caspase activation, and eventually apoptosis. In support of this idea, genistein has been shown previously to inhibit nonoxPPP ribose synthesis and cell proliferation in cancer cells (32,33). In healthy cells, stress induced stimulation of the proline cycle also drives the oxPPP via G6PDH recycling of $NADPH_2$, and G6PDH (and thereby the stress induced PL-PPP) can be inhibited by high $NADPH_2$ levels. In contrast, in tumorigenic cells G6PDH is dysfunctional (with activity decreased by up to 90%) and is no longer inhibited by high $NADPH_2$ (105,117,38). Therefore, while genistein stimulated PDH activity may be supported by $NADPH_2$ cycling between proline biosynthesis and G6PDH activity, a dysfunctional G6PDH enzyme may not allow the tumor cell to disengage the stress response mechanism in the presence of high levels of $NADPH_2$. Because the dysfunctional enzyme operates at such a low efficiency (~10%) and likely does not produce high $NADPH_2$ levels, there may not be enough $NADPH_2$ produced to support the anabolic demands of the antioxidant enzyme response system which would further hinder the tumor cell to defend itself against oxPHOS derived ROS.

Further, genistein possesses other biological activities that may help to promote apoptosis in tumor cells. Aside from the stimulation of caspase-3 which could result from mitochondrial PDH activation, genistein can inhibit NF-κB (13), which can block apoptosis (139).

9.4 SUMMARY AND IMPLICATIONS

Starvation is an efficient way to kill a living organism, even a diseased cell. An effective strategy to starve a tumor cell could be to potentiate a switch in energy metabolism from a dysfunctional and inefficient TCA/NADH linked mechanism to an even more dysfunctional alternative pathway (100). In tumor cells, mitochondrial ATP-generating activities are known to be negatively altered and to function inefficiently (103,104,140). If the energy metabolism could somehow be forced to revert back to the dysfunctional mitochondria, the inefficiency of the system may starve the tumor cell of chemical energy (ATP) for cellular activities (100). For a tumor cell to knowingly switch a core metabolism to favor a pathway or mechanism that would be detrimental to its survival seems unlikely, but it is possible that just such an action may occur through the activation of an underlying key stress response mechanism that functions in normal, healthy cells, for which activation triggers may still remain even after the normal cell transitions into a tumorigenic cell (100).

Here we describe a potential mechanism by which phenolic antioxidants such as the isoflavone genistein may act to promote tumor cell death by inducing the diseased cell to switch its energy metabolism from growth-promoting TCA cycle/ NADH linked system to a dysfunctional proline linked system through the activation of a key stress response mechanism involving the PL-PPP, whereby the cell is essentially starved of chemical energy (100). As the PL-PPP in plants can be stimulated by dietary (89,92,93,41), the same may be true in animal (e.g., human) cells. We have described how p53 mediated activation of PDH by genistein may create a metabolic demand for proline, therein activating the PL-PPP and causing a shift in energy metabolism to facilitate mitochondrial ATP generation, even though components of the stress response mechanism in tumor cells may be dysfunctional. A mitochondrial demand for proline might divert proline away from collagen biosynthesis, which may explain the interrupted tumor growth and proliferation activities observed in cancer cells treated with phenolic compounds. It is unknown whether or not tumor cell PDH is dysfunctional (100).

ROS accumulation precipitated by increased activity of leaky mitochondria should stimulate activity of antioxidant enzymes and the oxPPP for $NADPH_2$ to support reductant cycling systems (100). However, in tumor cells G6PDH is also dysfunctional and inefficient, such that the generation of adequate $NADPH_2$ levels to support the demands of both proline synthesis and the antioxidant enzyme response system is unlikely. Further, as the dysfunctional G6PDH is no longer inactivated by high $NADPH_2$, it is likely that once the PL-PPP mediated stress response mechanism is activated, the tumor cell may not be able to disengage it, and may become locked in its fate and thus prevented from returning to TCA cycle/ NADH linked energy metabolism (100).

Further, our hypothetical model helps to explain how phenolic antioxidants such as genistein could have both detrimental effects against tumor cells and beneficial effects on healthy cells. In healthy cells, addition of a phenolic antioxidant that can stimulate mitochondrial activity and the antioxidant enzyme response system (i.e., SOD, catalase, GPx) may aid in protection against oxidative damage and thus promote healthy cellular conditions (102,142). Furthermore, as normal G6PDH can be inactivated by high $NADPH_2$ levels, the PL-PPP-mediated stress response mechanism can be turned off as needed, something that a tumor cell may be unable to do and which may, ultimately, lead to the death of the diseased cell (100).

Numerous dietary phenolic antioxidants have been shown to inhibit cancer cell growth and proliferation, but until now no overall cellular mechanism that integrates the many observed phenolic linked activities has been put forth. We believe our hypothetical model for the chemopreventive actions of the soybean isoflavonoid genistein as described here offers a logical mode of action for dietary phenolic antioxidants based on key cellular metabolism linked to alternative energy and redox management. This model merits further experimental investigation in conjunction with genetic and signal transduction mechanisms associated with cancer biology (107). As various dietary phenolic compounds have been reported to possess activities that may facilitate apoptosis, our model provides a core mechanism by which many of these compounds could act, and a mechanism that could be supported by other such anticancer properties, such as NF-κB inhibition and related signal pathways.

REFERENCES

1. McCue, P., K. Shetty. Health benefits of soy isoflavonoids and strategies for enhancement: a review. *Crit. Rev. Food Sci. Nutr.* 44:1–7, 2004.
2. Wu, A.H., R.G. Ziegler, P.L. Horn-Ross, A.M. Nomura, D.W. West, L.N. Kolonel, J.F. Rosenthal, R.N. Hoover, M.C. Pike. Tofu and risk of breast cancer in Asian-Americans. *Cancer Epidemiol. Biomarkers Prev*. 5:901–906, 1996.
3. Zheng, W., Q. Dai, L.J. Custer, X.O. Shu, W.Q. Wen, F. Jin, A.A. Franke. Urinary excretion of isoflavonoids and the risk of breast cancer. *Cancer Epidemiol. Biomarkers Prev*. 8:35–40, 1999.
4. Anderson, J.W., B.M. Smith, C.S. Washnock. Cardiovascular and renal benefits of dry bean and soybean intake. *Am. J. Clin. Nutr*. 70(Suppl):464S–474S, 1999.
5. Yamakoshi, J., M.K. Piskula, T. Izumi, K. Tobe, M. Saito, S. Kataoka, A. Obata, M. Kikuchi. Isoflavone aglycone-rich extract without soy protein attenuates atherosclerosis development in cholesterol-fed rabbits. *J. Nutr.* 130:1887–1893, 2000.
6. Dai, Q., A.A. Franke, F. Jin, X.O. Shu, J.R. Hebert, L.J. Custer, J. Cheng, Y.T. Gao, W. Zheng, Urinary excretion of phytoestrogens and risk of breast cancer among Chinese women in Shanghai. *Cancer Epidemiol. Biomarkers Prev.* 11:815–821, 2002.
7. Jenkins, D.J., C.W. Kendall, C.J. Jackson, P.W. Connelly, T. Parker, D. Faulkner, E. Vidgen, S.C. Cunnane, L.A. Leiter, R.G. Josse. Effects of high- and low-isoflavone soyfoods on blood lipids, oxidized LDL, homocysteine, and blood pressure in hyperlipidemic men and women. *Am. J. Clin. Nutr*. 76:365–372, 2002.
8. Yamamoto, S., T. Sobue, M. Kobayashi, S. Sasaki, S. Tsugane. Soy, isoflavones, and breast cancer risk in Japan. *J. Natl. Cancer Inst*. 95:906–913, 2003.
9. Jacobsen, B.K., S.F. Knutsen, G.E. Fraser. Does high soy milk intake reduce prostate cancer incidence?: the Adventist health study (United States). *Cancer Causes Control* 9:553–557, 1998.
10. Lee, M.M., S.L. Gomez, J.S. Chang, M. Wey, R.T. Wang, A.W. Hsing. Soy and isoflavone consumption in relation to prostate cancer risk in China. *Cancer Epidemiol. Biomarkers Prev.* 12:665–668, 2003.
11. Darbon, J.M., M. Penary, N. Escalas, F. Casagrande, F. Goubin-Gramatica, C. Baudouin, B. Ducommun. Distinct Chk2 activation pathways are triggered by genistein and DNA-damaging agents in human melanoma cells. *J. Biol. Chem*. 275:15363–15369, 2000.
12. Lamartiniere, C.A. Protection against breast cancer with genistein: a component of soy. *Am. J. Clin. Nutr*. 71:1705S–1707S, 2000.

13. Xu, J., G. Loo. Different effects of genistein on molecular markers related to apoptosis in two phenotypically dissimilar breast cancer cell lines. *J. Cell. Biochem.* 82(1):78–88, 2001.
14. Lamartiniere, C.A., M.S. Cotroneo, W.A. Fritz, J. Wang, R. Mentor-Marcel, E. Elgavish. Genistein chemoprevention: timing and mechanisms of action in murine mammary and prostate. *J. Nutr.* 132:552S–558S, 2002.
15. Tanos, V., A. Brzezinski, O. Drize, N. Strauss, T. Peretz. Synergistic inhibitory effects of genistein and tamoxifen on human dysplastic and malignant epithelial breast cancer cells *in vitro*. *Eur. J. Obstet. Gynecol. Reprod. Biol.* 102:188–194, 2002.
16. Ohta, T., S. Nakatsugi, K. Watanabe, T. Kawamori, F. Ishikawa, M. Morotomi, S. Sugie, T. Toda, T. Sugimura, K. Wakabayashi. Inhibitory effects of *Bifidobacterium*-fermented soy milk on 2-amino-1-methyl-6-phenylimidazo- [4, 5-*b*]-pyridine-induced rat mammary carcinogenesis, with a partial contribution of its component isoflavones. *Carcinogenesis* 21: 937–941, 2000.
17. Wuttke, W., H. Jarry, T. Becker, A. Schultens, V. Christoffel, C. Gorkow, D. Seidlova-Wuttke. Phytoestrogens: endrocrine disrupters or replacement for hormone replacement therapy? *Maturitas* 44(1):S9-S20, 2003.
18. Safford, B., A. Dickens, N. Halleron, D. Briggs, P. Carthew, V. Baker. A model to estimate the oestrogen receptor mediated effects from exposure to soy isoflavones in food. *Regul. Toxicol. Pharmacol.* 38:196–209, 2003.
19. Coward, L., N. Barnes, K.D. Setchell, S. Barnes. Genistein, daidzein, and the β-glycoside conjugates: anti-tumor isoflavones in soybean foods from American and Asian diets. *J. Agric. Food Chem.* 41:1961–1967, 1993.
20. Chen, Z., W. Zheng, L.J. Custer, Q. Dai, X.O. Shu, F. Jin, A.A. Franke. Usual dietary consumption of soy foods and its correlation with the excretion rate of isoflavonoids in overnight urine samples among Chinese women in Shanghai. *Nutr. Cancer* 33:82–87, 1999.
21. De Kleijn, M.J., Y.T. van der Schouw, P.W. Wilson, H. Adlercreutz, W. Mazur, D.E. Grobbee, P.F. Jacques. Intake of dietary phytoestrogens is low in postmenopausal women in the United States: the Framingham study. *J. Nutr.* 131:1826–1832, 2001.
22. Peterson, T.G., G.P. Ji, M. Kirk, L. Coward, C.N. Falany, S. Barnes. Metabolism of the isoflavones genistein and biochanin A in human breast cancer cell lines. *Am. J. Clin. Nutr.* 68:1505S–1511S, 1998.
23. Rao, M., G. Muralikrishna. Evaluation of the antioxidant properties of free and bound phenolic acids from native and malted finger millet (Ragi, Eleusine coracana Indaf-15). *J. Agric. Food Chem.* 50:889–892, 2002.
24. Yuan, L., C. Wagatsuma, M. Yoshida, T. Miura, T. Mukoda, H. Fujii, B. Sun, J.H. Kim, Y.J. Surh. Inhibition of human breast cancer growth by GCP™ (genistein combined polysaccharide) in xenogeneic athymic mice: involvement of genistein biotransformation by β-glucuronidase from tumor tissues. *Mutat. Res.* 523-524:55–62, 2003.
25. Murota, K., S. Shimizu, S. Miyamoto, T. Izumi, A. Obata, M. Kikuchi, J. Terao. Unique uptake and transport of isoflavone aglycones by human intestinal caco-2 cells: comparison of isoflavonoids and flavanoids. *J. Nutr.* 132:1956–1961, 2002.
26. Setchell, K.D., N.M. Brown, L. Zimmer-Nechemias, W.T. Brashear, B.E. Wolfe, A.S. Kirshner, J.E. Heubi. Evidence for lack of absorption of soy isoflavone glycosides in humans, supporting the crucial role of intestinal metabolism for bioavailability. *Am. J. Clin. Nutr.* 76:447–453, 2002.
27. McCue, P., A. Horii, K. Shetty. Solid-state bioconversion of phenolic antioxidants from defatted soybean powders by *Rhizopus oligosporus*: role of carbohydrate cleaving enzymes. *J. Food. Biochem.* 27(6):501–514, 2003.
28. McCue, P., K. Shetty. Role of carbohydrate-cleaving enzymes in phenolic antioxidant mobilization from whole soybean fermented with *Rhizopus oligosporus*. *Food Biotechnol.* 17:27–37, 2003.
29. Okura, A., H. Arakawa, H. Oka, T. Yoshinari, Y. Monden. Effect of genistein on topoisomerase activity and on the growth of [Val 12]Ha-ras-transformed NIH 3T3 cells. *Biochem. Biophys. Res. Commun.* 157:183–189, 1988.

30. Akiyama, T., J. Ishida, S. Nakagawa, H. Ogawara, S. Watanabe, N. Itoh, M. Shibuya, Y. Fukami. Genistein, a specific inhibitor of tyrosine-specific protein kinases. *J. Biol. Chem.* 262:5592–5595, 1987.
31. Gong, L., Y. Li, A. Nedeljkovic-Kurepa, F.H. Sarkar. Inactivation of NF-κB by genistein is mediated via Akt signaling pathway in breast cancer cells. *Oncogene* 22:4702–4709, 2003.
32. Wang, S.Y., K.W. Yang, Y.T. Hsu, C.L. Chang, Y.C. Yang. The differential inhibitory effects of genistein on the growth of cervical cancer cells *in vitro*. *Neoplasma* 48:227–233, 2001.
33. Boros, L.G., S. Bassilian, S. Lim, W.N.P. Lee. Genistein inhibits non-oxidative ribose synthesis in MIA pancreatic adenocarcinoma cells: a new mechanism of controlling cancer growth. *Pancreas* 22:1–7, 2001.
34. Arora, A., M.G. Nair, G.M. Strasburg. Antioxidant activities of isoflavones and their biological metabolites in a liposomal system. *Arch. Biochem. Biophys.* 356(2):133–141, 1998.
35. Hollman, P.C., M.B. Katan. Health effects and bioavailability of dietary flavonols. *Free Radic. Res.* 31:S75-S80, 1999.
36. Barnes, S., B. Boersma, R. Patel, M. Kirk, V.M. Darley-Usmar, H. Kim, J. Xu. Isoflavonoids and chronic disease: mechanisms of action. *Biofactors* 12(1-4):209-215, 2000.
37. Cos, P., M. Calomme, J.B. Sindambiwe, T. De Bruyne, K. Cimanga, L. Pieters, A.J. Vlietinck, D. Van den Berghe. Cytotoxicity and lipid peroxidation-inhibiting activity of flavanoids. *Planta. Med.* 67(6):515–519, 2001.
38. Pike, A.C., A.M. Brzozowski, R.E. Hubbard, T. Bonn, A.G. Thorsell, O. Engstrom, J. Ljunggren, J.A. Gustafsson, M. Carlquist. Structure of the ligand-binding domain of oestrogen receptor beta in the presence of a partial agonist and a full antagonist. *EMBO J.* 18(17):4608–4618, 1999.
39. Hochanadel, G., J. Shifren, I. Zhdanova, T. Maher, P.A. Spiers. Soy isoflavones (phytoestrogens) in the treatment of the cognitive and somatic symptoms of menopause. *PCRS Abtracts* 71(4,1):20S, 1999.
40. Murkies, A.L., C. Lombard, B.J. Strauss, G. Wilcox, H.G. Burger, M.S. Morton. Dietary flour supplementation decreases post-menopausal hot flushes: effect of soy and wheat. *Maturitas* 21(3):189–195, 1995.
41. Dalais, F.S., G.E. Rice, A.L. Murkies, R.J. Bell, M.L. Wahlqvist. Effects of dietary phytoestrogens in postmenopausal women. *Maturitas* 27(1):214, 1997.
42. Albertazzi, P., F. Pansini, G. Bonaccorsi, L. Zanotti, E. Forini, D. De Aloysio. The effect of dietary soy supplementation on hot flushes. *Obstet Gynecol* 91(1):6–11, 1998.
43. Carusi, D. Phytoestrogens as hormone replacement therapy: an evidence-based approach. *Primary Care Update OB/GYNS* 7(6):253–259, 2000.
44. Arjmandi, B.H., R. Birnbaum, N.V. Goyal, M.J. Getlinger, S. Juma, L. Alekel, C.M. Hasler, M.L. Drum, B.W. Hollis, S.C. Kukreja. Bone-sparing effect of soy protein in ovarian hormone-deficient rats is related to its isoflavone content. *Am. J. Clin. Nutr.* 68(6):1364S–1368S, 1998.
45. S.M. Potter, J.A. Baum, H. Teng, R.J. Stillman, N.F. Shay, J.W. Erdman. Soy protein and isoflavones: their effects on blood lipids and bone density in postmenopausal women. *Am. J. Clin. Nutr.* 68(6):1375S–1379S, 1998.
46. J.J. Anderson, M.S. Anthony, J.M. Cline, S.A. Washburn, S.C. Garner. Health potential of soy isoflavones for menopausal women. *Public Health Nutr.* 2(4):489–504, 1999.
47. Scheiber, M.D., J.H. Liu, M.T. Subbiah, R.W. Rebar, K.D. Setchell. Dietary inclusion of whole soy foods results in significant reductions in clinical risk factors for osteoporosis and cardiovascular disease in normal postmenopausal women. *Menopause* 8(5):384–392, 2001.
48. Gallagher, J.C. Role of estrogens in the management of postmenopausal bone loss. *Rheum. Dis. Clin. North. Am.* 27:143–162, 2001.
49. Goodman, M.T., L.R. Wilkens, J.H. Hankin, L.C. Lyu, A.H. Wu, L.N. Kolonel. Association of soy and fiber consumption with the risk of endometrial cancer. *Am. J. Epidemiol.* 146(4):294–306, 1997.
50. Morishige, K., K. Matsumoto, M. Ohmichi, Y. Nishio, K. Adachi, J. Hayakawa, K. Nukui, K. Tasaka, H. Kurachi, Y. Murata. Clinical features affecting the results of estrogen

replacement therapy on bone density in Japanese postmenopausal women. *Gynecol. Obstet. Invest.* 52(4):223–226, 2001.

51. Lacey, J.V., P.J. Mink, J.H. Lubin, M.E. Sherman, R. Troisi, P. Hartge, A. Schatzkin, C. Schairer. Menopausal hormone replacement therapy and risk of ovarian cancer. *JAMA* 288(3):334–341, 2002.
52. Black, L.J., M. Sato, E.R. Rowley, D.E. Magee, A. Bekele, D.C. Williams, G.J. Cullinan, R. Bendele, R.F. Kauffman, W.R. Bensch, C.A. Frolik, J.D. Termine, H.U. Bryant. Raloxifene (LY139481 HCI) prevents bone loss and reduces serum cholesterol without causing uterine hypertrophy in ovariectomized rats. *J. Clin. Invest.* 93(1):63–69, 1994.
53. Ishimi, Y., C. Miyaura, M. Ohmura, Y. Onoe, T. Sato, Y. Uchiyama, M. Ito, X. Wang, T. Suda, S. Ikegami. Selective effects of genistein, a soybean isoflavone, on B-lymphopoiesis and bone loss caused by estrogen deficiency. *Endocrinology* 140(4):1893–1900, 1999.
54. Anderson, J.J., S.C. Garner. The effects of phytoestrogens on bone. *Nutr. Res.* 17(10):1617–1632, 1997.
55. Mühlbauer, R.C., F. Li. Frequency of food intake and natural dietary components are potent modulators of bone resorption and bone mass in rats. *Biomed. Pharmacother.* 51(8):360–363, 1997.
56. Barnes, S., T.G. Peterson, L. Coward. Rationale for the use of genistein-containing soy matrices in chemoprevention trials for breast and prostate cancer. *J. Cell. Biochem. Suppl.* 22:181–187, 1995.
57. Röhrdanz, E., S. Ohler, Q.H. Tran-Thi, R. Kahl. The phytoestrogen daidzein affects the antioxidant enzyme system of rat hepatoma H4IIE cells. *J. Nutr.* 132:370–375, 2002.
58. Su, S.J., N.H. Chow, M.L. Kung, T.C. Hung, K.L. Chang. Effects of soy isoflavones on apoptosis induction and G2-M arrest in human hepatoma cells involvement of caspase-3 activation, Bcl-2 and Bcl-xL down-regulation, and Cdc2 kinase activity. *Nutr. Cancer* 45:113–123, 2003.
59. Wilson, L.C., S.J. Baek, A. Call, T.E. Eling. Nonsteroidal anti-inflammatory drug-activated gene (NAG-1) is induced by genistein through the expression of p53 in colorectal cancer cells. *Int. J. Cancer* 105:747–753, 2003.
60. Cai, Q., H. Wei. Effect of dietary genistein on antioxidant enzyme activities in SENCAR mice. *Nutr. Cancer* 25:1–7, 1996.
61. Vissac-Sabatier, C., Y.J. Bignon, D.J. Bernard-Gallon. Effects of the phytoestrogens genistein and daidzein on BRCA2 tumor suppressor gene expression in breast cell lines. *Nutr. Cancer* 45:247–255, 2003.
62. Song, D., X. Na, Y. Liu, X. Chi. Study on mechanisms of human gastric carcinoma cells apoptosis induced by genistein. *Wei Sheng Yan Jiu* 32:128-130, 2003.
63. Andersson, C., Z. Servetnyk, G.M. Roomans. Activation of CFTR by genistein in human airway epithelial cell lines. *Biochem. Biophys. Res. Commun.* 308:518–522, 2003.
64. Yan, C., R. Han. Effects of genistein on invasion and matrix metalloproteinase activities of HT1080 human fibrosarcoma cells. *Chin. Med. Sci. J.* 14:129–133, 1999.
65. Naaz, A., S. Yellayi, M.A. Zakroczymski, D. Bunick, D.R. Doerge, D.B. Lubahn, W.G. Helferich, P.S. Cooke. The soy isoflavone genistein decreases adipose deposition in mice. *Endocrinology* 144:3315–3320, 2003.
66. Murakami, A., D. Takahashi, K. Hagihara, K. Koshimizu, H. Ohigashi. Combinatorial effects of nonsteroidal anti-inflammatory drugs and food constituents on production of prostaglandin E2 and tumor necrosis factor-alpha in RAW264.7 murine macrophages. *Biosci. Biotechnol. Biochem.* 67:1056–1062, 2003.
67. Ridges, L., R. Sunderland, K. Moernlan, B. Meyer, L. Astheimer, P. Howe. Cholesterol lowering benefits of soy and linseed enriched foods. *Asia. Pac. J Clin. Nutr.* 10:204–211, 2001.
68. Department of Health and Human Services. Food labeling: health claims, soy protein and coronary heart disease, final rule. Federal registers, Food and Drug Administration, 64:57700–57733, 1999.

69. Cheng, J., K. Kondo, Y. Suzuki, Y. Ikeda, X. Meng, K. Umemura. Inhibitory effects of total flavones of *Hippophae Rhamnoides* L. on thrombosis in mouse femoral artery and *in vitro* platelet aggregation. *Life Sci*. 72:2263–2271, 2003.
70. Dixon, R.A., C.L. Steele. Flavanoids and isoflavonoids – a gold mine for metabolic engineering. *Trends Plant Sci* 4(10):394–400, 1999.
71. Jung, W., O. Yu, S.M. Lau, D.P. O'Keefe, J. Odell, G. Fader, B. McGonigle. Identification and expression of isoflavone synthase, the key enzyme for biosynthesis of isoflavones in legumes. *Nat. Biotechnol*. 18(2):208–212, 2000.
72. Yu, O., W. Jung, J. Shi, R.A. Croes, G.M. Fader, B. McGonigle, J.T. Odell. Production of the isoflavones genistein and daidzein in non-legume dicot and monocot tissues. *Plant Physiol*. 124(2):781–794, 2000.
73. Liu, C.J., J.W. Blount, C.L. Steele, R.A. Dixon. Bottlenecks for metabolic engineering of isoflavone glycoconjugates in Arabidopsis. *Proc. Natl. Acad. Sci. USA* 99(22):14578–14583, 2002.
74. Dixon, R.A., L.W. Sumner. Legume natural products: understanding and manipulating complex pathways for human and animal health. *Plant Physiol.* 131:878–885, 2003.
75. Hwang, E.I., M. Kaneko, Y. Ohnishi, S. Horinouchi. Production of plant-specific flavanones by *Escherichia coli* containing an artificial gene cluster. *Appl. Environ. Microbiol.* 69(5):2699–2706, 2003.
76. Chassy, B.M. Food safety evaluation of crops produced through biotechnology. *J. Am. Coll. Nutr.* 21(3, Suppl):166S–173S, 2002.
77. Jwanny, E.W., M.M. Rashad, H.M. Abdu. Solid-state fermentation of agricultural wastes into food through *Pleurotus* cultivation. *Appl. Biochem. Biotechnol.* 50(1):71–78, 1995.
78. Okamura, T., T. Ogata, N. Minamimoto, T. Takeno, H. Noda, S. Fukuda, M. Ohsugi. Characteristics of wine produced by mushroom fermentation. *Biosci. Biotechnol. Biochem.* 65(7):1596–1600, 2001.
79. Sanchez, A., F. Ysunza, M.J. Beltran-Garcia, M. Esqueda. Biodegradation of viticulture wastes by *Pleurotus*: a source of microbial and human food and its potential use in animal feeding. *J. Agric. Food Chem.* 50:2537–2542, 2002.
80. Han, J. Solid-state fermentation of cornmeal with the basidiomycete *Hericium erinaceum* for degrading starch and upgrading nutritional value. Int. J. Food Microbiol. 80:61-66, 2003.
81. Kashyap, D.R., S.K. Soni, R. Tewari. Enhanced production of pectinase by *Bacillus* sp. DT7 using solid state fermentation. *Bioresource Technol*. 88:251–254, 2003.
82. Zheng, Z., K. Shetty. Solid-state bioconversion of phenolics from cranberry pomace and role of *Lentinus edodes* beta-glucosidase. *J. Agric. Food Chem.* 48(3):895–900, 2000.
83. Vattem, D.A., K. Shetty. Solid-state production of phenolic antioxidants from cranberry pomace by *Rhizopus oligosporus*. *Food Biotechnol*. 16:189–210, 2002.
84. Vattem, D.A., K. Shetty. Ellagic acid production and phenolic antioxidant activity in cranberry pomace mediated by *Lentinus edodes* using solid-state system. *Process. Biochem.* 39:367–379, 2003.
85. Esaki, H., R. Watanabe, H. Onozaki, S. Kawakishi, T. Osawa. Formation mechanism for potent antioxidative o-dihydroxyisoflavones in soybeans fermented with *Aspergillus saitoi*. *Biosci. Biotechnol. Biochem*. 63(5):851–858, 1999.
86. Kishida, T., H. Ataki, M. Takebe, K. Ebihara. Soybean meal fermented by *Aspergillus awamori* increases the cytochrome P-450 content of the liver microsomes of mice. *J. Agric. Food Chem*. 48:1367–1372, 2000.
87. Ohlsson, A.B., T. Berglund, P. Komlos, J. Rydstrom. Plant defense metabolism is increased by the free radical-generating compound AAPH. *Free Radic. Biol. Med.* 19(3):319–327, 1995.
88. Cantos, E., J.C. Espin, F.A. Tomas-Barberan. Effect of wounding on phenolic enzymes in six minimally processed lettuce cultivars upon storage. *J. Agric. Food Chem.* 49:322–330, 2001.
89. McCue, P., Z. Zheng, J.L. Pinkham, K. Shetty. A model for enhanced pea seedling vigor following low pH and salicylic acid treatments. *Process. Biochem.* 35:603–613, 2000.

90. Andarwulan, N., K. Shetty. Improvement of pea (*Pisum sativum*) seed vigor by fish protein hydrolysates in combination with acetyl salicylic acid. *Process. Biochem.* 35:159–165, 1999.
91. Zheng, Z., K. Shetty. Enhancement of pea (*Pisum sativum*) seedling vigor and associated phenolic content by extracts of apple pomace fermented with *Trichoderma* spp. *Process. Biochem.* 36:79–84, 2000.
92. Duval, B., K. Shetty. Stimulation of phenolics and antioxidant activity in pea (*Pisum sativum*) elicited by genetically transformed anise root extract. *J. Food Biochem.* 25:361–377, 2001.
93. McCue, P., K. Shetty. Clonal herbal extracts as elicitors of phenolic synthesis in dark-germinated mungbean for improving nutritional value with implications for food safety. *J. Food Biochem.* 26:209–232, 2002.
94. Randhir, R., K. Shetty. Light-mediated fava bean (*Vicia faba*) response to phytochemical and protein elicitors and consequences on nutraceutical enhancement and seed vigor. *Process. Biochem.* 38:945–952, 2003.
95. Randhir, R., Y.T. Lin, K. Shetty. Stimulation of phenolics, antioxidant and antimicrobial activity in dark-germinated mungbean sprouts in response to peptide and phytochemical elicitors. *Process. Biochem.* 39:637–646, 2004.
96. Randhir, R., Y.T. Lin, K. Shetty. Phenolics, antioxidant and antimicrobial activity in dark-germinated fenugreek sprouts in response to peptide and phytochemical elicitors. *Asia Pac. J. Clin. Nutr.* 2003.
97. McCue, P., K. Shetty. A biochemical analysis of mungbean (*Vigna radiata*) response to microbial polysaccharides and potential phenolic-enhancing effects for nutraceutical applications. *Food Biotechnol.* 16:57–79, 2002.
98. Shetty, P., M.T. Atallah, K. Shetty. Effects of UV treatment on the proline-linked pentose-phosphate pathway for phenolics and L-DOPA synthesis in dark-germinated *Vicia faba*. *Process. Biochem.* 37:1285–1295, 2002.
99. Strycharz, S., K. Shetty. Effect of *Agrobacterium rhizogenes* on phenolic content of *Mentha pulegium* elite clonal line for phytoremediation applications. *Process. Biochem.* 38:287–293, 2002.
100. McCue, P., K. Shetty. A hypothetical model for action of soybean isoflavonoids against cancer involving a shift to proline-linked energy metabolism through activation of the pentose-phosphate pathway. *Food Biotechnol.* 18:19–37, 2004.
101. Benzie, I.F. Evolution of antioxidant defense mechanisms. *Eur. J. Nutr.* 39:53–61, 2000.
102. Shi, D.Y., Y.R. Deng, S.L. Liu, Y.D. Zhang, L. Wei. Redox stress regulates cell proliferation and apoptosis of human hepatoma through Akt protein phosphorylation. *FEBS Lett.* 542:60–64, 2003.
103. Shetty, K., P. McCue. Phenolic antioxidant biosynthesis in plants for functional food application: integration of systems biology and biotechnological approaches. *Food Biotechnol.* 17(2):67–97, 2004.
104. Sang, S., K. Lapsley, W.S. Jeong, P.A. Lachance, C.T. Ho, R.T. Rosen. Antioxidative phenolic compounds isolated from almond skins (*Prunus amygdalus* Batsch). *J. Agric. Food Chem.* 50:2459–2463, 2002.
105. Middleton, E., C. Kandaswami, T.C. Theoharides. The effects of plant flavanoids on mammalian cells: implications for inflammation, heart dizease, and cancer. *Pharmacol. Rev.* 52:673–751, 2000.
106. Squier, T.C. Oxidative stress and protein aggregation during biological aging. *Exp. Gerontol.* 36:1539–1550, 2001.
107. Spitz, D.R., J.E. Sim, L.A. Ridnour, S.S. Galoforo, Y.J. Lee. Glucose deprivation-induced oxidative stress in human tumor cells. *Ann. NY Acad. Sci.* 899:349–362, 2000.
108. Simonnet, H., N. Alazard, K. Pfeiffer, C. Gallou, C. Beroud, J. Demont, R. Bouvier, H. Schagger, C. Godinot. Low mitochondrial respiratory chain content correlates with tumor aggressiveness in renal cell carcinoma. *Carcinogenesis* 23(5):759–768, 2002.
109. Savagner, F., B. Franc, S. Guyetant, P. Rodien, P. Reynier, Y. Malthiery. Defective mitochondrial ATP synthesis in oxyphilic thyroid tumors. *J. Clin. Endocrinol. Metab.* 86:4920–4925, 2001.

110. Dey, R., C.T. Moraes. Lack of oxidative phosphorylation and low mitochondrial membrane potential decrease susceptibility to apoptosis and do not modulate the protective effect of Bcl-x_L in osteosarcoma cells. *J. Biol. Chem.* 275(10):7087–7094, 2000.
111. Dominguez, J.E., J.F. Graham, C.J. Cummins, D.J. Loreck, J. Galarraga, J. Van der Feen, R. De La Paz, B.H. Smith. Enzymes of glucose metabolism in cultured human gliomas: neoplasia is accompanied by altered hexokinase, phosphofructokinase, and glucose-6-phosphate dehydrogenase levels. *Metab. Brain Dis.* 2(1):17–30, 1987.
112. Chesney, J., R. Mitchell, F. Benigni, M. Bacher, L. Spiegel, Y. Al-Abed, J.H. Han, C. Metz, R. Bucala. An inducible gene product for 6-phosphofructo-2-kinase with an AU-rich instability element: role in tumor cell glycolysis and the Warburg effect. *Proc. Natl. Acad. Sci. USA* 96(6):3047–3052, 1999.
113. Hasegawa, Y., T. Takano, A. Miyauchi, F. Matsuzuka, H. Yoshida, K. Kuma, N. Amino. Decreased expression of catalase mRNA in thyroid anaplastic carcinoma. *Jpn. J. Clin. Oncol.* 33(1):6–9, 2003.
114. Sander, C.S., F. Hamm, P. Elsner, J.J. Thiele. Oxidative stress in malignant melanoma and non-melanoma skin cancer. *Br. J. Dermatol.* 148:913–922, 2003.
115. Phang, J.M., S.J. Downing, G.C. Yeh. Linkage of the HMP pathway to ATP generation by the proline cycle. *Biochem. Biophys. Res. Commun.* 93:462–470, 1980.
116. Hagedorn, C.H., J.M. Phang. Transfer of reducing equivalents into mitochondria by the interconversions of proline and Δ^1-pyrroline-5-carboxylate. *Arch. Biochem. Biophys.* 225:95–101, 1983.
117. Phang, J.M. The regulatory functions of proline and pyrroline-5-carboxylic acid. *Curr. Top. Cell. Regul.* 25:91–132, 1985.
118. Nissler, K., H. Petermann, I. Wenz, D. Brox. Fructose 2,6-bisphosphate metabolism in Ehrlich ascites tumour cells. *Cancer Res. Clin. Oncol.* 121:739–45, 1995.
119. Le Goffe, C., G. Vallette, L. Charrier, T. Candelon, C. Bou-Hanna, J.F. Bouhours, C.L. Laboisse. Metabolic control of resistance of human epithelial cells to H_2O_2 and NO stresses. *Biochem. J.* 364(2):349–359, 2002.
120. Spolarics, Z. Endotoxemia, pentose cycle, and the oxidant/antioxidant balance in the hepatic sinusoid. *J. Leukoc. Biol.* 63:534–541, 1998.
121. Tuttle, S., T. Stamato, M.L. Perez, J. Biaglow. Glucose-6-phosphate dehydrogenase and the oxidative pentose phosphate cycle protect cells against apoptosis induced by low doses of ionizing radiation. *Radiat. Res.* 153:781–787, 2000.
122. Neumann, C.A., D.S. Krause, C.V. Carman, S. Das, D.P. Dubey, J.L. Abraham, R.T. Bronson, Y. Fujiwara, S.H. Orkin, R.A. Van Etten. Essential role for the peroxiredoxin Prdx1 in erythrocyte antioxidant defense and tumor suppression. *Nature* 424:561–565, 2003.
123. Friedmann, Y., I. Vlodavsky, H. Aingorn, A. Aviv, T. Peretz, I. Pecker, O. Pappo. Expression of heparanase in normal, dysplastic, and neoplastic human colonic mucosa and stroma. Evidence for its role in colonic tumorigenesis. *Am. J. Pathol.* 157:1167–1175, 2000.
124. Wilson, L.C., S.J. Baek, A. Call, T.E. Eling. Nonsteroidal anti-inflammatory drug-activated gene (NAG-1) is induced by genistein through the expression of p53 in colorectal cancer cells. *Int. J. Cancer* 105:747–753, 2003.
125. Seo, Y.R., M.R. Kelley, M.L. Smith. Selenomethionine regulation of p53 by a ref1-dependent redox mechanism. *Proc. Natl. Acad. Sci. USA* 99:14548–14553, 2002.
126. Brash, D.E., P.A. Havre. New careers for antioxidants. *Proc. Natl. Acad. Sci. USA* 99:13969–13971, 2002.
127. Donald, S.P., X.Y. Sun, C.A. Hu, J. Yu, J.M. Mei, D. Valle, J.M. Phang. Proline oxidase, encoded by p53-induced gene-6, catalyzes the generation of proline-dependent reactive oxygen species. *Cancer Res.* 61(5):1810–1815, 2001.
128. Maxwell, S.A., G.E. Davis. Differential gene expression in p53-mediated apoptosis-resistant vs. apoptosis-sensitive tumor cell lines. *Proc. Natl. Acad. Sci. USA* 97:13009–13014, 2000.
129. Maxwell, S.A., A. Rivera. Proline oxidase induces apoptosis in tumor cells, and its expression is frequently absent or reduced in renal carcinomas. *J. Biol. Chem.* 278:9784–9789, 2003.

130. Sun, Y. Free radicals, antioxidant enzymes, and carcinogenesis. *Free Radic. Biol. Med.* 8:583–599, 1990.
131. Cobb, C.S., D.S. Levi, K. Aldape, M.A. Israel. Manganese superoxide dismutase expression in human central nervous system tumors. *Cancer Res.* 56:3192-3195, 1996.
132. Iqbal, M., S.D. Sharma, Y. Okazaki, M. Fujisawa, S. Okada. Dietary supplementation of curcumin enhances antioxidant and phase II metabolizing enzymes in ddY male mice: possible role in protection against chemical carcinogenesis and toxicity. *Pharmacol. Toxicol.* 92:33–38, 2003.
133. Zheng, Q.S., Zhang, Y.T., Zheng, R.L. Ascorbic acid induces redifferentiation and growth inhibition in human hepatoma cells by increasing endogenous hydrogen peroxide. *Pharmazie*. 57 (11): 753–757, 2002.
134. Fahey, J.W., K.K. Stephenson. Pinostrobin from honey and Thai ginger (*Boesenbergia pandurata*): a potent flavanoid inducer of mammalian phase 2 chemoprotective and antioxidant enzymes. *J. Agric. Food. Chem.* 50:7472–7476, 2002.
135. Suzuki, K., H. Koike, H. Matsui, Y. Ono, M. Hasumi, H. Nakazato, H. Okugi, Y. Sekine, K. Oki, K. Ito, T. Yamamoto, Y. Fukabori, K. Kurokawa, H. Yamanaka. Genistein, a soy isoflavone, induces glutathione peroxidase in the human prostate cancer cell lines LNCaP and PC-3. *Int. J. Cancer* 99(6):846–852, 2002.
136. Chang, W.H., J.J. Liu, C.H. Chen, T.S. Huang, F.J. Lu. Growth inhibition and induction of apoptosis in MCF-7 breast cancer cells by fermented soy milk. *Nutr. Cancer* 43(2):214–226, 2002.
137. Hewitt, A.L., K.W. Singletary. Soy extract inhibits mammary adenocarcinoma growth in a syngeneic mouse model. *Cancer Lett.* 192(2):133–143, 2003.
138. Loreck, D.J., J. Galarraga, J. Van der Feen, J.M. Phang, B.H. Smith, C.J. Cummins. Regulation of the pentose phosphate pathway in human astrocytes and gliomas. *Metab. Brain Dis.* 2:31–46, 1987.
139. Javelaud, D., F. Besancon. NF-κB activation results in rapid inactivation of JNK in TNF α-treated Ewing sarcoma cells: a mechanism for the anti-apoptotic effect of NF-κB. *Oncogene* 20:4365–4372, 2001.
140. Stadtman, E.R., B.S. Berlett. Reactive oxygen-mediated protein oxidation in aging and dizease. *Drug Metab. Rev.* 30:225–243, 1998.
141. Andarwulan, N., K. Shetty. Stimulation of novel phenolic metabolite, epoxy-Psuedoisoeugenol-(2-Methylbutyrate) [EPB], in transformed anize (*Pimpinella anisum* L.) root cultures by fish protein hydrolysates. *Food Biotechnol.* 14:1–20, 2000.
142. Shetty, K., M.L. Wahlqvist. A model for the role of proline-linked pentose-phosphate pathway in phenolic phytochemical biosynthesis and mechanism of action for human health and environmental applications. *Asia Pac. J. Clin. Nutr.* 13(1): 1–24, 2004.

2.10

Functional Phytochemicals from Cranberries: Their Mechanism of Action and Strategies to Improve Functionality

Dhiraj A. Vattem and Kalidas Shetty

CONTENTS

10.1 INTRODUCTION: PHENOLIC PHYTOCHEMICALS

Phenolic compounds or phenolic phytochemicals are secondary metabolites of plant origin which constitute one of the most abundant groups of natural compounds and form an important component of both human and animal diets (1,2,3). These phenolic metabolites function to protect the plant against biological and environmental stresses and are therefore synthesized in response to pathogenic attack, such as fungal or bacterial infection, or high energy radiation exposure, such as prolonged UV exposure (4,5). Because of their important biological functions, phenolic phytochemicals are ubiquitous in plants and therefore find their place in almost all food groups. Common fruits such as apples, cranberries, grapes, raspberries, and strawberries, and fruit beverages like red wine and apple and orange juices, are rich sources of phenolic phytochemicals. In addition to fruits, vegetables such as cabbage and onion, and food grains such as sorghum, millet, barley, peas, and other legumes (6) are also described as important sources of phenolic phytochemicals. Varied biological functions of phenolic phytochemicals in plants have led to the evolution of diverse types of phenolic compounds. Depending on the evolutionary pressures experienced, plants have evolved a constituent profile of inducible phenolic phytochemicals, usually characteristic of a particular species of fruit or vegetable (6,7). For example, the most abundant phenolic compound in fruits and their products are flavonols. Cereals and legumes are rich in flavonoids, phenolic acids, and tannins (6,7). The major phenolic phytochemicals in wine include phenolic acids, anthocyanins, tannins, and other flavonoids (6,7).

10.2 CHEMICAL NATURE AND BIOSYNTHESIS OF PHENOLIC PHYTOCHEMICALS

The International Union for pure and applied chemists (IUPAC) defines phenol as hydroxybenzene. The term phenolic compounds refers to a relatively wide range of chemical compounds that contain at least one aromatic ring and usually one or more hydroxyl substituents (8). There are numerous different types of these phenolic phytochemicals, classified according to their ring structure and the number of carbon atoms substituting the ring and linking them together (Table 10.1).

Metabolic processing of phenolic phytochemicals in plants for their final biological function has led to chemical variations in basic phenolic structure. Differences in substituent groups and linkages have resulted in a wide variety of chemical structures having distinct properties. More than 8000 different phenolic structures, categorized into 10 classes, have been identified. They vary structurally from simple molecules (e.g., phenolic acids with a single ring structure), to biphenyls and flavonoids having 2–3 phenolic rings (9,10) (Figure 10.1). Another abundant group of phenolic phytochemicals in fruits and vegetables, often referred to as polyphenols, contain 12–16 phenolic groups and approximately 5–7 aromatic rings per 1000 relative molecular mass (Figure 10.1). These polyphenols are classified as condensed proanthocyanidins, tannins which include galloyl and hexahydroxydiphenoyl (or ellagoyl) esters and their derivatives, or phlorotannins (9,10) (Figure 10.1).

All the phenolic phytochemicals are derived from a common biosynthetic pathway, incorporating precursors from the shikimate or the acetate–malonate pathways or both (1,11) (Figure 10.2). Simple phenolic acids such as cinnamic acid and its derivatives are widespread in fruits and vegetables. They are derived primarily from the shikimate pathway via phenylalanine or tyrosine (1). Other phenolics of biological importance that are formed in this manner include coumaric acid and caffeic acid. Chemical modification of side chains by processes such as oxidation produce another group of compounds, such as protocatechuic

Table 10.1
The major classes of phenolic compounds in plants

Number of Carbon Atoms	Basic Skeleton	Class
6	C_6	Simple phenols
		Benzoquinones
7	C_6-C_1	Phenolic acids
8	C_6-C_2	Acetophenones
		Tyrosine derivatives
		Phenylacetic acids
9	C_6-C_3	Hydroxycinnamic acids
		Phenylpropenes
		Coumarins
		Isocoumarins
		Chromones
10	C_6-C_4	Naphthoquinones
13	C_6-C_1-C_6	Xanthones
14	C_6-C_2-C_6	Stilbenes
		Anthraquinones
15	C_6-C_3-C_6	Flavonoids
		Isoflavonoids
18	$(C_6$-$C_3)_2$	Lignans
		Neolignans
30	$(C_6$-C_3-$C_6)_2$	Biflavonoids
N	$(C_6$-$C_3)_n$	Lignins
	$(C_6)_n$	Catechol melanins
	$(C_6$-C_3-$C_6)_n$	Flavolans (Condensed Tannins)

acid and its positional isomer gentisic acid, which are referred to as benzoic acid derivatives (Figure 10.2). To increase the solubility and target the phenolics to specific parts of the plant, and to prevent their enzymatic and chemical degradation, phenolic phytochemicals are often esterified with sugars and other chemical components such as quinic acid through the hydroxyl groups of the phenolic ring. The esters are sugars and phenolic phytochemicals are called glycosides. A majority of glycosides contain glucose, but the glycosides of phenolics with galactose, sucrose, and rhamnose have also been described (11).

Flavonoids and biphenyls are another important class of phenolic phytochemicals that are especially rich in fruits and legumes. Flavonoids like quercetin constitute the most abundant group of phenolic phytochemicals. Their common structure consists of two aromatic rings linked through three carbons that usually form an oxygenated heterocycle; this type of chemical arrangement is often described as diphenylpropanes (C_6-C_3-C_6) (2,11). Structural variations within the rings resulting in an alteration in the extent of hydroxylation, methylation, isoprenylation, dimerization, and glycosylation (producing O- or C-glycosides) subdivide the flavonoids into several families: flavonols, flavones, isoflavones, antocyanidins and others (Figure 10.2).

10.2.1 Phenolic Phytochemicals from Berries

Phenolic phytochemicals are ubiquitous in nature and are especially rich in fruits (12,13). Fruits such as berries (14,15) are known to be good sources of phenolic antioxidants having protective health benefits. Fruit beverages such as wines have been shown to

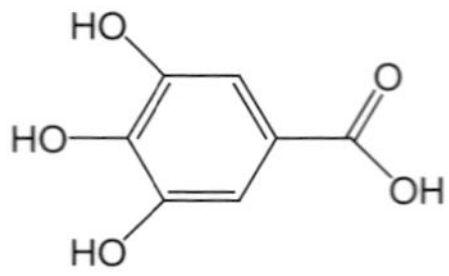

Gallic acid

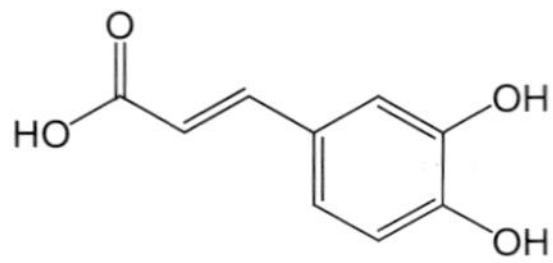

Caffeic acid

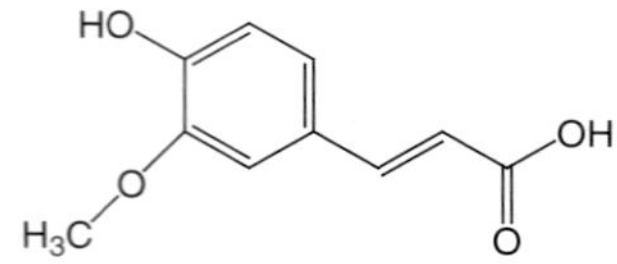

Ferulic acid

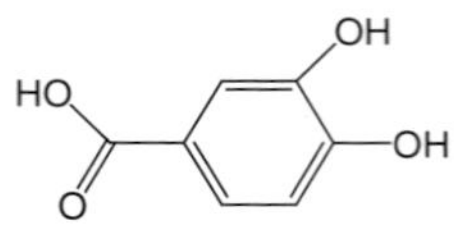

Protocatechuic acid

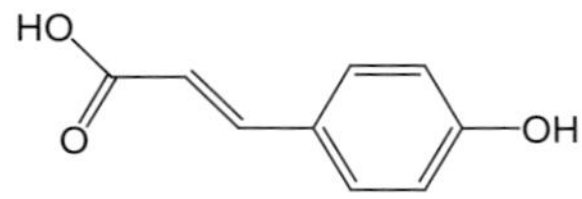

Coumaric acid

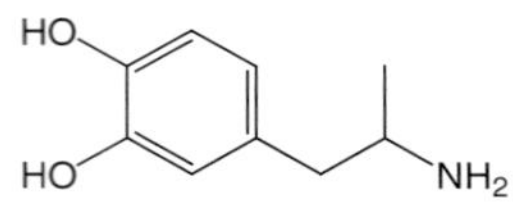

L-DOPA

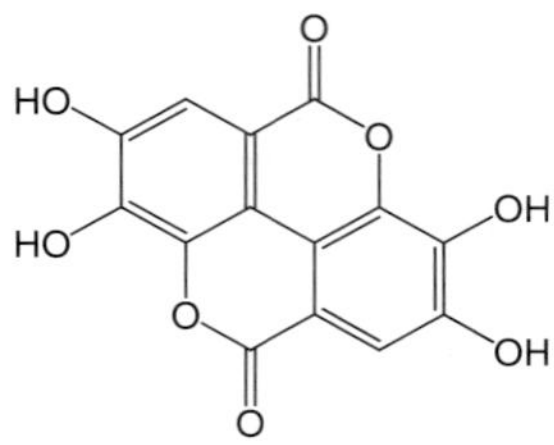

Ellagic acid

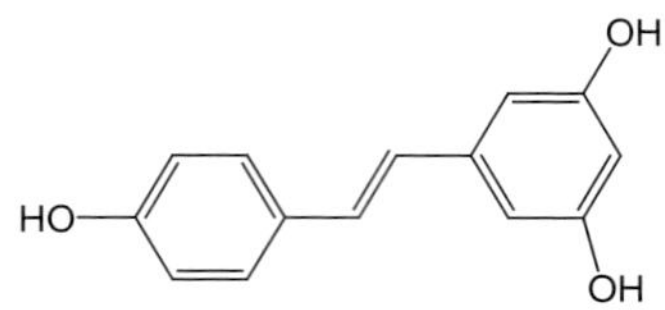

Resveratrol

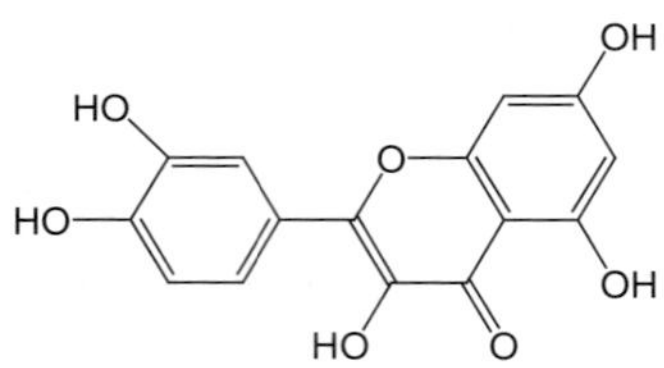

Quercetin

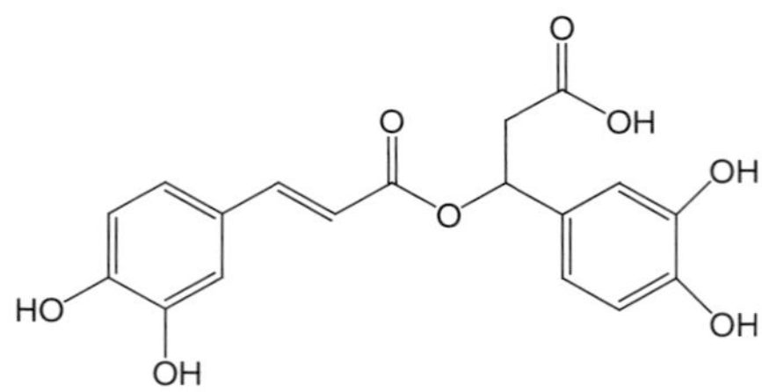

Rosmarinic acid

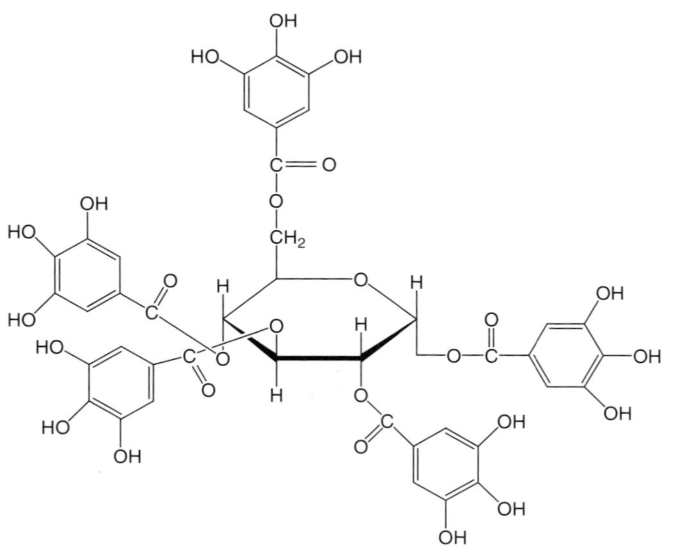

Figure 10.1 Common simple phenol, biphenyls, flavonoids, and tannins in plants.

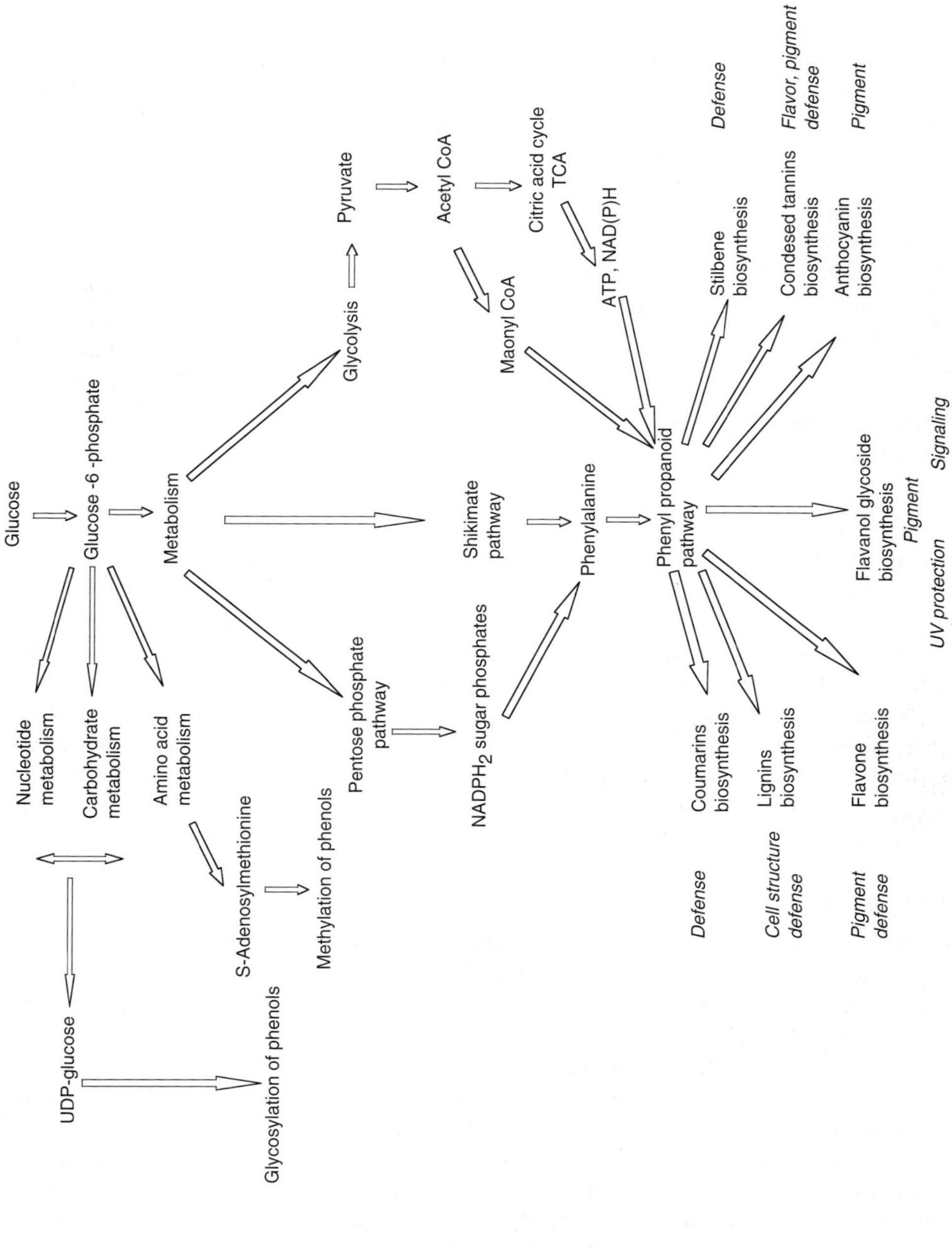

Figure 10.2 Biosynthesis of phenolic phytochemicals.

contain a variety of polyphenolic compounds, the most abundant being flavonoids such as anthocyanins (16,17,18).

Berries, like many other fruits, are rich in phenolic compounds, including diphenyls, flavonoids and phenolic acids, which exhibit a wide range of biological effects, including antioxidant (19,20,21) and anticarcinogenic properties (22). A high free radical scavenging activity of berry extracts toward chemically generated active oxygen species has been described in several studies (19–21). Increased fruit consumption in daily diets has shown to significantly reduce the incidence and mortality rates of cancer, cardiovascular disorders, and other degenerative diseases caused by oxidative stress (22–25). Epidemiological evidence suggests that high consumption of flavonoids, which are an important component of berries, may provide protection against coronary heart disease, cardiac stroke (26–28), lung cancer (22,28,29), and stomach cancer (30).

Berries are typically rich in flavonoids and diphenyls such as ellagic acid (Figure 10.1), and represent a large group of secondary plant metabolites (31). The diversity and complexity of the flavonoids found in berries depends on at least two factors: different variety of aglycones and the high number of glycosides, sometimes in acylated form; and condensation into complex molecules. Recent research has determined the antioxidant activity of different phenolic compounds (32–34) and attempted to define the structural characteristics which contribute to their activity (33,35). Phenolic acids present in berries are hydroxylated derivatives of benzoic acid and cinnamic acid (36). The other simple phenolics in berries include caffeic, chlorogenic, ferulic, sinapic, and *p*-coumaric, acids (37).

10.2.1.1 Cranberry

The cranberry *(Vaccinium macrocarpon),* also known as American cranberry, belongs to the family Ericaceae, which also includes blueberry (*V. angustifolium*) and bilberry (*V. myrtillus*). Even though cranberries have been historically associated with positive health benefits, scientific investigation into the positive health benefits of the cranberry has received little attention (38,39). Many studies have demonstrated high phenolic content in fruits such as grapes, apples, oranges, prunes, and berries (31). The American cranberry is a prominent agricultural food crop produced in Massachusetts, Wisconsin, Michigan, Canada, New Jersey, Oregon, and Washington. The crop size is approximately 500 million pounds annually and is processed into three basic categories: fresh (5%); sauce products, concentrate, and various applications (35%); and juice drinks (60%) (40).

10.2.1.2 Cranberry Pomace

Cranberry pomace is the byproduct of the cranberry juice processing industry with limited applications. Once the juice is extracted from the fruit, the remaining product is called cranberry pomace. Pomace is mainly composed of the skin, flesh, and seed of the fruit. It is rich in fiber and has relatively small amounts of protein and carbohydrates (41). Traditionally it has been used as an ingredient in animal feed, however, due to its low protein and carbohydrate content, it has little nutritive value. Its disposal into the soil or in a landfill poses considerable economic loss, and causes potential environmental problems due to its low pH (41).

Agricultural and industrial residues are attractive sources of natural antioxidants. Potato peel waste (42,43), olive peel (44), grape seeds, and grape pomace peels (45,46) have been studied for their use as cheap sources of antioxidants. Increased antioxidant activity in rat plasma after oral administration of grape seed extracts was reported recently (47). Identification of polyphenolic compounds from apple pomace (48) has also been reported. Phenolics are ubiquitous in plants, but seeds and skins are especially rich sources of phenolics (42,47,49–51), probably because of the role they play in protecting the fruit and the seed to ensure healthy propagation of the species. Pomace, which mainly consists of fruit skins and seeds, is also a rich source of phenolic compounds (47,48). However, several phenolics that are found in pomace and other plant products exist in conjugated forms either with sugars (primarily glucose), as glycosides, or as other moieties. This conjugation occurs via the hydroxyl groups of the phenolics, which reduces their ability to function as good antioxidants, because availability of free hydroxyl groups on the phenolic rings is important for resonance stabilization of free radicals. Lowered antioxidant capacity has direct implications on decreasing health functionality when these phenolics are ingested via food or as nutraceuticals. Therefore, if free phenolics are released from their glycosides or other conjugates, then the antioxidant, and thus the health functionality of these phytochemicals could be improved. Enzymatic hydrolysis of these phenolic glycosides appears to be an attractive means of increasing the concentration of free phenolic acids in fruit juice and wines to enrich taste, flavor, and aroma, also potentially increasing nutraceutical value (52–54).

10.2.2 Oxidative Stress Mediated Pathogenesis

Emerging evidence suggests that free radical mediated oxidative stress is responsible for the induction of pathogenicity in biological systems. The primary manner in which oxidative stress is mediated in cellular systems is by the generation of reactive oxygen species (ROS). Oxygen in its ground state is often described as a biradical, as it contains two electrons in its outer shell having the same spin. When one of the electron changes its spin, the oxygen is transformed into a singlet, and becomes a powerful oxidant and undergoes reduction. However, in biological systems, incomplete reduction of molecular oxygen occurs due to errors in metabolism, or due to controlled reduction of oxygen, resulting in the formation of what are now known as ROS. These ROS are highly reactive molecules, and include the hydroxyl radicals ($HO^{\cdot}$), superoxide anions ($O_2^{\cdot}$) and hydrogen peroxide (H_2O_2) (Figure 10.3). The superoxide anion is the precursor of most ROS in biological systems, and is formed by the one electron reduction of molecular oxygen. Spontaneous or enzymatic dismutation of superoxide anion produces the peroxide radical, which can undergo a complete reduction to form water or an incomplete reduction to form the highly reactive hydroxyl radical (55). Singlet oxygen can also react with nitric oxide (NO) in a diffusion limited reaction to form peroxynitrite which is a powerful

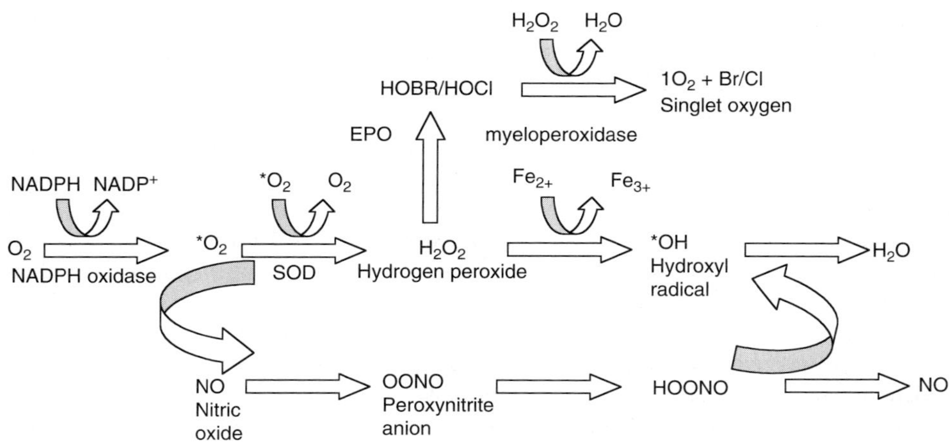

Figure 10.3 Biological mechanisms for ROS formation.

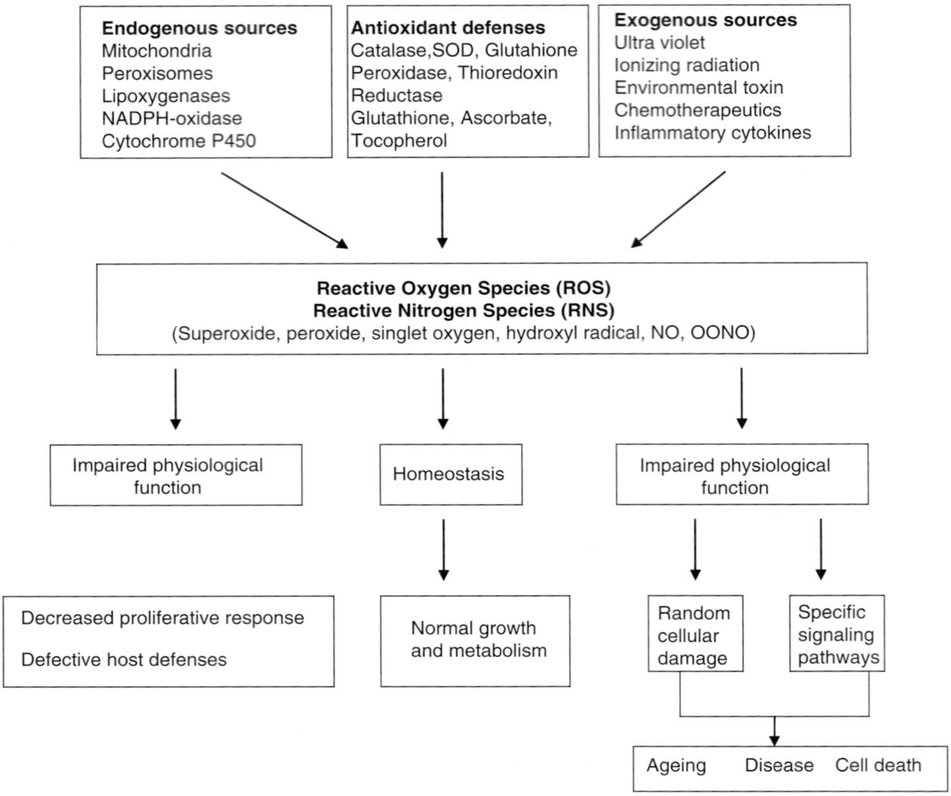

Figure 10.4 Reactive oxygen species mediated pathogenecity.

oxidant (56,57). The oxidants derived from NO are often referred to as reactive nitrogen species (RNS) (Figure 10.3).

Reactive oxygen species and reactive nitrogenspecies (RNS) are constantly produced in aerobic organisms both enzymatically and non-enzymatically. Many sources for the formation of ROS have been identified in cellular systems of living organisms (58,59). Superoxide is formed upon the one electron reduction of oxygen mediated by enzymes such as NADPH oxidase located on the cell membrane of polymorphonuclear cells, macrophages, and endothelial cells (60–62); from xanthine oxidase or from the respiratory chain (63). Superoxide radicals can also be formed from the electron cycling carried out by the cytochrome P450 dependent enzymes (64,65). Mitochondrial leakage of electrons or direct transfer of electron to oxygen via coenzymes or prosthetic groups such as flavins and Fe-S centers constitute another important source of ROS in cellular systems (58,59). The homolytic cleavage of water, or the breakdown of hydrogen peroxide in a high energy radiation (e.g., x-rays, UV) or metal catalyzed process, forms the hydroxyl radical which is the most reactive oxygen species (Figure 10.3, Figure 10.4). Another ROS, the hypochlorite ion, is formed by the macrophage myeloperoxidase, or related peroxidase activities, which catalyze the halide driven reduction of hydrogen peroxide to form the oxidant hypochlorous acid (HOCl) to kill invading microorganisms (63). Biological conversion of L-arginine to L-citrulline by nitric oxide synthase forms a cytotoxic oxygen species, nitric oxide (NO), which is involved in the cellular defense against malignant cells, invading fungi, and protozoa (56,57). Nitric oxide is also involved critically in signaling mechanisms in the vasculature controlling the vascular force in the blood vessels by vasodilation and inflammation (66).

Reactive oxygen species are ideally suited to be signaling molecules, as they are small, can diffuse small distances, and have several mechanisms of production that can be controlled and regulated. Low levels of ROS therefore have been implicated in many cellular processes, including intracellular signaling responsible for proliferation or apoptosis (67), for modulation of immune response (68), and for mounting a defense response against pathogens (69) (Figure 10.4). Even though the exact mechanism of action of ROS in effecting these physiological processes is not very well understood, there is growing evidence that ROS at some level are capable of activating or repressing many biological effector molecules (70). It has been shown to activate or repress transcription factors by directly activating them by oxidizing the sulphydryl groups present, or by modulating a complex array of kinases and phosphatases, which are important in signal transduction. Many ions that maintain the electrostatic balance of the cell required for many physiological processes such as cell growth and cell death are also now shown to be regulated by ROS. Another important mechanism by which ROS are now shown to be regulating the cell function is by altering the redox status of the cell, often by regulating the levels of oxidized and reduced glutathione (GSH/GSSG) (71).

10.2.2.1 Antioxidant Defense Systems

Redox regulated physiological processes are inevitably sensitive against excessive ROS production by any source. Such excessive levels of ROS may be generated either by overstimulation of the otherwise tightly regulated NAD(P)H oxidases or by other mechanisms that produce ROS in a nonregulated fashion, including the production of ROS by the mitochondrial electronic transport chain (ETC) or by xanthine oxidase. Diets rich in saturated fatty acids, and carbohydrates; and environmental factors such as exposure to high energy radiation, and ingestion of or exposure to toxins and carcinogens can also overstimulate metabolic systems resulting in the formation of ROS, which are beyond the cellular need for their regular functions (63–65) (Figure 10.4).

Several antioxidant systems are in place in the cell that can quickly remove the ROS from cellular systems. Biological defenses against ROS comprise a complex array of endogenous antioxidant enzymes, endogenous antioxidant factors including glutathione (GSH) and other tissue thiols, heme proteins, coenzyme Q, bilirubin and urates, and a variety of nutritional factors, primarily the antioxidant vitamins and phenolic phytochemicals from diet (72–74) (Figure 10.4, Figure 10.5).

Glutathione is the most abundant intracellular reductant in all aerobic cells. Tissue GSH and other tissue thiols exist at millimolar concentrations and are important systems to protect the cell against oxidative stress and tissue injury (71). Glutathione acts as a redox sensor and as a sulfhydryl buffer in cellular systems. Upon generation of ROS, GSH preferentially reacts with ROS in an almost sacrificial manner to form oxidized glutathione (GSSH) to prevent the oxidation of other biological molecules. Enzymes such as glutathione peroxidases (GPX) reduce soluble peroxides and membrane bound peroxides to the corresponding alcohols, at the expense of GSH, which is oxidized to GSSG and ascorbic acid, which is then oxidized to dehydroascorbic acid (71–73). Another class of proteins, called glutaredoxins (GRx), are reduced by GSH, and are capable of reducing protein disulfides from oxidative stress (Figure 10.4, Figure 10.5). However, for this antioxidant response to continue, cellular systems have to regenerate GSH constantly (74). Tissue GSH/GSSH ratios are maintained in the reduced state by the concerted action of antioxidant vitamins such as ascorbate and tocopherols using many antioxidant enzyme systems

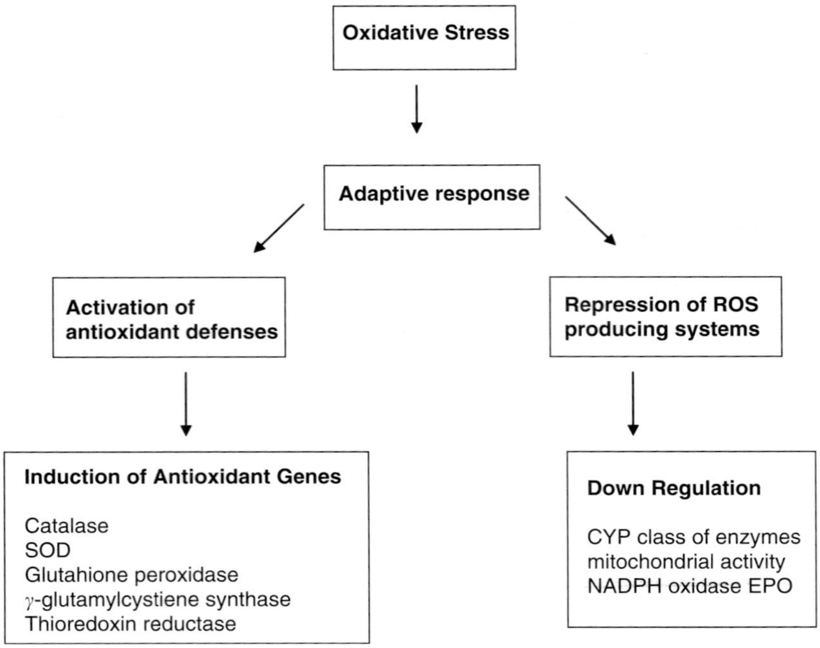

Figure 10.5 Biological adaptive responses to manage oxidative stress. (Morel, Y., R. Barouki, *Biochem. J.* 342(3): 481–496, 1999).

involving several oxidoreductases (72–74). The cascade of endogenous antioxidant enzymes requires energy to maintain the living system in the reduced state. Glutathione reductase and ascorbate peroxidase maintain the tissue GSH in the reduced state at the expense of $NADPH_2$ and $FADH_2$ (72–74) (Figure 10.6). Enzymatic reactions of the thioredoxin system contain a set of oxidoreductases with wide range substrate specificity that reduce the active site disulfide in thioredoxin (Trx) and several other substrates, directly under the consumption of $NADPH_2$. Reduced Trx is highly efficient in reducing disulfides in proteins and peptides including glutathione disulfide (GSSG).

Another biological defense against superoxide free radicals is the superoxide dismutase (SOD) enzyme, which is often considered the most effective antioxidant system.

Superoxide dismutase is an enzyme that catalyzes dismutation of two superoxide anions into hydrogen peroxide and molecular oxygen

$$2O_2^{\cdot -} + 2H^+ \rightarrow H_2O_2 + O_2 \quad (10.1)$$

The importance of SOD is that it is a very efficient enzyme system in removing superoxide radicals from biological systems. The function of SOD is so imperative for the protection of cells that it represents a substantial proportion of the proteins manufactured by the body (72,73). One of the contributing factors for the high efficiency of SOD is that its dismutation is always coupled to another enzyme system called catalase (CAT). Catalase is involved in removing hydrogen peroxide molecules, which are byproducts of the reactions created by SOD (72,73) (Figure 10.6). Catalase is also abundantly present in the body and is integrated into all the cellular systems that operate in an oxidant environment. Red blood cells, which transport oxygen to different cellular systems, have high levels of CAT and help to remove hydrogen peroxide from tissues to prevent both cellular damage and propagation

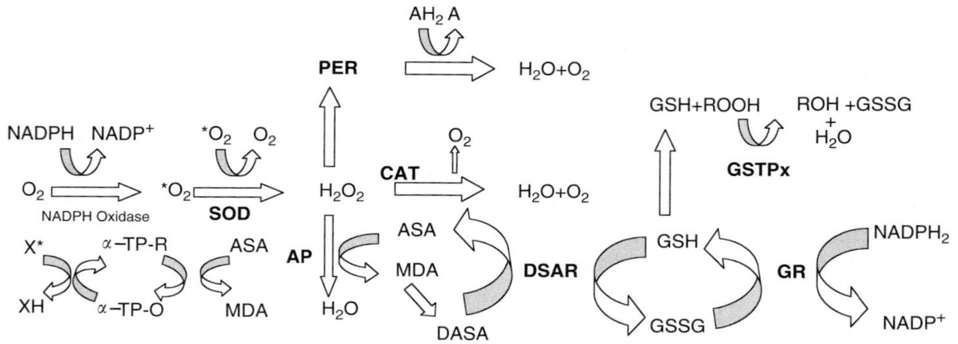

Figure 10.6 The antioxidant defense response of the cell carried out by enzymatic as well as the nonenzymatic antioxidants.

of free radicals. This natural interaction synergy between these two antioxidant enzymes constitutes the most effective system of free radical control (72,73).

Imbalance in the generation of ROS can occur, and deficiency of biological systems to control such an imbalance can result in conditions of oxidative stress, which is now believed to be a primary cause for the induction of many chronic diseases. Small fluctuations in the concentration of ROS may actually play a role in intracellular signaling (67). Uncontrolled increases in the concentrations of these oxidants lead to free radical mediated chain reactions that indiscriminately target proteins (75), lipids (76), and DNA (77,78), which can impair their normal biological functions (Figure 10.4).

Excessive generation of ROS can result in activation or repression of several key signaling pathways and transcription factors, which can impair normal physiological responses, resulting in impaired host defenses. These can lead to impaired cell growth and cell death, which have been shown to develop into cancer and diseases related to aging (67,70,79).

Pathogenesis mediated by virus and *Helicobacter pylori* are also now linked to oxidative stress and are associated with lower responses of CAT and SOD (80,81). Recent experimental findings suggest that overproduction of ROS and RNS lowered antioxidant defense; and alterations of enzymatic pathways in humans with poorly controlled diabetes mellitus can contribute to endothelial, vascular, and neurovascular dysfunction (82). Neurodegenerative disorders such as Alzheimer's disease, Parkinson's disease, and dementia have also been associated with the oxidation of carbohydrates and proteins by ROS, and reduced activities of the enzymes involved in the antioxidant defense response: GPX, SOD and CAT (83).

10.2.2.2 Phenolic Phytochemicals in Nutritional Management of Antioxidant Defense

Association of ROS with manifestation of several diseases has led to an increase in the management of oxidation linked diseases such as cancer, cerebrovascular disease, and diabetes by supporting the body's antioxidant defense system through consuming dietary antioxidants such as ascorbic acid, tocopherols, and carotenoids (82,84,85). Recent epidemiological studies have shown that diets rich in fruits and vegetables are associated with lower incidences of oxidation linked diseases such as cancer, cerebrovascular disease, and diabetes. These protective effects of fruits and vegetables are now linked to the presence of antioxidant vitamins and phenolic phytochemicals having antioxidant activity.

The ability of dietary antioxidants in managing diseases manifested by oxidative stress is not clearly understood. Most phenolic phytochemicals that have positive effects on health are believed to function by countering the effects of ROS species generated during cellular metabolism. Consumption of natural dietary antioxidants from fruits and vegetables has been shown to enhance the function of the antioxidant defense response mediated by GSH, ascorbate, SOD, CAT and GST interface (86–89). This was observed in systemic sclerosis and rheumatoid arthritis patients (90). It was observed that consumption of fruits and vegetables led to a direct increase in scavenging ROS, and prevented polymorphonuclear leukocytes from generating hydroxyl radicals (91).

10.2.3 Biological Function of Phenolic Phytochemicals

Chronic and infectious diseases have become the primary cause of mortality and are expected to become a major public health challenge. These chronic diseases such as cardiovascular disease, hypertension, diabetes mellitus, and some forms of cancer, have now been associated with changes in diet and lifestyle associated with calorie sufficiency. These include excessive dietary carbohydrate and fat intake, low intake of fruits and vegetables, smoking, lack of physical activity, and exposure to environmental toxins (92).

Compelling epidemiological and scientific evidence has led to an understanding that oxidative stress, as a consequence of an imbalance of prooxidants and antioxidants, is a key phenomenon in the manifestation of chronic diseases (90). Powerful strategies to control oxidative stress related pathogenicities are gaining prominence. Epidemiological evidence showing that populations consuming diets rich in fruits and vegetables have lower incidences of many chronic diseases such as cancer, cardiovascular diseases and diabetes has led to an interest in the use of diet as a potential tool for the control of these oxidative diseases (86–89). Recent *in vitro* and clinical studies have shown that diets rich in carbohydrates and fats induced oxidative stress, which was decreased by consuming fruits, vegetables, and their products (93). Among all the dietary components, fruits and vegetables have especially been shown to exert a protective effect (23–26). Phenolic phytochemicals with antioxidant properties are now widely thought to be the principle components in fruits and vegetables which have these beneficial effects. Phenolic phytochemicals exhibit a wide range of biological effects and can broadly be divided into two categories. The first and the most well described mode of action of these phenolic phytochemicals in managing oxidation stress related diseases is due to the direct involvement of the phenolic phytochemicals in quenching the free radicals from biological systems. It is well known that free radicals cause oxidative damage to nucleic acids, proteins, and lipids. Oxidation of biological macromolecules as a result of free radical damage has now been strongly associated with development of many physiological conditions which can develop into disease (67,70,79–83). Phenolic phytochemicals, due to their phenolic ring and hydroxyl substituents, can function as effective antioxidants due to their ability to quench free electrons. Phenolic antioxidants can therefore scavenge the harmful free radicals and inhibit their oxidative reactions with vital biological molecules (19).

The second and more significant mode of action is a consequence of their ability to modulate cellular physiology both at the biochemical or physiological level and at the molecular level. This mode of action is a result of the structure to function phenomenon linked to metabolic pathways. Because of their structural similarities with several key biological effectors and signal molecules, phenolic phytochemicals are able to participate in induction or repression of gene expression; or activation or deactivation of proteins, enzymes, and transcription factors of key metabolic pathways (67,71–73). They can critically modulate cellular homeostasis as a result of their physiochemical properties such as size, molecular weight, partial hydrophobicity, and ability to modulate acidity at biological pH through enzyme coupled reactions. As a consequence of many modes of action of phenolic phytochemicals they have been shown to have several different functions. Several studies have demonstrated anticarcinogenic properties of phenolic phytochemicals such as gallic acid, caffeic acid, ferulic acid, catechin, quercetin, and resveratrol (94–96). It is believed that phenolic phytochemicals might interfere with several of the steps that lead to the development of malignant tumors, including inactivating carcinogens and inhibiting the expression of mutant genes (97). Potential anticarcinogenic functions of phenolic phytochemicals have also been shown due to their ability to act as animutagens in the Ames test (94–96). Many studies have also shown that these phenolic phytochemicals can repress the activity of enzymes such as the CYP class of enzymes, involved in the activation of procarcinogens. The protective functions of the liver against carbon tetrachloride toxicity (98) have shown that phenolic phytochemicals also decrease the carcinogenic potential of a mutagen and can activate enzymatic systems (Phase II) involved in the detoxification of xenobiotics (2). Antioxidant properties of the phenolic phytochemicals can also prevent oxidative damage to DNA which has been shown to be important in the age related development of some cancers (99). Other phenolics such as caffeic and ferulic acids react with nitrite *in vitro* and inhibit nitrosamine formation *in vivo*. They inhibit the formation of skin tumors induced by 7,12-dimethyl-benz(a) anthracene in mice (100). Important biphenyls such as resveratrol (Figure 10.1) which is found in wine has been shown to inhibit the development of preneoplastic lesions in rat mammary gland tissue in cultures in the presence of carcinogens, and it was also shown to inhibit skin tumors in mice (101,102).

Another well described function of phenolic phytochemicals is the prevention of cardiovascular diseases (CVDs). The lower incidences of CVDs in populations consuming wine as part of their regular diet is well established, and is often referred to as the French paradox (103). Recent research into the beneficial effects of wine has led to an understanding that resveratrol, which is a phenolic phytochemical, is the active component in wine responsible for its beneficial effects. Resveratrol and other phenolic antioxidants have also been shown to prevent development of CVDs by inhibiting LDL oxidation *in vitro* (104) and preventing platelet aggregation. Phenolic phytochemicals have also been able to reduce blood pressure and have antithrombotic and antiinflammatory effects (105,106). Phenolic phytochemicals have also been shown to inhibit the activity of α-amylase and α-glucosidase, which are resposible for postprandial increase in blood glucose level, which has been implicated in the manifestation of type II diabetes (107,108).

In addition to managing oxidation, several studies have indicated the ability of phenolic phytochemicals to manage infectious diseases. Antibacterial, antiulcer, antiviral, and antifungal properties (109–112) of the phenolic extracts have been described. Immune modulatory activities of phenolic phytochemicals such as antiallergic properties as a result of suppressing the hypersensitive immune response have also been defined (113). Antiinflammatory responses mediated by suppression of the TNF-α mediated proinflammatory pathways have also been shown to be mediated by phenolic phytochemicals (114).

10.2.3.1 Biological Functionality of Cranberry Phenolics

Cranberries and their products have been part of North American and Western European cuisines for many centuries. Foods that contain cranberries and their products have been associated historically with many positive benefits on human health. For many decades, cranberry juice has been widely used, particularly in North America, as a folk remedy to treat urinary tract infections (UTIs) in women, as well as other gastrointestinal (GI) disorders (115,116). These infections have now been shown to be caused by the infections of the GI tract by *Escherichia coli* and other pathogens. Recent clinical studies have established a positive link in prevention of urinary tract infection with the consumption of cranberry juice (117). Cranberry, like other fruits and berries is rich source of many bioactive components including phenolic phytochemicals such as phenolic acids, flavonoids, anthocyanins and their derivatives (118). p-Hydroxybenzoic acid, a phenolic acid present at high concentrations in cranberry, was believed to be the primary bioactive component in preventing urinary tract infections (38). This was believed to be due to the bacteriostatic effect of hippuric acid which is formed from metabolic conversion of p-Hydroxybenzoic acid in the liver. Hippuric acid, when excreted into the renal system, causes acidification of urine and thus prevents the growth of *Escherichia coli* on the urinary tract (38). It is now well established that adherence of the pathogen to the host tissue is also one of the most important steps required for the colonization of the bacteria and their subsequent infection. A majority of infectious diseases, including UTIs that are caused by microorganisms, have now been shown to involve the adherence of the pathogen to the host tissue (119,120). Investigations into the mechanism of adherence to host tissue has led to an understanding that these are mediated by specific glycoprotein receptors called fimbriae or lectins on the bacterial cell surface which can specifically bind to sugars present on the mucosal or intestinal cell surfaces of the host tissue (119,120). Many soluble and nondigestible sugars and oligosaccharides, such as fructose and mannooligosaccharides, can act as decoy sugars, forcing the bacteria to bind to them instead of the host cell. Inability of the pathogen to bind to the cell surface causes the microorganism to be washed away by the constant peristaltic motion in the intestine. This type of binding however, occurs only via a specific type of fimbriae called type 1 (mannose sensitive) fimbriae (119,120). Recent investigations have shown that type P fimbriae [α-Gal(1$\rightarrow$4)β-Gal] mediated adhesion, which is mannose resistant, is also involved in bacterial adhesion. Components of fruit juices, including cranberry juice, have been proposed to inhibit bacterial adherence to the epithelial cells by competing to bind with both these fimbriae (119,120). In addition to the extensive studies done on the inhibition of the adherence of components of *Escherichia coli* to host mucus cells, recent *in vitro* studies indicate a high molecular weight component in cranberry to inhibit the siallylactose specific (S-fimbriae) adhesion of *Helicobacter pylori* strains to immobilized human mucus, erythrocytes, and cultured gastric epithelial cells. It is suspected that these high molecular weight components from cranberries can inhibit the adhesion of *Helicobacter pylori* to the stomach *in vivo* and therefore may have a potential inhibitory effect on the development of stomach ulcers (121,122). Certain high as well as low molecular weight preparations of cranberry juice were also effective in decreasing the congregation and salivary concetration of *Streptococcus mutans,* which causes tooth decay (123,124). The formation of catheter blocking *Proteus mirabilis* biofilms in recovering surgical patients was also significantly decreased by the consumption of cranberry juice (125). Adherence of *Fusobacterium nucleatum* to buccal cells was also reduced by the high molecular weight extract from cranberry juice (123,124). Low and high molecular weight components from cranberry are also suspected to have antiviral properties because of the ability of tannins and other polyphenols to form noninfectious complexes with viruses. Cranberry and its products are also known to inhibit many fungi belonging to *Candida* species, *Microsporum* species and *Trycophyton* species (126,127).

Recent studies have reported on the radical scavenging activities of the various flavonol glycosides and anthocyanins in whole cranberry fruit and their considerable ability to protect against lipoprotein oxidation *in vitro*. The flavonoid and hydrocinnamic acid derivatives in cranberry juice reduced the oxidation of LDL and LDL mobility (128). In an *in vitro* study cranberry samples significantly inhibited both H_2O_2 and TNFα induced vascular endothelial growth factor (VEGF) expression by the human keratinocytes (129). Matrigel assay using human dermal microvascular endothelial cells showed that edible cranberries impair angiogenesis (129). It is therefore believed that cranberry juice may also have beneficial effects on cardiovascular health (130,131).

Cranberry and cranberry extracts have been shown to have anticancer properties. Phenolic extracts from berries of *Vaccinium* species were able to modulate the induction and repression of ornithine decarboxylase (ODC) and quinone reductase that critically regulate tumor cell proliferation (132). Cranberry extracts showed *in vitro* antitumor activity by inhibiting the proliferation of MCF-7 and MDA-MB-435 breast cancer cells. Cranberry extracts also exhibited a selective tumor cell growth inhibition in prostate, lung, cervical, and leukemia cell lines (132,133). Solid-state bioprocessing of natural products including cranberry pomace had shown to enhance its functionality. The antioxidant activity of the cranberry pomace was improved significantly after solid-state bioprocessing with fungi. Bioprocessing of cranberry pomace was found to release phenolic aglycones and enhance the phenolic profile of the pomace with important functionally relevant diphenyls such as ellagic acid. The antimicrobial properties of the pomace extract against foodborne and human pathogens such as *Listeria monocytogenes, Eschereschia coli O157: H7, Vibrio parahemolyticus* and *Helicobacter pylori* significantly improved after solid-state bioprocessing (134).

10.2.3.2 Ellagic Acid

Ellagic acid is a naturally occurring phenolic lactone compound found in a variety of natural products (Figure 10.1). Ellagic acid is present in plants in the form of hydrolyzable tannins called ellagitannins as the structural components of the plant cell wall and cell membrane. Ellagitannins are esters of glucose with ellagic acid which, when hydrolyzed, yield ellagic acid. Ellagic acid is seen at high concentrations in many berries including strawberries, raspberries, cranberries, and grapes (38,39). Other sources of ellagic acid include walnuts, pecans (135), and distilled beverages (136). Recent studies have indicated that ellagic acid possesses antimutagenic, antioxidant, and antiinflammatory activity in bacterial and mammalian systems (137–140).

Ellagic acid has been shown to be a potent anticarcinogenic agent. One of the main mechanisms by which ellagic acid is proposed to have anticancer benefits is by modulating the metabolism of environmental toxins and therefore preventing initiation of carcinogenesis induced by these chemicals (141). It is also proposed to show antimutagenic activity by inhibiting the direct binding of these carcinogens to the DNA (142).

Ellagic acid was found to inhibit the mutagenesis induced by aflatoxin B1 in *Salmonella* tester strains TA 98 and TA 100 (143). On oral administration, ellagic acid exhibited hepatoprotective activity against carbon tetrachloride both *in vitro* and *in vivo* (144). Ellagic acid inhibited the DNA binding and DNA adduct formation of N-nitrosobenzylmethylamine (NBMA) in cultured explants of rat esophagus (145). Related studies have shown that ellagic acid inhibited both the metabolism of NBMA and the binding of NBMA metabolites to DNA (146). In human epithelial cells ellagic acid also inhibited the binding of carcinogenic benzo[a]pyrene metabolites to DNA (142), and dibenzo[a,l]pyrene-DNA adduction in human breast epithelial cell line MCF-7 (147). Smith et al. (148) also showed that ellagic acid resulted in substantially reduced (>70%) DNA binding of 7,12-dimethylbenz[a]anthracene (DMBA) and suggested that possible mechanisms for the observed adduct reduction include direct interaction of the chemopreventive agent with the carcinogen or its metabolite, inhibition of phase I enzymatic activity, or formation of adducts with DNA, thus masking binding sites to be occupied by the mutagen or carcinogen (142). Ellagic acid also significantly increased the GST enzyme activity, GST isozyme levels, and glutathione levels, and therefore is proposed to show strong chemoprotective effects by selective enhancement of members of the GST detoxification system in the different cancerous cells (142).

Ellagic acid was also found to significantly reduce the number of bone marrow cells with chromosomal aberrations and chromosomal fragments as effectively as alpha tocopherol (149). Moreover, administration of ellagic acid inhibited radiation induced DNA strand breaks in rat lymphocytes. Ellagic acid induced G1 arrest, inhibited overall cell growth, and caused apoptosis in tumor cells (150). One of the studies reported a better protection by ellagic acid than vitamin E against oxidative stress (151). Ellagic acid reduced cytogenetic damage induced by radiation, hydrogen peroxide, and mitomycin C in bone marrow cells of mice (149,152). Chen et al. (153) suggested that the antitumor promoting action of ellagic acid and other related phenolics may be mediated in part by inducing a redox modification of protein kinase C (PKC) which serves as a receptor for tumor promoters. Ellagic acid is also suggested to carry out its antimutagenic and anticarcinogenic effects through the inhibition of xenobiotic metabolizing enzymes (141), and by the induction of antioxidant responsive element (ARE) mediated induction of NAD(P)H: quinone reductase and glutathione S-transferase (GST) genes, which can detoxify carcinogens and reduce carcinogen induced mutagenesis and tumorigenesis (154,155). Wood et al. (140) showed that ellagic acid is a potent antagonist to the adverse biological effects of the ulimate carcinogenic metabolites of several polycyclic aromatic hydrocarbons, and suggested that this naturally occurring plant phenolic, normally ingested by humans, may inhibit the carcinogenicity of polycyclic aromatic hydrocarbons.

Studies have shown that ellagic acid is a potent inhibitor of DNA topoisomerases, which are involved in carcinogenesis. Structure and activity studies identified the 3,3′-hydroxyl groups and the lactone groups as the most essential elements for the topoisomerase inhibitory actions of plant phenolics (156). Some recent studies have shown that ellagic acid was found to be better than quercetin for chemoprevention (139). When the effect of both of these compounds on reduced glutathione (GSH), an important endogenous antioxidant, and on lipid peroxidation, was investigated in rats, both ellagic acid and quercetin caused a significant increase in GSH and decrease in NADPH dependent and ascorbate dependent lipid peroxidation. However, ellagic acid was found to be more effective in decreasing the lipid peroxidation and increasing the GSH (139). This suggested that it may be more effective in inducing the intracellular synthesis of GSH and may be more capable of regenerating the oxidized GSH. This may be one of the reasons for the well documented anticarcinogenic activity of ellagic acid compared to quercetin (139). When the ability of vitamin E succinate and ellagic acid to modulate 2,3,7,8-tetrachlorodibenzo-p-dioxin (TCDD)-induced developmental toxicity and oxidative damage in embryonic or fetal and placental tissues was compared in C57BL/6J mice (151), ellagic acid provided better protection than vitamin E succinate and decreased lipid peroxidation in embryonic and placental tissues.

10.2.3.3 Alternate Model for the Function of Ellagic Acid and Related Phenolic Phytochemicals from Cranberries

Recent research has shown that phenolic phytochemicals such as ellagic acid from cranberries and other fruits have several health benefits. Several studies have suggested many mechanisms as the mode of action of ellagic acid and several other related phenolics. Primarily the mechanism of action has been defined as being able to counter the negative effects of stress at late stages of pathogenecity by aiding the regeneration of cellular antioxidants such as GSH and ascorbate, and by activation or induction genes responsible for expressing enzymes such as superoxide dismutase (SOD), catalase (CAT), glutathione-S-transferase (GST), Quinone: NADPH oxidoreductase (QR), and others which are involved in managing the oxidative stress (141,155). Control of oxidative stress induced diseases is also believed to be brought about by repressing certain genes such as the cytochrome P450 dependent Phase I enzymes (157,158), inhibition of NADPH oxidase, and other systems that generate ROS (63). Though these proposed mechanisms are valid and justified with experimental findings, however, they do not explain the larger, more comprehensive, functions of

in Gram-negative bacteria such as *Eschereschia coli* which have an external lipopolysaccharide layer and additional minor membrane components besides an intact plasma membrane. This gives them potentially more buffering capacity and hydrophobicity and couls, therefore, create an unfavorable environment for simple phenolics to exert their hyperacidification effect.

10.3 INNOVATIVE STRATEGIES TO ENRICH FRUITS WITH PHENOLIC ANTIOXIDANTS

A large variety of phenolic phytochemicals that have several beneficial functions on human health are present in plants and especially in fruits. The phenolic phytochemicals are generally present in their glycosidic and nonglycosidic forms. The glycosides are mainly confined to hydrophilic regions in the cells, such as in vacuoles and apoplasts, probably because of their higher water solubility (197,198). Glycosylation of the hydroxyl groups on the phenolic ring of a phenolic phytochemical renders the molecule more water soluble and less reactive toward free radicals (6). Glucose is the most commonly involved sugar in glycoside formation, although phenolic glycosides of galactose, rhamnose, xylose, and arabinose; and disaccharides such as rutinose, have also been reported in plants (6). Polymeric phenolics such as tannins exist primarily as condensed tannins or proanthocyanidins and are formed biosynthetically by the condensation of single catechins and flavonols. They are present either as soluble tannins or bound to the cell wall. Hydrolysable tannins are esters of a sugar with either gallic acid (gallotannins) or ellagic acid (ellagitannins). Tannins, even though they have higher antioxidant properties than individual simple phenolics, are usually not bioavailable, and are to some extent antinutritive in their function because of their ability to bind and precipitate biological macromolecules such as proteins and carbohydrates (109). The total phenolic phytochemical content in plant foods also varies greatly. Their presence in plant foods is largely influenced by genetic factors and environmental conditions. Other factors, such as cultivar, variety, maturity, processing, and storage, also influence the content of plant phenolics (199–201). The effects of processing and storage on the changes and content of polyphenols in cranberry (202), plum (203), and grape juice (204) have been evaluated.

As a consequence of evidence that consumption of fruits and vegetables has been linked to decreased manifestation of chronic diseases, there has been a constant increase in the demand for diets rich in phenolic phytochemicals. Vast variation in the amounts of phenolic antioxidants available via diet (205) coupled with reduced bioavailability and functionality has led to an urgent need to develop innovative strategies to enrich diets with phenolics and specifically phenolic antioxidants with consistent phytochemical profile for enhanced health functionality.

Among many strategies, two are important to enrich phenolic antioxidants. The first is genetic modification of cultivars to produce plants that will yield fruits and vegetables with higher phenolic concentration. Currently, in terms of genetic improvement, breeding strategy coupled with micropropagation using tissue culture is being developed (206–208). These strategies, along with genetic modification, could be directed toward phytochemical enrichment and quality improvement. However, this method presents important issues, such as regulation of key metabolites by multiple genes and biochemical pathways, acceptance of genetically modified foods, and relative time and economic considerations that are involved (209). Another exciting strategy that can be used is the bioprocessing of botanicals using solid-state bioprocessing and synergies to generate phytochemical profiles with enhanced health functionality. This strategy can be used for juice and pulp as well as pomace that remains after the juice is extracted from the fruits. Fermentation of fruit juices, such as grape juice to wine, has already been shown to improve nutritive and health promoting activities (210–212). Solid-state bioprocessing done on the pulps using food grade fungi can result in enrichment of the pulps with phenolic antioxidants and functionally important phenolic phytochemicals, and also improve phytochemical profile consistency.

10.4 SOLID-STATE BIOPROCESSING

Fermented foods have been consumed by humans all over the world for centuries. Most fermentation processes are conducted with liquid nutrient broths. Well known examples in the food industry are the production of yogurt, beer, wine, lactic acid, and many food flavors (213). However, partial fermentation and aerobic microbial growth based bioprocessing has also been used for processing food and food wastes. Here, instead of a nutrient broth, moist solid nutrients with minimal water are used as a substrate for microbial growth. This process is referred to as solid-state bioprocessing. Microbial fermentation and aerobic microbial growth on foods in solid state, for preservation of food and flavor enhancement, has been done for centuries and some of the common examples for these processes include manufacture of cheese and bread (214). Other wellknown examples are the production of microbe laced cheeses such as Roquefort, and the production of fermented sausages. In Asia, solid-state bioprocessing has been used for food fermentation for over 2000 years, for the production of fermented foods such as tempeh, natto, and soy sauce (193). The preservation of fish and meat by solid-state bioprocessing has also been reported to be carried by early human civilizations (193,214). Fruit wastes have been extensively used as substrates for solid-state bioprocessing. These wastes have mostly been used for the production of fertilizers, animal feed, as a growth substrate for mushrooms, ethanol production, production of organic acids such as citric acid, tartaric acid, and lactic acid, and for the production of various kinds enzymes such as pectinases (41,213). Solid-state bioprocessing of fruit substrates has also been carried out for several decades to produce compounds like gallic acid and vinegar (193,214). Recent research has also shown that consumption of fermented foods, especially solid-state bioprocessed foods is beneficial for health (215,216). Solid-state bioprocessing of fruit wastes such as cranberry pomace using the food grade fungi *Rhizopus oligosporus* and *Lentinus edodes* has been shown to enrich phenolic antioxidants and improve phytochemical consistency. These studies have shown that during the course of solid-state growth, the antioxidant activity and phenolic content of the pomace extracts increase several fold (52,53). The process resulted in enrichment of functional phenolics to a level found usually in fresh fruits and their juice products (217). The antimicrobial activity of the extracts against pathogens such as *Escherichia coli O157:H7, Listeria monocytogenes, Vibrio parahemolyticus* and *Helicobacter pylori* of cranberry pomace was also enhanced by solid-state bioprocessing. It is suspected that the increase in phenolic and antioxidant activity could have been due to the production of various hydrolyzing enzymes by the fungi during the course of solid-state growth. These fungal hydrolases such as glucosidase and fructofuranosidases could possibly be hydrolyzing the glycosidic linkages between the phenolic moieties and sugars. A similar observation in the increase in phenolic aglycones was observed during the fermentation of soy milk for the production of tofu (218). The fungus, in adapting itself to utilize the fruit substrates, may produce various other types of hydrolases such as laccases and lignocellulases. The activity of these enzymes is suspected to responsible for the increase in the polymeric phenolics, and potentially to contribute to the enhanced bioavailability of such phenolic antioxidants. Enrichment of the solid-state bioprocessed fruit wastes such as cranberry pomace with ellagic acid after bioprocessing has been reported (52,53). This may have resulted due to the hydrolysis of ellagotannins by tannin hydrolyzing enzymes produced by the fungus. Further, it is suspected that phenolic enrichment could also occur through contribution from the growing fungal species. The endogenous phenolics present in the fruit wastes could be toxic to the growing fungus. In an attempt to adapt and utilize the substrate for growth, the fungus could be detoxifying the phenolics biochemically using a variety of enzymatic systems present in the fungus. The fungal detoxification can occur by a variety of mechanisms including methylating or demethylating the pheniolic ring, or by hydroxylation (219,220). Recent studies have shown methylated phenolic phytochemicals have excellent antibacterial properties against Gram-positive bacteria (221). Hydroxylation of the phenolic ring by the fungal system during its growth increases antioxidant properties (19) and therefore, phenolics resulting from biotransformation occurring during the solid-state bioprocessing may improve their functionality and be beneficial for human health. The advantage of this strategy is that the fungi, such as *Rhizopus oligosporus* and *Lentinus edodes* and other fungi used in this solid-state growth process are food grade and are generally recognized as safe (GRAS). This approach can easily be adapted to different substrates as well as be extended to liquid fermentation of juices to develop food and ingredients with enhanced functionality.

10.5 CRANBERRY SYNERGIES WITH FUNCTIONAL PHYTOCHEMICALS AND OTHER FRUIT EXTRACTS

Recent research has documented the evidence that whole foods, and not single compounds, have a better functionality in maintaining our health against many of the antioxidant diseases. Recent work on the effect of wine has shown that resveratrol is responsible for the decrease in atherogenesis in rats (101,102). However, when resveratrol was used as a supplement in diet such an effect was not seen. Other researchers have shown that the combination of resveratrol and quercetin exerts a synergic effect in the inhibition of growth and proliferation of human oral squamous carcinoma cells (221). Carbonneau et al. (223) during *in vivo* antioxidant assays with red wine, observed that different phenolics in wine could play a coantioxidant role, similar to that described for vitamin C, and a sparing role toward vitamin E, which increases due to supplementation with phenolics. Synergistic interactions between wine polyphenols, quercetin, and resveratrol were found to decrease the inducible nitric oxide synthase (iNOS) activity in cell culture systems (224). This suggests that the phytochemical profile in which the specific functional phenolic is present plays an important role in determining its functionality. Synergy can be defined as the ability of two or more functional components, such as antioxidants in a phytochemical background, to mutually enhance their functionalities. Typically, in a whole food background such as red wine, resveratrol is present in a background containing several simple as well as polyphenols, such as gallic acid, protocatcheuic acid and hydrolyzed tannins. Each phenolic phytochemical has its own mode of action against a particular target. These modes of actions could be due to their ability to function as classical antioxidants or because of their ability to modulate cellular physiology by disrupting membrane functions or by altering the redox balance and energy metabolism of the cell (160). However, when they are present together, their ability to function together rapidly improves the overall result of maintaining the cellular homeostasis, especially in eukaryotes, or killing the pathogen in prokaryotes. Also, it was observed that coadministration of coffee with ellagic acid enhanced the antigenotoxic effect compared with that of either coffee or the ellagic acid alone, suggesting that there is a significant synergistic interaction between coffee and the dietary constituents for antigenotoxic effects against different mutagens (225). The conditions created due to the mode of action of one phenolic significantly improves the chances of the other phenolics being able to function effectively, thereby reducing the overall dosage required to observe the desired positive effect. This may be one of the reasons why whole foods have a better functionality in maintaining human health compared to the consumption of supplements. This synergy concept can also be artificially duplicated, and this provides another way to significantly enhance the functionality of foods. Synergistic supplementation of foods such as fruit beverages with flavonoids and other functional phenolic phytochemical can drastically improve their functionality.

In a recent study the potential antimutagenic properties of cranberry phenolics, ellagic acid, rosmarinic acid, and their synergistic interactions on enhancing antimutagenic properties in *Salmonella typhimurium* tested the system against mutagens sodium azide and N-methyl-N′-nitro-N-nitrosoguanidine (MNNG) was investigated (226). Ability of these phytochemical treatments to prevent oxidative damage to DNA was also investigated, using the supercoiled DNA strand scission assay. Results showed that ellagic acid was most effective in inhibiting the mutations in *Salmonella typhimurium* system, whereas rosmarinic acid and ellagic acid were equally effective in protecting the DNA from oxidative damage. The antimutagenic functionality of cranberry powder made from juice extracts was significantly enhanced when 30% (w/w) of phenolics in cranberry powder were substituted with rosmarinic acid and ellagic acid possibly due to synergistic redox modulation which can influence mutagen function. It was suggested that the synergistic mixture of cranberry phenolics with rosmarinic acid could also be protecting the cell from mutations by modulating DNA repair systems.

Phenolic antioxidant and α-amylase inhibition activity of cranberry, wild blueberry and grape seed extracts and their synergistic mixtures were investigated in a recent study to develop an additional strategy to manage type II diabetes (227). The results indicated that all the extracts had α-amylase inhibition activity which correlated to the presence of specific phenolic phytochemicals such as chlorogenic acid, ellagic acid, and rosmarinic acid, suggesting a possible structure related inhibition of α-amylase. Among the fruit juice powders the cranberry powder had the highest α-amylase inhibition activity. A mixture containing 75% cranberry, 15% blueberry and 10% grape showed a synergistic mode of action and was the most optimal mixture to control α-amylase activity.

A similar study investigated the effect of cranberry, blueberry and grape seed extracts on inhibiting *H. pylori* (228). The results showed that the anti-*H. pylori* activity of cranberry juice extract was significantly improved by its synergistic blending with blueberry or grape seed extract. In both the studies, the lower efficacy of purified phenolics in inhibiting α-amylase or *H. pylori* compared with fruit powder at similar dosage levels suggests a synergistic mode of functionality of these individual phenolics in whole food background.

Consumption of blends of fruit juices with biologically active biphenyls or other fruit as well as herb extracts can impart unique functional attributes and could be a more effective strategy in developing diet based management of *H. pylori* infections, as well as other oxidation linked diseases, including diabetes and mutagen and DNA damage induced carcinogenesis.

10.6 CONCLUSIONS

Emerging epidemiological evidence is increasingly pointing to the beneficial effects of fruits and vegetables in managing chronic and infectious diseases. These beneficial effects are now believed to be due to the constituent phenolic phytochemicals having antioxidant activity. Cranberry, like other fruits, is also rich in phenolic phytochemicals such as phenolic acids, flavonoids and ellagic acid. Consumption of cranberry has been historically been linked to lower incidences of urinary tract infection and now has been shown to have a capacity to decrease peptic ulcer caused by *Helicobacter pylori*. Isolated compounds from cranberry have been shown to reduce the risk of CVD and cancer. Functional phenolic antioxidants from cranberry such as ellagic acid have been well documented to have antimutagenic and anticarcinogenic functionality. Even though many benefits have been associated with phytochemicals from cranberry, such as ellagic acid, their mechanism of action is still not very well understood. Emerging research exploring the mechanism of action of these phytochemicals from cranberry usually follows a reductionist approach, and is often focused on the disease or pathological target. These approaches to understanding the mechanism of action of phytochemicals have limitations as they are unable to explain the overall preventive mode of action of phenolic phytochemicals. The current proposed mechanisms of action of these phenolic phytochemicals have also overemphasized the antioxidant activity of phenolic phytochemicals associated with free radical quenching capacity. This mechanism of action is unable to explain the nonantioxidant (free radical scavenging) role, as well as sometimes contradictory modes of action of phenolic phytochemicals across different species, such as being protective in eukaryotes while being inhibitory to prokaryotes. The often promising results seen at laboratory scale have very rarely been successful at clinical levels in terms of seeing the beneficial effects of these phenolic phytochemicals. All these limitations of the present understanding strongly suggest the involvement of phenolic phytochemicals much earlier in the cellular response towards maintaining antioxidant stress response. A current line of investigation after a newly proposed integrated model suggests the involvement of phenolic phytochemicals in critically regulating energy metabolism of the cell by stimulating the PPP in order to supply the cell with reducing equivalents necessary for the efficient functioning of antioxidant response. A coupling of the proline biosynthesis with PPP has been suggested, which can further stimulate the PPP. Proline can also function as an alternative reductant to participate in the oxidative phosphorylation for ATP synthesis. This would effectively reduce the cellular needs for NADH, which could contribute to reducing oxidative stress as a result of reduced oxidative burst in the mitochondria. This model emphasizes the structure and function aspect of phenolic phytochemicals in their ability to cause hyperacidification, alter membrane permeability to ions, and interact with the membrane proteins and receptors to activate many signaling pathways which can be responsible for their beneficial mechanism of action in both prokaryotes and eukaryotes. Recent evidence suggests the ability of whole foods in being more effective in managing human health compared to individual phenolic phytochemicals. This suggests that the profile of phenolic phytochemicals determines the functionality of the phenolic phytochemical as a result of synergistic interaction of constituent phenolic phytochemicals. Solid-state bioprocessing using food grade fungi, as well as cranberry phenolic synergies with functional biphenyls such as ellagic acid and rosmarinic acid, and other fruit extracts have helped to advance these concepts. These strategies could be further explored to enhance cranberry and cranberry products with functional phytochemicals and further improve their functionality.

REFERENCES

1. Mann, J. *Secondary Metabolism*: *Oxford Chemistry Series*. Oxford: Clarendon Press, 1978.
2. Bravo, L. Phenolic phytochemicals: chemistry, dietary sources, metabolism, and nutritional significance. *Nutr. Rev*. **56**:317–333, 1998.
3. Crozier, A., J. Burns, A.A. Aziz, A.J. Stewart, H.S. Rabiasz, G.I. Jenkins, C.A. Edwards, M.E.J. Lean. Antioxidant flavonols from fruits, vegetables and beverages: measurements and bioavailability. *Biol. Res*. 33:79–88, 2000.

4. Shetty, K. Biotechnology to harness the benefits of dietary phenolics: focus on *Lamiaceae*. *Asia Pac. J. Clin. Nutr.* 6:162–171, 1997.
5. Briskin, D.P. Medicinal plants and phytomedicines: linking plant biochemistry and physiology to human health. *Plant Physiol.* 124:507–514, 2000.
6. Urquiaga, I., F. Leighton. Plant polyphenol antioxidants and oxidative stress. *Biol. Res.* 33:55–64, 2000.
7. Moure, A., J.M. Cruz, D. Franco, J.M. Domínguez, J. Sineiro, H. Domínguez, M.H. Núñez, J.C. Parajó. Natural antioxidants from residual sources. *Food Chem.* 72(2):145–171, 2001.
8. Bors, W., C. Michel. Chemistry of the antioxidant effect of polyphenols. *Ann. NY Acad. Sci.* 957:57–69, 2002.
9. Harborne, J.B. Plant phenolics. In: *Encyclopedia of Plant Physiology*, Vol. 8, Bella, E.A., B.V. Charlwood, eds., Heidelburg: Springer–Verlag. 1980, pp 329–395.
10. Haslam, E. *Practical Polyphenolics: From Structure to Molecular Recognition and Physiological Action*. Cambridge: Cambridge University Press, 1998.
11. Strack, D. Phenolic metabolism. In: *Plant Biochemistry*, Dey, P.M., J.B. Harborne, eds., San Diego: Academic Press, 1997, pp 387–416.
12. Wang, H., G. Cao, R.L. Prior. Total antioxidant capacity of fruits. *J. Agric. Food Chem.* 44:701–705, 1996.
13. Kalt, W., C.F. Forney, A. Martin, P.L. Prior. Antioxidant capacity, vitamin C, phenolics, and anthocyanins after fresh storage of small fruits. *J. Agric. Food Chem.* 47:4638–4644, 1999.
14. Abuja, P.M., M. Murkovic, W. Pfannhauser. Antioxidant and prooxidant activities of elderberry (*Sambucus nigra*) extract in low-density-lipoprotein oxidation. *J. Agric. Food Chem.* 46:4091–4096, 1998.
15. Heinonen, M., P.J. Lehtonen, A.L. Hopia. Antioxidant activity of berry and fruit wines and liquors. *J. Agric. Food Chem.* 46:25–31, 1998.
16. Sato, M., N. Ramarathnam, Y. Suzuki, T. Ohkubo, M. Takeuchi, H. Ochi. Varietal differences in the phenolic content and superoxide radical scavenging potential of wines from different sources. *J. Agric. Food Chem.* 44:37–41, 1996.
17. Lapidot, T., S. Harel, B. Akiri, R. Granit, J. Kranner. pH-dependent forms of red-wine anthocyanins as antioxidants. *J. Agric. Food Chem.* 47:67–70, 1999.
18. Larrauri, J.A., C. Sánchez-Moreno, P. Rupérez, F. Saura-Calixto. Free radical scavenging capacity in the aging of selected red spanish wines. *J. Agric. Food Chem.* 47:1603–1606, 1999.
19. Rice–Evans, C.A., N.J. Miller, G. Paganga. Antioxidant properties of phenolic compounds. *Trends Plant Sci.* 2:152–159, 1997.
20. Wang, S.Y., H. Lin. Antioxidant activity in fruits and leaves of blackberry, raspberry, and strawberry varies with cultivar and developmental stage. *J. Agric. Food Chem.* 48:140–146, 2000.
21. Kähkönen, M.P., A.I. Hopia, H.J. Vuorela, J.-P. Rauha, K. Pihlaja, T.S. Kujala, M. Heinonen. Antioxidant activity of plant extracts containing phenolic compounds. *J. Agric. Food Chem.* 47:3954–3962, 1999.
22. Knekt, P., R. Järvinen, R. Reppänen, M. Heliövaara, L. Teppo, E. Pukkala, A. Aroma. Dietary flavonoids and the risk of lung cancer and other malignant neoplasms. *Am. J. Epidemiol.* 146:223–230, 1997.
23. Gillman, M.W., L.A. Cupples, D. Gagnon, B.M. Posner, R.C. Ellison, W.P. Castelli, P.A. Wolf. Protective effect of fruits and vegetables on development of stroke in men. *J. Am. Med. Assoc.* 273:1113–1117, 1995.
24. Joshipura, K.J, A. Ascherio, J.E. Manson, M.J. Stampfer, E.B. Rimm, F.E. Speizer, C.H. Hennekens, D. Spiegelman, W.C. Willett. Fruit and vegetable intake in relation to risk of ischemic stroke. *J. Am. Med. Assoc.* 282:1233–1239, 1999.
25. Cox, B.D., M.J. Whichelow, A.T. Prevost. Seasonal consumption of salad vegetables and fresh fruit in relation to the development of cardiovascular disease and cancer. *Public Health Nutr.* 3:19–29, 2000.
26. Strandhagen, E., P.O. Hansson, I. Bosaeus, B. Isaksson, H. Eriksson. High fruit intake may reduce mortality among middle-aged and elderly men. The study of men born in 1913. *Eur. J. Clin. Nutr.* 54:337–341, 2000.

27. Hertog, M.G., E.J. Feskens, P.C. Hollman, M.B. Katan, D. Kromhout. Dietary antioxidant flavonoids and risk of coronary heart disease: the Zetphen Elderly Study. *Lancet* 342(8878):1007–1011, 1993.
28. Keli, S.O., M.G.L. Hertog, E.J.M. Feskens, D. Kromhout. Dietary flavonoides, antioxidant vitamins, and incidence of stroke: the Zetphen study. *Arch. Intern. Med.* 156:637–642, 1996.
29. Le Marchand, L., S.P. Murphy, J.H. Hankin, L.R. Wilkens, L.N. Kolonel. Intake of flavonoids and lung cancer. *J. Natl. Cancer Inst.* 92(2):154–160, 2000.
30. Garcia-Closas, R., C.A. Gonzalez, A. Agudo, E. Riboli. Intake of specific carotenoids and flavonoids and the risk of gastric cancer in Spain. *Cancer Causes Control* 10(1):71–75, 1999.
31. Zuo, Y., C. Wang, J. Zhan. Separation, characterization, and quantitation of benzoic and phenolic antioxidants in American cranberry fruit by GC-MS. *J. Agric. Food Chem.* 50(13):3789–3794, 2002.
32. Guo, C., G. Cao, E. Sofic, R.L. Prior. High-performance liquid chromatography coupled with coulometric array detection of electroactive components in fruits and vegetables: relationship to oxygen radical absorbance capacity. *J. Agric. Food Chem.* 45:1787–1796, 1997.
33. Cao, G., E. Sofic, R.L. Prior. Antioxidant and prooxidant behavior of flavonoids: structure-activity relationships. *Free Rad. Biol. Med.* 22:749–760, 1997.
34. Foti, M., M. Piattelli, M.T. Baratta, G. Ruberto. Flavonoids, coumarins, and cinnamic acids as antioxidants in a micellar system structure-activity relationship. *J. Agric. Food Chem.* 44:497–501, 1996.
35. Velioglu, Y.S., G. Mazza, L. Gao, B.D. Oomah. Antioxidant activity and total phenolics in selected fruits, vegetables, and grain products. *J. Agric. Food Chem.* 46:4113–4117, 1998.
36. Macheix. J.J., A. Fleuriet, J. Billot. *Fruit Phenolics*. Boca Raton, FL: CRC Press, 1990.
37. Larson, R.A. The antioxidants of higher plants. *Phytochemistry* 27:969–978, 1988.
38. Marwan, A.G., C.W. Nagel. Characterization of cranberry benzoates and their antimicrobial properties. *J. Food Sci.* 51:1069–1070, 1986.
39. Chen, H., Y. Zuo, Y. Deng. Separation and determination of flavonoids and other phenolic compounds in cranberry juice by high-performance liquid chromatography. *J. Chromatogr. A.* 13:913(1,2):387–395, 2001.
40. National Agricultural Statistics Service. *"Cranberries" Annual Reports, Fr Nt 4*, Agricultural Statistics board, Washington, DC, USDA, 2001.
41. Zheng, Z., K. Shetty. Cranberry processing waste for solid-state fungal inoculant production. *Proc. Biochem.* 33:323–329, 1998.
42. Rodríguez de Sotillo, D., M. Hadley, E.T. Holm. Potato peel waste, stability and antioxidant activity of a freeze-dried extract. *J. Food Sci.* 59:1031–1033, 1994.
43. Rodríguez de Sotillo, D., M. Hadley, E.T. Holm. Phenolics in aqueous potato peel extract, extraction, identification and degradation. *J. Food Sci.* 59:649–651, 1994.
44. Sheabar, F.Z., I. Neeman. Separation and concentration of natural antioxidants from the rape of olives. *J. Am. Oil Chem. Soc.* 65:990–993, 1988.
45. Yamaguchi, F., Y. Yoshimura, H. Nakazawa, T. Ariga. Free radical scavenging activity of grape seed extract and antioxidants by electron spin resonance spectrometry in an H_2O_2/NaOH/DMSO system. *J. Agric. Food Chem.* 47:2544–2548, 1999.
46. Koga, T., K. Moro, K. Nakamori, J. Yamakoshi, H. Hosoyama, S. Kataoka, T. Ariga. Increase of antioxidative potential of rat plasma by oral administration of proanthocyanidin-rich extract from grape seeds. *J. Agric. Food Chem.* 47:1892–1897, 1999.
47. Lu, Y., L.Y. Foo. The polyphenol constituents of grape pomace. *Food Chem.* 65:1–8, 1999.
48. Lu, Y., L.Y. Foo. Antioxidant and radical scavenging activities of polyphenols from apple pomace. *Food Chem.* 68:81–85, 2000.
49. Peleg, H., M. Naim, R.L. Rouseff, U. Zehavi. Distribution of bound and free phenolic acids in oranges (*Citrus sinensis*) and grapefruits (*Citrus paradisi*). *J. Sci. Food Agric.* 57:417–426, 1991.
50. Meyer, A.S., S.M. Jepsen, N.S. Sørensen. Enzymatic release of antioxidants for human low-density lipoprotein from grape pomace. *J. Agric. Food Chem.* 46:2439–2446, 1998.
51. Bocco, A., M.E. Cuvelier, H. Richard, C. Berset. Antioxidant activity and phenolic composition of citrus peel and seed extracts. *J. Agric. Food Chem.* 46:2123–2129, 1998.

52. Zheng, Z., K. Shetty. Solid-state bioconversion of phenolics from cranberry pomace and role of *Lentinus edodes* beta-glucosidase. *J. Agric. Food Chem.* 48:895–900, 2000.
53. Vattem, D.A., K. Shetty. Solid-state production of phenolic antioxidants from cranberry pomace by *Rhizopus oligosporus. Food Biotechnol.* 16(3):189–210, 2002.
54. Vattem, D.A., K. Shetty. Ellagic acid production and phenolic antioxidant activity in cranberry pomace (*Vaccinium macrocarpon*) mediated by *Lentinus edodes* using solid-state system. *Proc. Biochem.* 39(3):367–379, 2003.
55. Liochev S.I., I. Fridovich. The relative importance of HO* and ONOO- in mediating the toxicity of O*-. *Free Radic. Biol. Med.* 26(5,6):777–778, 1999.
56. Beckman J.S., W.H. Koppenol. Nitric oxide, superoxide, and peroxynitrite: the good, the bad, and ugly. *Am. J. Physiol.* 271(5,1):C1424–C1437, 1996.
57. Radi, R., A. Cassina, R. Hodara, C. Quijano, L. Castro. Peroxynitrite reactions and formation in mitochondria. *Free Radic. Biol. Med.* 33(11):1451–1464, 2002.
58. Barber, D.A., S.R. Harris. Oxygen free radicals and antioxidants: a review. *Am. Pharm.* NS34:26–35, 1994.
59. Betteridge, D.J. What is oxidative stress? *Metabolism* 49:3–8, 2000.
60. Vignais, P.V. The superoxide-generating NADPH oxidase: structural aspects and activation mechanism. *Cell Mol. Life Sci.* 59(9):1428–1459, 2002.
61. Babior, B.M., J.D. Lambeth, W. Nauseef. The neutrophil NADPH oxidase. *Arch. Biochem. Biophys.* 397(2):342–344, 2002.
62. Babior, B.M. The leukocyte NADPH oxidase. *Isr. Med. Assoc. J.* 4(11):1023–1024, 2002.
63. Parke, A., D.V. Parke. The pathogenesis of inflammatory disease: surgical shock and multiple system organ failure. *Inflammopharmacology* 3:149–168, 1995.
64. Parke, D.V. The cytochromes P450 and mechanisms of chemical carcinogenesis. *Environ. Health Perspect.* 102:852–853, 1994.
65. Parke, D.V., A. Sapota. Chemical toxicity and reactive oxygen species. *Int. J. Occ. Med. Environ. Health* 9:331–340, 1996.
66. Triggle, C.R., M. Hollenberg, T.J. Anderson, H. Ding, Y. Jiang, L. Ceroni, W.B. Wiehler, E.S. Ng, A. Ellis, K. Andrews, J.J. McGuire, M. Pannirselvam. The endothelium in health and disease: a target for therapeutic intervention. *J. Smooth Muscle Res.* 39(6):249–267, 2003.
67. Droge, W. Free radicals in the physiological control of cell function. *Physiol Rev.* 82(1):47–95, 2002.
68. Niess, A.M., H.H. Dickhuth, H. Northoff, E. Fehrenbach. Free radicals and oxidative stress in exercise: immunological aspects. *Exerc. Immunol. Rev.* 5:22–56, 1999.
69. Sculley, D.V., S.C. Langley-Evans. Salivary antioxidants and periodontal disease status. *Proc. Nutr. Soc.* 61(1):137–143, 2002.
70. Morel, Y., R. Barouki. Repression of gene expression by oxidative stress. *Biochem. J.* 342(3):481–496, 1999.
71. Lusini, L., S.A. Tripodi, R. Rossi, F. Giannerini, D. Giustarini, M.T. del Vecchio, G. Barbanti, M. Cintorino, P. Tosi, P. Di Simplicio. Altered glutathione anti-oxidant metabolism during tumor progression in human renal-cell carcinoma. *Int. J. Cancer 91* (1):55–59, 2001.
72. Mates, J.M., F. Sanchez-Jimenez. Antioxidant enzymes and their implications in pathophysiologic processes. *Front. Biosci.* 4:D339–345, 1999.
73. Mates, J.M., C. Perez-Gomez, I. Nunez de Castro. Antioxidant enzymes and human diseases. *Clin. Biochem.* 32(8):595–603, 1999.
74. Nordberg, J., E.S. Arner. Reactive oxygen species, antioxidants, and the mammalian thioredoxin system. *Free Radic. Biol. Med.* 31(11):1287–1312, 2001.
75. Stadtman, E.R., R.L. Levine. Protein oxidation. *Ann. NY Acad. Sci.* 899:191–208, 2000.
76. Rubbo, H., R. Radi, M. Trujillo, R. Telleri, B. Kalyanaraman, S. Barnes, M. Kirk, B.A. Freeman. Nitric oxide regulation of superoxide and peroxynitrite-dependent lipid peroxidation: formation of novel nitrogen-containing oxidized lipid derivatives. *J. Biol. Chem.* 269(42):26066–26075, 1994.
77. Richter, C., J.W. Park, B.N. Ames. Normal oxidative damage to mitochondrial and nuclear DNA is extensive. *Proc. Natl. Acad. Sci. USA* 85(17):6465–6467, 1988.

78. LeDoux, S.P., W.J. Driggers, B.S. Hollensworth, G.L. Wilson. Repair of alkylation and oxidative damage in mitochondrial DNA. *Mutat. Res.* 434(3):149–159, 1999.
79. Parke, A.L., C. Ioannides, D.F.V. Lewis, D.V. Parke. Molecular pathology of drugs: disease interaction in chronic autoimmune inflammatory diseases. *Inflammopharmacology* 1:3–36, 1991.
80. Schwarz, K.B. Oxidative stress during viral infection: a review. *Free Radic. Biol. Med.* 21(5):641–649, 1996.
81. Götz, J., C.I. va Kan, H.W. Verspaget, I. Biemond, C.B. Lamers, R.A. Veenendaal. Gastric mucosal superoxide dismutases in *Helicobacter pylori* infection. *Gut* 38:502–506, 1996.
82. Jakus, V. The role of free radicals, oxidative stress and antioxidant systems in diabetic vascular disease. *Bratisl. Lek. Listy.* 101(10):541–551, 2000.
83. Offen, D., P.M. Beart, N.S. Cheung, C.J. Pascoe, A. Hochman, S. Gorodin, E. Melamed, R. Bernard, O. Bernard. Transgenic mice expressing human Bcl-2 in their neurons are resistant to 6-hydroxydopamine and 1-methyl-4-phenyl-1,2,3,6- tetrahydropyridine neurotoxicity. *Proc. Natl. Acad. Sci.* 95:5789–5794, 1998.
84. Barbaste, M., B. Berke, M. Dumas, S. Soulet, J.C. Delaunay, C. Castagnino, V. Arnaudinaud, C. Cheze, J. Vercauteren. Dietary antioxidants, peroxidation and cardiovascular risks. *J. Nutr. Health Aging* 6(3):209–223, 2002.
85. Gerber, M., C. Astre, C. Segala, M. Saintot, J. Scali, J. Simony-Lafontaine, J. Grenier, H. Pujol. Tumor progression and oxidant-antioxidant status. *Cancer Lett.* 19:114(1,2):211–214, 1997.
86. Block, G., B. Patterson, A. Subar. Fruit, vegetables, and cancer prevention: a review of the epidemiological evidence. *Nutr. Cancer* 18:1–29, 1992.
87. Serdula, M.K., M.A.H. Byers, E. Simoes, J.M. Mendlein, R.J. Coates. The association between fruit and vegetable intake and chronic disease risk factors. *Epidemiology* 7:161–165, 1996.
88. Tapiero, H., K.D. Tew, G.N. Ba, G. Mathe. Polyphenols: do they play a role in the prevention of human pathologies? *Biomed. Pharmacother.* 56(4):200–207, 2002.
89. Duthie, G.G., P.T. Gardner, J.A. Kyle. Plant polyphenols: are they the new magic bullet? *Proc. Nutr. Soc.* 62(3):599–603, 2003.
90. Lundberg, A.C., A. Åkesson, B. Åkesson. Dietary intake and nutritional status in patients with systemic sclerosis. *Ann. Rheum. Dis.* 51:1143–1148, 1992.
91. Yoshioka, A., Y. Miyachi, S. Imamura, Y. Niwa. Anti-oxidant effects of retinoids on inflammatory skin diseases. *Arch. Dermatol. Res.* 278:177–183, 1986.
92. Wilks, R., F. Bennett, T. Forrester, N. Mcfarlane-Anderson. Chronic diseases: the new epidemic. *West Ind. Med. J.* 47(4):40–44, 1998.
93. Leighton, F., A. Cuevas, V. Guasch, D.D. Perez, P. Strobel, A. San Martin, U. Urzua, M.S. Diez, R. Foncea, O. Castillo, C. Mizon, M.A. Espinoza, I. Urquiaga, J. Rozowski, A. Maiz, A. Germain. Plasma polyphenols and antioxidants, oxidative DNA damage and endothelial function in a diet and wine intervention study in humans. *Drugs Exp. Clin. Res.* 25(2,3):133–141, 1999.
94. Yamada, J., Y. Tomita. Antimutagenic activity of caffeic acid and related compounds. *Biosci. Biotechnol. Biochem.* 60(2):328–329, 1996.
95. Mitscher, L.A., H. Telikepalli, E. McGhee, D.M. Shankel. Natural antimutagenic agents. *Mutat. Res.* 350:143–152, 1996.
96. Uenobe, F., S. Nakamura, M. Miyazawa. Antimutagenic effect of resveratrol against Trp-P-1. *Mutat. Res.* 373:197–200, 1997.
97. Kuroda, Y., T. Inoue. Antimutagenesis by factors affecting DNA repair in bacteria. *Mutat. Res.* 202(2):387–391, 1988.
98. Kanai, S., H. Okano. Mechanism of protective effects of sumac gall extract and gallic acid on progression of CC14-induced acute liver injury in rats. *Am. J. Chin. Med.* 26:333–341, 1998.
99. Halliwell, B. Establishing the significance and optimal intake of dietary antioxidants: the biomarker concept. *Nutr. Rev.* 57:104–113, 1999.
100. Kaul, A., K.I. Khanduja. Polyphenols inhibit promotional phase of tumorigenesis: relevance of superoxide radicals. *Nurt. Cancer* 32:81–85, 1998.

101. Clifford, A.J., S.E. Ebeler, J.D. Ebeler, N.D. Bills, S.H. Hinrichs, P.-L. Teissedre, A.L. Waterhouse. Delayed tumor onset in transgenic mice fed an amino acid-based diet supplemented with red wine solids. *Am. J. Clin. Nutr.* 64:748–756, 1996.
102. Jang, M., L. Cai, G.O. Udeani, K.V. Slowing, C. Thomas, C.W.W. Beecher, H.H.S. Fong, N.R. Farnsworth, A.D. Kinghorn, R.G. Mehta, R.C. Moon, J.M. Pezzuto. Cancer chemopreventive activity of resveratrol, a natural product derived from grapes. *Science* 275:218–220, 1997.
103. Ferrieres, J. The French paradox: lessons for other countries. *Heart* 90(1):107–111, 2004.
104. Frankel, E.N, J. Kanner, J.B. German, E. Parks, J.E. Kinsella. Inhibition of oxidation of human low-density lipoprotein by phenolic substances in red wine. *Lancet* 341:454–457, 1993.
105. Gerritsen, M.E., W.W. Carley, G.E. Ranges, C.-P. Shen, S.A. Phan, G.F. Ligon, C.A. Perry. Flavonoids inhibit cytokine-induced endothelial cell adhesion protein gene expression. *Am. J. Pathol.* 147:278–292, 1995.
106. Muldoon, M.F., S.B. Kritchevsky. Flavonoids and heart disease. *BMJ* 312(7029):458–459, 1996.
107. McCue, P.P., K. Shetty. Inhibitory effects of rosmarinic acid extracts on porcine pancreatic amylase *in vitro*. *Asia Pac. J. Clin. Nutr.* 13(1):101–106, 2004.
108. Andlauer, W., P. Furst. Special characteristics of non-nutrient food constituents of plants: phytochemicals: introductory lecture. *Int. J. Vitam. Nutr. Res.* 73(2):55–62, 2003.
109. Chung, K.T., T.Y. Wong, C.I. Wei, Y.W. Huang, Y. Lin. Tannins and human health: a review. *Crit. Rev. Food Sci. Nutr.* 38:421–464, 1998.
110. Ikken, Y., P. Morales, A. Martínez, M.L. Marín, A.I. Haza, M.I. Cambero. Antimutagenic effect of fruit and vegetable ethanolic extracts against N-nitrosamines evaluated by the Ames test. *J. Agric. Food Chem.* 47:3257–3264, 1999.
111. Vilegas, W., M. Sanomimiya, L. Rastrelli, C. Pizza. Isolation and structure elucidation of two new flavonoid glycosides from the infusion of *Maytenus aquifolium* leaves: evaluation of the antiulcer activity of the infusion. *J. Agric. Food Chem.* 47:403–406, 1999.
112. Abram, V., M. Donko. Tentative identification of polyphenols in *Sempervivum tectorum* and assessment of the antimicrobial activity of *Sempervivum* L. *J. Agric. Food Chem.* 47:485–489, 1999.
113. Noguchi, Y., K. Fukuda, A. Matsushima, D. Haishi, M. Hiroto, Y. Kodera, H. Nishimura, Y. Inada. Inhibition of Df-protease associated with allergic diseases by polyphenol. *J. Agric. Food Chem.* 47:2969–2972, 1999.
114. Ma, Q., K. Kinneer. Chemoprotection by phenolic antioxidants. Inhibition of tumor necrosis factor alpha induction in macrophages. *J. Biol. Chem.* 277(4):2477–2484, 2002.
115. Mowrey, D.B. *The scientific validation of herbal medicine*. New York: McGraw-Hill, 1990, pp 255–264.
116. Yan, X., B.T. Murphy, G.B. Hammond, J.A. Vinson, C. Neto. Antioxidant activities and antitumor screening of extracts from cranberry fruit (*Vaccinium macrocarpon*) *J. Agric. Food Chem.* 50:5844–5849, 2002.
117. Griffiths, P. The role of cranberry juice in the treatment of urinary tract infections. *Br. J. Comm. Nurs.* 8(12):557–561, 2003.
118. Vvedenskaya, I.O., R.T. Rosen, J.E. Guido, D.J. Russell, K.A. Mills, N. Vorsa. Characterization of flavonols in cranberry (*Vaccinium macrocarpon*) powder. *J. Agric. Food Chem.* 52(2):188-95, 2004.
119. Zafriri, D., I. Ofek, R. Adar, M. Pocino, N. Sharon. Inhibitory activity of cranberry juice on adherence of type 1 and type P fimbriated *Escherichia coli* to eukaryotic cells. *Antimicrob. Agents Chemother.* 33(1):92–98, 1989.
120. Sharon N, I. Ofek. Fighting infectious diseases with inhibitors of microbial adhesion to host tissues. *Crit. Rev. Food Sci. Nutr.* 42(3):267–272, 2002.
121. Burger, O., I. Ofek, M. Tabak, E.I. Weiss, N. Sharon, I. Neeman. A high molecular constituent of cranberry juice inhibits *Helicobactor pylori* adhesion to human gastric mucus. *FEMS Immunol. Med. Microbiol.* 29:295–301, 2000.

122. Burger, O., E. Weiss, N. Sharon, M. Tabak, I. Neeman, I. Ofek. Inhibition of *Helicobacter pylori* adhesion to human gastric mucus by a high-molecular-weight constituent of cranberry juice. *Crit. Rev. Food Sci. Nutr.* 42(3):279–284, 2002.
123. Weiss, E.I., R. Lev-Dor, Y. Kashamn, J. Goldhar, N. Sharon, I. Ofek. Inhibiting interspecies coaggregation of plaque bacteria with a cranberry juice constituent. *J. Am. Dent. Assoc.* 129(12):1719–1723, 1998.
124. Weiss, E.L., R. Lev-Dor, N. Sharon, I. Ofek. Inhibitory effect of a high-molecular-weight constituent of cranberry on adhesion of oral bacteria. *Crit. Rev. Food Sci Nutr.* 42(3):285–292, 2002.
125. Morris, N.S., D.J. Stickler. Does drinking cranberry juice produce urine inhibitory to the development of crystalline, catheter-blocking *Proteus mirabilis* biofilms? *BJU Int.* 88:192–197, 2001.
126. Swartz, J.H., T.F. Medrek. Antifungal properties of cranberry juice. *Appl. Microbiol.* 16(10):1524–1527, 1968.
127. Cavanagh, H.M., M. Hipwell, J.M. Wilkinson. Antibacterial activity of berry fruits used for culinary purposes. *Med. Food* 6(1):57–61, 2003.
128. Wilson, T., J.P. Porcari, D. Harbin. Cranberry extract inhibits low-density lipoprotein oxidation. *Life Sci.* 62:381–386, 1998.
129. Bagchi, D., C.K. Sen, M. Bagchi, M. Atalay. Anti-angiogenic, antioxidant, and anti-carcinogenic properties of a novel anthocyanin-rich berry extract formula. *Biochemistry* 69(1):75–80, 2004.
130. Reed, J.D., Krueger, C.G., Porter, M.L. Cranberry juice powder decreases low density lipoprotein cholesterol in hypercholesterolemic swine. *FASEB J.* 15(4,5):54, 2001.
131. Reed, J. Cranberry flavonoids, atherosclerosis and cardiovascular health. *Crit. Rev. Food Sci. Nutr.* 42:301–316, 2002.
132. Bomser, J., D.L. Madhavi, K. Singletary, M.A. Smith. *In vitro* anticancer activity of fruit extracts from *Vaccinium* species. *Planta Med.* 62(3):212–216, 1996.
133. Krueger, C.G., M.L. Porter, D.A. Weibe, D.G. Cunningham, J.D. Reed. Potential of cranberry flavonoids in the prevention of copper-induced LDL oxidation. *Polyphenols Communications*. 2:447–448, 2000.
134. Vattem, D.A., Y.-T. Lin, R.G. Labbe, K. Shetty. Phenolic antioxidant mobilization in cranberry pomace by solid-state bioprocessing using food grade fungus *Lentinus edodes* and effect on antimicrobial activity against select food borne pathogens. *Innovative Food Sci. Emerg. Technol.* 5(1):81–91, 2003.
135. Daniel, E.M., A.S. Krupnick, Y.H. Heur, J.A. Blinzler, R.W. Mims, G.D. Stoner. Extraction, stability and quantitation of ellagic acid in various fruits and nuts. *J. Food Comp. Anal.* 2:385–398, 1989.
136. Goldberg, D.M., B. Hoffman, J. Yang, G.J. Soleas. Phenolic constituents, furans and total antioxidant status of distilled spirits. *J. Agric. Food. Chem.* 47:3978–3985, 1999.
137. Loarca-Pina, G., P.A. Kuzmicky, E.G. de Mejia, N.Y. Kado. Inhibitory effects of ellagic acid on the direct-acting mutagenicity of aflatoxin B1 in the *Salmonella* microsuspension assay. *Mutat. Res.* 398(1,2):183–187, 1998.
138. Kaur, S., I.S. Grover, S. Kumar. Antimutagenic potential of ellagic acid isolated from *Terminalia arjuna. Ind. J. Exp. Biol.* 35:478–482, 1997.
139. Khanduja, K.L., R.K. Gandhi, V. Pathania, N. Syal. Prevention of N-nitrosodiethylamine-induced lung tumorigenesis by ellagic acid and quercetin in mice. *Food Chem. Toxicol.* 37(4):313–318, 1999.
140. Wood, A.W., M.T. Huang, R.L. Chang, H.L. Newmark, R.E. Lehr, H. Yagi, J.M. Sayer, D.M. Jerina, A.H. Conney. Inhibition of the mutagenicity of bay-region diol epoxides of polycyclic aromatic hydrocarbons by naturally occurring plant phenols: exceptional activity of ellagic acid. *Proc. Natl. Acad. Sci. USA* 79(18):5513–5517, 1982.
141. Zhang, Z., S.M. Hamilton, C. Stewart, A. Strother, R.W. Teel. Inhibition of liver microsomal cytochrome P450 activity and metabolism of the tobacco-specific nitrosamine NNK by capsaicin and ellagic acid. *Anticancer Res.* 13(6A):2341–2346, 1993.

142. Teel, R.W., M.S. Babcock, R. Dixit, G.D. Stoner. Ellagic acid toxicity and interaction with benzo[a]pyrene and benzo[a]pyrene 7,8-dihydrodiol in human bronchial epithelial cells. *Cell Biol. Toxicol.* 2(1):53–62, 1986.
143. Soni, K.B., M. Lahiri, P. Chackradeo, S.V. Bhide, R. Kuttan. Protective effect of food additives on aflatoxin-induced mutagenicity and hepatocarcinogenicity. *Cancer Lett.* 115(2):129–133, 1997.
144. Singh, K., A.K. Khanna, R. Chander. Hepatoprotective effect of ellagic acid against carbon tetrachloride induced hepatotoxicity in rats. *Ind. J. Exp. Biol.* 37:1025–1026, 1999.
145. Mandal S., N.M. Shivapurkar, A.J. Galati, G.D. Stoner. Inhibition of N-nitrosobenzyl-methylamine metabolism and DNA binding in cultured rat esophagus by ellagic acid. *Carcinogenesis* 9(7):1313–1316, 1988.
146. Mandal, S., G.D. Stoner. Inhibition of N-nitrosobenzyl-methylamine-induced esophageal tumorigenesis in rats by ellagic acid. *Carcinogenesis* 11:55–61, 1990.
147. Smith, W.A., J.W. Freeman, R.C. Gupta. Modulation of dibenzo[a,l]pyrene-DNA adduction by chemopreventive agents in the human breast epithelial cell line MCF-7 (meeting abstract). *Proc. Annu. Meeting Am. Assoc. Cancer Res.* 38:A2422, 1997.
148. Smith, W.A., U. Devanaboyina, R.C. Gupta. Use of a microsomal activation system as a potential screening method for cancer chemopreventive agents (meeting abstract). *Proc. Annu. Meeting. Am. Assoc. Cancer Res.* 36:A3555, 1995.
149. Thresiamma, K.C., J. George, R. Kuttan. Protective effect of curcumin, ellagic acid and bixin on radiation induced genotoxicity. *J. Exp. Clin. Cancer Res.* 17(4):431–434, 1998.
150. Narayanan, B.A, O. Geoffroy, D.W. Nixon. P53/p21 (WAF1/CIP1) expression and its possible role in G1 arrest and apoptosis in ellagic acid treated cancer cells. *Cancer Lett.* 136(2):215–221, 1999.
151. Hassoun, E.A., A.C. Walter, N.Z. Alsharif, S.J. Stohs. Modulation of TCDD-induced fetotoxicity and oxidative stress in embryonic and placental tissues of C57BL/6J mice by vitamin E succinate and ellagic acid. *Toxicology* 124(1):27–37, 1997.
152. Cozzi, R., R. Ricordy, F. Bartolini, L. Ramadori, P. Perticone, R. De Salvia. Taurine and ellagic acid: two differently acting natural antioxidants. *Environ. Mol. Mutag.* 26:248–254, 1995.
153. Chen, C., R. Yu, E.D. Owuor, A.N. Kong. Activation of antioxidant-response element (ARE), mitogen-activated protein kinases (MAPKs) and caspases by major green tea polyphenol components during cell survival and death. *Arch. Pharm. Res.* 23(6):605-612, 2000.
154. Barch, D.H., L.M. Rundhaugen. Ellagic acid induces NAD(P)H:quinone reductase through activation of the antioxidant responsive element of the rat NAD(P)H:quinone reductase gene. *Carcinogenesis* 15(9):2065–2068, 1994.
155. Barch, D.H., L.M. Rundhaugen, N.S. Pillay. Ellagic acid induces transcription of the rat glutathione S-transferase-Ya gene. *Carcinogenesis* 16(3):665–668, 1995.
156. Constantinou, A., G.D. Stoner, R. Mehta, K. Rao, C. Runyan, R. Moon. The dietary anti-cancer agent ellagic acid is a potent inhibitor of DNA topoisomerases *in vitro*. *Nutr. Cancer* 23(2):121–130, 1995.
157. Szaefer, H., J. Jodynis-Liebert, M. Cichocki, A. Matuszewska, W. Baer-Dubowska. Effect of naturally occurring plant phenolics on the induction of drug metabolizing enzymes by o-toluidine. *Toxicology* 186(1,2):67–77, 2003.
158. Jimenez-Lopez, J.M., A.I. Cederbaum. Green tea polyphenol epigallocatechin-3-gallate protects HepG2 cells against CYP2E1-dependent toxicity. *Free Radic. Biol. Med.* 36(3):359–370, 2004.
159. Shetty, K., P. McCue. Phenolic antioxidant biosynthesis in plants for functional food application: integration of systems biology and biotechnological approaches. *Food Biotechnol.* 17:67–97, 2003.
160. Shetty, K., M.L. Wahlqvist. A model for the role of proline-linked pentose phosphate pathway in phenolic phytochemical biosynthesis and mechanism of action for human health and environmental applications. *Asia Pac. J. Clin. Nutr.* 13(1):1–24, 2004.
161. Fabregat, I., J. Vitorica, J. Satrustegui, A. Machado. The pentose phosphate cycle is regulated by NADPH/NADP ratio in rat liver. *Arch. Biochem. Biophys.* 236(1):110–118, 1985.

162. Pfeifer, R., G. Karl, R. Scholz. Does the pentose cycle play a major role for NADPH supply in the heart? *Biol. Chem. Hoppe Seyler* 367(10):1061–1068, 1986.
163. Cabezas, H., R.R. Raposo, E. Melendez-Hevia. Activity and metabolic roles of the pentose phosphate cycle in several rat tissues. *Mol. Cell Biochem.* 201(1,2):57–63, 1999.
164. Owuor, E.D., A.N. Kong. Antioxidants and oxidants regulated signal transduction pathways. *Biochem. Pharmacol.* 64(5,6):765–770, 2002.
165. Chen, C., G. Shen, V. Hebbar, R. Hu, E.D. Owuor, A.N. Kong. Epigallocatechin-3-gallate-induced stress signals in HT-29 human colon adenocarcinoma cells. *Carcinogenesis* 24(8):1369–1378, 2003.
166. Hider, R.C., Z.D. Liu, H.H. Khodr. Metal chelation of polyphenols. *Methods Enzymol.* 335:190–203, 2001.
167. Yang, C.S., J.M. Landau, M.T. Huang, H.L. Newmark. Inhibition of carcinogenesis by dietary polyphenolic compounds. *Annu. Rev. Nutr.* 21:381–406, 2001.
168. Nicholson, C. Modulation of extracellular calcium and its functional implications. *Fed. Proc.* 39(5):1519–1523, 1980.
169. Brown, E.M. Physiology and pathophysiology of the extracellular calcium-sensing receptor. *Am. J. Med.* 106(2):238–253, 1999.
170. Rizzuto, R., P. Pinton, D. Ferrari, M. Chami, G. Szabadkai, P.J. Magalhaes, F. Di Virgilio, T. Pozzan. Calcium and apoptosis: facts and hypotheses. *Oncogene* 22(53):8619–8627, 2003.
171. Stout, C., A. Charles. Modulation of intercellular calcium signaling in astrocytes by extracellular calcium and magnesium. *Glia* 43(3):265–273, 2003.
172. Cohen, J.E., R.D. Fields. Extracellular calcium depletion in synaptic transmission. *Neuroscientist* 10(1):12–17, 2004.
173. Bellomo, G., H. Thor, S. Orrenius. Increase in cytosolic $Ca2^+$ concentration during t-butyl hydroperoxide metabolism by isolated hepatocytes involves NADPH oxidation and mobilization of intracellular $Ca2^+$ stores. *FEBS Lett.* 168(1):38–42, 1984.
174. Hagerman, A.E., L.G. Butler. The specificity of proanthocyanidin-protein interactions. *J. Biol. Chem.* 256(9):4494–4497, 1981.
175. Pan, C.Y., Y.H. Kao, A.P. Fox. Enhancement of inward $Ca(2^+)$ currents in bovine chromaffin cells by green tea polyphenol extracts. *Neurochem. Int.* 40(2):131–137, 2002.
176. Papadopoulou, A., R.A. Frazier. Characterization of protein–polyphenol interactions. *Trends Food Sci. Technol.* 5:(3,4)186–190, 2003.
177. Kim, H.J., K.S. Yum, J.H. Sung, D.J. Rhie, M.J. Kim, S. Min do, S.J. Hahn, M.S. Kim, Y.H. Jo, S.H. Yoon. Epigallocatechin-3-gallate increases intracellular $[Ca2^+]$ in U87 cells mainly by influx of extracellular $Ca2^+$ and partly by release of intracellular stores. *Naunyn. Schmiedebergs Arch. Pharmacol.* 369(2):260–267, 2004.
178. Choi, S.H., M.B. Gu. Phenolic toxicity: detection and classification through the use of a recombinant bioluminescent *Escherichia coli. Environ. Toxicol. Chem.* 20(2):248–255, 2001.
179. Tsuchiya, H. Biphasic membrane effects of capsaicin, an active component in *Capsicum* species. *J. Ethnopharmacol.* 75(2,3):295–299, 2001.
180. Tsuchiya H., M. Sato, Y. Kameyama, N. Takagi, I. Namikawa. Effect of lidocaine on phospholipid and fatty acid composition of bacterial membranes. *Lett. Appl. Microbiol.* 4(6):141–144, 1987.
181. Shetty, K. Role of proline-linked pentose phosphate pathway in biosynthesis of plant phenolics for functional food and environmental applications: a review. *Process. Biochem.* 39:789–803, 2004.
182. Mazumder, A., N. Neamati, S. Sunder, J. Schulz, H. Pertz, E. Eich, Y. Pommier. Curcumin analogs with altered potencies against HIV-1 integrase as probes for biochemical mechanisms of drug action. *J. Med. Chem.* 40 (19):3057–3063, 1997.
183. Durant, S., P. Karran. Vanillins: a novel family of DNA-PK inhibitors. *Nucleic Acids Res.* 31(19):5501–5512, 2003.
184. Wu, G. Intestinal mucosal amino acid catabolism. *J. Nutr.* 128(8):1249–1252, 1998.
185. Brosnan, J.T. Glutamate, at the interface between amino acid and carbohydrate metabolism. *J. Nutr.* 130(4S):988S-990S, 2000.

186. Newsholme, P., J. Procopio, M.M. Lima, T.C. Pithon-Curi, R. Curi. Glutamine and glutamate: their central role in cell metabolism and function. *Cell Biochem. Funct.* 21(1):1–9, 2003.
187. Sarkela, T.M., J. Berthiaume, S. Elfering, A.A. Gybina, C. Giulivi. The modulation of oxygen radical production by nitric oxide in mitochondria. *J. Biol. Chem.* 276(10):6945–6949, 2001.
188. Cadenas, E. Mitochondrial free radical production and cell signaling. *Mol. Aspects Med.* 25(1,2):17–26, 2004.
189. Phang, J.M. The regulatory functions of proline and pyrroline-5-carboxylic acid. *Curr. Topics Cell Reg.* 25:91–132, 1985.
190. Hagedorn, C.H., J.M. Phang. Transfer of reducing equivalents into mitochondria by the interconversions of proline and α-pyrroline-5-carboxylate. *Arch. Biochem. Biophys.* 225:95–101, 1983.
191. Shetty, K., R.L. Labbe. Food-borne pathogens, health and role dietary phytochemicals. *Asia Pac. J. Clin. Nutr.* 7:270–276, 1998.
192. Acamovic, T., C.S. Stewart. Plant phenolic compounds and gastrointestinal microorganisms. *AICR Proc.* 137–139, 1992.
193. Aidoo, K.E., R. Hendry, B.J.B. Wood. Solid state fermentation. *Ad. Appl. Microbiol.* 28:201–237, 1982.
194. McDonald, M., I. Mila, A. Scalbert. Precipitation of metal ions by plant polyphenols: optimal conditions and origin of precipitation. *J. Agric. Food Chem.* 44:599–606, 1996.
195. Kainja, C., L. Bates, T. Acamovic. The chelation of trace elements by tannins. In: *Toxic Plants and Other Natural Toxicants*, Garland, T., A.C. Barr, eds., Wallingford: CAB Intl., 1998, pp 111–114.
196. Muhammed, S.A. Anti-nutrient effects of plant polyphenolic compounds. PhD Thesis, University of Aberdeen, 1999.
197. McClure, J.W. Physiology and functions of flavonoids. In: *The Flavonoids,* Harborne, J.B., ed., New York: Academic Press, 1975, pp 45–77.
198. Wollenweber, E., V.H. Dietz. Occurrence and distribution of free flavonoid aglycones in plants. *Phytochemistry* **20**:869–932, 1981.
199. Kähkönen, M.P., A.I. Hopia, M. Heinonen. Berry phenolics and their antioxidant activity. *J. Agric. Food Chem.* 49:4076–4082, 2001.
200. Onyeneho, S.N., N.S. Hettiarachchy. Antioxidant activity, fatty acids and phenolic acids compositions of potato peels. *J. Sci. Food Agric.* 62:345–350, 1993.
201. Torres, A.M., T. Mau-Lastovicka, R. Rezaaiyan. Total phenolics and high-performance liquid chromatography of phenolic acids of avocado. *J. Agric. Food Chem.* 35:921–925, 1987.
202. Gil, M.I., D.M. Holcroft, A.A. Kader. Changes in strawberry anthocyanins and other polyphenols in response to carbon dioxide treatment. *J. Agric. Food Chem.* 45:1662–1667, 1997.
203. Raynal, J., M. Moutounet, J.M. Souquet. Intervention of phenolic compounds in plum technology, 1: changes during drying. *J. Agric. Food Chem.* 37:1046–1050, 1989.
204. Spanos, G.A., R.E. Wrolstad. Phenolics of ale, pear, and white grape juices and their changes with processing and storage: a review. *J. Agric. Food Chem.* 40:1478–1487, 1992.
205. Kris-Etherton, P.M., K.D. Hecker, A. Bonanome, S.M. Coval, A.E. Binkoski, K.F. Hilpert, A.E. Griel, T.D. Etherton. Bioactive compounds in foods: their role in the prevention of cardiovascular disease and cancer. *Am. J. Med.* 30(113)9B:71S–88S, 2002.
206. De Lumen, B.O., A.F. Galvez, M.J. Revilleza, D.C. Krenz. Molecular strategies to improve the nutritional quality of legume proteins. *Adv. Exp. Med. Biol.* 464:117–126, 1999.
207. Tabe, L.M., T. Wardley-Richardson, A. Ceriotti, A. Aryan, W. McNabb, A. Moore. Exceptional activity of ellagic acid. *Proc. Natl. Acad. Sci. USA* 79:5513–5517, 1982.
208. Viera Diaz, J. Genetic improvement of legumes. *Arch. Latinoam. Nutr.* 44(4,1):41S–43S, 1996.
209. Brar, D.S., T. Ohtani, H. Uchimiya. Genetically engineered plants for quality improvement. *Biotechnol. Genet. Eng. Rev.* 13:167–179, 1996.

210. Auger, C., B. Caporiccio, N. Landrault, P.L. Teissedre, C. Laurent, G. Cros, P. Besancon, J.M. Rouanet. Red wine phenolic compounds reduce plasma lipids and apolipoprotein B and prevent early aortic atherosclerosis in hypercholesterolemic golden Syrian hamsters (*Mesocricetus auratus*). *J. Nutr.* 132(6):1207–1213, 2002.
211. Mattivi, F., C. Zulian, G. Nicolini, L. Valenti. Wine, biodiversity, technology, and antioxidants. *Ann. NY Acad. Sci.* 957:37–56, 2002.
212. Waterhouse, A.L. Wine phenolics. *Ann. NY Acad. Sci.* 957:21–36, 2002.
213. Raimbault, M. General and microbiological aspects of solid substrate fermentation. *Electron. J. Biotechnol.* 1(3):1–15, 1998.
214. Pandey, R. Recent progress developments in solid-state fermentation. *Process Biochem.* 27:109–117, 1992.
215. Shekib, L.A. Nutritional improvement of lentils, chick pea, rice and wheat by natural fermentation. *Plant Foods Hum. Nutr.* 46(3):201–205, 1994.
216. Hadajini, S. Indegenous mucana tempe as functional food, *Asia Pac. J. Lin. Nutr.* 10(3):222–225, 2001.
217. Wang, S.Y., A.W. Stretch. Antioxidant capacity in cranberry is influenced by cultivar and storage temperature. *J. Agric. Food Chem.* 40:969–974, 2001.
218. Zheng, G.L., Y.G. Zhou, W.G. Gong. Isolation of soybean isoflavones from tofu wastewater. *ACTA Academiae Medicinae Zhejang* 8:23–25, 1997.
219. Morrissey, P., A.E. Osbourn. Fungal resistance to plant antibiotics as a mechanism of pathogenesis. *Microbiol. Mol. Biol. Rev.* 62(3):708–724, 1999.
220. Osbourn, A.E. Antimicrobial phytoprotectants and fungal pathogens: a commentary. *Fung. Genet. Biol.* 26:163–168, 1999.
221. Fukai, T., A. Marumo, K. Kaitou, T. Kanda, S. Terada, T. Nomura. Anti-*Helicobacter pylori* flavonoids from licorice extract. *Life Sci.* 71(12):1449–1463, 2002.
222. Elattar, T.M.A., A.A. Virji. The effect of red wine and its components on growth and proliferation of human oral squamous carcinoma cells. *Anticancer Res.* 19:5407–5414, 1999.
223. Carbonneau, M.A., C.I. Léger, B. Descomps, F. Michel, L. Monnier. Improvement in the antioxidant status of plasma and low-density lipoprotein in subjects receiving a red wine phenolics mixture. *J. Am. Oil Chem. Soc.* 75:235–240, 1998.
224. Chan, M.M., J.A. Mattiacci, H.S. Hwang, A. Shah, D. Fong. Synergy between ethanol and grape polyphenols, quercetin, and resveratrol, in the inhibition of the inducible nitric oxide synthase pathway. *Biochem. Pharmacol.* 60(10):1539–1548, 2000.
225. Abraham, S.K. Anti-genotoxic effects in mice after the interaction between coffee and dietary constituents. *Food Chem. Toxicol.* 34(1):15–20, 1996.
226. Vattem, D.A., H.-D. Jang, R. Levin, K. Shetty. Synergism of cranberry phenolics with ellagic acid and rosmarinic acid for antimutagenic and DNA-protection functions. *J. Food Biochem.*, 2004. [Submitted]
227. Vattem, D.A., Y.-T. Lin, R. Ghaedian, K. Shetty. Cranberry synergies for dietary management of α-amylase activity for type-II diabetes. [Unpublished results]
228. Vattem, D.A., Y.-T. Lin, R. Ghaedian, K. Shetty. Cranberry synergies for dietary management of *Helicobacter pylori* infections. *Process. Biochem.* 40(5): 1583–1592, 2004.

2.11

Rosmarinic Acid Biosynthesis and Mechanism of Action

Kalidas Shetty

CONTENTS

11.1 INTRODUCTION

Rosmarinic acid (RA) (Figure 11.1) is commonly found in substantial amounts in the family Lamiaceae, which has many important species such as oregano (*Origanum vulgare*), thyme (*Thymus vulgaris*), rosemary (*Rosmarinus officinalis*), holy basil (*Ocimum sanctum*), perilla or shiso (*Perilla frutescens*), spearmint (*Mentha spicata*), and several other species that have food and medicinal applications. This chapter highlights various sources of rosmarinic acid and their functional effects, RA biosynthesis in cell cultures and pathways associated with RA biosynthesis. In addition, investigations from my own research group have been summarized, where we have developed methods for RA biosynthesis through generation of high RA clonal lines using tissue cultures for field production and regulation of RA biosynthesis through a proposed critical control point (CCP), proline linked pentose–phosphate pathway. In terms of applications we have developed

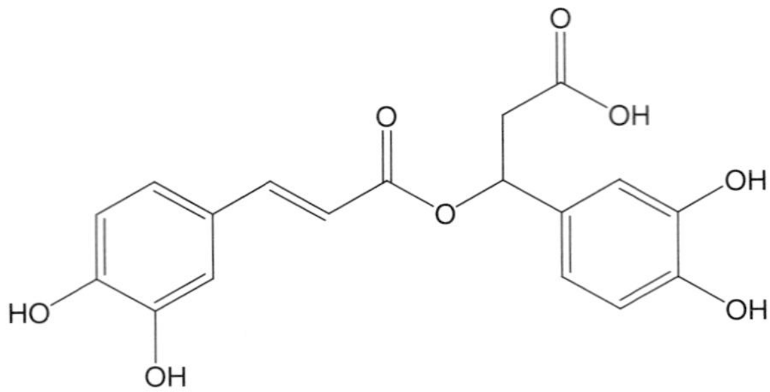

Figure 11.1 Rosmarinic acid

new applications for RA such as for amylase inhibition, which has potential for hyperglycemia management and antimicrobial effects of high rosmarinic acid-containing herbal extracts. Further, we have proposed a model of how RA can regulate human health relevant antioxidant response through the CCP, proline linked pentose–phosphate pathway of human cells.

11.2 ROSMARINIC ACID SOURCES AND FUNCTIONAL PHARMACOLOGICAL EFFECTS

Rosmarinic acid (RA) is an important caffeoyl ester (phenolic depside) with proven medicinal properties and well characterized physiological functions. Rosmarinic acid is found in substantial sources in the family Lamiaceae (1,2). *Salvia lavandulifolia* is used as choleretic, antiseptic, astringent, and hypoglycemic drug in Southern Europe and contains high levels of RA (3). Rosmarinic acid-containing *Ocimum sanctum* (holy basil) is used to reduce fevers and against gastrointestinal disorders in India and has antioxidant properties (2,4). In Mexico, high RA-containing *Hyptis verticillata* is widely used by Mixtec Indians against gastrointestinal disorders and skin infections (5). In Indonesia and several countries in Southeast Asia, RA-containing *Orthosiphon aristatus* is known for diuretic properties and is also used against bacterial infections of the urinary system (6). *Salvia cavaleriei*, a high RA-containing species is used in China for treatment of dysentery, boils, and injuries (7). The antioxidative, antimicrobial, and antiviral effects of *Prunella vulgaris* indicate its potential as a medicinal herb (8). Rosmarinic acid-containing *Origanum vulgare* (oregano), *Thymus vulgaris* (thyme), *Ocimum basilicum* (sweet basil), and *Rosmarinus officinalis* (rosemary) are important sources of antioxidants in food preservation (9–11) and for stability and enhancement of anthocyanin and related pigment color in berry based juices (12,13) and have potential health benefits as dietary amylase inhibitor in diabetes management (14).

Many pharmacological effects of RA are known. Rosmarinic acid inhibits several complement dependent inflammatory processes and has potential as a therapeutic agent for control of complement activation diseases (15,16). Rosmarinic acid has been reported to have effects on both the classical C3-convertase and on the cobra venom factor and ovalbumin/antiovalbumin mediated passive cutaneous anaphylaxis (15). Rosmarinic acid also inhibits prostacyclin synthesis induced by complement activation (17,18). Rosmarinic acid is also known to have complement independent effects, such as scavenging of oxygen free radicals and inhibiting elastase and is known to be safe (19). Other actions of RA are antithyrotropic activity in tests with human thyroid membrane preparations, inhibition of complement dependent components of endotoxin shock in rabbits, and the ability to react rapidly to viral coat proteins and so inactivate the virus (15). Rosmarinic acid also inhibits Forskolin induced activation of adenylate cyclase in cultured rat thyroid cells (20) and inhibits external oxidative effects of human polymorphonuclear granulocytes (21).

Recent research has indicated other benefits of RA containing *Perilla frutescens* on the reduction of lipopolysaccharide (LPS) induced liver injury in D-galactosamine sensitized mice (22). High RA *P. frutescens* also inhibited lung injury in mice induced by diesel exhaust particles (23) and also had antiallergic effect (24). The antiallergic titer of rosmarinic acid was more effective than tranilast, which is a widely used antiallergic drug (24). Rosmarinic acid also has potential antidepressive-like effect in mice based on a forced swimming test (25). Another interesting study has shown that RA inhibits calcium dependent pathways of T-cell antigen receptor mediated signaling (26). However, investigations so far on pharmacological effects of RA have not clarified the antiinflammatory effects but more evidence suggests RA's ability to block complement activation (27) and to inhibit cyclooxygenase (4).

Research findings on pharmacological potential of RA have substantially increased since 2000. RA synergistically inhibited LDL oxidation in combination with lycopene indicating its potential against atherosclerosis (28). In relation to HIV type 1, RA in addition to being an integrase inhibitor also inhibited reverse transcriptase (29,30). In mice studies, *Perilla frutescens* rich in RA reduced allergenic reactions using mice ear passive cutaneous anaphylaxis (PCA) reaction (24,31). In another mice model, RA in *Perilla* extract inhibited allergic inflammation induced by mite allergen (32). In a human clinical study related to allergy reduction, RA enriched *Perilla frutescens* proved to be an effective intervention for mild seasonal allergic rhino-conjunctivitis (SAR), partly through inhibition of polymorphonuclear leukocytes (PMNL) infiltration into the nostrils, which could contribute to reduction in treatment costs for allergic diseases (33). In mice studies, oral and intraperitoneal administration of RA had antidepressive effects and mechanism was suggested to not involve inhibition of monoamine transporters and monoamine oxidase (25,34). In relation to improvement of kidney related functions, RA has shown suppressive effects on mesangioproliferative glomerulonephritis in rats (35). Inhibitory effect of RA on the proliferation of cultured murine mesangial cells was previously reported (36). In this study, RA inhibited cytokine induced mesangial cell proliferation and suppressed PDGF and c-myc mRNA expression in PDGF mediated mesangial cells (36). Suppressive effects of RA enriched *Perilla frutescens* on IgA nephropathy in HIGA mice were also observed (37). In other pharmacological studies, RA reduced the defensive freezing behavior of mice exposed to conditioned fear stress (38). In relation to signal transduction, RA inhibited Ca2+ dependent pathways of T-cell antigen receptor mediated signaling by inhibiting the PLC-gamma 1 and Itk activity (26). RA is also known to inhibit TCR induced T cell activation and proliferation in an Lck dependent manner (39–41) and can also influence Lck dependent apoptotic activity (42). In other T cell studies, RA alone and in conjunction with currently used immunosuppressive drugs, inhibited *in vitro* splenic T-cell proliferation (43). RA enriched herb extract was also shown to have beneficial effects on suppression of collagen induced arthritis (44) and showed significant reduction in tumorigenesis in a murine two stage skin carcinogenesis model (45). Herbal extracts enriched in RA yielded higher inhibition of amylase than purified RA, suggesting RA in synergy with other phenolic compounds may contribute to amylase inhibition for potential modulation of hyperglycemia (14). Recently it was shown that oregano clonal extracts high in RA also have antimicrobial activity against *Listeria monocytogenes* (46) and *Helicobacter pylori* (47).

11.3 RA BIOSYNTHESIS IN CELL CULTURES

Rosmarinic acid has been targeted for production using undifferentiated cell suspension cultures in several species (6,48–56). Further, elicitors such as yeast extract and methyl jasmonate were used to stimulate RA content in cell cultures (6,51,57,58). The biosynthesis of RA were also evaluated and stimulated in hairy root cultures by elicitors (59). The main purpose of cell suspension cultures for production of RA is the potential for large scale production in bioreactors (60,61). Although large scale production in bioreactors is feasible for RA (62), undifferentiated cell suspension cultures are generally not practical for metabolites produced in differentiated structures (e.g., anethole in seeds of anise, curcumin in rhizomes of turmeric, eugenol in barks of cinnamon, and thymol in glandular cells of leaves). In a comparison of nodal shoot cultures and callus cultures of *Ocimum basilicum* L in airlift bioreactors, RA production in cell suspension cultures was 29 μg/g dry weight compared to 178 µg/g dry weight for shoot cultures (63). An additional disadvantage of undifferentiated callus based suspension cultures is that the DNA is more error prone and therefore more genetically unstable (64). Further, bioreactor based production requires high initial operating costs.

In terms of regulation of RA in cell cultures it is constitutively expressed in *Coleus blumei* without any medium manipulation (65). Plant cell cultures are known to accumulate 8–10% of their dry weight as RA, a content much higher than parent plants (53,66), contradicting the callus and shoot culture comparisons of the *Ocimum* study (63). The pathway of RA biosynthesis is through the aromatic amino acids, phenylalanine, and tyrosine (65). Cell suspension cultures of *C. blumei* (66), *Rosmarinus officinalis*, *Salvia officinalis* (67), and *Anchusa officinalis* (68,69) have been used to produce RA in cell suspension cultures. The influence of various macronutrients and growth regulators on growth of *A. officinalis* has been investigated (68,69). Concentrations of 3% sucrose, 15 mM nitrate, 3 mM phosphate, and 0.25 mM calcium were best for increasing both biomass and RA contents with yields in the range of 10–15% (68). Similar nitrogen, potassium, and phosphate optimizations proved useful for RA enhancement in *Lavandula vera* MM cell suspension cultures (70). When auxin effects were tested, they maintained growth and integrity of the cell suspensions, whereas cytokinins alone did not, suggesting that the culture of *A. officinalis* was auxin dependent (69). Among auxins, NAA had the most pronounced effect on RA content (69) and was confirmed in other recent studies where NAA at levels of 2 mg/l induced maximum RA level of 355 mg/l in cultures of *Zataria multiflora* (71). The kinetics of growth and RA production suggested that the increase in the final RA content and initiation of the period of biosynthesis was in the exponential, rather than the linear growth phase (69). Other studies have shown that increased sucrose concentrations stimulate RA content (72). The highest RA content reported so far was 36% of the cell dry weight in suspension cultures of *Salvia officinalis* at 5% sucrose (49), which is unusually high and never been reproduced.

11.3.1 Pathways Associated with Rosmarinic Acid Biosynthesis

The amino acids phenylalanine and tyrosine have been shown to be precursors of RA biosynthesis (52,73–75). Phenylalanine is transformed to an activated hydroxycinnamic acid by the enzymes of the general phenylpropanoid pathway, which are already well known for biosynthesis of flavonoid or lignin (75). Using radioactive phenylalanine and tyrosine, it was established that these two amino acids are incorporated into caffeic acid and 3,4-dihydroxyphenyllactic acid moieties, respectively (73). Steps in RA biosynthesis originating from phenylalanine and tyrosine have been characterized (Figure 11.2)

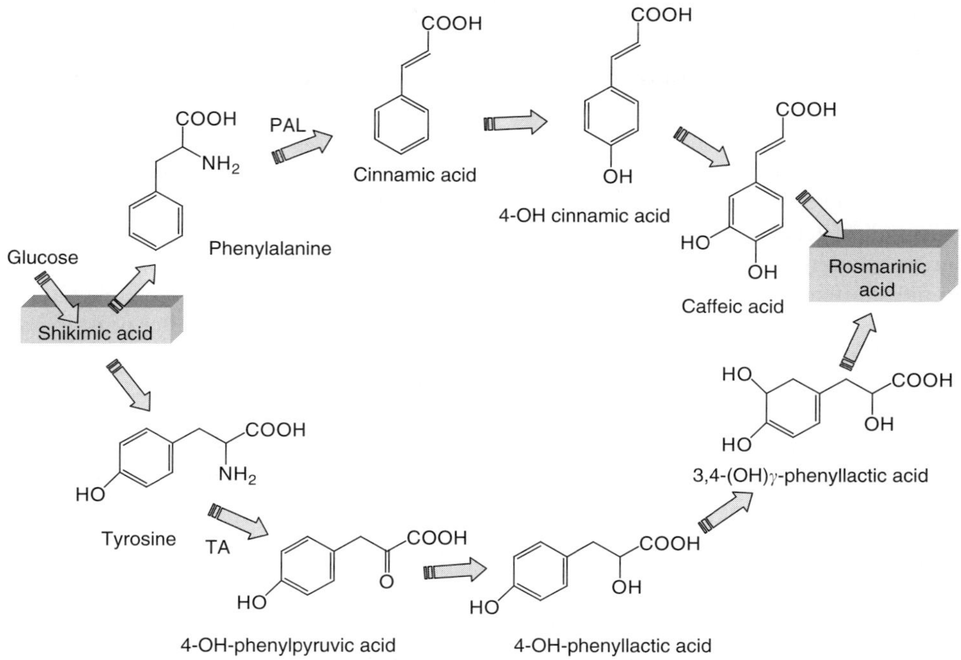

Figure 11.2 Rosmarinic acid biosynthesis from shikimic acid through tyrosine and phenylalanine

(6,52,65,76,77). In several cell suspension cultures, activity of phenylalanine ammonia-lyase (PAL) was correlated to RA (6,77). Basic characteristics of this enzyme were determined in protein preparations from suspension cell suspension cultures of *Coleus blumei* (78). The reaction, following formation of cinnamic acid, is the hydroxylation in the position 4 to 4-coumaric acid by cytochrome P450 monooxygenase cinnamate 4-hydroxylase (74). Generally, coenzyme A thioester, or glucose or chlorogenic acid have been shown to serve as donors of hydroxycinnamic acid moieties (75). Further, using *A. officinalis* cell suspension cultures, it was reported tyrosine amino transferase catalyzes the first step of the transformation of tyrosine to 3,4-dihydroxyphenyllactic acid. Several isoforms of tyrosine aminotransferase were found to be active in cell suspension cultures of *A. officinalis* (76,77). Prephenate aminotransferase in *A. officinalis* cell suspension cultures was found to be important, and its activity was affected by 3,4-dihydroxyphenyllactic acid (79). Other enzymes of late steps in the RA biosynthesis pathway, like hydroxyphenylpyruvate reductase and RA synthase (hydroxycinnamoyl-CoA, hydroxyphenyllactate, hydroxycinnamoyl transferase), were isolated and characterized in cell suspension cultures of *C. blumei* (80–82). Under the release of coenzyme A, the ester linkage is formed between carboxyl group of 4-coumaric acid and the aliphatic hydroxyl group of 4-coumaric acid and the aliphatic hydroxyl group of 4-hydroxyphenyllactate (75). Other studies have isolated microsomal hydroxylase, later confirmed as cytochrome P450 monooxygenases (74,83), whose activities introduce hydroxyl groups at position 3 and 3′-hydroxyphenyllactate to give rise to the aromatic rings of ester 4-coumaroyl-4′-hydroxyphenyllactate to give rise to RA (52). A number of cDNAs encoding cytochrome P450s, which can hydroxylate 4-coumaric acid or 4-coumaroyl moiety in an ester, have been isolated (84–86). This enzyme isolation led to the proposal that the complete biosynthetic pathway for RA biosynthesis originates from phenylalanine and tyrosine (52,75).

11.3.2 RA Biosynthesis and Generation of High RA Clonal Lines

High RA production can be achieved at low cost by incorporating superior varieties in traditional agronomic systems. Superior varieties can be isolated using tissue culture techniques using shoot based clonal lines of single seed origin (1). The major limitation of using dietary herbs for pharmacological applications from traditional wild collections and heterogeneous seeds is the inconsistency of phenolic phytochemicals such as RA due to the heterogeneity resulting from the cross pollinating nature of their breeding characteristics, and especially species in the family Lamiaceae (1). Plants which originate from different heterozygous seeds in a given pool of extract are phenotypically variable, resulting in the substantial phytochemical inconsistency, and therefore leads to unreliable clinical effects as well as inconsistent health benefits and functional value. In order to overcome the problem of phytochemical inconsistency due to genetic heterogeneity, plant tissue culture techniques have been developed to isolate a clonal pool of plants originating from a single heterozygous seed (87,88). A single elite clonal line with superior RA and phenolic profile can then be screened and selected based on tolerance to *Pseudomonas* sp. (89–91) and proline analogs (92–94). These elite clonal lines (each clonal line originating from a different heterozygous seed), following large scale clonal propagation (micropropagation) and evaluation of functionality, can be targeted as dietary sources of phenolics (with focus on RA compared in diverse total phenolic clonal backgrounds) for diverse food and pharmaceutical applications.

11.3.3 Role of Proline Linked Pentose–Phosphate Pathway in RA Biosynthesis in Clonal Systems

The hypothesis that synthesis of plant phenolic metabolites is linked to the critical control point (CCP), proline linked pentose–phosphate pathway (PLPPP) (1,95,96) (Figure 11.3) was developed based on the role of the PLPPP in regulation of purine metabolism in mammalian systems (97). Proline is synthesized by a series of reduction reactions from glutamate. In this sequence, P5C and proline known to be metabolic regulators function as a redox couple (97,98). During respiration, oxidation reactions produce hydride ions, which augment reduction of P5C to proline in the cytosol. Proline can then enter mitochondria through proline dehydrogenase (99) and support oxidative phosphorylation (alternative to NADH from Krebs/TCA cycle). This is important because shunting the TCA cycle toward proline synthesis likely deregulates normal NADH synthesis. The reduction of P5C in the cytosol provides $NADP^+$, which is the cofactor for glucose-6-phosphate dehydrogenase (G6PDH), an enzyme that catalyzes the rate-limiting step of the pentose–phosphate pathway. Proline synthesis is therefore hypothesized, and has been partly shown to both regulate and stimulate pentose–phosphate pathway activity in erythrocytes (100) and cultured fibroblasts (101) when P5C is converted to proline. This was shown to stimulate purine metabolism via ribose-5-phosphate, which affects cellular physiology and therefore function (97,102). Therefore, understanding the CCP, PLPPP is important for designing high RA clonal extracts isolated from single seed genetic origin among a heterogeneous seed population.

From the above insights Shetty (1,95,96) first proposed a model that CCP, PLPPP could stimulate shikimate and phenylpropanoid pathways and hypothesized that stress linked modulation of this pathway can lead to the stimulation of phenolic phytochemicals, including RA (1,95,96) (Figure 11.3). Using this model, proline, proline precursors, and proline analogs were effectively utilized to stimulate total phenolic content and RA in clonal shoot cultures (103,104). Further, it was shown that proline, proline precursors, and proline analogs stimulated somatic embryogenesis in anise, which correlated with increased soluble phenolic content (105). It was also established that during *Pseudomonas* mediated stimulation of total soluble phenolics, RA and proline content was stimulated in

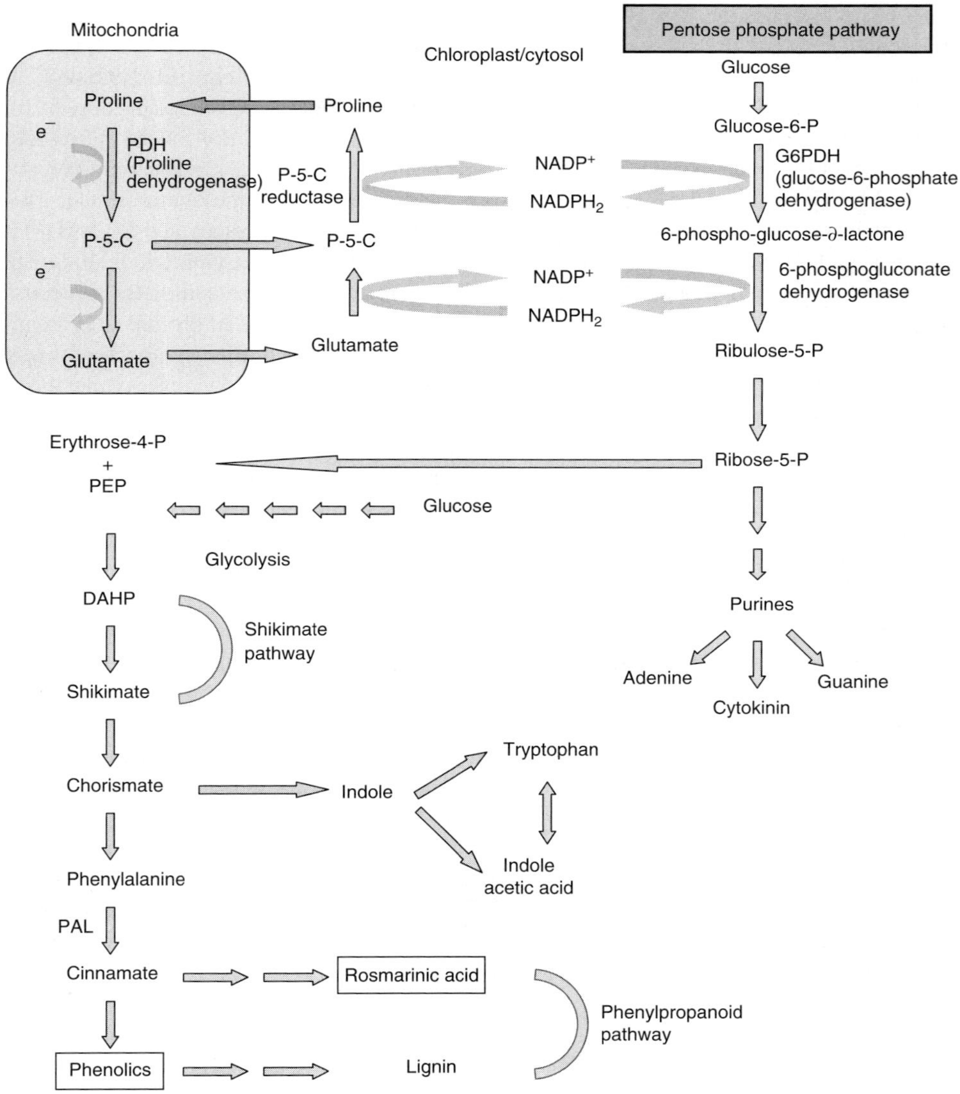

Figure 11.3 Model for RA biosynthesis through the proline linked pentose–phosphate pathway and its utilization to screen high RA and phenolic-producing clonal lines using microbial interaction and proline analogs (targeted at proline dehydrogenase-PDH)

oregano clonal shoot cultures (106). Therefore, it was proposed that $NADPH_2$ demand for proline synthesis during response to microbial interaction and proline analog treatment (1) may reduce the cytosolic $NADPH_2$/ $NADP^+$ ratio, which should activate G6PDH (107,108). Therefore, deregulation of the pentose–phosphate pathway by proline analog and microbial induced proline synthesis may provide the excess erythrose-4-phosphate (E4P) for shikimate and, therefore, the phenylpropanoid pathway leading to RA (95). At the same time, proline and P5C could serve as superior reducing equivalents (RE) reductant, an alternative to NADH (from Krebs/TCA cycle) to support increased oxidative phosphorylation (ATP synthesis) in the mitochondria during the stress response (97,98).

Therefore, proline analog, azetidine-2-carboxylate (A2C), and proline stimulating *Pseudomonas* interactions were used to isolate high RA clonal lines (1,92–95). A2C is an inhibitor of proline dehydrogenase (109) and is also known to inhibit differentiation in Leydig cells of rat fetal testis, which can be overcome by exogenous proline addition (110). Another analog, hydroxyproline, is a competitive inhibitor of proline for incorporation into proteins. According to the model of Shetty (1,96), either analog at low levels should deregulate proline synthesis from feedback inhibition, and stimulate proline synthesis (1). This would then allow the proline linked pentose–phosphate pathway to be activated for $NADPH_2$ synthesis, and concomitantly drive metabolic flux toward E4P for biosynthesis of shikimate and phenylpropanoid metabolites, including RA (Figure 11.4). Proline could also serve as a RE reductant for ATP synthesis via mitochondrial membrane associated proline dehydrogenase (99). Therefore, any high RA clonal line with a deregulated proline synthesis pathway should have an overexpressed pentose phosphate pathway which allows excess metabolic flux to drive shikimate and phenylpropanoid pathway toward total phenolic and RA synthesis. Similarly, such proline overexpressing clonal lines should be more tolerant to proline analog, A-2-C (95,96). At the same time if the metabolic flux to RA is overexpressed, it is likely to be stimulated in response to *Pseudomonas* sp (95). Therefore, such a clonal line is equally likely to be tolerant to *Pseudomonas* sp. Further, such a clonal line should also exhibit high proline oxidation and RA content in response to A2C and *Pseudomonas* sp. In addition, in the presence of A2C or *Pseudomonas* sp., increased activity of key enzymes G6PDH (pentose–phosphate pathway), P5C reductase (proline synthesis pathway), proline dehydrogenase (proline oxidation pathway), 3-deoxy-D arabino-heptulosonate-7-phosphate synthase (shikimate pathway), and phenylalanine ammonia-lyase (phenylpropanoid pathway) should be stimulated (96). As mentioned earlier, the rationale for the PLPPP model for RA biosynthesis is based on the role of the pentose–phosphate pathway in driving ribose-5-phosphate toward purine metabolism in cancer cells (97), differentiating animal tissues (110), and plant tissues (111). The success of this CCP, PLPPP strategy (89–94) provides ready access to critical interlinking metabolic pathways associated with RA biosynthesis and will allow more detailed analyses, which could lead to large scale greenhouse and Agronomic based field production systems for efficient RA biosynthesis. This strategy for investigation and production of RA can be the foundation for designing other dietary phenolic phytochemicals from cross pollinating, heterogeneous species (1,95,96).

11.4 MECHANISM OF RA ACTION THROUGH STIMULATION OF HOST ANTIOXIDANT RESPONSE

Investigations so far in food grade clonal herb systems (1) and legume sprouts (112–114) led to the development of the model that activity of CCP, proline linked pentose–phosphate pathway is important for stress induced phenolic biosynthesis such as RA and phenolics and that this stimulation of phenolics is likely closely linked to stimulation of antioxidant response pathways (Figure 11.4) (96,112–114). Further research has indicated that the proline biosynthesis pathway coupled to stress induced antioxidant response pathways could be also stimulated in legume sprouts using exogenous treatment of phenolic extracts from clonal oregano (113,115,116). Phenolic extracts from these clonal oregano lines have high free radical scavenging activity (47). Proline linked stimulation of antioxidant enzyme response pathways may also be stimulated by low pH and salicylic acid (117). Further, exogenous seed treatment with oregano phenolic antioxidant extracts enhanced endogenous phenolic content, GPX activity, and consequently, enhanced seeding vigor during germination (116). From these initial plant studies and plant PLPPP model (Figure 11.4), a human/mammalian cell PLPPP model has been developed (Figure 11.6) wherein a proton donation by phenolic antioxidants

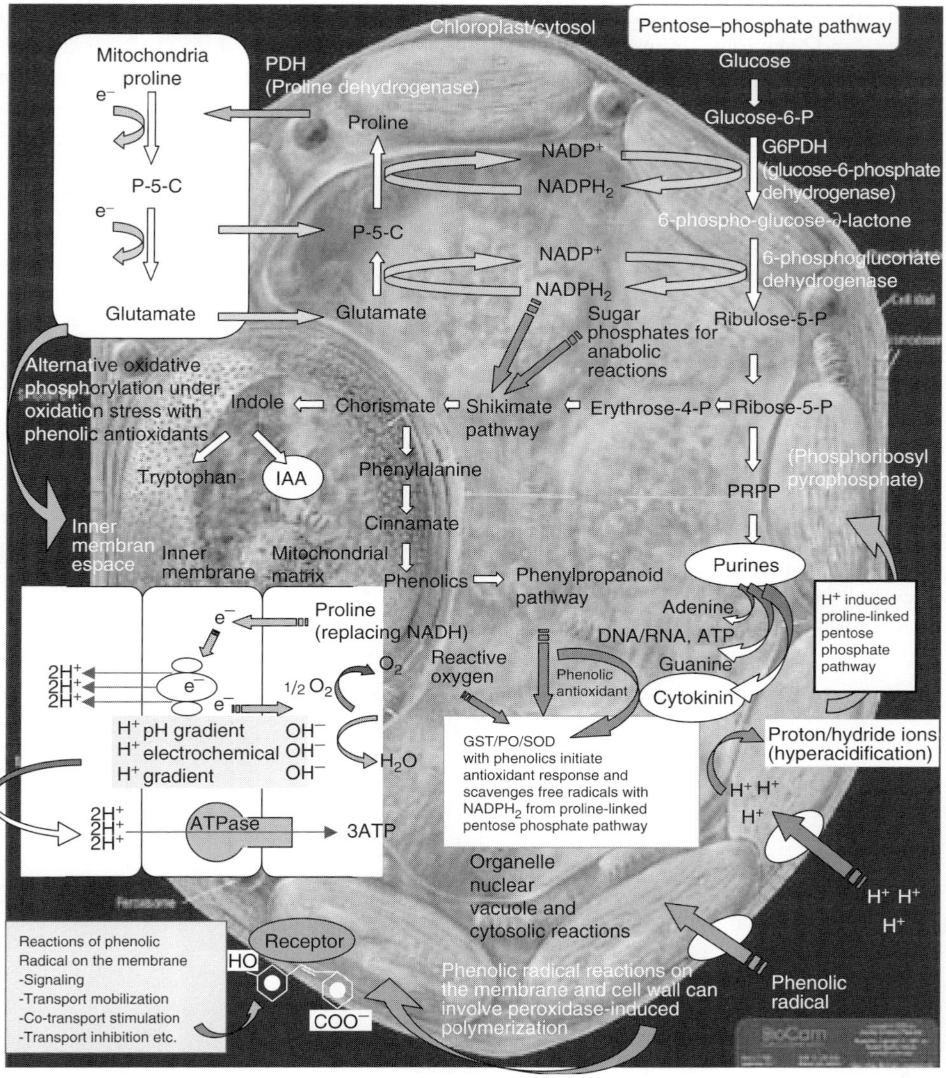

Figure 11.4 Expanded Model for the role of proline linked pentose–phosphate pathway in regulating phenolic biosynthesis, which also accommodates the mechanism of action of external phenolic phytochemicals such as RA from herbal extracts to trigger an endogenous antioxidant enzyme response. (Abbreviations: P5C: pyrroline-5-carboxylate; IAA: indole acetic acid; GST: Glutathione-s-transferase; PO: peroxidase; SOD: superoxide dismutase)

such as RA at the outer plasma membrane initiates a proton/hydride ion influx into the cytosol which activates an antioxidant response through the stimulation of the, CCP, proline linked pentose–phosphate pathway (95,96). Demand for $NADPH_2$ by stimulated proline biosynthesis also drives the production of precursors for phenolic (only in plants and fungi), purine and antioxidant pathways. In this host response PLPPP model, proline can also be used as a reducing equivalent (RE) reductant to support oxidative phosphorylation for ATP synthesis. Using this approach and rationale, we first developed several RA and phenolic overexpressing plant clonal systems for functional food and agro-environmental applications. Subsequently, PLPPP linked and optimized phenolic phytochemical clonal profiles can be used as sources of antioxidants and antimicrobials in biological systems based on host PLPPP response and have implications for human health and wellness.

Human health applications have been developed with additional insights based on the animal antioxidant response model (Figure 11.7) and the plant antioxidant response model (Figure 11.5). From these insights an innovative model for the mechanism of

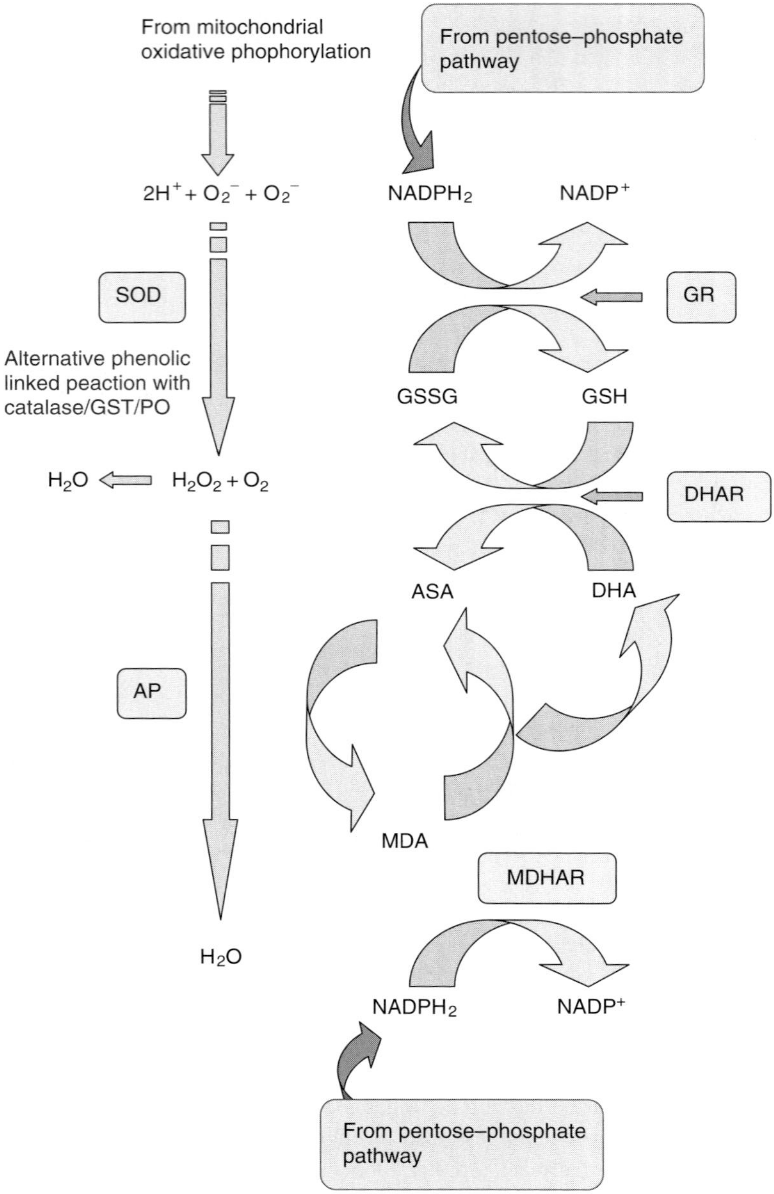

Figure 11.5 Model for specific steps in antioxidant enzyme response pathway in plants. (Abbreviations: SOD: superoxide dismutase; AP: ascorbate peroxidase; GR: glutathione reductase; GSSG: oxidized glutathione; GSH: reduced glutathione; DHAR: dehydroascorbate reductase; ASA: reduced ascorbate: DHA: dehydroascorbate; MDA: monodehydroascorbate; MDHA: monodehydroascorbate reductase)

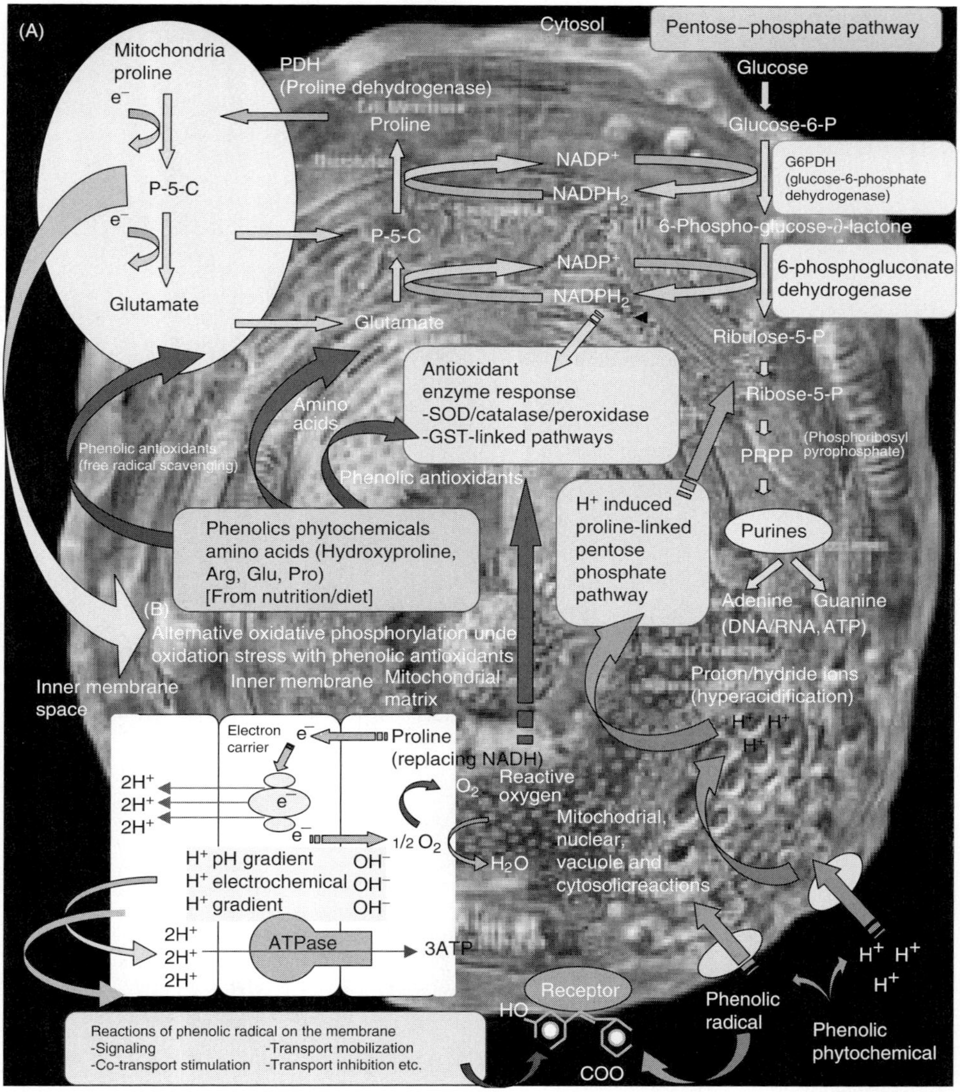

Figure 11.6 Extension of plant proline linked pentose–phosphate pathway model for the effect of external phenolic phytochemicals in human and mammalian systems. (Abbreviations: P5C: pyrroline-5-carboxylate; GST: Glutathione-s-transferase; SOD: superoxide dismutase)

action of phenolic antioxidants like RA for improving human health through protection against oxidation linked diseases involving the host proline linked pentose–phosphate pathway has been proposed (Figure 11.6). The major diseases afflicting humans today, related to excess calories, obesity, and environmental pollutant exposure, are oxidation linked chronic diseases such as cancer, cardiovascular disease, arthritis, cognition diseases, and diabetes. Oxidation linked immune dysfunction and the inability to fight pathogenic infection under a very low calorie and protein diet still remains a problem in several parts of the world and the challenge has to be addressed. Oxidation linked and infectious diseases involve free radical reactions. Free radicals are potential

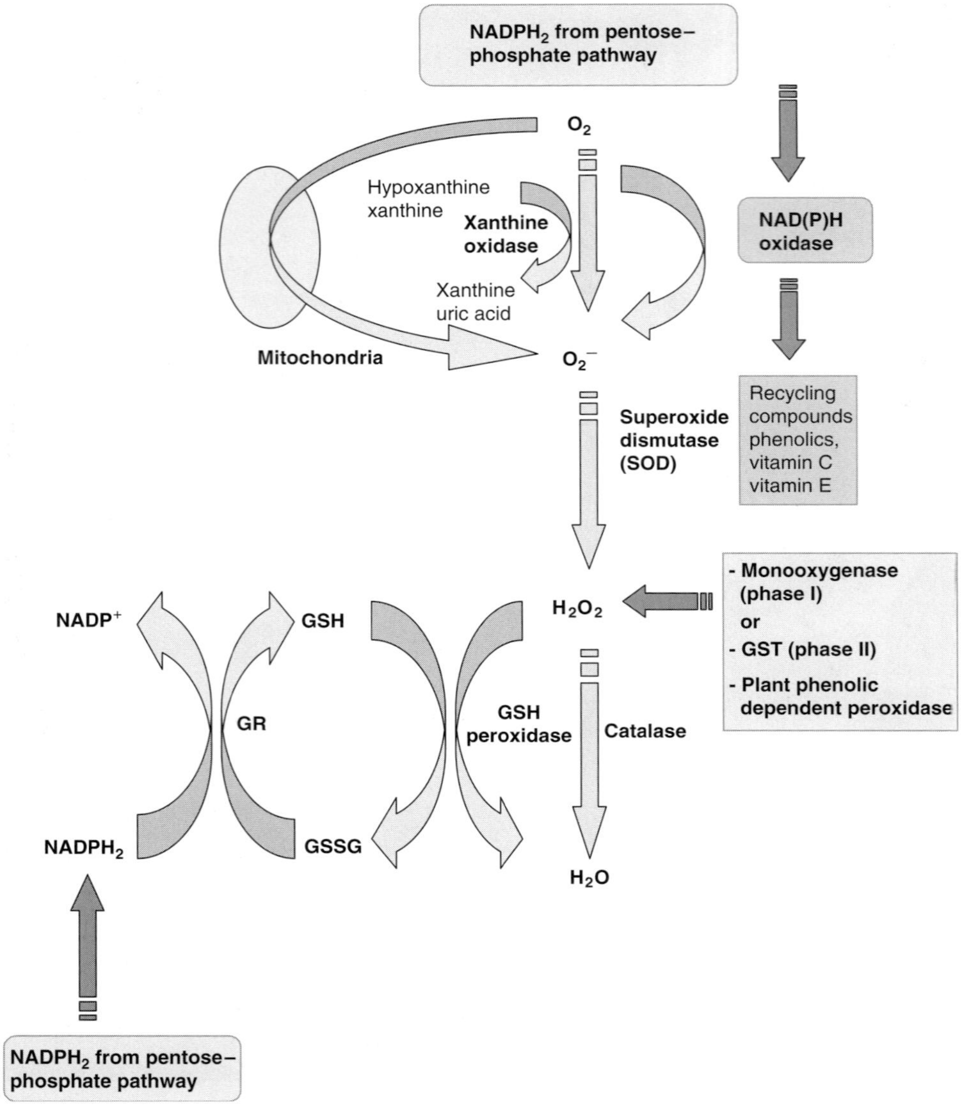

Figure 11.7 Model for specific steps in antioxidant response pathways in human and mammalian systems. (Abbreviations: SOD: superoxide dismutase; GST: glutathione-s-transferase; GR: glutathione reductase; PO: peroxidase; GP: glutathione peroxidase; GSSG: oxidized glutathione; GSH: reduced glutathione)

carcinogens because they can facilitate mutagenesis, tumor promotion and progression (118–120). For example, in the case of cardiovascular diseases, free radicals are implicated in the pathogenesis of atherosclerosis, which is characterized by the hardening of the arterial wall (118,121,122). Rheumatoid arthritis is a systemic autoimmune disease characterized by chronic joint inflammation with infiltration of macrophages and activated T cells. Production of free radicals at sites of inflammation is thought to contribute to the pathogenesis (118,123,124). Free radicals are implicated in the pathogenesis of Alzheimer disease (118,125). As significant amounts of lipid peroxidation in brain tissues have been observed and may help explain the progressive decline in cognition function and excessive neuronal loss in afflicted patients. The brain tissue shows numerous amyloid plaques. Free radicals have been implicated in diabetes mellitus (118,126,127). These oxidative stress linked disease conditions are associated with a prooxidative shift of glutathione redox state in the blood (118,128). Elevated glucose levels are also associated with increased production of free radicals by several different mechanisms (118,129,130).

Diet, environment, and lifestyle influence oxidation linked diseases and improvement of diet is an important part of the preventive management of these diseases (95,96). It is evident that a variety of plant based phenolic antioxidants have a positive impact on the prevention and modulation of various oxidation linked diseases (16,131–135) and must be considered important part of the dietary management of these diseases. Currently the modes of action and early stage effects of these phenolic antioxidants in positively modulating and preventing various diseases are not completely clear (95,96). Extensive research is underway to ascertain how free radicals modulate physiological control of cell function at the level of cell proliferation and deterioration (118) and at the level of gene expression (136). Phenolic antioxidants such as RA have been targeted to control the free radical linked cellular deterioration that can otherwise lead to major oxidation linked chronic diseases. We have proposed in this chapter that an understanding of RA biosynthesis through CCP and PLPPP could help to not only design the right RA profile through functional foods, but the host response to this RA profile could also operate effectively through an antioxidant enzyme response regulated by host (human) PLPPP.

Therefore, an important strategy to develop diet based interventions is through design of functional foods (conventional foods with clinically defined health promoting components) based on the understanding of phenolic antioxidant biosynthesis, such as RA in food plants (Figure 11.4) and the effect of RA in human and mammalian systems (Figure 11.6). In order for diet based interventions (through functional foods) to be effective, it is also important to understand the early stage modes of action of these functional compounds such as RA and how to deliver RA at consistent levels and with no toxicity problems. In the antioxidant enzyme response model for human health, proposed previously (95,96), a consistent and defined phytochemical profile of RA and in high phenolic background can be developed using clonal shoot systems using various dietary botanicals from the family Lamiaceae. In the human model (Figure 11.6), the early stage mode of action of RA in human cells has parallels to the models for plant systems (Figure 11.4), but within the scope of human cellular physiology, function, and diversity (95). By this model, plant extract enriched RA similarly initiate an inward proton flux at the outer human cell membrane, which increases the cytosolic proton/hydride ion concentration and activates the PLPPP. Some phenolic antioxidant radicals in the total phenolic background or hydrolyzed RA, depending on their size, may penetrate the plasma membrane along with the proton/hydride ion flux (cotransport) into the cytosol. The cytosolic proton/hydride ion flux then drives PLPPP generating $NADPH_2$, sugar phosphates for anabolic reactions and proline as an alternative RE reductant to generate ATP via oxidative pentose–phosphate pathway (95,96). The products of the pentose–phosphate pathway are important for purine biosynthesis and for stimulating antioxidant enzyme response pathways in conjunction with action of the dietary phenolic antioxidants (Figure 11.6). The control of free radicals that is likely associated with proline or TCA cycle generated NADH linked mitochondrial oxidative phosphorylation at this early stage could have a positive effect on any subsequent oxidation linked cellular deterioration and consequent oxidation linked chronic disease development. Other roles for RA or RA hydrolyzed phenolic radicals that penetrate the membrane could involve:

1. Stability and protection of organelle membranes and proteins from free radical damage
2. Participation in the antioxidant response pathway to quench super oxide and peroxide radicals
3. Protection of DNA and protein stability
4. Stimulation of proline linked pentose–phosphate pathway activity to satisfy demand for $NADPH_2$ in reactions involving penetrating phenolic radicals (95,96)

In specific cases where RA or phenolic radicals cannot normally penetrate the outer plasma membrane, other conceivable roles could include:

1. Stability and protection of the outer membranes and membrane proteins from free radical damage
2. Modulation of membrane transport
3. Inhibition of specific membrane proteins, including those involved in PMF and electron transport chain in Prokaryotes
4. Modulation of signal transduction
5. Modulation of membrane receptors
6. Co-transport with H^+, sugars or amino acids
7. Passive membrane transport through damaged membranes

An inward proton flux to the cytosol could be created even without phenolic radical penetration, which then could stimulate the proline linked pentose–phosphate pathway and couple its action to the various reactions and roles that may be initiated and modulated through interactions of phenolic radicals at the outer plasma membrane (95,96).

11.5 SUMMARY

It is clear that RA containing herb extracts have wide potential applications for functional food and pharmacological applications. In optimizing RA biosynthesis for such applications the potential role of microbial elicitation and proline linked pentose–phosphate pathway (PLPPP) has been exploited to develop clonal tissue culture systems of single seed genetic origin that can be grown in traditional and efficient greenhouse and agronomic systems. Such clonal systems allow the screening of phenotypic specific RA enriched herbal herbs extracts that offer potential for consistency for various applications and design of health specific functional foods. Further, using clues about the role of proline linked pentose–phosphate pathway as a CCP in relation to RA biosynthesis in herb clonal systems (Figure 11.4), a model has been developed for the mode of action of RA in mammalian and human systems (95,96) (Figure 11.6). In this model for animal systems, RA has been hypothesized to stimulate host antioxidant enzyme response and enhance free radical scavenging to counter oxidation pressure. Efficient operation of this free radical scavenging system requires the proper functioning of pentose phosphate pathway. Further, pentose phosphate pathway could be optimally stimulated by coupling it to proline biosynthesis. This coupled proline linked pentose–phosphate pathway in normal human cells could be stimulated by RA enriched herb extracts, thereby providing critical precursors like $NADPH_2$ for antioxidant enzyme response and therefore serving as a critical control point (CCP). Through this antioxidant response regulation and other specific RA related structure and function activity at the cellular gene and signal pathway levels, protective chemopreventive functions of RA can be potentially enhanced.

REFERENCES

1. Shetty, K. Biotechnology to harness the benefits of dietary phenolics: focus on Lamiaceae. *Asia Pac. J. Clin. Nutr.* 6:162–171, 1997.
2. Shetty, K. Biosynthesis and medical applications of rosmarinic acid. *J. Herbs Spices Med. Plant.* 8:161–181, 2001.
3. Canigueral, S., J. Iglesias, M. Hamburger, K. Hostettmann. Phenolic constituents of *Salvia lavdandulifolia* spp. lavdandulifolia. *Planta Medica* 55:92, 1989.
4. Kelm, M.A., M.G. Nair, G.M. Strasburg, D.L. Dewitt. Antioxidant and cyclooxygenase inhibitory phenolic compounds from *Ocimum sanctum* Linn. *Phytomedicine* 7:7–13, 2000.
5. Kuhnt, M.A., A. Probstle, H. Rimpler, R. Bauer, M. Heinrich. Biological and pharmacological activities and further constituents of *Hyptis verticilllata. Planta Medica* 61:227–232, 1995.
6. Sumaryono, W., P. Prokash, T. Hartmann, M. Nimtz, V. Wray. Induction of rosmarinic acid accumulation in cell suspension cultures of *Orthosiphon aristatus* after treatment with yeast extract. *Phytochemistry* 30:3267–3271, 1991.
7. Zhang, H.J., L.N. Li. Salvionolic acid I: a new depside from *Salvia cavaleriei. Plant Medica* 60:70–72, 1994.
8. Psotova, J., M. Kolar, J. Sousek, Z. Svagera, J. Vicar, J. Ulrichova. Biological activities of *Prunella vulgaris* extract. *Phytother. Res.* 17:1082–1087, 2003.
9. Kikuzaki, H., N. Nakatani. Structure and a new antioxidative phenolic acid from oregano (*Origanum vulgare L*). *Agric. Biol. Chem.* 53:519–524, 1989.
10. Madsen, H.L., G. Bertelsen. Spices as antioxidants. *Trends Food Sci. Tech.* 6:271–277, 1995.
11. Jayasinghe, C., N. Gotoh, T. Aoki, S.Wada. Phenolics composition and antioxidant activity of sweet basil (*Ocimum basilicum* L.). *J. Agric. Food Chem.* 16:4442–4449, 2003.
12. Eiro, M.J., M. Heinonen. Anthocyanin color behavior and stability during storage: effect of intermolecular co-pigmentation. *J. Agric. Food Chem.* 50:7461–7466, 2002.
13. Rein, M.J., M. Heinonen. Stability of berry juice color. *J. Agric. Food Chem.* 52:3106–3114, 2004.
14. McCue, P., K. Shetty. Inhibitory effects of rosmarinic acid extracts on Porcine amylase and implications for health. *Asia Pac. J. Clin. Nut.* 13:101–106, 2004.
15. Engleberger, W., U. Hadding, E. Etschenberg, E. Graf, S. Leyck, J. Winkelmann, M.J. Parnham. Rosmarinic acid: A new inhibitor of complement C3 – convertase with antiinflammatory activity. *Intl. J. Immunopharmac.* 10:729–737, 1988.
16. Peake, P.W., B.A. Pussel, P. Martyn, V. Timmermans, J.A. Charlesworth. The inhibitory effect of rosmarinic acid on complement involves the C5 convertase. *Int. J. Immunopharmac.* 13:853–857, 1991.
17. Bult, H., A.G. Hermann, M. Rampert. Modification of endotoxin-induced haemodynamic and haemolytical changes in rabbit by methylprednisolone. F(ab)2 fragments and rosmarinic acid. *Brit. J. Phamacol.* 84:317–327, 1985.
18. Rampart, M., J.R. Beetens, H. Bult, A.J. Herman, M.J. Parnham, J. Winklemann. Complement-dependent stimulation of prostacyclin biosynthesis: inhibition by rosmarnic acid. *Biochem. Pharmc.* 35:1397–1400, 1986.
19. Nuytinck, J.K.S., R.J.A. Goris, E.S. Kalter, P.H.M. Schillings. Inhibition of experimentally induced microvascular injury by rosmarinic acid. *Agents Actions* 17:373–374, 1985.
20. Kleemann, S., H. Winterhoff. Rosmarinic acid and freeze-dried extract of *Lycos virginicus* are able to inhibit Forskolin-induced activation of adenylate cyclase in cultured rat thyroid cells. *Planta Medica* 56:683–687, 1990.

21. Van Kessel, K.P., E.S. Kalter, J. Verhoef. Rosmarinic acid inhibits external oxidative effects of human polymorphonuclear granulocytes. *Agents Actions* 17:375–376, 1986.
22. Osakabe, N., A. Ysuda, M. Natsume, C. Sanbongi, Y. Kato, T. Osawa, T. Yoshikawa. Rosmarinic acid, a major polyphenolic component of *Perilla frutescens* reduces lipopolysaccharide (LPS)-induced liver injury in D-Galactosamine (D-GalN)-sensitized mice. *Free Rad. Biol. Med.* 33:798–806, 2002.
23. Sanbongi, C., H. Takano, N. Oskabe, N. Sasa, M. Natsume, R. Yanagisawa, K. Inoue, Y. Kato, T. Osawa, T. Yoshikawa. Rosmarinic acid inhibits lung injury by diesal exhaust particles. *Free Rad. Biol. Med.* 34:1060–1069, 2003.
24. Makino, T., Y. Furuta, H. Wakushima, H. Fujii, K. Saito, Y. Kano. Anti-allergic effect of *Perilla frutescens* and its active constituents. *Phytother. Res.* 17:240–243, 2003.
25. Takeda, H., M. Tsuji, M. Inazu, T. Egashira, T. Matusmiya. Rosmarinic acid produce antidepressive-like effect in the forced swimming test in mice. *Eur. J. Pharmcol.* 449:261–267, 2002.
26. Kang, M-A., S.-Y. Yun, J. Won. Rosmarinic acid inhibits Ca2+- dependent pathways of T-cell antigen receptor-mediated signaling by inhibiting the PLC-γl and Itk activity. *Blood* 101:3534–3542, 2003.
27. Sahu, A., N. Rawal, M.K. Pangburn. Inhibition of complement by covalent attachment of rosmarinic acid to activated C3b. *Biochem. Pharmacol.* 57:1439–1446, 1999.
28. Fuhrman, B., N. Volkova, M. Rosenblat, M. Aviram. Lycopene synergistically inhibits LDL oxidation in combination with vitamin E, glabridin, rosmarinic acid, carnosic acid or garlic. *Antioxid. Redox. Signal.* 2:491–506, 2000.
29. Hooker, C.W., W.B. Lott, D. Harrich. Inhibitors of humandeficiency virus type 1 reverse transcriptase target distinct phases of early reverse transcription. *J. Virol.* 75:3095–3104, 2001.
30. Tewtrakul, S., H. Miyashiro, N. Nakamura, M. Hattori, T. Kawahata, T. Otake, T. Yoshinaga, T. Fujiwara, T. Supavita, S. Yuenyongsawad, P. Rattanasuwon, S. Dej-Adisai. HIV-1 integrase inhibitory substances from *Coleus parivifolius. Phytother. Res.* 17:232–239, 2003.
31. Makino, T., A.S. Furuta, H. Fujii, T. Nakagawa, H. Wakushima, K. Saito, Y. Kano, Y. Effect of oral treatment of *Perilla frutescens* and its constituents on type-I allergy in mice. *Biol. Pharm. Bull.* 24:1206–1209, 2001.
32. Sanbongi, C., H. Takano, N. Osakabe, N. Sasa, M. Natsume, R. Yanagisawa, K.I. Inoue, K. Sadakane, T. Ichinose, T. Yoshikawa. Rosmarinic acid in perilla extract inhibits allergic inflammation induced by mite allergen, in a mouse model. *Clin. Exp. Allerg.* 34:971–977, 2004.
33. Takano, H., N. Osakabe, C. Sanbongi, R. Yanagisawa, K. Inoue, A. Yasuda, M. Natusume, S. Baba, E. Ichiishi, T. Yoshikawa. Extract of *Perilla frutescens* enriched for rosmarinic acid, polyphenolic phytochemical inhibits seasonal allergic rhinoconjunctivitis in humans. *Exp. Biol. Med.* 229:247–254, 2004.
34. H. Takeda, M. Tsuji, J. Miyamoto, T. Matsumiya. Rosmarinic acid and caffeic acid reduce the defensive freezing behavior of mice exposed to conditioned fear stress. *Psychopharmacology* 164:233–235, 2002.
35. Makino, T., T. Ono, N. Liu, T. Nakamura, E. Muso, G. Honda. Suppressive effects of rosmarinic acid on mesangioproloferative glomerulonephritis in rats. *Nephron* 92:898–904, 2002.
36. Makino, T., T. Ono, E. Muso, H. Yoshida, G. Honda, S. Sasayama. Inhibitory effects of rosmarinic acid on proliferation of cultured murine mesangial cells. *Nephrol. Dial. Transplant.* 15:1140–1145, 2000.
37. Makino, T., T. Ono, K. Matsuyama, F. Nogaki, S. Miyawaki, G. Honda, E. Muso. Suppressive effects of *Perilla frutescens* on IgA nephropathy in HIGA mice. *Nephrol. Dial. Transplant.* 18:484–490, 2003.
38. Takeda, H., M. Tsuji, T. Matsumiya, M. Kubo. Identification of rosmarinic acid in the leaves of *Perilla frutescens* Britton var. acuta Kudo (*Perilla Herba*). *Nihon Shinkei Seishin Yakurigaku Zasshi* 22:15–22, 2002.
39. Won, J., Y.G. Hur, E.M. Hur, S.H. Park, M.A. Kang, Y. Choi, C. Park, K.H. Lee, Y. Yun. Rosmarinic acid inhibits TCR-induced T cell activation and proliferation in an Lck-dependent manner. *Eur. J. Immunol.* 33:870–879, 2003.

40. Ahn, S.C., W.K. Oh, B.Y. Kim, D.O. Kang, M.S. Kim, G.Y. Heo, J.S. Ahn. Inhibitory effects of rosmarinic acid in Lck SH2 domain binding to a synthetic phosphopeptide. *Planta Med.* 69:642–646, 2003.
41. Park, S.H., S.H. Kang, S.H. Lim, H.S. Oh, K.H. Lee. Design and synthesis of small chemical inhibitors containing different scaffolds of lck SH domain. *Biorg. Med. Chem. Lett.* 13:3455–3459, 2003.
42. Hur, Y.G., Y. Yun, J. Won. Rosmarinic acid induces p56lck-dependent apoptosis in Jurkat and peripheral T cells via mitochondrial pathway independent from Fas/Fas ligand interaction. *J. Immunol.* 172:79–87, 2004.
43. Yun, S.Y., Y.G. Hur, M.A. Kang, J. Lee, C. Ahn, J. Won. Synergistic immunosuppressive effects of rosmarinic acid and rapamycin *in vitro* and *in vivo*. *Transplantation* 75:1758–1760, 2003.
44. Youn, J., K.H. Lee, J. Won, S.J. Huh, H.S. Yun, W.G. Cho, D.J. Paik. Beneficial effects if rosmarinic acid on suppression of collagen induced arthritis. *J. Rheumatol.* 30:1203–1207, 2003.
45. Osakabe, N., A. Yasuda, M. Natsume, T. Yoshikawa. Rosmarinic acid inhibits epidermal inflammatrory responses: anticarcinogenic effect of *Perilla frutescens* extracts in the murine two-stage skin model. *Carcinogenesis* 25:549–557, 2004.
46. Seaberg, A., R.L. Labbe, K. Shetty. Inhibition of *Listeria monocytogenes* by elite clonal extracts of oregano (*Origanum vulgare*). *Food Biotechnol.* 17:129–149, 2003.
47. Chun, S.-S., D.A. Vattem, Y.-T. Lin, K. Shetty. Phenolic antioxidants from clonal oregano (*Origanum vulgare*) with antimicrobial activity against *Helicobacter pylori. Process. Biochem.* 40:809–816, 2005.
48. De-Eknamkul, W., B.E. Ellis. Rosmarinic acid production and growth characterization of *Anchusa officinalis* cell suspension cultures. *Planta Medica* 50:346–350, 1984.
49. Hipplolyte, I.B., J.C. Marin, J.C. Baccou, R. Jonard. Growth and rosmarinic acid production in cell suspension cultures of *Salvia officinalis*. *Plant Cell Rep.* 11:109–112, 1992.
50. Lopez-Arnoldos, T., M. Lopez-Serano, A.R. Barcelo, A.A. Calderon, M.M. Zapata. Spectrophotometric determination of rosmarinic acid in plant cell cultures by complexation with Fe2++ ions. *Fresenius J. Anal. Chem.* 351:311–314, 1995.
51. Mizukami, H., Y. Tabira, B.E. Ellis. Methyl jasmonate-induced rosmarinic acid biosynthesis in *Lithospermum erythrorhizon* cell suspension cultures. *Plant Cell Rep.* 12:706–709, 1993.
52. Petersen, M., B. Hausle, B. Karwatzki, J. Meinhard. Proposed biosynthetic pathway of rosmarinic acid in cell cultures of *Coleus blumei* Benth. *Plant* 189:10–14, 1993.
53. Zenk, M.H., H. El-Shagi, B. Ulbrich. Production of rosmarinic acid by cell suspension cultures of *Coleus blumei. Naturwissenschaften* 64:585–586, 1977.
54. Nitzsche, A., S.V. Tokalov, H.O. Gutzeit, J. Ludwig-Muller. Chemical and biological characterization of cinnamic acid derivatives from cell cultures of lavender (*Lavandula officinalis*) induced by stress and jasmonic acid. *J. Agric. Food Chem.* 52:2915–2923, 2004.
55. Santos-Gomes, P.C., R.M. Seabra, P.B. Andrade, M. Fernandes-Ferreira. Determination of phenolic antioxidant compounds produced by calli and cell suspension of sage (*Salvia officinalis* L). *J. Plant Physiol.* 160:1025–1032, 2003.
56. Kintzios, S., O. Makri, E. Panagiotopoulos, M. Scapeti. *In vitro* rosmarinic acid accumulation in sweet basil (*Ocimum basilicum* L.). *Biotechnol. Lett.* 25:405–408, 2003.
57. Mizukami, H., T. Ogawa, H. Ohashi, B.E. Ellis. Induction of rosmarinic acid biosynthesis in *Lithospermum erythrorhizon* cell suspension cultures by yeast extract. *Plant Cell Rep.* 11:480–483, 1992.
58. Szabo, E., A. Thelen, M. Petersen. Fungal elicitor preparations and methyl jasmonate enhance rosmarinic acid accumulation in suspension cultures of *Coleus blumei. Plant Cell Rep.* 18:485–489, 1999.
59. Chen, H., F. Chena, F.C. Chiu, C.M. Lo. The effect of yeast elicitor on the growth and secondary metabolism of hairy root cultures of *Salvia miltiorrhiza*. *Enzyme Microb. Technol.* 28:100–105, 2001.

60. Kreis, W., E. Reinhard. The production of secondary metabolites by plant cells cultivated in bioreactors. *Plant Medica* 55:409–416, 1989.
61. Ulbrich, B., W. Wiesner, H. Arens. Large-scale production of rosmarinic acid from plant cell cultures of *Coleus blumei* Benth. In: *Primary and Secondary Metabolism of Plant Cell Cultures,* Neumann, K. Ed., Heidelberg: Springer-Verlag, 1985, pp. 292–303.
62. Georgiev, M., A. Pavlov, M. Ilieva. Rosmarinic acid production by *Lavandula vera* MM cell suspension: the effect of temperature. *Biotechnol. Lett.* 26:855–856, 2004.
63. Kinzios, S., H. Kollias, E. Straitouris, O. Makri. Scale-up micropropagation of sweet basil (*Ocimum basilicum* L.) in an air-lift bioreactor and accumulation of rosmarinic acid. *Biotechnol. Lett.* 26:521–523, 2004.
64. Phillips, R.L., S.M. Kaeppler, J. Olhoft. Genetic instability of plant tissue cultures: Breakdown of normal controls. *Proc. Natl. Acad. Sci. USA* 91:5222–5226, 1994.
65. De-Eknamkul, W., B.E. Ellis. Purification and characterization of tyrosine aminotransferase activities from *Anchusa officinalis* cell cultures. *Arch. Biochem. Biophys.* 257:430–438, 1987.
66. Razzaque, A., B.E. Ellis. Rosmarinic acid production in *Coleus* cell cultures. *Planta* 137:287–291, 1977.
67. Whitaker, R.J., T. Hashimoto, D.A. Evans. Production of secondary metabolites, rosmarinic acid by plant cell suspension cultures. *Ann. N.Y. Acad. Sci.* 435:363–368, 1984.
68. De-Eknamkul, W., B.E. Ellis. Effects of macronutrients on the growth and rosmarinic acid formation in cell suspension cultures of *Anchusa officinalis. Plant Cell Rep.* 4:46–49, 1985.
69. De-Eknamkul, W., B.E. Ellis. Effects of auxins and cytokinins on the growth and rosmarinic acid formation in cell suspension cultures of *Anchusa officinalis. Plant Cell Rep.* 4:50–53, 1985.
70. Pavlov, A.I., M.P. Ilieva, I.N. Panchev. Nutrient medium optimization for rosmarinic acid production by *Lavandula vera* MM cell suspension. *Biotechnol. Prog.* 16:668–670, 2000.
71. Mohagheghzadeh, A., M. Shams-Ardakani, A. Ghannadi, M. Minaeian. Rosmarinic acid from *Zataria multiflora* tops and *in vitro* cultures. *Fitoterapia* 75:315–321, 2004.
72. Gertolowski, C. M. Petersen. Influence of carbon source on growth and rosmarinic acid production in suspension cultures of *Coleus blumei. Plant Cell Tissue Org. Cult.* 34:183–190, 1993.
73. Ellis, B.E., G.H.N. Towers. Biogenesis of rosmarinic acid in *Mentha. Biochem. J.* 118:287–291, 1970.
74. Petersen, M. Cytochrome P-450-dependent hydroxylation in the biosynthesis of rosmarinic acid in *Coleus. Phytochemistry* 45:1165–1172, 1997.
75. Petersen, M., M.S.J. Simmonds. Molecules of interest: rosmarinic acid. *Phytochemistry* 62:121–125, 2003.
76. De-Eknamkul, W., B.E. Ellis. Tyrosine aminotransferase: the entry point enzyme of the tyrosine-derived pathway in rosmarinic acid biosynthesis. *Phytochemistry,* 26:1941–1946, 1987.
77. Mizukami, H., B.E. Ellis. Rosmarinic acid formation and differential expression of tyrosine aminotransferase isoforms in *Anchusa officinalis* cell suspension cultures. *Plant Cell Rep.* 10:321–324, 1991.
78. Petersen, M., E. Hausler, J. Meinhard, B. Karwatzki, C. Gertlowski. The biosynthesis of rosmarinic acid in suspension cultures of *Coleus blumei. Plant Cell Tissue Org. Cult.* 38:171–179, 1994.
79. De-Eknamkul, W., B.E. Ellis. Purification and characterization of prephenate aminotransferase from *Anchusa officinalis* cell cultures. *Arch. Biochem. Biophys.* 267:87–94, 1988.
80. Hausler, E., M. Petersen, A.W. Alfermann. Hydroxyphenylpyruvate reductase from cell suspension cultures of *Coleus blumei. Benth. Naturforsch.* 46C:371–376, 1991.
81. Petersen, M. Characterization of rosmarinic acid synthase from cell cultures of *Coleus blumei. Phytochemistry* 30:2877–2881, 1991.

82. Petersen. M., A.W. Alfermann. Two enzymes of rosmarinic acid biosynthesis from cell cultures of *Coleus blumei*: Hydroxyphenylpyruvate reductase and rosmarinic acid synthase. *Z. Naturfosch.* 43C:501–504, 1988.
83. Petersen, M. Cinnamic acid 4-hydroxylase from cell cultures of the hornwort *Anthoceros agrestis. Planta* 217:96–101, 2003.
84. Matsuno, M., A. Nagatsu, Y, Ogihara, B.E. Ellis, H. Mizukami. CYP98A6 from *Lithospermum erythrorhizon* encodes 4-coumaroyl-4′-hydroxyphenyllactic acid 3-hydroxylase involved in rosmarinic acid biosynthesis. *FEBS Lett.* 514:219–224, 2002.
85. Schoch, G., S. Goepfert, M. Morant, A. Hehn, D. Meyer, P. Ullmann, D. Werck-Reichhart. CYP98A3 from *Arabidopsis thaliana* is a 3′-hydroxylase of phenolic esters, a missing link in the phenylpropanoid pathway. *J. Biol. Chem.* 276:36566–36574, 2001.
86. Anterola, A.M., J.H. Jeon, L.B. Davin, N.G. Lewis. Transcriptional control of monolignol biosynthesis in *Pinus taeda*. Factors affecting monolignol ratios and carbon allocation in the phenylpropanoid metabolism. *J. Biol. Chem.* 277:18272–18280, 2002.
87. Shetty, K., O.F. Curtis, R.E. Levin, R. Witkowsky, W. Ang. Prevention of vitrification associated with *in vitro* shoot culture of oregano (*Origanum vulgare*) by *Pseudomonas* spp. *J. Plant Physiol.* 147:447–451, 1995.
88. Shetty, K., O.F. Curtis, R.E. Levin. Specific interaction of mucoid strains of *Pseudomonas* spp. with oregano (*Origanum vulgare*) clones and the relationship to prevention of hyperhydricity in tissue culture. *J. Plant Physiol.* 149:605–611, 1996.
89. Eguchi, Y., O.F. Curtis, K. Shetty. Interaction of hyperhydricity-preventing *Pseudomonas* spp. with oregano (*Origanum vulgare*) and selection of high rosmarinic acid-producing clones. *Food Biotechnol.* 10:191–202, 1996.
90. Shetty, K., T.L. Carpenter, D. Kwok, O.F. Curtis, T.L. Potter. Selection of high phenolics-containing clones of thyme (*Thymus vulgaris* L.) using *Pseudomonas* spp. *J. Agric. Food Chem.* 44:3408–3411, 1996.
91. Yang, R., O.F. Curtis, K. Shetty. Selection of high rosmarinic acid-producing clonal lines of rosemary (*Rosmarinus officinalis*) via tissue culture using *Pseudomonas* sp. *Food Biotechnol.* 11:73–88, 1997.
92. Al-Amier, H., B.M.M. Mansour, N. Toaima, R.A. Korus, K. Shetty. Tissue culture-based screening for selection of high biomass and phenolic-producing clonal lines of Lavender using *Pseudomonas* and azetidine-2-carboxylate. *J. Agric. Food Chem.* 47:2937–2943, 1999.
93. Al-Amier, H., B.M.M. Mansour, N. Toaima, R.A. Korus, K. Shetty. Screening of high biomass and phenolic-producing clonal lines of Spearmint in tissue culture using *Pseudomonas* and azetidine-2-carboxylate. *Food Biotechnol.* 13:227–253, 1999.
94. Al-Ameir, H., B.M.M. Mansour, N. Toaima, L. Craker, K. Shetty. Tissue culture for phenolics and rosmarinic acid in thyme. *J. Herbs Spices Med. Plant.* 8:31–42, 2001.
95. Shetty, K. Role of proline-linked pentose phosphate pathway in biosynthesis of plant phenolics for functional food and environmental applications: a review. *Process. Biochem.* 39:789–804, 2004.
96. Shetty, K., M.L. Wahlqvist. A model for the role of proline-linked pentose phosphate pathway in phenolic phytochemical biosynthesis and mechanism of action for human health and environmental applications: a review. *Asia Pac. J. Clin. Nutr.* 13:1–24, 2004.
97. Phang, J.M. The regulatory functions of proline and pyrroline-5-carboxylic acid. *Curr. Topics Cell. Reg.* 25:91–132, 1985.
98. Hagedorn, C.H., J.M. Phang. Transfer of reducing equivalents into mitochondria by the interconversions of proline and δ–pyrroline-5-carboxylate. *Arch. Biochem. Biophys.* 225:95–101, 1983.
99. Rayapati J.P., C.R. Stewart. Solubilization of a proline dehydrogenase from maize (*Zea mays* L.) mitochondria. *Plant Physiol.* 95:787–791, 1991.
100. Yeh, G.C., J.M. Phang. The function of pyrroline-5-carboxylate reductase in human erythrocytes. *Biochem. Biophys. Res. Commun.* 94:450–457, 1980.

101. Phang, J.M., S.J. Downing, G.C. Yeh, R.J. Smith, J.A. Williams. Stimulation of hexosemonophosphate-pentose pathway by δ-pyrroline-5-carboxylic acid in human fibroblasts. *Biochem. Biophys. Res. Commun.* 87:363–370, 1979.
102. Hagedorn, C.H., J.M. Phang. Catalytic transfer of hydride ions from NADPH to oxygen by the interconversions of proline and δ-pyrroline-5-carboxylate. *Arch. Biochem. Biophys.* 248:166–174, 1986.
103. Kwok, D., K. Shetty. Effect of proline and proline analogs on total phenolic and rosmarinic acid levels in shoot clones of thyme (*Thymus vulgaris* L.). *J. Food Biochem.* 22:37–51, 1998.
104. Yang, R., K. Shetty. Stimulation of rosmarinic acid in shoot cultures of oregano (*Origanum vulgare*) clonal line in response to proline, proline analog and proline precursors. *J. Agric. Food Chem.* 46:2888–2893, 1998.
105. Bela, J., K. Shetty. Somatic embryogenesis in anise (*Pimpinella anisum* L.): The effect of proline on embryogenic callus formation and ABA on advanced embryo development. *J. Food Biochem.* 23:17–32, 1999.
106. Perry, P.L., Shetty, K. A model for involvement of proline during *Pseudomonas*-mediated stimulation of rosmarinic acid. *Food Biotechnol.* 13:137–154, 1999.
107. Lendzian, K.J. Modulation of glucose-6-phosphate dehydrogenase by NADPH, NADP+ and dithiothreitol at variable NADPH/NADP+ ratios in an illuminated reconstituted spinach (*Spinacia oleracea* L.) choroplast system. *Planta* 148:1–6, 1980.
108. Copeland, L., J.F. Turner. The regulation of glycolysis and the pentose-phosphate pathway. In: *The Biochemistry of Plants*, Vol. 11, Stumpf, F., E.E. Conn, eds., New York: Academic Press, 1987; 107–125.
109. Elthon, T.E., C.R. Stewart. Effects of proline analog L-thiazolidine-4-carboxylic acid on Proline metabolism. *Plant Physiol.* 74:213–218, 1984.
110. Jost, A., S. Perlman, O. Valentino, M. Castinier, R. Scholler, S. Magre. Experimental control of the differentiation of Leydig cells in the rat fetal testis. *Proc. Natl. Acad. Sci. USA* 85:8094–8097, 1988.
111. Kohl, D.H., K.R. Schubert, M.B. Carter, C.H. Hagdorn, G. Shearer. Proline metabolism in N_2-fixing root nodules: energy transfer and regulation of purine synthesis. *Proc. Natl. Acad. Sci. USA* 85:2036–2040, 1988.
112. McCue, P., K. Shetty. A biochemical analysis of mungbean (*Vigna radiata*) response to microbial polysaccharides and potential phenolic-enhancing effects for nutraceutical applications. *Food Biotechnol.* 6:57–79, 2002.
113. McCue, P., K. Shetty. Clonal herbal extracts as elicitors of phenolic synthesis in dark-germinated mungbeans for improving nutritional value with implications for food safety. *J. Food Biochem.* 26:209–232, 2002.
114. Randhir, R., K. Shetty. Microwave-induced stimulation of L-DOPA, phenolics and antioxidant activity in fava bean (*Vicia faba*) for Parkinson's diet. *Process Biochem.* 39:1775–1784, 2004.
115. Randhir, R., P. Shetty, K. Shetty. L-DOPA and total phenolic stimulation in dark geriminated fava bean in response to peptide and phytochemical elicitors. *Process Biochem.* 37:1247–1256, 2002.
116. Randhir, R., K. Shetty. Light-mediated fava bean (*Vicia faba*) response to phytochemical and protein elicitors and consequences on nutraceutical enhancement and seed vigor. *Process Biochem.* 38:945–952, 2003.
117. McCue, P., Z. Zheng, J.L. Pinkham, K. Shetty. A model for enhanced pea seedling vigor following low pH and salicylic acid treatments. *Process Biochem.* 35:603–613, 2000.
118. Droge, W. Free radicals in the physiological control of cell function. *Physiol. Rev.* 82:47–95, 2002.
119. Dreher, D., A.F. Junod. Role of oxygen free radicals in cancer development. *Eur. J. Cancer* 32A:30–38, 1996.
120. Ha, H.C., A. Thiagalingam, B.D. Nelkin, R.A. Casero, Jr. Reactive oxygen species are critical for the growth and differentiation of medullary throid carcinoma cells. *Clin. Cancer Res.* 6:3783–3787, 2000.

121. Alexander, R.W. Theodore Cooper Memorial Lecture: hypertension and the pathogenesis of atherosclerosis: oxidative stress and the mediation of arterial inflammatory response: a new perspective. *Hypertension* 25:155–161, 1995.
122. Griendling, K.K., C.A. Minieri, J.D. Ollerenshaw, R.W. Alexander. Angiostatin II stimulates NADH and NADPH oxidase activity in cultured vascular smooth muscle cells. *Circ. Res.* 74:1141–1148, 1994.
123. Araujo, V., C. Arnal, M. Boronat, E. Ruiz, C. Dominguez. Oxidant-antioxidant imbalance in blood of children with juvenile rheumatoid arthritis. *Biofactors* 8:155–159, 1998.
124. Mapp, P.I., M.C. Grootveld, D.R. Blake. Hypoxia, oxidative stress and rheumatoid arthritis. *Br. Med. Bull.* 51:419–436, 1995.
125. Multhaup, G., T. Ruppert, A. Schlicksupp, L. Hesse, D. Beher, C.L. Masters, K. Beyreuther. Reactive oxygen species in Alzheimer's disease. *Biochem. Pharmacol.* 54:533–539, 1997.
126. Wolff, S.P. Diabetes mellitus and free radicals: free radicals, transition metals and oxidative stress in the aetiology of diabetes mellitus and complications. *Br. Med. Bull.* 49:642–652, 1993.
127. Baynes, J.W. Role of oxidative stress in development of complications in diabetes. *Diabetes* 40:405–412, 1991.
128. De Mattia, G., M.C. Bravi, O. Laurenti, M. Cassone-Faldetta, A. Armeinto, C. Ferri, F. Balsano. Influence of reduced glutathione infusion on glucose metabolism in patients with non-insulin-dependent diabetes mellitus. *Metabolism* 47:993–997, 1998.
129. Nishikawa, T., D. Edelstein, X.L. Du, S. Yamagishi, T. Matsumura, Y. Kaneda, M.A. Yorek, D. Beebe, P.J. Oates, H.P. Hammes, I. Giardino, M. Brownlee. Normalizing mitochondrial superoxide production blocks three pathways of hyperglycemic damage. *Nature* 404:787–790, 2000.
130. Van Dam, P.S., van Asbeck, B.S., Erkelens, D.W., Marx, J.J.M., Gispen, W.H. and Bravenboer, B. (1995) The role of oxidative stress in neuropathy and other diabetic complications. *Diabetes Metab. Rev.* 11:181–192.
131. Hertog, M.G.L., D. Kromhout, C. Aravanis, H. Blackburn, R. Buzina, F. Fidanza, S. Giampaoli, A. Jansen, A. Menotti, S. Nedeljkovic, M. Pekkarinen, B.S. Simic, H. Toshima, E.J.M. Feskens, P.C.H. Hollman, M.B. Kattan. Flavonoid intake and long-term rosk of coronary heart disease and cancer in the seven countries study. *Arch. Intern. Med.* 155:381–386, 1995.
132. Masuda, T., A. Jitoe. Antioxidative and anti-inflammatory compounds from tropical gingers: isolation, structure determination, and activities of cassumunins A, B and C, new complex curcuminoids from *Zingiber cassumunar*. *J. Agric. Food Chem.* 42:1850–1856, 1994.
133. Narayanan, B.A., G.G. Re. IGF-II down regulation associated cell cycle arrest in colon cancer cells exposed to phenolic antioxidant ellagic acid. *Anticancer Res.* 21:359-364, 2001.
134. Labriola, D., R. Livingston. Possible interactions between dietary antioxidants and chemotherapy. *Oncology* 13:1003–1008, 1999.
135. Freudenheim, J.L., J.R. Marshall, J.E. Vena. Premenupausal breast cancer risk and intake of vegetables, fruits and related nutrients. *J. Natl. Cancer Inst.* 88:340–348, 1996.
136. Morel, Y., R. Barouki. Repression of gene expression by oxidative stress. *Biochem. J.* 342:481–496, 1999.

2.12

Bioprocessing Strategies to Enhance L-DOPA and Phenolic Antioxidants in the Fava Bean *(Vicia faba)*

Kalidas Shetty, Reena Randhir, and Preethi Shetty

CONTENTS

12.1 INTRODUCTION AND ROLE OF PHENOLIC SECONDARY METABOLITES

In plants, primary products such as carbohydrates, lipids, proteins, photosynthetic components, and nucleic acids are common to all; they are involved in the primary metabolic processes of building and maintaining cells. In contrast, secondary metabolites do not appear to have such a vital biochemical role, but studies have indicated a role of these chemicals in defense and stress response of plants. Some of the most abundant stress induced secondary metabolites synthesized by plants are phenolics and their derivatives. Phenolic compounds include a large array of chemical compounds possessing an aromatic ring bearing one or more hydroxyl groups, together with a number of other side groups. Plant phenolics are a chemically heterogeneous group (1–3). These phenolics usually occur in conjugated or esterified form as glycosides (1,2,4). The diverse arrays of plant phenolics have many roles in plant growth and development. Therefore, emergence of dietary and medicinal applications for phenolic phytochemicals, harnessing especially their antioxidant and antimicrobial properties, for the benefit of human health and wellness is not altogether surprising. As stress damage at the cellular level appears similar among eukrayotes, it is logical to suspect that there may be similarities in the mechanism for cellular stress mediation between eukaryotic species. Plant adaptation to biotic and abiotic stress involves the stimulation of protective secondary metabolite pathways (5–7), resulting in the biosynthesis of phenolic antioxidants. Studies indicate that plants exposed to ozone responded with increased transcript levels of enzymes in the phenylpropanoid and lignin pathways (8). Increase in plant thermotolerance is related to the accumulation of phenolic metabolites and heat shock proteins that act as chaperones during hyperthermia (9). Phenolics and specific phenolic like salicyclic acid levels increase in response to infection, acting as defense compounds or serving as precursors for the synthesis of lignin, suberin, and other polyphenolic barriers (10). Antimicrobial phenolics, called phytoalexins, are synthesized around the site of infection during pathogen attack and, along with other simple phenolic metabolites, are believed to be part of a signaling process that results in systemic acquired resistance (5–7). Many phenylpropanoid compounds such as flavonoids, isoflavonoids, anthocyanins, simple phenolics, and polyphenols are induced in response to wounding (11), nutritional stress (12), cold stress (13), and high visible light (14). Ultraviolete (UV) irradiation induces light-absorbing flavonoids and sinapate esters in *Arabidopsis* to block radiation and protect DeoxyriboNucleic Acid (DNA) from dimerization and cleavage (15). In general, the initiation of the stress response arises from certain changes in the intracellular medium (16) that transmits the stress induced signal to cellular modulating systems, resulting in changes in cytosolic calcium levels, proton potential as a long distance signal (17), and low molecular weight proteins (18). Stress can also initiate free radical generating processes and shift the cellular equilibrium toward lipid peroxidation (19). It is believed that the shift in prooxidant antioxidant equilibrium is a primary nonspecific event in the development of the general stress response (20). Therefore, phenolic compounds are ubiquitous and have important roles in all vascular plants, and as a result are integral part of the human diet (21–23). These phenolic secondary metabolites that are synthesized through the shikimic acid pathway vary from simple phenolics such as the hydroxy benzoic acids and levodopa (L-DOPA) to biphenyls such as resveratrol and rosmarinic acid to large condensed tannins and hydrolysable tannins with high molecular weights (23,24). The polymers formed from plant phenolics in the cell wall provide structural support and form barriers to prevent moisture loss diffusion and pathogen encroachment. The phenolics also function in defense mechanisms with UV protectant, antifungal, antibacterial, antifeedant, and antimitotic properties, and in morphogenesis (25,26). When exposed to air, most phenolics readily undergo oxidation to colored quinone containing products. This response is frequently observed as a browning reaction of plant tissues as a part of a healing response. The oxidation of these compounds by polyphenoloxidase (PPO) has been suggested to be the main cause of apple browning (27). Therefore, protective phenolic metabolites involved in such secondary metabolite linked stress responses in food plant species can be targeted as a source of therapeutic and disease-preventing functional ingredients, especially in oxidation disease linked diets (diets containing foods with a high glycemic index and saturated fats) and environmentally (physically, chemically, and biologically) influenced chronic disease problems (23).

12.2 PLANT PHENOLICS IN HUMAN HEALTH AND AS ANTIOXIDANTS

It is evident that plant phenolic compounds constitute one of the most numerous and widely distributed groups of substances with more than 8000 phenolic structures currently known (28). In addition to stress linked phenolics coming only from the shikimate and phenylpropanoid pathways, a number of the phenolic compounds are found in plants, including the flavonoids that contribute to the characteristic flavor and fragrance of vegetables, fruits, tea, and wine. These compounds, which come from phenylpropanoid and polyketide (acetate–malonate) pathways, also have biological properties that are beneficial to human health.. Flavonoids such as quercetin and catechin and isoflavonoids, genistein for example, are being investigated for properties which may reduce the incidence of cancer(22,23). Flavonoids and isoflavonoids are a class of phenolic compounds that have appeared sequentially during plant evolution and are simple aromatic compounds generated from both the phenypropanoid and acetate–malonate (polyketide) pathways (24). From a functional health point of view, it is suggested that such phenolics, for example, through the consumption of tea, may provide protection against certain cancers; soybeans may provide protection against breast cancer and osteoporosis (22). The Japanese and Chinese frequently used plants rich in polyphenol tannins in their folk medicines for the treatment of inflammation, liver injury, kidney ailments, hypertension, and ulcers. Oregano extracts have been shown to inhibit lipid peroxidation by their flavonoid fractions such as flavone apigenin, the flavanone, eriodictoyl, dihydrokaempferol, and dihydroquercetin (29). Rosmarinic acid-containing *Ocimum sanctum* (holy basil), derived from the phenylpropanoid pathway, is commonly used in India to reduce fevers and gastrointestinal disease (30). Essential oils from thyme (*Thyme vulgaris* L.) have phenolic linked antioxidant properties, which may result from the presence of free radical scavengers in these oils (31–33). These diverse phenolic compounds in thyme are unusually derived from phenylpropanoid, polyketide, and terpenoid pathways. Many other phenolic compounds have the ability to inhibit platelet aggregation, block calcium influx, and protect low density lipoproteins (LDL) from oxidation (34). L-DOPA from the fava bean, that is the focus of this chapter, has been evaluated in Parkinson's management (35–37); it is derived from L-tyrosine of the phenylpropanoid pathway (Figure 12.1).

Among many functional roles, the most important function assigned to phenolics is their antioxidant activity. Antioxidants may be defined as substances, which when present in low concentrations compared with those of an oxidizable substrate, such as proteins, carbohydrates, and fats, delay or prevent the oxidation of the substrate (38). Phenolic antioxidants from dietary plants can be useful to counter reactive oxygen species related to human diseases. Reactive oxygen species (ROS) are able to oxidize cellular components such as DNA, lipids, and proteins (23). Dysfunction of oxidative phosphorylation at the mitochondrion has been recognized as a major physiological source of ROS (39). Lipid peroxidation damages the structural integrity of the mitochondria, which can result in organelle swelling, resulting in increased permeability to cations, decreased membrane potential, and damage to electron transfer activities (40). This form of tissue damage can ultimately lead to some of the major diseases such as cancer, cardiovascular disease, immune dysfunction, diabetes, and neurodegeneration (23). To deal with ROS, the biological systems, including human cellular systems, have an effective defense system, which includes enzymes such as superoxide dismutase (SOD), catalase (CAT), high molecular weight antioxidants, and an array of low molecular weight antioxidants such as ascorbic acid, α-tocopherol, β-carotene, and glutathione (41). The endogenous antioxidant responses, both enzymatic and nonenzymatic could also be enhanced by dietary intake of

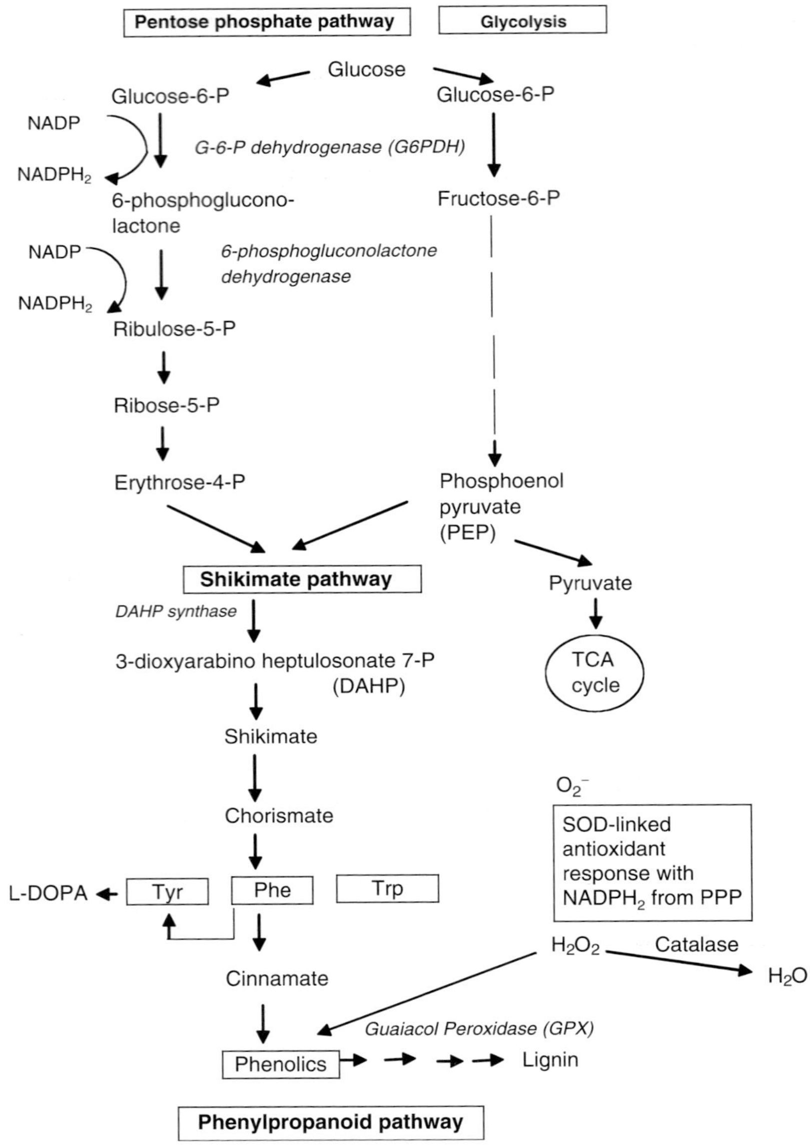

Figure 12.1 Pentose–phosphate pathway and L-DOPA biosynthesis in plants

phenolic antioxidants from plant foods (23,42). Therefore phenolic antioxidants, through design and use of functional foods, are receiving increasing interest from consumers and food manufacturers due to their synergistic roles as antioxidants, antimutagens, and scavengers of free radicals (23,43). They have the potential to function as antioxidants by trapping free radicals generated in the oxidative chemistry, which then normally undergo coupling reaction leading to polymeric or oligomeric products (43) or by enhancing host antioxidant enzyme response through a stimulation of antioxidant enzyme response through SOD, peroxidases, and CAT (23). Epidemiological studies have also suggested associations between the consumption of phenolic antioxidant rich foods or beverages and the prevention of diseases (44–49).

12.3 BIOSYNTHESIS OF PHENOLIC METABOLITES AND L-DOPA

Plant phenolics are synthesized using several different routes. This constitutes a heterogeneous group from a metabolic point of view. Two basic pathways are the shikimic acid pathway and the acetate–malonate (polyketide) pathway. The shikimic acid pathway represents the principal mode of accumulation of plant stress related phenolic compounds. The acetate–malonate pathway is also an important source of phenolic secondary products for biosynthesis of flavonoids and isoflavonoids that have many human disease protective chemoprevention properties.

The shikimic acid pathway requires substrates such as erythrose-4-phosphate (E4P) from the pentose phosphate pathway (PPP) and phosphoenol pyruvate (PEP) from glycolysis (Figure 12.1). The oxidative pentose phosphate pathway in plants is thought of as comprising two stages (50). The first is an essential irreversible conversion of glucose 6 phosphate (G6P) to ribulose-5-phosphate (Ril5P) by the enzymes glucose-6-phosphate dehydrogenase (G6PDH), 6-phosphogluconolactonase, and 6-phospho gluconolactonate dehydrogenase (6 PGDH). This oxidative stage provides reductant in the form of $NADPH_2$ for a wide range of anabolic pathways including the synthesis of fatty acids, the reduction of nitrite, and synthesis of glutamate (51,52). The second stage is the irreversible series of interconversions between phosphorylated carbon sugars. The function of this stage of the pathway is to provide a carbon skeleton for the shikimate pathway via erythrose-4-phosphate (E4P), nucleotide synthesis utilizing ribose-5-phosphate, as well as recycling of sugar phosphate intermediates for use in the glycolytic pathway (53).

The shikimate pathway is often referred to as the common aromatic biosynthetic pathway, even though in nature it does not synthesize all aromatic compounds by this route (54). The shikimic acid pathway converts the simple carbohydrate precursors to aromatic amino acids phenylalanine, tyrosine, and tryptophan. The flux from this pathway is critical for both auxin and phenylpropanoid synthesis (23,30). Auxins are plant hormones that regulate plant development (2). Up to 60% of the dry weight in some plant tissue consists of metabolites derived from the shikimate pathway. Activity of the distinct isoenzyme of 3-deoxy arabinose heptulosonate-7-phosphate (DAHP) synthase in the shikimate pathway is dependent on metabolite flux from E4P. This enzyme has been shown to be subject to feedback inhibition by L-phenylalanine, L-tyrosine, and L-tryptophan (54,55). Therefore, this enzyme controls the carbon flow into the shikimate pathway.

The most abundant class of secondary phenolic compounds in plants is derived from phenylalanine via the elimination of an ammonia molecule to form cinnamic acid. This reaction is catalyzed by phenylalanine ammonia lyase (PAL). This is the branch point between the primary (shikimate pathway) and the secondary (phenylpropanoid pathway) (6). Studies with several different species of plants have shown that the activity of PAL is increased by environmental factors such as low nutrient levels, light (through the effect of phytochrome), and fungal infection (11). Fungal invasion triggers the transcription of messenger RNA (mRNA) in the plant, which then stimulates the synthesis of phenolic compounds (6). Many phenylpropanoid compounds are induced in response to wounding or in response to microbial pathogens, insect pests, or herbivores. Anthocyanins increase in response to high visible light levels and it is thought that they attenuate the amount of light reaching the photosynthetic cells (6).

The product of PAL, *trans*-cinnamic acid, is converted to *para*-coumaric acid by the addition of a hydroxyl group on the aromatic ring in *para* position. Subsequent reactions lead to the addition of more hydroxyl groups and other substituents. These are simple phenolic compounds called phenylpropanoids because they contain a benzene ring and a three carbon side chain. Phenylpropanoids are building blocks for more complex phenolic compounds (1,3).

As previously discussed, phenolic compounds have wide ranges of functions. The synthesized phenolics can be either antioxidant in nature or they may function in lignification of the plant cells. Depending on the requirements, the type of the phenolics synthesized and their complexity vary from species to species in different environmental niches. Flavonoids, tannins, caffeic acids, curcumin, gallic acids, eugenol, rosmarinic acid, and many more have antioxidant properties (56,57). These ranges of phenolics provide plants with defense mechanisms and act as scavengers of free radicals as described earlier. Other functions of phenolics include their ability to provide structural stability to the plants by lignins and lignans. These are complex phenolics are formed from the polymerization of simple phenols (58). Figure 12.2 illustrates the origins of varied phenolic compounds from simple phenols.

Lignin is a polymer of aromatic subunits usually derived from phenylalanine. It serves as a matrix around the polysaccharide components of some plant cell walls, providing additional rigidity and compressive strength, as well as rendering the walls hydrophobic for water impermeability (59,60). The final enzymatic steps of lignin biosynthesis, the production of mesomeric phenoxy radicals from cinnamoyl alcohol, is catalyzed by peroxidase and must occur outside the cell to allow these short lived radicals to polymerize *in situ* (61). Phenol polymerization is catalyzed by the peroxidase enzyme (62) and a specific enzyme, Guaiacol peroxidase (GPX), is suggested to be important in the metabolic interconversion of phenolic antioxidants. The same phenylpropanoid pathway also supports the synthesis of L-DOPA, which is a simple phenolic compound found in many seeded legumes such as fava beans and velvet beans with relevance for Parkinson's therapy (37,63–66). This pathway

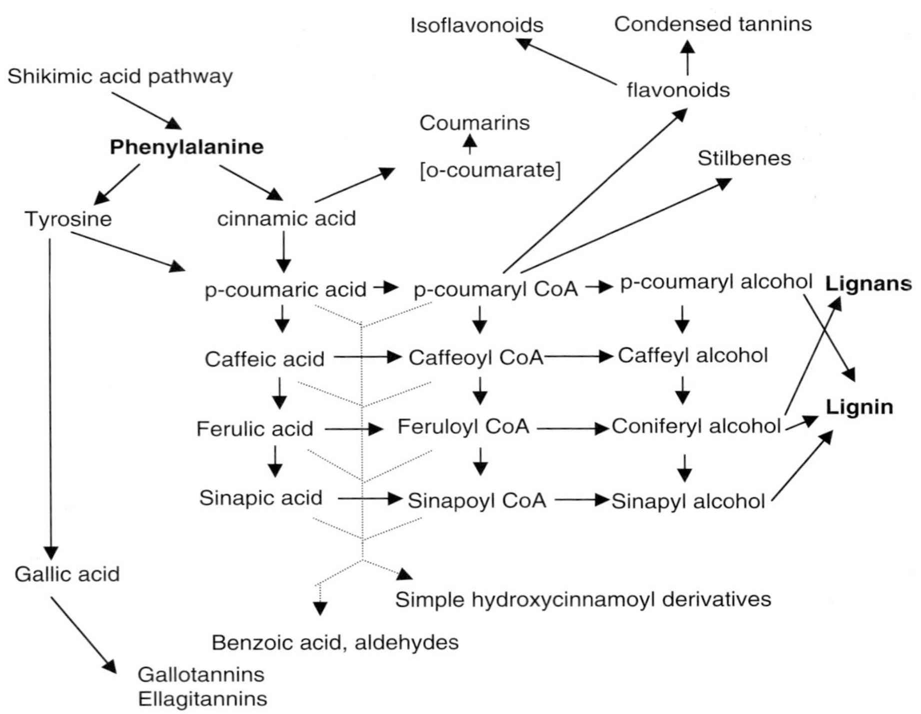

Figure 12.2 General phenolic flux through the phenylpropanoid pathway from the shikimic pathway

provides the precursor chorismate, which oxidizes to L-phenylalanine, before going through a hydroxylation step to form L-tyrosine and than L-DOPA (Figure 12.1).

12.4 L-DOPA AND PARKINSONIAN SYNDROME

The Parkinsonian syndrome has long held the interest of psychologists, psychiatrists, and other behavioral investigators. Lately, the field of cerebral monoamines has received increased attention; first with the serotonin hypothesis of Brodie, then the noradrenaline hypothesis of Schildrant, in relation to neuron function, is receiving particular attention. (67). Dopamine is postulated to be the key neurotransmitter responsible for the origin of human intelligence (67). Dopamine, the direct precursor for noradrenaline, has a specific distribution pattern within the brain (Figure 12.3). It is concentrated mainly in the striatum and substantia nigra. It is deficient in patients with Parkinson's diseases (PD). L-DOPA has been studied as a treatment for neurological disorders such as Parkinson's disease (68,69).

Parkinson's disease is a common neurological disorder, affecting 1% of the people over the age of 60 years; it is a disease with the signs of rigidity, resting tremor, postural instability, which is due to underproduction of the neurotransmitter dopamine (DA) (70). One of the factors leading to PD is the marked loss of melanized nigrostriatal dopamine neurons, one of the principal components responsible for the normal control of movement. Signs of PD do not appear until a large majority of the nigrostriatal DA neurons have been damaged. Another factor that may mitigate the loss of nigrostriatal DA neurons in PD is a reduced rate of DA activation (71). Also, mitochondrial swelling was noticed as a result of low concentrations of DA, causing a mitochondrial damage triggering neurodegenerative process and PD (72). DA receptors are expressed not only in the central nervous system (CNS), but also in several peripheral tissues including arteries, heart, thymus, and peripheral blood lymphocytes (73). Unconjugated DA represents only 20–40% of the total DA excreted in human urine, with the remainder being mainly excreted in the form of phenolic sulfate (74). It has been suggested that a low level of DA excretion is associated with hypokinesia and normal or excess rate of excretion with hyperkinesia and associated renal dysfunction (74). Overproduction of this catecholamine appears to be associated with the psychological disorder schizophrenia (68).

Introduction of L-DOPA has improved the clinical status of patients with Parkinson's disease (75). Levodopa (L-DOPA — synonym: levo-dihydroxyphenylalanine) therapy in Parkinson's disease was initiated after the suggestion of the role of dopamine in the neuronal system (71,76–81). Dopamine is also known to have a role in renal function, through its action on renal dopaminergic receptors and improved concentration of intrarenal area is dependent on L-DOPA availability (82). The biosynthesis dopamine is considered to start from tyrosine. obtained from dietary sources (82) (Figure 12.3). Blood borne tyrosine is taken into the brain to effect the functioning of dopaminergic neurons. When tyrosine has entered the neurons, L-DOPA (dihydoxyphenylalanine) is made using the cytsolic enzyme tyrosine hydroxylase (71). Subsequently, another cytosolic enzyme aromatic amino acid decarboxylase converts L-DOPA to dopamine (71). In the neurons, striatum is one of the main components of basal ganglia that are responsible for normal control of movement and role of dopamine is essential for this action (71). In Parkinson's disease, the importance of dopamine for striatum is clear due to the motor abnormalities seen in patients, which is characterized by a marked loss of melanized nigrostriatal dopamine neurons (71). Dopamine release in the striatum modulates activity in direct and indirect circuits, which are important for voluntary control of movement (71). Signs of Parkinson's

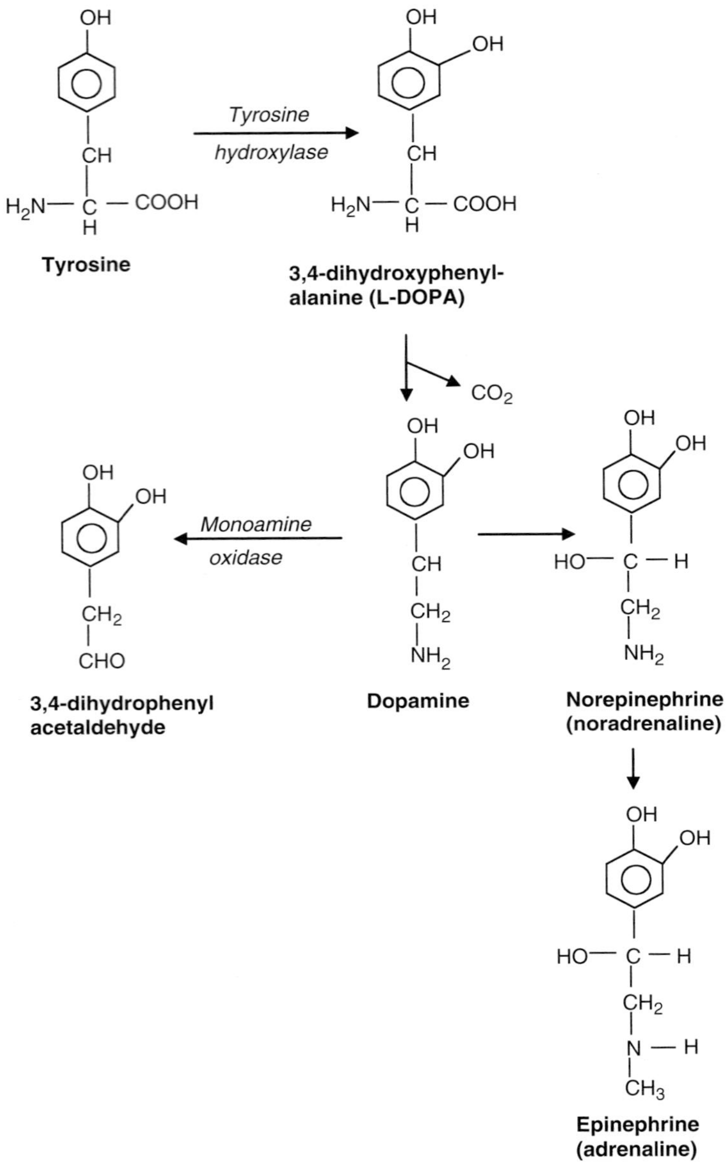

Figure 12.3 Dopamine pathway in neuronal cells

disease are observed after damage to majority of nigrostriatal dopamine neurons when 80% of striatal dopamine loss has occurred (71).

However, declining efficacy is observed over the course of L-DOPA treatment for Parkinson's (75). In addition, the L-DOPA therapy is associated with a variety of side effects such as dyskinesias, fluctuations in motor performance, confusion, hallucinations and sensory syndrome (37,75). To reduce the side effects, L-DOPA dosage is reduced, which limits the antiParkinsonian benefits (75). The benefits of lower dosage and short absence of L-DOPA intake to reduce side effects has had limited success (75) and therefore L-DOPA from fava beans is promising as an alternative Parkinson's therapy (35,37,69).

12.4.1 l-DOPA from Natural Sources

L-DOPA is present in certain food, and L-DOPA glycoside has been extracted from the broad bean (fava bean) *Vicia faba* (83). The fava bean is one of the best plant sources of L-DOPA; clinical studies have shown that seed sprouts have antiParkinson's effects without any of the side effects seen from the pure synthetic form (74). Fava beans are a widely cultivated legume and commonly consumed in the Mediterranean region (37). L-DOPA was identified in the seedling, pods, and beans of broad bean (fava bean) initially by Guggenheim in the early 1930s (37,69,74). Fava bean ingestion correlated with a significant increase in levels of L-DOPA in plasma and with an improvement in motor performance (35). As discussed earlier, exogenously synthesized L-DOPA is not effective in all PD patients and has side effects (37). The mechanism of efficiency of the natural source of L-DOPA may be due to the amino acid milieu generated in broad fava beans that may favor the selective transport of L-DOPA across the blood brain barrier (37). Fava beans seedling ingestion is useful in the treatment of PD patients, especially those in low income societies, where limited resources do not allow the purchase or manufacture of expensive drugs. Fava beans are available cheaply in all seasons and can be easily grown. Furthermore, fava beans serve as natriuretic agents (82). Other natural sources of levodopa, besides broad bean, have also been reported in *Stizolobium deeringianum* and *Mucuna pruriens*. *Mucuna pruriens* beans have been used as an efficacious herbal drug for PD treatment in India for many years (37).

12.4.2 l-DOPA from Fava bean

Studies have shown that ingestion of *Vicia faba* (fava bean) fruits (pods and seeds) improved substantially the clinical features in Parkinson's patients (35,36). In further studies the benefits from fava bean seedlings, which had 20-fold higher L-DOPA than fruit was even better, with higher plasma L-DOPA and substantial clinical improvement (69). In a limited clinical case study, fava beans as a source of L-DOPA, prolonged the "ON" periods in patients with Parkinson's disease who have "ON-OFF" fluctuations (37). In these studies, patients had been previously administered higher doses of L-DOPA, up to 800–100 mg per day, which failed to optimize the "ON" time and resulted in peak dose dyskinesias, but the fava bean showed beneficial effects by prolonging the "ON" time and shortened "OFF" time (37). Other studies using L-DOPA from *Mucana pruriens* beans have been used as an effective drug for Parkinson's disease in India for many years (84). In a clinical study of 60 patients over 12 weeks, an extracted powder from *M. pruriens* mixed with water significantly improved Parkinsonian motor scores, and from the studies authors speculated that the extract benefits may be a result of other antiParkinsonian compounds besides L-DOPA (Parkinson's study group for HP-200). The observation from these limited legume studies indicates that there may be additional complementary compounds that may work in conjunction with L-DOPA. Our hypothesis is that these additional complementary compounds could be other phenolics that have antioxidant benefits and amino acid cofactors. Therefore, the rationale for enhancing biosynthesis of L-DOPA and total phenolic profiles in various stages of fava bean seedlings following elicitation with physical and elicitor stress has merit. Further, because seedlings were shown to be the best source of L-DOPA (69), its biosynthesis is being investigated in light and dark germinated fava bean seedlings and correlated to total phenolic content and fava bean antioxidant activity at the metabolite and enzyme level. From this understanding of L-DOPA phenolic content and antioxidant response, the goal is to use various stages of the seedlings with different optimized levels of L-DOPA and total phenolics and to confirm that these ratios will affect the antioxidant enzyme response in a Parkinson's neuronal cell model system. This approach not only helps to optimize the best stage for L-DOPA biosynthesis in fava bean seedlings but will also confirm whether the additional factors that contribute to the benefits of L-DOPA from the fava bean is potentially the result of " phenolic antioxidant" and "amino acid" factors.

12.5 LINKING L-DOPA SYNTHESIS TO THE PENTOSE PHOSPHATE AND PHENYLPROPANOID PATHWAYS

L-DOPA in the seed is potentially derived from the phenylalanine. As mentioned earlier, phenylalanine is synthesized in the plant from the pentose phosphate pathway and the shikimate pathway, which forms the starting material of the phenylpropanoid pathway for the synthesis of phenolic acid (Figure 12.1). PAL catalyses its further conversion to cinnamate, which leads to the secondary metabolite synthesis. An alternative route for phenylalanine modification is the synthesis of tyrosine by phenylalanine hydroxylase. Tyrosine, in turn, is the precursor for L-DOPA production in the seed.

It is hypothesized that synthesis of free soluble phenolics and L-DOPA is regulated via the proline linked pentose–phosphate pathway (PLPPP), shikimate pathway, and phenylpropanoid pathway (23,30) (Figure 12.1). PPP is an alternate route for the breakdown of carbohydrates generating $NADPH_2$ for use in anabolic reactions and for providing erythrose-4-phosphate for the shikimate pathway (23,30). This route is vital for the biosynthesis of phenylpropanoid secondary metabolites, including L-DOPA (Figure 12.1). Glucose-6-phosphate dehydrogenase (G6PDH) catalyses the first committed and rate-limiting step of PPP (85). A putative correlation has been observed between proline levels and total soluble phenolics in thyme, oregano, seeds of *Pangium edule*, and pea, which suggests that proline accumulation may be linked to the regulation of the phenylpropanoid pathway (86–88). The stimulation of PPP, purine, (89) and soluble phenolic (23,30) synthesis is believed to occur through a redox cycle. Cytosolic pyrroline-5-carboxylate (P5C) is reduced to proline and $NADPH_2$ is oxidized to $NADP^+$ (89). The enzymes G6PDH and 6-phosphogluconate dehydrogenase utilize the generated $NADP^+$ as cofactors in PPP for their reactions. In many plants, free proline accumulates in response to the imposition of a wide range of biotic and abiotic stresses (90). The stimulation of L-DOPA through the proline linked pentose phosphate pathway (PLPPP) could also serve as the critical control point (CCP) for generating $NADPH_2$, not just for proline synthesis (as an alternative reductant for energy), but it could also meet the reductant needs of antioxidant enzyme response through superoxide dismutatse (SOD) and catalase (CAT) (Figure 12.4)(23,30).

Current investigations using fava bean sprouts indicate that under various biotic and abiotic elicitors, L-DOPA content is enhanced in the early stages of germination, which then gradually gets reduced and the total soluble phenolics steadily increases with germination, reaching highest synthesis in the later stages (63,64,66). Further, it is evident that stimulation of soluble phenolics in the late stages is regulated through the pentose phosphate pathway as reflected in the activity of glucose-6-phosphate dehydrogenase. This is likely coupled to proline synthesis (63,64). It is likely that as soon as germination takes place L-DOPA content is mobilized rapidly in early stages, and once germination leads to hypocotyl development, the phenolic flux is redirected for structural development with maximum demand for the pentose phosphate pathway because elicitor stress demand increases soluble phenolics in late stages (63,64,66). This late stage soluble phenolic stimulation in response to stress factors could be as important as

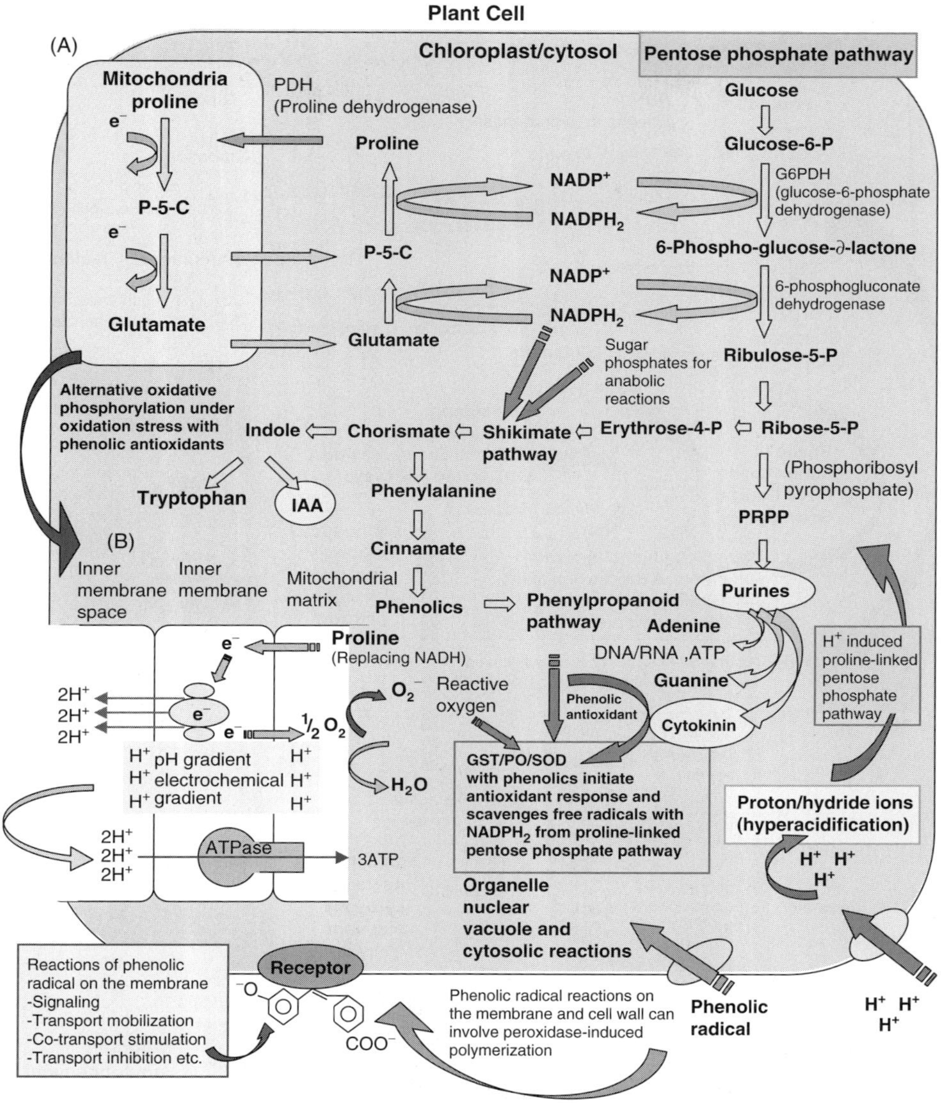

Figure 12.4 Improved model for the role of praline linked pentose–phosphate pathway in phenolic biosynthesis, which also accommodates the biosynthesis and mechanism of action of L-DOPA in a high phenolioc background from fava bean sprouts, and the effect of external elicitors like oregano phenolics. (Abbreviations: P5C: pyrroline-5-carboxylate; IAA: indole acetic acid: GST: Glutathione-s-transferase: PO: peroxidase; SOD: superoxide dismutase)

L-DOPA for Parkinson's management, as this could provide additional "phenolic antioxidant" and "amino acid" factors that could moderate the L-DOPA insensitivity often seen with pure synthetic L-DOPA treatments (37). These additional phenolic antioxidant and amino acid factors such as glutamic acid and proline could contribute soluble factors for proper redox management of neuronal cells as proposed for human and mammalian cells (23,91) (Figure 12.5).

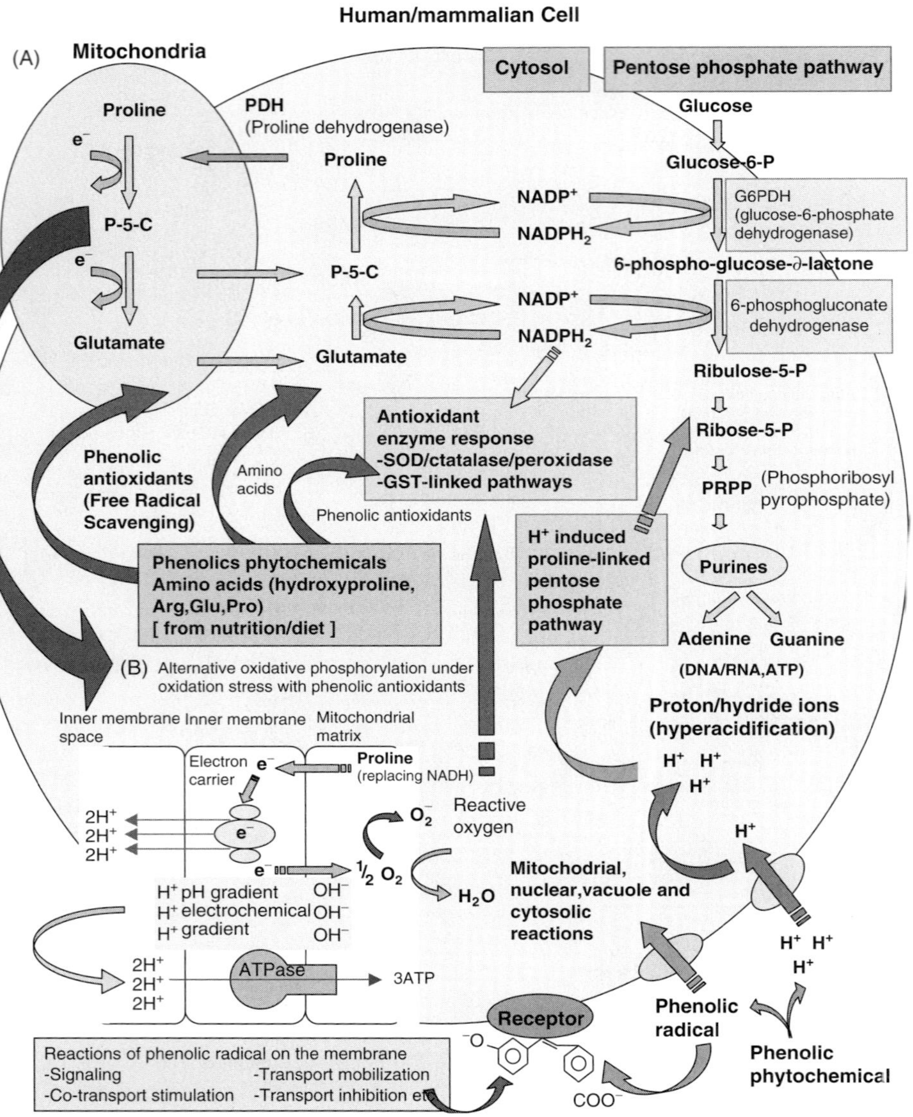

Figure 12.5 Extension of plant praline linked pentose–phosphate pathway model for the effect of external phenolic phytochemicals like L-DOPA in a high total phenolic background from fava bean sprouts or fermented extracts in human and mammalian systems. (Abbreviations: P5C: pyrroline-5-carboxylate; GST: Glutathione-s-transferase; SOD: superoxide dismutase)

12.6 RECENT PROGRESS ON ELICITATION LINKED BIOPROCESSING TO ENHANCE L-DOPA AND PHENOLICS IN FAVA BEAN SPROUTS

We have developed biotic and abiotic induced elicitation techniques to enhance L-DOPA, total phenolics, and antioxidant activity during sprouting of fava beans (63,64,66,92–94). The rationale for this approach is based on the understanding that pretreatment in the dry presoak stage or soak stage enhances critical stress responses as the seed germinates and that this will alter the stress related phenylpropanoid pathways linked to soluble phenolic and L-DOPA biosynthesis. This also likely mobilizes free amino acid from protein stored in the cotyledons. In our earliest work we investigated the stimulation of L-DOPA, total soluble phenolics, and related antioxidant activity in dark germinated fava bean seeds in response to bacterial polysaccharides from *Pseudomonas elodea* (gellan gum) and *Xanthomonas campestris* (xanthan gum) (92). Results indicated that gellan gum stimulated a ninefold increase in total soluble phenolics, compared to control. and this stimulation may be regulated via the pentose phosphate pathway based on the stimulation of glucose-6-phosphate dehydrogenase (92). The L-DOPA content was high in the initial days in the hypocotyl in all treatments and steadily reduced over the later stages. This approach allowed the development of two kinds of extracts, an early stage high L-DOPA with low soluble phenolics, and later stage medium L-DOPA with high soluble phenolics. The relative antioxidant activity based on the β-carotene assay did not change over the sprouting phase (92). Subsequent studies explored the stimulation of L-DOPA and total soluble phenolic in response to phytochemical and peptide elicitors in dark and light germinated conditions (93,94). Under dark germination conditions, the elicitors stimulated soluble phenolic content, with fish protein hydrolysates, lactoferrin, and oregano phenolic extracts stimulating the highest phenolic contents of 2.9–5.2 mg/g fresh weight (FW) during early stages of germination (93). The stimulation of primary metabolism through the pentose phosphate pathway occurred on day 5 for fish protein hydrolysates, and early stage of day 3 for lactoferrin and oregano phenolics (93). L-DOPA was stimulated from 20–40% for all elicitors immediately after treatment on day 1 and declined steadily. Under these conditions, fava extracts with any of the foregoing elicitations could be suitable for designing a dietary source of L-DOPA with high soluble phenolics between 2–5 days, depending on the elicitor. The same elicitors were tested for stimulation of light modulated seed vigor response and it was clear that L-DOPA content in seedlings was stimulated to over 80–100% by fish protein hydrolysate and oregano phenolics, which was followed by a change in flux toward higher soluble phenolics and enhanced antioxidant activity on day 20 (94).

In another study, UV treatment of seeds for 15 hours after overnight soaking stimulated highest L-DOPA content on day 1 and highest total soluble phenolic content day 6. The stimulation of soluble phenolic content was correlated to glucose-6-phosphate dehydrogenase, the first committed step in the pentose phosphate pathway. Further, proline content was slightly elevated compared non-UV treated control, indicating that soluble phenolic synthesis in late stages may be driven through the proline linked pentose–phosphate pathway (63). In another very exciting abiotic stress study, 30 seconds of microwave treatment of dry seed prior to overnight soaking and dark germination, resulted in over 50% stimulation of L-DOPA on day 1 and over 700 % stimulation of total soluble phenolics on day 7 (66). Each of these stages correlated to higher antioxidant activity both in terms of free radical scavenging assay as well as stimulation of superoxide dismutase (SOD) (66). Further, it was clear that late stage soluble phenolic stimulation correlated strongly to the stimulation of the pentose phosphate pathway (66). Additional studies using proline analog and proline combinations were used as elicitor treatments to confirm that the stimulation of soluble phenolics in the late stages of dark germination correlated with simultaneous stimulation of glucose-6-phosphate dehydrogenase and proline, further supporting the hypothesis that the proline linked pentose–phosphate pathway may be involved in the stress induced stimulation of phenolics (64). These studies support the model that PLPPP may be a critical control point (CCP) that not only stimulates the pentose phosphate pathway to support proline as an alternative reductant for energy through oxidative phosphorylations (alternative to NADH), but could also provide the critical $NADPH_2$ for supporting antioxidant enzyme response through SOD and CAT (Figure 12.4). The optimization of these steps during sprouting and seedling development could help to generate fava bean extracts with optimum L-DOPA with additional soluble phenolics and amino acid factors that could better modulate neuronal antioxidant enzyme response (Figure 12.5) and help to overcome any L-DOPA insensitivity seen with extended use of pure synthetic L-DOPA.

12.7 RECENT PROGRESS ON SOLID-STATE BIOPROCESSING OF THE FAVA BEAN TO ENHANCE L-DOPA AND PHENOLICS USING FOOD GRADE FUNGAL SYSTEMS

A solid-state bioconversion system using the food grade fungus *Rhizopus oligosporus* was developed to enrich the fava bean substrate with phenolic antioxidants and L-DOPA (65). The L-DOPA content in the fungal grown fava bean increased significantly to approximately twice that of control, accompanied by moderate soluble phenolic linked antioxidant activity based on free radical scavenging assay and higher fungal SOD activity during early stages of growth. This indicated that L-DOPA can be mobilized and formed from fava bean substrates by fungal bioconversion, conributing to the antioxidant functionality of such extracts (65). A high superoxide dismutase (SOD) activity during early and late growth stages indicates the likely oxidation stress of initial fungal colonization and during later growth stages due to nutrient depletion. High levels of soluble phenolics were observed during late growth stages. During the course of solid-state growth there was an increase in β-glucosidase activity, which correlated to an increase in total soluble phenolic content during the late stages. This suggests that the enzyme may play an important role in the release of phenolic aglycones from fava bean substrate, thereby increasing the soluble phenolic content and accompanying antioxidant activity (65). The implication from this study is that solid-state bioconversion of the fava bean by *R. oligosporus* can significantly improve the phenolic antioxidant activity and Parkinson's relevant L-DOPA content.

12.8 IMPLICATIONS AND SUMMARY

It is clear that elicitor induced sprout systems and fungal based solid-state bioprocessing systems can be used for specific stimulation of L-DOPA. In the case of sprouts, the L-DOPA increase was observed immediately after elicitor and soaking treatment on day 1, but was observed over an extended growth phase when using a fungal solid-state system, The contents were consistently higher over the middle growth stage from days 4–12. In contrast, the total soluble phenolics were substantially stimulated concurrent with high proline in late stages of sprout growth with general stimulation of the pentose phosphate pathway supporting this total soluble phenolic metabolite production. In the solid-state fungal bioprocessing system, soluble phenolic content was observed in the later stages. The extent of antioxidant activity based on free radical scavenging generally coincided with higher L-DOPA content in the early stage, and with total soluble phenolics in the late stage in both elicitor stimulated sprouts and fungal bioconversion system. These bioprocessing approaches provide avenues to optimize L-DOPA content in the fava bean system with optimum “phenolic antioxidant” and “amino acid” profiles that could be more effective than the current approaches observed for pure synthetic L-DOPA drug treatments. Insights provided by studies so far also indicate that the L-DOPA content and total soluble phenolic contents are likely stimulated through an alternative proline liked pentose–phosphate pathway (Figure 12.4). In this model under elicitor stress, soluble phenolic synthesis is enhanced to counter the stress, which likely involves both direct free radical scavenging using the produced phenolics as well as antioxidant enzyme response through SOD. Further, an alternative proline linked pentose–phosphate pathway more efficiently facilitates $NADPH_2$ and sugar phosphate flow for anabolic pathways, including phenolic and antioxidant response pathways, while proline meets the reductant need for oxidative phosphorylation replacing NADH from TCA Krebs cycle (Figure 12.4) (23). Extending this model into animal systems, we have hypothesized fava bean extracts with optimum total soluble phenolics, L-DOPA, and proline could better maintain the cellular homeostasis and function of neuronal cells by modulating host antioxidant response through a proline linked pentose–phosphate pathway (23) (Figure 12.5). This mode of alternative regulation could facilitate better management of cellular redox environment through dehydrogenases and antioxidant enzyme response through SOD and CAT under which L-DOPA can work effectively.

REFERENCES

1. Goodwin, T.W., E.I. Mercer. Plant phenolics. In: *Introduction to Plant Biochemistry*, New York: Pergamon Press, 1983, pp 528–566.
2. Taiz, L, E. Zeigler. Auxins. In: *Plant Physiology*, 2nd ed., Sunderland, MA: Sinauer Associates, 1998, pp 543–557.
3. Croteau, R., T.M. Kutchan, N.G. Lewis. Natural products: secondary metabolites. In: *Biochemistry and Molecular Biology of Plants*, Buchanan, B.B., W. Gruissem, R.L. Jones, eds., Rockville, MD: American Society of Plant Physiology, 2002, pp 1250–1318.
4. Walker J.R.L. *The Biology of Plant Phenolics*, 1st ed. London: Edward Arnold, 1975.
5. Dixon, R.A., M.J. Harrison, C.J. Lamb. Early events in the activation of plant defense responses. *Ann. Rev. of Phytopath.* 32:479–501, 1994.
6. Dixon, R.A., N. Paiva. Stress-induced phenylpropanoid metabolism. *Plant Cell*, 7:1085–1097, 1995.
7. Rhodes, J.M., L.S.C. Wooltorton. The biosynthesis of phenolic compounds in wounded plant storage tissues. In: *Biochemistry of Wounded Plant Tissues,* Kahl, G., ed, Berlin: Walter deGruyter, 1978, p 286.
8. Brooker, F.L., J.E. Miller. Phenylpropanoid metabolism and phenolic composition of soybean [*Glycine max.* (L) Merr.] leaves following exposure to ozone. *J. Exp. Bot.* 49:1191–1202, 1998.
9. Zimmerman, Y.Z., P.R. Cohill. Heat shock and thermotolerance in plant and animal embryogenesis. *New Biology* 3:641–650, 1991.
10. Yalpani, N., A.J. Enyedi, J. Leon, I. Raskin. Ultraviolet light and ozone stimulate accumulation of salicylic acid, pathogenesis related proteins and virus resistance in tobacco. *Planta.* 193:372–376, 1994.
11. Hahlbrock, K., D. Scheel. Physiology and molecular biology of phenylpropanoid metabolism. *Plant Mol. Biol.* 40:347–369, 1989.
12. Graham, T.L. Flavanoid and isoflavanoid distribution in developing soybean seedling tissue and in seed root exuddates. *Plant Physiol.* 95:594–603, 1991.
13. Christie, P.J., M.R. Alfenito, V. Walbot. Impact of low temperature stress on general phenylpropanoid and anthocyanin pathways: enhancement of transcript abundance and anthocyanin pigmentatation in maize seedlings. *Planta.* 194:541–549, 1991.
14. Beggs, C.J., K. Kuhn, R. Bocker, E. Wellmann. Phytochrome induced flavanoid biosynthesis in mustard (*Sinapsis alba* L.) cotyledons: enzymatic control and differential regulation of anthocyanin and quercitin formation. *Planta.* 172:121–126, 1987.

15. Lois, R., B.B. Buchanan. Severe sensitivity to ultraviolet light radiation in *Arabidopsis* mutant deficient in flavanoid accumulation: mechanisms of UV-resistance in *Arabidopsis. Planta*, 194:504–509, 1994.
16. Kurganova, L.N., A.P. Veselov, T.A. Goncharova, Y.V. Sinitsyna. Lipid peroxidation and antioxidant system protection against heat shock in pea *(Pisum sativum* L.*)* chloroplasts. *Fiziol. Rast.* (Moscow), (*Russ. J. Plant Physiol., Eng. Transl.*) 44:725–730, 1997.
17. Retivin, V., V. Opritov,S.B. Fedulina. (1997) Generation of action potential induces preadaptation of *Cucurbita pepo* L. stem tissues to freezing injury. *Fiziol. Rast.* (Moscow), *(Russ. J. Plant Physiol., Engl. Transl.)* 44:499–510.
18. Kuznetsov, V.V., N.V. Veststenko. Synthesis of heat shock proteins and their contribution to the survival of intact cucumber plants exposed to hyperthermia. *Fiziol. Rast* (Moscow), *(Russ. J. Plant Physiol., Engl. Transl.*) 41:374–380, 1994.
19. Baraboi, V.A. Mechanisms of stress and lipid peroxidation. *Usp. Sovr. Biol.* 11:923–933, 1991.
20. Kurganova, L.N., A.P. Veselov, Y.V. Sinitsina, E.A. Elikova. Lipid peroxidation products as possible mediators of heat stress response in plants. *Exp. J. Plant Physio.* 46:181–185, 1999.
21. Parr, A.J., G.P. Bolwell. Phenols in the plant and in man. The potential for possible nutritional enhancement of the diet by modifying the phenols content or profile. *J. Sci. Food and Agric.* 80:985–1012, 2000.
22. Scalbert, A., G. Williamson. Dietary intake and bioavailability of polyphenols. *Am. Soc. Nutr. Sci.* 130:2073S–2085S, 2000.
23. Shetty, K., M.L. Wahlqvist. A model for the role of proline-linked pentose phosphate pathway in phenolic phytochemical biosynthesis and mechanism of action of human health and environmental applications: a review. *Asia Pacific J. Clin. Nutrition* 13:1–24, 2004.
24. Driver, G.A.C., M. Bhattacharya. Role of phenolics in plant evolution. *Phytochemistry* 49:1165–1174, 1998.
25. Cowan, M.M. Plant products as antimicrobial agents. *Clin. Micro. Rev.* 12:564–582, 1999.
26. Lewis, N.G. Antioxidants in higher plants. In: *Plant Phenolics,* Boca Raton, FL: CRC Press, 1993, pp135–169.
27. Lu, Y., L.Y. Foo. Identification and quantification of major polyphenols in apple pomace. *Food Chem.* 59:187–194, 1997.
28. Bravo, L. Phenolic phytochemicals: chemistry, dietary sources, metabolism, and nutritional significance. *Nutr. Rev.* 56:317–333, 1998.
29. Vekiari, S.A., V. Oreopoulou, C. Taia, C.D. Thompson. Oregano flavonoids as lipid antioxidants. *J. Amer. Oil Chem. Soc.* 70:483–487, 1993.
30. Shetty, K. Biotechnology to harness the benefits of dietary phenolics: focus on *Lamiaceae. Asia Pac. J. Clin. Nutr.* 6:162–171, 1997.
31. Shetty, K., T.L. Carpenter, D. Kwok, O.F. Curtis, T.L. Potter. Selection of high phenolic containing clones of thyme (*Thymus vulgaris*) using *Pseudomonas* sp. *J. Agric. Food Chem.* 44:3408–3411, 1996.
32. Deighton, N., S.M. Glidewell, S.G. Deans, B.A. Goodman. Identification by EPR spectroscopy of carvacrol and thymol as the major sources of free radicals in the oxidation of plant essential oils. *J. Fd. Sci. Agric.* 63:221–225, 1993.
33. Deighton, N., S.M. Glidewell, B.A. Goodman, S.G. Deans. The chemical fate of the endogeneous plant antioxidants carvacrol and thymol during oxidative stress. *Proc. R. Soc. Edinburgh* 102B:247–252, 1994.
34. Frankel, E.N., A.L. Waterhouse, P.L. Teissedrel. Principal phenolic phytochemicals in selected California wines and their antioxidant activity in inhibiting oxidation of human low density lipoproteins. *J. Agric. Food Chem.* 43:890–894, 1995.
35. Rabey, J.M., Y. Vered, H. Shabtai, E. Graff, A.D. Korczyn. Improvement of Parkinsonian features correlate with high plasma levodopa values after broad bean (*Vicia faba*) consumption. *J. Neurosurg. Psychiatr.* 55:725–727, 1992.

36. Kempster, P.A., Z. Bogetic, J.W. Secombe, H.D. Martin, N.D.H. Balazs, M.L. Wahlqvist. Motor effects of broad beans (*Vicia faba*) in Parkinson's disease: single dose studies. *Asia Pacific J. Clin. Nutr.* 2:85–89, 1993.
37. Apaydin, H., S. Ertan, S. Ozekmekci. Broad bean (*Vicia faba*): a natural source of L-DOPA: prolongs "on" periods in patients with Parkinson's disease who have "on-off" fluctuations. *Movement Disorders.* 15:164–166, 2000.
38. Aruoma, O.I. Free radicals, antioxidants and international nutrition. *Asia Pac. J. Clin. Nutr.* 8:53–63, 1999.
39. Narayanswami, V., H. Sies. Oxidative damage to mitochondrial and protection by ebselen and other antioxidants. *Biochem. Pharm.* 40:1623–1629, 1990.
40. Bindoli, A. Lipid peroxidation in mitochondria. *Free Radical Bio. Med.* 5:247–261, 1998.
41. Whalley, C.V., S.M. Rankin, R.J. Hoult, W. Jessu, D.S. Leakes. Flavonoids inhibit the oxidative modification of low density lipoproteins by macrophages. *Biochem. Pharm.* 39:1743–1750, 1990.
42. Randhir, R., D.A. Vattem, K. Shetty. Antioxidant enzyme response studies in H_2O_2-stressed porcine muscle tissue following treatment with Oregano phenolic extracts. *Process Biochem.* 40:2123–2134, 2005.
43. Malaveille, C., A. Hautefeuille, B. Pignatelli, G. Talaska, P. Vineis, H. Bartsch. Antimutagenic dietary phenolics as antigenotoxic substances in urothelium of smokers. *Mutation Research.* 402:219–224, 1998.
44. Meyer, A.S., O.S. Yi, A. Pearson, A.L. Waterhouse, E.N. Frankel. Inhibition of human low-density lipoprotein oxidation in relation to consumption of phenolic antioxidants in grapes (*Vitis vinifera*). *J. Agric. Food Chem.* 45:1638–1643, 1997.
45. Jakus, V. The role of free radicals, oxidative stress and antioxidant systems in diabetic vascular disease. *Bratisl Lek Listy* 101:541–551, 2000.
46. Barbaste, M., B. Berke, M. Dumas, S. Soulet, J.C. Delaunay, C. Castagnino, V. Arnaudinaud, C. Cheze, J. Vercauteren. Dietary antioxidants, peroxidation and cardiovascular risks. *J. Nutr. Health Aging.* 6:209–23, 2002.
47. Block, G., B. Patterson, A. Subar. Fruit, vegetables, and cancer prevention: a review of the epidemiological evidence. *Nutr. Cancer.* 18:1–29, 1992.
48. Serdula, M.K., M.A.H. Byers, E. Simoes, J.M. Mendlein, R.J. Coates. The association between fruit and vegetable intake and chronic disease risk factors. *Epidemiology.* 7:161–165, 1996.
49. Yoshioka, A., Y. Miyachi, S. Imamura, Y. Niwa. Anti-oxidant effects of retinoids on inflammatory skin diseases. *Arch. Of Derm. Research.* 278:177–183, 1986.
50. Puskas, F., P. Gergely, K. Banki, P. Andras. Stimulation of the pentose phosphate pathway and glutathione levels by dehydroascorbate, the oxidized form of vitamin C. *FASEB J.* 14:1352–1361, 2000.
51. Appeldoorn, N.J.G., S.M. Bruijn, E.A.M. Koot Gronsveld, R.G.F. Visser, D. Vreughenhi, L.H.W. Van der Plas. Developmental changes in enzymes involved in the conversion of hexose phosphate and its subsequent metabolites during early tuberiztion of potato. *Plant Cell Environ.* 22:1085–1096, 1999.
52. Chugh, L.K., S.K. Sawhney. Effect of cadmium on activities of some enzymes of glycolysis and pentose phosphate pathway in pea. *Biologia Plantarum.* 42:401–407, 1999.
53. Debnam, P., M.J. Emes. Subcellular distribution of enzymes of the oxidative pentose phosphate pathway in root and leaf tissues. *J. Experimental Botany.* 50:1653–1661, 1999.
54. Herrmann, K.M. The shikimate pathway as an entry point to aromatic secondary metabolism. *Plant Physiol.* 107:7–12, 1995.
55. Shetty, K., D.L. Crawford, A.L. Pometto. Production of L-phenylalanine from starch by analog-resistant mutants of *Bacillus polymyxa. App. Environ. Micro.* 52:637–643, 1986.
56. Hollman, P.C.H. Evidence for health benefits of plant phenols: local or systemic effects? *J. Sci. Food Agric.* 81:842–852, 2001.
57. Rajalakshmi, D., S. Narashimhan. Food antioxidants: sources, and methods of evaluation. In: *Food Antioxidants: Technological, Toxicological and Health Perspectives.* Madhavi, D.L., S.S. Deshpande, D.K. Salunkhe, eds., New York: Marcel Dekker, 1995, pp 65–158.

58. Zheng, Z., U. Sheth, M. Nadiga, J.L. Pinkham, K. Shetty. A model for the role of proline-linked phenolic synthesis and peroxidase activity associated with polymeric dye tolerance in oregano. *Process. Biochem.* 36:941–946, 2001.
59. Whetten, R., R. Sederoff. Lignin biosynthesis. *The Plant Cell* 7:1001–1013, 1995.
60. Fukushima, R.S., R.D. Hatfield. Extraction and isolation of lignin for utilizing as a standard to determine lignin concentration using acetyl bromide spectrophotometric method. *J. Plant Physio.* 49:3133–3139, 2001.
61. McDougall, G.J. Cell wall associated peroxidases and lignification during growth of flax fibers. *J. Plant Physio.* 139:182–186, 1991.
62. Ebermann, R., H. Pichorner. Detection of proxidase catalyzed phenol polymerization induced by enzymatically reduced paraquat. *Phytochemistry* 28:711–714, 1989.
63. Shetty, P., M.T. Atallah, K. Shetty. Effects of UV treatment on the proline-linked pentose phosphate pathway for phenolics and L-DOPA synthesis in dark germinated *Vicia faba*. *Process Biochem.* 37:1285–1295, 2002.
64. Shetty, P., M.T. Atallah, K. Shetty. Stimulation of total phenolics, L-DOPA and antioxidant activity through proline-linked pentose phosphate pathway in response to proline and its analog in germination fava beans (*Vicia faba*). *Process Bichem.* 38:1707–1717, 2003.
65. Randhir, R., D. Vattem, K. Shetty. Solid-state bioconversion of fava bean by *Rhizopus oligosporus* for enrichment of phenolic antioxidants and L-DOPA. *Innovative Food Sci. and Emerging Technologies.* 5:235–244, 2004a.
66. Randhir, R., K. Shetty. Microwave-induced stimulation of L-DOPA, phenolics and antioxidant activity in fava bean (*Vicia faba*) for Parkinson's diet. *Process Biochem.* 39:1775–1784, 2004b.
67. Previc, F.H. Dopamine and the origin of human intelligence. *Brain and Cognition.* 41:299–350, 1999.
68. Lehninger, A.L., D.L. Nelson, M.C. Michael. *In Principles of Biochemistry.* Worth Publishers, 1993.
69. Vered Y., J.M. Rabey, D. Paleveitch, I. Grosskopf, A. Harsat, A. Yanowski, H. Shabtai, E. Graff. Bioavailability of levodopa after consumption of *Vicia faba* seedlings by Parkinsonian patients and control subjects. *Clin. Neuropharm.* 17:138–146, 1994.
70. Ahlskog, J.E. Treatment of Parkinson's disease, from theory to practice. *Postgrad. Med.* 95:55–69, 1994.
71. Elsworth, J.D., R.H. Roth. Dopamine synthesis, uptake, metabolism and receptors: relevance to gene therapy of Parkinson's Disease. *Exp. Neurol.* 144:4–9, 1997.
72. Boada , J., B. Cutillas, T. Roig, S. Bermudez, J. Ambrosio. MPP+ induced mitochondrial dysfunction is potentiated by dopamine. *Biochem. Biophys. Res. Comm.* 268:916–920, 2000.
73. Hiroi, T., S. Imaoka, Y. Funae. Dopamine formation from tyramine by CYP2D6. *Biochem. Biophys. Res. Comm.* 249:838–843, 1988.
74. Malhere, H.W., J.M. Van Buren. The excretion of dopamine metabolites in Parkinson's disease and the effect of diet thereon. *J. Lab. Clin. Med.* 74:306–318, 1969.
75. Kurlan, R., C.M. Tanner, C. Goetz, J. Sutton, D. Lichter, C. Deeley, L. Cui, C. Irvine, M.P. McDermott. Levodopa drug holiday versus drug dosage reduction in Parkinson's disease. *Clin. Neuropharm.* 17:117–127, 1994.
76. Blaschko, H. Metabolism and storage of biogenic amines. *Experientia* 13:9–12, 1959.
77. Carlsson, A. The occurrence, distribution and physiological role of catecholamines in the nervous system. *Pharmacol. Rev.* 11:490–493, 1959.
78. Ehringer, H., O. Hornykiewicz. Verteilung von Noradrenalin und Dopamin (3-Hydroxytyramin) im Gehirn des Menschen und ihr Vehalten bei Erkrankungen des extrapyramidalen Systems. *Wien. Klin. Wochenschr.* 38:1236–1239, 1960.
79. Barbeau, A. Biochemistry of Parkinson's disease. *Proc. 7th Intl. Cong. Neurology.* 2:925, 1961.
80. Birkmayer, W., O. Hornykiewicz. Der L-3,4-Dioxyphenylalanin (DOPA)-Effect bei der Parkinson-Akinese. *Wien. Klin. Wochenschr.* 73:787–788, 1961.

81. Cotzias, G.C., M.H. Van Woert, L.M. Schiffer. Aromatic amino acids and modification of parkinsonism. *N. Engl. J. Med.* 276:374–379, 1967.
82. Vered Y., I. Grosskopf, D. Palevitch, A. Harsat, G. Charach, M.S. Weintraub, E. Graff. Influence of *Vicia faba* (broad bean) seedlings on urinary sodium excretion. *Planta Medica.* 63:227–249.
83. Banwart, B., T.D. Miller, J.D. Jones, G.M. Tyce. Plasma dopa and feeding. *P.S.E.B.M.* 191:357–361, 1989.
84. Manyam, B.V., K.M. Patikh. HP-200: an herbal drug for treatment of Parkinson's disease. *Parkinson's Mag. Euro. Parkinson's Assoc.* 8:10–11, 1997.
85. Garrett R.H., C.M. Grisham. Nitrogen acquisition and amino acid metabolism. In: *Biochemistry,* New York: Harcourt Brace College Publishers, 1995, pp 826–870.
86. McCue, P., Z. Zheng, J.L. Pinkham, K. Shetty. A model for enhanced pea seedling vigour following low pH and salicylic acid treatment. *Process. Biochem.* 35:603–13, 2000.
87. Perry, P.L., K. Shetty. A model for improvement of proline during *pseudomonas*-mediated stimulation of rosmarinic acid levels in oregano shoot clones. *Food Biotech.* 13:137–154, 1989.
88. Andarwulan, N., K. Shetty. Improvement of pea (*Pisum sativu*) seed vigour response by fish protein hydrolysates in combination with acetyl salicylic acid. *Process Bichem.* 35:159–165, 1999.
89. Phang, J.M. The regulatory functions of proline-5-carboxylic acid. *Cur. Top. Cell Regul.* 25:91–132, 1985.
90. Hare, P.D., W.A. Cress. Metabolic implications of stress induced proline accumulation in plants. *Plant Growth Regulation* 21:79–120, 1997.
91. Shetty, K. Role of proline-linked pentose phosphate pathway in biosynthesis of plant phenolics for functional food and environmental applications: a review. *Process. Biochem.* 39:789–804, 2004.
92. Shetty, P., M.T. Atallah, K. Shetty. Enhancement of total phenolic, L-DOPA and proline contents in germinating fava bean (*Vicia faba*) in response to bacterial elicitors. *Food Biotech.* 15:47–67, 2001.
93. Randhir, R., P. Shetty, K. Shetty. L-DOPA and total phenolic stimulation in dark germinated fava bean in response to peptide and phytochemical elicitors. *Process. Biochem.* 37:1247–1256, 2002.
94. Randhir, R., K. Shetty. Light-mediated fava bean (*Vicia faba*) response to phytochemical and protein elicitors and consequences on nutraceutical enhancement and seed vigor. *Process. Biochem.* 38:945–952, 2003.

2.13

Phytochemicals and Breast Cancer Chemoprevention

Sallie Smith-Schneider, Louis A. Roberts, and Kalidas Shetty

CONTENTS

13.1 INTRODUCTION

13.1.1 Background

According to the National Cancer Institute, approximately 13% of American women will develop breast cancer during their lifetimes. There are approximately 216,000 newly diagnosed cases of breast cancer each year, and breast cancer is second only to lung cancer in cancer related deaths among women in the USA (American Cancer Society). Despite a gradual decline in deaths due to breast cancer (likely attributable to increased screening), there is a rise in the incidence of newly diagnosed breast cancer. Numerous pharmaceuticals exist to prevent breast cancer growth, and several drugs are used chemopreventatively to block the formation of new tumors in previously diagnosed patients. Several plant derived compounds have been identified that may offer promise as chemopreventative agents that can be used routinely by the general population. However, in order to develop improved strategies aimed at preventing breast cancer, a better understanding of the cellular and molecular mechanisms that contribute to breast cancer initiation and protect epithelial cells against initiation or progression to cancer is needed.

During normal development of the mammary gland, controlled cell growth and changes in differentiation that resemble the unchecked cell proliferation that is a hallmark of breast cancer occur in response to hormones (1,2). In particular, the signaling pathways used during development and cancer initiation or growth are the same, with the loss of regulation over normal signals representing the most significant difference in initiation of cancer. Early in fetal development an extended epidermal mammary ridge forms and subsequently regresses, producing one nipple per breast with a bud of epithelial cells that essentially remains dormant and unchanged in females until puberty. Pubertal hormones (i.e., estrogen) initiate rapid division, branching, and migration of epithelial cells to form a system of ducts which extend to the limits of the fat pad, with lobules surrounded by stroma and microvascularized connective tissue. The ends of the lobule are referred to as terminal end buds, which maintain the ability to proliferate when triggered by pregnancy hormones (including estrogen and progesterone). During pregnancy the cells of the terminal end buds rapidly divide, then differentiate during lactation into milk producing alveoli. Upon weaning, rapid apoptosis of the alveoli ensues (termed involution), and the breast is then virtually indistinguishable from those of nulliparous females. However, the epithelial terminal end buds retain the ability to proliferate, invade the stroma, and differentiate in response to hormones produced during pregnancy and lactation. Thus, mammary gland development is a highly ordered and controlled process that directs epithelial cell proliferation, differentiation, and apoptosis.

A vast majority of breast cancers initiate from the ductal and lobular epithelial cells already described. Initially, the cells that comprise ductal carcinoma *in situ* (DCIS) are still hormone responsive (express estrogen and progesterone receptors), but often display hormone independent growth upon progression to more metastatic tumors. Additionally, growth factor receptors are often overexpressed in aggressive metastatic tumors that further promote unrestrained proliferation, inhibition of apoptosis, and vascularization of the growing tumor (3). Many chemopreventative strategies are based upon promoting or restoring normal signaling mechanisms within these ductal and lobular epithelial cells to reduce the likelihood of breast cancer initiation. Interestingly, pregnancy exerts a protective effect on the breast, such that a full term pregnancy lowers breast cancer risk by ~50% (4). This effect appears to be mediated in part by sensitizing the tumor suppressor protein p53, and in part by altering growth factor signaling (5–9). Chemoprevention using dietary phytochemicals aims to mimic the protective effect of pregnancy hormones by modulating such signaling pathways (Figure 13.1).

13.1.2 Susceptibility and Chemoprevention: Molecular Pathways

There are numerous factors and molecular events that can increase or decrease the susceptibility toward developing breast cancer. Understanding the molecular basis of this disease often relies on identifying the signal transduction pathways whose activation (or inhibition) directly regulates the development of breast cancer or acts as a marker for breast cancer susceptibility or protection. Once these pathways are identified, development of chemopreventative strategies to decrease the risk of breast cancer and chemotherapeutic regimens to treat existing cancers will be greatly facilitated.

p53 is a tumor suppressor protein that when activated acts as a transcription factor to induce expression of a host of genes involved in responses to DNA damage (triggered by ultraviolet light, ionizing radiation, or carcinogenic chemicals) (6,10). The central function of p53 is to induce cell cycle arrest and decide if DNA damage can be fixed (arrest followed by resumption of the cell cycle), or is irreparable (activation of the apoptotic pathway). p53 thus plays a critical role in preventing genotoxic initiation of precancerous lesions. Mutation of the *p53* gene or loss of p53 responsivity is one of the most common events identified in human breast cancer (detected in ~50% of cases), and loss of the p53 allele in mouse models predisposes them to cancer. In fact, transformation of normal ductal epithelium to DCIS typically progresses to the formation of a malignant, invasive cancer of the breast, and is frequently linked to mutation of p53 (2). p53 can be sensitized, i.e., made more responsive, by the pregnancy hormones estrogen and progesterone (11) and many dietary phytochemicals (12). Sensitization of p53 minimizes cancer initiation by increasing the likelihood that cells with activated p53 will either repair damaged DNA, or will be removed by apoptosis.

The estrogen biosynthetic and signaling pathways have been very well studied in normal development of the breast (i.e., directing proliferation of ductal epithelial cells during puberty), as well as in breast cancer (13). The enzyme aromatase (itself a popular target of estrogen responsive breast cancer) (14,15) converts androgen into estrogen, produced distally by the ovaries and locally in the breast. Estrogen is highly mitogenic, and directs cell proliferation by binding to estrogen receptors and activating transcription. Estrogen also may ultimately lead to the release of epidermal growth factor (EGF), itself a mitogenic factor, and may be converted into hydroxylated estrogens which in turn can act to stimulate cell division. Inappropriate signaling through the estrogen response pathway can lead to unchecked cell proliferation, and breast cancer.

Upregulation of signaling through the aryl hydrocarbon receptor (AhR) has recently been closely linked with breast cancer initiation, likely by inducing hyperplasia in epithelial breast tissue. Environmental toxins such as dioxin bind to cytosolic AhR, which then acts as a transcription factor to induce the expression of several genes, including the monooxygenase CYP1A1. Protoxins are then metabolized into mutagenic substances by CYP1A1. With respect to breast cancer specifically, CYP1A1 is capable of generating hydroxylated estrogens (i.e., 16α-hydroxyestrone) with mitogenic activity, and variant or increased CYP1A1 activity is associated with an increased risk of breast cancer (16). Additionally, AhR activation leads to the upregulation of other genes that regulate the antioxidant response and the cell cycle.

Many phytochemicals have been shown to modulate AhR activation and CYP1A1 expression and activity, including polyphenols, flavonoids, and phytoalexins, and may have a role in breast cancer prevention.

Signaling by growth factors [such as insulin-like growth factor (IGF), vascular endothelial growth factor (VEGF), transforming growth factor beta (TGF-β), and EGF represents an important area of focus in cancer initiation and development. Increased signaling through IGF can inhibit apoptosis (likely by upregulating the protein kinase Akt and the transcription factor NF-κB) (17), and some correlation exists between high circulating blood levels of IGF and the risk of premenopausal breast cancer (18). Epithelial growth factor, signaling through its receptors epidermal growth factor receptor (EGFR) and HER-2/neu plays a significant role in stimulating cell proliferation in the mammary gland (19). Vascular endothelial growth factor expression is regulated by NF-κB as well as cyclooxygenase and lipoxygenase metabolites of arachidonic acid (AA), and signaling it increases the growth and migration of endothelial cells that leads to vascularization of growing tumors (20). In contrast, TGF-β acts as a negative regulator of cell growth, as disruption of TGF-β signaling is correlated with cancer development (21). Additionally, crosstalk amongst growth factor signaling networks exists. For example, both IGF and EGF are capable of regulating the expression of VEGF. Antagonistic effects are also seen. IGF activation can down regulate TGF-β signaling (22), while TGF-β can decrease the circulating levels of IGF. Modulation of growth factor signaling obviously represents a complex yet important area of research in the development of chemopreventative strategies.

The aberrant metabolism of arachidonic acid by cyclooxygenases (COX) and lipoxygenases (LO) has recently been shown to be a predictive marker for initiation and progression

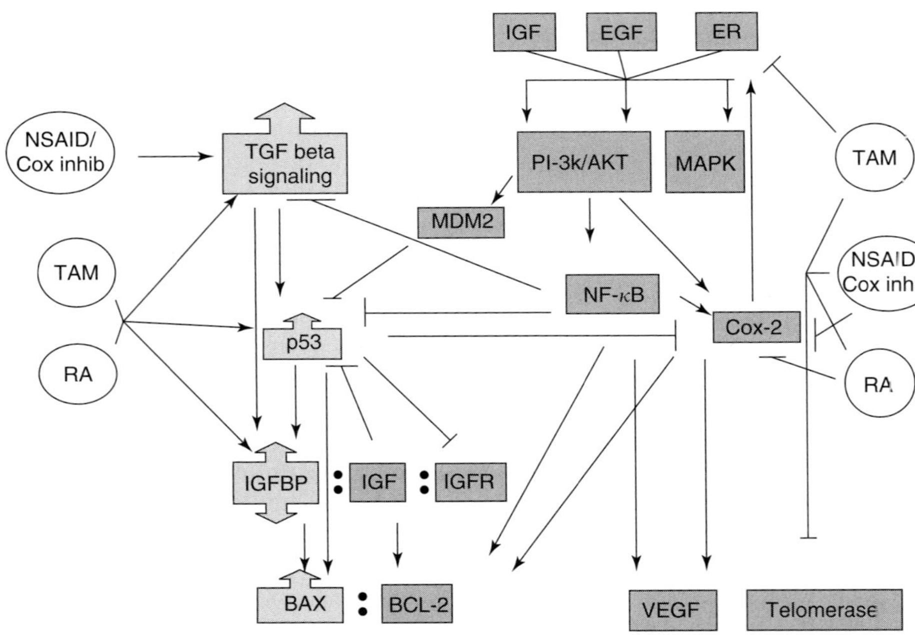

Figure 13.1 Signaling pathways which promote cell growth, survival, and metastasis and stimulate angiogenesis (dark grey) exist in a balance with signals that suppress growth, sensitize cell to apoptosis, and induce differentiation (light grey). Chemopreventive agents appear to have common overlapping roles in their ability to control growth signals through TGF beta family signaling, inhibit cyclooxygenase activity, regulate IGF, control oxidation, and sensitize or activate p53. (RA = retinoic acid, TAM = tamoxifen, NSAID = nonsteroidal antiinflammatory drug)

of a variety of cancers, and has been correlated with ultimate outcome in patients with breast cancer. COX-2 upregulation was first associated with a predisposition for colon cancer development, and COX-2 expression and activity are increased in breast cancer cells (23). The upregulation of COX-2 is indicative of, and is at least partially responsible for, increased cell growth and migration and decreased apoptotic rate. The AA metabolizing enzymes 5-lipoxygenase and 12-lipoxygenase are also overexpressed in many cancerous tissues, while decreased expression of 15-lipoxygenase is observed in many carcinomas. Recent studies have correlated increased expression of COX-2, 5-LO, and 12-LO with poor clinical outcome of breast cancer patients (24–26). Pharmacological COX inhibitors are currently in clinical trials as chemopreventative or chemotherapeutic agents (27), and the discovery and application of plant derived inhibitors of COX and LO represent an intense area of scientific and clinical research (28).

Many plant derived compounds either have been shown, or offer promise, to act as chemopreventative agents that block the development of breast cancer (29). These phytochemicals include but are not limited to the flavonoids, catechins, lignans, (poly)phenolics, and organosulfones. Interestingly, a subset of these phytochemicals act on multiple signaling pathways to coordinately suppress breast cancer initiation and progression, often by acting on a common signaling intermediate. Additionally, many of these plant derived compounds exert their protective effects by acting as antioxidants, which minimize oxidative stress and DNA damage. Several of the specific signaling pathways through which phytochemopreventative agents are known to act are detailed in the following (Figure 13.2).

	Estrogen signals	EGF signals	IGF signals	VEGF	COX2	TGF beta family signals	p53	Antioxidant enzyme response	AhR/ CYP1A1	LOX
Genistein	–		–	–		+	+		–	
Resveratrol	–					+	+		–	
Catechins	–	–	–	–	–					
Flaxseed lignans	–	–	–	–						
Ellagic acid							+			
Querciten	–		–				+		–	
Curcumin	–	–		–	–	+	+/–		–	–
Anthocyanidins	–	–			–		–			
PEITC							+			
Indole 3 carbinol	–	–							–	
Diallyl sulfide						+	+			
Perillyl alcohol						+	+			
Limonene										
Lycopene			–				–			
Baicalein					–				–	–

Figure 13.2 Numerous phytochemicals have been shown to regulate the balance of signals that promote cancer as well as the signals that control growth and sensitivity to DNA damaging agents. The (+) symbol indicates that in certain cell lines the phytochemical has been shown to activate or promote this pathway in some manner. The (−) symbol indicates that the phytochemical has been shown to have repressive or inhibitory effects on the signaling pathway at some level.

13.2 MOLECULAR PATHWAYS OF BREAST CANCER CHEMOPREVENTION

13.2.1 p53

p53 is a protein that activates transcription in response to a wide range of genotoxic stresses. Activation of p53-induced genes (i.e., p21, 14-3-3, bax, killer/DR5, GADD45, and NAG-1) results in execution of a cell cycle checkpoint important for the repair of DNA, cellular differentiation, or engagement of the apoptotic pathway. Loss of p53 activity due to mutation or viral inactivation results in immortalization of cells, and can represent one factor in the multihit process required for the progression to cancer.

The importance of proper control of p53 is underscored by its highly intricate regulation at numerous levels. The phosphorylation of p53 is complex and cooperative. Phosphorylation at specific residues are required for DNA binding, oligomerization, and transactivation. Ataxia telangiectasia mutated/ATM and Rad3 related kinase (ATM/ATR) are kinases which are believed to act as sensors of genotoxic stress. Ataxia telangiectasia mutated kinase appears to respond to double stranded DNA breaks, while ATR responds to stalled replication. These sensors transport signals of genotoxic damage through downstream targets such as Chk2 kinases, p53, and Brca. Activity of ATM/ATR is believed to be responsible for the phosphorylation of numerous N-terminal p53 residues such as serines 9, 15, and 46 as well as indirectly regulating the phosphorylation states of serines 20 and 376 (30). The phosphorylations of serines 15 and 20 control the binding of the ubiquitin ligase, MDM2, which regulates stability and cellular localization of p53 (31). Phosphorylated serine 15 (along with cooperative phosphorylations at serines 33, 37, 392) is believed to regulate the transcriptional activity of p53 to allow recruitment of general transcription factors (32–34). MAP kinase family members, in particular the stress activated members p38 and c-Jun N-terminal kinase (JNK), can also phosphorylate p53. c-Jun N-terminal kinase can phosphorylate threonine 81 to regulate p53 stability, transcriptional activity, and ability to initiate apoptosis (35), while p38 can phosphorylate serines 33 and 46 in response to UV radiation (36). The stability and cytoplasmic localization of the p53 protein is regulated by sequestration through interaction with the MDM2 protein, a ubiquitin ligase. The MDM2 protein can block p53 activity by binding to the N-terminus of p53 and either targeting p53 for degradation through ubiquitination, or by sequestering p53 in the cytoplasm. The regulation of MDM2 through the PI3k pathway is one mechanism by which deregulated signals through receptor tyrosine kinases can repress p53 activity and thus increase genetic instability (37,38). p53 can also be regulated by sumoylation and protein to protein interactions mediated through oxidative stress. The redox state of the cell determines the activity of thioredoxin reductase, which can control the activity of ref-1, which in turn can bind to and regulate transcriptional activity of p53 (39).

The mammary gland appears essentially normal in mice with a null deletion of p53, suggesting that p53 is not required for normal development. However, these mice are more prone to tumors in general and die at a relatively young age, mostly due to lymphomas. The loss of a single allele in the breast cancer susceptible mouse strain BALB/c results in 55% developing breast cancer by one year of age (40). A spectrum of p53 mutations has also been detected in cases of human breast cancer, and p53 appears to be inactivated by mutation in approximately 40% of human breast cancer cases (41,42).

A relatively new and interesting observation is that the mammary glands of virgin rodents are refractile to the activation of p53 by gamma radiation, while parous glands are sensitive (43). This offers a possible explanation for why the virgin rodent mammary gland is more susceptible to cancer induced by carcinogen exposure than the parous gland. The implications of pregnancy induced sensitization of p53 are obvious, as cells with DNA damage can be eliminated or repaired in the glands of parous mice, but may not be cleared or corrected in virgin mice. The importance of this pathway in humans remains to be determined; however, it is known that women who undergo pregnancy at an early age have a significantly reduced risk of breast cancer. Chemopreventative agents such as tamoxifen and retinoic acid have also been shown to sensitize p53 activity and the mammary gland (Tu, Schneider, and Jerry, unpublished). This suggests that the sensitization of p53 and the gland utilizes a common and important pathway in the protective response.

The ability to sensitize the mammary gland to repair and death via p53 represents an excellent target in the search for dietary phytochemicals which might act as strong chemopreventative agents. To date, numerous phytochemicals have been identified that can activate p53 and thus the apoptotic pathway in tumor cells. These agents include the organosulfur compounds phenethyl isothiocyanate (PEITC) (44) and diallyl disulfide (45); the phenolic compounds resveratrol (46), ellagic acid (47), and curcumin (48); and the plant isoflavonoids apigenin, kaempferol, luteolin (49), and genistein (50). While the mechanisms of p53 induction by most of these agents is not known, the flavonoids appear to induce p53 through the activation of ATM/ATR and the p38 stress related kinase; and some data suggest that genistein can alter Chk2 kinase activity which controls the activity of cdc25, a dual specificity kinase (49).

13.2.2 Estrogen Signaling

13.2.2.1 Background

Ovarian hormones have dichotomous effects when applied to the biology of the breast. On one hand, prolonged exposure through early menarche, belated menopause, or postmenopausal hormone replacement therapy increases the risk of breast cancer. On the other hand, an early full term pregnancy can significantly reduce breast cancer risk to humans (51), and in rodents pregnancy levels of estrogen and progesterone will impart a significant protection against carcinogen induced breast cancer (52,53). The context and duration of the signaling is likely to be the important determinant in the effects of estrogen. There is no doubt, though, that estrogen remains one of the critical hormones that regulate both normal and malignant development of the mammary gland.

13.2.2.2 Estrogen Regulation

Androgen is converted to estrogen by aromatase. Estrogen is produced primarily in the ovaries and released into the bloodstream, though aromatase activity has been identified in the stroma of the mammary gland. Estrogen is highly mitogenic, and regulates transcription of a variety of genes by activating several distinct signaling pathways within the cell. Estrogen specifically binds to estrogen receptors (ERα or ERβ) and activates transcription. Alternatively, recent data suggests estrogen may also bind a G-protein coupled plasma membrane receptor, which signals through Src, leading to matrix metalloproteinase (MMP) activation. Matrix metalloproteinase activavtion in turn cleaves heparin bound EGF such that EGF is released and can signal through EGF receptors (54). Finally, estrogen may also signal through metabolites generated by cytochrome P450 enzymes. 16a-hydroxyestrone has been shown to exert agonistic and proliferative effects upon mammary epithelial cells (55).

Estrogen is critical for the normal development of the mammary gland during puberty and pregnancy. Proper regulation of estrogen receptor (ER) signaling is important for directed proliferative events. During puberty proliferation of ductal epithelial cells increases dramatically in response to hormones, leading to extension of the ductal network through the stroma toward the limits of the fat pad. Mice with a targeted deletion of ERα exhibit a very prevalent phenotype in which the ducts are severely stunted. Estrogen is also partially required for differentiation and development of lobuloalveolar structures during pregnancy.

13.2.2.3 Protection and Susceptibility Duality of Estrogen Signaling

Ovariectomy in humans and mice (which reduces estrogen levels) substantially reduces the risk of breast cancer. A subset of potent chemopreventative and chemotherapeutic agents has the ability to inhibit certain ER activities (e.g., selective estrogen receptor modulators [SERM]) or inhibit estrogen synthesis (e.g., aromatase inhibitors). Tamoxifen (the best studied SERM) has partial ER antagonistic properties, and has been very popular as adjuvant therapy in humans to reduce the risk of recurrent breast cancer. Some of the documented successes of tamoxifen may be due to its pleiotropic effects. Tamoxifen decreases IGF signaling, increases TGF-β secretion, prevents oxidative damage, inhibits protein kinase C, and induces apoptosis. However, tamoxifen has agonistic effects on endometrial tissue, and because of its potential side effects long term use is discouraged. Raloxifene is a newer member of the SERM family which is also an ER antagonist, but lacks some of the side effects associated with tamoxifen. Preliminary studies with Raloxifene have yielded comparable preventative features. Furthermore, the aromatase inhibitor Letrozole decreases levels of estrogen and has also shown promising initial results.

Paradoxically, pregnancy or prolonged treatment with pregnancy levels of estrogen and progesterone (E+P) result in a significant reduction in the susceptibility to breast cancer. In humans, a full term pregnancy at a relatively young age results in approximately a 50% reduced risk of getting breast cancer (51,56). In rodents, a similar response is noted. Exposure to pregnancy levels of estrogen and progesterone for extended periods alters proliferative responses in untreated and carcinogen treated glands (5). Proliferation in the mammary glands of virgin estrogen and progesterone treated animals was overall lower than in untreated animals. Interestingly, the proliferation in ER positive cells was 10 times lower, suggesting that exposure to hormones that mimicked pregnancy altered important signaling pathways in a population with proliferative potential. The carcinogen n-nitroso methyl urea (NMU) induced a much more significant mitogenic response in virgin animals as compared to the hormone treated population. These hormones have also been shown to be the vital agent needed to sensitize p53 in ductal epithelial cells to make them responsive to DNA damage (5,7).

13.2.2.4 Estrogenic Phytochemicals

Epidemiological studies of populations with low versus high risk of breast cancer note that female Asian populations experience a much lower percentage of breast cancer cases (57). This protection is lost once the women relocate to the USA, suggesting that it is a difference in diet which imparts this protection. Analysis of the Asian diet indicates a higher amount of soy is consumed in comparison to a typical North American diet. Interestingly, soy contains significant amounts of estrogenic activity stemming from isoflavone components (specifically genistein and diadzein). Chemoprevention studies in rodents have indicated that soy can delay tumor development, and that administration of either diadzein or soy protein with isoflavones results in some protection (58,59). Genistein has been shown to have a chemopreventative effect on prepubertal rats (60). Besides the antiestrogenic ability of isoflavones, these compounds can enhance the antioxidant enzyme response, inhibit NF-kB and Akt activity, prevent angiogenesis, and reduce levels of IGF-I, ER-α, and PR (61,62).

Additional studies focusing on food benefits have identified a number of phytoestrogens, compounds found in plants that have both agonistic and antagonistic effects on the estrogen receptors. These phytoestrogens include the lignans found in flaxseed, the coumestans in alfalfa sprouts, and the polyphenolic catechins found in teas (63). Chemopreventative studies in rodents examining phytoestrogens indicate that at high concentrations, these agents can impart protection in the mammary gland, which can be hoped will translate into human chemoprotection.

13.2.3 CYP1A1 and Aryl Hydrocarbon Receptor (AhR)

The CYP1A1 p450 monooxygenase enzyme has recently been identified as a target of chemopreventative strategies. The protoxin dimethylbenz[a]anthracene (DMBA) is metabolized by CYP1A1 (and CYP1B1) into a DNA damaging agent that is a potent initiator of breast cancer (64). The CYP1A1 gene also catalyzes the conversion of estradiol to form hydroxylated estrogen; high estrogen levels are associated with an increased risk of breast and ovarian cancers. The CYP1A1 enzyme is capable of metabolizing arachidonic acid into 19-OH-AA (major product) and 14,15-EET (minor product) (65), but no link between these metabolites and breast cancer initiation or progression has been uncovered. Additionally, CPY1A1 expression is upregulated by a wide array of environmental toxins such as dioxins (most notably TCDD) and polychlorinated biphenyls (PCBs). The CYP1A1 gene is induced by the AhR, which is translocated to the nucleus upon ligand binding to act as a transcription factor. Dioxins act as ligands of AhR to induce expression of a class of genes (including CYP1A1) that possess a dioxin responsive element upstream of the coding sequence. In a screen of 90 vegetable derived compounds that may act as AhR antagonists, 37 were found to block dioxin mediated induction of AhR (16). The aryl hydrocarbon receptor also induces genes that regulate the cell cycle and the antioxidant response, two pathways that affect breast cancer development. Additionally, allelic variation or polymorphisms of the CYP1A1 gene may be correlated with an increased risk of breast cancer. For these reasons, identifying phytochemicals that modulate AhR/CYP1A1 signaling may be of significant value as chemopreventative agents.

The best characterized function of phytochemicals as chemopreventative agents is their role as antioxidants. Such plant derived compounds are divided into distinct classes: phenolics, flavonoids, and phytoalexins. Polyphenols are present at high levels in foods such as olives or olive oil (66) and fresh cut potatoes (67). Several polyphenolics (caffeic acid, syringic acid, and sinapic acid) have been shown to inhibit the growth of the human breast cancer cell line T47D (68). Caffeic acid was the most potent anti proliferative agent in these studies, and was capable of stimulating apoptosis of T47D cells. Induction of the apoptotic program in MCF-7 cells by the phenethyl ester of caffeic acid is mediated by inhibition of the transcription factor NF-κB and activation of Fas by a p53- and p38-dependent mechanism (69). Additionally, caffeic acid was able to bind to AhR antagonistically, and decrease CYP1A1 activity by inhibiting the activation of CYP1A1 transcription (68). Thus, polyphenols like caffeic acid are promising chemopreventative agents with respect to decreasing the likelihood of developing breast cancer.

The flavonoids have also received a great deal of attention for their potential as chemoprotective agents, especially with regards to their proestrogenic and antiestrogenic activity (70). Flavonoids are a diverse class of compounds that can act as either AhR agonists or antagonists, depending on the specific phytochemical and cell type. Focusing on the AhR responsive MCF-7 cell line, four compounds have been identified that act as AhR agonists by their induction of CYP1A1 gene expression: cantharidin (a lactone derived from insect extract); baicalein (an isoflavone that also is an inhibitor of 12-LO); chrysin (a flavonoid); and emodin (an herbal laxative) (71). Additionally, baicalein and emodin increase the protein level of CYP1A1, while baicalein and cantharidin increase the level of AhR protein. However, as dietary components it has not been shown the flavonoids would act as AhR agonists, due to their relatively low serum concentrations (16). Conversely, luteolin acts as a potent AhR antagonist in MCF-7 cells, while quercetin, kaempferol, and myricetin can act as mild AhR antagonists (71). Genistein and baicalein also are capable of inhibiting the conversion of environmental protoxins (like DMBA) into DNA damaging agents by blocking CYP1A1 activity (72).

Another class of plant derived compounds that modulate AhR/CYP1A1 functions and have potential applications as chemopreventative agents are the phytoalexins. These primarily stress induced compounds can act as antifungal agents in plants, but possess additional activities that make them attractive chemopreventative agents. For example, resveratrol (3,5,4′-trihydroxystilbene) is a polyphenolic phytoalexin found in red wine that has a wide spectrum of inhibitory properties (73). Resveratrol has been shown to induce apoptosis in established human breast cancer cell lines that have wild-type p53 tumor suppressor protein but not mutant p53 (46). In addition to its ability to inhibit cyclooxygenases, resveratrol is a competitive antagonist of AhR that causes nuclear translocation of AhR but does not induce CYP1A1 (74). Thus, resveratrol may act to prevent cancer initiation by inhibiting signaling through AhR and blocking the upregulation of CYP1A1.

In addition to these phytochemicals, other plant derived compounds affect the expression and activity of CYP1A1. Curcumin, a major component of the spice turmeric, binds to and activates AhR and can bind to the xenobiotic response element upstream of CYP1A1(75). However, several studies suggest that curcumin partially blocks DMBA induced CYP1A1 activity, as well as the conversion of procarcinogens into DNA damaging agents in breast and squamous cell carcinoma cells (75,76), and *in vitro* (77). Curcumin also inhibits 5-HETE (78) and prostaglandin formation (79), and has antioxidant properties; thus, its role as a tumor repressor or chemopreventative agent may be due to its effects on more than one single pathway. Indole-3-carbinol (derived from cruciferous vegetables) is also antitumorigenic, likely resulting from its ability to inhibit the activity of CYP1A1, blocking the hydroxylation of estradiol and resulting in the repression of estrogen signaling.

13.2.4 Growth Factor Signaling

13.2.4.1 Background

Growth factors are very important in the normal and malignant development of the mammary gland, and include epidermal growth factor (EGF), platelet derived growth factor (PDGF), hepatocyte growth factor (HGF), amphiregulin (AR), insulin-like growth factor (IGF), vascular endothelial growth factor (VEGF), fibroblast growth factor (FGF), transforming growth factor alpha (TGF-α), and transforming growth factor beta (TGF-β). These factors have diverse effects and work on different cell types, but ultimately their activities (in concert with hormones and other factors) regulate the architecture of the mammary gland. These growth factors bind receptor kinases on the surface of the plasma membrane. The deregulated expression or activity of these receptors often results in a susceptibility to breast cancer. Among them, EGF, IGF, TGF-β, and VEGF are particularly interesting with regards to chemoprevention due to their ability to regulate homeostasis in the mammary gland.

13.2.4.2 Epidermal Growth Factor (EGF)

Epidermal growth factor is secreted from cells and can act in an autocrine or paracrine fashion. Epidermal growth factor is one of approximately ten ligands (including amphiregulin and TGF-α) that bind to and activate a family of EGF receptors (HER1, 2, 3, and 4) found on the surface of the epithelial cells. These receptors are tyrosine kinases and play a strong role in the proliferation and differentiation of the mammary gland. The EGFR and HER-2 receptors are frequently upregulated or activated in breast cancer. New advances in breast cancer treatments of ER negative tumors target this family of receptors. Herceptin (Trastuzumab) is a humanized antibody to the HER-2 receptor which has been found to be successful in clinical trials of patients with very high expression of HER-2 (80,81).

Numerous dietary phytochemicals have been identified that can inhibit EGFR or HER-2/neu signaling. Typically these agents do not block the binding of the ligand to the receptor, but instead inhibit the autophosphorylation of the receptor and downstream signaling. Flaxseed has been shown to down regulate EGF receptor levels (82). Curcumin can inhibit EGFR activity and proliferation in A431 epidermoid carcinoma cells (83). Phase 1 clinical trials indicate curcumin is nontoxic up to 8 g/day, and patients with precancerous growths often saw stability of the disease or a histological improvement (84) with this curcumin regimen. However, trials with higher numbers of patients are needed.

Black and green teas have also been shown to have constituents which will decrease EGFR activity. Epigallocatechin-3-gallate (EGCG) or theaflavin-3,3′-digallate (TF-3) in particular can inhibit the autophosphorylation of the EGF receptor and inhibit proliferation. EGCG in prolonged treatment could enhance the apoptotic response by upregulating p53 (85), while TF-3 was better at blocking the binding of EGF to its receptor (86). Fruit derived anthocyanidins are structurally similar to EGCG and can also inhibit EGF receptor activation in A431 epidermoid carcinomas cells. Cyanidin and delphinidin have been shown to inhibit growth and block MAPK phosphorylation of ELK, and in turn inhibit EGF induced transcription though the ELK transcription factor (87).

13.2.4.3 Insulin-Like Growth Factor (IGF)

Insulin-like growth factor is a mitogen and a prosurvival factor which can act through autocrine and paracrine pathways. Signaling through the IGF axis is highly regulated at numerous levels, including expression of the IGF receptor (IGFR), the regulation of IGF binding protein (IGFBP) expression (which modulate the ability of IGF to bind to and activate receptors), regulation of IGFR autophosphorylation, and transmission of downstream signals through insulin receptor substrate-1 (IRS-1). The activation of the IGF receptor results in phosphorylation and activation of IRS, which in turn signals to the nucleus. The inositol triphosphate kinase (PI3k) pathway is one of the major pathways elicited by IGF in mammary epithelial cells.

Production of IGF occurs both systemically and locally. Growth hormone regulates the levels of IGF generated by the liver, the major site of IGF production. Increased serum levels of IGF are critical for the proper maintenance of terminal end bud structures in the mammary gland. The role of IGF in prevention of or susceptibility to breast cancer stems largely from correlative evidence in human studies and direct evidence in rodent studies. Epidemiological studies analyzing possible roles for IGF in breast cancer have indicated that premenopausal women with serum IGF levels in the highest quartile have an increased incidence of breast cancer. Conversely, well established protective therapies such as tamoxifen, fenretinide, and caloric restriction are associated with a decrease in IGF serum levels. However, the most convincing evidence that high IGF levels and signaling may be related to breast cancer is obtained from genetically defined rodent models with alterations in the IGF signaling pathway. Mice overexpressing IGF-I are sensitized to TPA induced tumor development (88). Mice that have a 25% reduction in serum IGF levels have a significantly delayed onset of breast tumor occurrence and a reduction in the number of tumors induced by C3(1)/SV40-LTA or in response to DMBA treatment (89). Interestingly, the reduced levels of IGF had no effect on the levels of hyperplasia induced in these models, but a significant difference in metaplasia, suggesting that the serum levels of IGF affect the progression, rather than the initiation, of breast cancer. Serum IGF levels also appeared to affect the growth of the tumors, as differences in tumor size were noted, especially in the early weeks after tumor detection (89).

The mechanism by which IGF predisposes mammary epithelial cells to tumorigenesis is unknown, but studies on various cell types indicate that signaling through this pathway counteracts most of the pathways important for the protective response. IGF has been shown to activate the PI3k pathway, which often leads to survival by activating Akt and NF-kB, or by phosphorylating and activating MDM2, which will degrade or sequester p53 in the cytoplasm (90). The PI3k pathway also can regulate the expression and localization of β-catenin, which will activate TCF/LEF signaling. In renal mesangial cells and colon carcinoma cells, IGF enhances cyclooxygenase activity (91,92). In certain cell types IGF has also been shown to down regulate the level of the progesterone receptor and TGF-β (22), suggesting pathways by which IGF might upset the balance of homeostasis. Furthermore, IGF has been shown to upregulate telomerase activity (93). Cancer cells have higher telomerase activity, which allows a bypass of naturally occurring senescence that limits the number of cell division cycles. Agents that inhibit telomerase thus represent a new potential target for cancer therapy.

Known chemopreventative agents appear to regulate IGF signaling at multiple levels. Retinoic acid (RA) inhibits growth of the MCF-7 breast cancer cells through inhibition of IRS and PI3k signaling (94), which in turn regulates c-fos expression (83). Retinoic acid also induces the secretion of IGFBPs in a cultured cancer cell line (95), which decreases the serum level of IGF-I. Data also suggests that in normal mammary epithelial cells RA will upregulate other binding proteins, such as IGFBP7 (96). Inhibitors of cyclooxygenase activity also upregulate IGFBP3 in immortalized breast epithelial cells, and tamoxifen and TGF-β inhibit the IGFBP3 protease to stabilize IGFBP3 in MCF-7 cells. In addition to decreasing plasma levels of IGF-I and upregulating IGFBP3, tamoxifen also decreases the levels of IRS-1 and Akt signaling in MCF-7 cells (97).

Dietary phytochemicals aimed at reducing IGF signaling have only recently been utilized in human chemoprevention. Flaxseed can inhibit growth of mammary tumors in rats, and has recently found favor in humans to prevent breast cancer. In addition to its antiestrogenic activity, flaxseed can inhibit IGF-I plasma levels in rats and IGF receptor levels in breast cancer cells (82,98). In prostate cancer cells genistein, biochaninA, quercetin, and kaempferol downregulated the activity and reduced phosphorylation of IRS as well as downstream Akt (99). In a different prostate cancer cell line black tea phenolics were also shown to reduce IGF signaling by decreasing the phosphorylation of the IGFR-I and Akt (100).

13.2.4.4 Transforming Growth Factor Beta (TGF-β)

The TGF-β growth factor is a cytokine with growth inhibitory properties for many cell types, including epithelial cells. Numerous members of the TGF-β family exist, with TGF-β1, 2, and 3 the most studied. Other family members include activin, BMP, and the more distantly related NAG-1 (also known as MIC-1, PLAB, PTGF-β, or GDF15). The three main forms of TGF-β have some overlapping properties and specific roles, and are expressed in response to different stimuli. All three (TGF-β 1, 2, and 3) are upregulated during involution, but TGF-β 3 expression is increased to the highest extent (101). Furthermore, mammary gland tissue from mice null for TGF-β 3 display a significant inhibition of involution, suggesting that this form is responsible for the apoptosis that occurs after weaning (102). The TGF-β growth factor is secreted in a latent form as part of a complex with other proteins, and is retained as such in the extracellular matrix until activation. Proteolytic cleavage is required to release the active form of TGF-β, and occurs in the mammary gland in response to estrogen and progesterone, as well as radiation(103,104). The active form of TGF-β then binds to one of three types of TGF-β receptors which either heterodimerize or homodimerize. Some receptors have higher affinities for particular TGF-β subtypes and some transduce signals more efficiently. Type I and type II receptors have cytoplasmic serine or threonine kinase domains responsible for downstream signaling events.

Cytokines of TGF-β are required for the normal development of the mammary gland. They are expressed at high levels in the stroma and can act in a paracrine fashion to control the growth and branching of ductal epithelium. Animal studies have also indicated that they have a role in regulating lobular alveolar development and involution. Pregnancy has been shown to upregulate the transcription of all three TGF-β isoforms. In rats, TGF-β 2 is increased late in pregnancy, and TGF-β 3 increases with milk stasis induced by weaning. Interestingly, TGF-β 3 remains at relatively high levels even after weaning, and is one of the few differences observed between the glands of virgin and parous females (9).

The loss of the TGF-β antiproliferative response is observed in many cases of breast cancer and appears to be an early hallmark of the immortalization of cells in culture. Intriguingly, the expression of TGF-β in tumor cells increases while the levels of TGF-β receptors are decreased. However, the lack of responsivity may be due to defects at multiple levels, such as loss of SMADs, mutation of receptors, or loss of other cooperative factors needed for growth arrest. Genetic manipulation of the genes encoding TGF-β and its receptor has demonstrated that modulation of this complex pathway within the mammary gland can impact the balance between a protective or susceptible phenotype. A constitutively active form of TGF-β crossed onto a mouse which has a TGF-α oncogene reduces breast cancer development (105). In mice that overexpress a dominant negative type II TGF-β receptor in the mammary gland, there is increased susceptibility to breast tumor development (21,106). These experiments suggest that properly regulated TGF-β can function as a tumor suppressor in the mammary gland.

Numerous chemopreventative agents regulate TGF-β expression. Natural and synthetic retinoids will induce TGF-β 1 or 2 in a number of cell types (107–109). Retinoic acid and tamoxifen will increase the expression of latent TGF binding proteins (LTBPs) (110). Tamoxifen upregulates the expression of TGF-β 2 (111,112) and activates TGF-β 1 in MCF-7 cells (113–115). The upregulation of TGF-β family members may be responsible for controlling the proliferation in response to growth factors and hormones, and may play a role in regulating cell death in response to ionizing radiation. Death by ionizing radiation requires TGF-β in concert with p53. It is also known that TGF-β inhibits telomerase activity (116).

The activation of TGF-β by dietary phytochemicals (such as perillyl alcohol) has been observed in mammary carcinomas (117). In addition, NAG-1 (NSAID activated gene) is a newly identified and more distant relative of TGF-β that is currently receiving significant attention. The NAG-1 gene was first identified as a factor that is expressed at high levels in the placenta, prostate, and brain(118,119); and is synthesized as a latent proenyzme which is cleaved to generate a 28kDa active form. Like other members of the TGF-β family, NAG-1 can inhibit tumor development and activate apoptosis. The upregulation of the NAG-1 protein is controlled by at least two different pathways: a p53-independent mechanism, which stabilizes Nag-1 RNA; or a p53-dependent transcriptional increase in expression. Expression of Nag-1 is induced by known chemopreventative agents like NSAIDs (120,121) and retinoids (122), but has also been shown to be upregulated by dietary phytochemicals such as diallyl disulfide, genistein, and resveratrol (120,121,123). These agents appear to induce NAG-1 through the p53-dependent mechanism. The role of NAG-1 in protection of the mammary gland is still under investigation, and its growth inhibitory and proapoptotic properties make it worthy of further study.

13.2.4.5 Vascular Endothelial Growth Factor (VEGF)

Oxygen is required for the sustained growth of tumors, and is delivered by an induced and directed growth of vascular tissue. Angiogenesis of tumors is controlled primarily by the tumor and growth factors released from the stoma, most notably VEGF and FGF. Vascular endothelial growth factor has been found to be released by breast tumor cells, and its expression is regulated by IGF, as well as by products of cyclooxygenase and lipoxygenase. In a synergistic fashion with FGF, VEGF binds receptors on the endothelial cells and stimulates growth and neovascularization. The targeting of angiogenesis has become a major focus of cancer research. In animal models significant inhibition of tumor development is observed; however, these results have not yet translated as successfully in human studies. Inhibitors that block VEGFR signaling (Neovastat), as well as monoclonal antibodies to VEGF are being examined. Drugs are also being developed that resemble endogenous angiogenesis inhibitors such as endostatin and angiostatin.

Phytochemicals may provide a partial defense against angiogenesis. Green tea, flaxseed, and numerous berry extracts decrease the expression of inducible VEGF. Green tea (in particular epigallocatechin gallate (EGCG) has been extensively studied in numerous cell types including the MDA-MB-231 breast cancer cell line, which constitutively expresses high levels of VEGF(124,125). EGCG ultimately blocks VEGF expression through the inhibition of the EGF receptor, which regulates NF-κB and STAT-3, two signaling proteins which have been demonstrated to impact the VEGF promoter. Green tea catechins in endothelial cell lines also appear to block VEGF induced tubule formation through the regulation of VE-cadherin and the inhibition of Akt activity (126). Fruits such as *Gleditsia sinensis* effectively control VEGF in the MDA-MB-231 breast cancer cell line (127), and other berries (cranberry, raspberry, strawberry, blueberry, elderberry, and bilberry) appear to significantly reduce VEGF expression in HaCaT cells (128). While most of the studies looking at phytochemical reduction of VEGF have used cell lines, a flaxseed study corroborated these findings *in vivo*. A diet that was supplemented with flaxseed reduced tumor formation in a breast cancer xenograft model and decreased the expression of VEGF (129).

13.2.5 Arachidonic Acid Metabolism

13.2.5.1 Background

Arachidonic acid and its metabolites play key roles in cell growth, signaling, and adhesion, and are key mediators of the inflammatory and immune responses. Arachidonic acid is released from plasma membrane phospholipids (i.e., phosphatidylcholine) by phospholipase A_2, where it can serve as a signaling molecule (autocrine or paracrine), or alternatively be metabolized by the monooxygenase (MO), lipoxygenase (LO), or cyclooxygenase (COX) enzymes. Natural product and pharmacological inhibitors of AA metabolism have been successfully employed to inhibit cell proliferation, transformation, and migration. Additionally, human breast cancer tissues overexpress 12-lipoxygenase and cyclooxygenase-2 (26). Thus, phytochemicals blocking AA and its metabolites offer great promise in chemopreventative strategies against a variety of cancers, including breast cancer (130,131). The COX and LO branches of AA metabolism are discussed in the following, and plant derived agents that block each of these pathways are presented.

13.2.5.2 Cyclooxygenase (COX)

Arachidonic acid can be metabolized by cyclooxygenase (COX) to produce prostaglandins (PGs), thromboxanes (TXs), and prostacyclins. These prostanoids have physiological roles in reproduction, pain response, fever, cell growth, and inflammation (23). There is much interest in prostaglandin E_2 (PGE_2), which is a potent mediator of the inflammatory response. PG synthesis can be achieved by two functionally distinct isoforms of cyclooxygenase: COX-1 and COX-2. Cyclooxygenase-1 is generally regarded as a constitutively expressed housekeeping gene used to synthesize PGs for necessary, routine functions, and is present in most cell types. In contrast, COX-2 is normally expressed only when induced by growth factors, cytokines, or upon viral transformation as an immediate early gene (132). Upregulation of COX-2 activity increases PGE_2 levels, which acts as a signaling molecule in an autocrine or paracrine manner to stimulate cell proliferation.

Constitutive expression of COX-2 (but not COX-1) and high PGE_2 levels have been detected in a variety of epithelial derived tumors, including cancers of the breast, colon, esophagus, and skin (23), and in human breast cancer cell lines (133). Several lines of evidence suggest that COX-2 overexpression may be important in breast cancer initiation. Cyclooxygenase-2 was typically overexpressed in invasive carcinomas (134), in adjacent DCIS (134,135), and in normal epithelium surrounding DCIS (135). Overexpression of COX-2 was sufficient to cause tumorigenesis in mice (136). Additionally, COX-2 overexpression was found to correlate with markers for premalignant breast cell transformation (137), and may be of prognostic significance in patient outcome (24,25). Prostaglandin E_2 has also been shown to be required for cell migration and metastasis by signaling through cAMP dependent protein kinase (PKA) (138,139), and along with TXA_2 and PGI_2 is involved in angiogenesis to supply blood to established tumors (140,141). Many oncogenes also stimulate COX-2 expression, including HER-2/neu (142), which directs EGF signaling leading to cell proliferation. Recently it has been determined that PGE_2 increases the expression of HER-2/neu, indicating that a positive loop drives overexpression of both genes in breast tumors (143). Increased COX-2 activity also has been shown to lead to a decrease in apoptosis, primarily through upregulation of the antiapoptotic factor Bcl-2 (144). The tumor suppressor protein p53 has also been shown to down regulate COX-2 expression (145). Thus, inhibition of COX-2 activity may serve to decrease breast tumor initiation and development.

The relationship between COX inhibition and cancer was first noted when links between frequent use of nonsteroidal antiinflammatory drugs (NSAIDs) that block COX activity, and decreased colon cancer incidence were identified (146). However, many of the first generation COX inhibitors used to treat pain and inflammation (e.g., aspirin, indomethacin) are nonselective; i.e., the activity of both COX isoforms are blocked, leading to many undesirable side effects, most frequently gastrointestinal discomfort or bleeding (23). Currently, there are many pharmacological selective COX-2 inhibitors (e.g., celeCOXib, SC-236, NS-398) available that possess potent antiinflammatory properties, lower colon cancer incidence when used regularly, and have greatly reduced side effects. There are many reported studies of NSAID use and breast cancer (147), most recently linking aspirin use to a reduced risk of hormone receptor positive breast cancer (148). However, more clinical studies are required to conclusively establish the use of aspirin or other NSAIDs in breast cancer chemoprevention.

Several scientific studies have demonstrated that inhibition of COX-2 activity and PGE_2 secretion may offer promise in breast cancer prevention and treatment. Using established human breast cell lines, indomethacin (nonselective COX inhibitor) is capable of inhibiting the proliferation of MCF-7 and MDA-MB-231 cancer cells and of normal human mammary epithelial cells (HMEC) (149,150). Treatment with the selective COX-2 inhibitor NS-398 was capable of inducing apoptosis in MDA-MB-231 cells (151). Treatment with NS-398 also increased sensitivity to radiation in COX-2 expressing rat intestinal epithelial cells (152). These scientific studies indicate a potential role may exist for COX-2 inhibitors in chemoprevention of breast cancer.

In addition to these pharmacological inhibitors of COX, many phytochemicals are capable of blocking COX activity. The best studied plant derived compound that inhibits COX is the phytoalexin polyphenol resveratrol, found in red wine, peanuts, and many herbs. Resveratrol induces growth arrest and apoptosis in several human cancer cell lines, including the breast cancer line MCF-7 (153). Treatment of mouse mammary glands with resveratrol inhibited the development of preneoplastic lesions (154), thus demonstrating resveratrol has chemopreventative functionality. Most of the initial research indicated that resveratrol was selective for inhibiting COX-1; however, it has since been shown in mammary epithelial cells to inhibit the both the expression and the activity of COX-2 (155,156). Resveratrol likely blocks COX-2 expression by a protein kinase C-dependent mechanism (155).

Several other phytochemicals have demonstrated the ability to inhibit COX-2 transcription and activity. Curcumin (a significant component of turmeric) prevents lipopolysaccharide and phorbol ester mediated upregulation of COX-2 gene expression, likely by blocking signaling through extracellular signal regulated kinase (ERK) and decreasing NF-κB and activator protein 1 (AP-1) binding to the COX-2 promoter (157,158). Curcumin also directly inhibits the enzymatic activity of COX-2 (159), indicating that curcumin acts on multiple levels to decrease PGE_2 levels. The phytochemical baicalein (flavone isolated from the roots of *Scutellaria baicalensis* that is also a 12-lipoxygenase selective inhibitor) is capable of inhibiting both cell growth and the production of PGE_2 in breast and prostate cancer cell lines (160). Additionally, various flavonoids such as EGCG (green tea) and apigenin (vegetables and fruits) block COX-2 gene expression by suppressing NF-κB (161).

13.2.5.3 Lipoxygenase (LO)

Three main branches of the lipoxygenase pathway give rise to specific products of AA, including leukotrienes (LTs) and hydroxyeicosatetraenonic acids (HETEs). Arachidonic acid can be metabolized by 5-lipoxygenase (5-LO) to LTs (key mediators of asthma and anaphylactic shock) or 5-HETE. 12-LO converts AA into 12-HETE. Metabolism of AA by 15-LO generates 15-HETE, which acts as a ligand for the growth controlling nuclear receptor PPARγ (162). Both 5-HETE and 12-HETE act as a mitogen to cultured breast and prostate cancer cells (163). Interestingly, studies with colon cancer cells indicate that unlike 5-LO and 12-LO, 15-LO activity is antiproliferative (164). Tissue samples taken from breast cancer patients indeed show that 15-LO expression is decreased in tissues taken from patients with breast cancer as compared to paired normal tissues. Therefore, the application of phytochemicals that block the activities of 5-LO or 12-LO or both in combination may be useful in the treatment of several cancers, including breast cancer. While 15-LO activity can be blocked by several flavonoids such as luteolin (165), the antiproliferative activity of this enzyme makes inhibition undesirable.

Increased metabolism of AA to 5-HETE by 5-LO has been observed in a variety of cancers (166), and overexpression of 5-LO in cancerous breast tissue has been correlated with poor clinical outcomes in patients with breast cancer (26). Application of exogenous 5-HETE to the breast cancer cell lines MCF-7 and MDA-MB-231 increased cell proliferation (167). Accordingly, treatment of these cell lines with the 5-LO inhibitor Rev-9501 induced apoptotic events, such as cytochrome *c* release and decreasing levels of antiapoptotic proteins. Several researchers have attempted to identify the signal pathways utilized by which 5-HETE exerts its mitogenic effect. In pancreatic cancer cell lines, 5-HETE was shown to induce tyrosine phosphorylation, and work through Erk and Akt to induce cell proliferation (168). Many pharmacological inhibitors of 5-LO activity exist (e.g., AA-861, MK-866, and the antiallergenic drugs montelukast, zafirlukast, and zileuton), but are not used for cancer chemoprevention. Natural phytochemicals that selectively inhibit 5-LO include the COX and CYP1A1 inhibitor curcumin (78), the prenylated flavanones sigmoidins A and B (isolated from *Erythrina sigmoidea hua*) (169), the dual 5-LO/12-LO inhibitor 2-(3,4-dihydroxyphenyl)ethanol from olive extracts (170) and hyperforin, a lipophilic component of *Hypericum perforatum* (St. John's wort) extract (171).

When normal breast tissue was compared to lobular and ductal carcinomas taken from the same sets of patients, cancerous tissues exhibited increased expression of 12-LO. Additionally, prognostic value of relating clinical outcome with the degree of increased 12-LO expression existed, where higher mortality rates were directly correlated with higher overexpression of 12-LO (26). Baicalein is a potent and selective inhibitor of 12-LO activity, and inhibits proliferation and induces apoptosis in MCF-7 and MDA-MB-231 cells, two well established breast cancer cell lines (160,167). Additionally, baicalein may be antiestrogenic in ER-alpha expressing MCF-7 cells and better at inducing apoptosis than the soy isoflavone genistein (172). Expression of 12-LO and levels of 12-HETE increase with progression of prostate cancer, and a similar correlation may also exist for breast cancer. Therefore, the use of baicalein as a chemopreventative agent against the development and progression of breast cancer may be of significant value as it inhibits both 12-LO and COX-2, two enzymes upregulated in breast cancer.

13.3 STRATEGIES FOR PHYTOCHEMICAL ENRICHMENT FOR CHEMOPREVENTION

13.3.1 Consistency of Phytochemicals

The major challenge in developing and characterizing phytochemicals for chemopreventative applications lies in the inconsistency of phytochemicals from botanicals. Additionally, in many cases a whole extract or food phytochemical profile provides more effective protection than individual phytochemicals in the profile. Soybeans are an example of this complication. Epidemiological studies have associated high soy intake with a reduced risk for certain types of cancer (173). Soybeans are a rich source of phenolic antioxidants known as isoflavonoids. The functionally active isoflavonoid found in soybeans is genistein, which has weak estrogenic activity that appears to antagonize the action of estradiol at the estrogen receptor in breast cells, and thus may protect against breast cancer development (174,175). While the chemopreventative properties of purified and synthetic genistein have been demonstrated (58,61,176–178), recent research has shown that fermented soy products performed better at reducing incidence of mammary tumor risk as compared to a similar mixture of soy constituent isoflavonoids (179). This suggests that the food background may play a positive role in the chemopreventative actions of its resident isoflavonoids (particularly genistein). The challenge posed by inconsistency of phytochemicals is even greater in highly cross pollinating botanical species, such as those belonging to *Lamiaceae*. These species are the source of rosmarinic acid (180), which in proper phenolic profiles has chemopreventative potential through inhibition of COX (181). However, each extract is likely to vary in content from source to source and batch to batch (180).

Development of consistent and optimized phytochemicals and whole extract profiles is critical for consistent dosage that is clinically relevant and safe for chemoprevention. Attempts to address this need include the development of approaches such as clonal screening from heterogeneous botanical sources, elicitor and stress induced sprouting of phytochemicals from seedlings, solid-state and liquid bioprocessing and fermentation of botanical extracts. By combining various optimized dietary phytochemicals, synergistic enhancement of chemoprevention can also be attempted.

13.3.2 Clonal Screening of Phytochemicals from a Heterogeneous Genetic Background

Rosmarinic acid is a phenolic biphenyl found in several species in the family *Lamiaceae* with many pharmacological effects. Rosmarinic acid inhibits several complement dependent inflammatory processes and thus has potential as a therapeutic agent for control of complement activation diseases (182,183). Rosmarinic acid has been reported to have effects on both the classical C3-convertase and on the cobra venom factor and ovalbumin and antiovalbumn mediated passive cutaneous anaphylaxis (182). Recent research has indicated other benefits of rosmarinic acid containing *Perilla frutescens* on the reduction of lipopolysaccharide (LPS)-induced liver injury in D-galactosamine sensitized mice (184). Additional evidence suggests that rosmarinic acid inhibits COX (181), and therefore represents a good candidate for chemoprevention of oxidation linked diseases, including cancer.

High rosmarinic acid producing, shoot based clonal lines originating from a single heterozygous seed amongst a heterogeneous bulk seed population of oregano, rosemary, lavender, spearmint, and thyme have been isolated. These lines were screened based upon tolerance to the proline analog A2C and a novel *Pseudomonas* species isolated from oregano (185–187). This strategy for screening and selection of high rosmarinic acid clonal lines is additionally based on the model that the proline linked pentose phosphate pathway is critical for driving metabolic flux (i.e., E4P) toward the shikimate and phenylpropanoid pathways (188). Any clonal line with a deregulated proline synthesis pathway should have increased pentose phosphate pathway activity, which would allow excess metabolic flux to drive the shikimate and phenylpropanoid pathways toward total phenolic and rosmarinic acid synthesis. This model hypothesizes that the same metabolic flux from increased activity of the proline linked pentose phosphate pathway also regulates the interconversion of ribose-5-phosphate to E4P, driving the shikimate pathway. Shikimate pathway flux is critical for both auxin and phenylpropanoid biosynthesis, including rosmarinic acid. High rosmarinic acid producing clonal lines selected by these approaches (185,186,189–191) are being targeted to produce consistent dietary phenolic phytochemicals from cross pollinating, heterogeneous species for functional foods and for chemoprevention strategies (180,191).

13.3.3 Stress and Elicitor Induced Sprouting

In many legumes such as soy (192), pea (193), fava bean (194), mung bean (195) and fenugreek (196), sprouting in combination with biological, chemical, and physical stress linked elicitation can stimulate phytochemicals relevant in chemoprevention. Preliminary results have provided empirical evidence for a link between proline biosynthesis and oxidation, as well as stimulation of G6PDH during phenolic phytochemical synthesis (180,188,197–199). In light mediated sprout studies in peas (*Pisum sativum)*, acetylsalicylic acid in combination with fish protein hydrolysates (a potential source of proline precursors) stimulated phenolic content and guaiacol peroxidase (GPX) activity during early germination, with corresponding higher levels of proline and G6PDH activity (200). In other light mediated studies in peas, low pH and salicylic acid treatments stimulated increased phenolic content and tissue rigidity, with concomitant stimulation of G6PDH and proline (199). This work supports the hypothesis that pentose phosphate pathway stimulation may be linked to proline biosynthesis, and that modulation of a proton linked redox cycle may also be operating through the proline linked pentose phosphate pathway. In dark germinated studies in peas, high cytokinin containing anise root extracts stimulated phenolic content and antioxidant activity, which correlated directly with increased proline content but inversely with G6PDH activity (193).

In dark germination studies in mung bean (*Vigna radiata*), dietary grade microbial polysaccharide, oregano phenolics and peptide elicitor treatments stimulated phenolic content and G6PDH and GPX enzyme activities compared to controls (193,195,201,202) with concomitant stimulation of proline content. In addition, specific elicitors xanthan gum, oregano phenolics, yeast extract, and yeast glucan stimulated antioxidant activity. The hypothesis that stimulation of proline linked pentose phosphate pathway would stimulate phenolic metabolism under elicitor and stress response was tested using dark germinated fava bean. In polysaccharide elicitor studies, gellan gum stimulated fava bean total phenolic content by ninefold in late stages of germination with a corresponding increase in proline content and GPX activity. However, the effect on antioxidant and G6PDH activity was inconclusive (203). In the same fava bean system, UV mediated stimulation of phenolic content in dark germinated fava bean sprouts indicated a positive correlation to G6PDH and GPX activities with a concomitant increase in proline content (204). It was further confirmed that the proline analog A2C also stimulated phenolic content in fava bean, with positive correlation to G6PDH and GPX activities as well as proline content (205). Similar to studies in clonal shoot cultures of thyme (197) and oregano (198), the proline analog mediated studies in fava bean confirmed that proline overexpression was not only possible, but involved stimulating G6PDH, and therefore likely diverted the pentose phosphate pathway toward phenylpropanoid biosynthesis. The late stage stimulation of phenolic content and GPX activity in response to microwave mediated thermal stress in dark germinated fava bean strongly correlated with stimulation of free radical scavenging activity of free phenolics as measured by the quenching of 1,1-diphenyl-2-picrylhydrazyl (DPPH) and stimulation of SOD activity (202). Recent studies have extended this concept of elicitor induced stimulation to fenugreek (196). These strategies and insights have allowed the development of a technology platform where stress induced elicitation can be used to capture and optimize the production of phytochemicals with chemopreventative potential in germinating seeds, which would not be otherwise possible. This approach could theoretically capture protective phytochemical profiles that are specifically produced under stress to protect the emerging young embryo and seedling, at a stage when respiration and oxidation stress during the sprouting process is high.

13.3.4 Solid-State Bioprocessing

Solid-state bioprocessing using dietary microbial systems under minimal water conditions has been used for over 2000 years in many parts of the world to develop specific foodstuffs such as soy based products like tempeh in Indonesia, tofu and related foods in East Asia, several legume products in India and cheese in Europe (206,207). In the context of chemoprevention, enhanced isoflavonoid content in soybean and soybean meal following bioprocessing by *Aspergillus* species has been reported (208,209). Dietary fungal bioprocessing of fruit and legume substrates to enrich aglycone phenolic antioxidants such as ellagic acid (cranberry), isoflavonoids (soybean) and L-DOPA (fava bean) have been similarly developed (210–215). Critical microbial enzymes that are involved in the optimal release of these phenolic phytochemicals have recently been identified (194,215–217). A better understanding of growth and enzyme regulation during phenolic phytochemical mobilization and release would help to optimize the desired phytochemical profile with maximum chemopreventative potential. In noncancer related chemoprevention investigations, results indicate that cranberry and soybean bioprocessing by solid-state methods not only stimulated free radical scavenging antioxidant activity, but also increased phenolic content in specific stages and inhibited foodborne pathogens (217–219) and the ulcer associated *Helicobacter pylori* (217,219). Such extracts are now being targeted for breast cancer chemoprevention and diabetes through modulation of the antioxidant enzyme response (207,220).

13.3.5 Liquid Fermentation

Recent research has shown that fermented soymilk performed better at reducing incidence of mammary tumor risk than a similar mixture of its constituent isoflavonoids, suggesting that the food background may play a positive role in the chemopreventative actions of soy based products (179). Fermented soymilk is rich in phenolic aglycones (which are more active and more readily taken up than their (β–glycosides), thus increasing the free phenolic content of soy based food through microbial bioprocessing may positively affect its medicinal and nutritional value (221–224). The role of lignin degrading enzymes in phenolic antioxidant mobilization during yogurt production from soymilk by active probiotic kefir cultures was investigated (225). Total soluble phenolic content and free radical scavenging antioxidant activity were measured every 8 hours for a 48 hour period. The activity of several enzymes (β-glucosidase, laccase, and peroxidase) associated with the microbial degradation of polymeric phenolics and lignin and previously linked to phenolic mobilization from soybean during solid-state bioprocessing by dietary fungi were also investigated. Soluble phenolic content initially increased with culture time and was strongly correlated to total peroxidase and laccase activity. However, phenolic content dropped sharply at 48 hours. Antioxidant activity increased with culture time and was strongly correlated to decreased soluble phenolic content over the same time period. This research has important implications for the optimization of functional phytochemicals in commercial soymilk based yogurts, which can be targeted for chemoprevention strategies (225).

REFERENCES

1. Wiseman, B., Z. Werb. Stromal effects on mammary gland development and breast cancer. *Science* 296:1046–1049, 2002.
2. Ronnov-Jessen, L., O. Petersin, M. Bissell. Cellular changes involved in conversion of normal to malignant breast: importance of the stromal reaction. *Physiol. Rev*. 78:69–125, 1996.
3. Beenken, S., K. Bland. Biomarkers for breast cancer. *Minerva Chirurgica* 57:437–448, 2002.
4. Rosner, B., G. Colditz, M. Willett. Reproductive risk factors in a prospective study of breast cancer: the nurses' health study. *Am. J. Epidemiol*. 139:819–835, 1994.
5. Sivaraman, L., S.G. Hilsenbeck, L. Zhong, J. Gay, O.M. Conneely, D. Medina, B.W. O'Malley. Early exposure of the rat mammary gland to estrogen and progesterone blocks co-localization of estrogen receptor expression and proliferation. *J. Endocrinol.* 171:75–83, 2001.
6. Sivaraman, L., O.M. Conneely, D. Medina, B.W. O'Malley. p53 is a potential mediator of pregnancy and hormone-induced resistance to mammary carcinogenesis. *Proc. Natl. Acad. Sci. USA* 98:12379–12384, 2001.
7. Kuperwasser, C., J. Pinkas, G.D. Hurlbut, S.P. Naber, D.J. Jerry. Cytoplasmic sequestration and functional repression of p53 in the mammary epithelium is reversed by hormonal treatment. *Cancer Res*. 60:2723–2729, 2000.
8. Medina, D. Breast cancer: the protective effect of pregnancy. *Clin. Cancer. Res.* 10:380S–384S, 2004.
9. D'Cruz, C.M., S.E. Moody, S.R. Master, J.L. Hartman, E.A. Keiper, M.B. Imielinski, J.D. COX, J.Y. Wang, S.I. Ha, B.A. Keister, L.A. Chodosh. Persistent parity-induced changes in growth factors, TGF-beta3, and differentiation in the rodent mammary gland. *Mol. Endocrinol.* 16:2034–2051, 2002.
10. Oren, M. Decision making by p53: life, death, and cancer. *Cell Death Diff.* 10:431–442, 2003.
11. Jerry, D., L. Minter, K. Becker, A. Blackburn. Hormonal control of p53 and chemoprevention. *Breast Cancer Res*. 4:91–94, 2002.
12. Singh, R., S. Dhanalakshmi, R. Agarwal. Phytochemicals as cell cycle modulators: a less toxic approach in halting human cancers. *Cell Cycle* 1:156–161, 2002.
13. Katzenellenbogen, B., J. Frasor. Therapeutic targeting in the estrogen receptor hormonal pathway. *Semin. Oncol*. 31:28–38, 2004.

14. Miller, W.R. Aromatase inhibitors: mechanism of action and role in the treatment of breast cancer. *Semin. Oncol.* 30:3–11, 2003.
15. Goss, P.E., K. Strasser-Weippl. Aromatase inhibitors for chemoprevention. *Best Pract. Res. Clin. Endocrinol. Metab.* 18:113–130, 2004.
16. Amakura, Y., T. Tsutsumi, K. Sasaki, T. Yoshida, T. Maitani. Screening of the inhibitory effect of vegetable constituents on the aryl hydrocarbon receptor-mediated activity induced by 2,3,7,8-tetrachlorodibeno-p-dioxin. *Biol. Pharmacol. Bull.* 26:1754–1760, 2003.
17. Zheng, W., S. Kar, R. Quirion. Insulin-like growth factor-1-induced phosphorylation of transcription factor FKHRL1 is mediated by phosphatidylinositol 3-kinase/Akt kinsase and role of this pathway in insulin-like growth factor-1-induced survival of cultured hippocampal neurons. *Mol. Pharmacol.* 62:225–233, 2002.
18. Sugumar, A., Y. Liu, Q. Xia, Y. Koh, K. Matsuo. Insulin-like growth factor (IGF)-I and IGF-binding protein 3 and the risk of premenopausal breast cancer: a meta-analysis of literature. *Int. J. Cancer* 111:293–297, 2004.
19. Darcy, K., A. Wohlhueter, D. Zangani, M. Vaughan, J. Russell, P. Masso-Welch, L. Varela, S. Shoemaker, E. Horn, P. Lee, R. Huang, M. Ip. Selective changes in EGF receptor expression and function during the proliferation, differentiation and apoptosis of mammary epithelial cells. *Eur. J. Cell Biol.* 78:511–523, 1999.
20. Ferrara, N. Molecular and biological properties of vascular endothelial growth factor. *J. Mol. Med.* 77:527–543, 1999.
21. Bottinger, E.P., J.L. Jakubczak, D.C. Haines, K. Bagnall, L.M. Wakefield. Transgenic mice overexpressing a dominant-negative mutant type II transforming growth factor beta receptor show enhanced tumorigenesis in the mammary gland and lung in response to the carcinogen 7,12-dimethylbenz-[a]-anthracene. *Cancer Res.* 57:5564–5570, 1997.
22. Huynh, H., W. Beamer, M. Pollak, T.W. Chan. Modulation of transforming growth factor beta1 gene expression in the mammary gland by insulin-like growth factor I and octreotide. *Int. J. Oncol.* 16:277–281, 2000.
23. Howe, L., K. Subbaramaiah, A. Brown, A. Dannenberg. Cyclooxygenase-2: a target for the prevention and treatment of breast cancer. *Endocrine Relat. Cancer* 8:97–114, 2001.
24. Ristimaki, A., A. Sivula, J. Lundin, M. Lundin, T. Salminen, C. Haglund, J. Joensuu, J Isola. Prognostic significance of elevated cyclooxygenase-2 expression in breast cancer. *Cancer Res.* 62:632–635, 2002.
25. Tan, K., W. Yong, T. Putti. Cyclooxygenase-2 expression: a potential prognostic and predictive marker for high-grade ductal carcinoma *in situ* of the breast. *Histopathology* 44:24–28, 2004.
26. Jiang, W., A. Douglas-Jones, R. Mansel. Levels of expression of lipoxygenases and cyclo-oxygenase-2 in human breast cancer. *Prostaglandins Leukotrienes Essential Fatty Acids* 69:275–281, 2003.
27. Arun, B., P. Goss. The role of COX-2 inhibition in breast cancer treatment and prevention. *Semin. Oncol.* 31:22–29, 2004.
28. Cline, J., C. Hughes. Phytochemicals for the prevention of breast and endometrial cancer. *Cancer Treatment Res.* 94:107–134, 1998.
29. Surh, Y. Molecular mechanisms of chemopreventative effects of selected dietary and medicinal phenolic substances. *Mutat. Res.* 428:305–327, 1999.
30 Saito, S., A.A. Goodarzi, Y. Higashimoto, Y. Noda, S.P. Lees-Miller, E. Appella, C.W. Anderson. ATM mediates phosphorylation at multiple p53 sites, including Ser(46), in response to ionizing radiation. *J. Biol. Chem.* 277:12491–12494, 2002.
31. Vargas, D.A., S. Takahashi, Z. Ronai. Mdm2: a regulator of cell growth and death. *Adv. Cancer Res.* 89:1–34, 2003.
32. Turenne, G.A., P. Paul, L. Laflair, B.D. Price. Activation of p53 transcriptional activity requires ATM's kinase domain and multiple N-terminal serine residues of p53. *Oncogene* 20:5100–5110, 2001.
33. Pise-Masison, C.A., M. Radonovich, K. Sakaguchi, E. Appella, J.N. Brady. Phosphorylation of p53: a novel pathway for p53 inactivation in human T-cell lymphotropic virus type 1-transformed cells. *J. Virol.* 72:6348–6555, 1998.

34. Kapoor, M., R. Hamm, W. Yan, Y. Taya, G. Lozano. Cooperative phosphorylation at multiple sites is required to activate p53 in response to UV radiation. *Oncogene* 19:358–364, 2000.
35. Buschmann, T., O. Potapova, A. Bar-Shira, V.N. Ivanov, S.Y. Fuchs, S. Henderson, V.A. Fried, T. Minamoto, D. Alarcon-Vargas, M.R. Pincus, W.A. Gaarde, N.J. Holbrook, Y. Shiloh, Z. Ronai. Jun NH2-terminal kinase phosphorylation of p53 on Thr-81 is important for p53 stabilization and transcriptional activities in response to stress. *Mol. Cell Biol.* 21:2743–2754, 2001.
36. Bulavin, D.V., S. Saito, M.C. Hollander, K. Sakaguchi, C.W. Anderson, E. Appella, A.J. Fornace, Jr. Phosphorylation of human p53 by p38 kinase coordinates N-terminal phosphorylation and apoptosis in response to UV radiation. *EMBO J.* 18:6845–6854, 1999.
37. Ogawara, Y., S. Kishishita, T. Obata, Y. Isazawa, T. Suzuki, K. Tanaka, N. Masuyama, Y. Gotoh. Akt enhances Mdm2-mediated ubiquitination and degradation of p53. *J. Biol. Chem.* 277:21843–21850, 2002.
38. Mayo, L.D., D.B. Donner. A phosphatidylinositol 3-kinase/Akt pathway promotes translocation of Mdm2 from the cytoplasm to the nucleus. *Proc. Natl. Acad. Sci. USA* 98:11598–11603, 2001.
39. Ueno, M., H. Masutani, R.J. Arai, A. Yamauchi, K. Hirota, T. Sakai, T. Inamoto, Y. Yamaoka, J. Yodoi, T. Nikaido. Thioredoxin-dependent redox regulation of p53-mediated p21 activation. *J. Biol. Chem.* 274:35809–35815, 1999.
40. Kuperwasser, C., G.D. Hurlbut, F.S. Kittrell, E.S. Dickinson, R. Laucirica, D. Medina, S.P. Naber, D.J. Jerry. Development of spontaneous mammary tumors in BALB/c p53 heterozygous mice: a model for Li-Fraumeni syndrome. *Am. J. Pathol.* 157:2151–2159, 2000.
41. Coles, C., A. Condie, U. Chetty, C.M. Steel, H.J. Evans, J. Prosser. p53 mutations in breast cancer. *Cancer Res.* 52:5291–5298, 1992.
42. Moll, U.M., G. Riou, A.J. Levine. Two distinct mechanisms alter p53 in breast cancer: mutation and nuclear exclusion. *Proc. Natl. Acad. Sci. USA* 89:7262–7266, 1992.
43. Minter, L.M., E.S. Dickinson, S.P. Naber, D.J. Jerry. Epithelial cell cycling predicts p53 responsiveness to gamma-irradiation during post-natal mammary gland development. *Development* 129:2997–3008, 2002.
44. Huang, C., W.Y. Ma, J. Li, S.S. Hecht, Z. Dong. Essential role of p53 in phenethyl isothiocyanate-induced apoptosis. *Cancer Res.* 58:4102–4106, 1998.
45. Hong, Y.S., Y.A. Ham, J.H. Choi, J. Kim. Effects of allyl sulfur compounds and garlic extract on the expression of Bcl-2, Bax, and p53 in non small cell lung cancer cell lines. *Exp. Mol. Med.* 32:127–134, 2000.
46. Laux, M., M. Aregullin, J. Berry, J. Flanders, E. Rodriguez. Identification of a p53-dependent pathway in the induction of apoptosis of human breast cancer cells by the natural product, resveratrol. *J. Alt. Compl. Med.* 10:235–239, 2004.
47. Narayanan, B.A., O. Geoffroy, M.C. Willingham, G.G. Re, D.W. Nixon. p53/p21(WAF1/CIP1) expression and its possible role in G1 arrest and apoptosis in ellagic acid treated cancer cells. *Cancer Lett.* 136:215–221, 1999.
48. Choudhuri, T., S. Pal, M.L. Agwarwal, T. Das, G. Sa. Curcumin induces apoptosis in human breast cancer cells through p53-dependent Bax induction. *FEBS Lett.* 512:334–340, 2002.
49. O'Prey, J., J. Brown, J. Fleming, P.R. Harrison. Effects of dietary flavonoids on major signal transduction pathways in human epithelial cells. *Biochem. Pharmacol.* 66:2075–2088, 2003.
50. Ye, R., A. Bodero, B.B. Zhou, K.K. Khanna, M.F. Lavin, S.P. Lees-Miller. The plant isoflavenoid genistein activates p53 and Chk2 in an ATM-dependent manner. *J. Biol. Chem.* 276:4828–4833, 2001.
51. Wigle, D.T. Breast cancer and fertility trends in Canada. *Am. J. Epidemiol.* 105:428–438, 1977.
52. Grubbs, C.J., D.L. Hill, K.C. McDonough, J.C. Peckham. N-nitroso-N-methylurea-induced mammary carcinogenesis: effect of pregnancy on preneoplastic cells. *J. Natl. Cancer Inst.* 71:625–628, 1983.
53. Russo, I.H., M. Koszalka, P.A. Gimotty, J. Russo. Protective effect of chorionic gonadotropin on DMBA-induced mammary carcinogenesis. *Br. J. Cancer* 62:243–247, 1990.

54. Razandi, M., A. Pedram, S.T. Park, E.R. Levin. Proximal events in signaling by plasma membrane estrogen receptors. *J. Biol. Chem.* 278:2701–2712, 2003.
55. Telang, N.T., A. Suto, G.Y. Wong, M.P. Osborne, H.L. Bradlow. Induction by estrogen metabolite 16 alpha-hydroxyestrone of genotoxic damage and aberrant proliferation in mouse mammary epithelial cells. *J. Natl. Cancer Inst.* 84:634–638, 1992.
56. Kelsey, J.L., M.D. Gammon, E.M. John. Reproductive factors and breast cancer. *Epidemiol. Rev.* 15:36–47, 1993.
57. Ketcham, A.S., W.F. Sindelar. Risk factors in breast cancer. *Prog. Clin. Cancer.* 6:99–114, 1975.
58. Constantinou, A.I., D. Lantvit, M. Hawthorne, X. Xu, R.B. van Breemen, J.M. Pezzuto. Chemopreventive effects of soy protein and purified soy isoflavones on DMBA-induced mammary tumors in female Sprague-Dawley rats. *Nutr. Cancer.* 41:75–81, 2001.
59. Gallo, D., S. Giacomelli, F. Cantelmo, G.F. Zannoni, G. Ferrandina, E. Fruscella, A. Riva, P. Morazzoni, E. Bombardelli, S. Mancuso, G. Scambia. Chemoprevention of DMBA-induced mammary cancer in rats by dietary soy. *Breast Cancer Res. Treat.* 69:153–164, 2001.
60. Cotroneo, M.S., J. Wang, W.A. Fritz, I.E. Eltoum, C.A. Lamartiniere. Genistein action in the prepubertal mammary gland in a chemoprevention model. *Carcinogenesis* 23:1467–1474, 2002.
61. Lamartiniere, C.A., M.S. Cotroneo, W.A. Fritz, J. Wang, R. Mentor-Marcel, A. Elgavish. Genistein chemoprevention: timing and mechanisms of action in murine mammary and prostate. *J. Nutr.* 132:552S–558S, 2002.
62. Sarkar, F.H., Y. Li. Soy isoflavones and cancer prevention. *Cancer Invest.* 21:744–757, 2003.
63. Goodin, M.G., K.C. Fertuck, T.R. Zacharewski, R.J. Rosengren. Estrogen receptor-mediated actions of polyphenolic catechins *in vivo* and *in vitro*. *Toxicol. Sci.* 69:354–361, 2002.
64. Guengerich, F. Characterization of roles of human cytochrome P450 enzymes in carcinogen metabolism. *Asia Pac. J. Pharmacol.* 5:327–345, 1990.
65. Schwarz, D., P. Kisselev, S. Ericksen, G. Szklarz, A. Chernogolov, H. Honeck, W. Schunck, I. Roots. Arachidonic and eicosapentaenoic acid metabolism by human CYP1A1: highly stereoselective formation of 17(R),18(S)-epoxyeicosatetraenoic acid. *Biochem. Pharmacol.* 67:1445–1457, 2004.
66. Boskou, D., F. Visioli. Biophenols in olive oil and olives. *Ind. Res. Signpost* 2003, pp 161–169.
67. Tudela, J., E. Cantos, J. Espin, F. Tomas-Barberan, M. Gil. Induction of antioxidant flavonol biosynthesis in fresh-cut potatoes: effect of domestic cooking. *J. Agric. Food Chem.* 50:5925–5931, 2002.
68. Kampa, M., V. Alexaki, G. Notas, A. Nifli, A. Nistikaki, A. Hatzoglou, E. Bakogeorgou, E. Kouimtzoglou, G. Blekas, D. Boskou, A. Gravanis, E. Castanas. Antiproliferative and apoptotic effects of selective phenolic acids on T47D human breast cancer cells: potential mechanisms of action. *Breast Cancer Res.* 6:R63–R74, 2004.
69. Watabe, M., K. Hishikawa, A. Takayanagi, N. Shimizu, T. Nakaki. Caffeic acid phenethyl ester induces apoptosis by inhibition of NFkappaB and activation of Fas in human breast cancer MCF-7 cells. *J. Biol. Chem.* 279:6017–6026, 2004.
70. Han, D., H. Tachibana, K. Yamada. Inhibition of environmental estrogen-induced proliferation of human breast carcinoma MCF-7 cells by flavonoids. In Vitro *Cell Develop. Biol. Anim.* 37:275–282, 2001.
71. Zhang, S., C. Qin, S. Safe. Flavonoids as aryl hydrdocarbon receptor agonists/antagonists: effects of structure and cell context. *Environ. Health Perspect.* 111:1877–1882, 2003.
72. Chan, H., L. Leung. A potential protective mechanism of soya isoflavones against 7,12-dimethylbenz(a)anthracene tumour initiation. *Br. J. Nutr.* 90:457–465, 2003.
73. Granados-Soto, V. Pleiotropic effects of resveratrol. *Drug News Perspect*, 73 16(5): 299–307, 2003.
74. Casper, R., M. Quesne, I. Rogers, T. Shirota, A. Jolivet, E. Milgrom, J. Savouret. Resveratrol has antagonist activity on the aryl hydrocarbon receptor: implications for prevention of dioxin toxicity. *Mol. Pharmacol.* 56:784–790, 1999.
75. Ciolino, H., P. Daschner, T. Wang, G. Yeh. Effect of curcumin on the aryl hydrocarbon receptor and cytochrome P450 1A1 in MCF-7 human breast carcinoma cells. *Biochem. Pharmacol.* 56:187–206, 1998.

76. Rinaldi, A., M. Morse, H. Fields, D. Rothas, P. Pei, K. Rodrigo, R. Renner, S. Mallery. Curcumin activates the aryl hydrocarbon receptor yet significantly inhibits (-)-benzo(a)pyrene-7R-trans-7,8-dihydrodiol bioactivation in oral squamous cell carcinoma cells and oral mucosa. *Cancer Res.* 62:5451–5456, 2002.
77. Thapliyal, R., G. Mar. Inhibition of cytochrome P450 isozymes by curcumins *in vitro* and *in vivo*. *Food Chem. Toxicol.* 39:541–547, 2001.
78. Flynn, D., M. Rafferty, A. Boctor. Inhibition of 5-hydroxy-eicosatetraenoic acid (5-HETE) formation in intact human neutrophils by naturally-occurring diarylheptanoids: inhibitory activities of curcuminoids and yakuchinones. *Prostaglandins Leukotriene Med.* 22:357–360, 1986.
79. Conney, A. Enzyme induction and dietary chemicals as approaches to cancer chemoprevention: the seventh DeWitt S. Goodman Lecture. *Cancer Res.* 63:7005–7031, 2003.
80. Vogel, C.L., M.A. Cobleigh, D. Tripathy, J.C. Gutheil, L.N. Harris, L. Fehrenbacher, D.J. Slamon, M. Murphy, W.F. Novotny, M. Burchmore, S. Shak, S.J. Stewart, M. Press. Efficacy and safety of trastuzumab as a single agent in first-line treatment of HER2-overexpressing metastatic breast cancer. *J. Clin. Oncol.* 20:719–726, 2002.
81. Baselga, J., D. Tripathy, J. Mendelsohn, S. Baughman, C.C. Benz, L. Dantis, N.T. Sklarin, A.D. Seidman, C.A. Hudis, J. Moore, P.P. Rosen, T. Twaddell, I.C. Henderson, L. Norton. Phase II study of weekly intravenous trastuzumab (Herceptin) in patients with HER2/neu-overexpressing metastatic breast cancer. *Semin. Oncol.* 26:78–83, 1999.
82. Chen, J., P.M. Stavro, L.U. Thompson. Dietary flaxseed inhibits human breast cancer growth and metastasis and downregulates expression of insulin-like growth factor and epidermal growth factor receptor. *Nutr. Cancer* 43:187–192, 2002.
83. Korutla, L., R. Kumar. Inhibitory effect of curcumin on epidermal growth factor receptor kinase activity in A431 cells. *Biochim. Biophys. Acta* 1224:597–600, 1994.
84. Cheng, A.L., C.H. Hsu, J.K. Lin, M.M. Hsu, Y.F. Ho, T.S. Shen, J.Y. Ko, J.T. Lin, B.R. Lin, W. Ming-Shiang, H.S. Yu, S.H. Jee, G.S. Chen, T.M. Chen, C.A. Chen, M.K. Lai, Y.S. Pu, M.H. Pan, Y.J. Wang, C.C. Tsai, C.Y. Hsieh. Phase I clinical trial of curcumin, a chemopreventive agent, in patients with high-risk or pre-malignant lesions. *Anticancer Res.* 21:2895–2900, 2001.
85. Sah, J.F., S. Balasubramanian, R.L. Eckert, E.A. Rorke. Epigallocatechin-3-gallate inhibits epidermal growth factor receptor signaling pathway: evidence for direct inhibition of ERK1/2 and AKT kinases. *J. Biol. Chem.* 279:12755–12762, 2004.
86. Liang, Y.C., Y.C. Chen, Y.L. Lin, S.Y. Lin-Shiau, C.T. Ho, J.K. Lin. Suppression of extracellular signals and cell proliferation by the black tea polyphenol, theaflavin-3,3′-digallate. *Carcinogenesis* 20:733–736, 1999.
87. Meiers, S., M. Kemeny, U. Weyand, R. Gastpar, E. von Angerer, D. Marko. The anthocyanidins cyanidin and delphinidin are potent inhibitors of the epidermal growth-factor receptor. *J. Agric. Food Chem.* 49:958–962, 2001.
88. Bol, D.K., K. Kiguchi, I. Gimenez-Conti, T. Rupp, J. DiGiovanni. Overexpression of insulin-like growth factor-1 induces hyperplasia, dermal abnormalities, and spontaneous tumor formation in transgenic mice. *Oncogene* 14:1725–1734, 1997.
89. Wu, Y., K. Cui, K. Miyoshi, L. Hennighausen, J.E. Green, J. Setser, D. LeRoith, S. Yakar. Reduced circulating insulin-like growth factor I levels delay the onset of chemically and genetically induced mammary tumors. *Cancer Res.* 63:4384 – 4388, 2003.
90. Heron-Milhavet, L., D. LeRoith. Insulin-like growth factor I induces MDM2-dependent degradation of p53 via the p38 MAPK pathway in response to DNA damage. *J. Biol. Chem.* 277:15600–15606, 2002.
91. Guan, Z., S.Y. Buckman, L.D. Baier, A.R. Morrison. IGF-I and insulin amplify IL-1 beta-induced nitric oxide and prostaglandin biosynthesis. *Am. J. Physiol.* 274:F673–F679, 1998
92. Di Popolo, A., A. Memoli, A. Apicella, C. Tuccillo, A. di Palma, P. Ricchi, A.M. Acquaviva, R. Zarrilli. IGF-II/IGF-I receptor pathway up-regulates COX-2 mRNA expression and PGE2 synthesis in Caco-2 human colon carcinoma cells. *Oncogene* 19:5517–5524, 2000.
93. Wetterau, L.A., M.J. Francis, L. Ma, P. Cohen. Insulin-like growth factor I stimulates telomerase activity in prostate cancer cells. *J. Clin. Endocrinol. Metab.* 88:3354–3359, 2003.

94. del Rincon, S.V., C. Rousseau, R. Samanta, W.H. Miller, Jr. Retinoic acid-induced growth arrest of MCF-7 cells involves the selective regulation of the IRS-1/PI 3-kinase/AKT pathway. *Oncogene* 22:3353–3360, 2003.
95. Adamo, M.L., Z.M. Shao, F. Lanau, J.C. Chen, D.R. Clemmons, C.T. Roberts, Jr., D. LeRoith, J.A. Fontana. Insulin-like growth factor-I (IGF-I) and retinoic acid modulation of IGF-binding proteins (IGFBPs): IGFBP-2, -3, and -4 gene expression and protein secretion in a breast cancer cell line. *Endocrinology* 131:1858–1866, 1992.
96. Swisshelm, K., K. Ryan, K. Tsuchiya, R. Sager. Enhanced expression of an insulin growth factor-like binding protein (mac25) in senescent human mammary epithelial cells and induced expression with retinoic acid. *Proc. Natl. Acad. Sci. USA* 92:4472–4476, 1995.
97. Guvakova, M.A., E. Surmacz. Tamoxifen interferes with the insulin-like growth factor I receptor (IGF-IR) signaling pathway in breast cancer cells. *Cancer Res.* 57:2606–2610, 1997.
98. Rickard, S.E., Y.V. Yuan, L.U. Thompson. Plasma insulin-like growth factor I levels in rats are reduced by dietary supplementation of flaxseed or its lignan secoisolariciresinol diglycoside. *Cancer Lett.* 161:47–55, 2000.
99. Wang, S., V.L. DeGroff, S.K. Clinton. Tomato and soy polyphenols reduce insulin-like growth factor-I-stimulated rat prostate cancer cell proliferation and apoptotic resistance *in vitro* via inhibition of intracellular signaling pathways involving tyrosine kinase. *J. Nutr.* 133:2367–2376, 2003.
100. Klein, R.D., S.M. Fischer. Black tea polyphenols inhibit IGF-I-induced signaling through Akt in normal prostate epithelial cells and Du145 prostate carcinoma cells. *Carcinogenesis* 23:217–221, 2002.
101. Faure, E., N. Heisterkamp, J. Groffen, V. Kaartinen. Differential expression of TGF-beta isoforms during postlactational mammary gland involution. *Cell Tissue Res.* 300:89–95, 2000.
102. Nguyen, A.V., J.W. Pollard. Transforming growth factor beta3 induces cell death during the first stage of mammary gland involution. *Development* 127:3107–3118, 2000.
103. Barcellos-Hoff, M.H., T.A. Dix. Redox-mediated activation of latent transforming growth factor-beta 1. *Mol. Endocrinol.* 10:1077–1083, 1996.
104. Ewan, K.B., R.L. Henshall-Powell, S.A. Ravani, M.J. Pajares, C. Arteaga, R. Warters, R.J. Akhurst, M.H. Barcellos-Hoff. Transforming growth factor-beta1 mediates cellular response to DNA damage *in situ*. *Cancer Res.* 62:5627–5631, 2002.
105. Pierce Jr., D.F., A.E. Gorska, A. Chytil, K.S. Meise, D.L. Page, R.J. Coffey, Jr., H.L. Moses. Mammary tumor suppression by transforming growth factor beta 1 transgene expression. *Proc. Natl. Acad. Sci. USA* 92:4254–4258, 1995.
106. Gorska, A.E., R.A. Jensen, Y. Shyr, M.E. Aakre, N.A. Bhowmick, H.L. Moses. Transgenic mice expressing a dominant-negative mutant type II transforming growth factor-beta receptor exhibit impaired mammary development and enhanced mammary tumor formation. *Am. J. Pathol.* 163:1539–1549, 2003.
107. Herbert, B.S., B.G. Sanders, K. Kline. N-(4-hydroxyphenyl)retinamide activation of transforming growth factor-beta and induction of apoptosis in human breast cancer cells. *Nutr. Cancer* 34:121–132, 1999.
108. Kishi, H., E. Kuroda, H.K. Mishima, U. Yamashita. Role of TGF-beta in the retinoic acid-induced inhibition of proliferation and melanin synthesis in chick retinal pigment epithelial cells *in vitro*. *Cell Biol. Int.* 25:1125–1129, 2001.
109. Jakowlew, S.B., H. Zakowicz, T.W. Moody. Retinoic acid down-regulates VPAC(1) receptors and TGF-beta 3 but up-regulates TGF-beta 2 in lung cancer cells. *Peptides* 21:1831–1837, 2000.
110. Weikkolainen, K., J. Keski-Oja, K. Koli. Expression of latent TGF-beta binding protein LTBP-1 is hormonally regulated in normal and transformed human lung fibroblasts. *Growth Factors* 21:51–60, 2003.
111. MacCallum, J., J.C. Keen, J.M. Bartlett, A.M. Thompson, J.M. Dixon, W.R. Miller. Changes in expression of transforming growth factor beta mRNA isoforms in patients undergoing tamoxifen therapy. *Br. J. Cancer* 74:474–478, 1996.

112. Kopp, A., W. Jonat, M. Schmahl, C. Knabbe. Transforming growth factor beta 2 (TGF-beta 2) levels in plasma of patients with metastatic breast cancer treated with tamoxifen. *Cancer Res.* 55:4512–4515, 1995.
113. Harpel, J.G., S. Schultz-Cherry, J.E. Murphy-Ullrich, D.B. Rifkin. Tamoxifen and estrogen effects on TGF-beta formation: role of thrombospondin-1, alphavbeta3, and integrin-associated protein. *Biochem. Biophys. Res. Commun.* 284:11–14, 2001.
114. Malet, C., F. Fibleuil, C. Mestayer, I. Mowszowicz, F. Kuttenn. Estrogen and antiestrogen actions on transforming growth factorbeta (TGFbeta) in normal human breast epithelial (HBE) cells. *Mol. Cell Endocrinol.* 174:21–30, 2001.
115. Perry, R.R., Y. Kang, B.R. Greaves. Relationship between tamoxifen-induced transforming growth factor beta 1 expression, cytostasis and apoptosis in human breast cancer cells. *Br. J. Cancer* 72:1441–1446, 1995.
116. Yang, H., S. Kyo, M. Takatura, L. Sun. Autocrine transforming growth factor beta suppresses telomerase activity and transcription of human telomerase reverse transcriptase in human cancer cells. *Cell Growth Differ.* 12:119–127, 2001.
117. Ariazi, E.A., Y. Satomi, M.J. Ellis, J.D. Haag, W. Shi, C.A. Sattler, M.N. Gould. Activation of the transforming growth factor beta signaling pathway and induction of cytostasis and apoptosis in mammary carcinomas treated with the anticancer agent perillyl alcohol. *Cancer Res.* 59:1917–1928, 1999.
118. Lawton, L.N., M.F. Bonaldo, P.C. Jelenc, L. Qiu, S.A. Baumes, R.A. Marcelino, G.M. de Jesus, S. Wellington, J.A. Knowles, D. Warburton, S. Brown, M.B. Soares. Identification of a novel member of the TGF-beta superfamily highly expressed in human placenta. *Gene* 203:17–26, 1997.
119. Strelau, J., M. Bottner, P. Lingor, C. Suter-Crazzolara, D. Galter, J. Jaszai, A. Sullivan, A. Schober, K. Krieglstein, K. Unsicker. GDF-15/MIC-1: a novel member of the TGF-beta superfamily. *J. Neural Transm. Suppl.* 273–276, 2000.
120. Baek, S.J., L.C. Wilson, T.E. Eling. Resveratrol enhances the expression of non-steroidal anti-inflammatory drug-activated gene (NAG-1) by increasing the expression of p53. *Carcinogenesis* 23:425–434, 2002.
121. Bottone Jr., F.G., S.J. Baek, J.B. Nixon, T.E. Eling. Diallyl disulfide (DADS) induces the antitumorigenic NSAID-activated gene (NAG-1) by a p53-dependent mechanism in human colorectal HCT 116 cells. *J. Nutr.* 132:773–778, 2002.
122. Newman, D., M. Sakaue, J.S. Koo, K.S. Kim, S.J. Baek, T. Eling, A.M. Jetten. Differential regulation of nonsteroidal anti-inflammatory drug-activated gene in normal human tracheobronchial epithelial and lung carcinoma cells by retinoids. *Mol. Pharmacol.* 63:557–564, 2003.
123. Wilson, L.C., S.J. Baek, A. Call, T.E. Eling. Nonsteroidal anti-inflammatory drug-activated gene (NAG-1) is induced by genistein through the expression of p53 in colorectal cancer cells. *Int. J. Cancer* 105:747–753, 2003.
124. Masuda, M., M. Suzui, J.T. Lim, A. Deguchi, J.W. Soh, I.B. Weinstein. Epigallocatechin-3-gallate decreases VEGF production in head and neck and breast carcinoma cells by inhibiting EGFR-related pathways of signal transduction. *J. Exp. Ther. Oncol.* 2:350–359, 2002.
125. Sartippour, M.R., Z.M. Shao, D. Heber, P. Beatty, L. Zhang, C. Liu, L. Ellis, W. Liu, V.L. Go, M.N. Brooks. Green tea inhibits vascular endothelial growth factor (VEGF) induction in human breast cancer cells. *J. Nutr.* 132:2307–2311, 2002.
126. Tang, F.Y., N. Nguyen, M. Meydani. Green tea catechins inhibit VEGF-induced angiogenesis *in vitro* through suppression of VE-cadherin phosphorylation and inactivation of Akt molecule. *Int. J. Cancer* 106:871–878, 2003.
127. Chow, L.M., C.H. Chui, J.C. Tang, F.Y. Lau, M.Y. Yau, G.Y. Cheng, R.S. Wong, P.B. Lai, T.W. Leung, I.T. Teo, F. Cheung, D. Guo, A.S. Chan. Anti-angiogenic potential of *Gleditsia sinensis* fruit extract. *Int. J. Mol. Med.* 12:269–273, 2003.
128. Roy, S., S. Khanna, H.M. Alessio, J. Vider, D. Bagchi, M. Bagchi, C.K. Sen. Anti-angiogenic property of edible berries. *Free Radic. Res.* 36:1023–1031, 2002.
129. Dabrosin, C., J. Chen, L. Wang, L.U. Thompson. Flaxseed inhibits metastasis and decreases extracellular vascular endothelial growth factor in human breast cancer xenografts. *Cancer Lett.* 185:31–37, 2002.

130. Cuendet, M., J. Pezzuto. The role of cyclooxygenase and lipoxygenase in cancer chemoprevention. *Drug Metab. Drug Interact.* 17:109–157, 2000.
131. Natarajan, R., J. Nadler. Role of lipoxygenases in breast cancer. *Frontiers Biosci.* 3:E81–E88, 1998.
132. Hershman, H. Prostaglandin synthase 2. *Biochem. Biophys. Acta* 1299:125–140, 1996.
133. Liu, X., D. Rose. Differential expression and regulation of cyclooxygenase-1 and -2 in two human breast cancer cell lines. *Cancer Res.* 56:5125–5127, 1996.
134. Half, E., X. Tang, K. Gwyn, A. Sahin, K. Wathen, F. Sinicrope. Cyclooxygenase-2 expression in human breast cancers and adjacent ductal carcinoma *in situ*. *Cancer Res.* 62:1676–1681, 2002.
135. Shim, V., M. Gauthier, D. Sudilovsky, K. Mantei, K. Chew, D. Moore, I. Cha, T. Tlsty, L. Esserman. Cyclooxygenase-2 expression is related to nuclear grade in ductal carcinoma *in situ* and is increased in its normal adjacent epithelium. *Cancer Res.* 63:2347–2350, 2003.
136. Liu, C., S. Chang, K. Narko, O. Trifan, M. Wu, E. Smith, C. Haudenschild, T. Lane, T. Hla. Overexpression of cyclooxygenase-2 is sufficient to induce tumorigenesis in transgenic mice. *J. Biol. Chem.* 276:18563–18569, 2001.
137. Crawford, Y., M. Gauthier, A. Joubel, K. Mantei, K. Kozakiewicz, C. Afshari, T. Tlsty. Histologically normal human mammary epithelia with silenced p16(INK4a) overexpress COX-2, promoting a premalignant program. *Cancer Cell* 5:263–273, 2004.
138. Stockton, R., B. Jacobson. Modulation of cell-substrate adhesion by arachidonic acid: lipoxygenase regulates cell spreading and ERK1/2-inducible cyclooxygenase regulates cell migration in NIH-3T3 fibroblasts. *Mol. Biol. Cell* 12:1937–1956, 2001.
139. Roberts, L., H. Glenn, R. Whitfield, B. Jacobson. Regulation of cell-substrate adhesion by the lipoxygenase and cyclooxygenase branches of arachidonic acid metabolism. *Adv. Exp. Med. Biol.* 507:525–529, 2002.
140. Gately, S., R. Kerbel. Therapeutic potential of selective cyclooxygenase-2 inhibitors in the management of tumor angiogenesis. *Progress Experim. Tumor Res.* 37:179–192, 2003.
141. Chang, S., C. Liu, R. Conway, D. Han, K. Nithipatikom, O. Trifan, T. Lane, T. Hla. Role of prostaglandin E2-dependent angiogenic switch in cyclooxygenase-2-induced breast cancer progression. *Proc. Nat. Acad. Sci. USA* 101:591–596, 2004.
142. Vadlamudi, R., M. Mandal, L. Adam, B. Steinbach, J. Mendelsohn, R. Kumar. Regulation of cyclooxygenase-2 pathway by HER2 receptor. *Oncogene* 18:305–314, 1999.
143. Benoit, V., B. Relic, X. Leval, A. Chariot, M. Merville, V. Bours. Regulation of HER-2 oncogene expression by cyclooxygenase-2 and prostaglandin E2. *Oncogene* 23:1631–1635, 2004.
144. Sheng, H., J. Shao, J. Morrow, R. Beauchamp, R. DuBois. Modulation of apoptosis and Bcl-2 expression by prostaglandin E2 in human colon cancer cells. *Cancer Res.* 58:362–366, 1998.
145. Subbaramaiah, K., N. Altorki, W. Chung, J. Mestre, A. Sampat, A. Dannenberg. Inhibition of cyclooxygenase-2 gene expression by p53. *J. Biol. Chem.* 274:10911–10915, 1999.
146. Thun, M. Aspirin use and reduced risk of fatal colon cancer. *New Eng. J. Med.* 325:1593–1596, 1991.
147. DuBois, R. Aspirin and breast cancer chemoprevention. *J. Am. Med. Assoc.* 291:2488–2489, 2004.
148. Terry, M., M. Gammon, F. Zhang, H. Tawfik, S. Teitelbaum, J. Britton, K. Subbaramaiah, A. Dannenberg, A. Neugut. Association of frequency and duration of aspirin use and hormone receptor status with breast cancer risk. *J. Am. Med. Assoc.* 291:2433–2440, 2004.
149. Earashi, M., M. Noguchi, K. Kinoshita, M. Tanaka. Effects of eicosanoid synthesis inhibitors on the *in vitro* growth and prostaglandin E and leukotriene F secretion of a human breast cancer cell line. *Oncology* 52:150–155, 1995.
150. Cunningham, D., L. Harrison, T. Shultz. Proliferative responses of normal human mammary and MCF-7 breast cancer cells to linoleic acid, conjugated linoleic acid and eicosanoid synthesis inhibitors in culture. *Anticancer Res.* 17:197–203, 1997.
151. Michael, M., M. Badr, A. Badawi. Inhibition of cyclooxygenase-2 and activation of peroxisome proliferator-activated receptor-gamma synergistically induces apoptosis and inhibits growth of human breast cancer cells. *Int. J. Molecular Med.* 11:733–736, 2003.

152. Pyo, H., H. Choy, G. Amorino, J. Kim, Q. Cao, S. Hercules, R. DuBois. A selective cyclooxygenase-2 inhibitor, NS-398, enhances the effect of radiation *in vitro* and *in vivo* preferentially on the cells that express cyclooxygenase-2. *Clin. Cancer Res.* 7:2998–3005, 2001.
153. Joe, A., H. Liu, M. Suzui, M. Vural, D. Xiao, I. Weinstein. Resveratrol induces growth inhibition, S-phase arrest, apoptosis, and changes in biomarker expression in several human cancer cell lines. *Clin. Cancer Res.* 8:893–903, 2002.
154. Jang, M., L. Cai, G. Udeani, K. Slowing, C. Thomas, C. Beecher, H. Fong, N. Farnsworth, A. Kinghorn, R. Mehta, R. Moon, J. Pezzuto. Cancer chemopreventative activity of resveratrol, a natural product derived from grapes. *Science* 275:218–220, 1997.
155. Subbaramaiah, K., W. Chung, P. Michaluart, N. Telang, T. Tanabe, J. Inoue, M. Jang, J. Pezzuto, A. Dannenberg. Resveratrol inhibits cyclooxygenase-2 transcription and activity in phorbol ester-treated human mammary epithelial cells. *J. Biol. Chem.* 273:21875–21882, 1998.
156. Banerjee, S., C. Bueso-Ramos, B. Aggarwal. Suppression of 7,12-dimethylbenz(a)anthra cene-induced mammary carcinogenesis in rats by resveratrol: role of nuclear factor-KB, cyclooxygenase-2, and matrix metalloprotease-9. *Cancer Res.* 62:2002.
157. Kang, G., P. Kong, Y. Yuh, S. Lim, W. Chun, S. Kim. Curcumin suppresses lipopolysaccharide-induced cyclooxygenase-2 expression by activator protein 1 and nuclear factor kappaB bindings in BV2 microglial cells. *J. Pharmacol. Sci.* 94:325–328, 2004.
158. Chun, K., Y. Keum, S. Han, Y. Song, S. Kim, Y. Surh. Curcumin inhibits phorbol ester-induced expression of cyclooxygenase-2 in mouse skin through suppression of extracellular signal-regulated kinase activity and NF-kappaB activation. *Carcinogenesis* 24:1515–1524, 2003.
159. Zhang, F., N. Altorki, J. Mestre, K. Subbaramaiah, A. Dannenberg. Curcumin inhibits cyclooxygenase-2 transcription in bile acid-and phorbol ester-treated human gastrointestinal epithelial cells. *Carcinogenesis* 20:445–451, 1999.
160. Ye, F., L. Xui, J. Yi, W. Zhang, D. Zhang. Anticancer activity of *Scutellaria baicalensis* and its potential mechanism. *J. Alt. Compl. Med.* 8:567–572, 2002.
161. Liang, Y., Y. Huang, S. Tsai, S. Lin-Shiau, C. Chen, J. Lin. Suppression of inducible cyclooxygenase and inducible nitric oxide synthase by apigenin and related flavonoids in mouse macrophages. *Carcinogenesis* 20:1945–1952, 1999.
162. Chen, G.G., H. Xu, J.F. Lee, M. Subramaniam, K.L. Leung, S.H. Wang, U.P. Chan, T.C. Spelsberg. 15-hydroxy-eicosatetraenoic acid arrests growth of colorectal cancer cells via a peroxisome proliferator-activated receptor gamma-dependent pathway. *Int. J. Cancer* 107:837–843, 2003.
163. Nie, D., M. Che, D. Grignon, K. Tang, K. Honn. Role of eicosanoids in prostate cancer progression. *Cancer Metastasis Rev.* 20:195–206, 2001.
164. Nixon, J., K. Kim, P. Lamb, F. Bottone, T. Eling. 15-lipoxygenase-1 has anti-tumorigenic effects in colorectal cancer. *Prostaglandins Leukotrienes Essential Fatty Acids* 70:7–15, 2004.
165. Sadik, C., H. Sies, T. Schewe. Inhibition of 15-lipoxygenases by flavonoids: structure-activity relations and mode of action. *Biochem. Pharmacol.* 65:773–781, 2003.
166. Ghosh, J., C. Myers. Arachidonic acid stimulates prostate cancer cell growth: critical role of 5-lipoxygenase. *Biochem. Biophys. Res. Commun.* 235:418–423, 1997.
167. Tong, W., X. Ding, T. Adrian. The mechanisms of lipoxygenase inhibitor-induced apoptosis in human breast cancer cells. *Biochem. Biophys. Res. Commun.* 296:942–948, 2002.
168. Ding, X., W. Tong, T. Adrian. Multiple signal pathways are involved in the mitogenic effect of 5(S)-HETE in human prostate cancer. *Oncology* 65f:285–294, 2003.
169. Njamen, D., J. Mbafor, Z. Fomum, A. Kamanyi, J. Mbanya, M. Recio, R. Giner, S. Manez, J. Rios. Anti-inflammatory activities of two flavanones, sigmoidin A and sigmoidin B, from *Erythrina sigmoidea*. *Planta Medica* 70:104–107, 2004.
170. Kohyama, N., T. Nagata, S. Fujimoto, K. Sekiya. Inhibition of arachidonate lipoxygenase activities by 2-(3,4-dihydroxyphenyl)ethanol, a phenolic compound from olives. *Biosci. Biotechnol. Biochem.* 61:347–350, 1997.
171. Albert, D., I. Zundorf, T. Dingermann, W. Muller, D. Steinhilber, O. Werz. Hyperforin is a dual inhibitor of cyclooxygenase-1 and 5-lipoxygenase. *Biochem. Pharmacol.* 64:1767–1765, 2002.

172. Po, L., Z. Chen, D. Tsang, L. Leung. Baicalein and genistein display differential actions on estrogen receptor (ER) transactivation and apoptosis in MCF-7 cells. *Cancer Lett.* 187:33–40, 2002.
173. Dai, Q., A.A. Franke, F. Jin, X.O. Shu, J.R. Hebert, L.J. Custer, J. Cheng, Y.T. Gao, W. Zheng. Urinary excretion of phytoestrogens and risk of breast cancer among Chinese women in Shanghai. *Cancer Epidemiol. Biomarkers Prev.* 11:815–821, 2002.
174. Adlercruetz, H., B. Goldin, S. Gorbach. Soy phytoestrogen intake and cancer risk. *J. Nutr.* 125:757S–770S, 1995.
175. Wiseman, H. *Phytochemicals: Epidemiological factors*. London: Academic Press, 1998, pp. 1549–1561.
176. Darbon, J.M., M. Penary, N. Escalas, F. Casagrande, F. Goubin-Gramatica, C. Baudouin, B. Ducommun. Distinct Chk2 activation pathways are triggered by genistein and DNA-damaging agents in human melanoma cells. *J. Biol. Chem.* 275:15363–15369, 2000.
177. C.A. Lamartiniere. Protection against breast cancer with genistein: a component of soy. *Am. J. Clin. Nutr.* 71:1705S–1707S; *discussion* 1708S–1709S, 2000.
178. Tanos, V., A. Brzezinski, O. Drize, N. Strauss, T. Peretz. Synergistic inhibitory effects of genistein and tamoxifen on human dysplastic and malignant epithelial breast cells *in vitro*. *Eur. J. Obstet. Gynecol. Reprod. Biol.* 102:188–194, 2002.
179. Ohta, T., S. Nakatsugi, K. Watanabe, T. Kawamori, F. Ishikawa, M. Morotomi, S. Sugie, T. Toda, T. Sugimura, K. Wakabayashi. Inhibitory effects of Bifidobacterium-fermented soy milk on 2-amino-1-methyl-6-phenylimidazo[4,5-b]pyridine-induced rat mammary carcinogenesis, with a partial contribution of its component isoflavones. *Carcinogenesis* 21:937–941, 2000.
180. Shetty, K. Biotechnology to harness the benefits of dietary phenolics; focus on *Lamiaceae*. *Asia Pac. J. Clin. Nutr.* 6:162–171, 1997.
181. Kelm, M.A., M.G. Nair, G.M. Strasburg, D.L. DeWitt. Antioxidant and cyclooxygenase inhibitory phenolic compounds from *Ocimum sanctum* Linn. *Phytomedicine* 7:7–13, 2000.
182. Engleberger, W., U. Hadding, E. Etschenberg, E. Graf, S. Leyck, J. Winkelmann, M. Parnham. Rosmarinic acid: a new inhibitor of complement C3-convertase with anti-inflammatory activity. *Int. J. Immunopharmacol.* 10:729–737, 1988.
183. Peake, P., B. Pussel, P. Martyn, V. Timmermans, J. Charlesworth. The inhibitory effect of rosmarinic acid on complement involves the C5 convertase. *Int. J. Immunopharmacol.* 13:853–857, 1991.
184. Osakabe, N., A. Yasuda, M. Natsume, C. Sanbongi, Y. Kato, T. Osawa, T. Yoshikawa. Rosmarinic acid, a major polyphenolic component of Perilla frutescens, reduces lipopolysaccharide (LPS)-induced liver injury in D-galactosamine (D-GalN)-sensitized mice. *Free Radic. Biol. Med.* 33:798-806, 2002.
185. Al–Amier, H., B.M. Mansour, N. Toaima, L. Craker, K. Shetty. Tissue culture for phenolics and rosmarinic acid in thyme. *J. Herbs Spices Med. Plants* 8:31–42, 2001.
186. Al-Amier, H., B.M. Mansour, N. Toaima, R.A. Korus, K. Shetty. Screening of high biomass and phenolic-producing clonal lines of spearmint in tissue culture using *Pseudomonas* and azetidine-2-carboxylate. *Food Biotechnol.* 13: 227–253, 1999.
187. Al-Amier, H., B.M. Mansour, N. Toaima, R.A. Korus, K. Shetty. Tissue culture based screening for selection of high biomass and phenolic producing clonal lines of lavender using pseudomonas and azetidine-2-carboxylate. *J. Agric. Food Chem.* 47:2937–2943, 1999.
188. Shetty, K. Role of proline-linked pentose phosphate pathway in biosynthesis of plant phenolics for functional food and environmental applications: a review. *Process Biochem.* 39:789–804, 2004.
189. Eguchi, Y., O. Curtis, K. Shetty. Interaction of hyperhydricity-preventing *Pseudomonas* spp. with oregano (*Origanum vulgare*) and selection of high rosmarinic acid-producing clones. *Food Biotechnol.* 10:191–202, 1996.
190. Yang, R., O. Curtis, K. Shetty. Selection of high rosmarinic acid-producing clonal lines of rosemary (*Rosmarinus officinalis*) via tissue culture using *Pseudomonas* sp. *Food Biotechnol.* 11:73–88, 1997.

191. Shetty, K., T. Carpenter, O. Curtis, T. Potter. Selection of high phenolics-containing clones of thyme (*Thymus vulgaris* L.) using *Pseudomonas* spp. *J. Agric. Food Chem.* 44:3408–3411, 1996.
192. McCue, P., K. Shetty. A hypothetical model for the action of soybean isoflavonoids against cancer involving a shift to proline-linked energy metabolism through activation of the pentose phosphate pathway. *Food Biotechnol.* 18:19–37, 2004.
193. Duval, B., K. Shetty. The stimulation of phenolics and antioxidant activity in pea (*Pisum sativum*) elicited by genetically transformed anise root extract. *J. Food Biochem.* 25:361–377, 2001.
194. Randhir, R., D. Vattem, K. Shetty. Solid-state bioconversion of fava bean by *Rhizopus oligosporus* for enrichment of phenolic antioxidants and L-DOPA. *Innovative Food Sci. Emerg. Technol.* 5:235–244, 2004.
195. McCue, P., K. Shetty. A biochemical analysis of mungbean (*Vigna radiata*) response to microbial polysaccharides and potential phenolic-enhancing effects for nutraceutical applications. *Food Biotechnol.* 16:57–79, 2002.
196. Randhir, R., Y.T. Lin, K. Shetty. Phenolics, their antioxidant and antimicrobial activity in dark germinated fenugreek sprouts in response to peptide and phytochemical elicitors. *Asia Pac. J. Clin. Nutr.* 13:295–307, 2004.
197. Kwok, D., K. Shetty. Effect of proline and proline analogs on total phenolic and rosmarinic acid levels in shoot clones of thyme (*Thymus vulgaris* L.). *J. Food Biochem.* 22:37–51, 1998.
198. Yang, R., K. Shetty. Stimulation of rosmarinic acid in shoot cultures of oregano (*Origanum vulgare*) clonal line in response to proline, proline analog, and proline precursors. *J. Agric. Food Chem.* 46:2888–2893, 1998.
199. McCue, P., Z. Zheng, J. Pinkham, K. Shetty. A model for enhancing pea seedling vigor following low pH and salicylic acid treatments. *Process Biochem.* 35:603–613, 2000.
200. Andarwulan, N., K. Shetty. Improvement of pea (*Pisum sativum*) seed vigor by fish protein hydrolysates in combination with acetyl salicylic acid. *Process Biochem.* 35:159–165, 1999.
201. McCue, P., K. Shetty. Clonal herbal extracts as elicitors of phenolic synthesis in dark-germinated mungbeans for improving nutritional value with implications for food safety. *J. Food Biochem.* 26:209–232, 2002.
202. Randhir, R., Y. Lin, K. Shetty. Stimulation of phenolics, antioxidant and antimicrobial activities in dark-germinated mung bean (*Vigna radiata*) sprouts in response to peptide and phytochemical elicitors. *Process Biochem.* 39:637–646, 2004.
203. Shetty, P., M. Atallah, K. Shetty. Enhancement of total phenolic, L-DOPA and proline contents in germinating fava bean (*Vicia faba*) in response to bacterial elicitors. *Food Biotechnol.* 15:47–67, 2001.
204. Shetty, P., M. Atallah, K. Shetty. Effects of UV treatment on the proline-linked pentose phosphate pathway for phenolics and L-DOPA synthesis in dark-germinated *Vicia faba*. *Process Biochem.* 37:1285–1295, 2002.
205. Shetty, P., M. Atallah, K. Shetty. Stimulation of total phenolics, L-DOPA and antioxidant activity through proline-linked pentose phosphate pathway in response to proline and its analog in germinating fava beans (*Vicia faba*). *Process Biochem.* 38:1707–1717, 2003.
206. Zheng, Z., K. Shetty. *Solid-State Fermentation and Value-Added Utilization of Fruit and Vegetable Processing By-Products.* New York: Wiley Publishers, 1999, pp 2165–2174.
207. McCue, P., K. Shetty. Health benefits of soy isoflavonoids and strategies for enhancement: a review. *Crit. Rev. Food Sci. Nutr.* 44:1–7, 2004.
208. Esaki, H., R. Watanabe, H. Onozaki, S. Kawakishi, T. Osawa. Formation mechanism for potent antioxidative o-dihydroxyisoflavones in soybeans fermented with *Aspergillus saitoi*. *Biosci. Biotechnol. Biochem.* 63:851–858, 1999.
209. Kishida, T., H. Ataki, M. Takebe, K. Ebihara. Soybean meal fermented by *Aspergillus awamori* increases the cytochrome P-450 content of the liver microsomes of mice. *J. Agric. Food Chem.* 48:1367–1372, 2000.

210. McCue, D., A. Horii, K. Shetty. Solid-state bioconversion of phenolic antioxidants from defatted powdered soybean fermented with *Rhizopus oligosprous*: role of carbohydrate-cleaving enzymes. *J. Food Biochem.* 27:501–514, 2003.
211. McCue, P., K. Shetty. Role of carbohydrate-cleaving enzymes in phenolic antioxidant mobilization from whole soybean fermented with *Rhizopus oligosprorus*. *Food Biotechnol.* 17:27–37, 2003.
212. Randhir, R., K. Shetty. Microwave-induced stimulation of L-DOPA, phenolics, and antioxidant activity in fava bean (*Vicia faba*) for Parkinson's diet. *Process Biochem.* 39:1775–1784, 2004.
213. Vattem, D., K. Shetty. Solid-state production of phenolic antioxidants from cranberry pomace by *Rhizopus oligosprous*. *Food Biotechnol.* 16:189–210, 2002.
214. Vattem, D., K. Shetty. Ellagic acid production and phenolic antioxidant activity in cranberry pomace mediated by *Lentinus edodes* using solid-state system. *Process Biochem.* 39:367–379, 2003.
215. Zheng, Z., K. Shetty. Solid-state bioconversion of phenolics from cranberry pomace and role of *Lentinus edodes* beta-glucosidase. *J. Agric. Food Chem.* 48:895–900, 2000.
216. McCue, P., A. Horii, K. Shetty. Mobilization of phenolic antioxidants from defatted soybean powders by *Lentinus edodes* during solid-state fermentation is associated with production of laccase. *Innovative Food Sci. Emerg. Technol.* 5:385–392, 2004.
217. McCue, P., Y. Lin, R. Labbe, K. Shetty. Sprouting and solid-state bioprocessing of *Rhizopus oligosporus* increase the *in vitro* antibacterial activity of aqueous soybean extacts against *Helicobacter pylori*. *Food Biotechnol.* 18:229–249, 2004.
218. Vattem, D., Y. Lin, R. Labbe, K. Shetty. Antimicrobial activity against select food-borne pathogens by phenolic antioxidants enriched cranberry pomace by solid-state bioprocessing using food-grade fungus *Rhizopus oligosporus*. *Process Biochem.* 39:1939–1946, 2004.
219. Vattem, D., Y. Lin, K. Shetty. Enrichment of phenolic antioxidants and anti-*Helicobacter pylori* properties of cranberry pomace by solid-state bioprocessing. *Food Biotechnol.* 19:51–68, 2004.
220. Shetty, K., M.L. Wahlqvist. A model for the role of the proline-linked pentose-phosphate pathway in phenolic phytochemical bio-synthesis and mechanism of action for human health and environmental applications. *Asia Pac. J. Clin. Nutr.* 13:1–24, 2004.
221. Izumi, T., M.K. Piskula, S. Osawa, A. Obata, K. Tobe, M. Saito, S. Kataoka, Y. Kubota, M. Kikuchi. Soy isoflavone aglycones are absorbed faster and in higher amounts than their glucosides in humans. *J. Nutr.* 130:1695–1699, 2000.
222. Rao, M., G. Muralikrishna. Evaluation of the antioxidant properties of free and bound phenolic acids from native and malted finger millet (Ragi, *Eleusine coracana* Indaf-15). *J. Agric. Food Chem.* 50:889–892, 2002.
223. Setchell, K.D., N.M. Brown, L. Zimmer-Nechemias, W.T. Brashear, B.E. Wolfe, A.S. Kirschner, J.E. Heubi. Evidence for lack of absorption of soy isoflavone glycosides in humans, supporting the crucial role of intestinal metabolism for bioavailability. *Am. J. Clin. Nutr.* 76:447–453, 2002.
224. Yuan, L., C. Wagatsuma, M. Yoshida, T. Miura, T. Mukoda, H. Fujii, B. Sun, J.H. Kim, Y.J. Surh. Inhibition of human breast cancer growth by GCP (genistein combined polysaccharide) in xenogeneic athymic mice: involvement of genistein biotransformation by beta-glucuronidase from tumor tissues. *Mutat. Res.* 523–524:55–62, 2003.
225. McCue, P., K. Shetty. Phenolic antioxidant mobilization during yogurt production from soymilk. *Process Biochem.* 40:1583–1592, 2004.

2.14

Biotechnology in Wine Industry

Moustapha Oke, Gopinadhan Paliyath, and K. Helen Fisher

CONTENTS

14.1 INTRODUCTION

Wine is a product of the alcoholic fermentation of grape juice, which is stored in a manner to retain its wine like properties (1). Clear evidence of winemaking dates from 5000 years ago, although archaeological records on wine date back to 6000 B.C. (2,3). It is believed that winemaking technology was first developed in Caucasia (currently part of Turkey, Iraq, Azerbaijan, and Georgia). The domestication of the commercial grape cultivar *Vitis vinifera* is also assumed to have originated in the same area (4). Some researchers believe that domestication of grapes may have independently occurred in Spain (5). From Caucasia, grape growing and winemaking spread toward Palestine, Syria, Egypt, Mesopotamia, and around the Mediterranean. In southern Spain however, there is evidence that extensive grape cultivation existed before the colonization of the region by the Phoenicians (6). It is not until the seventeenth century that modern winemaking technology became established. The quality of the wine was improved by the use of sulphur in barrels, glass bottles, and corks.

The grape growing regions of the world are located between 10°C and 20°C isotherms, (lines drawn around the earth connecting the average yearly temperatures). It is also possible to grow grapes in cooler and subtropical regions, but on those regions viticultural practices are limited because of less than ideal conditions. Currently, grapes are important fresh fruit crops globally, with production reaching over 60 million metric tons (second only to bananas, with a production level of 68.6 million metric tons in 2001) (7). Italy, France, and Spain produced more than 35% of the total world production of grapes. More than 80% of the grapes grown each year are fermented, and the rest are used for fresh consumption or dried for raisins. World wine production has varied from 250 to 330 million hectolitres since 1970, with the production level reaching 282 million hectolitres in 2001. Italy, France, and Spain produce about 50% of the world's wine. Other major wine producers include the USA, Argentina, South Africa, Germany, Australia, Portugal, and Chile.

Wine consumption varies in different regions of the world. Wine consumers are found primarily in European countries. For centuries, wine has been an integral part of daily regional food consumption and an important source of calories. The political conditions and religious beliefs of certain regions, and alcohol abuse in other countries, have resulted in increased restriction in alcohol consumption. It has long been accepted that excessive alcohol consumption is detrimental to human health, although recent studies support the beneficial effects of moderate wine drinking (8,9). There has been a noticeable decline in world wine consumption due possibly to the shift toward the use of less, but better quality, wine. The production of high quality wines involves the use of excellent quality grapes, and of ideal strains of yeasts called superyeasts. The objective of this chapter is to evaluate the current technologies for improving wine quality.

14.2 GRAPE CULTIVARS AND WINE TYPES

There is no unanimous classification for grapevine cultivars even though there are nearly 15,000 of them. Growers in France use ecogeographic associations based on ampelographic (structural) and physiological properties of grapes, while others use aroma profiles to assess the relationship between cultivars (10). Numerical taxonomy is another way of classifying grapevine cultivars (11). The common disadvantage of these methods is the subjectivity of data interpretation. Grouping of cultivars based on evolutionary evidence may be obtained using modern techniques such as DNA fingerprinting (restriction fragment length polymorphisms and amplified fragment length polymorphisms -RFLP and AFLP) (12). The application of microsatellite (SSR) markers is now available to *Vitis* species (13–18). Raisins and grape berries have been typed by microsatellites (19,20), and clonal line diversity has been investigated with such markers (21,22). Many authors have evaluated these markers for cultivar identification, and parentage studies (14,19,23–30). AmpeloCAD, a computer aided digitizing system for determining ampelographic measurements is another means of simplifying the identification (31).

Generally the classification of grape cultivars is based on their specific or interspecific origin. Currently there are four majors groups of grapevine cultivars that include the pure line of *Vitis vinifera*, the French–American hybrids, the American hybrids (*Vitis labrusca*), and some interspecific cultivars (1).

14.2.1 *Vitis vinifera* Cultivars

The genus *Vitis* is divided into two distinct subgenera, *Vitis* and *Muscadinia*. Except for *V. rotundifolia* and *V. popenoei*, all the rest of the species are members of the largest subgenus *Vitis*. The difference between the two genera is in the number of their chromosomes; 38 for *Vitis* against 40 for *Muscadinia* (32). Crossing between species of *Vitis* and *Muscadinia* is possible, but often result in nonfertile progenies, due to the imprecise pairing and unequal separation of the chromosomes during meiosis. The genetic imbalance creating infertility is probably caused by the synthesis of inhibitors such as quercetin glycosides in the pistil (33). However, interspecies crossings among *Vitis* and *Muscadinia* are relatively easy and often yield fertile progeny. This, on the other hand, complicates the classification of grapevine species, because the differences in shoot and leaf morphological characteristics such as hairiness are strongly dependent on environmental conditions.

Most commercial cultivars belong to the species *V. vinifera*. Their reputation comes from their fine wines with distinctive, balanced, and long aging potential. Their weakness, however, is the low productivity and demanding cultivation practices. *Vitis vinifera* cultivars are used to produce both white and red wines. There is room for improving quality attributes of *V. vinifera* cultivars using modern techniques such as genetic engineering.

14.2.1.1 White Cultivars

Chardonnay is the most prestigious white French cultivar, and does well in most wine producing regions of the world. It also yields one of the finest sparkling wines. Under optimal conditions, chardonnay produces wine with an aroma reminiscent of apple, peach, and melon. The cultivar is susceptible to powdery mildew and bunch rot.

Riesling (White Riesling or Johannesburg Riesling) is Germany's most appreciated variety, popular also in the U.S. and in Australia. Wines made from these varieties of grapes have the characteristics of a fresh, aromatic, and well aged wine, reminiscent of roses. Fruit clusters possess small to medium size berries. Although a cold hardy variety, the yield is moderate, and it is also susceptible to powdery mildew and bunch rot.

Parellada is a Spanish variety from Catalonia, producing white wine with apple to citrus like aromas, and sometimes with hints of licorice or cinnamon. Pinot Blanc and Pinot Gris are popular in France, Germany, and other cool regions of Europe, and are used mostly in the production of dry, botrytised, and sparkling wines. They yield wines with the fragrances of passionfruit and hard cheese, respectively.

Sauvignon Blanc is originally from Loire Valley in France, but is also grown in California, Eastern Europe, Italy, and New Zealand. Some clones possess a floral character, but most Sauvignon Blanc wines have an herbaceous aspect, and the aroma of green peppers. Clusters produce small size berries, resistant to bunch rot and downy mildew. Sauvignon Blanc is susceptible to powdery mildew and black rot.

Muller-Thurgau variety was developed by H. Muller-Thurgau in 1882, presumably as a cross between Riesling and Silvaner. However the DNA fingerprint showed that it was more likely a cross between Riesling and Chasselas de Courtillier (34). Muller-Thurgau is probably the best known modern *V. vinifera* in Germany, but is also extensively grown in cool regions of Europe and in New Zealand. It is a high-yielding cultivar, producing a light wine with fruity fragrance and mild acidity. The fruits are susceptible to powdery and downy mildew and bunch rot.

Traminer is an aromatic cultivar, grown throughout the cooler regions of the world. It is used to produce dry and sweet white wines. Gewürztraminer is a clone with an intense fragrance and litchi fruit aroma. Savagnin, in France, is a clone of Traminer with a mild aroma. The basic characteristic of Traminer is the relatively small size of the fruit and their tough skin. The modest clusters are susceptible to powdery mildew, bunch rot, and coulure.

Muscat Blanc is mostly used in the production of dessert wines. Muscat grapes are extensively grown throughout the world because of their marked and distinctive muscaty aroma. They also possess high levels of soluble proteins and flavanoids. Muscato Bianco is used for the production of sparkling wine in Asti, Italy.

Another well known white varietie is Viognier, from the Rhône Valley of France, which is characterized by the smaller size of the fruit and its Muscat character. The wine matures quickly and has a peach or apricot fragrance. Viura, cultivated in Rioja, Spain is another white cultivar, with excellent attributes. Generally, clusters are few and fruits are large. The wine has a floral aroma with aspects of citron when produced in cooler regions, and a rich butterscotch or banana fragrance, after aging in wood. Some *V. vinifera* cultivars that are regionally popular, but less well known are Fiano, Garganega, and Torbato, in Italy; Malvasia, in Spain; Arinto, in Portugal; Tãmîoasa Romîneascã, in Romania; and Furmint, in Hungary.

14.2.1.2 Red Cultivars

Cabernet Sauvignon is the most well known red variety, because of its association with Bordeaux, one of Europe's finest red wine producing areas. The wine has a black currant aroma under favorable conditions or a bell pepper aroma otherwise. The berries are small, seedy, and acidic, with darkly pigmented tough skin. The microsatellite study of Cabernet Sauvignon suggests a crossing between Cabernet Franc and Sauvignon Blanc (35). In order to accelerate the maturation of wine produced by Cabernet Sauvignon, it is blended with Merlot or Cabernet Franc. The cultivar is susceptible to several fungal diseases including powdery mildew, *Eurypa* and *Esca* wood decays, and phomopsis.

Merlot is a red variety grape, similar to Cabernet Sauvignon, which has the advantage of growing in cooler regions and moist soils. The ability to mature more quickly has made Merlot a popular substitute for Cabernet Sauvignon. The clusters are susceptible to coulure.

Pinot Noir is a famous French variety from Burgundy, suitable for the production of rosé and sparkling wines. Under favorable conditions it produces a wine with a distinctive aroma, which some researchers attribute to peppermint, beets, or cherries. Pinot Noir has a large number of clones such as Pinot Gris, Pinot Blanc, Meunier, Pinotage, (crossed with Cinsaut), and the California varietal Gamay Beaujolais (12,36). The clusters are compact with small to medium size fruits susceptible to bunch rot.

Barbera, ranked third after Sangiovese and Trebbiano in Italy, is also grown in California and in Argentina. The clusters possess long green stalks, and are susceptible to leaf roll and sometimes to bunch rot. Because of its high acidity, Barbera is often blended with some low acidic cultivars, but can also be used to make fruity wines by itself.

Gamay Noir à jus Blanc (Gamay with white juice), is mostly used in the production of Beaujolais wines. It produces fruit with medium size and tough skins. Gamay is a high yielding cultivar giving light red wines with little aroma. The fruity aroma of Beaujolais comes from the carbonic maceration technique. Gamay is susceptible to most fungal diseases.

Zinfandel is mostly grown in California and produces a wide range of wines varying from ports to light blush. The predominant aroma in rosé wines is reminiscent of raspberry. The full bodied wines possess berry flavors. The main problem with Zinfandel is that the fruits ripen unevenly.

Nebbiolo is basically grown in Northwestern Italy and adapts well in a wide range of soil pH and types. The wine produced from Nebbiolo is high in tannin and is acidic with a good aging potential. This cultivar is susceptible to powdery mildew and to bunch rot.

Tempranillo (Ull de Llebre) is the famous red variety in Spain, which is also grown in Argentina and in California under the name of Valdepeñas. Under optimal conditions it yields a fine wine with a complex berry jam fragrance with nuance of citrus. Tempranillo is sensitive to powdery and downy mildew.

Another important red variety of *V. vinifera* is Graciano, a Spanish variety with relatively low yield, but resistant to most fungal diseases and drought. Sangiovese is grown in central Italy and consists of a number of clones. It produces light to full bodied wines with nuances of citrus. The clusters are small to medium in size with oval fruits sensitive to bunch rot, powdery mildew and downy mildew. Syrah is a cultivar of the Rhône Valley in France, characterized by a low yield and high tannin content, allowing a good aging potential. It is also cultivated in Australia, where it is known under the name of Shiraz. The clusters are elongated with oval or small round fruits susceptible to drought, bunch rot, and berry moths.

Touriga National, mostly grown for the production of Port wine, is one of the famous Portugaese red cultivars. The clusters contain small berries and produce wine with a rich flavor and deep color. Ramisco, in Portugal, and Corvina and Dolcetto, in Italy, are nationally popular varieties with less international recognition.

14.2.2 French–American Hybrids

French–American hybrids are the second largest group of cultivars, and are derived from crosses between *V. vinifera* and one or more of the following: *V. riparia*, *V. rupestris*, and *V. aestivalis*. Primarily developed in France, their expansion was banned in the European Economic Community (EEC) in the late 1950s, because of their high yield and their non-traditional fragrance. Extensively grown in some South American countries and Asia as well, French–American Hybrids have been widely used by North American wine industries since 1960. These hybrids were developed in order to overcome the difficulties and high expenses of grafting *V. vinifera* onto *Phylloxera* resistant rootstocks. The hybrids obtained are easy to grow with high yielding characteristics, and show reduced sensitivity to most of leaf pathogens. These varieties are used in Europe by the breeding programs as sources of disease resistance. The most popular French–American hybrids are:

Baco Noir, resulting from a cross between Folle Blanche and *V. riparia,* yields wine with a specific flavor and a very good aging potential. This hybrid is unfortunately sensitive to bunch rot, and several soil borne viruses.

Marechal Foch, derived from a cross between *V. riparia* and *V. rupestris* selections, with Goldriesling (Riesling × Courtiller Musqué). This hybrid has characteristics such as early maturation, winter hardiness, high productivity, and resistance to downy mildew, which makes it popular in North America.

Chambourcin is a Johannes Seyve hybrid of unknown parentage, and is presumed to be one of the best French–American hybrids. The cultivar was popular in France in the 1960s and 1970s, and also in Australia, because of its long growing season. Chambourcin is characterized with resistance to downy and powdery mildews and produces wines with a rich flavor.

Vidal Blanc has excellent winemaking and viticultural properties including tough skin and long maturity, making it suitable for the production of high quality ice wines. Vidal Blanc yields wine with Riesling like character, under favorable conditions. It is a cold hardy cultivar, but susceptible to soil borne virus diseases, as are many other French–American hybrids (37).

De Chaunac, once extensively grown in Eastern North America, has lost its importance nowadays, due to its sensitivity to soil borne viruses and the neutral character of the produced wines. De Chaunac is a Seibel cross of unknown parentage.

Seyval Blanc is a Seyve-Villard hybrid of *V. vinifera*, *V. rupestris,* and *V. aestivalis lincecumii*. It is a cold hardy cultivar, which produces wine with a pomade fragrance and fairly bitter finish. Seyval Blanc is sensitive to bunch rot, but tolerant to many soil types.

Delaware, which is a cross between *V. labrusca, V. aestivalis,* and *V. vinifera*, is another important French–American hybrid In the past, it has been used in the production of sparkling wines. Delaware needs well drained soils to grow and tends to crack. It is susceptible to various fungal pathogens, such as *Phylloxera*. Duchess, which produces wine with a mild fruity aroma, is a cross between *Vitis labrusca* and *V. vinifera*, and is difficult to grow. Magnolia and Noble are self fertile strains of muscadine cultivars, which yield sweet bronze colored and dark red fruits, respectively.

14.2.3 American Hybrids

American hybrids are cultivars bred in North America, from hybrids between indigenous grapevines, and hybrids between these and *V. vinifera*, or both. Most important American hybrids are derived from *V. labrusca*. The major characteristics of American hybrids are low sugar content and high acidity, and abundant flavor, e.g., Niagara, Concord and Catawba. A method of reducing the intense labrusca flavor is early picking and cold fermentation. Norton and Cynthiana, derived from *Vitis aestivalis* (38), are grown in Arkansas and Missouri, and are resistant to indigenous diseases and pests. The group of American hybrids derived from *V. rotundifolia* have low sugar content, but are resistant to indigenous diseases, especially to Pierce's disease. American hybrids are extensively grown in Eastern North America, South America, Eastern Europe, and Asia, although their significance has decreased in North America.

Another group of hybrids are interspecific cultivars derived from crossing *V. vinifera* and species such as *V. amurensis*, *V. riparia*, *V. armata,* and *V. rotundifolia*,. Orion, Regent, and Phoenix are new hybrids developed in Germany, with winemaking properties equal or superior to most *V. vinifera*. Veeblanc was developed in Ontario from complex parentage and generates mild quality wine. Cayuga White from New York is a new hybrid with excellent winemaking attributes and resistance to fungal diseases. Its fragrance resembles apples and tropical fruits, with a rich mouth feel. The quality attributes of wine grapes can be improved by genetic engineering and clonal or somaclonal selection.

14.3 GENETIC ENGINEERING OF WINE GRAPES

The goal of grape breeding is to enhance agronomic properties of rootstock varieties and to improve winemaking attributes of fruit-bearing cultivars. There are a number of ways by which this can be accomplished, including simple cross and backcross between appropriate varieties. The improvement often involves the addition of novel qualities such as drought tolerance and disease resistance to established cultivars.

Standard breeding techniques are time consuming and expensive. Genetic engineering introduces the genes of choice without disrupting varietal characteristics of cultivars. The steps involved in genetic engineering include the isolation, amplification, and insertion of the target gene into the intended organism. An example is the insertion of a protein coat gene for the grapevine fanleaf virus (GFLV) into rootstock varieties in order to enhance their resistance (39). There are several laboratories around the world working on gene transfer technologies in grape vines. It has been reported that several transformed grape varieties are being evaluated in France (40) and the U.S. In Geneva, New York, a Thompson Seedless variety with an incorporated gene for resistance to tomato ringspot virus is being evaluated. Experiments in France include evaluation of rootstocks and Chardonnay carrying resistance genes to fanleaf virus; and another variety, Richter 1110, with an incorporated gene for resistance to chrome mosaic virus (CMV) (41). Attempts have also been made to introduce the chitinase gene into embryonic cultures of Merlot, Chardonnay, Pinot Noir, Concord, and Niagara varieties, through biolistic transformation. Increased chitinase activity may afford protection from fungal pathogens, such as *Botrytis*, which causes bunch rot, and *Uncinula necator*, which causes powdery mildew (42). These varieties are being evaluated for increased disease resistance. Superoxide dismutase is a key enzyme involved in the detoxification of active oxygen species and it is believed that a higher superoxide dismutase activity will confer added protection to grapevines during extreme cold conditions. A gene for superoxide dismutase was isolated from *Arabidopsis thaliana* and inserted into scion varieties in order to improve their cold resistance. A problem associated with these techniques is the uncertainty of the transfer and expression of traits under multigene control in another organism. Although genetic engineering requires isolated cells or tissues grown in tissue culture to develop into whole plants, the induction of embryos from grape tissue or cells is difficult. For transformed cells to be of value, they must be able to form embryos and differentiate into whole plants. The phenotypic variability of the progeny is another problem with propagating vines that originate from cells grown in tissue culture. The variability decreases with vine age or repeated propagation. Using buds from the fortieth node and onward from the stem apex, and grafting them onto desirable rootstocks, is a way to circumvent this problem (43). Progress has been made in both augmenting the proportion of embryonic tissue that develops into vines (44), and in understanding the origins of poor plantlet yield from grape cells (45).

14.3.1 Clonal Selection

Clones are forms of cultivars, derived vegetatively from a single parental plant, such that all derivatives are initially genetically identical (1). Clonal selection is a series of procedures designed to provide a premium stock of known characteristics. Basically, cuttings are multiplied and repeatedly assessed for their viticultural and fruit-bearing characteristics and their winemaking properties, as well as their resistance to systemic pathogens. Clonal selection is the primary means by which cultivar characteristics can be modified without significantly changing varietal attributes. Variations in monoclonal varieties consist of mutations accumulated since the origin of the cultivar which are reproduced unchanged in cuttings of the mutated clone, whereas in polyclonal varieties, variation occurs because the cultivar exists as a group of closely related strains. The difference between related clones may be phenotypical, such as color variants of Pinot Noir, Pinot Gris, and Pinot Blanc, or biotypical (almost genotypical), such as Sangiovese strains (46). Clones are grouped according to the combination of traits they possess. For example, among Pinot Noir cultivars, Pinot Fin, with trailing low yielding vines having small fruit clusters; Pinot Droit, having a high yielding vine with upright shoots; Pinot Fructifer, which has high yielding strains; and Mariafeld-type strains, with loose clustered, moderate yielding vines, show the variability in phenotypic characteristics (47). The main objective of clonal selection is the elimination of all systemic infectious organisms including pathogens. This is not always possible, as some systemic agents are not known to cause disease, and reinfection is always possible. The weakness of clonal selection is the resulting phenotypic variability, which is difficult to eliminate. In order to control the desirable variations, some breeders suggest planting several clones rather than just one (48).

Often, clonal instability is due to chimeric mutations, which occur in embryonic cells, and, as a result, cause vine tissues to differ from one another. Another important goal of clonal selection is the improvement in yield and grape quality. There is a negative association between yield and quality, but this can be overcome through rigorous viticultural practices such as more open canopy, increased plant density, and basal leaf removal. An exception to this rule is the Weinberg 29 clone of Riesling, which gives a high yield and shows a high level of sugar content in the fruits (49).

The assessment of quality is complex, time consuming, and costly. There is no unanimous criterion for quality. However, some parameters such as, ° Brix, pH, and acidity are used to assess maturity. Parameters such as the level of glycosyl glucose (G-G) in the must, the polyphenol content, and the color density and hue are commonly used as potential indicators of wine quality (50). The presence of specific flavor impact compounds such as 2-methyl-3-isobutylpyrazine in Cabernet Sauvignon is also another important characteristic. Unique varietal aromas can be integrated into clonal selection. Examples of this kind include the Muscat nuance of Chardonnay 77 and 809; the difference in wine making potential of Gamay 222 versus Gamay 509 in the production of Beaujolais cru vin de garde and vin nouveau (48). Because properties of a clone are dependent upon environmental conditions, some desirable traits may not be expressed, which requires vineyard growers to make their own assessment of a given clone.

14.3.2 Somaclonal Selection

Although clonal selection has reduced viral infection in grapes, more advances may come from incorporation of new genetic information through genetic engineering or induced mutations. Mutations can be created by exposing meristematic tissue or cells in tissue culture to mutagenic chemicals or radiation. Somaclonal selection enhances the expression and selection of clonal variation. It also involves the selective growth enhancement of cell lines. For example, isolation of transgenic vines or lines possessing tolerance for salinity or fungal toxins (51) is done by exposing cells to mutagenic agents, such as chemicals and radiation, during cell culture (39,52). However, the tolerance shown by tissue in culture may not always be expressed at the whole plant level (53).

14.3.3 Biotechnology in Viticulture

Biotechnology has been extensively applied in viticulture since the early work of breeders like F. Baco, B. Seyve, and A. Seibel. The main goal then was to develop cultivars with enhanced wine-producing characteristics of *Vitis vinifera* and the *Phylloxera* resistance of American species. Breeding has focussed on developing rootstocks resistant to drought, salt, lime, viruses, and nematodes. For example, *V. cinerea helleri* can donate resistance to lime induced chlorosis; *V. champinii* can supply tolerance to root knot nematodes; *V. vulpine* can provide drought tolerance under shallow soil conditions. Improvements have also been made in developing new pest and disease resistant fruit varieties. There is also a trend in producing grapevine varieties that produce fruits having increased nutritional value, such as antioxidant and vitamin contents, thereby improving the quality of the grapes and wine. Unfortunately, consumers are still sceptical of genetically modified organisms, which limits their acceptance. The benefits of such resistant varieties are their reduced production costs and use of pesticides. Their disadvantages however, are legal restrictions against interspecies crosses in the production of Appellation Contrôlée wines in most European countries. Recently 40 out of 60 Masters of Wine voted against any introduction of genetically modified (GM) vines or geneticallk modified organisms (GMOs) in any step of the wine making process (54). Despite these controversies, genetic engineering appears to be the most efficient way of successfully developing disease resistant cultivars.

Researchers are experimenting with the isolation and insertion of specific genes for resistance such as genes regulating the activities of chitinase, β-glucanase, and peroxidase isoenzymes.

14.4 GENETIC ENGINEERING OF YEAST FOR FERMENTATION

The two main organisms involved in fermentation are *Saccharomyces cerevisiae* and *Leuconostoc oenos*. Traditionally, indigenous yeasts conduct the fermentation, but induced fermentation with selected yeast strains is a standard procedure in most parts of the world (55). The reason for using specific yeasts is to avoid the production of off flavors sometimes associated with wild yeasts. New techniques such as mitochondrial DNA sequencing (56) and gene marker analysis (57) allow identification of strains responsible for fermentation. Regardless of yeast inoculation, red and white grape musts contain a significant amount of wild yeast, but their metabolic activity during induced fermentation is unknown. Thus, their significance in vinification is not clear. *Kloeckera apiculata* is the most frequently isolated wild yeast in grapes, believed to contribute greatly to the complexity of the wine. Other yeasts occasionally isolated are: *Brettanomyces*, *Candida*, *Debaryomyces*, *Hansenula*, *Kluyveromyces*, *Metschnikowia*, *Nadsonia*, *Pichia*, *Saccharomycodes,* and *Torulopsis* (58). *Saccharomyces cerevisiae* is the most important yeast species, because it may function as wine yeast, baker's yeast, distiller's yeast, and brewer's yeast. The original habitat of the yeast is uncertain (59), and only one healthy grape out of a thousand carries ideal wine yeasts (60). *Saccharomyces paradoxus* (*S. cerevisiae tetrasporus*), isolated from oak tree exudates, is assumed to be the ancestral form of *Saccharomyces cerevisiae*. Although *S. cerevisiae* is sometimes isolated from the intestinal track of fruit flies, the importance of insects in dispersal of *Saccharomyces* is not clear (59,61). *Saccharomyces* species such as *S. uvarum* and *S. bayanus* also can effectively conduct fermentations, and they are used in special applications; e.g., *S. uvarum* ferments well at temperatures down to 6°C and synthesizes desirable sensory components, whereas *S. bayanus* is well adapted for the production of sparkling wines and fine sherries. Other wild yeasts of sensory significance are *Candida stellata*, *Torulopsis delbruekii* and *Kloeckera apiculata. Candida stellata* persists in fermenting juice (62), and produces high concentrations of glycerol, increasing the mouth feel in wine. *Torulopsis delbruekii* positively influences sensory attributes of wine by producing low concentrations of acetic acid and succinic acid. *Kloeckera apiculata*, however, can produce up to 25 times the typical amount of acetic acid produced during induced fermentations (63), and is capable of producing above threshold amounts of 2-aminoacetophenone, giving a naphthalene like off odor to wines (64). *Kloeckera apiculata* may inhibit *S. cerevisiae* from completing fermentation (65). *Kloeckera apiculata* and *C stellata* normally produce only 4 to 10% alcohol, but can survive higher alcohol concentrations (66,67). Most of the other members of indigenous yeasts found on grapes are suppressed by low pH, high alcohol, oxygen deficiency, and sulphur dioxide concentrations. Most bacteria are inhibited by *S. cerevisiae*, except lactic acid bacteria. Commercially available strains of *S. cerevisiae* possess a wide range of winemaking characteristics such as low production of acetic acid, hydrogen sulphide (H_2S), and urea; the ability to produce fruit esters in abundance; and to enhance varietal flavorants. Yeasts can also be selected for their alcohol tolerance, fermentation speed, ability to restart stuck fermentation, production of specific wine styles (such as after carbonic maceration), resistance to killer factors, ability to ferment glucose or fructose selectively, fermentation at high pressure and at low temperature, and their ability to flocculate rapidly and completely after fermentation.

14.4.1 Ideal Yeast

Taking into account that each type of wine requires different types of yeast, it may be difficult to find an ideal type of yeast. However, the following are major characteristics of high quality wine yeasts:

1. Fermentation Speed: An ideal type of yeast should be able to initiate and complete the fermentation as quickly as possible after inoculation. The higher sugar content of the must, which creates an elevated osmotic pressure, is a limiting factor for most wine yeasts.
2. Alcohol tolerance: Yeast should be able to ferment all the sugars in the must in less than 3 weeks and be able to withstand up to 15% ethanol. Most fermentations stop at 1–2% sugar content creating conditions for spoilage. There are commercial strains of yeasts available that have the ability to restart stuck fermentations or to perform the second fermentation during sparkling wine production.
3. SO_2 Tolerance: Most of the yeasts are sensitive to SO_2, and sulphites are used as an antioxidant and a microbial inhibitor. The allowed concentration of SO_2 in the finished product is 70 parts per million in free state, and 350 parts per million in combined state, respectively. An ideal type of yeast should be able to dominate the fermentation even at this level of SO_2.
4. Cold tolerance: Cold tolerant yeasts are used especially for white wine production in order to keep the loss of the bouquet to a minimum. The fermentation temperature is often below 14°C.
5. Low foaming activity: Ideal yeasts should have a low foaming activity. High foaming yeasts spoil the quality of wine by producing off odors and off flavors.
6. Efficient conversion of sugar into alcohol: Ideal yeasts have the ability to convert sugar primarily into alcohol and produce low levels of other metabolites.
7. Production of desirable metabolites: Several secondary products of yeasts are components of the bouquet and flavor of the wine. The quality of a given wine is therefore dependent upon the production of the right amount of these metabolites. This is particularly important for those types of wines with low aging potential such as the Gamay Nouveau wines of Ontario, consumed within three months after production.
8. Low production of undesirable metabolites: The selection of yeasts is based on their ability to synthesize low levels of undesirable metabolites such as acetaldehyde, acetic acid, sulphur dioxide, hydrogen sulphide, and mercaptans.
9. Resistance to "Killer" yeasts: Most of the commercial yeasts are killer strains with the ability to ferment sugars completely and to resist toxins of contaminating killer yeasts. Killer yeasts are wild yeasts that possess the ability to synthesize and excrete proteins that kill other strain of yeasts. This is the reason why the yeast breeders breed yeasts containing the killer genes.
10. Flocculation: An ideal yeast strain should flocculate at the end of the fermentation, leaving the wine clear and requiring less rigorous filtration. Filtration negatively influences the complexity of wine.
11. Producing or reducing certain organic acids: Yeast strains are selected on the basis of their ability to deplete malic acid and other organic acids found in wine. This is very important taking into account the net difference in the acid content of grapes grown in hot and cool climate regions. Hot growing regions are ideal for the production of low acid grapes having high sugar content, whereas cool climate regions tend to produce high acid grapes.

14.4.2 Yeast Breeding and Wine Quality

The use of genetic engineering is very limited for improving wine yeasts, compared to yeasts used for other industrial fermentations. The reason for this is that more importance has been given to grape variety, fruit maturation, and fermentation temperature. Also, during genetic improvement, features controlled by one gene are easily influenced by other genes. For example, the expression of flocculation ability in yeasts is regulated by several genes, including epistatic genes and, possibly, cytoplasmic genetic factors (68). On the other hand, inactivating the gene that encodes sulphites reductase restricts the conversion of sulphites to sulphide (H_2S). Genetic improvement is attainable, although this may take time to achieve because many important enological attributes, such as the ability to ferment at low temperature and alcohol tolerance, are controlled by a combination of several genes. The consequences of changing the direction of a metabolic pathway on other characteristics of the yeasts are not very clear at present. For example increasing ester production could interfere with the alcohol tolerance ability of the strain (69). It is easier when the compound concerned is the end product of a pathway. There were no apparent undesirable side effects when genes for terpene synthesis were incorporated from a lab strain into a wine strain of *S. cerevisiae* (70). Rainieri and coworkers (71) found that thermotolerant *Saccharomyces cerevisiae* strains possess oenological potential and provide an important genetic resource for yeast improvement programs. In the search for a substitute for sulphite in winemaking, Suzzi and coworkers (72) selected a strain of *Saccharomyces cerevisiae* No 10278, able to form 30 to 80mg/L of sulphite, with an excellent stabilizing power. Techniques available for researchers for improving wine yeast qualities include selection and modification of the genetic makeup of the yeast. Simple selection is easy, if a selective medium is developed so that only cells containing the desirable trait can grow; otherwise, cells must be randomly isolated and individually studied for the presence of the desired trait. Modifying the genetic makeup involves procedures such as hybridization, mutagenesis, backcrossing, transformation, somatic fusion, and genetic engineering. To incorporate single traits into a desirable strain, backcross breeding is the preferred technique. Combined with a selection, backcross eliminates undesirable donor genes accidentally incorporated in the original cross.

Somatic fusion is used if a desirable feature missing in *S. cerevisiae* is found in another yeast species. This procedure requires enzymatic cell wall dissolution and mixing of protoplasts generated in the presence of polyethylene glycol, which enhances cell fusion. Instability is the major problem with somatic fusion, because fused cells often revert to one or the other of the original species.

On the other hand, the advantage of genetic transformation is that donor and recipient need not be closely related. The procedure involves bathing yeast protoplasts in a solution containing DNA from the donor organism. By this technique, the malolactic gene from *Lactococcus lactis* was transferred into *S. cerevisiae* along with the malate permease transport gene from *Schizosaccharomyces pombe* (73–76). Other genes of interest are pectic lyase to increase fruity flavor and β- (1-4)- endoglucanase to increase aroma. Boone and coworkers (77) integrated the yeast K1 killer toxin gene into the genome of laboratory and commercial yeasts by a gene replacement technology that generated recombinants containing only yeast DNA. The integration of the K1 killer gene into two K2 wine yeasts generated stable K1/K2 double killer strains, which have a wider spectrum of killing and a competitive advantage over other sensitive and killer strains of *Saccharomyces cerevisiae* in wine fermentations. Genetic engineering is useful if specific traits are desired, but not when multigenic characteristics are involved. However, the ethics of releasing genetically modified organisms that could enter the environment is questionable.

The main problem in breeding *S. cerevisiae* is its diploid character, which means each cell contain two copies of the genes. The diploid nature complicates improvement by masking the potentially desirable genes. More difficulties occur due to the low frequency of sexual reproduction, poor spore germination, and rapid return to the diploid form (75). Stability is the fundamental requirement for the yeast strain, which means that, for instance, flocculant strains should not loose their ability to flocculate.

14.4.3 Wine as a Functional Food

The beneficial effects of wine on human health may originate from their nonalcoholic components, such as flavanoids and phenols. Some phenolic compounds can be protective, whereas others have been found to be mutagenic at high doses in laboratory studies. Fazel and coworkers (78) found that quercetin can induce mutation during *in vitro* culture of animal cells, but is an anticarcinogen in whole animal dietary studies. This anomaly may be due to the differences in the concentrations of quercetin used and the low level of metal ions and free oxygen found in the animal body. Subbaramaiah and coworkers (79) observed that resveratrol, a member of the stilbene family found in wine, can inhibit the production of cyclooxygenase-2, thought to be important in carcinogenesis. As well, flavanols and flavones strongly reduce the action of the common dietary carcinogens such as the heterocyclic amines (80). In this regard, it is important to note that wines made from fruits such as cherry, blackberry and blueberry, and red grape show a very high complement of superoxide and hydroxyl radical scavenging ability due to the presence of several phenolic components (81). The flavanoids are also strong inhibitors of calcium second messenger function in animal systems, and the biological activities of wines may, to a large extent, result from this activity. Recently, during *in vitro* culture, the flavonoid wine components have been shown to be specifically cytostatic and cytotoxic to MCF-7 cells, which are estrogen receptor positive breast cancer cells. Under similar conditions, the normal human mammary epithelial cells were unaffected (82). Thus, by enhancing the components in the wine that afford health beneficial effects, the functional food value of the wine can be enhanced. Even though moderate wine consumption is believed to be beneficial to cardiovascular health, its effect on breast cancer development is still controversial. Several biotechnological approaches such as *in vitro* culture, genetic engineering of grapes for enhanced secondary metabolite biosynthetic pathways, and conditions of fermentation could ultimately enhance the functional food value of the wine.

14.5 CONCLUSIONS

It is possible to improve the quality attributes of wine using modern techniques such as selective breeding, hybridization, somatic fusion, transformation, and genetic engineering. They can be applied both to wine grapes and to yeasts. Although the safety of genetically modified food is an issue, the advantages of adapting these technologies are enormous; as they may help provide better yield, environmental protection and high quality wine products.

REFERENCES

1. Jackson, R.S. *Wine Science: Principles, Practices and Perception.* New York: Academic Press, 2000.
2. McGovern, P.E., D.L. Glusker, L.J. Exner, M.M. Voigt. Neolithic resinated wine. *Nature* 381:480–481, 1996.
3. Petrie, W.M.F. *Social Life in Ancient Egypt*. London: Methuen, 1923.
4. Zohary, D., M. Hopf. *Domestication of Plants in the Old World.* Oxford: Oxford University Press, 1988.

5. Nùñez, D.R., M.J. Walker. A review of paleobotanical findings of early Vitis in the Mediterranean and of the origins of cultivated grape-vines, with special references to prehistoric exploitation in the western Mediterranean. *Rev. Paleobot. Palynol.* 61:205–237, 1989.
6. Stevenson, A.C. Studies in the vegetational history of S.W. Spain, II: palynological investigations at Laguna de los Madres, Spain. *J. Biogeogr.* 12:293–314, 1985.
7. FAO. FAOSTAT database. http://apps1.fao.org/servlet/XteServlet?areas. September 2002.
8. Rimm, E.B., E.I. Giovannuchi, W.C. Willet, G.A. Colditz, A. Ascerio, B. Rosner, M.J. Stampfer. Prospecive study of alcohol consumption and risk of coronary disease in men. *Lancet* 338:464–468, 1991.
9. Soleas, G.J., E.P. Diamandis, D.M. Goldberg. Wine as a biological fluid: history, production and role in disease prevention. *J. Clin. Lab. Anal.* 11:287–313, 1997.
10. Lefort, P.L. Biometrical analysis of must aromagrams: application to grape breeding. In: *Proceedings of the 3rd International Symposium on Grape Breeding*, University of California, Davis, 1980, pp 120–129.
11. Fanizza, G. Multivariate analysis to estimate the genetic diversity of wine grapes (*Vitis vinifera*) for cross breeding in southern Italy. In: *Proceedings of the 3rd International Symposium on Grape Breeding*, University of California, Davis, 1980, pp 105–110.
12. Bowers, J.E., E.B. Bandman, C.P. Meredith. DNA fingerprint characterization of some wine grape cultivars. *Am. J. Enol. Vitic.* 44:266–274, 1993.
13. Bowers, J.E., G.S. Dangl, R. Vignani, C.P. Meredith. Isolation and characterization of new polymorphic simple sequence repeat loci in grape (*Vitis vinifera* L.). *Genome* 39:628–633, 1996.
14. Bowers, J.E., G.S. Dangl, C.P. Meredith. Development and characterization of additional microsatellite DNA markers for grape. *Am. J. Enol. Vitic.* 50(3):243–246, 1999.
15. Sefc, K.M., F. Regner, E. Turetschek, J. Glössl, H. Steinkellner. Identification of microsatellite sequences in *Vitis riparia* and their applicability for genotyping of different *Vitis* species. *Genome* 42:1–7, 1999.
16. Thomas, M.R., N.S. Scott. Microsatellite repeats in grapevine reveal DNA polumorphisms, when analysed as sequence-tagged sites (STSs). *Theor. Appl. Genet.* 86:173–180, 1993.
17. Thomas, M.R., S. Matsumoto, P. Cairn, N.S. Scott. Repetitive DNA of grapevine: classes present and sequences suitable for cultivar identification. *Theor. Appl. Genet.* 86:173–180, 1993.
18. Thomas, M.R., N.S. Scott, R. Botta, J.M.H. Kijas. Sequence-tagged site markers in grapevine and citrus. *J. Jpn. Soc. Hortic. Sci.* 67:1189–1192, 1998.
19. Lefort, F., K.K.A. Roubelakis-Angelakis. Genetic comparison of Greek cultivars of *Vitis vinifera* L. by nuclear microsatellite profiling. *Am. J. Enol. Vitic.* 52:101–108, 2001.
20. Sefc, K.M., S. Guggenberger, F. Regner, C. Lexer, J. Glössl, H. Steinkellner. Genetic analysis of grape berries and raisins with microsatellite markers. *Vitis* 37:123–125, 1998.
21. Silvestroni, O, D. Di Pietro, C. Intrieri, R. Vignani, I. Filipetti, C. Del Casino, M. Scali, M. Cresti. Detection of genetic diversity among clones of cv. Fortana (*Vitis vinifera* L.) by microsatellite DNA polymorphism analysis. *Vitis* 36:147–150, 1997.
22. Vignani, R., J.E. Bowers, C.P. Meredith. Microsatellite DNA polymorphism analysis of clones of *Vitis vinifera* 'Sangiovese.' *Scientia. Hort.* 65:163–169, 1996.
23. Botta, R., N.S. Scott, I. Eynard, M.R. Thomas. Evaluation of microsatellite sequence-tagged site markers for characterizing *Vitis vinifera* cultivars. *Vitis* 34:99–102, 1995.
24. Wilson, E.K. So many grapes, so little time. *Chem. Eng. News* 79:37–39, 2001.
25. Cipriani, G., G. Frazza, E. Peterlunger, R. Testolin. Grapevine fingerprinting using microsatellite repeats. *Vitis* 33:211–215, 1994.
26. Grando, M.S., U. Malossini, I. Roncador, F. Mattivi. Parentage analysis and characterization of some Italian *Vitis vinifera* crosses. *Acta Hort.* 528:183–187, 2000.
27. Sefc, K.M., H. Steinkellner, H.W. Wagner, J. Glössl, F. Regner. Application of microsatellite markers for parentage studies in grapevine. *Vitis* 36:79–183, 1997.
28. Sefc, K.M., F. Regner, J. Glössl, H. Steinkellner. Genotyping of grapevine and rootstock using microsatellite markers. *Vitis* 37:15–20, 1998.
29. Sefc, K.M., H. Steinkellner, J. Glössl, S. Kampfer, F. Regner. Reconstruction of grapevine pedigree by microsatellite analysis. *Theor. Appl. Genet.* 97:227–231, 1998.

30. Thomas, M.R., P. Cairn, N.S. Scott. DNA typing of grapevines: a universal methodology and database for describing cultivars and evaluating genetic relatedness. *Plant Mol. Biol.* 25:939–949, 1994.
31. Alessandri, S., N. Vignozzi, A.M. Vignini. AmpeloCADs (Ampelographic Computer-Aided Digitizing System): an integrated system to digitize, file, and process biometrical data from *Vitis*. leaves. *Am. J. Enol. Vitic.* 47:257–267, 1996.
32. Patel, G.I., H.P. Olmo. Cytogenetics of Vitis, I: The hybrid *V. vinifera* x *V. rotundifolia. Am. J. Bot.* 42:141–159, 1955.
33. Okamoto, G., Y. Fujii, K. Hirano, A. Tai, A. Kobayashi. Pollen tube growth inhibitors from Pione grape pistils. *Am. J. Enol. Vitic.* 46:17–21, 1995.
34. Sefc, K.M., H. Steinkellner, H.W. Wagner, J. Glossl, F. Regner. Application of microsatellite markers to parentage studies in grapevine. *Vitis* 36:179–184, 1997.
35. Bowers, J.E., C.P. Meredith. The parentage of a classic wine grape, Cabernet Sauvignon. *Nat. Genet.* 16:84–87, 1997.
36. Regner, F., A. Stadlbauer, C. Eisenheld, H. Kaserer. Genetic relationships among Pinots and related Cultivars. *Am. J. Enol. Vitic.* 51:7–14, 2000.
37. Alleweldt, G. Disease resistant varieties. In: *Proceedings of the 8th Australian Wine Industry Technology Conference*, Stockley, C.S., et al., eds., 1993, pp 116–119.
38. Reisch, I.B., R.N. Goodman, M.H. Martens, N.F. Weeden. The relationships between Norton and Cynthians, red wine cultivars derived from *Vitis aestivalis. Am. J. Enol. Vitic.* 44:441–444, 1993.
39. Mauro, M.C., S. Toutain, B. Walter, L. Pink, L. Otten, P. Coutos-Thevenot, A. Deloire, P. Barvier. High efficiency regeneration of grapevine plants transformed with the GFLV coat protein gene. *Plant Sci.* 112:97–106, 1995.
40. Carbonneau, A. The early selection of grapevine rootstocks for resistance to drought conditions. *Am. J. Enol. Vitic.* 36:195–198, 1985.
41. Kikkert, J.R., B.I. Reisch. Genetic engineering of grapevines for improved disease resistance. In: *Grape Research News*, 7, Goffinet, M., ed., Geneva, New York: New York State Agricultural Experiment Station, 1996.
42 Kikkert, J.R., G.S. Ali, M.J. Striem, M.-H. Martens, P.G. Wallace, L. Molino, B.I. Reisch. Genetic engineering of grapevine (*Vitis* sp) for enhancement of disease resistance. *Acta Hort.* 447:273–279, 1997.
43. Grenan, S. Multiplication *in vitro* et charactéristiques juveniles de la vigne. *Bull. O.I.V.* 67:5–14, 1994.
44. Reustle, G., M. Harst, G. Alleweldt. Plant regeneration of grapevine (*Vitis* sp.) protoplast isolated from embryogenic tissue. *Plant Cell Rep.* 15:238–241, 1995.
45. Maes, O., P. Coutos-Thevenot, T. Jouennem, M. Boulay, J. Guern. Influence of extracellular proteins, proteases and protease inhibitors on grapevine somatic embryogenesis. *Plant Cell. Tissue Org. Cult.* 50:97–105, 1997.
46. Calõ, A., A. Costacurta, G. Paludetti, M. Crespan, M. Giust, E. Egger, A. Grasselli, P. Storchi, D. Borsa, R. di Stephano. Characterization of biotypes of Sangiovese as a basis for clonal selection. In: *Proceedings of the International Symposium on Clonal Selection*, Rantz, J.M., ed., Davis, CA: American Society of Enology Viticulture, 1995, pp 99–104.
47. Wolpert, J.A. An overview of Pinot noir closes tested at UC, Davis. *Vineyard Winery Manage.* 21:18–21, 1995.
48. Boidron, R. Clonal selection in France: methods, organization, and use. In: *Proceedings of the International Symposium on Clonal selection*, Rantz, J.M., ed., Davis, CA: American Society of Enology Viticulture, 1995, pp 1–7.
49. Schöfflinger, H., G. Stellmach. Clone selection of grapevine varieties in Germany. *Fruit Var. J.* 50:235–247, 1996.
50. Somers, C. *The Wine Spectrum.* Adelaide, Australia: Winetitles, 1998.
51. Soulie, O., J.-P. Roustan, J. Fallot. Early *in vitro* selection of Eupypine-tolerant plantlets. Application to screening of *Vitis vinifera* cv. Ugni blanc somaclones. *Vitis* 32:243–244, 1993.

52. Kikkert, J.R., D. Hébért-Soulé., P.G. Wallace, M.J. Striem, B.I. Reisch. Transgenic plantlets of 'Chancellor' grapevine (*vitis* sp.) from biolistic transformation of embryonic cell suspensions. *Plant Cell Rep.* 15:311–316, 1996.
53. Lebrun, L., K. Rajasekaran, M.G. Mullins. Selection *in vitro* for NaCl-tolerance in *Vitis rupestris* Scheele. *Ann. Bot.* 56:733–739, 1985.
54. http://www.plant.uoguelph.ca/safefood/archives/agnet-archives.htm. Visited on July 29, 2002.
55. Barre, P., F. Vezinhet. Evolution towards fermentation with pure culture yeasts in wine making. *Microbiol. Sci.* 1:159–163, 1984.
56. Dubourdieu, D., A. Sokol, J.J. Zucca, P. Thalouarn, A. Dattee, M. Aigle. Identification des sources de levures isolées de vins par l'analyse de leur AND mitochondrial. *Connaiss. Vigne Vin.* 21:267–278, 1987.
57. Petering, J.E., P.A. Henschke, P. Langridge. The *Escherichia coli* β-glucuronidase gene as a marker for *Saccharomyces* yeast strain identification. *Am. J. Enol. Vitic.* 42:6–12, 1991.
58. Lafont-Lafourcade, S. Wine and Brandy. In: *Biotechnology Vol.5, Food and Feed Production with Microorganisms*, Reed, G., ed., Weinheim; Deerfield Beach, Florida; Basel: Verlag Chemie, 1983, pp 81–163. ISBN-0-89573-045-6.
59. Phaff, H.J. Ecology of yeast with actual and potential value in biotechnology. *Microb. Ecol.* 12:31–42, 1986.
60. Mortimer, R., M. Polsinelli. On the origin of wine yeast. *Res. Microbiol.* 150:199–204, 1999.
61. Wolf, E., I. Brenda. Qualität und resistenz, III. Das Futterwahlvermögen von *Drosophila melanogaster* gegenüber natürlichen Weinhefe-Arten und - Rassen. *Biol. Zentralbl.* 84:1–8, 1965.
62. Fleet, G.H., S. Lafon-Laforcade, P. Ribéreau-Gayon. Evolution of yeasts and lactic acid bacteria during fermentation and storage of Bordeaux wines. *Appl. Environ. Microbiol.* 48:1034–1038, 1984.
63. Ciani, M., F. Maccarelli. Oenological properties of non-*Saccharomyces* yeasts associated with winemaking. *World J. Microbiol. Biotechnol.* 14:199–203, 1998.
64. Sponholz, W.R., T. Hühn. Aging of wine: 1, 1, 6-Trimethyl-1, 2-dihycromaphthalene (TDN) and 2-aminoacetophenone. In: *Proceedings of the 4th International Symposium on Cool Climate Enology and Viticulture*, Henick-Kling, T., et al., eds., Geneva, New York: New York State Agricultural Experimental Station, VI:37–57, 1996.
65. Velázquez, J.B., E. Longo, C. Sieiro, J. Cansado, P. Calo, T.G. Villa. Improvement of the alcoholic fermentation of grape juice with mixed cultures of Saccharomyces cerevisiae wild strains; negative effects of *Kloeckera apiculata. World J. Microbiol. Biotechnol.* 7:485–489, 1991.
66. Gao, C., G.H. Fleet. The effects of temperature and pH on the ethanol tolerance of the wine yeasts, *Saccharomyces cerevisiae*, *Candida stellata* and *Kloeckera apiculata. J. Appl. Bacteriol.* 65:405–410, 1988.
67. Heard, G.M., G.H. Fleet. The effects of temperature and pH on the growth of yeast species during the fermentation of grape juice. *J. Appl. Bacteriol.* 65:23–28, 1988.
68. Teunissen, A.W.R.H., H.Y. Steensma. The dominant flocculation gene of *Saccharomyces cerevisiae* constitutes a new subtelomeric gene family. *Yeast* 11:1001–1013, 1995.
69. Guerzoni, M.E., R. Marchetti, P. Giudici. Modifications des composants aromatiques des vins obtenus par fermentation avec des mutants de *Saccharomyces cerevisiae. Bull. O.I.V.* 58:230–233, 1985.
70. Javelot, C., P. Girard, B. Colonna-Ceccaldi, B. Valdescu. Introduction of terpene productionability to a wine strain of *Saccharomyces cerevisiae. J. Biotechnol.* 21:239–252, 1991.
71. Rainieri, S., C. Zambonelli, V. Tini, L. Castellari, P. Giudici. The enological traits of thermotolerant *Saccharomyces* strains. *Am. J. Enol. Vitic.* 49:319–324, 1998.
72. Suzzi, G., P. Romano, C. Zambonelli. *Saccharomyces* strain selection in minimizing SO2, requirement during vinification. *Am. J. Enol. Vitic.* 36:199–202, 1985.
73. Bony, M., F. Bidart, C. Camarasa, L. Ansanay, L. Dulau, P. Barre, S. Dequin, Metabolic analysis of *Saccharomyces cerevisiae* strains engineered for malolactic fermentation. *FEBS Lett.* 410:452–456, 1997.

74. Volschenk, H., M. Viljoen, J. Grobler, F. Bauer, A. Lonvaud-Funel, M. Denayrolles, R.E. Subden, H.J.J. Van Vuuren. Malolactic fermentation in grape musts by a genetically engineered strain of *Saccharomyces cerevisiae*. *Am. J. Enol. Vitic.* 48:193–197, 1997.
75. Dharmadhikari, M.R., K.L. Wilker. Deacidification of high malate must with *Schizosac-charomyces pombe*. *Am. J. Enol. Vitic.* 49:408–412, 1998.
76. Dequin, S., E. Baptista, P. Barre. Acidification of grape musts by *Saccharomyces cerevisiae* wine yeast strains, genetically engineered to produce lactic acid. *Am. J. Enol. Vitic.* 50:45–56, 1999.
77. Boone, C., A.M. Sdicu, J. Wagner, R. Degré, C. Sanchez, H. Bussey. Integration of the yeast K1 killer toxin gene into the genome of marked wine yeasts and its effects on vinification. *Am. J. Enol. Vitic.* 41:3742–3748, 1990.
78. Fazel, F., A. Rahman, I. Greensill, K. Ainley, S.M. Hasi, J.H. Parish. Strand scission in DNA by quercetin and Cu(II): identification of free radical intermediates and biological consequences of scission. *Carcinogenesis* 11:2005–2008, 1990.
79. Subbaramaiah, K., P. Michaluart, W.J. Chung, A.J. Danneberg. Resveratrol inhibits the expression of cyclooxygenase-2 in human mammarian and oral epithelial cells. *Pharmaceut. Biol.* 36:35–43, 1998.
80. Kazanawa, K., T. Yamashita, H. Ashida, G. Anno. Antimutagenecity of flavones and flavonols to heterocyclic amines by specific and strong inhibition of the cytochrome P450 1A family. *Biosci. Biotechnol. Biochem.* 62:970–977, 1998.
81. Pinhero, R.G., G. Paliyath. Antioxidant and calmodulin inhibitory activities of phenolic components in fruit wines and its biotechnological implications. *Food Biotechnol.* 15:179–192, 2001.
82. Hakimuddin, F., G. Paliyath, K. Meckling. Selective cytotoxicity of a red grape wine flavonoid fraction against MCF-7 cells. *Breast Canc. Res. Treatment*. 85: 65–79, 2004.

2.15

Biotechnology of Nonnutritive Sweeteners

Reena Randhir and Kalidas Shetty

CONTENTS

15.1 INTRODUCTION

Consumer preference for sweetness in a diet with reduced calories has created a marketplace for sweeteners with few or no calories. Sweeteners are broadly classified as providing energy (nutritive) or not providing energy (nonnutritive). Nutritive sweeteners (e.g., sucrose, fructose) are generally recognized as safe (GRAS) by the Food and Drug Administration (FDA). The Code of Federal Regulations states that nonnutritive sweeteners are "substances having less than 2 percent of the calorie value of sucrose per equivalent unit of sweetening capacity" (1,2). By increasing the palatability of nutrient dense foods and beverages, sweeteners can promote health. Scientific evidence supports neither that intake of nutritive sweeteners by themselves increase the risk of obesity, nor that nutritive or nonnutritive sweeteners cause behavioral disorders. However, nutritive sweeteners increase risk of dental caries. High fructose intakes may cause hypertriglyceridemia and gastrointestinal symptoms in susceptible individuals (1,2).

Nonnutritive sweeteners are used to:

1. Expand food and beverage choices for those who want to control calorie, carbohydrate, or specific sugar intake
2. Assist weight control or reduction
3. Aid in the management of diabetes and hyperlipemia
4. Assist the control of dental caries
5. Enhance the usability of pharmaceuticals and cosmetics
6. Provide sweetness when sugar supply is limited
7. Assist the cost effective use of limited resources

These concerns are shaping the need for manufacturers and food and beverage processors to reconsider what additives they use in their products, and what alterations need to be made to meet consumer demand for something sweet, low calorie, and affordable (1,2). The search for noncarbohydrate nonnutritive sweeteners from natural sources has led to the discovery of many intensely sweet tasting substances. The characteristics of an ideal nonnutritive sweetener are:

1. A sweetness quality and profile identical to that of sucrose
2. Sensory and chemical stability under the relevant food processing and storage conditions
3. Compatibility with other food ingredients and stability toward other constituents in the food
4. Complete safety, shown as freedom from toxic, allergenic and other physiological properties
5. Complete freedom from metabolism in the body
6. High specific sweetness intensity (2)

Most sweet compounds, including all popular sweeteners, are small molecular weight compounds of different chemical nature, but there are also sweet macromolecules, both synthetic and natural, such as sweet proteins. Sweet molecules elicit their taste, in humans and other mammals, by interacting with the recently discovered T1R2–T1R3 taste receptors (3–5). The sequence of this protein indicates that it is a metabopromic 7 transmembrane G-protein coupled receptor with a high homology to the mGluR subtype 1 (4). The structure of the N-terminal part of the mGluR has been recently determined by x-ray diffraction and has been used as a template to build a homodimeric T1R3–T1R3 receptor model (4). It is very likely that small molecular weight sweet molecules occupy a pocket analogous to the glutamate pockets in the mGluR (5), possibly similar to the active site models predicted by indirect receptor mapping studies (6–8). These significant investigations have facilitated the understanding of sweetness perception in humans; and have also made it possible to modify genetically the sequence of the natural sweet proteins as necessary. The following review is a brief summary of the current nonnutritive sweeteners, namely Saccaharin, Acesulfame K, Sucralose, Aspartame, Neotame, Stevioside, Monellin, Brazzein, Thaumatin, and Manbinlin, which, although they are permitted food ingredients under different national statutory regulations, are not universally accepted. The evaluation of the sweetness of a given substance in relation to sucrose is made on a weight basis. Table 15.1 provides the relative sweetness of the nonnutritive sweeteners discussed in this chapter.

Table 15.1
Relative sweetness of nonnutritive sweeteners

Nonnutritive Sweetener	Approximate Sweetness (Sucrose = 1)
Saccharin	300
Acesulfame K	200
Sucralose	600
Aspartame	180–200
Neotame	8000
Stevioside	200
Monellin	1500–2000
Brazzein	2000
Thaumatin	1600
Mabinlin	100

15.2 SACCHARIN

Saccharin, commonly known as Sweet n Low™, was first discovered in 1879 by Remsen and Fahlberg and has been used commercially since then (9). It is 300 times as sweet as sucrose, produces no glycemic response in humans, synergizes the sweetening power of nutritive and nonnutritive sweeteners, and its sweetness is thermostable. It occurs as a white crystalline powder with a molecular formula of $C_7H_5NO_3S$ and molecular weight of 183.18 (Figure 15.1). It has been widely used to sweeten foods and beverages without calories or carbohydrates for over a century (10,11). It was widely used during the sugar shortages of the two world wars. Research also has shown that saccharin is beneficial to people with diabetes and obese people, and helps to reduce dental cavities (12). It is used in products such as soft drinks, tabletop sweeteners, baked goods, jams, chewing gum, canned fruit, candy, dessert toppings, and salad dressings. It is also used in cosmetic products, vitamins, and pharmaceuticals (10).

Although extensive research indicated that saccharin was safe for human consumption, there has been controversy over its safety. The basis for the controversy rests primarily on findings of bladder tumors in some male rats when fed high doses of sodium saccharin (13–18). Other saccharin research, however, indicates safety at human levels of consumption with no detectable metabolism using analytical techniques. The vast majority of the data on the biotransformation of saccharin demonstrate that this compound is excreted unchanged, predominantly in the urine of all lab animals tested (20–22). The production of saccharin is simple, inexpensive, and can be done in bulk. The production process uses the basic route described by Remsen and Fahlberg. Methyl anthranilate is diazotized by treatment with sodium nitrite and hydrochloric acid to form 2-carbomethoxybenzenediazonium chloride. Sulfonation of this compound produces 2-carbomethoxybenzenesulfinic acid, which is converted to 2-carbomethoxybenzenesulfonyl chloride with chlorine. Saccharin is synthesized by amidation of this sulfonychloride, followed by acidification. This is then treated with either sodium hydroxide or sodium bicarbonate to produce sodium saccharin (23).

15.3 ACESULFAME-K

Acesulfame-K, commonly known as Sunette™ or Sweet One™, was discovered by Clauss and Jensen, 1967 (24). It was approved by the FDA as tabletop sweetener and as an additive in a variety of desserts, confections, and alcoholic beverages. It is 200 times sweeter

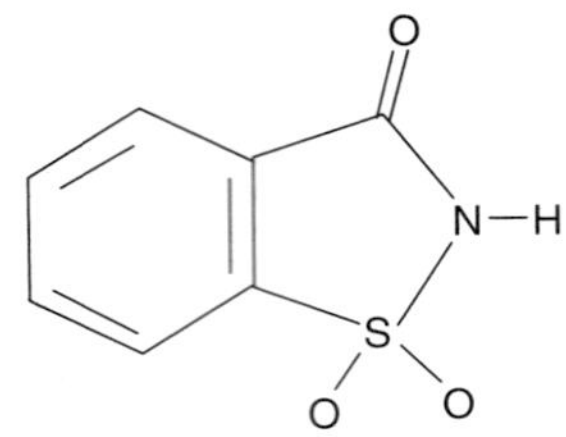

Figure 15.1 Chemical structure of saccharin

than sucrose, is noncarcinogenic and produces no glycemic response (25). Its chemical name is potassium salt of 5,6-methyl-1,2,3-oxathiazin-4(3H)-one-2,2-dioxide and its chemical formula is $C_4H_4KNO_4S$ (Figure 15.2). Its sweetening power is not reduced by heating, and can synergize the sweetening power of other nutritive and nonnutritive sweeteners. It does not provide any energy, is not metabolized in the body, and is excreted unchanged (12,26,27). However, toxicology research performed on albino mice for the genotoxic and clastogenic potential of acesulfame-K revealed chromosome aberrations in the bone marrow cells. In view of this significant *in vivo* mammalian genotoxicity data, it is advised that acesulfame-K should be consumed with caution (28).

Synthesis of acesulfame-K starts with the reaction of acetoacetic acid tert-butyl ester with fluorosulfonyl isocyanate (29). Both compounds form the intermediate of α–N-fluorosulfonylcarbamoyl acetoacetic acid tert-butyl ester. This compound is unstable, and by releasing CO_2 and isobutene is converted to N-fluorosulfonyl acetoacetic acid amides (30). This reaction can be accelerated by heating the compound slightly. In the presense of potassium hydroxide it cyclizes to the dihydrooxathiazinone dioxide ring system by separating out fluorides. Because these are highly acidic compounds, salts of this ring compound are formed in reactions with KOH, NaOH or $Ca(OH)_2$ (30).

15.4 SUCRALOSE

Sucralose is also known by the brand name Splenda™ and is 600 times sweeter than sucrose. It is the only low calorie sweetener made from sugar, with a chemical formula of $C_{12}H_{19}C_{13}O_8$ and molecular weight of 397.64 (Figure 15.3). It was discovered in 1976 and scientific studies conducted over a 20 year period have conclusively determined that sucralose is safe for everyone to consume (31,32). It was approved by the FDA, and by the Joint FAO/WHO Expert Committee on Food Additives (JECFA) in 1990. Sucralose has been approved by prominent regulatory authorities throughout the world, and has been consumed by millions of people internationally since 1991.

It can be used in place of sugar to eliminate or reduce calories in a wide variety of products, including beverages, baked goods, desserts, dairy products, canned fruits, syrups, and condiments. Heating or baking does not reduce its sweetening power. It has no calories, and the body does not recognize it as a carbohydrate (33). Clinical research showed that it produced no glycemic response in the human body; approximately 15% of sucralose was passively absorbed in the body, and the majority was excreted unchanged (34–36). The small amount that was passively absorbed was not metabolized and was eliminated within 24 hours. The FDA concluded that it does not pose a carcinogenic, reproductive, or neurological risk to humans, and this is supported by clinical research (35,37). However a few controversial investigations indicate the potential hazards of sucralose on human health (38).

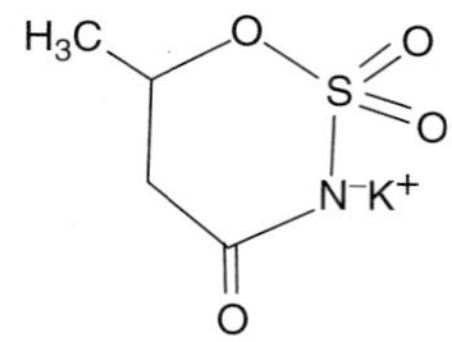

Figure 15.2 Chemical structure of Acesulfame-K

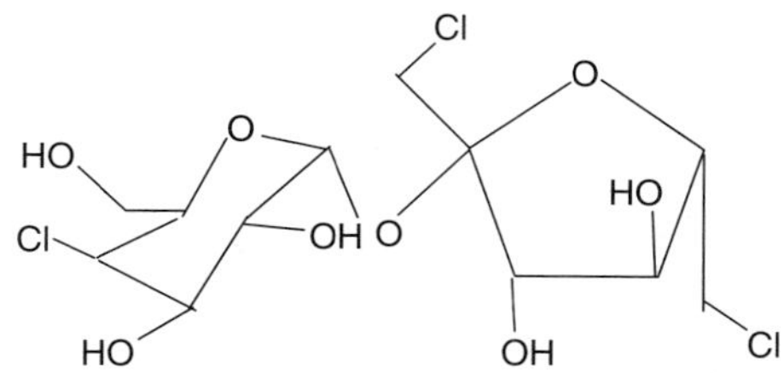

Figure 15.3 Chemical structure of Sucralose

Production of sucralose starts with a cane sugar molecule, then three hydrogen-oxygen groups are substituted with three tightly bound chlorine atoms, which makes it inert. A high yielding bioorganic synthesis of sucralose (4,1′,6′-trichloro-4,1′,6′-trideoxy-galactosucrose) involves the chemical chlorination of raffinose to form a novel tetrachloroRaffinose intermediate (6,4′,1″,6″-tetracyhloro-6,4′,1″,6″-tetradeoxygalactoraffinose; TCR) followed by the enzymatic hydrolysis of the α-1-6 glycosidic bond of the TCR to give sucralose and 6-chlorogalactose. The most active enzyme was produced by a strain of *Mortierella vinacea* and had a maximum rate of 118 μmol sucralose/g dry weight cells/hour, which was approximately 5% of the activity toward raffinose, and a Km of 5.8 mM toward TCR. The enzyme was used in the form of mycelial pellets in a continuous packed bed column reactor. Synthesis of raffinose was achieved from saturated aqueous solutions of galactose and sucrose using a selected α-galactosidase from *Aspergillus niger*. (39).

15.5 ASPARTAME

Aspartame was discovered by accident by James Schlatter, a chemist at G.D Searle Co. in 1965, when he was testing an antiulcer drug. Its chemical name is N-L- aspartyl-l-phenylalanine-1-methyl ester and marketed as Nutrasweet™, Equal™, Spoonful™, and Equal Measure™. It is an odorless, white crystalline powder that has a sweet taste (40,41). It is widely used as a flavor enhancer, and to sweeten foods and beverages (42). It provides the same energy as any protein (4 calories per gram) because it is a combination of phenylalanine and aspartic acid (two amino acids), which is then combined with methanol (Figure 15.4). It is 180–200 times sweeter than sucrose, so the small amount needed to sweeten products does not actually contribute a significant number of calories (40,41). Clinical studies on the safety of aspartame have been widely conducted on animals and humans (43,44). The data demonstrated a substantial margin of safety for aspartame and its metabolites at reasonable levels of consumption. However, the product is required to display a warning label about the possibility of Phenylketonuria, a genetic disease where the body cannot produce the enzyme necessary to use the amino acid phenylalanine, which is one of the breakdown metabolites of aspartame (45,46).

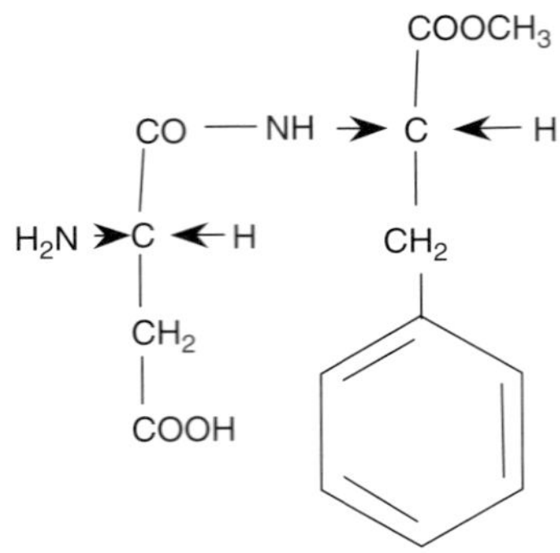

Figure 15.4 Chemical structure of Aspartame

The two primary components of aspartame (phenylalanine and aspartic acid) are chiral, which means that they have two isomers that are mirror images that cannot be superimposed upon each other. This means that the final aspartame molecule will have two stereogenic centers. If the wrong enantiomers are used, the aspartame molecule will not have the correct shape to fit the binding site of the sweetness receptors on the tongue (47). In the synthesis of aspartame, the starting materials are a racemic mixture (equal quantities of both enantiomers) of phenylalanine and aspartic acid. Only the L-enantiomer of phenylalanine is used: this is separated from the racemate by reacting it with acetic anhydride and sodium hydroxide. If the product of this reaction is then treated with the enzyme porcine kidney acylase, and an organic extraction with H^+ carried out, the L-enantiomer is found in the aqueous layer and the D-enantiomer remains in the organic layer (48,49).

Treatment of L-phenylalanine with methanol and hydrochloric acid esterifies the -CO_2H group, and this ester is then reacted with aspartic acid to give the final product. It is important that the amine group on aspartic acid be protected with carbobenzyloxy groups, and the acid group nearest the amine protected with benzyl groups, to prevent the L-phenylalanine reacting with these and giving unwanted byproducts. The acid group that is required to react is activated with Castro's reagent. Castro's reagent is displaced as L-phenylalanine is added, but the protective groups must be removed after the reaction. Carbobenzyloxy is removed by a reaction with hydrogen and platinum(IV) oxide with methanol and chloroform; benzyl is removed by reaction with hydrogen, palladium, and carbon, plus methanol and chloroform, completing the aspartame synthesis (48,49).

Another method for the synthesis of this high intensity sweetener is a very simple example of how proteases may be used in peptide synthesis. Most proteases show specificity in their cleavage sites, and may be used to synthesise specific peptide linkages (50–53). Aspartame is the dipeptide of L-aspartic acid with the methyl ester of L-phenylalanine (L-aspartyl-L-phenylalanyl-O-methyl ester). The chemical synthesis of aspartame requires protection of both the carboxyl group and the amino group of the L-aspartic acid. Even then, it produces aspartame in low yield and at high cost. If the carboxyl group is not protected, a cost saving is achieved, but about 30% of the isomer is formed and must subsequently be removed. When thermolysin is used to catalyze aspartame production, the regiospecificity of the enzyme eliminates the need to protect this carboxyl group, but the amino group must still be protected (usually by means of reaction with benzyl chloroformate to form the benzyloxycarbonyl derivative, i.e., BOC-L-aspartic acid) to prevent the synthesis of poly-L-aspartic acid. More economical racemic amino acids can also be used, as only the desired isomer of aspartame will be formed.

If stoichiometric quantities of L-aspartic acid and L-phenylalanine methyl ester are reacted in the presence of thermolysin, an equilibrium reaction mixture is produced giving relatively small yields of aspartame. However, if two equivalents of the phenylalanine methyl ester are used, an insoluble addition complex forms in high yield at concentrations above 1 M. The loss of product from the liquid phase due to this precipitation greatly increases the overall yield of this process. Later, aspartame may be released from this adduct by simply altering the pH. The stereospecificity of the thermolysin determines that only the L-isomer of phenylalanine methyl ester reacts but the addition product is formed equally well from both the D- and L-isomers. This allows the use of racemic phenylalanine methyl ester, the L-isomer being converted to the aspartame derivative and the D-isomer forming the insoluble complex shifting the equilibrium to product formation. D-phenylalanine ethyl ester released from the addition complex may be isomerised enzymically to reform the racemic mixture. The BOC-aspartame may be deprotected by a simple hydrogenation process to form aspartame (51).

15.6 NEOTAME

Neotame is the FDA approved sweetener developed by Monsanto Chemical Corporation, and was discovered by Nofri and Tinti. It is reported to be approximately 8,000 times sweeter than sugar. Neotame has approximately 40 times the sweetness potency of aspartame (54). Its chemical name is N-[N-(3,3-dimethylbutyl)-L-aspartyl]-L-phenylalanine 1-methyl ester (Figure 15.5). It was designed to overcome some of the problems with aspartame. It is more thermostable and does not break down during processing, which is a major drawback of aspartame. It is suggested for use as a tabletop sweetener, to sweeten frozen desserts, chewing gum and candy, baked goods, fruit spreads, and readymade cereals (55,56).

Neotame is a derivative of aspartame dipeptide made of amino acids: aspartic acid and phenylalanine (54). It is quickly metabolized and fully eliminated through normal biological processes. The dimethylbutyl part of the molecule is added to block the action of peptidases which are enzymes that break the peptide bond between the aspartic acid and phenylalanine. This reduces the availability of phenylalanine, eliminating the need for a warning on labels directed at people with phenylketonuria, who cannot properly metabolize phenylalanine (57,58).

15.7 STEVIOSIDE

Stevioside is a nonnutritive sweetener extracted from the leaves of the plant *Stevia rebaudiana*, which belongs to the Asteraceae family and was rediscovered by Bertoni in 1888 (59–61) (Figure 15.6). It is native to Brazil, Venezuela, Colombia, and Paraguay where the native Guarani tribes have used *caa-ehe* (stevia) for over 1500 years to sweeten otherwise unpalatable medicinal drinks. It is a herbaceous perennial, which is normally used as a natural herbal sweetener (59). The glycosides are found between the veins of the leaves which can be up to 200 times sweeter than sucrose. The sweetness is mainly attributed to the two compounds, namely stevioside (3–10% of dry leaf weight) and rebaudioside A (1–3%) (62–66). Research conducted at the College of Pharmacy at the University of Illinois found that when a bacterium was exposed to stevioside, the DNA of the bacteria was genetically altered (67). Mutagenic effects of steviol, the aglycone of stevioside, were

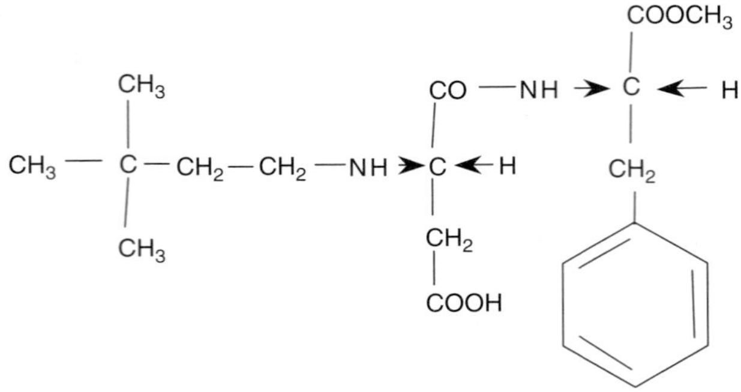

Figure 15.5 Chemical structure of Neotame

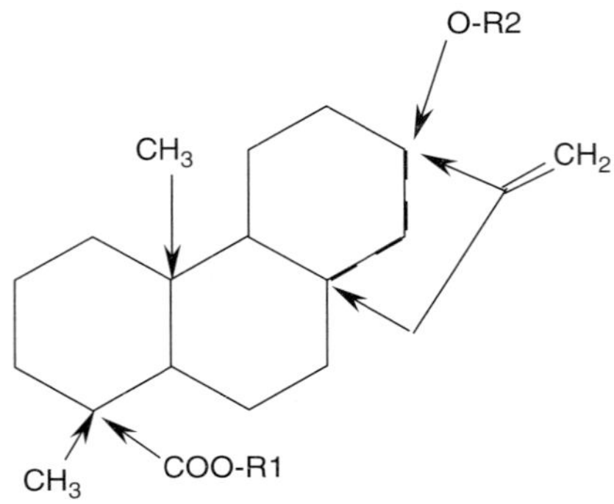

Figure 15.6 Chemical structure of Stevioside

also reported in *Salmonella typhimurium* TM677. After metabolic activation, it was shown that so far unknown steviol metabolites caused mutations in *Salmonella typhimurium* TM677, includiing transitions, transversions, duplications, and deletions at the guanine phosphoribosyltransferase (*gpt*) gene (67,68). Stevia, or stevioside, has not been granted GRAS status by the FDA, but the Dietary Supplement Act of 1994 allows stevia to be sold in the US as a dietary supplement. However, it has been used as a sweetener in South America for centuries, and in Japan for over 30 years.

The other uses for the plant and its extracts are in weight loss programs, because of its ability to reduce the cravings for sweet and fatty foods. The plant has also been used to treat diseases such as diabetes, hypoglycemia, candidiasis, high blood pressure, skin abrasions, and for inhibiting growth and reproduction of bacterialike plaque (69). The advantages of stevioside as a dietary supplement for human subjects are manifold: it is stable, it is noncalorific, it supports good dental health by reducing the intake of sugar, and opens the possibility for safe use by diabetic and phenylketonuria patients, and by obese persons. Stevioside has a few advantages over artificial sweeteners, in that it is stable at high temperatures (100°C) and a wide pH range (3–9), and does not darken with cooking. It is suggested in the use of sweetening soft drinks, ice cream, cookies, pickles, chewing gum, tea, and skin care products.

The biosynthesis of stevioside is via the recently discovered 2-C-Methyl-D-erythritol-4-phosphate pathway (MEP) due to the *ent*-kaurene skeleton of stevioside (69). The genes of the enzymes catalyzing the first two steps of this pathway, 1-deoxy-D-xylulose-5-phosphate synthase (DXS) and 1-deoxy-D-xylulose-5-phosphate reductoisomerase

(DXR) were cloned using reverse transcriptase polymerase chain reaction (PCR). The DXS and DXR from *Stevia* both contain an N-terminal plastid targeting sequence and show high homology to other known plant DXS and DXR enzymes. Furthermore, it was demonstrated through heterologous expression in *Escherichia coli* that the cloned cDNAs encode these functional proteins (70–73).

15.8 MONELLIN

Monellin is a sweet tasting protein isolated from the tropical African fruit *Dioscoreophyllum cumminsii*. Inglett and May first reported the isolation of a sweet substance from the berries in 1967. The unusual protein possesses the interesting property of having a very high specificity for the sweet receptors in the tongue (74). Monellin is made up of two dissimilar polypeptide chains, which are noncovalently associated. It consists of two peptide chains, the A chain of 44 residues and the B chain of 50 residues (75–80). These two chains, A and B, are linked by noncovalent interactions (Figure 15.7). Single chain mutants of monellins in, which the two chains are covalently linked, retain all their sweetening power and have greatly increased thermal stability (81). The first single chain monellin is obtained by joining the C-terminal residue of the B chain directly to the N-terminal residue of the A chain. The second single chain monellin is obtained by linking B and A chains *via* the Gly-Phe dipeptide. It has a molecular weight of 11,000 and is approximately 1500–2000 times sweeter than sugar on a molar basis and several thousand times sweeter on a weight basis. It does not contain carbohydrates or modified amino acids. Monellin is extracted from the fruit by removing the skin, leaching the fruit into water, and concentrating the product by salting out with ammonium sulphate. It shows little promise as a commercial sweetener because of taste qualities, stability, and difficulties in propagating the plant (74).

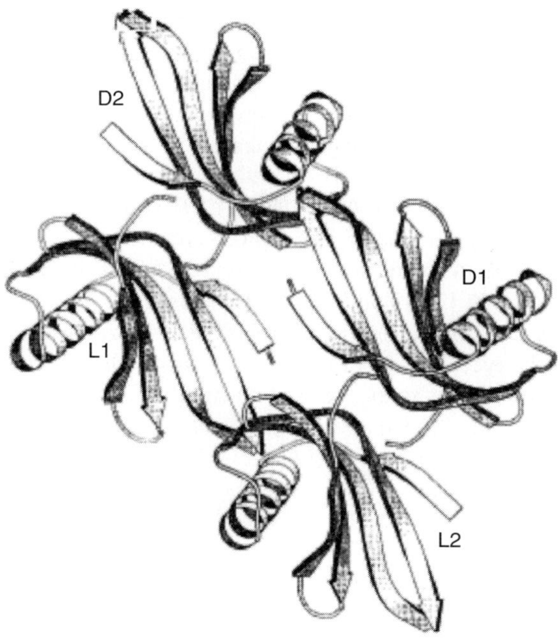

Figure 15.7 Schematic diagram of the spatial relationship of the two D-and two L-monellin molecules. Adapted from Hung L, M. Kohmura, Y. Ariyoshi, S. Kim. *J. Mol. Biol.* 285:311–321, 1999

The heterologous expression of the monellin gene in the yeast *Candida utilis* has been investigated. A single chain monellin gene was expressed under the control of the glyceraldehyde-3-phosphate decarboxylase gene promoter from *C. utilis* (80). A promoter deficient marker gene allowed high copy number integration of vectors into either the rDNA locus or the URA3 gene locus. Monellin was produced at a high level, accounting for 50% of the soluble protein. No significant decrease in the production level of monellin was detected in transformants after 50 generations of nonselective growth (80,81).

15.9 BRAZZEIN

Brazzein is a sweet tasting protein found in the fruit of the African plant *Pentadiplandra brazzeana* (82). It contains no carbohydrate, and its structure bears no structural resemblance to sucrose, small molecule chemical sweeteners, or the other sweet tasting proteins monellin and thaumatin. Moreover, members of the class of sweet proteins contain no conserved stretches of amino acids (83). It is the smallest, most heat stable and pH stable member of the set of proteins known to have intrinsic sweetness. These properties make brazzein an ideal compound for investigating the chemical and structural requirements of a sweet tasting protein. It is 2,000 times sweeter than sucrose and has a molecular mass of 6,473 Da (83,84).

Brazzein is a single chain polypeptide of 54 amino acid residues with four intramolecular disulfide bonds, no free sulfhydryl group, and no carbohydrate (82). It is rich in lysine, but contains no methionine, threonine or tryptophan. It exists in two forms in the ripe fruit. The major form contains pyroglutamate (pGlu) at its N-terminus; the minor form is without the N-terminal pGlu (des-pGlu1). Taste comparisons of chemically synthesized brazzein and des-pGlu1 brazzein revealed that the latter protein has about twice the sweetness of the former (85). It is highly water soluble with an isoelectric point ($pI = 5.4$) lower than those of other sweet proteins (82). It is remarkably heat stable, and its sweet taste remains after incubation at 80°C for 4 h (86). Chemical modification studies suggested that the surface charge of the molecule is important and led to the conclusion that Arg, Lys, Tyr, His, Asp, and Glu are important for brazzein's sweetness. The structure of brazzein was determined by nuclear magnetic resonance (NMR) spectroscopy in solution at pH 5.2 and 22°C (86,87). The study revealed that brazzein contains one short α-helix (residues 21–29) and three strands of antiparallel β-sheet structures (strand I, residues 5–7; strand II, residues 44–50; strand III, residues 34–39) held together by four disulfide bonds (Figure 15.8). It is proposed that the small connecting loop containing His31 and the random coil loop around Arg43 are the possible determinants of the molecule's sweetness (88,89). Brazzein can be synthesized by the fluoren-9-yl-methoxycarbonyl solid phase method which produces a compound identical to natural brazzein as demonstrated by high performance liquid chromatography, mass spectroscopy, peptide mapping, and taste evaluation. The D-enantiomer of brazzein, which is a mirror image of brazzein, was also synthesized, but was devoid of any sweetness and was essentially tasteless (85).

15.10 THAUMATIN

Thaumatin is a sweet tasting protein isolated from the arils of *Thaumatococcus daniellii* (Benth), a plant native to tropical West Africa (90). It is approved for use in many countries and has application as both a flavor enhancer and a high intensity sweetener. The

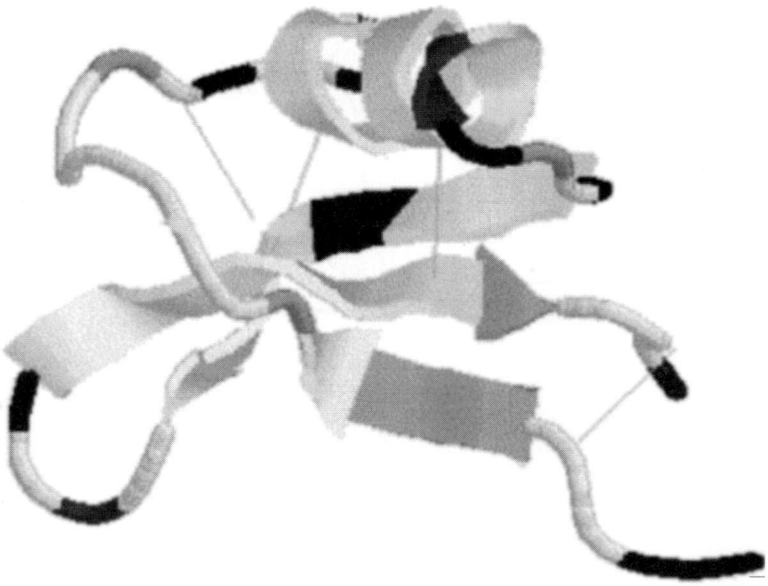

Figure 15.8 Diagram showing the three-dimensional backbone of brazzein. Adapted from Jin, Z., V. Danilova, P. Assadi, M. Fariba, D.J. Aceti, J.L. Markley, G. Hellekant. *FEBS Lett.* 544(3):33–37, 2003

availability of thaumatin of plant origin is very limited (91), and it is notoriously difficult to produce thaumatin by recombinant DNA methods (92). Production has been attempted with *Escherichia coli* (93), *Bacillus subtilis* (94), *Streptomyces lividans* (95), *Saccharomyces cerevisiae* (96), and *Aspergillus oryzae* (96). A synthetic gene for thaumatin with fungal codon usage has been synthesized, but expression in *Aspergillus niger* gave poor yields (97). Expression could be limited by a weak promoter, copy number, insertion location, inefficient processing of the prepropeptides, or bottlenecks in protein traffic and translocation through the membrane systems of the protein secretory pathway (98,99). Efficient production of a heterologous protein is usually achieved by increasing gene dosage, although overloading of the secretory pathway may result in abnormal folding and protein degradation.

Naturally occurring thaumatin consists of six closely related proteins (I, II, III, a, b, and c), all with a molecular mass of 22 kDa (207 amino acids) (90). Neither protein contains bound carbohydrate or unusual amino acids. The proteins have an isoelectric point of 12 (90). The sweet taste of thaumatins can be detected at threshold amounts 1600 times less than that of sucrose on a weight basis. The three dimensional structure of thaumatin I has been determined at high resolution (99,100), revealing that the protein consists of three domains: an 11 strand, flattened ß-sandwich (1–53, 85–127 and 178–207, domain I); a small, disulfide rich region (54–84, domain III); and a large disulfide rich region (128–177, domain II) (Figure 15.9). The five lysine residues, modification of which affected sweetness, are separate and spread over a broad surface region on one side of the thaumatin I molecule. These lysine residues exist in thaumatin, but not in nonsweet thaumatinlike proteins, suggesting that these lysine residues are required for sweetness. These lysine residues may play an important role in sweetness through a multipoint interaction with a putative thaumatin receptor (100,101).

Thaumatin was also secreted by the methylotrophic transgenic yeast *Pichia pastoris*. The mature thaumatin II gene was directly cloned from Taq polymerase amplified PCR

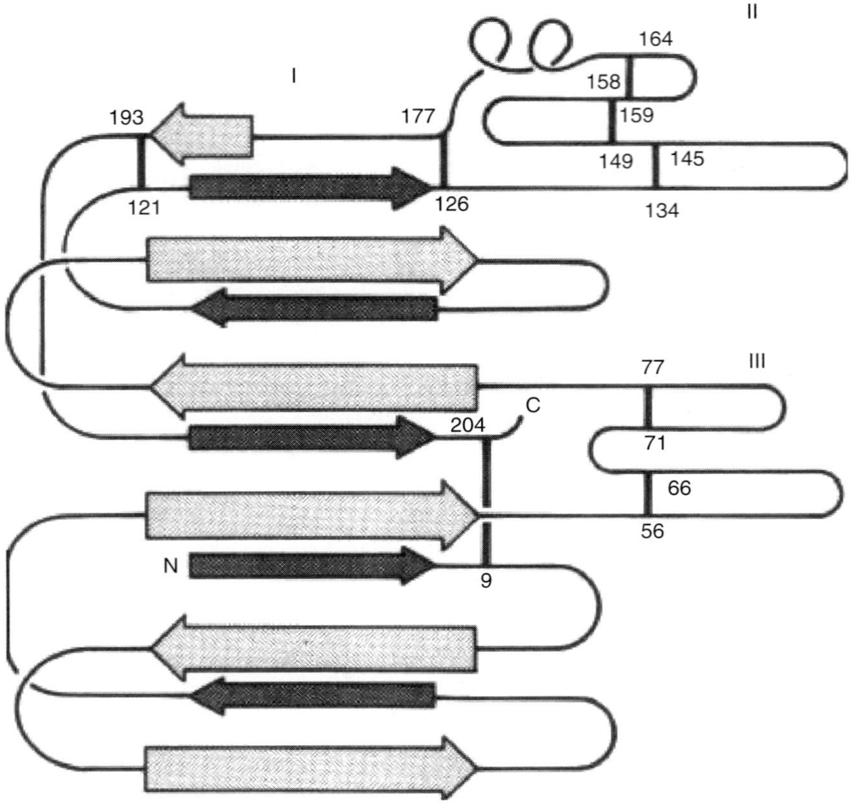

Figure 15.9 Topological structure of thaumatin. There are two beta sheets in the structure. The beta strands of the top sheet are indicated by broad arrows and those of the bottom sheet by narrow arrows. Also shown are the three domains of the protein and a crystallographic assignment of the disulphide bonds shown in black vertical bars. Figure adapted from: Vos, A.M., M. Hatada, H. Van der Wel, H. Krabbendam, A.F. Peerdeman, S.H. Kim. *Proc. Natl Acad Sci USA* 82(5):1406–1409, 1985

products by using cloning methods, and fused the pPIC9K expression vector that contains *Saccharomyces cerevisiae* prepro alpha mating factor secretion signal (101,102). Several additional amino acid residues were introduced at both the N- and C-terminal ends by genetic modification to investigate the role of the terminal end region for elicitation of sweetness in the thaumatin molecule. The secondary and tertiary structures of purified recombinant thaumatin were almost identical to those of the plant thaumatin molecule. Recombinant thaumatin II elicited a sweet taste just as native plant thaumatin II; its threshold value of sweetness to humans was around 50 nM, which is the same as that of plant thaumatin II. These results demonstrated that the functional expression of thaumatin II attained by *Pichia pastoris* systems and that the N- and C-terminal regions of the thaumatin II molecule did not play an important role in eliciting the sweet taste of thaumatin (102).

Research indicated that obtaining of thaumatin producing strains by transformation with an expression cassette containing a synthetic thaumatin gene (with an optimized codon usage) and the inactivation of a specific protease resulted in a significant increase of extracellular thaumatin (99). An alternative method for reduction of expression of a particular gene is the use of antisense RNA. Although the technique is simple, the effectiveness of the method is influenced by many factors (103). This technique has been successfully used to silence the *creA* gene in *Aspergillus nidulans* (104). It was, therefore used to silence the *pepB* gene in *A. awamori* by the antisense RNA technique, as a first approach to elimination of the negative effect of the presence of aspergillopepsin B on thaumatin accumulation. Research revealed that significant amounts of antisense RNA of the *pepB* gene are formed by using a strong fungal promoter, but that the aspergillopepsin B is not completely removed from the broths. On the other hand, *pepB* gene disruption by replacement using the double marker selection procedure led to the complete loss of aspergillopepsin B and to a 100% increase in thaumatin accumulation under optimal fermentation conditions (105).

15.11 MABINLIN

A new sweet tasting protein named mabinlin II was extracted from the seeds of *Capparis masaikai*. It was purified by ammonium sulfate fractionation, carboxymethylcellulose to Sepharose ion exchange chromatography, and gel filtration. The sweetness of mabinlin II was unchanged by at least 48 h incubation at nearly boiling temperature (106). Purified mabinlin II thus obtained gave a single band having a molecular mass of 14 kDa on SDS/PAGE (Sodium Dodecyl Sulphate-Polyacrylamide Gel Electrophoresis). In the presence of dithiothreitol, mabinlin II gave two bands having molecular masses of 4.6 kDa and 5.2 kDa on SDS/PAGE. Two peptides (A chain and B chain) were separated from reduced and S-carboxamidomethylated mabinlin II by HPLC. The amino acid sequences of the A chain and B chain were determined by the automatic Edman-degradation method. The A chain and B chain consist of 33-amino acid and 72-amino acid residues, respectively. The A chain is mostly composed of hydrophilic amino acid residues and the B chain also contains many hydrophilic residues. High similarity was found between the amino acid sequences of mabinlin II and 2S seed storage proteins, especially 2S albumin AT2S3 in *Arabidopsis thaliana* (106,107).

15.12 CONCLUSION

The pursuit for the ideal sweetener continues. The ideal sweetener is expected to taste like sucrose, and be colorless, odorless, noncariogenic, and nontoxic; and to present a pleasant, untainted taste without a delayed onset of persistence in sweetness; but at the same time be noncaloric and low cost. The long term safety of the nonnutritive sweeteners, which is of prime importance to consumers, has yet to be thoroughly investigated. However, the wide variety of sweeteners available today enables the development of a much wider range of new, good tasting, low calorie products to meet consumer demand. A variety of low calorie sweeteners also provides products with increased stability, improved taste, lower production costs, and more choices for the consumer. The search for noncarbohydrate sweeteners from natural sources has led to the discovery of many intensely sweet tasting substances. The occurrence of sweet tasting proteins has provided a new approach to the potential treatment of diabetes, obesity, and other metabolic disorders. Such choices allow a person to control their caloric intake in a manner best suited for their health conditions.

REFERENCES

1. Nabors, L.B., R.C. Gelardi. *Alternative Sweeteners: Food Science and Technology.* New York: Marcel Dekker, 1985.
2. Calorie and noncalorie sweeteners: per capita consumption, 1970–1982. In: *Sugar and Sweetener Outlook and Situatuion Report*, Washington: U.S. Department of Agriculture, Sept. 1983, p 29.
3. Nelson, G., M.A. Hoon, J. Chandrashekar, Y. Zhang, N.J. Ryba, C.S. Zuker. Mammalian sweet taste receptors. *Cell* 106:381–390, 2001.
4. Margolskee, R.F. Molecular mechanisms of bitter and sweet taste transduction. *J. Biol. Chem.* 277:1–4, 2002.
5. Li, X., L. Staszewski, X. Xu, K. Durick, M. Zoller, E. Adler. Human receptors for sweet and umami taste. *Proc. Natl. Acad. Sci. USA* 99:4692–4696, 2002.
6. Shallenberger, R.S., T. Acree. Molecular theory of sweet taste. *Nature* 216:480–482, 1967.
7. Kier, L.B. A molecular theory of sweet taste. *J. Pharm. Sci.* 61:1394–1397, 1972.
8. Temussi, P.A., F. Lelj, T. Tancredi. Three-dimensional mapping of the sweet taste receptor site. *J. Med. Chem.* 21:1154–1158, 1978.
9. Kaufman, G.B., P.M. Priebe. The discovery of saccharin: a centennial retrospect. *Ambix* 25:191–207, 1879.
10. Smith, J.C., A. Sclafani. Saccharin as a sugar surrogate revisited. *Appetite* 38(2):155–160, 2002.
11. Leclercq, C., D. Berardi, M.R. Sorbillo, J. Lambe. Intake of saccharin, aspartame, acesulfame K and cyclamate in Italian teenagers: present levels and projections. *Food Additives Contaminants* 16(3):99–111, 1999.
12. Ilback, N.G., S. Jahrl, H. Enghardt-Barbieri, L. Busk. Estimated intake of the artificial sweeteners acesulfame-K, aspartame, cyclamate and saccharin in a group of Swedish diabetics. *Food Additives Contaminants* 20(2):99–114, 2003.
13. Turner, S.D., H. Tinwell, W. Piegorsch, P. Schmezer, J. Ashby. The male rat carcinogens limonene and sodium saccharin are not mutagenic to male big blue rats. *Mutagenesis* 16(4):329–332, 2001.
14. Kurokawa, Y., T. Umemura. Risk assessment on an artificial sweetener, saccharin. *J. Food Hyg. Soc. Jpn.* 37(5):341–342, 1996.
15. Arnold, D.L. Long term toxicity study of orthotoluenesulfanamide and sodium saccharin in the rat. *Toxicol. Appl. Pharmacol.* 52:113–152, 1980.
16. Smith, J.C., W. Castonguayt, D.F. Foster, L.M. Bloom. Detailed analysis of glucose and saccharin in the rat. *Physiol. Behav.* 24(1):173–176, 1980.
17. Arnold, D.L., C.A. Moodie, H.C. Grice, S.M. Charbonneau, B. Stavric, B.T. Collins, P.F. McGuire. Long-term toxicity of toluenesulfonamide and sodium saccharin in the rat. *Toxicol. Appl. Pharmacol.* 52(1):113–152, 1980.
18. Arnold, D.L., C.A. Moodie, H.C. Grice, S.M. Charbonneau, B. Stavric, B.T. Collins, P.F. McGuire, I.C. Munro. The effect of toluene sulfonamide and sodium saccharin on the urinary tract of neonatal rats. *Toxicol. Appl. Pharmacol.* 51(3):455–464, 1979.
19. Rao, T.K., D.R. Stoltz, J.L. Epler. Lack of enhancement of chemical mutagenesis by saccharin in the *Salmonella typhimurium* assay. *Arch. Toxicol.* 43(2):141–146, 1979.
20. Kessler, I., J.P. Clark. Saccharin, cyclamate and human bladder cancer: a case-control study: no evidence of an association. *JAMA* 240:349–355, 1978.
21. Bailey, C.J., J.M.E. Knapper, S.L. Turner, P.R. Flatt. Antihyperglycaemic effect of saccharin in diabetic ob/ob mice. *Br. J. Pharmacol.* 120(1):74–78, 1997.
22. Chappel, C. A review and biological risk assessment of sodium saccharin. *Reg. Toxicol. Pharmacol.* 15(3):253–270, 1992.
23. Walter, G.J., M.I. Mitchell. Saccharin. In: *Alternative Sweeteners: Food Science and Technology*, New York: Marcel Dekker 17:15–41, 1985.
24. Clauss, K., H. Jensen. Oxathiazinon dioxides: a new group of sweetening agents. *Angew. Chem.* 85:965, 1973.

25. Peck, A. Use of acesulfame K in light and sugar-free baked goods. *Cereal Foods World* 39(10):743–745, 1994.
26. Abe, Y., Y. Takeda, H. Ishiwata, T. Yamada. Purity and content of a sweetener, acesulfame potassium, and their test methods. *Shokuhin Eiseigaku Zasshi.* 41(4):274–279, 2000.
27. Suami, T., L. Hough, T. Machinami, T. Saito, K. Nakamura. Molecular mechanisms of sweet taste 8: saccharin, acesulfame-K, cyclamate and their derivatives, *Food Derivatives* 63(3):391–396, 1998.
28. Mukherjee, J., J. Chakrabarti. *In vivo* cytogenetic studies on mice exposed to acesulfame-K, a nonnutritive sweetener. *Food Chem. Toxicol.* 35(12):1177–1179, 1997.
29. Arpe, H.J. Acesulfame-K: a new noncaloric sweetener. In: *Health and Sugar Substitutes, Proceedings of the ERGOB conference, Geneva, 1978,* p 178.
30. Wolfhard, G., R. Lipinski. Acesulfame-K. In: *Alternative Sweeteners: Food Science and Technology,* New York: Marcel Dekker, 1985, pp 89–102.
31. Stroka, J., N. Dossi, E. Anklam. Determination of the artificial sweetener Sucralose® by capillary electrophoresis. *Food Additives Contaminants* 20(6):524–527, 2003.
32. Knight, I. The development and applications of sucralose, a new high-intensity sweetener. *Can. J. Physiol. Pharmacol.* 72(4):435–439, 1994.
33. Finn, J.P., G.H. Lord. Neurotoxicity studies on sucralose and its hydrolysis products with special reference to histopathologic and ultrastructural changes. *Food Chem. Toxicol.* 38(2): S7–S17, 2000.
34. Grice, H.C., L.A. Goldsmith. Sucralose: an overview of the toxicity data. *Food Chem. Toxicol.* 38(2):S1–S636, 2000.
35. Baird, I.M., R.J. Merritt, G. Hildick-Smith. Repeated dose study of sucralose tolerance in human subjects. *Food Chem. Toxicol.* 38(2):S123–S129, 2000.
36. Sims, J., A. Roberts, J.W. Daniel, A.G. Renwick. The metabolic fate of sucralose in rats. *Food Chem. Toxicol.* 38(2):S115–S121, 2000.
37. Mann, S.W., M.M. Yuschak, S.J.G. Amyes, P. Aughton, J.P. Finn. A carcinogenicity study of sucralose in the CD-1 mouse. *Food Chem. Toxicol.* 38(2):S91–S98, 2000.
38. Goldsmith, L.A. Acute and subchronic toxicity of sucralose. *Food Chem. Toxicol.* 38(2): S53–S69, 2000.
39. Bennett, C., J.S. Dordick, A.J. Hacking, P.S.J. Cheetham. Biocatalytic synthesis of disaccharide high-intensity sucralose via a tetrachlororaffinose intermediate. *Biotechnol. Bioeng.* 39(2):211–217, 1992.
40. Cloninger, M.R., R.E. Baldwin. L-Aspartyl-L-phenylalanine methyl esther (aspartame) as a sweetener. *J. Food. Sci.* 39:347–349, 1974.
41. Prat, L.L., J.M. Oppert, F. Bellisle, G.B. Guy. Sweet taste of aspartame and sucrose: effects on diet-induced thermogenesis. *Appetite* 34(3):245–251, 2000.
42. Fellows, J.W., S.W. Chang, W.H. Shazer. Stability of aspartame in fruit preparations used in yogurt. *J. Food Sci.* 56(3):689–691, 1991.
43. Janssen, C. M., C.A. van der Heijden. Aspartame: review of recent experimental and observational data. *Toxicology* 50:1–26, 1988.
44. Bianchi, R.G., E.T. Muir, D.L. Cook, E.F. Nutting. The biological properties of aspartame: actions involving the gastrointestinal system. *J. Envir. Pathol. Toxicol.* 93:355–362, 1980.
45. Oyama, Y., H. Sakai, T. Arata, Y. Okano, N. Akaike, K. Sakai, K. Noda. Cytotoxic effects of methanol, formaldehyde, and formate on dissociated rat thymocytes: a possibility of aspartame toxicity. *Cell Biol. Toxicol.* 18(1):43–50, 2002.
46. Ishii, H. Incidence of brain tumors in rats fed aspartame. *Toxicol. Lett.* 7:433–437, 1981.
47. Duerfahrt, T., S. Doekel, P.L.M. Quaedflieg, M.A. Marahiel. Construction of hybrid peptide synthetases for the production of alpha-L-aspartyl-L-phenylalanine, a precursor for the high-intensity sweetener aspartame. *Eur. J. Biochem.* 270(22):4555–4563, 2003.
48. Isono, Y., M. Nakajima. Enzymatic synthesis of aspartame precursor in solvent-free reaction system with salt hydrates. *Nippon Shokuhin Kagaku Kogaku Kaishi* 49(12):813–817, 2002.
49. Ahn, J.E., C. Kim, C.S. Shin. Enzymic synthesis of aspartame precursors from eutectic substrate mixtures. *Process. Biochem.* 37(3):279–285, 2001.

50. Garbow, J.R., J.J. Likos, S.A. Schroeder. Structure, dynamics, and stability of beta-cyclodextrin inclusion complexes of aspartame and neotame. *J. Agric. Food Chem.* 49(4):2053–2060, 2001.
51. Li, J.P., Z.Y. He. Preparing process of aspartame. *Zhongguo YiyaoGongye Zazhi* 31(3):106, 2000.
52. Karikas, G.A., K.H. Schulpis, G. Reclos, G. Kokotos. Measurement of molecular interaction of aspartame and its metabolites with DNA. *Clin. Biochem.* 31(5):405–407, 1998.
53. Nakaoka, H., Y. Miyajima, K. Morihara. Papain-catalyzed synthesis of aspartame precursor: a comparison with thermolysin. *J. Ferment. Bioeng.* 85(1):43–47, 1998.
54. Nofre, C., J.M. Tinti. Neotame: discovery, properties, utility. *Food Chem.* 69(3):245–257, 2000.
55. E.J. Munson, S.A. Schroeder, I. Prakash, D.J.W. Grant. Neotame anhydrate polymorphs II: quantitation and relative physical stability. *Pharm. Res.* 19(9):1259–1264, 2002.
56. Dong, Z., V.G. Young, A. Sheth, E.J. Munson, S.A. Schroeder, I. Prakash, D.J.W. Grant. Crystal structure of neotame anhydrate polymorph G. *Pharm. Res.* 19(10):1549–1553, 2002.
57. Mayhew, D.A., C.P. Comer, W.W. Stargel. Food consumption and body weight changes with neotame, a new sweetener with intense taste: differentiating effects of palatability from toxicity in dietary safety studies. *Regul. Toxicol. Pharmacol.* 38(2): 124–143, 2003.
58. Flamm, W.G., G.L. Blackburn, C.P. Comer, D.A. Mayhew, W.W. Stargel. Long-term food consumption and body weight changes in neotame safety studies are consistent with the allometric relationship observed for other sweeteners and during dietary restrictions. *Reg. Toxicol. Pharmacol.* 38(2):144–156, 2003.
59. Kroyer, G.T. The low calorie sweetener stevioside: stability and interaction with food ingredients. *Lebensmittel Wissenschaft Technologie* 32(8):509–512, 1999.
60. Adduci, J., D. Buddhasukh, B. Ternai. Improved isolation and putrification of stevioside. *J. Sci. Soc. Thailand* 13(3):179–183, 1987.
61. Ogawa, T., M. Nozaki, M. Masanao. Total synthesis of stevioside. *Tetrahedron* 36(18):2641–2648, 1980.
62. Geuns, J.M.C. Stevioside. *Phytochemistry* 64(5):913–921, 2003.
63. Brandle, J.E., N. Rosa. Heritability for yield, leaf-stem ratio and stevioside content estimated from a landrace cultivar of *Stevia rebaudiana*. *Can. J. Plant Sci.* 72:1263–1266, 1992.
64. Crammer, B., R. Ikan. Sweet glycosides from the stevia plant. *Chem. Br.* 22:915–916, 1986.
65. Metivier, J., A.M. Viana. The effect of long and short day length upon the growth of whole plants and the level of soluble proteins, sugars and stevioside in leaves of *Stevia rebaudiana* Bert. *J. Exp. Bot.* 30:1211–1222, 1979.
66. Soejarto, D.D., C.M. Compadre, P.J. Medon, S.K. Kamath, A.D. Kinghorn. Potential sweetening agents of plant origin, II: field search for sweet-tasting *Stevia* species. *Econ. Bot.* 37:71–79, 1983.
67. Matsui, M., K. Matsui, Y. Kawasaki, Y. Oda T. Noguchi, Y. Kitagawa, M. Sawada, M. Hayashi, T. Nohmi, K. Yoshihira, T. Ishidate, M. Sofuni. Evaluation of the genotoxicity of stevioside and steviol using six *in vitro* and one *in vivo* mutagenicity assays. *Mutagenesis* 11:573–579, 1996.
68. Matsui, M., T. Sofuni, T. Nohmi. Regionally-targeted mutagenesis by metabolically-activated steviol: DNA sequence analysis of steviol-induced mutants of guanine phosphoribosyltransferase (gpt) gene of *Salmonella typhimurium* TM677. *Mutagenesis* 11:565–572, 1996.
69. Gregersen, S., P.B. Jeppesen, J.J. Holst, K. Hermansen. Antihyperglycemic effects of stevioside in type 2 diabetic subjects. *Metab. Clin. Exp.* 53(1):73–76, 2004.
70. Totté, N., L. Charon, M. Rohmer, F. Compernolle, I. Baboeuf, J.M.C. Geuns. Biosynthesis of the diterpenoid steviol, an *ent*-kaurene derivative from *Stevia rebaudiana* Bertoni, via the methylerythritol phosphate pathway. *Tetrahedron Lett.* 41:6407–6410, 2000.
71. Kim, K.K., H. Yamashita, Y. Sawa, H. Shibata. A high activity of 3-hydroxy-3-methylglutaryl coenzyme A reductase in chloroplasts of *Stevia rebaudiana* Bertoni. *Biosc. Biotech. Biochem.* 60:685–686, 1996.

72. Brandle, J.E., A. Richman, A.K. Swanson, B.P. Chapman. Leaf ESTs from *Stevia rebaudiana*: a resource for gene discovery in diterpene synthesis. *Plant Mol. Biol.* 50:613–622, 2002.
73. Kim, K.K., Y. Sawa, H. Shibata. Hydroxylation of ent-kaurenoic acid to steviol in *Stevia rebaudiana* Bertoni: purification and partial characterization of the enzyme. *Arch. Biochem. Biophys.* 332: 223–230, 1996.
74. Ogata, C., M. Hatada, G. Tomlinson, W.C. Shin, S.H. Kim. Crystal structure of the intensely sweet protein monellin. *Nature* 328(6132):739–742, 1987.
75. Kotlovyi, V., W.L. Nichols, L.F. Ten Eyck. Protein structural alignment for detection of maximally conserved regions. *Biophys. Chem.* 105(2,3):595–608, 2003.
76. Sung, Y.H., H.D. Hong, C. Cheong, J.H. Kim, J.M. Cho, Y.R. Kim, W. Lee. Folding and stability of sweet protein single-chain monellin: an insight to protein engineering. *J. Biol. Chem.* 276(47):44229–44238, 2001.
77. Kim, S.H., C.H. Kang, R. Kim, J.M. Cho, Y.B. Lee, T.K. Lee. Redesigning a sweet protein: increased stability and renaturability. *Protein Eng.* 2:571–575, 1989.
78. Tomic, M.T., J.R. Somoza, D.E. Wemmer, Y.W. Park, J.M. Cho, S.H. Kim. 1H resonance assignments, secondary structure and general topology of single-chain monellin in solution as determined by 1H 2D-NMR. *J. Biomol. NMR* 2:557–572, 1985.
79. Hung, L., M. Kohmura, Y. Ariyoshi, S. Kim. Structural differences in D- and L-monellin in the crystals of racemic mixture. *J. Mol. Biol.* 285:311–321, 1999.
80. Kondo, K., M. Yutaka, H. Sone, K. Kobayashi, H. Iijima. High-level expression of a sweet protein, monellin, in the food yeast *Candida utilis*. *Nat. Biotechnol.* 15(5):453–457, 1997.
81. Kim, H., K. Lim. Large-scale purification of recombinant monellin from yeast. *J. Ferment. Bioeng.* 82(2):180–182, 1996.
82. Ming, D., G. Hellekant. A new high-potency thermostable sweet protein from *Pentadiplandra brazzeana* B. *FEBS Lett.* 355(1):106–108, 1994.
83. Jin, Z., V. Danilova, P. Assadi, M. Fariba, D.J. Aceti, J.L. Markley, G. Hellekant. Critical regions for the sweetness of brazzein. *FEBS Lett.* 544(1–3):33–37, 2003.
84. Ishikawa, K., M. Ota, Y. Ariyoshi, H. Sasaki, M. Tanokura, M. Ding, J. Caldwell, F. Abildgaad. Crystallization and preliminary x-ray analysis of brazzein, a new sweet protein. *Acta Crystallogr. D. Biol. Crystallogr.* 52:577–578, 1996.
85. Izawa, H., M. Ota, M. Kohmura, Y. Ariyoshi. Synthesis and characterization of the sweet protein brazzein. *Biopolymers* 39(1):95–101, 1996.
86. Caldwell, J.E., F. Abildgaard, Z. Dzakula, D. Ming, G. Hellekant, J.L. Markley. Solution structure of the thermostable sweet-tasting protein brazzein. *Nat. Struct. Biol.* 5(6):427–431, 1998.
87. Somoza, J.R., F. Jiang, L. Tong, C.H. Kang, J.M. Cho, S.H. Kim. Two crystal structures of a potently sweet protein: natural monellin at 2.75 A resolution and single-chain monellin at 1.7 A resolution. *J. Mol. Biol.* 234(2):390–404, 1993.
88. Assadi-Porter, F.M., D.J. Aceti, J.L. Markley. Sweetness determinant sites of brazzein, a small, heat-stable, sweet-tasting protein. *Arch. Biochem. Biophys.* 376(2):259–265, 2000.
89. Assadi-Porter, F.M., D.J. Aceti, H. Cheng, J.L. Markley. Efficient production of recombinant brazzein, a small, heat-stable, sweet-tasting protein of plant origin. *Arch. Biochem. Biophys.* 376:252–258, 2000.
90. Van der Wel, H., K. Loeve. Isolation and characterization of thaumatin I and II, the sweet-tasting proteins from *Thaumatococcus daniellii* Benth. *Eur. J. Biochem.* 31(2): 221–225, 1972.
91. Kong, J.Q., Q. Zhao, L. Gao, X.T. Qui, Q.J. Yang. Sweet protein thaumatin and its genetic engineering. *Yichuan* 25(2):232–236, 2003.
92. Zemanek, E., B.P. Wasserman. Issues and advances in the use of transgenic organisms for the production of thaumatin, the intensely sweet protein from *Thaumatococcus daniellii*. *Crit. Rev. Food Sci. Nutr.* 35:455–466, 1995.
93. Faus, I., C. Patiño, J.L. del Río, C. del Moral, H. Sisniega, V. Rubio. Expression of a synthetic gene encoding the sweet-tasting protein thaumatin in *Escherichia coli*. *Biochem. Biophys. Res. Commun.* 229:121–127, 1996.

94. Illingworth, C., G. Larson, G. Hellekant. Secretion of the sweet-tasting plant protein thaumatin by *Bacillus subtilis*. *Biotechnol. Lett.* 10:587–592, 1988.
95. Illingworth, C., G. Larson, G. Hellekant. Secretion of the sweet-tasting plant protein thaumatin by *Streptomyces lividans*. *J. Ind. Microbiol.* 4:37–42, 1989.
96. Edens, L., I. Born, A.M. Ledeboer, J. Maat, M.Y. Toonen, C. Visser, C.T. Verrips. Synthesis and processing of the plant protein thaumatin in yeast. *Cell* 37:629–633, 1984.
97. Gwynne, D.I., M. Devchand. Expression of foreign proteins in the genus *Aspergillus*. In: *Aspergillus, the Biology and Industrial Applications,* Bennet, J.W., M.A. Klich, eds., London: Butterworth, 1992, pp 203–214.
98. Faus, I., C. del Moral, N. Adroer, J.L. del Río, C. Patiño, H. Sisniega, C. Casas, J. Bladé, V. Rubio. Secretion of the sweet-tasting protein thaumatin by recombinant strains of *Aspergillus niger* var. *awamori*. *Appl. Microbiol. Biotechnol.* 49:393–398, 1998.
99. Verdoes, J.C., P.J. Punt, C.A.M.J.J. van den Hondel. Molecular genetic strain improvement for the over-production of fungal proteins by filamentous fungi. *Appl. Microbiol. Biotechnol.* 43:195–205, 1995.
100. De Vos, A.M., M. Hatada, H. Van der Wel, H. Krabbendam, A.F. Peerdeman, S.H. Kim. Three-dimensional structure of thaumatin I, an intensely sweet protein. *Proc. Natl. Acad. Sci. USA* 82:1406–1409, 1985.
101. Kaneko, R., N. Kitabatake. Structure–sweetness relationship in thaumatin: importance of lysine residues. *Chem. Senses* 26:167–177, 2001.
102. Masuda, T., S. Tamaki, R. Kaneko, R. Wada, Y. Fujita, A. Mehta, N. Kitabatake. Cloning, expression and characterization of recombinant sweet-protein thaumatin II using the methylotrophic yeast *Pichia pastoris*. *Biotechnol Bioeng.* 85(7):761–769, 2004.
103. Agrawal, S., E.R. Kandimalla. Antisense therapeutics: is it as simple as complementary base recognition? *Mol. Med. Today* 6:72–81, 2000.
104. Bautista, L.F., A. Aleksenko, M. Hentzer, A. Santerre-Henriksen, J. Nielsen. Antisense silencing of the *creA* gene in *Aspergillus nidulans*. *Appl. Environ. Microbiol.* 66:4579–4581, 2000.
105. Berka, R.M., M. Ward, L.J. Wilson, K.J. Hayenga, K.H. Kodama, L.P. Carlomagno, S.A. Thompson. Molecular cloning and deletion of the gene encoding aspergillopepsin A from *A. awamori*. *Gene* 86:153–162, 1990.
106. Liu, X., S. Maeda, Z. Hu, T. Aiuchi, K. Nakaya, Y. Kurihara. Purification, complete amino acid sequence and structural characterization of the heat-stable sweet protein, mabinlin II.
107. Nirasawa, S., Y. Masuda, K. Nakaya, Y. Kurihara. Cloning and sequencing of a cDNA encoding a heat-stable sweet protein, mabinlin II. *Gene* 181(1,2):225–227, 1996.

2.16

Biotechnological Approaches to Improve Nutritional Quality and Shelf Life of Fruits and Vegetables

Reena Grittle Pinhero and Gopinadhan Paliyath

CONTENTS

16.1 INTRODUCTION

Fruits and vegetables, which can be consumed as fresh and as processed products, are important ingredients of a healthy diet. They are valuable sources of vitamins, minerals, antioxidants, and fiber. The important quality factors of fruits and vegetables are their color, flavor, texture, and nutritive value. Consumers always prefer to buy fruits and vegetables of high quality. As used by the industry, quality is a concept involving measurable attributes: degree of purity, firmness, flavor, color, size, maturity, condition, or any other distinctive attributes or characteristics of the product (1). The qualities of the produce bought by the consumer are influenced by many factors, such as the cultivar, the environmental conditions affecting growth, cultural practices, exposure to pests and diseases, time of harvesting, and postharvest and storage conditions used. Today, with the advancement of technology in several areas, the only factor on which the grower has no control of is the environment of the field conditions. Heredity (the identity of the cultivar) plays a major role in determining the quality of fruits and vegetables, as evidenced by the various varietal differences in quality. Even though traditional crop breeding is still being used as one means of crop improvement, continuing advances in knowledge and technology have dramatically expanded the biotechnological tools available for genetic improvement and production of vegetables and fruits. The term biotechnology is broad, encompassing a wide range of disciplines in science. In this chapter, the focus will be on plant transformation, where genes are modified or transferred by molecular means, and the resulting improvements in fruit and vegetable quality. Because this chapter is focused on genetic engineering, we will be discussing only the parameters and mechanisms that affect nutritional quality and shelf life (other than pests and diseases), and which can be improved or modified by genetic engineering. The discussion will also be based on two important crops; potatoes and tomato.

16.2 POTATOES

Potatoes are ranked fourth in production of all agricultural commodities in the world and yield more dry matter and protein per hectare than the major cereal crops (2). They are consumed as fresh and processed products, and used as raw material for many industrial purposes such as starch extraction. Potato chips and french fries are two of the most popular processed potato products. The consumer preference for these products is influenced by the color and crispness of these products. The primary problem associated with potato processing is the nonenzymatic browning of the product that occurs under the high temperature conditions used during frying, when reducing sugar levels are high in the tissue, a phenomenon known as Maillard reaction (3). The reducing sugars, glucose and fructose, combine with the α-amino groups of amino acids at the high temperatures used in frying operations, resulting in darker and more bitter flavored french fries and chips that are unacceptable to the consumer. The ideal content of reducing sugars is 0.1% of the tuber fresh weight; 0.33% is the upper acceptable limit (4). A four year study was conducted to determine the compositional differences during low temperature storage between low sugar accumulating and high sugar accumulating cultivars in relation to potato chip processing quality (5). Pearson analysis of the above data showed that chip color was most closely correlated with reducing sugar concentration. Multiple regression analysis revealed that the relative contribution of each of the parameters studied, such as sucrose, reducing sugars, nitrogen, protein, ascorbic acid and dry matter content, varied greatly among cultivars and selections evaluated and from season to season (5).

16.2.1 Factors Affecting Accumulation of Reducing Sugars in Potatoes

The factors that affect the sweetening or breakdown of starch into sucrose and its component reducing sugars glucose and fructose are drought, excess nitrogen during growth, high temperature at harvest, handling, aging, identity of the cultivar, anaerobic conditions, and low temperature during post harvest storage (6).

16.2.2 Low Temperature Sweetening in Potatoes

Low temperature sweetening (LTS) in potato tubers is a phenomenon that occurs when tubers are stored at temperatures below 10°C in order to minimize respiration and sprouting. LTS results in the accumulation of starch breakdown products, primarily sucrose and the reducing sugars glucose and fructose (7), which cause Maillard browning during potato chip frying operations (3,8). Fry color of Russet Burbank and Shepody potatoes has been shown to be more closely associated with glucose concentration than with fructose, total reducing sugars, sucrose, or total sugars (9). In order to avoid Maillard browning, processing potatoes are generally stored at temperatures around 10°C; but at this storage temperature potato tubers will sprout. To prevent sprouting during storage, the processing tubers are treated with chemical sprout inhibitors. However, due to health and environmental concerns, there is increasing pressure to reduce the use of chemical sprout inhibitors. The only solution to avoid this problem is using cultivars that are resistant to LTS.

Low temperature storage of potato tubers has many advantages, such as control of sprout growth and senescent sweetening, reduction of physiological weight loss due to decreased respiration and losses associated with bacterial and fungal pathogens, and extended marketability. Low temperature storage has several advantages, but the main drawback associated with it is the accumulation of reducing sugars and the resulting browning of processed products such as chips and fries.

The mechanism responsible for the initiation and subsequent regulation of LTS in potato tubers has not been fully elucidated. Many theories have been proposed to explain LTS based on starch metabolism, sucrose metabolism, glycolysis and the oxidative pentose phosphate pathway (PPP), and mitochondrial respiration (10,11), as well as membrane instability, lipid peroxidation, and electrolyte leakage (12–15). It has been suggested that in mature, cold stored potato tubers, the glycolytic or respiratory capacity plays a key role in the ability of potatoes to regulate their sugar concentration (16). Although LTS has not been elucidated at the molecular level, many factors may play a role in it. For example, chilling may influence compartmentation and membrane permeability by altering the phase transition of lipids in the bilayer, resulting in the leakage of key ions such as inorganic phosphates. This can alter the activity and synthesis of key enzymes involved in the metabolic pathways, ultimately resulting in LTS (17).

Many theories have been postulated and documented to explain LTS at the level of carbohydrate metabolism in stored potato tubers (7,11,12,17,18). The mechanism is complex and may involve the interaction of several pathways of carbohydrate metabolism and the genes that regulate these pathways. This discussion focuses on a theoretical model of the mechanisms involved in LTS based on information available about the roles of the major tuber carbohydrate metabolic pathways as well as changes in membrane stability (Figure 16.1).

16.2.3 Metabolism of Starch in Tuber

Starch is the major component in the main crop plants of the world, as well as an important raw material for many industrial processes. Potato tubers contain 60–80% starch, of which sugars represent only a small fraction (up to 3% on a dry weight basis) (19). There is evidence that the principal event in LTS is the cold induced synthesis of sucrose (7,12). The carbon needed for the synthesis of sucrose and reducing sugars for LTS is generally, but not always, provided by a net breakdown of starch. An increase in potato tuber sugar content occurs early during cold storage; over two to three months at storage temperatures of 1–3°C, tubers can convert as much as 30% of their starch content (20). In mature King Edward tubers stored at 2°C, the sugar content increased from 0.3 to 2.5% in three months, with the initial appearance of sucrose followed by reducing sugars (19). Coffin et al. (21) found that sucrose content increased within two days of 5°C storage for both mature and immature tubers of four cultivars, while fructose and glucose content increased more slowly. Pollock and ap Rees (22) reported an increase in both sucrose and reducing sugar content within 5 days in tubers stored at 2°C, and after 20 days storage, the sugar content was approximately

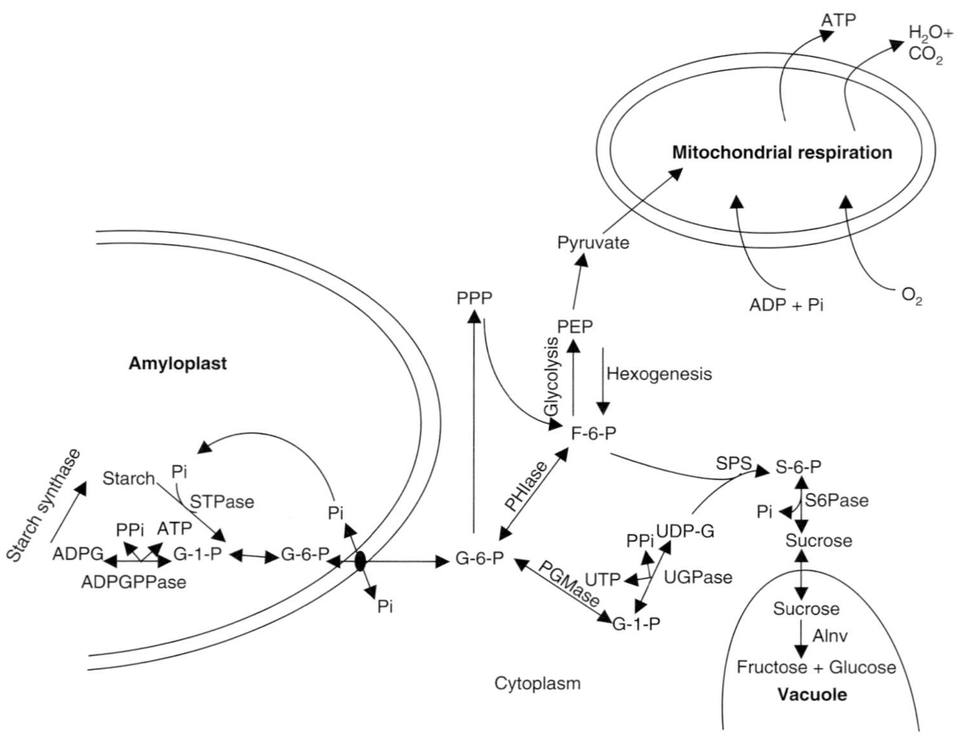

Figure 16.1 Starch-sugar interconversion in potato tubers. ADPG, ADP-glucose; ADPGPPase, ADP-glucose pyrophosphorylase; G-1-P, glucose-1-phosphate; G-6-P, glucose-6-phosphate; STPase, starch phosphorylase; F-6-P, fructose 6-phosphatase, SPS, sucrose-6-phosphate synthase; S-6-P, sucrose-6-phosphate; S6Pase, sucrose-6-P phosphatase; Ainv, acid invertase; PPP, pentose phosphate pathway; PEP, phosphoenolpyruvate (adapted from Sowokinos, 2001)

six times greater than at day 0. The sweetening response of tubers to low temperatures is fairly consistent, but is influenced by cultivar, locality, and conditions prior to cold storage. Isherwood (19) related energy requirements to possible biosynthetic pathways and concluded that sucrose was formed from starch when potato tubers were moved from 10° to 2°C and that starch was reformed when tubers were moved back from 2° to 10°C, although different metabolic pathways were involved. Reconditioning of tubers is sometimes used to improve chipping quality by decreasing the level of reducing sugars (20). After cold storage, potatoes are reconditioned at 18°C where sugar content decreases and starch content increases as sugars are resynthesized to starch. However, the response to reconditioning is neither consistent nor completely restorative, and tends to be cultivar dependent.

Preconditioning treatment has been used to lessen chilling injury in chilling sensitive plants. Storage at 10°C prior to cold storage can acclimatize tubers and lessen the LTS effect (8). Katahdin tubers preconditioned at 15.5°C for one to four weeks before 0°C storage did not show a change in sugar accumulation patterns or respiration rates (23). The use of intermittent warming (15.5°C for one week following 0°C for three weeks) decreased sugar levels and respiration rates to levels lower than those of tubers stored continuously at 0, 1 and 4.5°C, although sugar levels were not low enough for desirable chipping potatoes.

16.2.3.1 Starch Synthesis

Starch is synthesized in plastids (amyloplasts) upon tuber initiation, and both the number of starch grains and the grain size increase during tuber growth (Figure 16.1, Figure 16.2). Starch consists of two types of glucose polymers, the highly branched amylopectin, and relatively unbranched amylose. Potato starch is comprised of 21–25% amylose and 75–79% amylopectin (24). Starch is synthesized from ADPglucose by the concerted action of ADPglucose pyrophosphorylase (ADPGPase), starch synthase, and the starch branching enzymes (25). Following the conversion of sucrose into hexose phosphates in the cytosol, glucose-6-phosphate is transported into the amyloplast where it is converted into glucose-1-phosphate. A study involving antisense inhibition of plastidial phosphoglucomutase supported the theory that carbon from the cytosol was imported into potato tuber amyloplasts in the form of glucose-6-phosphate (26). Glucose-1-phosphate is subsequently converted to ADPglucose by ADPGPase. The starch synthases catalyze the polymerization of the glucose monomers into α-1,4-glucans using ADPglucose as a substrate, while the starch branching enzyme catalyzes the formation of the α-1,6-linkages of amylopectin (25).

ADPGPase is often referred to as a rate-limiting step in starch synthesis (25). It is subjected to allosteric activation by 3-P-glycerate and inhibition by inorganic phosphate (27). Strategies to alter the starch metabolism in tubers by genetic manipulation of ADPGPase may be helpful in reducing the accumulation of reducing sugars during LTS. It has been reported that transgenic tubers with an 80–90% reduction in ADPGPase activity have reduced starch content relative to wild type tubers (28,29). The reduction in ADPGPase activity resulted in a major reduction of carbon flux, with increased flux to sucrose and decreased flux to starch. Stark et al. (30) have reported that hexose accumulation was greatly reduced in cold stored tubers with overexpression of the mutated ADPGPase gene, glgC16, from *E. coli*. The glgC16 gene produces a mutant form of ADPGPase that is less responsive to allosteric effectors. It has been suggested that the observed decrease in hexose concentration could be due to the higher starch biosynthetic capacity of the transgenic tubers. Lorberth et al. (31) developed transgenic potatoes with decreased levels of R1 protein, a starch granule bound protein capable of introducing phosphate into starch-like glucans. By reducing the activity of the R1 protein using antisense technology, the phosphate content of starch was reduced, resulting in a starch that

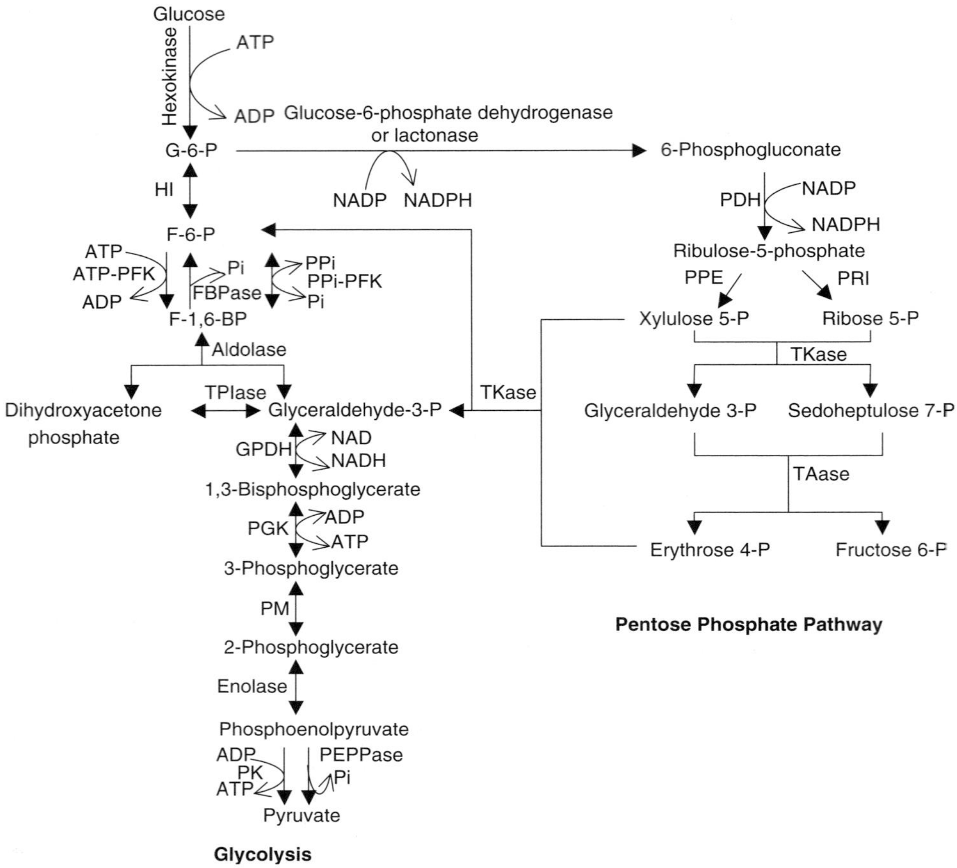

Figure 16.2 Interactions between glycolysis and pentose phosphate pathway. G-6-P, glucsoe 6-phosphate, HI, hexose phosphate isomerase, F-6-P, fructose 6-phosphate, ATP-PFK, ATP-dependent phosphofructokinase, FBPase, fructose 1,6-bisphosphatase, PPi-PFK, PPi-dependent phosphofructose kinase, F 1,6-BP, fructose 1,6 bisphosphate, TPlase, triose phosphate isomerase; GPDH, glyceraldehyde 3--phoshate dehydrogenase; PGK, phosphoglycerate kinase; PM, phosphoglycero mutase; PK, pyruvate kinase; PDH, 6-phosphogluconate dehydrogenase; PPE, phosphopentose epimerase; PRi, phoshoribo-isomerase; Tkase, transketolase; Taase, transaldolase.

was less susceptible to degradation at low temperatures relative to the starch of wild type tubers. It has been observed that after two months of storage at 4°C, the transgenic tubers contained up to ninefold lower concentrations of reducing sugars compared to the wild type. However, the commercial value of the modification of starch could not be assessed because the authors did not analyze the processing quality of transgenic tubers.

16.2.3.2 Starch Degradation

The differential response of potato cultivars to LTS may be the result of starch properties that affect the ability of enzymes to degrade it. The various starch properties ascribed are:

1. Chemical modifications of glucose units by attachment of covalently bound phosphate. Phosphate esters are attached to C3 or C6 glucosyl residues of amylopectin

in the larger interbranch chains and are absent around the branching points. This affects the cleavage sites and degradation product patterns (32).

2. Surface property alteration caused by the negative surface charge from surface phosphate, lipid, or protein can affect the properties of enzymes and other soluble compounds (33).
3. Association with starch-metabolizing enzymes such as endoamylase activity in cotton leaves (34) and starch synthase in potato tubers (35).
4. Physical characteristics of starch. Isolated starch grains from two cultivars differing in their sensitivity to LTS showed an increase in starch grain size over time with the disappearance of smaller starch grains while ND860-2, the resistant cultivar had a consistently smaller mean starch grain size (36,37). Higher levels of amylose in ND 860-2 were believed to be responsible for a more ordered crystallinity within the starch granule, decreased thermomechanical analysis swelling, increased resistance to gelatinization and decreased susceptibility to α-amylase hydrolysis.

In addition to the various properties of starch, which affect its degradation, other factors such as enzymes responsible for starch degradation during LTS have been studied. The pathway of starch breakdown during LTS is not well established. The degradation of starch is believed to occur in the amyloplast (38). The widespread distribution of α-glucan phosphorylase, α-amylase, β-amylase, and maltase (39) suggests that starch breakdown could be phosphoryltic, hydrolytic, or both. However it is assumed that starch breakdown in cold stored potato tubers is mainly conducted by starch phosphorylase, because sucrose is the first sugar to accumulate upon transfer of tubers to chilling temperatures (19). Amylase activities are too low at such cold temperature to catalyze the required rate of starch degradation (40), and no increase in either maltose or polymers of glucose larger than maltose, the common products of amylolytic starch degradation (41), have been observed during LTS (42).

Two types of potato phosphorylases are recognized based on glucan specificity, monomer size, and intracellular location. They are noninterconvertible proteins with different primary structures and different immunological properties (38). Type 1 isozyme, also known as type H, is localized in the cytoplasm, has a low affinity for maltodextrins, has a high affinity for branched polyglucans, and cross reacts with type H phosphorylase from potato leaves. Type 2 isozyme, also known as type L, is located in the amyloplast, has a high affinity for maltodextrins, has a low affinity for branched polyglucans, and cross reacts with type L enzyme from the leaf (38,43). Type L and type H isozymes do not cross react immunologically. The function of these isozymes in starch degradation and LTS is unknown. There are reports which suggest that starch breakdown during LTS is phosphorylitic. Kumar et al. (44) have demonstrated that although the activities of cytosolic and plastidic isozymes of starch phosphorylase were reduced by up to 70% in transgenic potatoes expressing antisense cDNA constructs of starch phosphorylase, this did not affect the accumulation of reducing sugars during 4°C storage.

Other studies suggest that starch breakdown in potato tubers during cold storage is not solely due to the activity of starch phosphorylase. Cochrane and coworkers, (45) using a modified amylase assay, found that α- and β-amylases and α-glucosidase activities were much higher in tubers stored at the colder temperature (4°C) than those stored at 10°C, and in cultivars known to be more susceptible to LTS. It was considered inappropriate to correlate reducing sugar content and amylase activity, because the formation of reducing sugars is influenced by many other cold labile processes in the tuber. Reducing sugar content and the activities of α- and β-amylases and debranching enzymes were measured by Cottrell et al. (46) over 139 days in five cultivars of potato tuber stored at 4 and 10°C. The activities of these enzymes were always greater at 4 than at 10°C, but cultivars that accumulated high levels of reducing sugars did not always display the greatest level of hydrolytic enzyme activity (46). It has been reported that the onset of sugar accumulation in tubers during low temperature storage coincided with an increase in the activity of one specific isoform of amylase, the β-amylase in the cultivar Desirée (47–49). β-Amylase activity was present at low levels in tubers stored at 20°C, and increased from four- to fivefold within 10 days of storage at 3°C. However, no specific role has been established for this cold induced β-amylase in LTS.

16.2.3.2.1 Effect of Inorganic Phosphates The intracellular compartmentalization of Pi has been suggested to influence carbon partitioning in nonphotosynthetic potato tubers in a manner similar to its role in photosynthetic tissues. It has been observed that increased inorganic phosphate in the amyloplast shifted metabolic activities toward starch degradation rather than accumulation (Figure 16.1) (17). Increased Pi concentration inhibits ADPGPase and enhances starch breakdown by α-glucan phosphorylase. A high concentration of Pi was found in tuber amyloplasts (50), and Pi in cold stored tubers was later found to be cleaved off from starch (51). Higher levels of Pi were found in Russet Burbank potatoes stored at 5.5 than at 15.5°C (52). A highly significant correlation was found between the Pi content and the accumulation of reducing sugars. Amyloplasts were found to have high concentrations of Pi, citrate, Cl^-, and K^+. It was suggested that Pi leaks from the amyloplast to the cytoplasm during cold storage and induces higher sugar concentration levels in tubers during LTS (53).

Another source of Pi in plant cells is the vacuole (54). The major portion of Pi is stored in potato tuber vacuoles where it is compartmentalized away from the cytoplasm. Loughman (55) examined the respiratory changes of potato tuber slices and found that the larger part of Pi in the cell was localized in the vacuole, and did not take part in the steady state metabolism of the cell. However, the Pi in the vacuole may become available for metabolism in the cytoplasm during cold storage when ionic pumps that utilize ATP in the tonoplast become unable to maintain ionic gradients (56). This scenario could happen by passive leakage of Pi into the cytoplasm or when the membrane becomes leaky due to changes in the properties of the lipid bilayer. Increased Pi in the cytoplasm could affect the metabolism by mobilizing carbon from the amyloplast into the cytoplasm in exchange for Pi transported into the amyloplast by the hexose phosphate translocator (Figure 16.1) (57). Inside the amyloplast stroma, Pi can serve as a substrate for the continued phosphorlysis of starch, via α-glucan phosphorylase, forming additional molecules of G-1-P.Elevated cytoplasmic levels of Pi initiated by leaky membranes, coupled with reduced levels of fructose 2,6,bisphosphate during cold stress, would direct carbons away from glycolysis and favor the buildup of hexose phosphates for gluconogenic reactions (17).

16.2.4 Starch–Sugar Balance

The close relationship between starch and sugar levels when potatoes are cooled from 10 to 2°C and then rewarmed from 2 to 10°C after an interval, has given rise to a misleading concept of a starch to sugar balance in which the overall change between the two compounds is seen as being reversible (19). All the available evidence suggests that sucrose is formed from starch by an irreversible pathway, and that starch is formed from sucrose by separate, but likewise irreversible, routes. The very close relationship between starch and sucrose in stored potatoes may be due to the fact that starch is the only possible source of carbon for sugar synthesis in the cold (7,11,12,17). There is a strong evidence to indicate that G-6-P is transported into the amyloplast of potato tubers to support starch synthesis (26). Thus the pathways of starch and sugar biosynthesis compete for the same pool of precursors.

A net flux of carbon from starch synthesis into sucrose occurs in cold stored tubers as evidenced from the use of radiolabels in experiments (4,58). The coexistence of the pathways of sucrose synthesis and starch breakdown in stored tubers may be regulated by fine control mechanisms. In potato discs incubated with ^{14}C glucose at 3°C and 15°C, a large proportion of label is recovered in starch. At low temperature, in a cold sensitive, high sugar-accumulating cultivar, the ratio of ^{14}C recovered in sucrose to that recovered in starch increased (4,58), but was unaffected in a cold tolerant, low sugar-accumulating cultivar (4,58). This suggests that genotypic variation in the capacity to maintain an active starch synthesizing system may help in alleviating the rate of sucrose accumulation.

16.2.5 Sucrose Metabolism

Sucrose is the first sugar to accumulate during LTS. Its accumulation in potato tubers has been recorded within hours of their placement at LTS inducing temperatures, with the accumulation of reducing sugars occurring a few days later (19). Sucrose synthesis occurs in the cytoplasm of the tuber either by sucrose 6-P synthase (SPS), or by sucrose synthase (SS) and the hexose phosphates required for this are transported from the amyloplast via a phosphate translocator (Figure 16.1) (26,59).

Pressey (60) reported that SS activities decreased after harvest, and continued to do so under low temperature storage conditions. SPS activity also decreased if tubers were held at warm temperatures but rapidly increased when tubers were held at low temperatures. This observation indicates that SPS is the enzyme responsible for sucrose synthesis at low temperatures. Pollock and ap Rees (22) reported that sucrose synthesis during LTS is catalyzed by SPS and not by SS. This was also confirmed by ^{13}C NMR studies (61).

The increase in sucrose synthesis upon transferring the tubers to low temperature has been associated with the increased expression of an isoform of SPS (SPS-1b, 127 kDa) (62). The cold induced increase in the SPS-1b isoform was found to correlate well with the change in the kinetic properties of the enzyme. The major isoform found in tubers stored at room temperature is a 125-kDa protein (SPS-1a). Reconditioning of the tubers at 20°C resulted in the disappearance of the cold induced SPS isoform after 2–4 days (49). An increase in the total amount of SPS transcript was observed at low temperature in each of these studies. SPS from potato tubers has been shown to be subject to fine regulation by allosteric effectors and protein phosphorylation (63). Potato tuber SPS is allosterically activated by G-6-P and inhibited by protein phosphorylation.

Antisense technology has been used as an effective tool to investigate the roles of enzymes that lead to the production of sucrose, as well as reducing hexoses such as glucose and fructose in LTS. Many researchers investigating LTS mechanisms have used this technology to substantiate the role of enzymes in the carbohydrate metabolic pathway. For example, in experiments involving transgenic tubers where the SPS activity was reduced by 70–80% either by antisense or cosuppression, cold sweetening was reduced by inhibiting the increase of the cold induced isoform of SPS (64). The authors also observed that the V_{max} of SPS was 50 times higher than the net rate of sugar accumulation in wild type tubers, and found that SPS is strongly substrate limited, particularly for UDP-G (Figure 16.1). These results indicate that the rate of cold sweetening in wild type tubers is not strongly controlled by the overall SPS activity or the overall amount of SPS protein. Alterations in the kinetic properties of SPS during cold temperature storage were more effective in stimulating sucrose synthesis than changes in SPS expression. The observation that changes in the kinetic properties of potato tuber SPS coincide with the onset of sugar accumulation points to the fact that the fine regulation of SPS may be more important than coarse regulation in controlling the ability of a cultivar to sweeten during cold storage.

However, it should be noted that SPS may not be the only candidate that regulates sugar accumulation during LTS, because other factors that affect the availability of hexose phosphates, such as glycolysis and the pentose phosphate pathway, may have key roles to play (Figure 16.1, Figure 16.2).

UDP-glucose pyrophosphorylase (UGPase) is a cytosolic enzyme that catalyzes the formation of UDP-G, one of the substrates required for the synthesis of sucrose (Figure 16.1). Depending on the physiological state of the tubers (i.e, growth or post harvest storage), the UGPase reaction may be directed toward the synthesis or degradation of starch (10). During the process of cold sweetening, it has been suggested that UDP-G and PPi have regulatory roles in directing carbon flux into glycolysis, starch synthesis, hexose formation, or a combination of the three (17,65). The activity of UGPase has been correlated with the amount of glucose that tubers of different cultivars accumulate in cold storage (12), leading to the assumption that this enzyme might be a control point for low temperature sweetening, as it regulates the rate of SPS and sucroneogenesis by controlling the levels of UDP-G (17,66).

Genetic manipulation to down regulate the expression of UGPase in potato tubers has resulted in contrasting results based on the physiological stage of the tubers. In two separate experiments in which the UGPase activity was reduced by 30–50% compared to their wild types, the transgenic tubers accumulated lower levels of sucrose during storage relative to wild type tubers at 4°C and 12°C (67) and at 6°C and 10°C (68). It has been suggested that by limiting the rate of UDP-G synthesis, UGPase may exert control over the flux of carbon toward sucrose during the cold storage of tubers. These observations are supported by the results of Hill et al. (47) who observed that following the initiation of cold sweetening, the concentration of UDP-G changed in parallel with the concentration of sucrose.

In contrast to the above results, Zrenner et al. (69) observed that carbohydrate metabolism of growing tubers was not affected when the transgenic plants had a 96% reduction in UGPase activity as compared with the wild type plants. No significant changes were observed in the levels of fresh mass, dry mass, starch, hexose phosphates, or UDP-G at harvest relative to the wild type tubers. It was reported that 4–5% of UGPase activity was still in considerable excess compared to the activity of other glycolytic enzymes in the tuber, and the antisense construct may have to reduce UGPase to negligible levels in transgenic potatoes before any phenotypic differences are noticeable (70).

It should be noted that the flow of carbon is different based on the physiological state of the tuber. In the growing tuber, most of the incoming sucrose is used for the synthesis of starch, while in the stored tuber the hexose–phosphate produced from starch degradation is converted into sucrose. This explains the different responses obtained by Zrenner et al. (69), Spychalla et al. (67), and Borovkov et al. (68). In cold stored tubers, when the rate of starch breakdown exceeds the rates of glycolysis and respiration, the conversion of G-1-P to UDP-G is the only means of controlling the level of hexose phosphates. Hence it is possible that a significant effect of reduced UGPase activity may be observed only in tubers acting in the direction of sucrose synthesis, such as during post harvest storage.

Two UGPase alleles have been identified in potato tubers: UgpA and UgpB (67). In a survey conducted on a number of American and European cultivars and selections stored at 4°C, it was observed that a relationship existed between the allelic polymorphism of UGPase and the degree of sweetening. The genotypes that resist sweetening during cold storage have demonstrated a predominance of the allele UgpA; the genotypes susceptible to sweetening have a predominance of the allele UgpB (68).

In order to assess the role of UGPase in LTS, Sowokinos (13) cloned UGPase from 16 American potato cultivars and selections that have varying degrees of cold sweetening ability during storage at 3°C. It was observed that cultivars that were resistant to LTS possessed a UgpA: UgpB allelic ratio of 4:0 or 3:1. The cultivars demonstrating LTS revealed a ratio of 1:3 or 0:4 in favor of the UgpB allele. Sowokinos (13) also observed that the cold sensitive potato cultivars expressed up to three acidic isozymes of UGPase (UGP1, UGP2, UGP3) with UGP3 being the most abundant. In addition to the three isozymes present in the sensitive cultivars, the cold resistant cultivars possessed another two isozymes, UGP4 and UGP5 that were more basic in nature. Sowokinos (18) studied the physicochemical and catalytic properties of the purified UGP4 and UGP5 isozymes, and suggested that the differences in sugar accumulation between the cultivars and selections that are either sensitive or resistant to LTS may be partially due to the unique nature of expression and catalytic properties of the isoforms in resistant lines, including pH optimum, substrate affinities for G-1-P and UTP, V_{max}, and the magnitude of product inhibition with UDP-G. The overall effect of these differences in isozyme expression is that it may decrease the rate of UDP-G formation, resulting in a lower accumulation of reducing sugars in the cold resistant clones.

16.2.5.1 Sucrose Degradation

Sucrose plays a pivotal role in plant growth and development because of its function in translocation and storage, and the increasing evidence that sucrose (or some metabolite derived from it) may play a nonnutritive role as a regulator of cellular metabolism, possibly by acting at the level of gene expression (71). As mentioned earlier, sucrose is the first sugar to form during LTS, and the source of glucose and fructose accumulation appears to be the degradation of sucrose (72). Sucrose is broken down by two types of enzymes in plants. By invertase action, it is hydrolyzed into glucose and fructose; whereas by the action of SS, it is converted into UDP-G and fructose in the presence of UDP (73).

Potato tubers are known to possess both alkaline and acid invertases. Acid invertase is localized in vacuoles, whereas alkaline (neutral) invertase is localized in the cytoplasm (72,74). Acid invertase isoforms that are ionically bound to the cell wall have also been identified (74). Alkaline invertases are sucrose specific, while acid invertases cleave sucrose at the fructose residue but can also hydrolyze other β-fructose containing oligosaccharides such as raffinose and stachyose (74).

Based on several observations of sucrose synthase and acid invertase activities in developing, mature, and cold stored tubers, and given the fact that sucrose is stored mainly in the vacuole, it is believed that sucrose synthase is responsible for sucrose degradation in developing tubers, whereas acid invertase is the principal enzyme responsible for the breakdown of sucrose into hexoses during LTS (60,70,73,75–77). Based on the widely established inverse correlation between sucrose content and vacuolar acid invertase activity, it is strongly believed that sucrose is broken down by acid invertase in the vacuole and the resulting glucose and fructose are transported into the cytosol for the formation of hexose phosphate by hexoskinase (72,78–80). It has been reported that glucose concentrations are frequently higher than fructose concentrations in stored potato tubers (81). Zrenner et al. (82) evaluated the glucose to fructose ratio of 24 different cultivars and found that the ratios were between 1.1 and 1.6, which is a strong indicator that invertase is the key enzyme responsible for the conversion of sucrose to hexose.

Zrenner et al. (82) studied the effect of soluble acid invertase activity in relation to the hexose to sucrose ratio in 24 different potato cultivars and observed a strong correlation between the hexose to sucrose ratio and the extractable soluble invertase activity. They also isolated a cold inducible acid invertase cDNA from potatoes and developed transgenic potatoes expressing the invertase cDNA in an antisense orientation. The subsequent 12–58% reduction of acid invertase activity compared to the wild type tubers resulted in an accumulation of sucrose and a decrease in the concentration of hexoses. The hexose to sucrose ratio was found to decrease with decreasing invertase activities; however, the total amount of soluble sugars did not significantly change. Based on these observations, it was concluded that invertases do not control the total combined amount of glucose, fructose, and sucrose in cold stored potato tubers, but are involved in the regulation of the ratio of hexoses to sucrose (82). Greiner et al. (83) strongly inhibited the activity of cold induced vacuolar invertase in potato plants by repressing the activity, or by the expression of a putative vacuolar invertase inhibitor from tobacco (Nt-inh), in potato plants under the control of the CaMV 35S promoter. It was possible to decrease the cold induced hexose accumulation up to 75% without affecting tuber yield. Although the concentration of sugar produced during cold induced sweetening was decreased, the level of hexose accumulated was still in excess of what is commercially acceptable for the production of potato chips and fries. The observation that antisense expression of acid invertase did not control the total amount of soluble sugars in cold stored potato tubers (82) indicates that other factors in the carbohydrate metabolism may influence the regulation of the total amount of sugars accumulated, and that acid invertase could be only one of the enzymes involved in starch–sugar conversion.

Metabolism is more rigorously regulated by intracellular compartmentalization in plants than in animals (84), and compartmentalization of the pathways of carbohydrate catabolism is realized to be a distinct feature of plant respiration (70). It is now believed that a "futile" cycling (simultaneous synthesis and degradation) of sucrose functions continuously to allow plants to respond rapidly to demand for carbon (85). This metabolic cycle may also be involved in LTS. For instance, it has been observed that in the first two weeks of 4°C storage, the initial rates of sucrose accumulation corresponded closely with the estimated rates of sucrose synthesis (47). The rate of total soluble sugar accumulation decreased with increasing duration of cold storage. It is suggested that sugar accumulation decreased because the rate of recycling equalled the rate of synthesis.

16.2.6 Glycolysis

The effects of cold exposure on the metabolism of potato tubers indicate that cold induced sweetening may at least in part be due to differential sensitivity to low temperature of the enzymes in the glycolytic pathway (86). The available data suggest that phophofructokinase (PFK) and pyruvate kinase are more sensitive to cold than are the other enzymes involved in the metabolism of hexose-6-phosphates (Figure 16.2), and by lowering the temperature, divert the latter to sucrose (Figure 16.1) (22,86,87). Another glycolytic enzyme that has been studied in relation to LTS is fructose-1,6-bisphosphatase (FBPase). Plants possess ATP dependent (ATP-PFK) and PPi dependent (PPi-PFK) phosphofructokinases (88).

PPi-PFK is a cytosolic enzyme and experiments to examine the role of PPi-PFK during the aging of tissue slices from potato tubers (starch-storing tissue) and carrot roots (sucrose-storing tissue) showed that both vegetables showed the same pattern of changes of phosphorylated metabolites and fructose 2,6-bisphosphate. But, the consumption of PPi by tubers and the production of PPi in carrots indicated PPi-PFK control of the glycolytic flux in tubers and catalysis of the opposite reaction in carrot roots (89). PPi-PFK is activated by fructose-2,6-bisphosphate which does not affect ATP-PFK (90). The activity of PPi-PFK is often equal to or exceeds that of ATP-PFK (72,90). It has been reported that the activity of PPi-PFK was ten times that of ATP-PFK in developing potato tubers, and hence it is suggested that glycolysis may proceed regardless of the activity of ATP-PFK (91). The maximum activity of PPi-PFK has also been shown to be greatly reduced in tubers stored at a low temperature of 5°C due to a decrease in PPi-PFK affinity for fructose 2,6-bisphosphate, an increase in sensitivity to fructose 2,6-bisphosphate as an activator, and a decrease in fructose 2,6-bisphosphate concentration at decreasing temperature (92). By contrast, in another study, no evidence was found for a cold induced inhibition of PPi-PFK in tubers stored at 2°C and 8°C (93). Hence it was postulated that PPi-PFK contributes to LTS by regulating the PPi concentration below inhibitory levels, facilitating the formation of UDP-G and subsequent synthesis of sucrose (Figure 16.1 and Figure 16.2).

ATP-PFK has been implicated in the regulation of LTS (94,95). It is responsible for the irreversible ATP dependent conversion of fructose-6-phosphate to fructose-1,6-bisphosphate. Potato tubers have been reported to possess four isozymes of ATP-PFK (96). It has been reported that the temperature coefficient (Q_{10}) of three of the four isoforms were higher at 2–6°C than at 12–16°C, indicating the cold lability of these isoforms and their roles in the accumulation of hexose phosphate and sucrose synthesis in LTS (95). This result supports the suggestion of Bryce and Hill (97) that ATP-PFK dominates the control of glycolysis, and thereby respiration, in plants. However, the observation that respiration of potato tubers increases concomitantly with the initial increase in sugar concentration (19,98), and the fact that the conversion of fructose-6-phosphate is also catalyzed by PPi-PFK (ure 16.2) indicate multiple regulatory controls in the biosynthesis of sugar phosphates.

Genetic manipulation of the two PFKs was carried out further to explore their roles in LTS. About 88–99% inhibition of PPi-PFK expression was obtained in stored tubers by antisense expression of PPi-PFK cDNA (99). Even though the transformation resulted in higher levels of hexose phosphates in transgenic tubers compared to their wild type tubers, no difference was observed between these tubers in the rates of sucrose and hexose accumulation, and the total amounts of sugars accumulated at 4°C. Besides, no change was observed in the maximum catalytic activities of ATP-PFK or other enzymes of glycolysis (pyruvate kinase) or sucrose breakdown (invertase and sucrose synthase) in the antisense tubers. This observation suggests that compensation occurs at the level of fine metabolic regulation rather than gene expression. The above results indicate that PPi-PFK may not control the rate of glycolysis at low temperatures, and that tubers possess excessive capacity to phosphorylate fructose-6-phosphate. The results are also not in agreement with the theory proposed by Claassen et al. (93), that PPi-PFK is involved in regulating the PPi concentration, as no evidence was observed to substantiate that the antisense and wild type tubers contained different PPi concentrations (99).

Expression of the *E. coli* pfkA gene in potato tubers resulted in a 14- to 21-fold increase in the maximum catalytic activity of ATP-PFK, without affecting the activities of other glycolytic enzymes (100). It was also found that no corresponding decrease in the concentration of hexose phosphate was observed, while the pool sizes of other glycolytic intermediates increased three- to eightfold. In another study, it was reported that a substantial increase in ATP-PFK activity did not affect the flux through glycolysis or a flux between glycolysis and the PPP (101). The above results suggest that ATP-PFK may not limit the rate of respiration of potato tubers. ATP-PFK is potently inhibited by phosphoenolpyruvate, and hence ATP-PFK activity may be dependent upon the activity of enzymes that metabolize phosphoenolpyruvate (PEP) such as pyruvate kinase and phosphoenolpyruvate phosphatase. The above contention is in agreement with the findings of Thomas et al. (102). Using metabolic control analysis (MCA) on tuber glycolysis, Thomas et al. (102) observed that ATP-PFK exerts little control over glycolytic flux, while far more control of flux resides in the dephosphorylation of PEP.

Fructose 1,6-bisphosphatase is localized in the plastids and in the cytosol. Cytosolic FBPase is involved in hexogenesis, converting fructose-1,6-bisphosphate to fructose-6-phosphate, which is used by SPS as one of the substrates for the production of sucrose-6-phosphate (Figure 16.1 and Figure 16.2). FBPase is potently inhibited by fructose-2,6-bisphosphate (90,103), a metabolite which is also a potent activator of PPI-PFK (104). In a study to investigate the role of FBPase in LTS, it was observed that there was a rapid increase in the levels of sucrose and reducing sugars in tubers stored at 2°C, but no change in FBPase activity, relative to 8°C storage.

In a study carried out to identify the regulatory steps in glycolysis, a decline in phosphoenolpyruvate and a rise in pyruvate were observed when potato tubers were stored under anoxic conditions (86). As this step is preceded by phosphofructokinase, pyruvate kinase cannot regulate glycolytic flux directly as it cannot control the entry of glucose-6-phosphate into glycolysis. However, pyruvate kinase could play a role in the regulation of the movement of carbon out of glycolysis and into the oxidative pentose phosphate pathway. It has been suggested that the cold lability of phosphofructokinase and pyruvate kinase could lead to a rapid reduction in hexose phosphate consumption, which could cause their diversion to sucrose (86).

The theory that the cold lability of enzymes in the glycolytic pathway diverts hexose 6-phosphate for sucrose production and thus to LTS cannot fully explain LTS, as it takes time for potatoes to sweeten fully (7). From the results of a study carried out by Marangoni et al. (16), by comparing LTS resistant (ND860-2) and LTS susceptible (Norchip) potato cultivars, it has been suggested that tubers with decreased invertase activity along with increased glycolytic or respiratory capacity, should be more tolerant to low temperature stress.

16.2.7 Oxidative Pentose Phosphate Pathway (PPP)

Although the PPP is usually depicted as being separate from glycolysis, the two pathways are intimately linked (Figure 16.2). They share the common intermediates glyceraldehyde 3-phosphate, fructose 6-phosphate and glucose 6-phosphate, and flow through either of the pathways will be determined by the metabolic needs of the cell. The main function of PPP is to generate NADPH for various biosynthetic reactions (105). It was proposed by Wagner et al. (106) that for low sugar accumulating cultivars, the PPP may provide a means of preventing the accumulation of high levels of sugars when tubers are stored below 10°C (by bypassing phosphofructokinase). However, no differences were observed in the specific activities of glucose 6-phosphate dehydrogenase and 6-phosphogluconate dehydrogenase among LTS resistant and LTS susceptible potato cultivars stored at 4°C and 12°C, respectively (107). It was observed that the LTS resistant cultivars exhibited higher activities of G6PDH and 6PGDH, relative to the LTS susceptible cultivars.

16.2.8 Mitochondrial Respiration

During storage of potato tubers below 5°C, in addition to changes in sugar accumulation patterns, respiration changes have also been observed (108). It has been reported that the cold resistant potato clone ND860-2 has shown a higher respiration rate throughout storage compared to the cold susceptible Norchip (109). Respiration decreases as storage temperature decreases, but at storage below 5°C, respiration is stimulated. There is a brief respiratory burst attributed to the combined effect of cyanide resistant (alternative pathway) and cytochrome mediated pathways (108), followed by a subsequent decrease in respiration rate to a new steady state (19,99).

It has been suggested that during chilling stress, an alternative oxidase pathway may play a protective role in the mitochondrion by preventing both an over reduction of the respiratory chain and the consequent production of reactive oxygen species that cause cellular damage (110). It has been suggested that the alternative pathway operates only during periods of high cellular energy charge, or when there is an imbalance between the supply of carbohydrates and the requirement for carbohydrates for structural growth, energy production, storage, and osmoregulation (111,112). There is also evidence which suggests that physical characteristics of the cellular membrane (i.e., mitochondrial membranes) may activate the alternative pathway (113). In a study carried out by Amir et al. (99) to study the relationship between respiration rate, sugar content, and ATP levels in cold stored tubers, an immediate decrease in respiration rate was observed upon storage at 4° C. The respiratory minimum was concomitant with an ATP maximum which is followed by a respiratory burst and a rapid decline in ATP content. This evidence suggests the presence of an active alternative pathway in cold stressed tubers. Expression of the alternative pathway is known to increase with decreasing temperatures (114). It has been suggested that sucrose formation could serve as an effective sink for excess ATP via the alternative pathway (113). In agreement with Solomos and Laties (113), it was observed that low O_2 levels, which inhibit the alternative pathway, were effective in suppressing sugar accumulation in tubers stored at 1°C (115) which suggests that LTS may be directly linked to the onset of cyanide resistant respiration.

16.2.9 Compartmentation and Stress Induced Membrane Changes

Membranes play an integral role in the response of plant tissues to chilling and freezing. It has been proposed that the thermotropic phase transition of membrane lipids might play an initiative role in the chilling sensitivity of plants (116–118). With further exposure to chilling, the phase separated biomembranes become incapable of maintaining ionic gradients and cellular metabolism becomes disrupted. The occurrence of phase separation as the initial event in chilling injury has been demonstrated in cyanobacterium *Anacystis nidulans* (119). It has been argued that such a phase separation would not occur in plant cells because they contain high levels of polyunsaturated fatty acids in their membranes. However, a positive correlation has been observed between chilling sensitivity of herbaceous plants and the level of saturated and transmonounsaturated molecular species of phosphatidylglycerol in thylakoid membranes (120,121).

It is likely that the regulation of starch breakdown and of sucrose synthesis is to some extent achieved by compartmentation. Therefore, it is possible that low temperature sweetening is, at least in part, due to effects of cooling on such compartmentation (7). In potatoes, studies have been performed on the effects of cold storage on lipid composition and membrane permeability (122,123), the associated biophysical changes of amyloplast membranes (124,125) and mitochondrial membranes (126), and lipid peroxidation (15,127–129). The results of these studies are described below.

16.2.9.1 Lipid Composition

Phospholipids and glycerolipids are the major potato lipids (130). It has been found that the combined proportion of polyunsaturated fatty acids (linoleic and linolenic) for all potato varieties examined consistently represents 70–76% of the total fatty acids, which help maintain membrane fluidity at lower temperatures (130).

In plants stressed by low temperature or other factors, the survival of the plant is based on the ability of the plant to maintain or reestablish membrane fluidity (131). Fatty acid desaturases play a central role in regulating the level of unsaturation of fatty acids in membrane lipids, which helps maintain membrane fluidity or refluidizes the membranes that have become rigid due to low temperature exposure (132). Bonnerot and Mazliak (133) reported cold induced oleyl-PC desaturase activity in microsomes from 16 h aged slices of potato tubers stored at 4°C for 3 months. Spychalla and Desborough (134) reported that the total amount of linoleic and linolenic acids remained constant, but the ratio of linolenic to linoleic increased over storage time for both tubers stored at 3°C and 9°C. Low temperature storage of potato tubers has been shown to increase the levels of

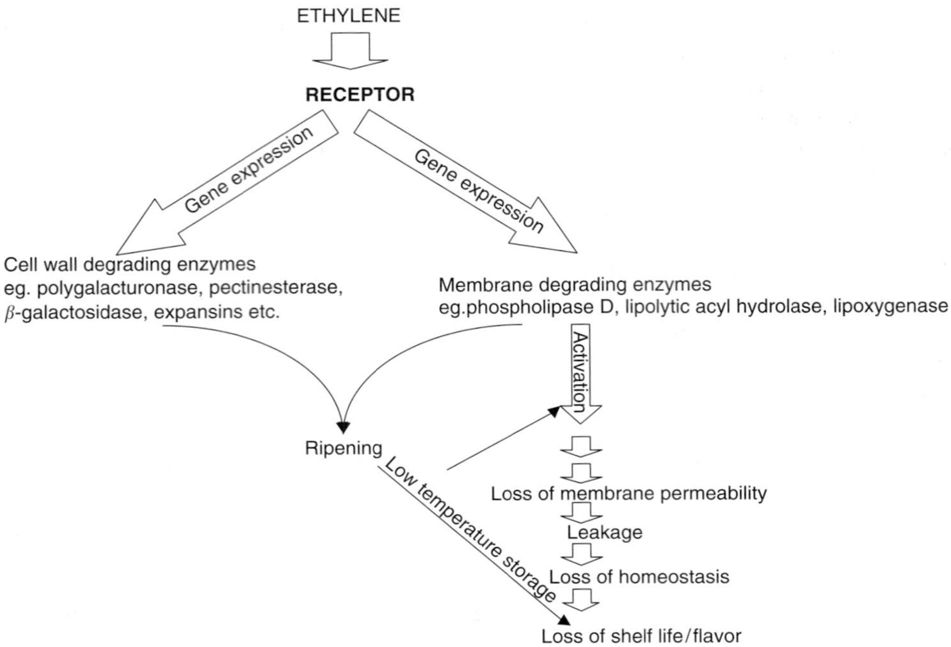

Figure 16.3 Schematic diagram illustrating the early events in the membrane and cell wall degradation during fruit ripening.

monogalactosyl diacylglycerol and digalactosyl diacylglycerol (135–137). Hence, it is possible that one or more of the low temperature induced changes in the lipid composition of the membrane may contribute to LTS or confer resistance to LTS.

16.2.9.2 Membrane Permeability

Several studies have reported that cold temperature damages membranes, resulting in the loss of compartmentation and homeostasis of the cell (138–140). As temperature decreases, membrane lipids undergo a phase transition from a liquid crystalline to a solid gel phase that results in cracks in the membrane, increases the membrane permeability and leakage of ions, and alters metabolism (141). The temperature range for the phase change in membranes is specific for each horticultural commodity and is a function of the heterogeneity of the lipid content, the ionic environment, and the presence of sterols and proteins (142).

From the earlier discussions, it is believed that LTS is caused by the effect of low temperature on many pathways of carbohydrate metabolism at the level of starch synthesis and breakdown, sucrose synthesis, hexogenesis, glycolysis, the PPP, and mitochondrial respiration. These pathways are also compartmentalized in the plant cell, and involve amyloplasts, the cytoplasm, vacuoles, and mitochondria. For the maintenance of homeostasis in normal cells, a tight control of the movement of the substrates or intermediates of these pathways is in place. However, when the plant, or a cell of the plant, experiences stress (such as low temperature stress as in the case of LTS), the normal metabolism of the cell will be lost, which might make the cell a candidate for readjustment at various levels. It is assumed that the changes caused by low temperature at the membrane level might be one of the factors contributing to LTS.

16.2.9.2.1 Tonoplast Membrane Permeability The "leaky membrane theory" of LTS suggests that the cause of LTS may be a leaky tonoplast membrane that allows Pi to be leaked into the cytoplasm from the vacuole (17). A high concentration of Pi in cytoplasm is believed to mobilize carbon from the amyloplast to the cytoplasm, while cytoplasmic Pi participates with G-1-P in a reversible exchange across the amyloplast membrane. Increased Pi concentration in the amyloplast favors the α-glucan phosphorylase activated starch breakdown and inhibits ADPGPase mediated starch synthesis (17). In addition Pi affects fructose 2,6,-bisphosphate, phosphofructokinase, sucrose synthase, and UGPase (see section 16.2.3.2.1). The leaky membrane theory is been supported by subtle changes in fatty acid composition of potato membranes as well as by increases in electrical conductivity, which is an indicator of electrolyte leakage and membrane permeability. It has been reported that the relative change in electrical conductivity of four cultivars of potato paralleled the increase in sugar concentration when temperature was dropped from 20°C to 0°C (143). A difference in electrical conductivity was noticed among cultivars that had accumulated similar amounts of sugars. It was concluded that the increased electrical conductivity was not due to the increase in the accumulation of sugars, particularly because respiration rates were found to increase before the increase in sugar concentration and electrical conductivity.

Knowles and Knowles (122) studied the relationship between electrolyte leakage and degree of saturation of polar lipids in Russet Burbank seed tubers stored at 4°C, and observed an inverse linear relationship between the double bond index (DBI) and electrical conductivity ($r = -0.97$). The DBI revealed that the proportion of unsaturated fatty acids in membranes decreased over storage time with an accompanying increase in electrical conductivity. The authors concluded that the ability to increase membrane lipid unsaturation in storage could confer resistance to electrolyte leakage by maintaining the fluidity of the membranes. This result was supported by another study, in which it was observed that tubers stored at 3°C had greater increase in sugar content, total fatty acid saturation, and membrane permeability, as compared to tubers stored at 9°C (123). These studies suggest that high initial or high induced levels of lipid unsaturation could prevent increased membrane permeability during low temperature storage. It has been reported that cyanobacteria transformed with *des*A gene, which encodes a 12 acyl-lipid desaturase in *Synechocystis* PCC6803, did not show any significant changes in photosynthetic activity below 10°C, whereas in the wild type cells, the photosynthetic activity was decreased irreversibly (132).

16.2.9.2.2 Amyloplast Membrane Permeability It has been suggested that the low temperature induced defects in amyloplast membrane allow α-glucan phosphorylase from the cytoplasm to enter the amyloplast and degrade starch, resulting in the accumulation of sugars during LTS (144,145). Electron spin resonance study, used to examine the amyloplast membrane in potato tubers stored at 5.5 and 15.5°C, showed a strong relationship between membrane permeability and starch to sugar conversion (146). Studies using spin labeled probes revealed that at low temperatures, membranes exhibited decreased lipid fluidity. O'Donoghue et al. (125) observed that membrane lipid phase transitions were higher for Norchip (an LTS susceptible cultivar) than ND 860-2 (an LTS resistant cultivar) at both 4 and 12°C storage, and Norchip amyloplast membranes were more ordered at 4 than 12°C. The drop in double bond index (DBI) was 93% for Norchip while only 70% drop was observed for ND 860-2 due to loss of linoleic and linolenic acid. It was suggested that low temperature caused Norchip membranes to undergo deterioration to a greater extent than ND 860-2 membranes and this could have contributed to LTS. By contrast, it has been reported that the amyloplast membrane breaks down during senescence but remains relatively intact during LTS (147–149). Based on TEM examination of membranes from LTS resistant and LTS susceptible cultivars of potatoes stored at 5 and 10°C, Yada et al. (149) concluded that LTS is not the result of amyloplast membrane breakdown. However, it is likely that changes in membrane can take place at the molecular composition or organizational level, which can affect the permeability or transport properties, or both, and can contribute to LTS.

16.2.10 Free radicals and Antioxidant Enzymes

Evidence from several lines of research suggests that a variety of toxic oxygen species such as superoxide radicals, hydrogen peroxide, and hydroxyl radicals are produced in plants exposed to various environmental stresses such as high and low temperatures, drought, light, and exposure to pollutants causing oxidative damage at the cellular level (150). Lipid peroxidation is considered to be one of the reasons for membrane deterioration during senescence and chilling injury (140), and results from the activity of lipoxygenase, resulting in the formation of lipid peroxides and free radicals (151). In a study to analyze the relationship between sugar accumulation and changes in membrane lipid composition associated with membrane permeability in early stages of LTS, sucrose cycling and accumulation were greatest for Norchip, a LTS susceptible cultivar at 4°C as compared to LTS tolerant ND 860-2 (15). No significant changes were observed in phospholipid, galactolipid, free sterol levels, or phospholipid to free sterol ratio. However, the double bond index obtained from the fatty acid profiles of the total lipid fraction decreased significantly (decreased unsaturation) for Norchip tubers at 4°C over time. Free fatty acid and diene conjugation values fluctuated and increased over time for both Norchip and ND 860-2 stored at 4°C and 12°C, with greater amplitude of fluctuations observed for Norchip stored at 4°C. From the results, it has been suggested that these effects may be due to the high levels of lipid acyl hydrolase and lipoxygenase found in potato tubers, and the observed peroxidation products could relate low temperature stress and the resultant LTS to chilling injury and drought stress (15). However, Fauconnier et al. (152) could not observe a correlation between cold sweetening and membrane permeability or lipid saturation status. They studied the effect of three storage conditions: at 4°C, at 20°C with sprout inhibitors, and at 20°C without sprout inhibitors, and observed that during storage at 20°C without sprout inhibitor, the increase in membrane permeability is inversely correlated to sucrose accumulation. It was also observed that lipoxygenase activity and gene expression are not correlated with the fatty acid composition of the membrane. It was also observed that the lipoxygenase activity and fatty acid hydroperoxide content are low in older tubers, irrespective of the storage conditions and the varieties. Spychalla and Desborough (134) studied the antioxidant potential of potato tubers stored at 3 and 9°C and observed that tubers stored at 3° C had higher superoxide dismutase activities than their 9°C counterparts and demonstrated time dependent increases in superoxide dismutase, catalase, and α-tocopherol during the 40 week storage period. They also observed that low sugar clones had significantly higher levels of superoxide dismutase and catalase than high sugar clones but significantly lower levels of α-tocopherol. The increased antioxidant responses could be due to increased free radical production as manifested by the higher levels of superoxide dismutase, catalase, and peroxidase activities in seed tubers stored at 4°C for 20 months as against those stored for 8 months (129).

From the above discussion, it can be concluded that LTS is not the result of a single cause. The sugar balance in potato is regulated by many intermediate carbohydrate metabolic pathways, which are subject to genetic and environmental control. It might be possible that the low temperature effects on enzymes involved in carbohydrate metabolism result in an imbalance in the normal metabolism combined with its effect on membrane fluidity, thus diverting or leaking the intermediates, or both, in the biosynthetic or respiratory pathway for sucrose and reducing sugar production. Hence, even though genetic engineering has great potential in manipulating or improving crop productivity and the quality of horticultural crops, because of the complexity of LTS, in depth research on various molecular and biochemical properties and their correlation to LTS need to be conducted before we can fully exploit that potential. The existence of cultivars resistant to LTS with several molecular, biochemical, or compositional characteristics, or combinations of these, might provide a better tool in understanding the mechanism that is responsible for resisting LTS in those cultivars. We can therefore be hopeful that the coordinated efforts of plant biochemists, molecular biologists and traditional plant breeders would help to better understand and control LTS, thus eliminating the use of chemical sprout inhibitors and their harmful safety issues.

16.3 TOMATO

Tomatoes rank second to potatoes in dollar value among all vegetables produced in the USA and in other parts of the world where they are grown (1). In terms of per capita consumption, processed tomato products lead all the other processed vegetables. The main factors in determining the postharvest deterioration of fruit and vegetable crops are the rate of softening of the fruit which influences quality, shelf life, wastage, infection by postharvest pathogens, and frequency of harvest, and which limit the duration of transportation and storage. Damage to the structure and function of the membrane affects the post harvest shelf life and quality of fruits, vegetables, and other food sources by causing leakage of ions from cellular storage compartments into the cytosol, thereby disrupting the homeostasis of the cell (153,154). A major problem faced by the fruit and vegetable fresh market and processing industry in the Northern Latitudes is the lack of a year round supply of high quality material. Even though cold storage can be used for long term storage of fruits and vegetables, in the case of tomatoes it is not effective due to the sensitivity of tomatoes to being chilled. Many factors affect the shelf life of tomato products, but our discussion will be based mainly on membrane changes during cold storage and the genetic manipulation to circumvent these factors.

16.3.1 Role of Membrane in Shelf life

The development of fruit in many plants can be interpreted as following a two step process. During the first phase, the ovary or hypothalamus within the flower expands and develops into a full sized fruit. During the second phase, the full sized fruit undergoes ripening, a complex set of molecular and physiological changes in the fruit. The ripening process brings dramatic changes to the fruit: softening, biosynthesis of pigments, and increase in sugar content, flavor, and aroma. In climacteric fruits such as tomatoes, and many other fruits, ripening begins with increased respiration and ethylene biosynthesis (155). Fruit ripening can be considered as the beginning of senescence of the fruit (156).

Senescence can occur at various levels, from cellular to whole plant levels, and is regulated by genetic, hormonal, and environmental factors (140,157). The plant hormone ethylene plays a major role in the ripening and senescence processes, and extensive work has been conducted in the past two decades on the role of ethylene in fruit ripening and signal transduction. Genetic manipulation to increase the shelf life of fruits, especially tomatoes, has been extensively undertaken, resulting in several new transgenic varieties with improved storage and quality characteristics (158–164).

Senescence is characterized by membrane deterioration resulting from the catabolism of membrane lipids and proteins. The pathway of the catabolism of phospholipids has been elucidated from several senescing systems and involves the sequential action of enzymes that include phospholipase D (PLD, phosphatidyl choline hydrolase, EC 3.1.4.4, PLD), phosphatidate phosphatase (3-*sn*-phosphatidate phosphohydrolase, EC 3.1.3.4), lipolytic acyl hydrolase and lipoxygenase (linoleate, oxygen oxidoreductase, EC 1.13.11.12) (140). In most systems studied, the first step in the lipid catabolic pathway is the conversion of phospholipid to phosphatidic acid by PLD, even though phosphatidylinositol and its phosphorylated forms may be acted upon by both phospholipase C and PLD (165,166). Phosphatidic acid does not accumulate, as it is immediately converted to diacylglycerol by phosphatidate phosphatase. Diacylglycerol is deacylated by lipolytic acyl hydrolase, liberating free fatty acids. Among the free fatty acids, unsaturated fatty acids with 1–4 pentadiene systems (18:2 and 18:3) serve as substrates for lipoxygenase, resulting in the formation of fatty acid hydroperoxides. Fatty acid hydroperoxides undergo a variety of reactions by virtue of their active unstable structure, including the generation of free radicals. The free radicals damage the protein as well as the membrane, giving rise to the characteristics features of senescence. The above reactions are deemed autocatalytic, as the reaction products increasingly contribute to the formation of gel phase and nonbilayer lipid structures resulting in the destabilization of the membrane and eventually in the loss of homeostasis. Similar changes occur in the membrane in response to chilling injury, but also involve the effect of low temperature on the catalytic activity of enzymes involved as well as the effect on the phase transition temperature of the lipids (see section 16.2.9). Here we will be emphasizing the role of PLD and lipoxygenase in enhancing the shelf life of tomatoes during cold storage.

16.3.2 Phospholipase D Gene Family

Phospholipids provide the backbone for biomembranes and serve as rich sources of signaling messengers. Phospholipase D (PLD, EC 3.1.4.4) catalyzes the hydrolysis of structural phospholipids to generate phosphatidic acid and a free head group. PLD has been grouped into three classes based on their requirements for Ca^{2+} and lipids in *in vitro* assays: the conventional PLD that is most active at 20 to 100 mM levels of Ca^{2+} ; the polyphosphoinositide (PI) dependent PLD that is most active at micromolar levels of Ca^{2+}; and the phosphatidylinositol (Ptdln) specific PLD that is Ca^{2+} independent (167). PLD has been cloned from a number of plants (168), animals (169), and fungi (170) and found to constitute a supergene family of many isoforms (171). The PLD isoforms from *Arabidopsis* have been divided into five groups: PLDα, β, γ, δ, and ε. (168). The PLDα gene product is responsible for the conventional PLD activity and differs physiologically from the PLDβ, and PLDγ isoforms based on Ca^{2+} requirements and pH (172). From tomatoes, three PLDα forms and two PLDβ classes have been cloned (173). Different expression patterns were observed in different plant tissues and organs for each PLD. In fruit, PLDα3 appeared to be transiently accumulated during early ripening, whereas PLDα2 was accumulated throughout fruit development and maturation (173).

16.3.2.1 Role of PLD in Senescence and Chilling

PLD activities have been observed in many cellular functions during seed germination, aging, various abiotic and biotic stresses, and senescence (140,167,174). In membranes of tomato fruit stored at low temperature, an accumulation of phospholipid catabolites occurs due to differential effects of reduced temperature on the activities of lipid degrading enzymes (175). Destabilization of membranes has been suggested as one of the causes of chilling injury (176). In ripening tomato fruits, rigidification of microsomal membranes has been reported to activate PLD and increase membrane catabolism (177). Increased PLD activities during and after chilling are suggested to result in chilling injury in maize (139) and cucumber fruits (178). In *Arabidopsis*, cold stress increased the expression of PLDα, but not PLDβ or PLDχ, implying that PLDα has a role in plant responses to low temperature stress (174). Recently the many roles of PLD in signal transduction have been reported (167). In tomatoes, PLDβ is suggested to have a role in signal transduction due to its low abundance, activation by micromolar concentration of Ca^{2+}, a concentration range that arises locally during signaling (173), whereas a metabolic role has been assigned for the various forms of PLDα. Our studies on the role of PLD during fruit ripening in cherry tomatoes showed that the soluble and membrane associated PLD activities increased during fruit development, which peaked at the mature green and orange stages (178a). It was reported by Jandus et al. (179) that PLD activity decreased slightly between the mature green and orange stages when the tomatoes ripened on the plants, but increased between the orange and red stages to values higher than at the mature green stage. However, when tomatoes were harvested at the mature green stage and left to ripen at room temperature, PLD activity decreased by about 40% between the mature green and red stages. It has been reported that phosphatidic acid increased as much as twofold while total phospholipids decreased about 20–25% during ripening of tomato pericarp (180,181). Treatment with lysophosphatidylethanolamine, which acts as a specific inhibitor of PLD activity (182) retarded senescence in tomato fruits and leaflets (183). From the observations on PLD activity at low temperature, ripening and senescence, it is conceivable that the quality of tomatoes, which is highly chilling sensitive, may be affected to a large degree by the modulation of PLD activity.

16.3.2.2 Regulation of PLD Activity

The activities of PLD are affected by a number of factors such as Ca^{2+} concentration, substrate lipid composition, pH changes, and mastoparan, a tetradecapeptide G-protein activator (167). Sequence analysis indicates that plant PLDs contain a Ca^{2+} to phospholipid binding fold, called the C2 domain at the N terminus. The C2 domains of PLDα and PLDβ have been demonstrated to bind Ca^{2+}, with PLDβ having a higher affinity for Ca^{2+}, whereas the Ca^{2+} requirement of PLDα is influenced by pH and substrate lipid composition (167). PLDα is active at near physiological, micromolar Ca^{2+} concentrations at an acidic pH of 4.5–5 in the presence of mixed lipid vesicles. PLDβ and PLDχ are optimally active under physiological micromolar concentration of Ca^{2+} concentrations at neutral pH and may play an active role in signal transduction. The relative distribution of PLD between the soluble and membrane fractions changes during development and in response to stress (184,185). It has been shown that Ca^{2+} binding increases the affinity of the C2 domains for membrane phospholipids (186). This shows that the C2 domain in PLD is responsible for mediating a Ca^{2+} dependent intracellular translocation between the cytosol and membranes. An increase in cytosolic Ca^{2+}, as well as a decrease in cytosolic pH, has been reported to occur in response to stress (187), which are favorable conditions for the activation of PLDα. PLDs associated with microsomal membranes are correlated with stress induced activation of PLD mediated hydrolysis (184,185). The increased association of preexisting PLD in the cell with membranes may represent a rapid and early step in PLD activation during stress responses (185).

16.3.2.3 Antisense Suppression of PLD Activity

In order to study the role of PLD in fruit ripening and senescence of fruits, we have developed transgenic tomatoes expressing antisense PLDα cDNA. The fruits from antisense Celebrity tomato (a fresh eating type) were smaller than the control fruits and showed a 30% decrease in PLD activity during development (178a). After storage for two weeks at room temperature, the control fruits developed wrinkles, indicative of senescence and dehydration, whereas the transgenic fruits appeared to be relatively normal. The transgenic fruits were also firmer, possessed a higher level of red pigmentation and increased level of soluble solids (178a). Transgenic Celebrity fruits showed a decrease in PLD expression as evidenced from Northern blot. Even though very few transcripts were detected at the mature green, orange and red stages in the antisense PLD celebrity fruits, PLD activity was present at these stages suggesting a very low turnover rate of PLD, and that PLD synthesized at young or intermediate stages remains functional even at the red stage (178a). These results suggest that for effective inhibition of PLD using antisense suppression, PLD expression has to be reduced at an early stage of fruit development using an appropriate fruit specific promoter. In our experiments we have used a constitutive promoter (CaMV 35S) for tomato transformation. The antisense Celebrity fruits also showed low levels of PLD activity during ripening, suggesting that the natural senescence process was retarded, which was translated into increased firmness in these fruits. This observation was contrary to earlier results in *Arabidopsis* where antisense suppression of PLDα resulted in retardation of ABA and ethylene promoted leaf senescence, without affecting the natural senescence of leaves (188). It has been reported that the phospholipid content of tomato fruit declines during ripening (180). This decrease in phospholipid content could be due to a high PLD activity. It was interesting to note that antisense suppression of PLDα in an ornamental cherry tomato cultivar, Microtom, did not show any significant reduction of PLD activity (178a). However, ethylene climacteric of the transgenic fruits was delayed by nearly six days, as compared to the control fruits. In Microtom, PLD activity declined during ripening in the control fruits, whereas transgenic fruits retained much higher levels of PLD activity. This may be related to the delayed climacteric in the transgenic fruits, indicating a slower rate of deterioration. *In situ* localization of PLD by immunolabeling followed by electron microscopy also supports this observation. These results suggest that fruits from different cultivars may differ in their pattern of senescence and the relative role of PLD may differ between fruits and leaves. It is unclear why PLD activity in transgenic Microtom was higher compared to control plants during ripening, as opposed to the observation in Celebrity. It has been reported that in PLDα suppressed *Arabidopsis*, the expression and activities of other PLD isoforms are not altered (189) which means that the other PLD members cannot compensate for the loss of PLDα.

16.3.3 Lipoxygenase

Lipoxygenases (LOX, EC 1.13.11.12) are a class of enzymes that catalyze the hydroperoxidation of *cis-cis*-1,4-pentadiene moieties in polyunsaturated fatty acids and that occur widely in both plant and animal kingdom. LOX pathways in higher plants and mammals are different in two main respects: in mammals the main LOX substrates are arachidonic and eicosapentaenoic acids whereas linoleate and α-linolenate are the most important LOX substrates in plants; and the hydroperoxide metabolizing enzymes are different in plants and mammals (190). Although LOX isoforms occur in most plant cells, the tissue specific expression level of LOX within a plant can vary substantially depending on developmental and environmental conditions (191). Plant lipoxygenases have been implicated as having a role in the loss of membrane integrity associated with senescence, flavor and odor formation, response to pest attack, and wounding (192). Products of the LOX pathway such as traumatin, jasmonic acid, oxylipins, and volatile aldehydes, are supposed to play a key role in signal transduction in response to wounding, as antimicrobial substances in host–pathogen interactions, as regulators of growth and development, and as aromatic compounds that affect food quality (191).

16.3.3.1 Role of Lipoxygenase in Membrane Deterioration

Lipid peroxidation is an inherent feature of senescence and generates a variety of activated oxygen species such as singlet oxygen and the alkoxy and peroxy radicals. Alkoxy and peroxy radicals are formed directly as decomposition products of lipid peroxides and singlet oxygen is formed through the interaction of lipid peroxy radicals (157). Lipid peroxidation is initiated either by the action of reactive oxygen species or enzymatically by the action of lipoxygenases (LOX). LOX appears to play an important role in the deterioration of membranes during senescence by initiating lipid peroxidation, and also by forming activated oxygen independently (193). Increasing data suggests that LOX is activated by various stresses such as wounding (194), water deficit (195), thermal stresses (196), and ozone stress (196). It is known that the primary site of action of various stresses is the biomembrane (151), which results in the liberation of free linoleic and linolenic acids (197). Some results also suggest the direct oxygenation of membrane lipids and biomembranes by LOX (195,198–200). The role of LOX in membrane deterioration is evident from its association with membrane in tomato fruits (192,201) and carnation petals (202) as a membrane bound LOX can attack the membrane lipids more readily than a soluble one. This is supported from the observation that LOX can oxygenate esterified polyenoic acids in complex lipids and biomembranes, in addition to free polyenoic fatty acids such as linoleic and linolenic acids (190). Yamauchi et al. (198) reported the oxygenation of dilinolenoyl monogalactosyldiacylglycerol in dipalmitoylphosphatidylcholine liposomes by a crude soybean LOX preparation containing all the three isozymes, thus demonstrating that plant LOX catalyzes the oxygenation of both free polyunsaturated fatty acids and monogalactolipids. Brash et al. (199) observed that soybean LOX-1 oxidizes fatty acid residues within phoshatidylcholine and other phospholipids such as phosphatidylinositol lipids and_phosphatidylethanolamine. Similar results were reported by Kondo et al. (200) from their studies on LOX action in soybean seedlings and Maccarrone et al. (195) on soybean LOX during water deficit. It has been proposed that the role of soybean cotyledon LOX during the early stages of seedling growth is the disruption of storage cell membranes, enhancing their permeability (203). These evidences corroborate the role of LOX in membrane deterioration. However, there is controversy regarding the role of LOX in senescence, as its activity increases during senescence in *Pisum sativum* foliage (204), whereas in soybean cotyledons, total LOX activity has been shown to decrease with advancing senescence (205). The occurrence of LOX activity in young and expanding tissues as well as the observation that soybean LOX is not induced in senescing tissues, argues against the role of LOX in senescence (191). However, soybean seedlings subjected to water stress showed an increase in specific activity of their major lipoxygenases, LOX-1 and LOX-2, which was paralleled by the increase of LOX content and mRNA, indicating that osmotic stress modulates the expression of LOX genes at the transcriptional level (195). Osmotic stress also increased the oxidative index of biomembranes by increasing the hydroperoxide content of the lipid ester fraction. Water deficit has been reported to impair cell membrane functioning (206). Based on the observed enhancement of both LOX activity and membrane oxidative state in response to water deficit, the authors suggested that their results corroborate the hypothesis of a role of LOX in plant membrane deterioration. Studies of the effects of thermal injury (heat shock and cold) and ozone treatment on LOX activity of soybean seedlings have shown that cold stress decreased the specific activities of LOX1 and LOX2, which is attributed to at least in part to a down regulation of gene expression at the translational level (196). Both heat shock and ozone treatment enhanced the LOX-specific activities, acting at the level of transcription of the genes. It is proposed that LOX-1 and LOX-2 are involved in the thermotolerance of soybeans and in the precocious aging induced by ozone. It has been suggested by the authors that cold, heat shock, and ozone can ultimately act on cell membranes. This corroborates the hypothesis for a major role of LOX in the control of membrane integrity. It has been observed that during ripening in tomato fruit pericarp, two distinct LOXs were identified based on their pH optima and their sensitivity to the LOX inhibitor nordihydroguaiaretic acid (207). Both these activities increased sharply during early ripening stages, and decreased after the fruit had ripened fully. It has been shown that the LOX require free fatty acids as their substrate and the timing and extent of peroxidative reactions initiated by LOX are determined by the availability of these substrates which are made available through the action of lipolytic acyl hydrolase (157,201). All these results suggest that a method of enhancing shelf life and quality of fruits could be by the regulation of LOX activity.

16.3.3.2 Tomato fruit Ripening and Lipoxygenases

Studies on LOX genes during tomato fruit ripening using a low ethylene producing fruit containing an ACC oxidase (*ACO1*) sense suppressing transgene, and tomato fruit ripening mutants such as *Never-ripe* (*Nr*), and *ripening-inhibitor* (*rin*), have demonstrated that expression of three LOX genes *TomloxA, TomloxB* and *TomloxC* is regulated differentially during fruit ripening, and that ethylene and a separate developmental component are involved (208). The expression of *TomloxA* declines during ripening and this is delayed in the *ACO1* transgenic low ethylene and *Nr* fruit, indicating that this phenomenon is ethylene regulated. Transcript abundance also declines during *rin* fruit development indicating that developmental factors also influence the expression of *TomloxA*. *TomloxB* expression increases during fruit ripening, which is also stimulated by ethylene. TomloxC gene expression is up regulated in the presence of ethylene and during ripening. The principal substrates of LOX in tomato fruit are the linoleic and linolenic acids, and the action of the 13-lyase on the 13-fatty acid hydroperoxide products of these substrates results in the production of hexanal and hexenal respectively. Release of these aldehydes following disruption of the tissue results in the production of the typical aroma characteristic of fresh tomatoes (209). As the ripening process continues, the thylakoid membranes break down as the chloroplasts are transformed into chromoplasts. It has been suggested that LOX may be the trigger for the chloroplast to chromoplast transition (210), and the polyunsaturated fatty acids in the thylakoid may be acted upon by LOX, and subsequently by a lyase, to release hexanal and hexenal, which in turn influence the flavor and aroma characteristics of the fruit (209). It has been suggested that the various LOX genes in tomato fruit may aid either in the defense mechanisms early in unripe fruit, or flavor and aroma generation and seed dispersal mechanism in the later developmental stages (208).

16.3.3.3 Regulation of LOX by Genetic Manipulation

With the genetic tools available today, it is possible to identify the role or function of a particular gene or gene product either by over expressing or down regulating the gene. Overexpression of LOX2 gene from soybean embryos fused with the enhancer of alfalfa mosaic virus under the control of a duplicated CaMV 35S promoter in transgenic tobacco increased the fatty acid oxidative metabolism as evidenced by a 50–529% increase in C_6 aldehyde production (211). The impact on C_6 aldehyde formation was greater than the effect on production of fatty acid hydroperoxides, which is consistent with other studies indicating the greater involvement of soybean embryo LOX2 in generating C_6 aldehydes than that of other well characterized LOX isozymes. To evaluate the role of LOX in the onset of plant defense, transgenic tobacco plants expressing the antisense tobacco LOX cDNA were developed which showed strongly reduced elicitor and pathogen induced LOX activity (212). A linear relationship was observed between the extent of LOX suppression and the size of the lesion caused by the fungus, *Phytophthora parasitica*. The antisense plants also showed enhanced susceptibility toward the compatible fungus *Rhizoctonia solani*. The authors suggested that their results demonstrate the strong involvement of LOX in the establishment of incompatibility in plant–microorganism interactions, consistent with its role in the defense of host plants. Antisense tomato plants were developed with

TomloxA under the control of fruit specific promoter 2A11 and ripening specific promoter of polygalacturonase (213). Reduced levels of endogenous *TomloxA* and *TomloxB* mRNA (2–20% of wild type) were detected in transgenic fruit containing 2A11 promoter compared to nontransformed plants, whereas the level of mRNA for *TomloxC* was unaffected. LOX enzyme activity was also reduced in these transgenic plants. However, no significant changes were observed in flavor volatiles. The transgenic plants with PG promoter were less effective in reducing endogenous LOX mRNA levels. The authors concluded that either very low levels of LOX are sufficient for the generation of C_6 aldehydes and alcohols, or a specific isoform *TomloxC* in the absence of *TomloxA* and *Tomlox*B is responsible for the production of these compounds. Transgenic tomato and tobacco plants were developed by transformation using the chimeric gene fusions of *TomloxA* and *TomloxB* promoter with β-glucuronidase (GUS) reporter gene. GUS activity in *tomloxA-gus* plants during seed germination peaked at day 5 and was enhanced by methyl jasmonate whereas no GUS activity was detected in *tomloxB-gus* seedlings (214). During fruit development, GUS expression in *tomloxA-gus* tobacco fruit increased 5 days after anthesis and peaked at 20 days after anthesis. In *tomloxB-gus* tobacco GUS activity increased at 10 days after anthesis and peaked at 20 days after anthesis. In *tomloxA-gus* tomato fruit, GUS activity was observed throughout fruit ripening, with highest expression at the orange stage, and the expression was localized to the outer pericarp during fruit ripening. In *tomloxB-gus* fruit, GUS activity was detected at the mature green stage, while expression was localized in the outer pericarp and columella. It has been shown that antisense transgenic potato plants with reduced levels of one specific 13-LOX isoform (LOX-H3) largely abolished the accumulation of proteinase inhibitors upon wounding, indicating that this LOX-H3 plays an important role in the regulation of wound induced gene expression (215).

The genetic manipulation studies on LOX show a specific role of LOX in various developmental processes and defense mechanisms. However, considering the presence of various isoforms in tomatoes and the roles they play, it is very important to characterize the specific role of each isoform before regulating its expression for a specific intent. For example, from the results of Beaudoni and Rothstein (214), it appears that antisense suppression of *TomloxA* may be helpful in enhancing the shelf life of the tomato fruit during storage without affecting the flavor of the fruit.

16.3.4 Cell Wall Metabolism and Fruit Softening

Tomato fruit ripening is a highly regulated developmental process requiring expression of a large number of gene products (216). Enzymes involved in the degradation of cell walls, complex carbohydrates, chlorophyll, and other macromolecules must be coordinately expressed with enzymes that make the fruit desirable nutritionally and aesthetically. Polyglalacturonases [PGs, poly (1→4-α-D-galacturonide) glycanohydrolases] are enzymes that catalyze the hydrolytic cleavage of galacturonide linkages in the cell wall, and are the most widely studied among cell wall hydrolases. PG has been implicated as an important enzyme in fruit softening based on its appearance during ripening, corresponding to the increase in fruit softening. In a number of cultivars, there is a correlation between levels of PG activity and the extent of fruit softening; it degrades isolated fruit cell walls *in vitro* in a manner similar to that observed during ripening and several ripening mutants that have been described with delayed or decreased softening are deficient in PG activity (217). Results on the genetic manipulation of PG has suggested that PG activity alone is not sufficient to affect fruit softening (218), and other enzymes such as pectin-methylesterase, β-galactosidase, and expansins are involved in fruit softening. This discussion is not intended to cover this area, as it has been reviewed recently by Brummell and Harpster (218) and through another chapter in this book.

16.4 CONCLUSION

Fruit color, texture, nutritional value, and flavors are the most important parameters that affect the quality of fruits and vegetables. The post harvest storage conditions of fruits and vegetables also affect these parameters as well as the shelf life, and the type of fruit and vegetables determines the extent of the effect. In the case of potatoes and tomatoes, it is apparent from the above discussion that chilling induced changes affect the above parameters even though preharvest factors and cultivar identity also contribute to this effect. While damages to membranes are attributed to be responsible for these changes in both crops, in the case of potatoes, carbohydrate metabolism also plays a major role. Our understanding of the genes involved in the membrane deteriorative processes as well as carbohydrate metabolism have resulted in the alleviation of these effects to a certain extent. Considering the roles of PLD and LOX in the membrane deteriorative pathway, it might be possible to enhance the shelf life of tomato by generating a double transgenic plant with suppressed activities of PLD and LOX.

REFERENCES

1. Gould, W.A. *Tomato Production, Processing and Technology*, 3rd ed. Baltimore (ISBN 0-930027-18-3): CTI Publications Inc., 1992, pp 3–5, 253–254.
2. Duplessis, P.M., A.G. Marangoni, R.Y. Yada. A mechanism for low temperature induced sugar accumulation in stored potato tubers: the potential role of the alternative pathway and invertase. *Amer. Potato J.* 73:483–494, 1996.
3. Roe, M.A., R.M. Faulks, J.L. Belsten. Role of reducing sugars and amino acids in fry colour of chips from potatoes grown under different nitrogen regimes. *J. Sci. Food Agr.* 52:207–214. 1990.
4. Davies, H.V., R. Viola. Regulation of sugar accumulation in stored potato tubers. *Postharvest News Inf.* 3(5):97N–100N, 1992.
5. Blenkinsop, R.W., L.J. Copp, R.Y. Yada, A.G. Marangoni. Changes in compositional parameters of tubers of potato (*Solanum tuberosum*) during low-temperature storage and their relationship to chip processing quality. *J. Agric. Food Chem.* 50:4545–4553, 2002.
6. Wismer, W.V. Sugar accumulation and membrane related changes in two cultivars of potato tubers stored at low temperature. PhD dissertation, University of Guelph, Guelph, Ontario, 1995.
7. ap Rees, T., W.L. Dixon, C.J. Pollock, F. Franks. Low temperature sweetening of higher plants. In: *Recent Advances in the Biochemistry of Fruits and Vegetables,* Friend, J., M.J.C. Rhodes, eds., Academic Press: New York, 1981, pp 41–61.
8. Burton, W.G. The sugar balance in some British potato varieties during storage, II: the effects of tuber age, previous storage temperature, and intermittent refrigeration upon low-temperature sweetening. *Eur. Potato J.* 12:81–95, 1969.
9. Prichard, M.K., L.R. Adam. Relationships between fry color and sugar concentration in stored Russet Burbank and Shepody potatoes. *Amer. Potato J.* 71:59–68, 1994.
10. Sowokinos, J.R. Postharvest regulation of sucrose accumulation in transgenic potatoes: role and properties of potato tuber UDP-glucose pyrophosphorylase. In: *The Molecular and Cellular Biology of the Potato,* Belkap, W., W.D. Park, M.E. Vadya, eds., Biotechnology in Agriculture, Series 12, Wallingford, UK: CABI Publishing, 1994, pp 81–106.
11. Wismer, W.V., A.G. Marangoni, R.Y. Yada. Low temperature sweetening in roots and tubers. *Hort. Rev.* 27:203–231, 1995.
12. Sowokinos, J.R. Stress-induced alterations in carbohydrate metabolism. In: *The Molecular and Cellular Biology of the Potato,* Vayda, M.E., W.D. Park, eds., Wallingford, UK: CAB Int, 1990, pp 137–158.

13. Sowokinos, J.R. Allele and isozyme patterns of UDP-glucose pyrophosphorylase as a marker for cold-sweetening resistance in potatoes. *Amer. J. Potato Res.* 78:57–64, 2001.
14. Spychalla, J.P., S.L. Desborough. Fatty acids, membrane permeability, and sugars of stored potato tubers. *Plant Physiol.* 94:1207–1213, 1990.
15. Wismer, W.V., W.M. Worthing, R.Y. Yada, A.G. Marangoni. Membrane lipid dynamics and lipid peroxidation in the early stages of low-temperature sweetening in tubers of *Solanum tuberosum. Physiologia Plant.* 102:396–410, 1998.
16. A.G. Marangoni, P.M. Duplessis, R.Y. Yada. Kinetic model for carbon partitioning in *Solanum tuberosum* tubers stored at 2°C and the mechanism for low temperature stress-induced accumulation of reducing sugars. *Biophys. Chem.* 65:211–220, 1997.
17. Sowokinos, J.R. Effects of stress and senescence on carbon partitioning in stored potatoes. *Amer. Potato J.* 67:849–857, 1990.
18. Sowokinos, J.R. Biochemical and molecular control of cold-induced sweetening in potatoes. *Amer. J Potato Res.* 78:221–236, 2001.
19. Isherwood, F. Starch-sugar interconversion in *Solanum tuberosum. Phytochemistry* 12:2579–2591, 1973.
20. Salunkhe, D.K., B.B. Desai, J.K. Chavan. Potatoes. In: *Quality and Peservation of Vegetables*, Eskin, N.A.M., ed., Boca Raton, FL: CRC Press, 1989, pp 2–52.
21. Coffin, R.H., R.Y. Yada, K.L. Parkin, B. Grodzinski, D.W. Stanley. Effect of low temperature storage on sugar concentrations and chip colour of certain processing potato cultivars and selections. *J. Food Sci.* 52:639–645, 1987.
22. Pollock, C.J., T. ap Rees. Effect of cold on glucose metabolism by callus and tubers of *Solanum tuberosum. Phytochemistry* 14:1903-1906, 1975.
23. Hruschka, H.W., W.L. Smith Jr., J.E. Baker. Reducing chilling injury of potatoes by intermittent warming. *Amer. Potato J.* 46:38-53, 1969.
24. Van Es, A., K.J. Hartmans. Starch and sugars during tuberization, storage and sprouting. In: *The Storage of Potatoes*, Rastovski, A., A. van Es, eds., Wageningen: The Netherlands Centre for Agricultural Publishing and Documentation, 1981, pp 79–113.
25. Smith, A.M., K. Denyer, C. Martin. The synthesis of the starch granule. *Annu. Rev. Plant Physiol. Plant Mol. Biol.* 48:67–87, 1997.
26. Tauberger, E., A.R. Fernie, M. Emmermann, A. Renz, J. Kossmann, L. Willmitzer, R.N. Trethewey. Antisense inhibition of plastidial phosphoglucomutase provides compelling evidence that potato tuber amyloplasts import carbon from the cytosol in the form of glucose-6-phosphate. *Plant J.* 23:43–53, 2000.
27. Sowokinos, J.R., J. Preiss. Pyrophosphorylase in *Solanum tuberosum* L, III: purification, physical and catalytic properties of ADP-Glucose pyrophosphorylase in potatoes. *Plant Physiol.* 69:1459–1466, 1982.
28. Müller-Röber, B.T., U. Sonnewald, L. Willmitzer. Inhibition of ADP-glucose pyrophosphorylase leads to sugar storing tubers and influences tuber formation and expression of tuber storage protein genes. *EMBO J.* 11:1229–1238, 1992.
29. Sweetlove, L.J., B.T. Müller-Röber, L. Willmitzer, S.A. Hill. The contribution of adenosine 5'-diphosphoglucose pyrophosphorylase to the control of starch synthesis in potato tubers. *Planta* 209:330–337, 1999.
30. Stark, D.M., G.F. Barry, G.M. Kishore. Engineering plants for commercial products and applications. *Ann. New York Acad. Sci.* 792:26–37, 1996.
31. Lorberth, R., G. Ritte, L. Willmitzer, J. Kossmann. Inhibition of a starch-granule-bound protein leads to modified starch and repression of cold sweetening. *Nature Biotechnol.* 16:473–477, 1998.
32. Takeda, Y., S. Hizukuri, Y. Ozono, M. Suetake. Actions of porcine pancreatic and *Bacillus subtillis* α-amylase and *Aspergillus niger* glucoamylase on phosphorylase (1-4)-α-D-glucan. *Biochim. Biophys. Acta* 749:302–311, 1983.
33. Marsh, R.A., S.G. Waight. Starch degradation. In: *The Biochemistry of Plants: A Comprehensive Treatise,* Vol. 14, Preiss, J., ed., San Diego: Academic Press, 1982, pp 103–128.

34. Chang, C.W. Enzyme degradation of starch in cotton leaves. *Phytochemistry* 21:1263–1269, 1982.
35. Vos-Scheperkeuter, G.H., W. de Boer, R.G.F. Visser, W.J. Feenstra, B. Witholt. Identification of granule-bound starch synthase in potato tubers. *Plant Physiol.* 82:411-416, 1986.
36. Barichello, V., R.Y. Yada, R.H. Coffin, D.W. Stanley. Low temperature sweetening in susceptible and resistant potatoes: starch structure and composition. *J. Food Sci.* 55:1054–1059, 1990.
37. Barichello, V., R.Y. Yada, R.H. Coffin. Starch properties of various potato (*Solanum tuberosum* L) cultivars susceptible and resistant to low temperature sweetening. *J. Sci. Food Agr.* 56:385–397, 1991.
38. Steup, M. Starch degradation. In: *The Biochemistry of Plants. A Comprehensive Treatise,* Vol. 14: *Carbohydrates*, Preiss, J., ed., San Diego: Academic Press, 1988, pp 103–128.
39. ap Rees, T. Pathways of carbohydrate breakdown in plants. *Biochemistry* 11:89–127, 1974.
40. Morrell, S., T. ap Rees. Control of the hexose content of potato tubers. *Phytochemistry* 25:1073–1076, 1986.
41. Preiss, J., C. Levi. Starch biosynthesis and degradation. In: *The Biochemistry of Plants*, Vol. 2, Preiss, J., ed., New York: Academic Press, 1980 pp 371–423.
42. Zhou, D., T. Solomos. Effect of hypoxia on sugar accumulation, respiration, activities of amylase and starch phosphorylase, and induction of alternative oxidase and acid invertase during storage of potato tubers (*Solanum tuberosum* cv. Russet Burbank) at 1°C. *Physiologia Plant.* 104:255–265, 1998.
43. Nakano, K., H. Mori, T. Fukui. Molecular cloning of cDNA encoding potato amyloplast α-glucan phosphorylase and the structure of its transit peptide. *J. Biochem.* 106:691–695, 1989.
44. Kumar, G.N.M., L.O. Knowles, N. Fuller, N.R. Knowles. Starch phosphorylase activity correlates with senescent sweetening but not low temperature-induced sweetening in potato. *Plant Physiol.* 123(suppl):126, 2000.
45. Cochrane, M.P., C.M. Duffus, M.J. Allison, R.G. Mackay. Amylolytic activity in stored potato tubers. 2: the effect of low-temperature storage on the activities of α- and β-amylase and α-glucosidase in potato tubers. *Potato Res.* 34:333–341, 1991.
46. Cottrell, J.E., C.M. Duffus, L. Paterson, G.R. Mackay, M.J. Allison, H. Bain. The effect of storage temperature on reducing sugar concentration and the activities of three amylolytic enzymes in tubers of the cultivated potato, *Solanum tuberosum* L. *Potato Res.* 36:107–117, 1993.
47. Hill, L.M., R. Reimholz, R. Schröder, T.H. Nielsen, M. Stitt. The onset of sucrose accumulation in cold-stored potato tubers is caused by an increased rate of sucrose synthesis and coincides with low levels of hexose-phosphates, an activation of sucrose phosphate synthase and the appearance of a new form of amylase. *Plant Cell Env.* 19:1223–1237, 1996.
48. Neilson, T.H., U. Deiting, M. Stitt. A β-amylase in potato tuber is induced by storage at low temperature. *Plant Physiol.* 113:503–510, 1997.
49. Deiting, U., R. Zrenner, M. Stitt. Similar temperature requirement for sugar accumulation and for the induction of new forms of sucrose phosphate synthase and amylase in cold-stored potato tubers. *Plant Cell Env*. 21:127–138, 1998.
50. Schwimmer, S., A. Bevenue, W.J. Weston. Phosphorus components of the white potato. *Agr. Food Chem.* 3:257–260, 1955.
51. Samotus, B., S. Schwimmer. Effect of maturity and storage on distribution of phophorus among starch and other components of potato tuber. *Plant Physiol.* 37:519–522, 1962.
52. V.V. Shekhar, W.M. Iritani. Starch to sugar interconversion in *Solanum tuberosum* L, 1: influence of inorganic ions. *Amer. Potato J.* 55:345–350, 1978.
53. Isherwood, F.A., M.G.H. Kennedy. The composition of the expressed sap from cold stored potatoes. *Phytochemistry* 14:83–84, 1975.
54. Bielski, R.L. Phosphate pools, phosphate transport, and phosphate availability. *Annu. Rev. Plant Physiol.* 24:245–252, 1973.
55. Loughman, B.C. Uptake and utilization of phosphate associated with respiratory changes in potato tuber slices. *Plant Physiol.* 35:418–424, 1960.

56. Graham, D., B.D. Patterson. Responses of plants to low, nonfreezing temperatures: proteins, metabolism and acclimation. *Annu. Rev. Plant Physiol.* 33:347–372, 1982.
57. Borchert, S., H. Grosse, H.W. Heldt. Specific transport of inorganic phosphate, glucose-6-phosphate, dihydroxyacetone phosphate and 3-phosphoglycerate into amyloplasts from pea roots. *FEBS Lett.* 253:183–186, 1989.
58. Viola, R., H.V. Davies. Effect of temperature on pathways of carbohydrate metabolism in tubers of potato (*Solanum tuberosum* L). *Plant Sci.* 103:135–143, 1994.
59. Kammerer, B., K. Fischer, B. Hilpert, S. Schubert, M. Gutensohn, A. Weber, U.I. Flügge. Molecular characterization of a carbon transporter in plastids from heterotrophic tissue: the glucose 6-phosphate/phosphate antiporter. *Plant Cell* 10:105–117, 1998.
60. Pressey, R. Changes in sucrose synthetase and sucrose phosphate synthetase activities during storage of potatoes. *Amer. Potato J.* 47:245–251, 1970.
61. Viola, R., H.V. Davies, A.R. Chudek. Pathways of starch and sucrose biosynthesis in developing tubers of potato (*Solanum tuberosum* L) and seeds of faba bean (*Vicia faba* L): elucidation by ^{13}C NMR spectroscopy. *Planta* 183:202–208, 1991.
62. Reimholz, R., M. Geiger, V. Haake, U. Deiting, K.-P. Krause, U. Sonnewald, M. Stitt. Potato plants contain multiple forms of sucrose phosphate synthase, which differ in their tissue distributions, their levels during development, and their responses to low temperature. *Plant Cell Env.* 20:291–305, 1997.
63. Reimholz, R., P. Geigenberger, M. Stitt. Sucrose-phosphate synthase is regulated via metabolites and protein phosphorylation in potato tubers, in a manner analogous to the enzyme in leaves. *Planta* 192:480–488, 1994.
64. Krause, K.-P., L. Hill, R. Reimholz, T.H. Nielson, U. Sonnewald, M. Stitt. Sucrose metabolism in cold-stored potato tubers with decreased expression of sucrose phosphate synthase. *Plant Cell Env.* 21:285–299, 1998.
65. Jelitto, T., U. Sonnewald, L. Willmitzer, M. Hajirezeai, M. Stitt. Inorganic pyrophosphate content and metabolites in potato and tobacco plants expressing *E. coli* pyrophosphatase in their cytosol. *Planta* 188:238–244, 1992.
66. Sowokinos, J.R., J.P. Spychalla, S.L. Desborough. Pyrophosphorylases in *Solanum tuberosum*, IV: purification, tissue localization, and physicochemical properties of UDP-glucose pyrophosphorylase. *Plant Physiol.* 101:1073–1080, 1993.
67. Spychall, J.P., B.E. Scheffler, J.R. Sowokinos, M.W. Bevan. Cloning, antisense RNA inhibition, and the coordinated expression of UDP-glucose pyrophosphorylase with starch biosynthesis genes in potato tubers. *J. Plant Physiol.* 144:444–453, 1994.
68. Borovkov, A.Y., P.E. McClean, J.R. Sowokinos, S.H. Ruud, G.A. Secor. Effect of expression of UDP-glucose pyrophosphorylase ribozyme and antisense RNAs on the enzyme activity and carbohydrate composition of field-grown transgenic potato plants. *J. Plant Physiol.* 147:644–652, 1996.
69. Zrenner, R., L. Willmitzer, U. Sonnewald. Analysis of the expression of potato uridine diphosphate-glucose pyrophosphorylase and its inhibition by antisense RNA. *Planta* 190:247–252, 1993.
70. ap Rees, T., S. Morrell. Carbohydrate metabolism in developing potatoes. *Amer. Potato J.* 67:835–847, 1990.
71. Jang, J.C., J. Sheen. Sugar sensing in higher plants. *Plant Cell* 6:1665–1679, 1994.
72. ap Rees, T. Hexose phosphate metabolism by nonphotosynthetic tissues of higher plants. In: *The Biochemistry of Plants*. Vol. 14, Preiss, J., ed., New York: Academic Press, 1988, pp 1–33.
73. Ross, H.A., H.V. Davies. Sucrose metabolism in tubers of potato (*Solanum tuberosum* L): effects of sink removal and sucrose flux on sucrose-degrading enzymes. *Plant Physiol.* 98:287–293, 1992.
74. Sturm, A. Invertases: primary structures, functions and roles in plant development and sucrose partitioning. *Plant Physiol.* 121:1-7, 1999.
75. Pressey, R. Potato sucrose synthase: purification, properties and changes in activity associated with maturation. *Plant Physiol.* 44:759–764, 1969.

76. Pressey, R., R. Shaw. Effect of temperature on invertase, invertase inhibitor and sugars in potato tubers. *Plant Physiol.* 41:1657–1661, 1966.
77. Avigad, G. Sucrose and other disaccharides. In: *Encyclopedia of Plant Physiology,* Vol. 13A, Loewus, F.A., W. Tanner, eds., Berlin: Springer–Verlag, 1982, pp 217–317.
78. ap Rees, T. Pathways of carbohydrate breakdown in plants. *MTP Int. Rev. Sci. Plant Biochem.* 11:89–127, 1974.
79. Davies, H.V., R.A. Jefferies, L. Scobie. Hexose accumulation in cold-stored tubers of potato (*Solanum tuberosum* L): the effect of water stress. *J. Plant Physiol.* 134:471–475, 1989.
80. Richardson, D.I., H.V. Davies, H.A. Ross, G.R. Mackay. Invertase activity and its relationship to hexose accumulation in potato tubers. *J. Exp. Bot.* 41:95–99, 1990.
81. Davies, H.V., R. Viola. Control of sugar balance in potato tubers. In: *The Molecular and Cellular Biology of the Potato,* 2nd ed., Belknap, W.R., M.E. Vayda, W.D. Park, eds., Wallingford: CAB Int., 1994, pp 67–80.
82. Zrenner, R., K. Schüler, U. Sonnewald. Soluble acid invertase determines the hexose-to-sucrose ratio in cold-stored potato tubers. *Planta* 198:246–252, 1996.
83. Greiner, S., T. Rausch, U. Sonnewald, K. Herbers. Ectopic expression of a tobacco invertase inhibitor homolog prevents cold-induced sweetening of potato tubers. *Nat. Biotechnol.* 17:708–711, 1999.
84. Dennis, D.T., J.A. Miernyk. Compartmentation of nonphotosynthetic carbohydrate metabolism. *Annu. Rev. Plant Physiol.* 33:27–50, 1982.
85. Geigenberger, P., M. Stitt. A 'futile' cycle of sucrose synthesis and degradation is involved in regulating partitioning between sucrose, starch and respiration in cotyledons of germinating *Ricinus communis* L seedlings when phloem transport is inhibited. *Planta* 185:81–90, 1991.
86. Dixon, W.L., T. ap Rees. Identification of the regulatory steps in glycolysis in potato tubers. *Phytochemistry* 19:1297–1301, 1980.
87. Dixon, W.L., T. ap Rees. Carbohydrate metabolism during cold-induced sweetening of potato tubers. *Phytochemistry* 19:1653–1656, 1980.
88. Huber, S.C. Fructose 2,6-bisphosphate as a regulatory metabolite in plants. *Annu. Rev. Plant Physiol.* 37:233–246, 1986.
89. Hajirezaei, M., M. Stitt. Contrasting roles for pyrophosphate: fructose 6-phosphate phosphotransferase during aging of tissue slices from potato tubers and carrot storage tissues. *Plant Sci.* 77:177–183, 1991.
90. Stitt, M. Fructose 2,6,-bisphosphate as a regulatory molecule in plants. *Annu. Rev. Plant Physiol. Plant Mol. Biol.* 41:153–185, 1990.
91. Morrell, S., T. ap Rees. Sugar metabolism in developing tubers of *Solanum tuberosum. Phytochemistry* 25:1579–1585, 1986b.
92. Trevanion, S.J., N.J. Kruger. Effect of temperature on the kinetic properties of pyrophosphate: fructose 6-phosphate phosphotransferase from potato tuber. *J. Plant Physiol.* 137:753–759, 1991.
93. Claassen, P.A.M., M.A.W. Budde, H.J. de Ruyter, M.H. van Calker, A. van Es. Potential role of pyrophosphate: fructose-6-phosphate phosphotransferase in carbohydrate metabolism of cold stored tubers of *Solanum tuberosum* cv. *Bintje. Plant Physiol.* 95:1243–1249, 1991.
94. Bredemeijer, G.M.M., H.C.J. Burg, P.A.M. Claassen, W.J. Stiekema. Phosphofructokinase in relation to sugar accumulation in cold-stored potato tubers. *J. Plant Physiol.* 138:129–135, 1991.
95. Hammond, J.B.W., M.M. Burrell, N.J. Kruger. Effect of low temperature on the activity of phosphofructokinase from potato tubers. *Planta* 180:613–616, 1990.
96. Kruger, N.J., J.B.W. Hammond, M.M. Burrell. Molecular characterization of four forms of phosphofructokinose purified from potato tuber. *Arch. Biochem. Biophys.* 267:690–700, 1988.
97. Bryce, J.H., S.A. Hill. Energy production in plant cells. In: *Plant Biochemistry and Molecular Biology,* Lee, P.J., R.C. Leegood, eds., Chichester: Wiley Europe Ltd, 1993, pp 1–21.
98. Amir, J., V. Kahn, M. Unterman. Respiration, ATP level, and sugar accumulation in potato tubers during storage at 4°C. *Phytochemistry* 16:1495–1498, 1977.

99. Hajirezaei, M., U. Sonnewald, R. Viola, S. Carlisle, D. Dennis, M. Stitt. Transgenic potato plants with strongly decreased expression of pyrophosphate: fructose 6-phosphate phosphotransferase show no visible phenotype and only minor changes in metabolic fluxes in their tubers. *Planta* 192:16–30, 1994.
100. Burrell, M.M., P.J. Mooney, M. Blundy, D. Carter, F. Wilson, J. Green, K.S. Blundy, T. ap Rees. Genetic manipulation of 6-phosphofructokinose in potato tubers. *Planta* 194:95–101, 1994.
101. Thomas, S., P.J. Mooney, M.M. Burrell, D.A. Fell. Finite change analysis of glycolytic intermediates in tuber tissue of lines of transgenic potato (*Solanum tuberosum*) overexpressing phosphofructokinose. *Biochem. J.* 322:111–117, 1997.
102. Thomas, S., P.J. Mooney, M.M. Burrell, D.A. Fell. Metabolic control analysis of glycolysis in tuber tissue of potato (*Solanum tuberosum*): explanation for the low control of coefficient of phosphofructokinose over respiratory flux. *Biochem. J.* 322:119–127, 1997.
103. Stitt, M. Frucotse 2,6-bisphosphate and plant carbohydrate metabolism. *Plant Physiol.* 84:201–204, 1987.
104. Van Schaftingen, E., B. Lederer, R. Bartons, H.-G. Hers. A kinetic study of pyrophosphate: fructose-6-phosphate phosphotransferase from potato tubers. *Eur. J. Biochem.* 129:191–195, 1982.
105. D.T. Dennis, Y. Huang, F.B. Negm. Glycolysis, the pentose phosphate pathway and anaerobic respiration. In: *Plant metabolism*, 2nd ed., Dennis, D.T., D.H. Turpin, D.D. Lefebvre, D.B. Layzell, eds. UK: Addison Wesley Longman Ltd., 1997, pp 105–123.
106. Wagner, A.M., T.J.A. Kneppers, B.M. Kroon, L.H.W. van der Plas. Enzymes of the pentose phosphate pathway in callus-forming potato tuber discs grown at various temperatures. *Plant Sci.* 51:159–164, 1987.
107. Barichello, V., R.Y. Yada, R.H. Coffin, D.W. Stanley. Low temperature sweetening in susceptible and resistant potatoes: starch structure and composition. *J. Food Sci.* 55:1054–1059, 1990.
108. Sherman, M., E.E. Ewing. Temperature, cyanide, and oxygen effects on the respiration, chip color, sugars, and organic acids of stored tubers. *Amer. Potato J.* 59:165–178, 1982.
109. Barichello, V., R.Y. Yada, R.H. Coffin, D.W. Stanley. Respiratory enzyme activity in low temperature sweetening of susceptible and resistant potatoes. *J. Food. Sci.* 55:1060–1063, 1990.
110. Purvis, A.C., R.L. Shewfelt. Does the alternative pathway ameliorate chilling injury in sensitive plant tissues? *Physiologia Plant.* 88:712–718, 1993.
111. Lambers, H. Cyanide-resistant respiration: a non-phosphorylating electron transport pathway acting as an energy overflow. *Physiologia Plant.* 55:478–485, 1982.
112. Elthon, T.E., C.R. Stewart, C.A. McCoy, W.D. Bonner, Jr. Alternate respiratory path capacity in plant mitochondria: effect of growth, temperature, the electrochemical gradient, and assay pH. *Plant Physiol.* 80:378–383, 1986.
113. Solomos, T., G.G. Laties. The mechanism of ethylene and cyanide action in triggering the rise in respiration in potato tubers. *Plant Physiol.* 55:73–78, 1975.
114. Lance, C. Cyanide-insensitive respiration in fruits and vegetables. In: *Recent Advances in the Biochemistry of Fruit and Vegetables,* Friend, J., M.J.C. Rhodes, eds., New York: Academic Press, 1981, pp 63–87.
115. Sherman, M., E.E. Ewing. Effects of temperature and low oxygen atmospheres on respiration, chip color, sugars, and malate of stored potatoes. *J. Amer. Soc. Hort. Sci.* 108:129–133, 1983.
116. Lyons, J.M. Chilling injury in plants. *Annu. Rev. Plant Physiol.* 24:445–466, 1973.
117. Raison, J.K. The influence of temperature-induced phase changes on kinetics of respiratory and other membrane-associated enzymes. *J. Bioenerg.* 4:258-309, 1973.
118. Shewfelt, R.L. Response of plant membranes to chilling and freezing. In: *A biophysical approach to structure, development and senescence,* Leshem, Y.Y., ed., Dordrecht: Kluwer, 1992, pp 192–219.
119. Murata, N., I. Nishida. Lipids of blue-green algae (cyanobacteria). In: *The Biochemistry of Plants*, Vol. 9. Orlando: Academic Press, 1987, 315–347.

120. Murata, N. Molecular species composition of phosphadtidylglycerols from chilling-sensitive and chilling-resistant plants. *Plant Cell Physiol.* 24:81–86, 1983.
121. Murata, N., N. Sato, N. Takahashi, Y. Hamazaki. Compositions and positional distributions of fatty acids in phospholipids from leaves of chilling-sensitive and chilling-resistant plants. *Plant Cell Physiol.* 23:1071–1079, 1982.
122. Knowles, N.R., L.O. Knowles. Correlations between electrolyte leakage and degree of saturation of polar lipids from aged potato (*Solanum tuberosum* L) tuber tissue. *Ann. Bot.* 63:331–338, 1989.
123. Spychalla, J.P., S.L. Desborough. Fatty acids, membrane permeability, and sugars of stored potato tubers. *Plant Physiol.* 94:1207–1213, 1990.
124. Shekhar, V.C., W.M. Iritani, J. Magnuson. Starch-sugar interconversion in *Solanum tuberosum* L, 11: influence of membrane permeability and fluidity. *Amer. Potato J.* 56:225–234, 1979.
125. O'Donoghue, E.P., R.Y. Yada, A.G. Marangoni. Low temperature sweetening in potato tubers: the role of the amyloplast membrane. *J. Plant Physiol.* 145(3):335–341, 1995.
126. Gounaris, Y., J.R. Sowokinos. Two-dimensional analysis of mitochondrial proteins from potato cultivars resistant and sensitive to cold-induced sweetening. *J. Plant Physiol.* 140:611–616, 1992.
127. Berkeley, H.D., T. Galliard. Lipids of potato tubers: effect of growth and storage on lipid content of the potato tuber. *J. Sci. Food. Agr.* 25:861–867, 1974.
128. Lojkowska, E., M. Holubowska. Changes of the lipid catabolism in potato tubers from cultivars differing in susceptibility to autolysis during the storage. *Potato Res.* 32:463–470, 1989.
129. Kumar, G.N.M., N.R. Knowles. Changes in lipid peroxidation and lipolytic and free-radical scavenging enzyme activities during aging and sprouting of potato (*Solanum tuberosum*) seed tubers. *Plant Physiol.* 102:115–124, 1993.
130. Galliard, T. Lipid of potato tubers, 1: lipid and fatty acid composition of tubers from different varieties of potato. *J. Sci. Food. Agr.* 24:617–622, 1973.
131. Thompson, G.A., Jr., K.J. Elnspahr, S. Ho Cho, T.C. Peeler, M.B. Stephenson. Metabolic responses of plant cells to stress. In: *Biological Role of Plant Lipids*, Biacs, P.A., K. Gruiz, T. Kremmer, eds., London: Plenum Publishing, 1989, pp 497–504.
132. Nishida, I., N. Murata. Chilling sensitivity in plants and cyanobacteria: the crucial contribution of membrane lipids. *Annu. Rev. Plant Physiol. Plant Mol. Biol.* 47:541–568, 1996.
133. Bonnerot, C., P. Mazliak. Induction of the oleoly-phosphatidylcholine desaturase activity during the storage of plant organs: a comparison between potato and jerusalem artichoke tubers. *Plant Sci. Lett.* 35:5–10, 1984.
134. Spychalla, J.P., S.L. Desborough. Superoxide dismutase, catalase, and α-tocopherol content of stored potato tubers. *Plant Physiol.* 94:1214–1218, 1990.
135. Jarvis, M.C., J. Dalziel, H.J. Duncan. Variations in free sugars and lipids in different potato varieties during low-temperature storage. *J. Sci. Food Agr.* 25:1405–1409, 1974.
136. Galliard, T., H.D. Berkeley, J.A. Mathew. Lipids of potato tubers, effects of storage temperature on total, polar, and sterol lipid content and fatty acid composition of potato tubers. *J. Sci. Food Agr.* 26:1163–1170, 1975.
137. Lijenberg, C., A.S. Sandelius, E. Selstam. Effect of storage in darkness and in light on the content of membrane lipids of potato tubers. *Physiologia Plant.* 43:154–159, 1978.
138. Phan, C.-T. Temperature effects on metabolism. In: *Postharvest Physiology of Vegetables,* Welchmann, J., ed., New York: Marcel Dekker, 1987, pp 173–180.
139. Pinhero, R.G., G. Paliyath, R.Y. Yada, D.P. Murr. Modulation of phospholipase D and lipoxygenase activities during chilling: relation to chilling tolerance of maize seedlings. *Plant Physiol. Biochem.* 36:213–224, 1998.
140. Paliyath, G., M.J. Droillard. The mechanisms of membrane deterioration and disassembly during senescence. *Plant Physiol. Biochem.* 30:789–812, 1992.
141. Lyons, J.M., J.K. Raison. Oxidative activity of mitochondria isolated from plant tissues sensitive and resistant to chilling injury. *Plant Physiol.* 45:386–389, 1970.

142. Quinn, P.J., W.P. Williams. Plant lipids and their role in membrane function. *Prog. Biophys. Mol. Biol.* 34:109–173, 1978.
143. Workman, M., A. Cameron, J. Twomey. Influence of chilling on potato tuber respiration, sugar, o-dihydroxyphenolic content and membrane permeability, *Amer. Potato J.* 56:277–288, 1979.
144. Schneider, E.M., J.U. Becker, D. Volkmann. Biochemical properties of potato phosphorylase change with its intracellular localization as revealed by immunological methods. *Planta* 151:124–134, 1981.
145. Brisson, N., H. Giroux, M. Zollinger, A. Camirand, C. Simard. Maturation and subcellular compartmentation of potato starch phosphorylase. *Plant Cell* 1:559–566, 1989.
146. Shekhar, V.C., W.M. Iritani, J. Magnuson. Starch-sugar interconversion in *Solanum tuberosum* L, II: influence of membrane permeability and fluidity. *Amer. Potato J.* 56:225–234, 1979.
147. Isherwood, F. Mechanism of starch-sugar interconversion in *Solanum tuberosum. Phytochemistry* 15:33–41, 1976.
148. Sowokinos, J.R., P.H. Orr, J.A. Knoper, J.L. Varns. Influence of potato storage and handling stress on sugars, chip quality, and integrity of the starch (amyloplast) membrane. *Amer. Potato J.* 64:213–226, 1987.
149. Yada, R.Y., R.H. Coffin, K.W. Baker, M.J. Lsezkowiat. An electron microscopic examination of the amyloplast membranes from potato cultivar susceptible to low temperature sweetening. *Can. Inst. Food Sci. Technol. J.* 23:145–148, 1990.
150. Allen, R.D. Dissection of oxidative stress tolerance using transgenic plants. *Plant Physiol.* 107:1049–1054, 1995.
151. McKersie, B.D., Y.Y. Leshem. *Stress and Stress Coping in Cultivated Plants.* Dordrecht, Kluwer, 1994, pp. 79–100.
152. Fauconnier, M.L., J. Rojas-Beltra, D. Delcarte, F. Dejaeghere, M. Marlier, P. Du Jardin. Lipoxygenase pathway and membrane permeability and composition during storage of potato tubers (*Solanum tuberosum* L) in different conditions. *Plant Biol.* 4:77–85, 2002.
153. Shewfelt, R.L., R.E. McDonald, H.O. Hultin. Effect of phospholipid hydrolysis on lipid oxidation in flounder muscle microsomes. *J. Food Sci.* 46:1297–1301, 1981.
154. Stanley, D.W. Biological membrane deterioration and associated quality losses in food tissues. *Crit. Rev. Food Sci. Nutr.* 30:487–553, 1991.
155. Giovannoni, J. Molecular biology of fruit maturation and ripening. *Annu. Rev. Plant Physiol. Plant Mol. Biol.* 52:725–749, 2001.
156. Sacher, J.A. Studies of permeability, RNA and protein turnover during aging of fruit and leaf tissue. In: *Aspects of the Biology of Aging*, Woolhouse, H.W., ed., New York: Academic Press, 1967 pp 269–303.
157. J.H. Brown, G. Paliyath, J.E. Thompson. Physiological mechanisms of plant sensescence. In: *Plant Physiology: A Treatise, X*, Steward, F.C., R.G.S. Bidwell, eds., New York: Academic Press, 1991, pp 227–275.
158. Lanahan, M.B., H.C. Yen, J.J. Giovannoni, H.J. Klee. The *Never Ripe* mutation blocks ethylene perception in tomato. *Plant Cell* 6:521–530, 1994.
159. Lashbrook, C.C., D.M. Tieman, H.J. Klee. Differential regulation of the tomato ETR gene family throughout plant development. *Plant J.* 15:243–252, 1998.
160. Payton, S., R.G. Fray, S. Brown, D. Grierson. Ethylene receptor expression is regulated during fruit ripening, flower senescence and abscission. *Plant Mol. Biol.* 31:1227–1231, 1996.
161. Tieman, D.M., M.G. Taylor, J.A. Ciardi, H.J. Klee. The tomato ethylene receptors NR and LeETR4 are negative regulators of ethylene response and exhibit functional compensation within a multigene family. *Proc. Natl. Acad. Sci. USA* 97:5663–5668, 2000.
162. Wilkinson, J., M. Lanahan, H. Yen, J.J. Giovannoni, H.J. Klee. An ethylene-inducible component of signal transduction encoded by Never-ripe. *Science* 270:1807–1809, 1995.
163. Yen, H., S. Lee, S. Tanksley, M. Lanahan, H.J. Klee, J.J. Giovannoni. The tomato Never-ripe locus regulates ethylene-inducible gene expression and is linked to a homologue of the *Arabidopsis* ETR1 gene. *Plant Physiol.* 107:1343–1353, 1995.

164. Zhou, D., P. Kalaitzis, A.K. Mattoo, M.L. Tucker. The mRNA for an ETR1 homologue in tomato is constitutively expressed in vegetative and reproductive tissues. *Plant Mol. Biol.* 30:1331–38, 1996.
165. Leshem, Y.Y. *Plant Membranes: A Biophysical Approach to Structure, Development and Senescence.* London: Kluwer Academic Publishers, 1992, pp 174–188.
166. Paliyath, G., D.P. Murr, J.E. Thompson. Catabolism of phosphorylated phosphatidylinositols by carnation petal microsomal membranes enriched in plasmalemma and endoplasmic reticulum. *Physiol. Mol. Biol. Plants* 1:141–150, 1995.
167. Wang, X. Plant phospholipases. *Ann. Rev. Plant Physiol. Plant Mol. Biol.* 52:211-231, 2001.
168. Wang, X. Multiple forms of phospholipase D in plants: the gene family, catalytic and regulatory properties, and cellular functions. *Prog. Lipid Res.* 39:109–149, 2000.
169. Hammond, S.M., Y.M. Alshuller, T. Sung, S.A. Rudge, K. Rose. Human ADP-ribosylation factor-activated phosphatidylcholine-specific phospholipase D defines a new and highly conserved gene family. *J. Biol. Chem.* 270:29640–29643, 1995.
170. Rose, K., S.A. Rudge, M.A. Frohman, A.J. Morris, J. Engebrecht. Phospholipase D signaling is essential for meiosis. *Proc. Natl. Acad. Sci. USA* 92:12151–12155, 1995.
171. Liscovitch, M., M. Czarny, G. Fiucci , X. Tang. Phospholipase D: molecular and cell biology of a novel gene family. *Biochem. J.* 345:401–415, 2000.
172. Wang, X. The role of phospholipase D in signaling cascades. *Plant Physiol.* 120:645–651, 1999.
173. Laxalt, A.M., B. ter Riet, J.C. Verdonk, L. Parigi, W.I.L. Tameling, J. Vossen, M. Haring, A. Musgrave, T. Munnik. Characterization of five tomato phospholipase D. cDNAs: rapid and specific expression of *LePLDβ1* on elicitation with xylanase. *Plant J.* 26:237–247, 2001.
174. Wang, C., C. Zien, M. Afitlhile, R. Welti, D.F. Hildebrand, X. Wang. Involvement of phospholipase D in wound-induced accumulation of jasmonic acid in *Arabidopsis*. *Plant Cell* 12:2237–2246, 2000.
175. Todd, J.F., G. Paliyath, J.E. Thompson. Effect of chilling on the activities of lipid degrading enzymes in tomato fruit microsomal membranes. *Plant Physiol. Biochem.* 30:517–522, 1992.
176. Marangoni, A.G., D.W. Stanley. Phase transitions in microsomal membranes from chilling-sensitive and chilling-resistant tomato plants and fruit. *Phytochemistry* 28:2293–2301, 1989.
177. McCormac, D.J., J.F. Todd, G. Paliyath, J.E. Thompson. Modulation of bilayer fluidity affects lipid catabolism in microsomal membranes of tomato fruit. *Plant Physiol.* 31:1–8, 1993.
178. Parkin, K.L., S.J. Kuo. Chilling-induced lipid degradation in cucumber (*Cucumis sativa* L. cultivar Hybrid C) fruit. *Plant Physiol.* 90:1049–1056, 1989.
178a. Pinhero, R.G., K.C. Almquist, Z. Novotna, G. Paliyath. Developmental regulation of phospholipase D in tomato fruits. *Plant Physiol. Biochem.* 41:223–240, 2003.
179. Jandus, J., O. Valentovä, J. Kas, J. Daussant, C. Thévenot. Phospholipase D during tomato fruit ripening. *Plant Physiol. Biochem.* 35:123–128, 1997.
180. Güçlü, J., A. Paulin, P. Soudain. Changes in polar lipid during ripening and senescence of cherry tomato (*Lycopersicon esculentum*): relation to climateric and ethylene increases. *Physiol. Plant.* 77:413–419, 1989.
181. Whitaker, B.D. Lipid changes in mature-green tomato fruit during ripening, during chilling, and after rewarming subsequent to chilling. *J. Amer. Soc. Hort. Sci.* 119:994–999, 1994.
182. Ryu, S.B., B.H. Karlsson, M. Ozgen, J.P. Palta. Inhibition of phospholipase D by lysophosphatidylethanolamine, a lipid-derived senescence retardation. *Proc. Natl. Acad. Sci. USA* 94:12717–12721, 1997.
183. Farag, K.M., J.P. Palta. Use of lysophosphatidylethanolamine, a natural lipid, to retard tomato leaf and fruit senescence. *Physiologia Plant.* 87:515–521, 1993.
184. Ryu, S.B., X. Wang. Activation of phospholipase D and the possible mechanism of activation in wound-induced lipid hydrolysis in castor bean leaves. *Biochim. Biophys. Acta.* 1303:243–250, 1996.
185. Wang X., C. Wang, Y. Sang, L. Zheng, C. Qin. Determining functions of multiple phospholipase Ds in stress response of *Arabidopsis*. *Biochem. Soc. Trans.* 28:813–816, 2000.

186. Zheng, L., R. Krishnamoorthi, M. Zolkiewski, X. Wang. Distinct Ca^{2+} binding properties of novel C2 domains of plant phospholipase Dα and β. *J. Biol. Chem.* 275:19700–19706, 2000.
187. Zocchi, G., J.B. Hanson. Calcium influx into corn roots as a result of cold shock. *Plant Physiol.* 70:318–319, 1982.
188. Fan, L., S. Zheng, X. Wang. Antisense suppression of phospholipase D alpha retards abscisic acid- and ethylene-promoted senescence of postharvest *Arabidopsis* leaves. *Plant Cell* 9:2183–2196, 1997.
189. Fan, L., S. Zheng, D. Cui, X. Wang. Subcellular distribution and tissue expression of phospholipase Dα, Dβ, and Dχ in *Arabidopsis*. *Plant Physiol.* 119:1371–1378, 1999.
190. Grechkin, A. Recent developments in biochemistry of the plant lipoxygenase pathway. *Prog. Lipid Res.* 37:317–352, 1998.
191. Rosahl, S. Lipoxygenase in plants: their role in development and stress response. *Z. Naturforsch* 51:123–138, 1996.
192. Droillard, M.J., M.A. Rouet-Mayer, J.M. Bureau, C. Lauriere. Membrane-associated and soluble lipoxygenase isoforms in tomato pericarp: characterization and involvement in membrane alterations. *Plant Physiol.* 103:1211–1219, 1993.
193. Lynch, D.V., J.E. Thompson. Lipoxygenase-mediated production of superoxide anion in senescing plant tissue. *FEBS Lett.* 173:251–254, 1984.
194. Mauch, F., A. Kmecl, U. Schaffrath, S. Volrath, J. Gorlach, E. Ward, J. Ryals, R. Dudler. *Plant Physiol.* 114:1561–1566, 1997.
195. Maccarrone, M., G.A. Veldink, A.F. Agro, J.F.G. Vliegenthart. Modulation of soybean lipoxygenase expression and membrane oxidation by water deficit. *FEBS Lett.* 371:223–226, 1995.
196. Maccarrone, M., G.A. Veldink, J.F.G. Vliegenthart. Thermal injury and ozone stress affect soybean lipoxygenase expression. *FEBS Lett.* 309:225–230, 1992.
197. Conconi, A., M. Miquel, J.A. Browse, C.A. Ryan. Intracellular levels of free linolenic and linoleic acids increase in tomato leaves in response to wounding. *Plant Physiol.* 111:797–803, 1996.
198. Yamauchi, R., M. Kojima, K. Kato, Y. Ueno. Lipoxygenase-catalyzed oxygenation of monogalactosyldilinolenoylglycerol in dipalmitoylphosphatidylcholine liposomes. *Agr. Biol. Chem.* 49:2475–2477, 1985.
199. Brash, A.R., C.D. Ingram, T.M. Harris. Analysis of specific oxygenation reaction of soybean lipoxygeanse-1 with fatty acids esterified in phospholipids. *Biochemistry* 26:5465–5471, 1987.
200. Kondo, Y., Y. Kawai, T. Hayashi, M. Ohnishi, T. Miyazawa, S. Itoh, J. Mizutani. Lipoxygenase in soybean seedlings catalyzes the oxygenation of phospholipid and such activity changes after treatment with fungal elicitor. *Biochim. Biophys. Acta.* 1170:301–306, 1993.
201. Todd, J.F., G. Paliyath, J.E. Thompson. Characteristics of a membrane-associated lipoxygenase in tomato fruit. *Plant Physiol.* 94:1225-1232, 1990.
202. M.A. Rouet-Mayer, J.M. Bureau, C. Lauriere. Identification and characterization of lipoxygenase isoforms in senescing carnation petals. *Plant Physiol.* 98:971–978, 1992.
203. B.A. Vick, D.C. Zimmerman. Oxidative systems for modification of fatty acids: the lipoxygenase pathway. In: *The Biochemistry of Plants*, Vol. 9, Stumpf, P.K., E.E. Conn, eds., Orlando: Academic Press, 1987, pp 53–90.
204. Grossman, S., Y. Leshem. Lowering of endogenous lipoxygenase activity in *Pisum sativum* foliage by cytokinins as related to senescence. *Physiologia Plant.* 43:359–362, 1978.
205. Peterman, T.K., J.N. Siedow. Behaviour of lipoxygeanse during establishment, senescence and rejuvenation of soybean cotyledons. *Plant Physiol.* 78:690–695, 1985.
206. Navari-Izzo, F., N. Vangioni, M.F. Quartacci. Lipids of soybean and sunflower seedlings grown under drought conditions. *Phytochemistry* 29:2119–2123, 1990.
207. Ealing, P.M. Lipoxygenase activity in ripening tomato fruit pericarp tissue. *Phtochemistry* 36:547–552, 1994.

208. Griffiths, A., C. Barry, A.G. Alpuche-Soils, D. Grierson. Ethylene and developmental signals regulate expression of lipoxygenase genes during tomato fruit ripening. *J. Exp. Bot.* 50:793–798, 1999.
209. R.G. Buttery, R. Teranishi, R.A. Flath, L.C. Ling. In: *Flavor Chemistry: Trends and Developments,* Teranishi, R., R.G. Buttery, eds., Washington: American Chemical Society, 1989, pp 213.
210. Ferrie, B.J., N. Beaudoin, W. Burkhart, C.G. Bowsher, S.J. Rothstein. The cloning of two tomato lipoxygenase genes and their differential expression during tomato fruit ripening. *Plant Physiol.* 106:109–118, 1994.
211. Deng, W., W.S. Grayburn, T.R. Hamilton-Kemp, G.B. Collins. Expression of soybean-embryo lipoxygenase 2 in transgenic tobacco tisue. *Planta* 187:203–208, 1992.
212. Rancé, L., J. Fournier, M.T. Esquerré-Tugayé.The incompatible interaction between *Phytophthora parasitica* var. nicotianaerace and tobacco is suppressed in transgenic plants expressing antisense lipoxygenase sequences. *Proc. Natl. Acad. Sci. USA* 95:6554–6559, 1998.
213. Griffiths, A., S. Prestage, R. Linforth, J. Zhang, A. Taylor, D. Grierson. Fruit-specific lipoxygenase suppression in antisense-transgenic tomatoes. *Postharvest Biol. Tech.* 17:163-173, 1999.
214. Beaudoin, N., S.J. Rothstein. Developmental regulation of two tomato lipoxygenase promoters in transgenic tobacco and tomato. *Plant Mol. Biol.* 33:835–846, 1997.
215. Royo, J., J, Léon, G. Vancanneyt, J.P. Albar, S. Rosahl, F. Ortego, P. Castañera, J.J. Sánchez-Serrano. Antisense-mediated depletion of a potato lipoxygenase reduces wound induction of proteinase inhibitors and increases weight gain of insect pests. *Proc. Natl. Acad. Sci. USA* 96:1146–1151, 1999.
216. Grierson, D., A.A. Kader. Fruit ripening and quality. In: *The Tomato Crop: A Scientific Basis for Improvement,* Atherton, J.G., J. Rudich, eds., London: Chapman & Hall, 1986, pp 241–280.
217. Kramer, M., R.E. Sheehy, W.R. Hiatt. Progress towards the genetic engineering of tomato fruit softening. *TIBTECH* 7:191–193, 1989.
218. Brummell, D.A., M.H. Harpster. Cell wall metabolism in fruit softening and quality and its manipulation in transgenic plants. *Plant Mol. Biol.* 47:311–340, 2001.

2.17

Genetic Modification of Peanut as a Solution to Peanut Allergy

Hortense Dodo, Koffi Konan, and Olga Viquez

CONTENTS

17.1 PREVALENCE AND IMPACT OF ALLERGIC DISEASES

Allergic diseases rank fourth among world diseases as recorded by the World Health Organization and they are becoming an increasing public health issue worldwide (1). Allergies are the 6th leading cause of chronic diseases in the US, costing the health care system, $18 billion per year (American Academy of Allergy, Asthma, and Immunology, The allergy report 1996–2001: Allergy Statistics http://www.aaaai.org/media/resources/media_kit/allergy_statistics.stm). Food allergy is a rapidly increasing phenomenon in industrial countries. Up to 25% increase every 10 years has been reported within children, as well as adult, populations. Food allergy is the leading cause of anaphylaxis treated in American hospital emergency rooms. It is estimated that in the US, food allergies affect 6 to 8% of the children and 1% of the adult population. In France, about 5% of the population is affected, and in the UK, about 2% (1). Food allergy is defined as an allergy that occurs when the immune system mistakenly responds defensively to a specific food protein that is not harmful to the body.

17.1.1 Peanut is One of the Most Potent Sources of Food Allergy

Peanut allergy is responsible for the majority of severe food induced anaphylaxis reactions. It is an immunoglobulin E (IgE) mediated hypersensitivity reaction (2) which is persistent, life threatening, and rarely outgrown, unlike allergy to egg and cow's milk (3). Reports of the exact prevalence of peanut allergy vary from countries to countries. Sicherer et al. (4) estimated that 1.1% of American children and adults (about 4.4 million people) suffer from allergy to peanut and tree nuts. Peanut allergy is characterized by more acute and severe symptoms and a higher rate of symptoms on minimal contact than other food allergies (5). Reports indicate that peanut allergies are responsible for half of the annual emergency room visits and approximately 63 to 67% of deaths due to anaphylaxis (6). Food anaphylaxis is defined as a life threatening sudden and severe allergic reaction to a foreign antigen mediated by IgE antibodies.

17.1.1.1 The Symptoms

The symptoms associated with peanut allergy may appear throughout the body. The most common sites are the mouth (swelling of the lips or tongue, itching lips), the digestive tract (stomach cramps, vomiting, diarrhea), the skin (hives, rashes, or eczema), and the airways (wheezing or breathing problems), and can progress to become systemic (dizziness, loss of consciousness, anaphylactic shock, coma) and sometimes deadly. As little as a trace amount of peanut can cause death. Anaphylaxis can quickly produce severe symptoms within as little as 5 to 15 minutes, although life threatening reactions may progress over hours.

17.1.1.2 Accidental Ingestion of Peanut Is Increasing

Accidental ingestion of peanuts in unsuspected food is increasing because of its ubiquitous use as an ingredient in processed foods, and the presence of peanut in the increasingly expanding diet in most industrial countries. One of Britain's most promising athletes died in June of 1999 at the age of 21, after a seizure triggered by accidentally ingesting peanuts in a chicken sandwich (*The Independent*, London, June 21, 1999). In Huntsville, AL, a high school student lost his life to peanut allergy. Despite the seriousness of peanut allergy, to date, there is no cure. Epinephrine and antihistamines are administered to reverse the symptoms. The most effective management strategy available to the allergic population is total avoidance of peanuts and peanut-containing products. However, avoiding peanuts in food products is difficult. In a recent study, 75% of patients with peanut allergy still failed to avoid peanuts in food products even if they were carefully trying to do so (John Weisnagel, 2002, http://www.allerg.qc.ca/peanutallergy.htm). This insecurity leads to severe psychological and emotional distresses for peanut allergic individuals and their families, and a considerable disruption of their daily activities and family relations (7).

17.1.1.3 Nutritional Impact of Peanut

Peanut (*Arachis hypogaea* L.) is not a nut as the name might imply but a legume and a member of the *Leguminosae* family just like soybean and garden peas. Peanut is a highly nutritious, popular, and inexpensive food worldwide. Peanut consumption is steadily increasing in industrialized countries. For example, peanut butter is an integral part of Americans' national culture and America's favorite protein rich sandwich. Peanut and peanut butter are protein powerhouses containing 15% of the Reference Daily Intake (RDI) of protein per serving.

Peanut is rich in proteins, monounsaturated and polyunsaturated fatty acids, carbohydrates, and fibers. Peanut is a good source of vitamins E, B1, B2, B6, folic acid, thiamin, niacin, and minerals such as copper, manganese, phosphorous, iron, magnesium, calcium, selenium, and zinc. Peanut has been linked to many health benefits. For example, it contains mostly heart healthy mono unsaturated fat, which has been shown to lower blood cholesterol and reduce the risk of heart diseases; fiber, which reduces the risk of some types of cancer, and controls blood sugar levels. Peanut is also a good source of folic acid, which helps prevent neural tube defects, and phytosterols which may offer protection from colon, prostate, and breast cancer. Resveratrol, another naturally occurring phytochemical found in peanut and in grape, is associated with reduced cardiovascular diseases and reduced cancer risks (www.peanut-institute.org).

17.1.1.4 Economic Impact of Peanut Allergy

Peanut is the #1 snack nut consumed in the US, accounting for two thirds of the total snack nut market. Half of the top ten selling candy bars in the US contain peanut or peanut butter. Peanut contributes over four billion dollars annually to the US economy. Americans eat more than 600 million pounds of peanuts and 700 millions pounds of peanut butter each year (www.nationalpeanutboard.com). About 70% of the total domestic consumption of peanut is in direct food use (peanut butter, snack peanut, peanut candy, salted peanut). Peanut is also valued when crushed, as high protein, and as vegetable oil, preferred for its long shelf life and cooking qualities. Domestic food use of peanut represents 2.34 billion pounds in 2001 and 2002 (Economic Research Service/USDA; Agricultural Outlook/March 2002). Although the trend of peanut consumption slightly rebounded for the years 1997 and 1998, the government purchases of peanut butter and roasted peanuts for its food programs have tremendously fallen. This is mainly due to the prevalence of peanut allergys (http://www.ces.uga.edu/Agriculture/agacom/outlook/peanuts/pnut2000.htm). In addition, there has been a reduction in household and institutional demands (e.g., by airlines and schools) of peanuts. For example, as of May, 2001, several major U.S. airlines, including

United, USAir, and TWA, have elected not to serve peanut snacks. (http://www.foodallergy.org/topics_archive/flying.html).

World peanut exports are approximately 1.3 million metric tons (shelled basis). The major peanut exporters are the US, China, and Argentina. The US exports about one fourth of the world peanut trade. Sixty percent of its raw peanut exports are destined to the European Union (EU). The major markets within the EU are the United Kingdom, the Netherlands, and Germany. The largest US export market for processed peanut butter is Saudi Arabia, followed by Canada, Japan, Germany, and Korea. Major snack nut markets are the Netherlands, Spain, the United Kingdom, France, and Germany (http://www.peanutsusa.com/trade/uspexp.html).

17.1.1.5 Peanut Allergy is a Liability to the Peanut and Food Industry

Peanut allergy has become a major problem for the peanut and the food industries because of the potential liability. The FDA has stepped up inspection of food manufacturing plants to make sure that no peanuts are getting into products that are not supposed to have them (8). Yet, the number of food allergy alerts and food recalls by manufacturers is increasing. For example, in the US, between May 10 and August 8, 2002, thirty-five foods were recalled for potential contamination with allergen containing foods, and 6 were due to undeclared peanuts. In 2002, in a landmark case, 10 million dollars were granted for the first time in a food allergy case to a boy who suffered permanent brain damage after going into shock while eating a candy bar laced with peanuts (9). This case will set a costly trend not only for the peanut industry and the American economy, but also for the peanut allergic individuals and their families.

17.1.2 Attempted Solutions to Peanut Allergy

Over the years several research teams have investigated and proposed a variety of ways to reduce or eliminate peanut allergy, including: thermal processing, immunotherapy, and anti-IgE and -DNA vaccines.

17.1.2.1 Thermal Processing

Studies conducted to eliminate allergens from peanut using thermal processing failed. Scientists at the USDA/ARS showed that peanut allergens are heat stable and that roasting and cooking enhanced the allergenic properties of peanuts due to protein cross linking and other nonenzymatic biochemical reactions (10–12).

17.1.2.2 Immunotherapy

Allergic individuals are desensitized with a series of injections of peanut extracts, which can sometime result in increased tolerance. However, such tolerance is not usually maintained (13–15). To date, immunotherapy is not recommended for peanut allergy because of the dangers involved during treatment. These dangers include severe anaphylactic reactions with reported deaths. In fact in one editorial comment to the *Canadian Journal of Allergy & Clinical Immunology,* Dr. Sussman (1998) warns that peanut immunotherapy may be an extremely dangerous procedure, which can create a false sense of security.

17.1.2.3 Anti-IgE Vaccine

This approach uses the recombinant protein technology to modify peanut proteins to prevent them from binding to IgE. They will still be able to stimulate an immune response when an unmodified peanut protein is introduced in the organism. The effect of anti-IgE therapy in patients with peanut allergy was studied. Leung et al. (16) concluded that TNX-901, a genetically engineered humanized IgG1 monoclonal antibody against IgE significantly increased the threshold of sensitivity to peanut, on oral food challenge, from half a peanut to almost nine peanuts. Another anti-IgE antibody, Xolair®, is also being tested for subcutaneous use in peanut allergy. (http://www.foodallergy.org/Research/antiigetherapy.html). A drawback to this approach, if successful, is the requirement for monthly injection of the drug which will affect the lifestyles of allergic patients.

17.1.2.4 DNA Vaccine

Researchers from Johns Hopkins School of Medicine have reported hope of effective prophylaxis against peanut allergy. Using a murine model they were able to induce tolerance using an oral vaccine made of DNA from the most offending peanut allergen Ara h 2 gene complexed with nanoparticles of chitosan, a natural polysaccharide, which acts as a vector for the DNA and protects it from digestion. They found that the chitosan-DNA vaccine did protect susceptible mice from developing an allergic response to Ara h 2. Control mice, fed nonencapsulated DNA, were not protected. The effect on peanut allergic individuals is not known, and the road between mice and human may be long (17).

It is important to stress that although encouraging, none of these approaches has led to a commercial solution for peanut allergy. To date, there is still no cure to peanut allergy and complete avoidance is the only proven management practice available to peanut allergic individuals. Therefore, a novel solution is offered the peanut allergic population via the genetic engineering of a hypoallergenic peanut.

17.1.3 Search for a Naturally Occurring Hypoallergenic or Allergen Free Peanut Variety

Before attempting to engineer an allergen reduced or hypoallergenic peanut plant, it was pertinent to ask if there was a naturally occurring hypoallergenic or allergen free peanut variety. A literature search made at the time revealed no data. We therefore proceeded to screen peanut germplasm in search of a null or hypoallergenic or allergen free peanut germplasm which could be used as the parent in peanut breeding programs and could also serve as a model plant to design genetic engineering experiments targeting at down regulating and silencing allergen genes in peanut.

Because this search for an allergen free Plant Introduction (PI) was the first of its kind it was critical to select a peanut germplasm which could provide a wide source of genetic diversity. It would be limiting to choose peanut varieties commercially grown in the US because the germplasm base of the US peanut cultivars is narrow. About half of the genes of all US cultivars are from two original cultivars (18). The ideal approach would be to screen peanut germplasm grown in its center of origin in Latin America. PI from countries such as Bolivia, Brazil, Paraguay, Peru, and Venezuela would provide a greater source of genetic diversity and variation. An indirect Enzyme Linked Immunosorbent Assay (ELISA) procedure was performed to detect the presence of peanut antigens or allergens in pooled sera of allergic patients (19).

The results of the ELISA revealed no naturally occurring allergen free peanuts. Two hypoallergenic germplasms, PI261942 and PI338386, were identified. Between the two, however, variation in the allergen content of the other peanut lines were not significantly different ($p < 0.05$) (Figure 17.1).

Hypoallergenic peanuts were defined as those PI which contain a significantly low ($p < 0.05$) amount of antigens capable of eliciting IgE production and thus, with a lower

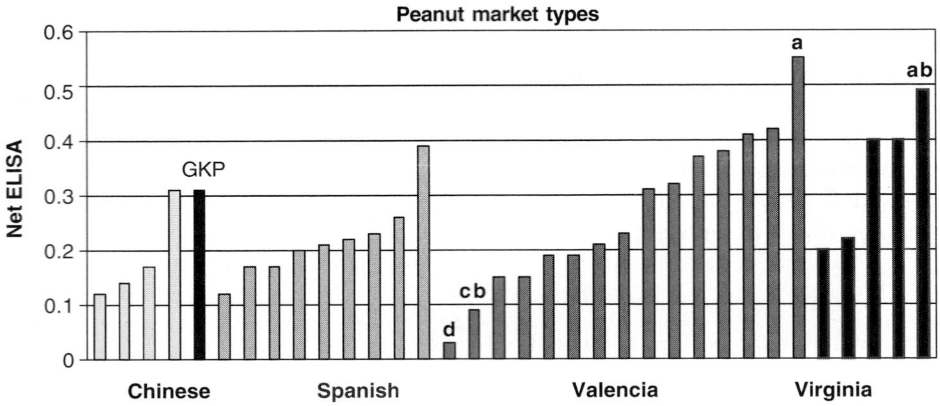

Figure 17.1 Distribution of allergen contents in four peanut market types. The PI are originating from peanut center of origin in South America. ELISA was performed using pooled sera of peanut allergic patients. Letters a, b, c and d are significantly different at $p \leq 0.05$.

likelihood of triggering an allergic reaction from peanut sensitive individuals, especially if the antigen present is a minor allergen. The significantly low optical density (OD) was set to be OD < 0.05 under the conditions of this experiment (19).

In addition, no significant difference was found in allergen content when comparing the four peanut market types used in the study (Figure 17.1). However, a wider range of variation was observed within the Valencia market type (0.020–0.550) compared to the Chinese (0.120–0.310) and Spanish market types (0.115–0.380). Breeding for a hypoallergenic or allergen free peanut variety might require the crossing of a null wild relative, essentially diploid ($2n = 2x = 20$), with a cultivated peanut which is an allotetraploid ($2n = 4x = 40$). The difference in ploidy level is an important criterion, making it difficult to successfully cross cultivated peanut with its closest wild relatives. In addition, plant breeding is long, tedious, laborious, and uncertain.

17.2 STRATEGY TO DEVELOP A GENETICALLY MODIFIED HYPOALLERGENIC PEANUT

Following advances made by the green revolution, genetic engineering will make it possible to improve agricultural crop in the areas of nutrition, medicine and environment. Genetic engineering holds one of the greatest potential to benefit global health. We embarked on this revolutionary projects with the goal to use a genetic engineering approach to dovelop a hypoallergenic and/or allergen-free peanut for the well being of peanut allergic patients and to boost the croded peanut Economy.

17.2.1 The Gene Silencing Principles

17.2.1.1 Posttranscriptional Gene Silencing (PTGS) Strategies to Produce a Hypoallergenic Peanut

Most of the allergen proteins in plants are present as isoforms encoded by genes which belong to multigene families (20,21). Isoforms for peanut allergens were identified for Ara h 1 (22,23), Ara h 2 (24,25), and Ara h 3 (26,27). Consequently any strategy that focuses on silencing single genes is unlikely to lead to a significant reduction in allergen production. Therefore, posttranscriptional gene silencing (PTGS) technologies, such as the use of antisense and RNA interference (RNAi) based constructs, which selectively inhibit gene expression by base-pairing to complementary regions or target sites on mRNA transcribed from the target genes, will provide the only effective approaches to eliminated allergens from peanut.

The RNA induced gene silencing phenomenon was first identified in the analysis of transgenic plants (28,29), in which transgenes were found to silence homologous endogenous genes. The phenomenon was initially called cosuppression to indicate that a transgene could not only suppress the expression of the homologous endogenous gene, but also silence itself. Cosuppression was later defined to happen at the transcriptional as well as posttranscriptional level (30). RNA directed transcriptional silencing involves histone H3 methylation and formation of heterochromatin (31), while posttranscriptional gene silencing (PTGS) is based on mRNA degradation.

The antisense RNA mediated gene silencing is a similar PTGS phenomenon based on hybridization of antisense sequence to complementary sequences in mRNA molecule coding for a specific protein. Both the cosuppression and the antisense methods have been extensively used to down regulate genes in plants. Of particular interest is the successful engineering of hypoallergenic varieties of important plants such as rice (32), soybean (33), and ryegrass (34).

17.2.1.2 Mechanism of PTGS

Meins (35) reviewed the speculative models for PTGS in which a double stranded RNA (dsRNA) molecule has been identified as the main silencing signal. This molecule has a dual role: it is able to trigger de novo methylation of homologous DNA sequences in the nucleus (36) and to induce the sequence specific degradation process responsible for PTGS (37). Sources of dsRNA molecule include: (1) pairing of transcripts transcribed from inverted repeats (IRs) transgenes, (2) pairing of the normal sense RNAs and antisense RNAs (asRNA) arising from aberrant transcription of the same trans (genes), (3) pairing of complementary regions of RNA degradation products, and (4) pairing of transcripts with asRNA produced by RNA directed RNA polymerase (RdRP), RNA viruses, transposons, or short endogenous hairpin RNAs (35,38).

Geley and Muller (39) divided the PTGS process into two distinct phases, (Figure 17.2) an initiation and an execution phase. The initiation phase involves processing of dsRNA into small RNA molecules, called small interfering RNAs (siRNA) by Dicer, a conserved member of the RNAse III gene family (40), which is found in many organisms including fission yeast, plants, *C. elegance*, Drosophila, mice, and human, suggesting all organisms use the same basic mechanism to initiate the PTGS pathway.

In the execution phase, Dicer generated siRNAs are incorporated into a large multiprotein complex called RNA induced silencing complex, or RISC (41,42). After unwinding of the siRNA duplex, a single RNA strand becomes incorporated into a RISC complex (43). In the siRNA mediated mRNA degradation pathway, the antisense strand of the siRNA is used to target the cognate mRNA for degradation (44). This process involves specific base pairing between the antisense strand of the siRNA and the target mRNA, endonucleolytic cleavage of the mRNA strand across the middle of the siRNA strand, and subsequent degradation of the now unprotected mRNA.

In a related pathway, short noncoding single stranded RNAs, called micro RNAs or miRNAs, which are derived from partially complementary dsRNA precursor molecules, regulate the translation rather than the stability of their complementary mRNA target. Micro RNAs are RNA precursors processed in the nucleus into pre miRNAs (45) of 60 to

70 nucleotides (nt) long. Pre miRNAs form stem loop structures, or hairpins. These noncoding pre miRNAs are then exported into the cytoplasm to be further processed by Dicer (46). Mature miRNAs are of similar size to siRNAs (20–25 nt), but single stranded, and they recognize sequences in the 3' UTR of their target mRNAs. Depending on the extent of homology to their target sequence, miRNAs can initiate mRNA degradation or block translation by an as yet poorly defined mechanism (47).

17.2.1.3 PTGS for Crop Improvement

Because of the dominant silencing properties of IR transgenes, they are being exploited for research purposes and practical applications. Phenotype suppression using IR transgenes, cosuppression, or antisense RNA can persist in vegetatively growing plants, and in some cases, it can be meiotically transmitted to progeny. However, silenced genes can return to an expressed state, indicating that the stable changes are potentially reversible (48).

For crop improvement, trait stability from one generation to the next is essential. Therefore, the gene silencing technologies have been improved for stable heritability of the silenced phenotype. Very few reports have described the stability of phenotype suppression by IR transgenes. This involved the use of hairpin RNA (hpRNA) transgene constructs (49). The authors reported that phenotype suppression by hpRNA transgenes was stably inherited at least as far as the T5 generation. Although no data are available beyond T5, the authors however expected the transgene to persist. In rice, the mutant line LGC-1 (Low Glutelin Content-1) was the first commercially useful cultivar produced by RNAi (50). It is low protein rice, and is useful for patients with kidney disease whose protein intake is restricted. This dominant mutation produced hpRNA from an inverted repeat for *glutelin*, the gene for the major storage protein glutelin, leading to lower glutelin content in the rice through RNAi. Interestingly, this mutant was isolated in the 1970s and the mutant trait appears to have been stable for over 20 generations. These examples suggest that the suppression of gene expression by hpRNA induced RNAi would be inherited stably.

hpRNA induced RNAi has also been shown to be much more efficient than antisense and cosuppression–mediated gene silencing (51). In an hpRNA producing vector, the target gene is cloned as an inverted repeat spaced with an unrelated sequence and is driven by a strong promoter such as the CaMV 35S promoter for dicots or the maize ubiquitin 1 promoter for monocots. When an intron is used as the spacer, which is essential for stability of the inverted repeat, the silencing efficiency becomes very high, almost 100% (52,53). Recently, Makoko (54), hypothesized that, hpRNA induced RNAi is inherited more stably than "conventional" PTGS because hpRNA induced RNAi does not require the generation of dsRNA mediated by RdRP for the suppression of the expression of the gene (Figure 17.3).

17.2.2 Application of the Gene Silencing Principles to Peanut

We designed genetic engineering protocol which incorporated several steps and included the isolation of allergen genes, cloning of the genes in carefully selected vectors, transformation of peanut tissues in culture via microprojectile bombardment or *Agrobacterium tumefaciens* and regeneration of transgenic issues to full maturity (Figure 17.4).

17.2.2.1 Step 1: Identification and Isolation of Peanut Allergens Genes

A prerequisite to silencing peanut allergen genes is the acquisition of DNA fragments encoding the target proteins and their use as transgenes in the construction of transformation cassettes. Peanut allergens are seed storage proteins and constitute about 5% of the total protein of a cell. Three major peanut allergens have been identified and characterized at the amino acid, cDNA, and genomic DNA levels. They are Ara h 1, Ara h 2, and Ara h 3

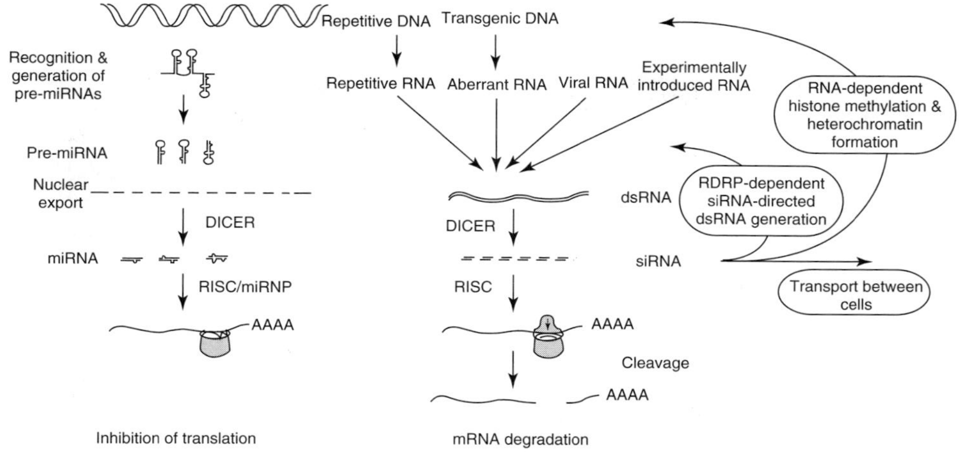

Figure 17.2 dsRNA-directed gene silencing mechanism adapted from Geley and Muller, 2004. Short dsRNA molecules can either be expressed by endogenous genes, invading viruses, or by experimental means, and are funneled into one of two different silencing mechanisms. The miRNA-dependant pathway, which mainly controls the translation of mRNAs, involves imperfect base pairing of the miRNA to its mRNA target, while siRNAs are perfectly complementary to their cognate mRNA species and induce their endonucleolytic cleavage and degradation.

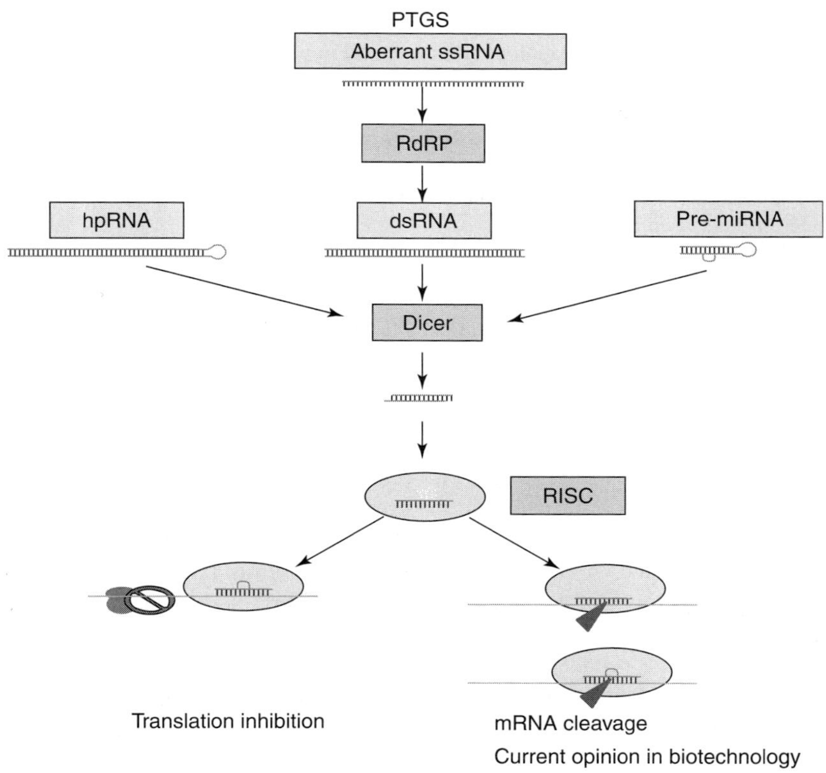

Figure 17.3 dsRNA, hpRNA, and pre-miRNA are processed by Dicer into 21 nt RNA duplexes and the unwound ssRNA is then incorporated into RISC (adapted from Makoto, 2004).

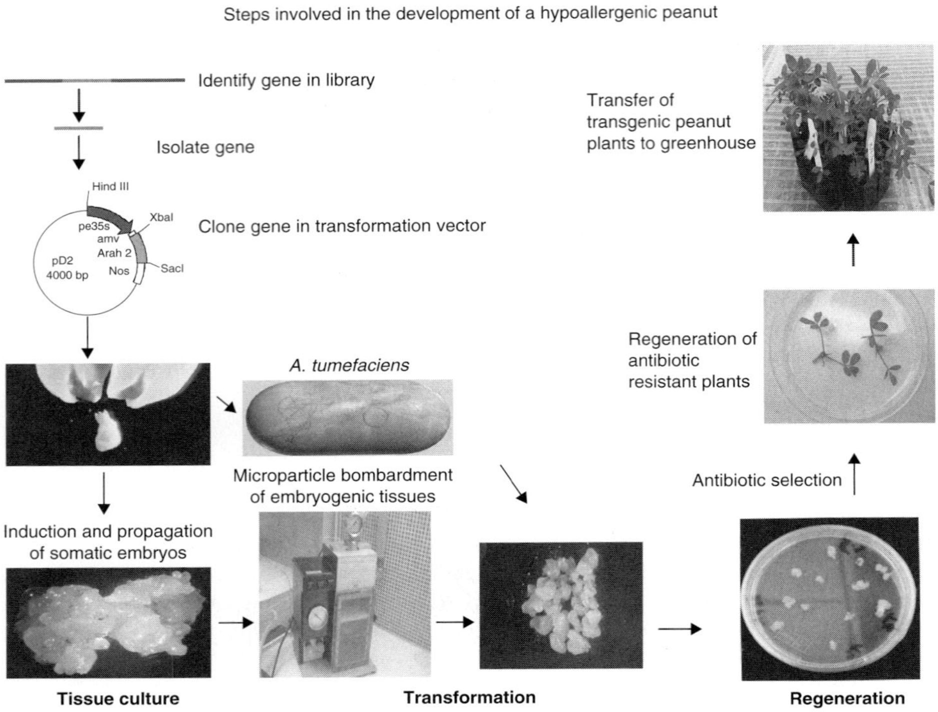

Figure 17.4 Schematic representation of steps involved in producing transgenic peanut via microprojectile bombardment or *Agrobacterium*–mediated transformation. The gene encoding peanut allergens are identified, isolated and cloned into a transformation vector. Somatic embryos initiated from peanut zygotic embryos are co-bombarded with plasmid DNAs containing the peanut allergen gene and an antibiotic selection marker gene. A binary vector containing the peanut allergan gene is mobilized into *A. tumerfaciens* for transformation in peanut tissues. Antibiotic-resistant cell lines are propagated for plant regeneration.

(23–25,55–57). Allergens are classified as "major" when they trigger a reaction in over 50% of the hypersensitive population. Although cDNA clones of major allergens were published, we made the decision to isolate and sequence the genomic clone of target allergens for several reasons. In addition to the basic information provided by a cDNA clone such as the size and coding region of the gene, a genomic clone will provide information on the features of the promoter, other regulatory regions, and the presence and sequences of introns. Introns can be used in the construction of intron hairpin RNA transformation cassette to facilitate dsRNA formation in the process of triggering RNAi. All of which are important contribution to the understanding of the structure of peanut allergen genes.

Ara h 1 encodes a 63.5 kDa protein recognized by serum IgE from over 90% of peanut sensitive individuals (55). At the protein level, it shares approximately 40% homology with vicilins from soybean and pea. The Ara h 1 transcript is about 2.3 kb and is an abundant message in mature peanut cotyledons. Ara h 1 contains 23 IgE binding epitopes that are evenly distributed along its linear sequence. A molecular model of Ara h 1 reveals that it is a homotrimer with IgE binding epitopes located in the area of monomer–monomer contact (58). This quaternary structure of Ara h 1 may be critical for its allergenicity as the epitopes appear to be structurally protected from digestive enzymes. The genomic clone of Ara h 1 was isolated and characterized in our Laboratory. Sequence analysis revealed a full length Ara h 1 gene with an open reading frame (ORF) of 2275 bp, and 3 introns (23).

Ara h 2 is the smallest of the three major allergen genes encoding a 17.5 kDa protein, and it is also the most potent allergen. At the DNA level it shares 39% amino acid homology with conglutinins from soybean and pea (57). Ara h 2 is recognized by serum IgE from 90% of allergic individuals tested. Ten antigenic (IgE binding) sites have been identified by screening overlapping peptides. Three out of these ten epitopes have been characterized as being immunodominant. A partial cDNA sequence for Ara h 2 (460 bp) has been published (55). The genomic clone of the Ara h 2 gene was isolated and sequenced by Viquez et al. (22). Sequence analysis revealed a full length genomic clone with an ORF of 622 nucleotides including a signal peptide and no intron. In the 5′ untranslated region (UTR) of the promoter, there is a TATA box TATTATTA. In the 3′ UTR there is a polyadenylation signal AATAAA (Figure 17.5). Two regions of sequence homologies were identified between major allergen Ara h 2 and two minor allergens Ara h 6 and Ara h 7 (59). The two regions present an opportunity to isolate the three genes together and also to silence them in concert during a single transformation event. Downregulating, or silencing, the most offending allergen Ara h 2 will considerably reduce the allergenicity of transgenic peanuts.

Ara h 3 encodes a 60 kDa preproglobulin (preprolegumin; arachin polypeptide); however, it is considered a minor allergen relative to Ara h 1 and Ara h 2 because it is recognized by less than 50% of peanut allergic individuals (56). The coding region of Ara h 3 is 1,530 nucleotides long coding for 510 amino acids that share 62% to 72% homology with glycinins (legumins) from soybean and pea. The protein contains four major epitopes represented by amino acid residues 33–47, 240–254, 279–293, and 303–317. One epitope, epitope 3, is considered to be immunodominant, being recognized by 100% of those individuals who are allergic to Ara h 3. The genomic Ara h 3 clone was also isolated in our laboratory. It is a 1614 bp clone with 3 introns (27).

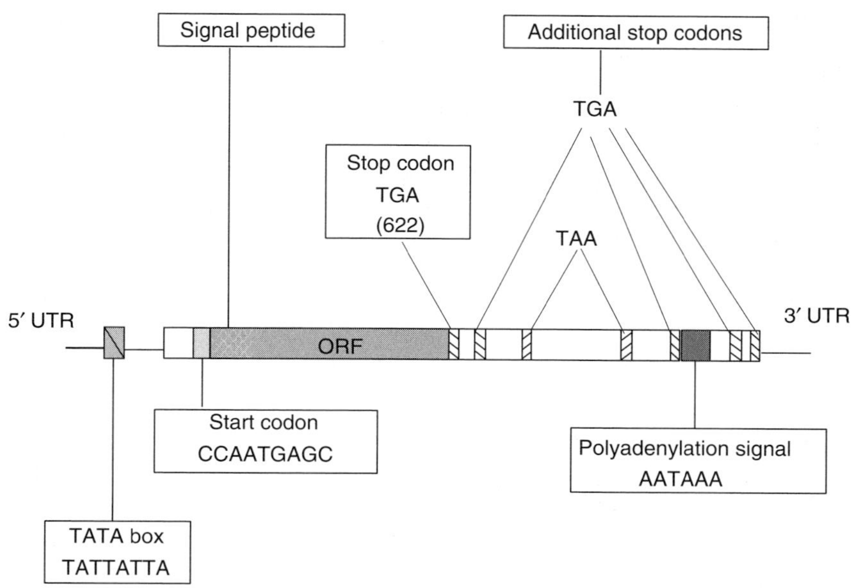

Figure 17.5 Genomic structure of peanut allergen Ara h 2 gene. The Open Reading Frame (ORF) of 622 bp is shown in grey and includes a signal peptide. In the 5′ UTR promoter region there is a typical TATA box and in the 3′ UTR a polyadenylation signal. Other features such as start and stop codons are listed.

17.2.2.2 Step 2: Construction of Peanut Ara h 2 Inverted Repeat Vectors

Ara h 2 is one of the most offending allergenic proteins in peanut (60). Silencing this major allergen is a critical step in the process of engineering an allergen free peanut variety. Peanut allergen Ara h 2 was therefore, selected as a model gene to be used to provide proof of the allergen gene silencing concept. The elimination of this protein alone will significantly reduce the allergenicity of the plant and result in the production of a hypoallergenic peanut.

According to the PTGS model, an inverted repeat construct made of Ara h 2 gene fragments targeting regions of sequence homologies with Ara h 6 and Ara h 7 would induce dsRNA for the specific degradation of Ara h 2, Ara h 6, and Ara h 7 mRNA. Two PCR fragments of a 430 bp region of the Ara h 2 gene were mounted as inverted repeats (IR) in the pUC18 vector which contains an enhanced version of the Cauliflower Mosaic Virus 35S promoter (CaMV 35S), and the Nopaline Synthase (NOS) terminator of *A. tumefaciens*. The IR transgene was fused to the 35S promoter and to the NOS term to construct plasmid pDK2 as shown in Figure 17.6.

17.2.2.3 Step 3: Induction of Peanut Somatic Embryos

The first protocol for successful transformation of peanut using particle bombardment of somatic embryos was developed by Ozias-Akins and coworkers (61). This protocol is today one of the most widely used because it presents the advantage of overcoming the genotype barrier, and allows the successful transformation of different peanut genotypes. Mature peanut embryonic axes were removed from seeds and cultured on Murashige and Skoog (MS) medium (62) containing 3 mg/l picloram to initiate somatic embryos. Embryogenic tissues were subcultured for three to four months for tissue increase, and to produce homogenous embryogenic tissues.

17.2.2.4 Step 4: Transformation and Regeneration of Transgenic Peanut

Plasmid pDK2 (Figure 17.6) was cobombarded with plasmid pCB13, a selection plasmid harboring the hygromycin resistance gene to transform peanut embryogenic tissues. Five independent bombardment experiments were performed, using embryogenic callus initiated from the peanut variety "Georgia green", a variety commonly used during particle bombardment mediated peanut transformation. Tissues were cobombarded by mixing at a 1:1 ratio plasmids pDK2 and pCB1, and bombarded tissues were subcultured for 2 to 3 months on the embryo induction medium supplemented with 10 mg/l hygromycin. Antibiotic resistant cell lines were independently propagated on the same medium for an additional 2 to 3 months for tissue increase. Transgenic embryos were produced on picloram free MS medium and germinated by supplementing the medium with 2 mg/l Naphthalene acetic acid

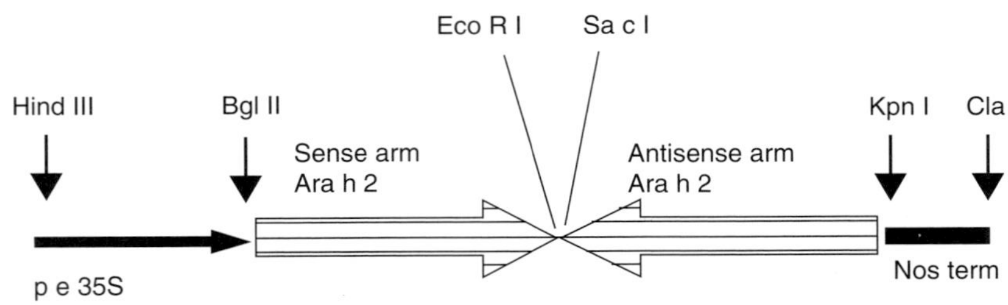

Figure 17.6 Inverted Repeats plasmid pDK2 constructed using Ara h 2 DNA fragments for transformation of peanut embryogenic tissues. Restriction sites are inserted for the restriction map of the transformation cassette.

(NAA) and 1 mg/l Benzyl Adenine Purine (BAP). Regenerated plants were transferred to the greenhouse.

No apparent phenotypic difference was detected between plants derived from somatic embryogenesis (transgenic as well as nontransgenic) and the control obtained from direct nontransgenic seed germination. However, the rate of flowering of most somatic embryo derived plants was lower and failed to seeds properly, while control plants from direct germination of nontransgenic seeds produced seeds within 3 months of transfer to greenhouse. These data suggest that the low level of flowering in tissue culture derived peanut plants resulted from the culture conditions, especially the lengthy time involved in the process. Therefore, a shorter tissue culture protocol was adopted.

17.2.3 Molecular Analyses

17.2.3.1 PCR Analysis

PCR analysis was performed to detect the presence of two foreign DNA fragments (the CaMV 35S promoter fragment, and the hygromycin gene) in transgenic peanut cells. PCR analyses were performed on all 48 hygromycin resistant callus lines produced. Two sets of primers were designed based on the nucleotide sequence of the DNA fragments to amplify the 35S promoter and the hygromycin selection gene which are DNA fragments not naturally present in peanut, but will be present in transgenic peanut. The primers amplified the expected 0.6 kb fragment of the promoter in 36 out of 48 (75%) of the callus lines. Using the same primers targeting the promoter, no amplification was observed in the two controls composed of the nontransgenic callus line, and the line transformed with the pCB13 selection plasmid only. To confirm the transgenic status of the control pCB13 callus line, amplification using the primers targeting the hygromycin gene was performed and the expected 0.7 kb fragment was generated.

17.2.3.2 Southern Hybridization

Southern hybridization was performed to confirm the stable integration of the transgene into the peanut genome. The target sequence was the CaMV 35S promoter. DNA was extracted from both transformed and nontransformed callus lines, and digested with *Hind*III/*Bgl*II to release the 0.6 kb promoter fragment. The hybridization probe was a ^{32}P labeled PCR product of the pe35S promoter. The hybridization bands (Figure 17.7) confirmed the stable integration of the transgene in the peanut genome (63).

17.2.3.3 Northern Analysis

Northern hybridization was performed to detect Ara h 2 inverted repeat transcripts in transgenic plants. Total RNA was extracted from callus, roots, petioles, and leaves of transgenic and nontransgenic plants, and hybridized with the 430 bp Ara h 2 transgene fragment. Data in Figure 17.8, show hybridization bands corresponding to the accumulation of the Ara h 2 transgene transcript is in the leaves and roots of transgenic peanut plants (Figure 17.8). These transcripts were also detected in the transgenic callus. All together, the data confirm (1) the constitutive expression of the pe35S promoter and (2) the stable and functional integration of the Ara h 2 transgene throughout the development of the transgenic plants (63).

17.2.2.4 Accumulation of Endogenous Ara h 2 Transcripts

In a parallel experiment total RNA extracted from different organs (roots, leaves, petioles, and seeds) of control nontransgenic peanut plant was hybridized with the 430 bp Ara h 2

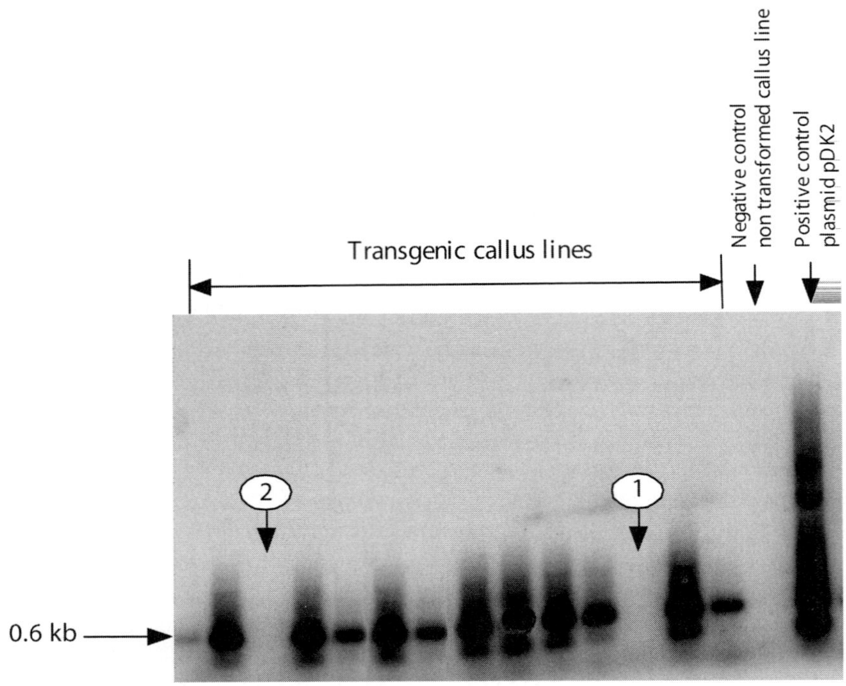

Figure 17.7 Southern hybridization in transgenic callus lines of peanut transformed with plasmid pDK2, showing the 0.6 kb band of the promoter region of the transgene. The probe was a ^{32}P labeled PCR product of the pe35S. Circled lanes 1 and 2 correspond to transgenic callus lines positive to hygromycin only.

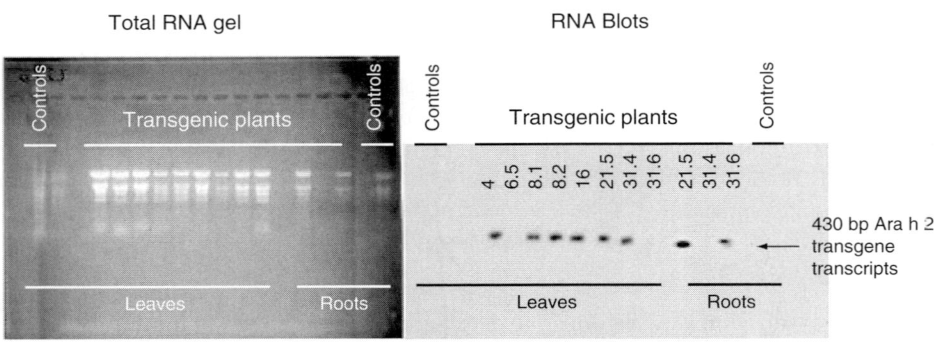

Figure 17.8 Northern analyses performed on total RNA extracted (A) from callus, and (B) from leaves and roots of mature peanut plants regenerated from transformed somatic embryos. Hybridization was performed using the 430 bp Ara h 2 transgene fragment labeled with ^{32}P.

transgene fragment to detect endogenous Ara h 2 transcripts. Figure 17.9, shows that endogenous Ara h 2 is expressed only in seed but not in vegetative tissues. These data suggest that although Ara h 2 transgene transcripts were accumulated in different organs of transgenic plants, the level of down regulation and silencing of Ara h 2 transcripts could not be assessed in the vegetative tissues but only in the seed (63). Therefore, experiments are ongoing to take the transgenic Ara h 2 hypoallergenic peanut to maturity. Nutritional

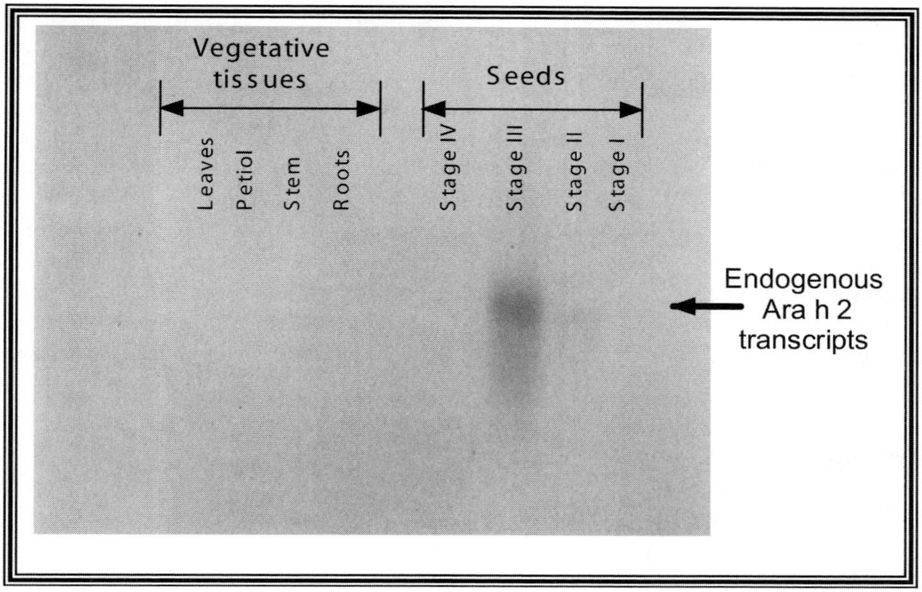

Figure 17.9 Northern analysis from leaves, roots, petioles, stems, and seeds of nontransformed peanut plants obtained by seed germination. Hybridization was performed using the 430 bp Ara h 2 transgene fragment labeled with ^{32}P.

content of the transgenic seeds will be evaluated and immunological analysis performed to determine the level of down regulation of allergen Ara h 2 in transgenic peanut seeds.

17.2.4 Current and Future Prospects: Production of Fertile Transgenic Peanut with Ara h Introns Hairpin RNA (ArahihpRNA) Constructs

Research groups using transgenic technologies for the development of improved peanut have experienced limited success mainly due to the increase sterility of transformed plants. However, only few articles have reported on this problem (64). Our personal experience and discussion with colleagues working on peanut tissue culture and transformation has confirmed that it is a common problem. The lengthy tissue culture and transformation procedures (12–17 months) currently used to produce transgenic peanut via somatic embryogenesis (61) is unanimously indexed as the major cause of sterility. These problems have prompted some researchers to adopt a nontissue culture based approach to generate transgenic peanut plants (65). Unfortunately, this procedure is inefficient ($<1\%$ transformation), labor consuming, and inadequate to produce a large number of independent transformants.

Alternative organogenesis-based protocols to produce fertile transgenic peanut plants were developed via *Agrobacterium tumefaciens* mediated transformation (66–69). However, one major disadvantage with these protocols are the problems associated with genotype and tissue specificity of *A. tumefaciens* strains. Cells which are susceptible to *A. tumefaciens* infection are restricted to certain positions within certain type of tissues and peanut susceptibility to *Agrobacterium* infection varied greatly among peanut genotypes (70).

Our preliminary experiments using *Agrobacterium* strain EHA105 to transform hypocotyls (66) and embryonic axe (67) transformation have shown good reproducibility. Therefore, future emphasis will be to optimize these transformation protocols for the production of fertile transgenic hypoallergenic peanut.

It is critical to understand that our goal in developing a hypoallergenic or allergen free peanut is not to encourage peanut allergic individuals to eat peanuts, but to promote the utilization of the transgenic peanuts in food products containing peanuts, such that in the event of an accidental ingestion of peanuts by an allergic individual, no severe allergic reactions will be triggered. In the unlikely event of an allergic reaction, it will at the most be very mild, without endangering human life. In addition the negative perception associated with peanut consumption, as well as the economic, emotional, and psychological impacts of peanut allergy, will be reduced and eventually dissipated. The development and marketing of a hypoallergenic peanut will give rise to a number of new enterprises including seed producers, processors, millers, and other allied businesses involved in the marketing of the value added products. Those who choose to market a safe hypoallergenic or allergen free peanut product for human consumption will have a competitive advantage.

REFERENCES

1. Molkhou, P. Épidémiologie de l'allergie alimentaire. *J. Puériculture* 17:249–253, 2004.
2. Taylor, S.L. Chemistry and detection of food allergens. *Food Technol.* 46:146–152, 1992.
3. Bock, S.A. The natural history of food sensitivity. *J Allerg. Clin. Immunol.* 69:173–177, 1982.
4. Sicherer, S.H., A. Munoz-Furlong, A.W. Burks, H.A. Sampson. Prevalence of peanut and tree nut allergy in the U.S. determined by a random digit dial telephone survey. *J. Allerg. Clin. Immunol.* 103:559–562, 1999.
5. Hourihane, J.O., S.A. Kilburn, P. Dean, J.O. Warner. Clinical characteristics of peanut allergy. *Clin. Exp. Allerg.* 27:634–639, 1997.
6. Bock, A.S., B.A. Munoz-Furlong, H.A. Sampson. Fatalities due to anaphylactic reactions to foods. *J. Allerg. Clin. Immunol.* 107:191–199, 2001.
7. Primeau, M.N., R. Kagan, L. Joseph, H. Lin, C. Dufresne, C. Duffy, D. Prhcal, A. Clarke. The psycological burden of peanut allergy as perceived by adults with peanut allergy and the parents of peanut allergic children. *Clin. Exp. Allerg.* 30:1135–1143, 2000.
8. Josephson, D. FDA target snack food industry over allergens. *BMJ* 322–883, 2001.
9. Steward, L. $10 M granted in first plaintiff victory in food allergy case. *The Legal Intelligence,* January 31, 2002.
10. Maleki, S.J., S.Y. Chung, E.T. Champagne, J.P. Raufman. The effects of roasting on the allergenic properties of peanut proteins. *J. Allerg. Clin. Immunol.* 106:763–768, 2000.
11. Maleki, S.J., O.M. Viquez, T. Jacks, H. Dodo, E.T. Champagne, S.Y. Chung, S. Laundry. The major peanut allergen, Ara h 2, functions as a trypsin inhibitor, and roasting enhances this function. *J. Allerg. Clin. Immunol.* 112:190–195, 2003.
12. Beyer, K., E. Morrow, L. Xiu-Min, L. Bardina, K. Ludmilla, G.A. Bannon, A.W. Burks, H.A. Sampson. Effects of cooking methods on peanut allergenicity. *J. Allerg. Clin. Immunol.* 107:1077–1081, 2001.
13. Oppenheimer, J.J., H.S. Nelson, A. Bock, F. Christensen, D.Y.M. Leung. Treatment of peanut allergy with rush immunotherapy. *J. Allerg. Clin. Immunol.* 90:256–262, 1992.
14. T Nelson, H.S., J. Lahr, R. Rule, A. Bock, D. Leung. Treatment of anaphylactic sensitivity to peanuts by immunotherapy with injections of aqueous peanut extract. *J. Allerg. Clin. Immunol.* 6:744–751, 1997.
15. Bock, S.A., H.A. Sampson, F.M. Atkins, R.S. Zeiger, S. Lehrer, M. Sachs, R.K. Bush, D.D. Metcalfe. Double-blind, placebo-controlled food challenge (DBPCFC) as an office procedure: a manual. *J. Allerg. Clin. Immunol.* 82:986–997, 1988.
16. Leung, D.Y., H.A. Sampson, J.W. Yunginger, A.W. Burks, L.C. Schneider, C.H. Wortel, F.M. Davis, J.D. Hyun, W.R. Shanahanr. Effect of anti-IgE therapy in patients with peanut allergy for the TNX-901 peanut allergy study group. *N. Engl. J. Med.* 348:986–993, 2003.

17. Roy, K., H.Q. Mao, S.K. Huang, K.W. Leong. Oral gene delivery with chitosan-DNA nanoparticles generates immunologic protection in a murine model of peanut allergy. *Nat. Med.* 5:387–391, 1999.
18. Isleib, T.G., J.C. Wynne. Use of plant introduction in peanut improvement. In: *Use of Plant Introduction to Cultivar Development*, Part 2, CSSA Spec. publication 20, Shands, H.L., L.E. Weisner, eds., Madison, WI: Crop Science Society of America, 1992.
19. Dodo, H., D. Marsic, M. Callender, E. Cebert, O. Viquez. Screening 34 peanut introductions for allergen content using ELISA. *Food Agric. Immunol.* 14:149–156, 2002.
20. Lowenstein, H., S.H. Sparholt, S.S. Klysner, H. Ipsen, J.N. Larsen. The significance of iso-allergenic variations in present and future specific immunotherapy. *Int. Arch. Allerg. Appl. Immunol.* 107:285–289, 1995.
21. Lagares, A., L. Puerta, L. Caraballo. Polymorphism in allergens. *Biomedical.* 22:51–62, 2002.
22. Becker W.M. Characterization of Ara h 1 by two-dimensional electrophoresis immunoblot and recombinant techniques: new digestion experiments with peanuts imitating the gastrointestinal tract. *Int. Arch. Allerg. Immunol.* 113:118–121, 1997.
23. Viquez, O.M., N.K. Konan, H.W. Dodo. Structure and organization of the genomic clone of a major peanut allergen gene, Ara h 1. *Mol. Immunol.* 40:565–571, 2003.
24. Viquez, O.M., C.G. Summer, H.W. Dodo. Isolation and molecular characterization of the first genomic clone of a major peanut allergen, Ara h 2. *J. Allerg. Clin. Immunol.* 107:713–717, 2001.
25. Chatel, J.M., H. Bernard, F.M. Orson. Isolation and characterization of two complete Ara h 2 isoforms cDNA. *Int. Arch. Allerg. Immunol.* 131(1):14–18, 2003.
26. Dodo, H.W., O.M. Viquez, S. Maleki, K. Konan. cDNA clone of a putative peanut (*Arachis hypogaea* L.) trypsin inhibitor has homology with peanut allergens Ara h 3 and Ara h 4. *J. Agric. Food Chem.* 10:1404–1409, 2004.
27. Viquez, O.M., N.K. Konan, H.W. Dodo. Genomic organization of peanut allergen gene, Ara h 3. *Mol. Immunol.* 41:1235–1240, 2004.
28. Napolis, C., C. Lemieux, R. Jorgensen. Introduction of a chimeric chalcone synthase gene into petunia results in reversible co-suppression of homologous genes in trans. *Plant.* 2:279–289, 1990.
29. van der Krol, A.R., L.A. Mur, M. Beld, J.N. Mol, A.R. Stuitje. Flavanoid genes in petunia: addition of a limited number of gene copies may lead to a suppression of gene expression. *Plant Cell.* 2(4):291–299, 1990.
30. Matzeke, M., A.J.M. Matzke, J.M. Kooter. RNA: guiding gene silencing. *Science.* 293:1080–1083, 2001.
31. Stevenson, D.S., P. Jarvis. Chromatin silencing: RNA in the driving seat. *Curr. Biol.* 13:13–15, 2003.
32. Tada, Y., M. Nakase, T. Adachi, R. Nakaruma, H. Shimada, M. Takahashi, T. Fujimura, T. Matsuda. Reduction of 14–16 kDa allergenic proteins in transgenic rice plants by antisense gene. *FEBS Lett.* 391:341–345, 1996.
33. Kinney, A.J., K.L. Stecco, R. Helm, E.M. Herman. Prospects for hypoallergenic soybean seeds. American Society of Plant Biology Annual meeting, Abs #158.
34. Bhala, P.L., I. Swoboda, M. Singh. Antisense-mediated silencing of a gene encoding a major ryegrass pollen allergen. *Proc. Natl. Acad. Sci.* 96:11676–11680, 1999.
35. Meins, F. RNA degradation and models for post-transcriptional gene silencing. Plant Molecular Biology. 43:261–273, 2000.
36. Mette, M.F., J. van der Winden, M.A. Matzke, A.J.M. Matzke. Production of aberrant promoter transcripts contributes to methylation and silencing of unlinked homologous promoters in trans. *EMBO J.* 18:241–248, 1999.
37. Fire, A. RNA- triggered gene silencing. *Trends Genet.* 15:358–363, 1999.
38. Hannon, G.J. RNA interference. *Nature.* 418:244–251, 2002.
39. Geley, S., C. Muller. RNAi: ancient mechanism with a promising future. *Exp. Gerontol.* 39:985–998, 2004.

40. Bernstein, E., A.A. Caudy, S.M. Hammond, G.J. Hannon. Role for a bidentate ribonuclease in the initiation step of RNA interference. *Nature*. 409:363–366, 2001.
41. Hammond, S.M., E. Berstein, D. Beach, G.J. Hannon. An RNA-directed nuclease mediates post-transcriptional gene silencing in drosophila cells. *Nature* 404:293–296, 2000.
42. Nykanen, A., B. Haley, P.D. Zamore. ATP requirements and small interfering RNA structure in the RNA interference pathway. *Cellular* 107:309–321, 2001.
43. Martinez, J., A. Patkaniowska, H. Urlaub, R. Luhrmann, T. Tuschl. Single-stranded antisense siRNAs guide target RNA cleavage in RNAi. *Cellular* 110:563–574, 2002.
44. Schwarrz, D.S., G. Hutvagner, B. Haley, P.D. Zamore. Evidence that siRNAs function as guides, not primer, in Drosophila and human RNAi pathways. *Mol. Cell.* 10:537–548, 2002.
45. Lee, Y., C. Ahn, J. Han, H. Choi, J. Kim, J. Yim, J.T. Lee, P. Provost, O. Radmark, S. Kim, V.N. Kim. The nuclear RNAse Drosha initiates microRNA processing. *Nature*. 425:415–419, 2003.
46. Lee, Y., K. Jeon, J.T. Lee, S. Kim, V.N. Kim. MicroRNA maturation: stepwise processing and subcellular localization. *EMBO J.* 21:4663–4670, 2002.
47. Doench, J.G., C.P. Petersen, P.A. Sharp. siRNAs can function as miRNAs. *Genes Dev.* 17:438–442, 2003.
48. Mitsuhara, I., S.N. Shirasawa, T. Iwai, S. Nakamura, R. Honkura, Y. Ohashi. Release from post-transcriptional gene silencing by cell proliferation in transgenic tobacco plants: possible mechanism for noninheritance of the silencing. *Genetics.* 160:343–352, 2002.
49. Stoutjesdijk, P.A., S.P. Singh, Q. Liu, C.J. Huristone, P.A. Waterhouse, A.G. Green. hpRNA-mediated targeting of the Arabidopsis FAD2 gene gives highly efficient and stable silencing. *Plant Physiol.* 129:1723–1731, 2002.
50. Kusaba, M., K. Miyahara, S. Lida, H. Fukuoka, T. Takano, H. Sassa, M. Nishimura, T. Nishio. Low glutelin content 1: a dominant mutation that suppresses the glutelin multigenee family via RNA silencing in rice. *Plant Cell.* 15:1455–1467, 2003.
51. Chuang, C.F., E.M. Meyerowitz. Specific and heritable genetic interference by double-stranded RNA in *Arabidopsis taliana. Proc. Natl. Acad. Sci. USA* 97:4985–4990, 2000.
52. Smith, N.A., S.P. Singh, M.B. Wang, P.A. Stoutjesdijk, A.G. Green, P.M. Waterhouse. Total silencing by intron-spliced haipin RNAs. *Nature* 407:310–320, 2000.
53. Wesley, S.V., C.A. Helliwell, N.A. Smith, M.B. Wang, D.T. Rouse, Q. Liu, P.S. Gooding, S.P. Sing, D. Abbott, P.A. Stoutjesdijk, S.P. Robinson, A.P. Gleave, A.G. Green, P.M. Waterhouse. Construct design for efficient, effective high-throughput gene silencing in plants. *Plant J.* 27:581–590, 2001.
54. Makoto, K. RNA interference in crop plants. *Curr. Opin. Biotechnol.* 15:139–143, 2004.
55. Burks, A.W., G. Cockrell, J.S. Stanley, R.M. Helm, G.A. Bannon. Recombinant peanut allergen Ara hI expression and IgE binding in patients with peanut hypersensitivity. *J. Clin. Invest.* 96:1715–1721, 1995.
56. Rabjohn, P., E.M. Helm, J.S. Stanley, C.M. West, H.A. Sampson, A.W. Burks, G.A. Bannon. Molecular cloning and epitope analysis of the peanut allergen Ara h 3. *J. Clin. Invest.* 103:535–542, 1999.
57. Stanley, J.S., N. King, A.W. Burks, S.K. Huang, H. Sampson, G. Cockrell, R.M. Helm, M. West, G.A. Bannon. Identification and mutational analysis of the immunodominant IgE binding epitopes of the peanut allergen Ara h 2. *Arch. Biochem. Biophys.* 342:244–253, 1997.
58. Shin, D.S., C.M. Compadre, S.J. Maleki, R.A. Kopper, H. Sampson, S.K. Huang, A.W. Burks, G.A. Bannon. Biochemical and structural analysis of the IgE binding sites on Ara h 1, an abundant and highly allergenic peanut protein. *J. Bio. Chem.* 273:13753–13759, 1998.
59. Kleber-Janke, T., R. Crameri, U. Appenzeller, M. Schlaak, W.M. Becker. Selective cloning of peanut allergens, including profilin and 2S albumins, by phage display technology. *Int. Arch. Allerg. Immunol.* 119:265–274, 1999.
60. Knol, E.F., M. Wensing, R. Vlooswijk, M. Ertmann, A.C. Knulst, J. Koppelman. Relevance of Ara h 1, Ara h 2, and Ara h 3 in peanut allergenic patients, as determined by IgE-Western-Blotting, basophil histamine release, and intracutaneous testing: Ara h 2 is the most important peanut allergen. *Suppl. J. Allerg. Clin. Immunol.* 111:194, 2003.

61. Ozias-Akins, P., J.A. Schnall, W.F. Anderson, C. Singsit, T.E. Clemente, M.J. Adang, A.K. Weissinger. Regeneration of transgenic peanut plants from stably transformed Embryogenic callus. *Plant Sci.* 93:185–194, 1993.
62. Murashige, T., F. Skoog. A revised medium for raid growth and bioassays with tobacco tissue culture. *Plant Physiol.* 15:473–497, 1962.
63. Konan, K.N., O.M. Viquez, H.W. Dodo. 2004. Towards the development of a hypoallergenic peanut through genetic transformation. *Appl. Biotechnol. Food Sci. Policy* 1(3):159–168, 2004.
64. Ozias-Alkins, P., R. Gill. Progress in the development of tissue culture and transformation methods applicable to the production of transgenic peanut. *J. Am. Peanut Res. Educ. Soc.* 28:123–131.
65. Rohini, V.K., K.S. Rao. Transformation of peanut (*Arachis hypogaea* L.): a non-tissues culture based approach for generating transgenic plants. *Plant Sci.* 150:41–49, 2000.
66. Egnin, M., A. Mora, C.S. Prakash. Factors enhancing *Agrobacterium tumefaciens*-mediated gene transfer in peanut (*Arachis hypogaea* L.). In Vitro *Cell. Dev. Biol. Plant.* 34:310–318, 1998.
67. Khandelwal, A., K.J.M. Vally, N. Geetha, P. Venkatachalam, M.S. Shaila, G. Lakshmi Sita. Engineering hemagglutinin (H) protein of rinderpest virus into peanut (*Arachis hypogaea* L.) as a possible source of vaccine. *Plant Sci.* 165:77–84, 2003.
68. Eapen, S., L. George. *Agrobacterium tumefaciens*-mediated gene transfer in peanut (*Arachis hypogaea* L.). *Plant Cell. Rep.* 13:582–586.
69. Sharma, K.K., V. Anjaiah. An efficient method for the production of transgenic plants of peanut (*Arachis hypogaea* L.) through *Agrobacterium tumefaciens*-mediated genetic transformation. *Plant Sci.* 159:7–19.
70. Mansur, E.A., C. Lacorte, V.G. de Freitas, D.E. de Oliveira, B. Timmerman, A.R. Cordero. Regulation of transformation efficiency of peanut (*Arachis hypogaea* L.) explants by *Agrobacterium tumefaciens*. *Plant Sci.* 89:93–99, 1995.

2.18

Recombinant Lipoxygenases and Oxylipin Metabolism in Relation to Food Quality

Rod Casey and Richard K. Hughes

CONTENTS

18.1 INTRODUCTION

Lipoxygenases (LOX) (EC 1.13.11.12) are nonheme, iron containing dioxygenases that catalyse the formation of *Z, E* conjugated hydroperoxides from polyunsaturated fatty acids (PUFAs), or any other molecules containing one or more *Z, Z*-1, 4-pentadiene structures.

LOXs are widely distributed throughout Nature. Their presence has been recorded in higher and lower plants and animals, including fungi, corals, algae, sea urchins, clams, and slime mold (Table 18.1). The presence of LOX has also been proposed in the brown alga *Laminaria angustata* (21), the diatom *Thalassiosira rotula* (22), the hydroid *Hydra vulgaris* (23), the barnacle *Balanus balanoides* (24), and the crab *Carcinus maenas* (25), but LOX sequences are absent from yeast and typical prokaryote genomes.

There are very many sequences within DNA databases that have been proposed as encoding LOX, but only a small number of these have been demonstrated to encode a functional, active enzyme. Despite considerable sequence diversity between and within plant and animal LOX sequences, the amino acids that are critical to catalysis, particularly the histidine residues that act as ligands to the active site iron, are generally highly conserved, and the amino acids that form part of the substrate (O_2 and PUFA) binding site show a greater degree of conservation than the rest of the sequence.

LOX are significant in food biotechnology for a number of reasons:

1. Their activity destroys PUFAs, which are essential components of human diet.
2. The fatty acid hydroperoxides that result from their activity can be further metabolized by other enzymes into compounds, such as six- or nine-carbon aldehydes and alcohols, that represent key flavor and aroma components of a wide range of fruits and vegetables.

Table 18.1
Examples of the recorded presence of LOX within lower organisms in Nature

Species
Fungi
Fusarium oxysporum (1,2)
Gäumannomyces graminis (3)
Saprolegnia parasitica (4)
Lagenidium giganteum(5)
Pityrosporum orbiculare (6)
Mushrooms
Psalliota campestris (7)
Psalliota bispora (7)
Agaricus bisporus (8)
Agaricus campetris (9)
Pleurotus ostreatus (10)
Others
Algae
Ulva lactuca (11)
Enteromorpha intestinalis (12)
Chlorella pyrenoidosa(13)
Miscellaneous
Plexaura homomalla (coral) (14)
Pseudoplexaura porosa (coral) (15)
Strongylocentrotus purpuratus (sea urchin) (16)
Pennaeus japonicus (shrimp) (17)
Dictyostelium discoideum (slime mold) (18)
Starfish and clam oocytes (19,20)

3. Some of the secondary products from fatty acid hydroperoxides, such as *n*-hexanal, impart off flavors to a number of vegetable derived, particularly legume derived, products (the so-called grassy-beany odor of soybean based products is a good example).
4. As a consequence of their reaction mechanisms, some LOX promote cooxidation processes that can have either positive or negative effects. An example of a positive effect is the improvement of bread dough rheology resulting from the LOX mediated formation of disulphide bonds in wheat gluten. The destruction of pigments, antioxidants, and vitamins by LOX mediated cooxidation would generally be regarded as negative, but pigment bleaching by cooxidation is a positive aspect in the production of white bread or pasta.
5. Plant LOX play important roles in defense against insect damage and fungal invasion. Although somewhat outside the scope of this article, plant health and the quality of foods derived from plants are obviously interrelated, and LOX activity clearly affects food quality in this more general sense.

The purpose of this chapter is to describe the nature, distribution, and activity of LOX, concentrating almost exclusively on plant LOX since these are most significant in food technology; and to describe how the production and analysis of recombinant LOX has led to a deeper appreciation of the action of the enzymes and their current and future roles in food biotechnology.

18.1.1 LOX Isoforms and Heterogeneity

Early studies of seed LOX assumed them to be a single molecular species. The realization that LOX can exist as a number of isoforms within a given plant organ came from work on soybean seed LOX (26,27), that demonstrated four isoforms: LOX-1, -2, -3a, and -3b. Subsequent studies, including analyses of genetic variants that lack LOX-3 (28), suggested that LOX-3a and -3b are derived from the same parent molecule and that there are actually only three major isoforms of soybean seed LOX. LOX-1 differs markedly from LOX-2 and -3 in a number of properties, including stability, but most notably in the pH for optimum activity, which is centred around 9.0 for LOX-1 and 6.5–7.0 for LOX-2 and -3. LOX-1 also shows a marked preference for charged fatty acids and little reactivity with esterified fatty acids. Most plant LOX reported to date are more similar to soybean LOX-2 and -3 in their properties; in this sense LOX-1, with its three dimensional structure and action mechanism which is known in some detail, is atypical of plant LOX.

Multiple forms of seed LOX have been recognized in several other species, including barley (29), broad beans (30), chickpeas (31,32), kidney beans (33), maize (34,35), peas (36,37), peanuts (38), rapeseed (39), rice (40), wheat (41), and winged beans (42). Isoforms are recognized on the basis of a number of parameters, including their amino acid sequence, their pH for optimal activity, differences in net surface charge that permit their separation by ion exchange chromatography, isoelectric focusing or chromatofocusing, sensitivity to Ca^{2+}, substrate preference, turnover numbers, apparent K_m values, the ability to cooxidize pigments and to generate carbonyl compounds, heat stability, and the nature of their products, which is a consequence of their regiospecificity (the position along the fatty acid chain where oxygen is inserted). Although it is clear from defense responses and from studies of developmental regulation that different vegetative LOX isoforms can play different physiological roles (43,44), it is not obvious why seeds should contain several LOX isoforms. Several mutants that lack one or more seed isoforms exist in soybean (28,45,46), pea (47), and rice (48) and, at least in the case of soybeans, would appear to have unimpaired agronomic performance (49,50), suggesting that some seed LOX may be dispensable for normal growth and development.

The genome of *Arabidopsis thaliana*, which is one of the smallest within the plant kingdom, contains six genes encoding LOX that fall into two classes; those with and those without N-terminal sequence extensions that suggest targeting to the chloroplast. In at least one instance, LOX has been shown to be targeted to the chloroplast . Those LOX that are associated with chloroplasts have markedly different sequences from those that lack such targeting peptides, and which are assumed to be located elsewhere within the cell. Shibata et al. (51) have classified those LOX that contain such a targeting sequence as *Lox2* gene products, and all others as *Lox1*.

Other plant LOX exhibit greater heterogeneity. Soybeans, as mentioned, have three major seed isoforms (LOX-1, -2, and -3); and at least three other vegetative (nonseed) isoforms (LOX-4, - 6) have been recognized. Peas have multiple LOX isoforms, with as many as five different seed forms, and other, different, LOX in leaves, stems, flowers, and roots (52). Furthermore, even a single isoform, such as pea seed LOX-2 or LOX-3, can comprise two or three very similar polypeptides from slightly divergent members of a small gene family. One consequence of this microheterogeneity is that the measured properties of pure preparations from plant sources will often be an average of the properties of the individual components, even for an apparently pure single isoform preparation; recombinant forms of such isoforms will, however, comprise just one type of polypeptide and may therefore differ slightly in properties from the corresponding plant derived preparation, but have many advantages.

18.1.2 The Value of Recombinant LOX

At present, the only large scale sources of LOX are crude sources such as enzyme active soybean flour; and these, as discussed, are limited because they contain a mixture of isoforms with different properties. The production of LOX as recombinant enzymes in microbial hosts such as *E. coli*, yeasts, or *Aspergillus* has several advantages: recombinant production systems from cloned enzymes are an effectively inexhaustible resource for enzyme production; cloned enzymes are a single type of recombinant protein, free of other isoforms; such systems produce very high biomass for high level production; and quality control over sterility and purity is simplified. Enzyme preparations from plant materials will have variable properties depending on the plant variety used as the enzyme source and the conditions under which the plant is grown. There are, for instance, plant varieties that lack particular LOX isoforms (53,54), and LOX enzyme levels can increase dramatically in response to the stresses encountered by plants growing in the field (55).

The benefits of using recombinant DNA technology for the industrial production of enzymes for use in food production are well established. A good example of its successful application is recombinant chymosin for vegetarian cheese. Production of enzymes in this way also has spinoff advantages, in that recombinant enzymes permit facile determination of kinetic and other properties on highly purified preparations, providing information that can be used in several ways. Purification, processing, and storage systems can be optimized, based on such information, but what is more important, enzyme properties can be improved (often dramatically) through the production of mutants, either by directed mutagenesis or by forced evolution. Thus, it becomes important to understand the properties of LOX through the use of recombinant enzymes in relation to their various uses in food biotechnology (56).

To date, a wide range of plant LOX have been produced, as whole or partial molecules, in microorganisms; some examples are given in Table 18.2. These have been produced for many different reasons: in some cases (65) to simply understand the nature of enzyme specificity, in others to establish a role in the aroma of foods (79), and in others still to produce LOX in sufficient amounts (gram quantities) for industrial use (56). In all

Table 18.2
Recombinant plant LOXs of known positional specificity. All positional specificities were determined with linoleate (18:2), except for potato 9-LOX, maize 13/9-LOX, potato 13-LOX-1 and potato 13-LOX-2 with linolenate (18:3), and rice 13-LOX RCI-1, which was with both. Figures shown are those published or calculated.

Source	Accession Number	Ref.	Positional Specificity[a]	Ref.	Designation
Pea seed	X07807	(57)	1:1-1:2	(58)	Pea 9/13-LOX
	X17061	(59)	7:1	(58)	Pea 13/9-LOX
Soybean seed	J02795	(60)	19:1	(61)	Soybean 13-LOX
	X06928	(62)	1:1	(63)	Soybean 9/13-LOX
Potato tuber	AF039651	(64)	1:3	(64)	Potato 9/13-LOX
	X95513	(65)	1:46	(65)	Potato 9-LOX
	Y18548	(66)	1:1	(66)	Potato 13/9-LOX
Potato leaf	X96405	(65)	31:1	(65)	Potato 13-LOX-1
	X96406	(65)	87:1	(65)	Potato 13-LOX-2
Lentil seedling	X71344	(67)	6:1	(68)	Lentil 13/9-LOX
Rice seedling	D14000	(69)	49:1	(70)	Rice 13-LOX-1
	X64396	(71,72)	11:1	(70)	Rice 13-LOX-2
	AB099850	(73)	1:24	(73)	Rice 9-LOX
	AJ270938	(74)	10:1(18:2) 99:1(18:3)	(74)	Rice 13-LOX RCI-1
Cucumber seedling	X92890	(75)	5:1	(76)	Cucumber 13/9-LOX
Barley leaf	U56406	(77)	9:1	(77)	Barley 13/9-LOX
Peanut seed	AF231454	(78)	2:1	(78)	Peanut 13/9-LOX
Almond seed	AJ418043	(79)	1:49	(79)	Almond 9-LOX
Maize leaves	AF271894	(80)	3:2	(80)	Maize 13/9-LOX

[a]Ratio of 13-*Z*,*E*-H(P)ODE:9-*E*,*Z*-H(P)ODE (no discrimination of *R* and *S* stereoisomers)

cases, there is an important interplay between understanding the structure and activity of plant LOX in relation to their uses in food biotechnology and the role in plant physiology for which they have naturally evolved.

18.2 THE STRUCTURE OF LOXs

All LOX so far examined are monomeric and, with the exception of the manganese-containing LOX from *Gäumannomyces graminis* (the "take all" fungus) (81), contain a single atom of nonheme, nonsulphur, high spin iron per polypeptide chain of Mr 75,000–95,000. In the resting or ground state, the catalytically inactive form in which LOX is usually prepared, the iron ion is ferrous [Fe(II)]; activation to the ferric [Fe(III)] ion is required to start the catalytic cycle, during which the iron undergoes redox shuffling.

Determinations of the crystallographic structures of soy LOX-1 (82,83) and LOX-3 (63) were conducted on the unliganded enzyme in the resting state. That of rabbit 15-LOX (84) was obtained in the presence of a bound inhibitor that had superficial similarity to arachidonic acid. The structure of the metastable product oxidized purple form of soy LOX-3 (85) has revealed an unexpected mode of fatty acid binding and identified amino acid residues that may play significant roles in catalysis, regiospecificity, and stereospecificity.

Recently, structures of soy LOX-3 complexed with (-)epigallocatechin gallate and curcumin have been determined (86,87). The soy LOX-1 and LOX-3 enzymes have two domains: an N-terminal β-barrel (146 residues in LOX-1) and a C-terminal helical bundle catalytic domain (693 residues in LOX-1). Mammalian LOX, such as rabbit 15-LOX, are considerably shorter than plant LOX, with 839 residues in soy LOX-1 compared to 662 in rabbit 15-LOX. Sequence comparisons, particularly of the highly variable N-domain, showed little similarity between plant and animal LOX, because of this, it was thought for some time that the N-terminal β-barrel of plant LOX was absent from the mammalian enzymes. The structure of rabbit 15- LOX (84), however, made it clear that the mammalian LOX has the same folds as its plant counterparts, achieving this through molecular parsimony (88), whereby the plant and mammalian enzymes have common core secondary structural elements but the mammalian form has fewer dispensable elements such as surface loops (Figure 18.1).

There are clear differences between soy LOX-1 and LOX-3, in and near the three cavities that connect the active site iron with the molecule surface (63), which form a theoretical basis for the difference in the profile of products from the two enzymes.

The iron in soy LOX-1 is coordinated to six ligands (Figure 18.2). Three of the coordination positions are occupied by histidine Nε atoms. A fourth ligand is oxygen from the carboxylate of the C-terminal isoleucine, and a fifth position is occupied by a water molecule. There is spectroscopic evidence to suggest that this water changes to a hydroxide upon oxidation, which is proposed to function as the active site base. The octahedral coordination sphere also includes an asparagine (Asn^{694}), which in the resting enzyme is involved in a number of labile polar interactions with other groups but is just beyond the maximum bonding distance to be a strong iron ligand; in rabbit 15-LOX this asparagine is a histidine, which has a larger side chain and is a stronger ligand. The distance between Fe

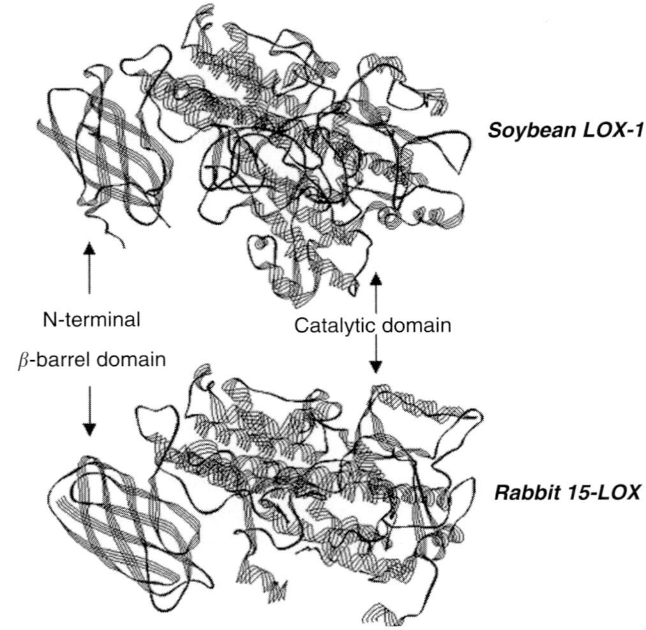

Figure 18.1 Diagram of the three dimensional structure of soybean LOX-1 and rabbit 15-LOX, showing the N-terminal β-barrel and the helical catalytic domain

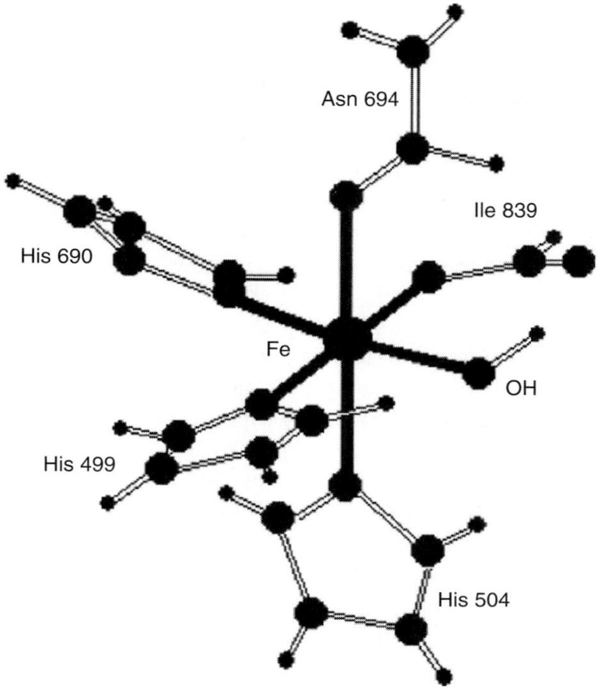

Figure 18.2 Digrammatic representation of the ligand geometry of the iron in the ferric (active) form of soybean LOX-1 (89)

and Asn^{694} depends on whether product is present or not — it is weak, or nonligand, in the Fe(II) form, giving 5-coordination geometry, but may shift position in the active, Fe(III) form to give 6-coordination geometry (Figure 18.2). Also, the bond to Wat^{842} is weak, but the strength and length of the Fe-O(Wat^{842}) bond changes noticeably on oxidation, becoming shorter and stronger as the character of the ligand changes from neutral water to the negatively charged OH group (89). The six ligands are referred to as the first coordination sphere (90). Asn^{694} is part of an important hydrogen bonding network that may provide a structural link between substrate binding and iron coordination (91). Glutamine residue Gln^{495}, which lines the putative substrate cavity wall, forms hydrogen bonds with Gln^{697}, which in turn is hydrogen bonded to Asn^{694}. This H-bond between Asn^{694} and Gln^{697}, is thought to play a key role in the modulation of the coordination flexibility of Asn^{694}, which in turn modulates activity (92). The two glutamines are referred to as the second coordination sphere. The hydrogen bond network plays a role in the hydrogen bond rearrangement of the enzymatic mechanism and steric interaction with the side chain of Gln^{495} is important in the proper positioning of the substrate (91).

18.3 LOX REACTION MECHANISM

There are several reasons to better understand the action mechanism and specificity of LOX in the context of food biotechnology. Substrate specificity will determine which PUFAs are preferentially destroyed by LOX activity in foods, and will influence the nature of the final metabolites from the LOX pathway; this in turn will influence aromas and tastes. Similarly, positional specificity — whether 9- or 13-hydroperoxides are produced — will determine the nature of subsequent taste and aroma compounds; 9-LOX activity will give rise to 9-carbon, and 13-LOX to 6-carbon, aldehydes, and alcohols, with their characteristic aroma properties. In addition, the ability of particular LOX to carry out cooxidation reactions has a major influence in a number of spheres of food biotechnology, including the stability of vitamins and antioxidants and the use of enzyme active soybean flour in bread making. Both of these characteristics — substrate and product specificity and cooxidation potential — depend on the exact nature of the LOX reaction, which differs according to the source of the enzyme, the specific LOX isoform involved, and the reaction conditions.

Isoform 1 of soybean seed LOX (soy LOX-1) is by far the most extensively studied and best understood LOX in terms of reaction mechanism, but with its high pH optimum and degree of thermal stability it is probably not typical of most LOX. Studies of soy LOX-3 as a product oxidized purple form (85) largely support the conclusions drawn from work on soy LOX-1, but differ in a number of details.

The substrate requirement of LOX is at least one 1,4-diene with at least one of the olefins in the *Z* geometry (93). In biological terms, this translates into polyunsaturated fatty acids (PUFAs) that contain *Z,Z*-pentadiene systems: linoleic (18:2), α- and γ-linolenic (18:3) acids in plants, and arachidonic (20:4) and higher order PUFAs in animals (Figure 18.3). With linoleic acid as substrate, the predominant reaction of soy LOX-1 involves formal abstraction of the *pro*-S hydrogen atom from methylene carbon C-11, a rearrangement of the double bond system, and the stereospecific addition of the peroxy group at C13 to form the 13-*S* (*Z,E*) product (Figure 18.4). With α-linolenic acid, hydrogen removal is in principle possible either at C-11 or C-14, or both, but it seems that plants

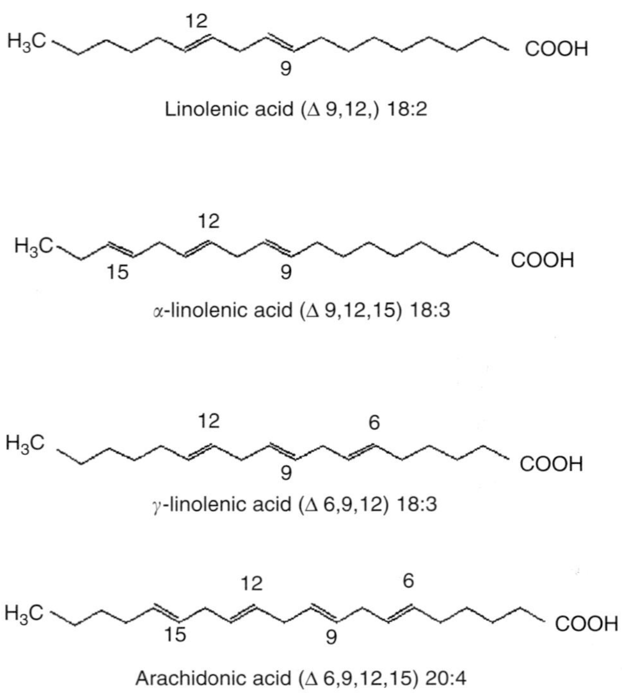

Figure 18.3 The major polyunsaturated fatty acids from plants (18:2, 18:3) and animals (20:4)

use only the one doubly allylic methylene group at C-11 in linoleic and α-linolenic acids (94). The enzyme activates the substrate PUFA for combination with molecular oxygen by this hydrogen abstraction reaction. The results of kinetic isotope effect experiments implicate a tunnelling mechanism for this key step (95), and steady state kinetics have shown that oxygen interacts with the enzyme only after hydrogen abstraction from the substrate (96). The water ligand may act as the Fe (III) bound hydroxide that serves as the effective base for hydrogen abstraction, and it has been suggested that an Fe(III)-OH^- active site base is formed as part of the mechanism, leading to a C-H bond cleavage transition state (Figure 18.5). This transition state requires that the substrate be positioned close enough to the Fe(III)-OH^- species to allow for tunnelling of the substrate hydrogen atom to the metal-bound hydroxide, leading to the fatty acid carbon based radical intermediate L˙ [plus Fe(II)-OH_2]. The radical intermediate reacts with O_2 to regenerate Fe(III)-OH and generate a peroxyl radical, which is subsequently reduced and protonated to form the final 13(*S*)-hydroperoxide product (Figure 18.6). As with most reactions between carbon radicals and O_2, the soy LOX-1 reaction is very rapid, but it is also stereospecific and regiospecific, forming almost exclusively the 13(*S*) product. The radical intermediate L˙ has generally been depicted as a pentadienyl radical, in which radical character is spread over C-9 - C-13, generating greater stability than for an allyl radical. This would make C9 and C13 equally reactive to O_2 attack, barring accessibility differences. However, the side

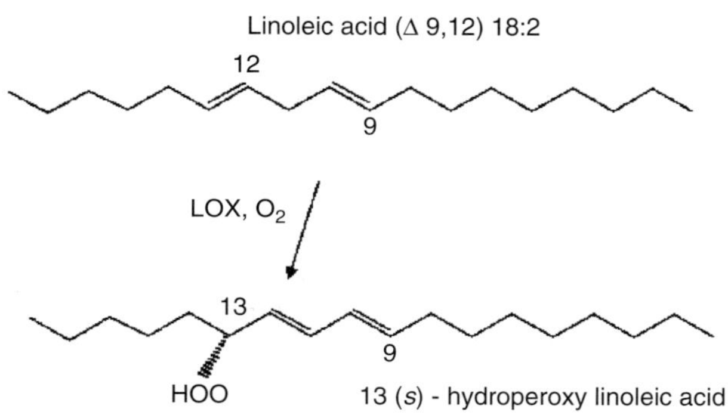

Figure 18.4 Formation of 13-(*S*)-hydroperoxy linoleic acid by soybean LOX-1

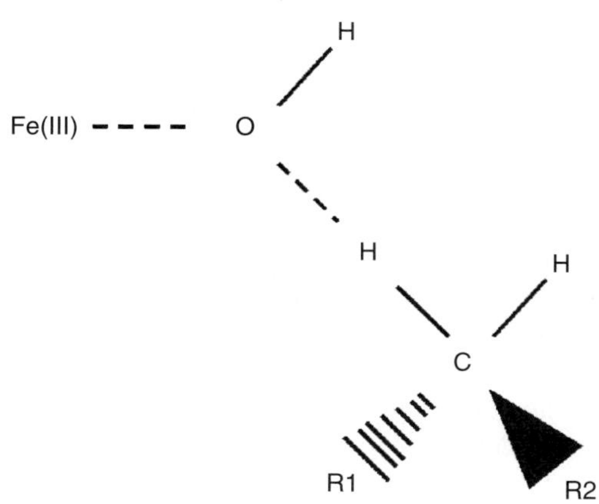

Figure 18.5 Postulated C-H bond transition state of soybean LOX-1 (91)

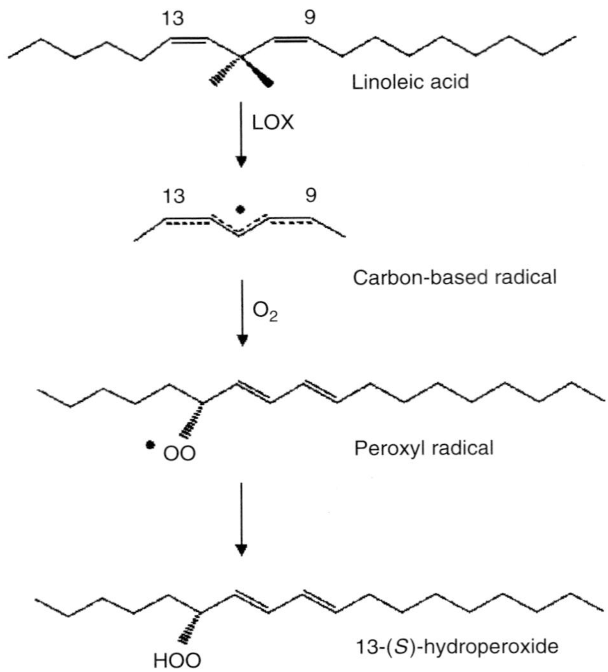

Figure 18.6 Generation of hydroperoxide from linoleic acid *via* carbon based and peroxyl radicals

chains of Ile[538] and Gln[495] in soy LOX-1 provide sufficient bulk to disfavor the planar pentadienyl intermediate, raising the possibility that L˙ is an ene-allyl radical. If the unpaired electron density of L˙ were localized near C13, in the form of a Δ^9-[11,12,13] allyl radical, then O_2 would be expected to selectively react at C13. However, electron paramagnetic resonance (EPR) investigations into the nature of L˙ suggested that radical character is localized near C-9. The Gln[495]→ Ala and Ile[538]→ Ala mutations (97) should each reduce the amount of strain that might favor a localized ene-allyl radical, but they have only modest effects on the rate of oxygenation, suggesting that either an ene-allyl radical is not formed during turnover, or there is rapid interconversion between the Δ^{12}-[9,10,11] and Δ^9-[11,12,13] radicals. There is, therefore, still some doubt over the exact nature of the carbon based radical that reacts with O_2.

The observed stereochemistry requires that hydrogen atom removal and oxygen attack occur antafaracially, on opposite faces of the pentadiene unit. There is some evidence that the enzyme engenders specificity by requiring a relatively specific approach of dioxygen through steric interference or other means, rather than the specific location of an Fe bound dioxygen (97). Mutation of soy LOX-1 Ile[553] → Phe greatly reduces the rate of oxygenation chemistry, while leaving k_{cat} (the overall rate of catalysis, which is probably limited by H abstraction) relatively unchanged. This suggests that Ile[553] → Phe impedes O_2 access to the carbon radical intermediate and supports the view that O_2 enters the active site through the channel that passes residue 553 (97). The studies of Knapp et al. (97) suggest that O_2 in soy LOX-1 requires a discrete, separate access channel, which is a departure from typical views of substrate-activating oxygenases. It seems that the primary role of the Fe^{3+} cofactor is to generate an enzyme bound radical, whereas the protein controls the stereochemistry and regiochemistry of O_2 encounters with this radical (98).

18.4 SUBSTRATE SPECIFICITY

Some LOX have quite specific substrate requirements, others less so. The two LOX (-2 and -3) from pea seeds, for instance, have markedly different substrate specificity despite having a high degree of sequence identity; pea seed LOX-3 is highly specific for linoleic acid, whereas pea seed LOX-2 is also active with linolenic acid and shows reasonable activity with arachidonic acid. Pea seed LOX-2 is almost equally active with linoleic acid and its methyl ester, whereas the latter is not a substrate for pea seed LOX-3 (58). The basis of such specificity is poorly understood, although Hornung et al. (99) used site-directed mutagenesis to change a single amino acid in cucumber lipid body LOX [which normally forms 13(*S*)-hydroperoxides from linoleic acid (100)] and generate a species that was active on γ-linolenic and arachidonic acids. This illustrates the value of recombinant enzymes to carry out such studies.

Although the majority of plant LOX strongly prefer free fatty acids as substrates, at least three – soy LOX-1, Arabidopsis and cucumber root LOX – are capable of using PUFAs esterified to phospholipids (101,102,103); and the LOX from cucumber lipid bodies and from barley seedlings are active with PUFAs that are esterified in neutral lipids such as triacylglycerols (104,76). It will be very relevant to biological function to discover how LOXs interact with substrates in these forms.

18.5 PRODUCT (REGIO- AND STEREO-) SPECIFICITY

As discussed, soy LOX-1 produces almost exclusively the 13(*S*)-hydroperoxide of linoleic acid under the appropriate conditions. Other LOX, such as those from potato tubers or tomato fruits (65,105), produce largely the 9-*S* hydroperoxide, whereas still others [soy LOX-3, pea seed LOX-3, (58)] produce mixtures of 9-*S*- and 13-*S*-hydroperoxides, in different ratios. There is, in addition, a class of LOX, from coral, sea urchin eggs, crabs, barnacles, marine and freshwater hydroids, and oocytes of starfish and clams, that naturally form (*R*)-hydroperoxides and are related in primary structure to the *S*-LOX (106).

It is not yet completely clear how the enzyme controls the regioselectivity and stereoselectivity of catalysis, or what factors govern the outcome of the peroxidation reaction and the formation of possible products. Such information may in principle be gleaned from comparisons of the structures and sequences of enzymes with different specificities. Soy LOX-1 and LOX-3 provide an interesting study in contrast, since their mechanisms lead to both single (LOX-1) and multiple (LOX-3) products of the 9/13 peroxidation of linoleic acid. A comparison of the structures of "native" soy LOX-3 and its 13-hydroperoxide product complex (85) shows few differences, with the only amino acid dislocated in a significant way from its original place being Gln^{514}, which in the native enzyme participates in a hydrogen bonding network which connects $His^{518}\rightarrow Gln^{514}$, $Gln^{514}\rightarrow Gln^{716}$, $Gln^{716}\rightarrow Asn^{713}$, and $Asn^{713}\rightarrow Leu^{773}$. Atoms C5-C6-C7 from the fatty acid occupy the place taken up by the Gln^{514} side chain in the native enzyme; in the complex with HPOD (hydroperoxy dienoic acid) this Gln^{514} side chain is disordered. This could be the basis for an explanation for the lack of regioselectivity in LOX-3 catalysis (85). If Gln^{514} were in its original hydrogen bonding network, the passage for the carboxyl end of the fatty acid would be blocked. If the COOH group of a fatty acid substrate could be stabilized by an interaction with the hydrogen bonding network around Gln^{514}, this would bring the C9, instead of the C13, atom from the fatty acid near the Fe^{3+} cofactor during the catalysed reaction. The binding site in general appears to be more flexible in soy LOX-3 than in LOX-1 as a consequence of amino acid differences in the aligned sequences: (LOX-1/LOX-3) Ser^{747}/Asp^{766}; Thr^{756}/Arg^{775}; Gln^{256}/Thr^{274}.

Although such comparisons give useful insights into the factors that may control specificity, they are limited and need further experimentation to check the hypotheses that they generate. Such experimentation is usually in the form of production of site directed mutants of recombinant enzymes; thus, the generation of recombinant LOX extends well beyond their utility as a production vehicle for biocatalysts, into the realms of improved understanding for the generation of improved activities. The production of site directed mutants, based on crystallographic structures and homology modelling, has been particularly useful in furthering our understanding of the basis of product specificity as well as the nature of the reaction mechanism.

Early mutagenesis experiments with human 15-LOX, based on sequence comparisons of mammalian 12- and 15-LOX (107), identified Met^{418} as a primary determinant of positional specificity. Mutagenesis of Met^{418} to Val converted the wild-type enzyme, which produces almost exclusively 15(*S*)-hydroperoxide from arachidonic acid, into a species that makes equal amounts of 12(*S*)- and 15(*S*)-hydroperoxides. It was proposed that this change in specificity was the consequence of a change in the bulk of the side chain at position 418, such that the substrate could move further into the active site, shifting the site of hydrogen abstraction from C13 to C10 and leading to 12-, rather than 15-, lipoxygenation. This particular amino acid became known as the Sloane determinant of mammalian LOX positional specificity. The suggestion that further translation of the fatty acid into the active site could lead to 5-LOX activity (84) was challenged (108) on the grounds that such further translation will not produce 5-LOX products with the correct stereospecificity of H^+ abstraction and oxygen addition. In addition, Borngräber et al. (109) made a strong argument that Phe^{353}, which also forms part of the binding pocket in 15-LOX, is a further key determinant of positional specificity of oxygenation, a residue that is referred as the Borngräber determinant (99). Borngräber et al. (110) extended their studies to include Ile^{593} of rabbit 15-LOX, amongst other residues, and suggested that enlargement of the substrate-binding pocket favored 12-lipoxygenation, whereas a smaller active site favored 15-lipoxygenation. They concluded that the overall size and shape of the substrate binding pocket is important to defining the specificity of the LOX reaction.

The equivalent position to mammalian 15-LOX Met^{418} is His^{608} in cucumber lipid body 13-LOX (100). Alteration of the bulky His to a less space-filling Val converted the enzyme to a 9-LOX. It was suggested that the alteration may unmask the positively charged guanidine group of Arg^{758}, which in turn may force an inverted head to tail orientation of the fatty acid substrate that leads to 9-, rather than 13-lipoxygenation. The equivalent mutation in pea seed LOX-3 and a specific potato tuber LOX, both of which have 9- and 13-LOX activity, however, did not alter positional specificity (66,111), suggesting that the determinants of specificity may vary from one LOX to another.

Overall, there are two hypotheses that rationalize the mechanistic differences between LOX with different positional specificities; the "orientation based" hypothesis and the "space based" hypothesis (Figure18.7). It is likely that each is correct for a particular circumstance, and the two, in any case, are not mutually exclusive in the sense that changes in the bulk of amino acid residues within the active site can influence both the degree of penetration of the substrate and its orientation.

Recent mutagenesis experiments with soy LOX-1 (97) suggest that steric restraints placed on the approach of oxygen to the enzyme bound substrate by the surrounding amino acid side chains culminate in the observed regioselectivity and stereoselectivity. Replacing bulky leucine residues with Ala at LOX-1 546 and 754 (= LOX-3 565 and 773), which created a constriction in the substrate cavity IIa, resulted in less 13-hydroperoxide, more 9-hydroperoxide, and a lower level of stereoselectivity at both positions. This indicated that the constriction in cavity IIa formed by Leu^{546} and Leu^{754} sterically prevents O_2 access to the C9 position of soy LOX-1 within the pentadiene radical, thus imparting selectivity.

18.6 COOXIDATION

The complete catalytic mechanism (Figure 18.6) involves the formation of a fatty acid peroxyl radical which is reduced, protonated, and released as hydroperoxide. Some LOX, including pea seed LOX-3, produce keto and hydroxy fatty acids that have been proposed (58) to be the consequence of dismutation of a prematurely released peroxyl radical (Figure 18.8). There are other routes to hydroxy fatty acids from hydroperoxides (94), but the production of hydroxy and keto fatty acids seems to be associated with the capacity of the enzyme to carry out cooxidation reactions such as the bleaching of β-carotene (112), which would be consistent with the premature release of a highly oxidizing peroxyl radical or its dismutation into singlet oxygen or both. It has been suggested that such LOX are "imperfect by design" in that the nature of their active site has a flexibility that allows both dual positional specificity and the premature release of a peroxyl radical. Pea seed LOX-3 is such an enzyme; whereas pea seed LOX-2 has a single positional specificity, does not produce hydroxy and keto fatty acids (58) (Figure 18.9) and does not efficiently cooxidize β-carotene. Whatever the mechanism, this ability to carry out cooxidation is highly significant to food quality.

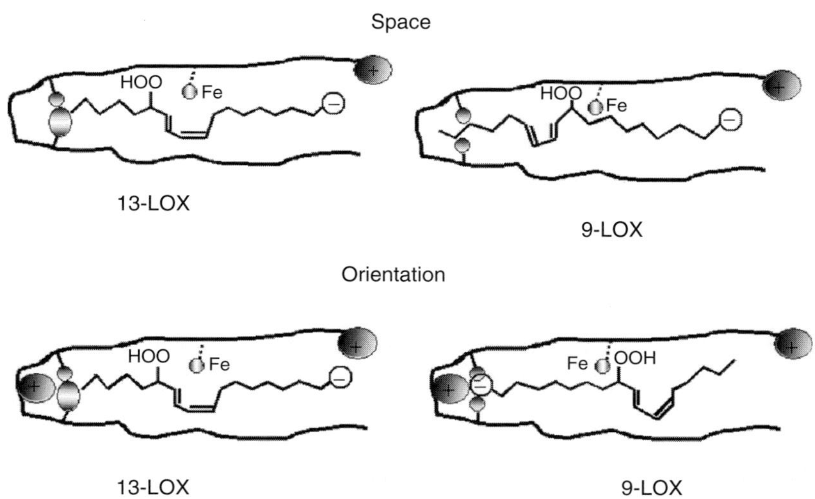

Figure 18.7 Representation of the "space" and "orientation" hypotheses of LOX positional specificity. In both cases the sizes of particular amino acids within the active site cavity play critical roles, but the underlying basis is different in the two hypotheses

$$2 \times LOO^{\bullet} \longrightarrow LOH + L{=}O + O_2\,(^1\Delta_g)$$

Figure 18.8 Dismutation of fatty acid peroxyl radical to produce hydroxy- and keto- fatty acids and singlet oxygen (58)

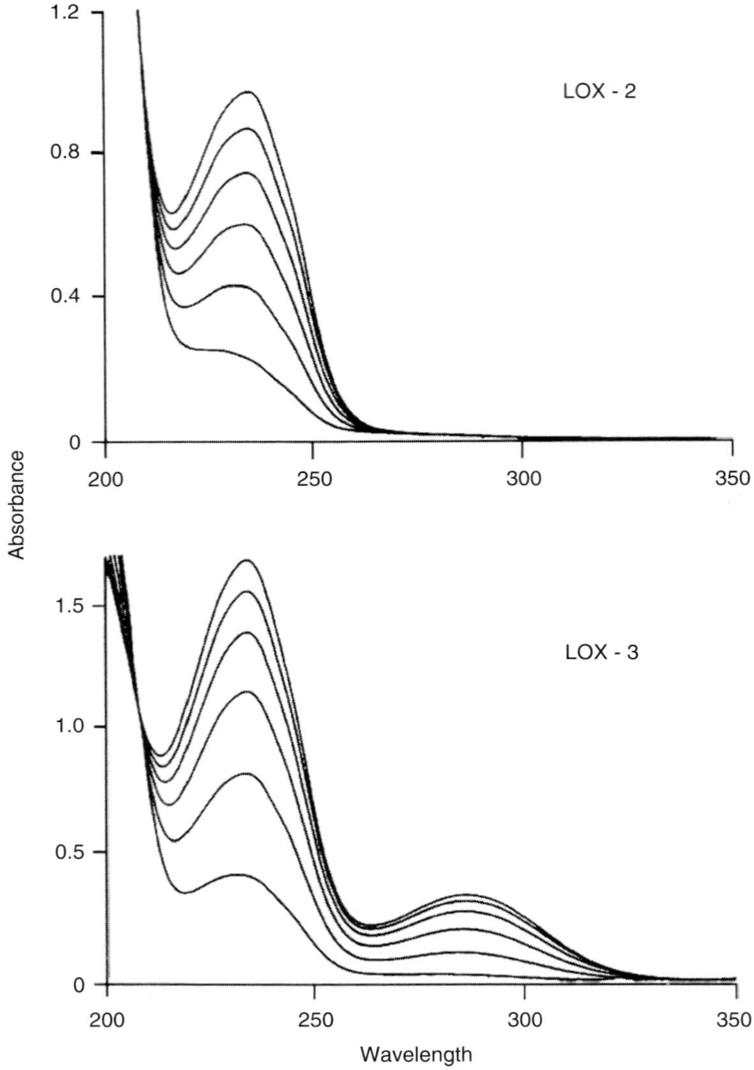

Figure 18.9 Pea seed LOX-3, but not LOX-2, generates carbonyl (keto) compounds that absorb at 280nm

18.7 APPLICATIONS OF LOX TO FOOD QUALITY

18.7.1 Flavors and Aromas

Fatty acid hydroperoxides are toxic compounds that are metabolized *in planta* by a range of enzyme activities (94), including several unusual members of the cytochrome P450 family known as CYP74s. These include allene oxide synthase (AOS), hydroperoxide lyase (HPL), and divinyl ether synthase (DES) (Figure 18.10). All three activities produce compounds that play direct or indirect roles in plant defense. AOS is the first committed step in the synthesis of jasmonate, an important signalling compound that activates defense gene expression upon wounding (113). DES in potato produces divinyl ether fatty acids,

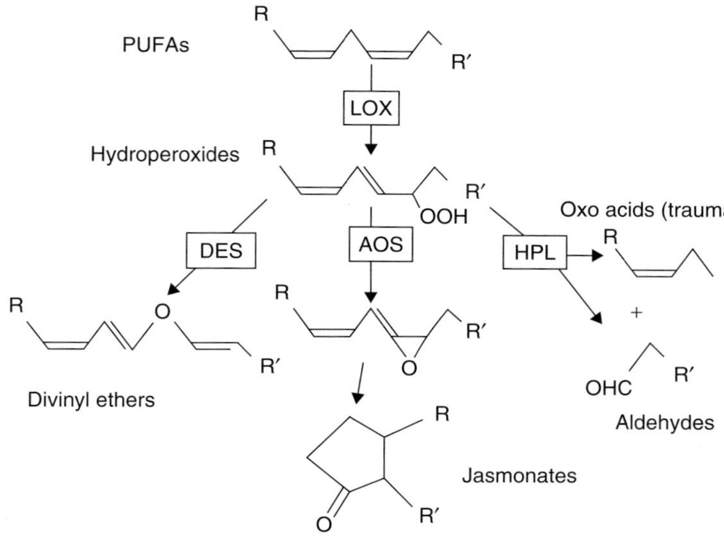

Figure 18.10 The formation of oxylipins from fatty acid hydroperoxide by CYP74 activity

which have been shown to act as protectants against *Phytophthora infestans*, the causative agent of late blight disease (114). HPL cleaves fatty acid hydroperoxides into 6- or 9-carbon volatile aldehydes, depending on enzyme specificity and, in the case of 13-HPL, acting on the 13-hydroperoxide of linolenic acid, traumatin (otherwise known as wound hormone). Traumatin plays a direct role in plant defense by accelerating cell division at wound sites, whereas volatile 6-C aldehydes can act directly as antimicrobial agents (115), or indirectly as defense signalling compounds (116). A number of plant HPL have been produced as recombinant molecules (Table 18.3) that may have a role to play in the production of fresh, green aromas (127).

Clearly, any compound which acts as a phytoprotectant will have an impact on food quality through reduced insect damage and diminished spoilage due to pathogen invasion. The volatile 6-C and 9-C aldehydes produced by HPL activity, and the alcohols derived from them by dehydrogenase action, however, have profound direct consequences for food quality in the form of taste and aroma compounds; the “green notes” (Figure 18.11) of fresh fruits and vegetables (53). They form the basis of the pleasant and distinctive aromas of tomatoes, potatoes, mushrooms, melons, cucumbers, bananas, and avocados, for instance. *E*-2-hexenal is one of the major flavor volatiles of fresh tomatoes (131) and is almost certainly derived from the action of a 13-HPL on the 13-hydroperoxide of linolenic acid. Aldehydes tend to provide green notes to tomato fruits, whereas their corresponding alcohols impart a more ripe aroma, and the balance between aldehyde and alcohol plays a major role in tomato flavor perception (132). Not all C-6 aldehydes are perceived as pleasant, however, a good example being *n*-hexanal, which is the grassy-beany flavor associated with soybean and other legume seed products. Thus the chemical difference between a pleasant and unpleasant aroma is very small (Figure 18.11).

Soybean seed has three major LOX isoforms, and genetic removal of LOX-2 reduces *n*-hexanal production and improves the consumer acceptability of soymilk and soy flour (133). LOX-2 is a 13-LOX and therefore a major source of *n*-hexanal via HPL activity, although subsequent studies have suggested that removal of both LOX-2 and LOX-3 has the greatest effect on off flavor reduction (134). Pea seeds contain two major LOX that are

Table 18.3

Recombinant plant HPLs of known substrate specificity. Functionality of all cDNA sequences were determined using purified enzymes after heterologous expression in *Escherichia coli*, with the exception of the potato sequence whose functionality was determined using crude extracts from untransformed and transgenic (antisense) plants.

Source	Accession Number	Preferred Substrate	Substrate Specificity[a] (13-HPOT:9-HPOT)	Designation	Ref.
Alfalfa seedling	AJ249245	13-HPOT	160:1	Alfalfa 13-HPL1	(117,118)
	AJ249246	13-HPOT	15500:1	Alfalfa 13-HPL2	(117)
	AJ249247	13-HPOT	60:1	Alfalfa 13-HPL3	(117)
Arabidopsis	AF087932	13-HPOT	19:1	Arabidopsis 13-HPL	(119,120,121)
Barley	AJ318870	13-HPOT	∞	Barley 13-HPL	(122)
Cucumber fruit	AF229811	9-HPOD	1:2	Cucumber 9/13-HPL	(123)
	AF229812	13-HPOD/T	∞	Cucumber 13-HPL	(123)
Melon fruit	AF081955	9-HPOT	1:3	Melon 9/13-HPL	(124)
Pepper fruit	U51674	n.d.	n.d.	Pepper HPL	(125,126,127)
Potato	AJ310520	13-HPOT	n.d.	Potato HPL	(128)
Tomato fruit	AF230372	13-HPOT	62.1	Tomato13-HPL1	(129)
	AF230372	13-HPOT	∞	Tomato13-HPL2	(130)

[a]Comparison of activity with 13-HPOT and 9-HPOT (derived from linolenic acid) or 13-HPOD and 9-HPOD (derived from linoleic acid).
n.d: not determined, or activity with 9-HPOD/T is unpublished, so designation as a 9-or 13-HPL was not possible.

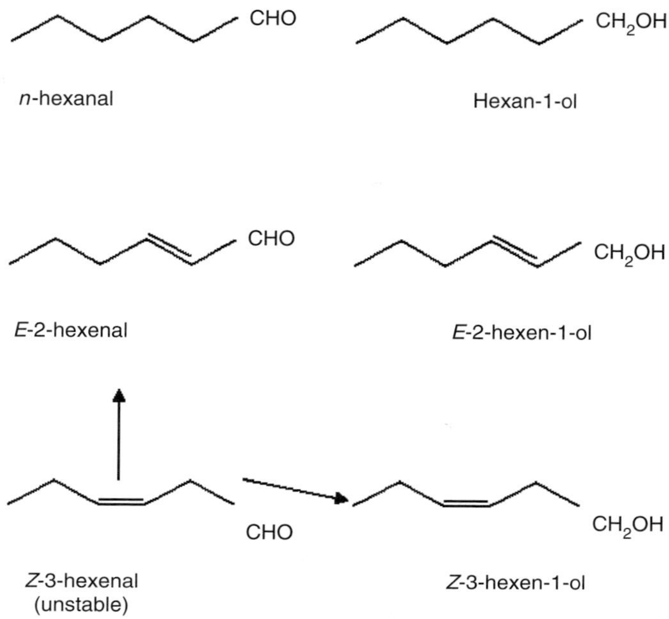

Figure 18.11 Some C-6 aldehydes and alcohols, formed by the combined action of LOX and HPL, that contribute to flavor aromas in fruits and vegetables

equivalent to LOX-2 and LOX-3 from soybean, and genetic variants have been produced that lack LOX-2. The properties of the LOX preparations from such seeds (54) are consistent with the loss of this particular isoform, based on our understanding of the properties of the two isoforms as recombinant enzymes (58). Thus, information gained from the study of recombinant enzymes can guide plant breeding processes. The recombinant enzymes can also be used as biocatalysts for the production of value added chiral chemicals; the advantage of such syntheses is control of stereochemistry. In the case of plant LOX, under the appropriate conditions virtually all of the hydroperoxide product will have (*S*) stereochemistry, whereas autoxidation products, for example, are a racemic mixture of (*S*) and (*R*) isomers, with implications for subsequent conversions. A combination of LOX and HPL can be used to produce aldehydes in industrial amounts to be used either as aroma additives or, in the case of *n*-hexanal, antimicrobial fumigants for postharvest crop protection.

18.7.2 Cooxidation and Food Quality

The ability of some LOX to prematurely release fatty acid peroxyl radicals is correlated with their capacity to carry out cooxidation reactions that catalyse the oxidation of sensitive molecules such as protein thiol groups, pigments, vitamins, and other anti-oxidants. The term "cooxidation" refers to the fact that the antioxidant or protein itself is not a direct LOX substrate, and cooxidation will not take place in the absence of PUFA, which is a mandatory part of the reaction. In general terms, cooxidation has a negative impact on food quality, destroying valuable antioxidants, decolorizing foods, and diminishing vitamin content.

18.7.2.1 Destruction of Vitamins

The adverse effect of LOX activity on antioxidant vitamins is well documented. In the presence of soybean LOX, vitamin E, vitamin C, and lutein are rapidly destroyed (135).

The antioxidants were scarcely affected by soybean extracts from lines in which the three seed LOX had been genetically removed, however, suggesting that soybeans that lack LOX could become a superior food ingredient which would improve not only the flavor of soybean products but also their nutritional quality and functionality (135). In frozen peas, the presence of LOX with cooxidizing potential may compromise vitamin C content, and genetic removal of pea seed LOX-2 (54), by conventional breeding using natural biodiversity may lead to an increased retention of vitamin C on processing (136). This principle of removing LOX to improve vitamin content can be applied generally to fruits and vegetables; the effect should be more pronounced if the LOX is a strong cooxidizer, and in this context it may be more effective to remove LOX-3, rather than (or as well as) LOX-2 from pea seeds because LOX-3 is better at cooxidation.

18.7.2.2 Pigment Bleaching

There are a large number of reports of the destruction of pigments as a consequence of LOX activity, including the loss of carotenoids from peppers (137), the bleaching of carotene and crocin by soybean LOX (138) and the cooxidation of β-carotene by potato (139) and pea (112) LOXs. The cooxidation process can be very rapid; LOX can quickly cooxidize β-carotene, even at 0°C in a predominantly organic solvent, to produce a range of breakdown products, some of which may themselves have flavor or aroma properties. The recombinant pea seed LOX-3 has a greater cooxidation to activity ratio than LOX-2 (112). Such pigment cooxidation obviously will have an impact on food quality, but this is not always negative. Commercial bread making includes the addition of enzyme active soybean flour for two reasons, both of which involve LOX. The cooxidation potential of LOX results in the bleaching of carotenoids and other pigments in the dough, leading to a brighter white loaf. The major application, however, is to brinng about the beneficial effect on dough rheology that results from the cooxidation of wheat flour proteins.

18.7.2.3 Bread-Making and Dough Rheology

It is common practice in commercial white bread making to add soybean LOX, as part of an enzyme active improver, to improve dough rheological and baking properties, by imparting tolerance to over mixing (140,141) (Figure 18.12), probably through the cooxidative formation of disulphide bonds between wheat flour proteins. The protein component of dough, gluten, is a complex mixture of glutenins, gliadins, and other proteins that interact through hydrogen bonding, disulphide bridges, and hydrophobic interactions (142). The high molecular weight (HMW) glutenin polypeptides (143) have a predicted β-spiral structure (144) that plays an important role in the viscoelastic behavior of gluten, possibly by acting as molecular springs. The HMW glutenin polypeptides have terminal cysteine residues and it is proposed (145) that the addition of LOX and PUFA, in the form of soybean flour, promotes the cooxidative formation of intermolecular disulphide bonds to generate a network of larger, stronger springs (Figure 18.13). Although wheat has endogenous grain LOX (41), they either are present in insufficient amounts, or have inappropriate effects, or both, to play this role in bread making.

There are several ways in which recombinant LOX can be used to provide the necessary cooxidation potential which are alternatives to the addition of soybean:

1. Engineer the yeast used in baking to produce and secrete a cooxidizing LOX during dough production

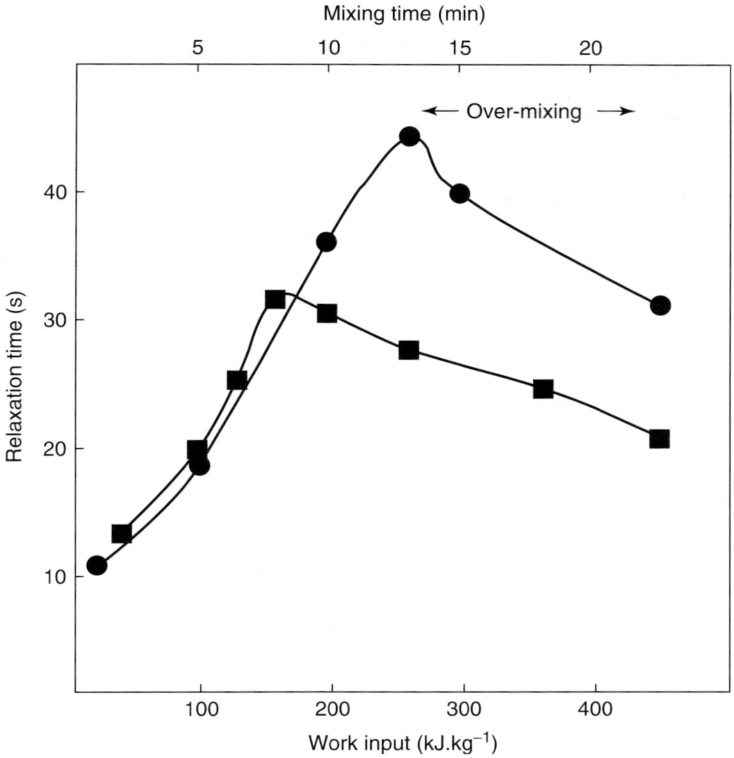

Figure 18.12 The effect of adding enzyme active soybean flour during high-power mechanical mixing of dough. Relaxation time equates to elasticity. ■, without and ●, with soybean flour (140)

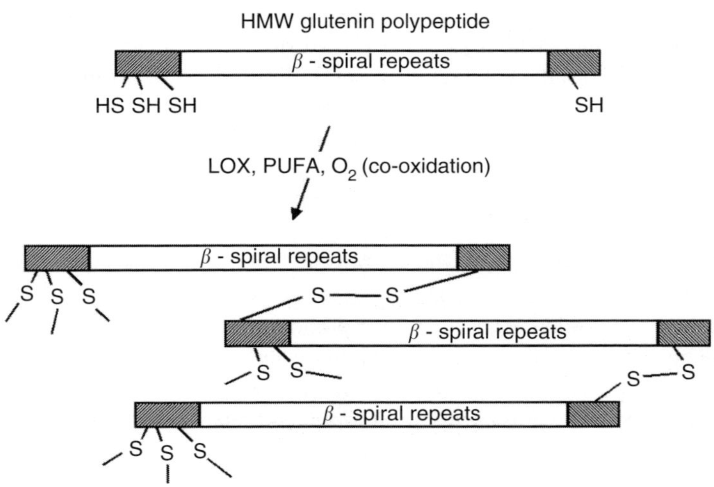

Figure 18.13 Diagrammatic representation of the cooxidative disulphide cross linking of HMW glutenin polypeptides

2. Use microbial fermentation in bacterial, fungal, or other, hosts to produce industrial amounts of cooxidizing LOX
3. Genetically modify wheat through recombinant DNA technology to produce a cooxidizing legume LOX within the developing wheat grain

18.7.2.3.1 Engineering Baker's Yeast to Secrete Recombinant LOX Active pea seed LOX-2 and -3 have been expressed in *Saccharomyces cerevisiae* using a constitutive phosphoglycerate kinase gene promoter (146), but secretion into the medium was not obtained, even with a yeast invertase signal sequence attached to the N-terminus of the enzyme. It appears that the signal sequence directed the pea seed LOX into the yeast secretory pathway, but potential cryptic vacuolar targeting sequences within the LOX coding sequence led to retention of the enzyme within the cell. The system has not been further exploited.

18.7.2.3.2 Production of Recombinant LOX by Fermentation in Microbes As indicated earlier, a large number of plant (and animal) LOX have been produced by over expression in *E. coli*, as tagged, fusion, or native proteins. For industrial use in the food industry, however, a number of criteria will need to be addressed. The enzymes will need to be produced in high yield and highly active form, and it is not clear whether LOX production by fermentation could ever be competitive with soybean products, unless it offers other advantages. Furthermore, if the recombinant enzyme is to be introduced directly into food, as opposed to being used as a biocatalyst for the production of food ingredients, then it should be identical to that from natural sources. This was the case for the high level expression of pea seed LOX in *E. coli*, which were produced as nonfusion proteins (147,56). The enzymes will need to have good cooxidation capacity if they are to be used in bread making, and there is considerable potential here for experimentation and improvement. Individual LOX vary a great deal in their cooxidation potential and their effects on the bread making process (148), and there are simple tests, such as carotenoid bleaching, that allow the screening of recombinant LOX and their site directed mutants for improved activity. Such high throughput assays also open up the possibility of forcing the evolution of LOX with greatly enhanced cooxidation potential. In addition, it is important that the enzyme activity does not generate aldehydes that impart off flavors to the product, which can easily be tested *in vitro*. The amount of soybean flour that can be added to the product is currently limited by the formation of off flavors (149), and recombinant activities that are effective cooxiders but do not generate the 13-hydroperoxides that give rise to C-6 aldehydes could be valuable. The stability of the enzyme preparations will be important to their production, purification and storage; pea seed LOX-3, which has the better cooxidation potential, is considerably more stable than LOX-2 (150).

18.7.2.3.3 Production of Recombinant Legume Lipoxygenase in Wheat Seeds The ability to transform wheat affords the possibility of expressing novel genes in wheat seeds, and this could certainly include a cooxidizing legume LOX. Other pea seed proteins have been successfully synthesized in transgenic wheat endosperm (151), the part of the seed in which the gluten proteins reside, using specific promoters to accomplish high level, seed specific expression. The *in planta* production of such recombinant LOX may or may not obviate the need for the addition of soybean improver during bread making, but it could widen the options available to plant breeders. The same considerations, in relation to cooxidation potential and the nature of the final LOX product, would apply to the production of recombinant LOX in wheat seeds under microbial fermentation, as was discussed.

18.8 FUTURE TRENDS AND APPLICATIONS OF RECOMBINANT LIPOXYGENASES AND OXYLIPIN METABOLISM

Recombinant enzymes offer immense opportunities for the development of new activities through site directed mutagenesis and forced evolution. As crystal structures for LOX become available, it becomes easier to understand the underlying bases of catalytic specificity and to test hypotheses through the production of site directed mutants. Many such experiments have been performed with plant and animal LOX, and have enhanced our understanding of regiospecificity of catalysis, but have not especially clarified the basis of substrate specificity, or the factors that influence cooxidation ability. The current lack of a structure with bound substrate is a disadvantage; that of a product bound complex (85) has proved enlightening, but will differ in detail from an enzyme substrate complex.

Site directed mutagenesis, based on three dimensional structures and predictive modelling, and aimed at altering catalytic behavior, nonetheless has some limitations. It necessarily makes assumptions about the role of specific amino acid residues in catalysis. Sequence differences within the active sites of similar enzymes with different specificities can be used as a basis for mutagenesis, as has been done for soy LOX-1 and LOX-3, but the outcomes of this may or may not be productive; in fact, often they are not. An alternative approach is to use forced evolution, or DNA shuffling, to generate altered activities. The original technique of forced evolution, pioneered by Stemmer (152), relied on the production and shuffling of PCR generated random mutants of single sequences and effective high throughput assays to select new molecules with altered, improved activities. The most effective way of carrying out forced evolution, however, is through the shuffling of naturally existing, closely similar sequences, such as LOX from soybean or pea, rather than random mutants generated by error prone PCR, because it takes advantage of sequence space that has already evolved to produce active molecules and avoids the wasteful production of nonactive mutants associated with random mutagenesis (153). Forced evolution makes no structural assumptions and, provided that the initial DNA shuffling has provided an appropriate substrate for subsequent evolution, guarantees changes that have a productive outcome in terms of altered catalytic behavior and can give great insights into the underlying basis of activity and specificity. The products of DNA shuffling or forced evolution do not necessarily reflect the properties of the parent molecules, which merely serve as sequence variants for recombination to force the emergence of new, positively selected, variants. Analysis of these variants can then be used to guide profitable, focussed site directed mutagenesis that can further the understanding of catalytic specificity or mechanism. Forced evolution can also provide new activities that are very unlikely to result from site directed mutagenesis.

There are a number of possible targets for the forced evolution of LOXs in the context of food biotechnology. Improvement of thermal stability is helpful and would benefit industrial scale purification and processing of recombinant enzymes; some LOX are much more thermostable than others, so there is potential not only for improvement through forced evolution, but also to better understand the underlying basis of the differences. Some LOX have relatively stringent requirements for substrate; forcing the evolution of activities that have a much broader specificity, or that will efficiently catalyse the lipoxygenation of novel substrates, could open up new opportunities in the production of aromas and flavors. Those LOX that catalyse cooxidation reactions are effectively mild, biological bleaching agents that require only a small amount of PUFA to decolorize specific pigments and break down antioxidants. It would be relatively straightforward to force the evolution of LOX that are much more efficient at cooxidation, and that may have application in the specific areas of decolorisation of food products. Such activities may also have potential in food applications that require cooxidation, such as bread making.

Although only loosely related to recombinant enzymes, the ability to modulate enzyme activity by plant genetic engineering has significant implications for food biotechnology, offering both a better understanding of the biochemistry that underlies a particular aspect and the opportunity to modify the organoleptic properties of foods. The LOX pathway plays an important role in the determination of the profile of PUFA derived volatiles (oxylipins) in fruits and vegetables (131), through the combined action of specific LOX, HPL, and alcohol dehydrogenase (ADH). To date, attempts to reduce the activity of the major tomato fruit LOX (TomloxA and -B) through antisense knockouts have not significantly altered the profile of C-6 volatiles (154), suggesting that either very low levels of LOX are sufficient for the generation of C-6 aldehydes and alcohols, or that a specific isoform such as TomloxC, in the absence of TomloxA and -B, is responsible for the production of these C-6 volatiles.

Alterations in the amounts of ADH in tomato fruits, on the other hand, have been more productive. ADH, which interconverts the LOX/HPL derived aldehydes and alcohols, has been shown to accumulate during tomato fruit ripening and is thought to play an important role in flavor development. Transgenic tomato fruits in which ADH activity had been increased or decreased showed alterations in the proportions of C-6 aldehydes and alcohols that were consistent with the changes in activity (132), albeit dependent on the method used to homogenize the fruits (155).

18.9 LIPOXYGENASES, OXYLIPIN METABOLISM, PLANT DEFENSE, AND FOOD QUALITY

There is no question that the lipoxygenase pathway plays a critical role in plant defense against pests and pathogens, including biting and sucking insects, bacteria, and fungi. Plant–insect interactions in turn can lead to virus infection. Any of these can seriously compromise food quality, through damage that diminishes product quality or even makes it unsaleable, severe reduction in yield, or reduced postharvest life. Experiments with transgenic plants in which specific components of the pathway of oxylipin metabolism have been knocked out have clearly shown a role for LOX in defense against fungal invasion (156) and damage by chewing insects (157,158), and for HPL in defense against aphids (128). In each case, reduction in enzyme activity led to increased susceptibility to the disease or pest. The corollary, in which increased expression leads to improved resistance in appropriate plant organs (and, by implication, potential improvement in food quality), has been demonstrated just once; the over expression of LOX in tobacco led to decreased susceptibility to virulent races of *Phytophthora parasitica nicotianae* (159).

The nature of the LOX pathway based defense mechanism varies from instance to instance. In the case of chewing insects, the response is through AOS, the synthesis of jasmonate and subsequent production of proteinase inhibitors (113) that interfere with insect digestive processes (157). In addition, HPL cleavage products play a dual role on wounding: the C-6 aldehydes protecting the wound site from infection and the C-12 products promoting wound healing. Divinyl ether fatty acids, from DES activity, are inhibitory to the growth of *Phytophthora infestans* and may, like the C-6 aldehydes, act as direct antimicrobial agents. All of these compounds are derived from CYP74-catalysed metabolism of fatty acid hydroperoxides, but there may be a role for the peroxides themselves in defense; the role of LOX in the incompatibility between certain cultivars of tobacco and particular races of *Phytophthora parasitica*, for instance (156), may have hydroperoxide-promoted local cell death as its underlying basis. Fatty acid hydroperoxides have been shown to be potent inducers of plant cell death (160), and may play an important role in restricting the spread of invading pathogens as part of a hypersensitive response (161).

18.10 CONCLUSIONS

LOX are widespread throughout plants, animals, and lower organisms. They, together with the enzyme activities that metabolize their oxylipin products, play key roles in defense but also have significant implications for food biotechnology. These latter include the production of flavors and aromas, the destruction of vitamins, pigments, and other antioxidants, and improvement of dough rheology during baking. Production of LOX and other enzymes of oxylipin metabolism in recombinant forms is important to food biotechnology in several respects: it provides large amounts of homogeneous biocatalysts for industrial processes; it leads to a better understanding of the mode of enzyme action; and it enables the production of new biocatalysts through directed mutagenesis or forced evolution.

REFERENCES

1. Matsuda, Y., T. Satoh, T. Beppu, K. Arima. Purification and properties of Co^{2+} requiring heme protein having lipoxygenase activity from *Fusarium oxysporum*. *Agric. Biol. Chem.* 40:963–976, 1976.
2. Bisakowski, B., S. Kermasha, M.-L. Klopfenstein. Partial purified lipoxygenase from *Fusarium oxysporum*: characterization and kinetic studies. *Process. Biotech.* 30:261–268, 1995.
3. Su, C., E.H. Oliw. Manganese lipoxygenase: purification and characterization. *J. Biol. Chem.* 273:13072–13079, 1998.
4. Hamberg, M., C.A. Herman, R.P. Herman. Novel biological transformations of 15-L_s-hydroperoxy-5,8,11,13-eicosatetraenoic acid. *Biochim. Biophys. Acta.* 877:447–457, 1986.
5. Simmons, C.A., J.L. Kerwin, R.K. Washino. Preliminary characterization of lipoxygenase from the entomopathogenic fungus *Lagenidium giganteum*. In: *The Metabolism, Structure and Function of Plant Lipids*, Stumpf, P.K., J.B. Mudd, W.D. Nes, eds., New York: Plenum Press, 1987, pp 421–423.
6. Nazzaro-Porro, M., S. Passi, M. Picardo, R. Mercantini, A. Breathnach. Lipoxygenase activity of *Pityrosporum in vitro* and *in vivo*. *J. Invest. Dermatol.* 87:108–112, 1986.
7. Grosch, W., M. Wurzenberger. Enzymic formation of 1-octen-3-ol in mushrooms. *Develop. Food Sci.* 10:253–259, 1984.
8. Mau, J.L., R.B. Beelman, G.R. Ziegler. 1-Octen-3-ol in the cultivated mushroom, *Agaricus bisporus*. *J. Food Sci.* 57:704–706, 1992.
9. Tressl, R., D. Bahri, K. Engel. Formation of eight-carbon and ten-carbon components in mushrooms *(Agaricus campetris)*. *J. Agric. Food Chem.* 30:89–93, 1982.
10. Kuribayashi, T., H. Kaise, C. Uno, T. Hara, T. Hayakawa, T. Joh. Purification and characterization of lipoxygenase from *Pleurotus ostreatus*. *J. Agric. Food Chem.* 50:1247–1253, 2002.
11. Kuo, J.-M., A. Hwang, D.-B. Yeh. Purification, substrate specificity and products of a Ca^{2+} stimulating lipoxygenase from sea algae (*Ulva lactuca*). *J. Agric. Food Chem.* 45:2055–2060, 1997.
12. Kuo, J.-M., A. Hwang, H.H. Hsu, B.S. Pan. Preliminary identification of lipoxygenase in algae (*Enteromorpha intestinalis*) for aroma formation. *J. Agric. Food Chem.* 44:2073–2077, 1996.
13. Zimmerman, D.C., B.A. Vick. Lipoxygenase in *Chlorella pyrenoidosa*. *Lipids* 8:264–266, 1973.

14. Boutaud, O., A.R. Brash. Purification and catalytic activities of the two domains of the allene oxide synthase-lipoxygenase fusion protein of the coral *Plexaura homomalla*. *J. Biol. Chem.* 274:33764–33770, 1999.
15. Bundy, G.L., E.G. Nidy, D.E. Epps, S.A. Mizsak, R.J. Wnuk. Discovery of an arachidonic acid C-8 lipoxygenase in the Gorgonian coral *Pseudoplexaura porosa*. *J. Biol. Chem.* 261:747–751, 1986.
16. Hawkins, D.J., A.R. Brash. Eggs of the sea urchin, *Strongylocentrotus purpuratus*, contain a prominent (11*R*) and (12*R*) lipoxygenase activity. *J. Biol. Chem.* 262:7629–7638, 1987.
17. Kuo, J.-M., B.S. Pan, H. Zhang, J.B. German. Identification of 12-lipoxygenase in the hemolymph of tiger shrimp (*Pennaeus japonicus* Bate). *J. Agric. Food Chem.* 42:1620–1623, 1994.
18. Brash, A.R. Lipoxygenases: occurrence, functions, catalysis and acquisition of substrate. *J. Biol. Chem.* 274:23679–23682, 1999.
19. Brash, A.R., M.A. Hughes, D.J. Hawkins, W.E. Boeglin, W.-C. Song, L. Meijers. Allene oxide and aldehyde biosynthesis in starfish oocytes. *J. Biol. Chem.* 266:22926–22931, 1991.
20. Meijer, L., A.R. Brash, R.W. Bryant, K. Ng, J. Maclouf, H. Sprecher. Stereospecific induction of starfish oocyte maturation by (8*R*)-hydroxyeicosatetraenoic acid. *J. Biol. Chem.* 261:17040–17047, 1986.
21. Boonprab, K., K. Matsui, Y. Akakabe, N. Yorsukura, T. Kajiwara. Hydroperoxy-arachidonic acid mediated *n*-hexanal and (*Z*|)-3- and (*E*)-2-nonenal formation in *Laminaria angustata*. *Phytochemistry* 63:669–678, 2003.
22. Pohnert, G. Phospholipase A_2 activity triggers the wound-activated chemical defense in the diatom *Thalassiosira rotula*. *Plant Physiol.* 129:103–111, 2002.
23. Di Marzo, V., L. De Petrocellis, C. Gianfrani, G. Cimino. Biosynthesis, structure and biological activity of hydroxyeicosatetraenoic acids in *Hydra vulgaris*. *Biochem. J.* 295:23–29, 1993.
24. E.M. Hill, D.L. Holland. Identification and egg hatching activity of monohydroxy fatty acid eicosanoids in the barnacle *Balanus balanoides*. *Proc. R. Soc. Lond. B*. 247:41–46, 1992.
25. Hampson, A.J., A.F. Rowley, S.E. Barrow, R. Steadman. Biosynthesis of eicosanoids by blood cells of the crab, *Carcinus maenas*. *Biochim. et Biophys. Acta*. 1124:143–150, 1992.
26. Christopher, J.P., E.K. Pistorius, B. Axelrod. Isolation of an isozyme of soybean lipoxygenase. *Biochim. et Biophys. Acta* 198:12–19, 1970.
27. Christopher, J.P., E.K. Pistorius, B. Axelrod. Isolation of a third isozyme of soybean lipoxygenase. *Biochim. et Biophys. Acta* 284:54–62, 1972.
28. Kitamura, K., C.S. Davies, K. Kaizuma, N.C. Nielson. Genetic analysis of a null-allele for lipoxygenase-3 in soybean seeds. *Crop Sci.* 23:924–927, 1983.
29. Doderer, A., I. Kokkelink, S. van der Veen, B.E. Valk, A.W. Schram, A.C. Douma. Purification and characterization of two lipoxygenase isoenzymes from germinating barley. *Biochim. et Biophys. Acta* 1120:97–104, 1992.
30. Nicolas, J., R. Drapron. Some physicochemical characteristics of horse bean lipoxygenase (*Vicia faba* L.). *Ann. Technol. Agric.* 26:119–132, 1977.
31. Borthakur, A., C.S. Ramadoss. Aerobic formation of ketodiene from linoleic acid catalysed by one of the two forms of lipoxygenase isolated from Bengal Gram (*Cicer arietinum*). *J. Agric. Food Chem.* 34:1016–1018, 1986.
32. Sanz, L.C., A.G. Perez, J.M. Olias. Purification and catalytic properties of chickpea lipoxygenase. *Phytochemistry* 31:2967–2972, 1992.
33. Sanz, L.C., A.G. Perez, J.J. Rios, J.M. Olias. Positional specificity of ketodienes from linoleic acid aerobically formed by lipoxygenase isoforms from kidney bean and pea. *J. Agric. Food Chem.* 41:696–699, 1993.
34. Belefant, H., F. Fong. Lipoxygenases in developing *Zea mays* kernels. *Plant Physiol. Biochem.* 29:99–104, 1991.
35. Jensen, A.B., E. Poca, M. Rigaud, G. Freyssinet, M. Pagés. Molecular characterization of L2 lipoxygenase from maize embryos. *Plant Mol. Biol.* 33:605–614, 1997.

36. Yoon, S., B.P. Klein. Some properties of pea lipoxygenase isoenzymes. *J. Agric. Food Chem.* 27:958–962, 1979.
37. Regdel, D., T. Schewe, H. Kuehn. Comparative characteristics of lipoxygenase isoenzymes from green pea seeds. *Biochem.* (Moscow) 60:715–721, 1995.
38. Sanders, T.H., H.E. Pattee, J.A. Singleton. Lipoxygenase isozymes of peanut. *Lipids* 10:681–685, 1975.
39. Kermasha, S., A. Khalyfa, I. Alli, B. Lee. Purification and characterization of lipoxygenase isozymes from canola (*Brassica napus* cv Westar) seed. *J. Food Biochem.* 15:219–238, 1991.
40. Ida, S., Y. Masaki, Y. Morita. The isolation of multiple forms and product specificity of rice lipoxygenase. *Agric. Biol. Chem.* 47:637–641, 1983.
41. Shiiba, K., Y. Negishi, K. Okada, S. Nagao. Purification and characterization of lipoxygenase isozymes from wheat germ. *Cereal Chem.* 68:115–122, 1991.
42. Truong, V.D., L.C. Raymundo, E.M.T. Mendoza. Winged bean lipoxygenase, part 2: physicochemical properties. *Food Chem.* 8:288–289, 1982.
43. Eiben, H.G., A.J. Slusarenko. Complex spatial and temporal expression of lipoxygenase genes during *Phaseolus vulgaris* (L.) development. *Plant J.* 5:123–135, 1994.
44. Saravitz, D.M., J.N. Siedow. The differential expression of wound-inducible lipoxygenase genes in soybean leaves. *Plant Physiol.* 110:287–299, 1996.
45. Hildebrand, D.F., T. Hymowitz. Inheritance of lipoxygenase-1 activity in soybean seeds. *Crop Sci.* 22:851–853, 1982.
46. Davies, C.S., N.C. Nielson. Genetic analysis of a null-allele for lipoxygenase-2 in soybean. *Crop Sci.* 26:460–463. 1986.
47. Wu, Z., D.S. Robinson, C. Domoney, R. Casey. High-performance liquid chromatographic analysis of the products of linoleic acid oxidation catalysed by pea (*Pisum sativum*) lipoxygenases. *J. Agric. Food Chem.* 43:337–342, 1995.
48. Suzuki, Y., T. Nagamine, A. Kobayashi, K. Ohtsubo. Detection of a new rice variety lacking lipoxygenase-3 by monoclonal antibodies. *JPN J. Breed.* 43:405–409, 1993.
49. Pfeiffer, T., D.F. Hildebrand, DM TeKrony. Agronomic performance of soybean lipoxygenase isolines. *Crop Sci.* 32:357–362, 1992.
50. Narvel, J.M., W.R. Fehr, G.A. Welke. Agronomic and seed traits of soybean lines lacking seed lipoxygenases. *Crop Sci.* 38:926–928, 1998.
51. Shibata, D., A. Slusarenko, R. Casey, D. Hildebrand, E. Bell. Lipoxygenases. *Plant Mol. Biol. Rep.* 12:S41–42, 1994.
52. Domoney, C., J.L. Firman, C. Sidebottom, P.M. Ealing, A. Slabas, R. Casey. Lipoxygenase heterogeneity in *Pisum sativum*. *Planta* 181:35–43, 1990.
53. Casey, R. Genetic manipulation of lipoxygenases for the agrifood industry. In: *Genetics and Breeding for Crop Quality and Resistance*, Mugnozza, G.T.S., E. Proceddu, M.A. Pagnotta, eds., Dordrecht: Kluwer, 1999, pp 259–269.
54. Forster, C., H. North, N. Afzal, C. Domoney, A. Hornostaj, D.S. Robinson, R. Casey. Molecular analysis of a null mutant for pea (*Pisum sativum* L.) seed lipoxygenase-2. *Plant Mol. Biol.* 39:1209–1220, 1999.
55. Casey, R., S.I. West, D. Hardy, D.S. Robinson, Z. Wu, R.K. Hughes. New frontiers in food enzymology: recombinant lipoxygenases. *Trends Food Sci. Technol.* 10:297–302, 1999.
56. Hughes, R.K., S.I. West, R. Casey. Recombinant plant lipoxygenases. *Ag. Biotech. Net.* 1:1–4, 1999.
57. Ealing, P.M., R. Casey. The complete amino acid sequence of a pea (*Pisum sativum*) seed lipoxygenase predicted from a new full-length cDNA. *Biochem. J.* 253:915–918, 1988.
58. Hughes, R.K., Z. Wu, D.S. Robinson, D. Hardy, S.I. West, S.A. Fairhurst, R. Casey. Characterization of authentic recombinant pea-seed lipoxygenases with distinct properties and reaction mechanisms. *Biochem. J.* 333:33–43, 1998.
59. Ealing, P.M., R. Casey. The cDNA cloning of a pea (*Pisum sativum*) seed lipoxygenase sequence comparisons of the 2 major pea and lipoxygenase isoforms. *Biochem. J.* 264:929–932, 1989.

60. Shibata, D., J. Stezcko, J.E. Dixon, M. Hermodson, R. Yazdanparast, B. Axelrod. Primary structure of soybean lipoxygenase-1. *J. Biol. Chem.* 262:10080–10085, 1987.
61. Berry, H., H. Debat, V. Larreta-Garde. Oxygen concentration determines regiospecificity in soybean lipoxygenase-1 reaction via a branched kinetic scheme. *J. Biol. Chem.* 273:2769–2776, 1998.
62. Kramer, J.A., K.R. Johnson, W.R. Dunham, R.H. Sands, M.O. Funk. Position 713 is critical for catalysis but not iron binding in soybean lipoxygenase 3. *Biochemistry* 33:15017–15022, 1994.
63. Skrzypczak-Jankun, E., L.M. Amzel, B.A. Kroa, M.O. Funk, Jr. Structure of soybean lipoxygenase L3 and a comparison with its L1 isoenzyme. *Proteins Struct. Funct. Genet.* 29:15–31, 1997.
64. Chen, X.Y., P. Reddanna, G.R. Reddy, R. Kidd, G. Hildenbrandt, C.C. Reddy. Expression, purification and characterization of a recombinant 5-lipoxygenase from potato tuber. *Biochem. Biophys. Res. Comm.* 243:438–443, 1998.
65. Royo, J., G. Vancanneyt, A.G. Pérez, C. Sanz, K. Stormann, S. Rosahl, J.J. Sánchez-Serrano. Characterization of three potato lipoxygenases with distinct enzymatic activities and different organ-specific and wound-regulated expression patterns. *J. Biol. Chem.* 271:21012–21019, 1996.
66. Hughes, R.K., S.I. West, A.R. Hornostaj, D.M. Lawson, S.A. Fairhurst, R.O. Sanchez, P. Hough, B.H. Robinson, R. Casey. Probing a novel potato lipoxygenase with dual positional specificity reveals primary determinants of substrate binding and requirements for a surface hydrophobic loop and has implications for the role of lipoxygenases in tubers. *Biochem. J.* 353:345–355, 2001.
67. Hilbers, M.P., A. Rossi, A. Finazzi-Agro, G.A. Veldink, J.F.G. Vliegenthart. The primary structure of a lipoxygenase from the shoots of etiolated lentil seedlings derived from its cDNA. *Biochim. Biophs. Acta* 1211:239–242, 1994.
68. Hilbers, M.P., A. Finazzi-Agro, G.A. Veldink, J.F.G. Vliegenthart. Purification and characterization of a lentil seedling lipoxygenase expressed in *E coli.* Implications for the mechanism of oxodiene formation by lipoxygenases. *Int. J. Biochem. Cell Biol.* 28:751–760, 1996.
69. Peng, Y.L., Y. Shirano, H. Ohta, T. Hibino, K. Tanaka, D. Shibata. A novel lipoxygenase from rice: primary structure and specific expression upon incompatible infection with rice blast fungus. *J. Biol. Chem.* 269:3755–3761, 1994.
70. Zhang, L.Y., M. Hamberg. Specificity of two lipoxygenases from rice: unusual regiospecificity of a lipoxygenase isoenzymes. *Lipids* 31:803–809, 1996.
71. Shirano, Y., D. Shibata. Low temperature cultivation of *Escherichia coli* carrying a rice lipoxygenase L-2 cDNA produces a soluble and active enzyme at a high level. *FEBS Lett.* 271:128–130, 1990.
72. Ohta, H., Y. Shirano, K. Tanaka, Y. Morita, D. Shibata. cDNA cloning of rice lipoxygenase L-2 and characterization using an active enzyme expressed from the cDNA in *Escherichia coli. Eur. J. Biochem.* 206:331–336, 1992.
73. Mizuno, K., T. Iida, A. Takano, M. Yokoyama, T. Fujimura. A new 9-lipoxygenase cDNA from developing rice seeds. *Plant Cell Physiol.* 44:1168–1175, 2003.
74. Schaffrath, U., F. Zabbai, R. Dudler. Characterization of RCI-1, a chloroplastic rice lipoxygenase whose synthesis is induced by chemical plant resistance activators. *Eur. J. Biochem.* 267:5939–5942, 2000.
75. Hohne, M., A. Nellen, K. Schwennesen, H. Kindl. Lipid body lipoxygenase characterized by protein fragmentation, cDNA sequence and very early expression of the enzyme during germination of cucumber seeds. *Eur. J. Biochem.* 241:6–11, 1996.
76. Feussner, I., A. Bachmann, M. Höhne, H. Kindl. All three acyl moieties of trili-nolein are efficiently oxygenated by recombinant His-tagged lipid body lipoxygenase *in vitro. FEBS Lett.* 43:433–436, 1998.
77. Voros, K., I. Feussner, H. Kuhn, J. Lee, A. Graner, M. Lobler, B. Parthier, C. Wasternack. Characterization of a methyljasmonate-inducible lipoxygenase from barley (*Hordeum vulgare* cv. Salome) leaves. *Eur. J. Biochem.* 251:36–44, 1998.

78. Burrow, G.B., H.W. Gardner, N.P. Keller. A peanut seed lipoxygenase responsive to *Aspergillus* colonization. *Plant Mol. Biol.* 42:689–701, 2000.
79. Mita, G., A. Gallo, V. Greco, C. Zasiura, R. Casey, G. Zacheo, A. Santino. Molecular cloning and biochemical characterization of a lipoxygenase in almond (*Prunus dulcis*) seed. *Eur. J. Biochem.* 268:1500–1507, 2001.
80. Kim, E.-S., E. Choi, Y. Kim, K. Cho, A. Lee, J. Shim, R. Rakwal, G.K. Agrawal, O. Han. Dual positional specificity and expression of non-traditional lipoxygenase induced by wounding and methyl jasmonate in maize seedlings. *Plant Mol. Biol.* 52:1203–1213, 2003.
81. Su, C., M. Sahlin, E.H. Oliw. Kinetics of manganese lipoxygenase with a catalytic mononuclear redox center. *J. Biol. Chem.* 275:18830–18835, 2000.
82. Boyington, J.C., B.J. Gaffney, L.M. Amzel. The three-dimensional structure of an arachidonic acid 15-lipoxygenase. *Science* 260:1482–1486, 1993.
83. Minor, W., J. Steczko, B. Stec, Z. Otwinowski, J.T. Bolin, R. Walter, B. Axelrod. Crystal structure of soybean lipoxygenase L-1 at 1.4 Å resolution. *Biochemistry* 35:10687–10701, 1996.
84. Gillmor, S.A., A. Villaseñor, R. Fletterick, E. Sigal, M.F. Browner. The structure of mammalian 15-lipoxygenase reveals similarity to the lipases and the determinants of substrate specificity. *Nat. Struct. Biol.* 4:1003–1009, 1997.
85. Skrzypczak-Jankun, E., R.A. Bross, R.T. Carroll, W.R. Dunham, M.O. Funk, Jr. Three-dimensional structure of a purple lipoxygenase. *J. Am. Chem. Soc.* 123:10814–10820, 2001.
86. Skrzypczak-Jankun, E., K. Zhou, J. Rankun. Inhibition of lipoxygenase by (-)-epigallocatechin gallate: X-ray analysis at 2.1 Å reveals degradation of EGCG and shows soybean LOX-3 complex with EGC instead. *Int. J. Mol. Med.* 12:415–422, 2003.
87. Skrzypczak-Jankun, E., K. Zhou, N.P. McCabe, S.H. Selman, J. Jankun. Structure of curcumin in complex with lipoxygenase and its significance in cancer. *Int. J. Mol. Med.* 12:17–24, 2003.
88. Funk, C.D., P.J. Loll. A molecular dipstick? *Nat. Struct. Biol.* 4:966–968, 1997.
89. Borowski, T., M. Król, M. Chruszcz, E. Broclawik. First principle calculations for the nonheme iron centers of lipoxygenases: geometrical and spectral properties. *J. Phys. Chem. B.* 105:12212–12220, 2001.
90. Holman, T.R., J. Zhou, E.I. Solomon. Spectroscopic and functional characterization of a ligand coordination mutant of soybean lipoxygenase-1: first coordination sphere analogue of human 15-lipoxygenase. *J. Am. Chem. Soc.* 120:12564–12572, 1998.
91. Tomchick, D.R., P. Phan, M. Cymborowski, W. Minor, T.R. Holman. Structural and functional characterization of second-coordination sphere mutants of soybean lipoxygenase. *Biochemistry* 40:7509–7517, 2001.
92. Schenk, G., M.L. Neidig, J. Zhou, T.R. Holman, E.I. Solomon. Spectroscopic characterization of soybean lipoxygenase-1 mutants: the role of second coordination sphere residues in the regulation of enzyme activity. *Biochemistry* 42:7294–7302, 2003.
93. Funk Jr., M.O., J.C. Andre, T. Otsuki. Oxygenation of trans polyunsaturated fatty acids by lipoxygenase reveals steric features of the catalytic mechanism. *Biochemistry* 26:6880–6884, 1987.
94. Feussner, I., C. Wasternack. The lipoxygenase pathway. *Annu. Rev. Plant Biol.* 53:275–297, 2002.
95. Rickert, K.W., J.P. Klinman. Nature of hydrogen transfer in soybean lipoxygenase-1: separation of primary and secondary isotope effects. *Biochemistry* 38:12218–12228, 1999.
96. Glickman, M.H., J.P. Klinman. Lipoxygenase reaction mechanism: demonstration that hydrogen abstraction from substrate precedes dioxygen binding during catalytic turnover. *Biochemistry* 35:12882–12892, 1996.
97. Knapp, M.J., F.P. Seebeck, J.P. Klinman. Steric control of oxygenation regiochemistry in soybean lipoxygenase. *J. Am. Chem. Soc.* 123:2931–2932, 2001.
98. Knapp, M.J., J.P. Klinman. Kinetic studies of oxygen reactivity in soybean lipoxygenase-1. *Biochemistry* 42:11466–11475, 2003.
99. Hornung, E., S. Rosahl, H. Kühn, I. Feussner. Creating lipoxygenases with new positional specificities by site-directed mutagenesis. *Biochem. Soc. Trans.* 28:825–826, 2000.

100. Hornung, E., M. Walther, H. Kühn, I. Feussner. Conversion of cucumber linoleate 13-lipoxygenase to a 9-lipoxygenating species by site-directed mutagenesis. *Proc. Natl. Acad. Sci. USA* 96:4192–4197, 1999.
101. Brash, A.R., C.D. Ingram, T.M. Harris. Analysis of a specific oxygenation reaction of soybean lipoxygenase-1 with fatty acids esterified in phospholipids. *Biochem.* 26:5465–5471, 1987.
102. Matsui, K., M. Nishioka, M. Ikeyoshi, Y. Matsumura, T. Mori, T. Kajiwara. Cucumber root lipoxygenase can act on acyl groups in phosphatidylcholine. *Biochim. Biophys. Acta* 1390:8–20, 1998.
103. B.A. Stelmach, A. Mülher, P. Hennig, S. Gebhart, M. Schubert-Zsilavecz, E.W. Weiler. A novel class of oxylipins, sn1-O-(12-OXO-phytodienoyl) – sn2-O-(hexadecatrienoyl) – monogalactosyl diglyceride, from *Arabidopsis thaliana*. *J. Biol. Chem.* 276:12732–12838, 2001.
104. Holtman, W.L., J.C. Vredenbregt-Heistek, N.F. Schmitt, I. Feussner. Lipoxygenase-2 oxygenates storage lipids in embryos of germinating barley. *Eur. J. Biochem.* 248:452–458, 1997.
105. Matthew, J.A., H.W.-S. Chan, T. Galliard. A simple method for the preparation of pure 9-hydroperoxides of linoleic acid and methyl linoleate based on the positional specificity of lipoxygenase in tomato fruit. *Lipids* 12:324–326, 1977.
106. Brash, A.R., W.E. Boeglin, M.S. Chang, B.-H. Shieh. Purification and molecular cloning of an 8*R*-lipoxygenase from the coral *Plexaura homonmalla* reveal the related primary structures of *R*- and *S*-lipoxygenases. *J. Biol. Chem.* 271:20949–20957, 1996.
107. Sloane, D.L., R. Leung, C.S. Craik, E. Sigal. A primary determinant for lipoxygenase positional specificity. *Nature* 354:149–152, 1991.
108. Prigge, S.T., B.J. Gaffney, L.M. Amzel. Relation between positional specificity and chirality in mammalian lipoxygenases. *Nat. Struct. Biol.* 5:178, 1993.
109. Borngräber, S., R.-J. Kuban, M. Anton, H. Kühn. Phenylalanine 353 is a primary determinant for the positional specificity of mammalian 15-lipoxygenases. *J. Mol. Biol.* 264:1145–1153, 1996.
110. Borngräber, S., M. Browner, S. Gillmor, C. Gerth, M. Anton, R. Fletterick, H. Kühn. Shape and specificity in mammalian 15-lipoxygenase active site: the functional interplay of sequence determinants for the reaction specificity. *J. Biol. Chem.* 274:37345–37350, 1999.
111. Hughes, R.K., D.M. Lawson, A.R. Hornostaj, S.A. Fairhurst, R. Casey. Mutagenesis and modelling of linoleate-binding to pea seed lipoxygenase. *Eur. J. Biochem.* 268:1030–1040, 2001.
112. Wu, Z., D.S. Robinson, R.K. Hughes, R. Casey, D. Hardy, S.I. West. Co-oxidation of β-carotene catalysed by soybean and recombinant pea lipoxygenases. *J. Agric. Food Chem.* 47:4899–4906. 1999.
113. Farmer, E.E., C.A. Ryan. Octadecanoid precursors of Jasmonic acid activate the synthesis of wound-inducible proteinase inhibitors. *Plant Cell* 4:129–134, 1992.
114. Weber, H., A. Chételat, D. Caldelari, E.E. Farmer. Divinyl ether fatty acid synthesis in late blight-diseased potato leaves. *Plant Cell* 11:485–493, 1999.
115. Croft, K.P.C., F. Juttner, A.J. Slusarenko. Volatile products of the lipoxygenase pathway evolved from *Phaseolus vulgaris* (L.) leaves inoculated with *Pseudomonas syringae* pv. *Phaseolicola*. *Plant Physiol.* 101:13–24, 1993.
116. Bate, N.J., S.J. Rothstein. C_6-volatiles derived from the lipoxygenase pathway induce a subset of defense-related genes. *Plant J.* 16:561–569, 1998.
117. Noordermeer, M.A., A.J.H. van Dijken, S.C.M. Smeekens, G.A. Veldink, J.F.G. Vliegenthart. Characterization of three cloned and expressed 13-hydroperoxide lyase isoenzymes from alfalfa with unusual N-terminal sequences and different enzyme kinetics. *Eur. J. Biochem.* 267:2473–2482, 2000.
118. Noordermeer, M.A., G.A. Veldink, J.F.G. Vliegenthart. Spectroscopic studies on the active site of hydroperoxide lyase; the influence of detergents on its conformation. *FEBS Letts.* 489:229–232, 2001.

119. Bate, N.J., S. Sivasankar, C. Moxon, J.C.M. Riley, J.E. Thompson, S.J. Rothstein. Molecular characterization of an *Arabidopsis* gene encoding hydroperoxide lyase, a cytochrome P-450 that is wound inducible. *Plant Physiol.* 117:1393–1400, 1998.
120. Matsui, K., J. Wilkinson, B. Hiatt, V. Knauf, T. Kajiwara. Molecular cloning and expression of *Arabidopsis* fatty acid hydroperoxide lyase. *Plant Cell Physiol.* 40:477–481, 1999.
121. Kandzia, R., M. Stumpe, E. Berndt, M. Szalata, K. Matsui, I. Feussner. On the specificity of lipid hydroperoxide fragmentation by fatty acid hydroperoxide lysase from *Arabidopsis thaliana*. *J. Plant Physiol.* 160:803–809, 2003.
122. Koeduka, T., M. Stumpe, K. Matsui, T. Kajiwara, I. Feussner. Kinetics of barley FA hydroperoxide lyase are modulated by salts and detergents. *Lipids* 38:1167–1172, 2003.
123. Matsui, K., C. Ujita, S. Fujimoto, J. Wilkinson, B. Hiatt, V. Knauf, T. Kajiwara, I. Feussner. Fatty acid 9- and 13-hydroperoxide lyases from cucumber. *FEBS Lett.* 481:183–188, 2000.
124. Tijet, N., C. Schneider, N.L. Mulle, A.R. Brash. Biogenesis of volatile aldehydes from fatty acid hydroperoxides: molecular cloning of a hydroperoxide lyase (CYP74C) with specificity for both the 9- and 13-hydroperoxides of linoleic and linolenic acids. *Arch. Biochem. Biophys.* 386:281–289, 2001.
125. Matsui, K., M. Shibutani, T. Hase, T. Kajiwara. Bell pepper fruit fatty acid hydroperoxide lyase is a cytochrome P450 (CYP74B). *FEBS Lett.* 394:21–24. 1996.
126. Psylinakis, E., E.M. Davoras, N. Ioannidis, M. Trikeriotis, V. Petrouleas, D.F. Ghanotakis. Isolation and spectroscopic characterization of a recombinant bell pepper hydroperoxide lyase. *Biochim. Biophys. Acta* 1533:119–127, 2001.
127. Delcarte, J., M.-L. Fauconnier, P. Jacques, K. Matsui, P. Thonart, M. Marlier. Optimisation of expression and immobilized metal ion affinity chromatographic purification of recombinant $(His)_6$-tagged cytochrome P450 hydroperoxide lyase in *Escherichia coli*. *J. Chromatogr. B.* 786:229–236, 2003.
128. Vancanneyt, G., C. Sanz, T. Farmaki, M. Paneque, F. Ortego, P. Castanera, J.J. Sánchez-Serrano. Hydroperoxide lyase depletion in transgenic potato plants leads to an increase in aphid performance. *Proc. Natl. Acad. Sci. USA* 98:8139–8144, 2001.
129. Matsui, K., C. Miyahara, J. Wilkinson, B. Hiatt, V. Knauf, T. Kajiwara. Fatty acid hydroperoxide lyase in tomato fruits: cloning and properties of a recombinant enzyme expressed in *Escherichia coli*. *Biosci. Biotechnol. Biochem.* 64:1189–1196, 2000.
130. G.A. Howe, G.I. Lee, A. Itoh, L. Li, A.E. DeRocher. Cytochrome P450-dependent metabolism of oxylipins in tomato. Cloning and expression of allene oxide synthase and fatty acid hydroperoxide lyase. *Plant Physiol.* 123:711–724, 2000.
131. Buttery, R.G., R. Teranishi, L.C. Ling. Fresh tomato aroma volatiles: a quantitative study. *J. Agric. Food Chem.* 35:540–544, 1987.
132. Speirs, J., E. Lee, K. Holt, K. Yong-Duk, N.S. Scott, B. Loveys, W. Schuch. Genetic manipulation of alcohol dehydrogenase levels in ripening tomato fruit affects the balance of some flavor aldehydes and alcohols. *Plant Physiol.* 117:1047–1058, 1998.
133. Davies, C.S., S.S. Nielson, N.C. Nielson. Flavor improvement of soybean preparation is by genetic removal of lipoxygenase-2. *J. Am. Oil Chem. Soc.* 64:1428–1433, 1987.
134. Evans, D.E., C. Tsukamoto, N.C. Nielson. A small-scale method for the production of soymilk and silken tofu. *Crop Sci.* 37:1463–1471, 1997.
135. Nishiba, Y., I. Suda. Degradation of vitamin E, vitamin C and lutein in soybean homogenate: a comparison of normal soybean and lipoxygenase-lacking (triple-null) soybean. *J. Agric. Food Chem.* 46:3708–3712, 1998.
136. Dörnenburg, H., K.J. Hunter, C. Davies. Pea lipoxygenase-2 (LOC-2) is a stress related enzyme and appears to be responsible for co-oxidation of ascorbate. Plant Protein Club Annual Symposium "Pathway Engineering in Plants", York, 1999, p 15.
137. Jarén-Galán, M., M.I. Mínguez-Mosquera. Effect of pepper lipoxygenase activity and its linked reactions on pigments of the pepper fruit. *J. Agric. Food Chem.* 47:4532–4536, 1999.
138. Weber, F., G. Laskawy, W. Grosh. Co-oxidation of carotene and crocin by soybean lipoxygenase isoenzymes. *Z. Lebensm. Unters. Forsch.* 155:142–150, 1974.

139. Azis, S., Z. Wu, D.S. Robinson. Potato lipoxygenase catalysed co-oxidation of β-carotene. *Food Chem.* 64:227–230, 1999.
140. Frazier, P.J., F.A. Leigh-Dugmore, N.W.R. Daniels, P.W.R. Eggitt, J.B.M. Coppock. The effect of lipoxygenase action on the mechanical development of wheat flour doughs. *J. Sci. Food Agric.* 24:421–436, 1973.
141. Frazier, P.J. Lipoxygenase action and lipid binding in breadmaking. *Bakers Dig.* 53:8–29, 1979.
142. Shewry, P.R., A.S. Tatham, F. Barro, P. Barcelo, P. Lazzeri. Biotechnology of breadmaking: unravelling and manipulating the multi-protein gluten complex. *Bio/Technol.* 13:1185–1190, 1995.
143. Flavell, R.B., A.P. Goldsbrough, L.S. Robert, D. Schnick, R.D. Thompson. Genetic variation in wheat HMW glutenin subunits and the molecular basis of bread-making quality. *Bio/Technol.* 7:1281–1285, 1989.
144. Shewry, P.R., J.A. Napier, A.S. Tatham. Seed storage proteins: structures and biosynthesis. *Plant Cell* 7:945–956, 1995.
145. Casey, R. Lipoxygenases and breadmaking. *Proceedings of the 1st European Symposium: Enzymes in grain processing*, Noordwijkerhout, TNO, Zeist, The Netherlands, 1997, pp 188–194.
146. Knust, B., D. von Wettstein. Expression and secretion of pea-seed lipoxygenase isoenzymes in *Saccharomyces cerevisiae*. *Appl. Microbiol. Biotechnol.* 37:342–351, 1992.
147. Casey, R., R.K. Hughes. From food to pharmaceuticals: the multipurpose lipoxygenases. *Biotech. LINK Newsl.* 4:4–5, 1998.
148. Cumbee, B., D.F. Hildebrand, J. Addo. Soybean flour lipoxygenase isozymes effects on wheat flour dough rheological and breadmaking properties. *J. Food Sci.* 62:281–294, 1997.
149. Hoover, W. Use of soy proteins in baked foods. *J. Am. Oil Chem. Soc.* 56:301–303, 1979.
150. Busto, M.D., R.K. Owusu-Apenten, D.S. Robinson, Z. Wu, R. Casey, R.K. Hughes. Kinetics of thermal inactivation of pea seed lipoxygenases and the effect of additives on their thermostability. *Food Chem.* 65:323–329, 1999.
151. Stöger, E., M. Parker, P. Christou, R. Casey. Pea legumin overexpressed in wheat endosperm assembles into an ordered paracyrstalline matrix. *Plant Physiol.* 125:1732–1742, 2001.
152. W.P.C. Stemmer. DNA shuffling by random fragmentation and reassembly: *in vitro* recombination for molecular evolution. *Proc. Natl. Acad. Sci. USA* 91:10747–10751, 1994.
153. Crameri, A., S.-A. Raillard, E. Bermudex, W.P.C. Stemmer. DNA shuffling of a family of genes from diverse species accelerates directed evolution. *Nature* 391:288–291, 1998.
154. Griffiths, A., S. Prestage, R. Linforth, J.L. Zhang, A. Taylor, D Grierson. Fruit-specific lipoxygenase suppression in antisense-transgenic tomatoes. *Postharvest Biol. Technol.* 17:163–173, 1999.
155. Prestage, S., R.S.T. Linforth, A.J. Taylor, E. Lee, J. Speirs, W. Schuch. Volatile production in tomato fruit with modified alcohol dehydrogenase activity. *J. Sci. Food Agric.* 79:1131–1136, 1999.
156. Rancé, I., J. Fournier, M.-T. Esquerré-Tugayé. The incompatible interaction between *Phytophthora parasitica* var. *nicotianae* race 0 and tobacco is suppressed in transgenic plants expressing antisense lipoxygenase sequences. *Proc. Natl. Acad. Sci. USA* 95:6554–6559, 1998.
157. Royo, J., J. León, G. Vancanneyt, J.P. Albar, S. Rosahl, F. Ortego, P. Castañera, J.J. Sánchez-Serrano. Antisense-mediated depletion of a potato lipoxygenase reduces wound induction of proteinase inhibitors and increases weight gain of insect pests. *Proc. Natl. Acad. Sci. USA* 96:1146–1151, 1999.
158. Ortego, F., C. Novillo, J.J. Sánchez-Serrano, P. Castanera. Physiological response of Colorado potato beetle and beet armyworm larvae to depletion of wound-inducible proteinase inhibitors in transgenic potato plants. *J. Insect. Physiol.* 47:1291–1300, 2001.

159. Mène-Saffrané, L., M.-T. Esquerré-Tugayé, J. Fournier. Constitutive expression of an inducible lipoxygenase in transgenic tobacco decreases susceptibility to *Phytophthora parasitica* var. *nicotianae*. *Mol. Breed* 12:271–282, 2003.
160. Knight, V.I., H. Wang, J.E. Lincoln, E.C. Lulai, D.G. Gilchrist, R.M. Bostock. Hydroperoxides of fatty acids induce programmed cell death in tomato protoplasts. *Physiol. Mol. Plant Pathol.* 59:277–286, 2001.
161. Jalloul, A., J.L. Montillet, K. Assigbetse, J.P. Agnel, E. Delannoy, C. Triantaphylides, J.F. Daniel, P. Marmey, J.P. Geiger, M. Nicole. Lipid peroxidation in cotton: *Xanthomonas* interactions and the role of lipoxygenases during the hypersensitive reaction. *Plant J.* 32:1–12, 2002.

2.19

Genetic Modification of Production Traits in Farm Animals

Vernon G. Pursel

CONTENTS

19.1 INTRODUCTION

As we begin the twenty-first century, those involved in food production may find it useful to think about the future. With the world's population increasing by more than 70 million people each year, it boggles the mind to think about how all these people are going to be fed. Most certainly, if this ever increasing population is going to avoid massive waves of starvation or wars over agricultural resources, we will need to harness all the advantages that modern agricultural methods have to offer; and plant and animal biotechnology will certainly be part of that methodology.

Some people will argue that as the world population continues to increase there will be no place for animal agriculture. However, the trend over several decades indicates quite the opposite. In some western countries animal products contribute about two thirds of the total protein in the food supply and provide almost three fourths of the eight essential amino acids (1). While people in the developed countries in recent years have attempted to reduce their consumption of animal protein, quite the reverse is seen in developing countries. It is well recognized that as per capita income increases so does the consumption of animal products; the same trend occurs as urbanization increases (2). Data from around the world show that on the whole the percentage of the population that can afford to consume animal products is increasing. Thus, unless this trend is reversed, we should expect to see the demand for animal products to increase tremendously during the coming decades. The use of genetic modification will be instrumental in providing a way for the animal industry to satisfy the increased demand for animal food and fiber products.

The following sections review the progress on genetic manipulation of productivity traits in farm animals as well as some of the areas that offer promise for manipulation in the future. These areas include stimulation of growth rate; increased efficiency of animal production; development of new or improved meat, milk, and fiber products; enhanced resistance to diseases; and reducing environmental pollution. At the outset it should be noted that progress on manipulation of animal traits is far slower than originally envisioned by the early proponents of this technology. The main reasons for this slow progress are: (1) Most economically important traits are controlled by multiple genes, which are largely unknown at this time, and are, understandably, not amenable to manipulation. (2) The low efficiency of gene transfer in farm animals makes research on transgenesis quite costly, so preliminary investigations are usually conducted in mice. Unfortunately, in many cases results obtained in mice are not directly applicable to farm animals. (3) The ability to regulate expression of transgenes is still far from adequate, even in mice, and frequently the regulation of expression obtained in transgenic mice is lacking when the same transgene is used in farm animals.

19.2 INCREASING GROWTH RATE AND FEED EFFICIENCY

Much of the earlier transgenic farm animal research was conducted using growth hormone (GH) or growth hormone releasing factor (GRF) genes because these peptide hormones regulate many of the physiological processes that influence growth rate and feed efficiency. This early work was fuelled by results that indicated transgenic mice expressing excess GH grew much faster and to a larger size than control mice (3). However, for farm animals, increased size or growth rate is not nearly as important as are improved efficiency of feed utilization or an increase in the proportion of lean body mass, which are economic traits that are altered when growth hormone is elevated.

19.2.1 GH and GRF Transgenics

A number of transgenic pigs and sheep were produced with human, bovine, rat, porcine, or ovine GH under the control of several gene promoters as is shown in Table 19.1. Vise and coworkers (14) reported that one pig expressing porcine GH (pGH) grew at an exceptionally rapid rate compared to littermates (1273 vs. 781 g/day), while Ebert and coworkers (6) reported two transgenic pigs expressing pGH grew at the same rate as littermates. In a study that compared two lines of transgenic pigs expressing bovine GH (bGH) with their sibling controls, Pursel and coworkers (17) found the transgenic pigs had an 11–14% faster growth rate and an 18% increase in feed efficiency. However, reduction in carcass fat as transgenic pigs approached market weight was the most consistent and dramatic effect of elevated GH, as is shown in Figure 19.1.

Unfortunately, pigs that continuously expressed excess GH exhibited several notable health problems, which included lameness, susceptibility to stress, gastric ulcers, parakeratosis, lethargy, anestrus in gilts, and lack of a libido in boars (10,17,18). In contrast, no increase in the incidence of these pathological conditions was observed in transgenic pigs carrying nonfunctional copies of GH transgenes or in transgenic pigs that expressed only low levels of bGH (10,22).

In contrast to the GH transgenic pigs, transgenic lambs expressing high levels of ovine GH (oGH) or bGH did not grow faster or utilize feed more efficiently than control lambs, but they were much leaner (11,12). In transgenic lambs, the lack of body fat may

Table 19.1
Growth-Related Genes Transferred into Genomes of Farm Animals

Gene Construct (Promoter-coding Sequence)	Abbreviation[1]	Animal	Ref.
Albumin-Growth Hormone Releasing Factor	mALB-hGRF	Pig, Sheep	(4,5)
Cytomegalovirus(LTR)-Growth Hormone	CMV-pGH	Pig	(6)
Mammary Tumor Virus(LTR)-Growth Hormone	MTV-bGH	Cattle	(7)
Metallothionein-Growth Hormone	mMT-hGH	Pig, Rabbit, Sheep	(8,9)
	mMT-bGH	Pig, Sheep	(10,11)
	oMT-oGH	Sheep, Pig	(12,13)
	hMT-pGH	Pig	(14,15)
Metallothionein-Growth Hormone Releasing Factor	hMT-hGRF	Pig	(16,17)
		Sheep	(11)
Metallothionein-Insulin-like Growth Factor-I	mMT-hIGF-I	Pig	(17)
Moloney Leukemia Virus (LTR)-Growth Hormone	MLV-rGH	Pig	(18)
	MLV-pGH	Pig	(6)
Mouse Sarcoma Virus (LTR)-cellular *SKI*	MSV-c*SKI*	Pig, Cattle	(19,20)
Phosphoenolpyruvate Carboxykinase-Growth Hormone	rPEPCK-bGH	Pig	(21)
Prolactin-Growth Hormone	bPRL-bGH	Pig	(22)
αSkeletal Actin-Estrogen Receptor	cASK-hER	Cattle	(23)
αSkeletal Actin-Insulin-like Growth Factor-I	cASK-hIGF-I	Cattle, Pig	(24,25)
Transferrin-Growth Hormone	mTF-bGH	Pig, Sheep	(26,5)

[1] Lower-case letters designate species from which DNA sequence was derived: b, bovine; c, chicken; h, human; m, murine; o, ovine; p, porcine; r, rat.

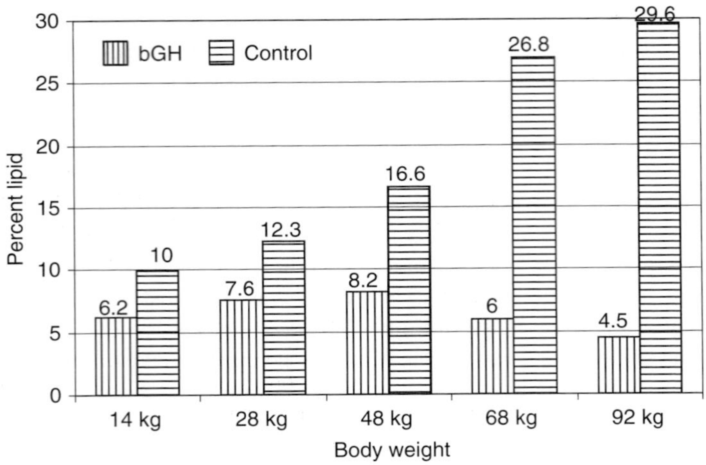

Figure 19.1 Least-squares mean total carcass lipid for four to six MT-bGH transgenic and sibling control pigs at each body weight. (From: Pursel, V.G., M.B. Solomon, *Food Rev. Int.* 9:423–439, 1993)

have been the result of hyperglycemia and glycosuria (5,11). Severely bowed front legs were a frequent anatomical abnormality observed in the transgenic lambs. More recent research with a modified oGH transgene that provided only a slight increase in circulating level of oGH resulted in lambs that grew faster and were leaner, but had significantly lower depth of eye muscle than their siblings (28).

Transfer of a GRF transgene is an alternative approach to obtain elevated GH that has been investigated in swine and sheep. Transgenic lambs that expressed GRF had continuously elevated GH and were phenotypically indistinguishable from GH transgenic lambs (11). In contrast, GH concentrations were not elevated in pigs that expressed GRF because the GRF in blood plasma was the 3–44 metabolite rather than the 1–44 native peptide (26).

It was recognized from the outset that regulation of transgene expression would be required to circumvent deleterious effects from continuous exposure of animals to elevated GH. While oGH transgenic mice expressed minimally without addition of zinc in their drinking water, the same construct produced high concentrations of constitutive expression in sheep (12) and swine (13). Similarly, expression of phosphoenolpyruvate carboxykinase-bGH (PEPCK-bGH) in mice could be regulated by altering the ratio of carbohydrate to protein in the diet (29). However, dietary manipulation did not control the level of GH expression when PEPCK-bGH was transferred into swine (27). Recently, tetracycline regulated and ecdysone regulated systems to switch transgene expression on and off in mice have been developed (30,31,32). These transgene switch systems show great promise, but are not yet perfected. When improved gene switch constructs become available there is little doubt that additional research with GH or GRF transgenes will be conducted.

19.2.2 IGF-I Transgenics

Recently transgenic pigs have been produced with a fusion gene composed of avian skeletal α-actin promoter, first intron and 3'-noncoding flanking regions, and human insulin-like growth factor-I (hIGF-I) coding region (33). Coleman and coworkers (34) previously reported that this fusion gene directs high levels of expression specifically in striated muscle in transgenic mice, and elicits myofiber hypertrophy. In the 12 founder transgenic pigs that were investigated, muscle insulin-like growth factor-I (IGF-I) concentrations varied from 20 to 1702 ng/g muscle compared to less than 10 ng/g muscle in nontransgenic control pigs (25). In marked contrast to previous experiences with GH transgenic pigs, definitive phenotypes for the IGF-I transgenic pigs were not detected, and no gross abnormalities, pathologies, or health related problems were encountered.

The founder transgenic pigs were then mated to nontransgenic pigs to produce G1 transgenic and sibling control progeny to investigate their growth potential and carcass composition. Neither average daily gain nor feed efficiency differed for transgenic and control pigs. At 120 kg body weight, 20 transgenic and 22 control pigs were sacrificed to evaluate carcass composition. The back fat measurements were 10% lower and the loin eye areas were 18% greater in transgenic than control pigs (35).

Subsequently, one of the transgenic founder boars with a hybrid damline background was mated to 12 nontransgenic gilts from two hybrid sire lines to test more fully the potential of the IGF-I transgene to function in a crossbreeding program. Pigs were provided feed ad libitum until they reached 120 kg body weight. Average daily gain was similar for transgenic and control pigs. At 120 kg body weight 33 transgenic and 42 sibling control pigs were sacrificed and their carcass composition evaluated. Transgenic gilts and barrows had significantly more carcass lean tissue, less total carcass fat, larger loin eye areas and lower average back fat thickness than sibling control pigs (Figure 19.2). While many of the carcass characteristics differed significantly for progeny of the two sire lines, IGF-I transgene expression had a beneficial effect on progeny of both lines. Based on these results the investigators concluded that enhancing IGF-I specifically in skeletal muscle had a positive effect on carcass composition of terminal cross market swine (36).

The enhanced carcass characteristics provided by the IGF-I transgene serves as an excellent prototype for a transgene that might be economically advantageous to the swine industry. Because society in general will not embrace use of a human gene for food products destined for consumption, porcine IGF-I gene should replace the human IGF-I gene used in these studies to make the transgene acceptable for public consumption. In addition, use of a gene switch that would enable IGF-I expression to decline to endogenous levels

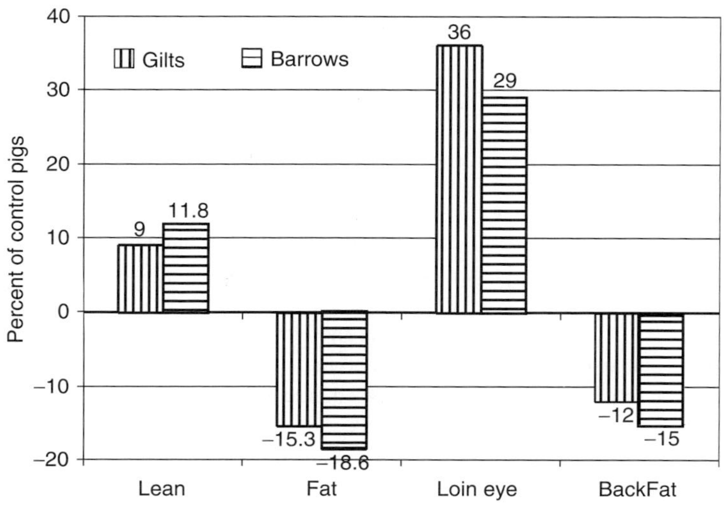

Figure 19.2 Percentage increases in carcass lean and loin eye area, and percentage decreases in carcass fat and back fat thickness for IGF-I transgenic barrows and gilts compared to sibling control barrows and gilts. (From Pursel, V.G., A.D. Mitchell, R.J. Wall, M.B. Solomon, M.E. Coleman, R.J. Schwartz, in: *Molecular Farming,* Toutant, J.P., E. Balazs eds., Paris: INRA, 2001, pp 77–86)

in striated muscle prior to harvest may be essential to overcome regulatory roadblocks that will accompany any transgenic animal that is destined for the marketplace.

19.3 MODIFICATION OF MILK COMPOSITION

Milk is widely considered one of nature's most perfect foods because of its high content of essential amino acids, essential fatty acids, vitamins, and bioavailable calcium. Because of this rich source of nutrients, dairy products provide about 20% of the protein in the diet of several western countries (37). Alteration of milk composition offers the dairy industry considerable potential for the future but has thus far received little research emphasis. A list of potential changes in milk components that have been suggested by various investigators is shown in Table 19.2.

About 80% of the protein in cow's milk is made up of the caseins (αS_1, αS_2, β and κ) while the whey fraction makes up the remainder. The caseins form the curds in cheese production, and for that reason they are extremely important for dairy manufacturing. In contrast, the whey proteins (β-lactoglobulin, α-lactalbumin, serum albumin and γ-globulin) make up only 20% of the milk proteins and represent a less valuable by product in the manufacture of cheese. Selective elimination of β-lactoglobulin, which makes up about one half of the whey,

Table 19.2
Some proposed modifications of milk constituents[1]

Change	Consequence
Increase α- and β-caseins	Enhanced curd firmness for cheese-making, improved thermal stability, and increased calcium content
Increase phosphorylation sites in caseins	Increased calcium content, improved emulsification
Introduce proteolytic sites in caseins	Increased rate of textural development to improve cheese ripening
Increase κ-casein concentration	Enhanced stability of casein aggregates, decreased micelle size, decreased gelation and coagulation
Eliminate β-lactoglobulin	Decreased high temperature gelation, improved digestibility, decreased allergenic response, decreased primary source of cysteine in milk
Decrease α-lactalbumin	Decreased lactose, increase market potential of fluid milk, decreased ice crystal formation, compromise osmotic regulation of mammary gland
Add human lactoferrin	Enhanced iron absorption, protect against gut infections
Add human lysozyme	Increase antimicrobial activity, reduce rennet clotting time, and increase cheese yield
Add proteolytic sites to κ-casein	Increased rate of cheese ripening
Decrease expression of acetyl CoA carboxylase	Decreased fat content, improved nutritional quality, reduce milk production costs
Express immunoglobulin genes	Protection against pathogens such as salmonella and listeria
Replace bovine milk proteins genes with human equivalents	Mimic human breast milk

[1]Adapted from Jimenez-Flores, R., T. Richardson, *J. Dairy Sci.* 71:2640–2654, 1985; H.-C. Yom, R.D. Bremel, *Am. J. Clin. Nutr.* 58:299–306, 1993; Maga, E.A., J.D. Murray, *Bio/Technology* 13:1452–1457, 1995.

would be beneficial to cheese production, because its presence in milk inhibits rennin's action on κ–casein (39). Because β-lactoglobulin is responsible for some of the allergic reactions to cow's milk, its elimination from fluid milk would also be advantageous for some consumers. Removal of β-lactoglobulin from cattle would require use of knockout technology that is now technically feasible by transfection of fetal fibroblasts *in vitro* (41,42,43).

19.3.1 Caseins

Because caseins are extremely important to the manufacture of cheese, modification of the relative amount or composition of the caseins may offer considerable economic value. One meritorious manipulation that has been suggested is to introduce extra copies of the αS_1-casein gene in cows to increase its proportion in milk. Jimenez-Flores and Richardson (38) suggest that such an alteration would enlarge the micelles and enhance textural development and curd characteristics. Alternatively, introduction of a chymosin sensitive region in the αS_1-casein gene by site specific mutagenesis may accelerate the rate of textural development (38).

Increasing the ratio of κ-casein to β-lactoglobulin might also be accomplished by inserting additional copies of the κ-casein gene. Such an alteration may be effective in reducing the size of casein micelles and improving the thermal stability of proteins in canned, sterilized milk products.

In addition, Jimenez-Flores and Richardson (38) suggest deletion of a phosphate group from β-casein would yield a softer cheese with higher moisture content, while addition of a phosphate group would yield a firmer cheese with less moisture and provide improved emulsifying properties. Each change would result in milk with characteristics that would be advantageous for specific market products.

19.3.2 Reduction in Butterfat

Nutrition conscious and diet conscious consumers are increasingly shunning products that have high fat levels. As a result of this declining demand, either surplus butterfat accumulates or the price declines sufficiently to stimulate consumption. Reduced butterfat production would provide a solution to this problem. Traditional genetic selection is not a suitable means of reducing fat content in milk because the genetic and phenotypic correlations for fat and protein yields are estimated at .86 and .93, respectively (44). Bremel, Yom and Bleck (45) suggest *de novo* fat synthesis from acetate might be reduced by blocking acetyl CoA carboxylase gene expression. The potential of this concept has been tested by transfecting preadipocytes with a ribozyme gene directed against acetyl CoA carboxylase mRNA (46). The rate of fatty acid synthesis in transfected cells was 30 to 70% that of the control adipocytes. If transfer of this ribozyme gene had a similar effect on butterfat production in the mammary gland, there would be the potential for a dramatic influence on the fat content of milk. Because a gram of fat has 2.25 times the amount of energy as a gram of carbohydrate or protein, reduction in the fat output in milk would greatly reduce the energy required to produce milk. Yom and Bremel (39) have suggested that reducing the fat content of milk from 3.8% to 2% would permit dairy farmers to increase the proportion of forage in the diet from about 60% to 83%. Consequently, the concentrate portion of the lactation ration would be reduced by 50%, and the farmers' feed bill for lactating cows would be reduced by 22%.

19.3.3 Lactose

Lactose is a major component of cow's milk that makes it an unacceptable food source for a sizable portion of the adult human population deficient in β-galactosidase and therefore unable to hydrolyze lactose (47). In addition, the low solubility of lactose is responsible for grittiness defects in ice cream. These problems might be alleviated by the partial inhibition of the α-lactalbumin gene, which is an essential cofactor in the synthesis of lactose. Because lactose is responsible for the movement of water through the secretory cells, a reduction in α-lactalbumin and lactose would partially reduce total milk volume, with no reduction in overall production of milk protein. Such an inhibition of lactalbumin expression might be accomplished by transfer of an antisense sequence of the α-lactalbumin gene ligated to a mammary specific promoter (39).

Conversely, Bleck and coworkers (48) have suggested that enhanced expression of α-lactalbumin in sow's milk may increase milk production and boost preweaning growth rates in piglets. An increase in α-lactalbumin should elevate the lactose synthase activity in the mammary gland and stimulate lactose content in sow milk. An increase in lactose content early in lactation might provide a valuable energy source for the newborn piglet. In addition, the increased lactose content should increase the total quantity of milk produced. Transgenic pigs overexpressing the milk protein bovine α-lactalbumin were subsequently developed at the University of Illinois. First parity α-lactalbumin gilts had higher milk lactose content in early lactation and 20–50% greater milk yield on days 3–9 of lactation than did nontransgenic gilts. Weight gain of piglets suckling α-lactalbumin gilts was greater (days 7–21 after parturition) than that of control piglets. Thus, transgenic overexpression of milk proteins provides a means for improving the lactation performance of pigs (49).

19.3.4 Human Milk Proteins

The introduction of several human milk protein genes into dairy cows, or replacement of these bovine genes with human genes, may one day play an important nutritional role in human infants that are now unable to receive milk from their mother. Milk from cows and humans differs considerably, and consequently cow's milk is not an ideal source of food for human infants. Total protein content of human milk is low (0.9%) and whey proteins predominate, while cow's milk contains about 3.3% protein with 80% of it being casein (50). In addition, bovine β-lactalbumin is quite allergenic to many human infants, and the concentration of lactoferrin in human milk is three times that of cow's milk.

Lactoferrin has long been known to provide antimicrobial activity to the human infant and may be an important source of bioavailable iron and other minerals (47). More recently, human lactoferrin has been reported to also have antifungal, antiendotoxin, and antiviral properties (51,52,53).

Therefore, enhancing the concentration of lactoferrin in cow's milk, or expression of human lactoferrin in cow's milk, may result in a milk product that is highly suitable for production of lactoferrin for the biopharmaceutical industry. Recently van Berkel and coworkers (54) reported the production of four lines of transgenic cattle that produce between 0.4 and 3.0 g of human lactoferrin per liter of milk. The iron binding and antibacterial activities of the lactoferrin protein were shown to be highly similar to that of natural human lactoferrin.

Human milk contains about 3,000 times as much lysozyme as is present in cow's milk. Although the biological function of lysozyme in human milk is not known, its antimicrobial activity is considered to be part of the natural defense against food microorganisms, and it may protect the mammary gland against bacterial infection (40). In addition, other research has showed that addition of lysozyme to cow's milk was beneficial for rennet clotting time, yield, and syneresis in cheese processing (55). Therefore, enhanced expression of lysozyme in milk may be beneficial for infant formulations and cheese production.

The antimicrobial properties of standard human lysozyme and the milk of transgenic mice expressing human lysozyme were investigated using bacterial strains important to the dairy industry (56). Milk from transgenic mice that secreted human lysozyme in their milk at an average concentration of 0.3 mg/mL was found to be bacteriostatic against *Pseudomonas fragi* and *Lactobacillus viscous*, which are cold spoilage organisms, and a mastitis causing strain of *Staphylococcus aureus*, but not against a pathogenic strain of *E. coli*. These results suggest that transgenic animals that produce human lysozyme in their milk may enhance the antimicrobial nature of milk.

19.4 IMPROVED WOOL PRODUCTION

Australia and New Zealand have large expanses of grazing lands that are used to convert forage to wool, which is a high quality textile fiber that is a valuable exportable commodity. The importance of wool to the economies of these countries has led investigators to attempt to improve the efficiency of wool production and enhance its characteristics as a textile fiber. Thus far three research approaches have been undertaken to improve wool production.

19.4.1 Synthesis of Cysteine

Production of high quality wool is dependent upon an abundant supply of amino acids in general and cysteine in particular. Keratins are the major structural proteins of the wool fiber, and a large amount of cysteine is required to form the extensive disulphide bridges that link the keratins as the fibers grow. Cysteine is the rate limiting amino acid for the production of wool. Addition of cysteine to the diet does not increase wool production because the rumen degrades proteins, and sulphur is lost as hydrogen sulphide. Bacterial genes are capable of synthesizing cysteine from the hydrogen sulphide and serine, both of which are available in the rumen. This limitation for wool production has led Australian investigators to attempt a genetic engineering approach to overcome this obstacle.

Serine transacetylase (*cys E*) and o-acetylserine sulfhydrylase (*cys M or K*) are required for the synthesis of cysteine from rumen substrates, hydrogen sulphide, and serine. Bacterial genes for these enzymes have been isolated, sequenced, and cloned for *Escherichia coli* (57) and *Salmonella typhimurium* (58). These linked bacterial genes (*cys E-cys M or K*) were used to form transgenes with three gene promoters that give broad expression, sheep metallothionein-Ia (oMT), Rous sarcoma virus long terminal repeat (RSVLTR) and mouse phosphoglycerate kinase-1(mPgk-1).

The *cys E-cys M or K* transgenes with oMT, RSVLTR and mPgk-1 promoters were first tested *in vitro* in tissue culture cells, and then in transgenic mice. While these transgenes expressed well in cultured mammalian cells (59), only the transgenic mice with the oMT promoter or the mPgk promoter expressed sufficient enzyme levels in stomach and intestines to support significant cysteine synthesis *in vitro* (60). Concurrently, these fusion genes were transferred into sheep zygotes and the resulting transgenic lambs were evaluated for expression of the enzyme genes (60,61). These three promoters were ineffective in directing expression of the bacterial enzymes in the rumen of the transgenic sheep. However, low levels of expression were detected in skeletal muscle of three of 23 RSVLTR lambs and one of 10 mPgk-1 lambs (60).

Subsequently, Rogers and coworkers (61) isolated a rumen specific gene from a sheep rumen cDNA library. This gene encodes a small proline rich protein (oSPR) that is highly expressed specifically in the ruminal epithelium. The *cys E-cys M or K* transgene with oSPR promoter was then transferred into sheep zygotes. Of seven transgenic sheep evaluated, only one expressed the transgene in the rumen at a low level at three months of age, and it lacked expression when reevaluated at six months of age.

19.4.2 Wool Keratin Proteins

The second approach to improved wool production that has been undertaken is to modify the protein composition of the wool fibers. The genes for many of the keratin proteins have been sequenced (61,62). Transgenic mice have been produced with three different wool keratin fusion genes expressed in the hair follicles. Two of these transgenes have produced visible phenotypic changes in the hair coat (63,64).

Subsequently, researchers at the University of Adelaide and the South Australian Research and Development Institute have produced a number of transgenic sheep in attempts to improve wool fiber quality by altering expression of wool fiber keratin and keratin associated protein genes in the wool follicle cortex (65). These investigations found: (1) Increasing the quantities of cysteine rich protein in the cortical karatin associated protein pool reduced the intrinsic strength of the wool fibers in transgenic sheep. (2) Expression of the human keratinocyte transglutaminase enzyme within the cortex increased isopeptide cross linkage but also weakened the fiber. (3) Enhanced expression of a glycine or tyrosine rich protein in the cortex also appeared to reduce the load bearing capacity of wool fibers. (4) The most promising results have been obtained with overexpression of a wool keratin intermediate filament gene in wool follicles. Using this approach, Bawden and coworkers (66) demonstrated that a high level of cortical specific expression of a wool type II intermediate filament keratin gene, K2.10, lead to marked alterations in both the microstructure and macrostructure of the wool fibers that yielded a higher luster and reduced crimp. Up to the present time only the sheep expressing the highest levels of the various transgenes have been evaluated. Transgenic sheep expressing low to moderate levels of these transgenes have been produced, but have not yet been fully evaluated (65).

19.4.3 IGF-I for Wool Follicle Growth

The third approach to improve wool production is to stimulate fiber growth by expression of insulin-like growth factor (IGF-I) in wool follicles. Several investigators reported that exogenous treatment of sheep with growth hormone stimulated wool growth (67,68,69). Subsequently, Harris and coworkers (70) reported that local infusion of sheep skin with IGF-I increased blood flow, oxygen uptake, and amino acid uptake. Because IGF-I mediates many of the actions of growth hormone and has mitogenic properties, study of its potential as a stimulator of follicle growth in transgenic sheep was undertaken in New Zealand.

Damak and coworkers (71) reported the production of transgenic sheep with a transgene composed of a mouse keratin promoter and ovine IGF-I cDNA. Two of five founder transgenic lambs expressed the IGF-I in their skin. The expressing transgenic ram was then bred to nontransgenic ewes, and 85 lambs were born of which 43 were transgenic. When shorn at 14 months of age, the clean fleece weight of the transgenic animals was 6.2% higher ($P = .028$) and the bulk was significantly higher ($P= .042$) than for the nontransgenic half sibs. Other fiber characteristics were unchanged, except that staple strength was lower for transgenic males than transgenic and nontransgenic females. Expression of the transgene did not have detectable negative physiological effect on these animals. Unfortunately, additional investigations with the second generation of transgenic animals failed to confirm the wool production enhancement that had been found in the first generation (72).

19.5 ENHANCED ANIMAL HEALTH

Economic losses from diseases of farm animals have been estimated to amount to 10 to 20% of the total production costs (73). Molecular biology and genetic manipulation may be useful in augmenting conventional breeding techniques to provide animals with improved resistance to some of these diseases, thereby reducing these costs and enhancing animal welfare. Several approaches under investigation include transfer of genes for providing naturally occurring disease resistance, preformed antibodies, viral envelope proteins, and antimicrobial peptides.

19.5.1 Naturally Occurring Disease Resistance

Only a few genes for disease resistance have thus far been identified (74). Mice carrying the autosomal dominant Mx1 allele are resistant to influenza viruses. Transfer of the Mx1 gene into mice that lacked the Mx1 allele was successful in conferring resistance to influenza-A viruses (75,76). Subsequently, three Mx1 fusion genes were transferred into swine to test their effectiveness (77). Two of five transgenic pigs harboring the Mx1 regulatory and structural sequences were found to respond to interferon induction of Mx1 mRNA. However, the response was insufficient to produce detectable amounts of Mx1 protein in the tissues. The other two fusion genes that were transferred into pigs were rearranged during integration, so they were not functional. In retrospect the investigators believe uncontrolled expression of Mx1 during embryogenesis was lethal, thus only animals with low levels of transgene expression of rearranged genes were produced (74). Once again the necessity of tight transgene regulation was crucial to the outcome, and results were not consistent for mice and pigs.

19.5.2 Preformed Antibodies

Genes encoding mouse α heavy and κ light chains from antibodies against phosphoryl choline (PC) were coinjected into ova to produce two transgenic pigs and three transgenic lambs (78). In the transgenic pigs, the mouse immunoglobulin A (IgA) was detected in the serum despite the failure of an intact mouse transgene to integrate. Transgenic progeny from both founders demonstrated high levels of serum mouse IgA starting at about five weeks of age. Average levels of mouse IgA were 630 μg/mL in one line and 1,293 μg/mL in the other. In both cases, IgA levels in progeny were higher than in the founders. However, mouse IgA showed little binding specificity for PC, presumably because secreted chimeric antibody included endogenous light chains with mouse heavy chains. In transgenic sheep, mouse IgA was detectable in peripheral lymphocytes but not in serum. These studies need to be expanded to obtain conclusive proof that the IgA transgene would be protective against pathogenic bacteria.

In a similar study, Brem and coworkers (79) produced two transgenic pigs and three transgenic rabbits that harbored mouse λ heavy and κ light chain transgenes from antibodies directed against the hapten 4-hydroxy-3-nitrophenylacetate. Titres of 100 to 300 μg IgG/mL in transgenic rabbits and up to 1,000 μg IgG/mL in one transgenic pig were present in the serum of founders and transgenic progeny. Further evaluation of the antibody composition indicated xenogenic antibodies had formed by association of light chains of rabbit and pig with heavy chains of mouse.

Most recently, Enjuanes and coworkers (80) generated 18 lines of transgenic mice secreting a recombinant monoclonal antibody (Mab) neutralizing transmissible gastroenteritis coronavirus (TGEV) into their milk. The genes encoding a chimeric Mab with the variable modules of the murine TGEV-specific Mab 6A.C3 and the constant modules of a human IgG, isotype Mab were expressed under the control of regulatory sequences derived from whey acidic protein (WAP), which is an abundant milk protein in mice. Antibody expression titers of 10^6 were obtained in the milk of transgenic mice that reduced TGEV infectivity by a factor of 10^6. The antibody was synthesized at high levels throughout lactation. These results suggest that generation of transgenic swine producing virus neutralizing antibodies in milk could provide an effective approach for protection against neonatal infections of the enteric tract.

19.5.3 Intracellular Immunization

A transgenic approach may be effective for producing farm animals that are genetically resistant to specific pathogenic viruses. Salter and Crittenden (81) produced transgenic chickens that were highly resistant to infection with the subgroup A avian leukosis virus (ALV) by introducing an ALV gene that encoded a viral envelope glycoprotein. Normally, ALV enters chicken cells by attachment of the envelope glycoprotein to cell membrane receptors. However, in the transgenic chickens the subgroup A ALV virus could not enter the cells because the membrane receptors were presumably occupied with envelope protein that had been produced by the transgene. These chickens were not resistant to infection by subgroup B ALV because a different receptor is used for entry.

A similar experiment was initiated in sheep with the envelope gene from visna virus. The sheep population is widely infected with visna virus, which is an ovine lentivirus similar to equine infectious anaemia virus, caprine arthritis encephalitis virus, and bovine, feline, simian, and human immunodeficiency virus. Visna viruses are usually transmitted to lambs in colostrum or milk, where they infect the macrophages or monocytes, and establish lifelong infections. The clinical disease in sheep is ovine progressive pneumonia, arthritis, mastitis, and occasionally paralysis (82). Immunizations with vaccines have not been effective in control of visna virus or other lentiviruses.

Three transgenic sheep were produced by microinjection of a visna virus fusion gene into pronuclei of sheep zygotes (83). The fusion gene was composed of a visna virus long term repeat (LTR) regulatory region fused to the coding region for visna virus envelope protein. All three lambs expressed the envelope glycoprotein in the macrophages as well as in fibroblasts isolated from the skin. These animals remained healthy and expression of the viral gene had no observable detrimental effect. Two of the three sheep produced antibodies to the envelope protein, which possibly indicated the viral gene was expressed relatively late in development and was not recognized as a self antigen. Transgenic progeny from these founders were subsequently challenged with an Icelandic strain of visna virus to test their abilities to resist infection. This challenge resulted in several of the transgenic lambs becoming infected with the virus in the same manner as the control sheep that were also challenged (C. E. Rexroad, Jr., personal communication, 1998). These results do not rule out the possibility that these transgenic sheep would have been resistant to infection if they had been challenged by natural exposure to visna virus under field conditions.

19.5.4 Anti Bacterial Peptides

As a first step toward enhancing mastitis resistance of dairy animals, researchers at the University of Vermont and the Beltsville Agricultural Research Center in Maryland generated transgenic mice that secrete a potent antistaphylococcal protein, lysostaphin, into milk (84). Lysostaphin is a peptidoglycan hydrolase normally produced by *Staphylococcus simulans* that is active against *Staphylococcus aureus* bacteria. *Staphylococcus aureus* is the major contagious mastitis pathogen, accounting for more than 15% of mastitis infections, and has proved difficult to control using standard management practices. Three lines of transgenic mice were produced with an ovine β-lactoglobulin gene directing the secretion of lysostaphin into milk. Progeny of these mice exhibited substantial resistance to an intramammary challenge of *S. aureus*, with the highest expressing line being completely resistant to infection. These results clearly demonstrated the potential of genetic engineering to combat the one of the most prevalent disease of dairy cattle, and as a consequence, the same transgene has now been used to produce transgenic dairy cattle that have not yet been fully evaluated.

19.6 REDUCING ENVIRONMENTAL POLLUTION

Researchers at the University of Guelph initially developed a transgenic mouse model to determine whether endogenous expression of phytase transgenes in the digestive tract of monogastric animals could increase the bioavailability of phytate, a major indigestible form of dietary phosphorus. Forsberg and coworkers (85) constructed a phytase transgene to provide expression in salivary glands and found that expression of salivary phytase in transgenic mice reduced fecal phosphorus by 11%. These results suggested that the introduction of salivary phytase transgenes into monogastric farm animals might offer a promising biological approach for relieving the requirement for dietary phosphate supplements and for reducing phosphorus pollution from animal agriculture.

Subsequently, these researchers used the phytase transgene to generate phytase transgenic pigs (86). The saliva of these pigs contained the enzyme phytase that allowed the pigs to digest the phosphorus in phytate, which is the most abundant source of phosphorus in the pig diet. Without this enzyme, phytate phosphorus passes undigested into manure to become the single most important manure pollutant of pork production. Their research showed that salivary phytase provided essentially complete digestion of dietary phytate phosphorus, relieved the requirement for inorganic phosphate supplements, and reduced fecal phosphorus output by up to 75%. These pigs offer a unique biological approach to the management of phosphorus nutrition and to reducing one of the major environmental pollutants generated on swine farms.

19.7 CONCLUSIONS

From the foregoing material it is clear that animal scientists are gradually making inroads toward harnessing the power of transgenic technology to improve economically important productivity traits of farm animals. In the past few years research on altering carcass composition, enhancing disease resistance, increasing milk production in sows, and reducing excretion of phosphate in pigs have shown the greatest progress. Modification of productivity traits in dairy cattle offers considerable potential, but much of this research awaits improved efficiencies of producing transgenic cattle through the use of nuclear transfer. For many productivity traits, further progress seems highly dependent on finding more effective regulatory systems that permit precise control of transgene expression. As people search for solutions to these problems, they will make unexpected discoveries, find new approaches, and expand fundamental knowledge. Then, at some time in the future, as superior plant and animal products resulting from genetic engineering become available, such items will be more appropriately called "Fantastic Foods" instead of "Frankenfoods".

REFERENCES

1. National Research Council. Current trends in consumption of animal products. In: *Designing Foods: Animal Product Options in the Marketplace.* Washington DC: National Academy Press, 1988, pp 18–44.
2. DeBoer, A.J., J.A. Yazman, N.S. Raun. Animal agriculture in developing countries: technology dimensions. In: *Developmental Studies Paper Series,* Seckler, K., ed., Morrilton, Arkansas: Winrock International, 1994, pp 1–43.
3. Palmiter, R.D., R.L. Brinster, R.E. Hammer, M.E. Trumbauer, M.G. Rosenfeld, N.C. Birnberg, R.M. Evans. Dramatic growth of mice that develop from eggs microinjected with metallothionein-growth hormone fusion genes. *Nature* 300:611–615, 1982.

4. Pursel, V.G., K.F. Miller, D.J. Bolt, C.A. Pinkert, R.E. Hammer, R.D. Palmiter, R.L. Brinster. Insertion of growth hormone genes into pig embryos. In: *Biotechnology of Growth Regulation*, R.B. Heap, C.G. Prosser, G.E. Lamming, eds., London: Butterworths, 1989, pp 181–188.
5. Rexroad, C.E., Jr., K.M. Mayo, D.J. Bolt, T.H. Elsasser, K.F. Miller, R.R. Behringer, R.D. Palmiter, R.L. Brinster. Transferrin- and albumin-directed expression of growth-related peptides in transgenic sheep. *J. Anim. Sci.* 69:2995–3004, 1991.
6. Ebert, K.M., T.E. Smith, F.C. Buonoma, E.W. Overstrom, M.J. Low. Porcine growth hormone gene expression from viral promoters in transgenic swine. *Anim. Biotech.* 1:145–159, 1990.
7. Roshlau, K, P. Rommel, L. Andreewa, M. Zackel, D. Roschlau, B. Zackel, M. Schwerin, R. Huhn, K.G. Gazaejan. Gene transfer experiments in cattle. *J. Reprod. Fertil. Suppl.* 38:153–160, 1989.
8. Brem, G., B. Brenig, H.M. Goodman, R.C. Selden, F. Graf, B. Kruff, K. Springman, J. Hondele, J. Meyer, E.-L. Winnaker, H. Krausslich. Production of transgenic mice, rabbits and pigs by microinjection into pronuclei. *Zuchthygiene* 20:251–252, 1985.
9. Hammer, R.E., V.G. Pursel, C.E. Rexroad, Jr., R.J. Wall, D.J. Bolt, K.M. Ebert, R.D. Palmiter, R.L. Brinster. Production of transgenic rabbits, sheep and pigs by microinjection. *Nature* 315:680–683, 1985.
10. Pursel, V.G., C.E. Rexroad, Jr., D.J. Bolt, K.F. Miller, R.J. Wall, R.E. Hammer, C.A. Pinkert, R.D. Palmiter, R.L. Brinster. Progress on gene transfer in farm animals. *Vet. Immunol. Immunopathol.* 17:303–312, 1987.
11. Rexroad, C.E., Jr., R.E. Hammer, D.J. Bolt, K.M. Mayo, L.A. Frohman, R.D. Palmiter, R.L. Brinster. Production of transgenic sheep with growth regulating genes. *Mol. Reprod. Dev.* 1:164–169, 1989.
12. Murray, J.D., C.D. Nancarrow, J.T. Marshall, I.G. Hazelton, K.A. Ward. The production of transgenic Merino sheep by microinjection of ovine metallothionein-ovine growth hormone fusion genes. *Reprod. Fert. Dev.* 1:147–155, 1989.
13. Pursel, V.G., R.J. Wall, M.B. Solomon, D.J. Bolt, J.D. Murray, K.A. Ward. Transfer of an ovine metallothionein-ovine growth hormone fusion gene into swine. *J. Anim. Sci.* 75:2208–2214, 1997.
14. Vize, P.D., A.E. Michalska, R. Ashman, B. Lloyd, B.A. Stone, P. Quinn, J.R.E. Wells, R.F.I. Seamark. Introduction of a porcine growth hormone fusion gene into transgenic pigs promotes growth. *J. Cell. Sci.* 90:295–300, 1988.
15. Nottle, M.B., H. Nagashima, P.J. Verma, Z.T. Du, C.G. Grupen, S.M. McIlfatrick, R.J. Ashman, M.P. Harding, C. Giannakis, P.L. Wigley, I.G. Lyons, D.T. Harrison, B.G. Luxford, R.G. Campbell, R.J. Crawford, A.J. Robins. Production and analysis of transgenic pigs containing a metallothionein porcine growth hormone gene construct. In: *Transgenic Animals in Agriculture,* Murray, J.D., G.B. Anderson, M.M. McGloughlin, A.M. Oberbauer, eds., Wallingford, UK: CAB International, 1999, pp 145–156.
16. Brem, G., B. Brenig, M. Muller, H. Krausslich, E.-L. Winnacker. Production of transgenic pigs and possible application to pig breeding. *Occasional Publ. Br. Soc. Anim. Prod.* 12:15–31, 1988.
17. Pursel, V.G., C.A. Pinkert, K.F. Miller, D.J. Bolt, R.G. Campbell, R.D. Palmiter, R.L. Brinster, R.E. Hammer. Genetic engineering of livestock. *Science* 244:1281–1288, 1989.
18. Ebert, K.M., M.J. Low, E.W. Overstrom, F.C. Buonomo, C.A. Baile, T.M. Roberts, A. Lee, G. Mandel, R.H. Goodman. A Moloney MLV-rat somatotropin fusion gene produces biologically active somatotropin in a transgenic pig. *Molec. Endocrinol.* 2:277–283, 1988.
19. Pursel, V.G., P. Sutrave, R.J. Wall, A.M. Kelly, S.H. Hughes. Transfer of c-SKI gene into swine to enhance muscle development. *Theriogenology* 37:278, 1992.
20. Bowen, R.A., M.L. Reed, A. Schnieke, G.E. Seidel, Jr., A. Stacey, W.K. Thomas, O. Kajikawa. Transgenic cattle resulting from biopsied embryos: expression of c-ski in a transgenic calf. *Biol. Reprod.* 50:664–668, 1994.

21. Wieghart, M., J.L. Hoover, M.M. McCrane, R.W. Hanson, F.M. Rottman, S.H. Holtzman, T.E. Wagner, C.A. Pinkert. Production of transgenic swine harboring a rat phosphoenolpyruvate carboxykinase - bovine growth hormone fusion gene. *J. Reprod. Fertil. Suppl.* 41:89–96, 1990.
22. Polge, E.J.C., S.C. Barton, M.H.A. Surani, J.R. Miller, T. Wagner, K. Elsome, A.J. Davis, J.A. Goode, G.R. Foxroft, R.B. Heap. Induced expression of a bovine growth hormone construct in transgenic pigs. In: *Biotechnology of Growth Regulation.* Heap, R.B., C.G. Prosser, G.E. Lamming, eds., London: Butterworth & Co., 1989, pp 189–199.
23. Massey, J.M. Animal production industry in the year 2000 A.D. *J. Reprod. Fertil. Suppl.* 41:199–208, 1990.
24. Hill, K.D., J. Curry, F.J. DeMayo, K. Jones-Diller, J.R. Slapak, K.R. Bondioli. Production of transgenic cattle by pronuclear injection. *Theriogenology* 37:222, 1992.
25. Pursel, V.G., R.J. Wall, A.D. Mitchell, T.H. Elsasser, M.B. Solomon, M.E. Coleman, F. DeMayo, R.J. Schwartz. Expression of insulin-like growth factor-I in skeletal muscle of transgenic swine. In: *Transgenic Animals in Agriculture*, Murray, J.D., G.B. Anderson, M.M. McGloughlin, A.M. Oberbauer, eds., Wallingford, U.K.: CAB International, 1999, pp 131–144.
26. Pursel, V.G., D.J. Bolt, K.F. Miller, C.A. Pinkert, R.E. Hammer, R.D. Palmiter, R.L. Brinster. Expression and performance in transgenic swine. *J. Reprod. Fertil. Suppl. 40*:235–245, 1990.
27. Pursel, V.G., M.B. Solomon. Alteration of carcass composition in transgenic swine. *Food Rev. Int.* 9:423–439, 1993.
28. Adams, N.R., J.R. Briegel, K.A. Ward. The impact of a transgene for ovine growth hormone on the performance of two breeds of sheep. *J. Anim. Sci.* 80:2325–2333, 2002.
29. McGrane, M.M., J. deVente, J. Yun, J. Bloom, E. Park, A. Wynshaw-Boris, T. Wagner, F.M. Rottman, R.W. Hanson. Tissue-specific expression and dietary regulation of a chimeric phosphoenolpyruvate carboxykinase/bovine growth hormone gene. *J. Biol. Chem.* 263:11443–11451, 1988.
30. Furth, P.A., L. St Onge, H. Boger, P. Gruss, M. Gossen, A. Kistner, H. Bujard, L. Hennighausen. Temporal control of gene expression in transgenic mice by a tetracycline-responsive promoter. *Proc. Natl. Acad. Sci. USA* 91:9302–9306, 1994.
31. Gossen, M., S. Freundlieb, G. Bender, G. Muller, W. Hillen, H. Bujard. Transcriptional activation by tetracyclines in mammalian cells. *Science* 268:1766–1769, 1995.
32. No, D., T.-P. Yao, R.M. Evens. Ecdysome-inducible gene expression in mammalian cells and transgenic mice. *Proc. Natl. Acad. Sci. USA* 93:3346–3351, 1996.
33. Coleman, M.E., V.G. Pursel, R.J. Wall, M. Haden, F. DeMayo, R.J. Schwartz. Regulatory sequences from the avian skeletal α-actin gene directs high level expression of human insulin-like growth factor-I cDNA in skeletal muscle of transgenic pigs. *J. Anim. Sci.* (1)73:145, 1995.
34. Coleman, M.E., F. DeMayo, K.C. Yin, H.M. Lee, R. Geske, C. Montgomery, R.J. Schwartz. Myogenic vector expression of insulin-like growth factor I stimulates muscle cell differentiation and myofiber hypertrophy in transgenic mice, *J. Biol. Chem.* 270:12109–12116, 1995.
35. Pursel, V.G., G. Bee, K.D. Wells, A.D. Mitchell, T. Elsasser, R.J. Wall, M.B. Solomon, M.E. Coleman, R.J. Schwartz. Expression of IGF-I in skeletal muscle of transgenic swine. *J. Anim. Sci.* (1)76:130, 1998.
36. Pursel, V.G., A.D. Mitchell, R.J. Wall, M.B. Solomon, M.E. Coleman, R.J. Schwartz. Transgenic research to enhance growth and lean carcass composition in swine. In: *Molecular Farming,* Toutant, J.P., E. Balazs eds., Paris: INRA, 2001, pp 77–86.
37. Gerrior, S.A., C. Zizza, Nutrient content of the U.S. Food Supply, 1909–90, *U.S. Department of Agriculture, Home Economics Research Report No. 52*, 1994, pp 1–120.
38. Jimenez-Flores, R., T. Richardson. Genetic engineering of the caseins to modify the behavior of milk during processing: a review. *J. Dairy Sci.* 71:2640–2654, 1985.
39. H.-C. Yom, R.D. Bremel. Genetic engineering of milk composition: modification of milk components in lactating transgenic animals. *Am. J. Clin. Nutr.* 58:299–306, 1993.

40. Maga, E.A., J.D. Murray. Mammary gland expression of transgenes and the potential for altering the properties of milk. *Bio/Technology* 13:1452–1457, 1995.
41. Denning, C., S. Burl, A. Ainslie, J. Bracken, A. Dinnyes, J. Fletcher, T. King, M. Ritchie, W.A. Ritchie, M. Rollo, P. de Sousa, A. Travers, I. Wilmut, A.J. Clark. Deletion of the alpha (1,3) galactosyl transferase (GGTA1) gene and the prion protein (PrP) gene in sheep. *Nat. Biotechnol.* 19:559–562, 2001.
42. Lai, L., D. Kolber-Simonds, K.W. Park, H.T. Cheong, J.L. Greenstein, G.S. Im, M. Samuel, A. Bonk, A. Rieke, B.N. Day, C.N. Murphy, D.B. Carter, R.J. Hawley, R.S. Prather. Production of alpha-1,3-galactosyltransferase knockout pigs by nuclear transfer cloning. *Science* 295:1089–1092, 2002.
43. Dai, Y., T.D. Vaught, J. Boone, S.H. Chen, C.J. Phelps, S. Ball, J.A. Monahan, P.M. Jobst, K.J. McCreath, A.E. Lamborn, J.L. Cowell-Lucero, K.D. Wells, A. Colman, I.A. Polejaeva, D.L. Ayares. Targeted disruption of the 1,3-galactosyltransferase gene in cloned pigs. *Nat. Biotechnol.* 20:251–255, 2002.
44. Bremel, R.D.. Potential role of transgenesis in dairy production and related areas. *Theriogenology* 45:51–56, 1996.
45. R.D. Bremel, H.-C. Yom, G.T. Bleck. Alteration of milk composition using molecular genetics. *J. Dairy. Sci.* 72:2826–2833, 1989.
46. Ha, J., K.H. Kim. Inhibtion of fatty acid synthesis by expression of an acetyl-CoA carboxylase-specific ribozyme gene. *Proc. Natl. Acad. Sci. USA* 91:9951–9956, 1994.
47. Mercier, J.C. Genetic engineering: some expectations. In: *Exploiting New Technologies in Animal Breeding,* Smith, C., J.W.B. King, J. McKay, eds. Oxford: Oxford University Press, 1986, pp 122–131.
48. Bleck, G.T., B.R. White, E.D. Hunt, L.A. Rund, J. Barnes, D. Bidner, M.B. Wheeler. Production of transgenic swine containing the bovine α-lactalbumin gene. *Theriogenology* 45:347, 1996.
49. M.S. Noble, S. Rodriguez-Zas, J.B. Cook, G.T. Bleck, W.L. Hurley, M.B. Wheeler. Lactational performance of first-parity transgenic gilts expressing bovine alpha-lactalbumin in their milk. *J. Anim. Sci.* 80:1090–1096, 2002.
50. Bounous, G., P.A. Kongshavn, A. Taveroff, P. Gold. Evolutionary traits in human milk proteins. *Med. Hypotheses* 27:133–140, 1988.
51. Soukka, T., J. Tenovuo, M. Lenander-Lumikari. Fungicidal effect of human lactoferrin against *Candida albicans*. *FEMS Microbiol. Lett.* 69:223–228, 1992.
52. Zhang, G.H., D.M. Mann, C.M. Tsai. Neutralization of endotoxin *in vitro* and *in vivo* by a human lactoferrin-derived peptide. *Infect. Immun.* 67:1353–1358, 1999.
53. Hasegawa, K., W. Motsuchi, S. Tanaka, S. Dosako. Inhibition with lactoferrin of *in vitro* infection with human herpes virus. *Jpn. J. Med. Sci. Biol.* 47:73–85, 1994.
54. van Berkel, P.H., M.M. Welling, M. Geerts, H.A. van Veen, B. Ravensbergen, M. Salaheddine, E.K. Pauwels, F. Pieper, J.H. Nuijens, P.H. Nibbering. Large scale production of recombinant human lactoferrin in the milk of transgenic cows. *Nat. Biotechnol.* 20:484–487, 2002.
55. Giangiacomo, R., F. Nigro, G. Messina, T.M. Cattaneo. Lysozyme: just an additive or a technological aid as well? *Food Addit. Contam.* 9:427–433, 1992.
56. Maga, E.A., G.B. Anderson, J.S. Cullor, W. Smith, J.D. Murray. Antimicrobial properties of human lysozyme transgenic mouse milk. *J. Food. Prot.* 61:52–55, 1998.
57. Ward, K.A., J.D. Murray, C.D. Nancarrow. The insertion of foreign genes into animal cells. In: *Biotechnology for Livestock Production,* FAO Animal Production and Health Division, ed., New York: Plenum Press, 1989, pp 17–28.
58. Rogers, G.E. Improvement of wool production through genetic engineering. *Trends Biotechnol.* 8:6–11, 1990.
59. Sivaprasad, A.V., E.S. Kuczek, C.S. Bawden, G.E. Rogers. Coexpression of the *cysE* and *cysM* genes of *Salmonella Typhimurium* in mammalian cells: a step towards establishing cysteine biosynthesis in sheep by transgenesis. *Transgenic. Res.* 1:79–92, 1992.

60. Bawden, C.S., A.V. Sivaprasad, P.J. Verma, S.K. Walker, G.E. Rogers. Expression of bacterial cysteine biosynthesis genes in transgenic mice and sheep: toward a new *in vivo* amino acid biosynthesis pathway and improved wool growth. *Transgenic. Res.* 4:87–104, 1995.
61. Powell, B.C., S.K. Walker, C.S. Bawden, A.V. Sivaprasad, G.E. Roger. Transgenic sheep and wool growth: possibilities and current status. *Reprod. Fertil. Dev.* 6:615–623, 1994.
62. Powell, B.C., G.E. Rogers. Hard keratin IF and associated proteins. In: *Cellular and Molecular Biology of Intermediate Filaments,* Goldman, R.D., P.M. Steinert, eds., New York: Plenum Press, 1990, pp 267–300.
63. Powell, B.C., G.E. Rogers. Cyclic hair-loss and regrowth in transgenic mice overexpressing an intermediate filament gene. *EMBO J.* 9:1485–1493, 1990.
64. Powell, B.C., A. Nesci, G.E. Rogers. Regulation of keratin gene expression in hair follicle differentiation. *Ann. NY Acad. Sci.* 642:1–20, 1991.
65. Bawden, C.S., C.J. McLaughlan, S.K. Walker, P.A. Speck, B.C. Powell, M.J. Huson, L.N. Jones, G.E. Rogers. Improvement of wool quality by transgenesis. In: *Molecular Farming,* Toutant, J.P., E. Balazs eds., Paris: INRA, 2001, pp 67–76.
66. Bawden, C.S., B.C. Powell, S.K. Walker, G.E. Rogers. Expression of a wool intermediate filament keratin transgene in sheep fiber alters structure. *Transgenic. Res.* 7:273–287, 1998.
67. Wagner, J.F., E.L. Veenhuizen. Growth performance, carcass composition and plasma hormone levels in wether lambs when treated with growth hormone and thyrotropin. *J. Anim. Sci.* (1)45:379, 1978.
68. Wolfrom, G.W., R.E. Ivy, C.D. Baldwin. Effects of growth hormone alone and in combination with Ralgro (Zeranol) in lambs. *J. Anim. Sci.* (1)60:249, 1985.
69. Heird, C.E., F.M. Hallford, R.A. Spoon, D.W. Holcombe, T.C. Pope, V.H. Olivares, et al. Growth and hormone profiles in fine-wool ewe lambs after long-term treatment with ovine growth hormone. *J. Anim. Sci.* (1) 66:201, 1988.
70. P.M. Harris, B.W. McBride, M.P. Gurnsey, B.R. Sinclair, J. Lee. Direct infusion of a variant of insulin-like growth factor-I into the skin of sheep and effects on local blood flow, amino acid utilization and cell replication. *J. Endocrin.* 139:463–472, 1993.
71. Damak, S., H.-Y. Su, N.P. Jay, D.W. Bullock. Improved wool production in transgenic sheep expressing insulin-like growth factor-1. *Bio/Technology* 14:185–188, 1996.
72. Su, H.-Y., N.P. Jay, T.S. Gourley, G.W. Kay, S. Damak. Wool production in transgenic sheep: results from first-generation adults and second-generation lambs. *Anim. Biotechnol.* 9:135–147, 1998.
73. Muller, M., G. Brem. Disease resistance in farm animals. *Experientia* 47:923–934, 1991.
74. Muller, M., G. Brem. Transgenic strategies to increase disease resistance in livestock. *Reprod. Fertil. Dev.* 6:605–613, 1994.
75. Arnheiter, H., S. Skuntz, M. Noteborn, S. Chang, E. Meier. Transgenic mice with intracellular immunity to influenza virus. *Cell* 62:51–61, 1990.
76. Kolb, E., E. Laine, D. Strehler, P. Staeheli. Resistance to influenza virus infection of Mx transgenic mice expressing Mx protein under the control of two constitutive promoters. *J. Virol.* 66:1709–1716, 1992.
77. Muller, M., B. Brenig, E.-L. Winnacker, G. Brem. Transgenic pigs carrying cDNA copies encoding the murine Mx1 protein which confers resistance to influenza virus infection. *Gene* 121:263–270, 1992.
78. Lo, D., V.G. Pursel, P.J. Linton, E. Sandgren, R. Behringer, C. Rexroad, Jr., R.D. Palmiter, R.L. Brinster. Expression of mouse IgA by transgenic mice, pigs and sheep. *Eur. J. Immunol.* 21:1001–1006, 1991.
79. Weidle, U.H., H. Lenz, G. Brem. Genes encoding a mouse monoclonal antibody are expressed in transgenic mice, rabbits and pigs. *Gene* 98:185–191, 1991.
80. Castilla, J., B. Pintado, I. Sola, J.M. Sanchez-Morgado, L. Enjuanes. Engineering passive immunity in transgenic mice secreting virus-neutralizing antibodies in milk. *Nat. Biotechnol.* 16:349–354, 1998.

81. Salter, D.W., L.B. Crittenden. Transgenic chickens: insertion of retroviral vectors into the chicken germline. *Theor. Appl. Genet.* 77:457–461, 1989.
82. Narayan, O., L.C. Cork. Lentiviral diseases of sheep and goats: chronic pneumonia, leukoencephalomyelitis and arthritis. *Rev. Infect. Dis.* 7:89–98, 1975.
83. Clements, J.E., R.J. Wall, O. Narayan, D. Hauer, R. Schoborg, D. Sheffer, A. Powell, L.M. Carruth, M.C. Zink, C.E. Rexroad, Jr. Development of transgenic sheep that express the visna virus envelope gene. *Virology* 200:370–380, 1994.
84. Kerr, D.E., K. Plaut, A.J. Bramley, C.M. Williamson, A.J. Lax, K. Moore, K.D. Wells, R.J. Wall. Lysostaphin expression in mammary glands confers protection against staphylococcal infection in transgenic mice. *Nat. Biotechnol.* 19:66–70, 2001.
85. Golovan, S.P., M.A. Hayes, J.P. Phillips, C.W. Forsberg. Transgenic mice expressing bacterial phytase as a model for phosphorus pollution control. *Nat. Biotechnol.* 19:429–433, 2001.
86. Golovan, S.P., R.G. Meidinger, A. Ajakaiye, M. Cottrill, M.Z. Wiederkehr, D.J. Barney, C. Plante, J.W. Pollard, M.Z. Fan, M.A. Hayes, J. Laursen, J.P. Hjorth, R.R. Hacker, J.P. Phillips, C.W. Forsberg. Pigs expressing salivary phytase produce low-phosphorus manure. *Nat. Biotechnol.* 19:741–745, 2001.

2.20

Enzyme Technology for the Dairy Industry

Kieran Kilcawley

CONTENTS

20.1 INTRODUCTION

The application of enzymes in dairy technology is well established and the use of enzymes to coagulate milk in cheese making is probably the earliest use of enzyme technology in the food industry (1). In comparison to other industries, the use of exogenous enzymes in the dairy industry is limited to a few major and some minor applications. However, enzymes have a role in many dairy applications as they are inherently present in milk or are incorporated through the addition of, or advantageous growth of, microorganisms. This chapter is a summary of the role enzymes play in the dairy industry, with emphasis on indigenous and exogenous enzymes.

Enzyme technology in the dairy industry can be divided into two main areas: indigenous enzymes in milk and their impact on dairy products, and the use of exogenous enzymes in the manufacture of dairy products.

20.2 INDIGENOUS ENZYMES OF BOVINE MILK

Bovine milk consists of mainly of water, protein, fat, lactose, salts, and microorganisms (the extent of which depend upon its quality). These microorganisms are a potential enzymatic source that can have positive or negative impacts in dairy applications. Most microorganisms do not survive pasteurization and therefore their potential is limited to products produced mainly from raw milk. The microflora of raw cold stored milk is dominated by Gram-negative, nonlactose fermenting psychrotrophic bacteria, with more than 70% of milk microflora consisting of *Pseudomonas* species. Only after prolonged storage or poor hygienic handling will bacterial enzymes exert an effect on the quality of pasteurized milk. The composition of the microflora of pasteurized cold stored milk is less predictable than that of cold raw milk. The microflora of pasteurized cold stored milk contains psychrotrophic thermoduric bacteria from the raw milk and from post pasteurization contamination, as well as psychrotrophic nonthermoduric post pasteurization bacteria (2).

Numerous indigenous enzymes are present in raw milk; approximately sixty have been reported to date. Indigenous milk enzymes arise mainly from three potential sources; blood, secretary cell cytoplasm, and the milk fat globule membrane. Most indigenous enzymes are inactive in milk due to inappropriate environmental conditions or lack of suitable substrates on which to act, and therefore do not present a major concern in milk itself; however, depending upon the application of the milk, these enzymes can exert an effect. Most indigenous milk enzymes and sources of enzymes, such as microorganisms, are destroyed by pasteurization to create a standardized product (3).

This section deals with those indigenous enzymes that can be exploited to aid certain dairy applications, or those that could have a major negative impact in dairy applications if not controlled.

20.2.1 Lipases

Tainting of milk can be a serious problem for its market acceptability. A main source of taints is hydrolytic rancidity. Lipolysis in milk has been widely studied and is due to the hydrolytic degradation of milk triacylglycerols or triglycerides, or mono- and diglycerides producing free fatty acids (FFAs). Due to their high volatility and low flavor thresholds, FFAs can impart specific flavors at low concentrations, Therefore, in certain instances, low levels of lipolysis can have a marked impact on flavor (4). In general the lower chain FFAs ($C_4 - C_8$) impart rancid flavors, while longer chain FFAs impart soapy flavors. The terms lipase and esterase are widely used to describe enzymes which degrade fat, and it is generally accepted that these enzymes are identical apart from the nature of their catalytic site which governs their selectivity (5). In summarizing esterase and lipase activities it is assumed that esterases act on soluble fat substrates with ester linkages of fatty acids of less than 8 carbon atoms, while lipases release fatty acids with ester linkages of more than ten carbon atoms in emulsified systems. The term lipase is used to represent both lipases and esterases in this chapter.

Milk has been described as an oil-in-water emulsion, where the fat globules are evenly dispersed in the serum phase. The fat globules are surrounded by a thin membrane called the milk fat globule membrane (MFGM), and it is this membrane which acts as a barrier between the lipolytic enzymes and the fat (Figure 20.1). The MFGM can be damaged by agitation, homogenization, foaming, or poor handling, which can result in the rapid interaction of enzyme and substrate, causing the development of off flavors (4).

20.2.1.1 Lipoprotein Lipase

The source of lipase or esterase activity in milk is mainly a milk lipase known as lipoprotein lipase (LPL) (EC 3.1.1.3). Indigenous LPL is present in milk via leakage through the

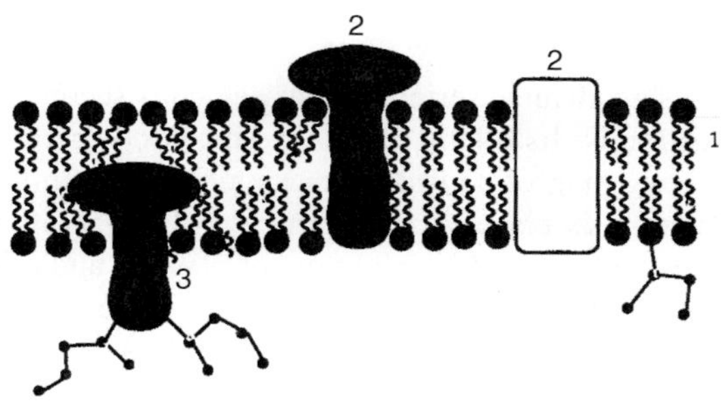

Figure 20.1 Schematic representation of a trilaminar cell membrance which is from the apical membrance of the mammary cell and forms the outer layer of the milk fat globule membrane following expression from the mammary cell, but which is more or less extensively lost on ageing.
1. Phospholipid/glycolipid
2. Protein
3. Glycoprotein

mammary cell membrane from the blood where it is involved in the metabolism of plasma triglycerols (6). LPL is activated by an apolipoprotein activator, apo-CII, and is present in high enough concentrations in milk to be of real danger in terms of its spoilage capability. LPL is mainly associated with casein and therefore is compartmentalized away from the fat by the MFGM. LPL has been shown to have 1, 3 regio-selectivity with little or no acyl specificity (7). LPL is susceptible to standard high temperature short time (HTST) pasteurization; however it is thought that it is not completely inactivated and may be active in some milk products (8). The fact that most LPL is inactivated by pasteurization is demonstrated by the fact that routine homogenization of milk immediately after or prior to pasteurization disrupts the MFGM and does not cause hydrolytic rancidity. The levels of LPL in milk are subject to variations, and are dependent upon levels in the blood serum in milk. It is thought that these levels vary in relation to the stage of lactation of the cow. For a more detailed description of LPL, see Olivecrona & Bengtsson-Olivecrona (1991) (6).

20.2.1.2 Bacterial Lipases

Extracellular lipases produced by psychrotrophic bacteria have the potential to cause hydrolytic rancidity in milk and milk products as these dominate in typical cold bulk milk storage conditions. Lipases of psychrotrophic bacteria have varying heat stability and some can survive pasteurization, however lipolytic spoilage rarely occurs in pasteurized milk (4). Studies have shown that lipases from psychrotrophs which survived pasteurization affected cheese quality if counts were reported to be greater than 10^6 cfu ml^{-1} in the raw milk (9).

20.2.1.3 Phospholipases

Phospholipase activity has been identified in milk LPL and in several bacteria. Phospholipases are potentially very important in milk as they degrade the phospholipids of the MFGM, thereby making the fat susceptible to hydrolysis. These enzymes are also quite heat stable and are not destroyed by pasteurization (4).

20.2.2 Proteinases

A few indigenous milk proteinases have been identified in bovine milk, but only two appear to be of importance in the dairy industry.

20.2.2.1 Plasmin or Milk Alkaline Proteinase

Plasmin (EC 3.4.21.7) is an alkaline serine proteinase and is secreted as plasminogen that is activated in blood and milk. Its role in blood is to break down blood clots and is a part of a complex system consisting of activators and inhibitors. In milk, there is about four times as much plasminogen as plasmin, and both, as well as plasminogen activators, are associated with the casein micelles. The inhibitors of plasmin and plasminogen activators are present in the milk serum. The concentration of plasmin and plasminogen in milk increases with advancing lactation, mastitic infection, and number of lactations (3).

Plasmin has a high specificity for peptide bonds with a carboxyl group supplied by lysine or arginine. It has an optimum activity at pH 7.5 and 35°C, and still exhibits about 20% activity at 5°C. It is active over a wide pH range from pH 4 to 9 and is quite heat stable. Its activity in milk increases after HTST pasteurization, probably due to inactivation of its inhibitors, and partly survives ultra high temperature (UHT) sterilization. Plasmin is associated with casein micelles in milk and degrades β-, α_{s1}-, and α_{s2}-caseins to γ-caseins, proteose–peptones, and possibly λ-caseins (3).

Plasmin can have a positive or negative effect on cheese making as it is involved in proteolysis during ripening, and is also linked to poor curd formation in cheese made from late lactation milks (8). Plasmin is also associated with the generation of "gluey" and "bitter" off flavors during storage of UHT milks and is a major factor affecting their shelf life (10). Plasmin aids in the initial hydrolysis of the major milk caseins, and recent studies suggest that the addition of urokinase to cheesemilk can accelerate cheese ripening by the activation of plasminogen to plasmin in the cheese (11).

For a more detailed review of plasmin activity in milk and dairy products see Fox and Stepaniak (1993); Fox (1992) and BastianBrown (1996) (8,12,13).

20.2.2.2 Cathepsin D or Acid Milk Proteinase

Cathepsin D (EC 3.4.23.5) is a lysozomal enzyme that has a similar specificity to chymosin, but has very poor milk clotting activity. It is relatively heat labile as it is inactivated by 70°C for 10 min, and may contribute to proteolysis in cheese, but not significantly (14).

20.2.3 Phosphatases

Milk contains a number of phosphatases, two of which are of technological significance.

20.2.3.1 Alkaline Phosphomonoesterase (Phosphatase)

Alkaline phosphatase (EC 3.1.3.1) occurs in the milks of all mammals, although the levels present may vary considerably. This enzyme is of technological importance to the dairy industry as the test for its presence in pasteurized milk is universally used as an index of the efficiency of HTST pasteurization. This is because alkaline phosphatase is always present in raw milk at easily measurable levels and its heat stability profile closely follows that required for adequate pasteurization (15). Alkaline phosphatase is concentrated in the fat globule membrane and has an optimum activity in the alkaline pH range of pH 9–10.5 at 37°C, and will hydrolyze most phosphate ester bonds, releasing inorganic phosphate including adenosine monophosphate, glycerophosphate, phosphates of glucose, serine, threonine, and phosphoproteins. The enzyme can be reactivated, particularly in the presence of magnesium and zinc ions, and this is of practical significance, because regulatory tests for pasteurization assume the absence of phosphatase activity. Generally bulk HTST milk does not show reactivation of phosphatase activity, however it is often observed in UHT treated milk. This problem is usually overcome by HTST pasteurization after UHT treatment. Overall alkaline phosphatase activity does not have a direct significance in milk or milk products (2).

20.2.3.2 Acid Phosphomonoesterase (Phosphatase)

Acid phosphatase (EC 3.1.3.2) is present in milk at approximately 2% of the level of alkaline phosphatase, but is quite heat stable, as 30 min of heating at 88°C is required for complete inactivation. The enzyme is a glycoprotein and has a lower optimum pH than that of alkaline phosphatase (pH 5). Acid phosphatase hydrolyses aromatic phosphomonoesters. Acid phosphatase has broad substrate specificity, and has a very different substrate specificity compared to milk alkaline phosphatase. The enzyme is active against caseins, although the major caseins act as competitive inhibitors, probably due to binding of the enzyme to the casein phosphate groups. Therefore, the effectiveness of the caseins as inhibitors is related to their phosphate content. The presence of this enzyme in milk and dairy products may be of significant importance as it affects casein micelle formation through dephosphorylation of casein, which destroys interchain bridging via phosphate groups between casein polypeptides and calcium ions. Because many of the technologically important properties of milk depend upon the presence of casein micelles and the maintenance of their integrity, the significance of acid phosphatase in the gelling of UHT milk, coagulation of milk by chymosin, curd structure, and cheese ripening have all been investigated (15). Overall, acid phosphatase has the potential to affect the quality of dairy products, but its effect has not been fully elucidated. Some dephosphorylated peptides have been found in certain cheese varieties, but they may be the result of action by phosphatases produced by bacteria. It may be that the low activity and ease of inhibition of acid phosphatase may impinge on its overall effect (3).

20.2.4 Lactoperoxidase

Peroxidases (EC 1.11.1.7) are present in high quantities in bovine milk and have been shown to be involved in antibacterial activity. Lactoperoxidase requires hydrogen peroxide (H_2O_2) and thiocyanate to cause bacterial inhibition by forming hypothiocyanite and other unidentified antibacterial agents. The mode of action of this system is well understood and has been used to cold pasteurize milk in situations where refrigeration or thermal pasteurization is unavailable. Lactoperoxidase is added to milk replacers for calves or piglets to reduce the incidence of enteritis. Lactoperoxidase inhibits starter cultures, which accumulate H_2O_2 through their oxygen metabolism, but has also been shown to stimulate some strains of Lactococci. Lactoperoxidase can cause nonenzymatic oxidation of unsaturated fatty acids (3,16).

20.2.5 Lysozymes

Milk contains an indigenous lysozyme, also know as muramidase (EC 3.2.1.17), which causes lysis of certain bacteria by hydrolyzing cell wall polysaccharides. Lysozyme hydrolyses the β(1-4)-linkage between muramic acid and N-acetylglucosamine of mucopolysaccharides of the bacterial cell wall, thereby causing cell lysis. Lysozyme is fairly heat stable in acid conditions, but heat labile at pH 7 or above. Lysozyme is effectively an antibacterial agent and is also used exogenously in cheese making. As levels of indigenous milk lysozyme are so low, it is thought not to have a significant effect on the shelf life of milk (3).

20.2.6 Xanthine Oxidase

Xanthine oxidase (EC 1.2.3.2) is a broad specificity oxido-reductase, concentrated in the MFGM where it is one of the principal proteins. Milk processing treatments can affect its activity, when it is transferred from the fat phase to the aqueous phase. All of the major milk proteins can act as either activators or inhibitors of xanthine oxidase, depending on their concentration, and may have some significance in the activation, inactivation, and reactivation of the enzyme. The pH optimum is about 8.5 and the enzyme requires flavin, adenine, dinucleotide, and other cofactors to work. The presence of xanthine oxidase in milk is associated with oxidative rancidity, as it can excite stable oxygen to from a prooxidant (3). Xanthine oxidase has also been implicated as having bacteriostatic properties, as it is thought to be secondarily involved in generating H_2O_2 for the lactoperoxidase system (17).

20.2.7 γ-Glutamyl Transpeptidase (Transferase)

γ-Glutamyl transpeptidase (EC 2.3.2.2) catalyses the transfer of γ-glutamyl residues from γ-glutamyl-containing peptides. It plays an important role in amino acid transport in the mammary gland. It may have a role in cheese making as γ-glutamyl peptides have been isolated from cheeses which have arisen through the activity of this enzyme (3).

20.2.8 Other Enzymes

A number of other enzymes have been identified in milk, but do not directly influence milk or milk products. For further information see Farkye (1991) (17).

20.3 EXOGENOUS ENZYMES USED IN THE DAIRY INDUSTRY

The use of enzymes in the dairy industry for commercial applications is, in general, fairly limited. Even then, the use of milk-clotting enzymes in the production of cheese and the use of lipases in cheese manufacture are easily the most widely used applications. However, the uses of enzymes in many minor applications in the dairy industry are significant. Most applications use proteinases or lipases because protein and fat are the major milk constituents.

20.3.1 Exogenous Enzymes – Role in Natural Cheese Making

20.3.1.1 Milk Clotting Enzymes (Rennet and Coagulants) – Role in Natural Cheese Making

Milk clotting enzymes for cheese manufacture have been used for thousands of years and go back to the earliest days of cheese making. Initially, unpurified crude preparations extracted from the stomachs of ruminants were used to coagulate milk. These were followed by the use of microbes and plant extracts. It is relatively unimportant in this context to go into any detail regarding the use of these crude preparations in cheese making. Suffice to say that it was probably the first recorded instance of the use of enzymes in the development of foods.

Some confusion has occurred over the years regarding nomenclature of milk clotting enzymes. Initially, milk clotting enzymes which were derived from the fourth stomach of calves were called chymosin, derived from "chyme," which means "gastric liquid" in Greek. In 1890 the word "rennet" was used and found wide acceptance in English speaking countries; however because of confusion with the proteolytic enzyme "renin" the enzyme was renamed chymosin in 1992 by the International Union of Biochemistry and Molecular Biology. However, as all cheeses are not produced using milk clotting enzymes from ruminant sources it was felt that milk clotting enzymes from different sources should be further classified. Therefore it is now generally accepted that the word "rennet" should be reserved for milk clotting enzyme preparations from ruminant stomachs, and that milk clotting enzymes from sources other than ruminant stomachs or genetically modified organisms (GMO) should be named "coagulants." Milk clotting enzymes derived from GMOs have had to be further categorized, and are now called "fermentation-produced chymosin" (FPC) (18).

Commercially available rennet has existed for many years, and a standardized chymosin was first introduced by Chr. Hansen A/S in Denmark in 1874 (18). Commercially available microbial and fermentation produced milk clotting enzymes have become widely available in recent years to meet increased demand. Other sources of milk clotting enzymes are used, but only in specialized cases. The use of plant sources of coagulants is generally limited to sources from dried cardoon flower heads, and is found mainly in Portugal. Other plant sources of coagulants exist but contain other proteinase activities, which impair milk coagulation. Some other nonbovine animal sources of rennet have also been used, but again their modern use is limited to specialized cheese production or used for religious reasons (19).

It is important to understand the molecular aspects of the different milk clotting enzymes to differentiate between them. All of these enzymes are aspartic proteinases (EC 3.4.23) and therefore have identical catalytic mechanisms, containing two aspartic residues in their catalytic site (20). Distinct differences exist between the various sources of milk clotting enzymes and have been well characterized to date. For a detailed review see Harboe and Budtz, (1999) (18).

The action of milk clotting enzymes is the first step in cheese making and results in the splitting of κ-casein which causes the destabilization of casein micelles and subsequently leads to the formation of a coagulum. Milk contains four major types of casein, α_{s1}-, α_{s2}-, β-, and κ-caseins which, along with calcium phosphate, form spherical structures called casein micelles which are colloidal aggregates and are stabilized by κ-casein. It is generally accepted that these micelles are formed by many submicelles held together by colloidal calcium phosphate bridges, hydrophobic interactions, and hydrogen bonding. α- and β-caseins are calcium sensitive and interact to form the core of the micelles through hydrophobic interactions with κ-casein, which binds calcium weakly on the surface of the micelles. κ-casein provides a stabilizing influence though the nature of its peptide sequence and consists of two parts, a hydrophobic part known as para-κ-casein (consisting of amino acids 1–105), and a hydrophilic part known as the glycomacropeptide or caseinomacropeptide (consisting of amino acids 106–169). Para-κ-casein is bound to the core proteins, with the caseinomacropeptide (CMP) projecting out of the micelle surface, which is negatively charged. This negative charge exerts a steric hindrance preventing micelles from coming together. It is at the junction between the para-κ-casein and the CMP that the enzyme chymosin acts, splitting κ-casein into its two components and thereby destabilizing the casein micelle structure by reducing intermicellar repulsive forces. Once approximately 80 to 85 % of κ-casein is hydrolysed, the caseins start to form a coagulum which retains fat and expels most whey proteins and some water, thereby ending the first phase of cheese making (Figure 20.2) (18,21). For a detailed account, see Fox & Grufferty (1991) and Guinee & Wilkinson (1992) (21,22).

20.3.1.1.1 Rennet – Milk Clotting Enzymes Derived from Animal Sources Bovine rennet consists of two main proteinases, the milk clotting enzyme chymosin and pepsin. The proportion of chymosin to pepsin is age dependent and varies between individual animals. For young calves, abomasum extracts can contain 80–90% chymosin, and adult animals can contain as much as 90% pepsin. Therefore, throughout the world the composition of rennet can vary markedly. Most commercial suppliers offer standardized rennets, which are available in liquid, powder, or paste form, and are stable for a period of months at given temperatures. The rennet is produced from either fresh, dried, or frozen stomachs, which are cut and minced. The rennet is extracted by water, buffer, or salts. The liquid component is separated by filtration or centrifugation and the extract consists of active enzymes and proenzymes. The proenzymes are activated by acid, usually at pH 2, and the pH is subsequently raised to solubilize protein. The proenzymes are then reclarified by centrifugation or filtration (18). Calf rennet is seen as the standard reference product to which all other milk clotting enzymes are compared, and is widely used in many countries. Lamb and kid-goat rennets are also used, but mainly in countries with traditional cheeses made from sheep or goat's milk. Lamb and kid-goat rennets are similar to calf rennet in activity and are produced by maceration and drying of the stomachs filled with milk to produce paste-type products.

Chymosin (EC 3.4.23.4) has a specific milk clotting activity and in calves is found in two allelic forms, A and B. Subtle differences exist between these enzymes, but for cheese making purposes they are all but identical. In adult animals, pepsin A (EC 3.4.23.1) is the predominant proteinase and has a broader specificity and higher pH dependency than

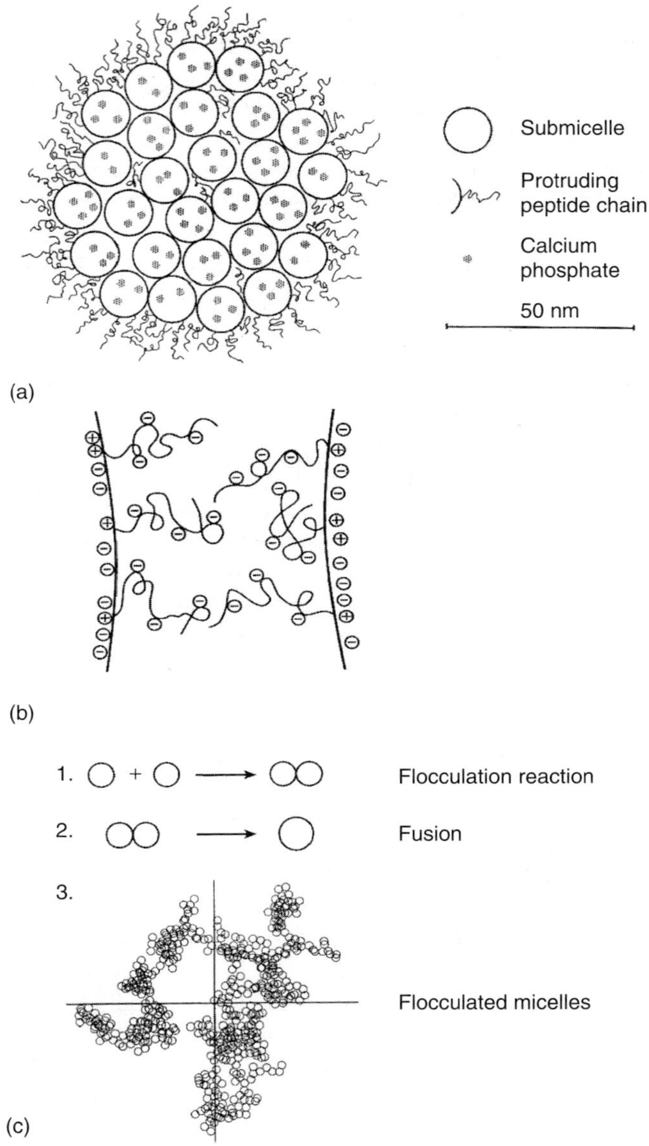

Figure 20.2 (a) Schematic of casein micelle aggregation, (b) Hypothetical diagram of electrostatic charges between two casein micelles, (c) Schematic model of a cross-section of a casein micelle.

chymosin. Two other proteinases exist in rennets and both are found in minor amounts in porcine stomachs: pepsin B (EC 3.4.23.2) and gastricin (EC 3.4.23.3), with the latter also found in adult bovine stomachs.

20.3.1.1.2 Coagulants – Milk Clotting Enzymes Derived From Microbes The declining supply of calves for slaughter and resulting chronic shortages and price increases fuelled the search for alternate sources of milk clotting enzyme (22). Various microorganisms were assayed for their suitability and approval for their use in cheese making was granted in the 1960s. Today three commercially available microbial coagulants exist which are all derived from fungal sources. The predominant coagulant is from

Rhizomucormiehei and exists as three distinct types; the other sources are from *Rhizomucor pusillus* and *Cryphonectria parasitica*. *Rhizomotor miehei* coagulants exist in three different types; the native type L which is heat stable and more proteolytic than calf rennet; the destabilized version type TL which is made by oxidation of the native enzyme and is heat labile and less proteolytic than type L; type XL which is the most heat labile and even less proteolytic. The *R. pusillus* coagulant is similar to the type L *R. miehei* product, with the *C. parasitica* coagulant being the most proteolytic and having a low pH dependency (18).

These coagulants are produced by fermentation where the proenzyme is converted to its active form during production by the slightly acidic environment. Production is typically by submerged batch fermentation, where a mother culture is propagated on selected media and grown to produce sufficient cells to inoculate a large scale culture. Production takes a number of days with process parameters closely controlled. The fungus is removed by filtration or centrifugation and the coagulant concentrated by ultrafiltration and clarified by filtration. The activity of these coagulants is standardized and stabilized to ensure product quality and consistency (19).

20.3.1.1.3 Fermentation Produced Chymosin – Milk Clotting Enzymes Produced Using Genetic Engineering The need for large amounts of standard milk clotting enzymes to meet the needs of the modern dairy industry brought about the development of fermentation produced chymosin (FPC) which is chymosin produced by fermentation brought about by an organism genetically modified with a gene for chymosin. Manufacturers claim that these enzymes give more consistent milk clotting and proteolytic activities and therefore offer a more standardized product. They are widely used in the USA and in some European countries; however some countries have continued to use standard calf rennet, presumably due to fear of affecting market share due to negative consumer feedback from the use of GMO products in foodstuffs. Two of these products are produced commercially containing chymosin from calf abomasum that has been expressed and produced in *Aspergillus niger* or *Kluyveromyces lactis*. A third product, containing calf chymosin produced by *Escherichia coli* is no longer commercially available. Another product commercially available contains the *R. miehei* coagulant proteinase expressed and produced in *Aspergillus oryzae*.

20.3.1.1.4 Effects of Milk Clotting Enzymes on Cheese Making Obviously some differences exist between the actions of different milk clotting enzymes; however in most circumstances they may have little influence on the overall cheese making process. Experiments with Cheddar cheese have shown that there was no advantage in terms of yield or quality of cheese produced with standard calf rennet and FPC (24). No significant differences were noted in yields between Cheddar cheeses made using identical procedures where the pH at whey drainage was 6.15 between standard calf rennet, the coagulants *R. miehei*, *R. pusillus*, and FPC products. (T. Guinee, personal communication). However, this does not support the conclusion that differences in yield or flavor and texture will not exist in cheese making with lower drainage pH or for different cheese varieties. An important aspect of differences between these milk clotting enzymes is their thermostablity, which differs considerably. Thermal stability is important when the whey is to be used in processing, as up to 90% of the enzyme can be lost to the whey during cheese making (18,25). Studies have shown that less than 2% of these enzymes remain in whey after normal low pasteurization at pH 6.0, which is acceptable for all uses of whey and whey products (18). The purity of milk clotting preparations is also an issue as some enzymatic side activities may survive pasteurization and act on whey proteins or in products containing whey proteins, which is particularly the case for starch degrading enzymes. The FPC coagulants do not have this inconvenience, because of their high purity. The cooking and scalding of curds in different cheese manufacturing procedures will also affect the activity of these enzymes in the curd. This is particularly relevant for Mozzarella and Swiss cheeses and is an issue which should be considered by the cheese maker, because of the range in thermostabilities of the different milk clotting enzymes available.

20.3.2 Lipase – Role in Natural Cheese Making

Lipases and esterases (EC. 3.1.1.3) are enzymes which hydrolyse tri-, di-, and monoglycerides at an oil-in-water interface. Lipases have different selectivity, which is dependent upon a number of factors such as source, concentration, and the environment in which they are active. Lipases can be grouped according to their selectivity: some are substrate specific in that they hydrolyse only monoacylglycerols, mono-and diacylglycerols, or just triacylglycerols. Others are regiospecific, i.e., are Sn-1, Sn-3 regioselective, or Sn-2 regioselective. Some are nonspecific, showing no preference for substrate or acyl position, while others show acyl specificity in that they have a preference for certain fatty acids (5).

Lipases are very important in the formation of aroma and flavor substances in certain cheese varieties, but uncontrolled lipolysis can also lead to major flavor defects and result in downgrading of cheese. The impact of lipolysis can be critical in natural cheese making and needs to be controlled in all cases. A major problem with lipolysis in cheese is because milk fat contains high levels of short chain fatty acids, which when in free form are very volatile and have low flavor thresholds. High levels of short chain FFAs have been implicated with the development of rancidity, while high levels of large chain FFAs have been implicated with soapy off flavors. The extent of lipolysis varies markedly between different cheese varieties as does its significance. Table 20.1 lists typical FFA levels in some well known cheese varieties.

Lipolysis in natural cheese originates traditionally from three sources: raw milk (containing indigenous milk lipases and microbial sources), rennet paste (pregastric esterases), and microorganisms (molds, yeasts, and bacteria).

20.3.2.1 Indigenous Milk Lipases – Role in Cheese Making

Indigenous milk lipase LPL has a role to play in cheese making, particularly in raw milk cheeses and to a much lesser extent in cheese made from pasteurized milk, as most are inactivated by pasteurization.

20.3.2.2 Rennet Paste – Role in Cheese Making

Rennet paste is almost exclusively used in the manufacture of Italian cheese varieties and is a combination of the milk clotting enzyme rennet and pregastric lipases, which contribute to the required flavor and aroma development. Traditional methods of manufacture consisted of feeding calves milk, then slaughtering them, removing the stomachs, drying, and macerating them. These pastes were often used in the manufacture of Italian cheese varieties in the USA, but in 1945 the FDA prohibited the use of imported rennet paste from Italy due food safety concerns. This brought about the development of purified lipases from ruminants, which were derived from salvia glands and pancreatic tissues (pregastric and gastric lipases). Oral or salvia lipase was found to be the lipase of interest in rennet paste and is now widely used, produced in powdered format. It can be used with rennet to provide a safe standardized alternative to traditional rennet pastes. Conversion to pregastric lipase brought about an improvement in the quality of Italian cheese varieties produced from them (19). The advantages of using pregastric lipase instead of rennet paste are improved microbiological quality, improved standardization of the cheese, and the ability of adding the lipase independently of the rennet. Traditional rennet pastes are still used in some regions of Italy today.

Table 20.1
Total Free Fatty Acid Content of Various International Cheeses

Cheese Type	Variety	FFA (mg/kg)
Extra-hard cheeses	Parmesan	4,993[a]
	Romano	6,754[b]
Hard cheeses	Cheddar	1,028[b]
	Cheshire	1,265[c]
	Idiazabal	5,557[a]
	Manchego	32,404[d]
	Roncal	8,178[a]
Semi-hard cheeses	Caephilly	1,253[c]
	Colby	550[b]
	Mahon	8,743[a]
	Majorero	20,794[a]
	Monterey Jack	736[b]
	Port Salut	700[b]
Cheeses with 'eyes'	Edam	356[b]
	Emmenttal	2,206[c]
	Gruyere	1,481[b]
Cheeses with internally ripened moulds	Blue (US)	32,230[b]
	Roqueforti	32,453[b]
Cheeses surfaced-ripened with moulds	Brie	2,678[b]
	Camembert	681[b]
Cheeses surface-ripened with bacteria	Brick	2,150[b]
	Limburger	4,187[b]
	Munster	6,260[e]
Pasta-filata cheeses	Mozzarella	363[b]
	Provolone	2,118[b]

[a] De la Feunte, M.A., J. Fontecha, M.Juarez. Fatty acid composition of the triglyceride and free fatty acid fractions in different cows-, ewes- and goats-milk cheeses. *Z. Lebensm. Unters. Forsch.* 196: 155–158, 1993.

[b] Woo, A.H., S.Kollodge, R.C. Lindsay. Quantification of major free fatty acids in several cheese varieties. *J. Dairy Sci* 67: 874–878, 1984.

[c] McNeill, G.P., J.F. Connolly. A method for the quantification of individual free fatty acids in cheese: application to ripening of cheddar-type cheeses. *Irish J. Food Sci. Technol.* 13: 119–128, 1989.

[d] Poveda, J.M., M.S. Perez-Coello, L.Cabezas. Evolution of the free fatty acid fraction of Manchego cheese during ripening. Milchwissenschaft 54. 685-687, 1999.

e de Leon-Gonzalez, L.P., W.L. Wendorff, B.H. Ingham, J.J. Jaeggi, K.B. Houck. Influence of salting procedure on the composition of Muenster-type cheese. *J. Dairy Sci* 83: 1396–1401, 2000.

20.3.2.3 Pregastric Esterases (PGE) – Role in Cheese Making

Semipurified sources of PGE are now widely used in making Italian-type cheeses, and are required to provide the sharp piquant flavor associated with Pecorino, Romano, and Provolone cheeses. This piquant flavor is primarily due to short chain FFAs arising from the action of PGE. PGEs are sometimes used in the production of Romano, Caciotta, Feta, and Manchego cheeses, which are typically made from ewe's or goat's milk. However, in shortages of ewe's and goat's milk, cow's milk is often added and the lipases are required to give the desired piquant flavor associated with the use of ewe's and goat's milk. The differences in flavor from ewe's and goat's milk in comparison to cow's milk arise from differences in their fatty acid composition (26). Varieties such as Fontina, Ras, and

Kopanisti, which were traditionally produced from raw milk, now have added lipases to replace indigenous milk lipases destroyed during pasteurization. Lipases can be added at very low levels to varieties such as, Mozzarella, Parmesan, Romi, and Cheddar cheeses to enhance flavor and aroma development (19).

PGEs are regioselective for the Sn1 and Sn3 positions of milk triglycerides and have a particular affinity for short chain fatty acids at the Sn 3 position. This ensures the release of the most volatile free fatty acids, which by their nature have the greatest impact on flavor and aroma. Slight differences exist between the specificity of PGEs from calves, kid-goats, and lambs and therefore give differences in the cheeses made from these sources. For a detailed review of PGE see Nelson, Jenson, and Pitas (1977) (27).

20.3.2.4 Microbial Lipases – Role in Cheese Making

Microbial lipases are lipolytic enzyme preparations, which are derived from yeasts, molds, and bacteria. In terms of natural cheese making, molds play a direct role in flavor and aroma development in some varieties, predominately through the action of lipases. Surface ripened and blue veined cheeses rely on lipases from molds added during their preparation to aid in the development of their characteristic appearance, flavor, and aroma. Extensive lipolysis occurs in many blue veined cheeses and not only does this contribute to flavor by the generation of FFA directly, but these also provide precursors of other flavor compounds, the most important of which are methyl ketones which give Blue cheeses their characteristic flavor. This reaction begins with the release of FFA from tri-acylglycerides by the lipase in the mold, which are oxidized to form β-ketonic acids, and subsequently decarboxylated to produce methyl ketones (28). The lipases originate from the mold *Penicillium roqueforti* used in the manufacture of the cheese, which is generally added to the milk. The cheeses are then pierced using rods, which enables carbon dioxide to be released from the curd and oxygen to penetrate. This together with the correct humidity, salt, and oxygen facilitates germination of the spores. Blue veined cheeses, such as Stilton, Roquefort, Gorganzola, and Danish Blue are cheeses typically made in this manner. The lipase of *P. roqueforti* is Sn-1, Sn-3 regiospecific with acyl specificity for short chain fatty acids (5). Other cheeses utilize surface mold for flavor development, such as Camembert, which uses a different mold, *Penicillium camemberti. Penicililum camemberti* lipase is substrate specific for triacylglycerols (5). For further details on the lipases activity of *P. roqueforti* and *P. camemberti* and their role in the development of different cheese varieties, see Kinsella and Hwang (1976) and Gripon (1993) (29,30).

Lactic acid bacteria (LAB) are essential for the manufacture of most cheeses. LAB are often called starter cultures as they initiate the start of the production of lactic acid, their primary role in natural cheese making. However, they also have a secondary role, which is to contribute to the cheese ripening process by the release of cell wall or intracellular enzymes. In the context of this section, the lipolytic activity of these bacteria is very low, but lipase activity has been identified in many LAB (31). Therefore, the role of this source of lipases in natural cheese making should be addressed. This is also the case for nonstarter lactic acid bacteria (NSLAB) which are predominant in the microbiological content of many cheese varieties during ripening and therefore may also have a role in the level of lipolysis in these cheeses (32).

Yeasts are often present on a variety of cheeses and are known to contribute to lipolysis levels in these cheeses. *Geotrichum candidum* is commonly found on some cheese varieties before salting and is very acylspecific for cis-9 unsaturated fatty acids (32). Other yeast species are present in different cheese varieties; however their lipolytic role in cheese has not been fully characterized.

Although many sources of microbial lipases are available commercially, vey few are used in natural cheese making. Some have been used in accelerated cheese ripening and are used in other dairy process applications.

20.3.3 Accelerated Cheese Ripening

One of the major costs of cheese production occurs with the ripening of cheese, which can take up to two years for some varieties. The high costs associated with ripening are involved in controlling the environment in which the cheeses are stored until ready for distribution to retail outlets. Also, money paid out at day zero is not recouped until months later at variable market dependent prices. In some varieties, costs associated with ripening are not a problem as these cheeses have short ripening periods. In others such as Cheddar, where a mature cheese is typically 12 to 18 months old, costs associated with ripening become a major issue. Therefore, the ability to reduce ripening time while maintaining quality has the potential to offer considerable cost savings. This chapter focuses on Cheddar cheese as it is one of the largest volume cheeses in the world. Various techniques described here to accelerate its ripening can also be applied to other varieties. Although the main driving force for accelerating the ripening of Cheddar cheese has and will remain cost reduction, over recent years other factors have also become important. The power of retail outlets to control the cost and type of cheese required has become an important factor for cheese producers, as the acquisition or loss of contracts to supply multinational retail chains can be the making of or the end of producers. Standards applied to the selection of cheeses by retailers have increased markedly over the years and are themselves in direct response to consumer demands. Consumers are not only interested in cost effective cheese but also in characteristic cheeses, which means that consistency of quality is vital. The development of characteristic Cheddar cheeses leads to brand loyalty, whereby the consumer associates a product from a certain manufacturer with a particular consistent flavor profile. Therefore the producer has to be able to produce a Cheddar cheese with characteristics that distinguish it from other brands, and which are acceptable to a wide range of consumers at a cost effective rate throughout the year, from a variable seasonal raw material, milk. This is not an easy task, considering the number of dynamics which come into play during production of Cheddar cheese.

Over the years, a wide variety of options has been used to accelerate ripening, all with various success rates, but none has yet proven to be the perfect solution (Table 20.2). In biochemical terms, cheese ripening is in itself an enzymatic process; therefore the solution to achieving the goals previously outlined involves, at least in part, an enzymatic system, whether involving the use of enzymes directly or indirectly through the use of LAB.

20.3.3.1 Nonenzymatic Methods For Accelerating Cheese Ripening

Nonenzymatic methods exist for accelerating cheese ripening and are mentioned here because they can be used in association with enzymatic methods. The first is elevated temperature ripening, which provides a better environment for the enzymes to work, thereby accelerating ripening. This mechanism has been favored in recent years, mainly because it is the most cost effective. Generally, Cheddar cheeses are ripened at 12 to 15°C using this method (as opposed to 8°C), however care needs to be taken to ensure microbial safety and to prevent textural defects. In commercial practice, cheese may be held at different temperatures throughout its ripening to make flavor development coincide with market demands at particular times of the year.

The second method is high pressure treatment, which still in its infancy, but may have the potential to accelerate cheese ripening. Work to date has shown conflicting

Table 20.2
Method for accelerated cheese ripening

Non-enzymatic method	Elevated ripening temperatures
	High Pressure
Enzymatic methods	Exogenous enzymes
	Attenuated starter cultures
	Adjuncts
	Genetic modification

results, with initial work by Yokoyama, Sawamura, and Motobayashi (1992) (33), reporting substantial accelerated ripening for Cheddar and Parmesan cheese. Later studies have not found any major differences between high pressure treatments and controls for other cheese varieties (34,35). However, it is too early to dismiss this as a possible mechanism for accelerating cheese ripening, as it may have a role in combination with other enzymatic technologies. A useful review of the use of high pressure treatment in cheese applications is given by O'Reilly, Kelly, Murphy, and Beresford (2001) (36).

20.3.3.2 Accelerating Cheddar Cheese Ripening Using Enzyme Technology

20.3.3.2.1 Exogenous Enzymes The addition of enzymes in the cheese making process has seen commercial application and has been tried as far back as 1978 (37). The potential of this system is very obvious as it involves adding increased amounts of enzymes during the cheese making process, enabling quicker hydrolysis of the major milk components and thereby reducing its ripening time. Achieving a balanced flavor and texture profile using this technology has proven difficult. Enhancing hydrolysis of the major casein proteins causes bitterness, losses in yield, and weakness in texture. These problems have become less critical with advances in understanding the proteolytic action of various enzymes in cheese flavor and texture development. Initial problems associated with this technology could be traced to uncontrolled protein hydrolysis. It was not until detailed investigations into the mode of action of different proteinases/peptidases were done that the most suitable enzymes were identified. Also, advances in understanding of why bitter peptides accumulate in cheese to cause bitterness, and how this could be alleviated through selection of the right types of peptidase activity, enabled bitterness to become less of an issue associated with this technology. The inclusion of exogenous lipases to accelerate Cheddar cheese ripening has not shown promise, as off flavors develop (38). However, other studies have shown lipolysis to be important in Cheddar cheese flavor development (39). A major difficulty in enhancing lipolytic flavor and aroma in Cheddar cheese is that small increases in the level of lipolysis can have major negative sensory impacts. Therefore the correct dosage level of lipase is critical for success.

Today blends of enzymes or enzyme cocktails are available to accelerate cheese ripening, which contain the necessary enzymes to accelerate ripening without negative effects. However, their use is not wholly widespread as problems can still occur with unbalanced ripening making it difficult to ensure a suitable shelf life of the cheese in the market place. Unbalanced ripening can occur when the enzymes have not been added to the curd uniformly, thereby creating zones in a cheese block where the rate of hydrolysis differs. In fact the point of addition of exogenous enzymes for the accelerating ripening of cheese is a major hurdle in the development of this technology. Differences in production techniques among different factories are also an issue, and manufacturers of these enzyme preparations work closely with individual cheese makers to try to alleviate problems that may occur due to differences in protocols and equipment. The issue of shelf life is important, as it may be possible to produce a cheese with typical characteristics of a cheese twice its age using exogenous enzymes, but stopping the rate of accelerated ripening, enabling the cheese to have a required shelf life, is also an issue that has not been wholly resolved. It appears that advances are being made in this area, and may involve changes in temperature and pH, and the use of starter culture technology, to control enzyme activity but as yet no detailed information has been published.

As mentioned, the point of addition of exogenous enzymes to the cheese making process has precluded its use in certain cheese manufacturing processes. Simply adding soluble enzymes directly to the milk is not feasible, as most of the enzyme would be lost in the whey, which would also adversely affect downstream applications of this product. Also, proteinases acting at the early stages of cheese making would affect the overall yield, as casein fragments would also be lost to the whey. An ideal system would enable enzymes to be added to the milk. The enzymes would then fully partition with the curd and not negatively affect yield. This system would be advantageous in that it would be easy to use, but also ensure even distribution of the enzyme in the curd. Work carried out by a number of groups in the 1980s and 1990s using liposome technology showed some promise. This involved encapsulating water soluble enzymes in the aqueous phase between phospholipid layers in liposome globules (Figure 20.3). The benefits of this system were even distribution of enzymes in the cheese curd, and early enzyme release during ripening as the liposome capsules break down. A number of problems existed with this technology and mainly revolved around maximizing enzyme encapsulation rates and eliminating high costs due to the expense of using soya lecithin in the production of liposomes. Recent work carried out in this area used a commercial phospholipid mixture (Pro-Lipo® S) from Lucas Meyer, France to encapsulate cell free extracts of LAB. This work has highlighted further limitations precluding the use of commercial phospholipid mixtures, such as increased moisture and changes in rheological characteristics of the cheese (40). It difficult to determine if this technology will find commercial application in the near future due to the limitations described, but also due to the fact that most soya lecithin is derived from genetically modified soya which further limits its use in certain countries.

Currently, adding exogenous enzymes at the salting stage (milling) of the cheese making process offers the best route for Cheddar cheese production, despite problems associated with uneven enzyme distribution in the curd. The enzymes may be preblended with the salt or added independently. Many different types of exogenous enzymes have been added to different cheeses, and most have shown an increase in the indices of ripening, although numerous defects were also reported. Most applications have involved Cheddar cheese and the use of commercial proteinase preparations, although quite a few have also involved the use of cell free extracts of various microbial cultures. For a more detailed description of the addition of exogenous enzymes during cheese making the reader is referred to Wilkinson (1993); Fox, Guinee, Cogan, and McSweeney (2000); Law (2001) (41,42,43).

20.3.3.2.2 Commercial Enzymes for Accelerating Cheese Ripening Very few commercially successful enzyme systems to accelerate cheese ripening exist. One which has survived the marketplace for years is the Accelase system (Rhodia Food). Accelase consists of a range of blends of enzyme systems (proteinase, peptidases, and lipases) tailored to suit different applications. They are purported to accelerate ripening, give particular flavor characteristics, and can be added to the milk or the curd (44). Other commercial systems exist (NaturAge, FlavourAge, DCA50, Enzobact, and ACR), but little information has been published regarding their use. For a more detailed review, see Wilkinson, Van den Berg, and Law (2002) (45).

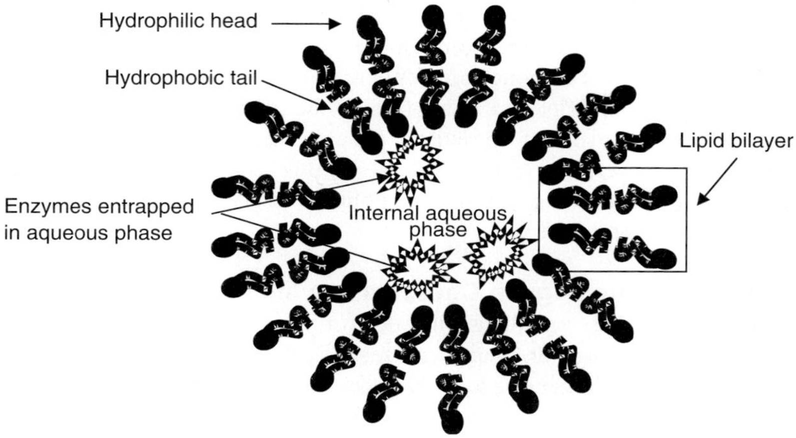

Figure 20.3 Schematic representation of 2-dimensional layer of a liposome with enzymes entrapped in the internal aqueous phase.

20.3.3.2.3 Attenuated Starter Cultures and Adjuncts Due to problems associated with the use of exogenous enzymes to accelerate cheese ripening, research has concentrated on using starter cultures for the same purpose. Starter cultures have been shown to contain the necessary pool of enzymes required to develop characteristic Cheddar flavor, as they have been used in cheese manufacture for years. They can also be added directly to the milk, and distribute evenly in the curd. Essentially, starter cultures are excellent vehicles for distributing enzymes into the curd to accelerate cheese ripening.

In natural Cheddar cheese, as in many other varieties, starter cultures are used to acidify the milk enabling curd formation, and to act as sources of enzymes necessary for the development of characteristic flavors. In commercial reality, starter cultures are selected for their phage resistance and their ability to grow well and produce acid over the temperature profiles used in cheese making. Those strains prone to phage attack, and those which do not produce good quality cheese, have been removed leaving the specific strains which produce good quality flavor (42). Cheese curd has the capacity to retain up to 3 times the level of starter typically used in cheese making, therefore opening the possibility of enabling extra starter culture, containing more enzymes, to be added to the curd to accelerate its ripening (43). To achieve this, the acid producing capability of the extra starter must be attenuated in order to prevent excess acid production, which would adversely affect the cheese making procedure. This can be achieved by using sublethal heat treatments or freeze-thaw treatments, which do not affect the enzyme activity of the starter bacteria. Lactose-negative strains may also be used, as these occur naturally, and because they cannot produce acid they can also be used to accelerate cheese ripening. An important facet of this technology is cell lysis, as many of the enzymes necessary to develop flavor are intracellular and therefore are not released until the cells lyse at death. Therefore the ability of a cell to lyse rapidly in the curd is important to enable the enzymes to work on the substrate as early as possible to accelerate ripening.

Another important aspect of cheese ripening in many varieties is the presence of nonstarter lactic acid bacteria (NSLAB), which predominate during the latter stages of ripening in some cheeses, such as Cheddar. Some of these cultures have been adopted and attenuated for use as adjunct cultures in conjunction with starter cultures in cheese making. They are generally used to impart certain characteristic flavors rather than accelerate ripening, but can be used to accelerate ripening (42).

The advent of genetic modification (GM) has enhanced the possibility of using cultures for the acceleration of cheese ripening. Not only has the taxonomy of cultures been advanced but also their enzyme complement, activities, bacteriophage resistance, and lysis potential. This technology has without doubt the potential to aid our understanding of cheese flavor development and provide a suitable mechanism for the acceleration of cheese ripening. However, the widespread use of this technology in the production of cheese has not occurred, due to consumer and political reservations about the safety of this technology, whether founded or not.

20.3.4 Enzyme Technology in Cheese as an Ingredient

Cheeses have been used as ingredients in foods for years with large volumes used in the retail, food service (catering), and industrial sectors. Natural cheese is widely used as an ingredient, but it also undergoes secondary processing into various products for a wide range of applications. Secondary processing of natural cheese includes shredding, grating, grinding, portioning, cooling, freezing, thawing, heating, or reheating; and cheese is still required to perform various rheological, visual, or sensory functions after such processes. A detailed review of the functionality of cheese as an ingredient is given by Guinee (2002) (46). The use of cheese as an ingredient has seen major growth in recent years, due to the increased demand for processed consumer foods and the demand for ethnic foods that use cheese in their preparation. Natural cheese is an essential ingredient in many products, providing flavor, aroma, color, texture, and functionality, but it can be quite expensive. This aspect has driven suppliers to develop cost effective alternatives to using natural cheese in ingredient formulations.

20.3.4.1 Fast Ripened Cheese

These cheese products have been loosely termed "ingredient-type cheeses" and are defined by their high moisture content and short ripening times, which can be weeks rather than months. They are developed to provide ingredient alternatives to semihard- and hard-type natural cheeses, which can take many months to ripen. Texture is not a factor with such cheeses as they are usually shredded after production and used solely as an ingredient cheese for further processing. The enzymes and starter cultures used in their production are similar to those outlined in the accelerated cheese ripening section and are added at levels higher than that found in natural cheese. These cheeses are high moisture cheeses, which are typically ripened at 15 to 20°C to increase the rate of enzyme reaction and thus hydrolysis. Once an acceptable flavor profile has been reached, the cheeses are typically shredded and stored chilled or frozen until required, and are solely used in the industrial sector. A variation of this type of cheese involves shredding the cheese and adding enzymes, then reblending it in a fashion similar to processed cheese, and ripening it at high temperatures. This cheese can be further processed into slices or grated depending upon the desired application. Virtually no information is published on this process, but it appears to offer potential to give the flavor, visual, and some functional characteristics of cheese to certain processed foods.

20.3.4.2 Enzyme Modified Cheese

Enzyme modified cheeses (EMCs) are defined as concentrated cheese flavors produced enzymatically from cheeses of various ages. They are principally used as a cheese flavor ingredient in processed foods, where they provide a cost effective alternative to natural cheese. Whereas natural cheese is consumed for its flavor, texture, appearance, and functional properties, either on its own or as an ingredient, EMCs are consumed almost exclusively as a flavor ingredient (47). EMCs can be used as the sole source of cheese flavor, to intensify an existing cheese taste, or to impart a specific cheese character to a blander product (1). The main applications of EMCs are in processed cheese, analogue cheese, cheese spreads, snack foods, soups, sauces, biscuits, dips, and pet foods. The main advantages of EMCs for use as an ingredient are low production costs, high flavor intensity, diverse flavor range, extended shelf life, and low storage costs (48). EMCs are generally perceived as being natural flavor providers as they are produced using the same flavor pathways that occur in natural cheese. The high degree of flavor intensity of EMCs enables them to be added to foods at very low levels (typically 0.1%, w/w) to achieve the desired flavor (47). This low addition level enables EMCs to be used in low fat foods and ensures that they have no adverse physical affect on the final product characteristics, which is particularly useful in frozen foods where they impart a smooth mouth feel. They can also be used cost effectively to replace a percentage of the natural cheese in a product where it is necessary to have some natural cheese for its physical characteristics. EMCs are available in a wide range of cheese flavors, which correspond to most natural cheese varieties. However, many different flavor types are also available for individual cheese types: e.g., Cheddar EMCs are available as a range of Cheddar-type EMC flavors. This highlights the fact that the relationship of EMCs to their corresponding natural cheese is at best unclear. EMC flavor is generated using the same pathways that occur in natural cheese ripening, i.e., proteolysis, lipolysis, and glycolysis. But, as opposed to natural cheese where the influence of each pathway on the flavor depends upon the variety (49), EMC flavor is principally derived through proteolysis and lipolysis with texture not recognized as a factor in its production. EMCs are used to provide the perception of cheese flavor and aroma to a product. The specific flavor and aroma characteristics they are expected to mimic typically resembles the main flavor and aroma character of a natural variety.

The basis of EMC production is the utilization of specific enzymes, under optimum conditions, to produce cheese flavors from dairy substrates, usually cheese curd. This technology has developed from studies by Kristoffersen, Mikolajeik, and Gould (1967) (50) where cheese flavor was produced rapidly using a curd slurry technique. The original process involved mixing fresh curd and a sodium chloride solution to make an emulsion of approximately 40% (w/v) solids. Enzymes and preservatives were added and the slurry was incubated at 30°C for 4–5 days with daily agitation, after which characteristic cheese flavors developed. This work demonstrated the potential of generating a range of intense cheese flavors in a short time from cheese substrates by modification of process parameters.

The fundamental procedure is similar today, where typically mature or immature cheese is incubated with specific exogenous enzymes or microorganisms under controlled conditions. The process is terminated by heat treatment and the final product standardized to a desired flavor intensity and composition. EMCs are generally manufactured by one of three enzymatic processes: an integrated approach, where a single substrate is hydrolysed simultaneously by proteolytic and lipolytic enzymes to the desired final flavor; a two step process where a single substrate is hydrolysed initially by proteolytic enzymes and subsequently by lipolytic enzymes to create the final product; and a component approach, where individual substrates are hydrolyzed independently by either proteinases or lipases and subsequently combined to create the desired flavor profile (Figure 20.4). (47,48,51,52).

Emulsification of the substrate is very important to optimize enzyme activity, as is consistency and quality of the curd substrate. Many EMC products are produced with the inclusion of flavor enhancers or potentiators, often called "top notes" in the industry, to create a greater diversity of products. The most commonly used flavor enhancers are monosodium glutamate, yeast extracts, and organic acids such as acetic acid, butyric acid, propionic acid, lactic acid, and diacetyl. The actual processing conditions used depend upon many factors, and directly relate to the enzymes being used. In general, EMCs are produced in 1–4 days at a temperature range of 25–45°C and with pH of 5–7 (48).

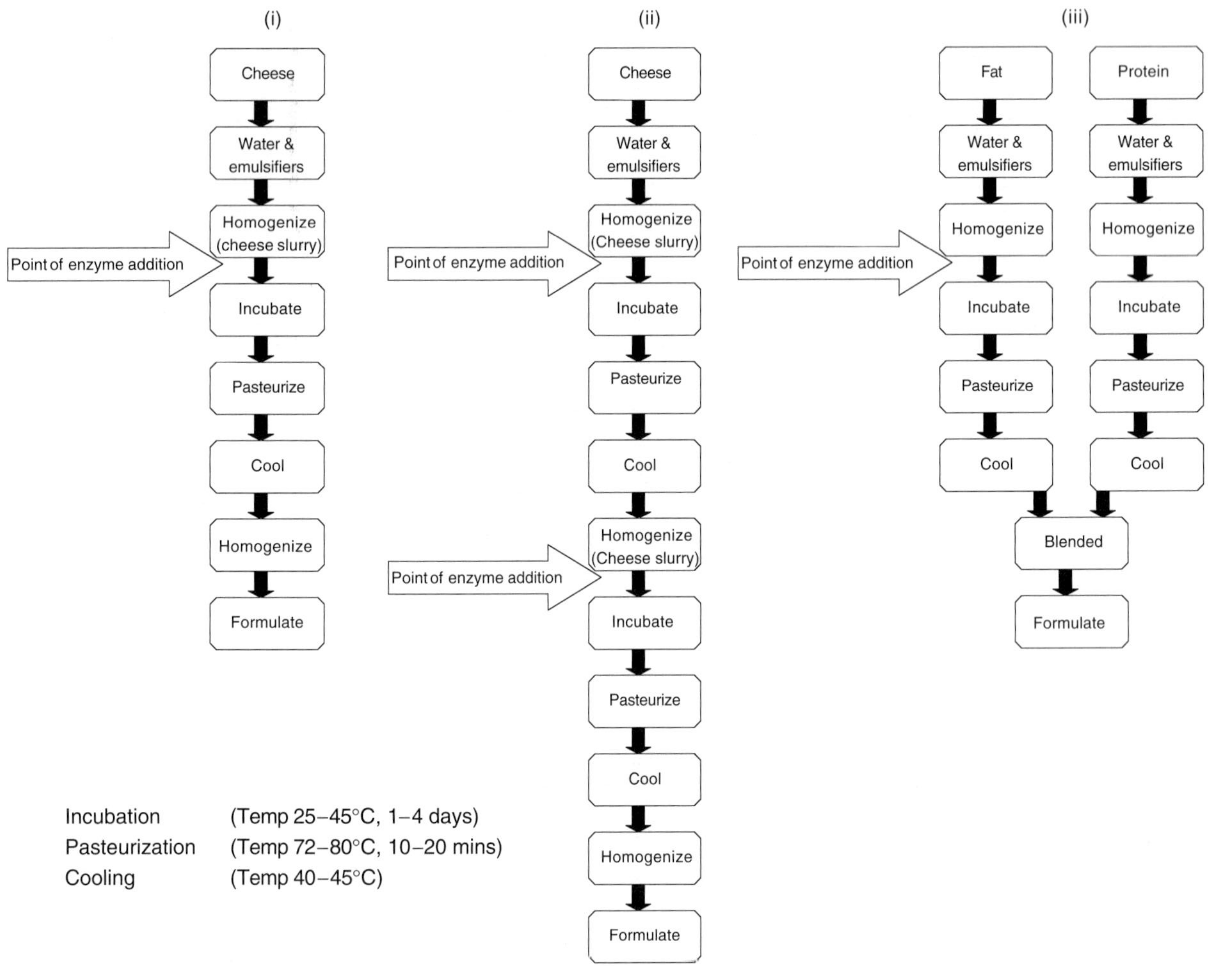

Figure 20.4 Methods of EMC production

The enzymes used in EMC manufacture are those that degrade dairy protein and fat into flavor compounds and precursors of flavor compounds (i.e., proteinases, peptidases, and lipase). A wide range of commercial enzymes are available, derived from plant, animal, or microbial sources. Most of the enzymes used in EMC production are derived from microbial or animal sources. Plant enzymes are not used as they are comparatively impure and are expensive because large volumes of crops are required to produce sufficient amounts of enzyme.

In the production of EMC it is necessary to rapidly develop a savory cheese flavor, so broad specificity proteinases are used widely. Quite a number of commercial proteinases are available for EMC production and all are microbial in nature. Most are derived from *Bacillus* or *Aspergillus* species and tend to be neutral or slightly acidic in nature. The choice of proteinase depends upon factors such as desired flavor, substrate, cost, and processing parameters. Peptidases are very important in EMC production, as they play an important role in flavor development but also in reduction of bitterness which can be a problem in EMCs due to the significant degradation of all the major caseins. Bitterness can be masked in EMC production by adding flavor enhancers and mature cheese, and is typically controlled by the use of specific peptidases. The most widely used commercially available peptidases are derived from LAB and *Aspergillus oryzae*. These products may contain a range of peptidase activities and, in some cases, proteinase activities. Commercial proteinase and peptidases for use in EMC production have been reviewed (53,54).

Hydrolysis of milk fat by lipases produces flavor and aroma compounds, which also act as precursors for additional flavor and aroma compounds. Lipolysis is very important in EMC production due to the generation of volatile compounds, which are important in providing cheese flavor to the product to which EMC is added. Lipases are available from two main sources, animal and microbial. The most significant animal lipases are isolated from bovine and porcine pancreatic tissues and the pregastric tissues of kid goat, lamb, and calf (27). Different PGEs produce characteristic flavor profiles. Calf PGE generates a "buttery" and slightly "peppery" flavor, kid PGE generates a sharp "peppery" flavor, often called "picante", whilst lamb PGE generates a "dirty sock" flavor, often called "pecorino" flavor (55). Microbial lipases are enzyme preparations derived from yeasts, molds, or bacteria, e.g., *Rhizomucor miehei*, *Rhizopus arrhizus*, *Aspergillus niger*, *Aspergillus oryzae*, *Geotrichum candidum*, *Penicillium roqueforti*, *Achromobacter lipolyticum*, *Pseudomonas* species, *Staphylococcus* species, and *Candida cylindracea* (56). Microbial lipases are generally cheaper than animal lipases because of lower production costs, and have the added advantage of being suitable for use in vegetarian and Kosher foods; and do not contain amylases which can cause problems in foods into which EMCs are added (57). Commercial lipase preparations for potential use in EMC manufacture have also been reviewed (54).

20.3.4.3 Dehydrated Cheese Products

These products can consist of spray dried natural cheese, dried grated cheeses, or extended cheese powders. The first two consist of natural cheese but the third, extended cheese powders, are produced by spray drying a slurry of cheese. Dried cheese powders are produced with varying low moisture contents for different applications. Some contain anticaking agents to prevent clumping, such as microcrystalline cellulose. Mature cheese is usually used because it has a better flavor profile, but also because it is easier to atomize and dry than young cheese. Young cheese has large amounts of intact casein, which produces viscous slurries making it difficult to dry. Extended cheese powders typically consists of mature cheese and enzyme modified cheese, flavorings, flavor enhancers, and emulsifying salts which are blended together under high shear, and then pasteurized and dried. Color, extenders (such as whey, skim milk, or milk fat), antioxidants, and carriers (maltodextrins) may be added depending upon the application and the drying procedure (58).

20.3.4.4 Substitute or Imitation Cheese

These are cheese products which, as opposed to natural cheese, are not made directly from milk. Two types of imitation cheese are produced in very high volumes: processed and analogue cheese. Neither imitation cheese uses enzymes directly in its manufacture, but either may contain ingredients produced using enzymes.

Cheese analogues are typically defined as products made by blending individual constituents, including nondairy fats or proteins, to produce a cheese-like product to meet specific requirements. They are cost effective as they employ a high proportion of cheaper nondairy fat and protein. This product is utilized in convenience foods , but is predominantly used as a source of cheese flavor in pizza topping to reduce the costs of using 100% natural cheese on a product. The major disadvantage of this product as a direct replacement to natural cheese is its lack of flavor. However, this has been overcome in certain applications by using enzyme modified cheese in addition to or in association with natural cheese.

Processed cheeses are further classified into different categories depending upon the ingredients used in their preparation. Different guidelines exist on labeling different processed cheese products and are dependent upon the countries where they are produced. Some processed cheese products also contain enzyme modified cheeses to impart characteristic flavors to the final product. For a detailed review of this area, see Fox, Guinee, Cogan, and McSweeney (2000) (59).

20.3.5 Lipolysed Cream and Butter Products

Lipases hydrolyze milk fat and catalyze the conversion of triglycerides to diglycerides, monoglycerides, and FFAs. These FFAs impart characteristic flavors and aromas, so the specificity of the lipase has a major role in determining the flavor of a product (60). Lipases are used to hydrolyze milk triacylglycerols to produce intensely flavored dairy products. As mentioned earlier, EMC technology uses this approach to create intensely flavored cheese for ingredient purposes. Lipolysed milk fats are used mainly in a wide range of confectionery and baking applications, but also find uses in nutritional products. A host of different microbial enzymes are available for these applications and flavors developed are directly related to the specificity of the lipases used in their manufacture. The specificity of commonly used lipases are detailed in Villeneuve and Fogilia, (1997); Balcao & Malcata, (1998) and Gunstone, (1999) (5, 61,62).

Production of lipase modified milk fat ingredients generally follows the following procedure (51):

1. Preparation of condensed milk or butter-oil substrate
2. Standardization of the lipase preparation
3. Contact of lipase and substrate
4. Homogenization to facilitate emulsion formation to enhance enzyme activity
5. Incubation of emulsion to achieve desired conversion of substrate to products
6. Enzyme inactivation in a manner that minimizes the loss of developed flavor volatiles
7. Final product standardization and formulation

Lipases can be used to produce pronounced dairy flavors in milk fat emulsions or in whole milk concentrate substrates. Such flavors are desired for the production of milk chocolate and account for the unique superiority of some milk chocolate flavors. These are produced in powdered form and are resistant to oxidative rancidity, which is of major advantage over their whole milk counterparts (63). Lipase modified milk fat products include milk chocolate, chocolate syrups, chocolate coatings, chocolate beverages, butter, margarine, butter creams and sauces, coffee whiteners, and creams. Lipolysed milk fat emulsions are also very effective carriers of synthetic fatty acids, diacetyl, butter esters, and lactones (61,64).

A lot of interest has been generated toward the development of structured lipids, which are basically lipids designed to have specific fatty acids attached to given positions on triglycerides. This approach can also be used to develop milk fat, in the formulation of confectionery fats, or to provide milk fat with enhanced nutritional qualities for specific applications. Milk fat can be modified by physical means, such as fractionation, or by chemical means, such as hydrogenation or interesterificiation. Interesterificiation is carried out by enzymes and involves the exchange and redistribution of acyl groups among milk fat triacylglycerols. Therefore, it is possible to enrich milk fat with certain fatty acids which could then be used in various flavor or nutritional applications, as well as imparting distinct physical properties for functional uses in food applications such as margarines (60,61).

20.3.6 Protein Hydrolysis

In the past, large scale generation of flavor during food processing was carried out by chemical or acid treatments and sometimes created unwanted byproducts or required very high energy demands, which has limited their potential uses. Recently natural product use has created a demand for enzyme produced flavors. Enzyme hydrolysis is generally seen as a noninvasive technology, and is an extrapolation of many natural processes.

Two milk proteins, casein and whey, are used in the food industry for specific flavor, functional, or nutritional applications. These proteins are hydrolysed to meet various applications, particularly in baby foods, baby milks, nutritionally fortified drinks, and other dietetic food formulations. A wide variety of commercial enzymes are useful in the preparation of hydrolysates that are very soluble, and stable at low pH and when heated (65). However, a major problem associated with protein hydrolysis, particularly for casein and whey proteins, is the development of bitterness, which is associated with the accumulation of hydrophobic peptides. Bitter peptides contain 3–15 amino acids and are characterized by the presence of the hydrophobic amino acid residues leucine, isoleucine, proline, valine, phenylalanine, tyrosine, and tryptophan. Bitterness can be alleviated by the degradation of these bitter peptides and is an area which has been widely studied over the years (66).

Generally, a number of key peptidase activities are important in bitterness reduction. General aminopeptidase activity, which cleaves single amino acids from the N-terminal of a peptide (Pep N and Pep C activity) is important. The removal of the amino acid proline from a peptide can greatly reduce bitterness, because it alters the peptide's three-dimensional structure, changing its solubility and increasing its susceptibility to hydrolysis (66,67). Key peptidase activities involved in removing proline are proline specific peptidases such as: postprolyl dipeptidyl dipeptidase (Pep X), which releases N-terminal dipeptides from prolyl peptides; proline aminopeptidase (Pep P), which exclusively hydrolyzes the N-terminal amino acid from peptides that have proline in the second position; proline iminopeptidase (Pep I) which removes unsubstituted N-terminal proline residues from tri-, oligo-, and polypeptides; prolinase (Pep R) which cleaves N-terminal Pro-X dipeptides; and prolidase (Pep Q) which hydrolyzes X-Pro dipeptides (68,69).

However it is not necessary to completely convert peptides to free amino acids to eliminate bitterness. In practice, the final level of bitterness will be the result of a tradeoff between the economics of bitterness control, desired functionality, and how tolerant the final application is to bitterness.

20.3.7 Lactose Hydrolysis

Lactose is hydrolysed by the enzyme β-galactosidase (EC 3.2.1.23) or lactase to produce glucose and galactose, and is used in a number of applications in the dairy industry. β-galactosidase is available from a number of animal, plant, or microbial sources and must be purified prior to use in dairy applications. The commonly used preparations are derived from yeasts and fungi and have different pH and temperature optima, so it is important to understand the conditions of a process prior to selecting the most suitable source.

β-galactosidase has found uses in certain dairy applications, particularly in products for people with lactose intolerance. Lactose intolerance is the inability to hydrolyse lactose into galactose and glucose in the small intestine due to a lack of β-galactosidase, and is a problem for a majority of the population (70). Symptoms include various intestinal disorders, including gas and diarrhea (28). As lactose is present in high levels in many dairy products, this is of major concern to processors, particularly in countries where a high proportion of the population are predisposed to lactose intolerance. β-galactosidase can be used to hydrolyse lactose to more digestible byproducts, enabling individuals with lactase intolerance to easily digest dairy products. Most of these products have a slightly sweeter taste as the byproducts of lactose, glucose and galactose, are sweeter than lactose. This technology has also been proposed for yogurt manufacture to make it sweeter without increasing its calorific content. It has also been shown to accelerate acid formation in yogurt (71).

β-galactosidase may stimulate starter and secondary starter activity and is thought to have a possible role in accelerating cheese ripening (22). However, studies were carried out on accelerating cheese ripening using β-galactosidase derived from the yeast *Kluyveromyces lactis*, which contained a proteinase that may have been responsible for the increased ripening rates (70).

Glucose–galactose syrups produced by the hydrolysis of lactose from whey are used for the production of syrups and sweeteners for the food industry. These are sweeter and more soluble than lactose. Such syrups have several applications in food products (71). Lactose can form insoluble crystals on drying and these can cause problems in some food products. Hydrolysis of milk lactose for its use in ice cream to prevent crystallization has been shown to be successful. The treatment of hydrolysed whey for animal feed removes problems associated with crystallization of lactose on drying and makes the whey more useful as a feed component (71).

20.3.8 Lysozymes

Lysozyme is used in some Dutch, Swiss, and Italian cheese varieties to prevent late gas blowing and off flavors caused by the growth of *Clostridium tyrobutyricum*. Most commercial lysozyme is derived from chicken egg white. Nitrate is also used in some countries for the same application. However because of possible nitrosamine production, its use is precluded or permitted levels reduced in other countries (8).

20.3.9 Transglutaminase

Transglutaminase (TGase) is an enzyme which catalyzes an acyl-transfer reaction introducing a covalent cross link between glutamine and lysine residues. TGase has found applications throughout the food sector particularly in improving functional properties of various products. The use of TGase in the dairy industry at present is generally limited to research on improving the gelation properties of yogurts and improving the rheological properties of whey proteins. However, some recent studies suggest that it has a role in masking bitterness in casein hydrolysates and in improving the curd firmness in imitation cheese. Other research has found that TGase has the potential to improve the heat stability of milk (72). It is quite likely that further dairy applications for this enzyme will be found in the future.

REFERENCES

1. Burgess, K., M. Shaw. Industrial application: dairy. In: *Industrial Enzymology: The Application of Enzymes in Industry*, Godfrey, T., J. Reichett, eds., Surrey: Macmillian Publishers Ltd, 1983, pp 260–283.
2. Sorhaug, T., L. Stepaniak. Microbial enzymes in the spoilage of milk and products. In: *Food Enzymology*, Fox, P.F., ed., London: Elsevier Applied Science, 1991, pp 169–270.
3. Fox, P.F., P.L.H. McSweeney. Enzymology of milk and milk products. In: *Dairy Chemistry and Biochemistry*, PF Fox, P.F., P.L.H. McSweeney, eds. London: Blackie Academic & Professional, 1998, pp 317–346.
4. Deeth, H.C., C.H. Fitz-Gerald. Lipolytic enzymes and hydrolytic rancidity in milk and milk products. In: *Advanced Dairy Chemistry, Vol. 2: Lipids*, 2nd ed., Fox, P.F., ed., London: Chapman & Hall, 1995, pp 247–308.
5. Villeneuve, P., T.A. Fogilia. Lipase specificities: potential application in liquid bioconversions. *Inform* 8:640–650, 1997.
6. Olivecrona, T., G. Bengtsson-Olivecrona. Indigenous enzymes in milk, part II: lipase. In: *Food Enzymology*, Fox, P.F., ed., London: Elsevier Applied Science, 1991, pp 62–78.
7. Olivecrona, T., S. Vilaro, G. Bengtsson-Olivecrona. Indigenous enzymes in milk. In: *Advanced Dairy Chemistry, Vol 1: Proteins*, 2nd ed, Fox, P.F., ed., London: Elsevier Applied Science, 1992, pp 292–310.
8. Fox, P.F., L. Stepaniak. Enzymes in cheese technology. *Int. Dairy J.* 3:509–530, 1993.
9. Fox, P.F., T.P. Guinee, T.M. Cogan, P.L.H. McSweeney. Bacteriology of cheesemilk. In: *Fundamentals of Cheese Science*, Fox, P.F., T.P. Guinee, T.M. Cogan, P.L.H. McSweeney, eds., Gaithersburg, MD: Aspen Publishers, Inc, 2000, pp 45–53.
10. Walstra, P., T.J. Geurts, A. Noomen, A. Jellema, M.A.J.S. van Boekel. Milk for liquid consumption. In: *Dairy Technology, Part III: Milk products*, Walstra, P., T.J. Geurts, A. Noomen, A. Jellema, M.A.J.S. van Boekel, eds., New York: Marcel Dekker, Inc, 1999, pp 383–404.
11. Barrett, F.M., A.L. Kelly, P.L.H. McSweeney, P.F. Fox. Use of exogenous urokinase to accelerate proteolysis in cheddar cheese during ripening. *Int. Dairy J.* 9:421–427, 1999.
12. Fox, P.F. Indigenous enzymes in milk: proteinases. In: *Advanced Dairy Chemistry*, Vol. 1: Proteins, Fox, P.F., ed., London: Elsevier Applied Science, 1992, pp 79-89.
13. Bastian, E.D., R.J. Brown. Plasmin in milk and dairy products: an update. *Int. Dairy J.* 6: 435–457, 1996.
14. McSweeney, P.L.H., M.J. Sousa. Biochemical pathways for the production of flavour compounds in cheeses during ripening: a review. *Lait* 80:293–324, 2000.
15. Andrews, A.T. Indigenous enzymes in milk, part IV: phosphatases. In: *Food Enzymology,* Fox, P.F., ed., London: Elsevier Applied Science, 1991, pp 90–99.
16. Bjorck, L. Indigenous enzymes in milk, Part V: lactoperoxidase. In: *Food Enzymology,* Fox, P.F., ed., London: Elsevier Applied Science, 1991, pp 100–106.
17. Farkye, N.Y. Indigenous enzymes in milk, part VI: other enzymes. In: *Food Enzymology,* Fox, P.F., ed., London: Elsevier Applied Science, 1991, pp 107–129.
18. Harboe, M., P. Budtz. The production, action and application of rennet and coagulants. In: *Technology of Cheesemaking,* Law, B.A., ed., Sheffield: Sheffield Academic Press, 1999, pp 33–65.
19. Wigley R.C. Cheese and Whey. In: *Industrial Enzymology*, T. Godfrey, S. West, eds., London: Macmillan Press Ltd, 1996, pp 133–154.
20. Flotmann, B.F. General and molecular aspects of rennets. In: *Cheese: Chemistry, Physics and Microbiology, Vol. 1: General aspects*, Fox, P.F., ed., London: Chapman & Hall, 1993, pp 37–68.
21. Fox, P.F., P.L.H. McSweeney. Chemistry and biochemistry of cheese and fermented milks. In: *Dairy Chemistry and Biochemistry*, Fox, P.F., P.L.H. McSweeney, eds., London: Blackie Academic & Professional, 1998, pp 379–436.
22. Fox, P.F., M.B. Grufferty. Exogenous enzymes in dairy technology. In: *Food Enzymology*, Fox, P.F., ed., London: Elsevier Applied Science, 1991, pp 219–269.

23. Guinee, T.P., M.G. Wilkinson. Rennet coagulation and coagulants in cheese manufacture. *J. Soc. Dairy Tech.* 45(4):94–104, 1992.
24. Banks, J.M. Yield and quality of cheddar cheese produced using a fermentation-derivied calf chymosin. *Milchwissenschaft* 47:153–156, 1992.
25. Fox, P.F., T.P. Guinee, T.M. Cogan, P.L.H. McSweeney. Enzymatic coagulation of milk. In: *Fundamentals of Cheese Science*, Fox, P.F., T.P. Guinee, T.M. Cogan, P.L.H. McSweeney, eds., Maryland: Aspen Publishers Inc, 2000, pp 98–137.
26. Christie, W.W. Composition and structure of milk lipids. In: *Advanced Dairy Chemistry, Vol. 2: Lipids*, 2nd ed., Fox, P.F., ed., London: Chapman & Hall, 1995, pp 1–36.
27. Nelson, J.H., R.G. Jenson, R.E. Pitas. Pregastric esterase and other oral lipases: a review. *J. Dairy Sci.* 60:327–362, 1977.
28. Desmazeaud, M., J. Cerning, J.C. Gripon. The use of enzymes in the dairy industry. *Proceedings of Food Ingredients Europe Conference*, Paris, 1989, pp 96–103.
29. Kinsella, J.E., D.H. Hwang. Enzymes of *Penicillium roqueforti* involved in the biosynthesis of cheese flavor. *Crit. Rev. Food Sci. Nutr.* 8:191–228, 1976.
30. Gripon, J.C. Mould-ripened cheeses. In: *Cheese: Chemistry, Physics and Microbiology, Vol. 2: Major cheese groups*, Fox, P.F., ed., London: Chapman & Hall, 1993, pp 111–136.
31. Gripon, J.C., V. Monnet, G. Lamberet, M.J. Desmazeaud. Microbial enzymes in cheese ripening. In: *Food Enzymology*, Vol 1, Fox, P.F., ed., London: Elsevier Applied Science, 1991, pp 131–168.
32. Meyers, S.A., S.L. Cuppett, R.W. Hutkins. Lipase production by lactic acid bacteria and activity on butter-oil. *Food Microbiol.* 13: 383–389, 1996.
33. Yokoyama, H., N. Sawamura, N. Motobayashi. Method for Accelerating Cheese Ripening. European Patent application EP 0 469 857 A1, 1992.
34. Messens, W., J. Estepar-Garcia, K. Dewettinck, A. Huyghebaert. Proteolysis of high-pressure treated Gouda cheese. *Int. Dairy J.* 9:775–782, 1999.
35. O'Reilly, C.E., P.M. O'Connor, P.M. Murphy, A.L. Kelly, T.P. Beresford. The effect of exposure to pressure of 50 Mpa on cheddar cheese ripening. *Innov. Food. Sci. Emerg. Technol.* 1:109–117, 2000.
36. O'Reilly, C.E., A.L. Kelly, P.M. Murphy, T.P. Beresford. High pressure treatment: applications in cheese manufacture and ripening. *Trends in Food Sci. Technol.* 12:51–59, 2001.
37. Law, B.A. The accelerated ripening of cheese by the use of non-conventional starters and enzymes: a preliminary assessment. *Int. Dairy Fed. Bull.* 108:40–48, 1978.
38. Law, B.A., A. Wigmore. Effect of commercial lipolytic enzymes on flavour development in cheddar cheese. *J. Soc. Dairy Tech.* 38:86–88, 1985.
39. Arbige, M.V., P.R. Freund, S.C. Silver, J.T. Zelko. Novel lipase for cheddar cheese flavor development. *Food Technol.* 40(4):91–98, 1986.
40. Laloy, E., J.C. Vuillemard, R. Simard. Characterisation of liposomes and their effect on the properties of cheddar cheese during ripening. *Lait* 78:401–412, 1998.
41. Wilkinson, M.G. Acceleration of cheese ripening. In: *Cheese: Chemistry, Physics and Microbiology,* Vol. 1, 2nd ed., Fox, P.F., ed., London: Chapman and Hall, 1993, pp 523–555.
42. Fox, P.F., T.P. Guinee, T.M. Cogan, P.L.H. McSweeney. Acceleration of cheese ripening. In: *Fundamentals of Cheese Science*, Fox, P.F., T.P. Guinee, T.M. Cogan, P.L.H. McSweeney, eds., Gaithersburg, MD: Aspen Publishers, Inc, 2000, pp 349–362.
43. Law, B.A. Controlled and accelerated cheese ripening: the research base for new technologies. *Int. Dairy. J.* 11:383–398, 2001.
44. Smith, M. Mature cheese in four months. *Dairy Ind. Int.* 62:7, 23–27, 1997.
45. Wilkinson, M.G., G. Van den Berg, B.A. Law. Technological properties of commercially-available enzyme preparations other than coagulants. *Bull. Int. Dairy Fed.* 371:16–19, 2002.
46. Guinee, T.P. The functionality of cheese as an ingredient: a review. *Aust. J. Dairy Tech.* 57:2, 79-92, 2002.
47. Moskowitz, G.J., S.S. Noelck. Enzyme modified cheese technology. *J. Dairy Sci.* 70:1761–1769, 1987.

48. Kilcawley, K.N., M.G. Wilkinson, P.F. Fox. Enzyme-modified cheese. *Int. Dairy. J.* 8:1–10, 1998.
49. Fox, P.F. Cheese: an overview. In: *Cheese: Chemistry, Physics and Microbiology*, Vol. 1, 2nd ed., Fox, P.F., ed., London: Chapman and Hall, 1993, 1–37.
50. Kristoffersen, T., E.M. Mikolajeik, J.A. Gould. Cheddar cheese flavour, 4: directed and accelerated ripening process. *J. Dairy Sci.* 50:292–297, 1967.
51. Kilara, A. Enzyme modified lipid food ingredients. *Process. Biochem* 20:35–45.
52. Kilara, A. Enzyme modified protein food ingredients. *Process. Biochem.* 20:149–157.
53. Kilcawley, K.N., M.G. Wilkinson, P.F. Fox. Properties of commercial microbial proteinase preparations. *Food Biotech* 16:1, 29–55, 2002.
54. Kilcawley, K.N., M.G. Wilkinson, P.F. Fox. Determination of key enzyme activities in commercial peptidase and lipase preparations from microbial or animal sources. *Enzyme Microb. Tech.* 31:310–320, 2002.
55. Birschbach, P. Pregastric lipases. *Bull. Int. Dairy. Fed.* 269:36–39, 1992.
56. Birschbach, P. Origins of lipases and their action characteristics. *Bull. Int. Dairy. Fed.* 294:7–10, 1994.
57. West, S. Flavour production with enzymes. In: *Industrial Enzymology*, 2nd ed., Godfrey, T., S. West, eds., London: Macmillan Press Ltd, 1996, pp 209–224.
58. McSweeney, P.L.H. Cheese as a food ingredient. In: *Proceedings of the 6th Cheese symposium*, Cogan, T.M., P.L.H. McSweeney, T.P. Guinee, eds., University College Cork and Teagasc, 2000, pp 176–182.
59. Fox, P.F., T.P. Guinee, T.M. Cogan, P.L.H. McSweeney. Processed cheese and substitute or imitation cheese products. In: *Fundamentals of cheese science*, Fox, P.F., T.P. Guinee, T.M. Cogan, P.L.H. McSweeney, eds., Maryland: Aspen Publishers Inc, 2000, pp 429–452.
60. Bosley, J.A. Enzymes used in oils and fats technology. In: *Ingredients Handbook Enzymes*, Rastall, R., ed., Surrey: Leatherhead Food RA Publishing, 1999, pp 79–95.
61. Balcao, V.M., F.X. Malcata. Lipase catalyzed modification of milkfat. *Biotech. Adv.* 16:2, 309–341, 1998
62. Gunstone, F.D. Enzymes as biocatalysts in the modification of natural lipids. *J. Sci. Food Agric.* 79:1535–1549, 1999.
63. Brown, R.J. Dairy Products. In: *Enzymes in Food Processing*, 3rd ed., Nagodawithana, T., G. Reed, eds., London: Academic Press, 1993, pp 347–361.
64. De Greyt, W., A. Huyghebaert. Lipase-catalysed modification of milk fat. *Lipid Tech*, 7(1):10–12, 1995.
65. Godfrey, T. Protein modification. In: *Industrial Enzymology*, 2nd ed., Godfrey, T., S. West, eds., London: Macmillan Press Ltd, 1996, pp 301–325.
66. Pawlett, D., G. Bruce. Debittering of protein hydrolysates. In: *Industrial Enzymology*, 2nd ed., Godfrey, T., S. West, eds., London: Macmillan Press Ltd, 1996, pp 177–186.
67. Alder-Nissen, J. Proteases. In: *Enzymes in Food Processing*, 2nd ed., Nagodawitha, T., G. Reed, eds., London: Academic Press, 1993, pp 159–203.
68. O'Cuinn, G., P.V. Jennings, I. Ni. Fhaolain, M. Booth, C.L. Bacon, M. McDonnell, M. Wilkinson, D.M. O'Callaghan, R. FitzGearld. The contribution of starter peptidases to flavour development in cheese. In: *Proceedings of the 4th Cheese Symposium*, Cogan, T.M., R.P. Ross, P.F. Fox, eds., Teagasc, Dublin, 1995, pp 68–71.
69. Bouchier, P.J. Hydrolytic and peptide debittering properties of purified Lactococcal aminopeptidases. Ph.D thesis, National University of Ireland, 1999.
70. Law, B.A., P.W. Goodenough. Enzymes in milk and cheese production. In: *Enzymes in Food Processing*, Tucker, G.A., L.F.J. Woods, eds., Glasgow and London: Blackie & Sons, 1991, pp 98–127.
71. Uhlig, H. Enzymes in the dairy industry. In: *Industrial Enzymes and Their Applications*, Uhlig, H., ed., New York: John Wiley & Sons, 1998, pp 323–337.
72. O'Sullivan, M.M., A.L. Kelly, P.F. Fox. Effect of transglutaminase on the heat stability of milk: a possible mechanism. *J. Dairy Sci.* 85:1–7, 2002.

2.21

Egg Yolk Antibody Farming for Passive Immunotherapy

Jennifer Kovacs-Nolan and Yoshinori Mine

CONTENTS

21.1 INTRODUCTION

Immunoglobulins, or antibodies, are large molecules formed by the immune system, developed by higher organisms to combat the invasion of foreign substances (1). In both humans and animals, the administration of specific antibodies is an attractive approach to establishing protective immunity against viral and bacterial pathogens, and has been prompted by the need to find alternatives to antibiotics in response to the increasing number of antibiotic resistant organisms, as well as organisms that are not responsive to traditional antibiotic treatment (2). There are also individuals who are unable to mount an active immune response against pathogens, including infants, children with congenital or acquired immunodeficiency syndromes, and those rendered immunodeficient by chemotherapy, malnutrition, or aging (3,4), who would benefit significantly from effective passive immunization techniques.

Passive immunization, a technique in which preformed pathogen specific antibodies are administered orally to individuals to prevent, and in some cases treat, infectious diseases, may be one of the most valuable applications of antibodies (5). It is proposed that the antibodies may exert a sort of antimicrobial activity against pathogens, by binding, immobilizing, and consequently reducing or inhibiting the growth, replication, or colony forming abilities of pathogens (5).

Traditionally, polyclonal antibodies have been produced in mammals such as mice, rats, rabbits, and goats. The antibodies are then collected from the serum of the animal (6). However, these antibodies cannot be prepared on the industrial scale that is required for passive immunization techniques, because of the limitation in the amount of blood that can be collected at any one time. Chickens have attracted considerable attention as an alternative source of antibodies for the prevention and treatment of infectious diseases (2,7,8). The serum immunoglobulin of chickens, referred to as immunoglobulin IgY, is transferred in large quantities to the egg yolk in order to give acquired immunity to the developing embryo (9), and can be readily extracted from the egg yolk.

The large scale potential of IgY, as well as its source, a normal dietary component, makes it ideally suited for passive immunization applications. Egg yolk antibodies, therefore, have significant potential for food fortification and nutraceutical applications for the reduction of the morbidity and mortality associated with infection by human and animal pathogens.

The following chapter details the structural and physical characteristics of IgY, the numerous advantages and immunotherapeutic applications of IgY, and future prospects in IgY research.

21.2 AVIAN EGG ANTIBODIES

21.2.1 Avian Immune System

Immunoglobulin production in chickens is significantly different from that of mammals. While the rearrangement of immunoglobulin genes is an ongoing process in mammals, in chickens, it occurs in one single wave during embryogenesis, limiting the possible number of antibodies to the number of B-cell precursors (10,11), estimated to be around 2–3 $\times$ 10^4 cells (12). In contrast to mammalian systems, gene rearrangement also contributes little to chicken antibody diversity (13). However, despite the fact that chickens have an extremely limited number of immunoglobulin genes compared to mammals, they have developed other methods of producing a wide range of immune responses and diverse antibody molecules (13), enabling their use for the production of specific antibodies.

Three immunoglobulin classes have been shown to exist in the chicken: IgA, IgM, and IgY. The IgA and IgM are similar to mammalian IgA and IgM. Chicken IgY is the functional equivalent of IgG, the major serum antibody found in mammals, and makes up about 75% of the total antibody population (2). The serum concentrations of IgY, IgA, and IgM have been reported to be 5.0, 1.25, and 0.61 mg/mL, respectively (14). In mammals, the transfer of maternal antibodies can take place after birth, however, in the chicken, the maternal antibodies must be transferred to the developing embryo to give acquired immunity to the chick (5,7). Antibody, specifically IgA and IgM, is incorporated into the egg white during egg formation. Serum IgY is selectively transferred to the yolk via a receptor on the surface of the yolk membrane which is specific for IgY translocation (15,16). Morrison et al. (17) identified the regions within the IgY molecule which are important for its uptake and specific translocation into the yolk. Egg white contains IgA and IgM at concentrations of around 0.15 and 0.7 mg/mL, respectively, whereas the yolk may contain up to 25 mg/mL of IgY (18). Mammalian equivalents of IgE and IgD have not been identified in chickens (13).

21.2.2 Molecular Properties of IgY

21.2.2.1 Structure of IgY

The structure of IgY is significantly different from that of mammalian IgG (7), even though they share a similar function. IgY contains two heavy (H) and two light (L) chains and has a molecular mass of 180 kDa, larger than that of mammalian IgG (159 kDa). IgY also possesses a larger molecular weight H chain (68 kDa) as compared to that in mammals (50 kDa). The H chain of IgG consists of four domains: the variable domain (V_H) and three constant domains (Cγ1, Cγ2 and Cγ3). The Cγ1 domain is separated from Cγ2 by a hinge region, which gives flexibility to the Fab fragments (the portion which contains the antigen binding activity). In contrast, the H chain of IgY does not have a hinge region, and possesses four constant domains (Cν1-Cν4) in addition to the variable domain (Figure 21.1). Sequence comparisons between IgG and IgY have shown that the Cγ2 and Cγ3 domains of IgG are closely related to the Cν3 and Cν4 domains of IgY; while the equivalent of the Cν2 domain is absent in the IgG chain, having been replaced by the hinge region (19). The content of β-sheet structure in the constant domains of IgY has been reported to be lower than that of IgG, and the flexibility between the Cν1 and Cν2 domains, corresponding to the hinge region of IgG, is less than that of IgG (20). Unlike IgG, IgY has two additional Cys residues, Cys331 and Cys 338, in the Cν2-Cν3 junction, which likely participate in the inter-ν chain disulfide linkages (19).

Both IgG and IgY are known to contain Asn linked oligosaccharides. However, the structures of oligosaccharides in IgY differ from those of any mammalian IgG, containing unusual monoglucosylated oligomannose-type oligosaccharides with $Glc_1Man_{7\text{-}9}GlcNAc_2$ structure (21,22).

Furthermore, the isoelectric point of IgY is lower than that of IgG (23), and it has been suggested that IgY is a more hydrophobic molecule than IgG (24). It also does not have the ability to precipitate multivalent antigens, except at high salt concentrations (25), possibly due to the steric hindrance caused by the closely aligned Fab arms. High salt concentrations may serve to release the Fab arms, thereby permitting precipitation (19).

21.2.2.2 Physicochemical Properties of IgY

IgY and IgG display differences in their stability when subjected to pH, heat, and proteolytic enzymes. Although the stability of both immunoglobulins is similar when subjected to alkaline conditions, IgY demonstrated much less stability than rabbit IgG to acid denaturation. Shimizu et al. (20,26) found that the activity of IgY was decreased by incubating at pH 3.5 or lower, and completely lost at pH 3. The rabbit IgG antibodies did not demonstrate

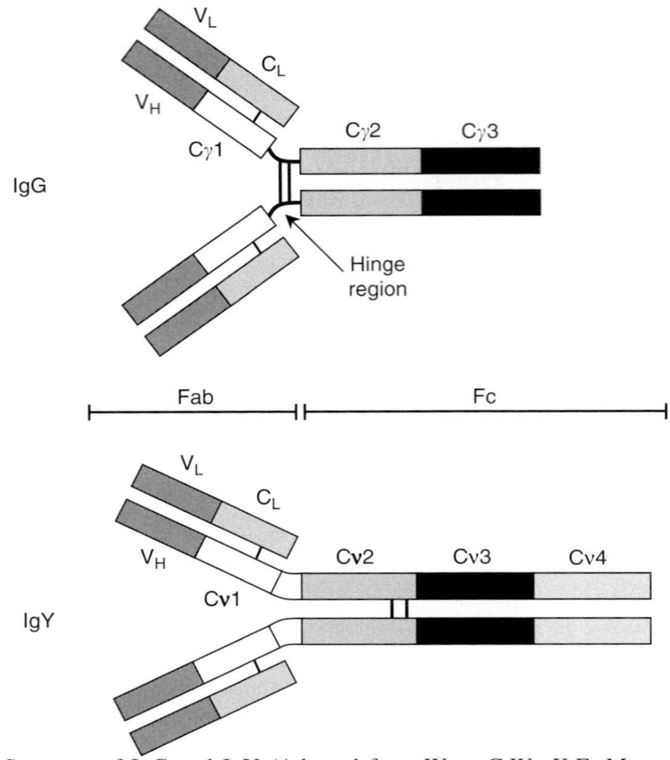

Figure 21.1 Structure of IgG and IgY (Adapted from Warr, G.W., K.E. Magor, D.A. Higgins, *Immunol. Today* 16:392–398, 1995.)

a loss of activity until the pH was decreased to 2, and even then some activity still remained. Similar results were also observed by Hatta et al. (27), using IgY produced against human rotavirus. The denaturing effects of acidic conditions on IgY were also found to be reduced by the addition of sucrose (28) or sorbitol (29). IgY is also more sensitive to heating than the rabbit IgG. Shimizu et al. (20) found that the activity of IgY was decreased by heating for 15 minutes at 70ºC or higher, whereas the activity of the IgG did not decrease until 75–80ºC or higher. The addition of stabilizers, such as sucrose, maltose, or glycerol to IgY preparations has been found to reduce the extent of thermal denaturation when heated at high temperatures (28,30). The addition of sugars may also increase IgY stability under various processing conditions, including high pressure (5000 kg/cm^2) (28) and freeze drying (31). It has also been observed that that IgY is more susceptible to digestion with proteolytic enzymes such as pepsin, trypsin, and chymotrypsin than IgG (32), but this will be discussed in more detail later.

Because immunoglobulins are large, complex molecules, the structural features causing the stability differences of the two immunoglobulins are unknown. Shimizu et al. (20) predicted that the lower content of β-structure in IgY may indicate that the conformation of IgY is more disordered, and therefore less stable, than mammalian IgG. The lack of a hinge region in IgY may be another factor affecting molecular stability. The lower flexibility of the Cν1 and Cν2 domains of IgY, as compared to the hinge region of IgG, may cause the rapid inactivation of the antibody by various treatments, because the flexibility of the hinge region is considered to influence the overall properties of immunoglobulin molecules (33).

21.2.3 Advantages of IgY

The production of polyclonal antibodies in chickens provides many advantages over the traditional method of producing antibodies in mammals (Table 21.1). The method of producing antibodies in hens is much less invasive, involving only the collection of eggs rather than blood, and is therefore less stressful to the animal (34). Animal care costs are also lower for chickens than for mammals such as rabbits (2). In contrast to mammalian serum, egg yolk contains only a single class of antibody, IgY, which can be easily purified from the yolk by precipitation techniques (35). The phylogenetic distance between chickens and mammals allows the production of antibodies in chickens against highly conserved mammalian proteins. This would not be possible in mammals, as demonstrated by Knecht et al. (36), who produced and compared antibodies against human dihydroorotate dehydrogenase in both rabbits and chickens. In contrast with mammalian systems, much less antigen is required to produce an efficient immune response in chickens (37), and sustained high titers reduce the need for frequent injections (35). Chicken antibodies will also recognize different epitopes than mammalian antibodies, giving access to a different antibody repertoire than with mammalian antibodies (7). Hens therefore provide a more hygienic, cost efficient, convenient, and plentiful source of antibodies, as compared to the traditional method of obtaining antibodies from mammalian serum (2,35).

In terms of safety, eggs are normal dietary components, so no risk of side effects from IgY consumption would be expected (2). In contrast to IgG, IgY does not activate mammalian complement or interact with mammalian Fc receptors that could mediate inflammatory response in the gastrointestinal tract (2). Even though IgY has been detected at low levels in the serum of calves and piglets after antibodies were orally administered, no absorption of intact antibodies has been shown in humans (2), so the risk of allergic reaction to purified IgY preparations would be expected to be low.

Nakai et al. (38) estimated that the productivity of antibodies in hens as nearly 18 times greater than that in rabbits based on the weight of antibody produced per head, which means that the use of hens can effect an overall reduction in animal use for the same level of antibody production (2). Because of the high yolk IgY concentrations, over 100 mg of IgY can be obtained from one egg (39). A laying hen produces approximately 20 eggs per month, therefore over 2 g of IgY per month may be obtained from a single chicken (7). Once purified, IgY has been found to be stable for years at 4°C (40). The large scale feeding, and automated collection and separation of eggs is a common practice (41), as is

Table 21.1
Comparison of the production and characteristics of IgG and IgY (Based on Schade, R., C. Pfister, R. Halatsch, P. Henklein. *ATLA* 19:403–419, 1991).

	Mammals (IgG)	Chicken (IgY)
Source of antibody	Blood serum	Egg yolk
Antibody sampling	Bleeding (invasive)	Egg collection (noninvasive)
Antibody amount	200 mg/bleed (40 ml blood)	50–100 mg/egg (5–7 eggs/week)
Amount of antibody per year	1400 mg	40 000 mg
Amount of specific antibody	~5%	2–10%
Protein A/G binding	Yes	No
Interference with rheumatoid factors	Yes	No
Activation of mammalian complement	Yes	No

vaccination of chicken flocks to control infections (42), making the large scale production and harvesting of antibodies in chicken eggs for therapeutic purposes quite feasible.

21.3 PASSIVE IMMUNIZATION WITH IGY

21.3.1 Immunotherapeutic Applications of IgY

Antibiotics have been widely used to treat infections. This method of treatment is being threatened by the growing numbers of immunosuppressed individuals, the emergence of new pathogens (including those which do not respond to antibiotics, such as viral pathogens), and the increasing resistance of organisms to antimicrobial drugs (43,44). This has prompted much research into the administration of specific antibodies as an alternative to antibiotics and antimicrobial chemotherapy to treat infections. Passive immunization also offers benefits in applications for which conventional vaccines may be less suitable, including providing immediate protection, controlled duration of activity, as well as selectivity of target organism (45).

The potential applications of IgY for the prevention and treatment of infections caused by pathogenic bacteria and viruses have been studied at length (Table 21.2).

21.3.1.1 Rotavirus

Human rotavirus (HRV) has been identified as the major causative agent of acute infantile gastroenteritis (46), infecting up to 90% of children under the age of three, and resulting in more than 600 000 deaths annually (47). Human rotavirus causes a shortening and atrophy of the villi of the small intestine (46), resulting in decreased water absorption, leading to severe diarrhea and vomiting and eventually death due to dehydration (48,49). Yolken et al. (3) found that orally administered antibodies isolated from the eggs of chickens immunized with three different serotypes of rotavirus (mouse, human, and monkey) were capable of preventing rotavirus induced diarrhea in mice infected with murine rotavirus, whereas IgY isolated from the eggs of unimmunized chickens failed to prevent rotavirus infection. Using an HRV infection model in suckling mice, Hatta et al. (27) reported that antiHRV IgY decreased the incidence of rotavirus induced diarrhea in mice, both when administered before and after HRV challenge, suggesting its use for both therapeutic and prophylactic applications. Similarly, Ebina (50) also observed the prevention of HRV induced symptoms in mice using antiHRV IgY. The production of specific IgY against recombinant HRV coat protein VP8*, a cleavage product of the rotavirus spike protein VP4 has been reported (51). VP4 has been implicated in several important functions, including cell attachment and penetration, hemagglutination, neutralization, and virulence (52–54). VP8 has been found to play a significant role in viral infectivity and neutralization of the virus. Kovacs-Nolan et al. (51) immunized chickens with recombinant VP8*, and found that the resulting antiVP8* IgY exhibited significant neutralizing activity against the Wa strain of HRV *in vitro*, indicating that antiVP8* IgY may be used for the prevention and treatment of HRV infection. The oral administration of antiHRV egg yolk to children infected with HRV resulted in a modest improvement in symptoms, but further studies are required (55).

Neonatal calf diarrhea caused by bovine rotavirus (BRV) is a significant cause of mortality in cattle (56). The passive protection of calves against BRV infection, using antiBRV IgY, has also been demonstrated (57).

21.3.1.2 Escherichia coli

Diarrhea due to enterotoxigenic *Escherichia coli* (ETEC) is a major health problem in humans and animals. ETEC is the most common cause of enteric colibacillosis encountered

Table 21.2
Specific IgY against various bacterial and viral antigens.

Pathogen	Effect of IgY	Ref.
Rotavirus	Prevented murine rotavirus in mice	3
	Prevented HRV-induced gastroenteritis in mice	27, 50
	Prevented HRV infection *in vitro*, using IgY against recombinant subunit protein VP8	51
	Protected calves from BRV-induced diarrhea	57
	Improved HRV-associated symptoms in children	55
Escherichia coli	Prevented K88+, K99+, 987P+ ETEC infection in neonatal piglets	63, 65
	Inhibited adhesion of ETEC K88 to piglet intestinal mucosa, *in vitro*	64
	Inhibited attachment of F18+ *E. coli* to porcine intestinal mucosa *in vitro*, and reduced diarrheal cases in infected piglets	66
	Protected neonatal calves from fatal enteric colibacillosis	67
	Prevented diarrhea in rabbits challenged with ETEC	68
	Inhibited growth of *E. coli* and colonization of cells by EPEC, using IgY against BfpA	69
	Protection of chickens against APEC-associated respiratory disease	70
Salmonella spp.	Inhibited SE and ST growth *in vitro*	72
	Inhibited adhesion of SE to human intestinal cells *in vitro*	73
	Protected mice challenged with SE or ST from experimental salmonellosis, using IgY against OMP, LPS and Fla	75, 81
	Protected mice challenged with SE from experimental salmonellosis, using IgY against SEF14	82
	Prevented fatal salmonellosis in neonatal calves exposed to ST or S. Dublin	83
Pseudomonas aeruginosa	Prevented pathogensis of *P. aeruginosa in vitro*	85
	Inhibited *P. aeruginosa* adhesion to epithelial cells *in vitro*	86
	Prolonged time before *P. aeruginosa* re-colonization, and prevented chronic infection in CF patients	84, 87
	Anti-*P. aeruginosa* IgY stable in saliva for 8 hrs	88
Streptococcus mutans	Reduced caries development in rats	91
	Prevented *S. mutans* adhesion *in vitro*, and *in vivo* in humans	93
	Prevented *S. mutans* accumulation and reduction of caries in rats using IgY against *S. mutans* GBP	94
	Reduced caries incidence in desalivated rats, using IgY against Gtf	89
Helicobacter pylori	Inhibited growth of *H. pylori* and reduced binding to human gastric cells, *in vitro*	97, 98
	Decreased gastric mucosal injury and inflammation in a gerbil model of *H. pylori*-induced gastritis	97, 98
Bovine coronavirus	Protected neonatal calves from BCV-induced diarrhea	101
Yersinia ruckeri	Protected rainbow trout against *Y. ruckeri* infection	103

(*Continued*)

Table 21.2 (*Continued*)

Pathogen	Effect of IgY	Ref.
Edwardsiella tarda	Prevented Edwardsiellosis in Japanese eels infected with *E. tarda*	104
Staphylococcus aureus	Inhibited production of *Staphylococcus aureus* enterotoxin-A	85
IBDV	Protected chicks from infectious bursal disease virus	106
PEDV	Protected piglets against porcine epidemic diarrhea virus	107
Cryptosporidium	Prevented cryptosporidiosis in mice	108

in children in developing countries and travelers to these countries, accounting for one billion diarrheal episodes annually (58). It is also a problem in neonatal calves (59) and piglets (60), causing significant economic losses to the pig industry from both mortality and reduced growth rates, killing 1.5–2.0% of weaned pigs (61).

The strains of ETEC which are associated with intestinal colonization and cause severe diarrhea are the K88, K99, and 987P fimbrial adhesins (62). The production of IgY against these fimbrial antigens has been described and was found to inhibit the binding of *E. coli* K88+, K99+, and 987P+ strains to porcine epithelial cells (63) and to porcine intestinal mucus (64), *in vitro*. The antifimbrial IgY was also found to protect piglets against infection with each of the three strains of *E. coli* in a dose dependent manner in passive immunization trials (63). In another feeding study, 21 day old pigs were challenged with ETEC. AntiK88 IgY was administered to the piglets in milk three times a day for 2 days. Control piglets developed severe diarrhea within 12 hrs and 30% of the pigs died. In contrast, the pigs that were fed IgY exhibited no signs of diarrhea 24 or 48 hrs after treatment (65). Imberechts et al. (66) produced IgY against *E.coli* F18ac fimbriae, and found that the IgY inhibited attachment of F18ac+ *E. coli* to porcine intestinal mucosa *in vitro*, and decreased incidences of diarrhea and death in animals infected with F18ac+ *E. coli*.

The passive protective effect of antiETEC IgY against fatal enteric colibacillosis in neonatal calves has also been examined (67). Calves fed milk containing IgY had only transient diarrhea, 100% survival and good body weight gain. O'Farrelly et al. (68) reported the use of IgY for the prevention of diarrhea in rabbits through the oral administration of antiETEC IgY. The use of IgY against enteropathogenic *E. coli* (EPEC), another major cause of childhood diarrhea, has also been examined. de Almeida et al. (69) produced IgY against a recombinant EPEC protein (BfpA), one of the factors required for EPEC pathogenesis, and found that the antiBfpA antibodies were capable of inhibiting the colonization of cells by EPEC, as well as inhibiting *E. coli* growth *in vitro*. Finally, IgY has also been suggested for the passive protection of chickens against respiratory disease caused by avian pathogenic *E. coli* (APEC) (70).

Because anti*E. coli* IgY has been shown to inhibit binding of *E. coli* both *in vitro* and *in vivo*, the clinical application of IgY for the passive protection against *E. coli* related diarrhea in humans is now being examined.

21.3.1.3 Salmonella species

Salmonella species are a significant cause of food poisoning, and it is estimated that 2 to 4 million cases of salmonellosis occur in the U.S. annually (71). Symptoms include fever, abdominal pain, headache, malaise, lethargy, skin rash, constipation, and changes in mental state. However, infants, the elderly, and immunocompromised individuals may develop more severe symptoms. In these cases, the infection may spread from the intestines to the blood stream and to other sites in the body, and can cause death. *Salmonella enteritidis* (SE) and *Salmonella typhimurium* (ST), in particular, are the major agents of salmonellosis (71).

Lee et al. (72) found that IgY against SE and ST was capable of inhibiting the growth of SE and ST *in vitro* when added to culture medium. They also suggested that the IgY bound to antigens expressed on the *Salmonella* surface, resulting in structural alterations on the bacterial cell. AntiSE IgY was also found to inhibit the adhesion of SE to human intestinal cells *in vitro* (73).

Salmonella has several surface components which are virulence related, including outer membrane protein (OMP) (74,75), lipopolysaccharides (LPS) (76,77), flagella (Fla), and in some strains fimbrial antigens (78,79). OMP is involved in pathogenicity determination (80), and has been used successfully for vaccine applications in both active and passive immunization studies (74,75). Lipopolysaccharides have also been shown to be highly immunogenic, producing large amounts of LPS specific IgY, and have demonstrated potential application for the inhibition of *Salmonella* adhesion (76,77). The ability of antiOMP, antiLPS or antiflagella (Fla) IgY to passively protect mice against experimental salmonellosis was examined. In mice challenged with SE, administration of specific IgY resulted in a survival rate of 80% (OMP specific IgY), 47% (LPS specific IgY), and 60% (Fla specific IgY), compared to only 20% in control mice. In the case of ST, the survival rate was 40%, 30%, and 20% using OMP, LPS or Fla specific IgY, while none of the control mice survived (81). A novel fimbrial antigen, SEF14, produced primarily by SE and *S. dublin* strains, has been described (82). Mice challenged with SE and treated with antiSEF14 IgY had a survival rate of 77.8%, compared to a 32% survival rate in the control mice.

Salmonella infection in calves is also a worldwide problem, with ST and *S. dublin* accounting for most salmonellosis within the first 2 weeks after birth. The passive protective effect of IgY was investigated by orally inoculating calves with SE or *S. dublin*, and administering IgY against SE or *S. dublin* orally 3 times a day for 7 to 10 days after challenge. All control calves died within 7–10 days. Calves given a low titer dose of IgY displayed 60–70% mortality, and calves treated with a high titer dose of IgY had only fever and diarrhea (83).

21.3.1.4 Pseudomonas aeruginosa

Cystic fibrosis (CF) is the most common fatal genetic disease in Caucasians. Colonization of the airways of CF patients by *Pseudomonas aeruginosa* (PA) is the main cause of morbidity and mortality, and once a chronic infection has been established it is very difficult to eliminate, even with the use of antibiotics (84).

It has been shown that specific IgY is capable of preventing the pathogenesis of PA (85). Carlander et al. (86) found that IgY raised against PA was capable of inhibiting adhesion of PA to epithelial cells *in vitro*, but did not inhibit bacterial growth, suggesting that specific IgY might be capable of interfering with the bacterial infection process and prevent colonization in CF patients.

In vivo studies with CF patients have demonstrated that antiPA IgY, when administered orally to CF patients on a continuous basis in the form of a mouth rinse, is capable of preventing, or prolonging the time before, PA recolonization (84,87). Kollberg et al. (84) found that the antiPA IgY prevented recolonization in 40% of CF patients, compared to only 14% with no recolonization in the control (untreated) group. As well, none of the IgY treated patients became chronically colonized with PA, in contrast to recolonization in 24% of the control patients. These results indicate that antiPA IgY may reduce the need for antibiotic treatment of CF patients.

The stability of antiPA IgY in saliva from healthy individuals, following a mouth rinse with an aqueous IgY solution, was examined (88). Antibody activity in the saliva remained after 8 hours. After 24 hours, the antibody activity had decreased significantly, but was still detectable in some subjects, suggesting the use of specific IgY for the treatment of various local infections, such as the common cold, and tonsillitis.

21.3.1.5 Streptococcus mutans

The role of *Streptococcus mutans* in the development of dental caries is well recognized (89). The molecular pathogenesis of *S. mutans* associated dental caries involves a series of binding events that eventually lead to the accumulation of sufficient numbers of these cariogenic bacteria to cause disease (90). IgY against *S. mutans* MT8148 or cell associated glucosyltransferases (CA-Gtf), one of the major virulence factors in the pathogenesis of dental caries (89), was prepared and tested against dental caries (91,92). Consumption of a cariogenic diet which contained more than 2% IgY containing yolk powder resulted in significantly lower caries scores (91), and effective passive protection against the colonization of *S. mutans* in the oral cavity, in a rat model of dental caries. It has also been reported that mouth rinse containing IgY specific to *S. mutans* was effective in preventing the dental plaque of humans *in vitro* and *in vivo* (93). Krüger et al. (89) found that the incidence of caries in desalivated rats fed antiCA-Gtf IgY was decreased in severity and number, indicating that IgY may be effective as prophylaxis for high caries risk patients.

Glucan-binding proteins (GBPs) are believed to be involved in *S. mutans* biofilm development, and antibodies against GBP-B appear to have the potential to modulate infection and disease caused by *S. mutans* (94). Smith et al. (94) found that rats treated with antiGBP-B IgY displayed a decrease in *S. mutans* accumulation, as well as a decrease in the overall amount of dental caries, as compared to control rats.

21.3.1.6 Helicobacter pylori

Helicobacter pylori is the most common cause of gastritis and gastric ulcers, and has been implicated in the development of gastric carcinomas. Around 25–50% of the population carry *H. pylori*, and this number is higher in developing countries (95). Antibiotic therapy is often employed to treat *H. pylori* infection, but this therapy is often undesirable due to increasing medical costs, low eradication rate, and the development of antibiotic resistant strains (29,96).

Anti*H. pylori* IgY was shown by Shin et al. (97) to significantly inhibit the growth of *H. pylori, in vitro*, and inhibit urease activity (a critical virulence factor), as well as reduce the binding of *H. pylori* to human gastric cells. The oral administration of specific IgY decreased gastric mucosal injury and inflammation *in vivo*, using a gerbil model of *H. pylori* induced gastritis (97,98), suggesting the use of anti*H. pylori* antibodies for the treatment of *H. pylori* associated gastric diseases.

Immunodominant *H. pylori* proteins were also identified, and used to produce highly specific anti*H. pylori* and antiurease IgY, to address concerns that antibodies produced against whole cell *H. pylori* might cross react with other bacteria, including normal human flora (96,99).

21.3.1.7 Coronavirus

Bovine coronavirus (BCV) is an important cause of neonatal calf diarrhea and acute diarrhea in adult cattle (100). Ikemori et al. (101) examined the protective effect of antiBCV antibodies in egg yolk in calves challenged with BCV. They found that control calves which received no antibodies experienced severe diarrhea and died within 6 days after infection. Calves fed milk containing egg yolk, however, all survived and had positive weight gain. The results indicate that orally administered egg yolk antibodies are capable of passively protecting calves against BCV infection.

21.3.1.8 Fish Pathogens

Yersinia ruckeri is the causative agent of enteric redmouth disease, a systemic bacterial septicemia, in salmonid fish (102). The persistence of *Y. ruckeri* in carrier fish for long periods and the shedding of bacteria in feces can present a continuing source of infection. The clearance of *Y. ruckeri* from carrier fish by orally administered anti*Y. ruckeri* antibody would provide a cost effective alternative to slaughtering fish which pose a health risk. The passive immunization of rainbow trout (*Oncorhynchus mykiss)* against *Y. ruckeri* infection has been examined (103). Groups of rainbow trout fed anti*Y. ruckeri* IgY prior to being challenged with *Y. ruckeri* had lower mortalities compared with fish receiving normal feed. The group fed IgY also appeared to have a lower number of infected fish, based on organ and intestine culture. The number of IgY fed fish carrying *Y. ruckeri* in intestine samples also appeared to be lower than the normal feed controls, regardless of whether the feeding was given before or after the challenge. The oral administration of specific IgY against fish pathogens could provide an alternative to the use of antibiotics and chemotherapy for prevention of bacterial diseases in fish farms, and would provide a novel approach for the prevention of viral infections in fish, as no medicine has been reported to be effective.

Edwardsiella tarda, another important fish pathogen, is spread by infection through the intestinal mucosa. Edwardsiellosis in Japanese eels is a serious problem for the eel farming industry, and egg yolk antibodies have been investigated for the prevention of this infectious disease, as treatment with antibiotics has been found to promote the growth of bacteria resistant strains (104). Eels were challenged with *E. tarda*, and anti*E. tarda* IgY was orally administered. Infected eels died within 15 days, whereas the eels given IgY survived without any symptoms of *E. tarda* infection (104,105).

21.3.1.9 Others

In addition, specific IgY has been shown to be effective at preventing and treating several other pathogens. Sugita-Konishi et al. (85) found that specific IgY was capable of inhibiting the production of *Staphylococcus aureus* enterotoxin-A *in vitro*. Its use has also been suggested for the passive protection of chicks against infectious bursal disease virus (IBDV) (106), for the protection against porcine epidemic virus (PEDV) in piglets (107), and for the prevention of cryptosporidiosis due to *Cryptosporidium* infection (108). Finally, IgY against *Bordetella bronchiseptica*, *Pasteurella multocida*, and *Actinobacillus pleuropneumoniae*, causative agents of swine respiratory diseases, has been proposed as an alternative method to control infectious respiratory diseases in pigs (109,110).

21.4 ANTIBODY STABILITY

21.4.1 Physiology of the Gastrointestinal Tract

The human gastrointestinal tract consists of the stomach, small intestine (which is divided into the duodenum, jejunum, and ileum), and the large intestine (111). Ingested proteins are subjected to denaturation by the acidic pH of the stomach, as well as to degradation by the proteolytic enzymes present in the stomach and small intestine (112).

The pH of the stomach ranges from 1 to 3.5 due to secretion of hydrochloric acid by the parietal cells (112–114). In neonates, however, the gastric fluid has a pH near neutral, which rapidly decreases to a pH below 3 within the first few days after birth (2). In both infants and adults, the pH of the gastric fluid will rapidly rise to about 4 or 5 soon after food reaches the stomach (111), but will return to low pH, occurring much more slowly in infants than in adults (115). It is in the stomach that proteins are subjected to digestion by pepsin, which is secreted by chief cells (116). Pepsin is optimally active at pH 1.8 to 3.5 (111), but will start to lose activity around pH 4 and is completely inactivated in alkaline conditions (111,117). The digestive activity of pepsin is terminated when the gastric contents are mixed with alkaline pancreatic juice in the small intestine (116). The extent of gastric protein digestion is determined by the physical state of the ingested protein, the activity of the pepsin in the gastric juice, and the length of time it stays in the stomach (111). The transit time in the stomach has been found to vary considerably, and has been reported to range from 0.5 to 4.5 (112,118), with a median time of 1 to 1.5 hours (114,118).

Trypsin and chymotrypsin continue digestion in the small intestine, where the pH ranges from around pH 6 to 7.4 (113,114). Transit through the small intestine has been reported to take from 1 to 4 hours (119), but may be as high as 7.5 hours (114). The pH rises to around 7.5 to 8 in the large intestine, where the transit time can be as long as 17 hours (112,114).

Along with the duration of exposure in each of the segments of the gastrointestinal tract, the ratio of enzyme to protein will also determine to what extent a protein is degraded (112).

21.4.2 Degradation and Stability of IgY

Several investigations have been carried out *in vitro* to study the effects of gastrointestinal enzymes and pH on the structure and activity of IgY, in order to predict its fate when administered orally (112). Antibodies were tested with pepsin at pH 2 to 4, and trypsin and chymotrypsin at a pH around neutral, simulating passage through the stomach and small intestine.

IgY has been found to be relatively resistant to trypsin and chymotrypsin digestion, but sensitive to pepsin digestion (120). It has, however, been found to be more susceptible to pepsin, trypsin, and chymotrypsin digestion than IgG (26,32). Using IgY raised against HRV, Hatta et al. (27) demonstrated that almost all of the IgY activity was lost following digestion with pepsin, but activity remained even after 8 hours incubation with trypsin or chymotrypsin. They found that digestion of IgY with pepsin at pH 2 resulted in complete hydrolysis of the antibody molecule, leaving only small peptides, however IgY digested with pepsin at pH 4 retained some of its activity, with heavy (H) and light (L) chains still present, along with smaller peptides. Shimizu et al. (120) also obtained similar results following the digestion of IgY against *Escherichia coli* with gastrointestinal enzymes.

In vivo results have also yielded similar results. Ikemori et al. (121) described the feeding of anti*E.coli* IgY to calves, and found that the activity of the antibodies was decreased significantly following passage through the stomach. Similarly, the feeding of specific IgY, also against *E. coli*, was carried out by Yokoyama et al. (122) using pigs. The morphology and physiology of the gastrointestinal tract of pigs has been found to be similar to that of humans, and therefore pigs represent a good animal model to study digestion in humans (123). Using piglets of varying ages, from 10 hours (neonatal) to 28 days old, it was observed that the IgY was degraded upon passage through the gastrointestinal tract, with some intact IgY detected in the intestine of neonatal pigs but to a lesser extent in the older pigs, possibly reflecting the increasing gastrointestinal maturity of the pigs.

Methods of encapsulating IgY, to provide resistance to the proteolytic and denaturing conditions of the stomach have been examined. Shimizu et al. (124) encapsulated IgY using lecithin or cholesterol liposomes, and found that the stability of the antibodies to *in vitro* gastric conditions was increased. Encapsulation of IgY in a water in oil in water (W/O/W) emulsion was examined by Shimizu and Nakane (125), who found that the antibody activity was significantly reduced by encapsulation alone, suggesting that denaturation of the IgY was occurring at the oil to water interface. Ikemori et al. (121) treated anti*E. coli* IgY with hydroxypropyl methylcellulose phthalate, an enteric coating substance used for drugs, and examined its activity following passage through the gastrointestinal tract of calves. They found that although the activity of the encapsulated IgY was decreased, it was not decreased to the same extent as the nonencapsulated antibodies. This research would indicate that the successful encapsulation of IgY would be an effective means for protecting IgY under gastric conditions.

21.5 BIOTECHNOLOGY OF AVIAN EGG ANTIBODIES

As demonstrated, the chicken is a useful animal for the production of specific antibodies for immunotherapeutic purposes. The many advantageous characteristics of IgY also make it ideal for use in diagnostic applications, where monoclonal antibodies have traditionally been used. Hybridomas can be cloned and grown in large quantity for indefinite periods of time, secreting high concentrations of monoclonal antibodies. The production of monoclonal IgY has also been examined, generated through the fusion of spleen cells from immunized chickens with chicken B cells, to produce monoclonal IgY-secreting hybridomas (126–131). These cells are capable of providing a consistent supply of antibody with a single and known specificity and homogeneous structure (1). Several researchers have described the production of single chain fragment variable region (scFv) monoclonal IgY via recombinant DNA technology (132–134), in order to improve upon the low levels of antibodies produced in the chicken hybridoma systems (133). Using RNA extracted from chicken hybridoma cells, Kim et al. (134) were able to express recombinant monoclonal IgY in *E. coli* and reported the production of 5–6 mg of IgY per litre of culture, suggesting that the production of monoclonal IgY on a large scale may be possible. Lillehoj and Sasaki (128) and Kim et al. (134) produced monoclonal IgY against *Eimeria* species, an intracellular parasite responsible for avian coccidiosis, in order to study the avian immune response to this parasite and to aid in vaccine development, as it was thought that monoclonal antibodies from mice would not adequately reflect the avian immune response due to the differences in antibody repertoire. Given that it is possible to produce antibodies against a vast array of antigens and epitopes using chickens, antibodies against any number of bacterial, viral, or biological antigens could be produced, suggesting the significant potential of avian antibodies for further use in immunodiagnostics and identification of disease markers, immunotherapy, and the treatment and prevention of disease, as well as affinity purification methods. Research is being continued to develop potential methods of production and application of egg yolk antibodies. Romito et al. (135) suggested the immunization of chickens with naked DNA, to elicit antigen specific IgY. Using DNA rather than a protein antigen could eliminate the protein expression and purification steps, and would allow the production of antibodies against pathogenic or toxic antigens. Not only are chickens useful for the production of specific IgY, but Mohammed et al. (136) demonstrated the deposition of recombinant human antibodies into the egg yolk of transgenic chickens, suggesting an extension of the production of specific IgY in eggs.

REFERENCES

1. Janeway, C.A., P. Travers. Structure of the antibody molecule and immunoglobulin genes. In: *Immunology: The immune system in health and disease.* Janeway, C.A., P. Travers, eds., London: Current Biology Ltd., 1996.

2. Carlander, D., H. Kollberg, P.-E. Wejaker, A. Larsson. Peroral immunotherapy with yolk antibodies for the prevention and treatment of enteric infections. *Immunol. Res.* 21:1–6, 2000.
3. Yolken, R.H., F. Leister, S.B. Wee, R. Miskuff, S. Vonderfecht. Antibodies to rotavirus in chickens' eggs: a potential source of antiviral immunoglobulins suitable for human consumption. *Pediatrics* 81:291–295, 1988.
4. Hammarström, L. Passive immunity against rotavirus in infants. *Acta Paediatr. Suppl.* 430:127–132, 1999.
5. Sim, J.S., H.H. Sunwoo, E.N. Lee. Ovoglobulin IgY. In: *Natural Food Antimicrobial Systems.* Naidu, A.S., ed., New York: CRC Press, 2000, pp 227–252.
6. Schade, R., C. Staak, C. Hendrikson, M. Erhard, H. Hugl, G. Koch, A. Larsson, W. Pollmann, M. van Regenmortel, E. Rijke, H. Spielmann, H. Steinbush, D. Straughan. The production of avian (egg yolk) antibodies: IgY. *ATLA* 24:925–934, 1996.
7. Carlander, D., J. Stalberg, A. Larsson. Chicken antibodies: a clinical chemistry perspective. *Ups. J. Med. Sci.* 104:179–190, 1999.
8. Sim, J.S., S. Nakai, W. Guenter. *Egg Nutrition and Biotechnology.* Oxon, UK: CAB Publishing, 1999.
9. Janson, A.K., C.I. Smith, L. Hammarström. Biological properties of yolk immunoglobulins. *Adv. Exp. Med. Biol.* 371:685–690, 1995.
10. Reynaud, C.-A., V. Anquez, J.-C. Weill. Somatic hyperconversion diversifies the single V_H gene of the chicken with a high incidence in the D region. *Cell* 59:171–183, 1989.
11. Reynaud, C.-A., A. Dahan, V. Anquez, J.-C. Weill. The chicken D locus and its contribution to the immunoglobulin heavy chain repertoire. *Eur. J. Immunol.* 21:2661–2670, 1991.
12. Pink, J.R.L., O. Vainio, A.-M. Rijnbeek. Clones of B lymphocytes in individual follicles of the bursa of Fabricius. *Eur. J. Immunol.* 15:83–87, 1985.
13. Sharma, J.M. The structure and function of the avian immune system. *Acta Vet. Hung.* 45:229–238, 1997.
14. Leslie, G.A., L.N. Martin. Studies on the secretory immunologic system of fowl, 3: serum and secretory IgA of the chicken. *J. Immunol.* 110:1–9, 1973.
15. Loeken, M.R., T.F. Roth. Analysis of maternal IgG subpopulations which are transported into the chicken oocyte. *Immunology* 49:21–28, 1983.
16. Tressler, R.L., T.F. Roth. IgG receptors on the embryonic chick yolk sac. *J. Biol. Chem.* 262:15406–15412, 1987.
17. Morrison, S.L., M.S. Mohammed, L.A. Wims, R. Trinh, R. Etches. Sequences in antibody molecules important for receptor-mediated transport into the chicken egg yolk. *Mol. Immunol.* 38:619–625, 2002.
18. Rose, M.E., E. Orlans, N. Buttress. Immunoglobulin classes in the hen's eggs: their segregation in yolk and white. *Eur. J. Immunol.* 4:521–523, 1974.
19. Warr, G.W., K.E. Magor, D.A. Higgins. IgY: clues to the origins of modern antibodies. *Immunol. Today* 16:392–398, 1995.
20. Shimizu, M., H. Nagashima, K. Sano, K. Hashimoto, M. Ozeki, K. Tsuda, H. Hatta. Molecular stability of chicken and rabbit immunoglobulin G. *Biosci. Biotech. Biochem.* 56:270–274, 1992.
21. Ohta, M., J. Hamako, S. Yamamoto, H. Hatta, M. Kim, T. Yamamoto, S. Oka, T. Mizuochi, F. Matsuura. Structure of asparagine-linked oligosaccharides from hen egg yolk antibody (IgY). Occurrence of unusual glucosylated oligo-mannose type oligosaccharides in a mature glycoprotein. *Glycoconj. J.* 8:400–413, 1991.
22. Matsuura, F., M. Ohta, K. Murakami, Y. Matsuki. Structures of asparagines linked oligosaccharides of immunoglobulins (IgY) isolated from egg-yolk of Japanese quail. *Glycoconj. J.* 10:202–213, 1993.
23. Polson, A., M.B. von Wechmar, G. Fazakerley. Antibodies to proteins from yolk of immunized hens. *Immunol. Commun.* 9:495–514, 1980.
24. Davalos-Pantoja, L., J.L. Ortega-Vinuesa, D. Bastos-Gonzalez, R. Hidalgo-Alvarez. A comparative study between the adsorption of IgY and IgG on latex particles. *J. Biomater. Sci. Polym. Ed.* 11:657–673, 2000.

25. Hersh, R.T., A.A. Benedict. Aggregation of chicken gamma-G immunoglobulin in 1.5 M sodium chloride solution. *Biochim. Biophys. Acta* 115:242–244, 1966.
26. Shimizu, M., H. Nagashima, K. Hashimoto. Comparative studies on molecular stability of immunoglobulin G from different species. *Comp. Biochem. Physiol.* 106B:255–261, 1993.
27. Hatta, H., K. Tsuda, S. Akachi, M. Kim, T. Yamamoto, T. Ebina. Oral passive immunization effect of anti-human rotavirus IgY and its behaviour against proteolytic enzymes. *Biosci. Biotech. Biochem.* 57:1077–1081, 1993.
28. Shimizu, M., H. Nagashima, K. Hashimoto, T. Suzuki. Egg yolk antibody (IgY) stability in aqueous solution with high sugar concentrations. *J. Food Sci.* 59:763–766, 1994.
29. Lee, K.A., S.K. Chang, Y.J. Lee, J.H. Lee, N.S. Koo. Acid stability of anti-Helicobacter pylori IgY in aqueous polyol solution. *J. Biochem. Mol. Biol.* 35:488–493, 2002.
30. Chang, H.M., R.F. Ou-Yang, Y.T. Chen, C.C. Chen. Productivity and some properties of immunoglobulin specific against *Streptococcus mutans* serotype c in chicken egg yolk (IgY). *J. Agric. Food Chem.* 47:61–66, 1999.
31. Jaradat, Z.W., R.R. Marquardt. Studies on the stability of chicken IgY in different sugars, complex carbohydrates and food materials. *Food Agric. Immunol.* 12:263–272, 2000.
32. Otani, H., K. Matsumoto, A. Saeki, A. Hosono. Comparative studies on properties of hen egg yolk IgY and rabbit serum IgG antibodies. *Lebensm. Wiss. Technol.* 24:152–158, 1991.
33. Pilz, I., E. Schwarz, W. Palm. Small-angle x-ray studies of the human immunoglobulin molecule. *Eur. J. Biochem.* 75:195–199, 1977.
34. Schade, R., C. Pfister, R. Halatsch, P. Henklein. Polyclonal IgY antibodies from chicken egg yolk: an alternative to the production of mammalian IgG type antibodies in rabbits. *ATLA* 19:403–419, 1991.
35. Gassmann, M., P. Thommes, T. Weiser, U. Hubscher. Efficient production of chicken egg yolk antibodies against a conserved mammalian protein. *FASEB J.* 4:2528–2532, 1990.
36. Knecht, W., R. Kohler, M. Minet, M. Loffler. Anti-peptide immunoglobulins from rabbit and chicken eggs recognize recombinant human dihydroorotate dehydrogenase and a 44-kDa protein from rat liver mitochondria. *Eur. J. Biochem.* 236:609–613, 1996.
37. Larsson, A., D. Carlander, M. Wilhelmsson. Antibody response in laying hens with small amounts of antigen. *Food Agr. Immunol.* 10:29–36, 1998.
38. Nakai, S., E. Li-Chan, K.V. Lo. Separation of immunoglobulin from egg yolk. In: *Egg Uses and Processing Technologies: New Developments*, Sim, J.S., S. Nakai, eds.,Wallingford, UK: CAB International, 1994, pp 94–105.
39. Akita, E.M., S. Nakai. Immunoglobulins from egg yolk: isolation and purification. *J. Food Sci.* 57:629–634, 1992.
40. Larsson, A., R. Balow, T.L. Lindahl, P. Forsberg. Chicken antibodies: taking advantage of evolution: a review. *Poult. Sci.* 72:1807–1812, 1993.
41. Cotterill, O.J., L.E. McBee. Egg breaking. In: *Egg Science and Technology*, 4th ed., Stadelman, W.J., O.J. Cotterill, eds., New York: The Haworth Press, Inc., 1995, pp 231–263.
42. Sharma, J.M. Introduction to poultry vaccines and immunity. *Adv. Vet. Med.* 41:481–494, 1999.
43. Casadevall, A., M.D. Scharff. Return to the past: the case for antibody-based therapies in infectious diseases. *Clin. Infect. Dis.* 21:150–161, 1995.
44. Wierup, M. The control of microbial diseases in animals: alternatives to the use of antibodies. *Int. J. Antimicrob. Agents* 14:315–319, 2000.
45. Zeitlin, L., R.A. Cone, T.R. Moench, K.J. Whaley. Preventing infectious disease with passive immunization. *Microbes Infect.* 2:701–708, 2000.
46. Kapikian, A.Z., R.M. Chanock. Rotaviruses. In: *Fields Virology,* 3rd ed., Fields, B.N., D.N. Knipe, P.M. Howley, R.M. Chanock, J.L. Melnick, T.P. Monath, B. Rolzman, S.E. Strauss, eds., New York: Raven Press, 1996, pp 1657–1708.
47. Clark, H.F., R.I. Glass, P.A. Offit. Rotavirus vaccines. In: *Vaccines*, 3rd ed., S.A. Plotkin, W.A. Orentstein, eds., Philadelphia: W.B. Saunders Company, 1999, pp 987–1005.
48. Ludert, J.E., A.A. Krishnaney, J.W. Burns, P.T. Vo, H.B. Greenberg. Cleavage of rotavirus VP4 *in vivo*. *J. Gen. Virol.* 77:391–395, 1996.

49. Hochwald, C., L. Kivela. Rotavirus vaccine, live, oral, tetravalent (RotaShield). *Pediatr. Nurs.* 25:203–204, 1999.
50. Ebina, T. Prophylaxis of rotavirus gastroenteritis using immunoglobulin. *Arch. Virol. Suppl.* 12:217–223, 1996.
51. Kovacs-Nolan, J., E. Sasaki, D. Yoo, Y. Mine. Cloning and expression of human rotavirus spike protein, VP8*, in *Escherichia coli. Biochem. Biophys. Res. Commun.* 282:1183–1188, 2001.
52. Mackow, E.R., R.D. Shaw, S.M. Matsui, P.T. Vo, M.-N. Dang, HB Greenberg. The rhesus rotavirus gene encoding protein VP3: location of amino acids involved in homologous and heterologous rotavirus neutralization and identification of a putative fusion region. *Proc. Natl. Acad. Sci. USA* 85:645–649, 1988.
53. Both, G.W., A.R. Bellamy, D.B. Mitchell. Rotavirus protein structure and function. In: *Rotaviruses.* Ramig, R.F., ed., New York: Springer-Verlag, 1994, pp 67–106.
54. Desselberger, U., M.A. McCrae. The rotavirus genome. In: *Rotaviruses.* R.F. Ramig, ed., New York: Springer-Verlag, 1994, pp 31–66.
55. Sarker, S.A., T.H. Casswall, L.R. Juneja, E. Hoq, E. Hossain, G.J. Fuchs, L. Hammarström. Randomized, placebo-controlled, clinical trial of hyperimmunized chicken egg yolk immunoglobulin in children with rotavirus diarrhea. *J. Pediatr. Gastroenterol. Nutr.* 32:19–25, 2001.
56. Lee, J., L.A. Babiuk, R. Harland, E. Gibbons, Y. Elazhary, D. Yoo. Immunological response to recombinant VP8* subunit protein of bovine rotavirus in pregnant cattle. *J. Gen. Virol.* 76:2477–2483, 1995.
57. Kuroki, M., Y. Ohta, Y. Ikemori, R.C. Peralta, H. Yokoyama, Y. Kodama. Passive protection against bovine rotavirus in calves by specific immunoglobulins from chicken egg yolk. *Arch. Virol.* 138:143–148, 1994.
58. Sack, R.B. Antimicrobial prophylaxis of travellers' diarrhea: a selected summary. *Rev. Infect. Dis.* 8:S160–S166, 1986.
59. Moon, H.W., S.C. Whipp, S.M. Skartvedt. Etiologic diagnosis of diarrheal diseases of calves: frequency and methods for detecting enterotoxic and K99 antigen produced by *Escherichia coli. Am. J. Vet. Res.* 37:1025–1029, 1976.
60. Morris, J.A., W.J. Sojka. *Escherichia coli* as a pathogen in animals. In: *The Virulence of Escherichia* coli. Sussman, M., ed., London: Academic Press, Inc., 1985, pp 47–77.
61. Hampson, D.J. Postweaning *Escherichia coli* diarrhoea in pigs. In: *Escherichia coli in Domestic Animals and Humans.* Gyles, C.L., ed., Oxon, UK: CAB International, 1994, pp 171–191.
62. Parry, S.H., D.M. Rooke. Adhesins and colonization factors of *Escherichia coli.* In: *The Virulence of Escherichia coli.* Sussman, M., ed., London: Academic Press, Inc., 1985, pp 79–159.
63. Yokoyama, H., R.C. Peralta, R. Diaz, S. Sendo, Y. Ikemori, Y. Kodama. Passive protective effect of chicken egg yolk immunoglobulins against experimental enterotoxigenic *Escherichia coli* infection in neonatal piglets. *Infect. Immun.* 60:998–1007, 1992.
64. Jin, L.Z., K. Samuel, K. Baidoo, R.R. Marquardt, A.A. Frohlich. *In vitro* inhibition of adhesion of enterotoxigenic *Escherichia coli* K88 to piglet intestinal mucus by egg yolk antibodies. *FEMS Immunol. Med. Microbiol.* 21:313–321, 1998.
65. Marquardt, R.R., L.Z. Jin, J.W. Kim, L. Fang, A.A. Frohlich, S.K. Baidoo. Passive protective effect of egg yolk antibodies against enterotoxigenic *Escherichia coli* K88+ infection in neonatal and early weaned piglets. *FEMS Immunol. Med. Microbiol.* 23:283–288, 1999.
66. Imberechts, H., P. Deprez, E. Van Driessche, P. Pohl. Chicken egg yolk antibodies against F18ab fimbriae of *Escherichia coli* inhibit shedding of F18 positive *E. coli* by experimentally infected pigs. *Vet. Microbiol.* 54:329–341, 1997.
67. Ikemori, Y., M. Kuroki, R.C. Peralta, H. Yokoyama, Y. Kodama. Protection of neonatal calves against fatal enteric colibacillosis by administration of egg yolk powder from hens immunized with K99-piliated enterotoxigenic *Escherichia coli. Am. Vet. Res.* 53:2005–2008, 1992.
68. O'Farrelly, C., D. Branton, C.A. Wanke. Oral ingestion of egg yolk immunoglobulin from hens immunized with an enterotoxigenic *Escherichia coli* strain prevents diarrhea in rabbits challenged with the same strain. *Infect. Immun.* 60:2593–2597, 1992.

69. de Almeida, C.M., V.M. Quintana-Flores, E. Medina-Acosta, A. Schriefer, M. Barral-Netto, W Dias da Silva. Egg yolk anti-BfpA antibodies as a tool for recognizing and identifying enteropathogenic *Escherichia coli. Scand. J. Immunol.* 57:573–582, 2003.
70. Kariyawasam, S., B.N. Wilkie, C.L. Gyles. Resistance of broiler chickens to *Escherichia coli* respiratory tract infection induced by passively transferred egg-yolk antibodies. *Vet. Microbiol.* 98:273–284, 2004.
71. Bell, C., A. Kyriakides. *Salmonella: A Practical Approach to the Organism and its Control in Foods.* New York: Blackie Academic & Professional, 1998.
72. Lee, E.N., H.H. Sunwoo, K. Menninen, J.S. Sim. *In vitro* studies of chicken egg yolk antibodies (IgY) against *Salmonella enteritidis* and *salmonella typhimurium. Poult. Sci.* 81:632–641, 2002.
73. Sugita-Konishi, Y., M. Ogawa, S. Arai, S. Kumagai, S. Igimi, M. Shimizu. Blockade of *Salmonella enteritidis* passage across the basolateral barriers of human intestinal epithelial cells by specific antibody. *Microbiol. Immunol.* 44:473–479, 2000.
74. Isibasi, A., V. Ortiz, M. Vargas, J. Paniagua, C. Gonzales, J. Moreno, J. Kumate. Protection against *Salmonella typhi* infection in mice after immunization with outer membrane proteins isolated from *Salmonella typhi* 9, 12, d, Vi. *Infect. Immun.* 56:2953–2959, 1988.
75. Udhayakumar, V., V.R. Muthukkaruppan. Protective immunity induced by outer membrane proteins of *Salmonella typhimurium* in mice. *Infect. Immun.* 55:816–821, 1987.
76. Sunwoo, H.H., T. Nakano, T. Dixon, J.S. Sim. Immune responses in chicken against lipopolysaccharides of *Escherichia coli* and *Salmonella typhimurium. Poultry Sci.* 75:342–345, 1996.
77. Mine, Y. Separation of *Salmonella enteritidis* from experimentally contaminated liquid eggs using a hen IgY immobilized immunomagnetic separation system. *J. Agric. Food Chem.* 45:3723–3727, 1997.
78. Thorns, C.J., M.G. Sojka, D. Chasey. Detection of a novel fimbrial structure on the surface of *Salmonella enteritidis* by using a monoclonal antibody. *J. Clin. Microbiol.* 28:2409–2414, 1990.
79. Thorns, C.J., M.G. Sojka, M. McLaren, M. Dibb-Fuller. Characterization of monoclonal antibodies against a fimbrial structure of *Salmonella enteritidis* and certain other serogroup D salmonellae and their application as serotyping reagents. *Res. Vet. Sci.* 53:300–308, 1992.
80. Galdiero, F., M.A. Tufano, M. Galdiero, S. Masiello, M.D. Rosa. Inflammatory effects of *Salmonella typhimurium* porins. *Infect. Immun.* 58:3183–3186, 1990.
81. Yokoyama, H., K. Umeda, R.C. Peralta, T. Hashi, F. Icatlo, M. Kuroki, Y. Ikemori, Y. Kodama. Oral passive immunization against experimental salmonellosis in mice using chicken egg yolk antibodies specific for *Salmonella enteritidis* and *S. typhimurium. Vaccine* 16:388–393, 1998.
82. Peralta, R.C., H. Yokoyama, Y. Ikemori, M. Kuroki, Y. Kodama. Passive immunization against experimental salmonellosis in mice by orally administered hen egg yolk antibodies specific for 14-kDa fimbriae of *Salmonella enteritidis. J. Med. Microbiol.* 41:29–35, 1994.
83. Yokoyama, H., R.C. Peralta, K. Umeda, T. Hashi, F.C. Icatlo, M. Kuroki, Y. Ikemori, Y. Kodama. Prevention of fatal salmonellosis in neonatal calves, using orally administered chicken egg yolk *Salmonella*-specific antibodies. *Am. J. Vet. Res.* 59:416–420, 1998.
84. Kollberg, H., D. Carlander, H. Olesen, P.-E. Wejaker, M. Johannesson, A. Larsson. Oral administration of specific yolk antibodies (IgY) may prevent *Pseudomonas aeruginosa* infections in patients with cystic fibrosis: a phase I feasibility study. *Pediatr. Pulmonol.* 35:433–440, 2003.
85. Sugita-Konishi, Y., K. Shibata, S.S. Yun, H.K. Yukiko, K. Yamaguchi, S. Kumagai. Immune functions of immunoglobulin Y isolated from egg yolk of hens immunized with various infectious bacteria. *Biosci. Biotech. Biochem.* 60:886–888, 1996.
86. Carlander, D.O., J. Sundstrom, A. Berglund, A. Larsson, B. Wretlind, H.O. Kollberg. Immunoglobulin Y (IgY): a new tool for the prophylaxis against *Pseudomonas aeruginosa* in cystic fibrosis patients. *Pediatr. Pulmonol.* 19:241 (Abstract), 1999.
87. Carlander, D., H. Kollberg, P.E. Wejaker, A Larsson. Prevention of chronic *Pseudomonas aeruginosa* colonization by gargling with specific antibodies: a preliminary report.

In: *Egg Nutrition and Biotechnology*. Sim, J.S., S. Nakai, W. Guenter, eds., New York: CABI Publishing, 2000, pp 371–374.

88. Carlander, D., H. Kollberg, A. Larsson. Retention of specific yolk IgY in the human oral cavity. *BioDrugs* 16:433–437, 2002.
89. Krüger, C., S.K. Pearson, Y. Kodama, A. Vacca Smith, W.H. Bowen, L. Hammarström. The effects of egg-derived antibodies to glucosyltransferases on dental caries in rats. *Caries. Res.* 38:9–14, 2004.
90. Hamada, S., H.D. Slade. Biology, immunology, and cariogenicity of *Streptococcus mutans*. *Microbiol. Rev.* 44:331–384, 1980.
91. Otake, S., Y. Nishihara, M. Makimura, H. Hatta, M. Kim, T. Yamamoto, M. Hirasawa. Protection of rats against dental caries by passive immunization with hen egg yolk antibody (IgY). *J. Dent. Res.* 70:162–166, 1991.
92. Hamada, S., T. Horikoshi, T. Minami, S. Kawabata, J. Hiraoka, T. Fujiwara, T. Ooshima. Oral passive immunization against dental caries in rats by use of hen egg yolk antibodies specific for cell-associated glucosyltransferase of *Streptococcus mutans*. *Infect. Immun.* 59:4161–4167, 1991.
93. Hatta, H., K. Tsuda, M. Ozeki, M. Kim, T. Yamamoto, S. Otake, M. Hirosawa, J. Katz, N.K. Childers, S.M. Michalek. Passive immunization against dental plaque formation in humans: effect of a mouth rinse containing egg yolk antibodies (IgY) specific to *Streptococcus mutans*. *Caries. Res.* 31:268–274, 1997.
94. Smith, D.J., W.F. King, R. Godiska. Passive transfer of immunoglobulin Y to *Streptococcus mutans* glucan binding protein B can confer protection against experimental dental caries. *Infect. Immun.* 69:3135–3142, 2001.
95. Dunn, B.E., H. Cohen, M.J. Blaser. *Helicobacter pylori*. *Clin. Microbiol. Rev.* 10:720-741, 1997.
96. Shin, J.-H., S.-W. Nam, J.-T. Kim, J.-B. Yoon, W.-G. Bang, I.-H. Roe. Identification of immunodominant *Helicobacter pylori* proteins with reactivity to *H. pylori*-specific egg-yolk immunoglobulin. *J. Med. Microbiol.* 52:217–222, 2003.
97. Shin, J.-H., M. Yang, S.W. Nam, J.T. Kim, N.H. Myung, W.-G. Bang, I.H. Roe. Use of egg yolk-derived immunoglobulin as an alternative to antibiotic treatment for control of *Helicobacter pylori* infection. *Clin. Diagn. Lab. Immunol.* 9:1061–1066, 2002.
98. Roe, I.H., S.W. Nam, M.R. Yang, N.H. Myung, J.T. Kim, J.H. Shin. The promising effect of egg yolk antibody (immunoglobulin yolk) on the treatment of *Helicobacter pylori*-associated gastric diseases. *Korean J. Gastroenterol.* 39:260–268, 2002.
99. Shin, J.-H., I.-H. Roe, H.-G. Kim. Production of anti-*Helicobacter pylori* urease-specific immunoglobulin in egg yolk using an antigenic epitope of *H. pylori* urease. *J. Med. Microbiol.* 53:31–34, 2004.
100. Kapil, S., A.M. Trent, S.M. Goyal. Excretion and persistence of bovine coronavirus in neonatal calves. *Arch. Virol.* 115:127–132, 1990.
101. Ikemori, Y., M. Ohta, K. Umeda, F.C. Icatlo Jr, M. Kuroki, H. Yokoyama, Y. Kodama. Passive protection of neonatal calves against bovine coronavirus-induced diarrhea by administration of egg yolk or colostrum antibody powder. *Vet. Microbiol.* 58:105–111, 1997.
102. Stevenson, R.M.W., D. Flett, B.T. Raymond. Enteric redmouth (ERM) and other enterobacterial infections of fish. In: *Bacterial Diseases of Fish*, Inglis, V., R.J. Roberts, N.R. Bromage, eds., Oxford: Blackwell Scientific Publications, 1993, pp 80–105.
103. Lee, S.B., Y. Mine, R.M.W. Stevenson. Effects of hen egg yolk immunoglobulin in passive protection of rainbow trout against *Yersinia ruckeri*. *J. Agric. Food Chem.* 48:110–115, 2000.
104. Hatta, H., K. Mabe, M. Kim, T. Yamamoto, M.A. Gutierrez, T. Miyazaki. Prevention of fish disease using egg yolk antibody. In: *Egg Uses and Processing Technologies: New Developments*. Sim, J.S., S. Nakai, eds., Oxon, UK: CAB International, 1994, pp 241–249.
105. Gutierrez, M.A., T. Miyazaki, H. Hatta, M. Kim. Protective properties of egg yolk IgY containing anti-*Edwardsiella tarda* antibody against paracolo disease in the Japanese eel, *Anguilla japonica* Temminck & Schlegel. *J. Fish Dis.* 16:113–122, 1993.

106. Eterradossi, N., D. Toquin, H. Abbassi, G. Rivallan, J.P. Cotte, M. Guittet. Passive protection of specific pathogen free chicks against infectious bursal disease by in-ovo injection of semi-purified egg-yolk antiviral immunoglobulins. *J. Vet. Med.* B44:371–383, 1997.
107. Kweon, C.-H., B.-J. Kwon, S.-R. Woo, J.-M. Kim, G.-H. Woo, D.-H. Son, W. Hur, Y.-S. Lee. Immunoprophylactic effect of chicken egg yolk immunoglobulin (IgY) against porcine epidemic diarrhea virus (PEDV) in piglets. *J. Vet. Med. Sci.* 62:961–964, 2000.
108. Cama, V.A., C.R. Sterling. Hyperimmune hens as a novel source of anti-*Cryptosporidium* antibodies suitable for passive immune transfer. *J. Protozool.* 38:42S–43S, 1991.
109. Ling, Y.S., Y.J. Guo, J.D. Li, L.K. Yang, Y.X. Luo, S.X. Yu, L.Q. Zhen, S.B. Qiu, G.F. Zhu. Serum and egg yolk IgG antibody titers from laying chickens vaccinated with *Pasteurella multocida*. *Avian Dis.* 42:186–189, 1998.
110. Shin, N.R., I.S. Choi, J.M. Kim, W. Hur, H.S. Yoo. Effective methods for the production of immunoglobulin Y using immunogens of *Bordatella bronchiseptica*, *Pasteurella multocida* and *Actinobacillus pleuropneumoniae*. *J. Vet. Sci.* 3:47–57, 2002.
111. Granger, D.N., J.A. Barrowman, P.R. Kvietys. *Clinical Gastrointestinal Physiology*. Toronto: W.B. Saunders Company, 1985.
112. Reilly, R.M., R. Domingo, J. Sandhu. Oral delivery of antibodies; future pharmacokinetic trends. *Clin. Pharmacokinet.* 4:313–323, 1997.
113. Evans, D.F., G. Pye, R. Bramley, A.G. Clark, T.J. Dyson, J.D. Hardcastle. Measurement of gastrointestinal pH profiles in normal ambulant human subjects. *Gut* 29:1035–1041, 1988.
114. Fallingborg, J., L.A. Christensen, M. Ingeman-Nielsen, B.A. Jacobsen, K. Abildgaard, H.H. Rasmussen, S.N. Rasmussen. Measurement of gastrointestinal pH and regional transit times in normal children. *J. Pediatr. Gastroenterol. Nutr.* 11:211–214, 1990.
115. Koldovsky, O. Development of human gastrointestinal functions: interaction of changes in diet composition, hormonal maturation, and fetal genetic programming. *J. Am. Coll. Nutr.* 3:131–138, 1984.
116. Castro, G.A. Digestion and absorption of specific nutrients. In: *Gastrointestinal Physiology,* 2nd ed., Johnson, E.R., ed., Toronto: The C.V. Mosby Company, 1981, pp 123–140.
117. Magee, D.F., A.F. Dalley. *Digestion and the Structure and Function of the Gut*. Basel, Switzerland: Karger, 1986.
118. Madsen, J.L. Gastrointestinal transit measurements, a scintographic method. *Dan. Med. Bull.* 41:398–411, 1994.
119. Davis, S.S. Small intestine transit. In: *Drug Delivery to the Gastrointestinal Tract*, Hardy, J.G., S.S. Davis, C.G. Wilson, eds., Toronto: John Wiley & Sons, 1989, pp 49–61.
120. Shimizu, M., R.C. Fitzsimmons, S. Nakai. Anti-*E. coli* immunoglobulin Y isolated from egg yolk of immunized chickens as a potential food ingredient. *J. Food Sci.* 53:1360–1366, 1988.
121. Ikemori, Y., M. Ohta, K. Umeda, R.C. Peralta, M. Kuroki, H. Yokoyama, Y. Kodama. Passage of chicken egg yolk antibodies treated with hydroxypropyl methylcellulose phthalate in the gastrointestinal tract of calves. *J. Vet. Med. Sci.* 58:365–367, 1996.
122. Yokoyama, H., R.C. Peralta, S. Sendo, Y. Ikemori, Y. Kodama. Detection of passage and absorption of chicken egg yolk immunoglobulins in the gastrointestinal tract of pigs by use of enzyme-linked immunosorbent assay and fluorescent antibody testing. *Am. J. Vet. Res.* 54:867–872, 1993.
123. Miller, E.R., D.E. Ullrey. The pig as a model for human nutrition. *Ann. Rev. Nutr.* 7:361–382, 1987.
124. Shimizu, M., Y. Miwa, K. Hashimoto, A. Goto. Encapsulation of chicken egg yolk immunoglobulin G (IgY) by liposomes. *Biosci. Biotechnol. Biochem.* 57:1445–1449, 1993.
125. Shimizu, M., Y. Nakane. Encapsulation of biologically active proteins in a multiple emulsion. *Biosci. Biotechnol. Biochem.* 59:492–496, 1995.
126. Nishinaka, S., T. Suzuki, H. Matsuda, M. Murata. A new cell line for the production of chicken monoclonal antibody by hybridoma technology. *J. Immunol. Methods* 139:217–222, 1991.
127. Asaoka, H., S. Nishinaka, N. Wakamiya, H. Matsuda, M. Murata. Two chicken monoclonal antibodies specific for heterophil hanganutziu-deicher antigens. *Immunol. Lett.* 32:91–96, 1992.

128. Lillehoj, H.S., K. Sasaki. Development and characterization of chicken-chicken B cell hybridomas secreting monoclonal antibodies that detect sporozoite and merozoite antigens of *Eimeria*. *Poult. Sci.* 73:1685–1693, 1994.
129. Nishinaka, S., H. Akiba, M. Nakamura, K. Suzuki, T. Suzuki, K. Tsubokura, H. Horiuchi, S. Furusawa, H. Matsuda. Two chicken B cell lines resistant to ouabain for the production of chicken monoclonal antibodies. *J. Vet. Med. Sci.* 58:1053–1056, 1996.
130. Matsushita, K., H. Horiuchi, S. Furusawa, M. Horiuchi, M. Shinagawa, H. Matsuda. Chicken monoclonal antibodies against synthetic bovine prion protein peptides. *J. Vet. Med. Sci.* 60:777–779, 1998.
131. Matsuda, H., H. Mitsuda, N. Nakamura, S. Furusawa, S. Mohri, T. Kitamoto. A chicken monoclonal antibody with specificity for the N-terminal of human prion protein. *FEMS Immunol. Med. Microbiol.* 23:189–194, 1999.
132. Yamanaka, H.I., T. Inoue, O. Ikeda-Tanaka. Chicken monoclonal antibody isolated by a phage display system. *J. Immunol.* 157:1156–1162, 1996.
133. Nakamura, N., Y. Aoki, H. Horiuchi, S. Furusawa, H.I. Yamamoto, T. Kitamoto, H. Matsuda. Construction of recombinant monoclonal antibodies from a chicken hybridoma line secreting a specific antibody. *Cytotechnology* 32:191–198, 2000.
134. Kim, J.K., W. Min, H.S. Lillehoj, S. Kim, E.J. Shon, K.D. Song, J.Y. Han. Generation and characterization of recombinant ScFv antibodies detecting *Eimeria acervulina* surface antigens. *Hybridoma* 20:175–181, 2001.
135. Romito, M., G.J. Viljoen, D.H. Du Plessis. Eliciting antigen-specific egg yolk IgY with naked DNA. *Biotechniques* 31:670–677, 2001.
136. Mohammed, S.M., S. Morrison, L. Wims, K. Ryan Trinh, A.G. Wildeman, J. Bonsellar, R.J. Etches. Deposition of genetically engineered human antibodies into the egg yolk of hens. *Immunotechnology* 4:115–125, 1998.

2.22

Application of Transgenic Fish Technology in Sea Food Production

Pinwen Peter Chiou, Jenny Khoo, and Thomas T. Chen

CONTENTS

22.1 INTRODUCTION

Fish into which a foreign gene or noncoding DNA fragment is artificially introduced and integrated into their genomes are called *transgenic fish*. Since 1985, a wide range of transgenic fish species have been produced mainly by microinjecting or electroporating homologous or heterologous transgenes into newly fertilized or unfertilized eggs, and sometimes sperm (11,21,27). To produce a desired transgenic fish, several factors should be considered. First, an appropriate fish species must be chosen. This decision would depend on the nature of each study and the availability of the fish holding facility. Second, a specific gene construct must be designed based on the special requirements of each study. For example, the gene construct may contain an open reading frame encoding a gene product of interest, and regulatory elements that regulate the expression of the gene in a temporal, spatial, or developmental manner, or some combination of the three. Third, the gene construct has to be introduced into sperm or embryos in order for the transgene to be integrated stably into the genome of the embryonic cells. Fourth, because not all instances of gene transfer are efficient, a screening method must be adopted for identifying transgenic individuals.

Since the development of the first transgenic fish in the mid 1980s, techniques to produce transgenic fish have improved tremendously, resulting in the production of many genetically modified (GM) fish species. In recent years, transgenic fish have been established as valuable models for different disciplines in biological research as well as human disease modeling. In addition, the application of transgenic fish technology to produce fish with beneficial traits, such as enhanced somatic growth and disease resistance, is also rising. In this chapter, we will review the progress of transgenic fish technology and its biotechnological applications.

22.2 METHODS OF TRANSGENIC FISH PRODUCTION

22.2.1 Types of Transgene Constructs

A transgene is a piece of nonself origin gene or DNA fragment that is introduced into fish for the production of transgenic fish. Transgenes are usually constructed in plasmids that contain appropriate promoter or enhancer elements for proper expression of transgenes in fish. A variety of promoters from nonfish species have been employed to control the expression of transgenes in fish (Table 22.1). Additionally, increasing number of fish promoters (Table 22.1) are available for the purpose of generating "all fish" expression cas-

Table 22.1
Examples of Promoters Used in Production of Transgenic Fish

	Origin Species	Promoter
Non-fish Origin:		
	Rat	GAP43[a] (Udvadia et al., 2001)
	Chicken	β-actin (Lu et al., 1992)
	Xenopus	elongation factor 1α (Amsterdam et al., 1995)
	Cytomegalovirus	CMV intermediate-early (Sarmasik et al., 2002)
	Moloney murine leukemia virus	MoMLV LTR (Lu et al., 1994)
Fish Origin:		
	Carp	β-actin (Lu et al., 2002)
	Goldfish	α1 tubulin[a] (Goldman et al., 2001)
	Medaka	elongation factor 1α (Wakamatsu et al. 2001a), β-actin (Hsiao et al., 2001)
	Zebrafish	H2A.F/Z (Pauls et al., 2001), α-actin[a] (Higashijima et al., 1997), heat shock protein 70[b] (Halloran et al., 2000)

[a]tissue-specific promoters- GAP43: nervous system; α1 tubulin: nervous system; α-actin: muscle (see Table 22.3 for more examples)
[b]Inducible promoters; LTR: long terminal repeat; H2A.F/Z: histone H2A.F/Z

settes, among which is the carp β-actin promoter, which has been shown to drive strong expression of transgenes in various fish cell types. Tissue specific promoters and inducible promoters such as zebra fish *heat shock protein 70* (*hsp70*) have also been used successfully to control the expression of transgenes in the intended tissue at the desired time. For example, activation of an *hsp70*-controlled transgene can be achieved by heat shock ;and, intriguingly, by focusing a sublethal laser microbeam onto individual cells that carry the transgene (28). The targeted cells appear normal after the treatment.

Depending on the purpose of the gene transfer studies, transgenes can be grouped into three major types: gain-of-function, reporter-function, and loss-of-function. The basic organization and the function of these transgenes are discussed below.

Gain-of-function transgenes are designed to add new functions to the transgenic individuals or to facilitate the identification of the transgenic individuals if the genes are expressed properly. The coding regions of the transgenes are usually homologous or heterologous genomic or cDNA sequences which encode polypeptide products (Figure 22.1A). The expression of the transgenes is usually driven by promoter or enhancer sequences of homologous or heterologous sources. If the transgenes contain functional signal peptide sequences, the gene products will be secreted out of the cells once they are synthesized. Transgenes containing the structural gene of mammalian or fish growth hormones (GH), or the respective cDNA fused to a functional promoter or enhancer of chicken or fish β-actin gene are examples of the gain-of-function transgene constructs. Expression of such transgenes in transgenic individuals has been shown to result in growth enhancement (12,16,52,84). Winter flounder antifreeze protein gene and chicken δ-crystalline gene, with their own promoter or enhancer sequences, are other commonly used gain-of-function transgenes used in various gene transfer studies (59,70).

Genes such as bacterial chloramphenicol acetyltransferase (CAT), β-galactosidase (β-gal), neomycin phosphotransferase II (*neo*^{R}), luciferase, and green fluorescent protein (GFP), are frequently used to produce gene construct with reporter-functions. These reporter genes serve as a convenient marker for indicating the success of gene transfer into the target cells, because the translational products of these genes can be easily monitored. Furthermore, because these gene products can be measured quantitatively, they are also routinely adopted for identifying the sequence of an undefined promoter or enhancer or measuring the relative activity of a promoter or enhancer element in question. A less frequently used reporter type is the one with target specificity. One such example is cameleon, a fluorescent calcium sensor, which was originally used to detect calcium transients in culture cells and pharyngeal muscle of *Caenorhabditis elegans*. Recently, cameleon and several other indicator genes have been used to establish transgenic zebra fish for monitoring neuronal activity (32). The prototypes of a reporter-function transgene are shown in Figure 22.1B.

Among the commonly used reporters, the jellyfish (*Aequorea victoria*) GFP and its variants have become the choice markers for transgenic fish because the detection of GFP signal is direct and easy, requiring no exogenous substrate (2). There are numerous GFP variants available, including those producing green, blue, and yellow fluorescence light. Together with the red fluorescence protein (RFP), these reporters can be easily detected in live transgenic fish such as zebra fish and medaka, whose embryos and body are transparent, without the need to sacrifice the fish (2,30). Additionally, these reporters can be used to study multiple transgenes simultaneously in the same fish. Since the first GFP transgenic fish in 1995 (2), GFP tagged transgenic embryos and fish have become the most convenient and widely used tools in basic biological research, particularly in the regulation of gene expression and morphogenetic movements during embryonic development. In many cases, similar applications can be achieved with the other reporter genes mentioned above. However, the detection of these reporters is indirect and involves in additional steps

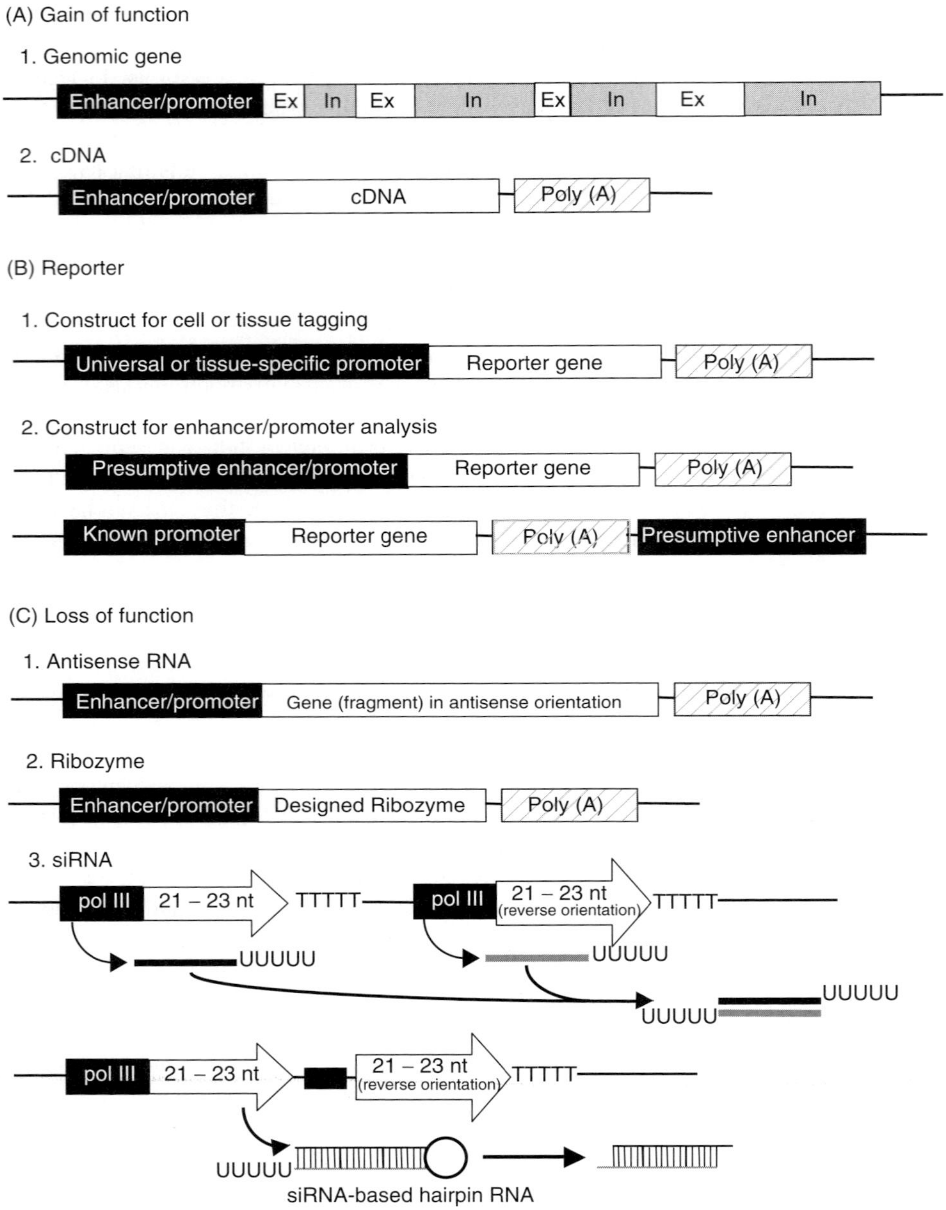

Figure 22.1 Prototypes of transgenes for fish transgenesis. Depending on the experimental purpose, the promoter used in the transgene construct can be either constitutive or inducible one. Ex: exon; In: intron; pol III: RNA polymerase III promoters.

such as enzymatic reaction with substrate and immunohistochemical staining with specific antibody to the reporter.

The loss-of-function transgenes are constructed for interfering with the expression of host genes at the transcriptional or translational level. Mechanistically, there are at least three groups of loss-of-function transgenes. The transgenes might encode antisense RNA or oligonucleotides that selectively hybridize to a target endogenous mRNA, resulting in degradation of the mRNA by RNase H, or interference with translation of the mRNAs via preventing the binding of ribosomes to mRNA. A variation of the antisense RNA approach is to introduce a group of chemically modified oligonucleotides known as morpholino phosphorodiamidate oligonucleotides (MOs) into fish embryos. In contrast to ordinary antisense molecules, MOs exhibit long term stability with less toxicity to living cells, and appear to function through translation blockage only. The second group of loss-of-function transgenes is catalytic RNA (ribozymes) that can cleave specific target mRNA at a specific site and thereby "knock down" the normal gene expression (14). The most recent development of the loss-of-function transgenes is to introduce a group of small RNAs known as small (short) interfering RNAs (siRNA) into cells. siRNA will guide a ribonuclease known as RISC (the RNA-induced silencing complex) to recognize mRNA containing a sequence homologous to the siRNA, resulting in degradation of the mRNA by the ribonuclease (19). The prototype of loss-of-function transgenes is shown in Figure 22.1C. Although there is currently no report of stable lines of transgenic fish carrying this type of transgenes, the feasibility of this approach in silencing genes has been demonstrated in numerous fish cell lines *in vitro*. Moreover, many recent studies also prove that MOs and siRNA molecules can silence their targets specifically and effectively when introduced into fish embryos, such as zebra fish, medaka, and rainbow trout (4,5,53).

22.2.2 Selection of Fish Species

Gene transfer studies have been conducted in several different fish species such as salmon, rainbow trout, common carp, goldfish, loach, channel catfish, tilapia, northern pike, walleye, zebra fish, and Japanese medaka (11,21,27). Depending on the purpose of the transgenic fish studies, the embryos of some fish species prove to be more suitable than others. In selecting a fish species for gene transfer studies, a series of parameters are considered. These parameters are: (1) length of the life cycle; (2) year round supply of eggs and sperm; (3) culture conditions; (4) size of the adult at maturity; and (5) availability of background information on genetics, physiology, and endocrinology of the fish species.

Japanese medaka (*Oryzias latipes*) and zebra fish (*Danio rerio*) are regarded as ideal fish species for conducting gene transfer studies. Both fish species have short life cycles (3 months from eggs to mature adults), produce hundreds of eggs on a regular basis without exhibiting a seasonal breeding cycle, and can be maintained easily in the laboratory for two to three years. Their eggs are relatively large, 1.0–1.5 mm in diameter, and possess very thin and semitransparent chorions. These features allow easy microinjection of DNA into the eggs if appropriate glass needles are used. Furthermore, inbred lines and various morphological mutants of both fish species are available. These fish species are suitable candidates for producing transgenic fish for: (1) studying developmental regulation of gene expression and gene action; (2) identifying regulatory elements that regulate the expression of a gene; (3) measuring the activities of promoters; and (4) producing transgenic models for human diseases and environmental toxicology. However, a major drawback of these two fish species is that their small body size makes them unsuitable for endocrinological or biochemical analysis.

The handicap of using small body size fish species such as medaka or zebra fish in gene transfer studies can be easily overcome by the use of fish species such as loach, killifish (*Fundulus*), goldfish, and tilapia. The body sizes of these fish species are big enough so that sufficient amounts of tissues can be isolated from individual fish for biochemical and endocrinological studies. Another important attribute of these fish species is their shorter maturation time compared to rainbow trout or salmon. In particular, tilapia is a very appropriate medium size fish species for gene transfer studies because it only requires about three months to complete its entire life cycle. It is possible to produce three generations of transgenic tilapia in one year. Unfortunately, the lack of a well defined genetic background and asynchronous reproductive behavior of these fish species present some problems for conducting gene transfer studies.

Rainbow trout, salmon, channel catfish, and common carp are commonly used large body size model fish species for transgenic fish studies. Since the knowledge of endocrinology, reproductive biology, and physiology of these fish species has been well worked out, they are well suited for studies on comparative endocrinology as well as aquaculture application. However, the maturation time for each of these fish species is relatively long. For example, rainbow trout and salmon require two to three years to reach reproductive maturity, and common carp or channel catfish about one to two years. Thus it will prolong inheritance studies. In addition, because these fish species have a single spawning cycle per year, the number of gene transfer attempts that can be conducted per year is restricted.

22.2.3 Methods of Gene Transfer

Techniques such as calcium phosphate precipitation, direct microinjection, retrovirus infection, electroporation, and particle gun bombardment have been widely used to introduce foreign DNA into animal cells, plant cells, and germ lines of mammals and other vertebrates. Among these methods, direct microinjection and electroporation of DNA into newly fertilized eggs have proven to be the most reliable methods of gene transfer in fish systems.

Microinjection of foreign DNA into newly fertilized eggs was first developed for the production of transgenic mice in the early 1980s (60). Since then, the technique of microinjection has been adopted for introducing transgenes into Atlantic salmon, common carp, catfish, goldfish, loach, medaka, rainbow trout, tilapia, and zebra fish (10,21). The gene constructs that were used in these studies include human or rat growth hormone (GH) gene, rainbow trout or salmon GH cDNA, chicken δ-crystalline protein gene, winter flounder antifreeze protein gene, *E. coli* β-galactosidase gene, and *E. coli* hygomycin resistance gene (11,21). In general, gene transfer in fish by direct microinjection is conducted as follows: Eggs and sperm are collected in separate, dry containers. Fertilization is initiated by mixing sperm and eggs, then adding water, with gentle stirring to enhance fertilization. Eggs are microinjected within the first few hours after fertilization. The injection apparatus consists of a dissecting stereo microscope and two micromanipulators, one with a glass micro needle for injection and the other with a micropipette for holding fish embryos in place. Routinely, about 10^6–10^8 molecules of a linearized transgene are injected into the egg cytoplasm. Following injection, the embryos are incubated in water until they hatch. Since natural spawning in zebra fish or medaka can be induced by adjusting photo period and water temperature, precisely staged newly fertilized eggs can be collected from the aquaria for gene transfer. If the medaka eggs are maintained at 4°C immediately after fertilization, the micropyle on the fertilized eggs will remain visible for two hours. The DNA solution can be delivered into the embryos by injecting through this opening during this time period. The parameters for microinjection are summarized in Table 22.2.

Depending on the species, the survival rate of injected fish embryos ranges from 35% to 80% while the rate of DNA integration ranges from 10% to 70% in the survivors (11,21). The tough chorions of the fertilized eggs in some fish species; e.g., rainbow trout and Atlantic salmon, can make insertion of glass needles difficult. This difficulty has been overcome by any one of the following methods: inserting the injection needles through the micropyle, making an opening on the egg chorions by microsurgery, removing the chorion by mechanical or enzymatic means, preventing chorion hardening by initiating fertilization in a solution containing 1 mM glutathione, or injecting the unfertilized eggs directly.

Electroporation is a successful method for transferring foreign DNA into bacteria, yeast, and plant and animal cells in culture (58,63,71). This method has become popular for

Table 22.2
Parameters of Fish Transgenesis by Microinjection and Electroporation

Parameters	Gene Transfer Method: Microinjection	Gene Transfer Method: Electroporation
Developmental stage	1 to 2 cells	1 to 2 cells
DNA size	<10 Kb	<10 Kb
DNA concentration	$10^{6\text{-}7}$ molecules/embryo	100 μg/ml
DNA topology	linear	linear
Chorion barrier	dechorionated/micropyle	intact chorion
Electrical field strength	N/A[a]	500 to 3,000 v
Pulse shape	N/A[a]	exponential/square
Pulse duration	N/A[a]	mS[c] to S
Temperature	RT[b]	RT[b]
Medium	PBS/saline	PBS/saline

[a]N/A, not applicable; [b]RT, room temperature (25°C); [c]mS, millisecond

transferring foreign genes into fish embryos or sperm in the past few years (52,64,72). Electroporation utilizes a series of short electrical pulses to permeate cell membranes, permitting the entry of DNA molecules into embryos or sperm. The patterns of electrical pulses can be emitted in a single pulse of exponential decay form (exponential decay generator) or high frequency multiple peaks of square waves (square wave generator). The basic parameters are summarized in Table 22.2. Studies conducted in our laboratory (52,64) and those of others (7) have shown that the rate of DNA integration in electroporated embryos is on the order of 20% or higher in the survivors. Although the overall rate of DNA integration in transgenic fish produced by electroporation was equal to or slightly higher than that of microinjection, the actual amount of time required for handling a large number of embryos by electroporation is orders of magnitude less than that required for microinjection. Several reports have also appeared in the literature describing successful transfer of transgenes into fish by electroporating sperm instead of embryos (72,75,78). With the success seen in electroporating sperm, electroporation can therefore be considered as an efficient and versatile mass gene transfer technology.

By microinjection or electroporation, plasmid encoded transgenes can be introduced at a satisfactory efficiency into many fish species to produce transgenic offspring. However, the resulting P_1 transgenic individuals are almost always mosaics as a result of delayed transgene integration. Furthermore, neither method is efficient or applicable in transferring foreign DNA into embryos of live bearing fish, marine fish, and invertebrates due to the restrictions imposed by the unique spawning behavior or anatomical structure of these species. Modification of transgenes has been adopted to increase the efficiency of gene transfer by microinjection or electroporation. Hsiao et al. (35) and Thermes et al. (76) have taken the following approaches: using the inverted terminal repeat DNA from adenoassociated virus to increase the stability of transgene integration; and employing *I-SceI* meganuclease to mediate high transportation efficiency of the transgene into the nucleus. Despite the improvement, these modified approaches can not be applied to the mentioned species whose transgenic offspring can not be produced by either microinjection or electroporation. Instead, transformation of mature or immature germ cells by infection with transgene encoded retroviral vectors may be an effective approach for transgenesis in these fish.

A replication defective pantropic retroviral vector containing the long terminal repeat (LTR) sequence of Moloney murine leukemia virus (MoMLV) and transgenes in a viral envelope with the G-protein of vesicular stomatitis virus (VSV), was developed by Burns et al. (8). Since entry of VSV into host cells is mediated by interaction of the VSV-G protein with a phospholipid component of the host cell membrane, this pseudotyped retroviral vector has a very broad host range and is able to transfer transgene into many different cell types. Using the pantropic pseudotyped defective retrovirus as a gene transfer vector, transgenes containing neo^R or β-galactosidase were transferred into zebra fish (48), medaka (50), and dwarf surf clams (51). More recently, Sarmasik et al. (66) and Chen et al. (unpublished results) also used the pantropic retroviral vector to transfer genes into the immature gonads of crayfish, desert guppy, and shrimp, respectively. They found that by using the pantropic retroviral vector as a gene transfer vector, the problem of transgene mosaicism in P_1 transgenic fish is eliminated, as the transgenes are introduced into immature male gonads by the pantropic retroviral vector.

Nuclear transplantation is a key technique in animal cloning. This method involves the transfer of nuclei from donor cells to enucleated eggs, stimulating eggs into cleavage and early phase embryonic development, and, in the case of mammals, the developing embryos are transferred into pseudopregnant females to complete the final stages of development. The first successful cloning of animals by nuclear transplantation was established for Northern Leopard Frog (*Rana pipiens*) by Briggs and King in 1952 (15). However, Dolly, the lamb, is the first mammal successfully cloned by the nuclear transplantation technique (83), and since then, many laboratories throughout the world have adopted this technique to clone other mammals (15). To date, production of diploid and fertile fish by the nuclear transplantation technique has been achieved in two small laboratory fish species, medaka and zebra fish (40,47,80). Three types of donor cells have been used for nuclear transfer: sperm, long term cultured somatic cells, and embryonic cells.

In the study reported by Wakamatsu et al. (80), embryonic cells from two transgenic medaka lines were used as donor cells to generate transgenic medaka. Blastoderm cells from the donor embryos in the mid-blastula stage were collected and dissociated into single cells. These were subsequently transplanted into the cytoplasm of the recipient enucleated eggs at the animal pole through the micropyle. In this study, 1 out of 588 eggs transplanted with the first donor and 5 out of 298 eggs transplanted with the second donor developed successfully to adulthood. The transgene of the donor nuclei was also transmitted to the F_1 and F_2 offspring in a Mendelian fashion.

Two versions of nuclear transfer technique have been established for zebra fish recently. Jesuthasan and Subburaju (40) reported the use of sperm nuclei as donor for nuclear transfer. Sperm collected from mature males were demembranated by treatment with lysolecithin, digitonin, or by freeze thawing. The demembranated sperm were then incubated with transgene DNA at room temperature prior to microinjection into the animal pole region of the unfertilized egg. This method could produce nonmosaic expression of transgene in all cells of the embryos if incubation of the demembranated sperm with DNA was prolonged to 20 minutes. Few of the treated eggs, however, developed into fertile adults and the transgene was shown to be transmitted through the germ line. In the report by Lee et al. (47), embryonic fibroblast cells from disaggregated embryos were used as donor cells. These cells were first modified by retroviral insertions expressing GFP as a reporter, and were cultured for at least 12 weeks before their nuclei were transplanted into enucleated, unfertilized eggs. The method resulted in 2% of the transplants developing into fertile, diploid offspring. This study clearly demonstrated that nuclei from slowly dividing cultured fish somatic cells could be reprogrammed to support rapid embryonic development and used as donor cells for nuclear transfer in fish. Moreover, the promising use of modified somatic cells provides an alternative choice of vehicle for gene targeting and fish cloning because fish ES cells are still unavailable. Overall, these studies have demonstrated the feasibility of using germinal and somatic cells from fish as donors for conducting nuclear transfer studies in different important fish species for both basic research and application purposes. In addition, the problem of transgene mosaicism in P_1 transgenic fish can be overcome by the nuclear transplantation method.

22.2.4 Characterization of Transgenic Fish

Identification of transgenic individuals is the most time consuming step in the production of transgenic fish. Traditionally, dot blot and Southern blot hybridization of genomic DNA were common methods used to determine the presence of transgenes in the presumptive transgenic individuals. These methods involve isolation of genomic DNA from tissues of presumptive transgenic individuals, digestion of DNA samples with restriction enzymes, and Southern blot hybridization of the digested DNA products. Although this method is expensive, laborious, and insensitive, it offers a definitive answer as to whether or not a transgene has been integrated into the host genome. Furthermore, it also reveals the pattern of transgene integration if appropriate restriction enzymes are employed in the Southern blot hybridization analysis. In order to handle a large number of samples efficiently and economically, a polymerase chain reaction (PCR) based assay has been adopted (12,52). The strategy of the assay involves the isolation of genomic DNA from a very small piece of fin tissue, PCR amplification of the transgene sequence, and Southern blot analysis of the amplified products. Although this method does not differentiate whether the transgene is integrated in the host genome or exists as an extrachromosomal unit, it serves as a rapid and sensitive screening method for identifying individuals that contain the transgene at the time of analysis. This method has been used in our laboratory as a routine preliminary screen for the presence of transgenes in thousands of presumptive transgenic fish.

Studies conducted in many fish species have shown that following injection of linear or circular transgene constructs into fish embryos, the transgenes are maintained as an extrachromosomal unit through many rounds of DNA replication in the early phase of the embryonic development. At a later stage of embryonic development, some of the transgenes are randomly integrated into the host genome while others are degraded, resulting in the production of mosaic transgenic fish (27). To determine the pattern of transgene integration, genomic DNA from PCR positive fish is digested with a series of restriction endonucleases, and the resulting products resolved in agarose gels for Southern blot analysis. In many fish species studied to date, it is found that multiple copies of transgenes are integrated in a head-to-head, head-to-tail, or tail-to-tail form, except in transgenic common carp and channel catfish where a single copy of the transgene was integrated at multiple sites on the host chromosomes (27,84).

Stable integration of transgenes is an absolute requirement for continuous vertical transmission to subsequent generations and establishment of a transgenic fish line. To determine whether the transgene is transmitted to subsequent generations, P_1 transgenic individuals are mated to nontransgenic individuals and the progeny are assayed for the presence of transgene by the PCR assay method described earlier (11,12,52,70). Although it has been shown that the transgene may persist in the F_1 generation of transgenic zebra fish as extrachromosomal DNA (73), detailed analysis of the rate of transmission of transgenes to F_1 and F_2 generations in many transgenic fish species indicates true and stable incorporation of gene constructs into the host genome (11,27). If the entire germ line of the P_1 transgenic fish is transformed, with at least one copy of the transgene per haploid genome, at least 50% of the F_1 transgenic progeny would be expected in a backcross involving a P_1 transgenic with a nontransgenic control. In many such crosses, only about 20% of the progeny are transgenic (12,52,70,73,74,84). However, when the F_1 transgenic is backcrossed with a nontransgenic control, at least 50% of the F_2 progeny are transgenics. These results clearly suggest that the germlines of the P_1 transgenic fish are mosaic as a result of delayed transgene integration during embryonic development.

An important aspect of gene transfer studies is the detection of transgene expression. Depending on the levels of transgene products in the transgenic individuals, the following methods are commonly used to detect transgene expression: RNA northern or dot blot hybridization; RNase protection assay; reverse transcription polymerase chain reaction (RT-PCR); immunoblotting assay; and other biochemical assays for determining the presence of the transgene protein products. Among these assays, RT-PCR is the most sensitive method and requires only a small amount of sample. The strategy of this assay is summarized in Figure 22.2. Briefly, it involves the isolation of total RNA from a small piece of tissue, synthesis of single stranded cDNA by reverse transcription, and PCR amplification of the transgene cDNA by employing a pair of oligonucleotides specific to the transgene product as amplification primers. The resulting products are resolved on agarose gels and analyzed by Southern blot hybridization using a radiolabeled transgene as a hybridization probe. Transgene expression can also be quantified by a quantitative RT-PCR method (3), or a quantitative real time RT-PCR (21).

22.3 BIOTECHNOLOGICAL APPLICATION OF THE TRANSGENIC FISH TECHNOLOGY

As techniques of producing various species of transgenic fish have become available in the last two decades, there has been a rapid boom in applying transgenic fish technology to different disciplines of basic research and biotechnological applications. Teleost fishes serve as extremely valuable models for basic research in vertebrate biology as they are functionally similar to mammalian species; and because they are sufficiently evolutionarily distant to mammals, they are well suited for comparative genomics. Since transgenic fish can be easily and economically produced and reared in large quantities, they are a better choice than transgenic micel, especially in studies requiring a large number of animals. In recent years, transgenic fish technology has made major contributions in studying vertebrate development, analyzing promoter or enhancer elements of genes, dissecting signal transduction pathways, and developing human disease models.

Transgenic fish technology research has also been driven toward biotechnological applications in recent years, with a hope to produce high economic returns for humans. Some of the advances will be reviewed below.

22.3.1 Growth Hormone Transgenic Fish

The initial drive of transgenic fish research came from attempts to increase production of economically important fish for human consumption. The worldwide supply of fishery products has traditionally depended upon commercial harvest of finfish, shellfish, and crustaceans from freshwater and marine sources. In recent years, while the world wide commercial catch of fish experienced a sharp reduction, the world wide demand for fish products has risen steeply. In order to cope with the demand for fish products, many countries have turned to aquaculture. Although aquaculture has the potential to meet the world demand for fish products, innovative strategies are required to improve its efficiency.

There are three aspects of fish growth characteristics that could be improved for aquaculture: increasing the initial growth rate of fry for an early head start; enhancing somatic growth rate in adults to provide larger market body size; and improving feed conversion efficiency of to achieve effective utilization of feeds. Among these three, enhanced somatic growth rates via manipulation of the GH gene show considerable promise. Studies conducted

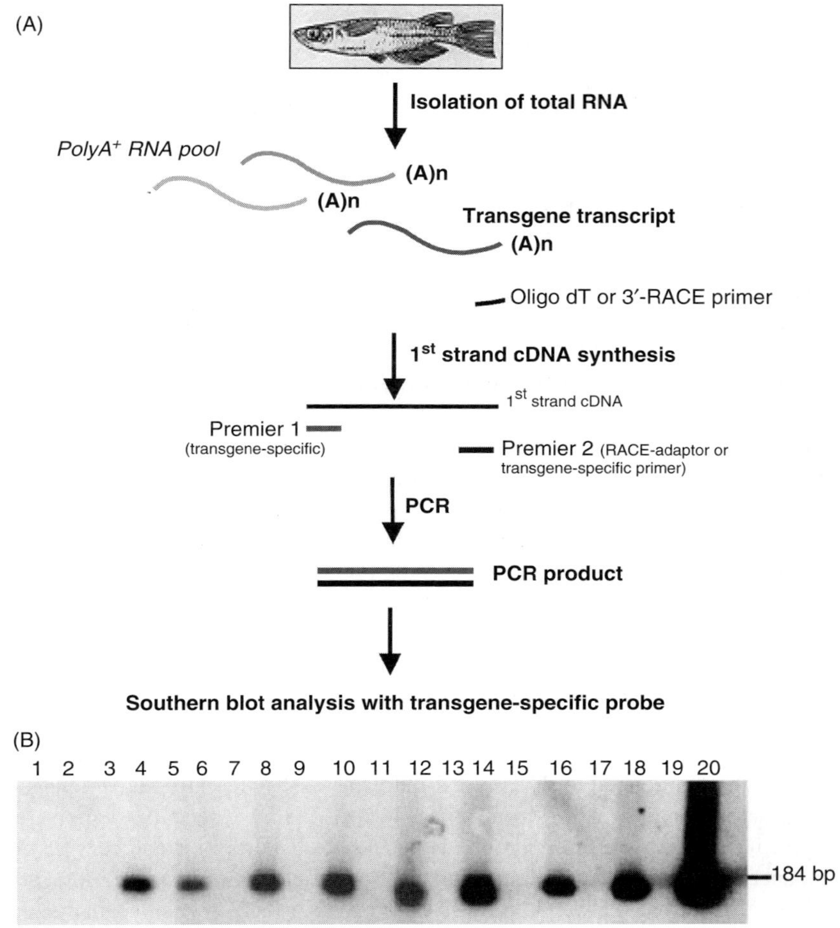

Figure 22.2 Determination of transgene expression by RT-PCR or RACE-PCR. (A) Strategy of RT-PCR and RACE-PCR. (B) An example of detecting cecropin transgene in transgenic medaka. Total RNA was isolated from fin tissues of F_2 transgenic medaka. Expression of cecropin transgene was detected by reverse transcribing poly A^+ RNA into complementary cDNA and PCR amplification of the cecropin cDNA by using cecropin-specific primers. The amplified product was further confirmed by Southern hybridization using a cecropin-specific probe. RT-PCR was conducted with mRNA from transgenic fish without prior reverse transcription (lanes 5, 7, 9, 11, 13, 15, 17, 19) or with 1st cDNA transcribed from mRNA isolated from the transgenic fish (lanes 4, 6, 8, 10, 12, 14, 16, 18). Lane 1: RT-PCR with no mRNA input; lane 2: RT-PCR with mRNA from negative control fish (c-mRNA); lane 3: PCR with c-mRNA without prior reverse transcription; lane 20: PCR with control plasmid containing cecropin gene. [From Sarmasik et al., 2002, with permission]

by Agellon et al. (1) and Paynter et al. (62) showed that treatment of yearling rainbow trout and oysters with recombinant rainbow trout growth hormone (GH) resulted in significant growth enhancement. Similar results of growth enhancement in fish treated with recombinant GH have been reported by many investigators (23,57,69). These results point to the possibility of improving the somatic growth rate fish by manipulating fish GH or its gene.

Zhu et al. (85) reported the first successful transfer of a human GH gene fused to a mouse metallothionein (MT) gene promoter into goldfish and loach. The F_1 offspring of these transgenic fish grew twice as large as their nontransgenic siblings. Since then, similar enhanced growth effect has been observed in the GH-transgenic fish of many other species, such as tilapia, carp, catfish, sea bream, and salmon (16,17,52,53,54,65,84). These studies have shown that the expression of a foreign human or fish GH gene could result in significant growth enhancement in the P_1, F_1 and F_2 transgenic fish. Additionally, Dunham et al. (18) demonstrated that transgenic common carp carrying rainbow trout GH transgene consistently display, in two consecutive generations, a favorable body shape, better dressout yield, and better quality of flesh compared with controls. These studies have demonstrated that the production of GH transgenic fish could be beneficial to worldwide aquaculture, and could help to alleviate starvation in many economically poor countries by providing more efficient, cheaper, and yet high quality protein sources.

22.3.2 Transgenic Fish with Enhanced Resistance to Pathogen Infection

Disease is one of the most severe bottlenecks in aquaculture. For the past few decades, efforts to control infectious diseases in commercially important teleosts have primarily focused on the development of suitable vaccines for fish and selection of fish strains with robust resistance to infectious pathogens. Although effective vaccines have been developed for several important fish pathogens, the current vaccination practice is expensive, laborious, and time consuming. Genetic selection based on traditional crossbreeding techniques is time consuming, and the outcome is frequently unpredictable (and sometimes unachievable due to the lack of desired genetic traits). More effective approaches for controlling fish disease in aquaculture are highly desirable.

Transgenic fish technology can facilitate the genetic selection process by directly modifying the undesirable genetic traits that confer vulnerability to pathogens or introducing specific genes that are related to disease resistance into fish. The introduced transgenes can be fish originated genes or genes from other species. We have recently demonstrated the feasibility of introducing the antimicrobial peptide genes, cecropin B and cecropin P1, into fish to enhance disease resistance. Cecropins, first identified in the *Cecropia* moth, are members of the antimicrobial peptide family that are evolutionarily conserved in species from insects to mammals (6). These peptides possess activities against a broad range of microorganisms, including bacteria, yeasts, and even viruses (6). *In vitro* studies conducted by Sarmasik et al. (67) showed that recombinant cecropin B inhibited the propagation of common fish pathogens such as *Pseudomonas fluorescens, Aeromonas hydrophila, and Vibrio anguillarum.* Furthermore, Chiou et al. (13) showed that cecropin and a designed analogue, CF-17 peptide, effectively inhibited the replication of fish viruses such as infectious hematopoietic necrosis virus (IHNV), viral hemorrhagic septicemia virus (VHSV), snakehead rhabdovirus (SHRV), and infectious pancreatic necrosis virus (IPNV) (Figure 22.3). More recently, our laboratory has shown that synthetic peptide CF-17 is also effective in inhibiting the propagation of insect baculovirus (Khoo et al. unpublished result). Additionally, Jia et al. (41) demonstrated the enhanced resistance to bacterial infection in fish which were continuously transfused with a cecropin–melittin hybrid peptide, CEME, and pleurocidin amide, a C-terminally amidated form of the natural flounder peptide. These results led to the hypothesis that production of disease resistant fish strains may be achieved by introducing into fish known antimicrobial peptide genes through transgenic fish technology.

To test this hypothesis, we introduced gene constructs containing prepro–cecropin, procecropin, mature cecropin B, and cecropin P1 into medaka embryos by electroporation (67). The resulting F_2 transgenic medaka were subjected to bacterial challenges at a LD_{50} dose with *Pseudomonas fluorescens* and *Vibrio anguillarum*, respectively (67). The resulting relative percent survival (RPS) of the tested transgenic F_2 fish ranged from 72 to 100%

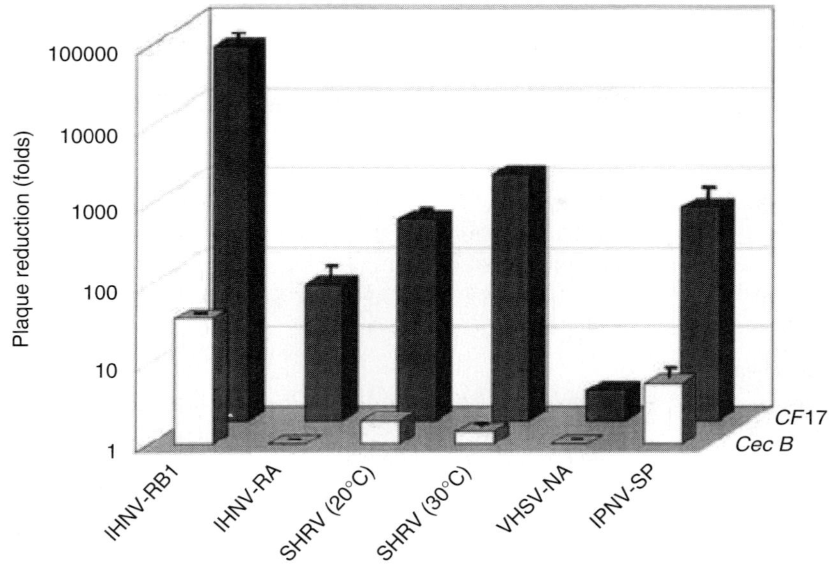

Figure 22.3 Effect of peptides cecropin and CF-17 on inhibition of fish virus propagation in vitro. The antiviral activity of the peptides cecropin B (Cec B) and CF17 was evaluated against one non-enveloped birnavirus, the SP isolate of infectious pancreatic necrosis virus (IPNV), and three enveloped rhabdoviruses: the RB1 (type2) and RA (type 1) isolates of infectious hematopoietic necrosis virus (IHNV), snakehead rhabdovirus (SHRV), and an avirulent North America isolate (NA) of viral hemorrhagic septicemia virus (VHSV). Viral infection was carried out in EPC cells for SHRV, and CHSE-214 cells for the rest viruses. [From Chiou et al., 2002, with permission]

against *P. fluorescens* and 25 to 75% against *V. anguillarum* (Table 22.4). These results clearly demonstrate the potential application of transgenic fish technology in producing fish with more robust resistance to infectious pathogens.

22.3.3 Transgenic Fish with Different Body Color

There are two main motives in producing fish with altered body color: to generate novel varieties of ornamental fish with rare colors for the purpose of rearing as pets, and to use the change of body color as indicators of environmental changes. GFP and several GFP variants are the most used genes for such purposes. Additionally, melanin-concentrating hormone (MCH) isolated from chum salmon has also been exploited to generate transgenic medaka fish with altered body color (43). So far, researchers have successfully produced several colorful transgenic zebra fish displaying whole body green, red, yellow, or orange fluorescent colors under daylight, dim light, or UV light (33,82). Some of these fish have been sterilized to avoid contaminating the wild population if they should accidentally be released from aquariums to the environments (33). For research purpose, mutant zebra fish and medaka with less interfering pigmentation for optical observation have also been developed. One such example is a line of see-through transgenic medaka that are transparent throughout their entire lives, thus allowing clear visualization of GFP that is introduced into the fish as a reporter (81).

22.3.4 Transgenic Fish as Environmental Biomonitors

Fish have long been used as models in environmental toxicology studies. Prompted by concerns of human health, tissues of wild caught fish have been monitored as indicators

Table 22.3
Examples of Cell Type and Tissue Type Specific Transgenic Zebrafish Lines With Gfp or Egfp Reporter

Specificity	Promoter	Ref.
Cell-specific		
germ cells	*Xenopus efl* α (*-vasa* 3'UTR)	Knaut et al., 2002
	zebrafish *versa*	KrØvel & Olsen, 2002
blood	zebrafish *gata 1*	Long et al., 1997
lymphoid cells	zebrafish *rag1*	Jessen & Willett, 1999
epithelia	zebrafish *krt8*	Gong et al., 2002
Tissue-specific		
muscle	zebrafish *α-actin*	Higashijima et al., 1997
pancreas	zebrafish *pdx-1*	Huang et al., 2001
	zebrafish *insulin*	
liver	zebrafish *l-fabp*	Her et al., 2003
intestine	zebrafish *i-fabp*	Her et al., 2004
vasculature	zebrafish *fli 1*	Lawson & Weinstein, 2002
heart	zebrafish *cmlc2*	Huang et al., 2003
nervous system	zebrafish *islet1*	Higigashijima et al., 2000
	zebrafish *huC*	Park et al., 2000
	goldfish *α1 tubulin*	Goldman et al., 2001

for the presence of dangerous pollutants such as polycyclic hydrocarbons, oxidants, and heavy metals in the waters or in the fish itself (9,68). Different assays have been employed to measure biological parameters that are impacted by toxic pollutants, including DNA damage, defense enzymes (e.g., glutathione peroxidase and superoxide dismutase), genes inducible by toxic chemicals (e.g., cytochrome P450 1A1 and 1A2), and factors that regulate redox potential (e.g., glutathione and ascorbic acid) (9). Although these assays are very sensitive, they require specialized techniques and equipment, and thus cannot be performed in the field when the samples are collected. In addition, the obtained data from wild caught fish samples give no indication of the exact time when those fish were exposed to the toxicants.

Transgenic fish can potentially serve as sensitive sentinels for aquatic pollution, with the benefit of avoiding the inconveniences mentioned above. The principle of such biomonitor fish is to introduce into the fish reporters that are easy to observe and are under the control of promoter or enhancer elements responsive to pollutants. GFP and its variants are thus the best reporters of choice for easy optical observation, whereas luciferase is best for applications where extreme sensitivity is needed. So far, researchers have made significant progress in establishing transgenic zebra fish carrying luciferase or GFP reporter under control of the aromatic hydrocarbon response elements (AHREs), electrophile response elements (EPREs), or metal response elements (MREs) (9). The specificity of each type of response elements is as follows: AHREs respond to numerous polycyclic hydrocarbons and halogenated coplanar molecules such as 2,3,7,8-tetrachlorodibenzo-p-dioxin (TCDD) and polychlorinated biphenyls; EPREs respond to quinones and numerous other potent electrophilic oxidants; MREs respond to heavy metal cations such as mercury, copper, nickel, cadmium, and zinc. In the zebra fish cell lineZEM2S, these three types of

Table 22.4
Relative Percent Survival (RPS) of F_2 Cecropin Transgenic Medaka Challenged with *P. fluorescens* and *V. anguillarum* [Adopted from Sarmasik et al., 2002, with permission]

Transgenic Lines	*P. fluorescene* RPS*	*V. anguillarum* RPS*
Preprocecropin B: CMV – pre – pro – cecropin B	100% (1)	50% (1)
Procecropin B: CMV – Ig – pro – cecropin B	72 – 100% (5)	25 – 50% (3)
Cecropin B: CMV – Ig – cecropin B	72 – 100% (2)	ND
Cecropin P1: CMV – Ig – cecropin P1	100% (2)	75% (1)

*RPS = [1 - % mortality of F_2 transgenic fish / % mortality of F_2 non-transgenic fish] × 100
CMV: cytomegalovirus promoter; Ig: catfish IgG signal peptide;
( ): number of F_2 families tested; ND: non-determined

elements can drive the expression of luciferase reporter in a dose dependent, chemical class specific manner in response to more than 20 environmental pollutants (9). The expression of GFP transgenes can be detected in the F_1 and sometimes the F_2 generations; however, the transgenes are eventually lost following that. There are still technical obstacles in generating transgenic biomonitor fish (9), but the problems can be possibly overcome in the near future.

22.3.5 Other Examples of Transgenic Fish

Other applications include the use of transgenic fish as a model for studying mechanisms of seawater adaptation (38) and generating cold resistant fish carrying antifreeze proteins (AFPs). AFPs are found in many cold water fish species, such as the winter flounder, ocean pout, and sea raven. Expression of *afp* genes permit these fish to survive in freezing seawater as low as -1.8°C (70). However, AFP transgenic Atlantic salmon does not demonstrate a measurable increase of cold tolerance (70).

Transgenic fish can potentially serve as bioreactors for producing therapeutic proteins beneficial to human health (26). The advantages of fish as bioreactors include: (1) relatively low cost as compared to mammals such as cows, (2) easy maintenance, (3) producing large quantity of progeny, and (4) most important, no known viruses or prions have been reported to infect both human and fish. Successful expression of bioactive human proteins such as coagulation factor VII in fish embryos (37) demonstrate this application of transgenic fish to produce therapeutic chemicals for humans in the future.

22.4 CONCERNS AND FUTURE PERSPECTIVES

The application of transgenic fish technology to produce fish with beneficial traits, such as environmental biomonitors and genetically modified (GM) food, is on the rise. There have been ground breaking discoveries made in the past few years based on transgenic fish models, and such applications are expected to grow in the future. The technique to produce transgenic fish has improved tremendously in the last two decades; nonetheless, there are still concerns about low efficiency, mosaic expression, nonfish components of the transgenes, and the long generation time for certain fish species. The following advances would thus be essential in developing the next generation transgenic fish: developing more efficient mass gene transfer technologies; developing targeted gene transfer technology such as embryonic stem cell gene transfer method; identifying more suitable fish origin promoters to direct the expression of transgenes at optimal levels and desired time; and developing methods to shorten the time required to generate homozygous offspring.

Despite the promising future of applying transgenic fish technology in food fish production, there are great concerns about the environmental impacts of such applications of transgenic fish technology to produce GM food or hobby pets. Based on mathematical modeling, a "Trojan Gene Effect" has been demonstrated in the male transgenic medaka carrying salmon growth hormone gene. Release of such fish into the environments could ultimately lead to the extinction of the wild type population, owing to their advantage in mating with the wild type female (34). The study provides a sound analysis of the potential impact of the accidental release of transgenic fish into the environment, reminding us of the delicate interaction between a newly introduced species and the native population and environment. However, we would reason that the potential environmental impact of transgenic fish is controllable and even avoidable. First of all, the Trojan Gene Effect might not necessarily exist for other types of transgenes or in other fish species. In practice, transgenic fish can be maintained in confined areas to avoid the escape of fish into the nearby waters. Moreover, as the desirable genetic traits can be introduced into fish, the undesirable behavior traits could be possibly removed from the transgenic population by screening the behavior pattern of each transgenic individual. Ultimately, establishment of sterile transgenic populations, such as a sterile triploid all female strain, could be applied to avoid the spreading of transgenes into the wild population. Ideally, such sterility should be reversible; otherwise it would mean the loss of the goose that lays golden eggs. Finally, from a historical perspective, the concerns about the environmental impacts in the early days of recombinant DNA technology have not been realized, and many lessons can be learned from this experience.

Acceptance of GM fish as food by society is a key factor for the development of transgenic fish as a new source of cheaper, high quality food. Largely, the concern is due to the public perception that GM food is unsafe for humans. Safety is the ultimate goal of GM fish development and relies on thorough characterization of the transgenes and safety assessment of the GM fish, resembling the development of new drugs. Unfortunately, it is a fact that there is always a subpopulation of people who are allergic, to different degrees, to many existing natural food sources, such as peanuts and shellfish. Therefore, the challenge is great in developing GM fish with absolutely no adverse effect on every single individual in the human population. A thorough evaluation process to fully analyze any health risk of GM fish to humans should be established; and the information should be clearly delivered to the public and well marked on the food product. Development of gene-inactivated transgenic fish may be more acceptable by people who are concerned of the safety of the "new" genes that are introduced into fish. One candidate of such GM fish is transgenic fish strains in which the myostatin gene has been inactivated. Inactivation of myostatin gene causes double muscling in animals, resulting in significant increase in mass of skeletal muscle. Natural mutation of myostatin gene has been found in the meaty Belgian Blue and Piedmontes cattle breeds; and the double muscling effect has been observed in mice whose myostatin gene is artificially inactivated (42,55). Such a gene-inactivation approach would be very valuable in producing transgenic fish with more flesh and higher food conversion rates, particularly for the economically important aquaculture species.

In summary, the development of transgenic fish has resulted in many valuable discoveries in basic biological science, any one of which could lead to the next breakthrough for new cures for human diseases. On the other hand, the potential of GM fish as bioreactors, or as new food sources, should not be deterred by the environmental and safety concerns. Instead, the concerns should be addressed with more research efforts to develop new GM species with beneficial traits that are also safe to human health and cause no harm to the environment.

ACKNOWLEDGMENT

Some results presented in this paper were supported by grants from NSF (IBN-0078067), USDA (CONTR # 58-1930-0-009) and Connecticut Sea Grant College (R/A 18) to TTC.

REFERENCES

1. Agellon, L.B., C.J. Emery, J.M. Jones, S.L. Davies, A.D. Dingle, T.T. Chen. Growth hormone enhancement by genetically engineered rainbow trout growth hormone. *Can. J. Fish. Aqua. Sci.* 45:146–151, 1988.
2. Amsterdam, A., S. Lin, N. Hopkins. The *Aequorea victoria* green fluorescent protein can be used as a reporter in live zebrafish embryos. *Dev. Biol.* 171:123–129, 1995.
3. Ballagi-Pordany, A., K. Funa. Quantative determination of mRNA phenotypes by polymerase chain reaction. *Anal. Biochem.* 196:88–94, 1991.
4. Boonanuntanasarn, S., G. Yoshizaki, Y. Takeuchi, T. Morita, T. Takeuchi. Gene knock-down in rainbow trout embryos using antisense Morpholino phosphorodiamidate oligonucleotides. *Mar. Biotechnol.* 4:256–66, 2002.
5. Boonanuntanasarn, S., G. Yoshizaki, T. Takeuchi. Specific gene silencing using small interfering RNAs in fish embryos. *Biochem. Biophys. Res. Commun.* 310:1089–1095, 2003.
6. Boman, H.G. Peptide antibiotics and their role in innate immunity. *Annu. Rev. Immunol.* 13:61–92, 1995.
7. Buono, R.J., F.D. Linser. Transient expression of RSVCAT in transgenic zebrafish made by electroporation. *Mol. Mar. Biol. Biotechnol.* 1:271–275, 1992.
8. Burns, J.C., T. Friedmann, W. Driever, M. Burrascano, J.K. Yee. VSV-G pseudotyped retroviral vector: concentration to very high titer and efficient gene transfer into mammalian and nonmammalian cells. *Proc. Natl. Acad. Sci. USA* 90:8033–8037, 1993.
9. Carvan, M.J. III, T.P. Dalton, G.W. Stuart, D.W. Nebert. Transgenic zebrafish as sentinels for aquatic pollution. *Ann. N.Y. Acad. Sci.* 919:133–147, 2000.
10. Chen, T.T., C.M. Lin, Z. Zhu, L.I. Gonzalez-Villasenor, R.A. Dunham, D.A. Powers. Gene transfer, expression and inheritance of rainbow trout growth hormone genes in carp and loach. In: *Transgenic Models in Medicine and Agriculture*, Church, R., ed., New York: Wiley Interscience, 1990, pp127–139.
11. Chen, T.T., D.A. Powers. Transgenic fish. *Trends Biotechnol.* 8:209–215, 1990.
12. Chen, T.T., D.A. Powers, C.M. Lin, K. Kight, M. Hayat, N. Chatakondi, A.C. Ramboux, P.L. Duncan, R.A. Dunham. Expression and inheritance of RSVLTR- rtGH1 cDNA in common carp, *Cyprinus carpio*. *Mol. Mar. Biol. Biotechnol.* 2:88–95, 1993.
13. Chiou, P.P., C.M. Lin, L. Perez, T.T. Chen. Effect of cecropin B and a synthetic analogue on propagation of fish viruses *in vitro*. *Mar. Biotechnol.* 4:294–302, 2002.

14. Cotten, M., P. Jennings. Ribozyme mediated destruction of RNA *in vivo*. *EMBO J.* 8:3861–3866, 1989.
15. Di Berardino, M.A., R.G. McKinnell, D.P. Wolf. The golden anniversary of cloning: a celebratory essay. *Differentiation* 71:398–401, 2003.
16. Du, S.J., G.L. Gong, M.A. Fletcheglr, M.A. Shears, M.J. King, D.R. Idler, C.L. Hew. Growth enhancement in transgenic Atlantic salmon by the use of an "all fish" chimeric growth hormone gene construct. *Bio. Technol.* 10:176–181, 1992.
17. Dunham, R.A., A.C. Ramboux, P.L. Duncan, M. Hayat, T.T. Chen, C.M. Lin, K. Kight, L.I. Gonzalez-Villasenor, D.A. Powers.Transfer, expression and inheritance of salmonid growth hormone genes in channel catfish, *Ictalarus punctatus*, and effects on performance traits. *Mol. Mar. Biol. Botechnol.* 1:380–389, 1992.
18. Dunham, R.A., N. Chatakondi, A.J. Nichols, H. Kucuktas, T.T. Chen, D.A. Powers, J.D .Weete, R.A. Cummins, R.T. Lovell. Effect of rainbow trout growth hormone complementary DNA on body shape, carcass yield, and carcass composition of F_1 and F_2 transgenic common carp (*Cyprinus carpio*). *Mar. Biotechnol.* 4:604–611, 2002.
19. Duxbury, M.S., E.E. Whang. RNA interference: a practical approach. *J. Surg. Res.* 117:339–344, 2004.
20. Fletcher, G.L., M.A. Shears, J.J. King, P.L. Davis, C.L. Hew. Evidence of antifreeze protein gene transfer in Atlantic salmon (*Salmo salar*). *Can. J. Fish. Aquat. Sci.*, 45:352–357, 1988.
21. Fletcher, G.L., P.L. Davis. Transgenic fish for aquaculture. In: *Genetic Engineering 13,* Setlow, J.K., ed.,New York: Plenum Press, 1991, pp 331–370.
22. Fronhoffs, S., G. Totzke, S. Stier, N. Wernert, M. Rothe, T. Bruning, B. Koch, A. Sachinidis, H. Vetter, Y. Ko. A method for the rapid construction of cRNA standard curves in quantitative real-time reverse transcription polymerase chain reaction. *Mol. Cell. Probes.* 16:99–110, 2002.
23. Gill, J.A., J.P. Stumper, E.M. Donaldson, H.M. Dye. Recombinant chicken and bovine growth hormone in cultured juvenile Pacific salmon. *Biotechnology* 3:4306– 4310, 1985.
24. Goldman, D., M. Hankin, Z. Li, X. Dai, J. Ding. Transgenic zebrafish for studying nervous system development and regeneration. *Transgenic Res.* 10:21–33, 2001.
25. Gong, Z., B. Ju, X. Wang, J. He, H. Wan, P.M. Sudha, T. Yan. Green fluorescent protein expression in germ-line transmitted transgenic zebrafish under a stratified epithelial promoter from keratin8. *Dev. Dyn.* 223:204–215, 2002.
26. Gong, Z., H. Wan, T.L. Tay, H. Wang, M. Chen, T. Yan. Development of transgenic fish for ornamental and bioreactor by strong expression of fluorescent proteins in the skeletal muscle. *Biochem. Biophys. Res. Commun.* 308:58–63, 2003.
27. PB Hackett, P.B. The molecular biology of transgenic fish. In: *Biochemistry and Molecular Biology of Fish, 2*, Hochachka, P.W., T.P. Mommsen, eds., Amsterdam: Elsevier Science B.V., 1993, pp 207–240.
28. Halloran, M.C., M. Sato-Maeda, J.T. Warren, F. Su, Z. Lele, H.P. Krone, J.Y. Kuwada, W. Shoji. Laser-induced gene expression in specific cells of transgenic zebrafish. *Development* 127:1953–1960.
29. Her, G.M., Y.H. Yeh, J.L. Wu. 435-bp liver regulatory sequence in the liver fatty acid binding protein (L-FABP) gene is sufficient to modulate liver regional expression in transgenic zebrafish. *Dev. Dyn.* 227:347–356, 2003.
30. Her, G.M., C.C. Chiang, J.L. Wu. Zebrafish intestinal fatty acid binding protein (I-FABP) gene promoter drives gut-specific expression in stable transgenic fish. *Genesis* 38:26–31, 2004.
31. Higashijima, S., H. Okamoto, N. Ueno, Y. Hotta, G. Eguchi. High-frequency generation of transgenic zebrafish which reliably express GFP in whole muscles or the whole body by using promoters of zebrafish origin. *Dev. Biol.* 192: 289–299, 1997.
32. Higashijima, S., Y. Hotta, H. Okamoto. Visualization of cranial motor neurons in live transgenic zebrafish expressing green fluorescent protein under the control of the islet-1 promoter/enhancer. *J. Neurosci.* 20:206–218, 2000.
33. Holden, C. That special glow (a news report in random samples), *Science* 300:1368, 2003.
34. Howard, R.D., J.A. DeWoody, W.M. Muir. Transgenic male mating advantage provides opportunity for Trojan gene effect in a fish. *Proc. Natl. Acad. Sci. USA* 101:2934–2938, 2004.

35. Hsiao, C.D., F.J. Hsieh, H.J. Tsai. Enhanced expression and stable transmission of transgenes flanked by inverted terminal repeats from adeno-associated virus in zebrafish. *Dev. Dyn.* 220:323–336, 2001.
36. Huang, H., S.S. Vogel, N. Liu, D.A. Melton, S. Lin. Analysis of pancreatic development in living transgenic zebrafish embryos. *Mol. Cell. Endocrinol.* 177:117–124, 2001.
37. Huang, C.J., C.D. Tu, C.D. Hsiao, F.J. Hsieh, H.J. Tsai. Germ-line transmission of a myocardium-specific GFP transgene reveals critical regulatory elements in the cardiac myosin light chain 2 promoter of zebrafish. *Dev. Dyn.* 228:30–40, 2003.
38. Inoue, K., Y. Takei. Asian medaka fishes offer new models for studying mechanisms of seawater adaptation. *Comp. Biochem. Physiol. Mol. Biol.* 136:635–645, 2003.
39. Jessen, J.R., C.E. Willett, S. Lin. Artificial chromosome transgenesis reveals long-distance negative regulation of rag1 in zebrafish. *Nat. Genet.* 23:15–16, 1999.
40. Jesuthasan, S., S. Subburaju. Gene transfer into zebrafish by sperm nuclear transplantation. *Dev. Biol.* 242:88–95, 2002.
41. Jia, X., A. Patrzykat, R.H. Devlin, P.A. Ackerman, G.K. Iwama, R.E. Hancock. Antimicrobial peptides protect coho salmon from *Vibrio anguillarum* infections. *Appl. Environ. Microbiol.* 66:1928–1932, 2000.
42. Kambadur, R., M. Sharma, T.P. Smith, J.J. Bass. Mutations in myostatin (GDF8) in double-muscled Belgian Blue and Piedmontese cattle. *Genome Res.* 7:910– 916, 1997.
43. Kinoshita, M., T. Morita, H. Toyohara, T. Hirata, M. Sakaguchi, M. Ono, K. Inoue, Y. Wakamatsu, K. Ozato. Transgenic medaka overexpressing a melanin-concentrating hormone exhibit lightened body color but no remarkable abnormality. *Mar. Biotechnol.* 3:536–543, 2001.
44. Knaut, H., H. Steinbeisser, H. Schwarz, C. Nusslein-Volhard. An evolutionary conserved region in the vasa 3'UTR targets RNA translation to the germ cells in the zebrafish. *Curr. Biol.* 12:454–466, 2002.
45. KrØvel, A.V., L.C. Olsen. Expression of a vas::EGFP transgene in primordial germ cells of the zebrafish. *Mech. Dev.* 116:141–150, 2002.
46. Lawson, N.D., J.W. Mugford, B.A. Diamond, B.M. Weinstein. Phospholipase C gamma-1 is required downstream of vascular endothelial growth factor during arterial development. *Genes Dev.* 17:1346–1351, 2003.
47. Lee, K.Y., H. Huang, B. Ju, Z. Yang, S. Lin. Cloned zebrafish by nuclear transfer from long-term-cultured cells. *Nat. Biotechnol.* 20:795–799, 2002.
48. Lin, S., N. Gaiano, P. Culp, J.C. Burns, T Friedmann, JK Yee, N Hopkins. Integration and germ-line transmission of a pseudotyped retroviral vector in zebrafish. *Science* 265:666–668, 1994.
49. Long, Q., A. Meng, H. Wang, J.R. Jessen, M.J. Farrell, S. Lin. GATA-1 expression pattern can be recapitulated in living transgenic zebrafish using GFP reporter gene. *Development* 124:4105–4111, 1997.
50. Lu, J.K., J.C. Burns, T.T. Chen. Retrovirus-mediated transfer and expression of transgenes in medaka. In: *Proceedings of the Third International Marine Biotechnology Conference*, Tromoso, Norway, 1994, p 72.
51. Lu, J.K., T.T. Chen, S.K. Allen, T. Matsubara, J.C. Burns, Production of transgenic dwarf surfclams, *Mulinia lateralis*, with pantropic retroviral vectors. *Proc. Natl. Acad. Sci. USA* 93:3482–3486, 1996.
52. Lu, J.K., C.L. Chrisman, O.M. Andrisani, J.E. Dixon, T.T. Chen. Integration expression and germ-line transmission of foreign growth hormone genes in medaka, *Oryzias latipes*. *Mol. Mar. Biol. Biotechnol.* 1:366–375, 1992.
53. Lu, J.K., B.H. Fu, J.L. Wu, T.T. Chen. Production of transgenic silver sea bream (*Sparus sarba*) by different gene transfer methods. *Mar. Biotechnol.* 4:328–337, 2002.
54. Martinez, R., M.P. Estrada, J. Berlanga, I. Guillen, O. Hernandez, E. Cabrera, R. Pimentel, R. Morales, F. Herrera, A. Morales, J.C. Pina, Z. Abad, V. Sanchez, P. Melamed, R. Lleonart, J. de la Fuente. Growth enhancement in transgenic tilapia by ectopic expression of tilapia growth hormone. *Mol. Mar. Biol. Biotechnol.* 5:62–70, 1996.

55. McPherron, A.C., S.J. Lee. Double muscling in cattle due to mutations in the myostatin gene. *Proc. Natl. Acad. Sci. USA* 94:12457–12461, 1997.
56. Boas Moav, Z., Y. Liu, W. Groll, P.R. Hackett. Selection of promoters for gene transfer into fish. *Mol. Mar. Biol. Biotechnol.* 1:338–345, 1992.
57. Moriyama, S., A. Takahashi, T. Hirano, H. Kawauchi. Salmon growth hormone is transported into the circulation of rainbow trout (*Oncorhynchus mykiss*) after intestinal administration. *J. Comp. Physiol.* B160:251–260, 1990.
58. Neumann, E., M. Schaefer-Ridder, Y. Wang, P.H. Hofschneider. Gene transfer into mouse lyoma cells by electroporation in high electric fields. *EMBO J.* 1:841–845, 1982.
59. Ozato, K., H. Kondoh, H. Inohara, T. Iwanatsu, Y. Wakamatusu, T.S. Okada. Production of transgenic fish: introduction and expression of chicken δ-crystallin gene in medaka embryos. *Cell Differ.* 19:237–244, 1986.
60. Palmiter, R.D., R.D. Brinster. Germ-line transformation in mice. *Ann. Rev. Genet.* 20:465–499, 1986.
61. Pauls, S., B. Geldmacher-Voss, J.A. Campos-Ortega. A zebrafish histone variant H2A.F/Z and a transgenic H2A.F/Z:GFP fusion protein for *in vivo* studies of embryonic development. *Dev. Genes Evol.* 211:603–610, 2001.
62. Paynter, K., T.T. Chen. Biological activity of biosynthetic rainbow trout growth hormone in the easter oyster, *(Crassostrea virginica)*. *Bio. Bull.* 181:459–462, 1991.
63. Potter, H., L. Weir, P. Leder. Enhancer-dependent expression of human *k* immunoglobulin genes introduced into mouse pre-B lymphocytes by electroporation. *Proc. Natl. Acad. Sci. USA* 81:7161–7165, 1984.
64. Powers, D.A., L. Hereford, T. Cole, K. Creech, T.T. Chen, C.M. Lin, K. Kight, R.A. Dunham. Electroporation: a method for transferring genes into gametes of zebrafish (*Brachydanio rerio)*, channel catfish (*Ictalurus punctatus*), and common carp (*Cyrinus carpio*). *Mol. Mar. Biol. Biotechnol.* 1:301–308, 1992.
65. Rahman, M.A., A. Ronyai, B.Z. Engidaw, K. Jauncey, G.-L. Hwang, A. Smith, E. Roderick, D. Penman, L.Varadi, N. Maclean. Growth and nutritional trials on transgenic Nile tilapia containing an exogenous fish growth hormone gene. *J. Fish Biol.* 59:62–78, 2001.
66. Sarmasik, A., I.-K. Jang, C.Z. Chun, J.K. Lu, T.T. Chen. Transgenic live-bearing fish and crustaceans produced by transforming immature gonads with replication-defective pantropic retroviral vectors. *Marine Biotechnol.* 3:465–473, 2001.
67. Sarmasik, A., G. Warr, T.T. Chen. Production of transgenic medaka with increased resistance to bacterial pathogens. *Mar. Biotechnol.* 4:310–322, 2002.
68. Schreurs, R.H., J. Legler, E. Artola-Garicano, T.L. Sinnige, P.H. Lanser, W. Seinen, B. Van der Burg. *In vitro* and *in vivo* antiestrogenic effects of polycyclic musks in zebrafish. *Environ. Sci. Technol.* 38:997–1002, 2004.
69. Sekine, S., T. Miizukzmi, T. Nishi, Y. Kuwana, A. Saito, M. Sato, H. Itoh, H. Kawauchi. Cloning and expression of cDNA for salmon growth hormone in *E. coli. Proc. Natl. Acad. Sci. USA* 82:4306–4310, 1985.
70. Shears, M.A., G.L. Fletcher, C.L. Hew, S. Gauthier, P.L. Davies. Transfer, expression, and stable inheritance of antifreeze protein genes in Atlantic salmon (Salmo salar). *Mol. Mar. Biol. Biotechnol.* 1:58–63, 1991.
71. Shigekawa, K., W.J. Dower. Electroporation of eukaryotes and prokaryotes: a general approach to introduction of macromolecules into cells. *Biotechniques* 6:742–751, 1988.
72. Sin, F.Y., S.P. Walker, J.E. Symonds, U.K. Mukherjee, J.G. Khoo, I.L. Sin. Electroporation of salmon sperm for gene transfer: efficiency, reliability, and fate of transgene. *Mol. Reprod. Dev.* 56(2 Suppl):285–288, 2000.
73. Stuart, G.W., J.V. McMurry, M. Westerfield. Replication, integration, and stable germ-line transmission of foreign sequence injected into early zebrafish embryos. *Development* 109:403–412, 1988.
74. Stuart, G.W., J.V. Vielkind, J.V. McMurray, M. Westerfield. Stable lines of transgenic zebrafish exhibit reproduction patterns of transgene expression. *Development* 109:293–296, 1990.

75. Symonds, J.E., S.P. Walker, F.Y.T. Sin. Development of mass gene transfer method in chinook salmon: optimization of gene transfer by electroporated sperm. *Mol. Mar. Biol. Biotechnol.* 3:104–111, 1994.
76. Thermes, V., C. Grabher, F. Ristoratore, F. Bourrat, A. Choulika, J. Wittbrodt, J.S. Joly. I-SceI meganuclease mediates highly efficient transgenesis in fish. *Mech. Dev.* 118:91–98, 2002.
77. Tomasiewicz, H.G., D.B. Flaherty, J.P. Soria, J.G. Wood. Transgenic zebrafish model of neurodegeneration. *J. Neurosci. Res.* 70:734–745, 2002.
78. Tseng, F.S., I.C. Lio, H.J. Tsai. Introducing the exogenous growth hormone cDNA into loach (*Misgurnus anguillicaudatus*) eggs via electroporated sperms as carrier. In: *3rd International Marine Biotechnology Conference Abstract*, Tromso, Norway, 1994, p 71.
79. Udvadia, A.J., E. Linney. Windows into development: historic, current, and future perspectives on transgenic zebrafish. *Dev. Biol.* 256:1–17, 2003.
80. Wakamatsu, Y., B. Ju, I. Pristyaznhyuk, K. Niwa, T. Ladygina, M. Kinoshita, K. Araki, K. Ozato. Fertile and diploid nuclear transplants derived from embryonic cells of a small laboratory fish, medaka *(Oryzias latipes). Proc. Natl. Acad. Sci. USA* 98:1071–1076, 2001.
81. Wakamatsu, Y., S. Pristyazhnyuk, M. Kinoshita, M. Tanaka, K. Ozato. The see-through medaka: a fish model that is transparent throughout life. *Proc. Natl. Acad. Sci. USA* 98:10046–10050, 2001.
82. Wan, H., J. He, B. Ju, T. Yan, T.J. Lam, Z. Gong. Generation of two-color transgenic zebrafish using the green and red fluorescent protein reporter genes gfp and rfp. *Mar. Biotechnol.* 4:146–154, 2002.
83. Wilmut, I., A.E. Schnieke, J. McWhir, A.J. Kind, K.H. Campbell. Viable offspring derived from fetal and adult mammalian cells. *Nature* 385:810–813, 1997.
84. Zhang, P., M. Hayat, C. Joyce, L.I. Gonzales-Villasenor, C.M. Lin, R.A. Dunham, T.T. Chen, D.A. Powers Gene transfer, expression and inheritance of pRSV- Rainbow Trout-GH-cDNA in the carp, *Cyprinus carpio* (Linnaeus). *Mol. Reproduc. Develop.* 25:3–13, 1990.
85. Zhu, Z., G. Li, L. He, S.Z. Chen. Novel gene transfer into the goldfish (*Carassius auratus* L 1758). *Angew Ichthyol.* 1:31–34, 1985.

2.23

The Production, Properties, and Utilization of Fish Protein Hydrolysates

Hordur G. Kristinsson

CONTENTS

23.1 INTRODUCTION

The situation of the world oceans is believed by many to be very close to experiencing major collapses in important fish stocks (1). Recent data on the world harvest estimates wild fish capture at close to 100 million metric tons, the estimated sustainable limit; if aquaculture is included, the value exceeds 120 million metric tons (2). This situation should encourage harvesters and processors to change their practices, leading to better utilization of the current harvest and directing efforts into developing new means to utilize underutilized species. Currently substantial amounts of protein rich material derived from aquatic animals are discarded or not utilized optimally. These materials are typically byproducts from seafood processing plants, such as frames, cut offs, and by catch from fisheries, such as shrimp fisheries (3). For example, it has been reported that possibly more than 60% of fish tissue remaining after processing is considered to be processing waste and not used as human food (4). This material is high in quality protein (≥10%) and other valuable compounds which could be utilized for human consumption. These values naturally can vary between species and the raw material in question. Furthermore, many species rich in quality protein and lipids are not harvested or utilized due to processing and stability problems of the final product (5). In many parts of the world, discarding byproducts is bound by strict regulations and comes at a high cost. Within the materials here listed are highly functional components (proteins, lipids, antioxidants, minerals) which could be converted into valuable products, provided the proper economic method exist and a market exists for the end products.

In the latter half of the last century there were great research efforts to find means to utilize the vast amount of fish byproducts and underutilized species. Despite this effort we have not come very far, in part due to economic obstacles of processing and generally low acceptance of the final products (3). These materials are typically rich in prooxidants, oxidative susceptible lipids, scales, skin, bones, connective tissue, and viscera. For this reason they have found very limited utilization. The great majority of utilizable byproducts, by catch, and underutilized species is transformed into fishmeal for animal feed and to some extent fertilizer, using harsh processing conditions that may adversely affect the function of the key components of the material. It has been estimated that about 30% of the world's catch is transformed into fishmeal for animal feed (6), and possibly more than 15% of the fish biomass harvested will not find any utilization (7).

There is an obvious need to better utilize processing byproducts, by catch, and underutilized species in light of the world fishing situation and the growing need for quality protein and lipids, both for humans and animals. Some exciting developments in the science and technology of fish utilization have developed recently which move us closer to this goal. To name one of the most important developments, a newly developed process utilizing highly alkaline or acidic conditions now allows for the effective recovery of stable functional proteins from underutilized and previously unutilized materials (8). The proteins are effectively separated from undesirable components in the raw material and can successfully be utilized for their gel forming ability (i.e., forming products similar to surimi) and other functional properties. The relatively pure proteins can also be further processed to modify their functionality for different applications. Another exciting process, enzymatic hydrolysis, although not new has, in the last few years, become more sophisticated and shown to be an excellent method to both recover protein from underutilized raw material and induce many desirable properties in the proteins. It has been shown by a number of researchers that, by applying controlled enzyme technology to recover and modify fish proteins present in byproducts and in underutilized species, it is possible to produce a broad spectrum of protein ingredients with a wide range of applications in the food, pharmaceutical, and specialized feed and fertilizer industries (3,7,9–14). These ingredients are collectively called fish protein hydrolysates (FPH), which can be defined as proteins that to a greater or lesser degree are chemically or enzymatically broken down into peptides of varying sizes (15). This promising approach could make significantly more use of fish proteins, increasing the margin of profit for the fishing industry as well as making it more environmentally friendly. Historically, the practice of fish protein hydrolysis extends back many decades. Much work on the hydrolysis of fish proteins was conducted in the 1960s and 1970s (16), and was directed into production of cheap, nutritious protein sources for rapidly growing developing countries, or toward animal feed production, primarily through production of fish protein concentrates (FPC), which employs chemical hydrolysis sometimes coupled with enzymatic hydrolysis. Most of these efforts were not successful due to a variety of reasons. In the past 20 years or so only scattered reports have emerged in the literature on FPH, but the past 3–5 years have seen a surge in publications, new developments, and novel uses of FPH. In this chapter we will give an overview of the production of FPH including their various properties, current uses, and future potential as functional ingredients in food, pharmaceutical, and feed and fertilizer products.

23.2 THE ENZYMATIC HYDROLYSIS PROCESS

Different techniques exist for extracting and or hydrolyzing protein from fish. These include using aqueous and organic solvents to extract proteins and concentrated acid or alkali to extract and hydrolyze proteins at elevated temperatures. These methods normally adversely affect the functional and organoleptic properties of FPH and may produce toxic byproducts (3). Biological processes using endogenous or exogenous enzymes are now far more frequently employed in industrial practices to make FPH, because enzyme hydrolysis is usually a mild process that results in products of high functionality, good organoleptic properties, and excellent nutritional value without the formation of toxic byproducts (3,17). Many different aquatic sources have been enzymatically processed into FPH as summarized by Kristinsson and Rasco (3,7). Commercial production of fish protein hydrolysates is growing, and has reached a sizable level in a few countries including Denmark, France, Iceland, Mexico, Norway, Japan, USA, Canada, and several Southeast Asian countries. Although enzymatic hydrolysis is preferred over chemical hydrolysis, it has some drawbacks such as: the potentially high cost of using large quantities of added commercial enzymes, which are not reusable; occasional difficulty in controlling the extent of the reaction (especially in mixed protease preparations), which can result in heterogeneous products consisting of fractions of varying molecular weight; lower protein recovery (depends on extent of hydrolysis); and the need to inactivate enzymes by high or low pH or heat treatment at the end of the reaction, which adds to the processing costs and may adversely affect some functional properties (3,18).

The process of protein hydrolysis is relatively simple, but many variations exist. Figure 23.1 outlines a fairly typical protein hydrolysis process as reported for Atlantic salmon muscle proteins (13). Going into the hydrolysis, the raw material has to be well homogenized and (ideally) diluted with water. The amount of water added can be an important variable in the hydrolysis process. Less water may not only slow down the hydrolysis due to insufficient access of the enzyme to the substrates, but also the high viscosity of the system may lead to less recovery of hydrolyzed protein. Slizyte and coworkers (19) reported that water addition was more important than enzyme type in the recovery and functionality of hydrolyzed proteins from cod byproducts. This was likely due to differences in the extent of hydrolysis and thus the properties of the peptides in the recovered fractions. Prior to hydrolysis, the starting material has to be thoroughly homogenized and diluted, so that the enzyme can have easy access to proteins. The protein substrate is then adjusted to a pH and temperature representing the optimal reaction conditions of the particular enzyme. The enzyme preparation of choice is then mixed into the material, and ideal pH and temperature are maintained. Initially, where a large number of peptide bonds are broken, hydrolysis is fast; followed by a slower rate due in part to enzyme inactivation or substrate limitation

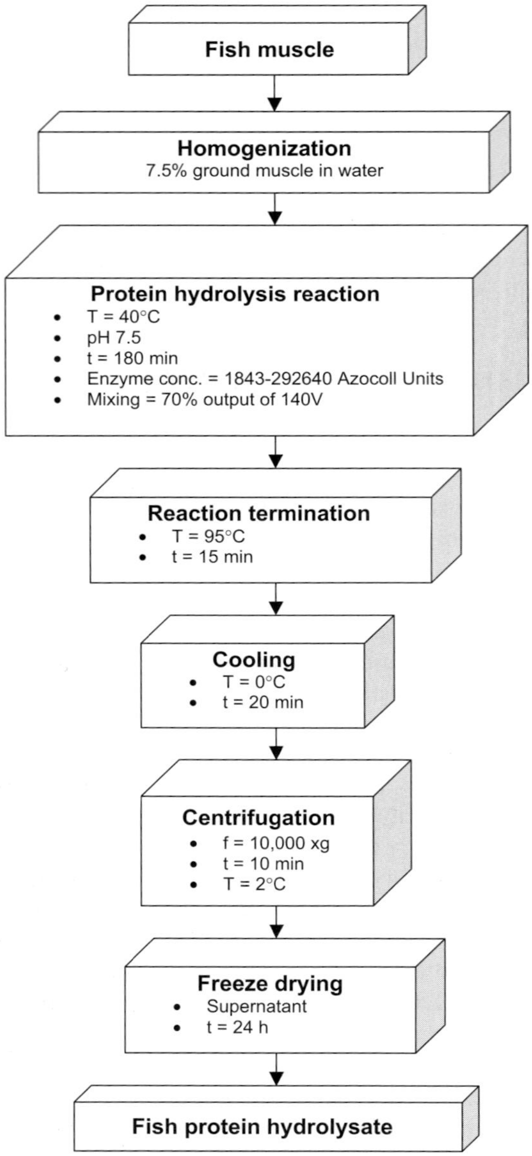

Figure 23.1 An example of a process to make fish protein hydrolysates from Atlantic salmon (adapted from Kristinsson and Rasco, *Proc. Biochem.* 36:131–139, 2000).

(6,10,13,20). Eventually the hydrolysis reaches a stationary phase. Because the extent of hydrolysis highly influences the recovery and the functionality of the proteins, it is imperative to control the hydrolysis if the goal is to produce a product with consistent specified molecular profile and functional properties. Many methods exist for monitoring and calculating the extent of hydrolysis, as summarized by Kristinsson and Rasco (3). The best and simplest method to monitor the degree of hydrolysis directly during the process is the pH-stat method (21), which is based upon maintaining constant pH during the reaction and recording the volume and molarity of base or acid used to maintain the pH. This method and other methods are discussed in detail by Kristinsson and Rasco (3). Unfortunately, control over the hydrolysis process is often lacking in the literature, and researchers often fail to specify the final degree of hydrolysis (%DH) of their FPH preparations, which makes it hard to compare properties of FPH from different studies. When hydrolysis has reached its desired end point in the process, it is terminated by inactivating the enzymes with high temperatures, or reducing or increasing pH, or a combination of both. The hydrolyzed proteins are then recovered by centrifugation or filtration, to separate them from insolubles; e.g., fat, scales, bones, connective tissue, and unhydrolyzed protein. A substantial amount of utilizable material, most notably proteins and lipids, could be recovered from the insoluble "sludge" fraction after hydrolysis and used for different applications than the FPH. If necessary the soluble fraction containing the hydrolyzed protein can be further treated by passing it through charcoal to deodorize and decolorize it. For long term storage and transportation, the hydrolysates can be stabilized with antimicrobial or antioxidant compounds and kept in liquid form (common for fertilizers), or dried by either freeze or spray drying depending on the use and value of the final product.

23.2.1 Importance of Raw Material Source

Although the process of protein hydrolysis is relatively simple, there are a number of factors that need careful consideration before, during, and after hydrolysis. The nature and quality of the raw material can have a great impact on the hydrolysis process and the quality and functionality of the final product. For example, it is important to distinguish between lean and fatty species, because the latter can lead to serious lipid oxidation problems in the final product, creating unpleasant flavors and odors (4,22). Furthermore, there are bound to be serious oxidative and color problems during the processing and in the final product, such as raw materials high in blood and dark muscle (common with pelagic species) due to high levels of membrane lipids and potent prooxidants (primarily heme proteins and to some extent transition metals). For materials rich in lipids, blood and dark muscle an alkaline processing pH would significantly reduce the development of oxidative and color problems (23), but could cause more microbial spoilage. Incorporating antioxidants and antimicrobial agents early in the processing could partially alleviate the problem, if the starting material is of decent quality. Other components of fish such as bones, skin, and connective tissue can also interfere with the hydrolysis process and contaminate the final product; and have adverse effects on its quality and function. The ideal scenario would be to start out with a "clean" and simple starting material, such as a skinned fillet. However, this is highly uneconomical. Recently a process was developed which aims at economically producing functional and essentially pure protein isolates from underutilized sources of fish, such as pelagic species and processing byproducts (5,24). Because both neutral and membrane lipids have been removed, along with some prooxidants, these proteins are of excellent quality. This revolutionary process would allow one to start out with a relatively clean protein product in the hydrolysis process, yielding a higher quality and more functional protein hydrolysate. This could also result in more protein recovery in the process as centrifugation and filtering methods would not be necessary at the end of hydrolysis.

23.2.2 The Choice of Enzyme and pH Conditions

The choice of enzyme is another very important factor in the hydrolysis process because different enzymes have different specificity, and therefore produce products of different molecular makeup and functional properties. The ratio of enzyme to substrate ratio has been investigated by several workers and an increase in this ratio (i.e., more enzyme) leads to a more rapid reaction and also more recovery of protein (25). However, excessive use of enzymes is costly and generally not necessary for an efficient hydrolysis process. The choice of enzyme, and its concentration, is usually based on a combination of the enzymes' effectiveness and cost as well as their substrate specificity (26–27). Proteolytic enzymes have either an endopeptidase or exopeptidase activity, or both, which has a dramatic impact on their speed of hydrolysis and the properties of the peptides formed. Endopeptidase enzymes are often very specific and cleave peptide bonds inside of the protein, usually leading to relatively small peptides but only small amounts of free amino acids. Generally, the less specific the enzyme is, the smaller the peptides are. The endopeptidases, on the other hand, cleave amino acids from the ends of the protein polypeptide chain resulting in mixtures of free amino acids and larger peptide fractions (depending on when the reaction is stopped). It is imperative to have a good understanding of the specificity and activity properties of the enzymes available for protein hydrolysis in order to control the extend of hydrolysis and properties of the end products. Commonly a combination of endo- and exopeptidase activities is favored in hydrolysis applications. With the above in mind, it becomes clear that using commercial or purified enzymes instead of endogenous enzyme mixtures (i.e., enzymes naturally present in the material being hydrolyzed) offers many advantages, because it allows good control of the hydrolysis and thereby the properties of the resultant products (10,13). According to the literature, a number of different commercial and purified enzymes have been used to hydrolyze fish protein. Table 23.1 lists some enzymes used for the hydrolysis of fish proteins (28). In early studies pepsin was found to be very efficient (29,30) but given that it is only active at acidic pH levels, it can adversely affect functional properties of the final product (31), as well as its oxidative quality and color, because at low pH lipid oxidation proceeds quickly (32). In part this is because heme proteins are rapidly denatured and become more active prooxidants (23). However, processing at low pH limits microbial growth, which positively contributes to the products' microbial stability. Enzymes active in the range of pH 6–8 have been more commonly employed in the past 20 years or so. One of the more popular enzymes, Alcalase (made by Novozymes, USA), an alkaline enzyme with both high endo- and exopeptidase activity, has repeatedly been reported to be one of the best enzymes used to produce functional FPH (13,33–36). Alcalase has also shown to give good protein recovery; between 70% and 88%, for a variety of species tested (10,12,13,33,36). Other enzymes have resulted in high recovery and protein content. The protease P "Amano" 3 gave ~87% recovery of proteins when used to produce hydrolysates from *Aristichtys noblis* (37). Kristinsson and Rasco (13) compared five different enzyme preparations (Alcalase 2.4L, Corolase PN-L, Corolase 7089, Flavourzyme 1000L, and a salmon visceral pyloric caeca extract) for their ability to produce protein hydrolysates from Atlantic salmon mince. All gave good protein recoveries and high protein contents of the final product (13) (Table 23.2, Table 23.3). Although Alcalase gave the highest protein content (>88%), Flavourzyme gave the highest protein recovery, with Alcalase and Corolase 7089 closely following. The protein recoveries were found to be highly dependent on the degree of hydrolysis (Table 23.3). The higher the %DH the more the protein recovery, while final protein content in the hydrolysate is less affected by %DH (13). This is very possibly due to the formation of more soluble peptides at higher degrees of hydrolysis which would stay suspended during the last centrifugation step.

It is also possible to perform the hydrolysis process using endogenous proteases to hydrolyze fish muscle proteins. This process depends on the action of endogenous digestive enzymes present primarily in the fish viscera (mainly trypsin and chymotrypsin from pyloric caeca and pepsin from stomach), and to some extent muscle tissues (lysosomal proteases). The end product is generally a fairly viscous liquid, rich in free amino acids and small peptides. Using visceral material to hydrolyze fish protein has an advantage over using commercial enzyme preparations in that processing costs can be reduced. However, a major drawback is that the viscera are a complex mixture of enzymes with different

Table 23.1
Some examples of enzymes used to hydrolyze fish proteins (adapted in part from Gilmartin and Jervis, *J. Agric. Food. Chem.* 50:5417–5423, 2002.)

Enzyme	pH Range	Temp. (°C) Range	Specificity	Source	Supplier
Alcalase 2.4L	6.5–8.5	55–70	Endo	*Bacillus licheniformis*	Novozymes
Flavourzyme 1000L			Endo & Exo	*Asperigillus oryzae*	Novozymes
Neutrase 0.5L	5.5–7.5	45–55	Endo	*Bacillus subtilis*	Novozymes
Protamex	5.5–7.5	35–60	Endo & Exo		Novozymes
Corolase 7089	5–7.5	<60	Endo	*Bacillus subtilis*	Röhm Enzymes
Corolase PN-L	5–8	<50	Endo & Exo	*Asperigillus sojae*	Röhm Enzymes
Corolase LAP	6–9	<70	Exo	*Asperigillus sojae*	Röhm Enzymes
Acid Protease II	3–5	40–50	N/A	*Rizopuz niveus*	Amano Enzymes
Protease P “Amano” 6	5–9	40–50	N/A	*Asperigillus melleus*	Amano Enzymes
Protease M “Amano”	3–6	40–60	N/A	*Asperigillus oryzae*	Amano Enzymes
Papain	5–7	65–80	Endo & Exo	Papaya fruit	e.g., Enzyme Development Corp.
Bromelain	3–9	50–60	Endo & Exo	Pineapple stem	e.g., Enzyme Development Corp.
Pepsin	1.5–2.5	35–50	Endo	Animal stomach	e.g., American Laboratories Inc.
Pancreatine	7.5–8.5	35–50	Endo & Exo	Animal pancreas	e.g., American Laboratories Inc.

Table 23.2
The recovery of proteins as a function of %DH and enzyme source during salmon muscle protein hydrolysis (adapted from Kristinsson and Rasco, *J. Agric. Food Chem.* 48:657–666, 2000)

Enzyme	5% DH	10% DH	15% DH
Alcalase	46.8	57.0	74.6
Flavourzyme	42.9	58.0	79.9
Corolase PN-L	41.6	56.3	67.4
Corolase 7089	52.0	61.8	74.5
Visceral serine protease extract (salmon)	40.7	48.6	68.6

Table 23.3
Protein content of freeze-dried salmon protein hydrolysates as a function of %DH and enzyme source (adapted from Kristinsson and Rasco, *J. Agric. Food Chem.* 48:657–666, 2000)

Enzyme	5% DH	10% DH	15% DH
Alcalase	86.9	88.4	88.1
Flavourzyme	84.3	82.7	79.0
Corolase PN-L	79.3	74.9	72.9
Corolase 7089	86.5	85.3	82.4
Visceral serine protease extract (salmon)	79.1	71.7	73.4

activity requirements, and the concentration of desirable enzymes can fluctuate greatly from season to season, as well as within and between species. This makes it very hard to control the hydrolytic process from one batch to another, resulting in hydrolysates with varying peptide and polypeptide molecular weights and varying functional properties. However, if the goal is extensive hydrolysis, variations in enzyme activity are not a major concern. For this reason extensive hydrolysis using endogenous proteases is widely used to produce hydrolyzed liquid products for human consumption, specifically fish sauces, fertilizers, and animal feeds.

Fish sauce is the major endogenously hydrolyzed fish product presently consumed in the world, with a reported annual production in Southeast Asia of about 250,000 metric tons (23). The raw material is typically small pelagic fishes, such as anchovies, that are mixed with concentrated (20–40%) salt solution at ambient temperatures (25–45°C). Hydrolytic breakdown via endogenous proteases proceeds under anaerobic conditions for 6–12 months. This slow but extensive breakdown results in a liquefied fish sauce composed predominantly of free amino acids. The goal of the process is not to improve the nutritive value of the protein per se, but to extend the shelf life of the protein along with improving its organoleptic characteristics (38). Microorganisms can also be used to produce liquefied fish products, via proteolysis and fermentation. For these processes, bacterial fermentation is typically initiated by mixing minced or chopped fish with a fermentable sugar. The fermentable sugar encourages the growth of lactic acid bacteria, which is advantageous because it produces acid and antibiotics that together limit the growth of competing spoilage bacteria (39).

Few research reports have been published on the utilization of endogenous enzymes to produce functional fish protein hydrolysates, especially for human food applications. Fish sauce is almost the only endogenously hydrolyzed fish product on the market in the USA and Europe. In order to produce a functional protein hydrolysate with specific properties, a good knowledge of the enzymes involved is crucial. The major limitation in being able to produce a functional hydrolysate with consistent properties (other than fish sauce) using endogenous enzymes is the lack of control over enzyme activity and specificity. Shahidi et al. (10) hydrolyzed ground capelin (*Mallotus villosus*) using endogenous enzymes, which yielded far less protein recovery (22.9%) than FPH produced with Alcalase (70.6%), a commercial enzyme mixture. Earlier Hale (10) reported a similar lower recovery of proteins from red hake muscle when it was hydrolyzed with fish viscera compared to Alcalase. Kristinsson and Rasco (13,14,27) however, reported high protein recovery and excellent functional properties using a preparation of endogenous visceral serine proteases to hydrolyze ground salmon muscle fillets and frames. Protein recovery was close to 70% at 15% degrees of hydrolysis (%DH), comparable to the recoveries obtained from using four commercial enzyme preparations (13). An added benefit of the endogenous mixture was that it had significantly higher proteolytic activities at ambient temperatures (20–22°C) compared to the commercial preparations, in addition to being less expensive (14,27). The salmon FPH produced with the endogenous enzymes also had better emulsifying properties than soy protein concentrate or FPH made from four commercially available enzymes (13). One of the reasons that the serine protease mixture gave significantly better results than were previously achieved using endogenous enzymes is very likely due to the fact that the mixture was a partially purified form of visceral enzymes with uniform enzyme properties, whereas other studies used ground visceral material, probably of highly variable enzymatic activities. This underlines how important it is to have a characterized enzyme preparation to be able to produce FPH with good and consistent functional properties.

23.3 PROPERTIES, APPLICATIONS, AND POTENTIAL USES OF FISH PROTEIN HYDROLYSATES

23.3.1 Food Functionality

Good functional properties of FPH are very important in order to successfully utilize it as a food ingredient. The size and chemical properties of the hydrolyzed proteins has a dramatic impact on their functionality, and thus control of enzyme specificity and %DH is critical if the goal is to produce hydrolysates with improved or different properties suitable for a wide array of products (13,41).

23.3.2 Solubility and Water-Binding Properties

Many functional properties of proteins are positively affected by increased solubility. Intact fish myofibrillar proteins have limited solubility at concentrations between ~0.3 mM and ~300 mM NaCl and in the pH range of most foods and therefore would find limited use as added ingredients in food products under normal conditions. Solubility of myofibrillar proteins is increased by enzymatic hydrolysis, as it leads to smaller peptides and newly exposed amino and carboxyl groups that enable more interactions with water (42,43). Sugiyama and coworkers (34) reported that solubility of sardine meal was greatly enhanced after treatment with enzymes. Quaglia and Orban (33,44) reported over 90% solubility for sardine hydrolysates produced with either Alcalase or papain. A recent study on salmon FPH made with Alcalase showed that it had between 96 and 100% solubility in 100 mM NaCl from pH 2 to 11. In the pH range ~4–8 (common to most food) intact myofibrillar proteins would have been largely insoluble in 100 mM NaCl. FPH thus has the advantage of being highly soluble over a wide range of pH and ionic strength (10,11,13) which could extend its use in many food systems. For example, FPH produced from Atlantic salmon with Alcalase exhibited 96–100% solubility in 100 mM NaCl from pH 2 to 11 (13). The good solubility and nutritive value of enzymatically hydrolyzed fish proteins has made them a feasible choice to produce milk replacers for weanling animals (6,45), which is presently being pursued in Japan and France (22). The good solubility of FPH could also make it a promising ingredient in nutritional drink formulas, provided that flavor is not adversely affected.

Due to good solubility over a wide range of pH, FPH is easily injected into muscle products for a variety of purposes, one of them being to aid in water uptake and retention. A number of commercially available hydrolyzed food proteins are used to control the water balance of food systems, to improve quality and increase yield and profits. The most common are soy and milk proteins. There is an advantage to using a muscle protein in a muscle product, in large part because of labeling issues. Limited research has been conducted on the potential of using FPH (or any muscle protein hydrolysate) as a water binder. In one of the earlier studies it was found that a significant increase in cooking yield was observed for a sardine hydrolysate when it was added to hamburgers (46). Shahidi et al. (10) reported that capelin protein hydrolysates incorporated into minced pork muscle increased its water holding capacity upon cooking. Similar findings were reported for shark protein hydrolysates mixed with comminuted pork, where cooking yield was increased in proportion to increased FPH addition (12). Kristinsson and Rasco (13) reported that salmon FPH incorporated into salmon mince patties prevented drip loss significantly better than egg albumin and soy proteins. In this study the type of enzyme used to make the hydrolysates was found to be more important than the %DH. Recent studies in the author's laboratory, injecting hydrolyzed fish protein isolates into fish fillets under certain conditions of pH and ionic concentration, have demonstrated a better overall performance in terms of water holding than by using salts and phosphates (Kristinsson, unpublished data). In the near future, therefore, hydrolysates could replace or in part substitute for phosphates, which are looked on unfavorably in several countries and by some consumers. Because the hydrolysates are a natural product, using them would help in creating a cleaner label for the food product. Fish protein hydrolysates could therefore be a great opportunity for the seafood industry to control the water balance of seafood products.

It is not well understood how FPH aids in the water retention characteristics of a food. The increase of terminal carboxyl and amino groups upon hydrolysis could increase the amount of water that can be bound compared to an intact protein, in part explaining the increased solubility. Free, small FPH peptides in the water phase of the mince or muscle could increase its osmotic character and thus slow water loss, explaining the decrease in drip loss. It is also possible that the peptides may aid in the gel forming mechanisms of the muscle proteins upon cooking. Recent studies report that fish hydrolysates may partially protect other proteins with which they are in contact. Khan et al. (47) found that FPH from five different species provided some stabilization of lizard fish myofibrillar proteins during drying, because more ATPase activity was retained. The FPH treated myofibrillar proteins also had a higher amount of bound unfrozen water, suggesting increased water binding to the myofibrils. The same study also reported similar effects of squid and krill protein hydrolysates on lizard fish myofibrillar proteins (48,49) and shrimp muscle proteins (48). These studies indicate that FPH could be a beneficial processing aid during drying of fish muscle or fish proteins to help retain more of the native protein conformation and function. Cryoprotective effects have also been reported for FPH (47,50). Khan and coworkers (47) reported that high gel forming ability and Ca-ATPase activity was retained after mixing FPH (5% w/w) into lizard fish surimi followed by freezing at -25°C. FPH added to products such as surimi could therefore be in part substituted for common ingredients like sucrose and sorbitol, and potentially give similar gel qualities. The mechanisms by which

FPH stabilizes proteins at freezing temperatures has not been well studied. Netto et al. (51) reported that FPH had similar effectiveness in lowering glass transition temperatures, and its effectiveness increased with lower molecular size. More research is needed in this area to understand the mechanisms of water retention, so that peptides can be tailor made to exhibit maximum water holding properties and be effective in protecting muscle proteins from denaturation.

23.3.3 Interfacial and Surface Properties

Fish protein hydrolysates with good interfacial and surface properties may find use as ingredients to aid in the formation and stabilization of oil in water or water in oil emulsions (e.g., margarine, dressing, and meat batters) and foams (e.g., mousse, whipped cream, and meringues).

Initial limited hydrolysis may expose more hydrophobic groups, thus increasing the surface and interfacial active properties of proteins. However, as the hydrolysis progresses, a decrease in surface and interfacial activity is observed (13). Even though small peptides may rapidly adsorb at the interface due to their high solubility, they are less able to produce good and stable emulsions as they become smaller (i.e., as the %DH becomes higher) (13). The specificity of the enzymes used is important here, because it can greatly influence the molecular size, topology, and hydrophobicity of the resulting polypeptides, all which influence the interfacial and surface properties of the FPH. Using enzymes with more narrow specificity and lower degree of hydrolysis is generally recommended, because larger and more amphiphatic peptides result. For good interfacial and surface properties, it has been suggested that peptides should not be smaller than 20 amino acid residues (52).

Results by Kristinsson and Rasco (13) showed that emulsifying capacity and stability of salmon FPH were greatly dependent upon the enzyme used. Using a serine protease mixture from salmon pyloric caeca produced FPH with the best emulsifying properties, and was found to be connected to the narrow specificity of the extract, which gave larger polypeptides than the other enzymes used (13). The salmon FPH had superior emulsifying capacity and stability compared to soy protein concentrates, but performed significantly more poorly than egg albumin. All salmon FPH demonstrated a decrease in emulsifying properties with an increase in %DH in agreement with previous finding about sardine hydrolysates (53). Sathivel and coworkers (54) reported a rather poor emulsifying capacity and stability for FPH made from different herring byproducts, when compared to egg albumin and soy protein concentrate. Liceaga-Gesualdo and Li-Chan (55) however, reported an increase in emulsifier formation and stability when herring muscle was hydrolyzed to 36% DH with Alcalase compared to unhydrolyzed herring muscle. Spinelli and coworkers (56,57) had reported similar findings earlier, that FPH from rockfish had improved emulsifying capacity compared to unhydrolyzed myofibrillar proteins. However, the same fish hydrolyzed with bromelain produced FPH with poor emulsifying capacity and stability (58). Jeon et al (59) studied the emulsifying activity and stability of cod protein hydrolysates, and reported that both were highly dependent on molecular size. Hydrolysates ultrafiltrated through a 10K Da membrane exhibited optimal emulsifying ability. Vieira and coworkers (11) reported poor emulsifying properties for lobster waste hydrolysates. Relatively poor emulsifying capacity and stability was also reported for capelin (10) and shark protein hydrolysates (12). The problem with many of the above studies is that the %DH of the FPH was not reported, which makes interpretations of the results impossible. Another complication with all of the above studies and other functional properties are that different methods are used to study them, making it very difficult to compare them.

Only a few studies have been conducted on the foaming properties of FPH. Shahidi and coworkers reported reasonably good whippability and foam stability for shark protein hydrolysates and capelin hydrolysates (10,12). However, compared to whey protein hydrolysates (60) and egg white proteins (61), the shark and capelin hydrolysates had significantly poorer foaming properties. Studies with herring FPH demonstrated that they had good foam forming properties but poor foam stability due to rapid breakdown, while unhydrolyzed proteins exhibited slightly lower foam forming properties, but far better stability (55). The decrease in peptide molecular weight as a result of hydrolysis leads to a decrease in foaming ability, because the small peptides do not have the ability to stabilize the air cells of the foam (55). Therefore low %DH is necessary to produce FPH with good foaming properties. This was demonstrated by Jeon et al. (59) for cod protein hydrolysates, where ultrafiltered samples had improved foaming characteristics in direct correlation with their particle sizes: 30KDa, > 10KDa > 5 KDa > 3KDa. Interestingly, the unfiltered hydrolysate had lower foamability and foam stability than the 30K, 10K and 5K filtrates, demonstrating the importance of molecular weight. Good foamability can, however, be a problem when it comes to other functional properties of FPH, especially if they are to be used as soluble and injectable protein ingredients. It is crucial to be able to produce a favorable balance of water solubility and good interfacial and surface properties for different applications, which is accomplished by proper selection of enzymes and processing conditions.

23.3.4 Lipid Binding

Many food products rich in lipids, e.g., meat batters (sausages) and confectionaries, require physical stabilization of the lipids. Although in part related to the ability of FPH to emulsify lipids, only a handful of studies have investigated directly the oil and fat binding property of FPH. Shahidi and coworkers (10,12) demonstrated the capacity of capelin and shark protein hydrolysates to bind oil, although no units were reported to express the amount of oil binding. Kristinsson and Rasco (13) reported that salmon FPH had significantly better oil binding than two common food protein ingredients, egg albumin and soy protein concentrate. This study revealed that oil binding was influenced by the enzyme used for the hydrolysis, with increased %DH leading to less oil binding regardless of enzyme used. At 5%DH oil binding varied from 4.48 to 7.07%; at 10% DH from 3.22 to 5.12%; and at 15% DH from 2.86 to 3.86%. In this study egg albumin and soy protein concentrate had oil binding of 2.36% and 2.90%. Sathivel and coworkers (54) also reported higher oil binding with different FPH from herring byproducts, although no %DH was specified.

23.3.5 Flavor Properties

One of the reasons behind the slow penetration of FPH into the food ingredient market is bitterness which often develops on hydrolysis. Bitterness in FPH is also of great importance for animal feed application utilizing FPH as a efficient protein source, because consumption of FPH can lead to increased fishy off flavors (62). There is a sensitive and complex balance between extent of hydrolysis, peptide profile, and bitterness development. Generally, limited hydrolysis eventually leads to the development of bitterness, because hydrophobicity of the peptides formed is increased over the parent protein. Hydrophobic peptides have found to play a major role in the bitterness sensation of protein peptides (63). From work reported by Yu and Fazidah (37) it could be seen that limited hydrolysis of *Aristichtys noblis* up to approximately 15%DH led to only limited bitterness, but between 15 and 20%DH bitterness developed rapidly. On the other hand, extensive hydrolysis yielding very small peptides or free amino acids generally leads to less bitterness. Sugiyama and coworkers (34) reported that sardine FPH with smaller peptides had significantly less bitterness than FPH with larger peptides. However, small basic tripeptides containing asparagine and lysine as the second and C-terminal residues, respectively, and with the N-terminal residue leucine or glycine, have been found to be responsible for bitter taste in fish protein hydrolysates (64). The relationship between peptide length and properties is therefore a complex one. Extensive hydrolysis of proteins from aquatic organisms may in some cases lead to a flavor enhancement effect, and the use of FPH as a seafood flavor or flavor enhancer is possibly one of its most promising applications. In an early study Fujimaki and coworkers (65) reported that fish protein concentrate treated with Pronase to make FPH had flavor enhancement effect similar to monosodium glutamate (MSG), but accompanied with some increased bitterness. The peptides responsible for this MSG-like flavor were found to be three dipeptides and five tripeptides (66). Vieira and coworkers (11) reported that extensive hydrolysis of lobster processing waste gave a product of superior quality and no bitterness. This is possibly due to large amounts of free amino acids and flavor enhancing nucleotides which lobster is rich in. Imm and Lee (57) prepared hydrolysates from several different species using enzymes specifically developed for maximizing flavor and minimizing bitterness (Flavourzyme and Savorase) and reported that increased levels of free amino acids led to a flavor enhancement effect (called the umami effect). These authors found that optimal flavor was obtained with a 6 hour hydrolysis. ChiaLing et al. (68) also reported an umami effect of hydrolysates made from proteins in tuna cooking juice. The authors concluded that the flavor and quality of these hydrolysates could make them successful as a condiment in the marketplace. There are several different seafood flavors or extracts produced commercially, and most of these are in fact hydrolyzed fish and shellfish proteins. There are multiple uses for these products such as sauces, seasonings, soups, bisques, frozen seafoods, fresh and prepared seafoods, appetizers, snacks, and imitation seafood products (e.g., crabsticks) (69).

To control bitterness when producing FPH, enzyme choice appears to be very important. Enzyme preparations with high exopeptidase activity and high preference for cleaving off hydrophobic amino acids in bitter peptides have been recommended (21,70). Several studies have also pointed out that a proper balance of endoprotease and exopeptidase activity produces less bitterness when compared to using each separately. Several enzyme preparations are now commercially available, with the goal of limiting the development of bitterness while at the same time obtaining good functional properties. Only a handful of studies have compared protease preparations for their effect on bitterness. Lalasidis and coworkers (71) hydrolyzed cod filleting offal which gave a product of bitter taste. Sugiyama and coworkers (34) reported that sardine FPH made with alkaline proteases were significantly less bitter than those prepared with neutral or acid proteases. Hoyle and Merritt (16) reported that herring FPH made with papain had more bitterness than FPH made with Alcalase. The papain samples were at a lower %DH than the Alcalase samples in that study. When the same substrate was defatted and then used to make FPH, much less bitterness developed. This suggests that lipid oxidation products may be a major source of much of the bitterness reported with FPH. This suggestion is strengthened by a recent study on autolytic hydrolysis of frigate mackerel byproducts which showed a good correlation of lipid oxidation and bitterness development during hydrolysis (72). This underlines the importance of substrate selection and quality of the starting material for FPH processes.

Several postprocessing techniques can be used to mask or reduce the bitterness of hydrolysates. Treating FPH with activated carbon may partly remove bitter peptides (10). It has also been reported that treating FPH with synthetic adsorbents led to a significant reduction in bitterness (34). Treating FPH with peptidases may also lower the bitterness of the product, as discussed above. Lalasidis and coworkers (71) found that when cod offal FPH was further treated with pancreatine, which is rich in exopeptidase activity, bitterness was essentially eliminated. Sugiyama and coworkers (34) also reported that bitterness was significantly reduced upon treating sardine FPH with different peptidases. Bitter peptides may also be extracted with solvents (71) and FPH may be debittered using ethyl alcohol

(73), both which may have a negative impact on the food functionality of the FPH. A milder and innovative approach to lower bitterness was described by Sugiyama et al (74). These workers found that when aldehydes were removed from FPH by yeast or acetic acid bacteria at low pH, bitterness was also reduced. Another unique approach to reduce bitterness is to rejoin the hydrolyzed peptide using certain proteases, with a process called the plastein reaction (22,75). These rejoined peptides have reduced exposed hydrophobic residues, i.e., less bitterness (22). Bitterness may also be masked by added ingredients. For example, Imm and Lee (67) reported that addition of 1.5% NaCl and 0.4% sodium tripolyphosphate masked some bitterness and off flavors that developed on hydrolysis of several fish species. A Japanese patent by Suzuki and coworkers (76) describes adding cyclodextrin to hydrolyzed proteins to eliminate bitter taste.

23.3.6 Antioxidant Properties

The potential of peptides to serve as antioxidants to limit lipid oxidation in food systems has recently received increased attention. Lipid oxidation is one of the major causes of quality deterioration in seafood products, especially those rich in lipids and potent prooxidants (77). Fish protein hydrolysates, which are rich in different types of peptides, have shown potential to be used to minimize oxidation. In an early study on the antioxidative potential of FPH, it was found that FPH prepared from sardine myofibrils showed some antioxidant activity which was substantially increased when it was combined with different commercial antioxidants, likely due to a synergistic effect (78). Shahidi and coworkers (10) reported that addition of capelin FPH at 0.5–3.0% level in minced pork reduced formation of thiobarbituric reactive substances (TBARS) by 17.7–60.4%. The authors proposed that this antioxidative property was likely due to chelation. Later work on capelin FPH showed that after fractionation with gel chromatography both prooxidative and antioxidative fractions could be identified (79,80). Presence of transition metals, heme, and heme proteins in the hydrolysates could contribute to the prooxidative effects of FPH. Chuang and coworkers (81) reported that mackerel FPH showed good antioxidant effect, because it decreased both lipoxygenase and hemoglobin mediated linoleic acid oxidation. The hydrolysate was, however, not as effective as the well known antioxidant ascorbic acid. The authors proposed that the antioxidant mechanism was likely due to histidine related dipeptides such as anserine and carnosine, which reportedly scavenge free radicals (81). Jeon and coworkers (59) also reported a good inhibition of cod FPH fractions below 5KDa in a linoleic acid model system. The antioxidative activity was reported to be equal to tocopherol. These antioxidative properties of FPH are likely highly dependent upon the amino acid composition and molecular size of the FPH peptides, as studies have shown that antioxidant activities of hydrolysates are highly dependent on the type of enzyme used (82). An elaborate study by Saito and coworkers (83) on synthesized tripeptides reported that the amino acid composition and sequence in a peptide greatly influences the level and type of antioxidative activity. More studies on molecular basis behind FPH potential to prevent oxidation and their antioxidative function in complex food systems is highly recommended.

23.4 NUTRITIONAL AND PHYSIOLOGICAL PROPERTIES

23.4.1 Animal Feed

It has been know for many centuries, if not thousands of years, that living organisms respond very positively to hydrolyzed fish material, although the reason for this was at that time unknown. A large amount of research in the past 50 years or so has shown that hydrolysates are excellent sources of protein due in part to their excellent amino acid balance, good digestibility, and rapid uptake. Hydrolyzed fish material (mostly in the form of silage, i.e., autolytically hydrolyzed fish) is widely used as feed for weaning pigs, minks, pigs, and poultry (3,22,84). Hydrolyzed fish products are also being increasingly used as pellets for pet food and in aquaculture feed formulations. The main problem with FPH in these applications has been bitterness, which can make the feed unpalatable for animals. Due to the excellent balance of essential amino acids, FPH have been suggested as a partial replacement for cereal proteins to improve the protein quality of cereal products (45,85). Enrichment of FPH with legume products has also been reported (85). Common Mexican dishes were prepared with cereals and legumes enriched with FPH, and were accepted by 70% of taste panelists even when FPH made up 35% of all protein (85). Fish protein hydrolysates have also been suggested as ingredients in bakery products (6). A French patent describes the use of dried skim milk (15%) and fish protein hydrolysate (14%) in a biscuit formulation designed as an energy food (86). Yu and Tan (87) reported that fried crackers made with tilapia protein hydrolysates were "highly acceptable" up to a 10% FPH addition level.

The protein quality and digestibility of FPH has been investigated by many. Bechtel and coworkers (88) recently reported that FPH from salmon heads and viscera had a calculated protein efficiency ratio (PER) of 2.6 and 2.8, respectively, and both FPH had over 94% protein digestibility. Both products were reported to be especially rich in lysine. Sugiyama and coworkers (34) reported an amino acid score of 100 for FPH from defatted sardine meal. This FPH was also found to have a PER of 3.2, a net protein ratio (NPR) of 5.2 and a biological value (BV) of 86 in rats. Digestibility and net protein utilization was 99% and 85%, respectively. This suggests that FPH would be an excellent protein source for feed and food applications. Good nutritional properties (both tested and predicted) of FPH have also been found for FPH made from tilapia (89,90), dogfish (35) and Atlantic cod (91). Rat trials have demonstrated that FPH inclusion in the diet led to more rapid growth and higher body weight than control proteins such as casein (89,92,93). Fish protein hydrolysates were compared to fish meal on their effect on the growth of rainbow trout (94). FPH diets had better feed conversion ratios than most of the fishmeal feeds and nitrogen retention was better for FPH feeds compared to fishmeal. Lian and Lee (95) reported that squid protein hydrolysates increased the survival rate of trout fingerlings and Atlantic salmon juveniles. The same hydrolysate was also found to serve as a feed attractant. Other studies have reported no or little benefit of using hydrolysates in feed formulations. Wilson and coworkers (96) found that catfish offal FPH was less effective than menhaden fish meal in catfish fingerling growth, feed, and protein efficiency, likely due to the deficiency of important amino acids in the FPH. Oliva-Teles and coworkers (97) reported that partially replacing fish meal with FPH, even at high levels, did not have a positive effect on growth or feed utilization of turbot juveniles. The size of the peptides, i.e., degree of hydrolysis, can play a major role in the effectiveness of FPH in feed formulations. Cordova-Murueta and Garcia-Carreno (98) reported that adding FPH to feed led to lower growth in shrimp than did adding krill protein hydrolysates (KPH). This was thought to be due to the large amount of low molecular weight peptides ($<$14 kDa) in the FPH compared to KPH, which can led to an imbalance of amino acid adsorption, i.e., lack of uptake of essential amino acids. This study suggested that properly controlling %DH and minimizing free amino acids in the feed is important for a successful application of FPH in animal feeds.

Some very interesting and novel developments have recently occurred in the use of FPH in aquaculture operations. FPH has been found to be a good supplemental protein source for juvenile fish (99). According to Gildberg (99) studies performed *in vitro* and *in vivo* suggest that low molecular weight fractions of cod stomach FPH (i.e., silage) makes fish more disease resistant and stimulates nonspecific immune response reactions.

Bogwald et al (100) reported that Atlantic cod FPH injected into Atlantic salmon led to a stimulation of salmon head kidney cells in a dose dependent manner. Gildberg and coworkers (101) reported that selected fractions (500–3000 Da) of cod stomach FPH could be used as an adjuvant in fish vaccine and as an immune stimulant for aquacultured fish. Fish protein hydrolysates could therefore not only be used to increase growth of fish, but also have a positive physiological impact on the animal.

23.4.2 Plant Growth

It is well known that using FPH as a fertilizer or a component of fertilizer positively affects crop growth and yield (3). Several commercial liquid FPH products are sold for this purpose. When sprayed directly on cranberry and cherry trees, FPH significantly increase crop yield (George Pigott, personal communication, 1997). Fish protein hydrolysates have been very successful as fertilizers on golf courses (Stephen D. Kelleher, personal communication, 1999). Novel applications taking advantage of this effect of FPH have been used to stimulate development of commercially important plant varieties. Eguchi and coworkers (102) compared FPH with proline, a known stimulator, to stimulate somatic embryogenesis in Anise (*Pimpinella anisum*) with good success. It has been proposed that FPH has the potential of becoming a proline and amino acid substitute in plant tissue culture applications and may find use in value added applications in the plant propagation industry (103). Research has shown that FPH has the potential to increase the production of valuable bioactive compounds in plant. Andarwulan and Shetty (104) found that mackerel FPH led to more production of rosemaric acid and phenolics in oregano when used as a component of the growth media. In a later study Andarwulan and Shetty (105) reported that mackerel FPH stimulated the growth of Epoxy-pseudoisoeugenol-(2-methylbutyrate) (EPB) in anise root cultures. This phenolic compound has the potential to regulate nutraceutical phytochemicals in plants. Fish protein hydrolysates were also found to increase the level of water extractable phenolics in cranberry pomace when added as a nitrogen source into the culture media (106). Fish protein hydrolysates contain a good amount of glutamic acid and proline, both which can be used to stimulate the pentose phosphate pathway and phenolic synthesis in plants (104). Using FPH can therefore stimulate expression of many commercially valuable phytochemicals which could find use as nutraceuticals in various food products or as antioxidants to extend shelf life of products, which is a new, and possibly very valuable arena for FPH applications.

23.4.3 Bioactivity

Several studies have reported various physiological functions of FPH. Oxidative processes result in various diseases, and the antioxidant potential of FPH has been reported, as previously discussed. Lipoxygenases have been implicated in low density lipoprotein oxidation, which is connected to hypertension and arteriosclerosis, and FPH has been shown to inhibit the activity of this enzyme (81). Angiotensin II is another compound connected to hypertension and it is produced from angiotensin I by the angiotensin I converting enzyme (ACE) (59). Fish protein hydrolysates from a variety of species have been found to have inhibitory effects on this enzyme and thus could possibly lead to less hypertension if consumed (107–110). This inhibition appears to be highly dependent on the peptide size of the hydrolysate, with smaller peptides generally being more inhibitory. Jeon and coworkers (59) found that the effectiveness of fractionated cod FPH to reduce the activity of ACE was in the following order 3kDa > 5kDa > 10 kDa > 30 kDa. Bordenave and coworkers (111) reported that sardine hydrolysate and cod head hydrolysate inhibited close to 30% of ACE activity while a shrimp hydrolysate inhibited 57% of ACE activity.

A commercial cod and mackerel hydrolysate (PC60 and its derivative, Stabilium 200) was found to reduce anxiety in humans (112) and improve learning performance and memory in humans (113) and rats (114). Furthermore Bernet and coworkers (115) reported the ability of the PC60 hydrolysates to reduce stress in laboratory rats, in a similar manner as the known drug diazepam (valium) does. Fish protein hydrolysates have also been found to enhance flow of red blood cells (81), and therefore could enhance blood flow *in vivo*. A stimulation of proliferation of white blood cells in humans was seen with peptides found in fish sauce, which is one form of FPH (116). In light of the popularity of functional foods FPH may find a niche in this market as a nutritional supplement or nutraceutical due to its positive effects *in vivo*.

23.4.4 Microbial Media

The rapid growth of the biotechnology industry has developed a great need for a variety of different growth media for microorganisms. There may be a niche market here for FPH because extensively hydrolyzed FPH are reportedly an excellent growth media for a variety of microorganisms (99,117,118). Fish peptone has performed better than commercial peptones for many microorganisms and has been successful in cultivating fish pathogens (99). FPH may also find use as growth factors for microorganisms in the food fermentation industry. Yugushi (119) found that FPH stimulated the growth of lactic acid bacteria in skim milk. The FPH was also found to lead to a decrease in curd tension during and after milk fermentation. FPH may also find increasing use as an alternative peptone source to beef peptone due to the advent of Mad Cow disease.

REFERENCES

1. Valdimarsson, G. Assuming responsibility for the worlds fisheries. *Seafood Int.* 69–75. 2003.
2. FAO. The State of World Fisheries and Aquaculture. Food and Agricultural Organization, Rome, Italy, 1998.
3. Kristinsson, H.G., B.A. Rasco. Fish protein hydrolysates: production, biochemical and functional properties. *CRC Crit. Rev. Food Sci. Nutr.* 32:1–39, 2000.
4. Mackie, I.M. Fish protein hydrolysates. *Proc. Biochem.* 17(1):26–32, 1982.
5. Hultin, H.O., S.D. Kelleher. Surimi processing from dark muscle fish. In: *Surimi and Surimi Seafood*, Park, J.W., ed., New York: Marcel Dekker, Inc., pp 59–77, 2000.
6. Rebeca, B.D., M.T. Pena-Vera, M. Diaz-Castaneda. Production of fish protein hydrolysates with bacterial proteases; Yield and nutritional value. *J. Food Sci.* 56:309–314, 1991.
7. Kristinsson, H.G., B.A. Rasco. Fish Protein Hydrolysates and Their Potential use in the Food Industry. In: *Recent Advances in Marine Biotechnology*, Vol 7, Fingerman, M., R. Nagabhushanam, eds., Enfield, NH: Science Publishers, Inc., 2002, pp 157–181.
8. Hultin H.O., H.G. Kristinsson, T.C. Lanier. Process for Recovery of Functional Proteins by pH Shifts. In: *Surimi and Surimi Seafood* 2nd ed., Park, J.W., ed., New York: Marcel Dekker Inc. 2004. In Press.
9. Baek, H.H., K.R. Cadwallader. Enzymatic hydrolysis of crayfish processing byproducts. *J. Food Sci.* 60:929–934, 1995.
10. Shahidi, F., X.-Q. Han, J. Synowiecki. Production and characteristics of protein hydrolysates from capelin (*Mallotus villosus*). *Food Chem.* 53:285–293, 1995.
11. Viera, G.H.F., A.M. Martin, S. Saker-Sampaiao, S. Omar, R.C.F. Goncalves. Studies on the enzymatic hydrolysis of Brazilian lobster (*Panulirus spp.*) processing wastes. *J. Sci. Food Agric.* 69:61–65, 1995.
12. Onodenalore, A.C., F. Shahidi. Protein dispersions and hydrolysates from shark (*Isurus oxyrinchus*). *J. Aquat. Food Prod. Technol.* 5:43–59, 1996.
13. Kristinsson, H.G., B.A. Rasco. Biochemical and functional properties of Atlantic salmon (*Salmo salar*) muscle proteins hydrolyzed with various alkaline proteases. *J. Agric. Food Chem.* 48:657–666, 2000.

14. Kristinsson, H.G., B.A. Rasco. Hydrolysis of salmon muscle proteins by an enzyme mixture extracted from Atlantic salmon (*Salmo salar*) pyloric caeca. *J. Food Biochem.* 24:177–187, 2000.
15. Skanderby, M. Protein hydrolysates: their functionality and applications. *Food Technol. Int. Eur.* 10:141–144, 1994.
16. Hoyle, N., J.H. Merritt. Quality of fish protein hydrolysates from herring (*Clupea harengus*). *J. Food Sci.* 59:76–79, 1994.
17. Gonzalez-Tello, P., F. Camacho, E. Jurado, M.P. Paez, E.M. Guadix. Enzymatic hydrolysis of whey proteins: II. Molecular-weight range. *Biotechnol. Bioeng.* 44:529–532, 1994.
18. Deeslie, W.D., M. Cheryan. Functional properties of soy protein hydrolsyates from a continous ultrafiltration reactor. *J. Agric. Food Chem.* 36(1):26–31, 1988.
19. Slizyte, R., E. Dauksas, E. Falch, T. Rustad. Functional properties of different fractions generated from hydrolysed cod (*Gadus morhua*) byproducts. In: Proceedings of the First Joint Trans-Atlantic Fisheries Technology Conference – TAFT, pp 301–303. June 10-14, 2003, Reykjavik, Iceland. Icelandic Fisheries Laboratories.
20. Mullally, M.M., D.M. O'Callaghan, R.J. FitzGerald, W.J. Donnelly, J.P. Dalton. Zymogen activation in pancreatic endoproteolytic preparations and influence on some whey protein characteristics. *J. Food Sci.* 60(2):227–233, 1995.
21. Adler-Nissen, J. *Enzymic Hydrolysis of Food Proteins*, Barking, UK: Elsevier Applied Science Publishers, 1986.
22. Gildberg, A. Enzymic processing of marine raw materials. *Process Biochem.* 28:1–15, 1993.
23. Kristinsson, H.G., N. Demir, B. Ingadottir, H.T. Petty. The functional and physical properties of protein ingredients made from muscles of warm water fish species. In: Proceedings of the First Joint Trans-Atlantic Fisheries Technology Conference – TAFT. pp 301–303. June 10–14, 2003, Reykjavik, Iceland. Icelandic Fisheries Laboratories.
24. Hultin, H.O., S.D. Kelleher, Process for isolating a protein composition from a muscle source and protein composition, US patent no. 6,005,073, 1999.
25. Liaset, B., R. Nortvedt, E. Lied, M. Espe. Studies on the nitrogen recovery in enzymic hydrolysis of Atlantic salmon (*Salmo salar*, L) frames by ProtamexTM protease. *Proc. Biochem.* 37:1263–1269, 2002.
26. Lahl, W.J., S.D. Braun. Enzymatic production of protein hydrolysates for food use. *Food Tech.* 58(10):68–71, 1994.
27. Kristinsson, H.G., B.A. Rasco. Kinetics of the hydrolysis of Atlantic salmon (*Salmo salar*) muscle proteins by alkaline proteases and a visceral serine protease mixture. *Proc. Biochem.* 36:131–139, 2000.
28. Gilmartin, L., L. Jervis. Production of cod (*Gadus morhua*) muscle hydrolysates. Influence of combinations of commercial enzyme preparations on hydrolysate peptide size ranges. *J. Agric. Food. Chem.* 50:5417–5423, 2002.
29. Hale, M.B. Relative activities of commercially-available enzymes in the hydrolysis of fish proteins. *Food Technol.* 23:107–110, 1969.
30. Liu, L.L. and G.M. Pigott. Preparation and use of inexpensive crude pepsin for enzyme hydrolysis of fish. *J. Food Sci.* 46:1569–1572, 1981.
31. Tarky, W., O.P. Agarwala, and G.M. Pigott, Protein hydrolysate from fish waste. *J. Food Sci.* 38:917–918, 1973.
32. Kristinsson H.G., H.O. Hultin. The effect of acid and alkali unfolding and subsequent refolding on the pro-oxidative activity of trout hemoglobin. *J. Agric. Food Chem.* 2004, in press.
33. Quaglia, G.B., E. Orban. Influence of the degree of hydrolysis on the solubility of the protein hydrolsyates from sardine (*Sardina pilchardus*). *J. Sci. Food Agric.* 38:271–276, 1987.
34. Sugiyama, K., M. Egawa, H. Onzuka, K. Oba. Characteristics of sardine muscle hydrolysates prepared by various enzymic treatments. *Nippon Suisan Gakkaishi* 57(3):475–479, 1991.
35. Diniz, F.M., D.M. Martin. Use of response surface methodology to describe the combined effects of pH, temperature and E/S ratio on the hydrolysis of dogfish (*Squalus acanthias*) muscle. *Int. J. Food Sci. Tech.* 31:419–426, 1996

36. Benjakul, B., M.T. Morrissey. Protein hydrolysates from Pacific whiting solid wastes. *J. Agric. Food Chem.* 45:3423–3430, 1997.
37. Yu, S.Y., S. Fazidah. Enzymic hydrolysis of proteins from *Aristichthys noblis* by protease P"Amano"3. *Trop. Sci.* 34:381–391, 1994.
38. van Veen, A.G., K.H. Steinkraus. Nutritive value and wholesomeness of fermented foods. *J. Agric. Food Chem.* 18(4):576–578, 1970.
39. Raa, J., A. Gildberg. Fish silage: a review. *CRC Crit. Rev. Food Sci. Nutr.* 14:383–419, 1982.
40. Hale, M.B. Making fish protein concentrate by enzymatic hydrolysis, *NOAA Technical Report NMFS SSRF-675*, U.S. Department of Commerce, Seattle, WA, USA, 1–31, 1972.
41. Mullally, M.M., D.M. O'Callaghan, R.J. FitzGerald, W.J. Donnelly, J.P. Dalton. Proteolytic and peptidolytic activities in commercial pancreatin protease preparations and their relationship to some whey protein hydrolysate characteristics. *J. Agric. Food Chem.* 42:2973–2981, 1994.
42. Gauthier, S.F., P. Paquin, Y. Pouliot, S. Turgeon. Surface activity and related functional properties of peptides obtained from whey proteins. *J. Dairy Sci.* 76(1):321–328, 1993.
43. Mahmoud, M.I., W.T. Malone, C.T. Cordle. Enzymatic hydrolysis of casein: effect of degree of hydrolysis on antigenicity and physical properties. *J. Food Sci.* 57(5):1223–1229, 1992.
44. Quaglia, G.B., E. Orban. Enzymic solubilisation of proteins of sardine (*Sardina pilchardus*) by commercial proteases. *J. Sci. Food Agric.* 38:263–269, 1987.
45. Yanez, E., D. Ballester, F. Monckeberg. Enzymatic fish protein hydrolyzate: chemical composition, nutritive value and use as a supplement to cereal protein. *J. Food Sci.* 41:1289–1292, 1976.
46. Vareltzis, K., N. Soultos, F. Zetou, F. Tsiaras. Proximate composition and quality of a hamburger type product made from minced beef and fish protein concentrate. *Lebensm. Wiss. u. Technol.* 23(2):112–115, 1990.
47. Khan, M.A.A, M.A. Hossain, K. Hara, K. Osatomi, T. Ishihara, Y. Nozaki. Effect of enzymatic fish protein hydrolysate from fish scrap on the state of water and denaturation of lizard fish (*Saurida wanieso*) myofibrils during dehydration. *Food Sci. Technol. Res.* 9:257–263, 2003.
48. Zhang, H., Y. Yamashita, Y. Nozaki. Effect of protein hydrolysate from Antartic krill meat on the state of water and denaturation by dehydration of lizard fish myofibrils. *Fish Sci.* 68:672–679, 2002.
49. Hossan M.A., T. Ishihara, K. Hara, K. Osatomi, M.A.A. Khan, Y. Nozaki. Effect of proteolytic squid protein hydrolysate on the state of water and dehydration-induced denaturation of lizard fish myofibrillar proteins. *J. Agric. Food Chem.* 51:4769–4774.
50. Sych, J., C. Lacroix, L.T. Adambounou, F. Castaigne. The effect of low- or non-sweet additives on the stability of protein functional properties of frozen cod surimi. *Int. J. Food Sci. Tech.* 26:185–197, 1991.
51. Netto, F.M., S.A. Desobry, T.P. Labuza. Effect of water content on the glass transition, caking and stickiness of protein hydrolysates. *Int. J. Food Prop.* 2:141–161, 1998.
52. Lee, S., W.M. Shimizu, S. Kaminogawa, K. Yamaguchi. Emulsifying properties of a mixture of peptides derived from the enzymatic hydrolysates of β-casein. *Agric. Biol. Chem.* 51:161–165, 1987.
53. Quaglia, G.B., E. Orban. Influence of enzymatic hydrolysis on structure and emulsifying properties of sardine (*Sardina pilchardus*) protein hydrolysates. *J. Food Sci.* 55(6):1571–1573, 1990.
54. Sathivel, S., P.J. Bechtel, J. Babbitt, S. Smiley, C. Crapo, K.D. Reppond, W. Prinyawiwatkul. Biochemical and functional properties of herring (*Clupea harengus*) byproduct hydrolysates. *J. Food Sci.* 68:2196–2200.
55. Liceaga-Gesualdo, A.M., E.C.Y. Li-Chan. Functional properties of fish protein hydrolysate from herring (*Clupea harengus*). *J. Food Sci.* 64(6):1000–1004, 1999.
56. Spinelli, J., B. Koury, R. Miller. Approaches to the utilization of fish for the preparation of protein isolates; Isolation and properties of myofibrillar and sarcoplasmic fish protein. *J. Food Sci. 37*:599–603, 1972a.

57. Spinelli, J., B. Koury, R. Miller. Approaches to the utilization of fish for the preparation of protein isolates; enzymic modifications of myofibrillar fish proteins. *J. Food Sci. 37*:604–608, 1972.
58. Miller, R., H.S Groninger. Functional properties of enzyme-modified acylated fish protein derivatives. *J. Food Sci.* 41:268–272, 1976.
59. Jeon, Y-J., H.-G. Byun, S.-E. Kim. Improvement of functional properties of cod frame protein hydrolysate using ultrafiltration membranes. *Proc. Biochem.* 35:471–478.
60. Althouse, P.J., P. Dinakar, A. Kilara. Screening of proteolytic enzymes to enhance foaming of whey protein isolates, *J. Food Sci.*, 60(5):1110–1112, 1995.
61. Ingadottir, B., H.G. Kristinsson. Acid and alkali unfolding and refolding strategies improve the foaming properties of egg albumen, IFT Annual Meeting, Chicago, IL, Abstract 42–4, 2003.
62. Wu, Y.C., R.O. Kellems, Z.A. Holmes, H.S. Nakaue. The effect of feeding four fish hydrolyzate meals on broiler performance and carcass sensory characteristics. *Poultry Sci.* 63:2414–2418, 1984.
63. Tamura, M., N. Mori, T. Miyoshi, S. Koyama, H. Kohri, H. Okai. Practical debittering using model peptides and related compounds. *Agric. Biol. Chem.* 54:41–50, 1990.
64. Hevia, P., H.S. Olcott. Flavour of enzyme-solubilized fish protein concentrate fractions. *J. Agric. Food Chem.* 25(4):772–775, 1977.
65. Fujimaki, M., S. Arai, M. Yamashita, H. Kato, M. Nogushi. Taste peptide fractionation from a fish protein hydrolysate. *Agric. Biol. Chem.* 37:2891–2895, 1973.
66. Noguchi, M., S. Arai, M. Yamashita, H. Kato, M. Fujimaki. Isolation and identification of acidic oligopeptides in a flavor potentiating fraction from a fish protein hydrolysate. *J. Agric. Food Chem.* 23(1):49–53, 1975.
67. Imm, J.Y., C.M. Lee. Production of seafood flavor from red hake (*Urophycis chuss*) by enzymatic hydrolysis. *J. Agric. Food Chem.* 47:2360–2366, 1999.
68. ChiaLing, J., K. WenChing, C.L. Jao, W.C. Ko. Utilization of cooking juice of young tuna processed into canned tuna as condiments: effect of enzymatic hydrolysis and membrane treatment. *Fish Sci.* 68:1344–1351, 2002.
69. In, T. Seafood flavourants produced by enzymatic hydrolysis. In: *Advances in Fisheries Technology and Biotechnology for Increased Profitability,* Voigt, M.N., J.R. Botta, eds., Lancaster, PA: Technomic Publishing Co. Inc, 1990, pp 425–436.
70. Petersen, B.R. The impact of the enzymatic hydrolysis process on recovery and use of proteins. In: *Enzymes and Food Processing*, Birch, G.G., N. Blakebrough, K.J. Parker, eds., London, UK: Elsevier Applied Science Publishers, 1981, pp 149–175.
71. Lalasidis, G., S. Bostrom, L.-B. Sjoberg. Low molecular weight enzymatic fish protein hydrolysates: chemical composition and nutritive value. *J. Agric. Food Chem.* 26(3):751–756, 1978.
72. Liu, C., K. Morioka, Y. Itoh, A. Obatake. Contributions of lipid oxidation to bitterness and loss of free amino acids in the autolytic extract from fish wastes: effective utilization of fish wastes. *Fisheries Sci.* 66:343–348, 2000.
73. Chakrabarti, R. A method of debittering fish protein hydrolysate, *J. Food Sci. Technol.* 20(4):154–158, 1983.
74. Sugiyama, K., K. Takada, I. Yamamoto, H. Onzuka, K. Oba. Removal of fish protein hydrolysate (FPH) aldehydes by microorganisms. *Nippon Nogeikagaku Kaishi* 64:1597–1602, 1990.
75. Montecalvo Jr., J., S.M. Constantinides, C.S.T. Yang. Enzymatic modification of fish frame protein isolate. *J. Food Sci.* 49:1305–1309, 1984.
76. Suzuki, Y., T. Fukuzaka, H. Shindai, K.K. Teijin. Method for removing the bitter taste of protein hydrolysates, Japanese Examined Patent no. 5,612,084, 1981.
77. Hultin, H.O. Oxidation of lipids in seafood. In: *Seafoods: Chemistry, Processing Technology and Quality*, Shahidi, F., J.R. Botta, eds., Glasgow: Blakie Academic and Professional, pp 49–74, 1994.
78. Hatate, H., Y. Numata, M. Kochi. Synergistic effect of sardine myofibril protein hydrolyzates with antioxidant. *Nippon Suisan Gakkaishi* 56(6):1011, 1990.

79. Shahidi, F. Marine oils and bioactive compounds as nutraceuticals and functional food ingredients: current status and future trends. In: *Proceedings of the First Joint Trans-Atlantic Fisheries Technology Conference – TAFT, June 10–14*, Reykjavik, Iceland: The Icelandic Fisheries Laboratories, 2003, pp 312–319.
80. Amarowicz, R., F. Shahidi. Antioxidant activity of peptide fractions of capelin protein hydrolysates. *Food Chem.* 58:355–359, 1997.
81. Chuang, W.-L., B. Sun Pan, J.-S. Tsai. Inhibition of lipoxygenase and blood thinning effects of mackerel protein hydrolysate. *J. Food Biochem.* 24:333–343. 2000.
82. Saiga, A., S. Tanabe, T. Nishimura. Antioxidant activity of peptides obtained from porcine myofibrillar proteins by protease treatment. *J. Agric. Food Chem.* 51:3661–3667, 2003.
83. Saito, K., D.-H. Jin, T. Ogawa, K. Muramoto, E. Hatakeyama, T. Yasuhara, K. Noki. Antioxidant properties of tripeptide libraries prepared by the combinatorial chemistry. *J. Agric. Food Chem.* 51:3668–3674, 2003.
84. Dong, F.M., W.T. Fairgrieve, D.I. Skonberg, B.A. Rasco. Preparation and nutrient analyses of lactic acid bacteria ensiled salmon viscera. *Aquaculture* 109:351–366, 1993.
85. Morales-de-Leon, J., A. Galvez-Mariscal, V. Tellez-Still. Preparation of fish protein isolate and hydrolyzate (*Mugil cephalus*) and their incorporation into Mexican foods. *Arch. Latinoamericanos Nutr.* 40:55–68, 1990.
86. Chevalier, J., B. Noel. Food product of high nutritional efficiency, French Patent Application no. FR 2,495,442 A1, 1982.
87. Yu, S.Y., L.K. Tan. Acceptability of crackers ('Keropok') with fish protein hydrolsyates. *Int. J. Food Sci. Technol.* 25(2):204–210, 1990.
88. Bechtel, P.J., S. Sathivel, A.C.M. Oliveira, S. Smiley, J. Babbitt. Properties of hydrolysates from pink salmon heads and viscera. In: Proceedings of the First Joint Trans-Atlantic Fisheries Technology Conference – TAFT. pp. 284–285. June 10–14, 2003, Reykjavik, Iceland. The Icelandic Fisheries Laboratories.
89. Atia, M., L. Shekib. Preparation of fish protein hydrolysate from bolti frame (*Tilapia nilotica*) and evaluation of its chemical composition. *Assiut. J. Agric. Sci.* 23:75–87, 1992.
90. Abdul-Hamid, A., J. Bakar, G. Hock Bee, H.B. Gan. Nutritional quality of spray dried protein hydrolysate from black tilapia (*Oreochrombis mossambicus*). *Food Chem.* 78:69–74, 2002.
91. Liaset, B., E. Lied, M. Espe. Enzymatic hydrolysis of byproducts from the fish-filleting industry; chemical characterisation and nutritional evaluation. *J. Sci. Food Agric.* 80(5):581–589, 2000.
92. Kienkas, I. Biological value of protein hydrolysate of cod muscle obtained by enzymic hydrolysis and dialysis. *Izvestiya Akademii Nauk Latviiskoi-SSR* 7:92–96, 1974.
93. Ballester, D., E. Yanez, O. Brunser, A. Stekel, P. Chadud, G. Castano, F. Mockenberg. Safety evaluation of an enzymatic fish protein hydrolysate: 10-month feeding study and reproduction performance in rats. *J. Food Sci.* 42:407–409, 1977.
94. Barrias, C., A. Oliva-Teles. The use of locally produced fish meal and other dietary manipulations in practical diets for rainbow trout Oncorhynchus mykiss (Walbaum). *Aquaculture Res.* 31:213–218, 2000.
95. Lian, P., C.M. Lee. Characterization of squid hydrolysates for its potential as aquaculture feed ingredient. In: Proceedings of the First Joint Trans-Atlantic Fisheries Technology Conference – TAFT. pp 379–380. June 10–14, 2003, Reykjavik, Iceland. The Icelandic Fisheries Laboratories.
96. Wilson, R.P., D.W. Freeman, W.E. Poe. Three types of catfish offal meals for channel catfish fingerlings. *Prog. Fish Cult.* 46:126–132, 1984.
97. Oliva-Teles, A., A.L. Cerqueira, P. Goncalves. The utilization of diets containing high levels of fish protein hydrolysate by turbot (*Scophthalmus maximus*) juveniles. *Aquaculture* 179:195–201.
98. Cordova-Murueta, J.H., F.L. Garcia-Carreno. Nutritive value of squid and hydrolyzed protein supplement in shrimp feed. *Aquaculture* 210:371–384, 2002.

99. Gildberg, A.J. Enzymes and bioactive peptides from fish waste related to fish silage: fish feed and fish sauce production. In: *Proceedings of the First Joint Trans-Atlantic Fisheries Technology Conference – TAFT, June 10–14*, Reykjavik, Iceland: The Icelandic Fisheries Laboratories, 2003, pp 328–331.
100. Bogwald, J., R.A. Dalmo, R. Leifson-McQueen, E. Stenberg, A. Gildberg. The stimulatory effect of a muscle protein hydrolysate from Atlantic cod, *Gadus morhua* L., on Atlantic salmon, *Salmo salar* l., head kidney leucocytes. *Fish Shellfish Immunol.* 6:3–16, 1996.
101. Gildberg, A., J. Bogwald, A. Johansen, E. Stenberg. Isolation of acid peptide fractions from a fish protein hydrolysate with strong stimulatory effect on Atlantic salmon (*Salmo salar*) head kidney leucocytes. *Comp. Biochem. Physiol. B* 114(1):97–101, 1996.
102. Eguschi, Y., J.S. Bela, K. Shetty. Simulation of somatic embryogenesis in Anise (*Pimpinella anisium*) using fish protein hydrolysates and proline. *J. Herbs and Spices* 5(3):61–68, 1997.
103. Milazzo, M.C., Z. Zheng, G. Kellett, K. Haynesworth, K. Shetty. Stimulation of benzyladenine-induced *in vitro* shoot organogenesis and endogenous proline in melon (*Cucumis melo* L.) by fish protein hydrolysates in combination with proline analogues. *J. Agric. Food Chem.* 47(4):1771–1775, 1999.
104. Andrawulan, N., K. Shetty. Influence of acetyl salicylic acid in combination with fish protein hydrolysates on hyperhydricity reduction and phenolic synthesis in oregano (*Origanum vulgare*) tissue cultures. *J. Food Biochem.* 23:619–635, 1999.
105. Andrawulan, N., K. Shetty. Stimulation of novel phenolic metabolite, epoxy-pseudoisoeugenol-(2-methylbutyrate) (EPB), in transformed anise (*Pimpinella anisum* L.) root cultures by fish protein hydrolysates. *Food Biotechnol.* 14(1–2):1–20, 2000.
106. Vattem D.A., K. Shetty. Solid-state production of phenolic antioxidants from cranberry pomace by Rhizopus oligosporus. *Food Biotechnol.* 16:189–210, 2002.
107. Kohama, Y., H. Oka, Y. Kayamori, K. Tsujikawa, T. Mimura, Y. Nagase, M. Satake. Potent synthetic analogues of angiotensin-converting enzyme inhibitor derived from tuna muscle. *Agric. Biol. Chem.* 55(8):2169–2170, 1991.
108. Ukeda, H., H. Matsuda, K. Osjima, H. Matufuji, T. Matsui, Y. Osjima. Peptides from peptic hyrolysate of heated sardine meat that inhibit angiotensin I converting enzyme. *Nippon Nogeikagaku Kaishi* 65(8):1223–1228, 1992.
109. Matsumura, N., M. Fujii, Y. Takeda, T. Shimizu. Isolation and characterization of angiotensin I-converting enzyme inhibitory peptides derived from bonito bowels. *Biosci. Biotechnol. Biochem.* 57(10):1743–1744, 1993.
110. Wako, Y., S. Ishikawa, K. Muramoto. Angiotensin I-converting enzyme inhibitors in autolysates of squid liver and mantle muscle. *Biosci. Biotechnol. Biochem.* 60(8):1353–1355, 1996.
111. Bordenave S, I. Fruitier, I. Ballander, F. Sannier, A. Gildberg, I. Batista, J.M. Piot. HPLC preparation of fish waste hydrolysate fractions. Effec on guinea pig ileum and ACE activity. *Prep. Biochem. Biotechnol.* 32:65–77, 2002.
112. Dorman, T., L. Bernard, P. Glaze, J. Hogan, R. Skinner, D. Nelson, L. Bowker, D. Head. The effectiveness of Garum armoricum (stabilium) in reducing anxiety in college students. *J. Adv. Med.* 8:193–200, 1995.
113. Le Poncin, M. Experimental study: stress and memory. *Eur. Neuropsychopharmacol.* 6:110–P10–2, 1996.
114. Le Poncin, M. Nutrient presentation of cognitive and memory performances. *Eur. Neuropsychopharmacol.* 6:187–P19–4, 1996.
115. Bernet, F., V. Montel, B. Noel, J.P. Dupouy. Diazepam-like effects of a fish protein hydrolysate (Gabolysat PC60) on stress responsiveness of the rat pituitary-adrenal system and sympathoadrenal activity. *Psychopharmacology* 149:34–40, 2000.
116. Thongthai, C., A. Gildberg. Asian fish sauce as a source of nutrition. In: *Asian Functional Foods*, Shi, J., F. Shahidi, C-T Ho, eds., Boca Raton, FL:CRC Press, 2005. (In Press).
117. Gildberg, A., I. Batista, E. Strom. Preparation and characterization of peptone obtained by a two-step enzymatic hydrolysis of whole fish. *Biotech. Appl. Biochem.* 11:413–423, 1989.

118. de la Broise, D., G. Dauer, A. Gildberg, F. Guerard. Evidence of positive effect of peptone hydrolysis rate on Escherichia coli culture kinetics. *J. Mar. Biotechnol.* 6(2):111–115, 1998.
119. Yugushi, H. Studies on utilization of the fish meat hydrolysate for fermented milk products, VII: Correlation of the chemical composition of the fish meat hydrolyate with the stimulatory activity for the growth of lactic acid bacteria and for the decrease of curd tension of fermented milk products. *Jpn. J. Dairy Food Sci.* 33:A81–A91.

2.24

Human Gut Microflora in Health and Disease: Focus on Prebiotics

Tamara Casci, Robert A. Rastall, and Glenn R. Gibson

CONTENTS

24.1 INTRODUCTION TO HUMAN GUT MICROFLORA AND FERMENTATION PROPERTIES

In humans, the colon receives dietary material, which has already been digested in the upper gut, from the ileum, and the contents are then mixed and retained for 6–12 hours in the cecum and right colon. Thereafter, digesta are ejected and pass through the transverse to the left colon for storage, absorption of water, and eventual excretion (1). The typical retention time for the colon is about 60 h in U.K. subjects (2). Thus, the large gut is a tubular system, with nutrients flowing into the cecum, and bacteria, their metabolic products, and undigested food being excreted in feces. The major end products of fermentation in humans are the SCFA (short chain fatty acids), acetate, propionate, and butyrate (3); gases H_2 and CO_2 (4,5); ammonia (6); amines (7); phenols (8,9); and energy, which bacteria use for growth and maintenance of cellular function. Only about 5% of total SFCA are excreted in feces. *In vivo*, SCFA contribute about 10% of the daily energy requirements of the host. Most SCFA produced by gut bacteria are absorbed and metabolized in the body (10).

A range of substrates, of dietary origin or produced by the host, is available for fermentation by the colonic microflora. From the diet, resistant starch is the most quantitatively important (2) and is readily fermented by gut bacteria including *Bacteroides*, *Bifidobacterium* spp., and *Fusobacterium* spp. (11). Nonstarch polysaccharides form the next largest contribution to the fermentable substrates and include plant derived substrates like pectin, cellulose, hemicelluloses, and chemically related substances (2). Oligosaccharides like lactose, lactulose, raffinose, stachyose, and fructooligosaccharides also escape absorption by the small intestine and are fermented in the colon. Mucin glycoproteins, which are produced by goblet cells in the colonic epithelium, are predominant endogenous substances fermented in the colon (2) and can be metabolized by a range of microorganisms, most importantly *Bifidobacterium*, *Ruminococcus* spp., and some *Bacteroides* (12–14). Proteins and peptides originating in the diet, in pancreatic secretions or produced by bacteria, are available in the colon (2,6), although to a lesser extent than are carbohydrates.

The principal polysaccharide degrading bacteria are thought to be the Gram-negative anaerobes belonging to the genus *Bacteroides* (15). They are able to synthesize a wide range of cell associated polysaccharide depolymerases and glycosidases that allows them to grow upon various polysaccharides (15,16). The breakdown of highly polymerized materials in the gut is a cooperative activity, with enzymes from many different species participating in the process (2).

Gibson and Roberfroid (17) defined prebiotics as "nondigestible food ingredients that beneficially affect the host by selectively stimulating the growth and/or the activity of one or a limited number of bacteria normally resident in the colon, and thus attempt to improve host health."

Prebiotic substances, such as lactulose, fructooligosaccharides (including inulin), and galactooligosaccharides, selectively stimulate the growth of *Bifidobacterium* spp. in the colon (18). Isomaltooligosaccharides, glucooligosaccharides, lactitol, xylitol, and xylooligosaccharides (XOS) are candidate prebiotics, but the specific enzymes for the degradation of these molecules have not yet been fully evaluated, so the full explanatory mechanism for any purported prebiotic effect is not yet evident (19).

Fermentation of prebiotic oligosaccharides by probiotic bacteria in the gastrointestinal tract has been shown to have a number of possible health implications including alleviation of lactose maldigestion, increased resistance to gut invasion by pathogenic bacteria, stimulation of the immune system, protection against cancer, effect on mineral bioavailability, and lowering of lipid levels. Potential health benefits may include reduction of the risk of intestinal infectious diseases, cardiovascular disease, noninsulin dependent diabetes, obesity, osteoporosis, and cancer (18).

Zubillaga et al. (18) reported on the link between probiotics and certain health conditions including ulceration, diarrhea, *H. pylori* infection, lactose intolerance, cancer, and immune system function. It was demonstrated that *L. acidophilus* and *B. bifidum* can act as an "ecological" therapy for gastritis and duodenitis, and that high intakes of fermented milk products were associated with reduced risk of ulceration.

In the context of diarrhea, it has been found that a fermented product containing *L. acidophilus* could inhibit growth of *S. dysenteriae*, *S. typhosa,* and *E. coli*. The beneficial effect has been attributed to antimicrobial substances produced by *L. acidophilus*, *in vivo*, which neutralize the enterotoxins of *E. coli* (20).

Helicobacter pylori infection was found to be associated with the deficiency of *Lactobacillus* species in the stomach (18). Moreover, more than one *in vitro* study has shown the ability of *L. salivarius* to inhibit growth of *H. pylori* (21–23).

Lactobacillus acidophilus is associated with an improvement of *in vitro* lactose fermentation (24). However, this is also known to be true for conventional yogurts. *Lactobacillus delbruekii bulgaricus, S. thermophilus, L. acidophilus,* and *Bifidobacterium* are thought to be antimutagenic and anticarcinogenic (25), and are potentially useful for inhibiting cancer. However, the evidence is still unclear and contradictory (18).

Spanhaak et al. (26) showed that consumption of milk fermented with *L. casei* strain Shirota could modulate the composition and metabolic activities of the intestinal flora. This agrees with the findings of Schiffrin et al. (27) who showed that *L. acidophilus* La1 and *B. bifidus* Bb 12 did not modify lymphocyte subsets in humans.

The gut microflora exert a considerable influence on host biochemistry including : enzymatic activity of intestinal contents, SCFA production in the lumen, oxidation reduction potential of luminal contents, host physiology, host immunology, and modification of host synthesized molecules (28). The SCFA and, to a lesser extent, gases primarily provide a source of energy and growth substrates to other gut microorganisms, salvage dietary energy (60–70% of epithelial cell energy is derived from butyrate), affect colonic mucosal cell proliferation and differentiation, control luminal pH, improve mineral ion absorption, and regulate metabolic pathways (29).

In vitro and *in vivo* studies show that the intermediate and end products of fermentation formed by colonic bacteria depend partly upon the chemical composition of the polysaccharide substrate. For example, starch fermentation yields high levels of butyrate, whereas with a more oxidized substrate such as pectin, more acetate is produced (30,31). Cummings and Macfarlane (2) suggested that this may be explained by the fact that different bacteria take part in fermenting different polysaccharides. The relative rate of depolymerization of complex carbohydrates in the colon may influence the types of fermentation product that are generated (2). A contribution is also given by the physiology and anatomical architecture of the gut (32,33). Transit time of digesta through the colon strongly influences the activities of the gut microflora (2) and fecal ammonia concentrations (34). Best estimates indicate that the amount of NSP (nonstarch polysaccharides) available for fermentation from the U.K. diet is in the range of 8–18 g/d (2,35). In persons consuming western type diets, the amount of fermented carbohydrate, other than starch and NSP, is unlikely to be more than 10 g/d (2). Principal sugars escaping absorption in the small intestine are lactose, raffinose, and stachyose (2).

In humans, acetate, propionate, and butyrate account for approximately 85–95% of total SCFA in all regions of the colon. Other acidic products of fermentation are found in the large gut, such as the branched chain fatty acids, isobutyrate, 2-methylbutyrate, and isovalerate, which are products of amino acid fermentation; while other organic acids, including the electron sink products lactate and succinate, accumulate to a lesser degree (36). Further acidic, neutral, or basic products of the hind gut fermentation include phenols, indoles, amines, and ammonia (7), which are generated during the catabolism of amino acids.

The precise mechanisms whereby SCFA are absorbed in humans are not known with certainty; however, the process is concentration dependent, and SCFA transport is associated with the appearance of bicarbonate ions and stimulation of sodium absorption, and is independent of bulk water flow (3,37,38).

24.2 GUT MICROFLORA IN HEALTH AND DISEASE

24.2.1 Acute Infections and Inflammatory Reactions

Disturbances in the normal intestinal microbial community structure can result in a proliferation of pathogens. Acute inflammatory reactions cause diarrhea and sometimes vomiting, and can be associated with a number of bacteria and viruses including *E. coli*, *Campylobacter* species, *Vibrio cholerae*, *S. aureus*, *B. cereus*, *Clostridium perfringens*, *Salmonella* species , *Shigella* species, *Yersinia* species, and protozoa, especially *Giardia lamblia*, *Entamoeba histolytica,* and *Cryptosporidium parvum* (39,40). Bacteria can also be linked to more chronic diseases in the colon. For example, *C. difficile* has been targeted as the primary causative agent of pseudomembranous colitis.

The GI tract functions as a barrier against antigens from microorganisms and food (41). Among the possible mechanisms of probiotic therapy is promotion of a nonimmunological gut defense barrier, which includes the normalization of increased intestinal permeability and altered gut microecology. Another possible mechanism is improvement of the intestinal immunological barrier; particularly through immunoglobulin A responses and alleviation of gut inflammatory responses, which produce a stabilising effect. Many probiotic effects are mediated through immune regulation, particularly from the balance control of proinflammatory and antiinflammatory cytokines (41).

Research has shown that *Bifidobacterium* spp. can exert powerful antagonistic effects toward *E. coli* and other pathogens like *Campylobacter* and *Salmonella*. The inhibitory effect was variable in species of *Bifidobacterium* spp., with *B. infantis* and *B. longum* exerting the greatest effect (42).

Several open studies suggest a beneficial role of *L. rhamnosus* GG, *S. boulardii,* and *L. plantarum* LP299v during *C. difficile* related infections (43–50). Some studies have shown promising effects of probiotics on inflammatory bowel disease in animals. Intracolonic administration of *L. reuteri* R2LC to rats decreased the disease, whereas *L. reuteri* HLC was ineffective (51).

While there is likely to be a more chronic component to Crohn's disease, *Eubacterium, Peptostreptococcus, Pseudomonas, Bacteroides vulgatus,* mycobacteria, and *C. difficile* have all been associated with its onset (52). In an open study, a 10 day administration of *L. rhamnosus* GG to 14 children with active or inactive Crohn's disease resulted in an increase in immunoglobulin A secreting cells to β-lactoglobulin and casein, which indicated an interaction between the probiotic and the local immune system (53). Various epidemiological and animal studies, from numerous groups, have agreed that ulcerative colitis has a microbial etiology. In particular, sulphate reducing bacteria are also thought to play an important role (54).

Studies have shown a probiotic preparation to be as effective as mesalazine in reducing and preventing relapse in ulcerative colitis (55,56). *Helicobacter pylori* survives in the mucosal layer of the gastric epithelium, and is thought to be involved in the etiology of ulcers, type B gastritis, and stomach cancer. It secretes ammonia as part of its normal metabolism, thereby allowing it to survive at the acidic pH in the stomach (57). Attempts to eradicate *H. pylori, in vivo* with a probiotic have mainly failed (58).

24.2.2 Antibiotic Associated Diarrhea (AAD)

Diarrhea occurs in about 20% of patients who receive antibiotics. Antibiotic associated diarrhea (AAD) results from microbial imbalance that leads to a decrease in the fermentation capacity of the colon. Invasion with *C. difficile*, *Klebsiella oxytoca,* and *C. septicum* are causes of AAD (39,59).

Studies have shown that oral administration of *S. boulardii* can decrease the risk of AAD (50,59). Other studies have shown the efficacy of *S. boulardii* (60), *Enterococcus faecium* SF68, and *L. rhamnosus* GG in shortening the duration of AAD (59).

Several trials have suggested a preventive effect of some fermented products on the risk of diarrhea in children (61,62). Saavedra et al. (63) showed that feeding *B. bifidum* and *S. thermophilus* to infants significantly reduced the risk of diarrhea and the shedding of rotavirus.

24.2.3 Immune Stimulation

In a human trial, administration of probiotic yogurt gave an increase in the production of γ-interferon (64). In animal models, probiotics have been shown to stimulate the production of antibodies (local and systemic), enhance the activity of macrophages, increase γ-interferon levels, and increase the concentration of natural killer cells (65).

An enhancement in the circulating IgA antibody secreting cell response was observed in infants supplemented with a strain of *L. casei*, and this correlated with shortened duration of diarrhea in the study group when compared to a placebo group (66). Other studies have reported an enhancement in the nonspecific immune phagocytic activity of granulocyte populations in the blood of human volunteers after consumption of *L. acidophilus* and *B. bifidum* (67–69). It is possible that stimulation of intestinal IgA antibody responses induced by lactic acid bacteria (LAB) may be explained partly by an effect on phagocyte cell functions. Ingestion of yogurt has been reported to stimulate cytokine production, including γ-interferon, in human blood mononuclear cells (70).

It has been shown in experimental animals that the postnatal maturation of small intestinal brush border membranes was associated with increased food protein binding capacity (71). The capacity of antigens to bind to epithelial cells is related to the rate and route of antigen transfer and shown to influence the intensity of mucosal immune responses (72). Oral introduction of *Lactobacillus* species can enhance nonspecific host resistance to microbial pathogens, thereby facilitating the exclusion of pathogens in the gut (73,74). Several strains of live lactic acid bacteria have been shown *in vitro* to induce the release of the proinflammatory cytokines tumor necrosis factor α, and interleukin 6, reflecting stimulation of nonspecific immunity (75).

Oral introduction of *L. casei* and *L. bulgaricus* activates the production of macrophages (73) and administration of *L. casei* and *L. acidophilus* activates phagocytosis in mice (74). Enhanced phagocytosis by *L. acidophilus* Lal was also reported in humans (67).

Probiotic bacteria have been shown to modulate phagocytosis differently in healthy and allergic subjects: in healthy persons there was an immunostimulatory effect, whereas in allergic persons, down regulation of the inflammatory response was detected (76).

Consumption of the adhesive strains *L. johnsonii* Lj1 and *B. lactis* Bb12 has been shown to increase the phagocytosis of *E. coli, in vitro* (67). A combination of both strains, together with *S. thermophilus*, has also been shown to possess adjuvant activity when consumed in combination with an attenuated *Salmonella typhimurium*. The preparation caused a significant increase in a specific serum IgA to the *Salmonella* (77).

In order to influence the immune system, a probiotic microorganism must activate the lymphoid cells of the gut associated lymphoid tissue, which are diffusely distributed amongst epithelial cells and populate the lamina propria and submucosa (78). Some probiotic strains are clearly antagonistic against *Helicobacter pylori, in vitro* although as mentioned earlier, experiments have not been so successful *in vivo* (59). A significant reduction of urease activity *in vivo* (which reflects the activity of *H. pylori*) has been reported in patients treated with a supernatant of *L. johnsonii* La1 (79).

Lactic acid bacteria have been observed to produce antimicrobial substances. The most widely produced antimicrobial substances are organic acids, especially lactic and acetic acids. Hydrogen peroxide and carbon dioxide are also widely produced by LAB. If probiotic LAB are metabolically active during their passage throughout the intestinal tract, it is very likely that some of these substances will be produced. A reduction in fecal pH confirms this. The production of other antimicrobial components, diacetyl, reuterin, pyroglutamic acid, and especially bacteriocins, is not certain under *in vivo* conditions. To date, no studies have been performed to investigate the production or efficacy of bacteriocins in the intestine. Evidence exists that they are produced and active in the oral cavity (80). Mechanisms implicated in the protective effects of probiotics may therefore include the production of acids, hydrogen peroxide, or antimicrobial substances; competition for nutrients or adhesion receptors; antitoxin actions; and stimulation of the immune system (59).

24.2.4 Lactose Intolerance

Lactose intolerance is a problem for approximately 70% of the world's population and is linked to a low intestinal β-galactosidase activity. Inadequate hydrolysis of dietary lactose leads to abdominal discomfort (81). Some lactic acid bacteria, including *L. acidophilus* and *B. bifidum,* produce β-D-galactosidase, the action of which increases tolerance to dairy products (81). There is good evidence for the alleviation of symptoms of lactose intolerance by specific probiotics and, recently, immune changes related to milk hypersensitivity have been shown to be down regulated (76).

It is well established that persons with lactose intolerance experience improved digestion and tolerance of the lactose contained in yogurt than of that contained in milk (82). At least two mechanisms have been reported to explain this: digestion of lactose in the gut lumen by lactase contained in the yogurt bacteria, and slower intestinal delivery or transit time of yogurt compared to milk (82–85). Kim and Gilliland (81) suggested that improved lactose digestion was not caused by lactose hydrolysis prior to consumption, but rather by the action of the enzyme in the digestive tract.

24.2.5 Colon Cancer

Bacterial links to colonic cancer are also evident. The colon is the second most common site of tumor formation in humans, and several bacterial products are carcinogenic (86). Development of colon cancer consists of a sequence of events that, although incompletely understood, occurs in definable steps. First is an initiating step, in which a carcinogen, produced by metabolic activation of a precursor, produces an alteration in DNA. At present, it is believed that several mutations must occur for a tumor to develop. The postinitiation steps are much less clear, but usually involve changes in signal transduction pathways. The next clear step is an overgrowth in colonic crypts, which can be seen morphologically in an aberrant crypt. Aberrant crypts, which are considered preneoplastic structures, are enlarged and elevated relative to normal crypts, and have a serpentine growth pattern. Aberrant crypts may occur singly or as groups of aberrant crypts within a single focus. A certain small but unknown fraction of these aberrant crypts will progress to polyps and eventually to tumors (87).

The carcinogenic effect of endogenous toxic and genotoxic compounds is probably influenced through the activity of the bacterial enzymes NAD(P)H dehydrogenase, nitroreductase, β-glucuronidase, β-glucosidase, and 7-α-dehydroxylase. Harmful and beneficial bacteria commonly found in the intestine differ in their enzyme activities (88). *Bifidobacterium* spp. and lactobacilli have negligible activities of xenobiotic metabolising enzymes, unlike bacteroides, clostridia, and enterobacteriaceae.

Ingestion of viable probiotics and prebiotics is associated with anticarcinogenic effects, one mechanism of which is the detoxification of genotoxins in the gut (89). Because of the complexity of cancer initiation, progression, and exposure in the gut, many types of interactions may be envisaged. It has been shown (89) that short lived metabolite mixtures isolated from milk that was fermented with strains of *L. bulgaricus* and *S. thermophilus,* were more effective in deactivating aetiological risk factors of colon carcinogenesis than cellular components of microorganisms. The underlying mechanisms are not known. After ingestion of resistant starch, the gut flora induces the chemopreventive enzyme glutathione transferase π (EC 2.5.1.18) in the colon. Together, these factors lead to a reduced load of genotoxic agents in the gut and increased production of agents that deactivate toxic components. Butyrate is one such protective agent and is associated with a lowering of cancer risk (89). The role of butyrate in modulation of nucleic acid metabolism is of particular interest, especially its effects upon regulation of gene expression and cell growth. Butyrate can reversibly alter the *in vitro* properties of human colorectal cancer cell lines by prolonging doubling time and slowing down growth rates (90). Low concentrations of this SCFA reduce DNA synthesis and suppress proliferation in a variety of cell types (91–93).

Ammonia is mainly formed through the deamination of amino acids (6,94). It is rapidly absorbed from the gut and detoxified by urea formation in the liver. Low concentrations can alter the morphology and intermediate metabolism of colonic epithelial cells (95,96), affect DNA synthesis, and reduce their lifespan. Thus, a high ammonia concentration in the colonic lumen may select for neoplastic growth.

Phenols and indoles do not appear to be carcinogenic alone, but act as cocarcinogens (97). Increased production of these metabolites has been found to occur in a variety of disease states (9,98,99). Phenolic and indolic compounds are produced by deamination of the aromatic amino acids, tyrosine, phenylalanine, and tryptophan. Phenols are not found in urine of germ free animals (8).

In colon cancer, the evidence of probiotic efficacy is still not certain. In animal models it was shown that a dietary intake of *B. longum* significantly suppressed development of azoxymethane (AOM) induced aberrant cryptic foci formation in the colon (100). This study was also confirmed by another using *B. longum* and inulin (101). Kinouchi et al. (102) showed that *L. acidophilus* and *B. adolescentis* suppressed ileal ulcer formation in animal models. *Lactobacilus acidophilus* fed to healthy volunteers significantly decreased β-glucuronidase, nitroreductase, and axoreductase activities (103).

Oral administration of *L. casei* strain Shirota has been shown to have inhibitory properties upon chemically induced tumors in rats (104). Also, it may prolong survival during specific cancer treatments (105). The beneficial effect of probiotic based cancer therapy has been associated with their antimutagenic properties and ability to modulate immune parameters including T-cell, natural killer (NK) cell, and macrophage activities, which are important in hindering tumor development (27,76,104). Some, but not all, epidemiological studies suggest that consumption of fermented dairy products may have certain protective effects against large colon adenomas or cancer (106).

Dietary administration of lyophilized cultures of *B. longum* strongly suppressed AOM induced colon tumor development. The inhibition of colon carcinogenesis was associated with a decrease in colonic mucosal cell proliferation and colonic mucosal and tumor ODC (ornithine decarboxylase) and *ras*-p21 oncoprotein activities (107). *Ras* activation represents one of the earliest and most frequently occurring genetic alterations associated with human cancers, especially that of the colon (108). Elevated levels of *ras*-p21 have been correlated with increased cell proliferation, histological grade, nuclear anaplasia, and degree of undifferentiation (109). The major conclusion from animal data is that there appears to be a synergistic effect of consumption of probiotics and prebiotics on the attenuation of development of colon cancer. The effect is often not large, but it is possible that it could be beneficial, in combination with other methods to reduce risk (87).

24.2.6 Cholesterol Lowering Effect and Lipid Metabolism

Prebiotics in the form of fermented milk are thought to have cholesterol lowering properties in humans. However, studies have given equivocal findings. Even recent studies fail to provide convincing evidence that "live" fermented milk products have any cholesterol lowering efficacy in humans (110).

In animals, dietary FOS caused a suppression of hepatic triglyceride and VLDL (very low density lipoprotein) synthesis, resulting in marked reductions in triglycerides, and to a lesser extent cholesterol, levels (110). Evidence for similar effects in humans is sparse and more studies are needed, particularly with respect to effects on postprandial triglyceride concentrations (110).

Propionate metabolism has also been studied in ruminants, where it is a major gluconeogenic precursor (111). In rats, it has been shown to lower cholesterol (112,113) either by inhibition of hepatic cholesterol synthesis or by redistribution of cholesterol from plasma to the liver (113). In humans, however, no change was seen in total cholesterol levels, although HDL (high density lipoprotein) cholesterol increased when a group of females were fed sodium propionate (114).

In CHD (coronary heart disease) there are variable data and no firm conclusions can be drawn (65). Schaafsma et al. (115) found that daily feeding of 125 mL of a test probiotic milk significantly lowered LDL (low density lipoprotein) cholesterol levels and total serum cholesterol.

24.3 STRUCTURE TO FUNCTION RELATIONSHIPS IN PREBIOTIC OLIGOSACCHARIDES

Eight NDOs have been licensed as having FOSHU (Foods for Specified Health Use) status by the Ministry of Health and Welfare in Japan: GOS (galactooligosaccharides), FOS (fructooligosaccharides), lactulose, XOS (xylooligosaccharides), soybean oligosaccharides, raffinose, lactosucrose, and isomaltooligosaccharides (116). The efficacy of prebiotics toward promoting human health is strongly related to their chemical structure. This section of the review therefore elucidates the interactions between oligosaccharide structure and prebiotic effects.

24.3.1 Fructooligosaccharides

"Fructooligosaccharides" (FOS) is a common name for fructose oligomers that are mainly composed of 1-kestose (GF_2), nystose (GF_3), and 1^F- fructofuranosyl nystose (GF_4) in which fructosyl units (F) are bound at the β-2,1 position of sucrose (GF) (117). FOS are produced from sucrose through the transfructosylating action of either β-fructofuranosidases (β-FFase, EC 3.2.1.26) or β-D-fructosyltransferases (β-FTase, EC 2.4.1.9) (119).

24.3.1.1 Fermentability of FOS

In vitro experiments (anaerobic batch culture fermenters inoculated with human fecal bacteria and a carbohydrate substrate) on the comparative fermentation of chicory inulin and FOS showed that both inulin and FOS were rapidly and completely metabolized by the microbial flora in the fermenters. The relative rate of fermentation was similar for both substrates. A more detailed analysis however, revealed that the rate of degradation of oligomers with a degree of polymerization (DP < 10) was approximately twice that of molecules with a higher DP. Also, GF_n type and F_m type components of inulin hydrolysates disappeared from the culture medium at a similar rate, and the two types are known to have similar bifidogenic effects (119). There is no difference between FOS obtained by partial enzymatic hydrolysis (containing fructan chains ending with a glucose moiety as well as fructan chains ending with a fructose moiety) and FOS obtained by enzymatic synthesis from sucrose (120).

The efficiency of FOS over sucrose and other sugars as substrates for most strains of *Bifidobacterium* spp. is due to the intracellular 2,1 β-D fructan-fructanhydrolase (E.C 3.2.1.7) activity (119,121), which liberates monomeric fructose molecules that are then transported into the bacterial cell (122). The fructose moiety is then metabolized in the specific "bifidus" pathway (19). The enzyme also gives the genus a competitive advantage in the human gut (121). The β(2-1) bond linking the fructose moieties in the FOS chains is central in the nondigestibility as well as the bifidogenic properties of FOS (123).

Lactobacillus plantarum and *Lactobacillus rhamnosus* were found to be capable of metabolising only tri- and tetrasaccharide fractions, not pentasaccharides (124). This suggests that there may be specific transport systems in these organisms for trisaccharides and tetrasaccharides (125). In general, feeding of FOS increases *Bifidobacterium* spp. and *Lactobacillus* species, increases SCFA concentrations, and decreases *Clostridium* spp., *Fusobacterium* spp., *Bacteroides*, and pH (29,126–128). The intake doses of FOS that elicit a bifidogenic effect in human studies has ranged from 4 to 15 g/d (128,129). That *Bifidobacterium* species selectively ferment the fructans in preference to other carbohydrate sources such as starch, fructose, pectin, and polydextrose was confirmed in a volunteer trial using fructooligosaccharides (127) and by other human studies (130,131).

Using continuous chemostat cultures inoculated with fecal slurries, Gibson and Wang (132) showed that after six turnovers, chicory derived FOS, but not glucose, were able to selectively stimulate *Bifidobacterium* spp. growth; the number of these bacteria was almost three orders of magnitude higher than *Bacteroides*. With glucose as the substrate, *Bacteroides* were two orders of magnitude higher than *Bifidobacterium* spp. (119). *In vitro* studies showed that the majority of *Bifidobacterium* species utilized FOS and low polymerized inulins, although twelve out of the thirty strains tested were not stimulated. Incorporation of FOS into the diet stimulated the proliferation of fecal *Bifidobacterium* spp. by 1.6 log cfu/g under *in vivo* conditions, whereas *B. animalis* KSp4 was not effective (133).

Administration of *Bifidobacterium* spp. together with the prebiotic improved the bifidogenic effect by around 1.4 log cfu/g of feces. Supplementation had almost no effect upon some groups of gut microflora (133). This indicates that utilization of inulin by

Bifidobacterium spp. depends on the degree of polymerization (DP) of fructooligomer chains and purity of the preparations. Among FOS, low DP inulin and high DP inulin, FOS was the most fermentable substrate and inulin the least.

It has also been suggested that fructans with higher DP are less rapidly fermented by *Bifidobacterium* (133). The results of this study indicate that the ability of *Bifidobacterium* spp. to metabolize FOS is a species dependent feature, and only to a small extent strain dependent. Muramatsu et al. (121) showed that the β-fructofuranosidase from *B. adolescentis* G1 has a unique substrate specificity to fructooligomers of DP 2-8 than to inulin and sucrose. All *L. acidophilus* strains tested fermented FOS, a result consistent with that recently reported by Sghir et al. (134). Such strains were: 33200 (ATCC), 837 (ATCC), DDS-1 (NC), and NCFM (NCSU), the latter two being widely promoted as probiotics. Three other commercial probiotic strains, *L. plantarum* MR240, *L. casei* MR191, and *L. casei* 685, also fermented FOS, whereas *Lactobacillus* strain GG was found to be a nonfermenter (despite being a well studied probiotic strain). Interestingly, most of the *L. bulgaricus* and *S. thermophilus* strains (the bacteria traditionally used in yogurt manufacture) were not FOS fermenters (124).

Perrin et al. (122) compared the physiological behavior of *B. infantis* ATCC 15697 growing on synthetic FOS or its components. The studies were carried out in batch cultures on semisynthetic media. Glucose, fructose, sucrose, and FOS as carbohydrate sources were compared. Glucose was the preferred substrate for growth and biomass production. Fructose was the best for lactate and acetate production. With FOS more metabolites were produced than with fructose. In a mixture of FOS, the shorter saccharides were used first [this has also been observed with other strains (135)] and fructose was released in the medium. Fructofuranosidase was inducible by fructose. The conclusion was that the glucose contained in FOS and sucrose might sustain growth and cell production while fructose may enable the production of major metabolites (122).

As a consequence of the metabolism of the FOS by fermentative bacteria, SCFA and lactic acid are produced. Both lead to a drop in the pH of the large intestine. This is beneficial for the organism as it constitutes an ideal medium for the development of the bifidogenic flora and, at the same time, limits the development of bacteria which are considered pathogens. The content of the metabolites produced by putrefactive bacteria decrease following the ingestion of FOS (136).

The addition of *L. acidophilus* 74-2 with FOS gave rise to an increase of *Bifidobacterium* spp. Major positive changes occurred in the production of organic acids: a strong upward trend was observed especially in the case of butyric and propionic acids. A noticeable increase of β-galactosidase activity was monitored while the activity of β-glucuronidase, generally considered undesirable, declined (137).

A variety of fermentation products influence the anatomy, physiology, and immunology of the host; and conversely the general constitution of the host affects the microbial community in the intestine (137). Gibson and Wang (132) have shown that *in vitro*, when *Bifidobacterium* spp. are given FOS as a substrate they secrete a peptide that is inhibitory to most of the common organisms causing acute diarrhea. In human healthy volunteer studies, oral dosing with 15 g of FOS daily led to *Bifidobacterium* spp. becoming the dominant species in the large bowel (127).

It was found that neither FOS nor inulin was, to a significant extent, fermented by *B. bifidum* (138). Bacteria other than the *Bifidobacterium* spp. can ferment FOS. These include *K. pneumoniae*, *S. aureus* and *epidermidis*, *E. faecalis* and *faecium*, *Bacteroides vulgatus*, *thetaiotamicron*, *ovatus* and *fragilis*, *L. acidophilus,* and *Clostridium* species, mainly *butyricum* (119). However, their ability to do so in mixed culture is repressed by *Bifidobacterium* spp. competition.

24.3.1.2 FOS Interactions with Lipid Metabolism

Evidence suggests that FOS acts as a soluble fiber and can be useful in increasing intestinal motility and transport, and in the reduction of high levels of plasma cholesterol (136). FOS can decrease triaglycerol in VLDL when given to rats. The triaglycerol lowering action of FOS is due to a reduction of *de novo* fatty acid synthesis in the liver through inhibition of all lipogenic enzymes (139).

Postprandial insulin and glucose concentrations are low in the serum of FOS fed animals and this could explain, at least partially, the metabolic effects of FOS. Some events occurring in the gastrointestinal tract after oligofructose feeding could be involved in the antilipogenic effects of this fructan: the production of propionate through fermentation, a modulation of the intestinal production of incretins, or modification of the availability of digestible carbohydrates. Recent studies have shown that the hypotriglyceridemic effect of fructans also occurs in humans (139).

The effects of inulin-type fructans on glycemia and insulinemia are not yet fully understood, and available data are often contradictory. Other nondigestible carbohydrates are known to modify the kinetics of absorption of carbohydrates, thus decreasing the incidence of glycemia and insulinemia (140,141).

24.3.1.3 FOS Interactions with Mineral Absorption

The production of a high concentration of short chain carboxylic acids in the colonic fermentation of nondigestible carbohydrates facilitates the absorption of minerals, Ca^{2+} and Mg^{2+} in particular (142). Roberfroid (143) argued that inulin-type fructans may improve mineral absorption and balance by an osmotic effect that transfers water into the large bowel, thus increasing the volume of fluid in which these minerals can dissolve. Also, acidification of the colon (as a consequence of carbohydrate fermentation) raises the concentration of ionized minerals, which favors passive diffusion. *In vivo* human studies have confirmed the positive effect of inulin and FOS upon absorption of calcium, but not of iron, magnesium, or zinc (144,145).

It has been observed that the effect of FOS on metabolic calcium balance, calcium content of bone, and prevention of ovariectomy induced loss of trabecular structure became more prominent when dietary calcium was high (146,147). It has been discussed that at least part of the stimulating effect of FOS on mineral absorption might be attributed to SCFA production (148).

24.3.1.4 FOS Interactions with Cancer

Prebiotics, in particular the chicory derived β(2-1) fructans, have been shown to exert cancer protective effects in animal models (149). A study was carried out to determine the effects of two chicory fructans: FOS (Raftilose P95, average DP 4) and long chain inulin (Raftiline HP, average DP 25) on apoptosis and bacterial metabolism in rats (149). Both substrates were shown to exert protective effects at an early stage in the onset of cancer.

It has been shown that inulin at dietary concentrations of 5–10% suppressed ACF (foci) in the rat colon (101,123,150).

NDOs may exert cancer protective effects at the cellular level following SCFA formation during fermentative bacterial metabolism. Butyrate, acetate, and propionate regulate colonic epithelial turnover and butyrate induces apoptosis in colon adenoma and cancer lines (151,152).

Rao et al. (150) and Rowland et al. (101) have reported protective effects of inulin at a postinitiation stage in the form of suppression of early preneoplastic lesions following treatment with a carcinogen. In this study, there was a trend toward increased colonic β-glucosidase activity and decreased ammonia concentration in the rats fed oligofructose or inulin. This was a positive response as ammonia is a known tumor promoter (96), so suppression of ammonia formation in the gut can be considered beneficial. β-glucuronidase is involved in DMH (1,2-dimethylhydrazine) and AOM metabolism, causing the release of MAM (methylazoxymethanol) from its biliary conjugate in the colon. Therefore, a decrease in β-glucuronidase may be protective (149). Large concentrations of butyrate can be produced as products of colonic fermentation by gut bacteria. Young (153) proposed that butyrate possessed antitumor properties, indicating the potential of probiotic strains such as *Lactobacillus* for preventing cancer. This study has shown a strong increase of β-galactosidase activity during administration of *L. acidophilus* 74-2 with FOS. After addition had stopped, the β-galactosidase activity again declined suggesting a causal link between the intake of a prebiotic product and increase of β-galactosidase activity, whereas β-glucuronidase activity declined (137).

24.3.2 Galactooligosaccharides

6' Transgalactosylated oligosaccharides (GOS, 154) are oligosaccharides produced by transgalactosylation of lactose using a β-galactosidase (β-Gal) (155). *In vitro,* GOS increased ATP, SCFA, and acetate concentrations, decreased pH and were 95% degraded from the fourth day of intake onward. They lead toward an increase in *Bifidobacterium* spp. species, while breath hydrogen excretion was reduced. However, methane excretion and levels of *Enterobacter* species are not affected (156).

GOS can be fermented by a number of intestinal bacteria including *Bifidobacterium* spp., *Bacteroides*, *Enterobacteria,* and *Lactobacillus* species (154,157,158).

A few human feeding studies investigating the prebiotic potential of GOS suggest that they may have bifidogenic potential (156,159). The bifidogenic nature of GOS has been related to a linkage specificity of the *Bifidobacterium* β-galactosidase, which cleaves β1-3 and β1-6 linkages, instead of β1-4 linkages (160).

Two probiotic strains, *Bifidobacterium lactis* DR10 and *Lactobacillus rhamnosus* DR20, were tested for their ability to utilize and grow upon GOS present in a commercial hydrolysed lactose milk powder (161). The results demonstrated that *B. lactis* DR10 preferentially utilized tri- and tetrasaccharides whereas *L. rhamnosus* DR20 preferred sugars with a lower degree of polymerization (di- and monosaccharides). Surprisingly, monosaccharides including glucose were the final sugars to be utilized by *B. lactis* DR10 (161). According to Rastall and Maitin (125) this suggests that *B. lactis* has a specific transport system for GOS that is not present in *L. rhamnosus*. A β-galactosidase that displays hydrolytic activity toward GOS but not lactose has been described in *B. adolescentis* (162). It has a preferential activity toward GOS with DP 2-3.

Fifty-four strains of lactic acid bacteria were studied for their ability to ferment lactose derived oligosaccharides. A good correlation was seen between the ability of a strain to utilize oligosaccharide and the presence of the lactose hydrolysing enzyme β-galactosidase (161). It has been suggested that GOS can be used more readily and selectively *in vitro* by *Bifidobacterium* spp. than other oligosaccharides such as lactulose and raffinose (163). However, the mechanism of utilization of this oligosaccharide by this group of bacteria is poorly understood. It is known that even though most *Bifidobacterium* spp. strains of human gut origin can readily use GOS, only a few strains from other genera, including *Lactobacillus* species, possess this ability (158,163). 4'-Galactosyllactose is selectively utilized by all the *Bifidobacterium* strains tested (157,158,164) compared with lactulose and raffinose whose specificity is less marked. Other studies also show that some strains of *Lactobacillus, Bacteroides* and *Clostridium* ferment GOS, and TD (transgalactosylated disaccharides) might even be better substrates for these bacteria (157,158). The bacterial species that ferment 4'- and 6'-GOS overlap, although some differences are found (116).

Physiological effects of GOS fermentation in the gut are:

1. Improvement of defecation. Deguchi et al. (165) showed that ingestion of 5g of GOS daily improved defecation frequency in a group of women.
2. Elimination of ammonia. In a human study, the ingestion of 4'-GOS at a dose of 3 g/d reduced the fecal ammonia concentration (166). The fact that other putrefactive products such as phenol, p-cresol, and indole were also decreased indicates that there was a dramatic change in metabolism upon ingestion of 4'-GOS.
3. Colon cancer prevention.
4. Stimulation of calcium absorption. Chonan et al. (167) found a stimulatory effect of GOS on calcium absorption and calcium content in femur and tibia with 0.5% calcium supplementation but not with 0.05%.

In a human volunteer trial, it was found that GOS increased *Bifidobacterium* spp. and *Lactobacillus* numbers while decreasing *Bacteroides* species and *Candida* species (159).

At a dose of 10 g/d, GOS given in the diet for 21 days decreased breath H_2 excretion and increased fecal *Bifidobacteria* without affecting *Enterobacteria* species or fecal pH (147). Similar results were found with inulin and lactose.

Van Laere et al. (162) reported on the effects of both reducing and nonreducing GOS comprising 2 to 8 residues on the growth of *B. adolescentis* DSM20083 and on the production of a novel β-galactosidase (β-GalII). In cells grown on GOS, in addition to the lactose degrading β-Gal (β-Gal I), another β-Gal (β-Gal II) was detected and it showed activity toward GOS but not toward lactose (162). During fermentation of GOS the di- and trisaccharides were first metabolized. β-Gal II was active toward β-galactosyl residues that were 1-4, 1-6, 1-3, and 1-1 linked, indicating its role in the metabolism of GOS by *B. adolescentis*, one of the predominant human fecal bacteria (162). This enzyme might allow *B. adolescentis* to utilize the oligosaccharides more efficiently than other microorganisms. Therefore, this GOS mixture, containing mainly higher molecular weight materials, might be an interesting prebiotic substrate. During the growth of *B. adolescentis* on GOS a large number of glycosidases are produced, including two arabinofuranohydrolases, which are involved in the degradation of arabinoxylooligosaccharides. This may offer an additional competitive advantage, since it allows the organism to scavenge the environment for a range of substrates and use the degradation products for growth (162). No endoglycanase activity could be detected in the cell extract, suggesting that *B. adolescentis* adopted a strategy aimed at utilizing polysaccharide degradation products generated by other microorganisms instead of taking part in the initial depolymerization stage of polysaccharides (162).

These results indicate that certain *Bifidobacterium* spp. together with prebiotics may be used for prophylaxis against opportunistic intestinal infections with antibiotic resistant pathogens (168).

24.3.3 Isomaltooligosaccharides

Isomaltooligosaccharides (IMO) are composed of glucose monomers linked by α1-6 glucosidic linkages. Isomalto-900, a commercial product, is produced by incubating α-amylase, pullulanase, and α-glucosidase with cornstarch. The major oligosaccharides in this mixture are isomaltose, isomaltotriose, and panose. IMO occur naturally in various fermented foods and sugars such as soybean sauce and honey (169).

IMO have been shown to be fermented by *Bifidobacterium* spp. and *Bacteroides fragilis*, but not by *E. coli* and other bacterial populations (169). IMO are efficient prebiotics in that they stimulate a lactic microflora as well as allow elevated production of butyrate, which is thought to be a desirable metabolite in the gut (65). Furthermore, the IMO mixture consisting of a high degree of branched oligosaccharides (due to α-GTase) is expected to be particularly effective at promoting *Bifidobacterium* spp. growth in the human intestine as growth of *Bifidobacterium* spp. is known to be enhanced in proportion to the degree of glucosidic polymerization of the IMO components (170). Kaneko et al. (171) reported that the IMO mixture containing trisaccharides and tetrasaccharides was more effective in enhancing the microflora than the IMO mixture of disaccharides.

Olano-Martin et al. (172) looked into the fermentability of dextran and novel oligodextrans (I, II, and III) by gut bacteria. In anaerobic batch culture, dextran and oligodextrans increased the number of *Bifidobacterium* spp. and butyrate concentration. Using a three stage continuous culture cascade system, it was shown that a low molecular weight oligodextran (IV) was the best substrate for *Bifidobacterium* spp. and lactobacilli. Moreover, dextran and oligodextrans stimulated butyrate production, which suggests that oligodextrans indeed have potential as modulatory agents for the gut microflora.

24.3.4 Lactulose

Lactulose (Galβ1,4 Fru) is formed by heating lactose syrup to caramelization (173). It was discovered by Montgomery and Hudson in 1930 (174) and its growth promoting effect for *Bifidobacterium* species was determined by Petuely (175).

It is difficult to crystallize in water solution, however, a new crystal form of lactulose was recently reported (173,176). It has an energy value of 2.2 and 2.0 kcal/g for miniature pigs and rats, respectively (177). The medical applications of lactulose include the treatment of hyperammonemia, hepatic encephalopathy, and chronic constipation. Lactulose is also thought to prevent renal (178) and liver failures, is expected to lower incidence of colorectal cancer, shows a therapeutic effect in salmonella carrier state, has an antiendoxin effect, and contributes toward a decreased response in diabetics (179). Lactulose prevents postsurgery complications of obstructive jaundice, activates the immune system, prevents contracting infections, including urinary tract infection and respiratory tract infections, and prevents the recurrence of colorectal adenomas (179). In a human study, total bacterial counts were not changed by administration of lactulose (180); however, the *Bifidobacterium* count increased significantly, and lectinase positive *Clostridium* spp., including *C. perfringens*, and *Bacteroides,* decreased significantly. As a result, the ratio of *Bifidobacterium* in total bacterial counts increased from 8.3% before intake to 47.4% after intake. Indole and phenol, metabolites of intestinal bacteria, decreased significantly during intake, as did ammonia. Enzymatic activities of β-glucuronidase, nitroreductase, and azoreductase in wet feces decreased during intake. Fecal pH decreased and fecal moisture content increased significantly during intake (179). Lactulose promoted the growth of a range of lactic acid bacteria, although their growth was more rapid on lactose, particularly for *B. bifidum* (181). When lactulose was incubated with various intestinally derived strains, *B. bifidum, L. casei,* and *Lactobacillus* species utilized the disaccharide well with no gas production, while *C. perfringens* metabolized lactulose with a large amount of gas production (182).

Lactulose has been known to exert an antiendotoxin effect. Liehr and Heine (183) reported that endotoxin mediated galactosamine hepatitis in rats, as induced by intravenous injection of galactosamine, developed necrosis of the hepatocytes, and the inflammatory reaction of the liver tissue was virtually prevented by oral (184) and intravenous (184) administration of lactulose beforehand. Iwasaki et al. (185) reported that when lactulose was administered to hepatic cirrhosis and hepatic cancer patients, improvement of endotoxemia was observed in 50% of the patients.

Chronic portal systemic encephalopathy is a metabolic disorder of the central nervous system and is characterized by psychiatric and neurological disturbances that may progress from mild mental aberration to coma. It is manifest when ammonia, which is produced in the intestine as a metabolite by intestinal putrefactive bacteria, escapes detoxication by hepatic function and is transported to the brain where it acts as a toxin. Lactulose reduces the ammonia level in blood and alleviates symptoms, because lactulose regulates intestinal activity, stimulates the growth of *Bifidobacterium* spp., and suppresses intestinal pH and therefore ammonia production. This lowers ammonia absorption in the intestine.

Organic acids and SCFA produced in the colon migrate to the blood and are metabolized in the liver. In one study, the production of C_4 and C_6 fatty acids was completely inhibited by lactulose in an *in vitro* incubation system. It was suggested that lactulose thus detoxified the profile of SCFA produced in the presence of proteins (186). Genovese et al. (187) observed that lactulose improved blood glucose response to an oral glucose test in noninsulin dependent diabetic patients by preadministration of lactulose. Also, lactulose activated the cell mediated immune system depressed during liver cirrhosis (188). A significant reduction in the number of patients contracting urinary infections and respiratory infections using lactulose therapy has been found (189).

Onishi et al. (190) studied the effect of lactulose on ammonia metabolism during dynamic exercise. Increases in blood ammonia in graded exercise were suppressed, and oxygen intake at moderate exercise was increased significantly through preadministration of lactulose. It was thus suggested that lactulose has some benefits when used in combination with exercise.

24.3.4.1 Colon cancer

In the context of colon carcinoma, lactulose administration to healthy volunteers lowered fecal concentrations of secondary bile salts (191,192). However Rooney et al. (193) did not confirm this observation. Roncucci et al. (194) reported that lactulose decreased the recurrence rate of colon adenomas. Lactulose suppressed DNA damage in the colon mucosa of rats treated with DMH (1,2 dimethylhydrazine)(195). Roncucci et al. (194) evaluated the effect of lactulose on the recurrence rate of adenomatous polyps. It was found that administration of lactulose reduced the recurrence rate of adenomas in the large bowel.

24.3.4.2 Mineral absorption

Lactulose significantly increased calcium absorption but only if the diet contained >0.3% Ca (196). Magnesium absorption from the cecum was affected by inulin but not by the background amount of dietary calcium (197). Suzuki et al. (198) observed that, in rats, lactulose intake promoted the absorption or retention in the body of Ca, Mg, Zn, Cu, and Fe. Igarashi and Ezawa (199) studied the effect of lactulose on bone strength. The results suggested that lactulose promoted calcium absorption in osteoporosis and increased the strength of bones in rats suffering osteoporosis.

24.3.4.3 Constipation

Lactulose has gained popularity in the treatment of constipation and chronic hepatic encephalopathy (147,200–203). Lactulose prevents constipation through its hyperosmotic effects, mainly based on unabsorbed SCFA and cations associated with it (204). Its effects are manifest through activation of intestinal peristaltic movement and intestinal acidification and osmotic reaction. However, this could be true of all prebiotics and may have more to do with all NDO structures than lactulose specifically (205).

24.3.5 Soya products

The primary oligosaccharides contained in soybeans are raffinose (Galα1,6Gluα1,2βFru) and stachyose (Galα1,6Galα1,6Gluα1,2βFru). They are growth promoters of *B. infantis, B. longum*, and *B. adolescentis* but do not stimulate *E. coli*, *E. faecalis*, or *L. acidophilus* (173,206). Also, raffinose effects a relatively short generation time of *B. infantis*, independent of the substrate upon which the organism has been previously cultured. In a study by Fooks et al. (65), all *Bifidobacterium* species tested fermented this carbohydrate with the exception of *B. bifidum,* while *L. salivarius*, *B. fragilis,* and *Mitsuokella multiacida* metabolized the substrate to a lesser degree. In another study, it was found that addition of a low concentration [0.1% (w/v)] of SOR (refined soybean oligosaccharides) to a two stage continuous culture of fecal bacteria (207) resulted in a threefold increase in the number of bacteria in the total bacterial count.

De Boever et al. (208) examined the survival of *L. reuteri* when challenged with glycodeoxycholic acid (GDCA), deoxycholic acid (DCA), and soygerm powder. Moreover the impact of *L. reuteri* on the bioavailability of isoflavones present in soygerm powder was examined. When 4 g/l soygerm powder was added, the *Lactobacillus* strain survived the bile salt burden better and the membrane damage in the haemolysis test decreased. The *Lactobacillus* strain cleaved β-glycosidic isoflavones during fermentation of milk supplemented with soygerm powder. Interactions between the bacteria and soygerm powder suggest that combining both in fermented milk could exhibit advantageous synbiotic effects. De Boever et al. (209) have shown that consumption of a soygerm powder may beneficially modulate the colonic fermentation. The soygerm powder contained constituents with putative health benefits such as β-glycosidic isoflavones. They can play an important role in the health benefits from soya food, such as lower incidence of estrogen related cancers, cardiovascular diseases, and osteoporosis (210). Saponins are also present in soygerm powder and these compounds may be responsible for the lowering of the serum cholesterol (211). In this study, soygerm powder and a *Lactobacillus reuteri* strain were combined. Because viability is essential for *L. reuteri* to exert its probiotic effects, the ability of soygerm powder to enhance the survival of *L. reuteri* when the latter was exposed to bile salts was investigated. Free isoflavones are only released from soygerm powder in the colon. A second objective was therefore to determine whether *L. reuteri* deconjugated β-glycosidic isoflavones during the fermentation of milk supplemented with soygerm powder, thereby increasing their bioavailability. Soygerm powder contains putative health promoting estrogenic isoflavones. These molecules occur as biologically inactive β-glycosides, which are poorly resorbed by humans. Isoflavones can only be absorbed in a biologically active form after cleavage of their β-glycosidic linkage. It is generally accepted that this does not occur until the compounds reach the colonic microbiota, which produce β-glycosidase (EC 3.2.1.21) (212). Day et al. (213), found that lactase phlorizin hydrolase (EC 3.2.1.108), a β-glycosidase present in the brush border of the mammalian small intestine, was capable of hydrolysing isoflavone glycosides thereby making the free isoflavones available for resorption. Nevertheless, a substantial loss of isoflavone glycosides into the large intestine can be expected because of a limited residence of food compounds in the small intestine and a possible decrease of lactase phlorizin hydrolase levels in adulthood (214). To take advantage of the putative beneficial isoflavones, Izumi et al. (215) suggested that a constant plasma concentration of isoflavones was needed. Therefore one should try to maximize the resorption of free isoflavones by consuming soya foods already containing free isoflavones or by releasing free isoflavones in the small intestine. The enzyme profile of the fermentations indicated that the presence of high levels of β-galactosidase released the isoflavones from the soygerm powder.

De Boever et al. (209) used the *in vitro* model SHIME (simulation of the human intestinal microbial ecosystem) to study the effect of a soygerm powder rich in β-glycosidic phytoestrogenic isoflavones on the fermentation pattern of the colon microbiota and to determine to what extent the latter metabolized the conjugated phytoestrogens. The addition resulted in an overall increase in bacterial marker populations (*Enterobacteriaceae*, coliforms, *Lactobacillus* species, *Staphylococcus* species and *Clostridium* species), with a significant increase in the *Lactobacillus* population. The SCFA concentration increased about 30% during the supplementation period; this was due mainly to a significant increase of acetic and propionic acids. Gas analyses revealed that the methane concentration increased significantly. Ammonia and sulphide concentrations were not influenced by soygerm supplementation. Use of an electronic nose apparatus indicated that odor concentrations decreased significantly during the treatment period. The β-glycosidic bonds of the phytoestrogenic isoflavones were cleaved under the conditions prevailing in the large intestine. The increased bacterial fermentation after addition of the soygerm powder was paralleled by substantial metabolism of the free isoflavones resulting in recovery of only 12–17% of the supplemented isoflavones (209).

Isoflavones have been shown to compete effectively with endogenous mammalian estrogens in binding to the estrogen receptor of mammalian cells preventing estrogen stimulated growth of cancerous cells (216,217). Microbiota play an important role in the generation of biologically active isoflavones, but also inactivate these bioactive compounds after further bacterial fermentation, resulting in a loss of their purported beneficial effects.

Much of the research concerning the health promoting effects of soya products has focused on the putative role of one dietary ingredient, the isoflavones. Soya is an important source of many other nutrients, however, including dietary fibre, oligosaccharides (which could classify as prebiotics), proteins, trace minerals, and vitamins, which could influence the host wellbeing (218). Soya products typically contain high concentrations of the α-glycosidic GOS, raffinose, and stachyose. These sugars are not absorbed in the upper part of the gastrointestinal tract or hydrolysed by human digestive enzymes (219). Upon delivery to the colon, the sugars are fermented by the colonic microbiota possessing α-galactosidase activity. Administration of soybean oligosaccharides to an *in vitro* culture of human gut microbiota has been reported to be bifidogenic (220). Therefore, Gibson and Roberfroid (17) suggested a potential prebiotic role for soybean oligosaccharides. Studies fully addressing this hypothesis have not been published.

Soygerm powder is marketed mainly because it contains a large amount of isoflavones, through which the product could relieve menopausal problems, reduce osteoporosis, improve blood cholesterol levels, and lower the risk of certain types of cancer (221,222). The oligosaccharides present in the soygerm powder consist mainly of indigestible raffinose (13 g/kg) and stachyose (87 g/kg), oligosaccharides which, therefore, serve as substrates for bacterial fermentation (219). During the treatment period, the concentration of all bacterial groups increased and the greatest increase was observed for *Lactobacillus* species. An increase in lactic acid bacteria is often driven by a decrease in luminal pH because the capacity to thrive in low pH environments gives these bacteria a selective advantage over other species (134). The pH reduction in a microniche could select for lactic acid producing bacteria in this environment, possibly excluding long term colonization by invasive pathogens (132). During the supplementation period, the fermentative capacity of the microbial community was increased, and this led to a general rise in SCFA. The largest increase was observed in vessel 3, indicating a high level of carbohydrate fermentation. A rise in SCFA concentration is a positive property because these acids, especially butyric acid, are the main energy source for colonocytes and influence colonic function by stimulating water and sodium absorption and modulating motility (223). Furthermore, butyric acid induces differentiation, stimulates apoptosis of cancerous cells *in vitro,* and thus arrests the development of cancer (224). During supplementation of soygerm powder, no effect on ammonia and hydrogen sulphide concentration was observed. Ammonia can alter the morphology and intermediary metabolism of intestinal cells, increase DNA synthesis, and promote tumorigenesis (225). Hydrogen sulphide selectively inhibits butyric acid oxidation in colonocytes and this may play a pathogenic role in inflammatory bowel diseases such as colitis. Hence, increases in the concentration of these compounds are considered to be potentially harmful to the host. Quantitative gas analyses revealed that methane concentration increased significantly in the last two colonic compartments during the supplementation period. Production of methane by the colonic microbiota is important because it permits a more complete fermentation, in that the removal of H_2 is energetically more favorable (226).

Phenolic and indolic compounds have been linked to a variety of disease states in humans, including initiation of cancer, malabsorption, and anaemia (227). Data revealed that during addition of the soygerm powder, the amount of odor significantly decreased. It is the first study in which the odor produced by the gut microbiota was monitored (209). This indicates that consumption of soygerm powder may influence the gut microbiota in a beneficial manner.

24.3.6 Lactosucrose

Lactosucrose is produced from a mixture of lactose and sucrose using the enzyme β-fructosidase. The fructosyl residue is transferred from the sucrose to the 1-position of the glucose moiety in the lactose, producing a nonreducing oligosaccharide (116) according to the following reaction (228):

$$\text{Gal}\beta 1 \rightarrow 4\text{Glc} + \text{Glc}\alpha 1 \leftrightarrow 2\beta\text{Fru} \rightarrow \text{Gal}\beta 1 \rightarrow 4\text{Glc}\alpha 1 \leftrightarrow 2\beta\text{Fru} + \text{Glc}$$

β-fructosidase is an artificial oligosugar that is indigestible and therefore low in calories. It enhances the growth of intestinal *Bifidobacteria.* Lactosucrose plays a major role in the refinement of sugars in the food industry (229).

In one study, lactosucrose was used by *B. infantis, B. longum,* and *B. adolescentis*, but not by *L. acidophilus, E. coli* or *E. faecalis* (206). *Bifidobacteria* species grew at a higher rate on lactosucrose than on raffinose (Galα-1,6Gluα-1,2βFru), and the former was utilized by a smaller number of *Enterobacteriaceae* strains than glucose, lactose or raffinose. Fujita et al. (230) found lactosucrose to be similar to FOS in its effects on intestinal bacteria in that it promoted *Bifidobacterium* spp. but few of the other organisms tested.

24.3.7 Xylooligosaccharides

Xylooligosaccharides (XOS) are chains of xylose molecules linked by β-1,4 bonds and mainly consist of xylobiose, xylotriose, and xylotetraose (231), and are found naturally in bamboo shoots, fruits, vegetables, milk, and honey (232). Xylose containing oligosaccharides can be XOS or WBO (wheat bran oligosaccharides). They are produced enzymatically by hydrolysis of xylan from birch wood (233), oats (234), or corncobs (235) by a xylanase from *Trichoderma* (233). WBO are prepared from wheat bran hemicellulose, consisting mainly of branched xylose containing and arabinose containing oligosaccharides such as arabinosylxylobiose, arabinosylxylotriose, arabinosylxylotetraose, and diarabinosylxylotetraose (236). XOS are poorly utilized by *Bifidobacterium* strains, but xylobiose was reported to be bifidogenic (231,237). Only a few *Bifidobacterium* spp. can ferment xylan (238,239). Few studies have been conducted on XOS. Okazaki et al. (237) carried out a small trial in which five male subjects were fed 1 g/d and five were fed 2 g/day XOS for three weeks. It was found that a dose of 1 g/d XOS was sufficient to elicit a bifidogenic response. In a study by Yamada et al. (236), although WBO could not stimulate all the *Bifidobacterium* strains tested, the substrate was concluded to be more selective than FOS or SOE (soybean oligosaccharide extract). A pure culture study by Jaskari et al. (234) concluded raffinose to be superior to xylooligomers and FOS in increasing *Bifidobacterium* populations without increasing potential pathogens. Van Laere et al. (240) concluded that FOS promoted *Lactobacillus* and other intestinal bacteria more than XOS. Hopkins et al. (231) concluded that overall GOS, Oligomate, and FOS were better bifidogenic substrates than XOS, inulin, and pyrodextrins.

24.4 STRUCTURE TO FUNCTION RELATIONSHIPS IN OTHER OLIGOSACCHARIDES

24.4.1 Lactitol

Lactitol (4-0-[β-galactopyranosyl]-D-glucitol) is a sweet tasting sugar alcohol derived from lactose by reduction of the glucose part of the disaccharide (241).

The fermentation of lactitol encourages the growth of saccharolytic bacteria and decreases the amount of proteolytic bacteria, which are responsible for the production of ammonia, carcinogenic compounds, and endotoxins (241). Most ingested lactitol reaches the colon because a suitable β-galactosidase is missing in the small intestine. Cell free extracts of *L. acidophilus* and *B. longum* showed high fermentation activity for both lactulose and lactitol. However, cell free extracts of *E. coli* and *B. fragilis* showed low activity with lactulose and lactitol indicating that they can aid the growth of saccharolytic bacteria in the colon (241). A 10% addition of lactitol to bacterial cultures reduces pH and gas formation (241). Growth of *Bifidobacterium* and *Lactobacillus* species was induced by lactitol (241). The addition of lactitol at 1, 2, 5, or 10% concentration caused inhibition of proteolytic, toxin producing strains and increased saccharolytic bacterial levels to an extent proportional to the lactitol concentration (241). It also caused a decline in pH and inhibition of NH_3 production (241). Lactitol is mainly fermented by anaerobic organisms (241). Scevola et al. (242) also described a reduction in growth of *E. coli* with lactitol. One possible explanation for this effect may be blockage of the receptor sites of the lectins by which the bacteria adhere to cell walls (241). Patients ingesting lactitol had a higher SCFA concentration, which resulted in a pH reduction. This resulted in increased solubility of minerals in the colon (241). Lactitol is used in the treatment of hepatic encephalopathy and other liver diseases because it aids the growth of saccharolytic bacteria such as *Bifidobacterium* and *Lactobacillus* species. Changing the microbial balance in favor of saccharolytic bacteria inhibits production of endotoxins in the colon and thereby lowers the amount absorbed in blood plasma (241). A marked decrease in the fecal cholesterol concentration and in the concentration of its metabolites, and in total fecal bile acid concentrations of humans has been observed upon lactitol addition to a normal diet (241). These results indicate that saccharolytic microflora were stimulated by lactitol, which has a positive effect on cholesterol.

24.4.2 Laevan Type Prebiotics

Laevan type exopolysaccharide (EPS) from *Lactobacillus sanfranciscensis*, laevan, inulin, and FOS were studied for their prebiotic properties (243). An enrichment of *Bifidobacterium* species was found for EPS and inulin but not for laevan or FOS. The bifidogenic effect of EPS was confirmed using a selective growth medium. The use of EPS and FOS resulted in enhanced growth of *Eubacterium biforme* and *Clostridium perfringens,* respectively (243). Numbers of *Bifidobacterium* spp. increased exclusively with EPS whereas those of *Lactobacillus* decreased concomitantly. With EPS, the coliform counts increased to a reduced extent in comparison to a control without added carbohydrate. Growth of *Clostridium* was stimulated not only by the two EPS but also by laevan (243). In another study, Korakli et al. (238) investigated the metabolism by *Bifidobacterium* spp. of EPS produced by *L. sanfranciscensis*. The prebiotic properties of bread, and metabolism by *Bifidobacterium* spp. of water soluble polysaccharides (WSP) from wheat and rye were also investigated (238). Cereal products are the most important staple food throughout the world. Cereal grains are predominantly composed of starch; nonstarch polysaccharides composed of glucose (β-glucan), fructose (polyfructan), xylose, and arabinose (arabinoglycan) are present (244). Some of these polysaccharides, like starch, are partially digested, and others are believed to be dietary fibre, such as arabinoxylan. In addition to polyfructan, wheat and rye flours contain kestose, nystose, and other FOS of the inulin type (233). Comparison of an intestinal microflora incubated in media with EPS and without a carbohydrate source showed that EPS favored the growth of *Bifidobacterium* spp. by more than two log values and the growth of *Clostridia* by less than one log (243). EPS of *L. sanfranciscensis* TMW1.392 is composed predominantly of fructans and is therefore probably not degraded under the conditions prevailing in the stomach or small intestine. Thus, EPS meets some requirements for its use as a prebiotic (239). Polysaccharides composed of glucose in WSP isolated from rye sourdough fermented with *L. sanfranciscensis* and from control rye dough could not be fermented by *B. adolescentis*. This can probably be attributed to the enzymatic degradation of a starch fraction during dough fermentation. Furthermore, polyfructan in flour was completely degraded by cereal enzymes (238).

In conclusion, EPS produced by *L. sanfranciscensis* TMW1.392 during sourdough fermentation was metabolized by *Bifidobacterium* spp. species. The WSP in wheat and rye were also degraded by *Bifidobacterium* spp. However, polyfructan and the starch fraction, which possess a bifidogenic effect, were degraded by cereal enzymes during dough fermentation while the EPS was retained. The stability of EPS should enable it to withstand the baking process. This suggests that EPS will improve the nutritional properties of sourdough fermented products (238).

24.4.3 Gentiooligosaccharides

Gentiooligosaccharides (GeOS) are glucose polymers of the form Gluβ1-6[Gluβ1-6]$_n$ where n=1-5. These are not hydrolyzed in the stomach or small intestine and should therefore reach the colon intact (235). Rycroft et al. (245) investigated the fermentation properties of GeOS as compared to FOS and maltodextrin in mixed fecal culture. Gentiooligosaccharides gave the largest significant increases in *Bifidobacterium, Lactobacillus,* and total bacterial numbers during *in vitro* batch culture incubations. However, FOS appeared to be a more selective prebiotic as it did not significantly stimulate growth of bacterial groups which were not probiotic in nature. Gentiooligosaccharides and maltodextrin produced the highest levels of SCFA. Lowest gas production was seen with GeOS and highest with FOS (245). Gentiooligosaccharides possessed bifidogenic activity *in vitro,* which was maintained over the 24-hour fermentation period (245). Large amounts of lactate were produced from GeOS and maltodextrin metabolism (21.60±7.26 mmol/L and 21.86± 10.58 mmol/L respectively compared to 11.44±3.21 mmol/L in FOS). Berggren et al. (246) had previously observed that barley β-glucans generated greater SCFA than FOS when fed to rats. This was also the case in this study, where GeOS were studied, not β-glucans. Highest butyrate production was recorded on FOS in accordance with previous batch fermentation studies (247,248), human trials (127), and rat studies (249,250). FOS also produced the lowest levels of propionate in the fermenters. Lowest gas production in the fermentations was detected with GeOS, whereas FOS produced the highest volume. This is an undesirable characteristic in a prebiotic (245).

24.4.4 Glucose Based Oligosaccharides

Glucose based oligosaccharides include cellooligosaccharides (β1,4), maltooligosaccharides (α1,4), gentiooligosaccharides (β1,6), isomaltooligosaccharides (α1,6) oligosaccharides, and those with mixed linkages such as glucooligosaccharides (α1,2, α1,4 and α1,6 linked), maltodextrin based oligosaccharides, and oligolaminarans.

Yazawa et al. (251) found that oligosaccharides of DP 3-5 from dextran hydrolysate promoted *B. infantis* but not *E. coli*, *E. faecalis* or *L. acidophilus*. Kohmoto et al. (169) observed that panose, isomaltose, isomaltotriose, and IMO900P were utilized as well as raffinose by all the *Bifidobacterium* spp. tested except *B. Bifidum,* which gave no growth on any of the substrates. *Bacteroides* species utilized all the sugars tested, but fewer *Clostridium* species grew on the IMO than the raffinose. Overall IMO900P appeared at least as selective, if not more so, for *Bifidobacterium* spp. than raffinose. IMO (NC) mixtures of different molecular weight distributions were incubated with fecal bacteria in batch cultures and complex gut models (172). The lower molecular weight products showed similar prebiotic activity to FOS with respect to increases in *Bifidobacterium* spp., decreases in *Bacteroides* and *Clostridium* spp., and the production of lactate and acetate. The higher molecular weight products generated higher butyrate concentrations. In the gut model, a validated three stage compound continuous culture system that represents the conditions typical of the human gastrointestinal tract (252), high molecular weight products stimulated *Bifidobacterium* spp. and *Lactobacillus* strains, while stimulating butyrate production, which persisted throughout the model and was greatest in vessel 3 (representing the distal colon). Once again this suggests that DP affected the prebiotic activity and fermentation of a molecule. In this case, a molecule with a low DP seemed to be more bifidogenic *in vitro*. IMO900P clearly had prebiotic potential *in vivo* and the minimum effective dose was 13–15 g/d IMO900 (253). The IMO3 fraction (trisaccharide) had a greater prebiotic effect than IMO2 (171). This may have been due to the greater hydrolysis of IMO2 than IMO3 by isomaltase in the jejunum so more substrate was available for the colonic microflora when IMO3 was administered.

In a limited pure culture study (254), while most *Bacteroides* strains, particularly *B. thetaiomicron*, degraded α-GOS, eight *Bifidobacterium* strains did not do so to the same extent, and three could not degrade it at all. Two *C. butyricum* strains degraded around 50% whereas most *Lactobacillus* strains hydrolysed less than 10% α-GOS, which appears not to have a prebiotic effect. IMO appeared to be prebiotic. Differences among glucooligosaccharides were due to the DP or linkage types of the molecules, suggesting that a low DP and α- rather than β- linkages favor fermentation and prebiotic activity.

24.5 CONCLUSIONS

Increasing consumer awareness for health and nutritional issues makes the market for prebiotics very promising; however, because of the high product costs and the still fragmented scientific backup of some of the health claims, this is still a disorganized area. Structure to function relationship studies, together with the development of cheaper manufacturing techniques, and more understanding of the physiological mechanism by which prebiotics bring about improved health, provide one answer to establishing the functional oligosaccharide market.

This chapter has reviewed research and conclusions on the complex effects of oligosaccharide structure and size on the human microflora. A large comparative study, as reported by Djouzi and Andrieux (250), involved feeding 4% (w/w) FOS, TOS (transgalactooligosaccharides), and α-GOS diets to rats for four weeks. TOS and FOS led to desirable changes in microflora composition but had little effect on enzyme activities. α-GOS did not affect microflora composition but favorably altered the metabolism. The desirable changes with α-GOS were reduced gas production and increased glycosidase activity. GOS was superior to FOS in the greater *Bifidobacterium* population and the increased β-galactosidase activity when GOS was administered. Galactooligosaccharides, however, gave higher hydrogen and methane excretions. Inulin had a similar effect to FOS although higher doses of the former were needed (127,131). This suggests that a lower molecular weight gives a higher prebiotic activity, in accordance with pure culture studies (251). Transgalactosylated disaccharides were not as effective prebiotics as TOS and Oligomate, as the former needed a higher dose (157,255,256). Therefore, longer chain GOS were more effective than their disaccharide counterparts. Although these findings are contradictory, they suggest that neither disaccharides nor polysaccharides are the most effective prebiotics, and that oligosaccharides of DP>3 would be most likely prebiotic candidates.

In human trials with soybean oligosaccharides (257,258) less SOE than raffinose was needed to gain a prebiotic effect, and since 3 times as much stachyose (tetrasaccharide) than raffinose is present in SOE, it is likely that stachyose has the best prebiotic activity of the two (220). In the case of IMO, a low DP was more bifidogenic *in vitro* fermentation studies (172). α-galactooligosacchride showed some prebiotic activity, MDO (maltodextrin based oligosaccharide) behaved like a polysaccharide, and IMO behaved like a prebiotic. So a low DP and α- rather than β- linkages favor fermentation and prebiotic activity (172,247,248,254). The complexity of structure to function studies is also due to population variation which accounts for many inconsistencies between studies, and the observation that very often the bifidogenic effect is dependent on the original level of the *Bifidobacterium* strains present (259). For this reason, there is a great need for an extensive comparative *in vitro* study of all prebiotics available, involving the same parameters and in the same controlled conditions, so as to allow a comparison of the same amounts. This will help to better inform the most efficacious prebiotics.

REFERENCES

1. Wiggins, H.S. Gastroenterological functions of dietary fibre. In: *Dietary Fibre*, Birch, G.G., K.J. Parker, eds., London: Applied Science, 1983, pp 205–219.
2. Cummings, J.H., G.T. Macfarlane. The control and consequences of bacterial fermentation in the human colon. *J. Appl. Bacteriol.* 70:443–459, 1991.
3. Cummings, J.H. Short chain fatty acids in the human colon. *Gut* 22:763–779, 1981.
4. Wolin, M.J., T.L. Miller. Carbohydrate fermentation. In: *Human Intestinal Microflora in Health and Disease*, Hentges, D.J., ed., London: Academic Press, 1983, pp 147–165.
5. Allison, C., G.T. Macfarlane. Effect of nitrate upon methane production by slurries of human faecal bacteria. *J. Gen. Microbiol.* 134:1397–1405, 1988.
6. Macfarlane, G.T., J.H. Cummings, C. Allison. Protein degradation by human intestinal bacteria. *J. Gen. Microbiol.* 132:1647–1656, 1986.
7. Drasar, B.S., M.J. Hill. *Human Intestinal Flora.* London: Academic Press, 1974.
8. Bakke, O.M., T. Midtvedt. Influence of germfree status on the excretion of simple phenols of possible significance in tumor promotion. *Experientia* 26:519–519, 1970.

9. E. Bone, A. Tamm, M. Hill. The production of urinary phenols by gut bacteria and their possible role in the causation of large bowel cancer. *Am. J. Clin. Nutr.* 29:1448–1454, 1976.
10. Macfarlane, G.T., G.R. Gibson. Carbohydrate fermentation, energy transduction and gas metabolism in the human large intestine. In: *Gastrointestinal Microbiology*, Vol. 1, Mackie, R.I., B.H. White, eds., London: Chapman & Hall, 1997, pp 269–318.
11. Macfarlane, G.T., H.N. Englyst. Starch utilisation by the human large intestinal microflora. *J. Appl. Bacteriol.* 60:195–201, 1986.
12. Hoskins, L.C., E.T. Boulding. Mucin degradation in human colon ecosystems. *J. Clin. Invest.* 67:163–172, 1981.
13. Roberton, A.M., R.A. Stanley. *In vitro* utilization of mucin by *Bacteroides fragilis*. *Appl. Environ. Microbiol.* 43:325–330, 1982.
14. Hoskins, L.C., M. Agustines, W.B. McKee, E.T. Boulding, M. Kriaris, G. Niedermeyer. Mucin degradation in human colon ecosystems. Isolation and properties of fecal strains that degrade ABH blood group antigens and oligosaccharides from mucin glycoproteons. *J. Clin. Invest.* 75:944–953, 1985.
15. Salyers, A.A., J.A.Z. Leedle. Carbohydrate utilization in the human colon. In: *Human Intestinal Microflora in Health and Disease,* Hentges, D.J., ed., London: Academic Press, 1983, pp 129–146.
16. Macfarlane, G.T., S. Hay, S. Macfarlane, G.R. Gibson. Effect of different carbohydrates on growth, polysaccharidase and glycosidase production by *Bacteroides ovatus*, in batch and continuous culture. *J. Appl. Bacteriol.* 68:179–187, 1990.
17. Gibson, G.R., M.B. Roberfroid. Dietary modulation of the human colonic microbiota: introducing the concept of prebiotics. *J.Nutr.* 125:1401–1412, 1995.
18. Zubillaga, M., R. Weill, E. Postaire, C. Goldman, R. Caro, J. Boccio. Effect of probiotics and functional foods and their use in different diseases. *Nutr. Res.* 21:569–579, 2001.
19. Gibson, G.R., R. Fuller. Aspects of *in vitro* and *in vivo* research approaches directed toward identifying probiotics and prebiotics for human use. *J. Nutr.* 130:391S–395S, 2000.
20. Rani, B., N. Khetarpaul. Probiotic fermented food mixture: possible applications in clinical anti-diarrhoea usage. *Nutr. Health* 12:97–105, 1998.
21. Aiba, Y., N. Suzuki, A.M. Kabir, A. Takagi, Y. Koga. Lactic acid-mediated supression of *Helicobacter pylori* by the oral administration of *Lactobacillus salivarius* as a probiotic in a gnotobiotic murine model. *Am. J. Gastroenterol.* 93:2097–2101, 1998.
22. Bazhenov, L.G., V.M. Bondarenko, E.A. Lykova. The antagonostic action of lactobacilli on *Helicobacter pylori. Zh. Mikrob. Epid. Immun.* 3:89–91, 1997.
23. Kabir, A.M., Y. Aiba, A. Takagi, S. Kamiya, T. Miwa, Y. Koga. Prevention of *Helicobacter pylori* infection by lactobacilli in a gnotobiotic murine model. *Gut* 41:49–55, 1997.
24. Jiang, T., D.A. Savaiano. *In vitro* lactose fermentation by human colonic bacteria is modified by *Lactobacillus acidophilus* supplementation. *J. Nutr.* 127:1489–1495, 1997.
25. Pool-Zobel, B.L., R. Munzner, W.H. Holzapfel. Antigenotoxic properties of lactic acid bacteria in the *S. typhimurium* mutagenicity assay. *Nutr. Cancer* 23:261–270, 1993.
26. Spanhaak, S., R. Hanevaar, G. Schaafsma. The effect of consumption of fermented by *Lactobacillus casei* strain Shirota on the intestinal microflora and immune parameters in humans. *Eur. J. Clin. Nutr.* 52:899–907, 1998.
27. Schiffrin, E.J., D. Brassart, A.L. Servin, F. Rochat, A. Donnet-Hughes. Immune modulation of blood leukocytes in humans by lactic acid bacteria: criteria for strain selection. *Am. J. Clin. Nutr.* 66:515S–520S, 1997.
28. Tannock, G.W. *Normal Microflora: An Introduction to Microbes Inhabiting the Human Body.* London: Chapman & Hall, 1995.
29. O' Sullivan, M.G. Metabolism of bifidogenic factors by gut flora: an overview. *Bull. IDF* 313:23–30, 1996.
30. Englyst, H.N., S. Hay, G.T. Macfarlane. Polysaccharide breakdown by mixed populations of human faecal bacteria. *FEMS Microbiol. Ecol.* 95:163–171, 1987.
31. Scheppach, W., C. Fabian, M. Sachs, H. Kasper. The effect of starch malabsorption on fecal short chain fatty acid excretion in man. *Scand. J. Gastroenterol.* 23:755–759, 1988.

32. Wiggings, H.S., J.H. Cummings. Evidence for the mixing of residue in the human gut. *Gut* 17:1007–1011, 1976.
33. Macfarlane, G.T., J.H. Cummings, S. Macfarlane, G.R. Gibson. Influence of retention time on degradation of pancreatic enzymes by human colonic bacteria grown in a 3-stage continuous culture system. *J. Appl. Bacteriol.* 67:521–527, 1989.
34. Cummings, J.H., M.J. Hill, E.S. Bone, W.J. Branch, D.J.A. Jenkins. The effect of meat protein and dietary fibre on colonic function and metabolism, part II: bacterial metabolites in feces and urine. *Am. J. Clin. Nutr.* 32:2094–2101, 1979.
35. Bingham, S.A., S. Pett, K.C. Day. NSP intake of a representative sample of British adults. *J. Hum. Nutr. Diet.* 3:339–344, 1990.
36. Cummings, J.H., E.W. Pomare, W.J. Branch, C.P.E. Naylor, G.T. Macfarlane. Short chain fatty acids in human large intestine, portal, hepatic and venous blood. *Gut* 28:1221–1227, 1987.
37. Argenzio, R.A., M. Miller, W. Von Engelhardt. Effect of volatile fatty acids on water and ion absorption from the goat colon. *Am. J. Physiol.* 229:997–1002, 1975.
38. Argenzio, R.A., S.C. Whipp. Interrelationship of sodium, chloride, bicarbonate and acetate transport by the colon of the pig. *J. Physiol.* 295:315–381, 1979.
39. Salminen, S., C. Bouley, M.-C. Boutron-Ruault, J.H. Cummings, A. Franck, G.R. Gibson, E. Isolauri, M.-C. Moreau, M. Roberfroid, I. Rowland. Functional food science and gastrointestinal physiology and function. *Br. J. Nutr.* 80(1):S147–S171, 1998.
40. Macfarlane, G.T., G.R. Gibson. Bacterial infections and diarrhoea. In: *Human Colonic Bacteria: Role in Nutrition, Physiology, and Pathology*, Gibson, G.R., G.T. Macfarlane, eds., Boca Raton, FL: CRC Press, 1995, pp 201–226.
41. Isolauri, E., Y. Sutas, P. Kankaanpaa, H. Arvilommi, S. Salminen. Probiotics: effects on immunity. *Am. J. Clin. Nutr.* 73:444S–50S, 2001.
42. Ziemer, C.J., G.R. Gibson. An overview of probiotics, prebiotics and synbiotics in the functional food concept: perspectives and future strategies. *Int. Dairy J.* 8:473–479, 1998.
43. Gorbach, S.L., T.W. Chang, B. Goldin. Successful treatment of relapsing *Clostridium difficile* colitis with *Lactobacillus* GG. *Lancet* 2:1519–1519, 1987.
44. Kimmey, M.B., G.W. Elmer, C.M. Surawics, L. McFarland. Prevention of further recurrence of *Clostridium difficile* colitis with *Saccharomyces boulardii*. *Dig. Dis. Sci.* 35:897–901, 1990.
45. Buts, J.P., G. Corthier, M. Delmee. *Saccharomyces boulardii* for *Clostridium difficile* associated enteropathies in infants. *J. Pediatr. Gastroenterol. Nutr.* 16:419–425, 1993.
46. Surawics, C.M., L. McFarland, G.W. Elmer, J. Chinn. Treatment of recurrent *Clostridium difficile* colitis with vancomycin and *Saccharomyces boulardii*. *Am. J. Gastroenterol.* 84:1285–1287, 1989.
47. Biller, J.A., A.J. Katz, A.F. Flores, T.M. Buie, S.L. Gorbach. Treatment of recurrent *Clostridium difficile* colitis with *Lactobacillus* GG. *J. Pediatr. Gastroenterol. Nutr.* 21:224–226, 1995.
48. Bennet, R.G., S.L. Gorbach, B.R. Goldin, T.-W. Chang, B.E. Laughon, W.B. Greenough III, J.G. Bartlett. Treatment of relapsing *C. difficile* diarrhea with *Lactobacillus GG. Nutr. Today* 31:35S–38S, 1996.
49. Levy, J. Experience with live *Lactobacillus plantarum* 299v: a promising adjunct in the management of recurrent *Clostridium difficile* infection. *Gastroenterology* 112:A379, 1997.
50. McFarland, L.V., C.M. Surawics, R.N. Greenberg, R. Fekety, G.W. Elmer, K.A. Moyer, S.A. Melcher, K.E. Bowen, J.L. Cox, Z. Noorani. A randomized placebo-controlled trial of *Saccharomyces boulardii* in combination with standard antibiotics for *Clostridium difficile*. *JAMA* 271:1913–1918, 1994.
51. Fabia, R., A. Ar'Rajab, M.L. Johansson, R. Willen, R. Anderson. The effect of exogeneous administration of *Lactobacillus reuteri* R2LC and oat fibre on acetic acid-induced colitis in the rat. *Scand. J. Gastroenterol.* 111:334–344, 1993.
52. Gibson, G.R., G.T. Macfarlane. Intestinal bacteria and disease. In: *Human Health: Contribution of Microorganisms,* Gibson, S.A.W., ed., London: Spinger-Verlag, 1994, pp 53–62.

53. Malin, M., H. Suomalainen, M. Saxelin, E. Isolauri. Promotion of IgA immune response in patients with Crohn's disease by oral bacteriotherapy with *Lactobacillus* GG. *Ann. Nutr. Metab.* 40:137–145, 1996.
54. Gibson, G.R., J.H. Cummings, G.T. Macfarlane. Growth and activities of sulphate-reducing bacteria in gut contents of healthy subjects and patients with ulcerative colitis. *FEMS Microbiol. Ecol.* 86:103–112, 1991.
55. Kruis, W., E. Schutz, P. Fric, B. Fixa, G. Judmaier, M. Stolte. Double-blind comparison of an oral *Escherichia coli* preparation and mesalazine in mantaining remission of ulcerative colitis. *Aliment. Pharmacol. Ther.* 11:853–858, 1997.
56. Rembacken, B.J., A.M. Snelling, P.M. Hawkey, D.M. Chalmers, A.T. Axon. Non-pathogenic *Escherichia coli* versus mesalazine for the treatment of ulcerative colitis: a randomised trial. *Lancet* 354:635–639, 1999.
57. Rolfe, R.D. Colonisation resistance. In: *Gastrointestinal Microbiology. Vol 2. Gastrointestinal microbes and host interactions,* Mackie, R.I., White, B.A., R.E. Isaacson, eds., New York: Chapman and Hall, 1997.
58. Bazzoli, F., R.M. Zagari, S. Fossi, M.C. Morelli, P. Pozzato, M. Ventrucci, G. Mazzella, D. Festi, E. Roda. *In vivo Helicobacter pylori* clearance failure with *Lactobacillus acidophilus. Gastroenterology* 102:(A38), 1992.
59. Marteau, P.R., M. de Vrese, C.J. Cellier, J. Schrezenmeir. Protection from gastrointestinal diseases with the use of probiotics. *Am. J. Clin. Nutr.* 73:430S–436S, 2001.
60. Adams, J., C. Barret, A. Barret-Bellet, E. Benedetti, A. Calendini, P. Daschen. Essais cliniques controlès en double insu de l'ultra-levure lyophilisèe. Etude multicentrique par 25 mèdecins de 388 cas. *Gazette Med. France* 84:2072–2078, 1977.
61. Marteau, P., P. Pochart, Y. Bouhnik, J.C. Rambaud. Fate and effects of some transiting microorganisms in the human gastrointestinal tract. *World Rev. Nutr. Diet.* 74:1–21, 1993.
62. Gibson, G.R., J.M. Saavedra, S. McFarlane, G.T. McFarlane. Probiotics and intestinal infections. In: *Probiotics 2: Applications and Practical Aspects,* Fuller, R., ed., New York: Chapman & Hall,1997, pp10–38.
63. Saavedra, J.M., N.A. Bauman, I. Oung, J.A. Perman, R.H. Yolken. Feeding of *Bifidobacterium bifidum* and *Streptococcus thermophilus* to infants in hospital for prevention of diarrhea and shedding of rotavirus. *Lancet* 344:1046–1049, 1994.
64. Halpern, G.M., K.J. Vruwink, J. Van der Water, C.L. Keen, M.E. Gershwin. Influence of long-term yogurt consumption in young adults. *Int. J. Immunother.* 7:205–210, 1991.
65. Fooks, L.J., R. Fuller, G.R. Gibson. Prebiotics, probiotics and human gut microbiology. *Int. Dairy J.* 9:53–61, 1999.
66. Kaila, M., E. Isolauri, E. Soppi, V. Virtanen, S. Laine, H. Arvilommi. Enhancement of the circulating antibody secreting cell response in human diarrhea by a human *Lactobacillus* strain. *Pediatr. Res.* 32:141–144, 1992.
67. Schiffrin, E.J., F. Rochat, H. Link-Amster, J.M. Aeschlimann, A. Donnet-Hughes. Immunomodulation of human blood cells following the ingestion of lactic acid bacteria. *J. Dairy Sci.* 78:491–497, 1994.
68. Marteau, P., M. Minekus, R. Havenaar, J.H.J. Huis in't Veld. Survival of lactic acid bacteria in a dynamic model of the stomach and small intestine: validation and the effects of bile. *J. Dairy Sci.* 80:1031–1037, 1997.
69. Marteau, P., J.P. Vaerman, J.P. Dehennin, S. Bord, D. Brassart, P. Pochart, J.F. Desjeux, J.C. Rambaud. Effects of intrajejunal perfusion and chronic ingestion of *Lactobacillus johnsonii* strain La1 on serum concentrations and jejunal secretions of immunoglobulins and serum propteins in healthy humans. *Gastroenterol. Clin. Biol.* 21:293–298, 1997.
70. Solis-Pereira, B., D. Lemonnier. Induction of human cytokines by bacteria used in dairy foods. *Nutr. Res.* 13:1127–1140, 1996.
71. Bolte, G., M. Knauss, I. Metzdorf, M. Stern. Postnatal maturation of rat small intestinal brush border membranes correlated with increase in food protein binding capacity. *Dig. Dis. Sci.* 43:148–155, 1998.

72. Van der Heijden, P.J., A.T.J. Bianchi, M. Dol, J.W. Pals, W. Stok, B.A. Bokhout. Manipulation of intestinal immune responses against ovalbumin by cholera toxin and its B subunit in mice. *Immunology* 72:89–93, 1991.
73. G. Perdigon, M.E. de Macias, S. Alvarez, G. Oliver, A.A. de Ruiz Holgado. Effect of perorally administered lactobacilli on macrophage activation in mice. *Infect. Immun.* 53:404–410, 1986.
74. Perdigon, G., M.E. de Macias, S. Alvarez, G. Oliver, A.P. de Ruiz Holgado. Systemic augmentation of the immune response in mice by feeding fermented milks with *Lactobacillus casei* and *Lactobacillus acidophilus*. *Immunology* 63:17–23, 1998.
75. Miettinen, M., J. Vuopio-Varkila, K. Varkila. Production of human tumor necrosis factor alpha, interleukin-6, and interleukin-10 is induced by lactic acid bacteria. *Infect. Immun.* 64:5403–5405, 1996.
76. Pelto, L., E. Isolauri, E.M. Lilius, J. Nuutila, S. Salminen. Probiotic bacteria down-regulate the milk-induced inflammatory response in milk-hypersensitive subjects but have an immunostimulatory effect in healthy subjects. *Clin. Exp. Allergy* 28:1471–1479, 1998.
77. Link-Amster, H., F. Rochat, K.Y. Saudan, O. Mignot, J.M. Aeschlimann. Modulation of a specific humoral immunoresponse and changes in intestinal flora mediated through fermented milk intake. *FEMS Immunol. Med. Microbiol.* 10:55–64, 1994.
78. Madara, L.J. The chameleon within: improving antigen delivery. *Science* 277:910–911, 1997.
79. Michetti, P., G. Dorta, P.H. Wiesel, D. Brassart, E. Verdu, M. Herranz, C. Felley, N. Porta, M. Rouvet, A.L. Blum, I. Corthesy-Theulaz. Effect of whey-based culture supernatant of *Lactobacillus acidophilus* (johnsonii) La1 on *Helicobacter pylori* infection in humans. *Digestion* 60:203–209, 1999.
80. Ouwehand, A.C., P.V. Kirjavainen, C. Shortt, S. Salminen. Probiotics: mechanisms and established effects. *Int. Dairy J.* 9:43–52, 1999.
81. Kim, H.S., S. Gilliland. *L. acidophilus* as a dietary adjunct for milk to aid lactose digestion in humans. *J. Dairy Sci.* 66:959–966, 1983.
82. De Vrese, M., A. Stegelmann, B. Richter, S. Fenselau, C. Laue, J. Schrezenmeir. Probiotics-compensation for lactase insufficiency. *Am. J. Clin. Nutr.* 73:421S–429S, 2001.
83. Marteau, P., B. Flourie, P. Pochart, C. Chastang, J.F. Desjeux, J.C. Rambaud. Effect of the microbial lactase activity in yogurt on the intestinal absorption of lactose: an *in vivo* study in lactase-deficient humans. *Br. J. Nutr.* 64:71–79, 1990.
84. Mahè, S., P. Marteau, J.F. Huneau, F. Thuilier, D. Tomè. Intestinal nitrogen and electrolyte movements following fermented milk ingestion in human. *Br. J. Nutr.* 71:169–180, 1994.
85. Lin, M., C.L. Yen, S.H. Chen. Management of lactose maldigestion by consuming milk containing lactobacilli. *Dig. Dis. Sci.* 43:133–137, 1998.
86. Rowland, I.R. Toxicology of the colon-role of the intestinal microflora. In: *Human Colonic Bacteria: Role in Nutrition, Physiology, and Pathology,* Gibson, G.R., G.T. Macfarlane, eds., Boca Raton, FL: CRC Press, 1995, pp 115–174.
87. Brady, L.J., D.D. Gallaher, F.F. Busta. The role of probiotic cultures in the prevention of colon cancer. *J. Nutr.* 130:410S–414S, 2000.
88. Mital, B.K., S.K. Garg. Anticarcinogenic, hypocholesterolemic, and antagonistic activities of *Lactobacillus acidophilus*. *Crit. Rev. Microbiol.* 21:175–214, 1995.
89. Wollowski, I., G. Rechkemmer, B. Pool-Zobel. Protective role of probiotics and prebiotics in colon cancer. *Am. J. Clin. Nutr.* 73:451S–455S, 2001.
90. Sakata, T. Stimulatory effect of short-chain fatty acids on epithelial cell proliferation in the rat intestine: a possible explanation for trophic effects of fermentable fibre, gut microbes and luminal trophic factors. *Br. J. Nutr.* 58:95–103, 1987.
91. Leder, A., P. Leder. Butyric acid, a potent inducer of erythroid differentiation in cultural erythroleukemic cells. *Cell* 5:319–322, 1975.
92. Hagopian, H.K., M.G. Riggs, L.A. Swartz, V.M. Ingram. Effect of *n*-butyrate on DNA synthesis in chick fibroplast and Hela cells. *Cell* 12:855–860, 1977.

93. Borenfreund, E., E. Schmid, A. Bendich, W.S. Franke. Constitutive aggregates of intermediate-sized filaments of the vimentin and cytokeratin type in cultured hepatoma cells and their dispersal by butyrate. *Exp. Cell Res.* 127:215–235, 1980.
94. Wrong, O.M., A.J. Vince, J.C. Waterlow. The contribution of endogenous urea to fecal ammonia in man, determined by ^{15}N labelling of plasma urea. *Clin. Sci.* 68:193–199, 1985.
95. Visek, W.J. Effects of urea hydrolysis on cell life-span and metabolism. *Fed. Proc.* 31:1178–1193, 1972.
96. Visek, W.K. Diet and cell growth modulation by ammonia. *Am. J. Clin. Nutr.* 31:S216–S220, 1978.
97. Dunning, W.T., M.R. Curtis, M.E. Mann. The effect of added dietary tryptophane on the occurrence of 2-acetylaminofluorene-induced liver and bladder cancer in rats. *Cancer Res.* 10:454–459, 1950.
98. Bryan, G.T. The role of urinary tryptophan metabolites in the etiology of bladder cancer. *Am. J. Clin. Nutr.* 24:841–847, 1971.
99. Duran, M., D. Ketting, P.K. DeBrec, C. van der Heiden, S.K. Wadman. Gas chromatographic analysis of urinary phenols in patients with gastrointestinal disorders and normals. *Clin. Chim. Acta* 45:341–347, 1973.
100. Kulkarni, N., B.S. Reddy. Inhibitory effect of *Bifidobacterium longum* cultures on the azoxymethane-induced aberrant crypt foci formation and fecal bacterial β-glucuronidase. *Proc. Soc. Exp. Biol. Med.* 207:278–283, 1994.
101. Rowland, I.R., C.J. Rumney, J.T. Coutts, L.C. Lievense. Effect of *Bifidobacterium longum* and inulin in gut bacterial metabolism and carcinogen-induced aberrant crypt foci in rats. *Carcinogenesis* 19:281–285, 1998.
102. Kinouchi, T., K. Kataoka, S. Ruo Bing, H. Nakayama, M. Uejima, K. Shimono, T. Kuwahara, S. Akimoto, I. Hiraoka, Y. Onishi. Culture supernatants of *Lactobacillus acidophilus* and *Bifidobacterium adolescentis* repress ileal ulcer formation in rats treated with a non-steroidal anti-inflammatory drug by suppressing unbalanced growth of aerobic bacteria and lipid peroxidation. *Microbiol. Immunol.* 42:347–355, 1998.
103. Goldin, B.R., S.L. Gorbach. Alterations of the intestinal microflora by diet, oral antibiotics, and *Lactobacillus*: decreased production of free amines from aromatic nitro compounds, azo dyes and glucorinides. *J. Natl. Cancer Inst.* 73:689–695, 1984.
104. Kato, I., K. Endo, T. Yokokura. Effects of oral administration of *Lactobacillus casei* on antitumor responses induced by tumor resection in mice. *Int. J. Immunopharmacol.* 16:29–36, 1994.
105. Okawa, T., H. Niibe, T. Arai, K. Sekiba, K. Noda, S. Takeuchi, S. Hashomoto, N. Ogawa. Effect of LC9018 combined with radiation therapy on carcinoma of the uterine cervix: a phase III, multicenter, randomized, controlled study. *Cancer* 72:1949–1954, 1993.
106. Rafter, J.J. The role of lactic acid bacteria in colon cancer prevention. *Scand. J. Gastroenterol.* 30:497–502, 1995.
107. Reddy, B.S. Prevention of colon cancer by pre- and probiotics: evidence from laboratory studies. *Br. J. Nutr.* 80(2):S219–S223, 1998.
108. Barbacid, M. *Ras* oncogenes: their role in neoplasia. *Eur. J.Clin. Invest.* 20:225–235, 1990.
109. Kotsinas, A., D.A. Spandidos, P. Romanowski, A.H. Wyllie. Relative expression of wild-type and activated Ki-*ras*2 oncogene in colorectal carcinomas. *Int. J. Oncol.* 3:841–845, 1993.
110. Taylor, G.R.J., C.M. Williams. Effect of probiotics and prebiotics on blood lipids. *Br. J. Nutr.* 80(2):S225–S230, 1998.
111. Bergman, D.N., W.E. Roe, K. Kon. Quantitative aspects of propionate metabolism and glucogenesis in sheep. *Am. J. Physiol.* 211:793–799, 1966.
112. Chen, W.-J.L., J.W. Anderson, D. Jennings. Propionate may mediate the hypocholesterolemic effects of certain soluble plant fibres in cholesterol-fed rats. *Proc. Soc. Exp. Biol. Med.* 175:215–218, 1984.

113. Illman, R.J., D.L. Topping, G.H. McIntosh, R.P. Trimble, G.B. Storer, M.N. Taylor, B.-Q. Cheng. Hypocholesterolaemic effects of dietary propionate studies in whole animals and perfused rat liver. *Ann. Nutr. Metab*. 32:97–107, 1988.
114. Venter, C.S., H.H. Vorster, J.H. Cummings. Effects of dietary propionate on carbohydrate and lipid metabolism in man. *Am. J. Gastroenterol*. 85:549–552, 1990.
115. Schaafsma, G., W.J.A. Meuling, W. van Dokkum, C. Bouley. Effects of a milk product, fermented by *Lactobacillus acidophilus* and with fructo-oligosaccharides added, on blood lipids in male volunteers. *Eur. J. Clin. Nutr*. 52:436–440, 1998.
116. Sako, T., K. Matsumoto, R. Tanaka. Recent progress on research and applications of non-digestible galacto-oligosaccharides. *Int. Dairy J*. 9:69–80, 1999.
117. Yun, J.W. Fructooligosaccharides: occurrence, preparation, and application. *Enzyme Microb. Technol*. 19(2):107–117, 1996.
118. Hidaka, H., M. Hirayama, N. Sumi. A fructooligosaccharide-producing enzyme from *Aspergillus niger* ATCC 20611 *Agric. Biol. Chem*. 52:1181–1187, 1988.
119. Roberfroid, M.B., J.A.E. Van Loo, G.R. Gibson. The bifidogenic nature of chicory inulin and its hydrolysis products. *J. Nutr*. 128(1):11–19, 1998.
120. Crittenden, R.G., M.J. Playne. Production, properties and applications of food-grade oligosaccharides. *Trends Food Sci. Technol*. 7:353–361, 1996.
121. Muramatsu, K., S. Onodera, S. Kikuchi, N. Shiomi. Substrate specificity and subsite affinities of β-fructofuranosidase from *Bifidobacterium adolescentis* G1. *Biosci. Biotech. Biochem*. 58:1642–1645, 1994.
122. Perrin, S., M. Warchol, J.P. Grill, F. Schneider. Fermentations of fructo-oligosaccharides and their components by *Bifidobacterium infantis* ATCC 15697 on batch culture in semi-synthetic medium. *J. Appl. Microbiol*. 90:859–865, 2001.
123. Rao, V.A. The prebiotic properties of oligofructose at low intake levels. *Nutr. Res*. 21:843-848, 2001.
124. Kaplan, H., R.W. Hutkins. Fermentation of fructooligosaccharides by lactic acid bacteria and *Bifidobacteria*. *Appl. Environ. Microbiol*. 66:2682–2684, 2000.
125. Rastall, R.A., V. Maitin. Prebiotics and synbiotics: towards the next generation. *Curr. Opin. Biotechnol*. 13:490–496, 2002.
126. Fuller, R., G.R. Gibson. Modification of the intestinal microflora using probiotics and prebiotics. *Gastroenterology* 32(222):28–31, 1997.
127. Gibson, G.R., E.B. Beatty, X. Wang, J.H. Cummings. Selective stimulation of *Bifidobacteria* in the human colon by oligofructose and inulin. *Gastroenterology* 108:975–982, 1995.
128. Gibson, G.R., A. Willems, S. Reading, M.D. Collins. Fermentation of non-digestible oligosaccharides by human colonic bacteria. *Proc. Nutr. Soc*. 55:899–912, 1996.
129. Roberfroid, M.B. Functional effects of food components and the gastrointestinal system: chicory fructooligosaccharides. *Nutr. Rev*. 54:S38–S42, 1996.
130. Buddington, R.K., C.H. Williams, S.C. Chen, S.A. Witherley. Dietary supplementation of neosugar alters the fecal flora and decreases activities of some reductive enzymes in human subjects. *Am. J. Clin. Nutr*. 63:709–716, 1996.
131. Kleesen, B., B. Sykura, H.J. Zunft, M. Blaut. Effects of inulin and lactose on fecal microflora, microbial activity and bowel habit in elderly constipated persons. *Am. J. Clin. Nutr*. 65:1397–1402, 1997.
132. Gibson, G.R., X. Wang. Regulatory effects of *Bifidobacteria* on the growth of other colonic bacteria. *J. Appl. Bacteriol*. 77:412–420, 1994.
133. Bielecka, M., E. Biedrzycka, A. Majkowska. Selection of probiotics and prebiotics for synbiotics and confirmation of their *in vivo* effectiveness. *Food Res. Int*. 35:125–131, 2002.
134. Sghir, A., J.M. Clow, R.I. Mackie. I: Continuous culture selection of *Bifidobacteria* and lactobacilli from human faecal samples using fructooligosaccharide as selective substrate. *J. Appl. Microbiol*. 85:769–777, 1988.
135. Hartemink, R, M.C.J. Quataert, K.M.J. Van Laere, M.J.R. Nout, F.M. Rombouts. Degradation and fermentation of fructo-oligosaccharides by oral streptococci. *J. Appl. Bacteriol*. 79:551–557, 1995.

136. Losada, M.A., T. Olleros. Towards a healthier diet for the colon: the influences of fructooligosaccharides and lactobacilli on intestinal health. *Nutr. Res.* 22:71–84, 2002.
137. Gmeiner, M., W. Kneifel, K.D. Kulbe, R. Wouters, P. De Boever, L. Nollet, W. Verstraete. Influence of a synbiotic mixture consisting of *Lactobacillus acidophilus* 74-2 and a fructooligosaccharide preparation on the microbial ecology sustained in a simulation of the human intestinal microbial ecosystem (SHIME reactor). *Appl. Microbiol. Biotechnol.* 53:219–223, 2000.
138. Wang, X. Comparative aspects of carbohydrate fermentation by colonic bacteria. PhD dissertation, University of Cambridge, Cambridge, U.K., 1993.
139. Delzenne, N.M., N. Kok. Effects of fructans-type prebiotics on lipid metabolism. *Am. J. Clin. Nutr.* 73:456S–458S, 2001.
140. Stanley, J.C., E.A. Newsholme. The effect of dietary guar gum on the activities of some key enzymes of carbohydrate and lipid metabolism in mouse liver. *Br. J. Nutr.* 53:215–222, 1985.
141. Leclère, C.J., M. Champ, J. Boillot, G. Guille, G. Lecannu, C. Molis, F. Bornet, M. Krempf, J. Delortlaval, J.P. Galmichem. Role of viscous guar gums in lowering the glycemic response after a solid meal. *Am. J. Clin. Nutr.* 59:914–921, 1994.
142. Roberfroid, M.B. Prebiotics and probiotics: are they functional foods? *Am. J. Clin. Nutr.* 71:1682S–1687S, 2000.
143. Ellegard, L., H. Andersson, I. Bosaeus. Inulin and oligofructose do not influence the absorption of cholesterol or the excretion of cholesterol, Ca, Mg, Zn, Fe, or bile acids but increases energy excretion in ileostomy subjects. *Eur. J. Clin. Nutr.* 51:1–5, 1997.
144. Coudray, C., J. Bellanger, C. Castiglia-Delavaud, C. Remesy, M. Vermorel, Y. Rayssignuier. Effect of soluble and partly soluble dietary fibres supplementation on absorption and balance of calcium, magnesium, iron and zinc in healthy young men. *Eur. J. Clin. Nutr.* 51:375–380, 1997.
145. Van den Heuvel, E.G.H.M., T. Muys, W. van Dokkum, G. Schaafsma. Oligofructose stimulates calcium absorption in adolescents. *Am. J. Clin. Nutr.* 69:544–548, 1999.
146. Scholz-Ahrens, K.E., J. van Loo, J. Schrezenmeir. Effect of oligofructose on bone mineralization in ovariectomized rats is affected by dietary calcium. *Am. J. Clin. Nutr.* 73:498S–498S, 2001.
147. Scholz-Ahrens, K.E., G. Schaafsma, E.G.H.M. van den Heuvel, J. Schrezenmeir. Effect of prebiotics on mineral metabolism. *Am. J. Clin. Nutr.* 73:459S–464S, 2001.
148. Ohta, A., M. Ohtsuki, S. Baba, T. Adachi, T. Sakata, E.I. Sagaguchi. Calcium and magnesium absorption from the colon and rectum are increased in rats fed fructooligosaccharides. *J. Nutr.* 125:2417–2424, 1995.
149. Hughes, R., I.R. Rowland. Stimulation of apoptosis by two prebiotic chicory fructans in the rat colon. *Carcinogenesis* 22(1):43–47, 2001.
150. Rao, C.V., D. Chou, B. Simi, H. Ku, B.S. Reddy. Prevention of colonic aberrant crypt foci and modulation of large bowel microbial activity by dietary coffee fiber, inulin and pectin. *Carcinogenesis* 10:1815–1819, 1998.
151. Hague, A., D.J.E. Elder, D.J. Hicks, C. Paraskeva. Apoptosis in colorectal tumor cells: induction by the short chain fatty acids butyrate, propionate and acetate and by the bile salt deoxycholate. *Int. J. Cancer* 60:400–406, 1995.
152. Hague, A., C. Paraskeva. The short chain fatty acid butyrate induces apoptosis in colorectal tumor cell lines. *Eur. J. Cancer Prev.* 4:359–364, 1995.
153. Young, G. Prevention of colon cancer: role of short chain fatty acids produced by intestinal flora. *Asia Pac. J. Clin. Nutr.* 5:44–47, 1996.
154. Matsumoto, K., Y. Kobayashi, N. Tamura, T. Watanabe, T. Kan. Production of galactooligosaccharides with β-galactosidase. *Denpun Kagaku* 36:123–130, 1989.
155. Van der Meer, R., I.M.J. Bovee-Oudenhoven. Dietary modulation of intestinal bacterial infections. *Int. Dairy J.* 8:481–486, 1998.
156. Bouhnik, Y., B. Flourie, L. D'Agay-Abensour, P. Pochart, G. Gramet, M. Durand, C.-J. Rambaud. Administration of transgalacto-oligosaccharides increases fecal *Bifidobacteria* and modifies colonic fermentation metabolism in healthy humans. *J. Nutr.* 127:444–448, 1997.

157. Tanaka, R., H. Takayama, M. Morotomi, T. Kuroshima, S. Ueyama, K. Matsumoto, A. Kuroda, M. Mutai. Effects of administration of TOS and *Bifidobacteria breve* 4006 on the human faecal flora. *Bifidob. Microfl.* 2:17–24, 1983.
158. Ishikawa, F., H. Takayama, K. Matsumoto, M. Ito, O. Chonan, Y. Deguchi, H. Kituchi-Hakayawa, M. Watanuki. Effects of β 1-4 linked galactooligosaccharides on human faecal microflora. *Bifidus* 9:5–18, 1995.
159. Ito, M., Y. Deguchi, K. Matsumoto, M. Kimura, N. Onodera, T. Yajima. Influence of galacto-oligosaccharides on human faecal microflora. *J. Nutr. Sci. Vitaminol.* 39:635–640, 1993.
160. Dumortier, V., C. Brassart, S. Bouquelet. Purification and properties of a β-galactosidase from *Bifidobacterium bifidum* exhibiting a transgalactosylation reaction. *Biotechnol. Appl. Biochem.* 19:341–354, 1994.
161. Gopal, P.K., P.A. Sullivan, J.B. Smart. Utilisation of galacto-oligosaccharides as selective substrates for growth by lactic acid bacteria including *Bifidobacterium lactis* DR10 and *Lactobacillus rhamnosus* DR20. *Int. Dairy J.* 11:19–25, 2001.
162. Van Laere, K.M.J., T. Abee, H.A. Schopls, G. Beldman, A.G.J. Voragen. Characterization of a novel β-galactosidase from *Bifidobacterium adolescentis* DSM 20083 active towards transgalactooligosaccharides. *Appl. Environ. Microbiol.* 66(4):1379–1384, 2001.
163. Tanaka, R., K. Matsumoto. Recent progress on probiotics in Japan, including galacto-oligosaccharides. *Bull. Int. Dairy Fed.* 336:21–27, 1998.
164. Ohtsuka, K., Y. Benno, A. Endo, H. Ueda, O. Ozawa, T. Ulchida, T. Mitsuoka. Effects of 4'galactosyllactose on human intestinal microflora. *Bifidus* 2:143–149, 1989.
165. Deguchi, Y., K. Matsumoto, T. Ito, M. Watanuki. Effects of β1-4 galactooligosaccharides administration on defecation of healthy volunteers with constipation tendency. *Jap. J. Nutr.* 55:13–22, 1997.
166. Tamai, S., K. Ohtsuka, O. Ozawa, T. Ulchida. Effect of a small amount of galactooligosaccharides on fecal *Bifidobacterium*. *J. Jap. Soc. Nutr. Food Sci.* 45:456–460, 1992.
167. Chonan, O., H. Takahashi, H. Yasui, M. Watanuki. Effects of beta 1→4 linked galactooligcsaccharides on use of magnesium and calcification of the kidney and heart in rats fed excess dietary phosphorus and calcium. *Biosci. Biotech. Biochem.* 60:1735–1737, 1996.
168. Asahara, T., K. Nomoto, K. Shimizu, M. Watanuki, R. Tanaka. Increased resistance of mice to *Salmonella enterica serovar Typhimurium* infection by synbiotic administration of *Bifidobacteria* and transgalactosylated oligosaccharides. *J. Appl. Microbiol.* 91:985–996, 2001.
169. Kohmoto, T., F. Fukui, H. Takaku, Y. Machida, M. Arai, T. Mitsuoka. Effect of isomalto-oligosaccharides on human fecal flora. *Bifidob. Microfl.* 7:61–69, 1988.
170. Lee, H.-S., J.-H. Auh, H.-G. Yoon, M.-J. Kim, J.-H. Park, S.-S. Hong, M.-H. Kang, T.-J. Kim, T.-W. Moon, J.-W. Kim, K.-H. Park. Cooperative action of α-glucanotransferase and maltogenic amylase for an improved process of isomaltooligosaccharide (IMO) production. *J. Agric. Food Chem.* 50:2812–2817, 2002.
171. Kaneko, T., Kohmoto, T., H. Kikuchi, M. Shiota, H. Iino, T. Mitsuoka. Effects of isomalto-oligosaccharides with different degrees of polymerisation on human faecal *Bifidobacteria*. *Biosci. Biotech. Biochem.* 58:2288–2290, 1994.
172. Olano-Martin, E., K.C. Mountzouris, G.R. Gibson, R.A. Rastall. *In vitro* fermentability of dextran, oligodextran and maltodextrin by human gut bacteria. *Br. J. Nutr.* 83:247–255, 2000.
173. Tamura, Y., T. Mizota, S. Shimamura, M. Tomita. Lactulose and its application to the food and pharmaceutical industries. *Bull. Int. Dairy Fed.* 289:43–53, 1993.
174. Montgomery, E.M., C.S. Hudson. Relations between rotatory power and structure in the sugar group: XXVII. synthesis of a new disaccharide ketose. *J. Am. Chem. Soc.* 52:2101–2106, 1930.
175. Petuely, F. Bifidusflora bei Flaschenkindern durch bifidogene Substanzen (Bifidusfacktor). *Z. Kinderheilkd.* 79:174–179, 1957.
176. Jeffrey, G.A., D. Huang, P.E. Pfeffer, R.L. Dudley, K.B. Hicks, E. Nitsch. Crystal structure and n.m.r. analysis of lactulose trihydrate. *Carbohydr. Res.* 266:29–42, 1992.

177. Bird, S.P., D. Hewitt, M.I. Gurr. Energy values of lactitol and lactulose as determined with miniature pigs and growing rats. *J. Sci. Food Agric.* 51:233–246, 1990.
178. Matsuda, Y., Y. Iematsu, T. Shibamoto, S. Oyama. Effect of lactulose on acute renal failure in rabbits. *J. Med.* 27:49–54, 1992.
179. Mizota, T. Lactulose as a growth promoting factor for *Bifidobacterium* and its physiological aspects. *Bull. Int. Dairy Fed.* 313:43–48, 1996.
180. Terada, A., H. Hara, M. Kataoka, T. Mitsuoka. Effect of lactulose on the composition and metabolic activity of the human faecal flora. *Microbiol. Ecol. Health Dis.* 5:43–50, 1992.
181. Smart, J.B. Transferase reactions of β-galactosidases: new product opportunities. *Bull. Int. Dairy Fed.* 289:16–22, 1993.
182. Sahota, S.S., P.M. Bramley, I.S. Menzies. The fermentation of lactulose by colonic bacteria. *J. Gen. Microbiol.* 128:319–325, 1982.
183. Liehr, H., W.D. Heine. Treatment of endotoxemia in galactosamine hepatitis by lactulose administered intravenously. *Hepato-Gastroenterology* 28:296–298, 1981.
184. Liehr, H., G. Emglish, U. Rasenack. Lactulose: a drug with antiendotoxin effect. *Hepato-Gastroenterology* 27:356–360, 1980.
185. Iwasaki, M., I. Maruyama, N. Ikejiri, M. Abe, T. Maeda, H. Nagata, H. Abe, K. Tanigawa. Liver disease and endotoxin (II): endotoxin in severe liver lesion. *Jap. J. Gastroenterol.* 77:386–393, 1980.
186. Mortensen, P.B., H.S. Rasmussen, K. Holtug. Lactulose detoxifies *in vitro* short-chain fatty acid production in colonic contents induced by blood: implications for hepatic coma. *Gastroenterology* 94:750–754, 1988.
187. Genovese, S., G. Riccardi, A.A. Rivellese. Lactulose improves blood glucose response to an oral glucose test in non-insulin dependent diabetic patients. *Diabetes Nutr. Metab.* 5:295–297, 1992.
188. Vendemiale, G., G. Palasciano, F. Cirelli, M. Altamura, A. De Vincentis, E. Altomare. Crystalline lactulose in the therapy of hepatic cirrhosis: evaluation of clinical and immunological parameters: preliminary results. *Drug Res.* 42(II):969–972, 1992.
189. Fulton, J.D. Infection limitation with lactulose therapy. *J. Clin. Exp. Gerontol.* 10(3&4):117–124, 1988.
190. Onishi, S., H. Yamazaki, C. Obayashi, Y. Suzuki. Effect of lactulose on ammonia metabolism during exercise. *Resp. Circ.* 38:693–697, 1990.
191. Nagengast, F.M., M.P. Hectors, W.A. Buys, J.H. van Tongeren. Inhibition of secondary bile acid formation in the large intestine by lactulose in healthy subjects of two different age groups. *Eur. J. Clin. Invest.* 18:56–61, 1988.
192. Owen, R.W. Faecal steroids and colorectal carcinogenesis. *Scand. J. Gastroenterol.* 32(222):76–82, 1997.
193. Rooney, P.S., L.M. Hunt, P.A. Clarke, K.A. Gifford, J.D. Hardcastle, N.C. Armitage. Wheat fibre, lactulose and rectal mucosal proliferation in individuals with a family history of colorectal cancer. *Br. J. Surg.* 81:1792–1794, 1994.
194. Roncucci, L., P.D. Donato, L. Carati, A. Ferrari, M. Perini, G. Bertoni, G. Bedogni, B. Paris, F. Svanoni, M. Girola, M.P. Leon. Antioxidant vitamins or lactulose for the prevention of the recurrence of colorectal adenomas. *Dis. Colon Rectum* 36:227–234, 1993.
195. Rowland, I.R., C.A. Bearne, R. Fischer, B.L. Pool-Zobel. Effect of lactulose on DNA damage induced by DMH in the colon of human flora-associated rats. *Nutr. Cancer Prev.* 26:37–47, 1996.
196. Brommage, R., C. Binacua, S. Antille, A.-L. Carriè. Intestinal calcium absorption in rats is stimulated by dietary lactulose and other resistant sugars. *J. Nutr.* 123:2186–2194, 1993.
197. Rèmèsy, C., M.-A. Levrat, L. Gamet, C. Demignè. Cecal fermentations in rats fed oligosaccharides (inulin) are modulated by dietary calcium level. *Am. J. Physiol.* 264:G855–G862, 1993.
198. Suzuki, K., M. Uehara, Y. Endo, S. Goto. Effect of lactose, lactulose and sorbitol on iron, zinc and copper utilization in rats. *J. Jap. Soc Nutr. Food Sci.* 39:217–219, 1986.

199. Igarashi, C., I. Ezawa. Effects of whey calcium and lactulose on the strength of bone in ovariectomized osteoporosis model rats. *Pharmacometrics* 42:245–253, 1991.
200. Van Loo, J., J. Cummings, N. Delzenne, H. Englyst, A. Franck, M. Hopkins, N. Kok, G. Macfarlane, D. Newton, M. Quigley, M. Roberfroid, T. van Vliet, E. van den Heuvel. Functional food properties of non-digestible oligosaccharides: a consensus report from the ENDO project (DGXII AIRII-CT94-1095). *Br. J. Nutr.* 81:121–132, 1999.
201. Wesselius de Casparis, A., S. Braadbaat, G.E. Bergh-Bolken, M. Mimica. Treatment of chronic constipation with lactulose syrup: results of a double-blind study. *Gut* 9:84–86, 1968.
202. Attar, A., M. Lemann, A. Ferguson, M. Halphen, M.C. Boutron, B. Flourie, E. Alix, M. Salmeron, F. Guillemot, S. Chaussade, A.M. Menard, J. Moreau, G. Naudin, M. Barthet. Comparison of a low dose polyethylene glycol electrolyte solution with lactulose for treatment of chronic constipation. *Gut* 44:226–230, 1999.
203. Clausen, M.R., P.B. Mortensen. Disaccharides and colonic flora. Clinical consequences. *Drugs* 53:930–942, 1997.
204. Teuri, U., R. Korpela. Galacto-oligosaccharides relieve constipation in elderly people. *Ann. Nutr. Metab.* 42:319–327, 1998.
205. Marteau, P. Prebiotics and probiotics for gastrointestinal health. *Clin. Nutr.* 20(1):41–45, 2001.
206. Minami, Y., K. Yazawa, Z. Tamura, T. Tanaka, T. Yamamoto. Selectivity of utilisation of galactosyl-oligosaccharides by *Bifidobacteria*. *Chem. Pharm Bull.* 31:1688–1691, 1983.
207. Saito, Y., T. Takano, I.R. Rowland. Effects of soybean oligosaccharides on the human gut microflora in *in vitro* culture. *Microbiol. Ecol. Health Dis.* 5:105–110, 1992.
208. De Boever, P., R. Wouters, W. Verstraete. Combined use of *Lactobacillus reuteri* and soygerm powder as food supplement. *Lett. Appl. Microbiol.* 33:420–424, 2001.
209. De Boever, P., B. Deplancke, W. Verstraete. Fermentation by gut microbiota cultured in a simulator of the human intestinal microbial ecosystem is improved by supplementing a soygerm powder. *J. Nutr.* 130:2599–2606, 2000.
210. Messina, M.J. Legumes and soybeans: overview of their nutritional profiles and health effects. *Am. J. Clin. Nutr.* 70:439S–450S, 1999.
211. Setchell, K.D.R. Phytoestrogens: the biochemistry, physiology, and implications for human health of soy isoflavones. *Am. J. Clin. Nutr.* 68:1333S–1346S, 1998.
212. King, R.A., D.B. Bursill. Plasma and urinary kinetics of the isoflavones daidzein and genistein after a single soy meal in humans. *Am. J. Clin. Nutr.* 67:867–872, 1998.
213. Day, A.J., F.J. Canada, J.C. Diaz, P.A. Kroon, R. Mclauchlan, C. Faulds, G.W. Plumb, M.R.A. Morgan, G. Williamson. Dietary flavonoid and isoflavone glycosides are hydrolysed by the lactase site of lactase phlorizin hydrolase. *FEBS Lett.* 468:166–170, 2000.
214. Harvey, C.B., Y. Wang, L.A. Highes, D.M. Swallow, W.P. Thurrell, V.R. Sams, R. Barton, S. Lanzomiller, M. Sarner. Studies on the expression of intestinal lactase in different individuals. *Gut* 36:28–33, 1995.
215. Izumi, T., M.K. Piskula, S. Osawa, A. Obata, K. Tobe, M. Saito, S. Kataoka, Y. Kubota, M. Kikuchi. Soy isoflavone aglycones are absorbed faster and in higher amounts than their glycosides in humans. *J. Nutr.* 130:1695–1699, 2000.
216. Brzezinski, A., A. Debi. Phytoestrogens: the "natural" selective estrogen receptor modulators? *Eur. J. Obstet. Gynaecol. Reprod. Biol.* 85:47–51, 1999.
217. Molteni, A., L. Brizio-Molteni, V. Persky. *In vitro* hormonal effects of soybeans isoflavones. *J. Nutr.* 125:751S-756S, 1995.
218. Slavin, J.L., M.C. Martini, D.R. Jacobs, L. Marquart. Plausible mechanims for the protectiveness of whole grains. *Am. J. Clin. Nutr.* 70:459S–463S, 1999.
219. Suarez, F.L., J. Springfield, J.K. Furne, T.T. Lohrmann, P.S. Kerr, M.D. Levitt. Gas production in humans ingesting a soybean flour derived from beans naturally low in oligosaccharides. *Am. J. Clin. Nutr.* 69:135–139, 1999.
220. Hayakawa, K., J. Mizutani, K. Wada, T. Masai, I. Yoshihara, T. Mitsuoka. Effect of soybean oligosaccharides on human fecal flora. *Microb. Ecol. Health Dis.* 3:293–303, 1990.

221. S. Watanabe, M. Yamaguchi, T. Sobue, T. Takahashi, T. Miura, Y. Arai, W. Mazur, K. Wahala, H. Aldercreutz. Pharmakokinetics of soybean isoflavones in plasma, urine and feces of men after ingestion of 60 g baked soybean powder (kinako). *J. Nutr.* 128:1710–1715, 1998.
222. Zhang, Y., G.J. Wang, T.T. Song, P.A. Murphy, S. Hendrich. Urinary disposition of the soybean isoflavones daidzein, genistein and glycitein differs among humans with moderate isoflavone degradation activity. *J. Nutr.* 129:957–962, 1999.
223. Cherbut, C., A.C. Aubè, H.M. Blottière, J.P. Glamiche. Effects of short-chain fatty acids on gastrointestinal motility. *Scand. J. Gastroenterol.* 32:52–57, 1997.
224. Scheppach, W., H.P. Bartram, F. Richter. Role of short-chain fatty acids in the prevention of colorectal cancer. *Eur. J. Cancer* 31(7-8):1077–1080, 1995.
225. Ichikawa, H., T. Sakata. Stimulation of epithelial cell proliferation of isolated distal colon of rats by continuous colonic infusion of ammonia or short-chain fatty acids is nonadditive. *J. Nutr.* 128:843–847, 1998.
226. Christl, S.U., P.R. Murgatroyd, G.R. Gibson, J.H. Cummings. Production, metabolism and excretion of hydrogen in the large intestine. *Gastroenterology* 102:1269–1277, 1992.
227. Macfarlane, G.T., S. Macfarlane. Human colonic microbiota: ecology, physiology and metabolic potential of intestinal bacteria. *Scand. J. Gastroenterol.* 32:3–9, 1997.
228. Gibson, G.R., P.B. Ottaway, R.A. Rastall. *Prebiotics*. Oxford: Chandos Publishing, 2000.
229. Pilgrim, A., M. Kawase, M. Ohashi, K. Fujita, K. Murakami, K. Hashimoto. Reaction kinetics and modeling of the enzyme-catalysed production of lactosucrose using β-fructofuranosidase from *Arthrobacter* sp. K-1. *Biosci. Biotech. Biochem.* 65(4):758–765, 2001.
230. Fujita, K., K. Hara, S. Sakai, T. Miyake, M. Yamashita, Y. Tsunetomi, T. Mitsuoka. Effect of 4^G-β-D-galactosylsucrose (lactosucrose) on intestinal flora and its digestibility in humans. *Denpun Kagaku* 38:249–255, 1991.
231. Hopkins, M.J., J.H. Cummings, G.T. Macfarlane. Interspecies difference in maximum specific growth rates and cell yields of *Bifidobacteria* cultured on oligosaccharides and other simple carbohydrate sources. *J. Appl. Microbiol.* 85:381–386, 1998.
232. Vasquez, M.J., J.L. Alonso, H. Dominguez, J.C. Parajo. Xylooligosaccharides: manufacture and applications. *Trends Food Sci. Technol.* 11:387–393, 2000.
233. Campbell, J.M., L.L. Bauer, G.C. Fahey, A.J.C.L. Hogarth, B.W. Wolf, D.E. Humter. Selected fructooligosaccharide (1-kestose, nystose, and 1^F-β-fructofuranosylnystose) composition of foods and feeds. *J. Agric. Food Chem.* 45:3076–3082, 1997.
234. Jaskari, J., P. Kontula, A. Siitonen, H. Jousimies-Somer, T. Mattila-Sandholm, K. Poutanen. Oat β-glucan and xylan hydrolysates as selective substrates for *Bifidobacterium* and *Lactobacillus* strains. *Appl. Microbiol. Biotech.* 49:175–181, 1998.
235. Playne, M.J., R. Crittenden. Commercially available oligosaccharides. *Bull. Int. Dairy Fed.* 313:10–12, 1996.
236. Yamada, H., K. Itoh, Y. Morishita, H. Taniguchi. Structure and properties of oligosaccharides from wheat bran. *Cereal Food World* 38:490–492, 1993.
237. Okazaki, M., S. Fujikawa, N. Matsumoto. Effects of xylooligosaccharide on growth of *Bifidobacteria*. *J. Jap. Soc. Nutr. Food Sci.* 43:395–401, 1990.
238. Korakli, M., M.G. Ganzie, R.F. Vogel. Metabolism by *Bifidobacteria* and lactic acid bacteria on polysaccharides from wheat and rye, and exopolysaccharides produced by *Lactobacillus sanfranciscensis*. *J. Appl. Microbiol.* 92:958–965, 2002.
239. Crociani, F., A. Alessandrini, M.M. Mucci, B. Biavati. Degradation of complex carbohydrates by *Bifidobacterium* spp. *Int. J. Food Microbiol.* 24:199–210, 1994.
240. Van Laere, K.M.J., R. Hartemink, M. Bosveld, H.A. Schols, A.G.J. Voragen. Fermentation of plant cell wall derived polysaccharides and their corresponding oligosaccharides by intestinal bacteria. *J. Agric. Food Chem.* 48:1644–1652, 2000.
241. Kummel, K.F., S. Brokx. Lactitol as a functional prebiotic. *Cereal Food World* 46(9):424–429, 2001.
242. Scevola, D., G. Bottari, L. Oberto, V. Monzillo. Intestinal bacterial toxins and alcohol liver damage. Effects of lactitol, a synthetic disaccharide. *Clin. Dietol.* 20:297–297, 1993.

243. Dal Bello, F., J. Walter, C. Hertel, W.P. Hammes. *In vitro* study of prebiotic properties of levan-type exopolysaccharides from lactobacilli and non-digestible carbohydrates using denaturing gradient gel electrophoresis. *Syst. Appl. Microbiol.* 24:232–237, 2001.
244. Belitz, H.D., W. Grosch. *Food Chemistry*. Berlin: Springer-Verlag, 1999.
245. Rycroft, C.E., M.R. Jones, G.R. Gibson, R.A. Rastall. Fermentation properties of gentiooligosaccharides. *Lett. Appl. Microbiol.* 32:156–161, 2001.
246. Berggren, A.M., I.M.E. Bjorck, E.M.G.L. Nyman, B.O. Eggum. Short-chain fatty acid content and pH in cecum of rats given various sources of carbohydrates. *J. Sci. Food Agric.* 63:397–406, 1993.
247. Michel, C., C. Benard, M. Lahaye, D. Formaglio, B. Kaeffer, B. Quemener, S. Berot, J.C. Yvin, H.M. Blottiere, C. Cherbut. Algal oligosaccharides as functional foods: *in vitro* study of their cellular and fermentative effects. *Sci. Aliment.* 19:311–332, 1999.
248. Flickinger, E.A., B.W. Wolf, K.A. Garleb, J. Chow, G.J. Leyer, P.W. Johns, G.C. Fahey. Glucose-based oligosaccharides exhibit different *in vitro* fermentation patterns and affect *in vivo* apparent nutrient digestibility and microbial populations in dogs. *J. Nutr.* 130:1267–1273, 2000.
249. Campbell, J.M., G.C. Fahey Jr., B.W. Wolf. Selected indigestible oligosaccharides affect large bowel mass, cecal and fecal short-chain fatty acids, pH and microflora in rats. *J. Nutr.* 127:130–136, 1997.
250. Djouzi, Z., C. Andrieux. Compared effects of three oligosaccharides on metabolism of intestinal microflora of rats inoculated with a human fecal flora. *Br. J. Nutr.* 78:313–324, 1997.
251. Yazawa, K., K. Imai, Z. Tamura. Oligosaccharides and polysaccharides specifically utilisable by *Bifidobacteria*. *Chem. Pharm. Bull.* 26:3306–3311, 1978.
252. Macfarlane, G.T., S. Macfarlane, G.R. Gibson. Validation of a three-stage compound continuous culture system for investigating the effect of retention time on the ecology and metabolism of bacteria in the human colon. *Microb. Ecol.* 35:180–187, 1998.
253. Kohmoto, T., F. Fukui, H. Takaku, T. Mitsuoka. Dose-response test of isomalto-oligosaccharides for increasing faecal *Bifidobacteria*. *Agric. Biol. Chem.* 55:2157–2159, 1991.
254. Djouzi, Z., C. Andrieux, V. Pelenc, S. Somarriba, F. Popot, F. Paul, P. Monsan, O. Szylit. Degradation and fermentation of α-gluco-oligosaccharides by bacterial strains from human colon: *in vitro* and *in vivo* studies in gnotobiotic rats. *J. Appl. Bacteriol.* 79:117–127, 1995.
255. Ito, M., Y. Deguchi, A. Miyamori, K. Matsumoto, H. Kikuchi, K. Matsumoto, Y. Kobayashi, T. Yajima, T. Kan. Effects of administration of galactooligosaccharides on the human fecal microflora, stool weight and abdominal sensation. *Microb. Ecol. Health Dis.* 3:285–292, 1990.
256. Ito, M., M. Kimura, Y. Deguchi, A. Miyamori-Watabe, T. Yajima, T. Kan. Effects of transgalactosylated disaccharides on the human intestinal microflora and their metabolism. *J. Nutr. Sci. Vitaminol.* 39:279–288, 1993.
257. Benno, Y., K. Endo, N. Shiragami, K. Sayama, T. Mitsuoka. Effects of raffinose intake on human faecal microflora. *Bifidob. Microfl.* 6:59–63, 1987.
258. Wada, K., J. Watabe, J. Mizutani, M. Tomoda, H. Suzuki, Y. Saitoh. Effects of soybean oligosaccharides in a beverage on human faecal flora and metabolites. *J. Agric. Chem. Soc. Japan* 66:127–135, 1992.
259. Rycroft, C.E. A comparative *in vitro* evaluation of the fermentation properties of potential prebiotic food ingredients: investigating structure-function relationships. PhD dissertation, University of Reading, Reading, UK, 2001.

2.25

Immunomodulating Effects of Lactic Acid Bacteria

Hanne Risager Christensen and Hanne Frøkiær

CONTENTS

25.1 INTRODUCTION

For decades, it has been believed that consumption of fermented dairy products is beneficial for health. In 1907, Eli Metchnikoff, the Nobel Prize winner and director of the Pasteur Institute, postulated in his book *The Prolongation of Life* that eating yogurt would extend the life span due to the health-improving influence of the lactic acid bacteria (LAB) involved in yogurt fermentation on the gut flora. Since that time, many studies have been conducted on gut-associated microorganisms, of which LAB are important members (1–3). These studies have led to the growing recognition today that the microflora colonizing the gastrointestinal tract of man and animal indeed play an important and essential role for the health of the host.

The gut flora comprises a highly evolved and extremely complex microbial ecosystem of hundreds of different microbial species. This metabolically highly active society is in close proximity with the gut epithelia through which it provides an array of signals and products that affect homeostasis of various physiologic processes of the body. Of these processes, the immune function constitutes a very important part; stimulating signals from the gut bacteria are critical for the development and maintenance of the mucosal and systemic immune compartments (4,5). A corollary, which has collected increasing evidence of support, is that the panel of stimulating signals bestowed by the gut flora is dynamically affected by the microbial composition, with some strains exerting stronger bioactivities than others (6–8). The gut flora are composed of a group of resident (autochthonous) bacterial species, and a variable group of transient (allochthonous) species that temporarily fills in functional gaps of the flora. In view of these facts, the concept of manipulating the composition of the gut flora in attempt to improve health of the host has emerged. That is, an increase in numbers but also potency of specific bacterial groups possessing particular health-promoting properties is desirable. One approach much investigated for this purpose is oral supplementation; e.g., via functional food products, with so called "probiotic" microorganisms that supply the gut flora with niches of beneficial functionality. The word probiotic is derived from the Greek meaning "for life." The definition of probiotics has changed over time; in 1989, Fuller defined probiotics as, "a live microbial food supplement which beneficially affects the host by improving the intestinal microbial balance" (9). However, to acknowledge emerging scientific data on proven effects, a broadening of the definition was recently proposed: "Probiotics are microbial cell preparations or components of microbial cells that have a beneficial effect on the health and well-being of the host" (10). In addition to the original concept of probiotics as acting solely via improvements of imbalances of the gut flora as a whole, this newly proposed definition embraces the bioactive effects of microbial compounds, not necessarily within the viable cell, but which are active also as nonviable whole cell or even as fractionated preparations. This broader concept gains ground in the recognition that supplementation of microbial preparations apart from viable cultures in some instances exerts a health promoting effect. Moreover, probiotic effects, e.g., on immune parameters, are not necessarily mediated via a preceding intervention with the gut flora balance (6,11,12).

The probiotic microorganisms currently used are predominantly, though not exclusively, lactic acid bacteria (LAB) such as lactobacilli or bifidobacteria (13). LAB are major components of the gut flora (1–3). Research has revealed these bacteria are far from just inert commensals, but may constitute one of the most significant microbial populations providing immunoregulatory signals (7,14–16).

In this chapter, we will discuss the current knowledge of how the gut flora as a whole affects the immune system, and then focus on the immunomodulating effects of LAB present either as a natural part of the gut flora or supplemented as probiotics. However, to provide a basis for understanding the functional influence conferred by the gut flora on the immune system, we will begin by giving a view of the organization and function of the intestinal immune system and its affiliation to the peripheral immune system.

25.2 THE GUT MUCOSA – A UNIQUE COMPARTMENT OF THE IMMUNE SYSTEM

The gut mucosal membranes form the interface between the highly controlled internal microenvironment of the body (basolateral side) and the external environment (gut lumen, apical side). To allow efficient absorption of nutrients, the surface area of the mucosal membranes is enormous, reaching around 400 m^2, which exceeds by far that of the skin (~2 m^2) (17). Along this huge area, the gut mucosa continually experiences a massive exposure to both pathogenic and harmless agents derived from the ingested food as well as the gut flora.

The immune compartment present in the gut mucosa — the gut associated lymphoid tissue (GALT) — is recognized as the largest immune organ of the body and plays a crucial role for keeping the body healthy. GALT constitutes a major part in the gut barrier function against pathogens and other harmful antigens entering the body through the gut. Equally important but oppositely directed, GALT is important in down regulating responses toward ingested harmless food antigens to permit entrance of nutritional food antigens into the body – a mechanism designated oral tolerance (18). In keeping with all this, GALT is not only responsible for local immune responses but also for systemic responses via primed immune cells that exit GALT and enter into peripheral circulation. As already touched on, appropriate development and balance maintenance of these rather distinct functions of GALT is affected by the microenvironment of the gut, most importantly the gut flora, including LAB, which is the core subject of the present chapter.

25.2.1 Organization of GALT

GALT is compartmentalized into inductive sites and effector sites; i.e., sites where priming and dissemination, respectively, of an immune response primarily takes place (17). This division is far from absolute, but is, however, very useful in understanding GALT functioning. GALT comprises the highly organized structures of the Peyer's patches and the mesenteric lymph nodes, and the more diffuse tissues consisting of a large number of lymphoid cells scattered throughout the lamina propria and intestinal epithelium (intraepithelial lymphocytes) (Figure 25.1).

Peyer's patches and the mesenteric lymph nodes are thought to be the main inductive sites and are found in all mammals. Peyer's patches are located primarily on the antimesenteric side of the small intestine. They contain a number of B cell rich zones with germinal centers called follicles, and, adjacent to these in the parafollicular areas, there are T cell rich zones. The subepithelial dome area overlying the follicles contains many antigen presenting cells, primarily dendritic cells. This area is extremely important for antigen processing and presentation for the initiation of an immune response such as IgA production. The dome area is covered by a follicle associated epithelium facing the gut lumen. In addition to conventional enterocytes, the follicle associated epithelium contains the very specialized microfold (M) cells. M cells lack microvilli, exhibit reduced enzymatic activity, and are adept at uptake and transport of intact luminal antigens (such as soluble food antigens, microorganisms, and viruses) into the Peyer's patch for further immunologic processing. M cells are particularly efficient in taking up particulate material such as microorganisms; in fact, particulate material seems to be almost exclusively taken up by M cells, making these cells the only portal for such antigens. Lymphoid structures similar to Peyer's patches are present also in the large intestine, particularly in the appendix and cecum. They consist only of isolated lymphoid follicles but are believed to function in a manner similar to Peyer's patches.

The mesenteric lymph nodes serve as secondary lymph nodes (Figure 25.1). Located in the center of the mesenteric tissue, they assemble all of the lymphatic vessels draining the gut mucosa as afferent (ingoing) lymphatic vessels into the nodes. Their efferent (outgoing) lymphatic vessels lead the lymph fluid to peripheral circulation via the thoracic duct. The mesenteric lymph nodes, therefore, seem to act as a crossover point between the peripheral and mucosal immune response.

The main effector site of intestinal immune responses is the lamina propria, which is the inner part of the intestinal villi (Figure 25.1). The lamina propria contains a large and heterogeneous group of scattered immune cells, including antigen presenting cells (dendritic cells and macrophages) and T and B cells highly differentiated into mature effector

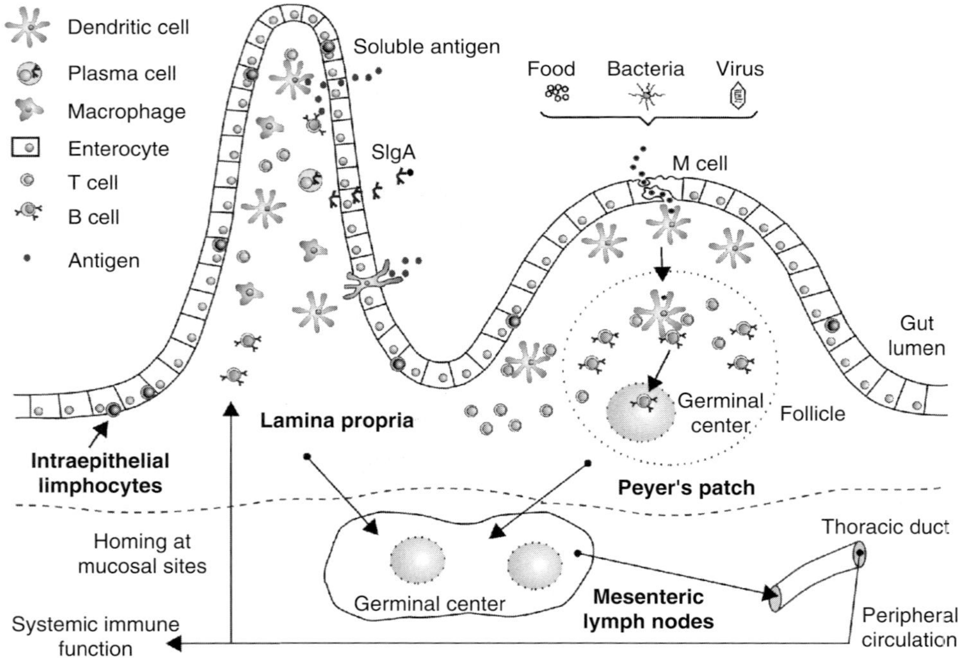

Figure 25.1 Schematic depiction of the gut-associated lymphoid tissue (GALT). The Peyer's patches and mesenteric lymph nodes are organized lymphoid tissues, while scattered immune cells are abundant in lamina propria and inside the epithelium (intraepithelial lymphocytes). Luminal antigen can enter either via M cells (primarily particulate antigen) overlying the Peyer's patch, through the enterocytes (soluble antigen) or potentially via capture by traversing dendritic cells. Immune cells travel from Peyer's patches and lamina propria to mesenteric lymph nodes from where they can enter the blood stream via the thoracic duct. Upon peripheral circulation, some immune cells, particularly IgA$^+$ B cells, home at effector sites primarily lamina propria.

cells, as will be discussed later. Many of the cells are primed in the Peyer's patches from where they migrate out and traffic through peripheral circulation to the lamina propria to exert immune responses. The most predominant type of response of the lamina propria is IgA production by plasma cells (differentiated B cells). Despite being a major effector site, certain immune responses may be induced in the lamina propria as well. Likewise, certain effector functions may disseminate in the Peyer's patches and mesenteric lymph nodes.

The important compartment of the intraepithelial lymphocytes can neither be grouped exclusively into an inductive or an effector site. The intraepithelial lymphocytes represent one of the largest lymphoid populations in the body and consist mainly of $CD8^+$ (cytotoxic) and $CD4^+$ (helper) T cells with different T cell receptor (TCR) phenotypes ($\alpha\beta$ TCR or $\gamma\delta$ TCR). The proximity of the intraepithelial lymphocytes to the enterocytes and the gut lumen makes them one of the first cellular mucosal immune components to respond to luminal antigens. They perform an effector function by exerting cytolytic activity, playing a role in the barrier function against pathogens; and an inductive function, as they seem to play a role in the induction of oral tolerance toward food antigens. Moreover, they appear very important in regulation of immune homeostasis including regulation of IgA production in the lamina propria.

As a second point of entry besides M cells, antigens can enter GALT through a population of intraepithelial dendritic cells (Figure 25.1) (19). Dendritic cells can take up and process antigens, and *in vitro* models suggest that dendritic cells can extend their dendrites between the enterocytes and into the lumen without disrupting the tight junctions that bind the enterocytes together (20).

Finally, it is important for the overview of GALT to mention the enterocytes. Besides their principal role as cells absorbing degraded luminal nutrients, enterocytes are also capable of transporting soluble intact antigens and thus make up a third point of antigen entry (21). Also, an increasing amount of data shows that these cells participate significantly in the initiation and regulation of immune responses via interaction with GALT, especially the intraepithelial cells (22). Enterocytes moreover, serve to transport IgA into the gut lumen.

25.2.2 Immune Responses of GALT and Peripheral Immune System

In GALT, the various immunocompetent cells such as epithelial cells, dendritic cells, T cells, B cells, and macrophages form a unique homeostatic immune network. During initiation of an immune response, the functionally distinct cells of GALT act coordinately to provide stimuli and share common regulatory elements. These processes are characterized by the production of a highly controlled set of molecules like cytokines, chemokines, and costimulatory molecules that alert the host to breach in the mucosal barrier and focus the immune response at the site of infection. Cytokines are soluble intercellular signaling molecules of principal importance for immune cell communication. They act to stimulate or suppress cell functions; and the chemokines, which constitute a cytokine subgroup, act specifically to attract immune cells.

The gastrointestinal surfaces, like the rest of the body, are protected against pathogens by both antigen nonspecific and specific mechanisms. The specific mechanisms implement immune cells committed specifically to act against a given antigen. The nonspecific mechanisms act without proceeding antigen specific commitment and include both nonimmune mechanisms as well as mechanisms exerted by immune cells. Thus, the nonspecific protective mechanisms are constantly in place to take action instantaneously, and therefore serve as the first line of defense. In contrast, the specific immune response takes time to establish, but afterward exerts a much more precisely targeted and effective response, especially toward reencountered antigens.

The gut flora and probiotic LAB affect both nonspecific and specific immune mechanisms. However, up till now most of the knowledge about the immune modulating effects of LAB was limited to only a few components of these immune mechanisms. To understand the relevance of such effects, an outline of the pertinent mechanisms of the nonspecific and specific immune system, particularly in the gut, is given in what follows.

25.2.2.1 Nonspecific Response Mechanisms

Most macromolecules are prevented from direct contact with the epithelial surface by mechanisms belonging to the nonimmune part of the nonspecific defense. The enterocytes have a folded apical surface of rigid structures (microvilli) that are coated with a thick layer of membrane anchored, negatively charged, mucin-like glycoproteins called glycocalyx. This layer forms a diffusion barrier and a highly degradative microenvironment. Bactericidal compounds such as lactoferrin, peroxidase, lysozyme, and defensins are produced by Paneth cells also situated in the epithelium. Another important nonimmune defense mechanism is the capability of the gut flora to compete with pathogens for nutrients and adhesion receptors and, moreover, to produce antimicrobial substances and acids that inhibit pathogenic colonization (23). These mechanisms are collectively called colonization resistance.

Interaction of microorganisms, such as pathogens, with epithelial cells, including enterocytes, triggers the epithelial cells to secrete factors, like cytokines and growth factors, that activate nonspecific mechanisms of the immune defense. The production of chemokines such as interleukin-8 (IL-8), monocyte chemoattractant protein-1 (MCP-1) and proinflammatory cytokines like IL-1β and tumor necrosis factor-α (TNF-α) by epithelial cells, stimulates activation and migration of phagocytic cells such as macrophages and neutrophils to the site of contact (infection) (24). Phagocytic cells bind and internalize antigens (a process called phagocytosis) through a repertoire of receptors, e.g., the antibody receptors (FcγRI, FcγRII, FcγRIII, FcαR and FcεRI), complement receptors (CR1 and CR3), and mannose receptor (MR). When phagocytosed, the microorganisms are killed by bactericidal compounds such as lysozyme or reactive oxygen species (ROS) — a process called the oxidative burst. Motile dendritic cells in Peyer's patches, lamina propria or within the epithelium are also capable of phagocytosing antigen, although much less efficiently. Upon stimulation by most pathogens, secretion of the cytokine IL-12 by macrophages and dendritic cells enhances cytotoxicity of natural killer (NK) cells. NK cells are another important part of the nonspecific immune system.

In addition to the importance for the initiation of the nonspecific immune defense, dendritic cells are of particular importance for the induction of a specific immune response. In this way, dendritic cells create a link between the nonspecific and the specific immune systems (25).

25.2.2.2 Specific Response Mechanisms – Immunity vs. Tolerance

An impressive feature of the protective mechanisms of GALT is the capacity to discriminate between harmless antigens of the diet or resident microflora, and antigens from potential pathogens. The outcome of exposure to foreign antigens can involve three types of antigen-specific mechanisms: mucosally induced tolerance, also termed oral tolerance; induction of mucosal IgA response; and induction of systemic response (18). It is essential for normal life that the intestinal immune system be exquisitely regulated to mobilize the right type of response to any encountered antigen; i.e., induction of tolerance versus immunity.

The mechanisms regulating the distinction between these responses in the intestine are far from completely understood. It is clear, however, that they depend on establishment and maintenance of homeostasis among the immunocompetent cells. It is increasingly believed that under well established homeostasis, antigenic molecules such as pure soluble antigens encountered by GALT are not intrinsically immunogenic but rather induce tolerance as the default response. In contrast to such "naked" antigens, antigens that are endowed with "danger" signals, such as microorganisms bearing pathogenic structures, drive the immune response away from the default response and instead induce active immunity (26). During imbalances in immune homeostasis, the immune system may, however, become either hyporesponsive to such danger signals and hence increase the propensity for development of infections or cancer; or it may become hyperresponsive, resulting in misinterpretations of weak signals from harmless antigens as being danger signals, which is the case for organ specific autoimmune diseases (reactions against "self" antigens) and hypersensitivity reactions such as allergy (e.g., reactions against food antigens).

When GALT, or other lymphoid tissues, encounter foreign antigens, via transport through M cells or enterocytes, the antigen is processed by professional antigen-presenting cells (macrophages or dendritic cells) (24). By binding to surface major histocompatibily molecules (MHC), antigen fractions are presented to T cells. T cells, primarily Thelper (Th) cells, that by chance have a receptor (TCR) specific to the antigen, bind to the antigen-presenting cells. During the subsequent intimate cell–cell interaction involving the MHC-antigen-TCR complex, costimulatory molecules, and cytokines, the T cells become activated and polarized to exert specific effector functions.

Activation and polarization of naïve Th cells are crucial for the type of immune response generated. Dendritic cells are the principal stimulators of naïve Th cells and, thus, the gatekeepers of an immune response (27). They perceive stimulating signals from microorganisms (pathogens and gut flora) through a panel of surface receptors (pattern recognition receptors) such as Toll-like receptors that recognize conserved structural patterns on the microorganisms; including lipopolysaccharide on Gram-negative bacteria, peptidoglycans on Gram-positive bacteria, and specific immunostimulatory DNA sequences (e.g., unmethylated CpG), carbohydrates, and glycolipids, occurring in many microbes (28). The dendritic cells, moreover, perceive signals such as cytokines coming from the epithelium and resident effector cells. Depending on the specific combination and timing of perceived signals, the dendritic cells mature and respond differentially whereby they control the type of immune response to be initiated (29–31). Dendritic cells accomplish this by driving differential polarization of naïve Th cells into active mature effector cells, producing different sets of cytokines that direct different types of immune responses as depicted in Figure 25.2. Three functionally discrete groups of T cells characterized by different cytokine profiles exist. Type 1 Th cells (Th1 cells) are characterized by secretion of IL-2 and IFN-γ, support cell-mediated immunity (cytotoxic T lymphocyte activation) and are immunopathologically associated with autoimmune diseases. In the gut, diseases like Crohn's disease are associated with a deleterious Th1 cell-driven response toward certain species of the gut flora. Th1 cells are induced in response to IL-12 produced by dendritic cells.

Type 2 Th cells (Th2 cells) produce IL-4, IL-5, IL-9, and IL-13 and support antibody production, and are immunopathologically associated with allergy such as food allergy, where a humoral IgE response is generated. A primary factor produced by dendritic cells for the initiation of Th2 cells is not identified, but absence of IL-12 and early presence of IL-4 (not produced by the dendritic cells) is important (32).

In contrast to Th1 and Th2 cells, mediating immunity, the third subpopulation of T cells, termed regulatory T (T_R) cells, are specialized in down regulating (suppressing) immune responses. T_R cells include two subtypes: Th3 and Tr1 cells, both of which are important effector cells in down regulating Th1 and Th2 cell activity and therefore play a central role in tolerance mechanisms and maintenance of immunologic homeostasis (33,34).

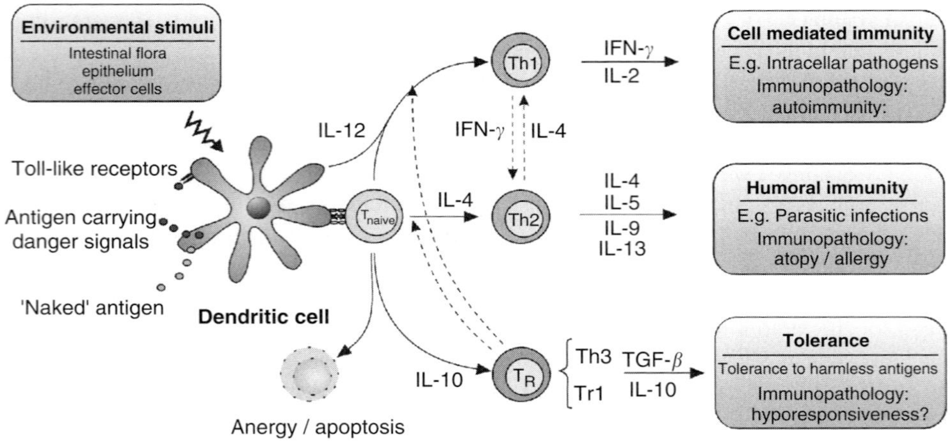

Figure 25.2 Schematic representation of polarized patterns of cytokine production and effector functions of differentiated T helper (Th) cells upon activation by dendritic cells. See text for further explanation.

T_R cells exert their immunoregulatory effect through expression of inhibitory cell surface molecules, or by secretion of immunoregulatory cytokines, most importantly TGF-β and IL-10, or both. Another response suppressive mechanism of the immune system is elimination of potential reactive (i.e., antigen specific) T cells either by "paralyzing" them (anergy) or by inducing apoptosis (deletion).

Many mechanistic details in the regulation of priming (Th1 or Th2) versus tolerance (T_R) are still incompletely understood. Induction of T_R cells or specific T cell anergy or deletion are the major mechanisms involved in the induction of oral tolerance, which depends on the dose of ingested antigen (35). An important prerequisite for tolerance induction is a preserved integrity of the mucosal barrier to avoid excessive antigen leakage into the mucosa (36). Even so, uptake and processing of intact soluble antigen by enterocytes are increasingly believed to play a major role in the induction of oral tolerance. This includes both presentation by the enterocytes themselves (MHC or CD1d molecules) to intraepithelial or lamina propria lymphocytes as well as basolateral releases of exosomes containing MHC-antigen complexes to be taken up by other cells, such as dendritic cells, present locally or at distant sites. Shortly after intake of antigen, antigen-loaded dendritic cells appear in Peyer's patches, lamina propria, and mesenteric lymph nodes, and whether tolerance or priming is the eventual outcome, antigen-specific activation of T cells takes place both in GALT and in the periphery soon after ingestion of antigen (37). Therefore, dissemination of activating or activated immune cells, or antigen bound to MHC or not, from inductive sites of GALT into peripheral tissues as well as into mucosal effector sites, creates an important link between GALT and the peripheral immune system.

It has become evident that enterocytes are not only capable of processing antigens, but, in fact, although they are not immune cells, they can respond actively to contact with luminal antigens, particular luminal bacteria such as LAB or pathogens (38). Enterocytes, like immune cells, express Toll-like receptors, through which they are stimulated by the gut flora and pathogens. The enterocytes respond by secreting a wide range of factors that affect intestinal immune cells, including macrophages, dendritic cells, and intraepithelial lymphocytes, which in turn influence the enterocytes by altering their functionality (39,40). In this way, the enterocytes, representing a predominant cell population, are intimately involved in the immunologic homeostasis and defense against pathogens, and also regulation of IgA production (41). As discussed later, several recent studies have focused on how the gut flora and LAB affect cellular interactions of enterocytes and intraepithelial cells and the resultant signal generation, important for regulation of tolerance vs. immunity. Impact on these interactions might be a key gate for the immunomodulating effect of LAB.

A last important hallmark of mucosal immunity is the secretory IgA (SIgA) response, which is neither a Th1, Th2, nor a T_R response. IgA is overwhelmingly the most important immunoglobulin in the intestine and other mucosal surfaces. Peyer's patches and mesenteric lymph nodes are important inductive sites for B cells to be committed for IgA production (Figure 25.1) (42). Antigen-loaded dendritic cells in Peyer's patches stimulate activation of B cells both through direct interaction with the B cells and via Th cell activation (IL-4-producing Th2 cells and TGF-β-producing Th3 cells). T cells colocalize with germinal center B cells, which are subsequently undergoing isotype switching from IgM to IgA. This occurs either in the Peyer's patch follicles or via cell migration to follicles of the mesenteric lymph nodes. Antigen-specific activated T and B cells then leave the inductive sites and migrate via the thoracic duct into peripheral circulation. Migrating cells finally home to mucosal effector sites, e.g., intestinal lamina propria or other mucosal membranes, where B cells, upon antigenic restimulation, complete their differentiation into IgA secreting plasma cells.

Production of IgA in lamina propria depends on the presence of IL-5, IL-6, IL-10 and TGF-β produced by effector T cells (43). Plasma cells synthesize IgA as a dimer of two IgA molecules interconnected by a J chain (44). Dimeric IgA is a ligand for the polymeric Ig receptor located on the basolateral surface of enterocytes. Here, bound IgA is taken up and transported across the cytoplasm to be released into the gut lumen by cleavage of the poly-Ig receptor molecule, leaving a so called secretory piece still bound to the IgA molecule. Due to the secretory piece, SIgA is resistant to the highly proteolytic environment in the gut.

The presence of SIgA in the gut lumen prevents infection of epithelial cells by coating pathogenic microorganism or virus and so forestalls binding to the cell surface. SIgA can likewise prevent absorption of antigen. By mediating transport of antigen across the

Table 25.1
Characteristics for Classical Immune Responses

Type of Response		Characteristics
Nonspecific responses		
Nonimmune:	Glycocalyx layer	Diffusion barrier
	Bactericidal compounds	Lysozyme, lactoferrin, defensins
	Gut flora	Competition/antimicrobial
Immune:	Phagocytosis	Neutrophils, macrophages, dendritic cells, IL-8[†], ROS
Specific responses		
Cell-mediated response		Th1, cytotoxic T cells, IL-1, IL-2, IL-12, IFN-γ, TNF-α, ROS
Systemic antibody response		Th2 and B cells, IgG, IgE, IL-4, IL-5, IL-9, IL-13
Tolerance		T_R (Th3, Tr1) cells suppressing Th1/Th2 cells, IL-10, TGF-β
Mucosal IgA response		(Th2), Th3 and B cells, IgA, TGF-β, IL-5, IL-6

[†] Abbreviations: IL: interleukin, IFN: interferon, TNF: tumor necrosis factor, ROS: reactive oxygen species, TGF: transforming growth factor

epithelium, SIgA can also neutralize, and carry back into the lumen, antigen that has leaked through the epithelial barrier. In this way, SIgA prevents pathogenic antigens from coming into contact with the body – a mechanism called immune exclusion. Due to the huge area of the gastrointestinal tract, great amounts of SIgA are necessary in performing its task. In man, approximately 3 grams of SIgA are delivered each day into the intestinal lumen; and 80–90% of all immunoglobulin-producing B cells (plasma cells) of the body, in fact, reside in the mucosa and exocrine glands (42,44).

Essential characteristics for these various immune responses, which are often subjects for measurement in evaluating immune modulation, are summarized in Table 25.1.

25.3 INFLUENCE OF THE GUT FLORA ON THE IMMUNE FUNCTION

It is well documented that the intestinal ecosystem features dynamic and reciprocal interactions among its microflora, the epithelium, and the immune system. Because LAB are a part of the gut flora, either naturally or by supplementation, an understanding of the mechanisms whereby LAB compel their immune modulating effects on the host requires an outline of what position LAB occupy in the gut flora and how the gut flora as a whole interact with the immune system of its host.

25.3.1 The Gut Flora

In man, greater than 75% of the fecal output is composed of bacterial cells derived from more than 400 different strains, totaling approximately 10^{14} viable bacteria, of which only about 40% are culturable (2). This enormous number of bacterial cells exceeds by a factor of 10 the number of eukaryotic cells constituting the human body. The diversity and numerical importance of the bacteria of the gastrointestinal tract vary in the different gut sections, and the

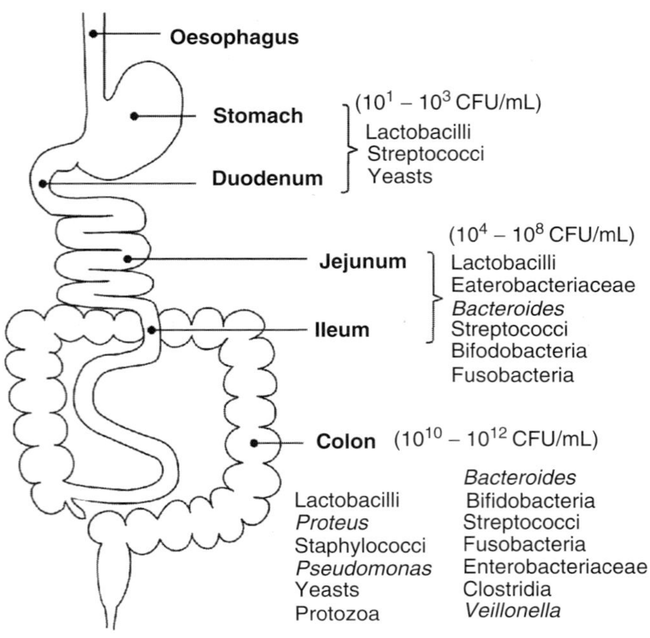

Figure 25.3 Microbial colonization of the human gut (175).

resident LAB are no exception (Figure 25.3). In adults, species of lactobacilli and streptococci as well as some other Gram-positive bacteria vastly dominate the flora of the sparsely populated stomach and duodenum. This is due to their ability to adhere to the epithelium and to resist acid, bile salts, and pancreatic secretions (45). At the distal part of the small intestine (jejunum and ileum), the microbial concentration and diversity increases, and here bifidobacteria species also enter the scene. In the colon which harbors a very diverse population of mostly strict anaerobic species, the levels of bifidobacteria reach 10^{10}–10^{11} cells/g whereas lactobacilli and enterococi species number only 10^{7}–10^{8} cells/g (46). In total, the population of bifidobacteria and lactobacilli in the gut of adults represent no more than 1–5% and <1%, respectively, of the fecal flora when analyzed by molecular analyses.

The predominant flora of adults is remarkably stable, whereas in children it is less stable, and in infancy highly unstable (46). In infancy, vast changes occur during development of the flora. Immediately after birth, the sterile gut of the fetus is rapidly colonized, and levels up to 10^{9} bacteria/g feces are reached within less than 48 hours. The composition of this initial flora represents a span of microbes originating from the environment, e.g., from the mother, but gives no indications of those likely to be retained in the tract (47). Subsequently, a selection of bacterial strains develops consistently along with maturation of the intestinal mucosa and under strong influence of both the microbial exposure and diet. Bifidobacteria constitute a major part of the flora of infants, ranging from 60 to 90%, whereas the values for Lactobacilli resemble those of adults (46). However, the flora varies greatly with neonatal feeding. In breast-fed infants, Bifidobacteria become dominant and the minor flora consist mainly of lactobacilli and streptococci. In contrast, formula-fed infants have similar amounts of bifidobacteria and bacteroides (~40%) and a minor flora frequently containing staphylococci, *Escherichia coli*, and clostridia (47). Upon weaning, the microbial profile becomes adult-like for both breast-fed and formula-fed with a significant increase of bacteroides.

Pioneer bacteria entering the gut in infancy can favor their own growth and inhibit growth of later introduced bacteria by inducing specific and lasting glycosylation of the glycocalyx as well as by modulating enterocyte gene expression. The initial colonization, therefore, widely shapes the final composition of the stable gut flora in adult life. As will be discussed later, increasing evidence shows that the development and maintenance of immunologic homeostasis is influenced by the composition of the gut flora, particularly the content of LAB. For example, distinct differences in the gut flora regarding bifidobacteria persist between atopic and nonatopic 2-year-olds (48). Thus, intake from birth or early infancy of probiotics may induce long term and enhanced efficacy of immune functions as well as other physiological conditions due to colonization as an integral part of the stable gut flora (autochthonous flora); contrasting probiotics ingested by older children and adults getting access only transiently (allochthonous flora) if at all, and thus demanding chronic administration of high concentrations.

In the gut flora of adults, the composition of the bifidobacteria population tends to be generally stable, while the composition of the lactobacilli tends to fluctuate more with only some persisting strains (46,49). Therefore, at least for the lactobacilli population, there seems to be a bacterial niche prone to alterations at a higher level than just bacteria in transit, e.g., through dietary changes or bacterial supplementation with probiotic lactobacilli.

Newly introduced harmless bacteria such as LAB stimulate the immune system including tolerance mechanisms and production of IgA. The produced IgA mediates immune exclusion whereby further contact between the bacteria and the immune system is attenuated as discussed in a previous section (50). Therefore, it has been suggested that for the allochthonous part of the flora, a high turnover of different strains of bacteria through the diet may participate in the continuous stimulation of GALT as a part of maintaining homeostasis.

25.3.2 Interactions Between the Gut Flora and the Host Immune System

In the elucidation of the impact of the gut flora on the immune system, germ-free (gnotobiotic) animals (pigs, mice, or rats) are very useful. Germ-free animals are obtained by delivering the pups through Caesarian section and hand-rearing them in sterile isolators. By comparing germ-free and conventional animals (animals with normal gut flora), or by deliberately colonizing the intestine of germ-free animals with defined microorganisms, it is possible to study the interaction between the gut flora, or gut flora components, and the host immune system. With this approach, species of LAB have often been investigated (51–53). A general observation from such studies is, however, that colonization with a single or a few gut flora-derived microbes is less effective in modifying GALT than complete conventionalization of the animals, revealing the importance of diverse bacterial stimuli.

The immune system of newborn conventional animals and humans matures along with colonization of the intestine. Such maturation does not occur in germ-free animals, leaving GALT development rudimentary and characterized by underdeveloped Peyer's patches and mesenteric lymph nodes lacking follicular germinal centers, reduced epithelial cell turnover, and low densities of immune cells both in gut mucosa and in peripheral lymph nodes (54). Upon microbial colonization, GALT and the systemic immune compartment undergo marked changes, of which the chief structural and functional effects on the immune system are listed in Table 25.2.

IgA production is one function highly influenced by the presence of intestinal microbes. Gut colonization instigates maturation of Peyer's patches, and consequently the numbers of IgA-producing plasma cells in the lamina propria increases (50,55). Within the compartment of intraepithelial lymphocytes, the number and cytolytic potential of $\alpha\beta$ T cells increases after microbial stimulation while the population of $\gamma\delta$ T cells is unaffected by the gut flora (56). As mentioned earlier, gut flora microbes can mediate phenotypic alterations of enterocytes affecting bacterial adhesion (expression of glycoconjugates) but

Table 25.2
Effect of Gut Flora on Structure and Function of the Immune System Observed in Studies of Germ-Free vs. Conventional Animals.

Compartment	Structural or Functional Change
Gut mucosa	Development of Peyer's patches
	Expansion of IgA-producing cells in lamina propria
	Expansion of intraepithelial T cells ($\alpha\beta$ TCR[†])
	Acquisition of cytotoxicity of T cells ($\alpha\beta$ TCR)
	Induction of MHC molecules on enterocytes
	Induction of glycoconjugates in enterocytes
	Enhanced cytokine production
Systemic	Increased number of immune cells in lymph nodes
	Enhanced cell-mediated immune response
	Enhanced cytokine production capacity of peritoneal cells
	Enhanced oxidative burst of macrophages and neutrophils
	Development of the capability to induce and maintain oral tolerance in antibody response

[†] Abbreviations: TCR: T cell receptor, MHC: major histocompatibility complex
Source: References 176 and 177.

also immune regulation; upon bacterial colonization, the expression of MHC class II molecules on enterocytes is provoked gradually (57). Thus, in contrast to the dietary antigens (ingested also by the germ-free mice), bacterial compounds are capable of stimulating the MHC class II expression, important for immunologic presentation of exogenous antigens. MHC class II expression is regulated by cytokines, and intestinal bacteria also greatly affect cytokine production in gut cells, which, in turn, conceivably affects the MHC class II expression (12).

The functional impact of diverse cellular changes may affect different immune functions. Besides enhancing the IgA production as previously mentioned, a major function of the gut flora is the impact on the host's capability to induce oral tolerance. Intake of dietary antigens by germ-free animals elicits an IgE antibody response along with increased levels of IL-4. This aberrant Th2 response, not induced in conventional mice, can, however, be corrected by reconstituting the gut flora of germ-free animals with *Bifidobacterium infantis* (a predominant gut bacteria of infants) at the neonatal stage, but not at a later age, signifying the importance of gut flora including LAB in the maturation of the neonatal immune system (51).

The neonate immune system of humans is immature with a poorly developed GALT. Intraepithelial lymphocytes are not expanded, and germinal centers are not formed, until sometime after birth, again reflecting the dependency of microbial stimuli (4,58). Newborns are generally Th2-skewed and have a reduced capability to induce Th1-mediated responses, and have poor IgA and IgG production capacity. The gut flora is believed to be a major Th2-counterbalancing (Th1/T_R-driving) force in the postnatal establishment of immune homeostasis (4). Many observations support the notion that most mucosal immune cells are competent even before birth, but need to undergo an activation process initiated by environmental signals, most significantly originating from the gut flora. Studies in animals (59) and in humans (60) have identified deficient antigen-presenting cell function as the chief explanation for immunologic immaturity; in neonates dendritic cells and macrophages are unable to deliver adequate costimulatory signals to activate naïve T cells. Thus, the effect of the gut flora, in particular the LAB dominating the flora of infants, may exert an effect on antigen-presenting cells, such as dendritic cells, either directly, or indirectly through interactions with other cells, e.g., enterocytes and intraepithelial lymphocytes, which, in turn, affect the antigen-presenting cells. As will be discussed in far more detail later, many LAB possess Th1- and T_R-inducing capacities and, indeed, interact with the gut epithelium (61,62). Such mechanisms of LAB may be crucial for the establishment and maintenance of immunologic homeostasis, and currently these phenomena have led to focused research into the immunomodulatory mechanisms of LAB as described in what follows.

25.4 IMMUNOMODULATORY MECHANISMS OF LAB

Numerous immunomodulating effects of LAB are observed in cellular mechanistic *in vitro* studies. Common to most of them is the absence of a mechanistic link to the immunomodulating effects observed *in vivo* and, one step further, to the observed "end point" effects on diseases or their prevention. This fact, in addition to the observation that a great deal of variation occurs among the different LAB even at the strain level, has caused probiotic research to be extremely multifarious and sometimes confusing. However, as more and more data accumulate some common characteristics come to light. In this section, we will go over some of the most prevailing immune mechanistic effects found for LAB as evident from either *in vitro* cell studies or *in vivo* studies in experimental animal models

Table 25.3
Summary of Immune Mechanistic Modulatory Effects of LAB[†]

Effect	Characteristics
Increased barrier integrity (Reduced permeability)	Reduced bacterial translocation and Ag leakage
Enhanced colonization resistance	Competitive LAB (adhesion sites, nutrients) Altered intestinal surface glycosylation (altered adhesion) LAB metabolites (antimicrobial substances, lactate)
Improved phagocytic cell function	Enhanced phagocytic and bactericidal capacity Reduced pathogen-induced apoptosis
Improved NK cell activity	Increased tumoricidal activity
Cytokine regulation	Enhanced production of pro-type 1 response cytokines Enhanced homeostasis (Th1/Th2/T_R cell balance)
Epithelial interaction	Counter-balance proinflammatory signals from G^- bacteria Stimulation of TGF-β production in enterocytes
Lymphocyte proliferation	Altered mitogen-induced T and B cell proliferation
Adjuvant capacity	Increased systemic and secretory Ag-specific Ab response
Secretory response	Nonspecific increase of IgA (incl. IgA^+ B cells in the gut) Enhanced Ag-specific IgA (e.g., against rotavirus)

[†] Abbreviations: LAB: lactic acid bacteria, Ag: antigen, Ab: antibody, G^-: Gram-negative

Table 25.4
Commonly Used Methods for Studying the Effects of Probiotic Bacteria on the Immune System

Method
Phagocytic capacity
Tumoricidal activity (NK cells)
Cell proliferation
Antigen-specific stimulation (primed cells)
Polyclonal mitogen stimulation
Cytokine determination
Supernatants from cell cultures
Serum
Cellular surface marker expression
Antibody determination
Antigen-specific
Isotype profile (mice: Th1: IgG2a, Th2: IgG1+IgE)

or humans. These mechanisms are summarized in Table 25.3. In the subsequent section we will go into the effect of LAB on diseases and the impact of viable vs. nonviable bacteria and effective dose.

25.4.1 Methods Used for Assessing Immunomodulatory Effect of LAB

Several methods have been employed for studying immune mechanisms as well as for evaluating clinical studies of immunomodulating effects of LAB, ranging from classical immune methods to sophisticated protocols for examining more unique parameters.

Table 25.4 summarizes the most commonly used immune methods, which will be discussed in brief in what follows.

Many different types of cells or cell mixtures are being investigated. When human cells are used for evaluating immune functions, whole blood or blood cell subgroups, such as peripheral blood mononuclear cells (PBMC) or polymorphonuclear cells are routinely used. Carcinoma cell lines of human enterocytes (Caco-2 and HT-29) are often used to study gut epithelial functions, sometimes in combination with PBMC to reflect the interaction between immune cells of GALT and enterocytes. With mice and rats, which are by far the most frequently used experimental animals to study immune functions, cells from the spleen, peritoneum, Peyer's patches, mesenteric lymph nodes, and lamina propria are widely used. Also, tissue samples for histological studies as well as secondary cultures (e.g., dendritic cells derived from bone marrow) and cell lines (macrophages, T, and B cells) are frequently used.

For the evaluation of nonspecific immune functions, cellular phagocytosis and NK cell activity are often determined, as these functions are important in combating microbial infection and growth of neoplastic cells. Phagocytosis is typically measured by cellular *in vitro* uptake of pathogenic bacteria like *E. coli* or *Staphylococcus aureus* following coculture of cell samples containing phagocytic cells such as whole blood, PBMC, or peritoneal lavages, and pathogens. Enumeration is either done by direct microscopic count or by flow cytometric analysis using fluorescence labeled bacteria. In humans, NK cell activity is normally measured as killing of dye-labeled or ^{51}Cr-labeled human tumor cells following coculture with PBMC. Phagocytic capacity and NK cell activity have been used to evaluate human clinical intervention studies.

Lymphocyte proliferation is a common method for measuring cellular response capacity toward *in vitro* stimulation either with antigen using *in vivo*-primed cells (immunization and vaccination), or nonspecifically with polyclonal mitogens. The most common mitogens applied to test lymphocyte proliferation are the T cell mitogens concanavalin A and phytohemagglutinin, and the B cell mitogens lipopolysaccharide and pokeweed mitogen. Although a commonly used method, the biological significance of mitogen stimulation is often questioned because of the broad nonspecific nature of the stimulation. Detection of cellular incorporation of [^{3}H]thymidine into DNA is commonly used for estimating changes in cell number.

Production of cytokines and other immune regulatory compounds are also methods frequently used in the assessment of immune mechanistic effects. Cytokines can be measured in supernatants of cell cultures or in whole blood samples with or without preceding stimulation with antigen or mitogen. Because many cytokines function at a focal level, e.g., in the synapse of interacting cells, and are often present only briefly and in low amounts, their detection is hampered and the biological significance of assessing their level in many *in vitro* experimental settings remains to be established. Cytokines are typically measured by enzyme-linked immunosorbent assay (ELISA) but can also be measured at the mRNA level.

Other measurements have been used to determine the effects of probiotics on immune status. For example, the number and type of immune cells such as blood or intestinal B and T cells can be determined by the use of fluorescence labeled antibodies directed against cell surface markers using flow cytometric analysis. Moreover, qualitative and quantitative determination of antibody production with or without preceding priming is used. Specific antibody responses to vaccination antigens or to bacterial antigens are used as an index for the adjuvant capacity of LAB; i.e., their capacity to potentiate a response to an antigen. The profile of the types of antibody produced indicates the type of response initiated, e.g., IgE in an allergic response.

25.4.2 Effects of LAB on Nonimmunologic Mucosa Barrier Functions

Many effects of LAB are very likely to be mediated through epithelial cell functions. For the nonimmunologic part, the intestinal epithelium provides a physical protective barrier, whose integrity is differentially influenced by various microbial signals. The integrity of the epithelial barrier is dependent on a tight cellular structure of enterocytes joined together by tight junctions. Defects in the barrier function can occur during invasion by pathogens, food intolerance reactions, or exposure to chemicals or radiation. Leakage through tight junctions or direct cell disruption results in increased permeability and consequently translocation of microbes and dietary antigens, which deteriorate the condition and further compromise barrier function (63). Factors such as proinflammatory cytokines, secreted by enterocytes and intraepithelial lymphocytes during contact with certain pathogenic bacteria, affect the integrity of the tight junctions and thus intestinal permeability, resulting in elevated transport of fluid and electrolytes eventually leading to diarrhea. Due to the same molecular mechanisms, beneficial microorganisms of the gut flora, including certain LAB, seem instead to enhance the integrity of the enterocytes tight junctions. For example, administration of the probiotic bacteria *L. rhamnosus* GG or *L. plantarum* 299v has been found to reverse increased intestinal permeability caused by cow's milk in suckling rats, or by *E. coli* infection in rats, respectively (64,65). Likewise, the probiotic product mixture VSL#3, which contains *B. longum*, *B. infantis*, *B. breve*, *L. acidophilus*, *L. casei*, *L. delbruekii* ssp. *bulgaricus*, *L. plantarum*, and *S. salivarius* ssp. *thermophilus*, was found to normalize colonic barrier integrity in parallel with a reduction of the proinflammatory cytokines TNF-α and IFN-γ in a colitis mouse model, where the permeability was otherwise increased by conventional gut colonization compared to germ-free animals (66).

Besides down regulation of proinflammatory factors, epithelial integrity could also be improved by LAB enhancing enterocyte proliferation and, thus increasing cellular turnover of injured and dysfunctional enterocytes. Such an effect has been seen in rats supplemented with *L. casei* (67). Alternatively or in addition, improved permeability could be mediated by LAB through pathogen exclusion due to enhanced resistance to pathogen colonization, which is another nonimmune defense mechanism promoted by certain LAB. Intestinal LAB such as strains of *Lactobacillus* are known to possess antimicrobial activities against pathogenic bacteria such as production of bactericidal peptides (68), and have been found to inhibit adhesion of pathogens to enterocytes *in vitro* (69). Bifidobacteria and some *Lactobacillus* strains, including many probiotics, adhere to enterocytes and are in this way believed to compete with pathogens for adhesion sites (70). The ability of certain bacteria to regulate surface expression of glycosylated compounds that mediate attachment of pathogens to the epithelial surface is another possible explanation for the improving effects of LAB on resistance to pathogen colonization (71). The capability of LAB to up regulate mucin production may also improve colonization resistance. *L. plantarum* 299v has been found *in vitro* to inhibit adhesion of *E. coli* to enterocytes and to up regulate mucin mRNA expression in the cells (72). Mucin up regulation may aid clearance of pathogens while also increasing attachment of LAB, as certain probiotic LAB are known to adhere to intestinal mucus (70). Much still needs to be confirmed and further clarified about the effect of LAB and the gut flora on the nonimmune related defense mechanisms.

25.4.3 Effects on Phagocytic Cells

Probiotic LAB possess the capability to enhance nonspecific immune functions of phagocytic cells (neutrophils and monocytes or macrophages). Although the exact mechanisms involved are largely unknown, the effect seems evident both from *in vitro* and *in vivo* studies, showing, however, a great variation among strains.

Increased phagocytic capacity is one important effect found for certain probiotic LAB: Ingestion of strains such as *L. johnsonii* La1, *L. rhamnosus* HN001, *B. lactis* Bb-12, and *B. lactis* HN019 significantly increases the phagocytic capacity of blood leukocytes or peritoneal cells (mice only), as shown in mice (73) and humans, including the elderly (6,74–77). In humans, this effect was shown to be stronger in individuals with poor phagocytic capacity prior to treatment (78). In contrast, ingestion of *L. casei* Shirota by humans was found not to influence the phagocytic capacity (79). Enhanced phagocytic capacity might be mediated through LAB-induced up regulation of phagocytic receptors, as observed for *L. rhamnosus* GG given to healthy humans, who showed an increased expression of CR1, CR3, FcγRI, and FcαR in neutrophils (80).

Increased phagocytic capacity mediated by probiotics is one mechanism that potentially accounts for enhanced resistance against pathogen infections. For example, mice challenged with pathogens like *Salmonella typhimurium* or *E. coli* 0157:H7 showed increased resistance to infection in conjunction with enhanced phagocytic capacity when fed probiotic LAB (*L. rhamnosus* HN001 or *B. lactis* HN019) (81,82).

Certain LAB are known to affect components involved in bactericidal activity of phagocytes, which is another mechanism potentially explaining increased resistance to infection. LAB enhance macrophage enzymatic activity when fed to mice (*L. casei* and *L. acidophilus*) (83) and have been found to enhance the oxidative burst (production of ROS) of phagocytic cells in humans (*L. johnsonii* La1) (14). Moreover, when exposed *in vitro* to macrophages, many strains of LAB readily induce production of nitric oxide and hydrogen peroxide, although the physiological relevance of such an *in vitro* effect remains unclear (84).

Pathogens such as *Salmonella typhimurium* have developed mechanisms to disseminate into deep tissues like the intestinal mucosa by inducing apoptosis (programmed cell death) of phagocytic and other immune cells. Recently, LAB strains (*L. delbruekii* ssp. *bulgaricus* and *S. thermophilus*) were found to inhibit such apoptosis induction and in so doing possibly enhance the capability of the phagocytic cells to kill the pathogen (85). This effect might, however, be mediated indirectly by an increased bactericidal capacity of the phagocytes that instigate killing of the pathogen before apoptosis is induced. Inhibition of pathogen-induced apoptosis of immune cells is yet another mechanism through which LAB might improve resistance to infection.

25.4.4 Effects on NK Cell Activity

NK cells are another group of cells playing an important role in the nonspecific immune defense against tumor cells, and viral and bacterial infections. A vast majority of NK cells reside in the liver, but they are also found at mucosal sites (86). Many studies, involving both animals and humans, have indicated that ingestion of certain LAB significantly enhances NK cell activity of PBMC observed as an increased tumoricidal activity (77,78,87,88). In some studies, a concomitant increase in the relative proportion of NK cells (cells expressing the CD56 surface molecule) among the isolated PBMC was observed, which could at least partly account for the observed increased tumoricidal activity. The increase in NK cell activity has been found to be positively correlated with age, and thus to benefit particularly elderly people of whom a high proportion have decreased NK cell activity (78). However, a direct link between LAB-induced increased NK cell activity and human health has yet to be documented.

NK cells are activated as an early event in an immune response by IL-12 produced by monocytes, macrophages, and dendritic cells. Activated NK cells produce IFN-γ that functions in the nonspecific immune response to further activate nonspecific immune mechanisms such as phagocytosis. NK cells themselves are also activated by IFN-γ.

Moreover, IFN-γ, like IL-12, is important in the promotion of specific Th1-polarized immune responses. The population of NK cells contained in PBMC is activated indirectly by LAB to produce IFN-γ through the effect of LAB on monocytes, which involves induction of IL-12 and costimulatory molecules (89). As will be discussed in what follows, many LAB induce IL-12 production in monocytes, macrophages, and dendritic cells (90–92), which potentially mediates the enhancing effect of LAB on NK cell activity.

25.4.5 Interactions Between LAB and Intestinal Epithelial Cells

Intestinal epithelial cells are endowed with several important immunoregulatory functions, as we have discussed, including antigen presentation and production of inflammatory and regulatory cytokines. Contact with pathogenic microorganisms such as *Salmonella typhimurium*, *E. coli* or *Helicobacter pylori*, activates enterocytes to secrete cytokines that induce migration of inflammatory cells like neutrophils into the mucosa. Such cytokines include the neutrophil attractant IL-8, and the proinflammatory cytokines IL-1β and TNF-α. Gram-positive versus Gram-negative bacteria appear to exert rather different effects on enterocytes (93). Gram-negative bacteria like *E. coli,* or *E. coli*-derived lipopolysaccharides, activate enterocytes to produce IL-8 and TNF-α through binding to the pattern recognition receptor CD14 (a cofactor for binding of several bacterial components to a Toll-like receptor). In contrast, Gram-positive bacteria including *L. johnsonii* La1 and *L. acidophilus* La10, as well as the cell wall glycolipid lipoteichoic acid derived thereof, inhibit the proinflammatory effect of *E. coli* (94). In line with this, *L. plantarum* 299v has been shown to inhibit *E. coli*-induced neutrophil transepithelial migration (95).

Migration of intraepithelial lymphocytes into the epithelial compartment depends on expression of adhesion molecules (αEβ7 integrin) on their surface, and these molecules are up regulated by TGF-β (96). In contrast to Gram-negative bacteria inducing inflammatory cytokines in epithelial cells, the probiotic *L. johnsonii* La1 was shown to induce TGF-β, potentially affecting the number of intraepithelial lymphocytes, and immunologic homeostasis, through the suppressive effect of TGF-β (61).

These observations suggest that LAB and other Gram-positive bacteria of the gut flora posses a capability to counterbalance the proinflammatory response of Gram-negative bacteria of the flora and so participate in the maintenance of immunologic homeostasis. The interaction of bacteria with enterocytes was found to be dependent on the presence of other immune cells, predominantly Th cells, and, in turn, to affect the activity of the immune cells, such as inducing immunosuppressive antigen-presenting cells (40,97). This provides evidence that luminal bacterial signaling is transduced to the host by a network of cross-talking immune cells in the gut, in which the gut epithelial cells participate considerably.

25.4.6 Effects on Cytokine Production

A large number of studies has clearly shown that LAB can influence cytokine production in immunocompetent cells of animals and humans. From *in vitro* studies with different types of immune cells such as human PBMC, murine spleen cells, dendritic cells or macrophage cell lines, a pattern for the type of cytokines typically induced following exposure to LAB has emerged: many lactobacilli, as well as other Gram-positive bacteria, are often capable of inducing Th1 response-related cytokines; IL-12, IL-18, IFN-γ, and TNF-α, and furthermore often induce IL-6 and IL-10 (87,90–92,98–100). In contrast to this, Gram-negative bacteria, e.g., nonpathogenic *E. coli*, are weak type 1 cytokine inducers but preferentially induce IL-10 (98,101,102). IFN-γ production by T cell or NK cells is induced through induction of the early response cytokine IL-12 in antigen presenting cells (monocytes, macrophages, or dendritic cells) (99,101,103). A large variation in the capacities to induce IL-12 occurs amongst species and strains of LAB; *L. casei* and *L. paracasei*, have, for instance, been repeatedly reported to include some of the strongest IL-12 inducing strains (91,92,99,104). IL-10 is, like IL-12, induced at different levels by different LAB, and strong IL-12-inducers appear often also to be strong IL-10 inducers. Induction of cytokines occurs in a dose dependent manner. In addition, the ratio between the induced cytokine levels varies not only with the strain of LAB but also with the concentration and duration of exposure, which adds further complexity to the phenomenon of bacterial signaling *in vivo* (92,102).

In addition to the capability of many LAB to be Th1-response potentiators *in vitro*, it has recently been found for a strain of *L. paracasei* (NCC2461), that, in addition to being a strong IL-10 and IL-12-inducer, this strain induced the development of a T_R-like Th cell population producing TGF-β and IL-10 in an *in vitro* system (104). Extension of such effect to *in vivo* conditions could explain some of the effects observed for LAB. In support of this, a recent *in vivo* study in mice showed that T_R cells with specificity toward the gut flora appeared to be dominant in the immune homeostasis of GALT (105).

In vivo cytokine production upon ingestion of LAB has been studied in both humans and animals. The results of these studies vary greatly and seem to depend on the LAB used and the experimental setting. As with the *in vitro* studies, animal feeding studies tend to show that some LAB strains favor Th1 responses. For example, *L. casei* CRL431 and *L. casei* Shirota fed to allergy-primed mice increased serum levels of IFN-γ and reduced IL-4 (106,107). However, in a study with healthy mice fed different LAB, no changes were observed for IFN-γ, TNF-α, or IL-6 (108). In another study, feeding yogurt to mice reduced cytokine expression, especially TNF-α (109). Moreover, in a study based on counting local cytokine-producing cells in the gut villi of healthy mice fed a range of lactobacilli, no changes were observed for IFN-γ, IL-4, or IL-10 (12). Yet in this study, significant induction of TNF-α and IL-2 was observed for some strains (*L. reuteri* and *L. brevis*), clearly showing a great variation between species and strains of LAB in their capacity to modulate the immune function at the site of delivery. Such local responses, however, may not necessarily be reflected systemically.

In human studies, administration of *L. rhamnosus* in combination with *L. reuteri* to patients with atopic disease did not alter IL-2, IL-4, IL-10, or IFN-γ production by PBMC, despite convalescence (110). However, in a similar study with atopic individuals ingesting *L. rhamnosus* GG, a transient up regulation of IL-10 was evident (111). Likewise, in patients suffering from intestinal inflammation, probiotics were shown to up regulate the tissue levels of IL-10 (112). In healthy individuals, intake of *L. brevis* ssp. *coagulans* was shown to increase a virus-induced IFN-α (an important cytokine in nonspecific cell mediated immunity) response. This effect followed a dose dependent pattern, with those subjects with initially lowest levels displaying the greatest increase (113). To date no studies have shown results on the effect of LAB on IL-12 production in humans. IL-12 and as other cytokines function focally between adjacent cells, which hampers factual *in vivo* assessment of changes of these cytokines.

On the whole, probiotic LAB are potentially capable of either improving homeostasis by inducing antiinflammatory cytokines like IL-10, or supporting Th1 responses by enhancing cytokines like IL-12. The pattern of LAB-induced modulation of cytokine production *in vivo* is likely to depend on the condition of the host; i.e., reflecting either immunologic imbalance due to disease or allergy, or proper immunologic balance when healthy.

25.4.7 Effects on Lymphocyte Proliferation

Lymphocyte proliferation responses to mitogens are used to examine T and B cell function upon treatment with immunosuppressive or immunoenhancing agents. The immunomodulating effect of ingested LAB on T and B cell proliferative capacity depends upon the species and strain of LAB involved. In some studies, ingestion of LAB such as *L. rhamnosus* HN001, *L. acidophilus* HN017, and *B. lactis* HN019 by mice was found to augment T and B cell function as evidenced by elevated proliferation response to concanavalin A, phytohemagglutinin, or lipopolysaccharide stimulation (73,114). However, in similar settings, strains of *L. casei, L. gasseri,* and *L. rhamnosus* inhibited mitogen-induced T and B cell proliferation, while *L. acidophilus* was without influence (115). Ingestion of LAB has, in some instances, been found to alter the proportion of T and B cells in lymphoid tissues, which may indirectly affect the proliferation response. Nevertheless, in cases with unchanged relative T cell populations, LAB were still found to alter the proliferation response (116). Cells from spleen, mesenteric lymph nodes and Peyer's patches have all exhibited altered proliferation response, showing, as for other analyses, that LAB exert immunomodulatory effects both at sites remote from and adjacent to the point of delivery. No explicit physiologic interpretation of altered proliferation response tested *in vitro* can be made. LAB inhibiting lymphocyte proliferation might be useful in the management of immune hyperresponsive diseases like allergy, autoimmunity, or intestinal inflammation, whereas LAB that enhance proliferation may be of general benefit to healthy individuals to improve the immune defense.

25.4.8 Effects on Specific Immune Responses – Adjuvant Capacity of LAB

A primary mechanism by which probiotic LAB mediate immune enhancement is through an adjuvant effect. An adjuvant is a compound that enhances the immune response toward a specific antigen when administered along with the antigen. Some examples of adjuvant capacity of certain LAB include studies where *L. rhamnosus*, *L. acidophilus*, *L. fermentum,* or *B. lactis* fed to healthy mice, significantly enhanced the formation of serum antibodies toward oral or parenterally administered antigens like cholera toxin, tetanus toxoid, or ovalbumin (12,73,117). Moreover, some LAB have been found to potentiate the antibody response against *E. coli* when cocolonized in germ-free animals (52,118). However, in this setting it was found that a large proportion of the antibody induced toward the LAB themselves cross reacted with the *E. coli*, which at least could partly account for the apparent adjuvant effect of LAB on the immune response against pathogens.

The LAB adjuvant effect has also been seen at the cytokine level. The *in vitro* antigen stimulated production of cytokines such as IFN-γ (Th1), IL-4, and IL-5 (Th2) in spleen and peritoneal cells of animals primed with ovalbumin and the Th2-driving adjuvant $Al(OH)_3$, was enhanced in animals fed *L. rhamnosus*, which indicates an adjuvant effect on both Th1 and Th2 cytokines (119). Feeding *L. casei* Shirota likewise enhanced Th1 cytokine production, but reduced Th2 cytokine production and, moreover, reduced the formation of ovalbumin-specific IgE (106). Hence, in addition to exhibiting an adjuvant capacity, different LAB strains seem to modulate cytokine regulation differentially, possibly resulting in the polarization of the response. Accordingly, a study in mice showed that intake of *L. acidophilus*, *L. delbruekii* ssp. *bulgaricus,* and *L. casei* all enhanced the IgG1 response (Th2-associated isotype) toward parenterally injected ovalbumin, whereas *L. acidophilus* was the only strain that also enhanced the IgG2a response (Th1-associated isotype), and was also the only strain inducing detectable IL-12-producing cells in lamina propria (117).

Whether such modulations of Th1 and Th2 cytokine production during an antigen-specific reaction are beneficial to the host still remains to be established. It is, however, hypothesized that mechanisms of LAB mediating suppression of Th2 responses, either by favoring Th1 responses or by enhancing tolerance induction mechanisms (in general improving Th1/Th2/T_R-homeostasis), may be effective in the antiallergy effect observed for certain probiotic LAB (62).

Adjuvant effects of LAB are also observed in human studies. For example, enhanced specific antibody responses to rotavirus are reported when patients who are infected with, or vaccinated against, rotavirus, ingest *L. rhamnosus* GG (120,121). However, in a study with subjects receiving attenuated *Salmonella typhi* as an oral vaccine, ingestion of *L. rhamnosus* GG did not enhance antibody formation, indicating significant variations between different experimental settings (122).

In many vaccination regimes and in the defense against pathogenic infections, a secretory IgA response is important. Like systemic antibody responses already discussed, certain LAB enhance secretory IgA responses, which is discussed in what follows.

25.4.9 Enhancement of IgA Production

Augmentation of IgA production following ingestion of LAB is one of the better documented effects of probiotic LAB. LAB enhance IgA production in both a specific and a nonspecific manner. From studies in mice, oral administration of a wide range of lactobacilli and bifidobacteria species are found to increase the number of IgA^+ B cells in the gut mucosa as well as total IgA production (16,53,123,124). Likewise, in humans, ingestion of LAB, such as *L. rhamnosus* GG or *L. johnsonii* La1, is found to increase the total IgA level in blood and feces (15,120,125).

On an antigen-specific level, many species of LAB also promote secretory IgA production as observed both in mice and humans. In mice, several studies show that along with an increased protection against pathogens, probiotic LAB increases pathogen-specific IgA (114,126). Moreover, oral administration of LAB is shown to enhance IgA against dietary antigens like the milk protein β-lactoglobulin or toxins (16,123,127). Such effects are also found in humans, showing that ingestion of *L. johnsonii* La1, bifidobacteria or *L. rhamnosus* GG increases the specific IgA formation to attenuated *Salmonella typhi* (125), poliovirus vaccinations (15), and dietary proteins (in patients with gut inflammation) (128), respectively. Increased rotavirus-specific IgA response in rotavirus patients given probiotic LAB such as *L. rhamnosus* GG has been demonstrated in several studies and is believed to be a contributing mechanism in the observed protection against reinfection with rotavirus (120,129,130).

The cytokine IL-6 is important for the maintenance of an IgA response; Th2 cytokines (IL-4 and IL-5) are more important during response induction (131). Thus, the ability of many LAB to induce IL-6 but not Th2 cytokines, as discussed in a previous section, may be linked to the capability of LAB to act as an adjuvant for IgA production.

25.5 EFFECTS OF LAB ON DISEASES

Scientific evidence is now being accumulated to support both therapeutic and prophylactic properties of probiotic LAB and fermented products. Many beneficial health effects have been claimed for probiotics, especially concerning their potential to prevent or ameliorate intestinal diseases. There are, however, fundamental differences between various LAB, and the clinical effects vary greatly, with only a few probiotic strains showing efficacy in randomized placebo-controlled clinical studies. The probiotic organism that has received by far the most clinical attention to date is *L. rhamnosus* GG. Several other organisms have, however, also been thoroughly studied, of which the most important and their effects are listed in Table 25.5. Most clinical studies have been performed using viable bacteria. However, in some instances, nonviable bacterial preparations have been shown also to be effective, which will be discussed at the end of this section.

Table 25.5
Examples of Probiotics with Reported Immunomodulating Effects in Humans

Bacteria	Effects	Ref.
L. rhamnosus GG	Prevention and treatment of diarrhea, vaccine adjuvant, antiallergy, increased IgA	120,121,128,135,161,162
L. johnsonii La1	Vaccine adjuvant, enhanced phagocytosis, antiinfection (*Helicobacter*), increased IgA	6,14,125,178
L. plantarum 299v	Antiinflammatory bowel disease	145
L. rhamnosus HN001	Enhanced phagocytosis and NK cell activity, antiinfection	77,81,82,179
B. lactis HN019	Enhanced phagocytosis and NK cell activity, antiinfection	76,114,132,180
L. reuteri DSM12246	Treatment of diarrhea, antiallergy	110,136
VSL#3 (mixture of LAB)	Antiinflammatory bowel disease, prevention of diarrhea	148,181,182
L. casei Shirota	Treatment of diarrhea, anticancer, antiallergy, antiinfection, enhanced NK cell activity	88,104,106,168,183,184
L. casei CRL431	Treatment of diarrhea, antiinfection, increased IgA	185–188
B. lactis Bb-12	Antiallergy, prevention of diarrhea vaccine adjuvant, increased IgA, enhanced phagocytosis	6,15,134,161

25.5.1 Effects of LAB on Intestinal Infections

Some LAB have proved effective in preventing gastrointestinal infections and in the recovery from infectious diarrhea due to miscellaneous causes including traveler's diarrhea and rotavirus.

Acute diarrhea caused by gastrointestinal infections is common throughout the world, especially among children. One aspect of the effect of LAB against infectious diseases in the gut is due to enhanced resistance to colonization, e.g., by production of antibacterial metabolites, and, thus, to a lesser extend related to a direct effect on the immune system. However, combating pathogenic infections has also been shown to be due to stimulation of the immune defense, particularly by increasing phagocytic capacity and the secretory antibody (IgA) response, as discussed previously. Studies in mice have demonstrated that in parallel with such immune stimulating effects, dietary supplementation with *L. rhamnosus* HN001 or *B. lactis* HN019 reduces the severity of infections with the pathogens *E. coli* O157:H7 and *Salmonella typhimurium* (81,82,114,132). In contrast, in a clinical study with healthy volunteers challenged with enterotoxigenic *E. coli*, administration of *L. acidophilus* and *L. bulgaricus* did not reduce the attack rate, incubation period, or duration of diarrhea (133).

Nevertheless, in studies concerning children with acute infectious diarrhea (viral or bacterial), probiotics such as *L. reuteri* DSM 12249, *L. rhamnosus* GG, or *B. bifidum* effectively ameliorated the diarrhea in both hospitalized and nonhospitalized patients (134–136). Probiotics have been found to be particularly effective against rotaviral infections. The effect of *L. rhamnosus* GG in the treatment of rotavirus diarrhea in infants and children is one of the best documented probiotic effects in human subjects (137). Mechanisms including normalization of gut flora and gut permeability, as well as enhancement of mucosal immune response, have been reported in several studies; e.g., clearance of virus after primary infection and subsequent protection against reinfection (seen for at least 1 year) correlates well with the production of mucosal and serum IgA, which are enhanced by probiotics (44).

Antibiotic-associated diarrhea is another form of acute diarrhea that occurs frequently. It is caused by a disruption of the ecosystem of the gut flora resulting in decreased colonization resistance and altered fermentation capacity (microbial imbalance). Under these conditions *Clostridium difficile* or *Klebsiella oxytoca* may colonize the gut and induce colitis. *L. rhamnosus* GG has shown to be effective in preventing antibiotic-associated diarrhea in children while being ineffective in adult patients. The probiotic microorganisms *Enterococcus faecium* SF68 and the yeast *Saccharomyces boulardii*, have also shown prophylactic effects in antibiotic-associated diarrhea (138).

Traveler's diarrhea is another common form of acute diarrhea. In several studies, probiotic LAB demonstrated no effect in preventing traveler's diarrhea, and data to support a prophylactic effect is limited. In one study, however, the incidence of diarrhea in tourists going to Egypt was reduced considerably in subjects given capsules of *S. thermophilus*, *L. bulgaricus*, *L. acidophilus,* and *B. bifidum*, indicating that the effect of probiotics in traveler's diarrhea not only depends upon the probiotic strain or strains, but also on the specific conditions, such as the type of microbial exposure, which can vary at different destinations (139).

Colonization of the gastric mucosa with *Helicobacter pylori* is associated with gastritis, ulcers, and some malignancies. Several probiotic LAB have demonstrated anti-*Helicobacter* activity in *in vitro* studies. *In vivo*, LAB strains including *L. johnsonii* La1 have also been found to exhibit anti-*Helicobacter pylori* activity by suppressing this organism, but do not appear to eradicate it (140).

In summary, probiotic LAB seem to possess the potential to be used in the management of several infectious diseases. In particular, probiotic bacteriotherapy of children and infants, to reduce the incidence and shorten the duration of acute infectious diarrhea, might be effective. This fact may be related to the ability of the bifidobacteria predominating in the flora of breast-fed infants to promote protection against pathogenic infections (141).

25.5.2 Effects on Inflammatory Bowel Disease

Inflammatory bowel disease (IBD), including ulcerative colitis, Crohn's disease, and pouchitis (inflammation of the ileal pouch arising after surgical resection of colon), refers to intestinal disorders of unknown cause that are characterized by chronic or recurrent intestinal inflammation (142). It appears that IBD, particularly Crohn's disease, develops due to genetically influenced dysregulation of mucosal immunity against components of the intestinal flora. Cytokine imbalance due to dysregulated Th cells appears to play an important role in the pathogenesis: Crohn's disease is mediated predominantly by Th1 response, whereas ulcerative colitis is usually associated with a Th2 response.

Some probiotics seem to be very efficient in the treatment of IBD. As discussed earlier in this chapter, IL-10 is a key regulatory cytokine for tolerance to luminal antigens, including the gut flora. IL-10 gene-deficient mice respond to their gut flora by developing spontaneous colitis resembling Crohn's disease and are, therefore, an effective experimental model for IBD (143). A gut flora deficient in LAB is a characteristic of IL-10 gene-deficient mice, and administration of probiotic LAB, such as *L. reuteri* and *L. plantarum* 299v, has been shown to be an effective treatment of these mice (144,145). As also mentioned earlier, bacterial immunostimulatory DNA sequences, especially unmethylated CpG motifs, are known to activate epithelial cells through interaction with Toll-like receptors (28). In different IBD mouse models, treatment with immunostimulatory DNA inhibited induction of colonic proinflammatory cytokines and chemokines (146). Likewise, genomic unmethylated DNA from the probiotic mixture VSL#3 ameliorated colitis in a similar model, whereas methylated DNA was ineffective, clearly demonstrating the importance of immunostimulatory DNA in probiotics effective against IBD and perhaps against other diseases involving an immune imbalance as well (147).

In humans, a number of clinical studies have now demonstrated an efficacy of probiotics against ulcerative colitis, Crohn's disease, or pouchitis. The VSL#3 product (*B. longum*, *B. infantis*, *B. breve*, *L. acidophilus*, *L. casei*, *L. delbruekii* ssp. *bulgaricus*, *L. plantarum*, and *S. salivarius* ssp. *thermophilus*) has proved very efficient in prophylaxis and reduction of recurrence of chronic relapsing pouchitis (148–150). Significantly, in IBD patients treated with VSL#3, the tissue levels of IL-10 were increased revealing an immunomodulatory mechanism elicited by the probiotics (112).

Other nonLAB microorganisms such as *E. coli* Nissle 1917 and *Saccharomyces boulardii* have also proved efficient in the treatment of IBD (151). Intense research is ongoing to elucidate further the efficacy of probiotic LAB in the treatment and prevention of IBD.

25.5.3 Antiallergy Properties of LAB

The prevalence of allergic diseases has progressively increased in Western industrialized countries (152). Allergy and atopy, featuring IgE production, are closely associated with a Th2 cell-driven response. Atopic eczema due to food allergy represents the earliest manifestations of allergic disease, starting from infancy.

A so called "hygiene hypothesis" was originally formulated by Strachan in 1989 (153), stating a possible explanation for the increasing prevalence of allergy. In essence, this hypothesis stated that the lower microbial exposure of the immune system due to increased hygiene causes an underdeveloped Th1 polarization (newborns are Th2-skewed) and, thus, an imbalance in the Th1/Th2 homeostasis, favoring the Th2-driven allergic responses. While this original hypothesis mainly implied the consequence of increased hygiene to be a reduction in the number of Th1-driving infections known to reduce allergic disorders, a revised hygiene hypothesis was proposed by Wold in 1998 (154), saying that the overly hygienic lifestyle in modern Western countries, rather than limiting pathogenic infections, has altered the normal intestinal colonization pattern in infancy leading to a failure to induce and maintain oral tolerance of innocuous antigens.

Indeed, there is data accumulating showing a clear tendency of atopic subjects to have an aberrant composition of the gut flora in comparison with nonatopic subjects: infants in whom atopy develops tend to have fewer bifidobacteria and more clostridia (8,155,156). Recent findings have, moreover, indicated a correlation between allergic disease and the composition of the intestinal bifidobacteria population, pointing toward allergic infants having a more adult-like bifidobacteria flora with high levels of *B. adolescentis* whereas healthy infants have high levels of *B. bifidum* (157,158). Interestingly, the adhesion capacity to human intestinal cells of the isolated fecal bifidobacteria from healthy infants has been found to be significantly higher than for allergic infants. Furthermore, when testing the capacity to induce cytokines in a macrophage cell line, bifidobacteria from allergic infants vs. healthy infants were found to induce more proinflammatory cytokines such as IL-1β, IL-12, and TNF-α, whilst inducing much less IL-10 with many strains not inducing IL-10 at all. This provides preliminary indications that although bifidobacteria are Th1 cytokine inducers and putatively counteract Th2 responses, the incapability of inducing the response suppressive IL-10 in combination with the TNF-α-inducing capacity (TNF-α is increased during allergic inflammation in the gut) of the gut flora derived bifidobacteria of allergic infants possibly plays a decisive role in allergic disease etiology (159).

A corollary of the documented immunologic repercussions to aberrant gut flora, as well as the general finding of LAB to improve immune homeostasis, is the investigation of the effect of supplementing LAB on allergy. As mentioned earlier, oral administration of *L. casei* Shirota to mice prior to Th2 response priming reduced IgE and IL-4 levels while increasing the Th1 cytokines IL-2 and IFN-γ (106). Likewise, in a mouse model of atopy, administration of *L. casei* CRL431 was shown to down regulate IgE and IL-4 synthesis by increasing IFN-γ. This was, however, only effective when administered before sensitization, indicating the impact of timing (107).

In humans, a number of studies have indicated the efficacy of *Lactobacillus* administration in the treatment and prevention of atopic eczema. In this regard, *L. rhamnosus* GG is the most investigated probiotic LAB strain and has proved efficient in several settings. In studies of infants with atopic eczema, supplementation with *L. rhamnosus* GG has been found to improve clinical symptoms, alleviate intestinal inflammation and enhance IL-10 production significantly (111,160,161). One of these studies also included *B. lactis* Bb-12, which demonstrated this strain to be equally efficient in reducing the severity of eczema (161). In a study involving older children (1–13 years) with chronic atopic eczema, oral treatment with a combination of *L. rhamnosus* 19070-2 and *L. reuteri* DSM12246, decreased the severity of eczema in more than half of the treated subjects (110). These studies substantiate the efficacy of some probiotics in the treatment of manifested allergy.

Clinical research has also aimed at the efficacy of probiotics to prevent development of allergy. The efficiency of *L. rhamnosus* GG to prevent the occurrence of atopic eczema in infants from atopic families has been studied (162). The bacteria were given to mothers during pregnancy, and mothers who were breast feeding continued to take the bacteria after delivery; otherwise the children were given bacteria for 6 months. At 2 years of age, the frequency of atopic eczema in the probiotic group was reduced by half (from 46% to 23%), demonstrating a potential of this probiotic LAB in prophylaxis of allergy. In line with this, administration of *L. rhamnosus* GG to the pregnant and lactating mothers increased the level of TGF-β in breast milk and reduced the risk of the children developing atopic eczema during the first two years of life (163).

There are indications that using probiotics in the management of allergy is most if not solely effective in allergies occurring in infancy or early childhood. For example, *L. rhamnosus* GG failed to prevent both birch pollen allergy and apple allergy in young adults (164). Supplementation of infants via formulas fortified with probiotics or supplementation of mothers during pregnancy and lactation are definite future perspectives for the applicability of probiotics. Research is ongoing, working to identify new probiotics and their mechanisms of action and optimal protocol of administration, especially regarding timing.

25.5.4 Antitumor Effect of LAB

Colorectal cancer is one of the most important causes of death from cancer in Western countries. Epidemiological studies show a diet associated risk, in particular of the low fiber "Western diet;" and many studies confirm the involvement of intestinal microflora in the onset of colon cancer. In animals, several studies have shown that bacteria of the bacteroides and clostridia genera increase the incidence and growth of colonic tumors, whereas other genera such as lactobacilli and bifidobacteria prevent tumorigenesis (165). Likewise, in a human study, high risk of colon cancer was found to be associated with presence of species of *Bacteroides*, whereas low risk was associated with presence of species of *Lactobacillus* (166). Thus, much attention has focused on decreasing cancer through diet alterations, in particular increasing intake of fiber and probiotics both intervening with the gut flora.

No direct experimental evidence has, yet, been established that consumption of LAB suppresses cancer in man; however, a vast amount of indirect evidence exists from laboratory experiments using experimental animal models (167). From such studies, several non-immunologic mechanisms of LAB have been suggested, such as decrease of fecal enzymes that may be involved in carcinogen formation, elimination of carcinogens, and production of antitumorigenic compounds. However, antitumor effects of LAB in experimental animals seem also to be mediated through immunologic mechanisms. As discussed in the previous section, LAB increase specific and nonspecific immune mechanisms, potentially protecting against tumor development. In different models of tumor bearing mice, injection of heat-killed *L. casei* Shirota (termed LC9018) to the site of tumor growth showed a potent inhibition of tumor growth and increased survival (106,168). This antitumor effect was mediated through production of the tumor-suppressing type 1 cytokines IL-12, IFN-γ, and most importantly TNF-α, as injection of neutralizing anticytokine antibodies abolished the antitumor effect of the bacteria. Studies on the effect of yogurt (*L. delbruekii* ssp. *bulgaricus* and *S. thermophilus*) showed that consumption of yogurt by mice inhibited chemically induced colorectal cancer in conjunction with an increase in IgA^+ B cells in the colon, and with IFN-γ and TNF-α production (169). In summary, these and other studies suggest that certain LAB impart an antitumor effect through modulation of the host's immune system, specifically cell-mediated immunity. However, such antitumor effects of LAB in humans remain to be determined.

25.5.5 Impact of Viability and Dose on the Effects of LAB

Besides strain-related primary selection criteria (adhesion, acid, and bile tolerance, not discussed in this chapter), some practical aspects have an impact on the immunomodulating effects of probiotics; most importantly the consequences of viable vs. nonviable bacterial preparations, and effective doses. Probiotics, as defined in the introduction, include also nonviable bacteria and bacterial components. However, research has shown that some effects demand the bacteria be alive whereas other effects are independent of viability. Effects exerted via prolonged epithelial adherence or (transient) colonization logically require live bacteria although there are indications that nonviable bacteria can, in fact, adhere to epithelial cells (170). A variety of immunomodulating activities, seem to be attributable to specific cell wall or intracellular bacterial components, and are exhibited by nonviable bacteria and even by the isolated components of the cell (100,147).

Only a few studies have focused specifically on testing the effects of viable vs. nonviable bacteria. Viable and heat-killed LAB were equally efficient in enhancing phagocytic capacity in mice, whereas only the viable bacteria enhanced IgA production (171,172). Likewise, only viable *L. rhamnosus* GG enhanced IgA production in patients with acute rotavirus diarrhea, while both viable and nonviable probiotics shortened the diarrhea (130). In an attempt to assess the efficacy of viable vs. heat-killed *L. rhamnosus* GG on atopic disease, the heat-killed preparation induced diarrhea and the study could not be completed, opening the question of safety of ingestion of large amounts of heat-killed bacteria (173). These and other studies clearly indicate that the importance of viable vs. nonviable bacteria depends on the effect in question, although viable probiotics tend to have more health effects than nonviable (113,174). The method of inactivation, e.g., inactivation by natural exhaustion in stored fermented products, which introduces cellular changes; inactivation by heat killing experimentally or due to pasteurization, which highly denatures cellular compounds; or inactivation by UV or gamma irradiation, which spares protein denaturation but affects DNA, may also play a role.

Another important factor involving the effects of probiotic LAB is effective dose. Typically, probiotic LAB are only effective if the dosage is sufficiently high. Accordingly, the minimum daily dose of viable *L. johnsonii* La1, which significantly increases phagocytic capacity and oxidative burst in humans, is 10^9 bacteria (14). In a study in mice, the minimum daily dose for a significant increase in blood cell phagocytic capacity was 10^7 with a clear dose-response effect up to 10^{11} bacteria per day (172). Most clinical studies apply doses at 10^9–10^{11} viable bacteria/day.

25.6 CONCLUDING REMARKS AND FUTURE PERSPECTIVES

Probiotics are just one aspect of the emerging field of functional foods, and have recently received renewed enthusiasm and far more rigorous scientific pursuit. Well documented beneficial effects on gastrointestinal disturbances are mediated by probiotics. The research worldwide is now focusing on the development of target specific probiotic products with well characterized microorganisms selected for their specific health enhancing properties. For this purpose, more research into the intestinal flora and its interaction with the host immune function is a prerequisite to enhance the scientific soundness of identifying novel probiotics. Currently, many questions remain to be answered as to how probiotic LAB exert their ostensibly wide spectrum of immunomodulating effects. It appears reasonable to assume that the multifactorial effects of LAB stem from merely one or a few primary effects, yet targeted at some early stages of the immune response and thus causing alterations in a variety of mechanisms along the response cascade. Accordingly, increasing amounts of data indicate that LAB greatly affect epithelial cells and antigen-presenting cells such as the dendritic cells important for the initiation and polarization of an immune response. Effects on these cells disseminate like ripples in a pond, potentially modulating multiple immune functions regulated by these cells. Equally important, such primary effects of probiotics can also evoke diversified outcomes depending on the immune status of the host; i.e., healthy, malfunctioning, or diseased.

Like many other areas of biology, regulation of the immune system is extremely complex. Metaphorically, the phenomenon of immune regulation could frankly be compared with jumping in a waterbed: when jumping down the bed, other parts of the bed inevitably go up, with some areas of the bed moving more vigorously than others; until stabilization, many areas move alternately up and down at different amplitudes and in an almost unpredictable pattern very closely linked to even small disparities in factors related to both the "jumper" and the bed — the "jumper" being the probiotic and the bed being the consumer.

ACKNOWLEDGMENT

The authors wish to thank Dr. Charlotte Madsen for critically reading through the manuscript.

REFERENCES

1. Molin, G., B. Jeppsson, M.L. Johansson, S. Ahrne, S. Nobaek, M. Stahl, S. Bengmark. Numerical taxonomy of *Lactobacillus* spp. associated with healthy and diseased mucosa of the human intestines. *J. Appl. Bacteriol.* 74:314–323, 1993.
2. Conway, P.L. Development of intestinal microbiota. In: *Gastrointestinal microbiology*, Mackie, R.I., B.A. White, R.E. Isaacson, eds., New York: Chapman & Hall, 1997, pp 3–38.
3. Ahrne, S., S. Nobaek, B. Jeppsson, I. Adlerberth, A.E. Wold, G. Molin. The normal *Lactobacillus* flora of healthy human rectal and oral mucosa. *J. Appl. Microbiol.* 85:88–94, 1998.

4. Bjorksten, B. The intrauterine and postnatal environments. *J. Allergy. Clin. Immunol.* 104:1119–1127, 1999.
5. Holt, P.G., C.A. Jones. The development of the immune system during pregnancy and early life. *Allergy* 55:688–697, 2000.
6. Schiffrin, E.J., D. Brassart, A.L. Servin, F. Rochat, A. Donnet-Hughes. Immune modulation of blood leukocytes in humans by lactic acid bacteria: criteria for strain selection. *Am. J. Clin. Nutr.* 66:515S–520S, 1997.
7. Blum, S., Y. Delneste, S. Alvarez, D. Haller, P.F. Perez, C.H. Bode, W.P. Hammes, A.M.A. Pfeifer, E.J. Schiffrin. Interactions between commensal bacteria and mucosal immunocompetent cells. *Int. Dairy J.* 9:63–68, 1999.
8. Kirjavainen, P.V., E. Apostolou, T. Arvola, S.J. Salminen, G.R. Gibson, E. Isolauri. Characterizing the composition of intestinal microflora as a prospective treatment target in infant allergic disease. *FEMS Immunol. Med. Microbiol.* 32:1–7, 2001.
9. Fuller, R. Probiotics in man and animals. *J. Appl. Bacteriol.* 66:365–378, 1989.
10. Salminen, S., A. Ouwehand, Y. Benno, Y.K. Lee. Probiotics: how should they be defined? *Trends Food Sci. Tech.* 10:107–110, 1999.
11. Hamann, L., V. El Samalouti, A.J. Ulmer, H.D. Flad, E.T. Rietschel. Components of gut bacteria as immunomodulators. *Int. J. Food. Microbiol.* 41:141–154, 1998.
12. Maassen, C.B.M., C. Holten-Neelen, F. Balk, J.H. Bak-Glashouwer, R.J. Leer, J.D. Laman, W.J.A. Boersma, E. Claassen. Strain-dependent induction of cytokine profiles in the gut by orally administered *Lactobacillus* strains. *Vaccine* 18:2613–2623, 2000.
13. Ouwehand, A.C., S. Salminen, E. Isolauri. Probiotics: an overview of beneficial effects. *Antonie Van Leeuwenhoek* 82:279–289, 2002.
14. Donnet-Hughes, A., F. Rochat, P. Serrant, J.M. Aeschlimann, E.J. Schiffrin. Modulation of nonspecific mechanisms of defense by lactic acid bacteria: effective dose. *J. Dairy Sci.* 82:863–869, 1999.
15. Fukushima, Y., Y. Kawata, H. Hara, A. Terada, T. Mitsuoka. Effect of a probiotic formula on intestinal immunoglobulin A production in healthy children. *Int. J. Food Microbiol.* 42:39–44, 1998.
16. Takahashi, T., E. Nakagawa, T. Nara, T. Yajima, T. Kuwata. Effects of orally ingested *Bifidobacterium longum* on the mucosal IgA response of mice to dietary antigens. *Biosci. Biotechnol. Biochem.* 62:10–15, 1998.
17. Mowat, A.M., J.L. Viney. The anatomical basis of intestinal immunity. *Immunol. Rev.* 156:145–166, 1997.
18. Strobel, S., A.M. Mowat. Immune responses to dietary antigens: oral tolerance. *Immunol. Today* 19:173–181, 1998.
19. Kelsall, B.L., W. Strober. Dendritic cells of the gastrointestinal tract. *Springer Semin. Immunopathol.* 18:409–420, 1997.
20. Rescigno, M., M. Urbano, B. Valzasina, M. Francolini, G. Rotta, R. Bonasio, F. Granucci, J.P. Kraehenbuhl, P. Ricciardi-Castagnoli. Dendritic cells express tight junction proteins and penetrate gut epithelial monolayers to sample bacteria. *Nat. Immunol.* 2:361–367, 2001.
21. Dearman, R.J., H. Caddick, D.A. Basketter, I. Kimber. Divergent antibody isotype responses induced in mice by systemic exposure to proteins: a comparison of ovalbumin with bovine serum albumin. *Food Chem. Toxicol.* 38:351–360, 2000.
22. Perdue, M.H. Mucosal immunity and inflammation, III: the mucosal antigen barrier: cross talk with mucosal cytokines. *Am. J. Physiol.* 277:G1–G5, 1999.
23. Kraehenbuhl, J.P., E. Pringault, M.R. Neutra. Intestinal epithelia and barrier functions. *Aliment. Pharmacol. Ther.* 11:3–8, 1997.
24. Kelsall, B., W. Strober. Gut-associated lymphoid tissue: antigen handling and T-lymphocyte responses. In: *Mucosal Immunology*, Ogra, P.L., J. Mestecky, M.E. Lamm, W. Strober, J. Bienenstock, J.R. McGhee, eds., San Diego: Academic Press, 1999, pp 293–317.
25. Palucka, K., J. Banchereau. Dendritic cells: a link between innate and adaptive immunity. *J. Clin. Immunol.* 19:12–25, 1999.
26. Strober, W., R.L. Coffman. Tolerance and immunity in the mucosal immune system: introduction. *Res. Immunol.* 148:489–490, 1997.

27. Banchereau, J., R.M. Steinman. Dendritic cells and the control of immunity. *Nature* 392:245–252, 1998.
28. Underhill, D.M., A. Ozinsky. Toll-like receptors: key mediators of microbe detection. *Curr. Opin. Immunol.* 14:103–110, 2002.
29. Vieira, P.L., E.C. de Jong, E.A. Wierenga, M.L. Kapsenberg, P. Kalinski. Development of Th1-inducing capacity in myeloid dendritic cells requires environmental instruction. *J. Immunol.* 164:4507–4512, 2000.
30. de Jong, E.C., P.L. Vieira, P. Kalinski, J.H.N. Schuitemaker, Y. Tanaka, E.A. Wierenga, M. Yazdanbakhsh, M.L. Kapsenberg. Microbial compounds selectively induce Th1 cell-promoting or Th2 cell-promoting dendritic cells *in vitro* with diverse Th cell-polarizing signals. *J. Immunol.* 168:1704–1709, 2002.
31. Shortman, K., Y.J. Liu. Mouse and human dendritic cell subtypes. *Nat. Rev. Immunol.* 2:151–161, 2002.
32. Banchereau, J., F. Briere, C. Caux, J. Davoust, S. Lebecque, Y.J. Liu, B. Pulendran, K. Palucka. Immunobiology of dendritic cells. *Annu. Rev. Immunol.* 18:767–811, 2000.
33. Roncarolo, M.G., R. Bacchetta, C. Bordignon, S. Narula, M.K. Levings. Type 1 T regulatory cells. *Immunol. Rev.* 182:68–79, 2001.
34. Singh, B., S. Read, C. Asseman, V. Malmstrom, C. Mottet, L.A. Stephens, R. Stepankova, H. Tlaskalova, F. Powrie. Control of intestinal inflammation by regulatory T cells. *Immunol. Rev.* 182:190–200, 2001.
35. Friedman, A., H.L. Weiner. Induction of anergy or active suppression following oral tolerance is determined by antigen dosage. *Proc. Natl. Acad. Sci. USA* 91:6688-6692, 1994.
36. Brandtzaeg, P. Current understanding of gastrointestinal immunoregulation and its relation to food allergy. *Ann. NY Acad. Sci.* 964:13–45, 2002.
37. Smith, K.M., J.M. Davidson, P. Garside. T-cell activation occurs simultaneously in local and peripheral lymphoid tissue following oral administration of a range of doses of immunogenic or tolerogenic antigen although tolerized T cells display a defect in cell division. *Immunology* 106:144–158, 2002.
38. Jung, H.C., L. Eckmann, S.K. Yang, A. Panja, J. Fierer, E. Morzyckawroblewska, M.F. Kagnoff. A distinct array of proinflammatory cytokines is expressed in human colon epithelial cells in response to bacterial invasion. *J. Clin. Invest.* 95:55–65, 1995.
39. Eckmann, L., H.C. Jung, C. Schurermaly, A. Panja, E. Morzyckawroblewska, M.F. Kagnoff. Differential cytokine expression by human intestinal epithelial cell lines regulated of interleukin-8. *Gastroenterology* 105:1689–1697, 1993.
40. Haller, D., C. Bode, W.P. Hammes, A.M.A. Pfeifer, E.J. Schiffrin, S. Blum. Non-pathogenic bacteria elicit a differential cytokine response by intestinal epithelial cell/leucocyte co-cultures. *Gut* 47:79–87, 2000.
41. Fujihashi, K., M.N. Kweon, H. Kiyono, J.L. VanCott, F.W. vanGinkel, M. Yamamoto, J.R. McGhee. A T cell/B cell epithelial cell Internet for mucosal inflammation and immunity. *Springer Semin. Immunopathol.* 18:477–494, 1997.
42. Brandtzaeg, P., I.N. Farstad, F.E. Johansen, H.C. Morton, I.N. Norderhaug, T. Yamanaka. The B-cell system of human mucosae and exocrine glands. *Immunol. Rev.* 171:45–87, 1999.
43. Kett, K., K. Baklien, A. Bakken, J.G. Kral, O. Fausa, P. Brandtzaeg. Intestinal B cell isotype response in relation to local bacterial load: evidence for immunoglobulin A subclass adaptation. *Gastroenterology* 109:819–825, 1995.
44. Macpherson, A.J., L. Hunziker, K. McCoy, A. Lamarre. IgA responses in the intestinal mucosa against pathogenic and non-pathogenic microorganisms. *Microb. Infect.* 3:1021–1035, 2001.
45. Dunne, C. Adaptation of bacteria to the intestinal niche: Probiotics and gut disorder. *Inflamm. Bowel Dis.* 7:136–145, 2001.
46. Vaughan, E.E., M.C. de Vries, E.G. Zoetendal, K. Ben Amor, A.D.L. Akkermans, W.M. de Vos. The intestinal LABs. *Antonie Van Leeuwenhoek* 82:341–352, 2002.
47. Harmsen, H.J.M., A.C.M Wildeboer-Veloo, G.C. Raangs, A.A. Wagendorp, N. Klijn, J.G. Bindels, G.W. Welling. Analysis of intestinal flora development in breast-fed and formula-fed infants by using molecular identification and detection methods. *J. Pediatr. Gastroenterol. Nutr.* 30:61–67, 2000.

48. Bjorksten, B., P. Naaber, E. Sepp, M. Mikelsaar. The intestinal microflora in allergic Estonian and Swedish 2-year-old children. *Clin. Exp. Allerg.* 29:342–346, 1999.
49. Tannock, G.W., K. Munro, H.J.M Harmsen, G.W. Welling, J. Smart, P.K. Gopal. Analysis of the fecal microflora of human subjects consuming a probiotic product containing *Lactobacillus rhamnosus* DR20. *Appl. Environ. Microbiol.* 66:2578–2588, 2000.
50. Shroff, K.E., K. Meslin, J.J. Cebra. Commensal enteric bacteria engender a self-limiting humoral mucosal immune response while permanently colonizing the gut. *Infect. Immun.* 63:3904–3913, 1995.
51. Sudo, N., S. Sawamura, K. Tanaka, Y. Aiba, C. Kubo, Y. Koga. The requirement of intestinal bacterial flora for the development of an IgE production system fully susceptible to oral tolerance induction. *J. Immunol.* 159:1739–1745, 1997.
52. Herias, M.V., C. Hessle, E. Telemo, T. Midtvedt, L.A. Hanson, A.E. Wold. Immunomodulatory effects of *Lactobacillus plantarum* colonizing the intestine of gnotobiotic rats. *Clin. Exp. Immunol.* 116:283–290, 1999.
53. Ibnou-Zekri, N., S. Blum, E. Schiffrin, T. von der Weid. Divergent patterns of colonization and immune response elicited from two intestinal *Lactobacillus* strains that display similar properties *in vitro*. *Infect. Immun.* 71:428–436, 2003.
54. Guarner, F., J.R. Malagelada. Gut flora in health and disease. *Lancet* 361:512–519, 2003.
55. Crabbe, P.A., H. Bazin, H. Eyssen, J.F. Heremans. Normal microbial flora as major stimulus for proliferation of plasma cells synthesizing IgA in gut-germ-free intestinal tract. *Int. Arch. Allerg. Appl. Immunol.* 34:362–367, 1968.
56. AbreuMartin, M.T., S.R. Targan. Regulation of immune responses of the intestinal mucosa. *Crit. Rev. Immunol.* 16:277–309, 1996.
57. Matsumoto, S., H. Setoyama, Y. Umesaki. Differential induction of major histocompatibility complex molecules on mouse intestine by bacterial colonization. *Gastroenterology* 103:1777–1782, 1992.
58. Brandtzaeg, P. Development and basic mechanisms of human gut immunity. *Nutr. Rev.* 56: S5–S18, 1998.
59. Ridge, J.P., E.J. Fuchs, P. Matzinger. Neonatal tolerance revisited: turning on newborn T cells with dendritic cells. *Science* 271:1723–1726, 1996.
60. Hunt, D.W.C., H.I. Huppertz, H.J. Jiang, R.E. Petty. Studies of human cord-blood dendritic cells: evidence for functional immaturity. *Blood* 84:4333–4343, 1994.
61. Blum, S., D. Haller, A. Pfeifer, E.J. Schiffrin. Probiotics and immune response. *Clin. Rev. Allerg. Immunol.* 22:287–309, 2002.
62. Cross, M.L. Immunoregulation by probiotic *Lactobacilli*: pro-Th1 signals and their relevance to human health. *Clin. Appl. Immunol. Rev.* 3:115–125, 2002.
63. Lewis, S.A., J.R. Berg, T.J. Kleine. Modulation of epithelial permeability by extracellular macromolecules. *Physiol. Rev.* 75:561–589, 1995.
64. Isolauri, E., H. Majamaa, T. Arvola, I. Rantala, E. Virtanen, H. Arvilommi. *Lactobacillus casei* strain GG reverses increased intestinal permeability induced by cow milk in suckling rats. *Gastroenterology* 105:1643–1650, 1993.
65. Mangell, P., P. Nejdfors, M. Wang, S. Ahrne, B. Westrom, H. Thorlacius, B. Jeppsson. *Lactobacillus plantarum* 299v inhibits *Escherichia coli*-induced intestinal permeability. *Dig. Dis. Sci.* 47:511–516, 2002.
66. Madsen, K., A. Cornish, P. Soper, C. McKaigney, H. Jijon, C. Yachimec, J. Doyle, L. Jewell, C. De Simone. Probiotic bacteria enhance murine and human intestinal epithelial barrier function. *Gastroenterology* 121:580–591, 2001.
67. Ichikawa, H., T. Kuroiwa, A. Inagaki, R. Shineha, T. Nishihira, S. Satomi, T. Sakata. Probiotic bacteria stimulate gut epithelial cell proliferation in rat. *Dig. Dis. Sci.* 44:2119–2123, 1999.
68. Jacobsen, C.N., N.V. Rosenfeldt, A.E. Hayford, P.L. Moller, K.F. Michaelsen, A. Paerregaard, B. Sandstrom, M. Tvede, M. Jakobsen. Screening of probiotic activities of forty-seven strains of *Lactobacillus* spp. by *in vitro* techniques and evaluation of the colonization ability of five selected strains in humans. *Appl. Environ. Microbiol.* 65:4949–4956, 1999.

69. Bernet, M.F., D. Brassart, J.R. Neeser, A.L. Servin. *Lactobacillus acidophilus* La1 binds to cultured human intestinal cell lines and inhibits cell attachment and cell invasion by enterovirulent bacteria. *Gut* 35:483–489, 1994.
70. Ouwehand, A.C., P.V. Kirjavainen, M.M. Gronlund, E. Isolauri, S.J. Salminen. Adhesion of probiotic micro-organisms to intestinal mucus. *Int. Dairy J.* 9:623–630, 1999.
71. Bry, L., P.G. Falk, T. Midtvedt, J.I. Gordon. A model of host-microbial interactions in an open mammalian ecosystem. *Science* 273:1380–1383, 1996.
72. Mack, D.R., S. Michail, S. Wei, L. McDougall, M.A. Hollingsworth. Probiotics inhibit enteropathogenic *E. coli* adherence *in vitro* by inducing intestinal mucin gene expression. *Am. J. Physiol.* 276:G941–G950, 1999.
73. Gill, H.S., K.J. Rutherfurd, J. Prasad, P.K. Gopal. Enhancement of natural and acquired immunity by *Lactobacillus rhamnosus* (HN001), *Lactobacillus acidophilus* (HN017) and *Bifidobacterium lactis* (HN019). *Br. J. Nutr.* 83:167–176, 2000.
74. Chiang, B.L., Y.H. Sheih, L.H. Wang, C.K. Liao, H.S. Gill. Enhancing immunity by dietary consumption of a probiotic lactic acid bacterium (*Bifidobacterium lactis* HN019): optimization and definition of cellular immune responses. *Eur. J. Clin. Nutr.* 54:849–855, 2000.
75. Gill, H.S., K.J. Rutherfurd. Probiotic supplementation to enhance natural immunity in the elderly: effects of a newly characterized immunostimulatory strain *Lactobacillus rhamnosus* HN001 (DR20 (TM)) on leucocyte phagocytosis. *Nutr. Res.* 21:183–189, 2001.
76. Gill, H.S., K.J. Rutherfurd, M.L. Cross, P.K. Gopal. Enhancement of immunity in the elderly by dietary supplementation with the probiotic *Bifidobacterium lactis* HN019. *Am. J. Clin. Nutr.* 74:833–839, 2001.
77. Sheih, Y.H., B.L. Chiang, L.H. Wang, C.K. Liao, H.S. Gill. Systemic immunity-enhancing effects in healthy subjects following dietary consumption of the lactic acid bacterium *Lactobacillus rhamnosus* HN001. *J. Am. Coll. Nutr.* 20:149–156, 2001.
78. Gill, H.S., K.J. Rutherfurd, M.L. Cross. Dietary probiotic supplementation enhances natural killer cell activity in the elderly: an investigation of age-related immunological changes. *J. Clin. Immunol.* 21:264–271, 2001.
79. Spanhaak, S., R. Havenaar, G. Schaafsma. The effect of consumption of milk fermented by *Lactobacillus casei* strain *Shirota* on the intestinal microflora and immune parameters in humans. *Eur. J. Clin. Nutr.* 52:899–907, 1998.
80. Pelto, L., E. Isolauri, E.M. Lilius, J. Nuutila, S. Salminen. Probiotic bacteria down-regulate the milk-induced inflammatory response in milk-hypersensitive subjects but have an immunostimulatory effect in healthy subjects. *Clin. Exp. Allerg.* 28:1474–1479, 1998.
81. Gill, H.S., Q. Shu, H. Lin, K.J. Rutherfurd, M.L. Cross. Protection against translocating *Salmonella typhimurium* infection in mice by feeding the immuno-enhancing probiotic *Lactobacillus rhamnosus* strain HN001. *Med. Microbiol. Immunol.* 190:97–104, 2001.
82. Shu, Q., H.S. Gill. Immune protection mediated by the probiotic *Lactobacillus rhamnosus* HN001 (DR20 (TM)) against *Escherichia coli* O157: H7 infection in mice. *FEMS Immunol. Med. Microbiol.* 34:59–64, 2002.
83. Perdigon, G., M.E.N. Demacias, S. Alvarez, M. Medici, G. Oliver, A.P.D. Holgado. Effect of a mixture of *Lactobacillus casei* and *Lactobacillus acidophilus* administered orally on the immune system in mice. *J. Food Prot.* 49:986–987, 1986.
84. Park, S.Y., G.E. Ji, Y.T. Ko, H.K. Jung, Z. Ustunol, J.J. Pestka. Potentiation of hydrogen peroxide, nitric oxide, and cytokine production in RAW 264.7 macrophage cells exposed to human and commercial isolates of *Bifidobacterium*. *Int. J. Food. Microbiol.* 46: 231–241, 1999.
85. Valdez, J.C., M. Rachid, N. Gobbato, G. Perdigon. Lactic acid bacteria induce apoptosis inhibition in *Salmonella typhimurium* infected macrophages. *Food. Agric. Immunol.* 13:189–197, 2001.
86. Pang, G., A. Buret, R.T. Batey, Q.Y. Chen, L. Couch, A. Cripps, R. Clancy. Morphological phenotypic and functional characteristics of a pure population of CD56+ CD16− CD3− large granular lymphocytes generated from human duodenal mucosa. *Immunology* 79: 498–505, 1993.

87. Desimone, C., B.B. Salvadori, R. Negri, M. Ferrazzi, L. Baldinelli, R. Vesely. The adjuvant effect of yogurt on production of gamma-interferon by Con-A-stimulated human peripheral blood lymphocytes. *Nutr. Rep. Int.* 33:419–433, 1986.
88. Nagao, F., M. Nakayama, T. Muto, K. Okumura. Effects of a fennented milk drink containing *Lactobacillus casei* strain *Shirota* on the immune system in healthy human subjects. *Biosci. Biotechnol. Biochem.* 64:2706–2708, 2000.
89. Haller, D., P. Serrant, D. Granato, E.J. Schiffrin, S. Blum. Activation of human NK cells by *Staphylococci* and *Lactobacilli* requires cell contact-dependent costimulation by autologous monocytes. *Clin. Diagn. Lab. Immunol.* 9:649–657, 2002.
90. Miettinen, M., S. Matikainen, J. Vuopio-Varkila, J. Pirhonen, K. Varkila, M. Kurimoto, I. Julkunen. *Lactobacilli* and *Streptococci* induce interleukin-12 (IL-12), IL-18, and gamma interferon production in human peripheral blood mononuclear cells. *Infect. Immun.* 66:6058–6062, 1998.
91. Hessle, C., L.A. Hanson, A.E. Wold. *Lactobacilli* from human gastrointestinal mucosa are strong stimulators of IL-12 production. *Clin. Exp. Immunol.* 116:276–282, 1999.
92. Christensen, H.R., H. Frokiaer, J.J. Pestka. *Lactobacilli* differentially modulate expression of cytokines and maturation surface markers in murine dendritic cells. *J. Immunol.* 168:171–178, 2002.
93. Lammers, K.M., U. Helwig, E. Swennen, F. Rizzello, A. Venturi, E. Caramelli, M.A. Kamm, P. Brigidi, P. Gionchetti, M. Campieri. Effect of Probiotic strains on interleukin 8 production by HT29/19A cells. *Am. J. Gastroenterol.* 97:1182–1186, 2002.
94. Vidal, K., A. Donnet-Hughes, D. Granato. Lipoteichoic acids from *Lactobacillus johnsonii* strain La1 and *Lactobacillus acidophilus* strain La10 antagonize the responsiveness of human intestinal epithelial HT29 cells to lipopolysaccharide and gram-negative bacteria. *Infect. Immun.* 70:2057–2064, 2002.
95. Michail, S., F. Abernathy. *Lactobacillus plantarum* inhibits the intestinal epithelial migration of neutrophils induced by enteropathogenic *Escherichia coli*. *J. Pediatr. Gastroenterol. Nutr.* 36:385–391, 2003.
96. Cepek, K.L., S.K. Shaw, C.M. Parker, G.J. Russel, J.S. Morrow, D.L. Rimm, M.B. Brenner. Adhesion between epithelial cells and T lymphocytes mediated by cadherin and the alpha(E)beta(7) integrin. *Nature* 372:190–193, 1994.
97. Haller, D., P. Serrant, G. Peruisseau, C. Bode, W.R. Hammes, E. Schiffrin, S. Blum. IL-10 producing CD14(low) monocytes inhibit lymphocyte- dependent activation of intestinal epithelial cells by commensal bacteria. *Microbiol. Immunol.* 46:195–205, 2002.
98. Pereyra, B.S., D. Lemonnier. Induction of human cytokines by bacteria used in dairy foods. *Nutr. Res.* 13:1127–1140, 1993.
99. Kato, I., K. Tanaka, T. Yokokura. Lactic acid bacterium potently induces the production of interleukin-12 and interferon-gamma by mouse splenocytes. *Int. J. Immunopharmacol.* 21:121–131, 1999.
100. Tejada-Simon, M.V., J.J. Pestka. Proinflammatory cytokine and nitric oxide induction in murine macrophages by cell wall and cytoplasmic extracts of lactic acid bacteria. *J. Food. Prot.* 62:1435–1444, 1999.
101. Haller, D., S. Blum, C. Bode, W.P. Hammes, E.J. Schiffrin. Activation of human peripheral blood mononuclear cells by nonpathogenic bacteria *in vitro*: evidence of NK cells as primary targets. *Infect. Immun.* 68:752–759, 2000.
102. Hessle, C., B. Andersson, A.E. Wold. Gram-positive bacteria are potent inducers of monocytic interleukin-12 (IL-12) while gram-negative bacteria preferentially stimulate IL-10 production. *Infect. Immun.* 68:3581–3586, 2000.
103. Shida, K., K. Makino, A. Morishita, K. Takamizawa, S. Hachimura, A. Ametani, T. Sato, Y. Kumagai, S. Habu, S. Kaminogawa. *Lactobacillus casei* inhibits antigen-induced IgE secretion through regulation of cytokine production in murine splenocyte cultures. *Int. Arch. Allerg. Immunol.* 115:278–287, 1998.
104. von der Weid, T., C. Bulliard, E.J. Schiffrin. Induction by a lactic acid bacterium of a population of CD4(+) T cells with low proliferative capacity that produce transforming growth factor beta and interleukin-10. *Clin. Diagn. Lab. Immunol.* 8:695–701, 2001.

105. Cong, Y., C.T. Weaver, A. Lazenby, C.O. Elson. Bacterial-reactive T regulatory cells inhibit pathogenic immune responses to the enteric flora. *J. Immunol.* 169:6112–6119, 2002.
106. Matsuzaki, T., J. Chin. Modulating immune responses with probiotic bacteria. *Immunol. Cell. Biol.* 78:67–73, 2000.
107. De Petrino, S.F., M.E.B. Bonet, O. Meson, G. Perdigon. The effect of *lactobacillus casei* on an experimental model of atopy. *Food Agric. Immunol.* 14:181–189, 2002.
108. Tejada-Simon, M.V., Z. Ustunol, J.J. Pestka. Effects of lactic acid bacteria ingestion of basal cytokine mRNA and immunoglobulin levels in the mouse. *J. Food Prot.* 62:287–291, 1999.
109. Ha, C.L., J.H. Lee, H.R. Zhou, Z. Ustunol, J.J. Pestka. Effects of yogurt ingestion on mucosal and systemic cytokine gene expression in the mouse. *J. Food Prot.* 62:181–188, 1999.
110. Rosenfeldt, V., E. Benfeldt, S.D. Nielsen, K.F. Michaelsen, D.L. Jeppesen, N.H. Valerius, A. Paerregaard. Effect of probiotic *Lactobacillus* strains in children with atopic dermatitis. *J. Allerg. Clin. Immunol.* 111:389–395, 2003.
111. Pessi, T., Y. Sutas, M. Hurme, E. Isolauri. Interleukin-10 generation in atopic children following oral *Lactobacillus rhamnosus* GG. *Clin. Exp. Allerg.* 30:1804–1808, 2000.
112. Ulisse, S., P. Gionchetti, S. D'Alo, F.P. Russo, I. Pesce, G. Ricci, F. Rizzello, U. Helwig, M.G. Cifone, M. Campieri, C. De Simone. Expression of cytokines, inducible nitric oxide synthase, and matrix metalloproteinases in pouchitis: effects of probiotic treatment. *Am. J. Gastroenterol.* 96:2691–2699, 2001.
113. Kishi, A., K. Uno, Y. Matsubara, C. Okuda, T. Kishida. Effect of the oral administration of *Lactobacillus brevis* subsp. *coagulans* on interferon-alpha producing capacity in humans. *J. Am. Coll. Nutr.* 15:408–412, 1996.
114. Shu, Q., H. Lin, K.J. Rutherfurd, S.G. Fenwick, J. Prasad, P.K. Gopal, H.S. Gill. Dietary *Bifidobacterium lactis* (HN019) enhances resistance to oral *Salmonella typhimurium* infection in mice. *Microbiol. Immunol.* 44:213–222, 2000.
115. Kirjavainen, P.V., H.S. El-Nezami, S.J. Salminen, J.T. Ahokas, P.F. Wright. The effect of orally administered viable probiotic and dairy *Lactobacilli* on mouse lymphocyte proliferation. *FEMS Immunol. Med. Microbiol.* 26:131–135, 1999.
116. Novakovic, S., I. Boldogh. *In vitro* TNF-alpha production and *in vivo* alteration of TNF-alpha RNA in mouse peritoneal macrophages after treatment with different bacterial derived agents. *Cancer Lett.* 81:99–109, 1994.
117. Perdigon, G., C.M. Galdeano, J.C. Valdez, M. Medici. Interaction of lactic acid bacteria with the gut immune system. *Eur. J. Clin. Nutr.* 56:S21–S26, 2002.
118. Herias, M.V., T. Midtvedt, L.A. Hanson, A.E. Wold. Increased antibody production against gut-colonizing *Escherichia coli* in the presence of the anaerobic bacterium *Peptostreptococcus. Scan. J. Immunol.* 48:277–282, 1998.
119. Cross, M.L., R.R. Mortensen, J. Kudsk, H.S. Gill. Dietary intake of *Lactobacillus rhamnosus* HN001 enhances production of both Th1 and Th2 cytokines in antigen-primed mice. *Med. Microbiol. Immunol.* 191:49–53, 2002.
120. Kaila, M., E. Isolauri, E. Soppi, E. Virtanen, S. Laine, H. Arvilommi. Enhancement of the circulating antibody secreting cell response in human diarrhea by a human *Lactobacillus* strain. *Pediatr. Res.* 32:141–144, 1992.
121. Isolauri, E., J. Joensuu, H. Suomalainen, M. Luomala, T. Vesikari. Improved immunogenicity of oral DxRRV reassortant rotavirus vaccine by *Lactobacillus casei* GG. *Vaccine* 13:310–312, 1995.
122. Fang, H., T. Elina, A. Heikki, S. Seppo. Modulation of humoral immune response through probiotic intake. *FEMS Immunol. Med. Microbiol.* 29:47–52, 2000.
123. Fukushima, Y., Y. Kawata, K. Mizumachi, J. Kurisaki, T. Mitsuoka. Effect of *Bifidobacteria* feeding on fecal flora and production of immunoglobulins in lactating mouse. *Int. J. Food Microbiol.* 46:193–197, 1999.
124. Perdigon, G., E. Vintini, S. Alvarez, M. Medina, M. Medici. Study of the possible mechanisms involved in the mucosal immune system activation by lactic acid bacteria. *J. Dairy Sci.* 82:1108–1114, 1999.

125. Link-Amster, H., F. Rochat, K.Y. Saudan, O. Mignot, J.M. Aeschlimann. Modulation of a specific humoral immune response and changes in intestinal flora mediated through fermented milk intake. *FEMS Immunol. Med. Microbiol.* 10:55–63, 1994.
126. Marteau, P., P. Seksik, R. Jian. Probiotics and health: new facts and ideas. *Curr. Opin. Biotechnol.* 13:486–489, 2002.
127. Tejada-Simon, M.V., J.H. Lee, Z. Ustunol, J.J. Pestka. Ingestion of yogurt containing *Lactobacillus acidophilus* and *Bifidobacterium* to potentiate immunoglobulin A responses to cholera toxin in mice. *J. Dairy. Sci.* 82:649–660, 1999.
128. Malin, M., H. Suomalainen, M. Saxelin, E. Isolauri. Promotion of IgA immune response in patients with Crohn's disease by oral bacteriotherapy with *Lactobacillus* GG. *Ann. Nutr. Metab.* 40:137–145, 1996.
129. Majamaa, H., E. Isolauri, M. Saxelin, T. Vesikari. Lactic acid bacteria in the treatment of acute rotavirus gastroenteritis. *J. Pediatr. Gastroenterol. Nutr.* 20:333–338, 1995.
130. Kaila, M., E. Isolauri, M. Saxelin, H. Arvilommi, T. Vesikari. Viable versus inactivated *Lactobacillus* strain GG in acute rotavirus diarrhea. *Arch. Dis. Child.* 72:51–53, 1995.
131. Husband, A.J., D.R. Kramer, S. Bao, R.M. Sutherland, K.W. Beagley. Regulation of mucosal IgA responses *in vivo*: cytokines and adjuvants. *Vet. Immunol. Immunopathol.* 54:179–186, 1996.
132. Shu, Q., H.S. Gill. A dietary probiotic (*Bifidobacterium lactis* HN019) reduces the severity of *Escherichia coli* O157: H7 infection in mice. *Med. Microbiol. Immunol.* 189:147–152, 2001.
133. Clements, M.L., M.M. Levine, R.E. Black, R.M. Robinsbrowne, L.A. Cisneros, G.L. Drusano, C.F. Lanata, A.J. Saah. *Lactobacillus* prophylaxis for diarrhea due to enterotoxigenic *Escherichia coli*. *Antimicrob. Agents Chemother.* 20:104–108, 1981.
134. Saavedra, J.M., N.A. Bauman, I. Oung, J.A. Perman, R.H. Yolken. Feeding of *Bifidobacterium bifidum* and *Streptococcus thermophilus* to infants in hospital for prevention of diarrhea and shedding of rotavirus. *Lancet* 344:1046–1049, 1994.
135. Guandalini, S., L. Pensabene, M. Abu Zikri, J.A. Dias, L.G. Casali, H. Hoekstra, S. Kolacek, K. Massar, D. Micetic-Turk, A. Papadopoulou, J.S. de Sousa, B. Sandhu, H. Szajewska, Z. Weizman. *Lactobacillus* GG administered in oral rehydration solution to children with acute diarrhea: a multicenter European trial. *J. Pediatr. Gastroenterol. Nutr.* 30:54–60, 2000.
136. Rosenfeldt, V., K.F. Michaelsen, M. Jakobsen, C.N. Larsen, P.L. Moller, M. Tvede, H. Weyrehter, N.H. Valerius, A. Paerregaard. Effect of probiotic *Lactobacillus* strains on acute diarrhea in a cohort of nonhospitalized children attending day-care centers. *Pediatr. Infect. Dis. J.* 21:417–419, 2002.
137. Szajewska, H., J.Z. Mrukowicz. Probiotics in the treatment and prevention of acute infectious diarrhea in infants and children: a systematic review of published randomized, double-blind, placebo-controlled trials. *J. Pediatr. Gastroenterol. Nutr.* 33:S17–S25, 2001.
138. Bergogne-Berezin, E. Treatment and prevention of antibiotic associated diarrhea. *Int. J. Antimicrob. Agents* 16:521-526, 2000.
139. Black, F.T., P.L. Andersen, J. Ørskov, F. Ørskov, K. Gaarslev, S. Laulund. Prophylactic efficacy of *Lactobacilli* on traveler's diarrhea. *Travel Med.* 7:333–335, 1989.
140. Michetti, P. *Lactobacilli* for the management of *Helicobacter pylori*. *Nutrition* 17:268–269, 2001.
141. Rubaltelli, F.F., R. Biadaioli, P. Pecile, P. Nicoletti. Intestinal flora in breast- and bottle-fed infants. *J. Perinat. Med.* 26:186–191, 1998.
142. Sartor, R.B. Pathogenesis and immune mechanisms of chronic inflammatory bowel diseases. *Am. J. Gastroenterol.* 92:S5–S11, 1997.
143. Madsen, K.L., D. Malfair, D. Gray, J.S. Doyle, L.D. Jewell, R.N. Fedorak. Interleukin-10 gene-deficient mice develop a primary intestinal permeability defect in response to enteric microflora. *Inflamm. Bowel Dis.* 5:262–270, 1999.
144. Madsen, K.L., J.S. Doyle, L.D. Jewell, M.M. Tavernini, R.N. Fedorak. *Lactobacillus* species prevents colitis in interleukin 10 gene-deficient mice. *Gastroenterology* 116:1107–1114, 1999.

145. Schultz, M., C. Veltkamp, L.A. Dieleman, W.B. Grenther, P.B. Wyrick, S.L. Tonkonogy, R.B. Sartor. *Lactobacillus plantarum* 299V in the treatment and prevention of spontaneous colitis in interleukin-10-deficient mice. *Inflamm. Bowel Dis.* 8:71–80, 2002.
146. Rachmilewitz, D., F. Karmeli, K. Takabayashi, T. Hayashi, L. Leider-Trejo, J.D. Lee, L.M. Leoni, E. Raz. Immunostimulatory DNA ameliorates experimental and spontaneous murine colitis. *Gastroenterology* 122:1428–1441, 2002.
147. Rachmilewitz, D., F. Karmeli, K. Takabayashi, E. Raz. Amelioration of experimental colitis by probiotics is due to the immunostimulatory effects of its DNA (abstr). *Gastroenterology* 122(1):T1004, 2002.
148. Gionchetti, P., F. Rizzello, A. Venturi, P. Brigidi, D. Matteuzzi, G. Bazzocchi, G. Poggioli, M. Miglioli, M. Campieri. Oral bacteriotherapy as maintenance treatment in patients with chronic pouchitis: a double-blind, placebo-controlled trial. *Gastroenterology* 119:305–309, 2000.
149. Gionchetti, P., F. Rizzello, A. Venturi, U. Helwig, C. Amadini, K.M. Lammers, F. Ugolini, G. Poggioli, M. Campieri. Prophylaxis of pouchitis onset with probiotic therapy: A double-blind, placebo controlled trial (abstr). *Gastroenterology* 118(1,2):1214, 2000.
150. Mimura, T., F. Rizzello, S. Schreiber, I.C. Talbot, R.J. Nicholls, P. Gionchetti, M. Campieri, M.A. Kamm. Once daily high dose probiotic therapy maintains remission and improved quality of life in patients with recurrent or refractory pouchitis: a randomised, placebo-controlled, double- blind trial (abstr). *Gastroenterology* 122(1):667, 2002.
151. Kruis, W., P. Fric, M.S. Stolte. Maintenance of remission in ulcerative colitis is equally effective with *Escherichia coli* Nissle 1917 and with standard mesalamine (abstr). *Gastroenterology* 120(1):680, 2001.
152. Bjorksten, B., D. Dumitrascu, T. Foucard, N. Khetsuriani, R. Khaitov, M. Leja, G. Lis, J. Pekkanen, A. Priftanji, M.A. Riikjarv. Prevalence of childhood asthma, rhinitis and eczema in Scandinavia and Eastern Europe. *Eur. Respir. J.* 12:432–437, 1998.
153. Strachan, D.P., Hay fever, hygiene, and household size. *Br. Med. J.* 299:1259–1260, 1989.
154. Wold, A.E. The hygiene hypothesis revised: is the rising frequency of allergy due to changes in the intestinal flora? *Allergy* 53:20–25, 1998.
155. Bottcher, M.F., E.K. Nordin, A. Sandin, T. Midtvedt, B. Bjorksten. Microflora-associated characteristics in faeces from allergic and nonallergic infants. *Clin. Exp. Allergy*. 30:1590–1596, 2000.
156. Kalliomaki, M., P. Kirjavainen, E. Eerola, P. Kero, S. Salminen, E. Isolauri. Distinct patterns of neonatal gut microflora in infants in whom atopy was and was not developing. *J. Allerg. Clin. Immunol.* 107:129–134, 2001.
157. He, F., A.C. Ouwehand, E. Isolauri, H. Hashimoto, Y. Benno, S. Salminen. Comparison of mucosal adhesion and species identification of *Bifidobacteria* isolated from healthy and allergic infants. *FEMS Immunol. Med. Microbiol.* 30:43–47, 2001.
158. Ouwehand, A.C., E. Isolauri, F. He, H. Hashimoto, Y. Benno, S. Salminen. Differences in *Bifidobacterium* flora composition in allergic and healthy infants. *J. Allerg. Clin. Immunol.* 108:144–145, 2001.
159. He, F., H. Morita, H. Hashimoto, M. Hosoda, J.I. Kurisaki, A.C. Ouwehand, E. Isolauri, Y. Benno, S. Salminen. Intestinal *Bifidobacterium* species induce varying cytokine production. *J. Allerg. Clin. Immunol.* 109:1035–1036, 2002.
160. Majamaa, H., E. Isolauri. Probiotics: a novel approach in the management of food allergy. *J. Allerg. Clin. Immunol.* 99:179–185, 1997.
161. Isolauri, E., T. Arvola, Y. Sutas, E. Moilanen, S. Salminen. Probiotics in the management of atopic eczema. *Clin. Exp. Allerg*. 30:1604–1610, 2000.
162. Kalliomaki, M., S. Salminen, H. Arvilommi, P. Kero, P. Koskinen, E. Isolauri. Probiotics in primary prevention of atopic disease: a randomised placebo-controlled trial. *Lancet* 357:1076–1079, 2001.
163. Rautava, S., M. Kalliomaki, E. Isolauri. Probiotics during pregnancy and breastfeeding might confer immunomodulatory protection against atopic disease in the infant. *J. Allerg. Clin. Immunol.* 109:119–121, 2002.

164. Herlin, T., S. Haahtela, T. Haahtela. No effect of oral treatment with an intestinal bacterial strain, *Lactobacillus rhamnosus* (ATCC 53103), on birch-pollen allergy: a placebo-controlled double-blind study. *Allergy* 57:243–246, 2002.
165. Horie, H., K. Kanazawa, M. Okada, S. Narushima, K. Itoh, A. Terada. Effects of intestinal bacteria on the development of colonic neoplasm: an experimental study. *Eur. J. Cancer Prevent.* 8:237–245, 1999.
166. Moore, W.E.C., L.H. Moore. Intestinal floras of populations that have a high risk of colon cancer. *Appl. Environ. Microbiol.* 61:3202–3207, 2003.
167. J. Rafter. Lactic acid bacteria and cancer: mechanistic perspective. *Br. J. Nutr.* 88:S89–S94, 2002.
168. Takahashi, T., A. Kushiro, K. Nomoto, K. Uchida, M. Morotomi, T. Yokokura, H. Akaza. Antitumor effects of the intravesical instillation of heat killed cells of the *Lactobacillus casei* strain *Shirota* on the murine orthotopic bladder tumor MBT-2. *J. Urol.* 166:2506–2511, 2001.
169. Perdigon, G., A.D. de LeBlanc, J. Valdez, M. Rachid. Role of yoghurt in the prevention of colon cancer. *Eur. J. Clin. Nutr.* 56:S65–S68, 2002.
170. Coconnier, M.H., M.F. Bernet, G. Chauviere, A.L. Servin. Adhering heat-killed human *Lactobacillus acidophilus,* strain LB, inhibits the process of pathogenicity of diarrhoeagenic bacteria in cultured human intestinal cells. *J. Diarrhoeal Dis. Res.* 11:235–242, 1993.
171. Neumann, E., M.A.P. Oliveira, C.M. Cabral, L.N. Moura, J.R. Nicoli, E.C. Vieira, D.C. Cara, G.I. Podoprigora, L.Q. Vieira. Monoassociation with *Lactobacillus acidophilus* UFV-H2b20 stimulates the immune defense mechanisms of germfree mice. *Braz. J. Med. Biol. Res.* 31:1565–1573, 1998.
172. Gill, H.S., K.J. Rutherfurd. Viability and dose-response studies on the effects of the immunoenhancing lactic acid bacterium *Lactobacillus rhamnosus* in mice. *Br. J. Nutr.* 86:285–289, 2001.
173. Kirjavainen, P.V., S.J. Salminen, E. Isolauri. Probiotic bacteria in the management of atopic disease: Underscoring the importance of viability. *J. Pediatr. Gastroenterol. Nutr.* 36:223–227, 2003.
174. Rayes, N., D. Seehofer, S. Hansen, K. Boucsein, A.R. Muller, S. Serke, S. Bengmark, P. Neuhaus. Early enteral supply of *Lactobacillus* and fiber versus selective bowel decontamination: a controlled trial in liver transplant recipients. *Transplantation* 74:123–128, 2002.
175. Holzapfel, W.H., P. Haberer, J. Snel, U. Schillinger, F.H.F. Huis in't Veld, V. Overview of gut flora and probiotics. *Int. J. Food. Microbiol.* 41:85–101, 1998.
176. Umesaki, Y., H. Setoyama. Structure of the intestinal flora responsible for development of the gut immune system in a rodent model. *Microbes Infect.* 2:1343–1351, 2000.
177. Elwood, C.M., O.A. Garden. Gastrointestinal immunity in health and disease. *Vet. Clin. N. Am. Small Anim. Pract.* 29:471–487, 1999.
178. Michetti, P., G. Dorta, P.H. Wiesel, D. Brassart, E. Verdu, M. Herranz, C. Felley, N. Porta, M. Rouvet, A.L. Blum, I. Corthesy-Theulaz. Effect of whey-based culture supernatant of *Lactobacillus acidophilus (johnsonii)* La1 on *Helicobacter pylori* infection in humans. *Digestion* 60:203–209, 1999.
179. Gill, H.S., M.L. Cross, K.J. Rutherfurd, P.K. Gopal. Dietary probiotic supplementation to enhance cellular immunity in the elderly. *Br. J. Biomed. Sci.* 58:94–96, 2001.
180. Arunachalam, K., H.S. Gill, R.K. Chandra. Enhancement of natural immune function by dietary consumption of *Bifidobacterium lactis* (HN019). *Eur. J. Clin. Nutr.* 54:263–267, 2000.
181. Delia, P., G. Sansotta, V. Donato, G. Messina, P. Frosina, S. Pergolizzi, C. De Renzis, G. Famularo. Prevention of radiation-induced diarrhea with the use of VSL#3, a new high-potency probiotic preparation. *Am. J. Gastroenterol.* 97:2150–2152, 2002.
182. Brigidi, P., B. Vitali, E. Swennen, G. Bazzocchi, D. Matteuzzi. Effects of probiotic administration upon the composition and enzymatic activity of human fecal microbiota in patients with irritable bowel syndrome or functional diarrhea. *Res. Microbiol.* 152:735–741, 2001.
183. de Waard, R., E. Claassen, G.C.A.M. Bokken, B. Buiting, J. Garssen, J.G. Vos. Enhanced immunological memory responses to *Listeria monocytogenes* in rodents, as measured by delayed-type hypersensitivity (DTH), adoptive transfer of DTH, and protective immunity, following *Lactobacillus casei Shirota* ingestion. *Clin. Diagn. Lab. Immunol.* 10:59–65, 2003.

184. Hori, T., J. Kiyoshima, H. Yasui. Effect of an oral administration of *Lactobacillus casei* strain *Shirota* on the natural killer activity of blood mononuclear cells in aged mice. *Biosci. Biotechnol. Biochem.* 67:420–422, 2003.
185. Perdigon, G., S. Alvarez, A.P.D. Holgado. Immunoadjuvant activity of oral *Lactobacillus casei*: influence of dose on the secretory immune response and protective capacity in intestinal infections. *J. Dairy Res.* 58:485–496, 1991.
186. Perdigon, G., M.E.N. Demacias, S. Alvarez, G. Oliver, A.A.P.D. Holgado. Prevention of gastrointestinal *infection using immunobiological methods with milk fermented with Lactobacillus* casei and *Lactobacillus acidophilus*. *J. Dairy Res.* 57:255–264, 1990.
187. Gonzalez, S., G. Albarracin, M.L.R. Pesce, M. Male, M.C. Apella, A.A.P.D. Holgado, G. Oliver. Prevention of infantile diarrhoea by fermented milk. *Microbiol. Aliment. Nutr.* 8:349–354, 1990.
188. Gonzalez, S.N., R. Cardozo, M.C. Apella, G. Oliver. Biotherapeutic role of fermented milk. *Biotherapy* 8:129–134, 1995.

2.26

Biochemical Markers for Antioxidant Functionality

Dhiraj A. Vattem and Kalidas Shetty

CONTENTS

26.1 INTRODUCTION

An imbalance of prooxidants and antioxidants, which leads to oxidative stress, is now believed to be a key contributing factor in the manifestation of chronic diseases such as cardiovascular disease, hypertension, diabetes mellitus, and some forms of cancer. (1,2). Recent research has shown that populations consuming diets rich in fruits and vegetables have lower incidences of many chronic diseases (3–6), which has encouraged the use of diet as a complementary strategy for the management of these oxidative diseases (7–10). Fruits and vegetables are rich sources of phytochemicals and highly antioxidant vitamins, which are now believed to be responsible for these beneficial effects. The function of antioxidants from fruits and vegetables has been categorized into two types depending on their mode of action. First, oxidation of biological macromolecules such as nucleic acids, proteins, and lipids as a result of free radical damage (oxidative stress) has now been strongly associated with development of many physiological conditions which can manifest as disease (11–17). The protective effect of antioxidants from fruits and vegetables, therefore, is believed to be due to their direct involvement in neutralizing free radicals in biological systems and preventing oxidative damage to cellular systems (18).

Second, reactive oxygen species (ROS) are naturally formed in the body as a result of many metabolic processes. Low levels of ROS therefore, have been implicated in many cellular processes, including intracellular signaling responsible for cell division (11,19,20), immune response (67–69), and via the activation and repression of many genes (11). However, excess formation of ROS can cause damage and therefore disturb cellular homeostasis, so cells have evolved several antioxidant enzyme coupled metabolic systems which can which can quickly remove the excess ROS from cellular systems (22,23). These antioxidant systems involved in the protection against ROS consist of many antioxidant factors including glutathione (GSH) and other tissue thiols, heme proteins, coenzyme Q, bilirubin and urates, and several antioxidant enzymes such as superoxide dismutase (SOD), catalase (CAT), glutathione peroxidase (GPX), and glutathione-S-transferase (GST) (Figure 26.1). Dietary antioxidants from fruits and vegetables and other functional foods are believed to modulate cellular physiology and participate in induction/repression of gene expression or activation/deactivation of proteins, enzymes, and transcription factors of key metabolic pathways (11,21–23) leading to activation of these biological antioxidant systems in the cell.

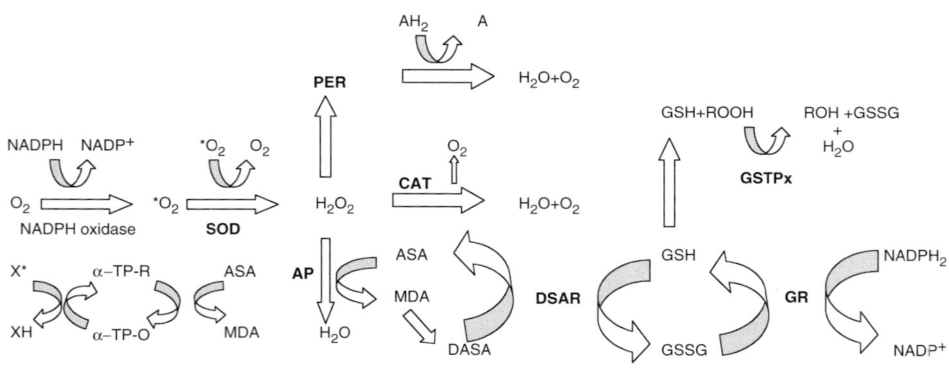

Figure 26.1 The antioxidant defense response of the cell carried out by enzymatic as well as the nonenzymatic antioxidants

The ability of fruits and vegetables to manage oxidation linked diseases has resulted in a strong interest in the development functional foods and other dietary strategies for management of oxidation diseases (3–10). These developments have led to the need for good biomarkers and analytical methodologies to determine how these diets work. Evaluation of the functional antioxidant efficacy of new phytochemicals or functional food products requires a wide range of specific, accurate, and sensitive biomarkers appropriately associated with oxidative stress and related diseases. The biomarkers help to characterize food and food products with respect to their antioxidant activity and associated functional properties; in *in vitro* and *in vivo* models as well as in clinical studies. Some well established biochemical biomarkers currently used for studying antioxidant function in the management of oxidative stress are described here.

26.2 DIRECT MEASUREMENT OF REACTIVE OXYGEN SPECIES

Molecular oxygen in biological system is relatively unreactive and plays an important role in the proper functioning of many biological processes (11). However, metabolic pathways in the cell can reduce oxygen incompletely to produce reactive oxygen species (ROS), which are unstable compared to molecular oxygen, and can react with a number of biological macromolecules and disturb the cellular homeostasis (11–17). ROS are constantly produced in aerobic organisms both enzymatically and nonenzymatically (Figure 26.2). One electron reduction of molecular oxygen in biological systems results in the formation of the superoxide anion which is the precursor of most ROS (24). Biological dismutation of the superoxide anion produces the peroxide radical, which can undergo a complete reduction to form water, or an incomplete reduction to form the highly reactive hydroxyl radical (24) (Figure 26.2). Mitochondrial leakage of electrons or direct transfer of electrons to oxygen via the electron transport chain is another important source of ROS in cellular systems (25,26). Many other sources for the formation of ROS have been identified in cellular systems of living organisms (25,26). Superoxide is also formed upon one electron reduction of oxygen mediated by enzymes such as NADPH oxidase located on the cell membrane of polymorphonuclear cells, macrophages, and endothelial cells; from xanthine oxidase or from the respiratory chain (27–30). Superoxide radicals can also be formed from the electron cycling carried out by the

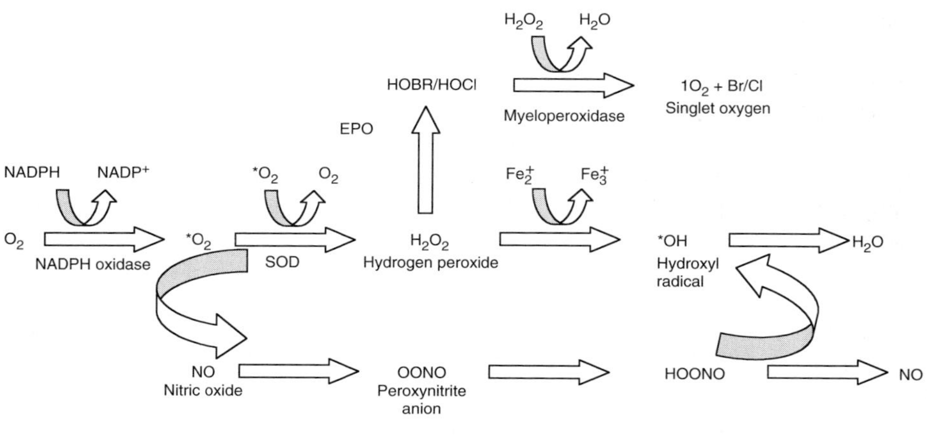

Figure 26.2 Biological mechanisms for ROS formation

cytochrome P450-dependent enzymes in detoxification reactions (31,32). Oxidation of biological molecules has been linked to manifestation of many diseases and therefore the presence of reactive oxygen species in biological systems is often considered as a marker for oxidative stress.

Many quantitative assays are available for the detection of ROS. The majority of the methods are based on the identification of the peroxide radical, as it more stable in aqueous solution at neutral pH (33–35). H_2O_2 and other ROS are powerful oxidizing agents; therefore the most common quantification of H_2O_2 has been based on their oxidative reaction with chemiluminescent or fluorescent dyes. The neutralization of the radicals from the H_2O_2 by the dyes results in their gain or loss of fluorescence which is monitored fluorometrically (36,37). Peroxidases are used to remove the H_2O_2 radicals from the reaction mixture, and enzyme activity is monitored by the change in the fluorescence. Dependence upon peroxidases to detect H_2O_2 and lack of absolute specificity for H_2O_2 (38) are some of the limitations of this method. Other methods that are also used for direct measurement of the ROS include the use of a Clark-type oxygen electrode. Here the oxygen released from H_2O_2 in the cell supernatant is measured after the addition of catalase (39,40). Superoxide radicals are measured by electron spin resonance spectroscopy (ESR) (40). $O_2^{\cdot -}$ in solution is indirectly detected by a spin trap method by using common trapping agents such as 5,5-dimethyl-1-pyrroline N-oxide (DMPO) (41). The ESR spectrum pattern obtained from DMPO is specific to $O_2^{\cdot -}$ and therefore enables its accurate detection (41). However, due to a lower rate constant between DMPO and $O_2^{\cdot -}$, a large amount of DMPO should be added to the solution, which increases the cost per assay. Another drawback with this method is the requirement of a relatively expensive ESR instrument (41,42).

26.3 ANTIOXIDANT ASSSAYS

The total antioxidant capacity of cell extract or blood plasma can give us an indication about the functionality of the compound or food of interest (39). A number of antioxidant assays are available to study the antioxidant activity.

26.3.1 Antioxidant Assays Based on Free Radical Neutralization

Many antioxidant assays are based on their ability to neutralize or quench free radicals. The two free radicals that have been most commonly used for assessing antioxidant activity are 1,1-diphenyl-2-picrylhydrazyl (DPPH)(43,44), and 2,2′-azinobis(3-ethyl-benzothiazoline-6-sulfonic acid) (ABTS)(45,46) (Figure 26.3).

The DPPH free radical is a stable radical with one electron delocalized over the molecule (Figure 26.3). This delocalization gives a deep purple color with absorption maxima at 517 nm in an ethanol solution (43,44). When an antioxidant capable of donating a hydrogen reacts with the DPPH radical it gives rise to a nonradical reduced form of DPPH which has a yellow color. The decrease in the absorption is measured spectrophotometrically and is compared with an ethanol control to calculate the DPPH free radical scavenging activity (43,44).

$$\text{DPPH (Purple)} + \text{AO} \rightarrow \text{DPPH (Yellow)} + \text{AO} \qquad (26.1)$$

This method is a very quick and simple method for the measurement of antioxidant activity. The antioxidant capacity in this assay depends on the chemical structure of the antioxidant (43,44). The reduction in the DPPH radical is dependent on the number of hydroxyl groups present in the antioxidant. This method, therefore, gives an indication of the structural dependence of the antioxidant functionality of many biological antioxidants (43,44).

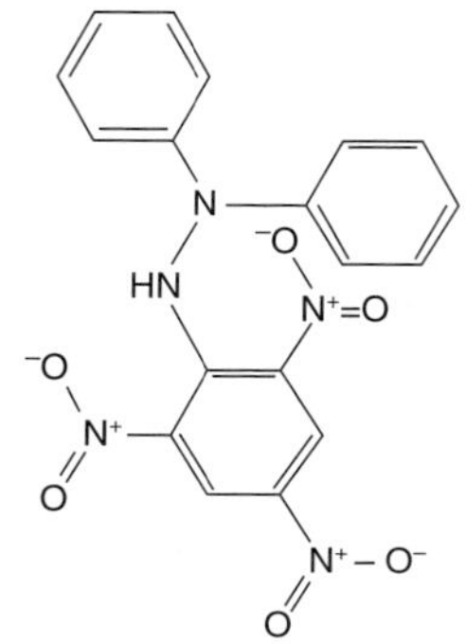

1-1-diphenyl-2-picryl-hydrazyl (free radical)

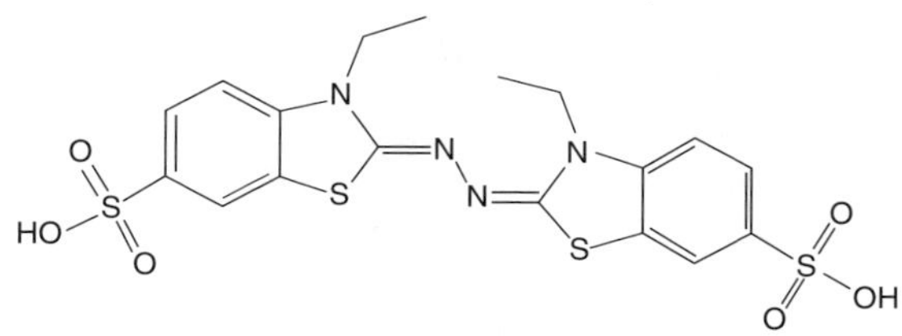

2,2′-azinobis(3-ethyl-benzothiazoline-6-sulfonic acid)

Figure 26.3 DPPH and ABTS — the two most commonly used indicators in antioxidant assays based on free radicals quenching

The absorption maxima of the DPPH radical in ethanol solution at 517 nm is considered as one of the main drawbacks of the assay as it precipitates out most of the proteins and reduces the activity of a majority of single valance phenolic hydroxides (47,48). Dependence on the number of hydroxyl groups also results in the different kinetic rates of reaction between the antioxidant and the DPPH radical and causes fluctuation in the reaction times significantly, suggesting the need for standardization of reaction time and kinetic modeling for different samples (43,44,47,48).

26.3.2 2,2'-Azinobis-(3-ethylbenzthiazoline-6-sulfonic acid)

The antioxidant assay based on the 2,2′-Azinobis-(3-ethylbenzthiazoline-6-sulfonic acid) (ABTS) is similar to the DPPH assay (Figure 26.3). ABTS can react with hydroxyl, peroxyl, alkoxyl and inorganic radicals to form a relatively stable radical cation (ABTS) with absorption maxima at 734 nm (45,46). The antioxidant, added to the reaction mixture prior to the addition of the ROS or the radical, scavenges the free radical and prevents the formation of the ABTS radical cation. This scavenging activity can be spectrophotometrically and quantitated by comparing the antioxidant activity of the sample, extract, or biological fluid with trolox (45,46).

$$\text{ABTS} + \text{ROS} \rightarrow \text{ABTS} \qquad (26.2)$$

Inhibition

$$\text{ROS} + \text{AO} \rightarrow \text{R{-}H} + \text{AO}$$

The ABTS assay can determine the antioxidant contribution from different components in the system as it is capable of reacting with hydroxyl, cysteinyl, and glutathione in biological systems, and therefore has been widely used to study the antioxidant activity of various compounds, both lipophilic and hydrophilic, food products and biological samples (45–48). However, in biological samples the antioxidant enzymes such as peroxidases have been shown to be able to promote the formation of ABTS radical therefore contributing to potential interference resulting in higher absorbance values. Due to the relatively low concentrations of hydrogen peroxide, this assay cannot accurately measure the free radical quenching, as the antioxidants can react with the ABTS radical cation directly instead of the hydrogen peroxide, which can make it look as if there is more antioxidant activity than there really is (45,46). The assay also measures the antioxidant activity of the sample in a fixed time, and does not take into account the time for the completion of the reaction, which may not be suitable as it overlooks the total antioxidant capacity of the sample (47,48).

26.3.3 Ferric Reducing Ability of Plasma (FRAP)

In the ferric reducing ability of plasma (FRAP), plasma or extract is allowed to react with ferric tripyridyltriazine (Fe-III-TPTZ) complex at low pH. Reduction of ferric iron (Fe III) in the complex to ferrous iron (Fe II) results in an intense blue color. This resulting complex can be measured spectrophotometrically at 593 nm, and the intensity of the blue color developed is directly proportional to the antioxidant capacity (49).

$$Fe^{3+}(TPTZ)_2 + AO \rightarrow Fe^{2+}(TPTZ)_2 \text{ (Blue)} + AO \quad (26.3)$$

The main advantages of the assay lie in its relative simplicity, which enables high throughput, and its cheaper cost (48,49). It is a fast assay, and reproducible results can be obtained from plasma and simple antioxidant solutions with minimal sample preparation. However, the reaction is nonspecific, as any compound with a lower redox potential than the standard redox potential of Fe(III) and Fe(II) (0.77 V) can reduce the complex and contribute to high FRAP values. The reaction also assumes that completion of the reduction is spontaneous. However, many biological antioxidants, especially phytochemicals such as caffeic acid and quercetin, continue to reduce the ferric complex even after the completion of reaction time (48,50). The antioxidant activity of any system depends on the pH of the medium in which it is present. The low pH of the reaction in the FRAP assay significantly inhibits one electron transfer from the antioxidant to the ferric ion, resulting in lower values which may not be accurate at physiologically relevant pH (pH 6.8–7.0) (47,48,50).

26.3.4 Oxygen Radical Absorbance Capacity (ORAC) Assay

The ORAC assay is very similar to the direct measurement ROS assay. In this assay, the time dependent decrease in the fluorescence of a fluorescent indicator is monitored at 37°C. The indicator often used is a protein called β-phycoerythrin (beta-PE) (50,51). 2-2′-azobis (2-amidinopropane)dihydrochloride (AAPH) is used to generate peroxyl radicals; oxidative damage of the beta-PE protein results in a decreased fluorescence, which is prevented by the antioxidant (50,51). The total area under the curve determines the antioxidant capacity of the compound or extract. This is a simple, sensitive, and reproducible method of measuring the antioxidant capacity of the natural extracts, biological fluids, and serum against peroxyl radicals (47,48). The method has also been adopted to measure antioxidant capacity against hydroxyl radicals, and the relative simplicity of the assay has resulted in the development of a microplate assay which can be used for screening large number of samples (50,51). Inconsistencies in the purity of the beta-PE protein, coupled with its interaction with many biological antioxidants, especially polyphenols, are one of the drawbacks of this method (48). Recently, fluorescein salt has been used in place of beta-PE to reduce some of the inconsistencies in the analysis (52) (Figure 26.4). One of the main limitations of this method is that it measures the ORAC capacity in comparison to a synthetic antioxidant, trolox, and the ORAC capacity is expressed as trolox equivalents (48). The assay also suggests comparison of the antioxidant activity to a synthetic antioxidant (trolox) which can give misleading information, because the antioxidant capacity of a biological sample depends on many factors such as structure, type of antioxidants, and the synergistic interaction among the constituent antioxidants (53–55). When copper is used as oxidant, trolox cannot be used, because it would act as a prooxidant in the presence of copper (47,56). In biological samples, especially plasma, interference in the scavenging of peroxyl radicals by protein thiol groups may occur; therefore protein needs to be removed from the extract before the assay (57). Nonspecificity in the reaction, especially in biological materials, is introduced by the presence of oxidizable compounds such as citric acid, which can react with the ROS and thus prevent the oxidative damage of the fluorescein and beta-PE protein, which reacts with oxygen radicals more slowly than most biological antioxidants such as

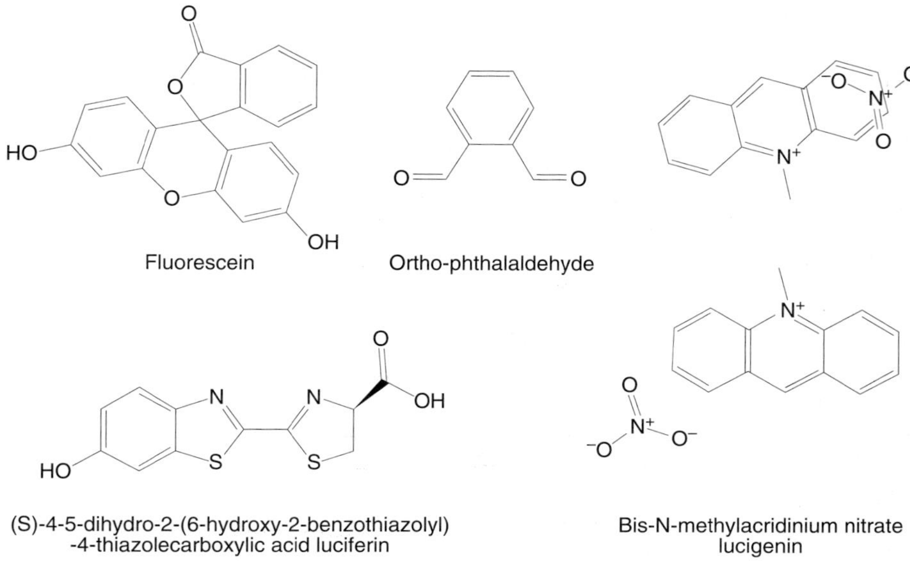

Figure 26.4 Some common fluorescent and chemiluminiscent dyes

thiols, uric acid, and ascorbate (48,52) which can cause a falsely high value for the antioxidant capacity.

26.3.5 β-carotene Oxidation (Bleaching) Assay

The antioxidant protection factor (APF) directly measures the ability of the antioxidant extract to prevent the H_2O_2 catalyzed oxidation of β-carotene. Many antioxidant methodologies based on oxidation of an indicator dye are time consuming, as they depend on the conditions used for the reaction and the oxidizability of the reagent (58–60). However, the β-carotene bleaching assay is a relatively fast method for analysis of antioxidant activity. The reaction is accelerated by carrying out the oxidation at 50°C for 30 min, and free radicals are directly generated using hydrogen peroxide (58,60). Because β-carotene is insoluble in aqueous media, it is introduced into the reaction mixture in the form of an emulsion. Many biological sites of ROS generation and oxidation are at a lipid water interface; therefore, it is believed that an antioxidant that is capable of preventing the bleaching of the β-carotene in the emulsion would be able to act efficiently at the lipid to water interface in biological systems, so the β-carotene bleaching assay will give a measure of the true antioxidant capacity (61,62). This method is sensitive and simple to perform, however, irregularities are introduced in the analysis due to the difficulties in keeping the emulsion particle size constant. The method also suffers for its nonspecificity as it is subject to interference from other naturally occurring oxidizing and reducing agents (60,63).

26.4 LIPID OXIDATION PRODUCTS

In aerobic organisms, reactive oxygen species (ROS) are formed by various endogenous and exogenous processes, such as the mitochondrial electron transport chain, cytochrome P-450 systems, NADPH oxidases, myeloperoxidases and xanthine oxidases, ultraviolet, and ionizing irradiation (11–17,64,65). Excessive formation of ROS, can lead to oxidative stress, which can damage cellular macromolecules including nucleic acids, proteins, and lipids

(11–17,64,65). Biological membranes contain unsaturated fatty acids which are especially prone to oxidation. This oxidation of lipids results in the formation of lipid peroxidation (LPO) products which further propagate free radical reactions. Lipid hydroperoxides (LOOH) are the primary oxidation products (66), and have been shown to oxidize other lipids and cellular proteins (67), and to form cytotoxic and genotoxic compounds (68,69). LPO has been therefore been implicated in the pathogenesis of a number of degenerative diseases, such as cancer (70), atherosclerosis (71–73), and Alzheimer's disease (74,75). It is now believed that primary and secondary lipid oxidation products can serve as good markers for oxidative stress and the functionality of antioxidants (70–75).

26.4.1 Malondialdehyde

Oxidation of lipids in biological systems by reactive oxygen species results in the formation malondialdehyde (MDA) which is a metabolite of lipid hydroperoxides (74–78). MDA is a secondary oxidation product of lipids and serves as a good marker for lipid oxidation and cell membrane injury (78). Recent research has shown that elevated levels of lipid peroxides are associated with cancer, heart disease, stroke, and aging. The MDA concentration in plasma can be determined by its reaction with thiobarbituric acid (TBA) (76). The reaction of TBA with MDA in biological materials can be detected spectrophotometrically at 532–535 nm. This method is simple and easy to perform, but suffers from lack of sensitivity as many compounds, especially the products of the side reactions, also have absorption maxima in the 532–535 nm range (73). Other products of lipid peroxidation, such as hydroperoxides and conjugated aldehydes, can also react with TBA and lower the sensitivity of the method (77). Reducing sugars and aldehydes that have a carbonyl group can also react with TBA to give higher readings (77). MDA in biological samples can also react with proteins and give abnormally low values in an oxidizing system (78). The method has been improved vastly by using diaminonapthalene (DAN) instead of TBA to form a DAN–MDA complex. (78). Unlike TBA, DAN reacts very specifically with MDA; other biological compounds such as sugars, aldehydes, and proteins cannot react with DAN. The DAN-MDA complex can also be easily measured by HPLC and has analytical sensitivity and specificity (78). The HPLC modification of the MDA assay has been extensively used to study lipid peroxidation in *in vivo* studies (78). Other methods of analysis include GC using N-methylhydrazine derivatization (79) and ELISA (80).

26.4.2 LDL Resistance to Oxidation

Lipid hydroperoxides are primary product of free radical oxidation of fatty acids and serve as substrates for the reactions that produce free radicals, which in turn form more lipid hydroperoxides (81). Due to their high reactivity and instability, measurement of LOOH *in vivo* is a very challenging (81). At present, the precise measurement of LOOH in plasma remains a difficult task because of low concentration, instability during extraction, and solvent evaporation, leading to both changes in LOOH composition and production of new hydroperoxides if chain reactions are initiated (82,83). Recent research has shown that peroxidative oxidation of LDL in biological systems is responsible for initiation of plaque formation and enhanced atherogenecity (81). Ability to reduce LDL oxidation by peroxidation is used as an important biomarker for studying the antiatherogenic functionality of antioxidant compounds (82,83). The LDL oxidation is induced by prooxidant metals such as copper, and other free radical generating systems such as 2,2′-azobis(2-amidinopropane) dihydrochloride (AAPH) (82,83). The presence and absence of antioxidants, signifying the lag phase before the oxidation of LDL , can be detected by measuring the rate of oxidation and total amount of dienes at 234 nm using the molar extinction coefficient for total lipid

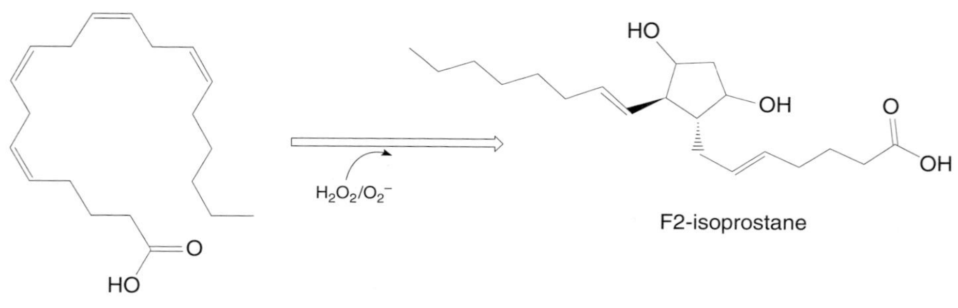

Figure 26.5 Nonenzymatic metabolism of arachidonic acid to F-2 Isoprostanes

hydroperoxides [ε_{234} = 29,500 (mol/L)$^{-1}$ cm^{-1}] (82,83). The total oxidation in the LDL can be measured spectrofluorometrically by extracting the lipid phase and drying it under nitrogen, then resuspending it in chloroform (84–86). This chloroform extract of the lipid phase is subjected to 360 nm excitation and 430 nm emission, and the results are expressed as units of relative fluorescence per milligram of LDL (84–86). The lipid peroxide content can be estimated fluorometrically by the formation of TBARS and expressed as nanomoles of malondialdehyde equivalents per milligram of LDL (84–86).

26.4.3 F2 Isoprostanes

Prostaglandin F2-like compounds (F2-isoprostanes) are formed by nonenzymatic free radical induced peroxidation of arachidonic acid (87) (Figure 26.5). These compounds are formed in phospholipids and then cleaved and released into the circulation before excretion in the urine as free isoprostanes (87). They have been extensively used as novel markers of endogenous lipid peroxidation and potential mediators of oxidant injury (88,89). Elevated levels of F2-isoprostanes have been seen in many diseases including diabetes, myocardial infraction, atherosclerosis, Alzheimer's, and in hepatic toxicity (90–94). Detection and quantification of F2-isoprostanes in biological samples including blood serum and plasma is now used as a tool in assessing oxidative stress status (95). The most common isoprostane formed is 8-isoprostaglandin F_2, which can be detected accurately and easily by using GC/MS or by HPLC/MS (95–97). The longer analysis time required for GC/MS and HPLC/MS has resulted in the development of many immunochemical assays (91,98). However, these immunochemical methods suffer crossreactivity from biological prostaglandins due to their close structural similarity(91).

26.5 DNA OXIDATION PRODUCTS

26.5.1 8-Oxo-7,8-dihydro-2′-deoxyguanosine

Free radicals have been shown to cause damage to lipids, DNA, and proteins. DNA oxidation in biological systems can occur by its interaction with ionic radiation, trace metals, and many carcinogens (11–17,64,65). DNA damage is caused by the oxidation and degeneration of nucleotide bases resulting in the formation of 2-OH-2′deoxyadenosine, 8-Oxo-7,8-dihydro-2′-deoxyguanosine, 5-hydroxymethyl-2′-deoxyuridine (99). However, 8-OHdG , formed from oxidative damage of guanine, is considered a typical marker of DNA damage and repair because guanine is the most fragile nuclear base (99,100). 8-Oxo-7,8-dihydro-2′-deoxyguanosine (8-OHdG), is formed from deoxyguanosine after reaction with a hydroxyl radical or

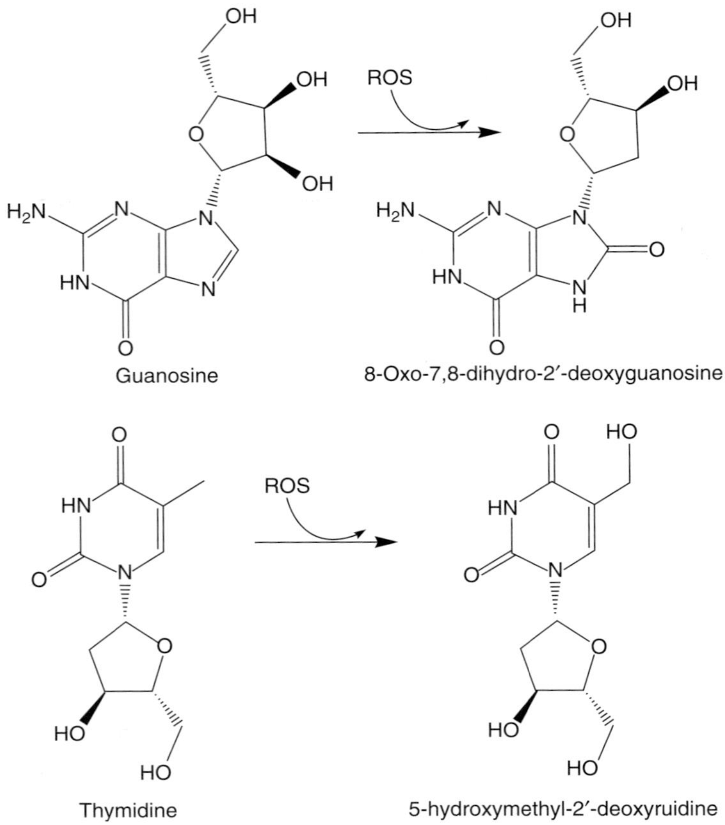

Figure 26.6 DNA oxidation by reactive oxygen species

singlet oxygen (Figure 26.6). 8-OHdG is not metabolized and is quantitatively excreted in urine independent of other metabolic processes, enabling it to be used as a noninvasive marker of oxidative DNA damage by ROS in various physiological and nutritional studies (99,100). Oxidative damage to DNA has now been shown to be responsible for aging, cardiovascular disease, degenerative neurological disease, and various types of cancer (11–17,64,65,99,100). 8-OHdG can be analyzed by many different assays. The most common methodology is high performance liquid chromatography using an electrochemical detector (HPLC-ECD) (99,100). ^{32}P-postlabeling assay using thin layer chromatography (TLC) (101), and gas chromatography (102) are also used in detecting 8-OHdG. Immunodetection using specific antibodies against 8-OHdG. Another possible biomarker, 5-hydroxymethyl-2′-deoxyuridine (HMdU) (103,104), which is a product of thymine oxidation, is gaining popularity because its higher sensitivity, specificity, and stability(105) (Figure 26.6).

26.5.2 Glutathione Reduced (GSH) and Oxidized (GSSG)

Glutathione (GSH, γ-glutamylcysteinylglycine) and its oxidized form (GSSG) (Figure 26. 7) together form the first line of defense against the cellular protection against reactive oxygen species (22,23,106,107). They carry out their protective functions by reducing ROS via their thiol groups, and therefore are involved in a variety of processes in the cellular systems that involve generation of free radicals such as detoxification of xenobiotics, reduction of hydroperoxides, and synthesis of leukotrienes and prostaglandins (106,107). The reducing power of glutathione is also involved in the maintenance of protein and membrane structure, and regulation of numerous enzyme activities and transcription factors that could be effected by oxidation as a result of reaction with ROS (11,106,107). The thiol groups are capable of participating in a number of chemical reactions including redox transitions, exchange of thiol groups, and radical scavenging (106,107). Recent research has also shown that the redox capability of thiol groups in the GSH/GSSG couple are also involved in gene regulation and intracellular signal transduction by critically regulating the functions of transcription factors and key cellular signaling molecules through oxidation and reduction (106). Because of the many protective functions of glutathione in the body, it is now considered one of the many biomarkers important to an understanding of the oxidative status of the cell (107). Many assays are based on conjugation of GSH with a chromophore or fluorophore, such as *o*-phthalaldehyde (OPA) (Figure 26.4). OPA does not fluoresce until it reacts with a thiol group in the presence of a primary amine. The OPA method for the detection of GSH has been limited due to the presence of other fluorogenic thiol and nonthiol components that react with OPA (107,108). Many biological constituents have been shown to quench fluorescence or inhibit the reaction of OPA with GSH, causing the underestimation of GSH levels or the overestimation of GSSG levels (107,108). Recent modifications in the OPA method by using a phosphate buffer at a lower pH, higher concentrations of OPA, and dithionite as the reductant for

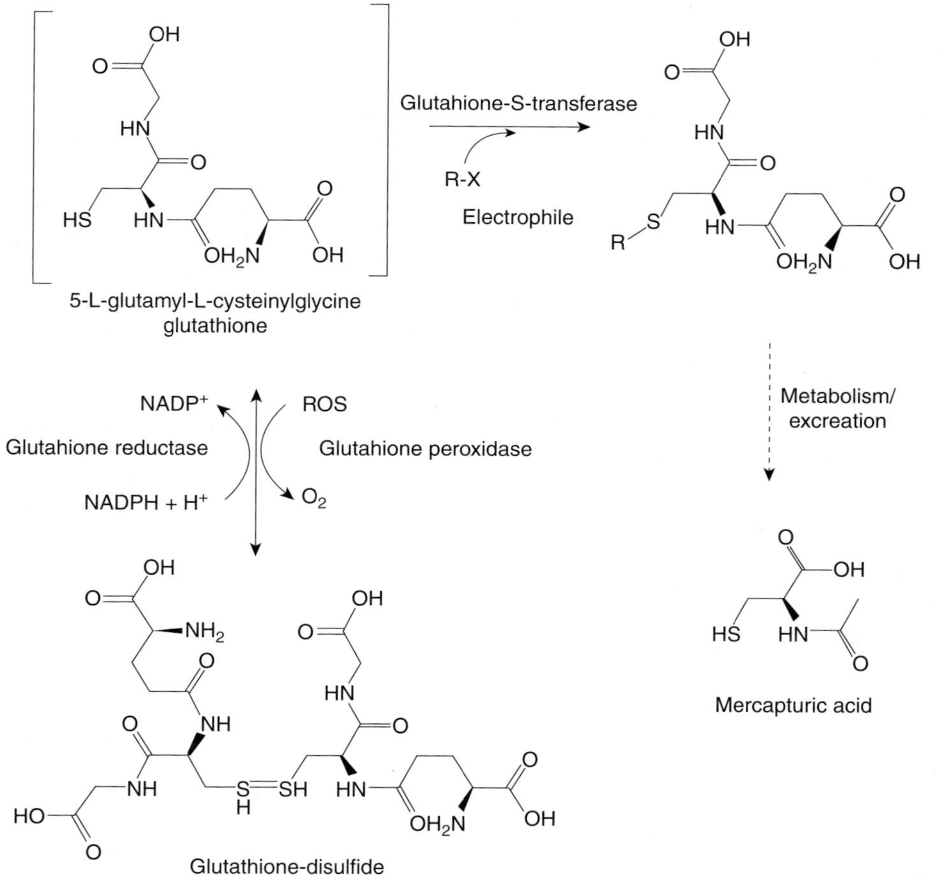

Figure 26.7 Metabolism of glutathione in eukaryotic cell

GSSG have led to rapid analysis, lower cost, and high sensitivity. These revised protocols have been able to estimate GSH and GSSG with accuracy comparable to established HPLC methods (108).

26.5.3 Glutathione Peroxidase

Glutathione peroxidase (GPX)(EC 1.11.1.9) is a selenium containing peroxidase involved in the cellular reduction of hydroperoxides. Two main types of GPX are found in the cells: mitochondrial peroxidase and phospholipid hydroperoxide glutathione peroxidase (22,23). Both types of GPX react directly with fatty acid hydroperoxides and hydrogen peroxide, and reduce them at the expense of glutathione (Figure 26.1, Figure 26.7). Depletion of GPX has now been observed in several forms of cancer, in heart disease, and in hypersensitive pathologies, and therefore is thought to make a good marker for studying the effects of certain types of foods and bioactive compounds on reducing the symptoms of certain types of heart disease, liver disease, and cancer (22,23). The most common assay method for measuring the activity of GPX is by glutathione reductase (GR) coupled assay. The reaction is initiated in the presence of a lipid hydroperoxide or hydrogen peroxide (109,110). The glutathione in the GPX reacts with GSH and oxidizes to GSSG which is reduced by GR back to GSH using $NADPH_2$. The rate of disappearance of $NADPH_2$ is monitored at 340 nm and is used to calculate the enzyme activity (109–111).

26.5.4 Glutathione-S-Transferases

The glutathione-S-transferases (GST) are important phase II detoxification enzymes found mainly in the cytosol, and are involved in the detoxification of drugs and harmful chemicals in the body (112). They are found in very high concentrations in the liver, where they conjugate glutathione to the electrophilic centers of lipophilic compounds and increase their solubility in order to facilitate their excretion (113,114) (Figure 26.7). GST posses a wide range of substrate specificities and can catalyze the reduction of breakdown products of macromolecules formed as a result of oxidative stress including reactive unsaturated carbonyls, oxidized DNA bases, and hydroperoxides, and therefore play a vital role in protecting tissues against oxidative damage and oxidative stress (112–115). As a result of the beneficial functions of GST it serves as an important biomarker for oxidative stress (115). Most of the methods to assay the GST activity involve measuring the conjugation of 1-chloro-2,4-dinitrobenzene (CDNB) with reduced glutathione (114). The conjugation of GSH with CDNB results in an increase in absorbance at 340 nm which is measured spectrophotometrically. The rate of increase is directly proportional to the GST activity in the sample. This assay can also be used to measure GST activity in plasma, cell lysates and tissue homogenates (114).

26.5.5 Superoxide Dismutase

The superoxide radicals are formed by the activity of enzymes like xanthine oxidase and NADPH-oxidase (22,23) (Figure 26.1). They are extremely reactive and have been shown to be able to oxidize biological macromolecules such as DNA, lipids and proteins (11,19,21–23). Superoxide dismutase (SOD) (EC 1.15.1.1) is involved in the dismutation of the superoxide radicals into hydrogen peroxide and molecular oxygen. In humans, there are four forms of SOD: cytosolic Cu, Zn-SOD, mitochondrial Mn-SOD, and extracellular-SOD (EC-SOD) (22,23).

The process of dismutation yields hydrogen peroxide, which is also a reactive oxygen species and causes oxidative damage to cellular systems. Therefore, in the body, activity of SOD is often coupled to another enzyme called "Catalase" (CAT) to remove hydrogen peroxide molecules which are byproducts of the reactions created by SOD (22,23). CAT is also abundant in the body and is prevents both cell damage and, more importantly, the formation of other, more toxic, free radicals (22,23).

Pathogenesis, mediated by virus and *Helicobacter pylori* (116,117), has been implicated in lowered antioxidant defense and alterations of enzymatic pathways; diabetes mellitus, endothelial, vascular, and neurovascular dysfunction. Neurodegenerative disorders such as Alzheimer's and Parkinson's disease are also now linked to oxidative stress and can be associated with lower responses of enzymes CAT and SOD (118,119). The activity of SOD and CAT is therefore considered as an important biomarker for these diseases (116–119).

The most common SOD detection method is based on spectrophotometric detection. Superoxide, which is a substrate of SOD, is generated by a xanthine to xanthine oxidase, which reacts with either cytochrome C (120) or nitroblue tetrazolium (NBT) (121), and changes color, which can be monitored spectrophotometrically. Detection of O_2^- by the cytochrome C reducing method is based on the color change to generate purple colored dye from reduced cytochrome C. The NBT method is based on the generation of water insoluble blue formazan dye (λmax: 560 nm) by a reaction with O_2^- (120,121). Recent methods developed use chemiluminescence probes such as lucigenin, and luciferin derivative (MCLA) for O_2^- detection (122,123). They react with the superoxide radical and produce a signal which can be used to quantitate SOD activity.

26.5.6 Catalase

Catalase (EC 1.11.1.6) is a ubiquitous oxidoreductase that is present in most aerobic cells. Catalase (CAT) is involved in the detoxification of hydrogen peroxide (H_2O_2), a (ROS), into molecular oxygen and two molecules of water (22,23). CAT can be assayed directly by measuring the rate of appearance of molecular oxygen using a Clark-type electrode (124). Here, the enzyme extract is allowed to react with a known concentration of hydrogen peroxide. The formation of oxygen changes the potential of the Clark-type electrode which is proportional to the enzyme activity (124). Catalase activity has also been determined spectrophotometrically at 240 nm by monitoring the disappearance of hydrogen peroxide at pH 7.0 (125). Catalase also has a peroxidative activity which has been used to assay its activity by reaction with methanol in the presence of an optimal concentration of H_2O_2 (126,127). The formaldehyde produced is measured spectrophotometrically with 4-amino-3-hydrazino-5-mercapto-1,2,4-triazole (Purpald) as the chromogen, in which low molecular weight alcohols can serve as electron donors. Fluorogenic dyes such as amplex red have also been recently used to measure CAT activity. Amplex red reacts with hydrogen peroxide to form a highly fluorescent resorufin which can be measured in a spectrofluoremeter (excitation and emission of 563/587 nm) (126,127).

26.5.7 Xanthine Oxidase and Uric acid

Xanthine oxidoreductase (XOR), a metalloflavoprotein, catalyzes the oxidation of the purine bases hypoxanthine and xanthine to uric acid. XOR is a site for ROS generation in the cell (128) (Figure 26.1). High levels of XOR activity resulting in ROS production has now been shown to play an important role in reperfusion injury; and in congestive heart failure (129). Xanthine oxidase and its reaction product uric acid are therefore now considered to be important biomarkers for oxidative stress and cardiovascular diseases (129,130). Uric acid in urine, serum, and plasma is measured using a spectrophotometric assay based on phosphotungstic acid reagent (131); however, for improved specificity and characterization, HPLC methods have been used recently (132,133). The most popular assay method for xanthine oxidoreductase involves the spectrophotometric measurement of urate production at 292 nm, using xanthine or hypoxanthine as the substrate (134). A spectrofluorometric assay utilizing pterin as the substrate for XOR, and the formation of fluorescent isoxanthopterin, is quantified for xanthine oxidase activity (135). Chemiluminescence methods are also used alternatively which assess the ROS production by xanthine oxidase which reacts with chemiluminescent dyes such as luminol or lucigenin (Figure 26.4) which are then measured and quantitated as enzyme activity (136).

26.5.8 NADPH Oxidoreductase

NADPH oxidase (NOX) is a membrane bound enzyme present in neutrophils and other phagocytic cells (138). This plasma membrane associated enzyme complex catalyzes the univalent reduction of oxygen using NADPH as an electron donor to produce superoxides, which are involved in the bactericidal activity of the neutrophils (137,138). Though NOX is involved in the generation of ROS for bactericidal activities, recent research has shown that a new family of NADPH oxidases are expressed in various cell types, including the epithelium, smooth muscle cells, and the endothelium (134,135). These enzymes produce superoxide and other ROS, including hydrogen peroxide (Figure 26.1). The activity of this enzyme in vascular cells is a main source of ROS in inflammation and is believed to be a cause for atherosclerosis and hypertension. NOX expression is also found to be high in certain forms of cancer and endocrine disorders as a result of superoxide and other ROS production (137–139). Elevated levels of ROS are also seen in patients with progressive neurological diseases like Alzheimer's, and Parkinson's disease (140). These recent discoveries about the many different functions of the nonphagocytic functions of NOX has made it an attractive target for therapeutic agents and an important biomarker for these diseases in different cell types (139,140). NOX is usually measured in cell free extracts by using them with $NADPH_2$ for superoxide radical generation, which is measured by following the reduction of ferric cytochrome-C at 550 nm (141,142) or by monitoring the inhibition of the formation of formazan blue from NBT at 560 nm (143). Use of a new tetrazolium based assay to study the production of superoxide radicals has also been done, with the NOX and $NADPH_2$ as a source for superoxide radical generation (143). Chemiluminescence assays based on oxidation of a chromophore such as lucigenin are also being used to measure NOX activity (144). In this assay, the chromophore becomes chemiluminescent after reacting with the superoxide radical generated from the NOX reaction. This is then measured and quantitated as NOX activity (145).

26.5.9 Myeloperoxidase

Myeloperoxidase (MPO) is a lysosomal enzyme that is found in white blood cells and neutrophils. In these cells the NADPH oxidase produces superoxide radicals which then dismutates to hydrogen peroxide. Myeloperoxidase uses hydrogen peroxide and chloride to produce another highly reactive ROX called hypochlorous acid (HOCl) (145,146). This potent oxidant then reacts with proteins, DNA, and lipids to cause cellular injury. The primary function of these enzymes is to mediate a defense response against invading pathogens. However, it has been shown recently that, due to the remarkably high oxidant potential of MPOs, they function as mediators of inflammatory pathologies, autoimmune diseases, and rheumatoid arthritis (145,146). The hypochlorous radical and tyrosyl radicals made by MPO have been shown to cause oxidation of LDL leading to atherosclerosis (148) and other cardiovascular diseases (145,146). MPO therefore has evolved into the most promising cardiac marker and a good inflammatory biomarker. The strong oxidant capacity of MPO is often exploited in assaying its activity. MPO is capable of oxidizing many different chromophores such as tyrosine (147) and 3′,5,5′-tetramethylbenzinide (TMB) (148) which are monitored spectrophotometrically or fluorometrically in cell free extracts. Possible interference from other peroxidases and ROS in the system has led to the development of many enzyme linked immunosorbent assays which are more specific to myeloperoxidases and their isoforms (149–151).

26.6 PLANT PHENOLIC DEPENDENT PEROXIDASES

Phenolic dependent peroxidases catalyze the oxidative coupling of phenolic compounds using hydrogen peroxide to make lignins, phenolic hemicellulose, and phenolic to tyrosine cross links to aid structural development in plant tissues (152,153). These phenolic dependent peroxidases, often referred to as guaiacol, peroxidases have been well characterized in plants and are known to be inducible enzymes, expressed in response to a need for structural growth and development (154,155). Plant peroxidases are induced in response to oxidative stress, exposure to environmental toxins, and high energy radiation (156–160). They protect the plants from these stresses by removing the reactive oxygen species, and by using them to oxidatively couple phenolic phytochemicals to make lignin and other cross-linked phenolics important in plant cell wall protection (154). The oxidative polymerization of phenolics also contributes to the increase in antioxidant activity which is believed to further protect the plant against oxidative stress (54,155). Recently, it has been shown that these peroxidases are also stimulated in response to exogenous phenolic elicitors in germinating fava bean sprouts and that the proxidases contributed to increased antioxidant activity observed during the course of germination (161). It has now been proposed that peroxidases similar to the phenolic dependent guaiacol peroxidase could be stimulated in human cells in response to exogenous phenolics for protection against oxidative stress (162). It is also believed that that these phenolic dependent peroxidases reduce oxidative stress by removing reactive oxygen species by oxidatively polymerizing dietary phenolics without affecting the cellular pools of glutathione, ascorbate, and other antioxidants (162). Stimulation of these peroxidases was recently observed in peroxide stressed porcine muscle models in response to treatment with phenolic extract from oregano (163).

The most common analytical method for assaying the activity of phenolic dependent peroxidases is based on oxidation of a substrate catalyzed by the enzyme in the presence of hydrogen peroxide, which can be followed spectrophotometrically (154,164). The time dependent oxidation of guaiacol to tertraguaiacol ($\varepsilon = 26.6\ mM^{-1}cm^{-1}$) at 470 nm, described by Laloue et al. (165), has been extensively used in characterization of these peroxidases. Other substrates that are used for assaying the activity of peroxidases include *o*-dianisidine, pyrogallol, syringaldazine and 2,2′-azinobis(3-ethylbenzothiazoline-6-sulphonate) (ABTS) (154).

REFERENCES

1. Lundberg, A.C., A. Åkesson, B. Åkesson. Dietary intake and nutritional status in patients with systemic sclerosis. *Ann. Rheum. Dis.* 51:1143–1148, 1992.
2. Wilks, R., F. Bennett, T. Forrester, N. Mcfarlane-Anderson. Chronic diseases: the new epidemic. *West Ind. Med. J.* 47(4):40–44, 1998.
3. Gillman, M.W., L.A. Cupples, D. Gagnon, B.M. Posner, R.C. Ellison, W.P. Castelli, P.A. Wolf. Protective effect of fruits and vegetables on development of stroke in men. *J. Am. Med. Assoc.* 273:1113–1117, 1995.
4. Joshipura, K.J, A. Ascherio, J.E. Manson, M.J. Stampfer, E.B. Rimm, F.E. Speizer, C.H. Hennekens, D. Spiegelman, W.C. Willett. Fruit and vegetable intake in relation to risk of ischemic stroke. *J. Am. Med. Assoc.* 282:1233–1239, 1999.

5. Cox, B.D., M.J. Whichelow, A.T. Prevost. Seasonal consumption of salad vegetables and fresh fruit in relation to the development of cardiovascular disease and cancer. *Public Health Nutr*. 3:19–29, 2000.
6. Strandhagen, E., P.O. Hansson, I. Bosaeus, B. Isaksson, H. Eriksson. High fruit intake may reduce mortality among middle-aged and elderly men: the study of men born in 1913. *Eur. J. Clin. Nutr*. 54:337–341, 2000.
7. Block, G., B. Patterson, A. Subar. Fruit, vegetables, and cancer prevention: a review of the epidemiological evidence. *Nutr. Cancer* 18:1–29, 1992.
8. Serdula, M.K., M.A.H. Byers, E. Simoes, J.M. Mendlein, R.J. Coates. The association between fruit and vegetable intake and chronic disease risk factors. *Epidemiology* 7:161–165, 1996.
9. Tapiero, H., K.D. Tew, G.N. Ba, G. Mathe. Polyphenols: do they play a role in the prevention of human pathologies? *Biomed. Pharmacother*. 56(4):200–207, 2002.
10. Duthie, G.G., P.T. Gardner, J.A. Kyle. Plant polyphenols: are they the new magic bullet? *Proc. Nutr. Soc*. 62(3):599–603, 2003.
11. Droge, W. Free radicals in the physiological control of cell function. *Physiol. Rev*. 82(1):47–95, 2002.
12. Sculley, D.V., S.C. Langley-Evans. Salivary antioxidants and periodontal disease status. *Proc. Nutr. Soc*. 61(1):137–143, 2002.
13. Parke, A.L., C. Ioannides, D.F.V. Lewis, D.V. Parke. Molecular pathology of drugs: disease interaction in chronic autoimmune inflammatory diseases. *Inflammopharmacology* 1:3–36, 1991.
14. Schwarz, K.B. Oxidative stress during viral infection: a review. *Free Radic. Biol. Med*. 21(5):641–649, 1996.
15. Götz, J., C.I. va Kan, H.W. Verspaget, I. Biemond, C.B. Lamers, R.A. Veenendaal. Gastric mucosal superoxide dismutases in *Helicobacter pylori* infection. *Gut* 38:502–506, 1996.
16. Jakus, V. The role of free radicals, oxidative stress and antioxidant systems in diabetic vascular disease. *Bratisl. Lek. Listy*. 101(10):541–551, 2000.
17. Offen, D., P.M. Beart, N.S. Cheung, C.J. Pascoe, A. Hochman, S. Gorodin, E. Melamed, R. Bernard, O. Bernard. Transgenic mice expressing human Bcl-2 in their neurons are resistant to 6-hydroxydopamine and 1-methyl-4-phenyl-1,2,3,6- tetrahydropyridine neurotoxicity. *Proc. Natl. Acad. Sci*. 95:5789–5794, 1998.
18. Rice-Evans, C.A., N.J. Miller, G. Paganga. Antioxidant Properties of Phenolic Compounds. *Trends Plant Sci*. 2:152–159, 1997.
19. Niess, A.M., H.H. Dickhuth, H. Northoff, E. Fehrenbach. Free radicals and oxidative stress in exercise-immunological aspects. *Exerc. Immunol. Rev*. 5:22–56, 1999.
20. Sculley, D.V., S.C. Langley-Evans. Salivary antioxidants and periodontal disease status. *Proc. Nutr. Soc*. 61(1):137–143, 2002.
21. Lusini, L., S.A. Tripodi, R. Rossi, F. Giannerini, D. Giustarini, M.T. del Vecchio, G. Barbanti, M. Cintorino, P. Tosi, P. Di Simplicio. Altered glutathione anti-oxidant metabolism during tumor progression in human renal-cell carcinoma. *Int. J. Cancer* 91(1):55–59, 2001.
22. Mates, J.M., F. Sanchez-Jimenez. Antioxidant enzymes and their implications in pathophysiologic processes. *Front Biosci*. 4:D339–345, 1999.
23. Mates, J.M., C. Perez-Gomez, I. Nunez de Castro. Antioxidant enzymes and human diseases. *Clin. Biochem*. 32(8):595–603, 1999.
24. Liochev, S.I., I. Fridovich. The relative importance of HO* and ONOO- in mediating the toxicity of O*-. *Free Radic. Biol. Med*. 26(5–6):777–778, 1999.
25. Barber, D.A., S.R. Harris. Oxygen free radicals and antioxidants: a review. *Am. Pharm*. NS34:26–35, 1994.
26. Betteridge, D.J. What is oxidative stress? *Metabolism* 49:3–8, 2000.
27. Vignais, P.V. The superoxide-generating NADPH oxidase: structural aspects and activation mechanism. *Cell Mol. Life Sci*. 59(9):1428–1459, 2002.
28. Babior, B.M., J.D. Lambeth, W. Nauseef. The neutrophil NADPH oxidase. *Arch. Biochem. Biophys*. 397(2):342–344, 2002.

29. Babior, B.M. The leukocyte NADPH oxidase. *Isr. Med. Assoc. J.* 4(11):1023–1024, 2002.
30. Parke, A., D.V. Parke. The pathogenesis of inflammatory disease: surgical shock and multiple system organ failure. *Inflammopharmacology* 3:149–168, 1995.
31. Parke, D.V. The cytochromes P450 and mechanisms of chemical carcinogenesis. *Environ. Health Perspect.* 102:852–853, 1994.
32. Parke, D.V., A. Sapota. Chemical toxicity and reactive oxygen species. *Int. J. Occ. Med. Environ. Health.* 9:331–340, 1996.
33. Baker, C.J., G.L. Harmon, J.A. Glazener, E.W. Orlandi. A Noninvasive Technique for Monitoring Peroxidative and H2O2-Scavenging Activities during Interactions between Bacterial Plant Pathogens and Suspension Cells. *Plant Physiol.* 108(1):353–359, 1995.
34. Low, P.S., J.R. Merida. The oxidative burst in plant defense: function and signal transduction. *Physiol. Plant* 96:533–542, 1996
35. Wojtaszek, P. Oxidative burst: an early plant response to pathogen infection. *Biochem. J.* 322(3):681–692, 1997.
36. Low, P.S., P.F. Heinstein. Elicitor stimulation of the defense response in cultured plant cells monitored by fluorescent dyes. *Arch. Biochem. Biophys.* 249(2):472–479, 1986.
37. Levine, A., R. Tenhaken, R. Dixon, C. Lamb. H_2O_2 from the oxidative burst orchestrates the plant hypersensitive disease resistance response. *Cell* 79:583–593, 1994.
38. Yoshiki, Y., T. Kahara, K. Okubo, K. Igarashi, K. Yotsuhashi. Mechanism of catechin chemiluminescence in the presence of active oxygen. *J. Biolumin. Chemilumin.* 11(3):131–136, 1996.
39. Halliwell, B., J.M.C. Gutteridge *Free Radicals In Biology And Medicine,* Third Ed., Oxford: Oxford University Press, 1999.
40. Able, A.J., D.I. Guest, M.W. Sutherland. Hydrogen peroxide yields during the incompatible interaction of tobacco suspension cells inoculated with Phytophthora nicotianae. *Plant Physiol.* 124(2):899–910, 2000.
41. Rosen, G.M., E.J. Rauckman. Spin trapping of superoxide and hydroxyl radicals. *Methods Enzymol.* 105:198–209, 1984.
42. Pou, S., G.M. Rosen, B.E. Britigan, M.S. Cohen. Intracellular spin-trapping of oxygen-centered radicals generated by human neutrophils. *Biochim. Biophys. Acta.* 991(3):459–464, 1989.
43. Bondet, V., W. Brand-Williams C. Berset. Kinetics and Mechanisms of Antioxidant Activity using the DPPH. Free Radical Method. *Lebensm.-Wiss. u.-Technol.* 30:609–615, 1997.
44. Molyneux, P. The use of the stable free radical diphenylpicrylhydrazyl (DPPH) for estimating antioxidant activity. *Songklanakarin. J. Sci. Technol.* 26(2):211–219, 2004.
45. Miller, N.J., C.A. Rice-Evans. Factors influencing the antioxidant activity determined by the ABTS.+ radical cation assay. *Free Radic. Res.* 26(3):195–199, 1997.
46. Re, R., N. Pellegrini, A. Proteggente, A. Pannala, M. Yang, C. Rice-Evans. Antioxidant activity applying an improved ABTS radical cation decolorization assay. *Free Radic. Biol. Med.* 26(9–10):1231–1237, 1999.
47. Sanchez-Moreno, C. Methods used to evaluate the free radical scavenging activity in foods and biological systems. *Food Sci. Tech. Int.* 8(3):121–137, 2002.
48. McAnalley, S., C.M. Koepke L. Le, R.A.C. Vennum, B. Bill McAnalley. *In vitro* methods for testing antioxidant potential: a review. *Glycoscience*. 4(2):1–9, 2003.
49. Benzie, I.F., J.J. Strain. The ferric reducing ability of plasma (FRAP) as a measure of "antioxidant power": the FRAP assay. *Anal. Biochem.* 239(1):70–76, 1996.
50. Cao, G., R.L. Prior. Comparison of different analytical methods for assessing total antioxidant capacity of human serum. *Clin. Chem.* 1998. 44(6 Pt 1):1309–15.
51. Cao, G., R.L. Prior. Measurement of oxygen radical absorbance capacity in biological samples. *Methods Enzymol.* 299:50–62, 1999.
52. Ou, B., M. Hampsch-Woodill, R.L. Prior. Development and validation of an improved oxygen radical absorbance capacity assay using fluorescein as the fluorescent probe. *J. Agric. Food Chem.* 49(10):4619–4626, 2001.
53. Rice-Evans, C.A., N.J. Miller, G. Paganga. Antioxidant Properties of Phenolic Compounds. *Trends Plant Sci.* 2:152–159, 1997.

54. Lotito, S.B., L. Actis Goretta, M.L. Renart, M. Caligiuri, D. Rein, H.H. Schmitz, F.M. Steinberg, C.L. Keen, C.G. Fraga. Influence of oligomer chain length on the antioxidant activity of procyanidins. *Biochem. Biophys. Res. Commun.* 276:945–951, 2000.
55. Vattem, D.A., H.-D. Jang, R. Levin, K. Shetty. Synergism of cranberry phenolics with ellagic acid and rosmarinic acid for antimutagenic and DNA-protection functions. *J. Food Biochem.* Submitted, 2004.
56. Frankel, E.N. A.S. Meyer. The problems of using one-dimensional methods to evaluate multifunctional food and biological antioxidants. *J. Sci. Food Agric.* 80:1925–1941, 2000.
57. Ghiselli, A., M. Serafini, G. Maiani, E. Azzini, A.A. Ferro-Luzzi. Fluorescence-based method for measuring total plasma antioxidant capability. *Free Rad. Biol. Med.* 18:29–36, 1995.
58. Marco, G.J. A rapid method for evaluation of antioxidants. *J. Am. Oil Chem. Soc.* 45:594–598, 1968.
59. Miller, H.E. A simplified method for evaluation of antioxidants. *J. Am. Oil Chem. Soc.* 48:91, 1971.
60. Lee, Y., L.R. Howard, B. Villalón. Flavonoids and antioxidant activity of fresh pepper (*Capsicum annuum*) cultivars. *J. Food Sci.* 60:473–476, 1995.
61. Vattem, D.A., K. Shetty. Solid-state production of phenolic antioxidants from cranberry pomace by *Rhizopus oligosporus. Food Biotechnol.* 16(3):189–210, 2002.
62. Vattem D.A., K. Shetty. Ellagic acid production and phenolic antioxidant activity in cranberry pomace (*Vaccinium macrocarpon*) mediated by *Lentinus edodes* using solid-state system. *Proc. Biochem.* 39(3):367–379, 2003.
63. Frankel, E.N., J. Kanner, J.B. German, E. Parks, J.E. Kinsella. Inhibition of oxidation of human low-density lipoprotein by phenolic substances in red wine. *Lancet* 20:341(8843):454–457, 1993.
64. Ames, B.N., M.K. Shigenaga, T.M. Hagen. Oxidants, antioxidants, and the degenerative diseases of aging. *Proc. Natl. Acad. Sci. USA* 90(17):7915–7922, 1993.
65. Beckman, J.S., W.H. Koppenol. Nitric oxide, superoxide, and peroxynitrite: the good, the bad, and ugly. *Am. J. Physiol.* 271(5,1):C1424–1437, 1996.
66. McLeod, L.L., A. Sevanian. Lipid peroxidation and modification of lipid composition in an endothelial cell model of ischemia and reperfusion. *Free Radic. Biol. Med.* 23(4):680–694, 1997.
67. Kawakami, A., A. Tanaka, T. Nakano, K. Nakajima, F. Numano. The role of remnant lipoproteins in atherosclerosis. *Ann. NY Acad. Sci.* 902:352–356, 2000.
68. Pryor, W.A., N.A. Porter. Suggested mechanisms for the production of 4-hydroxy-2-nonenal from the autoxidation of polyunsaturated fatty acids. *Free Radic. Biol. Med.* 8(6):541–543, 1990.
69. Esterbauer, H., R.J. Schaur, H. Zollner. Chemistry and biochemistry of 4-hydroxynonenal, malonaldehyde and related aldehydes. *Free Radic. Biol. Med.* 11(1):81–128, 1991.
70. Khanzode, S.S., M.G. Muddeshwar, S.D. Khanzode, G.N. Dakhale. Antioxidant enzymes and lipid peroxidation in different stages of breast cancer. *Free Radic. Res.* 38(1):81–85, 2004.
71. Yilmaz, M.I., K. Saglam, A. Sonmez, D.E. Gok, S. Basal, S. Kilic, C. Akay, I.H. Kocar. Antioxidant system activation in prostate cancer. Biol. Trace Elem. Res. 98(1):13–19, 2004.
72. Hennig, B., C.K. Chow. Lipid peroxidation and endothelial cell injury: implications in atherosclerosis. *Free Radic. Biol. Med.* 4(2):99–106, 1988.
73. Massy, Z.A., W.F. Keane. Pathogenesis of atherosclerosis. *Semin. Nephrol.* 16(1):12–20, 1996.
74. Pratico, D., M.Y. Lee V., J.Q. Trojanowski, J. Rokach, G.A. Fitzgerald. Increased F2-isoprostanes in Alzheimer's disease: evidence for enhanced lipid peroxidation *in vivo*. *FASEB J.* 12(15):1777–1783, 1998.
75. Montine, T.J., W.R. Markesbery, J.D. Morrow, L.J. Roberts, 2nd. Cerebrospinal fluid F2-isoprostane levels are increased in Alzheimer's disease. *Ann. Neurol.* 44(3):410–413, 1998.
76. Moore, K., L.J. Roberts, 2nd. Measurement of lipid peroxidation. *Free Radic. Res.* 28(6):659–671, 1998.

77. Karatas, F., M. Karatepe, A. Baysar. Determination of free malondialdehyde in human serum by high-performance liquid chromatography. *Anal. Biochem.* 1:311(1):76–79, 2002.
78. Steghens, J.P., A.L. van Kappel, I. Denis, C. Collombel. Diaminonaphtalene, a new highly specific reagent for HPLC-UV measurement of total and free malondialdehyde in human plasma or serum. *Free Radic. Biol. Med.* 31(2):242–249, 2001.
79. Wong, J.W., S.E. Ebeler, R. Rivkah-Isseroff, T. Shibamoto. Analysis of malondialdehyde in biological samples by capillary gas chromatography. *Anal. Biochem.* 220(1):73–81, 1994.
80. Khan, M.F., X. Wu, G.A. Ansari. Anti-malondialdehyde antibodies in MRL+/+ mice treated with trichloroethene and dichloroacetyl chloride: possible role of lipid peroxidation in autoimmunity. *Toxicol. Appl. Pharmacol.* 170(2):88–92, 2001.
81. Mayne, S.T. Antioxidant nutrients and chronic disease: use of biomarkers of exposure and oxidative stress status in epidemiologic research. *J. Nutr.* 133(3):933S–940S, 2003.
82. Jialal, I., C.J. Fuller. Effect of vitamin E, vitamin C and beta-carotene on LDL oxidation and atherosclerosis. *Can. J. Cardiol.* G:97G–103G, 1995.
83. Gaziano, J.M., A. Hatta, M. Flynn, E.J. Johnson, N.I. Krinsky, P.M. Ridker, C.H. Hennekens, B. Frei. Supplementation with beta-carotene *in vivo* and *in vitro* does not inhibit low density lipoprotein oxidation. *Atherosclerosis* 112(2):187–195, 1995.
84. Pierdomenico, S.D., F. Costantini, A. Bucci, D. De Cesare, F. Cuccurullo, A. Mezzetti. Low-density lipoprotein oxidation and vitamins E and C in sustained and white-coat hypertension. *Hypertension* 31(2):621–626, 1998.
85. Costantini, F., S.D. Pierdomenico, D. De Cesare, P. De Remigis, T. Bucciarelli, G. Bittolo-Bon, G. Cazzolato, G. Nubile, M.T. Guagnano, S. Sensi, F. Cuccurullo, A. Mezzetti. Effect of thyroid function on LDL oxidation. *Arterioscler. Thromb. Vasc Biol.* 18(5):732–737, 1998.
86. Princen, H.M., W. van Duyvenvoorde, R. Buytenhek, A. van der Laarse, G. van Poppel, J.A. Gevers Leuven, V.W. van Hinsbergh. Supplementation with low doses of vitamin E protects LDL from lipid peroxidation in men and women. *Arterioscler. Thromb. Vasc. Biol.* 15(3):325–333, 1995.
87. Morrow, J.D., L.J. Roberts, 2nd. The Isoprostanes: Novel Markers of Lipid Peroxidation and Potential Mediators of Oxidant Injury. *Adv. Prostaglandin Thromboxane Leukot. Res.* 23:219–224, 1995.
88. Morrow, J.D., L.J. Roberts, 2nd. The Generation and Actions of Isoprostanes. *Biochemistry* 121–135, 1997.
89. Halliwell, B. Establishing the significance and optimal intake of dietary antioxidants: The biomarker concept. *Nutr. Rev.* 57:104–113, 1999.
90. Gopaul, N.K., E.E. Anggard, A.I. Mallet, D.J. Betteridge, S.P. Wolff, J. Nourooz-Zadeh. Plasma 8-epi-PGF2a levels are elevated in individuals with non-insulin dependent diabetes mellitus. *FEBS Lett.* 368:225–229, 1995.
91. Janssen, L.J. Isoprostanes: Generation, pharmacology, and roles in free-radical-mediated effects in the lung. *Pulm. Pharmacol. Therap.* 13(4):149–155, 2000.
92. Pratico, D., Iuliano, L., A. Mauriello, L. Spagnoli, J.A. Lawson, J. Rokach, J. Maclouf, F. Violi, G.A. FitzGerald. Localization of distinct F2-isoprostanes in human atherosclerotic lesions. *J. Clin. Invest.* 100:2028–2034, 1997.
93. Mobert, J., B.F. Becker. Cyclooxygenase inhibition aggravates ischemia-reperfusion injury in the perfused guinea pig heart: involvement of isoprostanes. *J. Am. Coll. Cardiol.* 31:1687–1694, 1998.
94. Montine, T.J., M.F. Beal, M.E. Cudkowicz, H. O'Donnell, R.A. Margolin, L. McFarland, A.F. Bachrach, W.E. Zachert, L.J. Roberts, J.D. Morrow. Increased CSF F2-isoprostane concentration in probable AD. *Neurology* 52(3):562–565, 1999.
95. Morrow, J.D., T.M. Harris, L.J. Roberts, 2nd. Noncyclooxygenase oxidative formation of a series of novel prostaglandins: Analytical ramifications for measurements of eicosanoids. *Anal. Biochem.* 184:1–10, 1990.
96. Morrow, J.D., L.J. Roberts. Mass spectrometry of prostanoids: F2-isoprostanes produced by non-cyclooxygenase free radical-catalyzed mechanism. *Methods Enzymol.* 233:163–174, 1994.

97. Walter, M.F., J.B. Blumberg, G.G. Dolnikowski, G.J. Handelman. Streamlined F2-isoprostane analysis in plasma and urine with high-performance liquid chromatography and gas chromatography/mass spectroscopy. *Anal. Biochem.* 280(1):73–79, 2000.
98. Sasaki, D.M., Y. Yuan, K. Gikas, K. Kanai, D. Taber, J.D. Morrow, L.J. Roberts 2nd, D.M. Callewaert. Enzyme immunoassays for 15-F2T isoprostane-M, an urinary biomarker for oxidant stress. *Adv. Exp. Med. Biol.* 507:537–541, 2002.
99. Halliwell, B. Why and how should we measure oxidative DNA damage in nutritional studies? How far have we come? *Am. J. Clin. Nutr.* 72(5):1082–1087, 2000.
100. Kasai, H. Analysis of a form of oxidative DNA damage, 8-hydroxy-2′-deoxyguanosine, as a marker of cellular oxidative stress during carcinogenesis. *Mutat. Res.* 387:147–163, 1997.
101. Devanaboyina, U., R.C. Gupta. Sensitive detection of 8-hydroxy-2′deoxyguanosine in DNA by 32P-postlabeling assay and the basal levels in rat tissues. *Carcinogenesis* 17(5):917–924, 1996.
102. Teixeira, A.J., M.R. Ferreira, W.J. van Dijk, G. van de Werken, A.P. de Jong. Analysis of 8-hydroxy-2′-deoxyguanosine in rat urine and liver DNA by stable isotope dilution gas chromatography/mass spectrometry. *Anal. Biochem.* 226(2):307–319, 1995.
103. Santella, R.M. Immunological methods for detection of carcinogen-DNA damage in humans. *Cancer Epidemiol. Biomarkers Prev.* 8:733–739, 1999.
104. Frenkel, K., J. Karkoszka, T. Glassman, N. Dubin, P. Toniolo, E. Taioli, L.A. Mooney, I. Kato. Serum autoantibodies recognizing 5-hydroxymethyl-2′-deoxyuridine, an oxidized DNA base, as biomarkers of cancer risk in women. *Cancer Epidemiol. Biomarkers Prev.* 7:49–57, 1998.
105. Hu, J.J., C.X. Chi, K. Frenkel, B.N. Smith, J.J. Henfelt, M. Berwick, S. Mahabir, R.B. D'Agostino. Alpha-tocopherol dietary supplement decreases titers of antibody against 5-hydroxymethyl-2′-deoxyuridine (HMdU). *Cancer Epidemiol. Biomarkers Prev.* 8:693–698, 1999.
106. Dalton, T.P., H.G. Shertzer, A. Puga. Regulation of gene expression by reactive oxygen. *Annu. Rev. Pharmacol. Toxicol.* 39:67–101, 1999.
107. Senft, A.P., T.P. Dalton, H.G. Shertzer. Determining glutathione and glutathione disulfide using the fluorescence probe o-phthalaldehyde. *Anal. Biochem.* 280(1):80–86, 2000.
108. Floreani, M., M. Petrone, P. Debetto, P. Palatini. A comparison between different methods for the determination of reduced and oxidized glutathione in mammalian tissues. *Free Radic. Res.* 26(5):449–455, 1997.
109. Imai, H., K. Narashima, M. Arai, H. Sakamoto, N. Chiba Y. Nakagawa: Suppression of leukotriene formation in RBL-2H3 cells that overexpressed phospholipid hydroperoxide glutathione peroxidase. *J. Biol. Chem.* 273:1990–1997, 1998.
110. De Haan, J., C. Bladier, P. Griffiths, M. Kelner, R.D. O´Shea, N.S. Cheung, R. T. Bronson, M.J. Silvestro, S. Wild, S.S. Zheng, P.M. Beart, P.J. Herzog, I. Kola. Mice with a homozygous null mutation for the most abundant glutathione peroxidase, Gpx1, show increased susceptibility to the oxidative stress-inducing agents paraquat and hydrogen peroxide. *J. Biol. Chem.* 273:22528–22536, 1998.
111. Ding, L., Z. Liu, Z. Zhu, G. Luo, D. Zhao, J. Ni. Biochemical characterization of selenium-containing catalytic antibody as a cytosolic glutathione peroxidase mimic. *Biochem J.* 332:251–255, 1998.
112. Sheehan, D., G. Meade, V.M. Foley, C.A. Dowd. Structure, function and evolution of glutathione transferases: implications for classification of non-mammalian members of an ancient enzyme superfamily. *Biochem. J.* 360(1):1–16, 2001.
113. Smith, R.A., J.E. Curran, S.R. Weinstein, L.R. Griffiths. Investigation of glutathione S-transferase zeta and the development of sporadic breast cancer. *Breast Cancer Res.* 3(6):409–411, 2001.
114. Ito, N., S. Tamano, T. Shirai. A medium-term rat liver bioassay for rapid *in vivo* detection of carcinogenic potential of chemicals. *Cancer Sci.* 94(1):3–8, 2003.
115. Schwartz, J.L. Biomarkers and molecular epidemiology and chemoprevention of oral carcinogenesis. *Crit. Rev. Oral Biol. Med.* 11(1):92–122, 2000.

116. Schwarz, K.B. Oxidative stress during viral infection: a review. *Free Radic. Biol. Med.* 21(5):641–649, 1996.
117. Götz, J., C.I. va Kan, H.W. Verspaget, I. Biemond, C.B. Lamers, R.A. Veenendaal. Gastric mucosal superoxide dismutases in *Helicobacter pylori* infection. *Gut* 38:502–506, 1996.
118. Jakus, V. The role of free radicals, oxidative stress and antioxidant systems in diabetic vascular disease *Bratisl. Lek. Listy.* 101(10):541–551, 2000.
119. Offen, D., P.M. Beart, N.S. Cheung, C.J. Pascoe, A. Hochman, S. Gorodin, E. Melamed, R. Bernard, O. Bernard. Transgenic mice expressing human Bcl-2 in their neurons are resistant to 6-hydroxydopamine and 1-methyl-4-phenyl-1,2,3,6- tetrahydropyridine neurotoxicity. *Proc. Natl. Acad. Sci.* 95:5789–5794, 1998.
120. Oberley, L.W., D.R. Spitz. Assay of SOD activity in tumor tissue. In: *Methods in Enzymology*, Academic Press, 1984, 105:457–461.
121. Spychalla, J.P., S.L. Desborough. Superoxide dismutase, catalase and alpha tocopherol content of stored potato tubers. *Plant Physiol.* 94:1214–1218, 1990.
122. Antier, D., H.V. Carswell, M.J. Brosnan, C.A. Hamilton, I.M. Macrae, S. Groves, E. Jardine, J.L. Reid, A.E. Dominiczak. Increased levels of superoxide in brains from old female rats. *Free Radic. Res.* 38(2):177–183, 2004.
123. Yamaguchi, K., D. Uematsu, Y. Itoh, S. Watanabe, Y. Fukuuchi. *In vivo* measurement of superoxide in the cerebral cortex during anoxia-reoxygenation and ischemia-reperfusion. *Keio. J. Med.* 51(4):201–207, 2002.
124. Escobar, L., C. Salvador, M. Contreras, J.E. Escamilla. On the application of the Clark oxygen electrode to the study of enzyme kinetics in apolar solvents: the catalase reaction. *Anal. Biochem.* 184(1):139–144, 1990.
125. Beers, R., I. Sizer. A spectrophotometric method for measuring the breakdown of hydrogen peroxide by catalase. *J. Biol. Chem.* 195:133, 1952.
126. Johansson, L.H., L.A. Borg. A spectrophotometric method for determination of catalase activity in small tissue samples. *Anal. Biochem.* 174(1):331–336, 1988.
127. Wheeler, C.R., J.A. Salzman, N.M. Elsayed, S.T. Omaye, D.W. Korte Jr. Automated assays for superoxide dismutase, catalase, glutathione peroxidase, and glutathione reductase activity. *Anal. Biochem.* 184(2):193–199, 1990.
128. Fridovich, I. Quantitative aspects of the production of superoxide anion radical by milk xanthine oxidase. *J. Biol. Chem.* 245(16):4053–4057, 1970.
129. Berry, C.E., J.M. Hare. Xanthine oxidoreductase and cardiovascular disease: molecular mechanisms and pathophysiological implications. *J. Physiol.* 555(3):589–606, 2004.
130. Culleton, B.F., M.G. Larson, W.B. Kannel, D. Levy. Serum uric acid and risk for cardiovascular disease and death: the Framingham Heart Study. *Ann. Intern. Med.* 131(1):7–13, 1999.
131. Crowley, L.V. Determination of uric acid. An automated analysis based on a carbonate method. *Clin. Chem.* 10:838–844, 1964.
132. Kock, R., S. Seitz, B. Delvoux, H. Greiling. A method for the simultaneous determination of creatinine and uric acid in serum by high-performance-liquid-chromatography evaluated versus reference methods. *Eur. J. Clin. Chem. Clin. Biochem.* 33(1):23–29, 1995.
133. Czauderna, M., J. Kowalczyk. Quantification of allantoin, uric acid, xanthine and hypoxanthine in ovine urine by high-performance liquid chromatography and photodiode array detection. *J. Chromatogr. B. Biomed. Sci. Appl.* 744(1):129–138, 2000.
134. Harris, C.M., V. Massey. The reaction of reduced xanthine dehydrogenase with molecular oxygen: reaction kinetics and measurement of superoxide radical. *J Biol. Chem.* 272(13): 8370–8379, 1997.
135. Haining, J.L., J.S. Legan. Fluorometric assay for xanthine oxidase. *Anal. Biochem.* 21(3): 337–343, 1967.
136. Corbisier, P., A. Houbion, J. Remacle. A new technique for highly sensitive detection of superoxide dismutase activity by chemiluminescence. *Anal. Biochem.* 164(1):240–247, 1987.
137. Bokoch, G.M., U.G. Knaus. NADPH oxidases: not just for leukocytes anymore! *Trends Biochem. Sci.* 28(9):502–508, 2003.

138. Cheng, G., Z. Cao, X. Xu, E.G. van Meir, J.D. Lambeth. Homologs of gp91phox: cloning and tissue expression of Nox3, Nox4, and Nox5. *Gene* 269(1,2):131–134, 2001.
139. Cai, H., Griendling, K.K., Harrison, D.G. The vascular NAD(P)H oxidases as therapeutic targets in cardiovascular diseases. *Trends Pharmacol. Sci.* 24(9):471–478, 2003.
140. Shimohama, S., H. Tanino, N. Kawakami, N. Okamura, H. Kodama, T. Yamaguchi, T. Hayakawa, A. Nunomura, S. Chiba, G. Perry, M.A. Smith, S. Fujimoto. Activation of NADPH oxidase in Alzheimer's disease brains. *Biochem. Biophys. Res. Commun.* 273(1):5–9, 2000.
141. Cohen, H.J., P.E. Newburger, M.E. Chovaniec. NAD(P)H-dependent superoxide production by phagocytic vesicles from guinea pig and human granulocytes. *J. Biol. Chem.* 255(14):6584–6588, 1980.
142. Lopes, L.R., C.R. Hoyal, U.G. Knaus, B.M. Babior. Activation of the leukocyte NADPH oxidase by protein kinase C in a partially recombinant cell-free system. *J. Biol. Chem.* 274:15533–15537, 1999.
143. Able, A.J., D.I. Guest, M.W. Sutherland. Use of a new tetrazolium-based assay to study the production of superoxide radicals by tobacco cell cultures challenged with avirulent zoospores of phytophthora parasitica var nicotianae *Plant Physiol.* 117(2):491–499, 1998.
144. Schepetkin, I.A. Lucigenin as a substrate of microsomal NAD(P)H-oxidoreductases. *Biochemistry (Mosc)* 64(1):25–32, 1999.
145. Brennan, M.L., S.L. Hazen. Emerging role of myeloperoxidase and oxidant stress markers in cardiovascular risk assessment. *Curr. Opin. Lipidol.* 14(4):353–359, 2003.
146. Rutgers, A., P. Heeringa, J.W. Tervaert. The role of myeloperoxidase in the pathogenesis of systemic vasculitis. *Clin. Exp. Rheumatol.* 6(32):S55–63, 2003.
147. Savenkova, M.L., D.M. Mueller, J.W. Heinecke. Tyrosyl radical generated by myeloperoxidase is a physiological catalyst for the initiation of lipid peroxidation in low density lipoprotein. *J. Biol. Chem.* 269(32):20394–20400, 1994.
148. Haqqani, A.S., J.K. Sandhu, H.C. Birnboim. A myeloperoxidase-specific assay based upon bromide-dependent chemiluminescence of luminol. *Anal. Biochem.* 273(1):126–132, 1999.
149. Seim, S. Role of myeloperoxidase in the luminol-dependent chemiluminescence response of phagocytosing human monocytes. *Acta. Pathol. Microbiol. Immunol. Scand.* 91(2):123–128, 1983.
150. Cohen Tervaert, J.W. Association of autoantibodies to myeloperoxidase with different forms of vasculitis. *Arthritis Rheum.* 33:1264–1272, 1990.
151. Kallenberg, C.G.M. Antineutrophil cytoplasmic autoantibodies with specificity for myeloperoxidase. In: *Autoantibodies.* Peter, J.B., Y. Shoenfeld, eds., Amsterdam:Elsevier, 53–60, 1996.
152. Moorales, M., A.R. Barcelo. A basic peroxidase isoenzyme from vacuoles and cell walls of *Vitis vinifera*. *Phytochemistry* 45:229–232, 1997.
153. Barcelo, A.R. Peroxidase and not laccase is the enzyme responsible for cell wall lignification in the secondary thickening of xylem vessels in *Lupinus*. *Protoplasma* 186:41–44, 1995.
154. Wallace, G., S.C. Fry. Action of diverse peroxidases and laccases on six cell wall-related phenolic compounds. *Phytochemistry* 52:769–773, 1999.
155. Barcelo, A.R., F. Pomar. Oxidation of cinnamyl alcohols and aldehydes by a basic peroxidase from lignifying Zinnia elegans hypocotyls. *Phytochemistry* 57(7):1105–1113, 2001.
156. Kwak, S.S., S.K. Kim, I.H. Park, J.R. Liu. Enhancement of peroxidase activity by stress related chemicals in sweet potato. *Phytochemistry* 43:565–568, 1996.
157. Rao, M.V., G. Paliyath, D.P. Ormrod. Ultraviolet-B- and ozone-induced biochemical changes in antioxidant enzymes of Arabidopsis thaliana. *Plant Physiol.* 110(1):125–136, 1996.
158. Nicoli, M.C., S. Calligaris, L. Manzocco. Effect of enzymatic and chemical oxidation on the antioxidant capacity of catechin model systems and apple derivatives. *J. Agric. Food Chem.* 48(10):4576–4580, 2000.
159. Randhir, R., K. Shetty. Microwave-induced stimulation of L-DOPA phenolics and antioxidant activity in fava bean (*Vicia faba*) for Parkinson's diet. *Process Biochem.* 39(11):1775–1784, 2004.

160. Strycharz, S., K. Shetty. Effect of *Agrobacterium rhizogenes* on phenolic content of *Mentha pulegium* elite clonal line for phytoremediation applications. *Process Biochem.* 38:287–293, 2002.
161. Randhir, R., Y.-T. Lin, K. Shetty. Stimulation of phenolics, Antioxidant and antimicrobial activities in dark germinated mung bean (*Vigna radiata*) Sprouts in response to peptide and phytochemical elicitors. *Process Biochem.* 39:637–646, 2004.
162. Shetty, K. Role of proline-linked pentose phosphate pathway in biosynthesis of plant phenolics for functional food and environmental applications; A review. *Process Biochem.* 39:789–803, 2004.
163. Randhir, R., D.A. Vattem, K. Shetty. Antioxidant response studies on the effect of oregano phenolics on H_2O_2 stressed porcine muscle. *Process Biochem.* 39(11):1775–1784, 2004.
164. McDougall, G.J., D. Stewart, I.M. Morrison. Cell-wall-bound oxidases from tobacco (Nicotiana tabacum) xylem participate in lignin formation. Planta 194:9–14, 1994.
165. Laloue, H., F. Weber-Lofti, A. Lucau- Danila, P. Gullemat. Identification of ascorbate and guaiacol peroxidase in needle chloroplasts of spruce trees. *Plant Physiol. Biochem.* 35:341–346, 1997.

2.27

Enzymatic Synthesis of Oligosaccharides: Progress and Recent Trends

V. Maitin and R. A. Rastall

CONTENTS

27.1 OLIGOSACCHARIDES

Carbohydrates play an important role in numerous biological processes. However, until a few decades ago, the chemistry and biology of carbohydrates was a Cinderella field (1), an area with significant activity but lacking the glamor of genomes and proteins. In recent years, concomitant with the rapid establishment of the field of glycobiology based on a range of synthetic and analytical methods, the potential of carbohydrates in the maintenance of health and as therapeutic agents has been realized (2,3). Oligosaccharides consist of two to ten monosaccharide residues linked via glycosidic linkages; these can be liberated by depolymerization (4,5). Oligosaccharides composed of monosaccharides only are termed simple or true oligosaccharides and those linked to nonsaccharides such as peptides and lipids are termed conjugate oligosaccharides.

27.1.1 Biological Importance of Oligosaccharides

According to Varki (6), oligosaccharides either mediate specific recognition events or provide modulation of biological processes. The former activity includes the role of cell surface oligosaccharides and their conjugates as recognition sites for bacteria, viruses, toxins, antibodies, and hormones. The latter activity refers to their functions in cell recognition, cell growth, cell differentiation, cell adhesion, signal transduction, development, regulation, and other intercellular communications. Several workers seem to agree that the major function of oligosaccharides is in molecular recognition (2,3,7). The involvement of oligosaccharides in a wide range of specific recognition mechanisms results from their incredible structural diversity. This diversity arises from the fact that they can be highly branched, and the monomeric units may be connected to one another by many different linkage types.

27.1.2 Oligosaccharides as Drugs and Functional Foods

It has been established that many pathogens use carbohydrate binding proteins, or lectins, to attach to cells and initiate disease (8,9). Specific membrane bound oligosaccharides serve as receptors for the attachment of the pathogen onto the cell. One of the natural defense mechanisms of the body against infection involves decoy oligosaccharides present in the mucosal lining of epithelial cells and in saliva, tears, urine, sweat, and breast milk. This is especially significant in breast milk, where the presence of many oligosaccharides at high concentration (up to millimolar levels) protects infants from many infections (9,10). An invading pathogen binds to these decoy oligosaccharides instead of the host cell and the host is thus protected from infection. This is the basic concept behind use of oligosaccharides as antiinfective agents. Because these oligosaccharides prevent the adhesion of the pathogen to the host cell, the term antiadhesives is used to describe this class of carbohydrate drugs. They also have a possible antiinflammatory application against tumors or virally infected cells. Many pharmaceutical companies are now involved in research and development of carbohydrate drugs (2,11).

Two types of antiinfectives are under investigation: receptor oligosaccharides, for use against respiratory and gastrointestinal disease; and sialic acid analogues for use against influenza (10) and cholera. It is important that the antiadhesives used are safe, nontoxic, and nonimmunogenic. Short chain oligosaccharides, for example, human oligosaccharides about 1kDa in size are promising antiadhesive agents. This is because of their high water solubility, and resistance to heat and pH, making their delivery and administration more effective. Another important factor in their efficacy is the strength of binding to the bacterial lectins. For this reason, polyvalent oligosaccharides have been found to be much more effective than monovalent ones because they form a stable structure by binding to more than one adhesin (9). Several glycomimetics, such as glycodendrimers (12) and mixed type glycoclusters (13) have emerged as valuable tools in this regard. The concept of antiinfective therapy is an exciting one, considering its potential commercial impact and wide application. However, at present, it is in its developmental stages. There are practical limitations to its widespread use because of factors such as the lack of animal models for clinical trials, availability of appropriate oligosaccharides, and limited knowledge of the host surface glycoconjugates (9,10).

Apart from their role in recognition, there has been recent interest in the use of oligosaccharides as functional food ingredients to promote growth of probiotic or beneficial bacteria in the human gut (14). Oligosaccharides provide a combination of suitable physicochemical and physiological properties, which make them very good candidates for use as food ingredients. The 1991 legislation for Foods for Specified Health Use (FOSHU) of the Japanese government comprised 223 items, of which more than half incorporate oligosaccharides as the functional components (15). Food grade oligosaccharides are manufactured using simple methods, and avoiding expensive purification steps, in order to keep the prices competitive. As a result, they are generally impure products containing residual feedstock carbohydrate and monomer sugars (11). Pharmaceutical applications in most cases require pure compounds of a defined chemical structure, and synthesis of these oligosaccharides is complex and expensive.

The biological importance of oligosaccharides coupled with their potential as prophylactic agents and functional food ingredients has generated considerable research interest in this area about various aspects of their structure, function, and uses. The limiting factors in this research and application of the results are the limited quantity and range of available oligosaccharides, and their very high prices. It is therefore important to develop oligosaccharide synthesis methods that are efficient, cost effective, and easy to scale up, in order to increase the repertoire of oligosaccharides available.

There are two methods of oligosaccharide synthesis: chemical and enzymatic.

27.2 CHEMICAL SYNTHESIS

Traditionally, oligosaccharides are synthesized by chemical methods. Though well developed, chemical methods of synthesis are complicated and involve a number of steps. Because carbohydrates have several hydroxyl groups of similar reactivity, there are numerous protection and deprotection steps required in order to achieve regioselectivity (16). In spite of this, a mixture of isomeric oligosaccharides is often produced. Moreover, the number of steps increases with the size of the oligosaccharide (up to 7 steps needed for disaccharide synthesis and more than ten for a trisaccharide), making the synthesis of even simple oligosaccharides a lengthy process. The method needs expensive chemicals, yields are low, and scale up is difficult. For an overview of chemical oligosaccharide synthesis, the reader is referred to exhaustive reviews by Boons (17) and Davis (18). Several oligosaccharides of biological importance have been synthesized using chemical methods, including oligosaccharides corresponding to the capsular polysaccharide of *Cryptococcus neoformans* and outer cell wall of *Moraxella catarrhalis* (19), a high mannose nonasaccharide corresponding to part of the glycoprotein gp120 of the viral coat of HIV-1(20), myoinositol containing compounds (21), glycopeptide containing tumor associated carbohydrate antigens (22), α-1,2-linked disaccharide derivatives, (23) and recently, mixed type glycocluster oligosaccharide mimetics containing sugar derivatives from different sugar series (galactoside, mannoside and fucoside) (13) with potential as antiadhesives. Chemical methods have also been reported for synthesis of sugar nucleotides needed for enzymatic synthesis reactions using Leloir-transferases (24). New strategies to improve the chemical synthesis process are being developed and are summarized by Bartolozzi and Seeberger (25). A significant achievement has been that of programmable one pot synthesis. Based on difference in reactivity between glycosyl donors, a database, search engine, and computer program called Optimer was developed (26). The program enables the user to select from a list of the best reagent combinations for the synthesis of an oligosaccharide of interest, along with predicted yields (27). The program has been successfully employed for controlled synthesis of both linear and branched trisaccharides and tetrasaccharides, allowing the possibility of creating a library of oligosaccharides. To facilitate the synthesis of some linkages, which are difficult to synthesize by chemical methods, chemists are now using chemoenzymatic synthesis, wherein an enzyme is used to synthesize part of the oligosaccharide structure, in combination with chemical methods. One such example is the synthesis of β-mannosyl linkage of *N*-glycans; Watt et al. (28) used β -mannosyltransferase for the synthesis of the core of *N*-linked oligosaccharides.

In spite of these advancements, the area of chemical synthesis remains a challenge because of its technical complexity and absence of general procedures applicable to synthesis of a range of oligosaccharides, which makes their adaptation to industrial scale next to impossible. Their application is currently limited to synthesis of some bioactive oligosaccharides for research purposes; large scale manufacture for food and drug use remains impractical.

27.2.1 Enzymatic Synthesis

Because most of the complex heterooligosaccharides in nature are produced by enzymes, it was thought that it must be possible to use the synthetic potential of enzymes *in vitro* to bring about oligosaccharide manufacture (29). In recent years, enzymatic synthesis of oligosaccharides has become the method of choice for most researchers. It helps to overcome two major problems posed by chemical synthesis: selective protection and deprotection is not necessary, and the configuration of the newly formed anomeric centres can be controlled well. As a result, a mixture of isomers is not obtained. Other advantages include lack of byproducts, wider range of reaction conditions and regiospecificity and stereospecificity of the enzymes. Thus enzymes can be used to synthesize oligosaccharides by single step reactions and to synthesize novel oligosaccharides. Production of the enzymes by fermentation and immobilization to enable reuse can reduce the cost of the enzymes, making the process very economical.

Two main types of enzymes are used to catalyze oligosaccharide synthesis: glycosyltransferases (biosynthetic) and glycosidases (hydrolytic).

27.2.2 Glycosyltransferases

Glycosyltransferases are type II membrane bound glycoproteins consisting of a short N-terminal cytoplasmic domain, a transmembrane domain that anchors it to the cell membrane, and the C-terminal catalytic domain (30). They are currently classified into 60 families according to the sequence based classification of Coutinho and Henrissat (31). Glycosyltransferases catalyze the stereospecific and regiospecific transfer of a monosaccharide from a donor substrate (glycosyl nucleotide) to an acceptor substrate. They are classified according to the sugar transferred from the donor to the acceptor and by the acceptor specificity. For example, β 1, 4-galactosyltransferase (GalT) from bovine milk catalyzes the transfer of a galactose (Gal) unit from UDP-Gal to the 4-OH group of terminal N-acetylglucosamine (GlcNAc) β -R acceptors generating Gal β 1-4GlcNAc β -R (N-acetyllactosamine-R) structures. A completely different enzyme, β 1,3-GalT is required for synthesis of isomeric Gal β 1-3 GlcNAc β -R structures from the same substrates (32). The regiospecificity and stereospecificity, the high selectivity for the acceptor substrate, and the high yields that can be achieved are attractive features of these catalysts (16). The reaction can be represented by the following general equation:

$$\text{Nucleotide sugar} + \text{R-OH} \rightarrow \text{R-O-sugar} + \text{nucleotide} \quad (27.1)$$

where R is a free saccharide or a saccharide linked to a protein or lipid.

In vivo, the synthesis occurs via three fundamental steps: activation, transfer, and modification (33). In the first stage of the sequence, a monosaccharide is transformed into a sugar-1-phosphate (sugar-1-P) by a kinase enzyme. This then reacts with a nucleoside triphosphate (NTP) in a reaction catalyzed by nucleoside transferase, and forms a chemically activated nucleoside diphosphate sugar (NDP-sugar). The sugar unit is then transferred to a sugar acceptor by glycosyltransferases Each one is specific for a certain donor and acceptor and for the linkage position of the new glycosidic bond (29). Owing to this high specificity, they have been used in synthesis of oligosaccharides *in vitro*. However, such high specificity also means that a wide range of glycosyltransferases is required to enable the synthesis of a variety of desired oligosaccharides *in vitro*. At the moment, not many glycosyltransferases are commercially available [nine, as reported by Bastida et al. (34), of which only β-1,4-galactosyltransferase is available in amounts higher than 1 unit]. They are difficult to isolate and purify from their natural sources because they are usually membrane bound and present in low concentrations. Several glycosyltransferase genes, including those coding for glucosyltransferase (35,36), sialyltransferase (37,38), fucosyltransferase (34,39), mannosyltransferase (40,41), and galactosyltransferase (42–45) have been cloned, but many more need to be cloned and probably modified to make the enzymes more widely available and overcome problems of instability. In practice, the range of routinely used glycosyltransferases at the moment is mainly limited to β-1,4-galactosyltransferase, α-2,3-and α-2,6-sialyltransferases and α-1,3/4-fucosyltransferases (46,47).

Besides the availability of the actual enzyme, the availability of the sugar nucleotide donors was also a problem until recently, as they were very expensive and unstable. Eight sugar nucleotides: UDP-Glc, UDP-Gal, GDP-Fuc, CMP-NeuAc, UDP-glucuronic acid, GDP-Man, UDP-GlcNAc, and UDP-GalNAc act as donor substrates for most mammalian glycosyltransferases. With the exception of UDP-glucuronic acid, they are now commercially available, though still expensive. Because the initial report for regeneration of UDP-Glc and UDP-Gal by Wong et al. (48), numerous regeneration methods have been developed for all of these sugar nucleotides, which not only reduce their cost but also overcome the problem of product inhibition resulting from the released nucleoside monophosphates or diphosphates in the reaction (Figure 27.1). Representative papers include regeneration of UDP-Glu, UDP-Gal, CMP- NeuAc (49), and UDP-Gal, UDP-2-deoxyGal, UDP-Galactosamine (50) using multienzyme systems, regeneration of GDP-Man (40), and a simplified pathway for regeneration of UDP-Gal (51). Gram scale synthesis has been reported for UDP-Gal (52), and yields up to multikilogram levels have been achieved for some oligosaccharides such as SLe^x (32).

It must be realized, however, that the enzymes that are involved in these pathways are not recovered and recycled, which is essential for the process to be feasible at industrial scales. The obvious biotechnological solution was enzyme immobilization (53,54). Nishiguchi et al. (55) carried out the synthesis of a trisaccharide derivative Neu-5Acα (2$\rightarrow$6) Galβ(1$\rightarrow$4)GlcNAcβ-O-$(CH_2)_6$-NH_2 using recombinant β1,4-galactosyltransferase and α2,6-sialyltransferase immobilized by coupling reactions with activated Sepharose. The oligosaccharide was built up by glycosylation on a water soluble primer having GlcNAc residues and an α–chymotrypsin sensitive linker, for the release of final product by α-chymotrypsin treatment.

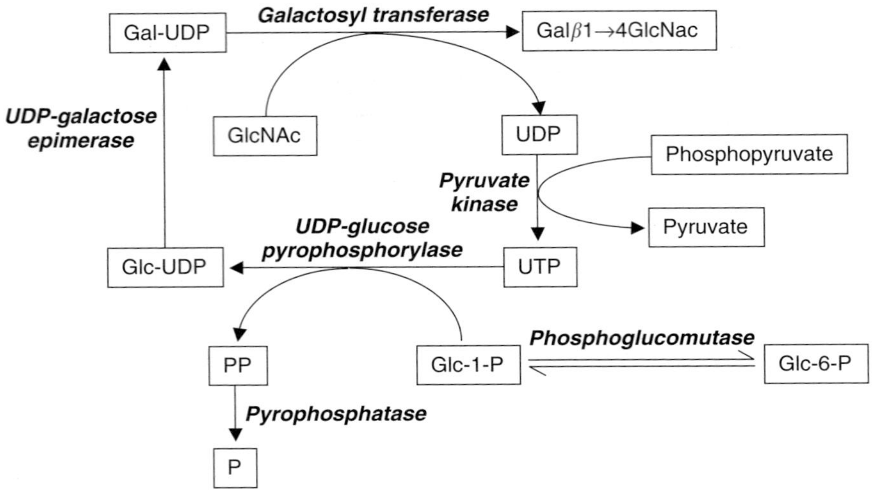

Figure 27.1 Multienzyme synthesis of *N*-acetyllactosamine using galactosyltransferase with cofactor regeneration

Several workers have also investigated the feasibility of using whole cells of recombinant bacteria as catalysts, as this approach obviates the need for enzyme isolation and in many cases tedious purification. Herrmann et al. (56) were able to synthesize disaccharides Manα1, 2Man, Manα1, 2ManOMe, and two dimannosylated glycopeptides using whole recombinant *E. coli* cells expressing cloned α-1,2-mannosyltransferase reported earlier (40). The group of Ozaki has produced a series of papers reporting the use of coupled recombinant or metabolically engineered bacterial strains: *Corynebacterium ammoniagenes* for NTP (re)generation, recombinant *E. coli* with NDP-sugar generation and regeneration cycle expression and recombinant *E. coli* with glycosyltransferase (57) : for large scale production of globotriose (188 g/L) (58), N-acetyllactosamine (107 g/L) (59), 3'-sialyllactose (33 g/L), (60) and sialyl-T_n epitope (45 g/L) (61). Chen et al. (62) have developed a superbug, an engineered *E. coli* strain containing a plasmid harboring an artificial gene cluster containing all the necessary genes for sugar nucleotide generation and regeneration as well as oligosaccharide accumulation for the synthesis of an α-Gal epitope. Another engineered *E. coli* strain (lacZ$^-$) has been used by Priem et al. (63) for the production of human milk oligosaccharides by efficient conversion from lactose. The lacZ$^-$ mutant accumulated lactose when grown on glycerol. The accumulated lactose acted as an acceptor for the heterologous β1,3-N-acetylglucosaminyltransferase enzyme expressed from a *Neisseria meningitidis* gene. The resultant trisaccharide acted as acceptor for the coexpressed *N. meningitidis* β1,4-galactosyltransferase enzyme. This system produced lacto-N-neotetraose and lacto-N-neohexaose at over 5 g/L yield. Similarly, a nanA$^-$ *E. coli* was able to produce sialyllactose.

Another particularly elegant regeneration methodology is sugar nucleotide regeneration beads or superbeads (57,64,65). The superbeads technology involves overexpression of individual enzymes along the sugar nucleotide biosynthetic pathway in a His_6-tagged form followed by their coimmobilization onto nickel nitrilotriacetate agarose beads. The beads can be employed for synthesis of specific oligosaccharide sequences by combination with a variety of glycosyltransferases. UDP-Gal regenerating superbeads were developed as a model system and in combination with β1,4-GalT and/or α 1,3/1,4-GalT, 100 mg-1.03 g scale synthesis of biologically important oligosaccharides was achieved (66). The first generation superbeads required four enzymes and the reactions were performed in stirred batch mode whereas the second generation superbeads (65) involve seven enzymes and the reactions performed by circulation of the reaction mixture through the superbeads in a packed bed column reactor configuration, proving to be more efficient.

The aforementioned advancements over the last decade have transformed the status of Leloir glycosyltransferases in oligosaccharide synthesis by alleviating the problems associated with them earlier: of low availability of enzymes and sugar nucleotide donors, product inhibition, and high reagent costs. These developments, together with their inherent regioselectivity and stereoselectivity should significantly increase their applicability.

27.2.3 Glycosidases

As the name suggests, the natural function of glycosidases is hydrolytic cleavage of glycosidic bonds. Glycosidases occur widely in nature; in viruses, microorganisms, and plant and animal cells (16), and consequently are more available than glycosyltransferases. They are also more stable, act on easily available substrates, and do not need cofactors.

Their main drawback is that if nonlinkage specific enzymes are used, the regioselectivity tends to be low, which results in the formation of a product which is a mixture of several different linkages (usually 1→6, 1→4, 1→3 and 1→2), posing problems of purification. Linkage specific glycosidases, however, generally produce products with very high regioselectivity. In spite of this, keeping in mind their availability, relative economy of use and simplicity of the reaction, glycosidases have become established as a popular choice for oligosaccharide synthesis. While the high specificity of glycosyltransferases can allow precise sequential construction of desired oligosaccharides, the wide acceptor specificity of glycosidases makes them useful for synthesis of novel oligosaccharides of unknown and potentially exploitable biological activities.

Glycosidases can be divided into two groups: the exoglycosidases, which cleave glycosidic bonds at the nonreducing end of the oligosaccharide, and the endoglycosidases, which cleave internal glycosidic bonds (33). In most instances, synthesis has been carried out using retaining exoglycosidases. According to Coutinho and Henrissat (31), glycosidases are currently classified into 90 families based on sequence similarities. Based on mechanism (67–71), they are classified as retaining or inverting glycosidases depending on whether the hydrolysis proceeds by net retention or inversion of the stereochemistry of the anomeric center. In either case, the reaction is acid base catalyzed through an oxocarbenium ionlike transition state, and involves two carboxylic groups at the active site. Inverting glycosidases operate via direct displacement of the leaving group by water. The two carboxylic groups are suitably positioned at the active site such that one provides base catalytic assistance to the attack of water, while the other provides acid catalytic assistance to cleavage of the glycosidic bond. Retaining glycosidases use a double displacement mechanism involving the formation of a covalent glycosyl enzyme intermediate. Unlike the inverting enzymes, where one residue functions as a general acid and the other as the general base, in retaining enzymes one residue serves both functions, acting as acid catalyst for the glycosylation step and base catalyst for the deglycosylation step. The second carboxylic group acts as a nucleophile and a leaving group. A significant difference between the two enzyme classes is the separation between the carboxylic acid side chains, which is approximately 11 Å in inverting enzymes and 5.5 Å in retaining enzymes.

Detailed studies have been carried out on the mechanism of retaining glycosidases using several strategies, details of which can be found in reviews by Withers (68,71). These include intermediate trapping by (1) modification of the substrate through fluorination of its 2-hydroxyl or 5-hydroxyl group by fluorosugar reagents which has led to the identification of the catalytic nucleophiles of several families of retaining glycosidases; and (2) mutation of the enzymes to decrease the turnover of the intermediates and enable determination of their three dimensional structure by crystallographic analysis. Crystallographic studies have also been performed on the enzyme itself (72) at various stages of reaction, providing snapshots of the free enzyme, enzyme substrate complex, covalent intermediate, and product complex.

The mechanism of efficient acid base catalysis by a single residue in retaining glycosidases was investigated using *Bacillus circulans* xylanase as a model. This was achieved by measuring the pK_a values of the two relevant carboxylic acid residues in the active site by ^{13}C NMR (73). While the pK_a of the catalytic residue (Glu172) was 6.7 in the first step, it dropped to 4.2 in the intermediate, allowing it to act as the base catalyst in the second step. Mutagenesis studies attribute the higher pK_a of 6.7 to the negative charge on the other residue Glu 78; in the absence of this charge, the pK_a drops to the lower value of 4.2. During the reaction, this shift in pK_a occurs when the charge on the catalytic nucleophile is removed upon ester formation. The carboxylic acid residue in question is thus able to operate as an acid or base catalyst depending on the charge on the second residue. This has been suggested as a general mechanism for retaining glycosidases.

There are two well established approaches for oligosaccharide synthesis using glycosidases: equilibrium controlled synthesis and kinetically controlled synthesis.

27.3 EQUILIBRIUM CONTROLLED SYNTHESIS

The equilibrium approach is based on the reversal of the catabolic role of enzymes in a thermodynamically controlled reaction (74). It involves a simple reversion of the glycoside hydrolysis reaction by combining a free monosaccharide and nucleophile by direct coupling (condensation). This is also referred to as direct glycosylation or reverse hydrolysis. The equilibrium constant of the reaction favors hydrolysis over glycoside formation and the yields therefore are usually low. In order to shift the reaction equilibrium toward product formation, various approaches have been used (33,46,74,75). The most common way is to use high concentrations of the products of the forward reaction (e.g., monosaccharides) and reduced concentrations of the reactants. This is achieved by incubating the enzyme in a highly concentrated solution of monosaccharide, of the order of 70–80% (w/w), that is, in conditions of lowered water activity (74). Second, to accelerate the otherwise low reaction rate, higher reaction temperatures of 50–60°C are used (74). Increased yields result from controlled and selective removal of the product oligosaccharides from the reaction mixture over the course of the reaction using molecular traps such as activated carbon (76,77). This makes reaction yields comparable to those achieved through transglycosylation. The greatest advantage of this method is the simplicity of the reaction procedure. Both homooligosaccharides and heterooligosaccharides can be synthesized using the equilibrium approach.

Mannooligosaccharides, both homo- and hetero-, have been obtained in good yields by reversal of the α-mannosidase reaction. There is interest in their synthesis because not only are they present in the high mannose type sugar chains of glycoproteins, but they have potential as antiadhesives. Different linkage isomers show varying degrees of inhibition at varying concentrations (78–80).The exceptionally high solubility of mannose in water has enabled synthesis to be carried out in mannose concentrations up to 85% (w/w) and maximum total yields of up to 70% mannooligosaccharide have been achieved using α-mannosidase from Jack bean (75,81). The same enzyme has been used to synthesize heteromannooligosaccharides by cocondensation with a range of acceptor sugars, in addition to homooligosaccharides (82,83). In general, the yields were higher at 70% total sugar concentration than at 50%, except in the case of lactulose, where the yield at 50% was more than double than at 70% total sugar. Higher yields were obtained for heteromannooligosaccharides using reverse hydrolysis compared to kinetic syntheses. A range of linkages, $\alpha(1\rightarrow1)$, $\alpha(1\rightarrow2)$, $\alpha(1\rightarrow3)$, $\alpha(1\rightarrow4)$, and $\alpha(1\rightarrow6)$, were obtained and the product spectrum showed a predominance of lower oligosaccharides, mainly disaccharides, trisaccharides, and tetrasaccharides in the case of homooligosaccharides. Partially purified α-mannosidases from almond meal and limpets were employed by Singh et al. (84) for synthesis of $\alpha(1\rightarrow2)$ and $\alpha(1\rightarrow3)$ linked mannose disaccharides in ratios of 65:35 and 57:43 respectively. Fungal α-mannosidase from *Aspergillus niger* (85) produced $\alpha(1\rightarrow2)$, $\alpha(1\rightarrow3)$ and $\alpha(1\rightarrow6)$ linked disaccharides and trisaccharides, but no $\alpha(1\rightarrow1)$ or $\alpha(1\rightarrow4)$ linkages were detected. The yield of individual regioisomers in equilibrium synthesis is correlated with the standard free energy. In the synthesis of oligosaccharides using α-glucosidase, the standard free energy of the $\alpha(1\rightarrow6)$ disaccharide was the lowest, followed by $\alpha(1\rightarrow2)$, $\alpha(1\rightarrow3)$, $\alpha(1\rightarrow4)$, and $\alpha(1\rightarrow1)$. The standard free energy was in inverse correlation to the oligosaccharide yields (76). Regioselective synthesis of $\alpha(1\rightarrow2)$ mannobiose and mannotriose was achieved (86) using specific 1,2-α-mannosidase from *Aspergillus phoenicis*. A novel α-1,6- mannosidase isolated from the same organism (87) resulted in synthesis of $\alpha(1\rightarrow6)$-linked mannobiose and mannotriose with absolute regioselectivity.

Other fungal enzymes used in synthesis by this approach include glucoamylase from *Aspergillus niger* (88) and β-glucanase from *Penicillium emersonii* (89). While homooligosaccharides were obtained in moderate yields (14–16%), the yields of heterooligosaccharides were rather low compared to the mannosidase enzymes. Using purified *A. niger* glucoamylase (90), a thorough investigation was carried out of its ability to form condensation products from aqueous mixtures of individual carbohydrates (arabinose, fructose, galactose, myoinositol, lyxose, mannose, ribose, and xylose) or their mixtures with glucose. Heterodisaccharides were produced with each of the eight sugars in combination with glucose. Interestingly, the enzyme was able to condense not only glucose but also galactose and mannose individually into disaccharides.

Cellulases (cellobiohydrolase I and endoglucanase I) from recombinant *Trichoderma reesei* (91), α-amylase from *Bacillus licheniformis* (92) and α-glucosidase from *Bacillus stearothermophilus* (93) provide some instances of use of bacterial glycosidases in reverse hydrolysis. Cello bio hydrolase (CBH) I gave yields in the order of 40% in the best case experimental conditions. Homooligosaccharides and heterooligosaccharides from DP 1-10 were obtained using α-amylase; however the products in this case as well in the case of CBH I were not characterized. The *B. stearothermophilus* α-glucosidase was able to produce isomaltose (51%), nigerose (25%), maltose (14%), and kojibiose (10%) in 50% glucose solution, and heterooligosaccharides were obtained both with mannose and xylose, with quantity being dependent on glucose:saccharide acceptor ratios.

Recently (94), a range of glycosidases were screened for their ability to synthesize thioglucosides, which in the context of glycobiology are useful as specific enzyme inhibitors and as glycosyl donors in oligosaccharide synthesis. Of the enzymes tested, β-glucosidase from almonds showed the highest activity in the reverse hydrolysis reaction and high yields of 1-propanethioglucoside were obtained (68% and 41%), based on 1-propanethiol and glucose respectively.

27.3.1 Kinetically Controlled Synthesis

This approach, also known as transglycosylation, employs an activated glycoside donor to form an active intermediate in high concentrations. This intermediate is then trapped by a second nucleophile (other than water) to yield a new glycoside (Figure 27.2) (33,46). The donor glycosides include oligosaccharides, substituted aryl glycosides and glycosyl fluorides.

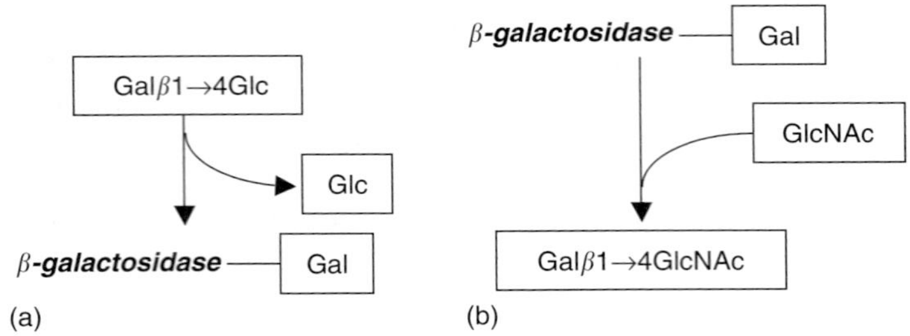

Figure 27.2 Synthesis of *N*-acetyllactosamine using *β*-galactosidase from *Bacillus circulans* (a) Activation to form a galactosyl-enzyme complex (b) Galactosyl transfer to *N*-acetyl glucosamine acting as an acceptor in competition with water

The reaction is efficiently controlled kinetically as the product is a potential substrate of the glycosidase enzyme (95). As a result, higher product yields using lower enzyme concentration can be obtained. However, the donor glycoside is used up during the reaction in greater than stoichiometric quantities and cannot be reused. This drives up the cost of synthesis.

Table 27.1 summarizes some recent glycosides synthesized using the kinetic approach.

27.4 INFLUENCE OF REACTION COMPONENTS AND CONDITIONS ON YIELD AND SELECTIVITY

27.4.1 Donor:Acceptor Sugar Ratio

In equilibrium controlled heterooligosaccharide synthesis, the proportions can be regulated by the ratio of the individual substrates in the reaction mixture. Usually, heterooligosaccharides are obtained exclusively at higher ratios of the saccharide acceptor to donor, but yields are invariably lower. Using glucoamylase from *A. niger*, heterooligosaccharides were produced exclusively at a fucose:glucose ratio of 85:15, with a yield of 1.6%; highest yield of 3% was at 65:35 ratio but comprised only 42% heterooligosaccharides (88). The same trend was observed in the case of α-glucosidase from *B. stearothermophilus* (93); at a xylose:glucose ratio of 90:10, heterodisaccharides were the major products, but their final concentration was less than half (22mM) of that obtained at a ratio of 50:50 (53mM).

In kinetic synthesis, the donor and acceptor determine the structure of the glycoside synthesized. Although the same glycoside can be synthesized with several different donor and acceptor combinations, the yields vary depending on their nature. For instance (Table 27.1), while α-L-Fuc-(1→3)-α-GlcNAc-OAll and α-L-Fuc-(1→3)-β-GlcNAc-OAll could be synthesized in 34% and 25% yields respectively using α-L-Fuc-O-NP as donor and α/β -GlcNAc-OAll as acceptors, the yield was only 8% for the former when α-L-Fuc-F was the donor instead, and the second glycoside could not be isolated (96). The low yield was attributed to the increased hydrolysis rate of the fluoride donor.

The donor:acceptor ratios need to be optimized for best yields. Equimolar ratios gave higher yields compared to ratios of 1:0.5 and 1:0.2, in a study of galactooligosacharide synthesis using β-galactosidases from three different sources (97); presumably due to higher transglycosylation rates and reduced secondary hydrolysis at 1:1 ratios. The nature of the glycosyl acceptor also influences the regiospecificity of the transfer. For instance, addition of α-Gal-O-p-NO_2Ph to α-Gal-O-Me gives predominantly 1→6 linked product, while the same addition to β-Gal-O-Me gives predominantly the 1→3 linked product (33).

Table 27.1
Synthesis of oligosaccharides using the kinetic approach

Enzyme	Source	Glycosyl Donor	Glycosyl Acceptor	Main Glycoside Product	Yield (%)	Ref.
N-acetylhexosaminidase	*A. oryzae*	p-NP- β-GlcNAc	GalNAc	β-GlcNAc-(1→6)-GalNAc	26	151
		p-NP- β-GalNAc	GalNAc	β-GalNAc-(1→6)-GalNAc	38	151
β -galactosidase	*B. circulans*	p-NP- β-Gal	β-GlcNAc-(1→6)-GalNAc	β-Gal-(1→4)- β-Gal(1→4)- β-GlcNAc-(1→6)-GalNAc	48	151
β -galactosidase	*B. circulans*	O-pNP-β-Gal	OAll- α/ β-GlcNAc	β-Gal-(1→4)-α/β-GlcNAc-OAll	α:66, β:30	152
			OAll- α/ β-Gal	β-Gal-(1→4)-α/β-Gal-OAll	α:55, β:49	
			Ph-S- β-Gal	β-Gal-(1→4)- β-Gal-S-Ph	63	
			O-pNP-β-Gal	β-Gal-(1→4)- β-Gal-O-pNP	23	
				β-Gal-(1→4)- β-Gal-(1→4)- β-Gal-O-pNP	10	
β -galactosidase	Porcine liver	o-NP-β-Gal	p-NP-α/β-Gal/GlcNAc/GalNAc	β-Gal-(1→6)-α/β-Gal-pNP	α:63.9, β:21.2	104
				β-Gal-(1→6)-α/β-GalNAc-pNP	α:78.7, β:28.1	
				β-Gal-(1→6)-α/β-GlcNAc-pNP	α:83.8, β:25.7	
β -galactosidase	*B. circulans*	o-NP-β-Gal	p-NP-α/β-Gal/GlcNAc/GalNAc	β-Gal-(1→3)-α-Gal-pNP	9.30	104
				β-Gal-(1→6)-α/β-Gal-pNP	α:22.8, β:11.7	
				β-Gal-(1→3)-α/β-GalNAc-pNP	α:75.9, β:20.9	
				β-Gal-(1→6)-α/β-GalNAc-pNP	α:3.20, β:18.6	
				β-Gal-(1→3)-α/β-GlcNAc-pNP	α:79.3, β:29.5	
				β-Gal-(1→6)-α/β-GlcNAc-pNP	α:0.30, β:6.0	
β -galactosidase	*A. oryzae*	o-NP-β-Gal	Xylose	Galactosyl-xylose	21	111
			Ethyleneglycol	Galactosyl-ethyleneglycol	70	
α-L- fucosidase	*P. multicolor*	α-L-Fuc-O-NP(1)	α-GlcNAc-OMe(3)	α-L-Fuc-(1→3)-α-GlcNAc-OMe (1+3)	29	96
		α-L-Fuc-F(2)	α/β -GlcNAc-OAll(α:4, β:5)	α-L-Fuc-(1→3)-α/β -GlcNAc-OAll (α:1+4, β:1+5)		
			GlcNAc(6)	α-L-Fuc-(1→3)-α-GlcNAc-OMe (2+3)	α:34, β: 25	
			Glc(7)	α-L-Fuc-(1→3)-α/β -GlcNAc-OAll (α:2+4, β:2+5)	13	
				α-L-Fuc-(1→3)-GlcNAc(2+6)	α:8, β: -7	
				α-L-Fuc-(1→3)-Glc (2+7)	34	

27.4.2 Temperature

The equilibrium synthesis reactions are typically carried out at temperatures of 50–60°C and kinetic synthesis reactions at 37°C, although temperatures ranging from room temperature to 45°C have been employed, depending on the enzyme source and thermal stability. The high substrate concentrations used in reverse hydrolysis exert a protective effect on the enzyme, thus allowing use of higher temperatures.

Statistical analysis was performed for effect of temperature (20, 30, 40, and 50°C) on lactose conversion by β-galactosidase from *B. circulans* with the aim of developing a kinetic model which could aid reactor design for the process (98). Reaction rate as well as yield increased at higher temperatures.

A recent study employing dextransucrase has reported surprising results where not only yield but also selectivity improved considerably with decreasing reaction temperature, best results being obtained in reactions in the frozen state (99). The yield of leucrose (0.1M sucrose as substrate) increased from 10% at 25°C to 65% at −4°C. Similarly, in reactions containing sucrose with isomaltulose as acceptor, overall yields (trisaccharide, tetrasaccharide, and pentasaccharide) at −4 to −10°C were 80−86%, double the yield of reactions in the 5–25°C range.

Temperature has been reported to have a bearing on the selectivity of β-glycosidases for donors differing in their glycon parts, as observed from an investigation with β-glucosidases and galactosidases from almond, a mesophilic (*K. fragilis*), and three thermophilic organisms (*C. saccharolyticum, S. solfataricus, P. furiosus*) (100). Selectivity, calculated as the specificity constant (V_{max}/K_m) or V_{max} ratio of glucoside to galactoside donor (p-NP or phenyl-β-glucoside/galactoside), decreased with increasing temperature in case of almond, *K. fragilis* and *C. saccharolyticum,* and remained constant for the other two thermophilic enzymes.

Klepsiella fragilis β-galactosidase became a β-glucosidase when the temperature was raised from 25 to 50°C.

The range of temperatures employed in enzymatic oligosaccharide synthesis combined with the high sugar concentrations has been reported in some instances (101,102) to limit optimum oligosaccharide yields owing to glycation induced enzyme inactivation. We have provided possible evidence for this (103) during synthesis of mannooligosacharides by reverse hydrolysis. Inhibitors of the Maillard reaction were partially able to alleviate these effects resulting in reduced loss of enzyme activity and increased oligosaccharide yield.

27.4.3 Enzyme Origin

Regioselectivity can vary with different enzymes, and careful selection of the enzyme can effect the formation of desired linkage. A major objective of researchers in this field is to build up a library of glycosidases capable of selectively catalysing the formation of any desired glycosidic bond (47).

Same donors and acceptors lead to a range of glycosides varying in linkage as well as yield (Table 27.1), when β-galactosidases from two different sources were used (porcine liver and *Bacillus circulans*). Only (1→6) linked products were obtained with the former, while the bacterial enzyme gave rise to (1→3) linkages as well (104). β-glucosidases and galactosidases from almond, *K. fragilis, C. saccharolyticum, S. solfataricus,* and *P. furiosus* differed in their substrate specificity and donor selectivity and in the extent to which it was influenced by temperature (100). Another study investigated β- galactosidases from *B. circulans, A. oryzae, K. lactis, and K. fragilis* (98). There were obvious differences with respect to the amount, size, and type of oligosaccharides produced by these enzymes. *Bacillus circulans* enzyme produced the highest yield and variety of oligosaccharides, and also largest saccharides (up to pentasaccharides). The *A. oryzae* enzyme also produced higher oligosaccharides but in smaller quantities. The enzymes from *Kluyveromyces* mainly produced trisaccharides, with negligible amounts of higher saccharides. Novel galactooligosaccharides with distinct spectra were synthesized with β-galactosidase from five different species of *Bifidobacterium* (105). Higher yields (48%) of β-galactooligosaccarides were obtained with thermostable β-galactosidase from *Sulfolobus solfataricus* (97) compared to those from *A. oryzae* (36%) and *E. coli* (32%). Maximum oligosaccharide yields were obtained fastest with the *A. oryzae* enzyme, were highest with the *S. solfataricus* enzyme, and lowest with the *E. coli* enzyme.

Partially purified α-mannosidases from almond meal and limpets ,when incubated with a high concentration of mannose in reverse synthesis reaction, gave rise to the same (1→2)- and (1→3)-mannose disaccharides, but in different ratios of 65:35 and 57:43 respectively (84). Taking advantage of the differences in oligosaccharide yields and specificities of mannosidases from different sources, we have recently devised a simple method of regioselective synthesis of α (1→3)-mannobiose (106). A commercial α-mannosidase from almond was highly efficient in reverse hydrolysis, and oligosaccharide yields of 45–50% were achieved. The products were a mixture of disaccharides (30.75%, w/w), trisaccharides (12.26%, w/w) and tetrasaccharides (1.89%, w/w) with 1→2, 1→3, and 1→6 isomers. α-1,2-linkage specific mannosidase from *P. citrinum* (106) and α-1,6-linkage specific mannosidase from *Aspergillus phoenicis* (87) were used in combination to hydrolyse the respective linkages from the mixture of isomers, resulting in α-(1→3)-mannobiose in 86.4% purity.

27.4.4 Use of Organic Solvents

Many workers have experimented with the use of nonaqueous media and organic cosolvents in enzymatic synthesis. While in reverse hydrolysis the aim is to reduce the water activity of the reaction systems and drive the reaction toward synthesis, in transglycosylation the aim is to reduce the extent of competing hydrolysis.

The feasibility of this approach depends to an extent on the enzyme source (107), as many enzymes undergo inactivation in such conditions and reduced yields or decreased reaction rates are observed. The effect of the solvent on enzyme activity under reaction conditions should be studied, and higher enzyme concentrations are generally required than in buffer. Using o-nitrophenyl galactoside as donor and 3-O-methyl Glc or GlcNAc as acceptor, β-galactosidase from *E. coli* synthesized 3-O-methyl *allo*lactose, Gal β(1→6) Gal and N-acetyl *allo*lactosamine in the presence of different concentrations of diglyme at a much reduced rate (55–144 h reaction time), compared to reaction time of only 6 h in aqueous buffer, and there was no improvement in yields. Conversely, with β-galactosidases from *K. fragilis* and *A. oryzae*, synthesis occurred only in the presence of organic solvents (trimethyl or ethyl phosphate, or tetraglyme in >60% v/v concentration) but the yields were not high. In aqueous buffer, these enzymes hydrolysed the donor and no transfer to acceptor was achieved.

Laroute and Willemot (108) investigated the effect of 66 different organic solvents on glucoamylase from *Rhizopus oryzae* and β-galactosidase from *Aspergillus flavus*. Both enzymes demonstrated similar activities in the range of solvents tested, residual activities after 24 h ranging from 0 (methanol, ethylene glycol, dimethyl sulfoxide, and amines) to 100% (1,4-butanediol, 1,5-pentanediol, water, and tributyl phosphate), being 65–100% for most ethers, alcohols, and esters. The exception was acetates, in which glucoamylase was denatured but β-galactosidase retained 60–70% activity. A protective effect against thermal denaturation was correlated with increasing carbon chain length and lower hydroxyl content. No clear correlation could be established between enzyme stability and hydrophobicity of the medium. In the case of ethers, higher half lives were obtained with water immiscible solvents. In the context of oligosaccharide synthesis, the study suggests that the study of enzyme stability is not sufficient to determine if a solvent would be suitable in synthesis, as synthesis failed to occur with glucoamylase in reaction mixtures containing tetrahydrofuran and tetrahydrofuranfurfuryl alcohol, though they did not cause enzyme denaturation. Most solvents allowed synthesis to occur and best yields were obtained with diethylene glycol diethyl ether (37%), 1-octanol (33%), and 1-hexanol (31%). Another study with glucoamylases in aqueous ether mixtures (109) showed that although the enzyme was stable, the final oligosaccharide concentration was lower compared to aqueous media due to limited solubility of glucose in organic rich media. Neither the product specificity nor the relative order of initial rates of disaccharide formation by glucoamylase was modified in aqueous ethers. Glucoamylase was considered unsuitable for oligosaccharide synthesis by the authors owing to its broad specificity.

The hydrophobicity of the solvent had a bearing on the size of oligosaccharides synthesized with thermophilic β-glycosidases from *S. solfataricus* and *P. furiosus* (110), synthesis of larger saccharides being favored in presence of a more hydrophobic solvent. Initial reaction rate increased as a function of log P. Toluene and nonane were found to be the best as organic phases, resulting in maximum yields equivalent to the aqueous solution.

Cosolvents DMF and acetone at 50% (v/v) decreased the yields of galactosyl and xylose compared to those in the absence of cosolvents, (111) but the rate of transglycosylation was doubled in the presence of 50% (v/v) acetone.

Improved synthesis of galactooligosaccharides was obtained from lactose by *A. oryzae* β-galactosidase in sodium bis (2-ethylhexyl) sulfosuccinate (AOT) isooctane reverse micelles (112). At 45% (w/v) lactose, a maximum yield of 42.5% (w/w) GOS was obtained in the AOT isooctane solution compared to 31% (w/w) in aqueous solution. The improved yield is a result of inhibition of the hydrolytic activity of the enzyme due to decreased water activity and increased lactose concentration in the core of the micelles.

Attempts to increase oligosaccharide synthesis using α-amylase from *B. licheniformis* in the presence of various organic solvents (ethanol, methanol, N-propanol, propanediol, N-butanol, dioxane, and dimethyl sulfoxide) were unsuccessful (92) and this could not be attributed to the loss of enzyme activity as the enzyme was stable and even stimulated in the presence of low concentrations of some solvents.

27.4.5 Substrate Concentration

As in the case of organic solvents, high substrate concentration drives the equilibrium reactions toward synthesis and decreases the hydrolytic side reaction in transglycosylation. High substrate concentrations also protect the enzyme against thermal denaturation. Substrate concentration is a major determinant of synthesis yields.

In reverse hydrolysis reaction using α-mannosidase from Jack bean, the ratio of initial reaction rate in 83% and 40% mannose was approximately two (81). Maximum yield was obtained at 70% mannose using α-mannosidases from Jack bean and *A. phoenicis* (83); reduced yields were obtained with the Jack bean enzyme at 85% mannose; and synthesis was completely inhibited at 80% mannose in the case of *A. phoenicis*. Synthesis using β-glucanase from *P. emersonii* gave maximum yield of 14–16% at 60% glucose concentration and yield fell sharply at 70% glucose. The initial lactose conversion rate and oligosaccharide production rate were independent of the initial lactose concentration during synthesis of oligosaccharides by thermostable β-glucosidase from *Pyrococcus furiosus* (108). However, the relative oligosaccharide yield increased slightly. Increase in concentration of xylose did not influence the rate of o-nitrophenyl-β-galactose (ONPG) conversion significantly in β-galactosidase catalysed synthesis of galactosyl xylose (106) but yield increased to 21% from 12% with increase in xylose concentration from 0.05M to 2.7M. In a study of oligosaccharide synthesis using β-galactosidases from four different sources (97), a higher lactose concentration increased yields in all the cases. Similarly, an increase in the maltose concentration (15–50%) led to an increase in the maximum trisaccharide (131– 498 mM) and tetrasaccharide (39–283 mM) concentrations using α-glucosidase from *B. stearothermophilus* and trisaccharide concentration (24–217 mM) for the same enzyme from brewer's yeast (109).

27.4.6 pH

pH of reaction has not been investigated as a variable in many synthesis studies. In most cases, the pH was chosen on the basis of the enzyme,e and variation of pH did not exert a significant influence on yields. In kinetic syntheses, pH can affect the proportion of hydrolysis and transglycosylation.

The influence of pH on the stability of Jack bean alpha mannosidase was studied (81) by measuring its residual activity in 83% mannose solution at pH 4.5, 5.5, and 6.5 after 24 h at 75°C and 80°C. At 75°C, the enzyme was completely stable at all pH values, whereas at 80°C, the residual activities were 35%, 65%, and 71% at pH 4.5, 5.5, and 6.5 respectively. At 75°C, the initial reaction rate at pH 4.5 and 5.5 was 0.42M disaccharides formed per hour and 0.23 M at pH 6.5. At 55°C, however, pH (4.5 or 5.5) did not influence reaction rates.

A pH to temperature correlation was also seen while optimization of galactooligosaccharide production from lactose using β-glycosidases from *S. solfataricus* and *P. furiosus* (110); optimal pH for synthesis differed with temperature. Optimal transglycosylation was obtained at pH 8.0 at 22°C and pH 5.0 at 90°C. This difference could be due to variable influence of pH on protein structure at different temperatures.

No effect of pH was found on the reaction rate between a pH range of 5.0–7.0 using *P. furiosus* β-glucosidase (113), reduced reaction rate was observed at pH 4.0. Relative yield was highest at pH 6.0 in the range tested.

27.4.7 Time

In reverse hydrolysis, the reaction time is the time taken to attain equilibrium, which is usually the point at which no increase in oligosaccharide yield is observed for a period of two or more days and is the point of maximum yield. This is influenced by many factors such as substrate concentration, enzyme source, activity and stability, reaction temperature, and pH. The time scales in equilibrium synthesis are generally in the order of 1–2 weeks although shorter times have been reported for mannooligosaccharide synthesis using Jack bean alpha mannosidase (81) and β-glucanase (89).

The time of reaction is more crucial in the case of transglycosylation than equilibrium synthesis and needs to be carefully monitored to prevent decreased yields due to secondary hydrolysis of products.

Using α-glucosidase from *B. stearothermophilus*, trisaccharides were rapidly synthesized up to a value of 27% on a molar basis from maltose within two days and their concentration declined thereafter (114). Tetrasaccharides reached their maximum concentration in 5 days and did not undergo appreciable hydrolysis. Only trisaccharides were formed by α-glucosidase from yeast, but they were stable over the time period of reaction (10 days). A time dependent modulation of product spectrum was found in the study; with α-(1→4) linked trisaccharides synthesized exclusively by the bacterial enzyme at 2 and 6 h intervals and a low proportion (<5%) of (1→3) and (1→6) linkages appearing after 20 h of reaction. All tetrasaccharides were α-(1→4) linked. The yeast glucosidase also produced mainly α-(1→4) linkages at 6 h, with 12% α-(1→3). The percentage of α-(1→3) linkage increased to 34% after 20 h accompanied by 7% α-(1→6).

27.4.8 Enzyme Immobilization

Synthesis using immobilized glycosidases drives the cost of synthesis down by enabling reuse of the catalyst. The yields are generally lower compared to synthesis using the free enzyme, but were improved in some cases. Immobilization is also able to modulate product spectrum, primarily due to steric effects on the immobilized enzyme. The reactor configuration and mode of operation (batch or continuous) also affects yield and spectrum.

A tetrasaccharide OS-1 (gal/glu 3:1) and a trisaccharide OS-2 (gal/glu 2:1) were synthesized in 32–35% total yield by free and immobilized β-galactosidase from *Thermus aquaticus* YT-1 (115). The enzyme was immobilized by cocrosslinking with bovine serum albumin followed by entrapment in agarose beads of about 2.0 mm diameter. The immobilized enzyme gave significantly higher lactose conversion rates, higher yields, and produced more OS-2 and less OS-1 than the free enzyme.

Using β-galactosidase from *Bullera singularis* immobilized by adsorption onto Chitopearl resin, galactooligosaccharides were produced in 55% yields with a productivity of 4.4 g/L/h from 100 g/L lactose solution in a continuous process in a packed bed reactor (116). In comparison, β-galactosidase from *A. oryzae* when immobilized on glutaraldehyde treated chitosan beads and used in a plug reactor (117) produced only 18% yield from 100 g/L lactose. Although the thermal stability of the immobilized enzyme was higher, yield of enzyme activity and galactooligosaccharides was lower than the free enzyme. High productivities of 80 and 106 g/L/h with corresponding yields of 21 and 26% (w/w) have been reported for the *A. oryzae* β-galactosidase immobilized on cotton cloth and operated in a plug flow reactor (118), from 200 g/L and 400 g/L lactose respectively.

A dramatic rise in yield (573%, oligomers mg/ml.mg of enzyme) was reported (119) for glucooligosaccharide synthesis using immobilized β-glucosidase from almond compared to the free enzyme at 7.5 M (1350 g/L) glucose concentration. Immobilization led to a decrease in hydrolytic activity and increase in the synthetic activity of the enzyme. It has been suggested that this might result from creation of a hydrophobic microenvironment in the region of the active site.

Immobilized α-glucosidase was employed in a column system comprising an immobilized α-glucosidase column and an activated carbon column sequentially (77). The disaccharides formed in the enzyme column were adsorbed in the carbon column preferentially, expelling glucose and making it available for recycling into the enzyme column. In this manner, formation of disaccharides occured by repeated condensation. This system resulted in a higher yield of α-(1$\rightarrow$4) linked disaccharide than α-(1$\rightarrow$6), which was different from the batch system. This was due to lower energy barrier of the α-(1$\rightarrow$4) linked disaccharide enzyme activation intermediate, resulting in its faster formation and immediate adsorption in the activated carbon column.

The method of immobilization influenced the regioselectivity of linkage specific α-mannosidase from *A. phoenicis* (120). The enzyme immobilized by entrapment in alginate beads gave rise predominantly to α-(1$\rightarrow$2) product, as in case of the free enzyme, but the predominant product was Man α-(1$\rightarrow$6) Man in the cases of enzyme immobilized on China clay and DE-52. Additionally, immobilization on China clay gave rise to α-(1$\rightarrow$3) linked disaccharide. The loss of regioselectivity has been correlated with the likely conformational freedom of the immobilized enzyme as the covalently crosslinked enzyme displayed the lowest regioselectivity followed by the noncovalently modified enzyme tightly bound to an ion exchange resin, and then by the enzyme entrapped in alginate beads. Total oligosaccharide yield was over 10% lower than that obtained with the free enzyme. Mannose at 80% concentration was used for synthesis with immobilized enzyme, but was inhibitory in synthesis with free enzyme.

27.5 RECOMBINANT AND ENGINEERED GLYCOSIDASES IN OLIGOSACCHARIDE SYNTHESIS

Recombinant glycosidases

The cloning of an increasing number of glycosidases is improving their availability, and many of these are being investigated for their synthetic potential.

Yield of cloned α-galactosidase from *Bifidobacterium adolescentis* expressed in *E. coli* (121) was 100 times higher than the native enzyme and could be purified by a single anion exchange chromatography step. The enzyme exhibited transglycosylating activity and α-galactooligosaccharides were synthesized from melibiose and stachyose.

Purified recombinant β-glycosidase from *Thermus thermophilus,* overexpressed in *E. coli,* had an optimum temperature of 88°C and performed transglycosylation at high temperature with yields of over 63% in transfucosylation reactions. The enzyme catalysed the hydrolysis of β-D-galactoside, β-D-glucoside, and β-D-fucoside. The enzyme specificity in decreasing order toward different linkages was β (1→3) (100%)> β (1→2) (71%)> β (1→4) (40%)> β (1→6) (10%). Potential application of the glycosidase was suggested in the synthesis of fucosyl adducts and fucosyl sugars.

β-glucosidase II from *Pichia etchellsii* expressed in *E. coli* (122) was able to carry out glucooligosaccharide synthesis both by reverse hydrolysis (glucobiose and triose, at 18 and 6 mmol/L from 167 mmol/L glucose) and transglycosylation (glucotriose and pentose at 4.5 and 2 mmol/L from 79 mmol/L cellobiose). High conversion of 25% was observed in transglycosylation reaction with sophorose. The enzyme exhibited low regioselectivity and β (1→2), β (1→3), and β (1→6)-linked glucobioses were obtained in equilibrium synthesis.

27.5.1 Mutant Glycosidases

27.5.1.1 Glycosynthases

Glycosynthases (123–126) are engineered mutant glycosidases constructed by mutation of the catalytic nucleophile of a retaining glycosidase to a small nonnucleophilic residue. These mutant enzymes are able to efficiently synthesize oligosaccharides, but lack hydrolytic activity due to their inability to form the glycosyl enzyme intermediate. Their hydrolytic inactivity obviates the main problem of secondary hydrolysis in transglycosylation and oligosaccharide synthesis using glycosynthases resulted in high yields of over 95% in some instances. The first reported glycosynthase was a Glu358Ala mutant of *Agrobacterium* species β-glucosidase (Abg) (123). The k_{cat} value was 10^7 times lower than the wild-type enzyme, but was restored 10^5-fold by addition of small anions such as azide or formate (127). In the case of azide, the reaction intermediate was identified as α-glucosyl azide indicating that the reaction proceeded via the direct attack of azide. This implies that the mutant underwent a shift in mechanism from retaining to inverting. The transfer of donor sugar to an acceptor is catalysed only from activated donors with a good leaving group such as fluoride and an opposite anomeric configuration to the normal substrate of the wild-type enzyme. The glycosyl fluoride substrate mimics the glycosyl intermediate and the mechanism of transfer is similar to that occurring during transglycosylation. Formate functions differently (124,128), acting as an assistant nucleophile in the formation of the intermediate from an activated β-substrate, termed as a retaining glycosynthase mechanism. During synthesis using Abg Glu 358 Ala, α-glucosyl fluoride proved to be a better donor, resulting in longer oligosaccharides and higher yields, than α-galactosyl fluoride, which catalysed only a single galactosyl transfer to give disaccharides. Synthesis of several valuable substrates and inhibitors of cellulases including those unattainable by the wild-type enzyme was achieved using this glycosynthase in gram scales.

The efficiency of the glycosynthase was increased dramatically with a 24-fold improvement in rate constant when serine instead of alanine was used to substitute the catalytic nucleophile, i.e., in an E358S mutant (129). The increased reaction rates resulted in higher yields. Also, the enzyme was able to glycosylate some glycosides like PNP-GlcNAc which were weak acceptors in the case of E358A. A specific stabilizing interaction such as hydrogen bonding between the Ser hydroxyl group and the anomeric fluorine of the α-glycosyl fluoride has been suggested as the reason for the increased glycosylation ability of Abg E358S. From a 20:1 culture, 3.5 g of the pure mutant enzyme could be obtained, which affords the possibility of its application in large scale oligosaccharide synthesis.

A β-mannosynthase constructed by Glu519Ser mutation of recombinant β-mannosidase from *Cellulomonas fimi* overexpressed in *E. coli* has been used for synthesis of β-mannosides using the donor α-D-mannosyl fluoride and various acceptors (130). Highest yield of 99% was obtained using PNP β-D-cellobioside as acceptor. The products contained a mixture of β-(1→3) and β-(1→4) linkages. The Glu519Ala mutant was both a poor hydrolase and mannosynthase and only 8% yields of disaccharides and trisaccharides were obtained.

Other glycosynthases include E387G mutant of *Sulfolobus solfataricus* β-glucosidase and those derived from endoglycosidases: E134A mutant of *B. licheniformis* glucanase and E197A mutant of *Humicola insolens* cellulase. The details of these enzymes are tabulated and discussed in reviews by Jakeman and Withers (125) and Williams and Withers (126). The two endoglycosynthases have been used in tandem in a one pot synthesis (131) of hexasaccharide substrates of 1,3/1,4- β-glucanases in 70–80% yields, useful in their kinetic and structural studies.

All of these glycosynthases give rise to β-(1→3) or β-(1→4) linked products. Jakeman and Withers (132) have generated a glycosynthase (Glu537Ser) from *E. coli* β-galactosidase capable of forming Gal- β- (1→6)-linkages. Product yields of 40–63% were obtained by reactions between α-D-galactosyl-fluoride (α-Gal-F) and β-D-glucopyranoside (pNPGlc), pNP β-cellobioside, and phenyl β-D-glucoside. A second mutation generated close to the enzyme's active site resulted in the double mutant Glu537Ser/Gly794Asp which was much more active and resulted in improved product yields of 80–85%. X-ray crystallography studies showed that the introduced Asp lies within the active site and might facilitate transglycosylation by interacting with the acceptor sugar.

A new endoglycosynthase has been generated by E231G, E231S, or E231A mutation of barley (1,3)-β-D-glucan endohydrolase (133). E231G exhibited the highest catalytic efficiency. The mutant enzymes had no activity on laminarin but activity could be restored partially with formate. The glycosynthase was able to bring about autocondensation of α-laminaribiosyl fluoride and heterocondensation of α-laminaribiosyl fluoride and 4'-nitrophenyl β-D-glucopyranoside to form polymeric products: linear (1,3)-β-D-glucans of DP 30−34.

The Abg E358S and E358G mutants of *Agrobacterium* species β-glucosidase have been used for solid phase oligosaccharide synthesis (134). A galactose moiety was transferred by these enzymes from α-Gal-F with >90% efficiency to acceptors linked to PEGA resin via a backbone amide linker (BAL) and the recovery of products was high.

The increasing repertoire of glycosynthases with a broadening donor and acceptor range, combined with their application in methodologies such as solid phase synthesis suitable for large scale use, make them attractive catalysts for oligosaccharide synthesis.

27.5.1.2 Mutant Enzymes other than Glycosynthases

The performance of β-glucosidase CelB from hyperthermophilic *Pyrococcus furiosus* in oligosaccharide synthesis was improved by two active site mutations (135): a phenylalanine to tyrosine (F426Y), and methionine to lysine (M424K). F426Y improved oligosaccharide yield from 40 to 45% compared to the wild type enzyme, and M424K increased the pH optimum of transglycosylation. The improvement in transglycosylation capacity and oligosaccharide yield was especially significant at low lactose concentrations of 10 to 20%. The F426Y/M424K double mutant gave 40% yield at 10% lactose, compared to only 18% for the wild-type enzyme. It also displayed a higher ratio of tetrasaccharides to trisaccharides. The increased yields were due to an increase in the transglycosylation:hydrolysis ratio. The high yields at low lactose concentrations are desirable characteristics for an industrial process.

β-galactosidase BIF3 from *Bifidobacterium bifidum* has a molecular mass of 188kDa and consists of a signal peptide, an N-terminal β-galactosidase region and a C-terminal galactose binding motif. The enzyme was converted from a hydrolytic to an efficient transgalactosylating enzyme by deletion of about 580 amino acid residues from the C-terminal including a putative galactose binding motif by deletion mutagenesis and termed a transgalactosylase (136). The truncated enzyme expressed in *E. coli* displayed a 9:1 ratio of trangalactosylating to hydrolytic activity during galactooligosaccharide synthesis over a range of lactose concentrations from 10 to 40% and yields of 37–44% were obtained.

27.5.1.3 Mutant Glycosidases Created by Directed Evolution

Directed evolution has been revolutionary in its approach to creating enhanced biocatalysts in that no structural information is required for the catalyst whose improvement is sought. The main requirements for successful directed evolution (137) are: the functional expression of the enzyme in a suitable microbial host, the availability of a screen (or selection) sensitive to the desired properties, and identifying a workable evolution strategy.

The techniques generally used for directed evolution were oligonucleotide directed mutagenesis and error prone PCR until the invention of a method called DNA shuffling by Stemmer (138). Since then, this has been the most widely used method for a range of enzymes. The method was further optimized by Zhao and Arnold (139) for improving the fidelity of recombination. Other methods of directed evolution include staggered extension process (140), random elongation mutagenesis (141), a phage display based method (142), cycling mutagenesis (143), and heteroduplex recombination (144).

Some examples of properties of biocatalysts improved by directed evolution can be found in Affholter and Arnold (145). Applications fall into a few major categories: improving function in nonnatural or extreme environments (where activities or stabilities are low), improving activity toward a new substrate, tuning specificity, and increasing functional expression in a heterologous host.

All of these general categories apply in the context of glycosidase catalysed synthesis of oligosaccharides and a few examples are available in recent literature, including improvement of regioselectivity.

Directed evolution of *E. coli* β-galactosidase carried out by DNA shuffling and colony screening in a plate based assay using chromogenic fucose substrates led to a 10- to 20-fold increase in K_{cat}/K_m for fucose substrates relative to the native enzyme (146). Thirteen base changes resulting in six amino acid changes were responsible for the change from galactosidase to fucosidase specificity.

Using staggered extension process, it has been attempted (147) to evolve the minor 1, 3 regioselectivity of α-galactosidase from *Bacillus stearothermophilus,* which is of greater biological and medical significance than the major 1, 6 selectivity. The initial screening of the mutants was based on their inability to grow on melibiose (1, 6-linked). Just one generation of random mutagenesis and one recombination of the best mutants led to the loss of 1, 6 hydrolytic and synthetic activity to a large extent. The mutant enzymes had preference for 1, 3 linkage ($R_{1,3/1,6}$=6.9, compared to 0.3 for the wild type), however no improvement in yield was obtained and it remained constant at about 5%. The increase in 1, 3 selectivity is thus primarily due to decrease in 1, 6 regioselectivity rather than an improvement in the 1, 3 regioselectivity.

The genes coding for β-glucosidase Cel B from *Pyrococcus furiosus* and β-glycosidase LacS from *Sulfolobus solfataricus* were subjected to evolution (148) by DNA family shuffling. Three rounds of screening led to isolation of two improved hybrids with 1.5-fold–3.5-fold and 3.5-fold–8.6-fold increases in lactose hydrolysis rates respectively. The study is an example of successful DNA shuffling between sequences of enzymes with limited homology.

Devising a suitable and efficient screen is, in fact, the most important step of a directed evolution experiment, and it is the first law of directed evolution that "You get what you screen for" (149). On this principle, a screening method has been developed by Mayer et al. (150) for identification of functional synthetic enzymes for oligosaccharide synthesis, in this case glycosynthases. The method was a plate based coupled enzyme screen employing two plasmids coexpressed in the same cell: one coding for the screening enzyme (releases a chromogenic product only from glycosynthase derived product and not the acceptor) and the other for the glycosynthase. The screen was used to detect glycosynthase activity in a library of random mutants generated by saturation mutagenesis of the catalytic nucleophile in *Agrobacterium* species β-glucosidase (Abg). In addition to the glycosynthases generated from this enzyme in previous studies (123,129), two new glycosynthases, E358G and E358C, were discovered using this assay. The E358G had a K_{cat}/K_m which was more than double than that of the previous best mutant, E358S.

27.6 CONCLUSIONS

It is evident that the use of biotechnology to manufacture oligosaccharides of food and medical importance has developed rapidly. We now have a range of synthetic approaches which can be employed to bring about the synthesis of a wide range of biologically important structures. The development of glycosynthases and the Superbead technology will increase the availability of complex mammalian oligosaccharides and facilitate drug development based on glycoconjugates. The use of glycosidases as synthetic catalysts is also maturing and we can now manufacture multigram quantities of several bioactive oligosaccharides without the need for derivatization. This approach is also a promising route to novel oligosaccharide structures. The application of directed evolution techniques to glycosidases can be expected to increase the range of glycosidases with desired specificities. Genetic engineering techniques can also be seen as a means for scaling up the manufacture of glycosidases from plant and mammalian sources and should, at least in principle, make them more readily available.

Developments in carbohydrate biotechnology are already facilitating efforts to understand the roles of complex carbohydrates in biological systems and they will also enable the exploitation of such knowledge for practical ends.

REFERENCES

1. Hurtley, S., R. Service, P. Szuromi. Cinderella's coach is ready. *Science* 291:2337–2337, 2001.
2. McAuliffe, J.C., O. Hindsgaul. Carbohydrate drugs: an ongoing challenge. *Chem. Ind.* 5:170–174, 1997.

3. Sharon, N., H. Lis. Carbohydrates in cell recognition. *Sci. Am.* 268:82–89, 1993.
4. El Khadem, S.H. Structure of oligosaccharides. In: *Carbohydrate Chemistry: Monosaccharides and their Oligomers*. San Diego: Academic Press, 1988, pp 191–222.
5. Pazur, J.H. Oligosaccharides. In: *The Carbohydrates: Chemistry and Biochemistry,* 2nd ed., Vol. IIA. Pigman, W., D. Horton, eds., New York: Academic Press, 1970, pp 69–137.
6. Varki, A. Biological roles of oligosaccharides: all the theories are correct. *Glycobiology* 3:97–130, 1993.
7. Dwek, R.A. Glycobiology: toward understanding the function of sugars. *Chem. Rev.* 96:683–720, 1996.
8. Ofek, I., N. Sharon. Adhesins as lectins: specificity and role in infection. *Curr. Top. Microbiol. Immunol.* 151:91–113, 1990.
9. Zopf, D., S. Roth. Oligosaccharides anti-infective agents. *Lancet* 347:1017–1021, 1996.
10. Karlsson, K.-A. Meaning and therapeutic potential of microbial recognition of host glycoconjugates. *Mol. Microbiol.* 29:1–11, 1998.
11. Playne, M.J., Glycoscience: oligosaccharides as drugs, functional foods, and receptors in the gut. *Aust. Biotechnol.* 12:35–37, 2002.
12. Röckendorf, N., T.K. Lindhorst. Glycodendrimers. *Top. Curr. Chem.* 217:201–238, 2001.
13. Patel, A., T.K. Lindhorst. Synthesis of "mixed type" oligosaccharide mimetics based on a carbohydrate scaffold. *Eur. J. Org. Chem.* 1:79–86, 2002.
14. Crittenden, R.G., M.J. Playne. Production, properties and application of food-grade oligosaccharides. *Trends Food Sci. Tech.* 7:353–361, 1996.
15. Nakakuki, T. Present status and future of functional oligosaccharide development in Japan. *Pure Appl. Chem.* 74:1245–1251, 2002.
16. Nilsson, K.G.I. Enzymatic synthesis of oligosaccharides. *Trends Biotech.* 6:256–264, 1988.
17. Boons, G.J. Strategies in oligosaccharide synthesis. *Tetrahedron* 52:1095–1121, 1996.
18. Davis, B.G. Recent developments in oligosaccharide synthesis. *J. Chem. Soc. Perkin Trans.* 1(14):2137–2160, 2000.
19. Ekelöf, K., P.J. Garegg, L. Olsson, S. Oscarson. Towards the 21st century: the emerging importance of oligosaccharide synthesis. *Pure Appl. Chem.* 69:1847–1852, 1997.
20. Grice, P., S.V. Ley, J. Pietruszka, H.M.I. Osborn, H.W.M. Priepke, S.L. Warriner. A new strategy for oligosaccharide assembly exploiting cyclohexane-1,2-diacetal methodology: an efficient synthesis of a high mannose type nonasaccharide. *Chem. Eur. J.* 3:431–440, 1997.
21. Garegg, P.J., P. Konradsson, D. Lezdins, S. Oscarson, R. Katinka, L. Öhberg. Synthesis of oligosaccharides of biological importance. *Pure Appl. Chem.* 70:293–298, 1998.
22. Allen, J.R., C.R. Harris, S.J. Danishefsky. Pursuit of optimal carbohydrate-based anticancer vaccines: preparation of a multiantigenic unimolecular glycopeptide containing the Tn, MBr1, and Lewis (y) antigens. *J. Am. Chem. Soc.* 123:1890–1897, 2001.
23. Wang, C.-C., J.-C. Lee, S.-Y. Luo, H.-F. Fan, C.-L. Pai, W.-C. Yang, L.-D. Lu, S.-C. Hung. Synthesis of biologically potent α-1,2-linked disaccharide derivatives via regioselective one-pot protection-glycosylation. *Angew Chem. Int. Ed.* 41:2360–2362, 2002.
24. Arlt, M., O. Hidsgaul. Rapid chemical synthesis of sugar-nucleotides in a form suitable for enzymatic oligosaccharide synthesis. *J. Org. Chem.* 60:14–15, 1995.
25. Bartolozzi, A., P.H. Seeberger. New approaches to the chemical synthesis of bioactive oligosaccharides. *Curr. Opin. Struct. Biol.* 11:587–592, 2001.
26. Ye, X.-S., C.-H. Wong. Anomeric reactivity-based one-pot oligosaccharide synthesis: a rapid route to oligosaccharide libraries. *J. Org. Chem.* 65:2410–2431, 2000.
27. Koeller, K.M., C.-H. Wong. Complex carbohydrate synthesis tools for glycobiologists: enzyme-based approach and programmable one-pot strategies. *Glycobiology* 10:1157–1169, 2000.
28. Watt, G.M., L. Revers, M.C. Webberley, I.B.H. Wilson, S.L. Flitsch. The chemoenzymatic synthesis of the core trisaccharide of N-linked oligosaccharides using a recombinant β -mannosyltransferase. *Carbohydr. Res.* 305:533–541, 1997.
29. Theim, J. Applications of enzymes in synthetic carbohydrate chemistry. *FEMS Microbiol. Rev.* 16:193–211, 1995.

30. Öhrlein, R. Glycosyltransferase-catalysed synthesis of non-natural oligosaccharides. *Top. Curr. Chem.* 200:227–254, 1999.
31. Coutinho, P.M., B. Henrissat. Carbohydrate-active enzymes server at URL: http://afmb.cnrs-mrs.fr/CAZY/index.html, 1999.
32. Palcic, M.M. Biocatalytic synthesis of oligosaccharides. *Curr. Opin. Biotechnol.* 10:616–624, 1999.
33. Toone, E.J., E.S. Simon, M.D. Bednarski, G.M. Whitesides. Enzyme-catalysed synthesis of carbohydrates. *Tetrahedron* 45:5365–5422, 1989.
34. Bastida, A., A. Fernandez-Mayoralas, R.G. Arrayas, F. Iradier, J.C. Carretero, E. Garcia-Junceda. Heterologous over-expression of alpha-1,6-fucosyltransferase from Rhizobium sp.: application to the synthesis of the trisaccharide beta-D-GlcNAc(1 -> 4)-(alpha-L-Fuc-(1 -> 6))-D-GlcNAc, study of the acceptor specificity and evaluation of polyhydroxylated indolizidines as inhibitors. *Chem. Eur. J.* 7:2390–2397, 2001.
35. Creeger, E.S., L.I. Rothfield. Cloning genes for bacterial glycosyltransferases. *Methods Enzymol.* 83:326–331, 1982.
36. Russel, R.R.B., M.L. Gilpin, H. Mukasa, G. Dougan. Characterization of glucosyltransferase expressed from a *Streptococcus sobrinus* gene cloned in *Escherichia coli*. *J. Gen. Microbiol.* 133:935–944, 1987.
37. Weinstein, J., E.U. Lee, K. McEntee, P.H. Lai, J.C. Paulson. Primary structure of beta-galctosidase alpha-2,6-sialyltransferase- conversion of membrane-bound enzyme to soluble forms by cleavage of the NH2-terminal signal anchor. *J. Biol. Chem.* 262:17735–17743, 1987.
38. Wen, D.X., B.D. Livingston, K.F. Medzihradszky, S. Kelm, A.L. Burlingame, J.C. Paulson. Primary structure of Gal-beta-1,3(4)GlcNAc alpha-2,3-sialyltransferase determined by mass-spectrometry sequence-analysis and molecular-cloning-evidence for a protein motif in the sialyltransferase gene family. *J. Biol. Chem.* 267:21011–21019, 1992.
39. Ernst, L.K., V.P. Rajan, R.D. Larsen, M.M. Ruff, J.B. Lowe. Stable expression of blood group-H determinants and GDP-L-fucose-beta-D-galactosidase 2-alpha-L-fucosyltransferase in mouse cells after transfection with human DNA. *J. Biol. Chem.* 264:3436–3447, 1989.
40. Wang, P., G.-J. Shen, Y.-F. Wang, Y. Ichikawa, C.-H. Wong. Enzymes in oligosaccharide synthesis: active domain overproduction, specificity study, and synthetic use of an α-1,2-mannosyltransferase with regeneration of GDP-Man. *J. Org. Chem.* 58:3985–3990, 1993.
41. Revers, L., R.M. Bill, I.B.H. Wilson, G.M. Watt, S.L. Flitsch. Development of recombinant, immobilised beta-1,4-mannosyltransferase for use as an efficient tool in the chemoenzymatic synthesis of N-linked oligosaccharides. *Biochim. Biophys. Acta Gen. Subjects* 1428:88–98, 1999.
42. Nakazawa, K., K. Furukawa, H. Narimatsu, A. Kobata. Kinetic study of human beta-1, 4-galactosyltransferase expressed in *Escherichia-coli*. *J. Biochem.* 113:747–753, 1993.
43. Malissard, M., L. Borsig, S. DiMarco, M.G. Grutter, U. Kragl, C. Wandrey, E.G. Berger. Recombinant soluble beta-1,4-galactosyltransferases expressed in *Saccharomyces cerevisiae*: purification, characterization and comparison with human enzyme. *Eur. J. Biochem.* 239:340–348, 1996.
44. Ju, T.Z., K. Brewer, A. D'Souza, R.D. Cummings, W.M. Canfield. Cloning and expression of human core beta 1,3- galactosyltransferase. *J. Biol. Chem.* 277:178–186, 2002.
45. Kudo, T., T. Iwai, T. Kubota, H. Iwasaki, Y. Takayma, T. Hiruma, N. Inaba, Y. Zhang, M. Gotoh, A. Togayachi, H. Narimatsu. Molecular cloning and characterization of a novel UDP-Gal: GalNAc alpha peptide beta 1,3-galactosyltransferase (ClGal-T2), an enzyme synthesizing a core 1 structure of O-glycan. *J. Biol. Chem.* 277:47724–47731, 2002.
46. Ichikawa, Y., G.C. Look, C.-H. Wong. Enzyme-catalyzed oligosaccharide synthesis. *Anal. Biochem.* 202:215–238, 1992.
47. Crout, D.H.G., G. Vic. Glycosidases and glycosyl transferases in glycoside and oligosaccharide synthesis. *Curr. Opin. Chem. Biol.* 2:98–111, 1998.

48. Wong, C.-H., S.L. Haynie, G.M. Whitesides. Enzyme-catalysed synthesis of N-acetyllactosamine with *in situ* regeneration of uridine 5'-diphosphate glucose and uridine 5'- diphosphate galactose. *J. Org. Chem.* 47:5416–5418, 1982.
49. Ichikawa, Y., J.L.C. Liu, G.J. Shen, C.H. Wong. A highly efficient multienzyme system for the one-step synthesis of a sialyl trisaccharide-insitu generation of sialic-acid and N-acetyllactosamine coupled with regeneration of UDP-Glucose, UDP-Galactose, and CMP-Sialic acid. *J. Am. Chem. Soc.* 113:6300–6302, 1991.
50. Wong, C.-H., R. Wang, Y. Ichikawa. Regeneration of sugar-nucleotide for enzymatic oligosaccharide synthesis: use of Gal-1-phosphate uridyltransferase in the regeneration of UDP-galactose, UDP-2-deoxygalactose, and UDP-galactosamine. *J. Org. Chem.* 57:4343–4344, 1992.
51. Bulter, T., L. Elling. Enzymatic synthesis of nucleotide sugars. *Glycoconjugate J.* 16:147–159, 1999.
52. Bulter, T., L. Elling. Enzymatic synthesis of UDP-galactose on a gram scale. *J. Mol. Catal. B Enzym.* 8:281–284, 2000.
53. Thiem, J., W. Treder. Synthesis of the trisaccharide Neu-5-Ac-Alpha(2-)6)Gal-Beta (1-)4)GlcNAc by the use of immobilized enzymes. *Angew. Chem. Int. Ed. Engl.* 25:1096–1097, 1986.
54. Lubineau, A., K. BassetCarpentier, C. Auge. Porcine liver (2->3)-alpha-sialyltransferase: substrate specificity studies and application of the immobilized enzyme to the synthesis of various sialylated oligosaccharide sequences. *Carbohydr. Res.* 300:161–167, 1997.
55. Nishiguchi, S., K. Yamada, Y. Fuji, S. Shibatani, A. Toda, S.-I. Nishimura. Highly efficient oligosaccharide synthesis on water-soluble polymeric primers by recombinant glycosyltransferases immobilized on solid supports. *Chem. Commun.* 19:1944–1945, 2001.
56. Herrmann, G.F., P. Wang, G.-J. Shen, C.-H. Wong. Recombinant whole cells as catalysts for the enzymatic synthesis of oligosaccharides and glycopeptides. *Angew. Chem. Int. Ed. Engl.* 33:1241–1242, 1994.
57. Nahalka, J., Z. Liu, X. Chen, P.G. Wang. Superbeads: immobilization in "sweet" chemistry. *Chem. Eur. J.* 9:372–377, 2003.
58. Koizumi, S., T. Endo, K. Tabata, A. Ozaki. Large-scale production of UDP-galactose and globotriose by coupling metabolically engineered bacteria. *Nat. Biotechnol.* 16:847–850, 1998.
59. Endo, T., S. Koizumi, K. Tabata, S. Kakita, A. Ozaki. Large-scale production of N-acetyllactosamine through bacterial coupling. *Carbohydr. Res.* 316:179–183, 1999.
60. Endo, T., S. Koizumi, K. Tabata, A. Ozaki. Large-scale production of CMP-NeuAc and sialylated oligosaccharides through bacterial coupling. *Appl. Microbiol. Biotechnol.* 53:257–261, 2000.
61. Endo, T., S. Koizumi, K. Tabata, S. Kakita, A. Ozaki. Large-scale production of the carbohydrate portion of the sialyl-Tn epitope, alpha-Neup5Ac-(2 -> 6)-D-GalpNAc, through bacterial coupling. *Carbohydrate Res.* 330:439–443, 2001.
62. Chen, X., Z.Y. Liu, J.B. Zhang, W. Zhang, P. Kowal, P.G. Wang. Reassembled biosynthetic pathway for large-scale carbohydrate synthesis: alpha-Gal epitope producing "superbug". *Chem. Biochem.* 3:47–53, 2002.
63. Priem, B., M. Gilbert, W.W. Wakarchuk, A. Heyraud, E. Samain. A new fermentation process allows large-scale production of human-milk oligosaccharides by metabolically engineered bacteria. *Glycobiology* 12:235–240, 2002.
64. Chen, X., J. Fang, J. Zhang, Z. Liu, J. Shao, P. Kowal, P. Andreana, P.G. Wang. Sugar nucleotide regeneration beads (Superbeads): a versatile tool for the practical synthesis of oligosaccharides. *J. Am. Chem. Soc.* 123:2081–2082, 2001.
65. Liu, Z.Y., J.B. Zhang, X. Chen, P.G. Wang. Combined biosynthetic pathway for *de novo* production of UDP-galactose: catalysis with multiple enzymes immobilized on agarose beads. *Chem. Biochem.* 3:348–355, 2002.
66. Zhang, J., B. Wu, Z. Liu, P. Kowal, X. Chen, J. Shao, P.G. Wang. Large-scale synthesis of carbohydrates for pharmaceutical development. *Curr. Org. Chem.* 5:1169–1176, 2001.

67. McCarter, J.D., S.G. Withers. Mechanisms of enzymatic glycoside hydrolysis. *Curr. Opin. Struct. Biol.* 4:885–897, 1994.
68. Withers, S.G. 1998 Hoffman La Roche Award Lecture: understanding and exploiting glycosidases. *Can. J. Chem.* 77:1–11, 1999.
69. Rye, C.S., S.G. Withers. Glycosidase mechanisms. *Curr. Opin. Chem. Biol.* 4:573–580, 2000.
70. Withers, S.G. Mechanisms of glycosyl transferases and hydrolases. *Carbohydr. Polym.* 44:325–337, 2001.
71. Vasella, A., G.J. Davies, M. Bohm. Glycosidase mechanisms. *Curr. Opin. Chem. Biol.* 6:619–629, 2002.
72. Davies, G.J., L. Mackenzie, A. Varrot, M. Dauter, A.M. Brzozowski, M. Schulein, S.G. Withers. Snapshots along an enzymatic reaction coordinate: analysis of a retaining beta-glycoside hydrolase. *Biochemistry* 37:11707–11713, 1998.
73. McIntosh, L.P., G. Hand, P.E. Johnson, M.D. Joshi, M. Korner, L.A. Plesniak, L. Ziser, W.W. Wakarchuk, S.G. Withers. The pK(a) of the general acid/base carboxyl group of a glycosidase cycle during catalysis: A C-13-NMR study of *Bacillus circulans* xylanase. *Biochemistry* 35:9958–9966, 1996.
74. Rastall, R.A., C. Bucke. Enzymatic synthesis of oligosaccharides. *Biotechnol. Genet. Eng. Rev.* 10:253–281, 1992.
75. Johansson, E., L. Hedbys, P.-O. Larsson, K. Mosbach, A. Gunnarsson, S. Svensson. Synthesis of mannose oligosaccharides via reversal of the α-mannosidase reaction. *Biotechnol. Lett.* 8:421–424, 1986.
76. Ajisaka, K., Y. Yamamoto. Control of regioselectivity in the enzymatic syntheses of oligosaccharides using glycosides. *Trends Glycosci. Glycotechnol.* 14:1–11, 2002.
77. Ajisaka, K., H. Nishida, H. Fujimoto. Use of an activated carbon column for the synthesis of disaccharides by use of a reverse hydrolysis activity of beta-galactosidase. *Biotechnol. Lett.* 9:387–392, 1987.
78. Firon, N., I. Ofek, N. Sharon. Carbohydrate specificity of the surface lectins of *Escherichia coli*, *Klebsiella pneumoniae*, and *Salmonella typhimurium*. *Carbohydr. Res.* 120:235–249, 1983.
79. Sharon, N. Bacterial lectins, cell-cell recognition and infectious disease. *FEBS Lett.* 217:145–157, 1987.
80. Pan, Y.T., B. Xu, K. Rice, S. Smith, R. Jackson, A.D. Elbein. Specificity of the high-mannose recognition site between *Enterobacter cloacae* pili adhesin and HT-29 cell membranes. *Infect. Immun.* 65:4199–4206, 1997.
81. Johansson, E., L. Hedbys, K. Mosbach, P.-O. Larsson, A. Gunnarsson, S. Svensson. Studies of the reversed α-mannosidase reaction in high concentrations of mannose. *Enzyme Microb. Technol.* 11:347–352, 1989.
82. Rastall, R.A., N.H. Rees, R. Wait, M.W. Adlard, C. Bucke. α-mannosidase-catalysed synthesis of novel manno-,lyxo-,and heteromanno-oligosaccharides: a comparison of kinetically and thermodynamically mediated approaches. *Enzyme Microb. Technol.* 14:53–57, 1992.
83. Suwasono, S., R.A. Rastall. Enzymatic synthesis of manno- and heteromanno-oligosaccharides using α-mannosidases: a comparative study of linkage-specific and non-linkage-specific enzymes. *J. Chem. Technol. Biotechnol.* 73:37–42, 1998.
84. Singh, S., M. Scigelova, D.H.G. Crout. Glycosidase-catalysed synthesis of mannobioses by the reverse hydrolysis activity of alpha-mannosidase: partial purification of alpha-mannosidases from almond meal, limpets and *Aspergillus niger. Tetrahedron Asymmetry* 11:223–229, 2000.
85. Ajisaka, K., I. Matsuo, M. Isomura, H. Fujimoto, M. Shirakabe, M. Okawa. Enzymatic synthesis of mannobioses and mannotrioses by reverse hydrolysis using α-mannosidase from *Aspergillus niger*. *Carbohydr. Res.* 270:123–130, 1995.

86. Suwasono, S., R.A. Rastall. A highly regioselective synthesis of mannobiose and mannotriose by reverse hydrolysis using specific 1,2-α-mannosidase from *Aspergillus phoenicis*. *Biotechnol. Lett.* 18:851–856, 1996.
87. Athanasopoulos, V.I., K. Niranjan, R.A. Rastall. Regioselective synthesis of mannobiose and mannotriose by reverse hydrolysis using a novel 1,6-α-D-mannosidase from *Aspergillus phoenicis*. *J. Mol. Catal. B Enzym.* 27:215–219, 2004.
88. Rastall, R.A., M.W. Adlard, C. Bucke. Synthesis of hetero-oligosaccharides by glucoamylase in reverse. *Biotechnol. Lett.* 13:501–504, 1991.
89. Rastall, R.A., S.F. Pikett, M.W. Adlard, C. Bucke. Synthesis of oligosaccharides by reversal of a fungal β-glucanase. *Biotechnol. Lett.* 14:373–378, 1992.
90. Pestlin, S., D. Prinz, J.N. Starr, P.J. Reilly. Kinetics and equilibria of condensation reactions between monosaccharide pairs catalyzed by *Aspergillus niger* glucoamylase. *Biotechnol. Bioeng.* 56:9–22, 1997.
91. Gama, F.M., M. Mota. Cellulases for oligosaccharide synthesis: a preliminary study. *Carbohydr. Polym.* 37:279–281, 1998.
92. Chitradon, L., P. Mahakhan, C. Bucke. Oligosaccharide synthesis by reversed catalysis using α-amylase from *Bacillus licheniformis*. *J. Mol. Catal. B Enzym.* 10:273–280, 2000.
93. Malá, Š., B. Králová. Heterooligosaccharide synthesis catalyzed by α-glucosidase from *Bacillus stearothermophilus*. *J. Mol. Catal. B Enzym.* 10:617–621, 2000.
94. Meulenbeld, G.H., B.M. Roode, S. Hartmans. Enzymatic synthesis of thioglucosides using almond β-glucosidse. *Biocatal. Biotrans.* 20:251–256, 2002.
95. Monsan, P., F. Paul. Enzymatic synthesis of oligosaccharies. *FEMS Microb. Rev.* 16:187–192, 1995.
96. Farkas, E., J. Thiem, K. Ajisaka. Enzymatic synthesis of fucose-containing disaccharides employing the partially purified α-L-fucosidase from *Penicillium multicolor*. *Carbohydr. Res.* 328:293–299, 2000.
97. Reuter, S., A.R. Nygaard, W. Zimmermann. β-galactooligosaccharide synthesis with β-galactosidases from *Sulfolobus solfataricus, Aspergillus oryzae,* and *Escherichia coli*. *Enzyme Microb. Technol.* 25:509–516, 1999.
98. Boon, M.A., A.E.M. Janssen, K. van't Riet. Effect of temperature and enzyme origin on the enzymatic synthesis of oligosaccharides. *Enzyme Microb. Technol.* 26:271–281, 2000.
99. Daum, B., K. Buchholz. High yield and high selectivity of reactions in the frozen state: the acceptor reaction of dextransucrase. *Biocatal. Biotrans.* 20:15–21, 2002.
100. Hansson, T., P. Adlercreutz. The temperature influences the ratio of glucosidase and galactosidase activities of β-glycosidases. *Biotechnol. Lett.* 24:1465–1471, 2002.
101. Bruins, M.E., E.W. Van Hellemond, A.E.M. Janssen, R.M. Boom. Maillard reactions and increased enzyme inactivation during oligosaccharide synthesis by a hyperthermophilic glycosidase. *Biotechnol. Bioeng.* 81:546–552, 2003.
102. Bruins, M.E., A.J.H. Thewessen, A.E.M. Janssen, R.M. Boom. Enzyme inactivation due to Maillard reactions during oligosaccharide synthesis by a hyperthermophilic glyclosidase: influence of enzyme immobilization. *J. Mol. Catal. B Enzym.* 21:31–34, 2003.
103. Maitin, V., R.A. Rastall. Enzyme glycation influences product yields during oligosaccharide synthesis by reverse hydrolysis. *J. Mol. Catal. B Enzym.* 30:195–202, 2004.
104. Zeng, X., R. Yoshino, T. Murata, K. Ajisaka, T. Usui. Regioselective synthesis of p-nitrophenyl glycosides of β-D-galactopyranosyl-disaccharides by transglycosylation with β-D-galactosidases. *Carbohydr. Res.* 325:120–131, 2000.
105. Rabiu, B.A., A.J. Jay, G.R. Gibson, R.A. Rastall. Synthesis and fermentation properties of novel galacto-oligosaccharides by β-galactosidases from *Bifidobacterium* species. *Appl. Environ. Microbiol.* 67:2526–2530, 2001.
106. Maitin, V., V. Athanasopoulos, R.A. Rastall. Synthesis of FimH receptor-active manno-oligosaccharides by reverse hydrolysis using α-mannosidases from *Penicillium citrinum, Aspergillus phoenicis* and almond. *Appl. Microbiol. Biotechnol.* 63:666–671, 2004.

107. Finch, P., J.H. Yoon. The effects of organic solvents on the synthesis of galactose disaccharides using β-galactosidases. *Carbohydr. Res.* 303:339–345, 1997.
108. Laroute, V., R.-M. Willemot. Effect of organic solvents on stability of two glycosidases and on glucoamylase-catalysed oligosaccharide synthesis. *Enzyme Microb. Technol.* 14:528–534, 1992.
109. Cantarella, L., Z.L. Nikolov, P.J. Reilly. Disaccharide production by glucoamylase in aqueous ether mixtures. *Enzyme Microb. Technol.* 16:383–387, 1994.
110. Hansson, T., P. Adlercreutz. Optimization of galactooligosaccharide production from lactose using β-glycosidases from hyperthermophiles. *Food Biotechnol.* 15:79–97, 2001.
111. Giacomini, C., G. Irazoqui, P. Gonzalez, F. Batista-Viera, B.M. Brena. Enzymatic synthesis of galactosyl-xylose by *Aspergillus oryzae* β-galactosidase. *J. Mol. Catal. B Enzym.* 19,20:159–165, 2002.
112. Chen, S.-X., D.-Z. Wei, Z.-H. Hu. Synthesis of galacto-oligosaccharides in AOT/isooctane reverse micelles by β-galactosidase. *J. Mol. Catal. B Enzym.* 16:109–114, 2001.
113. Boon, M.A., J. van der Oost, W.M. de Vos, A.E.M. Janssen, K. van't Riet. Synthesis of oligosaccharides catalysed by thermostable β-glucosidase from *Pyrococcus furiosus*. *Appl. Biochem. Biotechnol.* 75:269–278, 1998.
114. Malá, Š., H. Dvořáková, R. Hrabal, B. Králová. Towards regioselective synthesis of oligosaccharides by the use of α-glucosidases with different substrate specificity. *Carbohydr. Res.* 322:209–218, 1999.
115. Berger, J.L., B.H. Lee, C. Lacroix. Oligosaccharides synthesis by free and immobilized β-galactosidases from *Thermus aquaticus* YT-1. *Biotechnol. Lett.* 17:1077–1080, 1995.
116. H.-J. Shin, J.-M. Park, J.-W. Yang. Continuous production of galacto-oligosaccharides from lactose by *Bullera singularis* β-galactosidase immobilized in chitosan beads. *Process Biochem.* 33:787–792, 1998.
117. Sheu, D.-C., S.-Y. Li, K.-J. Duan, C.W. Chen. Production of galactooligosaccharides by β-galactosidase immobilized on glutaraldehyde-treated chitosan beads. *Biotechnol. Techniques* 12:273–276, 1998.
118. Albayrak, N., S.-T. Yang. Production of galacto-oligosaccharides from lactose by *Aspergillus oryzae* β-galactosidase immobilized on cotton cloth. *Biotechnol. Bioeng.* 77:8–19, 2002.
119. Ravet, C., D. Thomas, M.D. Legoy. Gluco-oligosaccharide synthesis by free and immobilized β-glucosidase. *Biotechnol. Bioeng.* 42:303–308, 1993.
120. Suwasono, S., R.A. Rastall. Synthesis of oligosaccharides using immobilized 1,2-α-mannosidase from *Aspergillus phoenicis*: immobilization dependent modulation of product spectrum. *Biotechnol. Lett.* 20:15–17, 1998.
121. Van den Broek, L.A.M., J. Ton, J.C. Verdoes, K.M.J. Van Laere, A.G.J. Voragen, G. Beldman. Synthesis of α-galacto-oligosaccharides by a cloned α-galactosidase from *Bifidobacterium adolescentis*. *Biotechnol. Lett.* 21:441–445, 1999.
122. Bhatia, Y., S. Mishra, V.S. Bisaria. Biosynthetic activity of recombinant *Escherichia coli*-expressed *Pichia etchellsii* β-glucosidase II. *Appl. Biochem. Biotechnol.* 102,103:367–379, 2002.
123. Mackenzie, L.F., Q. Wang, R.A.J. Warren, S.G. Withers. Glycosynthases: mutant glycosidases for oligosaccharide synthesis. *J. Am. Chem. Soc.* 120:5583–5584, 1998.
124. Moracci, M., A. Trincone, M. Rossi. Glycosynthases: new enzymes for oligosaccharide synthesis. *J. Mol. Catal. B Enzym.* 155–163, 2001.
125. Jakeman, D.L., S.G. Withers. Glycosynthases: new tools for oligosaccharide synthesis. *Trends Glycosci. Glycotechnol.* 14:13–25, 2002.
126. Williams, S.J., S.G. Withers. Glycosynthases: mutant glycosidases for glycoside synthesis. *Aust. J. Chem.* 55:3–12, 2002.
127. Wang, Q., R.W. Graham, D. Trimbur, R.A.J. Warren, S.G. Withers. Changing enzymatic reaction mechanisms by mutagenesis: conversion of a retaining glucosidase to an inverting enzyme. *J. Am. Chem. Soc.* 116:11594–11595, 1994.

128. Moracci, M., A. Trincone, B. Cobuzzi-Ponzano, G. Perugino, M. Ciaramella, M. Rossi. Enzymatic synthesis of oligosaccharides by two glycosyl hydrolases of *Sulfolobus solfataricus*. *Extremophiles* 5:145–152, 2001.
129. Mayer, C., D.L. Zechel, S.P. Reid, R.A.J. Warren, S.G. Withers. The E358S mutant of *Agrobacterium* sp. β-glucosidase is a greatly improved glycosynthase. *FEBS Lett.* 466:40–44, 2000.
130. Nashiru, O., D.L. Zechel, D. Stoll, T. Mohammadzadeh, R.A.J. Warren, S.G. Withers. β-mannosynthase: synthesis of β-mannosides with a mutant β-mannosidase. *Angew. Chem. Int. Ed.* 40:417–420, 2001.
131. Faijes, M., J.K. Fairweather, D. Hugues, A. Planas. Oligosaccharide synthesis by coupled endo-glycosynthases of different specificity: a straightforward preparation of two mixed-linkage hexasaccharide substrates of 1,3/1,4- β-glucanases. *Chem. Eur. J.* 7:4651–4655, 2001.
132. Jakeman, D.L., S.G. Withers. On expanding the repertoire of glycosynthases: mutant β-galactosidases forming β-(1,6)-linkages. *Can. J. Chem.* 80:866–870, 2002.
133. Hrmova, M., T. Imai, S.J. Rutten, J.K. Fairweather, L. Pelosi, V. Bulone, H. Driguez, G.B. Fincher. Mutated barley (1,3)-β-D-glucan endohydrolase synthesize crystalline (1,3)-β-D-glucans. *J. Biol. Chem.* 277:30102–30111, 2002.
134. Tolborg, J.F., L. Peterson, K.J. Jensen, C. Mayer, D.L. Jakeman, R.A.J. Warren, S.G. Withers. Solid-phase oligosaccharide and glycopeptide synthesis using glycosynthases. *J. Org. Chem.* 67:4143–4149, 2002.
135. Hansson, T., T. Kaper, J. van der Oost, W.M. de Vos, P. Adlercreutz. Improved oligosaccharide synthesis by protein engineering of β-glucosidase CelB from hyperthermophilic *Pyrococcus furiosus*. *Biotechnol. Bioeng.* 73:203–210, 2001.
136. Jorgensen, F., O.C. Hansen, P. Stougaard. High-efficiency synthesis of oligosaccharides with a truncated β-galactosidase from *Bifidobacterium bifidum*. *Appl. Microbiol. Biotechnol.* 57:647–652, 2001.
137. Kuchner, O., F.H. Arnold. Directed evolution of enzyme catalysts. *Trends Biotech.* 15:523–530, 1997.
138. Stemmer, W.P.C. DNA shuffling by random fragmentation and reassembly: *in vitro* recombination for molecular evolution. *Proc. Natl. Acad. Sci. USA* 91:10747–10751, 1994.
139. Zhao, H., F.H. Arnold. Optimization of DNA shuffling for high-fidelity recombination. *Nucleic Acids Res.* 25:1307–1308, 1997.
140. Zhao, H., L. Giver, Z. Shao, J.A. Affholter, F.H. Arnold. Molecular evolution by staggered extension process (StEP) *in vitro* recombination. *Nat. Biotechnol.* 16:258–261, 1998.
141. Matsuura, T., K. Miyai, S. Trakulnaleamsai, T. Yomo, Y. Shima, S. Miki, K. Yamamoto, I. Urabe. Evolutionary molecular engineering by random elongation mutagenesis. *Nat. Biotechnol.* 17:58–61, 1999.
142. Pedersen, H., S. Holder, D.P. Sutherlin, U. Schwitter, D.S. King, P.G. Schultz. A method for directed evolution and functional cloning of enzymes. *Proc. Natl. Acad. Sci. USA* 95:10523–10528, 1998.
143. Buchholz, F., P.-O. Angrand, A.F. Stewart. Improved properties of FLP recombinase evolved by cycling mutagenesis. *Nat. Biotechnol.* 16:657–662, 1998.
144. Volkov, A.A., Z. Shao, F.H. Arnold. Recombination and chimeragenesis by *in vitro* heteroduplex formation and *in vivo* repair. *Nucleic Acids Res.* 27, e18:1–6, 1999.
145. Affholter, J., F. Arnold. Engineering a revolution. *Chem. Br.* 35:48–51, 1999.
146. Zhang, J.-H., G. Dawes, W.P.C. Stemmer. Directed evolution of a fucosidase from a galactosidase by DNA shuffling and screening. *Proc. Natl. Acad. Sci. USA* 94:4504–4509, 1997.
147. Dion, M., A. Nisole, P. Spangenberg, C. Andre, A. Glottin-Fleury, R. Mattes, C. Tellier, C. Rabiller. Modulation of the regioselectivity of a *Bacillus* α-galactosidase by directed evolution. *Glycoconjugate J.* 18:215–223, 2001.
148. Kaper, T., S.J.J. Brouns, A.C.M. Geerling, W.M. De Vos, J. Van der Oost. DNA family shuffling of hyperthermostable b-glycosidases. *Biochem. J.* 368:461–470, 2002.

149. Arnold, F.H. Unnatural selection: molecular sex for fun and profit. *Eng. Sci.* 1/2:41–50, 1999.
150. Mayer, C., D.L. Jakeman, M. Mah, G. Karjala, L. Gal, R.A.J. Warren, S.G. Withers. Directed evolution of new glycosynthases from *Agrobacterium* β-glucosidase: a general screen to detect enzymes for oligosaccharide synthesis. *Chem. Biol.* 8:437–443, 2001.
151. Singh, S., M. Michaela Scigelova, G. Vic, D.H.G. Crout. Glycosidase-catalysed oligosaccharide synthesis of di-, tri- and tetra-saccharides using the N-acetylhexosaminidase from *Aspergillus oryzae* and β-galactosidase from *Bacillus circulans*. *J. Chem. Soc. Perkin Trans.* 1(16):1921–1926, 1996.
152. Farkas, E., J. Thiem. Enzymatic synthesis of galactose-containing disaccharides employing β-galactosidase from *Bacillus circulans*. *Eur. J. Org. Chem.* 11:3073–3077, 1999.
153. Ichikawa, Y. Glycozymes: tools for oligosaccharide synthesis and targets for drug development. *Trends Glycosci. Glycotechnol.* 9:S47–S59, 1997.

Section 3

Food Safety, Novel Bioprocessing, Traditional Fermentations, and Regulatory Issues

3.01

Molecular Evolution and Diversity of Food Borne Pathogens

Katy Windham, Kendra Kerr Nightingale, and Martin Wiedmann

CONTENTS

1.1 INTRODUCTION

An understanding of microbial evolution and diversity is critical to our ability to control food borne pathogens throughout the food chain and to reduce human food borne infectious diseases. While different phenotypic methods (such as multilocus enzyme electrophoresis and serotyping) have been used for many years to characterize the diversity of food borne pathogens, the advent of inexpensive, rapid, and increasingly easy to use molecular characterization methods (for example, polymerase chain reaction [PCR] and automated DNA sequencing) has revolutionized our ability to study the molecular diversity and evolution of food borne pathogens. Consequently, an extraordinary amount of information on the molecular diversity and evolution of food borne pathogens is currently available. Various molecular studies have not only provided a better understanding of the basic biological processes underlying the emergence and maintenance of virulence traits and the transmission dynamics of food borne pathogens, but have also yielded widely used tools to differentiate food borne pathogens in order to detect, track, and control outbreaks of food borne disease.

A sound understanding of the evolution of food borne pathogens can also allow for classification of these organisms into clonal groups, which share specific virulence and transmission characteristics. While we already possess a good understanding of the evolution of virulence and virulence associated traits for some food borne pathogens (e.g., *Escherichia coli*), our understanding of the evolution of other food borne pathogens is considerably more limited. Defining evolutionary related groups of food borne pathogens will be critical to allow for improved control of these pathogens throughout the food continuum, and to reduce the incidence of human food borne disease. For example, it can be crucial for the design and implementation of appropriate food processing steps and barriers to understand whether different evolutionary groups of a bacterial pathogen differ in heat or acid resistance. An understanding of clonal groups of a pathogen and their ability to cause disease, and infectious dose between clonal groups of a bacterial pathogen, may also have considerable implications for regulatory policies, intervention strategies, and risk assessments (1).

Food borne pathogens include viruses, protozoans, and bacteria. While some knowledge on the diversity and evolution of protozoan and viral food borne pathogens exists, the majority of research has been conducted on bacterial food borne pathogens. We will therefore focus our discussions in this chapter on bacterial food borne pathogens. This chapter provides an overview of mechanisms of evolution relevant to bacteria, and methods to infer evolutionary relatedness and other evolutionary processes in these organisms. In addition, this chapter includes an overview of molecular subtyping methods used to study the evolution and diversity of food borne bacterial pathogens, as well as brief summaries on our current knowledge of molecular diversity and evolution of selected food borne pathogens.

1.2 MECHANISMS OF BACTERIAL EVOLUTION

Evolution is the process of change in the genetic makeup of populations (Table 1.1). Bacteria demonstrate an amazing level of diversity in respect to cell structure, metabolism, and general way of life, considering the relatively small variation that is observed between the genome sizes of different bacterial species. More specifically, even within defined taxonomic groups such as enteric bacteria, of which several species are known to share a common ancestry, individual species show distinct affinity for defined ecological niches and ability to infect specific host species (2). The forces involved in the evolution

Table 1.1
Definitions of important terms

Term	Definition
Allele	Alternate forms of a gene at a given locus
Clone	A group of bacteria that descended from a common ancestor and represents a monophyletic branch on a phylogenetic tree
Evolution	The process of change in the genetic make-up of populations
Haplotype	Collective DNA polymorphism combinations based on several loci being studied
Homologous recombination-	Recombination between two DNA sequences that have the same or very similar nucleotide sequences
Horizontal gene transfer	Lateral transfer of genes from one genome to another
Intergenic recombination	Recombinational events between different genes
Intragenic recombination	Recombinational events within a given gene (e.g., between two repeats present within a given gene)
Intergenomic recombination	Recombinational event between two sequences present on a single DNA molecule or genome (e.g., between two rRNA operons of a bacterial chromosome)
Intragenomic recombination	Recombinational event between two sequences originating from two different genomes
Lineage	Evolutionary group that contains the ancestral sequence and its descendants
Linkage disequilibrium	Non-random assortment of alleles indicative of clonal population structure
Linkage equilibrium	Random association of alleles; presence of an allele at one locus is independent of presence or absence of alleles at other loci
Polymorphism	Presence of two or more genetic variants at a defined locus

of microorganisms include mutation, horizontal gene transfer, random genetic drift, migration, and natural selection. Random genetic drift can be defined as the change in allele frequency in a population that occurs by chance alone. Random genetic drift diminishes the genetic variation that is introduced by mutations in a population. The movement of organisms among subpopulations is known as migration. In essence, migration defines the limit of the amount of genetic diversity that can occur within a defined population (3). The two processes that are responsible for the majority of changes in genetic makeup that bacteria experience include mutation and horizontal gene transfer. Although mutations clearly contribute to the generation of prokaryotic diversity, they are more likely to cause subtle changes and gradual evolution; whereas horizontal transfer of genetic information provides a mechanism for rapid introduction of complex genetic traits, including those that can give rise to new pathogenic variants (4).

Bacteria multiply asexually by binary fission, in which two daughter cells are produced from a single mother cell. Because these daughter cells have a chromosome that is essentially identical to their mother cell, variation seen in the bacterial chromosome can be primarily attributed to changes in the genetic makeup of an organism that occur after reproduction. These mutations will then be passed on to descendents and, provided the mutations are successfully accumulated in the bacterial population, new lineages may ultimately emerge (5,6). The main focus of bacterial evolution studies for many years has been on mutations and the vertical transmission of genetic material between mother and daughter cells. Only more recently has it become apparent that the transfer of genetic information between unrelated cells by sexual means (horizontal or lateral transfer) plays an important role in the evolution of many bacterial food borne pathogens (6).

The advent of rapid and increasingly affordable whole genome sequencing strategies (7) has contributed considerably to our increased recognition of the importance and extent of horizontal transfer in bacterial food borne pathogens, as well as the impact this mechanism has on bacterial evolution. Thus, genome and large scale sequence data have challenged previous assumptions that evolution occurs as a treelike process. Recent evidence suggests that the relative contributions of mutations and recombination to microbial diversity and evolution differ considerably between bacterial genera, species, and possibly even genetic lineages (5). Some food borne pathogens (e.g., *Listeria monocytogenes*) appear to represent highly clonal populations, for which mutations have contributed considerably more than horizontal gene transfer to genetic diversity. Other food borne pathogens, such as *Staphylococcus aureus,* appear to have evolved predominantly by horizontal gene transfer, such that analysis of multilocus sequence data have revealed rates of recombination high enough to eliminate any detectable phylogenetic signal from mutations (5).

1.2.1 Mutations

While DNA sequences are normally copied exactly during the process of replication, rare errors in DNA repair and replication give rise to new DNA sequences. These erroneous changes are generally known as mutations, which can further be defined as the heritable unit of genetic change (8). Mutations may be classified into two broad categories by the length of DNA that is affected by these changes as either point mutations, which involve a single nucleotide, or segmental mutations, which affect several adjacent bases. More commonly, mutations are categorized by the type of change caused by the mutational event: nucleotide substitutions, deletions, insertions, inversions, and recombinations (9). Specific aspects of these different types of mutations and their importance for the evolution of bacterial food borne pathogens are discussed below.

1.2.1.1 Substitutions

Substitutions are often also referred to as point mutations. Substitutions involve the replacement ("substitution") of a single nucleotide base for another in a DNA sequence (10). Two types of nucleotide substitutions, transitions and transversions, may occur during the evolution of DNA sequences. Transitions are substitutions where a purine base (A or G) is replaced by another purine base or where a pyrimidine base (C or T) is replaced by another pyrimidine base. Transversions are substitutions where a purine base changes to a pyrimidine base or vice versa. Mutations that occur in coding sequences can also be classified as synonymous (silent), leading to no change in the amino acid sequence encoded, or as nonsynonymous changes, which lead to a change in the amino acid encoded. Nonsynonymous mutations that result in a stop codon, and therefore lead to expression of a truncated protein, are often referred to as "non-sense mutations" (9).

1.2.1.2 Insertions and Deletions

Insertion and deletions refer to the addition or removal of one or more bases in the DNA sequence of an organism. It is impossible to determine whether an actual insertion or deletion event caused an observed difference between sequences of interest. Therefore, when homologous sequences or genomes are compared, the presence or absence of one or more nucleotides compared is often referred to as an "indel". Insertions and deletions can occur by a variety of mechanisms, including unequal crossover, intrastrand deletion, and replication slippage or slipped strand mispairing (8). In coding regions, indels that are not multiples of three nucleotides incur a shift in the reading frame that results in numerous amino acid changes, and new, often premature, stop codons downstream of the indel. Mutations that disrupt the correct reading frame are thereby referred to as frameshift mutations (8).

In many bacteria (including in many food borne pathogens), insertions and deletions ranging from single nucleotides, through short repeats (as short as 3 nucleotides), to chromosomal virulence gene islands exceeding 10 kb in length have been observed. Both gene acquisition and gene loss clearly play an important role in the evolution and emergence of pathogens and bacterial lineages with novel phenotypic characteristics (11). While smaller indels generally involve intragenomic genetic events, insertions of larger sequences (e.g., virulence gene islands, antibiotic resistance genes), typically include horizontal (or lateral) gene transfer (intergenomic genetic events). Interestingly, it is not only the introduction of new genetic material that may cause important phenotypic changes and emergence of a more pathogenic lifestyle. Gene deletions have also been linked to the evolution of pathogenicity in food borne pathogens. For example, in *Shigella*, deletion of *cad*A results in the loss of lysine decarboxylase activity. This deletion has been identified as a pathoadaptive mutation (12) since cadaverine, the product of this lysine decarboxylase activity, acts as an enterotoxin inhibitor and thus decreases virulence of *Shigella* (13).

Duplication and deletion of repeated sequences (ranging from as few as 3 nucleotides to more than 300 nucleotides) represents a considerably more frequent genetic event than substitutions. Thus, detection of indels in repeat sequences has been employed for highly sensitive molecular subtyping of bacterial pathogens in different subtyping techniques, including a technique termed MLVA (multiple locus variable number tandem repeat [VNTR] analysis) (14).

1.2.1.3 Inversions

Inversion of DNA sequences within a chromosome or plasmid can occur by either chromosomal breakage and rejoining or by intrachromosomal crossover between two homologous segments oriented in opposite directions. While inversions of large chromosomal fragments have been observed in different bacterial species, including different *Salmonella* serotypes, the specific evolutionary importance of these inversions is often not clear. It has been hypothesized that large chromosomal inversions restore chromosomal balance after disruptions by other insertions, deletions, and inversions (15,16).

1.2.1.4 Recombination

The term "recombination" is generally used by geneticists to describe the process of exchanging two genetic loci that may be present on a single DNA molecule (intragenomic or intragenic recombination) or on two different DNA molecules (intergenomic recombination). Homologous or generalized recombination involves recombination between homologous sequences and can be classified as either crossing over (reciprocal recombination) or gene conversion (nonreciprocal recombination) (8). Both types of homologous recombination generally are thought to occur through an intermediate called the "Holliday junction". In contrast to homologous recombination, site specific recombination involves the exchange of a usually very short sequence by another typically long sequence that has no similarity to the original sequence (8). Site specific recombination is, for example, responsible for integration of phage genomes into bacterial chromosomes, a mechanism that has been shown to be important in the evolution of many food borne bacterial pathogens (e.g., introduction of shiga toxin encoding genes into certain strains of *E. coli)* (17).

In bacteria, homologous recombination can occur within sequences present in the genome of a given bacterial cell, which can be referred to as intragenomic recombination. For example, recombination between rRNA operons of a given genome has been shown to provide a mechanism for genomic rearrangement in eubacteria (18). Homologous recombination between a chromosomal sequence and a homologous sequence introduced into a recipient bacterial cell, usually from another closely related donor bacterium by horizontal gene transfer, can be referred to as intergenomic homologous recombination. Because relatively short sequence fragments generally participate in this intergenomic homologous recombination, this process differs in its evolutionary implication from the meiotic recombination of eukaryotes (3).

1.2.2 Horizontal Gene Transfer

Horizontal gene transfer in prokaryotes broadly involves the exchange of chromosomal DNA segments between different genomes, therefore introducing variability into the gene pool that may give some organisms an evolutionary advantage. Horizontal gene transfer can occur through three common mechanisms: transformation, conjugation, and transduction. These processes allow for unidirectional genetic exchanges that can introduce large DNA fragments (up to more than 10 kb) into a recipient cell. Conjugation, which usually requires a vector for information transfer, involves the physical transfer of host DNA (usually a plasmid or a transposon) from a donor to a recipient cell. Transformation involves the uptake of DNA from the environment by competent cells. Transduction represents transfer of DNA from one bacterial cell into another bacterial cell via a bacteriophage that packages donor DNA and injects the partial chromosome into a recipient cell.

The term "horizontal gene transfer" is generally used to describe lateral transfer of homologous genes between closely related bacteria (see Figure 1.1A) as well as to describe the transfer of nonhomologous sequences between sometimes highly divergent organisms (see Figure 1.1B). While the magnitude and level of bacterial horizontal gene transfer in natural populations is uncertain, sequence data have shown that these events are quite common in some species (2). While it has been hypothesized that horizontal transfer and

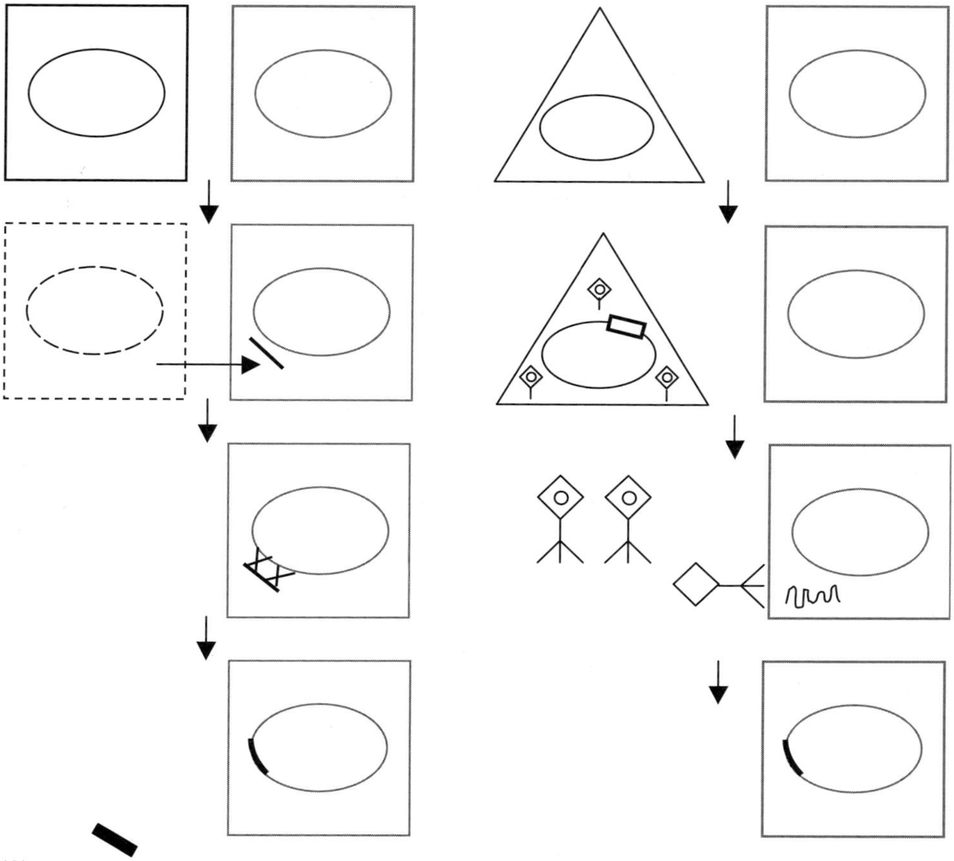

(A) (B)

Figure 1.1 Schematic representation of horizontal gene transfer of homologous gene sequences (Figure 1.1A) and horizontal gene transfer of non-homologous gene sequences (Figure 1.1B). Figure 1.1A show uptake of homologous gene sequences by transformation followed by integration into the host genome by homologous recombination. Figure 1.1B shows phage mediated transfer of a non-homologous gene sequence followed by site-specific recombination into the host genome. Donor cells are shown in black and recipient cells are shown in red.

subsequent recombination of homologous sequences plays a negligible role in ecological and phenotypic diversification of bacteria (2), lateral transfer of nonhomologous sequences from highly divergent organisms clearly has played an important role in the evolution of food borne pathogens.

1.2.2.1 Horizontal Gene Transfer of Homologous Gene Sequences

While it has become apparent that horizontal gene transfer of homologous gene sequences occurs commonly in many bacterial species, determining whether and to what extent horizontal gene transfer and recombination have occurred in bacterial isolates collected from natural sources can be challenging. A variety of mathematical and statistical methods have been developed over the years to probe for evidence for the exchange of genetic information (5). Multilocus enzyme electrophoresis (MLEE) data has traditionally been used to asses the presence of linkage equilibrium, which would indicate a considerable rate of horizontal gene transfer. Similarly, multilocus sequence data can be used to probe for the presence of linkage disequilibrium. However, analysis of linkage disequilibrium represents a highly insensitive method to probe for the presence of horizontal gene transfer, and Maynard Smith et al. (19) estimated that an allele must change at least 20 times more frequently by genetic exchange events than by point mutation to eliminate linkage disequilibrium within a population (5). In addition, other factors, such as sampling bias (5), could also cause an apparent linkage disequilibrium in bacterial populations with significant rates of horizontal gene transfer.

1.2.2.2 Horizontal Gene Transfer of Nonhomologous Gene Sequences

DNA sequence data have been used in different ways to identify lateral gene transfer of nonhomologous sequences. While bacterial species display considerable variation in their overall G+C content and codon usage, the genome of a particular species in general shows a very homogeneous base composition (2). Thus, sequences introduced through horizontal gene transfer from a divergent species can often be recognized by their distinct GC content or codon usage. However, amelioration of horizontally introduced nonhomologous gene sequences over time may make it more difficult to differentiate "foreign DNA" from ancestral genome sequences (20).

Horizontal gene transfer of nonhomologous gene sequences has been shown to critically contribute to the evolution of many food borne pathogens, in particular through transfer of antibiotic resistance genes and through transfer of so called pathogenicity islands. Pathogenicity islands (PAIs) are large genomic regions that range from 10 to 200 kilobases or more and carry genes that encode for one or more virulence factors. Typically, PAIs are found in the genomes of a given pathogenic species or lineage but are absent (or rarely seen) in closely related nonpathogenic organisms. While PAIs are generally found on the chromosome, they may also be part of bacteriophages, plasmids, or conjugative transposons (21). They often have a GC content that differs from the rest of the genome, are often flanked by repetitive DNA sequences, and frequently are associated with tRNA genes, which can act as sites for incorporation of foreign DNA (21). PAIs have been described in a variety of Gram-negative pathogens, but also have been identified in several Gram-positive genera such as *Listeria* and *Staphylococcus*.

Virulence factors such as adhesion molecules, invasion molecules, toxins, and various secretion systems can be encoded on these islands. In particular, type III secretion systems, which can inject bacterial proteins into host cells to facilitate the uptake of several Gram-negative food borne pathogens including *Salmonella, Shigella*, and *E. coli* by nonphagocytic host cells, are commonly associated with PAIs. Although type III secretion systems may be conserved between distantly related pathogens, the proteins that are secreted and injected into host cells differ greatly, resulting in a wide spectrum of observed virulence phenotypes between pathogens (22). Various numbers of PAIs have been identified in different pathogens. While in some organisms, such as *E. coli*, the acquisition of a single PAI appears to be sufficient to confer pathogenic potential; in other species, such as *Salmonella enterica*, as many as five pathogenicity islands have been identified in isolates of specific serotypes (11).

1.2.3 Natural Selection

While mutations and horizontal gene transfer are the mechanisms that are responsible for generating genetic diversity in bacterial populations, it is natural selection that determines whether distinct genotypes will survive and multiply in different environments and hosts. Natural selection is defined as the differential reproduction of divergent genotypes within a population (8). Fitness is a measure of an individual's ability to survive and multiply in a population. In most cases, new mutations will be deleterious and lower fitness; therefore, these mutations will be removed from a population by negative or purifying selection. Conversely, new mutations that confer an enhanced fitness to an individual are maintained and spread within a population through positive or adaptive selection (10).

Food borne pathogens are exposed to a variety of selective pressures in different environments and foods, as well as in human and often animal hosts. Evolution of food borne pathogens is thus driven by a variety of selective pressures, which need to be balanced to maximize pathogen transmission and multiplication. While external or environmental stressors encountered during food processing and preparation (i.e., pH, temperature, and osmotic pressure) often represent selective pressures different from those encountered in a host, sometimes similar selective pressures (e.g., acidic pH) may be present in certain foods and within a host, raising particular concerns as to whether changes in the food processing methods and environments may also select for pathogens with enhanced human virulence. Changes in selective pressures encountered by food borne pathogens within both human and animal hosts have also considerably impacted the evolution of these pathogens. One of the most obvious selective pressures has been the application of antibiotics in human and animal hosts, which has provided the selective pressure to facilitate the rapid emergence and spread food borne pathogens such as *Salmonella typhimurium* DT104 that are resistant to multiple antibiotics (23).

1.3 MOLECULAR SUBTYPING METHODS TO STUDY EVOLUTION OF FOOD BORNE PATHOGENS

In general, bacterial subtyping methods can be divided into phenotype based and molecular genetic or DNA based methods (24). While phenotype based methods have been used for many years to subtype food borne pathogens, molecular methods have revolutionized our ability to differentiate bacterial subtypes. Key aspects of selected and commonly used subtyping methods are summarized below; for a more comprehensive review the reader is referred to one of the many review articles on bacterial subtyping methods (24–26). Additional review articles and book chapters on the current state of subtyping methods for specific food borne pathogens such as *Campylobacter* (27), *Listeria monocytogenes* (28,29), *E. coli* O157:H7 (30), and *Salmonella* (31) are also available.

1.3.1 Phenotype Based Subtyping Methods

Commonly used phenotype based strain typing (or subtyping) methods for bacterial pathogens include serotyping, biotyping, phage typing, and MLEE. While a variety of shortcomings and concerns may be associated with different phenotype based strain typing methods (24), these methods are still regularly used and have some utility for characterization of bacterial food borne pathogens. Phenotype based methods may often lack discriminatory power and reproducibility (28). Furthermore, a considerable proportion of bacterial isolates may be untypable by some of these methods. To overcome these issues and to provide improved strain differentiation, molecular subtyping methods are increasingly applied for characterization of bacterial isolates, and it is likely that these molecular methods will ultimately replace phenotype based strain typing methods.

Among the phenotypic subtyping methods, phage typing, and particularly MLEE, have been instrumental tools for the development of an initial understanding of bacterial population genetics and evolution. For example, phage typing of *Salmonella* has contributed to the detection and description of the emergence of a multiresistant *Salmonella typhimurium* strain, which was termed "DT (definitive type) 104", based on its phage type (32,33). In France, routine application of phage typing to *L. monocytogenes* isolates from humans and foods led to the detection of at least one human listeriosis outbreak (34). Multilocus enzyme electrophoresis differentiates bacterial strains by variations in the electrophoretic mobility of different constitutive enzymes. Cell extracts containing soluble enzymes are separated by size in nondenaturing starch gels, and enzyme activities are determined in the gel by using color generating substrates (35). This method usually provides 100% typability, and MLEE has been widely used for studies on the population genetics of many bacterial pathogens (36). For example, MLEE data were critical for the development of an initial understanding of the evolution of *E. coli* O157:H7 (37,38).

1.3.2 DNA Based Subtyping Methods

The widespread development of different DNA based subtyping methods has dramatically improved our ability to differentiate subtypes of bacterial food borne pathogens and to study their evolution and population genetics. Commonly used DNA based subtyping approaches for bacterial isolates include plasmid profiling, Pulsed-Field Gel Electrophoresis (PFGE), ribotyping, Amplified Fragment Length Polymorphism (AFLP),and random amplification of polymorphic DNA (RAPD) as well as other PCR based subtyping methods (24,28). Most of these methods provide DNA banding based data outputs that can be used to classify bacteria into genetically similar subtypes. These methods have contributed considerably to our ability to characterize, often with high discriminatory power, genetically related subtypes within a bacterial species, and to develop a better understanding of the diversity of bacterial food borne pathogens. However, the data generated by these methods are generally not suitable for evolutionary analyses, since it is generally not feasible to infer the primary genetic changes leading to different DNA banding patterns. Methods that rely on DNA sequencing subtyping and characterization, such as multilocus sequence typing (MLST), on the other hand provide data suitable for evolutionary analyses and have provided considerable improved insight into the evolution of food borne pathogens. Even more recently, other DNA sequence based characterization and subtyping methods, such as MLVA and whole genome microarrays as well as comparative whole genome sequencing have further enhanced our ability to study the genome evolution of food borne pathogens.

1.3.2.1 DNA Sequencing Based Subtyping and Characterization Methods

Multilocus sequence typing refers to a molecular subtyping approach that uses DNA sequencing of multiple genes or gene fragments to differentiate bacterial subtypes, determine the genetic relatedness of isolates, and make inferences about population genetics parameters of interest. Sequencing of multiple housekeeping genes is often what MLST refers to (39), but sequencing of multiple virulence genes can also be used as a subtyping method (28). A major advantage of DNA sequencing based subtyping, as compared to other molecular subtyping methods is that sequence data are considerably less ambiguous (39) and easier to interpret than banding pattern based subtype data obtained by many other DNA based subtyping approaches (e.g., PFGE and ribotyping). Data from DNA sequencing also provide an opportunity to reconstruct ancestral relationships among bacterial isolates, allowing further insight into the evolution, epidemiology, and ecology of food borne pathogens.

The development of internet accessible databases for MLST information (such as the MLST database at http://www.mlst.net/) will also facilitate global, large scale surveillance and tracking of bacterial food borne pathogens, as well as large scale evolutionary studies (39). While sequencing of housekeeping genes provides an opportunity to probe the evolutionary relatedness of bacterial organisms without the confounding effects of adaptive selection, which often occurs among virulence genes, sequencing of virulence genes may provide critical information on the evolution of virulence characteristics among clonal groups of bacterial pathogens. In addition, studying the effects of adaptive selection in specific lineages or strains of food borne pathogens may help explain variations in the ability of these clonal groups to cause disease. As DNA sequencing continues to become cheaper and more broadly accessible and as DNA chip based DNA sequencing methods are developed, the use of DNA sequencing data for bacterial subtyping and evolutionary studies will continue to expand.

1.3.2.2 MLVA

Multiple locus variable number tandem repeat analysis (MLVA) utilizes the detection of length polymorphisms in short repeated nucleotide sequences (caused by insertions or deletions of single or multiple repeat elements) to classify bacterial isolates into subtypes (14). Typically, MLVA involves the use of multiple sets of PCR primers targeting different VNTRs; design of these primers generally requires the availability of a full genome sequence of the target organism to allow identifications of target VNTRs and PCR primers. VNTRs appear to be highly polymorphic (14) and MLVA consequently appears to show a higher discriminatory power than most other DNA based subtyping methods. Due to the apparent rapid evolution of VNTRs, MLVA seems to provide a suitable approach to study the short term evolution of bacterial pathogens, providing a unique tool to study the diversity and evolution of *Bacillus anthracis*, a highly monomorphic bacterial pathogen (14) and other bacterial pathogens, such as *Yersinia pestis*, that show limited diversity and appear to have emerged relatively recently (40).

1.3.2.3 Whole Genome Sequencing and Whole Genome Microarray Based Characterization Methods

As full genome sequences of a variety of bacterial food borne pathogens [e.g., *L. monocytogenes*, *Salmonella typhimurium*, and *E. coli* O157:H7 (41-43)] become available, our ability to study the genome evolution of food borne bacterial pathogens continues to increase. Comparative genomic studies on closely related pathogens and nonpathogens [e.g., *E. coli* O157:H7 (43) and *E. coli* K12 (44); *L. monocytogenes* and *L. innocua* (41)] have already provided some interesting insight in the evolution of virulence in the food borne pathogens *E. coli* O157:H7 and *L. monocytogenes*. While full genome sequencing of multiple strains of different bacterial pathogens may occur for some strains of selected pathogens, whole genome microarrays provide a more cost effective approach to characterize the entire genome content of different bacterial strains and to study genome evolution of bacterial pathogens. For example, Zhang et al. (45), used *L. monocytogenes* serotype 1/2a microarrays to gain further insight into the evolution of different lineages of the food borne pathogen *L. monocytogenes*. Combination of multilocus housekeeping and virulence gene sequencing with whole genome microarrays may currently provide the most cost effective method to probe comprehensively the evolution of food borne bacterial pathogens.

1.4 PHYLOGENETIC AND POPULATION GENETIC METHODS

Methods for analysis of phylogenetic and population genetics data have developed considerably over the past decades, and a variety of outstanding books on these topics are available. For a comprehensive overview of phylogenetic analysis readers are referred to “Inferring

Phylogenies" (46). Phylogeny is the study of evolutionary relationships between organisms. Molecular biology techniques are used to generate data that can then be used to reconstruct a phylogenetic tree that describes the degree of genetic relatedness and elucidates the genealogy among specific organisms of interest (47). The primary goals of phylogenetic analysis are to rebuild an accurate evolutionary relationship between organisms and estimate the time since organisms last shared a common ancestor (9). A general overview of some of the major concepts relevant to reconstruction of phylogenies and estimation of recombination and natural selection from DNA sequence data follows.

1.4.1 Models of DNA Substitution

All methods used to infer phylogeny make assumptions about the DNA substitution process, and phylogenetic analyses thus strongly depends on the model of DNA substitution used for a particular analysis. Therefore, all phylogenetic analyses should first determine which model of DNA substitution best fits a particular set of data (48). Models that are a poor fit for the data being analyzed can easily lead to inaccurate and biased phylogenies (49). Four major models of DNA substitution, including the Jukes and Cantor (JC69), Kimura-2-Parameter (K2P), Hasegawa, Kishino and Yano (HKY85), and General Time Reversible (GTR) models will be briefly discussed.

The JC69 model is the simplest model of DNA evolution and assumes equal base (A, T, G, and C) frequencies and an equal rate of substitution for all bases. The K2P model is similar to the JC69 model, except K2P allows a different rate of substitution for transitions and transversions. While both the JC69 and K2P models are easily computed, they are greatly simplified and do not reflect realistic DNA sequence evolution. The HKY85 model relaxes some of the assumptions of the JC69 and K2P models by allowing base frequencies and rates of substitution for transitions and transversions to be estimated from the actual sequence data. The JC69, K2P and HKY85 models are all reversible, that is, the probabilities of changing from A to T and from T to A are equal. The GTR model is a model in which one unit of time is the time in which we expect to see one substitution per base (46); in this model a different substitution rate for each of the six possible substitutions between the four bases is estimated from the sequence data. The GTR is the most general model of DNA evolution and all other models described above are nested within the GTR.

The likelihood ratio test is widely used to test the goodness of fit of different models of DNA substitution for a given set of DNA sequence data. Maximum likelihood methods to estimate phylogeny (discussed in more detail later) produce a log likelihood estimate for the most likely tree topology determined under a specific model of DNA substitution. The test statistic for the likelihood ratio test is calculated by multiplying by two the difference between the log likelihood estimate of the null model and alternative model. The test statistic is approximated by a chi-square test and the degrees of freedom are determined by the difference in the number of parameters between the nested models. The likelihood ratio test can thus be used to determine which model of DNA substitution best fits a given set of DNA sequence data.

Molecular studies indicate that all sites along a sequence do not change at the same rate and, in fact, a gamma distribution may explain among site substitution rate heterogeneity (50). The addition of a gamma distribution to explain substitution rate heterogeneity along DNA sequences to the most likely model of DNA evolution often significantly improves the ability to describe DNA sequence data for phylogenetic analysis. Furthermore, mutation rate variation may be described by assigning different weights to changes occurring at different positions in the codon; these models can also account for a proportion of invariable sites (51). The likelihood ratio test can also be utilized to evaluate the ability of these additional parameters to explain the heterogeneity of mutation rate along DNA sequences.

Posada and Crandall (48) developed the MODELTEST program, which allows an evaluation of 56 different nested DNA substitution models coupled with various other parameters (i.e., substitution rate heterogeneity). MODELTEST specifically uses log likelihood scores to select the model of DNA evolution and the additional parameters that best explain a given set of sequence data. However, it should be noted that this program uses Neighbor-joining trees to evaluate models of DNA substitution and evolutionary parameters. Therefore, the model and conditions determined by MODELTEST should be validated using the likelihood ratio test and a maximum likelihood tree to ensure the highest accuracy of inferring phylogeny.

1.4.2 Estimating Phylogenetic Trees

A phylogenetic tree consists of nodes and branches; furthermore, branches connect nodes, which are points where two or more branches diverge. While nodes signify taxonomic units, branches describe the descent and ancestry among taxa in the phylogenetic tree (52). Phylogenetic trees are composed of monophyletic groups known as clades, which share a more recent common ancestor than any other taxa on the tree (53). Trees may be presented either with a root (rooted tree) or without a root (unrooted tree), and the tree branching pattern is known as the tree topology. A rooted tree contains a specific node, known as the root, and the distance from the root to any other node on the tree is the same. Rooted trees should only be constructed for data that have been shown to follow a molecular clock, which implies that evolution occurs at a constant rate among lineages within a phylogeny; in these trees the root functions as the common ancestor for all taxa in the phylogeny. Although an unrooted tree describes the relationship between taxa, this phylogenetic presentation fails to define the course of evolution for the taxa being studied (Figures 1.2A, 1.2B). While most phylogenetic data follow a molecular clock, and can therefore be used for construction of rooted trees, phylogenetic tress for taxa that do not follow a molecular clock should be presented as an unrooted tree with an outgroup included. Any organism thought to have diverged earlier than the taxa being studied can be used as an outgroup to root an unrooted tree (9,54). Numerous approaches, including distance based, maximum parsimony, maximum likelihood, and Bayesian inference methods have been used for reconstructing phylogenetic trees, and are briefly described in the following sections.

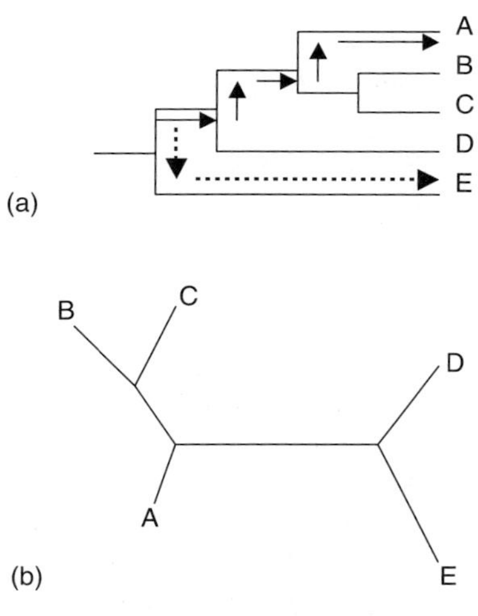

Figure 1.2 Rooted (a) and unrooted (b) five taxa phylogenetic trees. The evolutionary path for each taxa can be traced in the rooted tree as depicted by arrows.

1.4.2.1 Distance Based Methods to Infer Phylogeny

Distance based methods to estimate phylogeny calculate the difference in the number of nucleotide substitutions between all possible pairs of taxa and compress DNA sequence data to a distance matrix to estimate evolutionary distance. The major advantage of distance based methods is that they are very fast, but information is generally lost when DNA sequences are analyzed as a distance matrix, and an accurate estimate of the true phylogeny can be difficult to attain for diverse taxa. Neighbor-joining is the most popular distance based method, and Neighbor-joining trees can be constructed in several phylogenetic analysis programs, including PAUP* (50), PHYLIP (55) and MEGA (56). An unfavorable aspect of using distance based methods to construct phylogenetic trees is that, in some circumstances, these methods may yield two or more equally plausible tree topologies for a given data set. These trees must then be carefully evaluated to determine the reliability of branching patterns in order to prevent making incorrect inferences about the evolution of genes or populations represented by these alternative tree topologies (57). Neighbor-joining trees often serve as the starting point for more computationally intensive approaches to determine phylogeny (58).

1.4.2.2 Parsimony Methods to Infer Phylogeny

Unlike the distance based methods mentioned above, the phylogenetic trees estimated by maximum parsimony and maximum likelihood approaches illustrate the evolutionary relatedness of a group of taxa. Parsimony is a valuable approach for analyzing data without complex underlying model assumptions (59). Maximum parsimony conducts a site by site analysis to reconstruct a phylogenetic tree from sequence data; the number of nucleotide changes is summed over all sites to assign a parsimony score to each possible tree topology. Both the maximum parsimony and maximum likelihood phylogenetic approaches typically utilize what is known as a heuristic search to explore a large number of tree topologies (60). The tree with the least amount of changes is selected as the phylogenetic estimate of the data, and is termed the most parsimonious tree by the maximum parsimony method (61). More specifically, the shortest route or the fewest number of evolutionary changes that leads to the observed state of the sequence data is selected as the correct phylogeny. Maximum parsimony approaches are included in PAUP*, PHYLIP, and MEGA.

1.4.2.3 Maximum Likelihood Methods to Infer Phylogeny

Maximum likelihood selects the most likely phylogeny by assigning a maximum likelihood value to each character state arrangement within the sequences being studied and selects the tree topology with the highest maximum likelihood value. Specifically, phylogenetic trees generated under different models (which can also include additional free evolutionary parameters) are compared. A likelihood value is generated for each phylogenetic tree, and a heuristic search using these values determines the most likely tree topology under each hypothesized model and the parameters to select the most likely tree observed for a given data set (62). One of the main advantages of the maximum likelihood approach is that the likelihood ratio test can be used to define the model of DNA substitution and the additional parameters (e.g., mutation rate variation and molecular clock hypothesis) that best explain the observed DNA sequence data. Simulation studies (63,64) showed that the maximum likelihood approach determined the correct phylogeny more often than other methods including Neighbor-joining and parsimony methods. The most comprehensive and reliable phylogenetic program for utilizing maximum likelihood methods is PAUP*.

1.4.2.4 Measures of Confidence for Traditional Phylogeny Estimation Methods

Bootstrapping is a method to measure how strongly the sequence data actually support the phylogeny that has been inferred by traditional phylogenetic analysis methods. This tool has been widely used to assign a measure of confidence to each clade of a phylogenetic tree (65). The bootstrapping procedure works by randomly resampling (with replacement) the original data to generate replicate data sets, which are used to repeat the selected phylogenetic analysis on each replicate data set. The proportion of times that each clade observed in the original topology is recovered in replicate topologies is the bootstrap value (58). Bootstrap values have been criticized for being too conservative, and determining bootstrap values may be computationally intensive for phylogenies inferred by complex evolutionary models or for data sets with a large number of taxa (65).

1.4.2.5 Bayesian Methods

Bayesian methods are a close relative of maximum likelihood with the major exception being that Bayesian methods incorporate a prior distribution to explain the probability that a given tree topology is correct. Maximum likelihood relies on selecting the tree topology that shows the relationship of the taxa, branch lengths, and phylogenetic model parameter estimates that correspond with maximizing the probability of observing the DNA sequence data; while Bayesian statistics depend on the posterior probability of these characteristics to infer phylogeny (49). The model that maximizes the posterior probability is the most favorable one in the Bayesian perspective. The Bayesian approach relies on the Markov chain Monte Carlo (MCMC) search algorithm to sample the tree space (all possible tree topologies). One of the main advantages of Bayesian phylogenetics is that the primary analysis yields both a phylogenetic tree and a measure of confidence for grouping of taxa on that tree. Additionally, Bayesian methods can be used to estimate divergence times and to identify sites under recombination (58). Bayesian phylogenetic analysis can be performed in the MrBayes program (66).

1.4.3 Measuring Recombination

Due to the increasing availability of DNA sequence data for one or multiple bacterial genes or gene fragments from different isolates (e.g., from MLST analyses), analysis of DNA sequence data is now often used to probe for evidence of recombination among bacterial isolates (67). In recent years, it has become apparent that horizontal gene transfer and recombination of bacterial housekeeping genes occurs more commonly than originally anticipated (68). Traditional methods to infer phylogenetic relationships among organisms assume that recombination is absent. If recombination has occurred within a given DNA sequence data set, conclusions about the genealogy drawn from these sequences may therefore be incorrect. Different nonparametric and parametric methods have consequently been developed to probe for evidence of recombination and to estimate recombination rates based on DNA sequence data.

1.4.3.1 Nonparametric Methods to Detect Recombination

Several nonparametric methods have been developed to detect the presence of recombination in a set of DNA sequences without actually describing the rate at which recombination occurs. These nonparametric methods used to collect evidence of lateral gene transfer events can be divided into two classes: comparative, and phylogenetic dependent approaches. However, these comparative and phylogenetic based nonparametric methods are only useful to detect the presence of recombination in sequence data sets and so are limited due to their inability to estimate the rate of recombination in underlying populations (69).

Estimating the rate of recombination is crucial, though, to determine the validity of phylogenetic inferences and measure the diffusion rate of advantageous mutations and transfer of genetic information (70).

Comparative nonparametric methods attempt to recognize specific genes that appear to be atypical in comparison to the content of the complete genome (71). These methods specifically use multiple sequence alignments to detect regions within a sequence that show more similarity in base composition and codon usage than is expected by chance. These comparative methods have been used to identify evidence of horizontal gene transfer in bacterial species (20,72,73). The application of these methods to compare bacterial genomes to uncover acquisition of pathogenicity via incorporation of foreign genetic material is discussed in greater detail in the section on "Molecular evolution and diversity of selected food borne pathogens".

In addition, phylogenetic based methods can be used to identify uncharacteristic distribution of genetic information between organisms. Conflicting phylogenetic signals within a gene or between genes from the same organism provide evidence for recombination. For example, the split decomposition method for inferring phylogeny (74) does not force sequence data into a treelike structure. Instead, this method allows conflicting phylogenetic information between taxa to be visualized as a meshlike network. Generation of this meshlike network can thus be interpreted as evidence for conflicting phylogenies and may indicate recombination has occurred.

When determining the phylogenetic relationship of a group of taxa, DNA sequence data may be partitioned to give different weights to specific nucleotide changes (i.e., synonymous versus nonsynonymous changes), or sequence data from multiple genes may be combined to improve phylogenetic accuracy. Several parsimony based methods including the Templeton test, incongruent length difference test, and Rodrigo test evaluate the degree of incongruence between partitions in sequence data. These tests may be used to evaluate the accuracy of partitioned and combined data by determining whether the degree of incongruence observed for data partitions significantly affects the phylogenetic signal (75). If a dataset shows significantly different phylogenetic signals for different partitions, one might conclude that recombination occurred.

Several approaches have been utilized to assess the consistency of phylogenetic signal inferred along DNA sequences as a means to detect recombination. One approach implemented in the program called RETICULATE calculates a compatibility matrix based on pairwise comparison of all polymorphic sites in aligned sequences. Two sites are compatible if a phylogeny exists in which all nucleotide changes at the site can be inferred to have occurred only once; incompatible sites require multiple mutation events, indicating that these sites have experienced recombination or repeated mutation (76). The observed compatibility is then compared to the neighborhood similarity value obtained for a high number of randomized matrices (e.g., 10,000). A significant difference between the observed similarity value and the similarity and randomized matrix values may indicate that recombination has not occurred (77). Evaluation of the similarity matrix and a similarity matrix plot can also be used to identify likely regions of recombination, as shown by Reid et al (76).

Other approaches using sliding window analyses to detect evidence of recombination include the Dss (Difference in the sum of squares) method, which has been implemented in TOPAL. In this method, a sliding window is moved along the nucleotide alignment. The window is then split into two halves and a phylogenetic tree is estimated based on the first half of the window. This tree is used to compute the goodness of fit scores for both parts of the window; the Dss value represents the difference between these goodness of fit scores for each window (78). The same calculation is subsequently repeated with sliding windows moving in the opposite direction. The highest Dss value between the reverse run and the forward run becomes the Dss statistic for that window (78). These calculations yield a Dss value for each sliding window along the sequence alignment. Parametric bootstrapping (Monte Carlo simulation) is then used to simulate a distribution of Dss values under the null hypothesis that no recombination has occurred. The resulting 99% significance point of Dss obtained from the Monte Carlo simulation is used to identify observed Dss values above the threshold (99%) Dss, indicating regions with conflicting phylogenies, which may have been caused by recombination events. RECPARS [as detailed in Hein, 1993 (79)] represents another method that uses a parsimony based approach to probe for evidence of recombination in DNA sequences. This method overcomes the problems associated with using a finite window in a sliding window analysis. In this approach, the best phylogenies for different regions of the aligned sequences are defined; and a recombination event is postulated for sites that correspond to a transition from one tree topology into another topology.

Several methods, including the homoplasy test and the index of association, may be used to detect repeated recombination in a population. These tests recognize stretches of linked nucleotides with unique ancestry as potential recombination events. The homoplasy test implemented in the program HOMOPLASY TEST (80) builds a maximum parsimony tree for a sequence data set and the number of homoplasies (h) is calculated by subtracting the number of polymorphic sites from the number of steps required to build the tree. If the population is clonal then h is expected to equal zero, but if recombination has occurred then h is expected to be larger than 0. Another measure of recombination is the index of association (I_A), which has been used to analyze recombination in MLEE data. In this method, the variance of genetic diversity between pairs of isolates (observed variance) is divided by the variance obtained for a shuffled matrix (expected variance) to yield the I_A; if the I_A is greater than 1, the data is not greatly influenced by recombination (80–81).

Another approach, known as the informative sites test (implemented in PIST), evaluates the underlying assumption of clonality in DNA sequence data sets by analyzing the relationship between substitution rate heterogeneity and recombination. More specifically, the polymorphic site composition is expected to differ in data sets affected by recombination induced rate heterogeneity, while clonal data sets should maintain a more constant rate. The informative sites test compares the number of parsimony informative sites to an expected distribution of this number in clonal data sets generated by Monte Carlo simulation. An excess of parsimony informative sites indicates that null hypothesis of clonality is not valid (82).

1.4.3.2 Parametric Methods to Detect Recombination and Measure Rate of Recombination

Parametric methods rely on describing linkage disequilibrium within the haplotype structure of population genetics data. Specifically, these methods estimate population recombination rates from population genetic data by estimating the likelihoods for theta (effective population size multiplied by mutation rate) and r (per site recombination divided by per site mutation) rates under a coalescent based population genetics model (70,82,83,84). Several methods examine linkage disequilibrium between segregating sites to identify recombination. Linkage disequilibrium based methods offer the ability to make inferences about both the observed data and the underlying population, thus allowing estimates of the actual rate of recombination (69).

Husmeier and Wright (85) formulated an approximate Bayesian approach to discriminate between alternative sequence partitions that may have resulted from exchange of genetic information between taxa within a set of DNA sequence data. This method performed well when compared to nonparametric sliding windows methods like RECPARS (79) and TOPAL (86), but is limited by the number of taxa that can be analyzed.

The RECOMBINE program (84) uses maximum likelihood estimation under a neutral model of evolution via Metropolis Hastings sampling to search many recombinant genealogy reconstructions that are weighted by the posterior probability of the data and use different parameter values. A distinct advantage of this method is the ability to include other parameters such as population growth estimators [FLUCTUATE, (87)], and migration rate estimators [MIGRATE, (88)] in an analysis. By simultaneously estimating these population genetics parameters in the LAMARC package, the bias that could be introduced by estimating each parameter separately and by ignoring the effects of other important population parameters can be greatly reduced. Although this method is very powerful, it is also is very computationally intensive. Furthermore, the lack of ability to assess the validity of the likelihood estimate and to make statistical inferences represents limitation of this method (89).

In 2001, Hudson (89) suggested an approximate likelihood coalescent based method to estimate the recombination rate parameter in populations under a neutral mutation rate by assessing the level of linkage disequilibrium between sites. In contrast to the full multisite likelihood methods described previously, Hudson (89) used likelihood methods to analyze recombination in all pairwise combinations of segregating sites. Hudson reported that the presence of more than two linked polymorphisms within a given data set would lead to some loss of information in comparison to full likelihood methods. This method is intended to be used on data sets where full likelihood methods are not computationally practical.

McVean et al. (70) extended Hudson's method to accommodate bacterial and viral genomes that are expected to have high nucleotide substitution rates and contain sites that may have experienced multiple mutations. These patterns of genetic variation may be falsely categorized as recombination events, leading to bias in the estimation of the population recombination rate. McVean et al. (70) modified Hudson's method by implementing a finite sites mutation model and also suggested a likelihood permutation test to test the null hypothesis of no recombination. This group reported that both the likelihood permutation test and Hudson's method appeared to work well for data sets that contain recurrent mutations. Moreover, the likelihood permutation method was shown to be more powerful than the formerly described permutation methods and was not sensitive to specification of an incorrect mutation model (70).

1.4.4 Detecting Natural Selection and Models of Population Genetics

The strictly neutral theory of evolution (90) revolves around the concept that all mutations are either neutral or strongly deleterious. All alleles in a population evolving under the strictly neutral model have the same fitness; moreover, the estimation of many population genetic parameters assumes strict neutrality. Neutrality tests including Tajima's *D* test (91) and the HKA test (92) are designed to detect deviation from this strictly neutral model of evolution. Noncoding regions in DNA sequences (e.g., intergenic regions), as well as synonymous changes in coding regions, are assumed to not be influenced by selective constraints. The objective of Tajima's *D* test is to assess this assumption of neutrality by sampling sequences from a population and comparing the number of segregating sites and the average number of nucleotide differences (54). The HKA test is designed to test the hypothesis that the level of DNA variation is correlated both within and between populations, which is expected if the neutral assumption holds true (9). In 1973, Ohta (93) developed the "slightly deleterious model", which included slightly deleterious mutations but continued to disallow positive selection. Kimura (94) modified the strictly neutral model to describe negative mutations as following a gamma distribution and termed this model the effectively neutral model. In contrast, others (95–96) have suggested models of evolution that include positive selection (97).

1.4.4.1 Estimating Synonymous and Nonsynonymous Substitution Rates

The ability to identify adaptive evolution at the protein level is an important tool in evolutionary biology. Selection at the amino acid level can be detected by measuring the nonsynonymous:synonymous substitution rate ratio ($\omega = d_N/d_S$). Synonymous or silent mutations for the most part do not influence natural selection, while nonsynonymous or amino acid changing mutations may be under strong selective pressure. By evaluating the fixation rate of synonymous and nonsynonymous mutations, valuable insight into the evolution of specific protein coding sequences may be gained. An ω greater than 1 indicates evidence for positive selection; nonsynonymous mutations may thus have a distinct fitness advantage and become fixed in the population more rapidly than synonymous mutations. Values of $\omega = 1$ and $\omega < 1$ indicate neutral and purifying evolution, respectively. Codon substitution models that allow ω to vary among sites can be implemented to identify functionally constrained amino acid sites as well as to recognize specific amino acid sites that may be under positive selection. Analyzing the variation of ω among lineages can be used to determine if adaptive evolution is occurring in specific lineages of interest compared to other lineages in a data set (98–100).

1.4.4.2 Prediction of Amino Acid Sites Under Positive Selection

The ability to detect positive selection at specific amino acid sites has important implications in the biological and medical sciences. Several methods have been introduced for the purpose of assessing variability in selective pressures across amino acid sites. The Suzuki and Gojobori (SG) method is a parsimony based method that reconstructs the ancestral sequence state and counts the number of changes at each site along a tree to identify excessive nonsynonymous substitution events (101). This method is implemented in the ADAPTSITE program to test the null hypothesis of neutrality at each codon (102). Suzuki and Nei (103) conducted computer simulations to compare the SG parsimony method to likelihood methods. The authors indicated that likelihood methods were more prone to yield false positive results and parsimony seemed to be a more conservative method.

Nielsen and Yang developed codon substitution models of heterogeneous d_N/d_S ratios among sites using a maximum likelihood framework implemented in the PAML program (104). This maximum likelihood approach uses the likelihood ratio test to compare nested models, to determine whether or not a gene is undergoing positive selection. If the likelihood ratio test is significant and the null model, which does not allow for sites with $\omega > 1$, is rejected in favor of an alternative model that allows for sites with $\omega > 1$, then that gene is considered to be under positive selection. Once a gene is determined to be experiencing positive selection, an empirical Bayes approach is employed to predict which specific amino acid sites are positively selected for (105,106). More specifically, the posterior probability that a site fits in each ω class within the model given the data observed at that site is calculated, and sites with a high posterior probability of falling into the site class of $\omega > 1$ are considered to be under positive selection (107). In 2002, Yang and Swanson (108) refined these codon substitution models to allow sites to be partitioned into classes when prior information about specific sites indicates that those sites are expected to have different selective pressures. These models can be used to test the hypothesis that sites with an explicit biological function might be under diversifying selection or for combined analysis of several protein coding genes from the same taxa to evaluate substitution patterns.

1.4.4.3 Detecting Adaptive Evolution in Lineages

A lineage that experienced Darwinian selection would be expected to show a d_N/d_S ratio that is greater than one or is different from that of other lineages. Messier and Stewart (109) used both parsimony and likelihood approaches to reconstruct the ancestral state of lysozyme genes from extinct ancestors in a phylogeny. This reconstructed ancestral sequence, and the observed sequences, were used to determine d_N and d_S by making pairwise sequence comparisons along each branch or lineage in the primate phylogenetic tree. In 1998, Yang (110) raised concerns about the reliability of using parsimony methods to reconstruct the ancestral sequence and implemented likelihood methods to infer d_N and d_S from lineages in a phylogeny without using the reconstructed ancestral sequence state. In addition, the likelihood models developed by Yang (110) account for transition or transversion rate and codon usage bias that are not addressed in the pairwise comparison methods implemented by earlier researchers. Yang (110) developed several codon based models that permit ω to vary among lineages; however, these models assume that ω remains constant among sites. In addition, the codon substitution models described above assume that ω is invariable among lineages in the phylogeny. These nested models can then be used to perform the likelihood ratio test to establish whether or not ω is variable among lineages, and ultimately if specific lineages of interest demonstrate a d_N/d_S ratio that is greater than one, which would indicate that those lineages were under adaptive selection.

1.5 MOLECULAR EVOLUTION AND DIVERSITY OF SELECTED FOOD BORNE PATHOGENS

1.5.1 Evolution of *Escherichia coli*

Although most *E. coli* strains represent harmless commensals that are part of the normal flora of the gastrointestinal tract, several pathogenic *E. coli* groups that are associated with diverse human and animal diseases have been described. The acquisition of various sets of virulence genes has been critical in the evolution of these pathogenic *E. coli*. After its divergence from *Salmonella* around 100 million years ago, *E. coli* has undergone significant genetic change, and most of the horizontally transferred genetic material in *E. coli* is thought to have been acquired fairly recently (111). Genome comparisons have shown that about 25% of the genes found in *E. coli* do not appear to have homologs in the *Salmonella enterica* genome (112). Comparison of the complete genome sequences available for both *E. coli* K-12 (44), a nonpathogenic laboratory strain, and *E. coli* O157:H7, a pathogenic strain, (43) have shown that while 4.1 million base pairs represent genomic information conserved between the two genomes, 1.34 million base pairs of the *E. coli* O157:H7 genome were specific to this serotype. This analysis also identified some novel putative virulence genes unique to *E. coli* O157:H7 (113). It has been predicted that almost 18% of the *E. coli* K-12 chromosome was introduced from external sources and that the new genes were acquired in at least 234 separate events. (111). Horizontal gene transfer and loss of DNA are thus thought to have significantly contributed to the phenotypic differentiation of *E. coli* and *S. enterica* (111).

Within *E. coli* there are over 200 recognized O antigen serotypes and around 30 H antigen types (114). There are at least five recognized pathotypes of diarrheagenic *E. coli*, including enteropathogenic *E. coli* (EPEC), enterohemorrhagic *E. coli* (EHEC), enterotoxigenic *E. coli* (ETEC), enteroinvasive *E. coli* (EIEC), and enteroaggregative *E. coli* (EAEC). Each of these pathotypes is characterized by specific virulence traits. While strains belonging to EPEC, EHEC, ETEC, and EIEC have all been implicated in water or food borne outbreaks, it is still uncertain whether EAEC strains should be considered food borne pathogens (114). Enterohemorrhagic *E. coli* (EHEC), which includes *E.coli* O157: H7, have been an important cause of water and food borne outbreaks in various developed countries. First recognized as an emerging human food borne pathogen in the early 1980s, *E. coli* O157:H7 can cause severe hemorrhagic colitis as well as hemolytic uremic syndrome (HUS). The EHEC pathotype represents a subset of the Shiga toxin-producing *E. coli* (STEC), which includes all *E. coli* strains that share the ability to produce shiga toxins (Stx) (113). The ability to produce shiga toxins is one of the main characteristics that distinguishes EHEC from EPEC strains. The two main forms of the shiga toxin, Stx1 and Stx2, are encoded on bacteriophages (115).

Although most *E. coli* pathotypes fall into clonal lineages, EHEC and EPEC strains group into divergent lineages. Based on MLEE data, these two pathotypes have been divided into EPEC 1 and EPEC 2, as well as into EHEC 1 (*E. coli* O157:H7) and EHEC 2; EPEC 2 and EHEC 2 are the most closely related lineages (113). These data are consistent with a model that assumes the EHEC pathotype represents two distinct evolutionary emergence processes. Even though the Stx producing EHEC 2 group is quite divergent from *E. coli* O157:H7 strains, they do possess the same key virulence factors and could possibly become a significant emerging pathogen (113).

One feature that is common to almost all EHEC and EPEC is the presence of an approximately 35-kb pathogenicity island termed the locus of enterocyte effacement (LEE). This LEE island includes virulence genes that encode proteins responsible for bacterial adhesions to epithelial cells and the formation of the "attaching and effacing" (A/E) lesions characteristic for both EHEC and EPEC. One of the genes found in the LEE, *eae*A, encodes intimin, which is expressed in both EHEC and EPEC. Intimin is an outer membrane protein that shows similarity to proteins found in other pathogens (116). Tir, which serves as the host cell receptor for intimin, is also encoded in the LEE; this protein is translocated into host cells via the *E. coli* type III secretion systems. The intimin gene has shown high variability and has at least five antigenic types that usually correspond to particular lineages (117). Although there are some exceptions, α intimin is common to EPEC 1 strains, β intimin is mostly found in EPEC 2 and EHEC 2 strains, and γ intimin is generally associated with EHEC 1 strains (117). The genetic variations that are seen in intimin may be due to selective pressures that are often placed on outer membrane proteins in an effort to escape recognition by host immune responses (116). The LEE island was most likely acquired through horizontal gene transfer, because it is absent from nonpathogenic strains of *E. coli* and has a lower G+C content than the rest of the *E. coli* genome (113). Interestingly, the LEE has been found to be located at more that one chromosomal site; in *E. coli* with α or γ intimin it is inserted into *sel*C, but in strains with β intimin the insertion site is in *phe*U (76). These finding support the hypothesis that lateral acquisition of LEE has occurred at least twice in the evolution of *E. coli.* The *sel*C tRNA locus has also served as an integration site for other PAIs, including SPI-3 in *S. enterica* and SHI-2 in *Shigella flexneri* (2).

Even though O157:H7 falls into the EHEC category, this serotype is most closely related to *E. coli* O55:H7, an EPEC that does not produce shiga toxins but carries the LEE (118). Unlike most other *E. coli* strains, O157:H7 lacks the ability to ferment sorbitol (Sor-) and lacks β-glucuronidase activity (GUD-). However, one nonmotile (H-) *E. coli* O157 strain that was able to ferment sorbitol (Sor+) has been identified in an outbreak of HUS in Germany (113). Based on the various data available on the characteristics of different EPEC and EHEC strains, an evolutionary model of the emergence of *E. coli* O157:H7 has been proposed by Feng (118). An EPEC like *E. coli,* that expressed α-glucuronidase (GUD+) and fermented sorbitol (Sor+), and carried the LEE, served as an ancestor for

E. coli O55:H7. Subsequent phage mediated acquisition of Stx2 resulted in a shiga toxin producing *E. coli* O55:H7 strain, followed by a change in somatic antigens from O55 to O157 and the attainment of one or more EHEC plasmids, which encode various putative virulence factors including a hemolysin operon (113). Two lineages subsequently evolved from this hypothesized ancestor, one lineage lost the motility phenotype (converting to H-), but remained GUD+, Sor+ and Stx2 positive. The second lineage became GUD- and Sor- but acquired the Stx1gene, giving rise to the *E. coli* O157:H7 clone found worldwide today. If this hypothesis is correct, the German *E. coli* O157:H7 clone diverged early in the evolution of EHEC, since it retained some of the ancestral characteristics that are not seen in the *E. coli* O157:H7 isolated today (113).

In addition to the LEE pathogenicity island, the pathogenicity characteristics of EPEC strains are also associated with a plasmid know as the EAF (EPEC adherence factor) plasmid, which is present in most EPEC strains (119). The EAF plasmid includes the gene encoding the bundle forming pilus (Bfp), a specific virulence factor that facilitates bacterial adhesion to epithelial cell surfaces. Unlike EHEC strains, which lack the genes necessary to synthesize Bfp, EPEC strains have a characteristic adherence pattern known as localized adherence (120). Although the localized adherence phenotype typical for EPEC may not be exclusively due to Bfp, recent data has shown that Bfp are necessary for full virulence in EPEC strains (120).

Enteroinvasive *E. coli* strains have also been implicated in food borne disease cases and outbreaks (114), but they do not tend to group with EHEC or EPEC strains in phylogenetic trees (113). These strains are genetically similar to *Shigella* species and their virulence mechanisms are remarkably similar (121). Unlike most other *E. coli*, EIEC strains, as well as *Shigella* species lack lysine decarboxylase activity, which is encoded by *cadA*, which appears to have been lost in EIEC and *Shigella* species together with a large, approximately 90 kb region, of their genome. Loss of *cad*A gene leads to an inability to produce cadaverine, a molecule that acts as an enterotoxin inhibitor. This *cadA* deletion has thus been hypothesized to be responsible for an enhanced virulence of EIEC and *Shigella* species (13).

The ETEC pathotype includes *E. coli* strains that secrete one or both of the following entertoxins, LT, a heat labile toxin and ST, a heat stable toxin (119). The LT can be divided into the LT-I type, a toxin that is structurally and functionally similar to cholera entertoxins, and the LT-II type, a type that does not appear to play any role in human disease pathogenesis (121). The heat stable toxins also fall into two groups, STa and STb, with the genes encoding STa found on plasmids (119).

1.5.2 Evolution and Diversity of *Salmonella*

The genus *Salmonella* includes over 2,100 serovars, which are divided into two divergent species, *Salmonella enterica*, which contains seven subspecies (I, II, IIIa, IIIb, IV, VI, and VII), and *Salmonella bongori,* which was formerly classified as subspecies V (122). It has been estimated that the common ancestor for the genus *Salmonella* arose between 35 and 40 million years ago (123). Subspecies I, which includes *S. enterica typhimurium*, represents strains that are highly specialized to infect mammals and birds, and this subspecies accounts for more than 99% of enteric and systemic *Salmonella* infections in humans (124). Although the number of typhoid fever cases (caused by *Salmonella typhi*) has steadily decreased in the United States, the incidence of nontyphoidal salmonellosis has gradually increased to an estimated 1.4 million cases per year (125), with only a few serovars causing the majority of human cases (126).

While some *S. enterica* serovars have a broad host range (e.g., *typhimurium* and *enteriditis*), other serovars appear to be host adapted (e.g., *typh*i in humans). Based on

MLEE analysis, serovars that can infect a wide range of hosts have much greater genetic variability than serovars that are host specific (122). Although the acquisition of virulence genes may play a role in host adaptation, other factors such as selective pressure and competitive exclusion also play an important role in the evolution and ecology of *Salmonella*. For example, *S. enterica gallinarum*, an avian adapted serotype (O9 antigen, D1 serogroup) was eradicated from poultry in the US and Europe in an attempt to decrease economic losses. As long as *Salmonella gallinarum* was present in poultry flocks, other D1 serogroup strains may have been excluded from circulating in poultry, due to the presence of flock immunity against the O9 antigen. After the eradication of *Salmonella gallinarum* in the 1970s, the loss of flock immunity against the O9 antigen appears to have led to the availability of a new niche that was filled by another D1 serogroup, *S. enterica enteriditis*. The exclusion of *Salmonella gallinarum* from poultry thus may have led to the emergence of a new pathogen that triggered a human salmonellosis epidemic (127). This example clearly shows the complex nature of the evolution and population genetics of food borne pathogens, which is affected not only by pathogen factors, but also by host and environmental factors.

As in *E. coli*, horizontal acquisition of virulence genes and PAIs appears to have played a major role in the evolution of virulence characteristics in *Salmonella*. It has been specifically hypothesized that the fact that *S. enterica* serovar strains have the ability to affect a wide range of hosts and cause a variety of different disease manifestations, even though they appear to have a high genetic relatedness based on sequence data, may be largely due to lateral gene transfer of virulence genes (122). *Salmonella* has at least five known PAIs (termed *Salmonella* pathogenicity islands, SPI-1 through SPI-5) that encode an array of virulence factors ranging from epithelial cell invasion to systemic spread (128). These SPIs have not been found in related enteric species, e.g., *Shigella* or *E. coli*, and thus appear to have been introduced into the *Salmonella* genome rather recently, since *Salmonella*'s divergence from *E. coli* (129). SPI-1, which encodes genes that promote *Salmonella* invasion into human host epithelial cells (the *inv/spa* genes), is found in all *Salmonella* lineages, and is thus thought to be have been acquired early in the evolution of *Salmonella,* probably by an ancestor to all *Salmonella* (130). This 40 kb pathogenicity island (SPI-1) appears to have been acquired through horizontal gene transfer because it has a much lower G+C content (42%) than the rest of the genome. Also, several of the genes on SPI-1 are similar to invasion genes found on the *Shigella* virulence plasmid. Homologs of several of the *inv/spa* genes, which are known to aid in bacterial internalization into host cells, have also been found in various other bacterial pathogens even though their locations in the genome can vary (131). While most PAIs are usually flanked by tRNA genes, which are target sites for DNA integration, SPI-1 is bordered by two housekeeping genes, *fhl*A and *mut*S. The remaining SPIs are located near tRNA loci: SPI-2 is flanked by the *pyk*F gene and the *val*V tRNA gene; SPI-3 is located downstream of the *sel*C tRNA gene; SPI-4 is located downstream from an unspecified tRNA gene; and SPI-5 is located downstream of the *ser*T tRNA gene (128).

The SPI-2 island contains more than 40 genes and is divided into two segments, a 14.5 kb piece and a 25.3 kb piece, which may have been transferred into the genome in two distinct steps (132). SPI-2 was probably horizontally transferred into the *S. enterica* genome after its divergence from *S. bongori,* because several of the genes in SPI-2 are not found in *S. bongori* serotypes. The larger portion of the SPI-2 island, which is only found in *S. enterica,* was probably acquired more recently than the smaller segment of the island, which is found in both *Salmonella* species (132). SPI-2 also has a lower G+C content (41%) than the rest of the *Salmonella* genome, further supporting horizontal acquisition of this PAI. The genes found on SPI-2 appear to be specifically important for the ability of *Salmonella* to cause systemic disease (130).

The SPI-3 island, which has a mosaic structure, possesses several genes that are required for virulence and survival within the macrophage (122). Although SPI-3 is quite small (17 kb), the full island is not present in all *Salmonella* species, suggesting that it was acquired in multiple steps (128). The genes of the two remaining islands, SPI-4 and SPI-5, appear to play less of a role in virulence, with many of the genes functions still unknown (122).

Salmonellae also have various chromosomal fimbrial operons, which encode proteins that appear to be critical for host cell attachment and infection initiation. The specific distribution and expression of fimbrial operons among serovars may also affect the host range of different *Salmonella* strains, similar to what has been proposed for enteropathogenic *E. coli* (133). The *fim*, *lpf* and *agf* fimbrial operons were probably acquired early in the evolution of *Salmonella*, although *lpf* appears to have subsequently been lost from several lineages (133). In contrast, the distribution of the *ser* and *pef* fimbrial operons has been limited, probably due to recent acquisition by the *Salmonella* genome (133).

The acquisition of other mobile elements, such as plasmids, phages, and transposable elements, has also played a major role in the evolution of *Salmonella*. Phage genomes as well as phage remnants are widespread within the *Salmonella* chromosome, with many of them encoding important virulence factors. Most *Salmonella* serovars also carry a putative virulence plasmid bearing the *spv* genes. While these genes were initially thought to be important in the survival of *Salmonella* in the host, several groups could not find any evidence for a role of the *spv* gene products in intracellular survival. Further studies on these genes are thus necessary to clarify their function (132). Horizontal gene transfer of antibiotic resistance has also been of importance in the evolution of food transmitted *Salmonella* strains. For example, one strain of *S. enterica typhimurium*, known as definitive type (DT) 104, has emerged as a worldwide health concern because of its resistance to multiple antibiotics. It is resistant to at least five antibiotics including ampicillin, streptomycin, sulfonamides, tetracycline, and chloramphenicol (32). While the antibiotic resistance genes found in DT104 appear to be located on the chromosome, they are surrounded by plasmid and phagelike genes, supporting acquisition by lateral gene transfer and subsequent site specific recombination (122). Interestingly, although PFGE has been used quite successfully in the typing of *Salmonella* Typhimurium strains, DT104 has tested the discriminatory limits of PFGE with its genetic homogeneity (134). Because it is based on tandem repeats in DNA, VNTR has been shown to have a greater discriminatory power than PFGE for distinguishing among DT104 strains (134). These observations support the idea that DT 104 represents a recently emerged genetically homogeneous subtype, and also illustrate the potential for VNTR based subtyping methods for studies on the evolution and population genetics of genetically homogenous bacterial pathogen clones.

While we clearly already have acquired a considerable understanding of the evolution of *Salmonella*, the increasing availability of complete genome sequences for *Salmonella* serovars, including *S. enterica typhimurium* LT2 and *S. enterica typhi* CT18 (42,135) as well as other closely related organisms like *E. coli*, will provide the ability to further probe the genome evolution of these important human food borne pathogens.

1.5.3 Evolution and Diversity of *Listeria monocytogenes*

Of the different species within the genus *Listeria,* only *L. monocytogenes* is considered a human pathogen. Virulence characteristics of *L. monocytogenes* appear to be primarily associated with the presence of different pathogenicity islands, including the *prfA* virulence gene island and different virulence gene islands carrying genes of the internalin family (136). Recent comparative genomics studies facilitated by the availability of full genome sequences for *L. monocytogenes* and the closely related nonpathogenic species

L. innocua also identified additional virulence genes unique to *L. monocytogenes,* such as *bsh* which encodes a bile salt hydrolase (41). Interestingly, the *prfA* virulence gene island, which encodes most of the major *Listeria* virulence genes, has also been identified in the animal pathogen *L. ivanovii* and in the nonpathogenic species *L. seeligeri*. In *L. seeligeri*, the *prfA* virulence gene cluster appears to be improperly expressed apparently making this species incapable of causing human disease (136).

Classically, *L. monocytogenes* has been differentiated into 13 serotypes, with greater than 90% of human listeriosis cases caused by serotype 1/2a, 1/2b, and 4b strains. While the majority of human clinical infections occur as sporadic cases, human listeriosis can also occur in large epidemics. Most sporadic human listeriosis cases and large human food borne listeriosis epidemics have reportedly been caused by *L. monocytogenes* serotype 4b (28). Serotypes 1/2a and 1/2b are also responsible for significant numbers of sporadic cases of human illness, and a serotype 1/2a strain was responsible for a recent multistate human listeriosis outbreak in the USA (137). Serotyping data collected by the CDC in 1986 showed that serotypes 1/2a (30%), 1/2b (32%), 4b (34%) represented the majority of isolates from 144 human sporadic cases (138). Of 1,363 human isolates collected in the UK, 15% were 1/2a, 10% were 1/2b, and 64% were 4b (139). The remaining ten currently recognized *L. monocytogenes* serotypes have been only rarely linked to human disease. This apparent association between a few specific *L. monocytogenes* strains and most cases of human listeriosis raises the intriguing challenge of identifying unique characteristics enabling these strains to be more effective than others in causing human disease. Two hypotheses could explain the apparent predominance of serotype 4b strains in human epidemic listeriosis, and of 4b, 1/2a, and 1/2b strains in sporadic human cases: humans are more commonly exposed to these subtypes as compared to other *L. monocytogenes* serotypes, i.e., these strains are found in foods more frequently than other serotypes; or these subtypes have a unique pathogenic potential for humans.

Molecular subtyping methods have consistently grouped *L. monocytogenes* into two major lineages. Multilocus enzyme electrophoresis (MLEE), PFGE, ribotyping, and amplified restriction fragment length polymorphism analysis (AFLP) all show that *L. monocytogenes* can be separated into two major genetic groups (28). Allelic analyses of several virulence genes as well as ribotyping revealed a third phylogenetic lineage within *L. monocytogenes.* Specifically, a combination of virulence gene alleles and ribotype patterns allowed separation of *L. monocytogenes* strains into three distinct lineages, designated I, II, and III (28).

Evolutionary analysis of both DNA sequencing (140–141) and microarray data (45) confirmed the existence of two distinct *L. monocytogenes* lineages [previously termed lineages I, and II (140)]. These two lineages correlate with serotype groupings; lineage I comprised serotypes 1/2b, 3b, 3c and 4b; lineage II comprised serotypes 1/2a, and 1/2c (142). Preliminary microarray analyses also identified a considerable number of contigs present in lineage II strains, but absent from lineage I strains (45). Further analysis suggested a model in which the ancestor of the two lineages had the 1/2 somatic serotype, and the regions absent in the lineage I genome arose by loss of ancestral sequences (45). *L. monocytogenes* isolates previously designated as lineage III, which includes the rare *L. monocytogenes* serotypes 4a and 4c, appear to have a mosaic genome structure, based on DNA sequence analyses. Phylogenies based on nonvirulence genes further indicate that lineage III strains may classify into two distinct groups, including at least one group that appears to be closely related to *L. innocua.*

Characterization of a total of 42 human clinical isolates and 502 isolates from ready to eat foods collected in Maryland and California during 2000 and 2001 (143), as well as of 465 additional human clinical isolates collected throughout the USA between 1997 and

2002, was performed to probe associations between specific subtypes and human listeriosis (144). Genotypic analyses of isolates by automated *Eco*RI ribotyping and PCR-RFLP analysis of the *hly* gene (140) allowed assignment of isolates into one of 65 different *Eco*RI ribotypes and into one of three previously described genetic lineages. Statistical analyses showed that, while exclusive associations were rare, the majority of subtypes were significantly associated with isolation from either foods or humans. Using a large isolate set, this study did not only confirm previous data (145,146) that lineage I strains are more significantly associated with human listeriosis cases compared to both animal listeriosis cases and isolation from foods, but also provided specific data on the comparative prevalence of different ribotypes among human clinical and food isolates. Most strikingly, one specific *Eco*RI ribotype (DUP-1062A) was found to represent 30.1% of food isolates, but only 1.8% of human isolates, a highly significant difference in prevalence ($p<0.0001$). Overall these data indicate that the prevalence of certain subtypes among human listeriosis cases is not strictly a reflection of their prevalence in contaminated foods, but might rather reflect virulence differences among *L. monocytogenes* subtypes and clonal groups (28).

Evolutionary analyses of the stress response gene *sigB* and the virulence genes *actA* and *inlA* indicated that while evolution of *sigB* followed a molecular clock model, evolution of *actA* did not follow a clock model, and appeared to be under positive selection ($p<0.005$) as determined by using a likelihood ratio test. Using an empirical Bayes approach, 8 ActA amino acid sites with posterior probabilities $> 95\%$ of being positively selected were identified. Posterior probability plots revealed both highly conserved regions and regions with a significant frequency of positively selected amino acid sites in *actA* (141). These data indicate that active evolution of *actA* may play a role in the development of strain specific virulence characteristics.

In conclusion, currently available data indicate that *L. monocytogenes* is characterized by a predominant clonal population structure (particularly for lineages I and II), although strains previously characterized as lineage III show more diversity and indications for recombination. Both evolution by gene loss and active evolution of specific virulence genes (e.g., *actA*) appear to contribute to evolution of *L. monocytogenes*, including evolution of virulence related characteristics. Recent data (144,146) also support that the previously observed high prevalence of specific *L. monocytogenes* subtypes (e.g., serotype 4b strains) among human listeriosis cases and outbreaks appears to at least partially represent unique virulence characteristics of these subtypes rather then their abundance in contaminated food products.

1.6 CONCLUSIONS

Pathogenesis of food borne pathogens may incorporate complex interactions between the pathogen, diverse environments, and one or more host species. The ability of these pathogens to survive and compete in numerous environments may play an important role in the transmission and pathogenesis of food borne pathogens. Moreover, selective evolutionary pressures exerted by environmental stresses and the host may influence the evolution and emergence of virulence factors that allow specific strains to infect mammalian hosts. Understanding the molecular evolution of food borne pathogens will provide a more complete picture of these organisms' transmission dynamics, and generate essential knowledge required to control these pathogens and prevent their entry into the food system; ultimately reducing the risk of human food borne disease. Furthermore, a thorough understanding of the evolution of food borne pathogens is required to develop reliable subtyping methods for food borne pathogens and to correctly interpret subtyping data, e.g., in food borne disease outbreak investigations.

REFERENCES

1. Wiedmann, M. Subtyping of bacterial food borne pathogens. *Nutr. Rev.* 60:201–208, 2002.
2. Ochman,H., J.G. Lawrence, E.A. Groisman. Lateral gene transfer and the nature of bacterial innovation. *Nature* 405:299–304, 2000.
3. Hartl, D.L., A.G. Clark. *Principles of Population Genetics,* 3rd ed. Sunderland, MA: Sinauer Associates, Inc. 1997.
4. Lawrence, J.G. Gene transfer, speciation, and the evolution of bacterial genomes. *Curr. Opin. Microbiol.* 2:519–523, 1999.
5. Spratt, B.G., W.P. Hanage, E.J. Feil. The relative contributions of recombination and point mutation to the diversification of bacterial clones. *Curr. Opin. Microbiol.* 4:602–606, 2001.
6. Spratt, B.G., M.C. Maiden. Bacterial population genetics, evolution and epidemiology. *Phil. Trans. R. Soc. Lond. B. Biol. Sci.* 354:701–710, 1999.
7. Fraser, C.M., J.A. Eisen, S.L. Salzberg. Microbial genome sequencing. *Nature* 406:799–803, 2000.
8. Graur, D., W. Li. *Fundamentals of Molecular Evolution,* 2nd ed., Sunderland, MA: Sinauer Associates, Inc., 2000.
9. Li, W. *Molecular Evolution.* Sunderland, MA: Sinauer Associates, Inc. 1997.
10. Page, D.M., E.C. Holmes. *Molecular Evolution: A Phylogenetic Approach.* Oxford: Blackwell Scientific, 1988.
11. Ochman, H., N.A. Moran. Genes lost and genes found: evolution of bacterial pathogenesis and symbiosis. *Science* 292:1096–1098, 2001.
12. Day, W.A. Jr., R.E. Fernández, A.T. Maurelli. Pathoadaptive mutations that enhance virulence: genetic orgazation of the cadA regios of Shigella spp. *Infect. Imm.* 69:471–480, 2001.
13. Maurelli, A.T., R.E. Fernandez, C.A. Bloch, C.K. Rode, A. Fasano. "Black holes" and bacterial pathogenicity. a large genomic deletion that enhances the virulence of Shigella spp. and enteroinvasive Escherichia coli. *Proc. Natl. Acad. Sci. USA* 95:3943–3048, 1998.
14. Keim, P., L.B. Price, A.M. Klevytska, K.L. Smith, J.M. Schupp, R. Okinaka, P.J. Jackson, M.E. Hugh-Jones. Multiple-locus variable-number tandem repeat analysis reveals genetic relationships within Bacillus anthracis. *J. Bacteriol.* 182:2928–2936, 2000.
15. Liu, S., K. Sanderson. The chromosome of Salmonella paratyphi A is inverted by recombination between rrnH and rrnG. *J. Bacteriol.* 177:6585–6592, 1995.
16. Alokam, S., S. Liu, K. Said, K. Sanderson. Inversions over the terminus region in Salmonella and Escherichia coli: IS200s as the site of homologous recombination inverting the chromosome of Salmonella enterica serovar Typhi. *J. Bacteriol.* 184:6190–6197, 2002.
17. Boerlin, P. Evolution of virulence factors in Shiga-toxin-producing Escherichia coli. *Cell. Mol. Life Sci.* 56:735–741, 1999.
18. Teyssier, C., H. Marchandin, M.S. De Buochberg, M. Ramuz, E. Jumas-Bilak. Atypical 16S rRNA gene copies in Ochrobactrum intermedium strains reveal a large genomic rearrangement by recombination between rrn copies. *J. Bacteriol.* 185:2901–2902, 2003.
19. Smith, J. Maynard, N.H. Smith, M. O'Rourke, B.G. Spratt. How clonal are bacteria? *Proc. Natl. Acad. Sci. USA* 90:4384–4388, 1993.
20. Lawrence, J.G., H. Ochman. Amelioration of bacterial genomes: rates of change and exchange. *J. Mol. Evol.* 44:383–307, 1997.
21. Hacker, J., J.B. Kaper. Pathogenicity islands and the evolution of microbes. *Ann. Rev. Microbiol.* 54:641–679, 2000.
22. Hueck, C. Type III protein secretion systems in bacterial pathogens of animals and plants. *Microbiol. Mol. Biol. Rev.* 62:379–433, 1998.
23. Threfall, E.J. Antimicrobial drug resistance in Salmonella: problems and perspectives in food-and water-borne infections. *FEMS Microbiol Rev.* 26:141–148, 2002.
24. Olive, D.M., P. Bean. Principles and applications of methods for DNA-based typing of microbial organisms. *J. Clin. Microbiol.* 37:1661–1669, 1999.
25. de Boer, E., R. Beumer. Methodology for detection and typing of food borne microorganisms. *Int. J. Food Microbiol.* 50:119–130, 1999.

26. van Belkum, A., M. Struelens, A. de Visser, H. Verbrugh, M. Tibayrenc. Rose of genomic typing in taxonomy, evolutionary genetics, and microbial epidemiology. *Clin. Microbiol. Rev.* 14:547–560, 2001.
27. Wassenaar, T.M., D.G. Newell. Genotyping of Camplyobacter spp. *Appl. Environ. Microbiol.* 66:1–9, 2000.
28. Wiedmann, M. Molecular subtyping methods for Listeria monocytogenes. *J. Assoc. Off. Anal. Chem.* 85:524–531, 2002.
29. Graves, L.M., B. Swaminathan, S.B. Hunter. Subtyping Listeria monocytogenes. In: *Listeria, Listeriosis, and Food Safety,* Ryser, E., E. Marth, eds., New York: Marcel Dekker, 1999, pp 279–297.
30. Strockbine, N.A., J.G. Wells, C.A. Bopp, T.J. Barrett. Overview of detection and subtyping methods. In: *Escherichia coli O157-H7 and Other Shiga-toxin Producing E. coli Strains.* Kaper, J.B., A.D. O'brien, eds., Washington, DC: ASM Press, 1998, pp 331–356.
31. Threlfall, E., J. Frost. The identification, typing and fingerprinting of Salmonella: laboratory aspects and epidemiological applications. *J. Appl. Bacteriol.* 68:5–16, 1990.
32. Glynn, M.K., C. Bopp, W. Dewitt, P. Dabney, M. Mokhtar, F.J. Angulo. Emergence of multidrug-resistant Salmonella enterica serotype Typhimurium DT104 infections in the United States. *N. Engl. J. Med.* 338:1333–1338, 1998.
33. Threfall, E., J. Frost, L. Ward, B. Rowe, Increasing spectrum of resistance in multiresistant Salmonella typhimurium. *Lancet* 347:1053–1054, 1996.
34. Jacquet, C., B. Catimel, R. Brosch, C. Buchrieser, P. Dehaumont, V. Goulet, A. Lepoutre, P. Veit, J. Rocourt. Investigations related to the epidemic strain involved in the French listeriosis outbreatk in 1992. *Appl. Environ. Microbiol.* 61:2242–2246, 1995.
35. Selander, R., D. Caugant, H. Ochman, J.M. Musser, M.N. Gilmour, T.S. Whittam. Methods of multilocus enzyme electrophoresis for bacterial population genetics and systematics. *App. Environ. Microbiol.* 51:873–884, 1986.
36. Piffaretti, J.C., H. Kressebuch, M. Aeschbacher, J. Bille, E. Bannerman, J.M. Musser, B.K. Selander, J. Rocourt. Genetic characterization of clones of the bacterium Listeria monocytogenes causing epidemic disease. *Proc. Natl. Acad. Sci. USA* 86:3818–3822, 1989.
37. Whittam, T.S., I.K. Wachsmuth, R.A. Wilson. Genetic evidence of clonal descent of Escherichia coli 0157:H7 associated with hemorrhagic colitis and hemolytic uremic syndrome. *J. Infect. Dis.* 157:1124–1133, 1988.
38. Whittam, T.S., M.L. Wolfe, I.K. Wachmuth, F. Orskov, I. Orskov, R.A. Wilson. Clonal relationships among Escherichia coli strains that cause hemorrhagic colitis and infantile diarrhea. *Infect. Immun.* 61:1619–1629, 1993.
39. Spratt, B.G. Multilocus sequence typing: molecular typing of bacterial pathogens in an era of rapid DNA sequencing and the internet. *Curr. Opin. Microbiol.* 2:312–316, 1999.
40. Klevytska, A.M., L.B. Price, J.M. Schupp, P.L. Worsham, J. Wong, P. Keim. Identification and characterization of varible-number tandem repeats in the Yersinia pestis genome. *J. Clin. Microbiol.* 39:3179–3185, 2001.
41. Glaser, P., L. Frangeul, C. Buchrieser, C. Rusniok, A. Amend, F. Baquero, P. Berche, H.Bloecker, P. Brandt, T. Chakraborty, A. Charbit, F. Chetouani, E. Couvé, A. d. Daruvar, P. Dehoux, E. Domann, G. Domínguez-Bernal, E. Duchaud, L. Durant, O. Dussurget, K.D. Entian, H. Fsihi, F.G.D. Portillo, P. Garrido. L. Gautier, W. Goebel, N. Gómez-López, T. Hain, J. Hauf, D. Jackson, L.M. Jones, U. Kaerst, J. Kreft, M. Kuhn, F. Kunst, G. Kurapkat, E. Madueno, A. Maitournam, J.M. Vicente, E. Ng, H. Nedjari, G. Nordsiek, S. Novella, B. d Pablos, J.C. Pérez-Diaz, R. Purcell, B. Remmel, M. Rose, T. Schlueter, N. Simoes, A. Tierrez, J.A. Vásquez-Boland, H. Voss, J. Wehland, P. Cossart. Comparative genomics of *Listeria* species. *Science* 294:849–852, 2001.
42. McClelland, M., K.E. Sanderson, J. Spieth, S.W. Clifton, P. Latreille, L. Courtney, S. Porwollik, J. Ali, M. Dante, F. Du, S. Hou, D. Layman, S. Leonard, C. Nguyen, K. Scott, A. Hollmes, N. Grewal, E. Mulvaney, E. Ryan, H. Sun, L. Florea, W. Miller, T. Stoneking, M. Nhan, R. Waterston, R.K. Wilson. Complete genome sequence of Salmonella enterica serovar Typhimurium LT2. *Nature* 413:852–856, 2001.

43. Perna, N.T., G. Plunkett III, V. Burland, B. Mau, J.D. Glasner, D.J. Rose, G.F. Mayhew, P. S. Evans, J. Gregor, H.A. Kirkpatrick, G. Posfai, J. Hackett, S. Klink, A. Boutin, Y. Shao, L. Miller, E. J. Grotbeck, N.W. Davis, A. Lim, G. Yen, D.C. Schwartz, R.A. Welch, F.R. Blattner. Genome sequence of enterohaemorrhagic Escherichia coli 0157:H7. *Nature.* 409:529–533, 2001.
44. Blattner, F.R., G. Plunkett III, C.A. Bloch, N.T. Perna, V. Burland, M. Riley, J. Collado-Vides, J.D. Glasner, C.K. Rode, G.F. Mayhew, J. Gregor, N.W. Davis, H.A. Kirkpatrick, M.A. Goeden, D.J. Rose, B. Mau, Y. shao. The complete genome sequence of Escherichia coli. *Science* 277:1453–1462, 1997.
45. Zhang, C., M. Zhang, J. Ju, J. Nietfeldt, J. Wise, P.M. Terry, M. Olson, S.D. Kachman, M. Wiedmann, M. Samadpour, A.K. Benson. Genome diversification in phylogenetic lineages I and II of Listeria monocytogenes: idendification of segments unique to lineage II populations. *J. Bacteriol.* 185:5573–5584, 2003.
46. Felsenstein, J. *Inferring Phylogeny.* Seattle, WA: Department of Genome Sciences, University of Washington, 2003.
47. Li, W., D. Graur. *Fundamentals of Molecular Evolution.* Sunderland MA: Sinauer Associates, Inc., 1991.
48. Posada, D., K.A. Crandall. MODELTEST: testing the model of DNA substitution. *Bioinformatics.* 14:817–818, 1998.
49. Bollback, J.P. Bayesian model adequacy and choice in phylogenetics. *Mol. Biol. Evol.* 19:1171–1180, 2002.
50. Swofford, D.L. *Phylogenetic Analysis Using Parsimony (PAUP*),* version 4.0. Sunderland, MA: Sinauer Associates, 1998.
51. Whelan, S., P. Lio, N. Goldman. Molecular phylogenetics: state-of-the-art methods for looking into the past. *Trends in Genetics.* 17:260–262, 2001
52. Baldauf, S.L. Phylogeny for the faint of heart: a tutorial. *Trends in Genetics* 19:345–351, 2003.
53. Sanderson, M.J., A.C. Driskell. The challenge of construction large phylogenetic trees. *Trends in Plant Sci.* 8:374–379, 2003.
54. Nei, M., S. Kumar. *Molecular Evolution and Phylogenetics.* Oxford, New York: Oxford University Press, 2000.
55. Felsenstein, J. *PHYLIP (Phylogenetic Inference Package),* ver. 3.57. Seattle, WA: Dept. of Genetics, Univ. of Washington, 1995.
56. Kumar, S., K. Tamura, I.B. Jakobsen, M. Nei. MEGA2: molecular evolutionary genetics analysis software. *Bioinformatics.* 17:1244–1245, 2001.
57. Takezaki, N. Tie trees generated by distance methods of phylogenetic reconstruction. *Mol. Biol. Evol.* 15:772–737, 1998.
58. Holder, M., P.O. Lewis, Phylogenetic estimation: traditional and Bayesian approaches. *Nature Rev. Gen.* 4:275–284, 2003.
59. Steel, M., D. Penny. Parsimony, likelihood, and the role of models in molecular phylogenetics. *Mol. Biol. Evol.* 17:839–850, 2000.
60. Takahashi, K., M. Nei. Efficiencies of fast algorithms of phylogenetic inference under the criteria of maximum parsimony, minimum evolution, and maximum likelihood when a large number of sequences are used. *Mol. Biol. Evol.* 17:1251–1258, 2000.
61. Yang, Z. Phylogenetic analysis using parsimony and the likelihood methods. *J. Mol. Evol.* 42:294–307, 1996.
62. Lio, P., N. Goldman. Models of molecular evolution and phylogeny. *Genome Res.* 8:1233–1244, 1998.
63. Kuhner, M.K., J. Felsenstein. A simulation comparison of phylogeny algorithms under equal and unequal evolutionary rates. *Mol. Biol. Evol.* 3:459–468, 1994.
64. Huelsenbeck, J. The robustness of two phylogenetic methods: four-taxon simulations reveal a slight superiority of maximum likelihood over neighbor joining. *Mol. Biol. Evol.* 12:843–849, 1995.
65. Efron, B., E. Halloran, S. Holmes. Bootstrap confidence levels for phylogenetic trees. *Proc. Natl. Acad. Sci.* 93:7085–7090, 1996.

66. Ronquist, F., J.P. Huelsenbeck. MrBayes3: Bayesian phylogenetic inference under mixed models. *Bioinformatics.* 19:1572–1574, 2003.
67. Schierup, M.H., J. Hein. Consequences of recombination on traditional phylogenetic analysis. *Genetics.* 156:879–891, 2000.
68. Feil, E.J., E.C. Holmes, D.E. Bessen, M. Chan, N.P.J. Day, M.C. Enright, R. Goldstein, D.W. Hood, A. Kalia, C.E. Moore, J. Zhou, G.G. Spratt. Recombination within natural populations of pathogenic bacteria: short-term empirical estimates and long-term phylogenetic consequences. *Proc. Natl. Acad. Sci.* 98:182–187, 2001.
69. Awadalla, P. The evolutionary genomics of pathogen recombinations. *Nat. Rev. Genet.* 4:50–60, 2003.
70. McVean, G., P. Awadalla, P. Fearnhead. A coalescent-based method for detecting and estimating recombination from gene sequences. *Genetics.* 160:1231–1241, 2002.
71. Gogarten, J.P., W.F. Doolittle, J.G. Lawrence. Prokaryotic evolution in light of gene transfer. *Mol. Biol. Evol.* 19:2226–2238, 2002.
72. Lawrence, J.G., H. Ochman. Reconciling the many faces of lateral gene transfer. *Trends Microbiol.* 10:1–4, 2002.
73. Ragan, M.A. Detection of lateral gene transfer among microbial genomes. *Curr. Opin. Genet. Dev.* 11:620–626, 2001.
74. Bandelt, H.J., A.W.M. Dress. Split decomposition: a new and useful approach to phylogenetic analysis of distance data. *Mol. Phylogenet. Evol.* 1:242–252, 1992.
75. Cunningham, C.W. Can three incongruence tests predict when data should be combined? *Mol. Biol. Evol.* 14:733–740, 1997.
76. Reid, S.D., C.J. Herbelin, A.C. Bumbaugh, R.K. Selander, T.S. Whittam. Parallel evolution of virulence in pathogenic Escherichia coli. *Nature* 406:64–67, 2000.
77. Grassly, N.C., E.C. Holmes. A likelihood method for the detection of selection and recombination using nucleotide sequences. *Mol. Biol. Evol.* 14:239–247, 1997.
78. McGuire, G., F. Wright. TOPAL 2.0: improved detection of mosaic sequences withing multiple alignments. *Bioinformatics* 16:130–134, 2000.
79. Hein, J. A heuristic method to reconstruct the history of sequences subject to recombination. *J. Mol. Evol.* 36:396–405, 1993.
80. Smith, J. Maynard, N.H. Smith. Detecting recombination from gene trees. *Mol. Biol. Evol.* 15:590–599, 1998.
81. Smith, J. Maynard. The detection and measurement of recombination from sequence data. *Genetics.* 153:1021–1027, 1999.
82. Worobey, M, A novel approach to detecting and measuring recombination: new insights into evolution in viruses, bacteria, and mitochondria. *Mol. Biol. Evol.* 18:1425–1434, 2001.
83. Fearnhead, P., P. Donnelly, Estimating recombination rates from population genetic data. *Genetics* 159:1299–1318, 2001.
84. Kuhner, M.K., J. Yamato, J. Felsenstein. Maximum likelihood estimation of recombination rates from population data. *Genetics* 156:1393–1401, 2000.
85. Husmeier, D., F. Wright. A Bayesian approach to discriminate between alternative DNA sequence segmentations. *Bioinformatics* 18:226–234, 2002.
86. McGuire, G., F. Wright, M. Prentice. A graphical method for detecting recombination in phylogenetic data sets. *Mol. Biol. Evol.* 14:1125–1131, 1997.
87. Kuhner, M.K., J. Yamato, J. Felsenstein. Maximum likelihood estimation of population growth rates based on coalescent. *Genetics* 149:429–434, 1998.
88. Beerli, P., J. Felsenstein. Maximum likelihood extimation of migration rates and effective population numbers in two populations using a coalescent approach. *Genetics* 152:763–773, 1999.
89. Hudson, R.R. Two-locus sampling distributions and their application. *Genetics* 159:1805–1817, 2001.
90. Kimura, M. Evolutionary rate at the molecular level. *Nature* 217:624–626, 1968.
91. Tajima, F. Statistical method for testing the neutral mutation hypothesis by DNA polymorphism. *Genetics* 123:585–595, 1989.

92. Hudson, R.R., M. Krietman, M. Aguade. A test of neutral molecular evolution based on nucleotide data. *Genetics* 116:153–159, 1987.
93. Ohta, T. Slightly deleterious mutant substitutions in evolution. *Nature* 246:96–98, 1973.
94. Kimura, M. *The Neutral Theory of Molecular Evolution.* Cambridge: Cambridge University Press, 1983.
95. Fisher, R.A. The distribution of gene ratios for rare mutations. *Proc. R. Soc. Edinb. Sect. B.* 50:204–219, 1930.
96. Ohta, T. The nearly neutral theory of molecular evolution. *Annu. Rev. Ecol. Syst.* 23:263–286, 1992.
97. Nielsoen, R., Z. Yang. Estimating the distribution of selection of coefficients from phylogenetic data with applications to mitochondrial and viral DNA. *Mol. Biol. Evol.* 20:1231–1239, 2003.
98. Yang, Z., R. Nielsen, Codon-substitution models for detecting molecular adaptation at individual sites along specific lineages. *Mol. Biol. Evol.* 19:908–917, 2002.
99. Yang, Z., R. Nielsen. Estimating synonymous and nonsynonymous substitution rates under realistic evolutionary models. *Mol. Biol. Evol.* 17:32–43, 2000.
100. Yang, Z., R. Nielsen. Synonymous and nonsynonymous rate variation in nuclear genes of mammals. *J. Mol. Evol.* 46:409–418, 1988.
101. Suzuki, Y., T. Gojobori. A method for detecting positive selection at single amino acid sites. *Mol. Biol. Evol.* 16:1315–1328, 1999.
102. Suzuki, Y., T. Gojobori, M. Nei. ADAPTSITE: detecting natural selection at single amino acid sites. *Bioinformatics.* 17:660–661, 2001.
103. Suzuki, Y., M. Nei. Simulation study of reliability and robustness of statistical methods for detecticn positive selection at single amino acid sites. *Mol. Bio. Evol.* 19:1865–1869, 2002.
104. Yang, Z. *PAML (Phylogenetic Analysis by Maximum Likelihood) Version 3.0.* London: University College, 2000.
105. Nielsen, R., Z. Yang. Likelihood models for detecting positively selected amino acid sited and applications to the HIV-1 envelope gene. *Genetics* 148:929–936, 1998.
106. Yang, Z., R. Nielsen, N. Goldman. A.M.K. Pedersen. Codon=substitution modes of heterotgeneous selections pressure at amino acid sites. *Genetics* 155:431–449, 2000.
107. Anisimova, M., R. Nielsen, Z. Yang. Effect of recombination on the accuracy of the likelihood method for detecting positive selection at amino acid sites. *Genetics* 164:1229–1236, 2003.
108. Yang, Z., W.J. Swanson. Codon substitution models to detect adaptive evolution that account for heterogeneous selective pressure among site classes. *Mol. Biol. Evol.* 19:49–57, 2002.
109. Messier, W., C.B. Stewart. Episodic evolution of primate lysozymes. *Nature* 385:151–154, 1997.
110. Yang. Z. Likelihood ratio tests for detecting positive selection and application to primate lysozyme evolution. *Mol. Bio. Evol.* 15:568–563, 1998.
111. Lawrence, J.G., H. Ochman. Molecular archaeology of the Escherichia coli genome. *Proc. Natl. Acad. Sci. USA* 95:9413–9417, 1998.
112. Schaechter, M. and The View From Here Group. Escherichia coli and Salmonell 2000: the view from here. *Microbiol. Mol. Biol. Rev. Rev.* 65:119–130, 2001.
113. Donneberg, M.S., T.S. Whittam. Pathogenesis and evolution of virulence in enteropathogenic and enterohemorrhagic Escherichia coli. *J. Clin. Invest.* 107:539–548, 2001.
114. Jay, J.M. *Modern Food Microbiology,* 5th ed., Gaithersburg, MD: Aspen Publishers, 1998, pp 527–543.
115. Shaikh, N., P.I. Tarr. Escherichia coli 0157:H7 shiga toxin-encoding bacteriophages: integrations, excisions, truncations, and evolutionary implications. *J. Bact.* 185:3596–3605, 2003.
116. McGraw, E.A., J. Li, R.K. Selander, T.S. Whittam. Molecular evolution and mosaic structure of a,b, and y intimins of pathogenic Escherichia coli. *IMol. Biol. Evol.* 16:12–22, 1999.
117. Tarr, C., T.S. Whittam. Molecular evolution of the intimin gene in 0111 clones of pathogenic Escherichia coli. *J. Bacteriol.* 184:479–487, 2002.

118. Feng, P., K.A. Lampel, H. Karch, T.S. Whittam. Genotypic and phenotypic changes in the emergence of Escherichia coli 0157:H7. *J. Infect. Dis.* 177:1750–1753, 1998.
119. Salyers, A.A., D.D. Whitt. *Bacterial Pathogenesis: A Molecular Approach,* 2nd ed. Washington, DC: ASM Press, 2002, pp 381–421.
120. Clarke, S.C., R.D. Haigh, P.P.E. Freestone, P.H. Williams. Virulence of enteropathogenic Escherichia coli, a global pathogen. *Clin. Micro. Rev.* 16:365–378, 2003.
121. Nataro, J.P., J. B. Kaper. Diarrheagenic Escherichia coli. *Clin. Microbiol. Rev.* 11:142–201, 1998.
122. Porwollik, S., M. McClelland. Lateral gene transfer in Salmonella. *Microbes Infect.* 5:977–989, 2003.
123. Cotter, P.A., V.J. DiRita, Bacterial virulence gene regulation: an evolutionary perspective. *Annu. Rev. Microbiol.* 54:519–565, 2000.
124. Selander, R.K., J. Li, K. Nelson. In: *Escherichia coli and Salmonella: Cellular and Molecular Biology,* Neidhardt, F.C., R. Curtiss III, J.L. Ingraham, E.C.C. Lin, K.B. Low, B. Magasanik, W.S. Reznikoff, M. Riley, M. Schaechter, H.E. Umbarger, eds., Washington, DC: ASM Press, 1996, pp 2691–2707.
125. Mead, P.S., L. Slutsker, V. Dietz, L.F. McCaig, J.S. Bresee, C. Shapiro, P.M. Griffin, R.V. Tauxe. Food-related illness and death in the United States. *Emer. Infect. Dis.* 5:607–625, 1999.
126. Rabsch, W., H. Tschäpe, A.J. Bäumler. Non-typhoidal salmonellosis: emerging problems. *Microbes Infect.* 3:237–247, 2001.
127. Rabsch, W., B.M. Hargis, R.M. Tsolis, R.A. Kingsley, K. Hinz, H. Tschäpe, A.J. Bäumler. Competitive exclusion of *Salmonella Enteritidis* by *Salmonella Gallinarum* in poultry. *Emerg. Infect. Dis.* 6:443–449, 2000.
128. Groisman, E., A. Blanc-Potard, K. Uchiya. Pathogenicity islands and the evolution of Salmonella virulence. In: *Pathogenicity Islands and Other Mobile Genetic Elements,* Kaper, J., J. Hacker, eds., Washington, DC: ASM Press, 1999, pp 127–150.
129. Hacker, J., G. Blum-Oehler, I. Mühldorfer, H. Tschäpe. Pathogenicity islands of virulent bacteria: structure, function and impact on microbial evolution. *Mol. Microbiol.* 23:1089–1097, 1997.
130. Bäumler, A.J., R.M. Tsolis, T.A. Ficht, L.G. Adams. Evolution of host adaptation in Sallmonella enterica. *Infect. Immun.* 66:4579–4587, 1998.
131. Groisman, E., H. Ochman. How Salmonella became a pathogen. *Trends Microbiol.* 5:343–349, 1997.
132. Marcus, S.L., J.H. Brumell, C.G. Pfeifer, B.B. Finlay. Salmonella pathogenicity islands: big virulence in small packages. *Microbes Infect.* 2:145–156, 2000.
133. Bäumler, A.J., A.J. Gilde, R.M. Tsolis, A.W.M. van der Velden, B.M.M. Ahmer, F. Heffron. Contribution of horizontal gene transfer and deletion events to development of distinctive patterns of fimbrial operons during evolution of Salmonella serotypes. *J. Bact.* 179:317–322, 1997.
134. Lindstedt, B., E. Heir, E. Gjernes, G. Kapperud. DNA fingerprinting of Salmonella enterica subsp.enterica serovar Typhimurium with emphasis on phage type DT104 based on varible number of tandem repeat loci. *J. Clin. Microbiol.* 41:1469–1479, 2003.
135. Parkhill, J., G. Dougan, K.D. James, N.R. Thompson, D. Pickard, J. Wain, C. Churcher, K.L. Mungall, S.D. Bentley, M.T. Holden, M. Sebaihia, S. Baker, D. Basham, K. Brooks, T. Chillingworth, P. Connerton, A. Cronin, P. Davis, R.M. Davis, R.M. Davies, L. Dowd, N. White, J. Farrar, T. Feltwess, N. Hamlin, A. Haque, T.T. Hien, S. Holroyd, K. Jagels, A. Krogh, T.S. Larsen, S. Leather, S. Moule, P. O'Gaora, C. Parry, M. Quail, K. Rutherford, M.Simmonds, J. Skelton, K. Stevens, S. Whiteheat, B.G. Barrell. Complete genome sequence of a multiple drug resistant Salmonella enterica serovar Typhi CT18. *Nature* 413:848–852, 2001.
136. Kreft, J., J. Vasquez-Boland, E. Ng, W. Goebel. Virulence gene clusters and putative pathogenicity islands in Listeriae. In: *Pathogenicity Islands and Other Mobile Virulence Elements*, Kaper, J., J. Hacker, eds., Washington, DC: ASM Press, 1999, pp 219–232.

137. Hurd, S., Q. Phan, J. Hadler, B. Mackenzie, S. Lance-Parker, P. Blake, M. Deasy, J. Rankin, D. Frye, I. Lee, B. Werner, D. Vugia, S. Bidol, G. Stoltman, M. Boulton, M. Wiedmann, L. Kornstein, S. Reddy, B. Mojica, F. Guido, A. Huang, C. Vincent, A. Bugenhagen, J. Corby, E. Carloni, J. Corby, E. Carloni, M. Holcomb, S. Kondracki, R. Woron, S. Zansky, P. Smith, G. Dowdle, C. Nichols, F. Smith, D. Gerber, T. Jones, W. Moore, S. Ahrabi-Fard, J. Davis. Multistate outbreak of Listeriosis—United States, 2000. *MMWR Weekly* 49:1129–1130, 2000.
138. Schwartz, B., D. Hexter, C.V. Broome, A.W. Hightower, R.B. Hirschorn, J.D. Porter, P.S. Hayes, W.F. Bibb, B. Lorger, D.G Faris. Investigation of an outbreak of listeriosis: new hypotheses for the etiology of epidemic Listeria monocytogenes infections. *J. Infect. Dis.* 159:680–685, 1989.
139. McLauchlin, J. Distribution of serovars of Listeria monocytogenes isolated from different categories of patients with listeriosis. *Eur. J. Clin. Microbiol. Infect. Dis.* 9:210–213, 1990.
140. Wiedmann, M., J.L. Bruce, C. Keating, A.E. Johnson, P.L. McDonough, C.A. Batt. Ribotypes and virulence gene polymorphisms suggest three distinct Listeria monocytogenes lineages with differences in their pathogenic potential. *Infect. Immun.* 65:2707–2716, 1997.
141. Cai, S. *DNA Sequence-based Subtyping and Molecular Evolution of Listeria and Listeria Monocytogenes.* MS thesis, Cornell University, Ithaca, NY, 2002.
142. Nadon, C.A., D.L. Woodward, C. Young, F.G. Rodgers, M. Wiedmann. Correlations between molecular subtyping and serotyping of Listeria monocytogenes. *J. Clin. Microbiol.* 39:2704–2707, 2001.
143. Gombas, D.E., Y. Chen, R.S. Clavero, V.N. Scott. Survey of Listeria monocytogenes in ready-to-eat foods. *J. Food Prot.* 66:559–569, 2003.
144. Gray, M.J., R.N. Zadoks, M.R. Roma, E.D. Fortes, B. Dogan, S. Cai, Y. Chen, V.N. Scott, D.E. Gombas, K.J. Boor, M. Wiedmann. *Food and Human Isolates of Listeria Monocytogenes Form Distinct but Overlapping Populations.* Cornell University, Ithaca, NY, 2003.
145. Jeffers, G.T., J.L. Bruce, P. McDonough, J. Scarlett, K.J. Boor, M. Wiedmann. Comparative genetic characterization of Listeria monocytogenes isolates from human and animal listeriosis cases. *Microbiol.* 147:1095–1104, 2001.
146. Norton, D.M., J.M. Scarlett, K. Horton, D. Sue, J. Thimothe, K.J. Boor, M. Wiedmann. Characterization and pathogenic potential of Listeria monocytogenes isolates from the smoked fish industry. *Appl. Environ. Microbiol.* 67:646–653, 2001

3.02

Genetics and Physiology of Pathogenicity in Food Borne Bacterial Pathogens

Michael Gray and Kathryn J. Boor

CONTENTS

2.1 INTRODUCTION

Food borne diseases are estimated to be responsible for 76 million illnesses, 325,000 hospitalizations, and 5,000 deaths annually in the United States. The pathogens responsible for these illnesses include viruses, eukaryotic parasites, and bacteria. Bacteria are responsible for more than 70% of the deaths attributed to known food borne pathogens (1). Clearly, understanding these bacteria and controlling their presence in foods is a critical element of food safety.

At a physiological level, bacterial pathogens produce characteristic "virulence factors", which are proteins or protein complexes which contribute to the ability of an organism to cause disease in a host. Virulence factors may affect a host directly (e.g., toxins can disable or kill host cells through various mechanisms) or indirectly (e.g., flagella, which are comprised of a long filament and a set of spinning motor proteins, allow bacteria to efficiently move toward and colonize mucosal surfaces). For a pathogen to cause disease, virulence factors need to be produced and active at appropriate times and locations within the host organism. At the genetic level, virulence factors, like all other proteins, are encoded by genes on the bacterial DNA. Regulation of the expression of these genes constitutes the genetics of pathogenicity. In order to fully understand how a given organism causes disease, both the physiology and genetics of its pathogenesis must be deciphered.

Food borne bacterial diseases can be classified into two broad categories based on the virulence factors which are most important in causing the characteristic symptoms of the disease. Intoxications are illnesses with symptoms resulting from the effects of a toxin or toxins. Typically, the toxin alone can cause disease symptoms, even in the absence of live bacteria. Infections, on the other hand, are illnesses with symptoms resulting from growth of the bacteria inside the host. Most food borne pathogens primarily infect the gastrointestinal tract and therefore cause diarrheal symptoms, but some food borne pathogens can penetrate further into the host and colonize other organs. These invasive infections typically have more serious consequences for the host than gastrointestinal infections. In general, the virulence factors necessary for establishing an infection are more varied and complex than the factors involved in intoxications.

The study of the physiology and genetics of virulence has yielded many important insights into the mechanisms by which bacterial pathogens cause food borne illness. In order to obtain an overview of these insights, we will examine three examples of important food borne pathogens and the diseases they cause: *Clostridium botulinum*, which causes botulism, a food borne intoxication, *Escherichia coli* O157:H7, which causes a noninvasive infection, and *Listeria monocytogenes*, which causes listeriosis, an invasive infection. Each of these organisms poses serious human health concerns, and an understanding of how they cause disease is important in efforts to control them as well as to provide insight into the general mechanisms of pathogenesis in other bacteria.

2.2 *CLOSTRIDIUM BOTULINUM*

2.2.1 Introduction and History

Botulism is one of the oldest and deadliest of food borne diseases. One of the earliest reliably recorded outbreaks of what was then known as "sausage poisoning" occurred in Wildbad, Germany in 1793. Thirteen people became sick and six died from eating spoiled blood sausage, but the true cause of the disease was unknown at that time. It was not until a hundred years later, in 1897, that Dutch bacteriologist Emile Pierre Marie van Ermengem was able to isolate a pure culture of a bacterium he called *Bacillus botulinus* (after *botulus,* the Latin word for sausage) and demonstrate the essential facts about the organism responsible for food borne botulism (2).

2.2.1.1 Characteristics of Botulism

Food borne botulism is an intoxication. Ingestion of live bacteria is unnecessary for disease to develop; all that is required is consumption of a toxin produced by the bacteria. This toxin is among the most poisonous substances known, and as little as 0.1 to 1 μg may be lethal for a human (3), although some other species, such as dogs and chickens, are resistant to its effects (2). The botulism toxin causes paralysis by blocking neurotransmitter release. Paralysis generally starts with the nerves in the face and head, works its way downward, and is fatal if it reaches the heart or lungs (2). Botulism toxin is heat labile, so cooking food thoroughly will inactivate the toxin. Heating to 85°C for 5 minutes is enough to inactivate the most thermostable botulism toxin tested (4).

Before 1950, botulism had a very high mortality rate (approximately 60%), but improvements in supportive care and the availability of antitoxin have reduced the case fatality ratio to 15.5% (5). A vaccine, made from inactivated toxin, is available, but not commonly given to humans. This is mostly due to the expense and risk associated with production of large quantities of botulism toxin. Alternative methods for production of a botulism vaccine are being explored, but none have yet been fully developed (6).

On average, there are 23 reported cases of food borne botulism in the U.S. per year, mostly associated with consumption of improperly home canned or pickled foods, with approximately one death per year. *Clostridium botulinum* can also cause occasional wound infections, particularly associated with the abuse of injected drugs, and infant botulism, which results from the colonization of an infant's intestinal tract by *C. botulinum*, leading to production of toxin inside the child's intestine. Because commercial food processing is very effective at preventing food borne botulism, infant botulism is now the most common form of botulism in the U.S. (5).

2.2.1.2 Characteristics of Clostridium botulinum

Clostridium botulinum is a Gram-positive obligate anaerobic spore-forming bacterium (7). Gram-positive bacteria have a single cytoplasmic membrane with a thick peptidoglycan layer. As an obligate anaerobe, growth of *C. botulinum* is inhibited by oxygen. Spores are a stress tolerant resting state formed by *C. botulinum* and some other Gram-positive bacteria. The spores of *C. botulinum* are resistant to heat: boiling at 100°C for half an hour is only sufficient to reduce their numbers by 90% (a one log reduction, or one D value). Canning processes are designed to produce a 12D reduction in numbers of *C. botulinum* spores at temperatures of 121°C. This 12D process ensures a very low risk of *C. botulinum* survival and outgrowth in the anaerobic interior of a given can. Under most conditions, *C. botulinum* will not produce toxin at a pH below 4.6, so high acid foods (for example, many tomato products) require less stringent heat treatment than foods with a higher pH (2).

Historically, any Gram-positive, spore-forming anaerobe which produces botulism toxin has been classified as *C. botulinum* (3). Modern taxonomy methods, involving comparison of conserved DNA sequences, including the sequences encoding 16S rRNA, have shown that this group of bacteria is extremely diverse. Some researchers have proposed reclassifying *C. botulinum* into as many as 20 different genera and several families (8,9).

2.2.1.3 Classification of C. botulinum

Clostridium botulinum was first classified into different categories according to toxin type. In 1904, only seven years after Van Ermengem's isolation of *C. botulinum*, G. Landmann reported that the organism responsible for a botulism outbreak in Darmstadt, Germany produced a toxin with similar activity to Van Ermengem's strain, but antitoxin prepared against one would not cross protect against the other. These two types were later designated type A and type B. Over the course of the next 70 years, five more toxin types were discovered and given alphabetical designations (C–G) (2). Toxin types A, B, and E are associated with human disease, while types C and D are associated with animal disease (10).

Clostridium botulinum strains are also classified according to their physiological properties into four major Groups. Group I strains produce proteolytic enzymes and their spores are the most heat and acid tolerant. They produce toxins of type A, B, or F. Group II strains are nonproteolytic and much less heat tolerant than strains in the other groups, but they can grow at lower temperatures. For example, some Group II strains can grow at refrigeration temperatures (approximately 4°C). They produce toxins of type B, E, or F. Group III and IV strains are not associated with human disease, so are less well studied, but group III strains, associated with animal disease, are nonproteolytic and produce toxin types C or D. Group IV strains, which produce toxin type G, are not known to be associated with disease in either humans or animals (11).

2.2.1.4 Sources in the Environment

Clostridium botulinum is found in soils and sediments world wide, but different types are commonly found in different types of environments, which affects their prevalence in foods and the patterns of disease they cause. For example, strains producing type A and B toxins are both commonly found in relatively dry soils. In the U.S., type A strains are most common in soil west of the Rocky Mountains, with type B strains more common in the eastern U.S. This parallels the pattern of disease-causing strains in these regions. Similarly, *C. botulinum* in marine sediments commonly produce type E toxin, and botulism outbreaks caused by seafood are most frequently caused by type E strains (2).

2.2.2 Virulence Factors

2.2.2.1 The Botulism Toxin

The most important virulence factor of *C. botulinum* is the botulism toxin. As mentioned, this toxin is one of the deadliest poisons known (3). Because botulism is an intoxication, other properties of the organism are less important in the course of the disease. In fact, some strains of *C. botulinum* are phenotypically indistinguishable from other species of *Clostridium* (such as *C. novyi* or *C. sporogenes*) except for their ability to produce the botulism toxin (7).

Botulism toxin is released from *C. botulinum* by cell autolysis late in the growth cycle (3). The toxin is a zinc endopeptidase, a proteolytic enzyme which requires zinc and specifically cleaves synaptic proteins (VAMP, SNAP, or syntaxin, depending on the toxin type) in mammalian neurons, inhibiting release of the neurotransmitter acetylcholine at motor neuron synapses (10). As little as one toxin molecule can totally destroy function of a neuron (6). The result of this activity is paralysis, as the muscles controlled by the affected neurons become unable to contract. This is termed flaccid paralysis and may be fatal if the lungs or heart become paralyzed (10). As intensive care technology, especially artificial respiration, has advanced, the prognosis for victims of botulism poisoning has improved, although survivors are likely to have lingering neurological symptoms (5).

2.2.2.2 Toxin Structure

Botulism toxins are all similar in molecular structure. They are transcribed as a single polypeptide with a molecular weight of about 150,000 Daltons. This polypeptide has relatively low toxicity. Active toxin is produced by proteolytic cleavage of the polypeptide into a heavy (~100 kDa) and a light (~50 kDa) chain, linked by a single disulfide bond (11). Strains of *C. botulinum* which produce their own extracellular proteases (for example, the Group I strains) are more toxic than nonproteolytic strains, even if similar amounts of toxin protein are produced (2).

Botulism toxin belongs to a class of toxins called AB toxins. The light chain (or A subunit) is the active part of the toxin, and harbors the zinc endopeptidase activity. The heavy chain (or B subunit) is required for the light chain to cross the cell membrane into the host cell and is responsible for specific binding to neurons (10). A number of accessory proteins are also found bound to the botulism toxin protein to form large toxin complexes. These include HA (hemagglutinin) proteins and the large (~150 kDa) NTNH (nontoxic, nonhemolytic) protein. Depending on the specific strain, toxin type, and growth conditions of the culture, the final toxin complex may range in size from ~300 to 900 kDa (3). The accessory proteins are not required for toxicity, but seem to be involved in protection of the neurotoxin from proteolytic attack in the host environment (10).

2.2.3 Sporulation

Without toxin production, *C. botulinum* is avirulent, but the ability of *C. botulinum* to form heat resistant spores can also be considered an important virulence factor. Many harmless bacteria form spores, but in *C. botulinum* the ability to sporulate is important in how the organism causes disease. In order for botulism to occur, four conditions must be met:

1. *Clostridium botulinum* must be present in the food.
2. The bacteria must survive processing. Spores are critical in this stage, as they can survive conditions which kill normal bacterial cells.
3. Conditions must enable germination and outgrowth of spores, along with the associated production of toxin, i.e., anaerobic conditions within the food, a pH above about 4.6, high enough water activity to support growth ($a_w > 0.94$), and a temperature which allows the bacteria to reproduce (20°C − 45°C) (2).
4. The intact toxin must enter the body. If contaminated food is cooked thoroughly before consumption, the toxin will be destroyed (2).

One of the reasons botulism is so dangerous is because the spores can survive common cooking and preservative methods (condition 2). For example, a recent outbreak of botulism in Thailand resulted from consumption of bamboo shoots which had been canned and boiled for an hour. They were then stored at ambient temperature for 3–6 months (12). Without spores able to survive boiling, as well as storage conditions which encouraged the germination and growth of the spores, *C. botulinum* would not have been able to produce toxin in the food and cause disease.

Sporulation of *C. botulinum* begins in late exponential phase of growth (13). Spores develop at one end of the cell and have a complex structure. At the center of the spore is the protoplast which contains the tightly packed DNA. This is surrounded by several protective layers including the cytoplasmic membrane, the cortex (composed of peptidoglycan and diaminopimelic acid), and the inner and outer spore coats (which have a keratin-like structure and are rich in disulfide bonds). The spores are resistant to a number of stresses including heat, drying, radiation, alcohol, and acid (2).

2.2.4 Genetics and Regulation

2.2.4.1 Toxin Genes

The genes encoding botulism toxin and its various accessory proteins are found grouped together in the botulism toxin gene cluster (Figure 2.1). The exact genes in the cluster vary

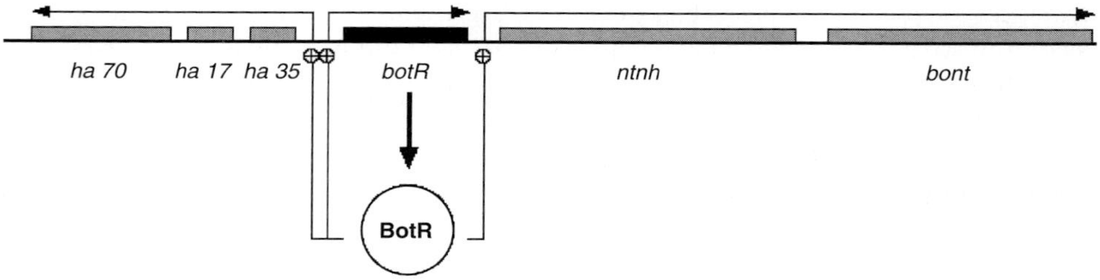

Figure 2.1 Regulation of virulence genes in *Clostridium botulinum*. The botulism toxin gene cluster shown is typical of a toxin type A strain. Gene products are as follows: *ha70*, *ha17*, and *ha35* = hemaglutinin proteins, *botR* = BotR regulatory protein, *ntnh* = nontoxic nonhemolytic protein, *bont* = botulism neurotoxin. BotR is a positive regulator at each of the three promoters indicated by +

among strains of different toxin types, but the toxin gene (*bont*) is always found associated with a gene immediately upstream which codes for nontoxic nonhemolytic protein (NTNH). The island also commonly contains genes for hemagglutinin proteins (*ha17*, *ha35*, and *ha70*) and a regulatory gene (*botR*) which will be discussed in more detail in what follows. Some toxin gene clusters also include other genes which encode proteins of unknown functions (3).

All of the different toxin types have had representative toxin gene clusters fully sequenced, and the degree of relatedness among them has been determined by DNA sequence homology. All of the botulism toxins have significant regions of similarity, suggesting that they are closely related to one another (11). The tetanus toxin of *C. tetani* is also related to the botulism toxins and has a similar structure and mode of action. The tetanus toxin, however, has different specificity and results in a spastic paralysis, in which the muscles affected contract and cannot relax, rather than the flaccid paralysis characteristic of botulism (10).

The genes for the botulism toxin are known to be located on pseudolysogenic bacteriophages for toxin types C and D and on large plasmids for toxin type G. A pseudolysogenic bacteriophage is a viral genome, integrated into the bacterial chromosome, which has lost some of the genes necessary to produce active phage particles. Plasmids are loops of DNA separate from the chromosome in the bacterial cytoplasm. In the absence of evidence for location on a mobile or extrachromosomal element, the genes for the other toxin types (A, B, E, and F) are assumed to be chromosomal. This has been shown to be the case for at least one type A strain of *C. botulinum* (3). The presence of some toxin genes on extrachromosomal elements suggests that it may be possible for these genes to be horizontally transferred from one bacterium to another, conveying the ability to produce botulism toxin to a previously nonpathogenic strain. In fact, strains of *C. butyricum* have been found which produce a type E neurotoxin which is 97% identical to that found in type E *C. botulinum* (11).

2.2.4.2 *Regulation of Toxin Production*

The genetic regulation of toxin production is still poorly understood, but several important points are known. The botulism toxin gene cluster in most toxin types contains a gene for a regulatory protein called BotR. BotR is required for toxin expression (at least in type A strains), has homology to known transcriptional regulators, such as TetR from *C. tetani* and UviA from *C. perfringens*, and has structural characteristics typical of this type of regulatory protein, including a DNA binding helix-turn-helix motif. BotR has been shown to bind to the promoter regions of NTNH and *ha35*, and overexpression of BotR results in overproduction of toxin and accessory proteins. BotR is therefore a positive regulator of toxin production. Some evidence has been found suggesting that there may also be negative regulation of toxin production. This might involve a regulatory protein or proteins which decrease expression of toxin genes, but the genes encoding this putative regulator or regulators have not yet been found. It is also unknown what upstream factors might influence the expression of BotR or botulism toxin (3).

2.2.4.3 Sporulation Genes

The regulation of sporulation is slightly better understood than the regulation of toxin production. While relatively few studies have directly examined the sporulation genes in any species of *Clostridium*, much less those of *C. botulinum*, it is clear from what has been observed that there is a fundamental similarity between the genes for sporulation in *Clostridium* and those found in the well studied genus *Bacillus* (13). In *Bacillus*, sporulation involves a complex regulatory network including a multi component phosphorelay chain and multiple alternative sigma factors. The phosphorelay is a signaling system which detects various stimuli and transmits a signal through the cell by transferring phosphate groups from one protein to another, ultimately activating transcriptional regulators. Sigma factors (symbolized by the Greek letter σ) confer specificity to RNA polymerase, and allow the bacterium to regulate transcription of large numbers of genes simultaneously (14).

Homologues of many of the sporulation specific sigma factors from *Bacillus* have been found in various *Clostridium* species. Genes encoding clostridial versions of the sigma factors σ^E, σ^G, and σ^K have all been found in *C. acetylbutylicum*, and the organization of these sporulation genes is very similar to that in *Bacillus*. Many of the phosphorelay genes also have homologues in *Clostridium*, including *spo0A*, *spoIIGA*, and *spoIID*. In addition, the promoter sequences for the clostridial sporulation genes match the expected sequences based on what is known from *Bacillus*. This suggests that the sporulation specific sigma factors also regulate genes involved in sporulation in *Clostridium* (13).

There are, however, distinct differences known in the regulation of sporulation between *Clostridium* and *Bacillus*. Sporulation is initiated by *Bacillus* species in response to starvation conditions, while *Clostridium* species cannot sporulate without an available carbon source. Sporulation in *Clostridium* is triggered by the cessation of growth or by exposure to oxygen. The mechanism and physiological rationale behind this difference are unknown (15), but it demonstrates that knowledge of sporulation in *Bacillus* is not sufficient to fully understand how sporulation is regulated in *Clostridium* species.

2.2.5 Summary

C. botulinum is the cause of botulism, a food borne intoxication. The botulism toxin (encoded by the *bont* gene) is a potent neurotoxin which can cause fatal paralysis. The toxin gene is located on a virulence gene cluster along with genes for several accessory proteins (NTNH, *ha17*, *ha35*, and *ha70*) and a transcriptional regulator called BotR (encoded by *botR*) which is a positive regulator of toxin production. *Clostridium botulinum* produces heat tolerant spores which contribute to pathogenicity by allowing the bacteria to survive some food processing methods. Regulation of sporulation in *C. botulinum* has some parallels with sporulation in *Bacillus*. Modern food processing techniques are very effective at preventing botulism, and this disease is now very rare in developed countries.

2.3 *ESCHERICHIA COLI* O157:H7

2.3.1 Introduction and History

In contrast to the long history of *C. botulinum*, *Escherichia coli* O157:H7 (classified as a strain of EHEC, enterohemorrhagic *E. coli*) has emerged quite recently as an important cause of food borne illness. The first outbreak of *E. coli* O157:H7 infection was described in 1982. Forty-seven people in two states were sick with severe abdominal pain and bloody diarrhea. The Centers for Disease Control and Prevention associated the disease with consumption of hamburgers at a restaurant chain by isolating and identifying *E. coli* O157:H7 from both human clinical cases and from uncooked ground beef patties. Prior to this outbreak, isolation of *E. coli* O157:H7 had been extremely rare (16–18). Pathogens which appear to be new or rapidly increasing in incidence are considered to be "emerging pathogens."

2.3.1.1 Characteristics of Disease

Since 1982, *E. coli* O157:H7 has been identified as the cause of increasing numbers of food borne illness (18). As of 1999, it was estimated to be responsible for approximately 73,000 illnesses and 61 deaths each year in the United States (1). *Escherichia coli* O157:H7 infection has a lower mortality rate than botulism (approximately 1% of cases), but this is still much deadlier than most food borne illnesses. Two to seven percent of infections result in a complication called hemolytic uremic syndrome (HUS). HUS occurs primarily in children under five and in the elderly, and can result in acute kidney failure. People who develop HUS may also suffer various long term complications including abnormal kidney function, high blood pressure, or seizures (17–19). Another rare but dangerous complication of *E. coli* O157:H7 infection is thrombotic thrombocytopenic purpura (TTP), which occurs mostly in adults and results in hemolysis, renal failure, neurological problems, and blood clots in the brain (18,19). TTP has been associated with a deficiency (either genetic or acquired) in an enzyme called von Willebrand factor-cleaving metalloprotease (20).

The majority of *E. coli* O157:H7 outbreaks have been associated with ground beef, but outbreaks have been linked to foods as diverse as Romaine lettuce, milk, apple cider, mayonnaise, coleslaw, and contaminated drinking water (18,21). Some of these foods (e.g., cider, mayonnaise) are unusual vectors for food borne disease due to their high acidity, but *E. coli* O157:H7 may have unusually high acid tolerance.

A major difference between *E. coli* O157:H7 and *C. botulinum* is that *E. coli* O157:H7 causes an infection, not an intoxication. In order for disease to develop, live *E. coli* O157:H7 cells need to be ingested and survive to colonize the host's large intestine and colon. *Escherichia coli* O157:H7 does not seem to penetrate the epithelial cells of the intestine in large numbers, so is considered generally noninvasive (17,22). The infective dose, or number of cells which must be ingested in order for the disease to develop, is remarkably low. Various studies have yielded estimates from 20 to 700 cells (17,18,23). In contrast, related strains of *E. coli* which cause milder forms of diarrheal disease (EPEC, enteropathogenic *E. coli*) require $> 10^8$ cells for an infection to become established (23,24).

2.3.1.2 Characteristics of E. coli O157:H7

Escherichia coli O157:H7, like other *E. coli* strains, is a Gram-negative, facultatively anaerobic bacterium (7). Gram-negative bacteria have a double cell membrane with a thin peptidoglycan layer between the two membranes. A facultative anaerobe can grow either in the presence or absence of oxygen (25). The designation O157:H7 is an example of a serotype. Serotyping allows differentiation of strains of bacteria based on the immunological properties of their surface proteins. For *E. coli*, there are three commonly typed surface antigens. The O antigen is the lipopolysaccharide (LPS) of the outer membrane, found in all Gram-negative bacteria. The H antigen is flagellar and the K antigen is used for strains which produce a polysaccharide capsule. Each separate antigen is numbered as new types are distinguished (18,26). In addition to serotype O157:H7, a number of other EHEC serotypes, including O111:H-, O26, and O103 have been associated with a similar diarrheal disease (17,19,23).

A relative of *E. coli* O157:H7, the nonpathogenic lab strain *E. coli* K-12, is an extensively studied and well understood model organism. Comparisons between *E. coli* K-12 and *E. coli* O157:H7 have been informative in determining what factors allow *E. coli* O157:H7 to cause severe disease (27). There are metabolic differences between the two strains, notably *E. coli* O157:H7's inability to ferment the sugar sorbitol or produce the enzyme β-glucuronidase, which have been useful for laboratory differentiation of *E. coli* O157:H7 from other *E. coli* strains (18,19,27).

The complete genomic DNA sequences of examples of both *E. coli* K-12 and *E. coli* O157:H7 have recently been determined (27). The two strains share 4.1 megabases (Mb) of very similar sequence, but large sections of each genome are unique to one strain or the other. *Escherichia coli* O157:H7 has 1387 genes (1.34 Mb of DNA) not found in *E. coli* K-12, while *E. coli* K-12 has 528 genes (0.53 Mb) not found in *E. coli* O157:H7 (27). The genes unique to *E. coli* O157:H7 have been examined to determine which of them are likely to contribute to virulence. The unique sequences include several bacteriophages integrated into the chromosome and clusters of genes encoding potential virulence factors such as secretion systems and adhesins. Among these the toxin-carrying bacteriophage Stx1 and Stx2 and a large pathogenicity island called the LEE (locus of enterocyte effacement) have been shown to make significant contributions to virulence. Many other potential virulence factors are as yet relatively uncharacterized (27).

2.3.1.3 Relatives of E. coli *O157:H7*

Escherichia coli belongs to a group of Gram-negative organisms known as enteric organisms, which are common inhabitants of the gastrointestinal tracts of animals, including humans, and include other species of food borne pathogens, including *Salmonella* and *Shigella* (7). The most important toxin produced by *E. coli* O157:H7 is called Shiga-like toxin, which is almost identical to the toxin produced by *Shigella dysenteriae*, the bacterium which causes dysentery (17,28). *Escherichia coli* which produce Shiga-like toxins (including *E. coli* O157:H7) are often grouped together as Shiga toxin producing *E. coli* (STEC) (17,23) or verocytotoxigenic *E. coli* (VTEC) (26). Not all types of *E. coli* which can produce Shiga-like toxin are implicated in causing human disease (17,23). The presence of other virulence factors, such as adhesins, secretion systems, or other toxins, appears to be necessary for efficient pathogenesis.

Strains of *E. coli* which do not produce Shiga-like toxins include harmless symbiotic bacteria as well as various pathogenic types. Examples of *E. coli* types which cause diarrheal disease are enteropathogenic *E. coli* (EPEC), enteroaggregative *E. coli* (EAggEC), enteroinvasive *E. coli* (EIEC), and enterotoxigenic *E. coli* (ETEC) (29). The diarrhea caused by these strains is more common [responsible for an estimated 160,000 illnesses per year in the U.S. (1)] and typically milder than that caused by *E. coli* O157:H7, although it is a significant cause of infant death in the developing world (25). Some of the virulence associated secretion systems and adherence factors found in these strains (particularly EPEC) are also found in *E. coli* O157:H7 (17,23).

2.3.1.4 Sources in the Environment

Escherichia coli are common commensal organisms found in the intestines of animals. Commonly, STEC are found among the colonic microflora of cattle (17,18). Cattle and other livestock are believed to be important reservoirs of *E. coli* O157:H7 and the presence of the bacteria in food is often associated with contamination by cattle feces. Ground beef is particularly vulnerable to this form of contamination during slaughter and processing, and more outbreaks are associated with foods of bovine origin than with any other source (17,18), but several outbreaks have been traced to the use of improperly treated manure as fertilizer for fruits or vegetables (19,30). About 4% of cattle herds in North America and the UK are positive for *E. coli* O157:H7 at any given time (17,18), but infection of any individual cow appears to be a very transient event (19,31). Cattle do not appear to naturally develop symptoms of STEC infection. The infectious dose in the laboratory for healthy adult cattle is extremely high ($>10^7$ cells), and experimental infections did not result in any diarrhea in cattle except in very young calves (19,31).

2.3.2 Virulence Factors

Escherichia coli O157:H7 has many virulence factors, not all of which have been fully characterized. This is in distinct contrast to botulism where a single, well understood factor (the botulism toxin) dominates the course of disease. In order for *E. coli* O157:H7 to cause its characteristic disease, the bacteria employ a number of adhesins, secretory systems, and toxins. Development of infection occurs in several distinct steps, each of which brings a different set of virulence factors into play.

2.3.2.1 Acid Tolerance and Low Infective Dose

The first step of *E. coli* O157:H7 infection is consumption of contaminated food. As mentioned, the infective dose of *E. coli* O157:H7 appears to be extremely low, and the organism can survive in relatively high acid foods like apple cider and fermented sausages. These characteristics may be related. A high acid tolerance might allow *E. coli* O157:H7 to survive passage through the acidic environment of the stomach. There are at least three known acid tolerance mechanisms in *E. coli*, but *E. coli* O157:H7 is only slightly more acid tolerant than the tested non-O157 *E. coli* strains (17,19,23). Whether acid tolerance contributes to the low infective dose of *E. coli* O157:H7 has not been clearly established.

2.3.2.2 Adhesion to the Intestinal Epithelium

Once *E. coli* O157:H7 reaches the large intestine, it adheres to the intestinal epithelium and produces characteristic lesions called attaching and effacing (A/E) lesions (17,18,24). Adhesion is an important stage of infection because it protects the bacteria from being flushed out of the gastrointestinal tract by peristaltic movements. Attachment of cells to the intestinal wall is well characterized in EPEC and EAggEC, in which specific proteins called adhesins mediate attachment. These adhesins include filamentous structures such as bundle-forming pili and fimbriae, as well as nonfilamentous outer membrane proteins (24,32). However, *E. coli* O157:H7 does not appear, in general, to possess these types of adhesins (17,23). The initial steps of *E. coli* O157:H7 adhesion to the intestines are not yet well understood.

2.3.2.3 A/E Lesions

The formation of the A/E lesion, on the other hand, is quite well characterized. A/E lesion formation has been most extensively studied in EPEC, but the mechanisms appear to be very similar in *E. coli* O157:H7 (17,24). In an A/E lesion, the bacterial cells form a strong, intimate attachment to the surface of the intestinal epithelium. The attached bacteria affect the host cell in such a way that the actin cytoskeleton of the host cells is extensively reorganized. The microvilli (projections on the surface of intestinal epithelial cells) disappear and are replaced by pedestal structures which support the *E. coli* cells (24). These pedestals can extend up to 10 μm above the surface of the host cell, are rich in host cell cytoskeletal components such as F-actin, α-actinin, and myosin, and move slowly across the surface of the host cell. Beyond their role in the strong attachment of *E. coli* cells to the cells surface, the function of the pedestals is unknown (33).

2.3.2.4 The Locus of Enterocyte Effacement (LEE)

All of the bacterial proteins necessary for the formation of A/E lesions are encoded by genes on the LEE (34). The LEE is a 35.6 kilobase (kb) chromosomal pathogenicity island which contains 5 operons and codes for many different virulence factors. It is found in *E. coli* O157:H7 as well as in some other pathogenic *E. coli*, notably EPEC strains (24). Transfer of the LEE from EPEC into nonpathogenic *E. coli* K-12 is sufficient to allow those bacteria to induce A/E lesions in cultured epithelial cells (35). Exactly how A/E lesions induce diarrhea is unknown, but evidence points toward the disruption of the tight junctions between intestinal cells, alterations in electrolyte transport across host cell membranes, and interactions with the host immune system, all of which could lead to fluid loss in the intestines (24,33).

2.3.2.5 Type III Secretion

Among many other genes, the LEE encodes a type III secretion system which is important in formation of the A/E lesion. Type III secretion systems are common among Gram-negative pathogens, including members of the genera *Yersinia*, *Salmonella*, *Pseudomonas*, and *Chlamydia*, as well as *E. coli*. They mediate contact dependent secretion of virulence factors, and can secrete toxins and other bacterial proteins directly into the host cell cytoplasm (34). The type III secretion apparatus is composed of approximately 10 proteins which form a "needle complex" spanning both the inner and outer membranes. The type III secretion mechanism can translocate proteins across both membranes in a single step (34,36). The needle complex structure appears to be related to the flagellar basal body (36). How the needle complex interacts with and translocates bacterial proteins into host cells is not fully understood (34).

The type III secretion system encoded by the LEE secretes a number of proteins. These include EspA, EspB, EspD, EspF (Esp stands for *E. coli* secreted protein), and Tir (34). All of these are encoded by genes in the LEE, and all except EspF are required for A/E lesion formation. The protein EspA forms filamentous structures on the surface of the bacterial cell and may form a channel between the bacterium and the host cell. Proteins EspB, EspD, and Tir are translocated into host epithelial cells. The exact functions of EspB and EspD are unknown, but they may form a pore in the host cell membrane, and may also act as effectors or toxins. Tir is translocated into the host cell membrane and serves as a receptor for the LEE encoded adhesin intimin. Intimin, a protein found in the *E. coli* outer membrane, binds tightly to Tir in the host cell membrane, and this interaction is required for the tight attachment of EPEC or EHEC to the epithelial cells and for pedestal formation (33,34).

Some differences in A/E lesion formation have been characterized between EPEC and *E. coli* O157:H7. In EPEC, Tir must be phosphorylated at a specific tyrosine residue by the host cell in order to be active. Tir from *E. coli* O157:H7 is not tyrosine phosphorylated (33). The *E. coli* O157:H7 LEE is not sufficient to confer the ability to form A/E lesions on *E. coli* K-12, unlike the LEE from EPEC (37). The reasons for these differences are not known.

2.3.2.6 Shiga-Like Toxin

A major difference between *E. coli* O157:H7 and EPEC strains is the presence of Shiga-like toxin, commonly referred to as Shiga toxin or verotoxin. Shiga toxins in *E. coli* O157:H7 are very closely related to the toxins produced by *Shigella dysenteriae*. Shiga toxin causes extensive damage to the intestinal epithelium, leading to bloody diarrhea (23).

Shiga toxins are, like botulism toxin, AB toxins, with an enzymatically active A portion and a B portion involved in binding to host cells. Active Shiga toxin is comprised of five 7.7 kDa B subunits and one 32 kDa A subunit. The A subunit, once internalized by a host cell and cleaved into A_1 and A_2 fragments, inhibits protein synthesis by cleaving an adenine residue from the 28S ribosomal RNA, leading to host cell death and lysis (17,28). The B subunit binds to specific receptors on host cell surfaces. Only host cells with the correct receptor (the cell membrane glycolipid Gb3) will be affected by the toxin. Cells with the correct glycolipid receptors include intestinal epithelial cells and renal endothelial cells (17). This may explain the specificity of Shiga toxins, in particular the severe kidney damage typical of HUS (17,28).

There are two main subtypes of Shiga toxin, called Stx1 and Stx2, which have slightly different properties. Stx2 is 1000 times more toxic to human kidney cells, and is more commonly associated with cases of HUS. Most *E. coli* O157:H7 isolates produce only Stx2, but many STEC produce Stx1, and some produce both types (23). Shiga toxins in *E. coli* strains are associated with bacteriophage sequences, and in *E. coli* O157:H7 both toxin types are encoded on lambdoid lysogenic prophage. Lysogenic prophages have integrated their genomes into the bacterial chromosome. The Stx phages are close relatives of the extensively studied phage lambda (λ). Under certain conditions, these prophages can enter a lytic cycle, where they excise themselves from the chromosome, replicate, and cause cell lysis, releasing phage particles which can infect other *E. coli* (38). The Shiga toxin genes are therefore highly mobile and can in theory be transferred easily among bacteria, for example from *Shigella* to *E. coli* or among *E. coli* strains.

2.3.2.7 Other Potential Virulence Factors

Escherichia *coli* O157:H7 has a variety of other genes whose importance to virulence has not been firmly established. Many of these have been identified by comparison of the *E. coli* O157:H7 and *E. coli* K-12 genomes. Genes found only in the *E. coli* O157:H7 genome may encode potential virulence factors (27). One, an extracellular serine protease called EspP, can cleave human coagulation factor V and may affect the blood clotting cascade (23). A hemolysin, a catalase/peroxidase, and a putative type II secretion system are encoded by the large plasmid pO157. This plasmid is almost always found in *E. coli* O157:H7 strains (39). Type II secretion is another system for exporting proteins from the cell to the outside environment which involves a two step translocation across the double cell membrane of Gram-negative bacteria (36). The pO157 plasmid also contains a gene for a protein with significant similarity to one of the cytoskeleton-altering toxins produced by *Clostridium difficile*. However, it is not known whether any of these genes contribute directly to virulence (23).

2.3.3 Genetics and Regulation

The regulation of virulence in *E. coli* O157:H7 is complex, and is accomplished through a combination of virulence factor specific regulatory systems and global regulatory mechanisms, which coordinately regulate expression of large sets of genes throughout the genome. Regulation of *E. coli* O157:H7 virulence genes is still incompletely understood, but regulation of the major virulence factors (the LEE and Shiga toxins) has been extensively studied (See Figure 2.2).

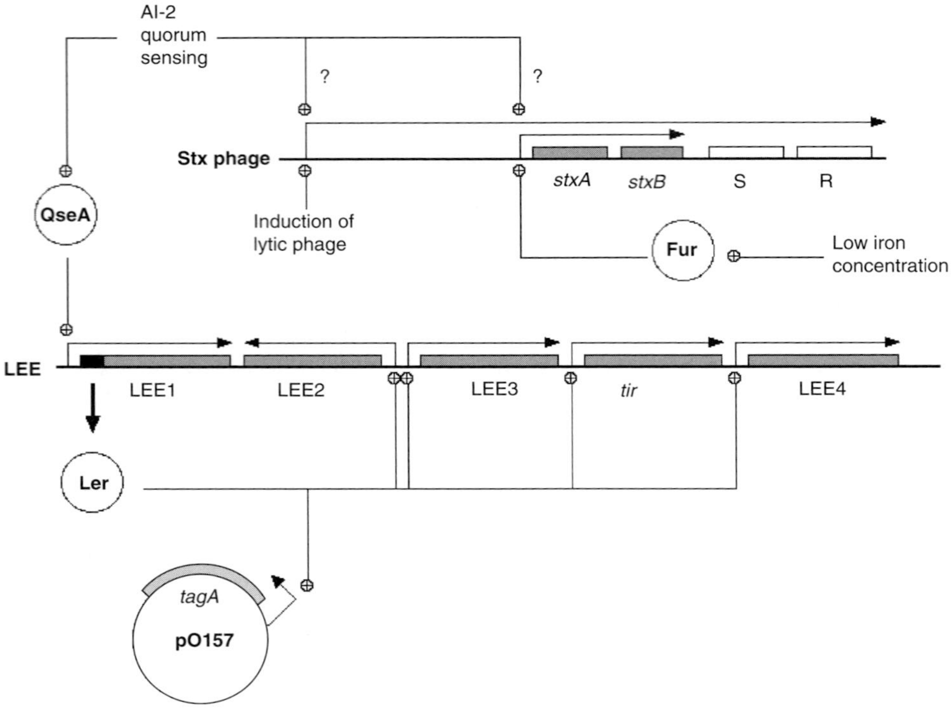

Figure 2.2 Regulation of virulence genes in *Escherichia coli* O157:H7. For simplicity, the multi-gene operons found in the locus of enterocyte effacement (LEE) are shown as single grey boxes. Gene products are as follows: *stxA* and *stxB* = Shiga toxin subunits, S and R = *Stx* phage late lytic phase proteins, LEE1 = includes *ler* (Ler regulatory protein) and type III secretion system components, LEE2 and LEE3 = type III secretion system components, *tir* = includes *tir* gene (Tir adhesin) and intimin gene, LEE4 = proteins secreted by type III secretion system, *tagA* = TagA protein on pO157 plasmid. The regulatory proteins QseA and Fur are shown without their respective genes, which are encoded at separate chromosomal loci. For each regulatory protein, positive regulation at their respective promoters is indicated by +

2.3.3.1 Structure of the LEE

The LEE is a pathogenicity island 35.6 kilobases in length, and is found in a variety of pathogens which cause A/E lesions. These include EPEC and *E. coli* O157:H7 as well as the animal pathogens *Citrobacter rodentium* and REPEC (rabbit enteropathogenic *E. coli*), among others (40). There are 41 open reading frames (ORFs) in the LEE, organized into 5 operons, termed LEE1, LEE2, LEE3, LEE4, and *tir*. The *tir* operon encodes Tir and intimin (34). The LEE1 operon encodes the regulatory protein Ler and some structural elements of the type III secretion system (*esc* genes – *E. coli* secretion). The rest of the type III secretion system is encoded by LEE2 and LEE3. LEE4 encodes proteins which are secreted by the type III secretion system (*esp* genes). There are several ORFs throughout the LEE which encode proteins of unknown function (40).

2.3.3.2 Regulation of the LEE

Expression of the LEE genes is dependent on several regulatory proteins. These include Ler, a regulatory protein specific to the LEE, as well as global regulatory mechanisms such as quorum sensing and alternative sigma factors.

2.3.3.3 LEE Encoded Regulator (Ler)

The first gene in the LEE1 operon is a gene called *ler* (LEE encoded regulator) (34). Ler protein is a transcriptional regulator whose DNA-binding domain is related to the H-NS/StpA family of DNA-binding proteins (37). Ler binds to the promoters of the other LEE operons (LEE2, LEE3, and *tir*) and activates their transcription (34,41). Activation of transcription of the LEE4 operon appears to be more complex. LEE4 is not expressed in a Δ*ler* *E. coli* O157:H7 mutant (which has had its *ler* gene deleted), but expression can be restored by addition of a plasmid carrying *ler*. However, when LEE4 is transferred into *E. coli* K-12, it is constitutively expressed at high levels in both the presence and absence of Ler. This suggests that a protein found in *E. coli* O157:H7, but not *E. coli* K-12, acts as a repressor of LEE4 transcription and that Ler acts to derepress LEE4 expression. The identity of this putative repressor is unknown (42). Efficient production of the type III secretion system and other components necessary for formation of A/E lesions (Tir, intimin) requires functional Ler protein in both EPEC and *E. coli* O157:H7 (41,42).

LEE encoded regulator (Ler) also regulates expression of genes outside the LEE in *E. coli* O157:H7. The *tagA* gene, located on the pO157 plasmid, is upregulated in the presence of Ler. The function of TagA is unknown, but it is very similar to TagA from *Vibrio cholerae*, the bacterium which causes cholera, a severe diarrheal disease. TagA in *V. cholerae* is also coregulated with virulence genes and has no known function (42). There is a large protein (110 kDa) which wild-type *E. coli* O157:H7 secretes, but which does not appear in cultures of Δ*ler* mutants. The identity of this protein is unknown. Finally, Δ*ler E. coli* O157:H7 mutants have long slender fimbriae (filamentous structures) which are not seen on the surface of wild-type cells (42).

2.3.3.4 Quorum Sensing

A global regulatory mechanism which affects expression of the LEE is called quorum sensing (34). Quorum sensing is a type of cell to cell communication which bacteria use to coordinate behavior at high cell densities. The bacteria produce a small molecule called an autoinducer. Once the concentration of autoinducer passes a certain threshold, transcription of the regulated genes is activated. Autoinducer concentration is essentially a measurement of bacterial cell density. Quorum sensing is found in both Gram-negative and Gram-positive bacteria, although the biochemical details differ (43).

Gram-negative bacteria may possess two known types of quorum sensing mechanisms. The first was originally discovered in the bioluminescent bacterium *Vibrio fischeri* and involves the genes *luxI* and *luxR*. In these systems, a LuxI type protein synthesizes an acylated homoserine lactone (AHL) autoinducer which diffuses through the cell membrane. Each LuxI type enzyme produces a distinct AHL molecule. The LuxR type protein is a transcriptional regulator which activates transcription of specific promoters when bound to its cognate AHL. LuxI/LuxR systems are widely distributed among Gram-negative bacteria (43).

A second type of quorum sensing in Gram-negative bacteria was discovered more recently. It uses a different type of autoinducer, called AI-2 (autoinducer 2), which is synthesized by an enzyme called LuxS. AI-2 has recently been discovered to be a furanosyl borate diester. Unlike the various LuxI/LuxR systems, which produce different and incompatible AHLs, the LuxS enzymes from a wide variety of species appear to produce the same autoinducer molecule (43). *Escherichia coli* O157:H7 has a *luxS* gene, and uses the AI-2 quorum sensing mechanism to regulate a variety of genes, including virulence genes (34,44,45).

In *E. coli* O157:H7, AI-2 activates expression of *ler*, which in turn activates expression of the rest of the LEE genes (45). Production of Shiga toxin is also upregulated in the presence of AI-2 (44). In experiments using genomic microarrays, AI-2 appears to regulate as much as 10% of the *E. coli* O157:H7 genome, including genes for cell division, flagella production, and 19 putative transcriptional regulators which could regulate even more genes. One of these, called QseA (quorum-sensing *E. coli* regulator A), has been examined in more detail. A Δ*qseA* mutant can still produce A/E lesions, but secretes significantly reduced amounts of Tir, EspA and EspB. QseA appears to enhance expression of *ler*, but does not have any effect on production of Shiga toxin or other virulence factors (46). The mechanism of Shiga toxin activation by AI-2 is unknown (17,44).

One interesting aspect of the quorum sensing activation of virulence genes in *E. coli* O157:H7 is the fact that identical AI-2 autoinducer is produced by a variety of different bacterial species. This appears to allow for a degree of interspecies communication. Presumably the levels of AI-2 in the large intestine, which has a dense population of bacteria, including *E. coli* and many other Gram-negative organisms, are quite high. One possibility is that *E. coli* O157:H7 detects the high levels of AI-2 in the large intestine and uses this as a signal to initiate infection. In this case, quorum sensing could activate virulence genes even without a dense population of *E. coli* O157:H7. This is consistent with the low infective dose of *E. coli* O157:H7 (17,45).

2.3.3.5 *Alternative Sigma Factors*

Another global regulatory mechanism used by bacteria involves alternative sigma factors, as discussed, for the regulation of sporulation in Gram-positive organisms. Housekeeping genes in *E. coli* are transcribed by RNA polymerase bound to σ^{70}. Many genes that are transcribed under stressful conditions, including environmental stresses (acid, heat) and entry into stationary phase growth, are transcribed by RNA polymerase bound to σ^{38} (encoded by a gene called *rpoS*) (47).

The promoter sequences for LEE1 and LEE2 are typical σ^{70} promoters, while the LEE3 and *tir* promoters more closely resemble σ^{38} promoters. In a Δ*rpoS* mutant of *E. coli* O157:H7, expression of LEE3 was lower than in the wild-type. It is likely that the genes in LEE3 (including *escV*, *escN*, and *escQ*, which encode genes for structural components of the type III secretion system) and probably *tir* are upregulated under stressful conditions (45). Differences in the *rpoS* genes of *E. coli* O157:H7 and other *E. coli* strains might also explain the slightly higher acid tolerance of *E. coli* O157:H7, but the importance of σ^{38} to virulence in *E. coli* O157:H7 has not yet been established (23). For two other enteric pathogens, *Salmonella enterica* and *Yersinia enterocolitica*, *rpoS* is known to influence the expression of virulence factors (48,49).

2.3.3.6 *Regulation of Shiga Toxin Production*

In addition to the activation of Shiga toxin production by quorum sensing discussed, there are a number of other factors which influence toxin production and release. As mentioned, Shiga toxins are encoded on integrated bacteriophage (Stx1 and Stx2) which are close relatives of phage λ (17). The regulation of gene expression in λ is one of the best understood systems in genetics, and an understanding of the λ lifecycle is invaluable in understanding expression of Shiga toxins.

2.3.3.7 *Bacteriophage Lambda* (λ)

Lambdoid phage (those closely related to λ) have both lytic and lysogenic phases to their lifestyle. In the lytic phase, the viral genome is actively replicated in the bacterial cytoplasm and the proteins which form the phage head are synthesized. Viral genomes are packaged into phage heads, and eventually the bacterium lyses, releasing intact phage particles into the environment. In the lysogenic phase, however, a single copy of the viral genome inserts into a specific site on the bacterial chromosome. An integrated bacteriophage is called a prophage. It is then carried by the bacterium and replicated each time the host cell divides. Although some prophage lose genes necessary for production of phage particles (becoming pseudolysogenic), an intact prophage can excise itself from the chromosome under the right conditions and reenter the lytic phase. In λ and its relatives, this transition is stimulated by the presence of single stranded DNA. In this way the phage can detect damage to the host DNA which may be fatal to the bacterium and produce phage particles, which are more likely to survive in some kinds of damaging environmental conditions than the host cell (47,50).

In many cases, the *stx* (Shiga toxin) genes of *E. coli* O157:H7 reside within fully functional lambdoid prophage. This means that under appropriate conditions, the Stx phage will enter the lytic phase. This has important implications for toxin production (17,38). Not only does it mean that the genes for toxin can be transferred to other bacteria via phage transmission, but production of toxin is coordinated with production of other phage proteins during the phage lifecycle.

2.3.3.8 Regulation of Stx Genes

The *stx* gene is inserted into the Stx phage genome near genes (called S and R) which are involved in the late stages of the lytic phase. It also has its own promoter, called p_{stx1}. The regulation of *stx* expression thus has two distinct aspects.

Expression from the p_{stx1} promoter is dependent on iron concentration. A small amount of toxin is produced under high iron conditions, but 10 times more is transcribed under low iron conditions. Iron dependent regulation of p_{stx1} is dependent on the regulatory protein Fur, which is a global regulator encoded on the *E. coli* chromosome and which is activated by the absence of available iron. Low iron availability is often used as a signal by pathogenic bacteria to determine whether they are in a host, because host environments are typically low in available iron. Under low iron conditions, Shiga toxin accumulates in the cytoplasm of *E. coli* O157:H7, but very little is detected in the surrounding medium (38).

The *stx* gene is also transcribed along with the lytic phase genes of the Stx phage. Any conditions which trigger the transition from lysogenic to lytic growth will also result in greatly enhanced (100-fold) expression of the *stx* gene. Among the known conditions which can induce lytic growth of Stx phage are damage to the bacterial DNA, the presence of some antibiotics, and attack by components of the host immune system (neutrophils). In contrast to the toxin produced by transcription from p_{stx1}, nearly all of the toxin produced under phage inducing conditions is found extracellularly (38).

Shiga toxins cause extensive damage to the intestinal epithelium and are thought to be responsible for the intestinal bleeding associated with *E. coli* O157:H7 infection (17). The induction of toxin production under low iron conditions in the large intestine and subsequent bleeding may allow more efficient growth of *E. coli* O157:H7, because it is able to utilize the iron bound to the hemoglobin in blood (23). Production of toxin under phage inducing conditions may also benefit the bacterium. When *E. coli* O157:H7 is under attack by the immune system, release of toxin from lysed cells may destroy immune cells in the vicinity. This mode of toxin production also helps to explain the observation that antibiotic treatment often results in the worsening of disease symptoms, consistent with more extensive toxin release upon phage induction (18,38).

2.3.4 Summary

Escherichia coli O157:H7 causes a noninvasive food borne infection with serious symptoms, including bloody diarrhea and kidney damage. It has many virulence factors. A large pathogenicity island called the LEE encodes a type III secretion system and other proteins responsible for tight attachment of the bacteria to the intestinal epithelium and formation of A/E lesions. The genes of the LEE are regulated by a specific transcriptional activator (called Ler) as well as by global regulatory mechanisms (which include quorum sensing, and alternative sigma factors). The other primary virulence factor of *E. coli* O157:H7 is the cytolytic Shiga-like toxin, which is encoded by *stx* genes on lysogenic bacteriophage. Production of Shiga-like toxin is also regulated by specific mechanisms (phage induction) and global regulatory mechanisms (quorum sensing and Fur). Both the LEE and the Stx bacteriophage appear to have been transferred more or less intact to *E. coli* O157:H7 from other species of bacteria, transforming a commensal organism into a dangerous pathogen.

2.4 *LISTERIA MONOCYTOGENES*

2.4.1 Introduction and History

Listeria monocytogenes causes a serious invasive disease called listeriosis. It is, like *E. coli* O157:H7, considered an emerging pathogen. However, *L. monocytogenes* is not a newly discovered organism. Murray, Webb, and Swann are generally credited with the first isolation and description of *L. monocytogenes* from the blood of infected rabbits in 1924 (51), but organisms which closely match the modern description of *L. monocytogenes* were isolated from rabbits in 1911, and from human cases of meningitis in 1915 and 1918 (52). The genus *Listeria* was named for the pioneering microbiologist Lord Joseph Lister, who is most famous for his invention of antiseptics for use in surgery (52,53).

During most of the twentieth century listeriosis was not a major human health concern. *Listeria monocytogenes* was known mostly as a pathogen of ruminants, especially sheep, although occasional human cases were seen. Then, in 1982, an outbreak of 41 cases of human listeriosis was linked to consumption of contaminated coleslaw (54). Since that time, recognition of *L. monocytogenes* infections has become more common, and numerous large outbreaks have been observed. *Listeria monocytogenes* is currently estimated to cause approximately 2500 illnesses and 500 deaths per year in the U.S. (1). These cases are usually associated with consumption of contaminated ready to eat foods like deli meats and soft cheeses (55,56). This increase in cases is particularly worrisome, because listeriosis has very serious symptoms unusual for a food borne disease, including central nervous system infection and abortion, and a 20–30% case fatality rate, even with early antibiotic treatment (57).

2.4.1.1 Characteristics of Listeriosis

Listeriosis usually begins with mild, flu-like symptoms, including chills, fatigue, headache, and muscle pain. In nonpregnant adults, this can be followed by increasingly severe symptoms. The symptoms which develop depend on the tissues infected most severely by the bacteria. Meningoencephalitis, which is the infection of the brain tissues and meninges, can cause movement disorders, changes in consciousness, and paralysis of cranial nerves. Systemic infection, or sepsis, has more generalized symptoms, but is equally fatal. In pregnant hosts, the mother is often asymptomatic or only has very mild symptoms, while the fetus becomes severely infected. This may lead to abortion, stillbirth, or sepsis in the newborn. Infected babies who survive birth are very likely to have neurological complications later in life. For unknown reasons, mothers who contract fetomaternal listeriosis are very rarely affected by meningoencephalitis or systemic infection. Infection by *L. monocytogenes* is much more common among certain high risk groups than in the general population. People at increased risk for listeriosis include the elderly, pregnant women, newborns, AIDS patients, or patients taking immunosuppressive medication (55,57).

Listeria monocytogenes has also been associated with cases of gastroenteritis, with symptoms including fever, vomiting, and diarrhea. These cases have generally been among healthy adults and appear to require extremely high doses of bacteria. Foods associated with outbreaks of *L. monocytogenes* gastroenteritis have contained 10^5 to 10^9 bacteria per gram. The same kinds of ready to eat foods are associated with this form of listeriosis as with invasive listeriosis. Some researchers suggest that many incidences of gastroenteritis with an unknown cause may actually be due to *L. monocytogenes* (58,59).

2.4.1.2 Characteristics of Listeria monocytogenes

Listeria monocytogenes is a Gram-positive, facultatively anaerobic bacterium which is motile at 10 to 25°C (7). The genus *Listeria* includes six species: *L. monocytogenes*, *L. ivanovii*, *L. seeligeri*, *L. innocua*, *L. welshimeri*, and *L. grayi*. Both *L. monocytogenes* and *L. ivanovii* are potential pathogens, but because *L. ivanovii* only rarely causes disease in humans, most research has focused on *L. monocytogenes*. The *Listeria* spp. are related to other Gram-positive organisms in the genera *Bacillus*, *Clostridium*, and *Enterococcus* (52).

Listeria monocytogenes is tolerant to a variety of environmental stresses. It will grow at temperatures from 1–45°C and at a pH between 6 and 9. It is also very salt tolerant, surviving in media containing as much as 20% NaCl. *Listeria monocytogenes* strains grow well in the presence of bile salts, which are lethal to most Gram-positive bacteria (7). High stress tolerance means that *L. monocytogenes* can survive many food processing methods designed to eliminate less resilient bacteria, although it does not form spores. *Listeria monocytogenes*'s ability to grow at refrigeration temperatures (approximately 4°C) is one of the most important factors in its emergence as a food borne pathogen. Shelf life extension of perishable processed foods often involves refrigerated storage for long periods of time before consumption. Extended refrigerated storage can function as a relative enrichment for *L. monocytogenes*, which can reproduce with reduced competition from mesophilic bacteria under those conditions (60).

Complete genome sequences have been published for *L. monocytogenes* and its nonpathogenic relative *L. innocua*. This allows the same kind of comparisons that have been made between *E. coli* O157:H7 and *E. coli* K-12, as discussed. Genes which are present in *L. monocytogenes* but not in *L. innocua* are good candidates for virulence factors. The genomes of the two *Listeria* species are very similar. Each is about 3 Mb in length, with about 3000 protein coding genes. There are 270 genes unique to *L. monocytogenes*, including some potential virulence genes and a 10kb virulence gene cluster, and 149 genes unique to *L. innocua* (61). In general, the genomes of *L. monocytogenes* and *L. innocua* are more similar to each other than are the genomes of *E. coli* O157:H7 and *E. coli* K-12.

2.4.1.3 Classification of Listeria monocytogenes

Many methods have been used to distinguish different strains of *L. monocytogenes* (62,63). Distinguishing between strains of *L. monocytogenes* is important for epidemiology and determination of the source of listeriosis outbreaks, but also has important implications for understanding virulence. Not all strains of *L. monocytogenes* are equally capable of causing disease (55,64), and various subtyping schemes have been developed to try and separate strains of differing virulence. For example, there are 13 serotypes of *L. monocytogenes*, three of which (1/2a, 1/2b, and 4b) account for 90% of human and animal cases of listeriosis. Serotype 4b alone accounts for 50% of listeriosis cases, and seems to be more frequently associated with fetomaternal disease than with systemic disease (55,63). However, despite the ability to separate *L. monocytogenes* strains into increasingly precise categories, no reliable predictor of virulence has been found.

For more discriminatory subtyping of *L. monocytogenes*, a wide variety of DNA based methods have been developed. One of the most commonly used is called ribotyping, and involves Southern hybridization using probes for the genes encoding ribosomes. Different alleles of these genes will produce different ribotype banding patterns on a gel. This method is standardized and allows researchers to distinguish different strains of *Listeria* (62,63).

Using a classification method based on ribotyping and allelic variation in virulence genes, *L. monocytogenes* can be subdivided into three lineages. Lineage I strains (serotypes 1/2b, 3b, 3c, and 4b) include all the strains responsible for human listeriosis outbreaks, along with human and animal isolates from sporadic cases. Lineage II (serotypes 1/2a, 1/2c, and 3a) contains both human and animal isolates, but no epidemic strains. Lineage III (serotypes 4a and 4c) is rare, but more common among animal isolates than among human isolates. This suggests that lineage I strains may have a greater potential to cause human disease, while lineage III strains may be better adapted to causing disease in animals (65,66). Regardless, almost any strain appears to be capable of causing isolated cases in humans or animals.

Three other common methods used for subtyping of *L. monocytogenes* are random amplified polymorphic DNA (RAPD) analysis, pulsed field gel electrophoresis (PFGE), and multiple enzyme electrophoresis (MEE) (62,67,68). RAPD is a polymerase chain reaction (PCR) based method which involves the use of short (approximately 10 nucleotide) arbitrary primers. PCR of genomic DNA with these primers under low stringency conditions results in banding patterns which vary depending on the genetic content of the bacterial strains (68). RAPD has been shown to be highly discriminatory for different isolates of *L. monocytogenes* from a variety of sources, although differences in methods can make comparison of RAPD types from different laboratories difficult (67–71). PFGE typing is a powerful discriminatory method which involves digesting genomic DNA with restriction enzymes and separating the fragments on a gel. The banding patterns produced by these DNA fragments differ from strain to strain. PFGE and RAPD typing give similar results for classifying and grouping *L. monocytogenes* strains (67,69,71). PFGE types cluster into two major divisions; division I, which correlates to serotypes 1/2a, 1/2c, 3a, and 3c, and division II, which correlates to serotypes 1/2b, 3b, 4b, 4d, and 4e (72). These correlate well with lineages I and II (62). PFGE types can also be correlated with virulence. For example, strains of PFGE type 1, which are commonly found in foods but are relatively rare in human cases, seem to be less invasive *in vitro* than other PFGE types (73). MEE is an older method which involves determining the electrophoretic mobility of 11 different cellular enzymes. Different banding patterns are designated as different electrophoretic types (ETs). Most ETs are in approximate agreement with PFGE types, although some ETs contain multiple PFGE types and vice versa (74). All of these methods allow differentiation among *L. monocytogenes* strains, but, like serotyping, ribotyping, and lineage determination, are of limited value in predicting the virulence of a particular isolate.

2.4.1.4 Sources in the Environment

Listeria spp are found in a wide variety of environmental sources, including soil, water, decaying vegetation, a wide variety of foods, and human and animal feces. Little is known about the life cycle of *L. monocytogenes* outside of an animal host, but it seems to be well suited to an independent, saprophytic existence. *Listeria* are most common in moist surface soils and decayed vegetation, and are detected in highest numbers in early spring (75). *Listeria monocytogenes* has been isolated from many species of mammals, birds, reptiles, amphibians, and insects, as well as being common in sewage and water (76). Because it is so ubiquitous in the environment, contamination of foods with *L. monocytogenes* can come from a variety of sources, and is therefore difficult to control.

2.4.2 Virulence Factors

In terms of number of virulence factors, *L. monocytogenes* falls between the single dominant virulence factor of *C. botulinum* and the elaborate network of interacting virulence factors in *E. coli* O157:H7. *Listeria monocytogenes* seems to be capable of infecting almost any animal host, and it uses the same relatively small set of virulence genes in all of them. As in *E. coli* O157:H7, at each stage in the course of infection *L. monocytogenes* brings a particular virulence factor to bear, the coordinated combination of which allows the bacterium to cause its characteristic disease. The important virulence factors of *L. monocytogenes* include internalins, membrane-lysing toxins, and an actin-nucleating protein called ActA (55).

2.4.2.1 Internalins

The first step in *L. monocytogenes* infection is attachment to and invasion of a susceptible host cell. The large family of proteins called internalins (Inl) are important in this process. There are two major groups of internalins, the large, membrane associated internalins (including InlA, InlB, InlC2, InlD, InlE, InlF, InlG, and InlH) and the small, secreted internalins (InlC in *L. monocytogenes* and a number of others in *L. ivanovii*) (55). The best studied of these are InlA and InlB, which are required for efficient invasion of host cells and formation of a close association between the bacteria and host cells (77). The internalin InlA binds to E-cadherin, a eukaryotic membrane protein found in epithelial cells, including intestinal cells, liver cells, dendritic cells, and placental epithelium; cell types which are particularly susceptible to invasion by *L. monocytogenes*. The internalin InlB has different specificity, and binds to several different host membrane proteins, including Met and gC1q-R, but not to E-cadherin (55,77). The receptors for the other internalins have not been determined. Binding of InlA or InlB to a host cell induces uptake of the *L. monocytogenes* cell. They do this by different molecular mechanisms, but the end result is the same: rearrangement of the host cell cytoskeleton and internalization of the bacteria within a vacuole by phagocytosis. Either InlA or InlB is sufficient for uptake into a cell with the appropriate receptor molecule, as demonstrated by experiments which observed internalization by host cells of latex beads that had been coated with the purified proteins (55,77). Different internalins are probably important for efficient invasion of different types of host cell, and may work synergistically to increase overall invasion efficiency (78).

2.4.2.2 Listeriolysin O (LLO)

Once *L. monocytogenes* has been internalized by a host cell, it must escape from the phagocytic vacuole into the cytoplasm. Any *L. monocytogenes* which fail to escape the vacuole are likely to be killed by acid and powerful oxidants. *Listeria monocytogenes* has several proteins involved in escape from the vacuole. The most important is listeriolysin O (LLO), also known as hemolysin, encoded by the *hly* gene. Pathogenic species of *Listeria* all produce LLO, while the nonpathogenic species do not (55). LLO is a cholesterol dependent secreted cytolysin, a member of a family of toxins produced by many Gram-positive pathogens, which act by forming pores in cholesterol containing cell membranes (79). Because bacterial membranes do not contain cholesterol, these toxins only affect host cell membranes. Most cholesterol dependent secreted cytolysins attack the cytoplasmic membrane of host cells and kill them, but LLO has very low activity at neutral pH, and is only active in the low pH environment of the phagocytic vacuole, where it lyses the vacuolar membrane and allows *L. monocytogenes* to escape into the cytoplasm without killing the host cell (79). Addition of the *hly* gene to nonpathogenic bacteria, such as *Bacillus subtilis*, is sufficient to allow those bacteria to escape from phagocytic vacuoles (80). *Listeria monocytogenes* reproduces readily in the eukaryotic cytoplasm, where it is protected from attack by the host's immune system.

2.4.2.3 Phospholipases

Listeria monocytogenes produces two enzymes with phospholipase activity which are also important for efficient escape from the phagocytic vacuole. They are called PI-PLC and PC-PLC, and even in the absence of LLO, their presence allows *L. monocytogenes* to escape from phagocytic vacuoles in some types of host cells (79). Deletion of the genes *plcA* (which encodes PI-PLC) and *plcB* (which encodes PC-PLC) individually have minor effects on virulence, but a double deletion decreases virulence markedly. *In vivo*, presumably LLO and the phospholipases work in concert to efficiently lyse vacuolar membranes (55).

Enzyme PI-PLC specifically cleaves phosphatidylinositol, a phospholipid which is found mostly on the inner side of eukaryotic membranes. Enzyme PC-PLC is less specific, and cleaves a number of other phospholipid components of eukaryotic membranes. Enzyme PI-PLC is most active at acidic pH, suggesting that, like LLO, it is active in phagocytic vacuoles (55). Enzyme PI-PLC seems to be more important in escaping the primary vacuole formed when *L. monocytogenes* is first taken up by a host cell, while PC-PLC seems to be more important for escaping the double membraned secondary vacuole formed when the bacteria are spreading from one host cell to another (55,79). The presence of three different membrane-lysing toxins with different specificities in *L. monocytogenes* reflects the importance of vacuolar escape to its pathogenic lifestyle, and may also be an adaptation to the variety of membrane structures it might encounter in different host species (55,79).

2.4.2.4 ActA and Actin Based Motility

Once *L. monocytogenes* escapes the phagocytic vacuole, it becomes mobile within the host cell cytoplasm (55,64,79). Outside of a host cell, *L. monocytogenes* use flagella for motility, but the cytoplasm is too viscous for flagella to be effective. Instead, *L. monocytogenes* stimulates polymerization of actin, a major protein of the eukaryotic cytoskeleton, into a tail like structure which pushes the bacterium through the cytoplasm (81,82). Using this rocket like mechanism, *L. monocytogenes* can attain speed of 0.3 μm/s inside a host cell (55). The main advantage of this form of intracellular motility is that it allows *Listeria* to spread from cell to cell. When an actin propelled bacterium reaches the host cell membrane it forms a long finger like protrusion into the adjacent host cell. Phagocytosis of this protrusion encases the *L. monocytogenes* in a double layered vacuole (81), which is lysed by LLO and the phospholipases, releasing the bacterium into a new host cell (55,79). *Listeria monocytogenes* can thereby spread through a tissue without being exposed to the host's immune system.

The process of actin based motility is mediated by a single bacterial protein, called ActA (82). Deletions of the *actA* gene result in *L. monocytogenes* which escape from vacuoles but cannot spread from cell to cell. ActA is a membrane bound protein which is localized at one pole of the rod shaped *L. monocytogenes* cell. It recruits host cell cytoskeletal proteins, including Ena/VASP proteins and the Arp2/3 complex, with the ultimate effect of causing actin to polymerize at that pole of the bacterium, forming the tail (55,79).

Insertion of the *actA* gene into *L. innocua* or *Streptococcus pneumoniae* is sufficient to give these bacteria the ability to form actin tails and spread intracellularly (55).

2.4.2.5 Other Virulence Factors

In addition to the virulence factors discussed, there are other genes which are important for virulence in *L. monocytogenes*. These include some housekeeping and stress response genes, as well as genes which seem to be more specialized for virulence functions.

Any food borne pathogen is exposed to attack by bile salts as it passes through the intestine of its host. *Listeria monocytogenes* is also commonly found growing in the livers of infected animals, where they are also exposed to bile. Gram-positive organisms are not ordinarily resistant to bile, but *L. monocytogenes* has a bile salt hydrolase enzyme (encoded by the *bsh* gene). The virulence of Δ*bsh* mutants is decreased in mouse and guinea pigs, and *bsh* is only found in pathogenic species of *Listeria* (*L. monocytogenes* and *L. ivanovii*), suggesting that bile salt hydrolase is a specific virulence factor and not a housekeeping gene (83).

Another recently discovered virulence gene in *L. monocytogenes* is *hpt*, which encodes a hexose phosphate transporter. Hexose phosphates (such as glucose-1-phosphate) are common in the cytoplasm of animal cells and appear to be an important carbon source for intracellular *L. monocytogenes*. Hpt is required for utilization of hexose phosphates and Δ*hpt* mutants grow much slower inside mammalian cells than wild-type bacteria. They are also attenuated for virulence in mice (84).

Evidence also points toward stress response genes having a role in virulence. The host environment, both in the intestine before invasion and inside the phagocytic vacuole, is a hostile one for *L. monocytogenes*, and the bacteria are exposed to a variety of stresses in that environment. These include exposure to acid, reactive oxygen species, and iron limitation, among others (55). General stress response is controlled in *L. monocytogenes* by an alternative sigma factor, σ^B, which regulates transcription of many stress response proteins (85,86). Mutants lacking σ^B are deficient in virulence, although the effects are relatively minor. Some evidence, however, suggests that regulation of stress response through σ^B may have a connection to regulation of some of the major virulence factors (85,87,88).

2.4.3 Genetics and Regulation

2.4.3.1 Organization of Virulence Genes

The genes for LLO (*hly*), both phospholipases (*plcAB*), and ActA (*actA*), along with a gene for a metalloprotease required for activation of PC-PLC (*mpl*) are all encoded on a 10 kb gene cluster (55). This cluster is only found in the two pathogenic species of *Listeria*, *L. monocytogenes* and *L. ivanovii*. *Listeria seeligeri* has the cluster, but with an insertion that blocks production of the virulence genes (89). The internalins are encoded separately, in a number of different loci. InlA and InlB are encoded together in the *inlAB* operon. Internalin InlC (the small secreted internalin of *L. monocytogenes*) and InlF are encoded as monocistronic genes, *inlC* and *inlF*, respectively. A fourth internalin locus in *L. monocytogenes* contains three genes, which may be *inlGHE* or *inlC2DE*, depending on the strain. These two operons are homologous, but not identical (55). The *hpt* and *bsh* genes are also encoded at separate loci on the *L. monocytogenes* chromosome (83,84) (Figure 2.3).

2.4.3.2 PrfA

The virulence gene cluster also contains the *prfA* (positive regulatory factor A) gene, which encodes a regulatory protein called PrfA. The protein PrfA is the master regulator of virulence genes in *L. monocytogenes*. Mutants lacking *prfA* are completely avirulent (90,91).

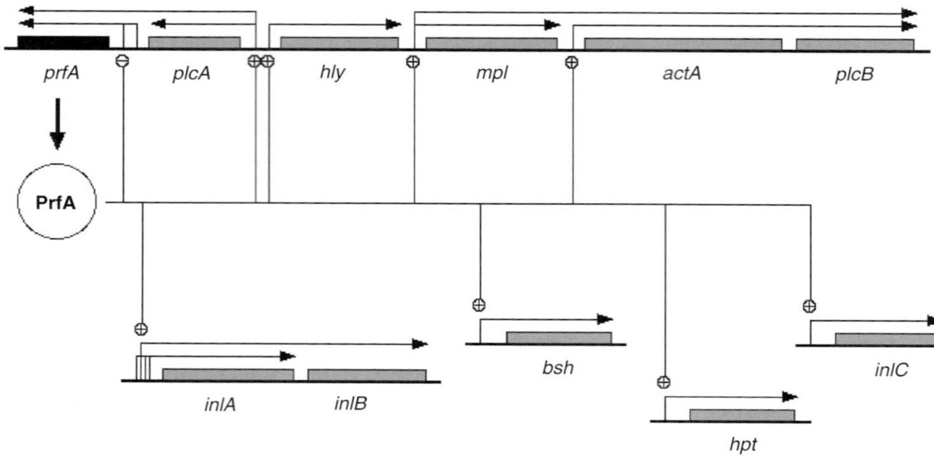

Figure 2.3 Regulation of virulence genes in *Listeria monocytogenes*. Gene products are as follows: *prfA* = PrfA regulatory protein, *plcA* = PI-PLC phospholipase, *hly* = listeriolysin O (LLO), *mpl* = metalloprotease, *actA* = ActA actin nucleating protein, *plcB* = PC-PLC phospholipase, *inlA* = internalin A, *inlB* = internalin B, *bsh* = Bsh bile salt hydrolase, *hpt* = Hpt hexose phosphate transporter, *inlC* = internalin C. PrfA is a positive regulator of promoters indicated by + and a negative regulator of the promoter indicated by −

Regulatory protein PrfA activates transcription of all the major virulence genes, including *hly*, *actA*, *plcAB*, *mpl*, and *prfA* itself. It also positively regulates expression of *bsh*, *hpt*, and the internalin genes *inlAB* and *inlC*. Interestingly, the other internalins (InlC2, D, E, F, G, and H), whose role in virulence appears to be relatively minor, are not regulated by PrfA (89).

Protein PrfA is related to a family of bacterial regulatory proteins called the Crp/Fnr family (92). The proteins Crp and Fnr are well studied regulatory proteins from *E. coli*; Crp is a regulator of catabolite repression, and responds to cyclic AMP (cAMP) levels, while Fnr regulates anaerobic growth factors. Although related, PrfA lacks several residues required in Crp for cAMP binding, and cAMP has no effect on PrfA activity. However, mutations in PrfA suggest that a domain homologous to the cAMP binding domain of Crp affects whether PrfA is in an active or an inactive conformation. Whether or not PrfA activity is affected by a small molecule cofactor is unknown (89). Two domains exist in PrfA that do not occur in other Crp/Fnr family regulators, a putative N-terminal helix-turn-helix domain and a putative C-terminal leucine zipper domain. Both are required for PrfA activity, but the exact roles of these domains are unknown (89,92).

2.4.3.3 Regulation of PrfA Expression

Because expression of virulence is dependent on activation by PrfA, regulation of expression of PrfA is of central importance in controlling pathogenicity in *L. monocytogenes*. Expression of PrfA is from three different promoters: P*plcA*, which produces a bicistronic transcript encoding both PI-PLC and PrfA, and two PrfA-specific promoters, P1 and P2, which produce monocistronic PrfA transcripts (93,94). The strong promoter P*plcA*, is dependent on PrfA, which can therefore upregulate its own expression. Expression from P1 is constitutive, but relatively weak, and translation of the P1 transcript is thermoregulated to prevent production of PrfA protein at temperatures below 37°C (95). Expression from P2 is also weak, and is dependent on the general stress sigma factor σ^B (87). The precise mechanisms and signals which *L. monocytogenes* uses to activate PrfA are not well understood, but conditions inside a host organism (high temperatures, low iron availability, acid stress) are believed to stimulate production of PrfA protein from P1 and P2, which in turn activates transcription from P*plcA*. This leads to an increase in the amount of PrfA in the bacterium (89). High concentrations of PrfA negatively regulate PrfA expression, ensuring that the autoregulatory loop does not overproduce PrfA, although this negative regulation does not appear to be required for virulence (94,96).

2.4.3.4 Regulation of Virulence Genes

As PrfA levels increase, transcription of virulence factors is activated. At low levels of PrfA, the virulence factors most important for escape from the phagocytic vacuole, LLO and PI-PLC, are produced (94). As the *L. monocytogenes* continue the infective process, beginning to spread from cell to cell, PrfA levels increase, and the other virulence factors important in this phase of infection are transcribed. These include ActA and PC-PLC, as well as Hpt and Bsh (83,84,97,98). The differential regulation of these genes depends on differences in their promoters. Promoters dependent on PrfA contain a sequence called the PrfA-box. As PrfA binds to the PrfA-box it activates transcription from these promoters. The more similar a PrfA-box is to the consensus PrfA-box sequence, the lower the concentration of PrfA protein is required to activate transcription. The promoters for PI-PLC and LLO have perfect PrfA-boxes, while the other virulence gene promoters have one or two mismatched bases in their PrfA-boxes, ensuring that they will not be transcribed until higher levels of PrfA protein have accumulated (83,84,89,92).

Regulation of internalin expression is more complex. Internalins A and B are required for invasion of most host cell types, but not for escape from a vacuole or cell to cell spread (55,77). The *inlAB* operon has four promoters (99). Expression of internalin is higher outside of host cells than when *L. monocytogenes* is growing intracellularly (100), but one of the *inlAB* promoters has a PrfA-box with two mismatches, suggesting that this promoter is activated during cell to cell spread (99). What appears to happen is that internalins are expressed at high levels outside of a host from the other three promoters, one of which is σ^B-dependent, and therefore activated by growth under stress (101). When *L. monocytogenes* is growing intracellularly, expression from these promoters is repressed by unknown mechanisms and relatively low levels of internalins are produced from the PrfA-dependent promoter. More work needs to be done to clarify the regulation of the internalins. The small secreted internalin InlC appears to be PrfA-dependent and expressed primarily intracellularly, suggesting it may have a role in cell to cell spread, but the function of this protein has not been established (102).

It appears that PrfA is also involved in negatively regulating a variety of genes in *L. monocytogenes*. These include *prfA* itself, some stress response genes, and genes for flagellar synthesis (94,96,103,104). In the case of the flagellar genes, this explains the loss of flagella by invading *L. monocytogenes* cells, which then become motile by means of actin polymerization. The rationale behind PrfA repression of some stress response genes is not clear.

2.4.4 Summary

Listeria monocytogenes causes an invasive food borne infection, which can cause sepsis, meningoencephalitis, abortion, and death. It primarily affects the immunocompromised. Major *L. monocytogenes* virulence factors include internalins, which stimulate phagocytosis and entry into host cells, membrane lysing toxins (LLO and phospholipases), which mediate escape from the phagocytic vacuole, and ActA, which allows intracellular motility and cell to cell spread. In addition, *L. monocytogenes* produces Hpt, which allows the use of hexose phosphates in intracellular growth, and Bsh, a bile salt hydrolase which is important for survival in the intestine and in liver cells. All of these virulence factors are regulated by the transcription factor PrfA. Regulation of virulence is also related to regulation of stress response. Regulation of virulence in *L. monocytogenes* is relatively well understood, but many details about how this bacterium senses changes in its environment and coordinates the multiple factors required for infection are still unknown. Why certain strains appear to be more pathogenic than others and whether it is possible to predict the virulence of an isolate remains unclear.

2.5 CONCLUSIONS

2.5.1 Common Themes in Bacterial Pathogenesis

Examining the regulation of virulence in three important food borne pathogens reveals that, while each organism has a different mode of pathogenesis depending on its particular complement of virulence factors, there are common themes which link all three. An important goal of the study of bacterial pathogens is to discover these common themes and general principles which govern the process of pathogenesis in all bacteria. In all three example organisms, pathogenesis requires coordination of virulence factors, which may include toxins, adhesins, secretion systems, or motility mechanisms. The regulation of virulence involves a complex interaction between the effect of virulence specific regulators (e.g., BotR, Ler, or PrfA) and the effects of more general global regulatory mechanisms (e.g., regulation of sporulation, quorum sensing, or general stress response) which regulate genes for both virulence factors and housekeeping proteins.

2.5.2 Pathogenicity Islands, Horizontal Gene Transfer, and the Evolution of Pathogens

Another common theme among bacterial pathogens is that virulence genes are generally found grouped together in virulence gene clusters. These virulence gene clusters include the botulism toxin gene cluster in *C. botulinum*, the LEE, and Stx phages in *E. coli* O157:H7, and the PrfA-dependent virulence gene cluster in *L. monocytogenes*. A region of DNA carrying several related virulence genes in close proximity is called a "pathogenicity island". Pathogenicity islands range in size from relatively small, like the *L. monocytogenes* PrfA-dependent virulence gene cluster (~10 kb, sometimes referred to as a pathogenicity "islet") through medium sized islands like the LEE from *E. coli* O157:H7 and EPEC (~36 kb) to very large islands (~200 kb) like those associated with virulence in uropathogenic *E. coli* (not a food borne pathogen) (105).

An important feature of pathogenicity islands is that they often contain sequences which indicate that they have or once had the ability to be mobilized, i.e., pathogenicity islands can be horizontally transferred from the genome of one bacterium into another. This is particularly clear in the case of *E. coli* O157:H7, which appears to have acquired the ability to cause disease by incorporating two pathogenicity islands, the LEE and the Stx phage (29). The LEE is widespread among Gram-negative enteric pathogens, especially among strains of *E. coli*, and, although the structure of the LEE remains consistent between strains, it is often found incorporated at different positions on the chromosome (106), which supports the idea that the LEE is mobile and can be spread from one strain to another. Stx phage are functional viruses capable of transducing the genes for Shiga-like toxin into any phage susceptible strain of bacteria. The Shiga-like toxin produced by *E. coli* O157:H7 is essentially identical to that produced by *Shigella dysenteriae*, but the *Shigella* toxin is encoded on the chromosome rather than on a bacteriophage. Stx phage may have acquired the toxin gene from *Shigella* and transferred it to *E. coli* (106). The botulism toxin gene cluster also shows evidence of horizontal transfer. Distantly related strains of *Clostridium* have similar gene clusters, and many are located on potentially mobile genetic elements like plasmids or prophage (107). There are no obvious mobilizing elements associated with the PrfA-dependent virulence gene cluster in *L. monocytogenes*, which has led to the suggestion that this pathogenicity islet may be very old. This hypothesis suggests that the PrfA-dependent gene cluster was once mobile, but lost that ability so long ago that evolution has completely eliminated any trace of the vestigial mobilizing elements (108).

Pathogenicity islands make it possible for previously innocuous bacteria to quickly acquire whole sets of virulence genes in a single step, which has enormous implications for the evolution and appearance of emerging pathogens. Current knowledge cannot fully explain how horizontally acquired genes become integrated into the global regulatory mechanisms of bacteria so that virulence genes can be correctly coordinated during infection.

2.5.3 Emerging Pathogens and Human Food Practices

The interaction between a pathogen and its host is only one factor in how a bacterium causes disease. The interaction between the bacteria and their environment is equally important, and for food borne pathogens, environmental factors are dependent on food processing and storage technology. Historically, one of the main goals of food processing is preservation or extension of shelf life. Most food processing technologies are intended to kill or slow the growth of the bacteria which are most likely to be a spoilage concern in that food product. Common preservative techniques include canning, pasteurization (with heat or electromagnetic radiation), drying, addition of salt, and refrigeration. However, sublethal applications of these treatments can select for organisms which can survive treatment, and produce an environment in which the survivors can multiply with minimal competition. If a new preservative technique provides an environment in which a potential pathogen can thrive, those organisms would then generally be considered emerging pathogens, because they were not a health hazard before the new technology gave them new opportunities to infect humans (109).

For example, fresh foods generally are not considered a botulism hazard. *Clostridium botulinum* typically only causes disease upon consumption of foods which contain anaerobic zones and have been stored for periods of time sufficient for outgrowth and toxin production. Food processing techniques, such as sausage making or canning, if improperly conducted could lead to these conditions and make botulism possible. Both sausage making and canning are intended to allow food to be stored for long periods of time by controlling or killing spoilage organisms, most of which are considerably less robust than *C. botulinum* spores. Food processing methods can be adjusted to minimize the risk of *C. botulinum* toxin production, by addition of nitrate in the case of sausage or through heat treatments designed to kill 12 logs of *C. botulinum* spores in the case of canning, which accounts for the relative rarity of botulism today. Most modern cases of food borne botulism are accounted for by inadequate home processing using methods that do not kill *C. botulinum* spores, but, instead, create the necessary conditions for bacterial growth and toxin production (2).

A somewhat more recent example of how food storage technology affects bacterial pathogens is the emergence of *L. monocytogenes* as an important food borne pathogen. Because *L. monocytogenes* can grow at refrigeration temperatures, which inhibit the growth of most spoilage bacteria, *L. monocytogenes* can grow with relatively little competition in food stored for long periods in a refrigerator. The increased demand for long shelf life, ready to eat food products has created a niche for these bacteria that probably has greatly increased human exposure to *L. monocytogenes*. More or less simultaneously, a growing population of immunocompromised and elderly persons supported by modern health care methods has appeared. These people are more susceptible to infection by *L. monocytogenes*. The combination of environmental and host factors accounts for how a once rarely encountered pathogen of sheep and rabbits has become, in the last 20 years, one of the deadliest food borne pathogens in the developed world (110).

How human food practices may have affected the development of *E. coli* O157:H7 is less clear. Unlike *C. botulinum* and *L. monocytogenes*, *E. coli* O157:H7 appears to have evolved relatively recently, by acquisition of the LEE and Stx phage, in contrast to being an "old" organism that is simply taking advantage of a newly available niche (111). However, the continued emergence of other Shiga-toxin producing serotypes of *E. coli* with similar properties and virulence suggests that some factor in the modern environment allows enterohemorrhagic *E. coli* to thrive and spread. What that factor or factors might be is unknown, but some possibilities present themselves. Food products are now often shipped to distant locations from a single processor. A single meat packing plant may supply ground beef to fast food restaurants across the country, which would allow a single isolated incidence of contamination to be spread widely. Another possible contributing factor is the development of very intensive, high density cattle farming, which might contribute to the carriage of *E. coli* O157:H7 among beef cattle (18,30).

For a bacterium to cause disease, several conditions must be met. The pathogen must possess genes for virulence factors which allow it to infect a host organism and the host must be susceptible to attack by that pathogen and its particular combination of virulence factors. In addition, the environment must allow exposure of the host to the pathogen, and for food borne pathogens, the most important environmental factors are dependent on how the food is produced, processed, stored, prepared, and consumed. Through development of new processing strategies and food handling systems, we have selected for many of the pathogens which potentially contaminate our foods.

2.5.4 Applications of Genetic and Molecular Understanding of Pathogenesis

Research into the genetic and physiological basis of virulence is critical in improving our ability to detect dangerous pathogens and prevent disease. When the genes for a particular pathogen's virulence factors are known, DNA based methods can be developed to quickly detect and identify that pathogen, to distinguish between virulent and avirulent strains of closely related bacteria, and to help develop vaccines specific to those virulence factors. Currently, DNA based methods have been most useful for diagnosis of infections caused by pathogens which are particularly difficult to culture (such as *Mycobacterium tuberculosis*, which causes tuberculosis, *Bordetella pertussis*, the cause of whooping cough, and a variety of viral diseases), but as automation and less expensive methods become more widely available, DNA based molecular diagnostics are expected to become increasingly valuable tools in identifying a wide range of pathogens (112).

Another important application of understanding the molecular mechanisms of pathogenesis is the development of DNA based subtyping methods. These methods allow rapid strain "fingerprinting" for epidemiological studies to assist in disease management and control. If an outbreak of disease is caused by a specific strain of a pathogen, and foods containing that specific strain can be found, it will allow epidemiologists to very precisely identify the source of the outbreak. Molecular subtyping has been an important tool used in tracking recent outbreaks of *L. monocytogenes* and *E. coli* O157:H7, as well as for a wide variety of other pathogens (26,62,112).

Finally, researchers hope that understanding the molecular mechanisms of pathogenesis will allow them to design new treatments or antibiotics which specifically attack those mechanisms. The better pathogens are understood at a molecular level, the better we are able to confront them at a clinical level. The continued emergence of new food borne pathogens and antibiotic resistant strains of known pathogens means that we will require innovative strategies to combat these bacteria and preserve the safety of the food supply. Studies of the genetic and molecular mechanisms of pathogenesis provide an important part of the information necessary to formulate these strategies.

2.6 RECOMMENDED READING

The following review articles and books may be useful to those interested in more in depth treatments of the organisms and concepts discussed in this chapter:

1. Regulation of virulence:
   a. Salyers, A.A., D.D. Whitt. *Bacterial Pathogenesis: A Molecular Approach*, 2nd ed. Washington, D.C.: ASM Press, 2002.
   b. Cotter, P.A., V.J. DiRita. Bacterial virulence gene regulation: an evolutionary perspective. *Ann. Rev. Microbiol.* 54:519–565, 2000.
2. *Clostridium botulinum*:
   a. Smith, L.D.S, H. Sugiyama. Botulism: the organism, its toxins, the disease. *American lecture series in clinical microbiology*, 2nd ed. Springfield, IL: Charles C. Thomas, 1988.
   b. Johnson, E.A., M. Bradshaw. Clostridium botulinum and its neurotoxins: a metabolic and cellular perspective. *Toxicon* 39:1703–1722, 2001.
3. *Escherichia coli* O157:H7:
   a. Donnenberg, M.S., ed. Escherichia coli: *virulence mechanisms of a versatile pathogen*. San Diego, CA: Academic Press, 2002.
   b. Duffy, G., P. Garvey, D.A. McDowell, eds. *Verocytotoxigenic* E. coli. Trumbull, CT: Food & Nutrition Press, 2001.
4. *Listeria monocytogenes*:
   a. Ryser, E.T., E.H. Marth, eds. *Listeria, listeriosis, and food safety*, 2nd ed. New York, NY: Marcel Dekker, 1999.
   b. Vázquez-Boland, J.A., M. Kuhn, P. Berche, T. Chakraborty, G. Domínguez-Bernal, W. Goebel, B. González-Zorn, J. Wehland, J. Kreft. *Listeria* pathogenesis and molecular virulence determinants. *Clin. Microbiol. Rev.* 14(3):584–640, 2001.
   c. Kreft, J., J.A. Vázquez-Boland. Regulation of virulence genes in *Listeria*. *Int. J. Med. Microbiol.* 291(2):145–157, 2001.
5. Evolution of pathogenesis:
   a. Hacker, J., J.B. Kaper. The concept of pathogenicity islands. In: Hacker, J., J.B. Kaper, eds., *Pathogenicity Islands and Other Mobile Virulence Elements*. Washington, DC: ASM Press, 1999, pp 1–11.
   b. Donnenberg, M.S., T.S. Whittam. Pathogenesis and evolution of virulence in enteropathogenic and enterohemorrhagic *Escherichia coli*. J. Clin. Invest. 107(5):539–548, 2001.

REFERENCES

1. Mead, P.S., L. Slutsker, V. Dietz, L.F. McCraig, J.S. Bresee, C. Shapiro, P.M. Griffin, R.V. Tauxe. Food-related illness and death in the United States. *Emerg. Infect. Dis.* 5(5):607–625, 1999.
2. Smith, L.D.S., H. Sugiyama. Botulism: the organism, its toxins, the disease. *American lecture series in clinical microbiology*, 2nd ed., Springfield, IL: Charles C. Thomas, 1988.
3. Johnson, E.A., M. Bradshaw. *Clostridium botulinum* and its neurotoxins: a metabolic and cellular perspective. *Toxicon* 39:1703–1722, 2001.
4. Woodburn, M.J., E. Somers, J. Rodriguez, E.J. Schantz. Heat inactivation rates of botulinum toxins A, B, E, and F in some foods and buffers. *J. Food Sci.* 44:1658–1661, 1979.
5. Anonymous. *Botulism in the United States,* 1899–1996: *Handbook for Epidemiologists, Clinicians, and Laboratory Workers.* Atlanta, GA: Centers for Disease Control and Prevention, 1998.
6. Byrne, M.P., L.A. Smith. Development of vaccines for prevention of botulism. *Biochimie* (Paris) 82:955–966, 2000.
7. Sneath, P.H.A., N.S. Mair, M.E. Sharpe, J.G. Holt, eds. *Bergey's Manual of Systematic Bacteriology*, Vol. 2. Baltimore: Williams and Wilkins, 1986.
8. Collins, M.D., P.A. Lawson, A. Willems, J.J. Cordoba, J. Fernandez-Garayzabal, P. Garcia, J. Cai, H. Hippe, A.E. Farrow. The phylogeny of the genus *Clostridium*: proposal of five new genera and eleven new species combinations. *Int. J. Syst. Bacteriol.* 44:812–826, 1994.
9. Stackebrandt, E., F.A. Rainey. Phylogenetic relationships. In: *The* Clostridia*: Molecular Biology and Pathogenesis*. Rood, J.I., B.A. McClane, J.G. Songer, R.W. Titball, eds., San Diego: Academic Press, 1997, pp 3–19.
10. Humeau, Y., F. Doussau, N.J. Grant, B. Poulain. How botulinum and tetanus neurotoxins block neurotransmitter release. *Biochimie* (Paris) 82:427–446, 2000.
11. Dodds, K.L., J.W. Austin. *Clostridium botulinum.* In: Doyle, M.P., L.R. Beuchat, T.J. Montville, eds. *Food Microbiology: Fundamentals and Frontiers.* Washington, DC: ASM Press, 1997, pp 288–304.
12. Anonymous. Foodborne botulism associated with home-canned bamboo shoots: Thailand, 1998. *Morb. Mortal. Wkly. Rep.* 48(21):437–439, 1999.
13. Sauer, U., J.D. Santangelo, A. Treuner, M. Buchholz, P. Dürre. Sigma factor and sporulation genes in *Clostridium. FEMS Microbiol. Rev*. 17:331–340, 1995.
14. Hoch, J.A. *spo0* genes, the phosphorelay, and the initiation of sporulation. In: Bacillus subtilis *and Other Gram-positive Bacteria: Biochemistry, Physiology, and Molecular Genetics.* Sonenshein, A.L., J.A. Hoch, R. Losick, eds., Washington, DC: ASM Press, 1993.
15. Woods, D.R., D.T. Jones. Physiological response of *Bacteroides* and *Clostridium* strains to environmental stress factors. *Adv. Microb. Physiol.* 28:1–64, 1986.
16. Riley, L.W., R.S. Remis, S.D. Helgerson, H.B. McGee, J.G. Wells, B.R. Davis, R.J. Hebert, E.S. Olcott, L.M. Johnson, N.T. Hargrett, P.A. Blake, M.L. Cohen. Hemorrhagic colitis associated with a rare *Escherichia coli* serotype. *N. Engl. J. Med.* 308(12):681–685, 1983.
17. Thorpe, C.M., J.M. Ritchie, D.W.K. Acheson. Enterohemorrhagic and other Shiga toxin-producing *Escherichia coli.* In: Escherichia coli: *virulence mechanisms of a versatile pathogen*, Donnenberg, M.S., ed., San Diego, CA: Academic Press, 2002, pp 119–155.
18. Parry, S., S. Palmer. *E. coli*: environmental and health issues of VTEC O157. In: Clay's Library of Health and the Environment, B Bassett, B., ed., London, UK: Spon Press, 2002.
19. Park, S., R.W. Worobo, R.A. Durst. *Escherichia coli* O157:H7 as an emerging foodborne pathogen: a literature review. *Crit. Rev. Microbiol.* 21(1):27–48, 2001.
20. Schmidt, H., H. Karch. Pathogenic aspects of STEC infection in humans. In: *Verocytotoxigenic* E. coli., Duffy, G., P. Garvey, D.A. McDowell, eds., Trumbull, CT: Food & Nutrition Press, 2001, pp 241–262.
21. Boer, E.D., A.E. Heuvelink. Foods as vehicles of VTEC infection. In: *Verocytotoxigenic* E. coli., Duffy, G., P. Garvey, D.A. McDowell, eds., Trumbull, CT: Food & Nutrition Press, 2001, pp 181–200.

22. Doyle, M.P., T. Zhao, J. Meng, S. Zhao. *Escherichia coli* O157:H7. In: *Food Microbiology: Fundamentals and Frontiers*, Doyle, M.P., L.R. Beuchat, T.J. Montville, eds.,Washington, DC: ASM Press, 1997, pp 171–191.
23. Law, D. Virulence factors of *Escherichia coli* O157 and other Shiga toxin-producing *E. coli*. *J. Appl. Microbiol*. 88:729–745, 2000.
24. Blank, T.E., J.-P. Nougayréde, M.S. Donnenberg. Enteropathogenic *Escherichia coli*. In: Escherichia coli: *Virulence Mechanisms of a Versatile Pathogen*, Donnenberg, M.S., ed., San Diego, CA: Academic Press, 2002, pp 81–118.
25. Salyers, A.A., D.D. Whitt. *Bacterial Pathogenesis: A Molecular Approach*, 2 ed. Washington, D.C.: ASM Press, 2002.
26. Thomson-Carter, F. General recovery, characterisation and typing protocols for VTEC. In: *Verocytotoxigenic* E. coli., Duffy, G., P. Garvey, D.A. McDowell, eds., Trumbull, CT: Food & Nutrition Press, 2001, pp 91–111.
27. Perna, N.T., J.D. Glasner, V. Burland, G. Plunkett. The genomes of *Escherichia coli* K-12 and pathogenic *E. coli*. In: Escherichia coli: *Virulence Mechanisms of a Versatile Pathogen*, Donnenberg, M.S., ed. San Diego, CA: Academic Press, 2002, pp 3–54.
28. Sandvig. K. Shiga toxins. *Toxicon* 39:1629–1635, 2001.
29. Johnson, J.R. Evolution of pathogenic *Escherichia coli*. In: Escherichia coli: *Virulence Mechanisms of a Versatile Pathogen*, Donnenberg, M.S., ed., San Diego, CA: Academic Press, 2002, pp 55–80.
30. Duffy, G., P. Garvey, D.A. McDowell, eds. *Verocytotoxigenic* E. coli. Trumbull, Connecticut: Food & Nutrition Press, Inc., 2001.
31. Blanco, J., M. Blanco, J.E. Blanco, A. Mora, M.P. Alonso, E.A. González, M.I. Bernárdez. Epidemiology of verocytotoxigenic *Escherichia coli* (VTEC) in ruminants. In: Duffy, G., P. Garvey, D.A. McDowell, eds. *Verocytotoxigenic* E. coli. Trumbull, Connecticut: Food & Nutrition Press, Inc., 2001, pp 113–148.
32. Nataro, J.P., T. Steiner. Enteroaggregative and diffusely adherent *Escherichia coli*. In: Escherichia coli: *Virulence Mechanisms of a Versatile Pathogen*, Donnenberg, M.S., ed., San Diego, CA: Academic Press, 2002, pp 189–208.
33. Celli, J., W. Deng, B.B. Finlay. Enteropathogenic *Escherichia coli* (EPEC) attachment to epithelial cells: exploiting the host cell cytoskeleton from the outside. *Cell Microbiol.* 2(1):1–9, 2000.
34. Crawford, J.A., T.E. Blank, J.B. Kaper. The LEE-encoded type III secretion system in EPEC and EHEC: assembly, function, and regulation. In: Escherichia coli: *Virulence Mechanisms of a Versatile Pathogen*, Donnenberg, M.S., ed., San Diego, CA: Academic Press, 2002, pp 337–360.
35. McDaniel, T.K., J.B. Kaper. A cloned pathogencity island from enteropathogenic *Escherichia coli* confers the attaching and effacing phenotype on *E. coli* K-12. *Mol. Microbiol.* 23(2):399–407, 1997.
36. Winstanley, C., C.A. Hart. Type III secretion systems and pathogenicity islands. *J. Med. Microbiol.* 50:116–126, 2001.
37. Elliott, S.J., J. Yu, J.B. Kaper. The cloned locus of enterocyte effacement (LEE) from enterohemorrhagic *Escherichia coli* O157:H7 is unable to confer the attaching and effacing phenotype upon *E. coli* K-12. *Infect. Immun.* 67:4260–4263, 1999.
38. Wagner, P.L., J. Livny, M.N. Neely, D.W.K. Acheson, D.I. Friedman, M.K. Waldor. Bacteriophage control of Shiga toxin 1 production and release by *Escherichia coli*. *Mol. Microbiol.* 44(4):957–970, 2002.
39. Burland, V., Y. Shao, N.T. Perna, G. Plunkett, H.J. Sofia, F.R. Blattner. The complete DNA sequence and analysis of the large virulence plasmid of *Escherichia coli* O157:H7. *Nucl. Acids Res.* 26(18):4196–4204, 1998.
40. Mellies, J.L., S.J. Elliott, V. Sperandio, M.S. Donnenberg, J.B. Kaper. The Per regulon of enteropathogenic *Escherichia coli*: identification of a regulatory cascade and a novel transcriptional activator, the locus of enterocyte effacement (LEE)-encoded regulator (Ler). *Mol. Microbiol.* 33(2):296–306, 1999.

41. Friedberg, D., T. Umanski, Y. Fang, I. Rosenshine. Hierarchy in the expression of the locus of enterocyte effacement genes of enteropathogenic *Escherichia coli. Mol. Microbiol.* 34(5):941–952, 1999.
42. Elliott, S.J., V. Sperandio, J.A. Girón, S. Shin, J.L. Mellies, L. Wainwright, S.W. Hutcheson, T.K. McDaniel, J.B. Kaper. The locus of enterocyte effacement (LEE)-encoded regulator controls expression of both LEE- and non-LEE-encoded virulence factors in enteropathogenic and enterohemorrhagic *Escherichia coli. Infect. Immun.* 68(11):6115–6126, 2000.
43. Bassler, B.L. Small talk: cell-to-cell communication in bacteria. *Cell* 109:421–424, 2002.
44. Sperandio, V., A.G. Torres, J.A. Girón, J.B. Kaper. Quorum sensing is a global regulatory mechanism in enterohemorragic *Escherichia coli* O157:H7. *J. Bacteriol.* 183(17):5187–5297, 2001.
45. Sperandio, V., J.L. Mellies, W. Nguyen, S. Shin, J.B. Kaper. Quorum sensing controls expression of the type III secretion gene transcription and protein secretion in enterohemorrhagic and enteropathogenic *Escherichia coli. Proc. Natl. Acad. Sci.* USA 96(26):15196–15201, 1999.
46. Sperandio, V., C.C. Li, J.B. Kaper. Quorum-sensing *Escherichia coli* regulator A: a regulator of the LysR family involved in the regulation of the locus of enterocyte effacement pathogenicity island in enterohemorrhagic *E. coli. Infect. Immun.* 70(6):3085–3093, 2002.
47. Snyder, L., W. Champness. *Molecular Genetics of Bacteria.* Washington, DC: ASM Press, 1997.
48. Iriarte, M., I. Stainier, G.R. Cornelis. The *rpoS* gene from *Yersinia enterocolitica* and its influence on expression of virulence factors. *Infect. Immun.* 63(5):1840–1847, 1995.
49. Ibanez-Ruiz, M., V. Robbe-Saule, D. Hermant, S. Labrude, F. Norel. Identification of RpoS (σ^S)-regulated genes in *Salmonella enterica* serovar typhimurium. *J. Bacteriol.* 182(20):5749–5756, 2000.
50. Friedman, D.I., D.L. Court. Bacteriophage lambda: alive and well and still doing its thing. *Curr. Opin. Microbiol.* 4:201–207, 2001.
51. Murray, E.G.D., R.A. Webb, M.B.R. Swann. A disease of rabbits characterised by a large mononuclear leucocytosis, caused by a hitherto undescribed bacillus *Bacterium monocytogenes* (n. sp.). *J. Pathol. Bacteriol.* 29:407–439, 1926.
52. Rocourt, J. The genus *Listeria* and *Listeria monocytogenes*: phylogenetic position, taxonomy, and identification. In: *Listeria, Listeriosis, and Food Safety*, Ryser, E.T., E.H. Marth, eds., New York, NY: Marcel Dekker, 1999, pp 1–20.
53. Brock, T.D., ed. *Milestones in Microbiology: 1546 to 1940.* Washington, DC.: ASM Press, 1961.
54. Schlech, W.F., P.M. Lavigne, R.A. Bortolussi, A.C. Allen, E.V. Haldane, A.J. Wort, A.W. Hightower, S.E. Johnson, S.H. King, E.S. Nicholls, C.V. Broome. Epidemic listeriosis - evidence for transmission by food. *N. Engl. J. Med.* 308(4):203–206, 1983.
55. Vázquez-Boland, J.A., M. Kuhn, P. Berche, T. Chakraborty, G. Domínguez-Bernal, W. Goebel, B. González-Zorn, J. Wehland, J. Kreft. *Listeria* pathogenesis and molecular virulence determinants. *Clin. Microbiol. Rev.* 14(3):584–640, 2001.
56. Ryser, E.T. Foodborne listeriosis. In: Listeria, *Listeriosis, and Food Safety*, Ryser, E.T., E.H. Marth, eds., New York: Marcel Dekker, 1999, pp 299–358.
57. Slutsker, L., A. Schuchat. Listeriosis in humans. In: Listeria, *Listeriosis, and Food Safety*, Ryser, E.T., E.H. Marth, eds., New York: Marcel Dekker, 1999, pp 75–96.
58. Frye, D.M., R. Zweig, J. Sturgeon, M. Tormey, M. LeCavalier, I. Lee, L. Lawani, L. Mascola. An outbreak of febrile gastroenteritis associated with delicatessen meat contaminated with *Listeria monocytogenes. Clin. Infect. Dis.* 35:943–949, 2002.
59. Sim, J., D. Hood, L. Finnie, M. Wilson, C. Graham, M. Brett, J.A. Hudson. Series of incidents of *Listeria monocytogenes* non-invasive febrile gastroenteritis involving ready-to-eat meats. *Lett. Appl. Microbiol.* 35:409–413, 2002.
60. Lou, Y., A.E. Yousef. Characteristics of *Listeria monocytogenes* important to food processors. In: Listeria, *Listeriosis, and Food Safety*, Ryser, E.T., E.H. Marth, eds., New York: Marcel Dekker, 1999, pp 131–224.

61. Glaser, P., L. Frangeul, C. Buchrieser, C. Rusniok, A. Amend, F. Baquero, P. Berche, H. Bloeker, P. Brandt, T. Chakraborty, A. Charbit, F. Chetouani, E. Couvé, A.d. Daruvar, P. Dehoux, E. Domann, G. Domínguez-Bernal, E. Duchaud, L. Durant, O. Dussurget, K.-D. Entian, H. Fsihi, F.G.-D. Portillo, P. Garrido, L. Gautier, W. Goebel, N. Gómez-López, T. Hain, J. Hauf, D. Jackson, L.-M. Jones, U. Kaerst, J. Kreft, M. Kuhn, F. Kunst, G. Kurapkat, E. Madueño, A. Maitournam, J.M. Vicente, E. Ng, H. Nedjari, G. Nordsiek, S. Novella, B.d. Pablos, J.-C. Pérez-Diaz, R. Purcell, B. Remmel, M. Rose, T. Schlueter, N. Simoes, A. Tierrez, J.A. Vázquez-Boland, H. Voss, J. Wehland, P. Cossart. Comparitive genomics of *Listeria* species. *Science* 294:849–852, 2001.
62. Wiedmann, M. Molecular subtyping methods for *Listeria monocytogenes*. *J. AOAC Int.* 85(2):524–531, 2002.
63. Graves, L.M., B. Swaminathan, S.B. Hunter. Subtyping *Listeria monocytogenes*. In: Listeria, *Listeriosis, and Food Safety*, Ryser, E.T., E.H. Marth, eds., New York: Marcel Dekker, 1999, pp 279–298.
64. Kuhn, M., W. Goebel. Pathogenesis of *Listeria monocytogenes*. In: Listeria, *Listeriosis, and Food Safety*, Ryser, E.T., E.H. Marth, eds., New York: Marcel Dekker, 1999, pp 97–130.
65. Wiedmann, M., J.L. Bruce, C. Keating, A.E. Johnson, P.L. McDonough, C.A. Batt. Ribotypes and virulence gene polymorphisms suggest three distinct *Listeria monocytogenes* lineages with differences in pathogenic potential. *Infect. Immun.* 65(7):2707–2716, 1997.
66. Nadon, C.A., D.L. Woodward, C. Young, F.G. Rodgers, M. Wiedmann. Correlations between molecular subtyping and serotyping of *Listeria monocytogenes*. *J. Clin. Microbiol.* 39(7):2704–2707, 2001.
67. Boerlin, P., E. Bannerman, F. Ischer, J. Rocourt, J. Bille. Typing *Listeria monocytogenes*: a comparison of random amplification of polymorphic DNA with 5 other methods. *Res. Microbiol.* 146:35–49, 1995.
68. Lawrence, L.M., J. Harvey, A. Gilmour. Development of a random amplification of polymorphic DNA typing method for *Listeria monocytogenes*. *Appl. Environ. Microbiol.* 59(9):3117–3119, 1993.
69. Giovannacci, I., C. Ragimbeau, S. Queguiner, G. Salvat, J.-L. Vendeuvre, V. Carlier, G. Ermel. *Listeria monocytogenes* in pork slaughtering and cutting plants: use of RAPD, PFGE, and PCR-REA for tracing and molecular epidemiology. *Int. J. Food. Microbiol.* 53:127–140, 1999.
70. Vogel, B.F., H.H. Huss, B. Ojeniyi, P. Ahrens, L. Gram. Elucidation of *Listeria monocytogenes* contamination routes in cold-smoked salmon processing plants detected by DNA-based typing methods. *Appl. Environ. Microbiol.* 67(6):2586–2595, 2001.
71. Vogel, B.F., L.V. Jørgensen, B. Ojeniyi, H.H. Huss, L. Gram. Diversity of *Listeria monocytogenes* isolates from cold-smoked salmon produced in different smokehouses as assessed by Random Amplified Polymorphic DNA analyses. *Int. J. Food. Microbiol.* 65:83–92, 2001.
72. Brosch, R., J. Chen, J.B. Luchansky. Pulsed-field fingerprinting of listeriae: identification of genomic divisions for *Listeria monocytogenes* and their correlation with serovar. *Appl. Environ. Microbiol.* 60(7):2584–2592, 1994.
73. Larsen, C.N., B. Nørrung, H.M. Sommer, M. Jakobsen. *In vitro* and *in vivo* invasiveness of different pulsed-field gel electrophoresis types of *Listeria monocytogenes*. *Appl. Environ. Microbiol.* 68(11):5698–5703, 2002.
74. Harvey, J., A. Gilmour. Characterization of recurrent and sporadic *Listeria monocytogenes* isolates from raw milk and nondairy foods by pulsed-field gel electrophoresis, monocin typing, plasmid profiling, and cadmium and antibiotic resistance determination. *Appl. Environ. Microbiol.* 67(2):840–847, 2001.
75. Fenlon. D.R. *Listeria monocytogenes* in the natural environment. In: Listeria, *Listeriosis, and Food Safety*, Ryser, E.T., E.H. Marth, eds., New York: Marcel Dekker, 1999, pp 21–38.
76. Wesley, I.V. Listeriosis in animals. In: Listeria, *Listeriosis, and Food Safety*, Ryser, E.T., E.H. Marth, eds., New York: Marcel Dekker, 1999, pp 39–74.

77. Cossart, P., J. Pizarro-Cerdá, M. Lecuit. Invasion of mammalian cells by *Listeria monocytogenes*: functional mimicry to subvert cellular functions. *Trends Cell. Biol.* 13(1):23–31, 2003.
78. Bergmann, B., D. Raffelsbauer, M. Kuhn, M. Goetz, S. Hom, W. Goebel. InlA- but not InlB-mediated internalization of *Listeria monocytogenes* by non-phagocytic mammalian cells needs the support of other internalins. *Mol. Microbiol.* 43(3):557–570, 2002.
79. Portnoy, D.A., V. Auerbach, I.J. Glomski. The cell biology of *Listeria monocytogenes* infection: the intersection of bacterial pathogenesis and cell-mediated immunity. *J. Cell. Biol.* 158(3):409–414, 2002.
80. Bielecki, J., P. Youngman, P. Connelly, D.A. Portnoy. *Bacillus subtilis* expressing a haemolysin gene from *Listeria monocytogenes* can grow in mammalian cells. *Nature* 345:175–176, 1990.
81. Tilney, L.G., D.A. Portnoy. Actin filaments and the growth, movement, and spread of the intracellular parasite, *Listeria monocytogenes*. *J. Cell. Biol.* 109:1597–1608, 1989.
82. Pistor, S., T. Chakraborty, K. Niebuhr, E. Domann, J. Wehland. The ActA protein of *Listeria monocytogenes* acts as a nucleator inducing reorganization of the actin cytoskeleton. *Mol. Microbiol.* 28:81–93, 1994.
83. Dussurget, O., D. Cabanes, P. Dehoux, M. Lecuit, European Listeria Genome Consortium, C. Buchrieser, P. Glaser, P. Cossart. *Listeria monocytogenes* bile salt hydrolase is a PrfA-regulated virulence factor involved in the intestinal and hepatic phases of listeriosis. *Mol. Microbiol.* 45:1095–1106, 2002.
84. Chico-Calero, I., M. Suarez, B. González-Zorn, M. Scortti, J. Slaghuis, W. Goebel, European Listeria Genome Consortium, J.A. Vázquez-Boland. Hpt, a bacterial homologue of the microsomal glucose-6-phosphate translocase, mediates rapid intracellular proliferation in *Listeria*. *Proc. Natl. Acad. Sci. USA* 99:431–436, 2002.
85. Wiedmann, M., T.J. Arvik, R.J. Hurley, K.J. Boor. General stress transcription factor σ^B and its role in acid tolerance and virulence of *Listeria monocytogenes*. *J. Bacteriol.* 180(14):3650–3656, 1998.
86. Becker, L.A., M.S. Çetin, R.W. Hutkins, A.K. Benson. Identification of the gene encoding the alternative sigma factor σ^B from *Listeria monocytogenes* and its role in osmotolerance. *J. Bacteriol.* 180(17):4547–4554, 1998.
87. Nadon, C.A., B.M. Bowen, M. Wiedmann, K.J. Boor. Sigma B contributes to PrfA-mediated virulence in *Listeria monocytogenes*. *Infect. Immun.* 70(7):3948–3952, 2002.
88. Milohanic, E., P. Glaser, J.-Y. Coppée, L. Frangeul, Y. Vega, J.A. Vázquez-Boland, F. Kunst, P. Cossart, C. Buchrieser. Transcriptome analysis of *Listeria monocytogenes* identifies three groups of genes differently regulated by PrfA. *Mol. Microbiol.* 47(6):1613–1625, 2003.
89. Kreft, J., J.A. Vázquez-Boland. Regulation of virulence genes in *Listeria*. *Int. J. Med. Microbiol.* 291(2):145–157, 2001.
90. Chakraborty, T., M. Leimeister-Wächter, E. Domann, M. Hartl, W. Goebel, T. Nichterlein, S. Notermans. Coordinate regulation of virulence genes in *Listeria monocytogenes* requires the product of the *prfA* gene. *J. Bacteriol.* 174:568–574, 1992.
91. Mengaud, J., S. Dramsi, E. Gouin, J.A. Vázquez-Boland, G. Milon, P. Cossart. Pleiotropic control of *Listeria monocytogenes* virulence factors by a gene that is autoregulated. *Mol. Microbiol.* 5:2273–2283, 1991.
92. Lampidis, R., R. Gross, Z. Sokolovic, W. Goebel, J. Kreft. The virulence regulator protein of *Listeria ivanovii* is highly homologous to PrfA from *Listeria monocytogenes* and both belong to the Crp-Fnr family of transcription regulators. *Mol. Microbiol.* 13:141–151, 1994.
93. Camilli, A., L.G. Tilney, D.A. Portnoy. Dual roles of *plcA* in *Listeria monocytogenes* pathogenesis. *Mol. Microbiol.* 8:143–157, 1993.
94. Freitag, N.E., L. Rong, D.A. Portnoy. Regulation of the *prfA* transcriptional activator of *Listeria monocytogenes*: multiple promoter elements contribute to intracellular growth and cell-to-cell spread. *Infect. Immun.* 61:2537–2544, 1993.
95. Johansson, J., P. Mandin, A. Renzoni, C. Chiaruttini, M. Springer, P. Cossart. An RNA thermosensor controls expression of virulence genes in *Listeria monocytogenes*. *Cell* 110:551–561, 2002.

96. Greene, S.L., N.E. Freitag. Negative regulation of PrfA, the key activator of *Listeria monocytogenes* virulence gene expression, is dispensible for bacterial pathogenesis. *Microbiol* 149:111–120, 2003.
97. Kocks, C., E. Gouin, M. Tabouret, P. Berche, H. Ohayon, P. Cossart. *L. monocytogenes*-induced actin assembly requires the *actA* gene product, a surface protein. *Cell* 68:521–531, 1992.
98. Vázquez-Boland, J.A., C. Kocks, S. Dramsi, H. Ohayon, C. Geoffroy, J. Mengaud, P. Cossart. Nucleotide sequence of the lecithinase operon of *Listeria monocytogenes* and possible role of lecithinase in cell-to-cell spread. *Infect. Immun.* 60:219–230, 1992.
99. Lingnau, A., E. Domann, M. Hudel, M. Bock, T. Nichterlein, J. Wehland, T. Chakraborty. Expression of the *Listeria monocytogenes* EGD *inlA* and *inlB* genes, whose products mediate bacterial entry into tissue culture cell lines, by PrfA-dependent and -independent mechanisms. *Infect. Immun.* 63:3896–3903, 1995.
100. Bubert, A., Z. Sokolovic, S.-K. Chun, L. Papatheodorou, A. Simm, W. Goebel. Differential expression of *Listeria monocytogenes* virulence genes in mammalian host cells. *Mol. Gen. Genet.* 261:323–336, 1999.
101. Kazmierczak, M., S. Mithoe, K.J. Boor, M. Wiedmann. The *Listeria monocytogenes* σ^B regulon includes stress response and virulence functions. *J. Bacteriol.* 185:5722–5734, 2003.
102. Engelbrecht, F., S.-K. Chun, C. Ochs, J. Hess, F. Lottspeich, W. Goebel, Z. Sokolovic. A new PrfA-regulated gene of *Listeria monocytogenes* encoding a small, secreted protein which belongs to the family of internalins. *Mol. Microbiol.* 21:823–837, 1996.
103. Michel, E., J. Mengaud, S. Galsworthy, P. Cossart. Characterization of a large motility gene cluster containing the *cheR, motAB* genes of *Listeria monocytogenes* and evidence that PrfA downregulates motility genes. *FEMS Microbiol. Lett.* 169:341–347, 1998.
104. Ripio, M.-T., J.A. Vázquez-Boland, Y. Vega, S. Nair, P. Berche. Evidence for expressional crosstalk between the central virulence regulator PrfA and the stress response mediator ClpC in *Listeria monocytogenes. FEMS Microbiol. Lett.* 158:45–50, 1998.
105. Hacker, J., J.B. Kaper. The concept of pathogenicity islands. In: *Pathogenicity islands and other mobile virulence elements*, Hacker, J., J.B. Kaper, eds., Washington, DC: ASM Press, 1999, pp 1–11.
106. Kaper, J.B., J.L. Mellies, J.P. Nataro. Pathogenicity islands and other mobile genetic elements of diarrheagenic *Escherichia coli*. In: *Pathogenicity islands and other mobile virulence elements*, Hacker, J., J.B. Kaper, eds., Washington, DC: ASM Press, 1999, pp 33–58.
107. Braun, V., C.v. Eichel-Streiber. Virulence-associated mobile elements in *Bacilli* and *Clostridia*. In: *Pathogenicity islands and other mobile virulence elements*, Hacker, J., J.B. Kaper, eds., Washington, DC: ASM Press, 1999, pp 233–264.
108. Kreft, J., J.A. Vázquez-Boland, E. Ng, W. Goebel. Virulence gene clusters and putative pathogenicity islands in listeriae. In: *Pathogenicity islands and other mobile virulence elements*, Hacker, J., J.B. Kaper, eds., Washington, DC: ASM Press, 1999, pp 219–232.
109. Buchanan, R.L. Identifying and controlling emerging foodborne pathogens: research needs. *Emerg. Infect. Dis.* 3(4):517–521, 1997.
110. Ryser, E.T., E.H. Marth, eds. Listeria, *Listeriosis, and Food Safety,* 2nd ed. 1999, Marcel Dekker: New York, NY.
111. Donnenberg, M.S., T.S. Whittam. Pathogenesis and evolution of virulence in enteropathogenic and enterohemorrhagic *Escherichia coli*. *J. Clin. Investig.* 107(5):539–548, 2001.
112. Gilbert, G.L. Molecular diagnostics in infectious diseases and public health microbiology: cottage industry to postgenomics. *Trends Mol. Med.* 8(6):280–287, 2002.

3.03

Biofilm Production by *Listeria monocytogenes*

William K. Shaw, Jr. and Lynne McLandsborough

CONTENTS

3.1 INTRODUCTION

In their natural environments, bacteria do not exist as isolated cells but grow and survive in organized communities (1). Simplistically, biofilms are microorganisms growing on a solid surface. In practice, however, biofilms are generally defined as matrix enclosed bacterial populations adhered to a surface or to each other, producing a dynamic environment in which the component microbial cells appear to reach homeostasis, optimally organized to make use of all available nutrients (2–6). Once colonizing the surface, microorganisms form a monolayer or multilayer of cells at the interfaces between solids and liquids, or air and liquids (1,5). Biofilms are formed by almost every type of microorganism under suitable conditions, including spoilage microorganisms such as *Pseudomonas* and pathogens of great concern to the food industry including the genera of *Bacillus, Vibrio, Listeria*, or *Salmonella*, (5,7,8). Throughout natural ecosystems, biofilms can be found on almost any surface exposed to water (9). In addition, biofilms are of great concern to the medical community, because biofilms play an important role in persistent human infections (4,10). In summation, biofilms can be found on almost any surface exposed to water (9) and, once established, are well documented to have a greater resistance to antimicrobial agents than free growing (planktonic) organisms (11–19). This chapter will discuss the basics of biofilm formation with an emphasis on *Listeria monocytogenes*.

3.2 BIOFILM FORMATION AND STRUCTURE

There are several steps in the formation of bacterial biofilms: transport, initial adhesion, substrate attachment, and microcolony formation or cell to cell adhesion (Figure 3.1). The first step in biofilm formation is transport of the organism to a solid surface. This can be via motility of the organism, diffusion of the organism through the environment, or convection of the system. Biofilm forming bacteria may use all of these mechanisms at one time or another. It is well documented that flagella mutants often have lower biofilm production under static conditions indicating that under these conditions flagella are involved in active cellular transport to surfaces. Under fluid flow greater deposition of bacteria has been observed when compared to static conditions (20), and it is speculated that turbulent flow may thrust bacterial cells onto the surface, enhancing adhesion and biofilm formation (17).

Once the organisms approach the surface, physicochemical forces are thought to influence the initial adhesion of the organisms. Physiochemical forces that can take place are: van der Waals interactions (>50 nm from the surface), repulsive electrostatic interactions (10–20 nm from the surface), a combination of repulsive and attractive electrostatic interactions (2–10 nm from the surface), and hydrophobic interactions (0.5–2 nm from the surface) (21). Van der Waals forces are brief dipole interactions between atoms in each body and are generally considered to be attractive forces (22). Electrostatic forces are due to the charge around each interacting body. Bacteria, as well as most natural solid surfaces,

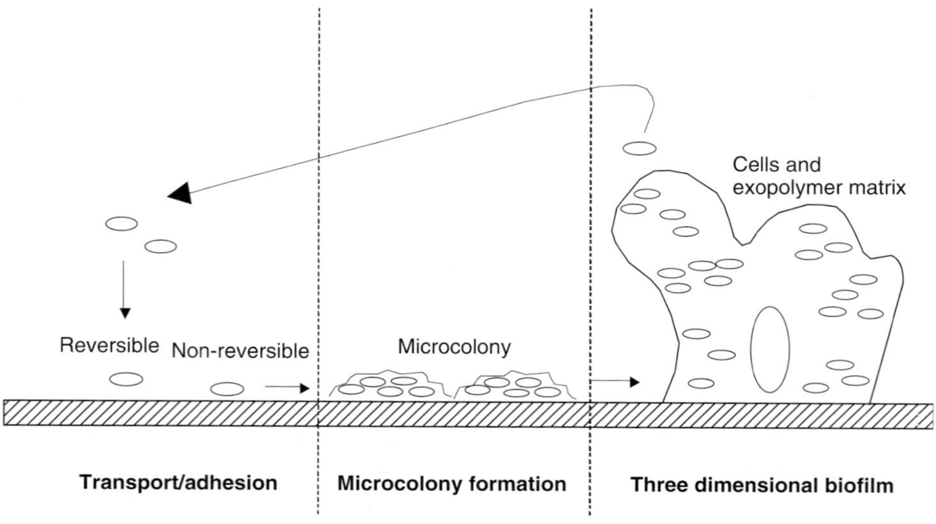

Figure 3.1 Schematic of steps involved in biofilm formation.

have an overall gross negative charge (23). Surface charge on biological materials and bacterial cells is due to ionization of surface groups, such as amino, carboxyl, phosphate, and less commonly, sulfate groups and capsular macromolecules. When immersed in an aqueous media, each surface becomes surrounded by oppositely charged ions, some of which are strongly attached to the surface, whereas the rest are distributed in a diffuse layer. The thickness of this diffuse layer depends on the ionic strength of the solution and the valences of the counter ions. The electrical interactions between particles (including bacteria) in solution are governed by the extension of the diffuse layer; when two particles are charged alike there is a repulsion between the two interacting bodies (23,24).

The Derjaguin, Landau, Verwey and Overbeek (DLVO) theory assumes that the interaction of the bacteria and the solid surface is the sum of the attractive van der Waals forces and electrostatic repulsive interactions (24,25). This interaction is dependent upon the distance of separation. Initially, van der Waals attractions are thought to predominate and pull the two surfaces closer together. As bacterial cells move closer to surfaces, the repulsive electrostatic force creates an energy barrier that must be overcome before the two interacting bodies can move closer together (25). At this point, the adhesion is considered reversible or in a secondary minimum (25,26). Hydrophobic or biological interactions are thought to be important in irreversible bacterial adhesion or primary minimum (22,27–30). Since the hydrophobic interactions in water are much stronger than van der Waals attraction at small separations (22), it has been suggested that hydrophobic interactions between the cell surface and the solid substratum may be one mechanism for overcoming the repulsive electrostatic force described by the DVLO theory, excluding water between the two surfaces and causing the bacterial to adhere irreversibly (31,32). Due to the complex nature of the bacterial cell surface, adherence may not always follow DVLO theory, because a single bacterial strain can adhere to a variety of surfaces with differing surface energies (21).

After initial adhesion occurs, bacteria form strong attachments to surfaces by synthesis of extracellular polymeric substances (EPS), facilitating irreversible bacterial attachment to a surface and maintaining the microcolony and biofilm structure (3,4,6,33,34). EPS has been shown to also enhance nutrient capture and resistance to environmental stress and antimicrobial agents (1,4–6,10,35). When mature, biofilms exist as a structured matrix with vertical and horizontal channels to allow liquid flow and dispersion of nutrients and waste. Within a microbial biofilm, the matrix consists of microbial cells (2–5%), polysaccharides (1–2%), and water (up to 97%), along with excreted cellular products (1–2%) (6). Although made up of these components, the actual structure of the biofilm matrix will vary depending on the types microbial cells present, their physiological status, the nutrients available, and the physical conditions present (6).

When growing in a biofilm, bacteria are known to have different growth rate, morphology, and physiology than their planktonic counterparts, and may exhibit varied physiological responses to nutrient conditions (9,36–38). Data show that biofilm bacteria receive less oxygen and fewer nutrients than cells in suspension, resulting in advantages in growth, altered physiology, and increased resistance to toxic agents compared with their planktonic forms (6,39,40). Sutherland proposes that these alterations dictate the placement of different organisms within the biofilm in order to take full advantage, matching environment to microorganism (6).

3.3 *LISTERIA MONOCYTOGENES*

Listeria monocytogenes is ubiquitous in the environment and resistant to diverse environmental conditions. It is associated with plant material (alive or dead), water, and soil.

Although exposure to *Listeria* is probably very common due to the wide environmental distribution on green plant material (some of which is consumed as food), listeriosis is a rare foodborne disease (41). It is the severity of the nonenteric disease (meningitis, septicemia, and abortion) in susceptible individuals (such as immunocompromised or pregnant woman) that warrants the alarm concerning this organism (41).

The very high mortality of listeriosis victims (20–30%) (41) has prompted the United States Department of Agriculture (USDA) to adopt a zero tolerance policy in processed foods. *Listeria monocytogenes* was the most frequent microbial agent associated with FDA food product recalls in the years 1994–1998 (61% of all food product recalls and 88% of all Class I food product recalls), with the next most frequent recall being due to *Salmonella* (11%) (42). Almost three fourths of the *Listeria* related recalls by the FDA occurred in dairy products, pastries, salads, or sandwiches (42). Other food products associated with outbreaks include hot dogs and ready to eat meats (43).

It is generally thought that biofilms within food processing environments are a major source of *L. monocytogenes* (5,44,45). The recent FSIS rule, which required that all establishments producing ready to eat meat and poultry products reformulate products to prevent growth of, and conduct environmental testing of food contact surfaces for the presence of *Listeria* species (43), acknowledges that transfer from the food processing environment is critical in the journey of this organism into processed foods after a lethal heat treatment. In processing plans, common sites of post processing *L. monocytogenes* contamination are filling or packaging equipment, conveyors, collators used for assembling product for packaging, racks for transporting product, hand tools, gloves, and freezers (5,44). One recall from a baking company involved 128 different products after it was discovered that the machine used to slice the items was contaminated with *L. monocytogenes* (42). The presence and persistence of this organism within the food processing environment continues to be a major challenge for the food industry (46).

3.3.1 Biofilm Formation by *Listeria monocytogenes*

3.3.1.1 Attachment and Biofilm Formation of L. monocytogenes to Surfaces

Research into the attachment by pure cultures of *L. monocytogenes* has focused on the influence of environmental conditions such as growth temperature, pH, and media composition (9,47–51); physiochemical forces of adhesion (51–53); and the influence of absorbed food components (or "conditioning layer") in *listerial* adhesion (54,55). In general, *L. monocytogenes* attaches better when grown in minimal or diluted rich media (48,56). Increased production of flagella and extracellular matrix components were observed at lower incubation temperatures (20°C or 4°C), although this may be due to increased incubation times required at lower temperatures (47,57). Neutral pH (7.0) appears to foster greater attachment than basic or acidic media (47,51).

Several different researchers have studied the involvement of physiochemical forces in adhesion. *Listeria monocytogenes* has been shown to attach tofood processing surfaces such as stainless steel, glass, polypropylene, and rubber with varying surface free energies (47,50,57). The surface charge of *L. monocytogenes* cells has been observed to be dependent upon growth temperature; as growth temperature decreased to 8°C, the electrophoretic condition of the cells became less negative (52). In addition, as the ionic strength of the suspending media decreased, the numbers of adherent *L. monocytogenes* cells increased, indicating that electrical interactions and Lewis acid base interactions are involved in adhesion to stainless steel (52). Others have measured cell surface hydrophobicity and calculated interfacial free energy of adhesion to stainless steel and rubber, and found these parameters alone did not explain the attachment capabilities of *L. monocytogenes* (50,53), suggesting another mechanism such as production of exopolymer or protein may be important for adhesion (50,53). This is further supported by the observation that the presence of the proteinase trypsin could reduce *listerial* adhesion to rubber and stainless steel by 99.9% (50). The presence of adsorbed proteins on surfaces generally reduced the adhesion of *L. monocytogenes* regardless of the surface composition or free energy (54,55,58,59). Flagella mutants of *L. monocytogenes* have been observed to attach to stainless steel at levels tenfold lower than wild-type cells under short incubation times (4 h) (40). With longer incubation periods, cell coverage by flagella negative mutants reach close to the same level as flagellated cells, indicating that the flagella are important for initial attachment (40).

3.3.1.2 Genetics and Proteomics of Biofilm Formation

Researchers have been investigating the genes needed for *L. monocytogenes* surface growth using both genetic and proteomic approaches. Using a transposon mutagenesis approach, *L. monocytogenes* mutants that were deficient in attached surface growth were isolated. The genes encoding (p)ppGpp synthase (*relA*), and a gene in the purine salvage pathway (hpt) were identified as responsible for the reduction in surface growth, and mutants were unable to accumulate (p)ppGpp and induce a stringent response (60). In addition, these researchers showed that the *relA* gene was induced after bacterial adhesion, suggesting that adhesive bacteria adapt physiologically to the early stages of biofilm formation (60). Others have used a proteomic approach to compare the proteins produced within a single strain growth in broth and within a biofilm, and have identified 31 proteins at variable levels between the two growth conditions (61). Twenty two proteins were up regulated while nine others were down regulated. Of the seven proteins identified and compared to the *L. monocytogenes* genome, the majority (5) were unknown in *Listeria*. Levels of superoxide dismutase and 30S ribosomal protein S2 were increased in biofilms and the level of flagellin protein decreased when compared to protein levels isolated from planktonically grown cells (61). In addition, Cabanes et al. (62) pointed out that *L. monocytogenes* has a gene homologue to Esp that was shown to be important in biofilm formation of *Enterococcus faecalis* (63).

Some have questioned whether *Listeria monocytogenes* has the ability to produce extensive growth as a biofilm (17,49,64). Hood and Zottola did not observe large increases in cell numbers after initial adhesion (49), and Kalmokoff et al. argue that because only one strain out of 36 tested produced a basic biofilm under high nutrient conditions, this organism may adhere to other organisms in the environment rather than produce an independent biofilm (64). Dolan questioned whether this organism could grow within a biofilm due to its fastidious growth requirements and inability to compete with other microflora (17).

The question remains: Does *Listeria monocytogenes* produce biofilms? In pure laboratory systems, *L. monocytogenes* has been observed to produce fibrils and extracellular materials when attached to a variety of surfaces (47,57,64,65). This growth has been shown to have a distinct three dimensional structure form: a honeycomb, or groups of aggregates surrounded by voids (56,65,66). The thickness of the biofilm varies, with higher cell density layers attached to the substratum and at the edge of the biofilm, and lower cell levels in the center (67). In addition, the growth rate in biofilms was slower that planktonic cells (67), the biofilm grown cells required a greater level of phosphate (9), and a stringent response was induced upon surface growth, which appears to be necessary for biofilm formation (60). This indicates that in laboratory systems *L. monocytogenes* has exhibited the three benchmark biofilm characteristics: extracellular material, three dimensional structure, and altered cellular physiology.

3.3.1.3 Influence of Lineages on Biofilm Formation

Many researchers have studied the variation in the ability of different strains to adhere and form biofilms on processing surfaces. This is because, simplistically, the strains that have the ability to produce persistent biofilms in the food processing environment would be more likely get into foods during processing and cause foodborne disease. However, the relationship between biofilms and postprocessing contamination may be more complex. Researchers have suggested that different isolates of *L. monocytogenes* may have different levels of pathogenicity (68,69). Theoretically, strains with differing levels of pathogenicity may also vary in their ability to form biofilms. Virulent strains that cannot form biofilms may be less likely to cause outbreaks in processed foods, because such organisms will not persist in the processing environment. On the other hand, a biofilm forming, but less virulent, strain of *L. monocytogenes* may also be less likely to cause foodborne outbreaks, due to differences in pathogenicity rather than ecological survival. Following in the same logic, the virulent strains that can form persistent biofilms would be most the likely to cause foodborne listeriosis, because of their ability to persist in the processing plant and their ability to cause human disease.

This potential relationship between pathogenesis and persistence has spurred a number of investigations into adhesion of biofilms produced by different strains of *L. monocytogenes*. Lunden et al. (70) observed that persistent strains had higher levels of adhesion to stainless steel than nonpersistent strains after two hours contact time, but this difference was not observed after increased incubation time (70). My lab adapted a commonly used rapid microtiter plate assay that utilizes crystal violet staining, as an indication of level of surface growth (56). We obtained thirty strains that have been characterized into three genetic lineages (69,71) from both clinical and food processing environments. The strains arrived were coded to remove any researcher bias. After testing using the biofilm microtiter plate assay, it was found that lineage 1 strains (predominately serotype 4b), which were generally associated with human disease, exhibited higher level of biofilm formation than the other two lineages. No correlation was found between environmental persistence and biofilm staining intensity. However, Boruchi et al. recently used this assay to compare a larger sample size (eighty isolates), and found that Division II strains (serotypes 1/2a and 1/2c) predominantly formed greater amounts of crystal violet stained films (72) than division I strains (serotypes 4b and 1/2b). They also compared a number of the same strains from the study by Djordjevic et al. (56) and found similar test results, indicating that differences between the two were not due to technique, but instead were related to sample sizes, and to the specific strains used in each study. This was confirmed by electron micrographs of highly stained biofilms' dense, multidimensional growth, while less stained strains showed loosely distributed cells adherent on a surface; once again indicating that strains of *L. monocytogenes* can vary widely in their biofilm production under laboratory test conditions (72). In addition, Boruchi et al. observed a statistical correlation between persistent strains from a dairy processing plant and greater biofilm formation (72).

A large variation in ability of pure cultures of *L. monocytogenes* to adhere and grow on surfaces under laboratory conditions has been shown (56,64,70,72,73). However, it is still unclear whether laboratory tests accurately evaluate the reasons behind strain persistence in the food processing environment. Recently Moretro and Langsrund proposed that the differences in persistence might be due to the ability of strains to adapt to the complex diversity of biofilms present in food processing environments (46) and if this is the case, monospecies evaluations may not be the best approach to give the answers behind ecological persistence.

3.3.2 *Listeria* in the Processing Environment

Jenkinson and Lappin-Scott describe a biofilm as a dynamic entity: cells continually enter or leave the community, promoting diversification or dispersal, and protozoal grazing and shear forces in flow systems lead to remodeling of the biofilm structure (1). Within the food processing environment, *Listeria monocytogenes* biofilms do not exist in pure culture, but as members of a diverse bacterial community. This commensal and mutual community allows *L. monocytogenes* to survive low nutrient and decreased temperature conditions on food processing surfaces, not as a pure culture but rather in conjunction with other microbes (1,9,74).

Several researchers have studied the interactions of *L. monocytogenes* with other bacteria. In one study by Sasahara and Zottola, *Listeria monocytogenes* levels in biofilms were greater when grown in mixed culture with *Pseudomonas fragi* (75). Other researchers have also observed enhanced biofilm formation with exopolysaccharide producing *Pseudomonas putida* (76). Jeong and Frank showed that *L. monocytogenes* grew with *Flavobacterium* sp. and *Pseudomonas* sp. in multispecies biofilms on stainless steel (77). There is some indirect evidence that *Listeria* may have favorable interaction with fungi and bacteria: because it can grow in the presence of trehalose and cellobiose, two carbohydrate sources commonly found and fungi (78) and enhanced biofilm formation has been observed with trahalose as the only carbon source (48) although this may be strain dependent (56). In addition, *Listeria* can utilize amino sugars glucosamine, N-acetylglucosamine, and N-acetylmuramic acid that are components of both bacterial and fungal cells walls.

In a recent project, in my laboratory in collaboration with R.E. Levin (University of Massachusetts) set out to evaluate levels of *L. monocytogenes* present in food processing biofilms using an enrichment most probable number (MPN) with polymerase chain reaction (PCR) confirmation. Floor drains in a seafood processing plant were sampled bimonthly over a course of 7 months, and two locations consistently were confirmed to harbor *L. monocytogenes*. The MPN had a detection limit of 0.03 MPN/cm^2 and when detected, the levels of *L. monocytogenes* varied between 9×10^{-1} and greater than 1×10^4 MPN/cm^2. The plate counts for heterotrophic bacteria for the same areas where consistently between 7×10^5 and 3×10^8 CFU/cm^2. This indicates that the population of *L. monocytogenes* within the biofilm was a vast minority of the population within a complex floor drain ecosystem (manuscript in preparation). It is important to point out that this data does not define how *Listeria* is interacting within the biofilm population. It may be randomly becoming entrapped and growing within the EPS of the diverse biofilm, specifically coaggregating with some members of the population, or possibly forming "microcolonies" or three dimensional compartments of *Listeria* within the biofilm. To date, limited knowledge is available with regard to the other organisms that are present in the processing environment, and the interaction of *Listeria* with these organisms.

The majority of biofilm research has been performed on single species biofilms under laboratory conditions. However, biofilms often exist as multispecies communities in the environment. One of the best studied areas of multispecies biofilms has been in the area of plaque formation in dental microbiology. In plaque, some species of bacterial cells are thought to be primary or early colonizers, adhering to the teeth and forming microcolonies and EPS. Secondary colonizing bacteria then adhere to the developing biofilm through a mechanism called coaggregation (79). In coaggregation, single cells or aggregated clumps of cells adhere to the primary colonizing cells through the presence of surface adhesions and receptors, thus developing a biofilm with a diverse bacterial population (79). Coaggregation has also been observed between bacterial species isolated from fresh water biofilms (80,81), and has been associated with lectin to polysaccharide interactions between species that were depending upon the growth phase of the cultures (81). Since *L. monocytogenes* appear to persist within floor drains in processing plants, it is a possibility that this organism may be interacting with fresh water biofilms.

3.3.3 Resistance of *L. monocytogenes* Biofilms to Sanitizing Agents

These dynamic communities pose great difficulties to food processors who attempt to remove them, because once the organisms attach to the surface, they become more resistant to chlorine, iodine, acid anionic, and quaternary ammonium sanitizers (74,82). *Listeria monocytogenes* has been reported to produce sanitizer resistant biofilms while attaching to glass, stainless steel, and buna-N rubber (14,83,84). Research has also determined that the processing surface used, concentration of sanitizer, and age of the biofilm affect the ease of removal (7). On stainless steel, for example, adherent cells in biofilms demonstrated significant resistance to conventional chemical sanitizers, germicides, and heat (9,14,57,83,85). With respect to age, entrapped cells grow and reproduce, producing multilayers within the biofilm and increasing resistance to chlorine with age (8,86).

An example of increased resistance of biofilms is shown in Figure 3.2. Colony biofilms were grown on filters placed on agar plates and exposed to sanitizer using a modification of the method of Anderl et al. (87). Colony biofilms growing at a solid to air interface exhibit a three dimensional structure similar to a biofilm grown at a liquid to solid interface [Figure 3.2 (a)]. The advantage of using colony biofilms is that they are easy to grow and easy to remove from the surface substrate. When grown on high nutrient media tryptic soy both (TSB) or minimal defined media (modified welshimers both (MWB) + 1.5% agar) (78), both colonies had similar growth patterns, reaching stationary phase within 20 hours and a final cell number of approximately 6×10^8 CFU/colony at 48h. When exposed to hypochlorine (200 ppm HOCl for six consecutive, ten minute exposures), cells grown under low nutrient conditions were more resistant (1.6 log reduction) than cells grown under high nutrient conditions (3.3 log reduction) [Figure 3.2 (b)]. However, when colonies are removed from the filter and then exposed to 200 ppm HOCl in suspension, the cells from both growth conditions exhibit a similar sensitivity, with a greater than 4 log reduction within 30 seconds. This is the same log reduction as seen with 48h stationary phase planktonically grown cells. This illustrates that the increased resistance of the biofilm is related to the spatial layout, because when the spatial layout is removed, the cells become sensitive to the sanitizer. However, the mechanism behind the increase in resistance to sanitizer when growing in biofilms is still somewhat ambiguous. In the case presented here, it is likely that the cellular components neutralized the HOCl solution as it diffused into the biofilm mass. Differences in high versus low nutrient conditions may be due to differences in cellular physiology or production of extracellular material.

It is likely the increased resistance of biofilms is due to a variety of factors that are either related to the structural form of the biofilm, or due to physiological differences in biofilm growing cells. Theories that support biofilm structure contributes to increased resistance suggest that it is due to reduced diffusion of chemicals through the biofilm mass, or chemical neutralization of sanitizers by the outer layers of biofilm, which would protect the inner cells (1,88). On the other hand, reduced growth rates and cellular physiology may also contribute to this resistance. Reduced growth rates in the biofilm reduce the susceptibility to select agents such as antibiotics that require active cellular component synthesis (such as protein translation or cell wall synthesis) (88). In addition, cell to cell communication may be more likely due to the close proximity of cells within a biofilm. Researchers have observed increased resistance to oxidative stress in biofilms due to quorum sensing in *Pseudomonoas aeruginosa* (89). Also in the *P. aeruginosa* system, others have hypothesized that the presence of a small population of "persister" cells with a higher resistance to antimicrobials were responsible for the increased resistance to antimicrobials in both biofilm and stationary phase cultures (90).

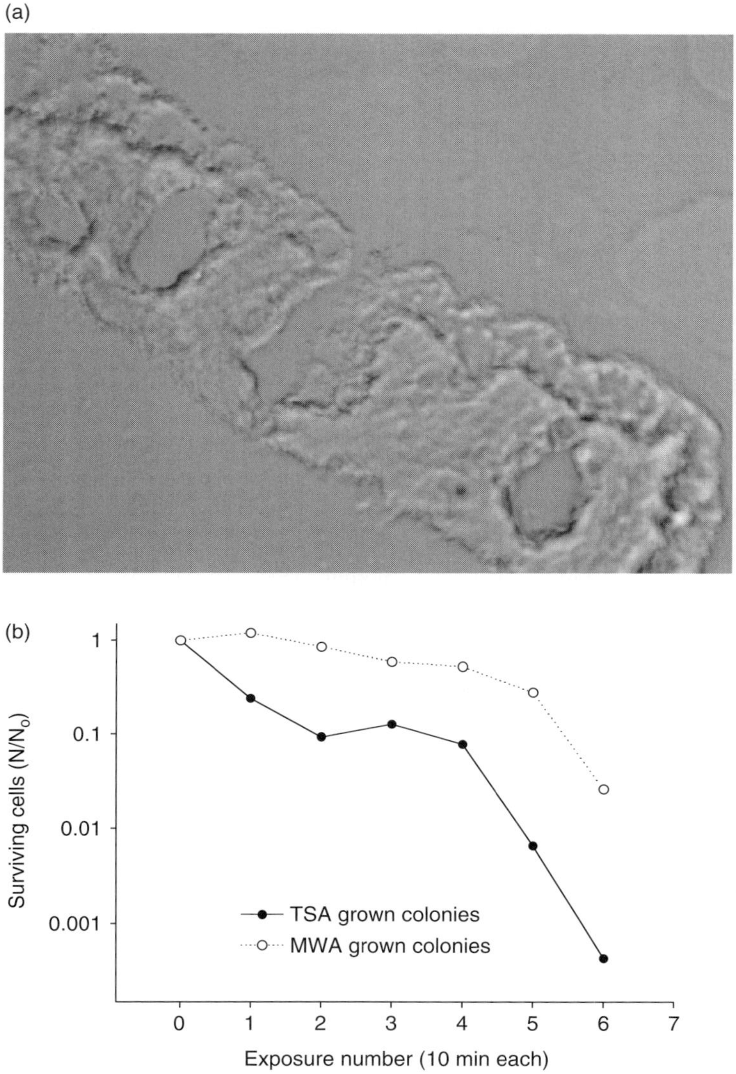

Figure 3.2 Colony biofilms were grown on 0.22 μm polycarbonate filters placed on solid 1.5% agar media with either high nutrients (tryptic soy broth) or low nutrients (modified Welshimers broth with 1.5% agar). Colonies were transferred to fresh agar plates every 24 h. (a) A cryosection of a 48 h colony grown on tryptic soy agar. (b) Colony biofilms were grown for 48 h at 32°C prior to consecutive 10 min exposures of 200 ppm HOCl solution pH 7.0. Every 10 min, surviving cell numbers were evaluated by removal from the filter in neutralization buffer and plated.

3.4 FUTURE DIRECTIONS

There are still many questions with regard to the ability of *L. monocytogenes* to grow and persist on food processing surfaces. Some genetic determinants needed for biofilm formation have been identified, and with the publication of the genome (91) it is likely that the genetic nature of monoculture biofilms will be unraveled in the future. At this time, it is unclear why certain strains of this organism persist within a processing environment and how it interacts and survives within a multispecies biofilm. Utilization of the published genome data along with the study of the microbial diversity of the food processing environment may be a start in answering these questions.

ACKNOWLEDGMENTS

Colony biofilm work was performed in the laboratory of Philip Stewart at the Center for Biofilm Engineering at Montana State University-Bozeman, Bozeman Montana with the assistance of Betsey Pitts. This material is based upon work supported by the Cooperative State Research Extension Education Service, U.S. Department of Agriculture, Massachusetts Agricultural Experiment Station and the Department of Food Science Under Project number MAS 00837 and Seafood Safety Special Research Grants 2000-34423-9131 and 2001-34423-1031.

REFERENCES

1. Jenkinson, H.F., H.M. Lappin-Scott. Biofilms adhere to stay. *Trends Microbiol.* 9:9–10, 2001.
2. An, Y.H., R.J. Friedman. Laboratory methods for studies of bacterial adhesion. *J. Microbiol. Meth.* 30:141–152, 1997.
3. Doyle, R.J. Microbial *Growth in Biofilms, Part A: Developmental and Molecular Biological Aspects.* Methods in Enzymology Vol. 336. San Diego: CA Academic Press, 1001.
4. O′Toole, G.A., H.B. Kaplan, R. Kolter. Biofilm formation as microbial development. *Annu. Rev. Microbiol.* 54:49–79, 2002.
5. Poulsen, L.V. Microbial biofilm in food processing. Lebensm.-Wiss. *u-Technol.* 32:321–326, 1999.
6. Sutherland, I.W. The biofilm matrix: an immobilized but dynamic microbial environment. *Trends Microbiol.* 9:222–227, 2001.
7. Marriott, N.G. *Principles of Food Sanitation,* 4th ed. Gaithersburg, MD: Aspen Publishers, 1999.
8. Sommer, P., C. Martin-Rouas, E. Mettler. Influence of the adherent population level on biofilm population, structure and resistance to chlorination. *Food Microbiol.* 16:503–515, 1999.
9. Kim, K.Y., J.F. Frank. Effect of nutrients on biofilm formation by *Listeria monocytogenes* on stainless steel. *J. Food Prot.* 58:24–28, 1995.
10. Costerton, J.W., K.J. Cheng, G.G. Geesey, T.I. Ladd, J.C. Nickel, M. Dasgupta, T.J. Marrie. Bacterial biofilms in nature and disease. *Annu. Rev. Microbiol.* 41:435–464, 1987.
11. Bower, C.K., M.A. Daeschel. Resistance responses of microorganisms in food environments. *Int. J. Food Microbiol.* 50:33–44, 1999.
12. LeChevallier, M.W., C.D. Cawthon, R.G. Lee. Inactivation of biofilm bacteria. *Appl. Environ. Microbiol.* 54:2492–2499, 1988.
13. Bolton, K.J., C.E.R. Dodd, G.C. Mead, W.M. Waites. Chlorine resistance of strains of *Staphylococcus aureus* isolated from poultry processing plants. *Lett. Appl. Microbiol.* 6:31–34, 1988.
14. Frank, J.F., R.A. Koffi. Surface-adherent growth of *Listeria monocytogenes* is associated with increased resistance to surfactant sanitizers and heat. *J. Food Prot.* 53:550–554, 1990.
15. Nickel, J.C., J.W. Costerton. Bacterial biofilms and catheters. A key to understanding bacterial strategies in catheter-associated urinary tract infection. *Can. J. Infect. Dis.* 3:619–624, 1992.
16. Reid, G., C. Tieszer, R. Foerch, H.J. Busscher, A.E. Khoury, A.W. Bruce. Adsorption of ciprofloxacin to urinary catheters and effect on subsequent bacterial adhesion and survival. *Coll. Surf. B. Biointerfaces.* 1:9–16, 1993.

17. Donlan, R.M., J.W. Costerton. Biofilms: survival mechanisms of clinically relevant microorganisms. *Clin. Microbiol. Rev.* 15:167–193, 2002.
18. Donlan, R.M. Biofilms: microbial life on surfaces. *Emerg. Infect. Dis.* 8:881–890, 2002.
19. Dunne, W.M. Bacterial adhesion: seen any good biofilms lately? *Clin. Microbiol. Rev.* 15:155–166, 2002.
20. Rijnaarts, H.H.M., W. Norde, E.J. Bouwer, J. Lyklema, A.J.B. Zehnder. Bacterial adhesion under static and dynamic conditions. *Appl. Environ. Microbiol.* 59:3255–3265, 1993.
21. Fletcher, M. Bacterial attachment in aquatic environments: a diversity of surfaces and adhesion strategies. In: *Bacterial Adhesion: Molecular and Ecological Diversity,* M. Fletcher, ed., New York: Wiley-Liss, 1996, pp 1–24.
22. Israelachvili, J.N. *Intermolecular and Surface Forces.* San Diego, CA: Academic Press Inc. 1992.
23. James, A.M. Charge Properties of Microbial Cell Surfaces. In: *Microbial Cell Surface Analysis: Structural and Physiochemical Methods.* Mozes, N., P.S. Handley, H.J. Busscher, P.G. Rouxhet, eds., New York: VCH Publishers, 1991, pp 221–262.
24. Olivera, D.R. Physio-chemical aspects of adhesion. In: *Biofilms: Science and Technology,* Melo, L.F., T.R. Bott, M. Fletcher, B. Capdeville, eds., Dordrecht: Kluwer, 1992.
25. McClements, D.J. Food emulsions: principles, practice and techniques. In: *Contemporary Food Science,* Clydesdale, F.M. ed., Boca Raton, FL: CRC Press, 1999.
26. Marshall, K.C., R. Stout, R. Mitchell. Mechanism of the initial events in the sorption of marine bacteria to surfaces. *J. Gen. Microbiol.* 68:337–348, 1971.
27. Busscher, H.J., M.M. Cowan, H.C.v.d. Mei. On the relative importance of specific and nonspecific approaches to oral microbial adhesion. *FEMS Microbiol. Rev.* 88:199–210, 1992.
28. Loosdrecht, M.C.M.V., J. Lyklema, W. Norde, G. Schraa, A.J.B. Zehnder. Electrophoretic mobility and hydrophobicity as a measure to predict the initial steps of bacterial adhesion. *Appl. Environ. Microbiol.* 53:1898–1901, 1987.
29. Loosdrecht, M.C.M.V., J. Lyklema, W. Norde, G. Schraa, A.J.B. Zehnder, The role of bacterial cell wall hydrophobicity in adhesion. *App. Environ. Microbiol.* 53:1893–1897, 1987.
30. Rosenberg, M., R.J. Doyle, Microbial cell surface hydrophobicity: history, measurement, and significance. In: *Microbial Cell Surface Hydrophobicity,* Doyle, R.J., M. Rosenberg, eds., Washington, D.C.: ASM, 1990.
31. Busscher, H.J., J. Sjollema, H.C.v.d. Mei. Relative importance of surface free energy as a measure of hydrophobicity in bacterial adhesion to solid surfaces. In: *Microbial Cell Surface Hydrophobicity.* Doyle, R.J., M. Rosenberg, eds., Washington, D.C.: ASM, 1990.
32. Busscher, H.J. A.H. Weerkamp. Specific and non-specific interactions in bacterial adhesion to solid substrata. *FEMS Microbiol. Rev.* 46:165–173, 1987.
33. Wimpenny, J., W. Manz, U. Szewzyk. Heterogeneity in biofilms. *FEMS Microbiol. Rev.* 24:661–671, 2000.
34. Wirtanen, G., T. Mattila-Sandholm. Epifluorescence image analysis and cultivation of foodborne biofilm bacteria grown on stainless steel surfaces. *J. Food. Prot.* 56:678–683, 1993.
35. Costerton, J.W. Overview of microbial biofilms. *J. Indust. Microbiol.* 15:137–140, 1995.
36. Hodgson, A.E., S.M. Nelson, M.R.W. Brown, P. Gilbert. A simple in vitro model for growth control of bacterial biofilms. *J. Appl. Bacteriol.* 79:87–93, 1995.
37. Kuchma, S.L., G.A. O′Toole. Surface-induced and biofilm-induced changes in gene expression. *Curr. Opin. Biotech.* 11:429–433, 2000.
38. O′Toole, G.A., R. Kolter. Initiation of biofilm formation in *Pseudomonas fluorescens* WCS365 proceeds via multiple, convergent signaling pathways: a genetic analysis. *Mol. Microbiol.* 28:449–461, 1998.
39. Frank, J.F., R.A.N. Chmielewski. Effectiveness of sanitation with quaternary ammonium compound or chlorine on stainless steel and other domestic food-preparation surfaces. *J. Food Prot.* 60:43–47, 1997.
40. Vatanyoopaisarn, S., A. Nazli, C.E.R. Dodd, C.E.D. Rees, W.M. Waites. Effect of flagella on initial attachment of *Listeria monocytogenes* to stainless steel. *Appl. Environ. Microbiol.* 66:860–863, 2000.

41. Rocourt, J., P. Cossart. *Listeria monocytogenes*. In: *Food Microbiology: Fundamentals and Frontiers*. Doyle, M.P., L.R. Beuchat, T.J. Montiville, eds., Washington, D.C.: ASM, 1997.
42. Wong, S., D. Street, S.I. Delgado, K.C. Klontz. Recalls of foods and cosmetics due to microbial contamination reported to the U.S. Food and Drug Administration. *J. Food. Prot.* 63:1113–1116, 2000.
43. FSIS/USDA., 9 CFR Part 430. Control of *Listeria monocytogenes* in ready-to-eat meat and poultry products, final rule. In: *Federal Register,* 2003, pp 34207–34254.
44. Tompkin, R.B., V.N. Scott, D.T. Bernard, W.H. Sveum, K.S. Gombas. Guidelines to prevent post-processing contamination from *Listeria monocytogenes*. *Dairy Food Environ. Sanit.* 19:551–562, 1999.
45. Reij, M.W., E.D.d. Aantrekker, I.E.R.A.i.M.T. Force. Recontamination as a source of pathogens in processed foods. *Int. J. Food Microbiol.* 91:1–11, 2004.
46. Moretro, T., S. Langsrud. *Listeria monocytogenes*: biofilm formation and persistence in food-processing environments. *Biofilms* 1:107–121, 2004.
47. Herald, P.J., E.A. Zottola. Attachment of *Listeria monocytogenes* to stainless steel surfaces at various temperatures and pH values. *J. Food Sci.* 53:1549–1552, 1988.
48. Kim, K.Y., J.F. Frank. Effect of growth nutrients on attachments of *Listeria monocytogenes* to stainless steel. *J. Food Prot.* 57:720–726, 1994.
49. Hood, S.K., E.A. Zottola. Adherence to stainless steel by foodborne microorganisms during growth in model food systems. *Int. J. Food Microbiol.* 37:145–153, 1997.
50. Smoot, L.M., M.D. Pierson. Influence of environmental stress on the kinetics and strength of attachment of *Listeria monocytogenes* Scott A to Buna-N rubber and stainless steel. *J. Food Prot.* 61:1286–1292, 1998.
51. Smoot, L.M., M.D. Pierson. Effect of environmental stress on the ability of *Listeria monocytogenes* Scott A to attach to food contact surfaces. *J. Food Prot.* 61:1293–1298, 1998.
52. Braindet, R., V. Leriche, B. Carpentier, M.N. Bellon-Fontaine. Effects of the growth procedure on the surface hydrophobicity of *Listeria monocytogenes* cells and their adhesion to stainless steel. *J. Food Prot.* 62:994–998, 1999.
53. Mafu, A.A., D. Roy, J. Goulet, L. Savoie. Characterization of physiochemical forces involved in adhesion of *Listeria monocytogenes* to surfaces. *Appl. Environ. Microbiol.* 57:1969–1973, 1991.
54. Al-Makhlafi, H., J. McGuire, M. Daeschel. Influence of preabsorbed milk proteins on adhesion of *Listeria monocytogenes* to hydrophobic and hydrophilic silica surfaces. *Appl. Environ. Microbiol.* 60:3560–3565, 1994.
55. Al-Makhlafi, H., A. Nasir, J. McGuire, M. Daeschel. Adhesion of *Listeria monocytogenes* to silica surfaces after sequential and competitive adsorption of bovine serum albumin and B-lactoglobulin. *Appl. Environ. Microbiol.* 61:2013–2015, 1995.
56. Djordjevic, D., M. Wiedmann, L.A. McLandsborough. Microtiter plate assay for assessment of *Listeria monocytogenes* to biofilm formation. *Appl. Environ. Microbiol.* 68:2950–2958, 2002.
57. Mafu, A.A., D. Roy, J. Goulet, P. Magny. Attachment of *Listeria monocytogenes* to stainless steel, glass, polypropylene and rubber surfaces after short contact times. *J. Food Prot.* 53:742–746, 1990.
58. Cunliffe, D., C.A. Smart, C. Alexander, V.E.N. Bacterial adhesion at synthetic surfaces. *Appl. Environ. Microbiol.* 65:4995–5002, 1999.
59. Barnes, L.-M., M.F. Lo, M.R. Adams, A.H.L. Chamberlain. Effect of milk proteins on adhesion of bacteria to stainless steel surfaces. *Appl. Environ. Microbiol.* 65:4543–4548, 1999.
60. Taylor, C.M., M. Beresford, H.A.S. Epton, D.C. Sigee, G. Shama, P.W. Andrew, I.S. Roberts. *Listeria monocytonenes rel A* and *hpt* mutants are impaired in surface-attached growth and virulence. *J. Bacteriol.* 184: 621–628, 2002.
61. Tremoulet, F., O. Duche, A. Namane, B. Martinie, T.E.L.G. Consortium, J.C. Labadie. Comparison of protein patterns of *Listeria monocytogenes* grown in biofilm or in planktonic mode by proteomic analysis. *FEMS Microbiol. Lett.* 210:25–31, 2002.
62. Cabanes, D., P. Dehoux, L. Frangeul, P. Cossart. Surface protein and the pathogenic potential of *L. monocytogenes*. *Trends in Microbiol.* 10:238–245, 2002.

63. Toledo-Arana, A., J. Valle, C. Solano, M.J. Arrizubieta, C. Cucarella, M. Lamata, B. Amorena, J. Leiva, J.R. Penades, I. Lasa. The enterococcal surface protein, ESP, is involved in *Enterococcus faecalis* biofilm formation. *Appl. Environ. Microbiol.* 67:4538–4545, 2002.
64. Kalmokoff, M.L., J.W. Austin, X.D. Wan, G. Sanders, S. Banerjee, J.M. Farber. Adsorption, attachment and biofilm formation among isolates of *Listeria monocytogenes* using model conditions. *J. Appl. Bacteriol.* 91:726–734, 200l.
65. Marsh, E.J., H. Luo, H. Wang. A three-tiered approach to differentiate *Listeria monocytogens* biofilm-forming abilities. *FEMS Microbiol. Lett.* 228:203–212, 2003.
66. Chavant, P., B. Martinie, T. Meylheuc, N.N. Bellon-Fontaine, M. Hebraud. *Listeria monocytogenes* L028: surface physiochemical properties and ability to form biofilms at different temperatures and growth phases. *Appl. Environ. Microbiol.* 68:728–737, 2002.
67. Chae, M.S. H. Schraft. Comparative evaluation of adhesion and biofilm formation of different *Listeria monocytogenes* strains. *Int. J. Food Microbiol.* 62:103–111, 2000.
68. Boelin, P., J.C. Piffaretti. Typing of human, animal, food, and environmental isolates of *Listeria monocytogenes* by multilocus enzyme electrophoresis. *Appl. Environ. Microbiol.* 57:1624–1629, 1991.
69. Wiedmann, M., J.L. Bruce, C. Keating, A.E. Johnson, P.L. McDonald, C.A. Batt. Ribotypes and virulence gene polymorphisms sugges three distinct *Listeria monocytogenes* lineages with differences in pathogenic potention. *Infect. Immun.* 65:2707–2716, 1997.
70. Lunden, J.M., M.K. Miettinen, T. Autio, H. Korkeala. Persistent *Listeria monocytogenes* strains show enhanced adherence to food contact times. *J. Food Prot.* 63:1204–1207, 2000.
71. Norton, D.M., J.M. Scarlett, K. Horton, D. Sue, J. Thimothe, K.J. Boor, M. Wiedmann. Characterization and pathogenic potential of *Listeria monocytogenes* isolates from the smoked fish industry. *Appl. Environ. Microbiol.* 67:646–653, 200l.
72. Borucki, M.K., J. D. Peppin, D. White, F. Loge, D.R. Call. Variation in biofilm formation among strains of *Listeria monocytogenes*. *Appl. Environ. Microbiol.* 69:7336–7342, 2003.
73. Norwood, D.E., A. Gilmour. Adherence of *Listeria monocytogenes* to stainless steel coupons. *J. Appl. Microbiol.* 86:567–582, 1999.
74. Blackman, I.C., J.F. Frank. Growth of *Listeria monocytogenes* as a biofilm on various food-processing surfaces. *J. Food Prot.* 59:827–831, 1996.
75. Sasahara, K.C., E.A. Zottola. Biofilm formation by *Listeria monocytogenes* utilizes a primary colonizing microorgansim in flowing systems. *J. Food Prot.* 56:1022–1028, 1993.
76. Hassan, A.N., D.M. Birt, J.F. Frank. Behavior of *Listeria monocytogenes* in a *Pseudomonas putida* biofilm on a condensate-forming surface. *J. Food Prot.* 67:322–327, 2004.
77. Jeong, D.K. J.F. Frank. Growth of *Listeria monocytogenes* at 10°C in biofilms with microorganisms isolated from meat and dairy processing environments. *J. Food Prot.* 57:576–586, 1994.
78. Premaratne, R.J., W.-J. Lin, E.A. Johnson. Development of an improved chemically defined medium for *Listeria monocytogenes*. *Appl. Environ. Microbiol.* 57:3046–3048, 1991.
79. Rickard, A.H., P. Gilbert, N.J. High, P.E. Kolenbrander, P.S. Handley. Bacterial coaggregation: an integral process in the development of multi-species biofilms. *Trends Microbiol.* 11:94–100, 2003.
80. Buswell, C.M., Y.M. Herlihy, P.D. Marsh, C.W. Keevil, S.A. Leach. Coaggregation amongst aquatic biofilm bacteria. *J. Appl. Microbiol.* 83:477–484, 1997.
81. Rickard, A.H., S.A. Leach, C.M. Buswell, N.J. High, P.S. Handley. Coaggregation between aquatic bacteria is mediated by specific-growth-phase-dependent lectin-saccharide interactions. *Appl. Environ. Microbiol.* 66:431–434, 2000.
82. Ren, T.J., J.F. Frank. Susceptibility of starved planktonic and biofilm *Listeria monocytogenes* to quaternary ammonium sanitizer as determined by direct viable and agar plate counts. *J. Food Prot.* 56:573–576, 1993.
83. Lee, S.H., J.F. Frank. Inactivation of surface-adherent *Listeria monocytogenes* hypochlorite and heat. *J. Food Prot.* 54:4–6, 1991.
84. Ronner, A., A.C.L Wong. Biofilm development and sanitizer inactivation of *Listeria monocytogenes* and *Salmonella typhimurium* on stainless steel and Buna-N rubber. *J. Food Prot.* 56:7500–7758, 1995.

85. Mustapha, A., M.B. Liewen. Destruction of *Listeria monocytogenes* by sodium hypochlorite and quaternary ammonium sanitizers. *J. Food Prot.* 52:306–311, 1989.
86. Bower, C.K., J. McGuire, M.A. Daeschel. The adhesion and detachment of bacteria and spores on food-contact surfaces. *Trends Food Sci. Technol.* 7:152–157, 1996.
87. Anderl, J.N., M.J. Franklin, P.S. Stewart. Role of antibiotic penetration limitation in *Klebsiella pneumoniae* biofilm resistance to ampicillin and ciprofloxacin. *Antimicrob. Agents Chemother.* 44:181–1824, 2000.
88. Cloete, T.E. Resistance mechanisms of bacteria to antimicrobial compounds. *Int. Biodeterior. Biodegrad.* 51:277–282, 2003.
89. Hassett, D.J., J.F. Ma, J.G. Elkins, T.R. McDermott, U.A. Ochsner, S.E.H. West, C.T. Huang, J. Fredericks, S. Burnett, P.S. Stewart, G. McFerts, L. Passador, B.H. Iglewski. Quorum sensing in *Pseudomonas aeruginosa* controls expression of catalase and superoxide dismutase genes and mediates biofilm susceptibility to hydrogen peroxide. *Mol. Microbiol.* 34:1082–1093, 1999.
90. Spoering, A.L., K. Lewis. Biofilms and planktonic cells of *P. aeruginosa* have similar resistance to killing by antimicrobials. *J. Bacteriol.* 183:6746–6751, 2001.
91. Glaser, P., L. Frangeul, C. Buchrieser, C. Rusniok, A. Amed, F. Baquer, P. Berche, H. Bloecker, P. Brandt, T. Chakrabroty, A. Charbit, F. Chetouani, E. Couve, A.D. Daruvar, P. Dehoux, E. Domann, G. Dominguez-Bernal, E. Duchaud, L. Durant, O. Dussurget, K.D. Entian, H. Fsihi, F.G.-D. Portillo, P. Garrido, L. Gautier, W. Goebel, N. Gomez-Lopez, T. Hain, J. Hauf, D. Jackson, L.-M. Jones, U. Kaerst, J. Kreft, M. Kuhn, F. Kust, G. Kurapkat, E. Madueno, A. Maitournam, J.M. Vicente, E. Ng, H. Nedjari, G. Nordsiek, S. Novella, B.d. Pablos, J.C. Perez-Diaz, R. Purcell, B. Remmel, M. Rose, T. Schlueter, N. Simoes, A. Tierrez, J.-A. Vazquez-Boland, H. Voss, J. Wehland, P. Cossart. Comparative genomics of *Listeria* species. *Science* 294:849–852, 2001.

3.04

Application of Microbial Molecular Techniques to Food Systems

Robert E. Levin

CONTENTS

4.1 THE POLYMERASE CHAIN REACTION (PCR)

4.1.1 Introduction

The Polymerase Chain Reaction (PCR) is one of the most powerful analytical techniques ever developed. It allows segments of minute amounts of double stranded DNA to be amplified several millionfold in several hours. Its most notable application to foods is for the detection of low numbers of food borne pathogenic and toxigenic bacteria in a wide variety of food products in addition to confirming the identification of such organisms isolated from food.

4.1.2 Requirements for the PCR

The polymerase chain reaction uses repeated temperature cycling involving template denaturation, primer annealing, and the activity of DNA polymerase for extension of the annealed primers from the 3′ ends of both DNA strands (Figure 4.1). This results in exponential amplification of the specific target DNA sequence. The availability of the thermostable *Taq* DNA polymerase, from the extreme thermophile *Thermus aquaticus* greatly facilitates repeated thermal cycling at ~95°C for template denaturation, without having to repeatedly add a less thermally stable DNA polymerase after each cycle. The notably high optimum temperature for *Taq* polymerase activity (75–80°C) allows high extension temperatures (72–75°C) which when coupled with a high annealing temperature (50–65°C) and denaturation at 95°C increases specificity, yield, and sensitivity of the PCR reaction (1). Polymerase chain reactions are usually performed in 0.5 mL or 0.2 mL thin walled polyethylene PCR tubes containing 50 μl total reaction volume. The availability of second generation thermal cyclers with heated lids has eliminated the previous need for overlaying the reaction volumes with 50 μl of mineral oil to prevent evaporation. The four deoxynucleotide triphophosphates (Table 4.1) are presently available commercially premixed. Variables that require optimization include components 5, 6, 7, 8, and 9 in Table 4.1. The concentration of $MgCl_2$ is particularly critical. Innis and Gelfand (2) have discussed the optimization of PCR in detail. Most thermal cycling of PCR encompasses 35 cycles; rarely are more than 35 cycles of benefit.

A typical thermal cycling protocol is given in Table 4.2. After an initial denaturation step at 95°C, steps 2, 3, and 4 are then sequentially performed for 35 cycles, followed by step 5 at 72°C to ensure that the final round of strand synthesis at high substrate concentration is completed. The 6th step, involving reduction of the temperature to 4°C, is used to terminate all reactions for convenient holding until agarose gel electrophoresis is performed. The time required to traverse from one temperature to another is referred to as the

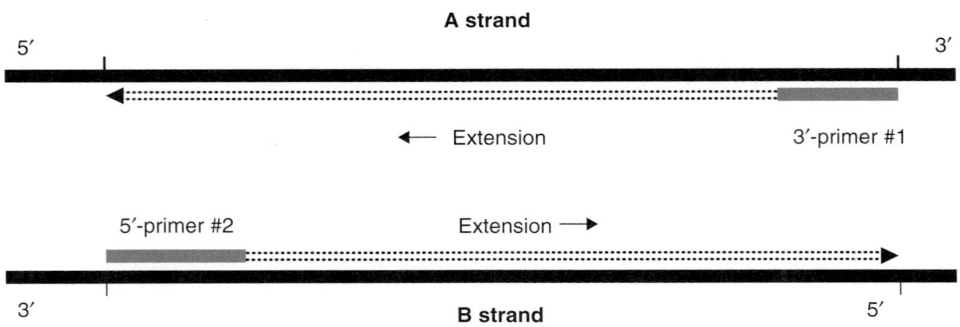

Figure 4.1 Amplification of a known target sequence with a set of two primers.

Table 4.1
Typical PCR reaction components.

1.	Template DNA	1-10 µl
2.	Tris-HCl (pH 8.3 @ 20°C)	20 mM
3.	KCl	25 mM
4.	Triton X-100	0.1%
5.	dNTP's (dATP, dCTP, dGTP, dTTP)	50 mM each
6.	*Taq* polymerase	1.0 unit
7.	$MgCl_2$	1.5 mM
8.	Primers	10 pmoles each

Table 4.2
Typical PCR thermal cycling protocol

Step#	Process	Temperature	Time
1	Initial denaturation	95°C	3 min
2	Denaturation	95°C	1 min
3	Annealing	60°C	1 min
4	Extension	72°C	1 min
5	Final extension step	72°C	4–7 min
6	Terminate reactions and hold	4°C	1–16 min

ramp time, and usually contributes significantly to the total thermal cycling time. Most 35 cycle amplification protocols for PCR are usually completed within 3 to 4 hrs. The total cycling time can often be significantly reduced to less than 90 min by lowering each holding time to 10–20 sec.

4.1.3 Sample Preparation Without Enrichment

A variety of components in foods and various tissues are capable of inhibiting PCR. Sample preparation is therefore one of the more critical steps that can adversely affect PCR. Meat and cheese products are particularly challenging. With ground beef, a 10 g sample is stomached with 90 mL of 0.01 M phosphate buffered saline (PBS, pH 6.0) in a Wirl-pak™ stomacher bag with a mesh insert and homogenized at normal speed for 60 s. A majority of the homogenate (~80 mL) is then centrifuged at low speed (160 g, or 1000 rpm) for 3 min at 4°C to pellet tissue debriss. Most of the supernatant (~75 mL) is then centrifuged at high speed (16,000 g or 10,000 rpm) for 10 min at 4°C to pellet bacterial cells. The supernatant is discarded and the pellet is resuspended in 1 mL of saline solution.

4.1.4 Lysing of Cells and Isolation of Bacterial DNA

Lysing solution (1 mL) (3) at 2X concentration (5.0 mg/ml of sodium azide and 4.0% Triton X-100 in 0.2 M Tris-HCl buffer, pH 8.0) containing proteinase K (0.4 mg/mL) is then added to the resuspended pellet from above. After incubation for 30 min at 37°C the sample is heated at 100°C for 10 min to inactivate the proteinase K and to lyse the target cells. The lysate is then centrifuged at 10,000 g for 10 min to remove cellular debriss.

Bacterial DNA is then isolated from the lysate with the "Wizard" DNA cleanup system (Promega Co, Madison, Wisconsin, USA) by mixing 0.45 mL of the lysate with 1.0 mL of the DNA cleanup resin. The mixture is then passed through a "Wizard" micro column, washed with 2.0 mL of 80% isopropanol, and dried by centrifugation (2 min at 10,000 g). DNA is then eluted from the column with 0.45 μl of sterile Milli-Q water. The eluate is then mixed with 1 μl of pellet paint (Novogen Co., Madison, Wisconsin, USA), which is a visible fluorescent dye labeled carrier to facilitate visualization of the ultimate DNA pellet, followed by the addition of 5 μl of 3M sodium acetate and 102 μl of ice cold 100% ethanol. The preparation is then stored at 4°C for 30 min and then centrifuged at 14,000 g at 4°C for 10 min. The supernatant is removed, the pellet air dried, and dissolved in 20 μl of sterile Milli-Q water, and the entire preparation incorporated into the PCR reaction.

4.1.5 PCR for Identification of Pure Cultures

Confirmation of the identity of pure cultures using PCR is most readily achieved by picking up a small visible amount of cells from an agar culture on the end of a needle and suspending the cells in 20 μl of Milli-Q water. Gene releaser (20 μl, Bioventure, inc., Murfreesboro, TN) is then added and the heating protocol of the manufacturer is used to lyse the cells. 1.0 μl of the resulting cell lysate is then added directly to the PCR reaction mixture.

4.1.6 Quantitative PCR

The author has found that the following methodology is ideally suited for quantitative assessment of target DNA incorporating an internal standard with the use of a conventional thermal cycler.

The operational assumption is that PCR yields an exponential rate of increase of the initial number of target DNA molecules. This is described by the following equation:

$$P = T\,(2^n) \tag{4.1}$$

where P is the number of PCR product molecules formed, T is the number of input target sequences, and n is the number of amplification cycles.

This exponential equation does not, however, apply to the entire amplification process. During the first few cycles, when the number of initial target molecules is very low, the rate of amplification can be anticipated to be low (4). During the last few cycles, when the ratio of PCR product molecules to unreacted primer and *Taq* polymerase molecules has greatly increased, the overall amplification rate can be expected to decrease significantly from an exponential rate. This is illustrated in Figure 4.2. It is only along the linear portion of the plot that a reliable quantitative relationship between the number of input molecules and final accumulated products can be derived. A direct approach to determine when amplification enters the plateau phase is to increase the number of input target molecules and measure the number of product molecules after cycling (Figure 4.2). With quantitative PCR assays, it is important that the initial number of target DNA molecules be low, because the linear range of amplification is represented by no more than about a two log increase in DNA. This will often result in a working range of input target colony forming units (CFU) of about $1.5 \times 10^2 - 1.5 \times 10^4$ (5). If an enrichment step (usually 4.5–6.0 hrs.) is used, then an appropriate dilution of the resulting CFU should be used for the PCR to ensure that the number of input target molecules falls within the linear PCR amplification range. There are several advantages to the use of nonselective enrichment cultivation of food samples: only viable CFU are increased, detection of the initial viable CFU is lowered to 0.5 to 1.0 CFU/g of sample, and dilution of PCR inhibitors in the food sample.

It is important to keep in mind that a 10% reduction in the efficiency of amplification will result in a reduction of PCR product accumulation by more than 95% (4). The inclusion of an internal standard involving an identical set of primers and a similar target can be used to eliminate variations in amplification efficiency, allowing reliable PCR quantitation.

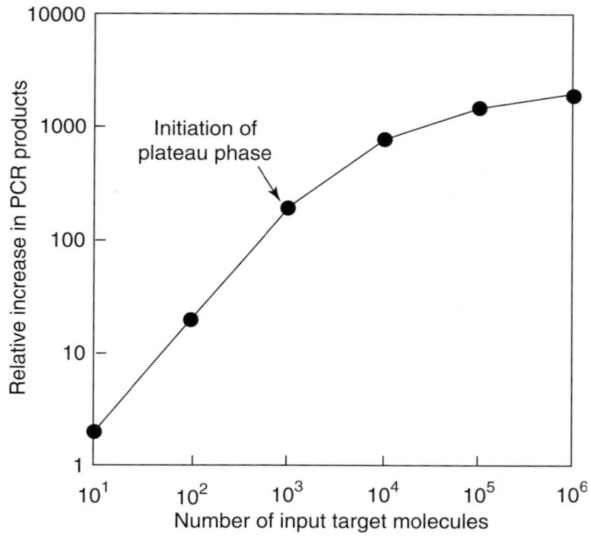

Figure 4.2 Dependence of PCR product accumulation on number of input targets from 35 amplification cycles.

4.1.7 Use of an Internal Standard for Quantitative PCR

The application of an internal standard is based on the coamplification of the target sequence and an internal DNA standard, which is amplified with the same efficiency as the target sequence by using the same primer pair. The internal standard contains the same primer binding sites as the target, and the two DNA pieces compete for reaction reagents to produce PCR products of different sizes, which can be separated in an agarose gel. The log ratio of intensities of amplified target DNA to internal standard is determined by the following equation given by Zarchar et al. (7):

$$- \text{Log } (N_{n1}/N_{n2}) = \log (N_{01}/N_{02}) + n \log (\text{eff}_1/\text{eff}_2) \tag{4.2}$$

Where N_{n1} is the number of resulting target molecules after amplification, N_{n2} is the number of resulting internal standard molecules after amplification, N_{01} is the number of initial target molecules, N_{02} is the number of initial internal standard molecules, n is a given cycle, eff_1 is the efficiency of amplification of target molecules, and eff_2 is the efficiency of amplification of internal standard molecules.

If the efficiencies of amplification ($\text{eff}_1/\text{eff}_2$) are equal, the ratio of amplified products (N_{n1}/N_{n2}) is dependant on the log ratio of starting reactants (N_{01}/N_{02}). Even if the efficiencies of the two reactions are not equal, the values for N_{n1}/N_{n2} still hold, assuming that $\text{eff}_1/\text{eff}_2$ is constant and amplification is in the exponential phase (7). With this technique, varying amounts of target DNA are coamplified with a constant amount of internal standard. The resulting log ratio of intensities of PCR products are plotted against the log of target CFU for construction of a standard curve which is used for determining the number of CFU per gram of food product.

4.1.8 Construction of an Internal Standard for Quantitative PCR

The ideal internal standard should consist of a nucleotide sequence derived from the target sequence, and should utilize the same two primers and primer binding sites so that when the internal standard is added to the PCR mix, amplification occurs in a competitive mode. An internal DNA standard can be synthesized by PCR using a hybrid primer and the reverse primer (8). The hybrid primer contains two components: a 5′-portion consisting of the original forward primer, and a 3′-portion consisting of a small DNA fragment derived from the target sequence. A 20 nucleotide sequence, about 200 to 300 nucleotides in from the 5′ terminal of the target sequence, is identified and a hybrid primer is synthesized (Figure 4.3) for use in the construction of an internal DNA standard. Complete details are given by Guan and Levin (6).

4.1.9 Construction of a Calibration Curve for Quantitative PCR

Varying amounts target DNA from a pure culture are coamplified with a constant amount of the internal standard (IS). The amplified co-PCR products are separated by electrophoresis in a 1.5% agarose gel (Figure 4.4). The gel images are then captured by an Electronic Archival System (Spectroline Model EAS-1000 Spectronics Corp., Westbury, New York, U.S.A.) and analyzed with NIH Image 1.61 software. To correct differences in the intensity of fluorescence of ethidium bromide stained PCR fragments of different sizes, the intensity of the internal standard is multiplied by the ratio of the number of nucleotides in the target sequence to that of the internal standard. The log of the ratio of fluorescence intensity of the amplified target sequence to that of the internal standard is then plotted against the log of CFU to establish a standard curve, using a constant amount of internal standard and varying the number of CFU (Figure 4.5). It is important to note that the calibration curve is operational over little more than a one log cycle of input target sequences (CFU).

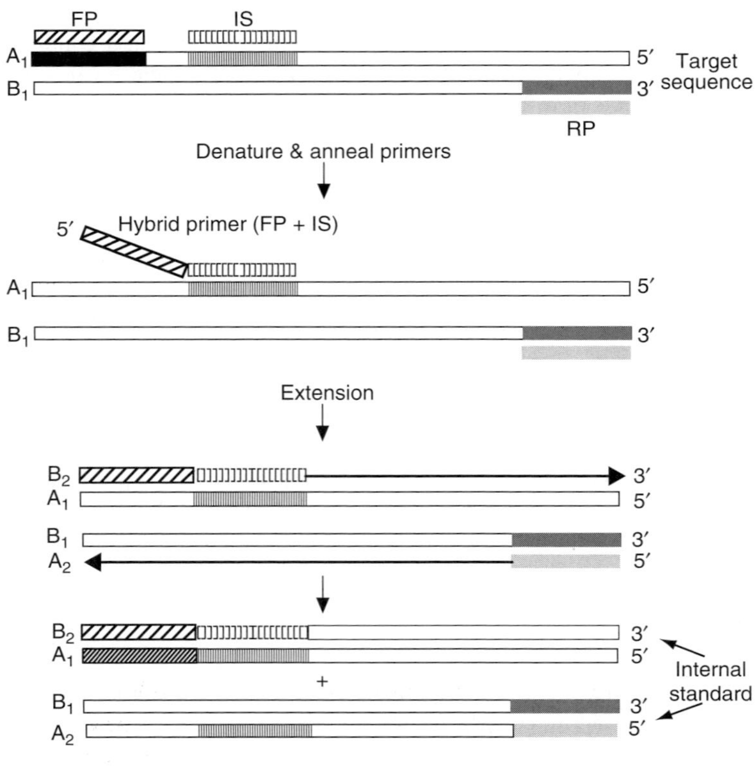

Figure 4.3 Construction of an internal DNA standard for quantitative PCR. FP: forward primer; RP: reverse primer; IS internal sequence.

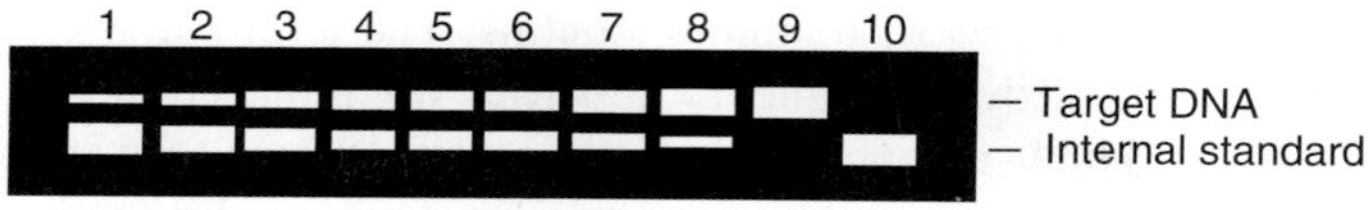

Figure 4.4 Idealized representation of competitive amplification of a constant amount of internal standard and varying amounts of target DNA. Lane 9, target DNA alone. Lane 10, internal standard alone.

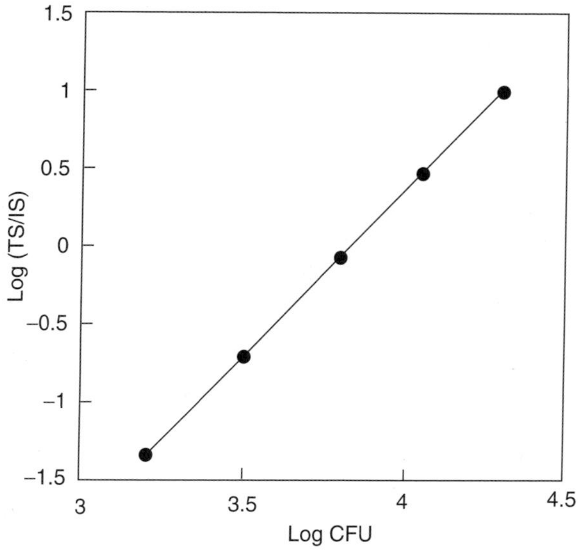

Figure 4.5 Standard curve for quantitative PCR.

4.1.10 The Use of Immunomagnetic Beads in Conjunction with PCR

The use of immunomagnetic beads coupled with PCR for the detection of specific pathogens in foods affords an additional element of selectivity if required (6). *Escherichia coli* O157:H7 produces two shigalike toxins, designated SLT-1 and SLT-2. Both of these toxins are also produced by other serotypes of *E. coli,* and by *Shigella shiga.* If DNA sequences derived from the genes coding for these toxins are used as the targets for amplification to detect this pathogen, the possibility exists that organisms other than *E. coli* O157:H7 may be mistaken for the target organism. The use of immunomagnetic beads coated with *E. coli* O157 specific antibody allows the specific target organism to be physically captured before PCR is undertaken. Immunomagnetic separation (IMS) of the target organism also allows its physical separation from PCR inhibitors in the food product or culture medium. Immunomagnetic separation is presently recognized as an efficient method to increase specificity and to concentrate a target organism from a food product.

A wide size range of immunomagnetic beads (IMB) is available. Generally, beads with a nominal mean diameter of 4.5 μ are used to capture bacterial cells. The magnetic beads are manufactured by precipitating microscopic grains of superparamagnetic iron oxide onto a monodispersion of polymer particles such as polystyrene. The particles may then be coated with a polymeric material such as polystyrene to facilitate subsequent surface binding of immunoglobulins. Alternatively, porous borosilicate glass particles are embedded with superparamagnetic iron oxide. Affinity purified immunoglobulin is covalently bound to the surface of the beads. Magnetic beads are also available for adsorption coating of immunoglobulins. The term paramagnetic refers to the ability of the beads to exhibit magnetic properties when placed in a magnetic field, without retaining residual magnetism when removed from the magnetic field. Superparamagnetic beads exhibit magnetic properties on all sides or throughout their surface. The beads may be irregular in shape or highly uniform spheres. The author has found that if the beads are not coated with polystyrene and they are added directly to a PCR reaction tube, they will inhibit the PCR, presumably due to leakage of Fe_2O_3 out of the beads. Paramagnetic beads are now commercially available, coated with sheep anti-rabbit IgG antibodies, to facilitate coating with any secondary IgG of rabbit origin. Beads are also available coated with sheep antimouse Ig to facilitate coating with any secondary monoclonal antibody. IgG coated beads for capture of *Listeria* species, *Salmonella* species, *E. coli* O157, and *Cryptosporidium parvum* are available at a concentration of at least 4×10^8 beads/mL.

The direct use of magnetic beads on a homogenized food, however, is encumbered by fibrous food products such as meats. With such products, a 4.5–6.0 hr nonselective enrichment is useful in that it allows subsequent dilution of the enrichment broth to eliminate interference of magnetic separation of the beads from tissue debris. Alternatively, without enrichment, differential centrifugation can be used with meat products involving the stomaching of 10 g of product with 90 mL of diluent such as PBS (pH 7.4), followed by initial low speed centrifugation for 3 min at 1,000 rpm (164 g) at 4°C to sediment tissue debris. The supernatant is then removed and subjected to high speed centrifugation at 10,000 rpm (16,300 g) for 10 min at 4°C to pellet the cells. The cell pellet is then suspended in 1.0 mL of PBS followed by the addition of 0.02 mL of IMB (2×10^6 total beads). The beads will capture the target organisms when the sample is subjected to slow intermittent agitation and usually requires 60 min for maximum binding of cells to the beads (6). Any magnet can be used to magnetically attract and pellet the beads with bound cells. We have found however, that the larger the magnet and its magnetic power, the more rapidly are the beads concentrated. We routinely use rectangular bar magnets (Harbor Freight Tools, Cammarillo, CA) with a lifting capacity of 150 lbs (6″ × 1″ × 1.75″) that magnetically pellet the beads within 2 min. The pelleted beads can be readily washed and resuspended several times with intermittent magnetic pelleting before the captured cells are lysed for PCR.

4.1.11 Nested PCR

The use of a pair of "nested primers" flanking a DNA sequence internal to the sequence encompassed by the external pair of primers allows a greater level of amplification than is normally achieved with a single set of primers. A target sequence is initially amplified by an external pair of primers for 20–40 cycles. A small aliquot of this reaction is then amplified a second time, for another 20–40 cycles using the internal or nested primer set (Figure 4.6). The inner pair of primers anneal to complementary sequences internal to the initially amplified product. This has been shown to result in greater amplification than reamplifying with the same initial pair of primers (9). The number of nucleotides flanked by the internally nested primers determines the final product size.

Nested PCR is of particular value with foods in the absence of enrichment cultivation, where inhibitors present from the food may significantly reduce the efficiency of amplification. In addition, the nested primers serve as a control for the specificity of the amplified external sequence and can therefore improve both sensitivity and specificity of DNA amplification (10). Nested PCR, however, introduces an additional risk of amplification of contaminating DNA when used routinely. This problem can be circumvented by using a "one tube nested PCR" involving the use of two primer pairs of

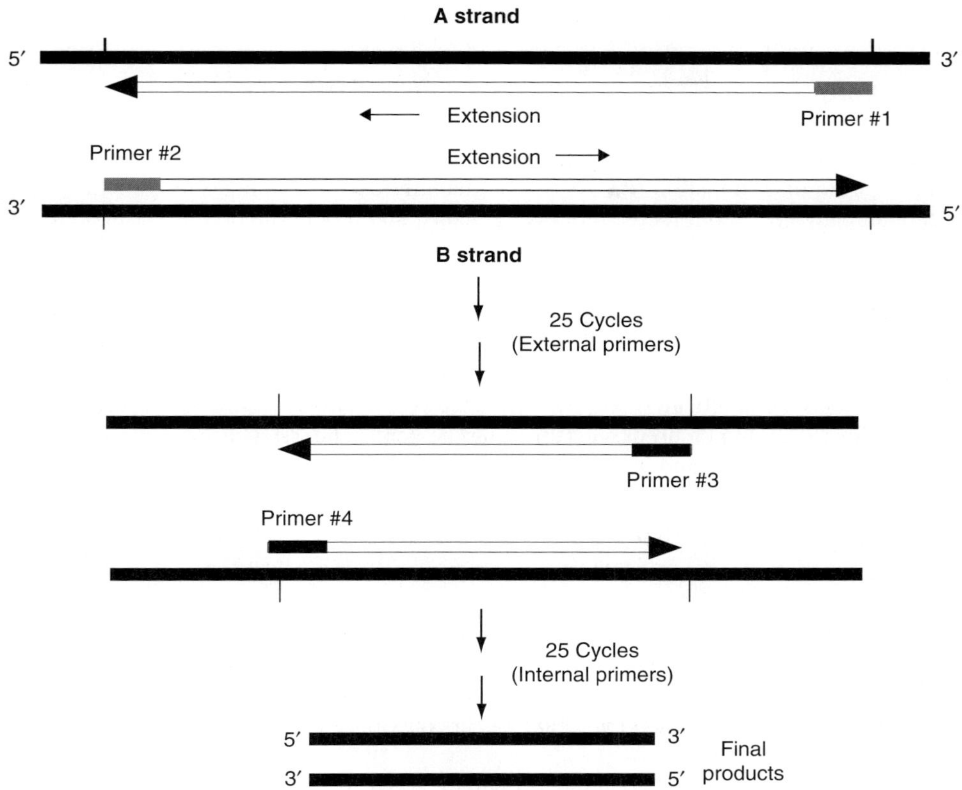

Figure 4.6 Two tube nested PCR. Amplification with external primers for 25 cycles followed by amplification of an aliquot for 25 cycles using internal nested primers.

different melting points (Tm values) in a single reaction mixture. Annealing of the nested primers during the first set of cycles is prevented by their lower Tm. The larger external fragment should ideally have a higher G + C content so that it will denature at a slightly higher temperature than the nested fragment. This difference in Tm values for the two amplified products has to be critically determined for both the larger and the nested fragment. A number of mathematical approaches have been developed for calculating the Tm based on either the G + C content of the fragments or slightly more accurate values based on nearest neighbor thermodynamics (11–13). Other calculations of Tm values for primers are based on formulas developed for nucleotides over 100 bp in length (14). The equation of Suggs et al. (15),

$$Tm = 2°C \times (A + T) + 4°C \times (G + C) \tag{4.3}$$

is widely used for its convenience and approximate determination of Tm values of primers. In addition, the GCG software program "Prime" will readily determine Tm values for primers.

Nested PCR has been applied to the detection of a variety of pathogens in a number of foods (16–20). The use of immunomagnetic capture of target cells prior to nested PCR has been found to increase the sensitivity of detection of *E. coli* O157:H7 from ground beef without enrichment cultivation from 110 CFU/10 g to 24 CFU/10 g (21) presumably due to magnetic separation of captured cells from PCR inhibitors.

4.2 RANDOM AMPLIFICATION OF POLYMORPHIC DNA

4.2.1 Introduction

Random Amplification of Polymorphic DNA (RAPD) has been found to be a rapid and valuable technique for distinguishing different strains of the same species (22) with a high level of strain discrimination (23). It is a particularly useful technique for genetic typing of human pathogens from foods, processing plants, and foodborne outbreaks. In addition, it has been found to be a powerful method can advantageously replace other more cumbersome typing methods such as serotyping, ribotyping, multilocus enzyme electrophoresis, restriction enzyme analysis, and phage typing, and has been found to be very efficient in differentiating strains while still allowing the clear recognition of clusters (24). Random Amplification of Polymorphic DNA has been found capable of distinguishing strains of a given species with identical 16s rDNA sequences (25). This high level of discrimination should allow RAPD to be used in establishing the persistence of a specific strain in foods and in processing plants and its distinction from transient strains of the same species.

4.2.2 Mechanism of RAPD

In conventional PCR, a known DNA sequence is amplified by using two primers, one that anneals to the 3′ end of the sequence on the A strand and is extended inward with *Taq* polymerase from the 3′ end of the primer. The second primer anneals to the 3′ of the B strand and is also extended inward from the 3′ end (Figure 4.1). After the first cycle and denaturation, four target sequences are then available for duplication. The sequence is usually highly specific for the target gene and occurs only once (monomorphic) in the genomic DNA of the organism.

With (RAPD) a single random primer of about 10 nucleotides is used, with no known target sequence being required, and the first round of amplification results in single strands having palindromatic termini (Figure 4.7). The single randomly chosen primer targets specific but unknown sites in the genomic DNA, which are polymorphic (repeating) with respect to the terminal sequences that the single primer anneals to. During subsequent cycles a number of different target sequences are amplified, many of which will be of differing base pair length so as to generate a variety of DNA agarose bands, resulting in a specific DNA banding pattern for each culture (Figure 4.8).

The ratio of primer to template in the RAPD reaction is critical. Template DNA concentration should be carefully titered against a standard concentration of primer so as to reveal the most reproducible amplified products (26). A hazy smear obscuring the amplified bands on the agarose gel is usually caused by failure to saturate the DNA template with primer. This is easily corrected by adjusting the ratio of primer to template DNA (26). Meuner and Grimont (27) expressed concern about the reproducibility of RAPD profiles, but concluded that reproducibility was excellent with standardized methodology. They found that reproducibility of banding patterns was dependant on the make of thermal cycler, with variations in patterns occurring with less rigid temperature control. Reproducibility of RAPD with whole cells is critically dependant on all strains being in the same stage of growth (24). Mazurier and Wernars (28) obtained reproducibility in RAPD profiling of strains of *L. monocytogenes* by using suspensions of washed whole cells grown overnight in broth that were adjusted to an absorbance at 600 nm of 1.5 yielding 7.5×10^6 CFU/ml. 5 μl of the cell suspension were then incorporated directly into a 50 μl PCR reaction volume without a prior cell lysis step. Establishing the same cell density among isolates being subjected to RAPD comparison is particularly important for reproducibility of faint bands.

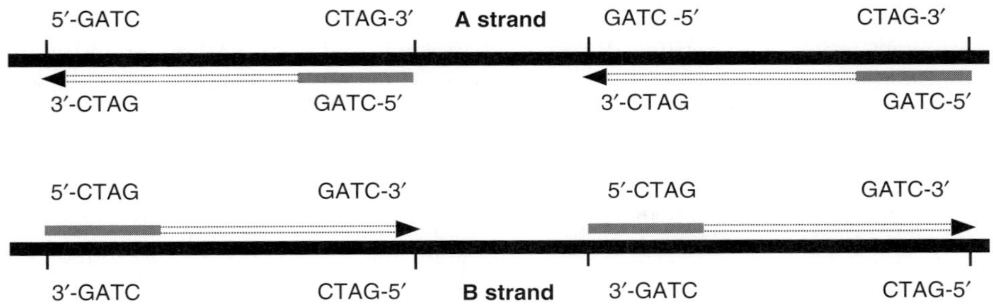

Figure 4.7 Random amplification of polymorphic sequences with a single random primer.

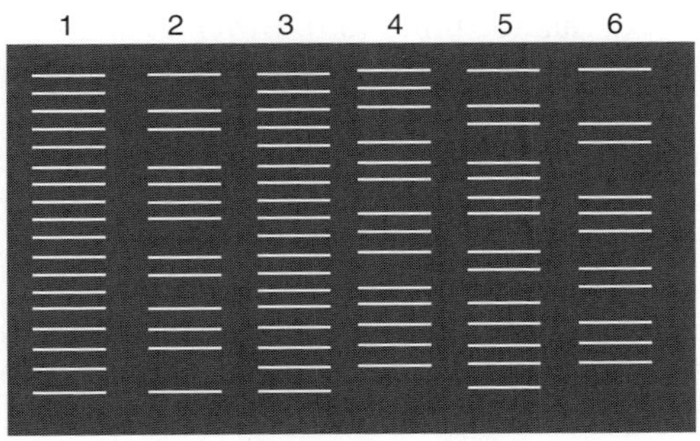

Figure 4.8 RAPD banding profiles of 6 bacterial isolates of the same species. Isolates corresponding to lanes 1 and 3 have identical DNA banding patterns and are therefore considered of identical clonal origin. Isolates corresponding to lanes 2, 4, 5, and 6 are distinguishable from each of the other 5 isolates on the basis of differences in the DNA banding patterns.

4.3 PULSED FIELD GEL ELECTROPHORESIS

4.3.1 Introduction

Pulsed Field Gel Electrophoresis (PFGE) is presently considered the method of choice for discriminating genetic differences and lineage among strains of the same bacterial pathogen. The method is particularly useful in epidemiology, and has also been applied to the detection of clonal strains of pathogens that have been found to persist in food manufacturing facilities (29–31).

4.3.2 Mechanism of PFGE

Pulsed Field Gel Electrophoresis is based on the use of low frequency restriction nucleases to generate a family of high molecular weight fragments derived from genomic DNA, and their subsequent resolution based on size using changes in the direction of the electric field during agarose electrophoresis. Large fragments stretch out linearly in the direction of the electric field. When the direction of the field changes, the fragments undergo an initial relaxation in conformation and then form multiple undulations or kinks in the direction of the new field, followed by linearization (32). The most widely adapted system presently is the contour clamped homogeneous electric field (CHEF) apparatus, involving a hexagonal distribution of electrodes that undergo periodic alternate uniform electric fields with an angle of 120° (Figure 4.9). Clamped homogeneous electric field instruments are available from Bio-Rad, Pharmacia, and BRL. A longer pulse time increases the size of the fragment that can be separated, but results in a decrease in resolution of fragments of similar size. Progressive increases in pulse time (pulse time ramping) can significantly increase resolution. A ramp from 5 s to 40 s will allow optimal separation of DNA fragments from 50 kb to 600 kb, while increasing the pulse time to 75 s will extend the separation to 1 Mb (33). Pulsed field gel electrophoresis is usually performed at 12–15°C for enhanced band resolution, and to prevent the development of temperature gradients in the gel. Agarose of high gel strength and low elecroendosmosis (EEO) is used. Fragments of DNA from 50 kb to 1 Mb are usually separated with 1.0% agarose. Reducing the agarose concentration will increase the pore size and allow separation of larger fragments. Higher concentrations of agarose (1.2–1.6%) reduces pore size, and will increase resolution and sharpness of bands, but reduces separation of larger fragments and results in longer running times (33). Birren et al. (34) have reported on optimized conditions for PFGE and the effects of a variety of variables on the resolution of DNA fragments.

The restriction nucleases used with PFGE are selected on the basis of the rarity of their recognition sequence in the target genome. Struelens et al. (33) and Tenover et al. (36) have listed restriction nucleases useful for PFGE analysis of a number of bacterial genera and species.

The usual methodology for DNA extraction is unsuitable for PFGE because the large size of genomic bacterial DNA (~5×10^9 Da) results in rapid shearing during pipeting to ~1×10^7 Da. To prevent such rupture of large DNA molecules, intact cells are embedded in an equal volume of 2.0% nuclease free, low melting temperature agarose plugs. The cells are then lysed *in situ* immersing the plugs first in lysozyme, and then proteinase, K, or other suitably lytic enzymes for the specific organism involved. To save time, lysozyme can be added to the agarose and cell mixture before solidification The agarose plugs are then placed in wells for PFGE. Uniform cell densities of $1 - 5 \times 10^9$ cells/mL are critical for valid comparison of resolved bands. Adjusting cell suspensions to a uniform A_{600} value is reliably effective and convenient, provided each culture is of the same age. To ensure adequate alignment and normalization of banding patterns, and accurate fragment size estimates, appropriate DNA ladders should be included in at least every fifth lane (33).

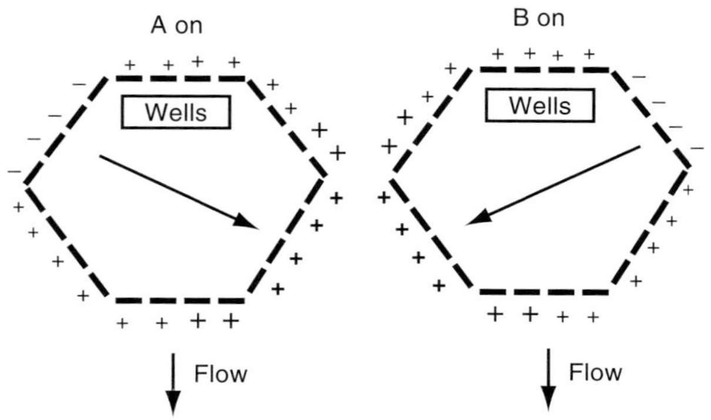

Figure 4.9 Voltage clamping by the CHEF-DR-II system. (a) Relative electrode potentials when +60° held vector is activated. (b) Relative electrode potentials when the −60° held vector is activated.

4.3.3 Interpretation of PFGE Banding Patterns

Distinctions in banding patterns are based on the size and shape of resolved bands. The problem of how minor differences in banding patterns should be interpreted has been dealt with in detail by Tenover et al. (36), with the establishment of well reasoned criteria based on the resolution of at least 10 distinct fragments derived from each culture being compared. The utility of strain typing using PFGE is based on the assumption that isolates derived from the same reference strain are of recent lineage from the original strain. A limitation occasionally arises when unrelated isolates may have similar or indistinguishable genotypes, especially if there is limited genetic diversity within a species or subtype (37). It should also be assumed that random genetic events, including end point mutations and deletions and insertions of nucleotides, will alter PFGE patterns of DNA from progeny derived from an original reference strain. An isolate is interpreted to be closely related to a reference strain if its PFGE pattern differs from the reference strain by changes reflecting a single genetic event, i.e., a point mutation or a frame shift mutation involving the insertion or deletion of one or more sequential nucleotides. Such changes usually result in two to three band differences. For example, a single spontaneous mutation that creates a new genomic restriction site will split one restriction fragment into two smaller ones. The loss of the original large fragment plus the appearance of two new smaller fragments will result in a three band difference between the reference strain and its direct progeny that has undergone such a single mutational event (36). In addition, variations of two to three bands have been reported in strains repeatedly cultured or isolated multiple times from the same patient (38, 39). Patterns that are distinctly different from an outbreak or reference strain by only two or three fragments should therefore be considered subtypes of the same clonal lineage. An isolate is considered to be possibly related to a reference strain if there are four to six different band differences. An isolate should be considered unrelated to a reference strain if its PFGE pattern differs by seven or more bands, reflecting three or more independent genetic events.

4.4 SUMMARY

The PCR is a powerful analytical method that can be used to rapidly identify microorganisms of public health significance. The method has been advanced to the state where it can be effectively used at present to detect and quantitate the presence of low numbers (~1 CFU/g) of specific organisms in foods and on processing equipment in near real time without the necessity for overnight cultivation, isolation, and subsequent immunological and time consuming metabolic identification. Those aspects of the food industry presently encountering massive and costly recalls are in critical need of near real time assays such as the PCR prior to shipment of product. When the PCR is coupled to immunomagnetic capture a high level of specificity can be achieved. The application of RAPD and PFGE for assessing clonal similarities and distinctions among strains of the same species derived from food processing equipment, foods, and foodborne outbreaks, is encountering widespread use and acceptance.

REFERENCES

1. Landre, P.A., D.H. Gelfand, R.M. Watson. The use of cosolvents to enhance amplification by the polymerase chain reaction. In: *PCR Strategies,* Innis, M.A., D.H. Gelfand, J.J. Sninsky, eds., New York: Academic Press, 1995, pp 3–16.

2. Innis, M.A., D.H. Gelfand. Optimization of PCRs. In: *PCR Protocols: a Guide to Methods and Applications*. Innis, M.A., D.H. Gelfand, J.J. Sninsky, T.J. White, eds., New York: Academic Press, 1990, pp 3–20.
3. Abolmaaty, A., C. Vu, J. Oliver, R.E. Levin. Development of a new lysis solution for releasing genomic DNA from bacterial cells for DNA amplification by polymerase chain reaction. *Microbios* 101:181–189, 2000.
4. Diaco, R. Practical considerations for the design of quantitative PCR assays. In: *PCR Strategies*. Innis, M.A., D.H. Gelfand, J.J. Sinsky, eds., New York: Academic Press, 1995, pp 84–108.
5. Guan, J., R.E. Levin. Sensitive and rapid detection of *Escherichia coli* O157:H7 in ground beef by nested PCR incorporating immunomagnetic separation. *J. Food Biotechnol.* 2002. (In press)
6. Guan, J., R.E. Levin. Quantitative detection of *Escherichia coli* O157:H7 in ground beef by immunomagnetic separation and competitive polymerase chain reaction. *J. Food Biotechnol.* 2002. (In press)
7. Zarchar, V.R., A. Thomas, A.S. Goustin. Absolute quantification of target DNA: a simple competitive PCR for efficient analysis of multiple samples. *Nucleic Acids Res.* 8:2917–2018, 1993.
8. Rupf, S., K. Merte, K. Eschrich. Quantification of bacteria in oral samples by competitive polymerase chain reaction. *J. Dent. Res.* 78(4):850–856, 1997.
9. Albert, J., E.M. Fenyo. Simple, sensitive and specific detection of human immunodeficiency virus type 1 in clinical specimens by polymerase chain reactions with nested primers. *J. Clin. Microbiol.* 28:1560–1564, 1990.
10. Jackson, D.P., J.D. Hayden, P. Quirke. Improving the sensitivity and specificity of PCR amplification. In: *PCR: A Practical Approach,* McPherson, M.J., P. Quirke, G.R. Taylor, eds., Oxford, England: Oxford Press, 1991, pp 42–50.
11. Rychlik, W., W.J. Spencer, R.E. Rhoads. Optimization of annealing temperature for DNA amplification *in vitro*. *Nucleic Acids Res.* 18:6409–6412, 1990.
12. Breslauer, K.J., F. Ronald, H. Blicker, L.A. Marky. Predicting DNA duplex stability from the base sequence. *Proc. Nat. Acad. Sci.* 83:3746–3750, 1986.
13. Freier, S.M., R. Kierzek, J.A. Jaeger, N. Sugimoto, M.H. Caruthers, T. Neilson, D.H. Turner. Improved free-energy parameters for predictions of RNA duplex stability. *Proc. Nat. Acad. Sci.* 83:9373–9377, 1986.
14. McConaughy, B.L., C.L. Laird, B.J. McCarthy. Nucleic acid reassociation in formamide. *Biochem.* 8:3289–3295, 1969.
15. Suggs, S.V., T. Hirose, E.H. Myake, M.J. Kawashima, K.I. Johnson, R.B. Wallace. Using purified genes. *ICN-UCLA Symp. Mol. Cell. Biol.* 23:683–693, 1981.
16. Ozbas, Z., A. Lehner, M. Wagner. Development of a multiplex and semi-nested PCR assay for detection of *Yersinia enterocolitica* and *Aeromonas hydrophila* in raw milk. *Food Microbiol.* 17(2):197–203, 2000.
17. Lindqvist, R., Detection of *Shigella* spp. in food with a nested PCR method: sensitivity and performance compared with a conventional culture method. *J. Appl. Microbiol.* 86:971–978, 1999.
18. Waage, A.S., T. Vardund, V. Lund, G. Kapperud. Detection of low numbers of *Salmonella* in environmental water, sewage and food samples by a nested polymerase chain reaction assay. *J. Appl. Microbiol.* 87:418–428, 1999.
19. Gilgen, M., P. Hubner, C. Hofelein, J. Luthy, U. Candrian. PCR-based detection of verotoxin-producing *Escherichia coli* (VTEC) in ground beef. *Res. Microbiol.* 149:145–154, 1998.
20. Kapperud, G., T. Vardund, E. Skjerve, E. Hornes, T.E. Michaelsen. Detection of pathogenic *Yersinia enterocolitica* in foods and water by immunomagnetic separation, nested polymerase chain reactions, and colorimetric detection of amplified DNA. *Appl. Environ. Microbiol.* 59:2938–2944, 1993.
21. Guan, J., R.E. Levin. Sensitive and rapid detection of *Escherichia coli* O157:H7 in ground beef by nested PCR incorporating immunomagnetic separation. *J. Food Biotechnol.* 2002. (In press)

22. Lawrence, L., J. Harvey, A. Gilmour. Development of a random amplification of polymorphic DNA typing method for *Listeria monocytogenes*. *Appl. Environ. Microbiol.* 59:3117–311, 1993.
23. Lawrence, L., A. Gilmour. Characterization of *Listeria monocytogenes* isolated from poultry products and from poultry-processing environment by random amplification of polymorphic DNA and multilocus enzyme electrophoresis. *Appl. Environ. Microbiol.* 61:2139–2144, 1995.
24. Boerlin, P., E. Bannerman, F. Ischer, J. Rocurt, J. Bille. Typing of *Listeria monocytogenes:* a comparison of random amplification of polymorphic DNA with 5 other methods. *Res. Microbol.* 146:35–49, 1995.
25. Czajka, J., N. Bsat, M. Piani, W. Russ, K. Sultana, M. Wiedman, R. Whitaker, C. Batt. Differentiation of Listeria monocytogens and *Listeria innocua* by 16S rRNA genes and intraspecies discrimination of *Listeria monocytogenes* strains by random amplified polymorphic DNA polymorphisms. *Appl. Environ. Microbiol.* 59:304–308, 1993.
26. del Tufo, J., S. Tingey. RAPD assay. In: *Protocols for Nucleic Acid Analysis by Nonradioactive Probes,* Isaac, P.G., ed., Totowa, NJ: Human Press, 1994, pp 237–241.
27. Meuner, J., P. Grimont. Factors affecting reproducibility of random amplified polymorphic DNA fingerprinting. *Res. Microbiol.* 144:373–379, 1993.
28. Mazurier, S., K. Wernars. Typing of *Listeria* strains by random amplification of polymorphic DNA. *Res. Microbiol.* 143:499–505, 1992.
29. Johansson, T., L. Rantala, L. Palmu, T. Honkanen-Buzalski. Occurrence and typing of *Listeria monocytogenes* strains in retail vacuum-packed fish products and in a production plant. *Int. J. Food Microbiol.* 47:111–119, 1999.
30. Autio, T., S. Hielm, M. Miettinen, A. SjÖberg, K. Aarnisalo, J. BjÖrkroth, T. Mattila-Sandhom, H. Korkeala. Sources of *Listeria monocytogenes* contamination in a cold-smoked rainbow trout processing plant detected by pulsed-field gel electrophoresis typing. *Appl. Environ. Mirobiol.* 65:150–155, 1999.
31. Brett, M., P. Short, J. McLauchlin. A small outbreak of listeriosis associated with smoked mussels. *Int. J. Food Microbiol.* 43:223–229, 1998.
32. Gurrieri, S., E. Rizzarelli, D. Beach, C. Bustamante. Imaging of kinked configurations of DNA molecules undergoing orthogonal field alternating gel electrophoresis by fluorescene microscopy. *Biochemistry* 29:3396–3401, 1990.
33. Struelens, M.J., R. De Ryck, A. Deplano. Analysis of microbial genomic macrorestriction patterns by pulsed-field gel electrophoresis (PFGE) typing. In: *New Approaches for the Generation and Analysis of Microbial Typing Data,* Diikshoorn, L., K. J. Towner, M. Struelens, eds., New York: Elsevier, 2001, pp 159–176.
34. Kim, Y., J.H. Jett, E.J. Larson, J.R. Penttila, B.L. Mrrone, R.A. Keller. Bacterial finger-printing by flow cytometry: bacterial species discrimination. *Cytometry.* 36:324–332, 1999.
35. Birren, B.W., E. Lai, S.M. Clark, L. Hood, M.I. Simon. Optimized conditions for pulsed field gel electrophoretic separations of DNA. *Nucleic Acids Res.* 16(15):7563–7582, 1988.
36. Tenover, F.C., R.D. Arbeit, R.V. Goering, P.A. Mickelsen, B.E. Murray, D.H. Persing, B. Swaminathan. Interpreting chromosomal DNA restriction patterns produced by pulsed-field gel electrophoresis: criteria for bacterial strain typing. *J. Clin. Microbiol.* 33:2233–2239, 1995.
37. Barrett, T. J., H. Lior, J.H. Green, R. Khakhria, J.G. Wells, B.P. Bell, K.D. Greene, J. Lewis, P.M. Griffin. Laboratory investigation of a multistate food-borne outbreak of *Escherichia coli* O157:H7 by using pulsed-field gel electrophoresis and phage typing. *J. Clin. Microbiol.* 32:3013–3017, 1995.
38. Arbeit, R.D., A. Slutsky, T.D.W. Barber, J.N. Maslow, S. Niemczyk, J.O. Falkinham III, G.T. O'Connor, C.F. von Reyn. Genetic diversity among strains of *Mycobacterium avium* causing monoclonal and polyclonal bacteremia in patients with AIDS. *J. Infect. Dis.* 167:1384–1390, 1993.
39. Sader, H.S., A.C. Pignatari, I.L. Leme, M.N. Burattini, R. Tancresi, R.J. Hollis, R.N. Jones. Epidemiologic typing of multiply drug-resistant *Pseudomonas aeruginosa* isolated from an outbreak in an intensive care unit. *Diag. Microbiol. Infect. Dis.* 17:13–18, 1993.

3.05

Control of Food Borne Bacterial Pathogens in Animals and Animal Products through Microbial Antagonism

Mindy M. Brashears, Alejandro Amezquita, and Divya Jaroni

CONTENTS

5.1 MICROBIAL ANTAGONISM

The concept of microbial antagonism or microbial interference has been known for a long time, even before other common food preservation methods (i.e., canning or refrigeration) were established. Microbial antagonism is the inhibition of undesired or pathogenic microorganisms caused by competition for nutrients, and by the production of antimicrobial metabolites (68,77,79,82,85,163). Pure cultures of lactic acid bacteria (LAB) have been used since the beginning of the twentieth century as starter cultures in fermented food products. Metabolism of these cultures may contribute in a number of different ways to the control of pathogens and the extension of the shelf life as well as modification of the sensory attributes of the food product (68,77,82,85). Antagonism between two different species or genera of microorganisms takes place when they compete for a common niche, or when one of the microorganisms produces an antagonistic extracellular agent or modifies the environment such that the other is inhibited (79,107,172).

The use of LAB as protective cultures rather than starter cultures has gained importance during recent years. Biopreservation is the extended storage life and enhanced safety of food products using their natural or controlled microflora and their antibacterial products (68,77,85,163). Additionally, the protective effects can inhibit food borne pathogens in a live animal prior to slaughter.

5.1.1 Lactic Acid Bacteria

The lactic acid bacteria constitute a group of Gram-positive bacteria that share similar morphologic, metabolic, and physiologic characteristics. They are nonsporeforming rods and cocci that ferment carbohydrates, forming lactic acid as the major end product (5), hence the denomination lactic acid bacteria. Depending on the metabolic pathways used to ferment carbohydrates and the resulting end products, lactic acid bacteria are divided into two major groups, homofermentative or heterofermentative. They are generally catalase negative, anaerobic, and nonmotile, and do not reduce nitrate (145). Lactic acid bacteria have complex nutritional requirements such as carbohydrates, amino acids, peptides, nucleic acid derivatives, fatty acids, salts, and vitamins for their growth (72). They are generally acid tolerant with different species having adapted to grow under widely different environmental conditions. They are widespread in nature, found wherever high concentrations of soluble carbohydrates, protein breakdown products, vitamins, and a low oxygen tension occur (146). Consequently they are common in milk and dairy products, fermented foods, intact and rotting vegetable material, silage, and the intestinal tracts and mucous membranes of humans and animals.

Phylogenetically, the lactic acid bacteria belong to the clostridial branch of the Gram-positive bacteria, which also includes genera such as *Clostridium, Bacillus, Listeria* and *Staphylococcus*, and is characterized by a low G + C DNA content (5). However, the term lactic acid bacteria has become commonly associated with the genera *Lactobacillus,*

Lactococcus, Leuconostoc, Pediococcus, Enterococcus, and *Streptococcus*. Members of this bacterial group are known to provide considerable benefits to humankind, some as natural inhabitants of the intestinal tract and others as fermentative bacteria that impart flavor and texture to a multitude of fermented foods. For many years there has also been widespread interest in the use of lactic acid bacteria in the biological preservation of foods. These organisms are particularly suitable as antagonistic microorganisms in food because of their ability to inhibit other food borne bacteria by a variety of means, including production of organic acids, hydrogen peroxide, or bacteriocins. In recent years their use as direct fed microbials for humans and animals has received increased attention (54,89). Their favorable effect on growth and health is thought to be due to the modulation of other bacterial growth through one or more of these antagonistic factors.

5.2 REDUCTION OF FOOD BORNE PATHOGENS IN FOODS OF ANIMAL ORIGIN

Biopreservation of foods in food and meat systems can be achieved by different mechanisms: (1) adding a pure culture of a bacteriocin-producing strain of LAB to favor *in situ* production of the bacteriocin, (2) adding only the bacteriocin, either as a crude preparation or as a purified compound to have more predictable results, or (3) adding strains of LAB that do not grow at refrigeration temperatures to serve as indicators of temperature abuse and to protect the products from the growth of pathogens under temperature abuse conditions (50,80,107,162).

Many strains and species of lactic acid bacteria are generally considered as food grade microorganisms, and are considered GRAS (Generally Regarded as Safe) because they have had a history of safe, widespread use in the manufacture of food prior to the 1958 Food Additives Amendment. The use of LAB cells in numerous food fermentations is a common practice. Cells of food grade LAB are added to foods mainly for three reasons: to produce safe, shelf stable, and desirable fermented foods; to serve as indicators of product abuse, especially temperature abuse of some refrigerated foods; and to control the growth of spoilage and pathogenic bacteria in refrigerated foods. The first two applications require the growth of LAB cells, whereas for the last application, the mere presence of cells is effective, and growth is undesirable (86,162).

5.2.1 Antimicrobial Substances Produced by Lactic Acid Bacteria During *In Vitro* Application

Reduction of pH and removal of large amounts of carbohydrates by fermentation are the primary preserving actions that LAB provide, mainly in fermented foods. However, it has also been recognized that LAB are capable of producing inhibitory substances other than organic acids, with inhibitory activity to different microorganisms. Additionally, these substances may be produced at refrigeration temperatures with no growth (8). These substances include hydrogen peroxide (38,89,137,141,175,180), diacetyl (85,128), reuterin (9,48), bacteriocins (27,32,91,127,163,164), and other low molecular weight metabolites (122).

5.2.1.1 Hydrogen Peroxide

Many fermentative bacteria, including LAB, produce hydrogen peroxide (H_2O_2) as a mechanism for protecting themselves against oxygen toxicity. Lactic acid producing bacteria, lack heme and thus do not utilize the cytochrome system (which reduces oxygen to water in respiratory metabolism) for terminal oxidation. Lactic acid producing bacteria utilize flavoproteins, which generally convert oxygen to H_2O_2. This mechanism, together with the absence of the heme protein catalase, generally results in the formation of H_2O_2 in amounts that are in excess of the capacity of the organism to degrade it. The H_2O_2 formed may inhibit or kill other members of the microbiota. Hydrogen peroxide is an effective antimicrobial due to its strong oxidizing effect on the bacterial cell; sulfhydril groups of cell proteins and membrane lipids can be oxidized (34,90,92,107,175,180). The formation and accumulation of H_2O_2 in growth media with a subsequent antagonistic effect was shown with *Staphylococcus aureus* (38) and *Pseudomonas* species (137). Hydrogen peroxide can react with other components to form inhibitory substances. In raw milk, for instance, hydrogen peroxide generated by LAB can react with endogenous thiocyanate that is catalyzed by lactoperoxidase to form intermediary oxidation products inhibitory to microorganisms. This mechanism, also known as the lactoperoxidase antibacterial system (LP system or LPS), has been well documented (34,98,135). Rodríguez et al. (141) studied the combined effect of the LP system and bacteriocin producing strains of LAB in milk at 4°C and 8°C to inhibit the growth of *Listeria monocytogenes*. These authors failed to demonstrate synergism between these two antimicrobial systems, given that the increment in the reduction levels of the pathogen was only 0.19 $\log_{10}$ cycles after 4 days at 8°C when the combined treatment was applied. The authors suggested that rather than synergism, the production of nisin by the evaluated strain of *Lactococcus lactis* subspecies *lactis* (strain ESI 515) was enhanced in activated in LP system milk.

5.2.1.2 Weak Organic Acids

Lactic acid bacteria are nonrespiring microorganisms, principally generating ATP by fermentation of carbohydrates coupled to substrate level phosphorylation. The two major pathways for the metabolism of hexoses are homofermentative or glycolysis (Embden-Meyerhof pathway), in which lactic acid is virtually the only end product; and heterofermentative (phosphoketolase pathway), in which other end products such as acetic acid, CO_2, and ethanol are produced in addition to lactic acid (9,93,184).

Weak organic acids are known to have strong antimicrobial activity. In solution, these acids exist in a pH dependent equilibrium between the undissociated and dissociated state. Their effectiveness as antimicrobials is greater at low pH because this favors the uncharged, undissociated state of the molecule, which is freely permeable across the cell membrane, because it is lipid soluble (36). Subsequently, upon encountering the higher pH inside the cell, the molecule will dissociate, resulting in the release and accumulation of charged anions and protons, which cannot cross the cell membrane (20,45,128).

Of the two major weak organic acids produced by lactic acid bacteria, acetic and lactic acid, the former is the strongest inhibitor due to its higher dissociation constant (pK_a=4.75) as compared to lactic acid (pK_a=3.08) at a given molar concentration and pH (45,77,128).

Ahamad and Marth (6) found that 0.2% (approximately 22 mM) lactic acid inhibited the growth of *Listeria monocytogenes* strains V7 and CA in tryptose broth at 7°C, but that a minimum of 0.3% (approximately 33 mM) was required for lethal effect.

5.2.1.3 Reuterin

Reuterin is a neutral, broad spectrum antimicrobial substance formed during anaerobic growth of *Lactobacillus reuteri* in the presence of glycerol (9). Reuterin is an equilibrium mixture of monomeric, hydrated monomeric, and cyclic dimeric forms of β-hydroxypropionaldehyde. The inhibitory effect of reuterin has been associated with its action on DNA synthesis by acting as an inhibitor of the substrate binding subunit of ribonucleotide reductase.

El-Ziney et al. (48) reported the antimicrobial activity of reuterin against *E. coli* O157:H7 and *Listeria monocytogenes* on the surface of cooked pork at 7°C. The initial levels of each pathogen on the surface of the product were approximately $\log_{10}$ 5 CFU/cm^2. After 15 seconds of application of reuterin (500 AU/ml), the numbers decreased by 0.45 and 0.3 $\log_{10}$ CFU/cm^2 for *E. coli* O157:H7 and *L. monocytogenes*, respectively. After 24 h of exposure to reuterin, the numbers decreased by 2.7 $\log_{10}$ CFU/cm^2 for *E. coli* O157: H7 and 0.63 $\log_{10}$ CFU/cm^2 for *L. monocytogenes*.

5.2.1.4 Diacetyl

Diacetyl is the compound responsible for the characteristic aroma and flavor of butter. It is produced by species of the genera *Lactobacillus*, *Leuconostoc*, *Pediococcus*, and *Lactococcus* (85). Diacetyl is formed from the metabolism of citrate via pyruvate (9,107). Jay (85) reported that this compound was more effective when the pH was lower than 7.0, but it was progressively ineffective at pH values higher than 7.0. This study was particularly extensive because it evaluated the antimicrobial effects of diacetyl against 40 different cultures, including 10 of LAB, 12 of Gram-positive non-LAB, 14 of Gram-negative bacteria, and 4 of yeasts. Diacetyl was found to be more active against Gram-negative bacteria, yeasts, and molds than against Gram-positive bacteria.

5.2.1.5 Bacteriocins

The biosynthesis, classification, mode of action and characterization of bacteriocins has been well reviewed and published (11,84,94,95,116,119,120,135,143,163,172,183). Bacteriocins are low molecular weight, heat stable, ribosomally synthesized and cationic proteinaceous compounds, produced by Gram-positive organisms, with antibiotic like functionality against closely related species, mainly Gram-positive bacteria, by adsorption to receptors on the target cells (84,95,96). Klaenhammer (95) proposed that bacteriocins be classified into four major groups, based on their biochemical properties. Class I or lantibiotics are small peptides ($<$ 5 kDa) and membrane active bacteriocins, containing the unusual dehydroamino acids and thioetheramino acids lanthionin and 3-methyllanthionine, respectively. Nisin, produced by *Lactococcus lactis* subspecies *lactis*, is the most widely studied Class I bacteriocin. Class II bacteriocins are subdivided into three subclasses, and in general are small, nonlanthionine containing and heat stable, membrane active peptides. Further subdivisions of Class II bacteriocins include class IIa or Listeria active bacteriocins; class IIb, poration complexes requiring two peptides; and class IIc, thiol activated peptides requiring cysteine residues. Examples of Class II bacteriocins are pediocin JD (IIa) produced by *Pediococcus acidilactici*, lactacin F (IIb) produced by *Lactobacillus johnsonii*, and lactococcin B (IIc) produced by *Lactococcus lactis* subspecies *cremoris*. Class III bacteriocins are large, heat labile proteins. Class IV includes complex bacteriocins, in which lipids and carbohydrates appear to be necessary for activity (95,128). In cases where the mode of action has been investigated, the cell membrane appears to be the site of action. There is enough evidence to conclude that bacteriocins produced by lactic acid bacteria act by the common mechanism of depleting proton motive force (PMF) (27).

Nisin is the best characterized LAB bacteriocin. The structure of nisin was first elucidated by Gross and Morell (70). Nisin dissipates the membrane potential ($\Delta\psi$) in cells of sensitive organisms (142), and causes PMF depletion of whole cells of *L. monocytogenes* (27) and *Clostridium sporogenes* (127). Other LAB bacteriocins such as lactococcin A, lactococcin B (174), and pediocin JD (32), share this common mechanism of action, which is the dissipation of the PMF in sensitive cells.

5.2.1.6 Low Molecular Weight Metabolites

Some species of LAB have been reported to produce some metabolites of low molecular weight, such as benzoic acid, mevalonolactone, and methylhydantoin, which exhibit inhibitory activity toward Gram-negative bacteria as well as some fungi (122). Niku-Paavola et al. (122) reported the production of compounds with a molecular weight lower than 700 Da that inhibited the growth of the Gram-negative *Pantoea agglomerans* when these substances were used in combination at a level of 10 ppm. Addition of 1% lactic acid produced a synergistic effect, inhibiting the growth of *P. agglomerans* by 100%.

5.2.2 Lactic Acid Bacteria and Inhibition of Food Borne Pathogens in Meat Products

Because much of the inhibitory activity of the lactobacilli is related to like species, much of the work has been focused on the Gram-negative pathogen, Listeria monocytogenes. The inhibition of Listeria monocytogenes in different types of meat products as well as in laboratory media by bacteriocin producing strains of LAB has been reported (8,16,25,28,29,39,41,43,79,87,121,123,135,138,149,150,177,181).

Different genera and species of LAB have been studied for antilisterial properties under different conditions. *Carnobacterium piscicola,* for instance, has been widely studied (28,29,43,123,135,150). In 1995, Pilet et al. (135) reported that two strains out of a total of 22 LAB isolates from fish products were selected for their antilisterial properties and identified as *Carnobacterium piscicola* and *Carnobacterium divergens*. These two isolates turned out to be bacteriocin producing strains, and their bacteriocins were heat resistant and sensitive to pronase E, proteinase K and trypsin. These bacteriocins, named piscicocin V1 and divercin V41, exhibited a bactericidal and nonbacteriolytic mode of action toward *L. monocytogenes*. When the pathogen was grown in the presence of cell free supernatants containing the bacteriocins, the numbers were reduced by approximately 2 $\log_{10}$ cycles after 120 hours of incubation at 30°C. A similar study reported by Duffes et al. (43) demonstrated that after screening 160 fish isolates of LAB, 23 were selected for antagonism toward *L. monocytogenes*, and 22 of this 23 belonged to the genera *Carnobacteria*. *Carnobacterium divergens* V41, *C. piscicola* V1 and *C. piscicola* SF668 were found to be the most inhibitory strains reducing the numbers of the pathogen from 10^3 to <10 CFU/mL after 11 days of storage at 4°C in the case of *C. divergens* V41, in a simulated cold smoked fish system (pH = 6.2, 4% NaCl). The mechanism of inhibition was strictly attributed to production of bacteriocin and not to nutritional competition between strains or to the production of other inhibitory substances such as acid or hydrogen peroxide.

Conversely, Buchanan and Bagi (28) reported that two bacteriocin producing strains of *Carnobacterium piscicola* inhibited the growth of *Listeria monocytogenes* in associative cultures in brain heart infusion mainly due to nutrient depletion rather than to the production of bacteriocins. These researchers suggested that *C. piscicola* is a microorganism of rapid growth and in associative cultures with *L. monocytogenes* may produce an insufficiency of a limiting nutrient for this pathogen. This was evident when coculturing the pathogen with *C. piscicola* in 1X, 3X and 6X brain heart infusion slightly increased the numbers of the pathogen after incubation at 12°C for over 10 days. This mechanism of inhibition was supported by another study that demonstrated that the reduction in the numbers of *L. monocytogenes* in cocultures with *C. piscicola* LK5 was not a function of acid production, or the accumulation of any other inhibitory extracellular substance. Instead, increasing levels of nutrients favored the growth of the pathogen, suggesting a mechanism involving nutrient depletion.

Nilsson et al. (123) added cells of *C. piscicola* to control the growth of *L. monocytogenes* on cold smoked salmon. These researchers found that out of 4 strains of *C. piscicola*, two of them produced bacteriocins that inhibited the pathogen in laboratory media, whereas the other two strains showed antilisterial activity that was independent of bacteriocin production. The bacteriocin producing strains of *C. piscicola* caused a long lag phase and reduced the number of *L. monocytogenes* from 10^3 to <10 CFU/mL after 32 days of storage at 5°C. Additionally, even the strains of *C. piscicola* that did not produce bacteriocins inhibited the growth of *L. monocytogenes*, suggesting that some other inhibitory compounds were produced during refrigerated storage. This study failed to demonstrate nutrient depletion as a mechanism of inhibition. These researchers selected for strains of *C. piscicola* with fast growth rates, because their objective was to enhance the production of bacteriocins at refrigeration temperatures, but did not take into account the increasing numbers of the LAB.

In a separate study, Schöbitz et al. (150) inoculated beef steaks with a partially purified bacteriocin like substance produced by *C. piscicola* L103, together with cells of *L. monocytogenes*, and evaluated the inhibitory activity at 4°C under vacuum package conditions. Numbers of *L. monocytogenes* decreased by 3 $\log_{10}$ cycles until complete inhibition of the pathogen after 14 days of storage when the substance was initially inoculated at a level of 100 AU/ml, whereas the numbers of the pathogen in the control treatment increased by 1 $\log_{10}$ cycle during the same time. The bacteriocin did not show any inhibitory effect toward natural LAB background flora.

Use of bacteriocin producing strains of *Pediococcus acidilactici* has also been studied to control the growth of *Listeria monocytogenes* in different meat products. Berry et al. (16) reported the use of bacteriocin producing *P. acidilactici* JD1-23 and its plasmid cured derivative JD-M for the control of *L. monocytogenes* in frankfurters at 4°C. The two strains of *P. acidilactici* exhibited a bacteriostatic activity toward the pathogen after 60 days of refrigerated storage under vacuum package conditions when the initial level of inoculation of the LAB was high (10^7 CFU/g). With low levels of *Pediococcus* organisms (10^3–10^4 CFU/g), *L. monocytogenes* grew, although the lag phase was increased. This study also reported the effect of package atmosphere on the growth of the pathogen in the presence of either strain of *Pediococcus*. Under aerobic conditions, *L. monocytogenes* was able to grow at 4°C in frankfurters inoculated with strain JD-M, whereas JD1-23 inhibited the growth for up to 30 days. Under temperature abuse conditions (storage at 15°C), the pathogen grew in the presence of either strain of *Pediococcus* under aerobic conditions, whereas the growth was inhibited under anaerobic conditions for up to 15 days.

Nielsen et al. (121) utilized a bacteriocin produced by a strain of *Pediococcus acidilactici* found in a commercial starter culture to inhibit the growth of *L. monocytogenes* in fresh meat. This study reported a strong antilisterial activity from this bacteriocin at concentrations ranging from 500 AU/mL to 5,000 AU/mL during the first 10 minutes of inoculation, reducing the levels of the pathogen by 1 to 2 $\log_{10}$ cycles depending upon the concentration of the bacteriocin. Bacteriocin production was reported to be stable after 28 days of refrigerated storage. However, when meat was treated with the bacteriocin and stored at 5°C for 28 days, only a 0.5 $\log_{10}$ reduction occurred after 7 days of storage; the numbers of *L. monocytogenes* actually increased in a similar fashion to the untreated control after 21 days of storage. These researchers suggested that the bacteriocin effect might have been diluted in the dripping fluid or exudate as a consequence of storage of meat for a long period of time. A study by Pucci et al. (138) also reported the use of bacteriocin PA-1 produced by *P. acidilactici* PAC 1.0 to inhibit the growth of *L. monocytogenes* in laboratory media and in different food systems including cottage cheese, half and half cream, and cheese sauce. The minimum inhibition concentration (MIC) for the inhibition of *L. monocytogenes* in a broth system was reported to be 54.7 AU/ml. Significant inhibition was detected at both 4°C and 32°C when the pH of the medium ranged from 5.5 to 7.0, and the initial inoculum of the pathogen was 1.2×10^2 CFU/ml. When the bacteriocin was applied to food systems, it reduced the numbers of *L. monocytogenes* during the first 7 days of storage at 4°C in non-acidic foods, but the pathogen was able to grow effectively after the first 7 days, indicating loss of activity. In acidic food systems, the combination of bacteriocin PA-1 and lactic acid present in the foods showed a synergistic effect controlling the growth of the pathogen for up to 14 days. Again, the effect of high pH values in the food systems favored the growth of *L. monocytogenes*, even in the presence of bacteriocins.

Several lactobacilli have also been reported to exhibit inhibitory activity toward *Listeria monocytogenes*. Juven et al. (1998) reported the use of the psychrotrophic strain FloraCarn L-2 of *Lactobacillus alimentarius* and its antibiotic resistant (streptomycin and rifampicin resistant) mutant SRL-2 to inhibit the growth of *L. monocytogenes* in ground beef (pH=5.4) at 4°C during a total storage time of 9 weeks. Both strains produced a 2-$\log_{10}$ reduction in the numbers of the pathogen during the storage period. The pH value in the samples inoculated with *L. alimentarius* L-2 decreased from 5.37 to 4.71, and the concentration of lactic acid increased from 90 to 142 mM after 9 weeks of storage at 4°C. The numbers of inoculated L-2 increased rapidly from 7 to 9.6 $\log_{10}$ CFU/g in the absence of *L. monocytogenes* and from 7 to 8.4 $\log_{10}$ CFU/mL in the presence of *L. monocytogenes* within 4 weeks of storage; progressive growth of strain SRL-2 was also detected during refrigerated storage. Inhibition of *L. monocytogenes* in this study was reported to be due to production of lactic acid and did not appear to be due to a bacteriocin. The concentrations of lactic acid produced by the two strains of *L. alimentarius* used in this study (approximately 50 mM) appeared to be high enough to explain the antilisterial effect.

Different strains of *Lactobacillus sake* have also been identified to have significant antilisterial properties. Bredholt et al. (25) isolated 5 strains of LAB from servelat sausage and ham for their antilisterial activity at temperatures below 8°C and their ability to grow at 3°C, in pH ranging from 5.8 to 6.2 with 3% NaCl added. The strains were identified as being *Lactobacillus sake*. Even if the five strains exhibited bacteriostatic activity toward *L. monocytogenes* in coinoculated ham samples at 8°C for 28 days of storage, the numbers of the LAB increased rapidly during the first three days of storage, from levels of 10^4-10^6 to levels of 10^8-10^9 CFU/g. Indigenous LAB background flora were also reported to increase from <100 CFU/g to 10^8-10^9 CFU/g in 4 weeks. Even with these increasing numbers of both inoculated LAB and indigenous flora, the authors reported nonsignificant sensory changes in the samples. Only 1 of the 5 strains produced bacteriocin. Nevertheless, all five strains exhibited the same level of inhibition toward *L. monocytogenes*, suggesting that mechanisms other than bacteriocin production were responsible; most likely it was competition for nutrients as the LAB strains were selected for their fast growth rates.

De Martinis and Franco (39) isolated a bacteriocin producing strain of *L. sake* from lingüiça, a Brazilian sausage, with antilisterial activity. In this study, the isolated strain was coinoculated in lingüiça with *L. monocytogenes* at a final concentration of 10^4 and 10^3 CFU/g respectively. The *L. sake* strain exhibited a bacteriostatic action over the pathogen during 4 weeks of storage of the sausage at 8°C. The counts of *L. monocytogenes* in the coinoculated treatment were 6 $\log_{10}$ cycles lower than the control treatment (*L. monocytogenes* only) after 4 weeks of storage, demonstrating effective diffusion of the bacteriocin in this particular food matrix. Conversely, Winkowski et al. (181) studied the inhibition of *L. monocytogenes* in beef cubes at 4°C and 10°C by using *Lactobacillus bavaricus* MN, a meat isolate which produces the bacteriocin bavaricin MN. The inhibition occurred only during the first week of storage at 10°C, but by the end of the fourth week of storage there were no differences among the coinoculated samples and the controls. However, when the beef cubes were held at the same temperature but immersed in gravy containing glucose, the numbers of *L. monocytogenes* cells were 5 $\log_{10}$ cycles lower than the controls at the end of a four week storage period. Inhibition of the pathogen in this study was not due to decrease in pH because the antilisterial effect was observed at pH values close to 6.1. Schillinger et al. (149) reported the production of sakacin A by *Lactobacillus sake* Lb 706, a bacteriocin with antilisterial properties. In MRS broth, the presence of sakacin A decreased the numbers of *L. monocytogenes* from 6×10^5 / mL to 4×10^3 /mL in 24 hours at 15°C. However, in minced beef stored at 8°C the viable numbers of *L. monocytogenes* cells remained unchanged for the first week, increasing by 1 $\log_{10}$ cycle after 9 days of storage, demonstrating that the bacteriocin effect was inactivated after 1 week.

Hugas et al. (80) reported another study in which a strain of *L. sake* was utilized to inhibit the growth of *L. monocytogenes* in meat products. These researchers utilized *L. sake* CTC494, which produces the bacteriocin sakacin K, to inhibit the growth of *Listeria* spp. on poultry breasts, cooked pork and raw minced pork samples under different conditions of packaging (oxygen permeable films, vacuum package, and modified atmosphere using 20% CO_2 and 80% O_2) at 7°C for 8 days. *Listeria innocua* and not *L. monocytogenes* was utilized as the target pathogen for this study because these two microorganisms show a similar behavior in meat packaged under modified atmospheres, and the former had shown to be more resistant to sakacin K in previous studies from the same research group. Inoculation of *L. sake* CTC494 or its bacteriocin sakacin K exhibited a bacteriostatic effect over *L. innocua* after 8 days of storage at 7°C in all of the studied samples. The greatest inhibition was observed in the vacuum packaged samples of poultry breasts and cooked pork, and in the modified atmosphere packaged samples of raw minced pork, suggesting that the type of meat and the levels of indigenous flora are important to sustain the growth of LAB.

The initial level of inoculation of protective cultures as it relates to the initial number of the pathogen present in the food matrix is important to ensure inhibitory activity. Buncic et al. (31) selected two LAB strains of a total of 210 isolates from chilled meat products due to their ability to produce antilisterial bacteriocins. The two selected strains were identified as *L. sake* 265 and *Lactobacillus casei* 52. Even if bacteriocin activity toward *L. monocytogenes* was detected in agar well diffusion tests, the authors failed to show inhibition of the pathogen in coinoculated experiments on vacuum packaged sausages at 4°C after 23 days of storage. The inability of these bacteriocin producing strains to inhibit the growth of the pathogen at refrigeration temperatures was attributed to the relatively low initial numbers of the selected strains (approximately 10^3 CFU/g). The authors suggested that when low initial numbers of LAB are applied on chilled meats there might be extensive inactivation of bacteriocins by binding to meat compounds and bacterial cells, and that depletion of the bacteriocin during storage may occur. Consequently, the concentrations of active bacteriocins present on the samples were too low to affect the growth of *L. monocytogenes*. The most effective strain reported in this study (*L. sake* 706) exhibited better inhibitory properties due to its ability to grow rapidly in the meat systems (from 10^3 /g to 10^8-10^9 /g within 3 weeks at 4°C). However, this rapidly increasing number of LAB produced undesirable sensory changes in the product. Similar results were obtained by Beumer et al. (17), who studied the interaction of two strains of lactic acid bacteria in coinoculated samples of sliced meat products (ham, chicken breast, and luncheon meats) with *L. monocytogenes*, and reported insufficient ability to inhibit the growth of the pathogen within 35 days of storage at 7°C due to a low initial inoculum level. The strains studied were *Lactobacillus curvatus* and nisin producing *L. lactis*. The initial level of the pathogen in the samples was approximately 10^1 CFU/g, and the initial numbers of lactic acid bacteria were in the order of 10^2 CFU/g.

Other genera and species of lactic acid bacteria have been reported to exhibit antagonistic activity toward *L. monocytogenes*. Harris et al. (73) used 14 different strains of LAB known to be bacteriocin producing microorganisms to inhibit the growth of *L. monocytogenes* in laboratory media. The 14 strains covered a wide range of genera including *Lactobacillus*, *Leuconostoc*, *Pediococcus* and *Lactococcus*. Out of the 14 strains studied, 7 of them showed inhibition toward the pathogen in agar spot tests (deferred antagonism), but only the supernatant of 3 of these strains exhibited antagonistic activity toward *L. monocytogenes* in well diffusion agar tests. The inhibitory strains were reported to be *Lactobacillus* UAL11, *Leuconostoc* UAL14 and *P. acidilactici* PAC 1.0. Dimitrijevic et al. (41) identified 5 different bacteriocin producing strains of LAB belonging to different genera and species, which were then evaluated for their antilisterial activity at 37°C for 24 h and at 4°C for 12 days in laboratory media. The strains evaluated were *Lactococcus* UW, *Lactobacillus sake* 148, *L. sake* 265, *Pediococcus* 347 and *Lactobacillus sake* 706. This study was particularly extensive because 98 strains of *L. monocytogenes* were evaluated, 50 human isolates and 48 strains of animal and food origin. By using well diffusion agar tests, these researchers reported that bacteriocins produced by *L. sake* 265, *L. sake* 706 and *Pediococcus* 347 exhibited antilisterial activity on almost every analyzed strain of the pathogen, with bacteriocins from *L. sake* 265 showing the greatest inhibition.

A study by Vignolo et al. (177) reported the use of *L. casei* CRL 705, and the bacteriocin produced by this microorganism, named lactocin 705, to control the growth of *L. monocytogenes* in a meat slurry. The effect of this bacteriocin producing microorganism was evaluated in sterile and nonsterile samples of ground beef at 20°C. Two different initial levels of *L. monocytogenes* (10^4 cells/mL and 10^7 cells/mL) and three different levels of bacteriocin (4200, 8400 and 16800 AU/ml) were studied. In all cases in which lactocin 705 was added, a bacteriostatic effect on the pathogen was observed and lasted over 24 hours of incubation; actually at the highest bacteriocin concentration there was a 2 $\log_{10}$ reduction in the numbers of the pathogen in autoclaved beef that was inoculated with approximately 10^4 cells of *L. monocytogenes*/mL. The use of cells of *L. casei* CRL 705 to produce the bacteriocin *in situ* had a bacteriostatic effect toward the pathogen over a period of 24 hours of incubation at 20°C. Numbers of *L. casei* CRL 705 utilized in this study as well as the change of these numbers during the incubation period were not reported.

Jeppesen and Huss (87) utilized one strain of *Leuconostoc* (V 6) and one strain of *L. plantarum* (LKE 5) to control the growth of *L. monocytogenes* in a shrimp model system (pH=6.0) at 5°C. Both lactic acid bacteria decreased the growth rate of the pathogen. After 15 days of storage, the maximum cell number was reduced from approximately 10^8 CFU/mL to approximately 10^6 CFU/mL when the coinoculated treatments were compared to the control treatment with no LAB. The greatest inhibitory effect was obtained with *Leuconostoc* (V 6) when the initial numbers of *L. monocytogenes* were approximately 10^2 CFU/mL and NaCl was added to the system at a level of 3% w/v; the effect observed was bacteriostatic because the numbers of the pathogen did not change significantly over 25 days of storage. However, there was a reduction of more than 3 $\log_{10}$ cycles compared to the control. The mechanism of inhibition was not clearly identified, although the authors suggested that hydrogen peroxide might have been involved in the antagonistic effect of *Leuconostoc*, and a proteinaceous substance might have been responsible for the inhibition exhibited by *L. plantarum*.

Schillinger et al. (149) utilized two bacteriocin producing strains, *L. sake* Lb 706, which produces sakacin A, and *Enterococcus faecium* BFE 900, which produces enterocin B, to control the regrowth of nisin resistant mutants of *L. monocytogenes*. In order to achieve this goal, these researchers selected for nisin resistant mutants of both LAB, which showed initially high sensitivity to low concentrations of nisin. The resulting mutants were able to grow at a concentration of 500 IU of nisin/mL and were still able to produce their respective bacteriocin. In the presence of 100 IU of nisin/mL, *L. monocytogenes* Li3 (nisin resistant) grew rapidly after a short lag phase to a high cell density at 30°C. In associative cultures with *L. sake* Lb 706-1a (nisin resistant), at the same temperature, almost no increase in viable counts of *L. monocytogenes* Li3 was observed after 24 h and the numbers decreased more than 2 $\log_{10}$ cycles after 48 h. *Enterococcus faecium* BFE 900-6a (nisin resistant) was less effective than *L. sake*, and the numbers of *L. monocytogenes* Li3 viable cells actually increased by almost 2 $\log_{10}$ cycles within 24 h. Both strains of LAB significantly decreased the numbers of a nisin sensitive strain of *L. monocytogenes* Scott A in a separate study. Hence, the authors demonstrated that a combination of nisin and a starter or protective culture producing a non-nisin bacteriocin might be effective in preventing the emergence of spontaneous nisin resistant subpopulations of *Listeria monocytogenes*.

Wessels and Huss (179) reported the use of the nisin producing strain *L. lactis* subspecies *lactis* ATCC 11454 to inhibit the growth of *L. monocytogenes* in laboratory media and in a cold smoked salmon infusion simulating conditions of light preservation. At 5°C in M17 broth, the *L. lactis* grew well and produced nisin. In associative cultures in brain heart infusion, this strain reduced the levels of *L. monocytogenes* by 5 $\log_{10}$ cycles at 30°C within 31 h. However, in the cold smoked salmon infusion at the same temperature, the microorganism did not grow. Also, at a level of NaCl of 4%, *L. lactis* grew well and produced nisin efficiently, while 5% NaCl resulted in very slow growth and no detectable nisin. Samples of sliced cold smoked salmon were coinoculated with an initial number of 10^4 CFU/g of *L. monocytogenes* and 10^6 CFU/g of *L. lactis*. The presence of the *L. lactis* cells led only to a decline in the pathogen count within the first 5 days of incubation at 10°C. After that period of time, the *Listeria* growth rate was only slightly depressed relative to the control (with no *L. lactis* inoculum). This study showed that under adverse conditions (5°C and high levels of salt), the use of nisin producing *L. lactis* ATCC 11454 does not prevent the growth of *L. monocytogenes* in cold smoked salmon; however, under temperature abuse conditions (10°C), this culture may be effective as a supplement to other preservative agents.

Amezquita and Brashears (8) isolated several species of LAB from refrigerated deli meats and hot dogs. They screened the isolates for their ability to inhibit *L. monocytogenes* during refrigeration without growing. Several isolates were obtained and one species of the LAB completely eliminated the growth of *L. monocytogenes* during refrigerated storage. Additionally, there were no increases in the number of LAB during the refrigerated storage. Sensory studies were conducted and there were no detectable differences in the products that contained the LAB compared to the controls.

While much of the research in meat products has focused on the inhibition of *Listeria monocytogenes*, other food borne pathogens can be inhibited in meat products during refrigerated storage. Brashears and Durre (23) reported that *L. lactis* completely eliminated the growth of *Salmonella* species and *E. coli* O157:H7 during growth and refrigerated storage. Brashears et al. (22) conducted a study to determine if four strains of lactic acid bacteria (LAB) inhibited *E. coli* O157:H7 and *Salmonella* species in ground beef at 5°C. In raw ground beef the cocktail combination of 4 LAB strains reduced *Salmonella* and *E. coli* O157 by 3–4 log cycles after 12 days of refrigerated storage. Similarly, Brashears et al. (24) reported that *E. coli* O157:H7 was significantly inhibited on raw chicken breasts during refrigerated storage when inoculated with *Lactobacillus* species. They reported that more than a 1 log cycle reduction occurred during a 7 day storage period.

5.3 USE OF LACTIC ACID BACTERIA FOR *IN VIVO* REDUCTION OF FOOD BORNE PATHOGENS

5.3.1 Definition of Probiotic and Direct Fed Microbial

The word probiotic, which means "for life," is derived from the Greek language and has been given several definitions over the years (59). It was first used in 1965 by Lilly and Stillwell (108) to mean substances produced by one protozoan that stimulated the growth of another. In 1974, Parker (130) defined it as "organisms and substances that have a beneficial effect on the host animal which contribute to intestinal microbial balance." However, this general definition was not precise enough, because the word substances includes chemical supplements such as antibiotics. In 1989, Fuller (58) revised the definition to "a live microbial feed supplement which beneficially affects the host animal by improving its intestinal microbial balance." This definition was less confusing, as it emphasized the importance of live cells as an essential component of an effective probiotic. The definition of probiotic has since been expanded to include treatments with live bacteria that are not administered by mouth and that do not exert their effect through the intestinal microflora. In 1991 Huis in't Veld and Havenaar (81) broadened the definition to "a mono- or mixed culture of live microorganisms which, applied to man or animal (e.g., as dried cells or as a fermented product), affects beneficially the host by improving the properties of the indigenous microflora."

Kmet et al. (99) defined ruminal probiotics as "live cultures of microorganisms that are deliberately introduced into the rumen with the aim of improving animal health or nutrition." Based on the various definitions, probiotic could possibly refer not only to microbial cultures, but also extracts and enzyme preparations. The Office of Regulatory Affairs of the Food and Drug Administration, as well as the Association of American Feed Control Officials, have recommended the term direct fed microbials (DFM) be used to describe feed products that contain a source of live, naturally occurring microorganisms. The term DFM will be used in the remainder of this text to describe live cultures of that are fed to animals to reduce food borne pathogens.

5.3.2 Use of Direct Fed Microbials in Farm Animals

5.3.2.1 Concept of Direct Fed Microbials

The gastrointestinal microflora, which are established immediately after birth, are considered very important for the performance of farm animals (58,124). The fetus *in utero* is sterile, but on passage through the reproductive tract during birth acquires microorganisms which are rapidly added to the gut (58). The gut microflora are also obtained from the immediate environment, which is heavily contaminated with bacteria from the mother (26,139). The final indigenous microflora that stabilize the gut are characteristic of the host species and forms a complex system of interrelationships between different microorganisms and between microorganisms and the host (58). This stable flora helps the animal to resist infections by creating a defense barrier against invading microorganisms or toxic substances in the diet (52,110). It also aids in digestion, especially in ruminants, where the metabolism of fibrous components in the diet is dependent on the fermentative action of bacteria in the rumen (147,148).

Although the protective flora in the gut are very stable, they can be influenced by dietary and environmental factors. Modern methods of animal rearing often restrict the access that the animal has to the mother, preventing it from acquiring the full complement of characteristic microbes (58). This in turn makes farm animals very susceptible to an imbalance in normal microflora, leading to inefficient utilization of feed and poor growth performance (124). Stress is another factor that affects gut flora, and can be induced by severe changes in the physical or emotional environment (165). Antibiotics have been used over the years as feed supplements to overcome these problems and promote the growth of the animal. However, there is growing concern that use of antibiotics is resulting in the development of resistant populations of harmful bacteria, which can make the use of antibiotic therapy difficult (75,109). Use of antibiotics also induces intestinal infections resulting in enteritis and diarrhea. In a study by Smith and Tucker (153), inclusion of subtherapeutic levels of antimicrobial growth promoters in the feed prolonged the excretion of *Salmonella* in the feces of chickens (52). A similar effect was obtained in mice dosed experimentally with antibiotics (52). Considering the effects produced by antibiotics, the Swann Committee in 1969 restricted the use of antibiotics as animal feed supplements to only those which were not used in the treatment of disease (58). On the other hand, the antiadditive lobby saw antibiotics as foreign substances that should not be present in the food chain (58). These conditions stimulated an interest in the use of more natural feed additives: Direct Fed Microbials.

Direct Fed Microbials that have been included in food for many years without any adverse effects are, according to the Food and Drug Administration (60), Generally Regarded As Safe (GRAS) and have been shown to have beneficial effects in the animal (58,89,105). The concept of direct fed microbials is now universally accepted, and a substantial amount of research is being directed towards formulation of mixtures of direct fed microbial strains that would have beneficial effects in the animal.

5.3.2.2 Microorganisms Used in Direct Fed Microbials

Intestinal strains of lactic acid bacteria and *Bifidobacteria* are most widely used as direct fed microbials, although yeasts such as *Saccharomyces, Aspergillus*, and *Torulopsis,* and microorganisms such as *Bacillus* and *Clostridium* have also been used (60,112,166,167). Among the lactic acid bacteria, strains of *Lactobacillus* species are most commonly used, and include *L. acidophilus, L. bulgaricus, L. helveticus, L. casei, L. lactis, L. plantarum, L. rhamnosus, L. reuteri, L. fermentum, L. brevis,* and *L. salivarius* (58,60,112). Other strains of lactic acid bacteria include *Enterococcus faecalis, E. faecium, Streptococcus salivarius thermophilus, S. lactis,* and *Pediococcus pentosaceus*. Most frequently used strains of *Bifidobacterium* include *B. bifidum, B. breve, B. thermophilus*, and *B. pseudolongum* (58,112). The composition of DFM preparations may vary from those containing a single strain of microorganisms to those containing multiple strains (58,60). There is, however, more interest in multiple strain preparations, as it is assumed that they are active against a wider range of conditions, including microbial infections and antibiotic relieved growth depression, and in a wider range of animal species (58,60).

5.3.2.3 Direct Fed Microbials and Microbial Angogonism in Animals

Since the first scientific explanation of the favorable effects of soured milk products in humans by Metchnikoff at the beginning of twentieth century (114), the most beneficial part of the intestinal flora has been considered to be lactic acid bacteria. Lactic acid bacteria are also the most common organisms used for commercial DFM preparations (8,170). The emphasis on the lactic acid bacteria stems from the fact that there is evidence that lactic acid bacteria occupy a central role in the gut flora that enables them to influence the composition of the flora to the benefits of the host. The stomachs of neonatal pigs are known to be colonized by *Lactobacillus* and *Streptococci* within 48 hours after birth (44). Similarly, in newborn calves one of the first groups of microorganisms in the rumen is lactic acid bacteria (24). Studies show that when the gut flora develop after birth, as the lactobacilli increase other components of the flora decrease (152). The claims made for probiotic effects of lactic acid bacteria in farm animals are many and varied, and are summarized in Table 5.1 (60).

The potential benefits of probiotic lactic acid bacteria can be placed into three broad categories: reduction of food borne pathogens, improved animal performance, and stimulation of immune response.

5.3.2.4 Pathogen Reduction

The most commonly identified beneficial effect of lactic acid bacteria as live feed supplements is their role in resistance to infection, particularly in the gastrointestinal tract. It has been proposed that these organisms can prevent infection through competitive exclusion (CE) or other mechanisms against pathogenic bacteria in the animal intestine (10,136). According to Bailey (10), competitive exclusion implies the prevention of entry of one entity into a given environment because that space is already occupied, the competing entity is better suited to establish and maintain itself in that environment, or the competing entity is producing a product hostile (toxic) to its competition. The phenomenon of CE was first described by Nurmi and Rantala (125) when they demonstrated that *Salmonella* colonization in the gut of a newly hatched chicken could be prevented by dosing it with a suspension prepared from gut contents of healthy adult chickens. The efficacy of the CE concept has since been demonstrated in several laboratories around the world (13,136,156).

5.3.2.4.1 Poultry Much of the work involving competitive exclusion by lactic acid bacteria has been done in poultry and has focused on the reduction of *Salmonella*. The most commonly studied organisms are strains of *Lactobacillus*, especially *L. acidophilus*. However, the results obtained from the studies to demonstrate the efficacy of DFM have been controversial. The original work by Nurmi and Rantala (125) to explain the concept was done with *Salmonella enteritidis*, and subsequent studies have shown that DFM can work against other strains of *Salmonella* such as *S. typhimurium, S. pullorum, S. salivarius*, and *S. blockley*. Jin et al. (88) demonstrated that a combination of *Lactobacillus* strains isolated from chicken intestine were able to inhibit growth of five strains of *Salmonella*: *S. enteritidis* 94/448, *S. typhimurium, S. pullorum, S. blockley,* and *S. enteritidis* 935/79, and three serotypes of *E. coli*: O1:K1, O2:K1 and O78:K88. Fuller (56) examined the effects of *Lactobacillus* on *E. coli* in the chicken crop and in associative cultures and observed that *E. coli* growth was suppressed both *in vitro* and *in vivo*. In a study by Gilliland and Speck (63), *Lactobacillus acidophilus* exerted antagonistic actions on growth of *S. typhimurium*, enteropathogenic *E. coli*, *Staphylococcus aureus,* and *Clostridium perfringens* when grown with each in associative cultures. Similar studies involving the

Table 5.1
Probiotic effects of lactic acid bacteria in farm animals

Effect	Ref.
Resistance to infectious diseases	(14) (61) (88)
Improved growth performance	(1) (37)
Improved feed efficiency	(12) (182)
Increased milk or egg production and quality	(115)

effect of lactic acid bacteria on *Salmonella, E. coli, S. aureus,* and *C. perfringens* have shown the inhibition of pathogens by lactic acid bacteria (14,178).

On the other hand, there are studies that have raised questions regarding the efficacy of DFM microorganisms in reducing colonization of pathogenic bacteria. In experiments conducted by Hinton and Mead (76), results showed that DFM products containing strains of *Lactobacillus* or *Enterococcus*, administered to day old chicks in feed or drinking water or by spraying on bird's feathers, did not reduce *Salmonella* in the ceca. Similarly, Adler and Da Massa (4) found no protection by *Lactobacillus* against *Salmonella* or *E. coli* colonization in the ceca of newly hatched chicks.

The controversy regarding the effectiveness of the DFM microorganisms can be explained partially by the use of defined and undefined cultures and also by using various strains and species to get the desired effects. Defined product comprises a known mixture of pure bacterial cultures derived from fecal and cecal contents of the bird, whereas undefined product consists of a homogenous mixture of known aerobic microorganisms and unknown, mainly anaerobic, microorganisms derived from the ceca of the bird (118). According to a review by Mulder et al. (118), results from studies on the effect of microflora consisting of 50 different pure cultures were less promising than those obtained after the administration of undefined microflora. Similarly, in a study by Stavric et al. (160), results with mixtures of pure cultures of *Lactobacillus* showed that the preparations were ineffective in reducing *Salmonella* carriage in chicks. On the other hand, undefined anaerobic culture prepared from feces of adult birds showed a significant reduction in the number of *Salmonella typhimurium* in chicks.

Even more important are the selection criteria used to obtain cultures to be used as DFM cultures. These criteria are discussed later in this chapter.

5.3.2.4.2 Pigs The work associated with the use of DFM in pigs is very limited, but the use of lactic acid bacteria has been growing in the past few decades. The most commonly tested lactic acid bacteria are the strains of *Lactobacillus* and *Enterococcus*, and the majority of the studies involve starter pigs, based on the assumption that adult pigs are more resistant to intestinal disorders. It has been demonstrated in several feeding trials that selected strains of lactic acid bacteria can be beneficial in reducing pathogenic bacterial count (14,40,129). A study by Barrow et al. (14) demonstrated that when two day old piglets, weaned to a sow's milk substitute diet, were given *L. fermentum* alone or in combination with *S. salivarius* in their milk, there was a significant decrease in the *E. coli* counts in the stomach and duodenum. Similarly, fecal coliform counts and hemolytic *E. coli* 0141:K85ab were reduced in piglets when treated with *E. faecalis* and *E. faecium* respectively (40,129). Underdahl et al. (171) also demonstrated that *E. faecium* reduced the number of pathogenic *E. coli* and the severity of illness associated with it in gnotobiotic piglets. However, contrasting results have also been reported by some researchers regarding the efficacy of DFM lactic acid bacteria in pigs. One such study involving the interaction between *Lactobacillus* species and *E. coli* K88 in gnotobiotic pigs showed that *Lactobacillus* was unable to prevent the adherence of *E. coli* to the intestinal mucosa (19). In spite of some negative results, the use of lactic acid bacteria in pigs holds considerable potential. The selection criteria for probiotic microorganisms are a big factor influencing the efficacy of a particular strain or mixture of strains, combined with the need for appropriate *in vitro* and animal models, sensitive and reproducible techniques, and repeated experimentation to validate the efficiency of probiotic lactic acid bacteria.

5.3.2.4.3 Cattle Since the recognition of cattle as the principal reservoir of *E. coli* O157:H7, the CE approach to reduce the carriage of pathogen in the animal has received tremendous attention. Unfortunately, the literature on the use of lactic acid bacteria as CE microorganisms in cattle is limited, because the probiotic products fed to cattle are primarily fungal cultures (78). Many of the studies examining the effects of probiotic lactic acid bacteria have been limited to calves (124). However, the number of studies done in feedlot animals is growing. As with other species, there is inconsistency in the results obtained from studies involving use of direct fed microbials in cattle. Nonetheless, use of DFM microorganisms in cattle is on an increase, and several studies have been conducted to understand the specific role of lactic acid bacteria in reducing the carriage of pathogenic bacteria in cattle.

One of the earlier studies, done by Ellinger et al. (46), reported a decrease in fecal coliforms when the liquid diet of newborn calves was supplemented with *Lactobacillus* cultures. In a similar study, Gilliland et al. (65) demonstrated that *Lactobacillus* strains isolated from cow were more effective than those isolated from pigs, suggesting the importance of host specificity of the probiotic strains. Ozawa and coworkers (129) tested the effect of *E. faecalis* BIO4R on intestinal flora of calves and found that the strain had an antagonistic effect on *Salmonella.*

Studies by Brashears et al. (21) indicate that *E. coli* O157:H7 can be reduced in feedlot age cattle. The use of Direct Fed Microbials (DFM) has been effective in significantly reducing the amount of *E. coli* O157:H7 detected in the feces and on the hides of beef feedlot cattle in two separate large scale studies conducted at Texas Tech University. In the first study, 180 beef feedlot cattle were separated into 3 treatment groups. One group received NP 51, one received NP 35, and one received a carrier of the DFM and served as a control. NP 51 and NP 35 contained two separate strains of *Lactobacillus acidophilus* as the DFM. The cultures were fed at a level of 1×10^9 cells/head/day for the last 60 days of the feeding period. Overall the reduction in shedding was 50% for the animals fed NP 51 compared to the control group (Figure 5.1). There were no significant reductions in those fed NP 35. At slaughter, the prevalence of *E. coli* on the hides was reduced from 20% in the controls to 3% and 0% in NP 51 and NP 35, respectively. In a follow up study conducted the following summer, the treated animals were fed NP 51 and a combination of other commercially available DFM cultures for the entire duration of the feeding period. At 7 days prior to slaughter and at slaughter, 27% of the fecal samples in the control animals tested positive for *E. coli* O157:H7 while the treated animals contained significantly fewer detectable numbers with only 13% being positive (Figure 5.2). Again there were significant reductions in the number of animals testing positive for *E. coli* O157:H7 on the hides with 14% of the control samples testing positive and only 5% of the treated samples testing positive. In all studies there was a trend toward improved performance characteristics.

5.3.2.5 *Improved Animal Performance*

Over the years, numerous claims have been made about the beneficial effects of lactic acid bacteria involving improved animal performance. Some of the potential benefits of feeding lactic acid bacteria as probiotic supplements, as suggested by Fuller (60), include increased growth rate, improved feed conversion, improved digestion and better absorption of nutrients, increased milk or egg production, and increased milk or egg quality. Several studies have shown that administration of lactic acid bacteria improved birth weight gain and mortality in piglets and calves (1,37) and also growth rate and egg production in chickens (42,115). An increase in daily weight gain and improvement in feed conversion was obtained by Baird (12) in separate experiments with feeder and growing finishing pigs using *Lactobacillus* supplements. In two studies reported by Wren (182) and Lee and Botts (103), higher feed intakes and weight gains were obtained in newborn claves when *Lactobacillus* species were supplemented with milk replacer diets. Dilworth and Day (42) evaluated the effects of adding *Lactobacillus* cultures to broiler diets and found a significant improvement in growth rate and efficiency. Miles et al. (115) reported significant increase in egg production when

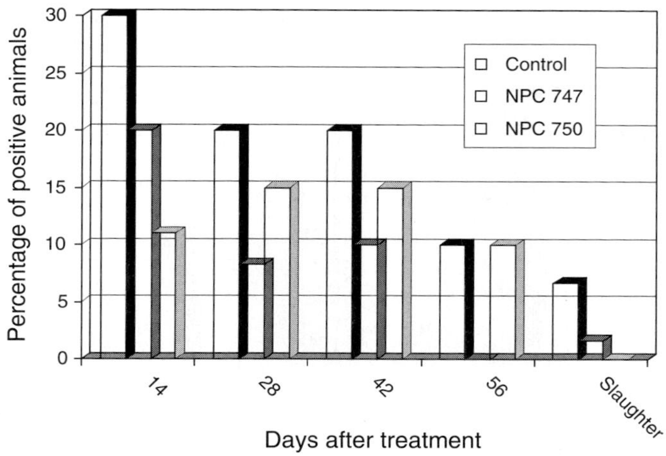

Figure 5.1 Prevalence of *E. coli* O157:H7 in 180 Cattle After Treatment with Direct Fed Microbials During the Last 60 Days of the Feeding Period

L. acidophilus cultures were fed to commercial layers. Although the statistical effects on growth, egg production, feed conversion, and daily weight gain have been reported, there is a lack of good, controlled field experiments and most of the positive results are obtained in studies involving young animals. There is therefore need for more research involving large scale field trials involving both young and mature animals.

5.3.2.6 Effect on Immune Response

The effect of probiotic lactic acid bacteria on the host immune response has been studied to some extent, and it is postulated that both mucosal and systemic immune responses can be affected by direct fed microbials. Bealmer et al. (15) demonstrated that conventional animals with complete gut flora have higher immunoglobulin levels and phagocytic activity than germ free animals. Roach and Tannock (140) suggested that a systemic effect was exerted by *E. faecium* that was established as a monoassociate in germ free mice, reducing *S. typhimurium* counts in the spleen. Similarly, *L. casei* was involved in the stimulation of phagocytic activity when administered perorally to mice in a study by Perdigon et al. (131). In order for a microorganism to affect systemic immunity it may be necessary for it to enter the systemic circulation. Bloksma et al. (18) showed that *Lactobacillus* species were able to survive in the spleen, liver, and lungs for several days. Saito and coworkers (144) showed that *L. casei*, given parenterally, stimulated phagocytic activity in mice. Serum IgA and IgG levels have been shown to be increased with administration of *Lactobacillus* in piglets and mice (106,132). These findings suggest that direct fed microbials have the potential to modulate immunity, and that their effect on systemic immune response can be used to overcome infections that occur in tissues away from the intestinal tract, caused by pathogens such as *Salmonella*.

5.3.3 Selection Criteria for Lactic Acid Bacteria to be Used as Direct Fed Microbials

Although lactic acid bacteria used as direct fed microbials are generally shown to have a positive effect *in vitro* and *in vivo*, the responses attained with the studies have been variable. One of the reasons for controversial results is the selection of strains for DFM use. There are

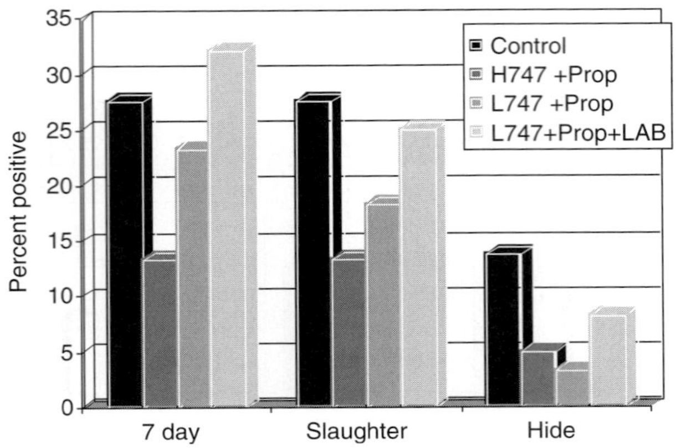

Figure 5.2 Prevalence of *E. coli* O157:H7 in 240 Cattle After Treatment with Direct Fed Microbials During the Entire Feedlot Feeding Period

certain attributes that an organism must possess in order to be functional or desirable as a probiotic microorganism. Following is a discussion of the selection criteria that are considered to produce suitable direct fed microbial bacteria.

5.3.3.1 Survival in the Gastrointestinal Tract

In order for a DFM to produce desirable effects, it must be able to survive and metabolize in the intestine. This means that the strain must be resistant to gastric juices, low pH, enzymes, organic acids, and bile and pancreatic secretions.

Most bacteria do not survive well at low pH values, and the severe acidic condition of the intestinal tract has an adverse effect on most microorganisms. It is therefore suggested that microbial cultures to be used as probiotic products should be screened for their resistance to acidity. In a study by Conway et al. (35) on the survival of lactic acid bacteria in the human stomach, strains showed variable survival at different pH conditions, demonstrating the importance of screening the strains for acid tolerance. Similarly resistance of lactic acid bacteria to bile is an important characteristic that enables them to survive and grow in the intestinal tract (64,66). Bile entering the duodenal section of the small intestine has been found to reduce survival of bacteria probably due to the fact that all bacteria have cell membranes consisting of lipids and fatty acids, which are very susceptible to destruction by bile salts. Gilliland et al. (1984) reported that when a diet supplemented with a more bile resistant strain of *L. acidophilus* was fed to newborn dairy calves, greater numbers of facultative lactobacilli were detected in the upper part of small intestine (jejunum) than when a strain with lower bile resistance was used. It was also suggested that the ability of *L. acidophilus* to cause significant increase in numbers of *Lactobacillus* in the intestinal tract may be critical for controlling growth of intestinal pathogens. Hence the success of probiotics also depends on the selected strain possessing bile resistant qualities.

5.3.3.2 Adhesion to Intestinal Epithelium

The adhesion properties of specific bacterial strains to intestinal cells enhance colonization and have become standard criteria for selecting new DFM strains (145). It is speculated that by attachment to the gut wall, the probiotic strain will occupy colonization sites and make them unavailable to other microorganisms, especially pathogens, thereby demonstrating competitive exclusion. The role of adhering bacteria in protection against enteric pathogens was recognized by Fuller (54), who suggested treating newly hatched chicks with pure cultures of adhering lactobacilli. Adhesion is also considered necessary for the microorganism to resist being washed away by contents of the stomach and intestine, and by peristalsis (60). In order for the probiotic organism to manifest its effect, ability to remain in the gut for a maximum amount of time is a crucial property (58) which can be attained by attachment. Another reason to use adhesion property as a selection criterion for strains is the variability obtained between strains of the same species. Mayra-Makinen et al. (113) reported that the degree of adherence varied greatly between the 13 strains of *Lactobacillus* that showed adherence to columnar epithelium of pigs and calves.

5.3.3.3 Host Specificity

In selecting strains for use as DFM supplements, the source of the organism is important because most of these organisms exhibit host specificity. For example, *Lactobacillus* isolated from a specific site of a specific animal source can only colonize epithelium of the same kind. Host specificity of bacterial strains is well recognized and documented (14,55). Barrow et al. (14) tested the attachment of lactic acid bacteria to gastric epithelium of pigs *in vitro* and found that with the exception of two strains of *Lactobacillus* isolated from the chicken gut, no isolates from animals other than domestic pigs and the closely related wild boar were able to adhere to pig squamous epithelium. Similarly, Fuller (54) demonstrated that *Lactobacillus* obtained from fowl crop only adhered to squamous epithelial cells of chicken intestine and not mouse, rat, or pigs.

5.3.3.4 Production of Antimicrobial Compounds

Lactic acid bacteria produce a wide variety of antimicrobial compounds such as bacteriocins, organic acids, hydrogen peroxide, and other low molecular weight metabolites. Production of these substances may give them a competitive advantage over pathogenic strains in the gastrointestinal tract and promote their colonization. However, production of antibacterial substances may not be an important criterion for selection, because several probiotic strains of lactic acid bacteria have been shown to exert beneficial effects in the intestinal tract without possessing this property.

5.3.3.5 Antibiotic Susceptibility

Antibiotic resistance in probiotic bacteria is an area of growing concern. According to a recent report, the FDA blocked the introduction of two DFM products for use in chickens on grounds that some of the microorganisms in the products were possibly antibiotic resistant, which could lead to humans contracting diseases not treatable by drugs (134). It is speculated that antimicrobial drugs used in food animals can promote emergence of resistant bacteria that may not necessarily be pathogenic to the animal, but may cause severe infections in humans (134). The use of antibiotics in food animals can also cause nonpathogenic bacteria to become resistant, which may directly or indirectly cause infections in humans.

In the past few years, the lactic acid bacteria most commonly associated with antibiotic resistance have been strains of the genera *Enterococcus*, especially *E. faecalis* and *E. faecium* (51). *Enterococcus* species have been used as direct fed microbials to maintain intestinal microbial balance and also as a treatment for gastroenteritis in humans and animals. However, the fact that these bacteria have acquired resistance toward most clinically used antibiotics, including the glycopeptide antibiotics vancomycin and teicoplanin, increases their threat as opportunistic pathogens (51). Resistance is acquired by gene transfer systems such as conjugative or nonconjugative plasmids, or transposons (133,168). Several antibiotic resistance plasmids from *Lactobacillus* species have also been detected. Ishiwa and Iwata (83) indicated plasmid linkage of tetracycline and erythromycin resistance in human isolates of *L. fermentum*. Morelli et al. (117) observed plasmid linked resistance for chloramhpenicol in *L. acidophilus* and *L. reuteri* isolated from poultry.

Plasmid associated antibiotic resistance is of special concern because of the possibility of resistance spreading to other, more harmful species and genera. Resistance can be transferred from nonpathogenic bacteria to pathogenic bacteria and from bacteria that are normally present in the intestinal tract of animals to those that cause infections in humans. *In vitro* studies have demonstrated that vancomycin resistance is transferable to other Gram-positive bacteria including *Listeria monocytogenes* and *S. aureus* (101). A chloramphenicol resistance plasmid from an *L. plantarum* strain isolated from raw ground pork (7) was shown to be transferred to other Gram-positive bacteria by the help of a wide host range (33). Of other concern is the potential risk of transfer of antibiotic resistance and associated virulence traits to other lactic acid bacteria in foods that could eventually lead to emergence of opportunistic pathogens, especially in immunocompromised individuals. Although *Streptococcus* and *Enterococcus* are the predominating lactic acid bacteria associated with human infections, other lactic acid bacteria have also been implicated in human infections despite their traditional GRAS status (61). New species and more specific strains of probiotic bacteria are constantly being sought, and these may not share the same historical safety of traditional strains. It therefore becomes important to carefully assess antibiotic resistance in new strains prior to incorporating them into probiotic or CE products.

5.3.3.6 Technological Properties

It was not until recently that technological properties were established as a selection criteria for DFM strains because they largely dictate the successful production and delivery of direct fed microbials. Direct fed microbials are fed to the animals directly or through their food in the form of pellets, capsules, paste, powder, or granules. Therefore, at the industrial level, DFM strains that are produced in large quantities have to undergo several processing steps before their use as feed supplements. The processing steps may involve separation by centrifugation or filtration, fermentation in cultured products, and freeze drying or lyophilization (97,124). Direct fed microbial bacteria should therefore be able to withstand stresses such as freezing, high pressure, and temperatures (60−80ºC for 5−10 minutes during pelleting); and should have a high growth rate and achievable cell mass. They should also be able to retain their viability under storage conditions. However, the stability of probiotic strains in continuous industrial culturing is a challenge for the industry (105). According to Nousiainen and Setala (124), most *Lactobacillus* strains do not tolerate pelletizing in an economically feasible way. Crowell et al. (97) showed a dramatic decrease in the viability of *L. acidophilus* in dried pellets held under refrigeration or at room temperature over a period of 12 months. Stability during culture propagation and storage may have an impact on the *in vivo* response toward the probiotic. It is therefore important, during the selection of a probiotic culture, to consider these production and stability criteria.

These selection criteria were used to select specific strains of lactic acid bacteria from cattle feces as potential candidates as direct fed microbials to reduce *E. coli* O157:H7 in the live animal (22). A total of 686 LAB were selected and screened for inhibition of *E. coli* O157:H7. One half of the 686 inhibited the pathogen. Further screening for acid and bile tolerance resulted in identification of 75 candidates. Well identified strains were subjected to inhibition studies in rumen fluid and manure during which only 2 LAB were identified as potential candidates. These candidates were used in *in vivo* trials and successfully reduced *E. coli* O157:H7 in the animal (22).

5.3.3.7 Possible In Vivo Mechanisms of Action

One of the reasons for the beneficial effects exhibited by lactic acid bacteria is a direct antagonistic action against harmful microorganisms. However, the exact mechanisms by which lactic acid bacteria affect the microflora of the intestinal tract are not clearly understood. Modes of action that have been proposed for competitive exclusion are production of antimicrobial metabolites, competition for nutrients and competition for adhesion sites on the gut epithelial surface.

Lactic acid bacteria are known to produce several antibacterial agents and primary metabolites, such as organic acids, hydrogen peroxide, enzymes, low molecular weight metabolites, and bacteriocins. Among the organic acids, lactic acid and acetic acid are the two major end products of carbohydrate fermentation by homofermentative and heterofermentative lactic acid bacteria respectively. The inhibition of pathogenic bacteria *in vitro* by production of organic acids is well documented, but the evidence for *in vivo* inhibition is not very convincing. In a study by Jin et al. (88) inhibition shown by *Lactobacillus* strains against pathogenic strains of *Salmonella* and *E. coli* was suggested to be due to the production of organic acids by *Lactobacillus* isolates. It is assumed that the primary reason for antagonism by lactic acid production is the reduction in pH, which inhibits the growth of many bacteria including Gram-negative pathogenic organisms (30,157). Although pH is the main factor in antagonism, it has also been demonstrated that lower pH values govern the activity of organic acids, because the undissociated forms are most bactericidal (2,157). The undissociated acid is easily diffused through the bacterial cell wall, thereby reducing the intracellular pH and slowing down metabolic activities of the bacteria (77). *Escherichia coli* is shown to be inhibited by lactic acid at a pH of 5.1 (71). Tramer (169) also showed that the inhibition of *E. coli* by *L. acidophilus* was due to the strong bactericidal effect of lactic acid at low pH. However, acetic acid, due to its higher dissociation constant, shows stronger inhibition than lactic acid at a given molar concentration and pH (77). These volatile acids are especially antimicrobial at low oxidation reduction potentials (144), which lactic acid bacteria help maintain in the intestine.

Hydrogen peroxide, which is produced in the presence of molecular oxygen together with lactate, pyruvate, and NADH by flavin enzymes (34,62,69,93), is one of the primary metabolites of lactic acid bacteria that may contribute to antagonism. It inhibits the growth of pathogens through its cytotoxic effect on the bacterial cells by generating highly reactive and toxic oxygen species, such as the hydroxyl radical that initiates oxidation of biomolecules (90). The antimicrobial activity of hydrogen peroxide is well recognized and documented. The formation and accumulation of hydrogen peroxide by *Lactobacillus* in growth media with a subsequent bactericidal effect on *S. aureus* has been reported by Dahiya and Speck (38). Gilliland and Speck (63) showed that *L. acidophilus* exhibited antagonistic activity against intestinal and food borne pathogens in associative cultures. In addition, hydrogen peroxide may also activate the lactoperoxidase system, indigenous to milk, causing the formation of hypothiocyanite and other antimicrobial products that may inhibit enteric.

Bacteriocin production by lactic acid bacteria is also recognized as one of the mechanisms used for antagonism against other microorganisms. Bacteriocins are defined as potent antimicrobial peptides or proteins produced by bacteria, which have a bactericidal action against other bacteria . The inhibitory spectrum of bacteriocins is restricted to closely related organisms, which implies that bacteriocins produced by lactic acid bacteria may not be active against Gram-negative pathogens and spoilage bacteria. A number of Gram-positive toxinogenic and pathogenic bacteria have been found to be inhibited by bacteriocins of certain lactic acid bacteria. For example, a strain of *Pediococcus acidilactici* was found to produce a bacteriocin (pediocin AcH) that inhibited several pathogenic bacteria associated with food including *Bacillus cereus, Clostridium perfringens, Listeria monocytogenes* and *S. aureus*. A bacteriocin like substance (lactocidin) produced by intestinal strains of *Lactobacillus* isolated from humans and several laboratory animals was found to be inhibitory against numerous genera including *Salmonella* species, *E. coli*, and *Staphylococcus* species.

Competition for adhesion sites on the intestinal epithelium, which prevents colonization of pathogens, is another mechanism involved in CE by lactic acid bacteria. A prerequisite for invasion by enteropathogens, including *E. coli*, is for the pathogen to have access to receptors on the host tissue (100,155). It is therefore believed that occupation of the receptor or attachment site by the native or protective intestinal flora is part of their protective role. The ability to adhere to mucosal surfaces has been suggested to be an important property of the bacterial strains used in probiotic products (111). Lactic acid bacteria, particularly *Lactobacillus* and *Streptococcus*, are known to be intimately associated with the nonsecretory squamous epithelial cells of pig stomach (14,57) and chicken crop (53). Mayra-Makinen et al. (113) demonstrated the adhesive capacity of *Lactobacillus* strains to columnar epithelial cells of calves and pigs *in vitro*. The role of adhering lactic acid bacteria in protection against enteric pathogens has been studied to some extent with mixed results. Barrow et al. (14) reported a statistically significant reduction in the numbers of *E. coli* in the stomach when strains of *Lactobacillus* and *Streptococci* were fed, alone or in combination, to artificially reared pigs. Stavric et al. (160) demonstrated that the microflora that remained attached to the cecal wall of chickens after four successive washes in buffered saline had a protective effect against *Salmonella*. On the other hand, in a trial by Spencer and Chesson (159) strongly adherent strains of *Lactobacillus* did not have any effect on the attachment of enterotoxigenic *E. coli* to porcine enterocytes under conditions of exclusion (*Lactobacillus* added to enterocytes before *E. coli*), competition (simultaneous addition of *Lactobacillus* and *E. coli*), or displacement (*E. coli* added before *Lactobacillus*). The inconsistencies in the results reported so far have made the mechanism of CE difficult to understand, because too many generalizations have been made about CE being able to work regardless of attention to individual pathogens' mechanism of adherence to host cells. Some researchers believe that the principle of exclusion by occupation of receptor sites is only applicable when both the lactic acid bacteria and the pathogens involved have the same attachment sites. It will not work with Gram-negative pathogens, because the mechanisms of attachment by lactic acid bacteria and Gram-negative bacteria are different. For example, adhesion of *E. coli* usually takes place by an interaction between the glycan component of host glycolipids and glycoproteins, which act as receptors for bacterial proteinaceous projections (fimbrial lectins) (126). The adherence of lactic acid bacteria, on the other hand, is a process mediated by extracellular components including carbohydrate, protein, or lipoteichoic acid polymers (173). However, another pool of scientists believe that adherent strains can mask pathogen and toxin receptors without necessarily binding to the same epitope, and thus limit the ability of a pathogen to colonize and infect (159). In view of the controversy associated with the exact mechanism that makes an adhering strain effective, it becomes necessary that the importance of adhesion in CE be evaluated thoroughly.

REFERENCES

1. Abe, F., N. Ishibashi, S. Shimamura. Effect of administration of bifidobacteria and lactic acid bacteria to newborn calves and piglets. *J. Dairy Sci.* 78:2838–2846, 1994.
2. Acheson, D. *Escherichia coli* II. *Food Quality* 1:54–56, 1999.
3. Adams, M.R., C.J. Hall. Growth inhibition of foodborne pathogens by lactic and acetic acids and their mixtures. *Int. J. Food Sci. Tech.* 23:287–292, 1988.
4. Adler, H.E., A.J. Da Massa. Effect of ingested lactobacilli on *Salmonella infantis* and *Escherichia coli* and on intestinal flora, pasted vents, and chick growth. *Avian Dis.* 24:868–878, 1980.
5. Aguirre, M., M.D. Collins. Lactic acid bacteria and human clinical infection. *J. Appl. Bacteriol.* 75:95–107, 1993.
6. Ahamad, N., E.H. Marth. Behavior of *Listeria monocytogenes* at 7, 13, 21, and 35°C in tryptose broth acidified with acetic, citric, or lactic acid. *J. Food Prot.* 52:688–695, 1989.
7. Ahn, C., D. Collins–Thompson, C. Duncan, M.E. Sites. Mobilization and location of the genetic determinant of chloramphenicol resistance from *Lactobacillus plantarum* caTC2R. *Plasmid* 27:169–176, 1992.
8. Amezquita, A., M.M. Brashears. Competitive inhibition of *Listeria monocytogenes* by lactic acid bacteria in ready to eat pork products. *J. Food Prot.* 65:316–325, 2000.
9. Anonymous. Probiotics: a review of some of the products currently available to compound feed manufacturers. *Feed Compd.* 5:58, 1990.
10. Axelsson, L., T.C. Chung, W.J. Dobrogosz, S.E. Lindgren. Production of a broad spectrum antimicrobial substance by *Lactobacillus reuteri*. *Microbiol. Ecol. Health Dis.* 2:131–136, 1989.
11. Bailey, J.S. Factors affecting microbial competitive exclusion in poultry. *Food Technol.* 88–92, 1987.
12. Barefoot, S.F., C.G. Nettles. Antibiosis revisited: bacteriocins produced by dairy starter cultures. *J. Dairy Sci.* 76:2366–2379, 1993.
13. Baird, D.M. Probiotics help boost feed efficiency. *Feedstuffs* 49:11–12, 1977.
14. Barnes, E.M., C.S. Impey, D.M. Cooper. Competitive exclusion of salmonellas from the newly hatched chick. *Vet. Rec.* 103:61, 1980.
15. Barrow, P.A., B.E. Brooker, R. Fuller, M.J. Newport. The attachment of bacteria to the gastric epithelium of the pig and its importance in the microecology of the intestine. *J. Appl. Bacteriol.* 48:147–154, 1980.
16. Bealmer, P.M., O.A. Holtermann, E.A. Mirand. Influence of the microflora on the immune response, I: general characteristics of the germ-free animal. In: *The Germ-Free Animal in Biomedical Research,* Coates, M.E., B.E. Gustafsson, eds., London: Academic Press, 1984, pp 335–346.
17. Berry, E.D., R.W. Hutkins, R.W. Mandigo. The use of bacteriocin-producing *Pediococcus acidilactici* to control post-processing *Listeria monocytogenes* contamination of frankfurters. *J. Food Prot.* 54:681–686, 1991.
18. Beumer, R.R., M.C. te Giffel, E. de Boer, F.M. Rombouts. Growth of *Listeria monocytogenes* on sliced cooked meat products. *Food Microbiol.* 13:333–340, 1996.
19. Bloksma, N., H. Ettekoven, F.M. Hothuis, L. Van Noorle-Jansen, M.J. De Reuver, J.G. Kreeflenberg, J.M. Willers. Effects of lactobacilli on parameters of non-specific resistance of mice. *Med. Microbiol. Immunol.* 170:45–53, 1981.
20. Bomba, A., R. Nemacova, R. Kastel, et al. Interactions of *Lactobacillus* spp. and enteropathogenic *Escherichia coli* under *in vitro* and *in vivo* conditions. *Vet. Medicina.* 41(5):155–158, 1996.
21. Booth, I.R., R.G. Kroll. The preservation of foods by low pH. In: *Mechanisms of Action of Food Preservation Procedures,* Gould, G.W., ed., New York, NY: Elsevier, 1989, pp 119–160.
22. Brashears, M.M., M.L. Galyean, J.E. Mann, K. Killinger-Mann, G. Loneragan. Reduction of *Escherichia coli* O157 and improvement in performance in beef feedlot cattle with a *Lactobacillus* direct fed microbial. *J. Food Prot.* 66:748–754, 2003.

23. Brashears, M.M., D. Jaroni, J. Trimble. Isolation, selection and characterization of lactic acid bacteria for a competitive exclusion product to reduce shedding of *E. coli* O157:H7 in cattle. *J. Food Prot.* 66(3):355, 2003.
24. Brashears, M.M., W.A. Durre. Inhibitory action of *Lactobacillus lactis* towards *Salmonella* spp. and *E. coli* O157:H7 during growth and refrigerated storage. *J. Food Prot.* 62:1336–1340, 1999.
25. Brashears, M.M., S.S. Reilly, S.E. Gilliland. Antagonistic action of cells of *Lactobacillus lactis* toward *Escherichia coli* O157:H7 on refrigerated raw chicken meat. *J. Food Prot.* 61(2):166–170, 1998.
26. Bredholt, S., T. Nesbakken, A. Holck. Protective cultures inhibit growth of *Listeria monocytogenes* and *Escherichia coli* O157:H7 in cooked, sliced, vacuum- and gas-packaged meat. *Int. J. Food Microbiol.* 53:43–52, 1999.
27. Bryant, M.P., N. Small. Observations on the ruminal microorganisms of isolated and inoculated calves. *J. Dairy Sci.* 43:654–667, 1960.
28. Bruno, M.E.C., A. Kaiser, T.J. Montville. Depletion of proton motive force by nisin in *Listeria monocytogenes* cells. *Appl. Environ. Microbiol.* 58:2255–2259, 1992.
29. Buchanan, R.L., L.K. Bagi. Microbial competition: effect of culture conditions on the suppression of *Listeria monocytogenes* Scott A by *Carnobacterium piscicola. J. Food Prot.* 60:254–261, 1997.
30. Buchanan, R.L., L.A. Klawitter. Effectiveness of *Carnobacterium piscicola* LK5 for controlling the growth of *Listeria monocytogenes* Scott A in refrigerated foods. *J. Food Safety* 12:219–236, 1992.
31. Burnett, G. S., J. Hanna. Effect of dietary calcium lactate and lactic acid on faecal *Escherichia coli* counts in pigs. *Nature* 197:815, 1963.
32. Bunčić, S. The incidence of *Listeria monocytogenes* in slaughtered animals, in meat, and in meat products in Yugoslavia. *Int. J. Food Microbiol.* 12:173–180, 1991.
33. Christensen, D.P., R.W. Hutkins. Collapse of the proton motive force in *Listeria monocytogenes* caused by a bacteriocin produced by *Pediococcus acidilactici. Appl. Environ. Microbiol.* 58:3312–3315, 1992.
34. Clewell, D., Y. Yagi, G. Dunny, S. Schults. Characterization of three plasmid DNA molecules from *Streptococcus faecalis*. Identification of a plasmid determining erythromycin resistance. *J. Bacteriol.* 117:283–289, 1974.
35. Condon, S. Responses of lactic acid bacteria to oxygen. *FEMS Microbiol. Rev.* 46:269–280, 1987.
36. Conway, P.L., S.L. Gorbach, B.R. Goldin. Survival of lactic acid bacteria in the human stomach and adhesion to intestinal cells. *J. Dairy Sci.* 70:1–12, 1987.
37. Cramer, J.A., J.H. Prestegard. NMR studies of pH-induced transport of carboxylic acids across phospholipid vesicle membranes. *Biochem. Biophys. Res. Com.* 75:295–301, 1977.
38. Cruywagen, C.W., I. Jordaan, L. Venter. Effect of *Lactobacillus acidophilus* supplementation of milk replacer on preweaning performance of calves. *J. Dairy Sci.* 79:483–486, 1996.
39. Dahiya, R.S., M.L. Speck. Hydrogen peroxide formation by lactobacilli and its effect on *Staphylococcus aureus. J. Dairy Sci.* 51:1568–1572, 1968.
40. De Martinis, E.C.P., B.D.G.M. Franco. Inhibition of *Listeria monocytogenes* in a pork product by a *Lactobacillus sake* strain. *Int. J. Food Microbiol.* 42:119–126, 1998.
41. Deprez, P., J. van den Branden, J. de Geest, E. Muylle. De invloed van *Streptococcus faecium* toediening op de excretie van *Echerichia coli* en het coorkomen van slingerzietke bij gespeende biggen. *Vlaams Diergeneeskde Tijdschr.* 58:113–117, 1989.
42. Dimitrijević, M., V. Teodorović, M. Baltić, M. Mirlović. Variations in the sensitivity of *Listeria monocytogenes* types to lactic acid bacteria bacteriocins. *Acta Veterinaria Beograd.* 49:49–56, 1999.
43. Dilworth, B.C., E.J. Day. Lactobacillus cultures in brooder diets. *Poultry Sci.* 57:1101, 1978.

44. Duffes, F., F. Leroi, P. Boyaval, X. Dousset. Inhibition of *Listeria monocytogenes* by *Carnobacterium* spp. strains in a simulated cold smoked fish system stored at 4°C. *Int. J. Food Microbiol*. 47:33–42, 1999.
45. Dulcuzeau, R. Implantation and development of the gut flora in the newborn piglet. *Pig News Inform.* 6:415, 1985.
46. Eklund, T. The antimicrobial effect of dissociated and undissociated sorbic acid at different pH levels. *J. Appl. Bacteriol.* 54:383–389, 1983.
47. Ellinger, D.K., L.D. Muller, P.J. Glantz. Influence of fermented colustrum and *Lactobacillus acidophilus* on fecal flora and selected blood parameters of young dairy calves. *J. Dairy Sci.* 61(Suppl. 1):126, 1978.
48. Elsinghorst, E.A., D.J. Kopecko. Molecular cloning of epithelial cell invasion determinants from enterotoxigenic *Escherichia coli*. *Infect. Immun.* 60:2409–2417, 1992.
49. El-Ziney, M.G., T. van den Tempel, J. Debevere, M. Jakobsen. Application of reuterin produced by *Lactobacillus reuteri* 12002 for meat decontamination and preservation. *J. Food Prot.* 62:257–261, 1999.
50. FDA. Office of Regulatory Affairs: Compliance Policy Guides. Sec. 689.100 Direct-Fed Microbial Products (CPG 7126.41). Available: http://www.fda.gov/ora/compliance_ref/cpg/cpgvet/cpg689-100.html. Accessed Jan. 29, 2003.
51. Foegeding, P. M., A. B. Thomas, D. H. Pilkington, T. R. Klaenhammer. Enhanced control of *Listeria monocytogenes* by *in situ*-produced pediocin during dry fermented sausage production. *Appl. Environ. Microbiol.* 58:884–890, 1992.
52. Franz, C.M.A.P., W.H. Holzapfel, M.E. Stiles. Enterococci at the crossroads of food safety? *Int. J. Food Microbiol*. 47:1–24, 1999.
53. Freter, R. Experimental enteric shigella and vibrio infection in mice and guinea pigs. *J. Exp. Med*. 104:411–418, 1956.
54. Fuller, R., Turvey, A. Bacteria associated with the intestinal wall of the fowl (*Gallus domesticus*). *J. Appl. Bacteriol*. 34:617–622, 1971.
55. Fuller, R. Ecological studies on the lactobacillus flora associated with the crop epithelium of the fowl. *J. Appl. Bacteriol.* 36:131–139, 1973.
56. Fuller, R. Nature of the detriment responsible for the adhesion of lactobacilli to chicken crop epithelial cells. *J. Gen. Microbiol.* 87:245–250, 1975.
57. Fuller, R. The importance of lactobacilli in maintaining normal microbial balance in the crop. *J. Poult. Sci.* 18:85–94, 1977.
58. Fuller, R. Epithelial attachment and other factors controlling the colonization of the intestine of the gnotobiotic chicken by lactobacilli. *J. Appl. Bacteriol.* 46:335–342, 1978.
59. Fuller, R. Probiotics in man and animals. *J. Appl. Bacteriol*. 66:365–378, 1989.
60. Fuller, R. *Probiotics: the Scientific Basis.* London: Chapman and Hall, 1992.
61. Fuller, R. Probiotics for farm animals. In: *Probiotics: A Critical Review,* Tannock, G.W., ed., Wymondham, U.K.: Horizon Scientific Press, 1999, pp 15–22.
62. Gasser, F. Safety of lactic acid bacteria and their occurrence in human clinical infections. *Bull. Inst. Pasteur*. 92:45–67, 1994.
63. Gilliland, S.E., M.L. Speck. Biological response of lactic streptococci and lactobacilli to catalase. *Appl. Microbiol*. 17(6):797–800, 1969.
64. Gilliland, S.E., M.L. Speck. Antagonistic action of *Lactobacillus acidophilus* toward intestinal and foodborne pathogens in associative cultures. *J. Food Prot*. 40:820–823, 1977.
65. Gilliland, S.E. Beneficial interrelationships between certain microorganisms and humans: candidate organisms for use as dietary adjuncts. *J. Food Prot*. 42:164–167, 1979.
66. Gilliland, S.E., B.B. Bruce, L.J. Bush, T.E. Stanley. Comparison of two strains of *Lactobacillus acidophilus* as dietary adjuncts for young calves. *J. Dairy Sci.* 63:964–972, 1980.
67. Gilliland, S.E., T.E. Stanley, L.J. Bush. Importance of bile tolerance of *Lactobacillus acidophilus* used as a dietary adjunct. *J. Dairy Sci*. 67:3045–3051, 1984.
68. Glass, K.A., M.P. Doyle. Fate of *Listeria monocytogenes* in processed meat products during refrigerated storage. *Appl. Environ. Microbiol*. 55:1565–1569, 1989.

69. Gombas, D.E. Biological competition as a preserving mechanism. *J. Food Safety* 10: 107–117, 1989.
70. Gotz, F., B. Sedewitz, E.F. Elstner. Oxygen utilization of *Lactobacillus plantarum*, I: oxygen consuming reactions. *Arch. Microbiol.* 125:209–214, 1980.
71. Gross, E., J.L. Morell. The structure of nisin. *J. Am. Chem. Soc.* 93:4634–4635, 1971.
72. Gudkow, A.V. Starters as a means of controlling contaminating organisms. *Milk Vital Force* 83–93, 1987.
73. Hardie, J. M., E. I. Garvie, O. Kandler, and N. Weiss. Gram-positive cocci. In: *Bergey's Manual of Systematic Bacteriology,* Sneath, P.H.A., N.S. Mair, M.E. Sharpe, J.G. Holt, eds., Baltimore: Williams and Wilkins, 1986, pp 1043–1209.
74. Harris, L.J., M.A. Daeschel, M.E. Stiles, T.R. Klaenhammer. Antimicrobial activity of lactic acid bacteria against *Listeria monocytogenes*. *J. Food Prot.* 52:384–387, 1989.
75. Harris, J.R., J. Mariano, J.G. Wells, B.S. Payne, H.D. Donnel, M.L. Cohen. Person-to-person transmission in an outbreak of enteroinvasive *Escherichia coli*. *Am. J. Epidemiol.* 122:245–252, 1985.
76. Hedges, A.J., A.H. Linton. Olaguinodox resistance in the coliform flora of pigs and their environment: an ecological study. *J. Appl. Bacteriol.* 64:429–444, 1988.
77. Hinton, M., and G. C. Mead. Salmonella control in poultry: the need for the satisfactory evaluation of probiotics for this purpose. *Lett. Appl. Microbiol.* 13:49–50, 1991.
78. Holzapfel, W. H., R. Geisen, and U. Schillinger. Biological preservation of foods with reference to protective cultures, bacteriocins and food-grade enzymes. *Intl. J. Food Microbiol.* 24:343–362, 1995.
79. Huber, J.T. Probiotics in cattle. In: *Probiotics 2: Applications and Practical Aspects,* Fuller, R., ed., London: Chapman and Hall, 1997, pp 162–186.
80. Hugas, M. Bacteriocinogenic lactic acid bacteria for the biopreservation of meat and meat products. *Meat Sci.* 49:S139–S150, 1998.
81. Hugas, M., F. Pagés, M. Garriga, J.M. Monfort. Application of the bacteriocinogenic *Lactobacillus sakei* CTC494 to prevent growth of *Listeria* in fresh and cooked meat products packed with different atmospheres. *Food Microbiol.* 15:639–650, 1998.
82. Huis in't Veld, J.H.J., R. Havenaar. Probiotics and health in man and animal. *J. Chem. Tech. Biotechnol.* 51:562–567, 1991.
83. Hurst, A. Microbial antagonism in foods. *Can. Inst. Food Sci. Technol. J.* 6:80–90, 1973.
84. Ishiwa, H, M. Iwata. Drug resistance plasmids in *Lactobacillus fermentum*. *J. Gen. Appl. Microbiol.* 26:71–74, 1980.
85. Jack, R.W., J.R. Tagg, B. Ray. Bacteriocins of Gram-positive bacteria. *Microbiol. Rev.* 59:171–200, 1995.
86. Jay, J.M. Antimicrobial properties of diacetyl. *Appl. Environ. Microbiol.* 44:525–532, 1982.
87. Jay, J.M. Microorganisms in fresh ground meats: the relative safety of products with low versus high numbers. *Meat Sci.* 43:S59–S66, 1996.
88. Jeppesen, V.F., H.H. Huss. Antagonistic activity of two strains of lactic acid bacteria against *Listeria monocytogenes* and *Yersinia enterocolitica* in a model fish product. *Int. J. Food Microbiol.* 19:179–186, 1993.
89. Jin, L.Z., Y.W. Ho, N. Abdullah, M.A. Ali, S. Jalaludin. Antagonistic effects of intestinal *Lactobacillus* isolates on pathogens of chicken. *Lett. Appl. Microbiol.* 23:67–71, 1996.
90. Juven, B.J., R.J. Meineresmann, N.J. Stern. Antagonistic effects of lactobacilli and pediococci to control intestinal colonization by human enteropathogens in live poultry. *J. Appl. Bacteriol.* 70:95–103, 1991.
91. Juven, B.J., M.D. Pierson. Antibacterial effects of hydrogen peroxide and methods for its detection and quantitation. *J. Food Prot.* 59(11):1233–1241, 1996.
92. Kanatani, K., M. Oshimura, K. Sano. Isolation and characterization of Acidocin A and cloning of the bacteriocin gene from *Lactobacillus acidophilus*. *Appl. Environ. Microbiol.* 61:1061–1067, 1995.

93. Kandler, O., N. Weiss. Genus *Lactobacillus*. In: *Bergey's Manual of Systematic Bacteriology,* Sneath, P.H., N.S. Mair, M.E. Sharpe, J.G. Holt, eds., Baltimore: Williams and Wilkins, 1986, pp 1209–1234.
94. Kandler, O. Carbohydrate metabolism in lactic acid bacteria. *Antonie van Leuwenhoek* 49:202–224, 1983.
95. Klaenhammer, T.R. Bacteriocins of lactic acid bacteria. *Biochimie* 70:337–349, 1988.
96. Klaenhammer, T.R. Genetics of bacteriocins produced by lactic acid bacteria. *FEMS Microbiol. Rev.* 12:39–86, 1993.
97. Klaenhammer, T.R. Genetics of intestinal lactobacilli. *Intl. Dairy J.* 5:1019–1058, 1995.
98. Klaenhammer, T.R. Functional activities of *Lactobacillus* probiotics: genetic mandate. *Intl. Dairy J.* 8:497–505, 1998.
99. Klebanoff, S.J. Myeloperoxidase-halide-hydrogen peroxide antibacterial system. *J. Bacteriol.* 95:2131–2138, 1968.
100. Kmet, V., H.J. Flint, R.J. Wallace. Probiotics and manipulation of rumen development and function. *Arch. Anim. Nutr.* 44:1–10, 1993.
101. Krogfelt, K.A. Bacterial adhesion: genetics, biogenesis and role in pathogenesis of fimbrial adhesions of *Escherichia coli*. *Rev. Infect. Dis.* 13:721–735, 1991.
102. Leclercq, R., E. Derlot, M. Weber, J. Duval, P. Courvalin. Transferable vancomycin and teicoplanin resistance in *Enterococcus faecium*. *Antimicr. Agent Chemother.* 33:10–15, 1989.
103. LeClerc, J.E., B. Li, W.L. Payne, T.A. Cebula. High mutation frequencies among *Escherichia coli* and *Salmonella* pathogens. *Science* 274:1208–1211, 1996.
104. Lee, R.W., R.L. Botts. Evaluation of single oral dosing and continuous feeding of *Streptococcus faecium* M74 (syntabac) on performance of incoming feedlot cattle. *J. Anim. Sci.* 66(1):460, 1988.
105. Lee, Y.K., S. Salminen. The coming age of probiotics. *Trends Food Sci. Technol.* 6:241–245, 1995.
106. Lee, Y.K., K. Nomoto, S. Salminen, S.L. Gorbach. *Handbook of probiotics*. NY: John Wiley & Sons, Inc. 1999.
107. Lessard, M., G.J. Brisson. Effects of a *Lactobacillus* fermentation product on growth, immune response, and fecal enzyme activity in weaned pigs. *Can. J. Anim. Sci.* 67:509–516, 1987.
108. Lindgren, P.E., W.J. Dobrogosz. Antagonistic activities of lactic acid bacteria in food and feed fermentations. *FEMS Microbiol. Rev.* 87:149–163, 1990.
109. Lilly, D.M., R.H. Stillwell. Probiotics: growth promoting factors produced by microorganisms. *Science*. 147:747–748, 1965.
110. Linton, A.H., A.J. Hedges, B.M. Bennet. Monitoring of resistance during the use of olaquinodox as a feed additive on commercial pig farms. *J. Appl. Bacteriol.* 64:311, 1988.
111. Lloyd, A.B., R.B. Cumming, R.D. Kent. Prevention of *Salmonella typhimurium* infection in poultry by pretreatment of chickens and poults with intestinal extracts. *Aus. Vet. J.* 53:82–87, 1977.
112. Lopez, E.L., M. Diaz, S. Grinstein, S. Devoto, F. Medilharzu, B.E. Murray, S. Ashkenazi, E. Rubeglio, M. Woloj, M. Vasquez, M. Turco, L.K. Pickering, T.G. Cleary. Hemolytic uremic syndrome and diarrhea in Argentine children: the role of Shiga-like toxins. *J. Infect. Dis.* 160:469–475, 1989.
113. Macfarlane, G.T., J.H. Cummings. Probiotics and prebiotics: can regulating the activities of intestinal bacteria benefit health? *Brit. Med. J.* 318:999–1003, 1999.
114. Mayra-Makinen, A., M. Manninen, H. Gyllenberg. The adherence of lactic acid bacteria to the columnar epithelial cells of pigs and calves. *J. Appl. Bacteriol.* 55:241–245, 1983.
115. Metchnikoff, E. *The Prolongation of Life*. London: Heinemann, 1907.
116. Miles, R.D., A.S. Arafa, R.H. Harms, C.W. Carlson, B.L. Reid, J.S. Crawford. Effects of a living non-freeze dried *Lactobacillus acidophilus* culture on performance, egg quality and gut microflora in commercial layers. *Poult. Sci.* 60:993–1004, 1981.

117. Montville, T.J., M.E.C. Bruno. Evidence that dissipation of proton motive force is a common mechanism of action for bacteriocins and other antimicrobial proteins. *Int. J. Food Microbiol.* 24:53–74, 1994.
118. Morelli, L., M. Vescovo, V. Bottazzi. Identification of chloremphenicol resistance plasmids in *Lactobacillus reuteri* and *Lactobacillus acidophilus. Int. J. Microbiol.* 1:1–5, 1983.
119. Mulder, R.W.A.W., R. Havenaar, J.H.J. Huis in't Veld. Intervention strategies: the use of probiotics and competitive exclusion microfloras against contamination with pathogens in pigs and poultry. In: *Probiotics 2: Applications and Practical Aspects,* Fuller, R., ed., London: Chapman and Hall, 1997, pp 187–207.
120. Nes, I.F., D.B. Diep, L.S. Håvarstein, M.B. Brurberg, V. Eijsink, H. Holo. Biosynthesis of bacteriocins in lactic acid bacteria. *Antonie van Leeuwenhoek* 70:113–128, 1996.
121. Nettles, C.G., S.F. Barefoot. Biochemical and genetic characteristics of bacteriocins of food-associated lactic acid bacteria. *J. Food Prot.* 56:338–356, 1993.
122. Nielsen, J.W., J.S. Dickson, J.D. Crouse. Use of bacteriocin produced by *Pediococcus acidilactici* to inhibit *Listeria monocytogenes* associated with fresh meat. *Appl. Environ. Microbiol.* 56:2142–2145, 1990.
123. Niku-Paavola, M.L., A. Laitila, T. Mattila-Sandholm, A. Haikara. New types of antimicrobial compounds produced by *Lactobacillus plantarum. J. Appl. Microbiol.* 86:29–35, 1999.
124. Nilsson, L., L. Gram, H.H. Huss. Growth control of *Listeria monocytogenes* on cold-smoked salmon using a competitive lactic acid bacteria flora. *J. Food Prot.* 62:336–342, 1999.
125. Nousiainen, J., H. Setälä. Lactic acid bacteria as animal probiotics. In: *Lactic Acid Bacteria,* S. Salminen, S., A. von Wright, A. eds., New York: Marcel Dekker, 1993, pp 315–356.
126. Nurmi, I.E., M. Rantala. New aspects of *Salmonella* infection in broiler production. *Nature* 241:210–211, 1973.
127. Ofek, I., N. Sharon. Adhesins as lectins: specificity and role in infection. *Curr. Topics Microbiol. Immun.* 151:91–113, 1990.
128. Okereke, A., T.J. Montville. Nisin dissipates the proton motive force of the obligate anaerobe *Clostridium sporogenes* PA 3679. *Appl. Environ. Microbiol.* 58:2463–2467, 1992.
129. Ouwehand, A.C. Antimicrobial components from lactic acid bacteria. In: *Lactic Acid Bacteria: Microbiology and Functional Aspects,* Salminen, S., A. von Wright, eds., New York: Marcel Dekker, 1998, pp 139–159.
130. Ozawa, K., K. Yabu-uchi, K. Yamanaka, Y. Yamashita, S. Nomura, I. Oku. Effect of *Streptococcus faecalis* BIO-4R on intestinal flora of weanling piglets and calves. *Appl. Environ. Microbiol.* 45:1513–1518, 1983.
131. Parker, R.B. Probiotics: the other half of the antibiotics story. *Ani. Nutr. Health.* 29:4–8, 1974.
132. Perdigon, G., M.E.N. De Macias, S. Alvarez, G. Oliver, A.A.P. de Ruiz Holgado. Effect of perorally administered lactobacilli on macrophage activation in mice. *Infect. Immun.* 53:404–410, 1986.
133. Perdigon, G., M.E. Nader de Macias, M.E. Roux, A.A.P. de Ruiz Holdaga. The oral administration of lactic acid bacteria increase the mucosal immunity in response to enteropathogens. *J. Food Prot.* 53:404–410, 1990.
134. Perreten, V., B. Kollöffel, M. Teuber. Conjugal transfer of the Tn916-like transposon TnF01 from *Enterococcus faecalis* isolated from cheese to other Gram-positive bacteria. *Syst. Appl. Microbiol.* 20:27–38, 1997.
135. Philip Brasher, 18 November 2000, *Bacteria vs Bacteria.* [Internet, E-mail to the author] Available as E-mail from the author, FSNet.
136. Pilet, M.F., X. Dousset, R. Barre, G. Novel, M. Desmazeaud, J.C. Piard. Evidence for two bacteriocins produced by *Carnobacterium piscicola* and *Carnobacterium divergens* isolated from fish and active against *Listeria monocytogenes. J. Food Prot.* 58:256–262, 1995.
137. Pivnick, H., E. Nurmi. The Nurmi concept and its role in the control of *Salmonella* in poultry. *In: Developments in Food Microbiology,* Vol. 1, Davis, R., ed., Barking, England: Applied Science, 1982, p 41.
138. Price, R.J., J.S. Lee. Inhibition of *Pseudomonas* species by hydrogen peroxide producing lactobacilli. *J. Milk Food Technol.* 33:13, 1970.

139. Pucci, M.J., E.R. Vedamuthu, B.S. Kunka, P.A. Vandenbergh. Inhibition of *Listeria monocytogenes* by using bacteriocin PA-1 produced by *Pediococcus acidilactici* PAC 1.0. *Appl. Environ. Microbiol.* 54:2349–2353, 1988.
140. Ratcliffe, B. The influence of the gut microflora on the digestive processes. In: *Proc. 3rd International Seminar on Digestive Physiology in the Pig*, Just, A., H. Jorgensen, J. A. Fernandez, eds., Beretning fra Statens Husdyrbrugsforsog, No. 580, 1985, pp 245–267.
141. Roach, S., G. W. Tannock. Indigenous bacteria that influence the number of *Salmonella typhimurium* in spleen of intravenously challenged mice. *Can. J. Microbiol.* 26:408–411, 1980.
142. Rodríguez, E., J. Tomillo, M. Nuñez, M. Medina. Combined effect of bacteriocin-producing lactic acid bacteria and lactoperoxidase system activation on *Listeria monocytogenes* in refrigerated raw milk. *J. Appl. Microbiol.* 83:389–395, 1997.
143. Ruhr, E., H.G. Sahl. Mode of action of the peptide antibiotic nisin and influence on the membrane potential of whole cells and on cytoplasmic and artificial membrane vesicles. *Antimicrob. Agents Chemother.* 27:841–845, 1985.
144. Sahl, H.G., R.W. Jack, G. Bierbaum. Biosynthesis and biological activities of antibiotics with unique post-translational modifications. *Eur. J. Biochem.* 230:827–853, 1995.
145. Saito, H., H. Tomioka, K. Sato. Enhanced resistance of *Lactobacillus* against *Listeria* infection in mice. *Med. Biol.* 102:273–277, 1981.
146. Salminen, S., M. Laine, A. von Wright, J. Vuopio-Varkila, T. Korhonen, T. Mattila-Sandholm. Development of selection criteria for probiotic strains to assess their potential in functional foods: a Nordic and European approach. *Biosci. Microflora.* 15:61–67, 1996.
147. Sandine, W.E. Roles of *Lactobacillus* in the intestinal tract. *J. Food Prot.* 42(3):259–262, 1979.
148. Savage, D.C. Microbial ecology of the gastrointestinal tract. *Ann. Rev. Microbiol.* 31:107–133, 1977.
149. Savage, D.C. Gastrointestinal microflora in mammalian nutrition. *Ann. Rev. Nutr.* 6:155–178, 1986.
150. Schillinger, U., M. Kaya, F. K. Lücke. Behaviour of *Listeria monocytogenes* in meat and its control by a bacteriocin-producing strain of *Lactobacillus sake*. *J. Appl. Bacteriol.* 70:473–478, 1991.
151. Schöbitz, R., T. Zaror, O. León, M. Costa. A bacteriocin from *Carnobacterium piscicola* for the control of *Listeria monocytogenes* in vacuum-packaged meat. *Food Microbiol.* 16:249–255, 1999.
152. Smith, L., J.E. Mann, M.M. Brashears. Competitive inhibition of *E. coli* O157:H7 and *Salmonella* spp. in ground beef. *J. Food Prot.*, 2003. (In Press)
153. Smith, H.W. The development of the flora of the alimentary tract in young animals. *J. Pathol. Bacteriol.* 90:495–513, 1965.
154. Smith, H.W., J.F. Tucker. The effect of feeding diets containing permitted antibiotics on the fecal excretion of *Salmonella typhimurium* by experimentally infected chicks. *J. Hyg.* 75:293–301, 1975.
155. Smith, H.W., P. Green, Z. Parsell. Vero cell toxins in *Escherichia coli* and related bacteria: transfer by phage and conjugation and toxic action in laboratory animals, chickens and pigs. *J. Gen. Microbiol.* 129:3121–3137, 1983.
156. Smith, H.W. Virulence determinants of *Escherichia coli*: present knowledge and questions. *Can. J. Microbiol.* 34:747–752, 1992.
157. Snoeyenbos, G.H., O.M. Weinack, A. Soerjadi. Competitive exclusion of some pathogens other than salmonella by native intestinal microflora of chickens. *Proc. 22nd World Vet. Congress,* Perth, Australia, 1978, p 191.
158. Sorrells, K.M., M.L. Speck. Inhibition of *Salmonella gallinarum* by culture filtrates of *Leuconostoc citrovorum*. *J. Dairy Sci.* 59:338–343, 1970.
159. Sorrells, K.M., D.C. Enigl, J.R. Hatfield. Effect of pH, acidulant, time and temperature on the growth and survival of *Listeria monocytogenes*. *J. Food Prot.* 52:571–573, 1989.

160. Spencer, R.J., A. Chesson. The effect of *Lactobacillus* spp. in the attachment of enterotoxigenic *Escherichia coli* to isolated porcine enterocytes. *J. Appl. Bacteriol.* 77:115–220, 1994.
161. Stavric, S., T.M. Gleeson, B. Blanchfield, P. Pivnick. Role of adhering microflora in competitive exclusion of *Salmonella* from young chicks. *J. Food Prot.* 50:928–932, 1987.
162. Stewart, P. J., W. Desormeaux, J. Chene, H. Lior. Hemorrhagic colitis in a home for the aged in Ontario. *Can. Dis. Week Rep.* 9:29–32, 1983.
163. Stiles, M.E. Biopreservation by lactic acid bacteria. *Antonie van Leeuwenhoek* 70:331–345, 1996.
164. Stiles, M.E., J.W. Hastings. Bacteriocin production by lactic acid bacteria: potential for use in meat preservation. *Trends Food Sci. Technol.* 2:247–251, 1991.
165. Tahara, T., M. Oshimura, C. Umezawa, K. Kantani. Isolation, partial characterization and mode of action of acidocin J 1132, a two-component bacteriocin produced by *Lactobacillus acidophilus* JCM 1132. *Appl. Environ. Microbiol.* 62:892–897, 1996.
166. Tannock, G.W. The effect of dietary and environmental stress on the gastrointestinal microbiota. In: *Human Intestinal Microflora in Health and Disease,* Hentges, D.J., ed., New York: Academic Press, 1983, pp 517–539.
167. Tannock, G.W. *Normal Microflora: An Introduction to Microbes Inhabiting the Human Body.* London: Chapman and Hall, 1995.
168. Tannock, G.W. Probiotic properties of lactic acid bacteria: plenty of scope for fundamental R & D. *Trends Biotechnol.* 15:270–274, 1997.
169. Teuber, M., V. Perreten, F. Wirsching. Antibiotikum-restente Bakterien: Eine neue Dimension in der Lebensmittel-mikrobiologie. *Lebensmitteltechnologie* 29:182–199, 1996.
170. Tramer, J. Inhibitory effect of *Lactobacillus acidophilus*. *Nature* 211:204–205, 1966.
171. Tuschy, D. Verwendung von "Probiotika" als Leistungsförderer in der Tierernährung. *Übers. Tierernährg.* 14:157, 1986.
172. Underdahl, N.R., A. Torres-Medina, A.R. Doster. Effect of *Streptococcus faecium* C63 in control of *Escherichia coli*-induced diarrhea in gnotobiotic pigs. *Am. J. Vet. Res.* 43:2227–2232, 1982.
173. Vandenbergh, P.A. Lactic acid bacteria, their metabolic products and interference with microbial growth. *FEMS Microbiol. Rev.* 12:221–238, 1993.
174. Vandervoorde, L., M. Vande Woestyne, B. Bruyneel, H. Christiaens, W. Verstraete. Critical factors governing the competition behavior of lactic acid bacteria in mixed culture. In: *The Lactic Acid Bacteria in Health and Disease,* Vol. 1. Wood, B.J.P., ed., London: Elsevier Applied Science, 1992, pp 447–475.
175. Venema, K., T. Abee, A.J. Haandrikman, K.J. Leenhouts, J. Kok, W.N. Konings, G. Venema. Mode of action of lactococcin B, a thiol-activated bacteriocin from *Lactococcus lactis*. *Appl. Environ. Microbiol.* 59:1041–1048, 1993.
176. Villegas, E., S.E. Gilliland. Hydrogen peroxide production by *Lactobacillus delbrueckii* subsp. *Lactis* I at 5°C. *J. Food Sci.* 63:1070–1074, 1998.
177. Vignolo, G., S. Fadda, M.N. De Kairuz, A.A.P. de Ruiz Holgado, G. Oliver. Effects of curing additives on the control of *Listeria monocytogenes* by lactocin 705 in meat slurry. *Food Microbiol.* 15:259–264, 1998.
178. Vignolo, G., S. Fadda, M.N. De Kairuz, A.A.P. de Ruiz Holgado, G. Oliver. Control of *Listeria monocytogenes* in ground beef by 'Lactocin 705', a bacteriocin produced by *Lactobacillus casei* CRL 705. *Int. J. Food Microbiol.* 29:397–402, 1996.
179. Watkins, B.A., B.F. Miller, D.H. Neil. *In vivo* inhibitory effects of *Lactobacillus acidophilus* against pathogenic *Escherichia coli* in gnotobiotic chicks. *Poultry Sci.* 61:1298–1308, 1982.
180. Wessels, S., H.H. Huss. Suitability of *Lactococcus lactis* subsp. *lactis* ATCC 11454 as a protective culture for lightly preserved fish products. *Food Microbiol.* 13:323–332, 1996.
181. Whittenbury, R. Hydrogen peroxide formation and catalase activity in the lactic acid bacteria. *J. Gen. Microbiol.* 35:13–26, 1964.

182. Winkowski, K., A.D. Crandall, T.J. Montville. Inhibition of *Listeria monocytogenes* by *Lactobacillus bavaricus* MN in beef systems at refrigeration temperatures. *Appl. Environ. Microbiol.* 59:3003–3010, 1993.
183. Wren, B. Probiotics: fact or fiction. *Large Anim. Vet.* 1:28–30, 1987.
184. Yang, R., B. Ray. Factors influencing productions of bacteriocins by lactic acid bacteria. *Food Microbiol.* 11:281–291, 1994.
185. Zourari, A., J.P. Accolas, M.J. Desmazeaud. Metabolism and biochemical characteristics of yogurt bacteria. *Lait* 72:1–34, 1992.

3.06

Bacteriocins: Antimicrobial Activity and Applications

A. Satyanarayan Naidu, Ragip Unal and Joseph Tulpinski

CONTENTS

6.1 INTRODUCTION

Bacteriocins (including colicins) are defined as antimicrobial proteins characterized by lethal biosynthesis, a very narrow range of activity, and adsorption to specific cell envelope receptors (1). The established association of bacteriocin biosynthesis with plasmids has been added to this definition, and it has been further modified to include the properties of bacteriocins produced by Gram-positive bacteria (2). To avoid confusion with therapeutic drugs, bacteriocins are not called "antibiotics." (3). These antimicrobial agents are polypeptide molecules with a narrow specificity of action against strains of the same or closely related species.

Bacteriocins from Gram-positive bacteria do not bind to specific receptors for adsorption, are generally of lower molecular weight (than colicins), and demonstrate a broader spectrum of antimicrobial activity (4). These antimicrobial agents also demonstrate different modes of release and cell transport, and possess leader sequences that cleave during maturation (5–7). These ribosomally synthesized polypeptides possess bactericidal properties and are rapidly hydrolyzed by proteases in the human digestive tract (8).

Lactic acid bacteria (LAB) produce a wide range of antagonistic factors that include metabolic products, antibiotic-like substances and bacteriocins (9,10). Several specific species of LAB are commonly used as starter cultures in the manufacturing of dairy, meat, vegetable and bakery products. In addition to health and nutritional benefits, LAB also contribute to the safety of fermented foods. Accordingly, these bacteria produce organic acids, diacetyl, hydrogen peroxide, and bacteriocins during fermentation (11–13). These compounds contribute to the desirable effect on food flavor and texture, but also inhibit undesirable microflora as well as extend product shelf life (14,15). Food applications of bacteriocins are limited, as efficacy and cost remain issues impeding the broad use of these antimicrobial compounds as biopreservatives. Hence, there is a continuous search for new and more effective bacteriocins as well as research and development activities ongoing for optimization of existing bacteriocins to overcome efficacy and economic limitations.

LAB produce bacteriocins either spontaneously or by induction. The genetic determinants for most bacteriocins are located on plasmids, with a few exceptions of chromosomally encoded ones (16). Secretion of bacteriocins into the milieu requires expression and activity of bacteriocin release proteins and the presence of detergent resistant phospholipase-A in the bacterial outer membrane of LAB (17,18). These antimicrobial agents are species specific and exert their lethal activity through adsorption to specific receptors located on the external surface of sensitive bacteria, followed by metabolic, biological, and morphological changes resulting in the killing of such bacteria. There is increasing evidence that bacteriocins from many bacterial genera share similar characteristics.

Currently, the term bacteriocins refers to a wide range of antimicrobial peptides or proteins produced by various bacteria. Molecular hydrolysis by proteolytic enzymes such as trypsin, α-chymotrypsin, and pepsin is a standard procedure to confirm the proteinaceous nature in the characterization of any new bacteriocin. Furthermore, any evaluation of bacteriocin for possible application as a food additive also requires consideration of protein stability at elevated temperatures given the widespread use of thermal processing in food production. Over the years, several publications have reviewed colicins and bacteriocins and their possible applications in foods, animal health, and medicine (3,5,19–25).

6.2 BIOSYNTHESIS AND FUNCTION

6.2.1 Genetics

Genes encoding bacteriocin production and immunity are linear ribosomally synthesized peptides generally organized in operon clusters (16–18,26). These peptides are post-translationally modified and contain unusual amino acid (dehydrate and lanthionine) residues. These are produced on the ribosome as a prepeptide and undergo extensive post-tranlational modification to form a bioactive peptide with a characteristic N-terminal leader (18,20,21,26). Synthesis of linear unmodified bacteriocins including plantaricins, carnobacteriocins, and sakacins is stimulated by specific inducing peptides or peptide pheromones located on the same gene cluster (27–29).

The best known and most characterized bacteriocin is nisin (18). There are 11 genes making the nisin cluster that code for nisin production, immunity, and externalization. The gene *nisA* encodes the nisin precursor molecule and is part of the gene cluster. Nisin A production and immunity genes are found on a 70 kb conjugative transposon (30). The transposons of nisin also encode *sacA*, *sacB* and *sacC* which are sucrose utilization genes. Nisin contains 34 amino acids arranged in a pentacyclic structure (16).

Bacteriocin gene clusters are located on the chromosome (i.e., subtilin and mersacidin); on plasmids (i.e., divergicin A, sakacin A); or on transposons (i.e., nisin, lacticin 481) (31–35). The operons for lantibiotic biosynthesis generally contain genes coding for the prepeptide (LanA - the abbreviation *lan* refers to homologous genes of different

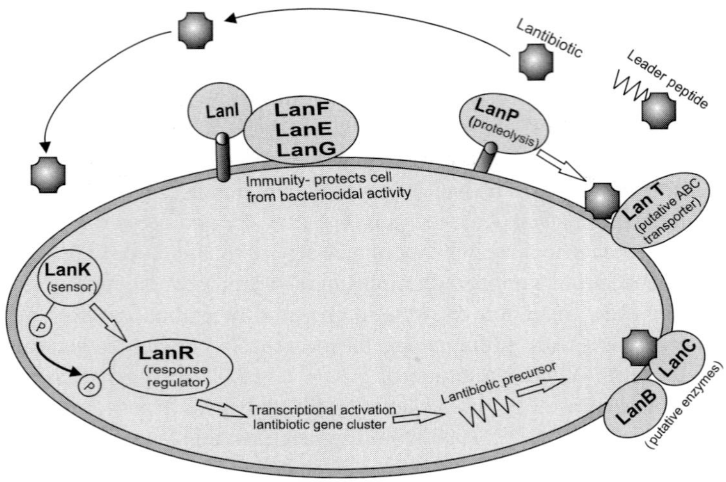

Figure 6.1 General scheme of lantibiotic biosynthesis consisting of pre-peptide formation, modification reactions, proteolytic cleavage of leader peptide, and translocation of modified pre-peptide or mature pro-peptide across cytoplasmic membrane.

lantibiotic gene clusters); enzymes for modification reactions (LanB,C/LanM); processing proteases for removal of the leader peptide (LanP), the ABC (ATP binding cassette), superfamily transport proteins involved in peptide translocation (LanT), regulatory proteins (LanR, K), and proteins involved in producer self protection (immunity) (LanI, FEG). These regulatory pathways have been established based on genetic analysis and biosynthesis of several lantibiotics including epidermin, nisin, subtilin, lacticin 481, and mersacidin (20,31,32,36–42).

Genetic regulation of class II bacteriocins, lactococcins A, B, and M; pediocin PA-1/AcH (pediocin PA-1 and AcH are the same molecule; the term pediocin PA-1 is more commonly used); and plantaricin A have been studied (43–50). Genes encoding the biosynthesis of these bacteriocins share many similarities in their organization. They consist of a structural gene that encodes a prepeptide containing a leader sequence with two glycine residues, followed immediately by a dedicated immunity gene and genes for an ABC transporter and an accessory protein, which are required for the externalization of the bacteriocins. In certain cases, regulatory genes have been identified. Furthermore, accessory proteins are essential for the export of class II bacteriocins, and such counterparts of these accessory proteins in lantibiotics are not reported (24,51).

6.2.2 Regulation

Most bacteriocins are synthesized as inactive prepeptides with an N-terminal leader sequence attached to the C-terminal propeptide. For lantibiotics, the serine, threonine, and cysteine residues in the propeptide region undergo extensive post-translational modification to form Lan/MeLan. The biosynthetic pathway of lantibiotics follows a general scheme with steps consisting of prepeptide formation, modification reactions, proteolytic cleavage of leader peptide, and translocation of modified prepeptide or mature propeptide across cytoplasmic membrane (Figure 6.1). Cleavage of the leader peptide could occur prior, during, or after export from the microbial cell. Based on the biosynthetic pathway, two categories of genetic organization of lantibiotics, Groups I and II, have been identified (16,26,52). During the synthesis of Group I lantibiotics (nisin, epidermin, subtilin, and Pep5), the dehydration reaction is presumably catalyzed by the LanB enzyme, while LanC

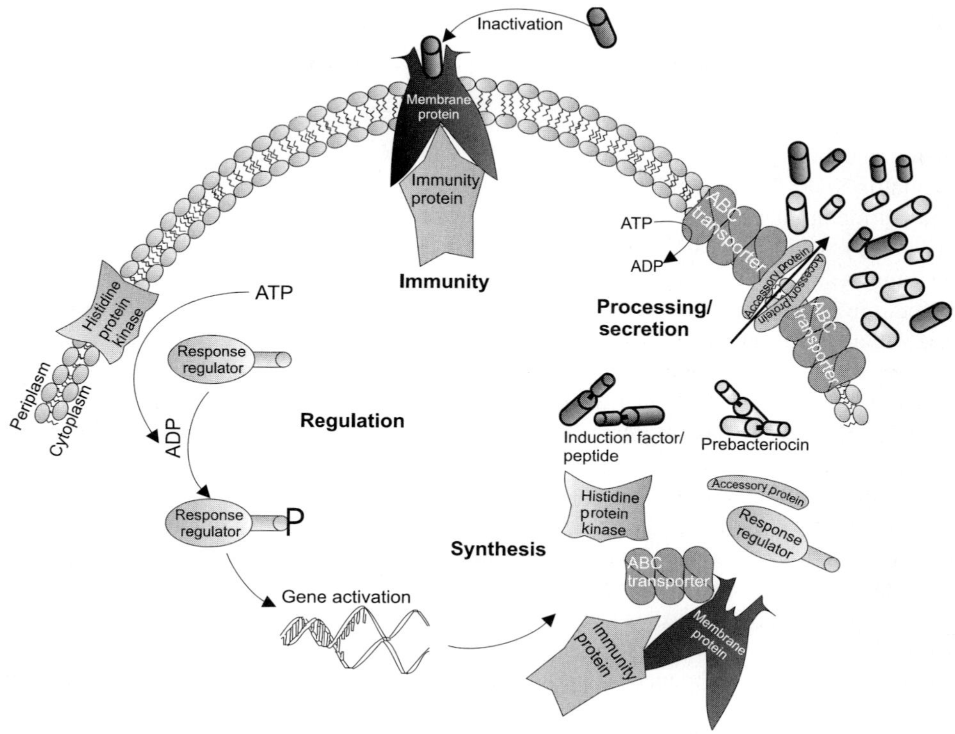

Figure 6.2 Pathways of class IIa bacteriocin biosynthesis. N-terminal leader peptide serves as a recognition site to direct the pre-peptide for maturation and transport proteins, self-immunity of producer strain and ensures conformational changes.

is involved in the thioether formation. The modified prepeptide is processed by a serine protease LanP and translocated through the ABC transporter LanT. In contrast, the Group II lantibiotics (cytolysin, lacticin 481, and mersacidin), are very likely modified by a single LanM enzyme and processing takes place concomitantly with transport by LanT(P) (16,53). Lactocin S is an exception to this group, because a single LanM enzyme modifies this bacteriocin and processing occurs prior to export (54).

Class II bacteriocins are synthesized as a prepeptide containing a conserved N-terminal leader and a characteristic double glycine proteolytic processing site, with the exception of class IIc bacteriocins, which are produced with a typical N-terminal signal sequence of the *sec*-type and processed and secreted through the general secretory pathway (33,55). Unlike the lantibiotics, class II bacteriocins do not undergo extensive post-translational modification. After formation, the prepeptide is processed to remove the leader peptide concomitant with export from the cell through a dedicated ABC transporter and its accessory protein (24,56).

The biosynthetic pathway of class II bacteriocins is shown in Figure 6.2. Several functions of the leader peptides have been proposed that may serve as a recognition site to direct the prepeptide toward maturation and transport proteins, protect the producer strain by keeping the bacteriocin in an inactive state while it is inside the producer, and interact with the propeptide domain to ensure a suitable conformation essential for enzyme to substrate interaction (16,51,57,58).

A three component system consisting of a histidine protein kinase, response regulatory, and induction factors has been shown to be responsible for most regulated Class IIa bacteriocins (56). An induction factor (IF) is required as a signal to induce the transcription of target genes (56,59). The IFs are small, heat stable, hydrophobic, cationic peptides that are produced as prepeptides with a double glycine leader sequence (24).

Regulation of lantibiotics responds to the extracellular bacteriocin with a two component regulatory system. This system is made up of two signal producing proteins, the membrane bound histidine kinase sensor protein NisK, and the cytoplasmic response regulator NisR. These regulatory systems as a class consist of two signal producing proteins, a membrane bound histidine protein kinase (HPK), and a response regulator (RR) (17,60,61). In the signal transduction pathway HPK autophosphorylates the conserved histidine residue in its intracellular domain when it senses a certain concentration of bacteriocin in the milieu. The phosphoryl group is subsequently transferred to the conserved aspartic acid residue on the RR receiver domain and the resulting intramolecular change triggers the response regulator to activate the transcription of the regulated genes, leading to the expression of the genes involved in immunity, synthesis, and post-translational modifications (62). These regulated genes include the structural gene, the export genes, the immunity genes, and in some cases, the regulatory genes themselves (59).

For nisin and subtilin, the bacteriocin molecule itself acts as an external signal to autoregulate its own biosynthesis via signal transduction (52,62). In contrast, most class II bacteriocins produce a bacteriocin-like peptide with no antimicrobial activity and use it as an induction factor (IF) to activate the transcription of the regulated genes. The IF is a small, heat stable, cationic, hydrophobic peptide that is first synthesized as a prepeptide with a double glycine leader sequence. A dedicated ABC transporter specifically cleaves the leader peptide of IF concomitant with export of the mature peptide from the cell. The secreted IF acts as an external signal that triggers transcription of the genes involved in bacteriocin production (24,56).

Other concepts for the bacteriocin-like peptides in the non-lantibiotic antimicrobial peptides is the peptide pheromone concept regulation of biosynthesis. This concept has been shown to be valid by several groups (27,63,64). Research has shown this concept valid for several bacteriocins including carnocin, plantaricin and sakacin.

6.2.3 Post-translational Events

A two step post-translational modification reaction of a pre-lantibiotic leading to formation of Lan/MeLan was first proposed by Ingram (65,66). The post-translational events begin with the dehydration of hydroxyl amino acids, serine and threonine, to yield 2,3-didehydroalanine and 2,3-didehydrobutyrine, respectively. Some of the dehydrated amino acids do not contain cysteine residues and are consistent in the mature peptide. Other dehydrated amino acids undergo an intramolecular addition reaction involving thiol groups of the neighboring cysteine residues and double bonds of the didehydroamino acids to form thioether bridges. The modified pre-lantibiotics undergo proteolytic processing to release the leader peptide for lantibiotic activation. For Group I lantibiotics, the leader peptide is removed by a serine protease, LanP. This event could occur before or after the peptide is exported from the producing cell via a dedicated ABC transporter, LanT, depending on the location of the protease LanP. For example, the proteases LanP of epicidin 280 and Pep5 are located intracellularly for proteolytic processing, however, the proteases of nisin and epidermin are extracellularly located, and lantibiotics are activated only after export by the ABC transporter (26,37,67). The ABC transporter contains 500 to 600 amino acids and is characterized by two membrane associated domains. The N-terminal domain consists of six membrane spanning helices that could recognize the substrate and form pathways across the membrane, while the cytoplasmic C-terminal domain contains two ATP binding domains with the conserved ATP binding or Walker motif. Hydrolysis of ATP, which likely occurs at the ATP binding domains, provides energy for the export process (16,68). The LanB and LanC enzymes, together with the LanT transporter, probably form a multimeric membrane associated complex (69,70). For Group II lantibiotics, which possess a conserved double glycine cleavage site, proteolytic processing takes place concomitantly with export through a hybrid ABC transporter. This unique ABC transporter possesses an N-terminal protease domain of ~150 amino acid residues that cleaves the double glycine leader (24,51).

Substantial similarities exist between the leader peptides of Class IIa and IIb and those of Group II lantibiotics. Both contain a characteristic double glycine cleavage site (24,56). Conservation of the cleavage site strongly suggests that the mechanism of processing and translocation of Class IIa and IIb bacteriocins is similar to Group II lantibiotics. Class IIc bacteriocins are processed by a signal peptidase during translocation across the cytoplasmic membrane.

6.2.4 Immunity and Resistance

Two systems of lantibiotic immunity in the producing microbial cell have been identified. Protection could be mediated by immunity proteins, LanI, and dedicated ABC transport proteins, LanFEG encoded on multiple open reading frames (16,71–74). These two immunity systems work synergistically to protect the producing microbial cells from their self bacteriocin (75,76). LanI, also known as PepI encoded by the *Pep5* operon, is a 69 amino acid protein containing a hydrophobic N-terminal domain and a hydrophobic C-terminal part. This suggests that the PepI protein is membrane associated and probably confers immunity to the producer microbial cells by antagonizing the pore formation by the bacteriocin (71). *LanF* codes the intracellular ATP binding domain whereas LanE and LanG represents the membrane spanning subunits. The collective LanFEG apparently acts by transporting bacteriocin molecules that have inserted into the membrane back to the surrounding medium, and thus keeping bacteriocin concentration below a critical level in the membrane.

For Class II bacteriocins, the immunity gene usually codes for a dedicated protein loosely associated with the cytoplasmic membrane. A major part of the immunity protein CbiB2 of carnobacteriocin B2 is found in the cytoplasm and a smaller proportion is associated with the membrane (77). Most of the immunity protein MesI of mesentericin Y105 is in the cytoplasm, with only a small proportion detected in the membrane (78). The immunity protein, which is cationic and ranges in size from 51 to 254 amino acids, provides total immunity against the bacteriocin. Interaction of immunity protein with the membrane could protect the producer against the self bacteriocin (79–81).

The presence of an antibacterial substance in a given milieu would eventually select for varieties of bacteria resistant to the antagonistic component. As found with therapeutic antibiotics in the environment, bacteriocin resistant mutants have been reported. Gravesen et al. (82) tested the responses of a number of strains of *L. monocytogenes* to pediocin PA-1 and nisin, and found a wide range of resistance to both bacteriocins. Influence of environmental stress (reduced pH, low temperature, and reduced salt) is bacteriocin specific; thus the frequency of nisin resistance is significantly reduced, whereas pediocin PA-1 is unaffected. Furthermore, stability of the phenotype of nisin resistance varies substantially, while resistance to pediocin is stable with continuous proliferation of *Listeria monocytogenes*.

6.3 ISOLATION AND CHARACTERIZATION

6.3.1 Purification Methods

Isolation and purification of bacteriocins from complex fermentation broth media is primarily based on the cationic and hydrophobic properties of these antimicrobial compounds (83).

Purification steps usually include an ammonium sulfate precipitation, followed by various combinations of ion-exchange and hydrophobic-interaction chromatography (HIC), with a final Reverse Phase (RP)-HPLC process. It is possible to replace chromatography with preparative isoelectric focusing (84). A rapid single step purification of nisin A using immunoaffinity chromatography has also been described (85). These procedures are not suitable for large commercial scale bacteriocin recovery and separation; however, these methods provide excellent yield and purity (86).

Several processes based on adsorption and desorption or on phase partitioning have been developed for large scale recovery and purification of bacteriocins. Bacteriocins could be recovered by adsorption on producer cells at pH 6.0–6.5, followed by cell separation and desorption at pH 2.0 with 0.1M NaCl. This method is effective for the separation of pediocin AcH, nisin, sakacin A, and leuconocin Lcm1, however, recovery could be limited for other strain and bacteriocin combinations (87,88). Vortex flow filtration systems could replace centrifugation and may be more amenable for large scale recovery of cells (89). However, when bacteriocin concentration is high, the ability of cells to adsorb nisin could be exceeded and result in partial recovery of bacteriocin from the cell fraction. The lantibiotics nisin and carnocin UI49 have been purified to homogeneity using a simple two-step protocol based on adsorption to HIC and cation exchange resins (90). Recovery is high (attributed to removal of interferences in the bioassay due to isolation procedures) and purification was 245- and 60-fold for carnocin and nisin, respectively. SepPak® C8 cartridges are excellent adsorbent materials for nisin Z. SepPak C8 cartridges have been used in an integrated fermentation system with a batch process coupled to microfiltration; permeate is circulated on the cartridge and returned to the fermenter (91). This process has resulted in an improved growth of the producer strain and a high yield of nisin.

Ingestible porous silica compounds such as Micro-Cel E® have also been used to separate bacteriocins (nisin, pediocin PO2, brevicin 286, and piscicolin 126) from fermentation broths (92). Adsorbed bacteriocins are eluted with 0.1% SDS and the final removal of SDS (by cold precipitation) from the eluting mixture was only partial (60–70%) (93). Bacteriocins isolated from this process are 110- to130-fold homogenous and showed an optimum inhibitory activity against target organisms.

A simple two-step purification system based on phase partitioning using the detergent Triton® 114, followed by an adsorption and desorption process on a cation exchange resin has been developed (94). Addition of 2% detergent to culture supernatants resulted in accumulation of divercin V41 in the detergent phase. Divercin V41 was recovered up to 95% purity after adsorption on a cation exchange resin followed by an elution with 0.7 M sodium chloride. This method is effective for the recovery of mesenterocin Y105 and nisin.

6.3.2 Quantitative Assays

A number of methods for detection and measurement of bacteriocin activity have been described (95–97). A common technique for screening bacteriocin activity in producing strains is the spot on lawn approach, which involves an agar overlay. In this method, the producing and indicator cultures are incubated concurrently before examination for zones of inhibition around the growth of producing strains (2,98).

In another procedure, the producing and indicator strains each are grown on different optimal media. The producing culture is spot inoculated on a spread plate; after growth, the agar mass is aseptically dislodged with a spatula from the petri dish bottom and transferred to the lid. A soft agar overlay seeded with the indicator is poured over the inverted agar. Following reincubation, bacteriocin positive cultures display a halo of lawn clearance around the original button of growth. This assay minimizes the effects of acids and bacteriophage, because the producing and indicator strains are physically separated by an agar layer (99).

Media composition is an important factor in plate assays; however, the limitation of these assays includes nutritional requirements of the producing and target organisms. The gelling agents used in solid media may interfere with bacteriocin diffusion and compromise the sensitivity of these screening methods (100). Other compounds could also interfere with or impede the results. For example, β-glycerol phosphate (a buffering agent) in M17 agar during the soft agar overlay of the Kekessy-Piguet methods could interfere with zones of inhibition by *Pediococcus acidilactici* PO2 compared to other agar media and buffering agents (101).

A matrix assisted laser desorption and ionization time of flight mass spectroscopy (MALDI-TOF MS) has been developed for rapid detection of pediocin, nisin, brochocins A and B, and enterocins A and B from culture supernatants (102). In this method, the interfering compounds in the supernatant are removed with a 30 s water wash. However, this washing step requires further improvement to eliminate other contaminants in foods that could potentially interfere with identification of bacteriocins.

Polymerase chain reaction (PCR) methods have been used to detect genes responsible for bacteriocin production and regulation in bacterial cultures. Rodriguez et al. (103) demonstrated the amplification of a 75-bp gene fragment of the lactocin S structural gene in seven bacteriocin positive strains of lactobacilli isolated from fermented sausages. Garde et al. (104) detected the genes necessary for synthesis of lacticin 481 and nisin using PCR techniques with specific probes in an isolate of *Lactococcus lactis* ssp. *lactis* from raw milk cheese.

Enterocin AS-48 from *Enterococcus faecium* isolates from milk and dairy products has been detected using dot blot and colony hybridization (105). Genes encoding for synthesis of AS-48 have been sequenced and a PCR technique has been developed for rapid detection of these genes in field isolates (106,107).

A rapid and sensitive flow cell method for detection of bacteriocins in fermentation broth has been described (108). Low concentrations of potassium ions released from bacteriocin sensitive indicator strains are directly measured and correlated to the concentrations of crude bacteriocin present in fermentation broth. This method demonstrated a high degree of correlation with conventional agar well diffusion assays.

6.4 CLASSIFICATION

Most bacteriocins produced by LAB are cationic, hydrophobic, or amphiphilic molecules composed of 20 to 60 amino acid residues (81). These bacteriocins are commonly classified into four groups that also include bacteriocins from other Gram-positive bacteria (19,24).

6.4.1 Class I: Lantibiotics

Lantibiotics are a group of ribosomally synthesized, post-translationally modified peptides containing unusual amino acids, such as lanthionine (Lan), α-methyllanthionine (MeLan), dehydroalanine, and dehydrobutyrine residues. The lantibiotics (from lanthionine containing antibiotic) are unique in that they are produced on the ribosome as a prepeptide which undergoes extensive posttranslational modification to form a bioactive peptide (20,21,26).

These bacteriocins are a heterogenous group of compounds with respect to molecular mass, structure, and mode of action. Based on chemical structure and antimicrobial mechanism, class I bacteriocins are further subdivided into types A and B lantibiotics (Table 6.1) (52,53,109).

Table 6.1
Class-I lantibiotic-types, their producing strains, modified residues and molecular mass

		Modified Residues				MW
Type	**Producing Strain**	**Lan**	**MeLan**	**Dha**	**Dhb**	**(Da)**
Type-A (I)						
Nisin A	*Lactococcus lactis* 6F3	1	4	2	1	3353
Nisin Z	*Lactococcus lactis* N8	1	4	2	1	3330
Subtilin	*B. subtilis* ATCC6633	1	4	2	1	3317
Epidermin	*Staph. epidermidis* Tu3298	2	1	0	1	2164
Gallidermin	*Staph. gallinarum* Tu3928	2	1	0	1	2164
Mutacin B-Ny266	*Strep. mutans*	2	1	1	1	2270
Mutacin 1140	*Strep. mutans* JH1000	2	1	1	1	2263
Mutacin III	*Strep. mutans*	2	1	1	1	2266
Pep5	*Staph. epidermidis* 5	2	1	0	2	3488
Epicidin 280	*Staph. epidermidis* BN280	1	2	0	1	3133
Epilancin K7	*Staph. epidermidis* K7	2	1	2	2	3032
Type-A (II)						
Lacticin 481	*Lactococcus lactis* CNRZ481	2	1	0	1	2901
Cytolysin A1	*E. faecalis* DS16		-	-	-	4164
Cytolysin A2	E. faecalis	-	-	-	-	2631
Lacticin 3147 A	*Lactococcus lactis* DPC3147	-	-	-	-	3322
Lacticin 3147 B	*Lactococcus lactis* DPC3147					2847
Staphylococcin C55α	*Staph. aureus* C55	-	-	-	-	3339
Staphylococcin C55β	*Staph. aureus* C55	-	-	-	-	2993
Salivaricin A	*Strep. salivarius* 20P3	-	-	-	-	2315
Lactocin S	*Lb. sake* L45	2	0	0	1	3764
Streptococcin A-FF22	*Strep. pyogenes* FF22	1	2	0	1	2795
Sublancin 168	*B. subtilis* 168	-	-	-	-	3877
Carnocin UI 49	Carnobacterium pisicola	-	-	-	-	4635
Variacin 8	*Micrococcus varians* MCV8	2	1	0	1	2658
Cypemycin	*Streptomyces* ssp.	0	0	0	4	2094
Type-B						
Cinnamycin	Streptomyces cinnamoneus	1	2	0	0	2042
Duramycin B	*Streptoverticillium* ssp.	1	2	0	0	1951
Duramycin C	Streptomyces griseoluteus	1	2	0	0	2008
Ancovenin	*Streptomyces* ssp.	1	2	1	0	1959
Mersacidin	*B. subtilis* HIL Y-85	0	3	1	0	1825
Actagardine	*Actinoplanes* ssp.	1	2	0	0	1890

Adapted from McAuliffe et al., *FEMS Microbiol. Rev.* 25:285–308, 2001; Guder et al., *Biopolymers* 55:62–73, 2000
Post-translational side chain modifications resulting in unusual amino acids.
Lan- ring formed by thioether amino acid lanthionine
MeLan- ring formed by thioether amino acid methyllanthionine
Dha- dehydrated amino acids 2,3- didehydroalanine
Dhb- dehydrated amino acids 2,3- didehydrobutyrine

6.4.1.1 Type A

These lantibiotics are elongated, flexible peptides with a net positive charge that could elicit bactericidal activity by pore formation in bacterial membranes. The most prominent member of this group is nisin, which is elaborated in Section 5.1.

6.4.1.2 Type B

These lantibiotics are small globular peptides due to their head to tail cross linkage. These peptides form a group of natural variants that carry a negative or no net charge. Their antimicrobial mechanism is by inhibition of specific enzyme functions. Bacteriocins such as ancovenin, duramycin A, B, C, and cinnamycin are members of this group.

6.4.2 Class II: Small Heat Stable Peptides (Non-lantibiotics)

These bacteriocins are ribosomally synthesized as inactive prepeptides. They are modified by post-translational cleavage of the N-terminal leader peptide, generally at a double glycine to release mature amphiphatic and thermostable cationic peptides (110). Potent anti-listerial activity is a common characteristic of this class of bacteriocins.

Classification of bacteriocins of this group is challenging. One of the criteria is based on the presence of a YGNGVXC amino acid motif near the N-terminus of the active peptide, and the presence of cysteine residues resulting in disulfide bridges. Bacteriocins consisting of the YGNGVXC motif frequently are referred to as pediocin-like or listeria-active. The function of this sequence, if any, is not known and both descriptions are misleading. The non-lantibiotic class II bacteriocins in this chapter are subdivided as class IIa to IIe (typical) and IIf (atypical) according to the scheme proposed by van Belkum and Stiles (110). The typical classes IIa to IId bacteriocins are listed in Table 6.2.

6.4.2.1 Class IIa

These are cystibiotics containing four cysteine residues that form two S-S bridges, one in the N-terminal half and the other in the C-terminal half of the molecule. Only three well characterized bacteriocins, pediocin PA-1/AcH, enterocin A, and divercin V41 fit this group. Reaction of pediocin PA-1 with dithiothreitol to maintain cysteine residues in the thiol form markedly decreases its activity (112). Pediocin from this subclass is described in Section 6.1.

6.4.2.2 Class IIb

These are the largest group of this class; cystibiotics containing one S-S bridge in the N-terminal half of the molecule. Many of these bacteriocins, including sakacins, contain a high degree of sequence homology in the N-terminal amino acid residues, whereas the C-terminal amino acids are relatively diverse (113). Sakacin from this subclass is described in Section 6.2.

6.4.2.3 Class IIc

These are cystibiotics which lack the YGNGVXC motif of class IIa and IIb, and the S-S bridge spans the N- and C-sections of the molecule. Carnobacteriocin A and enterocin B are included in this group. Their genetic arrangement is unusual and their structural and immunity genes are encoded on opposite strands of the DNA in the opposite orientation to one another.

6.4.2.4 Class IId

This is a small group of bacteriocins that do not contain the YGNGVXC motif and consists of only one or no cysteine residues. Lactococcins A and B are included in this group but lactococcin B is the only bacteriocin that has been characterized as a thiolbiotic (45).

Table 6.2
Amino acid sequence, molecular mass of CLASS-II a,b,c,d types of non-lantibiotic bacteriocins and their producing strains

Bacteriocin-Type	Producing Strain	Amino Acid (AA) Sequence	MW (Da)	AA (No.)
CLASS-IIa				
Pediocin PA-1/ACH	*Ped. acidilactici* H	KYYGNGVTCGKHSCSVDWGKATTCIINNGAMAWATGGHQGNHKC	4623	44
Enterocin A	*Ent. faecium* CTC492	TTHSGKYYGNGVYCTKNKCTVDWAKATTCIAGMSIGGFLGGAIPGKC	4829	43
Divercin V41	*Car. divergens* V41	TKYYGNGVYCNSKKCWVDWGQASGCIGQTVVGGWLGGAIPGKC	4509	43
CLASS-IIb				
Sakacin P	*Lb. sake* Lb674	KYYGNGVHCGKHSCTVDWGTAIGNIGNNAAANWATGGNAGWNK	4434	43
Sakacin A/Curvacin A	*Lb. sake* / *Lb. curvatus*	ARSYGNGVYCNNKKCWVNRGEATQSIIGGMISGWASGLAGM	4306	41
Leucocin A	*Leu. gelidum* UAL187	KYYGNGVHCTKSGCSVNWGEAFSAGVHRLANGGNGFW	3390	37
MesentericinY105	*Leu. mesenteroides* Y105	KYYGNGVHCTKSGCSVNWGEAASAGIHRLANGGNGFW	3868	37
Carnobacteriocin B2	*Car. piscicola* LV17B	VNYGNGVSCSKTKCSVNWGQAFQERYTAGINSFVSGVASGAGSIGRRP	4967	48
Carnobacteriocin BM1	*Car. piscicola* LV17B	AISYGNGVYCNKEKCWVNKAENKQAITGIVIGGWASSLAGMGH	4524	43
Enterocin P	*Ent. faecium* P13	ATRSYGNGVYCNNSKCWVNWGEAKENIAGIVISGWASGLAGMGH	4493	44
Bacteriocin 31	*Ent. faecalis*	ATYYGNGLYCNKQKCWVDWNKASREIGKIIVNGWVQHGPWAPR	5005	43
CLASS-IIc				
Carnobacteriocin A	*Car. piscicola* LV17A	DQMSDGVNYGKGSSLSKGGAKCGLGIVGGLATIPSGPLGWLAGAAGVINSCMK	5050	53
Enterocin B	*Ent. Faecium* BFE900	ENDHRMPNELNRPNNLSKGGAKCGAAIAGGLFGIPKGPLAWAAGLANVYSL	5230	51
Divergicin A	*Car. Divergens* LV13	AAPKITQKQKNCVNGQLGGMLAGALGGPGGVVLGGIGGAIAGGCFN	4221	46
CLASS-IId				
Lactococcin A	*Lactococcus lactis* subsp. *cremoris* 9B4	KLTFIQSTAAGDLYYNTNTHKYVYQQTQNAFGAAANTIVNGWMGG AAGGFGLHH	5775	54
Lactococcin A	*Lactococcus lactis* subsp. *cremoris* 9B4	KLTFIQSTAAGDLYYNTNTHKYVYQQTQNAFGAAANTIVNGWMGG AAGGFGLHH	5775	54
Lactococcin B	*Lac lactis* ssp. *cremoris* 9B4	SLQYVMSAGPYTWYKDTRTGKTICKQTIDTASYTFGVMAEGWGKTFH	5325	47

Adapted from Enhar et al., FEMS Microbiol. Rev. 24:85–106, 2000; van Belkum & Stiles, Nat. Prod. Rep. 17:323–335, 2000, and Graneau et al., Biochimie 84:577–592, 2002.

6.4.2.5 Class IIe

These are characterized by their antimicrobial activity, which is dependant on the functional combination of two peptides (Table 6.3). These two separate peptides are either type E (enhancing) in which one of the peptides has antibacterial activity that is enhanced by the other; i.e., thermophilin 13, lactacin F, and plantaracin S; or type S (synergistic) in which the peptides have little or no activity alone; i.e., lactococcin G, plantaricin A, and brochocin-C (110,114–116). Lactococcin G was the first two peptide bacteriocin described (117). This class also includes lactococcin M and plantaricins EF and JK (29). They do not contain cysteine residues and the antibacterial spectra are even more limited than the thiobiotics (5).

6.4.2.6 Class IIf – Atypical Class II Bacteriocins

These are cyclic and leaderless peptides; i.e., bacteriocin AS-48 which is better classified as a bacteriocin than a peptide antibiotic, widespread among *Enterococcus faecalis* and *E. faecium* strains (118). Some investigators even consider cyclic peptide bacteriocins as separate from the class II.

6.4.3 Class III: Large Heat Labile Proteins

This class of bacteriocins is not well characterized. This group consists of large (>30 kDa) heat labile proteins that are of limited interest to food scientists. This group includes Helvetin J produced by *Lactobacillus helveticus* and enterolysin produced by *E. faecium* (119,120).

6.4.4 Class IV: Complex Proteins

This group consists of either glycoproteins (lactocin 27) or lipoproteins (lacstrepcins) that require nonprotein moieties for antimicrobial activity (124,121). Plantaricin S also belongs to the group IV bacteriocins (19). Bacteriocins in this class have not been characterized adequately at the biochemical level, to the extent that the definition of this class requires additional descriptive information (16,122).

6.5 CLASS I: LANTIBIOTICS

Most of the class I bacteriocins have a fairly broad inhibitory spectrum. They not only inhibit closely related bacteria, such as species from the genera *Enterococcus*, *Lactobacillus*, *Lactococcus*, *Leuconostoc*, *Pediococcus*, and *Streptococcus*, but also inhibit several less closely related Gram-positive bacteria, such as *L. monocytogenes*, *Staphylococcus aureus*, *Bacillus cereus*, and *Clostridium botulinum*. Several bacteriocins in this class, such as nisin and thermophilin 13, prevent germination and proliferation of spores of *B. cereus* and *C. botulinum*. Interestingly, acidocin J1132 has a very narrow inhibitory spectrum and sensitive strains are limited to members of the genus *Lactobacillus*, while at the other extreme, plantaricin LP84 (produced by *Lb. plantarum* NCIM 2084) has a potent inhibitory activity against *Escherichia coli* (123).

6.5.1 *Type A:* Nisin

Nisin is a small, heat stable, antimicrobial peptide described as a class I type A bacteriocin, a group that comprises lantibiotics (123). In 1947, Mattick and Hirsch (124) first isolated nisin from culture broths of *Lactococcus lactis* ssp. *lactis*. Nisin is the first compound of this class type to be used in the food industry on a broad scale. Its discovery prompted the search for other antimicrobial peptide products from LAB, but nisin is still the only purified bacteriocin commercially available, and is unique in its acceptance worldwide as a food preservative. In 1953, the first industrial production of nisin (Nisaplin®) was launched by Aplin and Barrett, Ltd., UK.

Table 6.3
Amino acid sequence, molecular mass of different CLASS-II e (two-peptide) non-lantibiotic bacteriocins and their producing strains

Bacteriocin-Type		Producing Strain	Amino Acid (AA) Sequence	MW (Da)	AA (No.)
CLASS-IIe (two-peptide)					
ABP-118	*(Abp118α)*	*Lb. salivarius* UCC118	KRGPNCVGNFLGGLFAGAAAGVPLGPAGIVGGANLGMVGGALTCL	4097	45
		(Abp118β)	KNGYGGSGNRWVHCGAGIVGGALIGAIGGPWSAVAGGISGGFTSCR	4331	46
Brochocin-C	*(Brc-A)*	*B. campestris* ATCC43754	YSSKDCLKCIGKGIGAGTVAGAAGGGLAAGLGAIPGAFVGAHFGVIGGSAACIGGLLGN	5243	59
		(Brc-B)	KINWGNVGGSCVGGAVIGGALGGLGGAGGGCITGAIGSIWDQW	3945	43
Enterocin 1071	*(A)*	*Ent. faecalis* BFE1071	ESVFSKIGNNAVGPAATWILKGLGNMSDVTQADRINRKHN	4285	39
		(B)	GPGKWLPWLQPAYDFVTGLFAKGIGKEGNKNKWKN	3899	34
Enterocin L50	*(A)*	*Ent. faecium* L50	MGAIAKLVAKFGWPIVKKYYKQIMQFIGEGWAINKIIEWIKKHI	5187	44
		(B)	MGAIAKLVTKFGWPLIKKFYKQIMQFIGQGWTIDQIEKWLKRH	5175	43
Lactacin F	*(Laf-A)*	*Lb. johnsonni* VPI11088	NRWGDTVLSAASGAGTGIKACKSFGPWGMAICGVGGAAIGGYFGYTHN	4733	48
		(Laf-X)	RNNWQTNVGGAVGSAMIGATVGGTICGPACAVAGAHYLPILWTGVTAATGGFGKIRK	5598	57
Lactocin 705	*(705α)*	*Lb. casei* CRL505	MD-NLNK-FKKLSDNKLQATIGG	2334*	33
		(705β)	MESNKLEKFANISNKDLNKITGG	2550	33
Lactococcin G	*(α)*	*Lc. lactis* LMG 2081	GTWDDIGQGIGRVAYWVGKAMGNMSDVNQASRINRKKKH	4346	39
		(β)	KKWGWLAWVDPAYEFIKGNIGKGAIKEGNKDKWKN	4110	35

Lactococcin MN	*(LcnM)*	*Lc. lactis* ssp. *cremoris* 9B4	IRGTGKGLAAAMVSGAAMGGAIGAFGGPVGAIMGAWGGAVGGAMKYSI	4322	48
		(LcnN)	GSIWGAIAGGAVKGAIAASWTGNPVGIGMSALGGAVLGGVTYARPVH	4374	47
Leucocin H	*(Hα)*	*Leuconostoc* spp. MF215B	WXIGVTGAALGTG(Hyl)GV(Hyl)NVI	2538*	~40
		(Hβ)	WXAVFXNMAKYMFSQSKXVVIGFLVAS	3107	~40
Mutacin IV	*(NlmA)*	*Strep. mutans* UA140	KVS--------GGEAVAAIG-------ICATASAAIGGLAGATLVTPYCVGTWGLIRSH	4169	44
		(NlmB)	DKQAADTFLSAVG-----GAASGFTYC----ASNGVWHPYILAG-CAGVGAVGSVVFP	4826	49
Plantaracin EF	*(PlnE)*	*Lb. plantarum* C-11	FNRGGYNFGKSVRHVVDAIGSVAGIRGILKSIR	3543	33
		(PlnF)	VFHAYSARGVRNNYKSAVGPADWVISAVRGFIHG	3701	34
Plantaracin JK	*(PlnJ)*	*Lb. plantarum* C-11	GAWKNFWSSLRKGFYDGEAGRAIRR	2928	25
		(PlnK)	RRSRKNGIGYAIGYAFGAVERAVLGGSRDYNK	3501	32
Plantaracin S	*(α)*	*Lb. plantarum* LPCO10	RNKLAYNMGHYAGKATIFGLAAWALLA	2921	27
		(β)	KKKKQSWYAAAGDAIVSFGEGFLNAW	2871	26
Thermophilin 13	*(ThmA)*	*Strep. thermphilus* SFi13	YSGKDCLKDMGGYALAGAGSGALWGAPAGGVGAL-AHVGAIAGGFACMGGMIGNKFN	5776	62
		(ThmB)	QINWGSVVGHCIGGAIIGGAFSGGAAAGVGCLVGSGKAIINGL	3910	43

Adapted from van Belkum & Stiles, *Nat. Prod. Rep.* 17:323–335, 2000; and Garneau et al., *Biochimie* 84:577–592, 2002.

*-Molecular mass estimated by peptide sequence fragmentation modeling.

6.5.1.1 Mechanism of Action

The cytoplasmic membrane is the primary target site for both nisin A and nisin Z, where the activity results in pore formation, membrane perturbation and voltage dependent polarization of lipid bilayers (125,126). The flexible nature and amphiphilic properties of the nisin molecule are important for this mode of action.

The ability of nisin to inhibit N-acetylglucosamine synthesis in *B. subtilis*, and DNA, RNA and protein synthesis in *Micrococcus luteus* has prompted speculation that this bacteriocin interacts with the phospholipid components of the cytoplasmic membrane. Studies have indicated that isolated cytoplasmic membranes and phospholipid components of membranes show an antagonistic effect on nisin inhibition (127). Due to its cationic nature, nisin interacts with charged groups membrane phospholipids via ionic forces. The C-terminal region of the nisin molecule is responsible for the initial electrostatic interaction (128–131). Nisin could also bind to non-energized liposomes with a high affinity via the negatively charged phospholipids.

The nisin induced pore formation across the cytoplasmic membrane is non-selective and transient in nature with varying pore sizes. This suggests that nisin molecules could associate and dissociate with the pore in a dynamic process (5). The energy for formation and opening of the pores is accessed from the trans-membrane potential of the target organism (132,133). The lethal outcome with nisin is optimum against proliferating cells with actively energized membranes (133,134). However, nisin activity could be induced in the absence of a trans-membrane electric potential ($\Delta\Psi$) with high pH gradient across the membrane (ΔpH) (135).

Based on several lines of experimental evidence, a barrel stave mechanism of pore formation has been proposed for nisin lantibiotic mediated antimicrobial activity (Figure 6.3). According to this model, nisin monomers (the staves) bind to the membrane, insert into the membrane (dependent on the electric potential) and aggregate within the membrane to form a water filled pore (the barrel) (117,136). The nisin molecule orients perpendicularly into the membrane with its hydrophobic sides facing the membrane and its hydrophilic regions oriented into the lumen of the pore, forming the sides. If the membrane is sufficiently energized,

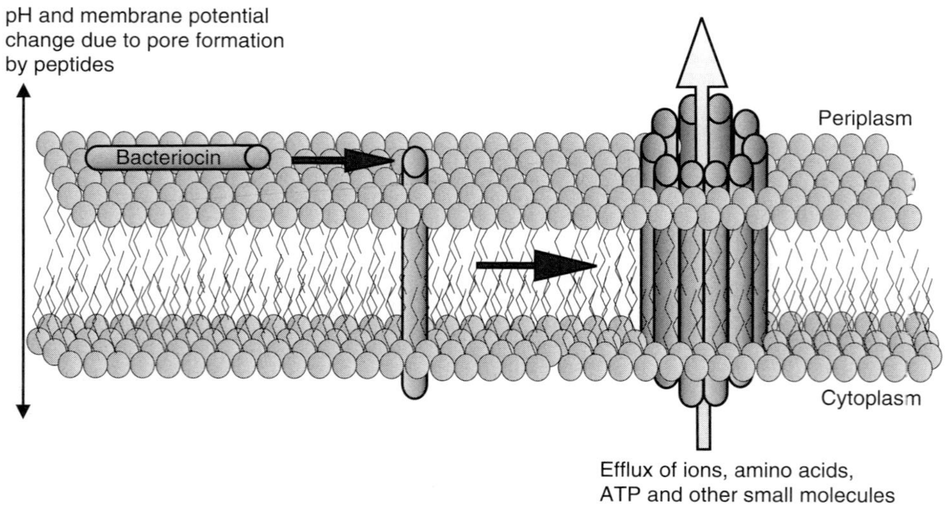

Figure 6.3 'Barrel-stave' model of pore formation by nisin. Nisin monomers ('staves') bind, insert and aggregate within the membrane to form a water-filled pore (the 'barrel').

nisin molecules travel through the membrane, bind to the surface, and penetrate across phospholipids. Binding of several nisin molecules at the same site results in the formation of a transient pore.

In another proposed wedge model, nisin molecules bind to anionic membrane surface by electrostatic interactions and disturb the lipid dynamics (130). This further leads to localized damage of the membrane that ultimately forces the nisin into the target cell membrane. This model suggests that multimeric nisin molecules form wedge-like pores and the electric potential propels the insertion of phospholipid surface bound nisin molecules into the cell membrane. Subsequently, the positively charged carboxyl terminal residues are drawn across the membrane in response to the electric potential, causing a structural defect in the phospholipid bilayer.

Nisin interaction and perturbation of membrane is dependent on an array of molecular properties such its amphipathic character, flexibility of its central segment, and overall negative surface charge of the membrane (128,131,137). Pore formation leads to leakage of small molecules such as K^+ ions, solutes, and metabolites such as ATP and amino acids. This phenomenon rapidly depletes the proton motive force, causing cellular energy deprivation and inhibition of vital biosynthetic processes (Figure 6.4) (26,125).

Other mechanisms of nisin activity against vegetative cells have been suggested (138). Interference of nisin with the murein biosynthesis of *B. subtilis* cell wall has been reported (127). Adsorption of nisin to teichoic acids in cell wall and activation of autolytic enzymatic degradation has also been suggested (139). These two mechanisms could be slower in effect than pore formation, but may contribute to an overall bactericidal effect.

6.5.1.2 Antimicrobial Spectrum

Nisin is a potent inhibitor of several species of Gram-positive bacteria, including the heat resistant endospore formers. The susceptibility of Gram-negative bacteria to nisin activity is limited due to their lantibiotic impermeable outer cell membrane. However, the cell wall deficient spheroplasts of *E. coli* and *Salmonella typhimurium* are susceptible to nisin (140). Furthermore, treatments that alter the outer membrane barrier functions could

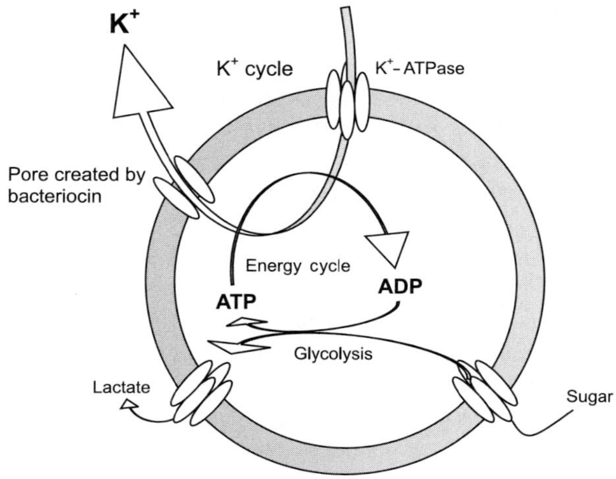

Figure 6.4 Cellular mechanism of pore-forming bacteriocin activity. Pore formation leads to depletion of proton motive force by causing cellular energy deprivation and inhibition of vital biosynthesis functions.

potentiate nisin activity in Gram-negative bacteria. Such treatments include hydrostatic pressure, heat, freezing and thawing and chelating agents such as EDTA or ethyl maltol (141–146). Yeasts, like the Gram-negative bacteria, are also insensitive to nisin, and any removal of their cell wall allows the bacteriocin access that leads to yeast cell lysis (147). However, the insensitivity of yeast to nisin has been exploited in fermentation applications with yeast based starter cultures and fermented products.

Nisin sensitivity could vary among strains even of the same species (148,149). About 10–100 times more nisin is needed to prevent the outgrowth of spores of *C. botulinum*, compared to *C. butyricum* (150). Spores and vegetative cells of thermophilic species such as *B. stearothermophilus* and *C. thermosaccharolyticum* are more sensitive to nisin than mesophilic spore forming species (151).

Comparative studies with nisin A, nisin Z, and pediocin showed that LAB, *Clostridium* and *Bacillus* (spores and cells) are more effectively inhibited by the nisin variants compared to pediocin (149). Antilisterial inhibition studies indicated that Nisin A is more efficacious than nisin Z, however, the effective dosages showed marked variations. Accordingly, nisin concentration required for complete inhibition of *L. monocytogenes* varies from 10 to 2500 μg/mL, depending on the culture medium and the target strain (152). The dosage limitation could be overcome in foods by using nisin in combination with a starter culture producing a nonlantibiotic antilisterial bacteriocin (153). In practice, however, varied listerial sensitivities are overcome by standard hurdle technology, where nisin acts in combination with other control measures in the food matrix, such as reduced water activity.

Spontaneous resistance to nisin has been reported in strains of *C. butyricum, L. lactis* ssp. *cremoris*, *L. monocytogenes* and *S. agalactiae* (154,155). Resistance has been linked to the production of an anionic phosphate polysaccharide with subunits of rhamnose and galactose (156).

The functional efficacy of nisin in a food matrix is affected by several factors including: changes in nisin solubility and charge, binding of nisin to food components (i.e., meat phospholipids), inactivation of nisin (by proteases or food ingredients), and changes in the cell envelope of the target organisms in response to environmental factors (157). The activity of nisin produced *in situ* may also be affected by food components.

6.5.1.3 Applications

Nisin is suitable for use in a wide range of foods: liquid or solid, canned or packaged, chilled or warm ambient storage (elaborated in Section 6.7). Based on target organisms, its usage falls into three broad categories: to prevent spoilage by Gram-positive endospore formers (particularly in heat processed food), to prevent spoilage by LAB and similar organisms such as *Brocothrix thermosphacta*, to kill or inhibit Gram-positive pathogens; i.e., *L. monocytogenes*, *B. cereus*, and *C. botulinum*.

Nisin provides advantages as an antimicrobial food additive in extending shelf life of both chilled and ambient storage temperatures (such as canned products in warm climates). Nisin also protect products, particularly chilled foods, from storage temperature abuse. Addition of nisin may contribute to the general preservation of a food so that thermal processing could be reduced. This has the double advantage of improving product quality (which may be impaired by harsh heat treatment) and reducing manufacturing costs, because both the time and temperature of thermal treatment could be diminished.

Nisin is best added as an aqueous solution, usually to the liquid portion of a product during processing. For instance, in canned foods nisin could be combined with a small quantity of canning brine followed by a thorough mixing into the bulk product. In dairy desserts and milks, nisin could be added to a small quantity of milk, followed by mixing into the bulk milk, then filled and processed (158). Nisin could also be added as a powder, provided a uniform dispersal throughout the food matrix is ensured. Nisin could also be used at higher concentrations as a spray or dip (processing aid) for surface decontamination.

The dosage and application of nisin depends on the food type, processing (heat, pH, or salts) and storage conditions as well as the shelf life expectations (Table 6.4). Nisin is often used in acidic foods, but is effective in products across a wide range of pH values (3.5–8.0). Nisin also seems to be an effective preservative in liquid egg, which generally has a pH of 7.3 to 7.8.

Several reports have indicated an effective application of nisin in packaging films and edible casings. Coatings of nisin on silicon surfaces and polyvinyl chloride films have been shown to demonstrate antimicrobial activity (92,159). Biodegradable films containing lysozyme, nisin and EDTA made from soy protein and corn zein prepared by heat

Table 6.4
Nisin applications as bio-preservative for various types of food products

Food Type	Nisin (μg/g)	Application
A) Dairy Foods		
1. Processed cheese	5 – 15	Inhibition of endospores from Bacillus spp. and Clostridium spp.
2. Cottage cheese	5 – 50	Inhibition of *L. monocytogenes* contaminants
3. Soft white fresh cheese	5	Preservation of minimal processed cheeses i.e., ricotta, panir and Latin-American cheeses queso blanco and queso fresco.
4. Natural cheese	5 – 20	Shelf life extension of Camembert and Feta cheeses.
5. Dairy desserts	1.25 – 25	Quality enhancement of crème caramel and cholcolates; anti-listerial protection of ice creams.
6. Yoghurts	0.5 – 1.25	Reduction of over-acidity in fermented milk products.
7. Milk and milk products	0.25 – 1.25	Extension of shelf life of pasteurised milk in developing countries
B) Beverages		
1. Fruit juice (pasteurized)	0.13 – 0.25	Alicyclobacillus acidoterrestris
2. Vegetable protein milks		Reduction in heat processing time and shelf life extension
3. Wine		Suppression of unwanted malolactic acid fermentation
4. Fruit brandy		Selective inhibition of lactic acid bacteria
5. Beer:		Inhibition of unwanted lactic acid bacteria i.e., Lactobacillus spp., Pediococcus spp. without affecting yeast
• pitching yeast wash	25 – 37.5	
• reduced pasteurization	0.25 – 1.25	
• during fermentation	0.63 – 2.5	
• post fermentation	0.25 – 1.25	
C) Other Foods		
1. Canned food	2.5 – 5	Destruction of clostridial and bacillary spores.
2. Dressings and sauces	1.25 – 5	Inhibition of bacterial, yeast and mold contaminants.
3. Liquid egg	1.25 – 5	Elimination of salmonella and endospore destruction
4. Fresh soups		Control of spoilage flora and shelf life extension
5. Crumpets	3.75 – 6.25	Inhibition of toxigenic *B. cereus* food poisonings.

Adapted from Thomas et al. (159)

press and casting methods exhibit a potent antimicrobial activity against *L. plantarum* and *E. coli* (160).

6.6 CLASS II: SMALL HEAT STABLE PEPTIDES

Compared to class I bacteriocins, most class IIa bacteriocins have a narrow spectrum of activity and inhibit closely related Gram-positive bacteria. In general, members of the genera *Enterococcus*, *Lactobacillus*, and *Pediococcus* are sensitive to Class IIa bacteriocins, and members of the genus *Lactococcus* are resistant. For example, pediocin PA-1 was found active against different species of *Enterococcus*, *Lactobacillus*, and *Pediococcus*; however, only 1 out of 11 *Lactococcus* strains tested (*Lactococcus lactis* LMG 2070) was sensitive to the bacteriocin (63). Certain class IIa bacteriocins, such as pediocin PA-1, demonstrate broad inhibitory spectra and could inhibit certain less closely related Gram-positive bacteria, such as *S. aureus* and vegetative cells of *Clostridium* species and *Bacillus* species. Other class IIa bacteriocins, such as mundticin from *Enterococcus mundtii*, even prevent the germination of spores of *C. botulinum*. Class IIa bacteriocins are generally active against *Listeria*. Different strains of *Listeria* including *L. monocytogenes*, *L. innocua*, and *L. ivanovii* were highly sensitive to four different class II bacteriocins (pediocin PA-1, enterocin A, sakacin P, and curvacin A) (63). The minimal inhibitory concentrations against *L. monocytogenes* for the four bacteriocins varied from 0.1 to 8 ng/mL; however, *L. monocytogenes* V7 and *L. innocua* LB1, have been found resistant to class II bacteriocins (enterocin A, mesentericin Y105, divercin V41, and pediocin AcH) (161).

6.6.1 *Class IIa:* Pediocins

Pediocin A, from *Pediococcus pentosaceus* is a 80 kDa bacteriocin and the pediocins of *P. acidilactici* strains are about 4.6 kDa peptides containing 44 amino acids as pediocin PA-1/AcH. Two *P. parvulus* strains and one *Lb. plantarum* strain also reportedly produce pediocin PA-1/AcH. In general, the bactericidal property of pediocins is relatively resistant to heat, storage conditions, pH, organic solvents, and hydrostatic pressure, but is destroyed by proteolytic enzymes and neutralized by certain anions. Pediocin PA-1/AcH by *P. acidilactici* strains is produced in large amounts in a non-buffered nutritional medium, especially at a low terminal pH. Studies have shown that pediocin PA-1/AcH, or the producer strains, could be effectively used in food systems to control *L. monocytogenes*. Under optimum processing conditions pediocins could also control Gram-positive and Gram-negative spoilage bacteria in foods.

6.6.1.1 Mechanism of Action

Early studies indicated that the antibacterial mechanism of pediocin PA-1/AcH against Gram-positive bacteria is a bactericidal and not a bacteriostatic effect (162). Subsequent studies showed that pediocin induced cell death by impairment of bacterial cytoplasmic membrane, which results in the leakage of cellular ions, small molecules, and UV absorbing materials (163). Furthermore, cationic pediocin molecule binds in a pH dependent manner to anionic lipoteichoic acids in the Gram-positive bacterial cell wall. In Gram-negative cells, due to the lack of teichoic acids, pediocin molecules are not absorbed (or adsorbed at very low concentrations, probably due to lipopolysaccharides which have a barrier function) and the inner membrane remains unaffected. However, Gram-negative bacteria, when stressed or injured, lose the barrier function of their outer membrane, and pediocin could then elicit a bactericidal effect (164).

The mechanism of bactericidal action of pediocin PA-1/AcH is unclear. Based on several lines of evidence, a hypothesis for the mechanism of action is elucidated (Figure 6.5) as follows:

1. The N-terminal half of the molecule adopts a β-sheet conformation, probably after membrane binding, which is stabilized by a disulfide bond between *Cys-9* and *Cys-14*.
2. Cationic amino acids (*Lys-11* and *His-12*) bind to the anionic phospholipids in the membrane through electrostatic interactions.
3. Protein receptors in the membrane are not essential for the binding of pediocin, but such a receptor, if present, could enhance the activity.
4. The C-terminal half also might have a random conformation in aqueous environment, but disulfide bridging between *Cys-24* and *Cys-44*, and β-sheet may help transform it into a defined α-helical amphiphilic structure following any contact with the lipid bilayer.
5. Pediocin molecules subsequently could assemble in the membrane, cause pore formation and lead to dissipation of proton motive force due to the collapse of pH gradient and membrane potential in a voltage independent manner.
6. Susceptible cells lose intracellular ions, UV absorbing materials, ATP, and could no longer transport amino acids or glucose.
7. These cellular events finally lead to cell death (165,166).

6.6.1.2 Antimicrobial spectrum

Compared to other bacteriocins in the class IIa family and to some lantibiotics, pediocin A of *P. pentosaceus* FBB61 and pediocins from *P. acidilactici* strains have a relatively wider bactericidal spectrum against Gram-positive bacteria. Pediocin A inhibits several strains of *P. pentosaceus, P. acidilactici, S. aureus, L. lactis, Lactobacillus* spp., *C. botulinum, C. perfringens* and *C. sporogenes* (167). Pediocin A has also been reported to inhibit six out of eight strains of *L. monocytogenes* (101).

Bhunia et al. (163) demonstrated that a partially purified fraction of pediocin PA-1/AcH effectively inhibited *B. cereus, C. perfringens, S. aureus* and several species and strains of *Lactobacillus* and *Leuconostoc*. In later studies, higher purity preparations of pediocin PA-1/AcH showed growth inhibition against strains of *L. monocytogenes, B. cereus, C. perfringens, C. botulinum, C. laramie*, and several *Lactobacillus, Leuconostoc, Enterococcus, Pediococcus* and *Lactococcus* species (168–170). Although bacterial spores are not sensitive to pediocin PA-1/AcH and other bacteriocins, no growth was observed following inoculation of spores of *C. laramie, C. perfringens* and *C. botulinum* in appropriate growth media containing pediocin. A later study showed that pediocin produced a bactericidal effect on the germinated and outgrowing spores. Studies have also indicated that resistant but stressed Gram-positive bacterial strains of *S. aureus, L. lactis* and Gram-negative bacterial strains of *Salmonella* sp., *E. coli* O157:H7, *Serratia liquefaciens* and *Pseudomonas* species turned susceptible to pediocin PA-1/AcH (142,143,171–173).

6.6.1.3 Applications

Pediocin A from *P. pentosaceus* FBB61 and pediocin PA-1/AcH from *P. acidilactici* strains as well as from *P. parvulus* strains and *L. plantarum* WHE92 are potent bactericidal agents against many normal as well as injured foodborne spoilage and pathogenic bacteria. Accordingly, these pediocins could be used as effective food biopreservatives (174).

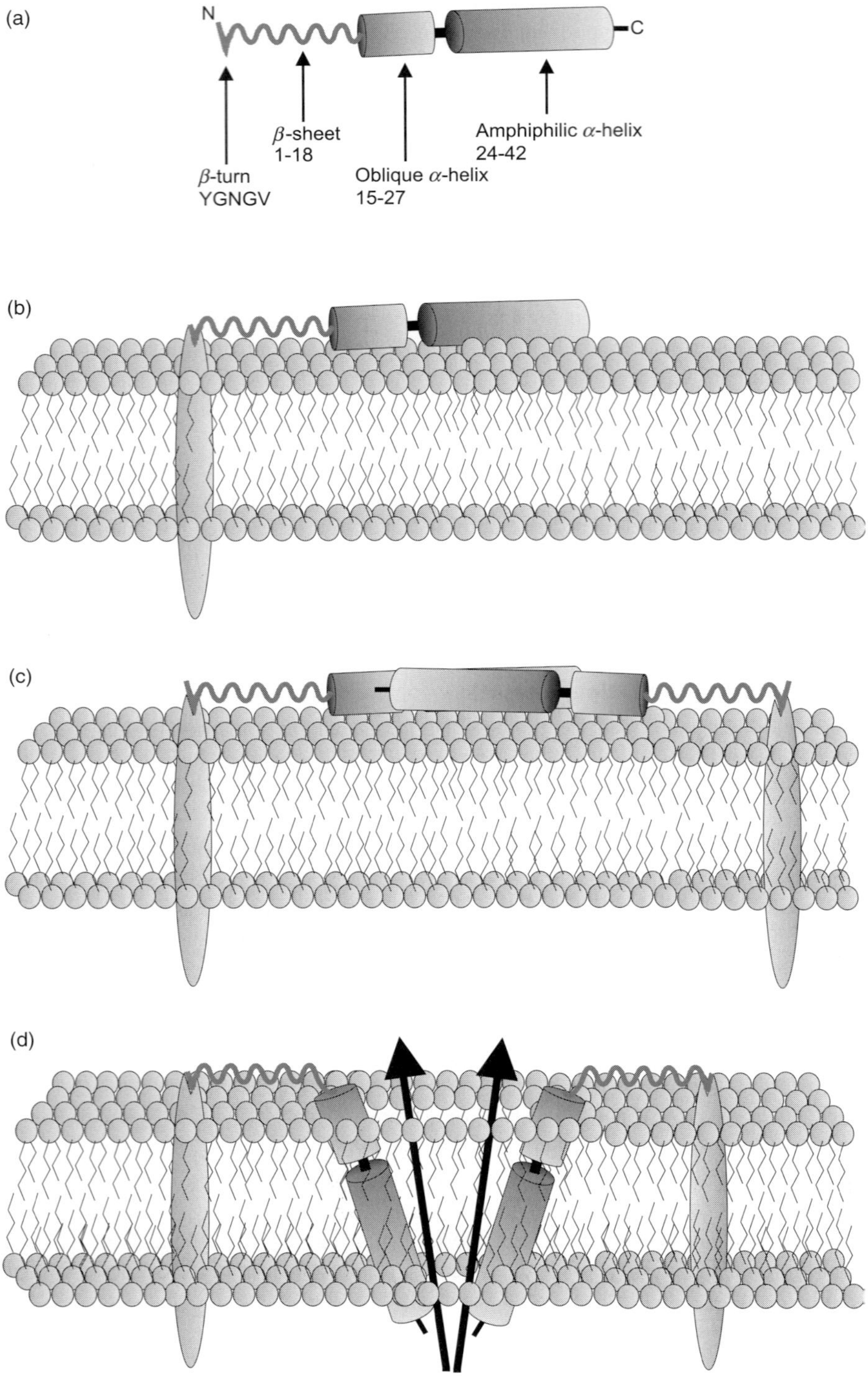

Figure 6.5 Membrane interaction model for class IIa bacteriocin with target cells. (a) Structure of the predicted domain of the bacteriocin. (b) Bacteriocin positioning by receptor (putative) recognition, electrostatic and hydrophobic interactions. (c) Avid binding of bacteriocin with the target site. (d) Bacteriocin aggregation and hydrophobic pore formation.

Other studies have suggested that pediocin A producing *P. pentosaceus* could reduce proteolysis and ammonia formation in caecal fermentation without interfering with the activity of normal caecal microflora (175). Furthermore, this strain could be used as a probiotic bacterium to maintain intestinal health. Pediocin producing *Pediococcus* strains could also control undesirable fermentation during silage fermentation (164).

6.6.2 *Class IIb:* Sakacins

Sakacins are class IIb bacteriocins produced by certain strains of *L. sake*. These small, cationic, hydrophobic peptides contain an N-terminal leader sequence of the double glycine type, and possess a strong antilisterial activity. Sakacin A was the first sakacin reported from *L. sake* Lb706, an isolate from raw meat (176). Subsequent sakacins discovered include sakacin M produced by *L. sake* 148, sakacin P by *L. sake* LTH673, sakacin 674 by *L. sake* Lb674 and *L. sake* Lb16, sakacin B by *L. sake* 251 and sakacin K by *L. sake* CTC 494 (177–182). All known sakacin producers of lactobacilli were predominantly isolated from meat sources.

6.6.2.1 Mechanism of Action

Bacterial cytoplasmic membrane is the primary target site for the antimicrobial activity of sakacins. The lethal action results from sakacin induced hydrophilic pore formation and membrane perturbation with a consequent leakage of vital cellular components as well as a depletion of proton motive force (5,23,183). Bactericidal effect is the antimicrobial outcome for sakacin types A, B, P and K, but only bacteriostatic for sakacin M (176–182). These antibacterial effects are influenced by several factors, including sakacin concentration and purity; testing buffer or broth; sensitivity of the indicator strain; and cell density of the target organism.

6.6.2.2 Antimicrobial Spectrum

Sakacins, as is common for class II bacteriocins, are highly inhibitory against *Listeria* species. (176–179,182,185,186). Furthermore, sakacins are characterized by a narrow inhibitory spectrum, being active against various strains of LAB, especially *Lactobacillus* species. (Table 6.5). The potent antilisterial activity and limited inhibitory spectrum toward LAB makes sakacin P, a promising bacteriocin to combat *Listeria* contamination of foods. Because sakacins do not inhibit the growth of meat starter cultures and certain strains of *Micrococci* and *S. carnosus*, they are suitable for practical use in the production of fermented meat products and to control meatborne pathogens *L. monocytogenes, E. faecalis* and *S. aureus* (176,177,179).

Table 6.5
Class-IIb Sakacin-types and their antimicrobial spectrum

Type	Susceptible Organism	Ref.
Sakacin-A	*L. monocytogenes, Lactobacillus* spp., *Pediococcus* spp., *Car. piscicola, Leu. paramesenteroides, Ent. faecium* and *Ent. faecalis*	(185)
Sakacin-B	*Lactobacillus* spp.and *Leuconostoc* spp.	(181)
Sakacin-K	*L. monocytogenes* and *L. innocua*	(182)
Sakacin-M	*L. monocytogenes, Staph. aureus, Lactobacillus* spp., *Leuconostoc* spp., *Carnobacterium* spp.	(177)
Sakacin-P	*L. monocytogenes, Leuconostoc, Carnobacterium, Ent. faecalis,* and *Br. thermosphacta*	(179)

6.6.2.3 Applications

The use of sakacin or sakacin producing LAB in foods, meat products in particular, has been under extensive evaluation (187–191). Besides the control of foodborne pathogens, incorporation of sakacins in foods could also inhibit spoilage bacteria such as the rope inducing, gas producing or flavor disturbing heterofermentative LAB. This characteristic could contribute to the overall sensory and organoleptic qualities of the food, when preserved with sakacins.

In meat technology, sakacin could be used as (partially) purified bacteriocin preparation or through *in situ* production by inoculating with sakacin producing strains. The latter application with nonfermented meat or meat products could result in suppression of growth of spoilage organisms or food pathogens (use as protective culture, for instance in fresh meat, high pH meat products, or minimally processed meat products like vacuum packed or modified atmosphere packaged meat). Furthermore, these sakacin producers may be used as starter cultures or cocultures in fermented meat products to control the fermentation process by inhibiting the growth of competing flora, including foodborne pathogens.

Applications of purified sakacin as food additive are currently not in practice due to the lack of approvals from regulatory agencies. Therefore, *in situ* applications as a food grade starter culture may be more suitable, because labeling is not necessary for sakacin producing LAB (192). Furthermore, it is likely that the starter flora of several fermented meat products already contain sakacin producers. Accordingly, there are several patents for the specific use of sakacin producing LAB as commercial starter cultures for meat fermentations (193,194).

6.7 FOOD APPLICATIONS

Biopreservation is defined as the use of antagonistic microorganisms and their metabolic products to inhibit or destroy undesired microorganisms in foods to enhance food safety and prolong shelf life. Bacteriocins of LAB are considered natural antimicrobial agents or biopreservatives that meet this definition. Biopreservation of foods with bacteriocins could be accomplished in three different ways: by inoculation of food with bacteriocin producing LAB; by direct addition of purified or semipurified bacteriocins to food; and by use of a product previously fermented with a bacteriocin producing LAB as an ingredient in another food.

6.7.1 Dairy Products

Nisin is the most commonly used bacteriocin in cheese manufacturing to inhibit the growth of both spoilage and pathogenic microorganisms, particularly *L. monocytogenes*. Addition of 2000 IU/g of nisin into cottage cheese increases the inactivation rate of *L. monocytogenes* by more than 3-log in 3 days (195). The growth of *L. monocytogenes* in ricotta-type cheeses is also significantly inhibited by addition of 100 IU/mL of nisin over 70 days or more at 6 to 8°C depending on the type of cheese (196). Shelf life of nisin containing pasteurized process cheese (301 and 387 IU nisin/g of cheese) is markedly enhanced over control cheese spreads (197). Nisin (100 and 300 IU/g) could decrease populations of *L. monocytogenes*, *S. aureus*, and *C. sporogenes* spores in cold processed and packed cheese spreads. Butyric acid fermentation by *Clostridium* species is one of the problems in cheese manufacturing; nisin could effectively control such spore forming *C. tyrobutyricum* (198).

Lacticin 3147 produced by *L. lactis* ssp. *lactis* DPC 3147 has a broad spectrum of antimicrobial activity for dairy applications. Ross et al. (199) described the control of cheddar cheese quality attributes by decreasing the population of nonstarter LAB at least 2-log after 6 months ripening with lacticin 3147 producing starter cultures. Nonstarter LAB failed to proliferate even after 6 months of ripening. Cottage cheese manufactured with *Lactobacillus lactis* DPC4275 reduced the population of *L. monocytogenes* by 3-log over a one week ripening period. Lacticin 3147 producing starter culture is an effective biopreservative for reducing listerial population on mold ripened cheese surfaces by 3-log (199). Nisin and lacticin 3147 are both produced by food grade LAB. However, the regulatory approval of the latter bacteriocin is still pending.

Dried preparation of pediocin PA-1/AcH from *P. acidilactici* PAC1.0 could effectively control *L. monocytogenes* growth in cottage cheese, half and half, and cheese sauce at 4°C for 14 d storage (200). Spraying of a pediocin PA-1/AcH producing cell suspension of *L. plantarum* WHE92 on Munster cheese prevented proliferation of *L. monocytogenes* during ripening for 21 d (201).

6.7.2 Meat Products

Naturally occurring microorganisms and their bacteriocins are widely used for enhancing food safety and shelf life of several meat products. Preservation of cooked ready to eat (RTE) foods and minimally processed refrigerated foods (MPRF) are the foremost candidates for such biopreservation.

Nisin (400 and 800 IU/mL) treatments alone and in combination with 2% NaCl were shown to inhibit growth of *L. monocytogenes* in minced raw buffalo meat (202). The inhibitory action of nisin was less pronounced when the storage temperature was increased to 37°C; however, the combination of 2% NaCl with nisin enhanced the bacteriocin efficacy.

Nisin has been used for sanitization of red meat carcass surfaces inoculated with *B. thermosphacta*, *C. divergens*, or *L. innocua* (203). Spray treatments with nisin (5000 IU/mL) decreased the populations 1.8 to 3.5 log CFU/cm^2 at day 0 and by 2.0 to 3.6-log CFU/cm^2 after one day storage at 4°C.

Combination of nisin (1000 or 10000 IU/ml) with modified atmosphere packaging (MAP) (100% CO_2, 80% CO_2 + 20% air) could effectively inhibit growth of *L. monocytogenes* and *Pseudomonas fragi* in cooked tenderloin pork (204). MAP/nisin combination was more effective at 4°C than at 20°C. The antilisterial activity of nisin A and pediocin AcH in fresh ground pork stored aerobically at 5°C has also been reported (205).

Nitrites are commonly used for stabilizing red meat color; inhibiting food spoilage and poisoning organisms, such as *C. botulinum*. However, nitrites and their derivatives the secondary amines, react with meats to form carcinogenic nitrosamines. Bacteriocins, as alternative additives, could reduce levels of nitrite for such meat applications. Combination of nisin (3000 IU/g) and nitrite (40 ppm) could totally inhibit growth of *C. botulinum* spores in meat slurries at 37°C for 56 days (206). Nisin (8000 IU/g) in combination with nitrite (60 ppm) also inhibits germination of clostridial spores in pork slurries at pH 5.5. Nisin (4000 IU/g) and nitrite (120 ppm) combination could delay toxin formation by *Clostridium* species in chicken frankfurter emulsions for 5 weeks (207).

Pediocin PA-1/AcH (1,350 AU/g or mL) could kill >99% of meat spoilage psychrotrophic *Leuconostoc mesenteroides* in ground beef, sausage mix, milk, and ice cream mix (208). Pediocin (3,000 AU/g) could also reduce the populations of *C. laramie*, *Lactobacillus* species and *Leuconostoc* species during storage at 4°C for 12 weeks (171,209). Incorporation of pediocin (from *P. acidilactici* PO_2) reduced the population and controlled growth of *L. cervatus* in a meat paste during 15 d storage at refrigerated temperature (170). The effectiveness of pediocin in controlling spoilage of salad dressing and salad caused by *L. Bifermentum,* and also in extending the shelf life of coleslaw, macaroni salad, and potato salad has been reported (210). A possible application of pediocin producing *P. parvatus* strains to control spoilage and pathogenic bacteria in fresh RTE vegetables has been suggested (211).

Efficacy of pediocin from *P. acidilactici* strains to control *L. monocytogenes* in processed meat products has been widely reported (208). In ground beef and sausage mix, pediocin could reduce populations of several strains of *L. monocytogenes* in a dose dependent manner. At a level of 1,350 AU/g the populations were reduced by 3.5 log cycles for strain $Ohio_2$, 3 log cycles for strain Scott A and 1.5 log cycles for strain CA. However, the resistant survivors multiplied during refrigerated storage. Pediocin application could also reduce populations of *L. monocytogenes* in fresh beef and beef byproducts (171,212). Pediocin (5,000 AU/g) could effectively reduce *S. aureus, L. monocytogenes, S. typhimurium* and *E. coli* O157:H7 in roast beef, Cotto salami, and summer sausage (172). Combination of this bacteriocin with hydrostatic pressure at 345 MPa at 50°C for 5 min further reduced the population compared to pressure treated products without pediocin (173,213).

Studies with pediocin producing strains of *P. acidilactici* strain JD 1-23 and PAC1.0 as starter cultures to produce dry fermented sausage showed potent antilisterial activity (214,215). Also, the pediocin PA-1/AcH producing *P. acidilactici* starter cultures when introduced in frankfurter packages along with *L. monocytogenes* cells and stored at 4°C for 60 days or more, caused marked reduction of this pathogen (216,217).

Sakacin K produced by *L. sake* CTC494 has been shown to inhibit *L. innocua* in vacuum packaged poultry breasts, cooked pork, and raw minced pork under MAP conditions (218). A bacteriocin-like substance produced by *C. piscicola* L103 completely inhibited the growth of *L. monocytogenes* in vacuum packaged meat after 14 days at 4°C (219). Lactocin 705 produced by *L. casei* CRL 705 inhibited the growth of *L. monocytogenes* in ground beef (220).

Other bacteriocins produced by LAB have also been examined to control the growth of *L. monocytogenes*. Enterocin CCM 4231 in dry fermented Hornád salami could decrease *L. monocytogenes* counts from 8 to 6.3-log immediately after addition of the bacteriocin (221).

6.7.3 Seafood Products

Several biopreservative applications for bacteriocins to protect various seafood products have been documented. Inhibitory effects of nisin and sakacin P against *L. monocytogenes* in vacuum packed cold smoked salmon stored at 10°C for 4 weeks has been reported (222). During a four week storage, addition of sakacin P-producing *L. sake* culture elicited a bacteriostatic effect against *L. monocytogenes.* However, a combination of purified sakacin P with the *L. sake* culture resulted in a potent listeriocidal outcome.

Efficacy of bacteriocin solutions (nisin and Microgard™) to reduce microbial counts, inhibit *L. monocytogenes*, and extend shelf life of cold smoked salmon was evaluated (223). Total aerobic counts on fresh chilled salmon were reduced by 2-log and the shelf life at 6°C was increased by 3 to 4 days with the combination of nisin and Microgard. This antimicrobial cocktail also reduced the growth of *L. monocytogenes* in frozen thawed salmon and increased the product shelf life (without affecting pH or color) from 5 to 10 days at 6°C.

Survival of *L. monocytogenes* in cold smoked salmon in the presence of nisin (500 or 1000 IU/g), CO_2 and low temperature was reduced by 1–2 logs (224). Antilisterial effect of a bacteriocin producing strain of *C. piscicola* A9b (bac$^+$) in model fish system and in vacuum packed cold smoked salmon has also been reported (120). The inhibition of *L. monocytogenes* seems due to deprivation of nutrients resulting from competitive growth of *C. piscicola* in the seafood product.

Efficacy of nisin Z and a preparation of crude bavaricin to prolong the shelf life of brined shrimp was reported (225). This bacteriocin combination has extended the seafood shelf life by 16 and 31 days. The addition of benzoate and sorbate solutions to the bacteriocins has further extended the shelf life of brined shrimp by 59 days.

The inhibition of *L. monocytogenes* and mesophilic aerobic bacteria in cold smoked rainbow trout by nisin (4000–6000 IU/mL) and sodium lactate (60%), alone or in combination has also been reported (226). Listerial counts in nisin and sodium lactate injected smoked trout samples decreased from 3.7 to 1.8 log CFU/g over 16 days at 8°C. Furthermore, the sensory characteristics of cold smoked rainbow trout remained unaffected by these treatments.

6.7.4 Canned Foods

Nisin is the only bacteriocin allowed in canned foods as a preservative. Nisin incorporation could shorten the heat processing time without compromising food safety and quality. Nisin has been used in canned potatoes, peas, mushrooms, soups, tomatoes, and cereal puddings. Amounts of nisin added to canned foods vary based on regulations of individual countries. In Europe, there is no specific limit for the amount of nisin for use in canned vegetables. In the United States, the amount of nisin in foods should not exceed 250 ppm (10000 IU/g). Because nisin is a relatively heat stable bacteriocin, it could be added to canned foods at levels of 100–200 IU/g to control thermophilic spore formers such as *B. stearothermophilus* and *C. tyrobutyricum* (227). For the control of acid tolerant spoilage microorganisms (i.e., *B. macerans* and *C. pasteurianum*), the enhanced activity of nisin at acidic pH makes it an ideal preservative in low pH foods such as canned tomatoes.

6.7.5 Beverages

Gram-negative bacteria, yeasts, and molds are insensitive to bacteriocins of LAB (228). Therefore, nisin could be advantageous in alcohol fermentations to control spoilage LAB without affecting growth and fermentative performance of brewing yeasts or compromising sensory characteristics of the product. Nisin has similar applications in wine making, except for those that require desirable malolactic fermentation. Malolactic fermentation by *Oenococcus oeni*, and various *Lactobacillus* and *Pediococcus* species converts L-malic acid to less sour L-lactic acid with CO_2 production. Nisin resistant strains of *O. oeni* with nisin could be used to control the progress of malolactic fermentation in wine making. Nisin is also used to reduce sulfite levels used in wine making to control spoilage bacteria (159).

6.8 "MULTIPLE HURDLE" APPLICATIONS

The hurdle concept comprises a combination of different preservation methods to inhibit or control microbial growth. In order to achieve the most effective combined treatments without adversely affecting the sensory and nutritional quality of food products, hurdle to hurdle interactions and careful integration of hurdles must be considered to avoid antagonism. Bacteriocins have been used as effective hurdles in combination with several food processing methods to improve the food safety (Table 6.6).

Table 6.6
Combination and synergistic application of nisin with other “hurdle” intervention systems

Combination System	Nisin	Effect	Ref.
Thermal processing			
55°C for 10 and 15 min exposure	100 IU/mL	Treatment for 6-h eliminated *E.coli* O157:H7	(229)
60°C for 5 min / 65°C for 2 min	25 mg/kg	Reduction of *L. monocytogenes* challenge by 3 to 5-logs in lobsters	(230)
55°C for 10 and 15 min exposure	500-2500 IU/mL	Enhanced inactivation of *Salmonella* Enteritidis PT4	(231)
Non-thermal processing			
High-pressure (155-400 MPa)	100 IU/mL	Enhanced bactericidal activity against selected foodborne pathogens	(232)
PEF (50 kV/cm; 32 pulse/2 µS)	10-100 IU/mL	Synergistic inhibition of *L. inocua* strains.	(233)
PEF (16.7 kV/cm; 50 pulse/2 µS)	0.06 µg/mL	Enhanced the inactivation of *Bacillus cereus*	(234)
Metal chelators			
EDTA (500 mg/kg)	500 mg/kg	Inhibited *E.coli O157:H7* and *Br. thermosphacta* in ham / bologna	(235)
20 mM EDTA at 37°C for 1-h	50 µg/mL	Reduction of salmonellae by 3.2 to 6.9-log with combination only.	(141)
EDTA/Citrate/Lactate	2000 IU/mL	Effective under MAP against *E.coli* O157:H7 and *S.* Typhimurium	(236)
Chemical interventions			
2% (w/w) Na-lactate or Na-citrate	500 IU/mL	Treatment for 15 min reduced *Arcobacter butzleri* from poultry skin	(237)
Monolaurin (250 µg/mL)	100 IU/mL	Inhibited *Bacillus cereus* growth in milk at 37°C for 5 days	(238)
5% Ethanol	10 IU/mL	Reduced *L. monocytogenes* by 3-logs at 5 to 37°C temp. ranges	(239)
Lactoperoxidase (LPO) and lysozyme			
LPO (10 enzyme units)	100 IU/mL	Effectively inhibited resident microflora in sardines	(240)
LPO (35 enzyme units)	100-200 IU/mL	Enhanced inhibitory activity against *L. monocytogenes*	(241)
LPO (0.2-0.8 ABTS units/mL)	10-100 IU/mL	Synergistic inhibition of *L. monocytogens* in UHT-processed milk	(242)
Lysozyme (25-125 µg/mL)	30 µg/mL	Synergistic inhibition of *E.coli O157:H* and *L. monocytogenes*	(243)
Plant extracts and essential oils			
0.1% carvacrol,/thymol/ eugenol	0.25 µg/mL	Synergistic inhibition of *L. monocytogens*	(244)
Bearberry ethanol extract (5 mg/mL)	500 µg/mL	Retarded the growth of *Brochothrix thermosphacta*	(245)
Garlic aqueous extract (0.5 mg/mL)	3 IU/mL	Synergistic inhibition of six *L. monocytogens* strains	(246)

6.8.1 Thermal Processing

Other preservative hurdles, such as heat treatments, low water activity, modified atmosphere, and low temperature could enhance the antimicrobial effects of nisin. Thermal synergism of nisin activity against *Lactobacillus* after exposure to a sequential treatment of mild heat (48–56°C) and then nisin (0.5 μg/mL) at 30°C has been observed (247). Combinations of low pH and high sodium chloride concentration have been shown to enhance nisin activity against *L. sake*, *S. aureus* and *L. monocytogenes* (248).

Duration of thermal inactivation time for bacteria could be reduced with bacteriocin combination resulting in improved food quality. The injury and inactivation kinetics of *S. enteritidis* PT4 in the presence of nisin (500 to 2500 IU/ml) in the media, liquid whole egg, or egg white could be reduced from the required pasteurization time by up to 35% (231). Nisin did not enhance the lethality of thermal processes and microbial cell injury was more severe in the egg white containing nisin, presumably as a result of bacteriocin interaction with antimicrobial factors in the egg white. Addition of nisin in cold pack lobster meat reduced the heat resistance of *L. monocytogenes* and significantly shortened the treatment time compared with thermal processing alone (230).

6.8.2 Nonthermal Food Processing

Food preservation with nonthermal food processing technologies, such as high pressure (HP) processing, ionizing radiation, pulsed electric field (PEF), and ultraviolet radiation is rapidly expanding with various product applications. Combination of these emerging food processing technologies with bacteriocins appears to be a promising antimicrobial hurdle mechanism. HP and PEF treatments induce sublethal injury to bacterial cells and enhance the activity of bacteriocins against *E. coli* O157:H7 and *S. typhimurium* (143). Furthermore, the HP inactivation of spores of *B. coagulans, B. subtilis*, and *C. sporogenes* is markedly enhanced with nisin treatment (249,250).

PEF has been successfully used with nisin for various food safety applications (234). Combination of low doses of nisin (0.06 mg/mL) and mild PEF (16.7 kV/cm, 50 pulses each of 2 mS duration) strongly reduced the numbers of *B. cereus* vegetative cells by more than the sum of reductions obtained with single treatments.

HP processing kills or injures microorganisms mainly by damaging the cell membrane, and such effect could potentiate the biopreservative activity of nisin that targets microbial membranes. A four dimensional process consisting of *Pressure x Time x Temperature x Antimicrobial* (such as nisin) has been suggested. Inclusion of nisin would allow lower pressure application with less impairment to food quality while maintaining a high kill level of food borne pathogens (164). *Listeria monocytogenes, E. coli,* and *S. typhimurium* when injured by these treatments turn more sensitive to nisin (143). Ultra HP inactivation of *E. coli* and *L. innocua* inoculated in liquid whole egg was even more effective by adding 1.25 and 5 mg/L nisin to the process (251). Hauben et al. (145) suggested that HP treated *E. coli* turned nisin sensitive due to pressure induced disruption of the bacterial outer membrane permeability. Evidence suggests that cells are least sensitive to HP between the temperature ranges of 20–30°C, however, bacteria turn highly sensitive at temperatures greater than 35°C due to phase transitions of membrane lipids (252–254). Increased cell death by HP processing (345 Mpa, 25 °C for 10 min) with combination of nisin and pediocin PA-1/ACH has been reported. Accordingly, *L. monocytogenes, S. aureus, E. coli* O157:H7 and *S. typhimurium* in the presence of such combined processing showed a total loss in viability ranging from 5–8 log cycles (255).

Combined effects of ultra HP treatment, heat, acidity, and nisin on *B. coagulans* spores has been reported (249). Nisin (0.8 IU/mL) in combination with 400 Mpa pressure in pH 4.0 buffer with a mild heat treatment (70 °C for 30 min) elicited ~6-log spore reduction.

6.8.3 Metal Chelators and Detergents

Resistance of Gram-negative bacteria to bacteriocins is largely due to the outer membrane as a permeability barrier. Combinations of two or more preservative systems, or hurdles, are better at preventing the emergence of less sensitive organisms, and reduce the extent of a single extreme system or preservative (256).

Nisin is effective against Gram-negative bacteria, if passed through the nisin impermeable outer membrane barrier, gaining access to its cytoplasmic targets. This could be achieved by disrupting the outer membrane with chelating agents (e.g., EDTA), organic acids, or treatments such as sublethal heat or freezing (141,142,144). Chelating agents remove divalent cations, such as Ca^{2+} and Mg^{2+}, from the Gram-negative cell walls, releasing phospholipids and lipoproteins and increasing cell wall permeability. Stevens et al. (257) found effective treatments that comprised 12–25μg nisin/g with 20 mM EDTA or citric monohydrate at 30–42°C. However, chelating agents are much less effective due to their preferential binding to free divalent ions present in the food matrix. Biochelators such as lactoferrin are milk antimicrobials that potentiate the activity of antimicrobials across outer membranes of Gram-negative bacteria (258).

A potent antimicrobial synergism between EDTA or lactoferrin with nisin against *L. monocytogenes, E. coli, S. enteritidis,* and *P. fluorescens* has been reported (243). Lactoferrin and EDTA enhanced the activity of nisin against two enterohemorrhagic *E. coli* strains. Combinations of nisin, lysozyme, and monolaurin with EDTA resulted in the killing of certain Gram-negative bacteria, whereas none of the antimicrobials alone demonstrated such bactericidal effect. Lactoferrin alone (2 mg/mL) did not inhibit any of the bacterial strains, but did enhance the nisin activity against *L. monocytogenes*.

A synergism between monolaurin and nisin against milk-borne *B. licheniformis* has been reported (238). A combination of nisin (0.75 μg/mL) and monolaurin (100 μg/mL) at pH 6.0 caused inhibition of spores and vegetative cells at 37°C for 7 days. A synergism between sucrose fatty acid esters (i.e., palmitate, stearate) and nisin was observed (259). Nisin activity could be stimulated by other neutral emulsifiers such as monolaurin, monooleate and Tween-80 (260).

6.8.4 Lactoperoxidase (LPO) System

Nisin synergism with the LPO system in the inactivation of *L. monocytogenes* was reported (242). Addition of nisin (10 or 100 IU/mL) to ultra high temperature (UHT) processed skim milk showed no effect on listerial counts after 24 h at 30°C. However, addition of LPO to this system resulted in 3-log reduction compared to control. Nisin combination with LPO elicited potent inhibition of *L. monocytogenes* by 5.6-log than the control. Such enhanced inactivation was also observed with Gram-negative bacteria, which otherwise are insensitive to nisin.

A novel method of using nisin against both Gram-negative and Gram-positive bacteria involves a treatment combining a solution of low concentration alkali metal orthophosphate (e.g., trisodium orthophosphate) with osmotic shock and lysozyme and nisin (261). Another synergistic combination, resulting in an antimicrobial effect against *Salmonella* species, has been described between nisin, LPO, thiocyanate, and hydrogen peroxide applied at 30–40°C at pH 3–5 has been described (262). The LPO thiocyanate hydrogen peroxide system occurs naturally in milk, and its antimicrobial effect is due to the generation of transient products of oxidation of thiocyanate by hydrogen peroxide catalyzed by LPO (263). Synergistic activity of LPO and nisin against *L. monocytogenes* in UHT skim milk has been reported to elicit a 5.6-log reduction of the pathogen compared to control (242).

6.8.5 Acid Interventions

Several commonly used acid preservatives have been reported to show synergism with nisin. Scannell et al. (264) found that 2% sodium lactate could potentiate nisin activity against *Salmonella kentucky* and *Staphylococcus aureus* in fresh pork sausages. In another study, an agar diffusion method was used to show that a combination of lactic acid and whey permeate fermented by a nisin producing *L. lactis* strain had a synergistic effect against fish isolates of *Micrococcus luteus*, *P. aeruginosa*, *P. fluorescens*, and *S. hominis* (265). Combinations of organic acids (citric acid, lactic acid, glucono delta lactone, or acetic acid) with nisin enhance inhibition of thermally stressed *Bacillus* spores, particularly under less acidic conditions and ambient storage (266). The combined treatment of a sorbate nisin mixture greatly reduced initial *L. monocytogenes* counts on vacuum packed and carbon dioxide packaged beef, and prevented growth during 4 weeks of refrigerated storage (267).

Synergistic effects between sodium diacetate and pediocin against *L. monocytogenes* in meat slurries have been reported (268). Listerial population increased from 4.5 to 8-log CFU/mL within 1 day at 25°C and within 14 days at 4°C in the control samples. A listeriocidal effect (~7-log CFU/mL difference compared to the control) was achieved in treatments containing pediocin (5000 AU/mL) with 0.5% diacetate at 25°C and pediocin with 0.3% diacetate at 4°C.

6.8.6 Plant Extracts and Essential Oils

Antibacterial activity of plant essential oils and their components has been recognized for a long time. Several applications have been developed using essential oils as natural food preservatives to control food spoilage organisms and pathogens (269). Combinations of nisin (0.25 μg/mL) and the essential oils of carvacrol, thymol, and eugenol (0.1%) have been shown to elicit a potent antimicrobial synergism to inhibit *L. monocytogenes* (244).

An antioxidant ethanolic extract of bearberry (*Arctostaphylos uva-ursi*) leaves when mixed with nisin demonstrated a strong inhibitory activity against 25 food related bacteria (245). This combination treatment with nisin (500 μg/mL) and bearberry extract (5 mg/mL) most notably has retareded the growth of *B. thermosphacta*, as compared to nisin alone.

Synergistic inhibition of six *L. monocytogenes* strains by nisin and aqueous extracts of garlic has also been reported (246). The synergistic interaction between nisin (3 IU/mL) and garlic extract (0.5 mg/mL) resulted in a strong bacteriostatic effect against *L. monocytogenes* in broth media, however, a potent bactericidal effect was observed at 4°C.

6.8.7 Bacteriocin Cocktails

Several mixtures of bacteriocins have also been tested in combination hurdles and found to enhance the biopreservation (270,271). The combined effects of nisin and leucocin F10 produced by *Leuconostoc carnosum* F10 against *L. monocytogenes* showed a potent inhibitory effect, and the addition of leucocin F10 seemed necessary to prevent the revival of *L. monocytogenes*.

Certain mixtures of antibiotics and nisin had an enhanced antibacterial effect in cheddar cheese (272). Nisin (12.5 μg/g curd), penicillin G (12 U/g curd), and streptomycin (12 mg/g curd) reduced nonstarter LAB by 4–5 log cycles compared with untreated control after 100–160 days.

6.9 FOOD PACKAGING APPLICATIONS

Incorporation of antimicrobial agents into food packaging systems is an innovative concept aptly undertaken in response to current consumer demands and market trends. Two methods are commonly used in the preparation of packaging films with bacteriocins: by direct incorporation into the polymer materials; and by coating the packaging films (273).

In the first method, nisin is incorporated into a polyethylene based plastic film commonly used in the vacuum packaging of beef carcasses. Nisin has retained its antimicrobial activity against *L. helveticus* and *B. thermosphacta* inoculated in carcass surface tissue sections (274). The population of *B. thermosphacta* was initially reduced by 2-log with nisin impregnated packaged beef within the first 2 days of storage at 4°C. *Brocothrix thermosphacta* populations from nisin impregnated plastic wrapped samples were significantly less than control after 20 days of refrigerated storage at 4 or 12°C.

Nisin was also incorporated into film forming edible hydroxypropylmethyl cellulose solution to inhibit the growth of *L. innocua* and *S. aureus* (275). However, stearic acid significantly reduced the inhibitory activity of nisin in this application due to the water vapor barrier properties of stearic acid. Further optimization is required to desorb nisin from the film and effectively diffuse the bacteriocin into the packaged food.

The second method for incorporation of bacteriocins into packaging films is to coat or adsorb bacteriocins onto polymer surfaces. Examples include methylcellulose coatings with nisin for poultry, and spray adsorption of nisin on various packaging materials such as polyethylene, ethylene vinyl acetate, polypropylene, polyamide, polyester, acrylics, and polyvinyl chloride was also reported (273).

Nisin coated onto silica surfaces seem to inhibit the growth of *L. monocytogenes* (276). Antimicrobial efficacy of cellulose based bioactive inserts and antimicrobial polyethylene and polyamide pouches incorporated with lacticin 3147 and nisin against *L. innocua* and *S. aureus* (along with *L. lactis* sp. *lactis*) has also been demonstrated (277). Nisin bound avidly to plastic film, and this bioactive film was stable for 3 months with or without refrigeration, and effective in reducing ~2-log bacteria in vacuum packed cheese in combination with MAP at refrigeration temperatures. Nisin incorporated cellulose based bioactive inserts, when placed between sliced products of cheese and ham under MAP, reduced growth of *L. innocua* by >3-log in cheese after 5 days at 4°C, and by ~1.5-log in sliced ham after 12 days; while *S. aureus* was reduced by 1.5- and 2.8-log in cheese and ham, respectively. The growth of *L. monocytogenes* in meats and poultry was completely inhibited by coating of pediocin onto cellulose casings and plastic bags during 12 weeks of storage at 4°C (278).

Nisin coated polymeric films such as PVC, linear low density polyethylene, and nylon inhibit proliferation of *S. typhimurium* on fresh broiler drumstick skin (279). Coating of solutions containing nisin, citric acid, EDTA, and Tween-80 on PVC, linear low density polyethylene, and nylon films reduced the counts of *S. typhimurium* in fresh broiler drumstick skin by 0.4 to 2.1-log after 24 h incubation at 4°C.

6.10 REGULATORY STATUS

Use of bacteriocin producing starter cultures as ingredients may not require special consideration in the USA if the microorganism is considered generally recognized as safe (GRAS) based on history of safe use prior to the 1958 Food Additives Amendment (190). For use of purified bacteriocin as a food preservative, the substance could be self affirmed as GRAS by the company according to the Code of Federal Regulations (U.S. Government

Printing Office, 1990), However, the US Food and Drug Administration (FDA) might require justification to such affirmation.

In 1969, based on the safety, tolerance, and efficacy data on nisin, the Food and Agriculture Organisation/World Health Organization (FAO/WHO) Expert Committee acknowledged that a level of 3.3×10^6 units/kg body weight had no adverse effect, and permitted an unconditional acceptable daily intake (ADI) to be set at 3.3×10^4 units/kg body weight (280). In 1988, The US FDA affirmed nisin as GRAS for use as a direct ingredient in human food. In the European Union, nisin is listed as E234, and may also be labeled as nisin preservative or natural preservative.

Pediocin PA-1, after nisin, is the most studied bacteriocin of LAB (174,281). Several scientific groups worldwide have recognized its potential as a biopreservative, especially for use in certain specific foods. However, pediocins have not yet been legally approved by the regulatory agencies in the USA, Europe, and other countries.

Several authors have outlined issues involved in the approval of new bacteriocins for food use, and the US FDA has published guidelines for the safety assessment of a new preservative (282). For approval to be granted, the bacteriocin must be chemically identified and characterized, and its use and efficacy must be proven. The manufacturing process must be described, and assays used for quantification and standardization of the peptide must be established. Finally, the toxicity data (safety and tolerance) and physiological turnover of the molecule after ingestion should be verified.

6.11 CONCLUSIONS

Several studies have advanced our knowledge on bacteriocins and their molecular properties, biosynthesis, isolation purification and mechanisms of antimicrobial activity. Genetic engineering or chemical modifications of bacteriocins to improve their functional properties have been the focus in the recent years, which has led to remarkable developments in bacteriocin technology. Accordingly, the solubility and stability of nisin Z has been improved by replacing *Asn-27* or *His-31* with lysine (283); the stability of pediocin PA-1 at 4°C and room temperature has been improved by replacing *Met-31* with alanine, isoleucine or leucine (284); and the development of a chimeric protein mutant of pediocin PA-1 displayed ~2.8-fold higher activity against an indicator strain, *L. plantarum* (166).

Despite isolation and characterization of a number of bacteriocins, only a few have demonstrated commercial potential in food application, and currently nisin is the only purified bacteriocin approved for food use in the USA. The use of pediocin PA-1 for food biopreservation has been commercially exploited and is covered by several US and European patents (56). A fermentate containing pediocin PA-1 (Alta™), is commercially available for use as a food preservative to inhibit pathogens, especially *L. monocytogenes*, in RTE meats (285). Lacticin 3147, which is active over a wider pH range than nisin, is another promising bacteriocin that might find applications in non-acid foods (199).

Because bacteriocins demonstrate relatively narrow spectrum of antimicrobial activity with limited effect against Gram-negative bacteria, these agents are now paving their way into hurdle concept technologies for food preservation. The simultaneous application of bacteriocins and non-thermal processing technologies, such as HP and PEF, to enhance food protection is highly attractive because products from these non-thermal interventions usually have better sensory and nutritional qualities. Because LAB and their metabolites have been consumed in high quantities by humans for centuries in cultured foods with no adverse effects, LAB continue to be the preferred source for food use bacteriocins, either in the form of purified compounds or as growth extracts. The potential application of bacteriocins from LAB, in particular, as natural food antimicrobial systems in minimally processed foods is steadily increasing with a high level consumer acceptance.

REFERENCES

1. Jacob, F., A. Lwoff, L. Siminovitch, E. Wallman. Definition de quelques termes relatifs a la Pysogenie. *Ann. Inst. Pasteur Paris* 84:222–224, 195.
2. Tagg, J.R., A.S. Dajani, L.W. Wannamaker. Bacteriocins of Gram-positive bacteria. *Bacteriol. Rev.* 40:722–756, 1976.
3. Cleveland, J., T.J. Montville, I.F. Nes, M.L. Chikindas. Bacteriocins: safe, natural antimicrobials for food preservation. *Int. J. Food Microbiol.* 71:1–20, 2001.
4. Gravesen, A., M. Ramnath, K.B. Rechinger, N. Andersen, L. Jansch, Y. Hechard, J.W. Hastings, S. Knochel. High-level resistance to class IIa bacteriocins is associated with one general mechanism in *Listeria monocytogenes*. *Microbiol.* 148:2361–2369, 2002.
5. Jack, R. W., J.R. Tagg, B. Ray. Bacteriocins of Gram-positive bacteria. *Microbiol. Rev.* 59:171–200, 1995.
6. James, R., C. Lazdunski, F. Pattus. *Bacteriocins, Micrococins and Lantibiotics*. New York: Springer-Verlag, 1992, p 519.
7. Riley, M.A. Molecular mechanisms of bacteriocin evolution. *Annu. Rev. Genet.* 32:255–278, 1998.
8. Joerger, R.D, D.G. Hoover, S.F. Barefoot, K.M. Harmon, D.A. Grinstead, C.G. Nettles-Cutter. Bacteriocins. In: *Encyclopedia of Microbiology,* Lederberg, J. ed., San Diego: Academic Press, 2000, pp 383–397.
9. Naidu, A.S., W.R. Bidlack, R.A. Clemens. Probiotic spectra of lactic acid bacteria (LAB). *Crit. Rev. Food Sci. Nutr.* 39:13–126, 1999.
10. Naidu, A.S., R.A. Clemens. Probiotics. In: *Natural Food Antimicrobial Systems*, Naidu, A.S., ed., Boca Raton, FL: CRC Press, 2000, pp 431–462.
11. Gilliland, S.E. Health and nutritional benefits from lactic acid bacteria. *FEMS Microbiol. Rev.* 87:175–178, 1990.
12. Sandine, W.E. Roles of bifidobacteria and lactobacilli in human health. *Contemp. Nutr.* 15:1, 1990.
13. Lindgren, S. E., W.J. Dobrogosz. Antagonistic activities of lactic acid bacteria in food and feed fermentations. *FEMS Microbiol. Rev.* 7:149–163, 1990.
14. Nettles, C.G., S.F. Barefoot. Biochemical and genetic characteristics of bacteriocins of food-associated lactic acid bacteria. *J. Food Prot.* 56:338–356, 1993.
15. Ray, B., M.A. Daeschel. *Food Biopreservatives of Microbial Origin*, Boca Raton, FL: CRC Press, 1992.
16. McAuliffe, O., R.P. Ross, C. Hill. Lantibiotics: structure, biosynthesis and mode of action. *FEMS Microbiol. Rev.* 25:285–308, 2001.
17. Nes, I.F., D.B. Diep, L.S. Havarstein, M.B. Brurberg, V. Eijsink, H. Holo. Biosynthesis of bacteriocins in lactic acid bacteria. *Antonie Van Leeuwenhoek* 70:113–128, 1996.
18. Papagianni, M. Robizomally synthesized peptides with antimicrobial properties, structure, function and applications. *Biotech. Adv.* 21:465–499, 2003.
19. Klaenhammer, T.R. Genetics of bacteriocins produced by lactic acid bacteria. *FEMS Microbiol. Rev.* 12:39–85, 1993.
20. de Vos, W.M., O.P. Kuipers, J.R. van der Meer, R.J. Siezen. Maturation pathway of nisin and other lantibiotics: posttranslationally modified antimicrobial peptides exported by Gram-positive bacteria. *Mol. Microbiol.* 17:427–437, 1995.
21. Sahl, H.-G., R.W. Jack, G. Bierbaum. Biosynthesis and biological activities of lantibiotics with unique posttranslational modifications. *Eur. J. Biochem.* 230:827–853, 1995.
22. Venema, K., G. Venema, J. Kok. Lactococcal bacteriocins: mode of action and immunity. *Trends Microbiol.* 3:299–304, 1995.

23. Abee, T. Pore-forming bacteriocins of Gram-positive bacteria and self-protection mechanisms of producer organisms. *FEMS Microbiol. Lett.* 129:1–10, 1995.
24. Nes, I.F., J.R. Tagg. Novel lantibiotics and their pre-peptides. *Antonie Van Leeuwenhoek* 69:89–97, 1996.
25. Chen, H., D.G. Hoover. Bacteriocins and their food applications. *Comp. Rev. Food Sci. Food Safety* 2:82–100, 2003.
26. Sahl, H-G., G. Bierbaum. Lantibiotics: biosynthesis and biological activities of uniquely modified peptides from Gram-positive bacteria. *Annu. Rev. Microbiol.* 52:41–79, 1998.
27. Quadri, L.E., M. Kleerebezem, O.P. Kuipers, W.M. de Vos, K.L. Roy, J.C. Vederas, M.E. Stiles. Characterization of a locus from *Carnobacterium piscicola* LV17B involved in bacteriocin production and immunity: evidence for global inducer-mediated transcriptional regulation. *J. Bacteriol.* 179:6163–6171, 1997.
28. Brurberg, M.B., I.F. Nes, V.G. Kijsink, J. Nissen-Meijer. Pheromone-induced production of antimicrobial peptides in *Lactobacillus*. *Mol. Microbiol.* 26:347–360, 1997.
29. Anderssen, E.L., D.B. Diep, I.F. Nes, V.G.H. Eijsink, J. Nissen-Meijer. Antagonistic activity of *Lactobacillus plantarum* C11: 2 new 2-peptide bacteriocins, plantaricins EF and JK, and the induction factor plantaricin A. *Appl. Environ. Microbiol.* 6:2269–2272, 1998.
30. Rauch, P.J.G., W.M. de Vos. Characterization of the novel nisin-sucrose conjugative transposon Tn5276 and its insertion in *Lactococcus lactis*. *J. Bacteriol.* 174:1280–1287, 1992.
31. Banerjee, S., J.N. Hansen. Structure and expression of a gene encoding the precursor of subtilin, a small protein antibiotic. *J. Biol. Chem.* 263:9508–9514, 1988.
32. Altena, K., A. Guder, C. Cramer, G. Bierbaum. Biosynthesis of the lantibiotic mersacidin: organization of a type B lantibiotic gene cluster. *Appl. Environ. Microbiol.* 66:2565–2571, 2000.
33. Worobo, R.W., M.J. van Belkum, M. Sailer, K.L. Roy, J.C. Vederas, M.E. Stiles. A signal peptide secretion-dependent bacteriocin from *Carnobacterium divergens*. *J. Bacteriol.* 177:3143–3149, 1995.
34. Axelsson, L., A. Holck. The genes involved in production of and immunity to sakacin A, a bacteriocin from *Lactobacillus sake* Lb706. *J. Bacteriol.* 177:2125–2137, 1995.
35. Dufour, A., A. Rince, P. Uguen, J.P. LePennec. IS1675, a novel lactococcal insertion element, forms a transposon-like structure including the lacticin 481 lantibiotic operon. *J. Bacteriol.* 182:5600–5605, 2000.
36. Bierbaum, G., F. Götz, A. Peschel, T. Kupke, M. van de Kamp, H.-G. Sahl. The biosynthesis of the lantibiotics epidermin, gallidermin, Pep5 and epilancin K7. *Antonie Van Leeuwenhoek* 69:119–127, 1996.
37. Geissler, S., F. Götz, T. Kupke. Serine protease EpiP from *Staphylococcus epidermidis* catalyzes the processing of the epidermin precursor peptide. *J. Bacteriol.* 178:284–288, 1996.
38. Buchmann, G.W., S. Banerjee, J.N. Hansen. Structure, expression, and evolution of a gene encoding the precursor of nisin, a small protein antibiotic. *J. Biol. Chem.* 263:16260–16266, 1988.
39. Mulders, J.W.M., I.J. Boerrigter, H.S. Rollema, R.J. Siezen, W.M. de Vos. Identification and characterization of the lantibiotic nisin Z, a structural nisin variant. *Eur. J. Biochem.* 201:581–584, 1991.
40. Klein, C., C. Kaletta, N. Schnell, K.D. Entian. Analysis of genes involved in biosynthesis of the lantibiotic subtilin. *Appl. Environ. Microbiol.* 58:132–142, 1992.
41. Rince, A., A. Dufour, P. Uguen, J.P. Le Pennec, D. Haras. Characterization of the lacticin 481 operon: the *Lactococcus lactis* genes lctF, lctE, and lctG encode a putative ABC transporter involved in bacteriocin immunity. *Appl. Environ. Microbiol.* 63:4252–4260, 1997.
42. Uguen, P., J.P. Le Pennec, A. Dufour. Lantibiotic biosynthesis: interactions between prelacticin 481 and its putative modification enzyme, LctM. *J. Bacteriol.* 182:5262–5266, 2000.
43. Holo, H., O. Nilssen, I.F. Nes. Lactococcin A, a new bacteriocin from *Lactococcus lactis* subsp. *cremoris*: isolation and characterization of the protein and its gene. *J. Bacteriol.* 173:3879–3887, 1991.

44. Stoddard, G.W., J.P. Petzel, M.J. van Belkum, J. Kok, L.L. McKay. Molecular analyses of the lactococcin A gene cluster from *Lactococcus lactis* subsp. *lactis* biovar *diacetylactis* WM4. *Appl. Environ. Microbiol.* 58:1952–1961, 1992.
45. van Belkum, M.J., J. Kok, G. Venema. Cloning, sequencing, and expression in *Escherichia coli* of lcnB, a third bacteriocin determinant from the lactococcal bacteriocin plasmid p9B4-6. *Appl. Environ. Microbiol.* 58:572–577, 1992.
46. Marugg, J.D., C.F. Gonzalez, B.S. Kunka, A.M. Ledeboer, M.J. Pucci, M.Y. Toonen, S.A. Walker, J.C. Zoetmulder, P.A. Vandenbergh. Cloning, expression, and nucleotide sequence of genes involved in production of pediocin PA-1, and bacteriocin from *Pediococcus acidilactici* PAC1.0. *Appl. Environ. Microbiol.* 58:2360–2367, 1992.
47. Motlagh, A.M., A.K. Bhunia, F. Szostek, T.R. Hansen, M.G. Johnson, B. Ray. Nucleotide and amino acid sequence of pap-gene (pediocin AcH production) in *Pediococcus acidilactici* H. *Lett. Appl. Microbiol.* 15:45–48, 1992.
48. Bukhtiyarova, M., R. Yang, B. Ray. Analysis of the pediocin AcH gene cluster from plasmid pSMB74 and its expression in a pediocin-negative strain. *Appl. Environ. Microbiol.* 60:3405–3408, 1994.
49. Venema, K., J. Kok, J.D. Marugg, M.Y. Toonen, A.M. Ledeboer, G. Venema, M.L. Chikindas. Functional analysis of the pediocin operon of *Pediococcus acidilactici* PAC 1.0: PedB is the immunity protein and PedD is the precursor processing enzyme. *Mol. Microbiol.* 17:515–522, 1995.
50. Diep, D.B., L.S. Havarstein, J. Nissen-Meyer, I.F. Nes. The gene encoding plantaricin A, a bacteriocin from *Lactobacillus plantarum* C11, is located on the same transcription unit as an agr-like regulatory system. *Appl. Environ. Microbiol.* 60:160–166, 1994.
51. Sablon, E., B. Contreras, E. Vandamme. Antimicrobial peptides of lactic acid bacteria: mode of action, genetics and biosynthesis. *Adv. Biochem. Eng. Biotechnol.* 68:21–60, 2000.
52. Guder, A., I. Wiedemann, H.-G. Sahl. Posttranslationally modified bacteriocins: the lantibiotics. *Biopolymers* 55:62–73, 2000.
53. van Kraaij, C., W.M. de Vos, R.J. Siezen, O.P. Kuipers. Lantibiotics: biosynthesis, mode of action and applications. *Nat. Prod. Rep.* 16:575–587, 1999.
54. Skaugen, M., C.I.M. Abildgaard, I.F. Nes. Organization and expression of a gene cluster involved in the biosynthesis of the lantibiotic lactocin S. *Mol. Gen. Genet.* 253:674–686, 1997.
55. Leer, R.J., J.M.B.M. van der Vossen, M. van Giezen, J.M. van Noort, P.H. Pouwels. Genetic analysis of acidocin B, a novel bacteriocin produced by *Lactobacillus acidophilus*. *Microbiology* 141:1629–1635, 1995.
56. Ennahar, S., T. Sashihara, K. Sonomoto, A. Ishizaki. Class IIa bacteriocins: biosynthesis, structure and activity. *FEMS Microbiol. Rev.* 24:85–106, 2000.
57. van der Meer, J.R., H.S. Rollema, R.J. Siezen, M.M. Beerthuyzen, O.P. Kuipers, W.M. de Vos. Influence of amino acid substitutions in the nisin leader peptide on biosynthesis and secretion of nisin by *Lactococcus lactis*. *J. Biol. Chem.* 269:3555–3562, 1994.
58. van Belkum, M.J., R.W. Worobo, M.E. Stiles. Double-glycine-type leader peptides direct secretion of bacteriocins by ABC transporters: colicin V secretion in *Lactococcus lactis*. *Mol. Microbiol.* 23:1293–1301, 1997.
59. Kuipers, O.P., P.G.G.A. de Ruyter, M. Kleerebezem, W.M. de Vos. Quorum sensing-controlled gene expression in lactic acid bacteria. *J. Biotechnol.* 64:15–21, 1998.
60. Stock, J.B., A.J. Ninfa, A.M. Stock. Protein phosphorylation and regulation of adaptive responses in bacteria. *Microbiol. Rev.* 53:450–490, 1989.
61. Parkinson, J.S. Signal transduction schemes of bacteria. *Cell* 73:857–871, 1993.
62. Kuipers, O.P., M.M. Beerthuyzen, P.G.G.A. de Ruyter, E.J. Luesink, W.M. de Vos. Autoregulation of nisin biosynthesis in *Lactococcus lactis* by signal transduction. *J. Biol. Chem.* 270:27299–27304, 1995.
63. Eijsink, V.G., M. Skeie, P.H. Middelhoven, M.B. Brurberg, I.F. Nes. Comparative studies of class IIa bacteriocins of lactic acid bacteria. *Appl. Environ. Microbiol.* 64:3275–3281, 1998.

64. Diep, D.B., L.S. Havarstein, I.F. Nes. A bacteriocin-like peptide induces bacteriocin synthesis in *L. plantarum* C11. *Mol. Microbiol.* 18:631–639, 1995.
65. Ingram, L.C. Synthesis of the antibiotic nisin: formation of lanthionine and β-methyllanthionine. *Biochim. Biophys. Acta.* 184:216–219, 1969.
66. Ingram, L.C. A ribosomal mechanism of synthesis for peptides related to nisin. *Biochim. Biophys. Acta.* 224:263–265, 1970.
67. van der Meer, J.R., J. Polman, M.M. Beerthuyzen, R.J. Siezen, O.P. Kuipers, W.M. de Vos. Characterization of the *Lactococcus lactis* nisin A operon genes nisP, encoding a subtilisin-like serine protease involved in precursor processing, and nisR, encoding a regulatory protein involved in nisin biosynthesis. *J. Bacteriol.* 175:2578–2588, 1993.
68. Fath, M.J., R. Kolter. ABC transporters: bacterial exporters. *Microbiol. Rev.* 57:995–1017, 1993.
69. Siegers, K., S. Heinzmann, K.D. Entian. Biosynthesis of lantibiotic nisin: posttranslational modification of its prepeptide occurs at a mutimeric membrane-associated lanthionine synthetase complex. *J. Biol. Chem.* 271:12294–12301, 1996.
70. Kiesau, P., U. Eikmanns, Z. Gutowski-Eckel, S. Weber, M. Hammelmann, K.D. Entian. Evidence for a multimeric subtilin synthetase complex. *J. Bacteriol.* 179:1475–1481, 1997.
71. Reis, M., M. Eschbach-Bludau, M.I. Iglesias-Wind, T. Kupke, H.-G. Sahl. Producer immunity toward the lantibiotic Pep5: identification of the immunity gene pepI and localization and functional analysis of its gene product. *Appl. Environ. Microbiol.* 60:2876–2883, 1994.
72. Siegers, K., K.D. Entian. Genes involved in immunity to the lantibiotic nisin produced by *Lactococcus lactis* 6F3. *Appl. Environ. Microbiol.* 61:1082–1089, 1995.
73. Peschel, A., F. Gotz. Analysis of the *Staphylococcus epidermidis* genes epi-F, -E, and -G involved in epidermin immunity. *J. Bacteriol.* 178:531–536, 1996.
74. Saris, P.E., T. Immonen, M. Reis, H.-G. Sahl. Immunity to lantibiotics. *Antonie Van Leeuwenhoek* 69:151–159, 1996.
75. Klein, C., K.D. Entian. Genes involved in self-protection against the lantibiotic subtilin produced by *Bacillus subtilis* ATCC6633. *Appl. Environ. Microbiol.* 60:2793–2801, 1994.
76. Abee, T., Krockel, L., C. Hill. Bacteriocins: modes of action and potentials in food preservation and control of food poisoning. *Int. J. Food Microbiol.* 28:169–185, 1995.
77. Quadri, L.E., M. Sailer, M.R. Terebiznik, K.L. Roy, J.C. Vederas, M.E. Stiles. Characterization of the protein conferring immunity to the antimicrobial peptide carnobacteriocin B2 and expression of carnobacteriocin B2 and BM1. *J. Bacteriol.* 177:1144–1151, 1995.
78. Abdel-Dayem, M., Y. Fleury, G. Devilliers, E. Chaboisseau, R. Girard, P. Nicolas, A. Delfour. The putative immunity protein of the Gram-positive bacteria *Leuconostoc mesenteroides* is preferentially located in the cytoplasm compartment. *FEMS Microbiol. Lett.* 138:251–259, 1996.
79. Nissen-Meyer, J., L.S. Havarstein, H. Holo, K. Sletten, I.F. Nes. Association of the lactococcin A immunity factor with the cell membrane: purification and characterization of the immunity factor. *J. Gen. Microbiol.* 139:1503–1509, 1993.
80. Venema, K., R.E. Haverkort, T. Abee, A.J. Haandrikman, K.J. Leenhouts, L. de Leij, G. Venema, J. Kok. Mode of action of LciA, the lactococcin A immunity protein. *Mol. Microbiol.* 14:521–532, 1994.
81. Nes, I.F., H. Holo. Class II antimicrobial peptides from lactic acid bacteria. *Biopolymers* 55:50–61, 2000.
82. Gravesen, A., A.M.J. Axelsen, J.M. da Silva, T.B. Hansen, S. Knochel. Frequency of bacteriocin resistance development and associated fitness costs in *Listeria monocytogenes*. *Appl. Environ. Microbiol.* 68:756–764, 2002.
83. Carolissen-Mackay, V., G. Arendse, J.W. Hastings. Purification of bacteriocins from lactic acid bacteria: problems and pointers. *Int. J. Food Microbiol.* 34:1–16, 1997.
84. Venema, K., M.L. Chikindas, J.F.M.L. Seegers, A.J. Haandrikman, K.J. Leenhouts, G. Venema, J. Kok. Rapid and efficient purification method for small, hydrophobic, cationic

bacteriocins: purification of lactococcin B and pediocin PA-1. *Appl. Environ. Microbiol.* 63:305–309, 1997.

85. Suarez, A.M., J.I. Azcona, J.M. Rodriguez, B. Sanz, P.E. Hernandez. One-step purification of nisin A by immunoaffinity chromatography. *Appl. Environ. Microbiol.* 63:4990–4992, 1997.
86. Cintas, L.M., P. Casaus, M.F. Fernández, P.E. Hernández. Comparative antimicrobial activity of enterocin L50, pediocin PA-1, nisin A and lactocin S against spoilage and food-borne pathogenic bacteria. *Food Microbiol.* 15:289–298, 1998.
87. Yang, R., M.C. Johnson, B. Ray. Novel method to extract large amounts of bacteriocins from lactic acid bacteria. *Appl. Environ. Microbiol.* 58:3355–3359, 1992.
88. Daba, H., C. Lacroix, J. Huang, R.E. Simard, L. Lemieux. Simple method of purification and sequencing of a bacteriocin produced by *Pediococcus acidilactici* UL5. *J. Appl. Bectroil.* 77:682–688, 1994.
89. Van't Hul, J.S., W.R. Gibbons. Concentration and recovery of the bacteriocin nisin from *Lactococcus lactis* subsp. *lactis*. *Biotechnol. Appl. Biochem.* 24:251–256, 1996.
90. Stoffels, G., H.-G. Sahl, A. Gudmundsdottir. Carnocin UI49, a potential biopreservative produced by *Carnobacterium piscicola*: large scale purification and activity against various Gram-positive bacteria including *Listeria* sp. *Int. J. Food Microbiol.* 20:199–210, 1993.
91. Chinachoti, N., T. Zaima, H. Matsusaki, K. Sonomoto, A. Ishisaki. Relationship between fermentative production and aeration condition using *Lactococcus lactis* IO-1. *J. Fac. Agric. Kyushu Univ.* 43:421–436, 1997.
92. Wan, J., J. Gordon, M.W. Hickey, R.F. Mawson, M.J. Coventry. Adsorption of bacteriocins by ingestible silica compounds. *J. Appl. Bacteriol.* 81:167–173, 1996.
93. Coventry, M.J., J.B. Gordon, M. Alexander, M.W. Hickey, J. Wan. A food-grade process for isolation and partial purification of bacteriocins of lactic acid bacteria that uses diatomite calcium silicate. *Appl. Environ. Microbiol.* 62:1764–1769, 1996.
94. Boyaval, P., P. Bhugaloo-Vial, F. Duffes, A. Metivier, X. Dousset, D. Marion. Production of concentrated bacteriocin solutions in high cell density bioreactors. *Lait.* 78:129–133, 1998.
95. Mayr-Harting, A., A.J. Hedges, R.C.W. Berkeley. Methods for studying bacteriocins. In: *Methods in Microbiology*, Vol. 7A, Norris, J.R., D.W. Ribbons, eds., New York: Academic Press, 1972, pp 315–422.
96. Piddock, L.J.V. Techniques used for the determination of antimicrobial resistance and sensitivity in bacteria. *J. Appl. Bacteriol.* 68:307–318, 1990.
97. Parente, E., A. Ricciardi. Production, recovery and purification of bacteriocins from lactic acid bacteria. *Appl. Microbiol. Biotechnol.* 52:628–638, 1999.
98. Sabine, D.B. An antibiotic-like effect of *Lactobacillus acidophilus*. *Nature* 199:811, 1963.
99. Kekessy, D.A., J.D. Piguet. New method for detecting bacteriocin production. *Appl. Microbiol.* 20:282–283, 1970.
100. Lindgren, S., G. Clevstrom. Antibacterial activity of lactic acid bacteria, 2: activity in vegetable silages, Indonesian fermented foods and starter cultures. *Swed. J. Agric. Res.* 8:67–73, 1978.
101. Hoover, D.G., K.J. Dishart, M.A. Hermes. Antagonistic effect of *Pediococcus* spp. against *Listeria monocytogenes*. *Food Biotechnol.* 3:183–196, 1989.
102. Rose, N.L., P. Sporns, L.M. McMullen. Detection of bacteriocins by matrix-assisted laser desorption/ionization time-of-flight mass spectrometry. *Appl. Environ. Microbiol.* 65:2238–2242, 1999.
103. Rodriguez, J.M., L.M. Cintas, P. Casaus, A. Suarez, P.E. Hernandez. PCR detection of the lactocin S structural gene in bacteriocin-producing lactobacilli from meat. *Appl. Environ. Microbiol.* 61:2802–2805, 1995.
104. Garde, S., E. Rodriguez, P. Gaya, M. Medina, M. Nunez. PCR detection of the structural genes of nisin Z and lacticin 481 in *Lactococcus lactis* subsp. *lactis* INIA 415, a strain isolated from raw milk Manchego cheese. *Biotechnol. Lett.* 23:85–89, 2001.

105. Rodriguez, E., M.I. Martinez, M. Medina, P.E. Hernandez, J.M. Rodriguez. Detection of enterocin AS-48-producing dairy enterococci by dot-blot and colony hybridization. *J. Dairy Res.* 65:143–148, 1998.
106. Martinez-Bueno, M., M. Maqueda, A. Galvez, B. Samyn, J. van Beeumen, J. Coyette, E. Valdivia. Determination of the gene sequence and the molecular structure of the enterococcal peptide antibiotic AS-48. *J. Bacteriol.* 176:6334–6339, 1994.
107. Joosten, H.M.L., E. Rodrieguez, M. Nunez. PCR detection of sequences similar to the AS-48 structural gene in bacteriocin-producing enterococci. *Lett. Appl. Microbiol.* 24:40–42, 1997.
108. Mugochi, T., M.P. Nandakumar, R. Zvauya, B. Mattiasson. Bioassay for the rapid detection of bacteriocins in fermentation broth. *Biotechnol. Lett.* 23:1243–1247, 2001.
109. Moll, G.N., W.N. Konings, A.J. Driessen. Bacteriocins: mechanism of membrane insertion and pore formation. *Antonie Van Leeuwenhoek* 76:185–198, 1999.
110. van Belkum, M. J., M.E. Stiles. Nonlantibiotic antibacterial peptides from lactic acid bacteria. *Nat. Prod. Rep.* 17:323–335, 2000.
111. Garneau, S., N.I. Martin, J.C. Vederas. Two-peptide bacteriocins produced by lactic acid bacteria. *Biochimie* 84:577–592, 2002.
112. Motlagh, A., M. Bukhtiyarova, B. Ray. Complete nucleotide sequence of pSMB 74, a plasmid encoding the production of pediocin AcH in *Pediococcus acidilactici*. *Lett. Appl. Microbiol.* 6:305–312, 1994.
113. Gallagher, N.L.F., M. Sailer, W.P. Niemczura, T.T. Nakashima, M.E. Stiles, J.C. Vederas. Three-dimensional structure of leucocin A in trifluoroethanol and dodecylphosphocholine micelles: spatial location of residues critical for biological activity in type IIa bacteriocins from lactic acid bacteria. *Biochemistry* 36:15062–15072, 1997.
114. Marciset, O., M.C. Jeronimus-Stratingh, B. Mollet, B. Poolman. Thermophilin 13, a nontypical antilisterial poration complex bacteriocin that functions without a receptor. *J. Biol. Chem.* 272:14277–14284, 1997.
115. Allison, G.E., C. Fremaux, T.R. Klaenhammer. Expansion of bacteriocin activity and host range upon complementation of 2 peptides encoded within the lactacin F operon. *J. Bacteriol.* 176:2235–2241, 1994.
116. Stephens, S.K., B. Floriano, D.P. Cathcart, S.A. Bayley, V.F. Witt, R. Jimenez-Diaz, P.J. Warner, J.L. Ruiz-Barba. Molecular analysis of the locus responsible for production of plantaricin S, a two-peptide bacteriocin produced by *Lactobacillus plantarum* LPCO10. *Appl. Environ. Microbiol.* 64:1871–1877, 1998.
117. Nissen-Meyer, J., H. Holo, L.S. Havarsteuin, K. Sletten, I.F. Nes. A novel bacteriocin whose activity depends on the complementary action of two peptides. *J. Bacteriol.* 174:5686–5692, 1992.
118. Martinez, M.I., E. Rodriguez, M. Medina, P.E. Hernandez, J.M. Rodriguez. Detection of specific bacteriocin-producing lactic acid bacteria by colony hybridization. *J. Appl. Microbiol.* 84:1099–1103, 1998.
119. Joerger, M.C., T.R. Klaenhammer. Characterization and purification of helveticin J and evidence for a chromosomally determined bacteriocin produced by *Lactobacillus helveticus* 481. *J. Bacteriol.* 167:439–446, 1986.
120. Nilsson, L., L. Gram, H.H. Huss. Growth control of *Listeria monocytogenes* on cold-smoked salmon using a competitive lactic acid bacteria flora. *J. Food Protect.* 62:336–342, 1999.
121. Oscariz, J. C., A.G. Pisabarro. Classification and mode of action of membrane-active bacteriocins produced by Gram-positive bacteria. *Int. Microbiol.* 4:13–19.
122. Jimenez-Diaz, R., J.L. Ruiz-Barba, D.P. Cathcart, H. Holo, I.F. Nes, K.H. Sletten, P.J. Warner. Purification and partial amino acid sequence of plantaricin S, a bacteriocin produced by *Lactobacillus plantarum* LPCO10, the activity of which depends on the complementary action of 2 peptides. *Appl. Environ. Microbiol.* 61:4459–4463, 1995.
123. Suma, K., M.C. Misra, M.C. Varadaraj. Plantaricin LP84, a broad-spectrum heat-stable bacteriocin of *Lactobacillus plantarum* NCIM 2084 produced in a simple glucose broth medium. *Int. J. Food Microbiol.* 40:17–25, 1998.

124. Mattick, A. T. R., A. Hirsch. Further observations on an inhibitory substance (nisin) from lactic streptococci. *Lancet* 2:5–7, 1947.
125. Ruhr, E., H.-G. Sahl. Mode of action of the peptide antibiotic nisin and influence on the membrane potential of whole cells and on cytoplasmic and artificial membrane vesicles. *Antimicrob. Agents Chemother.* 27:841–845, 1985.
126. Abee, T., F.M. Rombouts, J. Hugenholtz, G. Guibard, L. Letellier. Mode of action of nisin Z against *Listeria monocytogenes* Scott A grown at high and low temperatures. *Appl. Environ. Microbiol.* 60:1962–1968, 1994.
127. Henning, S., R. Metz, W.P. Hammes. Studies on the mode of action of nisin. *Int. J. Food Microbiol.* 3:121–134, 1986.
128. Breukink, E., C. van Kraaij, R.A. Demel, R.J. Siezen, O.P. Kuipers, B. de Kruijff. The C-terminal region of nisin is responsible for the initial interaction of nisin with the target membrane. *Biochemistry* 36:6968–6976, 1997.
129. Garcia-Garcera, M.J., M.G.L. Elferink, A.J.M. Driessen, W.N. Konings. *In vitro* pore-forming activity of the lantibiotic nisin. Role of proton motive force and lipid composition. *Eur. J. Biochem.* 212:417–422, 1993.
130. Driessen, A.J., H.W. van den Hoven, W. Kaiper, M. van den Kamp, H.S. Sahl, R.N.H. Konings, W.N. Konings. Mechanistic studies of lantibiotic-induced permeabilization of phospholipid vesicles. *Biochemistry* 34:1606–1614, 1995.
131. Demel, R.A., T. Peelen, R.J. Siezen, B. de Kruijff, O.P. Kuipers. Nisin Z, mutant nisin Z and lacticin 481 interactions with anionic lipids correlate with antimicrobial acitvity: a monolayer study. *Eur. J. Biochem.* 235:267–274, 1996.
132. Benz, R., G. Jung, H.-G. Sahl. Mechanism of channel formation by lantibiotics in black lipid membranes. In: *Nisin and Novel Lantibiotics*, Jung, G., H.G. Sahl, eds., Leiden: Escom, 1991, pp 359–372.
133. Sahl, H.-G. Pore formation in bacterial membranes by cationic lantibiotics. In: *Nisin and Novel Lantibiotics,* Jung, G., H.G. Sahl, eds., Leiden: Escom, 1991, pp 347–358.
134. Maisnier-Patin, S., N. Deschamps, S.R. Tatini, J. Richard. Inhibition of *Listeria monocytogenes* in camembert cheese made with a nisin-producing starter. *Lait* 72:249–263, 1992.
135. Moll, G.N., J. Clark, W.C. Chan, B.W. Bycroft, G.C.K. Roberts, W.N. Konings, A.J.M. Driessen. Role of transmembrane pH gradient and membrane binding in nisin pore formation. *J. Bacteriol.* 179:135–140, 1997.
136. Ojcius, D.M., J.D.E. Young. Cytolytic pore-forming proteins and peptides: is there a common structural motif? *Trends Biochem. Sci.* 16:225–229, 1991.
137. Giffard, C.J., H.M. Dodd, N. Horn, S. Ladha, A.R. Mackie, A. Parr, M.J. Gasson, D. Sanders. Structure-function relations of variant and fragment nisins studied with model membrane systems. *Biochemistry* 36:3802–3810, 1997.
138. Linnett, P. E., J.L. Strominger. Additional antibiotic inhibitors of peptidoglycan synthesis. *Antimicrob. Agents Chemother*. 4:231–236, 1973.
139. Bierbaum, G., H.-G. Sahl. Induction of autolysis of *Staphylococcus simulans* 22 by Pep5 and nisin and influence of the cationic peptides on the activity of the autolytic enzymes. In: *Nisin and Novel Lantibiotics,* Jung, G., H.G. Sahl, eds., Leiden: Escom, pp. 386–396, 1991.
140. Schved, F., Y. Henis, B.J. Juven. Response of spheroplasts and chelator-permeabilized cells of Gram-negative bacteria to the action of the bacteriocins pediocin SJ-1 and nisin. *Int. J. Food Microbiol.* 21:305–314, 1994.
141. Stevens, K.A., B.W. Sheldon, N.A. Klapes, T.R. Klaenhammer. Nisin treatment for the inactivation of *Salmonella* species and other Gram-negative bacteria. *Appl. Environ. Microbiol.* 57:3613–3615, 1991.
142. Kalchayanand, N., M.B. Hanlin, B. Ray. Sublethal injury makes Gram-negative and Gram-positive bacteria sensitive to the bacteriocins, pediocin AcH, and nisin. *Letts. Appl. Microbiol.* 15:239–243, 1992.
143. Kalchayanand, N., T. Sikes, C.P. Dunne, B. Ray. Hydrostatic pressure and electroporation have increased bactericidal efficiency in combination with bacteriocins. *Appl. Environ. Microbiol.* 60:4174–4177, 1994.

144. Delves-Broughton, J. The use of EDTA to enhance the efficacy of nisin towards Gram-negative bacteria. *Int. Biodet. Biodeg.* 32:87–97, 1993.
145. Hauben, K.J.A., E.C. Wuytack, G.C.F. Soontjens, C.W. Michiels. High pressure transient sensitization of *Escherichia coli* to lysozyme and nisin by disruption of outer membrane permeability. *J. Food Prot.* 59:350–355, 1996.
146. Schved, F., M.D. Pierson, B.J. Juven. Sensitization of *Escherichia coli* to nisin by maltol and ethyl maltol. *Lett. Appl. Microbiol.* 22:189–191, 1996.
147. Dielbandhoesing, S.K., H. Zhang, L.H.P. Caro, J.M. van der Vaart, F.M. Klis, C.T. Verrips, S. Brul. Specific cell wall proteins confer resistance to nisin upon yeast cells. *Appl. Environ. Microbiol.* 64:4047–4052, 1998.
148. Gupta, R.K., D.N. Prasad. Antibiotic activity of nisin in food preservation: a review. *Microbiol. Alim. Nutr.* 7:199–208, 1989.
149. Meghrous, J., C. Lacroix, R.E. Simard. The effects on vegetative cells and spores of three bacteriocins from lactic acid bacteria. *Food Microbiol.* 16:105–114, 1999.
150. Ramseier, H.R. The action of nisin on *Clostridium butyricum*. *Arch. Mikrobiol.* 37:57–94, 1960.
151. Jarvis, B. Resistance to nisin and production of nisin-inactivating enzymes by several *Bacillus* species. *J. Gen. Microbiol.* 47:33–48, 1967.
152. Benkerroum, N., W.E. Sandine. Inhibitory action of nisin against *Listeria monocytogenes*. *J. Dairy Sci.* 71:3237–3245, 1988.
153. Schillinger, U., H.-S. Chung, W.H. Holzapfel. Use of bacteriocinogenic lactic acid bacteria to inhibit spontaneous nisin-resistant mutants of *Listeria monocytogenes* Scott A. *J. Appl. Microbiol.* 85:657–663, 1998.
154. Hirsch, A. The assay of the antibiotic nisin. *J. Gen. Microbiol.* 4:70–83, 1950.
155. Harris, L.J., H.P. Fleming, T.R. Klaenhammer. Sensitivity and resistance of *Listeria monocytogenes* ATCC 19115, Scott A, and UAL500 to nisin. *J. Food Prot.* 52:384–387, 1991.
156. Breuer, B., F. Radler. Inducible resistance against nisin in *Lactobacillus casei*. *Arch. Microbiol.* 165:114–118, 1996.
157. Gänzle, M.G., S. Weber, W.P. Hammes. Effect of ecological factors on the inhibitory spectrum and activity of bacteriocins. *Int. J. Food Microbiol.* 46:207–217, 1999.
158. Fowler, G. G., B. McCann. The growing use of nisin in the dairy industry. *Austr. J. Dairy Technol.* 26:44–46, 1971.
159. Thomas, L.V., M.R. Clarkson, J. Delves-Broughton. Nisin. In: *Natural Food Antimicrobial Systems*, Naidu, A.S., ed., Boca Raton, FL: CRC Press, 2000, pp 463–524.
160. Padgett, T., I. Han, P.L. Dawson. Incorporation of food-grade antimicrobial compounds into biodegradable packaging films. *J. Food Prot.* 61:1330–1335, 1998.
161. Ennahar, S., N. Deschamps, J. Richard. Natural variation in susceptibility of *Listeria* strains to class IIa bacteriocins. *Curr. Microbiol.* 41:1–4, 2000.
162. Bhunia, A.K., M.C. Johnson, B. Ray. Purification, characterization, and antimicrobial spectrum of a bacteriocin produced by *Pediococcus acidilactici*. *J. Appl. Bacteriol.* 65:261–268, 1988.
163. Bhunia, A.K., M.C. Johnson, B. Ray, N. Kalchayanand. Mode of action of pediocin AcH from *Pediococcus acidilactici* H on sensitive bacterial strain. *J. Appl. Bacteriol.* 70:25–33, 1991.
164. Ray, B. *Pediococcus* in fermented foods. In: *Food Biotechnology Microorganisms*, Hui, Y.H., G.G. Khachatourians, eds., New York: VCH Publishers, 1995, pp 745–796.
165. Montville, T.J., Y. Chen. Mechanistic action of pediocin and nisin: recent progress and unresolved questions. *Appl. Microbiol. Biotechnol.* 50:511–519, 1998.
166. Miller, K.W., R. Schamber, O. Osmanagaoglu, B. Ray. Isolation and characterization of pediocin AcH chimeric protein mutants with altered bactericidal activity. *Appl. Environ. Microbiol.* 64:1197–2005, 1998.
167. Daeschel, M.A., T.R. Klaenhammer. Association of a 13.6-megadalton plasmid in *Pediococcus pentosaceus* with bacteriocin activity. *Appl. Environ. Microbiol.* 50:1538–1541, 1985.
168. Schved, F., A. Lalazar, Y. Henis, B.J. Juven. Purification, partial characterization and plasmid-linkage of pediocin SJ-1, a bacteriocin produced by *Pediococcus acidilactici*. *J. Appl. Bacteriol.* 74:67–77, 1993.

169. Jager, K., S. Harlander. Characterization of a bacteriocin from *Pediococcus acidilactici* PC and comparison of bacteriocin-producing strains using molecular typing procedures. *Appl. Microbiol. Biotechnol.* 37:631–637, 1992.
170. Coventry, M.J., K. Muirhead, M.W. Hickey. Partial characteristics of pediocin PO_2 and comparison with nisin for biopreservation of meat products. *Int. J. Food Microbiol.* 26:133–145, 1995.
171. Ray, B. Pediocin(s) of *Pediococcus acidilactici* as a food biopreservative. In: *Food Biopreservatives of Microbial Origin*, Ray, B., M.A. Daeschel, eds., Boca Raton, FL: CRC Press, 1992, pp 265–322.
172. Kalchayanand, N., A. Sikes, C.P. Dunne, B. Ray. Interaction of hydrostatic pressure, time and temperature of pressurization and pediocin AcH on inactivation of foodborne bacteria. *J. Food Prot.* 61:425–431, 1998.
173. Kalchayanand, N., A. Sikes, C.P. Dunne, B. Ray. Factors influencing death and injury of foodborne pathogens by hydrostatic pressure pasteurization. *Food Microbiol.* 15:207–214, 1998.
174. Ray, B., K.W. Miller. Pediocin. In: *Natural Food Antimicrobial Systems*, Naidu, A.S., ed., Boca Raton, FL: CRC Press, pp 525–566, 2000.
175. Piva, A., E. Meola, A. Panciroll. Effect of *Pediococcus pentosaceus* FBB61, pediocin A producer strain, in caecal fermentations. *J. Appl. Bacteriol.* 78:616–610, 1995.
176. Schillinger, U., F.K. Lucke. Antibacterial activity of *Lactobacillus sake* isolated from meat. *Appl. Environ. Microbiol.* 55:1901–1906, 1989.
177. Sobrino, O.J., J.M. Rodriguez, W.L. Moreira, M.F. Fernandez, B. Sanz, P.E. Hernandez. Antibacterial activity of *Lactobacillus sake* isolated from dry fermented sausages. *Int. J. Food Microbiol.* 13:1–10, 1991.
178. Sobrino, O.J., J.M. Rodriguez, W.L. Moreira, M.F. Fernandez, B. Sanz, P.E. Hernandez. Sakacin M, a bacteriocin-like substance from *Lactobacillus sake* 148. *Int. J. Food Microbiol.* 16:215–225, 1992.
179. Tichaczek, P.S., J.N. Meyer, I.F. Nes, R.F. Vogel, W.P. Hammes. Characterization of the bacteriocins curvacin A from *Lactobacillus curvatus* LTH1174 and sakacin P from *L. sake* LTH673. *Syst. Appl. Microbiol.* 15:460-468, 1992.
180. Holck, A., L. Axelsson, K. Hühne, L. Kröckel. Purification and cloning of sakacin 674, a bacteriocin from *Lactobacillus sake* Lb674. *FEMS Microbiol. Lett.* 115:143–149, 1994.
181. Samelis, J., S. Roller, J. Metaxopoulos. Sakacin B, a bacteriocin produced by *Lactobacillus sake* isolated from Greek dry fermented sausages. *J. Appl. Bacteriol.* 76:475–486, 1994.
182. Hugas, M., M. Garriga, M.T. Aymerich, J.M. Monfort. Inhibition of *Listeria* in dry fermented sausages by the bacteriocinogenic *Lactobacillus sake* CTC 494. *J. Appl. Bacteriol.* 79:322–330, 1995.
183. Montville, T.J., M.E. Bruno. Evidence that dissipation of proton motive force is a common mechanism of action for bacteriocins and other antimicrobial proteins. *Int. J. Food Microbiol.* 24:53–74, 1994.
184. Rodríguez, J.M., O.J. Sobrino, W.L. Moreira, M.F. Fernández, L.M. Cintas, P. Casaus, B. Sanz, P.E. Hernández. Inhibition of *Listeria monocytogenes* by *Lactobacillus sake* strains of meat origin. *Meat Sci.* 38:17–26, 1994.
185. Lewus, C.B., A. Kaiser, T.J. Montville. Inhibition of food-borne bacterial pathogens by bacteriocins from lactic acid bacteria isolated from meat. *Appl. Environ. Microbiol.* 57:1683–1688, 1991.
186. Motlagh, A.M., M.C. Johnson, B. Ray. Viability loss of foodborne pathogens by starter culture metabolites. *J. Food Prot.* 54:873–884, 1991.
187. Stiles, M.E., J.W. Hastings. Bacteriocin production by lactic acid bacteria: potential for use in meat preservation. *Trends Food Sci. Technol.* 2:247–251, 1991.
188. Eckner, K.F. Bacteriocins and food applications. *Dairy Food Environ. Sanit.* 12:204–209, 1992.
189. McMullen, L.M., M.E. Stiles. Potential use of bacteriocin-producing lactic acid bacteria in the preservation of meats. *J. Food Prot.* 55:64–71, 1996.
190. Muriana, P.M. Bacteriocins for control of *Listeria* spp. in food. *J. Food Prot.* 55:54–63, 1996.

191. Hugas, M. Biopreservation of meat and meat products. *Actes du Colloque Lactic 97*, Caen, France, 1997, pp 213–227.
192. Holzapfel, W.H., Geisen, R., U. Schillinger. Biological preservation of foods with reference to protective cultures, bacteriocins and food grade enzymes. *Int. J. Food Microbiol.* 24:343–362, 1995.
193. Hugas, M., M. Garriga, J.M. Monfort, J. Ylla. Bacteriocin from *Enterococcus faecium* active against *Listeria monocytogenes*, European Patent Office, EP0705843, 1996.
194. Tichaczek, P.S., S.B. Pohle, R.F. Vogel, W.P. Hammes. Use of bacteriocin-producing organisms to cure raw sausage. European Patent Office, EP0640291, 1995.
195. Ferreira, M.A.S.S., B.M. Lund. The effect of nisin on *Listeria monocytogenes* in culture medium and long-life cottage cheese. *Lett. Appl. Microbiol.* 22:433–438, 1996.
196. Davies, E.A., H.E. Bevis, J. Delves-Broughton. The use of the bacteriocin, nisin, as a preservative in ricotta-type cheeses to control the food-borne pathogen *Listeria monocytogenes*. *Lett. Appl. Microbiol.* 24:343–346, 1997.
197. Zottola, E.A., T.L. Yezzi, D.B. Ajao, R.F. Roberts. Utilization of cheddar cheese containing nisin as an antimicrobial agent in other foods. *Int. J. Food Microbiol.* 24:227–238, 1994.
198. Schillinger, U., R. Geisen, W.H. Holzapfel. Potential of antagonistic microorganisms and bacteriocins for the biological preservation of foods. *Trends Food Sci. Technol.* 7:158–164, 1996.
199. Ross, R.P., M. Galvin, O. McAuliffe, S.M. Morgan, M.P. Ryan, D.P. Twomey, W.J. Meaney, C. Hill. Developing applications for lactococcal bacteriocins. *Antonie Van Leeuwenhoek* 76:337–346, 1999.
200. Pucci, M.J., E.R. Vedamuthu, B.S. Kunka, P.A. Vandenbergh. Inhibition of *Listeria monocytogenes* by using bacteriocin PA-1 produced by *Pediococcus acidilactici* PAC1.0. *Appl. Environ. Microbiol.* 54:2349–2353, 1988.
201. Ennahar, S., D. Assobhel, C. Hasselmann. Inhibition of *Listeria monocytogenes* in a smear-surface soft cheese by *Lactobacillus plantarum* WHE92, a pediocin AcH producer. *J. Food Prot.* 61:186–191, 1998.
202. Pawar, D.D., S.V.S. Malik, K.N. Bhilegaonkar, S.B. Barbuddhe. Effect of nisin and its combination with sodium chloride on the survival of *Listeria monocytogenes* added to raw buffalo meat mince. *Meat Sci.* 56:215–219, 2000.
203. Cutter, C.N., G.R. Siragusa. Decontamination of beef carcass tissue with nisin using a pilot scale model carcass washer. *Food Microbiol.* 11:481–489, 1994.
204. Fang, T.J., L.W. Lin. Growth of *Listeria monocytogenes* and *Pseudomonas fragi* on cooked pork in a modified atmosphere packaging/nisin combination. *J. Food Prot.* 57:479–485, 1994.
205. Murray, M., J.A. Richard. Comparative study of the antilisterial activity of nisin A and pediocin AcH in fresh ground pork stored aerobically at 5°C. *J. Food Prot.* 60:1534–1540, 1997.
206. Rayman, K., B. Aris, A. Hurst. Nisin: a possible alternative or adjunct to nitrite in the preservation of meats. *Appl. Environ. Microbiol.* 41:375–380, 1981.
207. Taylor, S.L., E.B. Somers, L.A. Krueger. Antibotulinal effectiveness of nisin-nitrite combinations in culture medium and chicken frankfurter emulsions. *J. Food Prot.* 48:234-239, 1985.
208. Motlagh, A.M., S. Holla, M.C. Johnson, B. Ray, R.A. Field. Inhibition of *Listeria* spp. in sterile food systems by pediocin AcH, a bacteriocin produced by *Pediococcus acidilactici* H. *J. Food Prot.* 55:337–343, 1992.
209. Rozbeh, M., N. Kalchayanand, R.A. Field, M.C. Johnson, B. Ray. The influence of biopreservatives on the bacterial level of refrigerated vacuum-packaged beef. *J. Food Safety* 13:99–111, 1993.
210. Gonzalez, C.F. Methods for inhibiting bacterial spoilage and resulting composition. U.S. Patent 4883673, 1989.
211. Bennik, M.H.J., E.J. Smid, L.G.M. Gorris. Vegetable-associated *Pediococcus parvulus* produces pediocin PA-1. *Appl. Environ. Microbiol.* 63:2074–2076, 1997.
212. Nielsen, J.W., J.S. Dickson, J.D. Crouse. Use of bacteriocin produced by *Pediococcus acidilactici* to inhibit *Listeria monocytogenes* associated with fresh meat. *Appl. Environ. Microbiol.* 56:2142–2145, 1990.

213. Kalchayanand, N., B. Ray, A. Sikes, C.P. Dunne. Enhancement of safety of processed meat products by hydrostatic pressure in combination with temperature and bacteriocins. *Proceedings, 44th International Congress of Meat Science and Technology,* August 30 to September 4, 1998, Barcelona, Spain, Inst. Food Ag. Res. Technol. (IRTA), Spain, pp 522–523.
214. Berry, E.D., M.B. Liewen, R.W. Mandiog, R.W. Hutkins. Inhibition of *Listeria monocytogenes* by bacteriocin-producing *Pediococcus* during the manufacture of fermented semidry sausage. *J. Food Prot.* 53:194–197, 1990.
215. Foegeding, P.M., A.B. Thomas, D.H. Pilkington, T.R. Klaenhammer. Enhanced control of *Listeria monocytogenes* by *in situ*-produced pediocin during dry fermented sausage production. *Appl. Environ. Microbiol.* 58:884–890, 1992.
216. Berry, E.D., R.W. Hutkins, R.W. Mandigo. The use of bacteriocin-producing *Pediococcus acidilactici* to control postprocessing *Listeria monocytogenes* contamination of frankfurters. *J. Food Prot.* 54:681–686, 1991.
217. Degnan, A.J., A.E. Yousef, J.B. Luchansky. Use of *Pediococcus acidilactici* to control *Listeria monocytogenes* in temperature-abused vacuum-packaged wieners. *J. Food Prot.* 55:98–103, 1992.
218. Hugas, M., F. Pagés, M. Garriga, J.M. Monfort. Application of the bacteriocinogenic *Lactobacillus sakei* CTC494 to prevent growth of *Listeria* in fresh and cooked meat products packed with different atmospheres. *Food Microbiol.* 15:639–650, 1998.
219. Schöbitz, R., T. Zaror, O. León, M. Costa. A bacteriocin from *Carnobacterium piscicola* for the control of *Listeria monocytogenes* in vacuum-packaged meat. *Food Microbiol.* 16:249–255, 1999.
220. Vignolo, G., S. Fadda, M.N. de Kairuz, A.A.R. Holgado, G. Oliver. Control of *Listeria monocytogenes* in ground beef by 'Lactocin 705', a bacteriocin produced by *Lactobacillus casei* CRL 705. *Int. J. Food Microbiol.* 29:397–402, 1996.
221. Lauková, A., S. Czikková, S. Laczková, P. Turek. Use of enterocin CCM 4231 to control *Listeria monocytogenes* in experimentally contaminated dry fermented Hornád salami. *Int. J. Food Microbiol.* 52:115–119, 1999.
222. Katla, T., T. Møretrø, I.M. Aasen, A. Holck, L. Axelsson, K. Naterstad. Inhibition of *Listeria monocytogenes* in cold smoked salmon by addition of sakacin P and/or live *Lactobacillus sakei* cultures. *Food Microbiol.* 18:431–439, 2001.
223. Zuckerman, H., R.B. Abraharn. Quality improvement of kosher chilled poultry. *Poult. Sci.* 81:1751–1757, 2002.
224. Nilsson, L., H.H. Huss, L. Gram. Inhibition of *Listeria monocytogenes* on cold-smoked salmon by nisin and carbon dioxide atmosphere. *Int. J. Food Microbiol.* 38:217–227, 1997.
225. Einarsson, H., H.L. Lauzon. Biopreservation of brined shrimp (*Pandalus borealis*) by bacteriocins from lactic-acid bacteria. *Appl. Environ. Microbiol.* 61:669–676, 1995.
226. Nykänen, A., K. Weckman, A. Lapveteläinen. Synergistic inhibition of *Listeria monocytogenes* on cold-smoked rainbow trout by nisin and sodium lactate. *Int. J. Food Microbiol.* 61:63–72, 2000.
227. Rilla, N., B. Martinez, T. Delgado, A. Rodriguez. Inhibition of *Clostridium tyrobutyricum* in Vidiago cheese by *Lactococcus lactis* ssp. *lactis* IPLA 729, a nisin Z producer. *Int. J. Food Microbiol.* 85:23–33, 2003.
228. Soomro, A.H., T. Masud, K. Anwaar. Role of lactic acid bacteria (LAB) in food preservation and human health: a review. *Pakistan J. Nutr.* 1:20–24, 2002.
229. Lee, J.-I., H.-J. Lee, M.-H. Lee. Synergistic effect of nisin and heat treatment on the growth of *Escherichia coli* O157:H7. *J. Food Prot.* 65(2):408–410, 2002.
230. Budu–Amoako, E., R.F. Ablett, J. Harris, J. Delves-Broughton. Combined effect of nisin and moderate heat on destruction of *Listeria monocytogenes* in cold-pack lobster meat. *J. Food Prot.* 62:46–50, 1999.
231. Boziaris, I.S., L. Humpheson, M.R. Adams. Effect of nisin on heat injury and inactivation of Salmonella *Enteritidis* PT4. *Int. J. Food Microbiol.* 43:7–13, 1998.
232. Masschalck, B., R.V. Houdt, C.W. Michiels. High pressure increases bactericidal activity and spectrum of lactoferrin, lactoferricin and nisin. *Int. J. Food. Microbiol.* 64:325–332, 2001.

233. Calderón-Miranda, M.L., G.V. Barbosa-Cánovas, B.G. Swanson. Inactivation of *Listeria innocua* in skim milk by pulsed electric fields and nisin. *Int. J. Food Microbiol.* 51:19–30, 1999.
234. Pol, I.E., H.C. Mastwijk, P.V. Bartels, E.J. Smid. Pulsed-electric field treatment enhances the bactericidal action of nisin against *Bacillus cereus*. *Appl. Environ. Microbiol.* 66:428–430, 2000.
235. Gill, A.O., R.A. Holley. Inhibition of bacterial growth on ham and bologna lysozyme, nisin and EDTA. *Food Res. Int.* 33:83–90, 2000.
236. Cutter, C.N., G. Siragusa. Population reductions of Gram-negative pathogens following treatments with nisin and chelators under various conditions. *J. Food Prot.* 58:977-983, 1995.
237. Long, C., C.A. Phillips. The effect of sodium citrate, sodium lactate and nisin on the survival of *Arcobacter butzleri* NCTC 12481 on chicken. *Food Microbiol.* 20:495–502, 2003.
238. Mansour, M., J.B. Millière. An inhibitory synergistic effect of a nisin-monolaurin combination on *Bacillus* spp. vegetative cells in milk. *Food Microbiol.* 18:87–94, 2001.
239. Brewer, R., M.R. Adams, S.F. Park. Enhanced inactivation of *Listeria monocytogenes* by nisin in the presence of ethanol. *Lett. Appl. Microbiol. 34:18–21,* 2002.
240. Elotmani, F., O. Assobhei. *In vitro* inhibition of microbial flora of fish by nisin and lactoperoxidase system. *Lett. Appl. Microbiol. 38:60–65,* 2004.
241. Boussouel, N., F. Mathieu, A.M. Revol-Junelles, J.B. Milliere. Effects of combinations of lactoperoxidase system and nisin on the behaviour of *Listeria monocytogenes* ATCC 15313 in skim milk. *Int. J. Food Microbiol.* 61:169–175, 2000.
242. Zapico, P., M. Medina, P. Gaya, M. Nuñez. Synergistic effect of nisin and the lactoperoxidase system on *Listeria monocytogenes* in skim milk. *Int. J. Food Microbiol.* 40:35–42, 1998.
243. Branen, J.K., P.M. Davidson. Enhancement of nisin, lysozyme, and monolaurin antimicrobial activities by ethylenediaminetetraacetic acid and lactoferrin. *Int. J. Food Microbiol.* 90:63–74, 2004.
244. Yamazaki, K., T. Yamamoto, Y. Kawai, N. Inoue. Enhancement of antilisterial activity of essential oil constituents by nisin and diglycerol fatty acid ester. *Food Microbiol.* 21:283–289, 2004.
245. Dykes, G.A., R. Amarowicz, R.B. Pegg. Enhancement of nisin antibacterial activity by a bearberry (*Arctostaphylos uva-ursi*) leaf extract. *Food Microbiol.* 20:211–216, 2003.
246. Bhurinder, S., M.B. Falahee, M.R. Adams. Synergistic inhibition of *Listeria monocytogenes* by nisin and garlic extract. *Food Microbiol.* 18:133–139, 2001.
247. Ueckert, J.E., P.F. ter Steeg, P.J. Cole. Synergistic antibacterial action of heat in combination with nisin and magainin II amide. *J. Appl. Microbiol.* 85:487–494, 1998.
248. Blom, H., T. Katla, B.F. Hagen, L. Axelsson. A model assay to demonstrate how intrinsic factors affect diffusion of bacteriocins. *Int. J. Food Microbiol.* 38:103–109, 1997.
249. Roberts, C.M., D.G. Hoover. Sensitivity of *Bacillus coagulans* spores to combinations of high hydrostatic pressure, heat, acidity and nisin. *J. Appl. Bacteriol.* 81:363–368, 1996.
250. Stewart, C.M., C.P. Dunne, A. Sikes, D.G. Hoover. Sensitivity of spores of *Bacillus subtilis* and *Clostridium sporogenes* PA 3679 to combinations of high hydrostatic pressure and other processing parameters. *Innov. Food Sci. Emerg. Technol.* 1:49–56, 2000.
251. Ponce, E., R. Pla, E. Sendra, B. Guamis, M. Mor-Mur. Combined effect of nisin and high hydrostatic pressure on destruction of *Listeria innocua* and *Escherichia coli* in liquid whole egg. *Int. J. Food Microbiol.* 43:15–19, 1998.
252. Carlez, A., J.-C. Cheftel, J.P. Rosec, N. Richard, J.-L. Saldana, C. Balny. Effects of high pressure and bacteriostatic agents on the destruction of *Citrobacter freundii* in minced beef muscle. In: *High Pressure and Biotechnology,* Vol. 224, Balny, C., R. Hayashi, K. Heremand, P. Masson, eds., London: Colloque Inserm/John Libbey and Co., 1992, pp 365–368.
253. Ludwig, H., C. Bieler, K. Hallbauer, W. Scigalla. Inactivation of microoorganisms by hydrostatic pressure. In: *High Pressure and Biotechnology,* Vol 224, Balny, C., R. Hayashi, K. Heremand, P. Masson, eds., London: Colloque Inserm/John Libbey and Co., 1992, pp 365–368.
254. Kalchayanand, N., T. Sikes, C.P. Dunne, B. Ray. Viability loss kinetics of food spoilage and pathogenic bacteria at a moderate hydrostatic pressure. In: *Activities Report of the R & D*

Associates, San Antonio, TX: Research and Development Associates for Military Food and Packaging Systems, Inc., 1996, pp 331–341.
255. Kalchayanand, N., A. Sikes, C.P. Dunne, B. Ray. Interaction of hydrostatic pressure, time and temperature of pressurization and pediocin AcH on inactivation of foodborne bacteria. *J. Food. Prot.* 61:425–431, 1998.
256. Leistner, L. Basic aspects of food preservation by hurdle technology. *Int. J. Food Microbiol.* 55:181–186, 2000.
257. Stevens, K.A., B.W. Sheldon, N.A. Klapes, T.R. Klaenhammer. Effects of treatment conditions on nisin inactivation of Gram-negative bacteria. *J. Food Prot.* 55:763–766, 1992.
258. Naidu, A.S., R.R. Arnold. Lactoferrin interaction with salmonellae potentiates antibiotic susceptibility *in vitro*. *Diagn. Microbiol. Infect. Dis.* 20:68–75, 1994.
259. Thomas, L.V., E.A. Davies, J. Delves-Broughton, J.W. Wimpenny. Synergist effect of sucrose fatty acid esters on nisin inhibition of gram-positive bacteria. *J. Appl. Microbiol.* 85:1013–1022, 1998.
260. Jung, D.-S., F.W. Bodyfelt, M.A. Daeschel. Influence of fat and emulsifiers on the efficacy of nisin in inhibiting *Listeria monocytogenes* in fluid milk. *J. Dairy Sci.* 75:387–393, 1992.
261. Miles, R.J., C.A. Cassar, A.M. da Silva Carneiro de Melo. Bacterial decontamination method. *International Patent No.* WO 97/23136, 1997.
262. Bycroft, N.L., G.S. Byng, S.R. Good. Synergistic antimicrobial compositions. US Patent No. 5043176, 1991.
263. Naidu, A.S. Lactoperoxidase. In: *Natural Food Antimicrobial Systems*, Naidu A.S., ed. Boca Raton, FL: CRC Press, 2000, pp. 103-132.
264. Scanell, A.G.M., C. Hill, D.J. Buckley, E.K. Arendt. Determination of the influence of organic acids and nisin on shelf-life and microbiological safety aspects of fresh pork. *J. Appl. Microbiol.* 83:407–412, 1997.
265. Nykänen, A., S. Vesanen, H. Kallio. Synergistic antimicrobial effect of nisin whey permeate and lactic acids on microbes isolated from fish. *Lett. Appl. Microbiol.* 27:345–348, 1998.
266. Oscroft, C.A., J.G. Banks, S. McPhee. Inhibition of thermally-stressed Bacillus spores by combinations of nisin, pH and organic acids. *Lebensmittel Wissenschaft Technol.* 23:538–544, 1990.
267. Avery, S.M., S. Buncic. Antilisterial effects of a sorbate-nisin combination *in vitro* and on packaged beef at refrigeration temperature. *J. Food Prot.* 60:1075–1080, 1997.
268. Schlyter, J.H., K.A. Glass, J. Loeffelholz, A.J. Degnan, J.B. Luchansky. The effects of diacetate with nitrite, lactate, or pediocin on the viability of *Listeria monocytogenes* in turkey slurries. *Int. J. Food Microbiol.* 19:271–281, 1993.
269. Davidson, P.M., S. Naidu. Phyto-phenols. In: *Natural Food Antimicrobial Systems*, Naidu, A.S., ed., Boca Raton, FL: CRC Press, 2000, pp 265–294.
270. Hanlin, M.B., N. Kalchayanand, P. Ray, B. Ray. Bacteriocins of lactic acid bacteria in combination have greater antibacterial activity. *J. Food Prot.* 56:252–255, 1993.
271. Parente, E., M.A. Giglio, A. Ricciardi, F. Clementi. The combined effect of nisin, leucocin F10, pH, NaCl and EDTA on the survival of *Listeria monocytogenes* in broth. *Int. J. Food Microbiol.* 40:65–75, 1998.
272. Walsh, E.M., P.L.H. McSweeney, P.F. Fox. Use of antibiotics to inhibit non-starter lactic acid bacteria in cheddar cheese. *Int. Dairy J.* 6:425–431, 1996.
273. Appendini, P., J.H. Hotchkiss. Review of antimicrobial food packaging. *Innov. Food Sci. Emerg. Technol.* 3:113–126, 2002.
274. Siragusa, G.R., C.N. Cutter, J.L. Willett. Incorporation of bacteriocin in plastic retains activity and inhibits surface growth of bacteria on meat. *Food Microbiol.* 16:229–235, 1999.
275. Coma, V., I. Sebti, P. Pardon, A. Deschamps, F.H. Pichavant. Antimicrobial edible packaging based on cellulosic ethers, fatty acids, and nisin incorporation to inhibit *Listeria innocua* and *Staphylococcus aureus*. *J. Food Prot.* 64:470–475, 2001.
276. Bower, C., J. McGuire, M. Daeschel. Suppression of *Listeria monocytogenes* colonization following adsorption of nisin onto silica surfaces. *Appl. Environ. Microbiol.* 61:992–997, 1995.

277. Scannell, A.G.M., C. Hill, R.P. Ross, S. Marx, W. Hartmeier, E.K. Arendt. Development of bioactive food packaging materials using immobilised bacteriocins Lacticin 3147 and Nisaplin (R). *Int. J. Food Microbiol.* 60:241–249, 2000.
278. Ming, X., G. Weber, J. Ayres, W. Sandine. Bacteriocins applied to food packaging materials to inhibit *Listeria monocytogenes* on meats. *J. Food Sci.* 62:413–415, 1997.
279. Natrajan, N., B.W. Sheldon. Efficacy of nisin-coated polymer films to inactivate *Salmonella typhimurium* on fresh broiler skin. *J. Food Prot.* 63:1189–1196, 2000.
280. FAO/WHO Expert Committee on Food Additives. *Specifications for identity and purity of some antibiotics*. 12th Report, WHO Technical Report Series, No. 430, 1969.
281. Ray, B., D.G. Hoover. Pediocins. In: *Bacteriocins of Lactic Acid Bacteria*, Hoover, D.G., L.R. Steenson, eds., San Diego, California: Academic Press, Inc., 1993, pp 108–210.
282. Post, R.C. Regulatory perspective of the USDA on the use of antimicrobials and inhibitors in foods. *J. Food Prot.* (Suppl.):78–81, 1996.
283. Rollema, H.S., O.P. Kuipers, P. Both, W.M. de Vos, R.J. Siezen. Improvement of solubility and stability of the antimicrobial peptide nisin by protein engineering. *Appl. Environ. Microbiol.* 61:2873–2878, 1995.
284. Johnsen, L., G. Fimland, V. Eijsink, J. Nissen-Meyer. Engineering increased stability in the antimicrobial peptide pediocin PA-1. *Appl. Environ. Microbiol.* 66:4798–4802, 2000.
285. Rodriguez, J.M., M.I. Martinez, J. Kok. Pediocin PA-1, a wide-spectrum bacteriocin from lactic acid bacteria. *Crit. Rev. Food Sci. Nutr.* 42:91–121, 2002.

3.07

Genetic Characterization of Antimicrobial Peptides

Haijing Hu, Matthew M. Moake, and Randy W. Worobo

CONTENTS

7.1 INTRODUCTION

To survive in the natural environment and compete with other microorganisms for resources, bacteria produce antimicrobial compounds to inhibit or kill other competing strains. It is assumed that all bacteria produce some type(s) of antimicrobial compound(s), and that we need only find the right indicator strains and growth conditions to identify these compounds (1). The antimicrobial peptides represent just one of the many classes of antimicrobial compounds. These proteinaceous molecules often contain novel amino acid configurations and modifications when compared to normal ribosomally synthesized polypeptides (2). These compounds are typically broken up into two groups, the bacteriocins, and the peptide antibiotics, based upon their biosynthetic mechanisms.

Bacteriocins are ribosomally synthesized compounds produced by bacteria that are active against closely related bacteria (3). Most bacteriocins consist of 20–60 amino acids that may or may not undergo post translational modifications. As hydrophobic or amphiphilic compounds, most bacteriocins target cell membranes, while some also inhibit the biosynthesis of biopolymers (e.g., some colicins), or enzyme activities (e.g., duramycin).

The peptide antibiotics are nonribosomally synthesized and instead are formed through complex, stepwise condensation reactions that utilize large, nonribosomal peptide synthetases (NRPS) (4,5). These compounds often have nonprotein amino acids, including D-amino acids, hydroxy acids, or other unusual constituents and demonstrate a broad range of inhibition mechanisms and spectrums of activity (6).

Here, we will organize our coverage of the antimicrobial peptides into a slightly different scheme according to their applicability to the food industry, placing emphasis on those classes of compounds which are either applicable in foods or whose knowledge base provides insights into food applicable compounds. We will then review some of the current industrial applications of the antimicrobial peptides, and provide some insights into the exciting future prospectus of the field.

7.2 GRAM-POSITIVE BACTERIOCINS

Due to their natural occurrence in many types of foods, numerous Gram-positive bacteria species are considered to have food grade, or Generally Regarded as Safe (GRAS), status. Gram-positive bacteria produce a variety of bacteriocins, many of which have been biochemically and genetically characterized. Bacteriocins as a whole are generally sensitive to human intestinal proteases, which makes them promising resources for food preservation. The bacteriocins from two types of food grade bacteria, lactic acid bacteria (LAB) and *Propionibacteria*, have been intensely studied due to their potential applications in the food industry (7).

Bacteriocins produced by LAB are divided into four classes according to their structure and biochemical characteristics (8). This definition can also be applied to bacteriocins produced by other bacteria.

(1) Lantibiotics: small, membrane active peptides containing the unusual amino acid lanthionine.
(2) Small, heat stable, non-lanthionine-containing membrane active peptides characterized by a Gly-Gly-Xaa processing site in the precursor. This class can be further divided into three subclasses. 2(a) are *Listeria* active peptides with an N-terminal consensus sequence of –Tyr-Gly-Asn-Gly-Val-Xaa-Cys-. 2(b) are pore-forming complexes that require two peptides for activity. 2(c) are thiol activated peptides requiring reduced cysteine residues for activity.

(3) Large heat labile proteins (>30 KDa).
(4) Complex bacteriocins composed of protein plus one or more chemical moieties required for activity.

7.2.1 Class I Bacteriocins: Lantibiotics

Lantibiotics are a group of bacteriocins that, as their name suggests, contain the unique amino acid lanthionine. These bacteriocins also have other nonprotein amino acids, such as dehydroalanine and dehydrobutyrine, all of which are formed by post translational modifications. Based on their structural features and functional properties, lantibiotics can be grouped into three categories (9). Type A lantibiotics, also known as nisin-like or linear lantibiotics, like nisin and Pep5, have a linear structure consisting of around 34 amino acids with overall cationic character and some sequence homology. Their primary target is the cell membrane. Duramycins, which belong to type B, or duramycin-like lantibiotics, are smaller, consisting of 19 amino acids arranged in a globular conformation. The C-terminal residues of type B lantibiotics are involved in bridge formation within the globule. Type B lantibiotics inhibit phospholipases. Type C lantibiotics, like mersacidin, have structural features between those of types A and B and target the biosynthesis of the bacterial cell wall. Table 7.1 lists many of the naturally occurring lantibiotics, and Figure 7.1 shows several of their structures.

7.2.1.1 Nisin

Among the bacteriocins currently identified, nisin is the most intensely studied, and was the first one approved for use in the food industry on a commercial scale. It was discovered in 1928; the same year Fleming discovered penicillin (10,11). Five years later, Whitehead isolated nisin and identified its proteinaceous nature (12). Since then, many studies have been conducted on this antimicrobial compound and its producer strains. It has been previously shown that all the producer strains belong to serological group N of lactic streptococci. Nisin is named after its producer as "Group N Inhibitory Substance" (13). According to modern taxonomy, the producer strains are grouped to *Lactococcus lactis*.

7.2.1.1.1 Structure Nisin has two natural variants, nisin A and nisin Z. Both variants have 34 amino acids and similar sequences, except that nisin A has His27 while nisin

Table 7.1
Summary of several naturally occurring lantibiotics.

Lantibiotic	Structure Class	Producer	MW	Ref.
Nisin	Type A	*Lactococcus lactis*	3353	[179]
Subtilin	Type A	*Bacillus subtilis*	3317	[180]
Epidermin	Type A	*Staphylococcus epidermidis*	2165	[31]
Pep5	Type A	*Staphylococcus gallidermin*	3488	[181]
Gallidermin	Type A	*Streptococcus gallidermin*	2165	[182]
Duramycin A	Type B	*Streptomyces cinnamoneus*	2012	[183]
Duramycin B	Type B	*Streptomyces cinnamoneus*	2012	[184]
Duramycin C	Type B	*Streptomyces cinnamoneus*	2012	[184]
Ancovenin	Type B	*Streptomyces spp.*	1959	[185]
Cinnamycin	Type B	*Streptomyces cinnamoneus*	2041	[184]
Mersacidin	Type C	*Bacillus subtilis*	1825	[186]
Actagardine	Type C	*Actinoplanes sp.*	1890	[187]

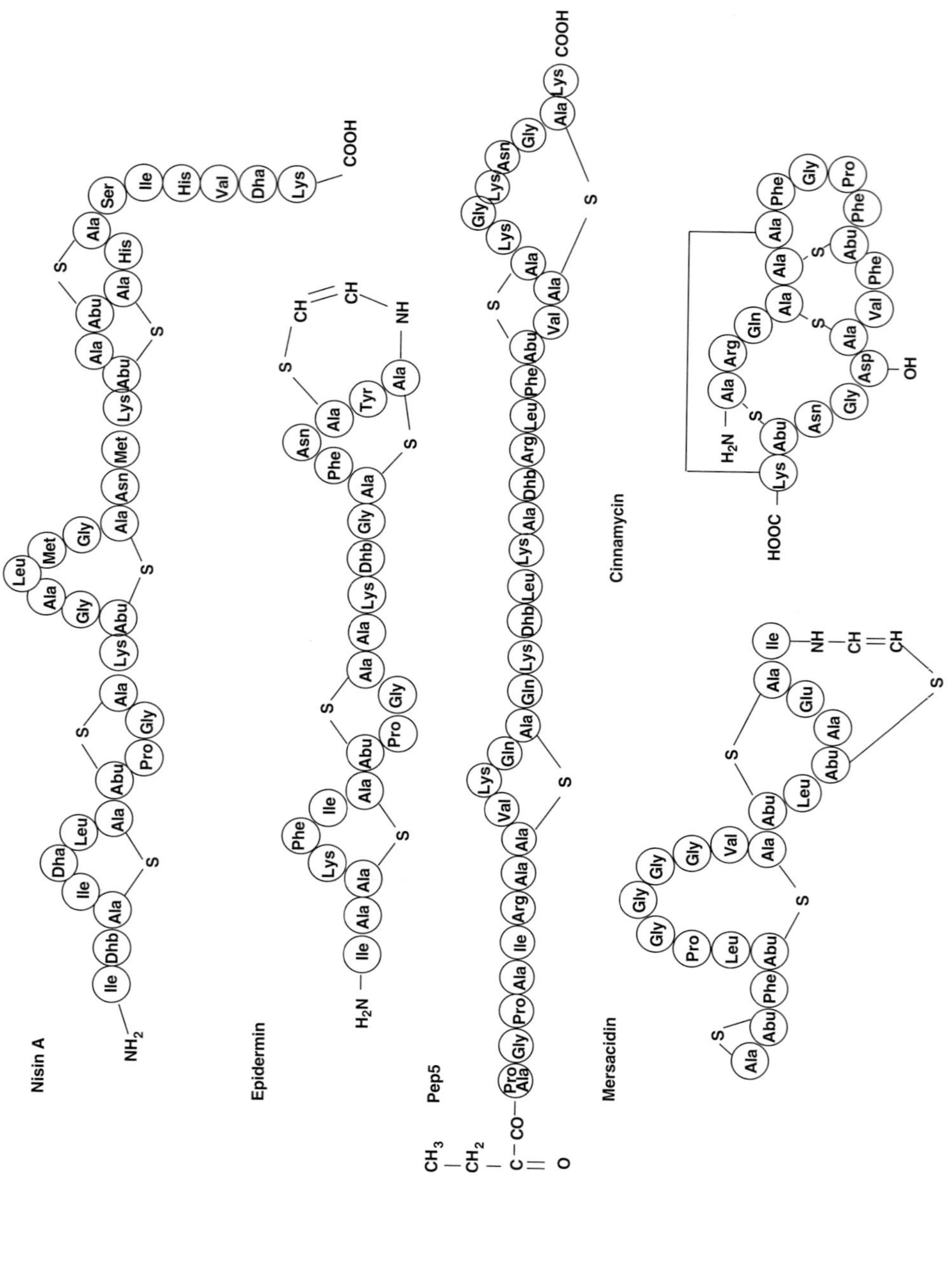

Figure 7.1 Structures of several lantibiotics. Dha: dehydroalanine; Dhb: dehydrobutyrine; A-S-A: lanthionine; Abu-S-A: 3-methyllanthionine.

Z has Gln27. Among these 34 amino acids exist several unusual dehydro amino acid residues including thioether bridged lanthionines. The primary structure of nisin A was proposed in 1971 (14). Figure 7.1 shows that nisin has a linear structure of 5 half rings. NMR spectra of nisin in aqueous solution and a membrane-mimicking environment confirmed and refined this structure (15,16). NMR data indicate that the nisin molecule has two amphipathic domains. The first three half rings on the N-terminal region (residues 3–19) form the first domain while the C-terminal domain (residues 22–28) consists of half rings D and E. The amino acid residues flanking these two structural domains have more variability. This is a typical membrane-acting structure.

7.2.1.1.2 Mode of Action Nisin is a membrane active bacteriocin (17,18) which destabilizes the phospholipid bilayer of the target cell, resulting in the rapid efflux of small molecules from the cytoplasm and loss of membrane potential. The most recent model for nisin interaction with the cell membrane includes three steps. First, the cationic nisin molecule binds to the cell membrane through electrostatic interactions with anionic membrane

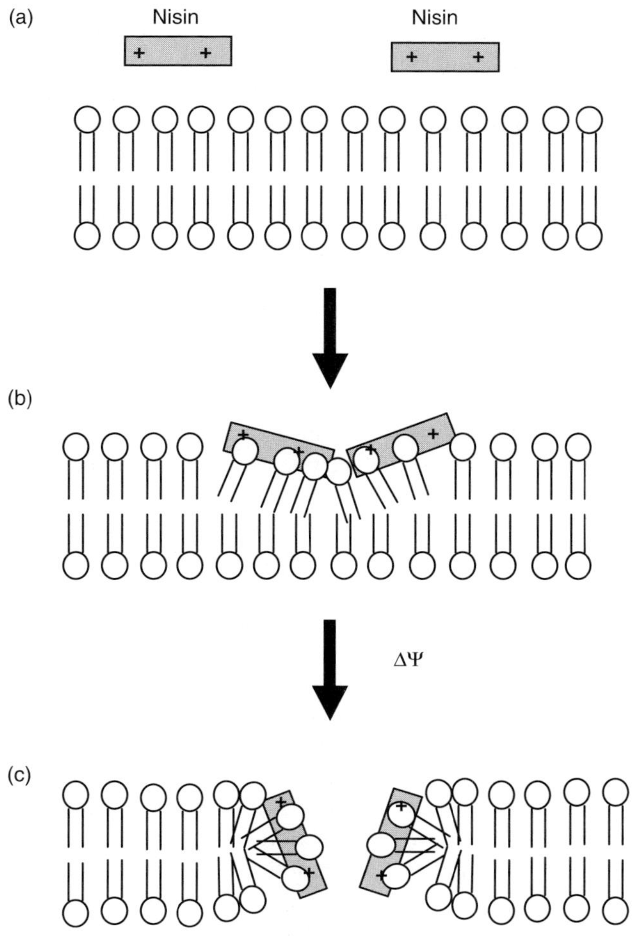

Figure 7.2 Nisin's pore-forming mode of action. (a) The cationic nisin molecule binds to the cell membrane through electrostatic interactions with anionic membrane lipids. (b) Nisin inserts into the cell membrane where its N-terminus penetrates into the outer leaflet of the membrane. (c) The C-termini of nisin molecules translocate across the membrane and form a transmembrane pore.

lipids. Nisin then inserts into the cell membrane where the N-terminus penetrates into the outer leaflet of the membrane. During the last step, the C-terminus of nisin translocates across the membrane (19). Several transmembrane molecules form a short lived transmembrane pore, and thus disturb the organization of the bilayer. This pore formation hypothesis is shown in Figure 7.2.

Nisin is able to inhibit both vegetative cell growth and the outgrowth of spores. The mechanisms of these inhibitions, however, appears to be different. Substitution of Dha at position 5 with Ala results in a loss of spore outgrowth inhibition, while the activity against vegetative cells remains the same (20).

Because nisin can react with liposomes and EDTA-treated Gram-negative strains, it is believed that this molecule interacts directly with the phospholipid bilayer (21). However, lipid II does act as a docking molecule during the attack. Through the interaction with lipid II on the cell wall, nisin inhibits peptidoglycan synthesis as well as interacts with the cell membrane. This combination of two killing mechanisms explains why nisin has a very low minimum inhibitory concentration (MIC) compared to other lantibiotics (22).

7.2.1.1.3 Nisin Biosynthesis The molecular mechanisms of nisin biosynthesis and regulation have been extensively investigated. It has been observed that nisin biosynthesis in *L. lactis* is inhibited by actinomycin, an inhibitor of RNA synthesis, but not mitomycin, an inhibitor of DNA synthesis (23,24). Protein inhibitors can also inhibit nisin production (25). These experiments provide solid data to demonstrate that nisin is ribosomally synthesized. Two questions remain unanswered at this point. (1) Do the nisin biosynthesis genes reside on a plasmid or on the chromosome? (2) What enzymes and mechanisms are responsible for the post translational modifications seen in nisin's unusual amino acids?

To answer this first question, a DNA degenerate probe based on the amino acid sequence of the peptide was designed to hybridize with plasmid DNA and digested chromosomal DNA of the nisin producer strain. The initial results obtained by different groups were conflicting. Some groups found that this probe hybridized with the plasmid DNA, while other groups showed hybridization with the chromosomal DNA (26–29). Researchers also found that nisin-producing strains are able to ferment sucrose, and that a correlation existed between nisin-production (Nip$^+$), nisin immunity (Nim$^+$) and sucrose fermentation (Suc$^+$) (27). The extra chromosomal genetic location has not been supported by plasmid transformation, while conjugation is successfully conducted between nisin producer and nonproducer. Most researchers now believe that the nisin biosynthesis gene cluster is located on a conjugative transposon within the chromosome.

The second question concerning nisin biosynthesis is how the unusual amino acids are formed after translation. The sequence of the structural gene, *nisA*, shows that the peptide product consists of 57 amino acid residues, none of which are unusual. The N-terminal region of this peptide product shows high homology to known leader peptides. Based on this information, a hypothesis for nisin maturation is proposed in Figure 7.3. A prenisin molecule is synthesized by translation, which includes the N-terminal leader peptide and several cysteine residues. Unusual amino acids are not present at this stage. The first step in modification is water elimination of serine and threonine residues, resulting in Dha and Dhb respectively. The next step is the formation of sulfide ring structures by nucleophilic addition between dehydro residues and cysteine residues. The molecule is now called pronisin. The last step is the cleavage of the leader peptide, and the active, mature molecule is secreted by the producer cells (30). This hypothesis is supported by later research on the nisin biosynthesis genes (26,31).

The organization of the nisin A biosynthesis gene cluster is shown in Figure 7.4. This gene cluster is around 13–15 kb and consists of 11 genes (21,32–35). Both nisin A and nisin Z have similar organization of their biosynthesis genes (34,36). Three promoters have

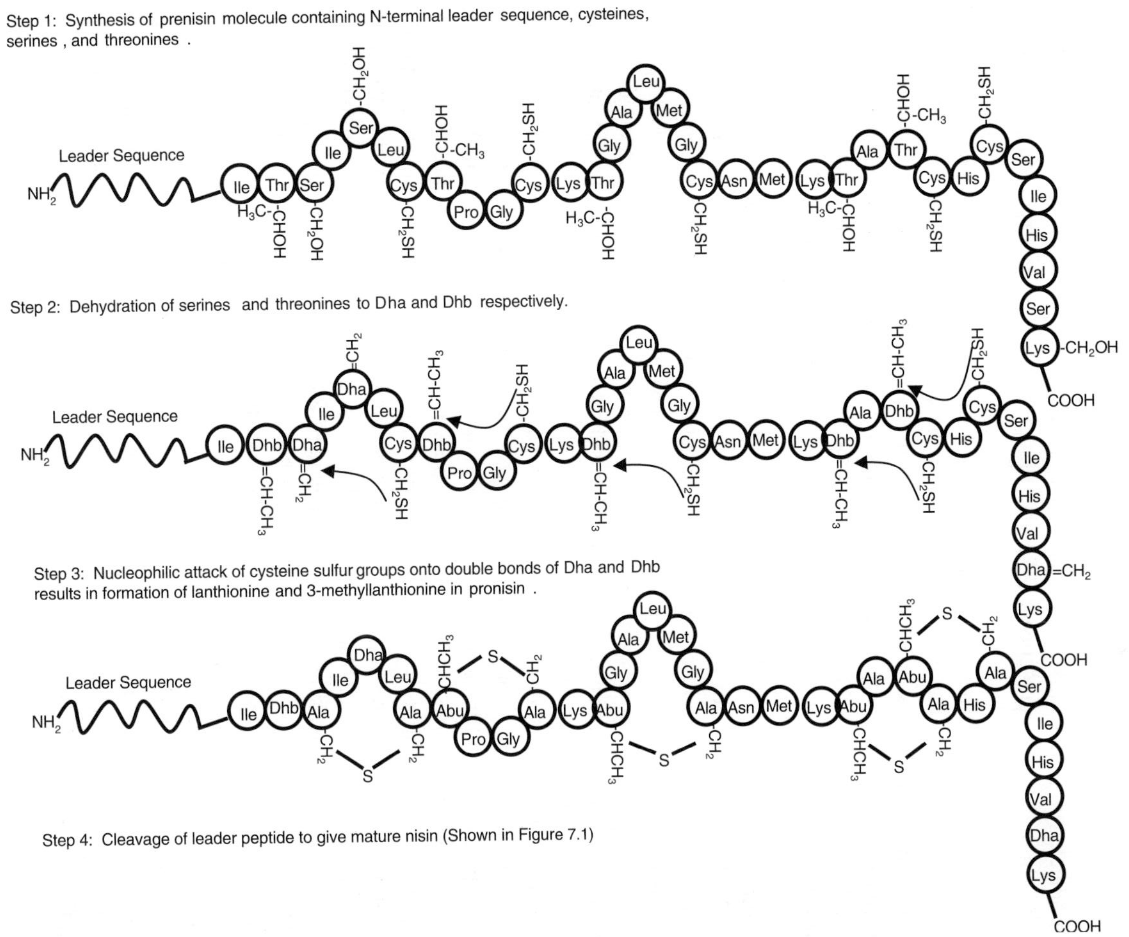

Figure 7.3 Posttranslational modification of nisin. Serine and threonine are dehydrated to form dehydroalanine and dehydrobutyrine respectively. Nucleophilic addition of the sulfhydryl of cysteine to the dehydro residues proceeds with stereo-inversion. Lanthionine and β-methyllanthionine result from the formation of thioether crosslinkages (30).

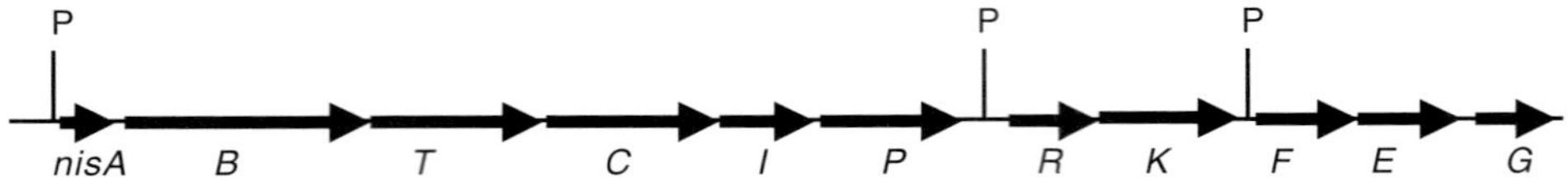

Figure 7.4 Organization of the nisin A biosynthesis gene cluster. Arrows indicate the open reading frames and their transcriptional directions. P represents a promoter site. Gene names and functions are covered in the text.

been detected in the nisin A gene cluster by primer extension, while only two have been detected in the nisin Z gene cluster (37). The first gene, *nisA*, is the structural gene encoding prenisin.

There is an inverted repeat sequence (IR) at the end of the *nisA* gene, suggesting a stem-loop structure on the transcript to control the expression rate of the structural and modification genes. The second gene and the fourth gene in the gene cluster are *nisB* and *nisC* respectively. While *nisB* encodes a protein of 993 amino acid residues, *nisC* encodes a protein of 414 amino acid residues. These two genes do not show any homology to other genes in the database outside of those involved in the biosynthesis of other lantibiotics. Mutational analysis and deletion experiments show that these two genes are necessary for the production of functional nisin and may be involved in dehydration and thioether formation (32,38). Disruption of these two genes abolishes the production of nisin and lowers the immunity of the producer cells (38). Computational analysis predicts several transmembrane alpha helix structures in *nis*B, which is consistent with experimental data showing that the *nis*B protein is membrane bound (34). Similar results have been obtained for *nis*C. Coimmunoprecipitation experiments have shown that anti *nis*C coprecipitates with the membrane fraction of cell extracts (39).

The *nisT* gene encodes a protein of 600 amino acid residues, which shows high homology to known ABC transporters (21,34). The ABC transporters are proteins of a super family that share common functional and structural characteristics (40). All contain ATP Binding Cassettes and are involved in membrane transport. It is reasonable to propose that *nis*T is involved in the secretion of nisin, and this process requires ATP. Experimental data has shown that mutation of the *nisT* gene results in accumulation of nisin within the cytoplasm (41). This same research also showed the *nisT* mutant to have decreased nisin immunity.

The yeast two hybrid system is a widely used tool for the analysis of *in vivo* protein interactions (42–44). Using this system, it has been demonstrated that nisin interacts with *nis*B and *nis*C; *nis*B interacts with *nis*C, and *nis*C interacts with *nis*T (39). Based on this information, a multiprotein membrane associated complex for nisin maturation has been proposed as shown in Figure 7.5.

The *nis*I gene is believed to be involved in immunity to nisin. Expression of *nis*I in a non-nisin-producing strain, *L. lactis* MG1614 conferred a low level of immunity to the host (45). Overexpression of this gene in *E. coli* under the control of the T7 promoter significantly enhances *Escherichia coli*'s resistance to nisin when treated with EDTA (21). The deduced protein sequence of this gene suggests that it is a lipoprotein of 245 amino acids, with its C-terminus anchored in the cell membrane (21).

The *nis*P gene encodes an extra cellular serine protease. This protein has an N-terminal signaling sequence, a serine protease domain, and a C-terminal membrane anchor (46). Expressed in *E. coli, nis*P can cleave the N-terminus of pronisin, releasing active nisin. This experiment also suggests that cleavage of the leader peptide is the last step in nisin maturation, which is confirmed by *nis*P disruption experiments. The *nis*P disrupted mutant secretes fully modified nisin with its N-terminal leader peptide and shows lowered levels of immunity (35). There is no direct evidence that *nis*P interacts with the multi metric complex of *nis*BCT.

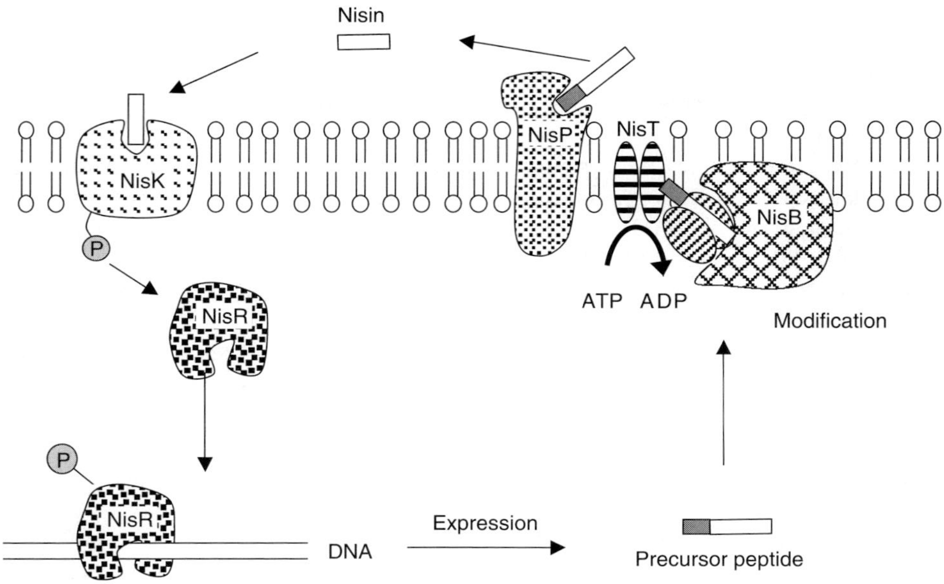

Figure 7.5 Nisin biosynthesis and regulation. (See text for details.)

The products of the *nis*R and *nis*K genes are involved in regulation of nisin biosynthesis. The *nis*R gene encodes a response regulator, and the *nis*K gene encodes a histidine protein kinase sensor (46). Nisin biosynthesis is an auto regulated process, and the mature nisin molecule acts as the signal for upregulation, as shown in Figure 7.5 (47).

7.2.1.1.4 Autoregulation of Nisin Biosynthesis Engelke et al. (1994) have found that an anti nisin antibody can reduce nisin expression levels (45). Disruption of *nis*K abolished the nisin expression in the producer strain. To study the biosynthesis of nisin, *L. lactis* NZ 9700 was constructed by mating the nisin-producing strain and a nonproducing strain *L. lactis* MG 1614, taking advantage of the fact that the nisin biosynthesis gene cluster was located on a conjugate. The resultant *L. lactis* NZ 9700 has only one copy of the nisin biosynthesis gene cluster. A 4 bp deletion in *nis*A was introduced into *L. lactis* NZ 9700 to obtain a nonproduction strain, *L. lactis* NZ 9800 (21). A *gus*A reporter gene driven by the *nis*A promoter was then introduced into *L. lactis* NZ 9800, allowing activation and expression of *gus*A by exogenously supplied native nisin but not with the unmodified nisin precursor. This result suggests that the mature nisin molecule, but not the precursors, activates the expression of its own synthesis genes. It is proposed that the binding of nisin to the membrane anchored *nis*K protein, which is a histidine sensor kinase, causes the phosphorylation of *nis*K. The activated *nis*K then transfers the phospho group to the corresponding response regulator, *nis*R. The transcription of genes, driven by both the *nis*ABTCP and *nis*FEG promoters, is activated by *nis*R–Pi (47). It is not clear whether other kinases are involved in this phosphorelay cascade.

Bacteria, as unicellular organisms, must communicate with each other so that they can behave in a synergistic, multi cellular way. In Gram-positive bacteria, modified oligopeptides are commonly found as signaling peptides (48). These peptides are secreted from the cells and then transported back into the cell to interact with receptors. As this regulatory mechanism is cell density or growth phase dependent, it is also called quorum-sensing regulation. In the case of nisin production, producer cells need to concert their biosynthetic efforts in order to ensure that enough antimicrobial peptide is produced to kill competing organisms. Meanwhile, the producer cells must also develop sufficient immunity before this high amount of nisin is produced. Studies have yet to examine the interaction between nisin and its membrane bound regulator *nis*K. One question to be solved by this potential research is whether this interaction occurs inside the cell or outside the cell.

A food grade expression system has been developed taking advantage of the nisin autoregulatory system. This is a controlled gene expression system which uses food grade nisin as an inducer. An MG1363 derivative with *nis*RK integrated onto the chromosome, *L. lactis* NZ 3900, is used as the host strain. The target gene is fused downstream of the *nis*A promoter and introduced into the host. High level expression is achieved by addition of nisin to the culture broth (49).

The last three genes in the nisin gene cluster, *nis*FEG, were discovered in 1995. The genes *nis*F and *nis*E appear to encode an ABC transporter, and *nis*G encodes a hydrophobic protein. Mutants with disruption in these genes can still produce nisin, but at decreased levels (35). This result suggests that *nisFEG* are not essential for nisin production. Some researchers have proposed that NisFEG are involved in nisin immunity together with *nis*I. Antisense *nis*EG and antisense *nis*G can effectively reduce nisin immunity, but the expression of *nis*F in *L. lactis* MG 1614 does not provide protection for the host (50). However, the ABC transporter may be indirectly involved in immunity via the nisin signal transduction process. It has been found that an ABC transporter is involved in some Gram-positive bacteria signal transduction processes by transporting the extracellularly modified peptide into the cell where it interacts with receptor proteins to regulate gene expression. An extra copy of *nis*RK-FEG can increase the production of nisin, which may be the result of this upregulation.

Genetic studies of nisin biosynthesis not only brought about the first food grade expression system, but have also provided powerful tools for biochemical studies. Protein engineering of nisin by site directed mutagenesis has been carried out to study and improve nisin properties (51). Introduction of lysine in nisin Z at N27 and H31 improves its solubility while the activity and inhibition spectrum remain similar. Substitution of Dha5 with Dhb gives higher stability toward chemical degradation but at the cost of decreased activity.

7.2.1.1.5 Nisin Resistance Before we start to discuss the mechanism of nisin resistance, it is necessary to clarify the difference between bacteriocin immunity and resistance. Immunity refers to the self protection mechanism by which the bacteriocin producer strains and cells can withstand high concentrations of their own bacteriocin. Most, but not all, bacteriocin producers discovered so far display immunity to their respective bacteriocins, and most of the time the immunity genes sit in the vicinity of the bacteriocin biosynthesis genes. The nisin immunity gene, *nisI*, as shown in Figure 7.4, is located inside the nisin biosynthesis gene cluster. Resistance on the other hand, is conferred by a mechanism developed by nonproducer strains to survive a specific bacteriocin. Two types of resistance are found for nisin. Resistance type Nis^{R} is encoded by a gene that is neither linked to nor located near the nisin biosynthesis gene. Also, Nis^{R} has been found on numerous plasmids (52–56). As nisin is a food grade bacteriocin, Nis^{R} is used as a food grade selective marker in cloning vectors (57). Another type of nisin resistance, Nis^{nr}, is caused by a spontaneous mutation in the sensitive cells. The mutation Nis^{m} has been reported by several research groups in different genera of nisin sensitive strains (53,54,58,59).

7.2.1.2 Biosynthesis of Other Lantibiotics

Nisin is the most prominent and most studied representative of the lantibiotic family, but many others also exist. Their producer strains include Bacillus, Streptococcus, Staphylococcus, Lactococcus, and Streptomyces (60). Several lantibiotic biosynthesis gene clusters from

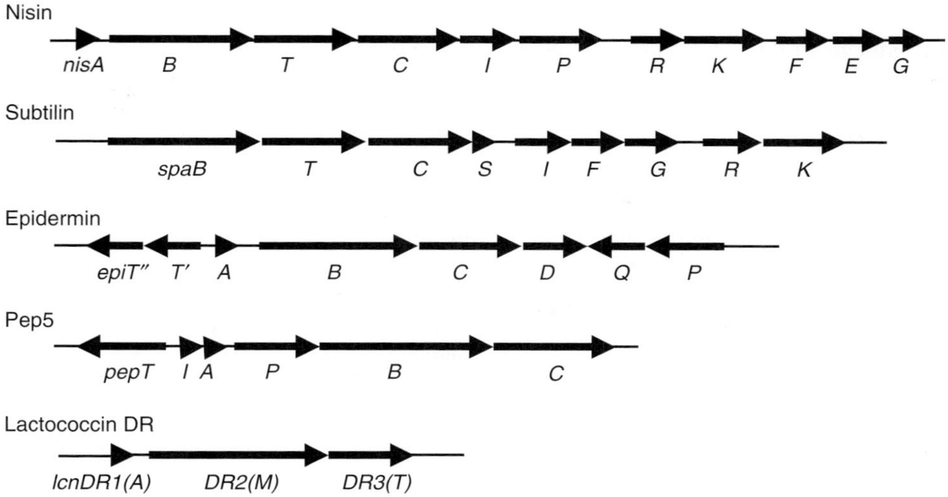

Figure 7.6 Summary of several lantibiotic gene clusters. Arrows indicate the open reading frames and their transcriptional directions.

producer strains such as these have been cloned. Figure 7.6 shows the genetic organization of these gene clusters, which show high homology to the nisin biosynthesis gene cluster. Genes involved in lanthionine formation are only found in the lantibiotic producer strains.

Worth special mention is the biosynthesis of subtilin. Genes involved in subtilin biosynthesis are located on the chromosomal DNA. A 40 kb region from the producer strain, *Bacillus subtilis* ATCC 6633, including the biosynthesis gene cluster, has been transferred to *B. subtilis* 168 (61). *Bacillus subtilis* 168 is a cloning strain, second only to *E. coli* when it comes to the amount of genetic information and tools available for genetic manipulation. Successful expression in this well known cloning strain has facilitated the protein engineering study of subtilin. Site directed mutagenesis was employed to construct a mutant subtilin, in which the Dha at position 5 was changed to alanine. Research examined the post translation processing by NMR, amino acid analysis, and Edman degradation, and found that this substitution did not interfere with the processing. This engineered peptide was fully active when reacted with vegetative *Bacillus cereus* cells, but failed to inhibit the outgrowth of spores. This result suggests that the mechanism by which subtilin attacks vegetative cells and spores may be different (62). Experiments for enhancing the properties of subtilin have also been performed by the same group with site directed mutagenesis. The Glu4 in subtilin was substituted by Ile. The design of this experiment is based on the observation that nisin is more stable than subtilin and has an Ile at position 4 in front of Dha5. Meanwhile, the chemical modification of Dha5 is coupled to a loss of activity (60). This change makes the peptide more stable than the natural one. This is the first example of mutagenesis demonstrating enhancement of a native molecule's properties (63).

7.2.2 Class II Bacteriocins: Small, Heat Stable, Non-Lanthionine-Containing Peptides

Class II bacteriocins are a group of heat stable antimicrobial peptides, which, unlike the lantibiotics, do not contain any modified amino acids. As mentioned before, they can be further categorized into 3 subgroups.

7.2.2.1 Class IIa Bacteriocins

Bacteriocins in this family are effective at killing *Listeria* spp. With 16 such antimicrobial peptides found to date, they form the largest group in the class II bacteriocin family. Table 7.2 shows the peptide sequences of the class IIa bacteriocins. They all share a common sequence on their N-terminus, YGNGV, with an overall sequence identity around 40–70% (64).

7.2.2.1.1 Mode of Action and Structure–Function Relationships Class IIa bacteriocins are membrane active peptides. The relationship between their structures and modes of action has been intensely studied. Bacteriocins in this family contain two modules in their primary structures. The N-termini of the peptides are the less hydrophobic modules that display high homology within the family and form β-sheet structures. The C-termini are more hydrophobic, more diverse, and have α-helical structures (65). Fimland et al. (1996) constructed hybrid class IIa bacteriocins by interchanging the N-terminal and C-terminal modules between different bacteriocins in this group. They reported that the hybrid products targeted the same bacteria as the natural bacteriocins from which the C-terminal modules were derived (65). The same research group also found that pediocin activity could be inhibited by a synthesized 15 mer fragment derived from its C-terminal module, while the synthesized fragments derived from the N-terminus did not show significant inhibition of activity (66). This is in agreement with the previous assumption that class IIa bacteriocins interact with receptors on the target membrane (67–69). However, the putative receptor protein has not been identified. Conflicting results have been obtained by other groups, which show that class IIa bacteriocins can interact with lipid vesicles. It may be that a docking molecule interacts with the N-terminus of the peptides. After the N-terminus of the bacteriocins recognize either the receptor or the docking molecule, hydrophobic interactions occur between the C-terminus of the peptides and the lipid acyl chain on the target cell membrane (66,69,70). Following this hydrophobic interaction, the peptides wedge into the membrane to create a more energy

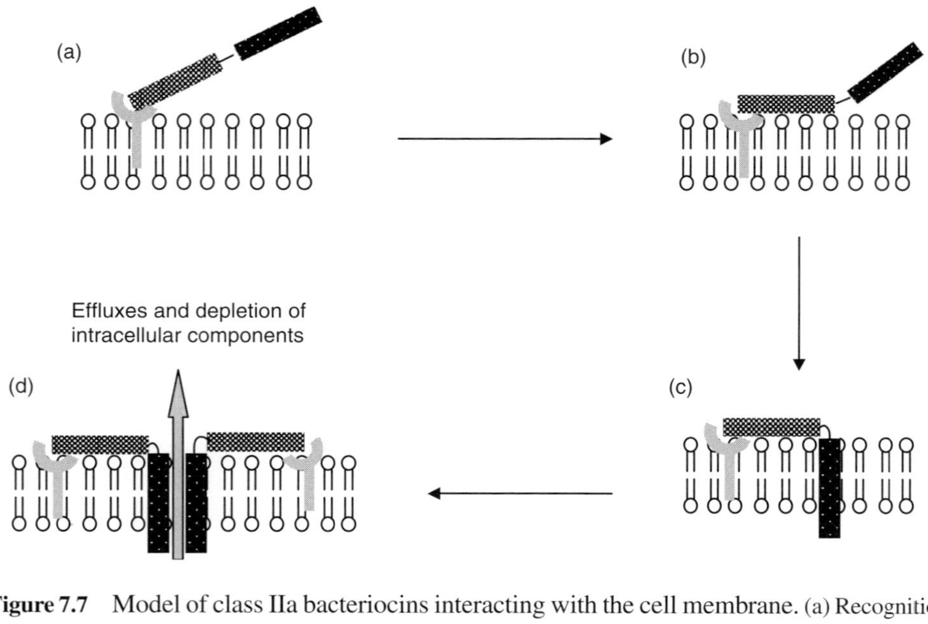

Figure 7.7 Model of class IIa bacteriocins interacting with the cell membrane. (a) Recognition between the bacteriocin and the putative receptor on the cell membrane. (b) Electrostatic interaction between the bacteriocin and the cell membrane. (c) Bacteriocin insertion into the cell membrane. (d) Aggregation of the bacteriocins and pore formation.

Table 7.2
Peptide sequences of the class IIa bacteriocins. Consensus sequences are underlined.

Class IIa Bacteriocin	Amino Acid Sequence	Ref.
Leucocin A	KYYGNGVHCTKSGCSVNWGEAFSAGVHRLANGGNGFW	[188]
Mesentericin Y105	KYYGNGVHCTKSGCSVNWGEAASAGIHRLANGGNGFW	[69]
Mundticin	KYYGNGVSCNKKGCSVDWGKAIGIIGNNSAANLATGGAAGWSK	[189]
Piscicolin 126	KYYGNGVSCNKNGCTVDWSKAIGIIGNNAAANLTTGGAAGWNKG	[190]
Bavaricin A	KYYGNGVHCGKHSCTQVDWGTAIGNIGNNAAANXATGXNAGG	[191]
Sakacin P	KYYGNGVHCGKHSCTVDWGTAIGNIGNNAAANWATGGNAGWNK	[192]
Pediocin PA-1	KYYGNGVTCGKHSCSVDWGKATTCIINNGAMAWATGGHQGWNKC	[72]
Bavarcin MN	TKYYGNGVYCNSKKCWVDWGQAAGGIGQTVVXGWLGGAIPGK	[70]
Divercin V41	TKYYGNGVYCNSKKCWVDWGQASGCIGQTVVGGWLGGAIPGKC	[193]
Enterocin A	TTHSGKYYGNGVYCTKNKCTVDWAKATTCIAGMSIGGFLGGAIPGKC	[194]
Enterocin P	ATGSYGNGVYCNNSKCWVNWGEAKENIAGIVISGWASGLAGMGH	[96]
Carnobacteriocin BM1	AISYGNGVYCNKEKCWVNKAENKQAITGIVIGGWASSLAGMGH	[195]
Sakacin A	ARSYGNGVYCNNKKCWVNRGEATQSIIGGMISGWASGLAGM	[196]
Carnobacteriocin B2	VNYGNGVSCSKTKCSVNWGQAFQERYTAGINSFVSGVASGAGSIGRRP	[195]
Lactococin MMFII	TSYGNGVHCNKSKCWIDVSELETYKAGTVSNPKDILW	[197]
Bacteriocin 31	ATYYGNGLYCNKQKCWVDWNKASREIGKIIVNGWVQHGPWAPR	[198]
Acidocin A	KTYYGTNGVHCTKRSLWGKVRLKNVIPGTLCRKQSLPIKQDLKILLGWATGAFGKYFH	[197]

stable orientation, which causes the destruction of the cell membrane. Figure 7.7 provides a model of this interaction.

7.2.2.1.2 Pediocin Class IIa bacteriocins are also referred to as pediocin-like bacteriocins, because pediocin is the most intensely studied of the family. Pediocin is produced by *Pediococcus acidilactici*, and the active peptide, pediocin PA-1, has been purified by several different groups (71–78). It has typical bacteriocin properties such as stability at 100°C for 60 minutes and sensitivity to protease, papain, pepsin, and α-chymotrypsin. Pediocin is active against Gram-positive bacteria including *Clostridium* spp., *Staphylococcus* spp., and *Listeria* spp. (79). Early biochemical characterizations based on partially purified bacteriocin samples revealed different molecular weights ranging from 2700 to 16500, and researchers tended to think that they were different molecules. Final identification by peptide sequencing and biosynthesis gene analysis showed that the bacteriocins isolated from different labs were actually the same molecule (72,80,81). The calculated molecular weight based on the peptide sequence is 4624, assuming that the cysteine residues form 2 disulfide bonds; a hypothesis which was confirmed by electrospray mass spectrometry (65). The molecular weight of 16500 may be a result of peptide aggregation.

7.2.2.1.3 Biosynthesis of Class IIa Bacteriocins Biosynthesis of the class IIa bacteriocins consists of two steps. First, a precursor peptide with an N-terminal leader sequence is synthesized on the ribosome. This precursor peptide is biologically inactive, just like the precursor of lantibiotics. This peptide is then cleaved at a specific site, and transported to the outside of the producer cell.

Biosynthesis genes of class IIa bacteriocins are organized in a gene cluster, either located on the chromosome or on a plasmid. Biosynthesis gene clusters of some representative class IIa bacteriocins are shown in Figure 7.8. The genes in the gene cluster can be transcribed in two different directions. In each gene cluster there is a structural gene encoding the bacteriocin peptide and the leader peptide. Although the length and amino acid sequence of the leader peptides are varied among different bacteriocins, most of them have two common glycine residues at the cleavage site of the leader peptide C-terminus. Table 7.3 shows some of the leader peptide sequences in this family.

The maturation process of class IIa bacteriocins consists only of leader peptide cleavage, as no amino acid modifications are involved. Formation of disulfide bonds is a spontaneous process. The leader peptide is cleaved during the secretion process. The cleavage and transport are actually accomplished by the same protein, which is an ABC transporter (82). The genes encoding the transporter proteins, sit either in the same operon as the structural gene, or in the neighboring operon, as shown in Figure 7.8. Deduced from the DNA sequence, ABC transporters have approximately 700 amino acid residues. The transporters, especially the C-terminal ATP-binding site, show significant homology amongst themselves and amongst other bacteriocin ABC transporters. The ABC transporters in the class IIa family have three domains. The 150 amino acids on the N-termini of the proteins are hydrophobic and have proteolytic activity. The ATP-binding sites are on the C-termini. A transmembrane domain is located between these two domains. It is proposed that the transporters form homodimers on the cell membrane. The N-terminal region may recognize the peptide precursor when ATP is bound to the C-terminal domain, and this binding triggers ATP hydrolysis, which in turn induces a conformational change in the protein. The leader peptide is removed, and the mature bacteriocin peptide is exported to the outside of the cell. The double glycine residues are believed to be involved in the precursor recognition (82,83).

Accessory proteins are involved in class IIa bacteriocin translocation. As in the lactococin A transport system, LcnC is the ABC transporter and LcnD is an accessory protein. Most of the accessory proteins consist of 460 amino acids. The pediocin accessory protein,

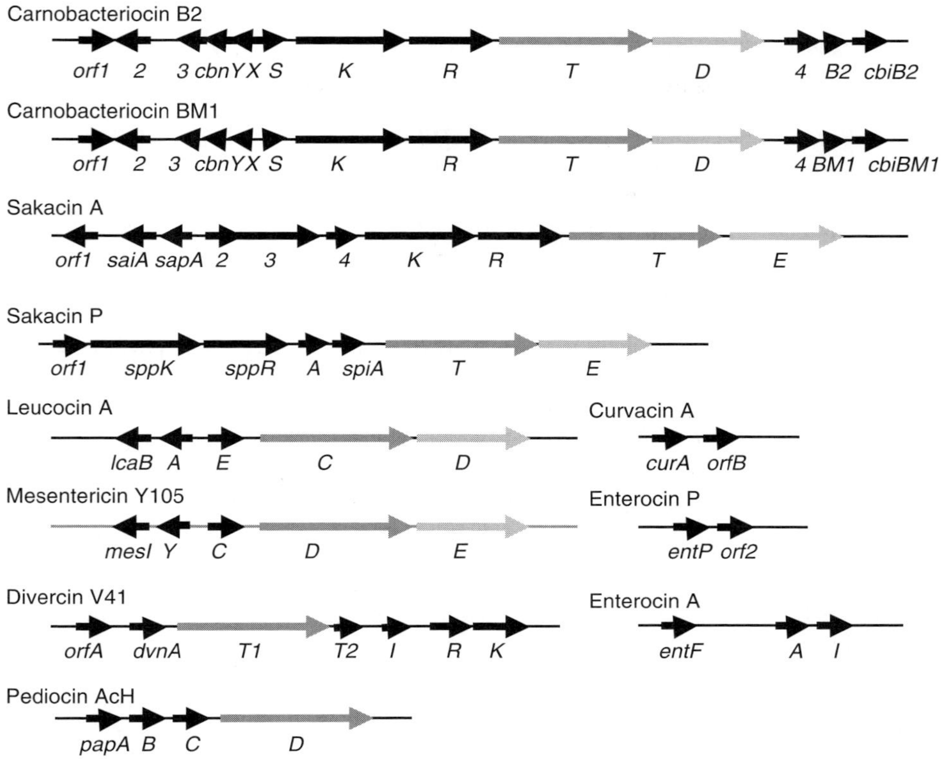

Figure 7.8 Organization of the genes involved in class IIa bacteriocin biosynthesis and immunity. Arrows indicate the open reading frames and their transcriptional directions. Structural genes are named after the corresponding bacteriocins, except that sakacin P structural gene is named A. I/i indicates the immunity gene; R represents the response regulator; K represents the sensor histine kinases. Dark gray arrows show the ABC transporter genes, and the light gray ones are the accessory protein genes. Functions of some ORFs are unknown.

however, is an exception, having only 174 amino acids. The accessory proteins have hydrophobic N-terminal regions and large hydrophilic regions. Experimental data shows that the accessory proteins are required for class IIa bacteriocin secretion (84,85). However, the exact function of this protein has yet to be elucidated.

By exploiting the general secretion pathway to secrete bacteriocin peptides, the producer strain decreases the amount of genetic information required for competition in the environment. Researchers have shown that class II bacteriocins can be transported by a general secretion pathway when their leader peptides are substituted with an appropriate signal peptide (86).

7.2.2.1.4 Class IIa Bacteriocin Immunity As shown in Figure 7.8, the immunity genes of the class IIa bacteriocins are located in the vicinity of the structural genes. In plasmid curing experiments involving the carnobacteriocin B2 producer strain, a loss of immunity was observed in conjunction with a loss of the plasmid containing the carnobacteriocin B2 biosynthetic operon. Homogeneous and heterogeneous expression of the immunity gene in the sensitive strains provides protection from carnobacteriocin B2 (87). Western blot analysis shows that the immunity protein, CbiB2, is mostly located in the cytoplasm. This 111 amino acid protein, deduced from the nucleotide sequence, does not react directly with the bacteriocin and does not provide protection for sensitive cells when added to the medium.

Table 7.3
Leader Peptide Sequences of Class IIa Bacteriocins.

Bacteriocin	Leader Peptide Sequence	Ref.
Sakacin A	MNNVKELSMTELQTITGG	[199]
Carnobacteriocin BM1	MKSVKELNKKEMQQIIGG	[89]
Carnobacteriocin BM2	MNSVKELNVKEMKQLHGG	[89]
Sakacin B	MEKFIELSLKEVTAITGG	[200]
Enterocin A	MKHLKILSIKETQLIYGG	[194]
Pediocin PA-1/AcH	MKKIEKLTEKEMANIIGG	[80, 201]
Acidocin A	MISMISSHQKTLTDKELALISGG	[197]
Divercin V41	MKNLKEGSYTAVNTDELKSINGG	[193]
Leucocin A	MMNMKPTESYEQLDNSALEQVVGG	[188]
Mesentericin Y105	MTNMKSVEAYQQLDNQNLKKVVGG	[202]
Enterocin P	MRKKLFSLALIGIFGLVVTNFGTKVDA	[96]
Bacteriocin 31	MKKKLVIGGIIGIGFTALGTNVEAAT	[198]

Producer strains of the class IIa bacteriocins show varying degrees of resistance to the noncognate bacteriocins from the same family (88). The immunity genes, however, do not show high homology. The immunity proteins range from 88 to 114 amino acids in size. There is no correlation between the homology of the bacteriocin sequence and the homology of the immunity protein sequence. Mesentericin Y105 and leucocin A differ in only 2 of 37 amino acids while their immunity proteins show only 74% identity. Sakacin A and carnobacteriocin BM1 show 67% sequence homology while their corresponding immunity proteins show only a 47% identity. However, curvacin A and acidocin, which only have a 34% sequence homology, have a 41.2% identity in their immunity proteins (64).

There is no experimental evidence showing how the immunity proteins provide protection. Based on the observations that the immunity proteins do not directly react with the bacteriocins, and that there may be receptors on the sensitive cells that interact with class IIa bacteriocins, it is reasonable to propose that the immunity proteins interact with the receptor on the producer cell membrane, so that the bacteriocins cannot attach to the same receptors. However, because sensitive cells gain no protection from externally supplemented immunity proteins, and furthermore, because most immunity proteins are localized within their respective host's cytoplasm, a model of external receptor competition seems unlikely. A second mechanism could be that the immunity protein may block the pores formed by the insertion of the bacteriocin molecules in the membrane. More research is needed to accurately determine this mechanism.

7.2.2.1.5 Biosynthetic Regulation The biosynthesis of class IIa bacteriocins, like that of the class I bacteriocins, is regulated by a peptide pheromone dependent quorum-sensing mechanism. Regulatory systems include a histidine protein kinase, a response regulator, and an induction factor (IF). In the nisin regulator described previously, the mature nisin molecule acts as the inducer. In the class IIa regulatory systems, either the bacteriocin or some specific inducer can be used to trigger the biosynthesis machinery, as shown in Figure 7.9. Induction factor (IF) accumulates during the growth of the bacteria. When the concentration reaches a certain critical level, the expression of bacteriocin starts (83).

In the systems using designated IFs, the IFs show some biochemical similarities to their corresponding bacteriocins. They are small, heat stable, cationic, and hydrophobic peptides. They are also synthesized as prepeptides with leader sequences, and are cleaved

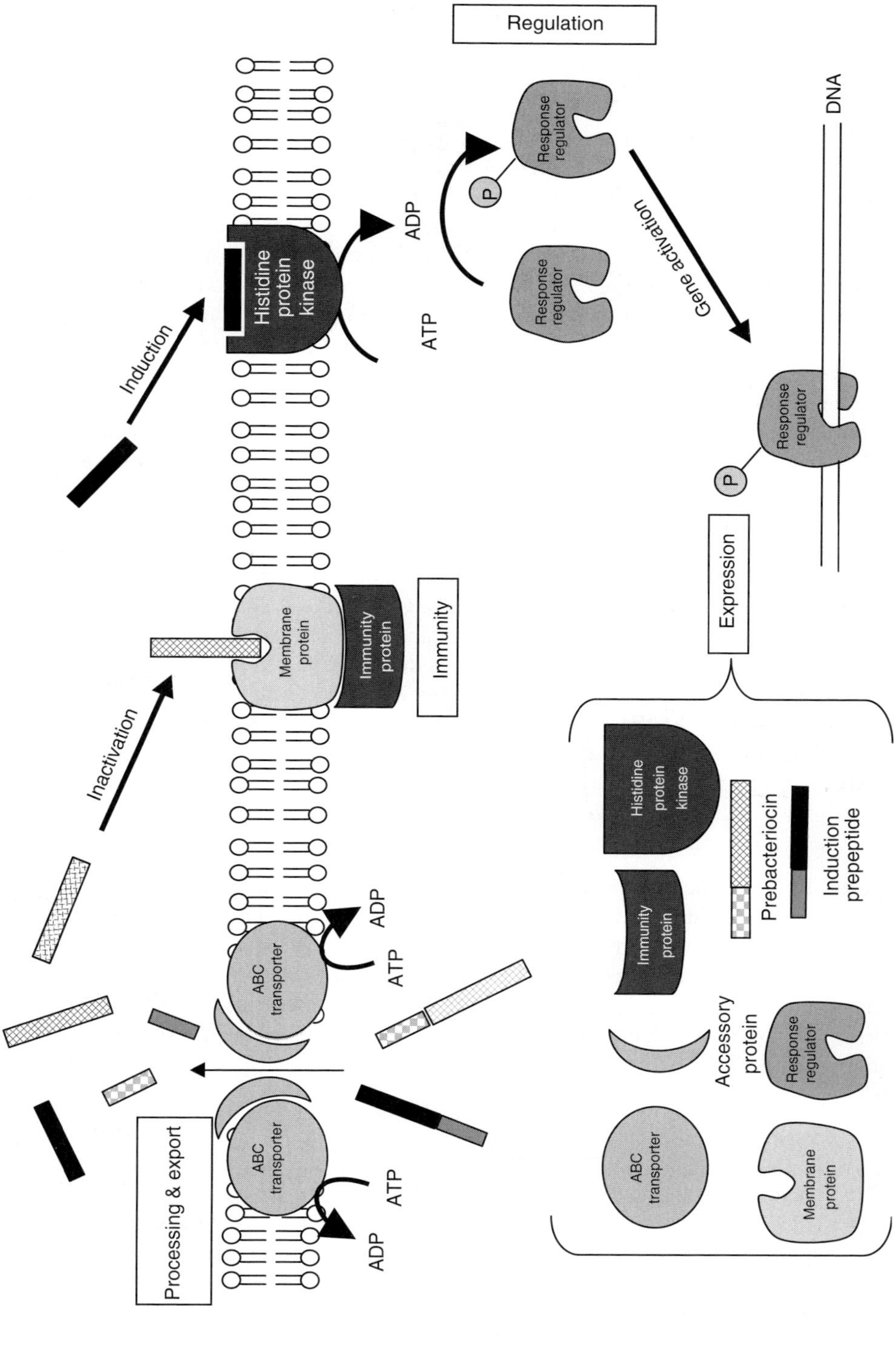

Figure 7.9 Class IIa bacteriocin biosynthetic mechanism. (See text for details.)

at the double glycine site. However, these IFs do not have high homology to the bacteriocins. Table 7.4 shows the sequence of several IFs.

The majority of class IIa bacteriocin biosynthesis is regulated by IFs. This system is also called a three component regulatory system. Divercin V41, produced by *Carnobacterium divergens* V41, also employs a second system in which, like nisin, the bacteriocin serves as the IF of its own expression. This system is also called a two-component system. The production of carnobacteriocin is therefore triggered by both systems. The bacteriocin itself as well as an inducer can trigger biosynthesis (89,90).

7.2.2.1.6 Other Class IIa Bacteriocins *Propionibacterum* spp., like lactic acid bacteria, has a long history of application in dairy fermentation. Several antimicrobial peptides produced by this genus have been found, such as jenseiin G, propionicin T1, and propionicin PLG-1 (91–93). Most of these bacteriocins have a narrow inhibition spectrum, only inhibiting *Propionibacteria* and *Lactobacillus*. Propionicin PLG-1, on the other hand, has a very broad spectrum, including Gram-positive bacteria, Gram-negative bacteria, and even some fungi. All of these peptides are around 10,000 Da and are heat stable. Recent genetic studies of these bacteriocins have revealed that they employ a sec dependent pathway for transportation. The structural genes of propionicin T1 and SM1 both encode prebacteriocins with a typical 27 amino acid signal peptide sequence (92,94). Another interesting aspect of these bacteriocins is that the producer strains may not have an immunity gene, and thus are sensitive to their own bacteriocins.

An antimicrobial peptide from *Propionibacterium jensenii* is found to be activated extracellularly by proteases. This protease activated antimicrobial peptide is encoded as a 225 aa propeptide including a 27 aa signal peptide on the N-terminus which is cleaved by a signal peptidase during export. The 64 aa on the C-terminus is activated by cleavage with proteinase K, proteinase A, trypsin, and subtilisin. None of these proteases are produced by the producer strain, and no immunity gene is detected (95).

The newly discovered enterocin P, from lactic acid bacteria, lacks the double glycine leader peptide typical of the class IIa bacteriocins, and is processed and exported by a general sec dependent pathway (96). Finally, a plasmid encoded bacteriocin from *Listeria innocua* is also a sec dependent bacteriocin (97).

7.2.2.2 Class IIb and IIc Bacteriocins

The class IIb bacteriocins, as summarized in Table 7.5, are a group of dipeptide bacteriocins. Individual peptides sometimes show modest activity, but the combination of paired peptides act synergistically to create a greatly heightened level of activity. Class IIb bacteriocins are membrane active peptides that have bactericidal effect (64). Figure 7.10 shows the mechanism of lactococcin G. When added to sensitive cells, different ratios of the two corresponding peptides yield different levels of activity, with a 1:1 ratio showing the best antimicrobial activity. Structures of lactococcin G, plantaricin E/F, and plantaricin J/K

Table 7.4
Sequence comparison of Induction Factors (IFs) involved in class IIa bacteriocins synthesis. ORF4 is the putative IF in sakacin A biosynthesis (178).

IF	Sequence
EntF	AGTKPQGKPASNLVECVFSLFKKCN
CbnS	SKNSQIGKSTSSISKCVFSFFKKC
ORF4	TNRNYGKPNKDIGTCIWSGFRHC

Table 7.5
Class IIb Bacteriocins. Shown here are the names of the active bacteriocins, their constituent dipeptides, and the sizes of each peptide.

Bacteriocin	Peptides	Amino Acids	Ref.
Lactococcin G	Peptide a and Pepdide b	39, 35	[203]
Lactacin F	Laf F and Laf X	57, 48	[204]
Lactococcin MN	LcnM and LcnN	58, 57	[205]
Plantaricin EF	PlnE and PlnF	33, 34	[100]
Plantaricin JK	PlnJ and PlnK	25, 32	[100]
Plantaricin S	Plc A and Plc B	27, 25	[206]
Enterocin 1071	Ent1071A and Ent1071B	39, 34	[207]
Thermophilin 13	ThmA and ThmB	43, 62	[208]
Acidocin J 1132	a and b (a + Gly)	6220 Da, 6289 Da	[209]
Brochocin C	brochocin A and brochocin B	59, 43	[210]

have been studied by circular dichroism (CD). It has been found that individual peptides are unstructured in aqueous solution and form some alpha helices in a lipid environment (98,99). It was also shown by these researchers that a 1:1 ratio of the two peptides enhances the formation of their dipeptide structure.

Genetic studies have revealed that the two paired peptides are encoded by two different genes that sit adjacent to each other in the same operon. Immunity genes and regulatory genes are also found close to the structural genes. A single immunity protein is encoded to provide protection against the dipeptide complex. The biosynthesis genes for bacteriocins in *Lactobacillus plantarum* C11 have been cloned and sequenced (100). This strain can produce four different bacteriocins: plantaricin A (Pln A), plantaricin EF, plantaricin JK, and another proposed single peptide bacteriocin. Each bacteriocin requires a biosynthetic operon which consists of the structural gene(s) and the immunity gene. Because all four bacteriocins have a double glycine site for cleavage, they share a single ABC transport system whose genes are located in the same region as the structural and immunity genes. The inducer regulating the expression of all the bacteriocins in the producer strain is Pln A.

Lactococcin B is one of the three bacteriocins produced by *Lactococcus lactis* subsp. *cremoris* 9B4 which also produces lactococcin A and M (101). It is a small, membrane-acting peptide with bactericidal effects. Lactococcin B is grouped into class IIc bacteriocins as a thiol activated antimicrobial peptide because previous research showed that a cysteine residue at position 24 must be in the reduced state for activity. This bacteriocin is inactive after oxidation with $HgCl_2$, and its activity is restored after Dithiothreotol (DTT) treatment. This result suggests that Cys24 and its corresponding oxidation state are critical to LcnB activity. However, site directed mutagenesis at Cys24 gave an opposite conclusion (102). Except for positively charged amino acid substitutions at position 24, the mutagenesis resulted in peptides that retained activity. The exact function of cysteine 24 during bacteriocin membrane interaction has therefore yet to be revealed conclusively. The structural and immunity genes of L-*canavanine* B (Lcn B) have been cloned and sequenced. All three, Lcn A, B, and M are encoded by a single plasmid. The Lcn B structural and immunity genes are of 68 and 91 codons respectively (103).

7.2.3 Class III Bacteriocins: Large, Heat Labile Proteins

Class III bacteriocins are large, heat labile proteins that include helveticin J (104), lacticin A and B (105), acidophilucin A (106), and caseincin 80 (107). Because of their sensitivity

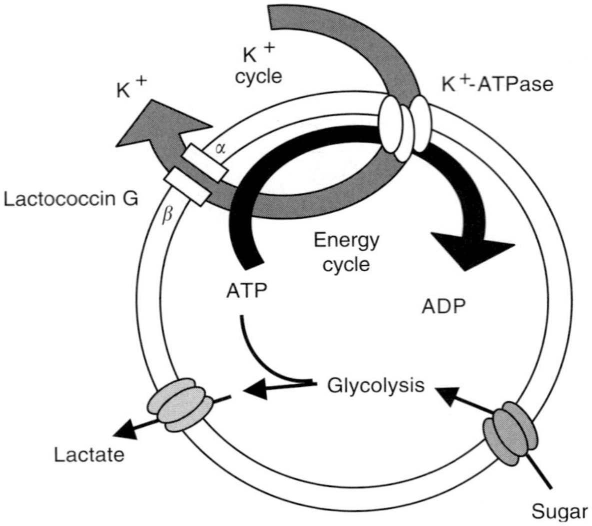

Figure 7.10 Lactococcin G mode of action. (See text for details.) (211).

to heat, they are not good candidates as food preservatives, and thus their biochemistry and genetics properties will be only briefly mentioned here.

7.2.3.1 Helveticin J

Helveticin J is a class III bacteriocin isolated from *Lactobacillus helveticus* 481. This bacteriocin inhibits closely related species, including *Lactobacillus delbrueckii*, *Lactobacillus bulgaricus*, and *Lactobacillus lactis* (104). Helveticin J is sensitive to proteases and heat treatment. Bacteriocin activity in the culture supernatant was inactivated by heating in a boiling water bath within 1 hour, which indicates that it has a high molecular weight.

Sodium Dodecyl Sulfate Polyacrylamide Gel Electrophoresis (SDS-PAGE) gives a molecular weight of around 37 kDa. The producer strain *L. helveticus* 481 harbors only one plasmid and plasmid curing experiments have shown that the biosynthetic operon of helveticin J is located on the chromosome. A genomic library of the producer strain was constructed, from which a 3.8 Kb fragment involved in biosynthesis was cloned (108). This fragment was introduced to *Lactobacillus acidophilus*, and the transformants showed antimicrobial activity similar to helveticin J. This result indicates that this 3.8 Kb fragment is sufficient for helveticin J production. Analysis of this fragment shows 5 putative open reading frames (ORFs). Figure 7.11 shows the organization of these ORFs, one of which, ORF3, encodes a putative hydrophilic protein of 37511 Da, which is very close to the molecular weight of helveticin J. This putative protein shows high glycine content on its N-terminal end. The function of ORF2 has not yet been elucidated. Because of the hydrophilic nature of helviticin J, it has been suggested that ORF2 encodes a protein involved in bacteriocin immunity and secretion. Whether other ORFs are involved in helveticin J production, and, if so, what their potential functions in this production may be, has not been investigated.

7.2.4 Class IV Bacteriocins: Complex, Modified Bacteriocins

Class IV bacteriocins are antimicrobial peptides carrying lipid or carbohydrate moieties. This group is well represented by Leuconocin S and Lactocin 27.

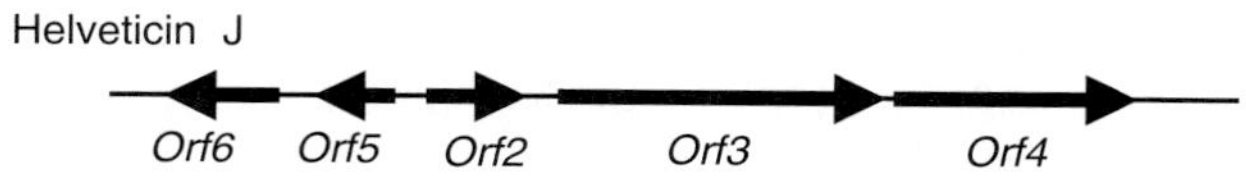

Figure 7.11 Organization of putative helveticin J biosynthesis genes. Arrows indicate the open reading frames and their transcriptional directions. Gene names and functions are covered in the text.

Leuconocin S is produced by *Leuconostoc paramesenteroides*, a strain isolated from retail meat (109). It is capable of inhibiting *Listeria monocytogenes*, *Aeromonas hydrophila*, *Staphylococcus aureus*, *Yersinia enterocolitica*, and some *Clostridium botulinum* strains (109,110). This peptide is resistant to lipase, but sensitive to α-amylase, trypsin, α-chymotrypsin, and proteinase K. Based on this information, this peptide has been determined to be a glycopeptide. Compared to bacteriocins of other families, leuconocin S is more heat sensitive. Approximately 50% of the activity is lost after heat treatment at 60°C for 60 minutes. The molecular weight of leoconocin S, as estimated by SDS-PAGE, is approximately 10 kDa. It is believed that leuconocin S is a membrane active molecule (109,111). Further information on its peptide sequence and biosynthesis pathway is not yet available.

Lactocin 27, produced by *Lactobacillus helveticus* LP27, is another example of a class IV bacteriocin. It is also a glycopeptide with a molecular weight of approximately 12 kDa (112). This bacteriocin shows bacteriostatic effects against closely related bacteria. It inhibits sensitive strains via inhibition of protein synthesis, Na^+ transport, and K^+ transport (113).

7.3 GRAM-NEGATIVE BACTERIOCINS AND BACTERIOCIN-LIKE COMPOUNDS

While the term bacteriocin is typically reserved for ribosomally synthesized antimicrobial peptides from Gram-positive species (114), it also appropriately includes similar compounds produced by Gram-negative species. The diversity of Gram-negative bacteriocins resembles that of the Gram-positive compounds and is best represented by the colicins and microcins. Because these compounds are produced by Gram-negative bacteria, which, due to their lack of natural occurrence in foods and typical association with fecal contamination, are considered to be nonfood grade, they are not directly applicable to food preservation. Still, despite their lack of direct food application, advances with Gram-negative bacteriocins and their respective modes of action, syntheses, post translational modifications, and bioregulatory mechanisms have proven to be much more rapid than the equivalent studies in Gram-positive species, and thus they provide good models for many of the same characteristics of food grade, Gram-positive bacteriocins. For instance, unlike their structural relatives the lantibiotics, the microcins have been synthesized *in vitro* (114). Furthermore, the membrane pore formation inhibition mechanism of some colicins has emerged as a model system not only for many bacteriocin modes of action, but also for numerous studies of the mechanisms and bioenergetics involved in various aspects of protein membrane interactions (115,116). These advances therefore warrant a brief review of the compounds here.

7.3.1 Microcins

Microcins are low molecular weight, ribosomally synthesized antibacterial peptides produced mainly by *Enterobacteriaceae* (117). They are highly modified and as such closely resemble the lantibiotics (114). Their molecular weights are generally below 5000 Da (118). Compared with colicins, microcins are less sensitive to proteases. This could be the result of their shorter length and of their amino acid modifications. Microcins are gene encoded and in most cases are plasmid encoded. The structural genes encoding microcins C7 and C51 each consists of 21 base pairs, probably the shortest coding sequences found to date (119,120).

Microcins vary in their structures, shown in Figure 7.12, and modes of action. They are divided into seven groups according to their cross immunity profiles (118,120). Presumably, microcins within the same cross immunity group have the same inhibition mechanism. These mechanisms include inhibition of metabolic enzymes, DNA replication, translation (118), amino acid transport (121), DNA biosynthesis (122), cell division (123), and cell membrane depolarization (124).

7.3.1.1 Microcin Biosynthesis

The biosynthesis and immunity genes of several microcins have been cloned and studied. Most of the known microcin genetic systems are located on plasmids, typically spanning a fragment of approximately 10 kb which includes both the biosynthesis and the immunity genes. As microcins undergo post translational modifications, the biosynthesis genes include both the structural genes encoding the peptides and the genes encoding enzymes responsible for modification. Figure 7.13 shows the biosynthesis gene clusters of several microcins.

The biosynthesis of microcin B17 (MccB17) has been the most thoroughly investigated. The translational product of the structural gene *mcbA*, preMccB17, consists of an N-terminal leader peptide and a C-terminal structural peptide. The gene products of *mcbB*, *mcbC*, and *mcbD* form a multimeric microcin synthetase complex which is responsible for

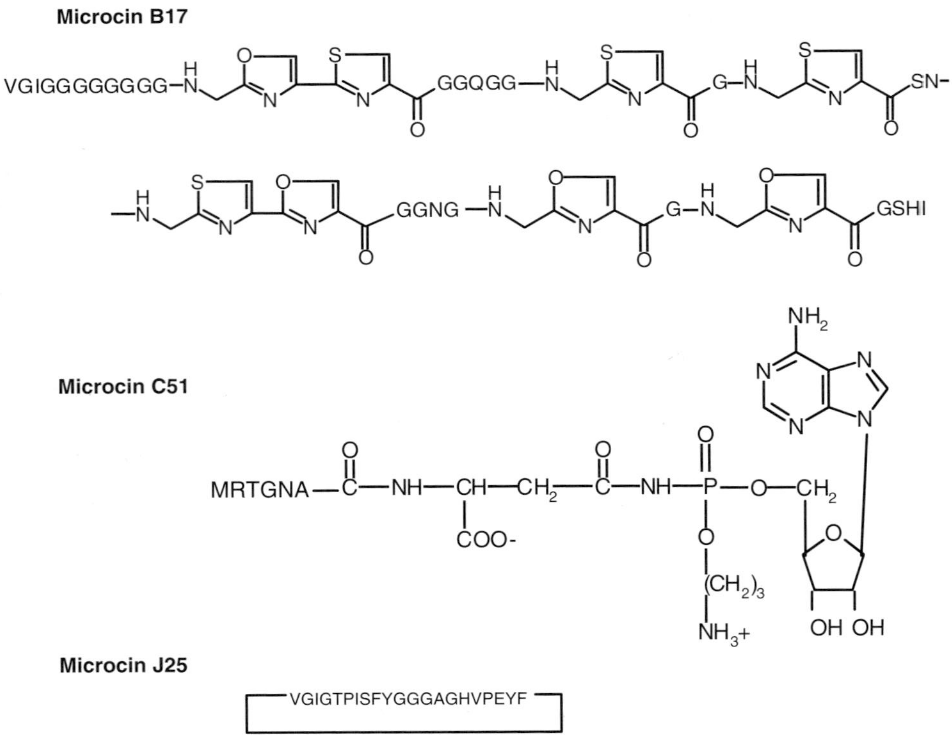

Figure 7.12 Structures of several microcins.

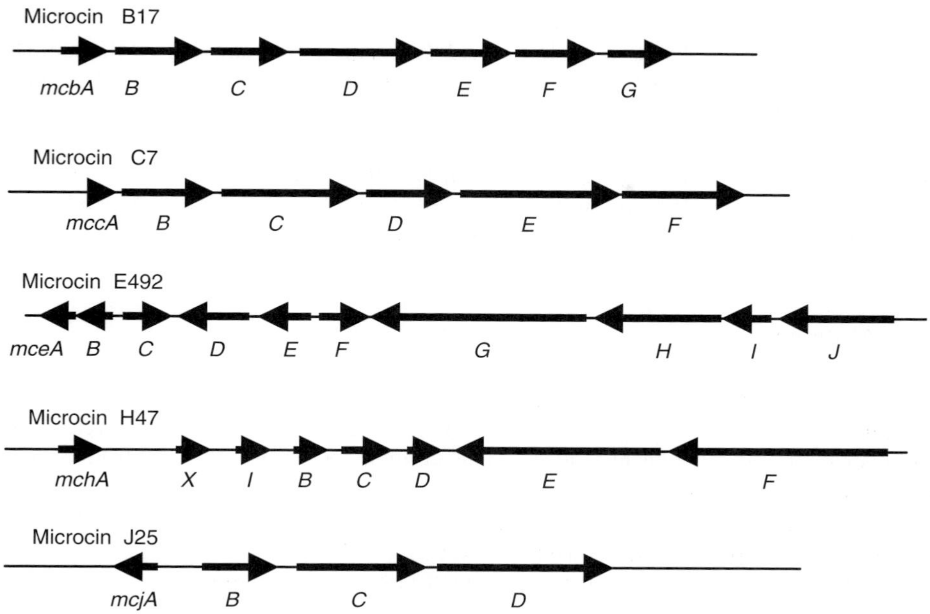

Figure 7.13 Biosynthesis gene clusters of several microcins. Arrows indicate the open reading frames and their transcriptional directions. Gene names and functions are covered in the text.

the formation of thiazoles and oxazoles (125). The leader peptide, along with ATP/GTP, is required for this process (126). The peptide is now called proMccB17. After removal of the leader peptide, the mature microcin B17 is secreted outside the producer cell. Both McbE and McbF are involved in transport of microcin B17.

The microcin E492 gene cluster is located on the chromosomal DNA. The entire gene cluster is around 13kb. The first gene in the cluster, *mceA*, encodes the precursor peptide with its corresponding leader sequence. The gene *mceB* encodes an immunity protein that is associated with the inner membrane (127). Expression of *mceA* and *mceB* is coordinated and occurs mainly in the exponential phase, with a decline during the stationary phase (124). Genes *mceC* and *mceI* show homology to glycosyl transferase and acyltransferase respectively. Gene *mceJ* does not show homology to any known gene. Genes MceC, MceI, and MceJ are necessary for production of active microcin, although microcin E492 does not contain any modified amino acid. Genes *mceG* and *mceH* encode an ABC transporter and its associated accessory protein, respectively (128).

Genes involved in the biosynthesis of microcin H47, which are located on the chromosomal DNA, have been cloned and sequenced. Eight ORFs of different transcriptional directions have been identified. It is proposed that *mchB* is the structural gene, while *mchA*, *C*, and *D* are involved in the maturation process. Genes *mchF* and *E* show high homology to other ABC transporter proteins. MchI is the immunity protein. There is a silent region between *mchA* and *mchI* with an ORF, *mchX,* of unknown function (129).

Four genes have been cloned from a low copy number plasmid that are related to microcin J25 synthesis and immunity. Gene *mcjA* is the structural gene, and *mcjD* is the immunity gene. Genes *mcjC* and *D* probably encode enzymes involved in cyclization of the peptide (130).

Although many of the microcin genes are located on plasmids, chromosomal genes are involved in the biosynthesis and secretory processes. A chromosomal gene *pmbA* is found to be involved in the maturation of microcins. This gene encodes a cytoplasmic protein that cleaves the leader peptide to yield the mature microcin (131). Microcin H47 is secreted by an ABC transporter, constituted by MchE, MchF, and the chromosomally encoded TolC (132).

The regulatory factors controlling the expression of microcins are encoded by the chromosomal DNA. Microcin B17 expression is activated by an integration host factor (IHF) and outer membrane protein OmpR (133). Expression of microcin C7 and C51 is regulated by cAMP/CRP and RpoS; the latter is an alternative subunit of RNA polymerase (120). Microcin J25 expression is controlled by ppGpp (134).

7.3.2 Colicins

Colicins are single, ribosomally synthesized polypeptides produced primarily by *Escherichia coli* which kill sensitive *E. coli* and closely related species including *Shigella* and *Salmonella*. Colicin production is not completely limited to *E.coli*, however, as several other Gram-negative bacteria also produce colicins, such as colicin U from *Shigella boydii* (135). The molecular weights of colicins range from 29000 to 89000 Da (136,137). This characteristic places them into the class III bacteriocins (colicin V is a small peptide that should be grouped with the class II bacteriocins). Compared with microcins, colicins are not modified except for the cleavage of an N-terminal methionine (115).

7.3.2.1 Structure and Mode of Action

The colicin mode of action is a sensitive cell suicide process which consists of three steps. First, the colicin molecules interact with receptor proteins on the outer membrane of the target cells. Next, the target cells actively uptake colicin into the cytoplasmic membrane or the cytoplasm, depending on the type of colicin. The colicin molecules then execute the killing process (116). Colicins consist of three functional domains which correspond to the three steps in the killing process. The N-terminal domains are important for uptake, the central regions interact with the receptor proteins, and the C-terminal domains are the toxic sites.

Colicins can be divided into two groups based upon their mode of action. The first group consists of the pore-forming colicins A, B, E1, Ia, Ib, K, and N (138). These colicins kill the target cells by forming a channel on the target membrane. Studies have shown that the channels formed by colicins can be up to 9 Å (139). The non-pore-forming colicins execute their killing task in the cytoplasm. Most of the colicins in this group have nuclease activity (115). Colicins E2, E7, E8, and E9 are endodeoxyribonucleases which degrade chromosomal DNA. Colicin E3, E4, E5, and E6 are endoribonucleases that inhibit protein biosynthesis by cleavage of the 16s rRNA (140). Colicin D inhibits protein synthesis by cutting tRNA-Arg (141). Colicin M blocks the biosynthesis of both peptidoglycan and O-antigen and causes autolysis of the cells (136).

Immunity proteins are found in the colicin-producing cells. Because of the differences among colicin modes of action, the protections provided by different immunity proteins are also different. They do, however, share one important characteristic, they all inhibit the colicin activity by binding to the active site of the corresponding colicins (142,143). This aspect makes colicins and their immunity proteins an excellent model system for the study of protein–protein interactions. For the pore-forming colicins, the corresponding immunity proteins are inserted in the membrane. For the non-pore-forming colicins, the immunity proteins are expressed and bound to the colicins inside the producing cells and are secreted with the colicins in their bound forms. The colicin M immunity protein is a membrane inserted protein, but it inactivates colicin M before the toxin reaches the cytoplasmic membrane (144).

7.3.2.2 Colicin Biosynthesis

Colicin-producing strains occur with high frequency among natural *E. coli* isolates (145). The genes encoding colicins have been found on plasmids. Biosynthesis genes for colicins A, D, E1 to E9, K, and N are located on small multi copy plasmids which are amplifiable but not self transmissible. Genes encoding colicins B, Ia, Ib, and M sit on large, low copy transferable plasmids that are not amplifiable (115). As shown in Figure 7.14, the colicin structural genes normally sit close to the immunity genes. Most of the nuclease colicins have their structural genes and their immunity gene in the same orientation, while the structural genes encoding the pore-forming colicins are typically in the opposite orientation of their corresponding immunity genes.

What is unique about colicin biosynthesis is the transport mechanism. Colicin secretions do not use the dedicated transport system as do lantibiotics or class IIa bacteriocins, nor do they utilize the universal transport system for other secreted proteins. A small lipopeptide is expressed coordinately with the colicin. This lipopeptide increases the permeability of the envelope, which causes the release of colicin as well as some other proteins.

The biosyntheses of colicins are under tight control because they are toxic to their own producer cells. Expression of colicin structural genes is driven by promoters that are repressed by the LexA protein. Upon UV irradiation or mitomycin C exposure, the resulting DNA damage signals the SOS response. This leads to the activation of RecA protease activity that degrades LexA, thereby allowing for the transcription of the colicin structural genes (146).

7.3.3 Other Gram-Negative Bacteriocins

Yersinia pestis produces pesticin, which is a muramidase (147). The pesticin gene cluster is located on a plasmid with the structural gene (*pst*) and the immunity gene (*pim*) having opposite transcriptional polarity. The expression of pesticin is, like that of the colicins, under the control of the LexA repressor of the SOS system. No lysis protein is detected near the gene cluster (148).

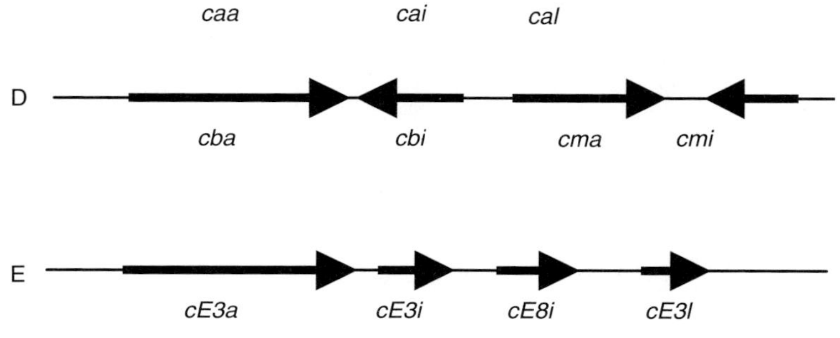

Figure 7.14 Organization of the colicin biosynthesis genes. The structural genes are called *cxa*, where x is the name of the colicin; the immunity genes are called *cxi*, and the lysis genes are *cxl*. The arrows indicate the direction of transcription.

Pseudomonas aeruginosa strains are frequently found to produce pyocin S1, S2, and AP41 (149). They are colicin E2-like bacteriocins with nuclease activity. They also have three functional domains: the receptor-binding domain, the translocation domain, and the DNAse domain (150). The biosynthesis genes of pyocin AP41 are located on a conjugative fragment on the chromosome (151). Pyocin AP41 expression is activated by DNA damage through the RecA pathway (152).

7.4 PEPTIDE ANTIBIOTICS

As previously mentioned, the peptide antibiotics refer to a group of antibiotics that have non-protein amino acids, including D-amino acids, hydroxy acids, or other unusual constituents (6). These unusual amino acids are not formed by post-translation modification as those in lantibiotics, but rather are synthesized along with the entire peptide on huge enzyme complexes. A number of peptide antibiotics have been discovered and studied in detail. Peptide antibiotics produced by *Bacillus* spp. include gramicidin, tyrocidine, bacitracin, and surfactin. The structures of some of these antibiotics are shown in Figure 7.15.

Paenibacillus (formally known as *Bacillus*) produces several different types of polymyxins. Polymyxin B was first isolated from *Paenibacillus polymyxa* (formally known as *Bacillus polymyxa*) in 1947. *Actinomycetes* spp. are known for the production of numerous different antibiotics. Some of these antibiotics also belong to the group of peptide antibiotics, such as actinomycin and bialaphos.

The biosynthesis of peptide antibiotics is quite different from that of ribosomally synthesized polypeptides. It has been found that the biosynthetic pathways of these peptide antibiotics have similar mechanisms that involve a head growth synthesis performed by

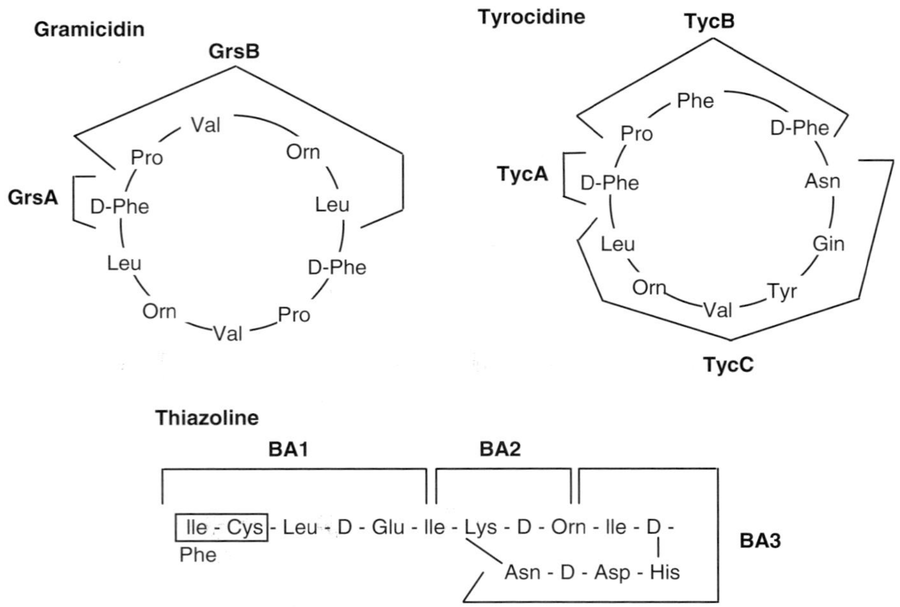

Figure 7.15 Structures of some peptide antibiotics produced by *Bacillus* spp. Indicated here are the peptide sequences of the respective antibiotics along with the NRPS responsible for their synthesis.

ATP consuming nonribosomal peptide synthetases (NRPSs) which utilize a thiotemplate multienzymatic mechanism (153–157).

Like bacteriocins, genes responsible for the biosynthesis of peptide antibiotics are usually organized in a gene cluster. Figure 7.16 shows the biosynthesis gene cluster of gramicidin, which inhibits bacteria by binding to nucleotides. Gramicidin S synthetase is a complex of two subunits. GrsA activates and racemizes L-Phe to D-Phe. GrsB activates Pro, Val, Orn, and Leu. The modules for activating and adding each amino acid are organized linearly on the synthetase GrsB, corresponding to the amino acid sequence of this peptide antibiotic.

Peptide antibiotics are considered secondary metabolites. Except for competitive exclusion, the functions of these antibiotics to their producer strains are not clear. The peptide antibiotics are widely used in the medical field to treat infection, and as such their use in the food industry is not allowed due to fears of propagating resistant strains of bacteria.

7.5 INDUSTRIAL APPLICATIONS OF ANTIMICROBIAL PEPTIDES

Humanity's perpetual coexistence with and codependence upon microbial life has lead to a high degree of necessity for being able to control the presence and actions of these microbes. Two areas where this need for control is particularly important are in the food and medical industries.

Microbes are involved in numerous aspects of the food industry, both good and bad. Many beneficial bacteria, namely lactic acid bacteria, are used to preserve food and to alter the taste content of foods to make them more appealing to consumers. Other, nonbeneficial bacteria, whether naturally occurring or contamination based, have negative effects on the food industry by causing food spoilage or diseases amongst consumers. To control these spoilage and pathogenic microbes, antimicrobial agents in foods are used as additives to prevent food spoilage and enhance food safety. Historically, these additives have been chemical compounds such as nitrites, sulfites, and organic acids (7). However, chemical preservatives such as these can sometimes be allergens, elicit nonallergic consumer sensitivities, alter the sensory properties of food, and have an overall negative impact upon consumers. Numerous alternatives exist to the use of chemical food preservatives, many of which make use of bacteria and their metabolites. Live probiotic bacteria, probiotic bacterial

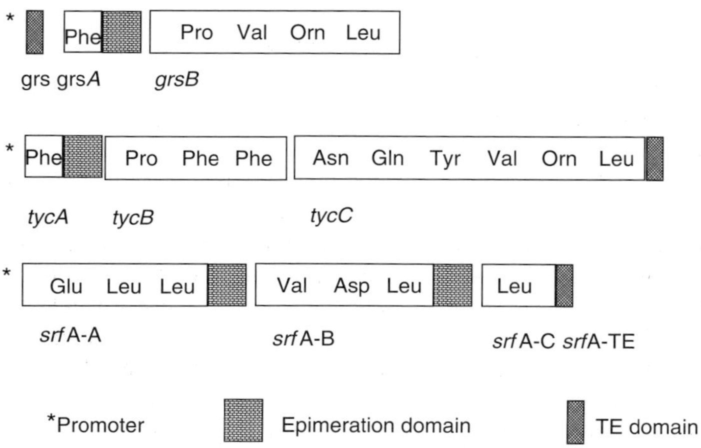

Figure 7.16 Organization of the peptide antibiotic gramicidin S biosynthesis genes.

fermentates, and purified bacterial metabolites have all found applications within the food industry. Amongst these products, antimicrobial peptides, with their selective antimicrobial activity and common sensitivity to harmless degradation by enzymes within the human intestinal tract, have emerged as a highly promising source of food preservatives. Two such compounds, nisin and pediocin, are currently in wide use today.

Nisin, as mentioned earlier, is a class I lantibiotic bacteriocin. It is a widely used antimicrobial peptide in the food industry and inhibits most Gram-positive bacteria, including *Bacillus, Clostridium*, and *Listeria*. It has a broad inhibition spectrum compared to other bacteriocins from Gram-positive bacteria. It can be degraded by several intestinal proteases, such as α-chymotrypsin and pancreatin (158–160), thus rendering it nontoxic to human beings. Its producer strain, *Lactococcus lactis*, is a food grade bacterium commonly found in milk and cheese products, and the purified nisin does not impart any off flavors to the finished products. These characteristics make nisin an excellent food grade preservative. As nisin is very stable at high temperature and low pH, it is perfect for low pH foods that are not suitable for heat processing. Nisin can be used for canned vegetables, high moisture hot plate bakery products, salad dressings, and various dairy products (161). Meanwhile, as nisin is not active against Gram-negative bacteria, it does not disturb the intestinal microflora of humans. More than 50 countries and regions have approved the use of nisin as a food preservative (161). In the United States, the FDA approved the use of nisin as a preservative for processed cheeses in 1989 (162).

Pediocin is a class IIa bacteriocin. It can be added to meat and meat products, milk and dairy products, and liquid whole eggs to inhibit food spoilage bacteria and pathogens (163). Experiments have been conducted to compare pediocin and its producer strains on their ability to control *Listeria monocytogenes*. Pediocin was observed to decrease the population of *L. monocytogenes* faster than the producer strain at first, but its activity was lost afterward. A lower final bacterial count was achieved with the producer strain (164,165). The addition of this peptide during fermentation of kimchi, a Korean fermented vegetable, decreased *L. monocytogenes* counts and inhibited its growth (166). Interestingly, pediocin is not active against *Lactococci*, a bacteria commonly used as a starter culture in dairy products. This makes pediocin a promising food preservative for fermented foods.

Pediocin has been commercially applied in the food industry as pediocin PA-1-containing fermentate (163). Due to the fact that the pediocin producer strains have GRAS status, they and their fermentates can be used in the food industry without special labeling (167). However, it is worthwhile to mention here that nisin is still the only bacteriocin licensed as a food grade preservative. Although pediocin has good *in vitro* activity against *L. monocytogenes*, its *in vivo* activity in food applications is much lower (168). This characteristic makes pediocin an additional safety factor when used together with other food protection methods.

Within the medical industry, the continued evolution of bacterial strains with increased antibiotic resistance has created an ever increasing need for new antimicrobial agents. Peptide antibiotics such as actinomycin, gramicidin, surfactin, and the polymyxins have found applications as both ingestible and topical antibiotics within the medical industry. Some bacteriocins have also found medical uses, many of which are nonantimicrobial in nature. Epidermine and gallidermin are used in the treatment of skin infections. Actagardine inhibits the biosynthesis of peptidoglycan. Ancovenin has a potential use in treatment of high blood pressure because it can inhibit angiotensin-converting enzymes. Cinnamycin shows antiviral activity against type I Herpes Simplex virus. Mersacidin has immunosuppressive activity. The medical applicability of nisin is currently under investigation.

7.6 FUTURE PROSPECTUS

Past and ongoing research surrounding antimicrobial peptides has shown that these compounds have a great deal of potential for applications in both the food and medical industries, and has also provided insights into key areas where the field might make, and should strive to make, strong advances in the future. Continued discoveries of new antimicrobial peptides and increased understanding of the biological systems involved in antimicrobial peptide synthesis, immunity, and regulation should allow for a great deal of advancement within the field, with an emphasis on practical applications to industry.

Bacteriocins, among all antimicrobial peptides, are of great interest to food scientists as potential food preservatives. While a great deal of research has been focused upon biochemical and genetic studies of these peptides, there are still many unrevealed fields yet to be explored. Various bacteriocins have been discovered during the past decades. Among them, only nisin and pediocin have found useful application in the food industry. While adept at inhibiting a wide range of Gram-positive pathogens, these two compounds do little on their own against Gram-negative pathogens. The most serious food borne illnesses, are, however, unfortunately caused by Gram-negative strains, such as pathogenic *E. coli* and *Salmonella* spp. These strains interfere with digestion and cause gastrointestinal problems, such as diarrhea and the more serious sequela. While, as mentioned earlier, nisin, in combination with some treatments such as EDTA or heat treatment can inhibit some Gram-negative bacteria, the processes involved are not always practical. A continued search for new antimicrobial peptides, particularly those with novel inhibition spectrums that include Gram-negative bacteria, as well as possibly yeast and mold, may yield novel compounds for use in the fight to control food spoilage and food pathogenic organisms.

Also, as our search for novel antimicrobial peptides continues, new sources for such compounds have begun to be revealed. While most of the classical research into novel antimicrobial peptides has involved strictly prokaryotic sources, ever increasing amounts of evidence suggests that almost all forms of life produce small peptides with antimicrobial activity. These include not only the bacterially produced bacteriocins, but also similar amphipathic proteinaceous compounds in eukaryotes such as the mammalian defensins and cathelicidins (169,170), as well as magainin, thanatin, and thionin from frogs, insects, and plants respectively. The widespread existence of these compounds suggests that they evolved fairly early in eukaryotic development, and further suggests that they exercise some sort of resistance proof mechanism of action upon susceptible bacteria. Current evidence indicates that this mechanism of action involves membrane pore formation similar to that induced by many bacteriocins (169). These recent data have spawned a great deal of interest into the discovery of both naturally occurring and synthetic antimicrobial peptides, and suggests that such peptides may provide the key to the continued problems of bacterial antibiotic resistance. Increased integration of these research advances with those in the microbial fields could well provide valuable insights for antimicrobial peptides.

Already, these types of discoveries have sparked large scale projects to synthesize a variety of peptides, particularly cyclic peptides, with the potential of possessing antimicrobial properties (171,172). The aims of these efforts might be greatly aided through advances in our understanding of antimicrobial peptide biosynthesis and modification. Such advances might allow for a similar synthesis in a biological setting through the use of protein engineering and interchangeable peptide coding cassettes, creating systems similar to those responsible for the creation and diversity of mammalian antibodies.

Next, there are a large number of known antimicrobial peptides that have been discovered but as of yet have not found a suitable application. More work should be done to investigate the applicability of these compounds, particularly to new areas of both the food and the medical industry. As previously mentioned, there are already a number of antimicrobial peptides that have shown various uses within the medical field. Along with those compounds already in use, many others are currently under investigation. An increasing wealth of knowledge suggests that the action of antimicrobial peptides can significantly decrease the prevalence of some cancers, particularly colon cancer (173). The mechanism for this prevention is not completely understood but appears to be due to the control of mutagenic compounds within the intestine, either by direct binding of the carcinogens or inhibition of the gut microbes that produce them (174). Also, some bacteriocins, including the colicins and pyocin S2 have even been shown to have anti tumor activities(175,176,177). Further research is needed to expand the knowledge surrounding these potential applications.

Finally, as the size of the antimicrobial peptide repertoire continues to expand, particularly to include compounds with novel inhibition spectrums, and our understanding of antimicrobial peptide immunity and biosynthesis increases, the potential for engineering improved food grade probiotic bacteria which are capable of producing multiple peptides, each of which is capable of inhibiting a specific subset of microbes, become more and more a realistic possibility. Such bacteria could have the potential to help preserve foods, keep them safe, and perhaps even reduce the risk of cancer.

While many of these suggestions are highly speculative, they demonstrate the fact that there is a great deal of potential surrounding antimicrobial peptides, and that more research into the area could very well produce results that could greatly impact and benefit both the food and medical industries.

ACKNOWLEDGMENTS

The authors of this chapter would like to thank Jason Amsel for his graphical assistance.

REFERENCES

1. Tagg, J.R. Bacteriocins of Gram-positive bacteria: an opinion regarding their nature, nomenclature and numbers. In: *Bacteriocins, Microcins and Lantibiotics*, James, R., B.A. Lazazzera, F. Pattus, eds., New York: Springer–Verlag, 1992, pp 33–35.
2. Zuber, P., M.M. Nakano, M.A. Marahiel. *Peptide antibiotics*. In: Bacillus Subtilis *and other Gram-Positive Bacteria: Biochemistry, Physiology and Molecular Genetics*, Sonenshein, A.L., J.A. Hoch, R. Losick, eds., Washington D.C.: ASM Press, 1993, pp 897–916.
3. Tagg, J.R., A.S. Dajani, L.W. Wannamaker. Bacteriocins of Gram-positive bacteria. *Bacteriol. Rev*. 40(3):722–756, 1976.
4. Marahiel, M.A., T. Stachelhaus, H.D. Mootz. Modular peptide synthetases involved in nonribosomal peptide synthesis. *Chem. Rev*. 97(7): 2651–2674, 1997.
5. Doekel, S., M.A. Marahiel. Biosynthesis of natural products on modular peptide synthetases. *Metab. Eng*. 3(1):64–77, 2001.
6. Kleinkauf, H., H. von Doehren. Nonribosomal biosynthesis of peptide antibiotics. *Eur. J. Biochem*. 192(1):1–15, 1990.
7. Stiles, M.E. Biopreservation by lactic acid bacteria. *Antonie van Leeuwenhoek* 70(2–4):331–345, 1996.
8. Klaenhammer, T.R. Genetics of bacteriocins produced by lactic acid bacteria. *FEMS Microbiol. Rev.* 12:39–86, 1993.
9. Sahl, H.-G. Gene-encoded antibiotics made in bacteria. *Ciba. Found. Symp*. 186:27–42, 1994.

10. Roger, L.A. The inhibiting effect of *Streptococcus lactis* on *Lactobacillus bulgaricus*. *J. Bacteriol.* 16:321–325, 1928.
11. Roger, L.A., E.A. Whittier. Limiting factors in lactic fermentation. *J. Bacteriol.* 16:211–214, 1928.
12. Whitehead, H.R. A substance inhibiting bacterial growth, produced by certain strains of lactic strepcocci. *Biochem. J.* 27:1793–1800, 1933.
13. Mattick, A.T.R., A. Hirsch. A powerful inhibitory substance produced by group N streptococci. *Nature* 154:551–552, 1944.
14. Gross, E., J.L. Morell. The structure of nisin. *J. Am. Chem. Soc.* 93:4634–4635, 1971.
15. van den Hooven, H.W., et al. Three-dimensional structure of the lantibiotic nisin in the presence of membrane-mimetic micelles of dodecylphosphocholine and of sodium dodecylsulphate. *Eur. J. Biochem.* 235(1):382–393, 1996.
16. van de Ven, F.J., et al. NMR studies of lantibiotics. The structure of nisin in aqueous solution. *Eur. J. Biochem.* 202(3):1181–1188, 1991.
17. Driessen, A.J., et al. Mechanistic studies of lantibiotic-induced permeabilization of phospholipid vesicles. *Biochemistry* 34(5):1606–1614, 1995.
18. Garcera, M.J., et al. *In vitro* pore-forming activity of the lantibiotic nisin: role of proton motive force and lipid composition. *Eur. J. Biochem.* 212(4): 417–422, 1993.
19. van Kraaij, C., et al. Pore formation by nisin involves translocation of its C-terminal part across the membrane. *Biochemistry* 37:16033–16040, 1998.
20. Chan, W.C., et al. Structure-activity relationships in the peptide antibiotic nisin: role of dehydroalanine 5. *Appl. Environ. Microbiol.* 62(8):2966–2969, 1996.
21. Kuipers, O.P., et al. Characterization of the nisin gene cluster nisABTCIPR of *Lactococcus lactis*: requirement of expression of the nisA and nisI genes for development of immunity. *Eur. J. Biochem.* 216(1):281–291, 1993.
22. Wiedemann, I., et al. Specific binding of nisin to the peptidoglycan precursor lipid II combines pore formation and inhibition of cell wall biosynthesis for potent antibiotic activity. *JBC* 276(1):1772–1779, 2001.
23. Hurst, A. Biosynthesis of the antibiotic nisin by whole *Streptococcus lactis* organisms. *J. Gen. Microbiol.* 44:209–220, 1966.
24. Hurst, A. Biosynthesis of the antibiotic nisin and other basic peptides by *Streptococcus lactis* grown in batch culture. *J. Gen. Microbiol.* 45:503–513, 1966.
25. Ingram, L. A ribosomal mechanism for synthesis of peptides related to nisin. *Biochim. Biophysi. Acta* 224:263–265, 1970.
26. Kaletta, C., K.-D. Entian. Nisin, a peptide antibiotic: cloning and sequencing of the nisA gene and posttranslational processing of its peptide product. *J. Bacteriol.* 171(3):1597–1601, 1989.
27. Donkersloot, J.A., J. Thompson. Simultaneous loss of N5-(carboxyethyl)ornithine synthase, nisin production, and sucrose-fermentation ability by *Lactococcus lactis* K1. *J. Bacteriol.* 172(7):4122–4126, 1990.
28. Steen, M.T., Y.J. Chung, J.N. Hansen. Characterization of the nisin gene as part of a polycistronic operon in the chromosome of *Lactococcus lactis* ATCC 11454. *Appl. Environ. Microbiol.* 57(4): 1181–1188, 1991.
29. Dodd, H.M., N. Horn, M.J. Gasson. Analysis of the genetic determinant for production of the peptide antibiotic nisin. *J. Gen. Microbiol.* 136(3):555–566, 1990.
30. Ingram, L. Synthesis of the antibiotic nisin: formation of lanthionine and beta-methyllanthionine. *Biochim. Biophys. Acta* 184(1):216–219, 1969.
31. Schnell, N., et al. Prepeptide sequence of epidermin, a ribosomally synthesized antibiotic with four sulphide-rings. *Nature* 333(6170):276–278, 1988.
32. de Vos, W.M., O.P. Kuipers. Maturation pathway of nisin and other lantibiotics: posttranslationally modified antimicrobial peptides exported by Gram-positive bacteria. *Mol. Microbiol.* 17(3):427–437, 1995.
33. de Vos, W.M., et al. Genetics of the nisin operon and the sucrose-nisin conjugative transposon Tn5276. *Dev. Biol. Stand.* 85:617–625, 1995.

34. Engelke, G., et al. Biosynthesis of the lantibiotic nisin: genomic organization and membrane localization of the NisB protein. *Appl. Environ. Microbiol.* 58(11):3730–3743, 1992.
35. Siegers, K., K.-D. Entian. Genes involved in immunity to the lantibiotic nisin produced by *Lactococcus lactis* 6F3. *Appl. Environ. Microbiol.* 61(3):1082–1089, 1995.
36. Immone, T., et al. The codon usage of the nisZ operon in *Lactococcus lactis* N8 suggests a non-lactococcal origin of the conjugative nisin-sucrose transposon. *DNA seq.* 5(4):203–218, 1995.
37. de Ruyter, P.G.G.A., et al. Functional analysis of promoters in the nisin gene cluster of *Lactococcus lactis*. *J. Bacteriol.* 178(12):3434–3439, 1996.
38. Ra, R., et al. Effects of gene disruptions in the nisin gene cluster of *Lactococcus lactis* on nisin production and producer immunity. *Microbiol.* 145(5):1227–1233, 1999.
39. Siegers, K., S. Heinzmann, K.-D. Entian. Biosynthesis of lantibiotic nisin: posttranslantional modification of its prepeptide occurs at a multimeric membrane-associated lanthionine synthetase complex. *J. Biol. Chem.*, 271(21):12294–12301, 1996.
40. Fath, M.J., R. Kolter. ABC transporter: bacterial exporters. *Microbiol. Rev.* 57(4):995–1017, 1993.
41. Qiao, M., P.E. Saris. Evidence for a role of NisT in transport of the lantibiotic nisin produced by *Lactococcus lactis* N8. *FEMS Microbiol. Lett.* 144:89–93, 1996.
42. Marykwas, D.L., H.C. Berg. A mutational analysis of the interaction between FliG and FliM, two components of the flagellar motor of *Escherichia coli*. *J. Bacteriol.* 178(5):1289–1294, 1996.
43. Bardwell, A.J., et al. Yeast DNA recombination and repair proteins Rad1 and Rad10 constitute a complex *in vivo* mediated by localized hydrophobic domains. *Mol. Microbiol.* 8(6):1177–1188, 1993.
44. Holt, K.H., et al. Phosphatidylinositol 3-kinase activation is mediated by high-affinity interactions between distinct domains within the p110 and p85 subunits. *Mol. Cell Biol.* 14(1):42–49, 1994.
45. Engelke, G., et al. Regulation of nisin biosynthesis and immunity in *Lactococcus lactis* 6F3. *Appl. Environ. Microbiol.* 60(3):814–825, 1994.
46. van der Meer, J.R., et al. Characterization of the *Lactococcus lactis* nisin A operon genes nisP, encoding a subtilisin-like serine protease involved in precursor processing, and nisR, encoding a regulatory protein involved in nisin biosynthesis. *J. Bacteriol.* 175(9):2578–2588, 1993.
47. Kuiper, O.P., et al. Autoregulation of nisin biosynthesis in *Lactococcus lactis* by signal transduction. *J. Biol. Chem.* 270(45):27299–27304, 1995.
48. Lazazzera, B.A. The intracellular function of extracellular signaling peptides. *Peptides* 22:1519–1527, 2001.
49. de Ruyter, P.G.G.A., O.P. Kuiper, W.M. de Vos. Controlled gene expression systems for *Lactococcus lactis* with the food-grade indecer nisin. *Appl. Environ. Microbiol.* 62(10):3662–3667, 1996.
50. Immone, T., P.E. Saris. Characterization of the nisFEG operon of the nisin Z producing *Lactococcus lactis* subsp. lactis N8 strain. *DNA seq.* 9(5,6):263–274, 1998.
51. Rollema, H.S., et al. Improvement of solubility and stability of the antimicrobial peptide nisin by protein engineering. *Appl. Environ. Microbiol.* 61(8):2873–2878, 1995.
52. Froseth, B.R., R.E. Herman, L.L. McKay. Cloning of nisin resistance determinant and replication origin on 7.6-kilobase EcoRI fragment of pNP40 from *Streptococcus lactis* subsp. diacetylactis DRC3. *Appl. Environ. Microbiol.* 54(8):2136–2139, 1988.
53. Klaenhammer, T.R., R.B. Sanozky. Conjugal transfer from *Streptococcus lactis* ME2 of plasmids encoding phage resistance, nisin resistance and lactose-fermenting ability: evidence for a high frequency conjugative plasmid responsible for abortive infection of virulent bacteriophage. *J. Gen. Microbiol.* 131:1531–1541, 1985.
54. McKay, L.L., K.A. Baldwin. Conjugative 40 megadalton plasmid in *Streptococcus lactis* subsp. diacetylactis is associated with resistace to nisin and baceriophage. *Appl. Environ. Microbiol.* 47(1):68–74, 1984.

55. von Wright, A., et al. Isolation of a replication region of a large lactococcal plasmid and use in cloning of a nisin resistance determinant. *Appl. Environ. Microbiol.* 56(7):2029–2035, 1990.
56. Duan, K., et al. Identification and characterization of a mobilizing plasmid, pND300, in *Lactococcus lactis* M189 and its encoded nisin resistance determinant. *J. Appl. Bacteriol.* 81(5):493–500, 1996.
57. Liu, C.Q., et al. Cloning vectors for lactococci based on a plasmid encoding resistance to cadmium. *Curr. Microbiol.* 33(1):35–39, 1996.
58. Harris, L.J., et al. Antimicrobial activity of lactic acid bacteria against *Listeria monocytogenes. J. Food Protect.* 52:384–387, 1989.
59. Daeschel, M.A., D.S. Jung, B.T. Waston. Controlling wine malolactic fermentation with nisin and nisin-resistant strains of *Leuconostoc oenos. Appl. Environ. Microbiol.* 57:601–603, 1991.
60. Hansen, J.N. Antibiotics synthesized by posttranslational modification. *Annu. Rev. Microbiol.* 47:535–564, 1993.
61. Liu, W., J.N. Hansen. Conversion of *Bacillus subtilis* 168 to a subtilin producer by competence transformation. *J. Bacteriol.* 173(22):7387–7390, 1991.
62. Liu, W., J.N. Hansen. The antimicrobial effect of a structural variant of subtilin against outgrowing *Bacillus cereus* T spores and vegetative cells occurs by different mechanisms. *Appl. Environ. Microbiol.* 59(2):648–651, 1993.
63. Liu, W., J.N. Hansen. Enhancement of the chemical and antimicrobial properties of subtilin by site-directed mutagenesis. *J. Biol Chem.* 267(35):25078–25085, 1992.
64. Nes, I.F., H. Holo. Class II antimicrobial peptides from lactic acid bacteria. *Biopoly* 55:50–61, 2000.
65. Fimland, G., et al. New biologically active hybrid bacteriocins constructed by combining regions from various pediocin-like bacteriocins: the C-terminal region is important for determining specificity. *Appl. Environ. Microbiol.* 62(9):3313–3318, 1996.
66. Fimland, G., et al. The bactericidal activity of pediocin PA-1 is specifically inhibited by a 15-mer fragment that spans the bacteriocin form the center toward the C terminus. *Appl. Environ. Microbiol.* 64(12):5057–5060, 1998.
67. Chikindas, M.L., et al. Pediocin PA-1, a bacteriocin from *Pediococcus acidilactici* PAC1.0, forms hydrophilic pores in the cytoplasmic membrane of target cells. *Appl. Environ. Microbiol.* 59(11):3577–3584, 1993.
68. Bhunia, A.K., M.C. Johnson, N. Kalchayanand. Mode of action of pediocin AcH from *Pediococcus acidilactici* H on sensitive bacterial strains. *J. Appl. Bacteriol.* 70:25–33, 1991.
69. Fleury, Y., et al. Covalent structure, synthesis, and structure-function studies of mesentericin Y 105(37), a defensive peptide from Gram-positive bacteria *Leuconostoc mesenteroides. J. Biol Chem.* 271(24):14421–14429, 1996.
70. Kaiser, A.L., T.J. Montville. Purification of the bacteriocin bavaricin MN and characterization of its mode of action against *Listeria monocytogenes* Scott A cells and lipid vesicles. *Appl. Environ. Microbiol.* 62(12):4529–4535, 1996.
71. Bhunia, A.K., M.C. Johnson, B. Ray. Direct detection of an antimicrobial peptide of *Pediococcus acidilactici* in SDS-PAGE. *J. Indust. Microbiol.* 2:319–322, 1987.
72. Henderson, J.T., A.L. Chopko, P.D. van Wassenaar. Purification and primary structure of pediocin PA-1 produced by *Pediococcus acidilactici* PAC-1.0. *Arch. Biochem. Biophys.* 295(1):5–12, 1992.
73. Nieto Lozano, J.C., et al. Purification and amino acid sequence of a bacteriocin produced by *Pediococcus acidilactici. J. Gen. Microbiol.* 138(9):1985–1990, 1992.
74. Daba, H., et al. Simple method of purification and sequencing of a bacteriocin produced by *Pediococcus acidilactici* UL5. *J. Appl. Bacteriol.* 77(6):682–688, 1994.
75. Ennahar, S., et al. Production of pediocin AcH by *Lactobacillus plantarum* WHE 92 isolated from cheese. *Appl. Environ. Microbiol.* 62(12):4381–4387, 1996.
76. Schved, F., et al. Purification, partial characterization and plasmid linkage of pediocin SJ-1, a bacteriocin produced by *Pediococcus acidilactici. J. Appl. Bacteriol.* 74:67–77, 1993.

77. Christensen, D.P., R.B. Hutkins. Collapse of the proton motive force in *Listeria monocytogenes* caused by a bacteriocin produced by *Pediococcus acidilactici*. *Appl. Environ. Microbiol.* 58(10):3312–3313, 1992.
78. Hoover, D.G., et al. A bacteriocin produced by *Pediococcus* species associated with a 5.5 megadalton plasmid. *J. Food Prot.* 59:29–31, 1988.
79. Taylor, S.L. Bacteriocins of lactic acid bacteria. In: *Food science and technology*, D.G. Hoover, L.R. Steenson, eds., San Diego, CA: Academic Press, Inc., 1993, pp 181-206.
80. Marugg, J.D., et al. Cloning, expression and nucleotide sequence of genes involved in production of pediocin PA-1, a bacteriocin from *Pediococcus acidilactici* PAC 1.0. *Appl. Environ. Microbiol.* 58:2360–2367, 1992.
81. Motlagh, A.M., et al. Nucleotide and amino acid sequence of pap-gene (pediocin AcH production) in *Pediococcus acidilactici* H. *Lett. Appl. Microbiol.* 15:45–48, 1992.
82. Havarstein, L.S., B.D. Diep, I.F. Nes. A family of bacteriocin ABC transporters carry out proteolytic processing of their substrates concomitant with export. *Mol. Microbiol.* 16(2): 229–240, 1995.
83. Ennahar, S., et al. Class IIa bacteriocins: biosynthesis, structure and activity. *FEMS Microbiol. Rev.* 24:85–106, 2000.
84. van Belkum, M.J., M.E. Stiles. Molecular characterization of genes involved in the production of the bacteriocin leucocin A from *Leuconostoc gelidum*. *Appl. Environ. Microbiol.* 61(10):3573–3579, 1995.
85. Venema, K., et al. Functional analysis of the pediocin operon of *Pediococcus acidilactici* PAC1.0: PedB is the immunity protein and PedD is the precursor processing enzyme. *Mol. Microbiol.* 17:515–522, 1995.
86. Worobo, R.W., et al. Characteristics and genetic determinant of a hydrophobic peptide bacteriocin, carnobacteriocin A, produced by *Carnobacterium piscicola* LV17A. *Microbiology* 140(3):517–526, 1994.
87. Quadri, L.E., et al. Characterization of the protein conferring immunity to the antimicrobial peptide carnobacteriocin B2 and expression of carnobacteriocins B2 and BM1. *J. Bacteriol.* 177(5):1144–1151, 1995.
88. Eijsink, V.G., et al. Comparative studies of class IIa bacteriocins of lactic acid bacteria. *Appl. Environ. Microbiol.* 64(9):3275–3281, 1998.
89. Quadri, L.E., et al. Characterization of a locus from Carnobacterium piscicola LV17B involved in bacteriocin production and immunity: evidence for global inducer-mediated transcriptional regulation. *J. Bacteriol.* 179(19):6163–6171, 1997.
90. Saucier, L., A. Poon, M.E. Stiles. Induction of bacteriocin in *Carnobacterium piscicola* LV17. *J. Appl. Bacteriol.* 78:281–91, 1995.
91. Grinstead, D.A., S.F. Barefoot. Jenseniin G, a heat-stable bacteriocin produced by *Propionibacterium jensenii* P126. *Appl. Environ. Microbiol.* 58(1):215–220, 1992.
92. Faye, T., et al. Biochemical and genetic characterization of propionicin T1, a new bacteriocin from *Propionibacterium thoenii. Appl. Environ. Microbiol.* 66(10): 4230–4236, 2000.
93. Lyon, W.J., J.K. Sethi, B.A. Glatz. Inhibition of psychrotrophic organisms by propionicin PLG-1, a bacteriocin produced by *Propionibacterium thoenii. J. Dairy Sci.*, 76(6):1506–1513, 1993.
94. Miescher, S., et al. Propionicin SM1, a bacteriocin from *Propionibacterium jensenii* DF1: isolation and characterization of the protein and its gene. *Syst. Appl. Microbiol.* 23(2):174–184, 2000.
95. Faye, T., et al. An antimicrobial peptide is produced by extracellular processing of a protein from *Propionibacterium jensenii. J. Bacteriol.* 184(13):3649–3656, 2002.
96. Cintas, L.M., et al. Biochemical and genetic characterization of enterocin P, a novel ssec-dependent bacteriocin from *Enterococcus faecium* P13 with a broad antimicrobial spectrum. *Appl. Environ. Microbiol.* 63:4321–4330, 1997.
97. Kalmokoff, M.L., et al. Identification of a new plasmid-encoded sec-dependent bacteriocin produced by *Listeria innocua* 743. *Appl. Environ. Microbiol.* 67(9):4041–4047, 2001.

98. Hauge, H.H., et al. Amphiphilic alpha-helices are important structural motifs in the alpha and beta peptides that constitute the bacteriocin lactococcin G: enhancement of helix formation upon alpha-beta interaction. *Eur. J. Biochem.* 251(3):565–572, 1998.
99. Hauge, H.H., et al. Membrane-mimicking entities induce structuring of the two-peptide bacteriocins plantaricin E/F and plantaricin J/K. *J. Bacteriol.* 181(3):740–747, 1999.
100. Diep, B.D., L.S. Havarstein, I.F. Nes. Characterization of the locus responsible for the bacteriocin production in *Lactobacillus plantarum* C11. *J. Bacteriol.* 178(15):4472–4483, 1996.
101. Venema, K., et al. Mode of action of lactococcin B, a thiol-activated bacteriocin from *Lactococcus lactis. Appl. Environ. Microbiol.* 59(4):1041–1048, 1993.
102. Venema, K., et al. Mutational analysis and chemical modification of Cys24 of lactococcin B, a bacteriocin produced by *Lactococcus lactis. Microbiology* 142(10):2825–2830, 1996.
103. Morgan, S., R.P. Ross, C. Hill. Bacteriolytic activity caused by the presence of a novel lactococcal plasmid encoding lactococcins A, B and M. *Appl. Environ. Microbiol.* 61(8):2995–3001, 1995.
104. Joerger, M.C., T.R. Klaenhammer. Characterization and purification of helveticin J and evidence for a chromosomally determined bacteriocin produced by *Lactobacillus helveticus* 481. *J. Bacteriol.* 167(2):439–446, 1986.
105. Toba, T., E. Yoshioka, T. Itoh. Lacticin, a bacteriocin produced by *Lactobacillus delbrueckii* subsp. lactis. *Lett. Appl. Microbiol.* 12:43–45, 1991.
106. Toba, T., E. Yoshioka, T. Itoh. Acidophilucin A, a new heat-labile bacteriocin produced by *Lactobacillus acidophilus* LAPT 1060. *Lett. Appl. Microbiol.* 12:106–108, 1991.
107. Rammelsberg, M., E. Muller, F. Radler. Caseincin 80: purification and characterization of a new bacteriocin from *Lactobacillus casei. Arch. Microbiol.* 154:249–252, 1990.
108. Joerger, M.C., T.R. Klaenhammer. Cloning, expression, and nucleotide sequence of the *Lactobacillus helveticus* 481 gene encoding the bacteriocin helveticin J. *J. Bacteriol.* 172(11):6339–6347, 1990.
109. Lewus, C.B., S. Sun, T.J. Montville. Production of an amylase-sensitive bacteriocin by a typical *Leuconostoc paramesenteroides* strain. *Appl. Environ. Microbiol.* 58(1):143–149, 1992.
110. Lewus, C.B., A.L. Kaiser, T.J. Montville. Inhibition of food-borne bacterial pathogens by bacteriocins from lactic acid bacteria isolated from meat. *Appl. Environ. Microbiol.* 57(6):1683–1688, 1991.
111. Waite, B.L., R.W. Hutkins. Bacteriocins inhibit glucose PEP:PTS activity in *Listeria monocytogenes* by induced efflux of intracellular metabolites. *J. Appl. Microbiol.* 85(2):287–292, 1998.
112. Upreti, G.C. Lactocin 27, a bacteriocin produced by homofermentative *Lactobacillus helveticus*. In: *Bacteriocins of Lactic Acid Bacteria: Microbiology, Genetics and Applications*, de Vuyst, L., E.J. Vandamme, eds., London: Blackie Academic & Professional, 1994.
113. Upreti, G.C., R.D. Hinsdill. Production and mode of action of lactocin 27: bacteriocin from a homofermentative Lactobacillus. *Antimicrob. Agents Chemother.* 7(2):139–145, 1975.
114. Jack, R., G. Jung. Lantibiotics and microcins: polypeptides with unusual chemical diversity. *Curr. Opin. Chem. Biol.* 4(3):310–317, 2000.
115. Braun, V., H. Pilsl, P. Gross. Colicins: structures, modes of action, transfer through membranes. *Arch. Microbiol.* 161:199–206, 1994.
116. Lazdunski, C., D. Cavard. Colicins: a minireview. *Toxicon* 20(1):223–228, 1982.
117. Martinez, J.L., J.C. Perez-Diaz. Cloning of the determinants for microcin D93 production and analysis of three different D-type microcin plasmids. *Plasmid* 23:216–225, 1990.
118. Baquero, F., F. Moreno. The microcins. *FEMS Microbiol. Lett.* 23:117-124, 1984.
119. Gonzalez-Pastor, J.E., et al. Structure and organization of plasmid genes required to produce the translation inhibitor microcin C7. *J. Bacteriol.* 177(24):7131–7140, 1995.
120. Khmel, I.A., et al. Microcins: new peptide antibiotics of Enterobacteria and genetic control of their synthesis. *Molekulyarnaya Biologiya* 33(1):96–102, 1999.

121. Duro, A.F., R. Serrano, C. Asensio. Effect of the antibiotic microcin 140 on the ATP level and amino acid transport of the *Escherichia coli*. *Biochem. Biophys. Res. Commun.* 88(1):297–304, 1979.
122. Martinez, J.L., J.C. Perez-Diaz. Isolation, characterization, and mode of action on *Escherichia coli* strains of microcin D93. *Antimicrob. Agents Chemother.* 29(3):456–460, 1986.
123. Blond, A., et al. The cyclic structure of microcin J25, a 21-residue peptide antibiotic from *Escherichia coli*. *Eur. J. Biochem.* 259:747–755, 1999.
124. Wilkens, M., et al. Cloning and expression in *Escherichia coli* of genetic determinants for production of and immunity to microcin E492 from *Klebsiella pneumoniae*. *J. Bacteriol.* 179(15):4789–4794, 1997.
125. Milne, J.C., et al. ATP/GTP hydrolysis is required for oxazole and thiazole biosynthesis in the peptide antibiotic microcin B17. *Biochemistry* 37:13250–13261, 1998.
126. Madison, L.L., et al. The leader peptide is essential for the post-translational modification of the DNA-gyrase inhibitor microcin B17. *Mol. Microbiol.* 23(1):161–168, 1997.
127. Lagos, R., J.E. Villanueva, O. Monasterio. Identification and properties of the genes encoding microcin E492 and its imminity protein. *J. Bacteriol.* 181(1):212–217, 1999.
128. Lagos, R., et al. Structure, organization and characterization of the gene cluster involved in the production of microcin E492, a channel-forming bacteriocin. *Mol. Microbiol.* 42(1):229–243, 2001.
129. Rodriguez, E., C. Gaggero, M. Lavina. The structural gene for microcin H47 encodes a peptide precursor with antibiotic activity. *Antimicrob. Agents Chemother.* 43(9):2176–2182, 1999.
130. Solbiati, J.O., et al. Genetic analysis of plasmid determinants for microcin J25 production and immunity. *J. Bacteriol.* 178(12):3661–3663, 1996.
131. Rodriguez-Sainz, M.C., C. Hernandez-Chico, F. Moreno. Molecular characterization of pmbA, an *Escherichia coli* chromosomal gene required for the production of the antibiotic peptide MccB17. *Mol. Microbiol.* 4(11):1921–1932, 1990.
132. Gaggero, C., F. Moreno, M. Lavina. Genetic analysis of microcin H47 antibiotic system. *J. Bacteriol.* 175(17):5420–5427, 1993.
133. Moreno, F., et al. *Escherichia coli* genes regulating the production of microcins MccB17 and MccC7. In: *Bacteriocins, Microcins and Lantibiotics*, James, R., C. Lazdunski, F. Pattus, eds., Berlin: Springer-Verlag. 1992, pp 3–13.
134. Chiuchiolo, M.J., et al. Growth-phase-dependent expression of the cyclopeptide antibiotic microcin J25. *J. Bacteriol.* 183(5):1755–1764, 2001.
135. Smajs, D., H. Pilsl, V. Braun. Colicin U, a novel colicin produced by *Shigella boydii*. *J. Bacteriol.* 179(15):4919–4928, 1997.
136. Harkness, R.E., T. Olschlager. The biology of colicin M. *FEMS Microbiol. Rev.* 8(1):27–41, 1991.
137. Frey, J., et al. Physical and genetic analysis of the ColD plasmid. *J. Bacteriol.* 166(1):15–19, 1986.
138. Lakey, J.H., S.L. Slatin. Pore-forming colicins and their relatives. *Curr. Top. Microbiol. Immunol.* 257:131–161, 2001.
139. Bullock, J.O., E.R. Kolen. Ion selectivity of colicin E1, III: anion permeability. *J. Membr. Biol.* 144(2):131–145, 1995.
140. Smarda, J., et al. Modes of action of colicins E4-E7: rates of basic biosyntheses inhibition. *Zentralbl. Bakteriol. Mikrobiol. Hyg.* 269(1):7–14, 1988.
141. Tomita, K., et al. A cytotoxic ribonuclease which specifically cleaves four isoaccepting arginine tRNAs at their anticodon loops. *Proc. Natl. Acad. Sci. USA* 97(15):8278–8283, 2000.
142. Kleanthous, C., D. Walker. Immunity proteins: enzyme inhibitors that avoid the active site. *Trends Biochem. Sci.* 26(10):624–631, 2001.
143. Cramer, W.A., et al. Structure-function of the channel-forming colicins. *Annu. Rev. Biophys. Biomol. Struct.* 24:611–641, 1995.
144. Gross, P., V. Braun. Colicin M is inactivated during import by its immunity protein. *Mol. Gen. Genet.* 251(3):388–396, 1996.

145. Riley, M.A., D.M. Gordon. A survey of Col plasmids in natural isolates of *Escherichia coli* and an investigation into the stability of Col-plasmid lineages. *J. Gen. Microbiol.* 138(7):1345–1352, 1992.
146. Spangler, R., et al. Colicin synthesis and cell death. *J. Bacteriol.* 163(1):167–173, 1985.
147. Vollmer, W., et al. Pesticin displays muramidase activity. *J. Bacteriol.* 179(5):1580–1583, 1997.
148. Rakin, R., E. Boolgakowa, J. Heesemann. Structural and functional organization of the *Yersinia pestis* bacteriocin pesticin gene cluster. *Microbiology* 142(12):3415–3424, 1996.
149. Kageyama, M., et al. Construction and characterization of pyocin-colicin chemeric proteins. *J. Bacteriol.* 178(1):103–110, 1996.
150. Sano, Y., M. Kobayashi, M. Kageyama. Functional domains of S-type pyocins deduced from chimeric molecules. *J. Bacteriol.* 175(19):6179–6185, 1993.
151. Sano, Y., M. Kageyama. Genetic determinant of pyocin AP41 as an insert in the *Pseudomonas aeruginosa* chromosome. *J. Bacteriol.* 158(2):562–570, 1984.
152. Matsui, H., et al. Regulation of pyocin genes in *Pseudomonas aeruginosa* by positive (prtN) and negative (prtR) regulatory genes. *J. Bacteriol.* 175(5):1257–1263, 1993.
153. Lipmann, F. The relation between the direction and mechanism of polymerization. *Essays Biochem.* 4:1–23, 1968.
154. Lipmann, F. Bacterial production of antibiotic polypeptides by thiol-linked synthesis on protein templates. *Adv. Microbiol. Physiol.* 21:227–266, 1980.
155. Turgay, K., M. Krause, M.A. Marahiel. Four homologous domains in the primary structure of GrsB are related to domains in a superfamily adenylated-forming enzymes. *Mol. Microbiol.* 6:529–546, 1992.
156. Cosmina, P., et al. Sequence and analysis of the genetic locus responsible for surfactin synthesis in *Bacillus subtilis. Mol. Microbiol.* 8(5):821–831, 1993.
157. Krause, M., et al. Molecular cloning of an ornithine-activating fragment of the gramicidin S synthetase 2 gene from *Bacillus brevis* and its expression in *Escherichia coli. J. Bacteriol.* 162(3):1120–1125, 1985.
158. Javis, B., R.R. Mahoney. Inactivation of nisin by alpha-chymotrypsin. *J. Dairy Sci.* 52:1448–1450, 1969.
159. Wilimowska-Pelc, A., et al. The use of gel-filtration for the isolation of pure nisin from commercial products. *Acta Microbiol. Pol.* 25:71–77, 1976.
160. Heinemann, B., R. Williams. The unactivation of nisin by pancreatin. *J. Dairy Sci.* 49:312–313, 1966.
161. Delves-Broughton, J., et al. Applications of the bacteriocin, nisin. *Antonie van Leeuwenhoek* 69:193–202, 1996.
162. Delves-Broughton, J. Nisin and its uses as a food preservative. *Food Technol.* 44(11):100, 102, 104, 108, 111–112, 117, 1990.
163. Rodriguez, J.M., M.I. Martinez, J. Kok. Pediocin PA-1, a wide-spectrum bacteriocin from lactic acid bacteria. *Crit. Rev. Food Sci. Nutr.* 42(2):91–121, 2002.
164. Yousef, A.E., et al. Behavior of *Listeria monocytogenes* in wiener exudates in the presence of *Pediococcus acidilactici* H or pediocin AcH during storage at 4 or 25 C. *Appl. Environ. Microbiol.* 57(5):1461–1467, 1991.
165. Degnan, A.J., A.E. Yousef, J.B. Luchansky. Use of *Pediococcus acidilactici* to control *Listeria monocytogenes* in temperature-abused cacuum-pack-aged wieners. *J. Food Prot.* 55:98–103, 1992.
166. Choi, S.Y., L.R. Beuchat. Growth inhibition of *Listeria monocytogenes* by a bacteriocin of *Pediococcus adicilactici* M during fermentation of kimchi. *Food Microbiol.* 11:301–307, 1994.
167. Holzapfel, W.H., R. Geisen, U. Schillinger. Biological preservation of foods with reference to protective cultures, bacteriocins and food-grade enzymes. *Int. J. Food Microbiol.* 24:343–362, 1995.
168. Muriana, P.M. Bacteriocins for control of *Listeria* spp. in food. *J. Food Protect.* 54–63, 1996.

169. Zasloff, M. Antimicrobial peptides of multicellular organisms. *Nature* 415(6870):389–395, 2002.
170. Nizet, V., et al. Innate antimicrobial peptide protects the skin from invasive bacterial infection. *Nature* 414(6862):454–457, 2001.
171. Fernandez-Lopez, S., et al. Antibacterial agents based on the cyclic D,L-alpha-peptide architecture. *Nature* 412(6845):452–455, 2001.
172. Ganz, T. Rings of destruction. *Nature* 412(6845):392–393, 2001.
173. Bures, J., et al. Colicinogeny in nonspecific intestinal inflammations and colorectal cancer. *Sb. Ved. Pr. Lek. Fak. Karlovy Univerzity Hradci Kralove Suppl.* 34(3):349–403, 1991.
174. de Roos, N.M. and M.B. Katan. Effects of probiotic bacteria on diarrhea, lipid metabolism, and carcinogenesis: a review of papers published between 1988 and 1998. *Am. J. Clin. Nutr.* 71(2):405–411, 2000.
175. Bures, J., et al. Colicinogeny in colorectal cancer. *Neoplasma* 33(2):233–237, 1986.
176. Watanabe, T., H. Saito. Cytotoxicity of pyocin S2 to tumor and normal cells and its interaction with cell surfaces. *Biochim. Biophys. Acta.* 633(1):77–86, 1980.
177. Farkas-Himsley, H., R. Cheung. Baterial proteinaceous products (bacteriocins) as cytotoxic agents of neoplasia. *Cancer Res.* 36(10):3561–3567, 1976.
178. Nilsen, T., I.F. Nes, H. Holo. An exported inducer peptide regulates bacteriocin production in *Enterococcus faecium* CTC492. *J. Bacteriol.* 180(7):1848-1854, 1998.
179. Buchman, G.W., S. Banerjee, J.N. Hansen. Structure, expression, and evolution of a gene encoding the precursor of nisin, a small protein antibiotic. *J. Biol. Chem.* 263(31):16260–16266, 1988.
180. Chung, Y.J., M.T. Steen, J.N. Hansen. The subtilin gene of *Bacillus subtilis* ATCC 6633 is encoded in an operon that contains a homolog of the hemolysin B transport protein. *J. Bacteriol.* 174(4):1417–1422, 1992.
181. Kaletta, C., et al. *Pep5,* a new lantibiotic: structural gene isolation and prepeptide sequence. *Arch. Microbiol.* 152(1):16–19, 1989.
182. Kellner, R., et al. Gallidermin: a new lanthionine-containing polypeptide antibiotic. *Eur. J. Biochem.* 177(1):53–59, 1988.
183. Navarro, J., et al. Interaction of duramycin with artificial and natural membranes. *Biochemistry* 24(17):4645–4650, 1985.
184. Fredenhagen, A., et al. Duramycins B and C, two new lanthionine containing antibiotics as inhibitors of phospholipase A2: structural revision of duramycin and cinnamycin. *J. Antibiot.* 43(11):1043–1012, 1990.
185. Kido, Y., et al. Isolation and characterization of ancovenin, a new inhibitor of angiotensin I converting enzyme, produced by *Actinomycetes. J. Antibiot.* 36(10):1295–1299, 1983.
186. Chatterjee, S., et al. Mersacidin, a new antibiotic from *Bacillus*: fermentation, isolation, purification and chemical characterization. *J. Antibiot.* 45(6):832–838, 1992.
187. Kettenring, J.K., et al. Sequence determination of actagardine, a novel lantibiotic, by homonuclear 2D NMR spectroscopy. *J. Antibiot.* 43(9):1082–1088, 1990.
188. Hastings, J.W., et al. Characterization of leucocin A-UAL 187 and cloning of the bacteriocin gene from *Leuconostoc gelidum. J. Bacteriol.* 173(23):7491–7500, 1992.
189. Bennik, M.H., et al. A novel bacteriocin with a YGNGV motif from vegetable-associated *Enterococcus mundtii*: full characterization and interaction with target organisms. *Biochim. Biophys. Acta* 1373(1):47–58, 1998.
190. Jack, R.W., et al. Characterization of the chemical and antimicrobial properties of piscicolin 126, a bacteriocin produced by *Carnobacterium piscicola* JG126. *Appl. Environ. Microbiol.* 62(8):2897–2903, 1996.
191. Larsen, A.G., F.K. Vogensen, J. Josephsen. Antimicrobial activity of lactic acid bacteria isolated from sour doughs: purification and characterization of bavaricin A, a bacteriocin produced by *Lactobacillus bavaricus* MI401. *J. Appl. Bacteriol.* 75(2):113–122, 1993.
192. Tichaczek, P.S., R.F. Vogel, W.P. Hammes. Cloning and sequencing of sakP encoding sakacin P, the bacteriocin produced by *Lactobacillus sake* LTH 673. *Microbiology* 140(2):361–367, 1994.

193. Metivier, A., et al. Divercin V41, a new bacteriocin with two disulphide bonds produced by *Carnobacterium divergens* V41: primary structure and genomic organization. *Microbiology* 144(10):2837–2844, 1998.
194. Aymerich, T., et al. Biochemical and genetic characterization of enterocin A from Enterococcus faecium, a new antilisterial bacteriocin in the pediocin family of bacteriocins. *Appl. Environ. Microbiol.* 62(5):1676–1682, 1996.
195. Quadri, L.E., et al. Chemical and genetic characterization of bacteriocins produced by Carnobacterium piscicola LV17B. *J. Biol. Chem.* 269(16):12204–12211, 1994.
196. Holck, A., et al. Purification and amino acid sequence of sakacin A, a bacteriocin from Lactobacillus sake Lb706. *J. Gen. Microbiol.* 138(12):2715–2720, 1992.
197. Kanatani, K., M. Oshimura, K. Sano. Isolation and characterization of acidocin A and cloning of the bacteriocin gene from Lactobacillus acidophilus. *Appl. Environ. Microbiol.* 61(3):1061–1067, 1995.
198. Tomita, H., et al. Cloning and genetic organization of the bacteriocin 31 determinant encoded on the Enterococcus faecalis pheromone-responsive conjugative plasmid pYI17. *J. Bacteriol.* 178(12):3585–3593, 1996.
199. Axelsson, L., A. Holck. The genes involved in production of and immunity to sakacin A, a bacteriocin from Lactobacillus sake Lb 706. *J. Bacteriol.* 177:2125–2137, 1995.
200. Huehne, K., et al. Analysis of the sakacin P gene cluster from Lactobacilllus sake Lb674 and its expression in sakacin-negative L. sake strains. *Microbiol.* 142:1437–1448, 1996.
201. Bukhtiyarova, M., R. Yang, B. Ray. Analysis of the pediocin AcH gene cluster from plasmid pSMB74 and its expression in a pediocin -negative Pediococcus acidilactici strain. *Appl. Environ. Microbiol.* 60:3405–3408, 1994.
202. Fremaux, C., Y. Hechard, Y. Cenatiempo. Mesentericin Y105 gene clusters in *Leuconostoc mesenteroides* Y105. *Microbiology* 141:1637–1645, 1995.
203. Nissen-Meyer, J., et al. A novel lactococcal bacteriocin whose activity depends on the complementary action of two peptides. *J. Bacteriol.* 174(17):5686–5692, 1992.
204. Allison, G.E., C. Fremaux, T.R. Klaenhammer. Expansion of bacteriocin activity and host range upon complementation of two peptides encoded within the lactacin F operon. *J. Bacteriol.* 176(8):2235–2241, 1994.
205. van Belkum, M.J., et al. Cloning of two bacteriocin genes from a lactococcal bacteriocin plasmid. *Appl. Environ. Microbiol.* 55(5):1187–1191, 1989.
206. Stephens, S.K., et al. Molecular analysis of the locus responsible for production of plantaricin S, a two-peptide bacteriocin produced by *Lactobacillus plantarum* LPCO10. *Appl. Environ. Microbiol.* 64(5):1871–1877, 1998.
207. Balla, E., et al. Characterization and cloning of the genes encoding enterocin 1071A and enterocin 1071B, two antimicrobial peptides produced by *Enterococcus faecalis* BFE 1071. *Appl. Environ. Microbiol.* 66(4):1298–1304, 2000.
208. Marciset, O., et al. Thermophilin 13, a nontypical antilisterial poration complex bacteriocin, that functions without a receptor. *J. Biol. Chem.* 272(22):14277–14284, 1997.
209. Tahara, T., et al. Isolation, partial characterization, and mode of action of Acidocin J1132, a two-component bacteriocin produced by *Lactobacillus acidophilus* JCM 1132. *Appl. Environ. Microbiol.* 62(3):892–897, 1996.
210. Garneau, S., et al. Purification and characterization of Brochocin A and Brochocin B(10-43): a functional fragment generated by heterologous expression in *Carnobacterium piscicola*. *Appl. Environ. Microbiol.* 69(3):1352–1358, 2003.
211. Moll, G., T. Urbink-Kok, H. Hildeng-Hauge, J. Nissen-Meyer, I.F. Nes, W.N. Konings, J.M. Driessen. Lactococcin G is a potassium ion-conducting two-component bacteriocin. *J. Bacteriol.* 178(3):600–605, 1996.

3.08

Phenolic Antimicrobials from Plants for Control of Bacterial Pathogens

Kalidas Shetty and Yuan-Tong Lin

CONTENTS

8.1 PHENOLIC ANTIMICROBIALS FROM PLANTS

Plants are excellent sources of phenolic metabolites. In particular, phenolic antioxidants from food grade plants have potential for long term chemo preventive and therapeutic applications against oxidation linked diseases (1–4) and increasingly have antimicrobial potential (2). Such phenolic metabolites are also broadly called phytochemicals. Phenolic phytochemicals have shown to be associated with antioxidative action in biological systems, acting as scavengers of singlet oxygen and free radicals (5,6). Much of this work has emphasized the role of phenolics from higher plants in relation to human health and in particular cancer and cardiovascular health (3,7,8). Phenolic phytochemicals (phenylpropanoids) serve as effective antioxidants (phenolic antioxidants) due to their ability to donate hydrogen from hydroxyl groups positioned along the aromatic ring to terminate free radical oxidation of lipids and other bio molecules (9). Phenolic antioxidants therefore short circuit a destructive chain reaction that ultimately degrades cellular membranes and in prokaryotes such phenolic antioxidants have potential for antimicrobial activity.

Plant phenolic extracts that impart flavor and aroma also have potential for inhibiting pathogenic microorganisms (10–12). These phenolic secondary metabolites are defensive antimicrobials produced against invading pathogens and stress and therefore the methods for exploiting them have to take this into account. In certain cases the induction is associated with action of diphenolic oxidases, and resulting modified compounds can have antimicrobial activity (13). In other situations dihydroxy phenolics are oxidized to quinones, which can interact with the proteins of the invading pathogens, forming melanoid polymers (13). Quinones are aromatic rings with 2 ketone substitutions and are highly reactive (10). These compounds are responsible for the enzymatic browning reaction of cut fruits and vegetables, and are an intermediate in melanin pigment production in humans (10). The quinones are sources of stable free radicals and complex irreversibly with nucleophilic amino acids and proteins (14) leading to inactivation of protein and loss of function (14). Therefore potential antimicrobial benefits of quinones are substantial (14). The potential targets for inhibition in the bacterial cells are surface adhesions; cell wall polypeptides and membrane bound enzymes (14).

In many herb and spice species investigated so far phenolic derivatives found in essential oils have been largely investigated for antimicrobial activity. Hydrophobic phenolics derived from the terpenoid pathway, thymol, and carvacrol, present in essential oil of thyme, oregano, savory, sage, and related species, have antimicrobial activity (11). The essential oil containing thymol can inhibit *Vibrio parahaemolyticus* (15). The addition of 0.05% of alcoholic extracts of thyme can inhibit the growth of *Staphylococcus aureus* (16). Sage extract was inhibitory to *Bacillus cereus* and *S. aureus* (17). Rosemary extract of 0.1% substantially inhibited the growth of *S. aureus* and *Salmonella typhimurium* (18). Among other types of phenolics, hydroxycinnamic acid derivatives such as caffeic acid, ferulic acid and p-coumaric acid derived from the phenylpropanoid pathway inhibit *E. coli, S. aureus* and *B. cereus* (19). Polymeric phenolics, such as tannins are inhibitory toward *Listeria monocytogenes, E. coli, S. aureus, Aeromonas hydrophila,* and *Streptococcus faecalis* (20). Hydroxylated phenols, such as catechol and pyrogallol, are known to be toxic to microorganisms (10). The site and number of hydroxyl groups is linked to the antimicrobial effect and, in some cases, more oxidized forms are more inhibitory (21,22). In other combination studies, growth of *E. coli, Salmonella spp., L. monocytogenes,* and *S. aureus* were inhibited by oregano essential oils in broth cultures and its effect in emulsion systems depended on pH, temperature, and other environmental factors (23,24). In meat studies essential oil of oregano inhibited *L. monocytogenes* and *S. typhimurium*

(25,26). Other diverse examples of antimicrobial effects of herbs and spices are documented well in Tassou et al. (27).

The mode of action of phenolics against bacterial pathogens has not been defined or clearly understood. It is suspected changes in membrane permeability through lipid compatible hydrophobic phenolics, like thymol from certain essential oils, and membrane localized hyperacidity from water and ethanol soluble phenolics like benzoic acid and rosmarinic acid, may affect proton motive force across the membrane and therefore energy depletion may take place (11,28,29). It is suggested as a result of damage to structural and functional properties of membranes, pH gradient and electrical potential of the proton motive force is disrupted (30,31). The disruption of the proton motive force and reduction of ATP pool lead to cell death (11,31,32). Further, leakage of ions, nucleic acid, and amino acids can occur (27,33–35). It is also proposed that enzyme inhibition by the oxidized compounds through reaction with sulfhydryl groups, or through nonspecific interactions with membrane proteins, may also be the reason for inhibition (10,36). Other proposed mechanisms include formation of Schiff's bases with membrane proteins by aldehyde groups of phytochemicals, which prevent cell wall biosynthesis (37,38), and interaction of ferrous iron with phenolic compounds, which can damage membranes by enhancing oxidative stress (39).

In spite of wide investigations around the world, using many diverse botanical sources, the results of antimicrobial efficacy of phytochemicals are mixed, to very poor, and mechanism of action is poorly understood. The three major reasons for this are:

1. Many phytochemicals have been derived from mixed heterogeneous genetic sources and consistency cannot be guaranteed from batch to batch and from source to source (1,2,40).
2. Many of the antimicrobial phytochemicals, and especially phenolics, are biotic or abiotic stress inducible and, if extracts are used from uninduced states, the efficacy is potentially low.
3. Investigations have largely focused on hydrophobic essential oils and focused solely on disruption of plasma membrane related functions.

Now, with the emergence of antibiotic resistance from overuse of single antibiotics from fungal and bacterial sources, new strategies using plant based sources are promising (10) and urgently needed (2,40). In line with this need, one of our major strategies focus on the hypothesis that high antioxidant phenolics from single seed origin clonal lines of herbs such as oregano, thyme, rosemary, and lavender, and genetically uniform sprouted legumes such as soybean, fava bean, fenugreek, chickpea, pea, and mung bean, which are self pollinating species would have excellent antimicrobial potential. Genetically uniform, high antioxidant phenolic profiles have been screened, evaluated for antioxidant efficacy and targeted to inhibit bacterial pathogens (41,42). In this screening strategy, both the concept of phytochemical consistency through clonal lines in herbs and stress (elicitor) based inducibility of phenolic antioxidants in self pollinating and genetically uniform legumes can be developed to explore the antimicrobial potential against bacterial pathogens. Further, we have targeted only water and ethanol extracted phenolic profiles from these systems that could work effectively and synergistically to inhibit biochemical targets at the cellular membrane and cytosol of the bacterial pathogen at the same time, thereby impacting multiple targets and offering fewer opportunities for resistance by the pathogen.

8.2 POTENTIAL OF *LAMIACEAE* AS SOURCE OF PHENOLIC ANTIMICROBIALS

Extracts from dietary herb species belonging to the family *Lamiaceae* (mint family) have been used by humanity as sources of complimentary medicine and food preservatives for over 4000 years. It is only recently that some of the bioactive components linked to medicinal and preservative function have been determined to be phenolic metabolites (4,43,44). Specific phenolic metabolites from Lamiaceae like rosmarinic acid (from rosemary, spearmint, thyme, lavender, and oregano) and thymol (from thyme and oregano) have anti inflammatory (4,45), antioxidant and antimicrobial properties (11,46–48), respectively. The antioxidant pharmacological functions from food grade dietary phytochemicals from various mint family species could contribute to long term prevention of oxidation influenced disease such as diabetes, cancer, CVD, cognition and inflammatory diseases (1,3,8). In general food processing applications oregano, rosemary, thyme, and lavender has been used widely as a source of flavors and potential food antioxidants (49–51). These herb species are native to the Mediterranean and are widely used in South European cuisine. The beneficial phenolic antioxidants that are being targeted for food and health applications in our research are biphenyl, rosmarinic acid that also has antioxidant, anti inflammatory, and antimicrobial properties (4,45,47), and several other simple soluble phenolic metabolites have antioxidant properties (49,51). The beneficial antioxidant activity from phenolic metabolites was also observed in many food preservation conditions such as in lard (50), salad dressing (52) and other food model systems. The phenolic antioxidants from these herb species also have potential antimicrobial properties (2,11,27,40,47,48) and, being partially ethanol and water soluble, could be effective at inhibiting targets on bacterial cell membrane as well as key dehydrogenase linked metabolic reactions in the cytosol, as opposed to being effective only in the hydrophobic regions as in case of wide variety of plant essential oils. These water and ethanol extracted phenolics are also less volatile and have fewer flavor related problems compared to volatile essential oils targeted to cell membranes. The water and ethanol extractable phenolics are also more compatible in a more diverse array of foods, are less volatile, and have fewer flavor problems.

8.2.1 Control of *Staphylococcus Aureus* by Phenolic Phytochemicals as a Model for Plant Based Antimicrobials

Staphylococcus aureus is an important pathogen in humans that has significant impact in the food chain (11) and high density communities and hospitals (53). The pathogen causes serious infections in open wounds, vascular tissues, bones, and joints (53,54). The importance of *S. aureus* as the etiological agent in catheter related and peritoneal dialysis patients is well known and often requires aggressive antibiotic treatment (55,56). Further it is well known that as a result of antibiotic resistance, it has been difficult to control infections ranging from abscesses, pneumonia, endocarditis, septicemia, and toxic shock syndrome. The one last line of defense is the antibiotic vancomycin (57–59). *Staphylococcus aureus* infections leading to complications from septic and toxic shock syndromes may result in organ failure (60,61). The existence of ecologically abundant hyper virulent clones has suggested that factors promoting ecological fitness of *S. aureus* also increase its virulence (62). The prospects of untreatable *S. aureus* infections have raised the need to search for alternative therapies, and development of a broadly protective vaccine based on an *in vivo* expressed antigen has some merit (63). In addition, consistent clonal profiles of plant phenolic metabolites from food grade herbs and inducible phenolics from legume sprouts can be an important complimentary therapy which can be used in conjunction with other therapies, including new microbial based antibiotics and vaccines.

Studies have indicated the antimicrobial potential of plant phenolic metabolites against *S. aureus*. The addition of 0.05% alcoholic extract of thyme can inhibit the growth of *Staphylococcus aureus* (16). Sage extract was also inhibitory to *S. aureus* (17). Rosemary extract of 0.1% substantially inhibited the growth of *S. aureus* and *Salmonella typhimurium* (18). Hydroxycinnamic acid derivatives such as caffeic acid, ferulic acid, and p-coumaric acid inhibited *S. aureus* (19). Polymeric phenolics, such as tannins, are inhibitory toward *S. aureus* (20). The hypothesis that elicitor inducible phenolic antioxidants from clonal herbs and legumes can also inhibit *S. aureus* holds much promise for improved efficacy and is being currently investigated. We also speculate that plant based phenolic antioxidants of consistent clonal origin have the potential to support human antioxidant enzyme response and therefore reduce human tissue damage from *S. aureus* toxin induced apoptosis (61) and improve host antimicrobial immune response and these investigations on mammalian cell responses are part of our future investigations.

8.2.2 Control of *Listeria Monocytogenes* by Oregano Clonal Extracts as a Model for Developing Herb Phenolics as Antimicrobials

Listeria monocytogenes is emerging to be a major food borne pathogen in post processing environments in meats and dairy foods. One of the many plants studied extensively for antimicrobial activity against *Listeria monocytogenes* is oregano (*Origanum vulgare*) (11,25,26,41,64). Oregano belongs to the family *Lamiaceae*, which also includes thyme and rosemary. Many studies indicated the antimicrobial activity of oregano and two of its major components, carvacrol and thymol, along with enhanced synergy with high rosmarinic acid extract (41). Kim et al. (65) reported that carvacrol was the most antibacterial phenolic compound they tested against five food borne pathogens (*Escherichia coli, E. coli* 0157:H7, *Salmonella typhimurium, L. monocytogenes, and Vibrio vulnificus*) using paper disk assays. Carvacrol was shown to have a larger zone of inhibition than did the other phenolic compounds at the same concentrations (citral, geraniol, terpineol, perillaldehyde, eugenol, linalool, and citronellal). Other studies have shown that nisin, when used in combination with carvacrol, inhibited the growth of *Bacillus cereus* and *L. monocytogenes* (66). One of the results found in this study was that at 20°C, growth of *L. monocytogenes* was completely inhibited in the presence of a concentration of nisin that was 16-fold lower (when used in combination with carvacrol) than when nisin was applied as the sole preservative. Thus, using carvacrol in combination with nisin did increase the effectiveness of the antimicrobial and allowed it to be used at much lower concentrations than it was used previously. Hurdle technology was also used in a study done by Karatzas et al. (67), which showed the combinations of carvacrol and high hydrostatic pressure inhibited *L. monocytogenes*. Other studies have also indicated that phenolic metabolites from oregano and other related herbs can be effective against *Listeria monocytogenes* (11,41).

Therefore the use of herbs and spices as a potential preservative for the food industry does have prospects for improving hurdle technology for controlling food borne pathogens. As discussed previously, due to their natural cross pollination nature, these herbs and spices are genetically heterogeneous (1,2). This heterogeneity results in a high degree of variability in the levels of phenolics in these plants (68). Furthermore, problems associated with climate and diseases can lead to even more quality variation in the plants. This produces many serious problems for the routine use of these plants as a hurdle to inhibit the growth of food borne pathogens in products developed by the food industry. Therefore tissue culture based clonal propagation was developed to develop consistent oregano extracts with antimicrobial activity against *Listeria monocytogenes* (41). Clonal propagation also allows us to obtain more detailed analysis of biosynthetic pathways of primary and secondary metabolites from uniform genetic material. This also allows us to study the effects of manipulating the pathways, such as adding specific precursors of a pathway or manipulating or deleting key enzymes (1,2).

In order to screen for high phenolic stimulated clonal lines with high antimicrobial activity, microbial elicitors, such as *Pseudomonas* sp., have been employed (69–72) (Figure 8.1). The rationale for this screening approach is that phenolic metabolites are stimulated in response to microbial elicitors (73) and, therefore, any clonal line tolerant to *Pseudomonas* must be a phenolic overexpressing phenotype (72). By using these techniques, it is now possible to obtain a reliable and genetically uniform plant source with excellent functionality for antioxidant and antimicrobial applications that can be used in the food industry. Using this approach we have shown that an elite clonal line of oregano was superior in inhibiting *Listeria monocytogenes* compared to commercial herbs extracts or individual terpenoid pathway derived phenolics normally found in essential oils of oregano (41). We are working further on obtaining *Listeria* specific inducible phenolics from specific clonal lines that could be more effective. In order to develop such *Listeria* specific herb clonal lines we have developed an innovative two layered plate assay to screen specific clonal lines that target

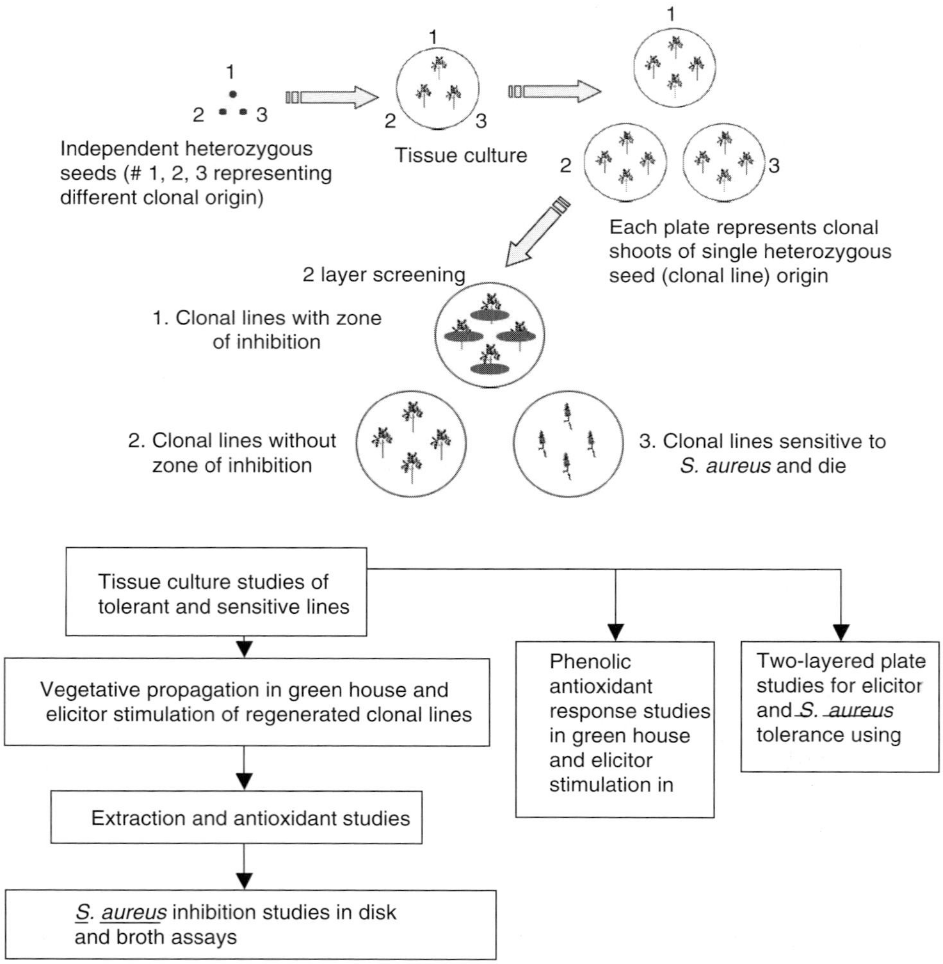

Figure 8.1 Steps for isolation of clonal lines of single seed origin heterozygous seed with potential for high phenolics with antimicrobial activity.

Listeria monocytogenes (Figure 8.2). Further, the hypothesis that elicitor inducible phenolic antioxidants from low flavored clonal lines can also inhibit *L. monocytogenes* holds much promise for improved efficacy and is being investigated. In addition, in our recent studies we have also found in meat systems that ethanol extracted phenolics from clonal oregano are more effective in inhibiting *Listeria monocytogenes* than individual hydrophobic phenolics such as carvacrol that are found in essential oils of oregano (41).

8.3 FILLING GAPS IN ANTIMICROBIAL STRATEGY WITH TISSUE CULTURE BASED SCREENING OF ELITE PHENOLIC PHYTOCHEMICAL-PRODUCING LINES

As mentioned, the major limitation of using dietary food grade herbs like oregano, thyme, and lavender for antioxidant and antimicrobial applications is the inconsistency of phenolic phytochemicals due to the heterogeneity resulting from the natural cross pollinating nature of their breeding characteristics (74). Plants which originate from different heterozygous seeds in a given pool of extract are phenotypically variable, resulting in the substantial phytochemical inconsistency, and therefore leads to unreliable clinical effects as well as inconsistent health benefits and functional value.

In order to overcome the problem of phytochemical inconsistency due to genetic heterogeneity, our laboratory has developed patented plant tissue culture techniques to isolate a clonal pool of plants originating from a single heterozygous seed (75,76) (Figure 8.1). A single elite clonal line with superior phenolic profile (a combination of several bioactive phenolics such as rosmarinic acid and simple phenolic acids) can then be screened and selected based on tolerance to *Pseudomonas* sp. (70–72). This screening strategy is being further developed to select and characterize several elite high antioxidant clonal lines of food grade herbs such as oregano (*Origanum vulgare*), thyme (*Thymus vulgaris*) and lavender (*Lavandula angustifolia*) based on response to microbial polysaccharide and bacterial pathogen targeted in a 2 layered plate assay (Figure 8.2) and based on tissue culture antioxidant response studies. These elite clonal lines (each clonal line originating from a different heterozygous seed), following large scale clonal propagation (micropropagation) in greenhouse systems are being evaluated for antioxidant profiles and functionality. High and low

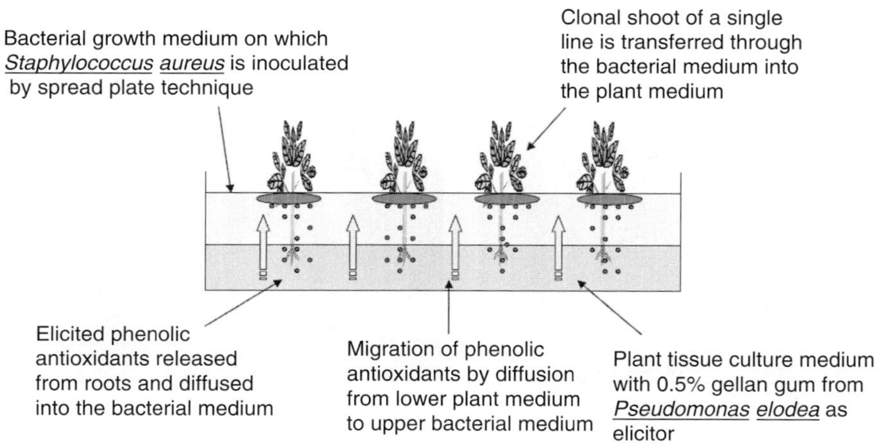

Figure 8.2 Innovative 2-layered plate assay from specific identification of clonal lines of herbs with potential of antimicrobial activity against the targeted bacterial pathogen.

antioxidant signature profiles of different clonal origin (different heterozygous seeds) are compared and targeted as dietary sources of phenolics that have high antioxidant efficacy and therefore potentially have antimicrobial benefits against food borne pathogens as evaluated *in vitro* disk (Figure 8.3) and broth assays. The advantages and importance of this approach is that it provides a rapid nontransgenic approach to obtain genetically consistent clonal lines with high rosmarinic acid as antioxidant in different total soluble phenolic and high antioxidant backgrounds from diverse heterogeneous gene pools that are generated due to natural cross pollination. Once the elite line is isolated by *in vitro* tissue culture methods it can be rapidly clonally propagated by vegetative cuttings for greenhouse and field production and therefore consistent phytochemical extracts of single seed genetic origin can obtained for potential antimicrobial applications against the pathogenic bacteria being targeted. Further, the clonal lines containing higher and lower total or specific phenolic content can be compared for synergistic contribution in different high total soluble phenolic and antioxidant clonal backgrounds. Therefore the approach envisioned in this chapter is excellent for generating consistent phytochemical antioxidant profiles for developing clinically or food safety relevant antimicrobial applications and for consistency of ingredients for food applications as a part of a hurdle technology. In addition, in future studies synergistic antimicrobial activity in combination with antimicrobials from multiple food grade sources can be evaluated against antibiotic resistant pathogens. Because antimicrobial phenolic profiles of single genetic origin are generated they offer consistency and yet have multiple phenolics that will make it difficult for pathogens to develop antimicrobial resistance. This synergy can be further enhanced in combinations with other dietary phenolic profiles. This strategy will be useful against hospital infections or food processing and packaging environments, especially infections from antibiotic resistant strains. Once a clonal profile antimicrobial efficacy is established, individual phenolic phytochemcials contributing to antimicrobial efficacy in specific clonal profiles can be investigated based on the model proposed on the mechanism of action of these phenolic phytochemicals. Such clonal screening is also ideal for screening for more ethanol and water soluble phenolic antimicrobials that can be effective in inhibiting or disrupting biochemical targets both in the plasma membrane and the cytosol.

8.3.1 Novel Model for Mechanism of Action of Phenolic Phytochemicals

Current theory and emerging data suggests that eukaryotes evolved from prokaryotes (77). Genetic evidence suggests that plant organelles like chloroplasts, mitochondria, and even vacuoles may have origins as free prokaryotes. The compartmentalized organization and

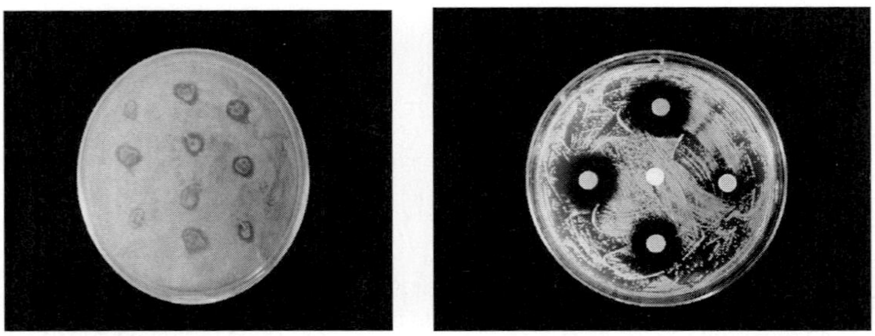

(a) (b)

Figure 8.3 (a) Example of antimicrobial zone of inhibition by specific oregano clonal line against *S.aureus* as outlined in Figure 8.2. (b) Dose-dependent inhibition of *S. aureus* by phenolic extracts from mature clonal plants developed from clonal screening as outlined in Figure 8.1.

cellular differentiation (into tissues) of plants apparently evolved in terrestrial environments about 500 million years ago. It is likely that through the millennia, plastic plant species constantly interacted with many microorganisms and environmental stresses, especially UV radiation and oxidative stress. The speculation is that constant environmental stress in many ways may have shaped the antimicrobial and phenolic antioxidant responses of plants through the evolutionary process. Natural selection of plants under various changing environmental conditions over millions of years, in many critical ways (directly or indirectly) may have shaped the current mammalian systems, including human nutrition, health, and dietary evolution (40).

It is from these assumptions that we are exploring the mechanism of antimicrobial action of plant phenolics on prokaryotic pathogens and the activation and maintenance of plant and eukaryotic antioxidant responses, at a similar phenolic concentration needed for bacterial inhibition (40). Plant phenolics and related synthetic food grade phenolics have been suggested to inhibit bacterial pathogens (particularly at low pH) by disrupting the proton motive force (PMF) (40) and through lipid membrane stacking with hydrophobic phenolics (11). This implies that regulation of H^+-ATPases and cofactors at the external membrane is critical, and may be a logical point to study the mechanism of inhibition by plant phenolics. Additionally, plant phenolics may disrupt the electron transport chain or destabilize the plasma membrane (10,11,40). At the same time it is important to investigate the role of plant H^+-ATPases at both the external membrane and organelle (chloroplast, mitochondria, and vacuolar tonoplast) membrane levels to determine their potential involvement in modulation of a phenolic radical and proton linked redox cycle (40). This cytosolic phenolic radical proton linked redox cycle (40) is hypothesized to activate the proline linked pentose–phosphate pathway, to utilize proline as an alternative RE for mitochondrial ATP synthesis, and to generate precursors and cosubstrates ($NADPH_2$ and sugar phosphates) for anabolic reactions, including a phenolic linked antioxidant response and purine synthesis (40). The genetic comparisons of H^+-ATPase alleles from bacterial and eukaryotes could provide clues toward understanding their susceptibility to plant phenolics in bacteria and the antioxidant response linked tolerance of plants and other eukaryotes (40). Such knowledge could facilitate the development of better structure function strategies for the development of better botanical phenolic profiles (i.e., elicitor induced herb and legume clonal extracts) that may provide multiple beneficial activities. A better conceptual understanding of phenolic antimicrobial effects in bacteria and antioxidant responses in eukaryotic hosts could also facilitate the development of diet based nutritional and functional food strategies to control bacterial pathogens with reduced potential for antibiotic resistance due to the concerted effect of a phenolic profile, instead of a single compound, as in case of antibiotics. Such an integrated approach involving phenolic phytochemicals has excellent potential to compliment current antimicrobial strategies (40).

Taking into account this rationale, a phenolic antimicrobial activity model (Figure 8.4) that incorporates the role of proline linked pentose–phosphate pathway provides a better perspective (40) than current putative antimicrobial models for phytochemical action (10). The phytochemical profiles that have the potential to inhibit pathogenic microorganisms (10–12) contain secondary metabolites that are defensive and inducible antimicrobials produced against both invading pathogens and stress. Therefore the methods for exploiting them must take this into account. In certain cases, the induction is associated with action of diphenolic oxidases and resulting modified compounds can have antimicrobial activity (13). In some cases, dihydroxy phenolics are oxidized to highly reactive quinones, which can interact with proteins of the invading pathogens and form melanoid polymers (10). As the compounds responsible for the enzymatic browning reaction of cut fruits and vegetables and an intermediate in melanin pigment production in humans (10), quinones are a

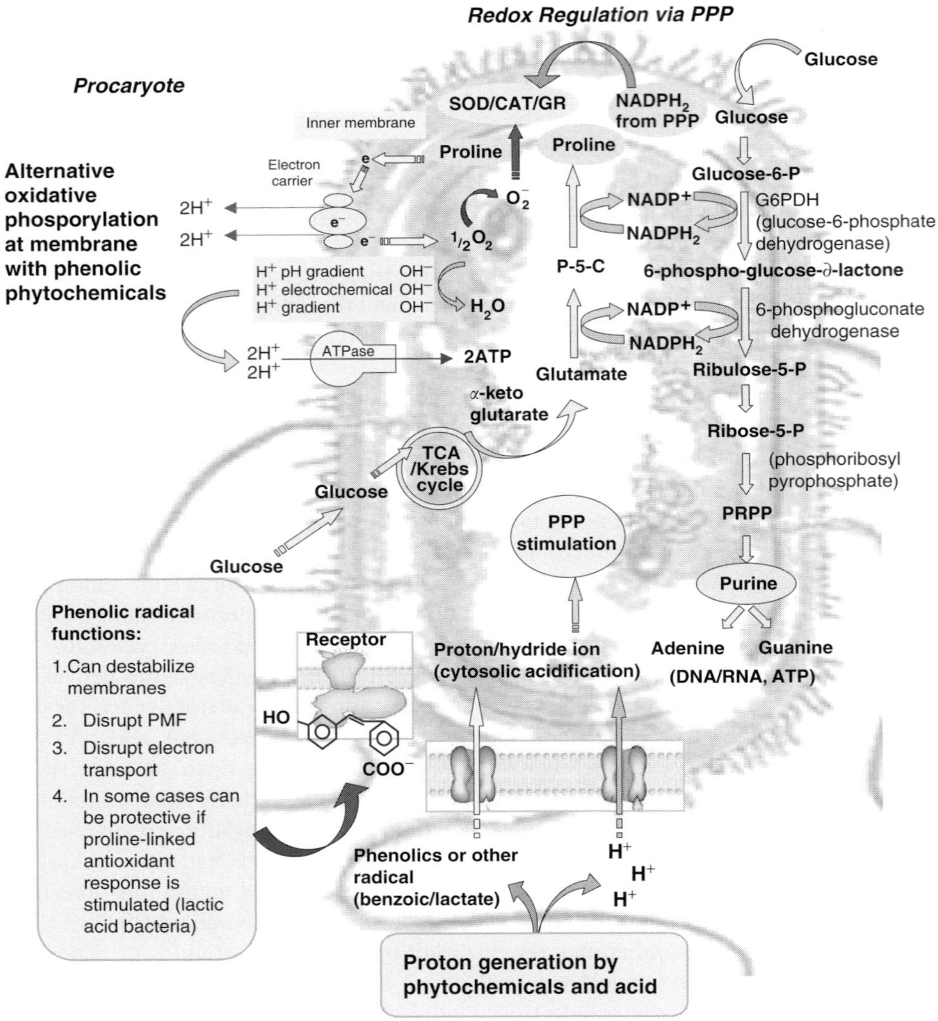

Figure 8.4 Model for inhibition of bacterial pathogens by phenolic phytochemical profiles (containing phenolics that are water soluble and effective in the cytosol as well as partially ethanol soluble phenolics that are effective at the membrane interface) based on membrane changes linked to energy (electron transport and proton flux changes) and cytosolic proton adjustment or lack of it that is linked to praline linked pathways and antioxidant response.

source of stable free radicals and complex irreversibly with nucleophilic amino acids (14) leading to inactivation of proteins and loss of function (10). Potential targets for inhibition in the bacterial cell are surface adhesions, cell wall polypeptides, and membrane bound enzymes (10). In herb species of the Lamiaceae family, phenolic derivatives are largely responsible for antimicrobial activity (2,41,42). Thymol, present in the essential oil of thyme, oregano, savory, sage, and related species, has good antimicrobial activity (11) but has flavor problems and is volatile. Other simple phenolics such as hydroxycinnamic acid derivatives such as caffeic acid, ferulic acid, and *p*-coumaric acid inhibited *E. coli, S. aureus* and *B. cereus* (19) could facilitate more hyper acidification in the cytosol and could be more effective with phenolics that are membrane targeted. Polymeric phenolics, such as tannins, were inhibitory toward *Listeria monocytogenes, E. coli, S. aureus, Aeromonas hydrophila,* and *Streptococcus faecalis* (20) and could be effective by stacking on the cell membrane and affecting energy function. Tannins could be made more potent in combination with cytosol targeted soluble phenolics. Hydroxylated phenols, such as catechol and pyrogallol, are known to be toxic to microorganisms (10) with a likely mechanism through membrane and cytosolic hyperacidification. The site and number of hydroxyl groups is linked to the antimicrobial effect and in some cases more oxidized forms are more inhibitory (21,22), which could be by inhibiting membrane related functions. In our laboratory we have shown that high rosmarinic acid enriched clonal profiles in ethanol and water extracts are effective in inhibiting *Listeria monocytogenes* (41) and *Helicobacter pylori* (42). We also have had similar success against the same pathogens using elicitor induced or fungal bio processed legume phenolics from soybean and other legumes.

From all the cited studies it is evident that the mode of action of phenolics against bacterial pathogens has not been clearly defined or understood and many modes of action have been suggested. Major mechanism models proposed have focused on the suspected changes in membrane permeability through hydrophobic linked lipid partitioning with certain phenolics to more membrane and cytosolic localized hyperacidity with soluble phenolics, which may affect PMF across the membrane resulting in energy depletion (11,28,29,31,32). Another model has proposed that enzyme inhibition by oxidized compounds through reaction with enzyme sulfhydryl groups or through nonspecific interactions with membrane proteins may explain the inhibition (10,27,36). Even with better models on the mechanism of action, another major limitation of using many phytochemicals is that they are derived from mixed heterogeneous genetic sources and consistency cannot be guaranteed (1,2). Now, with the emergence of antibiotic resistance from overuse of single antibiotics new strategies incorporating plant based antimicrobials are both promising (2,10,40). However, effective development of plant based antimicrobial strategies will require a better model to understand the mechanism of antimicrobial action involving consistent profiles of phenolic phytochemicals that inhibit bacterial pathogens at more than one biochemical steps. This could involve affecting cellular functions both at the membrane level with lipid or cellular interface compatible phytochemicals in conjunction with soluble phenolics that can cause cytosolic hyper acidification that will disrupt dehydrogenases and therefore disrupt the generation of proton motive force for ATP synthesis at an early cytosolic step along with reduced production of reductants for anabolic pathways (40).

An effective strategy proposed in this chapter focuses on the hypothesis that high antioxidant phenolics from single seed origin clonal lines of herbs, sprouted legumes, and fermented fruits would have excellent antimicrobial potential (40). Single seed high antioxidant phenolic profiles that are ethanol and water extracted have been screened, evaluated for antioxidant efficacy, and targeted to inhibit various food borne pathogens and chronic human infections, such as those by peptic ulcer-causing *Helicobacter pylori* (42) and *Listeria monocytogenes* (41). This strategy addresses both the concept of phytochemical consistency through the use of clonal lines and stress (elicitor) based inducible phenolic antioxidants from various developmental phases of growth and also the putative impact on antimicrobial potential. Therefore, using our strategy for developing consistent and inducible phytochemical profiles, an alternative and more robust model (Figure 8.4) for a mechanism of antimicrobial action incorporating the role of a critical control point (CCP) through proline linked pentose–phosphate pathway have been proposed. The mechanism of action of phenolic phytochemicals that may operate in prokaryotes involves the use hyperacidification (protons) from acids and phenolic metabolites and transport protons inside the cell (Figure 8.4), by more acid tolerant prokaryotes (lactic acid bacteria and some moderately acid tolerant Gram-negative pathogens like *H. pylori*, *Escherichia coli* and *Salmonella*) either passively or through H^+-transport membrane proteins (40). Such acid tolerant prokaryotes may stimulate redox cycling through the proline linked pentose–phosphate pathway and use proline (like the mitochondria of eukaryotes) as an alternative reductant for ATP synthesis at the single outer plasma membrane with oxygen as the terminal electron acceptor (40). As mentioned before, the process of proline biosynthesis could also be coupled to the pentose–phosphate pathway for $NADPH_2$ recycling and to make sugar phosphates for all anabolic needs. By this model, acid tolerant microorganisms could efficiently manage the ATP needs from stress increased oxidative phosphorylation through coupling to pentose–phosphate pathway activity, while recycling excess proton flux from hyperacidification. Proline linked oxidative phosphorylation could excrete excess protons outside the plasma membrane and augment the generation of PMF for ATP synthesis (40). This proline linked metabolism model may occur in acid and phytochemical tolerant lactic acid bacteria (Gram-positive) and in moderately acid and phytochemical tolerant Gram-negative bacteria but is less likely in other acid and phytochemical susceptible Gram-positive bacteria (40). The ethanol, water, and organic acid extracted phenolic antimicrobials would be highly compatible for the given model.

One way to prove this model would be to determine if proline overproducing mutants are more acid tolerant and if this tolerance is associated with increased generation of $NADPH_2$. An important difference in case of prokaryotes is that they have only one outer plasma membrane and no organelles for metabolic adjustment. Depending on the type of membrane modifications that occur between various bacterial species, phenolic radicals may negatively or positively affect membrane related functions, including transport, signaling, receptor modification, and energy metabolism. Membrane related modulation of metabolism could be closely linked to the cytosolic proton linked modulation of proline linked pentose–phosphate pathway. Based on this model, it is likely that Gram-positive bacteria (excluding lactic acid bacteria) would be most susceptible, followed by Gram-negative bacteria, and then by acid tolerant lactic acid bacteria, likely being more tolerant and actually protected by phenolic phytochemicals (as antioxidants) (40).

8.3.2 Control of *Helicobacter Pylori* by Phenolic Antimicrobials from Plants

We have extended our studies on recruiting dietary plant based phenolic antimicrobials in inhibiting food borne pathogens to control and manage ulcer and gastric cancer associated *Helicobacter pylori*. *Helicobacter pylori* induced gastrointestinal disease afflicts millions worldwide, in both developed and developing countries. It was long believed that bacteria could not live in the acidic environment of the stomach, but in 1984, Marshall and Warren (78) reported success in culturing an organism from the gastric mucosal layer of a clinical specimen and proposed that this bacterium, later named *Helicobacter pylori*, was the cause of gastritis and peptic ulcer (78). This proposal has since proved to be correct, as *H. pylori* has been associated with the presence of gastric inflammation (79) and gastric non-Hodgkin's lymphomas (80), and has been identified as a major cause of peptic ulcer disease (81) and of gastric cancer (82).

Helicobacter pylori is a spiral, microaerophilic, Gram-negative bacteria that mainly infects the gastric mucosal layer, but can also enter human primary cells from gastric epithelium, possibly to evade host defenses and antimicrobial treatments (83,84). It infects up to 50% of the global human population (85). *Helicobacter pylori* is associated with gastritis, peptic ulcer and ultimately gastric cancer (80,85,86–89). The infections are mostly asymptomatic, but in all cases involve inflammation of the gastric mucosa, and in about 10% of the infected cases develop duodenal or gastric ulcers leading to cancer of stomach

(90). In developing countries, *H. pylori* infection is as high as 80–90% of the population and this incidence correlates with higher rates of stomach cancers (90). *Helicobacter pylori* infection is currently the highest prevailing bacterial infection worldwide (90). Characteristic traits of *H. pylori* infection are chronic inflammation of the stomach and development of chronic gastritis, which may proceed to gastric cancer (91). An integral part of *H. pylori* infection is the induction of gastric inflammation, which is believed to aid in colonization of the mucosal surfaces (84,92). Pathogenesis of *H. pylori* has also been linked to the presence of an external urease that is embedded on the outer surface of the bacterium (93). Urease is an important enzyme that is produced up to 5–6% of the total protein of *H. pylori* and is linked to its infectivity (90). All clinical isolates of *H. pylori* produce a urease with characteristics unlike those of any other bacterial urease (93–96). Urease negative mutants cannot colonize the gastric tissue environment as evidenced by the failure of urease negative mutants to colonize mice and gnotobiotic piglets (97,98). It is thought that hydrolysis of urea by urease generates ammonia to counterbalance gastric acidity by forming a neutral microenvironment surrounding the bacterium within the gastric lumen (99). This hypothesis is supported by results showing that *H. pylori* survive low pH *in vitro* in the presence of urea and functional urease activity (100,101). A unique feature of *H. pylori* urease is that a significant fraction is found exclusively in the outer membrane (94,102). This urease becomes associated with the surface of *H. pylori* and the urease released as a result of autolysis of a part of bacteria becomes adsorbed to the remaining intact bacteria (99,102). Further research by the same group has indicated that cytoplasmic urease activity alone, though needed, is not sufficient to allow survival of *H. pylori* in acid and the activity of surface localized urease is essential for resistance of *H. pylori* to acid conditions (99). Recent advances, however, have indicated that acidity trigger cytoplasmic urease activity and deletion of the gene *urel* prevents activation of this enzyme (103). Urel protein is a membrane protein with 6 trans membrane segments and is needed for acid stimulated urea uptake triggers the activity of the cytoplasmic urease, which is active down to pH 2.5 (103). Another feature of urease action in addition to sustaining *H. pylori* is the prospect of physical injury in the gastric lumen from ammonia (104,105). Ammonia may also lead to inflammation due to host response or due to other mucosal damage (106). Other factors that may be involved in the pathogenesis of *H. pylori* and the colonization of mucosal surfaces are the adhesion and motility of *H. pylori* (91), the induction of IL-8 and ROS in host cells (83), and the induction of COX-2 in host cells (107). High rates of *H. pylori* infection are associated with low socioeconomic status and high densities of living and prevalence of infection increases with age worldwide (108–110).

8.3.3 Current Treatment Strategies for *H. Pylori* and Its Limitations

Before the link between *H. pylori* and gastrointestinal diseases was discovered, the main course of treatment for these diseases was to assuage the symptoms with acid (proton pump) inhibitors (PPI). Now that a link to *H. pylori* has been established, antibiotics are the current standard of care for patients with gastritis and peptic ulcers, with the rationale that elimination of *H. pylori* will eliminate a major cause of gastric disease and reduce the risk for subsequent development of gastric adenocarcinomas (111). Typical therapy for *H. pylori* infection consists of a first line combination of a PPI and a bismuth compound, and two antibiotics taken for at least 7 days, followed by a second line treatment of four PPIs taken for at least 7 additional days (112). When followed religiously, the antibiotic regimen has been shown to eradicate *H. pylori* in approx. 80–90% of the cases, although there is some debate as to whether *H. pylori* is actually being eradicated or merely suppressed (112).

Difficulties with current *H. pylori* therapies contribute to treatment failure at any age. Despite the high success rate of antibiotic–PPI combination therapy against *H. pylori*, low compliance due to negative side effects (i.e., diarrhea, drug hypersensitivity, antibiotic associated colitis), not to mention the risk of nephrotoxicity when bismuth is used, continues to be a problem (84,113,114). Cost and practicality of the treatments are also concerns. Primary and acquired resistance of *H. pylori* to the antimicrobial agents is a major reason for the failure of eradication therapies (115). Due to the increasing incidence of antibiotic resistant strains of *H. pylori*, there is still a need for new compounds or alternative phenolic phytochemical strategies to combat *H. pylori* as future therapies.

8.3.4 Dietary Phenolic Phytochemical Strategies for Control of *H. Pylori*

Much recent research has focused on compounds from dietary sources as new or alternative strategies to fight *H. pylori*. Studies have indicated the antimicrobial potential of medicinal herbs and food plant metabolites against *H. pylori*. Highly selective antimicrobial activity of novel alkyl quinolone alkaloids from Chinese herbal medicine, Gosyuyu, against *H. pylori in vitro* has been reported (116). Eradication potential of Chinese herbal medicine, without emergence of resistant colonies, has also been reported (117). Among plant food extracts and ingredients, tea catechins have been reported to have *in vitro* and *in vivo* activities against *H. pylori* (118). Among various catechins, epigallocatechin gallate showed the strongest activity (118). In studies on synergistic antimicrobial effects of garlic, the combination of garlic based on allicin content with omeprazole showed antimicrobial benefits indicating the need for clinical studies (119,120). Susceptibility of *H. pylori* to the antimicrobial activity of manuka honey has been reported (113). Honey is a traditional remedy for dyspepsia and use of phytochemical enriched honey is a traditional practice without a known rational basis (113). Therefore, evaluating the inhibition by honey of *H. pylori* that has been suggested as the probable cause of dyspepsia has merit and results are promising (113). Plant phenolic metabolites like capsaisin from diet are known to be associated with low rate of ulcers through inhibition of *Helicobacter pylori* (121). Another interesting food based inhibition of *H. pylori* is the *in vitro* antibacterial activity of Chilean red wines (122). The main active compound that was suggested to be the antimicrobial compound was resveratrol (122). In other studies, extracts from a yeast fermented soymilk yogurt (123), bee propolis (124), licorice (125), broccoli (126), and turmeric (127), and in our laboratory, clonal oregano (42) demonstrated an ability to inhibit the *in vitro* growth of *H. pylori*. Although the mechanism of action of each extract remains largely unknown, in majority of these cases antibacterial activity was linked to the presence of phenolic compounds.

8.4 DIETARY SOYBEAN APPROACH FOR CONTROL OF *H. PYLORI*

A strong inverse association between consumption of microbially fermented (bioprocessed) soybean (natto, tofu) and *H. pylori* infection (128) and incidence of death from stomach cancer (129) has been observed in Japanese populations. Fermented soybean curd (tofu) is a common source of vegetable protein in Japan (128). Sprouts with inducible phenolics and microbially fermented or bioprocessed soybean are rich in isoflavonoids and phenolics that have antioxidant activity (130–133). Many phenolic antioxidants (in eukaryotic cells) also possess strong antimicrobial activities against prokaryotic bacteria, therefore, phenolic extracts from sprouts and microbially bioprocessed soybean that possess antioxidant activity potentially may also possess antibacterial activity, specifically anti *H. pylori* activity. Therefore the main focus of our investigations is to explore the potential of phenolic rich extracts of sprouted soybean and solid-state bioprocessed defatted soybean by the dietary fungus *Rhizopus oligosporous* (used in tempeh production in

Indonesia and the U.S.) to inhibit the *in vitro* growth and pathogenesis related urease of *H. pylori*. Our preliminary results suggest:

LISTThat the phenolic rich, bioprocessed soybean extracts possess strong anti *H. pylori* activity.

That sprouting linked elicitor stimulation of phenolics and bioprocessing of soybean by *R. oligosporous* increases anti *H. pylori* activity in soybean extracts in a manner dependent upon the sprouting stage, elicitor concentration, and duration of solid-state growth of the dietary fungus.

Hint at a mechanism of action which may involve diphenolic and polymeric phenolic antioxidants, which may inhibit urease.

Therefore, the hypothesis that elicitor inducible phenolic antioxidants from sprouted and fermented soybean can also inhibit *H. pylori* holds much promise for improved efficacy and support dietary therapy using soybean phenolic extracts in food systems. Other studies have postulated that antioxidants such as vitamin C and astaxanthin may constitute supportive treatment for controlling *H. pylori* infections through support of host (eukaryotic) response (134). Our preliminary studies suggest that phenolic antioxidants from soybean may be better. They can not only, potentially, provide a supportive redox environment in the stomach (eukaryotic cell) through the antioxidant improved functionality, but more importantly, can be inhibitory to prokaryotic pathogens like *H. pylori,* a prokaryote with a single membrane is susceptible to the same phenolic antioxidants. Further, phenolic antioxidants could be potentially inhibitory against the pathogenesis linked urease and this will be evaluated using a novel plate assay described in this chapter. Through inhibition of urease by phenolics, hyperacidification at the membrane level can be maintained and as discussed for the general model (Figure 8.4), this proton efflux coupled to other functions of dietary phenolics can now be more effective in inhibiting *H. pylori* (Figure 8.5). Because dietary phenolics are offered as phenolic profiles it is likely to reduce the emergence of antibiotic resistance.

Extracts from dietary legumes and other phenolic phytochemicals can be important source of disease preventive phytochemicals which potentially can inhibit the target prokaryotic bacterial pathogens like *H. pylori*, which have a single plasma membrane that also contains the pathogenesis associated urease, and at the same time can be antioxidant, in function, in host eukaryotic cells that have organelle compartmentalization to manage a redox response linked to phenolics, triggering a protective response. Our strategy focuses on the hypothesis that inducible high antioxidant phenolics from clonal herbs screened from heterogeneous sources, seed sprouts of single genetic origin, and phenolic aglycones with high antioxidant activity from fermented dietary botanicals would have excellent antimicrobial potential. Our strategy also addresses the concept of phytochemical consistency through the mobilization of stress (elicitor) inducible phenolic antioxidants in various developmental phases of growth of soybean sprouts and bioprocessed extracts and evaluates the antimicrobial potential against *H. pylori.*

8.4.1 Sprouting of Soybean and Phenolic Antioxidant Stimulation Methods

We are exploring the use of high antioxidant sprout extracts of soybean and other legumes for antimicrobial benefit against gastric cancer linked *H. pylori* and therefore developed a diet based bacterial management strategy. Seeds of soybean are imbibed overnight in bacterial polysaccharide elicitors normally used in foods, such as gellan gum and xanthan gum (135). In earlier work on the legume mung bean (135), there was clear evidence that stress-modulating polysaccharide elicitors will stimulate phenolic content and antioxidant activity during the sprouting stages as a likely protective mechanism for the young

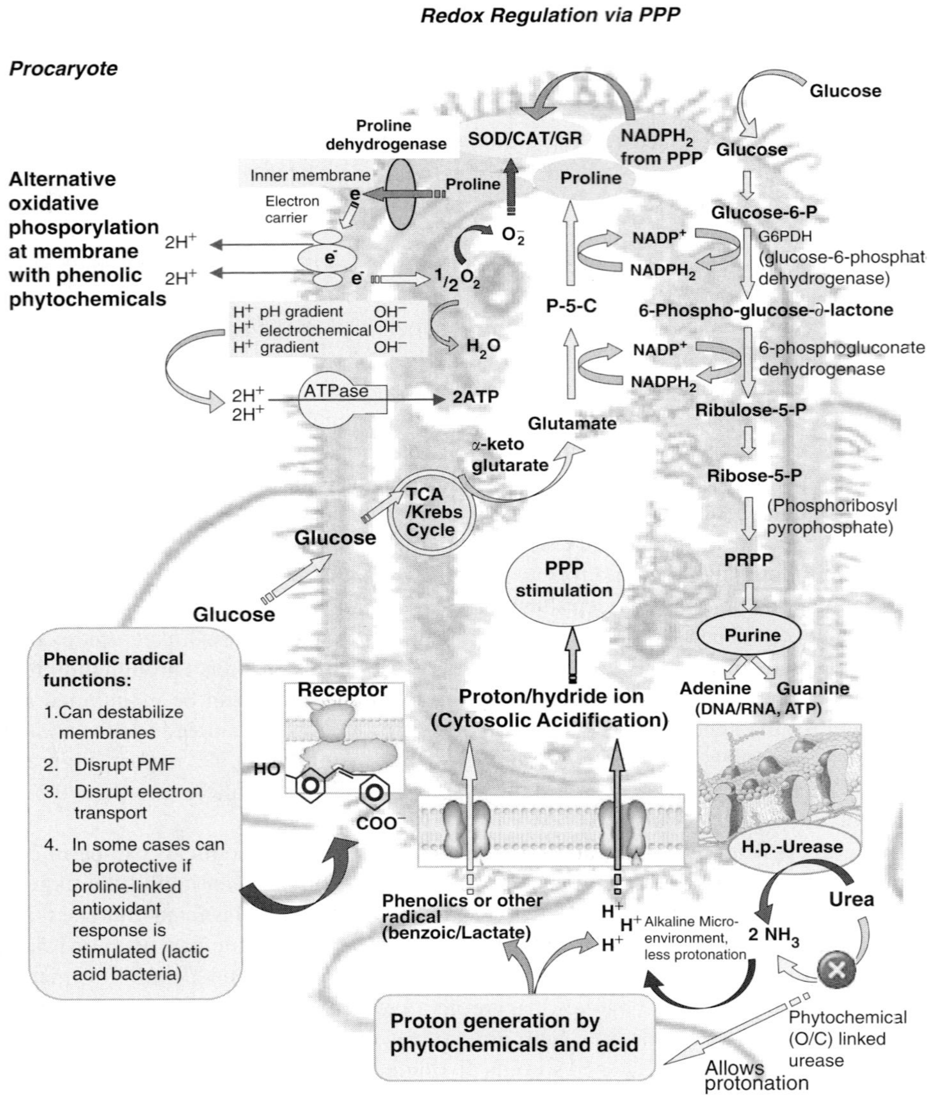

Figure 8.5 Model for inhibition of *Helicobacter pylori* by phenolic phytochemicals as in Figure 8.4 but also takes into account the role of urease.

seedling. Previously it has also been established that the percent of phenolic aglycones were the highest in soybean sprouts (136). The biological activity of the phenolics is stronger in the aglycone form than in the glycoside form. It is thought that isoflavonoids such as genistein and daidzein have a specific role during growth stages of soybean (130,136). Preliminary results from our laboratory have indicated that during dark germination, phenolic content was stimulated (inducible phenolics) by the given elicitors (137). Polysaccharide elicitor extracts from *Pseudomonas elodea* (gellan gum) are currently being tested in the concentration range of 0–500 mg/l of imbibed water. After elicitor treatment for 12 hours, sprouting through a seven day dark germination stage was done in specially designed plastic containers. Phenolic profiles, antioxidant activity and *H. pylori* inhibition studies was undertaken on extracts of various development stages. Optimized phenolic extracts was used for further *H. pylori* inhibition (137) and in the future for urease inhibition studies. Several soybean varieties from genetic stocks in U.S. and Asia are currently being tested. We have, currently, access to over 50 soy genetic accessions for testing. The basal uninduced phenolic content of soybean is around 1 to 3 mg/g of dry weight (DW) of sprouted and fermenting soybean, which is than stimulated to 4–5 mg/g DW.

8.4.2 Phenolic Response and Characterization of Soybean Extracts Inhibiting *H. Pylori* Urease

Initial preliminary efforts have so far focused on polysaccharide elicitor stimulated soybean seedling extracts, based on the rationale that if an individual clonal line (single seed origin variety soybean is self pollinating) has genetically evolved to produce high phenolics in a heterogeneous background, it may also have the genetic, biochemical and developmental mechanism to tolerate the same high level of phenolics and such high phenolics could have protective antioxidant benefits and antimicrobial effects against the bacterial pathogens or its elicitors. Further, we have also developed an innovative plate assay system to screen phenolic extracts that are inhibitory to *H. pylori* urease. In this assay system we have *H. pylori* medium at pH 5.5 with addition of 10 mM urea and bromophenol blue (0.1 g/L) as pH indicator. *Helicobacter pylori* will survive low pH conditions by using urea from the acidic environment and create a microenvironment of high pH by excreting urease mediated ammonia. When this happens, the bromophenol blue will turn blue as pH moves higher toward neutral. We have evidence that soybean extracts in a concentration dependent (100 μg (micro gram)/disk to 400 μg/disk) manner can inhibit urease activity as indicated the bromophenol blue remaining yellow at pH 5.5.

8.4.3 Solid-State Bioprocessing of Soybean Using *Rhizopus Oligosporus*

Rhizopus oligosporous used in soy Tempeh foods has been developed through solid-state bioprocessing of whole soy and would be a more suitable approach to enhance phenolic metabolites for inhibition of *Helicobacter pylori* (131,137). Using the methods developed in our laboratory (131), phenolic aglycones are being optimized from various varieties of defatted soy, and the fungal growth is carried out in Erlenmeyer flasks over a ten day period. Total soluble phenolics in water and ethanol extracts have been optimized with various specific phenolic ratios as in sprouts and using low, high, and medium phenolic categories of various soybean varieties.

8.5 RECENT PROGRESS AND OTHER STRATEGIES FOR DESIGNING PHENOLICS PHYTOCHEMICALS AS ANTIMICROBIALS

Using the clonal screening strategy we have isolated several high phenolic clonal herbs of oregano, thyme, rosemary, and spearmint of single seed origin. One superior oregano clonal line has been shown to be effective against *L. monocytogenes* in broth and meat system (41) and *H. pylori* in disk assay system (42). Further we have found that oregano extracts optimized on a phenolic basis and combined with cranberry phenolics shows synergistic effect in both broth and food systems against *L. monocytogenes* and *Vibrio parahaemolyticus* (64,138) and in disk assay system against *H. pylori* (139).

The elicitor induced spouting model in seed legumes was used to evaluate antimicrobial potential of various phenolic and antioxidant activity stages of sprouting against

H. pylori. Phenolic extracts of pea (140), chickpea (141), fenugreek (142), soy (137), and mung bean (143) were inhibitory against *H. pylori* in disk assays. Except for fenugreek, in all other sprout extract tested, the antimicrobial activity coincided with likely polymeric stage as indicated by high sprout peroxidase activity, when soluble phenolics were the lowest (137,140,141,143). In case of fenugreek, the highest antimicrobial activity coincided with the stage of highest soluble phenolics (142).

We have also developed solid-state bioprocessing systems using filamentous fungi, where under aerobic conditions phenolics from soy, cranberry, and pineapple substrates were mobilized (137,144–146). In case of soy bioprocessing with Tempeh fungus, *Rhizopus oligosporous* enhanced functional polymeric phenolics, as indicated by laccase activity at 2 days of growth, and had the highest anti *H. pylori* activity (137). In the same study, late stages, even though they had highest phenolics, had reduced antimicrobial activity. In the case of cranberry bioprocessing with *Rhizopus oligosporous* and *Lentinus edodes,* highest antimicrobial activity was found after bioprocessing, and effective antimicrobial activity varied for targeted bacterial food borne pathogens based on nitrogen sources and method of extraction (144,145). It was postulated that the type of phenolics mobilized during bioprocessing could have affect on the antimicrobial functionality (144,145). In other studies, pineapple wastes enriched with soybean and bioprocessed with *R. oligosporous* had effective anti *H. pylori* activity after 10 days of growth.

8.6 SUMMARY

Novel tissue culture techniques for screening high phenolic clonal lines and phenolic phytochemical bioprocessing approaches through sprouting from seed legumes and microbial solid-state mobilization from diverse dietary botanical substrates have been developed to isolate novel dietary and food grade phenolic metabolites as antimicrobials. In this effort only water and ethanol soluble phenolics which can likely create a localized hyper acidification at the cellular membrane interface and cytosol have been extracted and evaluated for antimicrobial effects. The goal is to design dietary phenolics that work at the cellular interface, and cytosol to disrupt the simpler prokaryotic cellular metabolism at membrane, energy, and cytosolic dehydrogenase level. In this model the functional property of dietary phenolics for localized cellular hyperacidification is coupled to a proposed critical control point (CCP), proline linked pentose–phosphate pathway (PLPPP). The presence and management of this PLPPP with membrane linked energy metabolism and cytosolic linked proton management through dehydrogenases serve as the central weak points for antimicrobial strategies. The same phenolics in eukaryotes, due to organelle compartmentalization, are likely to be managed through an antioxidant enzyme response and which in turn can positively support host response to bacterial pathogens (40).

REFERENCES

1. Shetty, K. Biotechnology to harness the benefits of dietary phenolics; focus on Lamiaceae. *Asia Pac. J. Clin. Nutr*. 6:162–171, 1997.
2. Shetty, K., R.L. Labbe. Food-borne pathogens, health and role dietary phytochemicals. *Asia Pac. J. Clin. Nutr.* 7:270–276, 1998.
3. Hertog, M.G.L., D. Kromhout, C. Aravanis, H. Blackburn, R. Buzina, F. Fidanza, S. Giampaoli, A. Jansen, A. Menotti, S. Nedeljkovic, M. Pekkarinen, B.S. Simic,, H. Toshima, E.J.M. Feskens, P.C.H. Hollman, M.B. Kattan. Flavonoid intake and long-term risk of coronary heart disease and cancer in the seven countries study. *Arch. Intern. Med.* 155:381–386, 1995.

4. Peake, P.W., B.A. Pussel, P. Martyn, V. Timmermans, J.A. Charlesworth. The inhibitory effect of rosmarinic acid on complement involves the C5 convertase. *Int. J. Immunopharmac.* 13:853–857, 1991.
5. Rice-Evans, C.A., N.J. Miller, P.G. Bolwell, P.M. Bramley, J.B. Pridham. The relative antioxidant activities of plant-derived polyphenolic flavonoids. *Free Rad. Res.* 22:375–383, 1995.
6. Jorgensen, L.V., H.L. Madsen, M.K. Thomsen, L.O. Dragsted, L.H. Skibsted. Regulation of phenolic antioxidants from phenoxyl radicals: an ESR and electrochemical study of antioxidant hierarchy. *Free Rad. Res.* 30:207–220, 1999.
7. Paganga, G., N. Miller, C.A. Rice-Evans. The polyphenolic contents of fruits and vegetables and their antioxidant activities: what does a serving constitute ? *Free Rad. Res.* 30:153–162, 1999.
8. Hertog, M.G.L., P.C.H. Hollman, M.B. Katan. Content of potentially anticarcinogenic flavonoids of 28 vegetables and 9 fruits commonly consumed in the Netherlands. *J. Agric. Food Chem.* 40:2379–2383, 1992.
9. Foti, M., M. Piattelli, V. Amico, G. Ruberto. Antioxidant activity of phenolic meroditerpenoids from marine algae. *J. Photochem. Photobiol.* 26:159–164, 1994.
10. Cowan, M.M. Plant products as antimicrobial agents. *Clin. Microbiol. Rev.* 12:564–582, 1999.
11. Burt, S. Essential oils: their antibacterial properties and potential applications in foods: a review. *Int. J. Food Microbiol.* 94:223–253, 2004.
12. Shelef, L.A. Antimicrobial effects of spices. *J. Food Safety* 6:29–44, 1984.
13. Walker, J.R.L. Antimicrobial compounds in food plants. In: *Natural Antimicrobial Systems and Food Preservation,* Board, R.G., V. Dillon, eds., Wallingford: CAB International, 1994, pp 181–204.
14. Stern, J.L., A.E. Hagerman, P.D. Steinberg, P.K. Mason. Phlorotannin-protein interactions. *J. Chem. Eol.* 22:1887–1899, 1996.
15. Beuchat, L.R. Sensitivity of *Vibrio parahaemolyticus* to spices and organic acids. *J. Food Sci.* 41:899–902, 1976.
16. Aktug, S.E., M. Karapinar. Sensitivity of some common food poisoning bacteria to thyme, mint and bay leaves. *Int. J. Food Microbiol.* 3:349–354, 1986.
17. Shelef, L.A., E.K. Jyothi, M.A. Bugarelli. Growth of enteropathogenic and spoilage bacteria in sage-containing broth and foods. *J. Food Sci.* 49:737–740, 1984.
18. Farbood, M.I., J.H. McNeil, K. Ostovar. Effect of rosemary spice extractive on the growth of microorganisms in meat. *J. Milk Food Technol.* 39:675–679, 1976.
19. Herald, P.J., P.M. Davidson. Antibacterial activity of selected hydroxycinnamic acids. *J. Food Sci.* 48:1378–1379, 1983.
20. Chung, K.T., C.A. Murdock. Natural systems for preventing contamination and growth of microorganisms in foods. *Food Microstruct.* 10:361–366, 1991.
21. Scalbert, A. Antimicrobial properties of tannins. *Phytochemistry* 30:3875–3883, 1991.
22. Urs, N.V.R.R., J.M. Dunleavy. Enhancement of the bactericidal activity of peroxidase system by phenolic compounds. *Phytopathology* 65:686–690, 1975.
23. Koutsomanis, K., K. Lambropoulou, G.J.E. Nychas. A predictive model for the non-thermal inactivation of *Salmonella enteritidis* in a food model supplemented with a natural antimicrobial. *Int. J. Food Microbiol.* 49:67–74, 1999.
24. Skandamis, P.N., G.J.E. Nychas. Development and evaluation of a model predicting the survival of *Escherichia coli* 0157: H7 NCTC 12900 in homemade eggplant salad at various temperatures, pHs, and oregano essential oil concentrations. *Appl. Environ. Microbiol.* 66:1646–1653, 2000.
25. Tsigarida, E., P. Skandamis, G.J.E. Nychas. Behavior of *Listeria monocytogenes* and autochthonous flora on meat stored under aerobic, vacuum and modified atmosphere packaging conditions with or without the presence of oregano essential oil at 5 C. *J. Appl. Bacteriol.* 89:901–909, 2000.
26. Skandamis, P., E. Tsigarida, G.J.E. Nychas. The effect of oregano essential oil on survival/death of *Salmonella typhimurium* in meat stored at 5°C under aerobic, vp/map conditions. *Food Microbiol.* 19:65–75, 2002.

27. Tassou, C.C., G.J.E. Nychas, P.N. Skandamis. Herbs and spices and antimicrobials. In: *Handbook of Herbs and Spices*, Peter, K.V., ed., Boca Raton, FL: CRC press, 2004, pp 23–40.
28. Conner, D.E., L.R. Beuchat, R.E. Worthington, D.A. Kautter. Effects of essential oils and oleoresins of plants on ethanol production, respiration and sporulation of yeasts. *Int. J. Food Microbiol.* 1:63–74, 1984.
29. Baranowski, J.D., P.M. Davidson, C.W. Nagel, A.L. Branen. Inhibition of *Saccharomyces cerevisiae* by naturally occurring hydroxycinnamates. *J. Food Sci.* 45:592–594, 1980.
30. Davidson, P.M. Chemical preservatives and natural antimicrobial compounds. In: *Food Microbiology Fundamentals and Frontiers*, Doyle, M.P., L.R. Beuchat, T.J. Montville, eds., New York: ASM Press, 1997, pp 520–526.
31. Ultee, A., M.H.J. Bennik, R. Moezellaar. The phenolic hydroxyl group of carvacrol is essential for action against the food-borne pathogen *Bacillus cereus. Appl. Environ. Microbiol.* 68:1561–1568, 2002.
32. Ultee, A., E.P.W. Kets, E.J. Smid. Mechanisms of action of carvacrol on the food-borne pathogen *Bacillus cereus. Appl. Environ. Microbiol.* 65:4606–4610, 1999.
33. Tahara, T., M. Ohsimura, C. Umezawa, K. Kanatani. Isolation and partial characterization and mode of action of acidocin J1132, a two-component bacteriocin produced by *Lactobacillus acidophilus* JCM 1132. *Appl. Environ. Microbiol.* 62:892–897, 1996.
34. Cox., S.D., J.E. Gustafson, C.M. Mann, J.L. Markham, Y.C. Liew, R.P. Hartland, H.C. Bell, J.R. Warmington, S.G. Wyllie. Tea tree oil causes K+ leakage and inhibits respiration in *Escherichia coli. Lett. Appl. Microbiol.* 26:355–358, 1998.
35. Tassou, C.C., K. Koutsoumanis, G.J.E. Nychas. Inhibition of *Salmonella enteritidis* and *Staphylococcus aureus* in nutrient broth by mint essential oil. *Food Res. Int.* 33:273–280, 2000.
36. Mason, T.L., B.P. Wasserman. Inactivation of red beet beta-glucan synthase by native and oxidized phenolic compounds. *Phytochemistry* 26:2197–2202, 1987.
37. Friedman, M. Chemistry nutrition and microbiology of D-amino acids. *J. Agric. Food Chem.* 47:3457–3479, 1999.
38. Patte, J. Biosynthesis of threonine and lysine. In: Escherichia coli *and* Salmonella, 2nd ed., Fredeci, M., F. Neidhardt, eds., Washington, D.C.: ASM Press, 1996, pp 528–541.
39. Friedman, M., G.A. Smith. Inactivation of quercetin mutagenicity. *Food Chem. Toxicol.* 22:535–539, 1984.
40. Shetty, K., M.L. Wahlqvist. A model for the role of proline-linked pentose phosphate pathway in phenolic phytochemical biosynthesis and mechanism of action for human health and environmental applications: a review. *Asia Pac. J. Clin. Nutr.* 13:1–24, 2004.
41. Seaberg, A., R.L. Labbe, K. Shetty. Inhibition of *Listeria monocytogenes* by elite clonal extracts of oregano (*Origanum vulgare*). *Food Biotechnol.* 17:129–149, 2003.
42. Chun, S.-S., D.A. Vattem, Y.-T. Lin, K. Shetty. Phenolic antioxidants from clonal oregano (*Origanum vulgare*) with antimicrobial activity against *Helicobacter pylori. Process Biochem*, 2004. [In press]
43. Deighton, N., S.M. Glidewell, S.G. Deans, B.A. Goodman. Identification by EPR spectroscopy of carvacrol and thymol as the major sources of free radicals in the oxidation of plant essential oils. *J. Food Sci. Agric.* 63:221–225, 1993.
44. Deighton, N., S.M. Glidewell, B.A. Goodman, S.G. Deans. The chemical fate of the endogeneous plant antioxidants carvacrol and thymol during oxidative stress. *Proceedings of the Royal Society of Edinburgh*, 102B:247–252, 1994.
45. Engleberger, W., U. Hadding, E. Etschenberg, E. Graf, S. Leyck, J. Winkelmann, M.J. Parnham. Rosmarinic acid: a new inhibitor of complement C3 – convertase with anti-inflammatory activity. *Intl. J. Immunopharmac.* 10:729–737, 1988.
46. Frankel, E.N., S.W. Huang, R. Aeschbach, E. Prior. Antioxidant activity of a rosemary extract and its constituents carnosic acid, carnosol and rosmarinic acid in bulk oil, and oil-in-water emulsion. *J. Agric. Food Chem.* 44:131–135, 1996.
47. Kuhnt, M., A. Probstle, H. Rimpler, R. Bauer, M. Heinrich. Biological and pharmacological activities and further constituents of *Hyptis verticillata. Planta Medica* 61:227–232, 1995.

48. Guggenheim, S., S. Shapiro. The action of thymol on oral bacteria. *Oral Microbiol. Immunol.* 10:241–246, 1995.
49. Kikuzaki, H., N. Nakazaki. Structure of a new antioxidative phenolic acid from oregano (*Origanum vulgare*). *Agric. Biol. Chem.* 53:519–524, 1989.
50. Economou, K.D., V. Oeropoulou, C.D. Thomoupoulos. Antioxidant activity of some plant extracts of the family Labiate. *J. Am. Oil Chem. Soc.* 68:109–113, 1991.
51. Madsen, H., G. Bertelsen. Spices as antioxidants. *Trends Food Sci. Tech.* 6:271–277, 1995.
52. Chipault, J.R., G. Mizuno, W.O. Lundberg. The antioxidant properties of spices in foods. *Food Technol.* 10:209–211, 1956.
53. Emori, T.G. and Gaynes, R.P. An overview of nosocomial infections, including the role of the microbiology laboratory. *Clin. Microbiol. Rev.* 6:428–442, 1993.
54. Steinberg, J.P., C.C. Clark, B.O. Hackman. Nosocomial and community-acquired *Staphylococcus aureus* bacteremias from 1980 to 1993: impact of intravascular devices and methicillin resistance. *Clin. Infect. Dis.* 23:255–259, 1996.
55. Vychytil, A., M. Loren, B. Schneider, W.H. Horl, M. Haag-Weber. New strategies to prevent *Staphylococcus aureus* infections in peritoneal dialysis patients. *J. Am. Soc. Nephrol.* 669–670, 1998.
56. Lowy, N. *Staphyloccocus aureus* infections. *New. Engl. J. Med.* 339:520–532, 1998.
57. Tenover, F.C., M.V. Lancaster, B.C. Hill, C.D. Steward, S.A. Stocker, G.A. Hancock. Characterization of stapylococci with reduced susceptibilities to vancomycin and other glycopeptides. *J. Clin. Microbiol.* 36:1020–1027, 1998.
58. Hiramatsu, K., et al. Dissemination in Japanese hospitals of strains of *Staphylococcus aureus* heterogeneously resistant to vancomycin. *Lancet* 350:1670–1675, 1997.
59. Smith,T.L., M.L. Pearson, K.R. Wilcox. Emergence of vancomycin resistance in *Staphylococcus aureus*. *N. Engl. J. Med.* 340:493–501, 1999.
60. Marrack, P., J. Kappler. The staphylococcal enterotoxins and their relatives. *Science* 240:705–711, 1990.
61. Bantel, H., B. Sinha, W. Domschke, G. Peters, K. Schulze-Osthoff, R.U. Janicke.α-Toxin is a mediator of *Staphylococcus aureus*-induced cell death and activates caspases via the intrinsic death pathway independently of death receptor signaling. *J. Cell. Biol.* 155:637–647, 2001.
62. Day, N.P.J., C.E. Moore, M.C. Enright, A.R. Berendt, J.M. Smith, M.F. Murphy, S. Peacock, B.G. Spratt, E.J. Feil. A link between virulence and ecological abundance in natural populations of *Staphylococcus aureus*. *Science* 292:114–116, 2001.
63. McKenney, D., K.L. Pouliot, Y. Wang, V. Murthy, M. Urlich, G. Doring, J.C. Lee, D.A. Goldmann, G.B. Pier. Broadly protective vaccine for *Staphylococcus aureus* based on an *in vivo*-expressed antigen. *Science* 284:1523–1527, 1999.
64. Lin, Y.-T., R.G. Labbe, K. Shetty. Inhibition of *Listeria monocytogenes* in fish and meat systems using oregano and cranberry synergies. *Appl. Environ. Microbiol.*, 2004. [Accepted for publication]
65. Kim, J., M.R. Marshall, C. Wei. Antibacterial activity of some essential oil components against five foodborne pathogens. *J. Agric. Food Chem.* 43:2839–2845, 1995.
66. Pol, I.E., E.J. Smid. Combined action of nisin and carvacrol on *Bacillus cereus* and *Listeria monocytogenes*. *Lett. Appl. Microbiol.* 29:166–170, 1999.
67. Karatzas, A.K., E.P.W. Kets, E.J. Smid, M.H.J. Bennik. The combined action of carvacrol and high hydrostatic pressure on *Listeria monocytogenes* Scott A. *J. Appl. Microbiol.* 90:463–469, 2001.
68. Fleisher, A, N. Sneer. Oregano spices and *Origanum* chemotypes. *J. Sci. Food Agric.* 33:441–446, 1982.
69. Al-Amier, H., B.M.M. Mansour, N. Toaima, R.A. Korus, K. Shetty. Tissue culture based screening and selection of high biomass and phenolic producing clonal lines of lavender using *Pseudomonas* and azetidine-2-carboxylate. *J. Agric. Chem.* 47:2937–2943, 2000.
70. Yang, R., O.F. Curtis, K. Shetty. Tissue-culture-based selection of high rosmarinic acid-producing clonal lines of rosemary (*Rosmarinus officinalis*) using hyperhydricity-reducing *Pseudomonas*. *Food Biotechnol.* 11:73–88, 1997.

71. Eguchi, Y., O.F. Curtis, K. Shetty. Interaction of hyperhydricity-preventing *Pseudomonas* sp. with oregano (*Origanum vulgare*) and selection of high phenolics and rosmarinic acid-producing clonal lines. *Food Biotechnol.* 10:191–202, 1996.
72. Shetty, K., T.L. Carpenter, D. Kwok, O.F. Curtis, T.L. Potter. Selection of high phenolics-containing clones of thyme (*Thymus vulgaris* L.) using *Pseudomonas* sp. *J. Agric. Food Chem.*, 40:3408–3411, 1996.
73. Dixon, R.A., N. Paiva. Stress-induced phenylpropanoid metabolism. *Plant Cell* 7:1085–1097, 1995.
74. Richards, A.J. Gynodioecy. In: *Plant Breeding Systems,* Richards, A.J. ed., London: George Allen and Unwin Ltd., 1986, pp 89–331.
75. Shetty, K., O.F. Curtis, R. E. Levin, R. Witkowsky, W. Ang. Prevention of vitrification associated with *in vitro* shoot culture of oregano (*Origanum vulgare*) by *Pseudomonas* spp. *J. Plant Physiol.* 147:447–451, 1995.
76. Shetty, K. Plant clones containing elevated secondary metabolites. U.S. patent # 5,869,340, 1996.
77. Margulis, L. Symbiosis everywhere and other chapters. In: *Symbiotic Planet (A New Look at Evolution)*, New York: Basic Books, 1998.
78. Marshall, B.J., J.R. Warren. Unidentified curved bacilli in the stomach of patients with gastritis and peptic ulceration. *Lancet* 1: 1311–1315, 1984.
79. Blaser, M.J. *Helicobacter pylori* and the pathogenesis of gastroduodenal inflammation. *J. Infect. Dis.* 161:626–633, 1990.
80. Parsonnet, J., S. Hansen, L. Rodriguez, A.B. Gelb, R.A. Warnke, E. Jellum, N. Orentreich, J.H. Vogelman, G.D. Friedman. *Helicobacter pylori* infection and gastric lymphoma. *N. Engl. J. Med.* 330:1267–1271, 1994.
81. NIH Consensus Conference. *Helicobacter pylori* in peptic ulcer disease. NIH Consensus Development Panel on *Helicobacter pylori* in Peptic Ulcer Disease. *JAMA* 272:65–69, 1994.
82. Uemura. N., S. Okamoto, S. Yamamoto, N. Matsumura, S. Yamaguchi, M. Yamakido, K. Taniyama, N. Sasaki, R.J. Schlemper. *Helicobacter pylori* infection and the development of gastric cancer. *N. Engl. J. Med.* 345:784–789, 2001.
83. Dunn, B.E., H. Cohen, M.J. Blaser. *Helicobacter pylori. Clin. Microbiol. Rev.* 10:720–741, 1997.
84. Rhen, M., S. Eriksson, M. Clements, S. Bergström, S.J. Normark. The basis of persistent bacterial infections. *Trends Microbiol.* 11:80–86, 2003.
85. Covacci, A., J.L. Telford, G. Del Giudice, J. Parsonnet, R. Rappuoli. *Helicobacter pylori* virulence and genetic geography. *Science* 284:1328–1333, 1999.
86. Marshall, B.J., C.S. Goodwin, J.R. Warren, R. Murray, E.D. Blincow, S.J. Blackboun, M. Phillips, T.E. Warers, C.R. Sanderson. Prospective double-blind trial of duodenal ulcer relapse after eradication of *Campylobacter pylori. Lancet* 2:1437–1442, 1988.
87. Rauws, E.A., G.N. Tytgat. Cure of duodenal ulcer associated with eradication of *Helicobacter pylori. Lancet* 335:1233–1235, 1990.
88. Parsonnet, J., G.D. Friedman, D.P. Vandersteen, Y. Chang, J.H. Vogelman, N. Orentreich, R.K. Sibley. *Helicobacter pylori* infection and the risk of gastric carcinoma. *N. Engl. J. Med. Med.* 325:1127–1131, 1991.
89. Forman, D., P. Webb, J. Parsonnet. *H.pylori* and gastric cancer. *Lancet* 34:243–244, 1994.
90. Labigne, A., H. De Reuse. Determinants of *Helicobacter pylori* pathogenicity. *Infect. Agents Dis.* 5:191–202, 1996.
91. Gerhard, M., R. Rad, C. Prinz, M. Naumann. Pathogenesis of *Helicobacter pylori* infection. *Helicobacter* 7(1):17–23, 2002.
92. Nedrud, J.G., S.S. Blanchard, S.J. Czinn. *Helicobacter pylori. Inflam. Immun. Helicobacter* 7(1):24–29, 2002.
93. Dunn, B.E., N.B. Vakil, B.G. Schneider, M.M. Miller, J.B. Zitzer, T. Puetz S.H., Phadnis. Localization of *Helicobacter pylori* urease and heat shock protein in human gastric biopsies. *Infect. Immun.* 65:1181–1188, 1997.

94. Dunn, B.E., G.P. Campbell, G.I. Perez-Perez, M.J. Blaser. Purification and characterization of urese from *Helicobacter pylori*. *J. Biol. Chem.* 265:9464–9469, 1990.
95. Mobley, H.L.T., M.J. Cortesia, L.E. Rosenthal, B.D. Jones. Chracterization of urease from *Campylobacter pylori*. *J. Clin. Microbiol.* 26:831–836, 1988.
96. Mobley, H.L.T., M.D. Island, R.P. Hausinger. Molecular biology of microbial ureases. *Microbiol. Rev.* 59:451–480, 1995.
97. Eaton, K.A., C.L. Brooks, D.R. Morgan, S. Krakowka. Essential role of urease in pathogenesis of gastritis induced by *Helicobacter pylori* in gnotobiotic piglets. *Infect. Immun.* 59:2470–2475, 1991.
98. Eaton, K.A., S. Krakowka. Effect of gastric pH on urease-dependent colonization of gnotobiotic piglets by *Helicobacter pylori*. *Infect. Immun.* 62:3604–3607, 1994.
99. Krishnamurthy, P., M. Parlow, J.B. Zitzer, N.B. Vakil, H.L.T. Mobley, M. Levy, S.H. Phadnis, B.E. Dunn. *Helicobacter pylori* containing only cytoplasmic urese is susceptible to acid. *Infect. Immun.* 66:5060–5066, 1998.
100. Marshall, B.J., L.J. Barret, C. Prakash, R.W. McCallum, R.I. Guerrant. Urea protects *Helicobacter pylori* from bactericidal effect of acid. *Gastroenterology* 99:697–702, 1990.
101. McGowan, C.C., T.L. Cover, M.J. Blaser. The proton pump inhibitor omeprazole inhibits acid survival of *Helicobacter pylori* by an urease-independent mechanism. *Gastroenterology* 107:1573–1578, 1994.
102. Phadnis, S.H., M.H. Parlow, M. Levy, D. Ilver, C.M. Caulkins, J.B. Connors, B.E. Dunn. Surface localization of *Helicobacter pylori* urease and a heat shock protein homolog requires bacterial autolysis. *Infect. Immun.* 64:905–912, 1996.
103. Weeks, D.L., S. Eskandari, D.R. Scott, G. Sachs. An H+ -Gated urea channel between *Helicobacter pylori* urease and gastric colonization. *Science* 287:482–485, 2000.
104. Desai, M.A., P.M. Vadgama. An *in vivo* study of enhanced H+ diffusion by urease action on urea: implications for *Helicobacter pylori*-associated peptic ulceration. *Scand. J. Gastroenterol.* 28:915–919, 1993.
105. Kawano, S.M., M. Tsujii, H. Fusamoto, N. Sato, T. Kamada. Chronic effect of intragastric ammonia on gastric mucosal structures in rats. *Dig. Dis. Sci.* 36:33–38, 1991.
106. Smoot, D.T. How does *Helicobacter pylori* cause mucosal damage?: direct mechanisms. *Gastroenterology* 113:S31–S34, 1997.
107. Sepulveda, A.R., L.G.V. Coelho. *Helicobacter pylori* and gastric malignancies. *Helicobacter* 7(1):37–42, 2002.
108. Mitchell, H., F. Mégraud. Epidemiology and diagnosis of *Helicobacter pylori* infection. *Helicobacter* 7(1):8–16, 2002.
109. Pilotto, A., N. Salles. *Helicobacter pylori* infection in geriatrics. *Helicobacter* 7(1):56–62, 2002.
110. Roma-Giannikou, E., P.L. Shcherbakov. *Helicobacter pylori* infection in pediatrics. *Helicobacter* 7(1):50–55, 2002.
111. Solnick, J.V., D.B. Schauer. Emergence of diverse *Helicobacter* species in the pathogenesis of gastric and enterohepatic diseases. *Clin. Microbiol. Rev.* 14(1):59–97, 2001.
112. Bazzoli, F., Pozzato, P., T. Rokkas. *Helicobacter pylori*: the challenge in therapy. *Helicobacter* 7(1):43–49, 2002.
113. Somal, N.A., K.E. Coley, P.C. Molan, B.M. Hancock. Susceptibility of *Helicobacter pylori* to the antibacterial activity of manuka honey. *J. Royal Soc. Med.* 87:9–12, 1994.
114. Tabak, M., R. Armon, I. Potasman, I. Neeman. *In vitro* inhibition of *Helicobacter pylori* by extracts of thyme. *J. Appl. Bacteriol.* 80:667–672, 1996.
115. Aldana, L.P., M. Kato, S. Nakagawa, M. Kawarasaki, T. Nagasako, T. Mizushima, H. Oda, J. Kodaira, Y. Shimizu, Y. Komatsu, R. Zheng, H. Takeda, T. Sugiyama, M. Asaka. The relationship between consumption of antimicrobial agents and the prevalence of primary *Helicobacter pylori* resistance. *Helicobacter* 7(1):306–309, 2002.
116. Hamasaki, N., E. Ishii, K. Tominaga, Y. Tezuka, T. Nagaoka, S. Kadota, T. Kuroki, I. Yano. Highly selective antibacterial activity of novel alkyl quinolone alkaloids from a Chinese

herbal medicine, Gosyuyu (Wu-Chu-Yu), against *Helicobacter pylori in vitro. Microbiol. Immunol.* 44:9–15, 2000.

117. Higuchi, K., T. Arakawa, K. Ando, Y. Fujiwara, T. Uchida, T. Kuroki. Eradication of *Helicobacter pylori* with a Chinese herbal medicine without emergence of resistant colonies. *Amer. J. Gastroenterol.* 94:119–120, 1999.
118. Mabe, K., M. Yamada, I. Oguni, T. Takahashi. *In vitro* and *in vivo* activities of tea catechins against *Helicobacter pylori. Antimicrob. Agents Chemother.* 43:1788–1791, 1999.
119. Jonkers, D., E. van den Broek, I. van Dooren, C. Thijs, G. Hageman, E. Stobberingh. Antibacterial effect of garlic and omeprazole on *Helicobacter pylori. J. Antimicrob. Chemother.* 43:837–839, 1999.
120. Canizares, P., I. Garcia, L.A. Gómez, C.M. de Argila, L. de Rafael, A. García. Optimization of *Allium sativum* solvent extraction for the inhibition of *in vitro* growth of *Helicobacter pylori*. *Biotechnol. Prog.* 18:1227–1232, 2002.
121. Jones, N.L., S. Shabib, P.M. Sherman. Capsaicin as an inhibitor of the growth of the gastric pathogen *Helicobacter pylori. FEMS Microbiol. Lett.* 146:223–227, 1997.
122. Daroch, F., M. Hoeneisen, C. Gonzalez, F. Kawaguchi, F. Salgado, H. Solar, A. Garcia. *In vitro* antibacterial activity of Chilean red wines against *Helicobacter pylori. Microbios* 104:79–85, 2001.
123. Oh, Y., M.S. Osato, X. Han, G. Bennett, W.K. Hong. Folk yogurt kills *Helicobacter pylori. J. Appl. Microbiol.* 93:1083–1088, 2002.
124. Boyanova, L., S. Derejian, R. Koumanova, N. Katsarov, G. Gergova, I. Mitov, R. Nikolov, Z. Krastev. Inhibition of *Helicobacter pylori* growth *in vitro* by Bulgarian propolis: preliminary report. *J. Med. Microbiol.* 52:417–419, 2003.
125. Fukai, T., A. Marumo, K. Kaitou, T. Kanda, S. Terada, T. Nomura. Anti-*Helicobacter pylori* flavonoids from licorice extract. *Life Sciences*, 71:1449–1463, 2002.
126. Fahey, J.W., X. Haristoy, P.M. Dolan, T.W. Kensler, I. Scholtus, K.K. Stephenson, P. Talalay, A. Lozniewski. Sulforaphane inhibits extracellular, intracellular, and antibiotic-resistant strains of *Helicobacter pylori* and prevents benzo[a]pyrene-induced stomach tumors. *Proc. Natl. Acad. Sci. USA* 99(11):7610–7615, 2002.
127. Mahady, G.B., S.L. Pendland, G. Yun, Z.Z. Lu. Turmeric (*Curcuma longa*) and curcumin inhibit the growth of *Helicobacter pylori*, a group 1 carcinogen. *Anticancer Res.* 22:4179–4182, 2002.
128. Shinchi, K., H. Ishii, K. Imanishi, S. Kono. Relationship of cigarette smoking, alcohol use, and dietary habits with *Helicobacter pylori* infection in japanese men. *Scand. J. Gastroenterol.* 32:651–655, 1997.
129. Nagata, C., N. Takatsuka, N. Kawakami, H. Shimizu. A prospective cohort study of soy product intake and stomach cancer death. *Br. J. Cancer* 87:31–36, 2002.
130. McCue, P., K. Shetty. A role for amylase and peroxidase-linked polymerization in phenolic antioxidant mobilization in dark-germinated soybean and implications for health. *Process. Biochem.* 39:1785–1791, 2004.
131. McCue, D.A., A. Horii, K. Shetty. Solid-state bioconversion of phenolic antioxidants from defatted powdered soybean by *Rhizopus oligosprous*: role of carbohydrate cleaving enzymes. *J. Food Biochem.* 27:501–514, 2003.
132. McCue, P., K. Shetty. Role of carbohydrate-cleaving enzymes in phenolic antioxidant mobilization from whole soybean fermented with *Rhizopus oligosprorus. Food Biotechnol.* 17:27–37, 2003.
133. Gyorgy, P., K. Murata, H. Ikehata. Antioxidants isolated from soybeans (tempeh). *Nature* 203:870–872, 1964.
134. Akyon, Y. Effet of antioxidants on the immune response of *Helicobacter pylori. Eur. Soc. Clin. Microbiol. Infect. Dis.* 8:438–441, 2002.
135. McCue, P., K. Shetty. A biochemical analysis of ming bean (*Vigna radiata*) response to microbial polysaccharides and poential phenolic-enhancing effects for nutraceutical applications. *Food Biotechnol.* 16:57–79, 2002.

136. Nakamura, Y., A. Kaihara, K. Yoshii, Y. Tsumura, S. Ishimitsu, Y. Tonogai. Content and composition of isoflavonoids in mature or immature beans and bean sprouts consumed in Japan. *J. Health Sci.* 47:394–406, 2001.
137. McCue, P., Y.-T. Lin, R.G. Labbe, K. Shetty. Sprouting and solid-state bioprocessing by Rhizopus oligosporus increase the *in vitro* antibacterial activity of aqueous soybean extracts against Helicobacter pylori. *Food Biotechnol.,* 2004. [In press]
138. Lin, Y.-T., R.G. Labbe, K. Shetty. Inhibition of *Vibrio parahaemolyticus* in seafood systems using oregano and cranberry phytochemical synergies and lactic acid. *Innovative Food Science and Emerging Technologies,* 2004. [Submitted]
139. Vattem, D.A., Y.-T. Lin, R. Ghaedian, K. Shetty. Cranberry synergies for dietary management of *Helicobacter pylori* infections. *Process Biochem,* 2004. [Accepted for publication]
140. Ho, C-Y., Y.-T. Lin, R.G. Labbe, K. Shetty. Inhibition of *Helicobacter pylori* by phenolic extracts of sprouted peas (*Pisum sativum* L.). *J. Food Biochem.*, 2004. [Submitted]
141. Ho, C.-Y., Y.-T. Lin, R.G. Labbe, K. Shetty. Phenolic extracts from sprouted chickpeas (Cicer arietinum L.) to inhibit ulcer-associated Helicobacter pylori. *Asia Pac. J. Clin. Nutr.,* 2004. [Submitted]
142. Randhir, R., Y.-T. Lin, K. Shetty. Phenolics, antioxidant and antimicrobial activity in dark germinated fenugreek sprouts in response to peptide and phytochemical elicitors. *Asia Pac. J. Clin. Nutr.*, 2004. [In press]
143. Randhir, R., Y.-T. Lin, K. Shetty. Stimulation of phenolics, antioxidant and antimicrobial activities in dark germinated mung bean (*Vigna radiata*) sprouts in response to peptide and phytochemical elicitors. *Process. Biochem.* 39:637–646, 2004.
144. Vattem, D.A., Y.T. Lin, R.G. Labbe, K. Shetty. Phenolic antioxidant mobilization in cranberry pomace by solid-state bioprocessing using food grade fungus *Lentinus edodes* and effect on antimicrobial activity against select food-borne pathogens. *Innovative Food Sci. Emerg. Technol.* 5:81–91, 2004.
145. Vattem, D.A., Y.T. Lin, R.G. Labbe, K. Shetty. Antimicrobial activity against select food-borne pathogens by phenolic antioxidants enriched cranberry pomace by solid-state bioprocessing using food-grade fungus *Rhizopus oligosporus*. Process Biochemistry, 2004. [In press]
146. Correia, R.T.P., P. McCue, D.A. Vattem, M.A.M. Margarida, G.R. Macedo, K. Shetty. Amylase and *Helicobacter pylori* inhibition by phenolic extracts of pineapple wastes bioprocessed by *Rhizopus oligosporus. J. Food Biochem.*, 2004. [Accepted for Publication]

3.09

Genetic Mechanisms Involved in Regulation of Mycotoxin Biosynthesis

Michael J. Miller and John E. Linz

CONTENTS

9.1 INTRODUCTION

9.1.1 Mycotoxins

Mycotoxins are a highly diverse family of secondary metabolites synthesized by molds as they grow on a variety of economically important food and feed crops including grains, oilseeds, and spices as well as fruits, dairy products, and meat products (1,2). Mycotoxins are biologically active compounds that can have mild to severe adverse health effects (including death) on animals that ingest contaminated feed (1,2,23,118). A significant body of evidence strongly suggests that mycotoxins have both acute and chronic effects on human health as well (1,2,23,118). Because of the potential health risks, the levels of aflatoxins and deoxynivalenol (DON, a trichothecene mycotoxin) are regulated in food and feed crops in the USA (2). Several countries have regulatory limits for additional mycotoxins including patulin, T-2 toxin, zearalenone, fumonisins, and ochratoxin A. The country specific regulatory limits for various mycotoxins can represent important barriers to international trade of contaminated commodities. The presence of mycotoxins in food and feed results in huge health and economic costs. Up to 25% of the world's crops are affected by mycotoxigenic molds each year, resulting in billions of dollars in losses (1). In the USA alone, losses due to aflatoxins, fumonisins, and deoxynivalenol are estimated at over $900 million annually (2).

Mycotoxin contamination of crops in the US occurs most frequently in specific regions and on specific crops (1) which strongly suggests that the growth environment (that is, moisture, temperature, and light), the genotype of the host plant, and the genotype of the fungal pathogen (or saprophyte) all play important roles in determining the degree of mycotoxin contamination. These factors directly or indirectly control the expression of genes involved in mycotoxin synthesis. An understanding of the genetic mechanisms that regulate mycotoxin gene expression will lead to novel control methods to reduce or eliminate contamination of food and feed crops. This chapter will provide the reader with current information regarding the genetic mechanisms that regulate mycotoxin gene expression in response to the growth environment. Although many mycotoxins are important from an economic and health standpoint, we provide examples primarily from the aflatoxins, trichothecenes, and fumonisins to illustrate key points, because a significant body of information has accumulated regarding the genes involved in their synthesis (23,118,142). Less is known about other groups of mycotoxins and these will be discussed only to a limited extent.

We begin with a summary of each major group of mycotoxins that includes the chemical structure, the primary food or feed crops affected, the mechanism of toxicity, and a brief overview of the biosynthetic pathway. This information provides the necessary framework for the detailed discussion of genetic mechanisms that follows. In this review, we will present fungal gene names by three lower case italicized letters followed by a capital letter locus designation (*aflR, pacC*). Protein names will be presented with a capital letter, followed by two lower case letters (no italics) and the capital letter locus designation (*aflR, pacC*). Exceptions to these conventions are clearly explained in the text.

9.1.1.1 Aflatoxins

Aflatoxins are a family of closely related, polyketide derived difuranocoumarins that are synthesized by several *Aspergillus* species (23,25). *Aspergillus flavus* and *A. parasiticus*

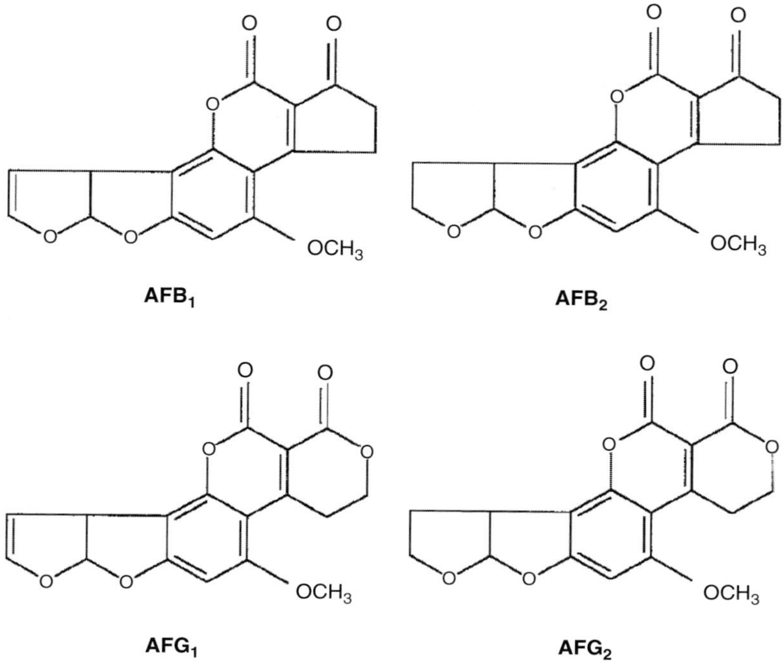

Figure 9.1 Molecular structure of the primary aflatoxins

are the predominant toxigenic species and, under appropriate environmental conditions, they can produce toxin on a variety of economically important host plants including corn, cottonseed, treenuts (almonds, pistachios, and walnuts), peanuts, figs, and spices (1,2). Aflatoxins B_1 (AFB_1), B_2, G_1, and G_2 are the primary aflatoxins observed in contaminated food and feed (Figure 9.1); AFB_1 is the most abundant, toxic, mutagenic, and carcinogenic of the naturally occurring aflatoxins (51).

Human aflatoxin exposure occurs primarily through ingestion of contaminated crops, by ingestion of products derived from animals given contaminated feed, or by inhalation of toxin containing materials (50). In the liver, aflatoxins can be metabolically detoxified by several different enzymes. However, aflatoxins can be converted to the 8,9 epoxide by a specific mixed function, cytochrome p450 monooxygenase (3A4) in the liver after ingestion (51). This highly reactive molecule then can form protein adducts (primarily at lysine) and DNA adducts (primarily at guanine) resulting in abnormalities in protein function and malfunctions in the replication, repair, and expression of DNA (51). The 8,9 epoxide is thought to intercalate into double stranded DNA, allowing reaction with ring nitrogen 7 in guanine (181). In animals, and presumably in humans as well, DNA damage (mutations) can result in development of hepatocellular carcinomas, presumably by activating or inactivating specific genes involved in signal transduction pathways that regulate cell growth (4,46). Another proposed mechanism of toxicity is the generation of reactive oxygen species (ROS) in cells exposed to AFB_1 (133) resulting in lipid peroxidation and cell injury (121). Experimental data clearly demonstrate that aflatoxins are carcinogenic in animals; a significant body of epidemiological evidence also suggests that aflatoxins, together with hepatitis B virus, are risk factors for human liver carcinoma (129).

At least 17 different enzyme activities catalyze AFB_1 biosynthesis (23,108,142,161). These activities are encoded by up to 25 genes clustered on one chromosome in *A. parasiticus*

Table 9.1
Summary of gene names, enzyme names, and enzyme activities in aflatoxin synthesis.

Gene Name	Enzyme Name (Reported)	Activity
fas-2 (aflA)	fatty acid synthase (α)	Converts acetate to hexanoate
fas1 (aflB)	fatty acid synthase (β)	Converts acetate to hexanoate
pksA (aflC)	polyketide synthase	Extends hexanoate to decaketide
nor-1 (aflD)	norsolorinic acid reductase	Converts norsolorinic acid to averantin
norA (aflE)		Converts norsolorinic acid to averantin
norB (aflF)		Converts norsolorinic acid to averantin
avnA (aflG)		Converts averantin to hydroxyaverantin
adhA (aflH)	alcohol dehydrogenase	Converts hydroxyaverantin to averufin
avfA (aflI)		Converts averufin to versiconal hemiacetal acetate (VHA)
estA (aflJ)	esterase	Converts VHA to versiconal
vbs (aflK)	versicolorin B synthase	Converts versiconal to versicolorin B (VerB)
verB (aflL)	desaturase	Converts VerB to versicolorin A (VerA)
ver-1 (aflM)	reductase	Converts VerA to demethylsterigmatocystin (DMST)
verA (aflN)		Converts VerA to DMST
omtB (aflO)	O-methyltransferase	Converts DMST to sterigmatocystin (ST)
omtA (aflP)	O-methyltransferase	Converts ST to O-methylsterigmatocystin (OMST)
ordA (aflQ)	oxidoreductase	Converts OMST to aflatoxin B1
aflR		positive transcription activator
aflJ (aflS)		putative transcription co-activator
aflT		putative toxin efflux pump
cypA (aflU)	p450 monooxygenase	function undetermined
cypX (aflV)	p450 monooxygenase	function undetermined
moxY (aflW)	monooxygenase	function undetermined
aflX	oxidase	function undetermined
aflY	open reading frame	function undetermined

Source: Yu, J-J., P-K Chang, J.W. Cary, D. Bhatnagar, T.E. Cleveland, G.A. Payne, J.E. Linz, C.P. Woloshuk, J.W. Bennett, Appl. Environ. Microbiol, 2003.

and *A. flavus* (153,168) (Table 9.1, Figure 9.2) and a similar gene cluster in *A. nidulans* (synthesizes sterigmatocystin, a late aflatoxin pathway intermediate) (15). During aflatoxin synthesis, ten molecules of acetyl CoA are polymerized head to tail by a specialized fatty acid synthetase (13,95,157) in association with a polyketide synthetase (32,153) to generate the first stable pathway intermediate, norsolorinic acid. The remainder of the biosynthetic pathway is dedicated to modification of this hydrophobic decaketide molecule to generate the highly reactive bisfuran ring structure and to modification of the coumarin moiety to carry either a cyclopentenone (B aflatoxins) or lactone (G aflatoxins) ring structure (108).

9.1.1.2 Trichothecenes.

Trichothecenes are a large family of sesquiterpene derived mycotoxins synthesized by several species of the filamentous fungus *Fusarium*, several other fungal genera (*Myrothecium*, *Stachybotrys*, *Trichothecium*) and even one plant genus (*Baccharis*) (118). There are over 180 unique trichothecenes which share a common core trichothecene structure (Figure 9.3)

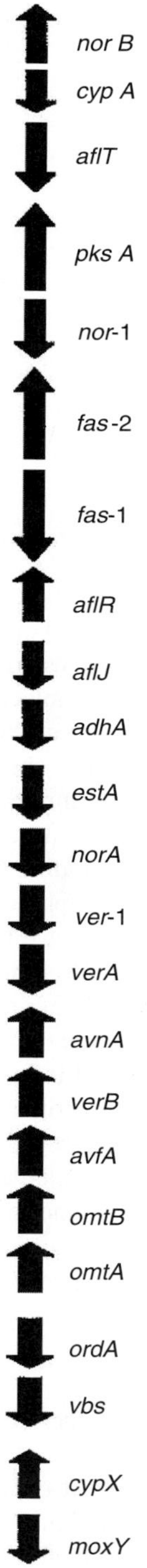

Figure 9.2 Genomic organization of the aflatoxin biosynthetic gene cluster in *Aspergillus parasiticus*. Arrowheads indicate the direction of transcription. Drawn approximately to scale

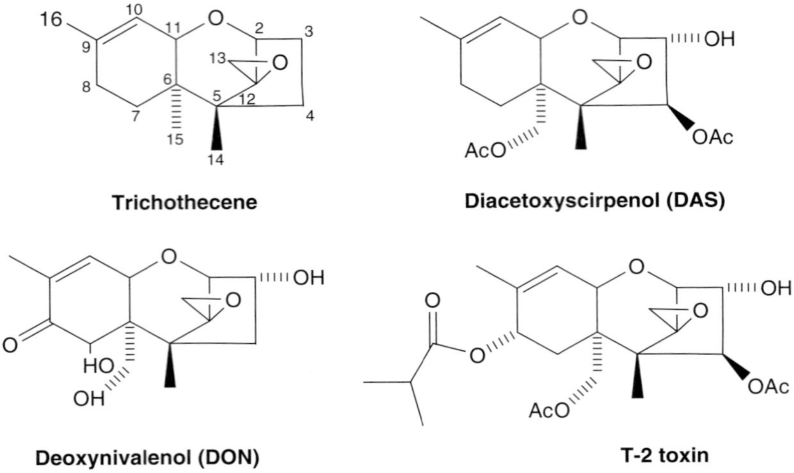

Figure 9.3 Chemical structures of trichothecene toxins

and generally all have an epoxide residue at carbons 12 (C-12) and 13 (C-13) and a double bond between C-9 and C-10. The trichothecenes are placed into four separate groups based on modifications to the core trichothecene molecule (154). Trichothecene contaminated grains, including wheat, barley, maize, and rye, represent the most important sources for human and animal exposure. Of the trichothecenes, DON is among the most commonly encountered in terms of frequency and concentration in grains (123).

Trichothecenes are potent protein synthesis inhibitors. This inhibition results in numerous acute toxic effects on animals including diarrhea, nausea, vomiting, dysfunction of the central nervous system, necrosis of lymphoid tissue and bone marrow, and death (1). Specific trichothecenes have been shown to have dose dependent effects on immune system function. For example, low doses of DON stimulate IgA synthesis in mice resulting in symptoms reminiscent of IgA nephropathy in humans (10). At higher doses, DON induces apoptosis in lymphocytes (163) and suppresses immune system function, resulting in increased susceptibility to bacterial and viral infection (10).

The trichothecenes are synthesized by polymerization of the isoprenoid intermediate farnesyl pyrophosphate by a trichodiene synthase (47). The remainder of the pathway involves a series of modifications of trichodiene to generate a variety of end products including DON (*F. culmorum*), T-2 (*F. sporotrichioides*), and diacetoxyscirpenol (*F. sporotrichioides*) (Figure 9.3). Like the aflatoxin pathway, the genes involved in trichothecene biosynthesis, called *tri* genes, are clustered and appear to be coordinately regulated (81,118).

9.1.1.3 Fumonisins

Fumonisins are aminopolyol mycotoxins synthesized by several *Fusarium* species including *F. moniliforme*, *F. anthophilum*, *F. dlamini*, *F. napiforme*, *F. nygamai*, *F. oxysporum*, and *F. proliferatum* (110,118). Fumonisins consist of a 20 carbon eicosane backbone with an amino group at C-2 and hydroxyl groups at C-3, C-5, or C-10 (Figure 9.4). Other modifications include methyl groups at C-12 and C-16, and propane-1,2,3 tricarballylic acid moieties esterified at C-14 and C-15. The fumonosins are placed into 2 major groups based on structure, group B and group A (amino group is acetylated). Fumonisin B_1, in group B, is the most abundant and is produced predominantly by *F. moniliforme* and *F. proliferatum* growing on

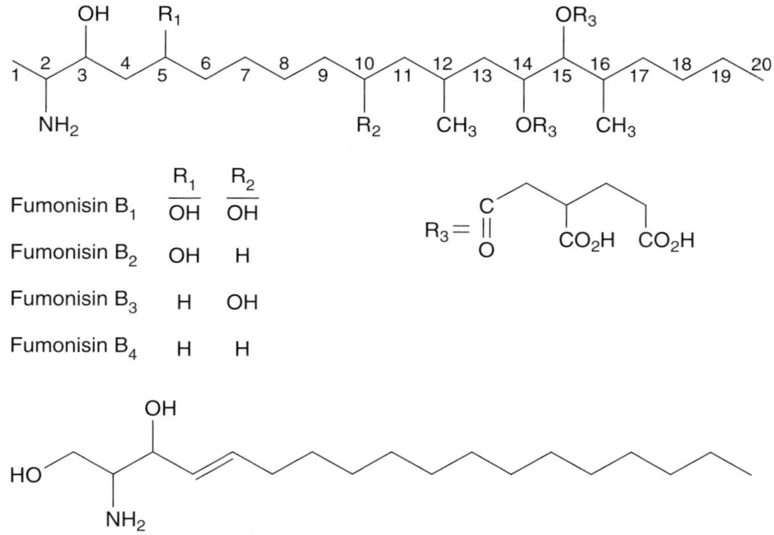

Sphingosine

Figure 9.4 Chemical structures of fumonisins

maize. Ingestion of fumonisin contaminated maize has been epidemiologically associated with human esophageal cancer in China and South Africa (96). In animals, fumonisin ingestion has been linked to cancer in rats, pulmonary edema in swine, and leukoencepahlomalacia in horses (96). This mycotoxin also causes apoptosis in a variety of cell types in culture and in kidney and liver cells in rats *in vivo* (150,151). It has been proposed that the toxic effects of fumonisins arise due to disruption of sphingolipid metabolism (126). In support of this hypothesis, fumonisins inhibit the activity of sphinganine n-acetyltransferase, which converts sphinganine to dihydroceramide and sphingosine to ceramide (156). The altered sphinganine/sphingosine balance presumably alters sphingolipid signaling which influences cell growth and death Following the pattern established for aflatoxins and trichothecenes, the currently identified genes involved in fumonisin biosynthesis, called *fum* genes, are coordinately regulated and clustered on a single chromosome (130). Although information on the details of the biosynthetic pathway is still limited, fumonisin synthesis is predicted to include at least five steps: (1) synthesis of a linear polyketide; (2) condensation of the polyketide with alanine; (3) reduction of a polyketide carbonyl group to hydroxyl; (4) hydroxylation of up to 4 polyketide carbons; and (5) esterification of tricarboxylic acid moieties to two hydroxyls (130). Cloning and functional analysis of several clustered genes encoding a polyketide synthetase (130), a cytochrome p450 monooxygenase, a type III alcohol dehydrogenase, a class II " amino transferase, and a dioxygenase (130) provide an important framework for developing a more detailed understanding of the synthesis of this polyketide derived mycotoxin.

9.2 MECHANISMS OF MYCOTOXIN GENE REGULATION

9.2.1 Transcriptional Regulation

Mycotoxins are secondary metabolites produced by several species of fungi. As secondary metabolites, mycotoxins are not constitutively synthesized. Rather, mycotoxin biosynthesis is a complex process that responds to developmental, environmental, and nutritional cues. One primary means that fungi utilize to regulate mycotoxin biosynthesis appears to be transcriptional regulation of the mycotoxin biosynthetic genes. Aflatoxin and trichothecene biosynthesis are the most characterized of the mycotoxins. Information gathered about aflatoxin and trichothecene biosynthesis has been helpful for researchers studying other mycotoxin biosynthetic pathways. In the following sections, the regulation of aflatoxin and trichothecene biosynthesis gene transcription is reviewed with brief mention of other mycotoxin pathways.

9.2.1.1 Aflatoxin Synthesis

9.2.1.1.1 AflR – Discovery Using parasexual analysis, a mutation in the *afl2* locus in *A. flavus* strain 650 was mapped to linkage group VII which contained nine other putative aflatoxin genes providing preliminary evidence for an aflatoxin gene cluster (8). The parental strain accumulated norsolorinic acid (Table 9.1) whereas *A. flavus* 650 did not accumulate this early pathway intermediate (8). Metabolite feeding studies and enzyme activity measurements of *A. flavus* 650 demonstrated that several aflatoxin biosynthetic enzymes were not present, suggesting that perhaps the mutation in *afl2* affected a regulatory gene and was responsible for the phenotype (116). The gene was later cloned in *A. flavus* by complementation and named *aflR* (116). Subsequently, *aflR* homologues were identified in *A. parasiticus* (31) and *A. nidulans* (169).

9.2.1.1.2 AflR – Function Several independent lines of biochemical evidence indicate a regulatory role for *aflR*. Metabolite feeding studies and enzymatic activity measurements with *A. flavus* 650 (*aflR* mutant) and wild-type *A. flavus* demonstrated that *aflR* is necessary for the aflatoxin biosynthetic enzyme activities to be detected (116). Later, studies demonstrated that *aflR* is required for aflatoxin biosynthetic gene transcript accumulation (169). Insertion of an additional copy of *aflR* into an *A. parasiticus* O-methylsterigmatocystin producing strain resulted in the overproduction of pathway intermediates including O-methylsterigmatocystin (31). In addition, insertion of an additional copy of *aflR* in *A. parasiticus* resulted in the transcription of pathway genes under aflatoxin noninducing conditions (34). Use of *aflR* linked with an inducible promoter in *A. nidulans* demonstrated that induction of *aflR* under aflatoxin noninducing conditions could activate genes in the biosynthetic pathway (169). Based on amino acid sequence identity, AflR was proposed to belong to a common class of fungal transcription factors called zinc binuclear cluster proteins that includes Gal4 (160). While the aminoterminus of AflR contains a DNA binding domain, the carboxyl terminus has a highly acidic region that functions as a transactivation domain (37). However, the total acidity in this region was not a major determinant in AflR transactivation (37).

DNA binding by AflR has also been investigated. Using methylation interference footprinting and electrophoretic mobility shift assays (EMSA), the palindromic AflR binding site was first identified as TCGNNNNNCGA in the *stcU* promoter in *A. nidulans* (62). The consensus AflR binding site was later defined as TCGSWNNSCGR (S = C/G, W = A/T, R = A/G) based on *in vitro* binding studies with recombinant AflR (61). Recombinant AflR binds to several aflatoxin biosynthetic gene promoters *in vitro* (52,54,62,105). AflR *cis*-acting sites have been shown to be necessary for transcriptional activation of several aflatoxin biosynthetic genes *in vivo* including *stcU* (62), *pksA* (55), *avnA* (26) and *nor-1* (116). Based on these and other data, AflR is proposed to be necessary for transcriptional activation of several, if not all, aflatoxin biosynthetic genes.

9.2.1.1.3 AflR – Regulation Aflatoxin synthesis is affected by environmental and nutritional factors such as pH, temperature, carbon source, and nitrogen source (93,94,107).

Because AflR is a pathway regulator, the simplest model for mediating the effects of these environmental and nutritional signals on aflatoxin gene transcription is directly through the *aflR* promoter. In support of this model, the timing of expression of *aflR* in response to these stimuli mimicked the expression of aflatoxin structural genes (93,107). *Aspergillus parasiticus* AreA (major nitrogen regulatory protein) was shown to bind to the AflR promoter *in vitro*, yet *in vivo* significance is unknown (38). A PacC (pH sensing) *cis*-acting site has also been identified in the *aflR* promoter using *in vitro* methods (52). A functional AflR binding site was located in the *aflR* promoter and is necessary for *aflR* transcription *in vivo,* suggesting autoregulation (52). Additional studies on the transcriptional regulation of *aflR* are needed in order to understand how environmental, nutritional, and even developmental stimuli affect aflatoxin biosynthesis.

9.2.1.1.4 Evidence for Other Transcription Factors Though AflR is necessary for transcriptional activation of several, if not all, aflatoxin biosynthetic genes (26,55,62,15), there is evidence for the involvement of additional transcription factors. Studies with the *pksA* promoter (55) provided evidence that the transcription factors PacC (pH sensing) and BrlA (sporulation) can impact *pksA* transcriptional regulation through *cis*-acting sites in the *pksA/nor-1* intergenic region (Figure 9.5). However, deletion of the consensus *cis*-acting sites for PacC and BrlA did not affect *nor-1* transcriptional regulation in *A. parasiticus* in submerged culture (105).

The *A. nidulans* gene *stcU* (*A. parasiticus ver-1* homologue) also appears to be regulated by additional transcriptional factors besides AflR. A full length *stcU* promoter fused to a glucuronidase (GUS) reporter resulted in only a two- to threefold increase in activity when transformed into a wild-type strain compared to transformation into an *aflR* mutant strain (62). In addition, substitution of both AflR binding sites in this promoter only resulted in an approximately fivefold reduction (not reduced to baseline expression levels) in GUS activity (62). Due to the relatively high activity in the *aflR* mutant strain and the moderate decrease resulting from binding site substitution, the possibility exists that other transcriptional activators besides AflR are involved in *stcU* transcriptional activation (62). For *avn*A in *A. parasiticus*, substitution of the AflR binding site resulted in a tenfold decrease in promoter activity (26). While there is no evidence of other transcriptional activators in the *avnA* promoter, a potential *cis*-acting site for a repressor of aflatoxin biosynthesis was located (26). Deletion of 78 bp upstream of an AflR binding site in an *avnA*::GUS reporter construct resulted in a threefold increase in reporter activity (26). In addition, protein extracts collected under nonaflatoxin inducing conditions demonstrated specific binding to this region (26).

The *nor-1* gene is perhaps the most studied of all the aflatoxin structural genes. A detailed analysis of the *nor-1* promoter has identified several *cis*-acting sites including AflR1, AflR2, a putative TATA box, CRE1 and NorL (116,117) (Figure 9.5, Figure 9.6).

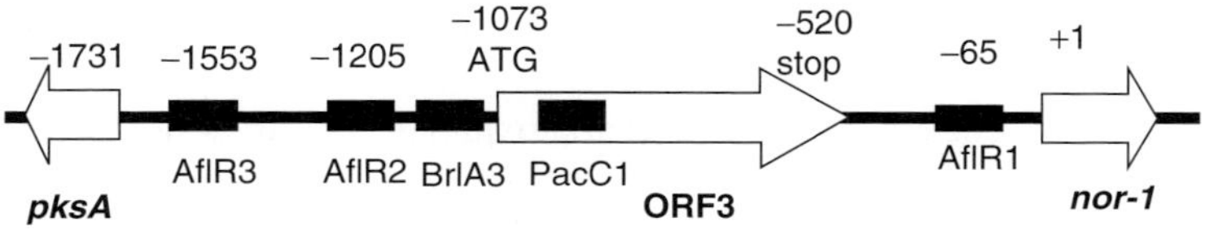

Figure 9.5 Schematic of the *nor-1/pksA* intergenic region in *A. parasiticus*. The numbers indicate the number of nucleotides included upstream from the primary transcriptional start site of *nor-1*. Several potential *cis*-acting sites are indicated including the AflR binding sites AflR1, AflR2 and AflR3 and PacC1 and BrlA3. The location of an open reading frame (ORF3) of unknown function is also shown (Adapted from Miller et al., 2003)

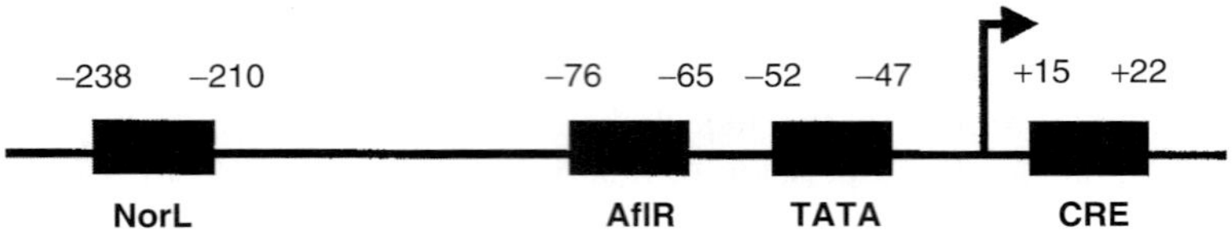

Figure 9.6 Identification of *cis*-acting sites in the *nor-1* promoter of *A. parasiticus*. The numbers indicate the number of nucleotides from the primary transcriptional start site of *nor-1*. (Adapted from Miller et al., 2003)

Although putative TATA boxes have been identified in aflatoxin biosynthetic promoters, the *nor-1* TATA box was the first to be rigorously tested for functional significance (106). Substitution of the TATA box in the context of a larger promoter resulted in nondetectable GUS activity (106). A novel *cis*-acting site (NorL) was identified that is necessary for maximum *nor-1* transcriptional activation *in vivo*. Using EMSA, specific protein binding to NorL was demonstrated (106). EMSA also identified another potential *cis*-acting site, CRE1, in the *nor-1* promoter (Roze et al., in preparation) (106,107). Both the NorL binding protein (NorLbp) and CRE1 binding protein (CRE1bp) appeared to rely on functional AflR for maximum DNA binding (106,107). Based on these observations, Miller proposed that the transcriptional regulation of *nor-1* involves additional proteins besides AflR (106).

9.2.1.1.5 Open Reading Frame 3 (ORF3), nor-1, and pksA Transcriptional Activation Reporter studies with *pksA*::GUS (55) and *nor-1*::GUS (105) both demonstrated that inclusion of the *cis*-acting site AflR2 (Figure 9.5) resulted in greater transcriptional activation. Miller and Linz proposed two alternative models to explain these data: (1) AflR2 works synergistically with AflR1 and AflR3 to mediate transcription of the *nor-1* and *pksA* promoters respectively; or (2) AflR2 mediates expression of ORF3 (potentially encodes a polypetide of approximately 300 amino acid residues) (Figure 9.5) directly downstream from AflR2 which directly or indirectly impacts transcription of the *nor-1* and *pksA* promoters (106). Identification of a cDNA corresponding to ORF3 in an *A. parasiticus* cDNA library strongly suggests it represents a functional gene. Interestingly, blast searches using ORF3 as a query sequence did not provide solid clues regarding potential function. Model 2 prompted two related predictions regarding *nor-1* and *pksA* promoter function (106): (1) Because fungal isolates carrying *nor-1*::GUS or *pksA*::GUS constructs with promoter fragments that include a functional AflR2 and ORF3 showed the highest GUS expression levels, accumulation of additional ORF3 protein in strains with two copies (native plus plasmid copy) overcomes a protein threshold resulting in extreme up regulation of *nor-1* and *pksA* promoter activity. (2) Loss of ORF3 function due to AflR2 deletion in the second copy accounts for down regulation (or lack of up regulation) of both *nor-1* and *pksA* expression. Similar results are seen with the insertion of an additional copy of *aflR* (31,34,98) suggesting a possible regulatory role for ORF3. Future experiments will determine the function of ORF3 and how AflR2 impacts *pksA* and *nor-1* transcriptional activation.

9.2.1.1.6 AflJ The function of *aflJ* in aflatoxin biosynthesis is still not entirely clear. The *aflJ* gene resides adjacent to *aflR* in the aflatoxin gene cluster with the two genes being divergently transcribed. An *A. flavus aflJ* knockout strain did not make aflatoxin, and lacked the ability to convert several aflatoxin intermediates to aflatoxin (103). However, the *aflJ* knockout strain did accumulate several aflatoxin biosynthetic transcripts under aflatoxin conducive conditions, suggesting that *aflJ* is not involved in the transcriptional regulation of aflatoxin biosynthesis (103). Sequence analysis of *aflJ* did not reveal any enzymatic function but did identify three potential membrane spanning domains and a putative microbody targeting signal (103). Consequently, Meyers et al. (103) proposed two hypotheses regarding the function of the AflJ protein: (1) AflJ is involved in either transmembrane transport of aflatoxin pathway intermediates through intracellular compartments; or (2) AflJ is involved in the localization of pathway enzymes to an organelle. In contrast to these studies, AflJ more recently has been shown to interact with AflR in a two hybrid assay (37). Insertion of an additional copy of *aflR* and *aflJ* into *A. parasiticus* resulted in greater aflatoxin biosynthesis than insertion of *aflR* alone, whereas insertion of *aflJ* alone had no affect on aflatoxin biosynthesis (30). Consequently, Chang et al. (30) described *aflJ* as being a transcriptional coactivator. Strong support for this idea was provided by a recent study by Chang (29). More work is needed to clearly define the activity of AflJ.

9.2.1.2 Trichothecene Synthesis

9.2.1.2.1 Tri6 – Discovery Following the identification of three trichothecene biosynthesis genes, *tri5* (74), *tri4* (75) and *tri3* (101), it was realized that they all resided in a 9 kb region and that the trichothecene biosynthesis genes might be clustered (77). The location of *aflR* within the aflatoxin biosynthetic gene cluster suggested the possibility that other mycotoxin gene clusters, including trichothecene, may be regulated by a gene present within the cluster (116). While the discovery of *aflR* relied on complementation of a regulatory mutant (31), the initial identification of *tri6* as a gene encoding a putative trichothecene biosynthesis transcriptional activator relied on nucleotide sequence analysis of the trichothecene gene cluster (120). Found immediately upstream of *tri5*, *tri6* encodes an open reading frame of 217 amino acids with regions similar to Cys_2His_2 zinc finger proteins (120). The Cys_2His_2 zinc finger is a common motif for transcription factors including BrlA from *A. nidulans* (3).

9.2.1.2.2 Tri6 – Function Several lines of evidence, in addition to sequence data, demonstrate that the Tri6 protein is a transcriptional activator for trichothecene biosynthesis. *tri6* expression mirrored the expression of other trichothecene genes including *tri3 and tri4* (120). Disruption of *tri6* resulted in a strain with greatly reduced trichothecene production, trichothecene enzyme activities, and trichothecene biosynthetic transcript steady states (120). Tri6 protein functioned as a transcriptional activator in *Saccharomyces cerevisiae* when fused to the DNA binding domain of the yeast transcription factor Gal4 (120). The Tri6 binding site was identified in *Fusarium sporotrichioides* as YNAGGCC using EMSA with Tri6 produced *in vitro* (76). The Tri6 binding site was confirmed *in vivo* using *F. sporotrichioides tri4* reporter strains (76). All of the trichothecene biosynthetic genes identified in the *F. sporotrichioides* and *F. graminearum* clusters have the Tri6 binding site located in their promoter regions except for *Tri10* (14,76).

9.2.1.2.3 Tri10 – Discovery Located upstream of *tri5* in the trichothecene gene cluster (Figure 9.7), expression of the gene *tri10* mirrored the timing of several trichothecene genes including *tri6*, *tri5* and *tri4* (144). A *tri10* disruption strain accumulated significantly less trichothecene biosynthetic transcripts (144). Conversely, transformants that had increased *tri10* expression resulted in significantly increased trichothecene gene expression (144). As noted, *tri10* lacks a consensus Tri6 binding site (14) and is significantly up regulated in the

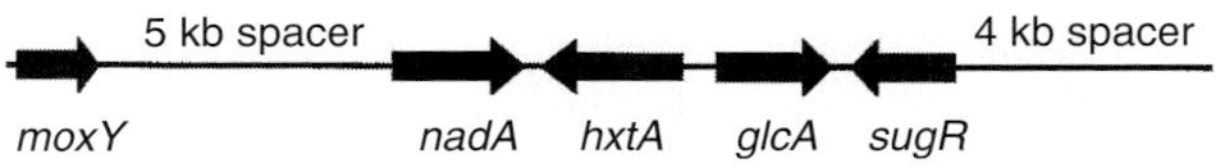

Figure 9.7 Schematic of the sugar utilization gene cluster in *A. parasiticus*. The *moxY* gene is at one end of the aflatoxin gene cluster. The spacer regions on either side of the sugar utilization cluster do not contain open reading frames. Map is roughly to scale (Adapted from Yu et al., 2000a)

tri6 disruptant strain (144). Tag et al. (144) postulated that *tri10* is not positively regulated by Tri6 but instead may be negatively regulated by unknown mechanisms when Tri6 is present. The impact of Tri10 extends to genes involved in the primary metabolic steps that generate trichothecene precursors including *Fpps* (144). While Tri10 protein lacks any known transcription factor motif, its function is clear as an essential regulator in trichothecene gene regulation (144). Future investigations are focused on determining the precise mode of action of Tri10 and on further identification of the regulatory circuits defined by Tri10 (144).

9.2.1.3 Other Mycotoxins

Although most information regarding mycotoxin gene regulation is centered on aflatoxin and trichothecene biosynthesis, other mycotoxin related transcription factors have been identified. In fact, information about aflatoxin and trichothecene biosynthesis pathways have aided in the analysis of different mycotoxin pathways. For example, a Tri6 homologue has been identified (MRTri6) in the distantly related trichothecene pathway of *Myrothecium roridum* (76). *Dothistroma pini* synthesizes dothistromin, a difuranoanthraquinone toxin similar to aflatoxin, with several analogous genes (11). Bradshaw et al. (11) hope to identify the dothistromin pathway regulator by using *A. parasiticus aflR* as a probe or by sequencing the entire dothistromin gene cluster. Sequencing of the paxilline biosynthesis cluster in *Penicillium paxilli* identified two predicted transcription factors, *paxR* and *paxS* (164). Like *aflR*, both *paxR* and *paxS* contain the uniquely fungal $Zn(II)_2Cys_6$ binuclear cluster DNA binding motif (164). The exact function of *paxR* and *paxS* in paxilline biosynthesis is currently unknown.

9.2.1.4 Activation and Inactivation of Transcription Factors

In the budding yeast *Saccharomyces cerevisiae*, a zinc finger transcription factor, Mig1p, in part mediates glucose signaling by binding to the promoters of glucose repressible genes (140). This carbon catabolite repression is released after phosphorylation of Mig1p by an Snf1p protein complex as part of the signaling cascade. Clues are beginning to emerge that phosphorylation of transcription factors also may play a role in mediating expression of genes involved in secondary metabolism. For example, phosphorylation of AflR is believed to inactivate this positive regulator of aflatoxin synthesis, and perhaps increase its susceptibility to proteolytic cleavage (N. Keller, personal communication). Similarly, phosphorylation of CRE1bp, a novel transcription factor proposed to mediate the cAMP, induced upregulation of at least one aflatoxin gene, *nor-1*, and appears to influence binding to a specific DNA binding motif in the *nor-1* promoter (124).

9.2.2 Posttranscriptional Regulation

Gene regulation studies in filamentous fungi have primarily focused on transcriptional mechanisms. Significantly less is known about posttranscriptional regulatory mechanisms in filamentous fungi (127). For example, recent data strongly suggest that regulation of aflatoxin synthesis may be mediated by at least three such mechanisms: (1) proteolytic cleavage; (2) protein localization; and (3) RNA processing.

Western blot analysis detected full length and proteolytically cleaved forms of OmtA in extracts of *A. parasiticus* SU-1 grown on yeast extract plus sucrose agar medium (YES; supports aflatoxin synthesis) (89); the proteolytically cleaved form of OmtA was observed in increased abundance in older cells within the colony (closer to the center of the center inoculated colony) as compared to younger cells (outer margin of the colony). The extent of cleavage was significantly greater in cells grown on solid medium than in cells grown in submerged liquid culture. In contrast, Western blot analysis of protein extracts of SU-1 grown in submerged liquid culture in YES liquid medium demonstrated that Nor-1 (173) and Ver-1 (92) proteins may be subject to at least one proteolytic cleavage under these culture conditions. However, Nor-1 and Ver-1 did not show a similar degree of proteolytic cleavage in colonies of SU-1 grown on YES agar medium. The specificity and timing of expression of the proteolytic enzymes and the role of cleavage in activation and inactivation of aflatoxin enzymes is clearly of interest for future studies.

In fungal tissues analyzed by immunofluorescence microscopy with highly specific polyclonal antibodies to Nor-1, Ver-1, and OmtA, the majority of the fluorescent signal was located in the cytoplasm of thin sections of paraffin embedded fungal tissue (89). Transmission electron microscopy using gold labeled antibodies confirmed the cytoplasmic location. In cells located near the basal surface (substrate surface) of the colony, OmtA antibodies generated an intense labeling in small and large organelles proposed to be vacuoles (175). Together these data suggest that specific aflatoxin enzymes may be targeted to specific organelles during synthesis. The protein composition and function of these organelles is one focus area for future work.

Certain strains of *Aspergillus parasiticus,* including SU-1, carry two copies of *aflR* (*aflR-1* and *aflR-2*) encoding an important positive regulator of aflatoxin synthesis (AflR). In analyzing the potential function of AflR-2 in transcriptional regulation of aflatoxin synthesis, Cary et al. (24) genetically disrupted *aflR-1*. The experiment demonstrated that AflR-2 alone is not effective as a positive regulator of aflatoxin synthesis. However, a basal level of transcription of *nor-1*, *ver-1*, and *omtA* was observed in the knockout strain using reverse transcriptase PCR (RT-PCR). Of particular interest, the transcripts in the knockout strain did not appear to be spliced to remove the introns, suggesting that an AflR dependent splicing factor was not expressed. Confirming the activity and specificity of the proposed splicing factor is a focus of future experimentation.

9.2.3 Overview of Regulatory Inputs

Filamentous fungi encounter a myriad of growth conditions in their natural environment, and have developed regulatory circuits to allow appropriate levels of gene expression in response to changes in their environment. Among these regulatory circuits are those that mediate alterations in mycotoxin production. As secondary metabolites, mycotoxins are synthesized predominantly during a transition from active growth to stationary phase by fungi. Although many factors are known to influence mycotoxin gene expression, this section will review those factors that have been shown to directly or indirectly modulate gene expression via the alteration of the expression or activity of positive or negative acting transcription factors that bind to specific *cis*-acting sites in mycotoxin gene promoters.

9.2.3.1 Nutrients

A vast array of chemical, nutritional, and environmental factors have been demonstrated to influence aflatoxin gene expression (23,42,49,56,68,115,152,170). Of these factors, most studies at the molecular level have focused on the effects of carbon, nitrogen, trace elements, and pH. Early on, it was demonstrated that transcripts for most, if not all, aflatoxin genes accumulate at the highest rate in a transition between active growth and stationary phase (liquid shake culture), suggesting that the aflatoxin genes are coordinately regulated (153). Based on pioneering work on the influence of carbon source on aflatoxin synthesis by Buchanan (16,17,159), Skory et al. (139) confirmed that *nor-1* and *ver-1* gene expression in *A. parasiticus* (liquid shake culture) was regulated at the level of transcript accumulation and that glucose appeared to be a key factor in mediating this influence. Using *A. flavus* and *A. parasiticus* strains carrying GUS (glucuronidase) reporter constructs, Flaherty et al. (65,179) and Liang et al. (92) showed that the regulatory influence on aflatoxin gene expression was, at least in part, at the transcriptional level. This set the stage for monitoring the interaction of specific transcription factors (such as AflR) with *cis*-acting sites on aflatoxin gene promoters and analyzing the influence of environment on transcription factor expression and activity.

9.2.3.1.1 Carbon Source It is well established that glucose and other simple sugars have an important regulatory influence on aflatoxin synthesis (17,18,159) although the molecular mechanisms that mediate this response are only now beginning to be understood. Glucose and other easily utilizable carbon sources are known to influence fungal gene expression via carbon catabolite repression (125). In *Aspergillus* species, carbon catabolite repression of genes involved in primary metabolism (e.g., *xlnA*, *alcA*) (87,114) requires CreA (48), a zinc finger protein that is reported to be homologous to Mig1 in *Saccharomyces cerevisiae*. A consensus binding site for CreA has been reported (5′-SYGGRG-3′) in the *alcA* promoter. A *creA* homolog, *cre1* has been shown to mediate carbon catabolite repression in the filamentous fungi *Trichoderma reesei* (141) and *Sclerotinia slerotiorum* (155) suggesting that the main components of this regulatory scheme are conserved among the filamentous fungi.

Possibly the most unusual feature of carbon regulation of aflatoxin synthesis is that glucose stimulates aflatoxin synthesis in contrast to the glucose mediated catabolite repression of another *Aspergillus* secondary metabolite, penicillin (12). Liu and Chu (93) reported that the level of AflR protein is increased in the presence of glucose. This positive regulator appeared at significantly lower levels in cell extracts from the fungus (liquid shake culture) when peptone was included in place of glucose (with ammonium sulfate as sole nitrogen source). Reduced levels of AflR protein were also observed with increased temperature, reduced levels of zinc, and when sodium nitrate was used to replace ammonium sulfate as the sole nitrogen source (similar culture conditions) suggesting that multiple nutrients and environmental factors influence AflR protein accumulation directly. This could influence aflatoxin synthesis indirectly by reducing AflR interaction with its *cis*-acting sites.

9.2.3.1.2 Nitrogen Source Like the carbon source, the nitrogen source has a large influence on the expression of specific fungal genes such as those involved in nitrate utilization (21,97). The effects of nitrogen are reported to be mediated by the function of a positive acting global nitrogen regulator encoded by *areA* in *Aspergillus* and by the homologous genes, *nit2* and *NRE*, in *Neurospora* and *Penicillium*, respectively. These GATA-type transcription factors bind specific consensus sites (so called GATA sites) in nitrogen regulated promoters. In addition, positive acting, pathway specific regulators such as those encoded by *nirA* in *Aspergillus* and *nit4* in *Neurospora* fine tune the expression of the nitrogen responsive genes. Early work on regulation of aflatoxin synthesis clearly demonstrated that the identity and concentration of the sole nitrogen source has a significant influence on the accumulation of aflatoxin and pathway intermediates in liquid shake culture (78,112). The nitrogen regulatory scheme in *A. flavus* and probably *A. parasiticus* is complicated by the fact that it is also strongly influenced by the pH of the growth medium (41) . This is important because growth of *A. flavus* in the presence of sodium nitrate (and with glucose or sucrose as the sole carbon source) results in a sharp increase in pH in liquid shake culture, while growth in the presence of ammonium sulfate results in a decrease in pH (78).

Western blot and Northern hybridization analyses and use of aflatoxin gene reporter constructs demonstrated that *nor-1* and *ver-1* are not expressed at detectable levels when sodium nitrate is used as the sole nitrogen source (with glucose as the sole carbon source), but these genes are expressed at normal levels with ammonium sulfate suggesting that this regulatory influence is mediated at the transcriptional level (105–107). This is likely in part explained by the observation that sodium nitrate did not support the accumulation of *aflR* transcript (105) or AflR protein (93). This regulatory scheme again is in direct contrast to the regulation of genes involved in penicillin biosynthesis in *Aspergillus*. Here the preferred nitrogen source, ammonium sulfate, represses penicillin gene expression while sodium nitrate stimulates expression (60). This allows one to predict that AreA protein, which is normally a positive global regulator, may act as a negative regulator of aflatoxin gene expression. A small quantity of available data appears, at least on the surface, to support this hypothesis.

Analysis of the intergenic region between *niaD* (nitrate reductase) and *niiA* (nitrite reductase) in *A. parasiticus* identified several HGATAR sequences that represent the consensus binding site for AreA, but also two copies of a palindrome, TCCGTGGA, which act as a binding site for the *N. crassa* Nit4 nitrogen global regulator (33). Later, Chang et al. (38) cloned and characterized the gene encoding AreA in *A. parasiticus*. The authors generated a recombinant AreA DNA binding domain fusion protein that bound cooperatively to single HGATAR sites in the *niaD* - *niiA* intergenic region. Of particular interest, this fusion protein also bound to the AflR - AflJ intergenic region which contains several consensus HGATAR sequences. However, binding to specific sites was not demonstrated. Because growth of *A. parasiticus* in the presence of sodium nitrate was shown to significantly reduce *aflR* expression at the level of protein (93) and transcript (106,107), these data combined are consistent with a negative regulatory role for nitrate in aflatoxin gene expression mediated by AreA. Clearly, followup will focus on the specificity of AreA binding and the regulatory consequences of binding to AflR/AflJ expression and activity.

An interesting wrinkle to the nitrogen regulatory scheme was presented by Feng and Leonard (61). They reported that sodium nitrate stimulated synthesis of sterigmatocystin (ST) and expression of the biosynthetic genes *stcA* (homologous to *pksA*), *stcE* (homologous to *nor-1*) and *stcD* in *A. nidulans* (liquid stationary culture) in contrast to expression of *pksA* and *nor-1* which were inhibited in *A. parasiticus*. Although these data suggest differential regulation of AF and ST biosynthesis with regards to nitrogen regulation, the use of stationary culture conditions (in contrast to liquid shake) may have had an influence on these observations.

Based on recent evidence, Chang et al. (34) proposed a model in which aflatoxin gene expression is regulated not only directly by influencing the level of AflR protein (perhaps via AreA), but also indirectly by interaction with a putative repressor protein that binds to existing AflR protein and blocks its interaction with AflR consensus sites in aflatoxin gene promoters. A second copy of *aflR* was introduced into the *A. parasiticus* SU-1 genome via transformation. One resulting transformant, SU-1N3(pHSP), was grown in minimal salts medium (liquid shake culture) with 1mM sodium nitrate as the sole nitrogen source. The accumulation of aflatoxin and pathway intermediates as well as the expression of *aflR*, *nor-1*, *ver-1*, and *omtA* were significantly increased in SU-1N3(pHSP) in comparison with SU-1, the parent strain. They hypothesized that excess levels of AflR produced in SU-1N3(pHSP) titrated a fixed level of negative acting repressor protein leaving unbound AflR to activate the aflatoxin gene promoters. Possibly in agreement with these data, high level constitutive expression of AflR in *A. flavus* relieved nitrate repression of aflatoxin gene expression (65). Nitrate did reduce the accumulation of aflatoxin in the overexpressing strain, suggesting that aflatoxin synthesis is not solely regulated at the level of transcription.

In support of this model, Chang et al. (36) generated a plasmid construct carrying the carboxyl terminus of AflR (AflRC) expressed from the *niaD* promoter. This plasmid was transformed into *A. parasiticus* SRRC 2043 and isolates with a copy of the construct located at both *aflR* and *niaD* were isolated. These dual integrants overproduced aflatoxin pathway intermediates regardless of nitrogen source. Increased aflatoxin intermediate production correlated with increased expression of the AflRC and native AflR suggesting that the excess AflRC protein titrated a negative regulator of AflR activity. Direct evidence for this repressor has yet to be presented. However, in *N. crassa*, an analogous regulatory mechanism to that proposed for aflatoxin synthesis has been experimentally demonstrated. In this organism, a negative acting repressor, called NMR, interacts directly with Nit2, the positive acting pathway specific regulator of nitrate metabolism (165). This interaction reduces the ability of Nit2 to bind to nitrogen regulated promoters and positively regulate the structural genes involved in nitrate metabolism.

Recent data now suggest that there is significant variability in nitrogen response between strains of *Aspergillus* (53). Two different groups of *A. flavus* isolates, called S_B (synthesize the aflatoxin B group only) and S_{BG} (synthesize both B and G group toxins) were grown in the presence of either sodium nitrate or ammonium sulfate as the sole nitrogen source. The growth medium was buffered over time to maintain a final pH range between 5.7 and 6.1, to reduce changes in pH that accompany N source metabolism and to eliminate pH as a variable. The quantity of aflatoxin synthesized, the dry weight, and the quantity of *aflR* and *aflJ* transcripts (adjacent and divergently transcribed genes in the aflatoxin gene cluster) were monitored under these culture conditions in the S_B and S_{BG} isolates (three isolates of each group). The authors concluded that nitrate reduced (repressed) aflatoxin B and G synthesis in the S_{BG} isolates but did not significantly reduce aflatoxin B synthesis in the S_B isolates. As a proposed explanation for these data, the ratio of *aflJ* transcript on ammonium versus *aflJ* transcript on nitrate, was approximately five-fold higher in the S_B isolates than in the S_{BG} isolates. In contrast, there was not a significant difference in *aflR* transcript ratio on the same media between the S_B and S_{BG} isolates. The authors observed that there was a larger number of HGATAR sites in the *aflR*/*aflJ* intergenic region in the S_B isolates (three consensus sites) than in the S_{BG} isolates (one consensus site), prompting them to propose that nitrate increased the level of AreA which had a larger inhibitory effect on expression of *aflJ* in the S_B isolates due to the larger number of *cis*-acting sites. The connection between increased *aflJ* transcript and increased aflatoxin accumulation is tentative and needs to be clarified. This interesting model also deserves a closer look by expanding on these observations and directly testing AreA quantities and promoter binding in both groups of *A. flavus* isolates.

Unlike aflatoxins, fumonisin synthesis appears to be regulated in *Gibberella fujikoroi* by a typical nitrogen catabolite repression pathway (134). At low levels of ammonium phosphate (1.25 or 2.5 mM), *G. fujikoroi* produced fumonisin B_1 as early as 15 hours in a defined medium (liquid submerged culture). Increasing the level of ammonium phosphate (10 or 20 mM) resulted in a delay in toxin synthesis for 75 to 125 hours. The specific effects of ammonium on fumonisin gene expression or the expression and activity of pathway specific transcription factors has not been reported to date.

9.2.3.1.3 Trace Elements Several studies clearly demonstrated that *A. parasiticus* requires zinc for aflatoxin synthesis to occur at optimum levels in culture (9,149). For example, addition of zinc sulfate at concentrations ranging from 50 to 400 mM resulted in several fold (greater than eightfold at 150 mM) increases in aflatoxin accumulation in liquid shake culture (149). Consistent with these observations, when zinc was omitted from the growth medium, no aflatoxin intermediates were detected in a mutant strain (accumulates versicolorin) of *A. parasiticus* in liquid shake culture (9). Furthermore, omission of zinc from the growth medium resulted in lower quantities of AflR protein detected in cell extracts of *A. parasiticus* (93). Although the mechanism for zinc regulation has not been clearly defined, these data suggest that zinc is required for AflR DNA binding activity in line with its proposed function as a positive acting zinc binuclear cluster transcription factor (160). Because AflR appears to up regulate its own expression (52) as well as the expression of most if not all pathway genes

(23), it appears likely that zinc limitation reduces aflatoxin synthesis via a reduction in AflR expression and DNA binding activity. Miller (106) also reported that zinc limitation results in limitation of growth and increased proteolysis. He proposed that these two factors likely play a role in reduction of aflatoxin synthesis under conditions of limited zinc.

9.2.3.2 Environment

9.2.3.2.1 pH The growth environment has been previously shown to have a great impact on the expression of genes involved in secondary metabolism in filamentous fungi. For example, pH strongly regulates penicillin biosynthesis in *Aspergillus* (58). One major regulator involved in pH mediated regulation is PacC, a zinc finger protein that acts as a positive regulator of alkaline expressed genes and a repressor of acid expressed genes at ambient alkaline pH in the growth medium. PacC is reported to require activation as a result of a signaling pathway encoded by the *pal* genes (58). Proteolytic cleavage of PacC appears to be in part responsible for its activator and repressor activity (113). Specific mutations in PacC obviate the requirement for pH signal transduction (146). The mechanism by which PacC can act as an activator and repressor on different sets of genes was recently clarified (64). In the simplest case, PacC *cis*-acting binding sites in alkaline expressed genes allow PacC interaction and activation of transcription. In acid expressed genes, such as *gabA* (involved in synthesis of gamma aminobutryrate), binding of PacC to double PacC binding sites blocks interaction of the pathway activator, IntA (mediates T amino acid induction), with its *cis*-acting site on the *gabA* promoter.

pH has also been shown to influence the level of aflatoxin synthesis (41) on a defined solid growth medium. A comparison was made between aflatoxin synthesis and sclerotial development on media containing either sodium nitrate or ammonium sulfate in buffered or unbuffered media (with glucose or sucrose as the sole carbon source). Metabolism of nitrate by *A. flavus* resulted in an alkaline pH, enhanced sclerotial development, and reduced aflatoxin synthesis, whereas metabolism of ammonium resulted in an acidic pH, reduced sclerotia, and high levels of aflatoxin synthesis. Buffering of the growth medium during growth to prevent a change from the initial pH (6.5) resulted in reduction of aflatoxin synthesis in ammonium and enhancement of aflatoxin synthesis in nitrate (relative to unbuffered levels in the same medium), suggesting that both nitrogen source and pH affect the levels of aflatoxin produced on solid medium.

In the closely related species *A. nidulans* and *A. parasiticus*, expression of homologous genes in the sterigmatocystin (*stcU*) and aflatoxin (*ver-1*) biosynthetic pathway was strongly influenced by pH (80). Transcripts for these genes accumulated at higher levels at acidic pH than at neutral or alkaline pH, and transcript levels correlated directly with the level of mycotoxin detected in culture. Mutation in the *A. nidulans* PacC gene (constitutive PacC activity) resulted in a tenfold decrease in sterigmatocystin as compared to the wild type (80). However, the level of *stcU* transcript was not affected in the PacC mutant, suggesting that PacC might not be solely responsible for the pH regulation of aflatoxin biosynthesis.

Indirect evidence for a role for PacC in aflatoxin gene expression in *A. parasiticus* was obtained using an *aflR* gene disruption strain (25). Transformation of the knockout strain with plasmid constructs carrying specific mutations in the *aflR* promoter identified three sites that were important for *aflR* expression; an AflR binding site, a PacC binding site, and a GA rich region near the transcriptional start site of *aflR*. Later, EMSA supported the functional significance of the PacC site by demonstrating that a protein in a nuclear extract of *A. parasiticus* interacted with this site in the *aflR* promoter. A PacC site also was shown to be functionally significant in a structural gene in the pathway called *pksA* (55) (encodes the polyketide synthase). In this study, the PacC site in the *pksA* promoter was replaced with nonrelated nucleotides resulting in a fivefold decrease in *pksA* promoter activity. These data are of particular interest because the pH of the growth medium in this study was acidic. These data suggest that PacC may operate as a positive regulator of genes involved in secondary metabolism in contrast with its usual positive activity at alkaline pH in regulation of genes involved in primary metabolism.

Although little is known about the molecular mechanisms that mediate the regulatory influence of pH on trichothecene synthesis, one study provided preliminary evidence that, in specific growth media (Fries Medium supplemented with corn steep liquor), higher levels of DON accumulate in liquid shake culture at pH above 8.0 suggesting a potential contrast in pH regulation of this biosynthetic pathway (117). However, maximum levels of fumonisins were observed in liquid shake culture under nitrogen limited conditions (82) between pH 3 and 4. When the growth medium was limited for oxygen, no fumonisins were produced, even at optimum pH.

9.2.3.2.2 Temperature *Aspergillus parasiticus* synthesizes aflatoxins at maximum levels in the temperature range between 25 and 30°C (75). Growth of an aflatoxin blocked mutant (accumulates the pathway intermediate versicolorin A, VA) at 37°C completely blocked VA accumulation on an aflatoxin inducing growth medium (138). As mentioned, high temperature down regulates AflR protein accumulation providing, at least in part, a potential explanation for a down regulation of toxin synthesis (93). Butchko et al. (20) reported a similar temperature sensitivity for sterigmatocystin synthesis in *A. nidulans* but the nonpermissive temperature was higher (42°C) than that observed for *A. parasiticus* (37°C).

9.2.4 Signal Transduction Pathways

Two pioneering studies stimulated interest in research on the role of signal transduction pathways in regulation of secondary metabolism. The first reported a positive influence of exogenous cAMP on AFB_1 accumulation in culture (145). The second reported an inverse relationship between ethylene synthesis and aflatoxin accumulation in *A. parasiticus* (132). With the development of molecular tools in filamentous fungi, progress has been made in discerning the molecular mechanisms that mediate these regulatory circuits. Understanding signaling pathway input into mycotoxin synthesis in filamentous fungi is of fundamental interest because it is currently unknown if the signaling pathways that control primary metabolism and secondary metabolism are the same, completely different, or modified during growth to accomplish different regulatory functions. At the applied level, signaling pathways present logical focus areas for study because they sense external signals at low concentration. In addition, external signals are often received by receptors in the cytoplasmic membrane, abrogating the need for potential inhibitors to enter the cell. These characteristics suggest that identification of molecules that interfere with signaling pathway function at the receptor level could be an important component in the development of novel and effective strategies to control toxin synthesis in food and feed products.

9.2.4.1 G-Protein and Protein Kinase A (PKA) Signaling

9.2.4.1.1 Coregulation of Toxin Synthesis and Sporulation in Aspergillus The filamentous fungus *A. nidulans* provides an excellent model system to study asexual sporulation at the genetic and molecular genetic level. As a result of early studies on this developmental process, a variety of mutant strains defective in the early stages of sporulation was generated. Hicks et al. (73) took advantage of this panel of early developmental mutants to provide the first strong evidence tying signal transduction pathways and regulation of both sporulation and mycotoxin synthesis. Biochemical and genetic data had previously suggested that mycotoxin synthesis and asexual sporulation (conidiation) are coregulated in *Aspergillus spp.* For example, mycotoxin synthesis was observed to begin concurrently with conidiophore formation and conidiation (91,173). In addition, conidiation mutants such as *fan*, *fluffy*, *fluP*, *fadA*^{G42R}, *fluG*, *flbA*, and DGP10 were also impaired in mycotoxin synthesis (7,25,71,73,79,173,174). Further, many compounds that inhibited conidiation also decreased mycotoxin synthesis. For example, fluoroacetic acid (200–400 mg/ml) completely inhibited conidiation in *A. parasiticus* and reduced aflatoxin synthesis with no detectable effect on hyphal growth (122). Similarly, a competitive inhibitor of ornithine decarboxylase, diaminobutanone (50 mM), completely blocked the onset of both conidiation and aflatoxin synthesis in *A. parasiticus* with little effect on hyphal growth (71).

9.2.4.1.2 cAMP, fadA, and PKA in A. nidulans and A. parasiticus Based on early studies performed in *A. nidulans*, Hicks et al. (73) proposed a model for signaling pathway regulation of growth, conidiation, and sterigmatocystin biosynthesis in *A. nidulans*. According to this model, Fa dA, a subunit of a heterotrimeric G protein, binds GTP and transduces a signal which promotes growth and inhibits conidiation and sterigmatocystin synthesis. FlbA, a GTPase activating protein homolog, negatively regulates FadA by stimulating GTP hydrolysis, blocking G protein signaling, decreasing growth, and triggers initiation of conidiation and sterigmatocystin synthesis. In support of this model, Shimizu and Keller (82) reported that an activated allele of *fad*A (*fad*A^{G42R}) down regulated aflatoxin synthesis and conidiation, resulted in a fluffy phenotype in *A. flavus* (personal communication), and decreased norsolorinic acid (an aflatoxin pathway intermediate) accumulation and conidiation in TJYP1-22, a *fad*A^{G42R} derivative of *A. parasiticus B62*.

Shimizu and Keller (136) recently reported genetic evidence that FadA/FlbA regulation of sterigmatocystin synthesis and conidiation is mediated by a cAMP dependent protein kinase (PKA) in *A. nidulans*. Overexpression of *pka*A (encodes one catalytic subunit of PKA) eliminated transcripts of two genes involved in sterigmatocystin synthesis (*stcU* and *aflR*) and sterigmatocystin accumulation; conidiation was reduced to levels below wild type. Deletion of *pkaA* increased conidiation, inhibited growth, and delayed expression of *aflR* and *stcU* on a minimal growth medium. Deletion of *pkaA* in a strain with an activated *fadA* allele (*fadA*^{G42R}) restored conidiation but not sterigmatocystin synthesis. These data prompted the authors to expand the previous model to propose that FadA up regulates PKA which then has a negative regulatory influence on AflR expression and conidiation. It was also proposed that additional signaling pathways and factors may influence sterigmatocystin synthesis and conidiation. Neither sterigmatocystin promoter activity nor intracellular cAMP levels were measured. In addition, no distinction was made between basal PKA activity (fraction of PKA activated by endogenous cAMP *in vitro*) and total PKA activity (PKA activated by exogenous cAMP *in vitro*). Here, we define activation as binding of cAMP to 2 PKA regulatory subunits resulting in the release of 2 active PKA catalytic subunits.

In contrast to these data reported for *A. nidulans*, Tice and Buchanan (145) suggested a positive regulatory role for cAMP signaling (theoretically via PKA) in regulation of aflatoxin synthesis in *A. parasiticus*. Addition of cAMP, but not cGMP, to the growth medium, resulted in a dose dependent stimulation of aflatoxin accumulation in stationary liquid culture; neither intracellular cAMP levels nor the effect of cAMP on conidiation were evaluated. In support of these data, Roze et al. (176) observed increased levels of aflatoxin accumulation and conidiation in *A. parasiticus* after addition of the PKA activators cAMP or DcAMP (a cAMP analog) to the growth medium.

To clarify these apparently contrasting data, Roze et al. (178) then studied the role of FadA, cAMP, and PKA in the regulation of conidiation and aflatoxin synthesis in *A. parasiticus*. *A. parasiticus* strain D8D3 (contains a *nor-1*::GUS reporter construct;

wild-type with respect to aflatoxin synthesis), and *A. parasiticus* TJYP1-22 (an isogenic strain containing an activated allele of *fadA*; *fadA*^{G42R}) were grown in the presence or absence of cAMP or DcAMP in liquid stationary culture or on GMS solid growth medium (with glucose and ammonium sulfate). Growth, conidiation, aflatoxin accumulation, intracellular cAMP levels, *nor-1* promoter activity, and basal and total PKA activity were measured. The data suggested that treatment of *A. parasiticus* D8D3 with exogenous cAMP or DcAMP increased intracellular cyclic nucleotides to nonphysiological levels. This down regulated total PKA activity and up regulated conidiation (fivefold) and AflR dependent aflatoxin synthesis (tenfold). Activation of FadA generated intermediate physiological intracellular cAMP levels and high basal and total PKA activity. This down regulated conidiation and aflatoxin synthesis. The studies provided direct evidence for involvement of FadA in regulation of intracellular cAMP levels and basal and total PKA activity, and prompted the authors to propose that FadA/PkaA regulation of mycotoxin synthesis occurs via similar regulatory mechanisms in *A. parasiticus* and *A. nidulans*. They also concluded that intracellular cAMP levels, at least in part, mediate a PKA dependent regulatory influence on conidiation and aflatoxin synthesis in *A. parasiticus*.

9.2.4.1.3 Identification of a cAMP Response Element Binding Protein (CRE1bp) EMSA was conducted to determine if specific transcription factors mediate the cAMP induced up regulation of aflatoxin synthesis and *nor-1* promoter activity (176). Using a subfragment of the *nor-1* promoter (*norR)* as probe and cell extracts from *A. parasiticus* strains D8D3 or TJYP1-22 (*fadA*^{G42R}) grown on GMS (glucose as sole carbon source; inducing conditions) or PMS (peptone as sole carbon source; noninducing conditions) agar medium, a single DNA and protein complex with similar mobility was observed in both extracts under the 2 different aflatoxin noninducing conditions (complex PS1 from D8D3 grown in PMS; complex FS1 from TJYP1-22 [*fadA*^{G42R}] grown in GMS). A complex of lower mobility (complex GS1) was generated by extracts of D8D3 grown on GMS agar (GS1; inducing conditions) but was not present under the noninducing conditions. Roze et al. (176) hypothesized that complexes PS1 and FS1 are structurally and functionally similar. Because TJYP1-22 (*fadA*^{G42R}) stimulated intracellular cAMP levels and down regulated AFB_1 synthesis and *nor-1* promoter function, they reasoned that a putative repressor was part of PMS1 and FS1, and this bound to a cAMP response element (CRE). Binding of additional proteins (or modification of bound proteins) converted the complex to GS1 in inducing conditions and up regulation of *nor-1* promoter activity. In support of this hypothesis, the authors identified a candidate CRE (named CRE1), *cis*-acting site (TGACATAA; nucleotide residues +15 to +22) in *norR* (−117 to +55) based on sequence similarity to a consensus CRE from yeast (TGACGTCA) (66,73,122). They then tested possible DNA binding activity at the CRE1 site *cis*-acting site using competition EMSA. A 250-fold excess of oligonucleotide containing the putative CRE1 site completely competed with both PS1 and FS1 complexes, while mutants containing 3 changed residues of this site (TG<u>TG</u>ATA<u>T</u>) or 2 changed residues (TG<u>TG</u>ATAA) were significantly less effective. A mutant containing 5 changed residues (CTAGCTAG) was totally ineffective at competing with these DNA/protein complexes. The consensus CRE from eukaryotic cells with flanking sequences from the *nor-1* promoter also completely competed with the shifted complexes, while oligonucleotides containing the AflR binding site and the TATA box could not compete. The fact that the pattern observed in competition EMSA was the same for complexes PS1 and FS1 strongly supported the hypothesis that they contained the same protein(s). Roze et al. (176) named this CRE binding protein, CRE1bp (<u>c</u>AMP <u>r</u>esponse <u>e</u>lement <u>b</u>inding <u>p</u>rotein). It now appears as though this CRE1 site has functional significance. A CRE replacement mutation in the *nor-1*::GUS reporter construct (2 nucleotide changes in the putative CRE1 site confirmed by sequence analysis) did not respond to exogenous cAMP to the same extent as wild type (about a threefold reduction).

Because a direct correlation was observed between PKA activity and complex FS1 in TJYP1-22 (*fadA*^{G42R}), Roze et al. (176) reasoned that phosphorylation of CRE1bp may be associated with DNA binding activity. They generated direct evidence for CRE1bp phosphorylation using supershift EMSA. Monoclonal antibodies to phosphoserine and phosphothreonine supershifted complex FS1 but phosphotyrosine did not. Clearly, followup is necessary to identify the amino acid residues that are phosphorylated and the influence on DNA protein complex formation and aflatoxin gene expression.

9.2.4.1.4 G protein signaling The effects of a constitutively activate *fadA* allele on penicillin biosynthesis in *A. nidulans* and trichothecene biosynthesis in *Fusarium sporotrichioides* were also analyzed (143). In contrast to aflatoxin and sterigmatocystin synthesis, the *fadA*^{G42R} allele stimulated expression of one gene in penicillin biosynthesis and increased accumulation of this antibiotic. Similarly, expression of the same allele in *Fusarium sporotrichioides* increased trichothecene gene expression and accumulation of this mycotoxin. These data strongly suggest that one signaling pathway can have diverse effects on different secondary metabolite pathways. The molecular mechanisms underlying this regulatory circuit are clearly a focus for future work.

9.2.4.1.5 Ras Proteins and Conidiospore Germination Because of the close regulatory association between mycotoxin synthesis and asexual sporulation in *Aspergillus,* it is worth mentioning a recent study that analyzed the effects of cAMP and *ras* signaling on conidial germination (63). Strains carrying mutations in adenylate cyclase (CyaA) or protein kinase A (PkaA) demonstrated altered kinetics of spore germination, together with several alterations in colony morphology. *Aspergillus nidulans* mutants overproducing a dominant activated form of RasA (a homologue of the human *ras* protooncogene) produced giant swollen spores with multiple nuclei that failed to germinate.

9.2.4.2 Lipids and Psi Factors

9.2.4.2.1 Aflatoxins and Lipids The effects of lipids on aflatoxin synthesis were reviewed recently (23). As a useful illustration of the phenomenon, Mellon et al. (102) demonstrated that cottonseed storage lipids were found to support growth and toxin synthesis by *A. flavus*. When the lipids (primarily triglycerides) were removed from the medium by ether extraction, aflatoxin production dropped by more than 800-fold. Keller and colleagues (19) provided a framework for integrating a large body of data related to the propensity of *Aspergillus* to produce mycotoxins on oilseed crops. Because *Aspergillus* species contain lipoxygenase (LOX) activities involved in synthesis of pheromones which regulate development, they proposed a theory which attempted to connect pheromone synthesis, development, and toxin synthesis on oil seed crops. *A. nidulans* contains LOX activities involved in synthesis of psi (precocious sexual inducers) factors, which regulate sexual and asexual sporulation in this organism (28). Deletion of *odeA* that encodes a delta 12 desaturase in *A. nidulans* resulted in a strain depleted in polyunsaturated fatty acids and psi factors derived from linoleic acid (22). The connection to the host plant was provided by the observation that host plants also contain LOX activities and the plant LOX metabolites strongly activate mycotoxin synthesis (43). Based on the existing evidence, it was reasonable to propose that peroxides of unsaturated fatty acids in oilseed formed during storage, or by host lipoxygenases during seed development, mimic the activity of fungal psi factors and alter fungal development and mycotoxin synthesis. Keller's group (19,67) has generated significant laboratory evidence in support of this theory, although more information is required to understand the importance of the lipids in regulation of mycotoxin synthesis in the field and during crop storage.

9.2.4.2.2 CP2 Inhibits Aflatoxin Gene Expression Support for the lipoxygenase theory was provided by studies on a naturally occurring compound identified as an aflatoxin synthesis inhibitor. Using a rapid two dimensional TLC assay, Annis et al. (6) discovered a compound (CP2) in black pepper extract (*Piper nigrum*) that down regulates *nor-1*::GUS expression at the level of transcription. CP2 also inhibited *nor-1* and *ver-1* transcript accumulation, suggesting that multiple aflatoxin promoters are affected. CP2 exhibited dose dependent inhibition of AFB_1 synthesis but did not inhibit expression of a β-tubulin::GUS reporter (housekeeping gene) nor accumulation of $-tubulin transcripts. CP2 appears to be relatively specific for the aflatoxin promoters. Gas chromatography mass spectroscopy (GC-MS) analysis revealed that CP2 is a C18 fatty acid that is structurally similar to the linoleic acid derived psi factors (27,28,99) that govern development of cleistothecia (structures for production of sexual spores) and conidiophores in *A. nidulans*. The authors proposed that CP2 acts via mimicry of the hormonal effects of psi factors and down regulates mycotoxin formation in *A. parasiticus*. Dr. Keller (19) has shown that hydroperoxide derivatives of linoleic acid can have opposing effects on sterigmatocystin synthesis in *A. nidulans*. A 13S hydroperoxy derivative of linoleic acid inhibited ST production, while a 9S hydroperoxy derivative did not inhibit ST synthesis but did stabilize mRNA transcripts derived from toxin genes. Of note, CP2 did not affect conidiation in *A. parasiticus,* which suggests that the association of psi factor to sexual development and toxin synthesis is complicated and requires further study.

9.2.4.3 Ethylene Signaling

9.2.4.3.1 Ethylene Inhibits AFB1 Synthesis Plants use ethylene signaling to regulate flowering and ripening of fruits and vegetables (131,137). Binding of ethylene to specific receptors influences the ethylene signaling pathway, initiating the ethylene response. Methyl cyclopropene (MCP) blocks ethylene/receptor interaction and inhibits the ethylene response (131,137). Roze et al. (177) showed that ethylene gas inhibits AFB1 biosynthesis and nor-1 promoter activity up to tenfold in a dose dependent manner, but the compound had no statistically significant effect on conidiation. Methyl cyclopropene added to A parasiticus enhanced AFB1 synthesis up to fivefold under 3 different growth conditions (193). Ethylene (0.5 PPM) also blocked the cAMP mediated stimulation of AFB1 synthesis but did not affect the number or viability of conidiospores. Based on these data, the authors hypothesized that A. parasiticus has ethylene receptors that mediate regulatory effects of ethylene on AFB1 synthesis; the group is currently attempting to clone the ethylene sensor or receptor based on presumed identity with ethylene receptors from plants (128).

9.2.4.4 Other Volatiles

Similar to ethylene, workers at the USDA in New Orleans have observed strong regulatory effects of a variety of plant volatiles, including alkenal and alkanal compounds on aflatoxin synthesis (69,162,171,172). Although such compounds hold potential in reducing toxin levels in food and feed, the mechanisms by which these compounds affect aflatoxin synthesis are currently not known.

9.2.4.5 Cyclins

9.2.4.5.1 Fumonisins and Cyclins Cyclins are a family of proteins involved in cell cycle control in eukaryotic organisms. Evidence for a role for one specific cyclin, FCC1, in regulation of fumonisin biosynthesis in *Fusarium verticelloides* was presented by Shim et al. (135). The authors observed that disruption of the gene, *FCC1*, reduced conidiation and expression of FUM5, which encodes a polyketide synthase in fumonisin biosynthesis on laboratory growth medium or corn kernels. The role of cyclin signaling and the exact mechanism by which the down regulation occurred are areas of focus for future work.

9.2.5 Gene Clusters and Positional Effects

While common in prokaryotes, the linkage of functionally related genes is relatively rare in higher eukaryotes. In filamentous fungi, however, the clustering of genes is a common feature for several metabolic pathways. Examples of gene clusters in secondary metabolism include aflatoxin (153), trichothecene (14), penicillin (45), fumonisin (130), dothistromin (11), paxilline (164), and several others. Several nutrient utilization pathways are also clustered in filamentous fungi including ethanol (64) and nitrate (182) utilization in *A. nidulans*. The physical linkage of related genes suggests two possible hypotheses: (1) linkage of metabolic pathway genes provides a means to regulate pathway gene transcription; and (2) linkage provides a means for genetic transfer of entire pathways between species.

9.2.5.1 Organization of Gene Clusters

9.2.5.1.1 Aflatoxin Gene Cluster Parasexual analysis of aflatoxin pathway mutants first established the likely clustering of at least some of the aflatoxin biosynthetic genes (8). Subsequently, the first two aflatoxin genes cloned, *nor-1* and *ver-1*, were found to be genetically linked on the same *A. parasiticus* cosmid (139). Analysis of overlapping cosmid and phage clones allowed the aflatoxin (sterigmatocystin) gene clusters to be mapped in *A. parasiticus* (153), *A. flavus* (168), and *A. nidulans* (15). As candidate genes in the aflatoxin cluster were identified, the timing of transcript accumulation from these cluster genes was observed to be consistent with their involvement in aflatoxin biosynthesis (15,153). To date, most cluster genes have been studied in detail and been confirmed to be involved in aflatoxin biosynthesis. Interestingly, the aflatoxin gene clusters of these three related organisms are not organized identically. For example, the distance between *aflR* and *ver-1* is 32 kb in *A. nidulans* but 8 kb in *A. parasiticus* and *A. flavus*. The maintenance of gene clusters despite changes in gene order suggests that gene clustering has functional significance for the fungi.

A 5 kb spacer region was identified at one end of the aflatoxin gene cluster (166,167). The spacer region contains no open reading frames and has a sugar utilization cluster located on the opposite side (Figure 9.8). Aflatoxin production is closely linked to the carbon source, with simple sugars like glucose and sucrose able to induce aflatoxin biosynthesis (17). Consequently, the localization of this sugar utilization cluster near the aflatoxin cluster suggests a regulatory connection between the two clusters. However, of the four genes in the sugar cluster, only *hxtA* (proposed hexose transporter protein) expression was shown to be concurrent with aflatoxin pathway cluster genes (166). It is unknown if there are any functional AflR *cis*-acting sites in the sugar cluster gene promoters. In addition, the border at the other end of the aflatoxin gene cluster has not yet been defined.

9.2.5.1.2 Trichothecene Gene Cluster Two overlapping cosmid clones were able to complement different trichothecene mutants, suggesting that the trichothecene genes are clustered in *Fusarium sporotrichioides* (77). More recently, a 23 kb trichothecene gene cluster was sequenced for both *F. sporotrichioides* and *F. graminearum*,which contains 12 genes (14). All of the 12 clustered genes studied so far have been shown to be involved in trichothecene biosynthesis [Figure 9.8 (a)] (14). Yet, the 12 genes in the identified trichothecene gene cluster are insufficient to account for all known trichothecene structures (14), leading

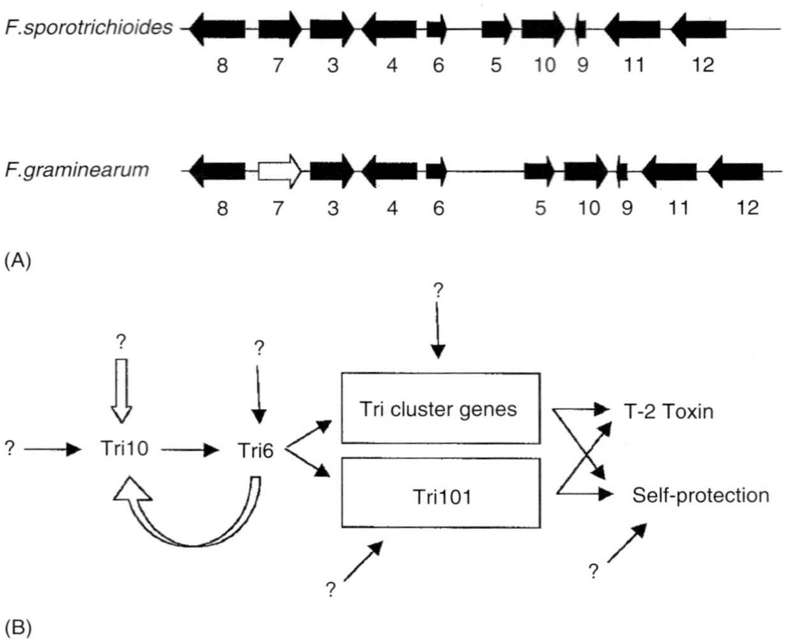

Figure 9.8 (A) Genomic organization of the trichothecene biosynthetic *tri* gene cluster of *Fusarium sporotrichioides* and *F. graminearum*. Arrowheads indicate the direction of transcription and the number underneath each arrow refer to the specific gene. Genes with the same number from both *Fusarium* species are homologues. *tri7* in *F. graminearum* is non-functional. Map is roughly to scale. (Adapted from Brown et al., 2001) (B) Proposed regulatory model for trichothecene biosynthesis. Solid arrows indicate positive activators while open arrows indicate inhibitory activities. Question marks indicate other proposed but unknown regulatory signals or factors (Adapted from Tag et al., 2001)

to two hypotheses: (1) additional trichothecene biosynthesis genes are located beyond the flanking sequence of *tri8* and *tri12*; and (2) all trichothecene genes are not located within the gene cluster. Because most, if not all, of the genes necessary for aflatoxin biosynthesis are located within the aflatoxin gene cluster, it would be reasonable to sequence beyond *tri8* and *tri12* to potentially identify additional trichothecene genes. In addition, it will be interesting to determine if there is an extended spacer region that forms a border of the trichothecene gene cluster as there is with the aflatoxin gene cluster.

tri101 exists outside of the 23 kb cluster in both *F. sporotrichioides* (100) and *F. graminearum* (85) and is the only known trichothecene gene to exist outside of the cluster. The genes on either side of *tri101* (UTP-ammonia ligase and phosphate permease) are not involved in trichothecene biosynthesis and *tri101* is at least 35 kb from either end of the identified trichothecene cluster in *F. graminearum* (84). Tri101 is a 3-O-acetyltransferase that is required for T-2 production in *F. sporotrichioides* (100). Because the enzymatic products of Tri101 are less toxic than the substrates, it was originally proposed that the purpose of Tri101 is to protect the fungus (83). Expression of *tri101* in yeast (83,100) and plants (109) resulted in increased tolerance for trichothecenes. Yet, a *F. sporotrichioides tri101* disruptant could both germinate and grow in the presence of trichothecenes, suggesting that *tri101* is not an essential self defense mechanism for *F. sporotrichioides* (100). The evolution of *tri101* is discussed in later sections.

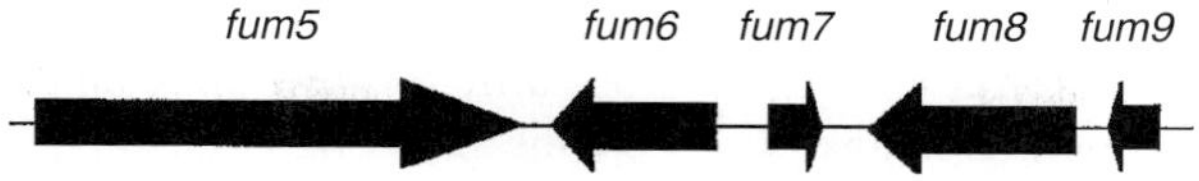

Figure 9.9 Schematic of the fumonisin biosynthetic gene (*fum*) cluster in *Fusarium verticillioides*. Arrowheads indicate the direction of transcription. Map is roughly to scale (Adapted from Seo et al., 2001)

9.2.5.1.3 Fumonisin Gene Cluster The identification of the fumonisin gene cluster utilized allele tests with three different *Fusarium monoliforme* fumonisin mutants (44). Subsequently, a polyketide synthase gene, *fum5*, was isolated and shown to be required for fumonisin biosynthesis (119). Based on the structure of fumonisin B1, at least 8 different enzymatic steps are needed for fumonisin biosynthesis (130). Sequencing downstream from *fum5* identified four additional ORFs, *fum6*, *-7*, *-8* and *-9*, whose expression is correlated with fumonisin production (Figure 9.9). Gene disruption analysis of *fum6* and *fum8* revealed that they are necessary for fumonisin biosynthesis (130). Nucleotide sequence analysis in the DNA regions flanking *fum5* and *fum9* may identify additional fumonisin biosynthetic genes. Based on aflatoxin and trichothecene gene clusters, it is reasonable to expect that a transcriptional activator specific for fumonisin biosynthesis located within the fumonisin gene cluster will be found with the additional sequencing.

9.2.5.2 Gene Cluster Dependent Regulation

Preliminary evidence for a role for clustering in aflatoxin gene regulation was reported by Liang et al. (92). The promoter of the aflatoxin biosynthesis structural gene *ver-1* was fused to GUS to generate a reporter plasmid (pHD6.6) that contained *niaD* (encodes nitrate reductase) as a selectable marker (92). pHD6.6 integrated predominantly at the *ver-1* or *niaD* locus via homologous recombination (92). Single copy integration of pHD6.6 at *niaD* resulted in a 500-fold reduction in *ver-1* promoter activity when compared with single copy integration at the *ver-1* locus; however, the temporal pattern of expression appeared to be similar at both loci (92). One explanation for reduced *ver-1* promoter function at the *niaD* locus is that location in the aflatoxin cluster results in positive, position dependent regulation of *ver-1* exression. An alternative hypothesis is that the expression of *ver-1* integrated at *niaD* is negatively influenced by *niaD* regulation. In the absence of preferred nitrogen sources and in the presence of nitrate, *niaD* is expressed. Under the rich growth conditions tested by Liang et al. (92), *niaD* may have been repressed, and therefore the lack of *ver-1* expression at the *niaD* site could have resulted from *niaD*-dependent regulation.

Subsequently, the promoter of the aflatoxin biosynthesis gene *nor-1* was fused to GUS to generate a reporter plasmid, pAPGUSNN-B, containing *niaD* as a selectable marker (40). Transformants with pAPGUSNN-B integrated at the *niaD* locus had no detectable GUS activity, while *nor-1* integrants had GUS activity (46). In addition, the cloned *nor-1* promoter functioned similarly to the native *nor-1* promoter when it integrated at the *nor-1* locus (40). Because *niaD* dependent regulation could account for the absence of expression at *niaD* for both *nor-1*::GUS (40) and *ver-1*::GUS (101), a third chromosomal location was analyzed using pAPGUSNP, which contained *nor-1*::GUS plus *pyrG* (encodes OMP decarboxylase) as a selectable marker (40). GUS expression was detectable only when pAPGUSNP integrated at *nor-1* and was not detectable at *pyrG*, even under growth conditions that required *pyrG* expression (40). While the mechanism is unknown, *nor-1* and perhaps other aflatoxin biosynthetic genes are susceptible to aflatoxin gene cluster dependent regulation.

The genome of some strains of *A. parasiticus* includes a partial duplication of the aflatoxin gene cluster (35). The region from *aflR* to *ver-1* plus *omtB* is duplicated in *A. parasiticus* SU-1 (35). It is unknown why the genes between *ver-1* and *omtB* are not part of the duplicated region. All of the duplicated genes appear to have mutations in them that would make them nonfunctional except for *aflR-2* and *aflJ-2* (35). Northern and RT-PCR analyses of RNA indicated that *aflR-2* is expressed at much lower levels than *aflR-1* (24). Nucleotide sequence analysis upstream of the *aflR-2* translational start codon revealed that the AflR binding site was intact and that there were few base changes (2%) compared to the corresponding region of *aflR-1* (24). Cluster dependent regulation may explain the poor expression of *aflR-2* (24,35). It is currently unclear if the other genes in the duplicated region are expressed at similar levels to their cluster counterparts. Reporter constructs that can integrate into the aflatoxin gene cluster and the duplicated region may provide valuable insight into cluster dependent mechanisms of regulation.

Trichothecene biosynthesis genes are also clustered (77), and there is some evidence for position dependent regulation with the pathway regulator *tri6* (39). Introduction of a plasmid containing a *tri5*::GUS fusion with a functional *tri6* resulted in 50- to 100-fold more GUS activity with *tri5* integration compared to ectopic integration (39). However, there were no differences in GUS activity between *tri5* and ectopic integration of a *tri5*::GUS plamid without a functional *tri6* (39). Integration of a *tri4*::GUS fusion into the *tri4* locus resulted in two to fivefold more GUS activity than ectopic integration (76). Additional experiments are necessary to clarify the significance of cluster dependent regulation in trichothecene biosynthesis.

It is currently unknown if cluster dependent regulation occurs with other mycotoxin gene clusters like fumonisin. However, cluster dependent regulation appears to be less significant with trichothecene biosynthesis than with aflatoxin biosynthesis. Except for aflatoxin biosynthesis, the description of mycotoxin gene cluster dependent regulation needs further study. While the construction of plasmids with mycotoxin promoters fused to reporter genes is useful in identifying *cis*-acting sites in the mycotoxin promoter, they also provide a means to determine differences between mycotoxin cluster integration and ectopic integration.

9.2.5.3 Significance of Gene Cluster Dependent Regulation

The targeting of reporter constructs to specific chromosomal locations for promoter analysis has been suggested by other investigators studying expression of fungal genes (72,148). The gene cluster dependent regulation seen with the aflatoxin gene cluster emphasizes the importance of determining the site of integration for mycotoxin reporter plasmids. In addition, screening transformants by genotype (site of integration) rather than phenotype (reporter activity) will prevent possible sampling bias. A rapid method for screening the site of integration for *A. parasiticus* has been described (40) that makes screening by genotype feasible.

Although the mechanisms of position dependent gene expression have not been fully elucidated in filamentous fungi, it has been hypothesized that enhancer elements may be responsible for the position dependent effect (86). Studies performed on the SpoC1 gene cluster in *A. nidulans* showed that clustered genes can be coordinately expressed during development, and that placement of cluster genes at ectopic chromosomal locations results in the loss of that coordination (104,147). The hypothesis that positive *cis*-acting factors have regional control over the transcription of aflatoxin genes is reasonable, because removal of aflatoxin reporter fusions from the aflatoxin gene cluster results in reduced GUS expression.

More recently, studies with the mammalian β-globin gene identified a region named a locus control region (LCR) (70). The β-globin LCR was identified as a region that was necessary to confer positional independence of a β-globin transgene (70). Subsequently, several additional LCRs have been identified (90). We hypothesize that aflatoxin gene expression is influenced by a locus control region. However, a locus control region has not been identified in fungal gene clusters. If an LCR does influence the regulation of *nor-1* transcription, it is located at least 3 kb upstream from the transcription initiation site in the 5′ *nor-1* region, at least 1.8 kb downstream of the transcription termination site in the 3′ region of *nor-1*, or within the *nor-1* coding region. Because similar position dependent expression is observed with the *ver-1*::GUS reporter construct (92), it is possible that the same LCR element is influencing the regulation of both *nor-1* and *ver-1*. Comparison of steady state levels of mRNA transcripts from genes present in the aflatoxin gene cluster and the duplicated region (i.e., *ver-1A* vs. *ver-1B*) may help narrow the search for the LCR(s). It will be interesting to determine if the duplicated region includes the LCRs necessary for cluster dependent regulation. In addition, the 5 kb spacer region at one border of the aflatoxin gene cluster is a tempting place to look for possible boundary elements. Boundary elements appear to provide a functional boundary for both accessible and inaccessible chromatin (88). The possible boundary element in the spacer region would prevent the LCRs from affecting transcriptional activation of genes in the sugar cluster or beyond.

While solid evidence for cluster dependent regulation has only been shown for aflatoxin biosynthesis, it is reasonable to predict that several other mycotoxin gene clusters will also be subject to cluster dependent regulation. Consequently, the proper use of reporter plasmids to study mycotoxin gene transcriptional regulation must include the identification of the site of integration of the plasmids to ensure proper data interpretation.

9.2.5.4 Evolution of Mycotoxin Gene Clusters

9.2.5.4.1 Origin of the Aflatoxin Gene Cluster The occurrence of fungal pathway gene clusters may result from horizontal gene transfer from prokaryotes where clustering of pathway genes is common (81). Perhaps the strongest case for the horizontal gene transfer of an entire secondary metabolic pathway from bacteria to fungi is with penicillin biosynthesis (5,158). The G+C content of the penicillin gene cluster is more similar to *Streptomyces* than the producing fungi and the fungal penicillin genes lack introns which is generally a characteristic of bacteria (81). Horizontal gene transfer is unlikely with the aflatoxin biosynthetic pathway due to the presence of introns and similar G+C content inside and outside of the aflatoxin cluster (15). Yet the changes in gene order within the aflatoxin gene cluster of different *Aspergilli* suggest that there is selective pressure to keep the aflatoxin genes clustered. *Dothistroma pini*, a fungal pine pathogen, produces the mycotoxin dothistromin that is structurally similar to versicolorin A, an intermediate in aflatoxin biosynthesis (11). Four genes identified in the *D. pini* dothistromin gene cluster all have high similarity with known aflatoxin biosynthesis genes (11). Detailed analysis of the dothistromin and aflatoxin gene clusters may provide clues to help understand the evolutionary history of the aflatoxin gene cluster.

9.2.5.4.2 Origin and Diversity of Trichothecene Gene Clusters The current literature provides few clues for the origin of the trichothecene gene cluster. The ten trichothecene genes identified so far in the cluster (*tri*3 through *tri*12) contain a total of 37 introns (14) suggesting that horizontal gene transfer from prokaryotes is unlikely. While there are many different trichothecenes that are produced by various fungi, *Fusarium* species appear to be the primary source of trichothecenes in agricultural products (76). The mechanism for how different Fusarium species produce different trichothecenes is unknown. The T-2 gene cluster in *Fusarium sporotrichioides* and the deoxynivalenol gene cluster in *F. graminearum* are very similar (Figure 9.7) in that the 23 kb region included 12 homologous genes (14). However, the Tri7 protein (acetylates the oxygen on C-4) is nonfunctional in *F. graminearum* (14) suggesting a possible mechanism for generating trichothecene structural diversity. The origin of the noncluster *tri101* gene may be different than the clustered trichothecene genes (85). Kimura et al. reasoned that a translocation event was unlikely to explain the location of *tri101* because it is flanked by essential primary metabolism genes (84). In addition, homologues of *tri101* were found in *Saccharomyces cerevisiae* and *Schizosaccharomyces pombe* (84). Consequently, Kimura et al. reasoned that it is feasible that trichothecene producers have acquired *tri101* through horizontal gene transfer (84).

9.2.6 Summary

This chapter has reviewed recent data related to molecular mechanisms that regulate mycotoxin gene expression in a variety of filamentous fungi. The overarching goals of this research are to increase our understanding the basic biology and chemistry of fungal secondary metabolism, and at the same time to identify practical methods to reduce mycotoxin contamination of food and feed. In this regard, we propose that research on the positive and negative regulators of mycotoxin gene expression and the signaling pathways that mediate environmental influences on these regulators hold considerable promise. Work on identification of food safe, natural plant compounds may have an important impact on mycotoxin contamination by down regulating toxin gene expression, and may move us closer to realizing a safer food supply. Clearly, the data reviewed illustrate important strides in achieving the overarching goals. However, a great deal more work remains to achieve a safe food supply.

REFERENCES

1. CAST. *Mycotoxins: Economic and Health Risks.* Ames, Iowa: Council for Agricultural Science and Technology, 1989, pp 1–91.
2. CAST. *Mycotoxins: Risks in plant, animal, and human systems.* Ames, Iowa: Council for Agricultural Science and Technology, 2003.
3. Adams, T.H., M.T. Boylan, W.E. Timberlake. *brlA* is necessary and sufficient to direct conidiophore development in *Aspergillus nidulans. Cell* 54:353–362, 1988.
4. Aguilar, F., S.P. Hussain, P. Cerutti. Aflatoxin B_1 induces the transversion of G-->T in codon 249 of the p53 tumor suppressor gene in human hepatocytes. *Proc. Natl. Acad. Sci. USA* 90:8586–8590, 1993.
5. Aharonowitz, Y., G. Cohen, J.F. Martin. Penicillin and cephalosporin biosynthetic genes: structure, organization, regulation, and evolution. *Annu. Rev. Microbiol.* 46:461–495, 1992.
6. Annis, S.L., L. Velasquez, H. Xu, R. Hammerschmidt, J. Linz, F. Trail. Novel procedure for identification of compounds inhibitory to transcription of genes involved in mycotoxin biosynthesis. *J. Agric. Food Chem.* 48:4656–4660, 2000.
7. Bennett, J.W., K.E. Papa. Genetics of aflatoxigenic *Aspergillus* species. *Adv. Plant Pathol.* 6:263–280, 1987.
8. Bennett, J.W., K.E. Papa. The aflatoxigenic *Aspergillus* spp. *Adv. Plant Pathol.* 6:263–280, 1988.
9. Bennett, J.W., P.L. Rubin, L.S. Lee, P.N. Chen. Influence of trace elements and nitrogen sources on versicolorin production by a mutant strain of *Aspergillus parasiticus. Mycopathologia* 69:161–166, 1979.
10. Bondy, G.S., J.J. Pestka. Immunomodulation by fungal toxins. *J. Toxicol. Environ. Health B. Crit. Rev.* 3:109–143, 2000.

11. Bradshaw, R.E., D. Bhatnagar, R.J. Ganley, C.J. Gillman, B.J. Monahan, J.M. Seconi. *Dothistroma pini*, a forest pathogen, contains homologs of aflatoxin biosynthetic pathway genes. *Appl. Environ. Microbiol.* 68:2885–2892, 2002.
12. Brakhage, A.A., P. Browne, G. Turner. Regulation of *Aspergillus nidulans* penicillin biosynthesis and penicillin biosynthesis genes *acvA* and *ipnA* by glucose. *J. Bacteriol.* 174:3789–3799, 1992.
13. Brown, D.W., T.H. Adams, N.P. Keller. *Aspergillus* has distinct fatty acid synthases for primary and secondary metabolism. *Proc. Natl. Acad. Sci. USA* 93:14873–14877, 1996.
14. Brown, D.W., S.P. McCormick, N.J. Alexander, R.H. Proctor, A. E. Desjardins. A genetic and biochemical approach to study trichothecene diversity in *Fusarium sporotrichioides* and *Fusarium graminearum*. *Fungal Genet. Biol.* 32:121–133, 2001.
15. Brown, D.W., J.H. Yu, H.S. Kelkar, M. Fernandes, T.C. Nesbitt, N.P. Keller, T.H. Adams, T.J. Leonard. Twenty-five co-regulated transcripts define a sterigmatocystin gene cluster in *Aspergillus nidulans*. *Proc. Natl. Acad. Sci. USA* 93:1418–1422, 1996.
16. Buchanan, R.L., S.B. Jones, W.V. Gerasimowicz, L.L. Zaika, H.G. Stahl, L.A. Ocker. Regulation of aflatoxin biosynthesis: assessment of the role of cellular energy status as a regulator of the induction of aflatoxin production. *Appl. Environ. Microbiol.* 53:1224–1231, 1987.
17. Buchanan, R.L., D.F. Lewis. Regulation of aflatoxin biosynthesis: effect of glucose on activities of various glycolytic enzymes. *Appl. Environ. Microbiol.* 48:306–310, 1984.
18. Buchanan, R.L., L.A. Ocker, H.G. Stahl. Effect of 2-deoxyglucose, alpha-methylglucoside, and glucosamine on aflatoxin production by *Aspergillus parasiticus*. *Arch. Microbiol.* 142:200–203, 1985.
19. Burow, G. B., T. C. Nesbitt, J. Dunlap, and N. Keller. Seed lipoxygenase products modulate *Aspergillus* mycotoxin biosynthesis. *Mol. Plant Microbe Int.* 10:380–387, 1997.
20. Butchko, R.A., T.H. Adams, N.P. Keller. *Aspergillus nidulans* mutants defective in *stc* gene cluster regulation. *Genetics* 153:715–720, 1999.
21. Caddick, M.X., D. Peters, A. Platt. Nitrogen regulation in fungi. *Antonie Van Leeuwenhoek* 65:169–177, 1994.
22. Calvo, A.M., H.W. Gardner, N. Keller. Genetic connection between fatty acid metabolism and sporulation in *Aspergillus nidulans*. *J. Biol. Chem.* 276:25766–25774, 2001.
23. Cary, J., D. Bhatnagar, J. Linz. Aflatoxins: biological significance and regulation of synthesis. In: *Microbial Foodborne Diseases: Mechanisms of Pathogenesis and Toxin Synthesis.* Cary, J., J. Linz, D. Bhantnagar, eds., Lancaster: Technomic Publishing Co., 2000, pp 317–361.
24. Cary, J.W., J.M. Dyer, K.C. Ehrlich, M.S. Wright, S.H. Liang, J.E. Linz. Molecular and functional characterization of a second copy of the aflatoxin regulatory gene, *aflR-2*, from *Aspergillus parasiticus*. *Biochim. Biophys. Acta* 1576:316–323, 2002.
25. Cary, J.W., K.C. Ehrlich, M. Wright, P.K. Chang, D. Bhatnagar. Generation of *aflR* disruption mutants of *Aspergillus parasiticus*. *Appl. Microbiol. Biotechnol.* 53:680–684, 2000.
26. Cary, J.W., B.G. Montalbano, K.C. Ehrlich. Promoter elements involved in the expression of the *Aspergillus parasiticus* aflatoxin biosynthesis pathway gene *avnA*. *Biochim. Biophys. Acta* 1491:7–12, 2000.
27. Champe, S.P., A.A. el Zayat. Isolation of a sexual sporulation hormone from *Aspergillus nidulans*. *J. Bacteriol.* 171:3982–3988, 1989.
28. Champe, S.P., P. Rao, A. Chang. An endogenous inducer of sexual development in *Aspergillus nidulans*. *J. Gen. Microbiol.* 133(5):1383–1387, 1987.
29. Chang, P.K. The *Aspergillus parasiticus* protein AflJ interacts with the aflatoxin pathway-specific regulator AflR. *Mol. Gen. Genom.* 268:711–719, 2003.
30. Chang, P. K., J. W. Bennett, and P. J. Cotty. Association of aflatoxin biosynthesis and sclerotial development in *Aspergillus parasiticus*. *Mycopathologia* 153:41–48, 2001.
31. Chang, P.K., J.W. Cary, D. Bhatnagar, T.E. Cleveland, J.W. Bennett, J.E. Linz, C.P. Woloshuk, G.A. Payne. Cloning of the *Aspergillus parasiticus apa-2* gene associated with the regulation of aflatoxin biosynthesis. *Appl. Environ. Microbiol.* 59:3273–3279, 1993.

32. Chang, P.K., J.W. Cary, J. Yu, D. Bhatnagar, T.E. Cleveland. The *Aspergillus parasiticus* polyketide synthase gene *pksA*, a homolog of *Aspergillus nidulans wA*, is required for aflatoxin B1 biosynthesis. *Mol. Gen. Genet.* 248:270–277, 1995.
33. Chang, P.K., K.C. Ehrlich, J.E. Linz, D. Bhatnagar, T.E. Cleveland, J.W. Bennett. Characterization of the *Aspergillus parasiticus niaD* and *niiA* gene cluster. *Curr. Genet.* 30:68–75, 1996.
34. Chang, P.K., K.C. Ehrlich, J. Yu, D. Bhatnagar, T.E. Cleveland. Increased expression of *Aspergillus parasiticus aflR*, encoding a sequence-specific DNA-binding protein, relieves nitrate inhibition of aflatoxin biosynthesis. *Appl. Environ. Microbiol.* 61:2372–2377, 1995.
35. Chang, P.K., J. Yu. Characterization of a partial duplication of the aflatoxin gene cluster in *Aspergillus parasiticus* ATCC 56775. *Appl. Microbiol. Biotechnol.* 58:632–636, 2002.
36. Chang, P.K., J. Yu, D. Bhatnagar, T.E. Cleveland. Repressor-AFLR interaction modulates aflatoxin biosynthesis in *Aspergillus parasiticus*. *Mycopathologia* 147:105–112, 1999.
37. Chang, P.K., J. Yu, D. Bhatnagar, T.E. Cleveland. The carboxy-terminal portion of the aflatoxin pathway regulatory protein AFLR of *Aspergillus parasiticus* activates *GAL1::lacZ* gene expression in *Saccharomyces cerevisiae*. *Appl. Environ. Microbiol.* 65:2508–2512, 1999.
38. Chang, P.K., J. Yu, D. Bhatnagar, T.E. Cleveland. Characterization of the *Aspergillus parasiticus* major nitrogen regulatory gene, *areA*. *Biochim. Biophys. Acta* 1491:263–266, 2000.
39. Chen, L., S.P. McCormick, T.M. Hohn. Altered regulation of 15-acetyldeoxynivalenol production in *Fusarium graminearum*. *Appl. Environ. Microbiol.* 66:2062–2065, 2000.
40. Chiou, C.H., M. Miller, D.L. Wilson, F. Trail, J.E. Linz. Chromosomal location plays a role in regulation of aflatoxin gene expression in *Aspergillus parasiticus*. *Appl. Environ. Microbiol.* 68:306–315, 2002.
41. Cotty, P.J. Aflatoxins and sclerotia production by *Aspergillus flavus*: influence of pH. *Phytopathology* 78:1250–1253, 1988.
42. Cotty, P.J. Agriculture, aflatoxins, and *Aspergillus*, In: *The Genus Aspergillus*. Powell, K.A., A. Fenwick, J.F. Peberdy, eds., New York: Plenum Press, 1994, pp 1–27.
43. DeLuca, C., S. Passi, A.A. Fabbri, C. Fanelli. Ergosterol oxidation may be considered a signal for fungal growth and aflatoxin production in *Aspergillus parasiticus*. *Food Addit. Contamin.* 12:445–450, 1995.
44. Desjardins, A.E., R.D. Plattner, R.H. Proctor. Linkage among genes responsible for fumonisin biosynthesis in *Gibberella fujikuroi* mating population A. *Appl. Environ. Microbiol.* 62:2571–2576, 1996.
45. Diez, B., S. Gutierrez, J.L. Barredo, P. van Solingen, L.H. van der Voort, J.F. Martin. The cluster of penicillin biosynthetic genes. Identification and characterization of the *pcbAB* gene encoding the alpha-aminoadipyl- cysteinyl-valine synthetase and linkage to the *pcbC* and *penDE* genes. *J. Biol. Chem.* 265:16358–16365, 1990.
46. Donnelly, P.J., T.R. Devereux, J.F. Foley, R.R. Maronpot, M.W. Anderson, T.E. Massey. Activation of K-ras in aflatoxin B1-induced lung tumors from AC3F1 (A/J x C3H/HeJ) mice. *Carcinogenesis* 17:1735–1740, 1996.
47. Doohan, F.M., G. Weston, H.N. Rezanoor, D.W. Parry, P. Nicholson. Development and use of a reverse transcription-PCR assay to study expression of Tri5 by *Fusarium* species *in vitro* and *in planta*. *Appl. Environ. Microbiol.* 65:3850–3854, 1999.
48. Dowzer, C.E., J.M. Kelly. Analysis of the *creA* gene, a regulator of carbon catabolite repression in *Aspergillus nidulans*. *Mol. Cell Biol.* 11:5701–5709, 1991.
49. Dutton, M.F. Enzymes and aflatoxin biosynthesis. *Microbiol. Rev.* 52:274–295, 1988.
50. Dvorackova, I. *Aflatoxins and human health*. Boca Raton, FL: CRC Press, 1990.
51. Eaton, D.L., E.P. Gallagher. Mechanisms of aflatoxin carcinogenesis. *Annu. Rev. Pharmacol. Toxicol.* 34:135–172, 1994.
52. Ehrlich, K.C., J.W. Cary, B.G. Montalbano. Characterization of the promoter for the gene encoding the aflatoxin biosynthetic pathway regulatory protein AFLR. *Biochim. Biophys. Acta* 1444:412–417, 1999.
53. Ehrlich, K.C., P.J. Cotty. Variability in nitrogen regulation of aflatoxin production by *Aspergillus flavus* strains. *Appl. Microbiol. Biotechnol.* 60:174–178, 2002.

54. Ehrlich, K.C., B.G. Montalbano, J.W. Cary. Binding of the C6-zinc cluster protein, AFLR, to the promoters of aflatoxin pathway biosynthesis genes in *Aspergillus parasiticus*. *Gene* 230:249–257, 1999.
55. Ehrlich, K.C., B.G. Montalbano, J.W. Cary, P.J. Cotty. Promoter elements in the aflatoxin pathway polyketide synthase gene. *Biochim. Biophys. Acta* 1576:171–175, 2002.
56. Ellis, W.O., J.P. Smith, B.K. Simpson, J.H. Oldham. Aflatoxins in food: occurrence, biosynthesis, effects on organisms, detection, and methods of control. *Crit. Rev. Food Sci. Nutr.* 30:403–439, 1991.
57. Espeso, E.A., J. Arst-HN. On the mechanism by which alkaline pH prevents expression of an acid-expressed gene. *Mol. Cell Biol.* 20:3355–3363, 2000.
58. Espeso, E.A., J. Tilburn, H.N. Arst, Jr., M.A. Penalva. pH regulation is a major determinant in expression of a fungal penicillin biosynthetic gene. *EMBO J.* 12:3947–3956, 1993.
59. Felinski, E.A., P.G. Quinn. The CREB constitutive activation domain interacts with TATA-binding protein-associated factor 110 (TAF110) through specific hydrophobic residues in one of the three subdomains required for both activation and TAF110 binding. *J. Biol. Chem.* 274:11672–11678, 1999.
60. Feng, B., E. Friedlin, G.A. Marzluf. A reporter gene analysis of penicillin biosynthesis gene expression in *Penicillium chrysogenum* and its regulation by nitrogen and glucose catabolite repression. *Appl. Environ. Microbiol.* 60:4432–4439, 1994.
61. Feng, G.H., T.J. Leonard. Culture conditions control expression of the genes for aflatoxin and sterigmatocystin biosynthesis in *Aspergillus parasiticus* and *A. nidulans*. *Appl. Environ. Microbiol.* 64:2275–2277, 1998.
62. Fernandes, M., N.P. Keller, T.H. Adams. Sequence-specific binding by *Aspergillus nidulans AflR*, a C6 zinc cluster protein regulating mycotoxin biosynthesis. *Mol. Microbiol.* 28:1355–1365, 1998.
63. Fillinger, S., M.K. Chaveroche, K. Shimizu, N. Keller, C. d'Enfert. cAMP and *ras* signaling independently control spore germination in the filamentous fungus *Aspergillus nidulans*. *Mol. Microbiol.* 44:1001–1016, 2002.
64. Fillinger, S., B. Felenbok. A newly identified gene cluster in *Aspergillus nidulans* comprises five novel genes localized in the *alc* region that are controlled both by the specific transactivator AlcR and the general carbon-catabolite repressor CreA. *Mol. Microbiol.* 20:475–488, 1996.
65. Flaherty, J.E., M.A. Weaver, G.A. Payne, C.P. Woloshuk. A beta-glucuronidase reporter gene construct for monitoring aflatoxin biosynthesis in *Aspergillus flavus*. *Appl. Environ. Microbiol.* 61:2482–2486, 1995.
66. Fox, M.E., T. Yamada, K. Ohta, G.R. Smith. A family of cAMP-response-element-related DNA sequences with meiotic recombination hotspot activity in *Schizosaccharomyces pombe*. *Genetics* 156:59–68, 2000.
67. Gardner, H.W., M.J. Grove, N. Keller. Soybean lipoxygenase is active on nonacqueous media at low moisture: a constraint to xerophilic fungi and aflatoxins. *JAOCS* 77:1801–1808, 1998.
68. Gourama, H., L.B. Bullerman. *Aspergillus flavus* and *Aspergillus parasiticus*: aflatoxigenic fungi of concern in food and feeds: a review. *J. Food Prot.* 58:1395–1404, 1995.
69. Greene-McDowelle, D. M., B. Ingber, M. Wright, H. J. Zeringue, D. Bhatnagar. The effects of selected cotton-leaf volatiles on growth, development, and aflatoxin production of *Aspergillus parasiticus*. *Toxicon* 37:883–893, 1999.
70. Grosveld, F., G. B. van Assendelft, D. R. Greaves, and G. Kolias. Position-independent, high-level expression of the beta globin gene in transgenic mice. *Cell* 51:975–985, 1987.
71. Guzman, d. P., H. J. Ruiz. Relationship between aflatoxin biosynthesis and sporulation in *Aspergillus parasiticus*. *Fungal. Genet. Biol.* 21:198–205, 1997.
72. Hamer, J. E., W. E. Timberlake. Functional organization of the *Aspergillus nidulans trpC* promoter. *Mol. Cell Biol.* 7:2352–2359, 1987.
73. Hicks, J. K., J. H. Yu, N. P. Keller, T. H. Adams. *Aspergillus* sporulation and mycotoxin production both require inactivation of the FadA G alpha protein-dependent signaling pathway. *EMBO J.* 16:4916–4923, 1997.

74. Hohn, T. M., A. E. Desjardins. Isolation and disruption of the Tox5 gene encoding trichodiene synthase in *Gibberella pulicaris*. *Mol. Plant Microbe Interact.* 5:249–256, 1992.
75. Hohn, T. M., A. E. Desjardins, S. P. McCormick. The Tri4 gene of *Fusarium sporotrichioides* encodes a cytochrome P450 monooxygenase involved in trichothecene biosynthesis. *Mol. Gen. Genet.* 248:95–102, 1995.
76. Hohn, T. M., R. Krishna, R. H. Proctor. Characterization of a transcriptional activator controlling trichothecene toxin biosynthesis. *Fungal Genet. Biol.* 26:224–235, 1999.
77. Hohn, T. M., S. P. McCormick, A. E. Desjardins. Evidence for a gene cluster involving trichothecene-pathway biosynthetic genes in *Fusarium sporotrichioides*. *Curr. Genet.* 24:291–295, 1993.
78. Kachholz, T., A. L. Demain. Nitrate repression of averufin synthesis and aflatoxin production. *J. Nat. Prod.* 46:499–506, 1983.
79. Kale, S. P., D. Bhantnagar, J. W. Bennett. Isolation and characterization of morphological variants of *Aspergillus parasiticus* deficient in secondary metabolite production. *Mycol. Res.* 98:645–652, 1994.
80. Keller, N., C. Nesbitt, B. Sarr, T. D. Phillips, G. B. Burow. pH regulation of sterigmatocystin and aflatoxin biosynthesis in *Aspergillus species*. *Postharvest Pathol. Mycotox.* 87:643–648, 1997.
81. Keller, N. P., T. M. Hohn. Metabolic pathway gene clusters in filamentous fungi. *Fungal. Genet. Biol.* 21:17–29, 1997.
82. Keller, S. E., T. M. Sullivan, S. Chirtel. Factors affecting the growth of *Fusarium proliferatum* and the production of fumonisin B1: oxygen and pH. *J. Ind. Microbiol. Biotechnol.* 19:305–309, 1997.
83. Kimura, M., I. Kaneko, M. Komiyama, A. Takatsuki, H. Koshino, K. Yoneyama, I. Yamaguchi. Trichothecene 3-O-Acetyltransferase protects both the producing organism and transformed yeast from related mycotoxins. *J. Biol. Chem.* 273:1654–1661, 1998.
84. Kimura, M., G. Matsumoto, Y. Shingu, K. Yoneyama, I. Yamaguchi. The mystery of the trichothecene 3-O-acetyltransferase gene: analysis of the region around Tri101 and characterization of its homologue from *Fusarium sporotrichioides*. *FEBS Lett.*435:163–168, 1998.
85. Kimura, M., Y. Shingu, K. Yoneyama, I. Yamaguchi. Features of Tri101, the trichothecene 3-O-acetyltransferase gene, related to the self-defense mechanism in *Fusarium graminearum*. *Biosci. Biotechnol. Biochem.* 62:1033–1036, 1998.
86. Kinsey, J., J. A. Rambosek. Transformation of *Neurospora crassa* with the cloned *am* (glutamate dehydrogenase) gene. *Mol. Cell Biol.* 4:117–122, 1984.
87. Kulmburg, P., M. Mathieu, C. Dowzer, J. Kelly, B. Felenbok. 1993. Specific binding sites in the *alcR* and *alcA* promoters of the ethanol regulon for the CREA repressor mediating carbon catabolite repression in *Aspergillus nidulans*. *Mol. Microbiol.* 7:847–857, 1993.
88. Labrador, M., V. G. Corces. Setting the boundaries of chromatin domains and nuclear organization. *Cell* 111:151–154, 2002.
89. Lee, L. W., C. H. Chiou, J. E. Linz. Function of native OmtA *in vivo* and expression and distribution of this protein in colonies of *Aspergillus parasiticus*. *Appl. Environ. Microbiol.* 68:5718–5727, 2002.
90. Li, Q., K. R. Peterson, X. Fang, G. Stomatoyannopoulos. Locus control regions. *Blood* 100:3077–3086, 2002.
91. Liang, S. H. The function and expression of the *ver-1* gene and localization of the Ver-1 protein involved in aflatoxin biosynthesis in *Aspergillus parasiticus*. PhD Dissertation, Michigan State University, 1996.
92. Liang, S. H., T. S. Wu, R. Lee, F. S. Chu, J. E. Linz. Analysis of mechanisms regulating expression of the ver-1 gene, involved in aflatoxin biosynthesis. *Appl. Environ. Microbiol.* 63:1058–1065, 1997.
93. Liu, B. H., F. S. Chu. Regulation of *aflR* and its product, AflR, associated with aflatoxin biosynthesis. *Appl. Environ. Microbiol.* 64:3718–3723, 1998.
94. Luchese, R. H., W. F. Harrigan. Biosynthesis of aflatoxin--the role of nutritional factors. *J. Appl. Bacteriol.* 74:5–14, 1993.

95. Mahanti, N., D. Bhatnagar, J. W. Cary, J. Joubran, J. E. Linz. Structure and function of *fas-1A*, a gene encoding a putative fatty acid synthetase directly involved in aflatoxin biosynthesis in *Aspergillus parasiticus*. *Appl. Environ. Microbiol.* 62:191–195, 1996.
96. Marasas, W. F. O. Fumonisins: history, world-wide occurrence, and impact. In: *Fumonisins in Foods*, Jackson, L.S., J. W. DeVries, L. B. Bullerman, eds., New York: Plenum Press, 1996, pp 1–17.
97. Marzluf, G. A. Genetic regulation of nitrogen metabolism in the fungi. *Microbiol. Mol. Biol. Rev.* 61:17–32, 1997.
98. Matsushima, K., P. K. Chang, J. Yu, K. Abe, D. Bhatnagar, T. E. Cleveland. Pre-termination in *aflR* of *Aspergillus sojae* inhibits aflatoxin biosynthesis. *Appl. Microbiol. Biotechnol.* 55:585–589, 2001.
99. Mazur, P., A. Nakanishi, E. El-Zayat, S. P. Champe. Structure and synthesis of sporogenic psi factors from *Aspergillus nidulans*. *J. Chem. Soc. Chem. Commun.* 20:1486–1487, 1991.
100. McCormick, S. P., N. J. Alexander, S. E. Trapp, T. M. Hohn. Disruption of TRI101, the gene encoding trichothecene 3-O-acetyltransferase, from *Fusarium sporotrichioides*. *Appl. Environ. Microbiol.* 65:5252–5256, 1999.
101. McCormick, S. P., T. M. Hohn, A. E. Desjardins. Isolation and characterization of Tri3, a gene encoding 15-O- acetyltransferase from *Fusarium sporotrichioides*. *Appl. Environ. Microbiol.* 62:353–359, 1996.
102. Mellon, J. E., P. J. Cotty, M. K. Dowd. Influence of lipids with and without other cottonseed reserve materials on aflatoxin B(1) production by *Aspergillus flavus*. *J. Agric. Food Chem.* 48:3611–3615, 2000.
103. Meyers, D. M., G. Obrian, W. L. Du, D. Bhatnagar, G. A. Payne. Characterization of *aflJ*, a gene required for conversion of pathway intermediates to aflatoxin. *Appl. Environ. Microbiol.* 64:3713–3717, 1998.
104. Miller, B. L., K. Y. Miller, K. A. Roberti, W. E. Timberlake. Position-dependent and -independent mechanisms regulate cell-specific expression of the SpoC1 gene cluster of *Aspergillus nidulans*. *Mol. Cell Biol.* 7:427–434, 1987.
105. Miller, M. J., C. S. Brown-Jenco, G. Obrian, G. A. Payne, J. Linz. *Cis*-acting sites, NorL, TATA box, and AflR1 play important roles in *nor-1* transcriptional activation in *Aspergillus* *Appl. Environ. Microbiol.*, 2003. (under review)
106. Miller, M. J. Transcriptional regulation of the *Aspergillus parasiticus* aflatoxin biosynthetic pathway gene *nor-1*. PhD dissertation, Michigan State University, 2003.
107. Miller, M., J.E. Linz. Promoter analysis of the aflatoxin biosynthesis gene *nor*-1 *Aspergillus parasiticus* using $-glucuronidase reporter constructs. *American Society of Microbiology National Meeting*, Los Angeles, CA, 2000.
108. Minto, R.E., C. A. Townsend. Enzymology and molecular biology of aflatoxin biosynthesis. *Chem. Rev.* 97:2537–2555, 1997.
109. Muhitch, M. J., S. P. McCormick, N. J. Alexander, T. M. Hohn. Transgenic expression of the Tri101 or PDR5 gene increases resistance of tobacco to the phytotoxic effects of the trichothecene 4,15 diacetoxyscirpenol. *Plant Sci.* 157:201–207, 2000.
110. Munkvold, G. P., A. E. Desjardins. Fumonisins in maize: can we reduce their occurrence? *Plant Dis.* 81:556–565, 1997.
111. Neely, L. A., C. S. Hoffman. Protein kinase A and mitogen-activated protein kinase pathways antagonistically regulate fission yeast *fbp1* transcription by employing different modes of action at two upstream activation sites. *Mol. Cell Biol.* 20:6426–6434, 2000.
112. Niehaus-WG, J., W. P. Jiang. Nitrate induces enzymes of the mannitol cycle and suppresses versicolorin synthesis in *Aspergillus parasiticus*. *Mycopathologia* 107:131–137, 1989.
113. Orejas, M., E. A. Espeso, J. Tilburn, S. Sarkar, J. Arst-HN, M. A. Penalva. Activation of the *Aspergillus* PacC transcription factor in response to alkaline ambient pH requires proteolysis of the carboxy-terminal moiety. *Genes Dev.* 9:1622–1632, 1995.
114. Orejas, M., A. P. MacCabe, J. A. Perez-Gonzalez, S. Kumar, D. Ramon. Carbon catabolite repression of the *Aspergillus nidulans xlnA* gene. *Mol. Microbiol.* 31:177–184, 1999.

115. Payne, G. A., M. P. Brown. Genetics and physiology of aflatoxin biosynthesis. *Annu. Rev. Phytopathol.* 36:329–362, 1998.
116. Payne, G. A., G. J. Nystrom, D. Bhatnagar, T. E. Cleveland, C. P. Woloshuk. Cloning of the *afl-2* gene involved in aflatoxin biosynthesis from *Aspergillus flavus. Appl. Environ. Microbiol.* 59:156–162, 1993.
117. Pestka, J. J., A. el Bahrawy, L. P. Hart. Deoxynivalenol and 15-monoacetyl deoxynivalenol production by *Fusarium graminearum* R6576 in liquid media. *Mycopathologia* 91:23–28, 1985.
118. Proctor, R. H. *Fusarium* toxins: trichothecenes and fumonisins, In: *Microbial Foodborne Disease: Mechanisms of Pathogenesis and Toxin Synthesis*, Cary, J., J. Linz, D. Bhatnagar, eds., Lancaster, PA: Technomic Publishing Co, 2000, pp 363–382.
119. Proctor, R. H., A. E. Desjardins, R. D. Plattner, T. M. Hohn. A polyketide synthase gene required for biosynthesis of fumonisin mycotoxins in *Gibberella fujikuroi* mating population A. *Fungal Genet. Biol.* 27:100–112, 1999.
120. Proctor, R. H., T. M. Hohn, S. P. McCormick, A. E. Desjardins. Tri6 encodes an unusual zinc finger protein involved in regulation of trichothecene biosynthesis in *Fusarium sporotrichioides. Appl. Environ. Microbiol.* 61:1923–1930, 1995.
121. Rastogi, R., A. K. Srivastava, A. K. Rastogi. Long term effect of aflatoxin B(1) on lipid peroxidation in rat liver and kidney: effect of picroliv and silymarin. *Phytother. Res.* 15:307–310, 2001.
122. Reiss, J. Development of *Aspergillus parasiticus* and formation of aflatoxin B1 under the influence of conidiogenesis affecting compounds. *Arch. Microbiol.* 133:236–238, 1982.
123. Rotter, B. A., D. B. Prelusky, J. J. Pestka. Toxicology of deoxynivalenol (vomitoxin). *J. Toxicol. Environ. Health* 48:1–34, 1996.
124. Roze, L., Linz, J. Stimulation of aflatoxin biosynthesis in *Aspergillus parasiticus* by cAMP analogs. *Aflatoxin Elimination Workshop 100*, 1999.
125. Ruijter, G. J., J. Visser. Carbon repression in Aspergilli. *FEMS Microbiol. Lett.* 151:103–114, 1997.
126. Riley, R. T.E. Wang, J.J. Schroeder, E.R. Smith, R.D. Plattner, H. Abbas, H-S. Yoo, A.H. Merrill. Evidence for disruption of sphingolipid metabolism as a contributing factor in the toxicity and carcinogenicity of fumonisins. *Natural Toxins* 4:3–15, 1996.
127. Sachs, M. S. Posttranscriptional control of gene expression in filamentous fungi. *Fungal Genet. Biol.* 23:117–124, 1998.
128. Salvadi-Goldstein, S. R. Fluhr. Signal transduction of ethylene perception. *Res. Probl. Cell. Differ.* 27:145–161, 2000.
129. Scholl, P., S. M. Musser, T. W. Kensler, J. D. Groopman. Molecular biomarkers for aflatoxins and their application to human liver cancer. *Pharmacogenetics* 5:S171–S176, 1995.
130. Seo, J. A., R. H. Proctor, R. D. Plattner. Characterization of four clustered and co-regulated genes associated with fumonisin biosynthesis in *Fusarium verticillioides. Fungal Genet. Biol.* 34:155–165, 2001.
131. Serek, M., E. C. Sisler, M. S. Reid. 1-Methylcylcopropene, a novel gaseous inhibitor of ethylene action, improves the life of fruits, cut flowers and potted plants. *Acta Horticulturae* 394:337–347, 1995.
132. Sharma, A., D. Padwal, Sr., G. B. Nadkarni. Possible implications of reciprocity between ethylene and aflatoxin biogenesis in *Aspergillus flavus* and *Aspergillus parasiticus. Appl. Environ. Microbiol.* 49:79–82, 1985.
133. Shen, H. M., C. N. Ong, C. Y. Shi. Involvement of reactive oxygen species in aflatoxin B1-induced cell injury in cultured rat hepatocytes. *Toxicology* 99:115–123, 1995.
134. Shim, W. B., C. P. Woloshuk. Nitrogen repression of fumonisin B1 biosynthesis in *Gibberella fujikuroi. FEMS Microbiol. Lett.* 177:109–116, 1999.
135. Shim, W. B., C. P. Woloshuk. Regulation of fumonisin B(1) biosynthesis and conidiation in *Fusarium verticillioides* by a cyclin-like (C-type) gene, FCC1. *Appl. Environ. Microbiol.* 67:1607–1612, 2001.

136. Shimizu, K., N. P. Keller. Genetic involvement of a cAMP-dependent protein kinase in a G protein signaling pathway regulating morphological and chemical transitions in *Aspergillus nidulans*. *Genetics* 157:591–600, 2001.
137. Sisler, E. C., M. Serek. Inhibitors of ethylene responses in plants at the receptor level; recent developments. *Physiol. Plantar.* 100:577–582, 1997.
138. Skory, C. D., P. K. Chang, J. Cary, J. E. Linz. Isolation and characterization of a gene from *Aspergillus parasiticus* associated with the conversion of versicolorin A to sterigmatocystin in aflatoxin biosynthesis. *Appl. Environ. Microbiol.* 58:3527–3537, 1992.
139. Skory, C. D., P. K. Chang, J. E. Linz. Regulated expression of the *nor-1* and *ver-1* genes associated with aflatoxin biosynthesis. *Appl. Environ. Microbiol.* 59:1642–1646, 1993.
140. Smith, F. C., S. P. Davies, W. A. Wilson, D. Carling, D. G. Hardie. The SNF1 kinase complex from *Saccharomyces cerevisiae* phosphorylates the transcriptional repressor protein Mig1p *in vitro* at four sites within or near regulatory domain 1. *FEB.* 77:1–18, 1999.
141. Strauss, J., R. L. Mach, S. Zeilinger, G. Hartler, G. Stoffler, M. Wolschek, C. P. Kubicek. Cre1, the carbon catabolite repressor protein from *Trichoderma reesei*. *FEBS Lett.* 376:103–107, 1995.
142. Sweeney, M. J., A. D. Dobson. Molecular biology of mycotoxin biosynthesis. FEMS *Microbiol. Lett.* 175:149–163, 1999.
143. Tag, A., J. Hicks, G. Garifullina, C. J. Ake, T. D. Phillips, M. Beremand, N. Keller. G-protein signaling mediates differential production of toxic secondary metabolites. *Mol. Microbiol.* 38:658–665, 2000.
144. Tag, A. G., G. F. Garifullina, A. W. Peplow, C. J. Ake, T. D. Phillips, T. M. Hohn, M. N. Beremand. A novel regulatory gene, Tri10, controls trichothecene toxin production and gene expression. *Appl. Environ. Microbiol.* 67:5294–5302, 2001.
145. Tice, G., R. L. Buchanan. Regulation of aflatoxin biosynthesis: effect of exogenously supplied cyclic nucleotides. *J. Food Sci.* 47:153–157, 1981.
146. Tilburn, J., S. Sarkar, D. A. Widdick, E. A. Espeso, M. Orejas, J. Mungroo, M. A. Penalva, J. Arst-HN. The *Aspergillus* PacC zinc finger transcription factor mediates regulation of both acid- and alkaline-expressed genes by ambient pH. *EMBO J.* 14:779–790, 1995.
147. Timberlake, W. E., E. C. Barnard. Organization of a gene cluster expressed specifically in the asexual spores of *A. nidulans*. *Cell* 26:29–37, 1981.
148. Timberlake, W. E., M. A. Marshall. Genetic regulation of development in *Aspergillus nidulans*. *Trends Genet.* 4:162–169, 1988.
149. Tiwari, R. P., V. Mittal, T. C. Bhalla, S. S. Saini, G. Singh, D. V. Vadehra. Effect of metal ions on aflatoxin production by *Aspergillus parasiticus*. *Folia Microbiol. Praha.* 31:124–128, 1986.
150. Tolleson, W. H., K. L. Dooley, W. G. Sheldon, J. D. Thurman, T. J. Bucci, P. C. Howard. The mycotoxin fumonisin induces apoptosis in cultured human cells and in livers and kidneys of rats. *Adv. Exp. Med Biol.* 392:237–250, 1996.
151. Tolleson, W. H., W. B. Melchior, Jr., S. M. Morris, L. J. McGarrity, O. E. Domon, L. Muskhelishvili, S. J. James, P. C. Howard. Apoptotic and anti-proliferative effects of fumonisin B1 in human keratinocytes, fibroblasts, esophageal epithelial cells and hepatoma cells. *Carcinogenesis* 17:239–249, 1996.
152. Trail, F., N. Mahanti, J. Linz. Molecular biology of aflatoxin biosynthesis. *Microbiology* 141:755–765, 1995.
153. Trail, F., N. Mahanti, M. Rarick, R. Mehigh, S. H. Liang, R. Zhou, J. E. Linz. Physical and transcriptional map of an aflatoxin gene cluster in *Aspergillus parasiticus* and functional disruption of a gene involved early in the aflatoxin pathway. *Appl. Environ. Microbiol.* 61:2665–2673, 1995.
154. Ueno, Y. General Toxicology. In: *Trichothecenes: Chemical, Biological and Toxicological Aspects*, Ueno, Y., ed., New York: Elsevier, 1983, pp 135–146.
155. Vautard-Mey, G., M. Fevre. Mutation of a putative AMPK phosphorylation site abolishes the repressor activity but not the nuclear targeting of the fungal glucose regulator CRE1. *Curr. Genet.* 37:328–332, 2000.

156. Wang, E., W. P. Norred, C. W. Bacon, R. T. Riley, A. H. Merrill, Jr. Inhibition of sphingolipid biosynthesis by fumonisins. Implications for diseases associated with *Fusarium moniliforme*. *J Biol. Chem.* 266:14486–14490, 1991.
157. Watanabe, C. M., D. Wilson, J. E. Linz, C. A. Townsend. Demonstration of the catalytic roles and evidence for the physical association of type I fatty acid synthases and a polyketide synthase in the biosynthesis of aflatoxin B1. *Chem. Biol.* 3:463–469, 1996.
158. Weigel, B. J., S. G. Burgett, V. J. Chen, P. L. Skatrud, C. A. Frolik, S. W. Queener, T. D. Ingolia. Cloning and expression in *Escherichia coli* of isopenicillin N synthetase genes from *Streptomyces lipmanii* and *Aspergillus nidulans*. *J. Bacteriol.* 170:3817–3826, 1988.
159. Wiseman, D. W., R. L. Buchanan. Determination of glucose level needed to induce aflatoxin production in *Aspergillus parasiticus*. *Can. J. Microbiol.* 33:828–830, 1987.
160. Woloshuk, C. P., K. R. Foutz, J. F. Brewer, D. Bhatnagar, T. E. Cleveland, G. A. Payne. Molecular characterization of *aflR*, a regulatory locus for aflatoxin biosynthesis. *Appl. Environ. Microbiol.* 60:2408–2414, 1994.
161. Woloshuk, C. P., R. Prieto. Genetic organization and function of the aflatoxin B1 biosynthetic genes. *FEMS Microbiol. Lett.* 160:169–176, 1998.
162. Wright, M. S., D. M. Greene-McDowelle, H. J. Zeringue, D. Bhatnagar, T. E. Cleveland. Effects of volatile aldehydes from *Aspergillus*-resistant varieties of corn on *Aspergillus parasiticus* growth and aflatoxin biosynthesis. *Toxicon* 38:1215–1223, 2000.
163. Yang, G. H., B. B. Jarvis, Y. J. Chung, J. J. Pestka. Apoptosis induction by the satratoxins and other trichothecene mycotoxins: relationship to ERK, p38 MAPK, and SAPK/JNK activation. *Toxicol. Appl. Pharmacol.* 164:149–160, 2000.
164. Young, C., L. McMillan, E. Telfer, B. Scott. Molecular cloning and genetic analysis of an indole-diterpene gene cluster from *Penicillium paxilli*. *Mol. Microbiol.* 39:754–764, 2001.
165. Young, J. L., G. Jarai, Y. H. Fu, G. A. Marzluf. Nucleotide sequence and analysis of NMR, a negative-acting regulatory gene in the nitrogen circuit of *Neurospora crassa*. *Mol. Gen. Genet.* 222:120–128, 1990.
166. Yu, J., P. Chang, D. Bhatnagar, T. E. Cleveland. Cloning of a sugar utilization gene cluster in *Aspergillus parasiticus*. *Biochim. Biophys. Acta* 1493:211–214, 2000.
167. Yu, J., P. K. Chang, D. Bhatnagar, T. E. Cleveland. Genes encoding cytochrome P450 and monooxygenase enzymes define one end of the aflatoxin pathway gene cluster in *Aspergillus parasiticus*. *Appl. Microbiol. Biotechnol.* 53:583–590, 2000.
168. Yu, J., P. K. Chang, J. W. Cary, M. Wright, D. Bhatnagar, T. E. Cleveland, G. A. Payne, J. E. Linz. Comparative mapping of aflatoxin pathway gene clusters in *Aspergillus parasiticus* and *Aspergillus flavus*. *Appl. Environ. Microbiol.* 61:2365–2371, 1995.
169. Yu, J. H., R. A. Butchko, M. Fernandes, N. P. Keller, T. J. Leonard, T. H. Adams. Conservation of structure and function of the aflatoxin regulatory gene *aflR* from *Aspergillus nidulans* and *A. flavus*. *Curr. Genet.* 29:549–555, 1996.
170. Zaika, L. L., R. L. Buchanan. Review of compounds affecting the biosynthesis of aflatoxin. *J. Food Prot.* 50:691–708, 1987.
171. Zeringue-HJ, J. Effect of C6 to C9 alkenals on aflatoxin production in corn, cottonseed, and peanuts. *Appl. Environ. Microbiol.* 57:2433–2434, 1991.
172. Zeringue, H.-J. J. Identification and effects of maize silk volatiles on cultures of *Aspergillus flavus*. *J. Agric. Food Chem.* 48:921–925, 2000.
173. Zhou, R. The function, accumulation, and localization of the Nor-1 protein involved in aflatoxin biosynthesis: the function of the *fluP* gene associated with sporulation in *Aspergillus parasiticus*. PhD Dissertation, Michigan State University, 1997.
174. Zhou, R., J. E. Linz. Enzymatic function of the *nor-1* protein in aflatoxin biosynthesis in *Aspergillus parasiticus*. *Appl. Environ. Microbiol.* 65:5639–5641, 1999.
175. Lee, Li-Wei, C-H. Chiou, K. Klomparens, J. Cary, J.E. Linz. Sub-cellular localization of aflatoxin biosynthetic enzymes, Nor-1 and Ver-1 in time-dependent fractionated colonies of Aspergillus parasiticus. *Arch. Microbiol.*, 2004. (in press)

176. Roze, L., M. Miller, N. Mahanti, J.E. Linz. A novel cAMP response element, CRE1, modulates expression of the *nor-1* promoter in *Aspergillus parasiticus*. *J. Biol. Chem.*, 2003. (under review)
177. Roze, L., A. Calvo-Byrd, R. Beaudry, J.E. Linz. Ethylene modulates development and toxin synthesis in *Aspergillus* likely via an ethylene sensor-mediated signaling pathway. *J. Food Protec.*, 2003. (in press)
178. Roze, L., N. Keller, R. Beaudry, J. E. Linz. Regulation of aflatoxin synthesis by FadA/cAMP/protein kinase A signaling in *Aspergillus parasiticus*. *Mycopathologia*, 2003. (in press).
179. Flaherty, J.E., G.A. Payne. Overexpression of aflR leads to upregulation of pathway gene transcription and increased aflatoxin production in *Aspergillus flavus*. *Appl. Environ. Microbiol.* 63:3995–4000, 1997.
180. Yu, J-J., P-K Chang, J.W. Cary, D. Bhatnagar, T.E. Cleveland, G.A. Payne, J.E. Linz, C.P. Woloshuk, J.W. Bennett. The clustered pathway genes in aflatoxin synthesis. Appl. Environ. Microbiol, 2003. (in press)
181. Bailey, G.S. Role of aflatoxin-DNA adducts in the cancer process. In: *The Toxicology of Aflatoxins*, Eaton, D.L., J.D. Groopman, eds., New York: Academic Press, 1994, pp 3–24.
182. Johnstone, I.L., P.C. McCabe, P.Greaves, S.J. Gurr, G.E. Cole, M.A. Brow, S.E. Unkles, A.J. Clutterbuck, J.R. Kinghorn, M.A. Innis. Isolation and characterization of the *crnAA-niia-niaD* gene cluster for nitrate assimilation in *Aspergillus nidulans*. *Gene* 90:181–192, 1990.

3.10

Application of ELISA Assays for Detection and Quantitation of Toxins in Foods

Robert E. Levin

CONTENTS

10.1 INTRODUCTION

ELISA assays by definition are enzyme linked immunosorbent assays and are presently used to detect and quantitate a wide variety of organic molecules such as hormones, pharmaceuticals, and toxins. The major advantages of ELISA assays involve a high level of specificity due to the immunological nature of the assays, extraordinary sensitivity, frequently in the picogram (pg) range of detection, and small sample size. All ELISA assays require at least one source of antibody (Ab). Most such antibodies involve the immunoglobulin G (IgG) fraction of antisera but are not necessarily restricted to the G fraction. Antisera are produced by the injection of suitable animals with the immunogenic molecule of interest. The rabbit is the conventional laboratory animal of choice for the small scale production of polyclonal antibodies (pAbs); mice are used for production of monoclonal Abs (mAbs) in conjunction with hybridoma methodology (1). Antisera to specific antigens are often additionally available commercially from larger animal sources such as the goat, cow, horse, sheep, cat, and dog. In general, the larger the animal, the less expensive is the antiserum. The rat, mouse, and poultry are additional sources of commercially available antisera. In addition, the industry produces a wide variety of anti immune sera, e.g., rabbit antisera that will react with all IgGs from the goat and vice versa. The availability of such a wide variety of anti immune sera greatly facilitates the development of antiglobulin sandwich ELISA assays.

10.2 HAPTENS AND THEIR CONJUGATION TO PROTEINS

Organic molecules having a molecular weight of less than 5000 are usually not immunogenic. Many of the organic toxins occurring in foods have molecular weights well below this value. The problem can be readily circumvented by covalently conjugating the small toxin molecule (hapten) to a large protein molecule or carrier. Bovine serum albumin (BSA) having a molecular weight of 67,000 Da is most frequently used as a protein carrier because of its high degree of purity, abundant availability, and low cost. Keyhole limpet hemocyanin (KLH, mol. wt. ~7 million kDa) and polylysine (mol. wt. 70,000–150,000 Da) are alternate protein carriers. Because of its large size, KLH may precipitate during cross linking which can create handling problems. BSA is readily soluble but is itself highly immunogenic. Ovalbumin (OVA, mol. wt. ~45,000 dalt.) is another suitable

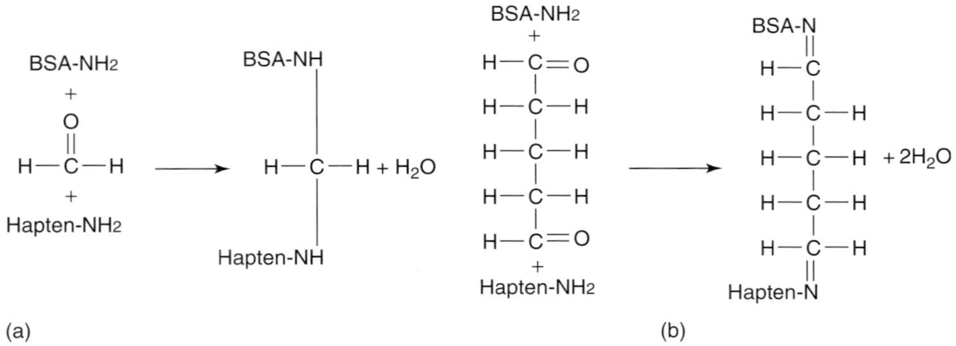

Figure 10.1 Aledyhdye conjugation reactions. (a) formaldehyde conjugation of hapten to BSA. (b) gluteraldehyde conjugation if hapten to BSA.

carrier and is ideal for a second carrier (as in antiglobulin ELISA sandwich assays) where BSA is used for primary antibody production.

If a free amino group is available on the hapten it can be readily conjugated to BSA using formaldehyde or gluteraldehyde as a covalent bridge (Figure 10.1). Reaction mechanisms are also available for conjugating through available –OH, -CHO, -COOH and -SH groups (2,3).

Injection of a mammal with such a purified conjugate will result in a family of Ab molecules, each directed against a specific domain of the conjugate. Some Ab molecules will be directed against overlapping domains of the BSA while others will be capable of recognizing the hapten. Such a family of Abs is referred to as polyclonal in that each Ab molecular species is produced by an individual B lymphocyte. Higher levels of immunological specificity can often be obtained with the use of monoclonal antibodies that are homogeneous, in that all the Ab molecules present are identical and will recognize the same domain of an antigen. The origin and methodology involving the production of monoclonal antibodies has been described in detail by Zola (1).

10.3 THE USE OF ENZYMES IN ELISA ASSAYS

The second critical component of all ELISA assays involves the use of an enzyme reaction to produce a visual or spectrophotometrically detectable indication of the presence or absence of the target antigen. The principle enzymes presently in use with ELISA assays are alkaline phosphatase, horse-radish peroxidase (HRP), and β–galactosidase (Table 10.1). The fastest response is usually with HRP because it has the highest turnover number (largest number of substrate molecules attacked per unit of time by a single enzyme molecule) and because it exhibits the highest level of heat resistance. This elevated heat resistance can be used to advantage with HRP by raising the incubation temperature for the enzyme reaction to 55°C so as to greatly accelerate the reaction rate and result in a colored reaction product within 5 to 10 min. HRP is cheaper than the other two enzymes but requires two substrates, H_2O_2 plus a proton donor such as ABTS, OPD, or TMB (Table 10.1). Among these three proton donors, TMB is preferred because it is nonmutagenic in the Ames assay (4). For an enzyme to be suitable for use in ELISA assays it should possess a high turnover number and have available a suitable colorless substrate that is converted to a stable, soluble, and colored end product. For quantitative ELISA assays, the product of the enzyme reaction must be soluble.

The enzymes are usually conjugated to specific or anti immune antibodies. Conjugation reactions that allow retention of enzyme activity usually involve the use of gluteraldehyde in a one step reaction (5) or with a two step reaction (6). The two step gluteraldehyde method results in enzyme and antibody being present in equal amounts in the conjugate resulting in enhanced sensitivity (7). Periodate conjugation (8) is also widely used.

10.4 THE USE OF MICROTITER PLATES IN ELISA ASSAYS

Most ELISA assays are presently performed using microtiter plates with 72 or 96 wells per plate, with each well having a capacity of 300 μl. The resulting miniaturized volume of ELISA assays results in a considerable saving of antibody preparations and other costly reagents. The multiple wells allow large numbers of samples to be simultaneously processed. The availability of microplate readers with variable wavelengths and optional internal incubation at a variety of temperatures greatly enhances the efficiency of processing multiple samples.

Table 10.1
Enzymes available for ELISA assays and their substrates

Enzyme	Substrate	Comments
Horseradish Peroxidase	2,2′ Azino-di[3-ethylben-thiazoline sulphonate] (ABTS)	Water soluble green product which absorbs light at 410nm and 650 nm. ABTS (1 mM) used with H_2O_2 (0.002% v/v) in phosphate / citrate buffer pH 4.3. Stop reaction with sodium azide (1.5 mM) in citric acid (0.1M).
	O-Phenylene diamine (OPD)	Water soluble absorbs light at 492nm. OPD (4mM) used with H_2O_2 (0.004% v / v) in citric acid (0.02M) Na_2HPO_4 (0.05M) buffer pH 5.0. Stop reaction with H_2SO_4 (0.1M).
	3,3′,5,5′-Tetramethylbenzidine (TMB)	Water soluble yellow product which absorbs light at 450nm. TMB (0.4mM) used with H_2O_2 (0.004% v / v) in 0.05M acetate buffer pH 6.0. Stop reaction with H_2SO_4 (2M).
Alkaline phosphatase	P-nitrophenyl phosphate (PNPP)	Water soluble yellow reaction product which absorbs lighter at 405nm. PNPP (2.5 mM), plus $MgCl_2$ (0.5mM) in diethanolamine buffer (10 mM) pH 9.5. Stop reaction with EDTA (0.1 M).
b-Galactosidase	O-nitrophenyl-b-D-galactopyranoside (ONPG)	Water soluble yellow product which absorbs light at 410nm. ONPG (3 mM), plus $MgCl_2$ (10 mM), 2-mercapto-ethanol (0.1M) in phosphate buffered saline, pH 7.5. Stop reaction with Na_2CO_3.
	4-methyl-umbelliferyl-β-D-galacto-pyranoside (MUG)	Water soluble product which fluoresces at 45nm. 4-MUG (0.1 mM) in NaCl (0.1 M), $MgCl_2$ (1 mM), 0.1% bovine serum albumin, sodium phosphate (10 mM), pH 7.0. Stop reaction with glycine (0.1 M), pH 10.3.

Polystyrene microtiter plates are most frequently used and harbor electrostatic binding sites on their surface that will bind proteins in a nonspecific manner. This is taken advantage of by the addition of diluted antiserum (or the hapten conjugated to a carrier protein) to wells (usually 100 µl) to achieve nonspecific binding of Abs or the hapten to the bottom of the well. Nonspecific binding of Abs to the wells often requires several hours and is most conveniently achieved by allowing binding to occur overnight in a refrigerator with the covers in place to prevent evaporation. The unbound Ab (or hapten-protein conjugate) molecules are then rinsed out and blocking solution (300 µl) is then added to completely fill the wells. Blocking solutions consist of any suitably inexpensive protein that will bind nonspecifically to the surface of the wells so as to prevent further nonspecific binding of ELISA reagents thereafter. An advantage of polystyrene plates is that blocking solution need only be added once. After 15 min., the blocking agent is rinsed out and thereafter only immunospecific binding will occur on the surface of the wells. Blocking agents consist of 0.1% BSA, 0.1% fat free skimmed milk (FFSM), or 0.1% gelatin. The latter two are readily available at food markets in powdered form. If FFSM is used it is first prepared as a 1.0% (w/v) solution in 0.1M phosphate buffer, pH 7.5 containing 8.5% NaCl. Boiling in the presence of phosphate salts solubilizes the casein. It is then diluted tenfold with distilled water for use. BSA should not be used as a blocking agent if the original hapten conjugate used for injection to produce Abs involved BSA as the protein carrier.

The sensitivity of ELISA assays can be significantly increased if the antibody portion of antiserum is purified before being bound to polystyrene wells or conjugated to an enzyme. This is readily achieved with the use of affinity chromatography involving protein A bound to agarose beads in miniature columns and frequently results in about a 100-fold purification of the IgG fraction of the antiserum.

It is important to keep in mind that all ELISA assays consist of a series of additions of reagents, and that after the addition of each reagent and the required incubation time, the wells are thoroughly washed with a washing solution consisting usually of 0.02% Tween 20 in 0.05M phosphate buffer (pH 7.5) containing 0.85% NaCl. In addition to Tween 20, Triton-X-100 and Nonidet-40 are also employed. These are nonionic detergents that are used to prevent the hydrophobic adsorption of proteins to the surface of the wells and to decrease the hydrophobic aggregation of proteins (9). Three commonly used ELISA formats using microtiter plates are presented in Figure 10.2. In the direct ELISA [Figure 10.2(a)], antigen or a hapten-protein conjugate is bound to the surface of the well followed by the addition of Ab-enzyme conjugate, and then substrate for the enzyme is added. The intensity of the color reaction after a predetermined incubation time is then directly related to the amount of antigen bound to the well. In the indirect ELISA [Figure 10.2(b)] antigen or a hapten-protein conjugate is first bound to the well followed by the addition of primary

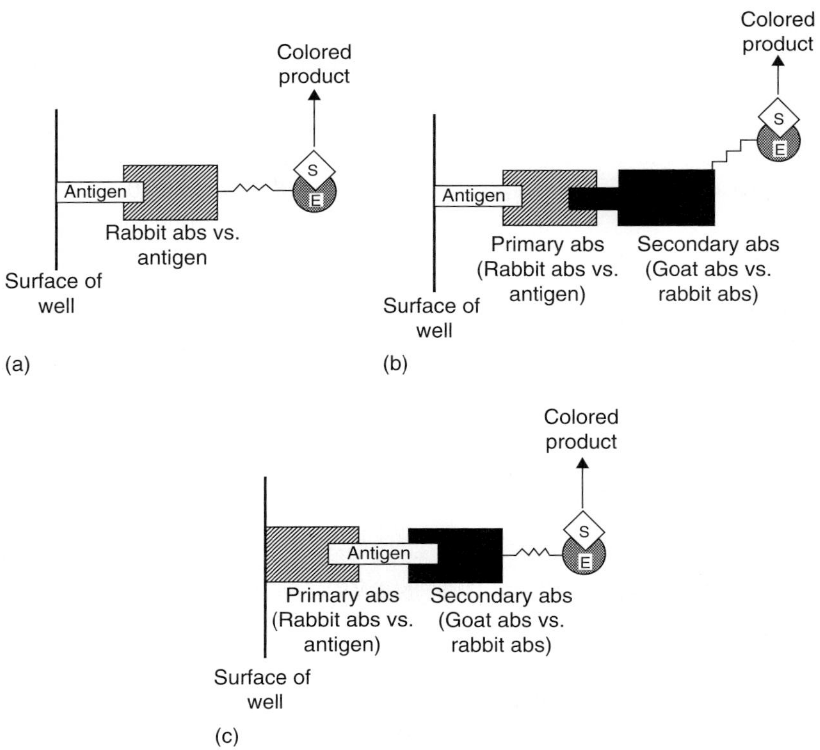

Figure 10.2 Fundamental ELISA techniques.

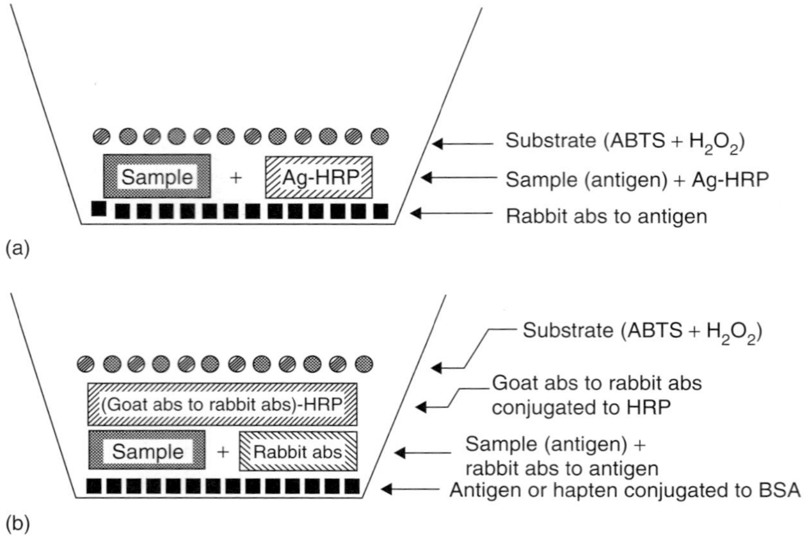

Figure 10.3 Direct and indirect competitive ELISA techniques.

antibody, then secondary antibody, then enzyme, and finally substrate. The advantage here involves the use of commercially available secondary Ab already conjugated to the enzyme. This mode is particularly useful if you are producing your own antibodies for research in rabbits or mice in that the primary antibody is not conjugated to the enzyme. The indirect ELISA assays are particularly useful for assaying antibody titers. The indirect sandwich ELISA [Figure 10.2(c)] is particularly useful if a hapten by itself is not capable of binding to the surface of the well. Here, primary Abs are bound initially to the well, followed by the addition of antigen (sample), then secondary Abs conjugated to the enzyme, and then the substrate. The competitive ELISA assays are the most sensitive and also facilitate the addition of samples to the wells. In the direct competitive ELISA [Figure 10.3(a)], Abs are first bound to the surface of the well followed by the simultaneous addition and incubation of the sample (antigen) and antigen-enzyme conjugate. These two components then compete for binding to the Abs bound to the well. The intensity of the final color developed is inversely related to the amount of antigen in the sample. In the indirect competitive ELISA [Figure 10.3(b)] antigen or hapten-protein conjugate is first bound to the surface of the well. This is then followed by the simultaneous addition and incubation of the sample (antigen) and primary Abs against the antigen. The primary Abs then compete for the soluble antigen in the sample and for the antigen bound to the well. After rinsing, secondary Ab-enzyme conjugate is then added followed by rinsing and the addition of substrate. The intensity of the final color developed is inversely related to the amount of antigen in the sample.

10.5 THE USE OF THE AVIDIN-BIOTIN SYSTEM IN ELISA ASSAYS

10.5.1 Nature of the Avidin-Biotin Complex

Avidin is a glycoprotein derived from egg white containing four identical subunits, having a combined mol. wt. of 66,000 kDa. Each subunit is capable of binding one biotin molecule.

One molecule of avidin therefore binds four molecules of biotin via a noncovalent interaction with an extremely low dissociation constant of 10^{-15}.

$$AB \xrightarrow{K} \frac{[A]\,[B]}{[AB]} = 10^{-15} \tag{10.1}$$

This extremely high binding affinity of avidin for biotin constitutes the strongest noncovalent bonding known. The avidin-biotin complex forms rapidly and spontaneously when both components are mixed in solution and greatly facilitates the formation of strong bonding complexes in ELISA assays without the need for a covalent reaction. The ability of avidin to bind four molecules of biotin is taken advantage of in greatly increasing the sensitivity of ELISA assays nearly 100-fold (10). Streptavidin is a molecule produced by *Streptomyces avidinii* which also binds four molecules of biotin with a similarly low dissociation constant. It is comprised of four subunits with a combined m. wt. of 47,000 kDa. Because avidin contains a carbohydrate moiety and has a pI = 10 it tends to bind nonspecifically. Streptavidin has a pI = 6.0 and therefore has almost no charge, and because it has no carbohydrate moiety it exhibits less nonspecific binding. An advantage of avidin is the higher enzyme activity of avidin-enzyme conjugates.

10.5.2 Biotinylation of ELISA Components and Conjugation of Avidin and Streptavidin to Proteins

Antibodies, antigens, and enzymes can be readily covalently bonded to biotin (biotinylation). Antibodies and antigens are usually biotinylated through their available amino groups although the sulfhydryl groups of both and the oligosaccharide components of Abs can also be used. The most common method used for biotinylating antibodies and protein antigens involves the use of the biotin derivative biotinyl N-hydroxysuccinimide (BNHS) which usually furnishes a satisfactory intervening spacer arm separating the biotin molecule from the reactive site of the protein. A detailed discussion of all commonly used methods for biotinylation of antibodies has been presented by Bayer and Wilchek (11). The conjugation of the proteins avidin and streptavidin to antibodies and enzymes involves the same conjugation chemistry as that of proteins in general. A number of methods for producing a variety of avidin derivatives and conjugates have been developed (12–15). Microtiter plates are also available precoated with streptavidin or poly D-lysine to facilitate the use of a wide variety of protocols.

There are three fundamental avidin-biotin techniques. The labeled avidin-biotin (LAB) technique involves adding an enzyme labeled avidin conjugate to a biotinylated primary (direct assay) or secondary antibody (indirect assay). Figure 10.4(a) illustrates the indirect (ILAB) reaction scheme. This method is commonly used for screening monoclonal antibodies from hybridoma cell lines. The bridged avidin-biotin assay (BRAB) results in amplification by using a sandwich technique where the avidin is not conjugated to the enzyme [Figure 10.4(b)]. Avidin is used as a bridge between a biotinylated antibody and the enzyme. This increases the amplification further by complexing more biotinylated enzyme than the ILAB assay. The avidin-biotin complex (ABC) technique is the most sensitive [Figure 10.4(c)]. Here, the biotinylated enzyme is preincubated with avidin, forming large complexes. When this solution is added to the wells, any remaining biotin-binding sites on the avidin will bind to immobilized biotinylated antibody. This leads to an increased number of enzyme molecules bound to the surface of the well resulting in maximum amplification.

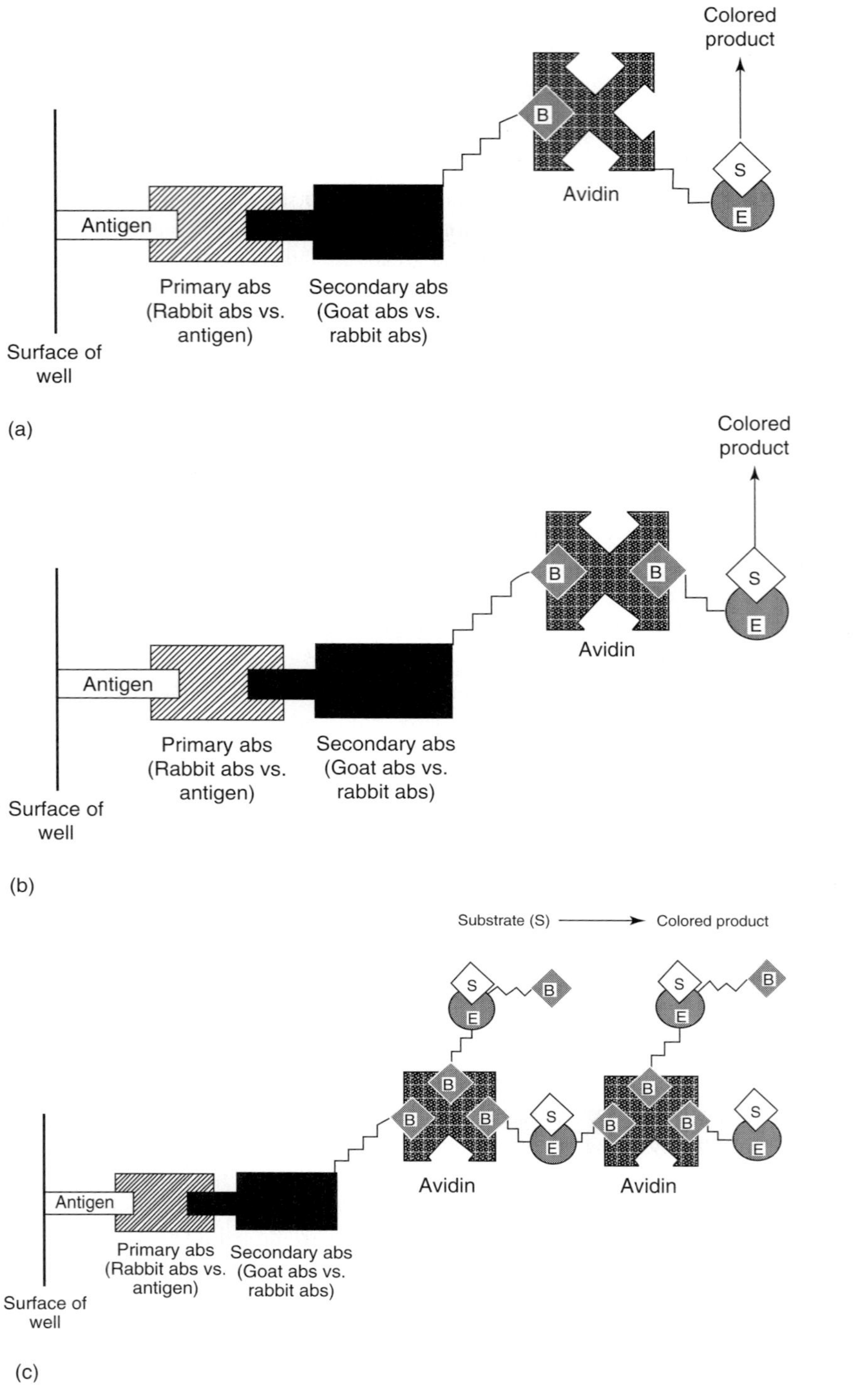

Figure 10.4 Fundamental avidin-biotin ELISA techniques.

10.6 ELISA DETECTION OF TOXINS IN FOODS

10.6.1 Seafood Toxins (Figure 10.5)

10.6.1.1 Saxitoxins

Saxitoxin (STX) is the most potent of a family of at least 18 known homologs (16) produced by single celled microscopic marine algae presently designated *Alexandrium tamarense* and *Alexandrium catenellum* and certain related organisms. These organisms are consumed by filter feeding shellfish such as clams, oysters, mussels, and other mollusks, and are the cause of red tides at sea that result from periodic blooms of these organisms. Consumption of shellfish containing elevated levels of saxitoxin and its homologs results in paralytic shellfish poisoning, hence the designation paralytic shellfish toxins (PST). These toxins resist boiling and function neurologically as sodium channel blocking agents, resulting in symptoms starting with paresthesia and leading to ataxia. In extreme cases, respiratory or cardiac failure has led to death. The lethal dose of STX for humans has been estimated to be between 0.5 to 4 mg (17). The U.S. Food and Drug administration has set the legal limit of STX in shellfish of 80 μg /100 g of edible tissue.

An indirect competitive ELISA for STX was developed using antibodies from rabbits generated by immunizing with an STX-BSA formaldehyde mediated conjugate (18). STX conjugated to BSA or polylysine was used to coat the wells, followed by simultaneous incubation with standard toxin and anti STX Abs. The amount of Ab bound to the solid phase was determined by incubation with goat anti rabbit Abs conjugated to peroxidase followed by H_2O_2 plus ABTS. The assay was found capable of detecting 2–10 pg of STX in a pure buffer solution (Figure 10.6). The detection limit for STX added to clam and mussel tissue was 50–100 ppb. There was little interference of the assay by clam tissue but mussel meat exhibited notable interference.

Davio et al. (19) conjugated STX to BSA, KLH, horse IgG, human IgG, and egg albumin with 1.4% formaldehyde. Only BSA and KLH bound significant quantities of STX and only KLH-STX induced STX binding antibodies in an ELISA assay. STX conjugated to KLH by formaldehyde with STX-BSA bound to the solid phase was used to develop a competitive indirect ELISA having a detection limit of 20 ppb (20).

Usleber et al. (21) coupled STX to HRP using a novel adaptation of the periodate reaction for the development of a direct ELISA assay that resulted in a detection limit for STX of 0.35 pg per assay and 3 ng per gram of shellfish tissue. This was about ten times more sensitive than the indirect ELISA (19) with the same antiserum due presumably to the absence of a methyl bridge resulting from formaldehyde coupling of STX to HRP. Only neo-STX showed cross reactivity (about 10% relative to STX) while several additional homologs of STX exhibited no cross reaction.

Kralovec et al. (22) precoated the wells of Maxisorp microtiter plates with BSA and obtained direct binding of STX in 0.01 M piperazine-glycylglycine buffer at pH 10.0. This yielded greater sensitivity with a competitive ELISA for detection of STX compared to binding of an STX conjugate. A protocol for the use of a direct competitive ELISA using either anti saxitoxin STX/STX-HRP or anti neo STX/STX-HRP pairs was found to detect 211 of 1450 (13.6%) naturally contaminated samples of shellfish as positive compared to 175 (11.3%) by the standard mouse bioassay (23).

A commercially available ELISA for detection of PST was found to exhibit unpredictable cross reactivities to a mixture of gonyautoxins (GTXs), and underestimated the toxicity of some naturally contaminated shellfish (24).

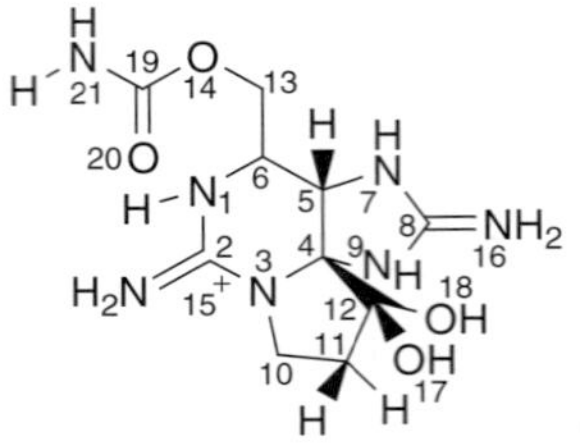

Saxitoxin

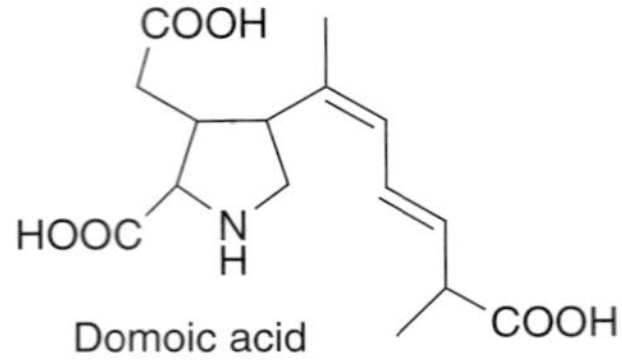

Domoic acid

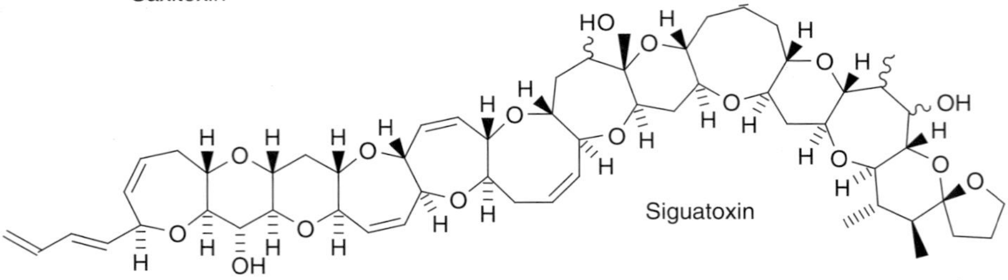

Siguatoxin

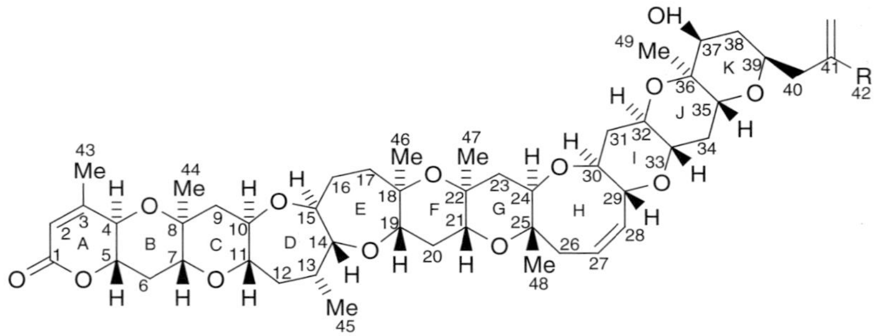

Brevitoxin type - 1

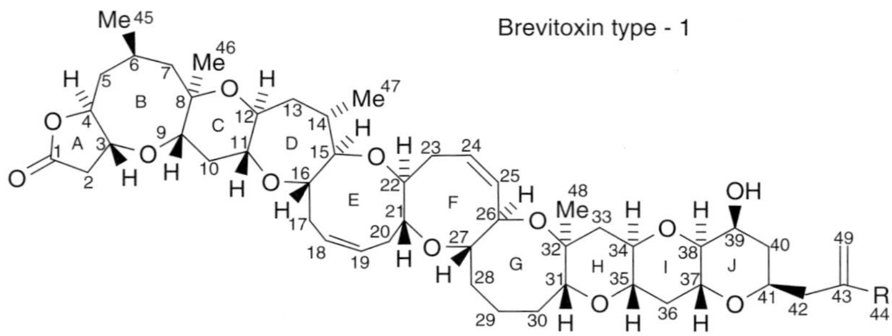

Brevitoxin type - 2

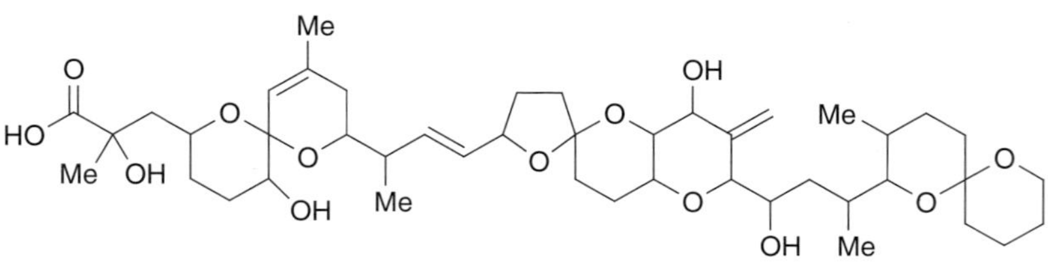

Okadaic acid

Figure 10.5 Structures of major marine food toxins.

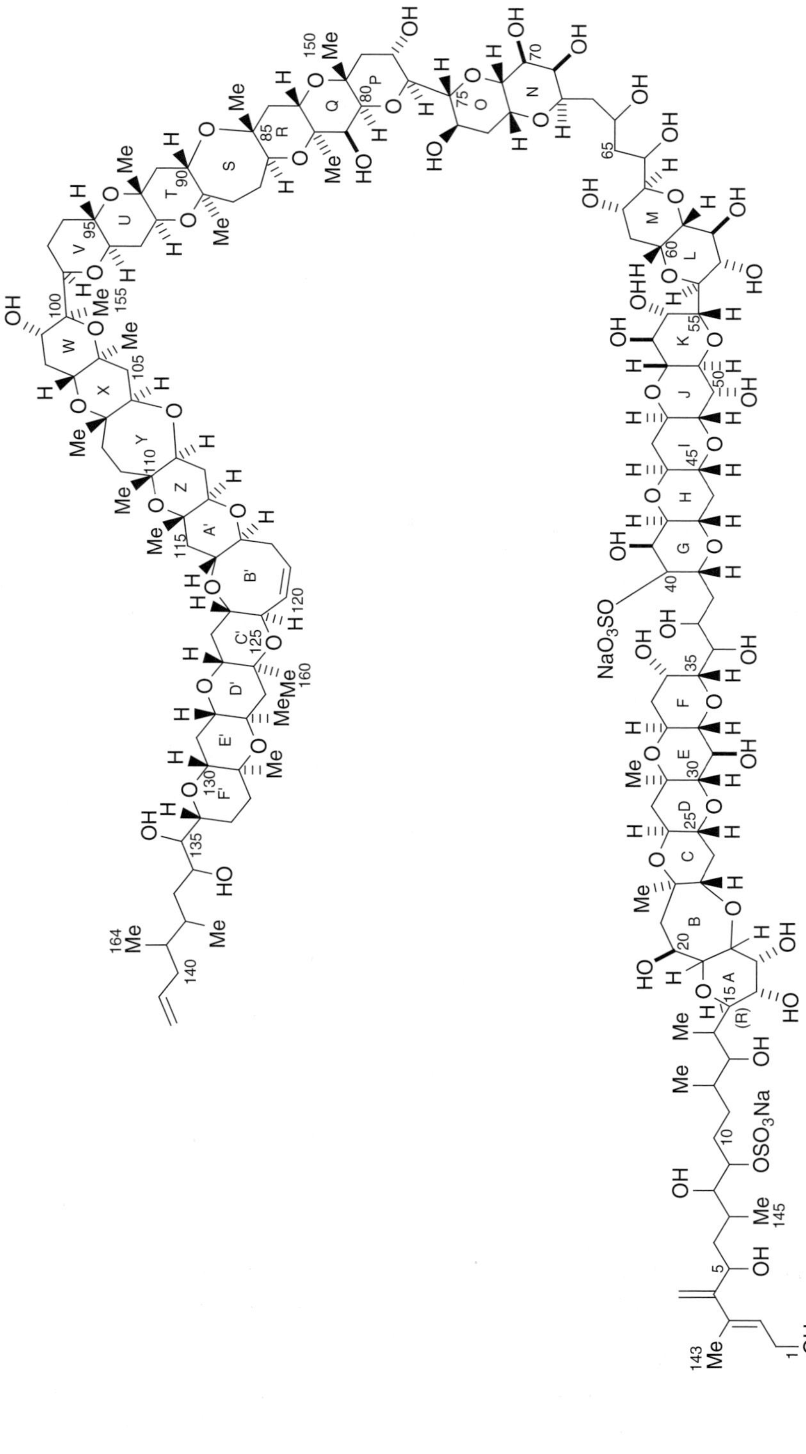

Maitotoxin

Figure 10.5 *(Continued)*

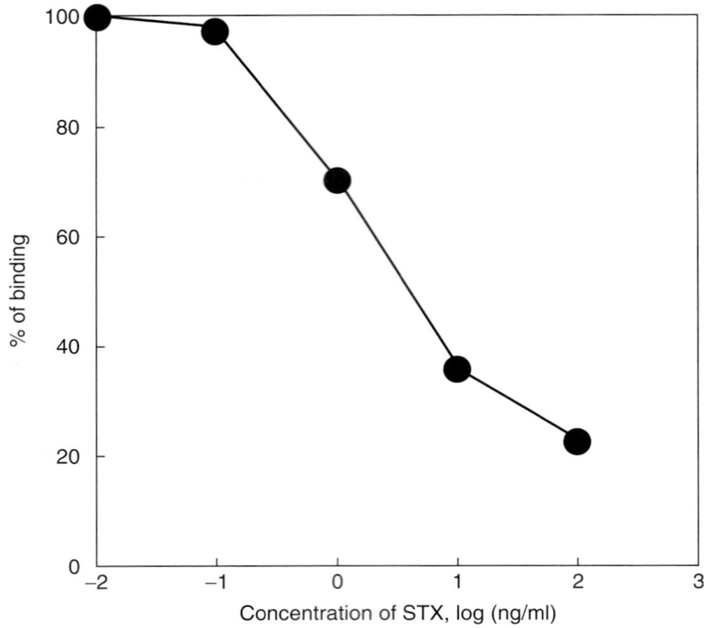

Figure 10.6 Standard curve for indirect competitive ELISA for STX. Redrawn from: Chu, F.S., T.S.L. Fan, *JAOAC* 68(1):13–16, 1985.

10.6.1.2 Ciguatoxin and Maitotoxin

Ciguatera poisoning refers to the neurological symptomology resulting from the ingestion of tropical fish contaminated with the polyether toxins ciguatoxin and maitotoxin. These fish include snappers, groupers, marine eels, jackfish, reef sharks, barracuda, sturgeon fishes, and occasionally others found in the Caribbean and tropical Pacific (25). Maitotoxin (MTX) is the predominant toxin in the tissues of such fish with ciguatoxin (CTX) occurring usually at one tenth the concentration of maitotoxin. Both toxins are produced by the dinoflagellate *Gambierdiscus toxicus*. MTX has a m. wt. of 3424 Da, contains 160 carbon atoms (26), and is the most potent of all marine toxins, having an LD_{50} for mice of 50 ng/kg body wt., I.P (27). MTX functions as a neuroexcitor by opening Ca^{++} channels causing an influx of calcium ions into neurons resulting in the release of the neuromuscular transmitter acetylcholine exciting both smooth and cardiac muscle tissue (28). Symptoms can last for many years. The major problem associated with the chemical detection and quantitation of maitotoxin is that it occurs at notably low concentrations in fish (28 μg/kg) (27) and requires initial extraction with methanol, solvent partitioning, and repetitive HPLC chromatography to achieve purification (26). Complete purification of ciguatoxin was achieved in 1989 (29).

Among a series of monoclonal antibodies derived from partially purified MTX and applied to an indirect ELISA assay, some reacted with various pigments present in the MTX extracts while others reacted in direct proportion to measured toxicity (30).

Mouse polyclonal antibodies derived from a CTX-1B fragment conjugated to BSA and OVA in indirect and direct competitive ELISA assays yielded similar antibody titers and a detection limit of 7 picomoles for CTX-1B in crude lipid extracts of fish tissue (31). Resulting antibodies reacted with CTX-1B but did not significantly cross react with brevitoxin-3 (PbTx-3) or other related polyether structures. Conjugation of the N-hydroxysuccinimide ester of the carboxylic acid CTX-1B fragment to the protein carriers was achieved using a standard carbodiimide condensation procedure.

10.6.1.3 Brevitoxins

The marine dinoflagellate *Ptychodiscus brevis* (formerly *Gymnodinium breve*) is responsible for production of a family of at least 10 polyether toxins referred to as brevitoxins (PbTx-1 to PbTx-910) (32–34). These toxins are grouped into two subgroups (type 1 and type 2) based on backbone structure. The organism is responsible for production of toxic red tides along the Gulf of Mexico and Texas (35) resulting in mass mortality of exposed fish. Consumption of toxic shellfish results in neurotoxic shellfish poisoning. Algal blooms along the shore have been found to produce an aerosol irritating to the respiratory tract (33). PbTx-3 functions by opening sodium channels resulting in depolarization of the axon membrane (36).

One of the earliest ELISA assays for brevitoxin was developed by Trainer and Baden (37). PbTx-3 was purified by HPLC and a carbodiimide derivative of PbTx-3 was prepared and conjugated to urease and to HRP for competitive ELISA assays. With the urease conjugate, the pH indicator bromcresol purple was used for observation of the color change from yellow to purple resulting from the release of ammonia by urease activity. Both systems yielded linear responses for quantitation with an operational range of 1 pmole to 10 nmoles of PbTx-3 per well. The urease conjugate however lacked enzyme stability and was abandoned in favor of the HRP conjugate.

Naar et al. (38) converted a minute amount of PbTx-3 (400 μg) to an hemisuccinate derivative and then conjugated to BSA and OVA in a reversed micellar medium for Ab production in mice. In competitive ELISA assays, both polyclonal and monoclonal Abs exhibited strong cross reactivity to other PbTx-2-type toxins (PbTx-2 and 9) but low or moderate cross reactivity to a PbTx-1-type toxin (PbTx-1). Monoclonal antibodies exhibited a low cross reactivity with okadaic acid but no significant cross reactivity was observed with two ciguatoxins (CTX-1B and CTX-3C). The monoclonal based assay resulted in a detection limit of 5 ng per well, with a working range of 8–150 ng per well. The assay appears to be a suitable alternative to the mouse bioassay for routine shellfish monitoring.

10.6.1.4 Domoic Acid

Domoic acid (DA) is the causative agent of amnesic shellfish poisoning and functions as a neuroexcitatory analog of glutamic acid causing lesions and neuronal necrosis in the hippocampus portion of the brain (39–41). Persistent short term memory loss is a frequent symptom. The pennate diatoms *Nitzschia pungens* (42) and *Nitzschia pseudodelicatissima* (43) have been found responsible for formation of domoic acid. A direct ELISA assay for domoic acid was developed by Kitts and Smith (44). Lower limits of detection in human plasma, urine, and milk were 0.25 mg/ml, 0.2 mg/ml, and 10 mg/ml respectively. The assay was also applicable for detection of domoic acid in shellfish tissue.

Garthwaite et al. (45) raised ovine antibodies against conjugates linked through the secondary amino group of domoic acid together with activated ester derived conjugates of domoic acid as a plate coater, for development of an indirect competitive ELISA. The assay had a working range for quantitation of 0.15–15 ng/ml of super(-1) DA, the lower limit being 500 times lower than the maximum permitted level in shellfish tissue in New Zealand.

10.6.1.5 Okadaic Acid

Okadaic acid (OA) and its seven homologs (46) are the causative agents of diarrheic shellfish poisoning (DSP) and are produced by dinoflagellated algae such as *Dinophysis* spp.

and *Prorocentrum* spp. An indirect competitive ELISA with electrochemical detection was developed for okadaic acid by Ramsay et al. (47). Soluble sample antigen competed with immobilized antigen for soluble antibody, followed by the addition of a secondary antibody labeled with alkaline phosphatase. The enzyme label hydrolyzed phenyl phosphate to yield phenol which was oxidized at +870 mV vs. Ag/AgCl and yielded a response inversely proportional to the soluble antigen concentration.

Total OA (OA plus methylOA) in clones of several dinoflagellates was determined by Morton and Tindall (48) using HPLC-fluorescence, the UBE ELISA test kit, and the Rougier Bio-Tech ELISA test kit, both of which involve use of monoclonal Abs. All three methods yielded consistent results for *P. hoffmannianum* which produces only OA. Results of the three methods however, were not consistent for *P. lima* which produces both OA and methylOA. The UBE ELISA demonstrated little or no cross reactivity with methyOA; whereas the Rougier ELISA demonstrated varying degrees of cross reactivity with that analog. Nonokadaic acid producing species yielded negative results. The authors concluded that both ELISA kits may underestimate total okadaic acid present in shellfish.

Nunez and Scoging (49) found that HPLC, a colorimetric protein phosphatase inhibition assay, and the mouse bioassay correlated well for detection of OA. In contrast, a commercial ELISA kit failed to accurately and consistently detect low levels of OA from the hepatopancreas of shellfish.

Lawrence et al. (50) using an indirect competitive ELISA found that a mAb detected with equal sensitivity OA, dinophysistoxin-4 (DTX-4) and dinophysistoxin-5 (DTX-5), and an OA diol ester.

Chin et al. (46) studied the affinity of the anti OA Ab from the Rugier test kit for OA and several of its homologs. Dinophysistoxins–2 and –1 could be detected by the assay but at concentration ranges 10- and 20-fold higher, respectively, than for OA. Dinophysistoxin-3, calyculin A, and brevitoxin-1 could not be detected. The working range of the assay for OA was 5 pg to 0.5 ng per well.

Usagawa et al. (51) developed an indirect competitive ELISA using mAb to OA derived by conjugating OA to ovalbumin for mouse injection. The assay involved binding OA-BSA to the surface of wells, followed by the simultaneous addition of AO containing samples and commercial alkaline phosphatase conjugated to rabbit anti mouse IgG+IgA+IgM (H+L) prior to myeloma cell fusion. Three mAbs were obtained against OA and were designated OA-1, OA-2, and OA-3. All three reacted with dinophysistotoxin-1 (DTX1) to the same extent as with OA , and failed to react with 7-*O*-acyl-OA but reacted slightly with Yessotoxin. Extraction of OA from shellfish tissue with higher than 45% methanol resulted in significant inhibition of the ELISA. Sensitivity was 0.25 ng per assay well.

Vale and Sampayo (52) found that a commercial ELISA test kit based on the methodology described by Usagawa et al. (50) was more sensitive, specific and faster than HPLC for determination of OA in shellfish tissue. No problems were encountered when using hydrolyzed semi purified tissue extracts to detect esters of OA. Sensitivity for OA was 0.25 ng per assay well. The assay kit also detected dinophysistotoxin-2 (DTX2) at a level of sensitivity about 50% below that for OA.

10.6.1.6 Seafood Allergens

Shiomi et al. (53) used an ELISA assay to asses the ability of sera from five fish sensitive patients to react with partially purified extracts from nine fish species. Allergenicity was found to vary among the fish species and individual patients. The major allergens in Japanese eel and bigeye tuna were found to be parvalbumins and higher molecular weight substances.

Ishikawa et al. (54) compared the allergens from fish species of shellfish using a competitive ELISA assay. Cross reactivity was recognized between shellfish allergens and the oyster allergen Cra g 1, suggesting that, irrespective of species, the major allergen in shellfishes is tropomyosin.

10.6.2 Fungal Toxins (Figure 10.7)

10.6.2.1 Aflatoxins

Aflatoxins are carcinogenic toxins that fluoresce brightly under long wavelength ultraviolet light. There are four major toxins in this group designated B1, B2, (fluoresce blue) and G1, G2 (fluoresce green yellow). B1 is the most potent carcinogen and mutagen of the group.

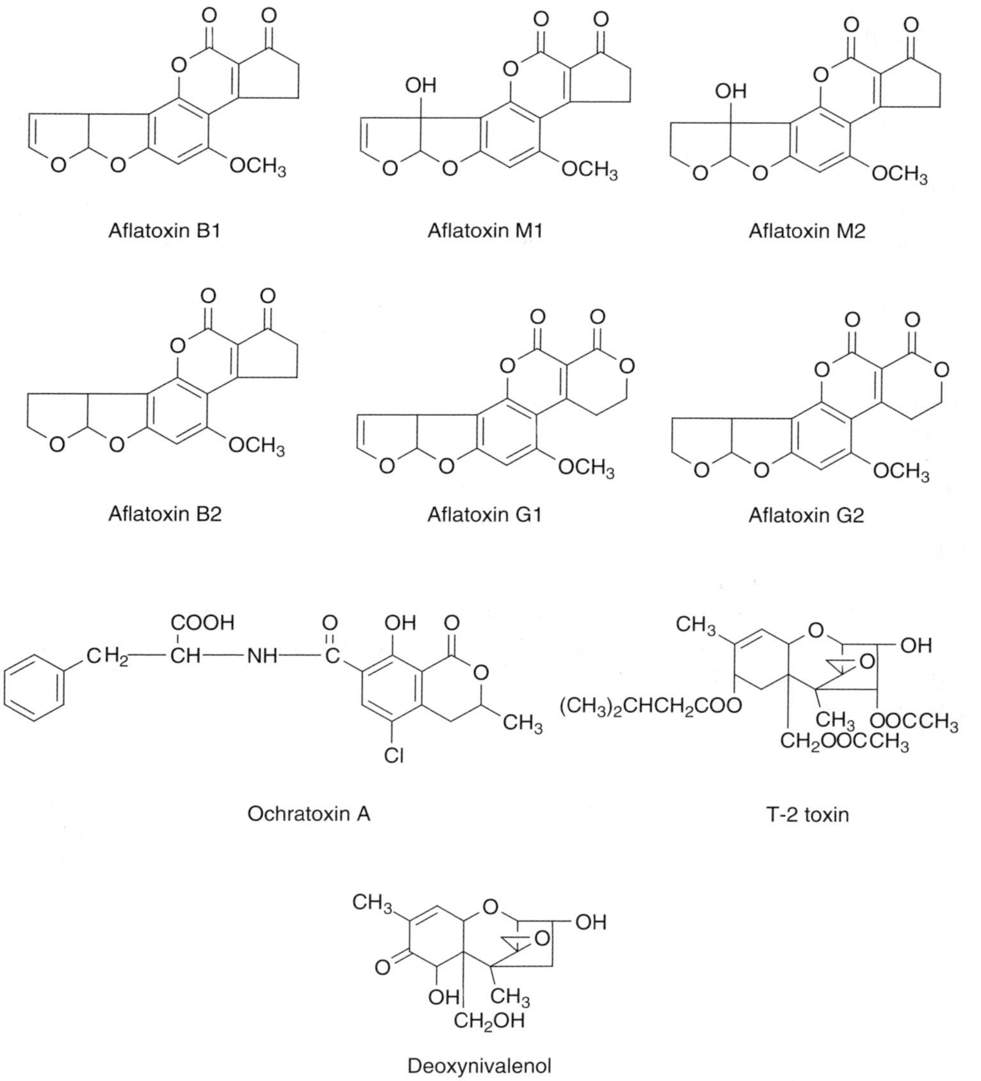

Figure 10.7 Structures of major fungal toxins.

M1 and M2 are produced by bovine tissue from B1 and can be found in milk. B2a and G2a are precursors. *Aspergillus flavus* and *Aspergillus parasiticus* are the major aflatoxin producing molds. These molds are typically green due to the color of their conidiospores. The attack of peanuts by these organisms prior to harvest as a result of rainfall and warm weather frequently occurs and may continue beyond the harvest period if sufficient moisture is present during storage. Peanut butter frequently contains detectable levels of aflatoxins. Corn, wheat, and other grains are also subject to attack by these molds with their B1 content being readily measured by ELISA (55). The limit in the U.S. for aflatoxins is 20 ppb in food, feeds, and nuts, and 0.5 ppb for milk (56).

Lawellin et al. (57) developed the first reported ELISA assay for aflatoxin B1. The assay was found capable of measuring between 100 to 1 pg of B1. Antiserum to B1-BSD conjugate was first bound to the surface of polystyrene tubes (16 x 125 mm). Samples containing free B1 were added and the B1 was allowed to bind. After rinsing, B1 conjugated to HRP was added, followed by rinsing and the addition of orthotolidine as substrate for the bound HRP. B2 resulted in complete inhibition, G1 and M1 resulted in 50% and 7.5% inhibition respectively.

Burkin et al. (58) used an indirect ELISA involving immobilization of a B1-BSA conjugate and polyclonal antibodies and obtained low relative cross reactivity against B2, G1, G2, M1, B2a, G2a and sterigmatocystin with a sensitivity of 0.04 ng per well or 4.0 ng/ml of sample.

Several interlaboratory collaborative studies have led to the acceptance of ELISA as an official method of quantitation of aflatoxins by the AOAC (59–61) and the development of commercial test kits. The development of monoclonal antibodies for aflatoxins has further increased the sensitivity of ELISA assays (62).

Devi et al. (63) reported that hybridomas that secreted antibodies for aflatoxin B1 were selected using two immunization protocols referred to as A and B. Protocol A is a standard immunization method and resulted in the selection of only two clones that produced monoclonal antibodies against aflatoxin B12. In protocol B, a unique immunization schedule was used which resulted in the generation of 10 hybridomas. Of the 10, one antibody was highly specific for B1, four antibodies reacted equally strongly with B1, and G1 but weakly with B2. Another four reacted strongly with B1 and weakly with B2 and G1. One clone reacted equally strongly with B1, G1, and B2. All of the 10 antibodies exhibited little or no cross reactivity with G2.

Monoclonal Ab (mAb) was found to be more accurate than pAb for determination of free and adducted B1 in urine samples (64). The ELISA had a sensitivity of 1 pg/ml compared to HPLC sensitivity of 0.1 ng/ml with a fluorescence detector and 4.5 ng with a UV detector.

Kim et al. (65) reported on the occurrence of M1 in Korean pasteurized milk and dairy products with the use of a direct competitive ELISA assay and HPLC. The limit of detection with ELISA was 2 pg/ml and with HPLC was 10 pg/ml. Among a total of 180 samples, the incidence of M1 in pasteurized milk, infant formula, powdered milk and yogurt was 76, 85, 75, and 83%, respectively, with mean concentrations of 18, 46, 200, and 29 pg/g, respectively.

An extremely unique approach was used by Devi et al. (66) for eliminating the requirement of a B1-BSA conjugate in an indirect competitive ELISA assay for B1 utilizing random peptide libraries displayed on bacteriophage. Peptides were fused to a coat protein of a suitable phage which also carried the encoding DNA for the peptide. Phage were then selected possessing a peptide capable of binding to mAb directed against B1. For two of the three mAbs tested, phage clones were isolated that bound specifically to mAb 24 (specific for B1 and G1) and mAb 13 (specific for B1, G1, and B2). The third mAb failed to react with any of the phage displayed peptides. Phage borne peptides that mimic antigen in binding antibodies are termed mimotopes. Their indirect competitive ELISA assay consisted of coating wells with phage harboring the requisite B1 mimotope. After washing and blocking, the wells were incubated simultaneously with mAb (to B1) plus various concentrations of B1 (representing samples). Following incubation and washing, goat anti mouse Ab conjugated to alkaline phosphatase was added, followed by incubation, washing and the addition of p-nitrophenyl phosphate as substrate.

Yong and Cousin (67) developed an ELISA assay to detect molds producing aflatoxins in maize and peanuts using an antibody to extracellular antigen from *A. parasiticus*. This antibody recognized species with phenotypic similarities to *A. parasitucus, A. flavus, A. sojae,* and *A. oryza.* For naturally contaminated maize samples, low and high levels of aflatoxin corresponded with low and high ELISA readings for mold antigens. *A. parasiticus* could be detected before the production of aflatoxins. The assay appears suitable for early detection of responsible molds before detectable aflatoxin is produced.

10.6.2.2 Ochratoxin

Ochratoxin is a toxic metabolite of certain strains of *Aspergillus ochraceu* and *Penicillium viridicatum* found on grains, legumes, peanuts, and green coffee, causing enteritis, renal necrosis (also known as Balkan nephropathy), and an increase in liver glycogen with an LD_{50} in rats of 20 mg/kg (68). The toxin is also carcinogenic.

A flow through membrane based direct competitive ELISA for the rapid detection of ochratoxin A in wheat was developed by De Saeger and Van Peteghem (69). An immunodyne 0.45 μ ABC membrane was first coated with rabbit anti mouse Abs and placed onto an absorbent pad, the membrane was blocked, followed by a competitive ELISA involving the addition of ochratoxin-mAb, washing, addition of ochratoxin (a standard solution or sample extract) plus ochratoxin A-HRP conjugate, washing, and the addition of HRP substrates. A portable colorimeter was used to confirm and quantify the color intensity. A wheat sample spiked with 4 μg/kg resulted in complete color suppression. Passage of 600 μL of sample extract through the membrane facilitated maximum binding of all ochratoxin present to the membrane in addition to the advantage of greatly reducing reaction times by bringing reactants into intimate contact in the matrix of the membrane by elimination of the dilution factor involved with reaction volumes. With coated membranes, the assay could be completed in less than 15 min. Sibanda et al. (70) subsequently applied this flow through ELISA assay to the detection of ochratoxin in samples of green coffee beens. The assay had a sensitivity of 8 ng/g and correlated with HPLC results.

Tirulamala-Devi et al. (71) developed an indirect competitive ELISA for ochratoxin A using pAbs capable of detecting 0.1 ng/ml or 0.06 ng of ochratoxin A per assay. The Abs did not react with ochratoxin B, coumarin, 4-hydroxycoumarin, L-phenylalanine, or aflatoxin B1.

10.6.2.3 Trichothecenes

Trichothecenes constitute a group of potent toxins that are esters of sesquiterpene alcohols having in common a trichothecene ring structure and an epoxide group. The trichothecenes are divided into three groups: the type A trichothecenes, which includes T-2 toxin, the type B trichothecenes, which includes deoxynivalenol, and the type C or macrocyclic trichothecenes, also known as satratoxins. Trichothecenes are responsible for severe outbreaks of mycotoxicosis, including alimentary toxic aleukia (ATA) and stachybotryotoxicosis in Russia, red mold poisoning in Japan and moldy corn toxicosis in the U.S. (72). Among the several dozen trichothecenes isolated to date, T-2 toxin, and deoxynivalenol are most frequently associated with human illness. Trichothecenes are produced by a number of fungal species and genera (73). The most important T-2 producer is *Fusarium tricinctum*. All members of the genus *Fusarium* produce characteristic sickle shaped conidiospores. T-2 is toxic to certain plants, fungi, brine shrimp insects, fish, and mammals. The i.p. LD_{50} in mice is 5.2 mg/kg (73). Toxic effects of T-2 result from inhibition of protein synthesis and with mammals include dermal necrosis, apoptosis, inflammation, and hemorrhaging of the gastrointestinal tract, edema, leucopenia, degeneration of the bone marrow, and death.

Attempts to produce mAbs to T-2 involving conjugating T-2 through the functional hydroxyl group at the C-3 position have been found to result in low levels of immunogenicity and low affinities for T-2 (74). The low level of immunogenicity was found to be due to the release of T-2 from the protein carrier with resulted toxicity to lymphoid cells. When T-2 was coupled at the C-8 position and the hemisuccinate of 3-acetyl-neosolaniol used for conjugation to BSA mAb the resulting conjugate was highly immunogenic and yielded mAb that was cross reactive with most group A trichothecenes (75). An indirect ELISA allowed the detection of 0.05 ng of T-2 (76).

Gendloff et al. (77) used the method of Chu et al. (78) for conjugating T-2 to BSA in the development of a competitive ELISA. The assay was used to screen for T-2 in *Fusarium sporotrichioides* infected corn. The assay detected T-2 in diluted methanol extracts of corn samples at a concentration of 0.05 ng/ml (detection limit of 1 pg per assay). HPLC and ELISA estimates were similar.

Nagayama et al. (79) developed an indirect mAb ELISA assay for the screening of T-2 toxin producing species of *Fusarium*. The minimum detection limit was 5 pg per assay.

10.6.3 Natural Plant Toxins

10.6.3.1 Ricin

Ricin is an extremely lethal heat labile toxin derived from the castor bean, and consists of two polypeptide chains designated A and B connected by a disulfide bridge. It is an irreversible inhibitor of protein synthesis. Both polypeptides have molecular weights of about 30,000 Da (80). The B chain, a lectin, with binding affinity for galactose, binds to the mammalian cell wall allowing entry of the A chain into the cell. The A chain is destructive to eukaryotic ribosomes by functioning as an N-glycohydrolase, which inactivates ribosomes by an enzymatic deadenylation of rRNA (81). A single molecule is considered sufficient to kill a mammalian cell and accounts for the notably low mouse MLD (i.p) of 0.001 μg of ricin D nitrogen per gram of body wt. In recent years, the A-chain coupled to antibodies has been used experimentally to target *in vivo* tumor cells. The tumor specificity is provided by the antibody of choice, while the cellular entry of the A-chain provides selective destruction of the tumor cells. The first ELISA assay for ricin was a sandwich method with a sensitivity of 6.8 ng/ml in buffer and was about tenfold less sensitive (60 ng/ml) if the ricin was assayed in rabbit body fluids (81). An ELISA for ricin, having a sensitivity of about 10 ng/ml, utilizing affinity purified rabbit antiserum, was developed for ricin assay in rat tissue after ricin injection (82). Later ELISA assays were reported with sensitivities of 25 pg/ml in plasma (83) and 200 pg/ml of tissue extracts (84). Poli et al. (85) developed an avidin/biotin alkaline phosphatase ELISA for ricin based on affinity purified goat antibodies. The minimum detection level was 1 ng/ml (100 pg/well). In this same study, the authors developed a chemiluminescence assay by replacing the substrate p-nitrophenyl-phosphate with Lumiphos-530 and were able to accurately quantitate the presence of 100 pg/ml (10 pg/well) of ricin. Lipps (86) injected mice with one half the predetermined lethal dose of ricin (1 µg), four times at two week intervals. An ELISA assay was used to assess the potency of the antitoxin. Because of the high antibody titers produced after three injections of this minute dose, the author concluded that ricin is a super antigen.

10.7 SUMMARY

ELISA assays for detection and quantititation of a variety of toxins in foods represents a reliable and highly sensitive technique that usually correlates closely with HPLC analysis. The use of mAbs affords an additional level of specificity and sensitivity to complement the use of pAbs, particularly when various homologs of a given toxin may be present. The present state of the immunological industry, and its wide range of products, serves to greatly facilitate the continued development and application of additional ELISA systems for the detection and quantitation of toxins in foods.

REFERENCES

1. Zolla, H. *Monoclonal Antibodies: a Manual of Techniques*. Boca Raton, FL: CRC Press, 1987, p 214.
2. Harlow, E., D. Lane. *Antibodies: a Laboratory Manual*. Cold Spring Harbor, NY: Cold Spring Harbor Laboratory Press, 1988, p 726.
3. E. Ishikawa. Labeling of antibodies and antigens. In: *Immunoassay,* Diamandis, E.P., T.K. Christopoulos, eds., New York: Academic Press, 1996, pp 191–204.
4. Bos, E.S., A.A. van der Doelen, N. van Rooy, A. Schuurs. 3,3',5,5'-tetramethylbenzidine as an Ames test negative chromogen for horse-radish peroxidase in enzyme immunoassay. *J. Immunoassay* 2(3,4):187–204, 1981.
5. Avrameas, S. Coupling of enzymes to proteins with gluteraldehyde. Use of conjugate for detection of antigens and antibodies. *Immunochemistry* 6:43–52, 1969.
6. Avrameas, S., T. Ternyck. Peroxidase labeled antibody and Fab conjugates with enhanced intracellular penetration. *Immunochemistry* 8:1175–1179, 1971.
7. Van Weemen, B.K., A. Schuurs. Immunoassay using antibody-enzyme conjugates. *FEBS Lett.* 43:215–218, 1974.
8. Nakane, P.K., A. Kawaoi. Peroxidase labeled antibody: a new method of conjugation. *J. Histochem. Cytochem.* 22:1084–1091, 1974.
9. Butler, J.E. Solid phases in immunoassay. In: *Immunoassay*, Diamandis, E.F., T.K. Christopoulos, eds., New York: Academic Press, 1996, pp 191–204.
10. Ternyck, T., S. Avrameas. Avidin-biotin system in enzyme immunoassays. *Methods Enzymol.* 184:469–481, 1990.
11. Bayer, E.A., M. Wilchek. The avidin-biotin system. In: *Immunoassay,* Diamandis, E.P., T.K. Christopoulos, eds., New York: Academic Press, 1996, pp 237–267.
12. Wilchek, M., E.A. Bayer. Avidin-biotin technology. *Methods Enzymol.* 184:746, 1990.
13. Savage, D., G. Mattson, S. Desai, G. Nielander, S. Morgensen, E. Conklin. *Avidin-Biotin Chemistry: A handbook.* Rockford, IL: Pierce Chemical Co. 1992, p 467.
14. Bayer, E.A., M. Wilchek. Avidin-biotin technology. In: *Immunochemical Protocols: Methods in Molecular Biology, Vol. 10.,* Manson, M.M. ed., Clifton, NJ: Humana Press, 1992, pp 137–148.
15. Bayer, E.A., M. Wilchek. Protein biotinylation. *Methods Enzymol.* 184:138–160, 1990.
16. Hall, S., G. Strichartz, E. Moczydlowski, A. Raavindran, P. B. Reichardt. The saxitoxins: sources, chemistry, and pharmacology. In: *Marine Toxins: Origin, Structure, and Molecular Pharmacology,* S. Hall and G. Strichartz, eds., Washington DC: American Chemical Society, 1990, pp 29–65.
17. Schantz, E.J. Phycotoxins from dinoflagellates. *Pure Appl. Chem.* 52:183–188, 1980.

18. Chu, F.S., T.S.L. Fan. Indirect enzyme-linked immunosorbent assay for saxitoxin in shellfish. *JAOAC* 68(1):13–16, 1985.
19. Davio, S.R., J.F. Hewetson, J.E. Beheler. Progress toward development of monoclonal antibodies to saxitoxin; antigen preparation and antibody detection. In: *Toxic Dinoflagellates*, Anderson, D.M., A.W. White, D.G. Baden, eds., 3rd. Int. Conf. On *Toxic Dinoflagellates*, St. Andrews, N.B., Canada, June 8–12, 1985, pp 343–348.
20. Renz, V., G. Terplan. Ein enzymimmunologischer Nachweis von saxitoxin. *Archiv. Fur Lebensmittelhyg* 39(2):30–33, 1988.
21. Usleber, E., E. Schneider, G. Terplan. Direct enzyme immunoassay in microtitration plate and test strip format for the detection of saxitoxin in shellfish. *Lett. Appl. Microbiol.* 13:275–277, 1991.
22. Kralovec, J.A., V. Laycock, R. Richards, E. Usleber. Immobilization of small molecules on solid matrices: a novel approach to enzyme-linked immunosorbent assay screening for saxitoxin and evaluation of anti-saxitoxin antibodies. *Toxicon* 34(10):1127–1140, 1996.
23. Chu, F.S., K. Hsu, X. Huang, R. Barrett, C. Alison. Screening of paralytic shellfish poisoning toxins in naturally occurring samples with three different direct competitive enzyme-linked immunosorbent assays. *J. Agric. Food Chem.* 44(12):4043–4047, 1996.
24. F. Kasuga, Y. Hara-Kudo, K. Machii. Evaluation of enzyme-linked immunosorbent assay (ELISA) kit for paralytic shellfish poisoning toxins. *J. Food Hyg. Soc. Japan* 37(6):407–410, 1996.
25. Li, K. Fish poisoning in the far east. *Far East Med. J.* 1:29–33, 1965.
26. Yasumoto, T., M. Murata. Polyether toxins involved in seafood poisoning. In: *Marine Toxins: Origin, Structure, and Molecular Pharmacology*, Hall, S., G. Strichartz, eds., Washington, DC: American Chemical Society, 1990, pp 120–132.
27. Yasumoto, T., M.S. Satake. Chemistry, etiology, and determination of ciguatera toxins. *J. Toxicol. Toxin Rev.* 15(2):91–107, 1996.
28. Ohizumi, Y., M. Kobayashi. Ca-dependant excitatory effects of maitotoxin on smooth and cardiac muscle. In: *Marine Toxins*. Hall, S., G. Strichartz, eds., Washington, DC: American Chemical Society, 1990, pp 133–143.
29. Murata, M., A. Legrand, Y. Ishinashi, T. Yasumoto. Structures of ciguatoxin and its congener. *J. Am. Chem. Soc.* 111:8929–8931, 1990.
30. Lundstrom, R.C., C. Martin, D.A. Adams, D.J. Rhoades, L.V. Sick. Progress in the development of monoclonal antibodies to semi-purified maitotoxin extracts from *Gambierdiscus toxicus*, *Proc. 12th Ann. Conf. Of the Tropical and Subtropical Fisheries Technol. Society of the Americas, Nov. 1987*, p 678.
31. Pauillac, S., M. Sasaki, J. Naar, M. Inoue, P. Branaa, P. Cruchet, M. Chinain, A.M. Legrand. Immunochemical methods for ciguatoxins detection in Pacific herbivorous and carnivorous fish, *Proc. 5th Indo-Pacific Fish Conference, Nouma (New Caledonia), Nov. 3–8, 1997*, pp 759–773.
32. Trainer, V.L., R.A. Edwards, A.M. Szmant, A.M. Stuart, T.J. Mende, D.G. Baden. Brevitoxins: unique activators of voltage-sensitive sodium channels. In: *Marine Toxins: Origin, Structure, and Molecular Pharmacology,* S. Hall and G. Strichartz, eds., Washington DC: American Chemical Society, 1990, pp 166–175.
33. Pierce, R.H., M.S. Henry, L.S. Proffitt, P.A. Hasbrouck. Red tide toxin (brevitoxin) enrichment in marine aerosol. In: *Toxic Marine Phytoplankton*, Granelli, E., B. Sundström, D.M. Anderson., eds., New York: Elsevier, 1990, pp 397–402.
34. Baden, D.G. Brevetoxins: unique polyether dinoflagellate toxins. *FASEB J.* 3(7):1807–1819, 1989.
35. Steidinger, K.A., D.G. Baden. Toxic Marine *Dinoflagellates*. In: *Dinoflagellates,* Specter, D. ed., New York: Academic Press, 1984, pp 201–261.
36. James, M., C. Huang, C.H. Wu. Pharmacological actions of brevitoxin from *Ptychodiscus brevis* on nerve membranes. In: *Red Tides: Biology, Environmental Science, and Toxicology,* Okaichi, T., D.M. Anderson, T. Nemoto, eds. *Proc. 1st Intl. Symp. On Red Tides, Nov. 10–14, 1987, in Takamatsu, Kagawa Prefecture, Japan,* New York: Elsevier, pp 379–382.

37. Trainer, V.L., D.G. Baden. Enzyme immunoassay of brevetoxins. In: *Toxic Marine Phytoplankton,* Granelli, E., B. Sundström, D.M. Anderson, eds., New York: Elsevier, 1990, pp 430–435.
38. Naar, J., P. Brannaa, M.-Y. Bottein-Dechraoui, M. Chinain, S. Pauillac. Polyclonal and monoclonal antibodies to PbTx-2-type brevetoxins using minute amounts of hapten-protein conjugates in a reversed micellar medium. *Toxicon* 39(6):869–878, 2001.
39. Novelli, A., J. Kispert, M.T. Fernandez-Sanchez, A. Torreblanca, V. Zitco. Domoic acid-containing toxic mussels produce neurotoxicity in neuronal cultures through a synergism between excitatory amino acids. *Brain Res.* 577:41–48, 1992.
40. Sutherland, R.J., J.M. Hoesing, I.Q. Whishaw. Domoic acid, an environmental toxin, produces hippocampal damage and severe memory impairment. *Neurosci. Lett.* 120:221–223, 1990.
41. Teitelbaum, J.S., R.J. Zatorre, S. Carpenter, D. Gendron, A.C. Evans, A. Gjede, N.R. Cashman. Neurological sequlae of domoic acid intoxication due to the ingestion of contaminated mussels. *N. Eng. J. Med.* 322:1781–1787, 1990.
42. Bates, S.S., C.J. Bird, A.S.W. de Freitas, R. Foxall, M. Gilgan, L.A. Hanic, G.D.R. Johnson, A.W. McCulloch, P. Odense, R. Poclington, M.A. Quilliam, P.G. Sim, J.C. Smith, D.V.S. Rao, E.C.D. Todd, J.A. Walter, J.L.C. Wright. Pennate diatom *Nitzschia pungens* as the primary source of domoic acid toxin in shellfish from Eastern Prince Edward Island, Canada. *Can. J. Fish Aquat. Sci.* 46:1203–1215, 1989.
43. Martin, J.L., K. Haya, L.E. Burridge, D.J. Wildish. *Nizchia pseudodelicatissima*: a source of domoic acid in the Bay of Fundy, eastern Canada. *Mar. Ecol. Prog. Ser.* 67:177–182, 1990.
44. Kitts, D.D., D.S. Smith. A serological method for the analysis of domoic acid in shellfish extracts and biological fluids. In: *Seafood Safety, Processing, and Biotechnology*, Shahidi, F., Y. Jones, D.D. Kitts, eds., Lancaster, PA: Technomic Publ. Co. Inc., 1997, pp 17–23.
45. Garthwaite, J., K.M. Ross, C.O. Miles, R.P. Hansen, D. Foster, A.L. Wilkins, N.R. Towers. Polyclonal antibodies to domoic acid and their use in immunoassays for domoic acid in seawater and shellfish. *Nat. Toxins* 6(3,4):93–104, 1998.
46. Chin, J.D., M.A. Quilliam, J.M. Fremy, S.K. Mohapatra, H.M. Sikorska. Screening for okadaic acid by immunoassay. *JAOAC* 78(2):508–513, 1995.
47. Ramsay, G., G.J. Ubrano, L.L. Nordyke, G.G. Guilbault. *Electrod Probes for the Rapid Assay of Seafood Toxicants: Phase 1 Report*. Metaire, LA: Universal Sensors, 1990, p 29.
48. Morton, S.L., D.R. Tindall. Determination of okadaic acid content of dinoflagellate cells: a comparison of the HPLC-fluorescent method and two monoclonal antibody ELISA test kits. *Toxicon* 34(8):947–954, 1996.
49. Nunez, P.E., A.C. Scoging. Comparison of a protein phosphatase inhibition assay, HPLC assay and enzyme-linked immunosorbent assay with the mouse bioassay for the detection of diarrhetic shellfish poisoning toxins of European shellfish. *Int. J. Food Microbiol.* 36(1):39–48, 1997.
50. Lawrence, J.E., A.D. Cembella, N.W. Ross, J.L.C. Wright. Cross-reactivity of an antiokadaic acid antibody to dinophsistoxin-4 (DTX-4), dinophsistoxin-5 (DTX-5), and an okadaic acid diol ester. *Toxicon* 36(8):1193–1196, 1998.
51. Usagawa, T., M. Nishimuara, Y. Itoh, T. Uda, T. Yasumoto. Preparation of monoclonal antibodies against okadaic acid prepared from the sponge *Halichondria okadai. Toxicon* 27(12):1323–1330, 1989.
52. Vale, P., M. Sampayo. Comparison between HPLC and a commercial immunoassay kit for detection of okadaic acid and esters in Portuguese bivalves. *Toxicon* 37(11):1565–1577, 1999.
53. Shiomi, K., Y. Hamada, K. Sekiguchi, K. Shimakura, Y. Nagashima. Two classes of allergens, parvalbumins and higher molecular weight substances, in Japanese eel and begeye tuna. *Fisheries Sci.* 65(6):943–948, 1999.
54. Isahikawa, M., Y. Nagashima, S. Kazuo. Immunological comparison of shellfish allergens by an enzyme-linked immunosorbent assay. *Fisheries Sci. Tokyo* 65(4):592–595, 1999.
55. El-Nakib, O., J.J. Pestka, F.S. Chu. Determination of aflatoxin B1 in corn, wheat and peanut butter by enzyme-linked immunosorbent assay and solid phase radioimmunoassay. *JAOAC* 64(5):1077–1082, 1981.

56. Labuza, T.P. Regulation of mycotoxins in food. *J. Food Prot.* 46:260–265, 1983.
57. Lawellin, D.W., D.W. Grant, B.K. Joyce. Enzyme-linked immunosorbent analysis for aflatoxin B1. *Appl. Environ. Microbiol.* 34:94–96, 1977.
58. Burkin, A.A., G.P. Konenko, N.A. Soboleva, E.V. Zotova. Immunoenzyme test system for detection of aflatoxin B1. *Prikladnaia Biokhimiia I Mikrobiologiia* 36(1):93–97, 2000.
59. Mortimer, D.N., M.J. Shepard, J. Gilbert, C. Clark. Enzyme-linked immunosorbent (ELISA) determination of aflatoxin B1 in peanut butter: collaborative trial. *Food Addit. Contam.* 5:601–608, 1988.
60. Park, D.L., B.M. Miller, L.P. Hart, G. Yang, J. McVey, S.W. Page, J.J. Pestka, L.H. Brown. Enzyme-linked immunosorbent assay for screening aflatoxin B1 in cottonseed products and mixed feed: collaborative study. *JAOAC* 72:326–332, 1989.
61. Patey, A.L., M. Sharman, J. Gilbert. Determination of total aflatoxin levels in peanut butter by enzyme-linked immunosorbent assay: collaboratve study. *JAOAC* 75(4):693–697, 1992.
62. Dixon-Holland, D.E., J.J. Pestka, B.A. Bidigare, W.L. Casale, R.L. Warner, B.P. Ram, L.P. Hart. Production of sensitive monoclonal antibodies to aflatoxin B1 and aflatoxin M1 and their application to ELISA of naturally contaminated foods. *J. Food Prot.* 51:201–204, 1988.
63. Devi, K.T., M.A. Mayo, K.L. Reddy, P. Delfosse, G. Reddy, S.V. Reddy, D.V. Reddy. Production and characterization of monoclonal antibodies for aflatoxin B1. *Lett. Appl. Microbiol.* 29(5):284–288, 1999.
64. Alvarez, M.T.D., M. Carvajal, F. Rojo, A. Escobar. Comparison between inhibitory indirect ELISA and HPLC methods to quantify free and adducted aflatoxins in human urine. *Nat. Toxins* 7(4):139–145, 1999.
65. Kim, E.K., D.H. Shon, D. Ryu, J.W. Park, H.J. Hwang, Y.B. Kim. Occurrence of aflatoxin M1 in Korean dairy products determined by ELISA and HPLC. *Food Addit. Contam.* 17(1):59–64, 2000.
66. Devi, K.T., J.S. Millere, G.K. Reddy, D.V.R. Reddy, M.A. Mayo. Phage-displayed peptides that mimic aflatoxin B1 in serological reactivity. *J. Appl. Microbiol.* 90:330–336, 2001.
67. Yong, R.K., M.A. Cousin. Detection of moulds producing aflatoxins in maize and peanuts by an immunoassay. *Intl. J. Food Microbiol.* 65(1,2):27–38, 2001.
68. Purchase, I.F.H., J.J. Theron. The acute toxicity of ochratoxin for rats. *Food Cosmet. Toxicol.* 6:479–483, 1968.
69. De Saiger, S., C. Van Petegham. Flow-through membrane-based enzyme immunoassay for rapid detection of ochratoxin A in wheat. *J. Food Prot.* 62(1):65–69, 1999.
70. Sibanda, L., S. De Sasiger, T.G.M. Bauteers, H.J. Nelis, C. Van Petegham. Development of a flow-through enzyme immunoassay and application in screening green coffee samples for ochratoxin A with confirmation by high-performance liquid chromatography. *J. Food Prot.* 64:1597–1602, 2001.
71. Thirulamala-Devi, K., M.A. Mayo, G. Reddy, S.V. Reddy, P. Delfosse, D.V. Reddy. Production of polyclonal antibodies against ochratoxin A and its detection in chiles by ELISA. *J. Food Agric. Food Chem.* 48(10):5079–5082, 2000.
72. Epply, R.M. Methods for the detection of trichothecenes. *JOAC* 58:906–908, 1975.
73. Beuchat, L.R. Food and Beverage Mycology. Westport, CT: AVI Publishing, 1978, p 527.
74. Hunter, K.W., Jr., A.A.L. Brimfield, M. Miller, F.D. Finkelman, S.F. Chu. Preparation and Characterization of monoclonal antibodies to trichothecene mycotoxin T-2. *Appl. Environ. Microbiol.* 49:168–172, 1985.
75. R.D. Wei, F.S. Chu. Production and characterization of a generic antibody against group A trichothecenes. *Anal. Biochem.* 160:399–408, 1987.
76. Fan, T.S.L., S.L. Schubring, R.D. Wei, F.S. Chu. Production and characterization of monoclonal antibody cross-reactive with most group A trichothecenes. *Appl. Environ. Microbiol.* 54:2959–2963, 1988.
77. Gendloff, E.H., J.J. Pestka, S.P. Swanson, L.P. Hart. Detection of T-2 toxin in *Fusarium sporotrichoides*–infected corn by enzyme-linked immunosorbent assay. *Appl. Environ. Microbiol.* 47:1161–1163, 1984.

78. Chu, F.S., S. Grossman, R. Wei, C.J. Mirocha. Production of antibody against T-2 toxin. *Appl. Environ. Microbiol.* 37:104–108, 1979.
79. Nagayama, S., O. Kawamura, K. Ohtani, J. Ryu, D. Latus, L. Sudheim, Y. Ueno. Application of an enzyme-linked immunosorbent assay for screening of T-2 toxin-producing *Fusarium* spp. *Appl. Environ. Microbiol.* 54(5):1302–1303, 1988.
80. R.B. Wellner, J.F. Hewetson, M.A. Poli. Ricin: mechanism of action, detection, and intoxication. *J. Toxicol. Toxin Rev.* 14:(4)483–522, 1995.
81. N. Koja, T. Shibata, K. Mochida. Enzyme-linked immunoassay of ricin. *Toxicon* 18:611–618, 1980.
82. Griffiths, G.D., H.V. Newman, D.J. Gee. Identification and quantification of ricin toxin in animal tissues using ELISA. *J. Forensic Sci. Soc.* 26:349–358, 1986.
83. A. Godal, O. Fodstad, K. Ingebrigtsen, A. Pihl. Pharmacological studies of ricin in mice and humans. *Cancer Chemother. Pharmacol.* 13:157–163, 1984.
84. Leith, A.G., G.D. Griffiths, M.A. Green. Quantification of ricin toxin using a highly sensitive avidin/biotin enzyme-linked immunosorbent assay. *J. Forensic Sci. Soc.* 28:227–236, 1988.
85. Poli, M.A., V.R. Rivera, J.F. Hewetson, G. Merrill. Detection of ricin by colorimetric and chemiluminescence ELISA. *Toxicon* 32:1371–1377, 1994.
86. Lipps, B.V. Production of polyclonal antibodies in mice against cobratoxin, botulinum toxin and ricin without altering their toxicity or use of adjuvant. *J.* Nat. Toxins 19(1):27–32, 2001.

3.11

Biosensors for Food Quality Assessment

Michele Del Carlo, Mihaela Nistor, Dario Campagnone, Bo Mattiasson, and Elisabeth Csöregi

CONTENTS

11.1 INTRODUCTION

11.1.1 Background

As agriculture and food technology have advanced, analytical and regulatory problems have become more and more complex. Food industries, operating in a highly competitive market are stimulated, on one side, to maintain and guarantee the characteristics of the existing products, and on another side, to innovate or introduce new products. Reliable data on the nutrient composition of foods consumed by people are critical in many areas: health assessment, the formulation of appropriate institutional and therapeutic diets, nutrition education, food and nutrition training, epidemiological research on relationships between diet and disease, plant breeding, nutrition labelling, food regulations, consumer protection, as well as for a variety of applications in trade, research, and development.

Improving the preservation, or increasing the yield, of agricultural products is a common aim for the producers, normally realized by using herbicides, insecticides, antibiotics or hormones. However, the use of these xenobiotics clearly represents a potential threat for consumers' health and people have always been concerned about the safety and quality of the food they eat. Additionally, there has always been a risk of fraud since food became a trade object, and despite the great progress of technology and preservation strategies, microbial contamination still remains a problem in food industry and distribution chain.

More recently, functional foods have been introduced on the market, when ingredients or components (e.g., vitamins, vaccine) are added to a product to tackle health and bad eating habit related issues, or from which possibly adverse components have been removed, thus introducing the need of monitoring an increasing number of various parameters.

Food quality comprises mainly three aspects:

1. Safety: A food product must not contain levels of contaminants, such as pathogens or toxins, likely to cause illness upon consumption
2. Shelf life: A food product must not contain contaminants at a level at which it becomes organoleptically spoiled in an unacceptably short time
3. Consistency: A food product must be of consistent quality with regard to both safety and shelf life, for example, the consumer will not accept products, which display large batch to batch variations in shelf life (1)

Obviously, the concern on the safety issue is a key aspect for food industry, regulatory agencies, and control laboratories. It is difficult to separate food quality from food safety, and the two aspects are, in fact, strictly related and the availability of analytical tools for assuring both is highly required. Regulatory agencies and the food industry are the two bodies most actively interested in determining and controlling the quality of food products. The regulatory authorities must fulfil their statutory responsibilities in order to protect the public from hazardous or inferior goods. The extent to which they are involved in food production and supply depends on the existing regulations and laws of the country in which they operate. Food stuff producing companies are equally interested, because association of their brand with products that are consistently good and safe will protect and enhance their good reputation on the market. The control is generally carried out by periodic chemical and microbiological analysis using standard reference procedures demonstrating the obvious need for highly sensitive, reliable, rapid, and affordable methods to determine an increasing number of compounds (2).

11.1.2 Food Products as Analytical Samples

The food industry is a process industry; suitable sensors for monitoring and control of key parameters are necessary not only for the characterization of the final product, but also for monitoring during the manufacturing process, both for process optimization and alarm in case of process failure. Food analysis comprises many aspects, including characterization of nutritional facts (product specific information on serving size, calories, content of fat, carbohydrates, proteins, vitamins, and minerals), microbial content (both pathogenic, i.e., *Salmonella*, *Listeria*, *Campylobacter*, *Escherichia coli O157*, *Yersinia*, *Shigella*, or *Vibrio* species, and spoilage, index or indicator organisms, i.e., *Bacillus cereus*, *Staphylococcus aureus*, *Faecal streptococci*, *Clostridium perfringens*, or *Pseudomonas sp.*), food additives (aspartame, saccharin, benzoate, sorbate, ascorbate, and sulphur dioxide), or pesticide and veterinary residues. Besides the characterization of the composition and properties of food products, the analytical methodology must also deal with possible chemical changes they might undergo during handling, processing, and storage. However, the difficulty of food analysis arises not only from this great number of components which need to be determined, but also by the complexity of the specific target food products, the analyte(s) of interest being always masked by a great number of matrix components, in a wide range of chemical classes and concentration levels. Moreover, food products might be liquids, pastes, or solids, the last two forms being incompatible with common existing analytical methodology, so that a suitable extraction method needs to be employed, in order to remove a great deal of matrix components and to bring the analyte(s) of interest in a liquid environment.

A successful analytical method requires the following characteristics: high selectivity and sensitivity; reliability; short assay times; stability; simplicity, low cost for operation and storage, and possibly a potential for miniaturization and facility of automation. Within the rapid method of analysis, biosensors are expected to play a prominent role in food processing, quality, and safety control, because they meet most of the criteria mentioned. The aim of this work is to review the development and application areas of biosensors in the food industry, their current situation, and future possibilities.

11.2 GENERAL ASPECTS OF BIOSENSORS

Biosensors are analytical devices incorporating a biological or biologically derived sensing element either integrated within or intimately associated with a physico-chemical transducer. Specific interactions between the target analyte and the complementary bio recognition layer produces a physico-chemical change that is detected and may be measured by the transducer (3,4).

The usual aim of using a biosensor is to produce a signal which is related to a single analyte or a related group of analytes.

The biological sensing elements used in biosensors are typically enzymes or binding proteins, such as antibodies, nucleic acids, bacteria and single cell organisms and even whole tissues or higher organisms. The choice of the biosensing element is decided by the substrate of interest (target analyte).

The physico-chemical transducers may take many forms depending upon the parameters being measured; electrochemical, optical, mass, and thermal changes are the most common. If the selectivity is mainly decided at the biosensing element, the transducer often determines the sensitivity of a biosensor (3,4) (See Figure 11.1).

Biosensors offer a number of advantages over traditional methods, such as high performance liquid chromatography and gas chromatography, which require high maintenance costs, expert operators, and long analysis times, making them less practical for food process monitoring. The measurements performed with a biosensor, are very *sensitive* and *selective*, due to the extraordinary nature of the biological interaction. The required equipment is generally simple, suitable for on site analysis and automation (5), and suitable for a quick food diagnosis.

11.3 POTENTIAL APPLICATIONS

Since the pioneering work of Clark and Lyons in 1962 (6), and Updike and Hicks in 1967 (7), there has been a phenomenal growth in the field of biosensors, with amperometrics dominating the literature. The application possibilities are very wide, medical and pharmaceuticals, food industry, and the environmental field being the three major categories. Table 11.1 presents some of the most important biosensors in the area of food analysis, described during the last six years (1999–2004).

11.3.1 Biosensors for Food Components Analysis

Application of biosensors to food component analysis has been recently reviewed covering the period 1997–2001 (49,50). In this section, the follow up is reported for the years 2001–2003 covering advancements in biosensor development and applications for the detection of carbohydrates, alcohols, phenols, carboxylic acids, and amino acids in food.

Carbohydrates (51–55) as well as alcohols, especially ethanol (23,56,57) still remain important targets for innovative biosensors. Both groups of analytes are key components in fermentation process industry, in addition to the prominent role of sugars in fruit juices, dairy products, soft drinks, and honey production. Within the biosensor area, the concept of multianalyte biosensor is gaining consideration and some attempts in this respect have been described for ethanol–glucose sensing (51,52). Most of the recently published reports are oriented to the use of chemically modified electrodes (CMEs), combined with enzymes. The aims of this chemical modification are: to reduce the working potential (23,51,52), to cast a diffusion membrane to reduce interferences (57) or to improve enzyme retention at the electrode surface (55).

Despite a certain interest in determining polyphenols using tyrosinase based biosensors during the late nineties (58,59), no recent papers have appeared on that topic, possibly due to the poor selectivity of the method; further development in that area is needed to address issues such as class specific determination, sensor stability, and selectivity.

Amino acid composition and total protein content are two important parameters that are studied in food component analysis and are related to food quality control. Total protein determination can be performed either using traditional methods such as Kjeldalh and its instrumental modification, or by commercially available photometric protein test kits. A recently developed biosensor for total protein composition based on the coupling of two enzymes, protease and L-amino acid oxidase (L-AAO) resulted in a rapid method (4 minutes/sample), requiring only 10 μl of sample and operating in the working range 0.17–1.0 mg/ml (60). A similar method, with comparable performance was developed by Kwan et al. in 2002. In addition, this method was applied to the determination of aspartame, by coupling L-AAO with pronase (61).

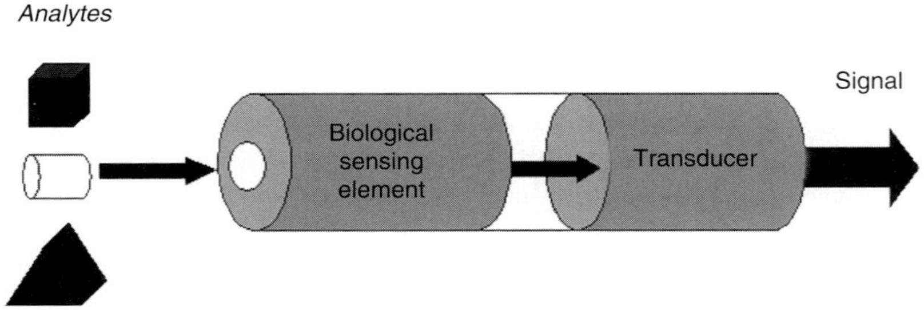

Figure 11.1 Schematic presentation of the biosensor principle

Table 11.1
Applications of the biosensors for food analysis.

Analyte	Biological Molecule	Transducer Type	Characteristics	Application	Ref.
Food components					
(Sugars, alcohols, phenols, carboxylic acids, aminoacids, biogenic amines, organic and inorganic ions, etc.)					
Glucose	Glucose oxidase	Amperometric E = +0.6 V vs. Ag/AgCl	S: 4.51 μA/mM LR: 0 – 0.5 mM	Mustard, Mayonnaise Soya souse	(8)
Glucose	Glucose oxidase	Amperometric E = −0.05V vs. Ag/AgCl	S: 3.9 mA/M cm^2 LR: 0.05 – 6 mM	Peer juice Peach juice Ananas juice	(9)
Glucose Ethanol Glycerol	Glucose dehydrogenase Alcohol dehydrogenase Glycerol dehydrogenase	Amperometric E = +0.2 V vs. Ag/AgCl	S_G: 87 mA/M cm^2 LR_G: 20 – 800 μM S_E: 220 mA/M cm^2 LR_E: 2.5 – 250 μM S_G: 32 mA/M cm^2 LR_G: 1 – 200 μM	Different types of wine	(10)
Glucose	Glucose oxidase	Optical (λ = 720 nm)	S: 0.036* U/mM LR: 0.05 – 2 mM	Apple juice Wine Coca-cola	(11)
Glucose	Glucose oxidase	Thermal	S: 1.4* mm/mM LR: 5 – 25 mM	Coca-cola Fruit Juice	(12)
Glucose Lactate	Glucose oxidase Lactate oxidase	Amperometric E = +0.6 V vs. Ag/AgCl	S_G: 9.9 nA/mM LR_G: 0.5 – 100 mM S_L: 7.2 nA/mM LR_L: 0.5 – 20 mM	Tomato juice	(13)
Glucose	Glucose oxidase	Amperometric E = +0.4 V vs. Ag/AgCl	S: 2.31 μA/mM LR: 0.1 – 9.6 mM	Soft drinks	(14)

(Continued)

Table 11.1 (*Continued*)

Analyte	Biological Molecule	Transducer Type	Characteristics	Application	Ref.
Sucrose Glucose	Invertase Mutarotase Glucose oxidase	Amperometric E = +0.7 V vs. Ag/AgCl	S_S: 55* nA/mM LR_S: 1 – 40 mM S_G: 33* nA/mM LR_G: 1 – 100 mM	Grape juice, Coca-cola, Orange juice, Kiwi juice	(15)
Glucose Ascorbic acid Citric acid	Glucose oxidase Horseradish peroxidase Urease	ISFET	LR_G: 1 – 10 mM LR_{AA}: 0.25 – 2 mM LR_{CA}: 5 – 100 mM	Fruit drinks	(16)
Lactose Glucose	*β*-Galactosidase Glucose oxidase	Manometric	S_L: 0.57 kPa/mM LR_L: 0 – 5 mM S_G: 0.95 kPa/mM LR_G: 0 – 5 mM	Milk	(17)
Lactulose Fructose	*β*-Galactosidase Fructose dehydrogenase	Amperometric E = +0.38 V vs. Ag/AgCl	S_L: 324 nA/mM LR_L: 0.01 – 5 mM S_F: 352 nA/mM LR_F: 0.001 – 5 mM	Milk	(18)
Fructose	D-Fructose dehydrogenase	Amperometric E = +0.2 V vs. SCE	S: 70 nA/mM LR: 0.05 – 1 mM	Jam, Honey, Chocolate, Biscuit, Juice, Wine	(19)
Sucrose	Sucrose phosphorylase Phosphoglucomutase Glucose-6-phosphate 1-dehydrogenase	Amperometric E = +0.15 V vs. Ag/AgCl	S: 1 nA/mM LR: 1 – 100 mM	Pineapple juice Orange juice Peach juice	(20)
Glucose Malic acid	Glucose oxidase Malic enzyme	Amperometric E = +0.0 or 0.2 V vs. Ag/AgCl	S_G: 125* nA/mM LR_G: 0 – 5 mM S_{MA}: 640 nA/mM LR_{MA}: 0.05 – 3 mM	Wine	(21)

Ethanol Methanol	Alcohol oxidase Horseradish peroxidase	Amperometric E = +0.0 V vs. Ag/AgCl	S_E: 259 nA/mM LR_E: 0.2 – 20 μM S_M: 5 nA/mM LR_M: 0.02 – 1.5 μM	Wine Beer Liquor	(22)
Ethanol	Alcohol dehydrogenase	Amperometric E = +0.3 V vs. Ag/AgCl	S: 336 mA/M cm^2 LR: 1 – 250 μM	Wine fermentation process	(23)
Polyphenols	Horseradish peroxidase DNA	Amperometric E = −0.05 V vs. Ag/AgCl	S: 181 nA/μM cm^2 LR: 1 – 50 μM	Vegetables	(24)
Polyphenols	Tyrosinase	Amperometric O_2-type Clark	S: 1* ΔDO(%)/μM LR: 0.5 – 6 ppm	Olive oil	(24)
Lactate Acetaldehyde	D-Lactate dehydrogenase Aldehyde dehydrogenase	Amperometric E = −0.15 V vs. Ag/AgCl	S_L: 589 nA/mM LR_L: 0.075 – 1 mM S_A: 1100 nA/mM LR_A: 0.01– 0.25 mM	Wine	(26)
Lactate Malate	L-Lactate dehydrogenase L-Malate dehydrogenase Diaphorase	Amperometric E = +0.3 V vs. SCE	S_L: 1.66* nA L/mg LR_L: 0.01 – 1.1 mM S_M: 1.16* nA L/mg LR_M: 0.01– 1.3 mM	Wine	(27)
Lactate	Lactate oxidase Peroxidase	Amperometric E = +0.0 V vs. Ag/AgCl	S: 424 nA/mM LR: 0.002 – 1 mM	Wine Yogurt	(28)
Acetate	Acetate kinase Pyruvate kinase Pyruvate oxidase	Amperometric E = −0.4 V vs. SCE	S: 0.07* μA/mM LR: 0.05 – 20 mM	Wine	(29)
Citric acid Pyruvic acid Oxaloacetic acid	Citrate lyase Oxaloacetate decarboxylase Pyruvate oxidase	Amperometric E = +0.65 V vs. Ag/AgCl	L_{CA}: 0 – 100 mM L_{PA}: 0 – 6 mM L_{OAA}: 0 – 6 mM	Synthetic samples	(30)
L-Ascorbic acid	Ascorbate oxidase	Amperometric O_2-type Clark	S:3*ΔDO(mg/L)/mM LR: 0.05 – 1.2 mM	Fruit Juices	(31)

(Continued)

Table 11.1 (*Continued*)

Analyte	Biological Molecule	Transducer Type	Characteristics	Application	Ref.
Folic acid	Antifolic acid antibody	Optical (SPR)	Accuracy 88 – 101 %	Milk powder, Cereal samples	(32)
Cholesterol	Cholesterol oxidase	Optical (oxygen transducer)	LR: 0.07 – 18 mM	Butter	(33)
Tryptophan Leucina Serina Valine	Amino acid oxidase	E = 0.0 V vs. Ag/AgCl	LR_T: 10 – 250 μM LR_L: 10 – 1200 μM LR_S: 50 – 500 μM LR_V: 25 – 500 μM	Grapes	(34)
Lysine	Lysine α-oxidase	Amperometric E = +0.1 V vs. Ag/AgCl	LR: 2-125 μM	Sample fermentation	(35)
Histamine Putrescine	Amine oxidase Horseradish peroxidase	Amperometric E = −0.05 V vs. Ag/AgCl	S_H: 73.7 mA/M cm^2 LR_H: 1 – 150 μM S_P: 194 mA/M cm^2 LR_P: 1 – 400 μM	Fish	(36)
Hypoxanthine	Xantine oxidase	Amperometric O_2-type Clark	LR: 1 – 400 μM DL: 0.8 μM	Fish	(37)
Phosphate	Alkaline phosphatase Polyphenol oxidase	Amperometric E = +0.6 V vs. SCE	S: 1.27 mA/M cm^2 DL: 2 μM	Drinking water	(38)
Sulfite	Sulfite oxidase	Amperometric E = +0.3 V vs. Ag/AgCl	LR: 4 – 750 ppm DL: 4 ppm	Water	(39)

Food contaminants					
(Bacteria, pesticides, antibiotics, heavy metals, etc)					
Escherichia coli O157:H7	Urease Anti *E. Coli* antibodies	Potentiometric (pH electrode)	DL: 10 cells/mL	Drinking water	(40)
Salmonella enteritidis	Anti *Salmonella* antibodies	PQC (10 MHz)	Specificity 92.9 %	Chickens, Eggs	(41)
Salmonella typhimurium	Anti *Salmonella* antibodies	SPR (Biacore)	DL: 1.7×10^3 CFU	Drinking water	(42)
Staphylococcal enterotoxin B	Anti *Staphylococcal enterotoxin* B antibodies	SPR (Biacore)	DL: 1 ng/mL	Milk, Mushrooms	(43)
Organophosphates	AChE	Photothermal	DL: 0.2 ng/mL	Salad, Onion	(44)
Carbamates	BChE	(488 nm, 120 mW)			
Paraoxon	AChE	Amperometric	DL_P: 0.1 nM	Water, Fruit juice	(45)
Carbofuran	BChE	E = +0.7 V vs. Ag/AgCl	DL_C: 0.01 nM		
Penicillin G	Protein with carboxypeptidase activity	SPR (Biacore)	DL: 2.6 μg/Kg	Milk	(46)
Ampicillin	Anti ampicillin antibodies	SPR (Biacore)	DL: 33 μg/L	Milk	(47)
Hg^{+2}, Zn^{+2}, Cu^{+2}, Ag^{+}, Mn^{+2}	Urease	Potentiometric (pH-electrode)	Regeneration of the enzyme activity in 10 min	Water	(48)

* calculated values.

Symbols: SCE-saturated calomel electrode, SPR-surface plasmon resonance, PQC-piezoelectric quartz crystal, AChE- Acetylcholinesterase, BChE- Butyrylcholinesterase.

More recently, a biosensor array based on photo patternable enzyme membranes for the simultaneous measurement of glucose, lactate, glutamate, and glutamine has been described and applied to fermentation broth (62). The microsystem consists of an upper glass chip with the integrated biosensor array, and a bottom part, comprised of a gold counter electrode, a 300 μm thick seal, and electrical interconnection lines. The flow device has a total internal volume of 2.1 or 6 μl when integrated with a mixer on the chip.

Finally, it is important to mention, that despite the number of relevant papers already present in the literature, new format biosensors for fish freshness evaluation are still under development, using electrochemical (63), or QCM (64), approaches.

Taking into account the extensive number of possible applications for biosensors in food analysis, the following section will discuss, in more detail, a few examples of biosensors developed in our laboratory.

11.3.1.1 Monitoring of Wine Quality: Biosensors for Ethanol, Glucose and Glycerol

Wine is a complex mixture of several hundred compounds, present simultaneously, at different concentrations. The dominants ones are water, ethanol, glycerol, sugars, organic acids, and various ions. Except ethanol and glycerol, other aliphatic and aromatic alcohols, amino acids, and phenolic compounds are present at much lower concentrations.

Three different PQQ-dehydrogenases [glucose dehydrogenase (GDH), alcohol dehydrogenase (ADH), and glycerol dehydrogenase, (GlDH)] newly isolated and purified from *Erwinia* sp. 34-1 or *Gluconobacter* sp.3-3 (65–67) have been used for development of biosensors for determination of key compounds in wine (10). The enzymes have been integrated in redox hydrogels by cross linking them to an Os complex modified nonconducting polymer (PVI_{13}dmeOs) using poly (ethyleneglycol)–diglycidyl ether (PEGDGE) as the cross linking agent. Due to the structural similarities of the studied enzymes, a common sensing mechanism could be proposed for the three biosensors (see Figure 11.2). Thus, the main enzyme substrate (glucose for GDH, glycerol for GlDH and ethanol for ADH) is firstly oxidized while the enzyme's cofactor is simultaneously reduced. The active form of the enzyme is regenerated via the interaction with the electrochemical mediator (Os modified redox polymer), which is maintained in its oxidized form by the positive potential applied at the electrode.

The developed biosensors were integrated in a single manifold flow injection line and the analytes were manually injected into the carrier flow and transported to the electrochemical cell equipped with the working, reference, and auxiliary electrodes respectively (10). The working potential of the enzyme modified electrodes have been controlled by a potentiostat. The electrodes were optimized with regard to their composition (enzyme/polymer cross linker ratio), flow rate, and pH. and evaluated first using standard solutions.

Table 11.2
Biosensors characteristics. Symbols: DL - detection limit (calculated as three times the signal-to-noise ratio), DR, DRwine - dynamic range, I_{max} and K_m^{app} were evaluated from Michaelis-Menten equation: $I = (I_{max}*[A]) / (K_m^{app}+[A])$, [A] - analyte concentration; S - sensitivity (calculated as I_{max} / K_m^{app}*Electrode surface)

Type of Sensor	Km (mM)	I_{max} (nA)	S (mA/M cm^2)	DL (μM)	DR (μM)	DR_{wine} (M)
Glucose	0.80	4963	88.62	9	20–800	0.002–1
Ethanol	0.60	9400	223.8	1.2	2.5–250	1–3.5
Glycerol	1.31	1797	19.5	1	1–200	0.01–0.1

Table 11.3
Content of ethanol, glucose and glycerol in Dunavár wine, Hungary, 1998.

Type of Sensor	Biosensor Analysis	Spectrophotometric Analysis
Ethanol	12.5 ± 0.3%	12%
Glucose	0.32 ± 0.01 g/L	0.30 g/L
Glycerol	7.8 ± 0.8 g/L	8.1 g/L

Next, the optimized electrodes were evaluated using real samples (wine) with declared content of the targeted analytes. The main characteristics of the developed biosensors are presented in Table 11.2.

The results obtained with the optimized biosensors were also compared with those yielded by other conventional analytical techniques. Thus, the results delivered by the developed biosensors have been validated and the applicability for the determination of these key compounds in wine samples could be successfully demonstrated (see Table 11.3).

Evaluation of the biosensors in terms of stability has been performed. While about 90% response of the ethanol sensor was kept after a continuous operation for 100 h, the other two electrodes were less stable, maintaining only 80% (glycerol sensor) and 60% (glucose sensor) of their initial signal after 20 h of continuous substrate injection. The storage stability was good for all three biosensor, less than 20% of their initial response being lost after one month of storage at 4°C (results not shown).

11.3.1.2 Monitoring of Fish Quality: Biosensors for Biogenic Amines

Biogenic amines are organic bases with aliphatic, aromatic, or heterocyclic structures that can be found in many foods such as meat, fish, cheese, wine, and milk, in which they are mainly produced by microbial decarboxylation of amino acids. Excessive consumption of these amines can be of health concern because they can generate different degrees of

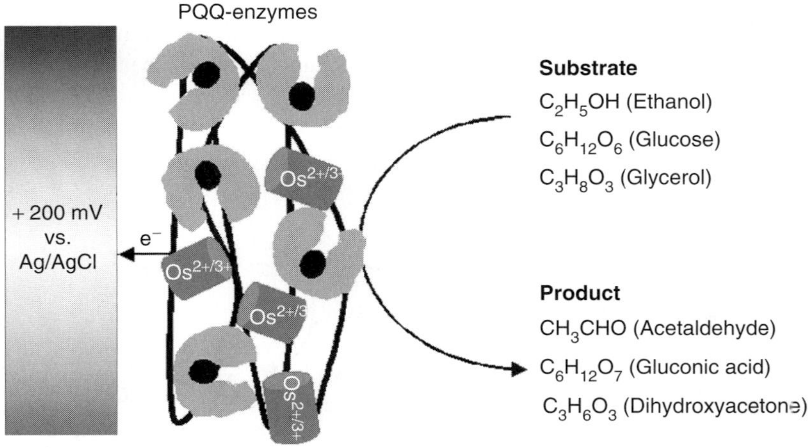

Figure 11.2 Proposed mechanism of bioelectroconversion of the substrate on PQQ-dehydrogenase modified electrodes; Electrodes composition: GDH-, ADH-, or GlDH-PQQ enzyme, $PVI_{13}dmeOs$ and PEGDGE

Table 11.4
Bioelectrochemical characteristics obtained for AO/HRP/ PVI_{13}-dmeOs/PEGDGE electrodes. Symbols as in Table 11.2.

Electrode Type	Analyte	K_m^{app} (μM)	I_{max} (μA)	S (mA/Mcm²)	DL (μM)	LR (μM)
AO/HRP/PVI_{13}-dmeOs/PEGDGE	Histamine	901 ± 85	4.85 ± 0.41	73.74 ± 1.7	0.33	1–150

diseases determined by their action on nervous, gastric, and intestinal systems and blood pressure (36,68).

The developed biosensor for the determination of biogenic amines was based on the direct coupling of two enzymes; amine oxidase (AO) and HRP (horseradish peroxidase) and their cross linking (PEGDGE) to a redox polymer (PVI_{13}-dmeOs). The operating potential of the resulting biosensor was 50 mV *vs.* Ag/AgCl. The electron transfer mechanism is outlined in Figure 11.3.

Different parameters, e.g., applied potential, flow rate, pH, and composition of the sensing film, have been optimized before testing the sensor in real samples. The kinetic parameters and the main biosensor characteristics obtained for the optimal electrodes are presented in Table 11.4.

The biosensors were also characterized with regard to their selectivity and response time, as well as operational and storage stability (36), showing a higher signal for aliphatic amines than for aromatic ones, a short response time (less than 1 minute) and a good operational stability (30% and 50% decrease in sensor activity for histamine and putrescine, respectively, after 10 h of continuous operation, at a sampling frequency of 30 injections h^{-1}) and long term stability (10% and 15% decrease in sensitivity for histamine and putrescine, respectively, after 10 days of storage).

Finally, the optimized biosensor was considered for monitoring of biogenic amines in real samples. The differentiation between the signals given by different amines is, however, not possible, only the total amine content of the sample was determined. Fish muscle samples, kept for 10 days at 4°C and 25°C were analysed by direct injection in the flow system after extraction in a phosphate buffer. The total amine content expressed in histamine equivalents is presented in Figure 11.4. The maximum accepted limit for total amine concentration in food products is 0.1 to 0.2 g kg^{-1} sample, and a concentration of 1 g kg^{-1}

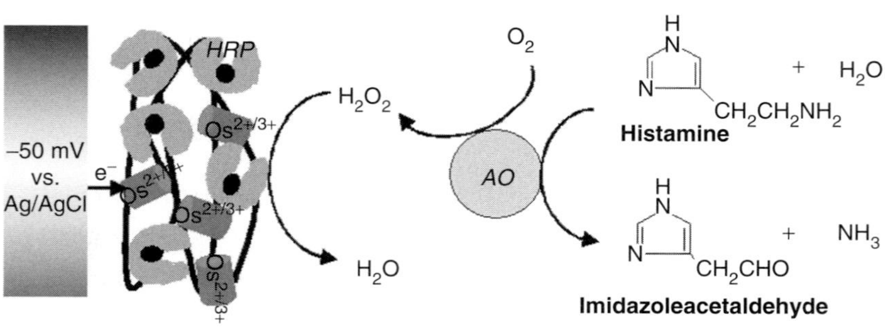

Figure 11.3 Proposed mechanism of bioelectroconversion of histamine on a bi-enzyme (amine oxidase (AO)/horseradish peroxidase (HRP)) electrode; Electrodes composition: AO, HRP, PVI_{13}dmeOs and PEGDGE

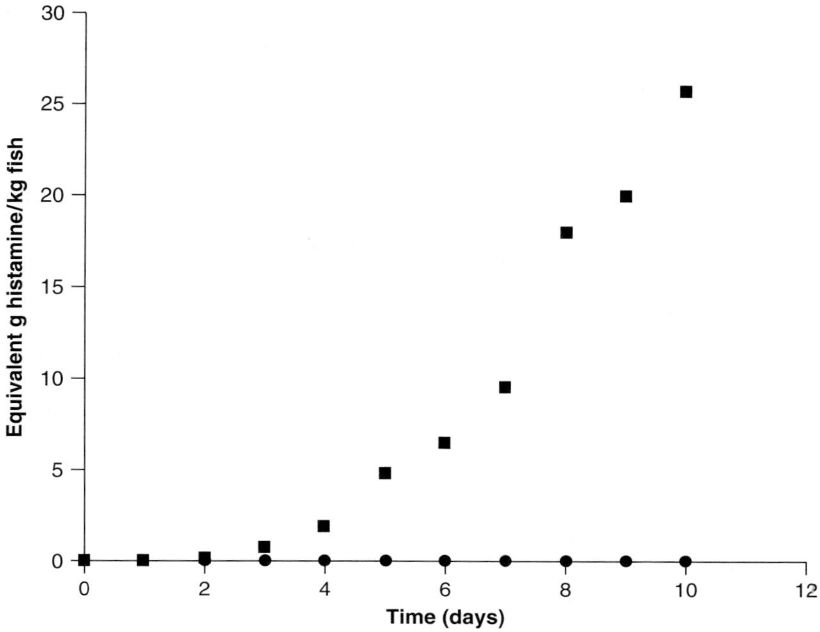

Figure 11.4 Monitoring of fish freshness using amine oxidase-based electrodes; Total amine content recorded for fish kept at 4°C (■) and 25°C (•), respectively; The total amine concentration was expressed in histamine equivalent units

is considered to be toxic. As seen, after 3 days of storage at room temperature, the fish became improper for consumption, while even after 10 days of storage at 4ºC there were no major changes observed in the total amine concentration.

11.3.2 Biosensors for Food Contaminants Analysis

Water is omnipresent in foodstuff and the surrounding atmosphere. Determination of water content is one of the most frequent analyses performed in food industry. The range of water concentrations in food is very broad, starting with a fraction of a per cent and reaching more than 98%. Fresh products and liquid food usually contain large amounts of water, whereas baked and dry products are poor in water. In many products the water is constitutive, but there are numerous food products in which water is added during processing.

The water present in food products affects their safety, stability, quality, and physical properties. Water is a vehicle for the transmission of many infectious agents, such as microorganisms, and continues to cause significant outbreaks of disease world wide.

11.3.2.1 Microorganisms in Food

Microorganisms, such as bacteria and viruses, are widely spread throughout the environment at any level such as marine and estuarine waters, soil, and the intestinal tracts of humans and animals, and they are present also in food. Many of these organisms have an essential function in nature, but certain potentially harmful microorganisms can have profound negative effects on both animal and human health. It is estimated that infectious diseases cause a significant number of total annual deaths worldwide, ranging between 15 and 75% depending on the geographical area (69). In addition, nonfatal infections afflict more than 200 million people each year (70). Food safety issues are continuously boosted by emerging concerns. Recently, recalls due to food poisoning bacteria such as *Listeria monocytogenes* (71,72) have evidenced the need for more rapid, sensitive, and specific methods of detecting these microbial contaminants to be used as early warning systems.

The detection and identification of bacteria is mainly based on specific microbiological and biochemical identification. Pathogens typically occur in low number in food. Some of the standard methods can be sensitive, inexpensive, and give both qualitative and quantitative information on the number and the nature of the microorganisms tested (73,74). Their main disadvantage, especially as warning systems, is the length of the procedures, which usually need the following steps: preenrichment, selective enrichment, biochemical screening, and serological confirmation. Other approaches are directed to identify and quantify microorganisms through the detection of characteristic DNA sequences using the polymerase chain reaction (PCR) (75–80). Nucleic acid based assays are effective identification tools, though they are limited, in that they will indicate the presence of a microorganism, but do not give any information on effective toxins or virulence (73). Finally, immunological techniques are used to detect microbes by their unique antigenic determinants. Polyclonal and monoclonal antibodies can be raised relatively quickly and cheaply (81). Immunological detection of microbial contamination has become more sensitive, specific, reproducible, and reliable with many commercial immunoassays available for the detection of a wide variety of microbes and their products such as biotoxins (82). It is also important to stress that antibody based techniques give only information on the presence of antigens regardless of whether the microorganims are viable or not. The Hazard Analysis Critical Control Point (HACCP) methodology requires, for its effective application, analytical devices that can be operated outside the laboratory by trained but not professionally skilled people. Particularly, the microbiological safety of food still needs rapid, user friendly and reliable tools. Biosensors offer the potential of detecting pathogens in near real time mode, but often this methodology still requires preenrichment in order to detect low numbers of pathogens.

11.3.2.1.1 Label Free Methods A label free method relies on the monitoring of analyte at the solid (i.e., sensor) or liquid (analyte solution) interface. The basis of this detection mode is the physical phenomena occurring on the transducer's surface during the biochemical reactions (e.g., change in optical or electrical features). Regardless of the large amount of literature produced on this topic, only a few articles report on the analysis of bacteria in real food matrices. A method based on the refractive index variation upon antigen antibody binding has been applied to the detection of *Salmonella typhimurium* in chicken carcass wash fluid (83). Based on the same assay principle, a method for the detection of *Staphylococcus enterotoxin A* in potato salad, canned mushroom, hot dog, and milk, has been described, quoting a total assay time of 4 minutes (84).

11.3.2.1.1.1 Optical Approaches for Microbial Detection Optical transducers are particularly attractive for label free detection of bacteria. Several optical techniques have been reported for the detection of bacteria including: monomode dielectric waveguides (85), surface plasmon resonance (86–89), ellipsometry (90), resonant mirror (91), and interferometry (92).

11.3.2.1.1.2 Bioluminescence Sensors The bioluminescence phenomenon is based on the ability of certain enzymes to emit photons as a product of enzymatic reactions. The basis of the assay scheme consists of the introduction of the gene encoding for the enzyme luciferase in the genome of a bacteriophage (a bacterial virus). When a host bacterium is infected, the luciferase encoding gene can be transferred to the bacteria, conferring bioluminescence to a previously nonbioluminescent organism. A wide range of microorganisms have been detected using this approach. Literature reports describe the detection of *Salmonella*

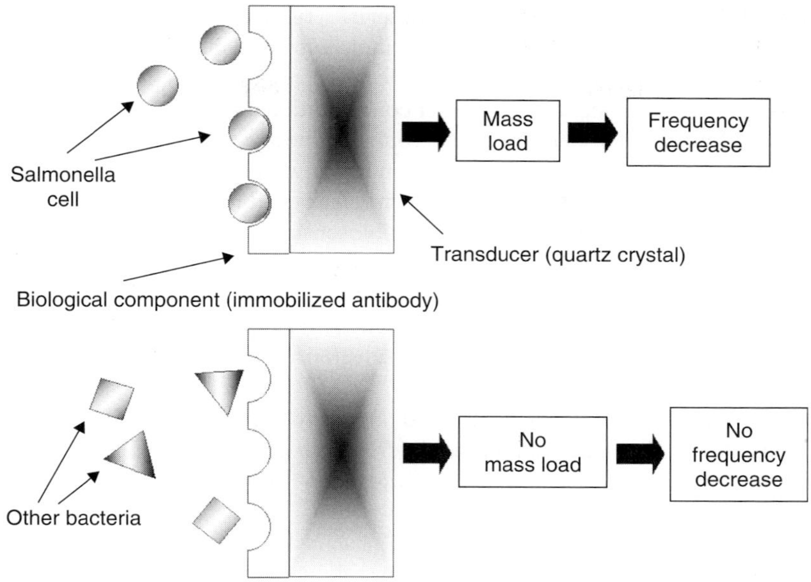

Figure 11.5 Working principle of the *Salmonella* piezoelectric biosensor

sp. (93,94), many *Listeria* strains, *Listeria monocytogenes* (95), *Salmonella newport*, and *E. coli* (96). The main advantage of this approach is its inherent ability to detect viable cells; the main limitation remains the poor sensitivity, which leads to lengthy protocols requiring a preenrichment step.

11.3.2.1.1.3 Piezoelectric Sensors Piezoelectric (PZ) biosensors are analytical tools that may be operated in a label free fashion.

The theoretical basis of piezoelectricity and the practical application to biosensors development are illustrated in a series of reports (97,98). The binding of the analyte to receptor (e.g., antibody) previously immobilized onto the surface, increases the mass on the sensor and determines a proportional decrease of the oscillation frequency of the piezoelectric crystal (Figure 11.5). Piezoelectric sensors surfaces can be functionalized with antibodies raised against the whole microbial cells, surface expressed antigens, microbial toxins, or even DNA specific sequences.

In the last 15 years, methods based on this approach (antibody) have been described for *Candida albicans* (99), *Salmonella typhimurium* (100,101) *Listeria monocytogenes* (102). Great efforts have been devoted toward sensors development to enhance and optimize analytical performance (103,104).

In one case, a flow injection set up was used for the detection of the *Staphylococcal Enterotoxin B* using polyclonal antibodies (105). Sensor surfaces have also been coated with specific antigens for the detection of serum antibodies and with microbial DNA probes for the detection of strain specific PCR amplified sequences (106). These sensors have also been used in the detection of several other microbial contaminants including *Candida albicans* (107), *L. monocytogenes* (108,109) and *Vibrio cholerae* (110). Most of the sensors have been applied to aqueous suspensions. Examples of the use of piezoelectric sensors in real matrices have been recently reported (111). The authors described the detection of *Salmonella enteriditis* in chicken and eggs using the biosensor as a diagnostic tool in veterinary medicine.

11.3.2.1.2 Indirect Detection of Bacteria Typically, the indirect biosensing of bacteria is based either on the use of a labelled molecule or on microbial metabolism tracking. The former approach is able to detect viable and nonviable cells, while the latter, using the cell metabolism, responds to viable cell concentrations. Among the indirect methods, electrochemical biosensors seem the most promising, because they are relatively low cost, operate even in turbid media, are amenable to miniaturization, and can be easily adapted to flow or flow injection systems. The sensors can be grouped in antibody based biosensors, DNA based biosensors, and metabolism based biosensors.

11.3.2.1.2.1 Antibody Based Biosensors Immunosensors, and immunochemical methods based on electrochemical detection, couple the specificity of the immunological reaction with the simplicity, rapidity, and sensitivity of the electrochemical detection. Immunosensors offer the potential to increase sensitivity and selectivity but lack a form of inherent signal amplification (112,113).

Different approaches have been used as assay scheme reflecting technological advancement. Mirhabibollahi et al. (114) described an enzyme linked immunosensor for the detection of *Staphylococcus aureus* and *Salmonella* in pure cultures using the two site immunometric approach (Figure 11.6).

An original improvement was introduced in 1995 by Kim et al. (115). They used the ability of hemolytic bacteria to disrupt a lipid membrane to design and develop a selective immunosensor for hemolytic organisms. Specific antibodies for *Listeria monocytogenes, Listeria welshimeri,* and *Escherichia coli* were immobilized on the surface of a platinum working electrode. After the capturing step, liposomes containing an electrochemical mediator were added to the medium. The presence of hemolytic organisms determined the rupture of the lipid bilayer of liposome, releasing the electrochemical mediator in the medium.

Another interesting approach for the detection of *Salmonella typhimurium*, consisted in the use of super paramagnetic material to perform antibody based separations. In this case, antibodies were immobilized onto magnetic beads and the secondary alkaline phosphatase labelled antibody for the analyte of interest formed the typical sandwich. Thereafter, with the aid of a magnet, the complex was placed onto the surface of the working electrode of a disposable electrochemical sensor where the assay medium was removed by aspiration and the enzyme substrate solution was added. Using this approach 8×10^3 cells/ml was detected in 80 minutes (112). Brewster and Mazenko (116) have developed an immunoelectrical sensor, coupled with filtration capture, for the rapid detection of *E. coli O157*. The sensor has a detection limit of 5000 cells/ml and an assay time of 25 min. The immunological complex is formed on the surface of a cellulose acetate membrane that is brought in contact with the electrode surface. Five min. after substrate addition, cells can be detected by the conversion of substrate (para-aminophenyl phosphate) to an electro active product (para-aminophenol).

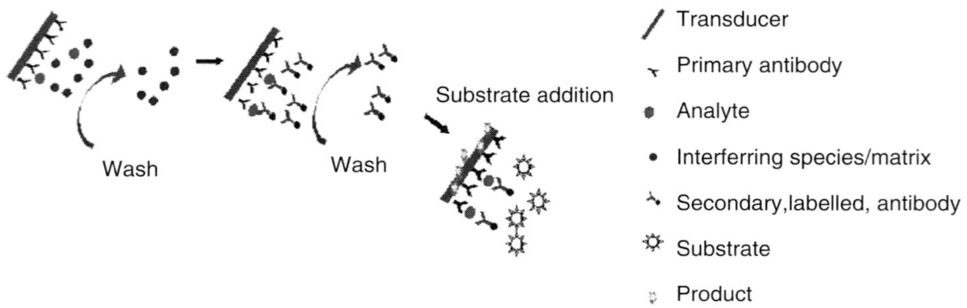

Figure 11.6 Schematic drawing of one of the "two site immunometric" immunosensors

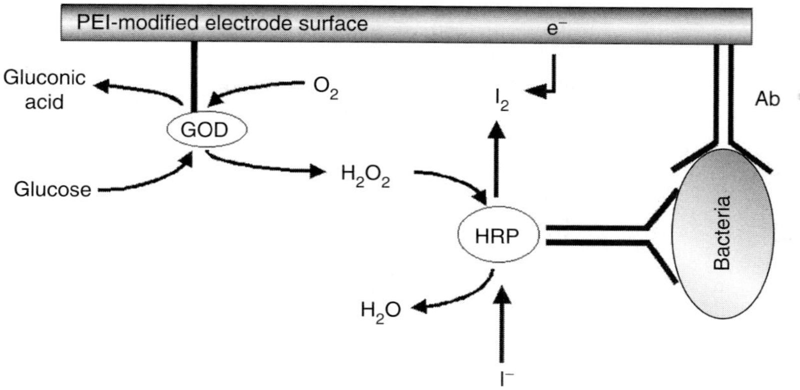

Figure 11.7 Separation-free, amperometric enzyme channeling immunoassay for *Staphylococcus aureus* with immobilized antibody and glucose oxidase (GOD) on the PEI-modified electrode surface

Rishpon and Ivnitski (117), have developed a separation free amperometric enzyme channelling immunosensor for the detection of *Staphylococcus aureus*. The immunosensor consists of a carbon electrode with anti Protein A antibodies, which detect *Staphylococcus aureus*, and the enzyme glucose oxidase immobilized on its surface. The test solution containing *Staphylococcus aureus*, a horseradish peroxidase (HRP) labelled anti Protein A antibody and iodide is then added. Upon antibody–antigen–antibody sandwich formation, hydrogen peroxidase is brought in contact with glucose oxidase at the electrode surface. After addition of glucose in solution, HRP catalyses the reaction of H_2O_2 and iodide ions. The reduction of iodide ions at the electrode surface allowed detection of *Staphylococcus aureus* to a limit of 1000 cells/ml of pure culture in 30 min. This assay format offers the advantage of being pseudo homogenous, i.e., without a washing step, as the polyethylenimine (PEI) membrane used discriminates between the analytical signal from the bound immunocomplex and background noise from HRP– H_2O_2 reacting in solution away from the electrode. Signal amplification is also facilitated through enzyme channelling (Figure 11.7).

Recently, Ivnitski et al. (118) have developed an amperometric immunosensor based on a supported planar lipid bilayer for the detection of *Campylobacter*. The artificial bilayer membrane, which served for the anti *Campylobacter* antibodies immobilization, was deposited on a stainless steel working electrode. Conformational changes, occurring when the antigen antibody is formed, allowed the ion to pass through the channel. The current produced by ions migration could be amperometrically measured. In addition to acting as a matrix for antibody, the lipid membrane also acts as a very thin electrical insulator and suppresses nonspecific binding of ligand. The biosensor shows a strong signal amplification effect, defined as the total number of ions transported across the bilayer. This is 10^{10} ions/s, which leads to a theoretical limit of detection of 1 cell in an assay time of just 10 min. The sensor may have potential for the detection of other microorganisms provided that high quality antibodies are available. Recently (119) conventional amperometry at an antibody-containing polypyrrole film electrode was found to be unsuccessful in detecting levels below 10^6 cells ml^{-1}. This sensor was capable of reproducibly detecting *Listeria* at levels of 10^5 cells ml^{-1} in 30min.

11.3.2.1.2.2 DNA/RNA Based Biosensors DNA electrochemical biosensors, coupling the high specificity of DNA hybridization with the inherent sensitivity of electrochemical transducer, hold great promise for the rapid screening of viral and bacterial pathogens in different samples matrices including food.

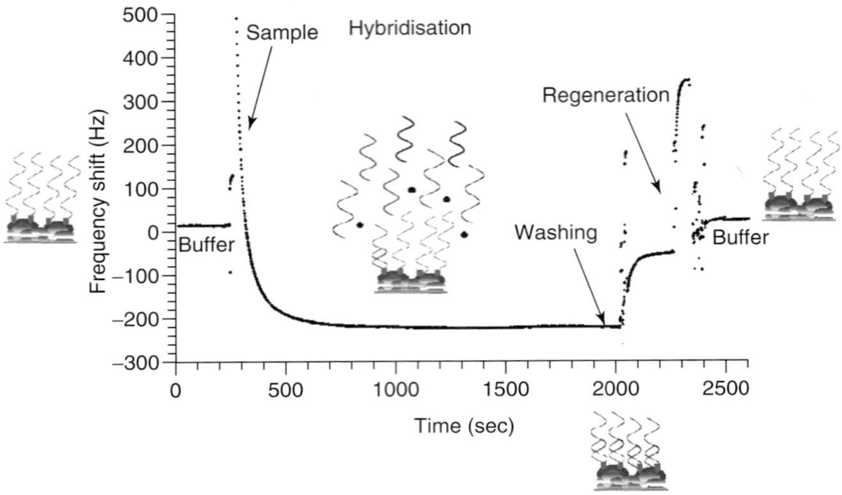

Figure 11.8 Typical response obtained for DNA hybridization detection using piezoelectric microbalance. The hybridization reaction generates a diminishment of the frequency of the oscillating that is proportional to the mass bound to the sensor surface. The sensor surface can be regenerated using appropriate washing buffers (Adapted from Mannelli, I., M. Minunni, S. Tombelli M. Mascini, *Biosens. Bioelectron.* 18:129–140, 2003)

A highly sensitive and specific RNA biosensor was developed for the rapid detection of viable *E. coli* as an indicator organism in water. The biosensor is coupled with protocols developed earlier for the extraction and amplification of mRNA molecules from *E. coli* (120). The biosensor was a membrane based DNA/RNA hybridization system using liposome amplification.

The biosensor resulted in a portable, inexpensive, and very easy to use device, which makes it an ideal detection system for field applications.

An excellent correlation to a much more elaborate and expensive laboratory based detection system was demonstrated, which can detect as few as 40 *E. coli* cfu/ml. Finally, the assay was tested regarding its specificity; no false positive signals were obtained from other microorganisms or from nonviable *E. coli* cells. Other recent application of DNA biosensors for bacteria detection have been described based on either fiber optic transduction mode (121,122) or piezoelectric microbalance (123) (Figure 11.8).

11.3.2.1.2.3 Metabolism Based Biosensors The detection of microorganism through the electrochemical monitoring of their metabolism is an interesting approach as it gives information on the viable cells. The main drawback which limits this approach is related to the poor selectivity which makes it applicable to well defined samples. Moreover the response is usually slow. The measurement scheme is based on the redox reactions of the metabolic pathways mediated via an electrochemical mediator. To date, 25 years of bacterial biosensing using this strategy led to a plethora of formats (124–127). The transducer can either detect the consumption of oxygen or of another electro active metabolite. Takayama et al. 1993 (125) showed that mediators such as $K_3Fe(CN)_6$ and dichlorophenol indophenol are able to accept electrons from glucose dehydrogenase (GDH) or from other enzymes of the respiratory chain, and that this mean can be used to detect viable bacteria.

11.3.2.2 Pesticides in Food

Since the end of World War II, in order to increase the production of both vegetal and animal products, chemicals have been extensively used both to control pests and microorganism infections. Today, over 500 compounds are registered worldwide as pesticides, or metabolites of pesticides.

A great number of pesticides are now used to raise crop productivity. It is calculated that the losses would double if pesticides were put out of use. However, the use of pesticides requires great care and control as they can pollute the environment and harm human health either directly or through the food chain.

Herbicides are by far the most commonly used pesticides, followed by insecticides and fungicides. This pesticide use, and additional environmental pollution caused by industrial emissions, has resulted in the occurrence of residues of these chemicals and their metabolites in food commodities, water, and soil. Legislations were enacted in the USA, the European Union (EU) and other countries to regulate pesticides in food products.

Multi residue methods (MRMs) of analysis, able to detect simultaneously more than one residue have been developed mainly based on chromatographic techniques. Two groups of MRMs are used: (1) multiclass MRMs that involve coverage of residues of various classes of pesticides, and (i2) selective MRMs, which concern multiple residues of chemically related pesticides (e.g., N-methyl carbamate pesticides (NMCs), carboxylic acids, and phenols). As foods are usually complex matrices, all of the preanalytical steps (matrix modification, extraction, and clean up), are often necessary.

Beside the instrumental methods, other analytical approaches, usually used as screening tests, are required because of the need of continuous monitoring of food commodities.

Organophosphate and carbamate pesticides are inhibitors of acetylcholinesterase, and therefore they are neurotoxic agents. The anticholinesterase activity of pesticides has been used as the basis to build a number of detection schemes for these classes of compounds. A large number of applications in pure standard solution or in environmental samples have been published (128) but only a few applications on real food samples have been reported. In spite of the significant demand for sensing of different analytes in food samples, it seems that the use of biosensors penetrate very slowly in the food industry.

Most of the inhibition bioassays or biosensors for organophosphate and carbamate pesticides are based on the amperometric detection of the enzymatic reaction product. Other detection techniques were also described, such as photo thermal analysis using excitation laser operating at 488 nm and 120 mW. This biosensor was tested for detection of carbofuran, propamocarb, oxydemeton–methyl, and parathion–ethyl in samples of salad, iceberg lettuce, and onion. The developed photo thermal biosensor offers a low cost mean to detect low concentrations of pesticides in vegetable samples with high throughput and little or no sample pretreatment (129).

Applications of amperometric biosensing strategies for pesticide detection in real or spiked food samples were recently reported. Most of the applications have been developed for vegetable matrices. Different formats of biosensors have been used: disposable screen printed choline oxidase biosensors (130) using acetylcholinesterase in solution, were utilized to detect pesticides in real samples of fruit and vegetable. Potato, carrot, and sweet pepper samples spiked with aldicarb, propoxur, carbaryl, carbofuran, and methomyl tested with this acetylcholinesterase biosensor, exhibited acceptable recoveries (79–96%) (131).

An interesting application on spiked (aldicarb, carbaryl, carbofuran, methomyl, or propoxur) fruit and vegetable samples was based on screen printed electrodes chemically modified with a carbon paste mix of cobalt(II) phtalocyanine, and acetylcellulose (132).

In this paper, a solventless extraction procedure of the spiked samples is reported. This was performed by mixing the fruits and passing the resulting juice through a sieve. No effect of matrices pH on the biosensor performance is reported. Screen printed sensor developed using photolithographic conducting copper track, graphite–epoxy composite and either acetylcholinesterase or butyrylcholinesterase were also used in the analysis of spiked (paraoxon and carbofuran) samples of tap water and fruit juices at subnanomolar concentration (133).

A recent application of a chemically modified electrode for the development of a biosensing device for chlorpyrifos methyl was developed by Del Carlo et al. (134). This method was demonstrated to detect the active molecule both in standard solution and in commercial products (Reldan® 22) with comparable sensitivity. The analytical protocol was then applied to vine and vine leaves samples in order to improve safety in wine making process (134) (Figure 11.9).

Other food samples that have been recently investigated with electrochemical biosensor based on acetylcholinesterase inhibition are infant foods. Considering the multitude of risks associated with pesticides intake by infants, the European Union has set a very low limit (10 μg/Kg) for pesticide in infant food (135). An amperometric biosensor which met the requirements set by the European Union, both for infant food and orange juice, with respect to detection limits has been developed. The method included an oxidation step of phosphorothioates pastiches to produce the oxygenated derivative, which represent the active pesticide molecule. Moreover the biosensors could be regenerated (i.e., the acetylcholinesterase activity could be recovered) through a chemical activation with pyridine-2-aldoxime methochloride (PAM). The biosensors performed well in solvent extract containing water, though it showed reduced recovery in food with lower water content (136).

Successive work by the same group addressed some of the major problems arising when electrochemical biosensors are in contact with food matrices: pH effect, and particle effect. Both problems were solved treating the biosensor surface with a Tween20®/phosphate buffer solution (pH 7.5) after the pesticide incubation. The treatment removed the

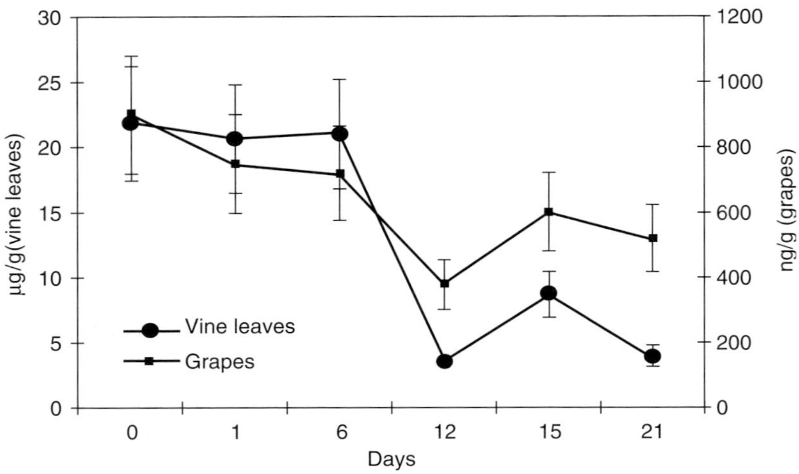

Figure 11.9 Analytical results obtained using an electrochemical method based on chemically modified screen printed electrodes for the detection of chlorpyriphos methyl in vine and vine leaves samples (redrawn from) Del Carlo, M., M. Mascini, A. Pepe, D. Compagnone, M. Mascini, *J. Agric. Food Chem.* 50:7206–7210, 2002.)

particulate and the correct pH was established while the pesticide enzyme inhibition was retained. A large number of samples (137) were analyzed and the results compared with reference standard methods.

As shown, there are only a few reports on cholinesterase based biosensors applied to food samples. Many other amperometric detection chemistries have been described that could be of potential use for food sample analysis, but the possibility to successfully apply these methods is closely related to their ability to overcome the matrix effect which leads either to complicated multistep sample preparation or to poor pesticide recovery.

11.3.2.3 Antibiotics in Food

The treatment of bacterial infections in cattle, sheep, pigs, and poultry with antibiotics, such as streptomycin and penicillin, is widely used in modern agriculture. Apart for their use in farming they have also been used in the protection of fruit trees. Public health concerns and perceptions of indiscriminate use of veterinary drugs have been supported by the presence of undesirable residues in some products.

The apprehension of the food industry on this topic is related to safety issues but also to technological problems that can arise from antibiotic contaminated products.

Residues of these antibiotics may present a potential hazard to the consumers in terms of toxicity, allergic reaction and the development of bacterial resistance. Moreover, in the dairy industry, milk contaminated with antibiotics can inhibit the starter cultures used in the production of yogurt and cheese thereby inflicting severe economic losses.

Regulatory authorities have established maximum residue limits (MRLs) for streptomycin and dihydrostreptomycin in milk, porcine kidney, and muscle, although none have yet been set for honey. Both the European Commission and the Codex Alimentarius MRLs have been set at 200μg/kg in both bovine and ovine milk, 1000μg/kg in porcine kidney, and 500μg/kg in porcine muscle. In contrast to the Commission MRLs, Codex MRLs are combined for streptomycin and dihydrostreptomycin. In the United States, a tolerance limit for dihydrostreptomycin has been set at 125μg/kg in milk, 2000μg/kg in porcine kidney, and 500μg/kg in porcine edible tissue.

Various analytical methods exist currently: high performance liquid chromatography (HPLC), gas chromatography mass spectroscopy (GC-MS) and liquid chromatography mass spectroscopy (LC-MS) have been described for confirmatory analysis, however, the long and tedious sample preparations involved prior to HPLC analysis also limit sample throughput.

Microbial inhibition assays, and immunoassays, are commonly employed as screening tests, however, recently the efficiency of agar diffusion techniques has been questioned.

A recent optical approach describes the use of white light interference for the detection of glycopeptides antibiotics in a fermentation process (138). Biosensor systems using the detection principle known as surface plasmon resonance (SPR) are an important addition to analytical instrumentation and have been shown to be versatile, robust, and capable of producing rapid and reliable results for the analysis of drug residues in complex matrices with minimal sample preparation (139–142). Among these approaches only reference 142 is applied to food samples. In particular, this was applied to accurately determine streptomycin and dihydrostreptomycin residues in milk, meat, and honey; results correlated well with a commercially available enzyme immunoassay (EIA) kit and a published confirmatory HPLC method.

11.3.2.4 Heavy Metals in Food

Heavy metals are a class of contaminants that can produce undesirable effects even if they are present in extremely minute quantities. All forms of ecological systems are affected to varying extents by heavy metals. They are the most insidious pollutants because of their nonbiodegradable nature and ability to persist for long periods.

Heavy metals from consumer's products may enter the environment either from natural sources or as discharge of the metallurgic industry, thus representing an almost ubiquitous environmental problem. They may contaminate fruit and vegetable constituents with different mechanisms, thus entering the food chain. Moreover, heavy metals are accumulated and stored in living organisms; especially in the marine organisms a very high bioaccumulation of heavy metals can take place. Finally they can be found in animal feedings. Recently European Community has established maximum admissible levels for metals in feeding stuff (European Directive 2002/32/CE), to prevent hazards to consumers health. Hence it is evident that, in a safety approach to food consumption, heavy metals must be monitored at various level of the food production process.

Various instrumental methods such as atomic absorption spectroscopy, inductively coupled plasma optical emission spectrometry, and inductively coupled plasma mass spectrometry (ICP-MS) are in wide use. These methods require sophisticated instrumentation and skilled personnel; therefore, there is a need for simpler methods of analysis. Classical electrochemical methods for metal ions determination include the use of ion selective electrodes, polarography, and other voltammetric techniques. The electrochemical methods are generally capable to selectively detect bioavailable heavy metal ions, that is, the free ion forms of metals. In this respect these methods deliver useful information on the potential hazards of metals.

Biosensors are promising analytical devices, and several different configurations have been described in the past for heavy metal detection. Three fundamental strategies can be outlined for biosensing of heavy metals: enzyme inhibition, cell based biosensor and chelating proteins as capture molecules.

11.3.2.4.1 Enzymes Inhibition Heavy metals are well known to inhibit the activity of enzymes. The main advantage of this inhibition mechanism for the determination of hazardous toxic elements consists in that the reduction of enzyme activity by a single inhibitor molecule can be large due to the amplification effect of the enzymatic reaction, leading to high sensitivity. Moreover, enzymes are often selective to the inhibitor and in many cases the inhibition effect of investigated pollutants is related to its biological toxicity. So the inhibition of enzymatic activity may offer a good choice as a simple and sensitive screen test for bioavailable metal ions. In these cases the choice of the enzyme leads to a certain degree of specificity with respect to the analyte. The postulated mechanism of enzyme inhibition by heavy metals is based on the interaction of metal ions with exposed thiol or methylthiol groups of protein amino acids often forming the active site of enzyme. The strongest interaction takes place in the case of mercury, copper and silver, and therefore those metals exhibit the largest inhibition effect.

Biosensors for mercury determination have been mainly developed, possibly due to its important toxicological profile. Numerous enzymes have been used for inhibition based determination of mercury traces, e.g., peroxidase (143,144), xanthine oxidase (145), invertase (146–148), glucose oxidase (149), butyrylcholinesterase (150), or isocitric dehydrogenase (151), but the most frequently applied is urease, as it is relatively cheap and easily available.

11.3.2.4.2 Cell Based Biosensors In cell based biosensing of heavy metals, two alternative strategies are prevalent: specific monitoring and the effect measurement. The former is obtained by means of one or more enzymes functionality in the cell on an inhibition basis; alternatively, a microbial strain that is deficient for a certain metal enzyme cofactor, is supplied with a medium containing the specific element. In this case the microbial metabolism strongly reacts. The latter is exemplified by Biological Oxygen Demand (BOD) measurements. Effects of heavy metal ions on cell metabolism may be quantified in such a manner.

11.3.2.4.3 Chelating Proteins as Capture and Transducer Element The use of metal–binding proteins can lead to the development of new generation of biosensors of varying specificity. It is believed that a large conformational change takes place when heavy metal ions bind to certain proteins. A suitable transducer should be able to detect this change directly. The use of strategies based on protein with distinct binding sites for heavy metals has been described. In this approach the conformational change induced by the selective binding of either Hg^{2+}, Zn^{2+}, Cd^{2+}, and Cu^{2+}, was used combined with a capacitive signal transducer which could detect the capacitance change in response to metal binding (152). All the metals could be detected as low as 10^{-15} M in standard solutions.

Despite the unanimously recognized need for the development and application of rapid, low cost, and easy to use methods for food analysis no application of biosensors to heavy metals detection has been so far described.

11.4 COMMERCIALLY AVAILABLE BIOSENSORS FOR FOOD ANALYSIS

Though a high number of publications, regarding the development and the applications of biosensors, appear every year, only very few devices are commercially available. Different studies on the biosensor market showed that 92% are sold in the clinical sector (90% being glucose sensors) while only 6% are commercialized in the food industrial sector (153). This high difference might be at least partially explained by the type of matrix in which the tests are carried out: medical biosensors are applied mainly on samples with a low variation of the matrix composition (blood or urine), whereas the biosensors in food industry must be adapted to a wide range of diverse matrices, each of them bearing particular difficulties in terms of matrix effects and concentration range of interest.

Biosensors for determination of food components, pathogens, toxins and pesticides are commercially available. Some examples and a summary of the biosensors characteristics are presented in Table 11.5.

The great majority of the biosensors available on the market are enzyme based, but antibody, receptor, or DNA based sensors are also commercialized. Among the first group, the working mechanism is generally based on a reaction catalyzed by an oxidase. Dehydrogenases are usually avoided, maybe because of the high cost and difficulty of immobilization of their cofactor, $NAD(P)^+$. Electrochemistry is the main principle of detection employed, with either the oxygen consumption or hydrogen peroxide formation, being amperometrically measured in combination with an immobilized oxidase. The principle of change in electrical conductance of the media caused by the growth of microorganisms has been exploited in the Malthus 2000 (Malthus Instruments Ltd) device for determination of coliform bacteria, yeasts, or fungi. The surface plasmon resonance principle of BIACORE® allows detection in real time of the specific interaction between an immobilized bio recognition element (antibody, receptor) and a variety of analytes (vitamins, hormones, and antibiotics). The biosensors are available in several forms, such as autoanalysers, manual laboratory instruments, and portable devices.

Even though the commercially available biosensors for determination of food components, pathogens and toxins are in continuous development, in regards to the principles of analysis, number of analytes that can be simultaneous measured, or sensors characteristics (sensitivity, selectivity, and robustness), the food industry is still not very receptive to biosensor technology. This limited acceptance might be due to the basic difference between classical analytical methods and bioassays in general, the latter requiring special conditions of operation, characterization, and data interpretation, due to the extraordinary

Table 11.5
Examples of commercial available biosensors used in food analysis.

Biosensor	Analytes	Biorecognition Element	Food Sample	Comments	Company	Ref.
ABD 3000 Biosensor Assay System	Ethanol	Alcohol oxidase	Beverages	- The enzyme is immobilized between two membrane layers; - O_2 or H_2O_2 sensors; - The biosensor assay is not affected by color, turbidity, particulates or pH and the sample requires little or no preparation; - Analysis time: 1-2 min; - < 100 μL sample size; - > 500 assays or 6 months; - Economical: ~10 cents per test; - Reproducibility: > 3 %.	Universal Sensor Inc., USA and Ireland	http://intel.ucc.ie/sensors/universal/
	Ascorbate	Ascorbate oxidase	Cereal products, Cookies			
	D-Glucose (Dextrose)	Glucose oxidase	Cereal products			
	Galactose	Galactose oxidase	Syrups, Cereals, Cheeses			
	Lactose	Galactose oxidase	Cheeses, Syrups, Cereal Products			
	L-Lysine	L-Lysine oxidase	Grains, Potatoes, Syrups			
	Sucrose	Invertase/Mutarotase/ Glucose oxidase	Potatoes, Cereals, Cookies			
SensAlyser-α	Alcohol	Alcohol oxidase	Beer and Wine production	- Analysis time: < 3min; - Compact and portable; - No reagents are required.	SensAlyse Ltd., UK	http://www.sensalyse.com
AM2 & AM3 Alcohol Analyser	Alcohol (Ethanol)	Alcohol oxidase	Cider, Beer, Wine, Spirit	- Analysis time: 20–25 sec; - Clark-type amperometric oxygen electrode; - Sample size: 5–10 μL; - Reagent stability: 9–15 months unopened and stored at 0–5 °C.	Analox Instruments, UK and USA	http://www.analox.com/

LM5 Lactate Analyser	Lactate	Lactate oxidase	Milk and Milk products			
GM10 Glucose Analyser	Dextrose (D-Glucose)	Glucose oxidase	Beverages			
	Lactose	β-Galactosidase/ Glucose oxidase				
	Sucrose	Invertase/Glucose oxidase				
YSI 2700 Select Food Analyser	Glucose	Glucose oxidase	Milk, Potatoes	- 10–25 μL sample size;	Yellow Springs Instruments, USA	http://www.ysi.com
	Lactose	Galactose oxidase	Cheese	- The enzyme is immobilized between two membrane layers, polycarbonate and cellulose acetate;		
	Galactose	Galactose oxidase	Fermentation process			
	Sucrose	Invertase/Mutarotase/ Glucose oxidase	Potatoes, Molasses, Cereal products, Ice cream			
	L-Lactate	L-Lactate oxidase	Lunch meat	- H_2O_2 formed is oxidized to a Pt electrode;		
	L-Glutamate	L-Glutamate oxidase	Tomatoes, Chicken broth			
	L-Glutamine	Glutaminase/ L-Glutamate oxidase	Pet foods	- Precision CV (n=10) ~2 %; - Analysis time: 1 min; - Number of samples which can be analysed automatically: 24;		
	Choline	Choline oxidase	n.a.			
	Ethanol	Alcohol oxidase	Beer			
	Methanol	Alcohol oxidase	Fermentation process			
	Starch	Amyloglucosidase/ Glucose oxidase	n.a.			

(*Continued*)

Table 11.5 (*Continued*)

Biosensor	Analytes	Biorecognition Element	Food Sample	Comments	Company	Ref.
SIRE® Biosensor P100	Ascorbic acid L-lactate Glucose Ethanol	Ascorbate oxidase Lactate oxidase Glucose oxidase Alcohol oxidase	Cereal products Baby food samples, Tomato paste n.a. Beverages	- Sensors with injectable recognition element; - Analysis time: 1 min; - The formed H_2O_2 gives an analyte concentration dependent of electrical signal - Repeatability: 1–3 %; - Reproducibility: 3–5 %;	Chemel AB, Sweden	http://www.chemel.com
SensLab 1a	Alcohols Glucose Lactate Ascorbate	Alcohol oxidase Glucose oxidase Lactate oxidase Ascorbate oxidase	Beverages Beverages n.a. n.a.	- Analysis time: 1min; - 5 μL sample size; - LR: 0–1 mM;	SensLab GmbH, Germany	http://www.senslab.de
The Answer 8000	Glucose	Horseradish peroxidase/ Glucose oxidase	Potatoes	- Automatic calibration; - Insulating cartridge filled with graphite, enzymes and ferrocene; - Analysis time: 20 sec.	Gwent Sensors Ltd., UK	http://www.g-s-l.co.uk/
OLGA	Glucose Sucrose Ascorbate Alcohol	Glucose oxidase n.a. Ascorbate oxidase Alcohol oxidase	Juice, Beer Juice, Beer n.a. Juice, Beer	- On-line general analyser based on principles of the sequential-injection mode; - Automatic calibration and dilution of sample; - Simultaneous analysis;	Sensolytics GmbH, Germany	http://www.iba-go.com/iba_us/olga/index.php?ol_intro.html

PerBacco 2000	Glucose Fructose	Glucose oxidase Fructose dehydrogenase	Must, Wine	- Solid binding matrix based composite transducers; - Filtration and dilution of the sample is recommended for red wines; - Remaining selectivity after 1 month of storage: > 90 %;	BioFutura S.R.L., Italy	http://www.biofutura.com
PerBacco 2002	Lactic acid Malic acid	Lactate dehydrogenase Malate dehydrogenase				
Biacore®Q	Folic acid Biotin Vitamin B12	Antibody, Enzyme, DNA or Receptor	Milk, Cereal, Juice, Flour, Jam, Pasta, Rice	-Surface plasmon resonance based biosensors; - Fully automated; - Quick and easy to change between different assays; - Analysis time: 2–10 min;	Biacore SA, Sweden	http://www.biacore.com
	Hormones Antibiotics	Antibody, Enzyme, DNA or Receptor	Meat and Milk production			
Freshness Meter KV-101	Adenosine Triphos-phate	n.a.	Fish, Meat	-Soluble enzymes and a Clark oxygen electrode; - Analysis time: 5–6 min;	Oriental Electric, Japan	(49,154)
Analyte 2000	E. coli 0157: H7, viruses, spores	Antibody	Hamburger, ham	- Evanescent-wave fluoroimmunoassay; - One sample per second;	Research International	http://www.resrchintl.com

Symbol: n.a. – not available.

nature of the bio recognition interaction. This attitude is, however, about to change, and can be directly related to numerous efforts on improving the biosensors robustness, by developing better stabilization and immobilization methods for the existing bio recognition elements, or by employing new artificially designed enzymes or antibodies.

11.5 CONCLUSIONS

Can we realistically imagine biosensors become widely applied analytical tools in food analysis?

In order to answer this question we have to consider two aspects: what is requested from the food industries, control agencies, and consumers, and how these needs will evolve in the next future, and secondly, the status of biosensors technological evolution, sample throughput, associated costs and acceptance by regulatory authorities.

Food industries have to play in a global market where the main characters are: costs, innovation, marketing, and fraudulent competition. Food industry needs analytical tools, characterized by ease of operation, low cost, and fast response to pursue safety and quality goals. Control Agencies operate as risk assessors, a duty of the risk managers, the politicians, from whom the consumers demand part of their request for control.

Consumers may act at two levels, asking the politicians to enforce warrant legislation and drive investment in control and safety, and exploiting actions of self assessment.

At present the main focus is certainly on safety, though it is difficult to separate this concept from quality. Recent food crisis, BSE in British beef, the tragic death of consumers as a result of *E. coli O157* in Scottish beef and the Belgian dioxin scandal in egg and chicken meat, which have been extensively reported by media and insufficiently coped with by authorities, had reduced consumer confidence toward food safety.

The controls cover all areas of chemical composition, hygienic aspects, and authenticity. In all of these, biosensors could possibly play an important role. In some cases, i.e., the electrochemical antibody based biosensors for *E. coli* O157 and *Campylobacter* (DETEX®) or others as described, biosensors are already a reality.

The potential value of biosensors technology in food safety and quality is basically related to their peculiar features: low cost per analysis can lead to increase the number of tests that industry can afford, with no consequence on the product price, fast response, and reduced sample preparation can increase the throughput of control laboratories enabling wide screening tests, ease of operation can enable consumers (distributors and end users) to perform their own tests (i.e., meat freshness or pesticide residue in vegetables or fruit).

Biosensor technology will have an impact on food safety and quality if the developed devices will be able to produce extensive data for anticipatory studies (i.e., emerging or new pathogens), regulatory (i.e., general toxicity index associated to group of contaminants), historical series (i.e., geographical area of risk).

An emerging area that will need appropriate regulations, in the next future, is that of the functional foods and nutraceuticals. These are foods where functional (i.e., antioxidants, probiotics) constituents or ingredients are present. Exaggerated or even fraudulent claims, in the form of marketing or labelling, are currently used by industry to drive consumer's choice. There is a trend to bind health issues with food consumption, orienting the market choice toward new food. This may represent a risk when the consumption of certain foods supposed to have beneficial effects on health is not supported by sound scientific evidences. In this area it is essential to gain exact knowledge of the functional constituents of food, its availability, and its metabolic fate before any health claims can be made.

In this respect, both catalytic and affinity biosensors can provide information on the activity of such constituents as the analytical response is produced by a biological reaction that can be adapted to gain functional information (i.e, protection with respect to an oxidising stress agent).

Possibly in the next years chemical sensor technology, and biosensors, will enter the HACCP procedures enabling an active approach to hazard analysis, creating the basis for an evolution of that approach to food safety and quality.

According to a survey published in 1999 when consumers were asked who they believe, the answer were: Environmental organizations (23%), Consumers organization (16%), Agricultural organization (16%), National authorities (4%), Industry (1%). And, even more notably, 75% believe that nobody can be fully trusted to speak the truth. It is evident that the poor trust toward national authorities and industry can be improved with a multilateral approach comprehensive of educational, communicative, and technical aspects. The availability of alternative and cheaper analytical tools can help to reduce the outbreaks of food related incidents and contribute to regain the consumers trust.

All of these actions require innovation (i.e., *new* analytes), improvement (i.e., selectivity, sensitivity), or simplification (i.e., sample pretreatment, analyte extraction) of the analytical protocols. Whether biosensor research will be able to tackle the challenges of the future depends on the ability of scientists to understand which technological aspects need further improvement and development.

It is a fact that biosensors are now increasingly becoming part of the main stream real time analytical measurement tools as detailed.

There is a clear need to develop biosensors for food analysis, both for screening a group of analytes (e.g., total amino acids and pesticides) and for identification of individual analyte within a group (e.g., different amino acids or single pesticide molecule) simultaneously. The technology to develop these types of class specific or single analytes detection schemes is available for many classes of compounds (e.g., mycotoxin, organophosphate and carbamate pesticide, polychlobiphenyls). However, for a biosensor, it is not only the new or enhanced transducer element that will form the basis of future sensors released onto the market, but also new sensing elements. Major limitations to the spreading of the use of biosensors, included the ones mentioned, are the instability of the (biological) sensing element and the cost that still remain noncompetitive. An important step forward is expected to come from the research on innovative biological recognition elements, such as engineered enzymes and abzymes (antibody possessing enzymic activity). In the affinity biosensors field, which seems the most promising with respect to commercial applications, the alternative physicochemical recognition elements, in their young age for sensing applications, are synthetic peptides and protein, nucleic acids, and molecular imprinted polymers (MIPs). Unlike antibodies or enzymes, the MIPs are inherently more stable and do not degrade under harsh environmental conditions; the active site can be regenerated many times without loss of recognition affinity and are relatively inexpensive to produce. Currently, however, their limitations include the use of high template concentrations in the production step leading to problems of incomplete template removal and hence, leaching of the target analyte causing high background noise. More research is needed before scale up procedures for the production of the MIPs can enter diagnostic companies.

Finally, there are at least two other issues that have to be considered to understand the potential impact of biosensors in the area of food analysis; automation and miniaturization. Nanotechnology (75–78) coupled with the laboratory on a chip (LOC) (72–74) concept appears to be able to change the entire way of considering chemical analysis and process control. Nanotechnology refers to exploitation of processes (e.g., physical, chemical, mechanical, and biological) to generate and utilize structures, components, and devices with a size range of about 0.1 nm (atomic and molecular scale) to about 100 nm (or larger in some cases) by control at atomic, molecular and macromolecular levels. It has been suggested that nanoscale sensors (e.g., monitoring of living cell metabolism at atomic level) and ultraminiaturized sensors (e.g., monitoring of a range of chemicals) could lead to the next generation of biotechnology based industries. In this context, a nanosensor (also referred to as an ion channel switch biosensor), which functions like a simplified cell membrane that is able of highly sensitive and rapid analytical measurements by monitoring changes in the electrical conductivity, has been reported (79). The concept of LOC entails miniaturization of all the essential components of analytical instrumentation (e.g., sample preparation, resolution of the components, reaction with appropriate reagents and detection) by microfabrication on a chip. Some of the components in LOC technology have already been released on the market (e.g., LabChip1 and GeneChip1 from Caliper Technologies and Af-matrix, respectively). A limitation to the full exploitation of lab-on-chip possibilities is sampling strategy and sample preparation. Going to the micronanoscale, in fact, gives the possibility to run several parallel assays in very small volumes but a careful sampling strategy is needed for food samples due to their extreme heterogeneity in most cases. Additionally, in many applications the sample preparation time, not the biosensor response is the rate determining step of the analytical process. Extreme pHs and ionic strength are noncompatible with most biosensors, thus a dilution or neutralising step is necessary. In the case of solid, or hydrophobic, food stuff an extraction procedure is required. Miniaturization of extraction procedure using microfluidic flow systems will enable to move the sample preparation closer to the biosensor in the LOC approach. In any case, new sampling and extraction procedures need to be pursued to fully utilize the potential of biosensors. Application of biosensors in organic solvents has been described and also the improvement of this possibility can enlarge the market for biosensors.

Six years ago, Luong (154) concluded his review on biosensors in food analysis on Tibtech with these words: "Biosensors must be inexpensive, reliable, robust enough to operate under realistic, and perhaps hostile, conditions, and possibly form part of highly integrated analytical system. This aspect is both technically challenging and mandatory for a successful commercialization of biosensors in this conservative and competitive industrial sector."

We have shown that in the period 1997–2003, the use of biosensors in the food area has received a considerable attention. Many devices have been developed and tested for application in food samples, some of which have entered the market. The results obtained in the challenges detailed, such as biocomponent stability, device robustness, LOC, and miniaturization development will determine the success of biosensor methods of analysis in the food sector over the next year.

ACKNOWLEDGMENTS

This work was financially supported by the European Commission (CRAFT, contract no. QLK1-CT-2002-70884 and NOVTECH, contract no. HPRN-CT-2002-00186), INTAS 00-0751, the Swedish Council for Forestry and Agricultural Research (SJSF), and The Swedish Research Council (contract no. 621-2002-4298).

REFERENCES

1. Adams, M.R., M.O. Moss. *Food Microbiology*, 2nd ed. Cambridge: Royal Society of Chemistry, 2000, pp 395–438.
2. Wagner, G., G.G. Guilbault. *Food biosensors analysis*. New York: Marcel Dekker, 1994, pp 11–103.
3. Cunningham, A.J. *Bioanalytical sensors*. New York: John Wiley & Sons, 1998, pp 1–11.
4. Scheller, F., F. Schubert. *Biosensors*. Amsterdam: Elsevier Science, 1992, pp 1–9.
5. Scott, A.O. *Biosensors for food analysis*. Cambridge: Royal Society of Chemistry, 1998, pp 1–9.
6. Clark, L., C. Lyons. *Ann. NY Acad. Sci.* 148:133–153, 1962.
7. Updike, S., G. Hicks. Enzyme electrode. *Nature* 214:986, 1967.
8. Adanyi, N., M. Toth-Markus, E.E. Szabo, M. Varadi, M.P. Sammartino, M. Tomassetti, L. Campanella. Investigation of organic phase biosensor for measuring glucose in flow injection analysis system. *Anal. Chim. Acta,* 2003. [In press]
9. Ricci, F., A. Amine, C.S. Tuta, A.A. Ciucu, F. Lucarelli, G. Palleschi, D. Moscone. Prussian blue and enzyme bulk-modified screen-printed electrodes for hydrogen peroxide and glucose determination with improved storage and operational stability. *Anal. Chim. Acta* 485:111–120, 2003.
10. Niculescu, M., R. Mieliauskiene, V. Laurinavicius, E. Csöregi. Simultaneous detection of ethanol, glucose and glycerol in wines using pyrroloquinoline quinone-dependent dehydrogenases based biosensors. *Food Chem.* 82:481–489, 2003.
11. Lenarczuk, T., D. Wencel, S. Glab, R. Koncki. Prussian blue-based optical glucose biosensor in flow-injection analysis. *Anal. Chim. Acta* 447:23–32, 2001.
12. Ramanathan, K., B.R. Jönsson, B. Danielsson. Sol-gel based thermal biosensor for glucose. *Anal. Chim. Acta* 427:1–10, 2001.
13. Palmisano, F., R. Rizzi, D. Centonze, P.G. Zambonin. Simultaneous monitoring of glucose and lactate by an interference and cross-talk free dual electrode amperometric biosensor based on electropolymerized thin films. *Biosens. Bioelectron.* 15:531–539, 2000.
14. Chia, J.L.L.S., N.K. Goh, S.N. Tan. Renewable silica sol-gel derived carbon composite based glucose biosensor. *J. Electroanal. Chem.* 460:234–241, 1999.
15. Surareungchai, W., W. Supinda, P. Sritongkum, M. Tanticharoen, K. Kirtikara. Dual electrode signal-subtracted biosensor for simultaneous flow injection determination of sucrose and glucose. *Anal. Chim. Acta* 380:7–15, 1999.
16. Volotovsky, V., N. Kim. Determination of glucose, ascorbic and citric acids by two-ISFET multienzyme sensor. *Sens. Actuators B* 49:253–257, 1998.
17. Jenkins, D.M., M.J. Delwiche. Adaptation of a manometric biosensor to measure glucose and lactose. *Biosens. Bioelectron.* 18:101–107, 2003.
18. Moscone, D., R.A. Bernardo, E. Marconi, A. Amine, G. Palleschi. Rapid determination of lactulose in milk by microdialysis and biosensors. *Analyst* 124:325–329, 1999.
19. Stredansky, M., A. Pizzariello, S. Stredanska, S. Miertus. Determination of D-fructose in foodstuffs by an improved amperometric biosensor based on a solid binding matrix. *Anal. Commun.* 36:57–61, 1999.
20. Maestre, E., I. Katakis, E. Dominguez. Amperometric flow-injection determination of sucrose with a mediated tri-enzyme electrode based on sucrose phosphorylase and electrocatalytic oxidation of NADH. *Biosens. Bioelectron.* 16:61–68, 2001.
21. Lupu, A., D. Companone, G. Palleschi. Screen-printed enzyme electrodes for the detection of marker analytes during winemaking. *Anal. Chim. Acta* 2003.
22. De Prada, A.G.V., N. Pena, M.L. Mena, A.J. Reviejo, J.M. Pingarron. Graphite-teflon composite bienzyme amperometric biosensors for monitoring of alcohols. *Biosens. Bioelectron.* 18:1279–1288, 2003.

23. Niculescu, M., T. Erichsen, V. Sukharev, Z. Kerenyi, E. Csöregi, W. Schuhmann. Quinohemoprotein alcohol dehydrogenase-based reagentless amperometric biosensor for ethanol monitoring during wine fermentation. *Anal. Chim. Acta* 463:39–51, 2002.
24. Mello, L.D., M.P. Taboada Sotomayor, L.T. Kubota. HRP-based amperometric biosensor for the polyphenols determination in vegetables extract. *Sens. Actuators. B* 96:636–645, 2003.
25. Capannesi, C., I. Palchetti, M. Mascini, A. Parenti. Electrochemical sensor and biosensor for polyphenols detection in olive oils. *Food Chem.* 71:553–562, 2000.
26. Avramescu, A., T. Noguer, M. Avramescu, J.L. Marty. Screen-printed biosensors for the control of wine quality based on lactate and acetaldehyde determination. *Anal. Chim. Acta* 458:203–213, 2002.
27. Katrlik, J., A. Pizzariello, V. Mastihuba, J. Svorc, M. Stredansky, S. Miertus. Biosensors for L-malate and L-lactate based on solid binding matrix. *Anal. Chim. Acta* 379:193–200, 1999.
28. Serra, B., A.J. Reviejo, C. Parrado, J.M. Pingarron. Graphite-teflon composite bienzyme electrodes for the determination of L-lactate: application to food samples. *Biosens. Bioelectron.* 14:505–513, 1999.
29. Mizutani, F., Y. Hirata, S. Yabuki, S. Iijima. Flow injection analysis of acetic acid in food samples by using trienzyme/poly(dimethylsiloxane)-bilayer membrane-based electrode as the detector. *Sens. Actuators B* 91:195–198, 2003.
30. Maines, A., M.I. Prodromidis, S.M. Tzouwara-Karayanni, M.I. Karayannis, D. Ashworth, P. Vadgama. An enzyme electrode for extended linearity citrate measurements based on modified polymeric membranes. *Electroanalysis* 12:1118–1123, 2000.
31. Akyilmaz, E., E. Dinckaya. A new enzyme electrode based on ascorbate oxidase immobilized in gelatin for specific determination of L-ascorbic acid. *Talanta* 50:87–93, 1999.
32. Caselunghe, M.B., J. Lindeberg. Biosensor-based determination of folic acid in fortified food. *Food Chem.* 70:523–532, 2000.
33. Wu, X.J., M.M.F. Choi. Hydrogel network entrapping cholesterol oxidase and octadecylsilica for optical biosensing in hydrophobic organic or aqueous micelle solvents. *Anal. Chem.* 75:4019–4027, 2003.
34. Dominguez, R., B. Serra, A.J. Reviejo, J.M. Pingarron. Chiral analysis of amino acids using electrochemical composite bienzyme biosensors. *Anal. Biochem.* 298:275–282, 2001.
35. Kelly, S.C., P.J. O'Connell, C.K. O'Sullivan, G.G. Guilbault. Development of an interferent free amperometric biosensor for determination of L-lysine in food. *Anal. Chim. Acta* 412:111–119, 2000.
36. Niculescu, M., C. Nistor, I. Frebort, P. Pec, B. Mattiasson, E. Csoregi. Redox hydrogel based amperometric bienzyme electrodes for fish freshness monitoring. *Anal. Chem.* 72, 2000.
37. Hu, S., C. Xu, J. Luo, D. Cui. Biosensor for detection of hypoxanthine based on xanthine oxidase immobilised on chemically modified carbon paste electrode. *Anal. Chim. Acta* 412:55–61, 2000.
38. Cosnier, S., C. Gondran, J.C. Watelet, W. Giovani, R.P.M. Furriel, F.A. Leone. A bienzyme electrode (alkaline phosphatase - polyphenol oxidase) for the amperometric determination of phosphate. *Anal. Chem.* 70:3952–3956, 1998.
39. Abass, A.K., J.P. Hart, D. Cowell. Development of an amperometric sulfite biosensor based on sulfite oxidase with cytochrome c, as electron acceptor, and a screen-printed transducer. *Sens. Actuators B* 62:148–153, 2000.
40. Ercole, C., M. Del Gallo, M. Pantalone, S. Santucci, L. Mosiello, C. Laconi, A. Lepidi. A biosensor for *Escherichia coli* based on a potentiometric alternating biosensing (PAB) transducer. *Sens. Actuators B.* 83:48–52, 2002.
41. Su, X., S. Low, J. Kwang, V.H.T. Chew, S.F.Y. Li. Piezoelectric quartz crystal based veterinary diagnosis for *Salmonella enteritidis* infection in chicken and egg. *Sens. Actuators B.* 75:29–35, 2001.
42. Bokken, G.C.A.M., R.J. Corbee, F. Knapen, A.A. Bergwerff. Immunochemical detection of *Salmonella* group B, D and E using an optical surface plasmon resonance biosensor. *FEMS Microbiol. Lett.* 222:75–82, 2003.

43. Nedelkov, D., A. Rasooly, R.W. Nelson. Multitoxin biosensor-mass spectrometry analysis: a new approach for rapid, real-time, sensitive analysis of staphylococcal toxins in food. *Int. J. Food Microb*. 60:1–13, 2000.
44. Pogacnik, L., M. Franko. Detection of organophosphate and carbamate pesticides in vegetable samples by a photothermal biosensor. *Biosens. Bioelectron*. 18:1–9, 2003.
45. Albareda-Sirvent, M., A. Merkoci, S. Alegret. Pesticide determination in tap water and juice samples using disposable amperometric biosensors made using thick-film technology. *Anal. Chim. Acta* 442:35–44, 2001.
46. Gustavsson, E., P. Bjurling, Å. Sternesjö. Biosensor analysis of penicillin G in milk based on the inhibition of carboxypeptidase activity. *Anal. Chim. Acta* 468:153–159, 2002.
47. Gaudin, V., J. Fontaine, P. Maris. Screening of penicillin residues in milk by a surface plasmon resonance-based biosensor assay: comparison of chemical and enzymatic sample pre-treatment. *Anal. Chim. Acta* 436:191–198, 2001.
48. Krawczyk, T.K., M. Moszczynska, M. Trojanowicz. Inhibitive determination of mercury and other metal ions by potentiometric urea biosensor. Biosens. Bioelectron. 15:681–691, 2000.
49. Mello, L.D., L.T. Kubota. Review of the use of biosensors as analytical tools in the food and drink industries. *Food Chem*. 77:237–256, 2002.
50. Csöregi, E., S. Gáspár, M. Niculescu, B. Mattiasson, W. Schuhmann. Amperometric enzyme-based biosensors for application in food and beverage industry. In: *Focus on Biotechnology Series: Physics and Chemistry Basis of Biotechnology*, de Cuyper, M., J.W.M. Bulte, eds., Dordrecht: Kluwer, 2001, pp 105–129.
51. Razumiene, J., V. Gureviciene, V. Laurinavicius, J.V. Gražulevicius. Amperometric detection of glucose and ethanol in beverages using flow cell and immobilised on screen-printed carbon electrode PQQ-dependent glucose or alcohol dehydrogenases. *Sens. Actuators B*. 78(1–3):243–248, 2001.
52. Razumiene, J., A. Vilkanauskyte, V. Gureviciene, V. Laurinavicius, N.V. Roznyatovskaya, Y.V. Ageeva, M.D. Reshetova, A.D. Ryabov. New bioorganometallic ferrocene derivatives as efficient mediators for glucose and ethanol biosensors based on PQQ-dependent dehydrogenases. *J. Org. Chem*. 668(1,2):83–90, 2003.
53. de Mattos, I.L., L.V. Lukachova, L. Gorton, T. Laurell, A.A. Karyakin. Evaluation of glucose biosensors based on Prussian Blue and lyophilised, crystalline and cross-linked glucose oxidases. *Talanta* 54(5):963–974, 2001.
54. Tkác, J., I. Voštiar, E. Šturdík, P. Gemeiner, V. Mastihuba, J. Annus. Fructose biosensor based on D-fructose dehydrogenase immobilised on a ferrocene-embedded cellulose acetate membrane *Anal. Chim. Acta* 439(1):39–46, 2001.
55. Rotariu, L., C. Bala, V. Magearu. Yeast cells sucrose biosensor based on a potentiometric oxygen electrode. *Anal. Chim. Acta* 458(1):215–222, 2002.
56. Pena, N., R. Tàrrega, A.J. Reviejo, J.M. Pingarròn. Reticulated vitreous carbon based composite bienzyme electrodes for the determination of alcohols in beer samples. *Anal. Lett*. 35(12):1931–1944, 2002.
57. Tkac, J., I. Vostiar, P. Gemeiner, E. Sturdik. Monitoring of ethanol during fermentation using a microbial biosensor with enhanced selectivity. *Bioelectrochemistry* 56(1,2):127–129, 2002.
58. Campo Dall'Orto, V., C. Danilowicz, I. Rezzano, M. Del Carlo, M. Mascini. Comparison between three amperometric sensors for phenol determination in olive oil samples. *Anal. Lett*. 231:124–131, 1999.
59. Romani, A., M. Minunni, N. Mulinacci, P. Pinelli, F.F. Vincieri, M. Del Carlo, M. Mascini. Comparison among differential pulse voltammetry, amperometric biosensor, and HPLC/DAD analysis for polyphenol determination. *J. Agric. Food Chem*. 48(4):1197–1203, 2000.
60. Setford, S.J., S.F. White, J.A. Bolbot. Measurement of protein using an electrochemical bi-enzyme sensor. *Biosens. Bioelectron*. 17(1,2):79–86, 2002.

61. Kwan, R.C.H., C. Chan, R. Renneberg. An amperometric biosensor for determining amino acids using a bienzymatic system containing amino acid oxidase and protease. *Biotechnol. Lett.* 24(14):1203–1207, 2002.
62. Moser, I., G. Jobst, G.A. Urban. Biosensor arrays for simultaneous measurement of glucose, lactate, glutamate, and glutamine. *Biosens. Bioelectron.* 17(4):297–302, 2002.
63. Nanjyo, Y., T. Yao. Rapid measurement of fish freshness indices by an amperometric flow-injection system with a 16-way switching valve and immobilized enzyme reactors. *Anal. Chim. Acta* 470(2):175–183, 2002.
64. Zhao, C., Y. Pan, L. Ma, Z. Tang, G. Zhao, L. Wang. Assay of fish freshness using trimethylamine vapor probe based on a sensitive membrane on piezoelectric quartz crystal. *Sens. Actuators B* 81(2,3):218–222, 2002.
65. Marcinkeviciene, L., I. Bachmatova, R. Semenaite, I. Lapenaite, B. Kurtinaitiene, V. Laurinavicius, R. Meskys. Characterization and application of glycerol dehydrogenase from *Gluconobacter* sp. 33. *Biologija* 2:49–51, 2000.
66. Marcinkeviciene, L., I. Bachmatova, R. Semenaite, R. Rudomanskis, G. Brazenas, R. Meskiene, R. Meskys. Purification and characterization of alcoholdehydrogenase from *Gluconobacter* sp. 33. *Biologija* 2:25–30, 1999.
67. Marcinkeviciene, L., I. Bachmatova, R. Semenaite, R. Rudomanskis, G. Brazenas, R. Meskiene, R. Meskys. Purification and characterization of PQQ-dependent glucose dehydrogenase from *Erwinia* sp.34-1. *Biotech. Lett.* 21:187–192, 1999.
68. Suzzi, G., F. Gardini. Biogenic amines in dry fermented sausages: a review. *Int. J. Food Microb.* 88:41–54, 2003.
69. Invitski, D., I. Abdel-Hamid, P. Atanasov, E. Wilkins. Biosensors for the detection of pathogenic bacteria. *Biosens. Bioelectron.* 14:599–624, 1999.
70. Anonymous. Waterborne pathogens kill 10M–20M people/year. *World Water Environ. Eng.* June 6, 1996.
71. Donnelly, C.W. Listeria monocytogenes: a continuing challenge. *Nutr. Rev.* 59:183–94, 2001.
72. Schlech,W.F., III. Foodborne listeriosis. *Clin. Infect. Dis.* 31:770–775, 2000.
73. De Boer, E., R.R. Beumer. Methodology for detection and typing of foodborne microorganisms. *Int. J. Food. Microbiol.* (50):119–130, 1999.
74. Artault, S., J.L. Blind, J. Delaval, Y. Dureuil, N. Gaillard. Detecting *Listeria monocytogenes* in food. *Int. Food. Hyg.* 12(3):23, 2001.
75. Toze, S. PCR and the detection of microbial pathogens in water and wastewater. *Water Res.* 33:3545–3556, 1999.
76. Bsat, N., M. Weidmenn, J. Czajka, F. Barany, M. Piani, C.A. Batt. Food safety applications of nucleic acid-based assays. *Food Technol.* 48:142–145, 1994.
77. Schwab, K.J., R. De Leon, M.D. Sobsey. Immunoaffinity concentration and purification of waterborne enteric viruses for detection by reverse transcriptase PCR. *Appl. Environ. Microbiol.* 63:4401–4407, 1996.
78. Jensen, M., J.A. Webster, N. Straus. Rapid identification of bacteria on the basis of polymerase chain reaction-amplified ribosomal DNA spacer polymorphisms. *Appl. Environ. Microbiol.* 59:945–52, 1993.
79. Rochelle, P.A., R. De Leon, M.H. Stewart, R.L.Wolfe. Comparison of primers and optimization of PCR conditions for detection of *Crypotosporidium parvum* and *Giardia lamblia* in water. *Appl. Environ. Microbiol.* 63:106–114, 1997.
80. Geary, T.G. Approaching helminth biology from the molecular direction. *Parasitol. Today* 12:373–375, 1996.
81. Köhler, G., C. Milstein. Continuous cultures of fused cells secreting antibody of predefined specificity. *Nature* 310:792–794, 1975.
82. Iqbal, S.S., M.W. Mayo, J.G. Bruno, B.V. Bronk, C.A. Batt, P. Chambers. A review of molecular recognition technologies for detection of biological threat agents. *Biosens. Bioelectron.* 15:549–578, 2000.
83. Hauck, S., C. Kosslinger, S. Drost, H. Wolf. *Lebensmittelchemie* 52:158–167,1998.
84. Rasooly, L., A. Rasooly. *Int. J. Food Microbiol.* 49:119–124, 1999.

85. Sloper, A.N., J.K. Deacon, M.T. Flannagan. A planar indium phosphate monomode waveguide evanescent field immunosensor. *Sens. Actuators B.* 1:285–297, 2001.
86. Pollard-Knight, D.V., E. Hawkins, D. Yeung, D.P. Paswhby, M. Simpson, A. McDougall, P. Buckle, S.A. Charles. Immunoassay and nucleic acid detection with a biosensor based on surface plasmon resonance. *Ann. Biol. Clin.* 48:642–646, 1990.
87. Karlsson, R., A. Michaelsson, L. Mattsson. Kinetic analysis of monoclonal antibody-antigen interactions with a new biosensor based analytical system *J. Immunol. Meth.* 145:229–240, 1991.
88. Brigham-Burke, M., J.R. Edwards, D.J. O'Shannessy. Detection of receptor ligand interactions using surface plasmon resonance: model studies employing the HIV-1 GP120/CD4 interactions. *J. Anal. Biochem.* 205:125–131, 1992.
89. Medina, M.B., L. Van Houten, P. Cooke, S.I. Tu. Real time analysis of antibody binding interactions with immobilized *E.coli* 0157:H7 cells using the BIAcore. *Biotechnol. Tech.* 11(3):173–176, 1997.
90. Fratamico, P.M., T.P. Strobaugh, M.B. Medina, A.G. Gehring. Detection of *Escherichia coli* 0157:H7 using a surface plasmon resonance biosensor. *Biotechnol. Tech.* 12(7):571–576, 1998.
91. Nakamura, N., A. Shigematsu, T. Matsunaga. Electrochemical detection of viable bacteria in urine and antibiotic selection. *Biosens. Bioelectron.* 6(7):575–580, 1991.
92. Schneider, B.H., J.G. Edwards, N.F. Hartman. Hartman interferometer: versatile integrated optic sensor for label-free, real time quantification of nucleic acids, proteins and pathogens. *Clin. Chem.* 43(9):1757–1763, 1997.
93. Turpin, P.E., K.A. Maycroft, J. Bedford, C.L. Rowlands, E.M.H. Wellington. A rapid luminescent phage-based MPN method for the enumeration of *Salmonella typhimurium* in environmental samples. *Lett. Appl. Microbiol.* 16:24–27, 1993.
94. Chen, J., M.W. Griffiths. *Salmonella* detection in eggs using lux bacteriophages. *J. Food Protect.* 59:908–914, 1996.
95. Loessner, M.J., C.E.D. Rees, G.S.A.B. Stewart, S. Scherer. Construction of luciferase reporter bacteriophage A511:lux AB for rapid and sensitive detection of viable *Listeria* cells. *Appl. Environ. Microbiol.* 62:1133–1140, 1996.
96. Blasco, R., M.J. Murphy, M.F. Sanders, D.J. Squirrell. Specific assays for bacteria using phage mediated release of adenylate kinase. *J. Appl. Microbiol.* 84:661–666, 1998.
97. Bao, L., L. Deng, L. Nie, S. Yao, W. Wie. Determination of microorganisms with a quartz crystal microbalance sensor. *Anal. Chim. Acta* 319:97–101, 1996.
98. Hobson, N.S., I. Tothill, A.P.F. Turner. Microbial Detection. *Biosens. Bioelectron.* 11(5):455–477, 1996.
99. Muramatsu, H., K. Kajiwara, E. Tamiya, I. Karube. Piezoelectric immunosensor for the detection of *Candida albicans* microbes. *Anal. Chim. Acta* 188:257–261, 1986.
100. Prusak-Sochaczewski, E., J.H.T. Luong, G.G. Guilbault. Development of a piezoelectric immunosensor for the detection of *Salmonella typhimurium. Enzyme Microb. Technol.* 12:173–177, 1990.
101. Babacan, S., P. Pivarnick, S. Letcher, A. Rand. Piezoelectric flow injection analysis biosensor for the detection of *Salmonella typhimurium. J. Food Sci.* 67(1):314–320, 2002.
102. Jacobs, M.B., R.M. Cater, G.J. Lubrano, G.G. Guilbault. A piezoelectric biosensor for *Listeria monocytogenes. Am. Lab.* 27(11):26–28, 1995.
103. Koenig, B., M. Gratzel. Detection of viruses and bacteria with piezoelectric immunosensors. *Anal. Lett.* 26(8):1567–1585, 1993.
104. Ye, J.M., S.V. Letcher, A.G. Rand. Piezoelectric biosensor for detection of *Salmonella typhimurium. J. Food Sci.* 62(5):1067–1072, 1997.
105. Harteveld, J.L.N., M.S. Nieuwenhuizen, E.R.J. Wils. Detection of *Staphylococcal Enterotoxin* B employing a piezoelectric crystal immunosensor. *Biosens. Bioelectron.* 12:661–667, 1997.
106. Tombelli, S., M. Mascini, C. Sacco, A.P.F. Turner. A DNA piezoelectric biosensor assay coupled with a polymerase chain reaction for bacterial toxicity determination in environmental samples. *Anal. Chim. Acta* 418:1–9, 2001.

107. Maramatsu, H., K. Kajiwara, E. Tamiya, I. Karube. Piezoelectric immunosensor for the detection of *Candida albicans* microbes. *Anal. Chim. Acta* 188:257–261, 1986.
108. Jacobs, M.B., R.M. Cater, G.J. Lubrano, G.G. Guilbault. A piezoelectric biosensor for *Listeria monocytogenes*. *Am. Lab.* 27(11):26–28, 1995.
109. Vaughan, R.D., C.K. O'Sullivan, G.G. Guilbault. Development of quartz crystal microbalance (QCM) immunosensor for the detection of *Listeria monocytogenes*. *Enzyme Microb. Tech.* 29:635–638, 2001.
110. Cater, R.M., J.J. Mekalanos, M.B. Jacobs, G.J. Lubrano, G.G. Guilbault. Quartz crystal microbalance detection of *Vibro cholerae* 0139 serotype. *J. Immunol. Methods* 187:121–125, 1995.
111. Su, X., S. Low, J. Kwang, V.H.T. Chew, S.F.Y. Li. Piezoelectric quartz crystal based veterinary diagnosis for *Salmonella enteritidis* infection in chicken and egg. *Sens. Actuators B.* 75:29–35, 2001.
112. Gehring, A.G., C.G. Crawford, R.S. Mazenko, L.J. Van Houten, J.D. Brewster. Enzyme-linked immunomagnetic electrochemical detection of *Salmonella typhimurium*. *J. Immunol. Methods* 195:15–25, 1996.
113. Tiefenauer, L.X., S. Kossek, C. Padeste, P. Thiébaud. Towards amperometric immunosensor devices. *Biosens. Bioelectron.* 12(3):213–223, 1997.
114. Mirhabibollahi, B., J.L. Brooks, R.G. Krool. A semi-homogeneous amperometric immunosensor for protein A-bearing *Staphylococcus aureus* in foods. *Appl. Microb. Biotechnol.* 34:242–247, 1990.
115. H.J. Kim, H.P. Bennetto, M.A. Halablab. A novel liposome based electrochemical biosensor for the detection of haemolytic microorganisms. *Biotechnol. Tech.* 9(6):389–394, 1995.
116. Brewster, J.D., R.S. Mazenko. Filtration capture and immunoelectrochemical detection for rapid assay of *Escherichia coli* 0157:H7. *J. Immunol. Methods* 211:1–8, 1998.
117. Rishpon, J., D. Ivnitski. An amperometric enzyme-channeling immunosensor. *Biosens. Bioelectron.* 12(2):195–204, 1997.
118. Ivnitski, D., E. Wilkins, H.T. Tien, A. Ottova. Electrochemical biosensor based on supported planar lipid bilayers for fast detection of pathogenic bacteria. *ELECOM* 2:457–460, 2000.
119. Minett, A.I., J.N. Barisci, G.G. Wallace. Coupling conducting polymers and mediated electrochemical responses for the detection of *Listeria*. *Anal. Chim. Acta* 475(1-2):37–45, 2003.
120. Baeumner, A.J., R.N. Cohen, V. Miksic, J. Min. RNA biosensor for the rapid detection of viable *Escherichia coli* in drinking water. *Biosens. Bioelectron.* 18(4):405–413, 2003.
121. Almadidy, A., J. Watterson, P.A.E. Piunno, S. Raha, I.V. Foulds, P.A. Horgen, A. Castle, U. Krull. Direct selective detection of genomic DNA from coliform using a fiber optic biosensor. *Anal. Chim. Acta* 461(1):37–47, 2002.
122. Slavík, R., J. Homola, E. Brynda. A miniature fiber optic surface plasmon resonance sensor for fast detection of staphylococcal enterotoxin B. *Biosens. Bioelectron.* 17(6-7):591–595, 2002.
123. Mannelli, I., M. Minunni, S. Tombelli M. Mascini. Quartz crystal microbalance (QCM) affinity biosensor for genetically modified organism (GMOs) detection. *Biosens. Bioelectron.* 18:129–140, 2003.
124. Ding, T., U. Bilitewski, R.D. Schmid, D.J. Korz, E.A. Sanders. Control of microbial activity by flow injection analysis during high cell density cultivation of *E. coli*. *J. Biotechnol.* 27:143–157, 1993.
125. Takayama, K., T. Kurosaki, T. Ikeda. Mediated electrocatalysis at biocatalyst electrode based on a bacterium luconobacter industrius. *J. Electroanal. Chem.* 356:295–301, 1993.
126. Gehring, A.G., C.G. Crawford, R.S. Mazenko, L.J. Van Houten, J.D. Brewster. Enzyme-linked immunomagnetic electrochemical detection of *Salmonella typhimurium*. *J. Immunol. Methods* 195:15–25, 1996.
127. Perez, F.G., M. Mascini, I.E. Tothill, A.P.F. Turner. Immunomagnetic separation with mediated flow-injection analysis amperometric detection of viable *E. coli* O157. *Anal. Chem.* 70:2380–2386, 1998.
128. Mulchandani, A., W. Chen, P. Mulchandani, J. Wang, K.R. Rogers. Biosensors for direct determination of organophosphate pesticides. *Biosens. Bioelectron.* 16:225–230, 2001.

129. Pogačnik, L., M. Franko. Detection of organophosphate and carbamate pesticides in vegetable samples by a photothermal biosensor. *Biosens. Bioelectron.* 18(1):1–9, 2003.
130. Palchetti, I., A. Cagnini, M. Del Carlo, C. Coppi, M. Mascini, A.P.F. Turner. Determination of anticholinesterase pesticides in real samples using a disposable biosensor. *Anal. Chim. Acta* 337:315–321, 1997.
131. Nunes, G.S., P. Skládal, H. Yamanaka, D. Barceló. Determination of carbamate residues in crop samples by cholinesterase-based biosensors and chromatographic techniques. *Anal. Chim. Acta* 362(1):59–68, 1998.
132. G.S. Nunes, D. Barceló, B.S. Grabaric, J.M. Díaz-Cruz, M.L. Ribeiro. Evaluation of a highly sensitive amperometric biosensor with low cholinesterase charge immobilized on a chemically modified carbon paste electrode for trace determination of carbamates in fruit, vegetable and water samples. *Anal. Chim. Acta*. 399(1,2):37–49, 1999.
133. Albareda-Sirvent, M., A. Merkoçi, S. Alegret. Pesticide determination in tap water and juice samples using disposable amperometric biosensors made using thick-film technology. *Anal. Chim. Acta* 442(1):35–44, 2001.
134. Del Carlo, M., M. Mascini, A. Pepe, D. Compagnone, M. Mascini. Electrochemical bioassay for the investigation of chlorpyrifos methyl in vine samples. *J. Agric. Food Chem.* 50:7206–7210, 2002.
135. Commission Directive 1999/50/EC of 25 May 1999 amending Directive 91/321/EEC on infant formulae and follow-on formulae, *Official Journal of the European Communities* No L 139, 02/06/1999, p 29.
136. Schulze, H., R.D. Schmid, T.T. Bachmann. Rapid detection of neurotoxic insecticides in food using disposable acetylcholinesterase-biosensors and simple solvent extraction. *Anal. Bioanal. Chem.* 372:268–272, 2002.
137. Schulze, H., E. Scherbaum, M. Anastassiades, S. Vorlova, R.D. Schmid, T.T. Bachmann. Development, validation, and application of an acetylcholinesterase biosensor test for the direct detection of insecticide residues in infant food. *Biosens. Bioelectron.* 17:1095–1105, 2002.
138. Tuennemann, R., M. Mehlmann, R.D. Suessmuth, B. Buehler, S. Pelzer, W. Wohlleben, H.P. Fiedler, K.H. Wiesmueller, G. Gauglitz, G. Jung. Optical biosensors: monitoring studies of glycopeptide antibiotic fermentation using white light interference. *Anal. Chem.* 73(17):4313–4318, 2001.
139. Cooper, M.A., M.T. Fiorini, C. Abell, D.H. Williams. Binding of vancomycin group antibiotics to D-Alanine and D-Lactate presenting self-assembled monolayers. *Bioorg. Med. Chem.* 8(11):2609–2616, 2000.
140. Gaudin, V., J. Fontaine, P. Maris. Screening of penicillin residues in milk by a surface plasmon resonance-based biosensor assay: comparison of chemical and enzymatic sample pre-treatment. *Anal. Chim. Acta* 436(2):191–198, 2001.
141. Gustavsson, E., P. Bjurling, Å. Sternesjö. Biosensor analysis of penicillin G in milk based on the inhibition of carboxypeptidase activity. *Anal. Chim. Acta* 468(1):153–159, 2002.
142. Ferguson, J.P., G.A. Baxter, J.D.G. McEvoy, S. Stead, E. Rawlings, M. Sharman. Detection of streptomycin and dihydrostreptomycin residues in milk, honey and meat samples using an optical biosensor. *Analyst* 127:951–956, 2002.
143. Dolmanova, I.F., T.N. Shekhovtsova, V.V. Kutcheryaeva. Assay of enzyme effectors. *Talanta* 34:201–205, 1987.
144. Shekhovtsova, T.N., S.V. Muginova, N.A. Bagirova. Determination of organomercury compounds using immobilized peroxidase. *Anal. Chim. Acta* 344:145–151, 1997.
145. Ghe, A.M., C. Stefanelli, D. Carati. Influence and role of metal ions in enzymatic catalysis with E.C. 1.2.3.2. xanthine oxidase: application to trace analysis. *Talanta* 31:241–247, 1984.
146. Mealor, D., A. Townshend. Applications of enzyme-catalysed reactions in trace analysis, I: determination of mercury and silver by their inhibition of invertase. *Talanta* 15:747–758, 1986.
147. Maslowska, J., J. Leszczynska. An enzyme-catalysed method for the determination of mercury traces in carbonated soft drinks, by the Hg^{+2} inhibition of β-fructofuranosidase. *Talanta* 32:883–886, 1985.

148. Amine, A., Cremisini, C., G. Palleschi. Detection of mercury compounds using invertase-glucose oxidase-based biosensor. In: *Environmental Monitoring and Hazardous Waste Remediation,* Vo-Dinh, T., R. Niebner, eds., *SPIE Ser.* 2504:209–220, 1995.
149. Donlan, A.M., G.J. Moody, J.D.R. Thomas. A prospect for the analysis of species acting as enzyme inhibitors as illustrated by heavy metal inhibition. *Anal. Lett.* 22:1873–1896, 1989.
150. Campanella, L., R. Cocco, M.P. Sammartino, M. Tomassetti. A new enzyme inhibition sensor for organophosphorus pesticides analysis. *Sci. Total Environ.* 123(124):1–16, 1992.
151. Kratochvil, B., S.L. Boyer, G.P. Hicks. Effect of metals on the activation and inhibition of isocitric dehydrogenase. *Anal. Chem.* 39:45–51, 1967.
152. Bontidean, I., C. Berggren, G. Johansson, E. Csöregi, B. Mattiasson, J.R. Lloyd, K.J. Jakeman, N.L. Brown. Detection of heavy metal ions at femtomolar levels using protein-based biosensors. *Anal. Chem.* 12(70):4162–4169, 1998.
153. http://www.cranfield.ac.uk
154. Luong, J.H.T., P. Bouvrete, K.B. Male. Developments and applications of biosensors in food analysis. *TIBTECH* 15:369–377, 1997.

3.12

Enzymatic Bioprocessing of Tropical Seafood Wastes

Rupsankar Chakrabarti

CONTENTS

ABSTRACT

A large quantity of fish in the tropical countries of the globe is being used for nonfood purposes. The tropical nations in South and South East Asia are the major exporters of aquaculture products to the developed nations in Europe, Asia, and the USA. Improper disposal of solid waste and untreated effluent or waste water from processing factories causes environmental hazards. The present status of enzymatic bioprocessing of seafood wastes along with underutilized unpopular species is discussed. The small and bony fish of South East Asia can be processed to popular fish paste and sauce using papain or bromelain to accelerate the enzymatic process. Unpopular fish species and the filleting waste of popular species can be used to prepare bitterness free fish protein hydrolysate using proteolytic enzymes in single preparations or in combinations. This product can be used as an ingredient in food or beverage formulations. The enzymatic hydrolysate from shellfish waste, and tuna head and viscera waste can be used as a flavoring agent and also as an ingredient in animal, fish, and shrimp feeds. Proteases can be used efficiently to remove protein in the extraction of (1) good quality chitin, (2) carotenoprotein from shell fish waste and (3) guanine crystals from fish scales. Technology transfer to industry is too slow due to the acceptability of new formulations over traditional products.

12.1 INTRODUCTION

With the stagnation of the global marine catch, aquaculture plays a key role in meeting the increasing global demand for fish products. Developed countries are the major importers of processed fish products; whereas the developing and underdeveloped nations are the major exporters. Tropical Asian countries namely, Taiwan, China, Thailand, India, Bangladesh, Malaysia, Singapore, and Indonesia are making rapid progress toward sustainable aquaculture, generating employment and export income. These countries contribute more than 50% of the world shrimp production and more than 70% by aquaculture (1). Other tropical nations, such as Saudi Arabia, are also trying to perfect aquaculture technology with available resources. There were only a few fish processing factories in tropical South and South East Asian nations before the mid twentieth century. With the growing demand for marine and aquaculture products, numerous export oriented fish processing factories were established in this region during the second half of the twentieth century. National laboratories in these countries are actively addressing the quality standard of finished products and the utilization of waste byproducts in an effort to maximize fishery resources for the human population needs.

Table 12.1 shows the nonfood use of total fish production by tropical countries (2). Central American countries such as Mexico, and Nicaragua reduced nonfood use (fish meal, feed and bait) of total fish production from 46% in 1980 to 26% in 1995, while their total catch and aquaculture remains almost stagnant. In South American countries the nonfood use of fish products increased from 60% in 1980 to 77% in 1995 due to increasing trends of total fish production. Since 1960, attempts have been made worldwide to use the total available catch as food for human consumption (3,4). For example, Table 12.2 shows

Table 12.1
Utilization of fish catch in tropical countries in the continents (tons)

	Africa		South America		Asia		Central America	
	Production	Nonfood Use	Production	Nonfood Use	Production	Nonfood Use	Production	Nonfood Use
1980	3687924	822758	7783347	4695405	10848353	1415775	1474000	684000
1985	4199333	621793	15934360	9018635	13208895	1451300	1521000	639000
1990	513741	535065	14453000	10870797	16514778	2061800	1541000	379000
1995	5539600	768064	19977310	15353049	20805199	2603600	1553000	407000

Nonfood use: Reduction to fish meal + used in feed and bait.

Source: *Fish and Fishery Products, FAO Fisheries Circular No.821 Revision 4***.** Rome: Food and Agriculture Organization, 1998, pp19–252

Table 12.2
Composition* of fish meat from 12 species of fish from East Coast of India

Fish	Body Weight (g)	Moisture (%)	Crude Protein (%)	Water Soluble Protein (%)	Fat (%)	Ash (%)	Calcium $mgkg^{-1}$	Potasium $mgkg^{-1}$	Iron $mgkg^{-1}$	Phosphorus $mgkg^{-1}$
Megalapis cordyla	100–150	75.1	20.1	5.8	1.9	1.4	465	1596	15	2315
Otolithes ruber	100–200	77.1	19.5	4.7	1.3	1.1	662	2404	12	1579
Nebia maculata	80–100	77.6	18.6	4.6	1.5	1.3	625	3021	22	3511
Arius dussumieri	130–150	77.5	18.5	6.6	1.2	1.1	632	1291	17	1677
Pricanthus hamrur	100–150	79.1	18.1	4.9	1	1.1	660	2370	21	2372
Decapterus sp.	40–50	75.1	19.3	5	2.9	1.4	958	4108	15	2845
Saurida tumbil	125–150	77.6	19.6	4.8	0.7	1.4	898	3595	17	3787
Atule mate	60–80	76.9	18.7	4	1.4	1.3	612	2892	35	2021
Dussumieri acuta	30–40	76.7	19.6	7	0.8	1.7	986	4434	18	5307
Upeneus vittatis	30–35	76.6	19.2	6.2	2.78	1.1	956	2361	20	2812
Sardinella dobsi	20–30	76.3	19.4	6.4	1.3	1.4	945	3231	31	2120
Sphyraena jello	200–500	78.4	20.1	5.6	0.7	1.3	682	2504	33	2058

* on wet weight basis ; all values average of 4 trials
Source: Chakrabarti, R., S. Gupta, *Fishery Technol.* 37(1):5–7, 2000.

the composition of meat from some unpopular species in the Indian Coast (5). However, the transfer of biotechnology is taking place at too slow a pace in some countries to provide a benefit.

Improper disposal of solid waste and untreated effluent or waste water from fish processing factories, including fish meal plants, causes an environmental hazard. Every nation has adopted regulatory standards for the disposal of fish processing waste. Several products have been developed in national laboratories from fish processing waste; but, again the transfer of biotechnological applications to industry is too slow. However, prospective entrepreneurs are being encouraged to start model environmentally friendly processing plants to use processing waste to make valuable products for human and animal consumption.

In the tropical countries, technologies of enzymatic processing of seafood wastes have been developed for the recovery of proteinaceous matter, pigment, flavorants, and chitosan (6–11). In other parts of the world, different research programs such as the Bay of Bengal Program were instigated to carry out work on tropical fishery products and tropical fishery waste (12–14). Research workers also reported using enzymes as food processing tools (for instance, proteases have been used in reduction of viscosity of stick water, a byproduct from fish meal, to improve drying rate). This chapter describes the role of enzymes in processing of seafoods into various products to be used either as human food or as an ingredient for preparation of human food.

12.2 TROPICAL FISHERY WASTE

12.2.1 Fishing Vessel Catch Waste

In ordinary pelagic fishing vessels, sorting of the catch is done as soon as the fish are taken on board in order to remove fish unsuitable for human consumption. Owing to the shortage of space in fishing vessels, these unsuitable fish are rarely brought to port for fishmeal manufacture. Gutting commences as soon as the catch comes on deck. The guts are thrown overboard, which could cause a serious pollution problem when the vessel is close to public beaches or inhabited areas. Big fishing vessels have separate and adequate facilities for freezing fish and reducing processing wastes into fishmeal on board. Table12.3 shows the weight loss during beheading, degutting, and trimming.

Shrimp trawling produces a large bycatch, that is, a catch of undesirable nontarget species, which comprises 80–90% of catch volume. Removal of shrimp heads at sea reduces the bacterial load on the shrimp product, and is more economical for icing and storage space. Weight loss due to beheading is 30–40% of the whole shrimp. Substantial quantities of shrimp bycatch and processing wastes are discarded at sea (15). High capital and operating costs mean that the operation is only profitable if revenues are high. Therefore, the industry focuses on shrimp and high value fish. The major portion of bulk catches of low priced fish is discarded at sea for the following reasons:

1 Limited ice and storage capacity
2. Lack of onboard processing facilities
3. Possible contamination of shrimp
4. Shortage of labor
5. Difficulty in selling large quantities of fish at port and thereby delay in transporting shrimps to processing facilities
6. Risk in transferring bycatch to another boat at mid sea

Table 12.3
Typical yield of edible flesh and organs as percentage of whole fresh fish

Species	Head (%)	Viscera* (%)	Skin (%)	Bones and Cartilages(%)	Flesh (%)
Horse mackerel	23–26	8–24	3–5	8–10	41–55
Jew fish	15–16	4.5–23	3.5–5	7–9	39–53
Cat fish	23–28	14–28	5–8	8–10	27–41
Tuna	18–26	8–23	4–6	8–10	60–65
Thread fin bream	14–18	13–22	3.5–5	10–12	27–49
Herring	12–18	6–21	3–5	6–8	42–60
Indian salmon	8–11	10–18	3–5	6–8	60–68
Seer	9–12	8–18	4–6	7–9	60–70
Carp (male)	15–17	8–10	4–6	10–12	42–49
Carp (Female)	15–17	24–26	3–5	8–10	33–36
Shrimp	30–40	-	15–20 (shell)	-	37–49
Crab	-	-	-	-	27–32
Cephalopods	-	-	-	-	70–78

*Viscera include gonads and liver.

Source: Clucas, I.J. *Fish Handling, Preservation and Processing in the Tropics, Part 2*. London: Tropical Development and Research Institute, 1982, pp.18–20; Zaitsev, V., I. Kizevetter, L. Lagunov, T. Makarova, L. Minder, V. Podsevalov. *Fish Curing and Processing*. Moscow: Mir Publishers, 1969, pp 460–485.

When lobsters are not landed alive, the tail and cephalothorax are separated as soon as they are brought on board. Whole lobsters deteriorate rapidly after death. If crabs are not landed alive, these crabs are cooked whole on board, or butchered by means of debacking and removal of viscera and gills. Enzymatic deterioration in dead crabs takes place very fast. The waste from crab processing may account for up to 70% of the original crab weight. In some big fishing vessels, cephalopods are immediately transferred to processing areas on board. After careful evisceration, the removal of eyes, mouth, skin, tentacles, and heads followed by deboning and trimming are common steps in processing of cuttlefish and squid. The edible portion is washed and stored in ice. The visceral content of cephalopods contains a large concentration of digestive enzymes and spoilage microorganisms. These will cause marked deterioration of quality if not removed in time. The trunk and tentacles, the edible portion, is only about 70–75% of the whole squid. The processing wastes of crabs and cephalopods are typically thrown overboard.

12.2.2 Waste During Handling at Landing Centers at Shore

Small and medium size fishing vessels engaged in short term voyages or daily fishing bring commercially important fish in on ice, while the remaining catch is not iced. The fresh fish are rejected if they are known to contain harmful, decomposed, or extraneous substances above acceptable levels. Diseased fish are discarded, and acceptable fresh fish are auctioned and transported to different processing units.

12.2.3 Waste from Filleting Units

Whether mechanical or manual filleting methods are used, the gutting of fish must be completed immediately on arrival in order to remove all pieces of gut, liver, blood, and backbone before the filleting operation. The fillets are processed into frozen products for domestic and export markets. All waste materials from filleting plants are collected immediately into suitable watertight and lidded containers, which are emptied regularly by manual or mechanical means.

12.2.4 Waste from the Freezing Units

Japan, the U.S. and European countries are the major importers of frozen shrimps. China, Hong Kong, Malaysia, and Singapore are the major importers of frozen fin fish from other developing nations of South and South East Asia. On average, India exports about 100 thousand tons of frozen shrimp, 120 thousand tons of frozen fish and 66 thousand tons of frozen cephalopods each year (16). Frozen fin fish include white pomfret, black pomfret, sea bream, and reef cod. Annually, fish and shrimp processing waste spread along the 8000 km coastline of India include 60–80 thousand tons of shellfish waste, 40–50 thousand tons of processing waste from fin fish and 30–35 thousand tons of cephalopod waste.

12.2.4.1 Waste During Processing of Fin Fish

The fresh or iced fin fish are degutted and the gills are removed. The dressed fish are thoroughly washed prior to freezing. The processing waste is 15–20% of the total fish weight. The waste washwater contains mainly blood and body slime.

12.2.4.2 Wastes from Peeling and Deveining of Shrimps and Lobsters

It is desirable to peel large shrimp by hand. When very small shrimp are peeled by machine, yields are reduced due to a high proportion of breakages. The wastes from heading, peeling, and deveining operations are processed for fertilizer, animal feed, and fish food. Heading, shucking, and deveining of raw or cooked lobster are performed thoroughly, so that no viscera are left to spoil the flavor of the final product. The processing waste may account for up to 60% of the original weight of raw materials. The dressed products are mainly processed into frozen products for export.

12.2.4.3 Waste from Processing Units of Crabs and Cephalopods

After butchering crabs, any remaining viscera and gills are removed by brushing and washing. Most crab species will throw off their legs and claws if placed directly in boiling water or steam. Back shells, viscera and other shell wastes (70% of original crab weight) are removed continuously from the processing area. Crab meat is picked by hand from dressed body parts and then frozen for export. During the dressing, gutting, and trimming of cephalopods all waste materials are collected immediately in suitable containers which are emptied regularly. The dressed cephalopods are graded and then frozen for export.

12.2.5 Wastes from Fish Canning Units

Only the edible portion of fish is used for canning. For large and medium sized fish, it is usual to cut off head, fins, and tail, and remove all viscera, and sometimes also the spinal column and skin. Small sized fish are headed, tailed, and partially gutted. Yield of dressed fish depends on the species, biological state, and size of fish. Table 12.3 shows the waste on the percentage of total fish weight. With small fish (sardines, small herring), the gut is pulled out without cutting open the belly; a cut is made at the neck and vertebra severed. When the head is pulled off, the pectoral fins and alimentary tract come away together with most of the guts. Preprocessing operations such as boiling in brine or frying in oil improve the taste, texture, and appearance of canned products through protein coagulation and partial extraction of moisture.

Protein coagulation and the breakdown of cells during boiling causes the release of fat, moisture, water soluble nitrogenous substances, vitamins, and minerals from fish flesh.

These losses are directly proportional to the duration and temperature of cooking, and inversely proportional to the fat content of the flesh. During canning of sardines in oil, processing losses amount to 35–38% (degutting and dressing) and 22–24% (steaming) on wet basis.

The yield of canned crabmeat varies from 17–28% of the wet crab, with cooking wastes of 11% on average. Crab processing produces 60–70% waste on a wet basis. Raw meat content in fresh shrimp varies from 24–41%; the yield of canned shrimp meat ranges from 18–32%. Shrimp waste ranges from 53–72% (cephalothorax and abdominal shell) of white shrimp.

The wet body weight of mollusks is 20–32% of the total weight. The edible parts are the fleshy foot, the adductor muscle, and the mantle (12–17% yield). When cooked, these mollusks lose 35–54% in weight. Mollusk wastes contain shells (60–80%) and inedible portions (8–12%). The shell is composed of calcium carbonate.

The edible parts (trunk, head, tentacles) of squid account for 68–75% of total weight. The squids are gutted and washed. After cooking (4% NaCl; 15 min), they are skinned, the eyes are removed, the mantle cavity is cleaned, and the appendages are separated, thoroughly washed, and put into the membrane (mantle) cavity for filling in cans. The processing waste includes the ink bag (6–10%), chitin plates (0.2–0.3%), liver (2–6%) and other organs (12–15%).

Octopi are dressed by slitting the mantle, body, and head and then removing the intestine, mouth apparatus, and eyes. With octopus larger than 2 kg, the fleshy parts of the tentacles are slit. Waste from dressing amounts to 11–20% of the wet weight.

12.2.6 Losses During Salting

Preserving fish in salt or brine begins with washing gutting, and splitting (for large fish), and ends with a second washing and mixing the fish with salt or brine in containers. The amount of protein and nitrogenous matter passed into brine depends on the salting conditions. When the fish are surrounded with a low brine concentration, more proteins and their fractions diffuse from the fish in larger quantities than with a high brine concentration. The proportion of nitrogenous matter in brine increases with the days of brining. Higher than ambient temperature increases enzyme activity and leads to increase losses of organic matter. Nitrogenous matter loss goes up to 10% of total amount of the available nitrogenous matter. Thus the food value of the product is appreciably reduced.

12.2.7 Losses During Reduction to Fishmeal

Table 12.1 shows the nonfood use of fish products, including fish from commercial catch and aquaculture. In South American countries 60–77% of total fish production is used for nonfood purposes (such as reduction to fish meal and use for feed and bait). In Central American tropical countries, 25–46% of total fish production is used for nonfood purposes. Nonfood use varies from 10–22% and 9–15% of total fish production in African and Asian tropical countries respectively. The raw materials used for nonfood use are divided into three main categories:

1. Fish caught for the sole purpose of fish meal production (e.g., anchovies in Peru, anchovies and pilchards in South Africa),
2. Low priced or unused fish from prawn bycatch, and
3. Fishery wastes from processing operations (e.g., damaged commercially important fish; trimmings from filleting operations; heads, guts, and scale from processing factories).

With expanding world populations and stagnation of fish catch in capture fishery, emphasis has been placed on increasing fish production through aquaculture; but water bodies suitable for aquaculture are limited. The search for development of new products from low valued, low priced, or wasted fish for human consumption is therefore increasing in importance.

12.2.8 Waste from Meat and Bone Separator Units

Many of the fish of shrimp bycatch are small and occur in relatively large quantities. There are a number of machines capable of flesh removal, generally known as meat and bone separators. Whole, ungutted fish produces nonwhite mince containing gut contents and blood. However, fish can be gutted and washed before separation in meat and bone separators to produce clean white meat. The wastes from these units include skin, scales, gut, gills, bones, and heads, which accounts for 50–60% of the whole fish. In developing countries, a major area of interest is minced meat and fish mince for the production of various value added products suited for internal and export markets.

12.2.9 Losses from Surimi Plants

Surimi is a minced and washed raw fish meat paste. The global demand for surimi based products, such as crab leg analogs and shrimp analogs, has increased every year. Losses of suspended solids and soluble solids, including soluble protein in washwater and presswater from surimi plants may accounts up to 35% of the solids entering the washing stage.

12.3 PRODUCT DEVELOPMENT FROM WASTE

Proteolytic enzymes or proteases are hydrolytic enzymes that catalyze degradation of specific peptide bonds in protein molecules. The applications of specific proteases in the processing of seafood waste are summarized in Table 12.4.

12.3.1 Products from Small and Bony Fish

12.3.1.1 Fish Protein Hydrolysate

The bycatch of shrimp trawlers consists of mostly small and bony varieties of miscellaneous unpopular fish. The use of proteolytic enzymes to liquefy fish protein has been found to be a feasible method to isolate protein easily from unused or underutilized fish. The enzymes can be from plants (papain, bromelain, or ficin), animals (trypsin or pancreatin), or microbial sources (pronase or alcalase). Sen et al. (17) defined digestion conditions using papain and found that at pH 7.0 maximum solubilization occurred in the first ten hours of hydrolysis. Hale (18) measured the relative activities of more than twenty commercially available enzymes for digestion of a washed and freeze dried fish protein substrate. The standard 24 hour test (concentration of enzyme required to effect 60% digestion in 24 hours) was carried out at near optimum conditions of temperature and pH for each enzyme. Hale (18) reported that pepsin, papain, and pancreatin in combination showed good activity at moderate cost (pronase exhibited greatest activity per unit weight). Mackie (19) used trypsin and bromelain for removal of fish protein from raw codfish. By conducting the digestion at relatively high temperature (50–60°C), it was possible to obtain a liquefied fish milk after a digestion of 30 minutes, which was long enough to dissolve the skin. Yanez et al. (20) employed hake as raw material for production of protein hydrolysate and used it for the supplementation of cereal protein. Iyer et al. (21) and Thankamma et al. (7) reported the preparation of good quality protein hydrolysate (even

Table 12.4
Application of proteases in seafood waste processing

S.No	Type of Waste	Purpose	Enzyme Used	Ref.
1	Head and viscera	Recovery of protein	Papain	Suryanarayana Rao et. al. (1978)
2	Red meat of tuna	Recovery of protein	Alcalase, papain	Thankappan et. al. (1998)
3	Filleting waste	Recovery of protein	Papain, bromelain	Nair et. al. (1985)
4	Shell fish waste	Recovery of protein and chitin	Trypsin Chymotrypsin Microbial proteases Alcalase Papain, trypsin, pepsin	Simpson and Haard (1994) Simpson et. al. (1994) Wang and Chio (1998) C.I.F.T (1998–99) Chakrabarti (2002)
5	Shellfish waste	Recovery of pigment and flavourant	Trypsin Trypsin Papain, trypsin, pepsin Corolase N	Simpson and Haard (1994) Cano Lapez (1987) Chakrabarti (2002) Haard and Simpson (1994)
6	Fish liver	Recovery of protein	Papain Papain, pepsin, trypsin	C.I.F.T (2000–01) Chakrabarti (personal comm.)
7	Stick water from fish meal	Reduction of viscosity	Papain / pepsin Alcalase	Seno (1974) Jacob and Rasmussen (1984)
8	Small fish as waste	Acceleration of protein hydrolysis in paste/sauce	Papain Pepsin	Canonizado (1978) Lee (1989)
9	Unpopular and unused fish	Solubilization of protein	Papain Papain Pepsin Papain	Sen et. al. (1962) Thankamma et. al. (1979) Venugopal and Lewis (1981) Warrier et. al, (1996)
10	Fish scales from carp, herring etc.	Recovery of guanine crystal	Pepsin	Zaitsev (1969)

bacteriological peptone grade) from miscellaneous trash fish such as thread fin bream, using papain at 55°C for 30 minutes. Venugopal and Lewis (8) reported maximum solubilization of low cost lean fish using pepsin. Warrier et al. (22) employed spray drying of the hydrolysate from deboned dhoma *(Johnius dussimeri)* meat using papain at 55°C for 2hrs, which yielded product with a creamy color and 90% protein. The process of the preparation of fish protein hydrolysate is outlined in Figure 12.1. During the process the loss of tryptophan, tryrosine, and lysine was reported by many workers (7,19).

Hevia (23) reported bitterness and glutamic acid taste in fish protein hydrolysate. He also reported that pronase hydrolysate was less bitter than that from bromelain or ficin. Bitter hydrolysate contained more basic n-terminal peptides residues (glycine, leucine, isoleucine, phenylalanine, and valine). Liquefied fish protein with negligible bitterness could be obtained using protease from *Aspergillus niger* (24). Lalasidis et al. (25) prepared fish protein hydrolysate with low molecular weight peptides, using bacterial endopeptidase: alcalase, followed by

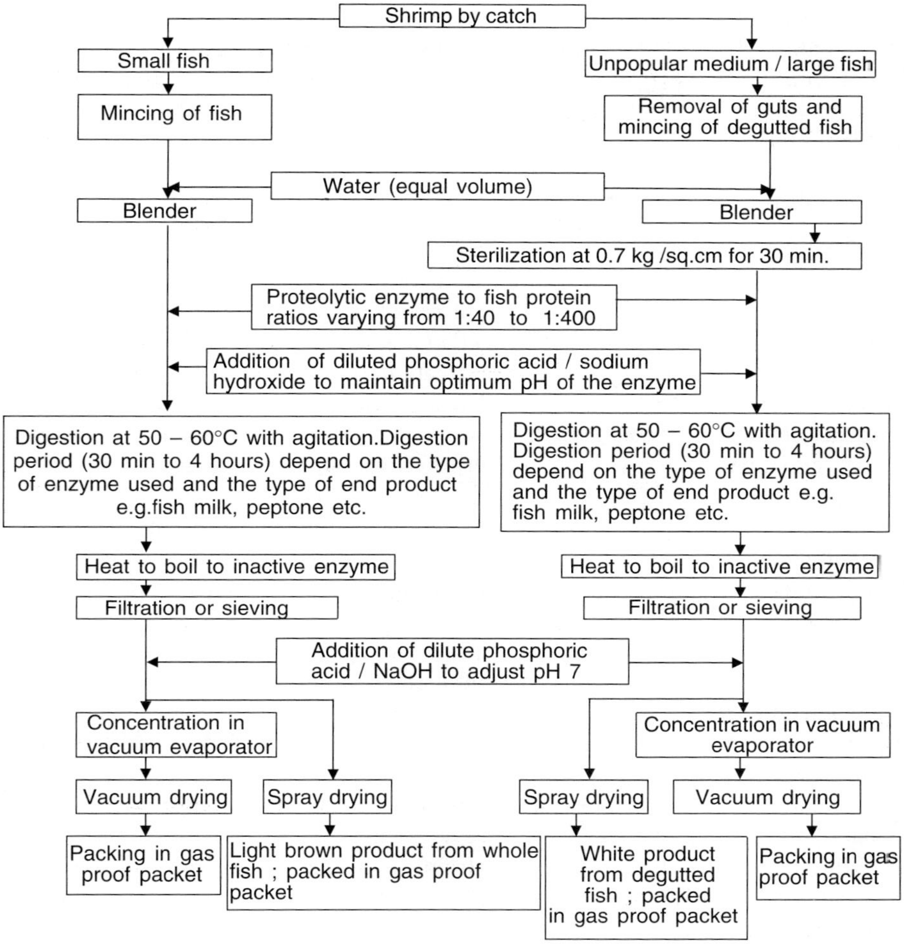

Figure 12.1 Outline of procedure for liquefying fish by proteolytic enzyme (Based on Mackie, 1974; Tankamma et al., 1979; Warrier et al., 1996)

pancreatine. This bitterness free hydrolysate had an average molecular weight of more than 1000. As an alternative method, Lalasidis and Sjoberg (26) reported the use of azeotropic secondary butyl alcohol (SBA) for removal of bitter compounds from bitter fish protein hydrolysate. An azeotropic mixture of 72.8% SBA and 27.2% water (v/v) was used, and when water levels reached 36% (v/v) a two phase system developed. SBA is not only a costly solvent, but its residue may also pose some human health problems. Chakrabarti (27) reported the use of food grade ethyl alcohol to remove bitter peptides from bitter fish protein hydrolysate. A solvent containing ethyl alcohol and water in the ratio 400:70 was used to remove bitter fractions from concentrated fish protein hydrolysate prior to vacuum drying. After debittering, the fish protein hydrolysate can be used in the wide range of food or beverage formulation. The solvent containing bitter fraction was concentrated in a vacuum evaporator and then vacuum dried. Chakrabarti (28) reported good growth of *Staphylococcus aureus, Escherichia coli* and *Bacillus subtilis* in media containing the bitter fraction of fish protein hydrolysate.

Quaglia and Orban (29) reported the sensory properties of fish protein hydrolysate prepared by alcalase were better than that prepared by papain. Kilara (30) reported bacterial neutral and alkaline proteases were about 140% more efficient than papain. Rebeca et al. (31) reported the use of several microbial proteases for the production of fish protein hydrolysate from mullet. Alkaline proteases (pH 6.5–11.0) from *Bacillus subtilis* were found to be more efficient than neutral proteases from the same bacterium, solubilizing more than 80% of the protein.

12.3.1.2 Fish Paste and Fish Sauce

12.3.1.2.1 Fish Paste Small and commercially unimportant fish, which can not be sold as fresh fish, are used as raw materials for preparation of fish paste. The common raw materials for fish paste production are round scad (*Sardinalla fimbriata*), herring (*Spiratelloides japonicus*), sardine (*Sardinella longicaps*), anchovy (*Stolephorus indicus*), small slipmouth (*Leioganathus*); tiny shrimps are used for bagoong, and belachan (6,32). In the Philippines and other South East Asian countries, fish paste or begoong is a semiliquid product obtained from the liquefaction of whole fish in a salt mixture. This is widely used as a condiment and also as a protein food for the poor section of the population. The method of preparation involves mixing fish and salt in the proportion of 1:3 or 2:7 (by weight). Fish proteins are broken down into their soluble constituents. Guevara et al. (33) reported an increased fermentation rate with the addition of 0.1–0.5% papain. They observed crude protein increase in the liquid portion of bagoong made from anchovy with the increase in enzyme concentrations. The hydrolysis was fastest at 45°C.

Autolytic enzymes are present at higher concentrations in the viscera and head than in other tissues (34,35). Thus their concentration during processing is influenced by the timing and completeness of beheading and evisceration, or by using whole uneviscerated fish if maximum enzymatic activity is required. The amount of indigenous enzymes can be reduced by beheading and evisceration. Commercial proteolytic enzymes, or fruit rich in enzymes (papaya or pineapple), can be used in controlled amounts to accelerate the enzymatic process, but their use is yet to be adopted by producers (6,13).

Some typical fish or shrimp pastes from South East Asian countries are ngapi of Mayanmar,, pra-hoc and mams of Cambodia, belachen of Malaysia and trassi of Indonesia. In general, fish or shrimps are pounded with 4–10% (w/w) of salt to form a paste, and this paste is subjected to varying periods of sun drying before being packed tightly in wooden tubs or boxes, excluding all air bubbles. The paste ferments upto 7 days and then is sun dried for 3–5 hours. The paste is then minced again and returned to the wooden tubs for an additional 30–90 days fermentation. It is then minced a third time and packed in blocks wrapped in cellophane or polyethylene paper. The moisture content and salt content in the paste vary from 35–50% and 13–18% respectively. The yield of paste is 40–50% in case of raw *Acetes* shrimps (14).

12.3.1.2.2 Fish Sauce In this process the fermentation of fish is carried out for a longer period than that in manufacture of fish paste. The classic fish sauces are nuoc-mam of Vietnam and Cambodia, nam-pla of Thailand and patis of Philippines. The best sauces are made from anchovies (*Stolephorus* species). The sauces are liquid containing a mixture of amino acids and other protein degradation products. A typical process is outlined in Figure 12.2. The use of uneviscerated fish ensures maximum autolytic enzyme activity; the combination of a salt concentration above 20% and anaerobiosis will totally inhibit any microbial growth during the production of sauce. In the production of most traditional high salt fish sauces, enzymes from external sources are not used. However, in certain parts of Vietnam nuoc mam is produced from eviscerated fish to which fresh pineapple juice is added, because of the lower content of the enzyme (34). In Japan koji (a preparation of cooked rice bearing growth of *Aspergillus orygae*) is used as a source of proteases in

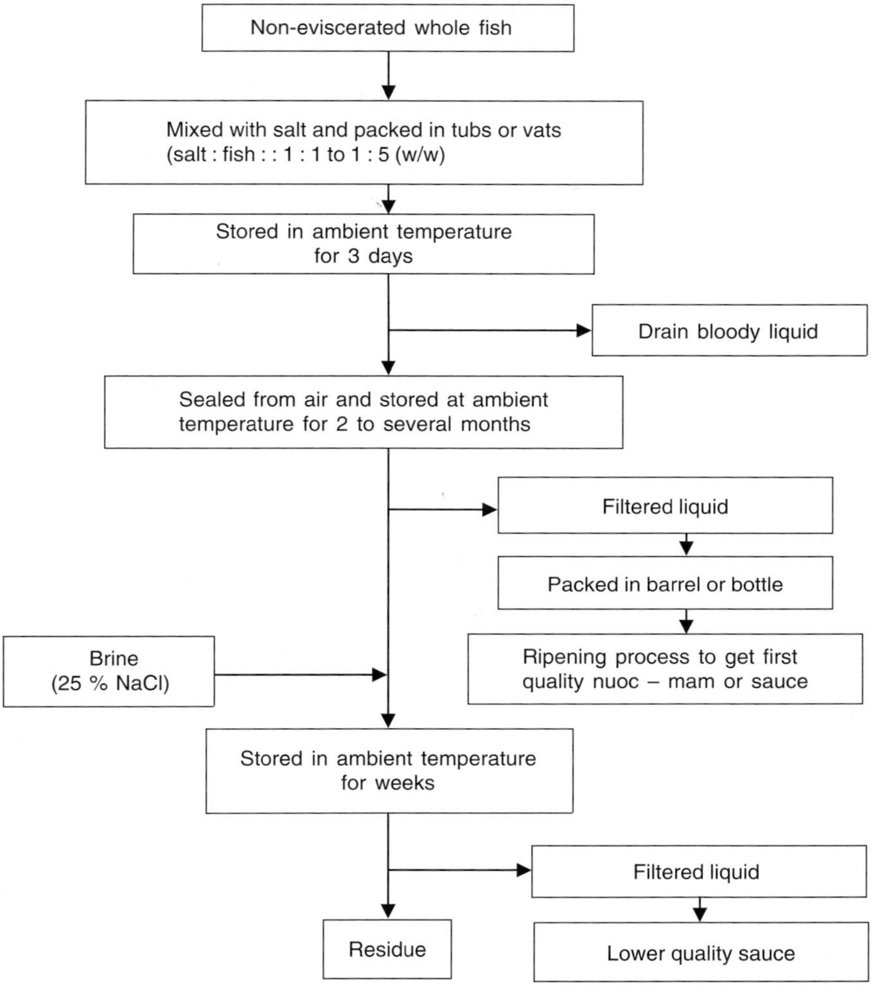

Figure 12.2 Flow diagram for production of sauce (Based on Clucas, 1982; Owen and Mendoza, 1985)

some traditional products (36). The cepalin fish sauce, prepared by using hepatopancreas as a source of proteases, is highly acceptable and preferable to commercial products from the Philippines (37).

12.3.2 Products from Filleting Waste

Filleting waste consists of heads, skins, skeletons with adhering flesh, and belly flaps. The waste is mostly discarded because of the difficulty in recovering the residual meat. Mackie (19) reported the use of trypsin and bromelain in hydrolysate preparation of minced raw filleting waste from cod. Hydrolyses were carried out with enzyme to protein ratios varying from 1:40 to 1:400 at 60°C for bromelain and 45°C for trypsin. Bromelain was found to be more effective than other enzymes at obtaining liquified product in 30 minutes, which was just long enough to disperse the proteins but not long enough to dissolve skin. Digestion was stopped by heating at 100°C for 10 minutes. The filtered protein suspension could be used either as sterilized liquid suspension or as spray dried powder, which was either white or light brown in color. The addition of lipid containing antioxidant is reported to be a very effective method of distributing antioxidaint rapidly throughout the homogenized tissue. The probable potential products are milk substitutes for both animals and humans.

Lalsidis et al. (25) prepared low molecular weight enzymatic fish protein hydrolysate from deboned cod fillet waste, by alcalase treatment at 50°C in the ratio 1:150 (enzyme:protein) followed by pancreatin at 37°C in the ratio 1:75 (enzyme:protein); the yields of soluble nitrogen were between 80 and 90%. Though the alcalase hydrolysis was continued for 3 hours, the pancreatin hydrolysis was continued for more than 8 hours to get product free from bitterness. The hydrolysate protein had an average molecular weight of over 1000.

Nair et al. (38) reported the use of papain and minced fresh pineapple as a source of bromelain, for hydrolysis of jewfish *(Juhnius* species*)* filleting waste. The filleting waste was homogenized with water (1:1), boiled for 5 minutes, and then cooled. The proteolysis was carried out separately with papain (activity 10 units/mg enzyme) at 55°C in the ratio 1:400 (enzyme:waste), and bromelain (minced fresh pineapples) at 30°C in the ratio 1:2 (mined pineapple:waste). Digestion was continued for two hours and then stopped by boiling for 10 minutes. The whole mixture was cooled, neutralized (using dilute sodium hydroxide) and filtered. The hydrolysate was concentrated on a water bath and dried in vacuum; the yield was 8–11%. The product prepared using papain was bitter and brown in color; but the product prepared using pineapple had better taste and better acceptability. The proximate compositions of jewfish waste and hydrolysate from waste are given in Table 12.5.

Ferreira and Hultin (39) used an acidic fungal protease (Newlase A from *Rhizopus niveus)* to recover protein from cod filleting waste. The enzyme hydrolysed the fish proteins at 40–50°C, at pH 3.5, within 75 minutes. Kim et al. (40) reported the use of a crude proteinase from tuna pyloric caeca for a period of 12 hours at 50°C to recover up to 80% of the protein from cod filleting waste.

12.3.3 Products from Head and Viscera Waste

Fishery wastes such as head and viscera, especially during on board processing, are discarded at sea, and the same wastes on land are largely unutilized or converted into manure. Mackie (19) reported the preparation of hydrolysate from cod offal using commercially

Table 12.5
Aproximate composition in percentage of jew fish waste and hydrolysate from waste.

		Hydrolysate	
Parameter	**Jew Fish Waste**	**Papain**	**Bromelain / Pineapple**
Moisture	74.8 ± 0.8	1.4 ± 0.4	5.3 ± 0.01
Fat (dry basis)	13.2 ± 0.4	Trace	Nil
Protein (dry basis)	66.1 ± 1.1	91.2 ± 2.3	60.2 ± 1.3
Ash (dry basis)	19.0 ± 0.2	8.6 ± 1.3	6.8 ± 0.7

All values are mean plus or minus standard deviation of three batches

Source: Nair, A.L., P.T. Mathew, P.V. Prabhu, in *Harvest and Post Harvest Technology of Fish*, Ravindran, K., N.U. Nair, P.A. Perigreen, P. Madhavan, A.G.G. Pillai, P.A. Panicker, M. Thomas, eds., Cochin: Society of Fisheries Technologists, 1985, pp 596–599 available proteolytic enzymes such trypsin and bromelain. He used 0.1 g enzyme to 200 g cod offal at 65°C for one hour. A 95% yield of nitrogen was recovered from the offal. In the offal hydrolysate the glycine and proline concentrations were correspondingly high at 569 and 323 mg/g total nitrogen respectively due to the greater amount of collagen and connective tissue present in the offal. Lalsidis et al. (25) reported the preparation of low molecular weight enzymatic bitter free hydrolysate from cod offal by treatment with alcalase followed by pancreatin. The yields of soluble nitrogen were between 80 and 90%. The hydrolysate had an average molecular weight of more than 1000 and a free amino acid content of 15–45%.

Rao et al. (41) reported the use of head and viscera of fresh water fish such as *Labeo rohita* and *Labeo calbasu*, and the viscera of mackerel for preparation hydrolysate to be used as bactopeptone. The process is outlined in Figure 12.3. The total yield of the dried

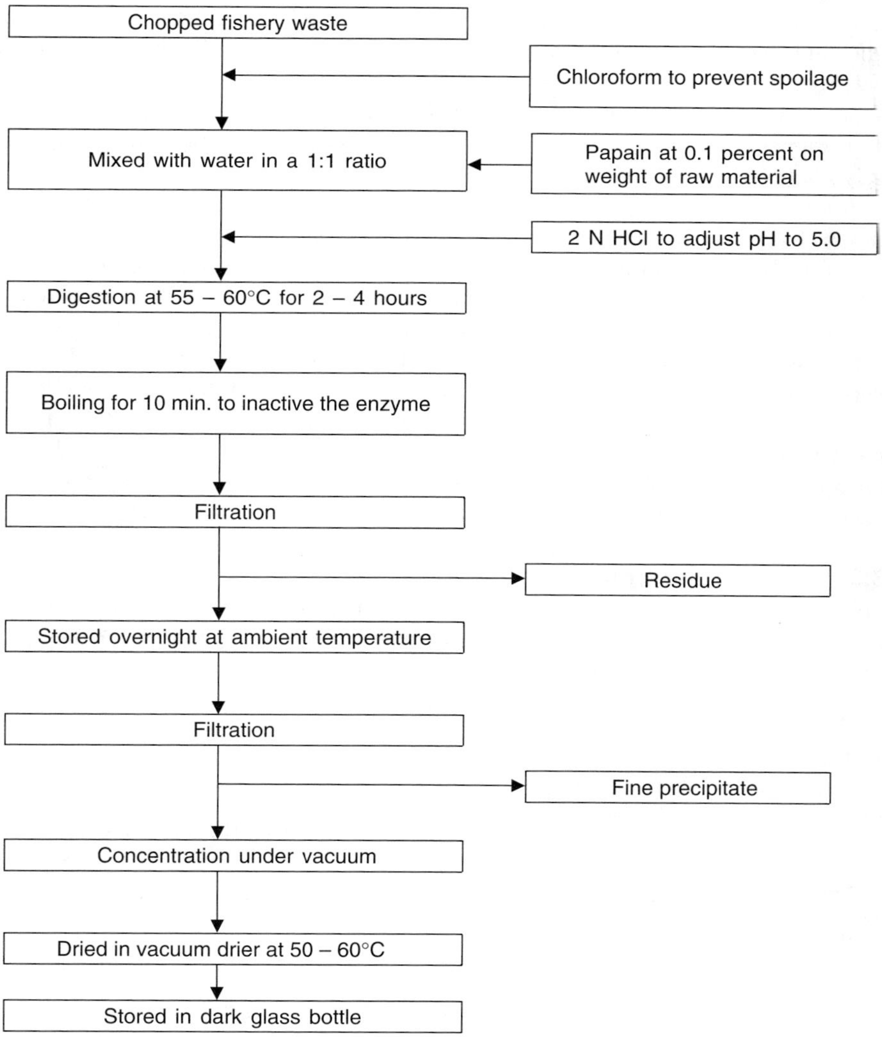

Figure 12.3 Flow diagram for production of bactopetone from head and viscera (Based on Rao et al., 1978)

product varied from 11–12.4%, whereas total nitrogen and ∝-amino nitrogen in the product varied from 12.3–12.6% and 24–34% of total nitrogen respectively. These pale yellow hygroscopic dried products were found to be equal to Oxoid bactopeptone in growth promotion of *Lactobacillus arabinosus* and *Streptococcus faecalis.*

Some tuna canning companies in Thailand manufacture tuna viscera powder (42) from enzymatic hydrolysis of tuna guts followed by dehydration and pulverization. It is typically dark brown in color, has the odor and flavor of tuna, and is highly hygroscopic. It is a soluble protein with components of water (<10%), protein (>60%), fat (8–18%) and ash (1%). It is used as a fish flavoring agent and also an ingredient in feed products for animals, fish, and shrimp.

12.3.4 Use of Red Meat from Tuna Waste

Tuna is abundantly available in tropical seawater. Commercially important and more available species are albacore, bigeye, and northern and southern bluefin and yellowfin species. Unlike other fishes, tuna contains a high proportion of red meat (about 11% of the body weight). Substantial quantities of white tuna meat are canned, while the red meat is discarded. Processing of tuna for canning gives 45–50% waste which can be used for preparing various products. Preparation of enzymatic hydrolysate from tuna waste was reported by Thankappan et al. (43) using alcalase. Approximate compositions of red meat and hydrolysate are given in Table 12.6.

After separation of the white meat, the remaining material was minced with potable water (1:2) in a meat mincer for 5 minutes. The pH of the slurry was adjusted to 8.0–8.5 for alcalase and 6.0–7.0 for papain in separate containers. The enzymes were added at a level of 0.5% of the material protein content. The enzymatic reaction was carried out for 60 min at 50–55°C with continuous stirring. The reaction was stopped by increasing the temperature to 85°C for 15 minutes for alcalase, and 95°C for 30 minutes for papain. The slurry was passed through filter cloth, and then filtered through Whitman No.1 filter paper. The filtrate was neutralized, concentrated, vacuum dried, and packed in sealed containers. The yield of hydrolysate using alcalase was 11.4%, and that using papain was 8.4%. Hydrolysate prepared by papain was bitter but that prepared by alcalase was less so.

12.3.5 Waste Water from Surimi Plants

Surimi making requires a large quantity of fresh water during the washing process. This ranges from 10–20 times the weight of the deboned meat, depending on the species. Presently, all the wash water is discharged as waste water. The waste water contains an

Table.12.6
Approximate composition in percentage of red meat of tuna and hydrolysate from red meat.

		Dried Hydrolysate	
Red Tuna Meat		**Alcalase Hydrolysis**	**Papain Hydrolysis**
Moisture	72.10	6.40	3.40
Protein	23.70	67.54	65.54
Fat	3.20	Nil	Nil
Ash	3.93	8.92	8.79

Source:Thankappan, T.K., G. George, K.G.R. Nair, in: *Technological Advancement in Fish*, Hameed, M.S., B.M. Kurup, eds., Cochin: University of Science and Technology, 1998, pp 329–333.

average of 3.4 g protein per liter; about 80% of the protein present is water soluble. Watanabe et al. (44) reported that the total protein loss accounted for approximately 30% of the deboned meat weight and varied from processing plant to plant. When correctly handled, the waste water is treated by an activated sludge method, alone or in combination with a dissolved air flotation method. Protein recovery by the air flotation method is aided by coagulants such as sodium polyacrylate and chitosan (45).

12.3.6 Product from Stick Water

Fish juice is a byproduct of the cooking procedure for canned products. Byproduct fish juice contains excess gelatin and has diluted fish flavor. Fish oil is removed from the byproduct juice and the stick water is concentrated to less than 50% moisture. Occasionally whole stick water is hydrolyzed to increase the capacity of the evaporation plant (for instance, addition of about 0.2% commercial preparation of proteolytic enzymes) (9).

Stick water is a byproduct obtained during the production of fish meal. The use of stick water including blood water from fishmeal plants is always recommended by regulatory authorities to avoid pollution. It may amount to 18–20% of the final fish meal solids. The protein content of stick water presents a problem during evaporation. Aguilera and Cortes (46) reported the use of enzymatically hydrolyzed whole stick water (containing blood water) to increase protein content and to depress water activity in fish meal. Hydrolyzed whole stick water concentrate was spray dried into a free flowing powder and then mixed with fish meal. Jacoben and Rasmussen (47) reported the addition of alcalase 0.6 L or neutrase 0.5 L at 0.2% to the stick water at neutral pH and 50°C for significantly reduced viscosity, which enabled further evaporative concentration.

12.3.7 Recovery of Chitin and Protein from Shellfish Waste

The shell fish waste is a good source of chitin. In India, the quantity of such waste from shrimp processing units amounts to 60 to 80 thousand tons annually. Wet shellfish waste contains 4–5% chitin. Recovery of chitin from the waste requires adhering proteinaceous matter removal followed by demineralization. Traditionally dilute alkali is used for deproteinization, which may cause a partial deacetylation and even polymer hydrolysis, resulting in a final product with inconsistent properties (48). Proteolytic enzymes or proteases can be used efficiently to remove protein from shellfish waste. Simpson et al. (49) used chymotrypsin for effective removal of protein from the waste. He maintained the enzyme to substrate ratio at 7:1000 (w/w) at pH 8, with a reaction time of 72 hours at 40°C. Wang and Chio (10) reported the use of a microbial protease from *Pseudomonas aeruginosa* for removal of protein from shrimp shell and crab shell powder. A solid-state fermentation of the materials for a period of 5–10 days removed protein in the range of 46–81%. Alcalase was found to be efficient in removal of 70% of the protein from shrimp shell waste. The product has a light color, less bitterness, and an acceptable appearance (50). The process of chitin and protein recovery from shellfish waste is outlined in Figure 12.4.

12.3.8 Extraction of Carotenoprotein and Protein Isolate from Shellfish Waste

The shrimp industry throughout the world produces enormous quantities of industrial waste in the form of shrimp heads and shells. Significant worldwide increase in crustacean aquaculture in the South and South East Asian countries has added tremendously to that waste. Shrimp shell waste ranges from 53–72% of whole shrimps. The major components of the shrimp shell waste are protein, chitin, flavorant protein, carotenoid protein, and minerals

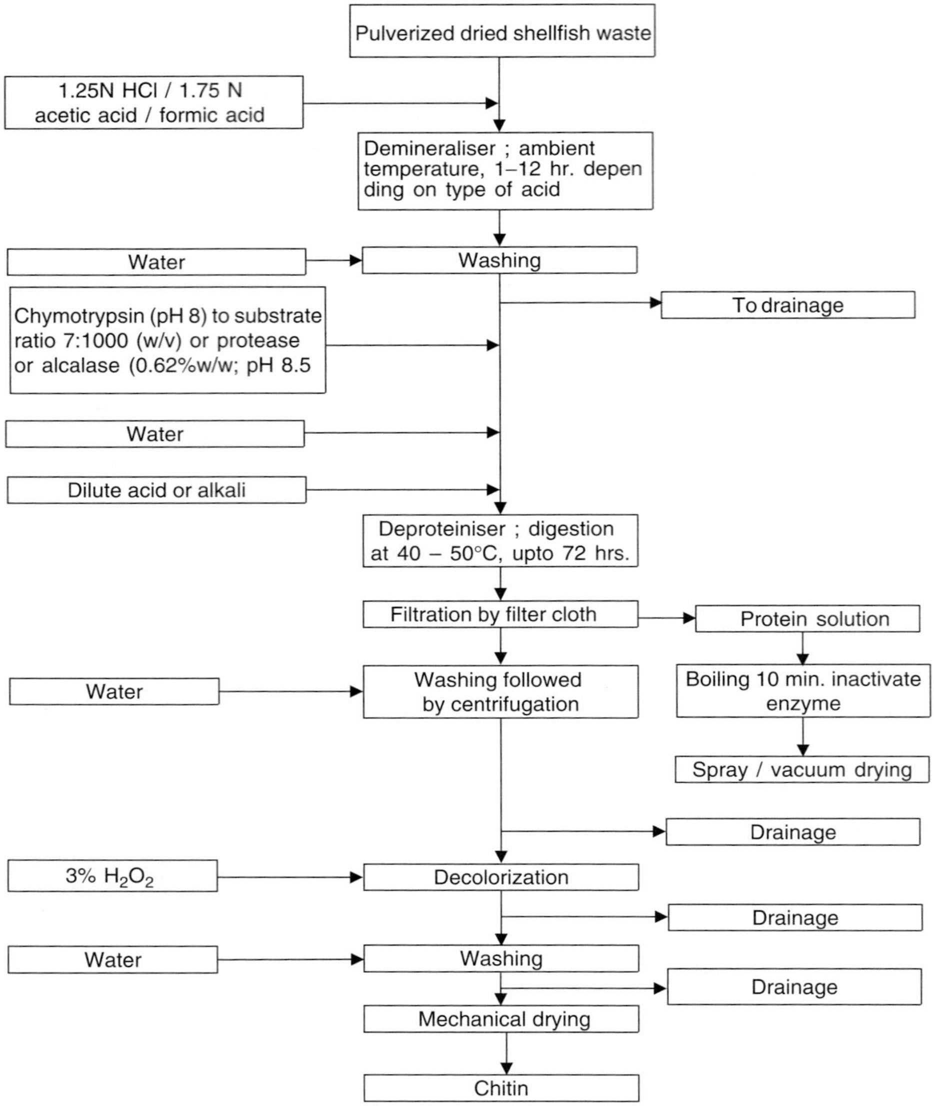

Figure 12.4 Flow diagram of recovery of chitin and protein from shellfish waste (Based on Simpson et al., 1994; CIFT, 1999; Madhavan and Nair, 1974)

(Table 12.7). The processing of shell waste not only resolves environmental problems, but also helps to produce valuable products such as carotenoid pigment, chitin, chitosan, and protein isolate. Simpson and Haard (51) reported the use of 25% bovine trypsin and 0.5 N ethylene diamine tetra acetic acid (EDTA) in an extraction medium at pH 7.7 and 4°C to recover 65% of the protein and 49% of the astaxanthin, a carotenoid pigment, from shrimp waste. He recovered the astaxanthin in the form of stable carotenoprotein complex. At higher temperatures the speed of carotenoprotein extraction increased and EDTA was not required. Cano-Lopez et al. (52) reported the use of cod trypsin at 4°C under identical conditions; the recovery of protein and astaxanthin increased to 81% and 64% of shrimp waste respectively. The study also indicated that only purified Atlantic cod trypsin was a more

Table 12.7
Approximate percentage composition of raw brown shrimp (*M.monoceros)* shell waste.

Component	Percentage
Moisture	72–76
Protein	8–10
Fat	2–3
Ash	7–9
Chitin	4.5–7
Carotenoids ($mgkg^{-1)}$	30–40

Source: Chakrabarti, R., *Food Biotechnol.* 16(1):81–89, 2002.

effective extraction aid than bovine trypsin for recovering carotenoprotein. However, the purified enzyme is too expensive. Haard and Simpson (53) reported the use of antioxidant in extraction media to stop oxidative activity and to retain the bright red orange color of carotenoprotein. Protein associated with astaxanthin is more resistant to oxidation and it is deposited in rainbow trout skin and flesh more efficiently than free pigment.

Chakrabarti (11) reported that trypsin showed maximum recovery (55%) of carotenoid pigment in 4 hours at 28±2°C from brown shrimp shell waste, but pepsin and papain showed about 50% recovery of the pigment during the same period. The yield and composition of carotenoprotein cake and protein paste are shown in Table 12.8. Though trypsin recovered the maximum amount of protein, the percentage of protein recovery by papain and pepsin was close to that of trypsin. The process is outlined in Figure 12.5. During the storage of the products at tropical ambient temperatures, Chakrabarti (11) reported the loss of carotenoid pigment from carotenoprotein cake with different enzymes was in the order of pepsin > papain > trypsin. He also mentioned that papain could be used for moderate recovery of stable carotenoprotein and good quality protein concentrate at lower cost from shrimp shell waste prior to its conversion to chitin or chitosan. The vacuum dried soluble protein paste, which contains good flavorants and concentrated protein (44–56%) can be used as an ingredient in food, feed, and beverage preparation.

An enzymatic method was developed to recover the carotenoid along with the protein from crab waste by Manu-Tawiah and Haard (54) using proteolytic enzyme. Wang and Chio (1998) reported the use of microbial proteases from *Pseudomonas aeruginosa* for protein removal, yielding up to 55% from crab shell powder during fermentation for a period of 7 days.

12.3.9 Production of Seafood Flavorings

Seafood flavor can be used as an additive in surimi based products such as crab analog and shrimp analog, and cereal based extrusion products. Seafood flavor can be extracted from shell and other raw materials by enzymatic reaction. The taste and volatile components of shrimp heads were recovered by enzymatic digestion using corolase N and Koji, or bacterial strain with high proteolysis rates followed by spray drying (53). The product contained 9–12% amino acids, mainly taurine, arginine, glycine, proline, and IMP.

In many fish processing procedures, the fish is boiled in brine or plain water. The water is drained and used as a food because it contains fishy flavor and some soluble nutrients. In Vietnam, a shrimp extract is prepared from dried shrimp heads and shells. The shrimp waste is boiled in water for several hours. Sugar is added and the liquid is concentrated to a thick syrup, producing a condiment. Similar condiment products are produced

Table 12.8
Yield and composition of the products from brown shrimp shell waste. All values are means ± standard deviations of four determinations.

		Caroteno Protein Cake			Protein Paste			Dried Solid Residue	
Enzyme	**Reaction Time (hr)**	**Yield* (%)**	**Carotenoids (mg kg^{-1})**	**Protein (g kg^{-1})**	**Ether Ext. (g kg^{-1})**	**Yield* (%)**	**Protein (g kg^{-1})**	**Yield* (%)**	**Protein (g kg^{-1})**
1. Pepsin	3	4.2 ± 0.25	345 ± 4.1	425 ± 3.3	30.3 ± 0.62	8.5 ± 0.45	562 ± 4.8	10.2 ± 0.56	101 ± 2.2
	4	4.4 ± 0.28	340 ± 3.5	371± 2.8	32.5 ± 0.68	12.3 ± 0.61	452 ± 4.0	7.2 ± 0.41	79 ± 1.8
Control PH 4.6	4	7.4 ± 0.31	159 ± 2.6	451± 5.0	16.2 ± 0.38	8.7 ± 0.41	322 ± 3.5	10.2 ± 0.62	127 ± 3.1
2. Papain	3	3.8 ± 0.21	370 ± 4.5	406 ± 3.2	32.5 ± 0.59	9.5 ± 0.52	450 ± 4.2	10.5 ± 0.52	81 ± 2.0
	4	4.1 ± 0.25	360± 4.2	383 ± 3.0	35.0 ± 0.72	11.6 ± 0.56	444 ± 3.8	10.0 ± 0.48	78 ± 1.8
Control pH 6.2	4	7.5 ± 0.42	156 ± 2.1	433 ± 4.3	15.0 ± 0.33	8.5 ± 0.38	343 ± 4.1	10.5 ± 0.55	138 ± 0.36
3.Trypsin	3	2.8 ± 0.22	505 ± 5.4	274 ± 2.4	40.0 ± 0.75	12.5 ± 0.58	443 ± 3.5	9.7 ± 0.51	76 ± 1.7
	4	3.1 ± 0.24	485 ± 5.1	245 ± 2.1	38.8 ± 0.68	14.0 ± 0.62	462 ± 4.1	8.5 ± 0.45	63 ± 1.4
Control pH 7.6	4	7.2 ± 0.35	164 ± 3.2	483 ± 4.5	18.3 ± 0.41	8.9 ± 0.42	303 ± 3.6	10.0 ± 0.45	118 ± 3.2

* Yield as a percentage of raw shrimp shell waste; moisture in cake and solid residue below 3%.
Source: Chakrabarti, R., *Food Biotechnol.* 16(1):81–89, 2002.

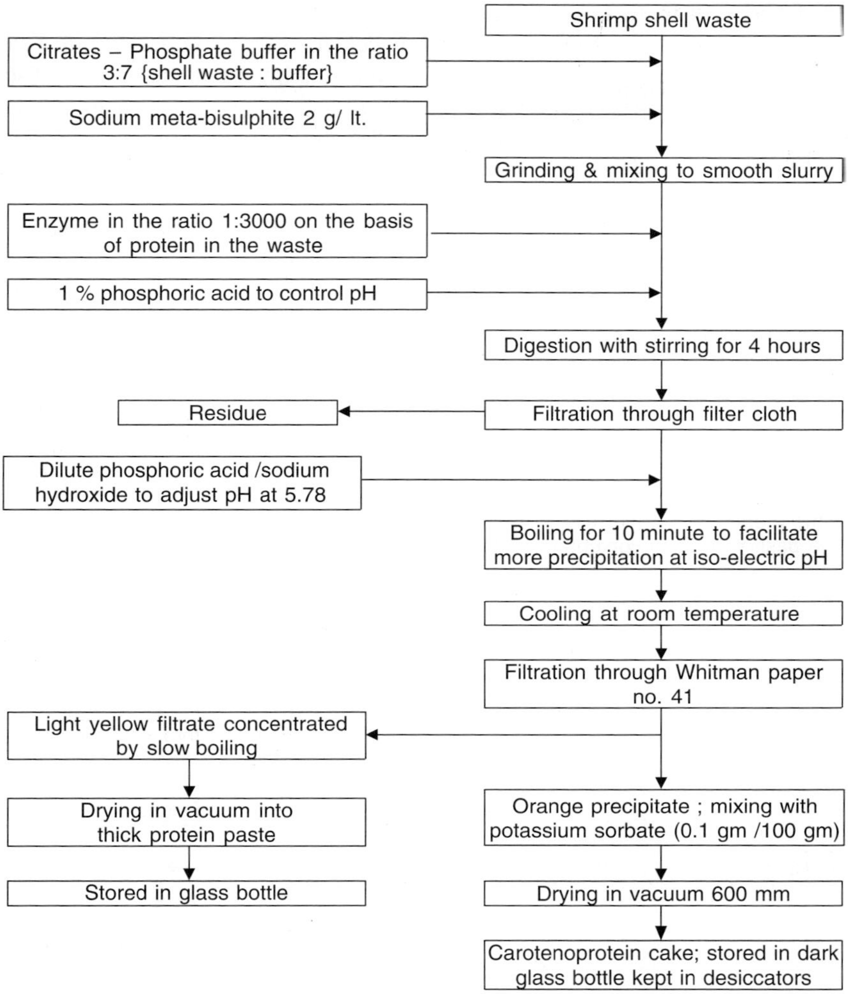

Figure 12.5 Flow diagram of production of carotenoprotein and protein paste from shrimp shell waste (Based on Chakrabarti, 2002).

in Hong Kong from dried oysters, in Indonesia from anchovies, and in the Minicoy islands of India from tuna (14).

12.3.10 Pearl Paste

Artificial pearls are made by coating glass beads with guanine crystal in lacquer. Pearl paste, a suspension of guanine crystals, is prepared from fish scales; preferably from various carps, herring, and other variety species of silvery fish. Scales are collected and preserved by storing in weak brine, but they must not be dried. The crude guanine crystals are removed by mechanical scrubbing and centrifuging. The crude guanine, which contains extraneous proteins, is treated with pepsin to digest the proteins. This process takes between 40–50 hours. The guanine is then washed in water, pressed, and freed from the oil by solvent extraction. The refined guanine is mixed with lacquer or organic solvent and sealed in glass containers (35).

12.3.11 Products from Liver Waste

Evisceration of fish either on board or at landing centers is the essential step for fish preservation and processing into various products. Liver is an important organ in the discarded viscera or waste in fish processing industries. The major components of fish liver are proteins, lipids, glycogen, carotenoids, vitamins, and minerals. Cod liver and shark liver are

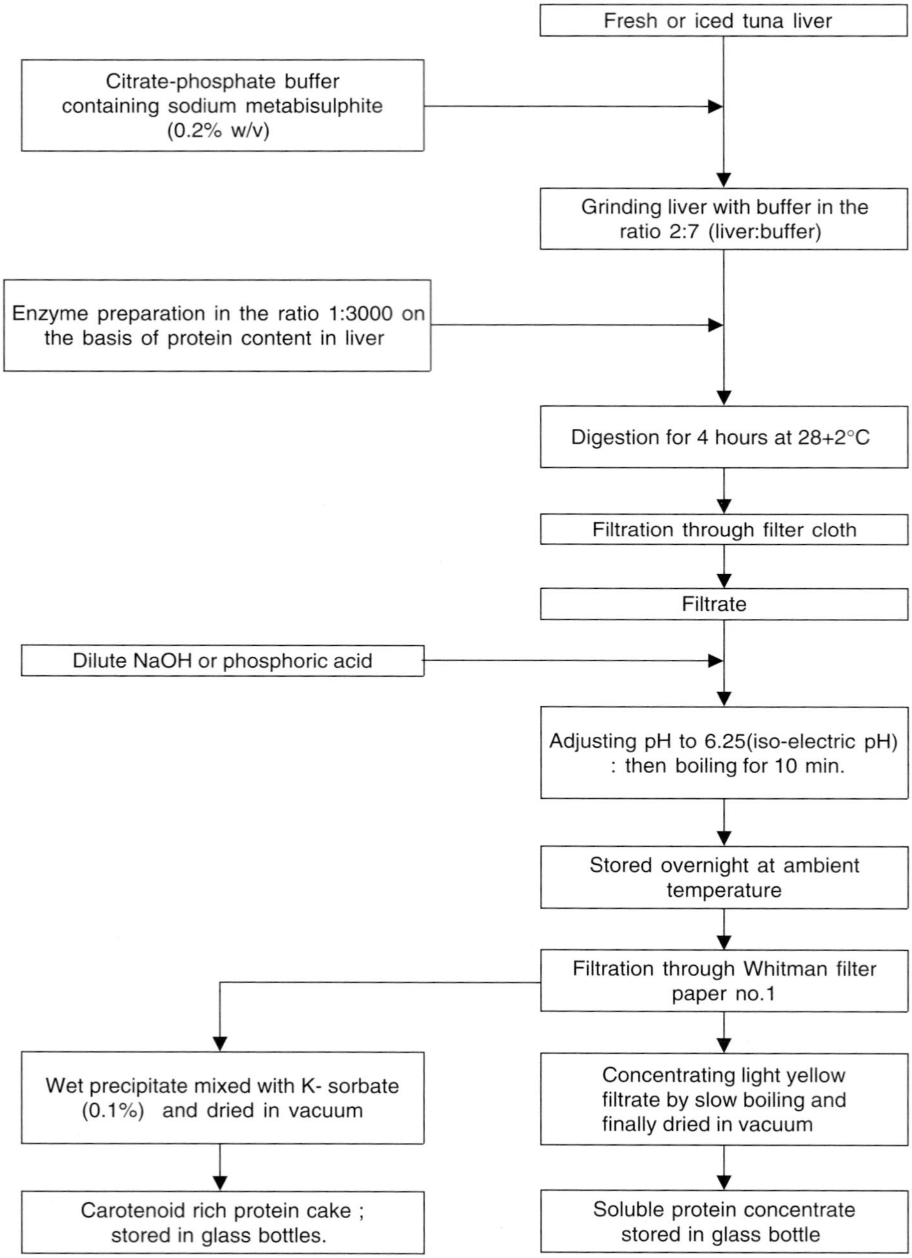

Figure 12.6 Flow diagram of processing of tuna liver to valuable products (Based on Chakrabarti, personal communication, 2002).

Table 12.9
Composition of raw tuna liver and products from livers

Parameter	Raw Tuna Liver	Carotenoids Rich Protein Cake		Soluble Protein Concentrate	
		Papain	Trypsin	Papain	Trypsin
Moisture (%)	71–74	2–3	2–3	3–5	3–5
Protein (%)	17–19	63–65	62–66	61–67	66–70
Fat (%)	5–7	17.8–18.4	17.5 18.5	Trace	Trace
Ash (%)	1.4–1.5	4–5	4–6	16–18	15–17
Carotenoid (mg kg^{-1})	10–15	140–150	135–152	–	–

Extraction time 3–4 hours
Source: R.Chakrabarti, personal communication, 2002.

widely used as a source of Vitamin A. However, the livers of other commercially important fish are either discarded with viscera or used as manure. Thankappan (43) reported tuna is abundantly available (24000 tons) in Indian Exclusive Economic Zone. Substantial quantities of tuna are canned and a portion is processed as masmin, a dark colored smoked product. CIFT (55) reported the use of tuna liver separated from tuna processing units waste in the preparation of carotenoid rich protein cake and soluble protein concentrate. Generally, the weight of liver in tuna is 1.1–2.0% of total body weight. Composition of raw tuna liver (*Enthynnus affinis*) is given in Table 12.9. The percentage of protein recovery from tuna liver by pepsin, trypsin, and papain was 91.0, 91.8 and 88.6% respectively, at the end of 4 hours at 28 ± 2°C. Carotenoids, which also include vitamin A and its isomers, can be recovered at 66 to 70% in 4 hours from tuna liver using the mentioned enzymes. The process is outlined in Figure 12.6 (R. Chakrabarti, personal communication, 2002). The vacuum dried carotenoid rich protein cake (62–70% protein; 108–152 ppm carotenoids) can be used as an ingredient in food and feed preparation. Soluble protein concentrate (63–71% protein; 4.3–5.6% ∝-amino nitrogen) can be used as an ingredient in the media preparation of sulphite reducing anaerobes such as *Clostridium* species. (56).

REFERENCES

1. Ferdouse, F. The international market for cultured shrimps. *INFOFISH Int.* Dec:16–22,1994.
2. FAO. *Fish and Fishery Products, FAO Fisheries Circular No.821 Revision 4*. Rome: Food and Agriculture Organization, 1998, pp 19–252.
3. Grantham, G.I. Minced fish technology: a review. *FAO Fish Technology paper No.216,* Rome: Food and Agriculture Organization, 1981, pp 1–51.
4. FAO. Fish by-catch: bonus from sea. *Report of a technical consultation on shrimp by-catch utilization held in Georgetown, Guyana, 27–30 October 1981*, pp. 43–163.
5. Chakrabarti, R., S. Gupta. Characterstics of gel from the meat of twelve species of fish from Visakhapatnam Coast. *Fishery Technol.* 37(1):5–7, 2000.
6. Canonizado, S.O. By-products technology and waste utilization. Proceeding of Indo Pacific Fishery Commission, Manila, 8–17 March, 1978, pp 507–509.
7. Thankamma, R., K. Gopakumar, A.L. Nair, A.V. Shenoy, M.A. James. Protein hydrolysate from miscellaneous fish. *Fishery Technol.* 16:71–74, 1979.
8. Venugopal, V., N.F. Lewis. Isolation of proteins from low priced fish. *Fleiwirtschaft* 61:1368–1370, 1981.

9. Seno, Y. Processing and utilisation of mackerel. In: *Fishery Products*, Kreuzer, R., ed., Farnham: FAO Fishing News Ltd., 1974, pp 147–151.
10. Wang, L., S.H. Chio. Deproteinization of shrimp and crab shell with protease of *Pseudomonas aeruginosa* K187. *Enzyme Microbiol. Technol.* 22:629–633, 1988.
11. Chakrabarti, R. Carotenoprotein from tropical brown shrimp shell waste by enzymatic process. *Food Biotechnol.* 16(1):81–89, 2002.
12. Gildberg, A. Fermented fish products in Thailand. *FAO Fisheries Report No.470 Supplement*, Rome: Food and Agriculture Organisation, 1992, pp 110–119.
13. Owens, J.D., L.S. Mendoza. Enzymatically hydrolysed and bacterially fermented fishery products. *J. Food Technol.* 20:273–293, 1985.
14. Clucas, I.J. *Fish Handling, Preservation and Processing in the Tropics*, Part 2. London: Tropical Development and Research Institute, 1982, pp. 18–20.
15. Gordon, A. *The By-Catch from Indian Shrimp Trawlers in the Bay of Bengal.* Madras: Bay of Bengal Programme, India, 1991, pp. 2–23.
16. Marine Product Export Development Authority. Seafood export exceeds records. *Seafood Export J.* 31(5):3–5, 2000.
17. Sen, D.P., N.V. Sripathy, N.L. Lahiry, A. Sreenivasan. Fish hydrolysates. *Food Technol.* 16(5):138–141, 1962.
18. Hale, M.B. Relative activities of commercially available enzymes in the hydrolysis of fish protein. *Food Technol.* 23:107–110, 1969.
19. Mackie, I.M. Potential production of powdered and liquid fish products for human consumption and animal feed. In: *Fishery Products,* Kreuzer, R., ed., Farnham: FAO, Fishing News Ltd., 1974, pp 136–140.
20. Yanez, E., D. Ballester, F. Monckeberg. Enzymatic fish protein hydrolysate; chemical composition, nutritive value and use as a supplement to cereal protein. *J. Food Sci.* 41:1289–1291, 1976.
21. Iyer, K.M., K. Gopakumar, A.V. Shenoy, M.A. James, M.R. Nair. Peptone from thread fin bream. *Proceeding of Indo Pacific Fishery Commission*, Manila, March, 8–17, 1978, pp 28–30.
22. Warrier, S.B., S.V. Ghadi, V. Ninjoor. Functional and nutritional properties of fish protein hydrolysate from dhoma (*Johnius dussumeiri*). *Fishery Technol.* 33:101–106, 1996.
23. Hevia, P., H.S. Olcott. Flavour of enzyme solubilized fish protein concentrate fractions. *J. Agric. Food Chem.* 25:772–775, 1977.
24. Kinumaki, T. Possibilities for the production and application of liquefied fish protein concentrate in the IPFC area. *Proceeding of Indo Pacific Commission*, Manila, March 8–17, 1978, pp 492–506.
25. Lalasidis, G., S. Bostrom, L.B. Sjoberg. Low molecular weight enzymatic fish protein hydrolysates: chemical composition and nutritive value. *J. Agric. Food Chem.* 26:751–753, 1978.
26. Lalasidis, G., L.B. Sjoberg. Two new methods of dibittering protein hydrolysate and fraction of hydrolysates with exceptionally high content of essential amino acids. *J. Agric. Food Chem.* 26:742–743, 1978.
27. Chakrabarti, R. A method of debittering fish protein hydrolysate. *J. Food Sci. Tech.* 20:154–156, 1983.
28. Chakrabarti, R. Suitability of bitter fraction separated from fish protein hydrolysate as bacteriological peptone. *Ind. J. Microbiol.* 24:167–170, 1984.
29. Quaglia, Q.B., E. Qrban. Enzymatic solubilization of protein of sardine (*Sardina pilchardus*) by commercial proteases. *J. Sci. Food Agric.* 38:263–267, 1987.
30. Kilara, A. Enzyme modified protein food ingredients. *Proc. Biochem.* 20:149–158, 1985.
31. Rebeca, B.D., M.T. Pena-Vera, M. Diaz-estaneda. Production of fish protein hydrolysates with bacterial proteases. *J. Food Sci.* 56:309–314, 1991.
32. Yeoh, O.L., Z. Merican. Processing of non-commercial low-cost fish in Malaysia. *Proceeding of Indo-Pacific Fishery Commission*, Manila, March 8–17, 1978, pp 572–580.
33. Guevara, G., V.C. Matias, P.D. Pena. Fish fermentation with the use of papain. *Proceeding of the Eight Annual convention, PAFT.* Manila, 1973, pp 103–107.

34. Westenberg, J. Fishery products of Indo-China. *Proceeding of Indo-Pacific Fisheries Council, 2nd meeting*, Australia, 1951, pp 125–150.
35. Zaitsev, V., I. Kizevetter, L. Lagunov, T. Makarova, L. Minder, V. Podsevalov. *Fish Curing and Processing*. Moscow: Mir Publishers, 1969, pp 460–485.
36. Tanikawa, E. *Marine Products of Japan*. Tokyo: Koseisha - Koseikaku Company; 1971, pp 1–507.
37. Raksakulthai, N., Y.Z. Lee, N.F. Haard. Effect of enzyme supplements on the production of fish sauce from male capelin (*Mallotus villosus*). *Can. Ins. Food Sci. Technol. J.* 19:28–33, 1986.
38. Nair, A.L., P.T. Mathew, P.V. Prabhu. Studies on hydrolysates from jew fish (*Johnius spp)*waste. In: *Harvest and Post Harvest Technology of Fish*, Ravindran, K., N.U. Nair, P.A. Perigreen, P. Madhavan, A.G.G. Pillai, P.A. Panicker, M. Thomas, eds., Cochin: Society of Fisheries Technologists, 1985, pp 596–599.
39. Ferreira, N.G., H.O. Hultin. Liquefying cod fish frames under acidic conditions with fungal enzymes. *J. Food Proc. Pres.* 18:87–101, 1994.
40 im, S.K., Y.J. Jeon, H.G. Byenn, Y.T. Kim, C.K. Lee. Enzyme recovery of cod frame proteins with crude proteinase from tuna pyloric caeca. *Fisheries Sci.* 63:421–427, 1997.
41. Rao, S.V.S., C.R. Saraswati, C.T. Dwarakanath. Studies on the utilization of fishery wastes for the production of microbiological media, *Proceeding of Indo Pacific Fishery Commission*, Manila, 1978, pp 364–368.
42. INFOFISH. Tuna Viscera powder. *INFOFISH International* 6/92:53, 1992.
43. Thankappan, T.K., G. George, K.G.R. Nair. Utilization of tuna and tuna waste. In: *Technological Advancement in Fish*. Hameed, M.S., B.M. Kurup, eds., Cochin: University of Science and Technology, 1998, pp 329–333.
44. Watanabe, H., R. Takai, A. Sekigawa, H. Hasegawa. An estimation of the amount of protein lost in the effluent from frozen surimi manufacture. *Bull. Jap. Soc. Sci. Fish* 48:869–871, 1982.
45. Lee, C.M. Surimi process technology. *Food Technol.* 38(11):69–80, 1984.
46. Aguilera, J.M., L. Cortis. Water sorption properties of fish meals and fish hydrolysates. In: *Seafood Science and Technology,* Bligh, E.G., ed., Victoria: Fishing News Books, 1992, pp 288–292.
47. Jacobsen, F., O.L. Rasmussen. Energy saving through enzymatic treatment of stickwater in fish meal industry. *Proc. Biochem.* 19:165–169, 1984.
48. Venugopal, V., R. Lakshmanan, S.N. Doke, D.R. Bongirwar. Enzymes in fish processing, biosensors and quality control: a review. *Food Biotechnol.* 14(1,2):21–77, 2000.
49. Simpson, B.K., N. Gogne, M. Simpson. Bio-processing of chitin and chitosan. In: *Fisheries Processing: Biotechnological Applications*, Martin, A.M., ed., New York: Chapman and Hall, 1994, pp 155–175.
50. CIFT. *Annual Report 1998–1999*. Kochi: Central Institute of Fisheries Technology, India, 1999, pp 19–20.
51. Simpson, B.K., N. Haard. The use of proteolytic enzymes to extract caroteno pigment from shrimp processing offal with the aid of trypsin. *J. Appl. Biochem.* 7:212–222, 1985.
52. Cano-Lopez, A., B.K. Simpson, N.F. Haard. Extraction of carotenoprotein from shrimp process waste with the aid of trypsin from Atlantic cod. *J. Food Sci*. 52(2):503–504, 1987.
53. Haard, N.F., B.K. Simpson. Proteases from aquatic organisms and their uses in the seafood industry. In: *Fisheries Processing: Biotechnological Applications,* Martin, A.M., ed., New York: Chapman and Hall, 1994, pp 132–149.
54. Manu-Tawiah, W., N.F. Haard. Recovery of carotenoprotein from exoskeleton of snow crab, *Chionectes opilio*. *Can. Inst. Food. Sci. Tech. J.* 20:31–33, 1987.
55. CIFT. *Annual Report 2000–2001*. Kochi: Central Institute of Fisheries Technology, 2001, pp 51–52.
56. Chakrabarti, R., B.M. Rao, P.G.V. Nair. Suitability of soluble protein concentrate from tuna liver for the growth of sulphite reducing *Clostridia*. *J. Food Sci Technol.* 41:405–408, 2004.

3.13

Cold Active Enzymes in Food Processing

Rajni Hatti-Kaul, Hákon Örn Birgisson, and Bo Mattiasson

CONTENTS

13.1 INTRODUCTION

Foods have been processed since ancient times using extracts of vegetables or microorganisms rich in enzymes. However, it was only in the nineteenth century that the various biological conversions that occurred during processing were ascribed to the action of enzymes, when isolation from malt of an enzyme complex "diastase" that converts gelatinized starch into sugars was reported. Soon afterward, the digestive enzyme pepsin was isolated from the calf stomach. The first major breakthrough for microbial enzymes in food industry occurred in the early 1960s with the launch of glucoamylase that allowed complete conversion of starch to glucose (1,2). The enzymatic methods have now become an important and indispensable part of the food industry (3). The specificity of their action and ability to function under mild conditions are the main advantages of using enzymes. The later part of the twentieth century has seen tremendous increases in enzyme production capacity of microorganisms as a result of advances in gene technology, and enzymatic processing has been extended to several novel applications (1). Of the estimated worldwide sales of $1.8 billion of the industrial enzymes in 2002, over 45% was accounted for by consumption in the food and feed industry, the main consumers being starch and dairy industries, while baking, brewing, fruit juice, animal feed, flavors, fats, oils, and protein hydrolysis, accounted for most of the rest (4).

The success of several processes, especially in the starch industry, has been due to the availability of enzymes with good activity and stability at high temperatures. In some cases, the operational stability of the enzymes was improved by means of immobilization, an important example being glucose isomerase. The enzymes from thermophilic and hyperthermophilic microorganisms that are naturally thermostable by way of adaptation to the high temperature environments was another focus (5). With the discovery of new enzymes, the temperature limits of the enzyme activity and stability reached unbelievable extremes. Protein engineering has also been used with a primary goal of increasing thermostability of the biomolecules.

In contrast to this, the interest in "cold activity" of enzymes is relatively recent, and has been promoted as a result of intense research activity on cold adapted organisms isolated from low temperature environments. One such attempt has been a rather large project sponsored by the EU Fourth Framework research program during the 1990s to examine enzymes from microorganisms isolated in Antarctica, both with respect to fundamental studies on cold adaptation and biotechnological applications. Simultaneously, laboratory evolution techniques have provided an alternative and powerful strategy to modify existing enzymes to function at lower temperatures. A number of reviews on cold active enzymes and mechanisms enabling low temperature activity have appeared in recent years (6–17). The applications of enzymes for low temperature processing may provide various benefits. These include:

lower energy requirements
protection of substrates or products from degradation
minimization of side reactions
prevention of the destruction of other substances associated with raw material
reduced risk of microbial spoilage
inactivation of the enzyme after completion of the process by a mild increase in temperature

13.2 COLD ADAPTED ORGANISMS AS A SOURCE OF COLD ACTIVE ENZYMES

The traditional means of obtaining cold active enzymes has been through screening of naturally occurring cold adapted prokaryotic and eukaryotic organisms for enzymatic activities of interest (18,19). Considering that a major part of Earth, including the vast extent of permanently cold environments such as the Antarctic, Arctic, and mountain regions, as well as the deep sea waters covering three quarters of the planet surface, is exposed to low temperatures (i.e., below 5°C) for a large part of the year, organisms inhabiting such environments should represent a significant portion of the living world (13,20). Some amazing examples of locations where microbial life has been found include super cooled cloud droplets in high altitude, and below 3500 meters of ice in the Antarctic subglacial Lake Vostok (21,22).

Most microorganisms that inhabit permanently cold environments are either psychrotolerant (also called as psychrotrophic) or psychrophilic. Psychrotolerant organisms grow well at temperatures close to the freezing point of water, but have optimal growth rates above 20°C while psychrophiles grow fastest at temperature of 15°C or lower but are unable to grow above 20°C. Metabolic fluxes displayed by such organisms are more or less comparable to those exhibited by mesophiles. A major challenge for such organisms is to counteract the reduction in chemical reactions due to low temperatures, and the best strategy to maintain sustainable activity under such conditions is to produce cold adapted proteins. Enzymes with up to tenfold higher specific activity are synthesized by psychrophiles to compensate for the low reaction rates. The specific activity is, however, generally lower than that of the mesophilic enzymes at 37°C. Furthermore, the temperature for apparent maximal activity is shifted toward the low temperatures, reflecting the weak stability of these proteins that are prone to unfolding and inactivation at moderate temperatures (17,23).

Besides cold habitats, other interesting sources of cold active enzymes include many fish, squids, and other aquatic creatures, which are adapted to environments that are permanently cold (24). For example, Greenland cod remains biologically active at about 0°C throughout the year living in sub-Arctic latitudes. Digestive enzymes from cold adapted fish possess unique activity and stability features for use as food processing aids at relatively low temperatures (25,26), and because they are obtained from edible animal tissue they have an advantage over microbial enzymes in that no toxicological testing of the raw material is needed to demonstrate their safety (24). A number of laboratories are developing commercial processes for isolating enzymes from fish processing offal. One company, KS Biotec-Mackzymal in Tromsö, Norway, produces such enzymes for use in food processing and other areas (27).

13.3 BASIS OF COLD ACTIVITY

Significant advances have been made during recent years in the study of proteins and enzymes from cold adapted species that have helped in providing insights into their molecular and functional adaptations to low temperature. This has been facilitated by the availability of recombinant enzymes for analysis by x-ray and biophysical techniques, and comparison with mesophilic and thermophilic homologues.

13.3.1 Kinetic and Thermodynamic Adaptations

In order for the cold active enzymes to maintain reaction rates at low temperature that are comparable to mesophilic enzymes at mesophilic temperatures, the loss of kinetic energy at low temperatures needs to be overcome. As revealed by the kinetic data available for the various enzymes from Antarctic or Arctic fish and by exoenzymes from Antarctic microorganisms, optimization of catalytic efficiency (k_{cat} /K_m) is the main physiological adaptation for the enzymes to be active at low temperatures (6,28). In the presence of high substrate concentration, optimization of k_{cat} should be the most relevant parameter, which is true for microbial extracellular enzymes (29–31). On the other hand, enzymes from marine or freshwater organisms secreted in a substrate poor medium should have highly optimized K_m values (6,32). Some enzymes can also improve both k_{cat} and K_m parameters as a way of adaptation (28).

In thermodynamic terms, the basis of increasing the reaction rate at a given temperature corresponds to reduction in activation free energy, ΔG, which has indeed been shown to be the case for enzymes from cold adapted species (33). The contribution of enthalpic and entropic energies to ΔG may further reflect adaptations to the low temperature environment in which these enzymes function ($\Delta G = \Delta H - T\Delta S$). The decrease in activation enthalpy (ΔH) seems to be the main adaptive feature of all cold enzymes studied so far, which reduces the dependence of k_{cat} on temperature and the inhibitory effect of low temperatures on enzyme reaction rates (according to Arrhenius' law) (17,34,35). Decrease in the number of enthalpy driven interactions also contributes to increased flexibility of the active site (17). The active site flexibility affects the substrate binding ability, providing a broader distribution of the ground state enzyme–substrate complex (ES), leading to increased entropy of this state as compared to the mesophilic enzymes (9,35). This feature of the cold enzymes requires lower activation energy for transforming the ground state ES complex to the activated form; i.e., k_{cat} will be increased at the expense of K_m. This is true for the cold adapted enzymes displaying the same substrate-binding site and active site makeup as the mesophilic enzymes (36). Greenland cod trypsin and psychrophilic α-amylase and subtilisin seem to make use of this strategy (6).

However, some enzymes seem to have undergone amino acid substitutions within the active site to maintain or improve the substrate binding affinity to meet their physiological requirements. The chitobiase from Antarctic bacteria was shown to display 8 times higher k_{cat} value, and its K_m for the substrate is 25 times lower than that of the mesophilic enzyme at 5°C, resulting in a nearly 200 times higher k_{cat}/K_m (28). The two tryptophan residues in the active site that serve as the substrate binding ligands in the mesophilic enzyme are replaced in the psychrophilic enzyme by polar residues that are able to perform stronger interactions as the temperature decreases.

13.3.2 Flexibility Versus Stability

Higher structural flexibility has been assumed to be a prerequisite for catalytic activity of the enzymes from cold adapted organisms at low temperatures (6,8). In a normal mesophilic enzyme, the molecular motions become restricted at lower temperatures, which interferes with catalysis. Higher flexibility in cold enzymes would overcome this restriction, but would lead to poor thermostability (37). The low stability has been demonstrated by a significant decrease in optimum temperature of activity and a higher susceptibility to unfolding at moderate temperatures (6,38). The difference between conformational stability of psychrophilic and mesophilic α-amylases has been estimated to be of the order of 30 kJ mol^{-1}. The balance between flexibility and stability represents one of the most crucial points in the adaptation of a protein to environmental temperature. However, the homologous enzymes from different temperature environments have nearly identical conformational plasticity under their respective optimal temperature working conditions (39).

13.3.3 Active Site of Cold Active Enzymes

The data available from increasing number of three dimensional structures of cold adapted bacterial and fish enzymes solved during past years (40–48) reveals that the cold adapted enzymes do not display any unusual structure, so that only minor structural modifications may be necessary to adapt a mesophilic or thermophilic homologue to cold. For example, the α-amylase from the Antarctic bacterium shares more than 40% identity with its mesophilic analogue from pig pancreas (49). All the active site residues involved in the reaction mechanisms are strictly conserved, suggesting that the fundamental reaction pathway is not modified among homologous enzymes adapted to different temperatures and that the molecular basis of adaptation lies elsewhere in the molecule. The structural changes necessary for cold adaptation often involve changes in the overall conformational properties and dynamics of the molecule (11,50).

The greater catalytic efficiency of cold active enzymes has been explained by better accessibility of the active site to the substrate, which may be achieved by one or more differences like small deletions in loops bordering the active site or replacement of bulky side chains for smaller groups at the entrance. (17,47). This may lead to the accommodation of substrates at lower energy costs, and hence reduce the activation energy required for the formation of the enzyme–substrate complex. Further, a relatively broad specificity and lower binding constants for the cold adapted enzymes may result, as shown in case of elastase from Atlantic salmon in contrast to mammalian elastases (12). Easier release of the product may also be facilitated from a larger active site. Cold adapted enzymes have further been shown to exhibit improved electrostatic potentials near the active site region when compared with their mesophilic or thermophilic counterparts, which would facilitate interaction with oppositely charged substrates (12,43,46,47,51–53).

13.3.4 Structural Adaptations

There is no consensus for a common paradigm of structural changes that confer cold activity. Structural models of cold active enzymes have been quite useful in revealing the specific strategies used in each enzyme, consisting of a set of structural adjustments to adapt to cold (6,17,52,54). These involve a decrease in proline residues in loops, arginine content, salt bridges, ion pairs, H-bonds and aromatic interactions, ion binding constants and hydrophobic interactions, changes in α-helix dipole interactions, increased clustering of glycine residues, and insertions (e.g., leading to extra surface charges) or deletions (resulting in weaker subunit interactions) of loops (47,48,52). The binding of stabilizing ions like calcium can be significantly weaker than in mesophiles. The compactness of the protein interior is often lowered due to smaller size and relative hydrophobicity of nonpolar residue clusters forming the protein core. Calculation of the solvent accessible area showed that some psychrophilic enzymes expose a higher proportion of nonpolar residues to the surrounding medium. This leads to increased destabilization, due to the reorganization of water molecules around the hydrophobic side chains (42,46,48). An excess of negative charges on the surface of the molecules, and more acidic isoelectric points, frequently noted in cold active enzymes as compared to that of mespophilic or thermophilic homologues, has been related to the improved interactions with the solvent, which could be of prime importance in rendering flexibility near zero degrees (55). Only a few of these structural alterations are used to acquire the required flexibility, but they do lead to a lowered stability of the protein molecule.

13.3.5 Mutational Analysis of Cold Activity

Comparison of crystal structures of extremophilic enzymes indicate a clear increase in the number and strength of all known weak interactions and structural factors involved in protein stability from psychrophiles to mesophiles and thermophiles. Very slight variations by site directed mutagenesis in a psychrophilic sequence can establish or mask the psychrophilic character of mutants that may affect the active site flexibility through changes in various conformational constraints (56). Site directed mutagenesis experiments on *Alteromonas haloplanktis* amylase to produce additional weak interactions found in its mesophilic homologue from pig pancreas, and analysis of the mutants, revealed that single mutations were not sufficient to modify the stability and catalytic parameters of the cold adapted enzyme drastically enough to obtain a true mesophilic α-amylase (50). However, accumulation of these mutations transformed the psychrophilic α-amylase into a true mesophilic enzyme. Even if the reaction pathway of the α-amylase has not been modified, the ability of the active site residues to perform catalysis has been altered, probably by modulating the plasticity of the whole catalytic cleft. Based on these studies, it has also been proposed that the lower temperature dependence of the activity in cold active enzymes arises from fewer interactions that need to be broken during the catalytic cycle, which is also responsible for a heat labile activity and higher entropic contribution (9).

Recent experimental results have indicated that relationships between stability, specific activity and flexibility is much more complex than expected. It seems that some elements of the protein structure control protein stability, while other regions are important in conferring flexibility, giving rise to an optimal catalytic efficiency at the environmental temperature. It is therefore probable that both a high specific activity and a good stability can be conferred to a cold adapted enzyme. This has even been supported by site directed mutagenesis experiments, where stabilities of enzymes were improved while the original catalytic properties were retained (30,57). This would support the hypothesis that organisms growing in stable cold environments would not encounter selective pressure to maintain enzyme stability at higher temperatures. So, although many cold active enzymes are thermolabile, this feature is not necessarily a prerequisite for the structural changes resulting in cold adaptation of the enzymes, but may simply be due to the lack of selective pressure (11,58,59).

13.4 DIRECTED EVOLUTION AS A TOOL FOR CREATION OF COLD ACTIVE ENZYMES

In contrast to selecting enzymes from a natural diversity of cold adapted organisms, laboratory evolution has contributed significantly toward modifying enzyme properties and functions. This evolution involves introduction of random mutations in a gene and screening the resulting library for the acquisition and improvement of a specific property. Cold adapted forms have been created from mesophilic enzymes by artificial evolution. A mesophilic subtilisin-like protease was converted into a cold active enzyme; only 4 amino acid substitutions were needed to obtain a mutant enzyme with a catalytic efficiency 9.6 times greater than that of its parent at 10°C (60). Even the catalytic activity of thermostable enzymes from hyperthermophilic organisms has been improved at low temperature by DNA shuffling and directed evolution (61,62). In contrast to the natural evolution, where selection takes place guided by the survival requirement and the protein properties are evolved to fit the existing environment, the selection conditions in the laboratory evolution are more defined. Mutated enzymes with high activity and stability can be selected from all combinations of activity and stability. Random mutagenesis thus proves that the fundamental characteristics of activity and stability are independent and not physically linked in proteins. However, mutations that simultaneously increase activity and stability are very rare and have proved to be extremely complex to obtain.

Evolutionary engineering based on the Darwinian sequential program of mutagenesis and selection has been used to produce cold active proteases with no change in thermal stability (63–66). The procedure consists of random mutagenesis for obtaining proteases with enhanced activities at low temperature, via multistep mutations with a combination of primary mutation causing activity loss and secondary mutation causing recovery of activity. Miyazaki et al. (58) combined random mutagenesis, saturation mutagenesis, and DNA shuffling to obtain variants of a psychrophilic subtilisin S41 with improved stability, that maintained the wild-type activity at 25°C.

13.5 COLD ACTIVE ENZYMES OF INTEREST FOR FOOD PROCESSING

Much of the efforts in research on cold active enzymes have so far gone into understanding the factors underlying cold activity. Hence, the use of cold active enzymes is not yet widespread in food processing or other applications. The exceptions are the digestive enzymes isolated from fish. The breakthrough for cold active enzymes in food processing relies on the applications involving heat sensitive foods and where residual enzyme activity is not desired due to potential changes in the product. Proteases and glycosidases are the main groups of enzymes with diverse applications, and several cold adapted candidates have been studied, some of which are described here.

13.5.1 Proteases

Enzymatic modification using proteases is often an attractive means of obtaining better functional and nutritional properties of food proteins, and for making a variety of new and valuable products (1,25,67). Table 13.1 lists many of the important applications of proteases in food processing, where low temperatures are preferred. Various proteases are widely used for manufacturing protein hydrolysates and extracts with improved functional properties from cheap vegetable, meat, and fish byproducts.

The fish industry uses proteases as gentle processing aids in a number of operations to yield products of high quality, and these proteases have in many cases replaced the alternative mechanical, chemical, or manual treatment procedures (Table 13.1) (24.68). For example, caviar produced using proteases has the advantages of better recovery, less damage to the eggs, cleanliness of products with no residues, less required labor, and good hygienic conditions compared to caviar produced mechanically. The enzymes used are the digestive enzymes isolated from fish, and the processing is invariably done at low temperatures. Experiments at Icelandic Fisheries Laboratories indicated that the scales from haddock could be removed without affecting the skin or flesh after incubating the fish with a protease at 0°C, followed by spraying with water (70).

In the dairy industry, rennets are used for coagulation during cheese production, while other proteases are used for accelerating cheese ripening, for modifying functional properties of some dairy products, and for modifying milk proteins to reduce the allergenic properties of cow milk products. Milk oxidation is an undesirable phenomenon in the dairy industry, and treatment of milk with trypsin improves its stability to oxidative rancidity, presumably through improved copper binding by the enzyme modified protein (73). Furthermore, enzymatic hydrolysis of milk proteins results in products with functionally important properties (Table 13.1). Specific bioactive peptides have been reported to have various positive effects on the human body, such as reduction of blood pressure, inhibiting effects on blood clotting, antibacterial effects, inducing effects on the immune system, and even reducing the craving for fat food (71,72,74,75).

Table 13.1
Food processing operations suitable for cold active proteases

Foods	Application	Ref.
Fish	Production of fish protein hydrolysates used to fortify the nutritional value of cereal-based foods, as dietetic and hypoallergenic agents, and as basic ingredient in milk substitutes for young animals	(24,68–70)
	Removal of skin from fish and squids that are difficult to de-skin by mechanical or manual means	
	Removal of scales without affecting the skin or flesh, e.g., from haddock	
	Removal of surface bacteria and extending the shelf life of stored fish	
	Removal of membranes and organs from seafood products	
	Production of caviar	
	Extraction of carotenoprotein pigment from shellfish processing discards for use as feed supplement in rations of farmed fish or as a colorant and flavorant for use in food products	
	Accelerating ripening of fermented products such as "maatjes" and fish silage	
Meats	Production of blood cell hydrolysates	(1)
	Production of meat extracts	
	Meat tenderization	
	Cleaning of bones	
	Upgrading of fresh bones to produce gelatin and meat protein hydrolysates	
Milk	Cheese production	(1,71,72)
	Hydrolysis of milk and whey proteins to produce non- and low allergenic cow milk products for baby foods as substitutes for mothers' milk	
	Production of various physiologically functional peptides such as opioid peptides, immunostimulating peptides and angiotensin I-converting enzyme (that play an important role in blood pressure regulation) inhibitors	
	Treatment of milk to protect against oxidative rancidity	
	Production of whey protein hydrolysates	
Vegetables	Limited hydrolysis of soy protein isolates to produce functional protein hydrolysates with better foam-expansion and –stability, good whipping properties, and potential use as egg white substitutes	(1)

Prevention of cloudiness or haze formation in beer during chilling that occurs as a result of the formation of complexes of peptides with polyphenolic compounds is among other interesting applications of proteases (1).

Most of the cold active proteases known so far originate from fish sources, while some enzymes developed by mutagenesis have also been obtained.

13.5.1.1 Cold Active Proteases from Fish

The digestive proteolytic enzymes can be recovered as byproducts from fish processing wastes. The cold active proteolytic enzymes like pepsin, trypsin, chymotrypsin, and elastase, have been isolated from the stomach of cold water species such as Atlantic cod (*Gadus morhua*), Polar cod (*Boreogadus saida*), Greenland cod (*Gadus ogac*), and Arctic fish, capelin. These enzymes invariably have a relatively lower optimum temperature of activity, higher catalytic efficiency, and much lower stability than their mammalian counterparts. Table 13.2 provides a comparison of the kinetic parameters of some of the cold adapted proteases with their homeothermic homologues. Pepsins isolated from the stomach of the fish species mentioned exhibit temperature optima of activity around 37–40°C and high instability above 40°C (27). Intestinal elastase isolated from Atlantic cod was shown to possess an optimal temperature of 40°C and about 3.5-fold higher k_{cat} and nearly fivefold higher catalytic efficiency than porcine elastase (79). Cod chymotrypsin also displayed higher activity at 3–15°C than the bovine enzyme (78). Trypsins from Antarctic fish (76), Greenland cod (77,80), Atlantic cod (81), and Atlantic salmon (82) are all shown to be catalytically efficient at all temperatures studied, but also in general much less stable at high temperatures, compared to other mammalian counterparts.

X-ray structures of Atlantic salmon trypsin and elastase and cod pepsin have been reported (40,44,45). Investigations on trypsins showed that the enhanced catalytic efficiency is mostly a result of lower K_m, indicating a better substrate affinity, which results in a 4- to 35-fold higher efficiency for the cold adapted fish trypsins (83). The increased substrate affinity is probably due to a lower electrostatic potential in the binding pocket and from the lack of five hydrogen bonds adjacent to the catalytic triad (52). Trypsins from fish and other species of marine origin are generally unstable at low pH values and stable under alkaline conditions, hence differing from mammalian proteases. The most notable difference in the cold trypsins with respect to their warm counterparts is a reduced hydrophobicity of internal residues accomplished by an apparently systematic substitution of smaller and less hydrophobic residues. In particular, a reduced number of isoleucines and a correspondingly increased number of valines is notable, which results in a less densely packed core and an increased rotational and vibrational freedom of the internal groups. The number of both charged and other polar amino acids is significantly higher in the cold trypsins. A difference in flexibility of a few loops, a higher number of methionine residues, and fewer interdomain hydrogen bonds and stabilizing interactions in helices are also observed (52). All these changes contribute to the low thermal stability of the cold trypsins; however, these substitutions do not fully explain the increased catalytic efficiency (83).

Cold active cod proteases are commercially available and are being applied in a number of processes. Preparation of fermented squid with supplements of Greenland cod trypsin has been found to be better than conventional process because of the higher molecular activity at low temperatures (4°C) during the early stages of fermentation. Cod trypsin was also more effective for the extraction of carotenoprotein than bovine trypsin at 4°C. Trypsin from cold adapted fish has an advantage in the treatment of milk for preventing oxidative rancidity, because thermal instability of the enzyme would avoid subsequent hydrolysis of milk proteins (25). Cod pepsins have been used as rennet substitutes in cheese production, in skin removal from fish, and as a digestive aid in fish feed (84). Pepsins from Atlantic cod (*Gadus morhua*) and orange roughly (*Hoplostecthus atlanticus*) have been used for caviar production from the roe of the same species in New Zealand (24,85).

Table 13.2
Kinetic parameters of some proteases from cold adapted fish as compared to the homeothermic homologues

Protease	Source	Substrate	Temp (°C)	K_m	k_{cat}	k_{cat}/K_m	Ref.
Trypsin	Antarctic fish (*Paranotothenia*	N-benzoyl-L-arginine-	5	121 μM	29 min^{-1}	0.24 $min^{-1}.\mu M^{-1}$	(76)
	magellanica)	p-nitroanilide	25	25	125	0.6	
	Bovine		5	170	5.4	0.03	
			25	2445	109	0.04	
	Greenland cod (*Gadus ogac*)	p-tosyl-L-arginine-	5	0.14 mM	5.2 U/μM	37 $min^{-1}.\mu M^{-1}$	(77)
		methyl ester	25	0.15	14.7	98	
	Bovine		5	0.05	1.7	35	
			25	0.05	9.1	182	
Chymotrypsin	Atlantic cod (*Gadus morhua*)	N-benzoyl-L-tyrosine	10	0.12 mM	117 s^{-1}	975 $s^{-1}.mM^{-1}$	(78)
	(isoenzyme A)	ethyl ester	25	0.14	207	1479	
			10	0.27	43	159	
	Bovine		25	0.27	100	370	
Elastase	Atlantic cod (*Gadus morhua*)	(Suc-Ala-Ala-Ala-	10	0.8 mM	25 s^{-1}	31 $s^{-1}.mM^{-1}$	(79)
		p-nitroanilide)	25	1.0	44	43	
	Porcine		10	0.7	12	16	
			25	0.7	16	22	

13.5.1.2 Proteases Made Cold Active by Mutagenesis

Mucor rennin, often used as a substitute for calf chymosin, possesses a high thermostability that causes survival of the enzyme activity after the curd is cooked, and thus may cause bad flavor in cheese during long maturating periods. Mutagenesis of the fungus *Mucor pusillus* was done to produce a mutant enzyme with decreased thermostability, which is desirable for the enzyme as a milk coagulant for cheese manufacturing (86). The marked decrease in thermostability was attributed to the mutation of Gly186 to Asp. Mutations involving exchange of Gly186 with various amino acids by site directed mutagenesis led to a decreased thermostability, indicating involvement of the residue to maintain conformation of the enzyme. Further, a double mutant, involving changes to Ala101Thr and Gly186Asp showed the lowest thermostability without decrease in enzymatic activity, as well as the best relative ratio of clotting to proteolytic activity.

Taguchi et al. (65,66) have used evolutionary engineering to double the activity of a cold active proteinase at 10°C by using sequential random mutagenesis and a screening program. A temperature dependent increase in the k_{cat} value was observed for the mutant enzyme.

13.5.2 Glycosidases

Although the starch industry accounts for a significant part of the applications requiring heat stable glycosidases (glycoside hydrolases), various other processes are potential targets for cold active enzymes.

The hydrolysis of lactose in milk and whey using β-galactosidase is a well known process to obtain better quality dairy products and ones that are more easily digested by lactose intolerant individuals. Another emerging application of the enzyme in the food industry is the production of galactooligosaccharides for use as prebiotics. β-galactosidase preparations such as Lactozyme from Novozymes (Bagsvaerd, Danmark) are commercially available. The use of a cold active enzyme would allow the treatment of milk in a refrigerated state.

Pectinases are now an integral part of fruit juice industries, being used to facilitate juice extraction and clarification. They can be used in combination with cellulases, hemicellulases, and amylases. In 1998, pectic enzymes alone accounted for about one quarter of the world's food enzyme production (87). Apple, orange, pear, grape, strawberry, raspberry, blackberry, lemon, mango, apricot, guava, papaya, pineapple, and banana are some examples of fruit juices that have been treated with pectinases (88). The enzyme treatment not only improves overall yields, but also reduces processing costs (87). For maceration of plant tissue during preparation of pulpy juices and nectars, baby foods, and ingredients for yogurts and puddings, enzymatic degradation of pectin after a mild mechanical treatment often improves product properties (88). Acidic pectic enzymes used in the fruit juice industry often come from fungal sources, especially *Aspergillus niger*. Some of the commercial enzyme preparations available are Pectinex from Novozymes (Bagsvaerd, Danmark) and Gammapect from Gamma Chemie (Darmstadt, Germany).

In bread making, the concentrations of starch, hemicelluloses, and gluten are very important for the quality of baking products. Amylases, along with xylanases and proteases, have therefore been applied in baking processes (1,89). These enzymes can be used to reduce the dough fermentation time and to improve the properties of dough and crumb, and can aid in the retention of aromas and levels of moisture. A clear advantage of cold active enzymes here is the ease of inactivating the enzymes by heat, so that the enzymatic activity can be controlled better and a prolonged action of the enzymes prevented, which otherwise can result in the crumb becoming too soft or too sticky (8,90). Fungal α-amylase is used, as it is less heat stable. Apart from the higher loaf volume, the enzyme treatment results in a softer crumb.

Glycosidases with antimicrobial activity have a clear potential for prolonging the shelf life of foods stored in cold or ambient temperatures. For example, lysozyme has been used as a food preservative in food products such as cheese and sausages because of its antibacterial activity (91).

Several glycosidases from cold adapted microorganisms have been isolated and characterized, and some even crystallized. Cold adapted features of a few glycosidases are listed in Table 13.3, where they are compared with other mesophilic and thermophilic counterparts. Optisize® Amylase and IndiAge® Super cellulase produced by Genencor are claimed to be active at low temperatures; however their source is not revealed (95).

13.5.2.1 Cold Active β-galactosidases

Several cold adapted organisms including *Arthrobacter* species (96–100), *Bacillus subtilis* (101), *Carnobacterium piscicola* (102), *Planococcus* sp. (103), and *Pseudoalteromonas haloplanktis* (34) with β-galactosidase activities have been reported.

Detailed characterization of the β-galactosidase from *P. haloplanktis* isolated from Antarctica has revealed significant similarities with the mesophilic enzyme from *Escherichia coli*, including the multimeric form and subunit mass, active site residues involved in catalysis, and strict requirement for divalent metal ions (34). However, the turnover (K_{cat}) of the enzyme toward its substrate is significantly higher than that of the *E. coli* enzyme in the temperature range of 0–40°C. As a result of increased k_{cat} and decreased K_m for its substrate lactose, the specificity constant k_{cat}/K_m at 25°C is increased 90-fold. The enzyme also exhibits a 10°C lower optimum temperature of activity and a weaker thermal stability as compared to the *E. coli* β-galactosidase. Increased flexibility of the cold-active β-galactosidase has been attributed to insertions of certain residues, lower Arg and Pro content, and stacking of Gly around the catalytic residues of the enzyme. A study on treatment of milk at 4°C with *P. haloplanktis* β-galactosidase led to 33% lactose hydrolysis after 50 min whereas only 12% lactose was hydrolysed under similar conditions by *Kluyveromyces lactis* enzyme (34). Application of the enzyme in an immobilized form has also been investigated (104).

The *Arthrobacter* isolates have been found to have two or more β-galactosidase isoenzymes (100,105,106). An isoenzyme of psychrotolerant *Arthrobacter* sp. C2-2 was shown to exhibit characteristics rather similar to that of the *P. haloplanktis* enzyme. It retained 20% of maximal activity at 10°C, and was further seen to catalyse not only lactose hydrolysis but also transglycosylation to produce galactooligosaccharides, predominantly trisaccharides. One of the isoenzymes of *Arthrobacter psychrolactophilus* β-galactosidase has a subunit mol wt of 71 kDa in contrast to 116 kDa typically encoded by *lac*Z family, and the gene encoding the enzyme differs from other members of the *lac*Z family but has regions similar to β-galactosidase isoenzymes from *Bacillus stearothermophilus* and *B. circulans*. It does not have a role in lactose utilization.

Carnobacterium piscicola β-galactosidase was shown to exhibit 49% amino acid identity with β-galactosidase from the thermophilic *B. stearothermophilus* but with an optimum temperature of activity at 30°C, while the *Planococcus* enzyme was most closely related to an enzyme from *B. circulans* (103). The latter enzyme is also a dimer with a mol wt of 155 kDa, no metal requirements, an optimum pH of 6.5 and optimum temperature of 42°C. It is active at NaCl or KCl concentrations up to 4 *M*. The enzyme is stable during storage at 5°C for several months, and undergoes irreversible inactivation after incubation for 10 min at 55°C.

Table 13.3
Activity and stability parameters of some cold active glycosidases in comparison with their mesophilic and thermophilic counterparts

Glycosidase	Organism	Opt. Temp (°C)	T_m(°C)	K_m	k_{cat}	k_{cat}/K_m	Ref.
β-galactosidase	*P. haloplanktis*	45	48	2.4 mM	33 s^{-1}	13.7 s^{-1}.mM^{-1}	(34)
	E. coli	55	56.5	13	2	0.15	
α-Amylase	*Alteromonas haloplanctis*	30	44	1.27 g/l	1363 s^{-1}	1075 s^{-1}.g^{-1}.l	(29,92,93)
	Porcine	50	65.6	1.12	326	292	
	Bacillus amyloliquefaciens	85	85.7				
Xylanase	*P. haloplanktis*	35	52.6	28 mg/ml	1247 s^{-1}	44.5 s^{-1}.mg^{-1}.ml	(35,94)
	Streptomyces sp. S38	62	63.1				
	Clostridium thermocellum	80	81.4				
Xylanase	*Cryptococcus adeliae*		48	3.8 g/l	46 s^{-1}	12 s^{-1}.g^{-1}.l	(6,31)
	Cryptococcus albidus		62	2	21	10.5	

Kinetic parameters determined at 25°C, using as substrates: lactose for β-galactosidase, soluble starch for α-amylase, birchwood xylan for bacterial xylanase and oatspelts xylan for yeast xylanase.

T_m: melting temperature determined by differential scanning calorimetry.

13.5.2.2 Cold Active Pectinases

Reports on pectinases from cold adapted organisms were until recently restricted to psychrotolerant spoilage bacteria such as different strains of *Pseudomonas fluorescens* (107). Endopolygalacturonases from psychrophilic fungi, *Sclerotinia borealis* and *Mucor flavus* have been reported which are active on both polygalacturonic acid and esterified pectin to different extents (108,109). Their molecular mass of ~40 kDa and optimum activity at low pH (pH 3.5–5.5) is similar to those from other fungal sources. Optimum temperature of their activity is around 40–50°C but the enzyme activity and stability is lowered rapidly at temperatures above 50°C. *S. borealis* polygalacturonase exhibited 30% of its optimal activity at 5°C and was very stable at this temperature for over 2 years (108). Recently, cold adapted polygalacturonase producing yeasts, *Cystofilobasidium lari-marini*, *Cystofilobasidium capitatum*, *Cryptococcus macerans* and *Cryptococcus aquaticus* have been reported (110). The enzymes were easily inactivated at 50°C.

Production of pectate lyases by the psychrotolerant bacteria *Chryseomonas liteola* and *P. haloplanktis* has been described (111,112). The latter organism isolated from seawater in Antarctica showed the presence of two pectate lyases, both of which showed maximal activity at 30°C, and 25–35% of activity at 5°C. One of the enzymes was stable at 30°C while the other had a half life of 4 h.

13.5.2.3 Cold Active α-amylase

The α-amylase from psychrophilic *A. haloplanctis* is an enzyme that has been examined in great detail with respect to its structure and thermodynamic properties (36,41,42,50,55,92, 93,113–115). It has the classical three domain structure of the other α-amylases, comprising an $(\alpha/\beta)_8$ barrel in the major domain A, a small β-pleated domain B, and a globular domain C made up of 8 β-strands (41,113). The catalytic triad of two aspartyl and one glutamyl residue, the 24 residues most intimately involved in forming the active site and the aromatic residues at the active site, are strictly conserved in the cold active α-amylase compared with the porcine enzyme (41,42). The amylase is among the most thermolabile enzymes known. The apparent temperature optimum of the psychrophilic enzyme is approximately 30°C lower than the mesophilic porcine enzyme, and at 4°C and 25°C its catalytic efficiency (k_{cat}/K_m) is 6.6 and 3.7 times greater, respectively. Both enzymes require a chloride ion for activity, which, however, is bound with tenfold lower affinity in the cold active enzyme and does not influence its thermostability in contrast to the porcine enzyme (92,114).

13.5.2.4 Cold Active Xylanase

Cold adapted xylanases belonging to families 8 and 10 from Antarctic psychrophilic bacteria *P. haloplanktis* and yeast *Cryptococcus adelia*, respectively, have been investigated (31,35,94). The enzymes exhibit high activity at low temperature, poor stability, and high flexibility, like the other cold adapted biocatalysts. The yeast family 10 xylanase displays lower activation energy and higher catalytic efficiency (threefold) in the range 0–20°C than its mesophilic homologue from *Cryptococcus albidus*, and loses 95% activity within 5 min at 50°C (31,116). The mature glycosylated enzyme secreted by the yeast shares 84% identity with its mesophilic counterpart. Analysis of computerized molecular models indicates that adaptation to cold consists of discrete changes in the tertiary structure. Among the 53 amino acid substitutions, 22 are presumably involved in the adaptation process. These changes lead to a less compact hydrophobic packing, the loss of one salt bridge, the loss of a charge dipole interaction, and to increased solvent interaction due to relative abundance of surface charge (31).

The crystal structure of the *P. haloplanktis* enzyme has recently been solved at 1.3 Å resolution (48). The parameters conferring cold-adapted properties were indicated to be fewer salt bridges, an increased exposure of hydrophobic residues, optimization of electrostatic potential at the active site, and an increased flexibility of the aromatic residues lining the subsites.

13.5.2.5 Cold Active Endoglucanases

Very few reports are available on cold active endoglucanases (cellulases) from cold adapted organisms. Two endoglucanases, CelG and EGD, from a ruminal bacterium *Fibrobacter succinogenes,* were found to have optimum temperature of activity at 25°C and 35°C, and to retain about 70% and 18% of the maximal activity, respectively, at 0°C (117). CelG has a 33-fold higher k_{cat} and a 73-fold higher k_{cat}/K_m value, and exhibits low thermal stability that has been ascribed to the presence of small amino acids around the putative catalytic residues. The presence of a cold active enzyme in a mesophile has been explained as a result of lateral gene transfer from a psychrophile or the origin of the *F. succinogenes* from the marine environment.

It was only very recently that a psychrotolerant *Clostridium* sp., isolated from a cold adapted cattle manure biogas digester, was shown to produce cellulolytic as well as xylanolytic enzymes, having temperature optima of 20°C (118). Also, crystallization of a cold active cellulase from the Antarctic psychrophile *P. haloplanktis* has been reported (119).

13.5.2.6 Cold Active Lysozyme

A cold active lysozyme-like ~11kDa enzyme designated chlamysin has been isolated from viscera of the marine bivalve *Chlamys islandica* that is a marine species located 20 m or deeper in the southern part of an arctic region having sea temperatures rarely exceeding 4°C (120). The protein was highly efficient in hydrolysing *Micrococcus luteus* cells only at low pH (pH 4.5–6.2) and at low temperature (4–35°C). Interestingly, the enzyme was completely stable during storage at room temperature for a month as well as after heating to 70°C for 15 min, suggesting high structural stability.

13.5.3 Lipase

Lipases are mainly used in the treatment of fats and oils for which activity at high temperature is normally desirable due to the solubility characteristics of these materials. Lipases are also used to some extent in cheese ripening. A lipase from an Antarctic psychrophilic bacterium *Psychrobacter immobilis* sp. has been described (121). When compared to the lipase from a mesophilic strain of *Pseudomonas aeruginosa*, it exhibits a limited thermal stability with a half life of 2 min at 60°C. Its apparent temperature optimum for activity is lowered by at least 30°C to around 35°C and its activation energy is reduced by a factor of 2. Several features characteristic of cold adapted enzymes have been revealed in the three dimensional model of the psychrophilic lipase, such as low proportion of arginines as compared to lysines, low proline content, a small hydrophobic core, fewer salt bridges, and aromatic–aromatic interactions.

13.5.4 Catalase

The enzyme catalase is important in the treatment of dairy and liquid egg products for the removal of residual hydrogen peroxide used for low temperature pasteurisation (122). The enzyme's own oxygen-generating capacity is also exploited in its application as a leavening agent for baked goods. So far, one psychrophilic bacterium exhibiting high catalase activity has been reported (123). The optimum temperature of the enzyme activity is 30°C, which is about 20°C lower than that for bovine catalase activity.

13.5.5 Other Potential Applications

There are other enzymes used in the processing of foods to achieve varying effects, but for which no cold active counterpart has been reported so far. Some examples include transglutaminase, α-L-rhamnosidase and laccase.

Transglutaminase is used to impart unique rheological and functional properties, such as increase in elasticity and firmness, texture improvement, gelation capability, and water holding capacity, by forming cross links between protein molecules (124). A bacterial (*Streptoverticillium* sp.) transglutaminase is widely used in seafood surimi (fish paste) products, meat products (sausage and ham), dairy products, and baked goods, and has potential in the manufacture of yogurt, cheese, and frozen dairy products with improved quality.

Alpha-L-rhamnosidases and laccases are applicable in the processing of fruit juices. The former enzyme degrades the bitter compound naringin in the juice to L-rhamnose and prunin – a compound three times less bitter than naringin (125,126). Alpha-L-rhamnosidases can also be used to dissolve hesperidin crystals, which sometimes form in fruit juices and make the juices cloudy (127). Laccases have antioxidant effects, as they facilitate removal of oxygen, thereby preserving the flavor and color of the juice (128).

13.6 ASPECTS ON PRODUCTION OF COLD ACTIVE ENZYMES

Application of cold active enzymes places demand on their cost effective production. Cloning and expression in *Escherichia coli* has provided sufficient quantities of the enzymes from cold adapted organisms to facilitate structural studies. It has been observed that recombinant psychrophilic enzymes may not be expressed or may be inactivated as a result of the high temperature (37°C) used for their expression (8). Furthermore, their properties seem to be altered as a result of misfolding or the occurrence of an inactivated population (115). It has, however, been possible to express recombinant psychrophilic enzymes in an active form, by growing the mesophilic host at the lowest possible temperature, even though its growth rate is drastically lowered (17). For example, significant amylase activity of recombinant *A. haloplanktis* cold active amylase could be found at growth temperatures for *E. coli* only below 25°C (115). When expressed at a sufficiently low temperature (15–18°C), a recombinant psychrophilic enzyme is indistinguishable from its wild-type parent molecule as regards kinetic parameters, folding properties, and three dimensional structure (42,115).

Some efforts are being made to develop cloning or expression systems in cold adapted bacteria that may circumvent the low stability of the product encountered during heterologous production in mesophilic hosts. Antarctic strains have mainly been tried as expression hosts (129,130). These organisms have been screened for characteristics like optimal temperature and ranges of growth temperature, survival after cryopreservation, natural resistance to antibiotics and transformation efficiency of broad host range plasmids. Some strains were shown to exhibit high gene transfer efficiencies, and induction levels up to 250-fold could be achieved at 4°C and 15°C (129). The cold active *P. haloplanktis* A23 α-amylase was produced and secreted in *P. haloplanktis* TAC 125 (which lacks the endogeneous amylase), using a cold adapted shuttle expression plasmid in which the gene coding for the amylase precursor gene was still subjected to the mesophilic lacZ promoter transcription regulation (130). For the purposes of producing enzymes for food processing, the choice of an expression host would also be dictated by reasons other than attaining high amounts by overexpression. For example, the GRAS (generally recognized as safe) status of the organism would need to be considered.

Subsequent to their production, strategies for purification of cold active enzymes need to be carefully evaluated because of their sensitive nature. Use of separation techniques allowing rapid and gentle purification are essential to minimize losses (104). Eventually, conditions for storage of the purified enzymes prior to their use would also need to be considered in specific cases.

In conclusion, it may be stated that cold active enzymes, even though having existed as long as the existance of life in cold environments, have been "discovered" relatively recently. Their potential in food processing is yet to be fully exploited. It is very likely that evolution in the laboratory will be used to assist the natural evolution or even gain preference in the design of cold active enzymes with required stability for the applications of interest.

REFERENCES

1. Olsen, H.S. Enzymes in food processing. In: *Biotechnology* Vol. 9, Reed, G., T.W. Nagodawithana, eds., Weinheim, Germany: VCH Publishers, 1995, pp 663–736.
2. http://www.novozymes.com
3. Whitehurst, R.J., B.A. Law, eds. *Enzymes in Food Technology*. Oxford: Blackwell Scientific, 2002.
4. http://www.bccresearch.com/editors/RC-147NA.html
5. Niehaus F., C. Bertoldo, M. Kähler, G. Antranikian. Extremophiles as a source of novel enzymes for industrial application. *Appl. Microb. Biotechnol.* 51:711–729, 1999.
6. Feller G., C. Gerday. Psychrophilic enzymes: molecular basis of cold adaptation. *Cell. Mol. Life. Sci.* 53:830–841, 1997.
7. Marshall, C.J. Cold-adapted enzymes. *Trends Biotechnol.* 15:359–364, 1997.
8. Gerday, C., M. Aittaleb, M. Bentahir, J.-P. Chessa, P. Claverie, T. Collins, S. D'Amico, J. Dumont, G. Garsoux, D. Georlette. Cold-adapted enzymes: from fundamentals to biotechnology. *Trends Biotechnol.* 18:103–107, 2000.
9. Lonhienne, T., C. Gerday, G. Feller. Psychrophilic enzymes: revisiting the thermodynamic parameters of activation may explain local flexibility. *Biochim. Biophys. Acta* 1543 1–10, 2000.
10. Russell, N.J. Toward a molecular understanding of cold activity of enzymes from psychrophiles. *Extremophiles* 4:83–90, 2000.
11. Sheridan, P.P., N. Panasik, J.M. Coombs, J.E. Brenchley. Approaches for deciphering the structural basis of low temperature enzyme activity. *Biochim. Biophys. Acta (BBA) - Protein Struct. Mol. Enzymol.* 1543:417–433, 2000.
12. Smalås, A.O., H.K. Leiros, V. Os, N.P. Willasen. Cold adapted enzymes. *Biotechnol. Annu. Rev.* 6:1–57, 2000.
13. D'Amico, S., P. Claverie, T. Collins, G. Feller, D. Georlette, E. Gratia, A. Hoyoux, M.-A. Meuwis, L. Zecchinon, C. Gerday. Cold-adapted enzymes: an unachieved symphony. In: *Protein Adaptations and Signal Transduction,* Storey, K.B., J.M. Storey, eds., New York: Elsevier, 2001, pp 31–42.
14. D'Amico, S., P. Claverie, T. Collins, D. Georlette, E. Gratia, A. Hoyoux, et al. Molecular basis of cold adaptation. *Phil. Trans. R. Soc. Lond. B.* 357:917–925, 2002.
15. Zecchinon, L., P. Claverie, T. Collins, S. D'Amico, D. Delille, G. Feller, et al. Did psychrophilic enzymes really win the challenge? *Extremophiles* 5:313–321, 2001.
16. Cavicchioli, R., K.S. Siddiqui, D. Andrews, K.R. Sowers. Low-temperature extremophiles and their applications. *Curr. Opin. Biotechnol.* 13:253–261, 2002.

17. Feller, G. Molecular adaptations to cold in psychrophilic enzymes. *Cell. Mol. Life Sci.* 60:648–662, 2003.
18. Brenchley, J.E. Psychrophilic microorganisms and their cold-active enzymes. *J. Ind. Microbiol.* 17:432–437, 1996.
19. Margesin, R., F. Schinner. *Biotechnological Applications of Cold-Adapted Organisms,* Heidelberg: Springer-Verlag, 1999.
20. Thomas, D.N., G.S. Dieckmann. Antarctic sea ice: a habitat for extremophiles. *Science* 295:641–644, 2002.
21. Price, P.B. A habitat for psychrophiles in deep Antarctic ice. *Proc. Natl. Acad. Sci. USA* 97:1247–1251, 2000.
22. Christner, B.C., E. Mosley-Thompson, L.G. Thomson, J.N. Reeve. Isolation of bacteria and 16S rDNAs from Lake Vostok accretion ice. *Environ. Microbiol.* 3:570–577, 2001.
23. Feller, G., E. Narinx, J.L. Arpigny, M. Aittaleb, E. Baise, S. Genicot, et al. Enzymes from psychrophilic organisms. *FEMS Microbiol. Rev.* 18:189–202, 1996.
24. Shahidi, F., Y.V.A. Janak Kamil. Enzymes from fish and aquatic invertebrates and their application in the food industry. *Trends Food Sci. Technol.* 12:435–464, 2001.
25. Simpson, B.K., N.F. Haard. Cold adapted enzymes from fish. In: *Food Biotechnology,* Knorr, D., ed., New York: Marcel Dekker, 1987, pp 495–527.
26. Stefansson, G. Enzymes in the fishing industry. *Food Technol.* 42(3):64–65, 1988.
27. De Vecchi, S., Z. Coppes. Marine fish digestive proteases: relevance to food industry and the south-west Atlantic region: a review. *J. Food Biochem.* 20:193–214, 1996.
28. Lonhienne, T., J. Zoidakis, C.E. Vorgias, G. Feller, C. Gerday, V. Bouriotis. Modular structure, local flexibility and cold-activity of a novel chitobiase from a psychrophilic Antarctic bacterium. *J. Mol. Biol.* 310:291–297, 2001.
29. Feller, G., F. Payan, F. Theys, M. Qian, R. Haser, C. Gerday. Stability and structural analysis of α-amylase from the Antarctic psychrophile *Alteromonas haloplanctis* A23. *Eur. J. Biochem.* 222:441–447, 1994.
30. Narinx, E., E. Baise, C. Gerday. Subtilisin from psychrophilic Antarctic bacteria: characterization and site-directed mutagenesis of residues possibly involved in the adaptation to cold. *Protein Eng.* 10:1271–1279, 1997.
31. Petrescu, I., J. Lamotte-Brasseur, J.-P. Chessa, P. Ntarima, M. Clayssens, B. Devreese, G. Marino, C. Gerday. Xylanase from the psychrophilic yeast *Cryptococcus adeliae. Extremophiles* 4:137–144, 2000.
32. Kim, S.Y., K.Y. Hwang, S.H. Kim, H.C. Sung, Y.S. Han, Y.J. Cho. Structural basis for cold adaptation: sequence biochemical properties and crystal structure of malate dehydrogenase from a psychrophile *Aquaspirillum arcticum. J. Biol. Chem.* 274:11761–11767, 1999.
33. Low, P.S., J.L. Bada, G.N. Somero. Temperature adaptations of enzymes: roles of free energy, the enthalpy and the entropy of activation. *Proc. Natl. Acad. Sci. USA* 70:430–432, 1973.
34. Hoyoux, A., I. Jennes, P. Dubois, S. Genicot, F. Dubail, J.M. Francois, E. Baise, G. Feller, C. Gerday. Cold-adapted β-galactosidase from the Antarctic psychrophile *Pseudoalteromonas haloplanktis. Appl. Environ. Microbiol.* 67:1529–1535, 2001.
35. Collins, T., M.A. Meuwis, C. Gerday, G. Feller. Activity, stability and flexibility in glycosidases adapted to extreme thermal environments. *J. Mol. Biol.* 328:419–428, 2003.
36. Aghajari, N., M. Roth, R. Haser. Crystallographic evidence of a transglycosylation reaction: ternary complexes of a psychrophilic alpha-amylase. *Biochemistry* 41:4273–4280, 2002.
37. Gerday, C., M. Aittaleb, J.L. Arpigny, E. Baise, J.-P. Chessa, G. Garsoux, I. Petrescu, G. Feller. Psychrophilic enzymes: a thermodynamic challenge. *Biochim. Biophys. Acta* 1342:119–131, 1997.
38. Somero, G.N. Proteins and temperature. *Annu. Rev. Physiol.* 57:43–68, 1995.
39. Jaenicke, R. Protein stability and molecular adaptation to extreme conditions. *Eur. J. Biochem.* 202:715–728, 1991.
40. Berglund, G.I., N.P. Willasen, A. Hordvik, A.O. Smalås. Structure of native pancreatic elastase from North Atlantic salmon at 1.61 Å resolution. *Acta Crystallogr. D.* 51:925–937, 1995.

41. Aghajari, N., G. Feller, C. Gerday, R. Haser. Crystal structures of the psychrophilic α-amylase from *Alteromonas haloplanctis* in its native form and complexed with an inhibitor. *Protein Sci.* 7:564–572, 1998.
42. Aghajari, N., G. Feller, C. Gerday, R. Haser. Structures of the psychrophilic *Alteromonas haloplanctis* α-amylase give insights into cold adaptation at a molecular level. *Structure* 6:1503–1516, 1998.
43. Aghajari, N., F. Van Petegem, V. Villeret, J.P. Chessa, C. Gerday, R. Haser, J. Van Beeumen. Crystal structures of a psychrophilic metalloprotease reveal new insights into catalysis by cold-adapted proteases. *Proteins: Struct. Funct. Genet.* 50:636–647, 2003.
44. Smalås, A.O., E.S. Heimstad, A. Hordvik, N.P. Willasen, R. Male. Cold adaptation of enzymes: structural comparison between salmon and bovine trypsins. *Proteins: Struct. Funct. Genet.* 20:149–166, 1994.
45. Karlsen, S., E. Hough, R.L. Olsen. The crystal structure and proposed amino acid sequence of a pepsin from Atlantic cod (*Gadus morhua*). *Acta Crystallogr. D.* 54:32–46, 1998.
46. Russell, R.J., U. Gerike, M.J. Danson, D.W. Hough, G.L. Taylor. Structural adaptations of the cold-active citrate synthase from an Antarctic bacterium. *Structure* 6:351–361, 1998.
47. Toyota, E., K.K. Ng, S. Kuninaga, H. Sekizaki, K. Itoh, K. Tanizawa, M.N. James. Crystal structure and nucleotide sequence of an anionic trypsin from chum salmon (*Oncorhyncus keta*) in comparison with Atlantic salmon (*Salmo salar*) and bovine trypsin. *J. Mol. Biol.* 324:391–397, 2002.
48. Van Petegem, F., T. Collins, M.A. Meuwis, C. Gerday, G. Feller, J. Van Beeuman. The structure of a cold-adapted family 8 xylanase at 1.3 Å resolution: structural adaptations to cold and investigations of the active site. *J. Biol. Chem.* 278:7531–7539, 2003.
49. D'Amico, S., C. Gerday, G. Feller. Structural similarities and evolutionary relationships in chloride-dependent alpha-amylases. *Gene* 253:95–105, 2000.
50. D'Amico, S., C. Gerday, G. Feller. Structural determinants of cold adaptation and stability in a large protein. *J. Biol. Chem.* 276:25791–25796, 2001b.
51. Aittaleb, M., R. Hubner, J. Lamotte-Brasseur, C. Gerday. Cold adaptation parameters derived from cDNA sequencing and molecular modeling of elastase from Antarctic fish *Notothenia neglecta*. *Protein Eng.* 10:475–477, 1997.
52. Schrøder Leiros, H.-K., N.P. Willassen, A.O. Smalås. Structural comparison of psychrophilic and mesophilic trypsins: elucidating the molecular basis of cold-adaptation. *Eur. J. Biochem.* 267:1039–1049, 2000.
53. Brandsdal, B.O., A.O. Smalås, J. Aqvist. Electrostatic effects play a central role in cold adaptation of trypsin. *FEBS Lett.* 499:171–175, 2001.
54. Gianese, G., P. Argos, S. Pascarella. Structural adaptation of enzymes to low temperatures. *Protein Eng.* 14:141–148, 2001.
55. Feller, G., D. D'Amico, C. Gerday. Thermodynamic stability of a cold-active α-amylase from the Antarctic bacterium *Alteromonas haloplanctis*. *Biochemistry* 38:4613–4619, 1999.
56. Tsigos, I., K. Mavromatis, M. Tzanodaskalaki, C. Pozidis, M. Kokkinidis, V. Bouriotis. Engineering the properties of a cold active enzyme through rational redesign of the active site. *Eur. J. Biochem.* 268:5074–5080, 2001.
57. van den Burg, B., G. Vriend, O.R. Veltman, G. Venema, V.G.H. Eijsink. Engineering an enzyme to resist boiling. *Proc. Natl. Acad. Sci.-Biol.* 95:2056–2060, 1998.
58. Miyazaki, K., P.L. Wintrode, R.A. Grayling, D.N. Rubingh, F.H. Arnold. Directed evolution study of temperature adaptation in a psychrophilic enzyme. *J. Mol. Biol.* 297:1015–1026, 2000.
59. Wintrode, P.L., F. H. Arnold. Temperature adaptation of enzymes: lessons from laboratory evolution. *Adv. Prot. Chem.* 55:161–225, 2000.
60. Wintrode, P.L., K. Miyazaki, F.H. Arnold. Cold adaptation of a mesophilic subtilisin-like protease by laboratory evolution. *J. Biol. Chem.* 275:31635–31640, 2000.
61. Merz, A., M.-C. Yee, H. Szadkowski, G. Pappenberger, A. Crameri, W.P.C. Stemmer, C. Yanofski, K. Kirschner. Improving the catalytic activity of a thermophilic enzyme at low temperature. *Biochemistry* 39:880–889, 2000.

62. Lebbink, J.H.G., T. Kaper, P. Bron, J. van der Oost, W.M. de Vos. Improving low-temperature catalysis in the hyperthermostable *Pyrococcus furiosus* β-glucosidase CelB by directed evolution. *Biochemistry* 39:3656–3665, 2000.
63. Tange, T., S. Taguchi, S. Kojima, K. Miura, H. Momose. Improvement of a useful enzyme (subtilisin BPN′) by an experimental evolution system. *Appl. Microbiol. Biotechnol.* 41:239–244, 1994.
64. Kano, H., S. Taguchi, H. Momose. Cold adaptation of a mesophilic serine protease, subtilisin, by *in vitro* random mutagenesis. *Appl. Microb. Biotechnol.* 47:46–51, 1997.
65. Taguchi, S., A. Ozaki, H. Momose. Engineering of a cold-adapted protease by sequential random mutagenesis and a screening system. *Appl. Environ. Microbiol.* 64:492–495, 1998.
66. Taguchi, S., A. Ozaki, T. Nonaka, Y. Mitsui, H. Momose. A cold-adapted protease engineered by experimental evolution system. *J. Biochem.* 126:689–693, 1999.
67. Haard, N.F., B.K. Simpson. Proteases from aquatic organisms and their uses in the seafood industry. In: *Fisheries Processing, Biotechnological Applications,* Martin, A.M., ed., London: Chapman & Hall, 1994, pp 132–154.
68. Rebeca, B.D., M.T. Pena-Vera, M. Díaz-Castaneda. Production of fish protein hydrolysates with bacterial proteases: yield and nutritional value. *J. Food Sci.* 56:309–314, 1991.
69. Raa, J. New commercial products from waste of the fish processing industry. In: *Making the Most of the Catch, Proceedings of the Seafood Symposium,* Bremner, A., C. Davis, B. Austin, eds., Hamilton, Qld:Qld Department of primary Industries, 1997, pp 33–36.
70. Stefansson, G., U. Steingrimsdottir. Applications of enzymes for fish processing in Iceland: present and future aspects. In: *Advances in Fisheries Technologies for Increased Profitability,* Voigt, M.N., J.R. Botta, eds., Lancaster, PA: Technomic Publication, 1990, pp 237–250.
71. Teschemacher, H., G. Koch, V. Brantl. Milk protein-derived opioid receptor ligands. *Biopolymers* 43:99–117, 1997.
72. Yamamoto, N., T. Takano. Antihypertensive peptides derived from milk proteins. *Nahrung-Food* 43:159–164, 1999.
73. Fox, P.E. Exogenous enzymes in diary technology. In: *Use of Enzymes in Food Technology,* Dupuy, P., ed., Paris, France: Symposium International, Versailles, 1982, pp 135–157.
74. Shau, H., A. Kim, S. Golub. Modulation of natural-killer and lymphokine-activated killer-cell cytotoxicity by lactoferrin. *J. Leuk. Biol.* 51:343–349, 1992.
75. Berkhout, B., J.L.B. van Wamel, L. Beljaars, D.K.F. Meijer, S. Visser, R. Floris. Characterization of the anti-HIV effects of native lactoferrin and other milk proteins and protein-derived peptides. *Antiviral Res.* 55:341–355, 2002.
76. Genicot, S., G. Feller, C. Gerday. Trypsin from Antarctic fish (*Paranotothenia magellanica* forster) as compared with trout (*Salmo gairdneri*) trypsin. *Comp. Biochem. Physiol. B.* 90:601–609, 1988.
77. Simpson, B.K., N.F. Haard. Purification and characterization of trypsin from the Greenland cod (*Gadus ogac*), 1: kinetic and thermodynamic characteristics. *Can. J. Biochem. Cell. Biol.* 62:894–900, 1984.
78. Asgeirsson, B., J.B. Bjarnasson. Structural and kinetic properties of chymotrypsin from Atlantic cod (*Gadus morhua*): comparison with bovine chymotrypsin. *Comp. Biochem. Physiol.* 99B:327–335, 1991.
79. Asgeirsson, B., J.B. Bjarnasson. Properties of elastase from Atlantic cod, a cold adapted proteinase. *Biochim. Biophys. Acta* 1164:91–100, 1993.
80. Simpson, B.K., N.F. Haard. Trypsin from Greenland cod, *Gadus ogac*: isolation and comparative properties. *Comp. Biochem. Physiol. B.* 79:613–622, 1984.
81. Asgeirsson, B., J.W. Fox, J.B. Bjarnasson. Purification and characterization of trypsin from poikilotherm *Gadus morhua*. *Eur. J. Biochem.* 180:85–94, 1989.
82. Outzen, H., G.I. Berglund, A.O. Smalås, N.P. Willasen. Temperature and pH sensitivity of trypsins from Atlantic salmon (*Salmo salar*) in comparison with bovine and porcine trypsin. *Comp. Biochem. Physiol. B.* 115:33–45, 1996.
83. Schrøder Leiros, H.-K., N.P. Willasen, A.O. Smalås. Residue determinants and sequence analysis of cold-adapted trypsins. *Extremophiles* 3:205–219, 1999.

84. Brewer, P., N. Helbig, N.F. Haard. Atlantic cod pepsin: characterization and use as a rennet substitute. *Can. Int. Food Sci. Technol. J.* 17:38–43, 1984.
85. Xu, R.A., R.J. Wong, M.L. Rogers, G.C. Fletcher. Purification and characterization of acidic proteases from the stomach of the deepwater finfish orange roughy (*Hoplostethus atlanticus*). *J. Food Biochem.* 20:31–48, 1996.
86. Yamashita, T., S. Higashi, T. Higashi, H. Machida, S. Iwasaki, M. Nishiyama, T. Beppu. Mutation of a fungal aspartic proteinase, *Mucor pusillus* rennin to decrease thermostability for use as a milk coagulant. *J. Biotechnol.* 32:17–28, 1994.
87. Alkorta, I., C. Garbisu, M. Llama, J. Serra. Industrial applications of pectic enzymes: a review. *Process. Biochem.* 33:21–28, 1998.
88. Kashyap, D., P. Vohra, S. Chopra, R. Tewari. Applications of pectinases in the commercial sector: a review. *Bioresour. Technol.* 77:215–227, 2001.
89. Hilhorst, R., H. Gruppen, R. Orsel, C. Laane, H.A. Schols, A.G.J. Voragen. Effects of xylanase and peroxidase on soluble and insoluble arabinoxylans in wheat bread dough. *J. Food Sci.* 67:497–506, 2002.
90. Monfort, A., A. Blasco, J.A. Prieto, P. Sanz. Combined expression of *Aspergillus nidulans* endoxylanase X24 and *Aspergillus oryzae* alpha-amylase in industrial baker's yeasts and their use in bread making. *Appl. Environ. Microbiol.* 62:3712–3715, 1996.
91. Ibrahim, H.R., S. Higashiguchi, Y. Sugimoto, T. Aoki. Antimicrobial synergism of partially-denatured lysozyme with glycine: effect of sucrose and sodium chloride. *Food Res. Inter.* 29:771–777, 1996.
92. D'Amico, S., C. Gerday, G. Feller. Temperature adaptation of proteins: engineering mesophilic-like activity and stability in a cold-adapted α-amylase. *J. Mol. Biol.* 332:981–988, 2003.
93. D'Amico, S., J.-C. Marx, C. Gerday, G. Feller. Activity-stability relationships in extremophilic enzymes. *J. Biol. Chem.* 278:7891–7896, 2003.
94. Collins, T., M.-A. Meuwis, I. Stals, M. Claeyssens, G. Feller, C. Gerday. A novel family 8 xylanase, functional and physicochemical characterization. *J. Biol. Chem.* 277:35133–35139, 2002.
95. http://www.genencor.com
96. Trimbur, D.E., K.R. Gutshall, P. Prema, J.E. Brenchley. Characterization of a psychrotrophic *Arthrobacter* gene and its cold-active β-galactosidase. *Appl. Env. Microbiol.* 60:4544–4552, 1994.
97. Gutshall, K., K. Wang, J.E. Brenchley. A novel *Arthrobacter* β-galactosidase with homology to eukaryotic β-galactosidases. *J. Bacteriol.* 179:3064–3067, 1997.
98. Loveland-Curtze, J., P.P. Sheridan, K.R. Gutshall, J.E. Brenchley. Biochemical and phylogenetic analyses of psychrophilic isolates belonging to the *Arthrobacter* subgroup and description of *Arthrobacter psychrolactophilus* sp. nov. *Arch. Microbiol.* 171:355–363, 1999.
99. Nakagawa, T., Y. Fujimote, M. Uchino, T. Miyaji, K. Takano, N. Tomizuka. Isolation and characterization of psychrophiles producing cold-active β-galactosidase. *Lett. Appl. Microbiol.* 37:154–157, 2003.
100. Karasova-Lipovova, P., H. Strnad, V. Spiwok, S. Mala, B. Kralova, N.J. Russell. The cloning, purification and characterisation of a cold-active β-galactosidase from the psychrotolerant Antarctic bacterium *Arthrobacter* sp. C2-2. *Enzyme Microb. Technol.* 33:836–844, 2003.
101. Rahim, K.A.A., B.H. Lee. Production and characterization of β-galactosidase from psychrotrophic *Bacillus subtilis* KL88. *Biotechnol. Appl. Biochem.* 13:246–256, 1991.
102. Coombs, J.M., J.E. Brenchley. Biochemical and phylogenetic analyses of a cold-active beta-galactosidase from the lactic acid bacterium *Carnobacterium piscicola* BA. *Appl. Environ. Microbiol.* 65:5443–5450, 1999.
103. Sheridan, P.P., J.E. Brenchley. Characterization of a salt-tolerant family 42 β-galactosidase from a psychrophilic Antarctic *Planococcus* isolate. *Appl. Env. Microbiol.* 66:2438–2444, 2000.
104. Fernandes, S., B. Geueke, O. Delgado, J. Coleman, R. Hatti-Kaul. β-Galactosidase from a cold-adapted bacterium: purification, characterization and application for lactose hydrolysis. *Appl. Microbiol. Biotechnol.* 58:313–321, 2002.

105. Loveland, J., K. Gutshall, J. Kasmir, P. Prema, J.E. Brenchley. Characterization of psychrophilic microorganisms producing β-galactosidase activities. *Appl. Environ. Microbiol.* 60:12–18, 1994.
106. Gutshall, K.R., D.E. Trimbur, J.J. Kasmir, J.E. Brenchley. Analysis of a novel gene and β-galactosidase isozyme from a psychrotrophic *Arthrobacter* isolate. *J. Bacteriol.* 177:1981–1988, 1995.
107. Schlemmer, A.F., C.F. Ware, N.T. Keen. Purification and characterization of a pectin lyase produced by *Pseudomonas fluorescens* W51. *J. Bacteriol.* 169:4493–4498, 1987.
108. Takasawa, T., K. Sagisaka, K. Yagi, K. Uchiyama, A. Aoki, K. Takaoka, K. Yamamato. Polygalacturonase isolated from the culture of the psychrophilic fungus *Sclerotinia borealis*. *Can. J. Microbiol.* 43:417–424, 1997.
109. Gadre, R.V., G. Van Driessche, J. Van Beeumen, M.K. Bhat. Purification, characterisation and mode of action of an endo-polygalacturonase from the psychrophilic fungus *Mucor flavus*. *Enzyme Microb. Technol.* 32:321–330, 2003.
110. Birgisson, H., O. Delgado, L. Arroyo, R. Hatti-Kaul, B. Mattiasson. Cold-adapted yeasts as producers of cold-active polygalacturonases. *Extremophiles* 7:185–193, 2003.
111. Laurent, P., L. Buchon, J.F. Guespin-Michel, N. Orange. Production of pectate lyases and cellulases by *Chryseomonas luteola* strain MFCL0 depends on growth temperature and the nature of the culture medium: evidence of two critical temperatures. *Appl. Environ. Microbiol.* 66:1538–1543, 2000.
112. van Truong, L., H. Tuyen, E. Helmke, L. Binh, T. Schweder. Cloning of two pectate lyase genes from the marine Antarctic bacterium *Pseudoalteromonas haloplanktis* strain ANT/505 and characterization of the enzymes. *Extremophiles* 5:35–44, 2001.
113. Aghajari, N., G. Feller, C. Gerday, R. Haser. Crystallization and preliminary X-ray diffraction studies of α-amylase from the Antarctic psychrophile *Alteromonas haloplanctis* A23. *Protein Sci.* 5:2128–2129, 1996.
114. Feller, G., O. le Bussy, C. Houssier, C. Gerday. Structural and functional aspects of chloride binding to *Alteromonas haloplanctis* α-amylase. *J. Biol. Chem.* 271:23826–23841, 1996.
115. Feller, G., O. le Bussy, C. Gerday. Expression of psychrophilic genes in mesophilic hosts: assessment of the folding state of a recombinant alpha-amylase. *Appl. Environ. Microbiol.* 64:1163–1165, 1998.
116. Gomes, J., I. Gomes, W. Steiner. Thermolabile xylanase of the Antarctic yeast *Cryptococcus adeliae*: production and properties. *Extremophiles* 4:227–235, 2000.
117. Iyo, A.H., C.W. Forsberg. A cold-active glucanase from the ruminal bacterium *Fibrobacter succinogenes* S85. *Appl. Env. Microbiol.* 65:995–998, 1999.
118. Akila, G., T.S. Chandra. A novel cold-tolerant *Clostridium* strain PXYL1 isolated from a psychrophilic cattle manure digester that secretes thermolabile xylanase and cellulase. *FEMS Microbiol. Lett.* 219:63–67, 2003.
119. Violot, S., R. Haser, G. Sonan, D. Georlette, G. Feller, N. Aghajari. Expression, purification, crystallization and preliminary x-ray crystallographic studies of a psychrophilic cellulase from *Pseudoalteromonas haloplanktis*. *Acta Crystallogr. D.- Biol. Crystallogr.* 59:1256–1258, 2003.
120. Nilsen, I.W., K. Øverbo, E. Sandsdalen, E. Sandaker, K. Sletten, B. Myrnes. Protein purification and gene isolation of chlamysin, a cold-active lysozyme-like enzyme with antibacterial activity. *FEBS Lett.* 464:153–158, 1999.
121. Arpigny, J.L., J. Lamotte, C. Gerday. Molecular adaptation to cold of an Antarctican bacterial lipase. *J. Mol. Catal. B.: Enzymatic* 3:29–35, 1997.
122. Dhaese, P. Catalase: an enzyme with growing industrial potential. *Chimica Oggi/Chemistry Today* 14:19–21, 1996.
123. Yumoto, I., H. Iwata, T. Sawabe, K. Ueno, N. Ichise, H. Matsuyama, H. Okuyama, K. Kawasaki. Characterization of a facultatively psychrophilic bacterium, *Vibrio rumoiensis* sp. nov., that exhibits high catalase activity. *Appl. Environ. Microbiol.* 65:67–72, 1999.
124. Kuraishi, C., K. Yamazaki, Y. Susa. Transglutaminase: its utilization in the food industry. *Food Rev. Int.* 17:221–246, 2001.

125. Chien, P., F. Sheu, Y. Shyu. Monitoring enzymatic debittering in grapefruit juice by high performance liquid chromatography. *J. Food Drug Anal.* 9:115–120, 2001.
126. Yadav, S., K. Yadav. Secretion of alpha-L-rhamnosidase by *Aspergillus terreus* and its role in debittering of orange juice. *J. Sci. Indust. Res.* 59:1032–1037, 2000.
127. Soria, F., G. Ellenrieder. Thermal inactivation and product inhibition of *Aspergillus terreus* CECT 2663 alpha-L-rhamnosidase and their role on hydrolysis of naringin solutions. *Biosci. Biotechnol. Biochem.* 66:1442–1449, 2002.
128. Minussi, R.C., G.M. Pastore, N. Duran. Potential applications of laccase in the food industry. *Trends Food Sci. Technol.* 13:205–216, 2002.
129. Remaut, E., C. Bliki, M. Iturizza, K. Keymeulen. Development of regulatable expression systems for cloned genes in cold-adapted prokaryotes. In: *Biotechnological Applications of Cold-Adapted Organisms,* Margesin, R., F. Schinner, eds., Berlin: Springer, 1999, pp 1–16.
130. Tutino, M.L., A. Duilio, R. Parrilli, E. Remaut, G. Sannia, G. Marino. A novel replication element from an Antarctic plasmid as a tool for the expression of proteins at low temperature. *Extremophiles* 5:257–264, 2001.

3.14

Biotransformations as Applicable to Food Industries

B. Suresh, T. Ritu and G. A. Ravishankar

CONTENTS

14.1 INTRODUCTION

Biotransformations are chemical reactions used to carry out specific conversion of complex substrates using plant, animal, or microbial cells or purified enzymes as catalysts. This is different from biosynthesis, where complex products are assembled from simple substrates by whole cells, organs or organisms. This is also different from biodegradation, in which complex substances are broken down to simple ones (1). Biotransformation has great potential to generate novel products or to produce known products more efficiently (2).

Biotransformation has the following goals:

1. Specific modifications of substrate structures via selective transformation reactions.

2. Partial degradation of substrates into desirable metabolites by means of controlled cultured cell reactions or reaction pathways.
3. Use of biosynthetic reactions to produce novel compounds

14.1.1 Potential of Microbial Cells and Plant Cells to Carry Out Biotransformation

Plant cell cultures exhibit a vast biochemical potential for the production of specific secondary metabolites. Some cultures may retain the ability to transform exogenous substrates into products of interest. The chemical compounds that can undergo biotransformation mediated by plant enzymes include phenyl propanoids, steroids, alkaloids, coumarins, terpenoids, lignans, and other molecular species. It is not necessary for the compounds to be natural intermediates in plant metabolism; they can be of synthetic origin also.

Plant cell cultures and enzymes have the potential to transform cheap and plentiful substances, such as industrial byproducts, into rare and expensive products. Multistep processes catalyzed by cell or organ culture often generate intermediary metabolites, which help to establish biosynthetic pathways (3).

Biotransformation capabilities of microorganisms and their enzymes for the production of a variety of fine chemicals are well known. Microbial systems are advantageous because biomass doubling time is short and production of biomass can be achieved quickly (4). In addition, methods for genetic manipulation of microbes are well established. Plant systems, on the other hand, produce a more limited range of enzymes, and undifferentiated plant cells have longer doubling times than microbial cells. The desired enzymes also are often produced in minute quantities (4). Despite these drawbacks, the plant kingdom contains some unique enzymes, which produce a variety of chemicals.

14.1.2 Reactions Involved in Biotransformation

Microorganisms employ both constitutive and inducible enzymes to degrade and synthesize great variety of chemical compounds, for not only their viability and reproduction, but also their secondary metabolism. A particular enzyme in a highly complex and well coordinated metabolic pathway catalyzes each reaction. In addition to their usual substrates, many of these enzymes accept other structurally related compounds and thus catalyze unnatural reactions when foreign substrates are added to the medium (Table 14.1). Reaction products, which are not further degraded, can usually be isolated from the fermentation medium.

Table 14.1
Reaction types mediated by microorganisms during biotransformation

Oxidations	Hydroxylation, epoxidation, oxidation of alcohols and aldehydes, oxidative degradation of alkyl, carboxyalkyl or ketoalkyl chains, oxidative removal of substitutes, oxidative deamination.
Reductions	Reduction of organic acids, aldehydes, ketones and hydrogenation of C=C bonds, reduction of hetero functions, dehydroxylation, reductive elimination of substituents
Hydrolysis	Hydrolysis of esters, amines, amides, lactones, ethers, lactams; hydration of C=C bonds and epoxides
Condensation	Dehydration, O- and N-acylation, glycosidation, esterification, lactonization, amination.
Isomerization	Migration of double bonds or oxygen functions, racemization, rearrangements

14.1.3 Advantages of Biotransformation

Biotransformation provides a number of advantages (1) such as:

1. Production of novel compounds.
2. Enhancement in the productivity of a desired compound.
3. Overcoming the problems associated with chemical analysis
4. Studies on biotransformation can lead to basic information to elucidate the biosynthetic pathway.
5. Catalysis can be carried out under mild conditions like ambient temperature and without the need of high pressure and extreme conditions, thus reducing undesired byproduct, energy needs, and cost.
6. Independence of geographical and seasonal variations and various environmental factors.
7. Rapidity of production.

So far, a few reports are available on scaleup of biotransformation. In the food area, the development of a process for making high fructose corn syrup using single step biotransformation, hydrolysis of lactose in milk using β-galactosidase from *Saccharomyces lactis* using a plug flow reactor, and production of L-aspartic acid from fumaric acid by using aspartase enzymes from *E. coli* for the synthesis of aspartame are some industrial examples.

14.1.4 Design of Biotransformation Processes

For a successful biotransformation, it is necessary that the substrate molecules come into contact with the enzymes such that the catalytic capability of microorganisms is not inactivated by the substrate or its product. Therefore, the ideal substrate should be soluble in the fermentation medium and able to pass the cell membrane without being toxic to the microorganism. Such a substrate is usually added to a growing culture as a sterile concentrated solution.

14.1.4.1 Selection of Microorganism

The first task in designing a biotransformation process is to find the microorganism or cell line that catalyzes the reaction of interest with the highest possible yield. Usually it is necessary to perform an extensive screening with a great number of pure cultures, which can be obtained from culture collections or isolated from natural sources. Sometimes it is desired or necessary to identify new strains of microorganisms that are capable of carrying out certain biotransformations (5).

14.1.4.2 Methods for Carrying Out Biotransformation Process

14.1.4.2.1 Biotransformation with Growing Culture The substrate is added to the fermentation medium at the time of inoculation or during a later phase of microbial growth. It is an easy method, and total time of incubation is relatively short because growth and biotransformation take place simultaneously.

14.1.4.2.2 Biotransformation with Resting Cells At first, the microorganisms are cultivated under conditions allowing optimum growth; then the biotransformation is performed in a second step. This method offers some advantages, such as:

1. Each step can be individually optimized.
2. Optimum conversion rate can easily be determined.
3. Biotransformation step is usually not susceptible to infection by contaminant microorganisms, and so can be conducted under nonsterile conditions.

14.1.4.2.3 Biotransformation with Immobilized Cells Immobilization can be done with different techniques:

1. Entrapment in a polymerous porous network (polyacrylamide, k-carrageenan, alginate, chitosan, collagen, agar, cellulose, or urethane).
2. Surface adsorption to a water insoluble, solid support (DEAE-cellulose, concanavalin A, ion exchange resins, or metal oxide).
3. Covalent attachment to a carrier material (e.g., carboxymethyl cellulose).
4. Induction of cell aggregation by physical or chemical cross linking (such as with gluteraldehyde).

For small scale laboratory procedures as well as for large scale industrial application, entrapment into polymerous matrix material has been most extensively applied. Biotransformation with immobilized cells has higher operational stability than cells free in solution, and the active cells are easily removed from the reaction mixture and can be used repeatedly. Furthermore, the product formation rate is high and the inhibitory influences are minimal.

On the other hand, the catalytic activity of immobilized cells is generally reduced due to damage occurring during the immobilization procedure. Increasing cell density can compensate for this loss of catalytic activity.

14.1.4.2.4 Biotransformation with Purified Enzymes In some cases, it may be advantageous or necessary to carry out a biotransformation with a purified enzyme. This is appropriate if

1. The membrane of intact cells prevents proper substrate or product permeation.
2. Further product degradation or undesirable side reactions take place due to the presence of other enzymes systems.
3. The enzyme of interest is excreted by the cells and can easily be purified from the medium after biomass removal.
4. Techniques for enzyme immobilization are well established and permit the repeated use of the biocatalyst and the opportunity to run continuous reactors.

Industrial applications of immobilized enzymes are mostly restricted to reactions like hydrolysis or isomerization (e.g., amino acylase, lactase, glucose isomerase, and glucoamylase). Immobilized enzyme technology is commercially used for lactose hydrolysis in milk by lactase from *Kluyveromyces* species. By far the largest industrial application is the isomerization of glucose from hydrolyzed corn starch to yield fructose, a sugar with increased sweetness (6).

14.1.4.2.5 Biotransformation with Multiphase Systems Biotransformation of substrates with low solubility in water has been successfully carried out with two liquid phase systems consisting of water and a poorly water miscible solvent (7). For this purpose, an ideal organic solvent should be practically immiscible in water, but should exhibit high solubility for the substrate and product. Furthermore, it should not exert an inhibitory effect on the biocatalyst. For reason of safety, nonflammable solvents are preferred. Organic solvents such as n-alkanes, cyclohexanes, toluene, benzene, carbon tetrachloride, chloroform, and methyl chloride have been used for biotransformation with intact cells or enzymes (8).

14.1.4.2.6 Combination of Two Sequential Biotransformation Steps Catalyzed by Different Microorganisms Biotransformations involving two (or several) sequential steps that are catalyzed by different microorganisms are favorably carried out as a one stage process using a mixture of both microorganisms in a single reactor. If the two microorganisms require different fermentation conditions, a two stage fermentation can be considered (9).

14.1.5 Improvement of Biotransformation Process

14.1.5.1 Optimization of Environmental Conditions

Enzyme formation and enzyme activity are heavily influenced by environmental conditions, such as medium composition (kind and concentration of nutrients; substrate product concentration) temperature, pH, and dissolved oxygen (depending upon aeration and agitation).

If the conditions for growth and associated enzymes synthesis are not optimal for the biotransformation, both processes must be separated and optimized individually, as in biotransformation with resting cells. For industrial processes, economic aspects (cheap raw material, short reaction time, increased substrate concentration, and low energy consumption) have to be considered (10).

14.1.5.2 Strain Improvement

Strain improvement is done for the amplification of beneficial properties or the elimination of detrimental properties of microorganisms involved in biotransformation. This can be done by inducing genetic modifications and selecting for mutants that improve the biotransformation by increasing the enzyme production, enhancing substrate or product tolerance, increasing membrane permeability, facilitating substrate transport into the cell and product efflux from the cell, as well as inactivation of the enzyme catalyzing side reaction or product degradation. The traditional method of strain development is from mutagenesis by commonly used agents (e.g., UV light or chemical mutagens) and subsequent selection for superior strains among survivors.

Screening efficiency can be considerably enhanced by adopting a screening strategy, which selects superior strains that give higher yields of the desired enzyme. Recombinant DNA technology provides new prospects of strain improvement. Plasmid or phage vectors can transfer any DNA fragment from eukaryotic, prokaryotic, or even synthetic origin into suitable recipient microorganisms (11).

14.1.5.3 Elimination of Side Reactions

Side reactions in a biotransformation process are observed if a substrate or its product is attacked by undesirable enzyme activities present in the cell. Suppression of such side reactions is desired in order to increase the yield of target product.

In order to eliminate side reactions selectively, cells have to be treated under conditions suppressing the activities of undesired enzymes while maintaining the desired biotransformation activity. Physical (heat) or chemical (pH shift) treatment of biomass can cause selective inactivation of enzymes that are responsible for side reactions.

A more laborious technique is mutagenesis and screening of mutants that have lost the ability to form enzyme systems responsible for side reactions.

14.1.6 Downstream Processing or Product Isolation

Usually biotransformation products are extracellular compounds of low or medium molecular weight. They are dissolved or suspended in the fermentation medium and can be isolated from the broth or (preferably) from a supernatant after removal of biomass (Table 14.2).

Table 14.2
Basic operations and common methods for the isolation of biotransformation product

	Basic Operation	Method
Fermentation broth	Removal of biomass and other solids (if necessary)	Centrifugation, Filtration, Sedimentation Sieving.
Raw Product	Concentration, Fractionation	Extraction, Ion exchange, Adsorption, Precipitation Distillation
Final product	Final purification	Chromatography, Distillation Decolorization Crystallization Drying

14.2 HIGH FRUCTOSE CORN SYRUP

Glucose is almost 65% as sweet as sucrose on a weight basis. Fructose is 120–180% as sweet, depending upon the conditions of testing. Glucose can be obtained from readily and cheaply available corn starch, either by acid or enzymatic hydrolysis (Figure 14.1). Further, it can be converted into high fructose syrup, which has more sweetness. Fructose plays an important role in the diet of diabetics, as it is only slowly reabsorbed by the stomach and intestinal tract, and does not influence the blood glucose level (12).

Until 1976 the main sweetener used in the world was sucrose, obtained from sugar beets (40%) and sugarcane (60%). In 1974, the price of sugar skyrocketed, making the cost of high fructose corn syrup (HFCS) production competitive. By the time the cost of sugar came down, the enzyme route production of HFCS was able to reduce cost enough to remain competitive (13).

14.2.1 Biotechnology Involved in Process

The possibility of using an enzyme to isomerize the dextrose in corn syrup into fructose has been recognized for some time. Marshall and Kooi (14) had reported the microbial enzyme capable of isomerizing dextrose into fructose. A number of different microorganisms from the genus *Streptomyces* have also been reported as being capable of producing glucose isomerase. Other microorganisms reported in literature include those from the genera *Pseudomonas, Aerobacter, Lactobacillus, Pasteurella* and *Bacillus*.

A process was developed for isomerizing dextrose into fructose using glucose isomerase from *Streptomyces* (Figure 14.2) (15) and for obtaining increased yields of glucose isomerase from *Streptomyces*.

14.2.2 Source of Enzyme for Production

The only commercially employed enzymes for HFCS production are D-xylose isomerase, and D-xylose ketol isomerase (EC 5.3.1.5). These enzymes have various advantages, such as fewer secondary reactions, high specific activity, a high temperature optimum (~80°C) and not needing cofactors (ATP, NAD^+).

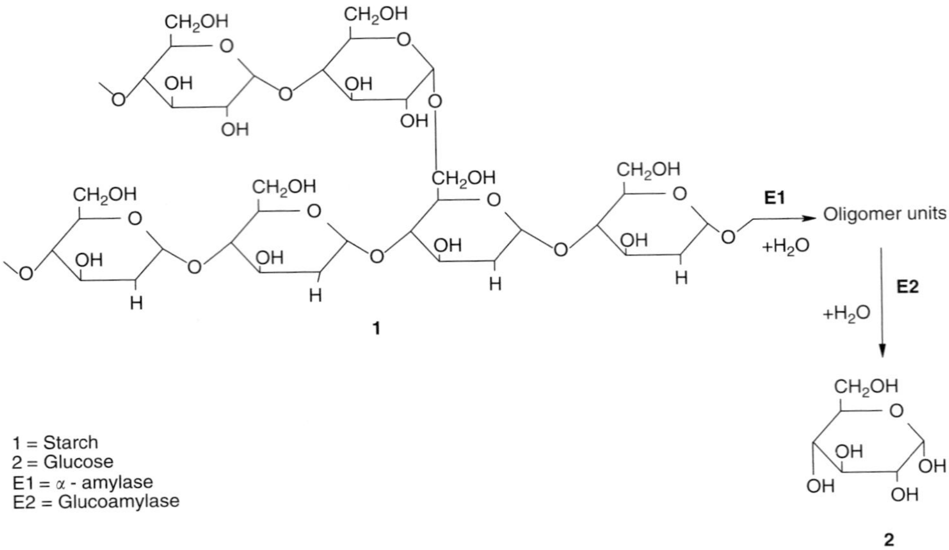

1 = Starch
2 = Glucose
E1 = α - amylase
E2 = Glucoamylase

Figure 14.1 Hydrolysis of starch to produce Glucose syrup

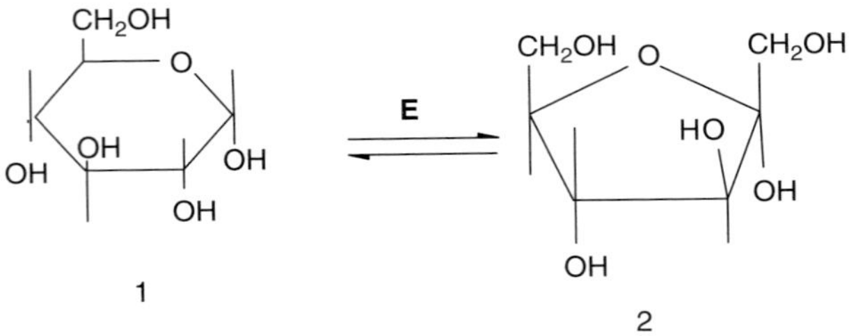

1 = Glucose
2 = Fructose

Figure 14.2 Isomerization of Glucose to Fructose

Glucose isomerizing enzyme has been shown in at least 50 different microorganisms (16). Among these strains with xylose isomerase activity, the following are of commercial interest today: *Bacillus coagulans, Arthrobacter* species, *Actinoplanes missouriensis, Streptomyces olivaceus,* and *Streptomyces olivochromogenes.*

14.2.3 Production of HFCS and Immobilization of Enzyme

As an intracellular enzyme, glucose isomerase is harvested after cultivation with the cells by centrifugation or filtration. The washed whole cells may be used for isomerization as such, after heat fixation of enzyme or immobilization. Enzyme immobilization can either be accomplished with the enzyme or with glucose isomerase bearing cells by the following procedures:

1. Covalent binding to an insoluble carrier
2. Adsorption to an insoluble carrier

3. Entrappment in a matrix
4. Immobilization within the cells

Commercially, glucose isomerases are obtained from different microorganisms, and vary in their properties (Table 14.3).

To prepare high fructose corn syrup, starch is first hydrolyzed into glucose using various hydrolytic enzymes (Table 14.4, Table 14.5, and Table 14.6). This syrup, containing 96% glucose, is subjected to conversion from aldose sugar to ketose sugar by the enzyme isomerase. The syrup is filtered through activated carbon and then passed through an ion exchange resin to remove soluble impurities before being evaporated to 40–45% dry solids. The pH of evaporated syrup is adjusted to 7.5–7.8 with NaOH, and the temperature is maintained at 55–60°C. The syrup is passed through a packed bed reactor containing immobilized glucose isomerase. After isomerization, the pH is adjusted to between 4.0 and 5.0. The syrup, containing 42% fructose, is again filtered and evaporated to 70–80% dry

Table 14.3
Comparison of commercial isomerase preparation

			Metal Required (M)	
Micro Organisms	**pH Optimum**	**Temp. Optimum (°C)**	**Cobalt**	**Magnesium**
Bacillus coagulans	7.0	75	0.001	0.1
Actinoplanes missouriensis	7.0–7.2	80–5	0.0003	0.003
Streptomyces olivaceus	8.5	60–70	+	+
Streptomyces albus	7.0	70	-	-
Streptomyces lavoviresn	6.8	70	0.25mM	0.02

Table 14.4
Reaction conditions of α amylase and glucoamylase

pH	liquefying : 6.0–6.5 / Glucosidation:4.2
Temperature	115°C–95°/60°C
Medium	Aqueous
Reaction-type	O-glucosyl-bond hydrolysis (Endo/exo)
Catalyst	soluble enzyme
Enzyme	1,4-α-D-Glucan glucanohydrolase(α-amylase, Glycogenase) / Glucan1,4-O-glucosidase (Glucoamylase, amylo glucosidase)
Strain	*Bacillus licheniformis / Aspergillus niger*

Table 14.5
Process parameters for the production of high fructose corn syrup.

Conversion	>95%
Reactor type	continuous operated stirred tank reactor
Residence time	2–3 h/48–72h
Downstream processing	Filtration

Table 14.6
Reaction conditions for xylose isomerase and glucose isomerase

pH	7.5–8.0
Temperature	50–60°C
Medium	Aqueous
Reaction type	Isomerization
Catalyst	Immobilized whole cells or isolated enzymes
Enzyme	D-xylose ketol-isomerase (xylose isomerase), Glucose (isomerase)
Strain	*Bacillus coagulans, Streptomyces rubiginosus, Streptomyces phaechromogenes*

solids. From this, HFCS 90 and HFCS 55,containing 90% and 55% fructose respectively, can be produced (17) (Figure 14.3).

14.2.4 Salient features of xylose isomerase and glucose isomerase

The process is a part of the production of high fructose corn syrup.

- This process provides an effective way for an important, low cost sugar substitute derived from grain.
- The corn kernels are softened by treatment with sulphur dioxide and lactic acid bacteria to separate oil, fiber and proteins. The enzymatic steps are cascaded to yield the product for the invertase process after liquefaction in continuous cookers, debranching and filtration.
- The process ends if all starch is completely broken down to limit the amount as oligomers of glucose and dextrins. Additionally, recombination of molecules has to be prevented.
- The thermostable enzymes can be used up to 115°C. The enzyme needs Ca^{2+} ions for stabilization and activation.
- Because several substances in corn can complex cations, the cation concentration is increased requiring a further product purification causing the necessity to refine the product.
- The commercially important varieties show superior affinity to xylose and are therefore classified as xylose isomerase.
- Because the isolation of the intracellular enzyme is very expensive, whole cells are used. In almost all cases the enzymes or cells are immobilized.
- Purified glucose (dextrose syrup) is the product from the saccharification stage.
- Isomerases belong to the group of metalloenzymes, as they require Co^{2+} and Mg^{2+} for their activity.
- To limit byproduct formation, the reaction time must be minimized. This can be done economically only by using high concentration of immobilized isomerase
- Several reactors are operated in parallel or in series, containing enzymes of different ages. The feed to a single reactor is controlled by the conversion of this reactor.
- The product has to be highly purified (by filtration, adsorption on charcoal, or ion exchange) to prevent fast deactivation and clogging of the catalyst bed.
- Plants producing more than 1,000 t of HFCS/day typically use at least 20 individual reactors.
- The product HFCS contain 42% fructose (53% glucose) or 55% fructose (41% glucose) as dry matter.

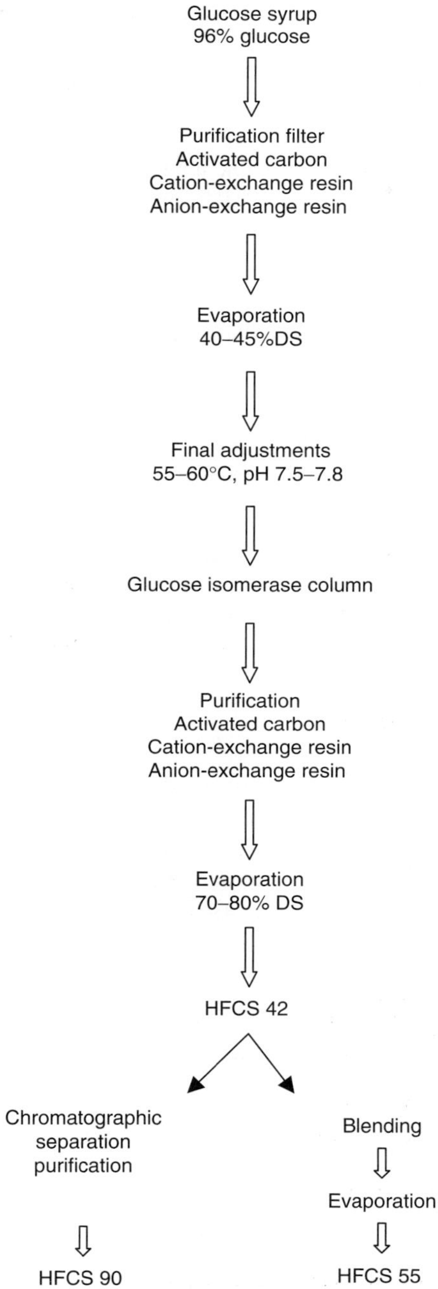

Figure 14.3 Isomerization of corn syrup

- Glucose isomerases have half lives of more than 100 days. To maintain the necessary activity, the enzyme is replaced after deactivation of about 12.5% of the original amount.
- The reaction temperature is normally above 55°C to prevent microbial infection, although enzyme stability is lowered.

According to Hartley et al. (18), current commercial glucose (xylose) isomerases can operate at 60°C and produce 45% fructose syrup. More thermally stable glucose isomerase would be useful for production of 55% fructose syrup at temperatures greater than 90°C. Glucose isomerase isolated from *Thermus thermophilus* (19) has a high temperature optimum (95°C) and can tolerate a neutral pH. A thermostable glucose isomerase isolated from thermophilic *Streptomyces* SK strain had a wide pH range with optima of 6.0 at 60°C and 6.2 at 90°C. It was optimally active at 95°C and completely stable at 80°C for at least 5.5 hours, with a half life of 5 hours at 90°C (20).

14.2.5 Process Parameters

The conversion of Glucose to fructose can be carried out under these conditions:

1. Reactor type – Continuous, fixed bed
2. Residence time – 0.17–0.33 hours
3. Downstream processing – 55% fructose

There are a number of companies producing xylose isomerase (Table 14.7). Several products based on HFCS are available (Table 14.8), and their list is growing. The importance of HCFS technology is evident from the number of patent application with innovation in the enzymatic processes and process engineering (Table 14.9).

14.3 OLIGOSACCHARIDES

Oligosaccharides are usually defined as glycosides that contain between three and ten sugar moieties (21). In general, food grade oligosaccharides are not pure products, but mixtures containing oligosaccharides of different degrees of polymerizaiton, the parent polysaccharide or disaccharide, and monomer sugars. Oligosaccharides provide several health benefits, which make their use as food ingredients particularly attractive.

Table 14.7
Companies which are producing xylose-isomerase

NOVO Nordisk	Denmark
Gist Brocades	Netherlands
Miles Kali-Chemie	Germany
Nagase	Japan

Table 14.8
Product application: HFCS can be used in preparation of many products some of which are listed in the table.

Carbonated beverages	table syrup
Fountain syrup	Apple sauce
Ketchup	Confections
Yeast raised baked goods	Toppings
Maraschino cherries	Pie fillings
Ice Cream	Pourable and spoonable salad dressings.

Table 14.9
Patent applications

Year	Patent No.	Authors	Topic
2001	US 6255084	Nielson et al.	Thermostable glucoamylase
1998	US 5731174	Deweer et al.	Process for the saccharification of starch
1999	US 5916789	Webbers et al.	Immobilized enzyme
1995	US 5384259	Lambeir et al.	Glucose isomerases with an altered pH optimum
1997	US 5656094	Peckous	Integrated process for producing crystalline fructose and a high fructose, liquid phase sweetener
1980	US 4199374	Dwivedi et al.	Process for preparing crystal fructose from high fructose corn syrup.

Worldwide, there are 12 classes of food grade oligosaccharides currently in commercial production. Oligosaccharide production was estimated at 35,000 t in 1990 and 65,000 t in 1995. Food grade oligosaccharides are manufactured using enzymatic processes, with the exception of soybean oligosaccharides (which are produced by direct extraction) and lactulose (which is produced using an alkali catalyzed reaction). They are either built up from simple sugars, such as sucrose or lactose, or by transglycosylation reactions; or formed by controlled hydrolysis of polysaccharides, such as starch, inulin, or xylan.

14.3.1 Classification of Oligosaccharides

Oligosaccharides can be divided into the following classifications:

1. Galactooligosaccharides: Produced commercially from lactose using the galactosyl transferase activity of β-galactosidase (EC 3.2.1.23), which dominates lactose hydrolysis at high lactose concentrations.
2. Lactulose: An alkali isomerization process is used to convert the glucose moiety in lactose to a fructose residue. The resulting disaccharide, lactulose, is not digested by humans and promotes the preferential growth of *Bifidobacteria* in the colon.
3. Lactosucrose: Lactosucrose is the third bifidogenic oligosaccharide that is produced using lactose as a raw material. This trisaccharide consists of a lactose molecule to which a fructose moiety is joined at the glucose residue by a $\beta(2\rightarrow1)$ glycosidic bond. It is manufactured from a mixture of lactose and sucrose using a transfructosylation activity of enzyme β-fructofuranosidase.
4. Palatinose (isomaltulose): Palatinose is produced from sucrose using an immobilized isomaltulaose synthase (EC 5.4.99.11).
5. Glycosyl sucrose (coupling sugar): The trisaccharide glycosyl sucrose (coupling sugar) is manufactured from the disaccharides maltose and sucrose using the enzyme cyclomaltodextrin glucanotransferase (EC 2.4.1.19).
6. Maltooligosaccharides: Maltooligosaccharides contain 2-10 α-D-glucose residues linked by α(1-4) or α(1-6) glycosidic linkages. They are produced commercially from starch by the action of α-amylase (EC 3.2.1.1).
7. Cyclodextins: Cyclodextrins are cyclic β(1-4) linked maltooligosaccharides consisting of 6–12 glucose units. They are formed from starch digest by the action of cyclomaltodextrin glucanotransferase.
8. Gentiooligosaccharides: Gentiooligosaccharides consist of several glucose residues linked by $\beta(1\rightarrow6)$ glycosidic bonds. They are produced from the glucose syrup by enzymatic transglucosylation.

9. Soybean oligosaccharides: Unlike other oligosaccharides, soybean oligosaccharides are extracted directly from the raw material and do not require enzymatic manufacturing processes.
10. Isomaltooligosaccharides: Like maltooligosaccharides, isomaltooligosaccharides are produced using starch as the raw material. Isomaltooligosaccharides consist of α-D-glucose residues comprising of α(1→6) linkage.
11. Xylooligosaccharides: The raw material for xylooligosaccharide synthesis is the polysaccharide xylan, which is extracted mainly from corncobs. The xylan is hydrolysed to xylooligosaccharide by the controlled activity of the enzyme endo-1, 4,β-xylanase (EC 3.2.1.8) to produce higher purity oligosaccharide products.

14.3.2 Fructooligosaccharides

Fructooligosaccharies represents one of the major classes of bifidogenic oligosaccharides in terms of their production volume. They are manufactured by two different processes, which result in slightly different end products.

In the first method, fructooligosaccharides are produced from the disaccharide sucrose (Figure 14.4) using the transfructosylation activity of the enzyme β-fructofuranosidase (EC 3.2.1.26) (22). A high concentration of the starting material is required for efficient transglycosylation. The fructooligosaccharides formed in this process contain between two and four β(1→2) linked fructosyl units linked to a terminal α-D-glucose residue. These are named 1-kestose (Glu-Fru$_{20}$), 1-nystose (Glu-Fru$_3$) and 1^F-fructosylnystose (Glu-Fru$_4$).

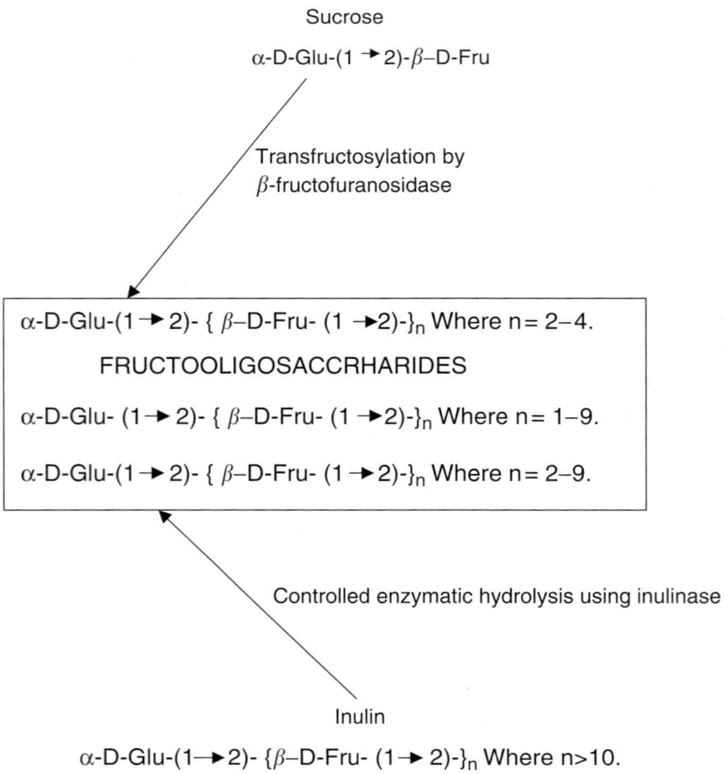

Figure 14.4 Production of Fructooligosaccharides

Glucose and small amounts of fructose are formed as byproducts in the reaction, as well as unreacted sucrose, and removal of these from the oligosaccharide mixture using chromatographic procedures is carried out to produce fructooligosaccharide products of high purity.

The second method used for fructooligosaccharide production is the controlled enzymatic hydrolysis of the polysaccharide inulin (inulin oligofructose) the fructooligosaccharide mixture formed by this process closely resembles the mixture produced from the transfructosylatin process (23). However, not all of the $\beta(1\rightarrow2)$ linked fructosyl chains end with a terminal glucose. Additionally, the oligosaccharide mixture is produced by the sucrose transfructosylation process.

14.3.3 Microbial Source for Enzyme Production

The industrial production oligosaccharides is a field that offers tremendous scope and opportunity. Meiji Seika (Japan) was the first company to meet success with commercial fructooligosaccharide production, with fructooligosaccharide production using fructosyl transferase from *Aspergillus* or *Aeureobasidium pullulans* M/S Cheils Food and Chemical Company (Korea) has been successful with Fructooligosaccharide production using immobilized cells of *A. pullulans*. Microbial production of oligosaccharides provides a convenient, cost effective alternative to chemical synthesis. The enzyme involved can be either intracellular or extracellular. Some microorganisms that can be used in fructooligosaccharide production are as follows.

1. *Aspergillus phoenicis:* Enzymes produced are sucrase-1^F-transferase (EC 2.4.1.99) under optimal conditions of 55°C; pH 8.0 for 8hrs, corresponding 60% (w/w) fructooligosaccharides i.e., 1-Ketose is obtained from sucrose (24).
2. *Aureobasidium pullulans*: Jung et al. (25) demonstrated that sucrose was an excellent carbon source for the production of intracellular fructosyl transferase. There was a tendency for the enzyme production to be increased as sucrose concentration is increased.
3. *Pencillium rugulosum*: Barthomeuf and Pourrot (26) used 70–85% sucrose in0.1M sodium acetate buffer at pH 5.0, 5.5 or 6.0, and an enzyme concentration between 2 and 10 Fructosyl transferase units/g.
4. *Aspergillus japonicus*: The reaction conditions of pH 5.0 and temperature 55°C; sucrose is the best carbon source and yeast extract is the best nitrogen source for the β-fructofuranosidase (EC 3.2.1.26) production (27).
5. *Penicillium citrium*: A novel syrup containing Neofructooligosaccharides was produced from sucrose (Brix 70) by whole cells of P. citrium. The efficiency of fructooligosaccharide production was more than 55%.
6. Mixed Enzyme System: A technique to produce high content fructooligosaccharide by the mixed enzyme system of fructosyltransferase and glucose oxidase has been studied. The mixed enzyme reaction has been carried out in a stirred tank reactor containing 40% (w/v) sucrose with 10 units of fructosyltransferase and 10 units of glucose oxidase per gram sucrose for 25h at 40°C and pH 5.5 Highly concentrated fructooligosaccharides (up to 90%) were obtained by the mixed enzyme system (28) (Table 14.10).

14.3.4 Properties of Oligosaccharides

Oligosaccharides are water soluble and mildly sweet, typically 0.3–0.6 times as sweet as sucrose. They exhibit low water activity, which is convenient in controlling microbial contamination. They are shown to be strong inhibitors of starch retrogradation. Many oligosaccharides are not digested by humans (29).

Table 14.10
Total Sugar Comparison of Fructooligosaccharides (FOS) and High content fructooligosaccharides (HCFOS)

Sugar	FOS % w/w	HCFOS % w/w
Fructose	1.0	1.31
Glucose	36.33	2.33
Sucrose	10.49	5.86
GF_2 (1-kestose)	32.32	50.13
GF_3 (Nystose)	24.78	38.19
GF_4 (fructofuranosylnystose)	1.68	2.18
$\Sigma^4{}_{n=2}GF_n$	52.18	90.4
Total	100	100

Table 14.11
Food grade oligosaccharides

Class of Oligosaccharides	Estimated Production in 1995(t)	Major Manufacturers	Trade Names
Fructooligosaccharides	12000	Meiji Seika Kaisha (Japan)	Meioligo
		Beghin-Meiji Industries (France)	Actilight
		Golden Technologies (USA)	NutraFlora
		Cheil Foods and Chemicls (Korea)	Oligo-Sugar
		ORAFTI (Belgium)	Raftilose and Raftiline
		Cosucra (Belgium)	Fibruline

14.3.5 Applications of Oligosaccharides

The major use of oligosaccharides is in the preparation of beverages. Oligosaccharides are being included in probiotic yogurts and yogurt drinks to produce symbiotic products. They are used as low cariogenic sugar substitutes in confectionery and baking, because they are not utilized by mouth microflora to form acid or polyglucans. Moreover, they can be used as bulking agent in desserts such as jellies and ice creams; bakery products including biscuits, breads, pastries; and also in infant milk formulas (29).

Some of the companies producing food grade oligosaccharides are given in Table 14.11 and the representative patents in Table 14.12.

14.4 VANILLA FLAVOR

Vanilla, which originated in Mexico, is a tropical orchid belonging to the family orchidaceae. About 110 species have been identified, but only three have been reported to be important in terms of commercial cultivation. *Vanilla fragens* or *Vanilla planifolia, Vanilla pompona* and *Vanilla tahitensis*. Among these, *V. planifolia* is the most valued for its flavor qualities, and is therefore widely cultivated and used for the production of food additives. The fully grown mature fruits of vanilla are called beans or pods.

Vanillin (4-hydroxy-3-methoxy benzaldehyde) is the characteristic organoleptic aroma compound of the cured vanilla pod, where it contributes to about 2% (w/w) of dry

Table 14.12
Some representative patents on food grade oligosaccharides

Year	Patent No.	Inventors	Topic
1981	UK patent Application 2072679 A	Adachi, T. and Hidaka, H.	A sweetener
1987	US 4681771	Adachi, T. and Hidaka, H.	Sweetener
1995	EP 0662479A2	Aga, H., Bhibuya, T, Sugimoto, T. and Miyake T.	Nonreducing oligosaccharide with neotrehalose structure and its production and uses.
1989	GB 2210545A	Beyts, P.K	Sweetening compositions
1989	FR2629985 AJ	Bitton,J. Gellf G. Michel,J.H. Paul F.B and Monsan,P.F	Used as food: oligosaccharides as low calorie sweeteners, and foods, dietetic products and beverages containing them.
1993	WO 93/02566A1	De soete,J and Frund.D	Reduced Calorie chocolate confectionary compositions
1986	EPO181988AJ	Gyllang, H.E, Ekblom, B.T. Wesshagen, V.V, and Ericson, K.Y	Beverage product
1974	JP 4940950	Hayashibara.K	Sweetening compositions
	EPO181988 AJ	Gyllang, H.E, Ekblom, B.T., Wesshagen, V.V, and Ericson, K.Y	Beverage product
1974	JP4940950	Hayashibara,K	Sweetening compositions
1989	EP0307523AJ	Kan,T, Kobayashi,Y., Sanoike,Y., Terashima T. and Mutai,M.	Liquid food
1992	FR2676164AJ	Landserbellani, M	Low Calorie chocolate
1995	EP0664299AJ	Plug,J.P	Oligosaccharides and use as food ingredient.
1995	US 5431929	Yatka et al.	Chewing gum products using oligofructose

matter. The proposed biosynthetic pathway of vanilla flavor metabolites formation from various phenyl propanoids is given in Figure 14.5.

14.4.1 Production of Vanillin by Biotransformation

The increasing trends towards natural foods ingredients have helped to maintain high demand for natural vanilla; therefore, a need was felt to develop alternative natural biological sources of vanilla flavoring. The production of vanilla flavor is possible from vanilla tissue culture; another possibility is biotransformation mediated production using either plant cell or microorganisms (30). Microbial transformation of natural precursors could provide the desired biovanillin, which may offer the natural status required by the European legislation (European Directive 88/388/CEE, Jo, No 184, 22nd of June 1988). The production of a natural biotechnological vanillin requires a natural precursor such as ferulic acid.

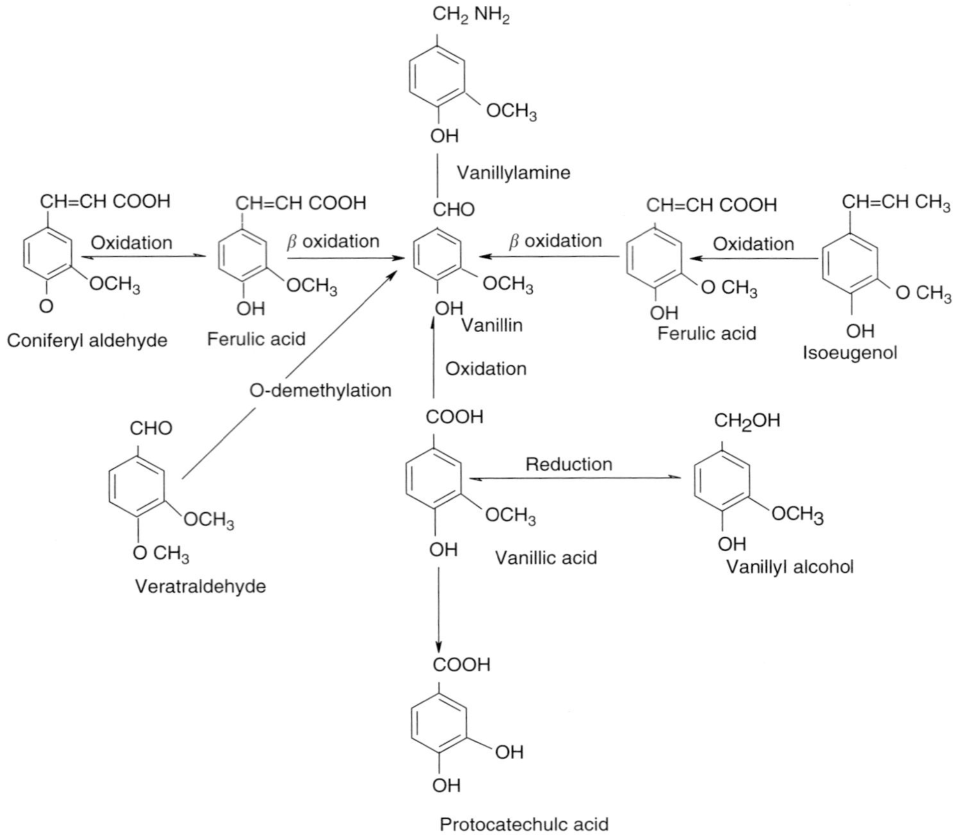

Figure 14.5 Proposed biosynthetic pathway of vanilla flavor metabolites formation from various phenylpropanoids

14.4.2 Vanilla Tissue Culture Studies

The rationale for the production of flavor compounds from cultured plant cells is based upon their unique biochemical and genetic capacity and on the totipotency of plant cells. Every cell of a vanilla plant contains the genetic information necessary to produce the numerous chemical compounds (or their precursors) that constitute natural vanilla flavor; feeding intermediates of the biosynthetic pathway can enhance the production of vanilla flavor metabolites by precursor biotransformation.

Feeding of cinnamic acid and ferulic acid to cell suspension cultures of *Vanilla planifolia* resulted in the formation of traces of p-hydroxy benzoic acid and vanillic acid (31). Processes have been reported (Table 14.13) for producing natural vanillin from ferulic acid precursors with aerial roots and charcoal as a product reservoir, developed by Westcott et al. (32). Organized aerial roots of *V. planifolia* plant detached and cultured in nutrient medium were used as biocatalysts in transforming ferulic acid to vanillin in the presence of charcoal.

14.4.3 Biotransformation of Lignin

Lignin is one of the most abundant natural sources of aromatic compounds. White rot fungi *Pycnoporus cinnabarinus* are able to delignify wood and decompose the lignin

Table 14.13
Biotransformation of Ferulic acid to vanilla flavor metabolites using microorganisms.

Micro-organism	Substrate	Product	Ref.
Pycnoporus cinnabarinus	Ferulic acid	Vanillin	36
Aspergillus niger	Ferulic acid	Vanillin	37
Pseudomonas acidovorans	Ferulic acid	Vanillin	38
Spirulina platensis	Ferulic acid	Vanillin	34
Haematococcus pluvialis	Ferulic acid	Vanillin	39
Pseudomonas fluorescenss	Ferulic acid	Vanillin	40
Escherichia coli	Ferulic acid	Vanillin	41
Alcaligenes paradoxus	Ferulic acid	Vanillin	42
Streptomyces setonii	Ferulic acid	Vanillin	43

Source: Ramachandra Rao, S., G.A. Ravishankar, *J. Food Sci. Agric.* 80:289–304, 2000.

polymer to yield vanillin, dehydrovanillin, vanillic acid, coniferyl aldehyde, ferulic acid, P-hydroxy cinnamyl aldehyde, and other compounds. The enzyme lignin peroxidase, which is responsible for lignin depolymerization, was isolated from *Phanerochaete chrysosporium*. However, in all reports of fungal degradation of lignin, vanillin has been detected only in trace amounts (33).

14.4.4 Biotransformation using Cell Culture of *Capsicum frutescens*

Production of vanilla flavor metabolites was observed when both freely suspended and immobilized cell cultures of *Capsicum frutescens* were treated with phenyl propanoid pathway intermediates like ferulic acid and vanillylamine. Vanillin is also produced by treating the cell cultures with protocatechuic aldehyde, caffeic acid, isoeugenol, coniferyl aldehyde, and veratraldehyde (30).

14.4.5 Biotransformation using Microorganisms

Various phenyl propanoid precursors, like ferulic acid, eugenol, isoeugenol, vanillyl alcohol, vanillyl amine, coniferyl alcohol, and veratryl alcohol can produce vanilla flavor metabolites by a large number of microbes such as bacteria, fungi, and yeast.

14.4.6 Biotransformation of Eugenol and Isoeugenol

Eugenol is the main constituent of the essential oil of the *Syzygium aromaticum* tree. It is a cheap commerciall available raw material for biotransformation. The first biotransformation of isoeugenol to vanillin was achieved with *Aspergillus niger* ATCC 9142 but it gave relatively low yield, because the efficiency of this transformation was only 10% due to further degradation of vanillin to vanillyl alcohol and vanillic acid. Strains of *Klebsiella, Enterobacter* and *Serratia* were used to transform eugenol and isoeugenol into vanillin. Biotransformation of isoeugenol and eugenol to vanillin and vanillic acid was shown in cultures of the microalga *Spirulina platensis* (34).

Recently a strain of *Bacillus subtilis* was isolated from soil that converts isoeugenol to vanillin with a molar yield of 12.4%. Rabenhorst (35) demonstrated with *Pseudomonas* species that they were capable of growing on eugenol as the sole carbon source and found vanillic acid, ferulic acid and coniferyl alcohol as metabolic intermediates. The conversion of isoeugenol to vanillin was also reported with cell cultures of *Capsicum frutescens*, however the yields were relatively low.

14.4.7 Biotransformation of Vanillic Acid

Vanillic acid is a main intermediate in lignin and ferulic acid degradation; in contrast to vanillin, it is often accumulated in remarkable amounts. With *Pseudomonas fluorescens* BF-13, a yield of 95% vanillic acid was obtained from ferulic acid within 5 hours.

The metabolism of vanillic acid has been intensively studied in *Sporotrichum pulverulentum* and *Pycnoporus cinnabarinus*. Vanillic acid is reduced to vanillin and vanillyl alcohol. Vanillic acid can be oxidatively decarboxylated to methoxy hydroquinone, which is the main problem with the vanillin production process using *P. cinnabarinus*. This problem was overcome by the addition of cellobiose prior to vanillic acid supplementation. The vanillin yield was further improved by use of high density cultures and different types of bioreactors.

14.4.8 Economic Considerations

The annual consumption of the world flavor market is about 6000 tons, out of which only about 0.2% originates from botanical sources. The main production is by chemical synthesis from guaiacol and lignin. The price of chemically synthesized "Nature identical" vanillin is very low (about U.S. $12/kg) compared to the price of cured vanilla pods (between U.S. $30/kg and U.S. $120/kg, which usually contains about 2% (w/w) vanillin. In the method described for biovanilla or biovanillin production by microbial biotransformation, the cost will be on the order of U.S. $1000/kg (30). The uses of vanilla flavor range from food to medicinal value.

14.4.9 Uses of Vanilla

Vanilla is the world's most popular flavoring for sweetened food.

Pure vanilla extracts are widely used as a flavorant for ice cream, soft drinks, chocolates, confections, candy, baked foods, pudding, cake, cookies, and liquors and as a fragrance ingredient in perfumery. It enhances the flavor of caramel, coffee, and some dairy products. It also has medicinal value as antimicrobial agent and antioxidant.

The demand for vanilla flavor has resulted in commercial companies establishing their own processes for the production of vanilla and vanilla flavor components from natural vanilla bean and from various substrates through biotransformation (Table 14.14). Similarly, there are increasing patent applications in the area of biotransformation for vanillin production (Table 14.15).

Table 14.14
Companies producing vanilla flavor

ESC Agenetics Corporation, San Carlos, CA, USA
David Michael Co Inc. Philadelphia, PA, USA
Nielsen- Massey Vanillas Inc. Waukegan, IL, USA
Premier Vanilla Inc, East Brunwick, NJ, USA
Virginia Dare, Brooklyn, NY, USA
Gernot Katzer, USA
Aust and Hachmann Ltd, Montreal, Canada
Food Research and Development Centre, Quebec, Canada
Haarmann and Reimer (H & R), Holzminden, Germany.

Source: Ramachandra Rao, S., G.A. Ravishankar, *J. Food Sci. Agric.* 80:289–304, 2000.

Table 14.15
Some Patent Applications on biotransformation to produce vanillin

Year	Patent No.	Inventors	Topic
1989	US 4874701	Cooper B	Preparation of coniferyl aldehyde by a micro organism.
1991	EP 0453368	Gross B, Asther M, Corrieu G	Production of vanillin by bioconversion of benzenic precursors
1992	US 5128253	Labuda et al.	Bioconversion process for the production of vanillin
1993	US 5262315	Gross et al.	Production of vanillin by bioconversion of benzenic precursors
1994	WO 9608576	Lesage-Heesen L, Haon M, Asther M	Procede d' obtention d' acide vanillique et de vanilline par bioconversion de micro organisms filamenteux
1994	US 5279950	Labuda et al.	Bioconversion process for the production of vanillin
1994	WO 9413614	Cheetam PSJ, Gradley ML, Sime JT	Flavor/aroma material and their preparation
1997	WO 9735999	Narbad A, Rhodes MJC Gasson MJ, Walton NJ	Production of vanillin
1998	US 59555137	Ago S, Kikuchi Y	Ferulic acid decarboxylase
1998	US 0885968	Muller B, Munch T Muheim A, Wetli M	Process for production of vanillin
1999	US 5866380	Lesage-Meessen et al.	Methods for bioconversion and ferulic acid to vanillinic acid or vanillin
2000	WO 0050622	Cheetam et al.	Production of vanillin
2000	US 6133003	Rabenhorst et al.	Process for production of vanillin and microorganisms suitable therefore
2001	US 6236607	Muheim et al.	Microbiological process for producing vanillin

14.5 MONOTERPENES

Monoterpenes are naturally occurring branched chain C-10 hydrocarbons formed from two isoprene units. They are widely distributed in nature and over 400 different naturally occurring monoterpenes have been identified. In particular, oxygenated derivatives of terpenes, commonly known as terpenoids, have strong and pleasant odors. Many of these terpenoids are considered as GRAS (Generally Recognized As Safe) and when obtained from natural sources, they can be added to foods without being considered as additives (44).

On the other hand, various available monoterpenes, such as (-)-α– pinene, (-)-β– pinene, and (+) – limonene are used in large quantities in the chemical industry for conversion into more valuable terpenoids. During the past decade, new information has become available on the purification and description of several enzymes involved in monoterpene degradation (Table 14.16). The genes encoding some of these enzymes have also been cloned and sequenced, opening up the possibility of doing single step biotransformation in a neutral background.

Table 14.16
Microbial transformation reactions involved in the bioconversion of monoterpenes

A. **Oxido reductases**
Monooxygenases catalyzing hydroxylation
Monooxygenases catalyzing epoxide formation
B. **Hydrolases**
Hydrolysis
Interesterification
Esterification
Ring closing and Ring opening reactions
Hydration
Allyl rearrangement
Racemization

14.5.1 Monoterpene Biotransformation

Problems encountered in Monoterpene biotransformation:

1. Chemical instability: Monoterpenes readily undergo spontaneous chemical autooxidation, cis-trans isomerization, racemization, hydration, cyclization, rearrangement, and polymerization reactions.
2. Low solubility: Monoterpenes are poorly soluble in water, which will affect the biotransformation rate.
3. Volatility: Monoterpenes are volatile in nature.
4. Toxicity: The problem associated with terpene biotransformation studies in the toxicity of monoterpenes to whole cells. Addition of the terpenes to the cell cultures has been observed to inhibit growth and the addition of high concentrations of monoterpenes can result in cell lysis.
5. Stability: Absence of product accumulation and product degradation.
6. Purity: Multiple metabolic pathways resulting in the formation of mixture of products.
7. Scarcity: Low product concentration and yields (e.g., in mg/L of culture media), resulting in high downstream processing costs.
8. Enzyme activity at times not detectable in cell extracts.
9. Long incubation times: Many bioconversions, especially those involving fungi, take many days to reach appreciable yield. The longer the fermentation time, the greater the cost and increased possibility of contamination by unwanted microorganisms.
10. Durability: Short lifetime of the biocatalyst.

14.5.2 Limonene

Limonene is monocyclic terpenoid, which besides pinene is the most widely distributed terpene in nature. The microbial transformation of limonene has been studied in detail by Dhavalikar et al. (45) and a large number of neutral and acidic metabolites have been isolated and identified. Several of these are of industrial importance such as carveol, carvone, dihydrocarvone, perillyl alcohol, and perillaldehyde. Dhavalikar et al. (45) proposed three pathways in the conversion of limonene.

An effort to increase the value of cheap (+)(–) limonene isolated from citrus was described by Bowen (5). Pure cultures of *Pencillium digitatum* and *P. italicum* were isolated

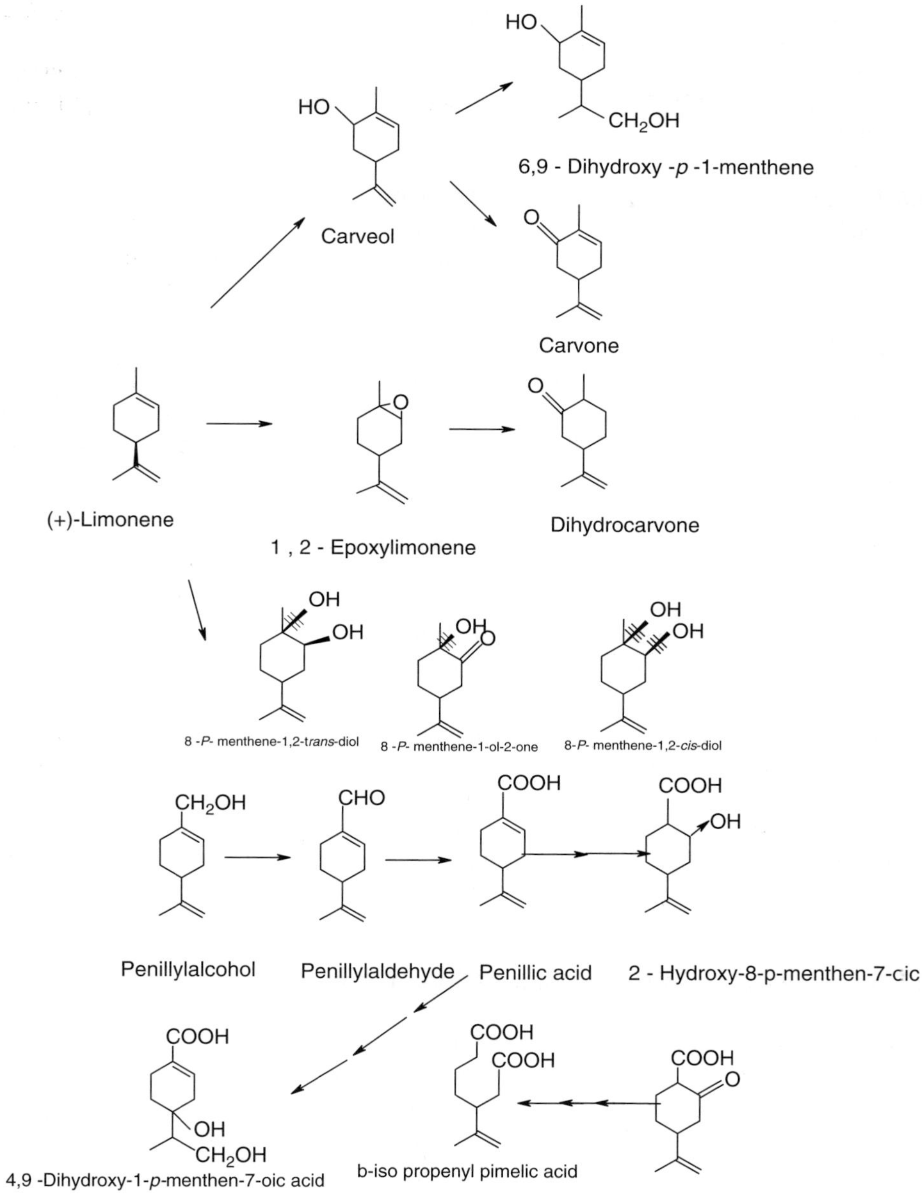

Figure 14.6 Pathway for degradation of (+) – Limonene in *Pseudomonas* sp. L

from overripe oranges by enrichment culture techniques on Czapek-Dox Broth containing 1% limestone as the sole carbon source. Transformation by *P. italicum* (Figure 14.6) yielded the following products: *Cis*- and *trans*-carveol (26%), carvane (6%), *cis*- and *trans*-p-menthe- 2.8-dien-1 – ol (18%), p-mentha-1,8-dien-4-ol (4%), penillyl alcohol (3%), and P-mentha-8-ene 1,2-diol (3%).

Noma et al. (46) used *Aspergillus cellulosae* M-77 for biotransformation of (+)(-) limonenes to major products like (-)– Perillyl alcohol, (-)– limonene-1, 2 trans-dilol and (-) – neodihydrocarveol. Biotransformation of limonene by *Xanthobacter* species C 20 into a novel limonene is described by Werf et al. (47).

Separate enantiomers of Limonene exhibit the characteristic Limonene odor profile: (R) – (+) with fresh, natural, citrus, orangelike odors, and (S) – (-) with harsh turpentine lemonlike notes.

14.5.3 Isolation and Identification of the Metabolites

The metabolites are separated and purified by a combination of column chromatography on silica gel and gas chromatography. The products are identified by comparison of the gas chromatography and mass spectra with those of authentic specimens (46).

14.5.4 Future Trends

The microbial transformation of terpenoid compounds has considerable potential for practical application in the flavor and fragrance industry. Nevertheless, microbial transformations of various monoterpenes have challenging industrial possibilities. There are three areas in the biotechnological production of monoterpene that can play an important role:

1. Formation of terpenoids from cheap and readily available monoterpenes such as (+) – limonene (−) -α- pinene.
2. Biotransformation of terpenoids with a high flavor impact that are not readily available from other biological or chemical sources.
3. Production of novel (natural) flavors with unique organoleptic properties.

In the future, genetic engineering techniques may provide modified strains catalyzing a single pathway to the desired product resulting in the production of pure aroma chemicals. As many of the characterized enzymes involved in monoterpene degradation appear to have broad substrate specificities, these systems might also be used for the production of closely related terpenoids. Furthermore, by means of genetic engineering, the control of biosynthetic pathways and the exploitation of the regulatory mechanisms can improve the yield of the processes.

14.5.5 Application of Biotransformation of Limonene

The transformation products of limonene yield (+)-cis-carveol and (+)-carvone, which are important constituents of caraway seed and dill seed oils, in addition to 1-p-menthene-6, 9-diol (-) – carvone which contributes to spearmint flavor. The other transformation products are perillyl alcohol, perillaldehyde, and perillic acid, constituents of various essential oils used in the flavor and fragrance industries. The selected patents in the area of monoterpene production are given in Table 14.17.

14.6 STRUCTURED LIPIDS

Interesterification is a process which is used in the oils and fat industry to modify the properties of triglyceride mixtures using enzyme or chemical catalysed reactions. Structured lipids (SL) will indicate any lipid that has been modified to change the fatty acid position or composition from the native stage. They are produced for commercial use by chemical hydrolysis of the natural oils and random reesterification, or by interesterification.

Structured lipids may be the most effective means of delivering designer triglycerols with desired fatty acids as nutraceuticals, functional foods, and medical foods to target specific diseases or metabolic conditions, and for optimal nutrition.

Table 14.17
Patent Applications

Year	Patent No.	Authors	Topic
2001	US 6200625	Beckett, sephnan Thomas	Preparation of chocolate production with limonene to reduce fat content
1998	US 5763237	Savithiry; N, Oriel P J	Method for production of monoterpene derivative of limonene
1996	US 5487988	Chang, HC; Oriel P J	Preparation of perillyl compounds using *Bccillus stereothermphilus*
1985	US 4551570	Johnson JR; Walter, E	Process for the isomerization of limonene to terpinolene
1997	US 5652137	Chang, HC; Oriel PJ	Process and bacterial cultures for the preparation of perillyl compounds

By exploitation of the specificity of the lipases, it is possible to produce useful glyceride mixtures that cannot be obtained by conventional chemical interesterification processes.

14.6.1 Enzymatic Synthesis

Lipases catalyze direct esterification and transesterification reactions at mild reaction conditions with little or no side products. Lipases can catalyze the incorporation of specific fatty acids and specific positions of the tricylglycerides.

Lipids with maximized functionality do not occur in nature at the proportions required for their functional use; therefore, enzymatic restructuring is a viable means of producing structured lipids with specific fatty acids at specific positions.

14.6.2 Substrate Specificity of Microbial Lipases

Extracellular microbial lipases (Glycerol ester hydrolases EC 3.1.1.3) are excreted by microorganisms into their growth medium to assist in the digestion of lipid material. They catalyze the hydrolysis of fats to give free fatty acids, partial glycerides, and glycerol.

The naturally occurring triglycerides of long chain fatty acids are water insoluble, and lipases are characterized by the ability to catalyze rapidly the hydrolysis of ester bonds at the interface between the insoluble substrate phase and aqueous phase, in which the enzyme is soluble.

The microbial lipases can be placed in three groups: nonspecific lipase, specific lipase and fatty acid specific lipase.

The first group shows no marked specificity. These lipases catalyze the complete breakdown of the triacylglycerol into fatty acid and glycerols, but diglycerides and monoglycerides appear as intermediates of reactions. Examples of this type are the lipases from *Candida cylindracae, Corynebacterium acnes* and *Staphylococcus aureus.*

The second group of lipases catalyzes the release of fatty acids specifically from the outer 1- and 3- positions of glycerides with these lipases. Triacyglycerides are hydrolysed to give free fatty acids, 1,2 (2,3)- diglycerides and 2-monoglycerides as reaction products. 1,3-specificity is common amongst microbial lipases and examples of enzymes from this group are the lipases from *Aspergillus niger, Mucor javanicus* and various *Rhizopus* species.

The third group of lipases catalyzes the specific release of a particular type of fatty acid from glyceride molecules. The lipase produced by *Geotrichum candidium* has been shown to possess a very marked specificity for the hydrolysis of esters of a particular type of long chain fatty acid.

14.6.3 Lipases as Interesterification Catalysts

Because lipase reactions are reversible, hydrolysis and resynthesis of glycerides occur when lipases are incubated with oils and fats. This hydrolysis and resynthesis causes acyl migration between glyceride molecules and gives interesterified products. Under conditions in which the amount of water in the reaction system is restricted, hydrolysis of the fat can be minimized so that lipase catalyzed interesterification becomes the dominant reaction (48).

If a nonspecific lipase is used to catalyse the interesterification of a triglyceride mixture, the triglycerides produced are similar to those obtained by chemical interesterification. However, with a 1, 3-specific lipase as catalyst, acyl migration is confined to the 1 and 3 positions, and a mixture of triglycerides that is unobtainable by chemical interesterification is produced. Mixtures of triglycerides and free fatty acid can also be used as reactants for lipase catalyzed reactions. In these cases, free fatty acid exchanges with the acyl groups of the triglycerides to produce new triglycerides enriched in an added fatty acid. With nonspecific lipases enrichment of all three glyceride positions occurs, but with 1,3-specific lipases the reaction is confined to the 1 and 3 positions of glycerides. If a fatty acid specific lipase is used, a particular fatty acid from a mixture of fatty acids can be selectively introduced.

The ability to produce novel triglyceride mixtures using specific lipases is of interest to the oils and fat industry because some of these mixtures have properties that make them valuable. For example, 1,3-specific lipase catalyzed interesterification of 1,3-dipalmitoyl-2- monooleine (POP), which is major triglyceride of the mid fraction of palm oil, with either stearic acid or tristearine gives products enriched in the valuable 1, (3) Palmitoyl-3(1) stearoyl -2-monooleine (P O St) and 1,3 distearoyl -2-monooleine (St O St). P O St and St O St are the main components of cocoa butter, and it is therefore possible by the interesterification reaction to produce a valuable cocoa butter equivalent from cheaper starting materials.

14.6.4 Interesterification Reaction Systems

The catalysts used for enzymatic interesterification are prepared by addition of a solvent such as acetone, ethanol, or methanol to a slurry of an inorganic particulate material such as kieselghur, hydroxylapatite, or alumina in buffered lipase solution. The precipitated enzyme coats the inorganic particles and the lipase coated particles are collected by filtration, then dried and stored. In the dried form the particles are almost inactive as interesterification catalysts, and to obtain high catalytic activity it is necessary to hydrate the particles by addition of up to 10% water prior to their use in the interesterification reaction systems.

14.6.5 Process for the Modification of Oil and Fats Via Enzymatic Interesterification

Most studies for the enzymatic oils and fats by lipase catalyzed interesterification have used the Stirred Tank Reactor process. Fuji oil used a Stirred Tank Reactor for the large scale production of cocoa butter equivalent by lipase catalyzed interesterification. The schematic arrangement of the process is shown in Figure 14.7.

Lipase catalyzed interesterification reactions can also be performed continuously using packed bed reactors. For operation of the packed bed reactors, the feed stock mixture, dissolved in petroleum ether, is treated to remove particulate materials and enzyme catalyst inhibitors and poisons, and then partially saturated with water prior to being pumped through a bed of hydrated catalyst particles. The lipase catalysts are reasonably stable under the conditions prevalent in the packed bed reactors, and it is possible to operate these reactors continuously for up to 600 hr with minimal loss of catalytic activity.

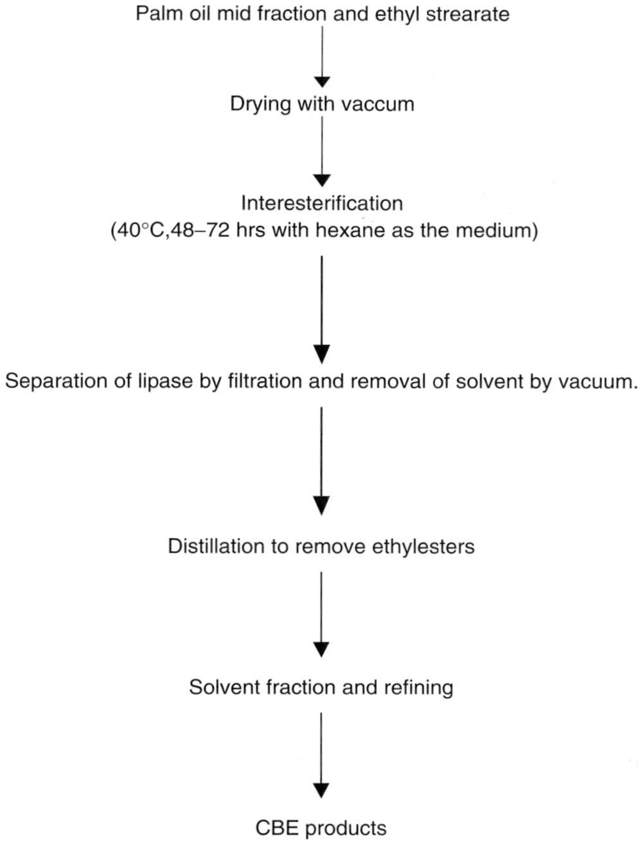

Figure 14.7 Process scheme for the production of cocoa butter equivalents (CBE) via lipase catalysed interesterification in a stirred tank reactor

14.6.6 Downstream Processing

At the end of reaction period, the catalyst is filtered off and a fraction containing the 1,(3)Palmitoyl-3(1)stearoyl-2-monooleine and 1,3 distearoyl-2-monooleine triglycerides isolated by conventional fat fractionation techniques such as countercurrent, liquid–liquid extraction, and crystallization from solvents. The catalyst is washed free of fatty material, dried and then reused in subsequent interesterification reactions. Use of the same catalyst particles in 10 successive batch interesterification reactions has proved to be possible. Some reported benefits of structured lipids are given in Table 14.18. A number of companies involved in the production of structured lipids are shown in Table 14.19. There are a number of patents in this area (Table 14.20) which are of great utility for the production of value added food products.

14.7 LACTOSE HYDROLYSIS

Lactose or milk sugar is a disaccharide found in milk. Lactose itself is less sweet and soluble than sucrose and it cannot be absorbed directly from the intestine. Lactase (β-galactosidase) hydrolyzes lactose into glucose and galactose (Figure 14.8), which have

Table 14.18
Some Reported benefits of Structured Lipids

*Enhanced absorption of the fatty acid at the Sn-2 position, e.g., 18 : 2n – 6 in cystic fibrosis patients.
*Improved absorption of other fats.
*Reduction in serum TAG, LD-Cholesterol and cholesterol
*Hypermetabolic response to thermal injury
*Reduced calorie fat
*Improved immunofunction
*Prevent thrombosis
*Reduced risk of cancer
*Lipid emulsions for parenteral feeding

Table 14.19
Some Companies producing structured lipids

Unilever, Black friars, London UK.
Arco Chem. Tech. Wilmington, DE, USA
Kraft General foods, Northfield, IL, USA
Raision Tehtaat, Finland
Procter and Gamble, USA.

a combined sweetening power of about 0.8 relative to sucrose; the hydrolyzed product is also three to four times more soluble than lactase, and the mono saccharides are easily and directly absorbed from the intestine (49). The hydrolysis of lactose to form glucose and galactose is of interest from several points of view:

1. A significant percentage of population of the world has β-galactosidase deficiency and are intolerant to lactose.
2. Due to its tendency to crystallize, its low solubility and compatible low sweetness, lactose would be impractical as a food sugar.
3. Prehydrolysis of milk with lactase would shorten yogurt and cheese production time by 20%, as well as increase the yield and sweetness.
4. During the hydrolysis of lactose, glucose and galactose are produced. The enzyme responsible for the reaction is β–D-galactoside galactohydrolase (Lactase, hydrolactase, β-galactosidase). The type of reaction is O-glycosyl bond hydrolysis. β-galactosidase is produced by a number of microorganisms which include yeast, fungi, and bacteria (Table 14.21).

14.7.1 Fermentation for Lactase Production

Fermentation methods include:

1. *Aspergillus niger*: The fermentation process can be solid-state or submerged fermentation. Solid-state fermentation can be done with wheat bran. Inoculation is done with *A. oryzae* spores, and the mixture is incubated for seven days at 30°C. Yields in solid-state fermentation are much higher than in submerged fermentation.

Table 14.20
Some selected patents on structured lipid production for developing value added food products.

Year	Patent No.	Authors	Topic
1998	EP882797	Bornscheuer, UT, Soumanou, MM; Schmid, RD	Preparation of symmetrical triglycerides comprises Reacting 2-monoglyceride with Reactant providing fatty acid moieties using 1,3-specific enzymes
1992	NL9002130	Sticht. Tech.Wetensch.	Cruciferous stearyol – ACP –Desaturate coding sequence
2000	EP103844AJ	Ozaka T, Yamadak;Nago A	Hard butter composition and its production
2000	FR2784116AJ	Hoang-le-chein	New process for manufacture of F.a.esters, protein flours, fibre and glycerol by direct transesterification of rapeseed erucic acid.
1996	WO96/07632AJ	Lamsa. M	Process for preparing a synthetic ester from a vegetable oil
1995	FR2711142AJ	Graille, JE Montel, DJL Ozenne, CL Lambert, GP	Process for enzymatic modification and TGS in a fat, especially milk fat
1992	US5288619	Brown, PH Caravallo, FD Thomas, C	Enzymatic method for preparing transesterification oils
1992	US5175323	Cooper CF	Preparation of esterified propoxylated glycerin
1991	GB2236537A	Macraejar Pandley, FB Chandler, IC	Transesterification
1990	US 4966876	Sankaran, V	Transesterification of triglycerides

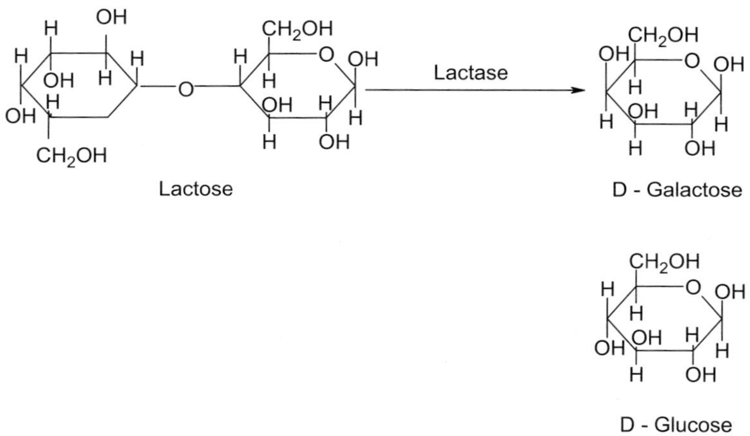

Figure 14.8 Hydrolysis of lactose to produce Glucose and Galactose

Table 14.21
Microorganisms producing lactases

Yeast
Kluyveromyces lactis, K.fragilis, Candida pseudotropicalis, Zygosaccharomyces lactis, Saccharomyces anamensis
Fungi
Aspergillus oryzae, A. niger, Mortierella vinacea, Penicillium multicolour, Curvularia inaequalis, Fusarium moniliforme, Mucor pusillus, Humicola grisea, Torula thermophila.
Bacteria
Bacillus coagulans, B. stearothermophilus, B. circulans, E. coli, Lactobacillus bulgaricus, L. thermophilus
Those microorganisms most important for commercial production are: *Kluyveromyces lactis, Kluyveromyces niger, Aspergillus oryzae, E. coli.*

2. *Escherichia coli*: Submerged fermentation is at pH 7.0. Enzymes can be isolated by centrifugation and precipitation.
3. *Bacillus stearothermophilus*: The lactases of *Bacillus stearothermophilus* are very thermostable. The organisms grow while aerated at 65°C in a batch fermented process or under continuous conditions.

A lactose free, low cost culture medium for the production of β–D Galactosidase by *Kluyveromyces macianus* was formulated in which sugar cane molasses is added as the carbon source.

14.7.2 Immobilization of Lactase

Lactase has been immobilized by a variety of procedures including adsorption, entrapment, and covalent linkage. Supports used are agar gel, alumina, carbon, cellulose, chitin, chitosan, collagen, glass beads, ion exchange resin, nickel, nylon, phenol-formaldehyde resin, polyacrylamide sephadex, and silk fibronin. (49). Immobilization is essential for reducing the cost of lactose hydrolysis as well as enabling changes of operational condition (pH, temperature, half life) (50).

Conditions for biotransformation: The conditions for biotransformation depend on the properties of microbial lactases. The optimal process conditions for the lactase derived from the dairy yeast *Kluyveromyces lactis* which is at present the most widely used commercial lactase preparation, is at 35–40°C temperature and pH 6.6–6.8, which are close to the natural temperature and pH of milk, and so it is most useful in the treatment of milk and sweet cheese whey. The properties of microbial lactases are given in Table 14.22.

14.7.3 Process Parameters

The synthesis of β–galactosidase by *K. lactis* has a conversion efficiency of 70–81%. The process can be operated as a fluidized bed in a column 3 inches in diameter. For industrial biotransformation, plug flow reactors are used, which have capacity of about 8000 L/d (at Central del Latte, Italy).

The first large scale pilot plant for the lactose hydrolyzing process was by Snam Progetti dairy plant in Milan, Italy. The second process for the production of lactase hydrolyzed whey has been developed by Corning Glass Works, USA.

Table 14.22
Properties of Microbial Lactases

Source	pH Optimum	pH Stability	Temperature Optimum °C	Km for Lactose	Cofactors Needed
A. niger	3.0–4.0	2.5–8.0	55–60	85	None
A. oryzae	5.0	3.5–8.0	50–55	50	None
K. fragilis	6.6	6.5–7.5	37	14	Mn2+, K+
K. lactis	6.9–7.3	7.0–7.5	35	12–17	Mn2+, Na+
E. coli	7.2	6-8	40	2	Na+, K+
Lactobacillus thermophilus	6.2	-	55	6	-
Leuconostoc citrovorum	6.5	-	60	7.8	None

Table 14.23
Companies using β–galactosidase

Sumitomo chemical industries, Japan
Snow brand milk products co. Ltd, Japan
Central del Latte, Italy
Rapidase Tokyo Tanaben, Japan
Miles Lab, Germany
Novo Industries, Denmark
Gist Brocades, Netherlands

Recently there are many companies that are producing lactose hydrolyzed whey, some of which are given in Table 14.23.

14.7.4 Application of Lactases

Lactases find their application in several areas, such as milk for people with lactose intolerance problems, especially infants; milk specific for cheese for cheese and yogurt making; whey or lactose for the production of sweeteners and soluble hydrolyzed whey syrups; concentrated milk products such as condensed milk; and for preventing the gritty texture of ice cream made with lactose.

Several patents on lactases have been obtained, and some representative ones are given Table 14.24.

14.8 D-TAGATOSE

A rare ketohexose, D-tagatose, is one of the D-galactose isomers. D-tagatose is a bulking agent in food, a noncalorific sweetener (sweetness 0.92), and tastes almost similar to sucrose. In addition, D-tagatose does not show a laxative effect, unlike other polyols. Zero energy or low energy alternative carbohydrate sweeteners play an important role in energy restricted diets. Alternative sweeteners are mainly used to control calorie intake in food and beverages, in the management of diabetes, to assist in the control of dental caries, and to enhance the usability of pharmaceuticals and cosmetics. Although D-tagatose occurs

Table 14.24
Representative patents on lactase

Year	Patent No.	Authors	Topic
1976	US 3, 957, 584	Kroinish et al.	Detection of beta-Galactosidase producing micro-organisms
1999	US 5, 962, 326	Shimada et al.	Hyper thermostable beta-galactosidase gene, enzyme encoded thereby and process for production
1995	US 5, 444, 161	Manning et al.	Substrates for beta-Galactosidase
1995	US 5, 403, 726	Wong et al.	Enzymatic process for producing a galactosyl. Beta 1, 3 glycal disaccharide using beta-galactosidase
1994	US 5, 283, 189	Takase et al.	Beta-galactosidase from *Thermus* sp
1993	US 5, 194, 382	Herrmann et al.	Method for increasing theenzymatic reactivity of beta-galactosidase by addition of a cynate, thiocyanate, azide or thiosulfate compound
1982	US 4, 332, 895	Griggiths et al.	Thermal stable beta galactosidase
1980	US 4, 234, 687	Bungard et al.	Beta-galactosidase
1980	US 4, 229, 539	Miwa, et al.	Beta-galactosidase and production there of

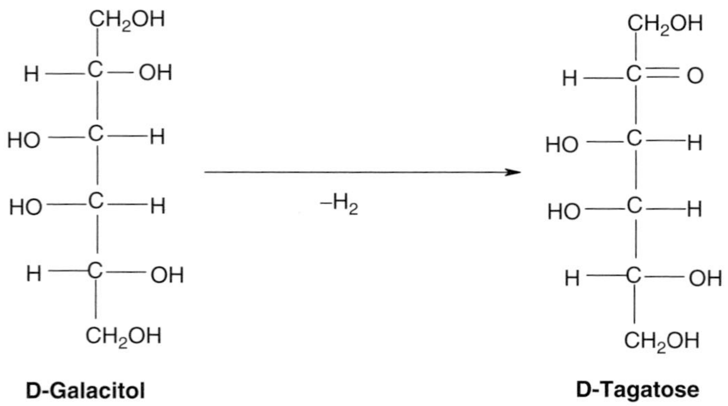

Figure 14.9 Oxidation of D-Galactital to D-Tagatose.

naturally, the amounts are too small for economic recovery. D-tagatose is a generally recognized as safe (GRAS) substance under FDA regulation (Table 14.25).

Several methods have been studied for the manufacture of D-tagatose. D-galactose could be converted into D-tagatose by a calcium catalyst. Although chemical synthesis is an economical process, this process also has the disadvantage of needing high temperature and high pressure. Biological production of D-tagatose has been studied intensely in recent years (51).

14.8.1 Reaction Involved in Biotransformation

D-tagatose production from galactitol using galactitol dehydrogenase is well known. Galactitol however, is more expensive than galactose and seems to have only little potential for commercial production (52). The D-tagatose is obtained by the biotransformation of D-galactitol and Bioconversion of D-galactose. During the biotransformation of D-Galactitol to D-tagatose the oxidation reaction takes place.

Table 14.25
Representative patents on D-Tagatose and its derivatives

Year	Patent No.	Authors	Topic
1985	US 4786722	Zehner, LR	D-tagatose as low calorie carbohydrate sweetener and bulking agent
1991	US 5002612	Beedle; James R Saunders; James Y Wajda, JR, Thomas J	Process for manufacturing D-tagatose
2000	US 6057135	Ibrahim OO, Sparadlin JE	Production of D-tagatose from whey or milk comprising production and isomerization of D-galactose with L-arabinose Isomerase
1987	JP-J62134095	Mitusi-sugar	Glycosyl tagatose and production thereof
1995	JP 07059584	Kyowa-Hakkg	Preparation of disaccharides and new disaccharides
1993	EP-552894	Kraf-Gen Foods	D-tagatose low calorie sweetener production
1985	P-J 60248196	Hayashibara EP 807682	Preparation of D-tagatose Biochem
1997		Szumori, K Tsusaki K	L-ribose isomerase enzyme
1992	US 507 8706	Beedle et al.	Process for manufacturing tagatose
1999	US 6015793	Levin, gilbert	Use of tagatose to enhance key blood factor
1994	US 5356879	Zehner LR	D-tagatose as antihyperglycemic agent

14.8.2 Microorganisms for Biotransformation to D-Tagatose

The first microorganism used was *Arthrobacter glabiformis* (53). The yield of D-tagatose, which accumulated in the medium from D-galactitol, was as high as 85%. Other microorganisms which can carry out the transformation are *Mycobacterium smegmatis, Enterobacter agglomerans* and *Gluconobacter oxydans* (Acetic acid bacteria) (53,54).

Bioconversion of galactose into tagatose can be achieved by the L-arabinose isomerase gene (*Ara A*) from *Escherichia coli, Bacillus subtilis,* and *Salmonella typhimurium* which could produce D-tagatose from D-galactose (52).

The cells of *Mycobacterium smegmatis* were grown on various sugar polyalcohols, but oxidation activity was found only in case of D-galactitol, and D-galactitol and D-tatitol. L-sorbose induces the enzymes for D-tagatose fermentation (53). Oxidation activity for D-galactiol (Figure 14.9) gives a yield about 60%. The intact cells and immobilized cells with calcium alginate have virtually the same activity with respect to D-tagatose production. The conversion rate from *Enterobacter agglomerans* was about 92% when 2% galactitol was used (54). At 5% substrate concentration, the conversion was about 86%. Better results have been obtained by adding galactitol to 24 hour old cultures of *Gluconobacter oxydans* the time corresponding to the stationary growth phase (51).

14.8.3 Downstream Processing

The reaction mixture containing D-tagatose is centrifuged to remove the bacterial cells. The supernatant fluid is deionized and then concentrated at 40°C under vacuum. After 2 days of standing in absolute ethyl alcohol at 5°C, white crystals of D-tagatose are obtained (51).

14.8.4 Status of D-Tagatose Development

Because D-tagatose is a full bulk noncalorific sweetener, it may play a significant dual role in helping people realize the health and longevity benefits of diet restriction and also in deterring the specific aging effect of glycosylation (55).

In future, the use of D-tagatose as a limited substitute for sucrose may prove feasible. Chocolates and chewing gum, both of which were successfully formulated by substituting D-tagatose for sucrose in each, are planned as initial products along with the use of D-tagatose as a tabletop sweetener for coffee and tea, and in single use packets.

14.9 CONCLUSION

The scope of biotransformation is wide and the use of biotransformation is beyond doubt. A number of products are already on the market, such as high fructose corn syrup, oligosaccharides, vanilla flavor, structured lipids, and lactose hydrolyzed milk. For a new product to be introduced, it needs safety data; and therefore although several reactions and biocatalytic agents have been identified, they may not be immediately commercialized.

The demand for natural, nature identical molecules is increasing. Therefore, the use of biotransformation processes for the production of novel compounds of utility value assumes greater importance. The advantage of utilizing transgenics will further enhance the efficiency of biocatalytic processes. Further, direct modification of enzymes using site directed mutagenesis would enhance the efficiency of the process to suit process conditions like thermostability and more affinity toward substrates.

The most important feature of biotransformation is to obtain highly desirable food constituents which are of use as ingredients in developing the targeted food or designer foods for humans such as functional foods and nutraceutical.

The number of patents which have been taken in recent times, and the investments that have been made by multinational companies, are indications of the commercial potential of biotransformation processes to produce desired products. Moreover, utilization of the products of biotransformation in traditional foods would be an attractive proposition. Hence from a futuristic point of view, the scenario is bright.

REFERENCES

1. Ramachandra Rao, S. Studies on biotransformation to produce phytochemicals of importance using plant cell cultures, Ph. D. Thesis, University of Mysore,1996.
2. Giri, A., V. Dhingra, C.C. Giri, A. Singh, P. Owen, Ward, M. Lakshmi Narasu. Biotransformations using plant cells, organ cultures and enzyme systems, current trends and future prospects. *Biotechnol. Adv.* 19:175–199, 2001.
3. Berlin, J., C. Mollenschott, S. Herminghaus, L.F. Fecker. Lysine decarboxylase transgenic tobacco root cultures biosynthesize novel hydroxycinnamoylcadaverines. *Phytochemistry* 48(1):79–84, 1998.
4. Ward, P.O., A. Singh. Enzymatic asymmetric synthesis by decarboxylases. *Curr. Opin. Biotechnol.* 11:520–526, 1991.
5. Bowen, E.R. Potential by products from microbial transformation of D-limonene. *Proc. Fla. State Hortic. Soc.* 88:304–308, 1975.
6. Veith, W.R., K. Venkatasubramanian. Process engineering with glucose isomerization by collagen-immobilized whole microbial cells. *Methods Enzymol.* 44:768–776, 1976.
7. M.D. Lilly. Two-liquid-phase biocatalytic reactions. *J. Chem. Tech. Biotechnol.* 32:162–169, 1982.

8. Buckland, B.C., P. Dunnill, M.D. Lily. The enzymatic transportation of water insoluble reactants in non-aqueous solvents: conversion of cholesterol to cholest-4-ene-3-one by a *Nocardia* sp. *Biotechnol. Bioeng*. 17, 815–826, 1975.
9. Takamatsu, S., I. Umemura, K. Yamamoto, T. Sato, T. Tosa, I. Chibata. Production of L-alanine from ammonium fumarate using two immobilized microorganisms. *Eur. J. Appl. Microbiol. Biotechnol.* 15:147–152,1982.
10. Banks, G.T. Scale-up of fermentation process. *Topics Enzyme Ferment. Biotechnol.* 3:170–226, 1970.
11. Puhler, A., W. Heumann. Genetic engineering. In: *Biotechnology*, Vol. I, Rehm, H.J., G. Reed, eds., Weinheim: VCH Verlagsge Sellschaft mbH, 1981, pp 331–354.
12. Crueger, A., W. Crueger. Carbohydrate biotransformation. In: *Biotechnology 6(a),* Rehm, H.J., G. Reed, eds., Weinheim: Verlag Chemie, 1984, pp 421–469.
13. Mermelstein, N.H. Immobilized enzymes produce high-fructose corn syrup. *Food Technol.* 29(6):20–26, 1975.
14. Marshall, R.O., E.R. Kooi. Enzymatic conversion of D-glucose to D-fructose. *Science* 125:648–649, 1957.
15. Bengston, B.L., W.R. Lamm. British patent 1368511, 1974.
16. Vaheri, M., V. Kauppinen. Improved microbial glucose isomerase production. *Process. Biochem.* 12:5–7, 1977.
17. Singh, T. Role of enzymes in food industry. *Ind. Food Ind.* 19(6):404–411, 2000.
18. Hartley, B.S., N. Hanlon, R.J. Jackson, M. Rangarajan. Glucose–isomerese: insights into protein engineering for increased thermostability. *Biochem. Biophy. Acta Protein Struct. Mol. Enzym.* 1543(2):294–335, 2000.
19. Dekker, K., H. Yamagata, K. Sakaguchi , S. Udaka. Xylose (glucose) isomerase gene from the Thermophile *Thermus thermophilus:* cloning, sequencing and comparison with other thermostable xylose isomerases. *J. Bacteriol.* 173:3078–3083, 1991.
20. Karima, S.B., B. Samir. A thermostable glucose isomerase having a relatively low optimum pH: study of activity and molecular cloning of the corresponding gene. *Biotechnol. Lett.* 20(6):553–556, 1998.
21. Crittenden, R.G., M.J. Playne. Production, properties and applications of food-grade oligosaccharides. *Trends Food Sci. Technol.* 7:353–361, 1996.
22. Patel, V., G. Saunders, C. Bucke. Production of fructooligosaccharides by *Fusarium oxysporum. Biotechnol. Lett.* 16:1139–1144, 1994.
23. Shiomi, N., J. Yamada, M. Izava. Isolation and identification of fructo-oligosaccharides in roots of *Asparagus* (*Asparagus officinallis* L.). *Agric. Biol. Chem.* 40:567–575, 1976.
24. Van Balken, J.A.M., J.G.M. Van Dooren, W.J.J. Vanden Tweel, J. Kamphuis, E.M. Meijer. Production of 1-Kestose with intact mycelium of *Aspergillus phoenicis* containing sucrose-1^F-fructosyltransferase. *Appl. Microbiol. Biotechnol.* 35:216–221, 1991.
25. Jung, K.H., J.Y. Li, S.J. Yoo, J.H. Lee, M.Y. Yoo. Production of fructosyl transferase from *Aureobasidium pullulans. Biotechnol. Lett.* 9(10):703–708, 1987.
26. Barthomeuf, C., H. Pourrat. Production of high-content fructo-oligosaccharides by an enzymatic system from *Pencillium rugulosum.* *Biotechnol. Lett.* 17(9):911–916, 1995.
27. Chen, W., C. Liu. Production of β-fructofuranosidase by *Aspergillus japonicus. Enzym. Micro. Technol.* 18:153–160, 1996.
28. Yun, J.W., S.K. Song. The production of high content fructo-oligosaccharides from sucrose by mixed enzyme system of fructosyltransferase and glucose oxidase. *Biotechnol. Lett.* 15:573–576, 1993.
29. Tomomatsu, H. Health effect of oligosaccharides. *Food Technol.* 48(3):61–65, 1994.
30. Ramachandra Rao, S., G.A. Ravishankar. Vanilla flavour: production by conventional and biotechnological routes. *J. Food Sci. Agric.* 80:289–304, 2000.
31. Funk, C., P. Brodelius. Phenyl propanoid metabolism in suspension cultures of *Vanilla planifolia* Andr. II effect of precursor feeding and metabolic inhibitor. *Plant Physiol.* 9:95–101, 1990.
32. Westcott, R.J., P.S.J. Cheetham, A.J. Barraclough. Use of organized viable vanilla plant aerial roots for the production of natural vanillin. *Phytochemistry* 35:135–138, 1994.

33. Priefert, H., J. Rabenhorst, A. Steinbuchel. Biotechnological production of vanillin. *Appl. Microb. Biotechnol.* 56:296–314, 2001.
34. Ramachandra Rao, S., G.A. Ravishankar, L.V. Venkataraman. An improved process for the preparation of vanillin. Indian patent 1022/DEL/96, 1996.
35. Rabenhorst, J. Pro duction of methoxyphenol type natural aroma chemicals by biotransformation of eugenol with a new *Pseudomonas* sp. *Appl. Microb. Biotechnol.* 46:470–474, 1996.
36. Falconnier, B., C. Lapierre, L. Lesage-Messen, G. Yonnet, P. Brunerie, B.C. Ceccaldi, G. Corrieu, M. Asther. Vanillin as a product of ferulic acid biotransformation by the white root fungus *Pycnoporus cinnabinus* I–37: identification of metabolic pathways. *J. Biotechnol.* 37:123–132, 1994.
37. Lessage-Messen, L., M. Delattre, M. Haon, J.F. Thibault, B.C. Ceccaldi, P. Brunerie, M. Asther. A two step bioconversion process for vanillin production from ferulic acid combining *Aspergillus niger* and *Pycnoporus cinnabarinus. J. Biotechnol.* 50:107–113, 1996.
38. Toms, A., J. Wood. The degradation of trans-ferulic acid by *Pseudomonas acidovarans. Biochemistry* 9:337–343, 1970.
39. Tripathi, U., S. Ramachandra Rao, G.A. Ravishankar. A process for the preparation of vanilla flavour metabolites through biotransformation. Indian patent 1193/DEL/99, 1999.
40. Andreoni, A., S. Bernasconi , G. Besetti. Biotransformation of ferulic acid and related compounds by mutant strains of *Pseudomonas fluorescens. Appl. Microbiol. Biotechnol.* 42:830–835, 1995.
41. Oruk, G. Degradation of ferulic acid by *Escherichia coli. J. Ferment. Technol.* 63:501–506, 1985.
42. Krishnamohan, S. Khanna. Metabolism of ferulic acid by *Alcaligenes paradoxus. Ind. J. Microbiol.* 34:303–306, 1994.
43. Sutherland, J.B., D.L. Crawford, A.L. Pomento. Metabolism of p-coumaric and ferulic acids by *Streptomyces setonii. Can. J. Microbiol.* 29:1253–1257, 1983.
44. Janssen, L., H.L. De Pooter, N.M. Schamp, E.J. Vandamme. Production of flavours by microorganisms. *Process. Biochem.* 27:195–215, 1992.
45. Dhavalikar, R.S., P.N. Rangachari, P.K. Bhattacharya. Microbiological transformations of terpenes, IX: pathways of degradation of limonene in a soil *Pseudomonad. Ind. J. Biochem.* 3:158–164, 1966.
46. Noma, Y., S. Yamasaki, Y. Asakawa. Biotransformation of Limonene and related compounds by *Aspergillus cellulosae. Phytochemistry* 31:2725–2727, 1992.
47. Werf, M.J.V., P.M. Keijzer, P.H. Schaft. *Xanthobacter* sp. C20 contains a novel bioconversion pathway for limonene. *J. Biotechnol.* 84(2):133–143, 2000.
48. Haumann, B.F. Tools: hydrogenation, interesterification. *Inform* 5(6):668–678, 1994.
49. Greenberg, N.A., R.R. Mahoney. Immobilization of lactase (β-galactosidase) for use in dairy processing. *Process. Biochem.* 2:8, 1981.
50. Rugh, S. Immobilization of β–galactosidase. *Appl. Biochem. Biotechnol.* 7:27–29, 1982.
51. Manzoni, M., M. Rollini, S. Bergomi. Biotransformation of D-galactitol to tagatose by acetic acid bacteria. *Process. Biochem.* 36:971–977, 2001.
52. Roh, H.J., P. Kim, C. Yong, H.J. Choi. Bioconversion of D-galactose into D-tagatose by expression of L-arabinose isomerase. *Biotechnol. Appl. Biochem.* 31:1–4, 2000.
53. Izumori, K., K. Tsuzaki, Production of D-tagatose from D-galactitol by *Mycobacterium smegmatis. J. Ferment. Technol.* 66(2):225–227, 1988.
54. Muniruzzaman, S., H. Tokunaga, K. Izumori. Isolation of *Enterobacter agglomerans* strains 221e from soil, a potent D-tagatose producer from Galactitol. *J. Ferment. Bioeng.* 78(2):145–148, 1994.
55. Levin, G.V., L.R. Zehner, J.P. Saunders, J.R. Beadle. Sugar substitutes: their energy values, bulk characteristics and potential health benefits. *Am. J. Clin. Nutr.* 62:1161–1169, 1995.

3.15

Solid-State Bioprocessing for Functional Food Ingredients and Food Waste Remediation

Kalidas Shetty

CONTENTS

15.1 INTRODUCTION

Solid-state bioprocessing (SSB) is an aerobic microbial metabolic process and is generally referred to as the process in which 1) microbial growth and product formation occur, 2) on suitable solid substrates, 3) in the near absence of free water or in low water that is absorbed within the solid substrate matrix (1–3).

Therefore, referring to this process as a solid-state fermentation (SSF) is a misnomer. Fermentation generally refers to growth with no oxygen, as in alcoholic or lactic acid fermentation. Solid-state bioprocessing deals with the utilization of water-insoluble materials for microbial growth and metabolite production (1). However, suitable materials are those that absorb well and have sufficient moisture to support aerobic growth (1). The SSB system has been widely used in Asia for over 2000 years for products like Koji for soy sauce and miso, as well as tempeh, and is being increasingly modernized and applied to production of other industrial products, including enzymes and pharmaceuticals (1). In the west, an SSB-like process is used in meat preservation, bread making, and cheese and mushroom production (4). This chapter will be limited to the discussion of certain unique applications of SSB for development of ingredients for functional foods, and some aspects of solid waste remediation for food processing wastes. A more detailed conceptual discussion of SSB-like process is covered elsewhere in the book.

15.2 MODERN APPLICATIONS

Since the late nineteenth and early twentieth century, SSB has been used for the production of many enzymes by microorganisms (4,5). It was also extended to the production of important metabolites, such as gluconic acid, citric acid, and gallic acid (4,5). Later, in the mid twentieth century, SSB was extended to the production of antibiotics and steroid transformation. Today it has many applications, especially the production of value added products from agricultural and food processing byproducts, which otherwise have very low value. Select examples of such applications, where SSB is required for initial aerobic breakdown, include the production of fertilizer and soil amendments (6,7), animal feed in the form of single cell proteins (8–12), mushrooms (13–15), ethanol (16–19), methane (bio) gas (20,21), citric acid (22–25), butanol (26), flavor and aroma volatiles (27), and enzymes (28–30). In terms of food applications, there is continuous contemporary improvement in the use of pure cultures and in metabolic processing for bread yeasts (31), cheese (32) and fermented sausage (33,34), Koji production for soy products (35), and tempeh (36,37), along with means of enhancing nutritional quality with such products as carotenoids (38), reducing toxins (39,40), and moderating antinutritional factors (41).

15.3 SSB FOR FOOD WASTE BIOPROCESSING AND APPLICATIONS

Solid-state bioprocessing (SSB) has been used for composting and sewage treatment since early in the twentieth century, because the early phases of waste breakdown requires the decomposition of polymeric substrates by aerobic processes (1–3). This concept has been extended to producing fertilizer and soil amendments (6,7) and to mushroom production (14,15). Another development of a value added product is the conversion of fruit processing wastes to fungal inoculants for agricultural and environmental applications (42–45). With suitable combinations of fruit processing wastes and fishery wastes, beneficial microorganism rich amendments can be developed for reclamation of marginal soils, and to provide proper addendum for efficient composting at landfills and the waste remediation of pollutants.

15.4 COMPLEXITY OF SSB SYSTEMS AND NEED FOR "SYSTEMS BIOLOGY" APPROACHES

Solid-state bioprocessing (SSB) systems, in almost all cases, involve complex microecosystems that have multiple components and variables; such complex systems require a more integrated "Systems Biology" approach. Systems Biology as described by Hood (46) is the study of the relationships of all elements in a biological system, rather than of each element one at a time. In the case of SSB systems, this can be extended to the consideration of a "microecosystem" with multiple substrates and a succession of microbial processes and metabolic interconversions. Further extending the concept of Systems Biology, Kitano (47) has looked at Systems Biology as being similar to understanding the overall workings of a machine or modern mechanical device, as a car or airplane. According to Kitano (47), to understand biology at the systems level one must examine the structure and dynamics of cellular and organismal function, rather than the characteristics of isolated parts. He elegantly provides four key properties from which systems level understanding of biological systems can be derived; these are:

1. Understanding systems structures such as genes, enzymes, and pathways
2. Understanding the dynamics of the pathways and interrelationships
3. Understanding control mechanisms that determine function
4. Understanding of control points

Once these four key properties are understood, one can design or redesign the overall function, the basic structure, and the functional relationships of each biological system (47). Various elements of this Systems Biology perspective, based on parallels to the workings of modern mechanical devices, have merits. However, the complexity and diversity of the process of designing and redesigning biological systems and biological and microbiological ecosystems for human applications require a more comprehensive perspective of evolutionary relationships across the mosaic of prokaryotes and eukaryotes, and the eco-evolutionary development and interrelationships that span two or three billion years. In particular, the perspective will have to start from fundamental energy production and the management of various biological systems individually and in an ecosystem. Further, the variations of energy and redox management from oxygen rich environments to low oxygen and anaerobic environments in complex ecosystems, such as are found in several SSB systems, have to be tackled before we can move on. From fundamental insights into metabolic critical points in energy and redox management across biological systems, we have a better framework to link and connect genetic, signal, enzyme, protein, and related metabolic critical points that drive the overall biological systems and the multiple biological systems in an ecosystem. This energy and redox management based Systems Biology rationale is a suitable way to determine critical control points (CCP) in SSB systems involving complex biological substrates and microbial systems, in order to better design and fine tune SSB systems for any applications.

15.5 SSB IN TRADITIONAL FOOD PRODUCTS AND IMPLICATIONS

To develop an energy and redox management based Systems Biology in SSB systems for food product design and functional food applications, it would make sound and logical sense to go back to over 2000 years of empirical SSB systems, and understand how they have been established in our modern food and cultural mosaic. In this book these concepts have been highlighted by insights into select traditional SSB systems from human development in Africa, China, India, and the Mediterranean. Examples of improvements in bioprocessing in these traditional SSB systems that have come with the knowledge of modern biology, especially microbiology, nutrition, the metabolic biology of energy management, and molecular biology, can provide the framework to adapt more integrated Systems Biology approaches.

15.6 APPROACH FOR DEVELOPMENT OF FUNCTIONAL FOODS AND INGREDIENTS USING SSB

Knowledge of traditional SSB systems can provide the principles, safe biological systems, and framework to develop functional foods (improvement of conventional foods with added health benefits) and ingredients. In line with this thinking, and based on the rationale of incorporating a more integrated Systems Biology approach, several innovative SSB systems have been developed for harnessing the potential of phenolic functional ingredients from safe and established dietary sources. In this chapter discussion will focus on the cultivation of soybeans, fava beans, and cranberries.

In these approaches, SSB systems have been adapted, over a period of more than 2000 years, to enrich phenolic metabolites from soybean, fava bean and cranberry crops for enhanced health benefits. In the case of cranberries, there is cross cultural integration of a traditional American Indian medicinal berry with Asian SSB based food processing using modern metabolic biology principles. In the case of soybeans and fava beans, we have used the traditional Asian tempeh SSB system and adapted it to enriching the beans with functional phenolics that have diverse health application potential. These strategies offer unlimited potential to capture the phytochemical potential of diverse dietary botanicals with well developed traditional microbial processing systems, using modern microbiology, metabolic biology, molecular biology, nutritional biochemistry, and analytical methods.

15.6.1 Soybeans

Epidemiological research has associated high soy intake with a lower risk for certain types of cancer (48,49). Soybean is a rich source of phenolic antioxidants with isoflavonoids being major components. The chief isoflavonoid found in soybean is genistein. Research has shown the chemopreventive properties of purified and synthetic genistein (50–54). However, more recent research has shown that fermented soymilk, developd through bioprocessing, performed better at reducing the incidence of mammary tumor risk than a similar mixture of its constituent isoflavonoids, suggesting that the food background may play a positive role in the chemopreventive actions of soy, in addition to the resident isoflavonoids like genistein (55). As fermented soymilk is rich in phenolic aglycones, which are more active and more readily taken up than their β-glycosides, increasing the free phenolic content of soy-based food through microbial bioprocessing may positively affect its medicinal and nutritional value (56–59). Thus, SSB of soybean substrates by a dietary fungus such as *Rhizopus oligosporus* or *Lentinus edodes* has been developed to mobilize free phenolic antioxidants (60–62). This concept was later extended to mobilization of phenolic antioxidants in soy kefir milk fermentation (63).

When SSB by *Rhizopus oligosporus* of whole soybean was undertaken based on the tempeh bioprocessing model, it was found that while total soluble phenolic content increased 120–135% in the extracts, increased antioxidant activity based on a free radical scavenging assay was limited to the early growth period (60). Higher antioxidant activity was linked to increased glucosidase and glucuronidase activity. Overall results based on phenolic antioxidant and enzyme activities suggested a possible involvement of lignin remobilization and degradation activities, potentially as a part of a detoxification pathway by *Rhiozpus oligosporus* (60).

Phenolic mobilization by SSB using *R. oligosporus* was further investigated, using defatted soybean powder. In this SSB system, phenolic content increased 41% in water extracts and 255% in ethanolic extracts after 10 days of growth (61). In this system beta-glucosidase was the main carbohydrate-cleaving enzyme that is required for efficient phenolic mobilization from defatted powdered soybean (61). This tempeh SSB model was extended to *Lentinus edodes*, which is a slow growing fungus used in shiitake mushroom (fruiting phase of *L. edodes*) production that grows only in the vegetative mycelial phase in soybean (62). It was determined that *L. edodes* SSB was effective in mobilizing phenolic antioxidants from defatted soybean powder (62). Postprocessing phenolic content had increased 90% to 11.3 mg/gram dry weight in water extract and 232 % to 5.8 mg/gram dry weight in ethanolic extracts. Furthermore, phenolic antioxidant mobilization by *L. edodes* was related to the activities of the carbohydrate-cleaving enzymes alpha-amylase and beta-glucosidase, as well as laccase, an enzyme involved in lignin biodegradation (62).

Further, in another study, phenolic mobilization by *L. edodes* was associated with manganese peroxidase and laccase (64). Taken together, these results so far indicate that adapting traditional tempeh fermentation using the same food grade fungus *R. oligosporus,* or exploiting the slower growing vegetative mycelial organism of Shiitake mushrooms, *L. edodes*, phenolic ingredients can be mobilized from soybean whole beans or defatted powders. The functional relevance of these phenolic metabolites were confirmed when it was clear that the phenolic mobilization by *R. oligosporus* on day 2 of SSB growth, that coincided with an increased laccase activity, also had the highest phenolic antioxidant activity and antimicrobial activity against ulcer bacteria *Helicobacter pylori* (65) and the food borne bacterial pathogen *Listeria monocyotogenes* (McCue et al., unpublished results). Results also indicate that phenolic mobilization by *L.edodes* vegetative mycelium enhanced DNA protection using a plasmid model system and had antihypertension potential (McCue et al., unpublished results).

15.6.2 Fava Beans and Other Legumes

Dopamine has several key physiological and biochemical roles in human health and well-being. The role of dopamine as a hormone for urinary and vasodilatory function has been postulated (66). The renal production of dopamine has been shown to be an important factor in the control of blood pressure (67). Further, the importance of dopamine for controlling normal motor functions is evidenced by the abnormality of the basal ganglia in Parkinson's disease, where marked loss of these dopamine linked neurons has been observed (68). A recent interesting article has proposed a general theory that attributes the origins of human intelligence to an expansion of dopaminergic systems in human cognition (69). According to Previc (69), "Dopamine is postulated to be the key neurotransmitter regulating six predominantly left hemispheric cognitive skills to human language and thought such as motor planning, working memory, cognitive flexibility, abstract reasoning, temporal analysis /sequencing and generativity."

L-Tyrosine is considered to be the initial precursor in the biosynthesis of dopamine via L-DOPA (68). Natural L-DOPA is found in significant quantities in fava beans. Studies have shown an antiParkinson effect of fava bean consumption (70–72). There is documentation of a substantial increase in plasma L-DOPA levels following consumption of fava beans, which correlated with substantial improvement in motor performance of patients (70). In a recent clinical case study, fava bean consumption prolonged on periods in patients with Parkinson's disease who had on–off fluctuations (73). This prolonged on period was not observed when higher levels (800–1000 mg) of pure, synthetic L-DOPA was given (73), suggesting that L-DOPA in the fava bean background is significant in the positive clinical benefits observed.

L-DOPA has been identified in seedlings, pods, and beans of the fava bean plant (74). A substantial clinical improvement in 6 patients with Parkinson's disease was observed after ingestion of 250 g of the fava bean fruit that included pods and seeds (70). A separate study has shown that the amount of L-DOPA in seedlings is up to 20 times higher than in seeds (75). Verad et al. (74) suggested that germinated seedlings have other advantages, such as ease of use and fewer side effects (i.e., flatulence, cramping). While a previous study (70) reported that consumption of boiled pods and seeds of the fava bean provided substantial improvement in the debilitating effects of Parkinson's disease, a later study (74) showed that consumption of germinated seedlings was even better, and that high plasma L-DOPA levels consistent with substantial clinical improvements were observed. The study is important, because germinated fava bean sprouts could be grown and used any time of the year and may be particularly useful for lower income populations. According to these studies, germinated sprouts of the fava bean offer an excellent and complimentary dietary approach to the management of Parkinson's disease, along with other prescribed treatments. In addition, we have explored the use of SSB fava bean systems using *Rhizopus oligosporus*, which offers more potential for enhanced functionality from phenolic enrichment by recruiting the complex metabolic interactions of fungal colonization of fava beans.

A fava bean SSB system was adopted from the soybean tempeh model using *Rhizopus oligosprous* (76). The L-DOPA content in the SSB fungal grown fava bean increased significantly, to approximately twice that of the control, accompanied by moderate phenolic linked antioxidant activity and high fungal superoxide dismutase (SOD) activity during the early stages of growth (76). This indicated that L-DOPA can be mobilized and formed from fava bean substrates by fungal SSB and this also contributed to the antioxidant functionality of such extracts. A high SOD activity during early and later growth stages indicates likely oxidation stress of the initial fungal colonization and in late stages is likely due to nutrient depletion, when levels of free soluble phenolics are high (76). High levels of soluble phenolics during late stages correlated to an increase in beta-glucosidase activity, indicating that the enzyme plays a role in the likely release of phenolic alycones and the high antioxidant activity of late stages. Therefore use of SSB for fava bean using *R. oligosporus* can improve antioxidant functionality and Parkinson's relevant L-DOPA content (76).

15.6.3 Cranberry

Pioneering investigation has shown that phenolic glycosides in cranberries can be hydrolyzed by enzyme glycosidases using food grade fungus *Lentinus edodes* during SSB (77). Further, preliminary results have shown that during the SSB there is a strong correlation between high antioxidant activity of the fruit extracts and glycosidase activity (78). Based on these preliminary results, the phenolic antioxidant content and activity is being investigated in fruit extracts and pomace byproducts of the cranberry using *R. oligosporus* and *L. edodes* (78,79). In this investigation, the food grade fungi *L. edodes* and *R. oligosporus* are used to mobilize the release of functional phenolic antioxidants linked to beta-glucosidase during SSB of cranberry fruit pulp extracts and pomace byproducts. This helps enhance the value of cranberries and other fruits as a source of food grade health ingredients that can be used and reconstituted in a wide array of processed food. Specifically, this approach has allowed the development of high ellagic acid enriched extracts, with enhanced antimicrobial activity against food borne pathogens (80,81) and ulcer associated *Helicobacter pylori* (82). This process not only enhances the economic value of fruits and fruit ingredients, it could provide an abundant source of health relevant phenolic antioxidants with nutraceutical value. This value added SSB approach has substantial implications for improving the functional value of food products and could be extended to a wide array of foods and food byproducts. We have expanded on these studies to regulate flavonoids and diphenyls in other important fruits, such as apples, raspberries, strawberries, and grapes.

15.6.4 Enzymes in Phenolic mobilization and Metabolic Biology of PLPPP

Phenolic mobilization from soybean substrates has been linked to the activity of carbohydrate-cleaving enzymes alpha amylase and beta-glucosidase (62), and to lignin degrading enzymes, such as laccase and peroxidase (64). Based on the low phenolics and the high antioxidant activity linked to peak laccase activity in early phase of *R. oligosporus* growth, it has been suggested that polymeric forms of phenolic may have more functional activity. It is likely that the polymeric form is the form with the most potent health benefits, as seen in anti*Helicobacter pylori* inhibition studies (65). These studies are now being extended to inhibition studies in breast cancer cell culture models (McCue et al., unpublished results). Based on carbohydrate mobilization and likely polymeric phenolic forms produced at early stages, it is likely that a growing fungus has to adjust to these conditions metabolically.

In this chapter a model is proposed that as free soluble sugars are enhanced during fungal growth and as polymeric phenolic forms are enhanced, the fungus likely switches to an alternative stress induced pathway. The polymeric phenolic forms may stack on the outer membrane of the fungal mycelia and some degraded phenolics may be have the ability to both enter the cytosol and donate protons to the cytosol through a localized hyperacidification and transport through H+-transporters (Figure 15.1). When these combinations of events occur the fungus switches to an alternative proline linked pentose–phosphate pathway (PLPPP).

By switching to a praline linked oxidative phosphorylation for ATP synthesis (away from a NADH based one that requires full operation of the tricarboxylic acid cycle), the fungus avoids entering a futile cycle, and at the same time supports the pentose phosphate pathway to supply the reductant $NADPH_2$ for anabolic pathways. In this way, $NADPH_2$ can not only meet the needs of proline, which can enter the mitochondria through proline dehydrogenase, but also meet the needs of an antioxidant enzyme response through superoxide dismuatase (SOD) and catalase (CAT). This is also likely linked to the glutathione redox flux (83). This PLPPP critical control point is hypothesized to be important in the adaptation of all fungal systems, including yeast based bioprocessing (both SSB and liquid). It may be especially important under high oxygen conditions, when oxygen pressure coupled to soluble sugar uptake is high. It could also be closely tied to the functional activities of phenolics from substrates on which these microorganisms grow. This concept is being investigated for fungal and yeast growth in the bioprocessing of soybeans, fava beans, cranberries, apples and black tea, and could be extended to any substrates where phenolic mobilization from botanical or other food substrates are involved.

15.7 EXTENSION OF PLPPP STRATEGIES FOR LIQUID YEAST FERMENTATION AND LACTIC ACID BACTERIA

Investigation has confirmed the involvement of lignin-degrading enzymes in phenolic antioxidant mobilization during yogurt production from soymilk by active probiotic kefir cultures (63). Kefir cultures contain a consortium of lactic acid bacteria and a yeast,

Saccharomyces fragilis (63). This approach was based on the rationale that fermented soymilk is rich in phenolic aglycones, which are more active and more readily taken up than their β-glycosides. Therefore, increasing the free phenolic content of soy based food through microbial bioprocessing may positively affect its medicinal and nutritional value (56–59). During the investigation, total soluble phenolic content and free radical scavenging antioxidant activity was measured every 8 h for 48 h. Further, the activity of several enzymes (β-glucosidase, laccase, peroxidase) associated with the microbial degradation of polymeric phenolics and lignin and previously linked to phenolic mobilization from soybean during solid-state bioprocessing by dietary fungi (62,64) were also investigated. Soluble phenolic content increased with kefir culture time and was strongly correlated to total peroxidase and laccase activity. However, phenolic content dropped sharply at 48 h. Antioxidant activity increased with kefir culture time and was strongly correlated to decreased soluble phenolic content over the same time period.

This research has important implications for the optimization of functional phytochemicals in commercial soymilk based yogurts, which can be targeted for disease chemoprevention strategies. This soymilk model can be extended to mobilization of phenolics in other liquid fermentation systems such as those for producing wine, beer, apple cider, buttermilk and any other process involving fermented yeast and acidic fermentative bacteria that utilizes an initial aerobic and subsequent low or no oxygen fermentation systems. In such systems the adjustment to initial oxygen stress and subsequent switch to low oxygen or no oxygen involves adjustment to ethanol or organic acids that are produced on the growing liquid substrates.

In the case of yeast, the fungal PLPPP model (Figure 15.1) described earlier, with modifications based on oxygen stress, may be involved. In the case of acid-producing fermentative bacterial systems which are prokaryotes, the fundamental energy metabolism and redox management takes place between the outer plasma membrane (like an independent mitochondria) and cytosol. It is conceivable that the acid-producing bacteria have evolved a high proton management cytosolic dehydrogenase system that can take the inward proton flux and protons generated from new lactic acid production under reduced oxygen and couple it to plasma membrane driven oxidative phosphorylations or substrate level phosphorylations linked to organic acid. It would make sense for this dehydrogenase linked energy and reductant management to occur through the same proline linked pentose–phosphate pathway (PLPPP), with the adjustment within the simpler prokaryotic coupling between the plasma membrane and cytosol (Figure 15.2). The ultimate efficiency and need for such a control point is the same as for mitochondrial linked systems in yeast or filamentous fungi. In these systems, energy metabolism through proline and reductant needs for anabolism through $NADPH_2$ from the pentose phosphate pathway can be provided during the initial crucial switch from high oxygen conditions to low and no oxygen conditions, when substrate level phosphorylation linked to alcohol or organic acids predominate.

It is also conceivable that organic acid-producing prokaryotes are more versatile in this redox switch, and therefore could have evolved a coupled mechanism to better survive high oxygen, high stress, and eventually high organic acids under much reduced oxygen. This rationale could be used in developing antimicrobial systems against pathogenic bacteria, where combinations of exogenous acids and hyperacidifying, proton-donating phenolic metabolites could be recruited to inhibit or kill pathogenic bacteria (83). Using the same logic, organic acid and phytochemical enriched SSB and liquid fermentation systems could be effective ways to design antimicrobials against pathogenic bacteria for improving food safety, as seen in many traditional fermented foods. In such a situation critical points at the level of energy and reductant management between the plasma membrane and cytosol become important, whether involving our hypothetical PLPPP or an alternative model.

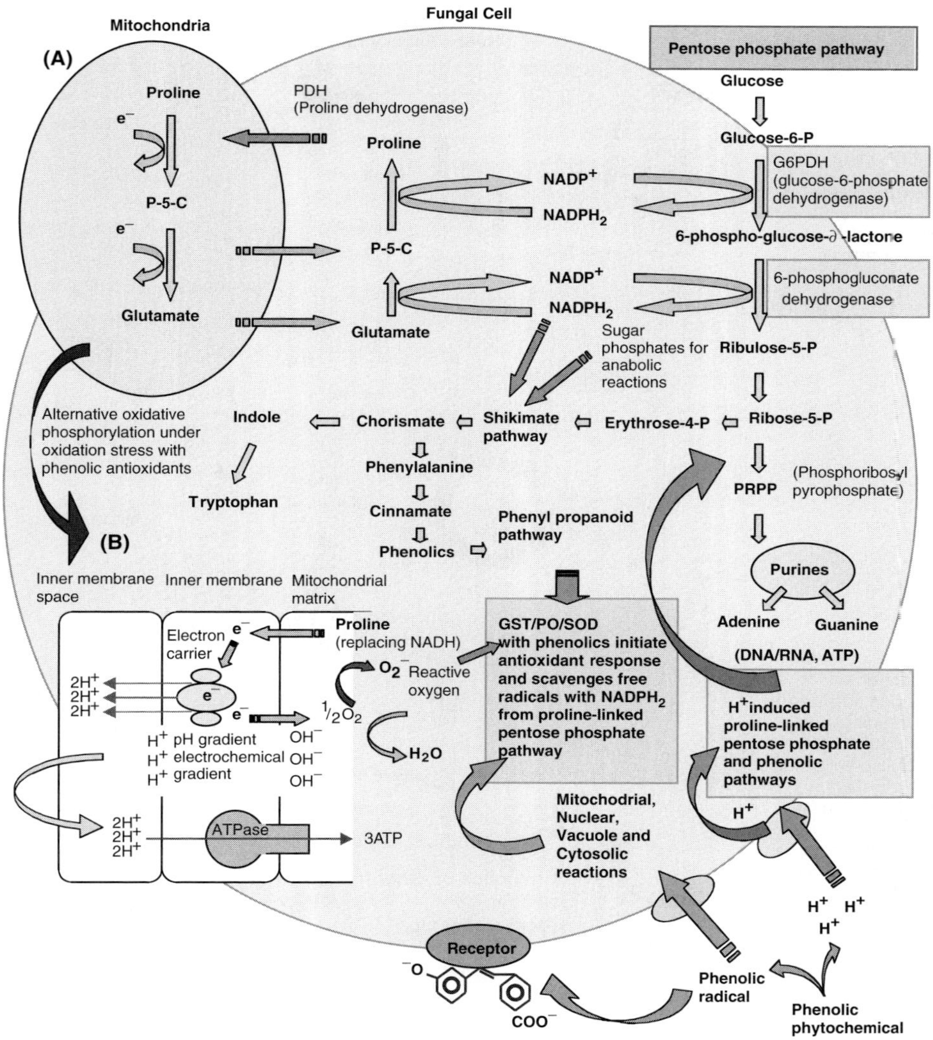

Figure 15.1 Extension of plant proline linked pentose–phosphate pathway model for the effect of external phenolic phytochemicals in yeast and fungal systems during solid-state bioprocessing under aerobic exposure when oxidation stress is maximum and sugar uptake in rapid. (Abbreviations: P5C; pyrroline-5-carboxylate, GST;Glutathione-s-transferase, PO;peroxidase, SOD;superoxide dismutase)

15.8 CONCLUDING PERSPECTIVES

More than 2000 years of microbial based bioprocessing (aerobic and fermentative–anaerobic) of foods and beverages in diverse food cultures around the world has provided us with many complex food substrate and microbial systems for developing functional foods and ingredients. In addition, value added food waste remediation systems have been developed. These food substrates and microbial systems can be further developed for human health benefits and environmental management using the Systems Biology approach of modern biology. These complex systems involve complex consortia of microbial ecosystems and metabolic biology acting on multiple food substrates. To understand these complex systems,

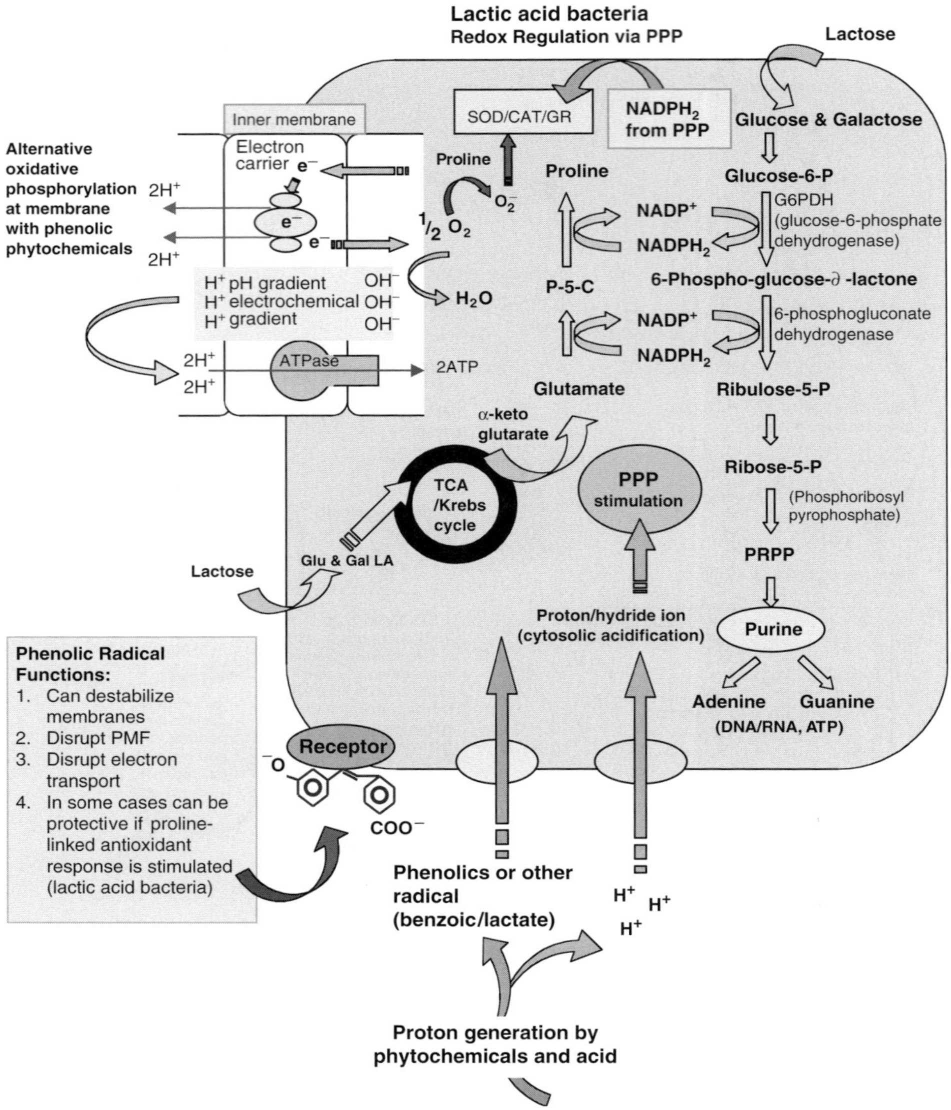

Figure 15.2 Extension of plant proline linked pentose–phosphate pathway model for the effect of external phenolic phytochemicals in prokaryotic organic acid producing bacterial systems, where released phenolics from fermenting substrates could contribute to a redox response and tolerance to acidity, unlike nonacid tolerant bacteria, which are inhibited by phenolics (Abbreviations: P5C;pyrroline-5-carboxylate, SOD;superoxide dismutase, CAT;catalase, GR;glutathione reductase, PPP;pentose phosphate pathway; Krebs {TCA-tricarboxylic acid} cycle), Glu; Glucose, Gal; Galactose, LA;lactic acid

a more integrated Systems Biology approach that first requires understanding of critical control points in energy and redox management across individual organisms to consortia is essential. An understanding of ecoevolutionary relationships of energy and redox metabolic biology across prokaryotes to eukaryotes from oxygen rich to low oxygen and to fermentative environments is also required. This energy- and redox-linked metabolic biology based platform can help us integrate exciting information from the genetic, signal, and enzyme–protein fields. Such an approach gives us a better understanding of the critical control points that define the functioning of organisms and ecosystems. The understanding of such control points can be utilized to better design functional food and ingredients using SSB systems.

Among many possible critical energy and redox-modulating pathways, we have focused on the relevance of the proline linked pentose–phosphate pathway (PLPPP) in SSB systems. Under the diverse stresses of high or low oxygen environments, organisms likely use PLPPP as an alternative pathway, with proline as a reductant, for oxidative phosphorylation at outer plasma membrane in prokaryotes and mitochondria in eukaryotic yeast and fungi. At the same time the pentose–phosphate pathway is utilized for redox management in the cytosol, to drive reductants such as NADPH2 for anabolism, including antioxidant enzyme response pathways involving superoxide dismutase and catalase. The phenolics mobilized from food substrates will, in many ways, define the behavior of the microbial energy and redox responses. The metabolic and cellular biology of these interrelationships can be harnessed to maximize enhancement and optimization of functional foods and ingredients.

REFERENCES

1. Zheng, Z., K. Shetty. Solid-state fermentation and value-added utilization of fruit and vegetable processing by-products. In: *Wiley Encyclopedia of Food Science and Technology,* 2nd ed., F.J. Francis, ed., New York: Wiley Interscience, 1999, pp 2165–2174.
2. Cannel, E., M. Moo-Young. Solid-state fermentation systems. *Process. Biochem.* 4:2–7, 1980.
3. Mudgett, R.E. Solid-state fermentations. In: *Manual of Industrial Microbiology and Biotechnology,* Demain, A.L., N.A. Solomon, eds., Washington, D.C.: American Society of Microbiology, 1986, pp 66–83.
4. Aidoo, K.E., R. Hendry, B.J.B Wood. Solid substrate fermentation. *Adv. Appl. Microbiol.,* 28:201–237, 1982.
5. Pandey, A. Recent progress in developments in solid-state fermentation. *Process. Biochem.* 27:109–117, 1992.
6. Logsdon, G. Pomace is a grape resource. *Biocyle* 33:40–41, 1992.
7. Hang, Y.D. Improvement of the nutritional value of apple pomace by fermentation. *Nutr. Rep. Int.* 38:207–211, 1988.
8. Rahmat, H., R.A. Hodge, G.J. Manderson. Solid-substrate fermentation of *Kloeckera apiculate* and *Candida utilis* on apple pomace to produce an improved stock-feed. *World J. Microbiol. Biotechnol.* 11:168–170, 1995.
9. Menezes, T.J.B. Protein enrichment in citrus wastes by solid-state fermentation. *Process. Biochem.* 24:167–171, 1989.
10. Chung, S.L., S.P. Meyer. Bioprotein from banana wastes. *Dev. Indust. Microbiol.* 20:723–731, 1979.
11. Enwefa, O. Biomass production from banana skins. *Appl. Microbiol. Biotechnol.* 36:283–284, 1991.
12. Fellows, P.J., J.T. Worgan. Growth of *Saccharromycopsis fibuliger* and *Candida utilis* in mixed culture in apple processing wastes. *Enzyme Microbiol. Technol.* 9:434–437, 1987.
13. Moyson, E., H. Verachtert. Growth of higher fungi on wheat straw and their impact on the digestability of the substrate. *Appl. Microbiol. Biotechnol.* 36:421–424, 1991.
14. Upadhyay, R.C., H.S. Sidhi. Apple pomace: a good substrate for the cultivation of edible mishrooms. *Curr. Sci.* 57:1189–1190, 1988.
15. Worrall, J.J., C.S. Yang. Shiitake and oyster mushroom production on apple pomace and sawdust. *Hort. Sci.* 27:1131–1133, 1992.
16. Hang, Y.D. Production of alcohol from apple pomace. *Appl. Environ. Microbiol.* 42:1128–1129, 1981.

17. Sandhu, D.K., V.K. Joshi. Solid-state fermentation of apple pomace for concomitant production of ethanol and animal feed. *J. Sci. Ind. Res. India* 56:86–90, 1997.
18. Saha, B.C., B.S. Dien, R.J. Bothast. Fuel ethanol production from corn fiber: current status and technical prospect. *Appl. Biochem. Biotechnol.* 70:115–125, 1998.
19. Roukas, T. Solid-state fermentation of carob pods for ethanol production. *Appl. Microbiol. Biotechnol.* 41:296–301, 1994.
20. Jewell, W.J., R.J. Cummings. Apple pomace energy and solids recovery. *J. Food Sci.* 48:407–410, 1984.
21. Knol, W., M.M. Van der Most, J. De Waart. Biogas production by anaerobic digestion of fruit and vegetable wastes. *J. Food Sci. Agric.* 29:822–827, 1978.
22. Hang, Y.D., E.E. Woodams. Apple pomace: a potential substrate for citric acid production by *Aspergillus niger. Biotechnol. Lett.* 6:763–764, 1984.
23. Tran, C.T., D.A. Mitchell. Pineapple waste: a novel substrate for citric acid production by solid-state fermentation. *Biotechnol. Lett.* 17:1107–1110, 1995.
24. Hang, Y.D., E.E. Woodams. Production of citric acid from corncobs by *Aspergillus niger. Bioresource Technol.* 64:251–253, 1998.
25. Roukas, T. Citric acid production from carob pod by solid-state fermentation. *Enzyme Microbial. Technol.* 24:54–59, 1999.
26. Voget, C.E., C.F. Mognone, R.J. Ertola. Butanol production from apple pomace. *Biotechnol. Lett.* 7:43–46, 1985.
27. Bramorski, A. Production of volatile compounds by edible *Rhizopus oryzae* during solid-state cultivation of tropical agro-industrial substrates. *Biotechnol. Lett.* 20:359–362, 1998.
28. Hours, R.A., C.E. Voget, R.J. Ertola. Apple pomace as raw material for pectinase production in solid-state fermentation. *Biol. Wastes* 23:221–228, 1988.
29. Rombouts, F.M., W. Pilnik. Enzymes in fruit and vegetable juice technology. *Process. Biochem.* 13:9–13, 1978.
30. Krishna, C., M. Chandrasekaran. Banana waste as substrate for alpha-amylase production by *Bacillus subtilis* (CBTK 106) under solid-state fermentation. *Appl. Microbiol. Biotechnol.* 46:106–111, 1996.
31. Reed, G. Production of baker's yeast. In: *Prescott and Dunn's Industrial Microbiology*, 4th ed., Reed, G. ed., Westport, CT: AVI, 1982, pp 593–633.
32. Pederson, C.S. *Microbiology of Food Fermentations,* 2nd ed., Westport, CT: AVI, 1979.
33. Selgas, M.D. Potential technological interest of *Mucor* strain to be used in dry fermented sausage production. *Food Res. Int.* 28:77–82, 1995.
34. Lucke, F.K. Fermented meat products. *Food Res. Int.* 27:299–307, 1994.
35. Wood, B.J. Technology transfer and indigenous fermented foods. *Food Res. Int.* 27:269–280, 1994.
36. Campbell-Platt, G. Fermented foods: a world perspective. *Food Res. Int.* 27:252–257, 1994.
37. Keuth, S., B. Bisping. Vitamin B12 production by *Citrobacter freundii* or *Klebsiella pneumoniae* during tempeh fermentation and proof of enterotoxin absence by PCR. *Appl. Environ. Microbiol.* 60:1495–1499, 1994.
38. Han, J.R. Sclerotia growth and carotenoid production of *Penicillium* sp. PT95 during solid-state fermentation of corn meal. *Biotechnol. Lett.* 20:1063–1065, 1998.
39. Steinkraus, K.H. Nutritional significance of fermented foods. *Food Res. Int.* 27:259–267, 1994.
40. Van-Veen, A.G., D.C.W. Graham, K.H. Steinkraus. Fermented peanut presscake. *Cereal Sci. Today* 13:96–99, 1968.
41. Reddy, N.R. Reduction of antinutritional and toxic compounds in plant foods by fermentation. *Food Res. Int.* 27:281–290, 1994.
42. Zheng, Z., K. Shetty. Cranberry processing waste for solid-state fungal inoculant production. *Process. Biochem.* 33:323–329, 1998.
43. Zheng, Z., K. Shetty. Solid state production of beneficial fungi on apple processing waste using glucosamine as the indicator of growth. *J. Agric. Food Chem.* 46:783–787, 1998.
44. Zheng, Z., K. Shetty. Effect of apple pomace-based *Trichoderma* inoculants on seedling vigor in pea (*Pisum sativum*) germinated in potting soil. *Process. Biochem.* 34:731–735, 1999.

45. Zheng. Z., K. Shetty. Enhancement of pea (*Pisum sativum*) seedling vigor and associated phenolic content by extracts of apple pomace fermented with *Trichoderma* spp. *Process. Biochem.* 36:79–84, 2000.
46. Hood, L. Systems biology: integrating technology, biology and computation. *Mech. Ageing Dev.* 124:9–16, 2003.
47. Kitano, H. Systems biology: a brief overview. *Sci.* 295:1662–1664, 2002.
48. McCue, P., K. Shetty. Health benefits of soy isoflavonoids and strategies for enhancement: a review. *Crit. Rev. Food Sci. Nutr.* 44:1–7, 2004.
49. Dai, Q., A.A. Franke, F. Jin, X.O. Shu, J.R. Hebert, L.J. Custer, J. Cheng, Y.T. Gao, W. Zheng. Urinary excretion of phytoestrogens and risk of breast cancer among Chinese women in Shanghai. *Cancer Epid. Biomarkers Prev.* 11:815–821, 2002.
50. Darbon, J.M., M. Penary, N. Escalas, F. Casagrande, F. Goubin-Gramatica, C. Baudouin, B. Ducommun. Distinct Chk2 activation pathways are triggered by genistein and DNA-damaging agents in human melanoma cells. *J. Biol. Chem.* 275:15363–15369, 2000.
51. Lamartiniere, CA. Protection against breast cancer with genistein: a component of soy. *Am. J. Clin. Nutr.* 71:1705S–1707S, 2000.
52. Xu, J., G. Loo. Differential effects of genistein on molecular markers related to apoptosis in two phenotypically dissimilar breast cancer cell lines. *J. Cell. Biochem.* 82:78–88, 2001.
53. Lamartiniere, C.A., M.S. Cotroneo, W.A. Fritz, J. Wang, R. Mentor-Marcel, E. Elgavish. Genistein chemoprevention: timing and mechanisms of action in murine mammary and prostate. J. Nutr. 132:552S–558S, 2002.
54. Tanos, V., A. Brzezinski, O. Drize, N. Strauss, T. Peretz. Synergistic inhibitory effects of genistein and tamoxifen on human dysplastic and malignant epithelial breast cancer cells *in vitro*. *Eur. J. Obstet. Gyn. Reprod. Biol.* 102:188–194, 2002.
55. Ohta, T., S. Nakatsugi, K. Watanabe, T. Kawamori, F. Ishikawa, M. Morotomi, S. Sugie, T. Toda, T. Sugimura, K. Wakabayashi. Inhibitory effects of *Bifidobacterium*-fermented soy milk on 2-amino-1-methyl-6-phenylimidazo[4,5-*b*]pyridine-induced rat mammary carcinogenesis, with a partial contribution of its component isoflavones. *Carcinogenesis* 25:937–941, 2000.
56. Izumi, T., M.K. Piskula, S. Osawa, A. Obata, K. Tobe, M. Saito, S. Kataoka, Y. Kubota, M. Kikuchi,. Soy isoflavone aglycones are absorbed faster and in higher amounts than their glucosides in humans. *J. Nutr.* 130:1695–1699, 2000.
57. Rao, M., G. Muralikrishna. Evaluation of the antioxidant properties of free and bound phenolic acids from native and malted finger millet (Ragi, Eleusine coracana Indaf-15). *J. Agric. Food Chem.* 50:889–892, 2002.
58. Setchell, K., N. Brown, L. Zimmer-Nechemias, W. Brashear, B. Wolfe, A. Kirschner, J. Heubi. Evidence for lack of absorption of soy isoflavone glycosides in humans, supporting the crucial role of intestinal metabolism for bioavailability. *Am. J. Clin. Nutr.* 76:447–453, 2002.
59. Yuan, L., C. Wagatsuma, M. Yoshida, T. Miura, T. Mukoda, H. Fujii, B. Sun, J.H. Kim, Y.J. Surh. Inhibition of human breast cancer growth by GCP™ (genistein combined polysaccharide) in xenogenetic athymic mice: involvement of genistein biotransformation by α-glucuronidase from tumor tissues. *Mutat. Res.* 523–524:55–62, 2003.
60. McCue, P., K. Shetty. Role of carbohydrate-cleaving enzymes in phenolic antioxidant mobilization from whole soybean fermented with *Rhizopus oligosporus*. *Food Biotechnol.* 17:27–37, 2003.
61. McCue, P, A. Horii, K. Shetty. Solid-state bioconversion of phenolic antioxidants from defatted soybean powders by *Rhizopus oligosporus*: role of carbohydrate cleaving enzymes. *J. Food Biochem.* 27:501–514.
62. McCue, P, A. Horii, K. Shetty, Mobilization of phenolic antioxidants from defatted soybean powders by *Lentinus edodes* during solid-state bioprocessing is associated with enhanced production of laccase. *Innovative Food Sci. Emerging Technol.* 2004. (In press)
63. McCue, P., K. Shetty. Phenolic antioxidant mobilization during yogurt production from soymilk. *Process. Biochem.* 40:1791–1797, 2005.

64. McCue, P., K. Shetty. A model for the involvement of lignin degradation enzymes in phenolic antioxidant mobilization from whole soybean during solid-state bioprocessing by *Lentinus edodes*. *Process. Biochem.* 40:1143–1150, 2005.
65. McCue P., Y.-T. Lin, R.G. Labbe, K. Shetty. Sprouting and solid-state bioprocessing by *Rhizopus oligosporus* increase the *in vitro* antibacterial activity of aqueous soybean extracts against *Helicobacter pylori*. *Food Biotechnol.* 18:229–249, 2004.
66. Vered, Y., I. Grosskopf, D. Palevitch, A. Harsat, G. Charach, M.S. Weintraub, E. Graff. The influence of *Vicia faba* (broad bean) seedlings on urinary sodium excretion. *Plant. Med.* 63:237–240, 1997.
67. Missale, C., S.R. Nash, S.W. Brown, M. Jaber, M.G. Caron. Dopamine Receptors: From Structure to Function. Physiological Reviews, 78:189–225, 1998.
68. Elsworth, J.D., R.H. Roth. Dopamine synthesis, uptake, metabolism, and receptors: relevance to gene therapy of Parkinson's disease. *Exper. Neurol.* 144:4–9, 1997.
69. Previc, F.H. Dopamine and the origins of human intelligence. Brain Cognit. 41:299–350, 1999.
70. Rabey, J.M., Y. Vered, H. Shabtai, E. Graff, A.D. Korczyn. Improvement of Parkinson's features correlate with high plasma levodopa values after broad bean (*Vicia faba*) consumption. *J. Neurol. Neurosurg. Psych.* 55:725–727, 1992.
71. Kempster, P.A., Z. Bogetic, J.W. Secombe, H.D. Martin, N.D.H. Balazss, M.L. Wahlqvist. Motor effects of broad beans (*Vicia faba*) in Parkinson's disease: single dose studies. *Asia Pac. J. Clin. Nutr.* 2:85–89, 1993.
72. Spengos, M., D. Vassilopoulos. Improvement of Parkinson's diseases after *Vicia faba* consumption. In: *Book of Abstracts, 9th International Symposium on Parkinson's Disease, 1988, p 46.*
73. Apaydin, H., S. Ertan, S. Ozekmekci. Broad bean (Vicia faba): a natural source of L-DOPA: prolongs "on" periods in patients with Parkinson's disease who have "on-off" fluctuations. *Movement Disorders* 15:164–166, 2000.
74. Vered, Y., J.M. Rabey, D. Palevitch, I. Grosskopf, A. Harsat, A. Yanowski, H. Shabtai, E. Graff. Bioavailability of levodopa after consumption of *Vicia faba* seedlings by Parkinsonian patients and control subjects. *Clin. Neuropharmacol.* 17:138–146, 1994.
75. Wong, K.P., B. Geklim. L-DOPA in the seedlings of *Vicia faba*: its identification, quantification and metabolism. *Biogen. Amines* 8:167–173, 1992.
76. Randhir, R., D. Vattem, K. Shetty. Solid-state bioconversion of fava bean by *Rhizopus oligosporus* for enrichment of phenolic antioxidants and L-DOPA. *Innovative Food Sci. Emerging Technol.* 5:235–244, 2004.
77. Zheng, Z., K. Shetty. Solid-state bioconversion of phenolics from cranberry pomace and role of *Lentinus edodes* beta-glucosidase. *J. Agric. Food Chem.* 48:895–900, 2000.
78. Vattem, D.A., K. Shetty. Solid-state production of phenolic antioxidants from cranberry pomace by *Rhizopus oligosporus*. *Food Biotechnol.* 16:189–210, 2002.
79. Vattem, D.A., K. Shetty. Ellagic acid production and phenolic antioxidant activity in cranberry pomace mediated by *Lentinus edodes* using solid-state system. *Process. Biochem.* 39:367–379, 2003.
80. Vattem, D.A., Y.T. Lin, R.G. Labbe, K. Shetty. Phenolic antioxidant mobilization in cranberry pomace by solid-state bioprocessing using food grade fungus *Lentinus edodes* and effect on antimicrobial activity against select food-borne pathogens. *Innovative Food Sci. Emerging Technol.* 5:81–91, 2004.
81. Vattem, D.A., Y.T. Lin, R.G. Labbe, K. Shetty. Antimicrobial activity against select food-borne pathogens by phenolic antioxidants enriched cranberry pomace by solid-state bioprocessing using food-grade fungus *Rhizopus oligosporus*. *Process. Biochem.* 39:1939–1946, 2004.
82. Vattem, D.A., Y.T. Lin, K. Shetty. Enrichment of phenolic antioxidants and anti-*Helicobacter pylori* properties of cranberry pomace by solid-state bioprocessing. *Food Biotechnol.* 19:51–68, 2005.
83. Shetty, K., M.L. Wahlqvist. A model for the role of proline-linked pentose phosphate pathway in phenolic phytochemical biosynthesis and mechanism of action for human health and environmental applications: a review. *Asia Pac. J. Clin. Nutr.* 13:1–24, 2004.

3.16

Fermentation Biotechnology of Traditional Foods of Africa

N. A. Olasupo

CONTENTS

16.1 INTRODUCTION

African countries require food processing technologies that will meet the challenges of the peculiar food security problem in the continent (1). Such technologies however, should be low cost to be affordable by the poor majority of the continent and should be the type that can address the perennial problems of food spoilage and food borne diseases, which are prevalent in the environment. Fermentation is one important food processing technology that meets these challenges. Fermentation is not new to the African countries, as most of the traditional foods in the community are fermented before consumption (2). The process of fermentation has been used in the preparation of many traditional foods in the countries of Africa for centuries. Fermentation as a means of improving the keeping quality of food has been used for more than 6000 years (3). Probably no other process has had such an impact on the nutrition habits and food culture of mankind. Fermentation performs a number of functions which assist the overall improvement of food quality and safety. First and foremost however, it serves to improve the shelf life and safety of foods and enables people in the moderate and cooler regions to survive winter seasons and drought periods. Moreover, it can be assumed that, because safe drinking water supplies were not always available in densely populated areas, fermented beverages with an extended shelf life, such as beer and wine (4), served as safe sources for consumption.

In addition to the preservation effect, fermentation serves as a means of improving sensory quality and acceptability of many raw materials to such an extent that many foods are preferred in a fermented state (e.g., iru, rather than the unfermented African locust bean; gari, rather than unprocessed cassava). Further attributes obtained from fermentation include reduction of toxic food components (e.g., linamarin in cassava and undesired antinutritional factors in legumes). Fermentation adds value and enhances nutritional quality and digestibility through nutritional enhancement. Furthermore, it provides dietary enrichment through aroma and flavor production. The production of fermented foods in Africa is largely home based and it is left to chance inoculation from the environment. The processes are not standardized, with little or no control involved in the processing. The fermentation vessels, utensils, and materials used for preparation or part of previous ferments usually serve as the source of inoculum for the initiation of the fermentation. The use of starter cultures is rare in the production of traditional foods in Africa but old stocks of previous ferments are used in some cases to initiate fermentation in new batches (a process called back slopping) (3). Because of the methods of preparation, indigenous fermented foods in Africa are faced with problems ranging from poor shelf life, doubtful hygiene status, and inconsistent quality to safety risk.

16.2 FERMENTED FOODS IN AFRICA

Many African foods are fermented before consumption. Over 50 different African fermented foods have been recorded. These foods can be classified into five groups (2):

1. Starchy root crops (fermentation)
2. Nonalcoholic cereal substrate (fermentation)

3. Alcoholic beverages
4. Fermented vegetable proteins
5. Fermented animal proteins

Detailed lists of the foods are shown in Table 16.1 to 16.5. The microorganisms involved in African food fermentation are restricted to a few groups of yeasts and bacteria.

Table 16.1
Fermented starchy root products (2,5,6)

Product	Area of Production	Substrate	Microorganisms Involved
Gari	West Africa	Cassava	*Streptococcus lactis* *Geotrichum candidum* *Corynebacterium manihot* Lactic acid bacteria
Lafun	Nigeria	Cassava	Yeast and lactic acid bacteria
Fufu	Nigeria	Cassava	Lactic acid bacteria
Chikawngue	Zaire	Cassava	Lactic acid bacteria and yeast
Cingwada	East and Central Africa	Cassava	Not known
Kocho	Ethiopia	Ensette (*Ensette Ventricosum*)	Lactic acid bacteria and yeast
Kivunde	Tanzania	Cassava	Lactic acid bacteria and yeasts
Agbelima	Ghana	Cassava	*Lactobacillus plantarum* *Bacillus sp.* *Candida tropicalis* *Geotrichum candidum* *Penicillium sp.*

Table 16.2
Fermented nonalcoholic cereal based foods (2,7,8)

Product	Area of Production	Substrate	Microorganisms Involved
Ogi	Nigeria-Benin	Maize, Sorghum, or Millet	*Lactobacillus* sp. and Yeast
Koko and Kenkey	Ghana	Maize, Sorghum, or Millet	*Lactobacillus* sp. and Yeast
Mahewu (Magou)	South Africa	Maize, Sorghum, or Millet	*Lactobacillus delbrueckii*, *L. bulgaris*
Uji	East Africa	Maize, Sorghum, or Millet	*Lactobacillus* sp.
Kisra	Sudan	Sorghum	Lactic acid bacteria
Injera	Ethiopia	Sorghum	*Candida guillermondii*
Ting	Botswana	Sorghum	Lactic acid bacteria
Obusera	Uganda	Millet	Lactic acid bacteria
Mawe	Benin	Maize	*Lactococcus lactis* *Lactococcus lactis*, *Pediococcus pentosaceus* *Lactobacillus plantarum*
Bogobe	Botswana	Sorghum	Unknown
Kunu-zaki	Nigeria	Millet, Sorghum	Lactic acid bacteria

Table 16.3
Fermented alcoholic beverages in Africa (2)

Product	Area of Production	Substrate	Microorganisms Involved
(1) From Sugary Substrates			
Palmwine	Nigeria (south)	Palm sap	Yeasts
Tej	Ghana	Honey	Yeasts
(2) From Starchy Substrates			
Pito	Nigeria Ghana	Guinea corn And maize	Molds, yeast and *Lactobacillus* sp.
Kaffir beer	South Africa	Kaffir corn or maize	*Lactobacillus* sp. and yeasts
Busaa (maize beer)	East Africa (Kenya)	Maize	Yeast and *Lactobacillus* sp.
Malawa beer	Uganda	Maize	*Candida Krusei*
Zambia opaque maize beer	Zambia	Maize	Yeasts
Merissa	Sudan	Sorghum	Lactic acid bacteria Acetic acid bacteria
Sheketch	Nigeria (south)	Maize	Unknown
Bouza	Egypt	Wheat (or Maize)	Unknown
Talla	Ethiopia	Sorghum	Unknown
Kishk	Egypt	Wheat or milk	*Lactobacillus* sp. yeasts and *Bacillus* sp.
(3) From Other Substrates			
Agadagidi (Plantain wine)	Southwestern Nigera	Plantain	Yeasts and Lactic acid bacteria
Cacao wine	Nigeria	Cacao	Yeasts
Alcoholic spirit	Kenya, Nigeria	Molasses or Cane sugar	Yeasts
Chang aa (Nubian gin)	Nigeria, Ghana	Palm wine	Yeasts
Ogogoro (Alpeteshi)	Nigeria, Ghana	Palm wine	Yeasts

Lactic acid bacteria (LAB) particularly *Lactobacillus* are involved in the fermentation of many African starchy foods. (Fermentation of foods involving lactic acid bacteria is very common in Africa.) Table 16.6 further corroborates the role of LAB in food fermentation.

16.2.1 Production Procedures of Foods

In Africa, the raw materials used for the production of indigenous fermented foods include cereals, root crops, (especially cassava) milk, and vegetables. However cereal and cassava based foods are usually large in number compared to food products from other raw materials (1). In Africa, fermentation of food is carried out through traditional, village art methods (2). Processing usually involves either soaking of raw materials in water contained in a fermenting vat, usually clay pots, for a period of time or an initial size reduction of the raw material by grating or milling in the wet form before being allowed to ferment. These two fermentation approaches have been described by Oyewole and Odunfa (9) as traditional submerged and solid-state fermentation methods.

Table 16.4
Fermented vegetable proteins in West Africa (2)

Product	Area of Production	Substrate	Microorganisms Involved
Iru	Southwestern Nigeria	Locust bean (*Parkia biglobosa*), Soybean (*Glycine max.*)	*Bacillus subtilis* *B. licheniformis*
Ogiri	Southwestern Nigeria	Melon seed (*Citrullus vulgaris*)	*Bacillus sp. (predomincnt)* *Proteus, Pediococcus*
Ogiri-nwan	Southeastern Nigeria	Fluted pumpkin bean (*Telfairia occidentalis*)	*Bacillus sp. (proteolytic)*
Ogiri-Igbo (same as ogiri-agbor)	Southern Nigeria	Castor oil seed (*Ricinus Communis*)	Various Bacillus sp; *B. subtilis, B. megaterium* *B. firmus*
Ogiri-Saro	Sierra Leone	Sesame seed (*Sesamum indicum*)	*Basillus* sp.
Ugba/Apara	Eastern Nigera (Igo); Southwestern Nigeria (Egabados)	Oil bean (*Pentaclethra macrophylla*)	*Bacillus subtilis* *Micrococcus* sp.
Kawai	Sudan	*Cassia obtusifolia*	*B. subtilis*

Table 16.5
Fermented animal proteins of Africa (2,8)

Product	Area of Production	Substrate	Microorganisms Involved
Maziwa Lala	East Africa	Milk	*Streptococcus lactis* *S. thermophilus*
Nono (Milk curd)	Northern part of West Africa	Milk	Lactic acid bacteria
Guedi	Senegal	Fish	Not known
Bonone (Stink fish)	Ghana	Fish	Not known
Leban (Sour milk)	Morocco	Milk	*Lactic streptococci* *Leuconostoc lactis* *L. cremoris*
Wara	West Africa	Milk	*Lactococcus lactis* *Lactobacillus* sp
Ergo	Ethiopia	Milk	*Lactobacillus* sp *Lactococcus sp*

There are numerous reports (2,10) on the methods of preparation of African fermented foods. Figures 16.1 to 16.6 show the flow charts for the preparation of some very common fermented foods in the continent of Africa.

16.2.2 Problems Associated with African Fermented Food

Because the production of fermented foods in Africa is usually a traditional family art, some problems are associated with the food products. The problems range from microbiology in processing, production environment, process control and nutritional to toxicological status (11).

Table 16.6
Food fermentations that use Lactic Acid Bacteria

Product	Typical LAB Involved
Cheeses and Fermented Milks	*Lactococcus lactis* subsp. *lactis*
	L. lactis subsp. *cremoris*
	L. lactis subsp. *lactis* var. *diacetylactis*
	Streptococcus thermophilus
	Lactobacillus helveticus
	Lb. acidophilus
	Lb. casei
	Lb. delbrueckii subsp. *bulgaricus*
	Leuconostoc cremoris
Meat Fermentations	*Lb. plantarum*
	Lb. pentosus
	Lb. sake
	Lb. curvatus
	Pediococcus pentosaceus
	P. acidilactici
Fermented Vegetables	*Lb. plantarum*
	Lb. bavaricus
	Lb. casei
	P. pentosaceus
	Leuconostoc mesenteroides
Wine	*Lc. oenos*
	Pediococcus
	Lb. delbrueckii
Soy Sauce	*Lb. delbrueckii*
	P. soyae
Probiotic Preparations	*Lb. acidophilus*
	Lb. casei
	Bifidobacterium bifidum
Bakery Products	
Soda Crackers	*Lb. plantarum*
	Lb. casei
	Lb. brevis
Sourdough	*Lb. sanfrancisco*
	Lb. brevis
	Lb. platarum
	Lb. fermentum
Indigenous Fermented Foods	Various LAB

16.2.2.1 Microbiology of Process

Much of the information on the microbiology of the process of production is faulty. The following factors make it difficult to assure a consistently contamination free environment for fermentation:

1. Unknown microbial composition of the inocula
2. Identification of type
3. Identity of the contaminating microbes

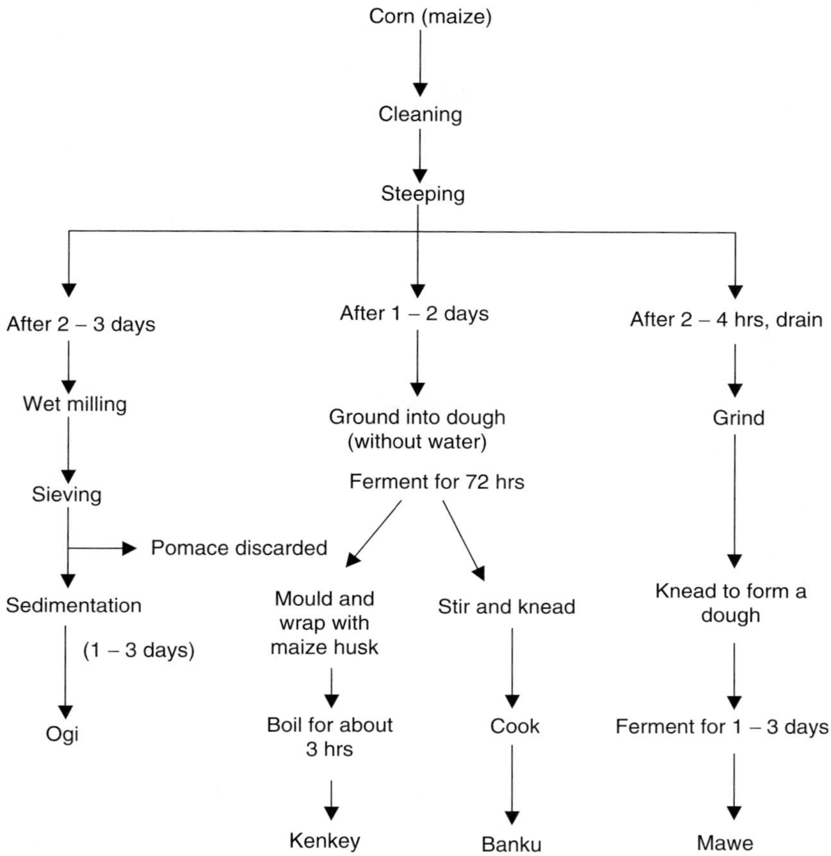

Figure 16.1 Flow chart for the preparation of Ogi, Kenkey, Banku, and Mawe

4. Age and purity of each culture
5. The predominant organisms
6. The conditions of pH
7. Temperature
8. Ionic strength
9. Information on food enhancing microbial performance in fermentation

16.2.2.2 Production Environment

Since the production of the foods is basically a traditional household practice, the processing environment is highly unpredictable because the equipment used is rudimentary (e.g., leaves, clay pot, calabash, baskets, earthenware, and cloth). The hygienic status of the handlers, equipment, and materials for preparation (especially water) is usually not checked. A tropical climate (especially the temperature and humidity) cannot be said to be optimum for fermentation and storage purpose.

16.2.2.3 Process Control

The practice of good process control is difficult to maintain because the processing procedures do vary from one ethnic group to another. Fermentation periods are sometimes

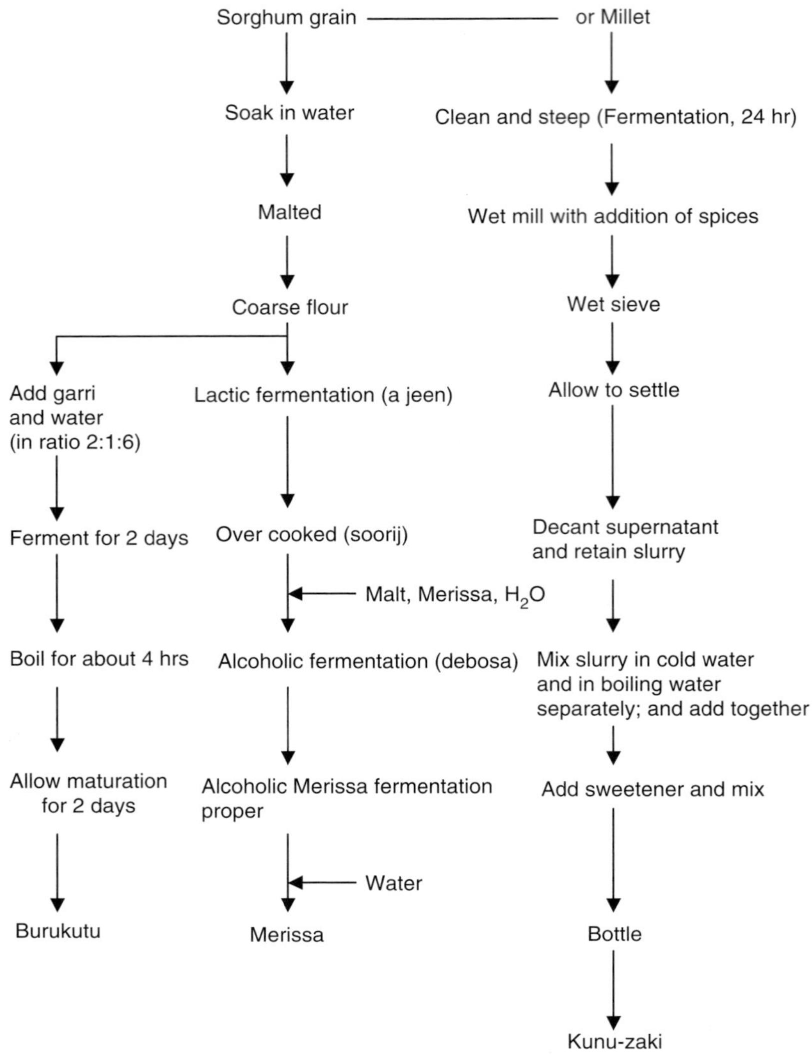

Figure 16.2 Flow chart for the preparation of Burukutu, Merissa, and Kunu-Zaki

chosen according to human judgement. The quality and quantity of water and substrate used are not usually regulated or standardized. Heat processes are not controlled or measured. The effect of all these factors is a finished product with inconsistent quality.

16.2.2.4 Nutritional and Toxicological Status

The dearth of information on the effect of fermentation on the nutritional and toxicological status of the traditional foods in Africa is a major problem, because the consumers are usually unaware of the actual nutritional quality and level of toxicity of the foods except the organoleptic attributes. Information guiding the consumers on the type and possible degree of danger posed by the consumption of fermented foods is not available.

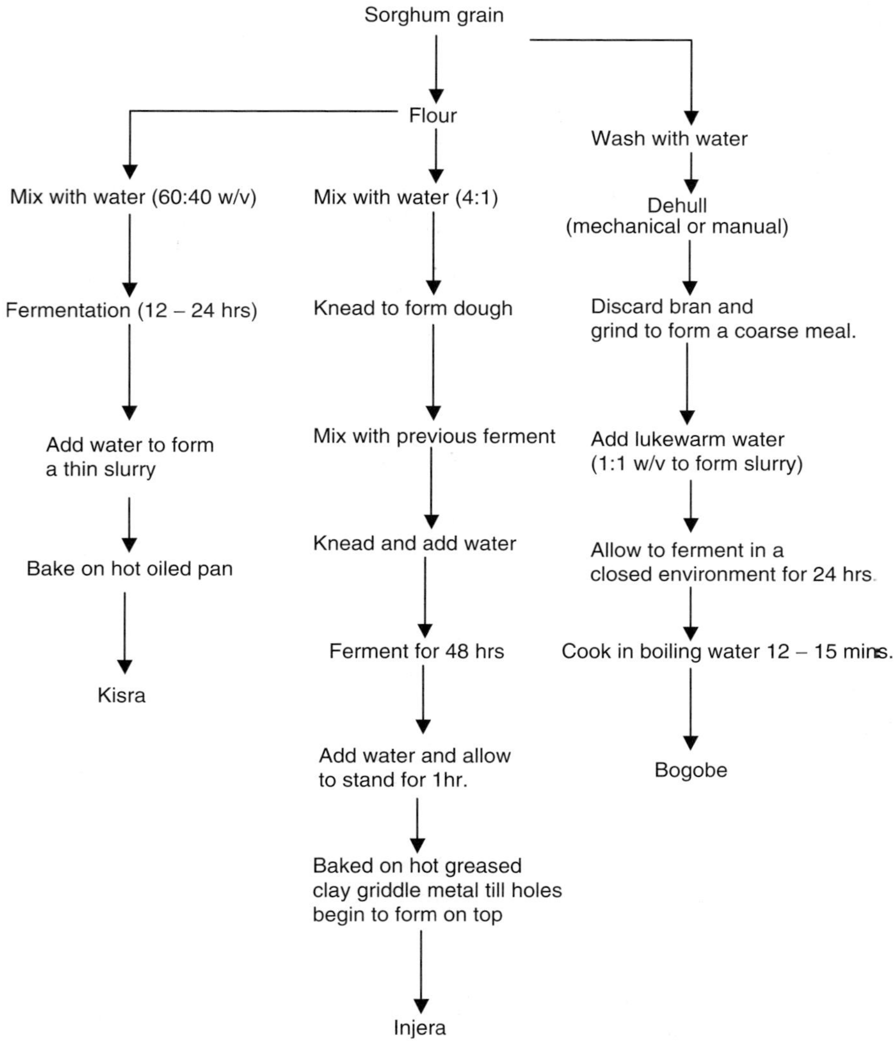

Figure 16.3 Flow chart for the preparation of Kisra, Injera, and Bogobe

16.3 TAXONOMY OF FERMENTING ORGANISMS

In the fermentation of indigenous foods, the organisms usually encountered are the lactic acid bacteria, yeasts, and molds. The type of predominant organisms present depends on the type of fermentation required for the production of the finished food product. If the food product is sour, the type of fermentation is usually lactic fermentation with LAB being the predominant fermenting organisms. However, if the final product is alcoholic in nature, the fermenting organisms usually include the yeasts.

The taxonomy of LAB has been well investigated (12). Several new genera have been added. Furthermore, the genera *Lactococcus, Leuconostoc, Lactobacillus,* and *Pediococcus*, as the classical groups of organisms involved in food fermentation are evolutionary well defined. The genera are gram positive and are located in the so called *Clostridium* branch

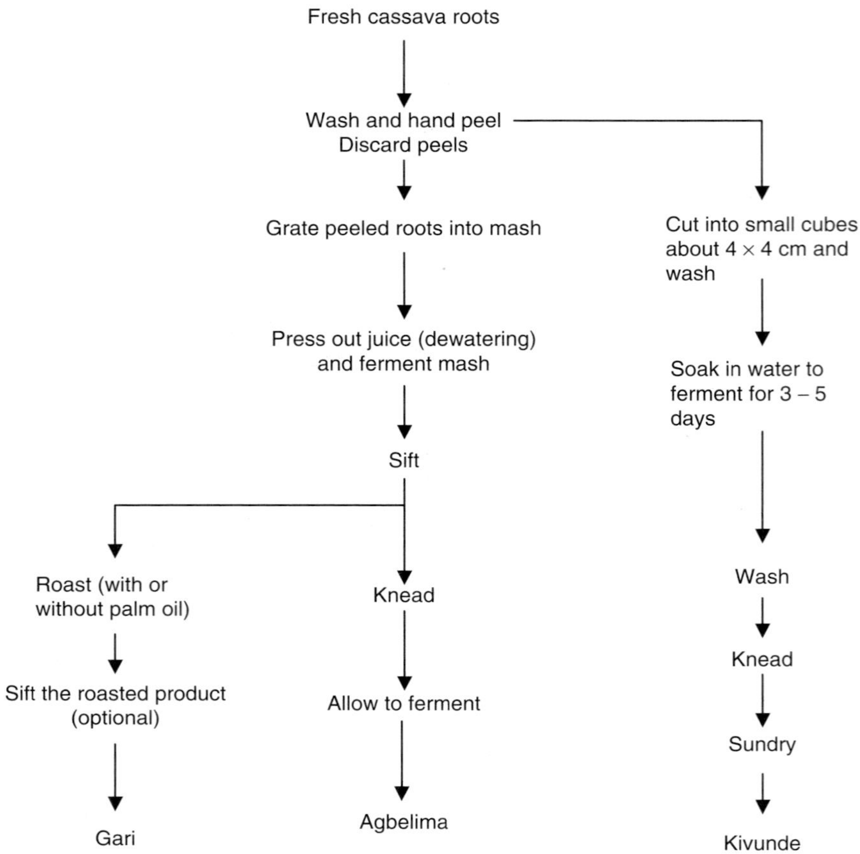

Figure 16.4 Flow diagram for the preparation of Gari, Agbelima, and Kivunde

of evolution (13), which is characterized by a low % G+C in their DNA (<55%). One group, the bifidobacteria, belongs to the high % G+C group of the *Actinomyces* branch. The LAB are characterized by a fermentative metabolism and derive their energy from the breakdown of sugars into metabolites, among which lactic acid is the most paramount. The homo fermentative LAB produce more than 85% of lactic acid while the hetero fermentative organisms produce less than 50%.

While information is available in Africa on the type and identity of organisms useful in traditional food fermentation, the methods for the identification of the organisms are usually inadequate and in most cases give wrong and confusing identity of strains. Very few works make use of more reliable and comprehensive identification methods to ensure correct species identification. Olasupo et al. (8) reported the identification of the lactic strains associated with the fermentation of some Nigerian fermented foods. The study showed that *Lactobacillus fermentum* was the predominant organism in the fermentation of wara, nono, iru, and kunun zaki. Similarly Amoa-Awua (6) identified different species of *Bacillus*, lactic acid bacteria, yeasts, and molds during the fermentation of Agbelima, a Ghanaian fermented cassava product.

In mawe, a Benin fermented maize product, Hounhouigan et al. (14) identified the predominant lactic organisms as *Lb. fermentum* (biotype cellobiosus), *Lb. fermentum,* and

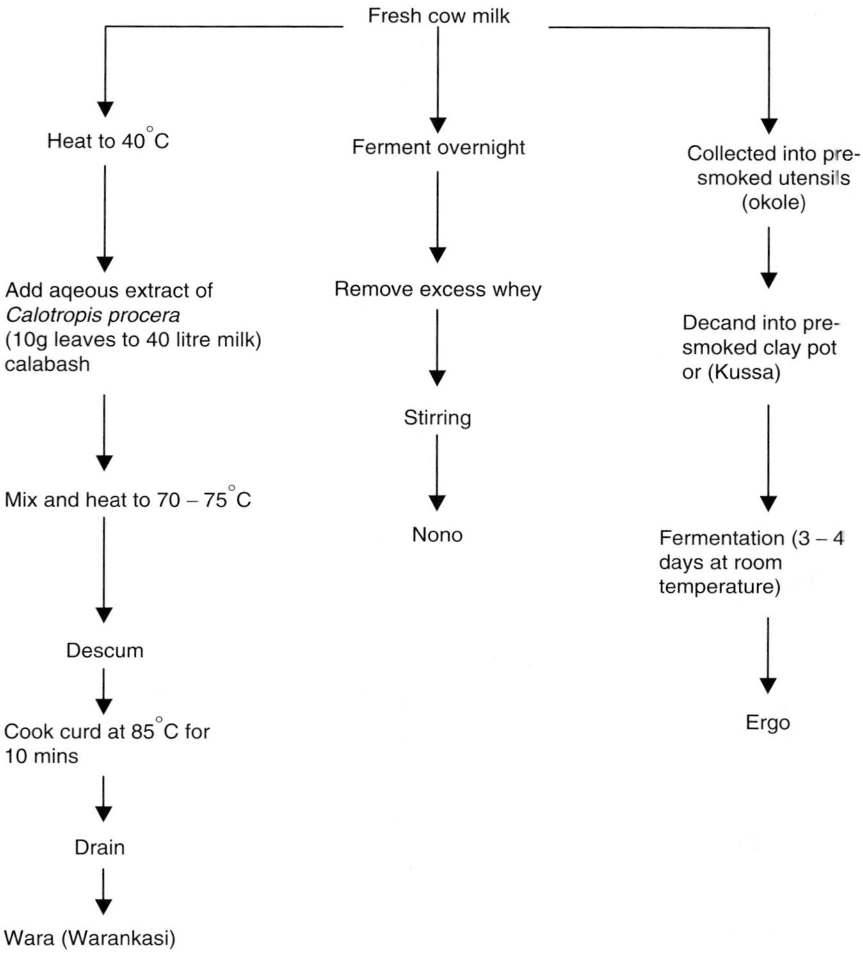

Figure 16.5 Flow Chart for the preparation of Nono, Ergo, and Wara

Lb. brevis while yeasts dominating the fermentation were *Candida* species which include *C. krusei* mainly, *C. kefyr* and *C. glabrata. C. Krusei* and *S. cerevisiae* were found together with lactic acid bacteria during the fermentation of busaa, a Kenyan opaque maize millet beer (15). Odunfa and Adeyele (16) found *Lactobacillus sp.* and *Lactococcus lactis* together with *C. krusei* and *Debaryomyces hansenni* during the fermentation of ogi-baba, a West African fermented sorghum gruel. Adegoke and Babalola (17) found *S. cerevisiae* together with *L. fermentum, L. brevis,* and *Enterococcus faecalis* in the fermentation of ogi, while Akinrele (18) found that corynebacteria *S. cerevisiae, E. cloacae,* and *L. plantarum* were predominant organisms in ogi. Halm et al. (19) found obligately hetero fermentative lactobacilli closely related to *L. fermentum,* and *L. reuteri*, in association with yeasts dominated by *Candida sp.* and *Saccharomyces sp.* in kenkey, a Ghanaian fermented maize dough.

The taxonomy of fermenting organisms in local food fermentation in the continent of Africa indicated close protobiotic association of lactic acid bacteria and yeasts, especially in cereal based foods.

African locust bean

↓

Boil for 12 – 24 hrs

↓

Dehulling by hand or pounding

↓

Boiling of cotyledons in water and kanwu (potash) for 1 – 2 hrs

↓

Spread cotyledons on calabash trays

↓

Wrapping of trays with coarse cotton cloth or banana leaves

↓

Ferment for 2 – 3 days at room temperature
(moist solid fermentation)

↓

Iru

Figure 16.6 Flow Chart for the preparation of iru

16.4 POTENTIAL OF FERMENTATION FOR THE REDUCTION OF MICROBIAL RISKS AND SHELF LIFE IMPROVEMENT

16.4.1 Antimicrobial Factors and Inhibition of Undesired Microorganisms

Food fermentation has been practiced and developed over several thousand years, first and foremost for the prevention of spoilage and the extension of shelf life of foods. Lactic fermented foods serve as special examples of safe and wholesome foods, and the production of lactic acid during fermentation results in acidification of the foods to pH values generally <4.2. This constitutes a major preservation and safety factor (3). In addition, several other antimicrobial metabolites (e.g., acetic acid, hydrogen peroxide, and bacteriocins) may also be produced during fermentation and thereby further contribute to the safety of lactic fermented foods (20). As is illustrated in Table 16.7, Gram-negative bacteria, including pathogens, putrefactive, and spoilage bacteria, are particularly inhibited by organic acids.

Bacteriocins of LAB and their possible role in food safety assurance have become a major focal point in studies toward food safety improvement. Nisin, a bacteriocin produced by some *Lactococcus lactis* strains and hitherto the one only approved for commercial use, is probably the best known and most studied bacteriocin. The possibility of using nisin producing strains of *Lact. lactis* subsp. *lactis* for protection of food against pathogenic organisms was first suggested by Hirsch et al. (21). They used nisin-producing strains that successfully prevented gas production ("late blowing") by clostridia in certain cheese varieties, and were able to show that nisin producing strains could also prevent the growth of *Staphylococcus aureus* in cheese. In addition, some other bacteriocinogenic LAB have been shown to effectively inhibit the growth of Gram-positive pathogens and toxinogens such as *Listeria monocytogenes, Staphylococcus aureus, Bacillus cereus,* and different *Clostridium* app. (see Table 16.8), even under *in situ* conditions (20). Information

Table 16.7
Metabolic products of lactic acid bacteria with antimicrobial properties (20, modified)

Product	Main Target Organisms
Organic acids	
Lactic Acid	Putrefactive and Gram-negative bacteria, some fungi
Acetic acid	Putrefactive bacteria, clostridia, some yeasts, and fungi
Hydrogen peroxide	Pathogens and spoilage organisms, especially in protein rich foods
Enzymes	
Lactoperoxidase system with H_2O_2	Pathogens and spoilage bacteria (milk and dairy products)
Low molecular weight Metabolites	
Reuterin (3-OH-Propionaldehyde)	Wide spectrum of bacteria, molds, and yeasts
Diacetyl	Gram-negative bacteria
Fatty acids	Different bacteria
Bacteriocins	
Nisin	Some LAB and Gram-positive bacteria, notably endospore formers
Other	Gram-positive bacteria, inhibitory spectrum according to producer strain and bacteriocin type

available in the literature generally shows a low frequency (from 0.6% to 22%) of bacteriocinogenic strains among food associated LAB (Table 16.9). A recent study by Olasupo et al. (22) indicates the frequency of bacteriocinogenic LAB in African fermented food within the same range (Table 16.9). A bacteriocin-producing strain of *Enterococcus faecium* isolated from wara, a traditional fermented diary product from West Africa, was found to produce a heat stable bacteriocin (enterocin) that showed antimicrobial activity against *Listeria monocytogenes* and *Enterococcus faecalis* (23). A *Lactococcus lactis* strain also isolated from wara, was found to produce a bacteriocin, structurally similar to nisin, the first to be confirmed for a traditional product from Africa (24). The bacteriocin inhibited notable food borne pathogens such as *Listeria monocytogenes* and *Bacillus cereus*.

Fermented dairy and meat products serve as examples of protein rich foods with extended shelf life and safety resulting from lactic fermentation. Olasupo et al. (25) reported extension of shelf life of agidi (solid form of ogi, the most common weaning food in Nigeria) from 6 days to 11 days at room temperature using bacteriocin-producing *Lactobacillus* starter in the fermentation of the food product.

Hounhouigan et al. (14) reported that lactic acid fermentation for the production of máwe (a fermented cereal of Bénin), reduced the Enterobacteriaceae population below the detection level ($<\log_{10}1.7$ cfu/g) after 24h of fermentation.

16.4.2 Food Borne Pathogens

Diarrheal diseases have frequently been indicated as a major cause of child morbidity and mortality in developing countries; the main aetiological agents of which include *Escherichia coli* (ETEC, EHEC, and EPEC strains), *Shigella, Salmonella, Campylobacter jejuni, Vibrio cholerae,* protozoa (e.g., *Entamoeba histolytica),* and rotavirus. These food borne infections often result from the fecal contamination of foods (26,27,28). The view of Metschnikoff, at the beginning of the century, that fermented foods such as yogurt could inhibit enteropathogens has been confirmed also for traditional lactic fermented foods.

Table 16.8
Bacteriocins of LAB with activity against Gram-positive food-borne pathogens (adapted from Holzapfel and Olasupo, in press)

		Activity Against:				
Genus	**Bacteriocin**	***Liseria monocytogenes***	***Staphylococcus aureus***	***Bacillus cereus***	***Clostridium botulinum***	***Clostridium perfringens***
Carnobacrium	Unnamed from					
	C.piscicola LK5	+	n.d.	n.d.	n.d.	n.d.
	Carnobacteriocins					
	A und B	+	n.d	n.d.	n.d.	n.d.
	Piscicolin 61	+	-	-	n.d	n.d.
Lactobacillus	Bavaricin A	+	-	-	n.d.	n.d.
	Bavaricin MN	+	-	n.d.	-	n.d.
	Curvacin A	+	(+)	-	n.d.	n.d.
	Curvaticin 13	+	+	+	n.d.	n.d.
	Plantaricin BN	+	-	n.d.	+	n.d.
	Plantarican NA	+	nd	n.d	n.d	n.d
	Sakacin A	+	-	n.d.	-	n.d.
	Sakacin M	+	(+)	-	(+)	(+)
	Sakacin P	+	-	-	n.d.	n.d.
Leuconostoc	Carnocin 44	+	-	-	n.d.	n.d.
	Leucocin UAL 187	+	-	n.d.	n.d.	n.d.
	Leuconocin S	+	+	n.d.	+	n.d.
	Mesenterocin 5	+	-	n.d.	n.d.	n.d.
	Mesenterocin 52	+	-	-	n.d.	n.d.
	Mesentericin Y 105	+	-	-	n.d.	n.d.
	From *Lc. Gelidum* IN 139	+	-	-	-	-
Pedicoccus	Pediocin A	+	+	n.d.	+	+
	Pediocin AcH	+	+	+	- (pores)	+
	Pediocin PA-1	+	-	n.d.	n.d.	n.d.
	Pediocin PC	+	-	-	n.d.	+
	Pediocin SJI	+	-	n.d.	n.d.	+

Table 16.9
Frequency of bacteriocin producers (Bac+) among *Lactobacillus* strains isolated from fermented foods in Africa (22)

Food	Number of Isolates	Bac^+
Wara	60	4
Kenkey	48	4
Ugba	36	2
Ogi	30	2
Kunu-zarki	12	0
Kufu	9	1
Iru	5	1
Total	200	14

The inhibition of food borne pathogens in lactic fermented cereal gruels have been highlighted by various workers (29,30,31,32). Fermented mixtures of water and whole flour of either maize or sorghum with a final pH of 3.8 and a lactic:acetic acid ratio of 9:1 were found to inhibit Gram-negative intestinal pathogens such as enterotoxigenic *Escherichia coli, Campylobacter jejuni, Shigella flexneri,* and *Salmonella typhimurium* (33). Lactic fermentation of cereal gruels was shown to inhibit *Salmonella typhimurium, Campylobacter jejuni, Yersinia enterocolitica,* ETEC, EPEC, *Shigella flexneri,* and *Shigella sonnei* (30,32,34,35). Other examples of inhibition of fermented foods include salmonellae, *E. coli, Shigella sonnei,* and *Staph. eureus* in rice based weaning food (36), and salmonellae, *E. coli,* and *Staph. eureus* in dairy products (37,38,39,40). Most of these antimicrobial activities may be termed 'unspecific' and are related to acidification and one or more of the factors summarized in Table 16.7. Antimicrobial activities against bacteria causing diarrhea have been related to LAB involved in fermentation of uji, a Kenyan indigenous fermented cereal gruel (41). Several studies have shown that pH values <4.0 will inhibit diarrhea causing pathogens in ready-to-eat fermented food products (42).

16.4.3 Inhibition of Toxin Production

Microbial toxins in a food usually result from growth and metabolic activities of toxinogenic bacteria and molds. Inhibition or elimination of these organisms either in the raw materials, or in the final product, may prevent the formation of these toxins. A number of reports indicate the feasibility of fermentation for reducing or preventing growth or metabolism of toxin producers.

Svanberg et al. (33) observed that a sustained inhibition of *Staphylococcus aureus* during fermentation of cereal gruels at pH<4.0 had been supported by an additional factor, probably a bacteriocin. Acid production and other antimicrobial metabolites have been suggested by Mbugua and Njenga (41) as inhibiting factors against *Staph. aureus* and some Gram-negative enteropathogens during uji fermentation. During lactic fermentation of cereal gruel, the level of *Staph. aureus* was strongly reduced at pH values <4.16 (30). Yusof et al. (36), showed the decisive importance of a high initial level of LAB for successful inhibition of *Staph. aureus* in a rice-based weaning food. Inhibition of *Staph. aureus* by LAB in typical fermented foods of industrialized countries has been sufficiently documented, both for fermented dairy products (37,39) and fermented meat products (43,44,45).

Even in nonfermented ready-to-eat foods such as salads, mere inoculation with *Lb. Plantarum* strains resulted in a decline of *Staph. aureus,* especially at refrigeration temperatures (46).

The inhibition of other bacterial toxinogens by LAB during fermentation has been particularly studied in fish and seafood with regard to *Bac. cereus* (34,46) and *Cl. botulinum* (48).

In contrast to antibacterial interactions of LAB, their antimycotic and antimycotoxinogenic potential – i.e., inhibition of either mold growth or metabolic activities resulting in mycotoxin production – has received relatively little attention thus far. Yet a number of reports strongly indicate the potential of *Lactobacillus* spp. (e.g., *Lb. acidophilus, Lb. delbrueckii ssp. bulgaricus, Lb. casei,* and *Lb. plantarum*), and *Streptococcus thermophilus* to inhibit mold growth and mycotoxin production. Special attention was given to inhibition of aflatoxin produciton and aflatoxinogenic molds such as *Aspergillus parasiticus* and *Aspergillus fumigatus* and, to a lesser extent, of *Aspergillus flavus* (49,50). Equivocal results were obtained with *Lactococcus lactis,* also suggesting stimulation effects on molds under particular conditions of pH and substrate composition (49). The mechanism of the inhibition of aflatoxin synthesis seems to be relatively complex and may not be explained only by low pH, depletion of nutrients, and microbial competition. It rather appears that lactic acid and other LAB metabolites (probably low molecular weight, heat stable compounds) play an important role (51). Relevant aspects have been excellently reviewed by Gourama and Bullerman (52).

16.5 THE FATE OF MICROBIAL TOXINS IN FERMENTATION

Food invaded by toxinogenic bacteria or fungi constitute a potential health risk. Fermentation may however contribute to toxin degradation (detoxification) during processing and thereby assist to reduce possible toxic safety risks. The role and potential of fermentation in reducing microbial toxins in food raw materials are discussed in what follows.

16.5.1 Toxins of Bacterial Origin

Because of their frequent production during fermentation, the biogenic amines (BA) probably constitute the most important bacterial metabolites that may pose a safety risk in traditional fermented foods. For this discussion they are considered together with typical bacterial toxins. Although the lactic fermented foods are generally considered safe, these fermentations are regularly accompanied by the production varying levels and types of BA, although generally at a low rate. The effect of the fermentation process on BA in such foods has been highlighted earlier (53,54). One of the few reports on biogenic amines in traditional fermented foods is the study by Nout et al. (53) on the presence of biogenic amines in kenkey, a Ghanaian fermented maize product. The study showed the level of biogenic amines in kenkey to be quite low (<60ppm). These levels however, increased up to tenfold when the food product was improved by addition of cowpeas. The addition of red cowpeas increased the level of biogenic amines in kenkey to a total concentration of ca.200ppm (mainly cadaverine and tyramine) and, with the addition of white cowpeas to ca.500ppm (mainly putrescine and tyramine). These increases were attributed to the presence of precursor amino acids in the respective cowpea varieties. In addition, polyphenols in the seed coats may have inhibited the activity of monoamine oxidases. Nout et al. (53) employed strains of *Lb. plantarum, Lb. confusus, Lb.brevis,* and *Ped. pentosaceus*, previously isolated from kenkey. However, no pure culture studies on biogenic amine production in traditional fermented foods have been conducted yet. Prolonging the cooking process during kenkey production may contribute only slightly to lowering of the amine levels (53).

Studies on the biogenic amine formation in Sauerkraut have showed that pure culture inoculation with *Lb. plantarum* significantly reduced putrescine formation during fermentation (54).

The report of Westby et al. (55) also showed histamine toxicity to be associated with poorly handled scombroid fish, but this has not been reported in fermented fish products.

Recent studies (56) have however showed the potential of some LAB strains to detoxify such BA by the action of monoamine oxidases.

Very little, if any, attention has thus far been given to the fermentative degradation of bacterial toxins. Their sensitivity to other factors such as heat, pH, and enzymes has however been the subject of extensive studies. This may be explained by the fact that, in contrast to mycotoxins, raw materials for fermentation are generally not contaminated with bacterial toxins.

16.5.2 Mycotoxins

Mycotoxins, especially aflatoxins, fumonisins, and citrinin, are a major risk factor in stored cereals and legumes, typical raw materials for fermented foods in Africa (3,55,57). Even with strict regulations on maximum tolerance levels, control mechanisms in developing countries are insufficient and very difficult to apply at the household and small scale levels to any significant extent. Increasing, although still controversial, reports indicate that some mycotoxins may be degraded or inactivated during fermentation of cereals and legumes. Recent information has been reviewed by Westby et al. (55) and is summarized in Table 16.10, indicating some of the effects of fermentation on the aflatoxin content of contaminated raw materials. The reduction of aflatoxin during the preparation of some traditional Nigerian fermented foods has been shown by Ogunsanwo et al. (58,59). They showed that, at low levels (16.4 ppb), aflatoxin B1 was completely destroyed in watermelon seeds during 4 days of fermentation in the preparation of ogiri (58). In addition, they found that fermentation of soya bean flour to produce soyogi reduced the initial aflatoxin B1 level of 0.3ppm (59). In a study by Adegoke et al. (60), >70% reduction was observed in the aflatoxin B1 level during the fermentation of maize and sorghum flour for the preparation of ogi, a popular weaning food in West Africa.

In contrast to the mentioned reports, Jespersen et al. (61) could not detect any decrease in the aflatoxin level during dough fermentation for kenkey preparation. Kpodo et al. (57) showed aflatoxins and citrinin to occur frequently in kenkey, with levels as high as 289 μg/kg for total aflatoxins and 584 μg/kg for citrinin, but only low levels of ochratoxin A and neither zearalenone nor α-zearalenol. Aflatoxin levels were reported to increase significantly during the initial stages of fermentation. However, traditional cooking of the maize dough for 3h was reported to result in a 80% reduction of aflatoxin B1 and G1 levels, 35% in the B2 and G2 levels, and a disappearance of citrinin.

Studies in the Institute of Hygiene & Toxicology in Karlsruhe, Germany (see Table 16.11) have shown the ability of single LAB strains isolated from Ghanaian aflata (for kenkey processing), to significantly reduce the amount of *Alternaria* toxins under defined conditions. In addition, some authetic LAB strains, and some isolated from traditional fermented Turkish foods, studied on semisynthetic medium, have been shown to play a significant role in the reduction of patulin concentration by more than 60% (62).

A reduction of patulin and ochratoxin A has also been reported during the fermentation of cider and beer (55). Other mycotoxins, such as zearalenone, ochratoxin A, T-2 toxin, deoxynivalenol, and diacetoxyscirpenol, have been shown to be degraded significantly particular fermented foods. In view of these reports, the potential contribution of fermentation to improved toxicological safety of some food products seems promising and should not be ignored. For some fermentation, metabolic action of microorganims has clearly been shown to contribute to the reduction of mycotoxins in some raw materials. The mechanisms of mycotoxin reduction by fermenting organisms are still not sufficiently understood. It may be partly explained by (hydrolytic) degradation, partly by inactivation and partly by

Table 16.10
Examples of the reported effects of fermentation on mycotoxins in raw material (Westby et al., 1997)

Toxin	Raw Material/Product	Type of Fermentation	Naturally Contaminated/Spiked	Extent of Reduction
Aflatoxin	Maize/Kenkey	Lactic acid		None
Aflatoxin	Sorghum/ogi	Lactic acid	Natural with B1	12–16%
Aflatoxin	Wheat/bread	Yeast (dough)	Spiked with B1	19%
Aflatoxin	Milk/yoghurt	Lactic acid	Natural with M1	None
	Milk/kefir	Lactic acid	As above	Decreased
Aflatoxin	Milk/yoghurt	Lactic acid	Spiked with B1, B2, G1, G2 and M1	None
Aflatoxin	Melon seed/*ogiri*	*Bacillus* spp.	Natural with B1 and G1	Complete removal after 4 days
Aflatoxin	Peanut press cake/ pure mould cultures	*Neurospora sitophila* and *Rhizopus oligosporus*	Not stated	50% and 70% respectively
Aflatoxin	Maize/Ogi Sorghum/Ogi	Lactic fermentation	Natural with B1	Greater than 70%
Alternariol and Alternariol-Monomethylether	Pure culture isolates from kenkey	Lactic acid bacteria	Spiked laboratory media	Reduction greater than 50% by all tested strains

Table 16.11
Degradation of Alternaria toxins by LAB strains isolated from Aflata (for Kenkey production)

	Reduction as% of Original	
Strains	**Alternariol**	**Alternariol-monomethylether**
Lb. plantarum 91	24	43
Lb. fermentum 85	26	38
Ped. acidilactici 100	47	29
Lb. fermentum DSM 20174	37	59

The purified toxins were added to basal MRS medium. Incubation was at 30ºC and the Alternaria toxins were determined by HPLC. All tested LAB strains showed reduction of alternariol and alternariol monomethyl ether by more than 50%.

adsorption of the mycotoxin to the cellular surface. The best long term approach to the mycotoxins problem in fermented food products would obviously be the prevention of the contamination of the raw materials used in the preparation of such foods (63). This goal may however not be achieved in the foreseeable future, and other measures, including controlled fermentation and the use of specially selected starter cultures, may contribute to an improvement in the reduction or elimination of mycotoxins in fermented foods.

16.6 NATURALLY OCCURING TOXINS AND ANTINUTRITIVE FACTORS IN FOOD FERMENTATION

16.6.1 Naturally Occurring Toxins

16.6.1.1 Cyanogenic Compounds in Cassava

Linamarin and lotaustralin are cyanogenic glucosides typically found in cassava, and may because of severe intoxications associated with the consumption of raw or unprocessed cassava (3). In Africa, methods of cassava processing usually involve peeling, soaking, heaping (a solid-state fermentation, with mainly molds involved) sun drying, and grating, typically followed by submerged (soaked) lactic fermentation. In addition to the endogenous linamarinase enzymes present in cassava, detoxification is mainly achieved by microbial action. Microorgnisms important in cassava fermentation include *Lactobacillus* (25,64), *Bacillus* (65,66), and yeasts and molds (66,67). The role of fermentation in the detoxification of cassava has been demonstrated by several workers with particular reference to traditionally processed fermented food products of Africa. Amoa-Awua (6) reported on significant detoxification of cassava during the processing of cassava into Ghanaian agbelima (Table 16.12). The cyanogenic glucosides in cassava (119.3 mg/kg) were completely removed by the fermentation process. Observations by Vasconcelos et al. (68) on gari preparation indicate the breakdown of linamarin in grated roots to be primarily due to the endogenous linamarinase activity. Cellular damage during grating was shown to release linamarase. Further clarification on the role of fermentation was provided by Westby and Choo (69), showing microbial growth to be essential for efficient cyanogen reduction in soaked cassava roots.

The use of selected strains of *Lb. plantarum* starter culture for controlled submerged fermentation of cassava into typical Tanzanian kivunde, was found to significantly enhance detoxification, (Table 16.13) as compared to spontaneous and back slopping operations

Table 16.12
Effect of inoculum on the cyanogenic glycosides content (mg/kg) of cassava during fermentation (6)

Sample	Cyanogenic Potential[b]	Cyanogenic Glucosides[c]	Cyanohydrins	Free Cyanide[d]
Fresh cassava	123,2	119,3	2,8	1,1
Start of fermentation				
Dough without inoculum	120,5	93,2	24,5	2,8
Dough with roasted inoculum	116,8	79,8	33,7	3,3
Dough with thatch inoculum	117,7	61,2	50,7	5,8
End of fermentation, 72h				
Dough without inoculum	93,7	0	88,9	4,8
Dough with roasted inoculum	49	0	41	8,1
Dough with thatch inoculum	22,5	0	22	0.5

[a] Average values for 2 samples
[b] Total cyanide content including cyanogenic glucosides and the hydrolysed products, Cyanohydrins and free cyanide
[c] Linamarin and lotaustralin
[d] Present as hydrogen cyanide

Table 16.13
Effect of fermentation on the cyanogenic glucoside content (mg/kg of dry weight) of cassava during proceesing into kivunde in Trial 3 (conducted at TIRDO, Dar-es-Salaam, Tanzania), as compared to Trials 1 and 2 (conducted at IHT, BFE, Karlsruhe, Germany); in each trial, the traditional procedure of submerged cassava fermentation for kivunde was followed, as given by Kimaryo et al. (2000)

Fermentation Period (day)	Fermentation Type								
	Spntaneous			Back-slopping			Starter Culture		
	3[a]	1[b]	2[c]	3[a]	1[b]	2[c]	3[a]	1[b]	2[c]
O (fresh cassava)	175.9	126.8	164.8	176.8	124.5	164.2	176.3	127.1	164.8
1	94.6	73.2	Nd	132.6	96.4	Nd	144.7	63.9	Nd
2	67.2	21.6	39.7	89.3	32.6	69.6	62.5	36.2	63.0
3	45.9	19.8	38.1	62.4	26.4	64.4	38.9	28.7	18.5
4	43.5	10.7	23.7	47.7	17.9	56.3	12.6	9.2	14.3
5	39.1	Nd	24.0	32.9	Nd	38.0	8.2	8.9	9.0
Dry kivunde	17.8	9.3	4.4	26.5	Nd	8.1	6.3	5.5	2.8

[a] Trial 3 conducted in Tanzania (TIRDO, Dar-es-Salaam)
[b] Trial 1 conducted in Germany (IHT, Karlsruhe)
[c] Trial 2 conducted in Germany (IHT, Karlsruhe)
Nd. No. data available because of spoilage or experimental error. "Spontaneously" fermented samples were spoiled after 3 days.

(5). Mechanisms of cyanogen reduction and the role of different microorganisms during fermentation of cassava, have been reviewed by several workers, and notably by Westby and coworkers (69,70,71,72).

The most important aspects regarding the different mechanisms and potential improvements are summarized in Table 16.14.

Table 16.14
Summary of mechanisms of cyanogen reduction during fermentation of cassava (Westby et al., 1997, modified)

Type of Fermentation	Role of Fermentation in Reducing Cyanogens	Mechanism of Cyanogen Reduction	Potential Improvements to Fermentation	Other Potential Improvements
Acidic fermentation of grated roots	None	-Mechanical grating brings linamarase into contact with linamarin -Acid stabilises cyanohydrin	-None required for linamarin hydrolysis -Organisms with hydroxynitrile lyase activity	-Ensure varieties have high linamarinase activity -Ensure efficiency of grating -Develop suitable post fermentation processing techniques to remove cyanohydrins plus HCN
Acidic fermentation of soaked roots	Major	Microorganisms cause cellular breakdown which facilitates leaching and probably contact between linamarin and endogenous linamarase	-Use of conditions favourable to growth of microorganisms with ability to degrade cellular structure and linamarin -Potential use of starter cultures with cell wall degrading activity -Use of microbes with hydroxynitrile lyase activity	-Ensure varieties have high linamarase activity -Ensure sufficient time is left for leaching
Solid substrate fermentation	Major	Microorganisms cause cellular breakdown which facilitates contact between linamarin and endogenous linamarase	-Use of conditions favourable to growth of microbes with ability to degrade cellular structure -Potential use of cultures with cell wall and linamarin degrading activity	-Ensure varieties have high linamarase activity -Ensure sufficient time for diffusion of HCN

The purified toxins were added to basal MRS medium. Incubation was at 30°C and the Alternaria toxins were determined by HPLC. All tested LAB strains showed reduction of alternariol and alternariol monomethyl ether by more than 50%.

16.6.2 Antinutritive Factors

The role of fermentation in the reduction of antinutritional factors in traditionally fermented foods from cereals and legumes, has been well documented (3,31,42,73,74,75,76). However, most of these reports indicate only a limited reduction of these undesired factors. Some of this information is highlighted in what follows.

16.6.2.1 Phytic Acid

Phytate (phytic acid) is a typical component of cereals, legumes, and some roots and tuber crops used in the preparation of such foods as garri, ogi, and iru. Its presence may reduce the bioavailability of iron and other minerals in foods and also result in reduced solubility and digestibility of proteins. The approximate phytate content of maize, millet, sorghum, and cowpeas on a dry matter basis (77) is presented in Table 16.15. Phytases hydrolyze phytate into lower inositol phosphates and are present in most cereals (78). These enzymes are activated during germination (e.g., during the soaking process in traditional household level processing) and fermentation. Phytate was shown to be completely hydrolyzed after fermentation of germinated white sorghum and as a result the amount of soluble iron was found to be strongly increased (79). The reduction of phytate by lactic acid fermentation has also been reported in maize (80), pearl millet (81,82), germinated finger millet (83), and in Indian idli (84). Similarly, Reddy and Salunkhe (85) observed almost complete elimination of phytate phosphorus within an 8 hr fermentation of rice (Table 16.16).

Information on the role of fermentation in phytate reduction for traditional African commodities is relatively sparse. Aderibigbe and Odunfa (86) reported on the contribution of fermentation to phytate reduction, but, did not find any phytase activity associated with the

Table 16.15
Approximate phytate content of sorghum, maize, millet, and cowpeas on a dry matter basis (77)

Product	Range (%)
Sorghum	0.57–0.96
Maize	0.44–1.2
Millet	0.85–1.1
Cowpeas	0.89–1.5

Table 16.16
Changes in the form of phosphorus during fermentation of rice (85)

Fermentation Period (h)	Total P (mg/g)	Phytate P (mg/g)	Phytate P Hydrolyzed (%)
Control, raw	1.89	0.97	0.00
Control, soaked In water 2h	1.96	0.65	32.99
Fermented:			
4h	1.75	0.07	92.78
8h	1.79	0.00	100.0

Bacillus spp. involved in the fermentation of African locust bean *Parkia biglobosa* for the preparation of iru or dawadawa. However, Eka (87) observed appreciable reduction in phytic acid during the fermentation of locust bean to produce iru (dawadawa) (see Table 16.17). Studies carried out at the Institute of Hygiene and Toxicology, Federal Research Centre for Nutrition in Karlsruhe, Germany on strains isolated from Ghanaian lactic fermented cereals and cereal–cowpea mixtures revealed this ability to be relatively rare among pure cultures. The HPLC analysis for quantification, showed that some *Lb. Plantarum* strains were, however, able to significantly reduce the phytic acid concentration in a synthetic medium containing sodium phytate after incubation at 37ºC for 120hrs. Enzymes of LAB responsible for such degradation may be acid phosphatases.

16.6.2.2 Tannins

As a component of cereals and legumes, tannis interact with proteins, and thereby inactivate digestive enzymes and decrease protein digestibility. Little has thus far been reported on their fermentative reduction in traditional African foods. Relative to other processing steps, (e.g., dehulling and cooking), microbial metabolism during fermentation does not appear to contribute significantly to tannin degradation. The underlying mechanisms, however, are still insufficiently understood (84).

16.6.2.3 Oligosaccharides (Flatus-producing Sugars)

Cereals and especially legumes contain oligosaccharides such as raffinose, stachyose, and verbascose, which may cause flatulence, diarrhea, and indigestion. Table 16.18 shows the typical levels of these oligosaccharides in legumes. These oligosaccharides, as members

Table 16.17
Vitamins and toxicant content of unfermented and fermented iru (dawadawa); Adapted from Odunfa (100)

Component (mg/100g)	Unfermented Dawadawa	Fermented Dawadawa
Thiamin	0.65	1.35
Riboflavin	0.45	1.30
Niacin	7.50	5.30
Oxalate	0.21	0.12
Phytic acid P	15.00	7.50

Table 16.18
Oligosaccharide content of some legumes (78)

Leguminose	Total Sugars (%)	Sucrose (%)	Raffinose (%)	Stachyose (%)	Verbascose (%)
Viciafaba	5.98	1.37	0.52	1.41	1.85
Lens culinaris (lentils)	5.13	1.14	0.45	1.65	0.62
Phaseolus vulgaris (garden beans)	4.9	1.25	0.45	1.8	0.25
Vinga sinensis (cowpeas)	6.05	1.51	0.77	3	0.3
Phaseolus aureus (mung beans)	6.11	1.39	0.39	1.67	2.66
Glycine max (soya beans)	12.02	6.42	1.26	4.34	traces

of the raffinose family, are usually resistant to cooking and other small scale processing steps, and possess α-galactosidic bonds which may be hydrolyzed by α-galactosidases produced by a number of molds and by bacteria associated with the digestive tract and with fermented foods. The role of fermentation in the degradation of these oligosaccharides has been well reported. According to Shallengerger et al. (88), stachyose was gradually hydrolyzed by fermenting molds, e.g., *Rhizopus oligosporus* during tempe production. However, the tempe molds could not utilize sucrose and raffinose (89).

The effect of fermentation of oligosaccharides has been well highlighted by the work of Odunfa (90). The study showed significant reduction or elimination of oligosaccharides such as sucrose, raffinose, and stachyose in fermented locust bean (iru) (see Table 16.19).The typical yogurt bacteria *(Lb. bulgaricus* and *Streptococcus thermophilis)* have been shown to reduce the stachyose content of yogurt produced from soy milk by almost 27% (91). In a related study on LAB strains isolated from fermented maize products in Ghana (92), showed the majority of the *Lb. plantarum* strains to ferment raffinose (Table 16.20).

Table 16.19
Changes in oligosaccharides during fermentation of locust bean (90)

	Oligosaccharide mg/g Dry Weight		
Fermentation Time (hr)	**Sucrose**	**Raffinose**	**Stachyose**
Unfermented (0 hr)	31.0 ± 4.0[a]	12.0 ± 2.0	28.0 ± 4.0
Fermented (24 hr)	5.5 ± 0.0	-	5.0 ± 0.8
Fermented (48 hr)	-	-	-

a, mean value with plus and minus standard variation.

Table 16.20
Raffinose fermentation by selected strains isolated from fermented maize products of Ghana (92)

Species	Strain Numbers[a]	Acid Production
Lactobacillus Plantarum	1,7,8,21,31,38,43,46,47,64,65,69	++
Lactobacillus Plantarum	37,39,44,53,55,70,91	+
Lactobacillus Plantarum	58,60,68	−
Pediococcus pentosaceus	90	−
Pediococcus acidilactici	100	−
Leuconostoc Mesenteroides ssp	92	−
Mesenteroides	DSM 10343	++

Acid production: −, negative; +, weak; ++, strong.
The qualitative fermentation tests were performed in MRS broth (pH 6.4) containing 1 % raffinose (instead of glucose) and 0.004% chlorophenolred as indicator. Incubation was at 30ºC for 3 days.

[a]Origin of strains:

Nos. 1–21, 92:	from fermenting maize (Aflata) during Kenkey production.
Nos. 31–47:	from fermenting maize: white cowpeas mixture (70:30).
Nos. 53–70:	from fermenting maize: red cowpeas mixture (70:30).
No. 90:	from Yakeyake after fermentation.
No. 91:	from Agbelima, after cooking.
No. 100:	from Aflata (for Ga-Kenkey) after fermentation.

16.6.2.4 Protease Inhibitors

Increased availability of essential amino acids including lysine, leucine, methionine, isoleucine, and tryptophan has been reported to be associated with lactic acid fermentation of cereal porridges and of Ghanaian kenkey (93,94). LAB strains isolated from lactic fermented foods in Ghana, showed varying abilities to degrade trypsin inhibitor under defined laboratory conditions, reaching up to 50% reduction by a *Leuconostoc* strain (3) (see Table 16.21). Furthermore, lactic acid fermentation has been reported to improve the *in vitro* protein digestibility of nontannin cereal grains (95,96) and of high tannin varieties (95,97). In children, the protein digestibility was reported to increase from 47% to 73% after lactic acid fermentation of whole grain nontannin sorghum flour that was prepared in Nasha, a traditional Sudanese fermented food for infants and young children (98). The effect of fermentation may be related to a reduction in proteinase inhibitors (e.g., trypsin inhibitor) in legumes, a reduction of tannins and of high levels of disulphide cross linkages in sorghum prolamine proteins (82,99).

16.7 FERMENTATION AND NUTRITIONAL IMPROVEMENT

Malnutrition and deficiency in micronutrients are highly prevalent and even increasing in different parts of African countries (101). It is estimated that more than 500 million people (mainly children) are underweight and even more are affected by micronutients deficiencies. Some reasons put forward to explain this situation include high population density, poverty, inadequate access to good sanitary and health facilities, and unavailability of adequate quantity and quality of food. Factors of immediate and direct influence on these nutritional problems are inadequate food consumption and diseases, which usually interact in a mutually reinforcing manner (Figure 16.7).

Several methods have been employed to improve the nutritional quality of cereals (73,76). These include genetic improvement, amino acid fortification, and supplementation with protein rich sources. Of equal importance is the application of various processing technologies to increase the bioavailability of indigenous nutrients in grains such as protein, starch, and minerals. However, numerous investigations have shown that one such household level technology that is widely used in many developing countries to improve the sensory and nutritional qualities of cereals and their products is fermentation.

The different ways by which fermentation process can affect the nutritional quality of foods include improving the nutrient density and increasing the amount of the bioavail-

Table 16.21
Degradation of trypsin inhibitor (TI) by lactic acid bacteria isolated from Aflata in Ghana (3)

Isolates	Reduction of TI (mg)	% Reduction
Lb. Plantarum 91	2.41	48.0
Lb. Fermentum 103	1.22	24.4
Pediococcus sp. 90	0.89	17.8
Pediococcus sp. 19	1.08	21.6
Leuconostoc sp. 106	2.68	53.6
Lactobacillus sp. 41	0.65	13.0
Lactobacillus sp. 17	1.86	37.2
Lactobacillus sp. 62	1.34	26.8

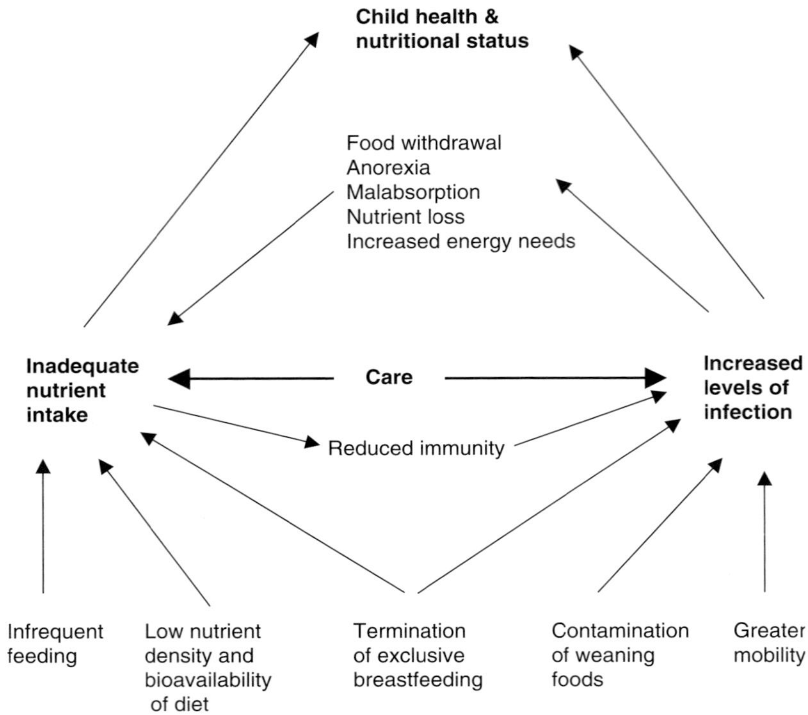

Figure 16.7 Interrelation of care, infections and nutrient intake and the health and nutritional status of infants and children (76)

ability of nutrients. The latter may be achieved by degradation of antinutritional factors, predigestion of certain food components, synthesis of promoters for adsorption, and by influencing the uptake of nutrients by the mucosa (76).

Fermentation of cereals with pure yeasts or lactic cultures has been shown to increase the protein, amino acid, and free sugar contents of the fermented products (102,103,104,105). In the production of ogi with *Lactobacillus* sp. starter (103), considerable increase in protein content was observed in ogi fermented with starter cultures over the one fermented naturally (Table 16.22). Similarly, Odunfa et al. (104) recorded up to a threefold increase in the lysine content of ogi using a *Lb. plantarum* starter. Furthermore, Teniola and Odunfa (106) reported significant increase in lysine and methionine content of ogi fermented with lactic and or yeasts starters. [Figure 16.8(a) and Figure 16.8(b)].

The changes in B-group vitamins of cereals and cereal–legume blends during natural fermentation have been investigated (107,108,109). Table 16.23 shows that thiamin, riboflavin, and niacin content increased significantly during natural fermentation of cereals: sorghum, rice, Pearl millet, and finger millet. Similarly increase in vitamins content during fermentation of African locust bean for iru production (see Table 16.17) was reported by Odunfa (90). The increase or decrease of vitamins varied with the nature of raw materials, temperature, nature of microflora, length of fermentation, and determination methods of these vitamins.

In a program to improve the digestibility of protein in iru (a Nigerian fermented proteinaceous food used as condiment), Aderibigbe and Odunfa (86) identified proteases producing strains of *Bacillus* with promising potential in the breakdown of protein.

Table 16.22
Estimation of protein and free sugars in ogi fermented naturally and that fermented with starter cultures (103)

	Concentrations of (mg/ml)*	
Sample	**Protein**	**Free Sugar**
Control	1.84 ± 0.017	0.41 ± 0.01
DK 52	2.63 ± 0.035	5.23 ± 0.053
DK77	2.28 ± 0.08	5.70 ± 0.083
DK 52/77	2.83 ± 0.28	4.02 ± 0.17

Control: naturally fermented ogi without the use of starter
DK 52: ogi fermented with *Lactobacillus acidophilus*
DK 77: ogi fermented with *Lactobacillus pentosus*
DK 52/77: ogi fermented with mixture of two starters
* mean value ± SD

Table 16.23
Changes in B-group vitamins during natural fermentation of cereals

	Vitamins (μg/g)		
Cereal	**Thiamin**	**Riboflavin**	**Niacin**
*Sorghum**			
Unfermented	20.2	0.90	37.9
Fermented	47.1	1.30	41.3
Rice †‡			
Unfermented	0.51	0.80	63.1
Fermented	0.35	1.35	41.6
Pearl millet			
Unfermented	0.37	0.19	-
Fermented	0.64	0.36	-
Finger millet			
Unfermented	0.30	0.16	-
Fermented	0.47	0.24	-

Adapted from: *Kazanas and Fields (107); † Tongnual and Fields (108); ‡ Aliya and Geervani (109).

The proteases are hydrophobic in nature which is an ideal property for solid-state fermentation.

16.8 SCOPE FOR IMPROVEMENT OF FERMENTED FOODS

Traditional fermented foods are produced on the household scale in a majority of African countries. Growing industrialization and urbanization trends in these countries will however dictate the need for larger scale production of fermented foods having consistent quality. Furthermore, variation of the quality attributes of fermented foods to meet the demands of the complicated and varied palates of industrialized communities will eventually be

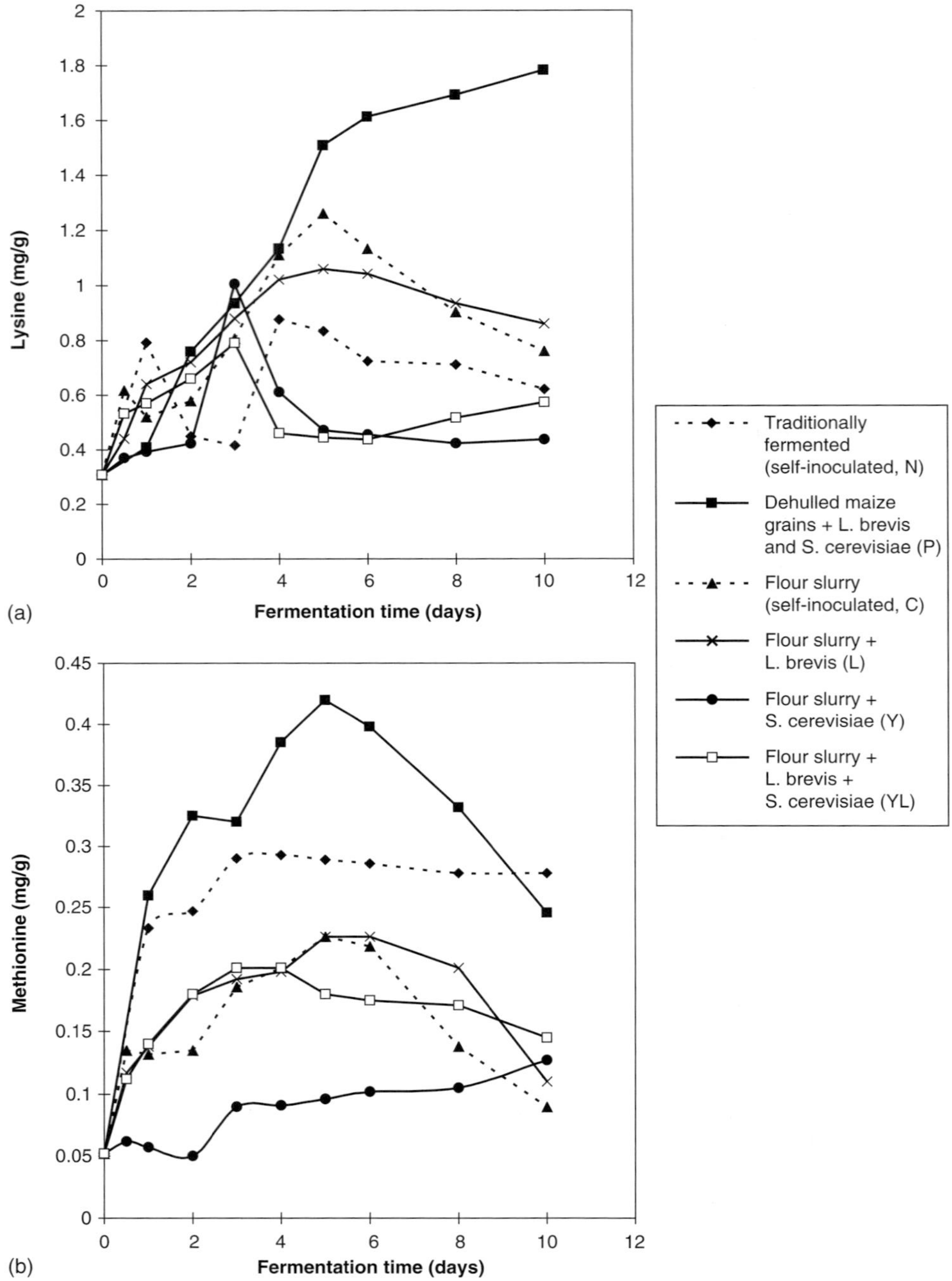

Figure 16.8 (a) Changes in lysine levels during ogi fermentation with different starter cultures. (b) Changes in methionine levels during ogi fermentation with different starter cultures

needed. Upgrading the production of fermented foods from the household to the industrialized level will necessitate several of the following critical steps (110):

1. Isolation and identification of the microorganisms associated with fermentation. Microorganisms associated with indigenous fermentations need to be isolated,

properly identified and preserved preferably in a recognized culture collection for future use.
2. Determination of the role(s) of the microorganism(s). The biochemical role(s) of microorganisms associated with food fermentations needs to be determined through chemical analysis of products released by the microorganisms under controlled laboratory conditions.
3. Selection and genetic improvement of microorganisms. Microorganisms responsible for effecting important changes in the food during fermentation should be selected and subjected to genetic improvement geared toward maximizing desirable quality attributes in the food and limiting any undesirable attributes.
4. Improvement in process controls for the manufacture of fermented foods. Improvements in the quality and quantity of fermented foods may be achieved by manipulating environmental factors such as temperature, moisture content, aeration, pH, and acidity, which influence the activity of microorganisms during the fermentation process.
5. Improvement in the quality of raw materials used in the production of fermented foods. Both the quality and the quantity of fermented foods may be improved by choosing raw materials other than those traditionally used for their production.
6. Laboratory simulation of the fermented foods. Prior to pilot scale production, and (ideally) after all the five stages have been well studied, fermented products may be produced under laboratory conditions. Laboratory simulation of fermented foods will involve the production of fermented foods by inoculating microbial isolates having desirable properties, into raw materials.
7. Pilot stage production. The pilot stage is the first clear departure from small scale production and should be based on the result of laboratory experiments.
8. Production or industrial plant stage. The production stage is the culmination of all the previous efforts and should lead to the availability of food of predictable and consistent quality on a larger scale.

16.9 CONCLUSION

Traditional small scale fermentation technologies offer considerable potential for stimulating development in the food industry of African countries in the light of their low cost, scalability contribution to food safety and nutrition, minimal energy and infrastructural requirements, and the wide acceptance of fermented products in these countries (111). In many traditional approaches, the advantages of some form of inoculation of a new batch, e.g., by back slopping or the repeated use of the same container (e.g., calabash) is appreciated and generally practiced. Still, the benefits of small scale starter culture application as a means of improved hygiene, safety, and quality control, in support of HACCP approaches, are not yet realized in small scale fermentation operations.

The introduction of appropriate starter culture techniques may constitute one major step toward improved safety, quality, and security of traditional small scale fermentation. The increasing scientific information on microbe substrate interactions and particularly on beneficial properties of LAB strains typically associated with fermentation, this potential is expected to be exploited to a much greater extent in the future.

ACKNOWLEDGMENTS

The author gratefully acknowledges Ford Foundation award through Lagos State University (LASU, Nigeria) toward the preparation of the manuscript.

REFERENCES

1. Oyewole, O.B. Lactic fermented foods in Africa and their benefits. *Food Contr.* 8:289–297, 1997.
2. Odunfa, S.A. African fermented foods. In: *Microbiology of fermented foods*. Wood, B.J., ed., London and New York: Elsevier Applied Science Publishers, 1985, pp151–191.
3. Holzapfel, W.H. Use of starter cultures in fermentation on a household scale. *Food Contr.* 8:241–258, 1997.
4. Campbell-Platt, G. Editorial: food control and its impact on food safety. *Food Contr.* 1:1–3, 1997.
5. Kimaryo, V.M., G.A. Massawe, N.A. Olasupo, W.H. Holzapfel. The use of starter culture in the fermentation of cassava for the production of kivunde, a traditional Tanzanian food product. *Int. J. Food Microbiol.* 56:179–190, 2000.
6. Amoa-Awua, W.K. The dominating microflora and their role in the fermentation of 'agbelima' cassava dough. Ph.D thesis, University of Ghana, 1996.
7. Ashworth, A., A. Draper. The potential of traditional technologies for increasing the energy density of weaning foods. WHO/CDD/EDP/92.4:7–19, 1992.
8. Olasupo, N.A., U. Schillinger, W.H. Holzapfel. Studies of some properties of predominant lactic acid bacteria isolated from Nigerian fermented foods. *Food Biotech.* 15:157–161, 2001.
9. Oyewole, O.B., S.A. Odunfa. Fermentation of cassava of lafun and fufu production in Nigeria. *Food Lab. News* 7:29–31, 1991.
10. Nout, M.J.R., P.K. Sakar. Lactic acid fermentation in tropical climates. *Antonie van Leeuwenhoek* 76:395–401, 1999.
11. Iwuoha, C.I., O.S. Eke. Nigerian indigenous fermented foods: their traditional process operation, inherent problems, improvements and current status. *Food Res. Int.* 29:527–540, 1996.
12. Schleifer, K.H., W. Ludwig. Phylogenic relationship of lactic acid bacteria. In: *The Lactic Acid Bacteria,* Vol, 2,Wood, B.J.B., W.H. Holzapfel, eds., New York: Elsevier, 1995, pp 7–18.
13. Stackebrandt, E., M. Teuber. Molecular taxonomy and phylogenetic position of lactic acid bacteria. *Biochime* 70:317–324, 1988.
14. D.J. Houngouigan, M.J.R. Nout, C.M. Nago, J.H. Houben, F.M. Rombouts. Starter cultures of lactobacilli and yeasts in fermentation of Mawé porridge. In: *Fermentation of Maize (Zea mays L.) Meal for Mawé Production in Bénin,* Hounhouigan, D.J., ed., Ph.D thesis, Agricultural University of Wageningen, The Netherlands. 1994.
15. Nout, M.J.R. Microbiological aspects of the traditional manufacture of Busaa, a Kenyan opaque maize beer. *Chem. Mikrobiol. Technol. Lebensm.* 6:137–142, 1980.
16. Odunfa, S.A., S. Adeyele. Microbiological changes during the traditional production of ogibaba, a West African fermented sorghum gruel. *J. Cereal Sci.* 3:173–180, 1985.
17. Adegoke, G.O., A.K. Babalola. Characteristics of microorganims of importance in the fermentation of fufu and ogi, two Nigerian foods. *J. Appl. Bacteriol.* 65:449–453, 1988.
18. Akinrele, I.A. Fermentation studies on maize during the preparation of a traditional African starch-cake food. *J. Sci. Food Agric.* 21:619–625, 1970.
19. Halm, M., A. Lillie, A.K. Sorensen, M. Jakobsen. Microbiological and aromatic characteristics of fermented maize doughs for kenkey production in Ghana. *Int. J. Food Microbiol.* 19:135–143, 1993.
20. Holzapfel, W.H., R. Geisen, U. Schillinger. Biological preservation of foods with reference to protective cultures, bacteriocins and food-grade enzymes. *Int. J. Food Microbiol.* 24:343–362, 1995.
21. Hirsch, A., E. Grinsted, H.R. Chapman, A.T.R. Mattick. A note on the inhibition of an anaerobic spore former in Swiss-type cheese by nisin-producing *streptococcus. J. Dairy Res.* 18:205–206, 1951.
22. Olasupo, N.A., D.K. Olukoya, S.A. Odunfa. Plasmid profiles of bacteriocin-producing *Lactobacillus* isolates from African fermented foods. *Folia. Microbiol.* 39:181–186, 1994a.

23. Olasupo, N.A., U. Schillinger, C.M.A.P. Franz, W.H. Holzapfel. Bacteriocin production by *Enterococcus faecium* NA01 from 'wara,' a fermented skimmed cow milk product from West Africa. *Lett. Appl. Microbiol.* 19:438–441, 1994.
24. Olasupo, N.A., U. Schillinger, A. Narbad, H. Dodd, W.H. Holzapfel. Occurrence of nisin Z production in *Lactococcus lactis* BFE 1500 isolated from wara, a traditional Nigerian cheese product. *Int. J. Food Microbiol.* 53:141–152, 1999.
25. Olasupo, N.A., D.K. Olukoya, S.A. Odunfa. Assssment of a bacteriocin-producing *Lactobacillus* strain in the control of spoilage of a cereal-based African fermented food. *Folia. Microbiol.* 42:31–34, 1997.
26. Black, R.E., M.H. Merson, A.S. Rahman, M. Yunus, A.R. Alim, I. Hug, R.H. Yolken, G.T. Curlin. A two-year study of bacterial, viral and parasitic agents associated with diarrhoea in rural Bangladesh. *J. Infect. Dis.* 142:660–664, 1980.
27. Huilan, S., L.G. Zhen, M.M. Mathan, M.M. Mathew, J. Olarte, R. Espejo, U. Maung, M.A. Ghafoor, M.A. Khan, Z. Sami, et al. Etiology of acute diarrhoea among children in developing countries: a multicentre study in five countries. *Bull. World Health Org.* 69:549–555, 1991.
28. Gomes, T.A.T. Enteropathogens associated with acute diarrhoeal diseases in urban infants in Sao Paulo, Brazil. *J. Inf. Dis.* 164:331–337, 1991.
29. Mensah, P.P.A., A.M. Tomkin, B.S. Drasar, T.J. Harrison. Effect of fermentation of Ghanaian maize dough on the survival and proliferation of 4 strains of *Shigella Flexneri. Trans. Roy. Soc. Trop. Med. Hyg.* 82:635–636, 1988.
30. Nout, M.J.R., F.M. Rombouts, A. Havelaar. Effect of accelerated natural lactic fermentation of infant food ingredients on some pathogenic microorganisms. *Int. J. Food Microbiol.* 8:351–361, 1989.
31. Lorri, W.S.M. Nutritional and microbiological evaluation of fermented cereal weaning foods. Ph.D thesis, Chalmers University of Technology, Göteborg, Sweden, 1993.
32. Olukoya, D.K., S.I. Ebigwei, N.A. Olasupo, A.A. Ogunjimi. Production of dogik: an improved ogi (Nigerian fermented weaning food) with potential for use in diarrhoeal control. *J. Trop. Ped.* 40:108–113, 1994.
33. Svanberg, U., E. Sjogren, W. Lorri, A.M. Svennerholm, B. Kaijser. Inhibited growth of common enteropathogenic bacteria in lactic fermented cereal gruels. *World J. Microbiol. Biotechnol.* 8:601–606, 1992.
34. Kingamkono, R., E. Sjogren, U. Svangerg, B. Kaijser. pH and acidity in lacctic- fermented cereal gruels: effects on viability of enteropathogenic micro-organisms. *World J. Microbiol. Biotechnol.* 10:664–669, 1994.
35. Mensah, P.P.A., A.M. Tomkin, B.S. Drasar, T.J. Harrison. Antimicrobial effect of fermented Ghanaian maize dough. *J. Appl. Bacteriol.* 70:203–210, 1991.
36. Yusof, R.M., J.B. Morgan, M.R. Adams. Bacteriological safety of a fermented weaning food containing L-lactate and nisin. *J. Food Protect.* 56:414–417, 1993.
37. Abdallah, O.M., P.M. Davidson, G.L. Christen. Survival of selected pathogenic bacteria in white picked cheese made with lactic acid bacteria or antimicrobials. *J. Food Protect.* 56:972–976, 1993.
38. Ashenafi, M. Fate of *Salmonella enteritidis* and *Salmonella typhimurium* during the fermentation of ergo, a traditional Ethiopian sour milk. *Ethiopian Med. J.* 31:91–98, 1993.
39. Isono, Y., I. Shingu, S. Shimizu. Identification and characteristics of lactic acid bacteria from Masai fermented milk in Northern Tanzania. *Biosci. Biotech. Biochem.* 58:660–664, 1994.
40. Arocha, M.M., M. McVey, S.D. Loder, J.H. Rupnow, L. Bullerman. Behaviour of hemorrhagic *Escherichia coli* 0157:H7 during the manufacture of cottage cheese. *J. Food Prot.* 55:379–381, 1992.
41. Mbugua, S.K., J. Njenga. The antimicrobial activity of fermented uji. *Ecol. Food Nutr.* 28:191–198, 1991.
42. Steinkraus, K.H. *Handbook of Indegenous Fermented Foods*, 2nd ed. New York: Marcel Dekker, Inc., 1996.
43. Barber, L.E., R.H. Deibel. Effect of pH and oxygen tension on staphylococcal growth and enterotoxin formation in fermented sausages. *Appl. Microbiol.* 24:891–898, 1972.

44. Daly, C., M. Lachance, W.E. Saandine, P.R. Elliker. Control of *Staphylococcus aureus* in sausage by starter cultures and chemical acidulation. *J. Food Sci.* 38:426–430, 1973.
45. Niskanen, A., E. Nurmi. Effect of starter culture on staphylococcal exterotoxin and thermonuclease production in dry sausage. *Appl. Environ. Microbiol.* 31:11–20, 1976.
46. Bonestroo, M.H., B.M.J. Kusters, J.C. de Wit, F.M. Rombouts. The fate of spoilage and pathogenic bacteria in fermented sauce-based salads. *Food Microbiol.* 10:101–111, 1993.
47. Holzapfel, W.H., N.A. Olasupo. Food safety in traditional food fermentation. In: *Quality System for Traditional Fermentation in Africa* , Jakobsen, M., A. Amoa-Awua, ed., Chapter 8 (In press)
48. Jeppesen, V.F. Biological preservation of seafood by lactic acid bacteria. *Infofish International* 5:29–34, 1993.
49. Nout, M.J.R. Fermented foods and food safety. *Food Res. Int.* 27:291–298, 1994.
50. Karunaratne, A., E. Wezenberg, L.B. Bullerman. Inhibition of mold growth and aflatoxin production by *Lactobacillus* spp. *J. Food Prot.* 53:230–236, 1990.
51. Gourama, H., L.B. Bullerman. Inhibition of growth and aflatoxin production of *Aspergillus flavus* by *Lactobacillus* species. *J. Food Prot.* 58:1249–1256, 1995.
52. Gourama, H., L.B. Bullerman. Antimycotic and antiaflatoxigenic effect of lactic acid bacteria: a review. *J. Food Prot.* 58:1275–1280, 1995.
53. Nout, M.J.R., P.F. Nche, P.C.H. Hohman. Investigation of the presence of biogenic amines and ethy carbamate in kenkey made from maize and maize-cowpea mixtures as influenced by process conditions. *Food Additiv. Contam.* 11:397–402, 1994.
54. Buchenhüskes, H.J., I. Sabatke, K. Gierschner. Zur Frage des Vorkommens biogener Amine in milchsauer fermentiertum Gemüse: citical review. *Industrielle Obst. Und. Gemüseverwertung* 1/92, 255–263, 1992.
55. Westby, A., P.J.A. Reilly, Z. Bainbridge. Review of the effect of fermentation on naturally occuring toxins. *Food Contr.* 8:329–339, 1997.
56. Bover-Cid, S.M., M. Izquierdo-Pulido, M.C. Vidal-Carou. Effect of proteolytic Starter Cultures of *staphylococcus* spp. on biogenic amine formation during the ripening of dry sauages. *Int. J. Food Microbiol.* 46:95–104, 1999.
57. Kpodo, K., A.K. Sorenson, M. Jabobsen. The occurrence of mycotoxins in fermented maize products. *Food Chem.* 56:147–153, 1996.
58. Ogunsanwo, B.M., O.O. Faboya, O.R. Idowu, T. Ikotun, D.A. Akano. The fate of aflatoxins during the production of 'ogiri:' a West African fermented melon seed condiment from artificially contaminated seeds. *Nahrung* 33:983–988, 1989.
59. Ogunsanwo, B.M., O.O. Faboya, O.R. Idowu, T. Ikotun, D.A. Akano. Fate of aflatoxins in soybeans during the preparation of 'soy-ogi'. *Nahrung* 33:485–487, 1989b.
60. Adegoke, G.O., E.J. Otumu, A.O. Akanni. Influence of grain quality, heat and processing time on the reduction of aflatoxin B level in tuwo and ogi: two cereal-based products. *Plant Food Hum. Nutr.* 45:113–117, 1994.
61. Jespersen, L., M. Halm, K. Kpodo, M. Jakobsen. Significance of yeast and moulds occuring in maize dough fermentation for 'kenkey' production. *Int. J. Food Microbiol.* 24:239–248, 1994.
62. Arici, M. Untersuchungen zum fermentativen Abbau von Patulin mit ausgewählten Milchsäurebakterien aus südeuropäischen Ländern, Dr.rer.nat. Dissertation, University of Karlsruhe (TH), 1997.
63. Smith, I.E., C.W. Lewis, J.G. Anderson, G.L. Solomons. Mycotoxins in human nutrition and health, European Union Directorate General, XII Report 16048 EN, 1994.
64. Amoa-Awua, W.K.A., F.E. Appoh, M. Jakobsen. Lactic acid fermentation of cassava dough into agbelima. *Int. J. Food Microbiol.* 31:87–98, 1996.
65. Ejiofor, M.A.N., N. Okafor. Comparison of pressed and unpressed cassava pulp for garri making. In: *Tropical Root Crops: Research Strategies for the 1980s,* Terry, E.R., K.A. Oduro, F. Vaceness, eds., Ottawa, Canada: International Development Research Centre, 1981, pp 154–158.

66. Essers, A.J.A. Removal of cyanogens from cassava roots: studies on domestic sun drying and solid substrate fermentation in rural Africa. Thesis, Landbouwuniversiteit Wageningen, The Netherlands, 1995.
67. Hahn, S.K. An overview of African traditional cassava processing and utilization. *Outlook Agric.* 18:110–118, 1989.
68. Vasconcelos, A.T., D.R. Twiddy, A. Westby, P.J.A. Reilly. Detoxification of cassava during gari preparation. *Int. J. Food Sci. Technol.* 25:198–203, 1990.
69. Westby, A., B.K. Choo. Cyanogen reduction during the lactic fermentation of cassava. *Acta Hort.* 376:209–215, 1994.
70. Westby, A. Importance of microorganisms in cassava processing. In: *Tropical Root Crops in a Developing Economy: Proc. 9th Symposium of the International Society for Tropical Root Crops, 20-26 October, 1991,* Ofori, F., S.K. Hahn, eds., Accra, Ghana: ISHS, pp 249–255, 1991.
71. Westby, A., D.R. Twiddy. Role of microorganisms in the reduction of cyanide during traditional processing of African cassava products. In: *Proceedings of a Workshop on Traditional African Foods: Quality and Nutrition,* Westby, A., P.J.A. Reilly, eds., International Foundation of Science, November 1991, pp 127–132.
72. Westby, A, D.R. Twiddy. Characterisation of gari and fufu preparation procedures in Nigeria. *World J. Microbiol. Biotechnol.* 8:175–182, 1992.
73. Chavan, J.K., S.S. Kadam. Nutritional improvement of cereals by fermentation. *Crt. Rev. Food Sci. Nutr.* 28:349–400, 1989.
74. Mbugua, S.K., R.H. Ahrens, H.N. Kigutha, V. Subramanian. Effect of fermentation, malted flour treatment and drum drying on nutritional quality of uji. *Ecol. Food Nutr.* 28:271–277, 1992.
75. Reedy, M.R., M.D. Pierson. Reduction in antinutritional toxic components in plant foods by fermentation. *Food Res. Int.* 27:281–290, 1994.
76. Svanberg, U., W. Lorri. Fermentation and nutrient availability. *Food Contr.* 8:319–327, 1997.
77. Abdel Gawad, A.S. Effect of domestic processing on oligosaccharide content of some dry legume seed. *Food Chem.* 46:25–31, 1993.
78. Irving, C.C.J. Phytase. In: *Inositol Phosphates: Their Chemistry, Biochemistry and Physiology,* Cosgrove, D.J., ed., Amsterdam: Elsevier,1980, p 85.
79. Svanberg, U., A.S. Sandberg. Improving iron availability of weaning foods through the use of germination and fermentation. In: *Improving Young Child Feeding in Eastern and Southern Africa: Household Level Food Technology: Proceedings of a Workshop, Nairobi, Kenya, October 1987*, D. Aldwick, S. Moses, O.G. Schmidt, eds., IDRC–265e, Ottawa, Ontario, Canada, 1988, pp 366–373.
80. Lopez, Y., D.T. Gordon, M.L. Fields. Release of phosphorus from phytate by natural fermentation. *J. Food Sci.* 48:953–954, 1983.
81. Mahajan, S., B.M. Chauhan. Phytic acid and entractable phosphorus of pearl millet flour as affected by natural lactic acid fermentation. *J. Sci. Food Agric.* 41:381–386, 1987.
82. Khetarpaul, N., B.M. Chauhan. Effect of fermentation of pure cultures of yeasts and lactobacilli on phytic acid and polyphenol content of pearl millet. *J. Food Sci.* 54:780–781, 1989.
83. Udaysekhara Rao, D., Y.G. Deosthle. *In vitro* availability of iron and zinc in white and coloured ogi: role of tannin and phytate. *Qual. Plant: Plant Food Hum. Nutr.* 38:35–41, 1988.
84. Reddy, M.R., M.D. Pierson, D.K. Salunkhe. *Legume–based fermented foods.* Boca Raton, FL: CRC Press, 1986.
85. Reedy, M.R., D.K. Salunkhe. Effect of fermentation on phytate phosphorus and mineral content in black gram, rice, and black gram and rice blends. *J. Food Sci.* 45:1708–1712, 1980.
86. Aderibigbe, E.Y., S.A. Odunfa. Growth and extracelluar enzyme production by strains of *Bacillus* species isolated from fermenting African locust bean, iru. *J. Appl. Bacteriol.* 69:662–671, 1990.
87. Eka, O.U. Effect of fermentation on the nutrient status of locust bean. *Food Chem.* 5:303–308, 1980.

88. Shallenberger, R.S., D.B. Hand, K.H. Steinkraus. *Changes in Sucrose, Raffinose and Starchyose during Tempeh Fermentation: Report of the 8th Bean Research Conference, Mellaire, MI.*, ARS 74–41, USDA, 1967, pp 68–71.
89. Sorensen, W.G., C.W. Hesseltine. Carbon and nitrogen utilization of *Rhizopus Oligosporus. Mycologia* 58:681–689, 1967.
90. Odunfa, S.A. Carbohydrate changes in fermenting African locust bean (*Parkia filicoidea*) during iru preparation. *Qual. Plant Food Hum. Nutr.* 32:3–10, 1983.
91. Buono, M.A., L.E. Erickson, D.Y.C. Fung. Carbonhydrate utilisation and growth kinetics in the production of yoghurt from soymilk, part II: experimental and parameter estimation results. *J. Food Process. Preserv.* 14:179–204, 1990.
92. Adam, M. Untersuchungen zum Abbau unerwünschter Oligosaccharide in der Nahrung durch ausgewählte Milchsäurebakterien. Diplomarbeit, Universität (TH) Karlsruhe, Germany, 1994.
93. Mbugua, S.K. The nutritional and fermentation characteristics of uji from dry milled maize flour (Unga Baridi) and whole wet milled maize. *Food Chem. Microbiol. Technol.* 10:154–161, 1986.
94. Nche, P.F. Innovation in the production of kenkey, a traditional fermented maize product of Ghana: nutritional physical and safety aspects. Ph.D thesis, Agricultural University of Wageningen, The Netherlands. 1995.
95. Khetarpaul, N., B.M. Chauhan. Effect of germination and fermentation on *in vitro* starch and protein digestibility of pearl millet. *J. Food Sci.* 55:883–884, 1990.
96. Lorri, W.S.M., U. Svanberg. Lactic fermented cereal gruels with improved *in vitro* protein digestibility. *Int. J. Food Sci. Nutr.* 44:29–36, 1993.
97. Back Knudsen, K.E., L. Munck, B.O. Eggum. Effect of cooking, pH and polyphenol level on carbohydrate composition and nutritional quality of a sorghum (*Sorgum bicolor* (L.) Moench) food, ugali. *Brit. J. Nutr.* 59:31–47, 1988.
98. Graham, G.G., W.C. MacLean Jr., E. Morales, B.R. Hamaker, A.W. Kireis, E.T. Mertz, J.D. Axtell. Digestibility and utilization of protein and energy from nasha, a traditional Sudanese fermented sorghum weaning food. *J. Nutr.* 116:978–984, 1986.
99. Hamaker, B.R., A.W. Kirieis, L.G. Butler, J.D. Axtell, E.T. Mertz. Improving the *in vitro* protein digestibility of sorghum with reducing agents. *Proc. Nat. Acad. Sci. USA* 84:626–628, 1987.
100. Odunfa, S.A., Dawadawa, I. *Legumes-Based Fermented Foods*. Boca Raton, FL: CRC Press, 1986, pp 179–189.
101. ACC/SCN. *Second Report on the World Nutrition Situation, Vol 1: Global and Reginal Results.* World Health Organisation, Geneva, 1992.
102. Wand, Y.D., M.L. Fields. Feasibility of home fermentation to improve the amino acid balance of corn meal. *J. Food Sci.* 43:1104, 1978.
103. Olukoya, D.K., S.I. Smith, N.A. Olasupo, A.A. Ogunjimi, A.M. Abaelu, A. Apena, R. Iyanda. Development of nutritionally improved cereal-based indigenous ogi. *Adv. Food Sci. (CMTL)* 22:77–80, 2000.
104. Odunfa, S.A., J. Nordstrom, S.A. Adeniran. Development of starter cultures for nutrient enrichment of ogi, a West African fermented cereal gruel: report submitted to HBVC research grants program. USAID, Washington, USA, 1994.
105. Odunfa, S.A., S.A. Adeniran, O.D. Teniola, J. Nordstrom. Evaluation of lysine and methionine production in some lactobacilli and yeasts from Ogi. *Int. J. Food Microbiol.* 63:159–163, 2001.
106. Teniola, O.D., S.A. Odunfa. The effects of processing methods on the levels of lysine, methionine and the general acceptability of ogi processed using starter cultures. *Int. J. Food Microbiol.* 63:1–9, 2001.
107. Tongnual, P.T., M.L. Fields. Fermentation and relative nutritive value of rice meal and chips. *J. Food Sci.* 44:1784, 1979.
108. Kazanas, N., M.L. Fields. Nutritional improvement of sorghum by fermentation. *J. Food Sci.* 46:819, 1981.

109. Aliya, S., P. Geervani. An assessment of the protein quality and vitamin B content of commonly used fermented products of legumes and millet. *J. Sci. Food Agric.* 32:837, 1981.
110. Odunfa, S.A. Cereal fermentation in African countries. In: *Fermented Cereals: A global perspective,* FAO Agricultural Services Bulletin No 138, Chapter 2, pp 1–17, 1999.
111. Rolle, R., M. Satin. Basic requirements for the transfer of fermentation technologies to developing countries. *Int. J. Food Microbiol.* 75:181–187, 2002.

3.17

Fermentation Biotechnology of Traditional Foods of China

Zuoxing Zheng, Changlu Wang, and Yang Zheng

CONTENTS

17.1 INTRODUCTION

Fermentation is one of the oldest, and remains one of the most economical, methods of producing and preserving traditional foods in China. These delicacy foods are made from ordinary agricultural products or byproducts through microbial fermentation to generate unique flavors and aromas. In China, there are many kinds of traditional fermented food products, which have a history that is centuries long. The main products include alcoholic beverages such as white wine and yellow wine, fermented condiments, and foods such as soy sauce, vinegar, cooking wine, Jiang, fermented tofu, Chinese bread, fermented meats and vegetables, as well as fermented tea. These fermented Chinese foods, in a broader term, have played a very important role in Chinese cuisines for thousands of years. They have not only contributed greatly to the unique palatability of Chinese foods, many of them are of beneficial to human's health and wellness. Today as the natural sciences and technology are rapidly advancing, the ancient methods of making fermented foods in China are also changing gradually through modern microbiology and biotechnology.

17.2 HISTORY OF FERMENTATION TECHNOLOGY IN CHINA

The earliest fermentation technology in China can be traced back to about 4,000 years ago when Jiu, a Chinese wine, was first made, followed by vinegar and Jiang (fermented soybean paste), which were originally made about 2,000 years ago (1,2). Today there are hundreds of fermented foods in China. The development history of fermentation technology in China can be divided into three stages: traditional, modern, and contemporary stages.

The traditional fermentation processes in China have two types: submerged fermentation and solid-state fermentation. The traditional oriental fermented foods are mostly made through solid-state fermentation processes. Examples of natural solid-state fermented foods include soy sauce, Chi (soybean paste), fermented tofu, red rice, and Mantou. Fermented vegetables can be made through either solid-state or submerged fermentation processes. Traditionally, such fermentation occurs naturally and the microorganisms come from local micro flora, therefore the fermented products often have very unique indigenous flavor characteristics. For example, a typical Chinese fermented vegetable is made by putting vegetables in a sealed tank containing salted water left at room temperature. The salt in the water inhibits the growth of many spoilage bacteria in vegetables while the lactic acid bacteria will grow and the resulting lactic acid further inhibits the growth of spoilage organisms, and eventually the lactic acid bacteria will outgrow and produce organic acids and flavor compounds in the products. Such natural fermentation processes are still used today in China.

The modern fermentation stage is characterized by controlled fermentation with selected pure cultures. Since about 100 years ago, as modern microbiology has rapidly advanced in the world, Chinese scientists have been able to study the mechanisms of fermentation and isolate microorganisms from natural fermented foods. Many pure cultures, including bacteria, mold, and yeast have been obtained and used commercially for mass production of traditional fermented foods in China. Some of the naturally isolated cultures have been undergone further improvement through natural breeding processes such as hybridization and mutation. Such improved strains are not genetically modified organisms, but are metabolically enhanced flavor producers or high yield producer strains. They have been widely used in food and fermentation industries in China. Mutation technology also can alter the natural metabolic pathways of microorganisms, thus certain natural metabolites can be inhibited while other natural metabolites are selectively "forced" to be overproduced. This technology is called metabolically controlled fermentation. Such technology has been widely used in China to commercially manufacture many industrial chemical products such as citric acid, polysaccharides, amino acid, vitamins, and ethanol. In addition to microbial isolation and improvement, the fermentation process has also been changed during this modern fermentation stage. For example, the fermentation medium, temperature, pH, moisture content, and fermentation time have all been optimized for maximal yield and best quality of targeted products. Some traditional fermentation processes have been changed from batch process to a continuous fermentation process.

Mechanical and analytical development has also made traditional fermentation process easier to control. Pure and improved culture and optimized process have significantly increased the productivity, quality, and safety of traditional fermented foods in China.

The contemporary fermentation technology is the result of the development in biotechnology, especially DNA technology. Since the mid twentieth century, the rapid advances in microbial physiology, cell infusion technology, immobilized cell and enzyme technology, genetic engineering, metabolic engineering, bioreactor design, and purification technology have dramatically promoted development of contemporary fermentation technology in China. Development in biosensors has made automation possible in fermentation process control to some extent. The fermentation products have been greatly expanded, and the concept of fermentation has also been changed. Fermentation is not limited to traditional fermented products or even food products any more; rather, many specialized products for applications in agricultural, food, chemical, and pharmaceutical industries have been commercially produced through microbial fermentation. Examples include various industrial enzymes such as amylases and proteases, organic acids such as lactic acid, acetic acid, propionic acid, malic acid, amino acids such as glutamic acid, phenylalanine and asparagines, nucleotides such as inosine and 5′-inosinic acid, vitamins such as ascorbic acid and vitamin B12, coenzyme A, glycerin, ethanol, and single cell protein. These fermentation products are well beyond the scope of traditional fermented Chinese foods, and the introduction of computerized control system to such fermentation and downstream process is becoming more and more popular in today's China.

The purpose of this article is mainly to introduce and review the process and technology development of some typical traditional Chinese fermented foods.

17.3 FERMENTED ALCOHOLIC BEVERAGES

Fermented alcoholic beverages in China are called "Jiu," which includes distilled and non-distilled spirits, wine, and beer. Jiu manufacturing has a very long history in China and is still a very important industry today. There are many varieties of the fermented alcoholic beverages in china, such as white wine, yellow wine, sweet wine, and cooking wine.

17.3.1 White Wine

Chinese white wine is recorded in Bencao Gangmu, a Chinese encyclopedia of medicinal herbs compiled by Shizhen Li during the Yuan Dynasty (2). There are different varieties of white wines according to different materials and processes used. They are classified as Big-Qu and Small-Qu wines. The Big-Qu wines are made with large usage level of Qu (starter culture) through solid-state fermentation, while the Small-Qu wines are made with smaller usage level of Qu through solid-state or submerged fermentation. They are also categorized into six major groups by the types of aroma: the light-scented type, the heavy-scented type, the fermented-sauce aroma type, the mixed-aroma type, the fermented-rice aroma type, and miscellaneous aroma type. The famous brand names of Fen, Luzhou Tequ, Maotai, Xifeng, Guilin Sanhua, and Dong represent each type, respectively.

The Chinese white wines are made from sorghum, wheat, and rice with Qu as the saccharification agent and fermentation starter culture through solid-state fermentation and distillation processes. The Qu is made from wheat inoculated with mother cultures obtained from naturally occurring wild type microorganisms. A typical Big-Qu making process is shown in Figure 17.1.

The Big-Qu has a history of several thousand years and it was originally used to make yellow wines. It was then used to make white wines. Many well known white wines,

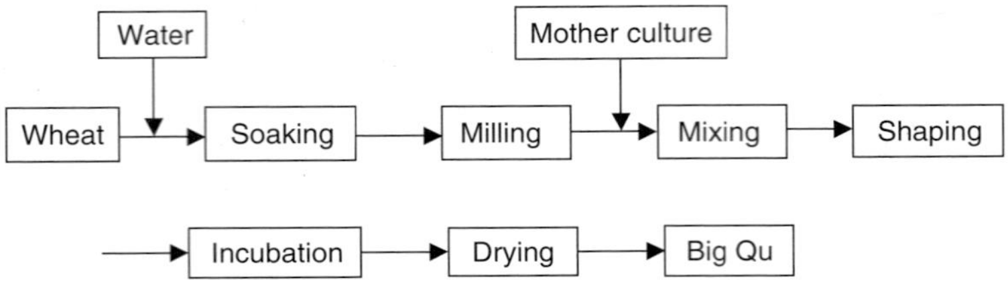

Figure 17.1 Flow chart of a typical Big Qu making

including Maotai, Fen Jiu, Luzhou Laojiao Tequ, Xifeng Jiu, Wuliang Ye, Lang Jiu, and Gujing Gongjiu are all made with Big-Qu.

Maotai is one of the best white wines made in China. Maotai Brewery is located in Maotai Village, Renhuai County in Guizhou Province. Maotai Jiu is made from sorghum through fermentation with Big-Qu and it involves repeated fermentation–distillation cycles for as many as 8 times. The Maotai brewing process is shown in Figure 17.2.

Because the repeated fermentation and distillation process, great quantities of Big Qu are required in Maotai making, almost equal to the amount of the raw material sorghum. The wine collected between the third and sixth distillations is the best, and the one collected last is the worst in quality. All collections will be mixed and adjusted to standardization prior to storage for at least one year before going to market (3–6).

Maotai wine has a strong yet elegant flavor which is believed to be generated from Big-Qu during fermentation. It is a typical example of Big-Qu wines with a soy sauce type aroma. The climate of Maotai Village in China is mild and humid, which provides ideal conditions for the growth of microorganisms and may contribute to the specific quality of the very famous brand wine.

The conversion rate, or the utilization efficiency, of starch used to produce white wines is relatively low in traditional processing. In order to increase productivity and conserve cereal grains, Chinese brewery scientists have conducted research on isolation and identification of fermentation microorganisms from traditional manufacturing environment and process samples. They obtained several pure cultures of *Aspergillus oryzae* and *Saccharomyces* species for fermentation. As a result of pure culture fermentation, the starch utilization efficiency was increased from 40–50% to 70–80% in wine fermentation process. Recently, wheat bran has been used to replace wheat in the traditional Big-Qu making process, and a submerged liquid culture was used in the fermentation process to increase the alcohol output. Although the productivity was increased, the aroma of the finished product was not as good as the one produced by the traditional methods.

To compensate for the quality, compromised by productivity increase, and improve the flavor of wine products, some new flavor enhancing techniques have been developed in China. Aroma stripping and aroma blending are the most commonly used methods. In aroma stripping process, a traditional fermented aroma concentrate called Xiangpei, is made separately, and is placed near the vapor outlet of the distillation apparatus during white wine process. When the alcohol vapor passes the aroma source Xiangpei, it strips some of the flavors and lets them cocondense in the product. This method still requires some traditional operations and it loses some alcoholic content when passing through the aroma source. Aroma blending method is a kind of artificial method. The alcohol used for blending is deodorized with active carbon before artificial flavorants are added. The ingredients of the artificial flavors used often contain acetic acid, butyric acid, lactic acid, citric acid, ethyl acetate, isopentyl alcohol, and glycerol. Usually, 10% traditional white wine is added to the product. Although the overall manufacturing cost of such wines is

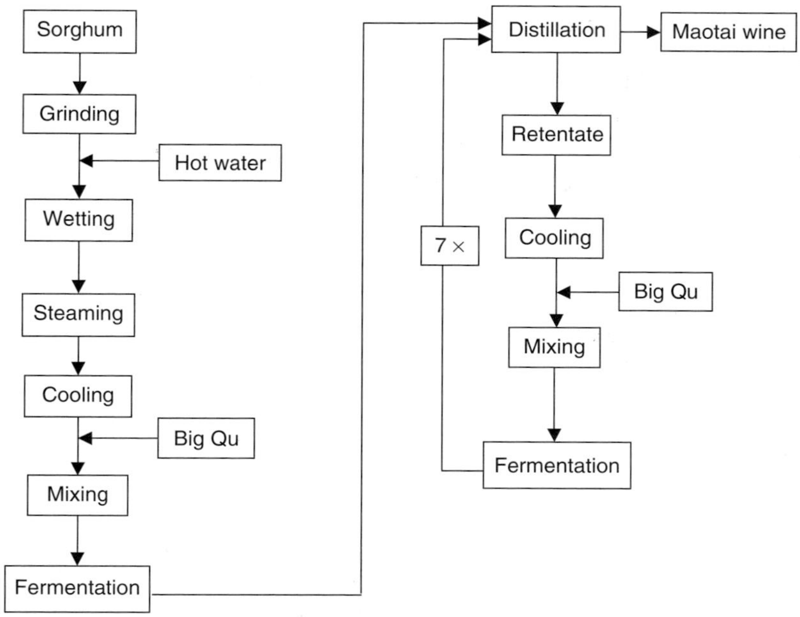

Figure 17.2 Flow chart of a Chinese white wine Maotai making

significantly reduced, the quality of such artificially blended wines is not as good as the traditional wines.

In 1975 the Mongolian Institute of Light Industry and Chemical Engineering isolated a caproic acid producing bacteria *Bacillus* species from the black soil of old wine-making cellars. This *Bacillus* isolate is strictly anaerobic and requires alcohol and sodium acetate as the major carbon source to produce caproic acid and other critical aroma compounds. This strain has been used commercially in white wine industries in China, and the resulting finished products contain high level of hexyl acetate and other flavors characteristic to the white wines. The sensory evaluation results for such products are fairly close to those of the strong-scented type of Big-Qu white wines.

17.3.2 Yellow Wine

Yellow wine is made from the fermentation of rice. It is called "yellow wine" because of the yellowish color of the finished product. It contains about 15% alcohol, and is rich in amino acids. The lysine content in yellow wine is much higher than that in beer, or grape wine. Routinely drinking certain amount of yellow wine is good for health. Sometimes it is used as extraction solvent of Chinese medicinal herbs for medical applications. More often however, yellow wine is used as a condiment in cooking for many Chinese cuisines. Yellow wine making is very popular in southern China where rice is the major cereal food. In yellow wine making, the three major materials, rice, Qu and water, are often referred as the "meat, bone, and blood" of yellow wine. The quality of these materials determines the quality of the wine. The ideal rice is rich in starch especially the branched starch, and low in protein and fat. Among many kinds of rice, the sticky rice, or glutinous rice, is considered the best. The water used for brewing should have a pH range of 6.8–7.2, with the hardness of 2–7°DH, and a very low level of minerals. Qu is made from fermentation of wheat, therefore, the wheat quality is important too. Sometimes 10–20% barley is added to wheat for Qu-making to improve aeration and microbial and enzymatic activities during fermentation. The amino acids, especially glutamic acid in the wheat, are the main source of the delicious taste in the yellow wine.

Just like white wine, yellow wine has many varieties. However, all varieties share some common characteristics. They are all made from rice or sorghum through steaming, saccharification, fermentation, press filtration, and storage processes. Qu and yeast are used as key saccharifying and fermenting agents, while many other natural microorganisms are also involved in the brewing process for the complicated flavor development. Saccharification and fermentation usually occur simultaneously at relatively lower temperatures. Starch is converted to sugars by Qu and the sugars are subsequently transformed to alcohol by yeast. Other microbial metabolites such as organic acids, amino acids, esters, and various alcohols are also formed during fermentation. They contribute significantly to the distinctive flavor of the finished product. Some well known brand names of yellow wines in China, such as Shaoxing Jianfan and Chengang Jiu are also found available on the market in the United States.

A typical process of manufacturing yellow wine is shown in Figure 17.3. The pasteurized yellow wine is hot filled in jugs and the jugs are usually stored for about 1 year of ripening before they enter the market. Some premium yellow wines are required to be aged for as long as 3 years. This storage period is similar to ripening processes in cheese making. During ripening, the fresh wine undergoes various chemical reactions to form esters and many other aroma compounds. The ripened yellow wine is very mild and balanced in aroma and taste. In general, fermented alcoholic beverages can enhance the pleasure of eating. Their nutritional roles include indirect contribution through subjective enhancement of appetite and a variety of other physiological effects (7).

17.4 FERMENTED CONDIMENTS AND FOODS

17.4.1 Soy Sauce

Soy sauce is probably man's oldest prepared seasoning and its exact origin is prehistoric and just like most other fermented foods it was initially used to prevent food spoilage and

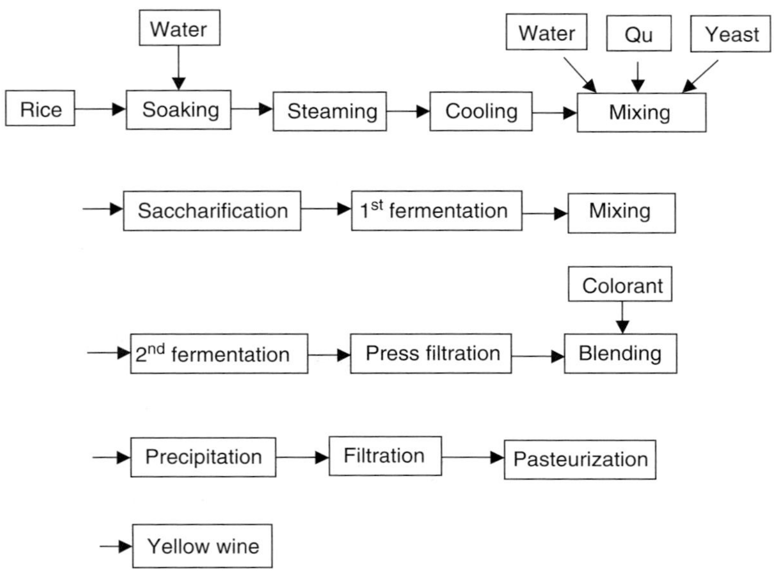

Figure 17.3 Flow chart of a typical yellow wine making process

preserve food. The origins of soy sauce, like other fermented foods, go back millennia. The first recorded reference to the soybean, from which soy sauce is made, is thought to be Shen Nan's *Materia Medica*, produced in China in 2838 B.C. (8). In the sixth century, Buddhism came to Japan from China. Along with this new religion came a vegetarian seasoning, a salty paste of fermented beans (9). This is the earliest direct ancestor of soy sauce. Soy sauce is known as Shoyu in Japan and Jiang-yiu in China. It is mainly used as all purpose seasoning in China and East Asia. Soy sauce has been regarded as one of the most popular condiments in the long history of China. It is estimated that, at present, more than 2 million tons of soy sauce are produced every year in China (2). Soy sauce is now well known in the western world, particularly in North America, where it has been popularized by communities with Chinese origins. It is also becoming popular in the U.S. as a sauce for barbecue, steaks, and hamburgers.

In China there are two types of soy sauce: fermented soy sauce and blended soy sauce. Fermented soy sauce is made from soybean, wheat, and wheat bran through natural fermentation. Blended soy sauce is made by blending 50% or more of fermented soy sauce with acid hydrolyzed vegetable proteins and other food additives (10,11). In addition to salt, soy sauce contains 18 amino acids and polypeptides, sugar, polysaccharides, organic acids, alcohols, aldehydes, esters, phenolic compounds, vitamins, and minerals (12).

Soy sauce fermentation process can be carried out in liquid state, semisolid state, or solid-state fermentation. In China, solid-state fermentation is most widely used in soy sauce manufacturing, and it accounts for more than 70% of total soy sauce produced in China. The typical manufacturing process is shown in Figure 17.4.

For thousands of years, Chinese soy sauce was produced by using soy sauce Qu, a natural fermented product, and the precursor of soy sauce. In 1930, a pure culture of

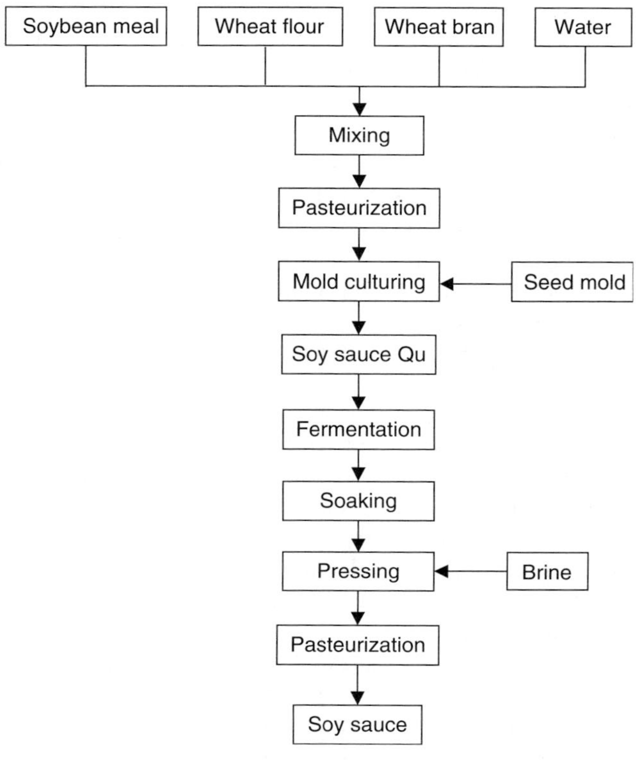

Figure 17.4 A typical manufacturing process of Chinese fermented soy sauce

Aspergillus oryzae was isolated from Qu and a new Qu fermentation method, using the pure culture, was developed for soy sauce manufacturing. The new pure culture fermentation significantly shortened the ripening time for soy sauce from 1 year to 2 months. Another invention of a no salt, soy sauce making process made it possible to reduce the fermentation time further down to 1–2 days, but the problem with temperature control has yet to be resolved. A low salt concentration and heat preservation method was then developed and fermentation time was about 10 days. This has become the most popular manufacturing method in the soy sauce industry in China today. Recently submerged liquid culture method has also shown some success and a promising future in soy sauce making.

To improve the quality of soy sauce, culture screening is one of the important activities in soy sauce research in China. Better species with a higher protease activity are still desired. The cytoplasmic fusion and *in vitro* gene recombination and other advanced techniques have been explored to screen and obtain a microorganism which possesses multifunctional enzymes, such as high activities of protease, glucoamylase, glutaminase and cellulase. The widely used pure culture of *Aspergillus oryzae* has a high protease activity, which hydrolyzes soy and wheat proteins to peptides and free amino acids including glutamic acid that are mainly responsible for soy sauce flavor and taste. Other characteristic flavor compounds found in soy sauce are the tautomers of 4-hydroxy-2-ethyl-5-methyl-3-furanone and 4-hydroxy-2-methyl-5-ethyl-3-furanone which exist in a ratio of approximately 3:2, and their content was found to correlate positively to the quality of soy sauce (13).

To increase the conversion rate and improve flavor in the product, Zheng (14) had tried to obtain a hybrid strain of *A. oryzae* and *A. niger* through protoplast infusion technology. The *A. niger* strain was a starter culture for vinegar production, and had a high amylase activity. The objective was to incorporate some properties of a vinegar culture into a soy sauce culture. The hybrid strain combined both high protease activity and high amylase activity from parent strains, and was able to shorten fermentation time and increase soy sauce yield as well as improve the flavor of soy sauce. Another approach to improving soy sauce quality is to use adjunct cultures such as yeast and lactic acid bacteria during fermentation to produce rich flavor in soy sauce.

Savory nucleotides such as 5′-inosine mono phosphate (IMP) and 5′-guanosine mono phosphate (GMP) are great flavor enhancers. The commercial production of 5′-nucleotides from enzymatic hydrolysis of poly nucleotides or through microbial fermentation has been successfully achieved (2,15). Addition of these nucleotides to soy sauce has created a product with superior flavor, which enjoys a good market. The conversion of 5′-xanthosine mono phosphate to 5′-IMP, and phosphorylation of inosine to 5′-IMP and the production of savory nucleotides directly by soy sauce fermentation, are among the efforts to optimize soy sauce processing and achieve premium quality.

17.4.2 Jiang

Jiang is fermented paste of soybean and wheat flour. The traditional process of Jiang making is similar to soy sauce making, except that the ripened fermented mash is consumed directly without brine treatment and drip-washing, which are necessary in soy sauce making. In the modern Jiang making, a pure culture of *Aspergillus oryzae* is used in the fermentation of the raw material simulating Qu making. Salt is added upon completion of the fermentation, followed by naturalization, which is a slow cooling process. A typical process is shown in Figure 17.5.

Because soybean Jiang making shares the same principle as soy sauce making, their processes both depend on the function of enzymatic saccharification and protocolization by microorganisms during fermentation. Large scale Qu can be made by solid-state fermentation or liquid fermentation. Both methods use the same culture of *Aspergillus oryzae*.

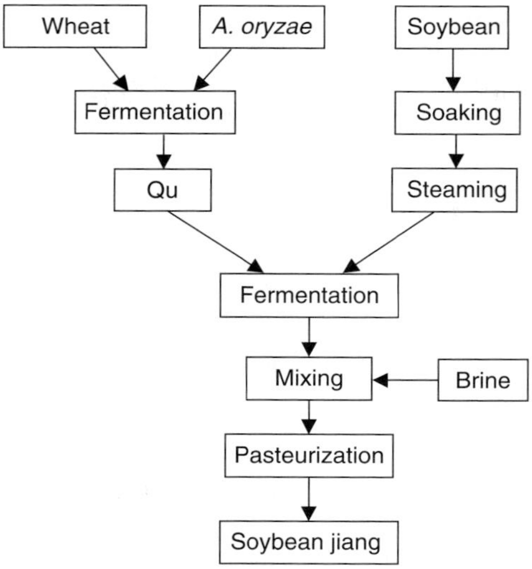

Figure 17.5 A typical manufacturing process of soybean Jiang

In solid-state fermentation, *A. oryzae* is introduced into powdered Qu, which is then used as a starter for Jiang making. The same starter can also be used in wheat Jiang making, in which the wheat flour is hydrated and steamed, then mixed thoroughly with powdered Qu and fermented at 45–50°C for a period of 10 days. During solid-state fermentation, the internal temperature could reach as high as 65°C on 8th day. This way, about 2 kg of Jiang can be made from every 1 kg of wheat flour. Although *A. oryzae* plays a key role in Jiang fermentation, many other microorganisms including molds, lactic acid bacteria, and yeasts are also involved during aging process to contribute to the overall flavor and taste of the finished product, Jiang (16).

17.4.3 Soybean Chi

Soybean Chi is a traditional fermented food in China and is widely distributed in most southern provinces such as Zhejinag, Fujian, Sichuan, Hunan, Hubei, Jiangsu, and Jiangxi, as well as some northern provinces in China. Although the exact origin of Chi is unknown, ancient Chinese literature indicates that Chi making existed before the Qin Dynasty (2). The traditional Chi making relies on natural fermentation of soybean. April and May are the best months for Chi making, followed by late July and August. Summer is not a good time for Chi making because it is difficult to control temperature.

In traditional Chi making, soybeans are sorted and steamed until they are soft. After cooling they are spread on mats on house floors. The thickness of the material and frequency of agitation depend on the temperature of the material. After a few days, yellow color mold will appear on the surface of the soybeans. After the mold spreads all over the soybeans, the material is mixed with salted water and is transferred to an earthen jar. The jar is tightly sealed and stored for a few months for ripening. The Chi produced in this way has a strong flavor similar to that of soybean Jiang.

Today, Chi is produced by fermentation with pure cultures, which can be either *Aspergillus, Mucor,* or bacteria, with *Aspergillus* predominating. Some selected *Aspergillus* cultures used in Chi industry today are obtained by natural screening or artificial mutation breeding. Such strains are usually able to produce better flavor and savoriness, with less off flavors.

17.4.4 Fermented Tofu

Fermented bean curd (Tofu) is called Doufuru or Furu in Chinese. It is one of the most famous Chinese fermented foods. It is flavorful, tasteful, and nutritious, and has been consumed for more than 1,000 years in China (2). Fermented tofu is produced all over the country in China and many provinces have their own characteristic Furu.

Traditionally, fermented Tofu is made by natural fermentation of Tofu. The process is still used in many places today. Tofu is nonfermented soybean curd and is very popular in China and Japan and is gaining popularity in the United States. Doufuru making consists of three major processes, i.e., tofu-making process, mold fermentation process, and brining and aging process, as shown in Figure 17.6. The soybeans are washed and soaked in water, and then milled into slurries. The slurries are diluted with warm water and pressed to obtain soymilk with 5–6% total solids. The soymilk is boiled and then a small amount of coagulant ($CaSO_4$ or $MgSO_4$) is added to the hot soymilk to precipitate the proteins. Immediately, the coagulated soymilk is placed in cloth bags and the bags are pressed in a mold. Stones or wooden planks are often used to squeeze off the water (soymilk whey). After the curd is formed it is sliced into different sized pieces. Recently, glucono-δ-lactone has been used widely as a new coagulant, instead of calcium sulfate or

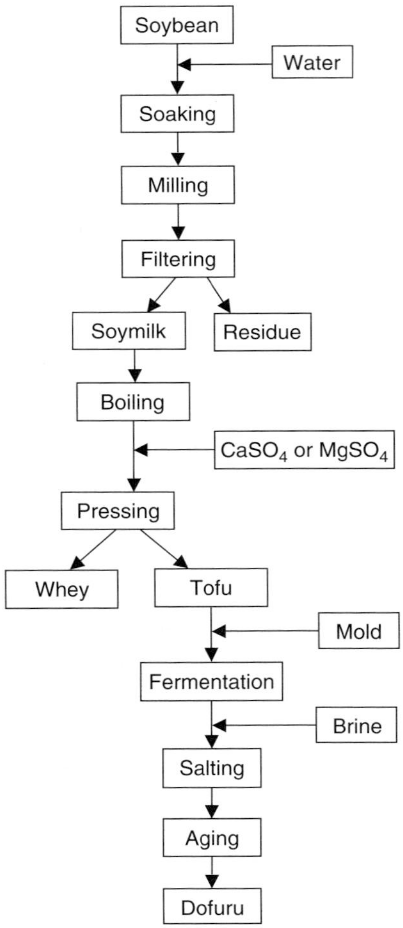

Figure 17.6 Manufacturing process of Doufuru

magnesium sulfate, because it needs less skill to form a high quality Tofu curd. In general Tofu contains 88% moisture, 6% proteins, and 3.5% oil. The regular Tofu curd is fragile and perishable and therefore is not suitable for Doufuru making. In Doufuru making, the Tofu is made so that its water content may be less than 70%, so that the texture of Tofu is much firmer than that of regular Tofu. The Tofu thus prepared is cut into 3 cm cubes and then heated. The purpose of heating is to pasteurize the cubes and to reduce moisture content on the cube surfaces, on which mold strains are inoculated. In mold fermentation, the molds belonging to the genus *Mucor* or *Actinomucor* are usually used, but the molds belonging to the genus *Rhizopus* are also used sometimes. For instance, *A. elegans, M. hiemalis, M. silvaticus, M. praini, M. subtilissimus, and R. chinensis* are used for the inoculation. The time of mold fermentation differs depending on the varieties of mold, and it normally takes about 7 days at 12°C for *R. chinensis*, 3 days at 24°C for *M. hiemalis* and *M. silvaticus*, and 2 days at 25°C for *M. praini* (17).

The last process of Doufuru making is salting and ripening in a dressing mixture. The freshly molded cubes are placed in various types of brining solution depending on the flavor desired (18). Normally the brining solution consists of salted fermented rice mash, soy sauce mash, and fermented soy paste. The time of ripening ranges from 1 to 12 months depending upon the varieties of brining solution. The salt content in the brining solution is very critical in ripening process. Salt inhibits the enzymatic ripening process for protein and lipid hydrolysis. Increasing the salt content will increase hardness and elasticity while reducing adhesiveness, whereas decreasing the salt content increases the free fatty acid content. In general, lowering the salt content will reduce ripening time, which may benefit manufacturers, but if the salt content is below 5%, the fermented tofu may become spoiled during ripening (19). Finally the product is bottled with the brine, sterilized and marketed as Doufuru.

Doufuru has a creamy cheese like texture, and has a strong flavor and taste and therefore it not only can be used as condiments in oriental cuisine, but also may be used in the same way as blue cheese is used in salad dressing.

17.4.5 Vinegar

A very detailed description of vinegar making was recorded in an ancient Chinese book, *Qimin Yaoshu*. Just like soy sauce, vinegar is another important condiment in Chinese cuisines. In China, Shanxi aged vinegar and Zhenjiang scented vinegar are considered the best among the traditional Chinese vinegars. Both of them were fermented using a mixed solid culture. Similar to wine making, the solid-state fermentation of vinegar employs Big-Qu fermentation and Small-Qu fermentation. Shanxi aged vinegar is a typical example of Big-Qu fermentation. Big-Qu is a fermentation starter, brick shaped, and is made of wheat, barley, or green peas. In such fermentation, large amount of Qu is used and a longer fermentation and aging time is needed. During fermentation, saccharification and alcoholic fermentation occur at a low temperature. The freshly obtained vinegar is then placed outdoors, heated in summer by the sun and frozen in the winter. The ice formed in winter is removed. After aging through a hot summer and a severe winter, the final product of Shanxi aged vinegar is ready, it is viscous in texture, dark purple in color, sweet in taste, and has a long shelf life. On the other hand, the Zhenjiang scented vinegar presents a typical example of Small-Qu fermentation. In the past, distilled yellow wine mash was used as the raw material. Regular rice or waxy rice is now used. The whole procedure includes alcoholic fermentation, fermented mash making, drip washing of the vinegar, and many minor steps. The whole process normally takes about 60 days. The finished product is famous for its delicate combination of color, fragrance, sourness, mellowness, and richness.

However, both Big-Qu and Small-Qu fermentation are a result of naturally mixed culture fermentation. They also suffer the similar disadvantage of low production rate because of a long, labor intensive fermentation process. Many modifications and improvements have been made to speed up the fermentation cycle, replace the manpower with machines, and increase the utilization efficacy of the raw materials. Figure 17.7 represents a typical process of vinegar making in today's vinegar manufacturing industry in China. Basically, it uses ground rice and rice bran as the raw materials for regular alcoholic fermentation. It also employs pure cultures for both alcoholic and acetic acid fermentation. In the fermentation tank, a removable perforated bottom is installed to facilitate the drip washing. The liquid washed off from the fermentation tank is the final product of vinegar, which contains 5% acetic acid, 1.32% total sugar, and 0.52% total nitrogen. The product has a reddish brown color and a sweet sour taste. The quality is comparable to Zhenjiang scented vinegar.

Submerged liquid fermentation was also developed to reduce the fermentation time and make it easier for automation in vinegar making. The alcoholic fermentation using ground rice as the raw material remains the same. The fermented alcoholic mash is then piped into a stainless steel fermentor, and the acetic acid fermentation starts with the addition of 10% acetobacter culture broth containing *Acetobacter lovaniensis*. The temperature is maintained at 32–35°C for 65–72 hours for the whole acetic acid fermentation process. The final product is obtained after the addition of salt followed by filtration, sterilization, and blending. A total of 3.5 kg of vinegar with acetic acid concentration of 5% can be obtained from 1 kg of rice. The advantages of submerged fermentation over solid-state

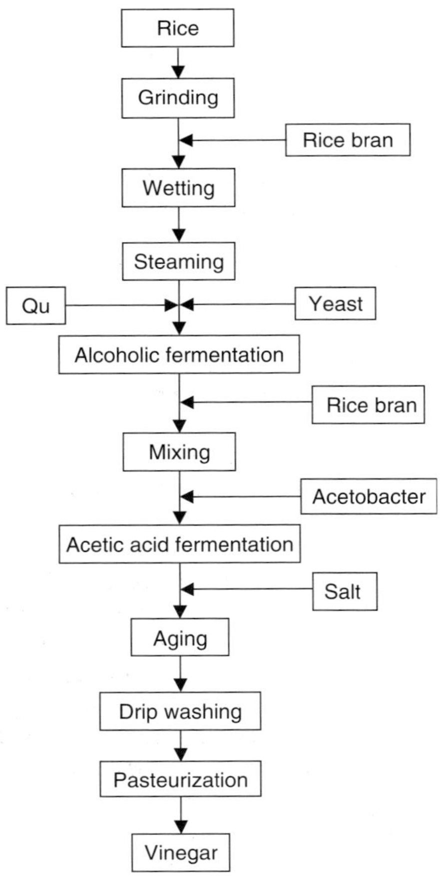

Figure 17.7 Vinegar manufacturing process

fermentation include shorter production cycle and higher extent of automation. However, the vinegar produced by submerged fermentation has fewer flavors than the one made by solid-state fermentation.

New technology development in vinegar making includes culture screening using protoplasmic fusion and genetic engineering techniques to obtain a hybrid which can resist high acetic acid concentrations up to 12% and which can produce good flavor comparable to that of established vinegar brands. Application of immobilized acetobacters for vinegar production has also been investigated in China.

17.4.6 Red Rice

Red rice, or Hong-Qu in Chinese, is a fermented product of rice with various fungal strains of *Monascus purpureus*. It was apparently first noted in the Yuan Dynasty (20). The detailed process information on red rice was first recorded in a Chinese famous ancient book Tiangong Kaiwu by Song Yingxing of the Ming Dynasty (1637 A.D.) (2).

Traditionally, red rice is made by natural fermentation. Rice is washed and soaked in water for 5–7 hours and then drained, steamed, and cooled on bamboo mats. The same procedure of steaming and cooling is repeated the next day. It is then mixed with Qu-starter, a milled and cream-like mash made from fermented rice, and moved into a fermentation room. The ambient conditions of the fermentation room are controlled at 30°C with moisture content of 45%. The inner temperature of the material is about 35°C. The mixture is turned over four to five times daily and the thickness of the layer is gradually reduced to 10–13 cm to compensate for the heat generated during fermentation. There are some little vent holes in the walls to adjust temperature of the windowless Qu-making room. On the third day, the material is taken out of the room and immersed in water for 10–15 minutes and then brought back to the room to continue the natural fermentation. On the fifth day, it is immersed in water again for 2–3 minutes. On the sixth day, it is sprayed with water and then transferred to an ordinary room, where the natural fermentation continues for 3 more days. It is dried in the sun on the ninth day. The dried fermented rice has a deep-red color on the surface and a reddish-white color inside. It can then be milled to powder and the red powder is called red rice or red-Qu. It contains *Monascus* and can be stored for a long time without deterioration.

The traditional method depends on the quality of the Qu-starter. The modern method simplified the whole process by using the pure culture of *Monascus* species. The pure culture is grown on rice to make high concentration starter. The starter is then used for inoculation in red rice fermentation. The fermentation time is shortened from about 8 days to 4 days.

The red rice is widely used as an enzymatic agent in alcoholic fermentation, as a condiment and colorant in cooking, and as a medicine in certain disease treatment in China. It is used in wine making in Fujian and Taiwan provinces, and the products are noted for their red color and mellow taste. The residue of the red rice alcoholic fermentation is called red Zao, and it is an ideal condiment in Chinese cuisine. The established Chinese dishes such as red Zao chicken, red Zao fish, and red Zao meat are among favorite menus in many Chinese restaurants.

Because the red rice powder is very rich in red pigment, it is often used as a natural food colorant. For example, it is used as a red colorant for fermented tofu (Doufuru). The red pigments obtained from *Monascus* are natural products and safe to be used as food grade color. Several high pigment-producing strains of *Monascus* species have been isolate for commercial production of food colorants (21,22).

The red rice also has antimicrobial activities against many spoilage microorganisms and therefore it is widely used for food preservation. Both the color and taste of the food can be well preserved by red rice. Fermented rice products including red rice have been used in traditional Chinese medicine too. A recent study demonstrated that the *Monascus purpureus* fermented red rice reduced serum total cholesterol and triglycerides in animal models of hypercholesterolemia and hyperlipidemia (23). It was also reported that an antibiotic, called Hongqu-mycin, was found in red rice (2). The antibiotic in the red rice is probably one of the key functional compounds in its medicinal applications, for its well known effectiveness in curing chronic enteritis and dysentery diseases in China. In addition, red rice is claimed to assist in digestion and blood circulation, and to benefit the spleen and stomach functions (2).

17.4.7 Chinese Bread

Chinese bread, called Mantou in Chinese, is white steamed yeast leavened dough. The making process of Mantou is very simple, and almost every Chinese family can make it during daily cooking. Traditionally, a Mantou starter (dried fermented dough, usually left from last Mantou preparation) is soaked in warm water until it is dissolved. Then it is mixed with wheat flour and water to make dough. The mixed dough is then placed in a warm place or in a water bath at a temperature of about 25–30°C for a few hours. During yeast fermentation, the dough is greatly expended in volume and it is mixed a few times to facilitate homogeneous fermentation inside the dough. A spongy dough texture and a distinct yeast flavor from the dough should be easily noticed. After the fermented dough is ready, it is then molded to an appropriate size and shape (usually hemisphere shape), and steamed for about 20–30 minutes. The steamed Mantou is ready to eat. Compared with bread, the Chinese bread or Mantou is bland in taste, because no sugar or additives are added to the dough. Usually a small portion of the fermented dough before molding is saved and dried in the air, and will be used as a Mantou starter for the next time preparation. Due to the nature of the natural fermentation, the Mantou starter contained yeast and many naturally occurring bacteria such as lactic acid bacteria. These bacteria are normally beneficial to improving the flavor and taste of the Mantou. However, sometimes they produce too much acid during dough fermentation, and in this case a small amount of soda powder is added to the fermented dough to neutralize the acid. The quality of the Mantou largely depends on the microbial content in the starter. Today dried pure yeast is readily available on the market, making it much easier to prepare Mantou at home.

17.5 OTHER FERMENTED PRODUCTS

Other fermented food products such as fermented vegetables, meats, and tea also have a long history and are still very popular in China. For example, according to Chinese history record, as early as 1200 A.D., pickles were consumed and loved by the King and the celebrities. Traditionally, the Chinese pickles are made by natural fermentation of vegetables soaked in a sealed earthen jar. The "starter" cultures are those native microbials naturally living on vegetables. The tightly sealed jar provided an anaerobic environment, and initially many anaerobic bacteria will grow but eventually the acid produced by *Lactobacillus* species such as *L. plantarum* will inhibit the growth of other bacteria and become dominant. As the pH drops, the growth of *L. plantarum* will slow down and other more acid tolerant lactic acid bacteria will continue to grow until the pH drops to as low as 3.6–3.8. Low pH and high acid in the pickles effectively prevent them from spoilage and they can be stored for months.

Fermented meats have a long history in China too. A well known Chinese fermented pork product called Jinhua Huotui is becoming increasingly popular in Asia. It is a traditional Chinese fermented food product made from aged, fermented ham, and is used in stock. Jinhua Huotui contains approximately 24% water, 24% protein, and 44% lipid. Comparing with commercial ham, although the total amino acid content in Jinhua Huotui is not different from that of ham, Jinhua Huotui contains more free amino acids, particularly glutamic acid, and high amounts of 5′-inosinic acid, which are responsible for its distinct delicious flavor and taste (24). The fermentation of Jinhua Huotui involves many kinds of mold, but the genus *Aspergillus* plays an important role in the decomposition of protein, and the genus of *Penicillium* is important in lipid degradation (25).

Tea growing originated in China and East Asia. In China, tea has been used as a beverage for 2,000–3,000 years, and it only later spread to other Asian countries, East Africa, the Middle East, and South America (26). The popularity of tea is derived from its delicate exotic aroma and possibly from the mild stimulating effects induced by caffeine. Some types of tea, such as green tea, are made by directly roasting tea leaves, while others such as black tea and oolong tea involve fermentation of tea leaves before drying. Green tea is roasted or steamed to stop enzymatic activity. For black tea, the leaves are allowed to ferment and this develops the aroma and color of the beverage. Oolong tea has the characteristics of both black tea and green tea, and the fermentation is only partial. Black tea processing involves four major steps:

1. Withering, in which tea shoots are placed in 30 cm deep beds or trays with forced circulation of warm air at a temperature of between 15 and 35°C for 4–18 hours until the leaf moisture is reduced from 75–80% down to 55–65%.
2. Leaf rolling or maceration, in which the leaf is broken into smaller particles and thus liberates its contents for the enzymes and catechins to mix and expose to the atmosphere resulting in enzymatic oxidation.
3. Fermentation, in which the sifted broken leaves are placed in fermentation drums or boxes with a constant supply of air for 30 minutes to 5 hours. During the natural fermentation, the leaf pieces darken and the flavor is developed further.
4. Drying, which stops the enzymatic and microbial activities. Either a fluidized bed system is employed or trays of leaves are placed in a hot air dryer at a temperature between 80 and 95°C. The final moisture of tea leaves is reduced to 3–5%. After drying, the finished product is graded and packaged for marketing. Fermented tea is now gaining more popularity in China, Japan, and many other countries.

This chapter discussed some major fermented foods in China. Of course there are many more fermented foods and beverages in China. Due to the long history of cultural evolution and huge geographic differences and large population in China, even the same fermented food can have many different varieties, and it is very difficult, if not impossible, to cover every aspect of the fermented food products in China. However, from the examples of the traditional fermented foods described in this chapter, one could draw a pretty good picture on an array of traditional Chinese fermented foods.

17.6 CONCLUSION

The ancient methods of making fermented foods are changing gradually through modern microbiology and other technologies. The fermentation technology passed down by ancient Chinese is a priceless treasure. However, some of the principles behind the technology still remain a mystery today. As a matter of fact, the making of Chinese fermented foods is really a combination of both science and art. Considerable progress has been made in the area of microbiology; however, some of the microorganisms involved in fermentation are yet to be isolated and identified. Unfortunately, today we understand that there are many microorganisms living in nature which are actually viable, but nonculturable by today's technology. This is probably one of the important reasons why the pure culture fermented products usually do not have as good flavor as the natural fermented products. Despite the sophisticated analytical techniques and equipment we have today, the functions and identities of many trace flavor compounds in the fermented foods remain unknown. Biochemical and biological interrelationships between different microorganisms are still beyond our present knowledge. In light of all these facts, it is not surprising to find that although the use of pure culture fermentation has greatly reduced the process time and increased the yield of the products, their qualities often are compromised. Therefore, even today, the best quality fermented foods such as Maotai wine and Luzhou Tequ are still produced by traditional processes in China.

The traditional process is a natural mixed fermentation in which some complicated biochemical and biological changes are made by multiple cultures. The superior quality thus produced must be a joint result of different metabolic and physiological processes by various microorganisms involved in the natural fermentation. To surpass the traditional processes not only in speed and efficiency but also in quality remains a great challenge and this area is still wide open for further research.

Generally speaking, traditional foods are closely related to the regional food cultures and eating habits and therefore some of their flavor might be considered to be unacceptable for the people who do not belong to these food cultures. However, these traditional foods are products which have been developed through thousands of years and had been selected through a long experience of the people. Thus, these flavors would be acceptable eventually for most of the people, even though unacceptable at the first trial. In fact, many oriental fermented foods have entered into the U.S. market and are gaining more and more popularity in the West. For example, the consumption of soy sauce in the United States is rapidly growing in recent years. The steady penetration of Chinese traditional fermented foods to the American population might be proof that a traditional fermented flavor in one area can also be accepted by most of the people in another area. Wider acceptance of Chinese fermented foods in the world will also stimulate research activities and promote technology development to finally benefit food industries and the consumers alike.

REFERENCES

1. Wong, Y, S. Zhang. Domestic and international development review in biotechnology. *Sheng-Wu Gong-Cheng Jin-Zhan* 5:1–10, 1998.
2. Chen, T., C.T. Ho. Past, present, and future of Chinese fermented food products. *Food Rev. Int*. 5:177–208, 1989.
3. Wang, F. *Modern Food Fermentation Technology*. Beijing: Chinese Light Industry Press, 1998.
4. Gu, L., W. Zhai. *Fermented Food Technology*. Beijing: Chinese Light Industry Press, 1998.
5. Kang, M. *Technology Manual of Chinese and International Well-Known Fermented Foods*. Beijing: Chinese Chemical Industry Press, 1997.
6. Zou, X., L. Wang, C. Shen. *Fermented Foods Preparation*. Beijing: Jindun Press, 2000.
7. Darby, W.J. The nutrient contributions of fermented beverages. In: *Fermented Food Beverages in Nutrition,* Gastineau, C.F., W.J. Darby, T.B. Turner, eds., New York: Academic Press, 1979, pp 61–74.
8. Rose, A.H. History and scientific basis of microbial activity in fermented foods. In: *Economic Microbiology Vol. 7: Fermented Foods*, Rose, A.H., ed., London: Academic Press, 1982, pp 1–13.

9. Nunomura, N., M. Sasaki. Soy sauce. In: *Legume-Based Fermented Foods,* Reddy, N.R., M.D. Pierson, D.K. Salunkhe, eds., Boca Raton, FL: CRC Press, 1986, pp 5–46.
10. Lin, Z.S. *Questions and Answers on Soy Sauce Brewing Technology*. Beijing: Chinese Light Industry Press, 2000.
11. Yang, T.Y. *Fermented Seasoning Technology*. Beijing: Chinese Light Industry Press, 2000.
12. Zhao, J.F. *Food Processing Technology,* 2nd ed. Beijing, Chinese Light Industry Press, 1999.
13. C.T. Ho, Y. Zhang, H. Shi, J. Tang. Flavor chemistry of Chinese foods. *Food Rev. Int.* 5:253–287, 1989.
14. Zheng, Z. Isolation of auxotrophic mutants of *Aspergillus oryzae* and *Aspergillus niger* and the preparation of their protoplasts. *Ind. Microbiol.* 24:18–20, 1994.
15. Zhang, K., Z. Zheng. Studies on the breeding of 5′-inosinic acid-producing strains and fermentation. *Food Ferment. Ind.* 5:15–21, 1989.
16. Chou, C.C., G.R. Hwang, F.M. Ho. Changes of microbial flora and enzyme activity during the aging of tou-pan-chiang, a Chinese fermented condiment. *J. Ferment. Technol.* 66:473–478, 1988.
17. Fukushima, D. Fermented vegetable protein and related foods of Japan and China. *Food Rev. Int.* 1:149–209, 1985.
18. Han, B.Z., R.R. Beumer, F.M. Rombouts, M.J.N. Robert. Microbiological safety and quality of commercial sufu: a Chinese fermented soybean food. *Food Contr.* 12:541–547, 2001.
19. Han, B.Z., J.H. Wang, F.M. Rombouts, M.J.R. Nout. Effect of sodium chloride on textural changes and protein and lipid degradation during the ripening stage of sufu, a Chinese fermented soybean food. *J. Sci. Food Agric.* 83:899–904, 2003.
20. Steinkraus, K.H. *Handbook of Indigenous Fermented Foods.* New York: Marcel Dekker, 1983, pp 547–553.
21. Lin, T.F., A.L. Demain. Effect of nutrition of *Monascus* sp. on the formation of red pigments. *Appl. Microbiol. Biotechnol.* 36:70–75, 1991.
22. Lotong, N., P. Suwanarit. Fermentation of ang-kak in plastic bags and regulation of pigmentation by initial moisture content. *J. Appl. Bacteriol.* 68:565–570, 1990.
23. Li, C., Y. Zhu, Y. Wang, J.S. Zhu, J. Chang, D. Kritchevsky. *Monascus purpureus*- fermented rice (red yeast rice): a natural food product that lowers blood cholesterol in animal models of hypercholesterolemia. *Nutr. Res.* 18:71–81, 1998.
24. Wagu, Y., T. Kakuta, H. Shindo, T. Koizumi. General components, amino acid, 5′-nucleotide and microorganism of Chinese fermented food Jinhua Huotui. *Nippon Shokuhin Kogyo Gakkaishi* 41:921–926, 1994.
25. Wagu, Y., T. Kakuta, H. Shindo, T. Koizumi. Identification and enzyme activity of molds isolated from Chinese fermented food Jinhua Huotui. *Nippon Shokuhin Kagaku Kaishi* 43:796–805, 1996.
26. Fowler, M.S., P. Leheup, J.L. Cordier. Cocoa, coffee and tea. In: *Microbiology of Fermented Foods*, 2nd ed., Wood, B.J.B., ed., London: Blackie Academic & Professional, 1998, pp 128–147.

3.18

Fermentation Biotechnology of Traditional Foods of the Indian Subcontinent

E. Rati Rao, S.V.N. Vijayendra, and M.C. Varadaraj

CONTENTS

18.1 INTRODUCTION

Mankind's eating habits have been a challenging area of scientific and technological developments. As advancements are made in food science and technology, man has made changes in his diet habits. In recent years, there has been a growing interest in traditional fermented foods, both at national and international levels. It is very well known that Asians and Africans are pioneers in utilizing fermentation as a means to preserve a wide variety of foods, including cereals, legumes, vegetables, milk, and meat. Early human civilizations, with no concept of microorganisms, were remarkable not only in developing the products, but also safely maintaining the cultures through centuries. These fermented foods are prized for the palatability and wholesomeness and are important components of diets in many parts of the world, especially in the Indian subcontinent.

Studies of the traditional fermented foods have established the multiple advantages as they pertain to nutritive values, therapeutic properties, and sensory attributes (1,2). In all these fermented foods, the process of fermentation is initiated mainly by lactic acid bacteria (LAB), and in a few of them, the fermentation is initiated by yeasts (3). In general, fermented foods are characterized by the accumulation of metabolites of microbial activities such as lactic acid, acetic acid, traces of ethanol, and other compounds collectively called as flavors. These secondary metabolites contribute to the overall acceptability of the final fermented product and also provide preservative effects.

India is a vast country on its own, and along with other countries of the subcontinent, has been the home of varied groups of people with and diverse living and dietary habits. This diversity accounts for a large number of traditional fermented foods. Nearly 90% of such traditional foods remain highly secretive in their preparation and tend to be regionalized, with most of these foods prepared at the household level. Very few of these traditional foods have been explored from the scientific and technological point of view. Some of these traditional foods and beverages have been studied, not only for their commercial value, but they have the potential to improve human health. These efforts have resulted in the preparation of products with uniform and consistent quality. This processing has the added benefit of providing healthy and safe foods for human consumption. The significance of the traditional fermented foods of India, an extensive review on the microbiological and biochemical aspects of selected fermented foods (4), as well as a comprehensive overview of the historical and technological developments of Indian foods is presented (5).

18.2 BIOTECHNOLOGICAL IMPETUS IN FOOD FERMENTATIONS

Fermented products are being produced throughout the world by converting desirable substrates into unique products through the action of desirable microorganisms, giving rise to acceptable products, both in terms of nutritional and sensory attributes. In many of these preparations, the mechanism of the conversion is not well defined nor understood. With the changes in eating patterns, wherein emphasis has been toward natural, healthy, and safe foods, the impetus has been on fermented foods, either of the ready to process or ready to eat nature. Therefore, demand is increasing for the production of traditional fermented foods on a large scale. As a result, problems of maintaining quality and safety of the products arise. This leads to a need for better understanding of the fermentation processes, through which the production processes can be made more reliable and predictable. In order to meet the demands of quality and safety, appropriate biotechnological developments, and an analysis of the basic microbiological parameters (6) is required.

Optimization of starter culture performance and elimination of the factors that can impede the fermentation process are the essential steps to improve the quality and safety of the finished products. Although considerable research has been carried out with regard to the improvement of starter cultures used in cheese preparation at the industrial level (7), very little effort has gone into the improvement of the starter cultures responsible for the preparation of quality traditional fermented products. The development of typical flavor in these foods is the major driving force behind the popularity of traditional foods. Under the prevailing circumstances, the biotechnological inputs toward obtaining consistent quality traditional fermented foods needs to be fully exploited.

Although manipulation of proteolytic system of certain lactic acid bacteria has been reported (8,9), the mechanism of flavor development has been less understood. To a certain extent, the role of proteases, peptides, and lipases, in flavor development has been established. Genetic engineering of the strains of lactic acid bacteria to produce derivatives with new attributes has been a challenge for food biotechnologists. In an era of scientific and technological advancements, there is definitely a need to look into the following aspects, which when achieved, can significantly place traditional fermented foods on the global map (10,11):

1. Desirable polysaccharides in enhanced levels to meet the quality parameters of body and texture of products
2. Desirable acid production in a shorter period
3. Increased therapeutic properties
4. Genetic stability and improved resistance to antibiotics and pesticides
5. Desirable production of dietary fiber
6. Threshold levels of desirable flavor components with stability during storage

Lactic acid bacteria have already been used as potent vehicles for vaccine delivery as a means to improve immunity in humans (12). Besides, genetic engineering studies with lactic acid bacteria have shown the approach to be positive in achieving the requisite attributes in traditional fermented foods.

18.3 FERMENTED FOODS OF THE INDIAN SUBCONTINENT

18.3.1 Milk Based Foods

Milk is the most important foodstuff for a mammal and has always been the first food of the newborn. Fermentation of milk, either knowingly or unknowingly, has occurred since early times, resulting in various fermented milk products. Fermented milk products are known for their taste, nutritive value, and therapeutic properties. The nature of these products has differed from region to region depending on the indigenous microflora, which in turn depends upon the surrounding environmental factors (13).

Although the origin of the cultured dairy products is not very clear, it dates back to the dawn of civilization. Milk from different species of domesticated animals has been used for the preparation of fermented products. The oldest traditional fermentations applied to the production of fermented milks depended on climatic and regional characteristics where human populations dwelled. The methods of fermentation were also influenced by old customs carried over from generation to generation. The most popular traditional fermented products of Indian subcontinent are *dahi, mishti doi, lassi* and *shrikhand*. Although these products are being prepared at the household level and in an unorganized dairy sector, the large scale production by industries is on a slow pace.

Traditionally, fermented milk products have been consumed either as beverages or as a part of meals. The preparation of these products, either at households or in unorganized sector, involves use of previous day's product as a starter, which invariably is a mixed culture, wherein there is no or little control on the fermentation process. On the contrary, the use of desired microorganisms as in the case of controlled fermentation would greatly enhance the chances of obtaining products with uniform and consistent quality products of acceptable attributes (1). Generally, the fermented milk products are prepared using freshly obtained milk of various milk animals. However, use of refrigerated milk stored up to four days have been found to be more suitable than fresh milk in obtaining fermented milks such as cultured milks, yogurt and probiotic yogurts with improved quality (14).

The increased activity of the lactic acid bacterial cultures in refrigerated milks has been attributed to heat stable enzymes and protein degradation products elaborated by psychrotrophic bacteria during refrigerated storage (15). The nutritional aspects (fermentation of lactose, hydrolysis of protein and lipids, vitamins, minerals) and therapeutic (physiological and health benefits) of indigenous peoples of the Indian subcontinent have been studied compared to the fermentation processes used by Western cultures (16). Use of genetically improved lactic cultures to improve the shelf life and safety of fermented milk products is envisaged in a few of the studies (17). Further attempts are being made to isolate lactic cultures having more health benefits. The work carried out on probiotic cultures in the Western world has influenced researchers in the Indian subcontinent. As a result, research has been focused on this angle to improve the nutritional and therapeutic quality of indigenous fermented milk products such as *dahi*, *lassi* and other cultured products, which have been documented in reviews (16,18).

18.3.1.1 Dahi

Dahi is one of the most popular fermented milk products of the Indian subcontinent, liked for its mild acidic taste and pleasant flavor. It can be consumed directly or along with other foods such as boiled rice and *chapathi*, a traditional wheat based pancake. Sushrut Samhita, the Indian medical treatise, describes *dahi* as a food promoting appetite and strength. It is prepared by lactic fermentation of pasteurized or boiled buffalo's or cow's milk through the action of single or mixed lactic cultures (19,20). The conventional method of *dahi* preparation is shown as a flow diagram in Figure 18.1.

A good quality *dahi* is of firm and uniform consistency with a sweet aroma and a clean acid taste. The surface is smooth and glossy and usually a cut surface is trim, free from cracks and gas bubbles. The nutritional and therapeutic values of *dahi* are rated as high when compared to that of the milk used for its preparation. It is easy to prepare, even at the household level. *Dahi* is easy to digest and it has been found to reduce the risk of cardiovascular problems and cancerous tumors, besides strengthening general immunity (21).

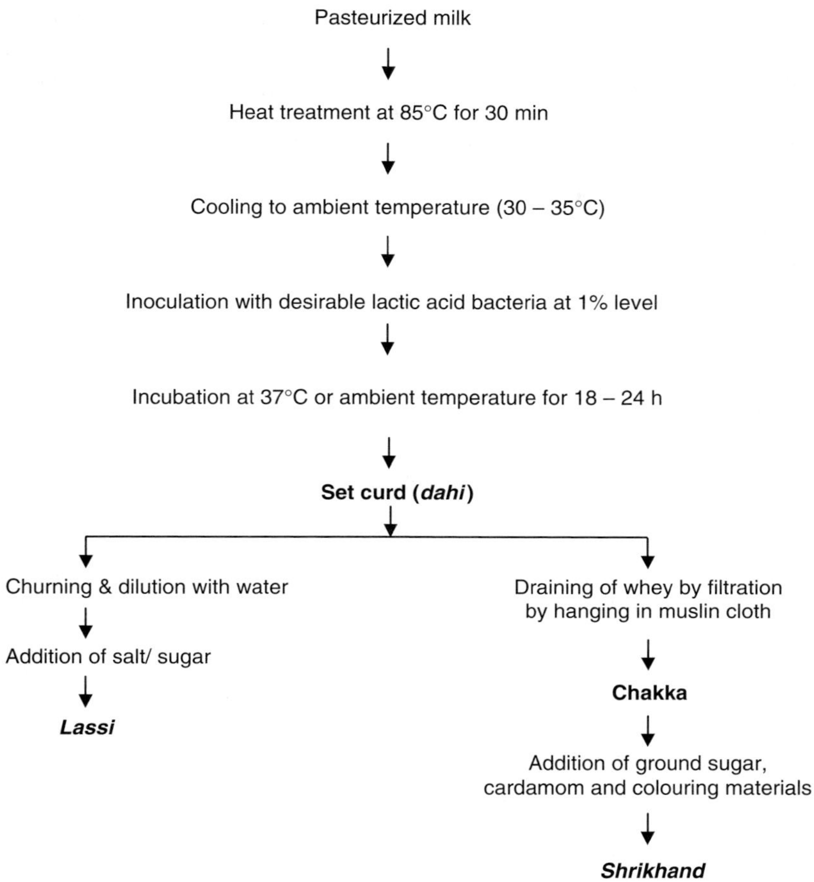

Figure 18.1 Traditional method of preparation of *dahi*, *lassi* and *shrikhand*

The composition and quality of *dahi* varies from place to place, as it is prepared under different domestic conditions. The earliest work (22) had established the average composition of *dahi* prepared from whole milk is as follows (on percentage basis): water 85–88%, fat 5–8%, protein 3.2–3.4%, lactose 4.6–5.2%, ash 0.7–0.75%, lactic acid 0.5–1.1%, calcium 0.12–0.14% and phosphorus 0.09–0.11%. The basic aspects, such as the quality of raw milk, processing parameters, starter cultures, biochemical and nutritive changes, shelf life and packaging have been well reviewed (1,23).

The commonly associated lactic acid bacterial cultures in the inoculum are strains of *Lactococcus lactis* ssp. *lactis, Lc. lactis* ssp. *cremoris, Lc. lactis* ssp. *diacetylactis, Leuconostoc cremoris, Lactobacillus delbrucecki* ssp. *bulgaricus, Lb. acidophilus,* and *Lb. helveticus.* Many studies have focused on several important aspects such as determining lactic microflora in homemade and marketed samples of *dahi*, isolation of lactic acid bacterial cultures from *dahi*, suitability of lactic cultures in preparing the product with acceptable quality with storage quality, and changes occurring during fermentation (22,24–26).

Selection of a good starter culture is important to obtain a good flavor with desirable characteristics. Use of both mesophilic and thermophilic starters in various combinations has been reported (27,28). Addition of probiotic cultures such as *Lb. acidophilus* and *Bifidobacterium bifidum,* along with the regular lactic cultures for *dahi* preparation, help to increase the therapeutic and nutritional value of *dahi* (29). *Dahi* was found to be a potential source of lactic acid bacteria, mainly species of *Lactobacillus,* active against food borne pathogenic and spoilage bacteria such as *Staphylococcus aureus, Bacillus cereus, B. licheniformis* and *Klebsiella* sp., and *Pseudomonas* sp. (30). Further, a strain of *Lb. delbruecki* ssp. *bulgaricus* producing a heat stable bacteriocin, having broad spectrum antibacterial activity with a potential for use as a food biopreservative, was isolated from *dahi* (31).

Presence of both spoilage and pathogenic microorganisms has been reported in market samples of *dahi* (32,33). Contrary to this, a good quality of *dahi* can be produced under varied incubation periods and temperatures such as 2–4 h at 40°C or 4–6 h at ambient temperature or 8–10 h at 25°C using a mixed culture obtained from a household (34). A shorter incubation was found to be sufficient to get a good quality *dahi*, as prolonged incubation may favor the growth of several undesirable microorganisms, which can affect the quality of *dahi*. This product was shown to have higher antibacterial activity against pathogens such as *Escherichia coli, Bacillus subtilis,* and *S. aureus.* Although many reports are available on genetic improvement of lactic cultures of dairy products, so far their use in the preparation of traditional fermented products has not been put into commercial use.

In India, very few cooperative milk federations are now taking up commercial scale production of *dahi*. The preparation of *dahi* by organized dairies has been on a slow pace, but it is making a brisk progress in the commercial market. However, the commercial success, to a large extent, is dependent upon producing uniform and consistent quality products. Heating of milk at 85°C for 30 min or 90°C for 5–15 min has been put into practice for the preparation of *dahi* on a commercial scale as a means to improve textural characteristics and water binding capacity due to denaturation of whey proteins and its interactions with casein fraction (19).

As a means of exploring alternatives or bringing in convenience, *dahi* powder has been made by spray drying of homogenized curd prepared from standardized buffalo milk of about 3.5% fat and 30% total solids (TS). Uniform quality *dahi* is prepared by reconstituting one part of *dahi* powder with three parts of lukewarm water and cooling for about one hour at refrigerated temperature (35).

In Sri Lanka, the traditional preparation of curd involves use of wide mouthed clay pots for production and distribution as well. The containers used are either wrapped with newspaper or with polyethylene, which cannot prevent contamination, resulting in a very poor quality of curd. Presence of coliforms, *E. coli, S. aureus* in high percentages has been reported (36).

Development of new innovative methods for rapid detection of emerging food pathogens such as *E. coli, Yersinia enterocolitica, Listeria monocytogenes*, and the absence of *Salmonella* spp. in dairy products is extremely important in obtaining a safe product. Implementation of quality management programs such as Hazard Analysis Critical Control Points can assure a uniform and consistent quality product with emphasis on food safety. Application of biotechnological approaches such as use of immunological assays, Deoxyribo Nucleic Acid (DNA) probe based assays, polymerase chain reaction (PCR) based pathogen detection systems, and use of biosensors in quality assurance of dairy products to improve the quality has also been shown to hold promise (37).

18.3.1.2 Mishti doi

Mishti doi is a popular variety of sweetened *dahi* available in the eastern region of India. It is also known as *lal dahi* or *payodhi*. It is generally characterized by light brown color and firm body with cooked or caramelized flavor. Two formulations for the preparation of *mishti doi* are identified (38). One combination involves use of *Streptococcus salivarius* ssp. *thermophilus, Lb. acidophilus* and *Lb. delbrueckii* ssp. *bulgaricus* and the other formula has *Lb. acidophilus, Lc. lactis* ssp. *lactis* and *Saccharomyces cerevisiae.* Microbiological examination of market samples revealed the presence of yeasts such as *Saccharomyces, Candida and Rhodotorula*, besides the LAB like *Lactobacillus, Lactococcus* and *Streptococcus.* The yeast count reported was 5×10^4 cfu/g of *mishti doi* and the coliform count was 46 cfu/g.

Proteose peptone (pp), a minor protein of milk, plays a very important role in imparting the brown color to milk products. The pp content changes significantly during the preparation of m*ishti doi.* A positive correlation was observed between pp content of fresh concentrated milk (before fermentation) and intensity of brown color in m*ishti doi*, with a correlation coefficient of 0.93 (significant at $p<0.01\%$ level). The pp content (mg/100g) of raw buffalo milk, after boiling for 30 and 40 min and m*ishti doi* are found to be 112, 290, 365 and 200.9, respectively (39).

18.3.1.3 Lassi

Lassi, also known as buttermilk, is another popular lactic fermented milk based beverage product of the Indian subcontinent, which is consumed mainly during summer months. It is a byproduct obtained during the preparation of ghee (country butter) from *dahi* by churning. It is also made by breaking the set *dahi* with the agitator and the addition of a required amount of water, sugar, or salt and flavor compounds. It is prepared by local vendors, in the small shops in market areas, and in households. The traditional method of lassi preparation is shown as a flow diagram in Figure 18.1.

The composition of *lassi* varies based on the type of milk used, the extent of dilution during churning, and the efficiency of fat removal. The average composition (on percentage basis) of *lassi* is as follows: water 96.2%, fat 0.8%, protein 1.29%, lactose 1.2%, lactic acid 0.44%, ash 0.4%, calcium 0.6%, and phosphorus 0.04% (40).

As the preparation and vending at many places are not done under hygienic conditions, high incidence of aerobic sporeformers such as *Bacillus subtilis, B. megaterium, B. cereus, B. licheniformis, B. coagulans. and B. pumilus* have been isolated from market samples of *lassi* (41). The shelf life of *lassi* has been extended by more than six days at 37°C by the addition of 0.03–0.35% sodium metabisulfite and the sulfur flavor imparted by this preservative could be masked by the addition of 0.07–0.09% crushed green ginger and 0.5–0.7% salt (42).

The preservative effect of nisin in enhancing the shelf life of *lassi* has been studied (43). The shelf life of lassi was up to 32–48 h at 30°C with the addition of Nisaplin at a concentration of 200–500 IU/ml, whereas, the stability was 8–10 days at refrigerated storage. To enhance the therapeutic value of *lassi*, a method has been standardized to prepare *lassi* using probiotic culture of *Lb. acidophilus* along with *Str. thermophilus* to get a desirable flavor in the final product (44). This product could be stored at refrigerated temperature for twenty-one days with acceptable sensory quality. Significant changes in the lactic counts during storage were not observed at the end of the storage period. The average lactobacilli count reported was 20.4×10^7 cfu/g, which is normally sufficient to be effective as a human dietary adjunct.

18.3.1.4 Shrikhand

Shrikhand is a sweetened lactic fermented product widely consumed in western and northern parts of India. It is a product having a refreshing taste with pleasant aroma, and smooth and homogenous texture and firm consistency. The traditional method of s*hrikhand* preparation is presented in Figure 18.1. *Shrikhand* preparation involves products of curd (*dahi*) by lactic fermentation of whole milk, either cow's or buffalo's milk, followed by the draining of whey from the curd through a suspended muslin cloth bag for 6–8 h. The resulting solid mass (known as *Chakka*) is uniformly mixed with ground sugar (44–45%) and made into a semi-solid mass to which flavoring substances such as cardamom and saffron are added. The product has a long shelf life of about 35–40 days at 5°C, while storage at ambient temperature results in a very short shelf life of 2–3 days.

The composition of *Shrikhand* is as follows (on a percentage basis): moisture 34.48–35.66%, fat 1.93–5.56%, protein 5.33–6.13%, reducing sugar 1.56–2.18% and nonreducing sugar 55.55–53.76% (1). Although it is prepared in households on at very small scale, successful attempts have been made to develop an industrial process for the manufacture of *Shrikhand* and it is available in the markets all over India. Buffalo milk is preferred to cow's milk for the manufacture of *chakka*, because of higher yield and consumer preference (45).

The lactic cultures used are of mesophilic lactics, the same as those involved in the preparation of *dahi*. A culture combination of LF 40 was found to be most suitable for *shrikhand* preparation (46). The technological and microbiological aspects of *shrikhand* preparation are reported in an earlier study (47). Preparation of dietetic *shrikhand* using buffalo skim milk fermented with a 2% combined culture of *Lb. acidophilus* (NDRI-AH1) and *Str. salivarius* ssp. *thermophilus* (NDRI-YHS) was reported to reduce the high fat content in the final product (48). During this process, the total solids recovered in the *chakka* were 54.46%.

To increase the yields (23% more than the traditional process) and for easy automation and process control for large scale production of *shrikhand*, a patented process has been developed to remove whey by an ultrafiltration technique (49). A comprehensive review on *shrikhand* preparation and its quality has been published (50). Fat was found to be the most important compositional parameter that affects the acceptability of *shrikhand*. It influences the flavor quality of the product. The fat and sugar contents between 7–9% and 33–39%, respectively, had higher acceptability (51).

Studies on the production of enterotoxin by *S. aureus* in *shrikhand* revealed that this organism, when occurring as a preprocessing contaminant in the milk, could grow and produce enterotoxin even in the presence of lactic cultures (52). More recently, the behavioral pattern of toxin producing pathogens like *E. coli* and *L. monocytogenes*, which can enter the product as a postprocessing contaminants in *shrikhand*, was predicted through response surface plots generated by using polynomial equations (53,54).

A change in the profile of mineral content from milk to the final product of *shrikhand* was studied. The fermentation of milk brought about highly significant changes in mineral content during transition from colloidal to soluble phase during *shrikhand* preparation. This process was faster and greater in buffalo's milk than in cow's milk (55).

Postproduction heat treatment (PPHT) of *shrikhand* at 70°C for 5 min enhanced the shelf life up to 15 days at 35°C and above 70 days at 8–10°C (56,57). The overall quality also improved, as the heat treatment decreased the microbial count. PPHT stabilized the product against microbial and biochemical changes with concomitant improvement in the shelf life of the product as revealed by the superior sensory scores of these samples. Dehydrated *shrikhand* has a shelf life of 90 days when stored at 30°C in gas packed containers.

Shrikhand powder was prepared by subjecting the product to spray drying at 160–170°C of inlet temperature and 100°C of outlet temperature (58). Similarly, standardization of *shrikhand* powder preparation was carried out using a colloid mill to grind the *chakka*, adding 25 kg of sugar per 100 kg of *chakka*, adjusting the total solids to 30%, and drying with inlet and outlet temperatures of 190°C and 95°C, respectively (59). For packaging of *shrikhand*, cups made of polypropylene or polystyrene form-fill-seal containers have been used successfully (60).

18.3.1.5 Sewsew

Sewsew is an anaerobically fermented yak's milk product produced by Sherpa communities of Nepal. It is used in making soup and consumed with the *dhindo* (a curd meal). The whole yak milk is boiled and set using lactic culture. The fermented milk is churned the next day. After removing butter, the buttermilk is boiled to precipitate the buttermilk solids, called *sher*. *Sher* is mixed with one third the amount of milk and kneaded. The kneaded mass is filled in clean, dry earthen pots and sealed by covering with leaves and plastering clay on top. The product could be consumed after a month or even after a year with a perception that older the product, the better the quality (61).

18.3.1.6 Churpi

Churpi is a traditional cheese-like product, which is commonly consumed by different ethnic groups of people in the Himalayan regions of the Darjeeling hills, Sikkim, and Ladakh in India, Nepal, Bhutan, and Tibet (62). It is available in two forms — soft and hard. Soft *churpi*, also known as *kachcha churpi* is prepared from cow's milk. The milk, either boiled or unboiled, is kept in a wooden vat at ambient temperature for 24 h. Often the cream is allowed to separate and the milk is curdled by boiling. The casein is wrapped tightly in a piece of muslin cloth and allowed to drain for 3–5 h. This is consumed as a condiment by mixing with it sliced radish or cucumber. It is also mixed with meats, vegetables and spices to prepare curry. The average per capita consumption of soft variety *chupri* is 6.9 g/day in the Darjeeling hills and 9.9 g/day in Sikkim (63).

The optimal process parameters for controlling the preparation of *churpi* have been documented (64). Also, the microbial population and predominant lactic acid bacteria profiles of the soft variety of *churpi* have been identified (65). The lactic acid bacteria, yeasts, and viable mesophilic microbial numbers ranged from 7.6–7.9, 7.0–7.4 and 7.7–8.0 $\log_{10}$ cfu/g, respectively. The predominant lactic microflora are *Lb. plantarum, Lb. curvatus, Lb. fermentum, Lb. paracasei* ssp. *pseudoplantarum and Leu. mesenteroides*. Most of these strains showed high hydrophobicity, which indicates the potential of adhesion to gut epithelial cells, one of the important features of a probiotic culture. The absence of proteinases (trypsin and chymotrypsin) and presence of high peptidase and esterase–lipase activities of the predominant organisms isolated from *churpi* was observed, which are the traits of desirable quality for their use in accelerated ripening of cheese and production of typical cheese flavors (66).

The hard variety of *churpi* is made from skimmed milk of cow or yak milk. The skimmed milk is boiled and curdled by adding whey. After filtration, the casein is wrapped tightly with a cloth and cured at ambient temperature (15–20°C) for 2–3 days under pressure of about 0.25 kg/cm^2 made with the aid of heavy stones. The cheese is sliced and allowed to dry in sunlight for 2–3 weeks. This type of *churpi* becomes very hard and can be stored for a number of years (62). The preparation of both varieties of *churpi* is shown in Figure 18.2.

18.3.2 Vegetable and Fruit Based Foods

Natural microbial flora of fruits and vegetables are capable of altering these foods for human consumption. Fruits are a rich source of sugars and are slightly acidic, which make them suitable for yeast growth leading to alcoholic fermentation, which is omitted from the preview of this chapter. Relatively few fruits with low sugar content undergo lactic fermentation. Lactic fermented fruits are usually consumed as condiments accompanying main dishes.

Vegetables, unlike fruits, have low sugar content, neutral pH and their composition is not favorable for spontaneous growth of lactic acid bacteria. However, over centuries people have traditionally developed lactic fermentation methods that stabilize and improve

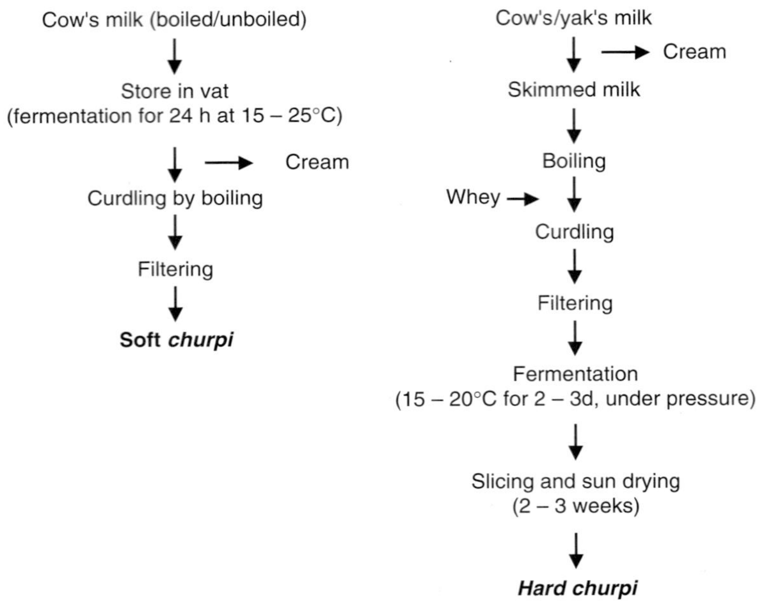

Figure 18.2 Flow sheet showing the differences in *churpi* preparation

nutritional quality of vegetables. Fermented vegetables represent an essential element of human diet worldwide. In Eastern and in many developing countries, fermented vegetables are staple foods, because they represent the traditional dietary element, which is easy to process. Lactic acid fermentation that enhances the organoleptic and nutritional quality of the vegetables has remained a household or small scale process. A few of the common lactic fermented products based on fruits and vegetables are shown in Table 18.1.

To promote lactic flora and to reduce initial microbial population from raw material, washing is essential. Natural inhibitors such as solanin in unripe tomatoes, phaseolin in beans, and oleuropein, the bitter glycoside in olives, needs to be eliminated. Choosing ripe tomatoes, blending the beans and adopting alkali treatment to olives prior to lactic fermentation eliminates the undesirable inhibitors. Certain ingredients are used in lactic fermentation such as sugar, mineral salts, and vitamins to promote the growth of desirable bacterial species. The lactic acid bacteria restrict other microbes by regulating the effect on pH or by producing inhibiting agents such as organic acids and bacteriocins during vegetable fermentation. Spices and herbs have a fine flavor determining role in fermented vegetables. Aromatic compounds in spices (including terpenes and polyphenols) often have antimicrobial effect, which selectively promote lactic bacteria. Certain spices such as garlic, cloves, chillies, — both unripe and red chillies — inhibit undesirable microflora. Many sulfur compounds (as in garlic) have antibacterial properties. Mustard seed contains allyl isothiocyanate, a volatile aromatic compound with antibacterial and fungal properties, which are known to inhibit yeasts. In India, ground mustard seed or oil is widely used in traditional lactic fermentation for prolongation of shelf life of the product. Chemical preservatives such as sorbic, benzoic acids and salts have sometimes been used in industrial production of cabbage, olives, and cucumbers (67).

Salting is an essential step in lactic fermentation. Salt concentration usually ranges from 20–80 g/l or up to 160 g/l as in the case of stored vegetables. Salt concentration induces plasmolysis in vegetables, thus promoting anaerobiosis for proliferation of lactic acid

Table 18.1
List of fruits and vegetables fermented by Lactic acid bacteria

Type	Example
Brassicaceous vegetables	cabbage, cauliflower, braceolis mustard vegetables.
Root vegetables	carrot, turnip, beetroot, radish
Vegetable fruits	cucumbers, olives, tomatoes, peppers, okra, beans, peas
Bulbs	onions, garlic
Vegetable juices	carrot, beetroot pepper, tomato pulp, mushrooms for sauces
Fruits	apples, pear, raw mango, immature plums, lemons, banana fruit pulp

Source: Montet, D., G. Loiseau, N. Zakhia, C. Mouquet, *Biotechnology: Food Fermentations, Vol. 2,* Joshi, V.K., A. Pandy, eds., New Delhi: Educational Publisher & Distributors, 1999, pp 951–969.

bacteria. Salting is done either by adding dry salt or by using brine. In India, *jaggery* (unrefined cane sugar called *gur*) is added during lactic fermentation of the sweet turnip. Adding 1–2 g% sucrose or gur to shredded turnip along with 6 g% each salt and mustard seed increases lactic flora (68). At higher concentration, sucrose (25 g%) acts as inhibitor of lactic fermentation. Traditionally, to increase lactic flora, cooked rice, flour, whey, or cereal brans have been used. Buffering of the medium with an acetic acid calcium acetate buffer system for vegetables (with high sugars) in order to stabilize pH of the medium during fermentation is useful. Calcium ions are released by acetate, complexed with pectin in vegetables, thus protecting them from enzymatic hydrolysis and enhance texture of the product.

The common fermented foods based on vegetables and fruits are presented in what follows, individually, to stress the importance of these foods in the diversified cultures of the Indian subcontinent.

18.3.2.1 Sauerkraut

In India, sauerkraut fermentation does not exist as a traditional art. However, several studies on lactic fermentation of cabbage for sauerkraut are available. Suitability of cabbage varieties for the preparation of sauerkraut has been identified (69,70). Six varieties of cabbage namely GoldenAcre, Sel 8, NL401, NL501, Alfaco, and Pusa synthetic were evaluated. Five kg heads of each variety were trimmed, cleaned, cored, cut into 2.5 mm thick shreds, salted at 2.25% (w/w), mixed thoroughly, filled into glass jars, covered, sealed with proper weight, and allowed to ferment at ambient temperature (25–30°C) for 7 days. The resultant sauerkraut was pasteurized for 5 min and preserved either with sodium benzoate (500 ppm) or potassium metabisulfite (350 ppm) at ambient and refrigeration temperature. A sauerkraut with a pH in the range of 3.5–3.1 and acidity values of 1.10–1.15% was attained in case of NL401 and NL501, which was low as compared to1.5–1.6 in case of quality sauerkraut. This could be attributed to high ambient temperature (25–30°C), which promotes wild yeast that oxidizes lactic acid. Sensory evaluation revealed that sauerkraut from NL 501, NL401, and Sel 8 were the best with regard to texture, flavor, and overall acceptability. The golden yellow color of kraut could be better retained by using potassium metabisulfite (KMS) as compared to sodium benzoate.

Pasteurized kraut could be stored at refrigerated temperature for more than six months. In cases of kraut stored at ambient temperature, browning was a problem (70). Temperature plays a vital role in production of good quality kraut. The effect of temperature on the quality of sauerkraut showed that kraut prepared at 18–25°C from either variety of cabbage (red cabbage or Pusa drum head) was superior in color, texture, flavor, and taste when compared to those prepared at 34°C or 12°C (71). The development of surface yeast called as *Mycoderma* during storage of sauerkraut results in the loss of acidity, development of off flavor and softening of the product. Accordingly, the tolerance limits of *Mycoderma* to various additives to prolong the storage life of kraut was studied (72). The study revealed that the wild yeast was tolerant to high salt (16%). In acidic slurry (1.5% lactic acid), its growth could be effectively controlled in low concentrations of salt (3%) in combination with low concentration of acetic acid (0.2%), sodium benzoate (300 ppm) or sorbic acid (100 ppm). Other (mustard-like) condiments and be added and they not only impart flavor and taste, but also act as a preservative. In India, mustard is extensively used in pickles. The addition of mustard powder, turnip, cabbage, and cucumber was shown to increase the rate of lactic fermentation (68,69).

18.3.2.2 Gundruk

Gundruk is a nonsalted fermented acidic vegetable product indigenous to the Himalayas (73). The fermenting substrate for *gundruk* is usually rayo (*Brassica campestris* L var *cumifolia* Roxb) leaves. Other leaves such as radish (*Raphanus sativus* L), shimrayo (*Cardamine hirsute* L var *sylvatica* Link), cauliflower (*Brassica oleracea* L var *botrytis* L) and mustard (*Brassica juncea* L) are also used. The word *gundruk* is derived from the newari word "gundru" (the Newaris being one of the ethnic groups of the Nepalese). *Gundruk,* though originally confined to Nepal, has become popular among all the ethnic groups of Darjeeling and Sikkim in India.

Gundruk is usually prepared during the months of December to February when the climate is less humid and an ample supply of vegetable is assured (62). The leaves are dried under sunlight (1–2 days), crushed mildly, soaked briefly in hot water, and hand pressed in perforated tin or earthen jars with heavy stones to remove excess water. The leaves are then kept in a warm, dry place for 15–22 days for fermentation. In the village process, a hole (about $1 m^3$) is dug in the ground and dried by fire. Banana or bamboo leaves are spread in the bottom (about 30 cm) on which the dried crushed leaves of the vegetables to be fermented are placed and covered on top by banana or bamboo leaves. Heavy stones are placed to compress the substrate. The holes are sometimes covered with a layer of cow dung. The leaves are allowed to ferment *in situ* until a fermenting odor develops (15–22 days). The *gundruk* is taken out and dried under sunlight for 2–4 days. It has a shelf life of one year.

An almost similar product known as *pani* (water) *gundruk* is commonly prepared and consumed in the Redong area in the Kalimpang subdivision of Darjeeling district of the West Bengal State of India. *Rayo* leaves are dried under sunlight for 2–3 days, crushed lightly, soaked in hot water, drained and kept in vats containing warm water (30–35°C) for only 3 days. They are then squeezed and dried under sunlight for about 7 days. Inhabitants of this region prefer *pani gundruk* for its sour taste, easier preparation, and greater resistance to spoilage microflora. *Gundruk* is typically used as a base for soup and as pickles (62). During *gundruk* fermentation *Lb. cellobiose* initiates the fermentation, followed by homofermentative *Pediococcus pentosaceus*, *Lb. casei* and finally by *Lb. plantarum*, producing lactic acid and acetic acid, which lower the pH of the substrates (74).

18.3.2.3 Sinki

Sinki is a unsalted traditional lactic acid fermented radish root product of Nepal. Recently, this fermented food is becoming popular in the Darjeeling district of West Bengal State and Sikkim of India, and in Bhutan. Like *gundruk*, *sinki* is prepared during the months of winter, when the climate is least humid and the supply of vegetables is assured (75). The method of preparation is similar to *gundruk*, except that the substrate is the taproot of radish (*Raphamus sativus* L) and the fermentation takes 30–40 days. In the traditional method, fresh tap roots of radish are washed, wilted by sun drying for 1–2 days until they become soft, shredded, washed, and placed tightly into earthen jars with the help of heavy wooden pestles. The jars are covered by radish leaves with earthen lids and kept in a warm, dry place for 15–30 days. In some places as in *gundruk* preparation, a pit (1 m^3 size) is dug in the ground, which is covered with dried leaves of bamboo, banana or radish. The shredded taproots are placed in it, pressed tightly and covered with dried leaves and heavy stones are placed for compression. The top of the pit is plastered with mud or cow dung. After 30–40 days, the fermented mass is removed and sun dried. *Sinki* has a shelf life of one year if dried in sunlight from time to time.

Studies (76) undertaken to assess the proximate composition and microflora of samples of *sinki* procured from markets of Darjeeling, Kalimpang and Gangtok showed the average moisture of 22.0, 20.4 and 21.3%, respectively. The protein, fat, and ash content are approximately 14.5, 2.5 and 11.5%, respectively. The pH of the products were 4.45, 4.28 and 4.40, with acidity at 0.65, 0.80 and 0.72%, respectively. The microbiological analysis of market samples showed the presence of *Lb. plantarum* and *Lb. brevis*. These isolates did not hydrolyze protein, fat, and starch, indicating that none of these substrates were utilized during fermentation. The development of acidity in *sinki* is attributed to the utilization of free sugars of the substrate by lactobacilli. Traditional process parameters optimization with respect to fermentation temperature (20, 30, and 40°C), utilization of different containers (glass jar, earthen jar, polythene bags) and the period of fermentation (7, 12, 20 d) have been studied (76). The optimum conditions of fermentation were at 30°C for a period of 12 days in glass jars. The product had a pH value of 3.0 and an acidity of 1.28% with highest organoleptic scores. The succession of microflora showed the initiation of fermentation by heterofermentative *Lb. fermentum,* followed by another heterofermentative *Lb. brevis* and finally succeeded by homofermentive *Lb. plantarum*. *Sinki* prepared under optimized condition could substantially reduce the fermentation time to 12 days, compared to the traditional method which takes about 30–40 days.

18.3.2.4 Khalpi

Khalpi is a fermented cucumber pickle of Sikkim and Himalayas. The ripe cucumber is wilted (1–2 days), peeled, sliced, and mixed with mustard oil; spices and salts were added and pressed tightly into a container, covered and fermented (5–7 days). *Khalpi* is a sour acidic pickle, which can be stored for several days (73).

18.3.2.5 Fermented Cucumber

A laboratory scale process for lactic fermentation of cucumber was standardized (77). Firm immature cucumbers were kept in glass jars with 5% brine. After 3 days, the brine was drained and replaced by 6% brine. Subsequently, at weekly intervals, brine concentration was raised to 10, 12 and 15% level. Salt stock cucumber has a storage life of several years. It could be used for the preparation of sweet, sour, sliced, or a mixed type pickle after desalting. Garg et al. (77) removed the cucumbers from the salt stock and preserved them in 2% acetic acid for 3 months at ambient temperature of 33–38°C as sour pickles, which could be consumed directly as pickles. However, reduction of acidity to 0.55% by treating with 5% $CaCl_3$ (overnight) gave desirable organoleptic characteristics, firm texture, and flavor for the pickles.

18.3.2.6 Kanji

In northern parts of India, a ready to serve traditional fermented beverage popularly known as *kanji* is prepared from roots of the black carrot (*Daucus carota* L). The traditional method for the preparation of *kanji* involves natural fermentation of longitudinally slit carrots suspended in an excess quantity of water (15–20 times the weight of the carrot) along with salt, chillies, and crushed mustard in an earthen pot. The product acquires acceptable taste and an attractive crimson color after seven days. The shelf life of the beverage is limited to about seven days and thereafter flavor degenerates (78).

Berry et al. (78) prepared *kanji* by both natural fermentation and controlled fermentation using *Lb. plantarum*. The fermentation medium for *kanji* contained the following: 1 kg of grated carrots mixed with 7 kg of water, 200 g of common salt, 40g of crushed mustard, and 8g of red chilli powder. One lot was subjected to natural fermentation; another lot was fermented using *Lb. plantarum*. The third lot also was subjected to fermentation with *Lb. plantarum* but without the crushed mustard. All three batches were fermented at ambient temperature (20–23°C) for 10 days and heat processed for 25 min in boiling water. The acidity increased from 0.18 to 0.43, 0.35 and 0.3%, respectively, with the total sugar at 1.94% in all the three batches. The bottled beverage was stable at ambient temperature (13–42°C) for six months. The natural fermented beverage was rated as the best, followed by the *Lb. plantarum* fermented batch containing crushed mustard. The beverage prepared using the dehydrated (0.7% moisture) black carrots, after three months storage, was also found to be satisfactory with respect to sensory quality attributes. Thus *kanji* may be made available to the consumer during off seasons when fresh carrots are not available.

Attempts have also been made to standardize the method of fermentation of *kanji* using preservatives such as sorbic acid or a combination of sodium benzoate and potassium metabisulfite. Fermentation of carrots in 2% brine with sorbic acid (0.1%) and calcium chloride (0.1%) at 30°C ± 2, for 6 days yielded an acceptable product in which 60% carotenoids were retained at the end of four months at ambient temperature of 30 ± 2°C (79).

A standard recipe for preparing concentrated fermented carrot juice using 3% brine, 1% mustard and sodium benzoate (0.015%) and potassium metabisulfite (0.01%) with a shelf life of 6 months at ambient is available (80). Microbiological and biochemical changes during fermentation of 1.5% sliced crimson colored carrots in 3 liter 3% brine with 4% mustard powder at ambient temperature (20 ± 2°C) for 15 days to obtain *kanji* was studied (81). *Leuconostoc mesenteroides* and *Leu. dextranicum* predominant during natural fermentation with occasional appearance of *Pediococcus* spp. were found. The total sugars increased up to 200 μg/ml. The pH dropped from 7.2 to 4.0 after four days of fermentation. The production of ethanol was 2–5% after 3 days, which dropped due to oxidation by acetic acid bacteria. At the end of 14 days, the *kanji* had an acidity of 0.3% and a pH of 4.0.

18.3.2.7 Mesu

Mesu is a salted fermented bamboo shoot product, traditionally consumed by the people in the bamboo growing regions of Nepal and in the Darjeeling hills of West Bengal and Sikkim in India (82). The Lepchas call it "sitit." Young edible shoots of bamboo (*Dendrocalamus humiltonii* Nees and Arnolt, *Bambusa tulda* Roob, and *Dendrocalamus sikkimesis* Gambli, locally known as "choya bans" or tama; "Karati bans" and bhalubans, respectively) are defoliated, chopped, pressed tightly with leaves of bamboo or other wild plants, and left at ambient temperature (20–25°C) to ferment for 7–15 days. The fermented product has typical *mesu* flavor, i.e., a strong ammoniacal flavor and sour taste.

Mesu is usually prepared in the months of June to September when bamboo shoots sprout. It is commonly used as a pickle by mixing it with salt, mustard oil, and green chillies. It is also used for preparing curry by frying it and mixing it with cooked meat. The fresh *mesu* has a shelf life of about 6 days, while its pickle can be stored for more than a year (82). Microbiology of *mesu* samples collected for different weekly markets of

Kalimpang (Darjeeling hills) and Gangtok (Sikkim) revealed *Lb. plantarum, Lb. brevis* and *Pediococcus pentosaceus*, among 327 strains of lactic acid bacteria from 30 samples. *Mesu* was dominated by *Lb. plantarum,* followed by *Lb. brevis*. Succession of microbial microflora in bamboo shoots during fermentation to produce *mesu* revealed dominance of *Ped. pentosaceus* in the early stages, followed by *Lb. brevis* and then again the domination of final product by *Lb. plantarum*. During fermentation, acidity increased from 0.04–0.95% and pH declined from 6.4–3.8.

18.3.2.8 Kachampulli

Kachampulli is a traditional souring concentrate made from fruits of malbar tamarind (*Garcinia cambogia*) used in culinary dishes in the Coorg district of Karnataka, India. These fruits are valued for their acidic and medicinal properties (83). The traditional method of making *kachampulli* involves softening of fruits (1–2 weeks) in specially made bamboo baskets. Softened fruits are pressed by wooden rollers to expel juice, which is filtered and concentrated by heating in earthen pots. Succession of microorganisms during *kachampulli* production is not studied. The modified process for *kachampulli* involves the use of the pectinase enzyme (0.5–1.0%) to soften the fruit, vacuum, or open pan evaporation of the juice to obtain *kachampulli*, a dark brown viscous concentrate of 50° Brix having 39% acidity (83). Fruit residues are used for byproducts. Juice from the seed and placenta were extracted using pectinase enzymes. Culture of *Saccharomyces cerevisiae* was used to complete fermentation (3 days) to yield low alcohol (5%) wine of pink color with a fruity aroma. The seeds were dried, cracked, and a low melting point (30–40°C) fat was extracted by boiling the seeds in water.

18.3.2.9 Indian Kinema

Fermentations involving highly alkaline products are generally safe if properly fermented. The organisms are proteolytic and proteins are hydrolyzed to peptides and amino acids. Ammonia is released and the pH rapidly reaches as high as 8.0 or higher. The combination of high pH and free ammonia along with very rapid growth of essential microorganisms at relatively high temperature (above 40°C) make it very difficult for other organisms to grow. Thus, the products are quite stable and well preserved especially when dried. These are safe foods even though they may be manufactured in an unhygienic environment (1).

Kinema is a soybean based fermented sticky, slightly alkaline product with slight ammonia-like flavor produced by bacterial fermentation. Tamang (84) has dealt in detail with the socio-economical aspects of *kinema*. It has been traditionally prepared in the Himalayan region of Nepal, India, and Bhutan for centuries. Soybean, locally known as "bahtmas" is a major summer crop grown under rain fed conditions in upland terraces up to an altitude of 1500 m. Indigenous local varieties of soybeans (*Glycine max*, L. Merrill) yellow cultivar, and dark brown cultivars are cultivated. The *kinema* may have originated from the "Limbu" (one of the castes of ethnic Nepalis) dialect kinambaa, "ki" meaning fermented and "nambaa" meaning flavor (74). In traditional process of *kinema* preparation, soybean is soaked in water overnight, cooked until it can be pressed easily, excess water is drained off, and cooked beans are transferred to wooden mortar and crushed moderately by a wooden pestle (locally called "okhali"). They are transferred to bamboo basket lined with jute bags and layers of fern (*Athyrium* sp.) leaves; about 1% wood ash is added to the soya bean grits before folding the fern leaves and sack clothes. The mixture is left to ferment naturally at ambient temperature (25–40°C) for 1–3 days above an earthen oven in the kitchen.

Kinema production has been confined to the rural household level. *Kinema* consumption in Sikkim is 2.2 g and in Darjeeling hills it is 3.3 g per day according to the estimate of 1997–1998 (85). The presence of *B. subtilis*, *Enterococcus faecium*, *Candida parapsilosis* and *Geotrichum candidum* in *Kinema* samples was reported (86). *Bacillus subtilis* was found to be the dominant microflora in *Kinema*. During *kinema* production, remarkable increase in water soluble nitrogen and trichloroacetic acid content was observed (87). Total amino acids, free amino acids and mineral contents increased and the ratio of free amino acid to total amino acids was 11.4% (88,89). A laboratory scale production of *Kinema* using *B. subtilis* KK-2:B10 (MTCC 2756) and GK-2:B10 (MTCC 2757) isolated from naturally fermenting *kinema* was optimized (90). The quality of *Kinema* was maintained with pure culture fermentation by *B. subtilis* KK-2:B 10 at 40°C for 20 h followed by maturation at 5°C for 1 day (91). A ready to use pulverized starter culture of *B. subtilis* was developed for *kinema* production (84).

18.3.2.10 Pickles

Cauliflower (*Bracica oleracea* var. *botrytis L*) is an important vegetable crop, which is popularly used for the preparation of pickles in India. The effect of mustard and its components on the fermentation of cauliflower was studied (92). In one study, whole mustard flour (fat free), mixed with total volatile oil of four species of mustard namely *Brassica juncea, B. nigra, B. campestris* var. *toria* and *B. campestries* var. *dichotoma* were studied for their suitability during lactic fermentation of cauliflower. Increasing the proportion of salt (2–12%) progressively decreased the rate of acid production and delayed spoilage by surface yeast. The maximum acidity was developed in 4% salt when used either alone or along with 2% mustard. On the other hand, increasing the proportion of mustard from 1 to 2% helped to increase the rate of acid production. The complement role of whole mustard was better than the residue flour or volatile oil. The mustard variety *B. nigra* showed better preservative action than *B. juncea "sarson"*or "*toria*."

In comparison to the previous example, the popular types of pickles consumed by the human population of India have been those based on unripe mangoes, gooseberries, lemons, *Decalepis hamitonii,* and a variety of mixed vegetables. In general, most of these pickles are prepared at the household level by subjecting the vegetable or fruit to natural fermentation. In one of the specific type of pickles, wherein preservation is achieved through high concentrations of salt and lactic acid fermentation, the raw materials are washed well, cut into suitably sized shapes and mixed with salt at a level of two times more than that of weight of the raw materials as well as a requisite spice powder. The specie powder mix mainly consists of chillies, mustard, and coriander seeds. The mouth of the containers are closed properly and kept for 7–8 days at ambient temperature (20–30°C) for the natural fermentation to take place *in situ*. During the storage period, acidity develops and a certain amount of aging provides the acceptable organoleptic attributes to the pickles. The amount of acid produced and time period is dependent upon the type of produce used in pickle preparation. Although not much research has been done into the nature of microflora and other attributes, it is thought that the microflora mainly comprises lactic acid bacteria and to a certain extent acetic acid bacteria. On similar lines, pickles are also prepared from a variety of vegetables.

In contrast to the previous discussion, wherein the final product is in the form of fermented vegetables or fruits immersed in liquid formed due to the presence of high salt and spice mixture, there is another specific type of pickle which is devoid of any liquid. In this case, the raw materials for pickle preparation are the same as described previously and preservation is through high concentration of salt, spice mix and edible oil. The raw materials, such as unripe mangoes, gooseberries, and lemons, either whole or cut pieces, are cleaned well and mixed with salt and spice mix as in the previously described pickle preservation.

This mixture is subjected to frequent mild frying with oil. After 3–4 similar treatments, the whole mixture is placed in a clean container and the mouth of container is closed properly. It is very essential to see that the pickle mixture is layered with sufficient quantity of edible oil, which is usually sesame oil or mustard oil depending upon the regions of preparation. In this specific type of preparation, fermentation occurs naturally. Generally, the lactic acid bacteria microflora is comprised mainly of those species that can survive and grow in presence of high salt concentrations.

Regardless of the type of pickle preparation, the shelf life is quite reasonable extending even up to periods of 6 months and beyond, if proper practices of hygiene and sanitation are taken. In the absence of any microbiological studies of the nature of pickle fermentation, from the product profile, it would appear that species of *Pediococcus* tend to predominate over the other lactic acid bacteria.

Traditionally, most of the vegetables, either in single or in combination with spices, are subjected to natural fermentation. This preparation has been more or less on a home scale operation, which contribute toward many health benefits without much scientific knowledgebase. Controlled fermentation of bitter gourd (*Momordica charantia*) and fenugreek leaves (*Trigonella foenum graecum*), individually, with an antagonistic culture of *Ped. pentosaceus* resulted in a more acceptable product with increased levels of pyridoxine, ascorbic acid and vitamin B_{12} (93).

18.3.3 Cereal and Legume Based

Cereal and legume based foods are a major source of economical dietary energy and nutrients, worldwide. Often, the regional specific cereals or legumes are subjected to natural or controlled fermentation to obtain desirable final products which are nothing but fermented foods. The involvement of desirable microorganisms, particularly those of lactic acid bacteria, yeasts, and fungi have been well documented (1,94,95).

A major proportion of the cereal and legume based foods are fermented by lactic acid bacteria, wherein these organisms possess the ability to increase palatability, keeping quality, safety, and nutritive value of the raw materials, either cereals or legumes or both. The successive growth phase of microorganisms in fermenting cereals and legumes also favors yeast growth, which often occurs as a component of mixed microflora and imparts specific characteristics for the product (3).

In general, the low pH resulting from lactic fermentation suppresses excess activity of alpha amylase and other enzymes during cereal fermentation. In addition, the fermentation process contributes to the bioavailability of minerals by phytate degradation (96,97,98). A good lactic acid fermentation process depends upon two important factors, namely soaking and cooking of raw materials (ingredients), and salting. The soaking and heating process causes solid losses, but it leaches out or inactivates certain factors that are detrimental to nutrition, such as phytic acid, flatus-causing oligosaccharides and trypsin inhibitors (99,100). The process also reduces the *in situ* microbial contaminants. Salting is a common step in almost all types of fermented foods. In addition to imparting a salty taste, sodium chloride acts as a preservative and bears a selective action on the microorganisms that prevail in the fermentation substrate.

One of the oldest human civilizations has been in the Indian subcontinent and in particular, India, being a vast country, has been the home of innumerable sects of human population. The diversity in cultures accounts for a large number of traditional fermented foods with the majority of them being region specific. Nearly 90% these cultures are highly secretive in their preparation of such fermented foods and tend to be regionalized, with many of the foods being prepared at the household level. Under these circumstances, very few of these traditional foods have been explored from scientific and technological angles. Most of these fermented foods are being prepared according to the traditional knowledge without any knowledge of technologically advanced methods.

Researchers have attempted to determine the nature of fermentation process, microflora involved, and the role played by these organisms in a variety of Indian fermented foods. The popular and widely consumed cereal and legume based fermented foods of the Indian subcontinent are presented in Table 18.2. A few of these fermented foods are presented with details relating to biotechnological aspects.

18.3.3.1 Idli

Among the closely related types of traditional fermented foods based on cereal and legume combination is that of *idli*. *Idli* is a white, fermented (acid leavened), steamed, soft and spongy texture product, widely popular and consumed in the entire South India. It has been documented that *idli* batter fermentation has been in use since 1100 AD (101). Recently, *idli* is also becoming popular throughout India, as well as in neighboring countries such as Sri Lanka and others. *Idli* is the resultant product from the naturally fermented batter made from washed and soaked milled rice (*Oryza sativus*) and dehulled Blackgram dhal (*Phaseolus mungo*). From a nutritional and health status point of view, *idli* appears to be an ideal human food for people of all ages and at all times. Considering the energy requirements of a diet for preschool children as well as women who lack proper diet intake, this traditional fermented food can play a vital role as a supplementary food.

Considering the importance of this traditional fermented food, exhaustive reviews have been published (1,4,102) covering all the aspects relating to preparation, microbiology, biochemistry and nutritional of *idli* batter fermentation. Investigations into the primary aspects of *idli* batter fermentation were initiated as early as 1955 at the Central Food

Table 18.2
Popular cereal and legume based fermented foods of the Indian subcontinent

Fermented Food	Region of Origin/ Popular/Consumed	Ingredients (Cereals/Legumes)	Microorganisms Involved
Ambali	India	Finger millet flour and rice cake	Lactic acid bacteria
Appamam/ Uttapamam	India, Sri Lanka	Blackgram dhal, rice, wheat flour	Lactic acid bacteria, Yeasts
Batura	India	Wheat flour	Yeasts, Lactic acid bacteria
Dhokla/Khaman	India	Bengalgram dhal, Blackgram dhal, rice	Lactic acid, bacteria Yeasts
Dosa	India	Blackgram dhal, rice	Lactic acid bacteria, Yeasts
Idli	India	Blackgram dhal, rice	Lactic acid bacteria, Yeasts
Jelaebi	India, Pakistan	Blackgram dhal	Lactic acid bacteria, Yeasts
Kadhi	India	Bengalgram flour, curds	Lactic acid bacteria
Kenima	Nepal, India	Soy beans	Yeasts and molds
Kulcha	India, Pakistan	Wheat flour	Yeasts
Papadam	India	Blackgram dhal rice, spices	Lactic acid bacteria
Poko	Nepal	Fermented rice	Yeasts, molds and Lactic acid bacteria
Punjabi warri	India	Blackgram dhal, spices	Lactic acid bacteria
Rabadi	India	Pearl millet flour	Lactic acid bacteria

Technological Research Institute, Mysore, India. These studies have not only been initiated by Indian researchers, but also by researchers in other developed countries for the purpose of scientific understanding and technological improvements.

The traditional method of preparation of *idli* is shown in Figure 18.3. *Idli* is prepared from a mixture of milled rice and Blackgram dhal with their proportions varying to a slight extent based on the regions of preparation. Several researchers have used different proportions of Blackgram cotyledons to rice, i.e., from 4:1 to 1:4 weight to weight (w/w) for making *idli* (102). Desikachar et al. (103) and Steinkraus et al. (104) used proportions of rice to Blackgram cotyledons of 2:1 to 4:1 and reported a preference for 2:1 and 3:1 over 4:1. The proportion of rice to Blackgram grains used in *idli* fermentation ranges from 1:4 to 4:1. Studies at the Central Food Technological Research Institute, Mysore have optimized the proportion of ingredients for *idli* batter fermentation during the years 1955–1960. Lewis et al. (105) tested rice to Blackgram ratios of 2:1 to 4:1 and preferred 2:1 and 3:1 over 4:1, where the rice flavor predominates. Radhakrishnamurthy et al. (106) also preferred a 2:1 ratio of rice and Blackgram dhal. The traditional method of *idli* preparation uses rice and Blackgram dhal in the proportion of 3:1.

Invariably the ingredients are soaked separately in water at ambient temperature for 6–8 h. Soaked ingredients are wet ground separately, mixed, and subjected to natural fermentation at ambient temperature ranging from 10–12 h, with overnight being the traditional time interval for *idli* and the like. Here also, the amount of water added to rice and Blackgram dhal batter as well as salt is variable. The amount of water required to prepare the batter of desirable and uniform consistency varies from 1:5 to 2:2 times the dry weight of the ingredients (103).

In traditional practice, rice and Blackgram dhal are soaked separately in water at ambient temperature for 5–10 h prior to grinding for batter preparation. The soaked ingredients

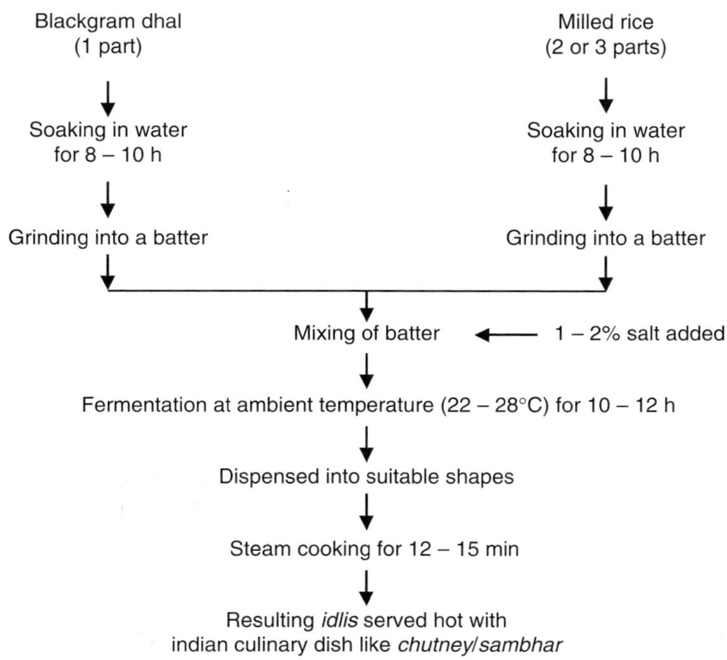

Figure 18.3 Preparation of *idli* by traditional method
Source: Yajurvedi, R.P., *Ind. Food Packer* 34: 33–38, 1980.

are ground wet, in a hand operated roller or stone grinder (107–109). In the laboratory, an electric grinder also can be used to provide the proper texture. In recent times, an electrically operated mixer or grinder is used in varied scale of operations, including home kitchens (1,108). As a means of convenience, in recent times instead of using milled rice, followed by soaking and grinding into a batter, requisite quantity of rice semolina (grits) is directly mixed into the Blackgram dhal batter prior to subjecting to natural fermentation.

In general, *idli* batter fermentation is carried out at ambient temperature. In the tropics, this generally means a temperature of 25–30°C, which is the optimum for a desirable fermentation (1,103). The fermented batter is prepared into suitable shapes and steamed for 10–15 min. The resultant product is known as *idli*, which is served hot along with compatible traditional culinary dishes such as chutney and sambhar.

Studies have demonstrated the optimum fermentation conditions for obtaining good *idlis* as well as the physiochemical and microbiological changes taking place over intermittent periods of incubation at varying temperatures (103,108,109). Usually, the microorganisms that develop during the initial and later soaking of the ingredients are sufficient to bring about the fermentation. The microbiological changes during the fermentation period have shown the involvement of varied genera and species of lactic acid bacteria and yeasts. The microflora associated with *idli* batter fermentation has been the subject of several investigations, both at the national and international levels (4,104,107,109–112). The bacterial organisms, identified as a part of the microflora, included *Leuconostoc mesenteroides*, *Lactobacillus delbrueckii, Lb. fermentum, Lb. lactis, Lb. brevis, Streptococcus faecalis* and *Pediococcus cerevisiae*, while the yeast flora comprised *Torulopsis candida, T. holmii*, *Candida cacaoi, C. fragicola, C. kefyr, C. tropicalis, Hansenula anomala,* and *Rhodotorula graminis*. Further, these studies have shown the predominance of major flora at various stages of fermentation in *idli*.

Studies (107) have shown that during *idli* batter fermentation, there is a rapid disappearance of the common aerobic contaminants during the initial soaking of the ingredients. Following this, *Leuconostoc mesenteroides* and *Streptococcus faecalis* develop concomitantly during soaking and continue to multiply following grinding and they play a significant role in *idli* fermentation. In the later stages, the number of these two microorganisms decreases and *Pediococcus cerevisiae* develops. A similar approach has been presented by Thyagaraja et al. (109).

Two significant changes occurring in *idli* fermentation are acidification and leavening of the batter. These two parameters are being used as the criteria for judging the progress of fermentation. Comprehensive studies on the various changes accompanying *idli* batter fermentation have shown that besides successive increase in microbial populations, the pH declined to 4.4–4.9 from an initial pH of 6.6 (4,109). Several attempts have been made to improve the *idli* fermentation by standardization of various physiochemical factors. An increasing rate of *idli* batter fermentation was observed to accompany a rise in temperature (103). Fortification of *idli* batter with glucose at 1% level showed a beneficial effect on gas production and leavening during fermentation (113).

Successive changes in microflora and accompanying biochemical changes during *idli* fermentation have been the subject of several studies (1,4,101,114,115). Results on the effects of fermentation on methionine and cystine content of the substrate mixture showed variable levels, with *idlis* prepared using rice and Blackgram dhal in 1:1 ratio resulted in an apparent increase in levels of methionine and cystine. Similarly, the relationship between the type of microflora and the biochemical attributes has also been studied, wherein there was an increase in the water soluble group B vitamins during *idli* fermentation (114).

Research investigations have been attempted to alter the preparation for the sake of convenience and for achieving uniform and consistent quality *idlis*. As a step toward convenience in the preparation, the concept of a dry blend mixture was put forth as early as 1960 by Desikachar et al. (103). A dry mix composition of one part of finely ground Blackgram flour (-52 + 56 B.S.I. mesh) and two parts of rice semolina (-18 + 20 B.S.I. mesh) and containing 2.8% sodium chloride were mixed together and stored. As and when required, the batter was prepared from this stock by uniformly dispersing it in the requisite quantity of water into a homogeneous consistency.

In order to obtain a more uniform and consistent quality *idlis*, the concept of controlled fermentation was attempted as early as 1960 (106). For the inoculua studies, Blackgram flour (120 g) was sterilized by autoclaving at 121°C for 60 min on two successive days, extracted with sterile water (800 ml), centrifuged twice at 2000 rpm and the centrifugate again autoclaved. Blackgram flour or rice semolina (120 g) was prefermented with 400 ml of distilled water for 5.5 hours at 30°C and agitated in a mechanical shaker for 30 min, after adding 400 ml more distilled water. The slurry was then centrifuged twice at 2000 rpm and the clear extract was used. For each study, 55 ml of the centrifugated mixture representing the microbial inoculua and the solubles that were extracted from 8.3 g of Blackgram flour or 16.6 g of rice semolina was taken.

In a similar approach, a process relating to an improved means of providing inoculua (lactic acid bacteria and yeast) in ready to use form *idli* fermentations has been developed (116). The novelty of the process is making available the inoculua for fermented foods in a ready to use form. The inoculua, which comprise microbial cultures, are made using a simple and easily affordable substrate. The growth of microbial cultures is achieved in the substrate by a series of steps which involve certain newer and simpler methods, as well as conditioning of the microbial rich substrate approach which does not require heating. This nonheating approach enables retention of high cell density of viable microbial cultures, which is lacking in other drying methods. The final ready to use product is a free flowing high volume with a high stability at low temperature storage.

In a direction towards reducing the fermentation time of *idli* batter and increasing the shelf life of fermented *idli* batter, an improved process for the preparation of shelf stable *idli* batter has been filed as an Indian patent (117). The novelty of the process lies in using optimal cell densities of defined microbial cultures, either to accelerate the fermentation time or to achieve fermentation of *idli* batter. It is then packed in polyester laminate pouches with a shelf life of 10 ± 2 days at 28–32°C and 3 ± 1 days at 40°C. Simultaneously, the flavor profile of such controlled fermented *idli* batter has shown the presence of desirable flavor compounds such as ketones, diols, and acids for a period of 8 days of storage. This flavor profile can be a reliable qualitiative and quantitative parameter for objective assessment (118).

In *idli* batters prepared by accelerated fermentation time period (6 h) using high cell densities of defined microbial cultures, a marginal increase in proteinase activity and a significant increase in alpha amylase activity was detected as opposed to the initial and 6 h naturally fermented batters. Similarly, there was an increased protein content in the extracted globulin from 6 h accelerated fermented *idli* batter and a reduced prolamin level (Varadaraj, unpublished data).

Under certain circumstances, the final product may not appear to be acceptable in terms of the quality attributes. This occurs due to the predominance of undesirable microbial contaminants over that of desirable lactic and yeast microflora. Studies on the inhibition of such undesirable bacterial species have revealed that plantaricin LP84, a bacteriocin produced by *Lb. planatrum* NCIM 2084, was able to retard the growth of the food borne pathogens such as *Bacillus cereus* F 4810, *Escherichia coli* D 21, and *Staphylococcus aureus* FRI 722 during *idli* batter fermentation. However, these pathogens occurring as contaminants in *idli* batter can survive and grow under conditions of natural fermentation (119).

In an effort to utilize certain nutritionally well balanced legumes such as soybeans in *idli* preparation, a few studies have shown its potential (101). *Idlis*, prepared replacing Blackgram with soybeans, has shown an increased roughage, amino nitrogen, and free sugars, while the resultant product had a somewhat different flavor and sticky surface (120). Soybean based *idli* batters have shown increased levels of biochemical attributes including B-group water soluble vitamins.

18.3.3.2 Dosa

Dosa is a fermented, thin, crisp, fried, pancake-like product widely consumed in South India, parts of Western India. It is becoming popular in other parts of India as well as in the Indian restaurants elsewhere in the world. The batter preparation for *dosa* has been almost the same as that of *idli* batter, except for the proportion of milled rice and Blackgram dhal. The product is prepared by soaking separately equal quantities of milled rice and dehulled Blackgram dhal for a period of 6–8 h at ambient temperature (25–30°C) and then grinding them together into a fine paste with requisite quantity of water (2.0–2.5 parts w/w). The batter is then allowed to undergo natural fermentation for a period of 10–14 h at ambient temperature (25–30°C). The fermented batter is then fried using oil into a thin, crisp, pancake-like product.

In almost similar to *idli* batter fermentation, traditional *dosa* batter fermentation has revealed the occurrence and role of lactic acid bacteria in combination with yeasts in bringing about various biochemical changes (121,122). The predominant species identified were *Leu. mesenteroides*, *Lb. fermentum*, *Lb. delbruecki*, *Ped. cerevisiae*, *Sacc. cerevisiae*, *Hansenula anomala,* and *Kluyveromyces* sp. Usually, these microorganisms come from the raw materials — rice and Blackgram dhal (4). *Dosa* batter fermentation also resulted in increased biochemical attributes, including that of water soluble group-B vitamins such as thiamine, riboflavin, and cyanocobalamin.

As in the case of *idli*, attempts were made to prepare *dosa*-like products by replacing Blackgram with other legumes like soybeans. Soybean-based *dosa* batter was found to be nutritious, but less preferred organoleptically (123).

18.3.3.3 Kadhi

Kadhi is a traditional fermented culinary food item with a mild acidic taste and a typical cooked flavor, widely consumed in most parts of India as a culinary food item. Traditionally, *Kadhi* is prepared by boiling stirred *dahi* (lactic fermented) with 5–8% (w/w) Bengalgram dhal flour (*besan*) as a thickening agent. The traditional method of *kadhi* preparation is presented in Figure 18.4. Microbiological studies in *kadhi* have been lacking.

Most of the traditional fermented foods are prone to contamination with food borne pathogenic and spoilage bacterial species. Such a situation is more frequently encountered due to the prevailing environmental factors aided by inconsistent quality raw materials. Considering the potential use of antagonistic LAB, studies have been undertaken to evaluate the effectiveness of antibacterial properties of selected LAB toward food borne pathogenic and spoilage bacterial species occurring in preprocessing and postprocessing contaminants in *kadhi* (124,125).

The potential use of antagonistic cultures of *Lb. delbruecki* ssp. *lactis* CFR 2023 and *Lb. delbruecki* ssp. *bulgaricus* CFR 2028 as biopreservatives in *kadhi* was established, wherein growth inhibition of *B. cereus*, *E. coli* and *S. aureus* (foodborne pathogens) occurred, when introduced, both as preprocessing and postprocessing contaminants in the product. In *kadhi* prepared using nonantagonistic cultures of *Lc. lactis* ssp. *lactis* CFR 2039 and market *dahi*, similar levels of growth inhibition of these pathogens were not

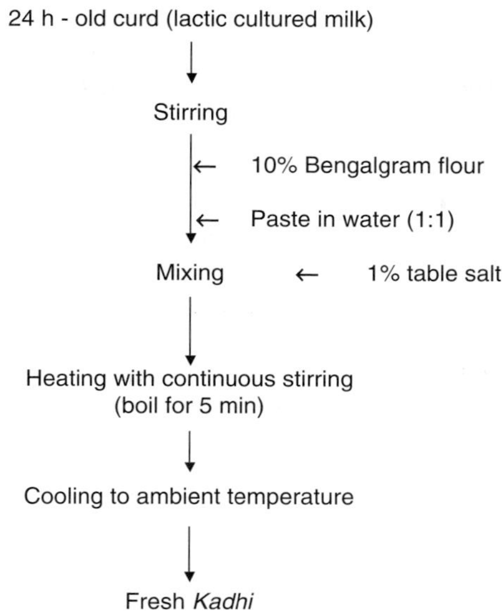

Figure 18.4 Flow sheet for *Kadhi* preparation (conventional method)
Source: Balasubramanyam, B.V., PhD Dissertation, Univ. Mysore, Mysore, 1996.

observed, but the increase in counts was very marginal. The study emphasized the benefits of using pure cultures of antagonistic LAB in the preparation of *kadhi*, with advantages of preservation and desirable quality attributes (124).

In a similar approach, the antagonistic cultures of *Lb. delbruecki* ssp. *bulgaricus* CFR 2028 and *Lb. delbruecki* ssp. *lactis* CFR 2023 were able to retard the growth of spoilage bacterial species such as *B. laterosporus*, *B. licheniformis,* and *B. subtilis* occurring as preprocessing and postprocessing contaminants in *kadhi*. However, these contaminants were able to grow and reach reasonable levels of population, when a nonantagonistic culture of *Lc. lactis* ssp. *lactis* CFR 2039 was used as the culture in *kadhi* preparation (125).

The efficacy of the antagonistic cultures of *Lactobacillus* species was evaluated in terms of microbiological and sensory parameters in *kadhi* during storage at ambient and refrigeration temperatures. *Kadhi* prepared individually with antagonistic cultures of *Lb. delbruecki* ssp. *bulgaricus* CFR 2028 and *Lb. delbruecki* ssp. *lactis* CFR 2023 were both (1) hot packed in laminates of polyester or aluminum foil or polyethylene, and (2) processed *in situ* using polypropylene pouches. The processed samples were then stored at 25–30°C and 6°C, separately. Stored samples of *kadhi* for 14 d at 25–30°C and 21 d at 6°C were stable from the microbiological point of view and in terms of sensory attributes, the products compared well with fresh *kadhi* (control) with no significant difference as evaluated by statistical analysis of mean scores recorded for individual sensory attributes (126).

Research studies have been attempted to enhance the nutritional status of ingredients that go into product preparation. Modification in the preparation of *kadhi* was achieved to provide better nutritional status through the use of lesser quantities of Bengalgram flour obtained from presoaked seeds. Considering the enhanced nutritional status of presoaked seeds, Bengalgram flour obtained from such seeds was used at lower concentrations relative to the level of flour obtained from resting seeds (*besan*) in *kadhi* preparation. On the basis of sensory attributes and the characteristic consistency of *kadhi*, the product prepared using Bengalgram flour (from presoaked seeds at 40 mg/g) was found to have better acceptance as apposed to the product prepared by conventional method using 100 mg/g of *besan*.

The alpha amylase and proteinase activities, as well as 11S globulin content (Table 18.3), varied in the samples of *Kadhi* prepared with Bengalgram flour obtained from soaked seeds using the strain of *Lb. delbruecki* ssp. *lactis* CFR 2125 and the sample prepared with commercially available *besan* (Varadaraj, unpublished data).

18.3.3.4 Dhokla

Dhokla, a steamed fermented product, is a traditional food popular in Western India, particularly in the States of Gujarat and Maharashtra. The product has a soft and spongy texture which is prepared from a mixture of Bengalgram dhal (*Cicer arietinum*), dehulled Blackgram dhal (*Phaseolus mungo*) and milled rice in the ratio of 2:1:1. The traditional method of *dhokla* preparation is presented in Figure 18.5. The mixture of legumes and cereal, after 6–8 h of soaking in water, is then ground into a grainy consistency. The resulting batter is mixed with curds in 1:1.5 (w/w) and allowed to ferment for 16–18 h. The

Table 18.3
Biochemical characteristics associated with *Kadhi* prepared using *Lb. delbruecki* ssp. *lactis* CFR 2125

Sample	Alpha Amylase Activity ($\times 10^2$ U/g Sample)	Acid Proteinase Activity (U/g Sample)	11S Globulin Content (mg/g Sample)
Kadhi A	12.3	1.6	1.8
Kadhi B	7.9	2.2	3.3

Kadhi A prepared with Bengalgram flour obtained from soaked seeds
Kadhi B prepared with commercially available besan

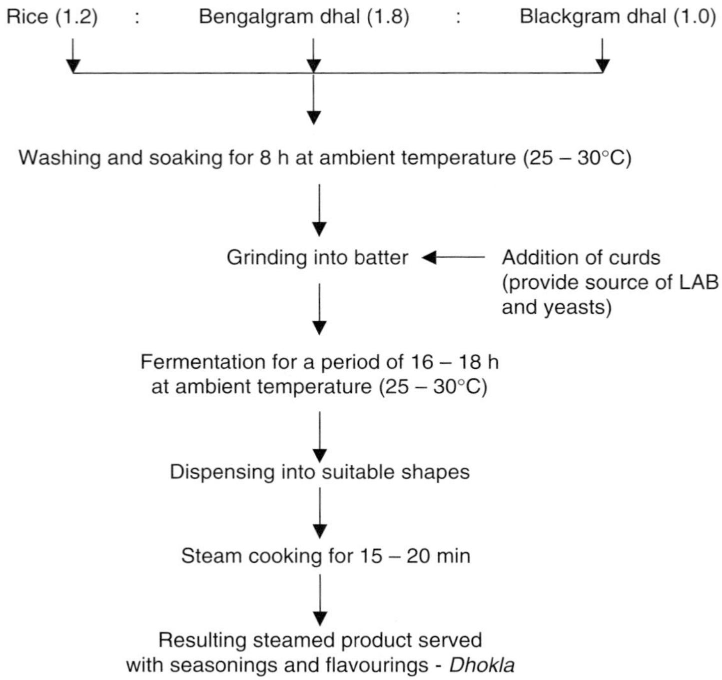

Figure 18.5 Flow sheet for *Dhokla* preparation (traditional method)
Source: Balasubramanyam, B.V., PhD Dissertation, Univ. Mysore, Mysore, 1996.

fermented batter is prepared into suitable shapes and steamed. The steamed product is then seasoned with oil, spices and greens (coriander leaves) prior to consumption (1,101).

In the background of limited studies, in one of the earlier studies (127) on the microbial and biochemical changes occurring during naturally fermented *dhokla* batter, it was established that the population levels of LAB and yeast increased during 0–18 h of fermentation period. The microflora usually comprised *Lb. fermentum, Leu. mesenteroides* and *Hansenula silvicola*. The increase in microbial counts was accompanied with the changes in pH, titratable, acidity and volatile fatty acids.

Studies with the use of antagonistic isolates of *Lactobacillus* spp. in the preparation of *dhokla* batter revealed that spoilage bacterial species, for example, *B. laterosporus, B. licheniformis,* and *B. subtilis* occurring as preprocessing contaminants were unable to increase in their population levels, while the same spoilage organisms were able to survive and increase in their population levels in *dhokla* batters prepared using nonantagonistic isolate of *Lactococcus* sp. (125).

In a similar type of study (126), the use of antagonistic cultures of *Lactobacillus* spp. to prepare *dhokla* batter resulted in the growth inhibition of food borne pathogenic bacterial species such as *S. aureus* and *B. cereus*. In an approach similar to *kadhi, dhokla* batters prepared with the antagonistic cultures of *Lactobacillus* spp. and heat sealed in laminate pouches revealed acceptable sensory attributes in a storage period of 7 d at 25–30°C and 14 d at 6°C. The stored batters were free from spoilage and pathogenic bacteria (126).

In the case of *dhokla*, alpha amylase and proteinase activities were on a higher side in batters prepared using the bacteriocinogenic culture of *Lb. delbruecki* subsp. *lactis* CFR 2125 as opposed to the one prepared from a commercial curd used as inoculua (Table 18.4). On the other hand, 11S globulin content was almost same in both the types of *dhokla* batters (Varadaraj, unpublished data).

18.3.3.5 Punjabi Warri

Punjabi warri, a legume based fermented and dried product is a very popular product in North India. *Warries* are spicy, hollow, brittle, fried balls of about 5–8 cm diameter each and are generally used as a condiment in cooking with vegetables, soup, or Indian *sambhar* (dhal based spicy liquid). The conventional method of *punjabi warri* preparation is presented in Figure 18.6. Dehulled Blackgram dhal (*Phaseolus mungo*), after being washed and soaked overnight in water, are ground to a batter of pasty consistency, which is then supplemented with a variety of spices, molded into small balls which are then subjected to fermentation and drying in open atmosphere for 4–8 d (128).

Microbiological and biochemical aspects associated with *punjabi warri* have been the subject of study in a few research investigations (4,128). These studies have established that the development and prevalence of microorganisms were affected by the seasons;

Table 18.4
Biochemical characteristics associated with *Dhokla* batters prepared using *Lb. delbruecki* ssp. *lactis* CFR 2125

Sample	Alpha Amylase Activity ($\times 10^2$ U/g Sample)	Acid Proteinase Activity (U/g Sample)	11S Globulin Content (mg/g Sample)
Dhokla A	3.9	6.7	0.4
Dhokla B	1.6	5.2	0.6

Dhokla A prepared with *Lb. delbruecki* ssp. *lactis* CFR 2125
Dhokla B prepared using curd of a commercial source

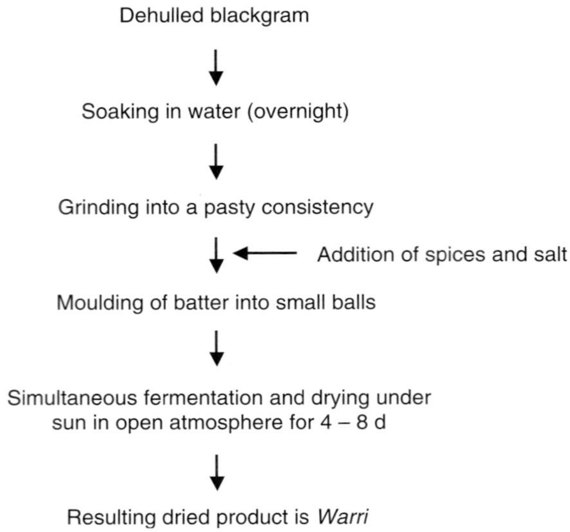

Figure 18.6 Traditional method of *Punjabi warri* preparation
Source: Balasubramanyam, B.V., PhD Dissertation, Univ. Mysore, Mysore, 1996.

summers being more favorable for bacteria, and winters for yeasts. These microbial types tend to increase significantly with the progress in fermentation. The microorganisms which bring about acidification and leavening in *warries* are those of *Leu. mesenteroides*, *Lb. fermentum,* and *Str. faecalis*. In addition, the yeast flora encountered were those of *Sacc. cerevisiae*, *Pichia membranaefaciens.* and *Trichosporon beigelii*. The microbiological changes have been associated with biochemical and nutritional changes. An increase in the amylase and proteinase activities was observed during warri fermentation, during which period there was also an appreciable increase in levels of water soluble B-vitamins such as thiamine, riboflavin, and cyanocobalamin.

The use of LAB as biopreservative in *punjabi warri* has been the subject of few studies. Although *warri* preparation involves the simultaneous process of natural lactic fermentation and drying, the conditions may not be ideal for the production of antibacterial factors by LAB. In order to enable the benefits of biopreservatives in *warries*, the product was prepared using antagonistic isolates of *Lb. delbruecki* ssp. *bulgaricus* CFR 2028 and *Lb. delbruecki* ssp. *lactis* CFR 2023. In these *warries*, the spoilage bacterial species of *B. laterosporus*, *B. licheniformis* and *B. subtilis* were retarded in their growth during the storage period of 10 d at 25–30°C, which happens to be the traditional method of simultaneous fermentation and drying (125). Similarly, the cultures of antagonistic *Lactobacillus* spp. resulted in the inhibition of selected food borne pathogenic bacterial species during *warri* fermentation (126).

18.3.3.6 Poko

Poko is a traditional rice based fermented food of Nepal characterized by a creamy color, soft and juicy, sweet and sour taste with slightly alcoholic and aromatic flavor. *Poko* is widely consumed by the rural populations in central Nepal, particularly during festivals and other celebrations. Nepalese people believe that consumption of *poko* promotes good health. Although these foods are commonly consumed, their production is confined to home or cottage scale operations. Traditionally, *poko* fermentation is brought about by *murcha* starters, the rice based starters known as *mana.* A few studies have revealed that mixed microflora consisting of the mucorales group of molds, yeasts, and LAB occur in *murcha* starters (75,129,130). The *murcha* starters are also used to prepare alcoholic beverages such as *jnard* and *rakshi*, respectively (131).

Preparation of *murcha* is regionalized and is kept as a close secret by certain sects of the Nepalese population and passed onto their next generation. The *murcha* starter cakes are of two types, namely *manapu* and *mana*. The *manapu* is prepared from rice flour and milled grains, while *mana* is prepared from wheat flakes. *Manapu* starters are prepared in the traditional style using rice flour or millet flour and other ingredients as shown in Figure 18.7. The traditional method of *mana* preparation from wheat flakes is illustrated in Figure 18.8. *Poko* preparation is presented as a flow chart in Figure 18.9.

Although *poko* has been a popular traditional fermented product, it has not been the subject of detailed microbiological studies. Research studies on microbiological and biochemical characteristics of *murcha* starters and *poko* fermentation revealed that LAB and yeasts are the predominant microbial groups, while a few fungal organisms do occur. In a few samples analyzed, there was the presence of *Escherichia coli* and *S. aureus*, but the samples were devoid of *B. cereus*. *Poko* fermentation was the resultant of microbial succession with the predominant microorganisms identified being *Sacc. cerevisiae*, *Candida versastilis*, *Lactobacillus* spp., *Pediococcus* spp., and *Rhizopus* spp. A quality *poko* product with desirable attributes of pH, acidity, reducing and total sugars and alcohol could be obtained in a fermentation period of 3 d at 30°C (132).

Considering the natural fermentation involved in *poko,* fermentation due to the use of starter culture cakes with mixed microflora, attempts have been made to use defined microbial organisms as starter cakes. Defined starter cakes for *poko* fermentation were prepared using a mixed inoculum of *Sacc. cerevisiae*, *Rhizopus chinensis,* and *Pediococcus pentosaceus*. The product prepared with the defined starter was evaluated

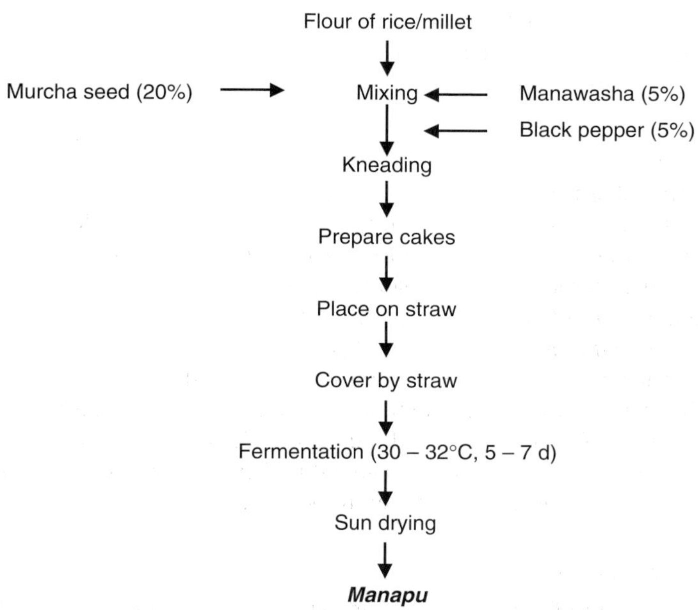

Figure 18.7 Traditional method of *manapu* preparation
Source: Shrestha, H., K. Nand, E.R. Rati., *Food Biotechnol.* 16:1–15, 2002.

Rice/Wheat
↓
Soaking in water
↓
Steaming for 30 min
↓
Washing in water
↓
Steaming
↓
Cooling by spreading over floor
↓
Cover by straw and straw mat
↓
Fermentation (30 – 32°C, 9 – 10 d)
↓
Green molds appear
↓
Sun drying
↓
Mana

Figure 18.8 Traditional method of *mana* preparation
Source: Shrestha, H., K. Nand, E.R. Rati., *Food Biotechnol.* 16:1–15, 2002.

for its performance against the traditional manapu starter. *Poko* prepared with the defined microbial starter had increased levels of vitamins including pyridoxine, thiamine, vitamin B_{12}, niacin, and folic acid. Additionally, the use of a defined starter ensured a better quality product with built in aspects of hygiene and safety (133).

18.3.3.7 Rabadi

Rabadi is a fermented beverage popular in north western parts of India, particularly the state of Rajasthan. The product is prepared from a mixture of pearl millet flour (*Pennisetum typhoideum*) or wheat flour and buttermilk, which is then placed in an earthen container and allowed to ferment for 4–6 h at ambient temperature. The fermented product is diluted with water, boiled and salted to taste. Due to the process of natural fermentation and lactic flora coming from buttermilk, lactobacilli count appears to increase (134).

Rabadi fermentation of freshly ground wheat millet flour brought about significant increase in HCl-extractability of calcium, iron, copper, zinc, manganese, and phosphorus (135). Consumption of such fermented foods may help to ameliorate the prevalent mineral deficiencies due to their limited bioavailability from such coarse grains and they may lead to better mineral status of the population (96–98).

18.3.3.8 Other Products

The diversified culture of the Indian subcontinent is well represented by the wide range of traditional fermented foods. The foods and their preparations previously discussed in this presentation have revealed the scientific knowledge and understanding of the

Rice
↓
Soaking in water for 24 h
↓
Addition of moisture (70 – 75%)
↓
Autoclaving at 121°C for 20 min
↓
Cooling to ambient temperature
↓
Addition of starter at 0.2%
↓
Incubation at 30 – 32°C for 5 d
↓
Fermented product
↓
Poko

Figure 18.9 Traditional method of *poko* preparation
Source: Shrestha, H., K. Nand, E.R. Rati., *Food Biotechnol.* 16:1–15, 2002.

biotechnological aspects involved in the major cereal and legume based fermented foods. Other popular food items are consumed widely by the Indian population, but scientific studies involving these foods are lacking. A few of these foods follow:

1. *Papadam* is a spiced condiment prepared from fermented Blackgram dhal paste. The prepared dough is slightly spiced as that of *punjabi warries* and is rolled into wafer-like pieces of about 10–20 cm diameter and 0.5 mm thick, which are then subjected to drying under sunlight for 6–8 h allowing for the simultaneous processes of fermentation and drying. The dried product is then deep fried in oil and consumed more as a culinary food item.
2. *Kulcha* is a popular product consumed in Pakistan and Northern India. The product is prepared from white wheat flour mixed with milk, sugar, salt, curds, dry yeast, baking powder, and water, with the ingredients kneaded well into soft stiff dough. The dough is subjected to fermentation for a period of 6–8 h at ambient temperature. The fermented dough is divided into equal size balls which are then rolled or flattened by hand to thick round *kulchas* of about six inches in diameter. The smoothly flattened dough is baked in a *tandoor* (metallic baking tray or special oven made of clay) to obtain a golden brown color. *Kulchas* are served hot.
3. *Batura* is a leavened bread, deep fried in oil, widely consumed in Northern India as snack item along with *masala channa* (*Cicer arietinum*). The product is prepared from a mixture of white wheat flour, baking powder, salt, curds, sugar, and water. The dough is prepared with little oil and covered with a hot cloth and kept for 1–2 h at ambient temperature. Then the dough is divided into equal sized balls and each one of them is rolled out into discs of five inches in diameter. The rolled out discs are deep fried in oil over medium flame, until the *baturas* reach light brown in color on both sides. The product is served hot.

4. *Naan* is a leavened flat bread baked in a clay oven called the *tandoor* and is widely consumed by the people of Northern India and Pakistan. The product is prepared from a mixture of white wheat flour, milk, egg, baking powder, salt, sugar, and curds. The dough is allowed to ferment at ambient temperature for 1–2 h. Then the dough is divided into equal sized medium balls, which are then rolled out into oblong shapes on a lightly flattened surface. The flattened dough is put on a wet cloth and then transferred onto the inner wall of preheated *tandoor* (clay oven) oven wall. The final product — *naan* — when fully cooked is crisp and brown on both sides, which is removed using skewers and served hot, topped with butter.
5. *Jelaebie* is a sweet savory item widely consumed in almost all parts of India and Pakistan on occasions of festivals and other celebrations. The product is prepared from a batter of wheat flour and water, fermented for 8–10 h. The fermented batter is squeezed through a thin cloth with a hole of 3–5 mm diameter, deposited as spirals into boiling vegetable oil and then deep fried. The light brown and crispy spirals are removed from the oil with a perforated ladle and immediately immersed in 70% sugar syrup. The crispy spirals, known as *jelaebie,* are removed from the sugar syrup and served hot as a delicious sweet savory item.

18.4 COMMERCIALIZATION OF TRADITIONAL FERMENTED FOODS

The review presented in this chapter has brought into focus the entire scenario of traditional fermented foods of the Indian subcontinent. In the background of a diversified Indian culture, it is quite natural that a large number of these fermented foods have their base in India. Although these foods are consumed almost on a daily basis by one section of population or the other, the approach toward commercialization is at the infancy stage. This slow progress may be due to the fact that these foods are still in the domain of home scale or at the most cottage scale operations. However, in recent times, with the changing social pattern, the consumer driven market is moving toward convenience foods, or more so fast foods, which encompasses a wide range of traditional fermented foods.

In the changing scenario, attempts have been made at the Central Food Technological Research Institute, Mysore, India to develop process know how for certain popular products. A few of them are:

1. Ready to use *idli* and *dosa* batter in retail packs, which afford a great deal of convenience with a shelf stability of 3–5 days at ambient temperature.
2. Preparation of pickles and *chutneys* which give a wider market, as these two food items are the regular items of lunch and breakfast menus.

In the light of an exhaustive biotechnological information about traditional fermented foods, in the future, food suppliers should look at aspects that would make these fermented foods achieve commercial outreach with built in concepts of convenience, shelf stability, hygiene, and safety toward improvement of human health.

ACKNOWLEDGMENT

The authors are thankful to Dr. V. Prakash, Director, CFTRI, Mysore for the support and interest in the area of traditional fermented foods.

REFERENCES

1. Steinkraus, K.H. *Handbook of Indigenous Fermented Foods*. New York: Marcel Dekker, 1995, pp 111–347.
2. Delaglio, F. Starters for fermented milks, I: taxonomy and metabolism. *Bull. Intl. Dairy Fed.* 227:7–18, 1988.
3. Salovaara, H. Lactic acid bacteria in cereal based products. In: *Lactic Acid Bacteria*, Salminen, S., Atte von Wright, eds., New York: Marcel Dekker, 1993, pp 111–126.
4. Soni, S.K., D.K. Sandhu. Indian fermented foods: microbiological and biochemical aspects. *Ind. J. Microbiol.* 30:135–157, 1990.
5. Achaya, K.T. *Indian Foods: A Traditional Companion*. Oxford: Oxford Univ. Press, 1994.
6. Caplice, E., G.F. Fitzgerald. Food fermentations: role of microorganisms in food production and preservation. *Intl. J. Food Microbiol.* 50:131–149, 1999.
7. Allison, G.E., T.R. Klaenhammer. Phage resistance systems in lactic acid bacteria. *Intl. Dairy J.* 8:207–226, 1998.
8. McGarry, A., J. Law, A. Coffey, C. Daly, P.F. Fox, G.F. Fitzgerald. Effects of genetically modifying the lactococcal proteolytic system on ripening and flavour development in Cheddar cheese. *Appl. Environ. Microbiol.* 60:4226–4233, 1994.
9. Mierau, I., E.R.S. Kunji, K.J. Leenhouts, M.A. Hellendoorn, A.J. Haandrikman, B. Poolman, W.N. Konings, G. Venema, J. Kok. Multiple peptidase mutants of *Lactococcus lactis* are severely impaired in their ability to grow in milk. *J. Bacteriol.* 178:2794–2803, 1996.
10. Mathur, B.N. Prospects for biotechnological applications in dairy processing industry for enhanced production. *Ind. Dairyman* 42:101–106, 1990.
11. de Vos, W.M. Metabolic engineering of sugar catabolism in lactic acid bacteria. *Antonie van Leeuwenhoek* 70:223–242, 1996.
12. Wells, J.M., K. Robinson, L.M. Chamberlain, K.M. Scholfield, R.W. LePage. Lactic acid bacteria as vaccine delivery vehicles. *Antonie van Leeuwenhoek* 70:317–330, 1996.
13. Tamime, A.Y., R.K. Robinson. Fermented milks and their future trends, part II: technological aspects. *J. Dairy Res.* 55:281–307, 1988.
14. Srinivas, H., R. Prabha, P.A. Shankar. Characteristics of cultured milks, yoghurt and probiotic yoghurt prepared from pre-refrigerated milks. *J. Food Sci. Technol.* 34:162–164, 1997.
15. Cousin, M.A., E.H. Marth. Changes in milk proteins caused by growth of psychrotrophic bacteria. *J. Dairy Sci.* 60(1):34, 1977.
16. Vaghela, M.N., A. Kilara. Nutritional and therapeutic aspects of indigenous and related Western fermented milk products: a review. *Ind. J. Dairy Sci.* 45:393–404, 1992.
17. Batish, V.K., S. Grover, S. Neelakantan. Improving shelf life and safety of fermented milk products through genetically improved microorganisms. *Ind. Dairyman* 45:51–56, 1993.
18. Vijayendra, S.V.N., R.C. Gupta. Therapeutic importance of bifidobacteria and *Lactobacillus acidophilus* in fermented milks. *Ind. Dairyman* 44:595–599, 1991.
19. BIS. *Specification for dahi – IS: 9617*. New Delhi: Bureau of Indian Standards, 1980.
20. Anonymous. *Prevention of Food Adulteration Act, 1955 (amendment-1988)*. Delhi: Delhi Law House, 1988.
21. Sinha, P.R., R.N. Sinha. Importance of good quality *dahi* in food. *Ind. Dairyman 52*:45–47, 2000.
22. Laxminarayana, H., V.K.N. Nambudripad, N. Lakshmi, S.N. Anantaramaiah, V. Sreenivasamurthy. Studies on *dahi*, II: general survey of the quality of market *dahi*. *Ind. J. Vety. Sci. Anim. Husb.* 22:13–25, 1952.
23. Misra, A.K. Commercial production of *dahi* by the dairy industry. *Ind. Dairyman* 44:501–503, 1992.
24. Srinivasan, M.R., B.N. Banerjee. Studies on the methods of preparation of *ghee*, II: souring process. *Ind. J. Vety. Sci. Anim. Husb.* 16:72–83, 1946.
25. Tiwari, M.P., Madan Lal, V.K.N. Nambudripad, H Laxminarayana. Studies on diacetyl producing *Streptococcus* isolated from *dahi*. *Ind. J. Dairy Sci.* 22:100–102, 1969.

26. Mohanan, K.R., P.A. Shankar, H. Lakshminarayana. Microflora of *dahi* prepared under household conditions of Bangalore. *J. Food Sci. Technol.* 21:45–56, 1984.
27. Baisya, R.K., A.N. Bose. Role of inoculating organisms on the physico-chemical changes in milk and final curd quality. *Ind. J. Dairy Sci.* 28:179–183, 1975.
28. Sharma, C.K., S.C. Jain. Effects of starter cultures and incubation (period and temperature) on the acidity of *dahi* (curd), part II. *J. Food Sci. Technol.* 12:81–83, 1975.
29. Vijayendra, S.V.N. Preparation, evaluation and use of bulk freeze-dried starter cultures in *dahi* and yoghurt manufacture. PhD dissertation, National Dairy Research Institute, Karnal, India, 1994.
30. Balasubramanyam, B.V., M.C. Varadaraj. *Dahi* as a potential source of lactic acid bacteria active against food borne pathogenic and spoilage bacteria. J. *Food Sci. Technol.* 31:241–243, 1994.
31. Balasubramanyam, B.V., M.C. Varadaraj. Cultural conditions for the production of bacteriocin by a native isolate of *Lactobacillus delbrueckii* ssp. *bulgaricus* CFR 2028 in milk medium. *J. Appl. Microbiol.* 84:97–102, 1998.
32. Mohanan, K.R., P.A. Shankar, H. Lakshminarayana. Growth and survival of microbial contaminants in *dahi*. *Ind. J. Dairy Sci.* 38:208–216, 1985.
33. Misra, R.K., K.L. Das, P.K. Roy. Bacteriological study of curd around Bhubaneswar. *Ind. J. Dairy Sci.* 45:309–310, 1993.
34. Neeraj, A., P.S. Panesar, S.S. Marwaha. Evaluation of different starter cultures for the production of quality *dahi* and their antibacterial activities. *Ind. Food Packer* 53:20–26, 1999.
35. Anonymous. *Dairy Handbook*. Karnal: National Dairy Research Institute, 1976.
36. Widanapathirana, S. Applications of lactic acid bacterial technology in industry: present status. Presented during International Workshop on Lactic Acid Bacteria, Central Food Technological Research Institute, Mysore, 1999.
37. Batish, V.K., S. Grover. Application of biotechnological approaches for quality assurance in dairy industry. *Ind. Food Ind.* 21:32–37, 2002.
38. Gupta, R.C., M. Bimlesh, V.K. Joshi, D.N. Prasad. Microbiological, chemical and ultra structural characteristics of *mishti doi* (sweetened *dahi*). *J. Food Sci. Technol.* 37:54–57, 2000.
39. Mann, B., V.K. Joshi. Proteose-peptone contents of *mishti doi* and their correlation with browning index. *J. Food Sci. Technol.* 34:425–426, 1997.
40. Rangappa, K.S., K.T. Achayya. *Indian Dairy products.* New Delhi: Asia Publishing House, 1974, pp 119–123.
41. Pillai, R.A.V., M.M.H. Khan, V.P. Reddy. Incidence of aerobic spore formers in *lassi*. *J. Food Sci. Technol.* 30:141–142, 1993.
42. Bhanumurthy, J.L., K.S. Trehan. Preservation and bottling of buttermilk. *Ind. Dairyman* 22:275–278, 1970.
43. Naresh, K., D.N. Prasad. Preservative action of nisin in *lassi* under different storage temperatures. *Ind. J. Anim. Sci.* 66:525–528, 1996.
44. Patidar, S.K., J.B. Prajapathi. Standardization and evaluation of *lassi* prepared using *Lb. acidophilus* and *S. thermophilus*. *J. Food Sci. Technol.* 35:428–431, 1998.
45. Kadam, S.J., D.N. Bhosle, I.G. Chavan. Studies on preparation of *chakka* from cow milk. *J. Food Sci. Technol.* 21:180–182, 1984.
46. Patel, R.S., M.H. Abdel-Salam. *Shrikhand*: an Indian analogue of western quarg. *Cult. Dairy Prod. J.* 21:6–7, 1986.
47. Reddy, K.K., M.P. Ali, B.V. Rao, T.J. Rao. Studies on production and quality of *shrikhand* from buffalo milk. *Ind. J. Dairy Sci.* 37:293–297, 1984.
48. Subramanian, B.S., C. Naresh Kumar, R. Narasimhan, A.M. Shanmugam, M.M.H. Khan. Selection of level and type of LAB starter in the preparation of dietetic *shrikhand*. *J. Food Sci. Technol.* 34:340–342, 1997.
49. Sharma, D.K. Ultrafiltration for manufacture of indigenous milk products: *chhana* and *shrikhand*. *Ind. Dairyman* 50:33–37, 1998.
50. Patel, R.S., B.K. Chakraborty. *Shrikhand*: a review. *Ind. J. Dairy Sci.* 41:109–115, 1988.

51. Nalawade, J.S., G.R. Patil, A.T. Sontakke, B.A. Hassan. Effect of compositional variables on sensory quality and consistency of *shrikhand*. *J. Food Sci. Technol.* 35:310–313, 1998.
52. Varadaraj, M.C., B. Ranganathan. Fate of *Staphylococcus aureus* in *shrikhand* prepared with *Lactobacillus acidophilus* and *L. bulgaricus*. *Ind. J. Dairy Sci.* 41:363–366, 1988.
53. Jagannath, A., M.N. Ramesh, M.C. Varadaraj. Predicting the behaviour of *Escherichia coli* introduced as a post processing contaminant in *shrikhand*, a traditional sweetened lactic fermented milk product. *J. Food Protect.* 64:462–469, 2001.
54. Jagannath, A., A. Ramesh, M.N. Ramesh, A.C. Chandrashekar, M.C. Varadaraj. Predictive model for the behaviour of *Listeria monocytogenes* Scott A in *shrikhand*, prepared with a biopreservative, pediocin K7. *Food Microbiol.* 18:335–343, 2001.
55. Boghra, V.R., O.N. Mathur. Physico-chemical status of major milk constituents and minerals at various stages of *shrikhand* preparation. *J. Food Sci. Technol.* 37:111–115, 2000.
56. Prajapati, J.P., K.G. Upadhyay, H.K. Desai. Comparative quality appraisal of heated *shrikhand* stored at ambient temperature. *Aust. J. Dairy Technol.* 47:18–22, 1993.
57. J.P. Prajapati, K.G. Upadhyay, H.K. Desai. Quality appraisal of heated *shrikhand* stored at room temperature. *Cult. Dairy Prod. J.* 28(14):16–17, 1993.
58. Mahajan, B.M., O.N. Mathur, D.C. Bhattacharya, M.R. Srinivasan. Production and shelf life of spray dried *shrikhand* powder. *J. Food Sci. Technol.* 16:9–10, 1979.
59. De, A., R.S. Patel. Technology of *shrikhand* powder. *Cult. Dairy Prod. J.* 25:21–28, 1990.
60. Goyal, G.K. Practice and research related to packaging of indigenous dairy products: a review. *Ind. Dairyman* 38:489–492, 1986.
61. Thapa, T.B. Technology of *sewsew* making in the high Himalayan region of Nepal. *J. Dairy. Sci.* 78(1):99, 1995.
62. Tamang, J.P., P.K. Sarkar, C.W. Hesseltine. Traditional fermented foods and beverages of Darjeeling and Sikkim: a review. *J. Sci. Food Agric.* 44:375–385,1988.
63. Yonzan, H., J.P. Tamang. Consumption pattern of traditional fermented foods in the Sikkim Himalaya, *J. Hill Res.* 11:112–115, 1998.
64. Pal, P.K., S.A. Hossain, P.K. Sarkar. Optimization of process parameters in the manufacture of *chhurpi*. *J. Food Sci. Technol.* 33, 219–223, 1996.
65. Tamang, J.P., S. Dewan, S. Thapa, N.A. Olasupo, U. Schillinger, A. Wijaya, W.H. Holzapfel. Identification and enzymatic profiles of the predominant lactic acid bacteria isolated from soft-variety *chhupri,* a traditional cheese typical of the Sikkim Himalayas. *Food Biotechnol.* 14:99–112, 2000.
66. Arora, G., B.H. Lee, M. Lamoureux. Characterization of enzyme profiles of *Lactobacillus casei* species by a rapid API-ZYM system. *J. Dairy Sci.* 73:264–273, 1990.
67. Montet, D., G. Loiseau, N. Zakhia, C. Mouquet. Fermented fruits and vegetables. In: *Biotechnology: Food Fermentations, Vol. 2,* Joshi, V.K., A. Pandy, eds., New Delhi: Educational Publisher & Distributors, 1999, pp 951–969.
68. Anand, J.C., D. Lakshmi. Effect of condiments on lactic fermentation in sweet turnip pickle. *J. Food Sci. Technol.* 8:143–145, 1971.
69. Mikki, M.S., Studies on sauerkraut. MSc dissertation, Indian Agricultural Research Institute, New Delhi, 1971.
70. Bawa, A.S., P.S. Ranote, A.S. Khatra. Drying and sauerkraut fermentation of cabbage. *Ind. Food Packer* 48:37–45, 1994.
71. Mikki, M.S., J.C. Anand. The effect of temperature on fermentation and quality of sauerkraut. *Ind. Food Packer* 5:9–11, 1974.
72. Mikki, M.S., J.C. Anand. Tolerance limits of mycoderma in sauerkraut to various additives. *Ind. Food Packer* 3:54–56, 1972.
73. Tamang, J.P. Traditional foods and beverages of Sikkim Himalayas in India: indigenous process and product characterization. *Proceedings of International Conference on Traditional Foods*, Mysore, 2000, pp 99–115.
74. Tamang, J.P. Indigenous fermented foods of the Sikkim Himalayas: socio-economic prospective. In: *Perspective for Planning and Development,* Raj, S.C., R.C. Sundriyal, E. Sharma, eds., Dehra Dun: Sikkim Science Society, 1998, pp 513–523.

75. Tamang, J.P., P.K. Sarkar. Traditional fermented foods and beverages of Darjeeling and Sikkim: a review. *J. Sci. Food Agric*. 44:375–385, 1988.
76. Tamang, J.P., P.K. Sarkar. A traditional lactic acid fermented radish root product. *J. Gen. Appl. Microbiol*. 39:395–408, 1993.
77. Garg, N., D.K. Tandan, S.K. Kalra. Lactic acid fermentation of cucumbers for pickling. *Beverage Food World* 20:17–18, 1993.
78. Berry, S.K., J.K. Mohanan, G.J. Joshi, A.K. Saxena, C.L. Kalra. Preparation and evaluation of ready to serve (RTS) black carrot beverage (*kanji*). *J. Food Sci. Technol*. 26:327–328, 1989.
79. Ramdas, A.R., P.R. Kulkarni. Fermentative preservation of carrots. *Ind. Food Packer* 41:40–48, 1987.
80. Sethi, V. Lactic fermentation of black carrot juice for spiced beverage. *Ind. Food Packer* 44:7–12, 1990.
81. Sura, K., S. Garg, F.C. Garg. Microbiological and biochemical changes during fermentation of *kanji*. *J. Food Sci. Technol*. 38:165–167, 2001.
82. Tamang, J.P., P.K. Sarkar. Microbiology of *mesu*, a traditional fermented bamboo shoot product. *Intl. J. Food. Microbiol*. 29:49–58, 1996.
83. Misra, M.C., S.P. Manjrekar. Production of *kachampulli*, a traditional product from Malabar tamarind (*Garcinia cambogia*). In: *Proceedings of International conference on Traditional Foods*, Mysore, 2000, pp 262–267.
84. Tamang, J.P. Case study on socio-economical prospective of *kinema*, a traditional fermented food. In: *Proceedings of International conference on Traditional Foods*, Mysore, 2000, pp 180–185.
85. Yonzan, H., J.P. Tamang. Consumption pattern of traditional foods in the Sikkim Himalaya. *J. Hill. Res*. 11:112–115, 1998.
86. Sarkar, P.K., J.P. Tamang, P.E. Cook, J.D. Owens. *Kinema*: a traditional soybean fermented food: proximate composition and microflora. *Food Microbiol*. 11:112–115, 1994.
87. Sarkar, P.K., J.P. Tamang. Changes in the microbial profile and proximate composition during natural and controlled fermentations of soybean to produce *kinema*. *Food Microbiol*. 12:317–325, 1995.
88. Tamang, J.P. Study of traditional fermented foods production in Darjeeling hills and Sikkim with emphasis on *kinema*. Postdoctoral report, United Nations University, Tokyo, 1995.
89. Nikkuni, S., T.B. Karki, K.S. Vilku, N. Okada. Mineral and amino acid contents of *kinema*, a fermented soybean food of Nepal. *Food Sci. Technol. Intl*. 1:107–111, 1995.
90. Tamang, J.P., S. Nikkuni. Selection of starter culture for production of *kinema*, a fermented soybean food of the Himalaya. *World J. Microbiol. Biotechnol*. 12:629–635, 1996.
91. Tamang, J.P., S. Nikkuni. Effect of temperature during pure culture fermentation of *kinema*. *World J. Microbiol. Biotechnol*. 14:847–850, 1998.
92. Sethi, V., J.C. Anand. Effect of mustard and its components on the fermentation of cauliflowers. *Ind. Food Packer* 38:41–46, 1984.
93. Gupta, U., E.R. Rati, R. Joseph. Nutritional quality of lactic fermented bitter guard and fenugreek leaves. *Intl. J. Food Sci. Nutr*. 49:101–108, 1998.
94. Campbell-Platt, G. *Fermented Foods of the World: A Dictionary and Guide*. London: Butterworths, 1987.
95. Wang, H.L., C.W. Hesseltine. Use of microbial cultures: legume and cereal products. *Food Technol*. 35:79–83, 1981.
96. Mahajan, S., B.M. Chauhan. Phytic acid and extractable phosphorus of pearl millet flour as affected by natural lactic acid fermentation. *J. Sci. Food Agric*. 41:381–386, 1987.
97. Khetarpaul, N., B.M. Chauhan. Improvement in HCl-extractability of minerals from pearl millet (*Pennisetum typhoideum*) by fermentation with yeasts and lactobacilli. *J. Sci. Food Agric*. 49:117–121, 1989.
98. Khetarpaul, N., B.M. Chauhan. Effect of pure and sequential culture fermentation by yeasts and lactobacilli on HCl-extractability minerals from food millet. *Food Chem*. 39:347–355, 1991.
99. Chang, R., S. Schwimmer, H.K. Burr. Phytate: removal from whole dry beans by enzymatic hydrolysis and diffusion. *J. Food Sci*. 42:1098–1101, 1977.

100. Wang, H.L., E.W. Swain, C.W. Hesseltine, H.D. Heath. Hydration of whole soya bean affects solid losses and cooking quality. *J. Food Sci.* 44:1510–1513, 1979.
101. Ramakrishnan, C.V. Studies on Indian fermented foods. *Baroda J. Nutr.* 6:1–7, 1979.
102. Reddy, N.R., S.K. Sathe, M.D. Pierson, D.K. Salunkhe. *Idli*, an Indian fermented food: a review. *J. Food Quality* 5:89–101, 1981.
103. Desikachar, H.S.R., R. Radhakrishnamurthy, G. Ramarao, S.B. Kadkol, M. Srinivasan, V. Subrahmanyan. Studies on *idli* fermentation, I: some accompanying changes in the batter. *J. Scient. Indu. Res.* 19C:168–172, 1960.
104. Steinkraus, K.H., A.G. van Veen, D.B. Thiebeau. Studies on *idli*: an Indian fermented black-gram-rice food. *Food Technol.* 27:110–113, 1967.
105. Lewis, Y.S., D.S. Johar, V. Subrahmanyan. Studies on process simplification in the preparation of a fermented type of foodstuff: *idli*. *Cent. Food Technol. Res. Inst. Bull.* 5:257–265, 1955.
106. Radhakrishnamurthy, R., H.S.R. Desikachar, M. Srinivasan, V. Subrahmanyan. Studies on *idli* fermentation, II: relative participation of blackgram flour and rice semolina in the fermentation. *J. Scient. Indu. Res.* 20:342–345, 1961.
107. Mukherjee, S.K., M.N. Albury, C.S. Pederson, A.G. van Veen, K.H. Steinkraus. Role of *Lecuconostoc mesenteroides* in leavening the batter of *idli*, a fermented food of India. *Appl. Microbiol.* 13:227–231, 1965.
108. Yajurvedi, R.P. Microbiology of *idli* fermentation. *Ind. Food Packer* 34:33–38, 1980.
109. Thyagaraja, N., H. Otani, A. Hosono. Studies on microbiological changes during the fermentation of *idli*. *Lebensmittel-Wissenschaft Technol.* 25:77–79, 1992.
110. Venkatasubbaiah, P., C.T. Dwarakanath, V. Sreenivasa Murthy. Microbiological and physicochemical changes in *idli* batter during fermentation. *J. Food Sci. Technol.* 22:59–63, 1984.
111. Soni, S.K. Studies on some Indian fermented foods: microbiological and biochemical aspects. PhD dissertation, Guru Nanak Dev Univ., Amritsar, 1987.
112. Thygaraja, N., H. Otani, A. Hosono. Microflora in *idli*, a traditional fermented cereal-pulse product from India. *Lebensmittel-Wissenschaft Technol.* 24:504–507, 1991.
113. van Veen, A.G., L.R. Hackler, K.H. Steinkras, S.K. Mukherje. Nutritive value of *idli*, a fermented food of India. *J. Food Sci.* 32:339–341, 1967.
114. Lakshmi, I. Studies on fermented foods. M.Sc. dissertation, MS University of Baroda, Baroda, 1978.
115. Soni, S.K., D.K. Sandhu. Microbial prevalence and succession associated with *idli* fermentation. *Ind. J. Microbiol.* 31:169–174, 1991.
116. Varadaraj, M.C., S.V.N. Vijayendra, E.R. Rati, R. Agrawal, M.S. Prasad, K. Nand. An improved process for the preparation of inoculum for fermented foods. Indian Patent Application 352/DEL/01, 2001.
117. Varadaraj, M.C., E.R. Rati, R. Agrawal, S.V.N. Vijayendra, M.S. Prasad, K. Nand. An improved process for the preparation of shelf stable *idli* batter. Indian Patent Application 1129/DEL/99, 1999.
118. Agrawal, R., E.R. Rati, S.V.N. Vijayendra, M.C. Varadaraj, M.S. Prasad, K. Nand. Flavour profile of *idli* batter prepared from defined microbial starter cultures. *World J. Microbiol. Biotechnol.* 16:687–690, 2000.
119. Jama, Y.H., M.C. Varadaraj. Antibacterial effect of plantaricin LP84 on foodborne pathogenic bacteria occurring as contaminants during *idli* batter fermentation. *World J. Microbiol. Biotechnol.* 15:27–32, 1999.
120. Sathe, S.K., D.K. Salunke. Fermentation of the great northern bean (*Phaseolus vulgaris* L.) and rice blends. *J. Food Sci.* 46:1374–1378, 1981.
121. Soni, S.K., D.K. Sandhu, K.S. Vilkhu. Studies on *dosa*: an indigenous Indian fermented food: some biochemical changes accompanying fermentation. *Food Microbiol.* 2:175–181, 1985.
122. Soni, S.K., D.K. Sandhu, K.S. Vilkhu, N. Kamra. Microbiological studies on *dosa* fermentation. *Food Microbiol.* 3:45–53, 1986.
123. Soni, S.K., D.K. Sandhu. Nutritional improvement of Indian *dosa* batters by yeast enrichment and blackgram replacement. *J. Ferment. Bioeng.* 68:52–55, 1989.

124. Balasubramanyam, B.V., M.C. Varadaraj. Antibacterial effect of *Lactobacillus* spp. on food borne pathogenic bacteria in an Indian milk-based fermented culinary item. *Cult. Dairy Prod. J.* 30:22–27, 1995.
125. Varadaraj, M.C., B.V. Balasubramanyam, R. Joseph. Antibacterial effect of *Lactobacillus* spp. on spoilage bacteria in selected Indian fermented foods. *Proceedings of Intl. Conference on Traditional Foods*, Central Food Technological Research Institute, Mysore, 2000, pp 210–226.
126. Balasubramanyam. B.V. Studies on the application of antagonistic lactic acid bacteria in the biopreservation of selected indigenous milk and cereal/pulse based foods. PhD Dissertation, Univ. Mysore, Mysore, 1996.
127. Neeta Joshi, S.H. Godbole, P. Kanekar. Microbial and biochemical changes during *dhokla* fermentation with special reference to flavour compounds. *J. Fd. Sci. Technol.* 26:113–115, 1989.
128. Sandhu, D.K., S.K. Soni. Micro flora associated with Indian *Punjabi warri* fermentation. *J. Fd. Sci. Technol.* 26:21–25, 1989.
129. Hesseltine, C.W., M.L. Ray. Lactic acid bacteria in *murcha* and *ragi*. *J. Appl. Bacteriol.* 64:395–398, 1988.
130. Tamang, J.P., P.K. Sarkar. Microflora of *murcha*: an amylolytic fermentation starter. *Microbios* 81:115–122, 1995.
131. Karki, T., H. Shrestha. Fermentation process of Nepal *murcha* starters. Presented at the National Science Conference Organized by Royal Nepal Academy of Science and Technology, Kathmandu, Nepal, 1999.
132. Shrestha, H., K. Nand, E.R. Rati. Microbiological profile of *murcha* starters and physico-chemical characteristics of *poko*, a rice based traditional fermented food product of Nepal. *Fd. Biotechnol.* 16:1–15, 2002.
133. Shrestha, H., E.R. Rati. Defined microbial starter formulation for the production of *poko*: a traditional fermented food product of Nepal. *Fd. Biotechnol.* 17:15–15, 2003.
134. Dankher, N., B.M. Chauhan. Technical note: preparation, acceptability and B vitamin content of *rabadi*: a fermented pearl millet food. *Intl. J. Food Sci. Technol.* 22:173–176, 1987.
135. Gupta, M., N. Khetarpaul, B.M. Chauhan. Effect of *rabadi* fermentation on HCl-extractability of minerals of wheat. *Food Chem.* 42:111–117, 1991.

3.19

Fermentation Biotechnology of Plant Based Traditional Foods of the Middle East and Mediterranean Region

Parthena Kotzekidou and Effie Tsakalidou

CONTENTS

19.1 HISTORY

Fermented foods of plant origin are popular in the Mediterranean and Middle East region. Traditional fermentation processes give rise to foods that are prized for their palatability and wholesomeness, but also enriched with essential amino acids and vitamins. Vegetables and fruits are preserved by lactic acid fermentation, producing pickled vegetables and table olives, which can be consumed without cooking. Alcoholic fermentation by yeast is generally less common and prevails mainly in bread making (1).

The Mediterranean food culture, rich in foods from plant materials, is linked to health and is thought to be associated with longer life expectancy. Throughout the Mediterranean Basin, a lower incidence of cardiovascular and cerebrovascular diseases, and certain cancers such as colon, breast, and other hormone dependent cancers, occur less frequently than in Northern Europe (2).

The objective of this chapter is to provide a view of the present state of knowledge of the fermentation of plant derived products like fermented vegetables, table olives, and bread in the Mediterranean and Middle East countries.

19.2 FERMENTED VEGETABLES

19.2.1 Introduction

The preparation of vegetables by lactic acid fermentation (pickling) on a small scale in the home has been applied in the Mediterranean and the Middle East region since antiquity. Nowadays, fermented vegetables such as cucumbers, green tomatoes, peppers, eggplants, carrots, okra, and mixed vegetables (including carrots, cauliflower, celery, green tomatoes, and cucumbers) serve as appetizers and are usually produced in salt brine by commercial processors. Vegetables for pickling are harvested while still immature, as fully ripened vegetables are too soft for most commercial uses. The fermentation is due primarily to the lactic acid bacteria, although yeasts and other microorganisms may be involved depending on the salt concentration and other factors. Salt influences the type and extent of microbial activity and prevents softening of the vegetable tissue (3).

The conventional method of fermenting vegetables used by industry is natural fermentation carried out by the indigenous microbial flora of the product. The microbial load of the fresh vegetables can, however, be highly variable, influenced by the ambient temperature during harvesting and other factors which are hard to control. This has numerous drawbacks that limit the yield of the process and affect the quality of the final product, making it difficult to anticipate the course of fermentation. To prevent defective fermentations, brines with high salt concentrations are sometimes used in order to regulate the type and extent of microbial growth, as well as to prevent softening of the product. Increased salt concentrations slow down the process, as the growth of lactic acid bacteria is inhibited, and the product may turn soft due to the growth of molds. In addition, subsequent removal of the salt is needed, which can result in leaching and substantial losses in vitamins and other nutrients, and the handling of highly polluting saline effluents is involved (4). Recent research is focused on the application of starter cultures in controlled fermentations to obtain a high quality end product as well as to reduce the amount of waste brines which can result in a significant environmental pollution problem (5). This review deals with fermented vegetables (with the exception of sauerkraut which is not a typical product of the Mediterranean and the Middle East Region) produced either by small scale producers who shelf their products in bulk, using refrigeration to assure stability, or by large scale producers after pasteurization at low temperature (such as an internal product temperature of 71°C with no holding time) to increase the shelf life. The microbiology of fresh pack pickles (pasteurized, unfermented pickles) is not included in the present review.

19.2.2 Microorganisms Involved in Fermentation

The microflora of fresh vegetables is mainly dominated by Gram-negative aerobic bacteria and yeasts, while lactic acid bacteria constitute a small portion in the initial population (6,7). Most vegetables, with the addition of salt brines (providing that salt concentration does not exceed 8%), under anaerobic conditions and appropriate temperature, undergo a spontaneous lactic acid fermentation during which lactic acid bacteria have a competitive advantage. The species of lactic acid bacteria most associated with the natural fermentation of vegetables are: *Pediococcus pentosaceus, Lactobacillus brevis,* and *Lactobacillus plantarum* (8). While *Pediococcus* species tolerate a minimum pH of 4.0 for growth, *Lactobacillus* species are the most acid tolerant, with a minimum pH of 3.8 (9). The growth of lactic acid bacteria depends on the available nutrients, salt concentration, pH value, and temperature of the environment. Changes of these parameters influence the dominant organisms. Organisms such as *Leuconostoc mesenteroides* are usually inhibited in the brine by the higher salt concentration, temperature, and rapid pH drop (5). The addition of salt and the rapid production of organic acid suppress the Gram-negative bacteria, which the lactic acid bacteria outnumber on the product. In addition, bacteriocins produced by lactic acid bacteria isolated from naturally fermented vegetables have a broad spectrum of activity and indicate that antimicrobial proteins have an important role in the ecology of fermented products (8,10–14).

Fermentation of vegetables occurs in four sequential stages (8). In the initial stage Gram-positive and Gram-negative microorganisms naturally present on the vegetables may grow. During the primary fermentation, growth of lactic acid bacteria occurs with or without fermentative yeasts. Secondary fermentation occurs if fermentable carbohydrates remain after the primary fermentation, which can be assimilated by fermentative yeasts while lactic acid bacteria are inhibited by low pH. Postfermentation is characterized by the depletion of fermentable carbohydrates and the absence of microbial growth under anaerobic conditions; while during the exposure of the brine to the atmosphere oxidative microorganisms grow on the surface. The proliferation of the naturally occurring lactic acid bacteria during initial and primary fermentation depends on their presence on the raw material and the physical and chemical conditions during fermentation.

During spontaneous fermentation of eggplants two stages appear: an initial stage (with a duration of 2 days) in which obligately heterofermentative *Lactobacillus fermentum* and *L. brevis* predominate, with a peak in the *L. fermentum* load on day 2; followed by a stage of vigorous sugar fermentation (with a duration of 5 days) dominated by *L. plantarum* (9). Similarly, during the spontaneous fermentation of okra, within the first 24 h of fermentation the heterofermentative *Lactobacillus cellobiosus* is dominant, while after 48 h of fermentation strains of *L. plantarum* predominate and the pH value of the brine is very low (pH 3.4) (15). *Lactobacillus plantarum* is the terminal organism in many natural lactic acid fermentations due to its high acid tolerance (16). Vegetables fermented with the homofermentative *L. plantarum* are microbiologically stable, provided all detectable sugars are removed, the pH is 3.8 or less, and they are stored in anaerobic conditions (17).

There have been only limited quantitative data obtained on the formation of products in vegetable fermentations by homofermentative or heterofermentative organisms (8). Carbon dioxide production is of major importance in the fermentation of vegetables. High levels of CO_2 are known to cause gassing or bloating of cucumber tissue, leading to undesirable defects (called "bloaters" or "floaters") (5). Although CO_2 is a major end product of heterolactic acid bacteria, CO_2 production by homofermentative lactic acid bacteria occurs by decarboxylation of malic and citric acids present in many vegetables as well as amino acids (18,19). In green bean juice fermented by the homofermentative *L. plantarum* only lactic acid is obtained as a fermentation product, whereas *Leuconostoc mesenteroides* and *Lactobacillus brevis* strains form typical heterofermentative products, i.e., acetic acid, in addition to lactic acid, ethanol, and mannitol, as well as a large amount of CO_2, are formed (20). In okra juice fermented by strains of *L. plantarum*, although the majority of the strains produced only lactic acid, three strains produced acetic acid and ethanol in addition to lactic acid, and showed low lactate dehydrogenase (LDH) activity, whereas the strains of *Lactobacillus cellobiosus* and *L. brevis* carried out a typical heterolactic fermentation (21).

19.2.3 Controlled Fermentation

The main advantages of controlled fermentation are the use of brines with low salt concentrations, the substantial shortening of the fermentation time, and the improvement of the quality, achieved by uniform sensory and nutritional quality attributes in the final product.

Controlled fermentation is commercially used in cucumber fermentation in a process thoroughly described by Fleming (3). The acidification of the brine to 4.5 with acetic acid or vinegar serves to virtually eliminate growth of the natural microbial flora. The concentration of salt required to prevent softening is 5–8%. Addition of calcium acetate or calcium chloride results in firmness retention of cucumbers at relatively low concentrations of sodium chloride (1.4–4%). The brine is purged with nitrogen to maintain an anaerobic headspace at the top of the tank as well as to remove dissolved CO_2 and prevent bloater formation. A pilot study showed that temperature (17 and 26°C) affects the rate of fermentation, and in combination with purging strongly affects the incidence of bloater damage (22). A 24 h buffering of the brine with sodium acetate serves to eliminate secondary fermentation by yeasts. With the pH at 4.7, the brine is inoculated with *L. plantarum* or *P. pentosaceus* or a combination. At temperatures between 26 and 29°C rapid growth of these starter cultures is favored (23). The inoculation with starter cultures is undertaken in order to complete the controlled fermentation as rapidly as possible to reduce purging costs (8).

19.2.4 Starter Cultures

To avoid the drawbacks of natural fermentation, addition of starter cultures and maintenance of anaerobiosis under controlled fermentation conditions is achieved. Although there are very few commercial starter cultures designed for vegetable fermentations (24), their benefits in vegetable fermentations are important. Starter cultures help to standardize the fermentation by controlling the competitive microbiota, especially in the cases when treatments are undertaken to rid the raw product of the natural epiphytic microflora that interfere with fermentation. This can be achieved either by blanching using hot water (66–80°C), chlorination and subsequent acidification (5,8), or lye treatment (lye concentrations 1–2% NaOH) (25). When mixed starter cultures are applied to the raw material, their selection is based mainly on the competitiveness between the starter and the natural flora, as well as on the sensory properties of the fermented product.

The commercial culture "Vege-Start 60™" (Chr. Hansen's Laboratorium, Horsholm, Denmark), composed of a mixture of *Lactobacillus* sp., primarily *L. plantarum,* was used for the fermentation of blanched Almagro eggplants (Appellation of Origin of a pickled product manufactured using a native variety of eggplant) in brine with a salt concentration not exceeding 6%, while the temperature in the fermenter was 32°C. A shortening of the fermentation process was observed; the fermented product was of good quality, though somewhat bitter due to the starter culture (26). In an effort to overcome the bitter taste, starter cultures composed of lactic acid bacteria isolated from spontaneous fermentations were used in order to select the most appropriate starter culture. By the combined action of the obligate heterofermentative species *L. brevis* and the facultative heterofermentative species *L. plantarum,* fermented eggplants with sensory characteristics preferred by panelists were obtained. Whereas *L. fermentum,* present in certain starter formulations tested, appeared to encounter difficulty growing during fermentation and exerted little influence on the sensory characteristics of the product (9,27).

During fermentation of okra a *Lactobacillus cellobiosus* strain (isolated from a spontaneous fermentation) could ferment okra mucilage, a polysaccharide composed of galacturonic acid, galactose, rhamnose, and glucose, lowering the pH and retaining the natural color of the product. When blanching preceded the fermentation improved quality characteristics of the fermented product in comparison to spontaneous fermentation were obtained. The commercial culture "Vege-Start" can antagonize the natural microbiota, lowering the pH of the raw product and improving the fermentation characteristics (28).

The addition of *L. plantarum* with pH control during fermentation of green beans, cucumbers, red and green bell peppers, and green tomatoes gives fermented products microbiological stablility during 12 months storage in hermetically sealed jars at approximately 24°C , providing all fermentable sugars are removed during fermentation and the product is stored at pH 3.8 or below (17). During fermentation of carrot, cabage, beet, and onion vegetable mixtures with cultures of *L. plantarum, Pediococcus acidilactici* and *L. mesenteroides* applied as a mixed starter culture, acceleration of the fermentation process and prevention of deterioration of the fermented product for up to 90 days is obtained. Although this mixed culture has a similar fermentation pattern to a spontaneous fermentation, the ethanol production is lower, and the acidification rate is faster, which helps to stabilize the final product as well as to reduce the production of gas during the fermentation and storage of the product (24). Peeled garlic blanched in hot water (90°C) for 15 min and fermented using a starter culture of *Lactobacillus plantarum* reaches a pH of 3.8 after 7 days, whereas the growth of the starter culture is inhibited in unblanched garlic. The blanched fermented garlic is microbiologically stable during storage at 30°C in acidified brine (containing 3% NaCl and pH 3.5 at equilibrium) and keeps the flavor (29).

19.3 TABLE OLIVES

19.3.1 Introduction

For centuries, the fruits of the cultivated olive tree *Olea europaea sativa*, a typical plant of the countries of the Mediterranean Basin, have been used for direct consumption and oil production. The first reference to the olive tree is reported in the Bible. In its first book, *Genesis*, the flight of the dove with the olive branch announcing the end of the flood is described. Nowadays, the cultivation of the olive tree is practiced in the Mediterranean area and the Middle East, as well as in parts of America and Australia. The Mediterranean region still serves as the major international olive growing area, accounting for almost 98% of the world's olive tree plantation (2). The main cultivars of table olives are presented in Table 19.1. Olive processing is the largest agroindustrial sector of the economy in some Mediterranean countries such as Spain, Italy, and Greece (32).

Table olives have been consumed since ancient times. For direct consumption, olives must first be processed in order to remove, at least partially, the natural bitterness of the fruit. As the bitterness is related to the stage of ripeness of the fruits, different processing methods have been used world wide. The methods of greatest importance are: Spanish

Table 19.1
Main table olive cultivars suitable for the preparation of green olives or black olives in brine in Mediterranean and Middle East region.

Country of Origin	Green Olives	Black Olives in Brine
Spain	Manzanilla	Holiblanca
	Sevillana or Gordal	Cacerena
		Lechin
Italy	Ascolana	Itrana
	Oliva di Spagna or Cerignola	Uovo di Piccione
	Sant' Agostino	
	Cucco	
Greece	Conservolea	Kalamon
	Chalkidiki	Conservolea
Turkey	Domat	Gemlik
	Memeli	Edencik
Morocco	Picholine	Picholine
	Zitoun	Zitoun
Algeria	Sigoise	Sigoise
	Azeradj	Azeradj
Tunisia	Meski	Meski
	Barouni	
Syria	Massabi	Dan
	Dan	Djlt
	Djlt	
Israel	Merhavia	Nuovo di Sicrone
Portugal	Carrasquenta	Galega Vulgar
	Redondil	

Data taken from Balatsouras, G. *The Table Olive*. Athens (*Monograph in Greek*), 1995, pp 40–66, and Garrido-Fermandez, A., M.J. Fernandez-Diez, M.R. Adams, *Table Olives,* London: Chapman & Hall, 1997, pp 10–45.

style green olives in brine; California style black olives in brine; and Greek style naturally black olives in brine (31). In the Spanish style and Californian style processes, bitterness is removed by an alkaline treatment, whereas in the case of the Greek style process, fruits are placed in brine and the debittering process is slow and sometimes partial.

Oleuropein, a secoiridoid glucoside that is commonly found in the leaves of the olive tree as well as in the fruits, is responsible for the natural bitterness. Hydrolysis of the oleuropein to more simpler, less bitter compounds, such as elenolic acid and β-3,4-dihydroxyphenyl ethyl alcohol (hydroxytyrosol), is obtained by olive processing in order to provide some form of debittering of the fruits (33,34). This can be achieved by alkaline hydrolysis (i.e., treatment of the olives with sodium hydroxide solutions before the fermentation; the product is referred to as treated olives in the international standards) and by further storage in brine or in dry salt. When there is no alkali treatment before fermentation and the olives are placed into brine or dry salted, bitterness may be also removed slowly and partially during the acid fermentation and the product is characterized as untreated. There are three main types of olives in brine according to the Unified Qualitative Standard Applying to Table Olives in the International Trade (35): green olives, turning color olives (untreated, as well as treated, including black olives; i.e., fruit harvested during the stage of turning color and blackened by oxidation in an alkaline medium), and naturally black olives.

There are numerous regional or national variations of the processing conditions applied (31) concerning:

1. The condition of the fruit (green color, naturally black color, turning color)
2. Treated or untreated (referred to the procedure applied to eliminate bitterness)
3. The preservation method (in brine or dry salting)

This review deals with table olive fermentation processing in the Mediterranean and the Middle East region; i.e., Spanish style green olives in brine (lye treated green olives in brine) and naturally black olives in brine (untreated naturally ripe black olives in brine).

The occurrence of biophenolic components in olives, and particularly oleuropein as well as compounds derived from the hydrolysis of the oleuropein (34), provides functional value to the Mediterranean food culture, owing to recognized antioxidant activities of these substances (36). Olive biophenols demonstrate antioxidant and free radical scavenger ability as well as antimicrobial activity, which are important for human health and well being (2). The natural bitter tasting oleuropein releases phytoalexins against pathogens. In vegetables, biophenols play an important role as the defense mechanism against pathogen and insect attack (37). In olives, the relevance of these bioactive molecules has been acknowledged for their contribution to flavor and to preventing rancidity (38). In the Spanish style green olives process oleuropein is completely degraded by the alkali treatment that is commonly used, yielding hydroxytyrosol and elenolic acid, with the consequence that lactic acid bacteria are the predominant microorganisms; whereas in undebittered olives phenolic substances inhibit *L. plantarum* strains (34,39) as well as mold growth and mycotoxin production (40). Although hydroxytyrosol has an inhibitory effect against *L. plantarum* ATCC 8014 inoculated in untreated olive brines, other *L. plantarum* strains are able to grow in alkali treated olive brines as a consequence of the different permeability of fruit for phenolic substances. It is suggested that proteins and amino acids of olives can mask the bactericidal effect by combining with the phenolic compounds, whereas in the case of untreated fruits the nutritional and neutralizing compounds take a long time to emerge from the fruits (39).

19.3.2 Microorganisms Involved in Fermentation

19.3.2.1 Spanish Style Green Olives in Brine

Spanish style green table olive processing consists of a treatment with NaOH solution (1.8–3.0%, w/v) in order to: hydrolyse the bitter glucoside oleuropein, promote alkaline hydrolysis of proteins in the flesh, and make the skin more permeable. In order to eliminate the lye inside the olives, one, two, or three washings follow. The lye treatment has the following consequences: the pH increases to values between 7.5 and 8.5 and the initial population of microorganisms is reduced, and during washing the nutrients and sugars from the olives are reduced. In order to partially neutralize the lye, food grade HCl or other strong acid can be added to the washing water, reducing the pH to approximately 6.0. The olives are then brined in a 10–12% w/v NaCl solution in which they undergo a typical lactic fermentation once a uniform NaCl concentration of 6–8% is reached. As a consequence of the alkaline treatment, brine pH values during the first days are well above neutrality; i.e., more than 10 in some cases. This has led to the recommendation of decreasing the pH artificially by bubbling CO_2, adding other acids, or waiting until the pH is around 6–7 as a result of natural microbial growth, with the risk of spoilage (41,42). Salt is added during the fermentation to maintain the NaCl concentration at 5–6%. The release into the brine of olive cell constituents (sugars, B-complex vitamins, and amino acids) controls the fermentation rate, as their diffusion from olive flesh into the brine depends on the permeability of the skin, temperature, salt concentration, and ratio of olive to brine (31). Later on in the fermentation vitamins may be excreted by autolysis of Gram-negative bacteria or yeasts occurring in the microflora. At the end of the fermentation, in order to avoid the growth of spoilage microorganisms, the level of NaCl is increased to 7%. When the fermentation is completed a pH below 3.5 coupled with a NaCl concentration greater than 5% is sufficient to preserve the fruit anaerobically (5).

The olive fermentation is described as occurring in three stages, though a fourth spoilage stage is described (5,43): During the initial stage of the fermentation Gram negative aerobic bacteria of the genera *Aeromonas, Flavobacterium* and *Pseudomonas,* as well as some molds, are suppressed under the normally anaerobic fermentation conditions. The facultative organisms of the coliform aerogenes group (*Enterobacter cloacae, Citrobacter* species, *Klebsiella aerogenes,* and *Escherichia coli*) predominate during the first 2 days of the fermentation, and gradually disappear during the second stage of the fermentation, as the pH decreases due to the increasing population of lactic acid bacteria of the genera *Pediococcus, Leuconostoc,* and *Lactococcus.* Lactic acid producing cocci occurring during the first and second stage of fermentation provide a good degree of acidity. However, some of these are heterofermentative producing low amounts of lactic acid. Gram-positive sporulating bacteria and yeasts occuring as contaminants may develop. Whereas sporulating Gram-positive bacteria disappear from the fourth to the fifth day, the yeasts develop quickly as the pH decreases. When the pH reaches 6.0 the second stage of the fermentation begins, which lasts until the pH reaches 4.5 and takes 10–15 days. During this stage the Gram-negative bacteria disappear, while lactobacilli reach the maximum population. The third stage of the fermentation lasts until the fermentable carbohydrates are depleted. *Lactobacillus plantarum* is the dominant microorganism occurring with fermentative and oxidative yeasts in reasonably high levels during this stage of fermentation. The yeasts produce a number of vitamins that may encourage the growth of *L. plantarum* (44). Although the contribution of individual yeasts to the fermentation is unclear, it is believed that they play a role in the organoleptic characteristics of fermented olives. A yeast population consisted of *Hansenula anomala, Candida krusei, Saccharomyces chevalieri, Candida parapsilosis* and *Hansenula subpelliculosa* dominate in some fermentations (43). From spontaneous fermentations of olive brines at different sites in Portugal ascomycetous species dominate the yeast flora, and among them the asporogenous are less numerous than the ascosporogenous yeasts. In the layter group *Pichia anomala* and *P. membranaefaciens* as well as its anamorphic state *Candida valida* and *P. pseudocactophila* are the predominant species. The species *Saccharomyces cerevisiae, Debaryomyces hansenii, Kluyveromyces lactis,* and *Rhodotorula mucilaginosa* are also present (45). The oxidative yeasts can be harmful because they metabolize lactic acid and raise the pH, which can lead to spoilage (5). During the period of storage the salt concentration is increased to avoid the growth of *Propionibacterium* during the fourth stage and the corresponding depletion of lactic acid. When the concentration of NaCl is ≤5% *Propionibacterium* may metabolize lactic acid with a consequent rise in pH (to 5) and the possibility of the development of species of *Clostridium* which lead to an unpleasant odor called "zapatera" (43). A yeast population composed of strains of *Saccharomyces cerevisiae, Pichia anomala, Candida etchellsii, Candida versatilis,* and *Rhodotorula glutinis* is isolated from boaters of fermented green table olives during storage (46). *In vitro* inhibition of these yeast species is achieved using laboratory media containing essential oil of garlic (47).

19.3.2.2 Naturally Black Olives in Brine

The process is practiced traditionally in Greece. The olives are harvested when they are completely ripe; i.e., when the skin has a color between violet and black, but before they are overripe. They are put into tanks and covered with 6–10% brine. The maintenance of strict anaerobiosis by the addition of olive oil is important to avoid the growth of film forming yeasts and surface growth of molds causing removal of sugars and acids and increasing the brine pH. As the olives are not treated by alkali, the diffusion of the soluble components through the skin is slow. A mixed population of Gram-negative bacteria of the coliform aerogenes group, lactic acid bacteria (mainly *Lactobacillus* sp.) and yeasts are present if the NaCl concentration is $\leq 7\%$. The growth of lactic acid bacteria throughout fermentation depends on various factors such as olive cultivars, NaCl concentration, and initial pH correction of brines (48). If the concentration of salt in the brine stays at 10%, the presence of yeasts is dominant. In this case an acetic fermentation caused by acid forming, salt tolerant yeasts occurs. When fermentative yeasts predominate, the final product has pH values between 4.5 and 4.8. Although the contribution of individual yeasts to the fermentation is unclear, it is believed that they contribute to the organoleptic characteristics of fermented olives. Yeast strains isolated from spontaneous naturally black olive fermentations are mainly *Saccharomyces oleaginosus, Torulaspora delbruekii,* and *Debaryomyces hansenii* (31,49,50). The main species of yeasts involved in fermentation of naturally black olives in brine in different Mediterranean countries are presented in Table 19.2. Scanning electron microscopy shows a yeast rich biofilm development on the epicuticular wax of the olive skin during fermentation. Especially, yeasts predominate in the stomatal openings, whereas bacteria are more numerous in intracellular spaces in the substomatal flesh (51).

Spoilage results from the growth of clostridia, *Propionibacterium,* and sulphate reducing bacteria. Because of the great maturity of the raw material different genera of microorganisms and enzymes present in the fruit are implicated in spoilage affecting the texture and the flavor of the product (5,43). During the whole fermentation process it is important to achieve anaerobiosis in order to avoid mold growth, as mycotoxin production is reported

Table 19.2
Main species of yeasts involved in fermentation of naturally black olives in brine in Mediterranean region.

Country of Origin	Species	Ref.
Greece	*Torulaspora delbrueckii*	(49)
	Debaryomyces hansenii	
	Cryptococcus laurentii	
Spain	*Saccharomyces oleaginosus*	(31)
	Torulopsis candida	
	Debaryomyces hansenii	
	Hansenula anomala	
	Candida diddensii	
	Pichia membranaefaciens	
Turkey	*Debaryomyces hansenii*	(50)
	Rhodotorula mucilaginosa	
	Rhodotorula glutinis	
	Pichia membranaefaciens	
	Saccharomyces cerevisiae	

from molds isolated from black olives in dry salt (52,53). Yeasts are involved in soft stem end spoilage, and include oxidative pink yeasts of the genus *Rhodotorula* and fermenting yeasts belonging to the genus *Saccharomyces* (5). Yeasts identified as *S. oleaginosus, S. kluyveri,* and *Hansenula anomala* var. *anomala* spoil the olives by a combination of gas pocket formation and softening (54). Strains of the "pink yeasts" *R. glutinis* var. *glutinis, R. minuta* var. *minuta,* and *R. rubra* grow and form pellicles in olive brines and produce polygalacturonases which cause a softening of olives held in storage (55).

In order to improve the quality of the table olives, that is, reducing shriveling and the salty taste of processed fruits, fermentation of naturally black olives under reduced salt concentrations is achieved. In spontaneous fermentations, the predominant microorganisms are yeasts, whereas in reduced salt concentrations lactic acid fermentation occurs in addition. A more active role of lactic acid bacteria is desirable to limit the negative effects of yeasts on the product quality, as some yeasts are responsible for deterioration; i.e., gas pocket formation and softening of the fruit (56).

19.3.3 Improved Fermentation

19.3.3.1 Controlled Fermentation

In traditional producing areas, physicochemical characteristics of the brines concerning initial NaCl content, pH, and the type of added acid are used to control the fermentations. In this case the spontaneous microflora are associated with *L. plantarum* and *L. pentosus* (57,58). The development of these species is essential to provide the large amount of lactic acid needed for olive preservation. Early in the fermentation *L. plantarum* and *L. pentosus* constitute a small proportion of the bacterial population, but within one or two weeks of the olives being placed in brine, *L. plantarum* and *L. pentosus* become dominant over Gram-negative bacteria and other lactic acid bacteria, and they coexist with a yeast population up to 3 months, until the end of the fermentation process (57–59). *Lactobacillus plantarum* plays an essential role during the fermentation process of Spanish style green olives, providing the large amount of lactic acid needed for olive preservation (43). Therefore, during the fermentation, the olives are handled in order to favor the growth of the small population of *L. plantarum* present on the fruit. However, in some cases, *L. plantarum* fails to grow and the undesirable, competing natural microflora predominates leading to wide variations in the flavor and quality of fermented olives and often to spoilage by a variety of microorganisms. Therefore, control procedures in combination with the use of suitable *L. plantarum* starter cultures are necessary to prevent spoilage and to provide a product of consistently high quality (43). A complete lactic acid fermentation can be achieved by the addition of glucose to the brine together with a *L. plantarum* starter culture. In addition, the complete control of fermentation results in a profound improvement of texture, color and other sensory qualities of the finished product (60) eliminating the initial microflora consisting of other spontaneously occurring lactic acid bacteria, Enterobacteriaceae, *Pseudomonas* species and micrococci (61).

In an effort to obtain green table olives with a lower sodium content the traditional method of using brine containing NaCl only, has been replaced in some instances with brines containing $CaCl_2$ and KCl. In the alternative brines, normal lactic fermentation is carried out allowing the olives to lose their bitter taste. The product obtained has low sodium content and good organoleptic characteristics (62).

19.3.3.2 Starter Cultures

Olive processing remains an empirical process and is far from being controlled. However, interest in developing starter cultures to be used in table olive fermentation is increasing.

In order to achieve an improved and more predictable fermentation process, an appropriate starter culture has to fulfill the following requirements:

1. Ability to dominate the indigenous microflora
2. Accelerated growth under fermentation conditions; i.e., growth in table olive brines, low temperature
3. Homofermentative metabolism
4. Tolerance to salt, acid, and polyphenols
5. Composed of strains specifically adapted to possible water soluble B complex vitamins deficiency
6. Oleuropein splitting capability
7. Ability to produce bacteriocins, providing the producer strain with a selective advantage over the competing and spoilage microflora which develops during fermentation

The successful use of *L. plantarum* as a starter culture in olive fermentations, in which this species dominates the epiphytic microflora, is due to the ability of some strains to produce bacteriocins which are active against the natural competitors of *L. plantarum* as well as against bacteria that can cause spoilage (63,64). *Lactobacillus plantarum* LPCO10, a strain originally isolated from a Spanish style green olive fermentation, produces two bacteriocins: plantaricin S (pls), a two peptide bacteriocin, and plantaricin T (plt) (63). The genes plsA and plsB encoding for production of plantaricin S, are commonly distributed among wild-type *L. plantarum* strains isolated from olive fermentations from different olive processing plants in South Spain (65). When *L. plantarum* LPCO10 is used as a starter culture, it does not completely eliminate the development of the indigenous microflora but has a substantial effect without reducing the quality of the product. The presence of bacteriocin in the brines at the end of the fermentation is useful in preventing the growth of spoilage microorganisms such as clostridia or propionibacteria, as both are sensitive to the antimicrobial compounds produced by *L. plantarum* LPCO10, and both occur at the late stage of fermentation (59). Production of plantaricin S by wild-type *L. plantarum* strains is a constitutive trait, for once the appropriate genes are present, bacteriocin production takes place. As plantaricin S is active against many other lactic acid bacteria present in the olive fermentation, its constitutive production provides the producer strain with an ecological advantage in order to dominate the fermentation process (65). Thermostable antibacterial proteins isolated from *L. plantarum* LB17.2b are active against other *L. plantarum* strains occurred naturally in olive fermentation brines as well as against *Weissella* strains and *E. faecalis* (64). Enterocin 900 produced by *E. faecium* BFE 900 isolated from black olives is active at pH values ranging from 2 to 10 and has antagonistic action toward *L. sakei, C. butyricum,* enterococci, and *Listeria* species, including *L. monocytogenes* (14). In Table 19.3 are presented the main bacteriocins produced by strains isolated from fermented olives.

The amount of inoculum is of importance when starter cultures are applied in food fermentations. In the case of olives the minimum inoculum size for significant improvement of the fermentation process ranges from 10^5 cfu/mL to 10^7 cfu/mL (59,66). In some cases, i.e., fermentation at low temperature (12°C), a high inoculum (4×10^7 cfu/mL) is necessary to carry out a practically normal Spanish style green olive fermentation using 3% NaCl and pH correction with HCl to 5.0 by the addition of three strains of *Lactobacillus plantarum* (LPC1, LPC2, and LP33) (67). An inoculum size necessary to achieve an initial population of *L. plantarum* or *L. pentosus* close to 10^6 cfu/mL, is considered the minimum for significant results in the case of inoculation at alkaline pH (above 9) of lye treated

Table 19.3
Bacteriocins produced by strains isolated from fermented olives.

Bacteriocin	Strain/ Product Isolated	Inhibitory Spectrum	Ref.
Plantaricin S & Plantaricin T	*L. plantarum* LPCO10 Spanish-style green olives	strains of lactobacilli, leuconostocs, pediococci, streptococci, propionibacteria, clostridia	(63)
Enterocin 900	*E. faecium* BFE 900 black olives	*Lactobacillus sakei* *Clostridium butyricum* Enterococci *Listeria* spp. *L. monocytogenes*	(14)
Thermostable antibacterial proteins	*L. plantarum* LB17.2b olive brine	*Lactobacillus* strains *Weissella* strains E. faecalis	(64)

green olives. A marked decrease of viability is noticed as a result of the unfavorable conditions, the survivors can withstand this hostile environment, and eventually grow and initiate lactic acid fermentation. The inoculum lethality is, on average, two log cycles when applied at day 0 and 1 log cycle at day 2. A previous pH correction is therefore not required, nor is waiting for pH reduction by other microorganisms necessary (42).

Starter cultures added to Spanish style green table olives (conservolea variety) declined the population of *Escherichia coli* O157:H7 inoculated in the fermentation brine, but the pathogen did not die out during fermentation. The main factor that governs the behavior of this pathogen during fermentation is the amount of acids produced (61).

19.3.3.3 Microbial Degradation of Oleuropein

Oleuropein is labile in alkaline conditions, yielding hydroxytyrosol and elenolic acid. So, during Spanish style green olive processing sodium hydroxide treatment is used to remove most of the bitterness of the fruits. However, the lye treatment causes many chemical and physical changes in the fruit, leading to the loss of soluble constituents and nutrients of the fruit. In order to overcome the problems associated with the lye treatment, alternative microbiological procedures to debitter table olives have been untertaken.

Oleuropein can also be hydrolyzed enzymatically by bacterial β-glucosidase and esterase in a two step process. In the first step the glycosidic linkage of the oleuropein is hydrolysed by β-glucosidase to glucose plus oleuropein aglycone, and in the second step the amount of aglycone formed is converted to hydroxytyrosol and elenolic acid by esterase activity (68). Lactobacilli have an important role in the biological debittering process of naturally ripened olives. Oleuropeinolytic *L. plantarum* strains (*L. plantarum* B17, B20, B21) isolated from the brine of naturally ripe olives not treated by alkali, grow at a pH 3.5, and tolerate 1% oleuropein and 8% NaCl in the growth medium. They hydrolyze oleuropein via β-glucosidase production to its aglycone; after the β-glycosidic linkage is completely hydrolyzed, as deduced from the disappearance of oleuropein, other compounds originate due to esterase activity that hydrolyzes the ester to β-3,4-dihydroxyphenylethanol and elenolic acid. Therefore, the hydrolysis of oleuropein during the fermentation process of olives not treated with alkali is attributed to the acid pH and to β-glucosidase produced by oleuropeinolytic microorganisms (33,34). The determination of the conditions for bacterial growth in olive brines is of great importance for the development of a new biotechnology for removing the bitter secoiridoid glucoside, oleuropein, from the olives replacing the traditional alkali treatment used for debittering of olives (68).

Oleuropein is also hydrolysed by β-glucosidase produced by *L. mesenteroides* DIP20 isolated from the brine of naturally ripened olives. The maximum amount of β-glucosidase is produced during the first 24 h of the exponential growth phase. The enzyme shows maximum catalytic activity at pH 8 and 55°C. In the presence of 20 U/mL of β-glucosidase extracted from the bacterium, fifty percent of the initial concentration of oleuropein (2.25 mM) is hydrolyzed in 6 h of incubation at pH 7 and 55°C (69).

19.4 BREAD MAKING

19.4.1 Introduction

Cereal based foods are a major source of inexpensive dietary energy and nutrients worldwide. Many cereal foods are made without any fermentation process, such as boiled or steamed rice, porridge, pasta, and cookies. When fermentation is applied it is often alcoholic fermentation by yeast that prevails, as in bread making and brewing. However, there are also a number of cereal based foods that are fermented by lactic acid bacteria, such as European sour rye bread, various Asian flat breads, and the numerous types of fermented sour porridges, dumplings, and nonalcoholic beers common in Africa and elsewhere. In these applications the lactic acid bacteria contribute technological and nutritional benefits, as well as affecting flavor and keeping properties of the products (70).

Addition of water to flour inevitably leads to acid and gas production in the resulting dough. This observation was made early in history and was exploited to produce leavened bread as a staple of the diet. Credit for the first technical innovation, the introduction of leavening, seems to belong to the Sumerians in Southern Mesopotamia and may date from approximately 6000 BC. At the very beginning, heated convex stones prepared the bread from flour or parched grain, with hot ashes sometimes thrown over the cake. In 3000–2000 BC the ancient Egyptians took over and carried this elementary art to high perfection. In the ancient world they were known as the "bread eaters". Bread was more than just food, it was a standard of social status; even salaries were paid in the form of bread. Herodotus says of them "they knead their dough with their feet, but their clay with their hands". A common shape for Egyptian bread revealed by ancient monuments or in excavated tombs in a small round loaf like a muffin or elongated roll, sprinkled on top with seeds. Ancient Greeks improved the taste of bread by adding other ingredients such as sesame, pieces of fruits and honey, and offering bread with a meal was an indication of hospitality. Bread making was initially a housewife's art, but in the fifth century BC the first public bake houses ("artopoieion") were established. Even then, in poor families the housewife continued to prepare bread. Bread was named according to the way of baking ("apopyrias" on charcoal, "esharitis" on grate), or the way of preparation ("zymitis" with yeast, "azymos" without yeast), or its shape (rectangular "vlomiaios", concave "mystili", flat "plakitis"). Rome continued the development of bread making. After the war with Perseus (171-168 BC) public bake houses came into use, although the Romans continued to regard bread making as a housewife's art. The ruins of Pompei reveal several home mills and ovens. About 100 AD, the Emperor Trajan founded the College of Pistores (bakers), leading Juvenal in his famous satire to make the remark that Romans only needed two things, "panem et circenses" (bread and circuses). The Romans built the first watermills, while windmills appeared in the middle ages. Even Leonardo Da Vinci worked on the improvement of mills, but the first iron made cylinder mills were constructed in the nineteenth century (71,72).

The superior palatability of yeast raised or leavened bread led to the almost universal adoption of the necessary bread making procedures in those countries where bread became the staple cereal based food. However, there were no further notable technological innovations for nearly 6000 years (73). Mechanization, which gradually took over toward the end of the nineteenth century, better wheat varieties, flour improvers taking over from atmospheric oxidation, and diastatic correctives ensuring satisfactory gas production during fermentation, all evolved the technology, but did not revolutionize it. On the other hand, it was not until the early twentieth century that the key role of yeasts and lactic acid bacteria in (sour)dough making was recognized (74,75). Wolffin in 1894 was the first to demonstrate the occurrence of lactic acid bacteria and yeasts in sourdough and to associate acidification with bacterial metabolism (76). However, the first defined starter culture for sourdough preparation, namely the San Francisco starter, was patented in the early 1970s (77). Recent research on sourdoughs has focused on the identification of the microflora, investigations of the carbohydrate and nitrogen metabolism of the yeasts and LAB involved, and of flavor generation during fermentation (78).

19.4.2 Microbiological and Technological Aspects

Conventionally, bread making starts with the preparation of dough, which is formed in the course of rapid kneading of flour, water, salt, and yeast. *Saccharomyces cerevisiae* is specialized for bread making, exhibiting maximal fermenting activity at about 30°C. The taste, aroma, and color of bread are the combined effect of substances that already exist in flour and those produced during fermentation and baking. During dough fermentation, yeasts ferment the free sugars of the flour (about 1%) to ethanol, carbon dioxide, and organic acids. During baking, the Maillard reaction and caramelization reaction that take place are responsible for the formation of aroma compounds such as aldehydes and (di)ketones. However, artisan bread production often employs sourdough processes or the use of preferments. This provides a wide, regional variety of breads and specialty products.

Sourdough is a mixture of wheat or rye flour and water that is fermented with lactic acid bacteria (LAB) and yeasts (79). Aeration of dough and bread probably was the principal reason for using sourdoughs from an historical perspective. Primarily fermenting yeasts causes the aeration occurring in a sourdough. Heterofermentative lactic acid bacteria may also have a significant leavening effect in sour rye dough. Sourdough fermentations improve dough properties, bread texture and flavor, retard the staling process, and prevent bread from mold and bacterial spoilage (80–82). Sourdough can be prepared in bakeries or obtained from commercial suppliers (83,84).

The benefits of lactic acid fermentation in bread making are more apparent in making rye bread. Sourness in white wheat bread induces a characteristic flavor not always acceptable. On the other hand, sour rye bread is favored over nonacidified rye bread. Acidification by the lactic fermentation also improves bread making properties of rye by suppressing the high endogenous activity of α-amylase and other enzymes (85,86). The water soluble rye flour proteins cannot form a proper network of hydrated gluten, which is crucial for retaining water as well as carbon dioxide during dough preparation and fermentation. In rye flour pentosans play the role of gluten and their action is enhanced when sourdough is used (87,88).

Due to their artisan and region dependent handling, sourdoughs are an immense source of diverse LAB and yeast species and strains. In sourdoughs the lactobacilli predominate although leuconostocs and pediococci have also occasionally been reported to be present. Yeasts are also present in sourdough unless the fermentation is carried out at an elevated temperature inhibiting yeast growth. In general, the spectrum of LAB involved in sourdough fermentations includes many different species of lactobacilli, in particular

L. sanfranciscensis and *L. brevis*. It seems, however, that the dominant microflora is mainly selected by the environmental conditions induced by the sourdough fermentation technology. Many researchers still report on the existence of nonidentifiable and perhaps new sourdough LAB species and strains (89). A precise identification of strains isolated from sourdoughs is often problematic because the properties of the isolated strains do not always fully comply with those of the model strains originating from other ecological environments. The use of modern identification methods shows earlier results in a new light (79,90,91). Molecular techniques including 16S rDNA sequencing, Randomly Amplified Polymorphic DNA (RAPD), and other polymerase chain reaction (PCR) based techniques are necessary for the taxonomic investigation of complex ecosystems (92). These polyphasic approaches have allowed the description of new species including *Lactobacillus pontis* (93), *Lactobacillus panis* (90), and *Lactobacillus frumenti* (94).

The yeasts present in sourdough are acid tolerant and form a stable combination of mutual benefit with the lactobacilli. The yeasts produce amino acids, vitamins, and other growth factors required by the lactic acid bacteria (95–97), whereas the acids produced by the lactic acid bacteria suppress the growth of other microbes. *Candida milleri* appears to be the most typical yeast species in sourdoughs, although baker's yeast *S. cerevisiae* and strains closely related to it are also commonly detected in sourdoughs. Many of the strains currently designated as *C. milleri* were earlier named as *Torulopsis holmii*, the asporogenous form of *Saccharomyces exiguous*. There is a peculiar symbiosis between *C. milleri* and *L. sanfrancisco* (identical to *L. brevis* var. *lindneri*), as was first described for the San Francisco sourdough French bread process by Kline and Sugihara (83). Later a corresponding relationship was also detected in sourdoughs from elsewhere, as in a German sourdough starter, a Sudanese Kirsa sourdough, and an Italian Panettone sourdough (98–100). An obvious key factor for this symbiosis and highly resistive relationship appears to be the strong preference of the lactobacilli for maltose, as clarified in details only recently (99–104). In these sourdoughs the *L. brevis* var. *lindneri* utilizes only maltose while releasing glucose that is used by the *C. milleri*, which is incapable of assimilating maltose.

According to the technology applied for their production, sourdoughs have been grouped into three types. Type I sourdoughs are traditional doughs sustained by continuous propagation at an ambient temperature of 20–30°C with a final pH of 3.9 (105). Mostly, traditional three stage fermentation processes are used. "Anstellgut" in Germany, "mother sponge" in San Francisco, "le chef" in France, "masa madre" in Spain, "madre" or "capolievito" in Italy and "prozimi" in Greece belong to this type (81,106). *Lactobacillus sanfranciscensis* and *L. pontis* are the predominant LAB in these doughs (107). Also, *Lactobacillus fructivorans*, *Lactobacillus fermentum*, *Lactobacillus brevis*, and *Lactobacillus farciminis* were found in some of these doughs (70,79). In contrast, leavening of type II sourdoughs is achieved by addition of bakers' yeast to the dough. This is essential, because type II doughs employ a less time consuming, one stage fermentation process at temperatures exceeding 30°C. Type II doughs are mostly used in industrial processes to produce bakery pre products. In most of the cases the fermentation broth is applied as a dough souring supplement and aroma carrier (104). Dominant LAB strains in these doughs are *L. panis*, *L. pontis*, *L. reuteri*, *L. johnsonii*, *L. sanfranciscensis*, *L. fermentum*, *L. delbruekii*, *L. acidophilus*, *L. brevis*, *L. amylovorus*, and *L. frumenti* (81,94). Type III doughs are dried preparations of doughs (81). They are made by (traditional) sourdough fermentation with subsequent water evaporation (86).

Most scientific literature on lactic acid bacteria in cereal based foods derives from studies on sour rye bread, which is typical for Central, Northern, and Eastern Europe. Much of this work was reviewed by Spicher and Stephan (108) in their book on sourdough technology. Recently, Loenner and Ahrne (109) reviewed applications of lactobacilli in

Table 19.4
Main microflora of traditional sourdoughs in Mediterranean and Middle East countries.

Country	Product	Yeasts	Lab	Ref.
Portugal	Traditional corn or rye sourdough	*I. cidentalis*	*L. brevis*	(120,121)
		I. orientalis	*L. curvatus*	
		K. marxianus	*L. lactis* subsp. *lactis*	
		P. anomala	*L. lactis* subsp. *lactis*	
		P. membranaefaciens	*E. casseliflavus*	
		S. cerevisiae	*E. durans*	
		S. kluyveri	*E. faecium*	
		T. delbreuckii	*S. constellantus*	
		C. pelliculosa	*S. equinus*	
Spain	Traditional wheat sourdough	*C. boidinii*	*E. faecium*	(109,110)
		C. guilliermondii	*L. brevis*	
		D. polymorphus	*L. cellobiosus*	
		H. anomala	*L. plantarum*	
		H. subpelliculosa	*L. mesenteroides*	
		O. margaritiferum		
		P. polymorpha		
		R. glutinis		
		S. cerevisiae		
		S. fructuum		
France	Traditional wheat sourdough	*C. holmii*	*L. acidophilus*	(110)
		C. tropicalis	*L. brevis*	
		S. cerevisiae	*L. buchneri*	
		T. delbreuckii	*L. casei*	
		L. delbreuckii		
		L. fermentum		
		L. plantarum		
		L. mesenteroides		
		P. cerevisiae		
		P. pentosaceus		
Italy	Traditional wheat sourdough	*C. holmii*	*L. acidophilus*	(109,110, 112,113)
		C. krusei	*L. alimentarius*	
		C. milleri	*L. brevis*	
		C. stellata	*L. brevis* var. *linderi*	
		H. anomala	*L. casei* subsp. *casei*	
		S. cerevisiae	*L. delbrueckii* subsp. *bulgaricus*	
		S. ellipsoideus	*L. farciminis*	
		S. exiguuus	*L. fermentum*	
		Torulopsis sp.	*L. fructivorans*	
			L. plantarum	
			L. sanfranciscensis	
			L. citreus	
			L. mesenteroides subsp. *mesenteroides* and *dextranicus*	

(*Continued*)

Table 19.4 (*Continued*)

Country	Product	Yeasts	Lab	Ref.
			L. lactis subsp. *lactis* *P. parvulus* *P. pentosaceus* *W. confusa*	
	Venetsiana, Colomba, Brioches, Cornetto, Panattone		*L. brevis* *L. delbrueckii* subsp. *bulgaricus* *L. plantarum* *L. sanfranciscensis*	(110,232)
Greece	Traditional wheat sourdough	*P. membranaefaciens* *S. cerevisiae* *Y. lopolytica* *D. hansenii*	*L. sanfranciscensis* *L. brevis* *L. paralimentarius* *P. pentosaceus* *E. faecium* *W. cibaria*	(72, 106, 116)
Egypt	Traditional wheat sourdough		*L. brevis* *L. fermentum*	(109)
Morocco	Traditional wheat sourdough	*C. milleri* S. cerevisiae	*L. plantarum* L. brevis	(127)

baking. Table 19.4 summarizes the existing literature data on the main microbial species found in traditional sourdoughs in Mediterranean and Middle East countries.

In Italy sourdough is used in more than 30% of bakery products (110), which include more than 200 types of sourdough breads (111). Most of these breads originate from very old traditions and differ in type of flour, other ingredients, type of sourdough, technology, and shelf life. These differences greatly influence the microbial composition of sourdoughs and, consequently, the characteristics of the baked end products. Ottogalli et al. (110) suggested two different microbial situations in Italian sourdoughs: the strict and the weak association. The strict association is characterized by the presence of microbial forms, one lactic rod, *L. sanfranciscensis* or *L. brevis* var. *linderi*, and one yeast, usually *S. exiguus* or *Candida holmii*. On the other hand, the weak association is characterized by a more heterogeneous microflora of yeasts, like *Saccharomyces, Candida, Hansenula, Pichia* and *Rhodotorula*, and lactic acid bacteria. The latter are rods belonging to obligately homofermentative, facultative heterofermentative, and obligately heterofermentative groups, and cocci belonging to the *Pediococcus*, *Leuconostoc,* and *Enterococcus* genera. For instance, Pugliese sourdoughs are composed of *L. plantarum*, *L. brevis*, *L. fermentum*, and *L. fructivorans*, and Foggia sourdoughs are composed of *L. brevis, L. sanfranciscensis, Leuconostoc citreum,* and *W. confusa* (110,112,113). *Lactobacillus sanfranciscensis* is the predominant bacterial species in Italian sourdoughs (110,112), and it is either associated with *L. plantarum* (112,114) or with *L. alimentarius* (113). The dominance of *L. paracasei* is rather uncommon, although this species has been found in pizza doughs (115).

Greek traditional wheat sourdoughs belong to type I and are prepared without the addition of baker's yeast. However, there is only very limited data on Greek sourdoughs. In a recent survey, it was found that the dominant yeast species were *Saccharomyces cerevisiae*, *Pichia membranaefaciens,* and *Yarrowia lipolytica*; it was actually the first time that the presence of *Y. lipolytica* in sourdough was reported (116). The isolated bacteria were classified into the species *L. sanfranciscensis*, *L. brevis*, *L. paralimentarius*, and *Weissella cibaria*, the last one never being previously isolated from sourdough. This consortium seems to be unique for Greek traditional wheat sourdoughs, while no *L. pontis* or *L. panis* strains were found (106). From an ecological point of view, the presence of *W. cibaria* is interesting and new for sourdough fermentations. The only *Weissella* species detected in previous studies is *W. confusa* (107,113). It may be assumed that isolates assigned to the latter species are misidentified and actually belong to the very recently described species *W. cibaria*, a species which is both genomically and phenotypically highly similar to *W. confusa* (117). A common characteristic for both species that allows distinguishing them from the other *Weissella* species is the ability to grow at 45°C, which may be in favor of type II sourdough fermentation processes. Both taxa have been isolated from fermented foods (117). A misidentification may also have occurred for sourdough isolates previously assigned to the species *L. alimentarius*. Recently, two phylogenetically highly related species, *L. paralimentarius* and *L. kimchii*, were described (118,119). In the description of the most recently described species *L. kimchii*, the taxon *L. alimentarius* is included for comparison, but not *L. paralimentarius*. However, according to De Vuyst et al. (106), it is shown that *L. kimchii* and *L. paralimentarius* are genomically related at a binding level close to the borderline of species delineation (68%). The observed instability of some phenotypic features of, for instance, *L. alimentarius*, included as a reference strain in both studies, may raise the question whether there was enough evidence to create two distinct species for these taxa.

In rural areas of Portugal a special type of corn and rye bread is prepared using as starter dough carried over from a previous making. According to Almeida and Pais (120), *S. cerevisiae* is the dominant yeast species in these doughs, followed by *Issatchenkia orientalis, Pichia membranaefaciens,* and *Torulaspora delbruekii*. In a recent and extended survey, more than 400 strains were isolated from sourdough as well as maize and rye flours from several geographic locations in Portugal and in two different periods within the agricultural year. The dominant groups were yeasts and lactic acid bacteria (LAB). The most frequently isolated yeasts were *Saccharomyces cerevisiae* and *Candida pelliculosa*. The most frequently isolated LAB were (heterofermentative) *Leuconostoc* species and (homofermentative) *Lactobacillus* species. *Lactobacillus brevis, L. curvatus*, and *L. lactis* ssp. *lactis* were the dominant species for the *Lactobacillus* genus; *Lactococcus lactis* ssp. *lactis* for lactococci; *Enterococcus casseliflavus, E. durans*, and *E. faecium* for enterococci; and *Streptococcus constellantus* and *S. equinus* for streptococci (121).

In Arab countries, North Africa, the Middle East, and elsewhere flat breads made from wheat are the staple food. Dough for the flat bread is often fermented in process closely similar to that used for sourdough rye bread in Europe. Differences exist between the protein requirements of pan bread and Arabic bread. In contrast to the linear relationship between protein content and loaf volume in pan bread, a parabolic relationship has been reported for the Arabic bread, with an optimum between 10–12% protein, within a wheat cultivar (122). These differences could be partially attributed to the differences in the dough characteristics and the time temperature combinations employed in the baking of these bread types. Pan bread doughs have relatively lower surface area:volume ratios especially as Arabic bread is baked from dough sheets, which are often 1–3 mm thick (123). Furthermore, pan bread requires longer baking time as compared to Arabic bread, which is fully baked after 15 or 90 s in an oven at 650 or 400°C, respectively (124). There are numerous local varieties of these breads, such as lavash, injera, idli and others described by Steinkraus (125), Chambell-Platt (126), and Quail (124). It has been shown that the yeasts *C. milleri* and *S. cerevisiae* and the lactics *L. plantarum* and *L. brevis* are the dominant microflora in traditional Moroccan sourdough (127). It has been suggested that the interactions between yeasts and lactic acid bacteria in spontaneously fermented sourdough constitute an important factor for the flavor and texture of Moroccan sourdough bread (128).

Many cereal based foods made by lactic acid fermentation are products other than sour bread. Such meals used to be staple foods in many countries, including many in the Mediterranean area, but nowadays they are common mainly in Africa, such as ogi and agidi in Nigeria, koko, akassa, and kenkey in Ghana, uji in East Africa, mawe in Benin, and mahewu in Southern Africa. These include beverages, gruels, and porridges, dumplings used in stews, and fried products. Maize, sorghum, millet, and other starchy materials are used for their preparation (126,129). Grains are first soaked in water for 1–3 days, usually at ambient temperature, softening for easy crushing or wet milling into slurry. Hulls, bran particles, and germ are removed by sieve procedures. During the soaking stage, mixed fermentations, including lactic acid fermentation, take place, and these go on in the slurry. This slurring stage has many similarities with the sourdough procedures used in traditional sourdough bread making. In the fully fermented slurry there may be up to 10^9 cfu/g of LAB, which equals cell densities found in a fermented sourdough. Mainly lactobacilli and leuconostocs are responsible for the souring process. When the fermentation is completed, the slurry is boiled with a necessary amount of water so that gelatinization of starch occurs and a product of desired consistency is obtained (130,131).

19.4.3 Biochemistry of the Fermentation Process

The main consideration on the effect of sourdough fermentation on the flavor and texture of wheat and rye bread is the carbohydrate metabolism by lactic acid bacteria (LAB) and yeasts. In the complex sourdough ecosystem, the metabolism of carbohydrates is affected either by the flour composition or by the LAB and yeast interactions. The concentration of soluble carbohydrates, namely maltose, sucrose, glucose, and fructose, of wheat flour varies with the type of flour from 1.55 to 1.85%, depending on the balance between starch hydrolysis by the flour and microbial enzymes, and microbial consumption (132). Maltose is the most abundant fermentable carbohydrate, and hence maltose catabolism is a key process during fermentation (133). The organic acids and ethanol produced contribute to the aroma of sourdoughs, whereas the production of acetic and lactic acid by sourdough LAB as well as specific inhibitors active against molds or rope forming bacilli, delay or prevent the growth of spoilage organisms during bread storage (80,134–136). Furthermore, they strongly affect the rheological properties of wheat doughs (87,137). Finally, biological acidification of the dough delays starch retrogradation effectively, and this effect is dependent on the selection of LAB used to initiate the fermentation (138).

In yeasts, carbohydrates enter the cell through facilitated diffusion by an either inducible or constitutive permease (139,140). Sucrose may be hydrolysed already outside the cell (141). Inside the cell, carbohydrates enter the glycolysis pathway to produce pyruvate. Under anaerobic conditions, pyruvate is decarboxylated to acetaldehyde, which is finally reduced to ethanol, serving in this way the necessary regeneration of NAD. Alternatively, NAD may be regenerated through the reduction of phosphodihydroxyacetone to 3-phospho-glycerol, which is finally dephosphorylated to glycerol (142,143). The formation of glycerol is essential for the recycling of the inorganic phosphate used in glycolysis (144,145), while accumulation of glycerol plays an important role in the swelling of doughs with high carbohydrate content (146,147). Among the by products of the alcoholic fermentation, trehalose, a storage disaccharide that is produced from 6-phospho-glucose, plays an important role in the protection of cell membranes (148,149). Its production is usually induced under stress conditions (150,151). Besides glycolysis, yeasts may also use the pentose phosphate pathway, which results in the production of

NADPH to be used in biosynthesis of fatty acids and amino acids. Depending on the stage of growth, *S. cerevisiae* may catabolize up to 20% of glucose via this pathway.

Regarding nitrogen source utilization, yeasts are able to use a wide range of inorganic and organic nitrogenous compounds, such as ammonium ions and amino acids, which are taken up by active transport systems (97,152,153). Most strains of *S. cerevisiae* use asparagine preferentially, then aspartate, glutamine, glutamate, and lysine, and to a much lower degree glycine, arginine, and proline (152,154,155).

Among lactic acid bacteria, those found in sourdough fermentations belong mainly to the heterofermentative species, which catabolize glucose via the pentose phosphate pathway. Lactic acid, ethanol, and carbon dioxide are the main end products of sourdough LAB cultivated under anaerobic conditions (85). Under microaerophilic conditions, both oxygen and fructose can be used as electron acceptors. This gives rise to the formation of additional metabolites such as acetate and mannitol (81). In sourdough, maltose is the most abundant fermentable carbohydrate, and hence maltose catabolism is a key process during fermentation. Microbial associations of maltose positive and maltose negative LAB strains are typical for sourdoughs dominated by *L. sanfranciscensis* (78). In *L. sanfranciscensis*, *L. reuteri* and *L. fermentum* a constitutive, intracellular maltose phosphorylase catalyses the phosphorolytic cleavage of maltose, yielding glucose-1-phosphate and glucose (93). Glucose-1-phosphate is then converted by phosphoglucomutase to glucose-6-phosphate, which is further metabolized via the pentose phosphate pathway (79,81). On the other hand, hexokinase activity, which catalyses the conversion of glucose to glucose-6-phosphate, is virtually absent in cells growing exponentially in maltose containing media, and thus the nonphosphorylated glucose is excreted in the medium in a molar ratio with maltose of about 1:1 (99,156). It has been shown, however, that no glucose accumulation occurred in the fermentation broth, and no maltose phosphorylase activity could be detected in cell extracts prepared from cells grown in the presence of both maltose and fructose, suggesting that in the presence of both maltose and fructose in the medium, induction of hexokinase activity does occur (106). Similarly, in experiments performed with growing cells of *L. sanfranciscensis*, no significant accumulation of glucose was observed in the medium as reported for resting cells of *L. sanfranciscensis*, *L. reuteri*, and *L. pontis* (101,103,157). It is also believed that hexokinase activity is induced in the presence of glucose or fructose in the medium (158).

Lactobacillus sanfranciscensis and *L. pontis* are able to use fructose as a carbon source; however, in the presence of maltose they use it mainly as an electron acceptor and fructose is reduced to mannitol (103,159,160). According to Röcken and Voysey, oxygen was proven to be the preferred hydrogen acceptor for the *L. sanfranciscensis* strains (86). When oxygen is depleted, fructose is used as an electron acceptor (102,103). Through the reduction of fructose to mannitol, extra ATP is produced via the acetate kinase reaction, and thus maltose fructose cometabolism yields a shorter lag phase and a higher growth rate and biomass production (161). It has been shown that at a molar ratio of 4:1 (fructose:maltose), acetic acid is the main product (102,103,162,163). *Lactobacillus sanfranciscensis* stoichiometrically converts fructose to mannitol, while *L. pontis* produces small amounts of lactic acid and ethanol (164).

It has been shown that several strains of *L. sanfranciscensis* are able to use citrate as an electron acceptor in the presence of maltose (103). On the other hand, cometabolism of maltose and citrate has not been observed for *L. pontis* (164). According to Gobbetti and Corsetti (165), during cometabolism of maltose and citrate, lactic acid and acetic acid are initially produced, but when citrate is exhausted lactic acid and ethanol are the main products. In all cases, maltose serves as a carbon source, while citrate as an electron acceptor. The production of small amounts of succinate from citrate has been also observed, indicating the presence of citrate lyase, malate dehydrogeanse, fumarase and succinate dehydrogenase.

During cometabolism of malate and maltose, higher amounts of lactic acid are produced by *L. sanfranciscensis*, and the lactic acid produced is stoichiometrically proportional to the malate consumed, indicating the presence of the malolactic enzyme (166). Fumarate is metabolized by *L. sanfranciscensis* to a lesser degree than malate. In *L. pontis*, it has been observed that malate and fumarate are converted to succinate via the reaction of succinate dehydrogenase and fumarase, while the presence of malate results in higher acetate production.

The acetate formed is of major importance for the development of flavor and the microbial stability of the bread. The molar ratio of lactate to acetate in bread is defined as the fermentation quotient (FQ), and it is considered optimum in the range 2.0 to 2.7 (81). The FQ of sourdough can be decreased by addition of fructose to the dough when using *L. brevis* as a starter culture, resulting in increased acetate titers (162,167).

Nitrogen is a limiting factor for lactic acid bacterial growth in various ecosystems. It has been shown that glutamate, isoleucine, and valine are essential amino acids for *L. brevis* subsp. *linderi* and *L. plantarum*. Arginine, methionine, and leucine enhance growth, while alanine, glycine, aspartate, lysine, histidine, cysteine, tyrosine, serine, threonine, and proline do not affect growth (97). In this context, the proteolytic system of lactic acid bacteria is very important. The proteolytic system of *Lactobacillus sanfranciscensis* has been studied (168,169), while the use of lactic acid bacteria able to assimilate arginine for the improvement of the organoleptic properties of bread is under examination (170–172).

The main issue during sourdough fermentation is the lack of balance between maltose consumption from the side of microorganisms and production of new maltose from the enzymatic hydrolysis of starch (173). This results in rapid exhaustion of soluble carbohydrates in wheat flour sourdough. This is not the case for rye flour sourdough due to the higher amylolytic activities (86). Generally, yeasts and lactic acid bacteria have a different carbohydrate uptake rate. Most yeasts take up hexoses and maltose via high affinity systems, while with lactic acid bacteria the uptake depends on the carbohydrate concentration and it is thus less efficient (78).

The competition for the main carbon sources is eventually the main prerequisite for the symbiosis of yeasts and lactic acid bacteria in sourdough. The lack of this competition between *L. sanfranciscensis* and *S. exiguous* for maltose is of essential importance for the traditional San Francisco sourdough (78). The glucose secreted by certain strains of *L. sanfranciscensis* during maltose fermentation can be used by yeasts such as *S. exiguous*, which cannot metabolize maltose, or it may restrain competitors such as *S. cerevisiae* via catabolite repression (101,156). When sucrose is used as carbon source, *L. plantarum* growth and lactic acid production is higher in the presence of *S. cerevisiae* and *S. exiguous* (156). Yeasts hydrolyze sucrose to fructose and glucose, which are used by *L. plantarum* more rapidly than sucrose itself. Sucrose hydrolysis by yeasts is by a factor of 200 more rapid than the consumption of the monosaccharides produced, leading this way to fast exhaustion of sucrose in the medium (132).

Concerning nitrogen source availability, coculture of *L. brevis* subsp. *linderi*, *L. plantarum, S. cerevisiae* and *S. exiguous* in a medium containing adequate amounts of carbon sources and vitamins promoted the growth of lactic acid bacteria due to the lack of competition over nitrogen sources (97). Yeasts prefer the uptake of ammonium salts compared to amino acids, and at the same time they release amino acids and small peptides during their growth or due to autolysis (97,174). In sourdough, *S. cerevisiae* releases mainly γ-aminobutyrate, proline, valine, sioleucine, glycine, and alanine, while lactic acid bacteria release glycine and alanine (175). The release of amino acids by the yeasts promotes the growth of *L. sanfranciscensis* even in media which did not initially contain the essential amino acids (valine and isoleucine) (78).

With respect to the production of aroma compounds, the heterofermentative lactic acid bacteria produce mainly ethylacetate and some alcohols and aldehydes, the homofermentative LAB produce mainly diacetyl and other carbonyl compounds, while yeasts produce various isoalcohols (176). The production of aroma compounds depends on strains and strain combinations. There are strain combinations which virtually did not affect the profile of aroma compounds of sourdough (177), while others yield a profile of high quality (78). Coculture of *L. sanfranciscensis* with other homofermentatative or heterofermentative lactic acid bacteria results in a balanced aromatic profile. Homofermentative lactic acid bacteria and *L. sanfranciscensis* promote yeasts metabolism, which are mainly responsible for the improvement of the organoleptic quality of the final product (178,179).

Special attention has been paid in recent years to the effect of the final metabolic products on the growth of sourdough microorganisms. Ethanol diffuses through the cytoplasmic membrane and accumulates intracellularly, independently the growth phase of yeast (180,181). The accumulation is species or even strain dependent (182). Medium ethanol concentrations (2–6% vol/vol) inhibit maltose phosphorylase inactivation due to catabolite repression in the presence of other fermentable carbohydrates (183). Higher ethanol concentrations limit growth and metabolic activity. Although the exact mechanism is not fully understood (184), experimental data indicate that this inhibition is mainly due to decreased water availability (185). Ethanol has also an impact on membrane structure and on enzymes, and it has been shown to inhibit glycolytic enzymes by *in vitro* experiments (186–188).

Lactic and acetic acid produced by lactic acid bacteria may inhibit yeast growth (189). At the pH of sourdough (3.6–3.9) both acids are in their undissociated form and thus can diffuse through the cytoplasmic membrane (190,191). Acetic acid is considered more important and its impact on yeast growth has been studied in more detail (192–195). Ethanol produced by yeasts and the heterofermentative lactic acid enhances the diffusion of acetic acid into the cytoplasm of yeasts (196,197). Inside the cell, acetic acid dissociates and inhibits several glycolytic enzymes (198,199).

19.4.4 Effect of Fermentation on the Nutritional Quality

Cereal grains are an important dietary source for the world's population, especially in the developing counties. They are estimated to directly provide 50% of human calories in the world and 70–80% of all food calories consumed in China and India. Worldwide, cereals are the direct or indirect source of about 75% of our dietary protein. At the same time, they are a significant source of some of the B group vitamins and minerals. Cereals contribute to 33.6% of thiamine, 22.7% of niacin, and 14.2% of riboflavin consumption in the U.S. diet. Wheat, which is mainly used for bread making, contains per 100 g fresh weight 454 mg potassium, 433 mg phosphorus, 183 mg magnesium, and 45 mg calcium. Furthermore, cereal based foods are a major source of dietary fiber in most diets (200).

Many cereal based foods, such as pasta and cookies, are made without any fermentation process. On the other hand, there are many others that are prepared through alcoholic and or lactic acid fermentation. These include bread, beer, and sour porridges. In these applications the microflora involved contribute to the improvement of the organoleptic properties, the shelf life of the final products and their nutritional profile. The reader may be reminded that phytic acid (myoinositol hexaphosphate) is considered an undesirable constituent of many cereal products as it may bind essential minerals (Zn, Fe, and Ca) and render them nutritionally unavailable. Furthermore, phytic acid seems to reduce protein nutritional quality in high protein wheat bran flour (201). It has been shown, however, that bread raised by yeast fermentation contains less phytic acid than "flat" bread. This destruction of phytic acid has been attributed to the indigenous flour phytase but also to the yeast phytase (202,203). However, some authors suggest that phytic acid may partially contribute to the beneficial effects attributed earlier exclusively to dietary fibers, such protective role in carcinogenesis and beneficial effects in heart disease (195,205).

The organic acids produced during sourdough fermentation enhance the perception of the other aroma compounds in bread, improve its physicochemical characteristics, and prolong its shelf life. Acetic acid at a concentration of 100–200 ppm increases the intensity of other aroma components in wheat bread, while at a fermentation quotient (FQ = the molar ratio of lactate to acetate) of 2.0–2.7 the taste and aroma of rye bread is more intense. It has been shown that acidification due to fermentation delays starch retrogradation more efficiently than chemical acidification (206,207). Lactic acid bacteria strains producing amylases and proteases but not pentosanases enhance that impact of biological acidification, although the exact mechanism is not understood (208–210).

Mold growth is the most frequent cause of bread spoilage. In unpreserved bread a shelf life of 3–4 days may be expected, especially if the hygiene during production is not sufficiently high. Besides the repelling sight of visible growth, molds are responsible for off flavor formation and production of mycotoxins and allergenic compounds, which may be formed even before growth is visible. In wheat bread *Penicillium commune*, *P. solium*, *P. corylophilum* and *Aspergillus flavis* dominates, whereas *P. roqueforti*, *P. corylophilum* and to some extend also *Euroticum* species are dominant in rye bread (211). The yeasts commonly known as chalk molds are also important spoilers of bread. The most important of these species is *Endomyces fibuliger* (212). The shelf life of bread inoculated with conidia of typical bread molds such as *Aspergillus niger*, *Cladosporium herbarum* and *Penicilium verrucosum* has been extended by increasing the amount of sourdough used. The fungistatic effect of sourdough has been attributed to lactic acid and, especially to acetic acid produced by heterofermentative lactic acid bacteria (213). However, other organic acids produced during sourdough fermentation such as caproic acid produced by *L. sanfranciscencis* (214), or phenyl lactate and p-hydroxyphenyl lactate produced by *L. plantarum* (136) seem also to play an important role in the prolongation of the shelf life of bread. Ropiness is the second most important spoilage of bread after molding and occurs particularly in summer when the climatic conditions favor growth of bacteria. Initially it is noticed as an unpleasant odor, followed by discolored, sticky and soft bread crumb, caused by the breakdown of starch and proteins and the production of extracellular slimy polysaccharides. It is mainly caused by *Bacillus subtilis*, but *B. licheniformis*, *B. megaterium*, and *B. cereus* have also been associated (215). The inhibition of *Bacillus* growth in sourdough bread has been attributed to the enhancement of spore thermal inactivation during baking by the presence of acetic acid (86), and the production of bacteriocins by the lactic acid bacteria involved in the fermentation (134,216–218).

It should be noticed, however, that most fermented cereal based foods are heat treated after fermentation, and thus the microflora are killed. However, in recent years there is an increased interest in cereal based functional foods that contain live probiotic bacteria. Fuller (219) and Havenaar and Huis in't Veld (220) provided earlier definitions for probiotic organisms, while Guarner and Shaafsman (221) refined the term to "living microorganisms, which upon ingestion in certain numbers, exert health benefits beyond inherent basic nutrition". In order for probiotic strains to exert their beneficial effect on the host, they have to be able to survive passage through the host's digestive tract, and colonize and metabolize in the lower intestine or colon. It was soon realized that in the presence of suitable dietary polysaccharides probiotic bacteria may be more effective. This has led to the prebiotic concept. Prebiotcs are nondigestible food ingredients that selectively increase the relative proportion of beneficial bacteria in the colon.

Carbohydrates of cereal origin and their hydrolysates are a potential source of prebiotics. Such polysaccharides must escape digestion in the upper gastrointestinal tract and they must be hydrolysable and fermentable by the gut flora. In cereal grains the most typical indigestible polysaccharides and dietary fiber components are cellulose, resistant starch, oligosaccharides, and polysaccharides, such as araboxylans or pentosans present predominantly in the cell wall of wheat and rye, and mixed linked β-glucans found in the cell walls of oats and barley (222–224). It has been shown that oligosaccharides isolated from wheat bran are selectively used by bifidobacteria (225), while cereal β-glucans were mostly converted to butyrate and probably metabolized by the colonic epithelial cells (226). The main end products of the intestinal fermentation of dietary fibers are short chain fatty acids (SCFA), such as acetate, propionate and butyrate, and gases such as H_2, CO_2, H_2S, and CH_4. The gases are either metabolized or excreted in breath and flatus (227). Fermentation intermediates like lactic acid are further converted by the colonic microflora to propionate. The SCFAs are readily metabolized and may contribute to many physiological beneficial effects. Butyrate is taken up and used as an energy source by colonic epithelial cells, while acetate and propionate enter the portal blood supply (227). There is a hypothesis that propionate and acetate present in the enterohepatic circulation may inhibit hepatic synthesis of cholesterol; however, this still needs to be confirmed (228). Although a lot of work has to be done in order to fully clarify the interactions between probiotic bacteria and prebiotic food components, some cereal based functional foods have recently entered the market (229–231).

REFERENCES

1. Steinkraus, K.H. Classification of fermented foods: worldwide review of household fermentation techniques. *Food Control* 8:311–317,1997.
2. Uccella, N. Olive biophenols: novel ethnic and technological approach. *Trends Food Sci. Technol.* 11:328–339, 2001.
3. Fleming, H.P. Fermented vegetables. In: *Economic Microbiology*: *Fermented Foods,* Vol. 7, Rose, A.H., ed., London: Academic Press, 1982, pp 227–258.
4. Jones, I.D. Some pigment changes in cucumbers during brining and brine storage. *Food Technol.* 3:324, 1975.
5. Harris, L.J. The microbiology of vegetable fermentations. In: *Microbiology of Fermented Foods.* 2nd ed., Vol. 1, Wood, B.J.B., ed., London: Blackie Academic & Professional, 1998, pp 45–72.
6. Mundt, J.O., J.L. Hammer. Lactobacilli on plants. *Appl. Microbiol.* 16:1326–1330, 1968.
7. Mundt, J.O. Lactic acid bacteria associated with raw plant food material. *J. Milk Food Technol.* 33:550–553, 1970.
8. Fleming, H.P., R.F. McFeeters, M.A. Daeschel. The lactobacilli, pediococci, and leuconostocs: vegetable products. In: *Bacterial Starter Cultures for Foods.* Gilliland, S.E., ed., Boca Raton, FL: CRC Press, 1985, pp 97–118.
9. Sanchez, I., L. Palop, C. Ballesteros. Biochemical characterization of lactic acid bacteria isolated from spontaneous fermentation of "Almagro" eggplants. *Int. J. Food Microbiol.* 59:9–17, 2000.
10. Daeschel, M.A., T.R. Klaenhammer. Association of a 13.6 megadalton plasmid in *Pediococcus pentosaceus* with bacteriocin activity. *Appl. Environ. Microbiol.* 45:1538–1541, 1985.
11. Anderson, R. Inhibition of *Staphylococcus aureus* and spheroplasts of Gram-negative bacteria by an antagonistic compound produced by a strain of *Lactobacillus plantarum. Int. J. Food Microbiol.* 3:149–160, 1986.
12. Daeschel, M.A., M.C. McKenney, L.C. McDonald. Bacteriocidal activity of *Lactobacillus plantarum* C-11. *Food Microbiol.* 7:91–98, 1990.
13. Uhlman, L., U. Schillinger, J.R. Rupnow, W.H. Holzapfel. Identification and characterization of two bacteriocin-producing strains of *Lactococcus lactis* isolated from vegetables. *Int. J. Food Microbiol.* 16:141–151, 1992.

14. Franz, C.M.A.P., U. Schillinger, W.H. Holzapfel. Production and characterization of enterocin 900, a bacteriocin produced by *Enterococcus faecium* BFE 900 from black olives. *Int. J. Food Microbiol.* 29:255–270, 1996.
15. Kotzekidou, P., T. Roukas. Characterization and distribution of lactobacilli during lactic fermentation of okra (*Hibiscus esculentus*). *J. Food Sci.* 51:623–625, 1986.
16. Fleming, H.P., R.F. McFeeters. Use of microbial cultures: vegetable products. *Food Technol.* 35(1):84–88, 1981.
17. Fleming, H.P., R.F. McFeeters, R.L. Thompson, D.C. Sanders. Storage stability of vegetables fermented with pH control. *J. Food Sci.* 48:975–981, 1983.
18. McFeeters, R.F., H.P. Fleming, R.L. Thompson. Malic and citric acids in pickling cucumbers. *J. Food Sci.* 47:1859–1861, 1865, 1982.
19. McFeeters, R.F., H.P. Fleming, R.L. Thompson. Malic acid as a source of carbon dioxide in cucumber juice fermentations. *J. Food Sci.* 47:1862–1865, 1982.
20. Chen, K.H., R.F. McFeeters, H.P. Fleming. Fermentation characteristics of heterolactic acid bacteria in green bean juice. *J. Food Sci.* 48:962–966, 1983.
21. Kotzekidou, P., T. Roukas. Fermentation characteristics of lactobacilli in okra *(Hibiscus esculentus)* juice. *J. Food Sci.* 52:487–488, 490, 1987.
22. Rodrigo, M., M.J. Lazaro, G. Garcia, F. Conesa, J. Safon. Pilot-study of cucumber fermentation: diffusing gases and bloater damage. *J. Food Sci.* 57:155–160, 1992.
23. McDonald, L.C., H.P. Fleming, M.A. Daeschel. Acidification effects on microbial-populations during initiation of cucumber fermentation. *J. Food Sci.* 56:1353–1356, 1359, 1991.
24. Gardner, N.J., T. Savard, P. Obermeier, G. Caldwell, C.P. Champagne. Selection and characterization of mixed starter cultures for lactic acid fermentation of carrot, cabbage, beet and onion vegetable mixtures. *Int. J. Food Microbiol.* 64:261–275, 2001.
25. de Castro, A., L. Rejano, A.H. Sanchez, A. Montano. Fermentation of lye-treated carrots by *Lactobacillus plantarum. J. Food Sci.* 60:316–319, 1995.
26. Ballesteros, C., L. Palop, I. Sanchez. Influence of sodium chloride concentration on the controlled lactic acid fermentation of "Almagro" eggplants. *Int. J. Food Microbiol.* 53:13–20, 1999.
27. Sesena, S., I. Sanchez-Hurtado, M.A.G. Vinas, L. Palop. Contribution of starter culture to the sensory characteristics of fermented Almagro eggplants. *Int. J. Food Microbiol.* 67:197–205, 2001.
28. Roukas, T., P. Kotzekidou. Lactic fermentation of okra (*Hibiscus esculentus*) with starter cultures for the improvement of quality of canned okra. In: *The Shelf Life of Foods and Beverages*. Charalambous, G., ed., Amsterdam: Elsevier Science, 1986, pp 629–640.
29. de Castro, A., A. Montano, A.H. Sanchez, L. Rejano. Lactic acid fermentation and storage of blanched garlic. *Int. J. Food Microbiol.* 39:205–211, 1998.
30. Balatsouras, G. *The Table Olive*. Athens (*Monograph in Greek*), 1995, pp 40–66.
31. Garrido-Fermandez, A., M.J. Fernandez-Diez, M.R. Adams. *Table Olives.* London: Chapman & Hall, 1997, pp 10–45.
32. Leone, F.G. The globalisation of the olive oil market and the competitive position of the sector in Italy: an international comparison. *Olivae* 83:10–14, 2000.
33. Ciafardini, G., V. Marsilio, B. Lanza, N. Pozzi. Hydrolysis of oleuropein by *Lactobacillus plantarum* strains associated with olive fermentation. *Appl. Environ. Microbiol.* 60:4142–4147, 1994.
34. Marsilio, V., B. Lanza, N. Pozzi. Progress in table olive debittering: degradation *in vitro* of oleuropein and its derivatives by *Lactobacillus plantarum. J. Am. Oil Chem. Soc.* 73:593–597, 1996.
35. International Olive Oil Council. *Unified Qualitative Standard Applying to Table Olives in International Trade*. Madrid, 1980.
36. Bianco, A., N. Uccella. Biophenolic components of olives. *Food Res. Int.* 33:475–485, 2000.
37. Bitonti, M.B., A. Chiappetta, A.M. Innocenti, I. Muzzalupo, N. Uccella. Biophenol functionality and distribution in *Olea Europea* L. drupes. *Olivo & Olio* 3:20–29, 2000.
38. Ryan, D., K. Robards. Phenolic compounds in olives. *Analyst* 123:31R–44R, 1998.

39. Ruiz-Barba, J.L., M. Brenes-Balbuena, R. Jimenez-Diaz, P. Garcia-Garcia, A. Garrido-Fernandez. Inhibition of *Lactobacillus plantarum* by polyphenols extracted from two different kinds of olive brine. *J. Appl. Bacteriol.* 74:15–19, 1993.
40. Gourama, H., B. Letutour, A. Tantaoui-Elaraki, M. Benbya, L.B. Bullerman. Effects of oleuropein, tyrosol, and caffeic acid on the growth of mold isolated from olives. *J. Food Protect.* 52:264–266, 1989.
41. Borbolla y Alcala, J.M.R., M.J. Fernandez-Diez, F. Gonzalez-Cancho. Influence of pasteurization and lye treatment on the fermentation of Spanish-style Manzanilla olives. *Appl. Microbiol.* 17:734–736, 1969.
42. Sanchez, A.-H., L. Rejano, A. Montano, A. de Castro. Utilization of high pH of starter cultures of lactobacilli for Spanish-style green olive fermentation. *Int. J. Food Microbiol.* 67:115–122, 2001.
43. Garrido-Fermandez, A., P. Garcia Garcia, M. Brenes Balbuena. Olive fermentations. In: *Biotechnology.* 2nd ed., Rehm, H.J., G. Reed, eds., New York: VCH, 1995, pp 593–627.
44. Ruiz-Barba, J.L., R. Jimenez-Diaz. Availability of essential B-group vitamins to *Lactobacillus plantarum* in green olive fermentation brines. *Appl. Environ. Microbiol.* 61:1294–1297, 1995.
45. Marquina, D., C. Peres, F.V. Caldas, J.F. Marques, J.M. Peinado, I. Spencer-Martins. Characterization of the yeast population in olive brines. *Lett. Appl. Microbiol.* 14:279–283, 1992.
46. Asehrau, A., C. Peres, D. Brito, M. Faid, M. Serhrouchni. Characterization of yeast strains isolated from bloaters of fermented green table olives during storage. *Grasas Aceites* 51:225–229, 2000.
47. Asehrau, A., S. Mohieddine, M. Faid, M. Serhrouchni. Use of antifungal principles from garlic for the inhibition of yeasts and molds in fermenting green olives. *Grasas Aceites* 48:68–73, 1997.
48. Quintana, M.C.D., C.R. Barranco, P.G. Garcia, M.B. Balbuena, A.G. Fernandez. Lactic acid bacteria in table olive fermentations. *Grasas Aceites* 48:297–311, 1997.
49. Kotzekidou, P. Identification of yeasts from black olives in rapid system microtitre plates. *Food Microbiol.* 14:609–616, 1997.
50. Borcakli, M., G. Ozay, I. Alperden, E. Ozsan, Y. Erdek. Changes in the chemical and microbiological composition of two varieties of olive during fermentation. *Grasas Aceites* 44:253–260, 1993.
51. Nychas, G.J.E., E.Z. Panagou, M.L. Parker, K.W. Waldron, C.C. Tassou. Microbial colonization of naturally black olives during fermentation and associated biochemical activities in the cover brine. *Lett. Appl. Microbiol.* 34:173–177, 2002.
52. Gourama, H., L.B. Bullerman. Mycotoxin production by molds isolated from "Greek-style" black olives. *Int. J. Food Microbiol.* 6:81–90, 1988.
53. Kivanc, M., A. Akgül. Mold growth on black table olives and prevention by sorbic acid, methyl-eugenol and spice essential oil. *Die Nahrung* 34:369–373, 1990.
54. Vaughn, R.H., K.E. Stevenson, B.A. Davé, H.C. Park. Fermenting yeasts associated with softening and gas-pocket formation in olives. *Appl. Microbiol.* 23:316–320, 1972.
55. Vaughn, R.H., T. Jakubczyk, J.D. MacMillan, T.E. Higgins, B.A. Davé, V.M. Crampton. Some pink yeasts associated with softening of olives. *Appl. Microbiol.* 18:771–775, 1969.
56. Özay, G., M. Borcakli. Effect of brine replacement and salt concentration on the fermentation of naturally black olives. *Food Res. Int.* 28:553–559, 1996.
57. Vaughn, R.H. The fermentation of olives. In: *Industrial Microbiology,* Reed, G., ed., Westport: AVI Publ, 1982, pp 207–236.
58. van den Berg, D.J.C., A. Smits, B. Pot, A.M. Ledeboer, K. Kerstens, J.M.A. Verbakel, C.T. Verrips. Isolation, screening and identification of lactic acid bacteria from traditional food fermentation processes and culture collections. *Food Biotechnol.* 7:189–205, 1993.
59. Ruiz-Barba, J.L., D.P. Cathcart, P.J. Warner, R. Jimenez-Diaz. Use of *Lactobacillus plantarum* LPCO10, a bacteriocin producer, as a starter culture in Spanish-style green olive fermentations. *Appl. Environ. Microbiol.* 60:2059–2064, 1994.

60. Balatsouras, G., A. Tsibri, T. Dalles, G. Doutsias. Effects of fermentation and its control on the sensory characteristics of Conservolea variety green olives. *Appl. Environ. Microbiol.* 46:68–74, 1983.
61. Spyropoulou, K.E., N.G. Chorianopoulos, P.N. Skandamis, G.J.E. Nychas. Survival of *Escherichia coli* O157:H7 during the fermentation of Spanish-style green table olives (conservolea variety) supplemented with different carbon sources. *Int. J. Food Microbiol.* 66:3–11, 2001.
62. Mule, R., A.S. Fodale, C.B. Bati, A. Tucci, A. Di Pisa. Preliminary results of a new processing in order to obtain green table olives with a low sodium content. *Ind. Aliment.* 39:844–847, 2000.
63. Jimenez-Diaz, R., R.M. Rios-Sanchez, M. Desmazeaud, J.L. Ruiz-Barba, J.-C. Piard. Plantaricins S and T, two new bacteriocins produced by *Lactobacillus plantarun* LPCO10 isolated from a green olive fermentation. *Appl. Environ. Microbiol.* 59:1416–1424, 1993.
64. Delgado, A., D. Brito, P. Fevereiro, C. Peres, J. Figueiredo Marques. Antimicrobial activity of *L. plantarum,* isolated from a traditional lactic acid fermentation of table olives. *Lait* 81:203–215, 2001.
65. Maldonado, A., J.L. Ruiz-barba, B. Floriano, R. Jimenez-Diaz. The *locus* responsible for production of plantaricin S, a class IIb bacteriocin produced by *Lactobacillus plantarum* LPCO10, is widely distributed among wild-type *Lact. plantarum* strains isolated from olive fermentations. *Int. J. Food Microbiol.* 77:117–124, 2002.
66. Etchells, J.L., A.F. Borg, I.D. Kittel, T.A. Bell, H.P. Fleming. Pure culture fermentation of green olives. *Appl. Microbiol.* 14:1027–1041, 1966.
67. Duran Quintana, M.C., P. Garcia Garcia, A. Garrido Fernandez. Establishment of conditions for green table olive fermentation at low temperature. *Int. J. Food Microbiol.* 51:133–143, 1999.
68. Marsilio, V., B. Lanza. Characterisation of an oleuropein degrading strain of *Lactobacillus plantarum.* Combined effects of compounds present in olive fermenting brines (phenols, glucose and NaCl) on bacterial activity. *J. Sci. Food Agric.* 76:520–524, 1998.
69. Ciafardini, G., B.A. Zullo. β-Glucosidase activity in *Leuconostoc mesenteroides* associated with fermentation of "Coratina" cultivar olives. *Ital. J. Food Sci.* 13:41–51, 2001.
70. Salovaara, H. Lactic acid bacteria in cereal-based products. In: *Lactic Acid Bacteria: Microbiology and Functional Aspects.* 2nd ed., Salminen, S., A. von Wright,eds., New York: Marcel Dekker, Inc, 1998, pp 115–137.
71. Spicer, A. The history of wheat and bread. In: *Bread: Social, Nutritional and Agricultural Aspects of Wheaten Bread.* Spicer, A., ed., London: Applied Science Publishers,1975, pp 1–4.
72. Paramythiotis, S. Symbiosis of lactic acid bacteria and yeasts in traditional sour dough. PhD Thesis, Agricultural University of Athens, Greece, 2002.
73. Chamberlain, N. Advances in bread making technology. In: *Bread: Social, Nutritional and Agricultural Aspects of Wheaten Bread*, Spicer, A., ed., Applied Science Publishers, London, 1975, pp 259–272.
74. Hollinger, W. Bakteriologische Untersuchungen ueber Mehlteiggaerung. *Zbl. F. Bakt. II* 9:305–312, 361–371, 473–483, 521–537, 1902.
75. Beccard, E. Beitraege zur Kenntnis der Saurteiggaering. *Zbl. F. Bakt. II Abt.* 112(54):465–471, 1921.
76. Wolffin, A. Bakteriologische und chemische Studien ueber die Saurteiggaerung. *Archiv. fuer Hygiene* 21:268, 1894.
77. Kline, L., T.F. Sugihara. Sourdough French bread. US patent No. 3, 734–743, 1973.
78. Gobbetti, M. The sourdough microflora: interactions of lactic acid bacteria and yeasts. *Trends Food Sci. Technol.* 9:267–274, 1998.
79. Vogel, R.F., R. Knorr M.R.A. Müller, U. Steudel, M.G. Gänzle, M. Ehrmann. Non-dairy lactic fermentations: the cereal world. *Antonie van Leeuwenhoek* 76:403–411, 1999.
80. Corsetti, A., M. Gobbetti, F. Balestrieri, F. Paoletti, L. Russi, J. Rossi. Sourdough lactic acid bacteria effects on bread firmness and staling. *J. Food Sci.* 63:347–351, 1998.

81. Hammes, W.P., M.G. Gänzle. Sourdough breads and related products. In: *Microbiology of Fermented Foods,* Vol. 1, Woods, B.J.B., ed, London: Blackie Academic & Professional, 1998, pp 199–216.
82. Rosenquist, H., A. Hansen. The antimicrobial effect of organic acids, sour dough and nisin against *Bacillus subtilis* and *B. licheniformis* isolated from wheat bread. *J. Appl. Microbiol.* 85:621–631, 1998.
83. Kline, L., T.F. Sugihara. Microorganisms of the San Francisco sour dough bread process, II: isolation and characterization of undescribed bacterial species responsible for the souring activity. *Appl. Microbiol.* 21:459–465, 1971.
84. Böcker, G., R.F. Vogel, W.P. Hammes. *Lactobacillus sanfrancisco* als stabiles Element in einem Reinzucht-Sauerteig-Präparat. *Getreide Mehl und Brot.* 44:269–274, 1990.
85. Seibel, W., J.M. Bruemmer. The sourdough process for bread in Germany. *Cereal Foods World* 36:299–304, 1991.
86. Röcken, W., P.A. Voisey. Sourdough fermentation in breadmaking. *J. Appl. Bacteriol. Symp. Suppl.* 79:38S–48S, 1995.
87. Wehrle, K., H. Grau, E.K. Arendt. Effects of lactic acid, acetic acid and table salt on fundamental rheological properties of wheat dough. *Cereal Chem.* 74:739–744, 1997.
88. Kaur, H., N. Sing. Effect of acetic acid and CMC on rheological and baking properties of flour. *J. Food Qual.* 22:317–327, 1999.
89. Rosenquist, H., A. Hansen. The microbial stability of two bakery sourdoughs made from conventionally and organically grown rye. *Food Microbiol.* 17:241–250, 2000.
90. Wiese, B.G., W. Strohmar, F.A. Rainey, H. Diekmann. *Lactobacillus panis* sp. nov. from sourdough with a long fermentation period. *Int. J. Syst. Microbiol.* 46:449–453, 1996.
91. Forschino, R., C. Arrigoni, C. Picozzi, D. Mora, A. Galli. Phenotypic and genotypic aspects of *Lactobacillus sanfranciscensis* strains form sourdoughs in Italy. *Food Microbiol.* 18:277–285, 2001.
92. Schleifer, K.H., M. Ehrmann, C. Beimfohr, E. Brockmann, W. Ludwig, R. Amann. Application of molecular methods for the classification and identification of lactic acid bacteria. *Int. Dairy J.* 5:1081–1094, 1995.
93. Vogel, R.F., G. Böcker, P. Stolz, M. Ehrmann, D. Fanta, W. Ludwig, B. Pot, K. Kersters, K.H. Schleifer, W.P. Hammes. Identification of lactobacilli from sourdough and description of *Lactobacillus pontis* sp. nov. *Int. J. Syst. Bacteriol.* 44:223–229, 1994.
94. Müller, M.R.A., M.A. Ehrmann, R.F. Vogel. *Lactobacillus frumenti* sp. nov., a new lactic acid bacterium isolated from rye-bran fermentations with a long fermentation period. *Int. J. Evolution Microbiol.* 50:2127–2133, 2000.
95. Ng, H. Factors affecting organic acid production by sourdough (San Francisco) bacteria. *Appl. Microbiol.* 23:1153–1159, 1976.
96. Nout, M.J.R. Ecology of accelerated natural lactic fermentation of sorghum based infant food formulas. *Int. J. Food Microbiol.* 12:217–224, 1991.
97. Gobbetti, M., A. Corsetti, J. Rossi. The sourdough microflora: interactions between lactic acid bacteria and yeasts: metabolism of amino acids. *World J. Microbiol. Biotechnol.* 10:275–279, 1994.
98. Hamad, S., G. Boecker, R.F. Vogel, W.P. Hammes. Microbiological and chemical analysis of fermented sorghum dough for Kirsa production. *Appl. Microbiol. Biotechnol.* 37:728–731, 1992.
99. Stolz, P., G. Böcker, R.F. Vogel, W.P. Hammes. Utilisation of maltose and glucose by lactobacilli isolated from sourdough. *FEMS Microbiol. Lett.* 109:237–242, 1993.
100. Gobbetti, M., M.S. Simonetti, A. Corsetti, F. Santinelli, J. Rossi, P Damiani. Volatile compound and organic acid production by mixed wheat sourdough starters: influence of fermentation parameters and dynamics during baking. *Food Microbiol.* 12:497–507, 1995.
101. Neubauer, H., E. Glaasker, W.P. Hammes, B. Poolman, W.N. Konings. Mechanism of maltose uptake and glucose excretion in *Lactobacillus sanfrancisco. J. Bacteriol.* 176:3007–3012, 1994.
102. Gobbetti, M., A. Corsetti, J. Rossi. Maltose-fructose co-fermentation by *Lactobacillus brevis* subsp. *lindneri* CB1 fructose-negative strain. *Appl. Microbiol. Biotechnol.* 42:939–944, 1995.

103. Stolz, P., G. Böcker, W.P. Hammes, R.F. Vogel. Utilization of electron acceptors by lactobacilli isolated from sourdough, I: *Lactobacillus sanfranciscensis*. *Z Lebensm Unters Forsch* 201:91–96, 1995.
104. Stolz, P., G. Böcker. Technology, properties and applications of sourdough products. *Adv. Food Sci.* 18:234–236, 1996.
105. Böcker, G., P. Stolz, W.P. Hammes. Neue Erkenntnisse zum Ökosystem Sauerteig und zur Physiologie des sauerteig-typischen Stämme *Lactobacillus sanfrancisco* und *Lactobacillus pontis*. *Getreide Mehl und Brot* 49:370–374, 1995.
106. De Vuyst, L., V. Schrijversi, S. Paramithiotis, B. Hoste, M. Vancanneyt, J. Swings, G. Kalantzopoulos, E. Tsakalidou, W. Messens. The biodiversity of lactic acid bacteria in Greek traditional wheat sourdoughs is reflected in both composition and metabolite formation. *Appl. Environ. Microbiol.* 68:6059–6069, 2003.
107. Vogel, R., M. Müller, P. Stolz, M. Ehrmann. Ecology in sourdoughs produced by traditional and modern technologies. *Adv. Food Sci.* 18:152–159, 1996.
108. Spicher, G., H. Stephan. *Handbuch Sauerteig: Biologie, Biochemie, Technologie.* 4th Aufage, Hamburg: Behr's Verlag GmbH & Co., 1993.
109. Loenner, C., S. Ahrne. *Lactobacillus* (Baking). In: *Food Biotechnology: Microorganisms.* Hui, Y.H., G.G. Khachatourians, eds., New York: VCH Publishers, 1995, pp 797–844.
110. Ottogalli, G., A. Galli, R. Forschino. Italian bakery products obtained with sourdough: characterization of the typical microflora. *Adv. Food Sci.* 18:131–144, 1996.
111. Barberis, C. G. Picchi. Atlante dei prodotti tipici: il pane. Angeli, F. ed., Roma: Istituto Nazionale di Sociologia Rulare, 1995, pp 13.
112. Gobbetti, M., A. Corsetti, J. Rossi, F. La Rosa, S. De Vincenzi. Identification and clustering of lactic acid bacteria and yeasts from wheat sourdoughs in central Italy. *Ital. J. Food Sci.* 1:85–94, 1994.
113. Corsetti, A., P. Lavermicocca, M. Morea, F. Baruzzi, N. Tosti, M. Gobbetti. Phenotypic and molecular identification and clustering of lactic acid bacteria and yeasts from wheat (species *Triticum durum* and *Triticum aestivum*) sourdoughs of Southern Italy. *Int. J. Food Microbiol.* 64:95–104, 2001.
114. Galli, A., L. Franzeti, M.G. Fortina. Isolation and identification of sourdough microflora. *Microbiol. Alim. Nutr.* 6:345–351, 1988.
115. Coppola, S., O. Pepe, P. Masi, M. Sepe. Characterization of leavened doughs for pizza in Naples. *Adv. Food Prot.* 18, 160–162, 1996.
116. Paramythiotis, S., M.R.A. Mueller, M.A. Ehrmann, E. Tsakalidou, H. Seiler, R.F. Vogel, G. Kalantzopoulos. Polyphasic identification of wild yeast strains isolated from Greek sourdoughs. *Syst. Appl. Microbiol.* 23:156–164, 2000.
117. Björkroth, K.J., U. Schillinger, R. Geisen, N. Weiss, B. Hoste, W.H. Holzapfel, H.J. Korkeala, P. Vandamme. Taxonomic study of *Weissella confusa* and description of *Weissella cibaria*, a novel species detected in food and clinical samples. *Int. J. Syst. Evol. Microbiol.* 52:141–148, 2002.
118. Cai, Y., H. Okada, H. Mori, Y. Benno, T. Nakase. *Lactobacillus paralimentarius* sp. nov. isolated from sourdough. *Int. J. Syst. Bacteriol.* 49:1451–1455, 1999.
119. Yoon, J.-H., S.-S. Kang, T.-I. Mheen, J.-S. Ahn, H.-J. Lee, T.-K. Kim, C.-S. Park, Y.H. Kho, K.H. Kang, Y.-H. Park. *Lactobacillus kimchii* sp. nov., a new species from kimchi. *Int J. Syst. Evol. Microbiol.* 50:1789–1795, 2000.
120. Almeida, M.J., C.S. Pais. Characterization of the yeast population from traditional corn and rye bread doughs. *Lett. Appl. Microbiol.* 23:154–158, 1996.
121. Rocha, J.M., F.X. Malcata. On the microbiological profile of traditional Portuguese sourdough. *J. Food Prot.* 62:1416–1429, 1999.
122. Quail, K.J., G.J. MacMaster, M. Wooton. Flour quality tests for selected wheat cultivars and their relationship to Arabic bread quality. *J. Sci. Food Agric.* 54:99–110, 1991.
123. Quail, K.J., G.J. MacMaster, J.D. Tomlinson, M. Wooton. Effect of baking temperature/time conditions and dough thickness on Arabic bread quality. *J. Sci. Food Agric.* 53:527–540, 1990.

124. Quail, K.J. *Arabic bread production.* St. Paul: American Association of Cereals Chemists, 1996, pp 73–97.
125. Steinkraus, K.H. *Handbook of indigenous fermented foods.* New York: Marcel Dekker, 1983.
126. Cambell-Platt, G. *Fermented Foods of the World: a Dictionary and a Guide.* London: Butterworths, 1987, p 291.
127. Boraam, F., M. Faid, J.P. Larpent, A. Breton. Lactic acid bacteria and yeasts associated with traditional sourdough Moroccan bread. *Sci. Alim.* 13:501–509, 1992.
128. Faid, M., F. Boraam, A. Achbab, J.P. Larpent. Yeast-lactic acid bacteria interactions in Moroccan sourdough bread fermenation. *Lebensm. Wiss. U. Technol.* 26, 443–446, 1993.
129. Hesselyine, C.W. Some important fermented foods in Mid-Asia, the Middle East and Africa. *J. Am. Oil Chem. Soc.* 56:367–374, 1979.
130. Adeyemi, I.A., A.T. Osunsami, M.A.B. Fakorede. Effect of corn varieties on ogi quality. *J. Food Sci.* 52:322–324, 1987.
131. Hounhouigan, D.J., M.J.R. Nout, C.M. Nago, J.H. Houben, F.M. Rombouts. Characterization and frequency distribution of species of lactic acid bacteria involved in the processing of mawe, a fermented maize dough from Benin. *Int. J. Food Microbiol.* 18:279–287, 1993.
132. Martinez-Anaya, M.A. Enzymes and bread flavour. *J. Agric. Food Chem.* 44:2469–2480, 1996.
133. Salovaara, H., T. Valjakka. The effect of fermentation temperature, flour type and starter on the properties of sour wheat bread. *Int. J. Food Sci.* 22:591–597, 1987.
134. Corsetti, A., M. Gobbetti, E. Smacchi. Antimicrobial activity of sourdough lactic acid bacteria: isolation of a bacteriocin-like inhibitory substance from *Lactobacillus sanfrancisco* C57. *Food Microbiol.* 13:447–456, 1996.
135. Gänzle, M.G., A. Höltzel, J. Walter, G. Jung, W.P. Hammes. Characterization of reutericyclin produced by *Lactobacillus reuteri* LTH 2584. *Appl. Environ. Microbiol.* 66:4325–4333, 2000.
136. Lavermicocca, P., F. Valerio, A. Evidente, S. Lazzaroni, A. Corsetti, M. Gobbetti. Purification and characterization of novel antifungal compounds from the sourdough *Lactobacillus plantarum* strain 21B. *Appl. Environ. Microbiol.* 66:4084–4090, 2000.
137. Wehrle, K., E.K. Arendt. Rheological changes in wheat sour dough during controlled and spontaneous fermentation. *Cereal Chem.* 74:739–744, 1998.
138. Corsetti, A., M. Gobbetti, B. De Marco, F. Balestrieri, F. Paoletti, L. Russi, J. Rossi. Combined effect of sourdough lactic acid bacteria and additives on bread firmness and staling. *J. Agric. Food Sci.* 48:3044-3051, 2000.
139. Ozcan, S., M. Johnston. Function and regulation of yeast hexose transporters. *Microbiol. Mol. Biol. Rev.* 63:554–569, 1999.
140. Stambuk, B.U., M.A. da Silva, A.D. Panek, P.S. de Araujo. Active α-glucosidase transport in *Saccharomyces cerevisiae. FEMS Microbiol. Lett.* 170:105–110, 1999.
141. Kotyk, A., D. Michaljanicova. Uptake of trehalose by *Saccharomyces cerevisiae. J. Gen. Microbiol.* 110:323–332, 1979.
142. Athenstaedt, K., S. Weys, F. Paltauf, G. Daum. Redundant systems of phosphatidic acid biosynthesis via acylation of glycerol-3-phosphate or dihydroxyacetone phosphate in the yeast *Saccharomyces cerevisiae. J. Bacteriol.* 181:1458–1463, 1999.
143. Remize, F., J.L. Roustan, J.M. Sablayrolles, P. Barre, S. Dequin. Glycerol overproduction by engineered *Saccharomyces cerevisiae* wine yeast strains leads to substantial changes in by-products formation and to a stimulation of fermentation rate in stationary phase. *Appl. Environ. Microbiol.* 65:143–149, 1999.
144. Thevelein, J.M., S. Hohmann. Trehalose synthetase: guard to the gate of glycolysis in yeast? *Trends Biochem. Sci.* 20:3–10, 1995.
145. Nevoigt, E., U. Stahl. Osmoregulation and glycerol metabolism in yeast *Saccharomyces cerevisiae. FEMS Microbiol. Rev.* 21:231–241, 1997.
146. Attfiled, P.V., S. Kletsas. Hyperosmotic stress response by strains of baker's yeast in high sugar concentration medium. *Lett. Appl. Microbiol.* 31:323–327, 2000.
147. Remize, F., J.M. Sablayrolles, S. Dequin. Re-assessment of the influence of yeast strain and environment factors on glycerol production in wine. *J. Appl. Microbiol.* 88:371–378, 2000.

148. Van Dijck, P., D. Colavizza, P. Smet, J.M. Thevelein. Differential importance of trehalose in stress resistance in fermenting and non fermenting *Saccharomyces cerevisiae* cells. *Appl. Environ. Microbiol.* 61:109–115, 1995.
149. Van Dijck, P., M.-F. Gorwa, K. Lemaire, A. Teunissen, M. Versele, S. Colombo, F. Dumortier, P. Ma, A. Tanghe, A. Loiez, J.M. Thevelein. Characterization of a new set of mutants deficient in fermentation-induced loss of stress resistance for use in frozen dough applications. *Int. J. Food Microbiol.* 55:187–192, 2000.
150. Cheng, L., J. Moghraby, P.W. Piper. Weak organic acid treatment causes a trahalose accumulation in low-pH cultures of *Saccharomyces cerevisiae*, not displayed by the more preservative resistant *Zygosaccharomyces bailii*. *FEMS Microbiol. Lett.* 170:89–95, 1999.
151. Birch, R.M., G.M. Walker. Influence of magnesium ions on heat shock and ethanol stress responses of *Saccharomyces cerevisiae*. *Enz. Microbiol. Technol.* 26:678–687, 2000.
152. Albers, E., C. Larsson, G. Liden, C. Niklasson, L. Gustafsson. Influence of the nitrogen source on *Saccharomyces cerevisiae* anaerobic growth and product formation. *Appl. Environ. Microbiol.* 62:3187–3195, 1996.
153. ter Schure, E.G., H.H.W. Sillje, E.E. Vermeulen, J.-W. Kalhorn, A.J. Verkleij, J. Boonstra, C.T. Verrips. Repression of nitrogen catabolites genes by ammonia and glutamine in nitrogen-limited continuous cultures of *Saccharomyces cerevisiae*. *Microbiology* 144:1451–1462, 1998.
154. Collar, C., A.F. Mascaros, C. Benedito de Barber. Amino acid metabolism by yeasts and lactic acid bacteria during bread dough fermentation. *J. Food Sci.* 57:1423–1427, 1992.
155. Thomas, D., Y. Surdin-Kerjan. Metabolism of sulfur amino acids in *Saccharomyces cerevisiae*. *Microbiol. Mol. Biol. Rev.* 61:503–532, 1997.
156. Gobbetti, M., A. Corsetti, J. Rossi. The sourdough microflora. Interactions between lactic acid bacteria and yeasts: metabolism of carbohydrates. *Appl. Microbiol. Biotechnol.* 41:456–460, 1994.
157. Stolz, P., R.F. Vogel, W.P. Hammes. Utilization of electron acceptors by lactobacilli isolated from sourdough, I: *Lactobacillus pontis*, *L. reuteri*, *L. amylovorus* and *L. fermentum*. *Z. Lebensm. Unters. Forsch.* 201:402–410, 1995b.
158. Stolz, P., W.P. Hammes, R.F. Vogel. Maltose-phosphorylase and hexokinase activity in lactobacilli from traditionally prepared sourdoughs. *Adv. Food Sci.* 18:1–6, 1996.
159. Hammes, W.P., M. Korakli, E. Schwarz, G. Wolf. Production of mannitol by *Lactobacillus sanfranciscensis*. *Book of Abstracts (p 22) of the 17th ICC Conference "Cereals across the Continents,"* Valencia, Spain, June 6–9, 1999.
160. Wolfrum, G., R.F. Vogel. Growth and metabolism of *Lactobacillus pontis* TMW 1.109 isolated from cereal fermentations described by differential equations. *Book of of Abstracts of the 17th ICC Conference "Cereals across the Continents,"* Valencia, Spain, 1999.
161. Antuna, B., M.A. Martinez-Anaya, W. Roecken. Modulation of acetic acid production by sourdough starters, *Book of Abstracts of the 17th ICC Conference "Cereals across the Continents,"* Valencia, Spain, 1999.
162. Martinez-Anaya, M.A., M.L. Llin, M.L. Macias, C. Collar. Regulation of acetic acid production by homo- and heterofermentative lactobacilli in whole-wheat sourdoughs. *Z. Lebensm. Unters. Forsch.* 199:186–190, 1994.
163. Gobbetti, M., P. Lavermicocca, F. Minervini, M. De Angelis, A. Corsetti. Arabinose fermentation by *Lactobacillus plantarum* in sourdough with added pentosans and a-L-arabinofuranosidase: a tool to increase the production of actetic acid. *J. Appl. Microbiol.* 88:317–324, 2000.
164. Hammes, W.P., P. Stolz, M. Gänzle. Metabolism of lactobacilli in traditional sourdoughs. *Adv. Food Sci.* 18:176-184, 1996.
165. Gobbetti, M., A. Corsetti. Co-metabolism of citrate and maltose by *Lactobacillus brevis* subsp. *linderi* CB1 citrate-negative strain: effect on growth, end-products and sourdough fermentation. *Z. Lebensm. Unters. Forsch.* 203:82–87, 1996.
166. Gobbetti, M., A. Corsetti, J. Rossi. *Lactobacillus sanfrancisco*, a key sourdough lactic acid bacterium: physiology, genetic and biotechnology. *Adv. Food Sci.* 18:167–175, 1996.
167. Röcken, W., M. Rick, M. Reinkemeier. Controlled production of acetic acid in wheat sourdoughs. *Z. Lebensm. Unters. Forsch.* 195:259–263, 1992.

168. Gobbetti, M., E. Smacchi, P. Fox, L. Stepaniak, A. Corsetti. The sourdough microflora. Cellular localization and characterization of proteolytic enzymes in lactic acid bacteria. *Lebensm. Wiss. U. Technol.* 29:561–569, 1996.
169. Gobbetti, M., E. Smacchi, A. Corsetti. The proteolytic system of *Lactobacillus sanfranciscensis* CB1: purification and characterization of a proteinase, a dipeptidase and an aminopeptidase. *Appl. Environ. Microbiol.* 62:3220-3226, 1996.
170. Liu, S.-Q., G.G. Pritchard, M.J. Hardman, G.J. Pilone. Arginine metabolism in wine lactic acid bacteria: is it via the arginine deaminase pathway or the arginine-urease pathway? *J. Appl. Bacteriol.* 81:486–492, 1996.
171. Liu, S.-Q., G.J. Pilone. Arginine metabolism in wine lactic acid bacteria and its practical significance. *J. Appl. Microbiol.* 84:315–327, 1998.
172. Ograbek, D., M.J. Brandt, W.P. Hammes. Arginine metabolism of *Lactobacillus sanfranciscenis* and *L. pontis* in sourdough, *Book of Abstracts of the 17th ICC Conference "Cereals across the Continents,"* Valencia, Spain, 1999.
173. Rouzaud, M., A. Martinez-Anaya. Effect of processing conditions on oligosaccharide profile of wheat sourdough. *Z. Lebensm. Unters. Forsch.* 197:434–439, 1993.
174. De Robichon-Szulmajester, H., Y. Surdin-Kerjan. Nucleic acid and protein synthesis and activity. In: *The Yeasts*, Rose A.H., J.S. Harrisson, eds., London: Academic Press, 1982, pp 336–418.
175. Collar, C. Biochemical and technological assessment of the metabolism of pure and mixed cultures of yeast and lactic acid bacteria in bread making applications. *Food Sci. Technol. Int.* 2:349–367, 1996.
176. Damiani, P., M. Gobbetti, L. Cossignani, A. Corsetti, M.S. Simonetti, J. Rossi. The sourdough microflora: characterization of hetero- and homofermentative lactic acid bacteria, yeasts and their interactions on the basis of the volatile compounds produced. *Lebensm. Wiss. U. Technol.* 29:63–70, 1996.
177. Martinez-Anaya, M.A., M.J. Torner, C. Beneditio de Barber. Microflora of wheat flour bread, XIV: changes in volatile compounds during fermentation of doughs prepared with pure organisms and their mixtures. *Z. Lebensm. Unters. Forsch.* 190:126–131, 1990.
178. Hansen, A., B. Hansen. Influence of wheat flour type on the production of flavour compounds in wheat sourdoughs. *J. Cereal Sci.* 19:185–190, 1994.
179. Hansen, A., B. Hansen. Flavour of sourdough wheat bread crumb. *Z. Lebensm. Unters. Forsch.* 202:244-249, 1996.
180. Loureiro, V., H.G. Ferreira. On the intracellular accumulation of ethanol in yeast. *Biotechnol. Bioeng.* 25:2263–2269, 1983.
181. Lebeau, T., T. Jouenne, G.-A. Junter. Diffusion of sugars and alcohols through composite membrane structures immobilizing viable yeast cells. *Enz. Microbiol. Technol.* 22:434–438, 1998.
182. Guijarro, J.M., R. Lagunas. *Saccharomyces cerevisiae* does not accumulate ethanol against a concentration gradient. *J. Bacteriol.* 160:874-878, 1984.
183. Lucero, P., E. Penalver, E. Moreno, R. Lagunas. Moderate concentrations of ethanol inhibit endocytosis of the yeast maltose transporter. *Appl. Environ. Microbiol.* 63:3831–3836, 1997.
184. Peres, M.F.S., C. Laluce. Ethanol tolerance of thermotolerant yeasts cultivated on mixtures of sucrose and ethanol. *J. Ferment. Bioeng.* 85:388–397, 1998.
185. Guerzoni, M.E., R. Lanciotti, M. Sinigaglia, M. Anese, C.R. Lerici. Influence of some selected ions on system water activity and on ethanol vapour pressure and its inhibitory action on *Saccharomyces cerevisiae*. *Can. J. Microbiol.* 40:1051–1056, 1994.
186. Hallsworth, J.E. Ethanol induced water stress in yeast. *J. Ferment. Bioeng.* 85:125–137, 1998.
187. Chi, Z., N. Arneborg. Relationship between lipid composition, frequency of ethanol-induced respiratory deficient mutants and ethanol tolerance in *Saccharomyces cerevisiae*. *J. Appl. Microbiol.* 86:1047–1052, 1999.
188. Guerzoni, M.E., M. Ferruzzi, F. Gardini, R. Lanciotti. Combined effects of ethanol, high homogenisation pressure and temperature on cell fatty acid composition in *Saccharomyces cerevisiae*. *Can. J. Microbiol.* 45:805–810, 1999.

189. Guldfeldt, L.U., N. Arneborg. Measurement of the effects of acetic acid and extracellular pH on intracellular pH of nonfermenting, individual *Saccharomyces cerevisiae* cells by fluorescence microscopy. *Appl. Environ. Microbiol.* 64:530–534, 1998.
190. Sousa, M.J., L. Miranda, M. Corte-Real, M. Leao. Transport of acetic acid in *Zygosaccharomyces bailii*: effects of ethanol and their implications on the resistance of the yeast to acidic environments. *Appl. Environ. Microbiol.* 62:3152–3157, 1996.
191. Paiva, S., S. Althoff, M. Casal, C. Leao. Transport of acetate in mutants of *Saccharomyces cerevisiae* defective in monocarboxylase permeases. *FEMS Microbiol. Lett.* 170:301–306, 1999.
192. Casal, M., H. Cardoso, C. Leao. Mechanisms regulating the transport of acetic acid in *Saccharomyces cerevisiae*. *Microbiology* 142:1385–130, 1996.
193. Stratford, M., P.A. Anslow. Comparison of the inhibitory action on *Saccharomyces cerevisiae* of weak-acid preservatives, uncouplers and medioum chain fatty acids. *FEMS Microbiol. Lett.* 142:53–58, 1996.
194. Kusumegi, K., H. Yoshida, S. Tomiyama. Inhibitory effects of acetic acid on respiration and growth of *Zygosaccharomyces rouxii*. *J. Ferment. Bioeng.* 85:213–217, 1998.
195. Sousa, M.J., F. Rodrigues, M. Corte-Real, M. Leao. Mechanisms underlying the transport and intracellular metabolism of acetic acid in the presence of glucose in the yeast *Zygosaccharomyces bailii*. *Microbiology* 144:665–670, 1998.
196. Fernanda Rosa, M., I. Sa-Correia. Intracellular acidification does not account for inhibition of *Saccharomyces cerevisiae* growth in the presence of ethanol. *FEMS Microbiol. Lett.* 135:271–274, 1996.
197. Mizoguchi, H., S. Hara. Permeability barrier of the yeast plasma membrane induced by ethanol. *J. Ferment Bioeng.* 85:25–29, 1998.
198. Imai, T., T. Ohno. The relationship between viability and intracellular pH in the yeast *Saccharomyces cerevisiae*. *Appl. Environ. Microbiol.* 61:3604–3608, 1995.
199. Pampoulha, M.P., M.C. Loureiro-Dias. Energetics of the effect of acetic acid on growth of *Saccharomyces cerevisiae*. *FEMS Microbiol. Lett.* 184:69–72, 2000.
200. Haard, N.F., G.W. Chism. Characteristics of edible plant tissues. In: *Food Chemistry,* 3rd ed., Fennema, O.R., ed., New York: Marcel Dekker, 1996, pp 943–1011.
201. Satterlee, L.D., R. Abdul-Kadir. Effect of phytate content on protein nutritional quality of soy and wheat bran proteins. *Lebensm. Wiss. U. Technol.* 16:8–14, 1983.
202. Nayini, N.R., P. Markakis. The phytase of yeast. *Lebensm. Wiss. U. Technol.* 17:24–26, 1984.
203. Lopez, H.W., V. Krespine, C. Guy, A. Messager, C. Demigne, C. Remesy. Prolonged fermentation of whole-wheat sourdough reduces phytate level and increases soluble magnesium. *J. Agric. Food Chem.* 49:2657–2662, 2001.
204. Graf, E., J.W. Eaton. Surpression of colonic cancer by dietary phytic acid. *Nutr. Canc. Int. J.* 19:11–19, 1993.
205. Graf, E., J.W. Eaton. Dietary surpression of colonic cancer: fiber or phytate? *Cancer* 56:717–718, 1995.
206. Armero, E., C. Collar. Texture properties of formulated wheat doughs: relationships with dough and bread technological quality. *Z. Lebensm. Unters. Forsch. A.* 204:136–145, 1997.
207. Gobbetti, M., M. De Angelis, P. Arnaut, P. Tossut, A. Corsetti, P. Lavermicocca. Added pentosans in breadmaking of derived pentoses by sourdough lactic acid bacteria. *Food Microbiol.* 16:409–418, 1999.
208. Martinez-Anaya, M.A., T. Jimenez. Functionality of enzymes that hydrolyse startch and non-starch polysaccharide in bread making. *Z. Lebensm. Unters. Forsch. A.* 204:209–214, 1997.
209. Gil, M.J., M.J. Callejo, G. Rodriguez, M.V. Ruiz. Keeping qualities of white pan bread upon storage: effect of selected enzymes on bread firmness and elasticity. *Z. Lebensm. Unters Forsch. A.* 208, 394–399, 1999.
210. Leuschner, R.G.K., M.J.A. O'Callaghan, E.K. Arendt. Moisture distribution and microbial quality of part baked breads as related to storage and rebaking conditions. *J. Food Sci.* 64:543–546, 1999.

211. Lund, F., O. Filtenborg, S. Westall, J.C. Frisvad. Associated mycoflora of rye bread. *Lett. Appl. Microbiol.* 23:213–217, 1996.
212. Nielsen, P.V., R. Rios. Inhibition of fungal growth on bread by volatile components from spices and herbs, and the possible application in active packaging, with special emphasis on mustrd essential oil. *Int. J. Food Microbiol.* 60:219–229, 2000.
213. Röcken, W. Applied aspects of sourdough fermentation. *Adv. Food Sci.* 18:212–216, 1996.
214. Corsetti, A., M. Gobbetti, J. Rossi, P. Damiani. Antimold activity of sourdough lactic acid bacteria: identification of a mixture of organic acids produced by *Lactobacillus sanfrancisco* CB1. *Appl. Microbiol. Biotechnol.* 50, 253–256, 1998.
215. Rosenquist, H., A. Hansen. Contamination profiles and characterization of *Bacillus* species in wheat bread and raw materials for bread production. *Int. J. Food Microbiol.* 26:353–363, 1995.
216. Larsen, A.G., F.K. Vogensen, J. Josephsen. Antimicrobial activity of lactic acid bacteria isolated from sourdoughs: purification and characterization of bavaricin A, a bacteriocin produced by *Lactobacillus bavaricus* MI401. *J. Appl. Microbiol.* 75:113–122, 1993.
217. Todorov, S., B. Onno, O. Sorokine, J.M. Chobert, I. Ivanova, X. Dousset. Detection and characterization of a novel antimicrobial substance produced by *Lactobacillus plantarum* ST31 isolated from sourdough. *Int. J. Food Microbiol.* 48:167–177, 1999.
218. Messens, W., L. De Vuyst. Inhibitory substances produced by *Lactobacilli* isolated from sourdoughs: a review. *Int. J. Food Microbiol.* 72:31–43, 2002.
219. Fuller, R. Probiotics in man and animals. *J. Appl. Bacteriol.* 66:372–378, 1989.
220. Havenaar, R., J.H.J. Huis in't Veld. Probiotics: a general view in the lactic acid bacteria. In: *The Lactic Acid Bacteria in Health and Disease,* Vol, 1, Wood, B.J.B. ed, New York: Chapman and Hall, 1992, pp 209-224.
221. Guarner, F., G.J. Shaafsma. Probiotics. *Int. J. Food Microbiol.* 39:237–238, 1998.
222. Olson, A.C., G.M. Gray, M.-C. Chiu, A.A. Betschart, J.R. Turnlund. Monosaccharides produced by acid hydrolysis of selected foods, dietary fibers and fecal residues from white and whole bread consumed by humans. *J. Agric. Food Chem.* 36:300–304, 1988.
223. Prosky, L. Inulin and oligofructose are part of the dietary fiber complex. *J. AOAC Int.* 82:223-226, 1999.
224. Thranathan, M., R.N. Tharanathan. Resistant starch in wheat-based products: isolation and characterization. *J. Cereal Sci.* 34, 73–84, 2001.
225. Yamada, H., K. Itoh, Y. Morishita, H. Taniguchi. Structure and properties of oligosaccharides from wheat bran. *Cereal Foods World* 38:490–492, 1993.
226. Bach Knudsen, K.E., B. Borg Jensen, I. Hansen. Oat bran but not a α-glucan enriched oat fraction enhances butyrate production in large intestine of pigs. *J. Nutr.* 123:1235–1247, 1993.
227. Chesson, A. Dietary fiber. In: *Food Polysaccharides and their Applications.* Stephen, A.M., ed., New York: Marchel Dekker, 1995, pp 547-576.
228. Gibson, G.R., M.B. Roberfroid. Dietary modulation of human colonic microbiota introducing the concept of prebiotics. *J. Nutr.* 125:1401–1412, 1995.
229. Molin, G., C.-E. Albertsson, S. Bengmark, K. Larsson. Nutrient composition and method for the preparation thereof. International Patent Application, PCT, WO 89/0840; US Patent No. 5,190,755, (1989/1993), 1993.
230. Salovaara, H. The time of cereal based functional foods is here: introducing Yosa®, a vellie. In: *Proccedings 26th Nordic Cereal Congress.* Skrede, G., E.M. Magnus, eds. Haugensund, 1996, pp. 195–202.
231. Salovaara, H., A. Kurka. Food product containing dietary fiber and method making said product. International Patent Application, PCT, WO 91/17672/ EP 0568530 (1991/1997), 1997.
232. Forschino, R., R. Terraneo, D. Mora, A. Galli. Microbial characterization of sourdoughs from sweet baked products. *Ital. J. Food Sci.* 11:19–28, 1999.

3.20

Fermentation Biotechnology of Animal Based Traditional Foods of the Middle East and Mediterranean Region

Effie Tsakalidou and Parthena Kotzekidou

CONTENTS

20.1 HISTORY

Fermented dairy and meat products have been known in the Mediterranean and Middle East region since ancient times. Fermentation is of great importance, as it contributes to the preservation of vast quantities of nutritious foods without significant fuel requirements, providing a wide diversity of flavors, aromas, and textures, and enriches food with vitamins and essential amino acids (1).

Lactic acid fermentation induced by lactic acid bacteria, which convert fermentable sugars primarily to lactic acid, are responsible for processing and preserving milk, producing sour milks, yogurts, and cheeses. Mixtures of yogurt and wheat undergoing lactic acid fermentation give rise to highly nutritional foods like Greek trahanas. These foods were originally developed empirically. Later, it was found that under certain treatments lactic acid bacteria occurring in raw materials improved keeping quality and safety as well as sensory characteristics of the fermented product, inhibiting the growth and survival of spoilage microorganisms and pathogens. Although acidity and the presence of undissociated organic acids are the most important inhibitory agents, other antimicrobial factors associated with lactic acid bacteria, such as bacteriocins, carbon dioxide, hydrogen peroxide, ethanol, and diacetyl are responsible for the inhibition of food spoilage and pathogenic bacteria (2).

Because of the recognition that fermentation plays an important, beneficial role in human nutrition and health as well as in food safety, the production of fermented products has shown a steady increase over recent years (3). Food fermentations that improve the balance of essential amino acids or their availability influence the nutritional value of fermented foods. In a similar way, fermentations that increase the content or availability of vitamins such as thiamine, riboflavin, niacin, or folic acid can have profound effects on the health of the consumers of such foods (1).

Mediterranean foods are produced through fermentation and other natural processes according to ancient recipes used locally. In the last few years there has been a tremendous back to nature emphasis, demanding a change in the market to naturally processed, additive free, and healthy products. The nutritional, sensory, and functional enhancement of traditional foods offer the potential for improving human health and wellbeing within the context of the Mediterranean diet (4).

The objective of this chapter is to provide food scientists, microbiologists, and nutritionists with a comprehensive and solid view of the present state of knowledge of the indigenous fermented foods of animal origin in the Mediterranean and Middle East countries, which since ancient times have played a vital role in feeding humans with nutritious and flavorsome foods in the context of a significant component of their cultural heritage.

20.2 DAIRY PRODUCTS

20.2.1 Introduction

Cheese making and fermented milk production are some of the oldest methods practiced by man for the conversion of a highly perishable and nutritional foodstuff (milk) into products with extended shelf life. It is commonly believed that cheese was developed in the Fertile Crescent between the Tigris and Euphrates some 8000 years ago. The Agricultural Revolution occurred here with the domestication of plants and animals. Cheese manufacture accompanied the spread of civilization throughout the Middle East, Greece, and Rome. There are several references to cheese in the Old Testament, such as Job (1520 BC) and Samuel (1170–1017 BC); in the tombs of Ancient Egypt; and in classical Greek literature, such as Homer (1184 BC), Herodotus (484–408 BC), and Aristotle (384–322 BC).

Movements of Roman armies spread the consumption of cheese throughout the known world of that time. After the fall of the Roman Empire, the great migrations of people throughout Europe promoted the further spread of cheese manufacture (5).

There are no records available regarding the origin of yogurt. However, the belief in its beneficial influence on human health and nutrition has existed in many civilizations for a long time. According to a Persian tradition, Abraham owed his fecundity and longevity to yogurt, while in more recent times Emperor Francis I of France was supposed to have been cured of a debilitating illness by consuming yogurt made out of goat's milk. It is likely, however, that the origin of yogurt was the Middle East. The evolution of this fermented product through the ages can be attributed to the culinary skills of the nomadic people living in that part of the world (6).

The first fermented dairy products were produced by an accidental combination of events: the ability of a group of bacteria, now known as lactic acid bacteria, to grow in milk and to produce just enough acid to reduce the pH of the milk to the isoelectric point of the caseins, at which these proteins coagulate. While lactic acid produced *in situ* is believed to have been the original milk coagulant, an alternative mechanism was also recognized from an early date. Many proteolytic enzymes can modify milk proteins, causing them to coagulate under certain circumstances. Before the development of pottery (ca. 5000 BC), the stomachs of slaughtered animals provided ready made, easily sealed containers. Stored in such containers, milk would extract coagulating enzymes from the stomach tissue, leading to its coagulation during storage.

Although the history of cheese and fermented milks can be traced back for thousands of years, cheese and yogurt making remained an art rather than a science until relatively recently. It is only since the turn of the twentieth century that serious attempts have been made to understand the associated microbiology. The result of this interest has been that many products are now manufactured with well defined starter cultures. Microbiological aspects of bovine cheeses manufactured worldwide, especially on an industrial scale and almost exclusively from pasteurized milk (e.g., Cheddar cheese), have been studied for decades. Not until recently have in depth studies focused on artisanal cheese manufactured in Mediterranean countries. Such studies deal with the characterization of those microbial groups related to ripening aspects and sanitary issues. It has always to be borne in mind that the production of fermented milks is a biological process, and the organisms involved are subject to the same variability as any other life form. Obviously, the selection pressure put upon a given species in a factory situation may be better defined than for species found in other environments. Even so, variation is still an active component of the system.

20.2.2 Microbiological and Technological Aspects

20.2.2.1 Cheese

Cheese is one of the most diverse food groups. According to Burkhalter (7) around 500 varieties are produced from cow's milk, while Kalantzopoulos (8) listed 500 more produced from sheep's or goat's milk. The variety of cheeses produced in the countries of the Mediterranean basin is probably the highest in the world. A major source of variation in the characteristics of cheese resides in the species from which the milk was produced. Although milk from several species is used in cheese manufacture, the cow is worldwide by far the most important, while sheep, goat, and buffalo are commercially important in certain areas. This is the case in the Mediterranean and Middle East countries, where herds of sheep are abundant and widely distributed, and the most representative cheese varieties are manufactured from ewe's milk. Mediterranean countries undoubtedly account for production of most ewe's and goat's milk worldwide. Almost all such milk, usually in raw form, is converted into cheese. Many of these cheeses are protected by Appellation d'Origin Protégée (AOP) (8).

The majority of the Mediterranean cheeses are produced using animal rennet as the coagulant. However, several cheeses in the Iberian Peninsula are traditionally produced with the use of plant rennet, namely an aqueous extract from the cardoon flowers (*Cyanara cadunuculus*). These include Serpa and Serra da Estrella in Portugal, and Casar de Caceres and La Serena in Spain. All these cheese varieties are produced from ewe's milk. Ewe's and goat's milk are used for the production of many other traditional varieties of cheeses in the Iberian Peninsula using animal rennet as coagulant. Although plant coagulants and chymosin have similar specificities in the cleavage sites, as they hydrolyse between hydrophobic amino acids, plant coagulants are considered more proteolytic than chymosin, and thus casein hydrolysis may result in lower gel firmness (9,10). In general, artisanal ewe's and goat's cheeses in the Iberian peninsula can be divided into four groups based on milk source and rennet type: (1) cheeses manufactured with raw ewe's milk and coagulated with plant rennet (e.g., Azetiao, Castelo Branco, Evora, Nisa, Serpa, Serra da Estrella, and La Serena), (2) cheeses made from raw ewe's milk and coagulated with animal rennet (e.g., Terrinco, Idiazabal, Manchego, Roncal and Zamorano), (3) cheeses manufactured with raw goat's milk and coagulated with animal rennet (e.g., Cabra Transmontano and Majorero), and (4) cheeses made from mixtures of raw ewe's and goat's milk and coagulated with animal rennet (e.g., Amarelo da Beira Baixa, Picante da Beira Baixa, and Rabacal). Most of these varieties are semihard or hard cheeses, apart from Sepra, Serra da Estrella, and La Serena, which are soft cheeses. A thorough review on the microbiology and biochemistry of these cheese varieties has been recently published by Freitas and Casalta (11). On the other hand, many important cheese varieties in Portugal and Spain are manufactured from cow's milk with animal rennet as coagulant. These include Pico (soft) and Sao Jorge (semihard) in Portugal, as well as Cabrales and Picon Bejes-Tresviso (blue veined), Cantabria and Cebreiro (soft), Mahon and Penamellera (semi hard) in Spain (12). Serra da Estrella is the most traditional cheese in Portugal, while Manchego is the most typical Spanish cheese. Traditionally, Serra da Estrella is produced without deliberately added starter cultures. At the end of ripening the cheese has a flat cylindrical shape; the rind is thin, uniform, smooth, and straw yellow; the cheese has no eyes; its color varies from ivory to white; and it exhibits a buttery texture, strong aroma and a clean, smooth and slightly acid flavor (13). Lactic acid bacteria, mainly *Lactococcus lactis*, *Enterococcus faecium* and *Leuconostoc mesenteroides*, and *Enterobacteriaceae* are the dominant fermentation organisms, while a large spectrum of yeasts, mainly *Sporobolomyces roseus* are also found (14,15). Although Manchego cheese is named after the La Mancha region, where the original product was traditionally made from raw ewe's milk, nowadays the manufacture of Manchego type cheeses is spread throughout Spain. When produced at industrial scale, *Lactococcus lactis* and *Lactococcus cremoris* are used as starters. Animal rennet is used for coagulation and the curd is heated to 37°C. The cheese must be ripened for at least 2 months after molding at 12–15°C and 75–86% RH. Mesophilic homofermentative lactobacilli, mainly *Lactobacillus plantarum* and *L. casei*; enterococci, mainly *E. durans*; *Leuconostoc* species, mainly *L. mesenteroides* spp. *dextranicum* and *L. paramesenteroides*; and *Pediococcus species*, mainly *P. pentosaceus,* are the most important groups (16,17,18). The most frequently isolated yeasts belong to the *Debaryomyces* and *Candida* genera, followed by *Yarrowia, Pichia, Saccharomyces* and *Tolusporosa* (19).

Many Mediterranean cheese varieties belong to the surface ripened cheeses. These include both the bacterial smear surface ripened and the mold surface ripened cheeses. The bacterial smear surface ripened cheeses can be broadly defined as those cheeses in which bacteria are present in large numbers on the surface of the cheese and play a significant role in determining the final characteristics and attributes of the cheese (20,21). French cheeses such as Beaufort, Comte, Gruyere, Saint Paulin, and Munster, and the Italian cheese Taleggio can be classified in this group. The pH on the surface of the cheese is about 5.0 and the cheese is generally ripened at temperatures of 12–16°C at a relative humidity of greater than 90%. These conditions result in rapid development of the smear. The microbiology of the smear is complex and its role in the ripening of the cheese is poorly understood. It is generally believed that yeasts grow during the first few days of ripening, oxidizing lactate fully to CO_2 and H_2O, and deaminating amino acids to the corresponding ketoacid and NH_3, thus leading to pH increase where the bacteria can grow (22). Yeasts of the genera *Candida*, *Debaryomyces*, *Kluyveromyces,* and *Rhodotorula* develop initially on the surface of Gruyere and Beaufort cheeses up to 10^9 cfu/g and then decrease to 10^5 for the rest of the ripening period (23). At the end of the ripening period, *Brevibacteria* species (mainly *B. linens*), coryneform bacteria, micrococci, and staphylococci are the dominant microorganisms (24,25,26). Compared with other types of cheese, the smear cheeses are characterized by a more rapid ripening process and by a more intense flavor due to a particular method of curing that involves the creation of conditions favorable for the optimal development of a complex microbiological community on the smear (27).

The mold surface ripened cheeses are mainly represented by the French varieties Camembert and Brie. Although during the initial stages of ripening their surface microflora may be similar to those of bacterial smear surface ripened cheeses, at the end of the ripening period, molds such as *Penicillium camemberti* are the dominant microorganisms (28). The composition and evolution of the microflora in these cheeses are complex, especially when raw milk is used, while the high biochemical activities of molds on the surface give these cheeses a very typical aroma and taste. In Camembert cheese, the starters used, mainly *Lactococcus lactis* spp. *lactis* and *Lactococcus lactis* spp. *cremoris*, dominate in the 24 h curd. At that point yeasts, mainly *Kluyveromyces lactis*, *Sacharomyces cerevisiae* and *Debaryomyces hansensi* grow on the surface. The mold *Geotrichum candidum* appears together with the yeasts, but its growth is limited by salting. A white felt of *P. camemberti* develops on the surface of Camembert cheese 6–7 days postmanufacture (29,30). *Penicillium camemberti* converts lactate to CO_2 and H_2O, and the surface is deacidified; and thus an aerophilic acid sensitive bacterial flora, consisting mainly of coryneform bacteria (*B. linens*) and micrococci becomes established (28). The French semihard cheeses Saint-Nectaire and Tome de Savoie are also white mold cheeses produced, however, in limited quantities. The sequential appearance of microorganisms on the rind and the curd of a Saint-Necatir type cheese has been documented (31).

Blue veined cheeses belong to the mold ripened cheeses. As the mold surface ripened cheeses, they are characterized by extensive proteolysis and lipolysis. However, in this case *Penicillium roqueforti* is grown not on the surface of the cheese but within the curd. The French Roquefort, the Italian Gorgonzolla and the Spanish Cabrales, Picon Bejes Tresviso and Valdeon belong to this group of cheeses. For the production of Roquefort cheese a water suspension of *P. roqueforti* spores is added to the milk just prior to setting, or spores are dusted onto the curd. In addition to *P. roqueforti* and lactic acid bacteria, the flora of Roquefort cheese, made from raw ewe's milk, also contains yeasts, micrococci, staphylococci and coliforms. *Lactococcus* species are always clearly dominant inside the cheese. *Lactobacillus* species, mainly *L. casei* and *L. plantarum*, reach a maximum just before salting, while yeasts and *Leuconostoc* species are also present from the beginning of the ripening (32–37). Cabrales is produced in Northern Spain from mixed raw cow's, ewe's, and goat's milk, with neither lactic starter nor *P. roqueforti* added. Its microflora during ripening has been investigated by Nunez (38). In the interior of the cheese, lactic streptococci, mainly *S. lactis*, predominate before and during salting (0–4 d).

Streptococcus lactis and lactic acid utilizing yeasts prevail during drying room ripening (5–15 d), while during cave ripening (16–120 d), molds, mainly *P. roqueforti*, and lactobacilli, mainly, *L. plantarum*, are the dominant species. On the cheese surface, *S. lactis* comprises the major flora before and during salting. *Streptococcus lactis*, micrococci, and *Leuconostoc* species are the main microbial groups during drying room ripening, and micrococci throughout cave ripening. Coliforms and coagulase positive staphylococci die off during the first month of maturation. Picon Bejes Tresviso is the second most important blue veined cheese in Spain after Cabrales. It is made primarily from cow's milk, which is sometimes mixed with small quantities of ewe's and goat's milk. The cheese is pierced on the 4th day after production and the growth of *P. roqueforti* is very abundant throughout the mass. In the first few days hardly any microbial growth on the surface is observed, but as the ripening process advances, a surface slime containing yeast and bacteria, which are very perceptible at the end of the ripening, develops on the cheeses. Once the ripening process is finished, and before covering the cheeses with *Acer pseudoplatanus* leaves, the cheeses are washed and the surface slime eliminated (39). Valdeon is a blue veined cheese produced in Spain from goat's milk. Nearly 95% of the aerophilic mesophiles on the cheese are lactic acid bacteria. *Enterococcus* and *Lactococcus* species are the dominant genera during the first stages of ripening, while *Lactobacillus* and *Leuconostoc* species replace the lactococci with proceeding ripening (40). Gorgonzolla is an Italian blue veined cheese produced from cow's milk. The use of natural lactic acid bacteria starters has been, in part, discontinued and selected strains of *L. delbruekii* spp. *bulgaricus* and *S. thermophilus*, along with *P. roqueforti*, have been introduced. According to Gobbetti et al. (41), *S. thermophilus*, *P. roqueforti*, lactococci, and surface grown micrococci, yeasts, and molds are present in the ripened cheese. As for other blue cheeses, the interactions between lactic acid bacteria and mold starters have been reviewed and a positive influence of the thermophilic lactic acid bacteria on the growth and the proteolytic activity of *P. roqueforti* has been reported (42). Due to the complex microflora and technology used to produce Gorgonzolla cheese, several studies also investigated the causes of some cheese defects, such as swelling by yeasts due to starter failure caused by phage (43). Kopanisti, a traditional Greek soft cheese prepared from cow's, ewe's, and goat's milk or mixtures of them has a strong flavor similar to that of blue veined cheeses. The procedure of manufacture and ripening promotes an abundant growth of a great variety of microorganisms, such as streptococci and lactobacilli, as well as yeasts. Kaminarides and Anifantakis (44) isolated and identified *Trichosporon cutaneum*, *Kluyveromyces lactis*, *Saccharomyces exiguus*, *Saccharomyces cerevisiae*, *Rhodotoula rubra*, *Trichosporon penicillium*, *Candida lusitanie* and *Debaryomyces hansenii*.

The French Emmental, Gruyere and Beaufort, and the Greek Graviera cheese belong to the so called Swiss-type cheeses (45,46). They are usually made with thermophilic starters (*S. thermophilus*, *L. helveticus* and *L. delbruekii* spp. *lactis*) and sometimes mesophilic lactic acid bacteria (*Lactococcus lactis* spp. *lactis*). The curd and whey mixture is always cooked between 50–56°C and the ripening time varies from 2 to 18 months. Often eyes are present due to the fermentation by propionic acid bacteria (*P. freudenreichii* spp. *shermani*) of lactate to propionate, acetate, and CO_2. The texture is slightly elastic and the flavor mild and slightly sweet, with intensity increasing with aging (47). However, these characteristics are not typical for all Swiss-type cheeses. For instance in Comte and Graviera, propionic acid bacteria are not intentionally added, and Beaufort has no openness. Some of these characteristics are found in other cheeses, like the high cooking temperature in Parmegiano-Regiano, in which no propionic acid fermentation occurs. The interactions between the different microbial groups of the Swiss-type cheeses microflora, as well as the genetic diversity within certain groups, have received much attention in recent years (48–51). The heterofermentative lactic acid fermentation provokes abundant eye formation but is of minor importance. Mixed acid fermentation may occur due to the growth of *Enterobacteriaceae* leading to an excess of eye formation. Butyric acid fermentation by *Clostridium butyricum* and *Clostridium tyrobutyricum* is undesirable, because lactate is converted to butyric and acetic acids, CO_2, and hydrogen, which causes the cheese loaf to blow (46). Some Swiss-type cheeses, such as Gruyere and Beaufort, undergo a smear ripening process. The surface of the cheese is first deacidified by yeasts, which provokes various salt tolerant microorganisms to develop, such as micrococci and *Brevibacteria* species, which then degrade the lactic acid and influence the flora and the color of the surface of cheese.

The grana cheeses, e.g., Parmigiano-Regiano and Grana Padano, are one of the most important characteristic groups of Italian cheeses from the technological point of view. The other characteristic groups are Pasta Filata and Pecorino Romano. They are prepared from partially skimmed cow's milk, using natural whey starters, while high scalding temperatures (54–55°C) and long ripening times (greater than 12 months) are applied. They are classified as extra hard because of the low moisture content (52). Lactic starters used in the production of grana cheeses are whey cultured mixed strains of homofermentative thermophilic lactobacilli and a small number of heterofermentative lactobacilli (53,54). Using randomly amplified polymorphic DNA (RAPD) fingerprinting, interactions between *L. helveticus*, *L. delbruekii* spp. *lactis* and *L. delbruekii* spp. *bulgaricus* were shown, and this was confirmed by comparison of growth kinetics and pH behavior in mixed and single strain cultures (55). The differences seen in bacterial species and strain composition between the whey starters and the curd and cheese are in agreement with the different biochemical and biophysical characteristics of the two ecosystems (56,57). A comprehensive study on the microbiological characteristics of milk, natural whey starter, and cheese during the first months of aging of Parmigiano Regiano showed that the thermophilic lactobacilli (between *L. helveticus*, *L. delbruekii* spp. *lactis* and *L. delbruekii* spp. *bulgaricus*) were derived from the natural whey starter, while S. *thermophilus* originated from the raw milk. Natural whey starter was also the source for *L. rhamnosus*, which was present throughout the entire period of cheese aging. The other components of the nonstarter microflora, mainly mesophilic lactobacilli and pediococci, were derived from the raw milk (58). The monitoring of propionibacteria, enterococci, coliforms, anaerobes and staphylococci has been also reported (59). In Grana cheese a key role is played by *L. helveticus,* which is the prevalent species in the natural whey starter cultures and in cheese in the early stages of ripening (54,60). Using polymerase chain reaction (PCR) fingerprinting it was shown that the community of *L. helveticus* dominating in Grana cheese is composed of different biotypes (61).

Pasta Filata cheeses are very popular in the Mediterranean countries. They include soft or semisoft varieties, typically consumed fresh or after only a brief period of aging (e.g., the Italian Mozzarella) and semihard or hard varieties, which are subjected to considerable aging before being consumed (e.g., the Italian Provolone and Caciocavallo). Similar to Caciocavallo is the Balkan variety Kashkaval, as well as Kasseri in Greece, Kasar peynir in Turkey and Romy in Egypt. Typical for the production of Pasta Filata cheeses is that after renneting, the curd is first cooked at 48–52°C and at a later stage it is kneaded in hot water (85–90°C) until it can be stretched (52,62). For cheese making, either natural whey starters, resulting from the incubation of cheese whey from the previous manufacture, or natural milk starters, obtained by incubation of heat treated raw milk, are used, in which thermophilic rods, like *L. delbruekii* spp. *lactis*, *L. delbruekii* spp. *bulgaricus* and *L. helveticus*, in addition to thermophilic cocci, like *S. thermophilis*, and mesophilic cocci, like *Lactococcus lactis* dominate. *Enterococcus* species, *Leuconostoc* species, and yeasts can also reach high numbers, while they are often contaminated by coliforms (63–67). Yeasts, in particular

Kluyveromyces and *Saccharomyces*, seem to affect the functional and sensory properties of Mozzarrella cheese (68). RAPD-PCR and 16S rRNA analysis are presently widely applied to determine the genetic diversity and to differentiate the biotypes in the microflora of Pasta Filata cheeses (67,69). A new *Streptococcus* species, *S. macedonicus*, has been recently isolated from traditional Greek Kasseri cheese (70). *Streptococcus macedonicus* seems to be also part of the microflora on hard Italian, semicooked (44–49°C) cheeses like Asiago d'Allevo, Fontina, Montasio, and Monte Veronese (71).

The subtropical climate in most of the Mediterranean countries imposed the production of white brined cheeses (white pickled cheeses). Brined cheeses share the practice of storage in brine (pickle), usually salted whey, for extended periods. They are of great importance for warm climates, because under these conditions the shelf life of milk is short and cheese deteriorates before it ripens. Possibly, these cheese varieties share the same origin, with various modifications to suit local conditions and needs, and they are therefore quite different in several aspects including the type of milk used and the manufacturing and storage conditions. The most well known cheeses of this class are Feta in Greece, Domiati in Egypt, Beyaz peynir in Turkey, and Halloumi in Cyprus (72,73). Feta is a soft white cheese made from ewe's milk, but goat's milk may be used as well, or mixtures of both. *Streptococcus thermophilus*, *L. bulgaricus* and *L. lactis* spp. *lactis* are commonly used as starters. Dry salting of the curd and slime formation are essential for the development of the characteristic flavor during ripening. In fresh Feta cheese, the predominant natural lactic microflora consists of the starter cultures, which are gradually replaced by salt resistant nonstarter lactic acid bacteria (NSLAB), consisting of mesophilic lactobacilli (mainly *L. plantarum*), pediococci (mainly *P. pentosaceus*), *Leuconostoc* species, and enterococci (mainly *E. faecium* and *E. durans*). The presence of yeasts (mainly *S. cerevisiae*), micrococci and coliforms in low numbers has also been reported (74–81). Domiati cheese is mainly made from buffalo's milk, cow's milk, or both, but also from ewe's and goat's milk. Large quantities of NaCl (8–15%) are added to milk before salting. This results in the need for more rennet and a longer coagulation time than for other cheeses, while the curd is usually weak. It is made either without or with the addition of starters, which are usually mixed cultures of lactococci, streptococci and lactobacilli (82). Enterococci (*E. faecium, E. faecalis* and *E. durans*), lactococci (*L. lactis* spp. *lactis* and *L. lactis* spp. *cremoris*), lactobacilli (*L. casei*, *L. plantarum*, *L. brevis* and *L. fermentum*) and streptococci are the most common microorganisms isolated from Domiati cheese, however enterococci are predominant among them. Lactococci dominate during the early stages of pickling and later lactobacilli (83,84). Several nonlactic acid bacteria have also been isolated, especially micrococci (*M. luteus*), yeasts (*Trichosporon, Saccharomyces, Pichia, Debaryomyces, Hansula, Torulopsis, Endomycopsis*, and *Cryptococcus*), and molds (*Aspergillus, Penicillium, Cladosporium* and *Geotricum*) (85–87). Turkish white cheese (Beyaz peynir) was manufactured initially from ewe's or goat's milk, but cow's milk or a combination of milks is now generally used for its production. In the traditional manufacture no starter cultures are used, and the milk may or may not be pasteurized, while today different mixed starters including thermophilic and mesophilic bacteria are used (88). At the beginning of ripening, *L. lactis* spp. *lactis* is the predominant species, while enterococci (*E. faecalis* and *E. faecium*) are the second most important group. *L. casei, L. plantarum, L. fermentum* and *L. brevis*, *Leuconostoc lactis* and *Leuconostoc mesenteroides* are also present. Lactoccoci decline during ripening, lactobacilli increase, and enterococci remain relatively constant, while staphylococci, micrococci and coliforms decrease (89,90). Halloumi, the traditional cheese of Cyprus, which appears to be a hybrid between the high salted Feta and the Pasta Filata varieties common in the Middle East, has received little attention to date. It is a semihard cheese preserved in brine and made from ewe's milk or mixtures of ewe's and goat's milk (91). In mature Halloumi cheese made either from bovine or ewe's milk, *E. faecium*, which dominated the microflora of the fresh ewe's cheese, was replaced by lactobacilli, including a new species, *Lactobacillus cypricasei* that was not found in the bovine samples. Yeasts counts in mature ewe's and bovine cheeses reached 2.3 – 2.8 10^5 cfu/g. Some of the yeasts were proteolytic or lipolytic and it was assumed that they were having a positive impact on the flavor of cheese (92).

Whey cheeses are produced from whey by coagulation via acidification or a combination of acid and heat. They are soft, unripened cheeses, ready for consumption once the manufacturing operations are complete (93). Ricotta in Italy, Mizithra, Anthotyro, and Manouri in Greece, Anari in Cyprus, and Karichee in Lebanon belong to this type of cheese. Most are produced on a small scale and enjoy a limited market, and published data on these cheese varieties are scarce.

20.2.2.2 Fermented Milks

According to Robinson and Tamime (94) fermented milks can be divided into three main categories based on the metabolic products: lactic fermentations (such as Yogurt, Laban, and Bulgarian buttermilk); yeast lactic fermentations (Kefir and Koumiss); and mold lactic fermentations (Villi). Closely related products are manufactured from fermented milks by dewheying to concentrate the product, which could resemble soft cheese (Labneh); drying of cereal fermented milk mixture (Kishk or Trahana); and freezing fermented milk to resemble ice cream (95,96). The organoleptic qualities of the fermented milks result from the multiple fermentations of milk constituents that take place. Bacteria, yeasts, and molds or combinations of these carry out fermentations. The main fermentation product is lactic acid from lactose. This is responsible for the sharp, refreshing taste of all fermented milks. It also serves as a background for the most distinctive flavors and aromas characteristic for each product. Fermented milks dominating in the Mediterranean basin are the lactic acid fermented milks, while cereal fermented milk mixtures are very popular in the Eastern part of the Mediterranean Sea.

Yogurt is the most important lactic acid fermented milk. Yogurt and yogurtlike products that appear in the Mediterranean basin include Cieddu, Mezzoradu, and Gioddu in Italy; Jiaourti in Greece; Jugurt and Ayran in Turkey; Leban or Laban in Lebanon; and Zabady in Egypt (97). Besides cow's milk, ewe's, goat's and buffalo's milk have also been used for the manufacture of yogurt. The fermentation of milk takes place at 40–45°C. The souring of milk results from the paradigmatic cooperation between *Lactobacillus delbruekii* spp. *bulgaricus* and *Streptococcus thermophilus*, two thermophilic lactic acid bacteria, which traditionally comprise the typical microflora of yogurt (96). Initially, production was confined to natural yogurt, and the market was limited to those who believed that yogurt is beneficial to health. Gradually however, attitudes towards yogurt changed and the advent of fruit yogurts during the 1950s gave the product an entirely fresh image, and it became a popular and inexpensive snack food or dessert. The method of producing yogurt has in essence changed little over the years. The fact that all commercial yogurts share a common core has lead to the word yogurt being applied to a whole range of products, such as Dried Yogurt, Frozen Yogurt and even Pasteurized Yogurt. The inclusion of these varieties under the banner of yogurt offends some people, because yogurt per se must by virtue of the process contain an abundance of viable bacteria originating from the starter culture. Presently, other lactic acid bacteria, mainly *Lactobacillus* spp. such as *L. casei*, *L. acidophilus* and *L. rhamnosus*, as well as *Bifidobacterium* species, such as *B. bifidum*, *B. infantis* and *B. longum*, or combinations of these two groups are widely used for the production of the so called probiotic yogurts.

Natural yogurt is prone to spoilage during storage, especially at ambient temperatures, because of the high content of water (85%). Traditional and advanced methods have been used to remove yogurt whey for the manufacture of strained yogurt. Traditionally, yogurt is strained in a special cloth bag, while modern methods include ultrafiltration and centrifugation (98,99). In the Middle East countries simple techniques are traditionally applied. Concentrated yogurt is produced either by the daily addition of milk to the yogurt hanging in goat's or sheep's skin, or by the addition of various quantities of salt. In Turkey it is called Torba, Kurut or Tulum, while in the Arab countries Labneh or Lebneh. In Egypt, yogurt is placed in earthenware vessels and condenses to Leben Zeer. In Lebanon, salted yogurt is made into small balls, placed in the sun to dry and then kept in either glazed earthenware pots or glass jars and covered with olive oil (Labneh Anbaris). Chanklichis similar to Labneh Anbaris, but it is enriched with herbs and spices, presumably to assist preservation. Both are consumed with bread and olive oil as appetizers. Production of smoked yogurt involves the heating of yogurt for a few hours over low fires of a special type of wood; the yogurt is then preserved over the winter months in jars covered either with olive oil or tallow (97,100). Drinking yogurt is also very popular in the Middle East countries. Yogurt is stirred, cooled, and finally diluted with an equal volume of water. Sometimes, it is flavored with salt or even flavoring extracts such as mint. In Turkey the product is called Ayran (97).

Dry forms of yogurt and cereal mixtures, known mainly as Kishk in the Arab countries and Trahana in Greece and Turkey, are very popular in the East Mediterranean countries. The methods employed for manufacture may differ from one region to another, because they are based on tradition. Milk from different species (cow, sheep, goat, buffalo, and camel) may be used for production of natural yogurt. The cereal fraction is crushed wheat, wheat flour, semolina, or parboiled wheat, known locally as bourghol (Arab countries) or bligur (Greece and Turkey). Natural or concentrated yogurt is mixed with the cereal fraction and the mixture is shaped into small rolls and placed in the sun to dry. It is sold either as rolls or in a ground form as flour. Kishk and Trahana as dishes are prepared by reconstituting the yogurt and cereal mixture with water and then simmering the mix gently over a fire. The consistency of this product, which is normally consumed with bread, is rather similar to porridge (97,101). There is no information in the literature on the starter cultures used in the preparation of Kishk or Trahana. It may be assumed that *L. bulgaricus* and *S. thermophilus* used for the production of yogurt constitute the major part of this microflora, while the indigenous microflora of milk and cereals may also play a significant role, depending on the technology applied. No improvement of the quality of Kishk was observed when *L. acidophilus* and *L. casei* were added to the yogurt starter culture (102). El-Sadek et al. (103) determined spore formers (*B. licheniformis, B. subtilis* and *B. megatherium*) to be the major part of the microflora in Egyptian Kishk, followed by lactic acid bacteria. Proteolytic and salt tolerant microorganisms were detected in appreciable numbers (3.9×10^6 cfu/g) in samples of Egyptian Kishk, along with *Bacillus subtilis*, *B. polymyxa*, *B. coagulans* and *B. cereus*, while yeasts and molds were also present ($9.0 \times 10^1 - 2.5 \times 10^3$ cfu/g) (104). Lactic acid bacteria isolated from Trahana were found to belong to the genera *Lactobacillus*, *Leuconostoc*, *Pediococcus* and *Enterococcus*. *Lactobacillus plantarum* was the most frequently found species, followed by *S. faecalis*, *L. mesenteroides* and *Lactobacillus brevis* (105).

20.2.2.3 Biochemistry of the Fermentation Process

The first essential step in milk fermentations is the catabolism of milk lactose by the lactic acid bacteria. Lactic acid bacteria as a group exhibit an enormous capacity to degrade different carbohydrates and related compounds. The main end product is lactic acid (more than 50% of sugar carbon). It should be noticed, however, that lactic acid bacteria adapt to various conditions and change their metabolism accordingly. This may lead to significantly different end product patterns.

Lactose fermentation is by far the most studied disaccharide metabolism in lactic acid bacteria. As shown for *Lactococcus lactis* and *Lactobacillus casei*, lactose is taken up via the phosphoenolpyruvate dependent phosphotransefase system (PTS) and enters the cytoplasm as lactose phosphate (106,107). Lactose phosphate is cleaved by phospho-β-D-galactosidase (P-β-gal) to yield glucose and galactose-6-phosphate, which are further catabolized. The enzyme systems of lactose PTS and P-β-gal are generally inducible, and repressed by glucose (108). An equally common way for lactic acid bacteria to metabolise lactose is by means of a lactose carrier (permease) and subsequent cleavage by β-galactosidase (β-gal) to yield glucose and galactose, which may again enter the major pathways (109,110).

For glucose fermentation two major pathways occur in lactic acid bacteria. The Embden-Meyerhof-Parnas pathway (glycolysis) is used by all lactic acid bacteria except leuconostocs, group III lactobacilli, oenococci, and weisellas. It is characterized by the formation of fructose-1,6-diphosphate (FDP), which is split by the FDP aldolase into dihydroxyacetone phosphate (DHAP) and glycerinaldehyde-3-phosphate (GAP). GAP (and DHAP via GAP) is then converted to pyruvate in a metabolic sequence including substrate level phosphorylation. One mole of glucose results in 2 moles of lactic acid and a net gain of 2 moles of ATP. The glycolysis pathway is used by the homofermentative lactic acid bacteria. The other main fermentation pathway is the pentose phosphate pathway. The key step is the phosphoketolase split of xylulose-5-phoshate to GAP and acetyl phosphate. GAP is then converted to lactate, while acetyl phosphate is converted to acetate and ethanol. This pathway is used by the heterofermentative lactic acid bacteria. The heterolactic fermentation gives 1 mole each of lactic acid, ethanol, and CO_2 and 1 mole of ATP per mole of glucose. It should be noted that glycolysis may lead to a heterolactic fermentation (meaning significant amounts of other end products besides lactic acid) under certain conditions, and some LAB, regarded as homofermentative, use the pentose phosphate pathway when metabolizing certain substrates (111). Regarding galactose, the galactose-6-phosphate formed by PTS is metabolized through the tagatose-6-phosphate pathway (112). The Leloir pathway is used by galactose fermenting LAB, which transport galactose with a permease and lack of galactose PTS (113).

Citrate, which is present in milk in concentrations around 9 mM, can also serve as an energy source for lactic acid bacteria. It is generally accepted that next to lactose, citrate metabolism plays an important role in milk fermentations. The ability of lactic acid bacteria to metabolize citrate is invariably linked to an endogenous plasmid that contains the gene encoding the transporter, which is responsible for citrate uptake from the medium (114). Within the cell, citrate is initially converted by the citrate lyase to acetate and oxaloacetate. Oxaloacetate in then decarboxylated into pyruvate. With involvement of the intracellular enzyme pool, pyruvate may be then converted to acetyl CoA (via the pyruvate dehydrogenase complex), which leads to acetate (via acetate kinase) and acetaldehyde or ethanol formation (via alcohol dehydrogenase); to formate (via pyruvate formate lyase); to α-acetolactate (via acetolactate synthase), which leads to acetoin (via acetolactate decarboxylate), and diacetyl and 2,3-butanediol (via diacetyl/acetoin reductase); and finally to lactate (via lactate dehydrogenase). The energy is mostly generated from the conversion of acetyl CoA to acetate, meaning that citrate acts as an electron acceptor, resulting in a higher production of acetate and ATP, probably via the acetate kinase pathway. Additional energy is produced during the initial breakdown of citrate into pyruvate (115). Furthermore, recent studies performed with *L. lactis* ssp. *lactis* var *diacetylactis* (114,116) and *Leuconostoc oenos* (117) indicated that the uptake of citrate is coupled to the generation of a proton motive force, which was shown to be strong enough to drive the additional ATP synthesis.

Some of the products of citrate catabolism, such as diacetyl, acetaldehyde, and acetoin, have very distinct aroma properties and influence significantly the quality of fermented foods. For instance, diacetyl determines the aromatic properties of fresh cheese, fermented milk, cream, and butter (118). The breakdown of citrate results as well in the production of carbon dioxide, which can add to the texture of some fermented dairy products (119). Not all lactic acid bacteria are able to metabolize citrate (120). Strains of *Lactococcus lactis* and *Leuconostoc* species have been extensively studied in respect to citrate metabolism and production of aroma compounds (119,121,122). Data also exist on citrate metabolism by *Enterococcus* strains (123–126). In many dairy products originated form the Mediterranean area enterococci comprise a major part of the fresh cheese microflora, and in some cases they are the predominant microorganisms in the fully ripened product.

The metabolism of lactate and sugars by propionic acid bacteria has been recently reviewed by Piveteau (127). Dairy propionic acid bacteria have the ability to utilize a variety of substrates for energy. Lactate, produced by lactic acid bacteria, is the main energy source for propionic acid bacteria in cheese. As early as the nineteenth century, the stoichiometry of the reaction resulting in the production of propionate from lactate was established by Fitz: 3 lactate = 2 propionate + 1 acetate + 1 CO_2 and 1 H_2O. However, in Swiss type cheeses, these theoretical equations are rarely found and the relative concentrations of propionate, acetate, and CO2 may be significantly different from the expected molar ration of 2:1:1 (128,129).

Proteolysis is considered the most complex of the three primary events during cheese ripening, the other two being lactose fermentation and lipolysis. It is important for flavor and texture development, and has therefore been a popular research subject. The main proteolytic agents in cheese manufacture are the rennet, the indigenous milk protease, and the proteolytic enzymes of starter and nonstarter microflora. The role of animal rennet, which consists of chymosin and pepsin, in milk clotting is well documented. The split of κ-casein at the Phe105 - Met106 bond to para-κ-casein and the glycomacropeptide is the first step for destabilizing casein micelles and subsequent milk coagulation. The specificity of calf chymosin on α_{s1}-, α_{s2}- and β-caseins has been established and the results can, largely, be extended to cheese (130,131). Gastric proteinases from calves, kids and lambs have been used traditionally as rennet. However, plant proteinases appear to have been used as rennet for hundreds of years as well. Extracts from dried flowers of *Cyanara carduncculus* have been employed successfully for many centuries in Portugal and in some regions in Spain for the production of cheeses like Serra da Estrella and La Serena. A few experiments have also been carried out pertaining to the manufacture of French cheeses like Camembert and Gruyere (132), and Italian cheeses like Grana, Provolone, and Bel Paese (133) using plant rennet. However, problems have been encountered, such as slightly more acid and bitter flavors and softer curd, with concomitant increased tendencies for loss of shape. *Cyanara carduncculus* proteinases acting upon casein show preference for bonds between bulky regions. They are claimed to be similar to chymosin in terms of specificity and kinetic parameters, but they produce higher levels of the ripening index than commercial animal rennet (10). Plasmin is the indigenous milk proteinase, and it is a trypsin like serine proteinase. Its complex consists of the active enzyme (plasmin), its zymogen (plasminogen) and plasminogen activators, all associated with the casein micelles, while the enzyme inhibitors are in the serum phase. The enzyme is active on all caseins, especially α_{s2}- and β-, but it shows very low activity on κ-casein. Plasmin activity in cheese is clearly indicated by the formation of γ-caseins from β-casein during ripening (130). Pasteurization increases plasmin activity in milk, probably because of the inactivation of plasmin inhibitors or by increasing the rate of plasminogen activation (134). In hard cooked cheeses like Swiss-type cheeses or the Italian grana cheeses, plasmin is mainly responsible for the primary hydrolysis of caseins, because rennet is almost completely inactivated during cooking and has only a weak action on caseins (135).

It is a general belief that lactic acid bacteria have limited abilities to synthesize amino acids, which are essential for their growth, and milk contains insufficient amounts of free amino acids and low molecular mass peptides to sustain growth (136,137). Although they are considered as weak proteolytic bacteria compared to other groups of microorganisms, it has been shown that lactic acid bacteria posses a complex proteolytic system capable of hydrolysing milk proteins to peptides and amino acids (138,139). Furthermore, it is generally accepted that their proteolytic system contributes to the degradation of milk proteins and hence to the texture, taste, and aroma of dairy fermented products (140–142). The most extensively studied proteolytic system is that of *Lactococcus lactis*, and it serves as a model for all lactic acid bacteria. An extracellular, membrane anchored serine proteinase (PrtP) is an essential component of this system. PrtP exists in at least two variants in lactococci with somewhat different specificities in the degradation of milk casein. The products of this initial casein degradation (amino acids and peptides) are transported into the cell by transport systems specific for amino acids (113), two di- and tripeptides (DtpT and DtpP) (143) and an oligopeptide transport system (Opp) accepting four to eight residue peptides (144). Inside the cell, several peptidases with a wide range of specificity complete the degradation (145). By analysing the peptide fraction in milk during growth, it was calculated that oligopeptides small enough to be accepted by the Opp system represent 98% of the nitrogen source for growth of lactococci in milk (146). Proteolytic activity by LAB is quite common, but only in a few cases has the system been thoroughly analysed (147). The gene for the PrtP-type proteinase of *Lactobacillus paracasei* has been cloned and sequenced (148). The derived amino acid sequence showed a very high degree of similarity to the lactococcal counterpart, indicating that the system in this species might be similar.

In recent years, it has become clear that a number of enzymes are involved in the conversion of amino acids to flavor components. These enzymes may catalyze reactions such as deamination, transamination, decarboxylation, and cleavage of the amino acid side chain. Branched chain amino acids can be transaminated to keto acids, which then undergo either spontaneous degradation or are enzymatically converted to the corresponding aldehydes or carboxylic acids (149–151). Most of these activities have been demonstrated *in vitro* for lactic acid bacteria, but how and whether they proceed *in vivo* is not always easy to establish. For instance, 3-methyl butanal originating from leucine has been characterized as an important volatile compound formed during the ripening of Parmesan cheese, which is responsible for a spicy cocoa flavor (152). It may have positive as well as negative effects on the sensory perception of cheese. Its contribution to the overall perception of cheese probably depends on the other volatiles present and the composition of the matrix.

The proteinases of propionic acid bacteria have been recently reviewed by Langsrud et al. (153), and they appear to be of modest importance in the breakdown of milk caseins. Their growth in cheese seems to depend on the primary hydrolysis of caseins by lactic acid bacteria. However, dairy propionibacteria have high peptidase activities, especially proline specific activities, which may be involved in the hydrolysis of the peptides produced from the initial breakdown of caseins (154–156).

Penicillium spp. play a major proteolytic role in mold ripened cheeses. Both *P. camemberti* and *P. roqueforti* synthesize a metalloproteinase, an aspartate proteinase as well as an acid carboxypeptidase and an alkaline aminopetidase. Moreover, *P. roqueforti* synthesizes one or more alkaline carboxypeptidases and some strains also produce an alkaline proteinase (28).

Yeasts are involved both directly and indirectly in the ripening of cheese (157,158). Assimilation of lactate, formation of alkaline metabolites, liberation of bacterial growth factors, fermentation of lactose, lipolysis, proteolysis, and formation of aroma compounds are some of the yeast activities which are considered important for the typical characteristics of the smear surface ripened cheeses (27,157,159).

Geotrichum candidum has properties of both yeast and molds and presently is considered to be a yeast. It is generally believed that *G. candidum* is found in all smear and mold ripened cheeses, although recent literature does not support this conclusion (22,160). The precise role of *B. linens* in the ripening of smear surface ripened cheeses is not known but the organism certainly makes a major contribution. *Brevibacterium linens* has an active proteolytic system, especially with respect to peptidase activity (161).

Lipolysis is among the principal events occurring during cheese ripening. The free fatty chain length can vary considerably depending on the type of milk fat and the hydrolytic agent present. Free fatty acids can be further converted to methyl ketones, lactones, thioesters, keto acids, and hydroxy acids, which contribute along with the free fatty acids to the flavor of the ripened product, while the volatile short chain fatty acids are responsible for the rancid flavor of milk (162). The main lipolytic agents in cheese include the indigenous milk lipoprotein lipase, lipases, and esterases produced by the starter and nonstarter bacteria, and depending on the cheese variety, enzyme preparations added during manufacturing. Milk lipase causes significant lipolysis in raw milk cheese, but it is inactivated during pasteurization at 76°C for 10 sec. It is selective for fatty acids at the sn-3 position, where most of the butyric acid in milk fat is esterified, explaining thus the unbalanced concentration of free butyric acid in cheese (161). In most cheese varieties relatively little lipolysis occurs, with the exception of mold ripened cheeses. The lipases produced by *P. roqueforti* and *P. camemberti* are the main contributors, while lipolytic yeasts and micrococci also play a role. Despite the interstrain variations, both *Penicillium* species produce very active extracellular lipases, which have been isolated and well characterized. Lipolysis in these cheeses does not, however, result in rancidity possibly due to the neutralization of the pH during ripening (21,28). The aroma and flavor of blue veined cheeses is dominated by methyl ketones, the main ones being 2-heptanone and 2-nonanone. They are formed via the β-oxidation pathway, and their concentration is proportional to lipolysis. Reduction of methyl ketones leads to the formation of the respective alcohols, which are present in appreciable concentrations, especially 2-pentanol, 2-heptanol, and 2-nonanol, in the blue veined cheeses (163). Another lipolytic agent used is the pregastric esterase (PGE), which is present in the rennet used for the production of some Italian cheeses (e.g., Provolone). It also shows high specificity for the short chain fatty acids esterified in the sn3 position and it is responsible for the characteristic piquant flavor of these cheeses (164). Lactic acid bacteria are generally considered low in lipolytic activity. It is true that their lipolytic system has not been extensively studied compared to the proteolytic system, although there are several reports on lipases and esterases produced by lactic acid bacteria (165–170). Esterases and lipases have been also been isolated and characterized from the cheese surface bacteria *B. linens* (171) and *Anthracobacter nicotiniae* (172), as well as from propionibacteria (173,174). Vital information on the cellular location of all these enzymes, their specificity, and their overall role during ripening remains to be determined.

20.2.2.4 *Effect of Fermentation on Nutritional Quality*

Milk and fermented dairy products are considered as foods of high nutritional value. The nutritional importance of dairy products arises from their high content of biologically valuable proteins. In cheese manufacture, the milk casein is incorporated in the cheese curd, while most whey proteins pass into the whey. Because the whey proteins are nutritionally superior to casein, which is deficient in sulfur amino acids, the biological value of the proteins in cheese is therefore lower than that of the total milk proteins. Concerning fat content, in recent years there has been worldwide a tendency for low fat milk and cheese, because they are perceived as healthy due to their low fat content. However, many consumers prefer high fat cheese because a high fat content contributes significantly to the flavor characteristics. However, the cholesterol content of cheese is rather low (up to 100 mg/100 g depending on cheese variety), and thus cheese cholesterol contributes only 3–4% to total cholesterol intake. Compared to milk, fermented products are superior for diets of persons suffering from lactose intolerance and diabetes. During cheese manufacture most of the milk lactose passes to the whey, and that retained in the cheese curd is partially or fully converted to lactic acid during cheese ripening. Cheese usually contains both lactic acid isomers, L(+) and D(-), depending on the type of starter used. During storage, L (+) lactate can be converted to D (-) by the nonstarter lactic acid bacteria. Humans can metabolize D (-) lactate only to a certain extent by the nonspecific mitochondrial enzyme D-2-hydroxy acid dehydrogenase. The World Health Organization (WHO) recommends D (-) lactic acid free diet for infants up to one year old, while there is no limit for adolescents or adults. Finally, milk and its products frequently comprise for humans the major source of minerals like calcium and phosphorus, and they are important sources for some vitamins of group B, especially riboflavin (vitamin B_2), while the fat soluble concentrated vitamins depend on the fat content (175,176).

Besides this classical positive impact of dairy products on human diet and thus health, presently a big discussion is taking place on the therapeutic properties of fermented milks. Lactic acid bacteria involved in dairy fermentations are known to contribute various desirable factors in addition to the physicochemical and organoleptic properties of the final product. These food grade and safe microorganisms have been used in food and feed preservation for centuries because they can produce a variety of antimicrobial agents, including organic acids like lactic, acetic and formic acid, and ethanol, carbon dioxide, diacetyl, and hydrogen peroxide (177). Additionally, several strains of lactic acid bacteria produce bacteriocins, which are peptides or proteins with an antibacterial activity against bacteria closely related to the producer strain. Furthermore, some of them inhibit spoilage bacteria and food borne pathogens including *Listeria monocytogenes*, *Staphylococcus aureus*, and *Clostridium botulinum*, and thus may be of great interest to the food industry as natural food preservatives (178,179). Problems concerning low production levels and instability of bacteriocins in certain food environments still need to be addressed. However, it is generally accepted that the use of bacteriocin producing cultures in food is of considerable advantage over using purified bacteriocin preparations. The latter application requires extensive and costly purification schemes, toxicology tests, and may suggest a nonnatural image of the additives, for instance with respect to the applied concentrations (180). No bacteriocin activity was detected during ripening of Feta cheese made with the antilisterial *E. faecium* FAIR-E 198 as adjunct starter (181). When crude enterocin CRL35 was added in goat's milk cheese deliberately inoculated with *L. monocytogenes,* the presence of the enterocin could not be detected in cheese during ripening; however, the growth of the pathogen was drastically inhibited (182). On the other hand, when a nisin producing *L. lactis* strain was used as starter in Camembert cheese contaminated with *L. monocytogenes*, the population of the latter increased upon prolonged ripening, especially on the surface (183). This was attributed to the probable nisin adsorption to fat and proteins, the lower nisin solubility at high pH values and the breakdown of the molecule by fungal proteases. It has also been reported that the use of a bactericinogenic *L. lactis* starter for Mozzarella cheese preparation resulted in the apparent death of *L. monocytogenes* after 24 h of storage, but after 7 days of storage revival and growth of this pathogen was observed (184). However, some successful examples of *in situ* bacteriocin production during cheese making have been described.

Giraffa et al. (185) showed that enterococci were suitable for producing bacteriocins in milk, in the presence of rennet, during an incubation period that simulated the first 55 h of Taleggio cheese making (186) and concluded that *E. faecalis* INIA 4 was able to produce enterocin in competition with a milk native microflora during the manufacture of Manchego cheese. Furthermore, an increase in enterocin activity was recorded in the cheese during the first week. During Taleggio cheese manufacture and ripening, Giraffa et al. (187) showed that bacteriocin production by *E. faecium* 7C5 started during whey drainage and generally reached a maximum amount just after cheese salting. The stability of the enterocin during ripening was not affected, although the heterogeneous microflora present in raw milk, as well as the enzymatic activity of rennet, were capable of inactivating the inhibitor. Furthermore, *E. faecium* 7C5 inhibited the growth *of L. monocytogenes* on the surface of Taleggio cheese (188).

The decarboxylation of free amino acids during cheese ripening produces amines. Biogenic amines, such as tyramine and histamine, exhibit biological activity, and the consumption of food containing high concentrations of these compounds may cause toxic effects for susceptible individuals. Biogenic amines are also of concern in relation to food spoilage. After fish, cheese is the next most commonly implicated food associated with histamine poisoning (189,190). The concentrations of individual amines in cheese show great variation and depend on the quality of the initial milk, the ripening period, the storage conditions, the level of proteolysis, the pH, the salt content, the microbial flora, and the synergistic effects between microorganisms (189,191–193). Among decarboxylase positive microorganisms, many strains of *Enterobacteriaceae* and certain lactobacilli, pediococci, and enterococci are particularly active (194).

Among biogenic amines, the predominant presence of tyramine has been reported for Idiazabal cheese (195), for Feta cheese (196), and for Beyaz peynir (197). Tyramine has also been found in Camembert, Parmesan, Romano, Roquefort, and Swiss-type cheeses (189). In Feta cheese, it appeared that the low pH and the high salt content did not create favorable conditions for amino acid decarboxylation, keeping the total level of biogenic amines relatively low (617 mg/kg in the 120 days mature cheese) (196). Storage of Azeitao cheese at 25°C promoted a significant increase in the tyramine and putrescine levels (198). In a survey of commercially available processed cheeses in Egypt, tyramine was detected in 70% of the samples at an average concentration of 58.7 mg/Kg, while no histamine was detected in any of the samples (199), but Ras cheese may contain appreciable amounts of biogenic amines (200,201).

In recent years, there has been much interest in the use of various strains of lactic acid bacteria as probiotics. Probiotics are broadly defined as "living microorganisms, which upon ingestion in certain numbers, exert health effects beyond inherent basic nutrition" (202). Most of the facts on probiotic bacteria are based on anecdotal reports and poorly controlled studies, which make the work inconclusive and recommendations difficult. However, evidence is currently accumulating from well designed, randomized and placebo controlled double blind studies indicating that a few well characterized lactic acid bacterial strains have documented health promoting effects when defined doses are administered. It should be noticed, however, that each probiotic strain should be tested on its own merit, because even closely related strains may behave differently in the human intestine (203).

Lactobacillus species (mainly *L. acidophilus* and *L. casei*), *Bifidobacterium* (*B. longum* and *B. infantis*) and *Enterococcus* (mainly *E. faecalis*) comprise the main genera to which the most commonly used probiotic bacteria belong. This is probably due to the perception that they are desirable members of the intestinal flora (204), while at the same time these bacteria have been traditionally used in the production of fermented dairy products and have the GRAS (generally regarded as safe) status (205). However, many probiotic organisms currently employed in the dairy food industry are not of human origin and, therefore, do not meet the criteria for the selection of probiotic microbes acceptable for human consumption (206).

Several publications have reviewed the area of physiological activity, mechanisms, and clinical studies (206–208). It seems that the key points of accepted and potential benefits are: aid in lactose intolerance, resistance to enteric pathogens, anticolon cancer effect, small bowel bacterial overgrowth, immune system modulation, allergy, blood lipids and heart disease, antihypertensive effects, and protection from urogenital infections, infections caused by *Helicobacter pylori,* and hepatic encephalopathy (206–208). Probiotic containing products appear in a variety of formats: conventional foods (probiotic containing yogurts, fluid milk, and cottage cheese, consumed primarily for nutritional purposes, but also for their probiotic benefits), food supplements or fermented milks [food formulations whose primary purpose is as a delivery vehicle for probiotic bacteria and their fermentation end products, consumed for health effects, in monoculture (Yakult, Japan) or mixed cultures (LC1, Nestle, CH; Actimel, Danone, F and others)], and dietary supplements (capsules and other formats designed to be taken by healthy individuals looking to enhance health) (209).

In the last two decades a number of studies have been performed on bioactive peptides present in the amino acid sequence of milk proteins. Although other animal and plant proteins contain potential bioactive sequences, milk proteins are currently the main source of a range of biologically active peptides. The structures of biologically active sequences were obtained from *in vitro* enzymatic and by *in vivo* gastrointestinal digestion of the appropriate precursor proteins (210,211). Bioactive peptides produced from milk proteins, especially from β-casein (β-casomorphins) are potential modulators of various regulatory processes in the body: Opioid peptides are opioid receptor ligands, which can modulate absorption processes in the intestinal tract, angiotensin-I-converting enzyme (ACE) inhibitory peptides are hemodynamic regulators and exert an antihypertensive effect; immunomodulating casein peptides stimulate the activities of cells of the immune system; antimicrobial peptides kill sensitive microorganisms; antithrombotic peptides inhibit aggregation of platelets; and caseinophosphopeptides may function as carriers for different minerals, especially calcium. Bioactive peptides can interact with target sites at the luminal side of the intestinal tract. Furthermore, they can be absorbed and then reach peripheral organs. Food derived bioactive peptides are claimed to be health enhancing components that can be used for functional food and pharmaceutical preparations (212). Recently, peptides inhibitory to ACE and endopeptidases and aminopetidases of lactic acid bacteria and *P. fluorescens* were isolated from Italian cheeses such as Mozzarella, Italico, Crescenza, and Gorgonzola (213).

20.3 FERMENTED MEATS

20.3.1 Introduction

Fermentation and drying of meat products are the most ancient ways of food preservation in the Middle East and Mediterranean region. The two processes are mentioned together, because in practice they are impossible to separate (214). The production of fermented meats is thought to have originated in the countries surrounding the Mediterranean Sea, as the climate, especially during the winter months, with fairly low temperatures and low relative humidities, is suitable.

We find proof of ancient sausage production in verse 43 of the eighteenth Rhapsody from Homer's The Odyssey. Sausages, described as "...goatpaunches...filled with blood and fat," were offered by Antinoos, son of Eupeithes, to the bravest suitors. Although the historical origin of fermented sausage still remains unknown, the Mediterranean sausage manufacturers were among the first to fill fresh meat into intestines and preserve the product by adding salt and drying. The Romans knew that ground meat with added salt, sugar, and spices turns into a palatable product with a long shelf life if prepared and ripened properly (215). Etymological investigations have shown that there are many linguistic modifications of the word sausage. The word salami may be derived either from the Greek island Salamis or the ancient city Salamis on the east coast of Cyprus; or it may be originated from the Italian word for salt, *sale* (216–218).

In early times people became aware of the preserving effects of salt and drying, and therefore over the centuries were able to develop many cured products. The term "curing" in the Mediterranean region means a long ripening (ageing) process, where complicated biochemical, proteolytic, and lipolytic modifications take place that are responsible for the distinctive flavor of the products (219,220). Fermented sausages are usually made from comminuted meat and fat mixed with salt, curing agents, sugar, and spices and filled into casings (215). Although there are too many variants, some characteristic examples are the Italian salami with all its variations, the French saucisson sec, the Spanish chorizo and salchichon, the Greek fermented sausage, the Turkish dry sausage (soudjuk or sucuk) and similar products in most Middle Eastern Countries. In most Mediterranean countries fermented sausages are produced with pork and beef, while Italian salamis are manufactured from pork only, and Turkish sucuk from cow and sheep meat. The products are often heavily spiced as in the case of sucuk, which contains various seasonings, i.e., garlic, black and red pepper, cumin, and cinnamon (221).

Dry cured ham is another cured meat product. The production of uncooked ham and similar cured products is confronted with a variety of extraordinary handicaps, which make it difficult to obtain a uniform quality product. There are national and local specialties as a consequence of variations in the use of raw materials, formulations, and manufacturing processes, which come from habits and customs of the different countries and regions (219). Some characteristic products are the Italian Parma and San Daniele hams, Spanish Iberian and Serrano hams, and French Bayonne and Corsican hams. A typical Mediterranean dry cured product made of beef is the Greek and Armenian Basturma or Turkish Pastirma or Pasterma; the meat is covered with a layer of a paste (3–5 mm thick) consisting of helba, garlic, salt, paprika, and pepper (222). Presently, in an effort to improve the safety and quality characteristics of the products, starter cultures are used.

Traditional intermediate moisture meat products are produced in the Middle East and Mediterranean region by salting and drying. The preservation of meat by salting and sun drying, a practice dating back several centuries in the Middle East, results in a product called kaddid. The successful preservation of this product is the result of lowering the water activity (a_w), salting, and spice flavoring (adding coriander, garlic, and pepper). The product is characterized by a strong flavor due to fat lipolysis and probably free fatty acid oxidation. Proteolysis may also help in flavor development by releasing amines and amino acids, which may have an influence on the organoleptic characteristics. Microbial populations of coliforms, enterococci, and staphylococci are reduced during processing, but staphylococci survive in the product, and the possibility of a foodborne outbreak exists (223).

20.3.2 Biochemical and Microbiological Changes in Fermented Meats

20.3.2.1 Fermented Sausage

The production of fermented meats is based on the enzymatic activity of certain microorganisms resulting in transformation and modification of the raw meat. Historically, the responsible microflora are accidentally introduced from the environment. Essentially two bacterial groups are responsible for the desired meat fermentations that lead to safe and stable products: the lactic acid bacteria and the catalase positive cocci (*Micrococcaceae*) (224). Gram-negative bacteria are usually present only in small numbers, which rapidly decline over time. Lactobacilli usually dominate the lactic acid bacteria microflora in naturally fermented sausages. The growth and metabolism of lactic acid bacteria is affected by the presence of oxygen. Usually the carbohydrates are metabolized via glycolysis. However, under certain conditions, the heterofermentative pathway is activated, resulting in undesirable flavor components, such as acetate (218). In the presence of oxygen, metabolites other than those found in anaerobic conditions may be observed. In addition to the formation of hydrogen peroxide, which may be formed during the aerobic metabolism of glucose, the yields of lactic acid, acetic acid, acetoin, and ethanol are also affected. *Lactobacillus plantarum,* which under anaerobic conditions mainly forms lactic acid from glucose, shows a dramatic increase in the production of acetic acid under aerobic conditions, together with small amounts of acetoin (224). Manganese is also required by lactic acid bacteria for various enzyme activities, including those of the Embden-Meyerhof pathway (225). In addition, some spices, such as red pepper, stimulate the rate of lactic acid formation by supplying manganese. Lactic acid bacteria also use manganese as a superoxide scavenger (which is presumed to be an intermediate in hydrogen peroxide formation) (108), which may explain the very high manganese requirements. During ripening of fermented sausage the main microbiological reactions that occur are the production of nitric oxide by nitrate and nitrite reducing bacteria, which is crucial for red color formation as a result of nitrosomyoglobin formed; and a decrease in the pH of the sausage emulsion via glycolysis by acidogenic microorganisms. When nitrate is used, it must first be reduced enzymatically to nitrite; the pH of the meat should be in the range of 6.0–5.4, because a stronger degree of acidity would prevent bacterial enzymatic activities. The competition of the microorganisms for the available nutrients and the pH dependent reduction of nitrate to nitrite are the two factors influencing these reactions. Nitric oxide formation and a decrease in pH are therefore the crucial points of sausage fermentation as the rate of nitric oxide myoglobin formation increases with falling pH and is therefore accelerated by the activity of the lactic acid bacteria in fermeted sausages. During nitrite inhibition, many homofermentative lactobacilli are found to release CO_2 as a result of the manganese increase in the cells and stimulation of the heterofermentative pathway (226). Under anaerobic conditions nitrates and nitrites serve as oxidizing agents for bacteria; catalase positive cocci occurring naturally in meat reduce nitrates into nitrites, affecting the formation of curing color in fermented sausages. However, peroxides formed by lactic acid bacteria in the presence of oxygen and peroxide groups present in fatty tissue can, at low pH values, attack the iron moiety in nitric oxide myochromogen resulting in discolorations (215). The nitrate reducing organisms, usually catalase positive cocci, make an important contribution to the flavor of the products and also produce catalase, which removes H_2O_2 produced by the lactobacilli that can cause color problems (227); and play an important role in forming and stabilizing the desired sensory properties of fermented sausages (215).

Relative humidity and air velocity in the drying chamber may affect the rate and intensity of acidification. With higher relative humidities and lower air velocities, rapid acidification may produce important defects because drying of the surface reduces the amount of free water available for bacterial growth. Good raw material and appropriate regulation of environmental conditions (temperature, relative humidity, and air velocity) are important factors affecting product quality. The fermentation of traditional meat products is based on the natural climatic conditions, and therefore the risk of faulty fermentations is rather high (218).

Typical Mediterranean-type products involve longer ripening periods, often without clear separation between fermentation and ageing. Smoke is not applied and acidulation to final pH values above 5 is slower (228). A slow curing process (up to 2 or 3 months) is given where nitrite is not usually used. Nitrate is added as sodium or potassium salts, which is transformed to nitrite by bacteria, with nitrate reductase activity naturally occurring in meat or added as starter cultures. Their flavor is directly related to the biochemical, proteolytic, and lipolytic mechanisms which take place during ripening (229). Shelf life is mainly determined by drying and lowered water activity (228).

The most important dry cured sausages in Spain are chorizo and salchichon (229). The fermentation process may be slow or fast, depending on the temperature used during ripening and the raw material. In the slow fermentation, temperatures of 15–16°C and nitrate for nitrification are used; nitrate is reduced to nitrite by bacterial nitrate reductase activity. The fast method involves the use of nitrite and exposure to high temperatures, between 22 and 25°C for 2 to 3 days depending on the diameter of the sausage, which favors development of the bacterial flora responsible for fermentation, consisting mainly of lactobacilli and micrococci in the case of chorizo (230). In modern industrial production of traditional products, an alteration of the traditional process using fast curing techniques results in a gradual loss of the typical sensory characteristics. The final result is a very intense acid taste, which is not acceptable to consumers (229).

20.3.2.2 Dry Cured Ham

A variety of processing practices have been developed throughout centuries for production of ham and similar cured products using different raw materials, ageing times, and manufacturing techniques, which have been adopted locally through individual experience and tradition. Traditional dry cured ham is produced by the rubbing on of salt, whereas the accelerated production processes include brine salting or brine injection pumping with the simultaneous addition of starter cultures (218). The main difference between these methods is the varying concentration of salt, the presence or absence of nitrate or nitrite, and the varying amounts of sugar and types of spices. Other distinctions are based on the degree to which the ham is smoked, aged, and dried in order to provide aroma and to ensure shelf life. In the long processed dry cured ham (the ripening period is 12 or more months, characterized as "slow process") the typical dry cured flavor is fully developed, whereas in the short processed dry cured ham (the ripening period has a minimum of 6 months, "rapid process") it is not (231). The lipolytic and oxidative degradation joined with the catabolism of amino acids produce volatile compounds which are responsible for the typical aroma of dry cured hams (220). The muscle aminopeptidases are the main contributors to the generation of free amino acids during the processing and contribute to flavor development (232). A common feature of Mediterranean raw ham manufacturing is a dry salting phase at low temperature (2–4°C) with a duration of 8 to 10 days followed by a post salting stage, during which the temperature is kept below 4°C for a period between 20 days and 2 months. The inhibition of spoilage microflora during the curing process makes limited storage possible (217). During the last ripening and drying stage, which has a duration from 6 to 24 months (for Serrano ham this stage takes about 9–12 months, while for Iberian ham it can be extended up to 18–24 months), the temperature is held between 14 and 20°C with the relative humidity decreasing from 90 to 70% (233). These temperatures set a limit to fermentation and the employment of starter cultures. During the last curing period the ham is frequently covered with lard in order to prevent excessive drying, whereas a natural mold covering can be developed (218).

The microbiological stability of dry cured ham is due to salt, curing agents, and the action of tissue enzymes with little contribution from microorganisms. In order to overcome the risk of growth of pathogens during the salting process, it is important to select meat pieces of normal pH ($\leq$ 5.8), otherwise injection of psychrotrophic lactic acid bacteria together with fermentable sugar is recommended (234). The large sizes of meat particles and the high salt concentrations have negative effects on growth and physiological activities of the microorganisms, but microorganisms play an important role in curing processes as they contribute to the appearance and flavor of raw hams by reducing nitrate and forming aroma precursors. *Micrococcaceae,* which comprise the predominant extrinsic flora, penetrate into hams, as do chlorides, during postsalting stages, and the technological characteristics of the process favor the development of *Micrococcaceae* on the deep muscular tissues of hams (235). Lactic acid bacteria, along with fermentable sugars, may be added to the injection brine in order to lower the pH value of the meat and to accelerate the salting process (236).

20.3.3 Starter Cultures

Although excellent fermented meat products can be produced without use of starter cultures (218), starters consisting of one (single starter) or several (mixed starter) strains are used today when manufacturing on a large industrial scale in order to standardize the product process and the quality of the product. Starter cultures are mixtures of lactic acid bacteria (*Lactobacillus*, *Pediococcus*), catalase positive cocci (*Staphylococcus, Kocuria* [formerly *Micrococcus*]), yeasts (*Debaryomyces*) and molds (*Penicillium*), depending on the final characteristics required, as they normally bring about and stabilize the desired sensory properties (237). Lactic acid bacteria are used less in Mediterranean countries because consumers reject the acid flavor and aroma (238); usually, they are used as mixed cultures in combination with catalase positive cocci. There are starter preparations for inoculating dry sausage emulsions as well as microorganisms for inoculating curing brines used for cured ham manufacture; and also for the surface inoculation of dry sausage and cured ham.

Metabolic activities of the microorganisms used as starters are essential for the desirable changes determining the particular characteristics of the fermented meat product. Lactic acid bacteria represent the most important group of starter organisms, as they are well adapted to the meat environment and are involved in all changes occurring during fermentation and ripening. The acidogenic bacteria transform carbohydrates into lactic acid through homofermentation. Lactic acid affects nitrate and nitrite reduction, which in turn influences meat color. Bacterial activity enhances gel formation, lending to a firmer consistency of the product as well as the development of aromatic compounds. Catalase positive cocci play an important role in sausage ripening by contributing to the development of the typical flavor as a consequence of their proteolytic and lipolytic activities. Catalase protects from color changes and rancidity, whereas nitrate and nitrite reductases aid in reddening and reduce residual nitrite content. The proteolytic activity of starter cultures affects the biogenic amine content of the product. Use of starter cultures guarantees safety (by control of pathogens or spoilage microorganisms), and is necessary to standardize product properties (consistent flavor and color, and improved texture).

Enterococcus strains isolated from different nonmeat origins and used as starter cultures in Spanish style dry fermented sausages improved the organoleptic properties of the product; so the enterococci may be suitable for addition to meat as cocultures to improve safety (239). *Enterococcus* strains may enhance sausage aroma and taste by their proteolytic activities, but may also compromise safety if opportunistic pathogenic strains proliferate. *Enterococcus* strains have recently assumed major importance in clinical microbiologigy as well, concerning their antibiotic resistance and their possible involvement in food borne illnesses due to the presence of virulence factors, such as the production of adhesins and aggregation substances (240).

Staphylococcus equorum is used in starter preparations with application in curing of raw ham, as the strains of this species grow well at temperatures below 10°C (241,242). *Halomonas elongata,* a psychrophilic and halophilic species that strongly reduces nitrate and nitrite, may be favorable in ensuring low nitrate concentration and in improving the sensory quality of ham (243).

Yeasts and molds traditionally play an important role in sausage fermentation, as well as maturation of hams, having an important role in the flavor of the product due to the enzymatic attack on the proteins and fats. The mycelial coat reduces moisture loss and facilitates uniform drying (244). Furthermore, mold protects the sausage from damage by intense light and high oxygen concentrations, and prevents rancidity by synthesizing catalase. Commercially available starter cultures of *Penicillium nalgiovense, P. chrysogenum, P. camemberti* may be used. The surface of the sausages is usually inoculated by dipping in a suspension of mold spores or by spraying. The cultivated molds avoid colonization by undesirable molds causing faults in the appearance and aroma of the products. Yeast starters such as *Debaryomyces hansenii* and *Candida famata* are used to produce a powdery surface or a fruity or alcoholic aroma (244). Strains of *Penicillium aurantiogriseum* and *P. camemberti* isolated from the surface of Spanish hams can give high quality characteristics to traditional meat products and can serve as a base for the production of starter cultures (245).

20.3.4 Microorganisms Occurring in Traditional Meat Products

20.3.4.1 Fermented Sausage

The microflora of naturally fermented sausages is usually dominated by lactobacilli (Table 20.1) and *Micrococcaceae* (Table 20.2). The microflora of Soppressata molisana, a traditional dry sausage produced in Southern Italy, is composed of lactobacilli and catalase positive cocci. The large majority of the lactobacilli isolated are *L. sake,* whereas a few strains are obligate heterofermentators belonging to *L. brevis.* Several isolates are hydrogen peroxide and acetoin producers (246). Among the catalase positive cocci occurring in the product during ripening, a predominance of micrococci (42%) and coagulase negative

Table 20.1
Main lactic acid bacteria involved in fermentation of traditional Mediterranean and Middle East meat products

Country of Origin Type of Sausage	Species	Ref.
Italy		
Soppressata molisana	*L. sakei, L. brevis*	(246)
Naples-type salami	*L.sakei, L. bavaricus*	(248)
Spain		
Chorizo	*L. sakei, L. curvatus, Pediococcus* sp.	(251)
Spanish-type sausage	*L. sakei, L. curvatus, L. bavaricus, L. plantarum*	(252)
Greece		
Greek-type sausage	*L. curvatus, L. sakei, L. plantarum, L. farciminis, C. piscicola, Weissella* sp., *E. faecium*	(259, 262)
Greek-type sausage	*L. sakei, L. curvatus, L. paracasei*	(260)
Turkey		
Sucuk	*L. sakei, L. curvatus, L. alimentarius, L. plantarum, L. brevis*	(264)

Table 20.2
Main *Micrococcaceae* involved in fermentation of traditional Mediterranean and Middle East meat products

Country of Origin Type of Sausage	Species	Ref.
Italy		
Soppressata molisana	*S. xylosus, M. kristinae*	(247)
Italian-style sausage	*S. xylosus*	(249)
Naples-type salami	a group between *S. xylosus* and *S.saprophyticus*	(248)
Spain		
Chorizo	*S. xylosus*	(253)
Greece		
Greek-type sausage	*S. saprophyticus, S. xylosus*	(261)
Greek-type sausage	*S. saprophyticus, S. carnosus, S. xylosus*	(263)

staphylococci (58%) is observed, with a prevalence of *Staphylococcus xylosus* and *Micrococcus kristinae* (247). Nine percent of staphylococci were coagulase positive and thermonuclease negative, but no strains of *Staphylococcus aureus* were isolated (247). Quite variable biochemical behavior is observed in the natural *Micrococcaceae* of the product. All strains are able to grow in the presence of 10% NaCl and a large majority are able to reduce nitrates to nitrites. Forty one percent of the micrococci and 16% of the staphylococci are able to hydrolyze pork fat, but only 2% of micrococci hydrolyze Tween 80. Micrococci showed a higher proteolytic activity in skim milk than staphylococci, whereas acetoin production was a strain specific trait (247). A natural fermentation therefore does not guarantee a consistently flavored product.

In Naples-type salami, manufactured without starter cultures, homofermentative lactobacilli constitute the predominant flora. Approximately 60% are identified as *L.sakei,* whereas 40% are *L. bavaricus* (248). *Micrococcaceae* constitute a fairly homogeneous group, which represents an intermediate variety between *S. xylosus* and *S. saprophyticus* and is isolated during the whole ripening process (248). Using a molecular method consisting of polymerase chain reaction amplification and denaturing gradient gel electrophoresis, *S. xylosus* proved to be the main species of the *Micrococcaceae* isolated during ripening of naturally fermented Italian sausages (249). The yeast population isolated from Naples-type salami mainly comprises *Debaryomyces* strains (248).

The dominant microflora in Spanish style dry cured sausage Chorizo Gallego is composed of lactic acid bacteria, but *Micrococcaceae,* enterococci, and bacilli also comprise a large portion of the total microflora of the product (250). The lactic acid bacterial microflora in different varieties of chorizo is composed of *L. sakei* (68.8%), *L. curvatus* (16.47%) and *Pediococcus* species (8.52%). Strains of *L. sakei* and *L. curvatus,* which ferment maltose but not lactose, are more dominant at the beginning and in the middle of the process, whereas lactose (or lactose and maltose) fermenting *L. sakei* and *L. curvatus* occur in higher numbers in the final product (251). The main lactobacilli isolated from dry fermented sausages in Spain are *Lactobacillus sakei* (55%), *L. curvatus* (26%), *L. bavaricus* (11%) and *L. plantarum* (8%); from the strains isolated, the most aciduric is *L. plantarum*. Although all *L. sakei* strains produce ammonia from arginine, in some strains the arginine deaminase pathway is induced only at a low oxygen level (252). *Staphylococcus xylosus* is the most predominant species among *Micrococcaceae* isolated from chorizo (95% of the strains isolated). Among such isolates, 38% are suitable as a starter culture as they show nitrate reductase and urease activity, low and moderate proteolytic and lipolytic activities, and do not produce acetoin (253).

The characteristics of the specific strains of lactobacilli isolated from meat origin have to be taken into account in order to select starter cultures in the manufacture of dry fermented sausages. From the evaluation as a starter culture of a mixture of *L. sakei, L. curvatus, L. bavaricus,* and *L. plantarum* strains isolated from naturally fermented Spanish sausages in comparison to a control (noninoculated batch) the following observations were made: a faster drop in pH was recorded in *L. curvatus* and *L. bavaricus* lots and correlated with a higher amount of nitrates. Some *L. sakei* strains gave rise to a product with the best sensory evaluation. *Lactobacillus plantarum* lots showed a high acidity, which was not well accepted by the panelists and was related to the highest amount of lactic acid. Especially dry sausages with a high D-lactate content had a very sour taste. One *L. curvatus* strain showed some important sensorial descriptors, in particular aged flavor, odor intensity, and typical flavor. The control lot showed suitable nitrate reduction, but the hydrogen sulfide odor detected, related to the higher content of *Enterobacteriaceae*, diminished its overall sensory acceptability (254).

Lactobacillus sakei CL 35 is a starter culture isolated from chorizo, which is able to inhibit Gram-positive and Gram-negative bacteria with the exception of the genus *Micrococcus.* When it is applied in combination with different concentrations of nitrite the microbiological quality of the product is improved (255). By use of the *Lactobacillus sakei* CL 35 as starter culture, a domination of the lactic acid bacteria (10^8–10^9 cfu/g) in the microflora is observed during ripening, followed by Micrococcaceae in high populations (10^5–10^6 cfu/g) and a decrease of the *Enterobacteriaceae.* This starter culture has a high acidifying capability and excretes antibioticlike substances ensuring the safety of the product (256).

In salchichon the bacterial flora is composed mainly of lactic acid bacteria and plays an important role in acidification. *Streptococcus, Microbacterium* and *Enterobacteriaceae* species are also involved, but only during exposure at high temperatures because their action is reduced as a consequence of growth of lactobacilli (257). During the fermentation stage of salchichon using a commercial mixture of *P. pentosaceus* and *Micrococcus varians,* proteolysis and lipolysis phenomena derived from high counts of lactic acid bacteria and *Micrococcaceae* occurred. The organoleptic characteristics of the final product are affected by an increase in the total free amino acid content, the peroxide index and the carbonyl compound content, as well as a decrease in protein solubility and the percentage of polyunsaturated free fatty acids (258).

In naturally fermented traditional Greek salami, the lactic acid bacterial species found are mainly *L. sakei* and *L. curvatus* followed by *L. plantarum* (259,260). *Carnobacterium* species are the second most frequent type of lactic isolates (about 10% of the isolated strains), identified as *C. piscicola* (261). Gas forming lactic acid bacteria of the genus *Weissella,* belonging to the species *W. hellenica, W. paramesenteroides, W. minor, W. halotolerans and W. viridescens,* occur as meat contaminants, due to poor hygienic precautions (262). The enterococci population, consisting mainly of *Enterococcus faecium* (259), increases during early fermentation and competes well in the fermented product comprising 4% of the isolated lactic acid bacteria (260). *Micrococcaceae* populations are poor competitors in the presence of actively growing aciduric bacteria, and their population is low consisting mainly of *Staphylococcus saprophyticus, S. carnosus* and *S. xylosus* (261,263). Although the role of *S. saprophyticus* in sausage aromatization is minor, *S. xylosus* is an important organism for dry sausage manufacture and is included in commercial starter culture preparations when a very aromatic product is intended (261).

In the microflora of Turkish sucuk, which is commonly fermented without starter culture, the predominant lactic acid organisms are *L. sakei, L. curvatus, L. alimentarius, L. plantarum* and *L. brevis* (264).

Although the role of lactobacilli in the production of traditional fermented meat products is well documented, there is little information available concerning lipases from *Lactobacillus* species. This is a consequence of a comparatively low production of lipase by these microorganisms, especially when compared with *Staphylococcus* species and *Micrococcus* species. Lactic acid bacteria lipases are involved in the development of flavor and aroma in fermented food products. A *L. plantarum* strain isolated from chourico, a traditional Portuguese dry fermented sausage, produces a highly thermostable extracellular lipase, which can be helpful for industrial utilization of lipases (265).

20.3.4.2 Dry Cured Ham

The natural microflora of lactic acid bacteria in dry cured hams is normally not very numerous (10–10^4 cfu/g) consisting of *Pediococcus pentosaceus,* the homofermentative lactobacilli *L. alimentarius* (43%), *L. curvatus* (16%) and *L. casei* var. *rhamnosus* (5%), and the heterofermentative *L. divergens* and atypical streptobacteria (266). Coagulase negative staphylococci are predominant in Spanish dry cured ham, probably because of their high salt tolerance, comprising 90% of the catalase positive cocci (267). *Staphylococcus xylosus* and *S. sciuri* are the main coagulase negative staphylococci growing in the product (267,268). In Spanish dry cured ham processed by both the slow and the fast process *S. xylosus* was the predominant microorganism among Gram-positive, catalase positive cocci isolated (269,270). A decrease in the number of *Enterobacteriaceae* species identified was observed as curing progressed during the two dry curing processes. At the end of the fast curing process, *Leclercia adecarboxylata* is the only species present, whereas at the end of the slow curing process, it is the most prevalent species, accompanied by *Klebsiella pneumoniae* and *Enterobacter agglomerans* (271). In Iberian dry cured ham, even after 16 months of maturing *S. xylosus* and *S. equorum* are the predominant organisms among the *Micrococcaceae* present in the product, and their metabolic activities contribute to the characteristics of the final product (241).

The Spanish hams are salted with marine salt alone or in combination with nitrate or nitrite. During the salting stage the microorganisms present in the salt develop several actions in the brine formed (reduction of nitrates to nitrites, proteolysis, and lipolysis) or penetrate the ham contributing to the sensorial properties. The salt during salting process of dry cured hams offers an ecosystem consisting of catalase positive cocci. Among 369 isolates, 60% belonged to the genus *Staphylococcus,* 25% to *Micrococcus,* 6% to *Kocuria,* 5.4% to *Dermacoccus* and 0.5% to *Stomatococcus.* The species most often isolated are *Staphylococcus xylosus, M. lylae, S. equorum* and *D. nishinomiyaensis* (272).

The good preservation of traditional Middle East and Mediterranean dry cured meat products like basturma is a result of lowering the water activity, salting, and spice flavoring (mainly the bactericidal effect of garlic) (273). The normal microflora of the product consists mainly of lactobacilli, staphylococci, and enterococci; whereas *Enterobacteriaceae* and sulfite reducing clostridia are present in low counts (222). From the 120 catalase positive cocci isolated, 42% were identified as *S. epidermidis,* 32% as *S. saprophyticus,* 12% as *S. simulans,* 4% as *S. carnosus,* 2% as *S. hyicus* subsp. *hyicus* and 7.5% as *K. varians* (274).

20.3.5 Quality Characteristics

20.3.5.1 Flavor Development

The typical flavor of fermented meat is a result of the interaction of numerous compounds resulting from the microbial and muscle enzyme metabolism of carbohydrates, proteins, and lipids which, in combination with spices and seasonings, create the specific profile for each product (219,275). In products where only curing agents are used, the flavor differs noticeably according to the use of nitrate or nitrite. Acidification plays a major part in these reactions, influencing not only the products produced but also the enzyme mechanisms of curing. With Mediterranean area products, curing with nitrate is considered indispensable because, among other factors, it directs the microbial development toward the growth of favorable microorganisms, especially lactic acid bacteria and catalase positive cocci that produce oxidized derivatives, aldehydes, and ketones, which together with nitrated and nitrosed compounds contribute decisively to the flavor (219). Acids derived from carbohydrate fermentation by staphylococci induce an acid taste (D-lactate) or aroma (acetic acid), whereas diacetyl and acetoin impart a buttery flavor (276). Because smoking is very seldom used in the Mediterranean area (277), the flavor of the products is directly related to the biochemical, proteolytic, and lipolytic mechanisms which take place during ripening (229). Although results are published indicating the importance of bacterial proteinases, the endogenous enzymes in meat play a crucial role in the development of flavor in fermented sausages (243). Bacterial metabolism of L-leucine, mainly by staphylococci, and to a less extent by *Lactobacilli,* is generally held responsible for the production of 3-methyl butanal, an important component of the dry sausage aroma (228). Bacterial lipases show very low activity under conditions found in fermented sausages, but lipid oxidation reflected in hexanal concentration and associated with a rancid aroma is important in Mediterranean sausages (276).

The flavor in the case of mold ripened dry sausages, which are normally ripened for a long time, is due to the surface flora consisting of molds and yeasts (219). Molds have an important role on the flavor of sausages as a consequence of proteolysis and lipolysis. A common type of enzymatic attack on amino acids is the oxidative deamination to ammonia and the corresponding alpha keto acid (278). The liberation of ammonia raises the pH of the sausages, which coincides with the abundant appearance of molds. The sausages produced have a mild flavor, which allows meaty notes to dominate the overall flavor profile (227). In dry cured Iberian ham proteolysis by *P. chrysogenum* contributes to the ripening and flavor of the product (279).

In salchichon the acid aroma is correlated with the acetic acid content, whereas the bitter flavor is correlated with the concentration of free amino acids: aspartic acid, glutamic acid, serine, glycine, asparagine, and threonine (280). It seems possible that controlling the level of free amino acids is a prospect of controlling the level of some aroma compounds important to dry sausage flavor.

20.3.5.2 Microbial Safety

20.3.5.2.1 Inhibition of Bacteria The most important mechanisms to ensure microbial safety of fermented meats are the formation of lactic acid, acetic acid, and possibly bacteriocins. Other metabolites of lactic acid bacteria, such as diacetyl and hydrogen peroxide, interfere with the sensory properties. Although the main acid formed is lactic acid, the antimicrobial effect of acetic acid should not be neglected, because at the same concentration and pH, it is more effective than lactic acid. The sensitivity of the different bacteria to acids depends on the action of other factors such as a_w and nitrite concentration. As a result, even small differences in acid concentrations have a major effect on acid sensitive microorganisms (237).

Bacteriocinogenic starter cultures are used to improve the competitiveness of the starter strains and to prevent the growth of food pathogens (243). Bacteriocin producing cultures of *L. curvatus, L. sakei,* and *L. plantarum* may control the meat microflora and inhibit foodborne pathogens during the production of fermented meat products (281). The levels of *Listeria monocytogenes,* particularly, can be reduced by one or two log cycles compared to nonbacteriocin producing cultures. The effect of bacteriocins on the safety of meat products is limited because of the resistance of Gram-negative bacteria to them, their inactivation in the meat, and the possibility of resistance development in the target organisms (237).

All genera of lactic acid bacteria found in fermented meats produce bacteriocins, which act against bacteria closely related to the producer organisms. Many of them also inhibit *Listeria* but only few are effective against *Bacillus, Clostridium,* and *Staphylococcus.* Bacteriocin resistant mutants are developed from sensitive bacteria. For instance, nisin and sakacin resistant mutants of *L. monocytogenes* have been reported, but nisin resistant variants of *L. monocytogenes* were sensitive to sakacin formed *in situ*; thus, combinations of bacteriocins may be more effective against *Listeria* than single bacteriocins (237). Gram-negative bacteria are sensitized to bacteriocins by treatment with chelating agents, and strains of lactic acid bacteria have been constructed that form and export colicin, but it is not known if such combinations are effective in meat systems (282).

The main factors affecting the effectiveness of bacteriocins in meat systems are the reduction of bacteriocin activity by the binding of bacteriocin molecules to food components, by the destabilizing action of proteases and other enzymes, and by an uneven distribution in the food matrix (283). The low carbohydrate content and low ripening temperature of fermented meat products may be responsible for a low bacteriocinogenic activity. Furthermore, *Listeria* inhibition is dependent on ecological factors prevailing in sausage fermentation such as pH and the concentration of sodium chloride, nitrite, and nitrate (284).

During Spanish style dry sausage fermentation, bacteriocin producing *Enterococcus* strains (CCM 4231, RZS C13) used as starter cultures strongly inhibited the growth of *Listeria* species (239). Enterocin P, produced by *E. faecium* P13, inhibited most of the tested spoilage and foodborne Gram-positive pathogenic bacteria, such as *L. monocytogenes, S. aureus, Clostridium perfringens,* and *Clostridium botulinum* (285). A different strain, *E. faecium* P21, produces two bacteriocins identical to enterocin A and enterocin B, which show a narrow antimicrobial activity against closely related lactic acid bacteria and strong antimicrobial activity against spoilage and foodborne Gram-positive bacteria, including *L. monocytogenes, S. aureus, C. perfringens,* and *C. botulinum* (286). A *Pediococcus acidilactici* strain isolated from commercial starter cultures used in the Spanish meat industry shows an inhibitory action against a wide range of Gram-positive bacteria including. *L. monocytogenes, E. faecalis, S. aureus,* and *C. perfringens* (287). Leucocin F10, a bacteriocin produced by a strain of *Leuconostoc carnosum* isolated from fermented sausages manufactured in Southern Italy, is inhibitory to *Listeria* species, *Leuconostoc* species, *E. faecalis,* and *L. sakei* (288). From Spanish fermented sausages the following bacteriocins were detected: plantaricin D, produced by *L. plantarum* CTC305 and active against different lactic acid bacteria and Gram-positive pathogens like *L. monocytogenes;* sakasin T, isolated from *L. sakei* CTC372 and active against *L. monocytogenes* and *S. aureus*; sakacin K, a heat stable proteinaceous antagonistic compound active against *L. monocytogenes* and isolated from *L. sakei* CTC494 (the first 30 amino acid residues from the N-terminus of sakacin K are identical to those of curvacin A in *L. curvatus* LTH1174 and sakacin A in *L. sakei* Lb706) (289). Sakacin B (with molecular weight estimated to 6.3 kDa) produced by *L. sakei* 251, a strain isolated from naturally fermented Greek dry sausage, is inhibitory toward various lactic acid bacteria of meat origin, but it is not effective against Gram-positive pathogenic bacteria. Thus, this bacteriocin is useful to control adventitious lactic flora in natural meat fermentations (290). The main bacteriocins produced by traditional fermented meat products of Mediterranean and Middle East region are presented in Table 20.3.

Bacteriocins produced *in situ* may contribute to the dominance of the producing strains over other lactic acid bacteria during fermentation as well as to the reduction of the levels of

Table 20.3
Bacteriocins produced by strains isolated from traditional fermented meat products of Mediterranean and Middle East region

Bacteriocin	Strain Isolated	Inhibitory Spectrum	Ref.
Enterocin P	*E. faecium* P13	*L. monocytogenes, S. aureus,*	(285, 286)
Enterocin A & Enterocin B	*E. faecium* P21	*C. perfringens* & *C. botulinum*	
Leucocin F10	*L. carnosum* F10	*Listeria* spp., *Leuconostoc* spp., *E. faecalis* & *L. sakei*	(288)
Plantaricin D	*L. plantarum CTC305*	*L. monocytogenes* & Lactic acid bacteria	(289)
Sakacin B	*L. sakei* 251	$Gram^+$ pathogens	(290)
Sakacin K	*L. sakei* CTC494	*L. monocytogenes*	(289)
Sakacin T	*L. sakei* CTC372	*L. monocytogenes* & *S. aureus*	(289)

the affected food pathogens. The use of *L. sake* is an adequate safety factor in chorizo, having a significant inhibitory effect on *Enterobacteriaceae* counts (256). Another strain of the same microorganism (*L. sakei* 121) proved to ensure safety in Soppressata Molisana (291).

Greek dry salami produced without starter cultures is free of *Salmonella* after formulation, while very low numbers of sulfite reducing clostridia and no coagulase positive staphylococci are present. *Listeria* species, including *L. monocytogenes,* occurring in raw meat are detected until the fourth day of fermentation, but disappear by the end of fermentation (261). Culture filtrates of *L. sake, L. curvatus* and *L. plantarum* strains isolated from Greek fermented sausage have antimicrobial activity against *L. monocytogenes* and *S. aureus* strains in well diffusion assays, whereas all isolated *S. xylosus* strains and a large majority of *S. carnosus* and *K. varians* strains exhibited antilisterial activity (260,263). Good hygienic conditions of the raw materials, a balanced amount of quickly and slowly fermentable carbohydrates, and a reasonably adjusted ripening program result in production of salamis of desired quality and safe. The addition of starter cultures is, however, of virtal importance in eliminating health risks (259,261). Strains selected as starter cultures can be followed during fermentation by rapid identification according to their enzymatic pattern (260,263).

In Turkish sucuk the probability of contamination with *L. monocytogenes* is deemed high (292); bacteriocinlike metabolites produced by lactic acid bacteria isolated from the product showed inhibitory activity against *L. monocytogenes* strains tested with agar spot tests and well diffusion assays (293). In laboratory scale the addition of 10^6–10^7 cfu/g of each one of *L. sake* ATCC 15521 and *P. acidilactici,* improved fermentation and controlled inoculated *Yersinia enterocolitica* O:3 (approximately 10^5 cfu/g), mainly due to lactic acid produced, whereas the inoculated foodborne pathogen is not eliminated in a natural fermentation (294). Starter cultures composed of *L. plantarum* with *S. carnosus* and *P. acidilactici* with *S. carnosus* improve the microbial safety and hygienic quality of sucuk (295,296).

In dry cured ham *Enterobacteriaceae* growing during the first nonrefrigerated steps of the curing process before the decrease of a_w can cause deep putrefaction of the product. Therefore, *Enterobacteriaceae* ought to be considered a microbial quality related hazard in the development of hazard analysis critical control point (HACCP) systems for dry cured hams (297). In general, spoiled hams have higher moisture content, and a lower content of sodium chloride, nitrates, and nitrites than the unspoiled product. Insufficient salt concentration during curing causes deep spoilage by *Serratia liquefaciens, Enterobacter cloacae* and *Proteus vulgaris.* These strains can grow in 7% NaCl at 5 °C, but the reduction of temperature to 2°C reduces the NaCl tolerance significantly (298). The most frequently isolated species in the spoiled hams during the fast dry curing process is *Serratia liquefaciens*, whereas in the slow dry curing process it is *Leclercia adecarboxylata.* In addition, strains of *S. aureus* were isolated from spoiled product with both processes (299).

In traditional Middle East and Mediterranean dry cured meat products like basturma, modifications of the original process or formula, such as increase of the final moisture content or decrease of the level of garlic or salt, introduced major safety hazards which caused an outbreak of salmonellosis in California in 1982 (300). Contamination of the product during processing with *Salmonella* species or *S. aureus* results in survival of the contaminating pathogens in the product for an extended period of time, and the occurence of a foodborne outbreak depends on the extent of initial contamination (222). In an effort to improve the quality characteristics and safety of the product, inoculation by starter culture combinations of *L. sakei, L. pentosus, L. curvatus, S. carnosus, S. xylosus and K. varians* has been used (301–303). In this case the inoculated bacteria are not spread evenly in the meat tissue and multiply in small groups (301) showing that a selection of starter cultures and the inoculation techniques is needed, because a homogeneous distribution of added microorganisms is important for their effect (304).

20.3.5.2.2 Biogenic Amines In long ripened meat products significant amounts of biogenic amines may be formed by the action of microbial decarboxylases on free amino acids, resulting from the proteolytic process that normally takes place during the ripening of fermented sausages. The microorganisms responsible for the decarboxylation reactions may be introduced as starter cultures in sausages or may be part of the natural microflora. Thus, *Enterobacteriaceae* are involved in the production of putrescine, histamine, and cadaverine, and *Pseudomonas* in the production of putrescine (305). From the lactic acid bacteria, all strains of *Carnobacterium* found in meat products and some strains of *L. curvatus* and *L. plantarum,* used as starter cultures, produce tyramine (306). *Micrococcaceae* and enterococci have been also involved in the production of biogenic amines (307). Spontaneous nonstarter lactic acid bacteria and *Enterococci* are indicated as responsible for tyramine formation during ripening in Italian sausages (305). Furthermore, wide variations have been observed in the amine content of different batches of the same commercial brand of fermented products. The presence of biogenic amines in foods is the consequence of a complex equilibrium between the physicochemical characteristics of the product (pH, a_w, and nutrients), the technological conditions (temperature, starter mediated fermentation, and use of raw materials with good hygienic quality) and the enzymatic activities of the microorganisms present. Especially, the availability of free amino acids, and the presence of microorganisms that can decarboxylate the amino acids, as well as the incidence of favorable conditions for their growth, affect the formation of amines (308). In sausages, histamine is often detected at low level, but can be occasionally produced in high amounts (309). The main amine found in Mediterranean sausages is tyramine, which is formed by tyramine forming lactic acid bacteria by decarboxylation of tyrosine (228). The amines putrescine and cadaverine can also be used as indicators of the microbial quality and safety of fermented meat products (308). In Turkish style sausage (sucuk) high concentrations of histamine and tyramine were observed in some samples and the addition of a commercial starter culture and high concentrations of additives reduced their formation (296). In fermented sausage produced in Egypt using *L. plantarum* and *P. acidilactici* as starter cultures, however, tyramine development was continued during ageing, although potassium sorbate was added at up to 0.06% (310). Two *Lactobacillus* strains, *L. curvatus* and *L. homohiochii,* isolated from a Portuguese traditional dry fermented sausage, showed increased proteolytic activity and were able to decarboxylate tyrosine to tyramine and, to a lesser extent, ornithine into putrescine (308). On the other hand, the use of *S. xylosus* S81, which can be isolated from fermented sausage in Southern Italy, as a starter culture influences the amount of histamine and other biogenic amines during ripening by influencing the level of the microbial amino oxidases and the amino acid decarboxylases (311). Starters (composed of *Micrococcus carnosus,* and *L. plantarum* or *P. pentosaceus*) seem to decrease biogenic amine formation but do not prevent it. The high background flora naturally present on the raw meat and pork lard seems to have a strong influence on biogenic amine formation during ripening (312). The use of a mixed starter culture composed of a tyramine producing *L. curvatus* and a proteolytic *S. xylosus* yielded lower amounts of tyramine than those produced by the wild microflora in the control batch, whereas the accumulation of the diamines putrescine and cadaverine in sausages was efficiently reduced (313).

Formation of carcinogenic nitrosamines constitutes an additional toxicological risk related to biogenic amines, especially in meat products that contain nitrite and nitrate as curing salts. It is important for the meat industry to produce fermented products with a high degree of safety with respect to biogenic amines, because of their possible involvement in various diseases such as migraines, headaches, gastric and intestinal ulcers, and allergic responses (314). Therefore, before being selected as starter cultures, strains should be screened for their proteolytic and decarboxylase activities. Biogenic amines in foods are of health concern, especially for people with low monoamine oxidase or diamine oxidase activities, due to medical treatment (311). Thus, the use of starter cultures exhibiting amine oxidase activity is of importance to prevent or reduce the accumulation of biogenic amines (311,315). In fact, biogenic amines are physiologically inactivated by amine oxidases, which are found in bacteria, fungi, plant, and animal cells and are able to catalyze the oxidative deamination of amines producing aldehydes, hydrogen peroxide, and ammonia (316).

20.3.5.3 Probiotic Properties

There is a world wide trend toward new probiotic functional foods, and meats may be another application area. The development of new probiotic functional foods, such as fermented meat products, presupposes the application of probiotic lactic acid bacteria with the following characteristics: resistance to sodium chloride and sodium nitrite; growth and activity during fermentation and ripening; resistance to acid environment; resistance to lysozyme; resistance to bile salts; and ability to colonize the human intestinal tract by means of some mechanisms for adhering or binding to the intestinal cells (317,318). There exists a need to select strains of lactic acid bacteria originating from the indigenous microflora of traditional products on the basis of these properties in order to obtain strains which are well adapted to the microenvironment which could dominate the microflora of the fermented meat product. *Lactobacillus curvatus* strains originating from commercial meat starter cultures have the capacity to survive 0.3% bile salts at pH 5 (319). Bile salts affect the cell membranes of the microorganisms, as they are composed of lipids and fatty acids. Some lactobacilli are able to hydrolyze bile salts with a bile salt hydrolase enzyme, weakening their detergent effect (319). The resistance to bile salts varies considerably between the *Lactobacillus* strains, and the mechanism is still unknown (319). A large majority of *L. sakei, L. curvatus* and *L. plantarum* strains isolated from Greek-style fermented sausage tolerate 0.3% bile salts (260). The probiotic properties of desirable bacteria are dependent on their ability to remain viable and to colonize the surface of human intestinal cells; a viable cell population of 10^7–10^9 cells per day is necessary in order for any beneficial effects to develop in humans (317). The application of probiotic cultures in fermented meats in conjunction with low fat and low salt content might have health promoting effects.

REFERENCES

1. Steinkraus, K.H. Classification of fermented foods: worldwide review of household fermentation techniques. *Food Control* 8:311–317, 1997.
2. Adams, M.R., L. Nicolaides. Review of the sensitivity of different foodborne pathogens to fermentation. *Food Control* 8:227–239, 1997.
3. Campbell-Platt, G. Fermented foods: a world perspective. *Food Res. Int.* 27:253–257, 1994.
4. Uccella, N. Olive biophenols: novel ethnic and technological approach. *Trends Food Sci. Technol.* 11:328–339, 2001.
5. Fox, P.F. Cheese: an overview. In: *Cheese Chemistry, Physics and Microbiology*, Vol. 1, Fox, P.F., ed., London: Chapman & Hall, 1993, pp 1–36.
6. Tamime, A.Y., R.K. Robinson. Introduction. In: *Yogurt: Science and Technology*, Tamime, A.Y., R.K. Robinson, eds., Oxford: Pergamon Press, 1985, pp 1–6.
7. Burkhalter, G. *IDF Catalogue of Cheeses, Document 141*. Brussels, Belgium: International Dairy Federation, 1981.
8. Kalantzopoulos, G.C. Cheeses from ewes and goats milk. In: *Cheese Chemistry, Physics and Microbiology, Vol. 2.* Fox, P.F., ed., London: Chapman & Hall, 1999, pp 507–543.
9. Esteves, C.L.C., J.A. Lucey, E.M.V. Pires. Rheological properties of milk gels made with coagulants of plant origin and chymosin. *Int. Dairy J.* 12:427–434, 2002.
10. Sousa, M.J., F.X. Malcata. Advances in the role of a plant coagulant (*Cyanara cardunculus*) *in vitro* and during ripening of cheeses from several milk species. *Lait* 82:151–170, 2002.
11. Freitas, C., F.X. Malcata. Microbiology and biochemistry of cheeses with Appelation d'Origin Protegee and manufacture in the Iberian Peninsula from ovine and caprine milks. *J. Dairy Sci.* 83:584–602, 2000.
12. Marcos, A., M.A. Esteban. Iberian cheeses. In: *Cheese Chemistry, Physics and Microbiology, Vol. 2,* Fox, P.F., ed., London: Chapman & Hall, 1999, pp. 173–219.
13. Macedo, A.C., M.L. Costa, F.X. Malcata. Changes in the microflora of Serra cheese: evolution throughout ripening time, lactation period and axial location. *Int. Dairy J.* 6:79–94, 1996.
14. Tavaria, F.K., F.X. Malcata. Microbiological characterization of Serra da Estrela cheese throughout its Appellation d'Origin Protegee region. *J. Food Prot.* 61:601–607, 1998.
15. Tavaria, F.K., F.X. Malcata. On the microbiology of Serra da Estrela cheese: geographical and chronological consideration. *Food Microbiol.* 17:293–304, 2000.
16. Nunez, M. Flora microbiana del queso Manchego, IV: Lactobaccilos. *Anal. INIA Ser. Gen.* 4:57–65, 1976.
17. Nunez, M. Flora microbiana del queso Manchego, V: Leuconostocs *Anal. INIA Ser. Gen.* 4:67–74, 1976.
18. Nunez, M. Flora microbiana del queso Manchego, VI : Pediococcus. *Anal. INIA Ser. Gen.* 4:75–81, 1976.
19. Serano, C.E., C. Garcia, L.M. Medina, E. Serrano. Caracterizacion de levaduras aisladas de queso Manchego con denominacion de origin. *Alimentaria* 7,8:65–67, 1996.
20. Bockelmann, W., T. Hoppe-Seyler. The surface flora of bacterial smear-ripened cheese from cow's and goat's milk. *Int. Dairy J.* 11:307–314, 2001.
21. Corsetti, A., J. Rossi, M. Gobbetti. Interactions between yeasts and bacteria in the smear surface-ripened cheeses. *Int. J. Food Microbiol.* 69:1–10, 2001.
22. Beresford, T.P., N.A. Fitzsimons, N.L. Brennan, T. Cogan. Recent advances in cheese microbiology. *Int. Dairy J.* 11:259–274, 2001.
23. Accolas, J.P., D. Melcion, L. Vassal. Study of the surface microflora of Gruyere and Beaufort cheeses, *Proceed. 20th Int. Dairy Cong.*, pp 762, 1978.
24. Gobbetti, M., S. Lowney, E. Smacchi, B. Battistotti, P. Damiani, P.F. Fox. Microbiology and biochemistry of Taleggio cheese during ripening. *Int. Dairy J.* 7:509–517, 1997.
25. Irlinger, F., A. Morvan, N. El Sohl, J.L. Bergere. Taxonomic characterization of coagulase negative staphylococci in ripening flora from traditional French cheeses. *Syst. Appl. Microbiol.* 20:319–328, 1997.

26. Irlinger, F., J.L. Bergere. Use of conventional biochemical tests and analyses of ripotype patterns for classification of micrococci isolated from dairy products. *J. Dairy Res.* 66:91–103, 1999.
27. Reps, A. Bacterial surface-ripened cheeses. In: *Cheese Chemistry, Physics and Microbiology, Vol. 2.* Fox, P.F., ed., London: Chapman & Hall, 1999, pp. 137–172.
28. Gripon, J.C. Mold-ripened cheeses. In: *Cheese Chemistry, Physics and Microbiology, Vol. 2.* Fox, P.F., ed. London: Chapman & Hall, 1999, pp 111–136.
29. Schmidt, J.L., J. Lenoir. Contribution a l'etudes de la flore levure du fromage de Camembert : son evolution au cours de la maturation, I. *Lait* 58:355–370, 1978.
30. Schmidt, J.L., J. Lenoir. Contribution a l'etudes de la flore levure du fromage de Camembert : son evolution au cours de la maturation, II. *Lait* 60:272–282, 1980.
31. Marsellino, N., D.R. Benson. Scanning electron and light microscopic study of microbial succession on Bethlem St. Nectaire cheese. *Appl. Environ. Microbiol.* 58:3448–3454, 1992.
32. Devoyond, J.J., G. Bret, J.E. Auclair. [Microbial flora of Roquefort cheese, I: its development during maturation and ripening of the cheese]. *Lait* 48:613–619, 1968.
33. Devoyond, J.J. [Microbial flora of Roquefort cheese, II: Staphylococci and micrococci]. *Lait* 49:20–39, 1969.
34. Devoyond, J.J. [Microbial flora of Roquefort cheese, IV: Enterococci]. *Lait* 49:637–650, 1969.
35. Devoyond, J.J., M. Muller. [Microbial flora of Roquefort cheese, III: Lactic streptococci and leuconostocs: influence at various contaminating microorganisms]. *Lait* 49:369–399, 1969.
36. Devoyond, J.J. [Microbial flora of Roquefort cheese, V : Lactobacilli]. Lait 50:277–284, 1970.
37. Devoyond, J.J., D. Sponem. [Microbial flora of Roquefort cheese, VI: Yeasts]. *Lait* 50:524–537, 1970.
38. Nunez, M. Microflora of Cabrales cheese: changes during maturation. *J. Dairy Res.* 45:501–508, 1978.
39. Prieto, B., I. Franco, J.M. Fresno, A. Bernardo, J. Carballo. Picon Bejes Tresviso blue cheese: an overall biochemical survey throughout the ripening process. *Int. Dairy J.* 10:159–167, 2000.
40. Lopez-Diaz, T.M., C. Alonso, C. Romain, M.L. Garcia-Lopez, B. Moreno. Lactic acid bacteria isolated from a hand-made blue cheese. *Food Microbiol.* 17:23–32, 2000.
41. Gobbetti, M., R. Burzigotti, E. Smacchi, A. Corsetti, M. De Angelis. Microbiology and biochemistry of Gorgonzola cheese during ripening. *Int. Dairy J.* 7:519–529, 1997.
42. Salvadori, P., B. Bianchi-Salvatori, G. Gower. Caratteristiche fisiologiche e colturali di associazioni microbiche constituite da batteri lattici thermophili e 'Penicillium roqueforti'. *L.' Industria del Latte* 10:23–37, 1974.
43. Cislaghi, S., P. Ceccarelli, S. Piravono, T. Sozzi. Difetto di produzione di Gorgonzolla causato dai batteriofagi. *Latte* 4:392–397, 1995
44. Kaminarides, S.E., E.M. Anifantakis. Evolution of the microflora of Kopanisti cheese during ripening: study of the yeast flora. *Lait* 69:537–546, 1989.
45. Moatsou, G.A., I.G. Kandarakis, A.K. Georgala, E.S. Alichanidis, E.M. Anifantakis. Effect of starters on proteolysis of Graviera Kritis cheese. *Lait* 79:303–315, 1999.
46. Steffen, C., P. Eberhard, J.O. Bosset, M. Ruegg. Swiss-type varieties. In: *Cheese Chemistry, Physics and Microbiology, Vol. 2,* Fox, P.F., ed., London: Chapman & Hall, 1999, pp 83–110.
47. Grappin, R., E. Beuvier, Y. Bouton, S. Pochet. Advances in the biochemistry and microbiology of Swiss-type cheeses. *Lait* 79:3–22, 1999.
48. Chamba, J.F. L'emmental : un ecosysteme complexe : consequences sur la selection et l'utilisation des ferments. *Sci. Alim.* 20:37–54, 2000.
49. Kerjean, J.R., S. Condon, R. Lodi, G. Kalantzopoulos, J.F. Chamba, T. Suomalainen, T. Cogan, D. Moreau. Improving the quality of European hard cheeses by controlling the interactions between lactic acid bacteria and propionibacteria. *Food Res. Int.* 33:281–287, 2000.
50. Berthier, F., E. Beuvier, A. Dasen, R. Grappin. Origin and diversity of mesophilic lactobacilli in Comte cheese, as revealed by PCR with repetitive and species-specific primers. *Int. Dairy J.* 11:293–305, 2001.

51. Bouton, Y., P. Guyot, E. Beuvier, P. Tailliez, R. Grappin. Use of PCR-based methods and PFGE for typing and monitoring homofermentative lactobacilli during Comte cheese ripening. *Int. J Food Microbiol.* 76:27–38, 2002.
52. Battistotti, B., C. Corradini. Italian cheese. In: *Cheese Chemistry, Physics and Microbiology,* Vol. 2. Fox, P.F., ed., London: Chapman & Hall, 1999, pp 221–243.
53. Bossi, F., V. Bottazzi, G.L. Scolari, B. Battistotti, F. Dellaglio. Lactic acid bacteria for Grana cheese production, I: technological characterization of thermophilic rod lactic acid bacteria. *Sci. Tech. Latt-Cas.* 41:105–136, 1990.
54. Neviani, E., S. Carini. Microbiology of Parmesan cheese. *Microbiol. Alim. Nutr.* 12:1–8, 1994.
55. Giraffa, G., P. De Vecchi, J. Reinheimer. Population dynamics of thermophilic lactobacilli in mixed starter whey cultures. *Food Res. Int.* 30:137–140, 1997.
56. Concoccelli, P., M.G. Parisi, L. Senini, V. Bottazzi. Use of RAPD and 16S rRNA sequencing for the study of *Lactobacillus* population dymanics in natural whey culture. *Lett. Appl. Microbiol.* 25:8–12, 1997.
57. Giraffa, G., L. Rossetti, G. Mucchetti, F. Addeo, E. Neviani. Influence of the temperature gradient on the growth of thermophilic lactobacilli used as natural starters in Grana cheese. *J. Dairy Sci.* 81:31–36, 1998.
58. Coppola, R., M. Nanni, M. Iorizzo, A. Sorrentino, E. Sorrentino, C. Chiavari, L. Grazia. Microbiological characteristics of Parmigiano Reggiano cheese during the cheesemaking and the first months of the ripening. *Lait* 80:479–490, 2000.
59. Thompson, T.L., E.H. Marth. Changes in Parmesan cheese during ripening: microflora-coliforms, enterococci, anaerobs, propionibacteria and staphylococci. *Milchwissenschaft* 41:201–205, 1986.
60. Torriani, S., M. Vescovo, G. Scolari. An overview of *Lactobacillus helveticus. Ann. Microbiol. Enzimol.* 44:163–191, 1994.
61. Giraffa, G., E. Neviani. Different *Lactobacillus helveticus* strain population dominate during Grana Padano cheese making. *Food Microbiol.* 16:205–210, 1999.
62. Caric, M. Ripened cheese varieties native to the Balkan countries. In: *Cheese Chemistry, Physics and Microbiology, Vol. 2.* Fox, P.F., ed., London: Chapman & Hall, 1999, pp 263–279.
63. Coppola, S., E. Parente, S. Dumontet, A. Le Peccerella. The microflora of natural whey cultures utilized as starters in the manufacture of Mozzarella cheese from water buffalo milk. *Lait* 68:295–310, 1988.
64. Villani, F., S. Copolla. Selection of enterococcal strains from water buffalo Mozzarella cheese manufacture. *Ann. Microbiol. Enzimol.* 44:97–105, 1994.
65. Parente, E., M.A. Rota, A. Ricciardi, F. Clementi. Characterization of natural starter cultures used in the manufacture of Pasta Filata cheese in Basilicata (southern Italy). *Int. Dairy J.* 7:775–783, 1997.
66. Moatsou, G., I. Kandarakis, E. Moschopoulou, E. Anifantakis, E. Alichanidis. Effect of technological parameters on the characteristics of Kasseri cheese made from raw or pasteurized ewes' milk. *Int. J. Dairy Technol.* 54:69–77, 2001.
67. Gobbetti, M., M. Morea, F. Baruzzi, M.R. Corbo, A. Matarante, T. Considine, R. Di Cagno, T. Guinee, P.F. Fox. Microbiological, compositional, biochemical and textural characterization of Caciocavallo Puglieses cheese during ripening. *Int. Dairy J.* 12:511–523, 2002.
68. Romano, P., A. Ricciardi, G. Salzano, G. Suzzi. Yeasts from water buffalo Mozzarella, a traditional cheese of the Mediterranean area. *Int. J. Food Microbiol.* 69:45–51, 2001.
69. Morea, M., F. Baruzzi, F. Cappa, P.S. Concocelli. Molecular characterization of the *Lactobacillus* community in traditional processing of Mozzarella cheese. *Int. J. Food Microbiol.* 43:53–60, 1998.
70. Tsakalidou, E., E. Zoidou, B. Pot, W. Ludwig, L. Wassil, L.A. Devriese, G. Kalantzopoulos, K.H. Schleifer, K. Kersters. Identification of streptococci from Greek Kasseri cheese and description of *Streptococcus macedonicus* sp. nov. *Int. J. Syst. Bacteriol.* 48:519–527, 1998.
71. Andrighetto, C., F. Borney, A. Barmaz, B. Stefanon, A. Lombardi. Genetic diversity of *Streptococcus thermophilus* strains isolated from Italian hard cheeses. *Int. Dairy J.* 12:141–144, 2002.

72. Abd El-Salam, M.H., E. Alichanidis, G.K. Zerfiridis. Domiati and Feta type cheeses. In: *Cheese Chemistry, Physics and Microbiology, Vol. 2.* Fox, P.F., ed., London: Chapman & Hall, 1999, pp 301–335.
73. Bintsis, T., P. Papademas. Microbiological quality of white-brined cheeses: a review. *Int. J. Dairy Technol.* 55:113–120, 2002.
74. Vafopoulou-Mastrojiannaki, A., E. Litopoulou-Tzanetaki, N. Tzanetakis. Effect of *Pediococcus pentosaceus* on ripening changes of Feta cheese. *Microbiol. Alim. Nutr.* 8:53–62, 1990.
75. Kaminarides, S.E., N.S. Laskos. Yeasts in factory brine of Feta cheese. *Austr. J. Dairy Techol.* 47:68–71, 1992.
76. Litopoulou-Tzanetaki, E., N. Tzanetakis. Microbiological study of white-brined cheese made from raw goat milk. *Food Microbiol.* 9:13–19, 1992.
77. Tzanetakis, N., E. Litopoulou-Tzanetaki. Changes in numbers and kinds of lactic acid bacteria in Feta and Teleme, two Greek cheeses from ewes' milk. *J. Dairy Sci.* 75:1389–1393, 1992.
78. Litopoulou-Tzanetaki, E., N. Tzanetakis, A. Vafopoulou-Mastrojiannaki. Effect of the type of lactic starter on microbiological, chemical and sensory characteristics of Feta cheese. *Food Microbiol.* 10:31–41, 1993.
79. Tzanetakis, N., A. Vafopoulou-Mastrojiannaki, E. Litopoulou-Tzanetaki The quality of white-brined cheese from goat's milk made with different starters. *Food Microbiol.* 12:55–63, 1995.
80. Sarantinopoulos, P., G. Kalantzopoulos, E. Tsakalidou. Effect of *Enterococcus faecium* on microbiological, physicochemical and sensory characteristics of Greek Feta cheese. *Int. J. Food Microbiol.* 76, 93–105, 2002.
81. Manolopoulou, E., P. Sarantinopoulos, E. Zoidou, A. Aktypis, E. Moschopoupou, I. Kandarakis, E. Anifantakis. Evolution of feta cheese microbial flora during traditional manufacture and ripening. *Int. J. Food Microbiol.* 82:153–161, 2003.
82. Abou-Donia, S.A. Egyptian Domiati soft white pickled cheese. *New Zealand J. Dairy Sci. Technol.* 21:167–195, 1986.
83. Naguib, M.M. Bacteriological and chemical studies on market Domiati cheese. MSc Thesis, Ain Ahams University, Cairo, Egypt, 1965.
84. Hemati, B., R. Suessmuth, N. Ezzat, H. El-Shafei, N. El Soda. Isolation and characterization of an *Enterococcus* starter from Egyptian Domiati cheese. Milchwissenschaft 53:198–202, 1998.
85. Ghoneim, N.A. Incidence of yeasts other than *Candida* species in Damietta cheese. *Milchwissenschaft* 23:482–484, 1968.
86. Mahmoud, S.A.Z., A.M. Moussa, G.N. Zein, K.M. Kamaly. Effect of adding some plant flavors to pickling solution on some microbiological properties of Domiati cheese, *Res. Bull. Fac. Agric.* 1033:22, 1979.
87. Seham, M., M.A. Sheleih, A.M. Saudi. Occurrence of yeasts in some Egyptian dairy products. *J. Egypt Ved. Med. Assoc.* 42:5–11, 1982.
88. Hayaloglu, A.A., M. Guven, P.F. Fox. Microbiological, biochemical and technological properties of Turkish white cheese 'Beyaz peynir'. *Int. Dairy J.* 12:635–648, 2002
89. Karakus, M., M. Borcakli, I. Alperden. Beyaz peynirin olgunlasma surecinde mikrobiyolojik ve kimyasal ozellikerindeki degismeler. *Gida Sanayi 6*:34–47, 1992.
90. Uraz, G., N. Gundogan. Beyaz peynirlerin mezofil mikroflorasinda kolifoerm, streptokok, LLP (laktobasil, lokonostok, pediokok) stafilokok ve basilussarin bulunm siklikari. *Gida* 23:391–401, 1998.
91. Phelan, J.A., J. Renaud. Some non-European cheese varieties. In: *Cheese Chemistry, Physics and Microbiology, Vol. 2,* Fox, P.F., ed., London: Chapman & Hall, 1999, pp 421–465.
92. Papademas, P., R.K. Robinson. A comparison of the chemical, microbiological and sensory characteristics of bovine and ovine Halloumi cheese. *Int. Dairy J.* 10:761–768, 2000.
93. Guinee, T.P., P.D. Pudia, N.Y. Farkye. Fresh acid-curd cheese varieties. In: *Cheese Chemistry, Physics and Microbiology.* Vol 2, Fox, P.F., ed., London: Chapman & Hall, 1999, pp 363–419.
94. Robinson, R.K., A.Y. Tamime. Microbiology of fermented milks. In: *Dairy Microbiology – The Microbiology of milk products. 2nd ed.*, Robinson, R.K., ed., London: Elsevier Applied Science, 1990, pp 291–343.

95. Kurmann, J.A., J.L. Rasic, M. Kroger. *Encyclopedia of Fermented Fresh Milk Products*. New York: Van Nostrand Reinbold, 1992, pp 81–287.
96. Tamime, A.Y., V.M.E. Marshall. Microbiology and technology of fermented milks. In: *Microbiology and Biochemistry of Cheese and Fermented Milk. 2nd ed.* Law, B.A., ed., London: Blackie Academic & Professional, 1997, pp 57–152.
97. Tamime, A.Y., R.K. Robinson. Traditional and recent developments in yogurt production. In: *Yogurt: Science and Technology.* 1st ed. Tamime, A.Y., R.K. Robinson, eds., Oxford: Pergamon Press, 1985, pp 234–275.
98. Tamime, A.Y., G. Davies, A.S. Chehade, H.A. Mahdi. The production of labneh by ultrafiltration: a new technology. *J. Soc. Dairy Technol.* 42:35–39, 1989.
99. Tamime, A.Y., M. Kalab, G. Davies, H.A. Mahdi. Microstucture and firmness of labneh (high solid yogurt) made from cow's, goat's and sheep's milk by a traditional method or by ultrafiltration. *Food Struct.* 10:37–44, 1991.
100. Nrgiz, C., A.K. Seckin. The losses of nutrients during the production of strained (Torba) yogurt. *Food Chem.* 61:13–16, 1998.
101. Tamime, A.Y., T.P. O'Connor. Kishk - a dried fermented milk/cereal mixture. *Int. Dairy J.* 5:109–128, 1995.
102. Abou-Donia, S.A., I.A. Attia, A.A. Khattab, Z. El-Shenawi. Formulation of dried cereal fermented milks with prolonged storage life. *Egypt J. Dairy Sci.* 19:283–299, 1991.
103. El–Sadek, G.M., M.R. Zawahry, S.A.Z. Mahmoud, L. Abd El-Motteleb. Chemical composition of Egyptian Kishsk. *Ind. J. Dairy Sci.* 11:67–75, 1958.
104. Atia, I.A., A.A. Khattab. Microbiological and chemical studies of Kishk. *Alex Sci. Exch.* 6:63–71, 1985.
105. Stephanopoulos, O., E. Litopoulou-Tzanetaki, N. Tzanetakis. La flora microbienne du trahana acide. *Ind. Alim. Agric.* 755–757, 1981
106. Thompson, J. Lactose metabolism in *Streptococcus lactis*: phosphorylation of galactose and glucose moieties *in vivo*. *J. Bacteriol.* 140:774–785, 1979.
107. Chass, B.M., C.A. Alpert. Molecular characterization of the plasmid encoded lactose – PTS of *Lactobacillus casei*. *FEMS Microbiol. Rev.* 63:157–166, 1989.
108. Kandler, O. Carbohydrate metabolism in lactic acid bacteria. *Antonie van Leeuwenhoek* 49:209–224, 1983.
109. McKay, L., A. Miller III, W.E. Sandine, P.R. Elliker. Mechanisms of lactose utilization by lactic streptococci: enzymatic and genetic analyses. *J. Bacteriol.* 102:804–809, 1970.
110. Bhowmik, T., E.H. Marth. β-Galactosidase of *Pediococcus* species: induction, purification and partial characterization. *Appl. Microbiol. Biotechnol.* 33:317–323, 1990.
111. Axelsson, L. Lactic acid bacteria: classification and physiology. In: *Lactic Acid Bacteria. Microbiology and Functional Aspects,* 2nd ed., Salminen, S., A. von Wright, eds., New York: Marcel Dekker, 1998, pp 1–72.
112. Bisset, D.L., R.L. Anderson. Lactose and D-galactose metabolism in group N streptococci: presence of enzymes for both the D-galactose-1-phosphate and D-tagatose-6-phosphate pathways. *J. Bacteriol.* 117:318–320, 1974.
113. Konings, W.N., B. Poolman, A.J.M. Driessen. Bioenergetics and solute transport in lactococci. *Crit. Rev. Microbiol.* 16:419–476, 1989.
114. Bandell, M., M.E. Lhotte, C. Marty-Teysset, A. Veyrat, H. Prévost, V. Dartois, C. Diviès, W.N. Konings, J.S. Lolkema. Mechanism of the citrate transporters in carbohydrate and citrate cometabolism in *Lactococcus* and *Leuconostoc* species. *Appl. Environ. Microbiol.* 64:1594–1600, 1998.
115. Hugenholtz, J. Citrate metabolism in lactic acid bacteria. *FEMS Microbiol. Rev.* 12:165–178, 1993.
116. Hugenholtz, J., L. Perdon, T. Abee. Growth and energy generation by *Lactococcus lactis* subsp. *lactis* biovar *diacetylactis* during citrate metabolism. *Appl. Environ. Microbiol.* 59:4216–4422, 1993.
117. Marty-Teysset, C., C. Posthuma, J.S. Lolkema, P. Schmitt, C. Divies, W.N. Konings. Proton motive force generation by citro-lactic fermentation in *Leuconostoc mesenteroides*. *J. Bacteriol.* 178:2178–2185, 1996.

118. De Figueroa, R.M., G. Cerutti de Guglielmone, I.L. Betino de Cárdenas, G. Oliver. Flavor compound production and citrate metabolism in *Lactobacillus rhamnosus* ATCC 7469. *Milchwissenschaft* 53:617–619, 1998.
119. Kimoto, H., M. Nomura, I. Suzuki. Growth energetics of *Lactococcus lactis* subsp. *lactis* biovar *diacetylactis* in cometabolism of citrate and glucose. *Int. Dairy. J.* 9:857–863, 1999.
120. Kennes, C., H.C. Dubourguier, G. Albagnac, E.-J. Nyns. Citrate metabolism by *Lactococcus plantarum* isolated from orange juice. *J. Appl. Bacteriol.* 70:380–384, 1991.
121. Drinan, D.F., S. Tobin, T.M. Cogan. Citric acid metabolism in hetero- and homofermentative lactic acid bacteria. *Appl. Environ. Microbiol.* 31:481–486, 1976.
122. Cogan, T.M. Co-metabolism of citrate and glucose by *Leuconostoc* spp.: effects on growth, substrates and products. *J. Appl. Bacteriol.* 63:551–558, 1987.
123. Coventry, M.J., A.J. Hillier, G.R. Jago. The metabolism of pyruvate and citrate in the thermoduric cheese starter *Streptococcus faecium* (*Streptococcus durans*). *Aust. J. Dairy Technol.* 33:148–154, 1978.
124. Freitas, A.C., A.E. Pintado, M.E. Pintado, F.X. Malcata. Organic acids produced by lactobacilli, enterococci, and yeasts isolated from Picante cheese. *Eur. Food Res. Technol.* 209:434–438, 1999.
125. Sarantinopoulos, P., G. Kalantzopoulos, E. Tsakalidou. Citrate metabolism by *Enterococcus faecalis* FAIR-E 229. *Appl. Environ. Microbiol.* 67:5482–5487, 2001.
126. Sarantinopoulos, P., L. Makras, F. Vaningelgem, G. Kalantzopoulos, L. De Vuyst, E. Tsakalidou. Growth and energy generation by *Enterococcus faecium* FAIR-E 198 during citrate metabolism. *Int. J. Food Microbiol.* 84:197–206, 2003.
127. Piveteau, P. Metabolism of lactate and sugars by propionibacteria: a review. *Lait* 79:23–41, 1999.
128. Hettinga, D.H., G.W. Reinbold. The propionic acid bacteria: a review, I: growth. *J. Milk Food Technol.* 35:295–301, 1972.
129. Hettinga, D.H., G.W. Reinbold. The propionic acid bacteria: a review. II. metabolism. *J. Milk Food Technol.* 35:358–372, 1972.
130. Fox, P.F., P.L.H. McSweeney. Rennets: their role in milk coagulation and cheese ripening. In: *Microbiology and Biochemistry of Cheese and Fermented Milk,* 2nd ed., Law, B.A., ed., London: Blackie Academic & Professional, 1997, pp 1–49.
131. Dalgelish, D.G. The enzymatic coagulation of milk. In: *Cheese Chemistry, Physics and Microbiology*, vol. 1, Fox, P.F., ed., London: Chapman & Hall, 1999, pp 69–100.
132. Barbosa, M., E. Valles, L. Vassal, G. Mocquot. L'utilization d'extrait de *Cynara cardunculus* L. comme agent coagulant en fabrication de fromages a pate molle et a pate cuite. *Lait* 56:1–17, 1976.
133. Barbosa, M., C. Corradini, B. Battistoni. Cheesemaking experiments carried out on some Italian cheeses with vegetable rennet from cardo (*Cynara cardunculus* L.). *Sci. Tec. Latt.-Cas.* 32:203–221, 1981.
134. Grufferty, M.B., P.F. Fox. Milk alkaline proteinase. *J. Dairy Res.* 55:609–630, 1988.
135. Baer, A. Influence of casein proteolysis by starter bacteria, rennet and plasmin on the growth of propionic acid bacteria in Swiss type cheeses. *Lait* 75:391–400, 1995.
136. Law, B.A., J. Kolstadt. Proteolytic systems in lactic acid bacteria. *Antonie van Leeuwenhoek* 49:225–245, 1983.
137. Thomas, T.D., G.G. Pritchard. Proteolytic enzymes of dairy starter cultures. *FEMS Microbiol. Rev.* 46:245-268, 1987.
138. Kunji, E.R.S., I. Mierau, A. Hagting, B. Poolman, W. Konings. The proteolytic systems of lactic acid bacteria. *Antonie van Leeuwenhoek* 70:187–221, 1996.
139. Mierau, I., E.R.S. Kunji, G. Venema, J. Kok. Casein and peptide degradation in lactic acid bacteria. *Biotechnol. Genet. Eng. Rev.* 14:279–301, 1997.
140. Adda, J., J.C. Gripon, L. Vassal. The chemistry of flavor and texture generation in cheese. *Food Chem.* 9:115–129, 1982.
141. Steele, J.L., G. Ünlü. Impact of lactic acid bacteria on cheese flavor development. *Food Technol.* 46:128–130, 1992.

142. McSweeney, P.L.H., M.J. Sousa. Biochemical pathways for the production of flavor compounds in cheeses during ripening: a review. *Lait* 80:293–324, 2000.
143. Foucaud, C., E.R.S. Kunji, A. Hahting, J. Richard, W.N. Konings, M. Desmazeaud, B. Poolman. Specificity of peptide transport system in *Lactococcus lactis*: evidence for a third system, which transports hydrophobic di- and tripeptides. *J. Bacteriol.* 177:4652–4657, 1995.
144. Tynkkynen, S., G. Buits, E. Kunji, J. Kok, B. Poolman, G. Venema, A. Haandrikman. Genetic and biochemical characterisation of the oligopeptide transport system of *Lactococcus lactis*. *J. Bacteriol.* 175:7523–7532, 1993.
145. Christensen, J.E., E.G. Dudley, J.A. Pederson, J. Steele. Peptidases and amino acid catabolism in lactic acid bacteria. *Antonie van Leeuwenhoek* 76:217–246, 1999.
146. Julliard, V., D. Le Bars, E.R.S. Kunji, W.N. Konings, J.C. Gripon, J. Richard. Oligopeptides are the main source of nitrogen for *Lactococcus lactis* during growth in milk. *Appl. Environ. Microbiol.* 61:3024–3030, 1995.
147. Pritchard, G.G., T. Coolbear. The physiology and biochemistry of the proteolytic system in lactic acid bacteria. *FEMS Microbiol. Rev.* 12:179–206, 1993.
148. Holck, A., H. Naes. Cloning, sequencing and expression of the gene encoding the cell-envelope associated proteinases from *Lactobacillus paracasei* subsp. *paracasei* NCDO 151. *J. Gen. Microbiol.* 138:1353–1364, 1992.
149. Gao, S., D.H. Oh, J. Steele. Aromatic amino acid catabolism by lactococci. *Lait* 77:371-381, 1997.
150. Smit, G., A. Verheul, R. van Kranenburg, E. Ayad, R. Siezen, W. Engels. Cheese flavor development by enzymatic conversion of peptides and amino acids. *J. Food Res. Int.* 33:153–160, 2000.
151. Tanous, C., A. Kieronczyk, S. Helinck, E. Chambellon, M. Yvon. Glutamate dehydrogenase activity: a major criterion for the selection of flavor-producing lactic acid bacteria strains. *Antonie van Leeuwenhoek* 82:271–278, 2002.
152. Barbieri, G., I. Bolzoni, M. Careri, A. Manglia, G. Parolari, S. Spangonoli, R. Virgili. Study of the volatile fraction of parmesan cheese. *J. Agric. Food Chem.* 42:1170–1176, 1994.
153. Langsrud, T., T. Sorhaug, G. Vegarud. Protein degradation and amino acid metabolism by propionibacteria. *Lait* 75:325–330, 1995.
154. Sahlstroem, S., C. Espinosa, T. Langsrud, T. Sorhaug. Cell wall, membrane, and intracellular peptidase activities of *Propionibacterium shermanii*. *J. Dairy Sci.* 72:342–350, 1989.
155. Quelen, L.C., C. Dupuis, P. Boyaval. Proline specific activities of *P. freudenreichii* subsp. *shermanii*. *J. Dairy Res.* 62:661–666, 1995.
156. Gagnaire, V., D. Molle, T. Sorhaug, J. Leonil. Peptidases of dairy propionic acid bacteria. *Lait* 79:43–57, 1999.
157. Fleet, G.H. Yeast in dairy products. *J. Appl. Bacteriol.* 68:199–211, 1990.
158. M.T. Wyder, Z. Puhan. Role of selected yeasts in cheese ripening: an evaluation in aseptic cheese curd slurries. *Int. Dairy J.* 9:117–124, 1999.
159. Noomen, A. The role of the surface flora in the softening of cheeses with a low initial pH. *Neth. Milk Dairy J.* 37:229–232, 1983.
160. Wouters, J.T.M., E.H.E. Ayad, J. Hugenholtz, G. Smit. Microbes from raw milk for fermented dairy products. *Int. Dairy J.* 12:91-109, 2002.
161. Fox, P.F., J. Law, P.L.H. McSweeney, J. Wallace. Biochemistry of cheese ripening. In: *Cheese Chemistry, Physics and Microbiology,* Vol. 1, Fox, P.F., ed., London: Chapman & Hall, 1999, pp 389–438.
162. El Soda, M., J. Law, E. Tsakalidou, G. Kalantzopoulos. Lipolytic activity of cheese related microorganisms and its impact on cheese flavor. In: *Food Flavors: Generation, Analysis and Process Influence,* Charalambous, G., ed., Amsterdam: Elsevier, 1995, pp 1823–1847.
163. Kinsella, J.E., D.H. Hwang. Enzymes of *Penicillium roqueforti* involved in the biosynthesis of cheese flavor. *Crit. Rev. Food. Sci. Nutr.* 8:191–228, 1976.
164. Nelson, J.H., R.G. Jensen, R.E. Pitas. Pregastric esterases and other oral lipases: a review. *J. Dairy Sci.* 60:327–362, 1977.

165. Tsakalidou, E., G. Kalantzopoulos. Purification and partial characterization of an intracellular esterase from *Lactococcus lactis* ssp. *lactis* strain ACA-DC 127. *Lait* 72:533–543, 1992.
166. Tsakalidou, E., I. Dalezios, G. Kalantzopoulos. Purification and partial characterization of an intracellular esterase from *Enterococcus faecium* ACA-DC 237. *J. Biotechnol.* 37:201–208, 1994.
167. Holland, R., T. Coolbear. Purification of tributyrin esterase from *Lactococcus lactis* subsp. *cremoris* E8. *J. Dairy Res.* 63:131–140, 1996.
168. Gobbetti, M., P.F. Fox, L. Stephaniak. Isolation and characterization of a tributyrin esterase from *Lactobacillus plantarum* 2739. *J. Dairy Sci.* 80:3099–3106, 1997.
169. Castilo, I., T. Requena, P. Fernandez de Palencia, J. Fontecha, M. Gobbetti. Isolation and characterization of an intracellular esterase from *Lactobacillus casei* subsp. *casei* IFPL731. *J. Appl. Microbiol.* 86:653-659, 1999.
170. Fernández, L., M.M. Beerthuyzen, J. Brown, R.J. Siezen, T. Coolbear, R. Holland, O.P. Kuipers. Cloning, characterization, controlled overexpression, and inactivation of the major tributyrin esterase gene of *Lactococcus lactis. Appl. Environ. Microbiol.* 66:1360–1368, 2000
171. Rattray, F.P., P.F. Fox. Purification and characterization of an intracellular esterase from *Brevibacterium linens* ATCC 9174. *Int. Dairy J.* 7:273–278, 1997.
172. Smacchi, E., M. Gobbetti, J. Rossi, P.F. Fox. Purification and characterization of an intracellular esterase from *Anthracobacter nicotiniae* 9458. *Lait* 80:255–265, 2000.
173. Dupuis, C., P. Boyaval. Esterase activity of dairy *Propionibacterium. Lait* 73:345–356, 1993.
174. Kakariari, E., M.D. Georgalaki, G. Kalantzopoulos, E. Tsakalidou. Purification and characterization of an intracellular esterase from *Propionibacterium freudenreichii* subsp. *freudenreichii* ITG 14. *Lait* 80:491–501, 2000.
175. Cheeseman, G.C. Milk as a food. In: *Therapeutic Properties of Fermented Milks,* Robinson, R.K., ed., Essex: Elsevier Applied Science, 1991, pp 1–21.
176. Renner, E. Nutritional aspects of cheese. In: *Cheese Chemistry, Physics and Microbiology,* Vol. 1, Fox, P.F., ed., London: Chapman & Hall, 1999, pp 357–580.
177. De Vuyst, L., E.J. Vandamme. *Bacteriocins of Lactic Acid Bacteria.* London: Blackie Academic & Professional, 1994.
178. Jack, R.W., J.R. Tagg, B. Ray. Bacteriocins of Gram-positive bacteria. *Microbiol. Rev.* 59:171–200, 1995.
179. Stiles, M.E. Biopreservation by lactic acid bacteria. *Antonie van Leeuwenhoek* 70:331–345, 1996.
180. De Vuyst, L. Technology aspects related to the application of functional starter cultures. *Food Technol. Biotechnol.* 38:105–112, 2000.
181. Sarantinopoulos, P., F. Leroy, E. Leontopoulou, M. Georgalaki, G. Kalantzopoulos, E. Tsakalidou, L. De Vuyst. Bacteriocin production by *Enterococcus faecium* FAIR-E 198 in view of its application as adjunct starter in Greek Feta cheese making. *Int. J. Food Microbiol.* 72:125–136, 2002.
182. Farias, M.E., R.N. Farias, A.P. de Ruiz Holgado, F. Sesma. Purification and N-terminal amino acid sequence of Enterocin CRL 35, a 'pediocin-like' bacteriocin produced by *Enterococcus faecium* CRL 35. *Lett. Appl. Microbiol.* 22:417–419, 1996.
183. Maisnier-Patin, S., N. Deschamps, S.R. Tatini, J. Richard. Inhibition of *Listeria monocytogenes* in Camembert cheese made with a nisin-producing starter. *Lait* 72:249–263, 1992.
184. Stecchini, M.L., V. Aquili, I. Sarais. Behavior of *Listeria monocytogenes* in Mozzarella cheese in presence of *Lactococcus lactis*. *Int. J. Food Microbiol.* 25:301–310, 1995.
185. Giraffa, G, E. Neviani, T.G. Tarelli. Antilisterial activity by enterococci in a model predicting the temperature evolution of Taleggio, an Italian soft cheese. *J. Dairy Sci.* 77:1176–1180, 1994.
186. Nuñez, M., J.L. Rodriguez, E. Garcia, P. Gaya, M. Medina. Inhibition of *Listeria monocytogenes* by enterocin 4 during the manufacture and ripening of Manchego cheese. *J. Appl. Microbiol.* 83:671–677, 1997.

187. Giraffa, G., N. Picchioni, E. Neviani, D. Carminati. Production and stability of an *Enterococcus faecium* bacteriocin during Taleggio cheesemaking and ripening. *Food Microbiol.* 12:301–307, 1995.
188. Giraffa, G., D. Carminati. Control of *Listeria monocytogenes* in the rind of Taleggio, a surface-smear cheese, by a bacteriocin from *Enterococcus faecium* 7C5. *Sci. Alim.* 17:383–391, 1997.
189. Stratton, J.E., R.W. Hutkins, S.L. Taylor. Biogenic amines in cheese and other fermented foods: a review. *J. Food Prot.* 54:460–470, 1991.
190. Santos, M.H.S. Biogenic amines: their importance in foods. *Int. J. Food Microbiol.* 29:213–231, 1996.
191. Joosten, H.M.L.J. Conditions allowing the formation of biogenic amines in cheese, 3: factors influencing the amounts formed. *Neth. Milk Dairy J.* 42:329–357, 1988.
192. Stratton, J.E., R.W. Hutkins, S.S. Sumner, S.L. Taylor. Histamine and histamine producing bacteria in retail Swiss and low-salt cheeses. *J. Food Prot.* 55:435–439, 1992.
193. Schneller, R., P. Good, M. Jenny. Influence of pasteurized milk and different ripening cultures on biogenic amine concentrations in semi-soft cheeses during ripening. *Zeitschrift Lebensmittel Untersuchung Forschung A.* 204:265–272, 1997.
194. Halasz, A., A. Barath, L. Simon-Sarkadi, W. Holzapfel. Biogenic amines and their production by microorganisms in food. *Trends Food Sci. Technol.* 5:42–49, 1994.
195. Ordonez, J.A., F.C. Ibanez, P. Torre, Y. Barcina. Formation of biogenic amines in Idiazabal ewe's milk cheese: effect of ripening, pasteurisation and starter. *J. Food Prot.* 60:1371–1375, 1997.
196. Valsamaki, K., A. Michaelidou, A. Polychroniadou. Biogenic amine production in Feta cheese. *Food Chem.* 71:259–266, 2000.
197. Durlu-Oezkaya, F., E. Alichanidis, E. Litopoulou-Tzanetaki, N. Tunail. Determination of biogenic amines content of Beyaz cheese and biogenic amine production ability of some lactic acid bacteria. *Milchwissenschaft* 54:680–682, 1999.
198. Pinho, O., I.M.P.L.V.O. Ferreira, E. Mendes, B.M. Oliveira, M. Ferreira. Effect of temperature on evolution of free amino acid and biogenic amine contents during storage of Azeitão cheese. *Food Chem.* 75:287–291, 2001.
199. El-Sayed, M.M. Biogenic amines in processed cheese available in Egypt. *Int. Dairy J.* 6:1079–1086, 1996.
200. Tawfik, N.F., A.R. Shalaby, B.A. Effat. Biogenic amine contents in Ras cheese and incidence of their bacterial producers. *Egypt J. Dairy Sci.* 20:219–225, 1992.
201. Kebary, K.M.K., A.H. El-Sonbaty, R.M. Badawi. Effects of heating milk and accelerating ripening of low fat Ras cheese on biogenic amines and free amino acids development. *Food Chem.* 64:67–75, 1999.
202. Guarner, F., G.J. Schaafsma. Probiotics. *Int. J. Food Microbiol.* 39:237–238, 1998.
203. Y.K. Lee, S. Salminen. The coming age of probiotics. *Trends Food Sci. Technol.* 6:241–245, 1995.
204. Berg, R.D. Probiotics, prebiotics or "conbiotics". *Trends Microbiol.* 6:89–92, 1998.
205. O'Sullivan, M.G., G.M. Thorton, G.C. O'Sullivan, J.K. Collins. Probiotic bacteria: myth or reality? *Trends Food Sci. Technol.* 3:309–314, 1992.
206. Dunne, C., L. Murphy, S. Flynn, L. O'Mahony, S. O'Halloran, M. Feeney, D. Morrissey, G. Thornton, G. Fitzgerald, C. Daly, B. Kiely, E.M. Quigley, G.C. O'Sullivan, F. Shanahan, J.K. Collins. Probiotics: from myth to reality: demonstration of functionality in animal models of disease and in human clinical trials. *Antonie van Leeuwenhoek* 76:279–292, 1999.
207. Tannock, G.W. *Probiotics. A critical review.* Wymondham: Horizon Press, 1999.
208. Ouwehand, A.C., S. Salminen, E. Isolauri. Probiotics: an overview of beneficial effects. *Antonie van Leeuwenhoek* 82:279–289, 2002.
209. Sanders, M.E., J. Huis in't Veld. Bringing a probiotic containing functional food to the market: microbiological, product, regulatory and labelling issues. *Antonie van Leeuwenhoek* 76:293–315, 1999.
210. Meisel, H. Overview on milk protein derived peptides. *Int. Dairy J.* 8:363–373, 1998.

211. Tanako, T. Anti-hypertensive activity of fermented dairy products containing biogenic peptides. *Antonie van Leeuwenhoek* 82:333–340, 2002.
212. Meisel, H., W. Bockelmann. Bioactive peptides encrypted in milk proteins: proteolytic activation and thropho-functional properties. *Antonie van Leeuwenhoek* 76:207–215, 1999.
213. Smacchi, E., M. Gobetti. Peptides from several Italian cheeses inhibitory to proteolytic enzymes of lactic acid bacteria, *Pseudomonas fluorescens* TACC 948 and to the angiotensin I-converting enzyme. *Enzym. Microbiol. Technol.* 22:687–694, 1998.
214. Zeuthen, P. Historical aspects of meat fermentations. In: *Fermented Meats.* Campbell-Platt, G., P.E. Cook, eds., London: Blackie Academic & Professional, 1995, pp 53–68.
215. Lücke, F.K. Fermented sausages. In: *Microbiology of Fermented Foods,* Vol. 2, 2nd ed., Wood, B.J.B., ed., London: Blackie Academic & Professional, 1998, pp 441–483.
216. Pederson, C.S. Fermented sausage. In: *Microbiology of Food Fermentations*, 2nd ed., Pederson, C.S., ed., Westport, CT: AVI Publishing, 1979, pp 210–234.
217. Liepe, H.U. Starter cultures in meat production. In: *Biotechnology: A Comprehensive Treatise in 8 Volumes,* Vol. 5, Rehm, H.J., G. Reed, eds., Weinheim: Verlag Chemie, 1983, pp 399–424.
218. Jessen, B. Starter cultures for meat fermentions. In: *Fermented Meats,* Campbell-Platt, G., P.E. Cook, eds., London: Blackie Academic & Professional, 1995, pp 130–159.
219. Flores, M., D.A. Ingram, K.L. Bett, F. Toldra, A.M. Spanier. Sensory characteristics of Spanish "Serrano" dry-cured ham. *J. Sens. Stud.* 12:169–179, 1997.
220. Gonzalez, C.B., H.W. Ockerman. Dry-cured Mediterranean hams: long process, slow changes and high quality: a review. *J. Muscle Food* 11:1–17, 2000.
221. Unluturk, A., F. Turantas. Fate of coliforms in Turkish soudjuk during ripening and storage. *J. Sci. Food Agric.* 57:399–404, 1991.
222. Kotzekidou, P., H.N. Lazarides. Microbial stability and survival of pathogens in an intermediate moisture meat product. *Lebensm. Wiss. U. Technol.* 24:419–423, 1991.
223. Bennani, L., M. Faid, A. Bouseta. Experimental manufacturing of kaddid, a salted dried meat product: control of the microorganisms. *Eur. Food Res. Technol.* 211:153–157, 2000.
224. Kröckel, L. Bacterial fermentation of meats. In: *Fermented Meats,* Campbell-Platt, G., P.E. Cook, eds., London: Blackie Academic & Professional, 1995, pp 69–109.
225. Raccach, M. Manganese and lactic acid bacteria. *J. Food Prot.* 48:895–898, 1985.
226. Collins-Thompson, D.L., I.Q. Thomson. Changes in manganese content in *Lactobacillus plantarum* during inhibition with sodium nitrite. *J. Food Prot.* 49:602–604, 1986.
227. Marchesini, B., A. Bruttin, N. Romailler, R.S. Moreton, C. Stucchi, T. Sozzi. Microbiological events during commercial meat fermentations. *J. Appl. Bact.* 73:203–209, 1992.
228. Demeyer, D., M. Raemaekers, A. Rizzo, A. Holck, A. De Smedt, B. ten Brink, B. Hagen, C. Montel, E. Zanardi, E. Murbrekk, F. Leroy, F. Vandendriessche, K. Lorentsen, K. Venema, L. Sunesen, L. Stahnke, L. De Vuyst, R. Talon, R. Chizzolini, S. Eerola. Control of bioflavor and safety in fermented sausages: first results of a European project. *Food Res. Int.* 33:171–180, 2000.
229. Flores, J., S. Bermell. Dry-cured sausages: factors influencing souring and their consequences. *Fleischwirtschaft* 76:163–165, 1996.
230. Mendoza, S., J. Flores, H. Silla. Influencia de la temperatura de estufado sobre las caracteristicas microbiologicas y quimicas del chorizo. *Rev. Agroquim. Tecnol. Alim.* 23:86–96, 1983.
231. Flores, J. Mediterranean vs. northern European meat products: processing technologies and main differences. *Food Chem.* 59:505–510, 1997.
232. Toldra, F., M.C. Aristoy, M. Flores. Contribution of muscle aminopeptidases to flavor development in dry-cured ham. *Food Res. Int.* 33:181–185, 2000.
233. Toldra, F., M. Flores, Y. Sanz. Dry-cured ham flavor: enzymatic generation and process influence. *Food Chem.* 59:523–530, 1997.
234. Hammes, W.P. Starterkulturen in der Fleischwirtschaft. *Chemie Mikrobiologie Technologie Lebensmittel* 9:131–142, 1986.

235. Cornejo, I., A.V. Carrascosa, M.E. Marin, P.J. Martin. Dry-cured ham consideration about the origin of microorganisms that grow on the deep muscular tissues of dry-cured Spanish hams during processing. *Fleischwirtschaft* 72:1422–1425, 1992.
236. Jaud, D., A. Fischer. Modellversuche zur Verbesserung der Pökelbereitschaft von DFD-Fleisch durch pH-Wert-Senkung. *Fleischerei* 44:545–548, 1993.
237. Lücke, F.K. Utilization of microbes to process and preserve meat. *Meat Sci.* 56:105–115, 2000.
238. Lücke, F.K., H. Hechelmann. Starter cultures for dry sausages and raw ham: composition and effect. *Fleischwirtschaft* 67:307–314, 1987.
239. Callewaert, R., M. Hugas, L. De Vuyst. Competitiveness and bacteriocin production of *Enterococci* in the production of Spanish-style dry fermented sausages. *Int. J. Food Microbiol.* 57:33–42, 2000.
240. Giraffa, G. Enterococci from foods. *FEMS Microbiol. Rev.* 26:163–171, 2002.
241. Rodriguez, M., F. Nunez, J.J. Cordoba, C. Sanabria, M.E. Bermudez, M.A. Asensio. Characterization of *Staphylococcus* spp. and *Micrococcus* spp. isolated from Iberian ham throughout the ripening process. *Int. J. Food Microbiol.* 24:329–335, 1994.
242. Rodriguez, M., F. Nunez, J.J. Cordoba, E. Bermudez, M.A. Asensio. Gram-positive, catalase-positive cocci from dry cured Iberian ham and their enterotoxiginenic potential. *Appl. Environ. Microbiol.* 62:1897–1902, 1996.
243. Hammes, W.P., C. Hertel. New developments in meat starter cultures. *Meat Sci.* 49(1): S125–S138, 1998.
244. Bacus, J.N. Fermented meat and poultry products. *Adv. Meat Res.* 2:123–164, 1986.
245. Ockerman, H.W., F.J.C. Sanchez, F.L. Crespo. Influence of molds on flavor quality of Spanish ham. *J. Muscle Foods* 11:247–259, 2000.
246. Coppola, R., B. Giagnacovo, M. Iorizzo, L. Grazia. Characterization of lactobacilli involved in the ripening of soppressata molisana, a typical southern Italy fermented sausage. *Food Microbiol.* 15:347–353, 1998.
247. Coppola, R., M. Iorizzo, R. Saotta, E. Sorrentino, L. Grazia. Characterization of micrococci and staphylococci isolated from soppressata molisana, a Southern Italy fermented sausage. *Food Microbiol.* 14:47–53, 1997.
248. Coppola, S., G. Mauriello, M. Aponte, G. Moschetti, F. Villani. Microbial succession during ripening of Naples-type salami, a southern Italian fermented sausage. *Meat Sci.* 56:321–329, 2000.
249. Cocolin, L., M. Manzano, D. Aggio, C. Cantoni, G. Comi. A novel polymerase chain reaction (PCR) - denaturing gradient gel electrophoresis (DGGE) for the identification of *Micrococcaceae* strains involved in meat fermentations. Its application to naturally fermented Italian sausages. *Meat Sci.* 58:59–64, 2001.
250. Dolazo, F., U. Steinhof, C. Ring, S. Pfeiffer. Microbiological status of the Spanish fermented sausage "Chorizo Gallego": sensory, chemical and chemo-physical properties. *Fleischwirtschaft* 78:1089–1092, 1998.
251. Santos, E.M., C. Gonzalez-Fernandez, I. Jaime, J. Rovira. Comparative study of lactic acid bacteria house flora isolated in different varieties of "chorizo". *Int. J. Food Microbiol.* 39:123–128, 1998.
252. Hugas, M., M. Garriga, T. Aymerich, J.M. Monfort. Biochemical characterization of lactobacilli from dry fermented sausages. *Int. J. Food Microbiol.* 18:107–113, 1993.
253. Garcia-Varona, M., E.M. Santos, I. Jaime, J. Rovira. Characterisation of *Micrococcaceae* isolated from different varieties of chorizo. *Int. J. Food Microbiol.* 54:189–195, 2000.
254. Garriga, M., M. Hugas, P. Gou, M.T. Aymerich, J. Arnau, J.M. Monfort. Technological and sensorial evaluation of lactobacillus strains as starter cultures in fermented sausages. *Int. J. Food Microbiol.* 32:173–183, 1996.
255. Fernandez, L.B., V.A. Diez. Antibacterial activity of lactobacilli isolated from "chorizo". *Fleichwirtschaft* 72:1005–1007, 1992.
256. Gonzalez, B., V. Diez. The effect of nitrite and starter culture on microbiological quality of "chorizo": a Spanish dry cured sausage. *Meat Sci.* 60:295–298, 2002.

257. Bucharles, C., J.P. Girard, J. Sirami, S. Pascal. Characteristics of a dry sausage showing excessive acidity. *Sci. Alim.* 4:137–143, 1984.
258. Lizaso, G., J. Chasco, M.J. Beriain. Microbial and biochemical changes during ripening of salchichon, a Spanish dry cured sausage. *Food Microbiol.* 16:219–228, 1999.
259. Samelis, J., F. Maurogenakis, J. Metaxopoulos. Characterization of lactic acid bacteria isolated from naturally fermented Greek dry salami. *Int. J. Food Microbiol.* 23:179–196, 1994.
260. Papamanoli, E., N. Tzanetakis, E. Litopoulou-Tzanetaki, P. Kotzekidou. Characterization of lactic acid bacteria isolated from a Greek dry-fermented sausage in respect of their technological and probiotic properties. *Meat Sci.* 65:859–867, 2003.
261. Samelis, J., J. Metaxopoulos, M. Vlassi, A. Pappa. Stability and safety of traditional Greek salami: a microbiological ecology study. *Int. J. Food Microbiol.* 44:69–82, 1998.
262. Collins, M.D., J. Samelis, J. Metaxopoulos, S. Wallbanks. Taxonomic studies on some *Leuconostoc*-like organisms from fermented sausages: description of a new genus *Weissella* for the *Leuconostoc paramesenteroides* group of species. *J. Appl. Bacteriol.* 75:595–603, 1993.
263. Papamanoli, E., P. Kotzekidou, N. Tzanetakis, E. Litopoulou-Tzanetaki. Characterization of *Micrococcaceae* isolated from dry fermented sausage. *Food Microbiol.* 19:441–449, 2002.
264. Gürakan, G.C., T.F. Bozoglu, N. Weiss. Identification of *Lactobacillus* strains from Turkish-style dry fermented sausages. *Lebensm Wiss. U. Technol.* 28:139–144, 1995.
265. Lopes, M.F.S., A.L. Leitao, M. Regalla, J.J.F. Marques, M.J.T. Carrondo, M.T.B. Crespo. Characterization of a highly thermostable extracellular lipase from *Lactobacillus plantarum. Int. J. Food Microbiol.* 76:107–115, 2002.
266. Silla, H., I. Molina, J. Flores, D. Silvestre. Study of the microbial flora in dry-cured ham, 1: isolation and growth. *Fleischwirtschaft* 69:1128–1131, 1989.
267. Molina, I., H. Silla, J. Flores, J.L. Monzo. Study of the microbial flora in dry-cured ham. *Fleischwirtschaft* 70:54–56, 1990.
268. Comi, G., C. Cantoni, F. Celori. Considerazioni sugli stafilococchi coagulasi negativi degli insaccati crudi stagionati. *Industrie Alimentari* 25:378–380, 1986.
269. Carrascosa, A.V., I. Cornejo. Characterization of *Micrococcaceae* strains selected as potential starter cultures to Spanish dry-cured ham processes, 2: slow process. *Fleischwirtschaft* 71:1207–1208, 1991.
270. Cornejo, I., A.V. Carrascosa. Characterization of *Micrococcaceae* strains selected as potential starter cultures to Spanish dry-cured ham processes,1: fast process. *Fleischwirtschaft* 71:99–101, 1991.
271. Marin, M.E., A.V. Carrascoda, I. Cornejo. Characterization of *Enterobacteriaceae* strains isolated during industrial processing of dry-cured hams. *Food Microbiol.* 13:375–381, 1996.
272. Cordero, M.R., J.M. Zumalacarregui. Characterization of Micrococcaceae isolated from salt used for Spanish dry-cured ham. *Lett. Appl. Microbiol.* 31:303–306, 2000.
273. El-Khateib, T., U. Schmidt, L. Leistner. Mikrobiologische Stabilität von türkischer Pastirma. *Fleischwirtschaft* 67:101–105, 1987.
274. Kotzekidou, P. Identification of staphylococci and micrococci isolated from an intermediate moisture meat product. *J. Food Sci.* 57:249–251, 1992.
275. Toldra, F. Proteolysis and lipolysis in flavor development of dry-cured meat products. *Meat Sci.* 49:S101–S110, 1998.
276. Montel, M.C., F. Masson, R. Talon. Bacterial role in flavor development. *Meat Sci.* 49(1): S111–S123, 1998.
277. Leistner, L. The essentials of producing stable and safe raw fermented sausages. In: *New Technologies for Meat and Meat Products,* Smulders, F.J.M., F. Toldra, J. Flores, M. Prieto, eds., Nijmegen: Audet Tijdschriften, 1992, pp 1–19.
278. Demeyer, D., A. Verplaetse, M. Gistelink. Fermentation of meat an integrated process. *Proceedings of 32nd European Meeting of Meat Research Workers*, Ghent, 1986, pp 241–247.
279. Rodriguez, M., F. Nunez, J.J. Cordoba, M.E. Bermudez, M.A. Asensio. Evaluation of proteolytic activity of micro-organisms isolated from dry cured ham. *J. Appl. Microbiol.* 85:905–912, 1998.

280. Beriain, M.J., G. Lizaso, J. Chasco. Free amino acids and proteolysis involved in "salchichon" processing. *Food Control* 11:41–47, 2000.
281. Campanini, M., I. Pedrazzoni, S. Barbuti, P. Baldini. Behaviour of *Listeria monocytogenes* during maturation of naturally and artificially contaminated salami: effect of lactic acid bacteria starter cultures. *Int. J. Food Microbiol.* 20:169–175, 1993.
282. McCormick, J.K., T.R. Klaenhammer, M.E. Stiles. Colicin V can be produced by lactic acid bacteria. *Lett. Appl. Microbiol.* 29:37–41, 1999.
283. Schillinger, U., R. Geisen, W.H. Holzapfel. Potential of antagonistic microorganisms and bacteriocins for the biological preservation of foods. *Trends Food Sci. Technol.* 7:158–164, 1996.
284. Hugas, M., M. Garriga, T. Aymerich, J.M. Monfort. Inhibition of *Listeria* in dry fermented sausages by the bacteriocinogenic *Lactobacillus sake* CTC494. *J. Appl. Bacteriol.* 79:322–330, 1995.
285. Cintas, L.M., P. Casaus, L. Havarstein, P.E. Hernandez, I.F. Nes. Biochemical and genetic characterization of enterocin P, a novel sec-dependent bacteriocin from *Enterococcus faecium* P13 with a broad antimicrobial spectrum. *Appl. Environ. Microbiol.* 63:4321–4330, 1997.
286. Herranz, C., P. Casaus, S. Mukhopadhyay, J.M. Martinez, J.M. Rodriguez, I.F. Nes, P.E. Hernandez, L.M. Cintas. *Enterococcus faecium* P21: a strain occuring naturally in dry-fermented sausages producing the class II bacteriocins enterocin A and enterocin B. *Food Microbiol.* 18:115–131, 2001.
287. J.C. Nieto-Lozano, J.I. Reguera-Useros, M.C. Pelaez-Martinez, A. Hardisson de la Torre. Bacteriocinogenic activity from starter cultures used in Spanish meat industry. *Meat Sci.* 62:237–243, 2002.
288. Parente, E., M. Moles, A. Ricciardi. Leucocin F10, a bacteriocin from *Leuconostoc carnosum. Int. J. Food Microbiol.* 33:231–243, 1996.
289. Aymerich, M.T., M. Garriga, J.M. Monfort, I. Nes, M. Hugas. Bacteriocin-producing lactobacilli in Spanish-style fermented sausages: characterization of bacteriocins. *Food Microbiol.* 17:33–45, 2000.
290. Samelis, J., S. Roller, J. Metaxopoulos. Sakacin B, a bacteriocin produced by *Lactobacillus sake* isolated from Greek dry fermented sausages. *J. Appl Bacteriol.* 76:475–486, 1994.
291. Coppola, R., B. Giagnacovo, M. Iorizzo, L. Grazia, F. Nazzaro, B. De Giulio, L. Maurelli. Influence of *Lactobacillus sakei* 121 on some microbiological and chemical parameters of "Soppressata Molisana" sausage. *Ind. Alim.* 40:1326–1330, 2001.
292. Con, A.H., M. Kaya, H.Y. Gökalp. Isolierung und Identifizierung von *Listeria monocytogenes* und weiteren listerienarten aus der türkischen rohwurst 'sucuk'. *Archiv für Lebensmittelhygiene* 47:65–66, 1996.
293. Con, A.H., H.Y. Gökalp, M. Kaya. Antagonistic effect on *Listeria monocytogenes* and *L. innocua* of a bacteriocin-like metabolite produced by lactic acid bacteria isolated from sucuk. *Meat Sci.* 59:437–441, 2001.
294. Ceylan, E., D.Y.C. Fung. Destruction of *Yersinia enterocolitica* by *Lactobacillus sake* and *Pediococcus acidilactici* during low-temperature fermentation of Turkish dry sausage (sucuk). *J. Food Sci.* 65:876–879, 2000.
295. Yurtyeri, A., B. Mutluer, I. Erol, G. Hildebrandt. Meat and sausage products from Turkey: constitutional and technological aspects of Turkish raw sausage. *Fleischerei* 44:3–8, 1993.
296. Bozkurt, H., O. Erkmen. Effects of starter cultures and additives on the quality of Turkish style sausage (sucuk). *Meat Sci.* 61:149–156, 2002.
297. Losantos, A., C. Sanabria, I. Cornejo, A.V. Carrascosa. Characterization of *Enterobacteriaceae* strains isolated from spoiled dry-cured hams. *Food Microbiol.* 17:505–512, 2000.
298. Paarup, T., J.C. Nieto, C. Pelaez, J.I. Reguera. Microbiological and physicochemical characterisation of deep spoilage in Spanish dry-cured hams and characterisation of isolated *Enterobacteriaceae* with regard to salt and temperature tolerance. *Eur. Food Res. Technol.* 209:366–371, 1999.
299. Marin, M.E., M.C. Delarosa, A.V. Carrascoda, I. Cornejo. Microbiological and physicochemical aspects of spoiled dry-cured Spanish hams. *Fleischwirtschaft* 72:1600–1605, 1992.

300. Genigeorgis, C., S. Lindroth. The safety of Basturma, an armenian-type dried beef with respect to *Salmonella, Proceedings of 30th European Meeting of Meat Research Workers,* Bristol, 1984, pp 211–218.
301. Katsaras, K., R. Lautenschläger, K. Boschkova. Das Verhalten von Mikroflora und Starterkulturen während der Pökelung, Trocknung und Lagerung von Pasterma. *Fleischwirtschaft* 76:308–314, 1996.
302. Aksu, M.I., M. Kaya. The effect of starter culture use in pastirma production on the properties of end product. *Turkish J. Vet. Anim. Sci.* 25:847–854, 2001.
303. Aksu, M.I., M. Kaya. Some microbiological and chemical properties of pastirma produced using potassium nitrate and starter culture. *Turkish J. Vet. Anim. Sci.* 26:125–132, 2002.
304. Katsaras, K., L. Leistner. Distribution and development of bacterial colonies in fermented sausages. *Biofoul* 5:115–124, 1991.
305. Ansorena, D., M.C. Montel, M. Rokka, R. Talon, S. Eerola, A. Rizzo, M. Raemaekers, D. Demeyer. Analysis of biogenic amines in northern and southern European sausages and role of flora in amine production. *Meat Sci.* 61:141–147, 2002.
306. Masson, F., R. Talon, M.C. Montel. Histamine and tyramine production by bacteria from meat products. *Int. J. Food Microbiol.* 32:199–207, 1996.
307. Halásy, A., A. Baráth, L. Simon-Sarkadi, W.H. Holzapfel. Biogenic amines and their production by microorganisms in foods. *Trends Food Sci. Technol.* 5:42–49, 1994.
308. Pereira, C.I., M.T. Barreto Crespo, M.V. San Romao. Evidence for proteolytic activity and biogenic amines production in *Lactobacillus curvatus* and *L. homohiochii. Int. J. Food Microbiol.* 68:211–216, 2001.
309. Parente, E., M. Martuscelli, F. Gardini, S. Grieco, M.A. Crudele, G. Suzzi. Evolution of microbial populations and biogenic amine production in dry sausages produced in Southern Italy. *J. Appl. Microbiol.* 90:882–891, 2001.
310. Shalaby, A.R., H.A.A. Elrahman. Effect of potassium sorbate on development of biogenic-amines during sausage fermentation. *Nahrung* 39:308–315, 1995.
311. Gardini, F., M. Martuscelli, M.A. Crudele, A. Paparella, G. Suzzi. Use of *Staphylococcus xylosus* as a starter culture in dried sausages: effect on the biogenic amine content. *Meat Sci.* 61:275–283, 2002.
312. Hernandez Jover, T., M. Izquierdo Pulido, M.T. Veciana Nogues, A. Marine Font, M.C. Vidal Carou. Effect of starter cultures on biogenic amine formation during fermented sausage production. *J. Food Prot.* 60:825–830, 1997.
313. Bover-Cid, S., M. Izquierdo-Pulido, M.C. Vidal-Carou. Effect of the interaction between a low tyramine-producing *Lactobacillus* and proteolytic staphylococci on biogenic amine production during ripening and storage of dry sausages. *Int. J. Food Microbiol.* 65:113–123, 2001.
314. Edwards, R.A., R.H. Dainty, C.M. Hibbard, S.V. Ramantanis. Amines in fresh beef of normal pH and the role of bacteria in changes in concentration observed during storage in vacuum packs at chill temperatures. *J. Appl. Bacteriol.* 63:427–434, 1987.
315. Leuschner, R.G.K., W.P. Hammes. Tyramine degradation by micrococci during ripening of fermented sausages. *Meat Sci.* 49:289–296, 1998.
316. Cooper, R.A. On the amine oxidases of *Klebsiella aerogenes* strain W70. *FEMS Microbiol. Lett.* 146:85–89, 1997.
317. Salminen, S., M. Deighton, Y. Benno, S. Gorbach. Lactic acid bacteria in health and disease. In: *Lactic Acid Bacteria,* Salminen, S., A. von Wright, eds., New York: Marcel Dekker, 1998, pp 211–254.
318. Sameshima, T., C. Magome, K. Takeshita, K. Arihara, M. Itoh, Y. Kondo. Effect of intestinal *Lactobacillus* starter cultures on the behaviour of *Staphylococcus aureus* in fermented sausage. *Int. J. Food Microbiol.* 41:1–7, 1998.
319. Erkkilä, S., E. Petäjä. Screening of commercial meat starter cultures at low pH and in the presence of bile salts for potential probiotic use. *Meat Sci.* 55:297–300, 2000.

3.21

Anaerobic Processes for the Treatment of Food Processing Wastes

Roger A. Korus

CONTENTS

21.1 INTRODUCTION

The effective treatment of wastewaters is crucial to the operation of food processing plants. Whereas aerobic processes effectively treat low strength wastes, anaerobic processes more effectively treat high strength wastes. These high strength wastes have chemical oxygen demand (COD) levels greater than 2,000 mg/L or biological oxygen demand (BOD) levels greater than 1,000 mg/L. However, these thresholds can be lowered by biomass retention and proper nutrient supplementation, and if the wastewater is warm and contains adequate alkalinity. Anaerobic processes treat high strength wastes without the high energy requirements and biomass production of aerobic processes. In addition methane rich biogas is produced as a renewable energy source. Up to 0.395 liters of methane (measured at 35°C and 1 atm) can be produced per gram of COD removed.

Anaerobic treatment is often a pretreatment method where effluent BOD is typically 100–500 mg/L and may be followed by aerobic treatment to achieve treatment objectives. However, there are examples of anaerobic treatment to less than 20 mg/l BOD (1). Anaerobic processes are typically conducted at approximately 35°C in the pH range of 6.8–7.4 and convert soluble organic carbon into carbon dioxide and methane. Toxic compounds can adversely affect anaerobic treatment, but toxins are not usually a major problem in food processing wastes. In order to assess the anaerobic treatability of a wastewater, a laboratory or pilot plant study is generally advisable. Especially with wastewaters that contain compounds of low solubility such as fats, oils, or greases (FOG) or complex carbohydrates, a treatability study is essential. Also, there is a wide range of anaerobic technologies to choose from. Fortunately, there are hundreds of anaerobic treatment facilities in operation and even more published research studies that guide in the selection of treatment options and feasibility determinations.

Over the past thirty years, the popularity of anaerobic wastewater treatment has greatly increased. In large part this has resulted from the development of high rate systems that overcome the slow growth rate of anaerobic microorganisms by retaining these organisms within the anaerobic reactor. With an increase in anaerobic biomass, much higher rates of anaerobic digestion can be obtained. Also, these systems are less susceptible to operational upsets, and start up can be accomplished within a few weeks using highly active anaerobic sludge from an existing facility.

21.2 TREATABILITY ASSESSMENTS

Initial laboratory tests can be used to determine the appropriateness of anaerobic treatment for a particular wastewater. These tests can determine alkalinity, rate of COD/BOD removal, waste toxicity, and supplemental nutrient requirements. Additional pilot plant tests can determine operational parameters. It is important in these tests to assess the need for acclimation of the anaerobic biomass to unfamiliar substrates and toxins. This acclimation period is not required if anaerobic biomass is obtained from a process treating a similar wastewater. If not, an acclimation period of 30–60 days may be required before treatability conclusions can be made.

The characterization of waste streams should not be neglected. Nearly always, some wastewater streams in a food processing plant will be more amenable to anaerobic treatment. The BOD and COD levels, pH, temperatures, and flow rates of the principal streams should be determined. Often the high level waste streams can be segregated, and treatment of these streams can achieve most of the treatment objectives. Once the wastewater streams to be treated are identified, representative samples can be taken for laboratory anaerobic tests. It is crucial that these samples be representative or time averaged samples that truly reflect the wastewaters to be treated.

Laboratory tests are conducted in well mixed batch reactors containing 50 mL to several liters of wastewater with a 10–20% inoculum. This inoculum can be obtained from an operating anaerobic reactor or a municipal anaerobic sludge digestor. The key measurement is the methane production rate. In this measurement it is important to measure only methane and not simply biogas production. The carbon dioxide can be removed by bubbling through a NaOH solution or can be determined by gas chromatography. This test should be run until the cessation of gas production, typically requiring 30–90 days with longer times required for acclimation to toxic or wastewaters of composition much different than the inoculum.

In laboratory tests temperature is typically held at 32–35°C. Alkalinity and pH are also measured during the test. The pH should not drop below 6, and if the alkalinity drops below 500 mg/L the dissolved CO_2 will cause the pH to drop below 6. Bicarbonate (HCO_3^-) will neutralize acid generated during anaerobic treatment by the following reaction:

$$HCO_3^- + H + \rightarrow H_2O + CO_2$$

As bicarbonate alkalinity is consumed by this reaction, there is reduced buffering. Supplemental alkalinity must be supplied often as Na_2CO_3 whenever tests indicate that alkalinity might be less than 2500–3000 mg/L during anaerobic treatment. Several titration methods are available for the determination of alkalinity and volatile acid concentrations (2,3). An alternative spectrophotometric method has recently been described (4). Alkalinity supplementation can be a major cost and can exceed in cost the value of methane produced.

Laboratory tests can determine whether supplemental nutrients can improve anaerobic treatment. Nitrogen and phosphorus are the major elements that might be limiting. For every 1000 parts BOD, 6–10 parts nitrogen and 1–2 parts phosphorus are required, much less than for aerobic treatment where the biomass production is much greater. Therefore, in anaerobic treatment nitrogen and phosphorus are seldom limiting. More likely it is trace metals that can be limiting. Speece recommends that a cocktail of 1 mg/L $FeCl_2$, 0.1 mg/L $CoCl_2$, and 0.1 mg/L $NiCl_2$ be added if high volatile acids persist in the effluent (1). There are many examples of iron, cobalt, and nickel supplementation increasing the rate and extent of anaerobic treatment (1). In addition, zinc, copper, manganese, molybdenum, selenium, tungsten, and boron have been reported to stimulate anaerobic activity (1). These trace elements can be quickly tested in laboratory assays where equal aliquots of biomass are placed in flasks and then supplemented with trace metals. Gas production can be used to determine if metal supplementation enhances the rate of gas production relative to the control. If methane yields are much less than the theoretical 0.395 liters per gram of COD, and if volatile fatty acid concentrations remain high after trace element additions, then the reasons for low methanogen activity should be investigated. Possible reasons are toxicity of the wastewater, or the washout, or low inoculum activity of the methanogens.

21.3 MONITORING AND PROCESS CONTROL

Monitoring and control are crucial to the successful operation of anaerobic process. The principal operational variables that must be controlled include flow rate of wastewater, pH and alkalinity, temperature, and foaming. Monitoring is essential to provide reliable indicators of operational problems and as the basis for control strategies that will maintain stable and efficient operation.

Volumetric loading rates of 1–25 kg BOD/m^3-day can be obtained in anaerobic digesters, but the higher loading rates require highly active, acclimated biomass concentrations of 10–50 g/L. These high biomass levels require either a high inoculum of acclimated biomass, or a long start up time, with proportionally low feed rates until the active biomass reaches a high concentration. These high biomass levels can only be reached in digesters that retain or recycle biomass. During startup and steady state operation, gas production, pH, alkalinity, volatile acid concentration, COD/BOD, and loading rate should be measured. These diagnostic tools will determine whether the digester is overloaded or if toxicity, nutrient limitations, or other reactor conditions are limiting performance.

The best monitoring gives the earliest indication of operational problems. The most common and serious problems usually result in a drop in pH preceded by a reduction in alkalinity. This can occur during startup if the digester is overloaded and during continuous operation if the methanogen activity is inadequate. Alkalinity is generally expressed as $CaCO_3$ equivalents. Given that the molecular weight of $CaCO_3$ is 100, 200 mM/L would correspond to an alkalinity of 2000 mg/L. Alkalinity measurements typically measure both partial alkalinity, a measure of bicarbonate concentration, and volatile fatty acid concentration. Partial alkalinity is a measure of buffering capacity, or the ability to maintain pH, as the concentrations of the volatile fatty acids increases or influent pH changes. Partial alkalinity or bicarbonate concentration has been found to be the most sensitive indicator for digestor pH upsets (2).

Poor digester performance is often attributed to a low concentration of methanogens. Of the two main groups of bacteria involved, acidogens and methanogens, the methanogens are slower growing and more susceptible to toxins. The most common biomass measurement, volatile suspended solids (VSS), gives no indication of the acidogen or methanogen populations. A characteristic of methanogens is their UV induced autofluorescence that permits counting by autofluorescence microscopy. A high correlation was evident between methane production and methanogenic population (5).

21.4 PROCESS TECHNOLOGIES

Many anaerobic processes and process configurations have been used for wastewater treatment. Several comparative studies have been conducted on high strength food processing wastewater. This review will focus on the anaerobic technologies most commonly used commercially, the Upflow Anaerobic Sludge Blanket (UASB), the anaerobic contact process, and the anaerobic filter, and will briefly discuss new hybrid and low rate technologies. There are approximately 800 UASB systems installed worldwide, nearly half in the food processing industries (6). Other technologies, which in total have fewer commercial installations than UASB installations, will be discussed in less detail.

21.4.1 Upflow Anaerobic Sludge Blanket (UASB) Process

The two leading companies in terms of the number of full scale anaerobic systems installed, are Paques and the Biothane Corporation, both with over 300 anaerobic installations. Paques and USFilter signed an exclusive license agreement in 2003 under which USFilter will design and supply Paques UASB anaerobic technologies for all industrial applications in the United States and Canada. Aerobic technologies often complement anaerobic treatment enabling treated water to be discharged directly into the environment.

The concept of UASB reactor design is simple (Figure 21.1). There are no moving parts within the UASB reactor. Pretreatment steps to remove suspended solids, fats, oils, and greases are required. Also, flow equalization is necessary for biomass retention. With such pretreatment in place, the UASB can convert achieve 70–95% BOD conversions. Wastewater is distributed into the tank through spaced inlets. The wastewater passes upward through a sludge bed composed of 0.5–2 mm microbial pellets. A high rate of anaerobic treatment results from retention of the anaerobic sludge bed within the reactor. The upward motion of gas bubbles causes hydraulic turbulence which provides reactor mixing. Baffles deflect the gas bubbles into a gas cap with a solid settling region above the gas outlet. The gas–liquid–solid three phase separator at the top of the UASB reactor is critical to effective operation. The UASB concept was first described in 1980 (7). The first pilot plant was installed at a beet sugar refinery in The Netherlands. Soon after, several full

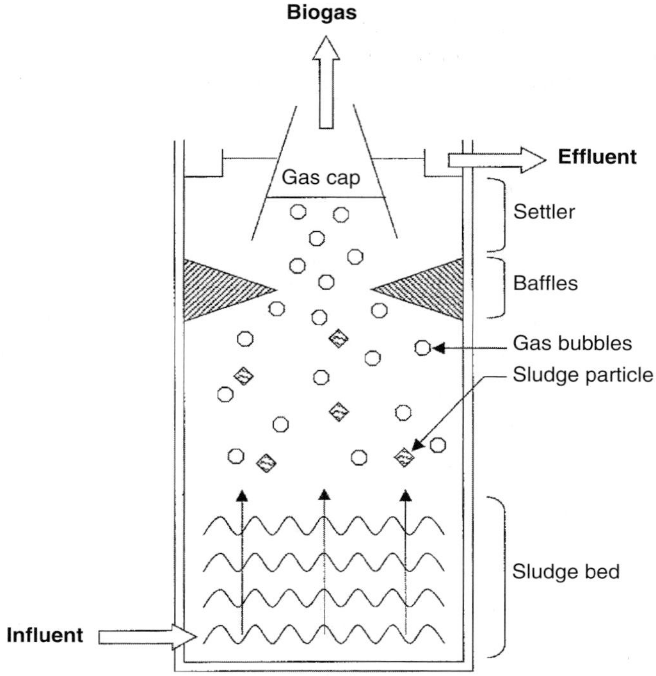

Figure 21.1 The upflow anaerobic sludge blanket (UASB) reactor

scale plants were installed at sugar refineries, recycle paper plants, potato starch plants, and other food industries.

UASB systems have a high organic loading capacity, 10–15 kg COD/m^3-day. A variation on the UASB concept, the expanded granular sludge bed (EGSB), uses a higher upward flow velocity that partially expands the granular sludge bed (8). This improves wastewater–sludge contact and allows higher organic loading in the range 15–35 kg COD/m^3-day. However, taller reactors or effluent recyclers must be used to achieve adequate wastewater treatment. The EGSB design is most appropriate for low strength wastewaters (less than 2000 mg COD/L) or for wastewaters that contain inert or poorly biodegradable suspended particles that should not be allowed to accumulate in the sludge bed.

The major consideration in UASB or EGSB operation is the development and maintenance of sludge granules. With the large number of existing UASB installations, large quantities of highly active anaerobic sludge are available, and start up of new installations can be made within a few weeks, sometimes within a few days. The maintenance and enhancement of biomass granulation is common to other high rate anaerobic systems where the retention of biomass within the reactor is crucial to efficient operation. The relatively large particle size granules have a higher settling velocity and higher specific activity than flocculent biomass.

Various forms of microbial conglomerates have been used in the form of flocs, pellets, and granules. Flocs have a loose structure and form a homogeneous layer after settling. Pellets and granules have a denser structure and a heterogeneous, well defined appearance after settling. UASB granules typically have a spherical form with diameters from 0.14–5 mm (9). Brewery-grown granules were observed to have a narrow size distribution of 1–2 mm, with a few 2–8 mm (10). However, only the outer 200–300 μm may be active because of diffusional limitations. Granules cultivated on acidified substrates, such as acetate, are generally smaller than granules grown on acidogenic substrates, such as glucose. Reported settling rates for UASB granules are in the range of 18–100 m/h (9). Densities of granules are 1.03–1.08 g/mL, but values up to 1.4 g/mL have been reported. The internal composition and organization of the granules varies with wastewater composition. Extracellular protein and polysaccharides in the granules are important for the structure and maintenance of the granules. This extracellular polymer content varies between 0.6–20% (9).

The UASB reactor operation is characterized by three zones (1). Granules grow mostly in the lower zone where volatile acid and hydrogen concentrations are high. Total COD is reduced to less than half its initial value, and total alkalinity decreases to a minimum as pH declines. In the middle zone total alkalinity and pH increase to stable values as the volatile acids are converted to methane and carbon dioxide. The upper zone is a settling zone where concentrations of soluble components are constant. The overall effect is an alkalinity requirement of 1.2–1.6 g alkalinity as $CaCO_3$/g COD for carbohydrate wastewaters. The function of alkalinity is primarily to control the pH decline in the lower zone.

Well developed, active granules with high physical strength and settling velocity are required for optimal operation of UASB reactors. The major advantages of granular sludge in high rate anaerobic reactors (11) are:

1. The superior settling characteristics of granular sludge. Settling velocities are commonly about 60 m/h in UASB reactors where upflow velocities are usually below 2 m/h. Solid retention times of over 200 days can be achieved with hydraulic retention times as low as 6 hours.
2. The high methanogenic activities of granular sludge. High volumetric rates in excess of 50 kg COD/m^3-day can be achieved with specific methanogenic activities of more than 2 kg COD/kg VSS.

Recent studies of UASB granules from fruit and vegetable cannery, two brewery, and slaughterhouse effluents illustrate the variations of granule structure on wastewater composition (10,12). Previous studies indicated that inferior granules were produced in UASB reactors treating protein-based, slaughterhouse wastewater. These recent studies verified that granule properties, including composition, size distribution, density, settling velocity, shear strength, and extra cellular polymeric substances (EPS), depend more on wastewater type than on reactor design or operating conditions such as pre acidification level. While the cannery and brewery-fed reactors had excellent shear strength, settling distribution, and density, the slaughterhouse granules had poor strength and settling velocity. Protein (slaughterhouse) wastewater was modeled using a biofilm model, and granules were predicted to have slow growth rates, low microbial density, and no trophic layers (12). The primary cause of this structure was the particulate nature of the wastewater rather than the presence of proteins. Some advantages of a well formed granular sludge are syntrophic associations with optimum distances between microbial partners, capability to withstand high gas and liquid shear rates, and resilience to shocks and toxins. As an example, an internal circulation (IC) reactor generates high levels of shear and requires rapid settling velocities but allows rapid reaction. Therefore, it would be ideal for treatment of cannery wastewater but would probably operate poorly when treating the wastewater from a complex protein source (10).

In efforts to improve granule formation, structure and activity, trace metal ion supplementation, especially iron, nickel, cobalt, and manganese, has been used to enhance UASB performance. Trace metal addition has promoted an increase in UASB specific activity to 2.6 g COD/g VSS-day (1). In a comparison study of pea and French fries wastewater, Fe, Co, and Ni supplementation increased the volatile suspended solids (VSS) form $3kg/m^3$ to 100 kg/m^3 and achieved COD reduction in excess of 95% at an organic loading rate of 10 kg/m^3-day.

Thickening agents have been used to enhance granule formation and to stabilize larger granules. This is especially important during the start up period of UASB reactors. Chitosan was shown to enhance the growth rate of granules from 19 μm/day in a control reactor to 56 μm/day with an ethanol–sucrose feed (13). Chitosan is a cationic natural polysaccharide polymer that was injected weekly at 25 mg/g suspended solids (SS) during a five week reactor operation. Synthetic cationic polymers have shown similar effects.

UASB reactors should be operated much differently during the start up period as opposed to the stable steady-state operation. Start up time can vary from less than a month to about a year. Reduced organic loading and monitoring of the sludge biomass are essential during start up. Sludge loading rates (SLR = g COD/g VSS-day) during start up have been proposed for several classes of wastewaters (14). For example, with a sugar molasses wastewater the proposed SLR during start up was 0.065–0.14 g COD/g VSS-day, while during stable operation SLR could be in the range of 0.30–0.78. At higher loading rates granules would not be retained in the UASB reactor. Also, higher levels of alkalinity and nutrients can improve start up performance. For example, by maintaining alkalinity:COD, N:COD, and P:COD at 1.06, 0.018 and 0.0028 during startup, granulation was observed after thirty days, and an average of 3.1 mm granules were obtained after ninety days of UASB operation (15). After this size granules were formed, addition of nutrients could be suspended and the alkalinity:COD ratio could be reduced to 0.4 without degrading process performance. Additionally, a high concentration of methanogens is advantageous. The inner core of granules typically consists of rod-shaped bacteria resembling *Methanosaeta* spp. The middle layer consists has a high content of syntrophic bacteria capable of producing and utilizing hydrogen, and the outer layer contains mostly acidogenic bacteria but also *Methanosarcina* spp. Granules were observed to grow at rates of 31, 21, 18, and 7μm/day in UASB reactors inoculated with syntrophic methanogenic consortia, *Methanosaeta* nuclei, *Methanosarcina* nuclei, and acidogenic floc, respectively (16).

Multivalent cations have been shown to enhance sludge granulation and reduce the time required for UASB start up (17). An optimum calcium concentration of 150–300 mg/L enhances biomass accumulation and granulation. The addition of Al^{3+} at a concentration of 300 mg/L reduced sludge granulation time in UASB reactors by approximately one month and produced larger granules compared to a control reactor.

Many of the operational considerations for UASB systems are valid for other high rate anaerobic processes. The formation of active granules with a high settling velocity is crucial to the retention of biomass and the operation with high VSS and organic loading rate.

21.4.2 Internal Circulation Reactor

The internal circulation (IC) process is based on the UASB concept of biomass granulation and the use of three phase separators. Its unique feature is the separation of produced biogas in two stages within a reactor with a large height–diameter ratio (Figure 21.2). The ratio of height–diameter is typically 4–8 with the reactor height up to 16–25m. The gas collected in the first stage is used to generate a gas lift that forces water up through a riser into the gas–liquid separator at the top of the reactor. Through this separator, the biogas leaves the reactor, while the water–sludge mixture returns through the downcomer back to the bottom of the system. The effluent from the first, lower reactor compartment is given additional treatment in the upper compartment. The upper level three phase separator collects the biogas formed in the upper compartment and the final effluent leaves the reactor

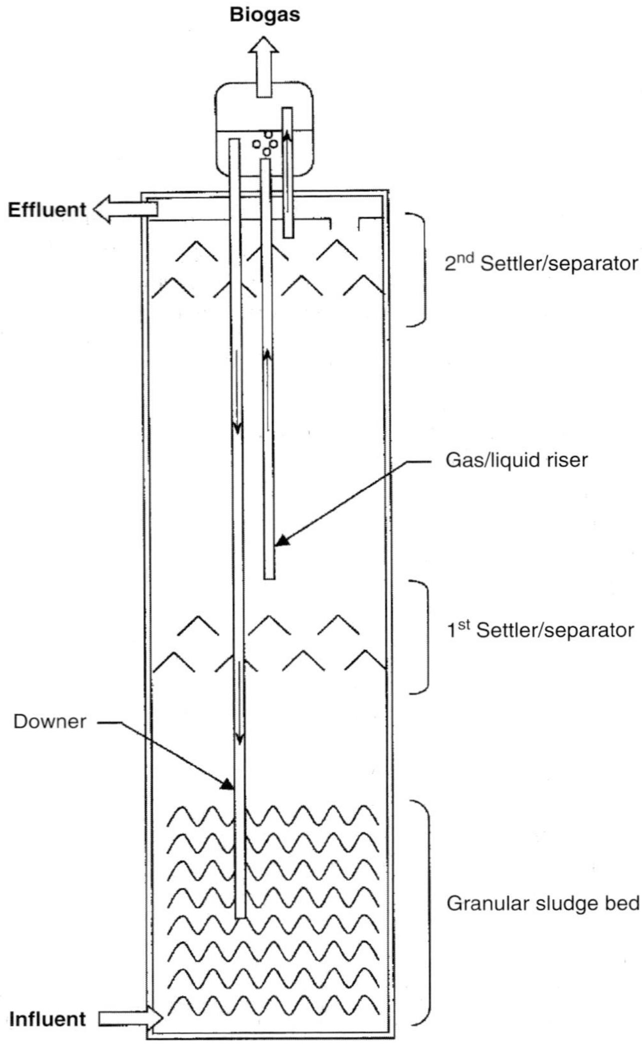

Figure 21.2 The internal circulation (IC) reactor

at the top. In the upper polishing section, there is a relatively long hydraulic retention time in a plug flow regime. This low superficial velocity and lower biomass concentration avoids solids loss especially during high peak loads.

The two stage design of the IC reactor allows up to three times higher volumetric loading rates and less than half the hydraulic retention times when compared to conventional UASB reactors (18). Hence the IC process is termed an ultra high rate process. A commercial IC process is offered by Paques as the BioPAQ IC system. This IC process is advertised to treat BOD in excess of 1000 mg/L with organic loading rates (OLR) in the range 20–35 kgCOD/m^3-day. Comparison tests showed that sludge granules from IC reactors have about two times higher methanogenic activities than UASB granules (18). Another advantage of the IC system is its very small footprint. The slim BioPAW IC systems are especially popular in densely populated areas such as South East Asia. More than 70 IC plants are being operated worldwide. Brewery and other food processing industries are the major users.

21.4.3 Anaerobic Contact and Anaerobic Filter Processes

After UASB systems, the anaerobic contact process and the anaerobic filter are the next most common anaerobic treatment systems with over 100 installations each (6). Historically, the first high rate anaerobic process, the anaerobic contact process, was developed in the 1950s. The anaerobic contact process consists of a completely well mixed digester and an external clarifier to separate and return the active biomass (Figure 21.3). The clarifier is required because of poor biomass retention within the digester. This problem is most acute during system start up. Mixing in the digester is achieved by mechanical stirrers or through the production and injection of biogas. The basis of this process is the maintenance of high sludge content in the digester for high rate organic removal and stable operation. The COD removal efficiency ranges from 65–98% depending on the type of wastewater. For efficient process control sludge loading in the digester and in the clarifier must not exceed capacity.

Anaerobic filters are usually operated in the upflow mode to minimize plugging (Figure 21.4). A major problem is clogging of the packing material. As with all high rate anaerobic processes, the limiting factor for efficient operation is the amount of active biomass that can be retained under high loading conditions. In anaerobic filters the space occupied by the packing material diminishes the overall sludge retention, but the high surface area provided by the packing encourages biofilm growth and biomass retention.

21.4.4 Hybrid Anaerobic Processes

There are several hybrid anaerobic processes that combine aspects of UASB, anaerobic filter, and other filtration processes. An example is the NewBio reactor that utilizes down flow with an intermittently backwashed sand filter to retain biomass and solids within the reactor (Figure 21.5). A major advantage of this hybrid technology is the capability of treating wastewaters containing high levels of solids, fats, oils, and greases. These wastewater components are difficult to treat with upflow processes as these components either wash out of the reactor or collect at the surface. The sand filter allows biomass and other water insoluble materials to be retained within the reactor. Clogging of the sand filter is prevented by periodically backwashing the filter bed to restore flow through the sand filter. The key to backwashing is a mixing blade that slowly rotates through the sand bed when pressure drop through the sand bed becomes excessive. Water is injected through nozzles in the front of the mixing blade to fluidize sand immediately in front of the blade. Water

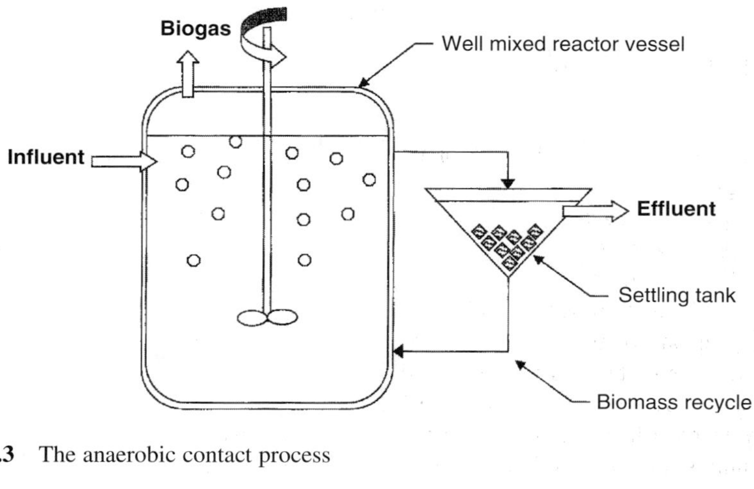

Figure 21.3 The anaerobic contact process

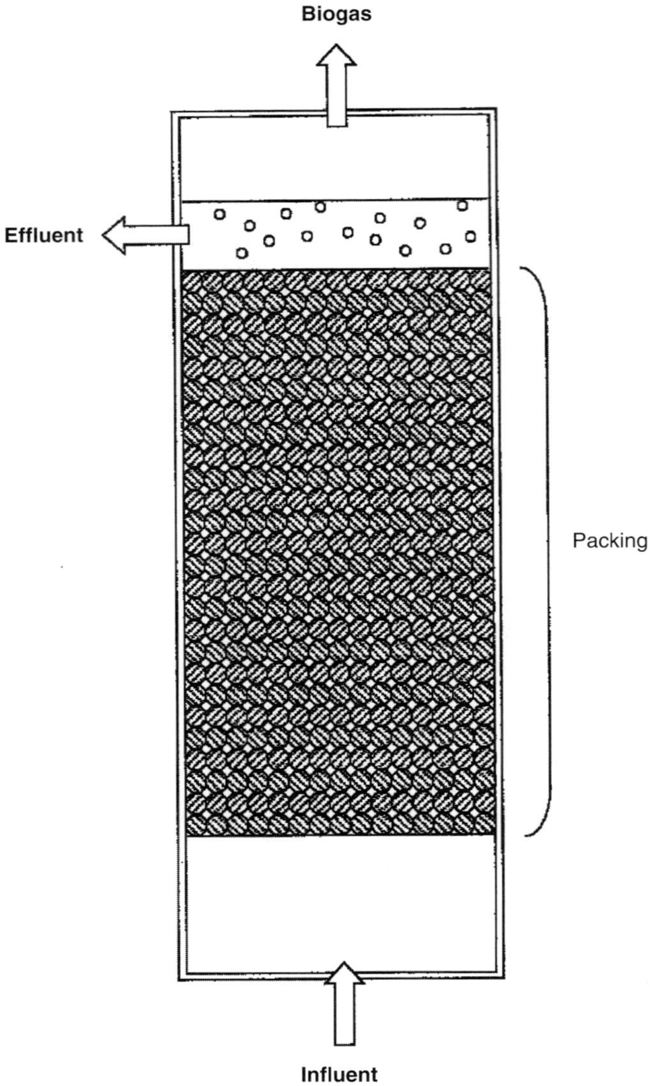

Figure 21.4 The anaerobic filter

pumped through nozzles in the rear of the blade provides a slow rotation of the mixing blade through the sand bed.

A series of pilot scale tests of the NewBio hybrid anaerobic process have been conducted with wastewater generated at two salad dressing production facilities (19). Operating at a hydraulic retention time of 12 days the system provided 98–99% BOD and total suspended solids (TSS) removals. Similar results were obtained in a full scale system shortly after startup at one of the salad dressing facilities. In addition, the computer controlled operation of these systems allows remote data acquisition and control.

In cases where granules cannot be formed from the wastewater treated and a fully packed anaerobic filter is too costly, an upflow hybrid reactor may be considered. This reactor contains an open section at the bottom and a packed section above. This packed

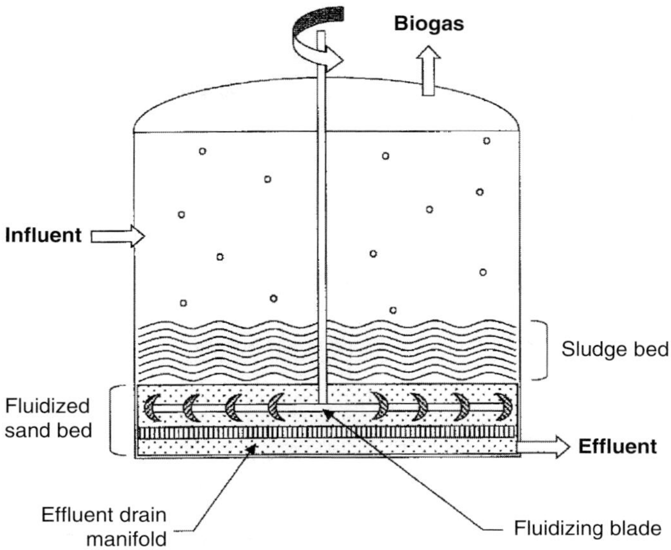

Figure 21.5 The NewBio hybrid anaerobic reactor

section improves the capture of the poor quality flocs escaping from the lower section. A column hybrid reactor with the bottom two thirds occupied by a sludge blanket and the upper one third by submerged clay rings was evaluated using slaughterhouse wastewater (20). COD removal was over 96% with loading rates up to 25 kg COD/m^3-day. The filter element of the reactor was highly efficient in retaining biomass; leading to a biomass accumulation yield coefficient of 0.029 g VSS/g COD higher than reported previously for either UASB or anaerobic filters operating independently. An example of a commercial system is the ADI-HYBRID reactor. Its major stated advantage is its stability and resilience against shock loadings. This system does not need a granular sludge because the overlying media intercepts solids, promotes flocculation, and returns solids to the sludge bed below. Loading rates of 10–20 kg COD/m^3-day are claimed.

21.4.5 Low Rate Process

An example of a low rate hybrid anaerobic process is the ADI bulk volume fermenter (BVF) process (Figure 21.6). This system combines features of the anaerobic contact process with intermittent mixing and sludge recycle. The reactor has two zones with a reaction zone at the inlet end and a clarification zone at the outlet end. The reactor can be an aboveground tank or a lined earthen basin with a floating insulated membrane cover for gas recovery and temperature and odor control (21). ADI-BVF systems have been installed at dairies, breweries, potato processing plants, corn processors, and meat and seafood processors. Over 80 BVF reactor installations worldwide are treating high strength organic wastewaters (22). These systems have the ability to digest high influent concentrations of fat, oil, and grease (FOG) and suspended solids (SS). An example is a 1.1 million gallon steel tank with a floating membrane, designed to treat 126,000 gal/day ice cream plant wastewaters, with 4,400 mg COD/L. COD and BOD are reduced by 85% and the effluent is discharged to the city sewer system.

The roof of the BVF reactor is typically a floating, insulated geomembrane cover that operates under a slight vacuum to collect and prevent escape of biogas into the environment (22). This low rate reactor has a relatively long retention time, typically seven

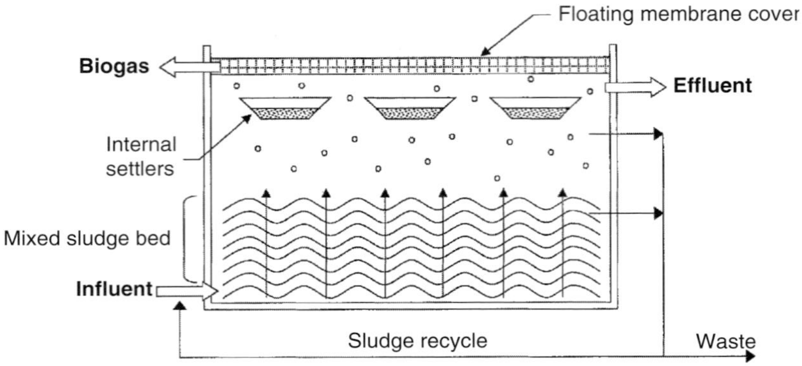

Figure 21.6 The ADI low rate bulk volume fermenter (BVF) process

days or more. The relatively large reactor size can serve to equalize flow from downstream processes as the floating membrane cover allows for expansion of the liquid within the reactor. When treating slow-degrading FOG or suspended solids, a scum layer can form at the liquid surface below the floating cover where it can gradually degrade without adversely affecting reactor operation. With the low upflow velocities and gas production rates, a low rate process can better retain sludge solids and maintain low effluent suspended solids concentrations. Internal settlers mounted in the upper portion of the BVF reactors enhance the liquid–solid separation.

21.4.6 Comparative Studies

Comparative studies illustrate the difficulty of ranking different high rate system performance. Performance is dependent on many factors such as properties of the inoculum, wastewater characteristics, and detailed design variables for each system. In a comparative study of an anaerobic filter, contact process, UASB system, and fluidized bed pilot plants each operating on ice cream wastewater, the upflow anaerobic filter gave the best results, and a full scale anaerobic filter was subsequently installed (23). However, poor biomass retention was observed in all reactors, possibly due to the high fat content of the wastewater. The UASB reactor gave the poorest performance because granulation of the biomass was not achieved. Several attempts at seeding the UASB with sewage sludge and sludge from an anaerobic digester treating creamery effluent were unsuccessful.

Laboratory scale anaerobic filter and fluidized bed with porous packing were compared for treating distillery wastewater (24). The highest organic loading rate, 32 kg COD/m^3-day, and a 96% COD removal at 5.88 kg COD/m^3-day was possible with the fluidized bed using SIRAN packing, an open pore sintered glass packing.

There are few comparative studies, and most food processors must make a comparison by visiting wastewater treatment facilities of different vendors and, if possible, conducting pilot plant studies with their wastewater. An example is Roger Sugar Ltd., which has been investigating options for treating sugarbeet-processing wastewater (25). An anaerobic–aerobic process was required to remove in excess of 99% of incoming BOD and to nitrify–denitrify the wastewater. At sugarbeet facilities the dominant anaerobic technologies are UASB and anaerobic contact systems. In addition, one German anaerobic fluidized bed process was reviewed. All processes operated with good stability. The anaerobic contact process found favor in the treatment of sugarbeet wastewaters where calcium concentrations can be greater than 600 mg/L, and calcium carbonate fouling is of less concern than with anaerobic fluidized bed or UASB processes. All exhibit good organic loading capacities, from 3–10 for anaerobic contact, 15 for UASB, and 25 kg/m^3-day for anaerobic fluidized bed systems. A major disadvantage of the fluidized bed is the need to discharge expensive carrier material. Differences among installed UASB systems were noted including shape, size, and concrete vs. steel materials, designs for distributing influent and recirculated flow, and separator designs at the top of the reactors. Generally phosphoric acid was fed to provide nutrient balance, antifoam, antiscalants, and weak acids, or chlorine to control foaming, calcium carbonate precipitation and to maintain clean heat exchange surfaces. A UASB system was selected after reviewing bids with performance specifications, performance guarantees, and material standards provided in the specification package.

21.5 APPLICATIONS

Food processing industries are usually situated in rural areas and provide a significant enhancement to the economies of many rural areas. These dairies, breweries, potato processors, corn processors, sugar factories, beverage industries, and many others are increasing limited by waste disposal requirements. Often these rural industries are forced to pay high disposal fees, and often their wastewater treatment requirements overwhelm the small local wastewater treatment facilities.

The anaerobic treatment of food processing wastes developed rapidly in European countries with a rapid growth in commercial installations during the late 1970s and 1980s, and more recently has developed in the United States. By 1990 there were over 600 anaerobic plants in Europe, most situated in the northern European countries where waste disposal costs were highest (26). By 1996, 106 anaerobic treatment plants had been installed at food processing plants in Germany (27). More than 850 agricultural biogas plants were reported in Germany in 2000 with volumes between 100 and 4,000 m^3, and 150 plants were planned for 2001 (28). The German government supported these plants by refunds based on electrical capacities. Government goals are to reduce the disposal of organic wastes, promote the recycling and utilization of wastes as a renewable energy source, and to decrease CO_2 emissions. The treatment of fat containing wastes from slaughterhouses or food industries has the highest biogas potential with gas yields up to 800 m^3 per ton of waste (28).

Anaerobic treatment produces 12×10^6 BTU as CH_4 per 1,000 kg of COD converted to CH_4 (1). Approximately 10,000 BTU are consumed in the generation of 1 kWhr of electricity. However, the biogas produced is 60–65% methane, with the remainder mostly CO_2. Methane, at standard temperature and pressure (0°C and 1 atm), has a heating value of 35,800 kJ/m^3 (960 Btu/ft^3). Natural gas has a heat content of 1,000 Btu/ft^3 because of small amounts of higher molecular weight hydrocarbons. Biogas will have proportionally less energy content than methane depending on the methane fraction.

Biogas from municipal anaerobic sludge digesters is routinely used for space heating and heating the digester. In larger plants, digester gas may be used as boiler fuel or for internal combustion engines, which can be used for generating electricity of operating pumps or blowers. A significant portion of the energy necessary to operate a wastewater treatment plant can be derived from the methane produced by anaerobic digestion. However, the conversion process requires expensive equipment and costly operation and maintenance. Most plants simply flare the excess methane.

The application of anaerobic treatment to food-processing waste management can best be discussed by focusing on specific food-processing industries. This review will focus on dominant food-processing industries where anaerobic treatment has been successfully applied in the treatment of wastewaters. Despite the differences among industries, many of the problems and solutions are common to all areas of food processing.

21.5.1 Fermentation and Beverage Industries

An area with increasing demand for high strength wastewater treatment has been the winery and microbrewery industries. Much of the research supporting anaerobic treatment of these wastewaters has been conducted in Europe and Australia. Very high organic loads characterize these fermentation wastewaters with additional problems associated with high suspended solids and phenolics and other byproducts of the fermentation processes that may be difficult to degrade. Before the introduction of high rate anaerobic reactors, aerobic processes or anaerobic sludge blanket systems treated these wastewaters.

Anaerobic processes have been widely applied for the treatment of brewery wastewaters. The ADI-BVD reactor has been used to treat this high carbohydrate, nutrient deficient wastewater. The high rate UASB and more recently IC reactors have been popular systems for treating brewery wastewaters. An anaerobic–aerobic wastewater treatment plant at a German brewery using a UASB first stage consistently removed greater than 80% COD at a loading rate of 8–10 COD/m^3-day (29). Using a large inoculum shortened start up time, and the recommended amount of inoculum was 10–20 kg volatile suspended solids (VSS)/m^3. No problems occurred when the UASB was restarted after standstill periods of several days or even weeks. Excessive scum layer problems have been observed in a field scale UASB, requiring shut down of the system for physical removal of the scum (30). Reducing the total suspended solids in the influent reduced, but did not eliminate, the scum layer accumulation.

An excellent case study for the application of the UASB process to the treatment of brewery wastewater is the seven year experience of Oriental Breweries in Korea (31). At the relatively low influent concentrations, averaging 1,620 mg COD/L, removal rates of approximately 80% were obtained, but at long hydraulic retention times. Gas produced by the digester was used to reduce operating costs by 30–45% to about \$.10/m^3. Good granule formation was difficult with this under loaded reactor, and this was attributed to excessive pre acidification. Later tests showed that a critical loading rate of 34–39 kg COD/m^3 total sludge bed volume-day (14–16 kg COD/m^3 reactor volume-day) was required to achieve 80% COD reduction, and the use of an upflow acidification reactor in series with a methanogenic UASB did not improve system performance (32). Field results (31) were only 20% of this critical rate.

Research and applications for winery wastewater treatment has largely focused on extensions of those anaerobic and aerobic methods successfully applied to the treatment of brewery wastewater. Little attention has been paid to the development of methods that successfully remove recalcitrant compounds that plague winery wastewaters such as phenols, tannins, and lignins. Tannins are the most common and crucial to the winemaking process, and tannins are most abundant in red wines. Tannins have been shown to inhibit microbial digestion (33).

UASB processes are most often described for winery wastewater treatment. UASB treatment of winery and cannery wastewaters has been used as the first step treatment followed by oxidation with ozone and ozone with hydrogen peroxide and granular activated carbon (34). With a hydraulic retention time of 24 hours, over 90% COD reduction was achieved. However, anaerobic treatment alone could not achieve the target COD effluent concentration of 75 mg/L, and the additional oxidations steps were required. Laboratory UASB reactors treated winery wastewater removed greater than 85% of the COD from wastewaters containing 1,000–17,000 mg COD/L at organic loading rates of 6.5–12.5 kg COD/m^3-day and hydraulic retention times of less than one day (35). There was a substantial decolorization of the wastewater with 45–67% reduction of polyphenol content.

An anaerobic sequencing batch reactor treated winery wastewater with a COD removal efficiency of greater than 98% operating at an organic loading rate of 8.6 kg COD/m^3-day and a hydraulic retention time of 2.2 days (36). Also, fluidized bed anaerobic reactors have been successfully used to treat winery wastewaters (37). Both laboratory and full scale of immobilized film technology reactors have successfully removed high percentages of COD. These reactors utilize carrier particles with a high surface area and low energy requirements for fluidization so that fluidization can be achieved at low upward flow rates.

Anaerobic processes have also been used to treat distillery wastewaters. For example, in whiskey production COD and suspended solids concentrations range between 10–60 and approximately 10 g/L, respectively. UASB granular bed anaerobic baffled reactor (GRABR), upflow anaerobic filter, and hybrid reactors have all been used to treat distillery wastewaters (38,39). Using the GRABR 90–96% COD reductions were obtained with the best performance at a hydraulic retention time of 4 days and a loading rate of 2.37 kg COD/m^3-day. The anaerobic filter with a support medium consisting of multi layer plates had a COD removal efficiency of 76% at 20 kg COD/m^3-day loading, removed approximately 70% of the total nitrogen by biological nitrification and denitrification, and converted approximately 70% of the organic phosphorus into soluble phosphate. A sequential anaerobic–aerobic laboratory system could achieve 99.5% COD removal with loadings as high as 39 kg COD/m^3-day in a two stage UASB (39). Nutrient supplementation had little effect on system performance.

An analysis of wastewater treatment at eleven Korean distilleries showed that barley, rice, tapioca, sweet potato, potato, and corn are used as raw materials (38). Difficulties in anaerobic treatment were reported only with the use of naked barley as the raw material. A thermophilic anaerobic reactor containing both a lower and an upper zone with modular blocks of cross flow media with a specific surface area of 102 m^2/m^3 was used to treat the naked barley wastewater with a soluble COD removal efficiency of 85–91% at loading rates of 5.45–11.52 kg/m^3-day.

Other applications of anaerobic wastewater treatment in the beverage industries include bottling plants and fruit juice processors. These are in general readily degradable wastewaters and high efficiencies of COD and BOD removal can be obtained. As with most food processing wastewaters, beverage wastewaters generally have low toxicity to anaerobic organisms and contain few refractory organic compounds. An exception is coffee, which is one of the most complex to degrade of the commonly encountered food commodities containing a high amount of lipids, 7.4 g/L tannin, and lignin (1). Also, in the roasting process a number of complex heterocyclic compounds are formed. Mesophilic anaerobic treatment of instant coffee wastewater has been difficult, but thermophilic operation has successfully treated the wastewater over a four month period.

21.5.2 Dairy Industries

High concentrations of fats are characteristic of wastewaters from dairy industries. These wastewaters are often very high strength, 40–80 g COD/L, and often contain high concentrations of proteins and lipids. Fat, oil, and grease (FOG) concentrations are typically in the 200–2000 mg/L range and can be even higher. Also, potentially inhibitory cleaning chemicals are often present due to the need to ensure continued disinfection of the production equipment. Because of the high organic content of these wastewaters, anaerobic digestion is essentially the only viable treatment method.

UASB reactors are very successful for wastewaters containing high concentrations of carbohydrates, but granule formation is more problematic with high fat wastewaters. When growing on these hydrophobic substrates, some bacteria produce extracellular polysaccharides that exhibit surface active properties that may inhibit granulation. After seeding a UASB reactor treating high strength ice cream wastewater with granular sludge, deterioration of granules can be observed within a short period of time (40). Also, emulsified triglycerides can cause severe flotation and washout of granular anaerobic sludge, and fat can agglomerate and float to the surface as a scum layer. For these reasons, packed bed reactors such as the anaerobic filter and hybrid reactors are more common for treating these wastewaters because these packed-bed reactors more effectively retain biomass.

An operational problem that must be considered when treating wastewaters high in protein and oil content is the tendency for foam formation. These substrates can lower the surface tension, and sudden bursts of gas production associated with increased loading can cause foam formation. Severe foaming can carry the biomass out of the reactor. Foam sensors can indicate when foam begins to form. Addition of an antifoam agent or a reduction in loading rate can reduce the potential for foaming.

A wide range of anaerobic reactors has successfully treated dairy wastewaters including low rate BVF reactors, UASB and hybrid UASB reactors, expanded granular sludge bed reactors, anaerobic filters, and other hybrid reactors. Operation at COD concentrations between 3–5 kg COD/m^3 is recommended to ensure high levels of biodegradation and to avoid flotation problems (41). Sludge flotation and washout in a UASB reactor was reported with a lipid loading rate exceeding 2 g COD/L-day.

There are approximately one dozen ADI-BVF reactors treating wastewaters from dairy and cheese processing facilities (22). This low rate technology is well suited for treating these high FOG and TSS-containing wastewaters. For six BVF reactors operating at cheese and multiproduct dairy plants, 87–94% COD removal rates were obtained for flows of 74,000–193,000 gpd, influent COD in the range 3,730–15,450 mg/L, FOG up to 1,470 mg/L, and TSS up to 4,170 mg/L (22). At a multiproduct cheese plant a 6.4 L Plexiglas reactor was used for bench scale testing. With a COD loading of 1.44 kg/m^3-day and an average hydraulic retention time of 10.6 days, 94% COD and 99.5% FOG removals were obtained. Based on the pilot testing no supplemental alkalinity, micronutrients, or macronutrients were required for the full scale system, and it was demonstrated that peak COD loadings of 3.36 kg/m^3-day would have little effect on digester performance. Also, tests showed that no primary treatment to remove solids or FOG was needed.

In many cases, alkalinity chemical addition is not required for low rate anaerobic systems treating dairy wastewaters, because there is often sufficient natural alkalinity present as a result of caustic cleaning chemicals used in the plant, and due to the digestion of protein in the anaerobic process which generates alkalinity (22). However, in the case of a multiproduct dairy that also produces nondairy products such as imitation sour cream and nondairy creamers, magnesium hydroxide addition was used to maintain alkalinity (22).

Addition of trace metals has been shown to enhance the anaerobic treatment of whey and other food processing wastewaters (1). Elevated levels (over 500 mg/L) of volatile acids in the reactor effluent indicate a need for supplemental metals. The addition of Co, Fe, and Ni can significantly enhance the COD removal rate. Whey can be difficult to treat with its high COD concentration, lack of alkalinity, foaming tendency, and the difficulty in obtaining granulation, and tendency for biomass washout (42, 46). Biochemical methane potential experiments showed that nutrient and trace metal supplementation is vial for the anaerobic treatment of cheese whey (42). With such supplementation, UASB reactors could treat undiluted cheese whey at hydraulic retention times of 2–3 days with COD removal efficiencies of 95–97% for influent COD concentrations of 42,700–55,100 mg/L. Similar results were obtained in a laboratory scale upflow hybrid reactor (43). Hydraulic retention times as low as two days were used to treat whey with influent concentrations of approximately 20 g/L. At COD loading rates up to 11.3 kg/m^3-day, removal efficiencies more than 95% were obtained when the alkalinity in the reactor was stabilized by controlling the influent pH.

To improve the retention of granular anaerobic sludge when treating cheese whey, a multiplate, compartmentalized reactor was designed (44). This 450m^3 reactor consists of four compartments with gas collected and evacuated through side outlets below each plate. With influent COD concentrations of 20–37 g/L, loading rates up to 14.7 kg COD/m^3-day, and hydraulic retention times of 55–68 hours, COD removal efficiencies of approximately 92% were obtained.

Ice cream wastewater is another challenge for anaerobic treatment. Ice cream wastewater is a complex mixture that includes high concentrations of milk proteins, lipids, and carbohydrates. High rate anaerobic processes are difficult to apply because granulation conditions may not be achieved. UASB reactors have been operated for three-year periods without granule production (1). Although granulation on ice cream wastewater was not achieved, commercially available UASB granules from a dairy reactor allowed stable operation of a laboratory UASB reactor for over a year (23). However, of four pilot scale digesters, the anaerobic filter operated at the highest loading rate of approximately 6 kg COD/m^3-day. The anaerobic filter was selected for full scale installation over UASB, fluidized bed and contact process designs. Poor biomass retention was observed in all reactors, possibly related to the fat content of the wastewater, and was the limiting factor in performance. For example, during shock loading experiments, where loading was increased to 14 kg COD/m^3-day, undegraded fat accumulated within the anaerobic filter bed (23).

A four compartment anaerobic baffled reactor removed 99% of the COD from ice cream wastewater at a loading rate of 15 kg COD/m^3-day, the highest loading rate examined (40). This compartment reactor could be classified as a series of UASB reactors with internal baffles retaining biomass in each compartment. The partial degradation of the fatty components of the ice cream wastewater in the first compartment may allow good granule development in subsequent compartments. Granulation was achieved in anaerobic baffled reactors within three months. With the addition of a cationic polyamide-epichlorohydrin polymer as a granulating agent, granules formed earlier and were larger and more compact (40).

Although the focus of this review has been on the removal of organics, often classified as COD, from food processing wastewaters, the removal of macronutrients such as nitrogen and phosphorus is of increasing concern. Increasingly, wastewater treatment processes will need to be designed for specific nutrient removals. As an example, three compartment pilot scale reactors, with anaerobic–anoxic–oxic zones, were used to treat dairy processing wastewater (45). Effluent was sent to a clarifier, and settled biomass was returned to the reactor. In the anaerobic phase, readily degradable carbon sources, such as volatile fatty acids, induce phosphate removing bacteria to take up the acids and release phosphate into solution. In the aerobic phase, luxury P uptake occurs with phosphorus removal at 75–98%. In the anoxic phase, both nitrogen and phosphate are removed. With hydraulic retention times of 1.2 days and inlet concentrations of 1,757 mg COD/L, 62 mg NH_4-n/L, and 57 mg P/L, removal efficiencies of 94%, 97% and 75% were obtained for COD, NH_4-N and P, respectively (45). A phosphate removal efficiency of 98% was obtained with a higher biomass recycle rate and longer retention time.

21.5.3 Meatpacking and Slaughterhouse Industries

Meat processing plants and slaughterhouses generate large volumes of high protein, high fat wastewaters with 0.4–3.1 m^3 wastewater produced per slaughtered animal (46). Nitrogen, phosphorus, and sulfur compounds are present in significant quantities. There is a large insoluble fraction consisting of colloidal and suspended matter in the form of fats, proteins, and cellulose, which degrade slowly in anaerobic reactors.

Wastewater COD levels vary with the difficult slaughterhouse processing and recovery methods. When blood and viscera are recovered, and pretreatment to recover fats and suspended matter is effectively used, COD can be approximately 1000 mg/L. Otherwise, COD levels can be as high as 12,000 mg/L. BOD levels are usually in the 1000–2400 mg/L range with the soluble fraction between 40–60% (46).

Anaerobic ponds have been used to reduce the BOD/COD levels of slaughterhouse wastewaters. However, these ponds degrade wastewaters at a low rate often with unpleasant odor generation. Performance is improved with effective pretreatment including screening and dissolved air flotation for the removal of fats and suspended solids. Covered anaerobic ponds reduce odor release and permit the capture of biogas. Retention times are typically 12–14 days and BOD loadings are approximately 0.33 kg/m^3-day (47). The biogas contains 0.2–0.7% H_2S that must be removed, for example by iron sponge filters, before use. BOD removals are typically 85–90%.

High rate anaerobic processes can be used to treat lagoon effluent or used independently. Sequencing batch, UASB, anaerobic contact, hybrid, and anaerobic filter reactors have been used to treat slaughterhouse wastewaters. Ultraviolet disinfection may be advantageous as a post treatment operation. Slaughterhouse wastewaters present two major problems for high rate anaerobic treatment:

1. The high fat and protein content of the wastewaters are much more resistant to anaerobic degradation than carbohydrates. The fats are sparingly soluble and tend to form scums, coat surfaces, and adsorb on biomass flocs and granules. Together with the high suspended solids content of the wastewater, granule formation is inhibited. The small flocs or granules that do form have a low settling rate.
2. The BOD/COD concentrations in the wastewater are relatively low for the successful operation of a high rate anaerobic process. These high rate processes operate better with BOD concentrations of 10,000 mg/L or more (47).

Because of the slow granule development, packed bed anaerobic reactors, such as the hybrid or filter reactors, and the anaerobic contact and sequencing batch reactors appear more suitable for these wastewaters. Pretreatment to remove fats and suspended solids is essential for optimum performance in all cases. The anaerobic contact process has been relatively popular for the treatment of slaughterhouse wastewaters with removal efficiencies of more than 84% BOD, 93% COD, and 75 % TSS in at a large plant with a COD loading of 3 kg/m^3-day (47). However, floating sludge problems that forced a plant in the UK to shut down has plagued other plants.

Large scale anaerobic filters have been installed in Europe, but little data concerning their performance is available (47). Reports list COD removals of 80–85% at COD loadings of 2–3 kg/m^3-day. Higher COD loadings appear to lead to poorer performance. Also, ineffective pretreatment resulting in a high fat and grease content of the wastewater caused unstable operation of an anaerobic filter treating slaughterhouse wastewater. A laboratory scale anaerobic filter was successfully used to treat wastewater from a fish meal factory (48). Fish meal processing wastewaters are characterized by a very high organic content (30–120 g COD/L) consisting of up to 70% protein. A reactor packed with PVC rings operating at a loading rate of 5 kg COD/m^3-day with a high recycle rate consistently operated at a COD removal rate of over 80%.

Other biofilm anaerobic reactors have been applied to the treatment of slaughterhouse wastewaters. A hybrid reactor was used to treat wastewater from a pig slaughterhouse at a loading rate of 2.7 kg COD/m^3-day. The main operating problems were carryout of granular sludge and fines, and loss of sludge during periods with high influent fat levels. The granular sludge required approximately two years to develop (12). A laboratory hybrid in which the bottom two thirds was occupied by a sludge blanket and the upper one third by clay rings operated with a 96% COD removal rate with slaughterhouse wastewater loading rates up to 25 kg COD/m^3-day (20).

The application of UASB reactors for the treatment of meat processing wastewaters is problematic because of the accumulation of suspended solids and floating fats and the poor quality of the granular sludge. This leads to biomass washout and low methanogenic activity. Large scale UASB reactors have been installed at slaughterhouses, but loss of sludge and lowered rates of BOD removal occurred because of the high fat concentrations (47). The key operating issue for UASB reactors is adequate pretreatment to remove suspended solids and fat to prevent their accumulation and subsequent loss of active sludge. UASB reactors often are forced to operating at low COD loading rates (0.3–1 kg/m^3-day), long hydraulic retention times (10–15 days), and poor COD removal (72–87%) (47). However, under optimum pretreatment conditions, UASB reactors can operate with COD loadings up to 11 kg/m^3-day with COD removal efficiencies of 80% or better (46,47).

Problems associated with granule formation in wastewaters high in protein and fat content have been examined in several studies. Anaerobic sequencing batch reactors were used to investigate the effects of pork fat particles on the rate of anaerobic digestion (49). The fat particle degradation was mainly controlled by the rate of free long chain fatty acid oxidation rather than the rate of fat hydrolysis. Also, these fatty acids adsorb on the biomass giving misleadingly low soluble COD values. Methane production was inhibited during the early phase of fat digestion. Low hydrogen partial pressure was required in order to favor fat oxidation, and high methanogen activity was essential for utilizing the hydrogen produced during fat oxidation. The particulate nature of slaughterhouse wastewater was predicted, and observed, to adversely affect granule formation (12). With high concentrations of sugar, a layered granule structure is predicted with fermenters in the outer layer, hydrogen utilizers and producers in the middle layer, and acetate utilizers in the center. High rate UASB processes treating slaughterhouse wastewater lack this structure and the resulting flocs have a low density of microorganisms. The tendency to form weak flocs is amplified at low substrate concentrations and low loading rates, which may be necessary in treating particulate wastewater.

In summary, the key requirements for the high rate anaerobic treatment of slaughterhouse wastewater are pretreatment to remove fats and suspended solids and to dampen flow fluctuations, and relatively low COD loadings (2–11 kg/m^3-day) (47).

REFERENCES

1. Speece, R.E. *Anaerobic Biotechnology.* Nashville, TE: Archae Press, 1996, pp 17–20, 221–244.
2. Anderson, G.K., G. Yang. Determination of bicarbonate and total volatile acid concentration in anaerobic digesters using a simple titration. *Water Environ. Res.* 64:53–59, 1992.
3. Buchaurer, K. A comparison of two simple titration procedures to determine volatile fatty acids in influents to waste-water and sludge treatment processes. *Water SA.* 24:49–56, 1998.
4. Jantsch, T.G., B. Mattiasson. A simple spectrophotometric method based on pH-indicators for monitoring partial and total alkalinity in anaerobic processes. *Environ. Technol.* 24:1061–1067, 2003.
5. Solera, R., L.I. Romero, D. Sales. Analysis of the methane production in thermophilic anaerobic reactors: use of autofluorescence microscopy. *Biotechnol. Lett.* 23:1889–1892, 2001.
6. Hulshoff, P.L., H. Euler, S. Schroth, T. Wittur, D. Grohganz. Promotion of anaerobic technology for the treatment of municipal and industrial wastes and wastewaters. In: *Proceedings 5th Latin American Workshop on Wastewater Anaerobic Treatment, Vina del Mar, Chile,* 1998, pp 27–30.

7. Lettinga, G., A.F.M. van Velsen, S.W. Hobma, W. De Zeeuw, A. Klapwijk. Use of upflow sludge blanket reactor concept for biological waste water treatment, especially for anaerobic treatment. *Biotechnol. Bioeng.* 22:699–734, 1980.
8. Kato, M., J.A. Field, P. Versteeg, G. Lettinga. Feasibility of the expanded granular sludge bed (EGSB) reactors for the anaerobic treatment of low strength soluble wastewaters. *Biotechnol. Bioeng.* 44:469–479, 1994.
9. Schmidt, J.E., B.K. Ahring. Granular sludge formation in Upflow Anaerobic Sludge Blanket (UASB) reactors. *Biotechnol. Bioeng.* 49:229–246, 1996.
10. Batstone, D.J., J. Keller. Variation of bulk properties of anaerobic granules with wastewater type. *Water Res.* 35:1723–1729, 2001.
11. Hulshoff Pol, L.W., S.I. de Castro Lopes, G. Lettinga, P.N.L. Lens. Anaerobic sludge granulation. *Water Res.* 38:1376–1389, 2004.
12. Batstone, D.J., J. Keller, L.L. Blackall. The influence of substrate kinetics on the microbial community structure in granular anaerobic biomass. *Water Res.* 38:1390–1404, 2004. (in press)
13. El-Mamouni, R., R. Leduc, S.R. Guiot. Influence of synthetic and natural polymers on the anaerobic granulation process. *Water Sci. Tech.* 38:341–347, 1998.
14. Singh, R.P., S. Kumar, C.S.P. Ojha. A critique on operational strategies for start-up of USAB reactors: effects of sludge loading rate and see/biomass concentration. *Biochem. Eng. J.* 1:107–119, 1998.
15. Gonzalez, J.S., A. Rivera, R. Borja, E. Sanchez. Influence of organic volumetric loading rate, nutrient balance and alkalinity: COD ratio on the anaerobic sludge granulation of an UASB reactor treating sugar cane molasses. *Int. Biodeterioration Biodegradation* 41:127–131, 1998.
16. El-Mamouni, R., R. Leduc, S.R. Guiot. Influence of the starting microbial nucleus type on the anaerobic granulation dynamics. *Appl. Microbiol. Biotechnol.* 47:189–194, 1997.
17. Riffat, R., K. Jahan, R. Tallent, K. Krongthamchat, S. Dararat. Anaerobic processes. *Water Environ. Res.* (CD) 74:1–134, 2002.
18. Gavrilescu, M. Engineering concerns and new developments in anaerobic waste-water treatment. *Clean Technol. Environ. Policy* 3:346–362, 2002.
19. Grismer, M.E., C.C. Ross, G.E. Valentine Jr., B.M. Smith, J.L. Walsh Jr. Food-processing wastes. *Water Environ. Res.* 74:377–384, 2002.
20. Borja, R., C.J. Banks, Z. Wang. Performance of a hybrid anaerobic reactor, combining a sludge blanket and a filter, treating slaughterhouse wastewater. *Appl. Microbiol. Biotechnol.* 43:351–357, 1995.
21. Eckenfelder, W.W., Jr. *Industrial Water Pollution Control,* 2nd ed. New York: McGraw-Hill, 1989, p 248.
22. Grant, S., R. Landine, D. Wilson, J. Molina, S. Norton, Z. Qiu, A. Cocci. Case studies of low-rate anaerobic treatment of dairy processing wastewaters. 8th Annual Industrial Wastes Technical and Regulatory Conference, Water Environment Federation, August 11–14, 2002.
23. Hawkes, F.R., T. Donnelly, G.K. Anderson. Comparative performance of anaerobic digesters operating on ice-cream wastewater. *Water Res.* 29:525–533, 1995.
24. Perez, M., L.I. Romero, D. Sales. Comparative performance of high rate anaerobic thermophilic technologies treating industrial wastewater. *Water Res.* 32:559–564, 1998.
25. Given, P.W., B. Vallance, J. With. Treatment technologies for food processing wastewater. www.ae.ca/about/papers/foodwater.html, 2004.
26. Wheatley, A.D. Anaerobic digestion: industrial waste treatment. In: *Anaerobic Digestion: A Waste Treatment Technology.* New York: Elsevier Applied Science, 1991, p 173.
27. Austermann-Haun, U., C.F. Seyfried, K.-H. Rosenwinkel. Full scale experiences with anaerobic pre-treatment of wastewater in the food and beverage industry in Germany. *Water Sci. Technol.* 36:321–328, 1997.
28. Weiland, P. Anaerobic waste digestion in Germany: status and recent developments. *Biodegradation* 11:415–421, 2000.

29. Austermann-Haun, U., K.-H. Rosenwinkel. Two examples of anaerobic pre-treatment of wastewater in the beverage industry. *Water Sci. Tech.* 36:311–319, 1997.
30. Laubscher, A.C.J., M.C. Wentzel, J.M.W. Le Roux, G.A. Ekama. Treatment of grain distillation wastewaters in an UASB system. *Water SA*. 27:433–444, 2001.
31. Ahn, Y., K. Min, R.E. Speece. Full scale UASB reactor performance in the brewery industry. *Environ. Technol.* 22:463–476, 2001.
32. Ahn, Y., K. Min, R.E. Speece. Pre-acidification in anaerobic sludge bed process treating brewery wastewater. *Water Res.* 35:4267–4276, 2001.
33. Sarni-Manchado, P., A. Deleris, S. Avallone, V. Cheynier, M. Moutunet. Analysis and characterization of wine condensed tannins precipitated by proteins. *Am. J. Enology Vitic.* 50:81–86, 1999.
34. Sigge, G.O., T.J. Britz, P.C. Fourie, C.A. Barnardt, R. Strydom. Combining UASB technology and advanced oxidation processes to treat food processing wastewaters. *Water Sci. Technol.* 45:329–334, 2002.
35. Kalyuzhnyi, S.V., M.A. Gladchenko, V.I. Sklyar, O.V. Kurakova, S.S. Shcherbakov. The UASB treatment of winery wastewater under submesophilic and psychrophilic conditions. *Environ. Technol.* 21:919–925, 2000.
36. Ruiz, C., M. Torrijos, P. Sousbie, J.L. Martinez, R. Moletta, J.P. Delgenes. Treatment of winery wastewater by an anaerobic sequencing batch reactor. *Water Sci. Technol.* 45:219–224, 2002.
37. Grismer, M.E., C.C. Ross, G.E. Valentine Jr., B.M. Smith, J.L. Walsh Jr. Food-processing wastes. *Water Environ. Res.* (CD) 73:1–29, 2001.
38. Nam, K.D., I. Chung, J.C. Young, W. Park. Treatment of naked-barley distillery wastewater using a thermophilic hybrid anaerobic filter. *J. Microbiol. Biotechnol.* 9:737–743, 1999.
39. Uzal, N., C.F. Gokcay, F.N. Demirer. Sequential (anaerobic/aerobic) biological treatment of malt whisky wastewater. *Proc. Biochem.* 39:279–286, 2003.
40. Uyanik, S., P.J. Sallis, G.K. Anderson. The effect of polymer addition on granulation in an anaerobic baffled reactor. *Water Res.* 36:933–943, 2002.
41. Vidal, G., A. Carvalho, R. Mendez, J.M. Lema. Influence of the content in fats and proteins on the anaerobic biodegradability of dairy wastewaters. *Bioresource Technol.* 74:2431–239, 2000.
42. Erguder, T.H., U. Tezel, E. Guven, G.N. Demirer. Anaerobic biotransformation and methane generation potential of cheese whey in batch and UASB reactors. *Waste Manage.* 21:643–650, 2001.
43. Calli, B., M.A. Yukselen. Anaerobic treatment by a hybrid reactor. *Environ. Eng. Sci.* 19:143–150, 2002.
44. Guiot, S.R., B. Safi, J.C. Frigon, P. Mercier, C. Mulligan, R. Tremblay, R. Samson. Performances of a full-scale novel multiplate anaerobic reactor treating cheese whey effluent. *Biotechnol. Bioeng.* 45:398–405, 1995.
45. Mulkerrins, D., E. O'Connor, B. Lawlee, P. Barton, A. Dobson. Assessing the feasibility of achieving biological nutrient removal from wastewater at an Irish food processing factory. *Bioresource Technol.* 91:207–214, 2004.
46. Caixeta, C.T., M.C. Cammarota, A.M.F. Xavier. Slaughterhouse wastewater treatment: evaluation of a new three-phase separation system in a UASB reactor. *Bioresource Technol.* 81:61–69, 2002.
47. Johns, M.R. Developments in wastewater treatment in the meat processing industry: a review. *Bioresource Technol.* 54:203–216, 1995.
48. Guerrero, L., F. Omil, R. Mendez, J.M. Lema. Treatment of saline wastewaters from fish meal factories in an anaerobic filter under extreme ammonia concentrations. *Bioresource Technol.* 61:69–78, 1997.
49. Masse, L., D.I. Masse, K.J. Kennedy, S.P. Chou. Neutral fat hydrolysis and long-chain fatty acid oxidation during anaerobic digestion of slaughterhouse wastewater. *Biotechnol. Bioeng.* 79:43–52, 2002.

3.22

International Aspects of the Quality and Safety Assessment of Foods Derived by Modern Biotechnology

John R. Lupien

CONTENTS

22.1 INTRODUCTION

Agriculture (including crops, fishery, forestry, and animal husbandry) must feed an increasing world population, forecast to reach 8000 million by 2020, of which 6700 million will be in developing countries (1). Although population growth rate is steadily decreasing, the increase in absolute numbers to be fed will require steady increased gains in productivity, often in countries where environmental degradation threatens decreased productivity.

To meet future needs and to be able to sustain agricultural production, agricultural research will have to use all available technologies, including the rapidly developing modern biotechnologies. The Food and Agriculture Organization of the United Nations (FAO) has recognized that the new techniques of biotechnology are powerful tools in agricultural development, and have great potential to benefit agriculture. At the same time, FAO has pointed out that there are a number of uncertainties and possible risks associated with their use (2) (Annex 1, FAO Statement on Biotechnology).

New technologies, such as modern biotechnologies, if properly focused, should provide solutions for some of the problems hindering sustainable rural development and the achievement of food security. Biotechnology may also offer a tool to resolve certain environmental problems, some of which derive from unsustainable agricultural and industrial practices.

22.2 BIOTECHNOLOGY IN FOOD

Biotechnology has a long history of use in food production and processing. It represents both traditional breeding techniques and the latest techniques based on molecular biology. Modern biotechnological techniques, in particular, open up very great possibilities of rapidly improving the quantity and quality of food available.

The 1992 United Nations Convention on Biological Diversity (CBD) defined biotechnology as "any technological application that uses biological systems, living organisms, or derivatives thereof, to make or modify products or processes for specific use (3)." In agriculture, biotechnology includes the application of tissue culture, immunological techniques, molecular genetics and recombinant DNA techniques in all facets of agricultural production and agro-industry.

Within the past few years, a variety of foods produced using biotechnology have been approved in many countries. Examples are crops such as maize, potatoes, soybeans, tomatoes, and oilseeds. The benefits of biotechnology are many and include providing resistance to crop pests to improve production and reduce chemical pesticide usage, thereby making major improvements in both food quality and nutrition. Biotechnology is also being used in a wide range of applications in fermentation techniques and in animals and plants for the production of food additives and pharmaceuticals.

It is important to consider any potential human health or environmental risks when foods are developed using biotechnology. It is vitally important to encourage worldwide efforts to develop and apply appropriate strategies and safety assessment criteria for food biotechnology research and to ensure the wholesomeness and safety of the food supply.

The increasing development of genetically modified organisms is accompanied by the need for all necessary controls related to their testing, release, use, and cross border movements. Adequate national legislation is necessary to protect the environment, biodiversity, and human health. There is also the need to consider how to carry out adequate levels of risk assessment and risk management of genetically modified organisms and their products and mechanisms and instruments for the application and control of biotechnology.

The Resumed Extraordinary Conference of the Parties to the Convention on Biological Diversity held in Montreal, Canada in 1998 adopted the Cartagena Protocol on Biosafety (4). The objective of the Protocol is to contribute to ensuring an adequate level of protection in the field of the safe transfer, handling and use of living modified organisms[1] (LMO) resulting from modern biotechnology[2] that may have adverse effects on the

[1] Cartagena Protocol on Biosafety definition: LMO means any living organism that possesses a novel combination of genetic material obtained through the use of modern biotechnology;

[2] Cartagena Protocol on Biosafety definition; "Modern Biotechnology" means the application of: (I) *In vitro* nucleic acid techniques, including recombinant deoxyribonucleic acid (DNA) and direct injection of nucleic acid into cells or organelles, or (ii) Fusion of cells beyond the taxonomic family that overcome natural physiological reproductive or recombination barriers and that are not techniques used in traditional breeding and selection

conservation and sustainable use of biological diversity, taking into account risks to human health, and specifically focussing on transboundary movements.

While there are a number of contradictions and imprecise statements in this new Protocol, the key concept seems to be the notion of advance informed agreement (AIA), which should be given by the country of import prior to the transboundary movement of an LMO. Many of the provisions of the Protocol need further discussion before they can be implemented and the Protocol contains provisions for such discussions. For example, it appears that the advance informed agreement procedure in the Protocol might not apply to LMOs intended for direct use as food or feed, or for processing. At the same time, the Protocol contains an article on the handling, transport, and identification of LMOs intended for "food, feed, or processing" (LMO/FFPs). Such contradictions require further discussion, preferably based on sound science.

The Protocol promotes a "precautionary approach" in assessing the usefulness and safety of LMOs and LMO/FFPs. The Protocol provides that risk assessments shall be carried out in a scientifically sound manner taking also into account risks to human health. The Protocol does not imply a change in the rights and obligations of a Party under any existing international agreement including WTO, SPS, and TBT Agreements. It is understood that the Protocol and trade agreements should be mutually supportive and none of them should be subordinated to the other.

This paper describes an overview of relevant instruments in the field of biotechnology that deal directly or indirectly with issues related to biotechnology, with particular reference to FAO and relevant bodies and ongoing processes within FAO, and outlines international mandates and capacity to advise UN member governments on matters and international regulations relevant to biotechnology and food and agriculture. Cooperation between FAO, the World Health Organization (WHO) and the World Trade Organization, and among other concerned UN agencies, is necessary to effectively carry out international work on biotechnology. To facilitate the quality and safety assessment of foods derived by means of modern biotechnology, action at the international level has been necessary to provide timely expert advice in this matter to all Member States.

22.3 INTERNATIONAL WORK ON BIOTECHNOLOGY

The United Nations and FAO have been involved for many years in biotechnology in order to advise and assist its Member Countries to adopt useful methodologies and monitor development in the area. In 1984, an FAO/International Atomic Energy Agency (IAEA) meeting discussed the role of relevant international organizations in biotechnology, and since then numerous sectoral and general meetings called by UN agencies have included specific aspects of biotechnology.

In 1991, the member countries of the FAO Council endorsed a draft "Code of Conduct for Biotechnology as It Affects the Conservation and Use of Plant Genetic Resources (5)." A draft was prepared, following a survey among 400 experts worldwide, requested by the Commission on Genetic Resources for Food and Agriculture (CGRFA). Its aim is to minimize possible negative effects of biotechnology on the overall conservation of plant genetic resources. Noting that CBD was considering the development of a biosafety protocol,[3] the Commission recommended that FAO participate in this work in

[3] The CBD definition of Biosafety "the safe and environmentally sustainable use of all biological products and applications for human health, biodiversity and environmental sustainability in support of improved global food security"

order to ensure that aspects of biosafety in relation to genetic resources for food and agriculture be appropriately covered.

A number of FAO publications — both meeting reports and technical bulletins — have addressed aspects of biotechnology, as mean of assisting member governments to acquire the technology, implications for agriculture, and trade-related issues, and many of these can be found on the FAO website (www.fao.org).

Particular biotechnology-related issues have been considered by a series of FAO/WHO expert consultations and workshops such as two Joint FAO/WHO Expert Consultations in 1990 and 1996 which addressed safety assessments of food derived by modern technology and outlined the procedures to be followed in establishing the quality and safety of such food.

The first Joint FAO/WHO Consultation on the Assessment of Biotechnology in Food Production and Processing as Related to Food Safety (November 1990) (6) reviewed the status of biotechnology as used in food production and processing and discussed foods derived from plant, animal and microbial sources. The consultation proposed safety assessment paradigms for each food source and recommended that safety assessment strategies should be based on the molecular, biological and chemical characteristics of the food to be assessed. It noted that traditional food safety assessment techniques, based on toxicological testing as used for food additives, for example, may not always apply to foods or food components produced by biotechnology.

A fundamental conclusion of the 1990 consultation concerning modern biotechnology was that "The use of these techniques does not result in food which is inherently less safe than that produced by conventional ones."

Another Joint FAO/WHO Consultation on Biotechnology and Food Safety held in 1996 recommended international guidelines for safety assessment of foods and food components which have been produced by techniques that change the heritable traits of an organism, such as recombinant DNA (rDNA) technology (7).

The 1996 Joint FAO/WHO Consultation established the concept of substantial equivalence, which is a dynamic, analytical exercise in the assessment of the safety of a new food relative to an existing food. This comparative approach was based on the possibilities that it may be possible to demonstrate that a genetically modified organism, or a food or food component derived from it, is substantially equivalent to a conventional counterpart already available in the food supply. Substantially equivalent in this context refers to both conventional nutritional equivalencies and to safety considerations.

If it is not possible to demonstrate substantial equivalence, it may be possible to demonstrate that a genetically modified organism, or food component derived from it, is substantially equivalent to its conventional counterpart apart from certain defined differences. It may not be possible to demonstrate substantial equivalence between the genetically modified organism or food component derived there from and a conventional counterpart, either because differences are not sufficiently well defined or because there is no appropriate counterpart with which to make a comparison.

While recognizing there may be limitations to the application of the substantial equivalence approach to safety assessment, the Consultation recommended that safety assessment based upon this concept be applied in establishing the safety of food products derived from genetically modified organisms to provide comparable or increased assurance of the safety of food products derived from biotechnology.

Three expert consultations, designed to provide expert advice to the Codex Ad Hoc Task Force and to FAO and WHO Member Countries, followed, in 2000 and 2001. In June 2000, FAO/WHO convened an Expert Consultation on Foods Derived from Biotechnology (8) that reviewed previous FAO/WHO recommendations and strongly reendorsed the concept of "substantial equivalence" as the best method of assessing the safety and suitability of foods derived from biotechnology.

In January 2001, FAO/WHO held a further Expert Consultation on Foods Derived from Biotechnology (9), concentrating on the possible allergenicity of food derived from biotechnology. This meeting pointed out that no problems with allergenicity have yet occurred with foods derived from biotechnology, mentioned that some food allergens might be removed from foods by future biotechnology developments, devised some suggested methods for testing new foods for possible allergenicity, and again endorsed the concept of "substantial equivalence."

In September 2001, FAO/WHO held a third Expert Consultation on Food Derived From Biotechnology (10), to consider safety and suitability aspects of genetically modified microorganisms. The report of this session has been published on the FAO website, as have been most of the previous reports. The meeting confirmed, in general, the recommendations of previous expert consultations, and pointed out some of the special features of genetically modified microorganisms in the human gut. It did not find any causes for immediate concern, recommended a system for evaluating genetically modified microorganisms, and again endorsed the concept of "substantial equivalence."

22.4 QUALITY AND SAFETY ASSESSMENT

The use of biotechnological processes, particularly genetic modification, is extremely important in devising new ways to increase food production, increase pesticide resistance and reduce use of pesticides, improve nutrient content, and provide better processing or storage characteristics. It follows that when new foods or food components are developed using biotechnology, there are both national legal requirements and consumer expectations for effective systems and procedures to assess the safety of the food and food component for consumption.

With regard to these needs, there is a series of instruments and international regulations in the field of food and agriculture in FAO that deal directly or indirectly with biosafety and biotechnology related issues which would be of relevance to the quality and safety assessment of foods derived by modern biotechnology.

22.5 FAO PROGRAMS ON FOOD QUALITY AND SAFETY

The Food Quality and Standards Service is a service within the Food and Nutrition Division of the Food and Agriculture Organization of the United Nations (FAO), located in Rome. The Secretariat of the Codex Alimentarius Commission is also located here. The Regular Program of the Food Quality and Standards Service provides the technical and scientific basis for FAO for all food quality matters, including food safety. This includes providing the secretariat for the Joint Expert Committee on Food Additives (JECFA) and participation in both the Joint Meeting of the FAO Panel of Experts on Pesticide Residues in Food and the Environment and the WHO Expert Group on Pesticide Residues (JMPR).

The Food Quality and Standards Service of FAO develops and publishes guidelines and manuals (including the FAO Food and Nutrition Series, Manuals of Food Quality Control), arranges expert consultations and conferences (examples of recent consultations include the Joint FAO/WHO Expert Consultation on Biotechnology and Food Safety, 30 September – 4 October 1996; the Joint FAO/WHO Expert Consultation on the Application of Risk Management to Food Safety Matters, 27–31 January 1997; the Joint FAO/WHO

Consultation on Food Consumption and Exposure Assessment to Chemicals, 10–14 February 1997; the FAO Consultation on Animal Feeding and Food Safety, 10–14 March 1997; the Joint FAO/WHO Expert Consultation on Risk Communication to Food Standards and Safety Matters, February 1998; and the Joint FAO/WHO Expert Consultation on Risk Assessment of Microbiological Hazards in Foods, March 1999) and has a major and continuing program of providing technical assistance related to food standards and food control to member countries, particularly developing countries and countries in transition from a centrally planned to a market economy.

JECFA and JMPR are expert committees that provide the independent scientific advice that forms the basis for the development of food quality and safety recommendations used in international trade. These committees are forums in which independent, invited experts assess the state of scientific knowledge of food additives, pesticide and veterinary drug residues in food, mycotoxins and other chemical contaminants in food, and make recommendations to member governments and to Codex on such matters.

FAO's Food Quality and Standards Service also develops and publishes Manuals of Food Quality Control which provide recommendations for the development and operation of food quality and safety systems. While aimed primarily at providing advice to developing countries, they document modern approaches including the development of quality control programmes throughout the food chain that are applicable to all countries. Such an approach is instrumental in facilitating international trade in food. Key titles in the series include Food Inspection, Food for Export, Management of Food Control Programmes, Imported Food Inspection and Quality Assurance in the Food Control Laboratory (see website www.fao.org.nutrition for entire list of these publications and source for ordering).

The program of technical assistance projects undertaken by FAO's Food Quality and Standards Service includes assistance in food quality control including safety and such projects have established or strengthened the food control systems in a number of developing countries. Typically, they assist in establishing the infrastructure for an enhanced food control programme, assessing laboratory service requirements, providing guidance to develop legislation and procedural manuals, setting up reputable inspection and certification systems and providing training and staff development. In these assistance projects, the standards established by the Codex Alimentarius Commission are basic guides to international requirements.

22.6 CODEX ALIMENTARIUS

The Codex Alimentarius Commission (CAC) was formed by FAO and WHO in 1962 to implement the Joint FAO/WHO Food Standards Program. The objectives of the Program are to ensure consumers health and fair practices in the food trade. The CAC is an intergovernmental statutory body of FAO and WHO. Its current membership is 165 countries and the Commission member countries are empowered by the CAC statutes and rules to adopt and recommend for approval to the governing bodies of FAO and WHO standards, guidelines and recommendations related to the protection of health and the promotion of fair international trade of foods (11,12).

The scope of Codex Standards includes all food safety considerations, description of essential food hygiene and quality characteristics, labelling, methods of analysis and sampling, and systems for inspection and certification. Codex Standards, guidelines and recommendations are based on current scientific knowledge including assessments of risk to human health. As mentioned, risk assessments are carried out by FAO/WHO expert panels of independent scientists selected on a worldwide basis, and the results of their review and deliberations are provided to Codex for use in Codex work, and to FAO/WHO Member Countries. The range of standards developed by the CAC covers all foods whether processed, semi processed or raw, intended for sale to the consumer or for intermediate processing. Over 200 standards, 45 Codes of Practice and 2,000 Maximum Limits for residues of agricultural and veterinary chemicals have been established.

In 1999, the Codex Alimentarius Commission established the Ad Hoc Intergovernmental Codex Task Force on Foods Derived from Biotechnology to develop standards, guidelines, or other recommendations on foods derived from biotechnology. The first Session of the Ad Hoc Intergovernmental Codex Task Force was held in Chiba, Japan from 14 to 17 March, 2000. The second Session was held in Chiba from 26–30 March 2001.

This Task Force is elaborating standards, guidelines, or recommendations, as appropriate, for foods derived from biotechnology or traits introduced into foods by biotechnology, on the basis of scientific evidence, risk analysis and having regard, where appropriate, to other legitimate factors relevant to the health of consumers and the promotion of fair trade practices. It must take full account of existing work carried out by national authorities, FAO, WHO, other international organizations and other relevant international for a, with close coordination and collaboration with appropriate Codex Committees within their mandate as relates to foods derived from biotechnology.

In its second meeting in 2001 the Task Force prepared "Proposed Draft Principles for the Risk Analysis of Foods Derived from Modern Biotechnology" and a "Proposed Draft Guideline for the Conduct of Food Safety Assessment of Foods Derived from Recombinant DNA Plants." Both of these documents were discussed and adopted by Codex at Step 5, of the 8 Step, Codex procedure at the July 2001 24th Session of the Codex Alimentarius Commission.

A further discussion of these documents of the 3rd Session of the Task Force in 2002 resulted in agreement on all aspects of the texts. The Task Force recommended final approval at Step 8 of the Codex procedures at the next Commission Session. The final texts adopted by the Task Force can be found in Annex 2 and Annex 3 to this Chapter. A Codex Draft Annex on the Assessment of Possible Allergenicity, Annex 4, was also recommended for Step 8 final Commission approval and a proposed Draft Guideline for the Conduct of Food Safety Assessment of Food Produced using Recombinant DNA Microorganisms, Annex 5, was approved by the Task Force and advanced to Step 5 of the Codex process (13–16).

Other Codex Committees such as the Codex Committee on Food Labeling, and the Codex Committee on Food Import Certification and Inspection Systems, are discussing biotechnology related topics such as voluntary or mandatory labeling systems, and means of assuring "Traceability" of foods and ingredients for foods (and feeds) derived from modern biotechnology. All of this Codex work, when applied at the national level by FAO/WHO member countries should assist greatly in assuring that harmonized procedures are used in the assessment and regulatory control of food derived from biotechnology.

The Uruguay Round of Multilateral Trade Negotiations established a new World Trade Organization (WTO) and included negotiations on reducing non tariff barriers to international trade in agricultural products and included Agreements on the Application of Sanitary and Phytosanitary Measures (the SPS Agreement) and on Technical Barriers to Trade (the TBT Agreement) (17). Both Agreements have implications for the work of the Codex Alimentarius Commission.

The SPS Agreement confirms the right of WTO Member countries to apply measures necessary to protect human, animal and plant life and health provided that "such measures are not applied in a manner which would constitute a means of arbitrary or unjustifiable discrimination between countries where the same conditions prevail, or a disguised restriction on international trade."

With respect to food safety, the SPS references the standards, guidelines and recommendations established by the Codex Alimentarius Commission relating to food additives, residues of veterinary drugs and pesticides, contaminants, methods of sampling and analysis, and codes and guidelines of hygienic practice.

Therefore, measures need to be taken with respect to foods to ensure to the Codex maximum levels or guidelines levels for contaminants, and to the Codex maximum residue limits (MRLs) for pesticide and veterinary drugs. Measures also need to be taken to ensure that the appropriate hygienic practices are followed at all stages of the animal feeding chain to prevent, eliminate, or reduce potential hazards in the food.

The objective of the TBT Agreement is to prevent the use of national or regional technical requirements, or standards in general, as unjustified technical barriers to trade. It covers all types of standards, including all aspects of food standards, other than those related to SPS measures, and includes a very large number of measures designed to protect the consumer against deception and economic fraud. The aspects of food standards it covers relate specifically to quality provisions, nutritional requirements, labelling, and methods of analysis. The TBT Agreement basically provides that all technical requirements and regulations must have a legitimate purpose and that the impact or cost of implementing the measure must be proportional to the purpose of the measure. It also places emphasis on international standards.

Codex standards, guidelines and other recommendations are not binding on Member States, but are a point of reference in international law (General Assembly Resolution 39/248; Agreement on the Application of Sanitary and Phytosanitary Measures; Agreement on Technical Barriers to Trade).

However, increased scientific, legal and political demands are being made on the standards, guidelines and recommendations elaborated by Codex. This is in part due to increased consumer interest in food safety, the WTO's SPS and TBT Agreements, harmonization initiatives, calls for increased scientific rigor, the need for transparency, and shrinking national regulatory resources.

The CAC is considering the development of a general standard that would apply basic food safety and food control disciplines to foods derived from biotechnology. The advice of prior FAO/WHO expert consultations in this area will be used as guidance for the conditions required for foods prepared from biotechnology.

22.7 OTHER FAO INSTRUMENTS THAT DEAL WITH ISSUES PERTAINING TO BIOSAFETY

22.7.1 The Commission on Genetic Resources for Food and Agriculture

The Commission on Plant Genetic Resources was established by the FAO Conference in 1983 (18). The mandate of the Commission was broadened to include all genetic resources that pertain to food and agriculture in 1995. The current Membership of the Commission on Genetic Resources for Food and Agriculture is 158 countries and the European Community.

The Commission has developed the following international agreements relevant to the biosafety protocol and to CBD:

1. The International Undertaking on Plant Genetic Resources, adopted by the FAO Conference in 1983. There are 113 countries that have adhered to the Undertaking. The revision of the Undertaking in harmony with the Convention on Biological Diversity is currently being negotiated by countries through the Commission.

2. The international Code of Conduct for Plant Germplasm Collecting and Transfer, which was adopted by the FAO Conference in 1993.

In 1989 and 1991 the Commission considered reports on technical and policy issues regarding Biosafety, within the context of biotechnology in general as explained previously.

22.7.2 Code of Conduct for Responsible Fisheries

The FAO Code of Conduct for Responsible Fisheries was adopted in 1995 by the 28th Session of the FAO Conference and provides a framework for the sustainable use and conservation of aquatic biodiversity (19). The code was created through negotiations with member Countries, NGOs and IGOs and contains articles on:

1. General Principles
2. Fisheries Management
3. Fishing Operations
4. Aquaculture Development
5. Integration of Fisheries into Coastal Area management
6. Post harvest Practices and Trade
7. Fisheries Research

Although the Code is voluntary, parts of it are based on relevant rules of international law, including those reflected in the United Nations Convention on the Law of the Sea.

Aquaculture is a primary means for the purposeful introduction of aquatic alien species, as well as the main motivation for the use of living modified aquatic organisms. Therefore, Article 9 on Aquaculture Development deals with these topics, specifically: Article 9.2 on the "responsible development of aquaculture including culture based fisheries within transboundary aquatic ecosystems" and Article 9.3 on the "use of aquatic genetic resources for the purpose of aquaculture, including culture-based fisheries."

22.7.3 The International Plant Protection Convention (IPPC)

Some of the potential environmental risks concern plant pests. The inclusion of pest resistance in plants should be carefully evaluated for potential development of resistance in pests and possible side effects on beneficiary organisms.

The IPPC is an international treaty for cooperation in plant protection, deposited with FAO and administered by FAO through the Secretariat for the IPPC. The purpose of the Convention is "*to secure common and effective action to prevent the spread and introduction of pests of plants and plant products, and to promote appropriate measures for their control.*"

The Convention had its beginnings in 1951 and came into force in 1952. It is recognized as the primary instrument for international cooperation in the protection of plant resources from harmful pests.[4] There are currently 106 governments that are contracting parties to the IPPC.

The role of the Convention with respect to trade has changed significantly as a result of the SPS Agreement. This is reflected in substantial amendments found in the New Revised Text approved by FAO Conference in 1997 (20).

[4] "Pests" are defined in the IPPC to be "any species, strain, or biotype of plant, animal or pathogenic agent injurious to plants or plant products"

The IPPC calls for phytosanitary measures to be based on a pest risk analysis, which covers both economic and environmental factors including possible detrimental effects on natural vegetation. The Convention also allows for the prohibition or restriction of the movement of biological control agents and other organisms of phytosanitary concern claimed to be beneficial into the territories parties.

22.8 CONCLUDING REMARKS

Adequate biosafety regulations, risk assessment of quality and safety of foods derived from biotechnology, mechanisms and instruments for monitoring use, and compliance are necessary to ensure that there will be no harmful effects on the environment and the health of people.

International organizations such as FAO, WHO, IAEA at the service of Member States by assisting in building capacities and providing technical advice and assistance in priority assignment, resource allocation, and international regulations.

REFERENCES

1. U.N. Fund for Population Assistance. The state of world population 2001. NewYork: UNFPA, 2001.
2. FAO. Statement on biotechnology. Rome: FAO, March, 2000.
3. U.N. Secretariat of the Convention on Biodiversity. 1992 Convention on biodiversity. Montreal, Canada: U.N. Secretariat of the Convention on Biodiversity, 1992.
4. U.N. Secretariat of the Convention on Biodiversity. Cartagena protocol on biosafety. Montreal, Canada: U.N. Secretariat of the Convention on Biodiversity, January, 2000.
5. FAO. FAO code of conduct for biotechnology as it affects the conservation and use of plant genetic resources. Rome: FAO, April 1999.
6. FAO. FAO/WHO joint consultation on the assessment of biotechnology in food production and processing as related to food safety. Rome: FAO, 1990.
7. FAO/WHO. FAO/WHO joint consultation on biotechnology and food safety. Rome: FAO, Geneva: WHO, 1996.
8. FAO/WHO. FAO/WHO joint consultation on foods derived from biotechnology. Rome: FAO, Geneva: WHO, 2000.
9. FAO/WHO. FAO/WHO joint consultation on foods derived from biotechnology. FAO, Rome, WHO, Geneva, January 2001.
10. FAO/WHO. FAO/WHO joint consultation on foods derived from biotechnology. Rome: FAO, Geneva, WHO, September 2001.
11. FAO: *Codex Alimentarius Commission Procedural Manual*, 12th Ed. Rome: FAO, 2001.
12. FAO: Understanding the Codex Alimentarius, Rome: FAO, 1999.
13. FAO: *Codex Proposed Draft Principles for the Risk Analysis of Foods Derived From Modern Biotechnology*. Rome: FAO, 2002.
14. FAO. *Codex Proposed Draft Guideline for the Conduct of Food Safety Assessment of Foods Derived from Recombinant-DNA Plants*. Rome: FAO, 2002.
15. FAO. *Codex Draft Annex on the Assessment of Possible Allergenicity*. Rome: FAO, 2002.
16. FAO. *Codex Draft Guideline for the Conduct of Food Safety Assessment of Food Produced Using Recombinant-DNA Micro-organisms*. Rome, FAO, 2002.
17. World Trade Organization. *The Results of the Uruguay Round of Multilateral Trade Negotiations: The Legal Texts*. Geneva: WTO 1994.
18. FAO. *International Undertaking on Plant Genetic Resources*. Rome: FAO, 1983.
19. FAO. Code of conduct for responsible fisheries. Rome: FAO, 1995.
20. FAO: *International Plant Protection Convention*, new revised text. Rome: FAO, 1997.

3.23

Patenting Inventions in Food Biotechnology

R. Stephen Crespi

CONTENTS

23.1 INTRODUCTION

Patent law is a system for recognizing technological innovation and protecting the inventor for a limited period against the unauthorized use of the patented invention by others. It has existed for centuries and, despite criticism, it survives as the most useful type of intellectual property for protecting technological advance. Patent law has to respond continually to the needs of new and rapidly developing technologies and among these none has made more demands on it than biotechnology.

Patent law is based on the bargain theory of providing legal protection in return for full disclosure of inventions, as the alternative to a policy of industrial and trade secrecy. Open disclosure will appeal to researchers in academic and similar institutions, who will usually seek scientific publication in parallel with patenting. Some of the pitfalls to be avoided by inventors of this kind have been outlined previously (1). For industry, the choice between these alternatives will always be made, case by case, as matters of commercial judgement. Coca cola and Chartreuse are the most cited successful examples of products made by secret processes. From limited experience in the food industry, I well remember the panels of tasters who tried to divine the process by which our competitors managed to make their chocolate taste so good. But for a novel food product which can be analysed and the manufacturing process reverse engineered, patenting might be the preferred option. In the past, the alcoholic beverage and food industries have tended to avoid using the patent system and have preferred to rely on secrecy. The same can be said for the microbial strain improvement programs followed by industry for the production of antibiotics and other useful metabolites. However, considerable use has been made of patents where it is possible to protect key intermediates and final products of a fermentation process.

23.2 BASIC PRELIMINARIES

23.2.1 The Nature of Patents

A patent is a property right granted by state authority which excludes others from the use or benefit of the patented invention without the consent of the patent holder. A patent does not confer a positive right to use an invention because freedom of use may be dependent on prior rights. For example a patent for an improvement on a basic product or process will often be subject to a prior patent for the basic product or process.

23.2.2 The Patent Application

To obtain a patent, an application must be filed with the relevant national authority (patent office) and will be examined for compliance with legal requirements. Separate patent applications are usually necessary in each country where protection is required but a single application in the European Patent Office (EPO) can cover any selection, or all, of the 30 European countries which are member states of the European Patent Convention (EPC) up to the point at which rights are granted.

There is no such thing as a world patent, although a so-called "international patent application" can be filed as a single patent application under an international convention (the Patent Cooperation Treaty, PCT) to which 126 member states are signatories. The early stages of a PCT application are managed by the World Intellectual Property Organization (WIPO), based in Geneva, and the specifications of these pending applications are published by WIPO. Eventually a PCT application splits into separate national applications for those countries which the applicant has designated for protection and these are then handled by the national patent offices of the designated states.

23.2.3 Patentability

The principal legal requirements for patentability are that the invention is new, involves an inventive step, and has an industrial or other useful capability. Also the patent application must include a specification of the invention which contains adequate instruction to enable the skilled person to produce or perform the invention. The invention is defined in the "claims" which are listed at the end of the patent document. A claim is a verbal formula which sets out what the patent protects (the scope of the patent), and it is therefore usual to include a series of claims directed to the various aspects of the invention. Common forms of claim are those directed to an apparatus or device, a process or product of manufacture, and a method of treatment, testing, or use.

23.2.4 Official Examination

The Patent Office will carry out a search of previously published documents including the scientific and patent literature to determine what is already known that may be relevant to the invention (the "prior art"). The application will then be examined in the light of the search results and the Examiner will raise objections if appropriate. Examination of the application usually involves extended argument about the specification, especially the scope of the claims, and may take considerable time to settle.

23.2.5 Opposition or Reexamination

Even after acceptance by the patent office, a patent application or patent can, in most countries, be opposed by third parties who may raise objections and prior art similar to or additional to those already overcome by the applicant. This is usually termed Opposition and involves argument between the applicant or patentee and an opponent, who have equal status as contending parties. U.S. patent law does not provide for opposition in this sense but allows a third party to request official reexamination of the patent in the light of prior art which has not already been considered. The third party can also participate up to a point in the reexamination procedure.

23.2.6 Special Features of U.S. Patent Applications

U.S. patent practice is unique in allowing an initial patent application to be refiled in the form of a "continuation" application which reproduces the text of the original application and retains its original priority date. In effect, this attractive facility extends the time to complete the process of argument with the Examiner over objections which may prove difficult to resolve speedily. A series of such continuing applications may be filed if desired and, provided the identical original text is adhered to in the refiled application, it will be possible subsequently to employ further arguments or scientific evidence to support the applicant's case. It will also be possible to make changes in the claims to improve the chances of reaching agreement with the Examiner as to their form and scope. This practice is used extensively by applicants for U.S. patents.

If any added data or other matter is added on refiling, the application is then designated as a "continuation-in-part" application (CIP) and, for this, the original priority date will be preserved only for what was originally present; the added matter will only be entitled, for priority, to the date of filing the CIP. Any intervening publications may therefore be citable as prior art against claims based on the added matter. This may also have consequences for disputes over priority, as will be apparent from the next section.

23.2.7 Conflicting Patent Applications

When two or more inventors are independently seeking a patent for the same invention, i.e., their claims cover the same ground, the U.S. Patent and Trade Mark Office (USPTO) has to decide which application has priority over the other(s). Provided the dates of filing of their respective applications are close to one another, the USPTO will declare "Interference." This is a procedure (too complex to attempt a summary here) based on examining laboratory notebook records and other evidence to determine the dates on which each party made the invention. This reflects the fact that U.S. patent law is a "first-to-invent" system. Other countries operate a "first-to-file" system, according to which the application with the earliest filing date will usually prevail, assuming that it is effective as a proper "enabling disclosure" of the invention.

23.2.8 Duration of a Patent

The term of a patent differs from country to country. In the U.S. and most European countries, the term is 20 years from the application date. The payment of annual official renewal fees is required in most countries to maintain a patent in force.

23.2.9 Enforcing Patents

A patent gives the patent holder no positive right to do anything except seek from a court a remedy against unauthorized use of the invention by others. A patent holder who takes such action has to face the possibility that the defendant will counterclaim that the patent is invalid and should therefore be revoked. Instead of going to law, a patent holder will often choose to regulate the situation by licensing the patent to the other party on reasonable terms.

23.2.10 Licensing Patents

The patent holder may wish to be the sole provider of the product or service covered by the patent and, subject to certain safeguards, this is permitted. Alternatively, the patent holder may license the patent to others for appropriate payment, either to one other party only (an exclusive license) or to more than one party (nonexclusive licenses). Where the patent holder is not an industrial or commercial organization, and does not wish to create a start up company to commercialize the invention, licensing to the established industry is the most effective way of securing a financial return on the investment in research.

23.3 MAIN TYPES OF BIOTECHNOLOGY PATENTS

To be patentable, all biological materials, including biopharmaceuticals, agrobiotechnology products, food products, and their methods of manufacture or application, have to meet the same three legal tests of novelty, inventiveness, and practical utility as do the simpler chemical compounds and preparative methods of synthetic chemistry. Biotechnology inventions must also be capable of adequate description and definition in the patent specification, the all important document upon which rest all questions of patent scope (how much the patent covers) and patent validity (whether it is enforceable).

23.3.1 Food Products

In the past, the patent laws of many countries refused patents on food products as a matter of principle, although claims to inventive processes for producing them were allowable. The same applied to chemical products and medicines. In some countries, while food products were not automatically excluded from patentability, it was not possible to patent mere mixtures of known ingredients unless some special effect was involved. The underlying basis for this was that, if the properties of the mixture were simply the additive effect of the ingredients, no inventiveness was involved in "putting two and two together." In the parallel rule for medical products, evidence of a synergistic effect was often sufficient to overcome the objection. This type of exclusion persists in a few countries as, for example, Australia, which refuses patents for "Substances capable of being used as foods or medicines (whether for human beings or animals...), which are a mere mixture of known ingredients, and processes of producing such substances by mere admixture." However, in most other countries, this is dealt with under the heading of lack of inventiveness.

Novel food additives such as antioxidants, flavor enhancers, enzymes, and preservatives would be patentable as products in their own right.

23.3.2 Biological Products

23.3.2.1 Naturally Occurring Substances

Substances present as components of complex mixtures of natural origin, can in principle be patented where they are isolated, identified, and made practically available for the first time, and a process is developed for producing them so that they can be put to a useful purpose. This applies to inanimate substances and also to living matter. In appropriate circumstances such substances are not ruled out as mere "discoveries" but are considered as inventions by the legal authorities. For example, it is legitimate to isolate the true causative agents of folk remedies so that these may perhaps be synthesized and used in a more scientific way in therapy or for nutritional purposes, and there are many examples of patents based on this kind of research.

23.3.2.2 Microorganism Patents

These are now obtainable in most industrially developed countries, following the landmark decision of the U.S. Supreme court in 1980 that the living nature of microorganisms does not preclude them from patentability.

23.3.2.3 Plant Patents

These are also obtainable in U.S., Europe, Japan, Australia, and some other countries. Patent rights must be distinguished from the separate legal system of Plant Variety rights (also called plant breeders' rights). From the early 1960s, new plant varieties have been afforded protection under national laws of Plant Variety rights and an International Convention (UPOV). To avoid legal confusion, patent law in Europe excluded plant varieties from patentability, e.g., in the prototype provision of the European Patent Convention (EPC) Article 53(b) which excludes patents for "plant and animal varieties" and "essentially biological processes for the production" of plants and animals.

It is incorrect and confusing to refer to plant variety rights as "plant patents." The U.S. law provides for a special form of "plant patent" for asexually propagated plants, e.g., rose trees, but the normal U.S. "utility" patent can be used to protect other types of plant innovation, especially genetically modified plants. The U.S. has a Plant Variety Protection Act for sexually propagated plants and this corresponds to the plant variety rights system in other countries. These topics are more fully covered in specialist works on IPR in Agricultural Biotechnology (2).

Plant genetic improvement is more suitably protected under patent law especially if the invention is generically applicable, i.e., not limited to use with a single variety. The European Union Directive 98/44/EC makes this clear for the European Community. Also, the enabling methodology for this type of improvement can only be protected under patent law. Resolution of the apparent problem of interface between patent law and plant variety right law in Europe has been outlined previously (2).

23.3.2.4 Animal Breeds

Animal breeding by traditional methods has no legal system for protection comparable to plant breeder's rights. U.S. patents may in principle be obtained for nonnaturally occurring nonhuman multi cellular living organisms including animals. The first transgenic animal patent issued in 1988 to Harvard University with claims covering the "onco-mouse," one in which an onco-gene has been introduced to make the animal more susceptible to cancer and therefore more sensitive in testing possible carcinogens. Transgenic animal patents are available in European countries also. A European patent was granted for the onco-mouse but was formally opposed by animal rights groups.

23.3.2.5 DNA and RNA Sequences

Patents for nucleic acids may also be obtained in industrially developed countries. "Gene patents" are controversial and require more discussion than is feasible in this brief summary. The patent authorities in countries where such patents have been challenged consider that the gene in its natural state is unpatentable but that a patent can be granted when the gene is isolated and made available for a practical industrial or other useful purpose. Genes therefore constitute a special case of the general rule for naturally occurring substances. The common objections to gene patents have been discussed previously (3).

23.4 PATENTING IN BIOTECHNOLOGY

The special problems of biotechnology patents began to emerge with the development, after the Second World War, of microbiological processes for producing antibiotics, amino acids, enzymes, and other microbial products of industrial importance. To approach these problems it was inevitable that inventors and patent experts would draw upon a century or more of experience in the patenting of chemical inventions and apply the established principles to the new situations. On the whole this has been successful, due to the fair degree of parallelism that exists between the two technologies. But the significant differences between living cells and the more manageable inanimate molecules have brought about a reexamination of traditional ideas of patent law to adapt it to deal effectively with biological systems and so stimulate innovation in this new field as it has done for older technologies.

23.4.1 Microbiological Inventions

Microbiological inventions follow the chemical pattern to a large extent, and the usual categories are shown in Table 23.1.

The microbiological equivalents of chemical products will be microbially produced products, or new microorganisms themselves or some other type of biological material including cell lines and plant or animal cells.

The early difficulties which appeared to be an obstacle to the granting of patents for living organisms have now been overcome intellectually and there is a clear trend toward the allowance of such patents at least in the major industrialized countries. Undoubtedly the decision of the U.S. Supreme Court in 1980, in the Chakrabarty case, to allow the following claim to a microorganism, was of prime influence in this connection.

"A bacterium from the genus *Pseudomonas* containing therein at least two stable energy-generating plasmids, each of said plasmids providing a separate hydrocarbon degradative pathway."

23.4.2 The Enabling Description of Microorganisms

It is a fundamental requirement of patent law that in return for legal protection, an inventor must disclose his invention in a manner sufficiently clear and complete to enable others of ordinary skill in the art to repeat or reproduce the process or product for which the patent is granted. Where the invention consists of or depends on a specific microorganism or other kind of biological material, this must be identified in the patent application to provide the "enabling disclosure" the law requires.

As compared with the chemical case, where a written description of the process is usually adequate, the requirement to provide a description which is reliably repeatable presents greater difficulty when living material is involved. Living material does not admit of a total description. But, more importantly, a total description would not by itself guarantee the means of producing the material. Where the microorganism is known and already available to the

Table 23.1
Categories of microbiological inventions

MICROBIOLOGICAL INVENTIONS
Production of microorganism (PROCESS)
New Microorganism (per se) (PRODUCT)
End Products of biosynthesis (PRODUCT)
Formulations of microorganism (COMPOSITION)
Use of microorganism (PROCESS) to produce
biomass
by-product of microbial growth
an extracted product
an improved substrate
a biotransformation product.

skilled person and the invention resides, for example, in the discovery of some new property or use of practical value, it is usually sufficient to refer to the microorganism by name. But for a new microorganism, say a newly isolated or developed strain of a known species, the skilled person who attempts to repeat the procedure described in the patent specification will in most cases need not only a description of the organism but also a means of access to it.

Now if the patent application gives reliable instructions how to reisolate, rediscover, or reconstruct the new organism, this will be a sufficient disclosure. However, in most cases this cannot be achieved with certainty. Patent law has solved this problem by making use of the Culture Collection deposit system which the scientific community had created much earlier for its own needs. The practice has developed internationally, both through case law and in the express obligations written into modern patent laws, in many countries. The International Convention dealing with this requirement is *The Budapest Treaty on the International Recognition of the Deposit of Microorganisms for the Purposes of Patent Procedure*. It came into force at the end of 1980 and now has 60 Member States. The Treaty provides for the recognition of culture collections as International Depositary Authorities (IDAs) in any one of which a new strain of microorganism (or other type of biological material) can be deposited for the purposes of a patent application in any member state. At present there are 36 IDAs distributed among 21 of the Member States.

The maxim that what cannot be fully described must be deposited has therefore become part of patent law. This development has added a new dimension to patent law and practice for which no parallel exists in chemistry or other fields of technological innovation. Thus, the deposit of the organism supplements the written description and fulfils other important functions. First, it provides a reference material for resolving any dispute over the alleged novelty of the organism. Secondly, its reference function may be called upon to decide whether any third party is infringing the patent by using the same organism without a licence from the patentee. Finally, the deposit provides an available source material to enable others to make use of it when they are legally free to do so, i.e., when the patent is allowed to lapse or expires at the end of its normal term.

23.4.3 Availability of the Deposit

As to when a deposited culture can become available to the applicant's competitors and other third parties there is a difference between both the U.S. and Japanese patent systems, which allow access to the deposited culture only after an enforceable right has been granted, and the corresponding laws in European countries which allow access to the deposited culture upon first publication of the European or National Patent application (at 18 months after the filing date or priority date).

Loss of control of the new strain, at least for competitive research purposes, is mitigated under European patent practice by the option to elect for the so-called independent expert solution in the interim period between publication of the application and eventual grant of the patent. The function of the independent expert is to act for other persons or firms, e.g., the applicant's competitors, and to test, experiment with, and generally evaluate the invention on their behalf. The independent expert, though acting for third parties, must not pass the strain on to them. Under this alternative a measure of control on the use of the organism is provided for the applicant before he obtains an enforceable right.

Eventually, there must be full disclosure to the public and the skilled person must be put in a position of reproducing the inventive process or product, including access to deposited biological material. From then on policing problems vary with the nature of the patent protection obtained. When the microbiological invention leads to new products which can be the subject of product patents then detection of infringement is relatively straightforward. Where the products are not new, however, and novelty lies only in the strain used or in some other parameter of the new microbiological process, then policing becomes more difficult. These factors can usually be fully anticipated before the decision to proceed by the patent route is first taken. One crumb of comfort may lie in the fact that the laws of some countries allow infringement actions to proceed on the basis of a strong *prima facie* case that the defendant is using the patented process. If the record shows that the defendant has obtained a sample of the patentee's deposited strain from the culture collection this might be sufficient to put the burden of proof of nonuse on the defendant.

23.4.4 Recombinant DNA Patents

In a recombinant DNA patent it will be common to have claims of various types, as shown in Table 23.2. For example, a possible claim might involve diagnostic methods and kits. Novel diagnostic methodology and reagents (or reagent combinations) for testing for contaminants, in food and other products, provide a growth area for patenting. Such patents usually have general application to products of many kinds in which microbial contaminants may be found but those specific for food products are also possible.

23.5 PRACTICAL EXAMPLES

Some remarks on selected patents in food biotechnology are given as illustrative of these principles. Being a random selection, it cannot be comprehensive as to the topics covered. Patents of interest are identified by patent numbers, from which copies can be obtained from the respective patent offices. Many will have lapsed by now either through nonpayment of renewal fees or by expiry at the end of their normal term.

23.5.1 Mycoprotein (*Fusarium* Strains)

The patenting of certain strains of *Fusarium graminearum* and foodstuffs based thereon began in mid 1970, at which time patents on living organisms were controversial in some countries. In Great Britain, two patents were obtained on this early work without much difficulty. UK patent 1,346,061 covered the various strains, a typical claim being "*Fusarium graminearum* Schwabe deposited with the Commonwealth Mycological Institute and assigned the number IMI 145425 and variants and mutants thereof".

Separate claims were also included for "fungal cultures" of these organisms and various aspects of the methodology.

For commercial reasons, and in view of the uncertainty as to the validity of microorganism per se patents at this date, the cultivation process for producing the edible mycelial protein was covered separately in UK patent 1,346,062. This also contained claims to the end product of the defined process, the "edible protein-containing substance" itself.

Table 23.2
Recombinant DNA patent claims

(1)	Recombinant protein products (and alleles, variants, derivatives)
(2)	DNA sequences coding for the products of (1)
(3)	Vectors containing the DNA sequences of (2)
(4)	Microorganisms, cell lines, and other organisms transformed with vectors (3)
(5)	Processes of producing products as in (1) by expression of DNA sequences (2) in a recombinant host organism

In the USA, protection for the strains by means of the ordinary "utility patent" was not acceptable to the U.S. Patent Office, so the applicants resorted to the unusual measure of protecting them under the plant patent law, which extends to fungi. Plant patent 4,347 covers the commercially important isolate. Three U.S. patents were apparently necessary to protect all aspects of this development, including the basic process (U.S. 3,937,654), the fungal cultures (U.S. 4,294,929) and the "Article of Manufacture" claims (U.S. 4,061,781). Distribution of the various claims to the technology over separate patents is usually required by official search and examination criteria but is sometimes done tactically by the applicant for licensing and other business reasons.

All these early patents have expired but later patents may still be in force covering subsequent developments of this commercially successful product (QUORN, Registered Trade Mark).

23.5.2 Mushroom (Mutant Strains)

In view of the Chakrabarty decision in 1980 it was no longer possible for the U.S. Patent Office to refuse patents for novel microorganisms, although in many instances it was necessary to restrict the claim to a "biologically pure culture" of the novel strain, newly isolated from nature or obtained by a mutagenic and selection process. U.S. patent 4,608,775 is a good example of a claim to certain fungicide resistant strains of *Agaricus Bisporus* produced by irradiation of commercially available strains followed by a special selection procedure. The derived strains were deposited at the Commonwealth Mycological Institute (England) and the patent claims refer to their accession numbers, this being a common practice for mutants.

23.5.3 Yeast Strains

Novel yeast strains were among the first types of microorganism to be patented. The earliest known patent of this kind is U.S. 141,072, granted to Louis Pasteur in 1873 for "yeast free from organic germs of disease, as an article of manufacture." Nowadays this would be claimed as "a biologically pure culture" in U.S. patent practice.

In the UK, patents were frequently obtained for brewers' and bakers' yeasts developed by breeders for some special advantage. As clearly useful in food or beverage manufacture, these aroused no controversy.

In more recent times, yeasts have been patented in terms of their possession of specific properties, as meeting some manufacturing requirement. For example, in UK patent 1,539,211, the yeast is claimed in terms of the volumes of gas released in certain established tests (A'3 and A'4), these being the criteria discovered by the patentee as indicative of suitability for fermenting sweetened dough. This patent was upheld in Litigation.

In European patent 8554, a similar approach to claim drafting succeeded. The claim defines the strains by reference to the amounts of gas released in four specified tests, A'1 to A'4. In the patent description these strains were produced by derivation from a parent strain α 1217, itself on deposit with two Culture Collections. The patent was opposed on the ground that the claim was too broad because no instructions were given for deriving the yeasts from any other strain than the deposited strain. This argument failed and the patent was maintained without limitation.

23.5.4 Trehalose

The disaccharide trehalose is an attractive food sweetener and has other useful properties for use in foods. It is produced by fermentation or enzymatic conversion methods.

U.S. patent 5,858,735 is directed to producing trehalose in high yields by cultivation of *Pleurotus eryngii*. The claims as originally filed also covered the use of some other named species of *Pleurotus*, but the finally allowed claims were limited to use of *P. eryngii* in view of the unexpectedly high yields obtained with this species.

Genetic modification of various organisms to produce increased yields of trehalose and to benefit transformed organisms also in other ways is described in U.S. patent 5,422,254. Genes encoding the long and short chains of yeast trehalose synthase were isolated, sequenced, cloned, and used to transform yeast. This patent also indicates how the methodology can be extended to transform crop plants. The patent claims the isolated DNA fragments defined by nucleotide sequence, vectors containing the DNA, transformed yeasts, and the corresponding fermentation process technology. It was necessary to distinguish this invention from an earlier published European patent application 0451896 which had referred to a trehalose 6-phosphate synthase gene but without disclosing its isolation or sequence. This lack of "enabling disclosure" in a cited prior reference frequently provides a way of discounting any relevance it might be thought to have as prior art.

23.5.5 Extracted Proteins

23.5.5.1 Chocolate Flavor

Two patents have been granted in the U.S., Europe, and UK for proteins extracted from cocoa beans (*Theobroma cacao*) which are in part responsible for chocolate flavor. These claim, respectively, an extracted (or extractable) 23kD protein and a 67kD protein, both of defined amino acid sequence or partial sequence. The 67kD protein is thought to be the primary translation product which is processed *in vivo* to produce 47kD and 31kD polypeptides.

These patents also claim the isolated or recombinant nucleic acid encoding these proteins, and the chemical synthesis or microbial biosynthesis of the proteins by which large amounts of product are obtainable. The ultimate commercial purpose of this technology is at present uncertain. The patents are:

- 23kD
  - U.S. 5,668,007
  - GB 2,260,327
  - EP 586372
- 67kD
  - U.S. 5,770,433
  - GB 2,260,328
  - EP 535053.

23.5.5.2 Brazzein Sweetener

Heat resistant protein sweeteners, with long lasting taste, isolated from *Pentadiplandra brazzeana Baillon,* or produced recombinantly from the DNA of defined sequence, are covered in U.S. patents 5,326,580, 5,346,998, and 5,527,555. Patents of this type give rise to debates over bioprospecting (termed by some as "biopiracy").

23.5.6 Flavorsavor Tomato

The basic principle of antisense regulation of gene expression in plants was covered by U.S. patent 5,107,065 (Calgene Inc.) which has a broad claim to inserting DNA into a plant cell which is transcribed into RNA complementary to indigenous RNA for the gene concerned. This covers application to all plants. A specific claim to the use of the method for tomatoes, by reducing expression of the polygalacturonase gene, is covered in Calgene's U.S. 5,453,566; this also has product claims to the resulting tomato plant cells and plants. An interference between Calgene and ICI on this development was settled out of court. In Europe, ICI obtained EP patent 0271988 for the use of those parts of the gene for which they had the earlier priority date. This patent is also relevant for use with processed tomatoes and the production of tomato puree.

For various reasons, these technologies are not currently in commercial use.

A later patent for achieving ethylene mediated response in the tomato by targeting the E8 gene is covered in U.S. 5,545,815 (Regents of the University of California).

23.5.7 Soybean Cultivar

New plant varieties produced by traditional plant breeding methods, e.g., involving crossing parental varieties and backcrossing to one of the parents, are unpatentable in Europe. Such methods are categorized as essentially biological and are excluded on this ground. The U.S. patent law is more generous and allows for patents for varieties of this type provided the method or product is assessed as inventive in some respect.

One example is U.S. patent 5,750,856 for a soybean cultivar having resistance to Roundup™ herbicide and high yield potential. The variety is designated 93149194423034 and seeds have been deposited with the American Type Culture Collection (ATCC) under Accession No. 209546. The patent claims seeds, plants, pollen, ovules, tissue cultures, and regenerated plants of the claimed variety, as well as the methodology to produce hybrids of this and other varieties. This type of claim structure is usual in such cases.

It appears that no corresponding patents have been applied for in other countries. This is not unexpected for reasons given earlier.

23.5.8 Golden Rice™

Technology has been developed (4) to raise levels of Vitamin A or the Vitamin A precursor, beta-carotene, in rice, and thereby to respond to the nutritional needs of peoples for whom rice is the most basic food crop.

This strategy, initiated by I. Potrykus and P. Beyer, is described in published International Patent Application WO 00/53768. It entails the insertion of 3 genes into the rice plant to complete the beta-carotene biosynthetic pathway. This application has not yet matured into granted patents in the many countries designated in the application, but the claims contemplated by the applicants can be seen in the published application.

One of the claims in WO 00/53768 is directed to:

> An isolated DNA molecule comprising a nucleotide sequence providing one or more expression cassettes capable of directing production of one or more enzymes specific for the carotenoid biosynthesis pathway selected from the group consisting of: 1) phytoene synthase derived from plants, fungi, or bacteria, 2) phytoene desaturase derived from plants, fungi, or bacteria, 3) carotene desaturase derived from plants, and 4) lycopene cyclase derived from plants, fungi, or bacteria, under the proviso that an expression cassette capable of directing production of phytoene synthase alone is excluded.

Additional claims cover vector systems containing the DNA and transgenic plants containing the DNA.

To achieve this strategy, and in addition to the proprietary genes, the methodology involves the use of a number of plant transformation vectors, promoters, and antibiotic resistance markers which are the subject of previous patent applications and granted patents held by various owners or covered by material transfer agreements. These are listed in the ISAAA Brief. ISAAA refrained from recording conclusions as to the relevance of each and every one of these prior rights, but they point to alternative strategies for international organizations concerned with facilitating the introduction of the genetically improved rice varieties into developing country agriculture.

This case is a striking example of the complications sometimes encountered in achieving the desired objective. But, as pointed out by ISAAA, patents are territorially restricted. Not all of these prior patents may be granted (or may have even been applied for) in the rice-growing countries, in which case public research bodies would be able to exploit the technology without undue restriction in these countries. One of the relevant patent holders (Monsanto, 35S promoter) has been reported as willing to make its technology freely available to those wishing to promote gm rice (*Science*, Vol. 289, 11 August 2000).

23.5.9 Bovine Somatotrophin

For more than 50 years it has been known that the administration of bovine somatotropin to dairy cows will increase milk yield and cause other desirable effects. Possible commercial exploitation of this finding came under active consideration in the late 1970s but appeared prohibitively costly at that time. A perfunctory patent search carried out for the purposes of this chapter shows that it was not until the 1980s, when recombinant DNA technology came into its own and was being applied to human growth hormone that work began seriously on the use of BST in beef and dairy farming. Some perspective on the development of this technology may be obtained by a study of selected examples from the extensive patent literature on this subject, listed in roughly chronological order.

The increase in the mammary parenchyma caused by administration of growth hormone between the onset of puberty and the first parturition was disclosed in U.S. patent 4,521,409, applied for in October 1983 (Cornell Research Foundation). Although very brief in terms of detail, it mentions the possibility of using the recombinant hormone, by then "known in the art." Ensuring the proper conformation of the protein by treatment with 2-amino-2-methyl 1-propanol is the subject of U.S. 4,975,529 (Monsanto Co).

Subsequent patents issued for specific formulations of the hormone which gave improvements of various kinds. For example, U.S. patent 4,977,140 first applied for in August 1985 (Eli Lilly & Co) refers to somatotropin produced in various genetically modified organisms and formulated with a carrier composed of wax and oil. U.S. patent 5,004,728 is directed to obtaining an improvement in milk yield by the coterminous administration of the hormone and a long chain fatty acid.

A major advance came with the work of the Monsanto Co. in the development of an oil-based formulation designed for prolonged release of the hormone. This is represented by a series of patent applications originating as far back as October 1984, and prosecuted as continuation applications, maturing into patents granted from 1991 onward. Thus U.S. 5,013,713 contains method claims directed to the use of a nonaqueous composition comprising from 10% or more of BST in a continuous phase of a biocompatible oil. A later patent issuing from this series in 1997 is U.S. 5,595,971, with product claims directed to the composition itself, in which BST, which may be in metal-associated form, is dispersed in a biocompatible oil containing an antihydration agent which retards penetration of the oil by aqueous body fluids after injection. These two significant patents are especially relevant to the commercial product produced by the patent owner.

An interesting diagnostic test for animals that have been treated with BST is disclosed in U.S. patent 5,135,401, based on the fact that treated cows have lower levels of phosphorylation of fatty acid binding protein in milk fat globules.

According to U.S. 5,106,631, improved cheese quality is obtained from milk in which the BST concentration is maintained less than 0.2 mg/l in order to regulate the plasmin level.

As might be expected, patents soon began to issue on variant forms of somatotropin made possible by protein engineering. Thus, U.S. 5,089,473 applied for in August 1988 (Monsanto Co), is directed to the use of site-specific mutagenesis of the starting gene to replace asparagine by glutamine in the region of residues 95–101 thus improving the stability, and hence recovery, of the recombinant product. Related to this (but separate from it) is U.S. 5,130,422 which covers the corresponding encoding DNA. Another patent of this type is U.S. 5,631,227 (The Upjohn Co) filed in June 1985 but, again, stemming from a series of more than ten earlier related applications. This specifies a replacement of Asparagine 99 by aspartic or glutamic acid. On the front page of this patent, where prior scientific literature and prior patents are cited as "prior art" the list includes the preceding example (U.S. 5,130,422).

Stabilization of BST to minimize aggregation by reacting the 183 and 191 cysteine SH groups with ethyleneoxy-type derivatizing reagents is covered by U.S. 5,951,972. A similar approach is found in U.S. 6,010,999. In U.S. 5,891,840, stability is enhanced by altering or eliminating at least one of the 4 cysteine residues by replacement, modification, deletion, or chemical derivatization.

23.5.10 Detecting Microbial Contaminants

A considerable body of patent literature exists on methods of detecting pathogenic microorganisms in clinical samples, water systems, foods and other materials. Molecular biological methods utilize nucleic acid probes specific for many bacterial species and presented in various forms to hybridize to products obtained after performing polymerase chain reaction (PCR) or other methods of amplification of DNA or RNA in the test sample. Method claims are common in this field but novel probes can be patented as products where appropriate, as well as the ingenious ways in which probes are displayed on support structures.

One method of detection, designed to identify a number of different species in a single test, is described in International patent application WO 00/52203. The method targets a specific segment of 23 S ribosomal DNA common to a wide variety of bacteria and uses defined "universal" primers to amplify this sequence, following which the resulting mixture of amplicons is probed with a set of organism specific oligonucleotide probes immobilized on a hybridization membrane. Although originally invented in a hospital context, the method has many other possible applications.

U.S. patent 5,498,525 makes use of bacteriophage, specific to target organisms, and is said to enable organisms such as Salmonella and Listeria to be assayed in hours rather than days. Phage added to the test sample is taken up by the specific bacteria, if present, following which any nonabsorbed phage is inactivated or destroyed. Incubation is continued and phage is released from the bacteria and can be assayed. A convenient assay makes use of reporter bacteria which have been genetically modified to express a suitable signal when triggered by phage infection, e.g., luciferase. This invention is distinguished from the previously known and allegedly less convenient use of phage which has itself been genetically modified to express the signal directly.

23.6 CONCLUDING REMARKS

The documents cited are exemplary of the wide variety of subject matter that can be protected by patents and they have been selected from an extensive patent literature in their respective fields. It is common in many such areas of research to find that only a minority of the patents granted are key patents in terms of commercial importance, and only a minority of patent holders are in sustained commercial development. In the early stages of any project it is often not clear what will prove essential to the technology finally adopted for commercial exploitation and so, in the interim, a fairly deep purse may be needed to maintain all the likely candidate patent applications and patents in force. Fortunately, patent costs are not all upfront costs and the major burden can be spread over extended periods during which the value of specific items can be assessed. Continual review of patent portfolios against agreed budgets will therefore be essential to all who make use of the potential of patent law to promote their interests, be they industry or public research organisations.

REFERENCES

1. Crespi, R. Stephen. Patenting for the research scientist: bridging the cultural divide. *Trends in Biotechnol.* 16:450–455, 1998.
2. Intellectual property rights. In: *Agricultural Biotechnology, Biotechnology in Agriculture Series No. 20*, 2nd ed., Erbisch, F.H., K.M. Maredia, eds., Oxon, UK: CABI Publishing.
3. Crespi, R. Stephen. Patents on genes: can the issues be clarified? *Biosci. Law Rev.* 5:199–204, 1991–2000.
4. The intellectual and technical property components of pro-vitamin A rice (GoldenRice™). In: *International Service for the Acquisition of Agri-Biotech Applications*, Ithaca, NY: ISAAA Brief No. 20, 2000, publications@isaaa.org.

Index